Advanced Optical and Wireless Communications Systems

Ivan B. Djordjevic

Advanced Optical and Wireless Communications Systems

Second Edition

Springer

Ivan B. Djordjevic
Department of Electrical
and Computer Engineering
University of Arizona
Tucson, AZ, USA

ISBN 978-3-030-98493-9 ISBN 978-3-030-98491-5 (eBook)
https://doi.org/10.1007/978-3-030-98491-5

This Springer imprint is published by the registered company Springer Nature Switzerland AGThe registered company address is: Gewerbestrasse 11, 6330 Cham, Switzerland

To my family

Preface

The purpose of this book is to introduce the reader to most advanced topics of (i) wireless communications, (ii) free-space optical (FSO) communications, (iii) indoor optical wireless (infrared, IR) communications, and (iv) fiber-optics communications. So far, these different types of communication systems have been considered as separate disciplines. However, the fundamental concepts, such as propagation principles, modulation formats, channel coding, diversity principles, MIMO signal processing, multicarrier modulation, equalization, adaptive modulation and coding, detection principles, and software-defined transmission, are common. The key idea of this book is to consider these different types of communication systems in a unified fashion. The fundamental concepts, listed above, are described first, followed by detailed description of each particular system. The book is self-contained and structured to provide straightforward guidance to readers looking to capture fundamentals and gain theoretical and practical knowledge about wireless communications, free-space optical communications, and fiber-optics communications, which can be readily applied in their research and practical applications.

This book unifies wireless, free-space optical, IR, and fiber-optics communications. Unique features of the book include the following:

- This book integrates wireless communications, free-space optical communications, indoor IR (optical wireless) communications, and fiber-optics communications' technologies.
- Use of the book does not require prior knowledge in communication systems.
- Use of the book does not require any prerequisite material except basic concepts on vector algebra at undergraduate level.
- This book offers in-depth exposition on propagation effects in different media (free-space, fiber-optics, atmospheric turbulence channels), channel impairments in these media, noise sources, key components and modules enabling wireless and optical communications, channel capacity studies, advanced modulation and multiplexing techniques, advanced detection and channel compensation techniques, OFDM for wireless and optical communications, diversity techniques, MIMO techniques, advanced coding and coded modulation techniques, spread spectrum techniques, CDMA systems, multiuser detection, ultra-wideband communications, and physical-layer security, to mention few.
- Multiple courses can be offered: advanced wireless communications, advanced fiber-optics communications, advanced free-space optical communications, advanced indoor IR (wireless optical) communications, and advanced wireless and optical communication systems.

This book is intended for very diverse group of readers in communications engineering, optical engineering, wireless communications, free-space optical communications, optical wireless communications, mathematics, physics, communication theory, information theory, devise electronics, photonics, as well as computer science.

The book comprises 11 chapters. Each chapter, except introductory one, contains more than 20 problems to help the reader get a deeper understanding of the corresponding text material. In the introductory chapter (Chap. 1), we provide historical perspective of both wireless and optical communications systems, describe their fundamental concepts, identify various noise sources and channel impairments, and briefly explain how to deal with them, with full details provided in the chapters that follow. Chapter 2 is devoted to the detailed description of propagation effects, noise sources, and channel impairments for wireless communications, free-space optical communications, and fiber-optics communications. In Chap. 3, the basic components, modules, and subsystems relevant in both optical and wireless communications are described. Chapter 4 is devoted to both wireless channel and optical channel capacities. Both memoryless and channels memory are considered. Chapter 5 is related to advanced modulation and multiplexing techniques suitable for both wireless and optical communication systems. Chapter 6 is related to advanced detection and channel impairments compensation techniques applicable to both wireless and optical communication systems. Chapter 7 is devoted to OFDM fundamentals and applications to wireless and optical communications. In Chap. 8, various diversity and MIMO techniques are described, capable of improving single input single output system performance, applicable to both wireless and optical communication systems. Chapter 9 is related to the advanced channel coding and coded modulation techniques, relevant to optical and wireless communication systems. Chapter 10 is devoted to spread spectrum (SS), CDMA, and ultra-wideband (UWB) communication systems. Finally, Chap. 11 provides detailed description of the physical-layer security concepts applied to wireless and optical channels. The detailed description of chapters is provided in the Introduction.

I would like to thank my former students, postdocs, and collaborators for the research we performed together.

Finally, special thanks are extended to Mary E. James of Springer, USA, for her tremendous effort in organizing the logistics of the book, in particular promotion and edition, which is indispensable to make this book happen.

Tucson, AZ, USA Ivan B. Djordjevic

Contents

About the Author

Ivan B. Djordjevic (Fellow, IEEE) is Professor of Electrical and Computer Engineering and Optical Sciences at the University of Arizona, director of the Optical Communications Systems Laboratory (OCSL) and Quantum Communications (QuCom) Lab, and co-director of the Signal Processing and Coding Lab. He is both IEEE Fellow and the OSA (Optica) Fellow. He received his PhD degree from the University of Nis, Yugoslavia, in 1999.

Dr. Djordjevic has authored or co-authored 10 books, more than 550 journal and conference publications, and he holds 54 US patents. Dr. Djordjevic serves as an area editor/associate editor/member of editorial board for the following journals: *IEEE Transactions on Communications, OSA (Optica)/IEEE Journal of Optical Communications and Networking, Optical and Quantum Electronics*, and *Frequenz*. He was serving as editor/senior editor/area editor of *IEEE Communications Letters* from 2012 to 2021. He was serving as editorial board member/associate editor for *IOP Journal of Optics* and *Elsevier Physical Communication Journal* both from 2016 to 2021.

Prior to joining the University of Arizona, Dr. Djordjevic held appointments at the University of Bristol and the University of the West of England in UK, Tyco Telecommunications in the USA, National Technical University of Athens in Greece, and State Telecommunication Company in Yugoslavia.

Introduction

1

Abstract

In this introductory chapter, we first describe the historical perspective related to both wireless and optical communication systems and describe various generations, from early beginning until today. After that we describe the fundamentals of optical systems and networks, including WDM network architecture, typical optical networking architecture, bit rates for synchronous/asynchronous optical communication systems, and wavelength regions for fiber-optics communications. In the same section, we briefly describe various multiplexing schemes such as time-division multiplexing (TDM), wavelength-division multiplexing (WDM), frequency-division multiplexing (FDM), subcarrier multiplexing (SCM), code-division multiplexing (CDM), polarization-division multiplexing (PDM), and spatial-division multiplexing (SDM). We also identify the main noise sources and optical channel impairments, in particular of importance for fiber-optics communications. After that the key optical devises and modules are identified and briefly described. In section related to wireless communications fundamentals, we describe a typical wireless communication point-to-point link and identify key devices/subsystems with brief description of operational principles. The multiaccess communications are then described with key problems being identified. Further, various effects affecting the received power are described including path loss, shadowing, and multipath fading. We also introduce various approaches to deal with multipath fading including orthogonal frequency-division multiplexing (OFDM), diversity, multiple-input multiple-output (MIMO) signal processing, equalization, and adaptive modulation and coding. The MIMO operation principle is briefly described next. Finally, the cellular concept is introduced. In section on organization of the book, the detailed description of chapters is provided. Final section provides some relevant concluding remarks.

1.1 Historical Perspective of Optical and Wireless Communication Systems

The earliest optical and wireless communications systems consisted of fire or smoke signals, signaling lamps, and semaphore flags to convey a single piece of information [1–4]. For example, a relatively sophisticated ancient communication system, along the Great Wall of China, was composed of countless beacon towers [5]. In this ancient communication system, the number of lanterns or the color of smoke was used as a means to inform the size of an invading enemy, which represents a crude form of simultaneously digital communication and multilevel signaling. By using the beacon towers, with the guards in each tower, positioned at regular distances along the Great Wall, a message could be transmitted from one end of the Great Wall to the other, more than 7300 km, in slightly more than 1 h [5]. Therefore, this ancient communication system has many similarities with today's relay or regeneration systems, in which the beacon towers can be considered as relays. Relay or regeneration systems were further studied by Claude Chappe in 1792 to transmit coded messages optically over distance of 100 km [1]. In next two subsections, we describe the evolution of both wireless communication systems [2, 33–37] and optical communication systems [1, 3–5, 38–42].

I. B. Djordjevic, *Advanced Optical and Wireless Communications Systems*, https://doi.org/10.1007/978-3-030-98491-5_1

1.1.1 Wireless Communication Historical Perspective

Electromagnetic wave propagation theory is developed in the 1860s and summarized by Maxwell and demonstrated later by Hertz. The first radio transmission was demonstrated in 1893 by Tesla, followed by Marconi in 1896. At that time, however, the systems were of low frequency, high power, huge size, expensive, and mostly mechanical. Interestingly enough, the telegraphic systems were digital, based on Morse code (1837), and the early steps of electronic age were driven by the need of wireless communications, rather than wired systems. Unfortunately, with the invention of amplitude-modulated (AM) radio by Fessenden and de Forest as well as invention of frequency-modulated (FM) radio by Armstrong and de Forest, the era of analog communication started instead. The development of communication systems was first boosted by the invention of vacuum tubes in 1906 by de Forest, which can be considered as the start of electronic age. Communication systems with vacuum tubes were huge, power hungry, and not stable. The real engine of electronic age started with the invention of transistor by William Shockley during 1948–1951, when the communication hardware get revolutionized for the second time. Another relevant revolution came in the same time, namely, the information theory was developed by Claude Shannon in 1948, and the golden age of information theory was from 1948 to 1960. Many sophisticated military radio systems were developed during and after WW2. The third revolution for communication hardware is the invention of integrated circuit (IC) in 1958 by Jack Kilby (independently by Robert Noyce in 1959), which was developed in the 1960s and enabled the exponential growth since then.

Another driving force of wireless systems was the invention of mobile phone, with the first trial being in 1915, between New York and San Francisco. The commercial networks started in 1946, while the takeoff was in the 1980s with the introduction of the cellular concept. Many wireless communication systems were proposed in the 1990s, with great failures around 2000. The *first generation (1G)* of cellular systems was introduced in the 1980s and was based on analog, employing the frequency-division multiplexing access (FDMA). The most popular 1G cellular systems were [2] advanced mobile phone service (AMPS) in North America and Australia; total access communication system (TACS) in the UK; Nordic mobile telephone (NMT) in the Nordic countries, Switzerland, the Netherlands, Eastern Europe, and Russia; and C-450 in West Germany, Portugal, and South Africa, to mention a few.

The *second generation (2G)* of cellular systems was introduced in 1991, based on digital communication principles. In addition to voice, they supported the short message service (SMS) as well as low data rate transmissions (including multimedia messaging service or MMS). The 2G technologies can be classified into two broad categories, depending on the multiplexing method: time-division multiple access (TDMA)-based and code-division multiple access (CDMA)-based standards. The most popular standards in 2G are (i) global system for mobile communications (GSM), which is TDMA-based, originally used in Europe but later in most of the world except North America; (ii) IS-95, which was CDMA-based, invented by Qualcomm and later adopted by the Telecommunications Industry Association (TIA); (iii) Japanese digital cellular (JDS or PDC), which was TDMA-based; and (iv) IS-136 (also known as digital AMPS or D-AMPS), TDMA-based. Regarding the data rates supported, with (1) general packet radio service (GPRS), there is a theoretical maximum data rate of 50 kbit/s, and (2) in enhanced data rates for GSM evolution (EDGE), there is a theoretical maximum data rate of 1 Mbit/s.

The first *3G networks* were introduced in 1998, and 3G technology is based on a set of standards that comply with the International Mobile Telecommunications-2000 (IMT-2000) specifications by the International Telecommunication Union (ITU). To meet the IMT-2000 standards, a 3G system is required to provide peak data rates of at least 200 kb/s. However, many 3G provide higher data rates than the minimum technical requirements for a 3G service. In particular, 3.5G and 3.75G releases provide mobile broadband access of several Mb/s to smartphones and laptop mobile modems. The most relevant 3G standards are (1) Universal Mobile Telecommunications Service (UMTS), standardized by 3GPP, used primarily in Europe, Japan, and China as well as other regions predominated by GSM 2G system infrastructure, and (2) the CDMA2000 system, standardized by 3GPP2, used in North America and South Korea, sharing infrastructure with the IS-95 2G standard. The UMTS can support the data rates up to 2 Mb/s for stationary/walking users and 348 kb/s for moving vehicles. The High-Speed Packet Access (HSPA) technology specifications, representing an intermediate phase toward 4G, initially supported increased peak data rates of up to 14 Mb/s in the downlink and 5.76 Mb/s in the uplink. This technology contains two mobile protocols High-Speed Downlink Packet Access (HSDPA) and High-Speed Uplink Packet Access (HSUPA). The predecessors of 4G technology are (i) 3GPP Long-Term Evolution (LTE) offering 100 Mbit/s in the downlink and 50 Mbit/s in the uplink over 20 MHz, (ii) mobile WiMAX (IEEE 802.16e-2005) offering up to 128 Mb/s in the downlink and 56 Mb/s in the uplink over 20 MHz of channel bandwidth, (iii) UMB (formerly EV-DO Rev. C), and (iv) fast low-latency access with seamless handoff OFDM (also known as FLASH-OFDM). The bandwidth as well as location information available to 3G devices gives rise to some new applications including global positioning system (GPS), location-based services, mobile TV, telemedicine, video conferencing, and video on demand.

The *fourth generation (4G)* of mobile communications technologies provides a comprehensive and secure all-IP-based solution, with current/potential applications including mobile web access, IP telephony, gaming services, high-definition mobile TV, video conferencing, 3D television, and cloud computing. ITU-R specified in Mar. 2008 the set of requirements for 4G standards, named International Mobile Telecommunications-Advanced (IMT-Advanced) specification, setting peak speed requirements for 4G service at 100 Mbit/s for high-mobility communication links (such as from trains and cars) and 1 Gbit/s for low-mobility communication links (such as pedestrians and stationary users). IMT-2000 compliant 4G standards are LTE Advanced and WirelessMAN-Advanced (IEEE 802.16 m). The LTE Advanced is essentially an enhancement to LTE, with the coordinated multipoint transmission that allows for higher system capacity and helps handling the enhanced data speeds. The peak download data rate is 1 Gb/s, while the peak upload data rate is 500 Mb/s. On the other hand, the IEEE 802.16 m represents the evolution of 802.16e offering 1 Gb/s data rate for stationary reception and 100 Mb/s for mobile reception. Principal 4G technologies can be summarized as follows: (i) multi-antenna and multi-user (MIMO) concept; (ii) frequency-domain equalization and frequency-domain statistical multiplexing; (iii) turbo codes which are used as channel codes to reduce the required SNR at Rx side; (iv) channel-dependent scheduling which is used to utilize the time-varying channel; (v) link adaptation which is used based on adaptive modulation and error-correcting codes; and (vi) relaying which is used including fixed relay networks (FRNs) and the cooperative relaying concept.

The *fifth generation (5G)* wireless systems go beyond the 4G IMT-Advanced standards. The 5G standard should fulfill the following Next Generation Mobile Networks Alliance requirements [6, 7]: (1) supporting tens of thousands of users with data rates of tens of Mb/s; (2) for metropolitan areas, data rates of 100 Mb/s should be supported; (3) the workers located on the same floor should get 1 Gb/s simultaneously; (4) for wireless sensor networks, we should be able to establish several hundreds of thousands of simultaneous connections; (5) significantly higher spectral efficiency compared to 4G; (6) the coverage must be improved; (7) latency must be reduced significantly compared to LTE; and (8) the signaling efficiency must be improved. Toward these goals, the FCC recently approved the new spectra for 5G, including the 28 GHz, 24 GHz, 37 GHz, 39 GHz, and 47 GHz bands [8, 9]. The FCC is working on freeing 2.75 GHz for 5G applications in the 26 and 42 GHz bands and has plans for additional millimeter wave spectrum in the 70/80/90 GHz bands [9]. FCC also considers allocating 600 MHz bandwidth in 2.5 GHz, 3.5 GHz, and 3.7–4.2 GHz bands [9]. Key enabling technologies for 5G include [18] (i) the employment of a wireless software-defined network (WSDN), (ii) the use of network function virtualization (NFV), (iii) introduction of massive MIMO, (iv) much higher network density, (v) the utilization of the millimeter wave spectrum, (vi) the breakthroughs in big data and mobile cloud computing, (vii) high mobility device-to-device connections, (viii) better energy-efficiency communications, (ix) the introduction of new radio access technologies, and (x) scalable Internet of Things techniques. The main applications, as specified by the ITU-R, for the enhanced capabilities of 5G are Enhanced Mobile Broadband (eMBB), Ultra Reliable Low Latency Communications (URLLC), and Massive Machine Type Communications (mMTC). The ITU's IMT-2020 standard requires a peak download speed of 20 Gb/s with spectral deficiency up to 30 bits/s/Hz and 10 Gb/s upload speed with spectral efficiency up to 15 bits/s/Hz [10, 11]. The target values for the user experienced data rates in downlink and uplink are 100 Mb/s and 50 Mb/s, respectively. The user plane latency minimum requirements are 4 ms for eMBB and 1 ms for URLLC [10]. Another relevant standard air interface for the 3GPP 5G networks is the 5G NR (New Radio) [12]. The 5G NR employs two frequency bands: frequency range 1 (FR1) 410 MHz–7.125 GHz and frequency range 2 (FR2) 24.25–52.6 GHz. Early FR1 deployments employ 5G NR software on 4G hardware, which are slightly better than newer 4G systems, and essentially represent 4.5G technology. The 5G-Advanced standard is currently under development by 3GPP [13].

The *6th generation (6G)* wireless systems and standards are currently under development. It is expected that the 6G networks will be more diverse than the 5G networks and these networks will be able to support various applications beyond current mobile applications including virtual and augmented reality, remote surgery, holographic projection, pervasive intelligence, the IoT, and instant communications, to mention few [14–17]. The peak data rates in 6G could be up to 10 Tb/s for downlink, with much larger spectral efficiencies compared to 5G. The user plane latencies should go down to 0.1 ms. To achieve those goals, in addition to mm-wave technologies, the THz and free-space optical (FSO) communications must be exploited.

Regarding the *wireless LANs (WLANs)*, the most relevant is the IEEE 802.11 standard family, representing the set of MAC and PHY standards implementing WLAN computer communications in 0.9, 2.4, 3.6, 5, and 60 GHz frequency bands [19–24]. The first wireless networking standard from this family was 802.11–1997; however, the 802.11b was the widely accepted one, which was followed by 802.11a, 802.11 g, 802.11n, and 802.11 ac. Other standards in this family (such as c–f, h, j) represent the service amendments to the existing ones that that are used to extend the scope, or eventually the amendments represent the corrections to the previous specification. The IEEE 802.11 g PHY standard is very similar to 801.11a, except that it is related to the 2.4 GHz unlicensed ISM band (2.4–2.497 GHz).

To increase the aggregated data rates in IEEE 802.11n and 802.11 ac PHY standards, the MIMO-OFDM concept, explained in Chap. 8, is used. Namely, through MIMO concept, several independent data streams can be simultaneously transmitted increasing the aggregate data rates. The IEEE 802.11n PHY standard is applicable to both 2.4 and 5 GHz bands while 802.11 ac to 5 GHz band only. In IEEE 802.11n standard, the available channel bandwidths are 20 MHz and 40 MHz, while in 802.11 ac (Wi-Fi 5), the available channel bandwidths are 20, 40, 80, and 160 MHz. The highest data rate in 802.11 ac standard, for 160 MHz bandwidth, is 780 Mb/s. Recent 802.11ax (Wi-Fi 6E) standard, also known as a high-efficiency Wi-Fi, adopted in 2020, is designed to operate in license-exempt bands (1–7.125 GHz), including commonly used 2.4 and 5 GHz as well as the 6 GHz band (5.925–7.125 GHz in the USA) [25]. The main goal of Wi-Fi 6E standard is to enhance throughput/area for high-density scenarios such as shopping malls, corporate offices, and dense residential areas. The modulation formats range from BPSK to 1024-QAM, with code rates from 1/2 to 5/6. The data rates range from 135 Mb/s (for 20 MHz channels) to 1.201 Gb/s (for 120 MHz channels).

The *WiMAX* represents one of the most popular broadband wireless access (BWA) technologies, aiming to provide high-speed bandwidth access for wireless MANs [21, 26–28]. The WiMAX is based on IEEE 802.16 set of standards, in particular IEEE 802.16–2004 (fixed-WiMAX) and IEEE 802.16e-2005 (mobile-WiMAX) standards, which provide multiple PHY and MAC options. The WiMAX represents the last mile wireless broadband access alternative to the cable and DSL services, and it is a direct competitor to the LTE Advanced standard. The IEEE 802.16e-2005 provides several improvements compared to the IEEE 802.16–2004 including [26–28] (1) providing support for mobility by enabling the soft and hard handover between base stations; (2) introducing the concept of scalable OFDM (SOFDM) by scaling the FFT to the channel bandwidth to ensure that the carrier spacing is constant across different channel bandwidths, ranging from 1.25 MHz through 5 MHz and 10 MHz all the way to 20 MHz and thus improving the spectral efficiency; (3) introducing the advanced antenna diversity schemes and hybrid automatic repeat request (HARQ); (4) introducing the adaptive antenna systems (AAS) and MIMO signal processing; (5) providing the denser sub-channelization and therefore improving the indoor penetration; (6) introducing downlink sub-channelization, in order to trade the coverage for capacity or vice versa; (7) introducing the LDPC coding to improve the error correction strength; and (8) introducing an extra QoS class for VoIP applications. In September 2018 the IEEE 802.16–2017 standard was released, relevant for combined fixed and mobile point-to-multipoint BWA systems providing multiple services [29]. The IEEE 802.16 Working Group ceased their activities on 9 March 2018 [30].

1.1.2 Optical Communication Historical Perspective

Thanks to the success of telegraphy, telephony, and radio communications in the first half of the twentieth century, the optical communication systems were actually forgotten. However, in the late twentieth century, different communication systems came to saturation in terms of reach and capacity. For instance, a typical coaxial cable-based transport system operating at 155 Mb/s requires the regeneration at approximately every 1 km, which is costly to operate and maintain. The natural step was to study the optical communication systems, which can dramatically increase the total capacity. The research in optical communication was boosted upon demonstration of a laser principle [31]. The first step was to fabricate an appropriate optical transmission medium. Kao and Hockman [32] proposed to use the optical fiber as the medium, although at the time it had unacceptable fiber loss. Their argument was that attenuation mostly was coming from impurities, rather than any fundamental physical effect such as Rayleigh scattering, which could be reduced by improving the fabrication process. Their prediction was that an optical fiber with attenuation of 20 dB/km should be sufficient for telecom applications, which surprisingly was developed within 5 years since initial proposal, by researchers from Cornell. This invention opens up opportunities for development of fiber-optics communication systems. Since then several generations of optical communication systems were developed. The optical communication evolution path can be summarized as follows.

The *first generation* appeared in the 1970s, and the operating wavelengths were 0.8 μm–0.9 μm with data rates ranging from 45 Mb/s to 500 Mb/s. Repeater spacing was 10 km, which was much greater than that for comparable coax systems. Lower installation and maintenance costs resulted from fewer repeaters.

The second generation, which was focused on a transmission near 1.3 μm to take advantage of the low attenuation (< 1 dB/km) and low dispersion, was deployed during the early 1980s. Sources and detectors were developed that use InGaAsP semiconductor. The bit rate of these systems was limited to <100 Mb/s due to dispersion in multimode fibers (MMFs). *Single-mode fiber* (SMF) was then incorporated. By 1987 the second-generation systems were operating at 2.5 Gb/s at 1.3 μm with repeater spacing of 50 km.

The third-generation systems were based on the use of 1.55 μm sources and detectors and dominated in the first half of the 1990s. At this wavelength the attenuation of fused silica fiber is minimal. The deployment of these systems was delayed however due to the relatively large dispersion at this wavelength. Two approaches were proposed to solve the dispersion problem. The first approach was to develop single-mode lasers, and the second was to develop dispersion-shifted fiber (DSF) at 1.55 μm. In 1990, 1.55 μm systems operating at 2.5 Gb/s were commercially available and were capable of operating at 10 Gb/s for distances of 100 km [1, 3–5, 38–40]. The best performance was achieved with DSFs in conjunction with single-mode lasers. A *drawback* of these systems was the need for *electronic regeneration* with repeaters typically spaced every 60–70 km. Coherent optical detection methods were investigated in late 1980s and early 1990s to increase receiver sensitivity. However, this approach was superseded by the development of the optical amplifier.

The *3.5th-generation* systems are based on the use of *optical amplifiers* to increase repeater spacing and *wavelength-division multiplexing* (WDM) to increase the aggregate bit rate. Erbium-doped fiber amplifiers (EDFAs) were developed to amplify signals without electronic regeneration during the 1980s [1, 3–5, 38–40]. In 1991 signals could be transmitted 14,300 km at 5 Gb/s without electronic regeneration [1, 3–5, 38–40]. The first transpacific commercial system went into operation sending signals over 11,300 km at 5 Gb/s, and other systems are being deployed. System capacity is increased through use of WDM. Multiple wavelengths can be amplified with the same optical amplifier. In 1996 20 × 5 Gb/s signals were transmitted over 9100 km providing a total bit rate of 100 Gb/s and a bandwidth-length (B-L) product of 910 (Tb/s)-km. In these broadband systems, dispersion becomes an important issue to be addressed.

In the *3.75th-generation* systems, the effort is primarily concerned with the fiber dispersion problem. Optical amplifiers solve the loss problem but increase the dispersion problem since dispersion effects accumulate over multiple amplification stages. An ultimate solution is based on the novel concept of *optical solitons* [1, 43, 44]. These are pulses that preserve their shape during propagation in a loss less fiber by counteracting the effect of dispersion through fiber nonlinearity. Experiments using stimulated Raman scattering as the nonlinearity to compensate for both loss and dispersion were effective in transmitting signals over 4000 km [1, 43, 44]. EDFAs were first used to amplify solitons in 1989 [1, 43, 44]. By 1994 a demonstration of soliton transmission over 9400 km was performed at a bit rate of 70 Gb/s by multiplexing seven 10 Gb/s channels [1, 43, 44]. In parallel, dispersion compensating fibers (DCFs) were invented to deal with chromatic dispersion, and various dispersion maps were proposed [1, 43, 44]. The WDM channel count increased to maximum 128, with data rates per single wavelength ranging from 2.5 Gb/s to 10 Gb/s. The operating wavelength region ranges from 1530 nm to 1560 nm. Regarding the networking aspects, WDM multiplexers/demultiplexers as well as optical add-drop multiplexers (OADMs) have been used. Regarding the optical networking protocols, the following protocols have been used: SONET/SDH/OTN, IP, ATM, ESCON/FICON, and Ethernet. The readers interested learning more about various optical networking protocols are referred to [45–59].

In the *fourth-generation* systems, the efforts have been directed toward realizing greater capacity of fiber systems by multiplexing a large number of wavelengths. These systems are referred to as dense wavelength-division multiplexing (DWDM) systems, with corresponding channel spacing being either 100 GHz or 50 GHz. Controlling wavelength stability and the development of wavelength demultiplexing devices are critical to this effort. The maximum DWDM channel count is 160, with the data rates per single wavelength being 10 Gb/s and/or 40 Gb/s, and the corresponding lasers are tunable. In addition to EDFAs, the Raman amplifiers have been used as well. Regarding the networking aspects, DWDM multiplexers/demultiplexers, reconfigurable OADMs (ROADMs), and optical cross-connects (OXCs) have been used. Additionally, the layered control plane is introduced. Regarding the optical networking protocols, the following protocols have been used: SONET/SDH/OTN, IP, MPLS, and Ethernet.

The *fifth-generation systems* are related to optical transmission systems with data rates per wavelength ranging from 100 Gb/s to 1 Tb/s, by employing various multilevel modulation and channel coding schemes, polarization multiplexing, digital signal processing (DSP), and coherent optical detection. The orthogonal frequency-division multiplexing (OFDM) appears to be an excellent candidate to deal with chromatic dispersion and polarization mode dispersion (PMD). Additionally, the MIMO signal processing is applied as well. Regarding the optical networking protocols, the following protocols have been used: OTN, IP, MPLS, and Ethernet. Regarding the standardization activities, ITU-T, IEEE 802.3ba, and OIF have completed the work on 100 Gb/s Ethernet (100 GbE) [60, 61], while the activities for adoption of 400 Gb/s Ethernet, 1 Tb/s Ethernet, and beyond are currently underway.

The systems with channel rates beyond 10 Tb/s over SMFs, such as those described in [62], can be placed into the *sixth generation*, and these systems are currently in research phase. Finally, the systems based on spatial-division multiplexing (SDM) [62, 63] (and references therein), employing various SDM fibers including few-mode fiber (FMF), few-core fiber (FCF), and few-mode-few-core fiber (FMFC), to mention a few, can be placed into the *seventh generation*.

1.2 Optical Communications Systems and Networks Fundamentals

A generic WDM optical network, which can be used to identify the key optical components, concepts, and system parameters, is provided in Fig. 1.1. The end-to-end optical transmission involves both electrical and optical signal paths. To perform conversion from electrical to optical domain, the optical transmitters are used, while to perform conversion in the opposite direction (optical-to-electrical conversion), the optical receivers are used. The singe-mode fiber (SMF) serves as a foundation of an optical transmission system because the optical fiber is used as medium to transport the optical signals from the source to destination. The optical fibers attenuate the signal during transmission, and someone has to use optical amplifiers, such as EDFAs, Raman amplifiers, or parametric amplifiers, to restore the signal level. Unfortunately, the amplification process is accompanied with the noise addition. For better exploitation of enormous bandwidth of SMF, the WDM concept is introduced, which corresponds to the scheme with multiple optical carriers at different wavelengths that are modulated by using independent electrical bit streams, as shown in Fig. 1.1, and then transmitted over the same SMF. During transmission of WDM signals, occasionally several wavelengths have to be added/dropped, which is performed by the reconfigurable optical add-drop multiplexer (ROADM), as shown in Fig. 1.1. The optical networks require the switching of information among different fibers, which is performed by the optical cross-connect (OXS). To combine several distinct wavelength channels into composite channel, the wavelength multiplexers are used. On the other hand, to split the composite WDM channel into distinct wavelength channels, the wavelength demultiplexers is used. To impose the information signal and perform electro-optical conversion, the optical modulators are used. The optical modulators are commonly used in combination with semiconductor lasers.

The optical transmission systems can be *classified* according to different criterions. When *bit rate* is used as classification criteria, the optical transmission systems can be classified as low-speed (tens of Mb/s), medium-speed (hundreds Mb/s), high-speed (Gb/s), and ultra-high-speed (tens of Gb/s). From *application perspective* point of view, the systems can be either power budget (loss) limited or bandwidth (transmission speed) limited. If *transmission length* is used for classification, we can identify very short reach (hundreds of meters), short reach (from several kilometers to several tens of km), long reach (hundreds of kilometers), and ultra-long reach (thousands of kilometers) optical transmission systems.

To provide a global picture, we describe a typical optical network shown in Fig. 1.2. We can identify three ellipses representing the *core network*, the *edge network*, and the *access network* [3, 4]. The long-haul core network interconnects big cities, major communications hubs, and even different continents by means of submarine transmission systems. The core networks are often called the wide area networks (WANs) or interchange carrier networks. The edge optical networks are deployed within smaller geographical areas and are commonly recognized as metropolitan area networks (MANs) or local exchange carrier networks. The access networks represent peripheral part of optical network and provide the last-mile access or the bandwidth distribution to the individual end users.

The ultimate goal of an optical signal transmission system is usually defined as achieving desired bit error rate (BER) performance between two end users or between two intermediate nodes in network reliably and at affordable cost. In order to achieve so, an optical transmission system needs to be properly designed, which includes the management of key optical communication systems engineering parameters. These parameters can be related to power, time, and wavelength or be interrelated. The parameters related only to power are power level, fiber loss, insertion loss, and extinction ratio (the ratio of powers corresponding to bit "1" and bit "0"). The parameters related only to time are jitter, first-order PMD, and bit/data rate. The parameters related to wavelength include optical bandwidth and wavelength stability. The parameters, signal

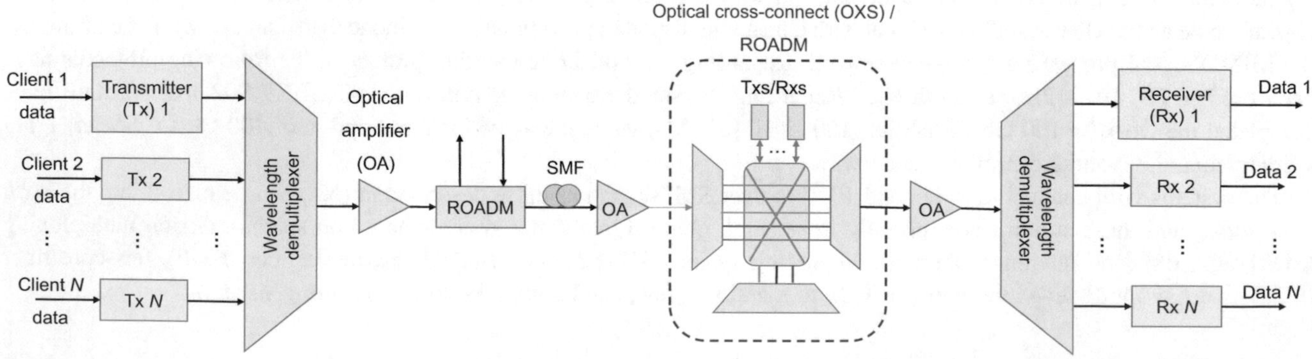

Fig. 1.1 A generic WDM optical network identifying key optical components, concepts, and parameters

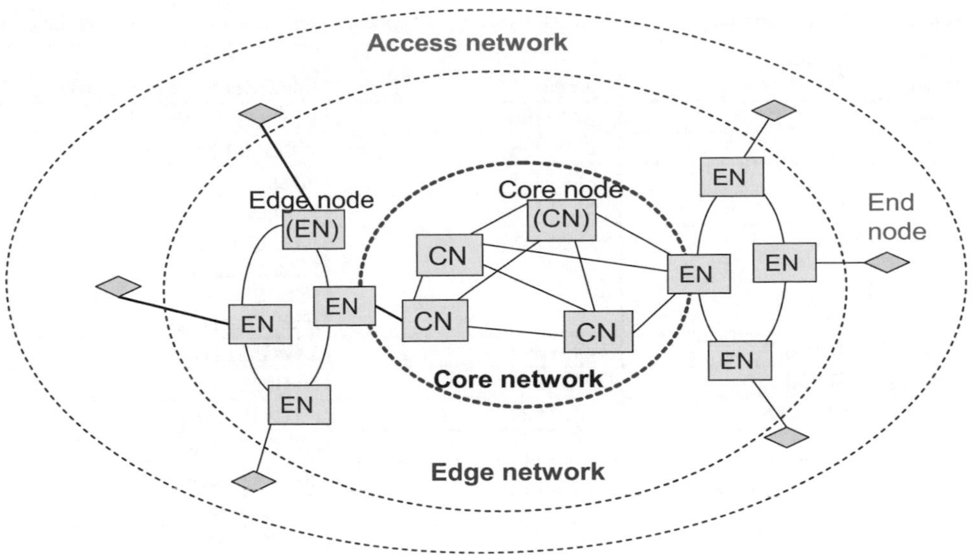

Fig. 1.2 A typical optical networking architecture

impairments, and additive/multiplicative noise sources related to both power and wavelength are optical amplifier gain, optical noise (such as amplified spontaneous emission (ASE) noise), different crosstalk effects, four-wave mixing (FWM), and stimulated Raman scattering (SRS). The parameters related to time and wavelength are laser chirp, second-order PMD, and chromatic dispersion. The parameters related to power and time are BER, modulation format, polarization-dependent loss (PDL), and quantum noise. Finally, the channel impairments related to time, power, and wavelength simultaneously are self-phase modulation (SPM), cross-phase modulation (CPM), and stimulated Brillouin scattering (SBS). Any detection scenario must include electronic noise, such as thermal noise, which is associated with receiver design. These different parameters, noise sources, and channel impairments are subject of investigation in Chap. 2.

Different high-speed optical transmission *enabling technologies* can be related to the usage of novel/better *devices*, such as Raman and parametric amplifiers, PMD and chromatic dispersion compensators, and modulators, or be related to the novel *methods*, such as advanced modulation formats (various multilevel modulation schemes with both direct and coherent optical detections and OFDM), forward error correction (FEC), coded modulation, constrained (modulation/line) coding, advanced detection schemes including maximum likelihood sequence detection/estimation (MLSD/E) and maximum a posteriori probability (MAP) detection (BCJR algorithm-based equalizers), and various multiplexing schemes including polarization multiplexing, optical time-division multiplexing (OTDM), subcarrier multiplexing (SCM), code-division multiplexing (CDM), OFDM, and spatial-division multiplexing. These various enabling technologies will be described in incoming chapters of the book.

An important concept to be introduced here is related to the so called *lightwave path*, which can be defined as the trace that optical signal passes between the source and destination without experiencing any opto-electrical-opto (O-E-O) conversion [3, 4]. Generally speaking, the lightwave paths may differ in lengths and in the information capacity that is carried along and can traverse though different portions of an optical network. The lightwave path can be considered as bandwidth wrapper for lower speed transmission channels, which form virtual circuit services [3, 4]. The time-division multiplexing (TDM) technique is applied to aggregate the bandwidth of virtual circuits before it is wrapped in the lightwave path. TDM of virtual circuits can be either *fixed* (each circuits receives a guaranteed amount of the bandwidth, a bandwidth pipe) or *statistical* (in packet switching the data content is divided into data packets, which can be handled independently). The fixed multiplexing of virtual circuits is defined by SONET/SDH standards. Bit rates of different bandwidth channels, for both synchronous and asynchronous transmission, are given in Table 1.1 based on [3, 4] (and references therein).

Optical fiber is the key ingredient of an optical transmission system because it has much wider available bandwidth, lower signal attenuation, and smaller signal distortions compared to any other wired or free-space physical media. The total bandwidth is approximately 400 nm, or around 50 THz, when related to the wavelength region with fiber attenuation being below 0.5 dB/km. The usable optical bandwidth is commonly split into several wavelength bands, as summarized in Table 1.2 based on refs. [3, 4, 40]. The bands around the minimum attenuation point, usually referred to as C and L bands, are the most suitable for high channel count DWDM transmission. The wavelength region around 1300 nm is less favorable for optical signal transmission because signal attenuation is higher than attenuation in S, C, and L bands. On the other hand, it is quite suitable for CATV signals, and the course-WDM (CWDM) technique is usually employed in this region.

Table 1.1 Bit rates for different synchronous/asynchronous optical communication systems (both introduced and envisioned)

TDM/synchronous channels	Bit rate	Data/asynchronous channels	Bit rate
DS-1	1.544 Mb/s	10-BaseT Ethernet	10 Mb/s
E-1	2.048 Mb/s	100-BaseT Ethernet	100 Mb/s
OC-1	51.84 Mb/s	FDDI	100 Mb/s
OC-3 ⇔ STM-1	155.52 Mb/s	ESCON	200 Mb/s
OC-12 ⇔ STM-4	602.08 Mb/s	Fiber Channel-I Fiber Channel-II Fiber Channel-III Fiber Channel IV	200 Mb/s 400 Mb/s 800 Mb/s 4 Gb/s
OC-48 ⇔ STM-16	2.488 Gb/s	1 Gb Ethernet	1 Gb/s
OC-192 ⇔ STM-64	9.953 Gb/s	10 Gb Ethernet	10 Gb/s
OC-768 ⇔ STM-256	39.813 Gb/s	40 Gb Ethernet	40 Gb/s
		100 Gb Ethernet	107–125 Gb/s
		400 Gb Ethernet	428–500 Gb/s
		1 Tb Ethernet	1.07–1.25 Tb/s
		4 Tb/10 Tb Ethernet.	4 Tb/s 10 Tb/s

Table 1.2 The wavelength bands for fiber-optics communications

Wavelength band	Descriptor	Wavelength range (nm)
O-band	Original	1260–1360
E-band	Extended	1360–1460
S-band	Short	1450–1530
C-band	**Conventional**	**1530–1565**
L-band	Long	1565–1625
U-band	Ultra-long	1625–1675

The key optical components, which will be described in next chapter, can be classified as follows: (i) semiconductor light sources including light-emitting diodes (LEDs) and semiconductor lasers (Fabry-Pérot (FP), distributed feedback (DFB), distributed Bragg reflector (DBR), vertical cavity surface emitting (VCSEL), tunable lasers (external cavity laser, mutilaser chip, three-section tunable)); (ii) optical modulators including both direct optical modulators and external modulators (Mach-Zehnder modulator (MZM), electro-absorption modulator, and I/Q modulator); (iii) optical fibers (MMFs and SMFs); (iv) optical amplifiers such as semiconductor optical amplifier (SOA), EDFA, Raman amplifiers, and parametric amplifiers; (v) photodiodes including PIN, avalanche photodiodes (APDs), and metal-semiconductor-metal (MSM) photodetectors; and (vi) various optical components (optical isolators, optical circulators, optical filters, optical couplers, optical switches, and optical multiplexers/demultiplexers).

A monochromatic electromagnetic wave, which is commonly used as a signal carrier, can be represented through its electric field as $\boldsymbol{E}(t) = \boldsymbol{p}A\cos(\omega t + \phi)$ (A-amplitude, ω-frequency, ϕ-phase, $\boldsymbol{p}$-polarization orientation), for which each parameter can be used to impose the message signal. If the message signal is analog, corresponding modulation formats are amplitude modulation (AM), frequency modulation (FM), phase modulation (PM), and polarization modulation (PolM). On the other hand, when the modulating signal is digital, then the carrier signal duration is limited to symbol duration, and corresponding modulation formats are amplitude shift keying (ASK), frequency shift keying (FSK), phase shift keying (PSK), and polarization shift keying (PolSK).

In order to better utilize the enormous bandwidth of the optical fiber, we have to transmit simultaneously many channels over the same bandwidth through *multiplexing*. The commonly used methods of multiplexing in optical communications are given below as follows:

- *Wavelength-division multiplexing* (WDM) is already introduced above.
- *Time-division multiplexing* (TDM), in which many lower-speed signals are time-interleaved to generate a high-speed signal. The multiplexing can be performed either in electrical domain, when is known as electrical TDM (ETDM), or in optical domain, when is known as optical TDM (OTDM).

- *Frequency-division multiplexing* (FDM), in which continuous-wave (CW) modulation is used to translate the spectrum of the message signal into a specific frequency slot of the passband of optical channel. The optical version of FDM is commonly referred to as WDM.
- *Orthogonal frequency-division multiplexing* (OFDM) is a particular version of FDM in which the orthogonality among subcarriers is obtained by providing that each sub-carrier has exactly an integer number of cycles in the symbol interval. The number of cycles between adjacent subcarriers differs by exactly one.
- *Subcarrier-multiplexing* (SCM) is again a particular version of FDM in which different independent data streams are first microwave multiplexed and then transmitted using the same wavelength carrier.
- *Code-division multiplexing* (CDM) in which each message signal is identified by a unique *signature sequence* ("code"), with signature sequences being orthogonal to each other.
- *Spatial-division multiplexing* (SDM) in which independent signals excite different spatial modes in FMFs/FCFs.

During the transmission over an optical fiber, the transmitted signal is impaired by various noise sources and channel impairments. The noise sources can be additive in nature (dark current noise, thermal noise, ASE noise, and crosstalk noise) or be multiplicative in nature (mode partition noise (MPN), laser intensity noise (RIN), modal noise, quantum shot noise, and avalanche shot noise). Different channel impairments can be related to fiber attenuation, insertion loss, dispersion effects, or fiber nonlinearities. Fiber attenuation originates from material absorption, which can be intrinsic (ultraviolet, infrared) or extrinsic (water vapor, Fe, Cu, Co, Ni, Mn, Cr, various dopants: GeO_2, P_2O_5, B_2O_3), Rayleigh scattering and waveguide imperfections (Mie scattering, bending losses, etc.). The dispersion effects can originate from intermodal (multimode) dispersion (in MMFs), chromatic dispersion (material and waveguide dispersion effects present in SMF), PMD, and polarization-dependent loss (PDL). The fiber nonlinearities can originate from non-elastic scattering effects (SBS, SRS) or Kerr nonlinearities (SPM, XPM, FWM). Various noise sources and optical channel impairments are described in Chap. 2.

1.3 Wireless Communications Systems Fundamentals

A typical, generic, point-to-point single-input single-output (SISO) wireless communication system, used in various wireless applications, is provided in Fig. 1.3. The binary data sequence is first channel encoded [see Fig. 1.3(a)], and the encoder output is written into interleaver. Typical channel codes include convolutional codes, concatenated convolutional and RS codes, turbo codes, or LDPC codes [3, 35, 64–66]. For turbo coding s-random interleaver can be used, while for LDPC coding,

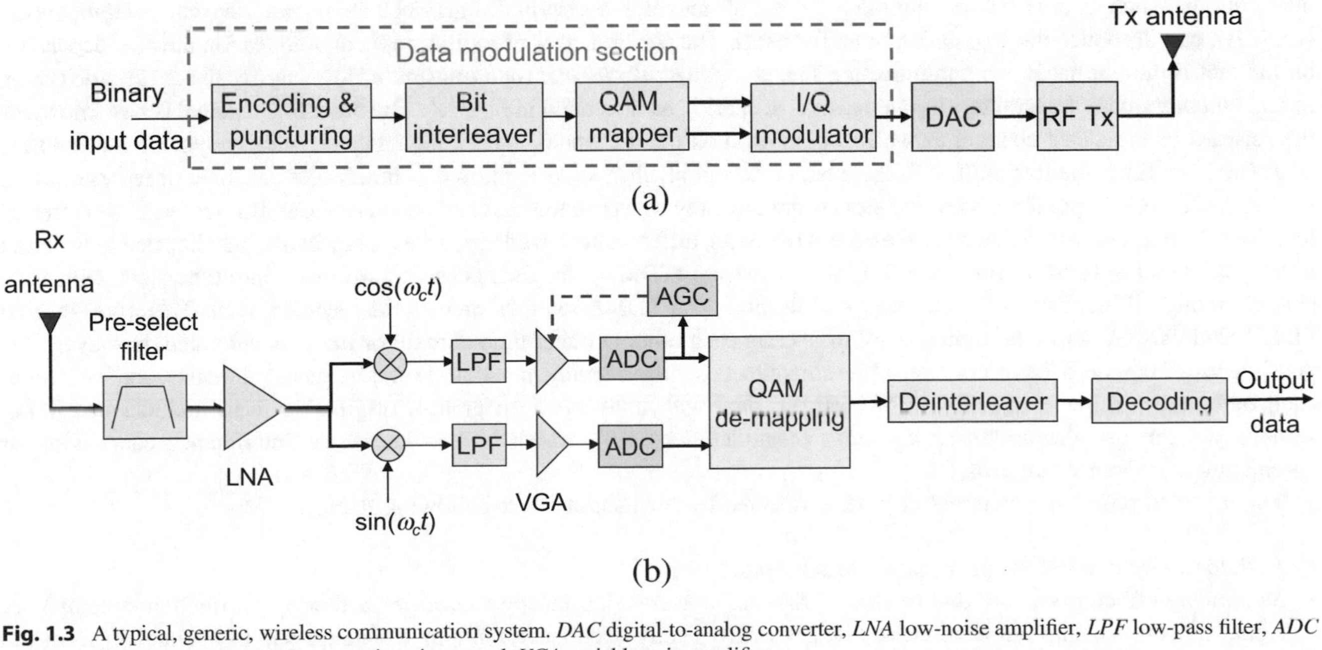

Fig. 1.3 A typical, generic, wireless communication system. *DAC* digital-to-analog converter, *LNA* low-noise amplifier, *LPF* low-pass filter, *ADC* analog-to-digital converter, *AGC* automatic gain control, *VGA* variable-gain amplifier

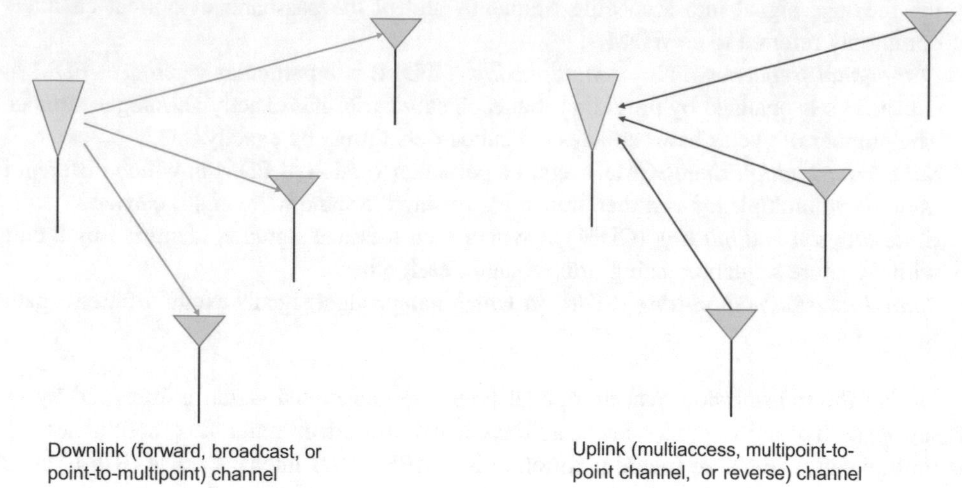

Fig. 1.4 Illustration of downlink and uplink channels

properly designed block interleaver can be used [67]. For block interleaver, $\log_2(M)$, where M is signal constellation size, codewords are written in row-wise fashion into interleaver. Notice that arbitrary two-dimensional (2D) constellation can be used. Then $\log_2(M)$ bits are taken into from the interleaver and used to select a point from 2D constellation such as QAM, by applying appropriate mapping rule. After digital-to-analog conversion (DAC), the corresponding RF signal is upconverted, amplified, and transmitted by using a transmit antenna.

On receiver side, shown in Fig. 1.3(b), the signal from receive antenna is pre-filtered to select the desired frequency band, followed by low-noise amplification, down conversion, and analog-to-digital conversion (ADC). The automatic gain control (AGC) ensures that the signal level at the output of variable gain amplifier (VGA) is constant. After that the QAM de-mapping takes place, followed by de-interleaver and channel decoder. The purpose of interleaving process is to spread the burst errors due to fading effects so that the channel code deals only with random errors. When soft-decision decoding is employed, such as in either turbo or LDPC coding, the QAM de-mapping block is in fact the a posteriori probability (APP) de-mapper, where symbol log-likelihood ratios (LLRs) are calculated, followed by bit LLRs calculation that gets passed to the soft-decision decoder, being either turbo decoder or LDPC decoder [3, 35, 64–67].

In wireless communications, multiple users can share the same spectrum, which is commonly referred to as *multiaccess communications*. Multiaccess communications suffer from *multiaccess interference*, which can be both intentional and unintentional, such as crosstalk or multipath fading. Moreover, the system design, such as in *code-division multiple access* (CDMA), can introduce the crosstalk among the users. The performances of multiaccess communication are also dependent on the fact if the multiaccess communication link is a *downlink channel* (one transmitter to many receivers scenario) or an *uplink channel* (many transmitters to one receiver scenario), as illustrated in Fig. 1.4. The downlink channel is also known as the forward or broadcast channel as well as point-to-multipoint channel, and all signals are typically synchronous as they originate from the same transmitter. In contrast, in the uplink channel, also known as multiaccess channel or reverse link as well as multipoint-to-point channel, the signals are typically asynchronous as they originate from different users at different locations. In frequency-division multiple access (FDMA), different non-overlapped frequency bands are allocated to different users. On the other hand, in time-division multiple access (TDMA), the time scale is partitioned into time slots, which are shared among different users in a round-robin fashion. The TDMA requires precise time synchronism. Therefore, in both FDMA and TDMA, users are required not to overlap each other in either time domain or frequency domain. However, it is possible users' signals to be orthogonal while simultaneously overlapping in both time and frequency domains, and we refer to such multiaccess scheme as CDMA [68, 69]. To deal with multiuser interference, originating from multiple interfering signals, we can use the *multiuser detection (demodulation)* [69], which is also known as interference cancelation or co-channel interference suppression.

The received power in a wireless channel is affected by attenuation due to following factors:

- *Path loss*, representing the propagation loss average.
- *Shadowing* effect, occurring due to slow variations in attenuation mostly caused by scattering obstruction structures. A signal transmitted through a wireless channel will typically experience random variations due to attenuation from the objects in the signal path, giving rise to random the variations of the received power at a given distance.

- *Multipath fading*, representing rapid random fluctuations in received power due to multipath propagation and Doppler frequency shift. In urban/indoor environment, a radio signal transmitted from a fixed source will encounter multiple objects that produce reflected, diffracted, or scattered copies of transmitted signal, known as *multipath signal components*.

Because of multipath propagation effect, when a single pulse is transmitted over the multipath channel, the received signal will appear as a pulse train rather than a single pulse, so that the receiver will experience variations in received power. The number of multipath components is random, while the *i*-th multipath component can be characterized by random amplitude $a_i(t)$, random delay $\tau_i(t)$, random Doppler frequency ν_{Di} [Hz], and random phase shift $\phi_i(t) = \omega_c\tau_i(t)$- $2\pi\nu_{Di}t$ -ϕ_O. This time-varying nature of a linear wireless channel is described by the *time-varying impulse response*, denoted as $h(\tau, t)$, and here τ is so called delay spread and t is the time instance. The *delay spread* is defined as the time difference in signal arriving times of the last and first arriving (line of sight, LOS) multipath components.

To deal with various fading effects, in particular multipath fading, we will describe in incoming chapters the multicarrier modulation, in particular orthogonal division multiplexing (OFDM) [2–5, 20, 70, 71], diversity [2, 37], multi-input multi-output (MIMO) signal processing [2–5, 35, 72–74], adaptive modulation and coding [2–5], and channel equalization [2–4], to mention a few. The key idea behind the multicarrier modulation is to divide the transmitted bit stream into many lower speed substreams, which are typically orthogonal to each other, and send these over many different subchannels. The number of substreams is chosen to insure that each subchannel has a bandwidth smaller than the characteristic channel parameter known as the coherence bandwidth, so that each subchannel will experience the flat fading. This is an efficient way to deal with frequency-selective fading effects. The key idea behind the diversity-combining techniques is to use several independent realizations of that same transmitted signal to improve tolerance to channel impairments. The independent signal copies transmitted over different paths or degrees of freedom will have a low probability of experiencing the same level of channel distortion. In wireless communications, the probability that all independent fading paths are in deep fade will be low. The independent realizations of the same signal will be properly combined to improve the overall reliability of transmission. In MIMO systems, multiple transmitters and receivers can exploit the knowledge about the channel to determine the rank of so-called channel matrix $\boldsymbol{H}$, which is related to the number of independent data streams that can be simultaneously transmitted. The basic concept of wireless MIMO communication is illustrated in Fig. 1.5, where the number of transmit antennas is denoted as M_{Tx} and the number of receive antennas is denoted as M_{Rx}.

The 2D symbol transmitted on the *m*–th transmit antennas is denoted by x_m ($m = 1, 2,\ldots, M_{\mathrm{Tx}}$), while the received symbol on the *n*-th received antenna is denoted as y_n. The channel coefficient relating the received symbol on *n*-th receive antenna and transmitted symbol on the *m*-th transmit antenna is denoted as h_{mn}. With MIMO signal processing, we can improve the signal-to-noise ratio (SNR) of SISO system, improve the symbol error rate slope (vs. SNR), and improve the overall data rate multiple times. In adaptive modulation techniques, we adapt the transmitter to time-varying wireless channel conditions to enable reliable and high spectral-efficient transmission. Various wireless communication system's parameters can be adapted including the data rate, power, code rate, and error probability. In adaptive coding, we adapt error correction strength (code rate) depending on time-varying wireless channel conditions. When channel conditions are favorable, the weaker (higher rate)

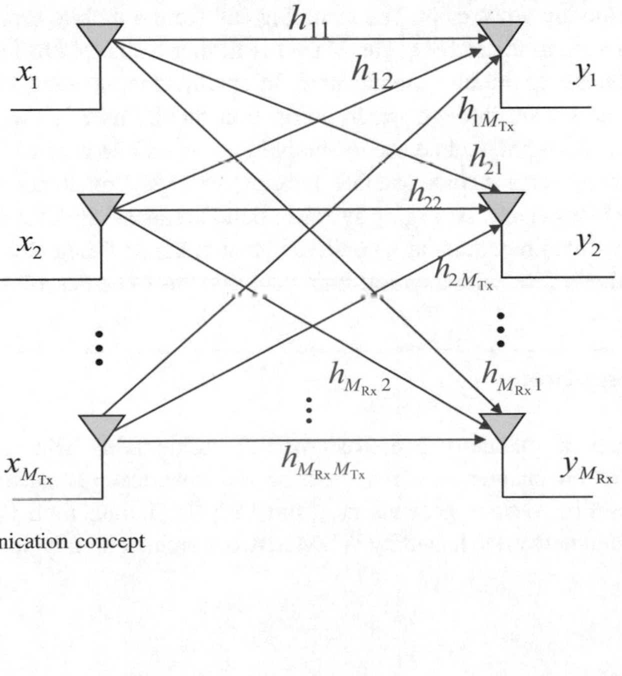

Fig. 1.5 The wireless MIMO communication concept

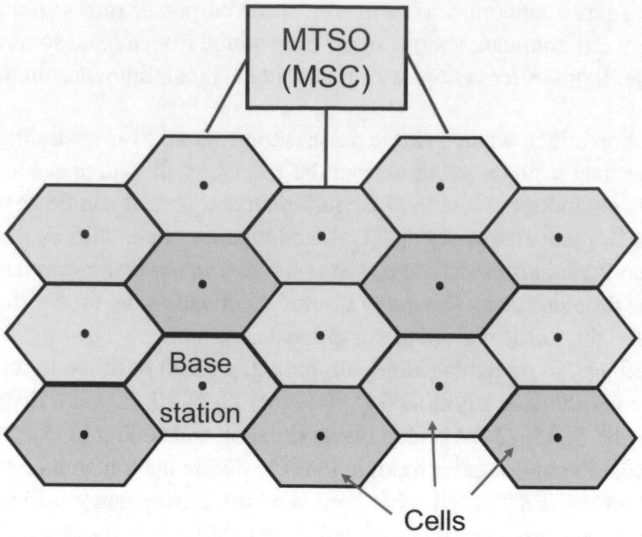

Fig. 1.6 Illustration of the cellular concept

channel code is used. On the other hand, when channel conditions deteriorate, the stronger (lower code rate) channel code is used. In adaptive modulation and coding (AMC), the channel code and modulation formats are simultaneously adapted.

Before concluding this section, we briefly describe the *cellular concept*, which is illustrated in Fig. 1.6. A cellular network consists of large number of mobile users (subscribers) and fixed number of base stations, arranged in particular way to provide the coverage to the mobile users. The area covered by a particular base station (BS) is called the cell. For better utilization of resources, the cells are of hexagonal shape. However, in practice the base station is typically arranged irregular, which is dictated by the configuration of terrane, location of building, and so on. The key idea of cellular concept is to reuse the channels, to improve the efficiency. The cells denoted by the same color in Fig. 1.6 employ the same frequency in FDMA, the same time slot in TDMA, or the same signature sequence (code) in CDMA. However, the channel reuse introduces the interference known as intercell interference.

The spatial separation of the cells that use the same channel, known as the reuse distance, should be as small as possible but should not exceed the maximum tolerable intercell interference threshold. The different base stations, in a given geographic area, are connected to the *mobile telephone switching office (MTSO)*, also known as the mobile switching center (MSC). Base stations and MTSOs coordinate handoff (of a user from one BS to the other) and control functions. Shrinking cell size increases the capacity, but, on the other hand, it increases the networking burden. The MTSO is connected by the high-speed link to the public switched telephone network (PSTN), typically by SMF, coaxial cable, or microwave link. The MTSO can provide the internet access or route the voice calls. The incoming call from a mobile user is connected to the base station, which is on the other hand connected to the MTSO. The MTSO is further connected to the wired network. The destination could be an ordinary wire telephone or another mobile user. In another mobile user, the corresponding MTSO will be identified, and in MTSO, the base station through which destination mobile user is supposed to be connected to will be identified as well. Finally, the call from one mobile use to another remote mobile user will be established.

Other wireless communication systems include satellite systems; wireless broadband access systems such as WiMAX, providing fixed and fully mobile internet access; paging systems; radio broadcast systems (AM radio, FM radio, analog TV, digital audio broadcasting, digital video broadcasting); cordless phone systems; Bluetooth; wireless LAN (Wi-Fi); and ultra-wideband (UWB) radio, to mention a few, with the most important systems to be describe in incoming chapters.

1.4 Organization of the Book

Each chapter of the book, except introductory one, has over 20 problems to help reader gain deeper knowledge in corresponding fields. In introduction chapter, we first describe the historical perspective of both wireless and optical communication systems and describe various generations, from early beginning until today. After that we describe the fundamentals of optical systems and networks, including WDM network architecture, typical optical networking architecture, bit rates for synchronous/asynchronous optical communication systems, and wavelength regions for fiber-optics communications. In the same section, we briefly describe various multiplexing schemes such as TDM, WDM, FDM, SCM, CDM, PDM, and SDM. We also identify the main noise sources and optical channel impairments, in particular for fiber-optics communications. After that the key optical devises and modules are identified and briefly described. In section related to wireless communications fundamentals, we describe a typical wireless communication point-to-point link and identify key devices/subsystems with brief description of operational principles. The multiaccess communications are then described with key problems being identified. Further, various effects affecting the received power are described including path loss, shadowing, and multipath fading. We also describe various approaches to deal with multipath fading including OFDM, diversity, MIMO, equalization, and adaptive modulation and coding. The MIMO operation principle is briefly described next. Finally, the cellular concept is introduced. In section on organization of the book, the detailed description of chapters is provided.

Chapter 2 discusses the propagation effects, noise sources, and channel impairments for wireless communications, free-space optical communications, and fiber-optics communications. The key idea of this chapter is to derive the wave equation for a given medium from Maxwell's equation and determine its solutions subject to corresponding boundary conditions. In the same chapter, we study vectorial nature of the light, derive Snell's law of refraction from Maxwell's equations, study total internal reflection, and explain the light confinement in step-index optical fibers. We also describe how arbitrary polarization state can be generated by employing tree basic polarizing elements, namely, polarizer, phase shifter (retarder), and rotator. We also introduce the electromagnetic potentials to simplify the study of generation of electromagnetic waves, followed by antenna theory basics. After that we discuss interference, coherence, and diffraction effects in optics. Following the detailed description of laser beam propagation effects over the atmospheric turbulence channels, we describe the simulation models suitable to study atmospheric turbulence effects. In the section devoted to wireless communication channel effects, we describe path loss, shadowing, and multipath fading propagation effects. Both ray theory-based and statistical multipath wireless communication channel models are described. The focus is then moved to signal propagation effects in optical fibers including fiber attenuation and insertion loss, chromatic dispersion effects, PMD effects, and fiber nonlinearities. Regarding the fiber nonlinearities, both Kerr nonlinear effects (self-phase modulation, cross-phase modulation) and nonlinear scattering effects, such as stimulated Raman scattering, are described. The generalized nonlinear Schrödinger equation is then introduced to study the propagation effects over both single-mode and few-mode fibers, as well as the corresponding split-step Fourier algorithms to solve it. The following noise sources related to optical communication are further described: mode partition noise, relative intensity noise, laser phase noise, modal noise, thermal noise, spontaneous emission noise, noise beating components, quantum shot noise, dark current noise, and crosstalk. Finally, in section devoted to indoor optical wireless communications, the channel models for both infrared and visible light communications are described.

In Chap. 3, basic components, modules, and subsystems relevant in both optical and wireless communications are described. The first half of the chapter is related to key components, modules, and subsystems for optical communications. After describing optical communication basic, focus is changed to optical transmitters. The following classes of lasers are introduced: Fabry-Perot, distributed feedback (DFB), and distributed Bragg reflector (DBR) lasers. Regarding the external modulators, both Mach-Zehnder modulator and electro-absorption modulator are described. Regarding the external modulator for multilevel modulation schemes, phase modulator, I/Q modulator, and polar modulator are described. In subsection on optical receivers, after the photodetection principle, the p-i-n photodiode, avalanche photodiode (APD), and metal-semiconductor-metal (MSM) photodetectors are briefly described. In the same subsection, both optical receiver high-impedance front end and transimpedance front end are described. In section on optical fibers, both single-mode fiber (SMF) and multimode fiber (MMF) are introduced. In subsection on optical amplifiers, after describing optical amplification principles, the following classes of optical amplifiers are described: semiconductor optical amplifier (SOA), Raman amplifier, and Erbium-doped fiber amplifier (EDFA). In subsection on optical processing components and modules, the following components/modules are described: optical couplers, optical filters, and WDM multiplexers/demultiplexers. After that, the principles of coherent optical detection are introduced, followed by the description of optical hybrids. In subsection on coherent optical balanced detectors, both homodyne and heterodyne optical balanced detectors for both binary and multilevel (QAM, PSK, APSK) modulations are described.

In the second half of Chap. 3, basic components and modules of relevance to modern wireless communications are described. First, the DSP basics, including various discrete-time realizations, are provided. Then the direct digital synthesizer (DDS), relevant in discrete-time (DT) wireless communications systems, is described. In subsection on multirate DSP and resampling, various upsampling, downsampling, and resampling schemes/modules are described. In the same subsection, the polyphase decomposition of the system (transfer) function is introduced and employed in resampling. In subsection on antenna array antennas, the linear array antennas are described. Finally, in subsection on automatic gain control (AGC), the AGC schemes operating in passband and baseband are described.

Chapter 4 is devoted to both wireless channel and optical channel capacities. The chapter starts with definitions of mutual information and channel capacity, followed by the information capacity theorem. We discuss the channel capacity of discrete memoryless channels, continuous channels, and channels with memory. Regarding the wireless channels, we describe how to calculate the channel capacity of flat-fading and frequency-selective channels. We also discuss different optimum and suboptimum strategies to achieve channel capacity including the water-filling method, multiplexed coding and decoding, channel inversion, and truncated channel inversion. We also study different strategies for channel capacity calculation depending on what is known about the channel state information. Further, we explain how to model the channel with memory and describe McMillan-Khinchin model for channel capacity evaluation. We then describe how the forward recursion of the BCJR algorithm can be used to evaluate the information capacity of channels with memory. The next topic is related to the evaluation of the information capacity of fiber-optics communication channels with coherent optical detection. Finally, we describe how to evaluate the capacity of hybrid free-space optical (FSO)-RF channels, in which FSO and RF subsystems cooperate to compensate for shortcoming of each other.

Chapter 5 is devoted to advanced modulation and multiplexing techniques suitable for both wireless and optical communication systems. The chapter starts with signal space theory concepts applied to wireless communication systems. After the geometric representations of signals, we describe various multidimensional modulators and demodulators suitable for wireless communications applications, in particular the Euclidean distance, correlation, and matched filter-based detectors, followed by the description of frequency-shift keying (FSK). Both continuous-time (CT) and discrete-time (DT) implementations for pulse amplitude modulation schemes, suitable for wireless communications, are described as well. After that the focus is moved to multilevel schemes suitable for both wireless and optical communications applications, including M-ary PSK, star-QAM, square-QAM, and cross-QAM. Regarding the optical communications, the transmitter for M-ary PSK, star-QAM, and square/cross-QAM are described in detail. The next topic in the chapter is related to multicarrier modulation, including description multicarrier systems with both non-overlapping and overlapping subcarriers as well as introduction of various approaches to deal with fading effects at subcarrier level. The concept of OFDM is also introduced; however, details are provided in Chap. 7. The MIMO fundamentals are then provided, including the description of key differences with respect to diversity scheme, as well as the introduction of array, diversity, and multiplexing gains. The parallel MIMO channel decomposition is briefly described. The details on MIMO signal processing are postponed for Chap. 8. In section on polarization-division multiplexing (PDM) and four-dimensional (4D) signaling, we describe key differences between PDM and 4D signaling and describe how both types of schemes can be implemented in both wireless and optical communications. The focused is the moved to the spatial-division multiplexing (SDM) and multidimensional signaling. We describe how SDM can be applied in wireless communications first. Then we describe how various degrees of freedom including amplitude, phase, frequency, polarization states, and spatial modes can be used to convey the information in optical domain. In the same sections, the SDM concepts for fiber-optics communications are described as well. The section concludes with SDM and multidimensional signaling concepts applied to free-space optical (FSO) communications. The next topic is devoted to the signal constellation design, including iterative polar modulation (IPM), signal constellation design for circular symmetric optical channels, energy-efficient signal constellation design, and optimum signal constellation design (OSCD). The final section of the chapter is devoted to the nonuniform signaling, in which different signal constellation points are transmitted with different probabilities.

Chapter 6 is related to advanced detection and channel impairments compensation techniques applicable to both wireless and optical communication systems. To better understand the principles behind advanced detection and channel impairments compensation, we start the chapter with fundamentals of detection and estimation theory including optimum receiver design in symbol error probability sense and symbol error probability derivations. In section on wireless communication systems performance, we describe different scenarios including outage probability, average error probability, and combined outage-average error probability scenarios. We also describe the moment generating function-based approach to average error probability calculation. We also describe how to evaluate performance in the presence of Doppler spread and fading effects. In section on channel equalization techniques, after short introduction, we first describe how zero-forcing equalizers can be used to compensate for ISI, chromatic dispersion, and polarization mode dispersion (PMD) in fiber-optics communications; ISI and multipath fading effects in wireless communications; and ISI and multipath effects in indoor optical wireless communications. After that we describe the optimum linear equalizer design in the minimum mean-square error sense as well as Wiener filtering concepts. The most relevant post-compensation techniques, applicable to both wireless and optical communications, are described next including feedforward equalizer, decision-feedback equalizer, adaptive equalizer, blind equalizers, and maximum likelihood sequence detector (MLSD) also known as Viterbi equalizer. The turbo equalization technique is postponed for Chap. 9. The next section is devoted to the relevant synchronization techniques. In section on adaptive modulation techniques, we describe how to adapt the transmitter to time-varying wireless/optical channel conditions to enable reliable and high spectral-efficient transmission. Various scenarios are described including data rate, power, code rate, error probability adaptation scenarios, as well as combination of various adaptation strategies. In particular, variable-power variable-rate modulation techniques and adaptive coded modulation techniques are described in detail. Next, the Volterra series-based equalization to deal with fiber nonlinearities and nonlinear effects in wireless communications is described. In section on digital backpropagation, we describe how this method can be applied to deal with fiber nonlinearities, chromatic dispersion, and PMD in a simultaneous manner. In section on coherent optical detection, we describe various balanced coherent optical detection schemes for two-dimensional modulation schemes; polarization diversity and polarization demultiplexing schemes; and homodyne coherent optical detection based on phase-locked loops, phase diversity receivers, as well as the dominant coherent optical detection sources including laser phase noise, polarization noise, transimpedance amplifier noise, and amplifier spontaneous emission (ASE) noise. In section on compensation of atmospheric turbulence effects, we describe various techniques to deal with atmospheric turbulence including adaptive coding, adaptive coded modulation, diversity approaches, MIMO signal processing, hybrid free-space optical (FSO)-RF communication approach, adaptive optics, and spatial light modulators (SLM)-based backpropagation method. In particular, linear adaptive optics techniques are described in detail.

Chapter 7 is devoted to OFDM fundamentals and applications to wireless and optical communications. The chapter starts with basic of OFDM including the generation of OFDM signal by inverse FFT. After that several basic manipulations on OFDM symbols, required to deal with multipath effects in wireless communications and dispersion effects in fiber-optics communications, including guard time and cyclic extension as well as the windowing operation, are described. Next, the bandwidth efficiency of OFDM is discussed. The next topic in the chapter is devoted to the OFDM system design, description of basic OFDM blocks, and OFDM-based parallel channel decomposition, including multipath fading radio and dispersion-dominated channels decomposition. Following the description of coherent optical OFDM (CO-OFDM) principles, the basic estimation and compensation steps, common to both wireless and optical communication systems, are described including DFT windowing, frequency synchronization, phase estimation, and channel estimation. Regarding the channel estimation, both pilot-aided and data-aided channel estimation techniques are described. Following the differential detection in OFDM description, the focus is on various applications of OFDM in wireless communications including digital audio broadcasting (DAB), digital video broadcasting (DVB), Wi-Fi, LTE, WiMAX, and ultra-wideband communication (UWB). Next, optical OFDM applications are discussed starting with description of basic OFDM systems types such as CO-OFDM and direct detection optical OFDM (DDO-OFDM) systems. Regarding the high-speed spectrally efficient CO-OFDM systems, the polarization-division multiplexed OFDM systems and OFDM-based superchannel optical transmission systems are described. Before, the concluding remarks, the application of OFDM in multimode fiber links is described.

Chapter 8 is devoted to various diversity and MIMO techniques capable of improving single-input single-output system performance. Through MIMO concept it is also possible to improve the aggregate data rate. The diversity and MIMO techniques described are applicable to wireless MIMO communications as well as free-space optical and fiber-optics communications with coherent optical detection. Various diversity schemes are described including polarization diversity, spatial diversity, and frequency diversity schemes. The following receive diversity schemes are described: selection combining, threshold combining, maximum-ratio combining, and equal-gain combining schemes. Various transmit diversity schemes are described, depending on the availability of channel state information on transmitter side. The most of the chapter is devoted to wireless and optical MIMO techniques. After description of various wireless and optical MIMO models, we describe the parallel decomposition of MIMO channels, followed by space-time coding (STC) principles. The maximum likelihood (ML) decoding for STC is described together with various design criteria. The following classes of STC are described: Alamouti code, orthogonal designs, linear space-time block codes, and space-time trellis codes. The corresponding STC decoding algorithms are described as well. After that we move our attention to spatial-division multiplexing (SDM) principles and describe various BLAST encoding architectures as well as multi-group space-time coded modulation. The following classes of linear and feedback MIMO receiver for uncoded signals are subsequently described: zero-forcing, linear minimum MSE, and decision-feedback receivers. The next topic in the chapter is related to suboptimum MIMO receivers for coded signals. Within this topic we first describe linear zero-forcing (ZF) and linear MMSE receivers (interfaces) but in context of STC. Within decision feedback and BLAST receivers, we describe various horizontal/vertical (H/V) BLAST architecture including ZF and MMSE ones. The topic on suboptimum receivers is concluded with diagonal- (D-)-BLAST and iterative receiver interfaces. The section on iterative MIMO receivers starts with brief introduction of concept of factor graphs, following with the description of factor graphs for MIMO channel and channels with memory. We then describe the sum-product algorithm (SPA) operating on factor graphs, followed by description of SPA for channels with memory. The following iterative MIMO receivers for uncoded signals are described: ZF, MMSE, ZF H/V-BLAST, and LMMSE H/V-BLAST receivers. After brief description of factor graphs for linear block and trellis codes, we describe several iterative MIMO receivers for space-time coded signals. The section on broadband MIMO describes how frequency selectivity can be used as an additional degree of freedom, followed by description of MIMO-OFDM and space-frequency block-coding principles. The focus is then moved to MIMO channel capacity calculations for various MIMO channel models including deterministic, ergodic, and non-ergodic random channels, as well as correlated channel models. The concepts of ergodic and outage channel capacity are described. The section on MIMO channel capacity ends with MIMO-OFDM channel capacity description. In MIMO channel estimation section, the following MIMO channel estimation techniques are described: ML, least squares (LS), and linear minimum MSE MIMO channel estimation techniques. In massive MIMO communications section, the concepts and various detection schemes are described.

Chapter 9 is related to the advanced channel coding and coded modulation techniques. After linear and BCH codes fundamentals, we provide the trellis description of linear block codes and describe the corresponding Viterbi decoding algorithm. After describing the fundamentals of convolutional, RS, concatenated, and product codes, we describe coding with interleaving as an efficient way to deal with burst of errors and fading effects. Significant space in the chapter is devoted to codes on graphs, in particular. Turbo, turbo product, and LDPC codes are described together with corresponding decoding algorithms. Regarding LDPC codes, both binary and nonbinary (NB) LDPC codes are introduced, and corresponding decoding algorithms as well as their FPGA implementation are described. Additionally, LDPC codes design procedures are described, followed by rate adaptation. Rate adaptive FGPA implementations of LDPC and generalized LDPC codes are described as well. The next portion of the chapter is devoted to coded modulation (CM) and unequal error protection (UEP). After providing the coded modulation fundamentals, we describe trellis-coded modulation (TCM), multilevel coding and UEP, bit-interleaved-coded modulation (BICM), turbo TCM, and various hybrid multidimensional-coded modulation schemes suitable for ultra-high-speed optical transmission including multilevel nonbinary LDPC-coded modulation. After coded modulation sections, the focus of the chapter is on multidimensional turbo equalization. The following topics are described: nonlinear channels with memory, nonbinary MAP detection, sliding-window multidimensional turbo equalization, simulation study of multidimensional turbo equalization, and several experimental demonstrations including time-domain 4D-NB-LDPC-CM and quasi-single-mode transmission over transoceanic distances. In section on optimized signal constellation design and optimized bit-to-symbol mapping-based coded modulation, we describe multidimensional optimized signal constellation design (OSCD), EXIT chart analysis of OSCD mapping rules, nonlinear OSCD-based coded modulation, and transoceanic multi-Tb/s transmission experiments enabled by OSCD. Finally, in adaptive coding and adaptive coded modulation section, we describe adaptive coded modulation, adaptive nonbinary LDPC-coded multidimensional modulation suitable for high-speed optical communications, and adaptive hybrid free-space optical (FSO)-RF coded modulation.

Chapter 10 is devoted to spread spectrum (SS), CDMA, and ultra-wideband (UWB) communication systems. After the introduction of SS systems, both direct sequence spread spectrum (DS-SS) and frequency-hopping spread spectrum (FH-SS) systems are described in detail. Regarding the DS-SS systems, after typical transmitter and receiver configurations are described, we describe in detail the synchronization process including acquisition and tracking stages. To deal with the multipath fading effects, the use of RAKE receiver is described. The section on DS-SS systems concludes with the spreading sequences description, in particular pseudo-noise (PN) sequences. In this section, after introduction of the basic randomness criterions, we introduce maximal-length sequences (m-sequences) and describe their basic properties. In the same section, the concept of nonlinear PN sequences is introduced as well. Regarding the FH-SS systems, both slow-frequency hopping (SFH) and fast-frequency hopping (FFH) systems are introduced. We provide the detailed description of FH-SS systems, including both transmitter and demodulator as well as various time acquisition schemes. In the same section, we describe how to generate the FH sequences. After the introduction of CDMA systems, we describe various signature waveforms suitable for use in DS-CDMA systems including Gold, Kasami, and Walsh sequences. We then describe relevant synchronous and asynchronous CMDA models, with special attention devoted to discrete-time CDMA models. In section on multiuser detection (MUD), we describe how to deal with interference introduced by various CDMA users. The conventional single-user detector is used as the reference case. The various MUD schemes are described then, with special attention being paid to the jointly optimum MUD. The complexity of jointly optimum MUD might be prohibitively high for certain applications, and for such applications, we provide the description of the decorrelating receiver, which can compensate for multiuser interference but enhance the noise effect. To solve for this problem, the linear minimum mean-square error (MMSE) receiver is introduced. The section on MUD concludes with description of the nonlinear MUD schemes employing decisions on bits from other users (be preliminary or final), with description of successive cancelation and multistage detection schemes. In section on optical CDMA (OCDMA), we describe both incoherent and coherent optical detection-based OCDMA schemes. Regarding the incoherent optical detection-based schemes, we first describe how to design various unipolar signature sequences, known as optical orthogonal codes (OOC), followed by the description of basic incoherent OCDMA schemes including time-spreading, spectral amplitude coding (SAC), and wavelength hopping (WH) schemes. Related to coherent optical detection-based OCDMA systems, we describe both pulse laser-based and CW laser-based OCDMA systems. Further, the hybrid OFDM-CDMA systems are described, taking the advantages of both OFDM and CDMA concepts into account, in particularly in dealing with frequency-selective fading and narrowband interference while at the same time supporting multiple users. The following broad classes of hybrid OFDM-CDMA systems are described: multicarrier CDMA (MC-CDMA), OFDM-CDMA, and OFDM-CDMA-FH systems. In section on UWB communications, we describe the concepts and requirements as well as transmitter and receiver configurations. Further, various modulation formats, suitable for UWB communications, are described, categorized into time-domain category (pulse-position nodulation and pulse-duration modulation) and shape-based category [on-off keying, pulse-amplitude modulation, bi-phase modulation (BPM) or BPSK, and orthogonal pulse modulation]. Regarding the pulse shapes suitable for orthogonal pulse modulation (OPM), the modified Hermite polynomials, Legendre polynomials, and orthogonal prolate spheroidal wave functions (OPSWF) are described. Regarding the UWB channel models suitable for indoor applications, the model due to Saleh and Valenzuela is described. Both types of UWB systems are described: the impulse radio UWB (IR-UWB), typically DS-CDMA-based, and multiband OFDM (MB-OFDM).

Chapter 11 is devoted to the physical layer security (PLS) for wireless and optical channels. The chapter starts with discussion on security issues, followed by the introduction of information-theoretic security and comparison against the computational security. In the same section, various information-theoretic security measures are introduced, including strong secrecy and weak secrecy conditions. After that, the Wyner's wiretap channel model, also known as the degraded wiretap channel model, is introduced. In the same section, the concept of secrecy capacity is introduced as well as the nested wiretap coding. Further, the broadcast channel with confidential messages is introduced, and the secrecy capacity definition is generalized. The focus is then moved to the secret key generation (agreement), the source and channel-type models are introduced, and corresponding secret key generation protocols are described. The next section is devoted to the coding for the physical layer security systems, including both coding for weak and strong secrecy systems. Regarding the coding for weak secrecy systems, the special attention is devoted to two-edge-type LDPC coding, punctured LDPC coding, and polar codes. Regarding the coding for strong secrecy systems, the focus is on coset coding with dual of LDPC codes and hash functions/extractor-based coding. The attention is then moved to information reconciliation and privacy amplification. In wireless channels PLS section, the following topics are covered: MIMO fundamentals, wireless MIMO PLS, and secret key generation in wireless networks. In section on optical channels PLS, both PLS for spatial-division multiplexing (SDM)-fiber-based systems and free-space optical (FSO) systems are discussed.

1.5 Concluding Remarks

The historical perspective of both wireless and optical communication systems has been described in Sect. 1.1, together with various generations, from early beginning until today. After that, in Sect. 1.2, the fundamentals of optical systems and networks have been described, including WDM network architecture, typical optical networking architecture, bit rates for synchronous/asynchronous optical communication systems, and wavelength regions for fiber-optics communications. In the same section, various multiplexing schemes have been described such as TDM, WDM, FDM, SCM, CDM, PDM, and SDM. The main noise sources and optical channel impairments have been described, in particular for fiber-optics communications. After that the key optical devises and modules have been identified and briefly described. In section related to wireless communications fundamentals, Sect. 1.3, we have described a typical wireless communication point-to-point link and identified key devices/subsystems with brief description of operational principles. The multiaccess communications have been then described with key problems being identified. Further, various effects affecting the received power have been described as well including path loss, shadowing, and multipath fading. The various approaches to deal with multipath fading have been then introduced including OFDM, diversity, MIMO, equalization, and adaptive modulation and coding. The MIMO operation principles have been briefly described next. Finally, the cellular concept has been introduced. In section on organization of the book, Sect. 1.4, the detailed description of chapters is provided.

References

1. G.P. Agrawal, *Fiber-Optic Communication Systems*, 4th edn. (Wiley, Hoboken, NJ, 2010)
2. A. Goldsmith, *Wireless Communications* (Cambridge University Press, Cambridge, UK/New York, 2005)
3. I. Djordjevic, W. Ryan, B. Vasic, *Coding for Optical Channels* (Springer, New York, 2010)
4. M. Cvijetic, I.B. Djordjevic, *Advanced Optical Communications and Networks* (Artech House, Boston/London, 2013)
5. W. Shieh, I. Djordjevic, *OFDM for Optical Communications* (Elsevier/Academic, Amsterdam/Boston, 2010)
6. A. Osseiran, F. Boccardi, V. Braun, K. Kusume, P. Marsch, M. Maternia, O. Queseth, M. Schellmann, H. Schotten, Scenarios for 5G mobile and wireless communications: The vision of the METIS project. IEEE Commun. Mag. **52**(5), 26–35 (2014)
7. A. Osseiran, J.F. Monserrat, P. Marsch, *5G Mobile and Wireless Communications Technology* (Cambridge University Press, Cambridge, UK, 2016)
8. T. Wheeler, *Leading Towards Next Generation "5G" Mobile Services* (Federal Communications Commission, 2015)
9. America's 5G Future, FCC Initiatives, available at https://www.fcc.gov/5G
10. Minimum requirements related to technical performance for IMT-2020 radio interface(s), available at: https://www.itu.int/pub/R-REP-M.2410
11. M.2150: Detailed specifications of the terrestrial radio interfaces of International Mobile Telecommunications-2020 (IMT-2020), available at: https://www.itu.int/rec/R-REC-M.2150/en
12. 5G NR (Rel-15), available at: https://www.3gpp.org/lte-2
13. 3GPP Release 18, available at: https://www.3gpp.org/release18
14. Dohler M et al., Internet of skills, where robotics meets AI, 5G and the Tactile Internet. In Proc. 2017 European Conference on Networks and Communications (EuCNC), (2017), pp. 1–5
15. W. Saad, M. Bennis, M. Chen, A vision of 6G wireless systems: Applications, trends, technologies, and open research problems. IEEE Netw. **34**(3), 134–142 (2020)
16. H. Yang, A. Alphones, Z. Xiong, D. Niyato, J. Zhao, K. Wu, Artificial-intelligence-enabled intelligent 6G networks. IEEE Netw. **34**(6), 272–280 (2020)
17. R. Chataut, R. Akl, Massive MIMO systems for 5G and beyond networks-overview, recent trends, challenges, and future research direction. Sensors (Basel) **20**(10), 2753 (2020)
18. I.F. Akyildiz, S. Nie, S.-C. Lin, M. Chandrasekaran, 5G roadmap: 10 key enabling technologies. Comput. Netw. **106**, 17–48 (2016)
19. IEEE Standard for Information technology--Telecommunications and information exchange between systems Local and metropolitan area networks--Specific requirements Part 11: Wireless LAN Medium Access Control (MAC) and Physical Layer (PHY) Specifications, in *IEEE Std 802.11–2012 (Revision of IEEE Std 802.11–2007)*, pp.1–2793, March 29 2012, https://doi.org/10.1109/IEEESTD.2012.6178212
20. J. Heiskala, J. Terry, *OFDM Wireless LANs: A Theoretical and Practical Guide* (Sams Publishing, 2002)
21. Hanzo L, Akhtman Y, Wang L, Jiang M, MIMO-OFDM for LTE, WiFi, and WiMAX,. IEEE/Wiley, 2011
22. M.S. Gast, *802.11 Wireless Networks: The Definitive Guide*, 2nd edn. (O'Reilly Media, Inc, Sebastopol, CA, 2005)
23. M.S. Gast, *802.11ac: A Survival Guide* (O'Reilly Media, Inc, Sebastopol, CA, 2013)
24. M.S. Gast, *802.11an: A Survival Guide* (O'Reilly Media, Inc, Sebastopol, CA, 2012)
25. Unlicensed Use of the 6 GHz Band; Expanding Flexible Use in Mid-Band Spectrum Between 3.7 and 24 GHz, available at: https://docs.fcc.gov/public/attachments/FCC-20-51A1.pdf
26. L. Nuaymi, *WiMAX: Technology for Broadband Wireless Access* (Wiley, Chichester, UK, 2007)
27. IEEE 802.16–2004, IEEE Standard for Local and Metropolitan Area Networks, Air Interface for Fixed Broadband Wireless Access Systems, October 2004
28. IEEE 802.16e, IEEE Standard for Local and Metropolitan Area Networks, Air Interface for Fixed Broadband Wireless Access Systems, Amendment 2: Physical and Medium Access Control Layers for Combined Fixed and Mobile Operation in Licensed Bands and Corrigendum 1, February 2006
29. IEEE Standard for Air Interface for Broadband Wireless Access Systems. IEEE Std 802.16–2017 (Revision of IEEE Std 802.16–2012), pp.1–2726, 2 March 2018
30. IEEE 802.16 Working Group on Broadband Wireless Access Standards, see https://grouper.ieee.org/groups/802/16/
31. T.H. Maiman, Stimulated optical radiation in ruby. Nature **187**(4736), 493–494 (1960)
32. K.C. Kao, G.A. Hockman, Dielectric-fiber surface waveguides for optical frequencies. Proc. IEEE **113**(7), 1151–1158 (1966)
33. D. Tse, P. Viswanath, *Fundamentals of Wireless Communication* (Cambridge University Press, Cambridge, UK, 2005)
34. S. Benedetto, E. Biglieri, *Principles of Digital Transmission with Wireless Applications* (Kluwer Academic/Plenum Publishers, New York, 1999)
35. E. Biglieri, *Coding for Wireless Channels* (Springer, New York, 2005)
36. T.M. Duman, A. Ghrayeb, *Coding for MIMO Communication Systems* (Wiley, Chichester, UK, 2007)
37. G.L. Stüber, *Principles of Mobile Communication*, 3rd edn. (Springer, New York, 2012)
38. L. Kazovsky, S. Benedetto, A. Willner, *Optical Fiber Communication Systems*, vol 1996 (Artech House, Boston, 1996)
39. G. Keiser, *Optical Fiber Communications*, 3rd edn. (McGraw-Hill, Boston, 2000)
40. R. Ramaswami, K. Sivarajan, *Optical Networks: A Practical Perspective*, 2nd edn. (Morgan Kaufman, San Francisco, CA, 2002)
41. Z. Ghassemlooy, W. Popoola, S. Rajbhandari, *Optical Wireless Communications: System and Channel Modelling with MATLAB* (CRC Press, Boca Raton, FL, 2013)
42. S. Arnon, J.R. Barry, S.R. Karagiannidis, M. Uysal, *Advanced Optical Wireless Communication Systems* (Cambridge University Press, New York, 2012)
43. G.P. Agrawal, *Nonlinear Fiber Optics*, 4th edn. (Elsevier/Academic Press, Amsterdam/Boston, 2007)
44. G.P. Agrawal, *Lightwave Technology: Telecommunication Systems* (Wiley, Hoboken, NJ, 2005)
45. F.E. Ross, FDDI – a tutorial. IEEE Commun. Mag. **24**(5), 10–17 (1986)
46. M.W. Sachs, A. Varma, Fiber channel and related standards. IEEE Commun. Mag. **34**(8), 40–49 (1996)

47. Telcordia Ericsson, Synchronous Optical Network (SONET) Transport Systems: Common Generic Criteria, GR-253-CORE (2009)
48. P. Loshin, F. Kastenholz, *Essential Ethernet Standards: RFCs and Protocols Made Practical (Internet Standards)* (Wiley, New York, 1999)
49. S.A. Calta, J.A. deVeer, E. Loizides, R.N. Strangwayes, Enterprise systems connection (ESCON) architecture-system overview. IBM J. Res. Dev. **36**(4), 535–551 (1992)
50. S.V. Kartalopoulos, *Understanding SONET/SDH and ATM* (IEEE Press, Piscataway, NJ, 1999)
51. A. Farrel, I. Bryskin, *GMPLS: Architecture and Applications* (Morgan Kaufmann Publishers, San Francisco, CA, 2006)
52. P. Yuan, V. Gambiroza, E. Knightly, The IEEE 802.17 media access protocol for high-speed metropolitan-area resilient packet rings. IEEE Netw. **18**(3), 8–15 (2004)
53. ITU-T, Rec. G. 681: Functional characteristics of interoffice and long-haul line systems using optical amplifiers, including optical multiplexing, ITU-T (10/96), 1996
54. ITU-T, Rec G.704: Synchronous frame structures used at 1544, 6312, 2048, 8448 and 44 736 kbit/s hierarchical levels, ITU-T (10/98), 1998
55. ITU-T, Rec G.7041/Y.1303: Generic framing procedure (GFP), ITU-T (12/01), 2001
56. ITU-T, Rec G.7042/Y.1305: Link capacity adjustment scheme (LCAS) for virtual concatenated signals, ITU-T (11/01), 2001
57. ITU-T, Rec G.694.2: Spectral grids for WDM applications: CWDM wavelength grid, ITU-T (06/02), 2002
58. ITU-T, Rec G.8110/Y.1370.1: Architecture of transport MPLS layer network, ITU-T (11/06), 2006
59. ITU-T, Rec. G.798: Characteristics of Optical Transport Network Hierarchy Functional Blocks", ITU-T (02/01), 2009
60. S. Gringery, E.B. Basch, T.J. Xia, Technical considerations for supporting bit rates beyond 100 Gb/s. IEEE Commun. Mag. **50**(2), 21–30 (2012)
61. Djordjevic IB, Shieh W, Liu X, Nazarathy M (2014), Advanced digital signal processing and coding for multi-Tb/s optical transport. IEEE Sig. Proc. Mag. 31 (2),. pages 15 and 142. (From the Guest Editors)
62. I.B. Djordjevic, M. Cvijetic, C. Lin, Multidimensional signaling and coding enabling multi-Tb/s optical transport and networking. IEEE Sig. Proc. Mag. **31**, 104–117 (2014)
63. I.B. Djordjevic, On advanced FEC and coded modulation for ultra-high-speed optical transmission. IEEE Commun. Surv. Tutorials **18**(3), 1920–1951 (2016)
64. S. Lin, D.J. Costello, *Error Control Coding: Fundamentals and Applications*, 2nd edn. (Pearson Prentice Hall, Upper Saddle River, NJ, 2004)
65. C. Heegard, S.B. Wicker, *Turbo Coding* (Springer, New York, 1999)
66. W.E. Ryan, S. Lin, *Channel Codes: Classical and Modern* (Cambridge University Press, Cambridge, UK/New York, 2009)
67. D. Zou, C. Lin, I.B. Djordjevic, FPGA-based LDPC-coded APSK for optical communication systems. Opt. Express **25**(4), 3133–3142 (2017)
68. S. Glisic, B. Vucetic, *Spread Spectrum CDMA Systems for Wireless Communications* (Artech House, Inc, Boston/London, 1997)
69. S. Verdú, *Multiuser Detection* (Cambridge University Press, Cambridge, UK/New York, 1998)
70. R. Van Nee, R. Prasad, *OFDM Wireless Multimedia Communications* (Artech House, Boston/London, 2000)
71. R. Prasad, *OFDM for Wireless Communications Systems* (Artech House, Boston/London, 2004)
72. E. Biglieri, R. Calderbank, A. Constantinides, A. Goldsmith, A. Paulraj, H.V. Poor, *MIMO Wireless Communications* (Cambridge University Press, Cambridge, UK, 2007)
73. J.R. Hampton, *Introduction to MIMO Communications* (Cambridge University Press, Cambridge, UK, 2014)
74. A. Pulraj, R. Nabar, D. Gore, *Introduction to Space-Time Wireless Communications* (Cambridge University Press, Cambridge, UK, 2003)

Propagation Effects in Optical and Wireless Communications Channels, Noise Sources, and Channel Impairments

2

Abstract

This chapter discusses the propagation effects, noise sources, and channel impairments for wireless communications, free-space optical communications, and fiber-optics communications. The key idea of the chapter is to derive the wave equation for a given medium from Maxwell's equation and determine its solutions subject to corresponding boundary conditions. After the introduction of the vector derivatives in orthogonal curvilinear, cylindrical polar, and spherical coordinate systems, we study vectorial nature of the light, derive Snell's law of refraction from Maxwell's equations, study total internal reflection, and explain the light confinement in step-index optical fibers. We also describe how arbitrary polarization state can be generated by employing three basic polarizing elements, namely, polarizer, phase shifter (retarder), and rotator. We then introduce the electromagnetic potentials to simplify the study of generation of electromagnetic waves, followed by antenna theory basics. We than discuss interference, coherence, and diffraction in optics. After that the detailed description of laser beam propagation effects over the atmospheric turbulence channels is provided, including the description of simulation models. In the section on wireless communication channel effects, we study path loss, shadowing, and multipath fading propagation effects. Both ray theory-based and statistical multipath wireless communication channel models are described. We then move to signal propagation effects in optical fibers and describe (i) fiber attenuation and insertion loss, (ii) chromatic dispersion effects, (iii) polarization mode dispersion effects, and (iv) fiber nonlinearities. Regarding the fiber nonlinearities, both Kerr nonlinear effects (self-phase modulation, cross-phase modulation) and nonlinear scattering effects, such as stimulated Raman scattering, are described. The generalized nonlinear Schrödinger equation is then described for both single-mode and few-mode fibers, as well as the corresponding split-step Fourier algorithms to solve it. The following noise sources related to optical communication are described: mode partition noise, relative intensity noise, laser phase noise, modal noise, thermal noise, spontaneous emission noise, noise beating components, quantum shot noise, dark current noise, and crosstalk. In section on indoor optical wireless communications, the channel models for both infrared and visible light communications are described. The set of problems is provided after the concluding remarks.

2.1 Electromagnetic Field and Wave Equations

Maxwell's equations are used to describe the change of electric field $\boldsymbol{E}$ and magnetic field $\boldsymbol{H}$ in space and time [1–5]:

$$\nabla \times \boldsymbol{E} = -\partial \boldsymbol{B}/\partial t \tag{2.1}$$

$$\nabla \times \boldsymbol{H} = \boldsymbol{J} + \partial \boldsymbol{D}/\partial t \tag{2.2}$$

$$\nabla \cdot \boldsymbol{D} = \rho \tag{2.3}$$

$$\nabla \cdot \boldsymbol{B} = 0 \tag{2.4}$$

I. B. Djordjevic, *Advanced Optical and Wireless Communications Systems*, https://doi.org/10.1007/978-3-030-98491-5_2

where $\boldsymbol{B}$ denotes the magnetic flux density, $\boldsymbol{D}$ is the electric flux density, and $\boldsymbol{J} = \sigma \boldsymbol{E}$ is the current density. The flux densities are related to the field vectors by constitutive relations:

$$\boldsymbol{D} = \varepsilon_0 \boldsymbol{E} + \boldsymbol{P} \tag{2.5}$$

$$\boldsymbol{B} = \mu_0 (\boldsymbol{H} + \boldsymbol{M}) \tag{2.6}$$

where $\boldsymbol{P}$ and $\boldsymbol{M}$ denote the induced electric and magnetic densities, $\varepsilon_0 = [4\pi \times 9{\cdot}10^9]^{-1}$ F/m is the permeability in the vacuum, and $\mu_0 = 4\pi \times 10^{-7}$ H/m is the permittivity in the vacuum. Given that there is no electric charge in free-space or fiber media, the volume density of electric charge $\rho = 0$, while the conductivity of either free-space or silica is extremely low $\sigma \approx 0$; the Maxwell's equations can be simplified. Moreover, the silica and free-space are nonmagnetic material, so $\boldsymbol{M} = \boldsymbol{0}$. $\boldsymbol{P}$ and $\boldsymbol{E}$ are mutually connected by:

$$\boldsymbol{P}(\boldsymbol{r}, t) = \varepsilon_0 \int_{-\infty}^{\infty} \chi(\boldsymbol{r}, t - t') \boldsymbol{E}(\boldsymbol{r}, t') dt', \tag{2.7}$$

where χ denotes the linear susceptibility, which is generally speaking a second-rank tensor. However, this tensor becomes scalar for an isotropic medium (where $\boldsymbol{P}$ and $\boldsymbol{E}$ are collinear). In free-space $\boldsymbol{P} = \boldsymbol{0}$, so that Maxwell's equations become:

$$\nabla \times \boldsymbol{E} = -\mu_0 \partial \boldsymbol{H} / \partial t \tag{2.8}$$

$$\nabla \times \boldsymbol{H} = \varepsilon_0 \partial \boldsymbol{E} / \partial t \tag{2.9}$$

$$\nabla \cdot \boldsymbol{E} = 0 \tag{2.10}$$

$$\nabla \cdot \boldsymbol{H} = 0 \tag{2.11}$$

We are in position now to separate electric and magnetic fields as follows. By taking the curls of Eqs. (2.8) and (2.9), we obtain:

$$\nabla \times (\nabla \times \boldsymbol{E}) = -\mu_0 \partial \left(\underbrace{\nabla \times \boldsymbol{H}}_{\varepsilon_0 \partial \boldsymbol{E} / \partial t} \right) / \partial t = -\varepsilon_0 \mu_0 \partial^2 \boldsymbol{E} / \partial t^2 \tag{2.12}$$

$$\nabla \times (\nabla \times \boldsymbol{H}) = \varepsilon_0 \partial \left(\underbrace{\nabla \times \boldsymbol{E}}_{-\mu_0 \partial \boldsymbol{H} / \partial t} \right) / \partial t = -\varepsilon_0 \mu_0 \partial^2 \boldsymbol{H} / \partial t^2 \tag{2.13}$$

By employing the following vector identity:

$$\nabla \times (\nabla \times \ ) = \nabla (\nabla \cdot \ ) - \nabla^2 (\ ), \tag{2.14}$$

and the divergence conditions (2.10), (2.11), we obtain the *wave equations*:

$$\nabla^2 \boldsymbol{E} = \frac{1}{c^2} \frac{\partial^2 \boldsymbol{E}}{\partial t^2}, \ \nabla^2 \boldsymbol{H} = \frac{1}{c^2} \frac{\partial^2 \boldsymbol{H}}{\partial t^2}, \tag{2.15}$$

where

$$c = (\varepsilon_0 \mu_0)^{-1/2} = 299792458 \text{ m/s} \tag{2.16}$$

is the speed of light. Therefore, the changes of electric and magnetic fields in vacuum propagate with a speed equal to the speed of light.

Maxwell's equations for electric and magnetic fields in nonconducting isotropic media are the same as those for vacuum, except that constants ε_0 and μ_0 for vacuum need to be replaced by the corresponding constant of medium ε and μ. The same observation is applicable to the wave equations. As a consequence, the speed of propagation v of electromagnetic fields in a medium is determined by:

$$v = (\varepsilon\mu)^{-1/2} \tag{2.17}$$

The medium constants ε and μ are related to the vacuum constants ε_0 and μ_0 as follows:

$$\varepsilon = \varepsilon_0\varepsilon_r, \quad \mu = \mu_0\mu_m, \tag{2.18}$$

where ε_r is called dielectric constant or the relative permittivity, while μ_m is called the relative permeability. The ratio of the speed of light in vacuum c and the speed of light in medium is known as *index of refraction n* and can be written as:

$$n = c/v = \left(\frac{\varepsilon\mu}{\varepsilon_0\mu_0}\right)^{1/2} = \left(\frac{\varepsilon_0\varepsilon_r\mu_0\mu_m}{\varepsilon_0\mu_0}\right)^{1/2} = (\varepsilon_r\mu_m)^{1/2}. \tag{2.19}$$

Transparent optical media are typically nonmagnetic ($\mu_m = 1$) so that the index of refraction is related to the dielectric constant by $n = (\varepsilon_r)^{1/2}$. As an illustration, the indices of refraction of the yellow light in the air [at 101.325 kPa (1 atm)] and water are 1.0002926 and 1.33, respectively. The index of refraction in the glass ranges from 1.5 to 1.7, depending on the type of glass.

2.1.1 Vector Derivatives (grad, div, and curl) in Orthogonal Curvilinear Coordinate System

At this point is convenient to introduce the orthogonal *curvilinear* coordinate system [5, 6]. Any point in orthogonal curvilinear coordinate system (u, v, w) is determined by the intersection of three warped planes: $u = \text{const}$, $v = \text{const}$, and $w = \text{const}$. Let $u = u(x,y,z)$, $v = v(x,y,z)$, and $w = u(x,y,z)$ be three independent (uniquely) differentiable functions of Cartesian coordinates (x,y,z). Then arbitrary point $P(x,y,z)$ in Cartesian coordinates get mapped by these functions into point $P(u, v, w)$ in orthogonal curvilinear coordinates system u, v, w. The infinitesimal increment in Cartesian coordinates $\mathrm{d}\boldsymbol{r} = (\mathrm{d}x, \mathrm{d}y, \mathrm{d}z)$ can be represented in curvilinear coordinate system as:

$$d\boldsymbol{r} = \frac{\partial \boldsymbol{r}}{\partial u}du + \frac{\partial \boldsymbol{r}}{\partial v}dv + \frac{\partial \boldsymbol{r}}{\partial w}dw = h_u du\widehat{\boldsymbol{u}} + h_v dv\widehat{\boldsymbol{v}} + h_w dw\widehat{\boldsymbol{w}}, \tag{2.20}$$

where Lamé coefficients h_u, h_v, and h_w are defined, respectively, as:

$$h_u = \left|\frac{\partial \boldsymbol{r}}{\partial u}\right| = \sqrt{\left(\frac{\partial x}{\partial u}\right)^2 + \left(\frac{\partial y}{\partial u}\right)^2 + \left(\frac{\partial z}{\partial u}\right)^2}, h_v = \left|\frac{\partial \boldsymbol{r}}{\partial v}\right| = \sqrt{\left(\frac{\partial x}{\partial v}\right)^2 + \left(\frac{\partial y}{\partial v}\right)^2 + \left(\frac{\partial z}{\partial v}\right)^2},$$

cropenup6pt

$$h_w = \left|\frac{\partial \boldsymbol{r}}{\partial w}\right| = \sqrt{\left(\frac{\partial x}{\partial w}\right)^2 + \left(\frac{\partial y}{\partial w}\right)^2 + \left(\frac{\partial z}{\partial w}\right)^2}, \tag{2.21}$$

and $(\widehat{\boldsymbol{u}}, \widehat{\boldsymbol{v}}, \widehat{\boldsymbol{w}})$ are unit-length vectors along coordinate axes in curvilinear coordinate system. The line element can now be defined as:

$$dl = \sqrt{d\boldsymbol{r} \cdot d\boldsymbol{r}} = \sqrt{(h_u du)^2 + (h_v dv)^2 + (h_w dw)^2}. \tag{2.22}$$

The volume element in curvilinear coordinate system is simply:

$$dV = (\widehat{\boldsymbol{u}} \cdot d\boldsymbol{r})\widehat{\boldsymbol{u}} \cdot \{[(\widehat{\boldsymbol{v}} \cdot d\boldsymbol{r})\widehat{\boldsymbol{v}}] \times [(\widehat{\boldsymbol{w}} \cdot d\boldsymbol{r})\widehat{\boldsymbol{w}}]\} = h_u h_v h_w du dv dw. \tag{2.23}$$

The surface elements perpendicular to each coordinate curve are simply:

cropenup8pt
$$
\begin{aligned}
dS_u = (h_v dv)(h_w dw) = h_v h_w dv dw,\ dS_v = (h_u du)(h_w dw) = h_u h_w du dw, \\
dS_w = (h_u du)(h_v dv) = h_u h_v du dv.
\end{aligned} \tag{2.24}
$$

Del or nabla operator in orthogonal curvilinear coordinate system is defined as:

$$
\nabla = \frac{1}{h_u}\frac{\partial}{\partial u}\widehat{\boldsymbol{u}} + \frac{1}{h_v}\frac{\partial}{\partial v}\widehat{\boldsymbol{v}} + \frac{1}{h_u}\frac{\partial}{\partial w}\widehat{\boldsymbol{w}}. \tag{2.25}
$$

The gradient, divergence, and curl in curvilinear coordinates can be determined as follows. The *gradient* of a scalar function φ is a vector:

$$
\text{grad}\,\varphi = \nabla\varphi = \frac{1}{h_u}\frac{\partial\varphi}{\partial u}\widehat{\boldsymbol{u}} + \frac{1}{h_v}\frac{\partial\varphi}{\partial v}\widehat{\boldsymbol{v}} + \frac{1}{h_u}\frac{\partial\varphi}{\partial w}\widehat{\boldsymbol{w}}. \tag{2.26}
$$

The *divergence* of a vector function $\boldsymbol{A}$ is a scalar:

$$
\text{div}\boldsymbol{A} = \nabla\boldsymbol{A} = \frac{1}{h_u h_v h_w}\left[\frac{\partial(h_v h_w A_u)}{\partial u} + \frac{\partial(h_u h_w A_v)}{\partial v} + \frac{\partial(h_u h_v A_w)}{\partial w}\right]. \tag{2.27}
$$

The *curl* of a vector function $\boldsymbol{A} = A_u\widehat{\boldsymbol{u}} + A_v\widehat{\boldsymbol{v}} + A_w\widehat{\boldsymbol{w}}$ is a vector:

$$
\text{curl }\boldsymbol{A} = \nabla\times\boldsymbol{A} = \frac{1}{h_u h_v h_w}\begin{vmatrix} h_u\widehat{\boldsymbol{u}} & h_v\widehat{\boldsymbol{v}} & h_w\widehat{\boldsymbol{w}} \\ \dfrac{\partial}{\partial u} & \dfrac{\partial}{\partial v} & \dfrac{\partial}{\partial w} \\ h_u A_u & h_v A_v & h_w A_w \end{vmatrix}. \tag{2.28}
$$

Finally, the *Laplacian* $\Delta = \nabla^2 = \nabla\cdot\nabla$ of a scalar function φ is a scalar:

$$
\text{div grad}\,\varphi = \nabla^2\varphi = \frac{1}{h_u h_v h_w}\left[\frac{\partial}{\partial u}\left(\frac{h_v h_w}{h_u}\frac{\partial\varphi}{\partial u}\right) + \frac{\partial}{\partial v}\left(\frac{h_u h_w}{h_v}\frac{\partial\varphi}{\partial v}\right) + \frac{\partial}{\partial w}\left(\frac{h_u h_v}{h_w}\frac{\partial\varphi}{\partial w}\right)\right]. \tag{2.29}
$$

Now by using the div and curl Eqs. (2.27) and (2.28), we are in position to solve Maxwell's equations (2.8, 2.9, 2.10, and 2.11), subject to the appropriate boundary conditions. On the other hand, by using the Laplacian Eq. (2.29), we can solve the wave equations (2.15).

In *Cartesian coordinate system*, the generalized coordinates are $u = x$, $v = y$, and $w = z$ ($\widehat{\boldsymbol{u}} = \widehat{\boldsymbol{x}}, \widehat{\boldsymbol{v}} = \widehat{\boldsymbol{y}}, \widehat{\boldsymbol{w}} = \widehat{\boldsymbol{z}}$) so that Lamé coefficients h_u, h_v, and h_w are all equal to 1. The corresponding vector derivatives are:

$$
\text{grad}\,\varphi = \nabla\varphi = \frac{\partial\varphi}{\partial x}\widehat{\boldsymbol{x}} + \frac{\partial\varphi}{\partial y}\widehat{\boldsymbol{y}} + \frac{\partial\varphi}{\partial z}\widehat{\boldsymbol{z}}, \tag{2.30}
$$

$$
\text{div}\boldsymbol{A} = \nabla\boldsymbol{A} = \frac{\partial A_x}{\partial x} + \frac{\partial A_y}{\partial y} + \frac{\partial A_z}{\partial z}, \tag{2.31}
$$

cropenup3pt
$$
\begin{aligned}
\text{curl }\boldsymbol{A} = \nabla\times\boldsymbol{A} &= \begin{vmatrix} \widehat{\boldsymbol{x}} & \widehat{\boldsymbol{y}} & \widehat{\boldsymbol{z}} \\ \dfrac{\partial}{\partial x} & \dfrac{\partial}{\partial y} & \dfrac{\partial}{\partial z} \\ A_x & A_y & A_z \end{vmatrix} \\
&= \left(\frac{\partial A_z}{\partial y} - \frac{\partial A_y}{\partial z}\right)\widehat{\boldsymbol{x}} + \left(\frac{\partial A_x}{\partial z} - \frac{\partial A_z}{\partial x}\right)\widehat{\boldsymbol{y}} + \left(\frac{\partial A_y}{\partial x} - \frac{\partial A_x}{\partial y}\right)\widehat{\boldsymbol{z}},
\end{aligned} \tag{2.32}
$$

$$\text{div grad}\,\varphi = \nabla^2\varphi = \frac{\partial^2\varphi}{\partial u^2} + \frac{\partial^2\varphi}{\partial v^2} + \frac{\partial^2\varphi}{\partial w^2}. \tag{2.33}$$

The vector derivatives (grad, div, curl, and Laplacian) in cylindrical polar and spherical coordinate systems are described in incoming section.

2.1.2 Vector Derivatives (grad, div, and curl) in Cylindrical Polar and Spherical Coordinate Systems

The *cylindrical polar coordinates* correspond to the polar (ρ,ϕ) coordinates with an additional z-coordinate directed out of the xy-plane. The cylindrical polar coordinates (ρ,ϕ,z) are related to Cartesian coordinates (x,y,z) as follows:

$$x = \rho\cos\phi, y = \rho\sin\phi, z = z \tag{2.34}$$

The Lamé coefficients h_ρ, h_ϕ, and h_z are determined as follows:

$$\begin{aligned}
h_\rho &= \sqrt{\left(\frac{\partial x}{\partial\rho}\right)^2 + \left(\frac{\partial y}{\partial\rho}\right)^2 + \left(\frac{\partial z}{\partial\rho}\right)^2} = \sqrt{(\cos\phi)^2 + (\sin\phi)^2 + 0 = 1} \\
\textit{cropenup3pt}\; h_\phi &= \sqrt{\left(\frac{\partial x}{\partial\phi}\right)^2 + \left(\frac{\partial y}{\partial\phi}\right)^2 + \left(\frac{\partial z}{\partial\phi}\right)^2} = \sqrt{(-\rho\sin\phi)^2 + (\rho\cos\phi)^2 + 0 = \rho} \\
h_z &= \sqrt{\left(\frac{\partial x}{\partial z}\right)^2 + \left(\frac{\partial y}{\partial z}\right)^2 + \left(\frac{\partial z}{\partial z}\right)^2} = \sqrt{0+0+1} = 1
\end{aligned} \tag{2.35}$$

The unit vectors in cylindrical polar coordinate system are related to the unit vectors in Cartesian coordinate system by:

$$\hat{\boldsymbol{\rho}} = \cos\phi\hat{\boldsymbol{x}} + \sin\phi\hat{\boldsymbol{y}}, \hat{\boldsymbol{\phi}} = -\sin\phi\hat{\boldsymbol{x}} + \cos\varphi\hat{\boldsymbol{y}}, \hat{z} = \hat{z} \tag{2.36}$$

By replacing Lamé coefficients from Eq. (2.35) into Eqs. (2.26, 2.27, 2.28, and 2.29), the operators of space differentiation become:

$$\text{grad}\,\varphi = \nabla\varphi = \frac{\partial\varphi}{\partial\rho}\hat{\boldsymbol{\rho}} + \frac{1}{\rho}\frac{\partial\varphi}{\partial\phi}\hat{\boldsymbol{\phi}} + \frac{\partial\varphi}{\partial z}\hat{z}, \tag{2.37}$$

$$\text{div}\boldsymbol{A} = \nabla\boldsymbol{A} = \frac{1}{\rho}\left[\frac{\partial(A_\rho\rho)}{\partial\rho} + \frac{\partial A_\phi}{\partial\phi} + \frac{\partial A_z}{\partial z}\right], \tag{2.38}$$

$$\text{curl}\,\boldsymbol{A} = \nabla\times\boldsymbol{A} = \frac{1}{\rho}\begin{vmatrix} \hat{\boldsymbol{\rho}} & \rho\hat{\boldsymbol{\phi}} & \hat{z} \\ \frac{\partial}{\partial\rho} & \frac{\partial}{\partial\phi} & \frac{\partial}{\partial z} \\ A_\rho & \rho A_\phi & A_z \end{vmatrix}, \tag{2.39}$$

$$\begin{aligned}
\text{div grad}\,\varphi = \nabla^2\varphi &= \frac{1}{\rho}\left[\frac{\partial}{\partial\rho}\left(\rho\frac{\partial\varphi}{\partial\rho}\right) + \frac{\partial}{\partial\phi}\left(\frac{1}{\rho}\frac{\partial\varphi}{\partial\phi}\right) + \frac{\partial}{\partial z}\left(\rho\frac{\partial\varphi}{\partial z}\right)\right] \\
&= \frac{1}{\rho}\frac{\partial}{\partial\rho}\left(\rho\frac{\partial\varphi}{\partial\rho}\right) + \frac{1}{\rho^2}\frac{\partial^2\varphi}{\partial\phi^2} + \frac{\partial^2\varphi}{\partial z^2}.
\end{aligned} \tag{2.40}$$

The *spherical coordinates* (ρ,θ,ϕ) are related to the Cartesian coordinates (x,y,z) through the following relations:

$$x=\rho\cos\phi\sin\theta, y=\rho\sin\phi\sin\theta, z=\rho\cos\theta \tag{2.41}$$

The Lamé coefficients h_ρ, h_θ, and h_ϕ are determined as follows:

$$\begin{aligned} h_\rho &= \sqrt{\left(\frac{\partial x}{\partial\rho}\right)^2+\left(\frac{\partial y}{\partial\rho}\right)^2+\left(\frac{\partial z}{\partial\rho}\right)^2}=1 \\ h_\theta &= \sqrt{\left(\frac{\partial x}{\partial\theta}\right)^2+\left(\frac{\partial y}{\partial\theta}\right)^2+\left(\frac{\partial z}{\partial\theta}\right)^2}=\rho \\ h_\phi &= \sqrt{\left(\frac{\partial x}{\partial\phi}\right)^2+\left(\frac{\partial y}{\partial\phi}\right)^2+\left(\frac{\partial z}{\partial\phi}\right)^2}=\rho\sin\theta \end{aligned} \tag{2.42}$$

By replacing Lamé coefficients from Eq. (2.42) into Eqs. (2.26, 2.27, 2.28, and 2.29), the operators of space differentiation become:

$$\mathrm{grad}\,\varphi=\nabla\varphi=\frac{\partial\varphi}{\partial\rho}\widehat{\boldsymbol{\rho}}+\frac{1}{\rho}\frac{\partial\varphi}{\partial\theta}\widehat{\boldsymbol{\theta}}+\frac{1}{\rho\sin\theta}\frac{\partial\varphi}{\partial\phi}\widehat{\boldsymbol{\phi}}, \tag{2.43}$$

$$\mathrm{div}\boldsymbol{A}=\nabla\boldsymbol{A}=\frac{1}{\rho^2\sin\theta}\left[\sin\theta\frac{\partial\left(A_\rho\rho^2\right)}{\partial\rho}+\rho\frac{\partial(\sin\theta A_\theta)}{\partial\theta}+\rho\frac{\partial A_\phi}{\partial\phi}\right], \tag{2.44}$$

$$\mathrm{curl}\ \boldsymbol{A}=\nabla\times\boldsymbol{A}=\frac{1}{\rho^2\sin\theta}\begin{vmatrix}\widehat{\boldsymbol{\rho}} & \rho\widehat{\boldsymbol{\theta}} & \rho\sin\theta\widehat{\boldsymbol{\phi}}\\ \dfrac{\partial}{\partial\rho} & \dfrac{\partial}{\partial\theta} & \dfrac{\partial}{\partial\phi}\\ A_\rho & \rho A_\theta & \rho\sin\theta A_\phi\end{vmatrix}, \tag{2.45}$$

$$\begin{aligned}\mathrm{div\ grad}\,\varphi=\nabla^2\varphi &=\frac{1}{\rho^2\sin\theta}\left[\sin\theta\frac{\partial}{\partial\rho}\left(\rho^2\frac{\partial\varphi}{\partial\rho}\right)+\frac{\partial}{\partial\theta}\left(\sin\theta\frac{\partial\varphi}{\partial\theta}\right)+\frac{1}{\sin\theta}\frac{\partial^2\varphi}{\partial\phi^2}\right]\\ &=\frac{1}{\rho^2}\frac{\partial}{\partial\rho}\left(\rho^2\frac{\partial\varphi}{\partial\rho}\right)+\frac{1}{\rho^2\sin\theta}\frac{\partial}{\partial\theta}\left(\sin\theta\frac{\partial\varphi}{\partial\theta}\right)+\frac{1}{\rho^2\sin^2\theta}\frac{\partial^2\varphi}{\partial\phi^2}.\end{aligned} \tag{2.46}$$

2.2 Propagation of Electromagnetic Waves

2.2.1 Propagation of Plane Waves

If we resolve the vector wave equations for electric and magnetic fields, Eq. (2.15), into Cartesian coordinates, we will notice that each component will satisfy the following scalar equation:

$$\nabla^2U=\frac{\partial^2U}{\partial x^2}+\frac{\partial^2U}{\partial y^2}+\frac{\partial^2U}{\partial z^2}=\frac{1}{v^2}\frac{\partial^2U}{\partial t^2},\ U\in\left\{E_x,E_y,E_z,H_x,H_y,H_z\right\} \tag{2.47}$$

It can easily verified by substitution that the following three-dimensional (3D) *plane harmonic wave function* satisfies the Eq. (2.47):

$$U(x,y,z,t)=U_0e^{\,j(\omega t-\boldsymbol{k}\cdot\boldsymbol{r})}, \tag{2.48}$$

where $\boldsymbol{r}$ is the position vector given by $\boldsymbol{r}=x\widehat{\boldsymbol{x}}+y\widehat{\boldsymbol{y}}+z\widehat{\boldsymbol{z}}$, ω is the carrier angular frequency, and $\boldsymbol{k}$ is the *propagation (wave) vector* represented in Cartesian coordinates as $\boldsymbol{k}=k_x\widehat{\boldsymbol{x}}+k_y\widehat{\boldsymbol{y}}+k_z\widehat{\boldsymbol{z}}$ (j is the imaginary unit). The magnitude of propagation vector is the (angular) wavenumber $k = |\boldsymbol{k}| = \omega/v = 2\pi/\lambda$, where λ is the *wavelength* (the propagation distance over which the wave function goes through one cycle). The *surfaces of constant phase*, also known as *wavefronts*, are determined by:

$$\omega t - \boldsymbol{k}\cdot\boldsymbol{r} = \omega t - k_x x - k_y y - k_z z = \text{const}, \tag{2.49}$$

and the corresponding *phase velocity* is given by:

$$v = \frac{\omega}{k} = \frac{\omega}{\sqrt{k_x^2 + k_y^2 + k_z^2}}. \tag{2.50}$$

Clearly, the equiphase surfaces of plane wave travel at phase velocity speed in direction of propagation vector $\boldsymbol{k}$, which is normal to the plane wave surfaces.

Let us now observe two harmonic wave traveling slightly different angular frequencies denoted, respectively, as $\omega - \Delta\omega$ and $\omega + \Delta\omega$. The propagation constants can be slightly different as well: $k - \Delta k$ and $k + \Delta k$. By assuming that these two waves propagate along z-axis, by employing the superposition principle, we can write:

$$U = U_0 e^{\,j[(\omega-\Delta\omega)t-(k-\Delta k)z]} + U_0 e^{\,j[(\omega+\Delta\omega)t-(k+\Delta k)z]} = 2\cos(\Delta\omega t - \Delta k z)U_0 e^{\,j(\omega t+kz)}. \tag{2.51}$$

Therefore, in this case, the phase travels at phase velocity, while the envelope travels at the *group velocity* v_g, defined as:

$$v_g = \frac{d\omega}{dk} = v - \lambda\frac{dv}{d\lambda} = v\left(1 - \frac{k}{n}\frac{dn}{dk}\right). \tag{2.52}$$

In media in which signals at different wavelengths travel at different speeds, the *dispersion* effect occurs.

The action of the time derivative on the harmonic wave, given by Eq. (2.48), is simply multiplication by $j\omega$. On the other hand, the action of del operator ∇ is multiplication with $-j\boldsymbol{k}$. By taking these two properties into account, the Maxwell's equations for plane harmonic waves become:

$$\boldsymbol{k} \times \boldsymbol{E} = \omega\mu\boldsymbol{H} \tag{2.53}$$

$$\boldsymbol{k} \times \boldsymbol{H} = -\omega\varepsilon\boldsymbol{E} \tag{2.54}$$

$$\boldsymbol{k}\cdot\boldsymbol{E} = 0 \tag{2.55}$$

$$\boldsymbol{k}\cdot\boldsymbol{H} = 0 \tag{2.56}$$

From harmonic wave Maxwell's equations, we conclude that electric and magnetic fields are perpendicular to one another and they are simultaneously perpendicular to the direction of propagation. The magnitudes of electric and magnetic fields are related as:

$$H = \varepsilon \underbrace{\frac{\omega}{k}}_{v} E = \varepsilon v E = n\frac{E}{Z_0}, Z_0 = \sqrt{\frac{\mu_0}{\varepsilon_0}} = 120\pi \cong 377\ \Omega \tag{2.57}$$

where Z_0 is commonly referred to as the *impedance of the free space*. The impedance of the medium, characterized by dielectric constant ε and magnetic permittivity μ, will be then $Z = \sqrt{\mu/\varepsilon}$.

The flow of electromagnetic energy per unit area is given by the *Poynting vector*, defined as the cross product of the electric and magnetic field vectors:

$$\boldsymbol{S} = \boldsymbol{E} \times \boldsymbol{H}^*. \tag{2.58}$$

For plane harmonic waves, the electric and magnetic fields components have the form given by Eq. (2.48) so that the Poynting vector becomes:

$$\boldsymbol{S} = \boldsymbol{E} \times \boldsymbol{H}^* = \boldsymbol{E}_0 e^{j(\omega t - \boldsymbol{k}\cdot\boldsymbol{r})} \times \boldsymbol{H}_0^* e^{-j(\omega t - \boldsymbol{k}\cdot\boldsymbol{r})} = \boldsymbol{E}_0 \times \boldsymbol{H}_0^* = \underbrace{E_0 H_0}_{I} \underbrace{\frac{\boldsymbol{k}}{k}}_{\widehat{\boldsymbol{n}}} = I\widehat{\boldsymbol{n}}, \tag{2.59}$$

where

$$I = E_0 H_0 = |E_0|^2 / \underbrace{(Z_0/n)}_{Z} = |E_0|^2 / Z$$

is known as *irradiance* and $\widehat{\boldsymbol{n}}$ is a unit vector in the direction of propagation.

2.2.2 Vectorial Nature of the Light, Snell's Law of Refraction, Reflection Coefficients, and Total Internal Reflection

2.2.2.1 Polarization

The electric/magnetic field of *plane linearly polarized waves* is described as follows [1, 7–10]:

$$\boldsymbol{A}(\boldsymbol{r}, t) = \boldsymbol{p} A_0 \exp\left[j(\omega t - \boldsymbol{k}\cdot\boldsymbol{r})\right], \ \boldsymbol{A} \in \{\boldsymbol{E}, \boldsymbol{H}\} \tag{2.60}$$

where $\boldsymbol{E}$ ($\boldsymbol{H}$) denotes electric (magnetic) field, $\boldsymbol{p}$ denotes the polarization orientation, $\boldsymbol{r}$ is the position vector, and $\boldsymbol{k}$ denotes the wave vector. For the x-polarization waves ($\boldsymbol{p} = \widehat{\boldsymbol{x}}, \widehat{\boldsymbol{k}} = k\widehat{\boldsymbol{z}}$), the Eq. (2.60) becomes:

$$\boldsymbol{E}_x(z, t) = \widehat{\boldsymbol{x}} E_{0x} \cos(\omega t - kz), \tag{2.61}$$

while for y-polarization ($\boldsymbol{p} = \widehat{\boldsymbol{y}}, \widehat{\boldsymbol{k}} = k\widehat{\boldsymbol{z}}$), it becomes:

$$\boldsymbol{E}_y(z, t) = \widehat{\boldsymbol{y}} E_{0y} \cos(\omega t - kz + \delta), \tag{2.62}$$

where δ is the relative phase difference between the two orthogonal waves. The resultant wave can be obtained by combining (2.61) and (2.62) as follows:

$$\boldsymbol{E}(z, t) = \boldsymbol{E}_x(z, t) + \boldsymbol{E}_y(z, t) = \widehat{\boldsymbol{x}} E_{0x} \cos(\omega t - kz) + \widehat{\boldsymbol{y}} E_{0y} \cos(\omega t - kz + \delta). \tag{2.63}$$

The *linearly polarized* wave is obtained by setting the phase difference to an integer multiple of 2π, namely, $\delta = m \cdot 2\pi$:

$$\boldsymbol{E}(z, t) = (\widehat{\boldsymbol{x}} E \cos\theta + \widehat{\boldsymbol{y}} E \sin\theta) \cos(\omega t - kz); \ E = \sqrt{E_{0x}^2 + E_{0y}^2}, \theta = \tan^{-1}\frac{E_{0y}}{E_{0x}}. \tag{2.64}$$

By ignoring the time-dependent term, we can represent the linear polarization as shown in Fig. 2.1(a). On the other hand, if $\delta \neq m \cdot 2\pi$, the *elliptical polarization* is obtained. From Eqs. (2.61) and (2.62), by eliminating the time-dependent term, we obtain the following equation of ellipse:

$$\left(\frac{E_x}{E_{0x}}\right)^2 + \left(\frac{E_y}{E_{0y}}\right)^2 - 2\frac{E_x}{E_{0x}}\frac{E_y}{E_{0y}} \cos\delta = \sin^2\delta, \tag{2.65}$$

which is shown in Fig. 2.1(b). By setting $\delta = \pm\frac{\pi}{2}, \pm 3\frac{\pi}{2}, \ldots$, the equation of ellipse becomes:

$$\left(\frac{E_x}{E_{0x}}\right)^2 + \left(\frac{E_y}{E_{0y}}\right)^2 = 1. \tag{2.66}$$

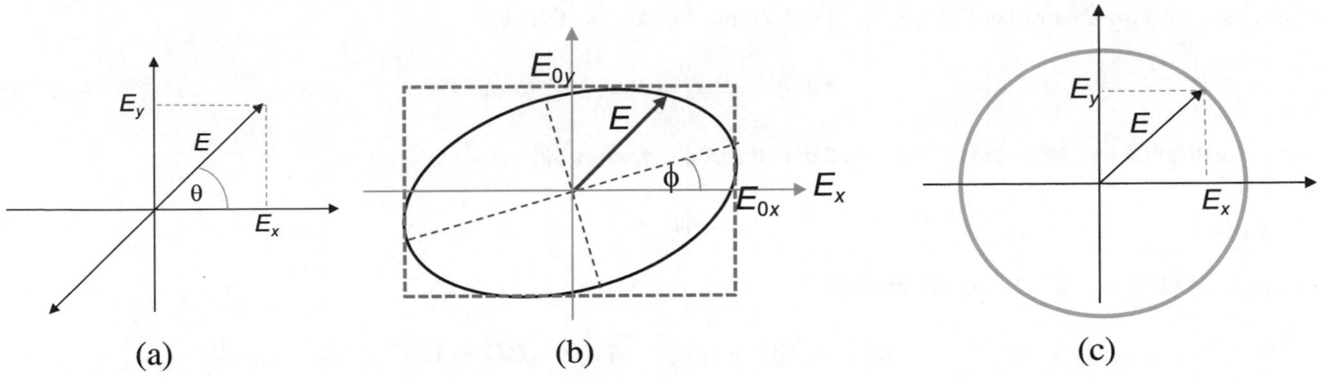

Fig. 2.1 Polarization forms: (**a**) linear, (**b**) elliptic, and (**c**) circular polarizations

By setting further $E_{0x} = E_{0y} = E_0$; $\delta = \pm\frac{\pi}{2}, \pm 3\frac{\pi}{2}, \ldots$, the equation of ellipse becomes the circle:

$$E_x^2 + E_y^2 = 1, \tag{2.67}$$

and corresponding polarization is known as *circular polarization* (see Fig. 2.1c). *Right circularly polarized* wave is obtained for $\delta = \pi/2 + 2m\pi$:

$$\boldsymbol{E} = E_0[\widehat{\boldsymbol{x}} \cos(\omega t - kz) - \widehat{\boldsymbol{y}} \sin(\omega t - kz)], \tag{2.68}$$

Otherwise, for $\delta = -\pi/2 + 2m\pi$, the polarization is known as left circularly polarized.

Very often, the *Jones vector representation* of polarization wave is used:

$$\boldsymbol{E}(t) = \begin{bmatrix} E_x(t) \\ E_y(t) \end{bmatrix} = \begin{bmatrix} E_{0x} \\ E_{0y}e^{j\delta} \end{bmatrix} e^{\,j(\omega t - kz)}, \tag{2.69}$$

with complex phasor term being typically omitted in practice. If we normalize the Jones vector by dividing by $\sqrt{E_{0x}^2 + E_{0y}^2}$, we can represent x-polarization state, denoted as $|x\rangle$, and y-polarization state, denoted as $|y\rangle$, respectively, by:

$$|x\rangle = \begin{bmatrix} 1 \\ 0 \end{bmatrix}, |y\rangle = \begin{bmatrix} 0 \\ 1 \end{bmatrix}. \tag{2.70}$$

On the other hand, right- and left-circular polarization states, denoted respectively as $|R\rangle$ and $|L\rangle$, can be written as:

$$|R\rangle = \frac{1}{\sqrt{2}}\begin{bmatrix} 1 \\ j \end{bmatrix}, |L\rangle = \frac{1}{\sqrt{2}}\begin{bmatrix} 1 \\ -j \end{bmatrix}. \tag{2.71}$$

In Eqs. (2.70 and 2.71), we used Dirac notation to denote the column vectors [7]. In Dirac notation, with each column vector ("ket") $|\psi\rangle$, defined as

$$|\psi\rangle = \begin{bmatrix} \psi_x \\ \psi_y \end{bmatrix}, \tag{2.72}$$

we associate a row vector ("bra") $\langle\psi|$ as follows:

$$\langle\psi| = \begin{bmatrix} \psi_x^* & \psi_y^* \end{bmatrix}. \tag{2.73}$$

The *scalar (dot) product* of ket $|\phi\rangle$ bra $\langle\psi|$ is defined by "bracket" as follows:

$$\langle\phi|\psi\rangle = \phi_x^*\psi_x + \phi_y^*\psi_y = \langle\psi|\phi\rangle^*. \tag{2.74}$$

The normalization condition can be expressed in terms of dot product by:

$$\langle\psi|\psi\rangle = 1. \tag{2.75}$$

Based on (2.70 and 2.71), it is evident that:

$$\langle x|x\rangle = \langle y|y\rangle = 1, \qquad \langle R|R\rangle = \langle L|L\rangle = 1.$$

Because the vectors $|x\rangle$ and $|y\rangle$ are *orthogonal*, their dot product is zero:

$$\langle x|y\rangle = 0, \tag{2.76}$$

and they form the *basis*. Any polarization state vector $|\psi\rangle$ can be written as a *linear superposition* of basis kets as follows:

$$|\psi\rangle = \begin{pmatrix} \psi_x \\ \psi_y \end{pmatrix} = \psi_x|x\rangle + \psi_y|y\rangle. \tag{2.77}$$

Another interesting representation is *Stokes vector representation*:

$$\mathbf{S}(t) = \begin{bmatrix} S_0(t) \\ S_1(t) \\ S_2(t) \\ S_3(t) \end{bmatrix}, \tag{2.78}$$

where the parameter S_0 is related to the optical intensity by:

$$S_0(t) = |E_x(t)|^2 + |E_y(t)|^2. \tag{2.79}$$

The parameter $S_1 > 0$ is related to the preference for horizontal polarization, and it is defined by:

$$S_1(t) = |E_x(t)|^2 - |E_y(t)|^2. \tag{2.80}$$

The parameter $S_2 > 0$ is related to the preference for $\pi/4$ SOP:

$$S_2(t) = E_x(t)E_y^*(t) + E_x^*(t)E_y(t). \tag{2.81}$$

Finally, the parameters $S_3 > 0$ is related to the preference for right-circular polarization, and it is defined by:

$$S_3(t) = j\left[E_x(t)E_y^*(t) - E_x^*(t)E_y(t)\right]. \tag{2.82}$$

The Stokes vector for the elliptically polarized light, given by Eq. (2.69), can be expressed as follows:

$$\mathbf{S} = \begin{bmatrix} S_0 \\ S_1 \\ S_2 \\ S_3 \end{bmatrix} = \begin{bmatrix} E_{0x}^2 + E_{0y}^2 \\ E_{0x}^2 - E_{0y}^2 \\ 2E_{0x}E_{0y}\cos\delta \\ 2E_{0x}E_{0y}\sin\delta \end{bmatrix}. \tag{2.83}$$

The parameter S_0 is related to other Stokes parameters by:

$$S_0^2(t) = S_1^2(t) + S_2^2(t) + S_3^2(t). \tag{2.84}$$

The *degree of polarization*, denoted as *DOP*, is defined by:

$$DOP = \frac{\left[S_1^2 + S_2^2 + S_3^2\right]^{1/2}}{S_0}, \qquad 0 \le DOP \le 1 \tag{2.85}$$

For $DOP = 1$ the light is completely polarized. The *natural light* is *unpolarized* as it represents *incoherent superposition* of 50% of light polarized along x-axis and 50% of light polarized along y-axis. The Stokes vector for unpolarized light is given by:

$$\boldsymbol{S}_{\text{unpolarized}} = \begin{bmatrix} S_0 \\ 0 \\ 0 \\ 0 \end{bmatrix} = S_0 \begin{bmatrix} 1 \\ 0 \\ 0 \\ 0 \end{bmatrix}. \tag{2.86}$$

The *partially polarized light* represents the mixture of the polarized light and unpolarized light so that the corresponding Stokes vector is given by:

$$\boldsymbol{S} = (1 - DOP) \begin{bmatrix} S_0 \\ 0 \\ 0 \\ 0 \end{bmatrix} + DOP \begin{bmatrix} S_0 \\ S_1 \\ S_2 \\ S_3 \end{bmatrix}. \tag{2.87}$$

The x-polarization (linear horizontal polarization) is defined by $E_{0y} = 0$, while the *y*-polarization (linear vertical polarization) by $E_{0x} = 0$ so that the corresponding Stokes representations are given, based on Eq. (2.83), by:

$$\boldsymbol{S}_x = S_0 \begin{bmatrix} 1 \\ 1 \\ 0 \\ 0 \end{bmatrix}, \boldsymbol{S}_y = S_0 \begin{bmatrix} 1 \\ -1 \\ 0 \\ 0 \end{bmatrix}. \tag{2.88}$$

The right-circular polarization is defined by $E_{0y} = E_{0x} = E_0$ and $\delta = \pi/2$, while left-circular polarization is defined by $E_{0y} = E_{0x} = E_0$ and $\delta = -\pi/2$ so that the corresponding Stokes representations are given, based on Eq. (2.83), by:

$$\boldsymbol{S}_R = S_0 \begin{bmatrix} 1 \\ 0 \\ 0 \\ 1 \end{bmatrix}, \boldsymbol{S}_L = S_0 \begin{bmatrix} 1 \\ 0 \\ 0 \\ -1 \end{bmatrix}. \tag{2.89}$$

After the *normalization* of Stokes coordinates S_i with respect to S_0, the normalized Stokes parameters are given by:

$$s_i = \frac{S_i}{S_0}; i = 0, 1, 2, 3. \tag{2.90}$$

If the normalized Stokes parameters are used, the polarization state can be represented as a point on a *Poincaré sphere*, as shown in Fig. 2.2. The points located at the opposite sides of the line crossing the center represent the orthogonal polarizations. The normalized Stokes (Cartesian) coordinates are related to the sphere coordinates, as defined in Fig. 2.2, by:

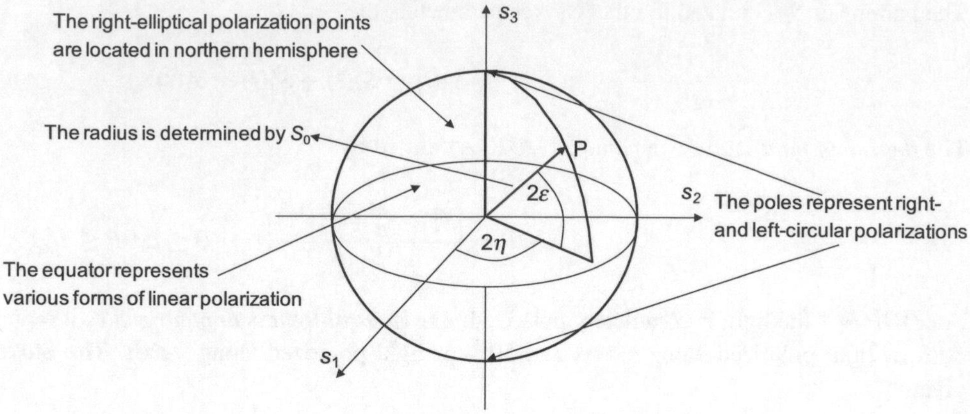

Fig. 2.2 Representation of polarization state as a point on a Poincaré sphere

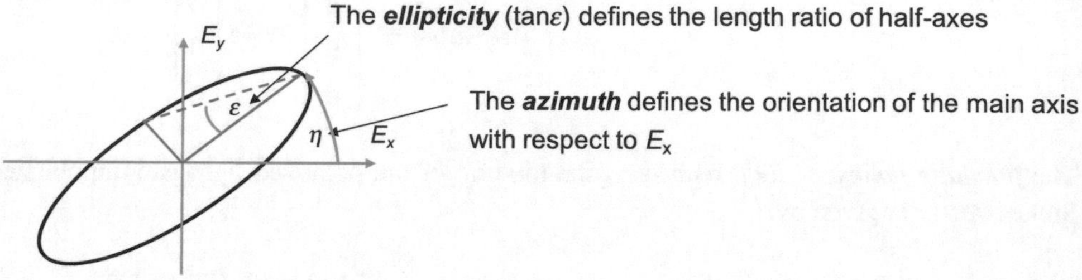

Fig. 2.3 The ellipticity and azimuth of the polarization ellipse

$$s_1 = \cos(2\varepsilon)\cos(2\eta), s_2 = \cos(2\varepsilon)\sin(2\eta), s_3 = \sin(2\varepsilon). \tag{2.91}$$

The *polarization ellipse* is very often represented in terms of the ellipticity and the azimuth, which are illustrated in Fig. 2.3. The ellipticity tanε is defined by the ratio of half-axes lengths. The corresponding angle is called *ellipticity angle* and denoted by ε. The small ellipticity means that polarization ellipse is highly elongated, while for zero ellipticity, the polarization is linear. For $\varepsilon = \pm\pi/4$, the polarization is circular. For $\varepsilon > 0$ the polarization is right-elliptical. On the other hand, the azimuth angle η defines the orientation of the main axis of ellipse with respect to E_x (x-axis). The Stokes parameters are related to the azimuth (orientation) angle η and ellipticity angle ε as follows:

$$S_1 = S_0\cos(2\varepsilon)\cos(2\eta), S_2 = S_0\cos(2\varepsilon)\sin(2\eta), S_3 = S_0\sin(2\varepsilon). \tag{2.92}$$

The azimuth angle η and ellipticity angle ε are related to the parameters of the polarization ellipse, given by Eq. (2.65), by

$$\tan 2\mu = \frac{2E_{0x}E_{0y}\cos\delta}{E_{0x}^2 - E_{0y}^2}, \ \sin 2\varepsilon = \frac{2E_{0x}E_{0y}\sin\delta}{E_{0x}^2 + E_{0y}^2}. \tag{2.93}$$

Let us now observe the *polarizer-analyzer ensemble*, shown in Fig. 2.4. When an electromagnetic wave passes through the polarizer, it can be represented as a vector in the xOy plane transversal to the propagation direction, as given by Eq. (2.63), where the angle θ depends on the filter orientation. By introducing the unit vector $\widehat{p} = (\cos\theta, \sin\theta)$, the Eq. (2.63) can be rewritten as:

$$\boldsymbol{E} = E_0\widehat{p}\cos(\omega t - kz). \tag{2.94}$$

If $\theta = 0$ rad, the light is polarized along x-axis, while for $\theta = \pi/2$ rad, it is polarized along y-axis. After the analyzer, whose axis makes an angle ϕ with respect to x-axis, which can be represented by unit vector $\widehat{n} = (\cos\phi, \sin\phi)$, the output electric field is given by:

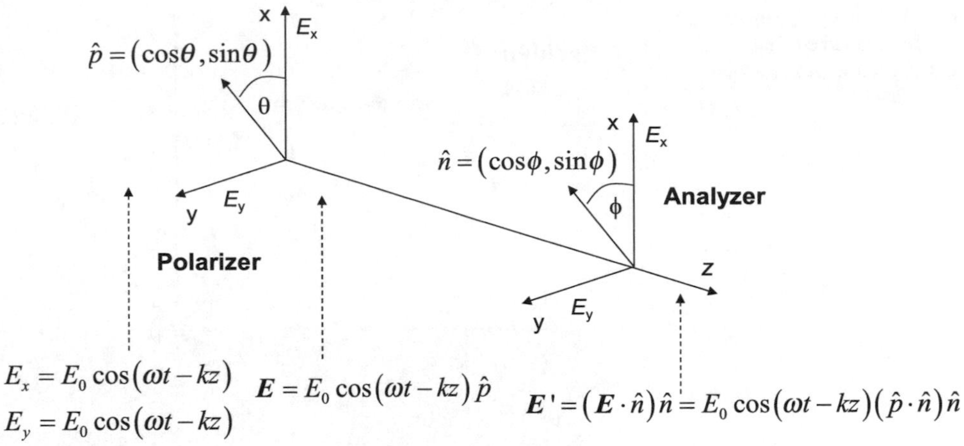

Fig. 2.4 The polarizer-analyzer ensemble for study of the photon polarization

$$\begin{aligned} \boldsymbol{E}' &= (\boldsymbol{E}\cdot\widehat{n})\widehat{n} = E_0\cos(\omega t - kz)(\widehat{p}\cdot\widehat{n})\widehat{n} \\ &= E_0\cos(\omega t - kz)[(\cos\theta, \sin\theta)\cdot(\cos\phi, \sin\phi)]\widehat{n} \\ &= E_0\cos(\omega t - kz)[\cos\theta\cos\phi + \sin\theta\sin\phi]\widehat{n} \\ &= E_0\cos(\omega t - kz)\cos(\theta - \phi)\widehat{n}. \end{aligned} \tag{2.95}$$

The intensity of the analyzer output field is given by:

$$I' = |\boldsymbol{E}'|^2 = \underbrace{|E_0|^2}_{I}\cos^2(\theta - \phi) = I\cos^2(\theta - \phi), \tag{2.96}$$

which is commonly referred to as *Malus law*.

2.2.2.2 Snell's Law of Refraction, Reflection Coefficients, and Total Internal Reflection

We now study space-time dependence of the incident $\boldsymbol{k}$, reflected $\boldsymbol{k}_r$, and transmitted $\boldsymbol{k}_t$ waves at the plane boundary as illustrated in Fig. 2.5. The space-time dependencies of electric/magnetic fields of these waves are given by:

$$\begin{aligned} \boldsymbol{A} = \boldsymbol{A}_0\exp[j(\omega t - \boldsymbol{k}\cdot\boldsymbol{r})],\ \boldsymbol{A}_r = \boldsymbol{A}_0\exp[j(\omega t - \boldsymbol{k}_r\cdot\boldsymbol{r})], \\ \boldsymbol{A}_t = \boldsymbol{A}_0\exp[j(\omega t - \boldsymbol{k}_t\cdot\boldsymbol{r})]. \end{aligned} \tag{2.97}$$

At the boundary, since the time terms are already equal, the space terms must be equal as well:

$$\boldsymbol{k}\cdot\boldsymbol{r} = \boldsymbol{k}_r\cdot\boldsymbol{r} = \boldsymbol{k}_t\cdot\boldsymbol{r}, \tag{2.98}$$

indicating that the propagation vectors are coplanar, laying in the same plane commonly referred to as the *incident plane*, and that their projections on the boundary (xz) plane are equal:

$$k\sin\theta = k_r\sin\theta_r = k_t\sin\phi. \tag{2.99}$$

Given that the incident and reflected wave propagate in the same medium corresponding wavenumbers are equal $k = k_r$ so that the first equation in (2.99) reduces to:

$$\theta = \theta_r, \tag{2.100}$$

which is known as the *reflection law*.

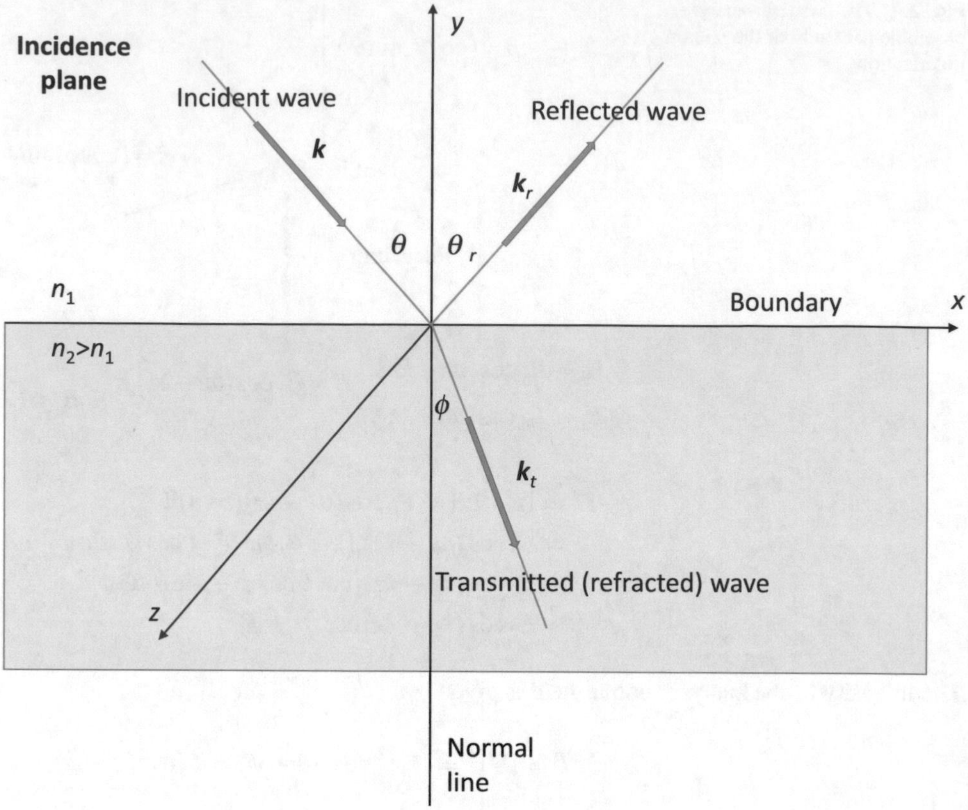

Fig. 2.5 Space dependence of incident, reflected, and transmitted waves at a plane boundary

On the other hand, because the ratio of wavenumbers of incident and transmitted waves can be written as:

$$\frac{k}{k_t} = \frac{\omega/v}{\omega/v_t} = \frac{1/v}{1/v_t} = \frac{c/v}{c/v_t} = \frac{n_1}{n_2}, \tag{2.101}$$

from Eq. (2.99), we obtain:

$$n_1 \sin\theta = n_2 \sin\phi, \tag{2.102}$$

which is known as *Snell's law* of refraction.

From Maxwell's curl Equations (2.53) and (2.54), we can determine the spatial dependence of propagation, electric field, and magnetic field vectors as illustrated in Fig. 2.6, where two cases of interest are observed. In Fig. 2.6(a), the incident electric field vector is parallel to the boundary plane, that is, perpendicular to the incidence plane. This case is commonly referred to as the *transverse electric* or *TE polarization.* The dot in the circle indicates that the electric field vector is perpendicular to the plane of incidence with direction towards the reader. In Fig. 2.6(b), the incident magnetic field vector is parallel to the boundary plane, that is, perpendicular to the incidence plane. This case is commonly referred to as the *transverse magnetic* or *TM polarization*. The cross in the circle indicates that the electric field vector is perpendicular to the plane of incidence with direction form the reader.

Now we apply well-known boundary conditions from electromagnetics [5, 8, 11] requiring that tangential components of the electrical and magnetic fields must be continuous as we cross over the interface of two media (the boundary plane). From Fig. 2.6(a), we conclude that for TE polarization $E + E_r = E_t$. On the other hand, from Fig. 2.6(b), we conclude that for TM polarization $H–H_r = H_t$, which is equivalent to $kE–kE_r = k_tE_t$, since H is proportional to kE. The magnetic field tangential components must satisfy boundary condition for TE polarization, which leads to:

$$-H\cos\theta + H_r\cos\theta = -H_t\cos\phi \Leftrightarrow -kE\cos\theta + kE_r\cos\theta = -k_tE_t\cos\phi \tag{2.104}$$

Fig. 2.6 Spatial dependence of wave vectors for: (**a**) TE polarization and (**b**) TM polarization

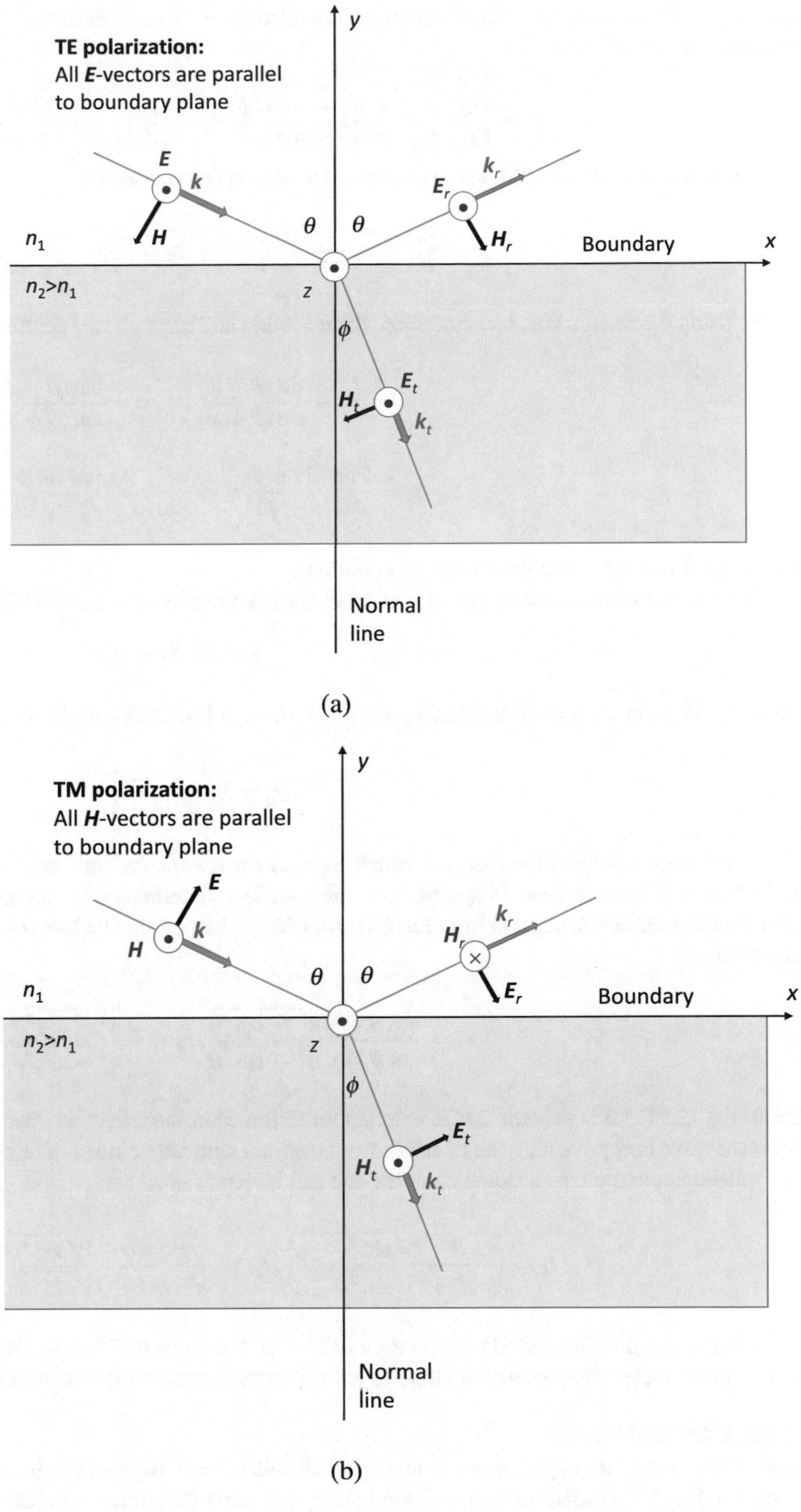

In equivalent representation of (2.104), we used the fact that H is proportional to kE. The electric field tangential components must satisfy boundary condition for TM polarization, which leads to:

$$E\cos\theta + E_r\cos\theta = E_t\cos\phi. \tag{2.105}$$

Based on the boundary conditions' equations provided above, we can determine the *coefficients of reflections* for TE and TM polarizations as follows:

$$r_s = \left[\frac{E_r}{E}\right]_{TE} = \frac{\cos\theta - n\cos\phi}{\cos\theta + n\cos\phi}, r_p = \left[\frac{E_r}{E}\right]_{TM} = \frac{\cos\phi - n\cos\theta}{\cos\phi + n\cos\theta}, n = \frac{n_2}{n_1}. \tag{2.106}$$

In similar fashion, the *coefficients of transmission* can be determined as:

$$t_s = \left[\frac{E_t}{E}\right]_{TE} = \frac{2\cos\theta}{\cos\theta + n\cos\phi}, t_p = \left[\frac{E_t}{E}\right]_{TM} = \frac{2\cos\theta}{\cos\phi + n\cos\theta}. \tag{2.107}$$

By using the Snell's law, $n = \sin\theta/\sin\phi$, the reflection and transmission coefficients can be rewritten in the following form:

$$r_s = -\frac{\sin(\theta - \phi)}{\sin(\theta + \phi)}, r_p = -\frac{\tan(\theta - \phi)}{\tan(\theta + \phi)}. \tag{2.108}$$

$$t_s = \frac{2\cos\theta\sin\phi}{\cos(\theta + \phi)}, t_p = \frac{2\cos\theta\sin\phi}{\sin(\theta + \phi)\cos(\theta - \phi)}, \tag{2.109}$$

and these forms are known as *Fresnel's equations*.

The fraction of reflected energy of incident light is known as *reflectance* and can be calculated as:

$$R_s = |r_s|^2, R_p = |r_p|^2. \tag{2.110}$$

As an illustration for normal incidence, when $\theta = 0$, from Eq. (2.98), we find that:

$$R_s = R_p = \left[\frac{n-1}{n+1}\right]^2. \tag{2.111}$$

So far we considered the case for which $n_2 > n_1$ (or $n > 1$), and this case is known as external reflection. In external reflection, the incident wave is approaching the boundary (interface) from the medium with smaller refraction index. If we eliminate the refractive angle ϕ from Eq. (2.106), with the help of Snell's law, $n = \sin\theta/\sin\phi$, the reflection coefficients can be rewritten as:

$$r_s = \frac{\cos\theta - \sqrt{n^2 - \sin^2\theta}}{\cos\theta + \sqrt{n^2 - \sin^2\theta}}, r_p = \frac{\sqrt{n^2 - \sin^2\theta} - n^2\cos\theta}{\sqrt{n^2 - \sin^2\theta} + n^2\cos\theta}. \tag{2.112}$$

From Eq. (2.112) it is evident that in external reflection case, the reflection coefficients are always real. However, when the incident wave is approaching the interface from medium with larger index of refraction, $n < 1$, for incident angles $\theta > \sin^{-1}n$, the reflection coefficients become complex and can be represented as:

$$r_s = \frac{\cos\theta - j\sqrt{\sin^2\theta - n^2}}{\cos\theta + j\sqrt{\sin^2\theta - n^2}}, r_p = \frac{-n^2\cos\theta + j\sqrt{\sin^2\theta - n^2}}{n^2\cos\theta + j\sqrt{\sin^2\theta - n^2}}, \theta > \tan^{-1}n. \tag{2.113}$$

This case is commonly referred to as *internal reflection*. The angle $\tan^{-1}n$ is known as the *critical angle*. The reflectance when the incident angle is larger than or equal to critical angle is equal to $R = 1$, indicating that *total reflection* occurs.

2.2.2.3 Optical Fibers

One of the important applications of total internal reflection is the propagation of light in *optical fibers*. The optical fiber, shown in Fig. 2.7, consists of two waveguide layers, the *core* (of refractive index n_1) and the *cladding* (of refractive index n_2), protected by the jacket (the buffer coating). The majority of the power is concentrated in the core, although some portion can spread to the cladding. There is a difference in refractive indices between the core and cladding ($n_1 > n_2$), which is achieved by mix of dopants commonly added to the fiber core. The refractive index profile for step-index fiber is shown in Fig. 2.7(a),

Fig. 2.7 Optical fibers: (**a**) refractive index profile for step-index fiber and (**b**) the light confinement in step-index fibers through the total internal reflection

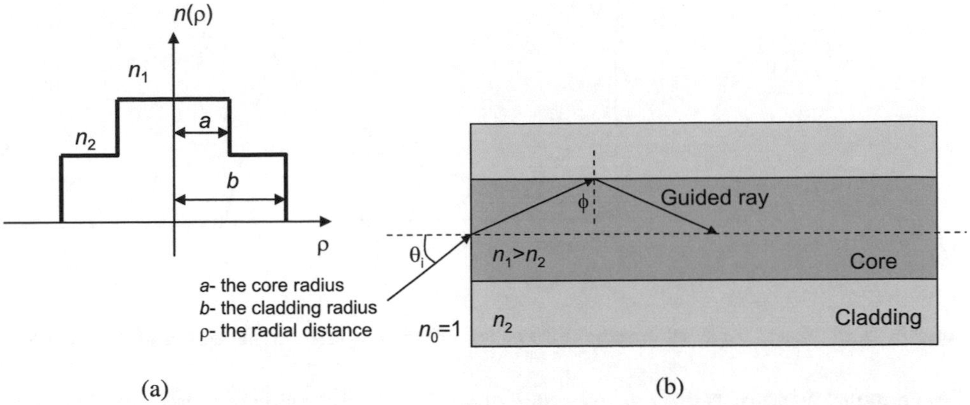

while the illustration of light confinement by the total internal reflection is shown in Fig. 2.7(b). The ray will be totally reflected from the core-cladding interface (a guided ray) if the following condition is satisfied: $n_0 \sin\theta_\text{i} < \sqrt{n_1^2 - n_2^2}$, where θ_i is the angle of incidence. The maximum semi-angle of the vertex of a cone of rays entering the fiber, known as the *acceptance angle*, can be determined by $\alpha = \sin^{-1}\sqrt{n_1^2 - n_2^2}$. The parameter max($n_0\sin\theta_\text{i}$), therefore, defines the light gathering capacity of an optical fiber, and it is called the *numerical aperture* (*NA*):

$$NA = \sqrt{n_1^2 - n_2^2} \approx n_1\sqrt{2\Delta}, \quad \Delta \ll 1 \tag{2.114}$$

where Δ is the *normalized index difference* defined as $\Delta = (n_1 - n_2)/n_1$. Therefore, from the geometrical optics point of view, light propagates in optical fiber due to series of total internal reflections that occur at the core-cladding interface. The smallest angle of incidence ϕ (see Fig. 2.7(b)) for which the total internal reflection occurs is called the critical angle and equals $\sin^{-1} n_2/n_1$.

2.2.2.4 Polarizing Elements

With the following three polarizing elements polarizer, phase shifter (retarder), and rotator, any elliptical polarization state can be obtained [9]. The action of polarizing element on polarization vector (state) given by Eq. (2.69) [Eq. (2.72)] can be described by *Jones operator (matrix)*:

$$|\psi\rangle = J|\psi_\text{in}\rangle = \begin{bmatrix} J_{xx} & J_{xy} \\ J_{yx} & J_{yy} \end{bmatrix} |\psi_\text{in}\rangle, |\psi_\text{in}\rangle = \begin{bmatrix} \psi_{x,\text{in}} \\ \psi_{y,\text{in}} \end{bmatrix} \tag{2.115}$$

where $|\psi_\text{in}\rangle$ is the input polarization state. *Linear polarizer* changes the amplitude of polarization vector, and it is characterized by absorption coefficients along x- and y-axis, denoted as p_x and p_y, respectively. The action of linear polarizer can be described by the following Jones matrix:

$$J_\text{polarizer} = \begin{bmatrix} p_x & 0 \\ 0 & p_y \end{bmatrix}; \ 0 \le p_i \le 1, i \in \{x, y\}. \tag{2.116}$$

Ideal linear horizontal polarizer is specified with $p_x = 1$ ($p_y = 0$), while an ideal vertical polarizer with $p_y = 1$ ($p_x = 0$).

The *wave plates* (phase shifters, retarders) are polarizing elements that introduce the phase shift $\phi/2$ along x axis (the fast axis) and the phase shift $-\phi/2$ along y-axis (the slow axis), and corresponding Jones matrix is given by:

$$J_\text{wave plate} = \begin{bmatrix} e^{j\phi/2} & 0 \\ 0 & e^{-j\phi/2} \end{bmatrix} \Leftrightarrow J_\text{wave plate} = \begin{bmatrix} 1 & 0 \\ 0 & e^{-j\phi} \end{bmatrix}. \tag{2.117}$$

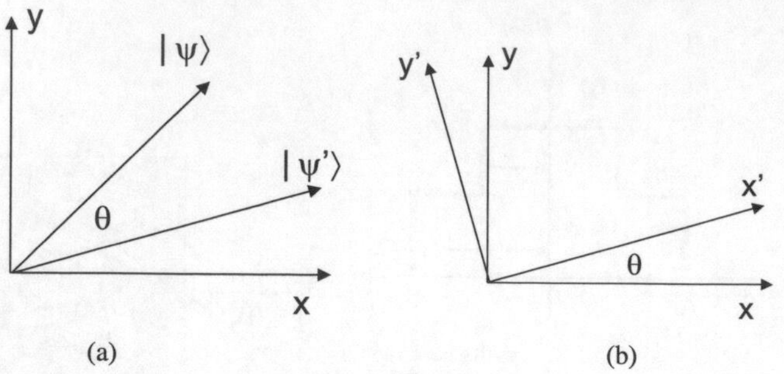

Fig. 2.8 Illustration of the transformation of the photon polarization basis: (**a**) the rotation of the state vector, (**b**) the rotation of a coordinate system

Two popular wave plats are *quarter-wave plate* (QWP), for which $\phi = \pi/2$, and *half-wave plate* (HWP), for which $\phi = \pi$. Their Jones matrix representations are as follows:

$$J_{\mathrm{QWP}} = \begin{bmatrix} 1 & 0 \\ 0 & -j \end{bmatrix}, J_{\mathrm{HWP}} = \begin{bmatrix} 1 & 0 \\ 0 & -1 \end{bmatrix}. \tag{2.118}$$

For instance, if the input state is linear ±45° state, denoted as $|\pm 45°\rangle = [1 \pm 1]^{\mathrm{T}}/\sqrt{2}$, after passing though QWP, it gets converted to:

$$J_{\mathrm{QWP}}|\pm 45°\rangle = \frac{1}{\sqrt{2}}\begin{bmatrix} 1 & 0 \\ 0 & -j \end{bmatrix}\begin{bmatrix} 1 \\ \pm 1 \end{bmatrix} = \frac{1}{\sqrt{2}}\begin{bmatrix} 1 \\ \mp j \end{bmatrix}, \tag{2.119}$$

in other words, left−/right-circular polarization state.

To determine the Jones matrix representation of *rotator*, let us observe the rotation operation as illustrated in Fig. 2.8. In Fig. 2.8(a), the vector state $|\psi\rangle$ is rotated for θ in a clockwise direction, which is equivalent to the rotation of coordinate system in opposite (counterclockwise) for the same angle (θ), as shown in Fig. 2.8(b).

The state $|\psi\rangle$ can be represented in original $\{|x\rangle, |y\rangle\}$ basis by:

$$|\psi\rangle = |x\rangle\langle x|\psi\rangle + |y\rangle\langle y|\psi\rangle, \tag{2.120}$$

where $\langle x|\psi\rangle$ ($\langle y|\psi\rangle$) denotes the projection along x-polarization (y-polarization). By multiplying (2.120) by $\langle x'|$ and $\langle y'|$ from the left side, we obtain:

$$\langle x'|\psi\rangle = \langle x'|x\rangle\langle x|\psi\rangle + \langle x'|y\rangle\langle y|\psi\rangle, \langle y'|\psi\rangle = \langle y'|x\rangle\langle x|\psi\rangle + \langle y'|y\rangle\langle y|\psi\rangle, \tag{2.121}$$

or by rewriting the Eq. (2.121) into matrix form, we obtain:

$$|\psi'\rangle = \begin{pmatrix} \langle x'|\psi\rangle \\ \langle y'|\psi\rangle \end{pmatrix} = \underbrace{\begin{pmatrix} \langle x'|x\rangle & \langle x'|y\rangle \\ \langle y'|x\rangle & \langle y'|y\rangle \end{pmatrix}}_{J(\theta)} \begin{pmatrix} \langle x|\psi\rangle \\ \langle y|\psi\rangle \end{pmatrix}. \tag{2.122}$$

From Fig. 2.8(b) it is clear that old basis $\{|x\rangle, |y\rangle\}$ is related to the new basis $\{|x'\rangle, |y'\rangle\}$ by:

$$|x\rangle = \cos\theta|x'\rangle - \sin\theta|y'\rangle, \quad |y\rangle = \sin\theta|x'\rangle + \cos\theta|y'\rangle. \tag{2.123}$$

The corresponding projections are:

$$\langle x'|x\rangle = \cos\theta, \langle x'|y\rangle = \sin\theta, \langle y'|x\rangle = -\sin\theta, \langle y'|y\rangle = \cos\theta, \tag{2.124}$$

so that the state vector in new basis can be expressed as follows:

$$|\psi'\rangle = J_{\text{rotation}}(\theta)|\psi\rangle, \tag{2.125}$$

where $J_{\text{rotation}}(\theta)$ is the rotation operator, which can be represented in matrix form, based on (2.122) and (2.124), as follows:

$$J_{\text{rotation}}(\theta) = \begin{pmatrix} \cos\theta & \sin\theta \\ -\sin\theta & \cos\theta \end{pmatrix}. \tag{2.126}$$

In other words, the Eq. (2.114) can be written as:

$$|\psi'\rangle = \left(\left\langle x'|\psi\rangle\langle y'|\psi\rangle\right)\right. = \begin{pmatrix} \cos\theta & \sin\theta \\ -\sin\theta & \cos\theta \end{pmatrix} \underbrace{\left(\langle x|\psi\rangle\langle y|\psi\rangle\right)_{|\psi\rangle}}. \tag{2.127}$$

The Jones matrix representation of a rotated polarizing element J, polarizer or wave plate, rotated by θ from the x-axis, is given by:

$$J(\theta) = J_{\text{rotation}}(-\theta) \cdot J \cdot J_{\text{rotation}}(\theta). \tag{2.128}$$

By interpreting TM and TE polarizations as x- and y-polarizations, respectively, the *reflection matrix R* and *transmission matrix T* are given, respectively, by:

$$R = \begin{bmatrix} -r_p & 0 \\ 0 & r_s \end{bmatrix}, \quad T = \begin{bmatrix} t_p & 0 \\ 0 & t_s \end{bmatrix}. \tag{2.129}$$

For the normal incidence case ($\theta = \phi = 0$), from Eq. (2.106), we conclude that $r_s = r_p = (1 - n)/(1 + n)$, so that the reflection matrix is given by:

$$R = \frac{1-n}{1+n}\begin{bmatrix} -1 & 0 \\ 0 & 1 \end{bmatrix} = \frac{n-1}{n+1}\begin{bmatrix} 1 & 0 \\ 0 & -1 \end{bmatrix}, n > 1. \tag{2.130}$$

So if the incident polarization state is $|L\rangle$, the reflected polarization state is:

$$|\psi\rangle = R|L\rangle = \frac{n-1}{n+1}\begin{bmatrix} 1 & 0 \\ 0 & 1 \end{bmatrix}\frac{1}{\sqrt{2}}\begin{bmatrix} 1 \\ j \end{bmatrix} = \frac{n-1}{n+1}\underbrace{\frac{1}{\sqrt{2}}\begin{bmatrix} 1 \\ j \end{bmatrix}}_{|R\rangle} = \frac{n-1}{n+1}|R\rangle. \tag{2.131}$$

To describe the action of polarizing elements in Stokes coordinates, the *Mueller matrix representation* can be used:

$$\underbrace{\begin{bmatrix} S_0' \\ S_1' \\ S_2' \\ S_3' \end{bmatrix}}_{\boldsymbol{S}'} = \underbrace{\begin{bmatrix} M_{00} & M_{01} & M_{02} & M_{03} \\ M_{10} & M_{11} & M_{12} & M_{13} \\ M_{20} & M_{21} & M_{22} & M_{23} \\ M_{30} & M_{31} & M_{32} & M_{33} \end{bmatrix}}_{\boldsymbol{M}} \underbrace{\begin{bmatrix} S_0 \\ S_1 \\ S_2 \\ S_3 \end{bmatrix}}_{\boldsymbol{S}} \Leftrightarrow \boldsymbol{S}' = \boldsymbol{M} \cdot \boldsymbol{S}. \tag{2.132}$$

The Mueller operator (matrix) representation for the *linear polarizer* is given by:

$$\boldsymbol{M}_{\text{polarizer}} = \frac{1}{2}\begin{bmatrix} p_x^2 + p_y^2 & p_x^2 - p_y^2 & 0 & 0 \\ p_x^2 - p_y^2 & p_x^2 + p_y^2 & 0 & 0 \\ 0 & 0 & 2p_x p_y & 0 \\ 0 & 0 & 0 & 2p_x p_y \end{bmatrix}, \tag{2.133}$$

where p_x and p_y are the absorption coefficients along x- and y-axes, respectively. The Mueller operator (matrix) representation for the *wave plates* is given by:

$$\boldsymbol{M}_{\text{waveplates}} = \begin{bmatrix} 1 & 0 & 0 & 0 \\ 0 & 1 & 0 & 0 \\ 0 & 0 & \cos\phi & -\sin\phi \\ 0 & 0 & \sin\phi & \cos\phi \end{bmatrix}, \tag{2.134}$$

where ϕ is the phase shift between x- and y-polarization states. The Mueller matrix representation of a *rotator* is given by:

$$\boldsymbol{M}_{\text{rotator}} = \begin{bmatrix} 1 & 0 & 0 & 0 \\ 0 & \cos(2\theta) & \sin(2\theta) & 0 \\ 0 & -\sin(2\theta) & \cos(2\theta) & 0 \\ 0 & 0 & 0 & 1 \end{bmatrix}, \tag{2.135}$$

where θ is the rotation angle. The *rotation of a polarizing element*, described by Mueller matrix $\boldsymbol{M}$, can be described by the following Mueller matrix:

$$\boldsymbol{M}(\theta) = \boldsymbol{M}_{\text{rotator}}(-2\theta)\cdot\boldsymbol{M}\cdot\boldsymbol{M}_{\text{rotator}}(2\theta). \tag{2.136}$$

2.2.3 Electromagnetic Potentials and Electromagnetic Waves

2.2.3.1 Electromagnetic Potentials

The problem of electromagnetic wave propagation over isotropic, linear, and homogenous media can be simplified when *electrical scalar potential* φ and *magnetic vector potential* $\boldsymbol{A}$ are first determined, which are related to electric field vector $\boldsymbol{E}$ and magnetic field vector and magnetic flux density vector $\boldsymbol{B}$ as follows [5, 8]:

$$\boldsymbol{E} = -\text{grad}\varphi - \frac{\partial \boldsymbol{A}}{\partial t} \quad \text{and } \boldsymbol{B} = \text{curl}\boldsymbol{A}. \tag{2.137}$$

Given that curl grad $\varphi = 0$ and div curl $\boldsymbol{A} = 0$, while spatial and time differentiation operators commute, Maxwell's equations get satisfied. In isotopic media the constitutive relations are given by $\boldsymbol{D} = \varepsilon\boldsymbol{H}$ and $\boldsymbol{B} = \mu\boldsymbol{H}$ so that from Maxwell's equation (2.2) we obtain:

$$\text{curl}\,\boldsymbol{B} = \mu\boldsymbol{J} + \varepsilon\mu\frac{\partial \boldsymbol{E}}{\partial t}. \tag{2.138}$$

By substituting Eqs. (2.137) into (2.138), we obtain:

$$\text{grad div}\boldsymbol{A} - \vec{\Delta}\boldsymbol{A} = \mu\boldsymbol{J} - \varepsilon\mu\,\text{grad}\frac{\partial\varphi}{\partial t} - \varepsilon\mu\frac{\partial^2\boldsymbol{A}}{\partial t^2}, \tag{2.139}$$

where $\vec{\Delta}\boldsymbol{A}$ denotes the Laplacian vector operator, defined as:

$$\vec{\Delta}\boldsymbol{A} = \text{grad div}\boldsymbol{A} - \text{curl}\,\text{curl}\boldsymbol{A}, \tag{2.140}$$

which in Cartesian coordinates becomes simply:

$$\vec{\Delta}\boldsymbol{A} = \Delta A_x\widehat{x} + \Delta A_y\widehat{y} + \Delta A_z\widehat{z}.$$

Divergence of vector potential can be arbitrarily chosen, and we chose it as follows:

$$\text{div}\boldsymbol{A} = -\varepsilon\mu\frac{\partial\varphi}{\partial t}, \tag{2.141}$$

so that the gradient terms in Eq. (2.139) disappear, which now becomes the *vector wave equation*:

$$\vec{\Delta}\boldsymbol{A} - \varepsilon\mu\frac{\partial^2\boldsymbol{A}}{\partial t^2} = -\mu\boldsymbol{J}. \tag{2.142}$$

The Eq. (2.141) is often called the *Lorentz gauge*. By substituting the electrical field from Eq. (2.137) into Maxwell's equation (2.3), with the help of (2.141), we obtain the *scalar wave equation*:

$$\Delta\varphi - \varepsilon\mu\frac{\partial^2\varphi}{\partial t^2} = -\frac{\rho}{\varepsilon}. \tag{2.143}$$

The wave equations for the monochromatic state, when $\partial/\partial t$ get replaced by jω, are known as *Helmholtz equations*.

2.2.3.2 Concept of Antenna

We now study the propagation of electromagnetic waves from the source or *transmit antenna*, as illustrated in Fig. 2.9. The volume V is composed of homogeneous material of dielectric constant ε and magnetic permittivityμ. The volume density of charges within this volume V is denoted by $\rho = \rho(\boldsymbol{r}', t)$, while the density of conduction currents is denoted by $\boldsymbol{J} = \boldsymbol{J}(\boldsymbol{r}', t)$, where $\boldsymbol{r}'$ denotes position vector of a source volume element dV'. We are interested in determining the electric in magnetic fields at point P, outside of source, and the position vector of this point is denoted by $\boldsymbol{r}$. In order to do so, we apply the *superposition principle* [5, 8], which claims that the electromagnetic field of a complicated system can be determined as a summation of electromagnetic fields originating from individual parts. By assuming that the distance from point P to the source element dV' is much larger than the dimension of the source, the source element dV' can be interpreted as the point source. The charge of the source element is then given by d$q = \rho$dV', and the corresponding electric scalar potential at point P is given by:

$$d\varphi_P = \frac{\rho(\boldsymbol{r}', t - R/v)}{4\pi\varepsilon R}dV', \quad R = |\boldsymbol{r} - \boldsymbol{r}'|, \tag{2.144}$$

where v is the speed of electromagnetic waves, namely, $v = (\varepsilon\mu)^{-1/2}$. By integrating over volume V, we obtain the electric scalar potential, originating from all charges within the volume V, as follows:

$$\varphi_P(\boldsymbol{r}, t) = \frac{1}{4\pi\varepsilon}\int_V \frac{\rho(\boldsymbol{r}', t - R/v)}{R}dV'. \tag{2.145}$$

It is straightforward to show that the electric scalar potential satisfies the scalar wave equation (2.143). By close inspection of the scalar wave and vector wave equations, we conclude that the i-th component, $i\in\{x,y,z\}$, of the magnetic vector potential

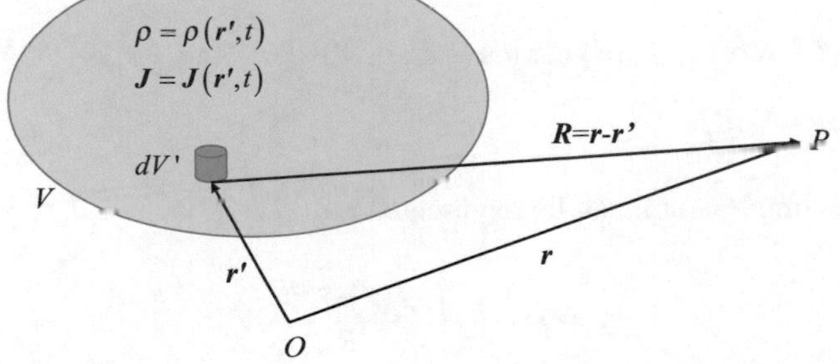

Fig. 2.9 Illustration of the source of electromagnetic waves (transmit antenna)

can be obtained from the scalar wave equation by replacing ε with 1/μ and ρ with J_i. Therefore, the *i*-th component of the magnetic vector potential can be obtained as follows:

$$A_{i,P}(\boldsymbol{r},t)=\frac{\mu}{4\pi}\int_V\frac{J_i(\boldsymbol{r}',t-R/v)}{R}dV',i\in\{x,y,z\} \tag{2.146}$$

The corresponding vector representation of the magnetic vector potential is given by:

$$\boldsymbol{A}_P(\boldsymbol{r},t)=\frac{\mu}{4\pi}\int_V\frac{\boldsymbol{J}(\boldsymbol{r}',t-R/v)}{R}dV' \tag{2.147}$$

The Eqs. (2.145) and (2.147), when the time-dependence is omitted, are known as the *Helmholtz integrals*. The potentials given by Eqs. (2.145) and (2.147) in electrodynamics are known as *retarded potentials*.

Clearly, the retarded potentials for electromagnetic field are generated by time-varying charge density or electric current. As the electromagnetic field propagates at finite speed, the speed of light in a given medium, denoted as v, the retarded time at a given point in space $\boldsymbol{r}$ is given by $t_r=t-|\boldsymbol{r}-\boldsymbol{r}'|/v=t-R/v$..

By substitution (2.145) and (2.147) into (2.135), with the help of constitutive relationship $\boldsymbol{H}=\mathbf{B}/\mu$, we obtain the following expressions for electric and magnetic field vectors:

$$\begin{aligned}\boldsymbol{E}_P(\boldsymbol{r},t)=&\frac{1}{4\pi\varepsilon v}\int_V\frac{\partial\rho(\boldsymbol{r}',t-R/v)}{\partial t}\frac{\widehat{\boldsymbol{R}}}{R}dV'+\frac{1}{4\pi\varepsilon}\int_V\rho(\boldsymbol{r}',t-R/v)\frac{\widehat{\boldsymbol{R}}}{R^2}dV'\\&-\frac{\mu}{4\pi}\int_V\frac{\partial\boldsymbol{J}(\boldsymbol{r}',t-R/v)}{\partial t}\frac{1}{R}dV',\end{aligned} \tag{2.148}$$

and

$$\boldsymbol{H}_P(\boldsymbol{r},t)=\frac{1}{4\pi v}\int_V\frac{\partial\boldsymbol{J}(\boldsymbol{r}',t-R/v)}{\partial t}\times\frac{\widehat{\boldsymbol{R}}}{R}dV'+\frac{1}{4\pi}\int_V\boldsymbol{J}(\boldsymbol{r}',t-R/v)\times\frac{\widehat{\boldsymbol{R}}}{R^2}dV', \tag{2.149}$$

where $\widehat{\boldsymbol{R}}=\boldsymbol{R}/R$.

When the *volume density charge is sinusoidal*:

$$\begin{aligned}&\rho(\boldsymbol{r}',t-R/v)=\rho_m(\boldsymbol{r}')\cos[\omega t-kR+\theta(\boldsymbol{r}')]=\mathrm{Re}\left\{\underline{\rho}(\boldsymbol{r}')e^{j(\omega t-kR)}\right\},\\&\underline{\rho}(\boldsymbol{r}')=\rho_m(\boldsymbol{r}')e^{j\theta(\boldsymbol{r}')}\end{aligned} \tag{2.150}$$

the electrical scalar potential in complex form can be represented as:

$$\underline{\varphi}_P(\boldsymbol{r})=\frac{1}{4\pi\varepsilon}\int_V\frac{\underline{\rho}(\boldsymbol{r}')e^{-jkR}}{R}dV'. \tag{2.151}$$

When the *current density is sinusoidal*:

$$\begin{aligned}&\boldsymbol{J}(\boldsymbol{r}',t-R/v)=\boldsymbol{J}_m(\boldsymbol{r}')\cos[\omega t-kR+\theta(\boldsymbol{r}')]=\mathrm{Re}\left\{\underline{\boldsymbol{J}}(\boldsymbol{r}')e^{j(\omega t-kR)}\right\},\\&\underline{\boldsymbol{J}}(\boldsymbol{r}')=\boldsymbol{J}_m(\boldsymbol{r}')e^{j\theta(\boldsymbol{r}')}\end{aligned} \tag{2.152}$$

the magnetic vector potential in complex form can be represented as:

$$\underline{\boldsymbol{A}}_P(\boldsymbol{r})=\frac{\mu}{4\pi}\int_V\frac{\underline{\boldsymbol{J}}(\boldsymbol{r}')e^{-jkR}}{R}dV' \tag{2.153}$$

2.2.3.3 Spherical Waves

In points located far away from the source, $r >> r'$, in the calculation of the electric scalar and magnetic vector potentials, the following two approximations can be made:

(1) The coarse approximation in amplitude:

$$R^{-1} = |\boldsymbol{r} - \boldsymbol{r}'|^{-1} \cong r^{-1}, \tag{2.154}$$

(2) The fine approximation in phase:

$$\begin{aligned} kR &= k|\boldsymbol{r} - \boldsymbol{r}'| = k\sqrt{r^2 - 2\boldsymbol{r}\boldsymbol{r}' + \boldsymbol{r}'^2} \overset{r>>r'}{\cong} k\sqrt{r^2 - 2\boldsymbol{r}\boldsymbol{r}'} \\ &= kr\sqrt{1 - \frac{2\boldsymbol{r}\boldsymbol{r}'}{r^2}} \cong kr - k\widehat{\boldsymbol{r}}\boldsymbol{r}', \widehat{\boldsymbol{r}} = \boldsymbol{r}/r. \end{aligned} \tag{2.155}$$

Using these two approximations, the magnetic vector potential expressions (2.153) become:

$$\underline{\boldsymbol{A}}_P \cong \frac{\mu}{4\pi} \underbrace{\int_V \underline{\boldsymbol{J}}(\boldsymbol{r}') e^{jk\widehat{\boldsymbol{r}}\boldsymbol{r}'} dV'}_{\underline{\boldsymbol{N}}} \cdot \frac{e^{-jkr}}{r} = \underbrace{\frac{\mu}{4\pi}\underline{\boldsymbol{N}}}_{\underline{\boldsymbol{A}}_0} \frac{e^{-jkr}}{r} = \underline{\boldsymbol{A}}_0 \frac{e^{-jkr}}{r}, \tag{2.156}$$

where the vector $\underline{\boldsymbol{N}}$ in (2.156) is known as the *radiation vector*. The radial dependence of the magnetic vector potential is of the form:

$$\Phi = \frac{e^{-jkr}}{r}. \tag{2.157}$$

Such electromagnetic waves, whose amplitude changes with distance as 1/r, while the phase is a linear function of distance, are commonly referred to as the *spherical electromagnetic waves*. At distances sufficiently far away for the source (or transmit antenna), regardless of the shape or size, the electromagnetic waves are spherical. This space region is called the *far-field* region or *Fraunhofer* region. The propagation function of spherical waves satisfies the following *radiation condition*:

$$\frac{\partial \Phi}{\partial r} = -\frac{1 + jkr}{r}\Phi. \tag{2.158}$$

In spherical coordinate system (r,θ,ϕ), because of $1/r$ dependence of radial component, in the far-field, only angle components of magnetic vector potential exist and are given by:

$$\underline{A}_\theta = \underline{A}\frac{e^{-jkr}}{r}\widehat{\theta}, \underline{A}_\theta = \underline{A}\frac{e^{-jkr}}{r}\widehat{\phi}. \tag{2.159}$$

By using the approximations (2.154) and (2.155), the electrical scalar potential (2.151) becomes:

$$\underline{\varphi}_P(\boldsymbol{r}) = \frac{1}{4\pi\varepsilon} \int_V \frac{\underline{\rho}(\boldsymbol{r}') e^{-jkR}}{R} dV'. \tag{2.160}$$

Given that the action of the time derivative on the harmonic wave is multiplication by $j\omega$, from Eq. (2.141), we obtain:

$$\underline{\varphi}_P = \frac{j}{\omega\varepsilon\mu} \text{div}\underline{\boldsymbol{A}}, \tag{2.161}$$

and by substituting (2.156) into (2.161), the scalar potential becomes:

$$\begin{aligned}\underline{\varphi}_P &= -j\underbrace{\frac{1}{\omega\varepsilon\mu}\frac{\mu}{4\pi}}_{Z/4\pi k}\int_V \underline{\boldsymbol{J}}(\boldsymbol{r}')\widehat{\boldsymbol{r}}\frac{1+jkr}{r}\frac{e^{-jkr}}{r}e^{\,jk\widehat{\boldsymbol{r}}\boldsymbol{r}'}dV' \\ &= -j\frac{Z}{4\pi k}\int_V \underline{\boldsymbol{J}}(\boldsymbol{r}')\widehat{\boldsymbol{r}}\frac{1+jkr}{r}\frac{e^{-jkr}}{r}e^{\,jk\widehat{\boldsymbol{r}}\boldsymbol{r}'}dV', Z=\sqrt{\frac{\mu}{\varepsilon}}.\end{aligned} \tag{2.162}$$

The electric field vector can be now determined as:

$$\begin{aligned}\underline{\boldsymbol{E}} &= -\text{grad}\ \varphi - j\omega\underline{\boldsymbol{A}} = -j\frac{\omega\mu}{4\pi}\int_V \underline{\boldsymbol{J}}(\boldsymbol{r}')\frac{e^{-jkr}}{r}e^{\,jk\widehat{\boldsymbol{r}}\boldsymbol{r}'}dV' \\ &+j\frac{Z}{4\pi k}\int_V\left\{\frac{1+jkr}{r^2}\underline{\boldsymbol{J}}(\boldsymbol{r}')+\underline{\boldsymbol{J}}(\boldsymbol{r}')\widehat{\boldsymbol{r}}(k^2r^2-3-jkr)\frac{\widehat{\boldsymbol{r}}}{r^2}\right\}\frac{e^{-jkr}}{r}e^{\,jk\widehat{\boldsymbol{r}}\boldsymbol{r}'}dV'\end{aligned} \tag{2.163}$$

The magnetic field vector can be obtained as:

$$\boldsymbol{H} = \frac{1}{\mu}\text{curl}\,\underline{\boldsymbol{A}} = \frac{1}{4\pi}\int_V\left[\underline{\boldsymbol{J}}(\boldsymbol{r}')\times\widehat{\boldsymbol{r}}\right]\frac{1+jkr}{r}\frac{e^{-jkr}}{r}e^{\,jk\widehat{\boldsymbol{r}}\boldsymbol{r}'}dV'. \tag{2.164}$$

In far-field, the components in (2.163) and (2.164), the components whose amplitudes decay as 1/r, dominate so that the electric and magnetic field vectors can be written as:

$$\underline{\boldsymbol{E}} = j\frac{Zk}{4\pi}\frac{e^{-jkr}}{r}\int_V\left\{\left[\underline{\boldsymbol{J}}(\boldsymbol{r}')\widehat{\boldsymbol{r}}\right]\widehat{\boldsymbol{r}}-\underline{\boldsymbol{J}}(\boldsymbol{r}')\right\}e^{\,jk\widehat{\boldsymbol{r}}\boldsymbol{r}'}dV' \tag{2.165}$$

The magnetic field vector can be obtained as:

$$\boldsymbol{H} = \frac{1}{\mu}\text{curl}\,\underline{\boldsymbol{A}} = \frac{1}{4\pi}\int_V\left[\underline{\boldsymbol{J}}(\boldsymbol{r}')\times\widehat{\boldsymbol{r}}\right]\frac{1+jkr}{r}\frac{e^{-jkr}}{r}e^{\,jk\widehat{\boldsymbol{r}}\boldsymbol{r}'}dV'. \tag{2.166}$$

Similarly to magnetic vector potential, because of $1/r$ dependence of radial component, in the far-field only angle components of electric and magnetic fields exist and are given, respectively, by:

$$\begin{aligned}\underline{E}_\theta &= -j\omega\underline{A}_\theta,\ \underline{H}_\theta = \underline{E}_\theta/Z, \\ \underline{E}_\phi &= -j\omega\underline{A}_\phi,\ \underline{H}_\phi = -\underline{E}_\phi/Z.\end{aligned} \tag{2.167}$$

It can be straightforwardly shown that the electric and magnetic field vectors are mutually perpendicular and simultaneously perpendicular to the radial direction:

$$\boldsymbol{E}\cdot\boldsymbol{H}=0, \boldsymbol{E}\cdot\widehat{\boldsymbol{r}}=0, \boldsymbol{H}\cdot\widehat{\boldsymbol{r}}=0. \tag{2.168}$$

The Poynting vector has only the radial component, in the far-field, as shown below:

$$\boldsymbol{S}=\boldsymbol{E}\times\boldsymbol{H}^*=S\widehat{\boldsymbol{r}}, S=\underline{E}_\theta\underline{H}_\phi^*+\underline{E}_\phi\underline{H}_\theta^*=\frac{\underline{E}_\theta\underline{E}_\theta^*+\underline{E}_\phi\underline{E}_\phi^*}{Z}=\frac{\omega^2}{Zr^2}\left(\underline{A}_\theta\underline{A}_\theta^*+\underline{A}_\phi\underline{A}_\phi^*\right) \tag{2.169}$$

The flux of the Poynting vector over a sphere of radius r, containing the transmit antenna/source, represents the radiated power:

$$P=\int_{\theta=0}^{\pi}\int_{\phi=0}^{2\pi}\frac{\omega^2}{Z}\left(\underline{A}_\theta\underline{A}_\theta^*+\underline{A}_\phi\underline{A}_\phi^*\right)d\theta d\phi. \tag{2.170}$$

2.2.3.4 Plane Waves

When the spherical wave are observed within the small region of dimensions much smaller than the distance from the transmit antenna/source, the variations in amplitude can be neglected. Such electromagnetic waves, whose amplitudes do not change, while the phases are linear function of distance, are commonly referred to as the *plane waves*. The propagation function of the plane waves is given by:

$$\Phi = e^{-jkr}, \tag{2.171}$$

and satisfies the following radiation condition:

$$\frac{\partial \Phi}{\partial r} = -jk\Phi. \tag{2.172}$$

Strictly speaking, the plane waves do not exist in nature, as the antenna to emit them cannot be manufactured. Namely, such antenna would be composed of infinitely large plane with the surface current density J_s and emission frequencyω. By expressing the current density of such "antenna" as:

$$\widehat{\underline{J}} = \underline{J}_S \delta(x)\widehat{\boldsymbol{x}}, \tag{2.173}$$

the magnetic vector potential will have only x-component, only z-dependent, so that the wave equation is given by:

$$\frac{d^2\underline{A}}{dz^2} + k^2\underline{A} = -\mu\underline{J}_S, \tag{2.174}$$

The corresponding solution is given as:

$$\underline{A} = -\frac{j\mu\underline{J}_S}{2k} e^{-jk|z|}. \tag{2.175}$$

The electric scalar potential is, based on Eq. (2.161), $\varphi = 0$. The components of electric field are then obtained as follows:

$$\underline{E}_x = -j\omega\underline{A} = -\frac{\omega\mu\underline{J}_S}{2k} e^{-jk|z|}, E_y = E_z = 0. \tag{2.176}$$

On the other hand, the components of magnetic field are obtained as:

$$\underline{H}_z = \begin{cases} -\frac{\underline{J}_S}{2} e^{-jkz}, z > 0 \\ \frac{\underline{J}_S}{2} e^{jkz}, z < 0 \end{cases}, H_x = H_y = 0. \tag{2.177}$$

Cleary, the nonzero electric and magnetic fields components (2.176), (2.177) have constant complex amplitudes, while the phases are a linear function of the distance z, indicating that they describe the plane wave. Even though that plane waves strictly speaking do not exist, they represent a nice approximation in the far-field, when the observation volume is small compared to the distance, and as such they have been widely used.

2.2.3.5 Hertzian Dipole

Let us observe the current element of momentum $\underline{I}\delta l\widehat{z}$, representing the *straight-wire antenna* with $\delta l \to 0$, located in the origin (of the coordinate system) as illustrated in Fig. 2.10, which is also known as the *Hertzian dipole*. The magnetic vector potential has only z-component:

$$\underline{\boldsymbol{A}} = \underline{A}\widehat{z}, \underline{A} = \frac{\mu}{4\pi}\underline{I}\delta l\frac{e^{-jkr}}{r}, \tag{2.178}$$

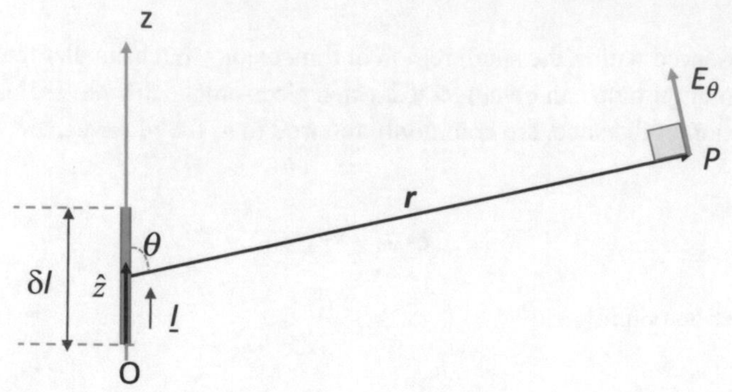

Fig. 2.10 Illustration of the Hertzian dipole

and the spherical components are given by:

$$\underline{A}_r = \underline{A}\widehat{z}\widehat{r} = \underline{A}\cos\theta, \underline{A}_\theta = \underline{A}\widehat{z}\widehat{\theta} = -\underline{A}\sin\theta, \underline{A}_\phi = 0. \tag{2.179}$$

The components of electric and magnetic fields in the far-field can now be determined as:

$$\begin{gathered} \underline{E}_\theta = -\mathrm{j}\omega\underline{A}_\theta = \mathrm{j}\frac{Z}{4\pi}\underline{I}k\delta l\frac{e^{-jkr}}{r}\sin\theta = ZH_\phi, \\ E_r = 0, E_\phi = -ZH_\theta = 0, H_r = 0. \end{gathered} \tag{2.180}$$

The electric field is angle θ dependent and exhibits emission directivity properties. The minimum radiation is for $\theta = 0, \pi$, while the maximum emission is achieved for $\theta = \pi/2$. The following function:

$$F(\theta) = \frac{k\delta l}{2}\sin\theta \tag{2.181}$$

can be used to describe the directivity properties of this transmit antenna, and it is commonly referred to as the *radiation pattern*. Since the radiation pattern is independent of ϕ, we say that this antenna is *omnidirectional*. The θ-component of electric field can now be expressed in terms of radiation function as follows:

$$\underline{E}_\theta = \mathrm{j}\frac{ZI}{2\pi}\frac{e^{-jkr}}{r}F(\theta). \tag{2.182}$$

The very often, the normalized radiation pattern is used:

$$f(\theta) = \frac{F(\theta)}{\max F(\theta)} = \frac{F(\theta)}{F(\theta = \pi/2)}. \tag{2.183}$$

The beamwidth of the transmit antenna is defined as the width of the main lobe (the lobe containing maximum power direction), usually specified by the *half power beam width*. In other words, the angle encompassed between the points on the side of the main lobe where the power has fallen to half (−3 dB) of its maximum value:

$$\frac{1}{\sqrt{2}} \leq f(\theta) \leq 1. \tag{2.184}$$

For Hertzian dipole, the antenna beamwidth is $\pi/2$.

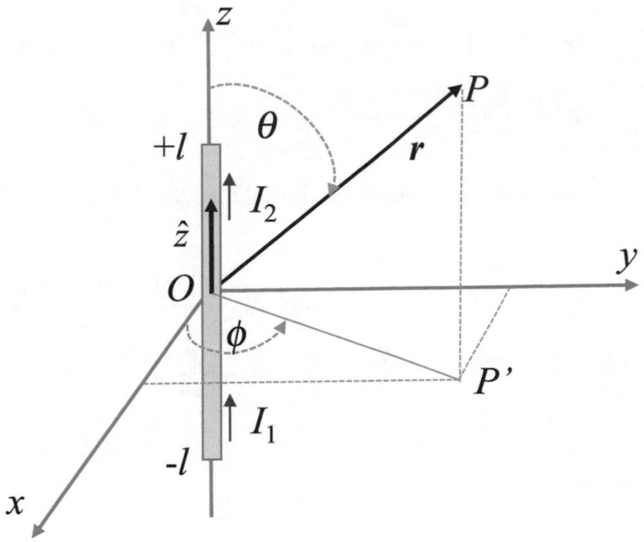

Fig. 2.11 Center-driven straight-wire antenna

2.2.3.6 Straight-Wire Antenna

Let us now consider the straight-wire antenna, driven from the center, extending from $z = -l$ to $z = +l$, as shown in Fig. 2.11.

The current density is given by:

$$\boldsymbol{J} = \delta(x)\delta(y)f(z)\hat{z}, \tag{2.185}$$

while the corresponding current is integral over cross section of this wire:

$$\boldsymbol{I}(z) = \int \boldsymbol{J}dxdy = f(z)\hat{z}. \tag{2.186}$$

Since the antenna is center-drive, the unknown function $f(z)$ must satisfy the following conditions:

$$f(-l) = f(l) = 0. \tag{2.187}$$

Additionally, the unknown function $f(z)$ must satisfy one-dimensional Helmholtz equation:

$$\frac{d^2f(z)}{dz^2} + k^2f(z) = 0. \tag{2.188}$$

There exist two solutions for this differential equation, e^{jkz} and e^{-jkz}. The corresponding current of upper and lower segments can be written as:

$$\begin{aligned} I_1(z) &= A_1e^{jkz} + B_1e^{\ jkz}, \ -l \le z \le 0; \\ I_2(z) &= A_2e^{jkz} + B_2e^{-jkz}, 0 \le z \le l. \end{aligned} \tag{2.189}$$

By using the continuity condition $I_1(0) = I_2(0)$ and Eq. (2.187), it can be shown that the overall solution is the standing wave:

$$I(z) = I_0 \sin k(l - |z|). \tag{2.190}$$

The magnetic vector potential is given by Eq. (2.156):

$$\underline{\mathbf{A}} = \frac{\mu}{4\pi}\int_V \underline{\mathbf{J}}(\mathbf{r}')e^{jk\widehat{\mathbf{r}}\mathbf{r}'}dV' \cdot \frac{e^{-jkr}}{r} = \frac{\mu}{4\pi}I_0\int_{-l}^{l} e^{jkz'\cos\theta}\sin k(l-|z|)dz' \cdot \frac{e^{-jkr}}{r}\widehat{z}$$
$$= -\frac{\mu}{4\pi}2I_0\frac{\cos(kl\cos\theta)-\cos kl}{k\sin^2\theta}\frac{e^{-jkr}}{r}\widehat{z}. \tag{2.191}$$

The nonzero components of electric and magnetic field in the far-field are given by:

$$\underline{E}_\theta = -\mathrm{j}\omega\underline{A}_\theta \overbrace{=}^{\underline{A}_\theta=\underline{A}\widehat{z}\widehat{\theta}=-\underline{A}\sin\theta} -\mathrm{j}\omega\sin\theta\frac{\mu}{4\pi}2I_0\frac{\cos(kl\cos\theta)-\cos kl}{k\sin^2\theta}\frac{e^{-jkr}}{r}$$
$$= -j\frac{Z}{2\pi}I_0\frac{e^{-jkr}}{r}\frac{\cos(kl\cos\theta)-\cos kl}{\sin\theta} = -j\frac{Z}{2\pi}I_0\frac{e^{-jkr}}{r}F(\theta),$$
$$H_\phi = E_\theta/Z, \tag{2.192}$$

where the radiation pattern is given by:

$$F(\theta) = \frac{\cos(kl\cos\theta)-\cos kl}{\sin\theta}. \tag{2.193}$$

Two illustrative examples of radiation patterns are shown in Fig. 2.12. For $kl > \pi$ in addition to the main lobe, two side lobes appear. The radial component of Poynting vector is given by:

$$S_r = \frac{1}{2}\left(\underline{E}_\theta\underline{H}_\phi^* + \underline{E}_\phi\underline{H}_\theta^*\right) = \frac{Z}{8\pi^2}\frac{I_0^2}{r^2}F^2(\theta). \tag{2.194}$$

An important parameter of the antenna is so-called directivity gain, denoted by g_d, defined as follows [8]:

$$g_d = \frac{4\pi^2 S_{r,\max}}{\int_0^{4\pi} S_r(r,\theta,\phi)r^2\sin\theta d\theta d\varphi}, \tag{2.195}$$

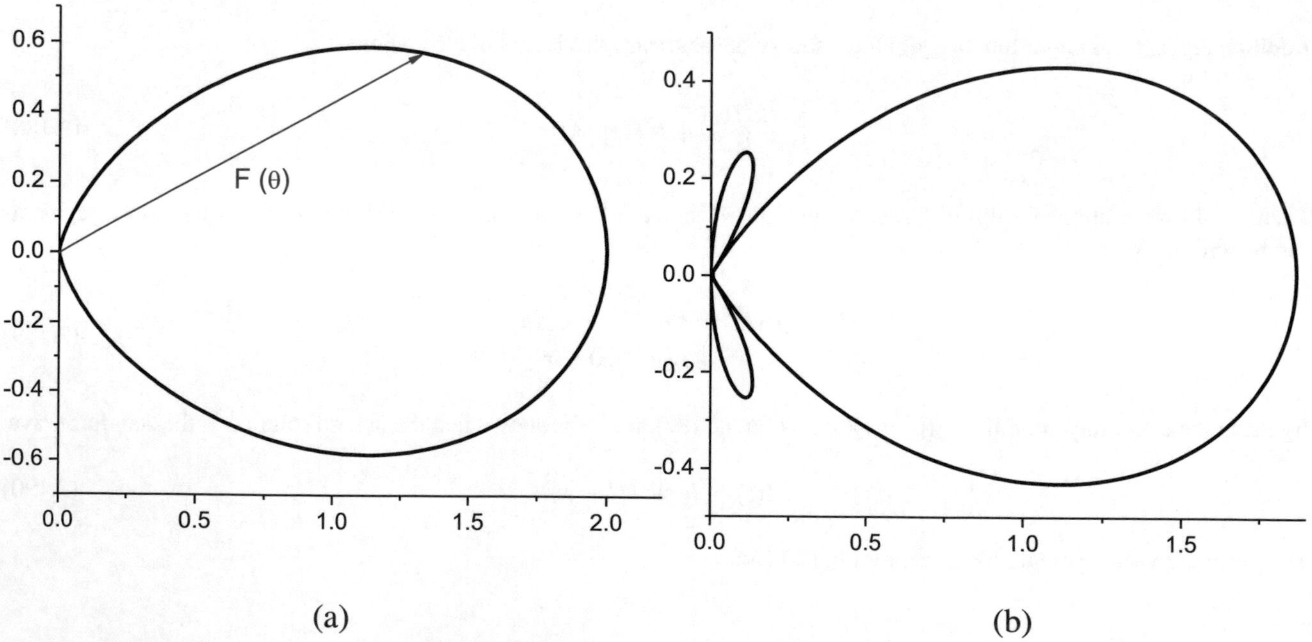

Fig. 2.12 Radiation patterns for center-driven straight-wire antenna: (**a**) $kl = \pi$ and (**b**) $kl = 7\pi/6$

where Sr, max is the major lobe maximum of the Poynting vector in the far-field and r is the far-field sphere radius. Clearly, the directivity gain of an isotropic antenna is $g_d = 1$, while the of directive antennas is larger than 1. The denominator in (2.195) represents the time-average power radiated by the antenna:

$$P = \int_0^{4\pi} S_r(r, \theta, \phi) r^2 \sin\theta d\theta d\varphi. \tag{2.196}$$

The directivity gain for the center-driven straight-wire antenna from Fig. 2.11 can be calculated by:

$$g_d = \frac{4\pi^2 \max\limits_{\theta} \left[\frac{\cos(kl\cos\theta) - \cos kl}{\sin\theta}\right]^2}{2\pi \int_0^{\pi} \frac{[\cos(kl\cos\theta) - \cos kl]^2}{\sin\theta} d\theta}. \tag{2.197}$$

For the half-wave dipole ($kl = \pi/2$), it is straightforward to show that directivity gain is $g_d = 1.64$.

2.2.3.7 Cylindrical Waves

Let us now consider the current element whose current density is given as:

$$\underline{J} = \frac{I}{2r\pi} \delta(r) z, \tag{2.198}$$

the magnetic vector potential will only have the axial component, namely, $\underline{A} = A\hat{z}$. The corresponding wave equation is given by:

$$\frac{\partial}{\partial r}\left(r \frac{\partial A}{\partial r}\right) + k^2 r \underline{A} = -\frac{\mu I}{2\pi} \delta(r). \tag{2.199}$$

The solutions of this wave equation are Bessel functions of the first and second kind, $J_0(kr)$ and $Y_0(kr)$. In the absence of any obstacle, only the incident wave will propagate; therefore, the Hankel functions of the second kind will be used as a solution of the wave equation [5]:

$$H_0^{(2)}(kr) = J_0(kr) - jY_0(kr). \tag{2.200}$$

The corresponding solution for the magnetic vector potential becomes:

$$\underline{A} = \frac{\mu I}{4\pi}(-j\pi) H_0^{(2)}(kr) = -j\frac{\mu I}{4} H_0^{(2)}(kr). \tag{2.201}$$

As a generalization, for sufficiently high frequencies, the solid conducting cylinder of sufficient length can be used as a more realistic transmit antenna. Thanks to the strong skin effect, only the surface current of density $\boldsymbol{J_S}$ will exist, and the corresponding magnetic vector potential can be written as:

$$\underline{A} = -j\frac{\mu}{4} \oint_C \boldsymbol{J}_S(\boldsymbol{r}') H_0^{(2)}(k|\boldsymbol{r} - \boldsymbol{r}'|) dl', \tag{2.202}$$

where C is the contour defining the cross section of the conducting cylinder. In the far-field, $r >> r'$, so that the following approximation of the Hankel function can be used:

$$H_0^{(2)}(kr) \overset{kr\to\infty}{\rightarrow} \sqrt{\frac{2}{\pi kr}} e^{-j(kr - \pi/4)}. \tag{2.203}$$

After substituting the Eq. (2.203) into (2.202), we obtain the following expression for the magnetic vector potential in the far-field:

$$\underline{A} = -j\frac{\mu}{4}e^{-j\pi/4}\underbrace{\sqrt{\frac{2}{\pi k}\frac{e^{-jkr}}{\sqrt{r}}}}_{\Phi}\oint_C \underline{J}_S(r')e^{jk\widehat{rr'}}\,dl'. \tag{2.204}$$

Clearly, the propagation function Φ has the following form:

$$\Phi = \frac{e^{-jkr}}{\sqrt{r}}, \tag{2.206}$$

and satisfies the following radiation condition:

$$\frac{\partial \Phi}{\partial r} = -\left(\mathrm{j} + \frac{1}{2kr}\right)\Phi. \tag{2.207}$$

The electric field will only have axial component $\underline{E}_z$ and magnetic field only angle $\underline{H}_\theta$ component, given, respectively, as:

$$\underline{E}_z = -j\omega\underline{A},\ \underline{H}_\theta = -\underline{E}/Z. \tag{2.208}$$

The electromagnetic waves, whose propagation function has the form given by Eq. (2.206), are commonly referred to as the *cylindrical waves*.

2.2.3.8 Paraxial Approximation and Gaussian Beam

Before concluding this section, we briefly return back to the spherical wave, whose propagation function is given by (2.157). Based on Fig. 2.9, we conclude that the position vector magnitude of observed point $P(x,y,z)$ with respect to the origin $O(x_0,y_0,z_0)$ can be written as:

$$r = \sqrt{(x-x_0)^2 + (y-y_0)^2 + (z-z_0)^2} = \Delta z\sqrt{1 + \left(\frac{x-x_0}{\Delta z}\right)^2 + \left(\frac{y-y_0}{\Delta z}\right)^2}, \quad \Delta z = z - z_0. \tag{2.209}$$

In *paraxial approximation*, when the region of interest is the space close to the optical axis, assuming that the propagation direction is aligned with the positive z-direction, the following approximation for square root can be used $\sqrt{1+u} \cong 1 + u/2$, and from previous equation, we obtain:

$$r \cong \Delta z\left[1 + \frac{1}{2}\left(\frac{x-x_0}{\Delta z}\right)^2 + \frac{1}{2}\left(\frac{y-y_0}{\Delta z}\right)^2\right]. \tag{2.210}$$

Corresponding paraxial spherical wave approximation becomes:

$$\Phi \cong \frac{e^{jk\Delta z}}{\Delta z}e^{j\frac{k}{2\Delta z}\left[\left(\frac{x-x_0}{\Delta z}\right)^2 + \left(\frac{y-y_0}{\Delta z}\right)^2\right]}. \tag{2.211}$$

Another relevant electromagnetic wave, in particular in optics, is the *Gaussian beam wave*, which has a Gaussian amplitude profile, but the spherical wavefront and corresponding propagation function can be written as:

$$\Phi_G = \frac{1}{q(z)}e^{jk\frac{x^2+y^2}{2q(z)}}, \tag{2.212}$$

where the complex beam parameter $q(z)$ is defined as:

$$q(z) = z - jz_R,\ z_R = \frac{\pi w_0^2}{\lambda} \tag{2.213}$$

and z_R is the so-called Rayleigh distance (*range*), the distance at which the waist w is √2 time larger than that at the beam waist w_0 (at $z = 0$). Another popular representation for $q(z)$ is given in term of reciprocal of $q(z)$ as follows:

$$\begin{aligned}\frac{1}{q(z)} &= \frac{1}{z - jz_R} = \frac{z}{z^2 + z_R^2} + j\frac{z_R}{z^2 + z_R^2} = \frac{1}{\underbrace{z\left[1 + \left(\frac{z_R}{z}\right)^2\right]}_{R(z)}} + j\frac{1}{z_R\underbrace{\left[1 + \left(\frac{z}{z_R}\right)^2\right]}_{w^2(z)/w_0^2}} \\ &= \frac{1}{R(z)} + j\frac{1}{z_R}\frac{1}{w^2(z)/w_0^2}\end{aligned} \tag{2.214}$$

$$R(z) = z\left[1 + \left(\frac{z_R}{z}\right)^2\right],\ w^2(z) = w_0^2\left[1 + \left(\frac{z}{z_R}\right)^2\right]$$

where $R(z)$ denotes the radius of the curvature of the beam's wavefronts at distance z and $w(z)$ denotes the radius at which the field amplitude falls to $1/e$ of their axial value, at the plane z along the beam.

2.2.4 Interference, Coherence, and Diffraction in Optics

2.2.4.1 Interference

According to the superposition principle, the electric field vector $\boldsymbol{E}$ at the observed point of space $\boldsymbol{r}$ can be obtained as the vector sum of individual field vectors $\boldsymbol{E}_i$ originating from different sources:

$$\boldsymbol{E} = \sum_{i=1}^{N} \boldsymbol{E}_{(i)}. \tag{2.215}$$

If the density of conduction currents $\underline{J}(r',t)$ from Fig. 2.9 can be represented as the sum of component conduction currents:

$$\boldsymbol{J}(\boldsymbol{r}',t) = \sum_{i=1}^{N} \boldsymbol{J}_{(i)}(\boldsymbol{r}',t), \tag{2.216}$$

wherein each of the satisfies the vector wave equation:

$$\vec{\Delta}\boldsymbol{A}_{(i)} - \varepsilon\mu\frac{\partial^2\boldsymbol{A}_{(i)}}{\partial t^2} = -\mu\boldsymbol{J}_{(i)}, \tag{2.217}$$

then the overall magnetic vector potential can be represented, based on (2.147), as follows:

$$\begin{aligned}\boldsymbol{A}(\boldsymbol{r},t) &= \frac{\mu}{4\pi}\int_V \frac{\sum_{i=1}^{N}\boldsymbol{J}_{(i)}(\boldsymbol{r}',t - R/v)}{R}dV' = \sum\nolimits_{i=1}^{N}\underbrace{\frac{\mu}{4\pi}\int_V \frac{\boldsymbol{J}_{(i)}(\boldsymbol{r}',t - R/v)}{R}dV'}_{\boldsymbol{A}_{(i)}}, \\ &= \sum\nolimits_{i=1}^{N}\boldsymbol{A}_{(i)},\end{aligned} \tag{2.219}$$

and clearly satisfies the overall vector wave equation (2.142). Let the charge density of the volume V_i in which the i-th conduction current is located be denoted as ρ_i. Based on Eq. (2.145), the overall scalar potential can be represented as:

$$\varphi(\boldsymbol{r},t)=\frac{1}{4\pi\varepsilon}\int_V\frac{\sum_{i=1}^{N}\rho_{(i)}(\boldsymbol{r}',t-R/v)}{R}dV'=\sum_{i=1}^{N}\underbrace{\frac{1}{4\pi\varepsilon}\int_V\frac{\rho_{(i)}(\boldsymbol{r}',t-R/v)}{R}dV'}_{\varphi_{(i)}(\boldsymbol{r},t)}$$
$$=\sum_{i=1}^{N}\varphi_{(i)}(\boldsymbol{r},t). \tag{2.220}$$

By substituting (2.219) and (2.220) into (2.137), we prove the superposition principle given by Eq. (2.215).

As an illustrative example, let us observe two plane linearly polarized waves of the same angular frequency:

$$\boldsymbol{E}_{(1)}=\boldsymbol{E}_1e^{j(\omega t-\boldsymbol{k}_1\cdot\boldsymbol{r}+\phi_1)},\boldsymbol{E}_{(2)}=\boldsymbol{E}_2e^{j(\omega t-\boldsymbol{k}_2\cdot\boldsymbol{r}+\phi_2)}. \tag{2.221}$$

When the phase difference $\phi_1-\phi_2$ is constant, we say that two sources are *mutually coherent*. As we have seen in Sect. 2.2.1, the irradiance function at the given point $\boldsymbol{r}$ is proportional to the magnitude square of the electric field, so that after applying the superposition principle given by Eq. (2.215) for $N=2$, the irradiance function, after ignoring the constant term, is obtained as follows:

$$\begin{aligned}I&=|\boldsymbol{E}|^2=\boldsymbol{E}\cdot\boldsymbol{E}^*=\left(\boldsymbol{E}_{(1)}+\boldsymbol{E}_{(2)}\right)\left(\boldsymbol{E}_{(1)}^*+\boldsymbol{E}_{(2)}^*\right)\\&=\underbrace{|\boldsymbol{E}_{(1)}|^2}_{I_1}+\underbrace{|\boldsymbol{E}_{(2)}|^2}_{I_2}+\underbrace{\boldsymbol{E}_{(1)}\boldsymbol{E}_{(2)}^*+\boldsymbol{E}_{(1)}^*\boldsymbol{E}_{(2)}}_{2\boldsymbol{E}_1\cdot\boldsymbol{E}_2\cos(\boldsymbol{k}_2\cdot\boldsymbol{r}-\boldsymbol{k}_1\cdot\boldsymbol{r}+\phi_1-\phi_2)}\\&=I_1+I_2+2\boldsymbol{E}_1\cdot\boldsymbol{E}_2\cos(\boldsymbol{k}_2\cdot\boldsymbol{r}-\boldsymbol{k}_1\cdot\boldsymbol{r}+\phi_1-\phi_2)\end{aligned} \tag{2.222}$$

The last term in the bottom line of (2.222) is commonly referred to as the *interference term*. Since the angle $\boldsymbol{k}_2\cdot\boldsymbol{r}-\boldsymbol{k}_1\cdot\boldsymbol{r}+\phi_1-\phi_2$ is the position vector $\boldsymbol{r}$ dependent, the periodic spatial variations in the intensity occur, and these variations are responsible for the interference fringes when two mutually coherent light beams are combined. On the other hand, when two mutually incoherent light beams are combined, the interference term is averaged out, and the interference effect does not occur.

Even when two mutually coherent beams are monochromatic and with constant amplitudes, if they propagate different paths before the interference occur, the instantaneous light flux at a given point $\boldsymbol{r}$ in space fluctuates, and we need to define the irradiance by using the time-average operation:

$$\begin{aligned}I&=\left\langle|\boldsymbol{E}|^2\right\rangle=\langle\boldsymbol{E}\cdot\boldsymbol{E}^*\rangle=\left\langle\left(\boldsymbol{E}_{(1)}+\boldsymbol{E}_{(2)}\right)\left(\boldsymbol{E}_{(1)}^*+\boldsymbol{E}_{(2)}^*\right)\right\rangle\\&=\underbrace{\left\langle|\boldsymbol{E}_{(1)}|^2\right\rangle}_{I_1}+\underbrace{\left\langle|\boldsymbol{E}_{(2)}|^2\right\rangle}_{I_2}+\underbrace{\left\langle\boldsymbol{E}_{(1)}\boldsymbol{E}_{(2)}^*+\boldsymbol{E}_{(1)}^*\boldsymbol{E}_{(2)}\right\rangle}_{2\,\mathrm{Re}\left\langle\boldsymbol{E}_{(1)}\boldsymbol{E}_{(2)}^*\right\rangle}\\&=I_1+I_2+2\,\mathrm{Re}\underbrace{\langle E_1(t)E_2^*(t+\tau)\rangle}_{\Gamma_{12}(\tau)}=I_1+I_2+2\,\mathrm{Re}\,\Gamma_{12}(\tau),\end{aligned} \tag{2.223}$$

2.2.4.2 Coherence

The cross-correlation function:

$$C_{12}(\tau)=\Gamma_{12}(\tau)=\langle E_1(t)E_2^*(t+\tau)\rangle \tag{2.224}$$

in optics is also known as the *mutual coherence function* (MCF). On the other hand, the following autocorrelation function:

$$C_{11}(\tau)=\Gamma_{11}(\tau)=\langle E_1(t)E_1^*(t+\tau)\rangle, \tag{2.225}$$

in optics is known as the *self-coherence function*. Clearly, we assumed that the random processes $E_1(t)$ and $E_2(t)$ are *wide-sense stationary*, meaning that the mean values are constant over the time and that autocorrelation functions are solely dependent of the time difference at which the process is sampled. The sharp brackets in (2.223, 2.224 and, 2.225) are used to denote the time-average function:

$$\langle f \rangle = \lim_{T\to\infty} \frac{1}{T}\int_{-T/2}^{T/2} f(t)dt. \tag{2.226}$$

The normalized cross-correlation function in optics is known as the *degree of partial coherence*:

$$\gamma_{12}(\tau) = \frac{\Gamma_{12}(\tau)}{\sqrt{\Gamma_{11}(0)\Gamma_{22}(0)}} = \frac{\Gamma_{12}(\tau)}{\sqrt{I_1 I_2}}. \tag{2.227}$$

Based on (2.227), the interference term in (2.223) can be expressed as $2\,\mathrm{Re}\,\Gamma_{12}(\tau) = 2\sqrt{I_1 I_2}\,\gamma_{12}(\tau)$, so that the irradiance equation can be rewritten as:

$$I = I_1 + I_2 + 2\sqrt{I_1 I_2}\,\gamma_{12}(\tau). \tag{2.228}$$

The magnitude of thc degree of partial coherence can be used to clarify different types of coherence into three types: (i) complete coherence when $|\gamma_{12}| = 1$, (ii) complete incoherence when $|\gamma_{12}| = 0$, and (iii) partial coherence, otherwise. Another relevant parameter in optics is the *fringe visibility*, defined as:

$$\mathcal{V} = \frac{I_{\max} - I_{\min}}{I_{\max} + I_{\min}}. \tag{2.229}$$

and based on irradiance Eq. (2.228), the fringe visibility can be expressed as:

$$\mathcal{V} = \frac{I_1 + I_2 + 2\sqrt{I_1 I_2}\,|\gamma_{12}| - \left(I_1 + I_2 - 2\sqrt{I_1 I_2}\,|\gamma_{12}|\right)}{I_1 + I_2 + 2\sqrt{I_1 I_2}\,|\gamma_{12}| + I_1 + I_2 - 2\sqrt{I_1 I_2}\,|\gamma_{12}|} = \frac{2\sqrt{I_1 I_2}\,|\gamma_{12}|}{I_1 + I_2}. \tag{2.230}$$

In particular, when $I_1 = I_2 = I$:

$$\mathcal{V} = \frac{2\sqrt{I^2}\,|\gamma_{12}|}{2I} = |\gamma_{12}|, \tag{2.231}$$

the fringe visibility is equal to the magnitude of the degree of partial coherence. For the complete coherence ($|\gamma_{12}| = 1$), the fringe visibility will have the maximum contrast of unity. On the other hand, for the compete incoherence ($|\gamma_{12}| = 0$), there are no interference fringes at all as the contrast is zero.

Let us now observe the *quasimonochromatic* field:

$$E(t) = E_0 e^{\,j[\omega t + \phi(t)]}, \tag{2.232}$$

where the phase is constant for the duration of time interval τ_c and changes rapidly for $\tau > \tau_c$. For simplicity, let us assume that the phase changes uniformly every τ_c seconds. The degree of partial coherence in this case is just the correlation coefficient:

$$\gamma(\tau) = \frac{\langle E(t)E^*(t)\rangle}{\left\langle |E|^2 \right\rangle} = \left\langle e^{-j\omega\tau} e^{\,j[\phi(t)-\phi(t+\tau)]} \right\rangle = e^{-j\omega\tau} \lim_{T\to\infty}\frac{1}{T}\int_0^T e^{\,j[\phi(t)-\phi(t+\tau)]}dt. \tag{2.233a}$$

For each time interval of duration τ_c, the phase difference $\phi(t) - \phi(t + \tau)$ is nonzero for t between $\tau_c - \tau$ and τ_c so that:

$$\gamma(\tau) = \frac{1}{\tau_c} e^{-j\omega\tau}\int_0^{\tau_c-\tau} e^{\,j0}dt + \underbrace{e^{-j\omega\tau}E\left[\frac{1}{\tau_c}\int_0^{\tau_c-\tau} e^{\,j\Delta\phi}dt\right]}_{=0} = e^{-j\omega\tau}\left(1 - \frac{\tau}{\tau_c}\right), \tau < \tau_c \tag{2.233b}$$

where we used the statistical average operator $E\{\cdot\}$ to perform the statistical averaging with respect to the random phase $\Delta\phi$. Clearly, we assumed that the process is ergodic so that the statistical and time-averaging operators can be interchangeably used. Therefore, the fringe visibility is known:

$$\mathcal{V} = |\gamma_{12}| = \begin{cases} 1 - \frac{\tau}{\tau_c}, \tau < \tau_c \\ 0, \quad \tau \geq \tau_c \end{cases} \tag{2.234}$$

Evidently, the fringe visibility drops to zero when the time difference between two observations τ is larger than the characteristic time interval τ_c, which is commonly referred to as the *coherence time*. Equivalently, for the nonzero fringe visibility, the path difference between two beams must be small than:

$$l_c = c\tau_c, \tag{2.235}$$

and we refer to the path difference l_c as the *coherence length*.

We are concerned now with the coherence between two fields at different points in space, in other words with *spatial coherence*. Let us observe quasimonochromatic point source S. If the two observation points lie in the same direction from the source, the corresponding spatial coherence is known as the *longitudinal spatial coherence* of the field. As long as the distance between these two points is smaller than l_c, the coherence between the fields is high. On the other hand, if the two observation points P_1 and P_2 are located at the same distance from the source S, but in different directions, then the normalized MCF will measure the so-called transverse spatial coherence of the field. The partial coherence between the observation points P_1 and P_2 occurs when the source S is extended source. If the extended source is composed of two mutually incoherent point sources S_1 and S_2, as illustrated in Fig. 2.13, the corresponding electric fields in observation points P_1 and P_2 are given as:

$$E_1 = E_{P_1S_1} + E_{P_1S_2}, E_2 = E_{P_2S_1} + E_{P_2S_2}. \tag{2.236}$$

The normalized cross-correlation function is given by:

$$\begin{aligned}
\gamma_{12}(\tau) &= \frac{\langle E_1(t)E_2^*(t+\tau)\rangle}{\sqrt{I_1I_2}} = \frac{\left\langle [E_{P_1S_1} + E_{P_1S_2}]\left[E^*_{P_2S_1} + E^*_{P_2S_2}\right]\right\rangle}{\sqrt{I_1I_2}} \\
&= \underbrace{\frac{\left\langle E_{P_1S_1}E^*_{P_2S_1}\right\rangle}{\sqrt{I_1I_2}}}_{\gamma(\tau_{S_1})/2} + \underbrace{\frac{\left\langle E_{P_1S_1}E^*_{P_2S_2}\right\rangle}{\sqrt{I_1I_2}}}_{=0} + \underbrace{\frac{\left\langle E_{P_1S_2}E^*_{P_2S_1}\right\rangle}{\sqrt{I_1I_2}}}_{=0} + \underbrace{\frac{\left\langle E_{P_1S_2}E^*_{P_2S_2}\right\rangle}{\sqrt{I_1I_2}}}_{\gamma(\tau_{S_2})/2}, \\
\tau_{S_1} &= \tau + \frac{r_{P_2S_1} - r_{P_1S_1}}{c}, \tau_{S_2} = \tau + \frac{r_{P_2S_2} - r_{P_1S_2}}{c}
\end{aligned} \tag{2.237}$$

where we used the fact that point sources are mutually incoherent, while $\gamma(\tau)$ is given by Eq. (2.233). The fringe visibility between the observation points is related to [10]:

$$|\gamma_{12}(\tau)| \approx \sqrt{\frac{1 + \cos\left[\omega(\tau_{S_2} - \tau_{S_1})\right]}{2}\left(1 - \frac{\tau_{S_1}}{\tau_c}\right)\left(1 - \frac{\tau_{S_2}}{\tau_c}\right)} \tag{2.238}$$

and it has the periodic spatial dependence. The mutual coherence drops to zero, at either side of the central line, when:

$$\omega(\tau_{S_2} - \tau_{S_1}) = 2\pi\frac{c}{\lambda}\left[\frac{r_{P_2S_2} - r_{P_1S_2}}{c} - \frac{r_{P_2S_1} - r_{P_1S_1}}{c}\right] = \pi. \tag{2.239}$$

Let s denote the distance between point sources, l denote the distance between observation points, and r denote the mean distance between the sources and observation points.

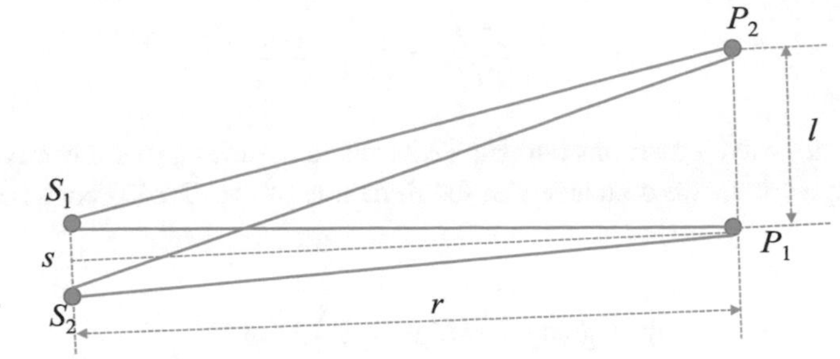

Fig. 2.13 Illustration of the lateral coherence from two mutually incoherent sources, assuming $r >> s, l$

This case is commonly referIllustration of the lateral coherence from two mutually incoherent sources, assuming $r>>s, l$. From Fig. 2.13, we see that $r_{P_1S_2} \approx r_{P_1S_1}$, so that [10]:

$$l_t \cong \frac{r}{s}\lambda. \tag{2.240}$$

The angle separation of the point sources as seen from the observation point P_1 is given by $\tan\theta_s \simeq \theta_s \simeq s/r$ so that we can write:

$$l_t \cong \lambda/\theta_s. \tag{2.241}$$

Since l_t represents the separation of receiving points for which mutual coherence is relevant, it can be called the *transverse coherence width*. When the source is extended circular, we have to modify the previous equation as follows [10]:

$$l_t \cong 1.22\lambda/\theta_s. \tag{2.242}$$

2.2.4.3 Diffraction Theory

The *divergence theorem* [5, 6], also known as *Gauss's theorem* or *Ostrogradsky's theorem*, represents a relevant theorem in vector calculus, which relates a flux of vector field through a surface to the behavior of the vector field in the volume inside of that surface. Let $\boldsymbol{F}$ be a differentiable vector field defined on the neighborhood of the volume $\mathcal{V}$ bounded by the closed surface S. The *divergence theorem* can be formulated as follows: the flux of the vector field $\boldsymbol{F}$ over the closed surface S is equal to the divergence of the vector filed $\boldsymbol{F}$ inside the volume $\mathcal{V}$. Mathematical representation of this theorem is given by:

$$\oint_S \boldsymbol{F} d\boldsymbol{S} = \int_{\mathcal{V}} \mathrm{div}\,\boldsymbol{F}\, d\mathcal{V}, d\boldsymbol{S} = \hat{\boldsymbol{n}} dS, \tag{2.243}$$

where $\hat{\boldsymbol{n}}$ is the outward surface normal. By setting $\boldsymbol{F} = U\nabla V - V\nabla U$, previous equation can be rewritten:

$$\oint_S \underbrace{(U\nabla V - V\nabla U)\hat{\boldsymbol{n}}}_{U\,\mathrm{grad}_{\hat{n}} V - V\,\mathrm{grad}_{\hat{n}} U} dS = \int_{\mathcal{V}} \underbrace{\nabla(U\nabla V - V\nabla U)}_{U\nabla^2 V - V\nabla^2 U} d\mathcal{V}, \tag{2.244}$$

and by multiplying the both sides of equation with -1, we obtain:

$$\oint_S \left(V\,\mathrm{grad}_{\hat{n}} U - U\,\mathrm{grad}_{\hat{n}} V\right) dS = \int_{\mathcal{V}} \left(V\nabla^2 U - U\nabla^2 V\right) d\mathcal{V}, \tag{2.245}$$

where we used $\mathrm{grad}_{\hat{n}} U$ to denote the normal component of the gradient at the surface S. When both U and V are the wave functions, they both satisfy the corresponding wave equations:

$$\nabla^2 U = \frac{1}{v^2}\frac{\partial^2 U}{\partial t^2},\ \nabla^2 V = \frac{1}{v^2}\frac{\partial^2 V}{\partial t^2}. \tag{2.246}$$

By substituting the left sides of the wave equations into Eq. (2.245) and assuming that the wave functions contain harmonic time dependence of the form exp.($\pm j\omega t$), we conclude that the right side of Eq. (2.245) becomes zero so that the Eq. (2.245) reduces to:

$$\oint_S \left(V \operatorname{grad}_{\hat{n}} U - U \operatorname{grad}_{\hat{n}} V \right) dS = 0. \tag{2.247}$$

Let us now suppose that wave function V is spherical and contains the harmonic time dependence as follows:

$$V = V_0 \frac{e^{j(\omega t + kr)}}{r}. \tag{2.248}$$

We are interested in determining the wave function U at point P, where $r = 0$. For $r = 0$, the V becomes infinitely large, and in integration given by Eq. (2.247), we have to exclude this point from the integration. This can be achieved by subtracting an integral over an infinitesimal sphere of radius $\rho \to 0$, with the point P being in its center, as follows:

$$\begin{aligned}&\oint_S \left(\frac{e^{jkr}}{r} \operatorname{grad}_{\hat{n}} U - U \operatorname{grad}_{\hat{n}} \frac{e^{jkr}}{r} \right) dS - \\ &\lim_{\rho \to 0} \oint_{S:r=\rho} \left[\frac{e^{jkr}}{r} \frac{\partial}{\partial r} U - U \underbrace{\frac{\partial}{\partial(-r)} \frac{e^{jkr}}{r}}_{-\left(jk - \frac{1}{r}\right)\frac{e^{jkr}}{r}} \right]_{r=\rho} r^2 d\Omega = 0,\end{aligned} \tag{2.249}$$

where $d\Omega = \sin\theta dr d\theta d\phi$ represents the solid angle in the spherical coordinates (r,θ,ϕ) on the sphere of radius ρ centered at P and r^2 $d\Omega$ is the corresponding area. On the interior infinitesimal sphere, enclosing the point P, the normal vector points inward, so that $\operatorname{grad}_{\hat{n}} = -\partial/\partial r$. In the limit U in the second term becomes U_P, while the second integral is simple $\oint -U_p e^{jk0} \frac{1}{\rho^2} \rho^2 d\Omega = -4\pi U_P$, and from Eq. (2.249), we derive the following solution for U_P:

$$U_P = -\frac{1}{4\pi} \oint_S \left(U \operatorname{grad}_{\hat{n}} \frac{e^{jkr}}{r} - \frac{e^{jkr}}{r} \operatorname{grad}_{\hat{n}} U \right) dS, \tag{2.250}$$

and this equation is commonly referred to as the *Kirchhoff integral theorem*.

Now we apply this equation to the problem related to the *diffraction of light*. The diffraction of light represents the deflection of light on sharp edges, narrow slits, and small apertures. Our problem at hands is to determine the optical disturbance at the point of interest P, originating from the source S, as illustrated in Fig. 2.14. The position of the point in the aperture observed from the source is denoted as $\boldsymbol{r}'$, while the position of the same point in the aperture from the point P is denoted as $\boldsymbol{r}$.

To simplify the problem, we employ the *Saint-Venant's hypothesis* claiming that (i) U in the aperture is the same as if there was no aperture at all and (ii) $U = 0$ on the screen and at very large distances (outside of the shaded space). The wave function of the point in the aperture at the position $\boldsymbol{r}'$ is simply:

$$U = U_0 \frac{e^{j(kr' - \omega t)}}{r'}. \tag{2.251}$$

By substituting this wave function into Kirchhoff integral, we obtain:

$$U_P = \frac{1}{4\pi} U_0 e^{-j\omega t} \oint_S \left(\frac{e^{jkr}}{r} \operatorname{grad}_{\hat{n}} \frac{e^{jkr'}}{r'} - \frac{e^{jkr'}}{r'} \operatorname{grad}_{\hat{n}} \frac{e^{jkr}}{r} \right) dS, \tag{2.252}$$

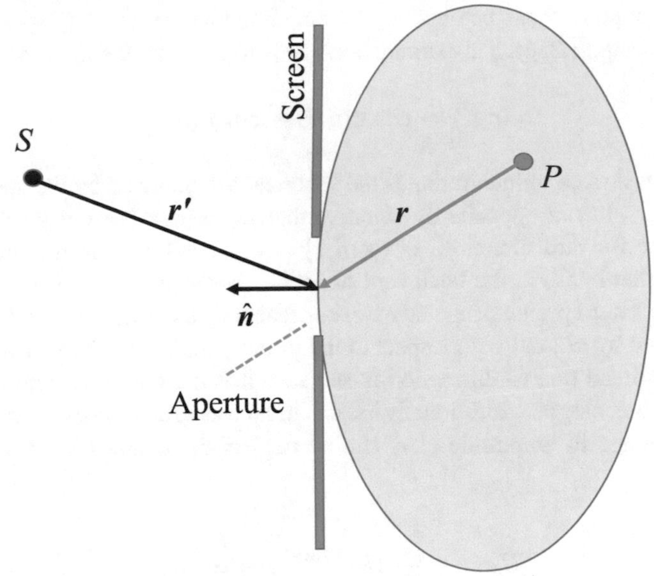

Fig. 2.14 Illustration of the of the diffraction problem on the narrow slit

where the integration is performed over the surface of aperture opening. Given that $\mathrm{grad}_{\widehat{n}}$ denotes the normal component of the gradient at the aperture opening, the corresponding gradients are evaluated as:

$$\begin{aligned}\mathrm{grad}_{\widehat{n}}\frac{e^{jkr'}}{r'} &= \cos\measuredangle(\widehat{\boldsymbol{n}},\boldsymbol{r}')\frac{\partial}{\partial r'}\frac{e^{jkr'}}{r'} = \cos\measuredangle(\widehat{\boldsymbol{n}},\boldsymbol{r}')\left(\frac{jke^{jkr'}}{r'}-\frac{e^{jkr'}}{r'^2}\right),\\ \mathrm{grad}_{\widehat{n}}\frac{e^{jkr}}{r} &= \cos\measuredangle(\widehat{\boldsymbol{n}},\boldsymbol{r})\frac{\partial}{\partial r}\frac{e^{jkr}}{r} = \cos\measuredangle(\widehat{\boldsymbol{n}},\boldsymbol{r})\left(\frac{jke^{jkr}}{r}-\frac{e^{jkr}}{r^2}\right),\end{aligned} \tag{2.253}$$

where the terms decaying as $1/r^2$ can be neglected when r and r' are much larger than the wavelength of the source.

After substitution of (2.253), the Kirchhoff integral (2.252) becomes:

$$U_P = -\frac{jk}{4\pi}U_0 e^{-j\omega t}\oint_S \frac{e^{jk(r+r')}}{rr'}[\cos\measuredangle(\widehat{\boldsymbol{n}},\boldsymbol{r}) - \cos\measuredangle(\widehat{\boldsymbol{n}},\boldsymbol{r}')]dS, \tag{2.254}$$

and this equation is commonly referred to as the *Fresnel-Kirchhoff integral formula*. By introducing the following substitution:

$$U_S = U_0\frac{e^{jkr'}}{r'}, \tag{2.255}$$

Fresnel-Kirchhoff integral formula can be rewritten as:

$$U_P = -\frac{jk}{4\pi}\oint_S U_S\frac{e^{j(kr-\omega t)}}{r}[\cos\measuredangle(\widehat{\boldsymbol{n}},\boldsymbol{r}) - \cos\measuredangle(\widehat{\boldsymbol{n}},\boldsymbol{r}')]dS, \tag{2.256}$$

which represents the mathematical statement of the *Huygens' principle*. The U_S represents the complex amplitude of the incident primary spherical wave at the aperture (of surface S). From this primary spherical wave, each element dS of the aperture gives rise to the secondary spherical wave at point P:

$$U_S\frac{e^{j(kr-\omega t)}}{r}dS, \tag{2.257}$$

and the total optical disturbance at point P can be obtained by summing up the secondary spherical waves form each aperture surface element according to the Eq. (2.256). This summation needs to consider the *inclination (obliquity) factor*:

$$I_F(\boldsymbol{r},\boldsymbol{r}') = \cos \measuredangle(\widehat{\boldsymbol{n}},\boldsymbol{r}) - \cos \measuredangle(\widehat{\boldsymbol{n}},\boldsymbol{r}'). \tag{2.258}$$

When the aperture is circular, the surface of integration is the spherical cap bounded by the aperture, and the normal vector is aligned with the positon vector $\boldsymbol{r}'$, but in opposite direction so that $\cos \measuredangle(\widehat{\boldsymbol{n}},\boldsymbol{r}') = \cos\pi = -1$, and the inclination factor becomes $\cos \measuredangle(\widehat{\boldsymbol{n}},\boldsymbol{r}) + 1$. In the forward direction, $\cos \measuredangle(\widehat{\boldsymbol{n}},\boldsymbol{r}) = \cos 0 = 1$ so that the inclination factor is equal to the maximum value of 2. On the other hand, in the backward direction, $\cos \measuredangle(\widehat{\boldsymbol{n}},\boldsymbol{r}) = \cos\pi = -1$, and the inclination factor becomes 0, indicating that no backward progressing wave can be created by the original wavefront. The factor –j indicates that diffracted waves are phase shifted by $\pi/2$ rad with respect to the primary incident spherical wave.

In the *far-field*, the angular spread due to diffraction is small so that the obliquity factor does not vary much over the aperture and can be taken out of the integral. Additionally the spherical wave function exp.(jkr')/r' can be considered constant over the aperture, while the change in amplitude ($1/r$) can be neglected, so that the Fresnel-Kirchhoff integral formula reduces to:

$$U_P = -\frac{jk}{4\pi r} U_0 e^{-j\omega t} \frac{e^{jkr'}}{r'} I_F \oint_S e^{jkr} dS, \tag{2.259}$$

indicating that the diffracted waves are effectively plane waves. Such a diffraction is commonly referred to as the *Fraunhofer (far-field) diffraction*. On the other hand, when either the receiving point or source point is close to the aperture so that the curvature of the wavefront is relevant, such diffraction is known as the *Fresnel (near-field) diffraction*. There is no strict distinction between two cases; however, the following criteria can be used. Let the size of the aperture opening (diameter) be denoted by D, while the distances of the receiving point P and the source point S are located at distances d and d' from the diffracting aperture plane, respectively, as illustrated in Fig. 2.15. The wave is effectively plane over the aperture when the curvature measure defined below is much smaller than the operating wavelength:

$$\frac{1}{2}\left(\frac{1}{d} + \frac{1}{d'}\right) D^2 \ll \lambda. \tag{2.260}$$

For $d \approx d'$ we are in the Fraunhofer region when $d >> D^2/\lambda$.

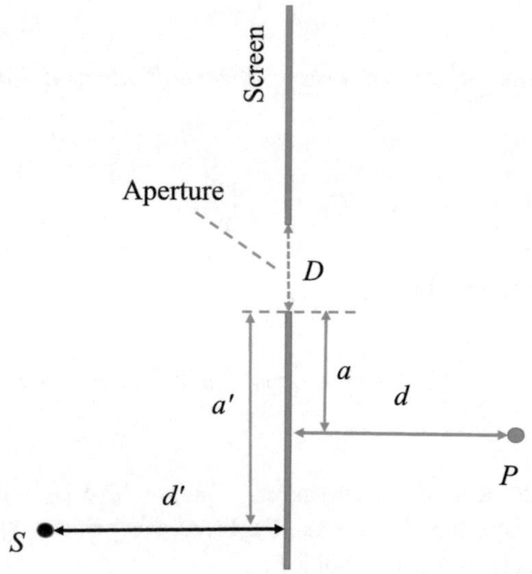

Fig. 2.15 Distinguishing between Fraunhofer and Fresnel diffractions

The variation in $r + r'$ from one edge of the aperture to the other, based on Fig. 2.15, can be determined as:

$$\begin{aligned}\Delta(r+r') &= \sqrt{d'^2+(a'+D)^2}+\sqrt{d^2+(a+D)^2}-\sqrt{d'^2+a'^2}-\sqrt{d^2+D^2}\\ &= (\frac{a'}{d'}+\frac{a}{d})+\frac{1}{2}(\frac{1}{d}+\frac{1}{d'})D^2+\ldots\end{aligned} \tag{2.261}$$

Clearly, the second term in Taylor expansion represents the curvature that was used in Eq. (2.260). In terms of the Fresnel number, the Fraunhofer condition becomes $D^2/(\lambda d) << 1$.

The inclination factor can also be expressed as follows:

$$\begin{aligned}I_F &= \cos\measuredangle(\hat{\boldsymbol{n}},\boldsymbol{r})-\cos\measuredangle(\hat{\boldsymbol{n}},\boldsymbol{r}') = \frac{\hat{\boldsymbol{n}}\cdot\boldsymbol{r}}{\underbrace{|\hat{\boldsymbol{n}}|}_{=1}|\boldsymbol{r}|}-\frac{\hat{\boldsymbol{n}}\cdot\boldsymbol{r}'}{|\hat{\boldsymbol{n}}||\boldsymbol{r}'|}\\ &= \hat{\boldsymbol{n}}\cdot\hat{\boldsymbol{r}}-\hat{\boldsymbol{n}}\cdot\hat{\boldsymbol{r}}';\hat{\boldsymbol{r}}=\boldsymbol{r}/|\boldsymbol{r}|,\hat{\boldsymbol{r}}'=\boldsymbol{r}'/|\boldsymbol{r}'|\end{aligned} \tag{2.262}$$

so that the *Fresnel-Kirchhoff integral formula* can be represented in the following alternative form:

$$U_P = -\frac{jk}{4\pi}U_0e^{-j\omega t}\oint_S\frac{e^{jk(r+r')}}{rr'}[\hat{\boldsymbol{n}}\cdot\hat{\boldsymbol{r}}-\hat{\boldsymbol{n}}\cdot\hat{\boldsymbol{r}}']dS, \tag{2.263}$$

which is valid when both receiving point P and source point S are a few wavelengths away from the aperture, in so-called wave zone ($kr>>1$).

Assuming that the inclination factor is nearly constant over the aperture, and replacing $U_0e^{-j\omega t}e^{jkr'}/r'$ with $U_{\text{inc}}(\boldsymbol{r}')$, the Eq. (2.263) can be rewritten as:

$$U_P = -\frac{jk}{4\pi}I_F\oint_S U_{\text{inc}}(\boldsymbol{r}')\frac{e^{jkr}}{r}dS, \tag{2.264}$$

and this equation is applicable to the wavefront U_{inc} generated by an *extended source* (a collection of point sources) as well. By representing the aperture by a transmission function $\tau(r')$, describing the changes in amplitude and phase, we come up with more general expression as follows:

$$U_P = -\frac{jk}{4\pi}I_F\oint_S U_{\text{inc}}(\boldsymbol{r}')\tau(\boldsymbol{r}')\frac{e^{jkr}}{r}dS, \tag{2.265}$$

As an illustration, let us determine the transmission function of a *thin plano-convex lens* [12, 13], shown in Fig. 2.16. The optical path between the ray shown in figure and one aligned along z-axis is $kn(t-\delta)\approx kn[t-\rho^2/(2R)]$, where t is the thickness of the lens and n is the refractive index of lens material.

Let us assume that incident beam is the plane wave $U_{\text{inc}}=\exp.(jkz)$ and then transmitted beam is given by, for ray shown in Fig. 2.16, as follows:

$$U_{\text{trans}} = \exp\left\{jk\left[z+\underbrace{\frac{\rho^2}{2R}}_{\delta}+n\left(t-\frac{\rho^2}{2R}\right)\right]\right\}. \tag{2.266}$$

The *transmission function for the plano-convex lens* will be then:

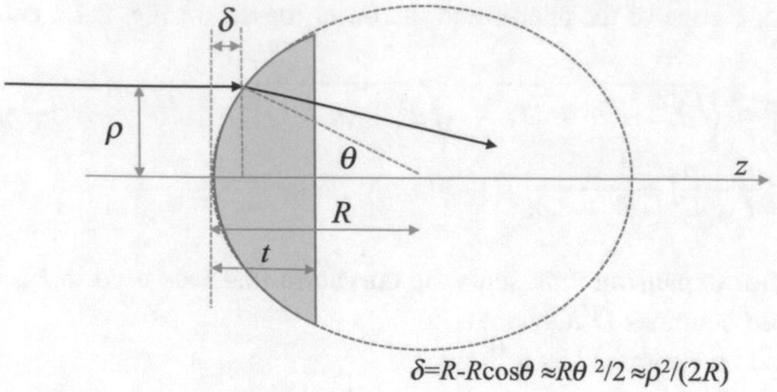

Fig. 2.16 Thin plano-convex lens

$$\tau(\rho) = U_{\text{trans}}/U_{\text{inc}} = e^{jknt}e^{-jk(n-1)\frac{\rho^2}{2R}}. \tag{2.267}$$

For a *biconvex lens*, with corresponding curvatures being R_1 and $-R_2$, the transmission function will be:

$$\tau(\rho) = e^{jkn(t_1+t_2)}e^{-jk(n-1)\frac{\rho^2}{2}\left(\frac{1}{R_1}-\frac{1}{R_2}\right)} = e^{jkn(t_1+t_2)}e^{-jk\frac{\rho^2}{2f}}, \frac{1}{f} = (n-1)\left(\frac{1}{R_1}-\frac{1}{R_2}\right), \tag{2.268}$$

where f, known as the *focal length*, is related to the net curvature of the wavefront.

2.2.4.4 Fresnel and Fraunhofer Approximations

We now discuss the *Fresnel and Fraunhofer approximations* with more details. For this purpose, in Fig. 2.17, we provide the coordinate systems for source, screen, and observed planes. The point source in source plane is denoted with S, one particular point in aperture is denoted by Q, and the receiving point is denoted by P. Since we have changed the direction of some position vectors to facilitate explanations, the Eq. (2.265) is rewritten below to match the notations in Fig. 2.16:

$$U_P = \frac{jk}{4\pi} I_F U_0 \oint_S \frac{e^{jkr}}{r}\tau(r')\frac{e^{jks}}{s}dS. \tag{2.269}$$

The position vector of the aperture point Q with respect to the source point S is denoted by $\boldsymbol{r}$, while $\boldsymbol{r}'$ denotes the origin in screen plane with respect to the source point P. The position of the aperture point Q in screen plane is denoted by $\boldsymbol{\rho}$. These three vectors are related as:

$$\boldsymbol{r} = \boldsymbol{r}' + \boldsymbol{\rho}, \tag{2.270}$$

so the length of vector r can be written as:

$$r = \sqrt{\boldsymbol{r}\cdot\boldsymbol{r}} = \sqrt{(\boldsymbol{r}'+\boldsymbol{\rho})\cdot(\boldsymbol{r}'+\boldsymbol{\rho})} = \sqrt{r'^2 + 2\boldsymbol{r}'\cdot\boldsymbol{\rho} + \rho^2} = r'\left[1 + \frac{2\boldsymbol{r}'\cdot\boldsymbol{\rho}}{r'^2} + \frac{\rho^2}{r'^2}\right]^{1/2}. \tag{2.271}$$

By using the binomial expansion, $(1 + x)^{1/2} \cong 1 + x/2 - x^2/8 + x^3/16 - \cdots$ and by keeping the first three terms the following approximation is obtained:

$$r \cong r' + \frac{\boldsymbol{r}'\cdot\boldsymbol{\rho}}{r'} + \frac{\rho^2}{2r'} - \frac{(\boldsymbol{r}'\cdot\boldsymbol{\rho})^2}{2r'^3} = r' + \hat{\boldsymbol{r}}'\cdot\boldsymbol{\rho} + \frac{\rho^2}{2r'} - \frac{(\boldsymbol{r}'\cdot\boldsymbol{\rho})^2}{2r'^3}. \tag{2.272}$$

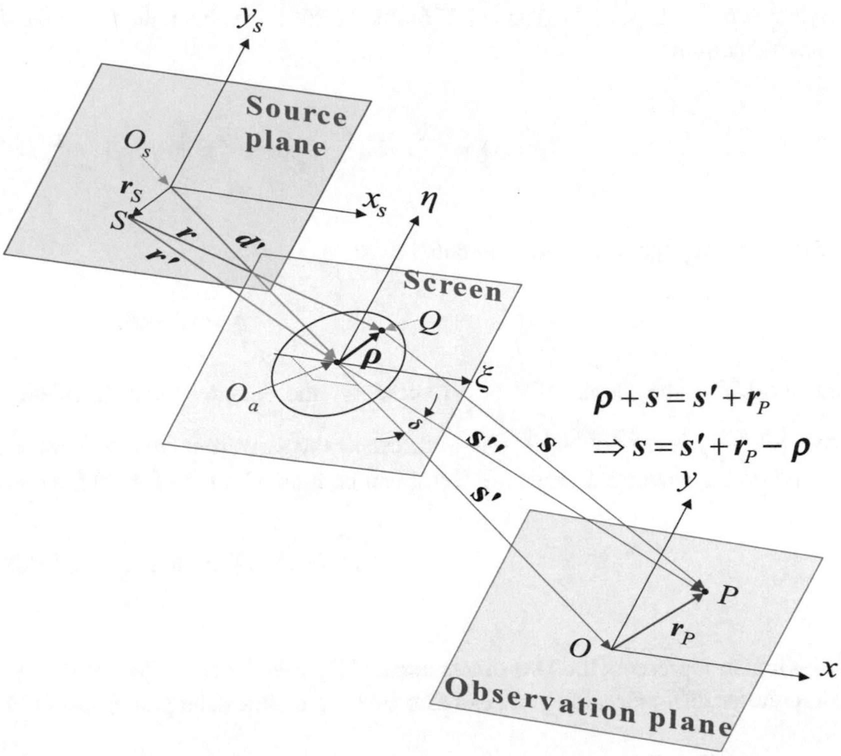

Fig. 2.17 Coordinate systems and geometry for Fresnel and Fraunhofer approximations

By close inspection of Fig. 2.17, we conclude that projection of vector $\boldsymbol{\rho}$ on $\boldsymbol{r}'$ can be determined as $\boldsymbol{r}'\cdot\boldsymbol{\rho}$. On the other hand, the projection of vector $\boldsymbol{\rho}$ on the normal to the $\boldsymbol{r}'$ is given by $\sqrt{\rho^2-(\boldsymbol{r}'\cdot\boldsymbol{\rho})^2}$. By applying the Pythagorean theorem, we establish the following connection between r and r':

$$r=\sqrt{(r'+\boldsymbol{r}'\cdot\boldsymbol{\rho})^2+\rho^2-(\boldsymbol{r}'\cdot\boldsymbol{\rho})^2}=\sqrt{r'^2+2\boldsymbol{r}'\cdot\boldsymbol{\rho}+\rho^2}=r'\left[1+\frac{2\boldsymbol{r}'\cdot\boldsymbol{\rho}}{r'^2}+\frac{\rho^2}{r'^2}\right]^{1/2}, \quad (2.273)$$

which is the same expression as given by Eq. (2.271). By applying the following two *Fresnel approximations*:

1. The coarse approximation in amplitude: $1/r \approx 1/r'$.
2. The fine approximation in phase:

$$kr \cong kr' + k\widehat{\boldsymbol{r}}'\cdot\boldsymbol{\rho}+k\frac{\rho^2-(\widehat{\boldsymbol{r}}'\cdot\boldsymbol{\rho})^2}{2r'}, \quad (2.274)$$

the spherical wave propagation function becomes:

$$\frac{e^{jkr}}{r}\cong\frac{e^{jkr'}}{r'}e^{jk\widehat{\boldsymbol{r}}'\cdot\boldsymbol{\rho}}e^{jk\frac{\rho^2-(\widehat{\boldsymbol{r}}'\cdot\boldsymbol{\rho})^2}{2r'}}, \quad (2.275)$$

where $\exp(jkr')/r'$ term represents the spherical wave at the aperture origin, $\exp(jk\boldsymbol{r}'\cdot\boldsymbol{\rho})$ corresponds to the plane wave in direction of $\widehat{\boldsymbol{r}}'$, and $\exp\{jk[\rho^2-(\widehat{\boldsymbol{r}}'\cdot\boldsymbol{\rho})^2]/(2r')\}$ represents the lowest order phase approximation of the spherical wavefront. By applying this phase approximation to the spherical wave $\exp(jks)/s$, we obtain:

$$\frac{e^{jks}}{s}\cong\frac{e^{jks'}}{s'}e^{jk\widehat{\boldsymbol{s}}'\cdot(\boldsymbol{r}_P-\boldsymbol{\rho})}\exp\left[jk\frac{\overbrace{|\boldsymbol{r}_P-\boldsymbol{\rho}|^2}^{(\boldsymbol{r}_P-\boldsymbol{\rho})(\boldsymbol{r}_P-\boldsymbol{\rho})}-[\widehat{\boldsymbol{s}}'\cdot(\boldsymbol{r}_P-\boldsymbol{\rho})]^2}{2s'}\right]. \quad (2.276)$$

By substituting Eqs. (2.275) and (2.276) into (2.269), we obtain the following *diffraction integral of a point source in Fresnel approximation*:

$$U_P(\boldsymbol{r},\boldsymbol{s}) = \frac{jk}{4\pi} I_F U_0 \frac{e^{jk(r'+s')}}{r's'} e^{jk\widehat{\boldsymbol{s}'}\cdot\boldsymbol{r}_P} \oint_S \tau(\boldsymbol{r}) e^{j\boldsymbol{q}\cdot\boldsymbol{\rho}} e^{jk\left\{\frac{\rho^2-\left(\widehat{\boldsymbol{r}'}\cdot\boldsymbol{\rho}\right)^2}{2r'}+\frac{(\boldsymbol{r}_P-\boldsymbol{\rho})^2-\left[\widehat{\boldsymbol{s}'}\cdot(\boldsymbol{r}_P-\boldsymbol{\rho})\right]^2}{2s'}\right\}} dS, \tag{2.277}$$

where the propagation vector $\boldsymbol{q}$ is defined as:

$$\boldsymbol{q} = k\widehat{\boldsymbol{r}'} - k\widehat{\boldsymbol{s}'}. \tag{2.278}$$

In (2.277), the term $e^{jk\widehat{\boldsymbol{s}'}\cdot\boldsymbol{r}_P}$ represents the plane wave incident at receiving point P, while the term $\exp\left\{jk\frac{\rho^2-\left(\widehat{\boldsymbol{r}'}\cdot\boldsymbol{\rho}\right)^2}{2r'}+\frac{(\boldsymbol{r}_P-\boldsymbol{\rho})^2-\left[\widehat{\boldsymbol{s}'}\cdot(\boldsymbol{r}_P-\boldsymbol{\rho})\right]^2}{2s'}\right\}$ represents the *wavefront curvature correction factor*.

When the wavefront curvature factor can be ignored, in the far-field, the *Fraunhofer diffraction integral* is obtained:

$$U_P(\boldsymbol{r},\boldsymbol{s}) = (\text{constant}) \oint_S \tau(\boldsymbol{r}) e^{j\boldsymbol{q}\cdot\boldsymbol{\rho}} dS, \tag{2.279}$$

which also represents the two-dimensional (2D) *Fourier transform of the aperture transmission function* $\tau(\boldsymbol{r})$. Fresnel and Fraunhofer diffraction integrals can also be used to study the propagation of the light beam in the free-space optical (FSO) links.

Fraunhofer diffraction integral for a rectangular aperture of dimensions $2a$ (in ξ-coordinate) × $2b$ (in η-coordinate) yields to:

$$\begin{aligned} U_P(\boldsymbol{r},\boldsymbol{s}) &= \underbrace{\frac{jk}{4\pi} I_F U_0 \frac{e^{jk(r'+s')}}{r's'} e^{jk\widehat{\boldsymbol{s}'}\cdot\boldsymbol{r}_P}}_{(\text{constant})} \int_{-a}^{a} \int_{-b}^{b} e^{j(q_x\xi+q_y\eta)} d\xi d\eta \\ &= (\text{constant}) 2\text{sinc}\,(q_x a)\text{sinc}\,(q_y b), \end{aligned} \tag{2.280}$$

where sinc(x) = sin(πx)/(πx). The corresponding intensity is given by:

$$I(\boldsymbol{q}) = I(0)\text{sinc}^2(q_x a)\text{sinc}^2(q_y b), I(0) = 4|(\text{constant})|^2. \tag{2.281}$$

Fraunhofer diffraction integral for a circular aperture of radius a is given by:

$$\begin{aligned} U_P(\boldsymbol{r},\boldsymbol{s}) &= \underbrace{\frac{jk}{4\pi} I_F U_0 \frac{e^{jk(r'+s')}}{r's'} e^{jk\widehat{\boldsymbol{s}'}\cdot\boldsymbol{r}_P}}_{(\text{constant})} \int_0^{2\pi} \int_0^{a} e^{jq\rho\cos\phi} \rho d\rho d\phi \\ &= (\text{constant})\pi a^2 \frac{2J_1(qa)}{qa}, \end{aligned} \tag{2.282}$$

where $J_1(x)$ is the Bessel function of the first kind and the first order [14]. The corresponding intensity, given by:

$$I(\boldsymbol{q}) = I(0)\left[\frac{2J_1(qa)}{qa}\right]^2, I(0) = |(\text{constant})|^2\left(\pi a^2\right)^2, \tag{2.283}$$

is commonly referred to as the *Airy pattern*. For a small diffraction angle θ, the following is valid $q \cong k\theta$. The first zero of $J_1(x)$ is 1.22π, so that $qa \cong (2\pi/\lambda)\theta_0 a = 1.22\pi$, and we can write:

$$\theta_0 \cong 1.22\underbrace{\frac{\lambda}{2a}}_{d} = 1.22\frac{\lambda}{d}, d = 2a, \tag{2.284}$$

which is the same as Eq. (2.242).

2.2.5 Laser Beam Propagation Over the Atmospheric Turbulence Channels

2.2.5.1 Paraxial Wave Equation

In this section we are concerned with the laser beam propagation over the random inhomogeneous medium, namely, the atmospheric turbulence channels. Let us introduce the slow-varying envelope of the electric field vector $\boldsymbol{E}(\boldsymbol{r},t)$ as follows:

$$\boldsymbol{E}(\boldsymbol{r},t) = \boldsymbol{e}\underline{E}(\boldsymbol{r})e^{j\omega t - jkz}, \tag{2.285}$$

where $\boldsymbol{e}$ is the polarization vector. After substitution of (2.285) into the wave equation and upon canceling the common $\boldsymbol{e}$ exp.($j\omega t$) term, we obtain:

$$\nabla^2\left[\underline{E}(\boldsymbol{r})e^{-jkz}\right] + k^2 n^2(\boldsymbol{r})\underline{E}(\boldsymbol{r})e^{-jkz} = 0, \tag{2.286}$$

where the refractive index now is a function of the position vector $\boldsymbol{r}$. By applying the definition of Laplacian and canceling the common exp($-jkz$) term, we obtain:

$$\left(\underbrace{\frac{\partial^2}{\partial x^2} + \frac{\partial^2}{\partial y^2}}_{\nabla_\perp^2}\right)\underline{E}(\boldsymbol{r}) - 2jk\frac{\partial}{\partial z}\underline{E}(\boldsymbol{r}) + \frac{\partial^2}{\partial z^2}\underline{E}(\boldsymbol{r}) + k^2\left[n^2(\boldsymbol{r}) - 1\right]\underline{E}(\boldsymbol{r}) = 0, \tag{2.287}$$

where we use $\nabla_\perp^2 = \partial^2\backslash\partial x^2 + \partial^2\backslash\partial y^2$ to denote the *transverse part of the Laplacian*. In the *paraxial approximation*, when $\measuredangle(\boldsymbol{k},\hat{z}) \ll 1$, the derivative of $\underline{E}$ is the slow-varying function of z so that the following is valid:

$$\left|\frac{\partial^2}{\partial z^2}\underline{E}(\boldsymbol{r})\right| << \left|2k\frac{\partial}{\partial z}\underline{E}(\boldsymbol{r})\right|, \tag{2.288}$$

and the third term in (2.287) can be neglected so that the wave equation reduces to:

$$\nabla_\perp^2\underline{E}(\boldsymbol{r}) - 2jk\frac{\partial}{\partial z}\underline{E}(\boldsymbol{r}) + k^2\left[n^2(\boldsymbol{r}) - 1\right]\underline{E}(\boldsymbol{r}) = 0, \tag{2.289}$$

which is commonly referred to as the *paraxial wave equation*. By replacing $n^2(\boldsymbol{r})$ with $n^2(\boldsymbol{r})/n_0^2$, since in the air $n_0 \cong 1$, we can represent $\frac{n^2}{n_0^2} - 1 = \frac{\overbrace{(n-n_0)}^{\delta n}\overbrace{(n+n_0)}^{2n_0}}{n_0^2} \cong 2\delta n$ so that the paraxial wave equation can be rewritten as:

$$2jk\frac{\partial}{\partial z}\underline{E}(\boldsymbol{r}) = \left[\nabla_\perp^2 + 2k^2\delta n(\boldsymbol{r})\right]\underline{E}(\boldsymbol{r}), \tag{2.289}$$

providing that $\delta n < <1$.

2.2.5.2 Split-Step Beam Propagation Method

To solve the paraxial wave equation, the finite-difference method can be used [15]; however, the complexity of such approach is high. The most common method to study the propagation over the atmospheric turbulence channels is the Fourier *split-step beam propagation method* [16–24]. The key idea behind this algorithm is to split the atmospheric turbulence link into sections of length Δz each and perform integration of the paraxial wave Eq. (2.289) on each section. The parabolic Eq. (2.289) on section of length Δz can be solved by separation of variables:

$$\frac{\partial \underline{E}(\boldsymbol{r})}{\underline{E}(\boldsymbol{r})} = \frac{1}{2jk}\left[\nabla_{\perp}^2 + 2k^2\delta n(\boldsymbol{r})\right]\partial z. \tag{2.290}$$

By performing the integration over the *i*-th section, from z_i to z_{i+1}, we obtain:

$$\underbrace{\int_{z_i}^{z_{i+1}} \frac{d\underline{E}(\boldsymbol{r})}{\underline{E}(\boldsymbol{r})} dz}_{\ln\left[\underline{E}(\boldsymbol{r})\right]\Big|_{z_i}^{z_{i+1}}} = \underbrace{\int_{z_i}^{z_{i+1}} \frac{-j}{2k}\left[\nabla_{\perp}^2 + 2k^2\delta n(\boldsymbol{r})\right]dz.}_{\frac{-j}{2k}\left[\nabla_{\perp}^2\Delta z + 2k^2\int_{z_i}^{z_{i+1}}\delta n(\boldsymbol{r})dz\right]} \tag{2.291}$$

The Eq. (2.291) can also be written as:

$$\underline{E}(x,y,z_{i+1}) = \exp\left\{\frac{-j}{2k}\left[\Delta z\nabla_{\perp}^2 + 2k^2\int_{z_i}^{z_{i+1}}\delta n(\boldsymbol{r})dz\right]\right\}\underline{E}(x,y,z_i). \tag{2.292}$$

By introducing the *refraction operator* $\mathcal{R}$ and the *diffraction operator* $\mathcal{D}$ as follows, respectively:

$$\mathcal{R}(z_i, z_{i+1}) = -jk\int_{z_i}^{z_{i+1}}\delta n(\boldsymbol{r})dz, \mathcal{D}(\Delta z) = \frac{-j}{2k}\Delta z\nabla_{\perp}^2, \tag{2.293}$$

the Eq. (2.292) can be represented as:

$$\underline{E}(x,y,z_{i+1}) = \exp\left\{\mathcal{D}(\Delta z) + \mathcal{R}(z_i, z_{i+1})\right\}\underline{E}(x,y,z_i). \tag{2.294}$$

This notation is consistent with that used in fiber-optics communications [2, 4]. The Taylor expansion can be used to represent the operators in exponential form:

$$\exp\left[\mathcal{D}(\Delta z) + \mathcal{R}(z_i, z_{i+1})\right] = \sum_{n=0}^{\infty}\left[\mathcal{D}(\Delta z) + \mathcal{R}(z_i, z_{i+1})\right]^n/n! \tag{2.295}$$

Instead of Taylor expansion, the following two methods known as *symmetric split-step* method (SSSM) and *asymmetric split-step* method (ASSM) are commonly used in practice:

$$\underline{E}(x,y,z_{i+1}) = \begin{cases} e^{\mathcal{D}(\Delta z/2)}e^{\mathcal{R}(z_i,z_{i+1})}e^{\mathcal{D}(\Delta z/2)}\underline{E}(x,y,z_i), \text{forSSSM} \\ e^{\mathcal{D}(\Delta z)}e^{\mathcal{R}(z_i,z_{i+1})}\underline{E}(x,y,z_i), \text{forASSM} \end{cases} \tag{2.296}$$

To study the accuracy of these two methods, we employ the *Baker-Hausdorff formula* from quantum mechanics [7, 25]:

$$e^A e^B = e^{A+B+\frac{1}{2}[A,B]+\frac{1}{12}[A,[A,B]]-\frac{1}{12}[B,[A,B]]-\ldots}, \tag{2.297}$$

where [*A*,*B*] is the commutator of operators *A* and *B*, defined as [*A*,*B*] = *AB*-*BA*. By applying Baker-Hausdorff formula on ASSM and SSM, by interpreting $A = \mathcal{D}$ and $B = \mathcal{R}$, we obtain:

$$\begin{aligned} e^{\mathcal{D}(\Delta z)}e^{\mathcal{R}(z_i,z_{i+1})} &\cong e^{\mathcal{D}(\Delta z)+\mathcal{R}(z_i,z_{i+1})+\frac{1}{2}[\mathcal{D}(\Delta z),\mathcal{R}(z_i,z_{i+1})]+\ldots}, \\ e^{\mathcal{D}(\Delta z/2)}e^{\mathcal{R}(z_i,z_{i+1})}e^{\mathcal{D}(\Delta z/2)} &\cong e^{\mathcal{D}(\Delta z)+\mathcal{R}(z_i,z_{i+1})+\frac{1}{6}[\mathcal{R}(z_i,z_{i+1})+\mathcal{D}(\Delta z/2)[\mathcal{R}(z_i,z_{i+1}),\mathcal{D}(\Delta z/2)]]+\ldots} \end{aligned} \tag{2.298}$$

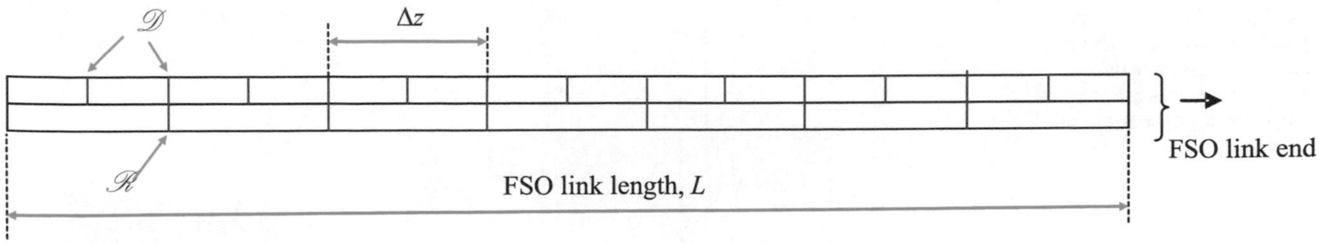

Fig. 2.18 Illustration of symmetric split-step method. *FSO* free-space optical

For sufficiently small step Δz, both diffraction and refraction operators are proportional to Δz indicating that the accuracy of ASSM is in the order of $(\Delta z)^2$, while the accuracy of SSSM is in the order of $(\Delta z)^3$. Therefore, the accuracy of SSSM is better. The SSSM can be illustrated as shown in Fig. 2.18. The algorithm proceeds by first applying the diffraction operator $\mathcal{D}$ for one half step length ($\Delta z/2$). Then the refraction operator $\mathcal{R}$ is applied for one step length Δz. The linear operator is again applied for the full step-size Δz. Therefore, the diffraction operator is applied at every half step ($\Delta z/2$), while the refraction operator at every full step (Δz).

The diffraction and refraction operators can also be defined in exponential form [20, 21]. If the *refraction operator* $\mathcal{R}$ and the *diffraction operator* $\mathcal{D}$ in exponential form are defined as follows:

$$\mathfrak{R}(z_i, z_{i+1}) = \exp\left[-jk\int_{z_i}^{z_{i+1}} \delta n(\boldsymbol{r})dz\right], \mathfrak{D}(\Delta z) = \exp\left[\frac{-j}{2k}\Delta z \nabla_\perp^2\right], \tag{2.299}$$

the SSSM can be described as:

$$\underline{E}(x, y, z_{i+1}) = [\mathfrak{D}(\Delta z/2)\mathfrak{R}(z_i, z_{i+1})\mathfrak{D}(\Delta z/2)]\underline{E}(x, y, z_i). \tag{2.300}$$

2.2.5.3 Paraxial Fresnel and Fraunhofer Diffraction Approximations

Let us now return to the Fig. 2.17 and observe the normal incidence case for which the inclination factor is $I_F = 2$. If the aperture is so large, that beam can be considered as a plane so that $\tau(\boldsymbol{r}') = 1$. For this case Eq. (2.269) becomes:

$$U_P = j\frac{2\pi/\lambda}{4\pi}2U_0\oint_S \frac{e^{jk(r+s)}}{rs}dS = \frac{j}{\lambda}U_0\oint_S \frac{e^{jk(r+s)}}{rs}dS. \tag{2.301}$$

By applying the Fresnel approximations:

$$1/r \approx 1/r', 1/s \approx 1/s', jk(r+s) \cong jk\left[r' + s' + \frac{(\boldsymbol{\rho} - \boldsymbol{r}_S)^2}{2r'} + \frac{(\boldsymbol{r}_P - \boldsymbol{\rho})^2}{2s'}\right], \tag{2.302}$$

the Eq. (2.301) can be represented as:

$$U_P = \frac{j}{\lambda}U_0\oint_S \frac{e^{jk(r'+s')}}{r's'}e^{jk\left[\frac{(\boldsymbol{\rho}-\boldsymbol{r}_S)^2}{2r'} + \frac{(\boldsymbol{r}_P-\boldsymbol{\rho})^2}{2s'}\right]}dS. \tag{2.303}$$

and this equation describes the free-space propagation from a point source to the screen plane and from the screen plane to the receiving point P. By denoting the terms incident to the screen plane as U_{inc}, the Fresnel diffraction integral describing the propagation from the screen plane to the observation plane can be written as:

$$U_P = \frac{j}{\lambda}\frac{e^{jks'}}{s'}\oint_S U_{\text{inc}}(\boldsymbol{\rho})e^{jk\frac{(\boldsymbol{r}_P-\boldsymbol{\rho})^2}{2s'}}dS. \tag{2.304}$$

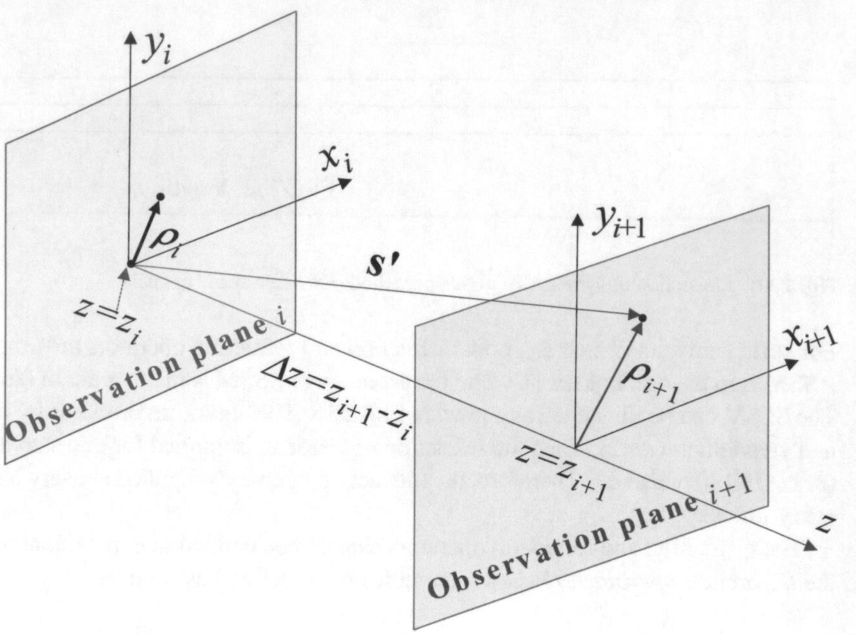

Fig. 2.19 Illustration of paraxial approximation to describe free-space propagation between observation planes z_i and z_{i+1}. The $\boldsymbol{\rho}$ denotes the transversal position

For *paraxial approximation*, as shown in Fig. 2.19, we can see that the Eq. (2.304) can be represented in the following form:

$$U(\boldsymbol{\rho}_{i+1}) = \frac{j}{\lambda} \frac{e^{jk(z_{i+1}-z_i)}}{z_{i+1}-z_i} \oint_{S_i} U(\boldsymbol{\rho}_i) e^{jk\frac{(\boldsymbol{\rho}_{i+1}-\boldsymbol{\rho}_i)^2}{2(z_{i+1}-z_i)}} dS_i, \tag{2.305}$$

where $z_{i+1} - z_i$ is the spacing between the observation planes i and $i + 1$, while $\boldsymbol{\rho} = x\hat{x} + y\hat{y}$ is the transversal radius vector. This equation is widely used in the existing literature [21, 26]. The Fresnel diffraction integral for paraxial approximation in Cartesian coordinates can be straightforwardly obtained from Eq. (2.305):

$$U(x_{i+1}, y_{i+1}) = \frac{j}{\lambda} \frac{e^{jk\Delta z}}{\Delta z} \int_{-\infty}^{\infty} \int_{-\infty}^{\infty} U(x_i, y_i) e^{jk\frac{(x_{i+1}-x_i)^2+(y_{i+1}-y_i)^2}{2\Delta z}} dx_i dy_i. \tag{2.306}$$

The Eq. (2.306) essentially represents the action of the diffraction operator on electrical field envelope in the i-th section of the ASSM, from observation plane i to the observation plane $i + 1$. For the SSM, in the i-th section, we would need to apply Eq. (2.306) twice, once to describe FSO propagation from z_i to $z_i + \Delta z/2$, followed by the application of refraction operator at $z_i + \Delta z$, and the second time to describe FSO propagation from $z_i + \Delta z/2$ to $z_i + \Delta z(=z_{i+1})$. Unfortunately, the form given by Eq. (2.306) is not suitable for efficient simulation of diffraction effects. If we apply the binomial formula, and take out the terms independent of integrand variables, we obtain the following form:

$$\begin{aligned}
U(x_{i+1}, y_{i+1}) &= \frac{j}{\lambda} \frac{e^{jk\Delta z}}{\Delta z} e^{jk\frac{x_{i+1}^2+y_{i+1}^2}{2\Delta z}} \int_{-\infty}^{\infty} \int_{-\infty}^{\infty} U(x_i, y_i) e^{jk\frac{x_i^2+y_i^2}{2\Delta z}} e^{-j\frac{2\pi}{\lambda}\frac{x_i x_{i+1}+y_i y_{i+1}}{\Delta z}} dx_i dy_i \\
&= \frac{j}{\lambda} \frac{e^{jk\Delta z}}{\Delta z} e^{jk\frac{x_{i+1}^2+y_{i+1}^2}{2\Delta z}} \int_{-\infty}^{\infty} \int_{-\infty}^{\infty} U(x_i, y_i) e^{jk\frac{x_i^2+y_i^2}{2\Delta z}} \exp\left[-j2\pi\left(\underbrace{\frac{x_{i+1}}{\lambda\Delta z}}_{f_{x_{i+1}}} x_i + \underbrace{\frac{y_{i+1}}{\lambda\Delta z}}_{f_{y_{i+1}}} y_i\right)\right] dx_i dy_i \\
&= \frac{j}{\lambda} \frac{e^{jk\Delta z}}{\Delta z} e^{jk\frac{x_{i+1}^2+y_{i+1}^2}{2\Delta z}} \int_{-\infty}^{\infty} \int_{-\infty}^{\infty} U(x_i, y_i) e^{jk\frac{x_i^2+y_i^2}{2\Delta z}} e^{-j2\pi\left(f_{x_{i+1}} x_i + f_{y_{i+1}} y_i\right)} dx_i dy_i
\end{aligned} \tag{2.307}$$

where $f_{x_{i+1}} = x_{i+1}/(\lambda\Delta z)$ and $f_{y_{i+1}} = y_{i+1}/(\lambda\Delta z)$ are *spatial frequencies* along x- and y-axes, respectively. The Eq. (2.307) can be interpreted as the two-dimensional (2D) Fourier transform (FT) of $U(x_i, y_i)\exp\left[jk\left(x_i^2 + y_i^2\right)\backslash(2\Delta z)\right]$. The 2D FT employing 1D fast Fourier transform (FFT) algorithms can be used for this purpose as explained in [27]. Alternatively, efficient multidimensional FFT algorithms can be used [28]. We can also interpret the Eq. (2.307) as the *convolution operation*:

$$U(x_{i+1}, y_{i+1}) = -U(x_i, y_i) \otimes \frac{e^{jk\Delta z}}{j\lambda\Delta z} e^{jk\frac{x_i^2+y_i^2}{2\Delta z}}, \tag{2.308}$$

where we used the operation ⊗ to denote the spatial-domain convolution. The operator * will be reserved to denote the time-domain convolution throughout the book. By employing the convolution theorem, claiming that FT of the convolution of two functions is equal to the product of corresponding FTs, we can rewrite the Eq. (2.308) as follows:

$$U(x_{i+1}, y_{i+1}) = -FT^{-1}\{FT[U(x_i, y_i)]FT[\frac{e^{jk\Delta z}}{j\lambda\Delta z} e^{jk\frac{x_i^2+y_i^2}{2\Delta z}}]\}. \tag{2.309}$$

Given that FT of the second term in (2.308) can be determined in closed form by:

$$H(\boldsymbol{f}_i) = FT[\frac{e^{jk\Delta z}}{j\lambda\Delta z} e^{jk\frac{x_i^2+y_i^2}{2\Delta z}}] = e^{jk\Delta z} e^{-j\pi\lambda\Delta z\left(f_{x_i}^2 + f_{y_i}^2\right)}, \boldsymbol{f}_i = \left[\, f_{x_i} \;\; f_{y_i}\right]^T, \tag{2.310}$$

the Eq. (2.309) can be written in the form more suitable for numerical computation:

$$U(x_{i+1}, y_{i+1}) = -FT^{-1}\{FT[U(x_i, y_i)]e^{jk\Delta z} e^{-j\pi\lambda\Delta z\left(f_{x_i}^2 + f_{y_i}^2\right)}\}, \tag{2.311}$$

and this method is known in the literature as the *angular-spectral propagation* method (see [21] and references therein).

When the distance between two observation planes Δz is sufficiently large, the quadratic term $k\left(x_i^2 + y_i^2\right)/(2\Delta z)$ in (2.307) becomes negligible, and paraxial approximation becomes the paraxial Fraunhofer diffraction approximation:

$$\begin{aligned} U\left(x_{i+1}, y_{i+1}\right) &= \frac{j}{\lambda}\frac{e^{jk\Delta z}}{\Delta z} e^{jk\frac{x_{i+1}^2+y_{i+1}^2}{2\Delta z}} \int_{-\infty}^{\infty}\int_{-\infty}^{\infty} U(x_i, y_i)\exp\left[-j2\pi\left(\underbrace{\frac{x_{i+1}}{\lambda\Delta z}}_{f_{x_{i+1}}} x_i + \underbrace{\frac{y_{i+1}}{\lambda\Delta z}}_{f_{y_{i+1}}} y_i\right)\right] dx_i dy_i \\ &= \frac{j}{\lambda}\frac{e^{jk\Delta z}}{\Delta z} e^{jk\frac{x_{i+1}^2+y_{i+1}^2}{2\Delta z}} \int_{-\infty}^{\infty}\int_{-\infty}^{\infty} U(x_i, y_i) e^{-j2\pi\left(f_{x_{i+1}} x_i + f_{y_{i+1}} y_i\right)} dx_i dy_i. \end{aligned} \tag{2.312}$$

2.2.5.4 Rytov Method

Another approach to study the diffraction in free space is to employ the *method due to Rytov* [29] (see also [20]) as follows. Let us represent the complex envelope of the electric field in the form of Fourier series [20]:

$$\underline{E}(x, y, z) = \sum_{m=-N/2+1}^{N/2} \sum_{n=-N/2}^{N/2} E_{mn}(z) e^{j\left[\frac{2\pi}{L}(mx+ny)\right]}, \tag{2.313.1}$$

where the expansion coefficients are determined by:

$$E_{mn}(z) = \frac{1}{4\pi^2 L^2} \int_0^L\int_0^L \underline{E}(x, y, z) e^{-j\left[\frac{2\pi}{L}(mx+ny)\right]} dx dy, \tag{2.313.2}$$

wherein L is the size of the expansion domain and N is the number of terms in each dimension. By substituting the Fourier series expansion (2.312) into the paraxial wave equation (2.289) (by ignoring the refraction term) and after cancelation common terms, we obtain:

$$2jk\frac{\partial}{\partial z}E_{mn} = -\frac{4\pi^2}{L^2}\left(m^2+n^2\right)E_{mn}, \tag{2.314}$$

which can be solved in closed form by separation of variables:

$$E_{mn}(z) = E_{mn}(0)e^{\,j\frac{4\pi^2}{2kL^2}\left(m^2+n^2\right)z}. \tag{2.315}$$

This approach appears to be conceptually simpler and requires the Fourier expansion coefficients to be calculated only once before transmission takes place at $z = 0$. On the other hand, we need to compute N^2 expansion coefficients.

2.2.5.5 Modeling Atmospheric Turbulence Effects

To solve for the refraction problem in the i-th layer of SSSM, we need to determine the variation of the refractive index $\delta n(\boldsymbol{r})$, and the refraction effect can be described by passing the beam through the phase screen as follows:

$$\varphi_i(\boldsymbol{\rho}) = -k\int_{z_i}^{z_{i+1}}\delta n(\boldsymbol{\rho}, z)dz, \tag{2.316}$$

and the refraction operator can be determined based on Eq. (2.299) as:

$$\mathscr{R}(z_i, z_{i+1}) = \exp\left[-jk\int_{z_i}^{z_{i+1}}\delta n(\boldsymbol{r})dz\right] = \exp\left[j\varphi_i(\boldsymbol{\rho})\right]. \tag{2.317}$$

This approach can be used to study the thermal blooming of the high-power laser beams as described in [20]. Here we are concerned with a fundamental problem that occurs during free-space propagation of the laser beam, even under clear weather conditions, namely, so-called atmospheric turbulence [4, 20, 21, 30–40]. A commonly used turbulence model assumes that the variations of the medium can be understood as individual cells of air or eddies of different diameters and refractive indices. In the context of geometrical optics, these eddies may be observed as lenses that randomly refract the optical wavefront, generating a distorted intensity profile at the receiver of a communication system. The intensity fluctuation is known as scintillation and represents one of the most important factors that limit the performance of an atmospheric FSO communication link. The most widely accepted theory of turbulence is due to Kolmogorov [41–44]. This theory assumes that kinetic energy from large turbulent *eddies* (pockets of air, vortices), characterized by the parameter known as the outer scale L_0, is transferred without loss to the eddies of decreasing size down to sizes of a few millimeters characterized by the inner scale parameter l_0. The inner scale represents the cell size at which energy is dissipated by viscosity. The refractive index varies randomly across the different turbulent eddies and causes phase and amplitude variations to the wavefront. Turbulence can also cause the random drifts of optical beams – a phenomenon usually referred to as wandering – and can induce the beam defocusing. Clearly, the longitudinal variation in the refractive index is in the scale of l_0, which is typically much smaller than the longitudinal discretization step Δz, indicating that the Eq. (2.316) cannot be determined in the closed form. Instead, the statistical methods must be used to properly describe the turbulent fluctuations of the refractive index. The *correlation function* of the random phase fluctuation appears to be a good starting point [20]:

$$R_\varphi(\boldsymbol{\rho}) = \langle\varphi(\boldsymbol{\rho}_0)\varphi(\boldsymbol{\rho}_0+\boldsymbol{\rho})\rangle = k^2\int_{\Delta z}\int_{\Delta z}\langle n(\boldsymbol{\rho}_0, z_i+z')n(\boldsymbol{\rho}_0+\boldsymbol{\rho}, z_i+z'')\rangle dz'dz''. \tag{2.318}$$

Based on Eq. (2.318), the spectral density of random process φ, denoted as Φ_φ, given as the FT of the correlation function $R\varphi$, can be related to the 3D spectral density of refractive index, denoted as Φ_n. For von Karman spectrum of atmospheric turbulence, to be described below, it has the following form [20]:

$$\Phi_\varphi(\boldsymbol{\kappa}) = FT\{R_\varphi(\boldsymbol{\rho})\} = 2\pi k^2 \Delta z \Phi_n(\boldsymbol{\kappa}_\perp, \kappa_z = 0, L_0) K(\boldsymbol{\kappa}, L_0/\Delta z), \tag{2.319}$$

where $\boldsymbol{\kappa}$ is the angular spatial frequency vector, defined as $\boldsymbol{\kappa} = 2\pi\left(f_x\widehat{x} + f_y\widehat{y} + f_z\widehat{z}\right)$, and $K(\cdot)$ is the correction factor, which is close to 1 when the ratio $L_0/\Delta z$ is sufficiently small. The refraction problem in either SSSM or ASSM can be solved by generating the random phase screens satisfying the statistics described by Eq. (2.319). In situations when the random phase φ is a stationary process, we can generate the samples from a Gaussian process and colored them according to Eq. (2.319) by passing the Gaussian process though a properly designed filter. Before we return to this problem, we provide additional details on *Kolmogorov theory of turbulence* [41–43].

2.2.5.6 Kolmogorov Theory of Turbulence and Corresponding Generalizations

To describe the turbulent flow, which represents a nonlinear process, Navier-Stokes equations [46] should be used. *Navier-Stokes equation for incompressible fluid* of constant density ρ is given by [46]:

$$\frac{\partial \boldsymbol{u}}{\partial t} + (\boldsymbol{u}\nabla)\boldsymbol{u} = -\frac{\nabla p}{\rho} + \nu\nabla^2\boldsymbol{u}, \tag{2.320}$$

where $\boldsymbol{u}$ is the velocity vector of the fluid pocket, p is the pressure, and ν is the kinematic viscosity equal to $1.51{\cdot}10^{-5}$ m^2/s for the air (and $1.01{\cdot}10^{-6}$ m^2/s for water). To solve the Navier-Stokes equations in turbulence regime on eddies level, it would be quite challenging. Instead the ensemble average properties should be studied as suggested by Kolmogorov in 1941 [41]. The turbulence can be considered as the collection of eddies of different sizes embedded in each other, changing in a random fashion. Even under laboratory conditions, it would be quite challenging to reproduce the detailed velocity's distribution. The turbulent motions span wide range of eddy sizes r, ranging from macroscale at which the energy is supplied all the way down to the microscale at which the energy is dissipated by viscosity. The average size of the largest eddies, denoted as L_0, is called the *outer scale*. On the hand, the average size of smallest eddies, denoted as l_0, is called the *inner scale*. The range for which $l_0 << r << L_0$ is known as the *inertial range*. Finally, the range for which $r < l_0$ is known as the *viscous dissipation range*. A parameter that can be used as a control parameter of the flow is known as *Reynolds number*, denoted as *Re*, and defined as [41]:

$$Re = \frac{LU}{\nu}, \tag{2.321}$$

where L is the characteristic geometric size (length) of the system and U is the characteristic velocity. The Reynolds number, for a turbulent flow to occur, must be very large so that the kinetic energy of the source, such as wind, is sufficiently high to generate so-called energy cascade. According to this model, the kinetic energy of the external source excites the largest possible eddies L_0, and the energy from these eddies is gradually passed to smaller eddies, all the way down to the inner scale l_0 when the energy is dissipated by the viscosity. As an illustration, for height of 1 km above ground and a typical wind speed of 10 m/s, the Reynolds number is huge $Re = 6.6{\cdot}10^8$, so that the conditions for turbulence to occur are not that difficult to satisfy. If the state of turbulence is *statistically steady*, then the rate of energy transfer, denoted as ϵ, from one scale of eddies to the next must be the same for all scales so that eddies within the same scale see the same total energy. In this scenario, the energy rate transferred from the source to the largest eddies (L_0) is the same as the energy dissipated at the shortest scale (l_0). The unit of the energy supplied to fluid per unit mass and time is $\epsilon_u =$ kg m^2 s^{-2} / (kg s) $=$ m^2 s^{-3}. The characteristics of eddies of scale r depend on angular speed u and energy cascade rate ϵ. Since the unit of speed is $u_u =$ m/s, while unit of scale is $r_u =$ m, the only dimensionally acceptable relationship among u, r, and ϵ is:

$$u(r) \sim \epsilon^{1/3} r^{1/3} \tag{2.322}$$

So the larger the energy transfer rate, the larger the speed of eddies is. The implication of this is that the largest eddies (with scale L_0) contain the bulk of kinetic energy while the smallest eddies (with scale l_0) have the lowest speeds. Additionally, greater energy transfer will generate stronger eddies. Clearly, the largest eddies cannot be larger than the geometrical dimension of the system, so $L_0 \leq L$. So the outer scale can be estimated as the cubic root of the volume of the system. On the other hand, the inner scale is set up by viscosity, because the smaller the eddy size, the stronger velocity shear and more important the viscosity is. The unit of kinetic viscosity is $\nu_u =$ m^2/s, as already indicated above. Therefore, the inner scale l_0

must be dependent on kinetic viscosity and the energy transfer rate, so that the only dimensionally acceptable relationship is as follows:

$$l_0 \sim \epsilon^{-1/4} \nu^{3/4}, \tag{2.323}$$

and often we refer to the inner scale as the *Kolmogorov's length*. Given the energy cascade discussion, the span of eddies' sizes is related to the Reynolds number. Based on Eq. (2.322) $U = u(L_0) \sim \epsilon^{1/3} L^{1/3}$, and from the definition of Reynolds number, we conclude that the energy supply/dissipation rate can be estimated as:

$$\epsilon \sim U^3/L. \tag{2.324}$$

Now based on Eqs. (2.323 and 2.324) as well as definition of Reynolds number, we conclude that:

$$l_0 \approx \frac{L}{Re^{3/4}}. \tag{2.325}$$

For the example above, we calculate that inner scale is ~0.24 mm and energy dissipation rate is ~1 m^2/s^3.

Another important characteristics of the turbulent channel is the *energy spectrum*, and it gives us the distribution of the energy per mass across the various eddies' sizes. Since the wave number is related to the wavelength by $k = 2\pi/\lambda$, we conclude that the wave number is reversely proportional to the scale r. Therefore, the extreme values for the wave number are $k_{\min} \sim \epsilon^{1/4} \nu^{-3/4}$ and $k_{\max} = \pi/L_0$. The kinetic energy per mass of fluid, denoted as E, has the unit E_u = kg m^2 s^{-2}/ kg = m^2/s^2. The fraction of eddies with wave numbers between k and $k + dk$ is given by:

$$dE = E(k)dk. \tag{2.326}$$

Since the unit of $E(k)$ is m^3/s^2, we conclude that only dimensionally correct relationship between $E(k)$, ϵ, and k is:

$$E(k) = C\varepsilon^{2/3} k^{-5/3}, \ 1/L_0 << k << 1/l_0 \tag{2.327}$$

and we commonly refer to this equation as *Kolmogorov's 5/3 law* or Kolmogorov-Obukhov 5/3 law. It has been found experimentally that this law is valid in *inertial range*. The constant C in (2.327) can be determined from:

$$\int_{k_{\min}}^{\infty} E(k)dk = \frac{U^2}{2}. \tag{2.328}$$

Common way, inspired by Kolmogorov's theory, to study the velocity fields in a turbulent channel is by means of flow velocity increments:

$$\boldsymbol{\delta u}(\boldsymbol{r}) = \boldsymbol{u}(\boldsymbol{x} + \boldsymbol{r}) - \boldsymbol{u}(\boldsymbol{x}), \boldsymbol{x} = \begin{bmatrix} x_1 \\ x_2 \\ x_3 \end{bmatrix} = \begin{bmatrix} x \\ y \\ z \end{bmatrix}. \tag{2.329}$$

Under isotropic assumption, in Kolmogorov's sense, in other words, when statistical properties are invariant with respect to rotations and reflections of the original coordinate system, the flow increments (in statistical sense) depend only on $|\boldsymbol{r}|$, and the velocity increments emphasize the effects of scale of the order of separation $|\boldsymbol{r}|$. Based on Eq. (2.322), we conclude that the n-th moment of the velocity increment is given by:

$$\langle [\boldsymbol{\delta u}(\boldsymbol{r})]^n \rangle = C^n \varepsilon^{n/3} r^{n/3}, \ l_0 << r << L_0. \tag{2.330}$$

The second moment of the velocity increment is commonly referred to as the *structure function*:

$$D_u(r) = C_u^2 r^{2/3}, \; l_0 << r << L_0, \tag{2.331}$$

and the constant $C_u^2 = C^2\varepsilon^{2/3}$ is known as the velocity structure parameter. For laminar flow, when $r << l_0$, the dependence in scale r is quadratic:

$$D_u(r) = C_u^2 l_0^{-4/3} r^2, \; 0 \le r << l_0. \tag{2.332}$$

Obukhov has shown in [45] that the potential temperature, defined as $\theta = T + a_h h$ (h is the height and a_h is the adiabatic rate of temperature decrease with height), structure parameter is similar to that of the velocity. The adiabatic rate is negligible for small heights, close to the ground. The refractive index in point $\boldsymbol{r}$ in the atmosphere can be represented as:

$$n(\boldsymbol{r}) = \underbrace{\langle n(\boldsymbol{r})\rangle}_{\cong 1} + \delta n(\boldsymbol{r}) = 1 + \underbrace{a\left(1 + \frac{b}{\lambda^2}\right)\frac{p(\boldsymbol{r})}{T(\boldsymbol{r})}}_{\delta n(\boldsymbol{r})} \tag{2.333}$$

where p denotes the pressure and T denotes the absolute temperature (in K). The constants a and b at optical frequencies are found to be [21, 30]: $a = 77.6 \cdot 10^{-6}$ and $b = 7.52 \cdot 10^{-3}$. The total differential of the refractive index is given by:

$$dn = a\left(1 + \frac{b}{\lambda^2}\right)\left[\frac{dP}{T} - \frac{dT}{T^2}\right], \tag{2.334}$$

and since it is directly proportional to dT, we expect the same behavior for the refractive index structure parameter:

$$D_n(r) = \begin{cases} C_n^2 r^{2/3}, & l_0 << r << L_0 \\ C_n^2 l_0^{-4/3} r^2, & 0 \le r << l_0 \end{cases} \tag{2.335}$$

where the refractive index structure parameter C_n^2 [m$^{-2/3}$] is given by:

$$C_n^2 = C_T^2\left[a\left(1 + \frac{b}{\lambda^2}\right)\frac{p}{T}\right]^2. \tag{2.336}$$

The typical values for the refractive index structure parameter $C_n^2$are in the range [10–17] m$^{-2/3}$.

Based on Eq. (2.333), we conclude that the refractive index can be represented as the sum of mean value and fluctuating part, so that the corresponding covariance function points $\boldsymbol{r}_0$ and $\boldsymbol{r}_0 + \boldsymbol{r}$ are given by:

$$C_n(\boldsymbol{r}_0, \boldsymbol{r}_0 + \boldsymbol{r}) = \langle [n(\boldsymbol{r}_0 + \boldsymbol{r}) - \langle n(\boldsymbol{r}_0 + \boldsymbol{r})\rangle][n(\boldsymbol{r}_0) - \langle n(\boldsymbol{r}_0)\rangle]\rangle. \tag{2.337}$$

Under homogeneity assumption (in Kolmogorov's sense), meaning that statistical properties are independent on initial location but dependent only on the separation $\boldsymbol{r}$ between two points, the covariance function is only dependent on $\boldsymbol{r}$, and by taking the Eq. (2.333) into account, we can rewrite previous equation as:

$$C_n(\boldsymbol{r}) = \langle \delta n(\boldsymbol{r}_0 + \boldsymbol{r})\delta n(\boldsymbol{r}_0)\rangle. \tag{2.338}$$

The correlation function and the spectral density represent the Fourier transform pair, according to Wiener-Khintchine theorem [48, 49]. Notice that in optics and physics the definitions of Fourier transform (FT) and inversed Fourier transform (IFT) [2, 30, 33, 47] are reversed compared with corresponding definitions in electrical and computer engineering [48–50]. To be consistent with reach literature on atmospheric turbulence effects [2, 12, 26, 30, 32–34, 47, 51], in this section we adopt corresponding definitions used in optics. The spectral density $\Phi_n(\boldsymbol{\kappa})$ is related to the covariance function $C_n(\boldsymbol{r})$ as follows:

$$\Phi_n(\boldsymbol{\kappa}) = \frac{1}{(2\pi)^3}\int C_n(\boldsymbol{r})e^{-j\boldsymbol{\kappa}\cdot\boldsymbol{r}}d^3\boldsymbol{r} = \frac{1}{(2\pi)^3}\int C_n(\boldsymbol{r})\cos(\boldsymbol{\kappa}\cdot\boldsymbol{r})d^3\boldsymbol{r} \tag{2.339}$$

On the other hand, the covariance function is related to the spectral density by:

$$C_n(\boldsymbol{r}) = \int \Phi_n(\boldsymbol{\kappa})e^{j\boldsymbol{\kappa}\cdot\boldsymbol{r}}d^3\boldsymbol{\kappa} = \int \Phi_n(\boldsymbol{\kappa})\cos(\boldsymbol{\kappa}\cdot\boldsymbol{r})d^3\boldsymbol{\kappa} \tag{2.340}$$

Under isotropic assumption (in Kolmogorov's sense), the covariance function depends only on the distance between two observation points, but not on the orientation of the line passing through them; we conclude that $C_n(\boldsymbol{r}) = C_n(r)$, $r = |\boldsymbol{r}|$, and $\Phi_n(\boldsymbol{\kappa}) = \Phi_n(\kappa)$. By using the spherical coordinate system, as defined in Sect. 2.1.2 , and by moving to the spherical coordinates $\boldsymbol{\kappa} = (\kappa,\theta,\phi)$, given that $d^3\boldsymbol{\kappa} = \kappa^2\sin\theta\, d\theta d\phi d\kappa$ and $\boldsymbol{\kappa}\cdot\boldsymbol{r} = \kappa r\cos\theta$, it is straightforward to show that Eqs. (2.339) and (2.340) can be simplified to:

$$\Phi_n(\boldsymbol{\kappa}) = \frac{1}{2\pi^2\kappa}\int C_n(r)\sin(\kappa r)rdr, \tag{2.339}$$

and

$$C_n(\boldsymbol{r}) = \frac{4\pi}{r}\int \Phi_n(\kappa)\sin(\kappa r)\kappa d\kappa. \tag{2.340}$$

The refractive index structure function is given by:

$$\begin{aligned} D_n(\boldsymbol{r}) &= \left\langle [\delta n(\boldsymbol{r}_0 + \boldsymbol{r}) - \delta n(\boldsymbol{r}_0)]^2 \right\rangle \\ &= 2[C_n(0) - C_n(\boldsymbol{r})] = 2\int \Phi_n(\boldsymbol{\kappa})[1 - \cos(\boldsymbol{\kappa}\cdot\boldsymbol{r})]d^3\boldsymbol{\kappa}. \end{aligned} \tag{2.341}$$

By using the Eq. (2.340), we can relate the refractive index structure function to the spectral density as follows:

$$\begin{aligned} D_n(r) &= 2\left[4\pi\int_0^\infty \Phi_n(\kappa)\underbrace{\lim_{r\to 0}\left[\frac{\sin(\kappa r)}{\kappa r}\right]}_{\lim_{r\to 0}\left[\frac{\kappa\cos(\kappa r)}{\kappa}\right]=1}\kappa^2 d\kappa - \frac{4\pi}{r}\int_0^\infty \Phi_n(\kappa)\sin(\kappa r)\kappa d\kappa\right] \\ &= 8\pi\int_0^\infty \Phi_n(\kappa)\left[1 - \frac{\sin(\kappa r)}{\kappa r}\right]\kappa^2 d\kappa]. \end{aligned} \tag{2.342}$$

The term $[1-\sin(\kappa r)/(\kappa r)]$ has the filtering properties and severely attenuations spatial frequencies for which $\kappa < r^{-1}$; in other words, the spatial frequencies with periods that are larger than the spacing between observation points.

To avoid for this problem, we can employ the *Tatarskii's approach* [51]. By taking the Laplacian of (2.341), we obtain:

$$\Delta D_n(\boldsymbol{r}) = \nabla(\nabla D_n(\boldsymbol{r})) = \nabla\{2\int \Phi_n(\boldsymbol{\kappa})\sin(\boldsymbol{\kappa}\cdot\boldsymbol{r})\boldsymbol{\kappa}d^3\boldsymbol{\kappa}\} = 2\int \Phi_n(\boldsymbol{\kappa})\cos(\boldsymbol{\kappa}\cdot\boldsymbol{r})\underbrace{\boldsymbol{\kappa}\boldsymbol{\kappa}}_{\kappa^2}d^3\boldsymbol{\kappa} \tag{2.343}$$

By inverting the Eq. (2.343), we obtain:

$$\begin{aligned} \frac{1}{(2\pi)^3}\int \cos(\boldsymbol{\kappa}\cdot\boldsymbol{r})\Delta D_n(\boldsymbol{r})d^3r &= 2\frac{1}{(2\pi)^3}\int \cos(\boldsymbol{\kappa}\cdot\boldsymbol{r})\int \Phi_n(\boldsymbol{\kappa})\cos(\boldsymbol{\kappa}\cdot\boldsymbol{r})\kappa^2 d^3\kappa d^3r \\ &= 2\frac{1}{(2\pi)^3}\int \underbrace{\left[\int \cos(\boldsymbol{\kappa}\cdot\boldsymbol{r})\Phi_n(\boldsymbol{\kappa})d^3r\right]}_{C_n(r)}\cos(\boldsymbol{\kappa}\cdot\boldsymbol{r})\kappa^2 d^3\kappa = 2k^2\Phi_n(\boldsymbol{\kappa}) \end{aligned} \tag{2.344}$$

Therefore, from (2.344), we obtain:

$$\Phi_n(\boldsymbol{\kappa}) = \frac{1}{16\pi^3\kappa^2}\int \cos\,(\boldsymbol{\kappa}\cdot\boldsymbol{r})\Delta D_n(\boldsymbol{r})d^3r. \tag{2.345}$$

Since for locally isotropic medium in spherical coordinates the Laplacian is defined as:

$$\Delta D_n(\boldsymbol{r}) = \frac{1}{r^2}\frac{\partial}{\partial r}\left[r^2\frac{\partial D_n(\boldsymbol{r})}{\partial r}\right],$$

the previous equation becomes:

$$\Phi_n(\kappa) = \frac{1}{4\pi^2\kappa^2}\int_0^\infty \frac{\sin\,(\kappa r)}{\kappa r}\frac{d}{dr}\left[r^2\frac{d}{dr}D_n(r)\right]dr, \tag{2.346}$$

where $D_n(r) = D_n(\boldsymbol{r})$.

The following covariance function and spectral density represent the Fourier transform pair [51]:

$$\begin{aligned} C(r) &= \frac{a^2}{2^{\upsilon-1}\Gamma(\upsilon)}\left(\frac{r}{r_0}\right)^\upsilon K_\upsilon\left(\frac{r}{r_0}\right), C(0) = a^2 \\ \Phi(\kappa) &= \frac{\Gamma(\upsilon+3/2)}{\pi\sqrt{\pi}\Gamma(\upsilon)}\frac{a^2r_0^3}{\left(1+\kappa^2r_0^2\right)^{\upsilon+3/2}}, \end{aligned} \tag{2.347}$$

where $\Gamma(a)$ is the gamma function, defined as $\Gamma(a) = \int_0^\infty t^{a-1}e^{-t}dt$. The $K_\upsilon(x)$ denotes the modified Bessel function of the second kind and order υ (also known as Macdonald function) defined as $K_\upsilon(u) = \{\pi/[2\,\sin\,(\upsilon\pi)]\}[I_{-\upsilon}(u) - I_\upsilon(u)]$, which is related to the modified Bessel function of the first kind by $I_\upsilon(u) = (1/\pi)\int_0^\pi \cos\,(\upsilon\vartheta)\exp\,(-u\cos\vartheta)d\vartheta$. The corresponding structure function is given by:

$$D(r) = 2[C(0) - C(r)] = 2a^2\left[1 - \frac{1}{2^{\upsilon-1}\Gamma(\upsilon)}\left(\frac{r}{r_0}\right)^\upsilon K_\upsilon\left(\frac{r}{r_0}\right)\right]. \tag{2.348}$$

By using the following approximation for the modified Bessel function $K_\upsilon(u) \approx \{\pi/[2\,\sin\,(\upsilon\pi)]\}\{[1/\Gamma(1-\upsilon)](z/2)^{-\upsilon} - [1/\Gamma(1+\upsilon)](z/2)^\upsilon\}$ which is valid for $r << r_0$, we obtain the following approximation for the structure function:

$$D(r) \cong 2a^2\frac{\Gamma(1-\upsilon)}{\Gamma(1+\upsilon)}\left(\frac{r}{2r_0}\right)^{2\upsilon}, r << r_0. \tag{2.349}$$

By introducing the substitutions $C^2 = 2a^2\frac{\Gamma(1-\upsilon)}{\Gamma(1+\upsilon)(2r_0)^{2\upsilon}}, \mu = 2\upsilon$, we can rewrite the Eq. (2.349) as follows:

$$D(r) \cong C^2r^\mu, \tag{2.350}$$

which is consistent with the refractive index structure function (2.335), by setting μ to either 2/3 or 2. Since the approximation (2.349) is valid for $r << r_0$, or equivalently $\kappa >> 1/r_0$, the corresponding approximation for spectral density, based on (2.347), becomes:

$$\Phi(\kappa) \cong \frac{\Gamma(\upsilon+3/2)}{\pi\sqrt{\pi}\Gamma(\upsilon)}\frac{a^2r_0^3}{(\kappa r_0)^{2\upsilon+3}} = \Gamma(\mu+2)\frac{\sin\,(\pi\mu/2)}{4\pi^2}C^2\kappa^{-(\mu+3)}. \tag{2.351}$$

The refractive index spectral density based on (2.347) and (2.351) can be written as:

$$\Phi(\kappa) = \begin{cases} \overbrace{\frac{\Gamma(2/3+2)\sin(\pi/3)}{4\pi^2}}^{0.033005391} C_n^2 \kappa^{-11/3}, & 2\pi/L_0 << \kappa << 2\pi/l_0 \\ \frac{\Gamma(4)}{4\pi^2} C_n^2 l_0^{-4/3} \kappa^{-5}, & \kappa >> 2\pi/l_0 \end{cases}$$
$$= \begin{cases} 0.033005391 C_n^2 \kappa^{-11/3}, & 2\pi/L_0 << \kappa << 2\pi/l_0 \\ 0.151981775 C_n^2 l_0^{-4/3} \kappa^{-5}, & \kappa >> 2\pi/l_0 \end{cases} \tag{2.352}$$

and this spectral density is commonly referred to as *Kolmogorov refractive index spectral density*.

To include the effects related to both inner and outer scales of turbulence, the various modification to the Kolmogorov spectrum have been introduced so far [20, 21, 30, 32, 33, 51, 62]. The *modified von Kármán spectral density* is given by:

$$\Phi(\kappa) = 0.033005391 C_n^2 \left(\kappa^2 + \kappa_0^2\right)^{-11/6} e^{-\kappa^2/\kappa_m^2}, \kappa_0 = 2\pi/L_0, \kappa_m = 5.92/l_0 \tag{2.353}$$

The spectrum is given by Eq. (2.353), for which $\kappa_0 = 0$ is known as the Tatarskii spectrum [62]. Clearly, by setting $l_0 = 0$, and $L_0 \to \infty$, the Kolmogorov spectral density is obtained. Another relevant spectral density is *Andrews spectral density*, representing the generalization of the modified von Kármán spectral density [52] and providing nice agreement with experimentally verified Hills spectrum:

$$\Phi(\kappa) = 0.033 C_n^2 \left[1 + 1.802(\kappa/\kappa_l) - 0.254(\kappa/\kappa_l)^{7/6}\right] \left(\kappa^2 + \kappa_0^2\right)^{-11/6} e^{-\kappa^2/\kappa_l^2}, \kappa_l = 3.3/l_0 \tag{2.354}$$

As an illustration, in Fig. 2.20 we plot spectral density for common models of atmospheric turbulence: Kolmogorov, modified von Kármán, Andrews, and Tatarskii spectra. Clearly, in inertial range we cannot see much difference among different models. For high spatial frequencies, Tatarskii spectrum behaves the same as modified von Kármán spectrum. On the

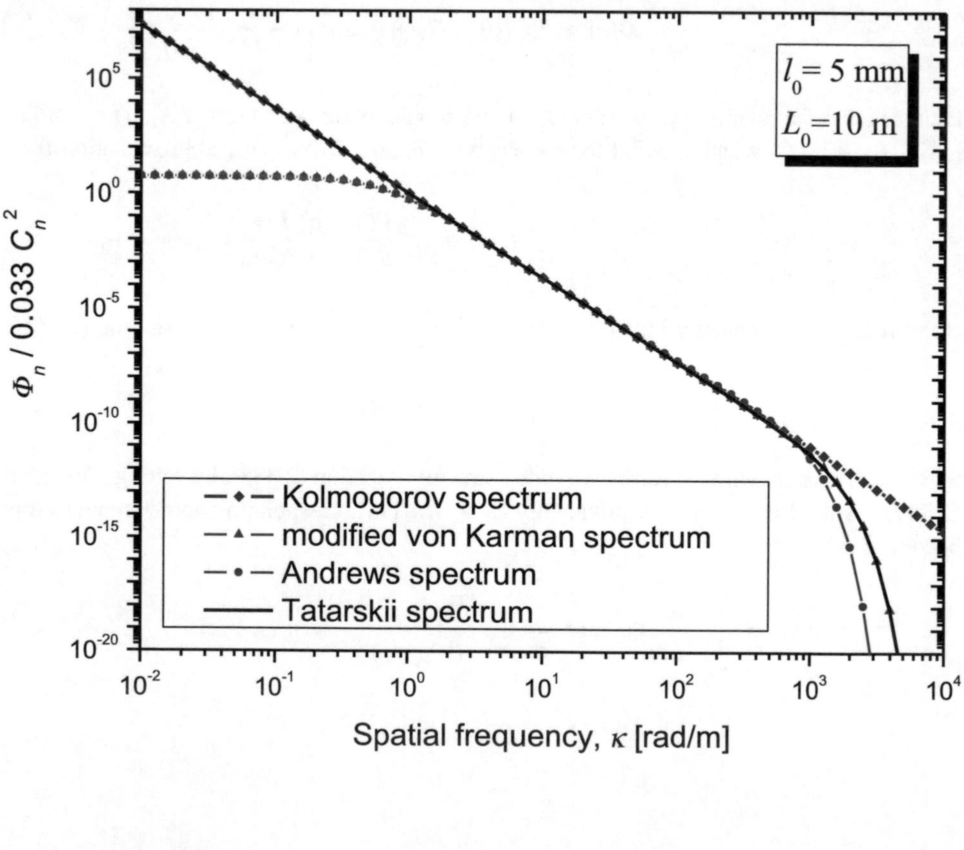

Fig. 2.20 Spectral densities for common models of atmospheric turbulence

other hand, Tatarskii spectrum behaves the same as Kolmogorov spectrum at low spatial frequencies. Andrews spectrum decays faster than modified von Kármán spectrum in high spatial frequencies region.

Now by using the Eq. (2.319), we can establish the connection between spectral densities of atmospheric turbulence-induced random phase and spectral density of refractive index, which is locally a correct relationship. Since the refractive index structure parameter is propagation distance dependent, we have to perform the integration over z so that the following relationship is obtained for Kolmogorov spectral density:

$$\Phi_\varphi(\boldsymbol{\kappa}) = 2\pi k^2 \cdot 0.033005391\kappa^{-11/3}\int_0^{\Delta z} C_n^2(z)dz. \tag{2.355}$$

By taking the zenith-angle ζ into account, the Eq. (2.355) can also be written in more common form as follows:

$$\begin{aligned}\Phi_\varphi(\boldsymbol{\kappa}) &= \underbrace{\frac{2\pi \cdot 0.033005391}{0.423}}_{0.490258}\kappa^{-11/3}\left\{\underbrace{\left[0.423k^2 \sec(\zeta)\int_0^{\Delta z} C_n^2(z)dz\right]^{-3/5}}_{r_0}\right\}^{-5/3} \\ &= 0.490258\kappa^{-11/3}r_0^{-5/3},\ r_0 = \left[0.423k^2 \sec(\zeta)\int_0^{\Delta z} C_n^2(z)dz\right]^{-3/5}\end{aligned} \tag{2.356}$$

where r_0 is commonly referred to as *Fried parameter* [53]. Fried defined atmospheric turbulence diameter r_0 in a different context, as the scale length over which the phase errors (due to atmospheric turbulence) in the wavefront are of the order of 1 rad. In other words, the phase standard deviation over an aperture of size (diameter) r_0 is approximately 1 rad. The Fried parameter in Eq. (2.352) is valid for plane waves; the corresponding expression for spherical waves must be modified with a correction factor $z/\Delta z$ to obtain [30]:

$$r_0^{(\text{spherical waves})} = \left[0.423k^2 \sec(\zeta)\int_0^{\Delta z} C_n^2(z)(z/\Delta z)^{5/3}dz\right]^{-3/5}. \tag{2.357}$$

By following the similar derivation, as for Kolmogorov spectral density, the phase spectral density for the modified von Kármán refractive index is given by:

$$\Phi_\varphi^{(\text{modifiedvonKarman})}(\boldsymbol{\kappa}) = 0.490258r_0^{-5/3}\left(\kappa^2+\kappa_0^2\right)^{-11/6}e^{-\kappa^2/\kappa_m^2}. \tag{2.358}$$

Let the Fourier transform of the atmospheric turbulence perturbed phase $\varphi(\boldsymbol{\rho}) = \varphi(x,y)$ be denoted as $\widetilde{\varphi}(f_x, f_y)$. Therefore, it is related to $\varphi(x,y)$ as follows:

$$\varphi(x,y) = \int_{-\infty}^{\infty}\int_{-\infty}^{\infty}\widetilde{\varphi}(f_x, f_y)e^{j2\pi(f_x x + f_y y)}df_x df_y. \tag{2.359}$$

The total power in the phase screen can be determined by employing the Parseval's theorem by:

$$P_{\text{total}} = \int_{-\infty}^{\infty}\int_{-\infty}^{\infty}|\varphi(x,y)|^2 dxdy = \int_{-\infty}^{\infty}\int_{-\infty}^{\infty}\Phi_\varphi(f_x, f_y)df_x df_y. \tag{2.360}$$

By using the trapezoidal rule, we can write Eq. (2.360) as:

$$P_{\text{total}} = \int_{-\infty}^{\infty}\int_{-\infty}^{\infty} \Phi_\varphi(f_x, f_y) df_x df_y \cong \sum_m \sum_n \Phi_\varphi(m\Delta f_x, n\Delta f_y)\Delta f_x \Delta f_y, \tag{2.362}$$

where Δf_x and Δf_y are frequency spacing along f_x- and f_y-axes. On the other hand, by applying the trapezoidal rule on Eq. (2.359), we obtain:

$$\varphi(x, y) \cong \sum_m \sum_n \tilde{\varphi}(m\Delta f_x, n\Delta f_y) e^{j2\pi(f_x x + f_y y)} \Delta f_x \Delta f_y. \tag{2.363}$$

From Eqs. (2.357), (2.358), we conclude that:

$$\left\langle |\tilde{\varphi}(m\Delta f_x, n\Delta f_y)|^2 \right\rangle \Delta f_x \Delta f_y = \Phi_\varphi(m\Delta f_x, n\Delta f_y)\Delta f_x \Delta f_y. \tag{2.364}$$

To generate the random phase screen, we can use the Gaussian random generator with zero-mean and unit variance and multiply the samples with the square root of variance given by Eq. (2.364). On such a way we can generate the samples for 2D random phase screen. Now we have everything needed to implement the symmetric split-step method. The main drawback of this method random phase screen generator is the low accuracy of the trapezoidal rule. To improve the accuracy, Zernike polynomial-based approach can be used [20, 54]. Alternatively, the non-uniform sampling can be employed and sample low spatial frequencies more precisely [55, 56]. Finally, the subharmonic method can be used [57] (see also [21]). In *subharmonics' method*, the low spatial frequency phase screen is a sum of Np phase screens with different accuracy levels [21, 57]:

$$\varphi_{\text{LSF}}(x, y) \cong \sum_{p=1}^{N_p} \sum_{m=-1}^{1} \sum_{n=-1}^{1} \tilde{\varphi}\left(m\Delta f_{x_p}, n\Delta f_{y_p}\right) e^{j2\pi(f_x x + f_y y)} \Delta f_{x_p} \Delta f_{y_p}, \tag{2.365}$$

where the frequency spacing for the p-th frequency grid is set to $\Delta f_{x_p} = \Delta f_{y_p} = \Delta f / b^p$, with Δf being the frequency spacing used for medium and high spatial frequency ranges, and $b \times b$ denotes the size of grid in the p-th frequency grid. As it has been shown in [21, 57], the subharmonics' method can significantly improve the accuracy of random phase screen generation.

We turn our attention now back to the correlation function of the random phase correlation, given by Eq. (2.318).

$$R_\varphi(\boldsymbol{\rho}) = k^2 \int_{\Delta z}\int_{\Delta z} \langle n(\boldsymbol{\rho}_0, z_i + z') n(\boldsymbol{\rho}_0 + \boldsymbol{\rho}, z_i + z'')\rangle dz' dz'' = k^2 \int_{\Delta z}\int_{\Delta z} R_n(\boldsymbol{\rho}, z'' - z') dz' dz''. \tag{2.366}$$

When Δz is much larger than the correlation length of the refractive index, the autocorrelation function of the refractive index can be written as $R_n(\boldsymbol{\rho}, z'' - z') = R_n(\boldsymbol{\rho}, z'')\Delta z \delta(z'' - z')$, and after extending the limits of integration and applying the sifting property, we obtain:

$$R_\varphi(\boldsymbol{\rho}) = k^2 \Delta z \int_{-\infty}^{\infty} R_n(\boldsymbol{\rho}, z) dz. \tag{2.367}$$

Given that the $\langle \delta n \rangle = 0$, we conclude that $C_n(\boldsymbol{\rho}, z) = R_n(\boldsymbol{\rho}, z)$ and consequently $C_\varphi(\boldsymbol{\rho}) = R_\varphi(\boldsymbol{\rho})$. The structure function of the random phase can now be written as:

$$\begin{aligned} D_\varphi(\boldsymbol{\rho}) &= 2[C_\varphi(0) - C_\varphi(\boldsymbol{\rho})] = k^2 \Delta z \int_{-\infty}^{\infty} 2[C_n(0, z) - C_n(\boldsymbol{\rho}, z)] dz \\ &= k^2 \Delta z \int_{-\infty}^{\infty} \left\{ \underbrace{2[C_n(0,0) - C_n(\boldsymbol{\rho}, z)]}_{D_n(\boldsymbol{\rho}, z)} - \underbrace{2[C_n(0,0) - C_n(0, z)]}_{D_n(0, z)} \right\} dz \end{aligned} \tag{2.368}$$

Since for Kolmogorov model $D_n(r) = C_n^2 r^{2/3}$, the Eq. (2.368) becomes:

$$D_\varphi(\boldsymbol{\rho}) = k^2 \Delta z C_n^2 \int_{-\infty}^{\infty} \left\{ \left(\rho^2 + z^2\right)^{1/3} - z^{2/3} \right\} dz = 2.914 k^2 \Delta z C_n^2 \rho^{5/3}. \tag{2.369}$$

Since the refractive index structure parameter is propagation distance dependent, we have to perform the integration over z, and after taking the zenith-angle into account, we obtain:

$$D_\varphi(\boldsymbol{\rho}) = 2.914 k^2 \sec(\zeta) \rho^{5/3} \int_0^{\infty} C_n^2(z) dz = 6.889 \left(\frac{\rho}{r_0}\right)^{5/3}. \tag{2.370}$$

Another parameter related to atmospheric turbulence, relevant in astronomy, is so-called isoplanatic angle. Namely, in optical systems in which the system characteristics are not shift-invariant, we say that such systems exhibit the anisoplanatism. To measure the severity of the anisoplanatism, the starting point would be the Eq. (2.369), which after inclusion of the zenith-angle can be written as:

$$D_\varphi(\rho) = 2.914 k^2 \Delta z \sec(\zeta) C_n^2 \rho^{5/3}. \tag{2.371}$$

A stellar separation of θ rad corresponds to the optical path from the stars to the telescopes through the single layer of atmospheric turbulence at the height Δh above the telescope, and we can write $\rho = \theta\, \Delta h \sec\zeta$. By substituting this value of ρ into (2.366), we obtain:

$$D_\varphi(\rho) = 2.914 k^2 \Delta z \sec(\zeta) C_n^2 (\theta \Delta h \sec\zeta)^{5/3} = 2.914 k^2 \Delta z (\Delta h)^{5/3} \sec^{8/3}(\zeta) C_n^2 \theta^{5/3}. \tag{2.372}$$

By integrating over the entire atmosphere, we obtain:

$$\begin{aligned} D_\varphi(\rho) &= 2.914 k^2 \sec^{8/3}(\zeta) \theta^{5/3} \int_0^\infty C_n^2(z) z^{5/3} dz = \left(\frac{\theta}{\theta_0}\right)^{5/3}, \\ \theta_0 &= \left[2.914 k^2 \sec^{8/3}(\zeta) \int_0^\infty C_n^2(z) z^{5/3} dz\right]^{-3/5}, \end{aligned} \tag{2.373}$$

where θ_0 represents the isoplanatic angle. Clearly, the isoplanatic angle is angle for which $D_\varphi(\rho) = 1 \text{ rad}^2$.

2.2.5.7 Rytov Method of Small Perturbations

One important method to study the laser beam propagation over atmospheric turbulence channels is the *Rytov method of small perturbations* [58–62] (see also [30, 47, 51]), in which the optical field in cylindrical coordinates $\boldsymbol{r} = (\boldsymbol{\rho}, L)$ (L-the transmission distance) is represented as:

$$U(\boldsymbol{\rho}, L) = U_0(\boldsymbol{\rho}, L) \exp[\psi(\boldsymbol{\rho}, L)], \psi(\boldsymbol{\rho}, L) = \psi_1(\boldsymbol{\rho}, L) + \psi_2(\boldsymbol{\rho}, L) + \ldots \tag{2.374}$$

where $U_0(\boldsymbol{r})$ is the vacuum solution, while $\psi(\boldsymbol{r})$ is used to denote the complex phase perturbation. In (2.369), we use $\psi_i(\boldsymbol{r})$ to denote the i-th order perturbation. These successive perturbations are used to determine the statistical moments of $\psi(\boldsymbol{r})$. The mean value of the optical field is given by $\langle U(\boldsymbol{\rho}, L)\rangle = U_0(\boldsymbol{\rho}, L)\langle \exp[\psi(\boldsymbol{\rho}, L)]\rangle$ The *mutual coherence function (MCF)* is essentially the second moment of the optical field:

$$\Gamma(\boldsymbol{\rho}_1, \boldsymbol{\rho}_2, L) = \langle U(\boldsymbol{\rho}_1, L) U^*(\boldsymbol{\rho}_2, L)\rangle. \tag{2.375}$$

The normalized magnitude of the MCF is often called the modulus of the complex *coherence factor* (or just the coherence factor):

$$\gamma(\boldsymbol{\rho}_1, \boldsymbol{\rho}_2, L) = \frac{|\Gamma(\boldsymbol{\rho}_1, \boldsymbol{\rho}_2, L)|}{[\Gamma(\boldsymbol{\rho}_1, \boldsymbol{\rho}_1, L)\Gamma(\boldsymbol{\rho}_2, \boldsymbol{\rho}_2, L)]^{1/2}}. \tag{2.376}$$

The coherence factor over homogeneous and isotropic turbulence is only function of a displacement, $\rho = |\boldsymbol{\rho}| = |\boldsymbol{\rho}_2 - \boldsymbol{\rho}_1|$, and for plane wave, we obtain:

$$\gamma(\rho, L) = \exp\left\{-4\pi^2 k^2 \int_0^L \int_0^\infty \Phi_n(\kappa, z)[1 - J_0(\kappa\rho)]dzd\kappa\right\}. \tag{2.377}$$

When Kolmogorov spectral density is used, the coherence factor simplifies to:

$$\gamma^{(\text{Kolmogorov})}(\rho, L) = \exp\left\{-1.457k^2\rho^{5/3}\int_0^L C_n^2(z)dz\right\}. \tag{2.378}$$

The *spatial coherence radius* of an optical wave can be defined as the e^{-1} of the coherence factor:

$$\rho_0 = \left[1.457k^2 \int_0^L C_n^2(z)dz\right]^{-3/5}. \tag{2.379}$$

The complex phase perturbation $\psi(\boldsymbol{r})$ can be expressed as follows:

$$\psi(\boldsymbol{\rho}, L) = \log\left[\frac{U(\boldsymbol{\rho}, L)}{U_0(\boldsymbol{\rho}, L)}\right] = \chi + jS, \tag{2.380}$$

where χ is the log-amplitude fluctuation and S is the corresponding phase fluctuation. In the weak turbulence regime, it is reasonable to assume χ and S to be Gaussian random processes, by invoking the central limit theorem. The irradiance I of the optical field at distance L can be determined as:

$$I = |U|^2 = \underbrace{|U_0|^2}_{I_0}\exp(2\chi) = I_0 \exp(2\chi). \tag{2.381}$$

Let the variance of the log-amplitude fluctuation be obtained as:

$$\sigma_\chi^2 = \langle\chi^2\rangle - \langle\chi\rangle^2. \tag{2.382}$$

We are concerned to determine the distribution of new random variable I, related to the old random variable χ by Eq. (2.381). In general, when the transformation of random variable X has the form $Y = f(X)$, the distribution for Y is given by $f_Y(y) = f_X(x)|dx/dy|$. By applying this theory, we obtain the following distribution of the irradiance I:

$$\begin{aligned} f(I) &= f(\chi)\left|\frac{d\chi}{dI}\right| = f(\chi)\underbrace{\left|\frac{d}{dI}\left(\overbrace{\frac{1}{2}\log\left(\frac{I}{I_0}\right)}^{\chi}\right)\right|}_{1/(2I)} = \frac{1}{\sigma_\chi\sqrt{2\pi}}e^{-\frac{(\chi-\langle\chi\rangle)^2}{2\sigma_\chi^2}}\frac{1}{2I} = \frac{1}{2\sigma_\chi I\sqrt{2\pi}}e^{-\frac{\left[\frac{1}{2}\log\left(\frac{I}{I_0}\right)-\langle\chi\rangle\right]^2}{2\sigma_\chi^2}} \\ &= \frac{1}{2\sigma_\chi I\sqrt{2\pi}}e^{-\frac{\left[\log\left(\frac{I}{I_0}\right)-2\langle\chi\rangle\right]^2}{8\sigma_\chi^2}}. \end{aligned} \tag{2.383}$$

Clearly, the distribution of irradiance is the *log-normal*. This distribution is valid only when the first perturbation term in Rytov approximation dominates, in other words, in so-called weak turbulence regime (to be precisely defined below).

The coherence factor can be used to determine the *wave structure function*, defined as follows:

$$D(\rho_1,\rho_2,L) = -2\log\gamma(\rho_1,\rho_2,L) = D_\chi(\rho_1,\rho_2,L) + D_S(\rho_1,\rho_2,L), \tag{2.384}$$

where we use D_χ and D_S to denote the log-amplitude and phase structure parameters, respectively. The phase spectral density is now related to the phase structure function, based on Eq. (2.346), as follows:

$$\Phi_S(\kappa) = \frac{1}{4\pi^2\kappa^2}\int_0^\infty \frac{\sin(\kappa r)}{\kappa r}\frac{d}{dr}\left[r^2\frac{d}{dr}D_S(r)\right]dr. \tag{2.385}$$

Fluctuations in irradiance are governed by the fourth-order moment of the optical field given as follows:

$$\Gamma_4(\rho_1,\rho_2,\rho_3,\rho_4,L) = \langle U(\rho_1,L)U^*(\rho_2,L)U(\rho_3,L)U^*(\rho_4,L)\rangle. \tag{2.386}$$

From this equation, the second moment of irradiance $\langle I^2(\rho_1, L)\rangle$ is obtained from Eq. (2.383) by setting $\rho_4 = \rho_3 = \rho_2 = \rho_1$, and it is used to define so-called scintillation index by:

$$\sigma_I^2 = \frac{\langle I^2\rangle - \langle I\rangle^2}{\langle I\rangle^2} = \frac{\langle I^2\rangle}{\langle I\rangle^2} - 1. \tag{2.387}$$

The scintillation index of an unbounded plane wave in a weak turbulence regime (defined as the regime for which the scintillation index is less than unity) based on a Kolmogorov spectral density is commonly referred to as the *Rytov variance* [4, 30]:

$$\sigma_R^2 = 1.23\ C_n^2\ k^{7/6}\ L^{11/6}. \tag{2.388}$$

Even though it is derived under the first-order perturbation assumption, which is valid in weak turbulence only, the Rytov variance is commonly used as a measure of optical turbulence strength. The weak fluctuations are associated with $\sigma^2_R < 1$, the strong with $\sigma^2_R > 1$, and the saturation regime is defined by $\sigma^2_R \to \infty$ [4, 30].

From probability theory we know that when X has the normal distribution with mean value μ and variance σ^2, the distribution of random variable $Y = \exp.(X)$ is log-normal with the first two moments given by $\mu_1 = \exp.(\mu + \sigma^2/2)$ and $\mu_2 = \exp.(\mu + \sigma^2)$. By using this fact, the scintillation index in weak turbulence regime, where the distribution of irradiance is given by Eq. (2.383), can be related to the log-amplitude variance by:

$$\sigma_I^2 = \frac{\langle I^2\rangle}{\langle I\rangle^2} - 1 = \frac{e^{2\left(2\langle x\rangle + 4\sigma_\chi^2\right)}}{\left[e^{2\langle x\rangle + 4\sigma_\chi^2/2}\right]^2} - 1 = e^{4\sigma_\chi^2} - 1. \tag{2.389}$$

2.2.5.8 Gamma-Gamma Distribution of Irradiance

We are now concerned with extending the Rytov approximation to strong turbulence. By close inspection on modified von Kármán spectral density given by Eq. (2.353), we notice that Tatarskii's term $\exp.(-\kappa/\kappa_m)$ corresponds to the large-scale filter function spatial frequency cutoff κ_m. On the other hand, the term $\left(\kappa^2 + \kappa_0^2\right)^{-11/6}$ corresponds to the small-scale filter function with spatial frequency cutoff κ_0. By denoting the large-scale and small-scale cutoff spatial frequency by κ_x and κ_y, respectively, we can generalize the Kolmogorov spectrum as follows:

$$\Phi_n(\kappa) = \underbrace{0.033005391C_n^2\kappa^{-11/3}e^{-\kappa^2/\kappa_x^2}}_{\Phi_{n,X}(\kappa)} + \underbrace{0.033005391C_n^2\left(\kappa^2 + \kappa_y^2\right)^{-11/6}}_{\Phi_{n,Y}(\kappa)}. \tag{2.390}$$

The small-scale effects are contributed by eddies of size smaller than small scale size l_{ss} defined by Eq. (2.391). On the other hand, the large-scale effects are contributed by eddies of size larger than L_{ls}, defined as:

$$l_{ss} = \min\left(\rho_0, \sqrt{L/k}\right), \quad L_{ls} = \max\left(\sqrt{L/k}, L/(k\rho_0)\right). \tag{2.391}$$

In (2.391), ρ_0 denotes the spatial coherence radius as defined by Eq. (2.379), and $\sqrt{L/k}$ is related to the first Fresnel zone radius and $L/(k\rho_0)$ to the scattering disk size. The addition operation in spectral density of random phase indicates the product of random variable X and Y in irradiance I, namely, $I = XY$. If we assume that large-scale fluctuations, describing the refraction properties, described by random variable X and the small fluctuations, describing the diffraction properties, described by random variable Y are mutually independent, in statistical sense, the scintillation index can be written as follows:

$$\sigma_I^2(\rho, L) = \langle X^2\rangle\langle Y^2\rangle - 1 = \left(1+\sigma_X^2\right)\left(1+\sigma_Y^2\right) - 1 = \sigma_X^2 + \sigma_Y^2 + \sigma_X\sigma_Y. \tag{2.392}$$

By expressing the variances of random variables X and Y in terms of log-amplitudes' variances, we can rewrite the scintillation index as follows:

$$\sigma_I^2(\rho, L) = \exp\left(\sigma_{\log X}^2\right)\exp\left(\sigma_{\log Y}^2\right) - 1. \tag{2.393}$$

By assuming that both large-scale and small-scale irradiance functions are governed by gamma distributions [63] (see also [30]), with different parameters:

$$\begin{aligned} f_X(x) &= \frac{\alpha(\alpha x)^{\alpha-1}}{\Gamma(\alpha)} e^{-\alpha x}; \alpha > 0, X > 0 \\ f_Y(y) &= \frac{\beta(\beta y)^{\beta-1}}{\Gamma(\beta)} e^{-\beta y}; \beta > 0, Y > 0 \end{aligned} \tag{2.394}$$

the conditional probability density function (PDF) $f_{I|X}(I)$ is given by:

$$f_{I|X}(I) = \frac{\beta(\beta I/x)^{\beta-1}}{\Gamma(\beta)} e^{-\beta I/x}. \tag{2.395}$$

By averaging out over random variable X in conditional PDF (2.395), we obtain the following PDF of irradiance I:

$$\begin{aligned} f(I) &= \int_0^\infty f_{I|X}(I) f_X(x) dx = \int_0^\infty \frac{\beta(\beta I/x)^{\beta-1}}{\Gamma(\beta)} e^{-\beta I/x} \frac{\alpha(\alpha x)^{\alpha-1}}{\Gamma(\alpha)} e^{-\alpha x} \\ &= \frac{2(\alpha\beta)^{\frac{\alpha+\beta}{2}}}{\Gamma(\alpha)\Gamma(\beta)} I^{\frac{\alpha+\beta}{2}-1} K_{\alpha-\beta}\left(2\sqrt{\alpha\beta I}\right), \end{aligned} \tag{2.396}$$

which is commonly referred to as *gamma-gamma distribution* [4, 30, 35–39, 63]. Given that the expected value of irradiance is 1, the second moment of irradiance for gamma-gamma distribution is given by:

$$\langle I^2\rangle = (1 + 1/\alpha)(1 + 1/\beta), \tag{2.397}$$

so that the scintillation index is given by:

$$\sigma_I^2 = \frac{1}{\alpha} + \frac{1}{\beta} + \frac{1}{\alpha\beta}. \tag{2.398}$$

By close inspection of Eqs. (2.393) and (2.397), we conclude that $1 + 1/\alpha = \exp\left(\sigma^2_{\log X}\right)$ and $1 + 1/\beta = \exp\left(\sigma^2_{\log Y}\right)$, and after rearranging we obtain:

$$\alpha = \left[\exp\left(\sigma^2_{\log X}\right) - 1\right]^{-1} = \sigma_X^{-2}, \beta = \left[\exp\left(\sigma^2_{\log Y}\right) - 1\right]^{-1} = \sigma_Y^{-2}. \tag{2.399}$$

Therefore, parameters α and β in gamma-gamma distribution can be related to large-scale and small-scale scintillation effects, respectively.

To completely specify the gamma-gamma distribution, we need to derive the log-amplitude variances. For this purpose, we employ the method of small perturbations as described by Tatarskii [51] (see also [30, 47]). We return back to the Helmholtz equation:

$$\nabla^2 U + k^2 n^2(\boldsymbol{r}) U = 0, \tag{2.400}$$

and expand the optical field U into a series of terms:

$$U = U_0 + U_1 + U_2 + \ldots; \quad |U_{m+1}| << |U_m|, m = 0, 1, 2, \ldots \tag{2.401}$$

Let us observe the first two terms in equation above, initially $U \cong U_0 + U_1$, and after the substitution into (2.400), we the following two equations:

$$\nabla^2 U_0 + k^2 U_0 = 0, \; \nabla^2 U_1 + k^2 U_1 = -2k^2 \delta n U_0, \tag{2.402}$$

where we obtained these equations by setting to zero all terms of the order δn in the approximation of the square refractive index Eq. (2.333):

$$n(\boldsymbol{r}) = [1 + \delta n(\boldsymbol{r})]^2 \cong 1 + 2\delta n(\boldsymbol{r}), \delta n(\boldsymbol{r}) << 1 \tag{2.403}$$

By interpreting the term $2k^2 \delta n U_0$ as the excitation, by employing the Green's theorem [5], we can determine U_1 as the 3-D convolution of Green's function (spherical wave) $G(\boldsymbol{r},\boldsymbol{r}') = (4\pi)^{-1}\exp(jk|\boldsymbol{r} - \boldsymbol{r}'|)/|\boldsymbol{r} - \boldsymbol{r}'|$ and the excitation $2k^2 \delta n U_0$:

$$U_1(\boldsymbol{r}) = \frac{1}{4\pi} \int_V \frac{e^{jk|\boldsymbol{r}-\boldsymbol{r}'|}}{|\boldsymbol{r} - \boldsymbol{r}'|} 2k^2 \delta n(\boldsymbol{r}') U_0(\boldsymbol{r}') d^3\boldsymbol{r}'. \tag{2.404}$$

Now by observing only U_m and U_{m+1} terms, we can establish the identical relationship:

$$U_{m+1}(\boldsymbol{r}) = \frac{1}{4\pi} \int_V \frac{e^{jk|\boldsymbol{r}-\boldsymbol{r}'|}}{|\boldsymbol{r} - \boldsymbol{r}'|} 2k^2 \delta n(\boldsymbol{r}') U_m(\boldsymbol{r}') d^3\boldsymbol{r}'; m = 0, 1, 2, \ldots \tag{2.405}$$

In cylindrical coordinates we can write $\boldsymbol{r} = (\boldsymbol{\rho}, z)$. Let us now apply the coarse approximation in amplitude $|\boldsymbol{r} - \boldsymbol{r}'| \approx z - z'$ and fine approximation in phase as follows:

$$k|\boldsymbol{r} - \boldsymbol{r}'| = k(z - z')[1 + \frac{(\boldsymbol{\rho} - \boldsymbol{\rho}')^2}{2(z - z')^2} + \frac{(\boldsymbol{\rho} - \boldsymbol{\rho}')^4}{8(z - z')^4} + \ldots] \tag{2.406}$$

By substituting (2.405) into (2,404), we obtain.

$$U_{m+1}(\boldsymbol{r}) \cong \frac{k^2}{2\pi} \int_V e^{jk(z-z')} \frac{\exp\left[jk\frac{(\boldsymbol{\rho}-\boldsymbol{\rho}')^2}{2(z-z')^2}\right]}{z - z'} \delta n(\boldsymbol{r}') U_m(\boldsymbol{r}') d^3\boldsymbol{r}'; m = 0, 1, 2, \ldots \tag{2.407}$$

By normalizing the perturbations with $U_0(\boldsymbol{r})$, we can rewrite the Eq. (2.406) as follows:

$$\frac{U_{m+1}(\boldsymbol{r})}{U_0(\boldsymbol{r})} \simeq \frac{k^2}{2\pi}\int_V e^{jk(z-z')}\frac{\exp\left[jk\frac{(\rho-\rho')^2}{2(z-z')^2}\right]}{z-z'}\delta n(\boldsymbol{r}')\frac{U_m(\boldsymbol{r}')}{U_0(\boldsymbol{r})}d^3\boldsymbol{r}';m=0,1,2,\ldots \tag{2.408}$$

For unbounded plane wave, $U_0 = \exp.(jkz')$ and $m = 1$, and Eq. (2.408) simplifies to:

$$\frac{U_1^{(\text{plane wave})}(\boldsymbol{r})}{U_0(z)} = \frac{k^2}{2\pi}\int_V \frac{\exp\left[jk\frac{(\rho-\rho')^2}{2(z-z')^2}\right]}{z-z'}\delta n(\boldsymbol{r}')d^3\boldsymbol{r}'. \tag{2.409}$$

By taking only the first two terms in expansion (2.401), the log-amplitude expression can be written as:

$$\psi = \chi + jS = \log\left(\frac{U_0+U_1}{U_0}\right) = \log\left(1+\frac{U_1}{U_0}\right) \cong_{\substack{|U_1|<<|U_0| \\ \log(x)\approx x - x^2/2}} \frac{|U_1|}{|U_0|} + j\arg\left\{\frac{U_1}{U_0}\right\}. \tag{2.410}$$

Based on Eqs. (2.407) and (2.410), we conclude that:

$$\chi \cong \frac{k^2}{2\pi}\int_V \frac{\cos\left[k\frac{(\rho-\rho')^2}{2(z-z')^2}\right]}{z-z'}\delta n(\boldsymbol{r}')d^3\boldsymbol{r}', S = \frac{k^2}{2\pi}\int_V \frac{\sin\left[k\frac{(\rho-\rho')^2}{2(z-z')^2}\right]}{z-z'}\delta n(\boldsymbol{r}')d^3\boldsymbol{r}' \tag{2.411}$$

Under assumption that refractive index variation is only in the plane perpendicular to the propagation direction, we can represent the refractive index variation as a 2D Fourier-Stieltjes transform [51] (see also [47]):

$$\delta n(\boldsymbol{r}) = \int e^{j\boldsymbol{\kappa}\cdot\boldsymbol{\rho}} dv(\boldsymbol{\kappa}, z), \tag{2.412}$$

where the random amplitude of the refractive index fluctuations has the following correlation function:

$$\begin{aligned}\langle dv(\boldsymbol{\kappa}, z')dv^*(\boldsymbol{\kappa}, z'')\rangle &= \delta(\boldsymbol{\kappa}-\boldsymbol{\kappa}')F_n(\boldsymbol{\kappa}', z'-z'')d^2\boldsymbol{\kappa}d^2\boldsymbol{\kappa}', \\ F_n(\boldsymbol{\kappa}', z'-z'') &= \int_{-\infty}^{\infty} \Phi_n(\boldsymbol{\kappa}', \kappa_z)\cos\left[\kappa_z(z'-z'')\right]d\kappa_z.\end{aligned} \tag{2.413}$$

By substituting (2.412) in (2.411), and after performing the integration with respect to ρ, we obtain the following solutions [51] (see also [47]):

$$\chi = \int e^{j\boldsymbol{\kappa}\cdot\boldsymbol{\rho}} \int_0^{k_z} \sin\left[\frac{\kappa^2(z-z')}{2k}\right]dz'dv(\boldsymbol{\kappa}, z'), S = \int e^{j\boldsymbol{\kappa}\cdot\boldsymbol{\rho}} \int_0^{k_z} \cos\left[\frac{\kappa^2(z-z')}{2k}\right]dz'dv(\boldsymbol{\kappa}, z'). \tag{2.414}$$

The covariance function of χ is defined by:

$$C_\chi(\boldsymbol{\rho}, z) = \langle \chi(\boldsymbol{\rho}_1+\boldsymbol{\rho}, z)\chi^*(\boldsymbol{\rho}_1, z)\rangle, \tag{2.415}$$

and after substitution of (2.414) into (2.415), we obtain:

$$\begin{aligned}C_\chi(\boldsymbol{\rho}, z) &= \\ k^2 &\iint e^{j\boldsymbol{\kappa}\cdot(\boldsymbol{\rho}_1+\boldsymbol{\rho})-j\boldsymbol{\kappa}'\cdot\boldsymbol{\rho}} \int_0^z\int_0^z \sin\left[\frac{\kappa^2(z-z')}{2k}\right]\sin\left[\frac{\kappa'^2(z-z'')}{2k}\right]dz'dz''\langle dv(\boldsymbol{\kappa}, z')dv^*(\boldsymbol{\kappa}', z'')\rangle \\ &= \int e^{j\boldsymbol{\kappa}\cdot\boldsymbol{\rho}} \int_0^z\int_0^z \sin\left[\frac{\kappa^2(z-z')}{2k}\right]\sin\left[\frac{\kappa^2(z-z'')}{2k}\right]F_n(\boldsymbol{\kappa}, z'-z'')dz'dz''d^2\boldsymbol{\kappa}\end{aligned} \tag{2.416}$$

For isotropic assumption we have that $F_n(\kappa, z' - z'') = F_n(\boldsymbol{\kappa}, z' - z'')$, and after introducing the following substitutions $z' - z'' = \xi$, $z' + z'' = 2\eta$ (whose Jacobian is 1), we can write the amplitude spectrum as follows [47]:

$$F_\chi(\kappa, 0) = k^2 \int_0^L \int_{\xi/2}^{L-\xi/2} \left[\cos\left(\frac{\kappa^2 \xi}{2k}\right) - \cos\left(\frac{\kappa^2 (L-\eta)}{2k}\right)\right] F_n(\kappa, \xi) d\xi d\eta \tag{2.417}$$

Under assumptions $\xi \cong L$, $\kappa^2\xi/(2k) \cong L$, for sufficiently long propagation distance L, we have that $\int_0^\infty F_n(\kappa, \xi) d\xi = \pi\Phi_n(0, \kappa) = \pi\Phi_n(\kappa)$ so that the Eq. (2.417) simplifies to:

$$F_\chi(\kappa, 0) = \pi k L^2 \left[1 - \sin\left(\kappa^2 L/k\right)/\left(\kappa^2 L/k\right)\right]\Phi_n(\kappa). \tag{2.418}$$

For isotropic assumption, it is convenient to move from Cartesian coordinates $\boldsymbol{\kappa} = (\kappa_x, \kappa_y)$ to polar coordinates (κ, ϕ), so that the covariance function at distance $z = L$ becomes [47]:

$$C_\chi(\rho) = 2\pi \int_0^\infty \underbrace{\frac{1}{2\pi}\int_0^{2\pi} e^{j\kappa\rho\cos\phi} d\phi}_{J_0(k\rho)} F_\chi(\kappa, 0) d\kappa = 2\pi \int_0^\infty J_0(k\rho) F_\chi(\kappa, 0) d\kappa, \tag{2.419}$$

where $J_0(\cdot)$ is the zero-order Bessel function of the first kind. By substituting (2.418) into (2.419), we obtain the following expression for covariance function of χ:

$$C_\chi(\rho) = 2\pi^2 k^2 L \int_0^\infty \left[1 - \frac{\sin\left(\kappa^2 L/k\right)}{\kappa^2 L/k}\right] J_0(\kappa\rho)\Phi_n(\kappa)\kappa d\kappa, \tag{2.420}$$

The variance of χ for plane wave can be determined as:

$$\sigma_\chi^2 = C_\chi(0) = 2\pi^2 k^2 L \int_0^\infty \left[1 - \frac{\sin\left(\kappa^2 L/k\right)}{\kappa^2 L/k}\right] \Phi_n(\kappa)\kappa d\kappa. \tag{2.421}$$

For Tatarskii spectral density, given by [62]:

$$\Phi_n(\kappa) = 0.033 C_n^2 \kappa^{-11/3} \exp\left(-\kappa^2/\kappa_m^2\right), \tag{2.422}$$

we obtain the following expression for variance of χ:

$$\sigma_\chi^2 \cong 0.033 \frac{6\pi^2}{91} L^3 C_n^2 \kappa_m^{7/3} = 0.214745 L^3 C_n^2 \kappa_m^{7/3}. \tag{2.423}$$

In this derivation, the following series expansions have been used: $\sin(x) \approx x - x^3/6$ and $\exp(-x) \approx 1 - x$. In weak turbulence regime, the cutoff spatial frequency κ_m is related to the first Fresnel zone radius by $\kappa_m = 1.166336\sqrt{k/L}$, and after substitution in (2.423), we obtain:

$$\sigma_\chi^2 \approx 0.3075 C_n^2 k^{7/6} L^{11/6}. \tag{2.424}$$

Based on Eq. (2.389), we obtain the following expression for scintillation index:

$$\sigma_I^2 = e^{4\sigma_\chi^2} - 1 = \exp\left(1.23 C_n^2 k^{7/6} L^{11/6}\right) - 1. \tag{2.425}$$

In weak turbulence regime, the following series expansion can be used exp(x) ≈ 1 + x, so that the Eq. (2.424) simplifies to:

$$\sigma_I^2 \approx 1.23 C_n^2 k^{7/6} L^{11/6} = \sigma_R^2, \tag{2.426}$$

which represents the Rytov variance given by Eq. (2.388).

To extend this theory to all turbulence regimes, we need to use the spatially filtered spectral density as given by Eq. (2.390), and after substitution into (2.421), we obtain:

$$\begin{aligned}\sigma_\chi^2 &= 2\pi^2 k^2 L \int_0^\infty \left[1 - \frac{\sin\left(\kappa^2 L/k\right)}{\kappa^2 L/k}\right] \left[\Phi_{n,X}(\kappa) + \Phi_{n,Y}(\kappa)\right] \kappa d\kappa \\ &= 2\pi^2 k^2 L \left\{ \int_0^\infty \left[1 - \frac{\sin\left(\kappa^2 L/k\right)}{\kappa^2 L/k}\right] 0.033 C_n^2 \kappa^{-11/3} e^{-\kappa^2/\kappa_x^2} \kappa d\kappa \right. \\ &\quad \left. + \int_0^\infty \left[1 - \frac{\sin\left(\kappa^2 L/k\right)}{\kappa^2 L/k}\right] 0.033005391 C_n^2 \left(\kappa^2 + \kappa_y^2\right)^{-11/6} \kappa d\kappa \right\}.\end{aligned} \tag{2.425}$$

After solving the integrals in (2.421) (for details an interested reader is referred to [30]), we obtain the following solution for scintillation index, assuming the zero inner scale case [30]:

$$\begin{aligned}\sigma_I^2 &= \exp\left(4\sigma_\chi^2\right) - 1 = \exp\left(\sigma_{\log X}^2 + \sigma_{\log Y}^2\right) - 1 \\ &\cong \exp\left[\underbrace{\frac{0.49\sigma_R^2}{\left(1 + 1.11\sigma_R^{12/5}\right)^{7/6}}}_{\sigma_{\log X}^2} + \underbrace{\frac{0.51\sigma_R^2}{\left(1 + 0.69\sigma_R^{12/5}\right)^{5/6}}}_{\sigma_{\log Y}^2}\right] - 1.\end{aligned} \tag{2.426}$$

Based on Eqs. (2.399) and (2.426), we derive the following expressions for parameters in gamma-gamma distribution in terms of Rytov variance, for zero inner scale:

$$\alpha = \left\{ \exp\left[\frac{0.49\sigma_R^2}{\left(1 + 1.11\sigma_R^{12/5}\right)^{7/6}}\right] - 1 \right\}^{-1}, \; \beta = \left\{ \exp\left[\frac{0.51\sigma_R^2}{\left(1 + 0.69\sigma_R^{12/5}\right)^{5/6}}\right] - 1 \right\}^{-1}. \tag{2.427}$$

As an illustration, in Fig. 2.21, we provide the gamma-gamma probability density function with different turbulence strengths, represented in terms of Rytov standard deviation, for zero inner scale scenario. The case with Rytov standard deviation $\sigma_R = 0.2$ (Rytov variance 0.04) corresponds to the weak turbulence regime. The corresponding parameters of gamma-gamma distribution are $\alpha = 51.913$ and $\beta = 49.113$, while the scintillation index is 0.041. In weak turbulence regime, gamma-gamma distribution resembles the log-normal distribution. The case with Rytov standard deviation $\sigma_R = 1$ corresponds to the medium turbulence regime. The corresponding parameters of gamma-gamma distribution are $\alpha = 4.3939$ and $\beta = 2.5636$, while the scintillation index is 0.70644. Clearly, as the turbulence strength increases, the PDF skews towards lower values of the irradiance. The case with Rytov standard deviation $\sigma_R = 3$ (Rytov variance 9) corresponds to the strong turbulence regime. The corresponding parameters of gamma-gamma distribution are $\alpha = 5.485$ and $\beta = 1.1156$, while the scintillation index is 1.2421. The distribution function in medium-to-strong turbulence resembles the Rayleigh

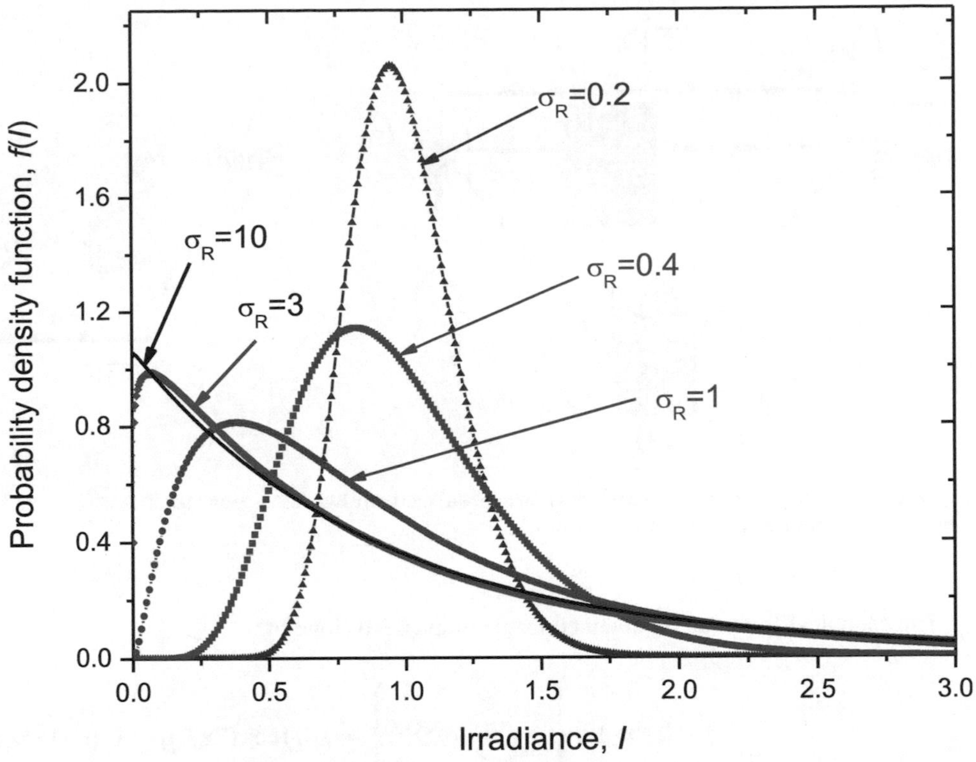

Fig. 2.21 Gamma-gamma probability density function for different turbulence strengths

distribution. Finally, the case with Rytov standard deviation $\sigma_R = 10$ (Rytov variance 100) corresponds to strong-to-saturation regime. The corresponding parameters of gamma-gamma distribution are $\alpha = 14.11$ and $\beta = 1.0033$, while the scintillation index is 1.1382. In saturation regime, clearly, the distribution resembles the exponential distribution.

2.3 Propagation Effects in Wireless Communication Channels

In this section we study various signal propagation effects in wireless channels [64–68]. We also describe various wireless channel models.

2.3.1 Path Loss and Shadowing Propagation Effects in Wireless Communication Channels

The received power in a wireless channel, as illustrated in Fig. 2.22, is affected by attenuation due to following factors:

(i) The *path loss*, representing the propagation loss average.
(ii) The *shadowing* effect, occurring due to slow variations in attenuation mostly caused by scattering obstruction structures.
(iii) The *multipath fading*, representing rapid random fluctuations in received power due to multipath propagation and Doppler frequency shift.

Transmitted signal (passband) representation in canonical (standard) form:

$$s(t) = s_I(t)\cos(2\pi f_c t) - s_Q(t)\sin(2\pi f_c t), \tag{2.428.1}$$

where $s_I(t)$ and $s_Q(t)$ represent in-phase and quadrature signals, while f_c is the carrier frequency. The corresponding complex envelope (low-pass) representation is given by:

$$\underline{s}(t) = s_I(t) + j s_Q(t). \tag{2.428.2}$$

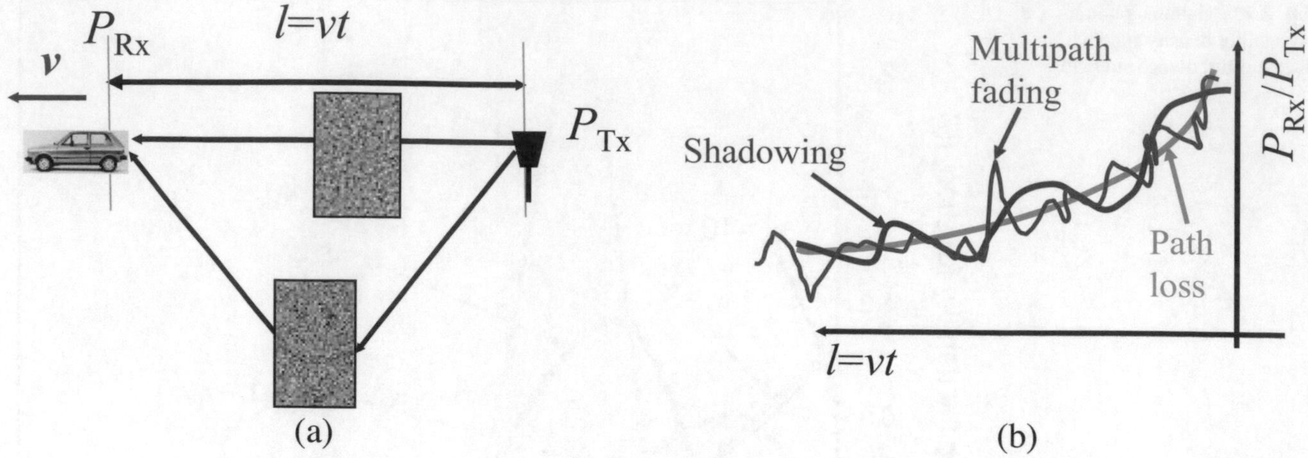

Fig. 2.22 Time-varying nature of the wireless channel: (**a**) one particular instance and (**b**) illustration of path loss, shadowing, and multipath fading effects against the distance traveled l

The canonical form can be obtained from complex envelope by:

$$s(t) = \mathrm{Re}\left\{ \underbrace{s(t)}_{s_I(t)+js_Q(t)} e^{j2\pi f_c t} \right\} = s_I(t)\cos(2\pi f_c t) - s_Q(t)\sin(2\pi f_c t), \tag{2.428.3}$$

The received signal, when the transmitter and receiver are in close proximity of each other, can be written as:

$$r(t) = \mathrm{Re}\left\{s(t) * h(t)e^{j2\pi f_c t}\right\} + n(t), \tag{2.429}$$

where $n(t)$ is the additive white Gaussian noise (AWGN) and $h(t)$ is the channel impulse response.

2.3.1.1 Free-Space Path Loss and Doppler Shift

Path loss definitions in bot linear- and dB-scales are given as:

$$P_L = \frac{P_{\mathrm{Tx}}}{P_{\mathrm{Rx}}}, \quad P_L[\mathrm{dB}] = 10\log_{10}\left(\frac{P_{\mathrm{Tx}}}{P_{\mathrm{Rx}}}\right). \tag{2.430.1}$$

where P_{Tx} is the transmit power and P_{Rx} is the received power averaged out for variations due to shadowing effect. The *path gain* is related to the path loss by:

$$P_G = -P_L = -10\log_{10}\left(\frac{P_{\mathrm{Tx}}}{P_{\mathrm{Rx}}}\right). \tag{2.430.2}$$

When transmitter or receiver is moving, the received signal will have a *Doppler shift* associated with it, as illustrated in Fig. 2.23. The Doppler shift is caused by the fact that transmitter/receiver movement introduces a distance difference dl that transmitted signal travels between two observations, which can be evaluated by $dl = vdt\cos\phi$, where ϕ is the angle of arrival of the transmitted signal relative to the direction of motion. The corresponding phase shift is given by $d\varphi = \underbrace{k}_{2\pi/\lambda} dl = (2\pi/\lambda)vdt\cos\phi$. The Doppler shift frequency can be determined as the first derivative of the phase:

$$\nu_D = \frac{1}{2\pi}\frac{d\varphi}{dt} = \frac{v}{\lambda}\cos\phi. \tag{2.431}$$

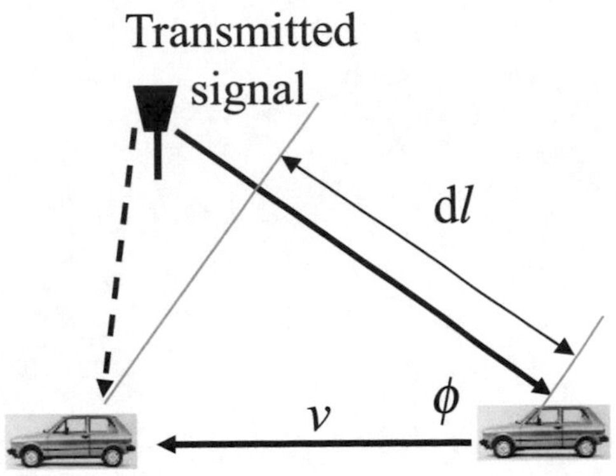

Fig. 2.23 Illustration of the Doppler effect

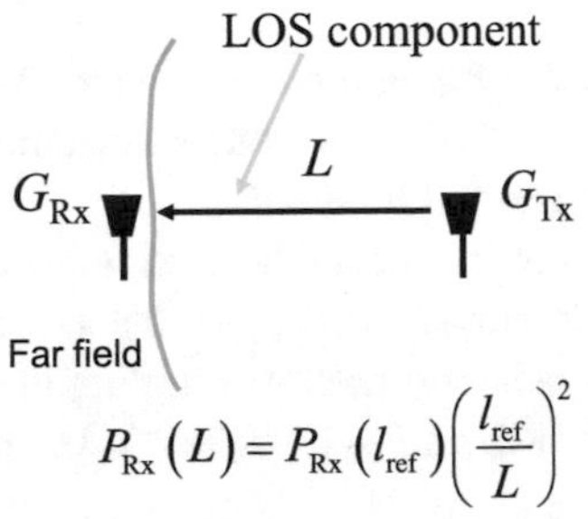

Fig. 2.24 Illustration of the illustration of the LOS component and corresponding power law in the far-field. l_{ref}: the reference distance

Clearly, the receiver will experience either increase or decrease in frequency depending on the angle of arrival. Received signal at distance L from transmitter:

$$r(t) = \mathrm{Re}\left\{\frac{\lambda}{4\pi}\sqrt{G_L}\frac{e^{-j\frac{2\pi}{\lambda}L}}{L}s(t)e^{j2\pi f_c t}\right\}, \tag{2.432}$$

where $\sqrt{G_L}$ is the product of transmit antenna (Tx) and receive (Rx) antenna radiation patterns in the line-of-sight (LOS) direction, as illustrated in Fig. 2.24. Receive-to-transmit power ratio is given by:

$$\frac{P_{Rx}}{P_{Tx}} = \left[\frac{\sqrt{G_L}}{4\pi L}\lambda\right]^2, \tag{2.433}$$

and this equation is commonly referred to as the *Friis free-space equation*. Received power can be expressed in dB-scale as follows:

$$\begin{aligned} P_{Rx}\,[\mathrm{dBm}] &= 10\log_{10}P_{Rx} \\ &= P_{Tx}[\mathrm{dBm}] + 10\log_{10}(G_L) + 20\log_{10}(\lambda) - 20\log_{10}(4\pi) - 20\log_{10}(L). \end{aligned} \tag{2.434}$$

Free-space path loss in dB-scale:

$$P_L[\mathrm{dB}] = 10\log_{10}\frac{P_{Tx}}{P_{Rx}} = -10\log_{10}\frac{G_L\lambda^2}{(4\pi L)^2}. \tag{2.435}$$

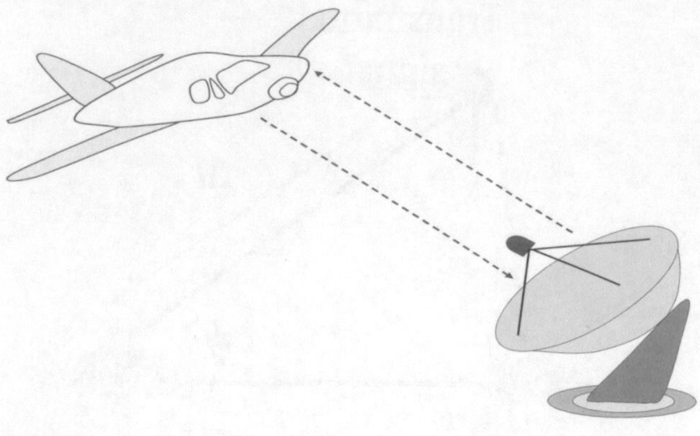

Fig. 2.25 Illustrating the monostatic radar concept

2.3.1.2 Radar Range Equation

At this point, it is convenient to derive the *radar range equation*, which will be used later on in this section. The basic operation of the monostatic classical radar, see Fig. 2.25, is to send an electromagnetic (EM) wave towards the target, which get reflected by the target, and radar return is measured by the receiver.

The range to the target can be determined by $R = c\Delta t/2$, where Δt is the transit time (time interval that elapses from the transmission of the pulse until its return upon reflection). Let G_{Tx} denote the gain of transmit antenna and P_{Tx} be the power of transmitter. The power density from uniformly radiating antenna emitting the spherical wave at distance R is given by $P_{\mathrm{TX}}/(4\pi R^2)$. Therefore, the power density at the target will be $P_{\mathrm{TX}}G_{\mathrm{Tx}}/(4\pi R^2)$. The power of reflected signal from the target will be:

$$P_t = \frac{P_{Tx}G_{Tx}}{4\pi R^2}\sigma, \tag{2.436}$$

where σ is the radar cross section (RCS) representing the measure of energy that target reflects back towards the radar. The power density of the reflected signal at receiver side will be:

$$W_{\mathrm{Rx}} = \frac{P_{Tx}G_{Tx}}{4\pi R^2}\frac{\sigma}{4\pi R^2}. \tag{2.437.1}$$

The power of the reflected signal at receiver side will be:

$$P_{\mathrm{Rx}} = W_{\mathrm{Rx}}A_{\mathrm{Rx}} = \frac{P_{Tx}G_{Tx}}{4\pi R^2}\frac{\sigma A_{\mathrm{Rx}}}{4\pi R^2} = \frac{P_{Tx}G_{Tx}\sigma A_{\mathrm{Rx}}}{(4\pi)^2 R^4}, \tag{2.437.2}$$

where A_{Rx} is the effective area of receive antenna. This equation is commonly referred to as the *radar range equation*. If the medium (air) is not transparent due to presence of clouds and/or fog, we need to introduce the atmospheric attenuation F, defined as $F = \exp(-\alpha R/2)$, with α being the attenuation coefficient, as follows:

$$P_{\mathrm{Rx}} = \frac{P_{Tx}G_{Tx}\sigma A_{\mathrm{Rx}}}{(4\pi)^2 R^4}F^4. \tag{2.437.3}$$

The RCS can be defined more formally by [69, 70]:

$$\sigma = \lim_{R\to\infty} 4\pi R^2 \frac{|\mathbf{E}_{Rx}|^2}{|\mathbf{E}_t|^2}, \tag{2.438}$$

where $\mathbf{E}_{\mathrm{t}}$ represents the incident electric filed at the target, while $\mathbf{E}_{\mathrm{Rx}}$ denotes the scattered electric field measured at the receiver side.

2.3.1.3 Ray Tracing Approximation

In urban/indoor environment, a radio signal transmitted from a fixed source will experience multiple objects that produce reflected, diffracted, or scattered copies of transmitted signal, known as *multipath signal components*, as illustrated in Fig. 2.26.

Ray tracing approximation employs the principles of *geometrical optics* to describe the wireless channel propagation effects. Geometry determines received signal from each signal component. This approximation typically includes reflected rays, scattered rays, and diffracted rays. It requires the knowledge of the site parameters such as the geometry and dielectric properties of objects. In *two-ray model*, the received signal consists of two components: the LOS ray and reflected ray, as shown in Fig. 2.27. The ground reflected ray approximately cancels LOS path above after some characteristic distance, known as the critical distance l_c (to be defined below). The power dependence law is proportional to l^{-2}, i.e., at −20 dB/decade, when the distance between transmit and receive antennas l is larger than the transmitting antenna height h_{Tx} but smaller than the critical distance l_c. For distances larger than the critical distance, the power dependence is proportional to l^{-4}, i.e., at −40 dB/decade.

The received signal for the two-ray model (by ignoring surface wave attenuation effect):

$$r(t) = \mathrm{Re}\left\{\frac{1}{2k}\left[\sqrt{G_L}\frac{e^{-jkL}}{L}s(t) + r_{\mathrm{refl}}\sqrt{G_{\mathrm{refl}}}\frac{e^{-jk(l+l')}}{l+l'}s(t-\tau)\right]e^{j\omega_c t}\right\}, k = \frac{2\pi}{\lambda} \tag{2.439}$$

where the frits term corresponds to the LOS ray and the second term to the reflected ray. The *delay spread*, representing the difference in arrival times for the LOS and reflected rays, based on Fig. 2.27, is determined by $\tau = (l + l' - L)/c$ (c is the speed of the light in the air). The corresponding phase shift is given by $\Delta\phi = (2\pi/\lambda)(l + l' - L)$. In Eq. (2.439), we use r_{refl} to denote the ground *reflection coefficient*, which based on Eqs. (2.122), and can be written in the following form:

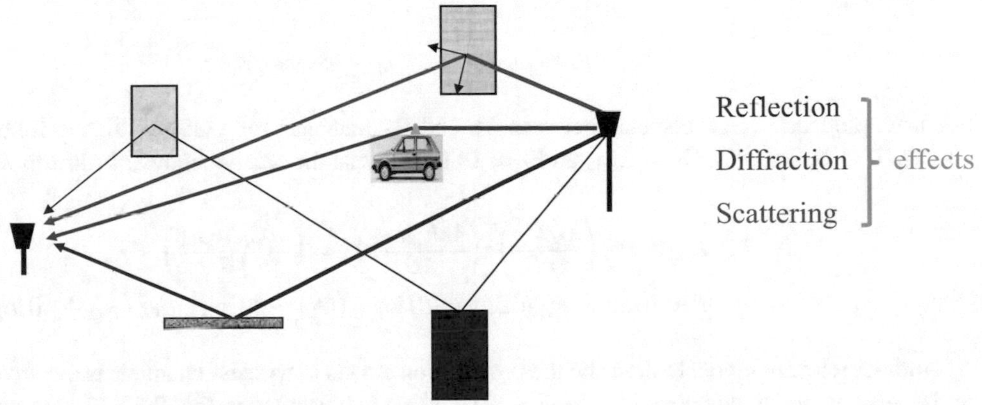

Fig. 2.26 Illustration of reflection, diffraction, and scattering effects

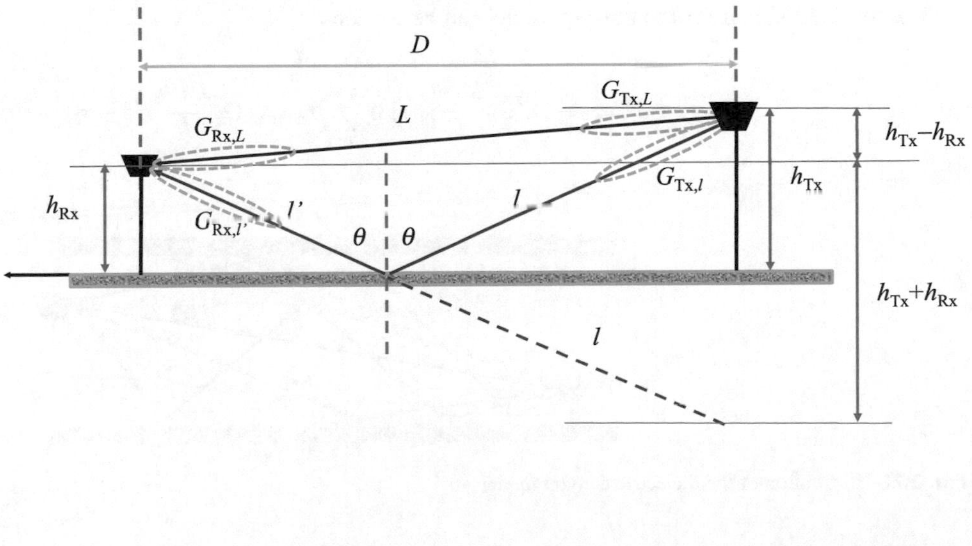

Fig. 2.27 Illustration of the two-ray channel model. h_{Tx} transmit antenna height, h_{Rx} receive antenna height

$$r_{\text{refl}} = \pm\frac{\cos\theta - Z}{\cos\theta + Z},\ Z = \begin{cases} \sqrt{\varepsilon_r - \sin^2\theta}, \text{ TE} \\ \sqrt{\varepsilon_r - \sin^2\theta/\varepsilon_r}, \text{ TM} \end{cases} \tag{2.440}$$

where ε_r is the dielectric constant of the round, related to the refractive index by $\varepsilon_r = n^2$. The sign + corresponds to TE (vertical) polarization while the sign − to the TM (horizontal) polarization state. The gain of LOS ray is determined by $G_L = G_{\text{Tx},L}G_{\text{Rx},L}$, while the gain of reflected ray is given by $G_{\text{refl}} = G_{\text{Tx},l}G_{\text{Rx},l'}$.

For narrowband transmitted signals, the delay spread is much smaller than the reciprocal of the signal bandwidth ($\tau << 1/B_s$), and we can write $s(t) \approx s(t - \tau)$. Under this assumption, the received power is related to the transmit power by:

$$P_{Rx} = P_{Tx}\left(\frac{\lambda}{4\pi}\right)^2\left|\frac{\sqrt{G_L}}{L} + \frac{r_{\text{refl}}\sqrt{G_{\text{refl}}}e^{\,j\Delta\phi}}{l+l'}\right|^2,\qquad \Delta\phi = k(l + l' - L). \tag{2.441}$$

By using the Pythagorean theorem twice (see Fig. 2.27), we determine the LOS and reflected ray paths difference by:

$$l + l' - L = \sqrt{(h_{Tx} + h_{Rx})^2 + D^2} - \sqrt{(h_{Tx} - h_{Rx})^2 + D^2}, \tag{2.442}$$

where D is the horizontal separation of antennas. When $D >> h_{Tx} + h_{Rx}$, we can use the following approximation $(1 + x)^{1/2} \approx 1 + x/2$ in (2.442) to obtain:

$$\Delta\phi \approx k\frac{2h_{Tx}h_{Rx}}{D}. \tag{2.443}$$

When $\Delta\phi = \pi$ the LOS and reflected ray are out of phase, and the distance for which this happens is known as the critical distance:

$$l_c = 4h_{Tx}h_{Rx}/\lambda. \tag{2.444}$$

For asymptotically large distance between Ts and Rx antennas for LOS ray L, the following approximations are valid: $l + l' \approx L \approx D$, $\theta \approx \pi/2$, $G_L \approx G_{\text{refl}}$, and $r \approx 1$ (TE), so that the received power in linear- and dB-scales can be written as:

$$\begin{aligned} P_{Rx} &\approx P_{Tx}\left(\frac{\lambda\sqrt{G_L}}{4\pi D}\right)^2\left(\frac{4\pi h_{Tx}h_{Rx}}{\lambda D}\right)^2 = \left(\frac{\sqrt{G_L}h_{Tx}h_R}{D^2}\right)^2 P_{Tx}, \\ P_{Rx}[\text{dBm}] &= P_{Tx}[\text{dBm}] + 10\log_{10}(G_L) + 20\log_{10}(h_{Tx}h_{Rx}) - 40\log_{10}(D). \end{aligned} \tag{2.445}$$

Another relevant model to describe the propagation effects in wireless channels is the *dielectric canyon model*, also known as the ten-ray model, developed by Amitay [71], which is illustrated in Fig. 2.28. This model incorporates all rays with one, two, or three reflections: line-of-sight (LOS), ground-reflected (GR), single-wall (SW) reflected, double-wall (DW) reflected, triple-wall reflected (TW), wall-ground (WG) reflected, and ground-wall (GW) reflected rays.

The received signal for the ten-ray model can be written as:

$$r(t) = \text{Re}\left\{\frac{1}{2k}\left[\sqrt{G_L}\frac{e^{-jkL}}{L}s(t) + \sum_{i=1}^{9} r_{\text{refl},i}\sqrt{G_{l_i}}\frac{e^{-jkl_i}}{l_i}s(t-\tau_i)\right]e^{\,j\omega_c t}\right\}, k = \frac{2\pi}{\lambda} \tag{2.446}$$

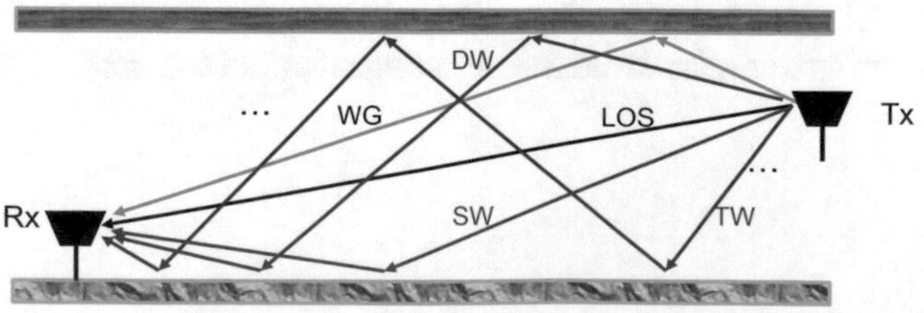

Fig. 2.28 Illustration of the dielectric canyon model

where $r_{\text{refl},i}$ is the overall reflection coefficient of the i-th reflected ray, $\sqrt{G_{l_i}}$ is the product of the transmit and receive radiating patterns corresponding to the i-th reflected ray, and l_i is the distance traveled by the i-th reflected ray. The delay spread of the i-th reflected ray relative to the LOS ray is given by $\tau_i = (l_i - L)/c$. Under the narrowband assumption ($\tau << 1/B_s$), $s(t) \approx s(t - \tau_i)$ so that the received power in dielectric canyon model can be written as:

$$P_{Rx} = P_{Tx}\left(\frac{1}{2k}\right)^2 \left| \frac{\sqrt{G_L}}{L} + \sum_{i=1}^{9} \frac{r_{\text{refl},i}\sqrt{G_{l_i}}e^{\,j\Delta\phi_i}}{l_i} \right|^2, \quad \Delta\phi_i = k(l_i - L) \tag{2.447}$$

The power roll-off with distance is proportional to l^{-g}, where $g \in [2 \text{ to } 6]$.

2.3.1.4 General Ray Tracing (GRT) Model

Another relevant model is so-called general ray tracing (GRT) model [72, 73]. GRT model can be used to predict the electric field strength and delay spread of any building configuration and antenna placement, providing that building database (height, location, and dielectric properties) and Tx and Rx locations are accurately specified. The methods of geometric optics, such as reflection, diffraction, and scattering, as illustrated in Fig. 2.29(a), are used to trace the propagation of different multipath components. Diffraction occurs on "sharp edges" and is commonly modeled by *Fresnel knife-edge diffraction model*, as shown in Fig. 2.29(b).

Here we derive the attenuation due to diffraction from a building or a mountain by modeling the obstruction as the diffracting knife-edge. Figure 2.29(b) has similarities with the Fig. 2.15. The additional distance traveled of diffracted ray with respect to the LOS ray can be determined from Fig. 2.29(b), by employing the Pythagorean theorem, as follows:

$$\begin{aligned} \Delta l &= D_{Tx} + D_{Rx} - (d_{Tx} + d_{Rx}) = \sqrt{d_{Tx}^2 + h_m^2} + \sqrt{d_{Rx}^2 + h_m^2} - d_{Tx} - d_{Rx} \\ &= d_{Tx}\sqrt{1 + \frac{h_m^2}{d_{Tx}^2}} + d_{Rx}\sqrt{1 + \frac{h_m^2}{d_{Rx}^2}} - d_{Tx} - d_{Rx} \\ &\overset{(1+x)^{1/2}\approx 1+x/2}{\simeq} d_{Tx}\left(1 + \frac{1}{2}\frac{h_m^2}{d_{Tx}^2}\right) + d_{Rx}\left(1 + \frac{1}{2}\frac{h_m^2}{d_{Rx}^2}\right) - d_{Tx} - d_{Rx} \\ &= \frac{h_m^2}{2}\left(\frac{1}{d_{Tx}} + \frac{1}{d_{Rx}}\right). \end{aligned} \tag{2.448}$$

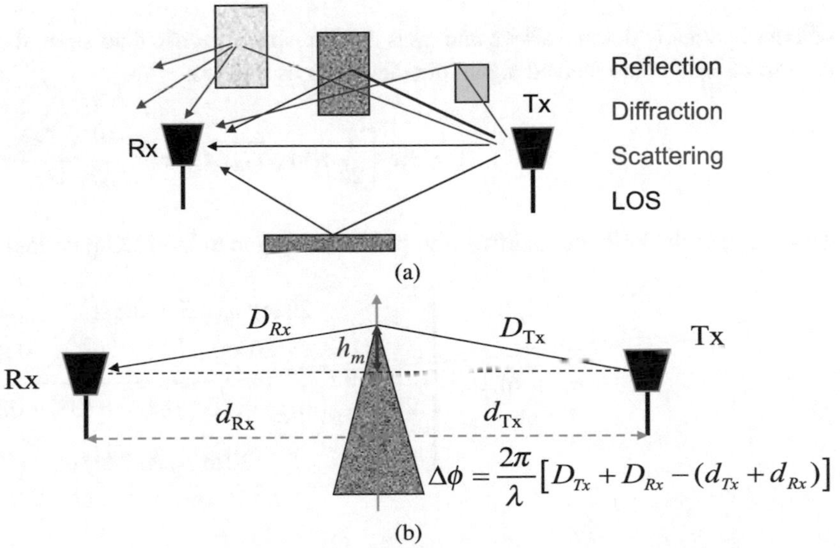

Fig. 2.29 (**a**) The methods of geometric optics are used in GRT model. (**b**) The knife-edge diffraction model

The corresponding phase shift difference is given by:

$$\Delta\phi = k\Delta l \simeq k\frac{h_m^2}{2}\left(\frac{1}{d_{Tx}}+\frac{1}{d_{Rx}}\right) = \frac{\pi}{2}v^2, v = h_m\sqrt{\frac{2}{\lambda}\left(\frac{1}{d_{Tx}}+\frac{1}{d_{Rx}}\right)} \tag{2.449}$$

where v is so-called Fresnel-Kirchhoff diffraction parameter.

From Sect. 2.2.4 we learned that the wave function at receiving point P can be approximated in the Fresnel regime as:

$$\psi_P \approx \int t(x,y)e^{jk\frac{x^2+y^2}{2R}}dS, \quad \frac{1}{R} = \frac{1}{d_{Tx}}+\frac{1}{d_{Rx}}, \tag{2.450}$$

where $t(x,y)$ is the aperture transmission function (it can also be called the electric field transfer function), which is 1 above the knife-edge. The wave function in our case is:

$$\psi_P \approx \int_{y_1}^{y_2} e^{jk\frac{y^2}{2R}}dy \sim \int_{w_1}^{w_2} e^{\frac{j\pi u^2}{2}}du. \tag{2.451}$$

By using the Fresnel integral definition:

$$F(w) = \int_0^w e^{\frac{j\pi u^2}{2}}du = C(w) + jS(w), \tag{2.452}$$

we can rewrite the Eq. (2.451) as follows:

$$\psi_P \sim \int_{w_1}^{w_2} e^{\frac{j\pi u^2}{2}}du = C(w_2) - C(w_1) + j\{S(w_2) - S(w_1)\}. \tag{2.453}$$

Therefore, the electric field in the shadow behind the half-plan can be written as:

$$E_{\text{diff}} = \frac{E_0}{2}(1+j)F(v), \tag{2.454}$$

where v is already defined above and E_0 is the free-space electric field strength. Alternatively, an approximate method can be used to describe the received signal diffracted ray as follows:

$$r(t) = \text{Re}\left\{\frac{1}{2k}\alpha(v)\sqrt{G_d}s(t-\tau)\frac{e^{-jk(D_{Tx}+D_{Rx})}}{D_{Tx}+D_{Rx}}e^{\,j\omega_c t}\right\}, k = \frac{2\pi}{\lambda} \tag{2.455}$$

where $\alpha(v)$ is the knife-edge diffraction path loss relative to the LOS path loss, which can be estimated as [74]:

$$\alpha(v)\ [\text{dB}] = \begin{cases} 20\log_{10}(0.5-0.62v), & -0.8 \le v < 0 \\ 20\log_{10}\left(0.5e^{-0.95v}\right), & 0 \le v < 1 \\ 20\log_{10}\left(0.4-\sqrt{0.1184-(0.38-0.1v)^2}\right), & 1 \le v \le 2.4 \\ 20\log_{10}(0.225/v), & v > 2.4 \end{cases} \tag{2.456}$$

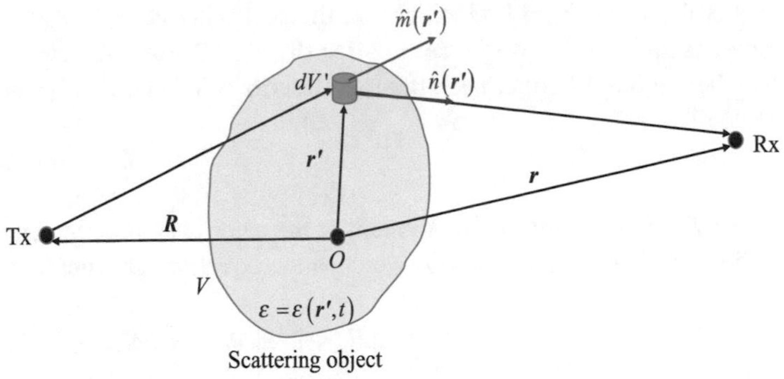

Fig. 2.30 Illustration of the scattering effect

2.3.1.5 Scattering Effects

The *scattering of electromagnetic waves* [51, 75, 76] occurs due to the interaction of the incident wave with the obstacle matter, which results in the remission of the energy of the incident wave into other directions different from the incident wave direction. The energy of incident wave is not absorbed but rather redistributed to different directions of propagation. The scattering effects are contributed to the inhomogeneities in a dielectric constant of the obstacle. A simple model to study the scattering effects is shown in Fig. 2.30. The origin is located inside of the scattering object of volume V. The scattering object is inhomogeneous in dielectric constant $\varepsilon = \varepsilon(\boldsymbol{r}',t)$. The position vectors of the transmitter Tx and receiver Rx are denoted by $\boldsymbol{R}$ and $\boldsymbol{r}$, respectively. The position of an arbitrary point inside the scattered object is denoted by $\boldsymbol{r}'$. The unit vector in direction of the incident electric field vector is denoted by $\widehat{m}(\boldsymbol{r}')$, while the unit vector pointing towards the receiver is denoted by $\widehat{n}(\boldsymbol{r}')$.

Let $\boldsymbol{E}$ and $\boldsymbol{H}$ be slow-varying complex electric and magnetic field amplitudes. The corresponding curl and divergence Maxwell's equation can be written as:

$$\nabla\times\boldsymbol{E} = jk\boldsymbol{H} - \frac{1}{c}\frac{\partial\boldsymbol{H}}{\partial t}, \nabla\times\boldsymbol{H} = -jk\varepsilon\boldsymbol{E} + \frac{1}{c}\frac{\partial\varepsilon\boldsymbol{E}}{\partial t}, \quad \nabla(\varepsilon\boldsymbol{E}) = 0. \tag{2.457}$$

By taking the curl of $\nabla\times\boldsymbol{E}$ and substituting $\nabla\times\boldsymbol{H}$ into such obtained equation, we arrive at:

$$\nabla\times(\nabla\times\boldsymbol{E}) = k^2\varepsilon\boldsymbol{E} + 2\mathrm{j}\frac{k}{c}\frac{\partial\varepsilon\boldsymbol{E}}{\partial t} - \frac{1}{c^2}\frac{\partial^2\varepsilon\boldsymbol{E}}{\partial t^2}. \tag{2.457}$$

By employing the identity $\nabla\times(\nabla\times\boldsymbol{E}) = \nabla(\nabla\boldsymbol{E}) - \Delta\boldsymbol{E}$ and by rewriting the divergence equation in (2.457) as $\nabla\boldsymbol{E} = -\boldsymbol{E}\frac{\nabla\varepsilon}{\varepsilon} = -\boldsymbol{E}\nabla(\log\varepsilon)$, the Eq. (2.457) becomes:

$$\Delta\boldsymbol{E} + k^2\varepsilon\boldsymbol{E} = -\nabla(\boldsymbol{E}\nabla(\log\varepsilon)) - 2\mathrm{j}\frac{k}{c}\frac{\partial\varepsilon\boldsymbol{E}}{\partial t} + \frac{1}{c^2}\frac{\partial^2\varepsilon\boldsymbol{E}}{\partial t^2}. \tag{2.458}$$

Assuming that dielectric constant in the scattering object changes slowly, we can write:

$$\varepsilon = \langle\varepsilon\rangle + \delta\varepsilon, \tag{2.459}$$

where $\delta\varepsilon$ is slowly varying perturbation around mean value $\langle\varepsilon\rangle$. For the propagation through atmosphere, we can set the mean value to be 1, and since $|\delta\varepsilon| < < \langle\varepsilon\rangle$, we can use the following approximation $\log(1 + \delta\varepsilon) \approx \delta\varepsilon$, so that the Eq. (2.458) becomes:

$$\Delta\boldsymbol{E} + k^2(1+\delta\varepsilon)\boldsymbol{E} = -\nabla(\boldsymbol{E}\nabla\delta\varepsilon) - 2\mathrm{j}\frac{k}{c}\frac{\partial\varepsilon\boldsymbol{E}}{\partial t} + \frac{1}{c^2}\frac{\partial^2\varepsilon\boldsymbol{E}}{\partial t^2}. \tag{2.460}$$

Given that ε is function of the location in the scattering object, the last two terms in (2.460) can be neglected so that this equation simplifies to:

$$\Delta\boldsymbol{E} + k^2\boldsymbol{E} = -k^2\delta\varepsilon\boldsymbol{E} - \nabla(\boldsymbol{E}\nabla\delta\varepsilon). \tag{2.461}$$

Let the wave $\boldsymbol{E}_i(\boldsymbol{r})$ be incident on the scattering object volume element as illustrated in Fig. 2.30, and this electric field clearly satisfies the divergence equation div $\boldsymbol{E}_i = 0$. Since the dielectric constant is slowly varying function of position, we can use the method of small perturbation, described in Sect. 2.2.5, and represent the electric field vector as:

$$\boldsymbol{E} = \boldsymbol{E}_i + \boldsymbol{E}_s, \tag{2.462}$$

where $\boldsymbol{E}_s$ is the scattering component of the electric field. After substituting (2.462) into (2.461), by retaining only term linear in $\delta\varepsilon$, we obtain the following two separate equations for incident field and scattered field:

$$\Delta\boldsymbol{E}_s + k^2\boldsymbol{E}_s = -k^2\delta\varepsilon\boldsymbol{E}_i - \nabla(\boldsymbol{E}_i\nabla\delta\varepsilon),\ \Delta\boldsymbol{E}_i + k^2\boldsymbol{E}_i = 0. \tag{2.463}$$

By interpreting the term $k^2\delta\varepsilon\boldsymbol{E}_i + \nabla(\boldsymbol{E}_i\nabla\delta\varepsilon)$ as the perturbation, the solution of the scattered field component can be found as the convolution of the perturbation and the Green function $G(\boldsymbol{r}, \boldsymbol{r}') = (1/4\pi)\exp(jk|\boldsymbol{r}-\boldsymbol{r}'|)/(jk|\boldsymbol{r}-\boldsymbol{r}'|)$:

$$\boldsymbol{E}_s(\boldsymbol{r}) = \frac{1}{4\pi}\int_V \frac{\exp(jk|\boldsymbol{r}-\boldsymbol{r}'|)}{|\boldsymbol{r}-\boldsymbol{r}'|}\left[k^2\delta\varepsilon(\boldsymbol{r}')\boldsymbol{E}_i(\boldsymbol{r}') + \nabla(\boldsymbol{E}_i(\boldsymbol{r}')\nabla\delta\varepsilon(\boldsymbol{r}'))\right]d^3\boldsymbol{r}' \tag{2.464}$$

By employing the Gauss theorem $\int_V u \text{ grad } v\, dV + \int_V v \text{ grad } u\, dV = \int_S uv d\boldsymbol{S}$, by interpreting u as $\exp(jk|\boldsymbol{r}-\boldsymbol{r}'|)/(jk|\boldsymbol{r}-\boldsymbol{r}'|)$ and v as $\boldsymbol{E}_i\nabla\delta\varepsilon(\boldsymbol{r}')$, we can rewrite the Eq. (2.464) as follows:

$$\begin{aligned}\boldsymbol{E}_s(\boldsymbol{r}) &= \frac{1}{4\pi}\int_V \frac{\exp(jk|\boldsymbol{r}-\boldsymbol{r}'|)}{|\boldsymbol{r}-\boldsymbol{r}'|}k^2\delta\varepsilon(\boldsymbol{r}')\boldsymbol{E}_i(\boldsymbol{r}')d^3\boldsymbol{r}' \\ &-\frac{1}{4\pi}\int_V(\boldsymbol{E}_i(\boldsymbol{r}')\nabla\delta\varepsilon(\boldsymbol{r}'))\text{grad}_{r'}\left[\frac{\exp(jk|\boldsymbol{r}-\boldsymbol{r}'|)}{|\boldsymbol{r}-\boldsymbol{r}'|}\right]d^3\boldsymbol{r}' \\ &+\frac{1}{4\pi}\int_S\frac{\exp(jk|\boldsymbol{r}-\boldsymbol{r}'|)}{|\boldsymbol{r}-\boldsymbol{r}'|}\boldsymbol{E}_i(\boldsymbol{r}')\nabla\delta\varepsilon(\boldsymbol{r}')d^3\boldsymbol{S},\end{aligned} \tag{2.465}$$

where S is the surface enclosing the volume V and d$\boldsymbol{S}$ is the vector of magnitude dS directed along the outer normal line to the surface. The grad term in equation above can be determined as:

$$\begin{aligned}\text{grad}_{r'}\left[\frac{\exp(jk|\boldsymbol{r}-\boldsymbol{r}'|)}{|\boldsymbol{r}-\boldsymbol{r}'|}\right] &= \frac{-jk\exp(jk|\boldsymbol{r}-\boldsymbol{r}'|)|\boldsymbol{r}-\boldsymbol{r}'| + \exp(jk|\boldsymbol{r}-\boldsymbol{r}'|)}{|\boldsymbol{r}-\boldsymbol{r}'|^2}\underbrace{\frac{\boldsymbol{r}-\boldsymbol{r}'}{|\boldsymbol{r}-\boldsymbol{r}'|}}_{\widehat{n}(\boldsymbol{r},\boldsymbol{r}')} \\ &= \frac{-jk\exp(jk|\boldsymbol{r}-\boldsymbol{r}'|)|\boldsymbol{r}-\boldsymbol{r}'| + \exp(jk|\boldsymbol{r}-\boldsymbol{r}'|)}{|\boldsymbol{r}-\boldsymbol{r}'|^2}\widehat{n}(\boldsymbol{r},\boldsymbol{r}').\end{aligned} \tag{2.466}$$

In the far-field, $k|\boldsymbol{r}-\boldsymbol{r}'| >> 1$ so that the last term in (2.466) can be neglected as it decays as $|\boldsymbol{r}-\boldsymbol{r}'|^{-3}$. The surface term in (2.465) is negligible compared to volume terms for sufficiently large scattering object so that after substitution of (2.466) into (2.465), we obtain:

$$\begin{aligned}\boldsymbol{E}_s(\boldsymbol{r}) &\cong \frac{1}{4\pi}\int_V \frac{\exp(jk|\boldsymbol{r}-\boldsymbol{r}'|)}{|\boldsymbol{r}-\boldsymbol{r}'|}k^2\delta\varepsilon(\boldsymbol{r}')\boldsymbol{E}_i(\boldsymbol{r}')d^3\boldsymbol{r}' \\ &+\frac{1}{4\pi}\int_V(\boldsymbol{E}_i(\boldsymbol{r}')\nabla\delta\varepsilon(\boldsymbol{r}'))jk\frac{\exp(jk|\boldsymbol{r}-\boldsymbol{r}'|)}{|\boldsymbol{r}-\boldsymbol{r}'|}d^3\boldsymbol{r}'.\end{aligned} \tag{2.467}$$

However, we are still left with another grad term that can be transformed by using Gauss theorem, and after applying div $\boldsymbol{E}_i = 0$ and neglect the terms decaying faster than $|\boldsymbol{r}-\boldsymbol{r}'|^{-3}$, we arrive at the following solution for the scattered electrical field:

$$\boldsymbol{E}_s(\boldsymbol{r}) \cong \frac{k^2}{4\pi}\int_V \frac{\exp(jk|\boldsymbol{r}-\boldsymbol{r}'|)}{|\boldsymbol{r}-\boldsymbol{r}'|}\delta\varepsilon(\boldsymbol{r}')\{\boldsymbol{E}_i(\boldsymbol{r}') - \widehat{n}(\boldsymbol{r},\boldsymbol{r}')(\widehat{n}(\boldsymbol{r},\boldsymbol{r}')\cdot\boldsymbol{E}_i(\boldsymbol{r}'))\}d^3\boldsymbol{r}'. \tag{2.468}$$

By using the cross-product identity $\boldsymbol{a} \times (\boldsymbol{b} \times \boldsymbol{c}) = \boldsymbol{b}(\boldsymbol{a} \cdot \boldsymbol{c}) - \boldsymbol{c}(\boldsymbol{a} \cdot \boldsymbol{b})$, we can rewrite the Eq. (2.468) in term of cross-products as follows:

$$\boldsymbol{E}_s(\boldsymbol{r}) \cong \frac{k^2}{4\pi}\int_V \frac{\exp\left(jk|\boldsymbol{r}-\boldsymbol{r}'|\right)}{|\boldsymbol{r}-\boldsymbol{r}'|}\delta\varepsilon(\boldsymbol{r}')\{\widehat{n}(\boldsymbol{r},\boldsymbol{r}') \times [\boldsymbol{E}_i(\boldsymbol{r}') \times \widehat{n}(\boldsymbol{r},\boldsymbol{r}')]\}d^3\boldsymbol{r}'. \tag{2.469}$$

Clearly, the evaluation of scattered electric field requires the knowledge of the spatial variation of dielectric constant, which is not easy to determine. As suggested by Tatarskii [51], the covariance function of the dielectric constant can be used instead, defined as:

$$\begin{aligned} C_\varepsilon(\boldsymbol{r}',\boldsymbol{r}'') &= \left\langle \left[\delta\varepsilon(\boldsymbol{r}') - \underbrace{\langle\delta\varepsilon(\boldsymbol{r}')\rangle}_{=0}\right]\left[\delta\varepsilon(\boldsymbol{r}'') - \underbrace{\langle\delta\varepsilon(\boldsymbol{r}'')\rangle}_{=0}\right]\right\rangle = \langle\delta\varepsilon(\boldsymbol{r}')\delta\varepsilon(\boldsymbol{r}'')\rangle \\ &= C_\varepsilon(\boldsymbol{r}'-\boldsymbol{r}'') = C_\varepsilon(\boldsymbol{\rho}), \end{aligned} \tag{2.470}$$

which is clearly spatially homogeneous. The covariance function C_ε is related to the spectral density of the dielectric constant fluctuations, denoted as $\Phi_\varepsilon(\boldsymbol{\kappa})$, by the following Fourier transform:

$$C_\varepsilon(\boldsymbol{\rho}) = \iiint \Phi_\varepsilon(\boldsymbol{\kappa})e^{j\boldsymbol{\kappa}\boldsymbol{\rho}}d^3\boldsymbol{\rho}. \tag{2.471}$$

The element of scattered energy in direction of $\widehat{n}'$ in a solid angle $d\Omega$ can be determined as:

$$dW = S_i(\boldsymbol{r}')\sigma_i(\boldsymbol{r}')dV'd\Omega, \tag{2.472}$$

where S_i is the incident wave Poynting vector magnitude and $\sigma_i(\boldsymbol{r})$ is the *effective scattering cross section* from a unit volume into a unit solid angle in the direction $\widehat{m}$ (see Fig. 2.30), determined as [62]:

$$\sigma_i(\boldsymbol{r}) = \frac{\pi}{2}k^4\Phi_\varepsilon(k\widehat{m}(\boldsymbol{r}) - k\widehat{n}(\boldsymbol{r}))\sin\gamma(\boldsymbol{r}), \tag{2.473}$$

where $\gamma(\boldsymbol{r})$ is the angle between $\boldsymbol{E}_i(\boldsymbol{r})$ and $\widehat{n}(\boldsymbol{r})$. The energy flux density of energy dW scattered in solid angle $d\Omega$ is given by [62]:

$$\frac{dW}{|r-r'|^2 d\Omega} = \frac{S_i(\boldsymbol{r}')\sigma_i(\boldsymbol{r}')dV'd\Omega}{|r-r'|^2 d\Omega} = \frac{S_i(\boldsymbol{r}')\sigma_i(\boldsymbol{r}')dV'}{|r-r'|^2}, \tag{2.474}$$

where $|r - r'|^2 d\Omega$ is the area of scattered energy distribution. The average Poynting vector at receiver Rx side can be determined as follows:

$$\langle \boldsymbol{S}(\boldsymbol{r})\rangle = \int_V \widehat{n}(\boldsymbol{r}')\frac{S_i(\boldsymbol{r}')\sigma_i(\boldsymbol{r}')}{|\boldsymbol{r}-\boldsymbol{r}'|^2}d^3\boldsymbol{r}'. \tag{2.475}$$

The total received scattered power can be estimated as $P_{Rx} \approx |\langle \boldsymbol{S}(\boldsymbol{r})\rangle|^2$. The scattering theory discussed above is applicable to both wireless communications and free-space optical communications.

Clearly, the calculation of electrical field strength and power of scattered wave on Rx side is not a trivial task. To simplify this study, we can proceed as follows. Let the incident wave be a spherical wave $\boldsymbol{E}_i(\boldsymbol{r}) = \boldsymbol{E}_i \exp\left(jk|\boldsymbol{R}-\boldsymbol{r}|\right)/|\boldsymbol{R}-\boldsymbol{r}|$, and after substitution into (2.468), we obtain:

$$\begin{aligned} \boldsymbol{E}_s(\boldsymbol{r}) &\cong \frac{k^2}{4\pi}\int_V \frac{e^{jk|\boldsymbol{r}-\boldsymbol{r}'|}}{|\boldsymbol{r}-\boldsymbol{r}'|}\frac{e^{jk|\boldsymbol{R}-\boldsymbol{r}'|}}{|\boldsymbol{R}-\boldsymbol{r}'|}\delta\varepsilon(\boldsymbol{r}')\{\boldsymbol{E}_i - \widehat{n}(\boldsymbol{r},\boldsymbol{r}')\underbrace{(\widehat{n}(\boldsymbol{r},\boldsymbol{r}')\cdot\boldsymbol{E}_i)}_{E_i\sin\gamma(\boldsymbol{r}')}\}d^3\boldsymbol{r}', \\ &= \frac{k^2}{4\pi}\int_V \frac{e^{jk|\boldsymbol{r}-\boldsymbol{r}'|}}{|\boldsymbol{r}-\boldsymbol{r}'|}\frac{e^{jk|\boldsymbol{R}-\boldsymbol{r}'|}}{|\boldsymbol{R}-\boldsymbol{r}'|}\delta\varepsilon(\boldsymbol{r}')\{\boldsymbol{E}_i - \widehat{n}(\boldsymbol{r},\boldsymbol{r}')E_i\sin\gamma(\boldsymbol{r}')\}d^3\boldsymbol{r}', \end{aligned} \tag{2.476}$$

wherein $\gamma = \measuredangle(\widehat{\mathrm{n}}, \boldsymbol{E}_\mathrm{i})$.

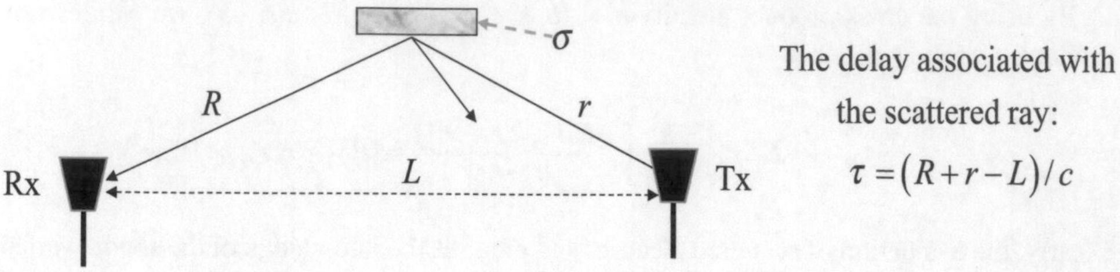

Fig. 2.31 Simplified scattering model

In the far-field zone, we can use coarse approximation in amplitudes $|\boldsymbol{r} - \boldsymbol{r}'| \approx r$, $|\boldsymbol{R} - \boldsymbol{r}'| \approx R$ and finer approximation in phases $k|\boldsymbol{r} - \boldsymbol{r}'| \approx k\boldsymbol{r} - k\boldsymbol{r}'\widehat{\boldsymbol{r}}$, $k|\boldsymbol{R} - \boldsymbol{r}'| \approx k\boldsymbol{R} - k\boldsymbol{r}'\widehat{\boldsymbol{R}}$ so that the previous equation simplifies to:

$$\boldsymbol{E}_s(\boldsymbol{r}) \cong \frac{k^2}{4\pi}\frac{e^{\,jk(r+R)}}{rR}\int_V e^{-jk(\boldsymbol{r}'\widehat{\boldsymbol{r}}+\boldsymbol{r}'\widehat{\boldsymbol{R}})}\delta\varepsilon(\boldsymbol{r}')\{\boldsymbol{E}_i - \widehat{n}(\boldsymbol{r},\boldsymbol{r}')E_i \sin\gamma(\boldsymbol{r}')\}d^3\boldsymbol{r}'. \tag{2.477}$$

By taking the power gain antennas definitions G_{Tx} and G_{Rx} as well the effective ("radar") cross-sectional area of the scattering σ object into account, based on radar range Eq. (2.437.2) and Eq. (2.477), the received signal can be written into an approximate form as follows:

$$r_{Rx}(t) \cong \text{Re}\left\{\frac{1}{2k}\frac{1}{\sqrt{4\pi}}\frac{e^{jk(r+R)}}{rR}s(t-\tau)\sqrt{\underbrace{G_{Tx}G_{Rx}\sigma}_{G_s}}e^{\,j\omega_c t}\right\}, k = \frac{2\pi}{\lambda} \tag{2.478}$$

where $s(t)$ is the transmitted signal and τ is delay associated with scattered ray, relative to the LOS ray, given by $\tau = (R + r - L)/c$. This simplified scattering model is illustrated in Fig. 2.31. This equation is consistent with the radar technique [76] (see also [64]). It was found in [77] that typical values for effective cross-sectional area in large German cities range from −4.5 to 55.7 dBm^2.

The total received signal in GRT model, obtained as a superposition of a LOS ray, N_{refl} reflected rays, N_{diff} diffracted rays, and N_{scatt} diffusely scattered rays, can be written as:

$$\begin{aligned} r(t) = \text{Re}\Bigg\{ crBigg\frac{1}{2k}\Bigg[& \sqrt{G_L}\frac{e^{-jkL}}{L}s(t) + \sum_{i=1}^{N_{\text{refl}}} r_{\text{refl},i}\sqrt{G_{l_k}}\frac{e^{-jkl_i}}{l_i}s(t-\tau_i) \\ & + \sum_{m=1}^{N_{\text{diff}}}\alpha_m(v)\sqrt{G_{d_m}}s(t-\tau_j)\frac{e^{-jk(D_{Tx,m}+D_{Rx,m})}}{D_{Tx,m}+D_{Rx,m}} \\ & + \sum_{n=1}^{N_{\text{scatt}}}\frac{\sqrt{G_{s_n}\sigma_n}}{(4\pi)^{1/2}}s(t-\tau_n)\frac{e^{-jk(R_n+r_n)}}{R_n r_n}\Bigg] e^{\,j\omega_c t}\Bigg\}, k = \frac{2\pi}{\lambda} \end{aligned} \tag{2.479}$$

2.3.1.6 Simplified Path Loss and Multislope Models

The *simplified path loss model*, which can be used for *general trade-off analysis* of various system designs, is suitable for use when the path loss dominated by reflections. The received power in this model is given by:

$$P_{Rx} = P_{Tx}C\left(\frac{l_{\text{ref}}}{l}\right)^s, \quad l > l_{\text{ref}} \tag{2.480}$$

where $s \in [2 \text{ to } 8]$ is the path loss exponent ("the slope"), whose exact value needs to be determined empirically. The referent distance for antenna far-field l_{ref} is typically 1–10 m for indoor applications and 10–100 m for outdoor applications [64], with exact value to be determined empirically. The corresponding dB-scale form is given by:

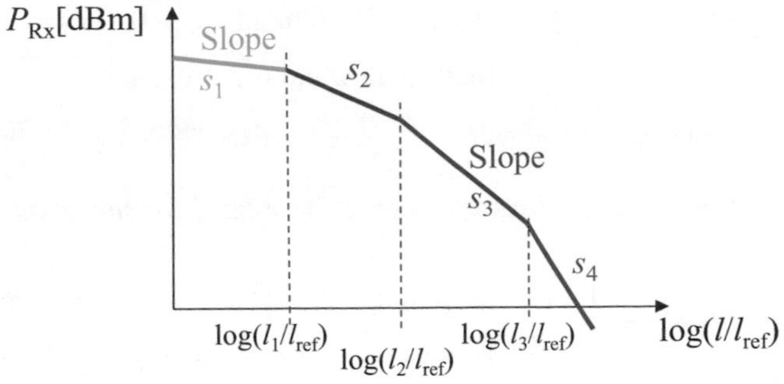

Fig. 2.32 Illustration of the multislope model

$$P_{Rx}[\text{dBm}] = P_{Rx}[\text{dBm}] + C[\text{dB}] - 10s\log_{10}\left(\frac{l}{l_{\text{ref}}}\right), \tag{2.481}$$

where C[dB] is the free-space path gain given by $C[\text{dB}] = 20\log_{10}[1/(2kl_{\text{ref}})]$. The two-ray model discussed earlier belongs to this category.

A better empirical model is a *multislope (piecewise linear) model*, which is for four slopes illustrated in Fig. 2.32. A multislope model has multiple segments with the slope of the i-th segment being s_i ($i = 1,2,\ldots$). For multislope model with N segments, the $N - 1$ breaking points $l_1,\ldots,l_{N-1}$ must be specified. The simplest model from this category is the *dual-slope model*, for which the received power in dB-scale is given by:

$$P_{Rx}(l)\,[\text{dBm}] = \begin{cases} P_{Tx}(l)\,[\text{dBm}] + C - 10s_1\log_{10}(l/l_{\text{ref}}), l_{\text{ref}} \le l \le l_1 \\ P_{Tx}(l)\,[\text{dBm}] + C - 10s_1\log_{10}(l_1/l_{\text{ref}}) - 10s_2\log_{10}(l/l_{\text{ref}}), l > l_1 \end{cases} \tag{2.482}$$

Two-ray model is in fact a dual-slope model with break point being the critical distance $l_1 = l_c$ and attenuation slopes being $s_1 = 20$ dB/decade and $s_2 = 40$ dB/decade. A useful approximation for the dual-slope model, in linear-scale, is given by [78] (see also [64]):

$$P_{Rx} = \frac{P_{Tx}C}{\alpha(l)}, \quad \alpha(l) = \left(\frac{l}{l_{\text{ref}}}\right)^{s_1}\left[1 + \left(\frac{l}{l_1}\right)^{-(s_2-s_1)q}\right]^{1/q}, \tag{2.483}$$

where q is the path loss smoothness parameter to be determined empirically.

2.3.1.7 Empirical Path Loss Models

The GRT model is quite difficult to develop in urban areas, in particular in large cities. To solve for this problem, the *empirical path loss models* have been developed such as Okumura model [79], Hata model [80], Hata model extension (COST 231) [81], Hata-Davidson model [82], and Walfisch-Ikegami model [83]. In the empirical models, the path loss at a desired distance $P_L(l)$ is determined by averaging out the local mean attenuation (LMA) measurement results in a given environment (a city, suburban area, or office building). The purpose of the averaging is to remove the multipath and shadowing components present in measurements. Okumura model is empirically based on site–/frequency-specific model, applicable to frequency ranges 150–1500 MHz and distances 1–100 km [79]. The Hata model represents the analytical approximation to Okumura model, and it is valid for distances ranging from 1 km to 20 km. The empirical path loss under Hata model is given by [79, 83]:

$$\begin{aligned} P_{L,\text{Hata}}(l)\,[\text{dB}] &= 69.55 + 26.16\log_{10}(f_c) - 13.82\log_{10}(h_{Tx}) - a(h_{Rx}) \\ &+ (44.9 - 6.55\log_{10}(h_{Tx}))\log_{10}(l) - b(f_c, \text{environment}), \end{aligned} \tag{2.484}$$

where h_{Tx} (h_{Rx}) is the height of transmit (receive) antenna as before, f_c [MHz] is the carrier frequency, l [km] is the link length, and $a(h_{\text{Rx}})$ is the receive antenna, carrier frequency, and environment-specific attenuation function:

$$a(h_{Rx}) = \begin{cases} (1.1\log_{10}(f_c) - 0.7)h_{Rx} - (1.56\log_{10}(f_c) - 0.8)\text{ dB}, \\ \qquad \text{small to medium-size cities} \\ 3.2(\log_{10}(11.75h_{Rx}))^2 - 4.97\text{ dB},\ \text{large cities } (f_c > 300\text{ MHz}) \end{cases} \tag{2.485}$$

The function $b(f_c$, environment) represents the correction factor to be applied for *suburban and rural areas* [79, 83]:

$$b(f_c, \text{environment}) = \begin{cases} 2[\log_{10}(f_c/28)]^2 + 5.4, \text{suburbanareas} \\ 4.78[\log_{10}(f_c)]^2 - 18.33\log_{10}(f_c) + C, \text{rural areas} \end{cases} \tag{2.486}$$

where C ranges from 35.94 in countryside to 40.95 for desert areas [64]. The ITU-R extended the Hata model for distances ranging from 20 km to 100 km as follows [83]:

$$\begin{aligned} P_{L,\text{Hata-ITU-R}}(l)\ [\text{dB}] &= 69.55 + 26.16\log_{10}(f_c) - 13.82\log_{10}(h_{Tx}) - a(h_{Rx}) \\ &+ (44.9 - 6.55\log_{10}(h_{Tx}))(\log_{10}(l))^g - b(f_c, \text{environment}), \end{aligned} \tag{2.487}$$

where g is the exponent correction factor [83]:

$$g = \begin{cases} 1, l < 20\text{ km} \\ 1 + \left(0.14 + 0.000187f_c + 0.00107\dfrac{h_{Tx}}{1 + 7\cdot 10^{-6}h_{Tx}^2}\right)\left(\log_{10}\left(\frac{l}{20}\right)\right)^{0.8}, l \geq 20\text{ km} \end{cases} \tag{2.488}$$

The Hata model was extended by the European Cooperation in the Field of Scientific and Technical Research (EURO-COST) to 2 GHz [81] by adding the correction factor of +3 dB in Eq. (2.484) for metropolitan centers and restricting the validity of model to 1.5 GHz $< f_c <$ 2GHz, 1 km $\leq l \leq$ 10 km, 30 m $\leq h_{Tx} \leq$ 200 m, 1 m $\leq h_{Rx} \leq$ 10 m.

The Telecommunications Industry Association (TIA) recommended the modification of the Hata model, now known as *Hata-Davidson model*, as follows [83]:

$$P_{L,\text{Hata-Davidson}}(l)\ [\text{dB}] = P_{L,\text{Hata}}(l)\ [\text{dB}] - A(h_{Tx}, l) - S_1(l) - S_2(h_{Tx}, l) - S_3(f_c) - S_4(f_c, l), \tag{2.489}$$

where

$$\begin{aligned} A(h_{Tx}, l) &= \begin{cases} 0, l < 20\text{ km} \\ 0.62137(l - 20)\left[0.5 + 0.15\log_{10}\left(\dfrac{h_{Tx}}{121.92}\right)\right], 20\text{ km} \leq l < 300\text{km} \end{cases} \\ S_1(l) &= \begin{cases} 0, l < 64.38\text{ km} \\ 0.174(l - 64.38), 64.38\text{ km} \leq l < 300\text{km} \end{cases} \end{aligned} \tag{2.490}$$

and

$$\begin{aligned} S_2(h_{Tx}, l) &= 0.00784|\log_{10}(9.98/l)|(h_{Tx} - 300), h_{Tx} > 300\ m \\ S_3(f_c) &= f_c/250\log_{10}(1500/f_c), \\ S_4(f_c, l) &= [0.112\log_{10}(1500/f_c)](l - 64.38), l > 64.38\text{ km} \end{aligned} \tag{2.491}$$

2.3.1.8 Shadowing Effects

A signal transmitted through a wireless channel will typically experience random variations due to attenuation from the objects in the signal path, giving rise to random the variations of the received power at a given distance, as illustrated in Fig. 2.33. This effect is commonly referred to as the *shadowing effect*. Attenuation through the building of total wall depth l is given by $a(l) = \exp.(-\alpha l)$, where α is the attenuation coefficient. Attenuation through multiple buildings can be described as:

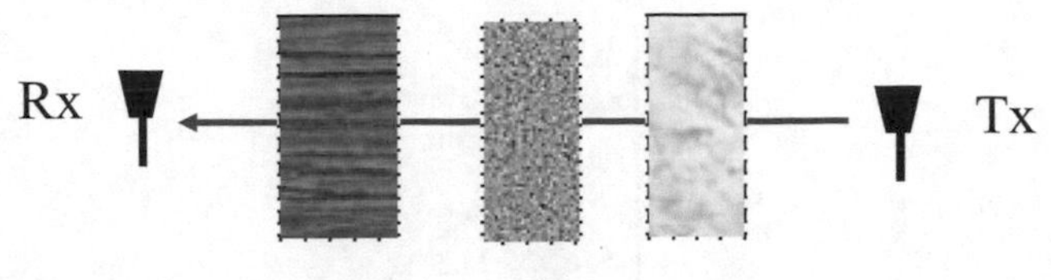

Fig. 2.33 Illustration of the shadowing effect

$$a(l_t) = e^{-\sum_i \alpha_i l_i} = e^{-\alpha_t l_t},\ l_t = \sum_i l_i, \tag{2.492}$$

where α_i is the attenuation coefficient of the i-th building and l_i is the correspond total wall depth. The corresponding attenuation in Np-scale is given by:

$$\log a(l_t) = -\sum_i \alpha_i l_i = -\alpha_t l_t\ [\text{Np}]. \tag{2.493}$$

Since different buildings have different dielectric constant and wall depth would be different as well, according to the central limit theorem, the overall distribution of attenuation in dB-scale is Gaussian:

$$p(\psi_{\text{dB}}) = \frac{1}{\sqrt{2\pi}\sigma_{\psi_{\text{dB}}}} \exp\left[-\frac{\left(\psi_{\text{dB}} - \mu_{\psi_{\text{dB}}}\right)^2}{2\sigma^2_{\psi_{\text{dB}}}}\right],\ \psi_{\text{dB}} > 0. \tag{2.494}$$

where $\psi_{\text{dB}} = 10\log_{10}\psi$, $\psi = P_{TRx}/P_{Rx}$, with standard deviation $\sigma^2_{\psi_{\text{dB}}} \in [4\,\text{dB}, 13\,\text{dB}]$ [64, 84, 85]. The mean value is contributed to the path loss, so that we set $\mu_{\psi_{\text{dB}}} = 0$.

By transformation of random variables, we determine that the distribution of ψ is log-normal:

$$p(\psi) = \frac{10/\log 10}{\sqrt{2\pi}\sigma_{\psi_{\text{dB}}}\psi} \exp\left[-\frac{\left(10\log_{10}\psi - \mu_{\psi_{\text{dB}}}\right)^2}{2\sigma^2_{\psi_{\text{dB}}}}\right],\ \psi > 0 \tag{2.495}$$

An important parameter in in shadowing is the distance over which the auto-covariance drops down to e^{-1} of its maximum valuc, and we refer to this parameter as the *decorrelation distance* L_c. In other words, from the following equation:

$$A(d) = \left\langle \left(\psi_{\text{dB}}(l) - \mu_{\psi_{\text{dB}}}\right)\left(\psi_{\text{dB}}(l+d) - \mu_{\psi_{\text{dB}}}\right)\right\rangle = \sigma_{\psi_{\text{dB}}} e^{-d/L_c}, \tag{2.496}$$

we can determine the decorrelation distance L_c, which for outdoor systems ranges from 50 to 100 m [64, 86].

The path loss and shadowing effects can be combined as illustrated in Fig. 2.34(a). The very slow variation is associated with the path loss, while slow fluctuations are associated with the shadowing effect.

The receive-to-transmit power ratio in dB-scale can be determined as:

$$\frac{P_{Rx}}{P_{Tx}}(dB) = 10\log_{10}C - 10s\log_{10}\left(\frac{l_{\text{ref}}}{l}\right) + \psi_{dB}, \tag{2.497}$$

where the path loss decreases linearly with the $\log_{10}(l)$, s corresponds to the path loss exponent, and ψ_{dB} is a zero-mean random variable with variance $\sigma^2_{\psi_{\text{dB}}}$. The corresponding linear-scale model is described as:

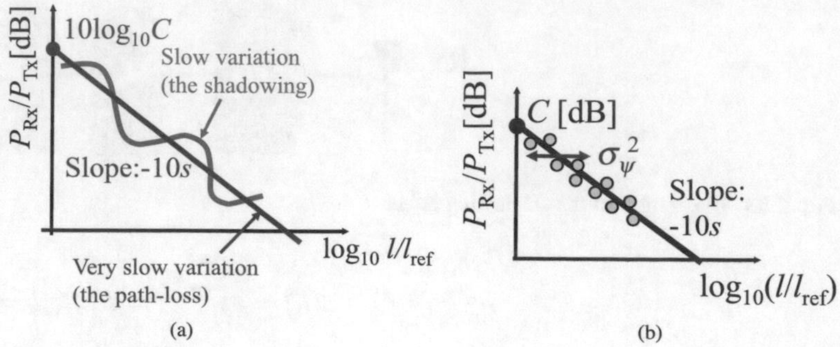

Fig. 2.34 (**a**) Illustration of the combined path loss and shadowing effect vs. normalized distance. (**b**) Determination of parameters from the empirical measurements

$$\frac{P_{Rx}}{P_{Tx}} = C\left(\frac{l_{\mathrm{ref}}}{l}\right)^s \psi, \tag{2.498}$$

where transmit-to-receive power ratio ψ is log-normally distributed (2.495). It is possible to determine the parameters of combined model from empirical measurements as illustrated in Fig. 2.34(b). "Best fit" straight line through measured attenuation data gives the slope -10s, from which we determine the path loss exponent. The intersection of the line with y-axis gives us the parameter C [dB]. Form the fluctuation of the measurements points form the straight line, we determine the variance $\sigma^2_{\psi_{\mathrm{dB}}}$ of the shadow model.

2.3.1.9 Outage Probability and Cell Coverage Area

An important characteristic of the combined path loss and shadowing effects is so-called outage probability. The *outage probability*, denoted as $P_{\mathrm{outage}}(P_{\mathrm{min}},l)$, under the combined path loss and shadowing, is defined as the probability that the received power at an observed distance l, $P_{\mathrm{Rx}}(l)$, falls below the minimum tolerable received power P_{min}:

$$\begin{aligned} P_{\mathrm{outage}}(P_{\mathrm{min}}, l) &= \Pr(P_{Rx}(l) < P_{\mathrm{min}}) \\ &= 1 - Q\left(\frac{P_{\mathrm{min}} - (P_{Tx} + 10\log_{10}C - 10s\log_{10}(l/l_{\mathrm{ref}}))}{\sigma_{\psi_{\mathrm{dB}}}}\right), \end{aligned} \tag{2.499}$$

where the Q-function is defined as:

$$Q(u) = \Pr(X > u) = \int_u^\infty \frac{1}{\sqrt{2\pi}} \exp\left(-y^2/2\right) dy, \tag{2.500}$$

and it is related to the complementary error function by:

$$Q(u) = \frac{1}{2}\mathrm{erfc}\left(\frac{u}{\sqrt{2}}\right), \mathrm{erfc}(u) = \frac{2}{\sqrt{\pi}}\int_u^\infty \exp\left(-y^2\right) dy. \tag{2.501}$$

Another relevant characteristic of the combined path loss and shadowing effect model is so-called cell coverage area, illustrated in Fig. 2.35. In the figure, the contours of constant received power corresponding to both combined path loss and average shadowing and combined path loss and random shadowing are shown. Clearly, the contour of constant received power for combined path loss and average shadowing is the circle of radius R, while the corresponding contour for combined path loss and random shadowing is an amoeba-like, which changes its shape over the time. The *cell coverage area* in a cellular system is defined as the expected percentage of locations within a cell, in which the received power is above the minimum tolerable received power P_{min}. Let P_{dS} denote the probability received power is larger than minimum power in an surface element dS, that is, $P_{dS} = \Pr(P_{\mathrm{Rx}}(r) > P_{\mathrm{min}}$ in $dS) = \Pr(P_{\mathrm{Rx}}(r) - P_{\mathrm{min}} > 0$ in $dS)$. By introducing the indicator function as follows $I(x)= 1$, for $x > 0$, we can write this probability in terms of the indicator function as:

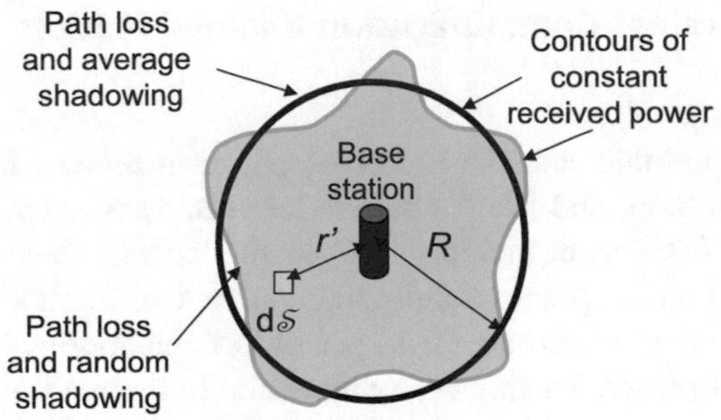

Fig. 2.35 Illustration of the cell coverage area

$$\begin{aligned} P_{dS} &= \Pr(P_{Rx}(r') - P_{\min} > 0 \text{ in } dS) = \langle I(P_{Rx}(r') - P_{\min} > 0 \text{ in } dS) \rangle \\ &= 1 - \Pr(P_{Rx}(r') - P_{\min} < 0 \text{ in } dS) = 1 - P_{\text{outage}}(P_{\min}(r'), l) \\ &= Q\left(\frac{P_{\min} - (P_{Tx} + 10 \log_{10} C - 10 s \log_{10}(r'/l_{\text{ref}}))}{\sigma_{\psi_{\text{dB}}}} \right). \end{aligned} \tag{2.502}$$

Now the cell coverage area can be calculated as:

$$\begin{aligned} C &= \left\langle \frac{\int_{\text{cell area}} I[P_{\text{Rx}}(r') > P_{\min} \text{ in } dS] dS}{\pi R^2} \right\rangle = \frac{\int_{\text{cell area}} \langle I[P_{\text{Rx}}(r) > P_{\min} \text{ in } dS] \rangle dS}{\pi R^2} \\ &= \frac{\int_{\text{cell area}} P_{dS} dS}{\pi R^2} = \frac{\int_0^{2\pi} \int_0^{R} P_{dS} r' dr' d\theta}{\pi R^2} = \frac{\int_0^{2\pi} \int_0^{R} [1 - P_{\text{outage}}(P_{\min}(r'), l)] r' dr' d\theta}{\pi R^2} \\ &= \frac{2\pi}{\pi R^2} \int_0^R Q\left(\frac{P_{\min} - (P_{Tx} + 10\log_{10} C - 10 s \log_{10}(r'/l_{\text{ref}}))}{\sigma_{\psi_{\text{dB}}}} \right) r' dr' \end{aligned} \tag{2.503}$$

After introducing the following notation [64]:

$$\overline{P}_{Rx}(R) = P_{Tx} + 10 \log_{10} C - 10 s \log_{10}(R/l_{\text{ref}}), c_1 = \frac{P_{\min} - \overline{P}_{Rx}(R)}{\sigma_{\psi_{\text{dB}}}}, \quad c_2 = \frac{10 s \log_{10}(e)}{\sigma_{\psi_{\text{dB}}}}, \tag{2.504}$$

we can rewrite Eq. (2.503) in a compact form:

$$C = \frac{2}{R^2} \int_0^R Q\left(c_1 + c_2 \log \frac{r'}{R} \right) r' dr'. \tag{2.505}$$

As an illustration, when the average power at distance R is equal to the minimum tolerable power, the outage probability will be 0.5 and $c_1 = 0$, so that the cell coverage area becomes $C = 0.5 + \exp\left(2/c_2^2\right) Q(2/c_2)$. The large cell coverage area of a given cell will mean that this cell will introduce the interference to the neighboring cells, and in practice we need to compromise between the cell coverage area and the interference.

2.3.2 Statistical Multipath Wireless Communication Channel Models

2.3.2.1 Multipath Channel Models

The term fading is used to describe the fluctuations of received power in time. It is caused by the combined defects of multipath propagation effect, see Fig. 2.36, and relative motion between transmitter and receiver, which introduces time-varying variation in received power. Because of multipath propagation effect, when a single pulse is transmitted over the multipath channel, the received signal will appear as a pulse train rather than a single pulse. In this particular example, the receiver moves at speed v, and at different time instances, the position of neighboring buildings changes, so that the receiver will experience variations in received power. This time-varying nature of the multipath channel is illustrated in Fig. 2.37. Therefore, the number of multipath components is random, while the i-th multipath component can be characterized by random amplitude $a_i(t)$, random delay $\tau_i(t)$, random Doppler frequency ν_{Di} [Hz], and random phase shift $\phi_i(t) = 2\pi f_c\tau_i(t) - 2\pi\nu_{Di}t - \phi_O$. This time-varying nature of a linear wireless channel is described by the *time-varying impulse response*, denoted as $h(\tau,t)$; here τ is so-called delay spread, and t is the time instance. The *delay spread* is defined as the time difference in signal arriving times of the last and first arriving (LOS) multipath components, which is illustrated in Fig. 2.36.

Let $\underline{s}(t)$ represent the low-pass representation of the passband signal, as discussed in previous section. The corresponding passband representation of the transmitted signal is determined by $s(t) = \text{Re}\{\underline{s}(t)e^{j\omega_c t}\} = s_I(t)\cos(\omega_c t) - s_Q(t)\sin(\omega_c t)$, where ω_c is the angular carrier frequency [rad/s], while s_I and s_Q are in-phase and quadrature components. The signal at receiver side originating from the i-th multipath component can be written as $\underline{s}(t-\tau_i)a_i(t)\exp.[-j\phi_i(t)]$. The received signal $r(t)$, originating from all multipath components, can be determined as:

$$r(t) = \text{Re}\{\underline{x}(t)e^{j\omega_c t}\} = \text{Re}\left\{\left[\sum_{i=0}^{N(t)} a_i(t)\underline{s}(t-\tau_i(t))e^{-j\phi_i(t)}\right]e^{j\omega_c t}\right\} = \text{Re}\{[\underline{h}(\tau,t)*\underline{s}(t)]e^{j\omega_c t}\}, \tag{2.506}$$

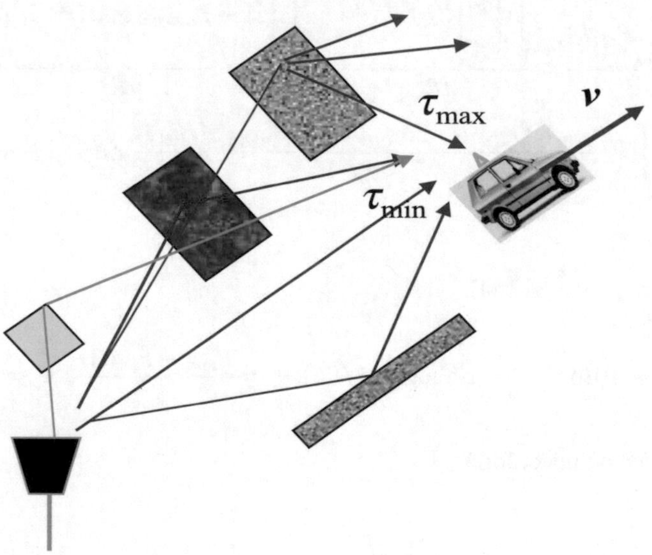

Fig. 2.36 Illustration of the multipath fading (multipath propagation effect and Doppler shift) and delay spread

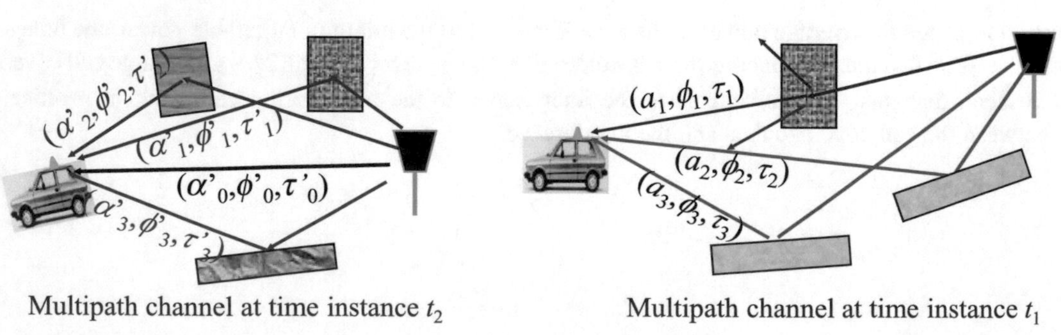

Multipath channel at time instance t_2 Multipath channel at time instance t_1

Fig. 2.37 Illustration of the time-varying nature of multipath channel

where the *time-varying impulse response* is given by:

$$\underline{h}(\tau, t) = \sum_{i=0}^{N(t)} a_i(t) e^{-j\phi_i(t)} \delta(t - \tau_i(t)), \tag{2.507}$$

and we say that channel is *time-selective* and *frequency-selective*. The time selectivity indicates that impulse response changes in time. Given that the output signal can be written in terms of convolution of the input signal and time-varying impulse response, as given by Eq. (2.506), different frequency (multipath) components will experience different fading, and we say that such channel is frequency-selective channel. Based on Eq. (2.506), we derive the multipath channel model, shown in Fig. 2.38(a).

When $f_c\tau_i(t) >> 1$, which is typical in both indoor and outdoor wireless communication systems, a small change in the path delay $\tau_i(t)$ will lead to a significant phase change $2\pi f_c \tau_i(t) - \phi_{Di} - \phi_0$. Since the phase changes rapidly, a uniform phase

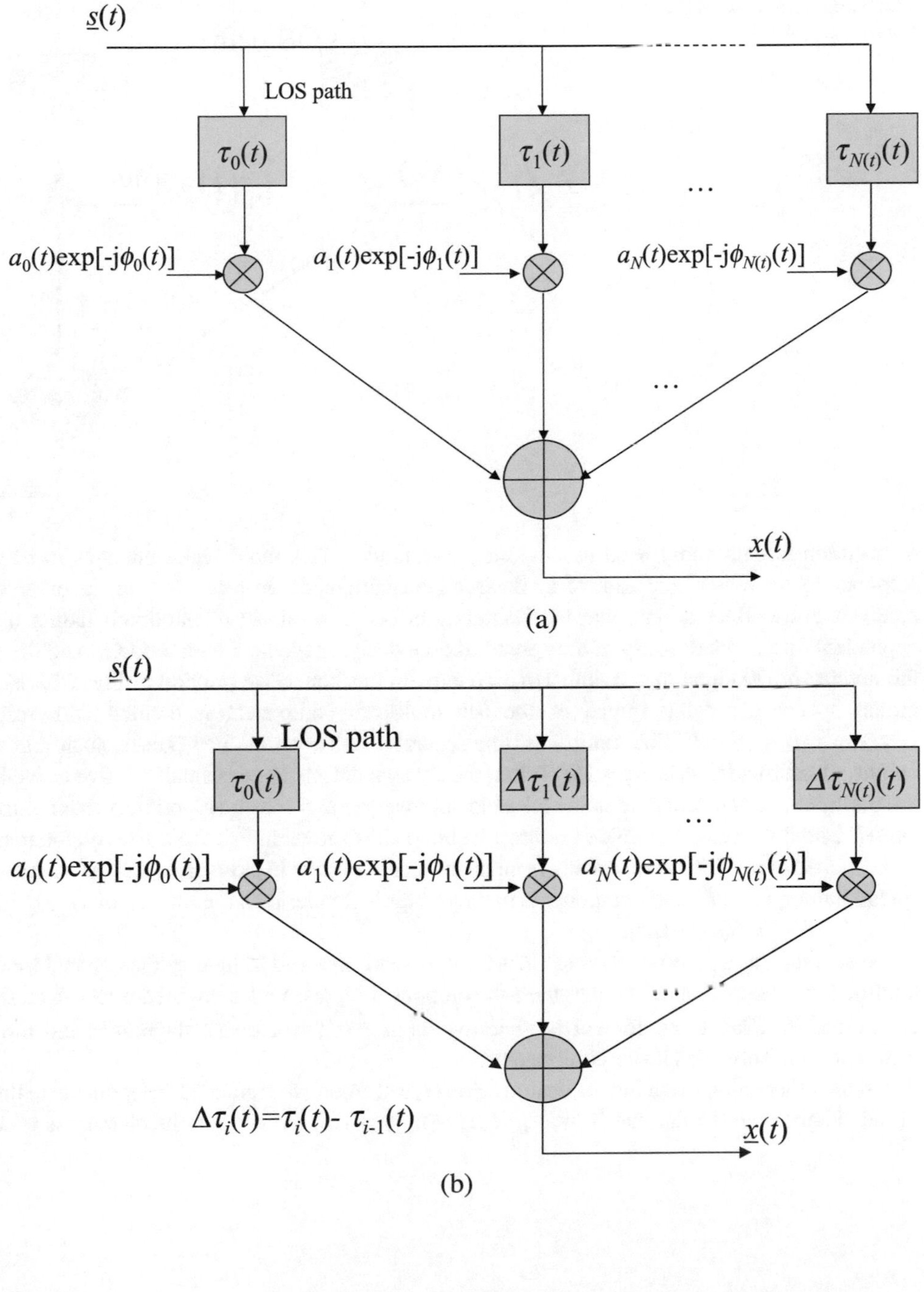

Fig. 2.38 (**a**) The multipath channel model. (**b**) The equivalent multipath channel model

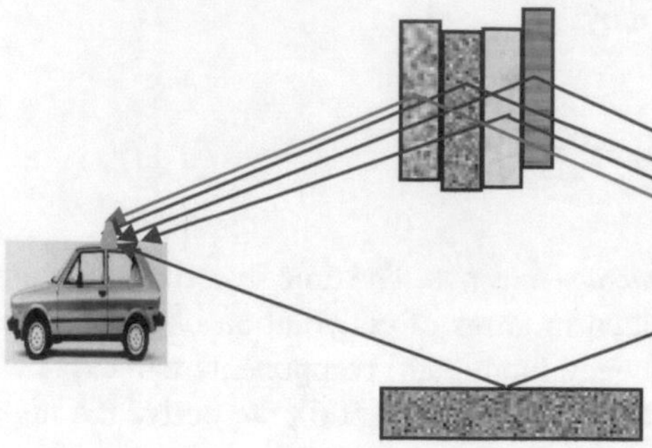

Fig. 2.39 An illustrative example of a nonresolvable path

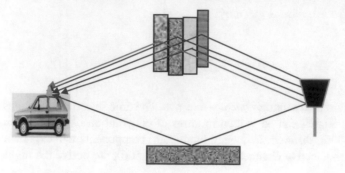

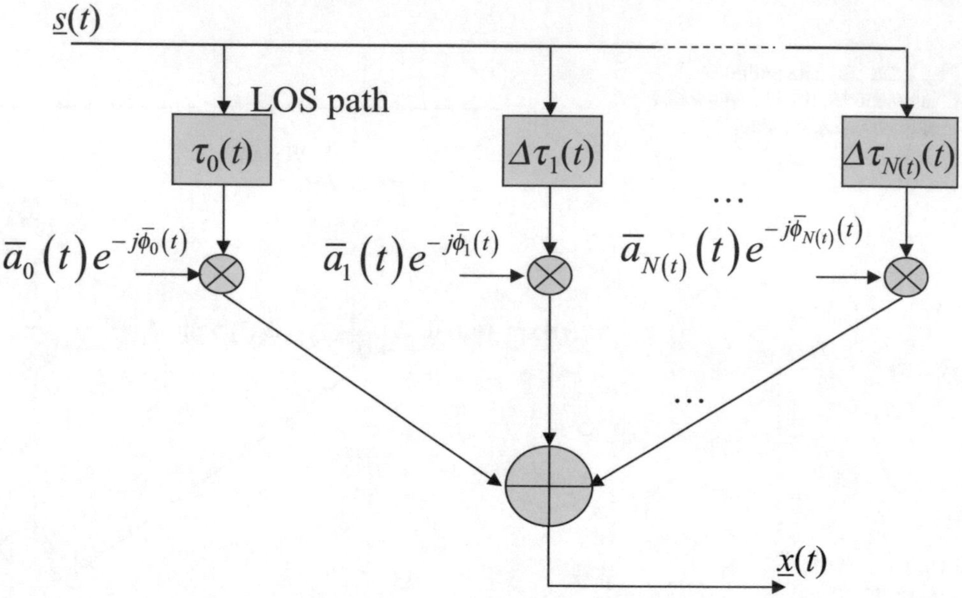

Fig. 2.40 The modified multipath channel model accounting for nonresolvable cluster components

distribution is commonly used as random phase model. This rapid phase changes in different multipath components are responsible for either constructive or destructive interference on receiver side; in other words, this effect introduces the received power fluctuations, that is, the fading effect. The impact of multipath fading on the received signal is strongly dependent on the relationship of the spread of time delays associated with the LOS and different multipath components with the inverse of the signal bandwidth $1/B_s$. To explain this better, we provide in Fig. 2.38(b) an equivalent multipath channel model, where the delay spread of the i-th multipath component is defined with respect to the $(i - 1)$-th one by $\Delta\tau_i(t) = \tau_i(t) - \tau_{i-1}(t)$. The minimum time separation between the two pulses, such that they can be resolved at the filter output, of bandwidth B, is $\Delta\tau > 1/B$. When the channel delay spread is small relative to the inverse of signal bandwidth, then the multipath components are nonresolvable, and we say that the corresponding channel model is the *narrowband channel model*. One illustrative example of nonresolvable path is shown in Fig. 2.39. Cleary, the nonresolvable multipath component is composed of reflected components from the cluster of objects in close proximity of each other. Namely, the maximum delay spread among the reflected components in the cluster is smaller than the inverse of the signal bandwidth so that we can write $\max_{m,n}|\tau_n(t) - \tau_m(t)| << 1/B_s$.

To account for the nonresolvable cluster component, we need to modify the channel model as shown in Fig. 2.40. In this modified multipath channel model, the i-th component represents a combined effect from multiple nonresolvable paths, and corresponding channel coefficient is given by $\overline{a}_i(t)e^{-j\overline{\phi}_i(t)}$, where the magnitude and the phase are averaged out among different nonresolvable cluster components.

On the other hand, when the maximum delay spread among the reflected components in the cluster is larger than the inverse of the channel bandwidth, that is, $\max_{m,n}|t_n(t) - t_m(t)| > 1/B$, we say that the channel is *wideband channel*.

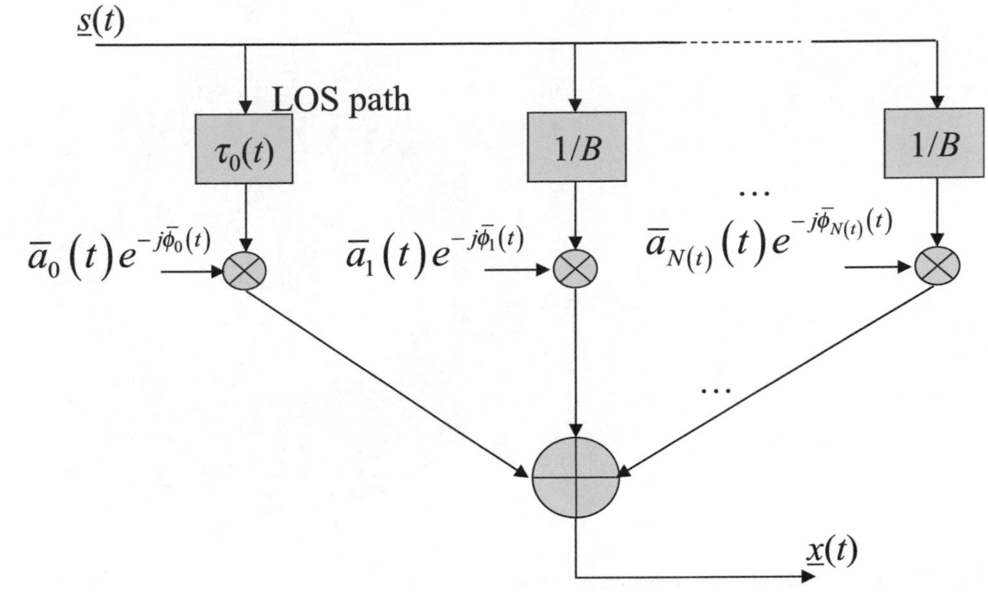

Fig. 2.41 The narrowband multipath channel model

2.3.2.2 Narrowband Fading Model

In *narrowband channel model*, the maximum delay spread between any two components is smaller than the inverse of signal bandwidth, and we can use the following approximation $\underline{s}(t) \approx \underline{s}(t - \tau_i(t))$, for all i. The received signal is now given by:

$$r(t) = \text{Re}\left\{\underline{x}(t)e^{j\omega_c t}\right\} = \text{Re}\left\{\underline{s}(t)\left[\sum_{i=0}^{N(t)} a_i(t)e^{-j\phi_i(t)}\right]e^{j\omega_c t}\right\}, \tag{2.508}$$

and corresponding narrowband channel model is shown in Fig. 2.41. Since the spreading in time is negligible in narrowband model, there is no channel distortion by intersymbol interference as different pulse from multipath fading stay in the same pulse duration. In the extreme narrowband case, we can even write that $\underline{s}(t) \cong 1$. The received signal in the extreme narrowband case can be written as:

$$r(t) = \text{Re}\left\{\underline{x}(t)e^{j\omega_c t}\right\} = \text{Re}\left\{\left[\sum_{i=0}^{N(t)} a_i(t)e^{-j\phi_i(t)}\right]e^{j\omega_c t}\right\} = r_I(t)\cos(\omega_c t) - r_Q(t)\sin(\omega_c t), \tag{2.509}$$

where the in-phase and quadrature components are given as:

$$r_I(t) = \sum_{i=0}^{N(t)} a_i(t)\cos\phi_i(t), r_Q(t) = -\sum_{i=0}^{N(t)} a_i(t)\sin\phi_i(t); \phi_i(t) = \omega_c\tau_i(t) - \phi_{D_i} - \phi_0. \tag{2.510}$$

If the number of multipath components $N(t)$ is large, and magnitude $a_n(t)$ and phase $\phi_n(t)$ are independent, then $r_I(t)$ and $r_Q(t)$ are jointly Gaussian random processes according to the central limit theorem. The *autocorrelation function of in-phase component* is given by:

$$\begin{aligned}
R_{r_I}(t, t+\tau) &= \langle r_I(t)r_I(t+\tau)\rangle = \sum_i \langle a_i^2\rangle\left\langle \cos\underbrace{\phi_i(t)}_{2\pi f_c\tau_i - 2\pi f_{D_i}t - \phi_0} \cos\underbrace{\phi_i(t+\tau)}_{2\pi f_c\tau_i - 2\pi f_{D_i}(t+\tau) - \phi_0}\right\rangle \\
&= \sum_i \langle a_n^2\rangle\left[\frac{1}{2}\langle\cos(2\pi f_{D_i}\tau)\rangle + \frac{1}{2}\underbrace{\langle\cos(4\pi f_c\tau_i - 4\pi f_{D_i}t - 2\phi_0)\rangle}_{=0}\right] \\
&= \frac{1}{2}\sum_i \langle a_i^2\rangle\langle\cos(2\pi f_{D_i}\tau)\rangle = \frac{1}{2}\sum_i \langle a_i^2\rangle\cos\left(2\pi\frac{v}{\lambda}\tau\cos\theta_i\right).
\end{aligned} \tag{2.511}$$

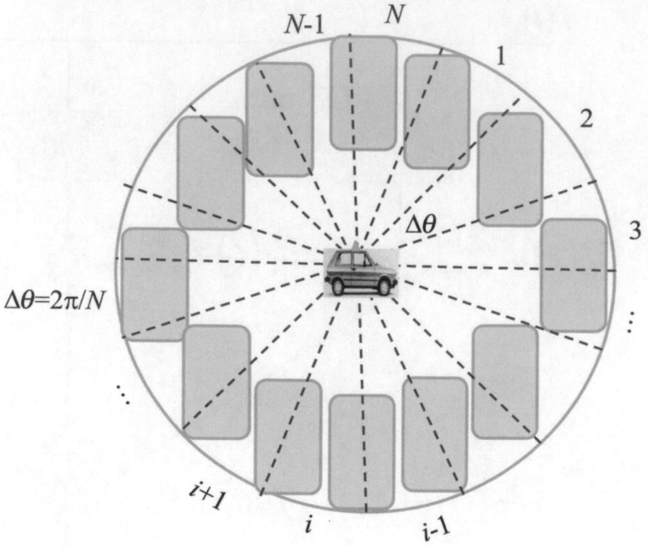

Fig. 2.42 Clark-Jakes uniform scattering (US) model

Clearly, the autocorrelation function of the in-phase component is not dependent on time origin t, and we can write $R_{r_I}(t, t+\tau) = R_{r_I}(\tau)$, meaning that the process is wide-sense stationary (WSS). It can be shown in similar fashion that the autocorrelation function of the quadrature component is equal to that of the in-phase component. The *cross-correlation function of the in-phase and quadrature components* is given by:

$$\begin{aligned} R_{r_I,r_Q}(t,t+\tau) &= R_{r_I,r_Q}(\tau) = \langle r_I(t) r_Q(t+\tau)\rangle = -\frac{1}{2}\sum_i \langle a_i^2\rangle \sin\left(2\pi\frac{v}{\lambda}\tau\cos\theta_n\right) \\ &= -\langle r_Q(t) r_I(t+\tau)\rangle = -R_{r_Q,r_I}(\tau). \end{aligned} \tag{2.512}$$

Finally, the autocorrelation function of the passband signal is given as follows:

$$R_r(t,t+\tau) = R_r(\tau) = \langle r(t) r(t+\tau)\rangle = R_{r_I}(\tau)\cos(2\pi f_c\tau) + R_{r_I,r_Q}(\tau)\sin(2\pi f_c\tau). \tag{2.513}$$

2.3.2.3 Clark-Jakes Model

To simplify the WSS model further, the Clark [87] proposed so-called uniform scattering (US) model, shown in Fig. 2.42, which was further developed by Jakes [84]. In this model, which we will refer to as *Clark-Jakes model*, the channel consists of many densely packed scattering objects, with the angle of arrival of the i-th multipath component being $\theta_i = i\Delta\theta$, where $\Delta\theta = 2\pi/N$ and the corresponding average power per scattering object being $\langle a_i^2\rangle = 2P_{Rx}/N$, where P_{Rx} is the total received power. After substituting the expression for the average power per scattering object into Eq. (2.511), we obtain the following expression for the autocorrelation function of the in-phase signal:

$$R_{r_I}(\tau) = \frac{P_{Rx}}{N}\sum_i \cos\left(2\pi\frac{v}{\lambda}\tau\cos(i\Delta\theta)\right) \overset{N=2\pi/\Delta\theta}{\widehat{=}} \frac{P_{Rx}}{2\pi}\sum_{i=1}^{N} \cos\left(2\pi\frac{v}{\lambda}\tau\cos(i\Delta\theta)\right)\Delta\theta. \tag{2.514}$$

In the limit as $N \to \infty$, the summation in (2.514) becomes integral:

$$R_{r_I}(\tau) = \frac{P_{Rx}}{2\pi}\int_0^{2\pi} \cos\left(2\pi\frac{v}{\lambda}\tau\cos\theta\right)d\theta = P_{Rx}J_0(2\pi\nu_D\tau), \tag{2.515}$$

where $J_0(x) = (1/\pi)\int_0^\pi e^{-jx\cos\theta} d\theta$ is the Bessel function of the first kind and zero order, as introduced earlier in this chapter. The cross-correlation function of the Clark-Jakes US model is clearly zero, that is:

$$R_{r_I,r_Q}(\tau) = \frac{P_{Rx}}{2\pi}\int \sin\left(2\pi\frac{v}{\lambda}\tau\cos\theta\right)d\theta = 0. \tag{2.516}$$

Since the first zero of $J_0(x)$ is 2.4048, we conclude that the in-phase component decorrelates after $\nu_D\tau = 2.4048/2\pi \cong 0.382$. In other words, if we separate two receive antennas by approximately 0.4λ, they can be considered as independent of each other.

The power spectral densities (PSDs) are obtained by calculating the Fourier transforms of corresponding autocorrelation functions. The PSD of in-phase and quadrature components is given by:

$$PSD_{r_I}(f) = PSD_{r_Q}(f) = FT\{R_{r_I}(\tau)\} = \begin{cases} \frac{2P_{Rx}}{\pi f_D}\frac{1}{\sqrt{1-(f/f_D)^2}}, |f| \le f_D \\ 0, \text{otherwise} \end{cases} \tag{2.517}$$

Clearly, as f tends to f_D, the PSD tends to infinity, but this never occurs in practice. To solve for this problem, the PSD should be truncated for $f/f_D \to 1$. The PSD of the passband signal is related to the PSD of in-phase component as follows:

$$\begin{aligned} PSD_r(f) &= \frac{1}{2}[PSD_{r_I}(f - f_c) + PSD_{r_I}(f + f_c)] \\ &= \begin{cases} \frac{P_{Rx}}{\pi f_D}\frac{1}{\sqrt{1-(|f \pm f_c|/f_D)^2}}, |f \pm f_c| \le f_D \\ 0, \text{otherwise} \end{cases} \end{aligned} \tag{2.518}$$

The PSDs are suitable to build a simulator based on Clark-Jakes US model, by using zero-mean unit variance Gaussian random generator. By setting the double-sided PSD of an additive white Gaussian noise (AWGN) by $N_0/2 = 2P_{Rx}/(\pi f_D)$, we need to determine the filter of transfer function $H(f)$ such that the PSD of the filter output is equal to (2.517), in other words:

$$\frac{2P_{Rx}}{\pi f_D}|H(f)|^2 = \frac{2P_{Rx}}{\pi f_D}\frac{1}{\sqrt{1-(f/f_D)^2}}. \tag{2.519}$$

Clearly, the magnitude of transfer function of such filter should be chosen as:

$$|H(f)| = \left[1-(f/f_D)^2\right]^{-1/4}. \tag{2.520}$$

2.3.2.4 Envelope and Power Distributions in Narrowband Models

Fluctuations in the envelope of the received signal are study of the fluctuations of the power of the received signal, in other words the study of fading effects. The received signal for narrowband channel model can be represented in either Cartesian or polar coordinates as follows:

$$r(t) = r_I(t)\cos(\omega_c t) - r_Q(t)\sin(\omega_c t) = \rho(t)\cos[\omega_c t + \psi(t)], \tag{2.521}$$

where the envelope and phase in polar coordinates are related to the in-phase and quadrature components by:

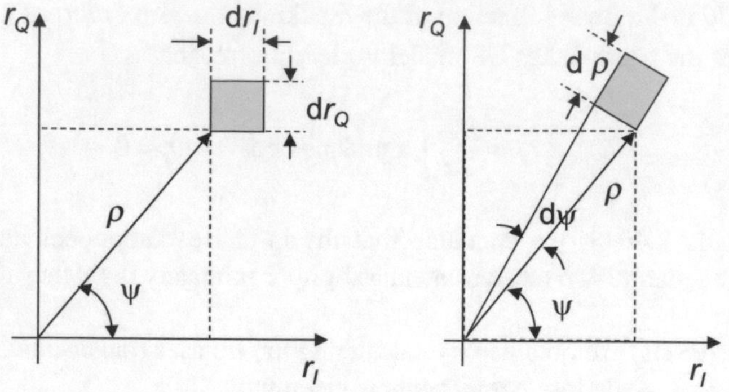

Fig. 2.43 Transformation of in-phase and quadrature components into envelope and phase coordinates

$$\rho(t) = |r(t)| = \sqrt{r_I^2(t) + r_Q^2(t)}, \qquad \psi(t) = \tan^{-1}\left[\frac{r_Q(t)}{r_I(t)}\right]. \tag{2.522}$$

Since the samples in-phase and quadrature components are independent and Gaussian distributed, the corresponding joint distribution will be:

$$f_{R_I,R_Q}(r_I, r_Q) = \frac{1}{2\pi\sigma^2}\exp\left[-\frac{r_I^2 + r_Q^2}{2\sigma^2}\right]. \tag{2.523}$$

Now we can use the method of transformation of random variables, illustrated in Fig. 2.43, to determine the joint distribution of envelope and phase. After the following substitutions $r_I = \rho\cos\psi$, $r_Q = \rho\sin\psi$, the joint distribution of envelope and phase is given as:

$$f_{R,\Psi}(\rho,\psi) = f_{R_I,R_Q}(r_I, r_Q)|J(\rho,\psi)|, \tag{2.524}$$

where the Jacobian is determined by:

$$|J(\rho,\psi)| = \begin{vmatrix} \dfrac{\partial r_I}{\partial \rho} & \dfrac{\partial r_I}{\partial \psi} \\ \dfrac{\partial r_Q}{\partial \rho} & \dfrac{\partial r_Q}{\partial \psi} \end{vmatrix} = \rho$$

which after substitution in (2.524), it yields to:

$$f_{R,\Psi}(\rho,\psi) = \frac{\rho}{2\pi\sigma^2}\exp\left[-\frac{\rho^2}{2\sigma^2}\right] = \underbrace{\frac{1}{2\pi}}_{f_\Psi(\psi)} \cdot \underbrace{\frac{\rho}{\sigma^2}\exp\left[-\frac{\rho^2}{2\sigma^2}\right]}_{p_R(\rho)=\frac{\rho}{\sigma^2}\exp\left(-\frac{\rho^2}{2\sigma^2}\right)}. \tag{2.524}$$

Therefore, the distribution of phase is uniform:

$$f_\Psi(\psi) = \int_0^\infty f_{R,\Psi}(\rho,\psi)d\rho = \frac{1}{2\pi}, \quad 0 \le \psi \le 2\pi, \tag{2.525}$$

while the distribution of envelope is *Rayleigh*:

$$f_R(\rho) = \int_0^{2\pi} p_{R,\Psi}(\rho,\psi)d\psi = \frac{\rho}{\sigma^2}\exp\left(-\frac{\rho^2}{2\sigma^2}\right). \tag{2.526}$$

Since there is no dominant LOS component, the total power in multipath components is given by $\overline{P}_{Rx} = \sum_i \langle a_i^2 \rangle = 2\sigma^2$, and we can represent the Rayleigh distribution as follows:

$$p_R(\rho) = \frac{2\rho}{\overline{P}_{Rx}} \exp\left(-\frac{\rho^2}{\overline{P}_{Rx}}\right), \rho \geq 0. \tag{2.527}$$

The corresponding power distribution is exponential, obtained after the substitution $p = \rho^2(t) = |r(t)|^2$, as given below:

$$p_{R^2}(x) = \frac{1}{2\sigma^2} \exp\left(-\frac{p}{2\sigma^2}\right) = \frac{1}{\overline{P}_{Rx}} \exp\left(-\frac{p}{\overline{P}_{Rx}}\right), p \geq 0. \tag{2.528}$$

In the presence of a strong LOS component, denoted as $s\cos\omega_c t$, the received signal can be represented as:

$$r(t) = s\cos(\omega_c t) + r_I(t)\cos(\omega_c t) - r_Q(t)\sin(\omega_c t) = [s + r_I(t)]\cos(\omega_c t) - r_Q(t)\sin(\omega_c t). \tag{2.529}$$

The corresponding envelope and phase are given by:

$$\rho(t) = |r(t)| = \sqrt{[s + r_I(t)]^2 + r_Q^2(t)}, \quad \psi(t) = \tan^{-1}\left[\frac{r_Q(t)}{s + r_I(t)}\right]. \tag{2.530}$$

By following the similar procedure of transformation of random variable, we found that the distribution of the phase is again uniform, while the distribution of envelope is *Rician*:

$$f_R(\rho) = \frac{\rho}{\sigma^2} \exp\left(-\frac{\rho^2 + s^2}{2\sigma^2}\right) I_0\left(\frac{\rho s}{\sigma^2}\right), I_0(x) = \frac{1}{2\pi} \int_0^{2\pi} \exp(x\cos\psi)d\psi, \tag{2.531}$$

where $I_0(x)$ is the modified Bessel function of the first kind and zero order. As an illustration, in Fig. 2.44 we show normalized Rician distribution for different values of parameter $a = \rho/\sigma$. Clearly, for $a = 0$, Rician distribution reduces down to Rayleigh. When parameter a is large, the distribution is similar to Gaussian.

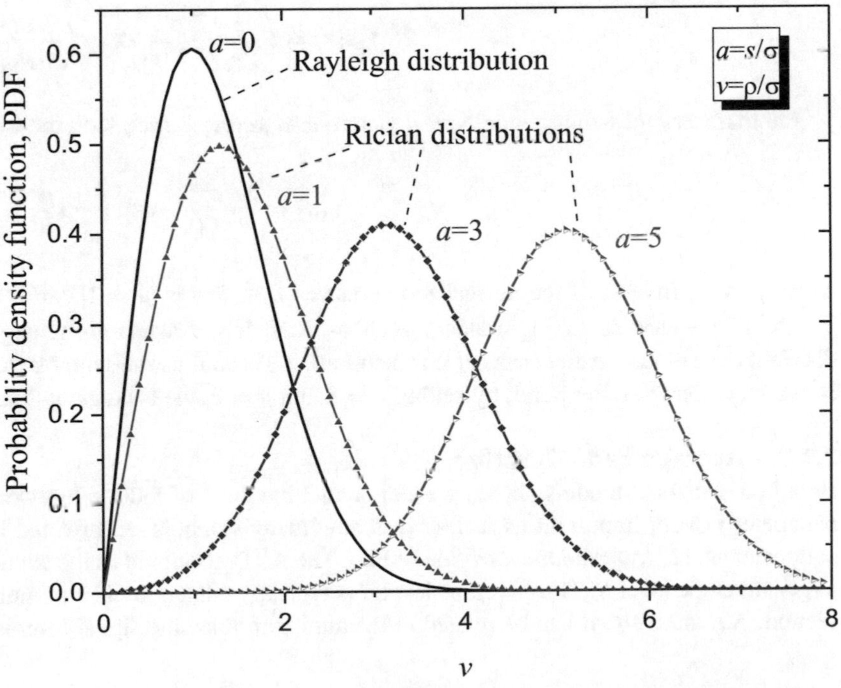

Fig. 2.44 Normalized Rayleigh and Rician distributions

The total power in multipath component different from the LOS component is $2\sigma^2 = \sum_{i,i\neq 0}\langle a_i^2\rangle$, while the total power is given by:

$$\overline{P}_{Rx} = \int_0^{\infty} \rho^2 f_R(\rho) d\rho = s^2 + 2\sigma^2. \tag{2.532}$$

The Rician distribution can also be expressed in terms of *Rician parameter K*, defined by $K = s^2/(2\sigma^2)$, as follows:

$$f_R(\rho) = \frac{2\rho(K+1)}{\overline{P}_{Rx}} \exp\left(-K - \frac{(K+1)\rho^2}{\overline{P}_{Rx}}\right) I_0\left(2\rho\sqrt{\frac{K(K+1)}{\overline{P}_{Rx}}}\right), \rho \geq 0. \tag{2.533}$$

The Rician parameter can be used as a measure of the fading severity: the small K implies severe fading, a large K implies mild fading, and when K tends to infinity, there is no fading at all.

More general fading distribution is the *Nakagami distribution*, given by??:

$$f_R(\rho) = \frac{2m^m \rho^{2m-1}}{\Gamma(m)\overline{P}_{Rx}^m} \exp\left(-\frac{m\rho^2}{\overline{P}_{Rx}}\right), m \geq 0.5 \tag{2.534}$$

The parameter m in Nakagami distribution represents the fading figure, defined as the receive power-to-variance of z^2 ratio. For integer values of m, $f_R(\rho)$ is the PDF of the following random variable $Y = \left(\sum_{i=1}^m X_i^2\right)^{1/2}$, where X_i are independent Rayleigh-distributed random variables. The special cases of Nakagami fading are:

- $m = 1$: Rayleigh fading,
- $m = (K + 1)^2/(2\,K + 1)$: Rician fading,
- $m = 1/2$: single-sided Gaussian distribution,
- $m\text{-} > \infty$: No fading, the constant fading.

The power distribution for Nakagami fading:

$$f_{R^2}(x) = \left(\frac{m}{\overline{P}_{Rx}}\right)^m \frac{p^{m-1}}{\Gamma(m)} \exp\left(-\frac{mp}{\overline{P}_{Rx}}\right). \tag{2.535}$$

Another relevant fading model is *α-μ generalized fading model*, with the following distribution [88–91]:

$$f_R(\rho) = \frac{\alpha\mu^\mu \rho^{\alpha\mu-1}}{\widehat{\rho}^{\alpha\mu}\Gamma(\mu)} \exp\left(-\mu\frac{\rho^\alpha}{\widehat{\rho}^\alpha}\right), \tag{2.536}$$

where μ is the inverse of the normalized variance of ρ^α, that is, $\mu = 1/\text{Var}(\rho^\alpha)/\langle\rho^\alpha\rangle^2$, and $\widehat{\rho}$ is the α-root mean value of ρ^α: $\widehat{\rho} = \langle\rho^\alpha\rangle^{1/\alpha}$. Various fading models such as Rayleigh, Nakagami-m, exponential, Weibull, and one-sided Gaussian distributions are just spatial cases of this distribution. As an illustration, by setting $\alpha = 2$ and $\mu = 1$, the Rayleigh distribution is obtained. On the other hand, by setting $\alpha = 2$ and $\mu = 2$, the Nakagami-2 distribution is obtained.

2.3.2.5 Average Fade Duration

Based on statistical models above, we can predict the level of fading, but we cannot predict the duration of the fade. This parameter is very important as it describes how many symbols get affected by the deep fade effect, and we refer to this parameter as the *average fade duration* (AFD). The AVD is defined as the average time for which the signal is below a given target envelope level R. The envelope level R is typically related to the minimum tolerable power P_{min}, discussed in previous section. Alternatively, it can be related to the minimum tolerable signal-to-noise ratio. Let T_{obs} denote the observation time interval. The time intervals for which the signal is below target envelop value R are denoted as t_i ($i = 1,2,..$). The probability that the envelope is below target value R can be determined as:

$$\Pr(\rho(t) < R) = \frac{1}{T_{\text{obs}}} \sum_i t_i. \tag{2.537}$$

In order to calculate the AFD, we need first to determine the level R crossing rate, denoted as L_R [number of crossings/s]. The *level crossing rate* is defined as the expected rate at which the signal envelope crosses the envelope level R with negative slope. The AFD can now be defined mathematically as:

$$\bar{t}_R = \frac{\frac{1}{T_{\text{obs}}} \sum_{i=1}^{L_R T_{\text{obs}}} t_i}{L_R} = \frac{\Pr(\rho(t) < R)}{L_R}. \tag{2.538}$$

In order to determine the level crossing rate, the first step is to determine the joint distribution of envelope ρ and its derivative dρ/dt, denoted as $f(\rho, \dot{\rho}), \dot{\rho} = d\rho/dt$. For details of derivation of this joint distribution, an interested reader is referred to [92–94]. The level crossing rate can now be determined as the expected value of the envelope slope, where averaging is performed over negative slopes:

$$L_R = \int_{-\infty}^{0} \dot{\rho} f(\rho, \dot{\rho}) d\dot{\rho}. \tag{2.539}$$

Number of crossing in observed interval T_{obs} is simply $N_R = L_R T_{\text{obs}}$. Under US assumption, the envelope and envelope slope are independent, and the distribution of envelope slope is zero-mean Gaussian:

$$f(\dot{\rho}) = \frac{1}{\sigma_{\dot{\rho}} \sqrt{2\pi}} e^{-\frac{\dot{\rho}^2}{2\sigma_{\dot{\rho}}^2}}, \tag{2.540}$$

where the variance of the slope is given by [93] $\sigma_{\dot{\rho}}^2 = (2\pi\nu_D)^2 \sigma^2/2$. The level crossing rate for Rician distribution would be then:

$$\begin{aligned} L_R &= \int_{-\infty}^{0} \dot{\rho} f(\dot{\rho}) f(\rho) d\dot{\rho} \\ &= \frac{1}{2} \underbrace{\int_{-\infty}^{0} \frac{1}{\sigma_{\dot{\rho}} \sqrt{2\pi}} e^{-\frac{\dot{\rho}^2}{2\sigma_{\dot{\rho}}^2}} d\dot{\rho}^2}_{\sigma_{\dot{\rho}}/\sqrt{2\pi}} \frac{2\rho(K+1)}{\overline{P}_{Rx}} e^{-K - \frac{(K+1)\rho^2}{\overline{P}_{Rx}}} I_0\left(2\rho \sqrt{\frac{K(K+1)}{\overline{P}_{Rx}}}\right) \\ &= \sqrt{2\pi(K+1)} \nu_D \frac{\rho}{\sqrt{\overline{P}_{Rx}}} \exp\left(-K - \frac{(K+1)\rho^2}{\overline{P}_{Rx}}\right) I_0\left(2\rho \sqrt{\frac{K(K+1)}{\overline{P}_{Rx}}}\right), \end{aligned} \tag{2.541}$$

which after substitution $\rho_n = \rho/\sqrt{\overline{P}_{Rx}}$ can be rewritten in compact form as follows:

$$L_R = \sqrt{2\pi(K+1)} \nu_D \rho_n \exp\left(-K - (K+1)\rho_n^2\right) I_0\left(2\rho_n \sqrt{K(K+1)}\right). \tag{2.542}$$

For Rayleigh distribution, $K = 0$, and the level crossing rate is simply:

$$L_R = \sqrt{2\pi} \nu_D \rho_n \exp\left(-\rho_n^2\right). \tag{2.543}$$

The cumulative distribution function (CDF) for Rayleigh fading is given by:

$$\mathrm{CDF}(\rho) = \Pr(\rho < R) = 1 - e^{-\rho^2/\overline{P}_{Rx}} = 1 - e^{-\rho_n^2}, \tag{2.544}$$

so that the average fade duration becomes:

$$\bar{t}_R^{(\mathrm{Rayleigh})} = \frac{\Pr(\rho(t) < R)}{L_R} = \frac{1 - \exp\left(-\rho_n^2\right)}{\sqrt{2\pi}\nu_D \rho_n \exp\left(-\rho_n^2\right)} = \frac{\exp\left(\rho_n^2\right) - 1}{\sqrt{2\pi}\nu_D \rho_n}. \tag{2.545}$$

On the other hand, the CDF for Rician fading is given by:

$$\mathrm{CDF}(\rho) = \Pr(\rho < R) = \int_0^R f_R^{(\mathrm{Rician})}(\rho) d\rho = 1 - Q\left(\sqrt{2K}, \sqrt{2(K+1)\rho_n^2}\right), \tag{2.546}$$

where $Q(a,b)$ is the Marcum Q-function, defined as:

$$Q(a,b) = 1 - \int_0^b z \exp\left(-\left(z^2 + a^2\right)/2\right) I_0(az) dz. \tag{2.547}$$

After substitution of (2.546) into (2.538), we obtain the following expression for AFD of Rician fading:

$$\bar{t}_R^{(\mathrm{Rician})} = \frac{\Pr(\rho(t) < R)}{L_R} = \frac{1 - Q\left(\sqrt{2K}, \sqrt{2(K+1)\rho_n^2}\right)}{\sqrt{2\pi(K+1)}\nu_D \rho_n e^{-K-(K+1)\rho_n^2} I_0\left(2\rho_n\sqrt{K(K+1)}\right)}. \tag{2.548}$$

When AFD is much longer than a bit duration, a long burst of errors will be introduced, indicating that fading is slow. Therefore, AFD can be used to determine the number of bits affected by a deep fade. Error correction codes developed for AWGN channels will most probably fail as they need to deal with both random and burst errors. To solve for this problem, interleaving is typically used so that the burst errors get spread to multiple codewords. When AFD is roughly equal to a bit duration, a single bit error gets introduced by fading. In this case, a regular error correction code, designed for random errors, is applicable. Finally, when AFD is much shorter than a bit duration, the fading is averaged out, which indicates that in this case fading is fast and the Doppler effect may be beneficial in this case.

2.3.2.6 Discrete-Time (DT) Model

When all multipath components are resolvable, the *discrete-time (DT) model due to Turin* can be used [95]. The N multipath components are resolvable if for $m \neq n$, the following is $|\tau_n(t) - \tau_m(t)| >> 1/B$, where B is the transmission bandwidth. This model is particularly suitable for spread-spectrum systems. The continuous-time output of such a channel is given by:

$$\underline{x}(t) = \sum_{i=0}^{N} \alpha_i(t) e^{-j\phi_i(t)} \underline{s}(t - \tau_i). \tag{2.549}$$

By sampling at $t = nT_s$ (T_s is the sampling interval), we obtain:

$$\underline{x}(nT_s) = \sum_{i=0}^{N} \underbrace{\alpha_i(nT_s) e^{-j\phi_i(nT_s)}}_{\alpha_i(n)} \underbrace{\underline{s}(nT_s - iT_s)}_{\underline{s}(n-i)} = \sum_{i=0}^{N} \alpha_i(n) e^{-j\phi_i(n)} \underline{s}(n - i). \tag{2.550}$$

Based on previous expression, we conclude that DT impulse response is given by:

$$\underline{h}(n) = \sum_{i=0}^{N} \underline{h}(i,n)\delta(n-i), \quad \underline{h}(i,n) = \alpha_i(n) e^{-j\phi_i(n)}. \tag{2.551}$$

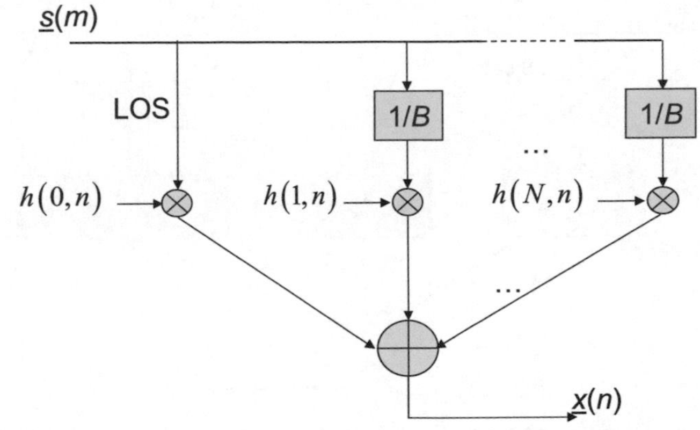

Fig. 2.45 Illustration of the discrete-time model

The corresponding DT model is illustrated in Fig. 2.45. The DT model can be further simplified as follows [95]. The channel use is discretized into MT intervals of duration T each, such that MT is longer than average delay spread. The binary indicator function is then used in each interval of duration T. Whenever the binary indicator function output is 1, the random amplitude and phase for a given multipath component are generated from an empirical distribution. This empirical distribution may belong to the same family of fading distributions but with different parameters.

2.3.2.7 Finite-State Markov Channel (FSMC) Model

Another relevant channel model is the *finite-state Markov channel (FSMC) model* due to Wang and Moayeri [96], in which the fading is approximated as a discrete-time Markov process with a time being discretized by a symbol duration. The FSMC model divides the range of all possible fading powers p into discrete regions as follows $RG_i = \{p| A_i \leq p < A_{i+1}\}$ ($i = 1,2...$), where A_i and A_{i+1} are boundaries of the i-th region. The stationary probabilities of the regions are determined by:

$$\pi_i = \Pr(p \in RG_i) = \Pr(A_i \leq p \leq A_{i+1}), \tag{2.552}$$

and can be straightforwardly derived from the cumulative distribution function of the power for each particular fading model. The transition probabilities between regions are determined by [96]:

$$p_{i,i+1} = \frac{L_{i+1}T}{\pi_i}, \quad p_{i,i-1} = \frac{L_i T}{\pi_i}, \quad p_{i,i} = 1 - p_{i,i+1} - p_{i,i-1}, \tag{2.553}$$

where L_i is the level A_i crossing rate and T is the symbol duration. Clearly, according to the FSMC model, from the Region RG_i, we can stay in the same region with probability $p_{i,i}$, move to one region below with probability $p_{i,i-1}$, or move one region above the observed region $p_{i,i+1}$. Since we can move only to neighboring regions, this model is accurate for sufficiently small normalized Doppler frequency $\nu_D T$.

2.3.2.8 Wideband Channel Models

In *wideband channel models*, the components in a given cluster are resolvable. Let τ_{MDS} denote the multipath delay spread (MDS), defined as $\tau_{MDS} = \tau_{max} - \tau_{min}$ (τ_{max} is the maximum delay spread and τ_{min} is the minimum delay spread). If B_s is transmitted bandwidth, the maximum number of resolvable components is given by $\tau_{MDS}B_s$. Now since the delay between any two components in a cluster is much longer than reciprocal of the signal bandwidth, then different multipath components from a pulse transmitted in a given time slot arrive in neighboring time slots introducing the intersymbol interference. In other words, the pulse of duration T results in a received pulse of duration $T + \tau_{MDS}$. Since the components with each cluster are resolvable, the received signal can be represented as follows:

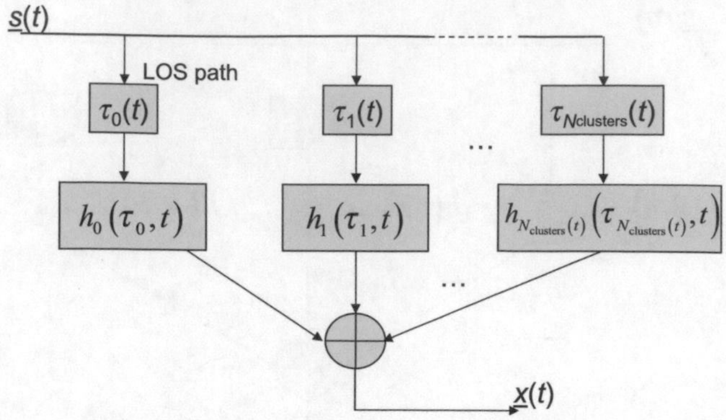

Fig. 2.46 The wideband channel model in which the impulse response of the *i*-th cluster is denoted as $h_i(\tau_i,t)$

$$\begin{aligned} r(t) = \mathrm{Re}\Bigg\{\Bigg[& \underbrace{s(t-\tau_0(t))\sum_{i\in C_0} a_i(t)e^{-j\phi_i(t)}}_{h_0(\tau_0,t)} + \underbrace{s(t-\tau_1(t))\sum_{i\in C_1} a_i(t)e^{-j\phi_i(t)}}_{h_1(\tau_1,t)} \\ & +\ldots + \underbrace{s(t-\tau_{N_{clussters}}(t))\sum_{i\in C_{N_{\text{clusters}}}} a_i(t)e^{-j\phi_i(t)}}_{h_{N_{\text{clusters}}}(\tau_{N_{\text{clusters}}},t)}\Bigg] e^{\,j2\pi f_c t}\Bigg\}, \end{aligned} \tag{2.554}$$

where we use C_i to denote the *i*–th cluster ($i = 0,1,\ldots,N_{\text{clusters}}$), while $h_i(\tau_i,t)$ is the impulse response of the *i*-th cluster. The overall impulse response of the wideband channel, based on previous equation, is given by:

$$h(\tau,t) = \sum_{i=0}^{N_{\text{clusters}}(t)} h(\tau_i,t)\delta(\tau-\tau_i), \tag{2.555}$$

and the corresponding *wideband channel model* is shown in Fig. 2.46.

Based on wideband channel model, we conclude that there are two time variables, the deal spread τ and time instance t. Corresponding variables in frequency domain are frequency f and Doppler spread ν, respectively. Since the impulse response is time-varying, it makes sense to define the *autocorrelation function of the impulse response* as follows:

$$R_h(\tau_1,\tau_2;t,t+\Delta t) = \langle h^*(\tau_1,t)h(\tau_2,t+\Delta t)\rangle. \tag{2.556}$$

Under the wide-sense stationary (WSS) assumption, the autocorrelation functions is not a function of the time origin, and we can write:

$$R_h(\tau_1,\tau_2;\Delta t) = \langle h^*(\tau_1,t)h(\tau_2,t+\Delta t)\rangle. \tag{2.557}$$

When the WSS assumption is extended by uniform scattering (US), we obtain the following simplified model for the autocorrelation function, introduced by Bello [97]:

$$\langle h^*(\tau_1,t)h(\tau_2,t+\Delta t)\rangle = R_h(\tau_1,\Delta t)\delta(\tau_1-\tau_2) \stackrel{\Delta}{=} R_h(\tau,\Delta t), \tag{2.558}$$

which indicates that the channel responses associated with different multipath components are mutually uncorrelated.

2.3.2.9 Power Delay Profile and Coherence Bandwidth

The WSS + US autocorrelation function of impulse response for which $\Delta t = 0$ is commonly referred to as the *power delay profile (PDP)*, denoted as $R_h(\tau)$, and gives the information how the total received power is distributed among different multipath components. The PDP can be used to determine the average delay spread:

$$\mu_{\tau_{\text{MDS}}} = \frac{\int_0^\infty \tau R_h(\tau)d\tau}{\int_0^\infty R_h(\tau)d\tau}, \tag{2.559}$$

and root-mean-squared (rms) delays spread:

$$\sigma_{\tau_{\text{MDS}}}(\tau) = \sqrt{\frac{\int_0^\infty \left(\tau - \mu_{T_m}\right)^2 R_h(\tau)d\tau}{\int_0^\infty R_h(\tau)d\tau}}. \tag{2.560}$$

The Fourier transform of $h(\tau,t)$ with respect to the delay spread τ, at particular time instance t:

$$H(f;t) = \int_{-\infty}^{\infty} h(\tau;t)e^{-j2\pi f\tau}d\tau, \tag{2.561}$$

is also time-varying, so that it makes sense to define the autocorrelation function of time-varying transfer function $H(f,t)$ as follows:

$$\begin{aligned} R_H(f_1, f_2; \Delta t) &= \langle H^*(f_1, t)H(f_2, t + \Delta t)\rangle \\ &= \left\langle \int_{-\infty}^{\infty} h^*(\tau_1;t)e^{j2\pi f_1\tau_1}d\tau_1 \int_{-\infty}^{\infty} h(\tau_2;t+\Delta t)e^{-j2\pi f_2\tau_2}d\tau_2 \right\rangle \\ &= \int_{-\infty}^{\infty}\int_{-\infty}^{\infty} \langle h^*(\tau_1;t)h(\tau_2;t+\Delta t)\rangle e^{j2\pi f_1\tau_1}e^{-j2\pi f_2\tau_2}d\tau_1 d\tau_2, \end{aligned} \tag{2.562}$$

which under WSS + US assumption simplifies to:

$$R_H(f_1, f_2; \Delta t) = \int_{-\infty}^{\infty} R_h(\tau, \Delta t)e^{-j2\pi(f_2 - f_1)\tau}d\tau = R_H(\Delta f; \Delta t), \tag{2.562}$$

and becomes the function on frequency difference Δf. For the fixed observation time ($\Delta t = 0$), we conclude that the PDP and autocorrelation function of transfer function represent a Fourier transform pairs:

$$R_H(\Delta f) = \int_{-\infty}^{\infty} R_h(\tau)e^{-j2\pi\Delta f\tau}d\tau. \tag{2.563}$$

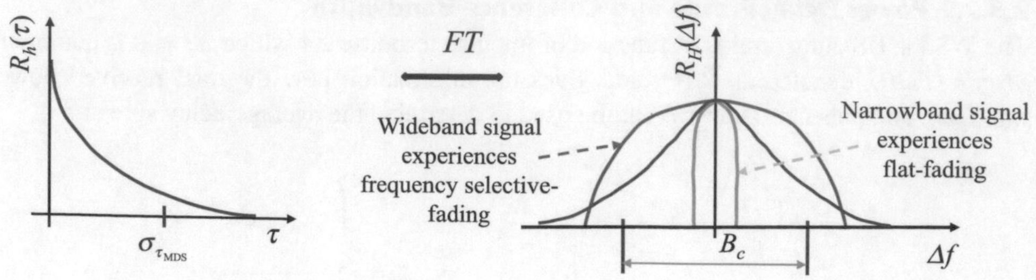

Fig. 2.47 The relationship between DPD and autocorrelation function of frequency response $H(f,t)$ (for $\Delta t = 0$) and the definition of the coherence bandwidth B_c

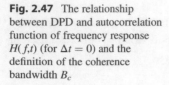

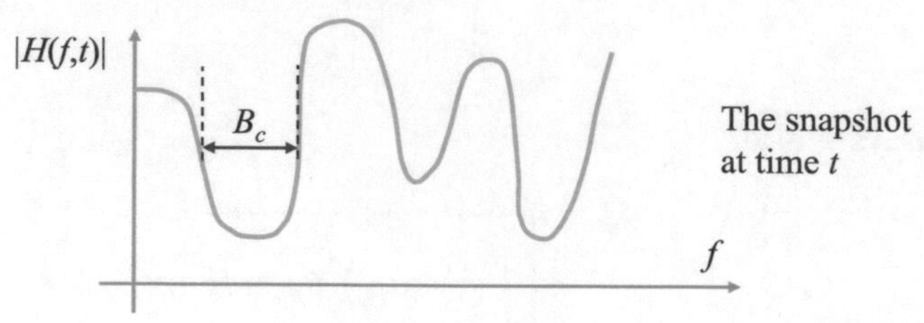

Fig. 2.48 The illustration of frequency-selective fading for one snapshot of the frequency response

The range of frequencies for which $R_H(\Delta f)$ is nonzero is known as the *coherence bandwidth* B_c. The definition of coherence bandwidth and its relationship to the PDP and rms delay spread is illustrated in Fig. 2.47. An example of the narrowband signal as well as an example of wideband signal is shown to illustrate how the fading affects the signal. Clearly, for the narrowband signal, in this particular example, different frequency components experience roughly the same fading amplitude. On the other hand, for the wideband signal, different frequencies experience different fading and refer to this type of fading as *frequency-selective fading*. Therefore, the coherence bandwidth can be used to classify the fading into two categories: flat-fading and frequency-selective fading. When the bandwidth of the signal is much smaller than coherence bandwidth, $B_s << B_c$, the fading is *flat-fading*. In other words, for the flat-fading, the multipath delay spread is much smaller than the symbol duration. On the other hand, in *frequency selective fading*, as illustrated in Fig. 2.48, the bandwidth of the signal is much larger than the coherence bandwidth, $B_s >> B_c$. In other words, in frequency-selective fading, the MDS is much longer than the symbol duration.

2.3.2.10 Channel Coherence Time and Fading Channels Classification

The time variations of the channel arising from transmitter and/or receiver motion introduce a Doppler shift in the received signal. This Doppler effect can be characterized by taking the Fourier transform of $R_H(\Delta f;\Delta t)$ with respect to Δt:

$$S_H(\Delta f, \nu) = \int_{-\infty}^{\infty} R_H(\Delta f, \Delta t) e^{-j2\pi\nu\Delta t} d\Delta t. \tag{2.564}$$

In order to characterize Doppler effect at a single Doppler frequency, we have to set $\Delta f = 0$ and define the *Doppler spectrum* $S_H(\nu) = S_H(0;\, \nu)$ as follows:

$$S_H(\nu) = \int_{-\infty}^{\infty} R_H(\Delta t) e^{-j2\pi\nu\Delta t} d\Delta t. \tag{2.565}$$

The range of time intervals over which the $R_H(\Delta t)$ is approximately nonzero is commonly referred to as the *coherence time* T_c, which is illustrated in Fig. 2.49. The time-varying channel decorrelates after T_c seconds. The coherence time can be used to determine how many symbols get affected by the fading. The maximum ν value for which the Doppler spectrum magnitude is larger than zero, $|S_H(\nu)| > 0$, is known as the *Doppler spread* of the channel, denoted as B_D. Based on coherence time, we can classify fading channels as *slow-fading* and *fast-fading* channels. For *slow-fading channels*, the symbol duration T_s is much shorter than the coherence time, $T_s << T_c$. In other words, in slow-fading channels, the signal bandwidth is much larger than

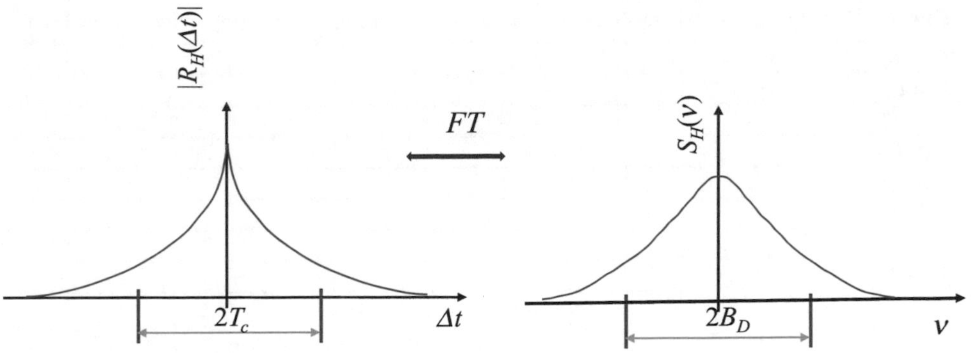

Fig. 2.49 The illustration of definitions of coherence time T_c and Doppler spread B_D

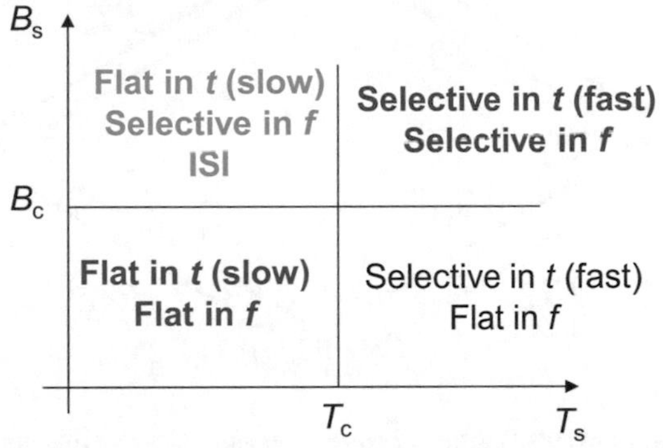

Fig. 2.50 Classification of wireless channels based on coherence bandwidth B_c and coherence time T_c

the Doppler spread, $B_s >> B_D$. On the other hand, in *fast-fading channels*, the symbol duration T_s is much longer than the coherence time, $T_s >> T_c$. In other words, in slow-fading channels, the signal bandwidth is much smaller than the Doppler spread, $B_s << B_D$. The classification of wireless channels based on coherence bandwidth and coherence time is summarized in Fig. 2.50.

When the symbol duration is much longer than coherence time, we also say that the channel is *ergodic*, as all possible channel conditions (states) can take place. On the other hand, when $T_s << T_c$ we say that the channel is *nonergodic*.

Finally, the factor $T_c B_c \approx 1/\tau_{\mathrm{MDS}} B_{\mathrm{D}}$ is known as the *spread factor* [67]. When $\tau_{\mathrm{MDS}} B_{\mathrm{D}} < 1$ (equivalently $B_c T_c > 1$), we say that the channel is *underspread*, and such channel can easily be estimated. On the other hand, when $\tau_{\mathrm{MDS}} B_{\mathrm{D}} > 1$ (equivalently $B_c T_c < 1$), we say that the channel is *overspread*.

The classification of wireless channels based on coherence bandwidth, coherence time, and spread factor is summarized in Table 2.1.

Another relevant function in characterizing the wireless channels is the *scattering function* $S_h(\tau; \nu)$, which is related to the autocorrelation function $R_h(\tau;\Delta t)$ by the Fourier transform:

$$S_h(\tau, \nu) = \int_{-\infty}^{\infty} R_h(\tau, \Delta t) e^{-j2\pi\nu\Delta t} d\Delta t. \tag{2.566}$$

The scattering function describes how the average output power is distributed among multipath delay τ and Doppler frequency ν. The scattering function $S_h(\tau; \nu)$ and the autocorrelation function $R_H(\Delta f;\Delta t)$ are related by:

$$S_H(\tau, \nu) = \int_{-\infty}^{\infty} \int_{-\infty}^{\infty} R_H(\Delta f, \Delta t) e^{-j2\pi\nu\Delta t} e^{\,j2\pi\tau\Delta f} d\Delta t d\Delta f. \tag{2.567}$$

Different correlation functions and scattering functions are mutually related as summarized in Fig. 2.51.

Table 2.1 Classification of wireless channels based on coherence bandwidth, coherence time, and spread factor.

$B_s \ll B_c$	**Frequency-flat fading**
$B_s \geq B_c$	**Frequency-selective channel**
$T_s \ll T_c$	**Time-flat (slow) fading**
$T_s \geq T_c$	**Time-selective (fast) fading**
$B_c T_c > 1$	Underspread channel
$B_c T_c \ll 1$	Overspread channel
$T_s \ll T_c$	**Nonergodic channel**
$T_s \gg T_c$	**Ergodic channel**

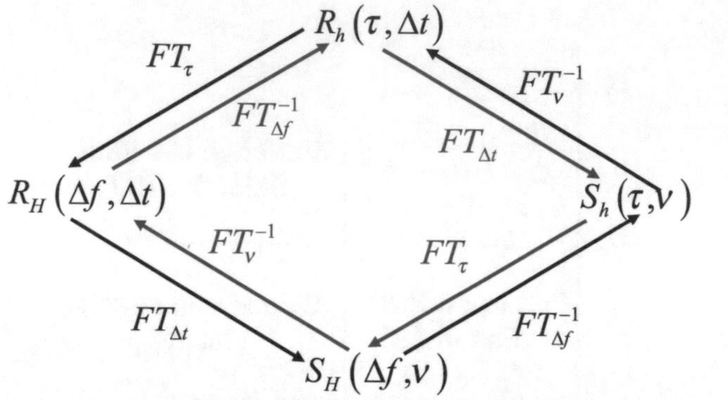

Fig. 2.51 The Fourier transform (FT) relationship among different correlation and scattering functions

2.4 Signal Propagation in Optical Fibers and Corresponding Channel Impairments

In this section, we describe basic signal and noise parameters in fiber-optics communications and major channel impairments including chromatic dispersion, multimode dispersion, polarization mode dispersion (PMD), and fiber nonlinearities [1–4, 98–104].

2.4.1 Fiber Attenuation and Insertion Losses

Fiber attenuation can be described by the general relation $dP/dz = -\alpha P$, where α is the *power attenuation coefficient per unit length*. If P_{in} is the power launched into the fiber, the power remaining after propagating a length L within the fiber P_{out} is $P_{out} = P_{in}\exp(-\alpha L)$. The absorption coefficient varies with wavelength as many of the absorption and scattering mechanisms vary with λ. For instance, *Rayleigh scattering* in fiber is due to microscopic variations in the density of glass (density fluctuation sites < λ) and varies as [1–4]:

$$\alpha_R = C/\lambda^4, \quad C = 0.7 - 0.9 \quad (\text{dB/km})\text{-}\mu\text{m}^4 \tag{2.658}$$

Mie scattering is caused by imperfections (scattering) in the core-cladding interface that are larger than λ.

Intrinsic absorption can be identified as (i) *infrared absorption* (in SiO_2 glass, vibrational modes of Si-O bonds cause an absorption peak at $\lambda > 7$ μm, which has a tail extending to the $\lambda = 1.55$ μm range) and (ii) *ultraviolet absorption* (which is due to the electronic resonances occurring in ultraviolet region ($\lambda < 0.4$ μm)).

Extrinsic absorption results from the presence of impurities: (i) *transition metal impurities* (Fe, Cu, Co, Ni, Mn, and Cr) absorb strongly in 0.6–1.6 μm, and the loss level is reduced below 1 dB/km if their amount is kept below 1 part per billion; and (ii) *residual OH-* ions cause peaks near $\lambda = 0.95$ μm, 1.24 μm, and 1.39 μm

Fiber loss is not the only source of optical signal attenuation along the transmission lines. The *fiber splices* and *fiber connectors* also cause the signal attenuation. The fiber splices can be fused or joined together by some mechanical means, with typical attenuation being 0.01–0.1 dB per fused splice and slightly above 0.1 dB per mechanical splice.

Optical connectors are removable and allow many repeated connections and disconnections, with typical insertion loss for high-quality SMF not above 0.25 dB. To minimize the *connector return loss* (a fraction of the optical power reflected back into the fiber at the connector point), the angled fiber-end surfaces are commonly used.

The number of optical splices and connectors depends on transmission length and must be taken into account, unless the total attenuation due to fiber joints is distributed and added to the optical fiber attenuation.

2.4.2 Chromatic Dispersion Effects

Dispersion problem can be described as follows. Short optical pulses entering a dispersive channel such as an optical fiber get spread out into a much broader temporal distribution. Both the *intermodal dispersion* and *chromatic dispersion* cause the distortion in multimode optical fibers, while chromatic dispersion is the only cause of the signal distortion in SMF. The intermodal dispersion can be specified through the *optical fiber bandwidth* B_{fib} that is related to a 1-km-long optical fiber, and this parameter is specified by the manufactures and commonly measured at wavelength around 1310 nm (chromatic dispersion is negligible in this region compared to intermodal dispersion) [1, 4, 98]:

$$B_{fib,L} = \frac{B_{fib}}{L^{\mu}}, \quad \mu = 0.5 - 1 \tag{2.569}$$

Single-mode optical fibers do, however, introduce another signal impairment known as the chromatic dispersion. *Chromatic dispersion* is caused by difference in velocities among different spectral components within the same mode and has two components: (i) *material dispersion* and (ii) *waveguide dispersion* component. The material dispersion is caused by the fact that the refractive index is a function of wavelength, defined by Sellmeier equation [2, 4, 98]:

$$n(\lambda) = \left[1 + \sum_{i=1}^{M} \frac{B_i \lambda^2}{\lambda^2 - \lambda_i^2}\right], \tag{2.750}$$

with typical B_i and λ_i parameters for pure silica being:

$$\begin{aligned} B_1 &= 0.6961663 \ \text{ at } \lambda_1 = 0.0684043 \ \mu\text{m} \\ B_2 &= 0.4079426 \ \text{ at } \lambda_2 = 0.1162414 \ \mu\text{m} \\ B_3 &= 0.8974794 \ \text{ at } \lambda_3 = 9.896161 \ \mu\text{m} \end{aligned}$$

The group index n_g can be determined by using the Sellmeier equation and the following definition expression:

$$n_g = n + \omega dn/d\omega = n - \lambda dn/d\lambda.$$

Waveguide dispersion is related to the physical design of the optical fiber. Since the value of Δ is typically small, the refractive indices of the core cladding are nearly equal, and the light is not strictly confined in the fiber core, and the fiber modes are said to be *weakly guided*. For a given mode, say fundamental mode, the portion of light energy that propagates in the core depends on wavelength, giving rise to the pulse-spreading phenomenon known as waveguide dispersion. By changing the power distribution across the cross-sectional area, the overall picture related to the chromatic dispersion can be changed. The distribution of the mode power and the total value of waveguide dispersion can be manipulated by *multiple cladding layers*. This is commonly done in optical fibers for special application purposes, such as *dispersion compensation*. Different refractive index profiles of single-mode fibers used today are shown in Fig. 2.52. The conventional SMF index profile is shown in Fig. 2.52(a). The nonzero dispersion-shifted fiber (NZDSF), with refractive index profile shown in Fig. 2.52(b), is suitable for use in WDM systems. Large effective area NZDSFs, with index profile shown in Fig. 2.52(c), are suitable to reduce the effect of fiber nonlinearities. Dispersion compensating fiber (DCF), with index profile shown in Fig. 2.52(d), is suitable to compensate for positive chromatic dispersion accumulated along the transmission line.

Because the source is non-monochromatic, different spectral components within a pulse will travel at different velocities, inducing the pulse broadening. When the neighboring pulses cross their allocated time slots, the ISI occurs, and the signal bit

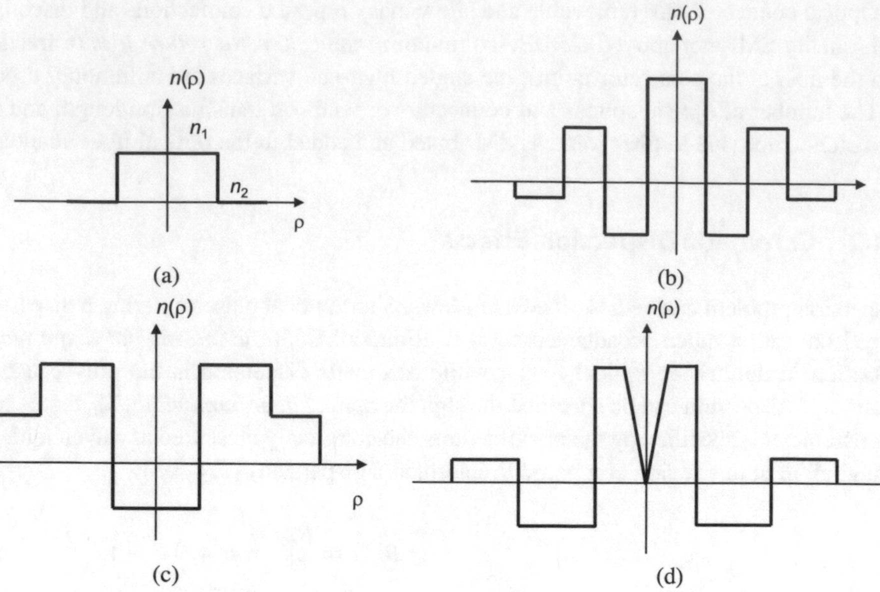

Fig. 2.52 Refractive index profiles for different SMF types: (**a**) standard SMF, (**b**) NZDSF with reduced dispersion slope, (**c**) NZDSF with large effective area, and (**d**) DCF

rates that can be effectively transmitted can be severely limited. A specific spectral component, characterized by the angular optical frequency ω, will arrive at the output end of the fiber of length L after some *delay* τ_g, known as the group delay:

$$\tau_g = \frac{L}{v_g} = L\frac{d\beta}{d\omega} = \frac{L}{c}\frac{d\beta}{dk} = -L\frac{\lambda^2}{2\pi c}\frac{d\beta}{d\lambda}, \tag{2.751}$$

where β is the propagation constant introduced earlier and $v_g = [d\beta/d\omega]^{-1}$ is the group velocity, the speed at which the energy of an optical pulse travels. As a result of the difference in time delays, the optical pulse disperse after traveling a certain distance and the *pulse broadening* ($\Delta\tau_g$) can be characterized by:

$$\Delta\tau_g = \frac{d\tau_g}{d\omega}\Delta\omega = \frac{d\tau_g}{d\lambda}\Delta\lambda, \tag{2.752}$$

where $\Delta\omega$ represents the frequency bandwidth of the source and $\Delta\lambda$ represents the wavelength bandwidth of the source. By substituting Eqs. (2.29) into (2.30), we get:

$$\Delta\tau_g = \frac{d\tau_g}{d\omega}\Delta\omega = \frac{d\left(L\frac{d\beta}{d\omega}\right)}{d\omega}\Delta\omega = L\frac{d^2\beta}{d\omega^2}\Delta\omega = L\frac{d^2\beta}{d\omega^2}\left(-\frac{2\pi c}{\lambda^2}\Delta\lambda\right) = LD\Delta\lambda, \tag{2.753}$$

where D [ps/nm-km] represents the chromatic dispersion parameter defined by:

$$D = -\frac{2\pi c}{\lambda^2}\frac{d^2\beta}{d\omega^2} = -\frac{2\pi c}{\lambda^2}\beta_2, \quad \beta_2 = \frac{d^2\beta}{d\omega^2} \tag{2.754}$$

where β_2 denotes the previously introduced group velocity dispersion (GVD) parameter. The chromatic dispersion can be expressed as the sum of two contributing factors:

$$D = \frac{d\left(-\frac{\lambda^2}{2\pi c}\frac{d\beta}{d\lambda}\right)}{d\lambda} = -\frac{1}{2\pi c}\left(2\lambda\frac{d\beta}{d\lambda} + \lambda^2\frac{d^2\beta}{d\lambda^2}\right) = D_M + D_W,$$
$$D_M = -\frac{\lambda^2}{2\pi c}\frac{d^2\beta}{d\lambda^2} \qquad D_W = -\frac{\lambda}{\pi c}\frac{d\beta}{d\lambda} \tag{2.755}$$

where D_M represents the material dispersion and D_W represents the waveguide dispersion.

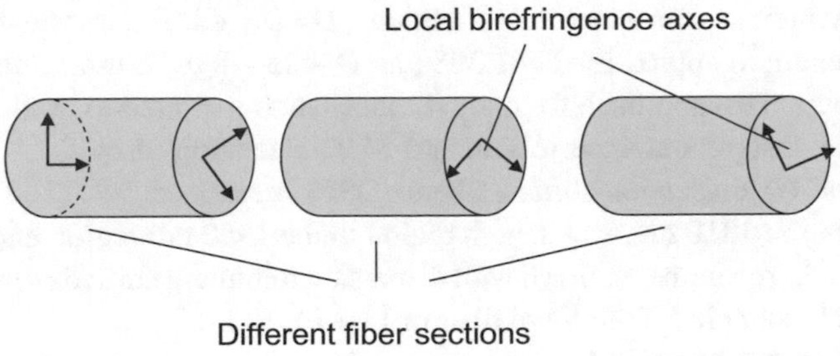

Fig. 2.53 A simplified PMD model

The *material dispersion* arises due to wavelength dependence of the refractive index on the fiber core material, which causes the wavelength dependence of the group delay. The wavelength dependence of the refractive index $n(\lambda)$ is well approximated by the *Sellmeier equation*, introduced earlier. The material dispersion is related to the slope of $n_g = n - \lambda dn(\lambda)/d\lambda$ by $D_M = (1/c)dn_g/d\lambda$. For pure silica fiber, the following approximation is valid [2]:

$$D_M \approx 122(1 - \lambda_{ZD}/\lambda), \quad 1.25\mu\text{m} < \lambda < 1.66\mu\text{m} \tag{2.756}$$

where $\lambda_{ZD} = 1.276$ μm is the zero-dispersion wavelength. We can see that for wavelengths larger than zero-dispersion wavelength, the material dispersion coefficient is positive, while the GVD is negative, and this regime is known as the *anomalous* dispersion region.

The *waveguide dispersion* occurs because the propagation constant is a function of the fiber parameters (core radius and difference between refractive indices in fiber core and cladding) and at the same time is a function of wavelength. Therefore, the propagation paths for a mode due to different boundary matching conditions are slightly different at different wavelengths. Waveguide dispersion is related to the physical design of the optical fiber. Since the value of Δ is typically small, the refractive indices of the core cladding are nearly equal, and the light is not strictly confined in the fiber core. For a given mode, say fundamental mode, the portion of light energy that propagates in the core depends on wavelength: the longer the wavelength, the more power in the cladding. The effect of waveguide dispersion on pulse spreading can be approximated by assuming that refractive index of material is independent of wavelength. To make the result independent of fiber configuration, we should express the group delay in terms of the normalized propagation constant b, defined as:

$$\beta \approx n_2 k(b\Delta + 1), \quad \Delta \ll 1 \tag{2.757}$$

The group delay due to waveguide dispersion can be found as [1]:

$$\tau_W = \frac{L}{c}\frac{d\beta}{dk} \simeq \frac{L}{c}\left(n_2 + n_2\Delta\frac{d(Vb)}{dV}\right), \quad \frac{d(Vb)}{dV} = b\left[1 - \frac{2J_m^2(pa)}{J_{m+1}(pa)J_{m-1}(pa)}\right] \tag{2.758}$$

where p is the core parameter introduced earlier ($p^2 = n_1^2k^2 - \beta^2$). The pulse broadening due to waveguide dispersion can be determined using Eq. (2.758) by:

$$\Delta\tau_W = \frac{d\tau_W}{d\lambda}\Delta\lambda = \frac{d\tau_W}{dV}\frac{dV}{dk}\frac{dk}{d\lambda}\Delta\lambda = L\frac{n_2\Delta}{c}\frac{d^2(Vb)}{dV^2}\frac{V}{k}\left(-\frac{k}{\lambda}\right)\Delta\lambda = LD_W\Delta\lambda, \tag{2.759}$$

where the waveguide dispersion parameter D_W is defined by:

$$D_W \simeq -\frac{n_2\Delta}{c}\frac{1}{\lambda}\left[V\frac{d^2(Vb)}{dV^2}\right], \tag{2.760}$$

and it is negative for the normalized frequencies between 0 and 3.

The total dispersion can be written as sum of two contributions, $D = D_M + D_W$. The waveguide dispersion (i.e., negative) shifts the zero-dispersion wavelength to approximately 1.308 μm. D ~ 15–18 ps/(km-nm) near 1.55 μm. This is the low loss region for fused silica optical fibers. This approach for mutual cancellation was used to produce several types of single-mode optical fibers that are different in design compared to standard SMF, standardized by ITU-T. In addition to standard SMF, there are two major fiber types: (i) dispersion-shifted fibers (DSF) (described in ITU-T recommendation G.653) with dispersion minimum shifted from the 1310 nm wavelength region to the 1550 nm region and (ii) nonzero-dispersion-shifted fibers (NZDSF) (described in G.655 recommendation), with dispersion minimum shifted from 1310 nm window to anywhere within C or L bands (commercial examples: TrueWave fiber or LEAF).

Dispersion effects do not disappear completely at zero-dispersion wavelength. Residual dispersion due to higher-order dispersive effects still exists. Higher-order dispersive effects are governed by the *dispersion slope parameter* $S = dD/d\lambda$. Parameter S is also known as differential dispersion parameter or second-order dispersion parameter, and it is related to the GVD parameter and second-order GVD parameter $\beta_3 = d\beta_2/d\omega$ by:

$$S = \frac{dD}{d\lambda} = \frac{d}{d\lambda}\left(-\frac{2\pi c}{\lambda^2}\beta_2\right) = \frac{4\pi c}{\lambda^3}\beta_2 + \left(\frac{2\pi c}{\lambda^2}\right)^2\beta_3. \tag{2.761}$$

The *geometrical optics* (or *ray theory*) approach, used earlier to describe the light confinement in step-index fibers through the total internal reflection, as shown in Fig. 2.7 and text related to it, is valid when the fiber has a core radius a much larger than the operating wavelength λ. Once the core radius becomes comparable to the operating wavelength, the propagation of light in step-index fiber is governed by Maxwell's equations describing the change of electric $\boldsymbol{E}$ and magnetic $\boldsymbol{H}$ fields in space and time, as discussed at the beginning of the chapter. Since the silica is nonmagnetic material $\boldsymbol{M} = \boldsymbol{0}$, so that the induced electric and magnetic densities, $\boldsymbol{P}$ and $\boldsymbol{E}$, are mutually connected by Eq. (2.7), which is clearly the convolution integral. By applying the convolution theorem of the Fourier transform (FT), the Eq. (2.7) becomes:

$$\begin{aligned}&\underline{\boldsymbol{P}}(\boldsymbol{r}) = \varepsilon_0 \underline{\chi}(\boldsymbol{r})\underline{\boldsymbol{E}}(\boldsymbol{r}),\\ &\underline{\boldsymbol{P}}(\boldsymbol{r},\omega) = FT\{\boldsymbol{P}(\boldsymbol{r},t)\}, \underline{\boldsymbol{E}}(\boldsymbol{r},\omega) = FT\{\boldsymbol{E}(\boldsymbol{r},t)\}, \underline{\chi}(\boldsymbol{r},\omega) = FT(\chi(\boldsymbol{r},t)).\end{aligned} \tag{2.762}$$

The constitutive relation (2.5) now becomes:

$$\begin{aligned}\underline{\boldsymbol{D}} &= \varepsilon_0\underline{\boldsymbol{E}} + \underline{\boldsymbol{P}} = \varepsilon_0\underline{\boldsymbol{E}} + \varepsilon_0\underline{\chi}(\boldsymbol{r},\omega)\underline{\boldsymbol{E}} = \varepsilon_0\left[\underbrace{1+\underline{\chi}(\boldsymbol{r},\omega)}_{\underline{\varepsilon}(\boldsymbol{r},\omega)}\right]\underline{\boldsymbol{E}},\\ &= \varepsilon_0\underline{\varepsilon}(\boldsymbol{r},\omega)\underline{\boldsymbol{E}},\end{aligned} \tag{2.763}$$

where the dielectric permittivity frequency response $\underline{\varepsilon}(\boldsymbol{r},\omega)$ is related to the refractive index $n(\boldsymbol{r},\omega)$ by:

$$\begin{aligned}&\underline{\varepsilon}(\boldsymbol{r},\omega) = [n(\boldsymbol{r},\omega) + j\alpha(\boldsymbol{r},\omega)]^2;\\ &n(\boldsymbol{r},\omega) = \sqrt{1+\mathrm{Re}\,\underline{\chi}(\boldsymbol{r},\omega)}, \alpha(\boldsymbol{r},\omega) = \frac{\omega}{cn(\boldsymbol{r},\omega)}\mathrm{Im}\underline{\chi}(\boldsymbol{r},\omega);\end{aligned} \tag{2.764}$$

where $\alpha(\boldsymbol{r},\omega)$ is the attenuation coefficient. By applying the curl of Maxwell equation (2.1), and after substitution of (2.2) into (2.1), we obtain:

$$\nabla\times(\nabla\times\boldsymbol{E}) = -\mu_0\partial\underbrace{\nabla\times\boldsymbol{H}}_{\nabla\times\boldsymbol{H}=\partial\boldsymbol{D}/\partial t}/\partial t = -\mu_0\partial^2\boldsymbol{D}/\partial t^2. \tag{2.765}$$

Now after applying the FT on (2.765) and by substituting the constitutive relation (2.763) into (2.765), we obtain:

$$\nabla\times(\nabla\times\underline{\boldsymbol{E}}) = \underbrace{-\mu_0\varepsilon_0}_{1/c^2}\underline{\varepsilon}(\boldsymbol{r},\omega)\omega^2\underline{\boldsymbol{E}} = \frac{\underline{\varepsilon}(\boldsymbol{r},\omega)\omega^2}{c^2}\underline{\boldsymbol{E}}. \tag{2.766}$$

By employing the cross-product identity, we can rewrite the previous equation as follows:

$$\nabla\times(\nabla\times\underline{E})=\nabla(\nabla\underline{E})-\underbrace{\nabla\nabla}_{\Delta}\underline{E}=\frac{\underline{\varepsilon}(\boldsymbol{r},\omega)\omega^2}{c^2}\underline{E}. \tag{2.767}$$

Based on Maxwell equation (2.1), we conclude that $\nabla\underline{E}=0$, and since the typical attenuation in modern SMS is below 0.2 dB/km, we can neglect the attenuation in (2.764) so that $\underline{\varepsilon}(\boldsymbol{r},\omega)\cong n^2(\boldsymbol{r},\omega)$. The spatial dependence of refractive index can be neglected, and after substituting $\underline{\varepsilon}(\boldsymbol{r},\omega)\cong n^2(\omega)$ and $\nabla\underline{E}=0$ into (2.767), we obtain the following *wave equation*:

$$\Delta\underline{E}+\frac{n^2(\omega)\omega^2}{c^2}\underline{E}=0. \tag{2.768}$$

An alternative form of the wave equation is given by:

$$\Delta\underline{E}+n^2(\omega)k_0^2\underline{E}=0, \tag{2.769}$$

where $k_0=\omega/c=2\pi/\lambda$ is the free-space wave number. A specific solution of the wave equation, subject to appropriate boundary conditions, is commonly referred to as an *optical mode*. The spatial field distributions, $\boldsymbol{E}(\boldsymbol{r},\omega)=\boldsymbol{E}_0(\rho,\phi)\exp[j\beta(\omega)z-j\omega t]$ and $\boldsymbol{H}(\boldsymbol{r},\omega)=\boldsymbol{H}_0(\rho,\phi)\exp[j\beta(\omega)z-j\omega t]$, of a mode do not change as the mode propagates along z-axis except for a multiplicative factor exp.$[j\beta(\omega)z]$, with $\beta(\omega)$ being the *propagation constant* of a mode. The fiber modes may be classified as *guided modes*, *leaky modes*, and *radiation modes*. Different modes in fiber propagate with different values of propagation constant, in other words with different speeds, causing the pulse spreading. Therefore, it is desirable to design the fiber which supports only one mode-*fundamental mode*. Such a fiber is called *single-mode fiber* (SMF). If the *weak guidance condition* ($\Delta<<1$) is not satisfied, the conventional modes TE, TM, EH, and HE can be found. The cylindrical symmetry of fiber suggests the use of *cylindrical coordinate system* (ρ,ϕ,z). The wave equation is to be solved for six components: E_ρ,E_ϕ,E_z and H_ρ,H_ϕ,H_z. We can solve the wave equation for axial components only (E_z and H_z) and use the system of Maxwell's equations to express the other components as functions of axial ones. Based on Eq. (2.40), the wave equation for E_z component can be written as follows:

$$\frac{\partial^2E_z}{\partial\rho^2}+\frac{1}{\rho}\frac{\partial E_z}{\partial\rho}+\frac{1}{\rho^2}\frac{\partial^2E_z}{\partial z^2}+n^2k_0^2E_z=0,\quad n=\begin{cases}n_1, & \rho\le a\\ n_2, & \rho>a\end{cases} \tag{2.770}$$

where n_1 is the refractive index of the core of radius a and n_2 is the refractive index of the cladding, as shown in Fig. 2.7. The wave equation for E_z component can easily be solved using the *method of separation of variables*, leading to the following overall solution [1–4]:

$$E_z(\rho,\phi,z)=\begin{cases}AJ_m(p\rho)e^{jm\phi}e^{j\beta z}, & \rho\le a\\ BK_m(q\rho)e^{jm\phi}e^{j\beta z}, & \rho>a\end{cases} \tag{2.771}$$

where $J_m(x)$ and $K_m(x)$ are corresponding Bessel functions of mth order, introduced earlier (A and B are constants to be determined from the boundary conditions). The similar equation is valid for H_z. The other four components can be expressed in terms of axial ones by using the Maxwell's equations, and in the core region, we obtain [2–4, 98]:

$$\begin{aligned}E_\rho&=\frac{j}{p^2}\left(\beta\frac{\partial E_z}{\partial\rho}+\mu_0\frac{\omega}{\rho}\frac{\partial H_z}{\partial\phi}\right) & E_\phi&=\frac{j}{p^2}\left(\frac{\beta}{\rho}\frac{\partial E_z}{\partial\phi}-\mu_0\omega\frac{\partial H_z}{\partial\rho}\right)\\ H_\rho&=\frac{j}{p^2}\left(\beta\frac{\partial H_z}{\partial\rho}-\varepsilon_0n^2\frac{\omega}{\rho}\frac{\partial E_z}{\partial\phi}\right) & H_\phi&=\frac{j}{p^2}\left(\frac{\beta}{\rho}\frac{\partial H_z}{\partial\phi}+\varepsilon_0n^2\omega\frac{\partial H_z}{\partial\rho}\right)\end{aligned} \tag{2.772}$$

where $p^2=n_1^2k_0^2-\beta^2$. The similar equations can be obtained for cladding by replacing p^2 with $-q^2$, where $q^2=\beta^2-n_2^2k_0^2$. By satisfying the boundary conditions, the homogeneous system of equation is obtained, which has the nontrivial solution only if corresponding determinant is zero, which leads to the following characteristic (eigenvalue) equation [2–4, 98]:

$$\left[\frac{J'_m(pa)}{paJ_m(pa)}+\frac{K'_m(qa)}{qaK_m(qa)}\right]\left[\frac{J'_m(pa)}{paJ_m(pa)}+\frac{n_2^2}{n_1^2}\frac{K'_m(qa)}{qaK_m(qa)}\right]$$
$$=m^2\left[\frac{1}{(pa)^2}+\frac{1}{(qa)^2}\right]\cdot\left[\frac{1}{(pa)^2}+\frac{n_2^2}{n_1^2}\frac{1}{(qa)^2}\right] \tag{2.773}$$

where *pa* and *qa* are related by the following equation:

$$(pa)^2+(pq)^2=V^2. \tag{2.774}$$

For given fiber parameters n_1, n_2, and a and the operating wavelength, we can determine the normalized frequency V. For fixed V and a given m, there exist multiple solutions $n = 1,2,3,..$ that lead to different modes-*propagating modes*. The case $m = 0$ corresponds to meridional rays (the rays that periodically intersect the center axis of the fiber), the electric/magnetic components are independent of ϕ, and the corresponding modes are classified as TE_{0n} ($E_z = 0$) and TM_{0n} ($H_z = 0$) modes. The case $m \neq 0$ corresponds to skew rays, the electric and magnetic field components are functions of ϕ, and the corresponding modes are classified as HE_{mn} (H_z dominates over E_z) and EH_{mn} (E_z dominates over H_z) modes. Once the mode is identified (m and n are fixed) from Eq. (2.773) and by using the definition expressions for p and q above, we can easily determine the propagation constant of the mode β_{mn}.

The propagating light pulse is a composite optical signal containing a number of monochromatic spectral components. Each spectral component behaves differently in a dispersion medium, such as optical fiber, leading to the light pulse distortion. Each axial component of the monochromatic electromagnetic wave can be represented by its complex electric field function [2–4, 98]:

$$E(z,t)=E_a(z,t)\exp\left[j\beta(\omega)z\right]\exp\left[-j\omega' t\right]. \tag{2.775}$$

The light pulse distortion, which is observed through the pulse broadening along the fiber, can be evaluated by knowing the frequency dependence of the propagation constant $\beta = \beta(\omega)$ at a specific distance z along the fiber. Each spectral component will experience a phase shift proportional to $\beta(\omega)z$. The amplitude spectrum observed at point z along the fiber length, in the frequency domain, is given by:

$$\underline{E}_a(z,\omega)=\underline{E}_a(0,\omega)\exp\left[j\beta(\omega)z\right]. \tag{2.776}$$

The behavior of the pulse envelope during the propagation process can be evaluated through the inverse Fourier transform of previous equation, which is very complicated to calculate, unless additional simplifications are made, for example, by expressing the propagation constant in terms of a Taylor series:

$$\beta(\omega)\approx\beta(\omega_c)+\left.\frac{d\beta}{d\omega}\right|_{\omega=\omega_c}(\omega-\omega_c)+\frac{1}{2}\left.\frac{d^2\beta}{d\omega^2}\right|_{\omega=\omega_c}(\omega-\omega_c)^2+\frac{1}{6}\left.\frac{d^3\beta}{d\omega^3}\right|_{\omega=\omega_c}(\omega-\omega_c)^3+\ldots \tag{2.777}$$

where $\beta_1 = (d\beta/d\omega)|_{\omega=\omega_c}$ is related to the group velocity v_g by $\beta_1 = 1/v_g$, $\beta_2 = \left(d^2\beta/d\omega^2\right)\big|_{\omega=\omega_c}$ is the group velocity dispersion (GVD) parameter, and $\beta_3 = \left(d^3\beta/d\omega^3\right)\big|_{\omega=\omega_c}$ is the second-order GVD parameter. Introducing the concept of slow-varying amplitude $A(z,t)$ as follows [2–4, 98]:

$$E(z,t)=E_a(z,t)\exp\left[j\beta(\omega)z\right]\exp\left(-j\omega_c t\right)=A(z,t)\exp\left(j\beta_c z-j\omega_c t\right) \tag{2.778}$$

we get:

$$A(z,t)=\frac{1}{2\pi}\int_{-\infty}^{\infty}\widetilde{A}(0,\omega)\exp\left[j\left(\beta_1\Delta\omega+\frac{\beta_2}{2}\Delta\omega^2+\frac{\beta_3}{6}\Delta\omega^3\right)z\right]\exp\left(-j(\Delta\omega)t\right)d(\Delta\omega). \tag{2.779}$$

By taking partial derivative of $A(z,t)$ with respect to propagation distance z, we obtain:

$$\frac{\partial A(z,t)}{\partial z} = \frac{1}{2\pi} \int_{-\infty}^{\infty} j\left(\beta_1 \Delta\omega + \frac{\beta_2}{2}\Delta\omega^2 + \frac{\beta_3}{6}\Delta\omega^3\right) \tilde{A}(0,\Delta\omega) e^{j\left(\beta_1\Delta\omega + \frac{\beta_2}{2}\Delta\omega^2 + \frac{\beta_3}{6}\Delta\omega^3\right)z} e^{-j(\Delta\omega)t} d(\Delta\omega) \tag{2.780}$$

Finally, by taking the inverse Fourier transform of (2.780), we derive basic propagation equation describing the *pulse propagation* in single-mode optical fibers:

$$\frac{\partial A(z,t)}{\partial z} = -\beta_1 \frac{\partial A(z,t)}{\partial t} - j\frac{\beta_2}{2}\frac{\partial^2 A(z,t)}{\partial t^2} + \frac{\beta_3}{6}\frac{\partial^3 A(z,t)}{\partial t^3}. \tag{2.781}$$

2.4.3 Polarization Mode Dispersion (PMD)

The polarization unit vector, representing the state of polarization (SOP) of the electric field vector, does not remain constant in practical optical fibers; rather, it changes in a random fashion along the fiber because of its fluctuating birefringence [105–108]. Two common birefringence sources are (i) *geometric birefringence* (related to small departures from perfect cylindrical symmetry) and (ii) *anisotropic stress* (produced on the fiber core during manufacturing or cabling of the fiber). The degree of birefringence is described by the difference in refractive indices of orthogonally polarized modes $B_m = |n_x - n_y| = \Delta n$. The corresponding difference in propagation constants of two orthogonally polarized modes is $\Delta\beta = |\beta_x - \beta_y| = (\omega/c)\Delta n$. Birefringence leads to a periodic power exchange between the two polarization components, described by *beat length* $L_B = 2\pi/\Delta\beta = \lambda/\Delta n$. Typically $Bm \sim 10^{-7}$, and therefore $L_B \sim 10$ m for $\lambda \sim 1$ μm. Linearly polarized light remains linearly polarized only when it is polarized along one of the principal axes; otherwise, its state of polarization changes along the fiber length from linear to elliptical, and then back to linear, in a periodic manner over the length L_B.

In certain applications it is needed to transmit a signal through a fiber that maintains its SOP. Such a fiber is called a *polarization-maintaining fiber* (PMF). (PANDA fiber is a well-known PMF.) Approaches to design PMF having high but constant birefringence are (i) the *shape birefringence* (the fiber having an elliptical core) and (ii) *stress birefringence* (stress-inducing mechanism is incorporated in fiber). Typical PMF has a birefringence $B_m \sim 10^{-3}$ and a beat length $B_m \sim 1$ mm, resulting in a polarization crosstalk smaller than −30 dB/km. Unfortunately, the loss of PMFs is high ($\alpha \sim 2$ dB/km).

The *modal group indices* and *modal group velocities* are related by:

$$\overline{n}_{gx,y} = \frac{c}{v_{gx,y}} = \frac{c}{1/\beta_{1x,y}}, \quad \beta_{1x,y} = \frac{d\beta x, y}{d\omega} \tag{2.782}$$

so that the difference in time arrivals (at the end of fiber of length L) for two orthogonal polarization modes, known as the *differential group delay* (DGD), can be calculated by [2, 4, 98]:

$$\Delta\tau = \left|\frac{L}{v_{gx}} - \frac{L}{v_{gy}}\right| = L\left|\beta_{1x} - \beta_{1y}\right| = L\Delta\beta_1. \tag{2.783}$$

DGD can be quite large in PMF (~ 1 ns/km) due to their large birefringence. Conventional fibers exhibit much smaller birefringence, but its magnitude and polarization orientation change randomly at scale known as the *correlation length* l_c (with typical values in the range 10–100 m). The analytical treatment of PMD is quite complex due to statistical nature. A simple model divides the fiber into a large number of segments (tens to thousands of segments), with both the degree of birefringence and the orientation of the principal states being constant in each segment but change randomly from section to section, as shown in Fig. 2.53. Even though the fiber is composed of many segments having different birefringence, there exist *principal states of polarization* (PSP). If we launch a pulse light into one PSP, it will propagate to the corresponding output PSP without spreading due to PMD. Output PSPs are different from input PSPs. Output PSPs are frequency independent up to the first-order approximation. A pulse launched into the fiber with arbitrary polarization can be represented as a superposition of two input PSPs. The fiber is subject to time-varying perturbations (such as temperature and vibrations) so that PSPs and DGD vary randomly over time.

DGD is *Maxwellian distributed* random variable with the mean-square DGD value [2, 4, 98]:

$$\left\langle (\Delta T)^2 \right\rangle = 2(\Delta\beta_1)^2 l_c^2 \left[\exp\left(-\frac{L}{l_c}\right) + \frac{L}{l_c} - 1 \right], \tag{2.784}$$

where the parameters in (2.784) are already introduced above. For long fibers, $L >> l_c$, the root-mean-squared (RMS) DGD follows:

$$\left\langle (\Delta T)^2 \right\rangle^{1/2} = \Delta\beta_1 \sqrt{2 l_c L} = D_p \sqrt{L}, \tag{2.785}$$

where D_p[ps/√km] is the PMD parameter. Typical values for PMD parameter are in the range 0.01–10 ps/(km)$^{1/2}$, in new fibers $D_p < 0.1$ ps/(km)$^{1/2}$. The first-order PMD coefficient, D_p, can be characterized again by Maxwellian PDF [105–108]:

$$f(D_p) = \sqrt{\frac{2}{\pi}} \frac{D_p^2}{\alpha^3} \exp\left(-\frac{D_p^2}{2\alpha^2}\right), \tag{2.786}$$

where α is the coefficient with a typical value around 30 ps [105, 106]. The mean value $<D_p>$ determined from Maxwellian PDF: $<D_p> = (8/\pi)^{1/2}\alpha$. The overall probability $\Pr(D_p)$ that coefficient D_p will be larger than pre-specified value can be determined by:

$$CDF(D_p) = \int_0^{D_p} f(D_p) dD_p, \tag{2.787}$$

The value $CDF(3 < D_p >)$ is about $4 \cdot 10^{-5}$, so that the practical expression to characterize the first-order PMD is [98]:

$$\Delta T = 3 \langle D_p \rangle \sqrt{L}. \tag{2.788}$$

The *second-order PMD* occurs due to frequency dependence of both DGD and PSP and can be characterized by coefficient D_{P2} [105]:

$$D_{P2} = \sqrt{\left(\frac{1}{2}\frac{\partial D_{P1}}{\partial \omega}\right)^2 + \left(\frac{D_{P1}}{2}\frac{\partial |s|}{\partial \omega}\right)^2}, \tag{2.789}$$

where the first term describes the frequency dependence of DGD and the second term describes the frequency dependence of Stokes vector s (describing the position of PSP) [106]. The statistical nature of second-order PMD coefficient in real fiber is characterized by PDF of D_{P2} [107]:

$$f(D_{P2}) = \frac{2\sigma^2 D_{P2}}{\pi} \frac{\tanh(\sigma D_{P2})}{\cosh(\sigma D_{P2})}. \tag{2.790}$$

The probability that the second-order PMD coefficient is larger than $3 < DP2 >$ is still not negligible. However, it becomes negligible for larger values of $5 < D_{P2} >$, so that the total pulse spreading due to the second-order PMD effect can be expressed as [98]:

$$\Delta\tau_{P2} = 5 \langle D_{P2} \rangle L. \tag{2.791}$$

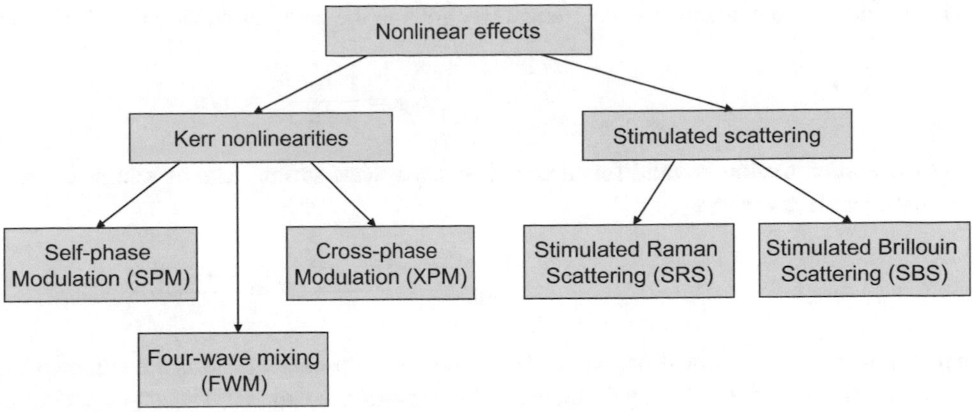

Fig. 2.54 Fiber nonlinearities classification

2.4.4 Fiber Nonlinearities

The basic operational principles of optical transmission can be explained assuming that optical fiber medium is *linear*. The linearity assumption is valid if the launched power does not exceed several mW in a single channel system. In modern WDM technology, high-power semiconductor lasers and optical amplifiers are employed, and the influence of fiber nonlinearities becomes important. Moreover, in some special cases, the fiber nonlinearities may be used to enhance the fiber transmission capabilities (e.g., soliton transmission).

2.4.4.1 Fiber Nonlinearities Classification

There are two major groups of fiber nonlinearities related either to nonlinear refractive index (*Kerr effect*) or to *nonlinear optical scattering*, which is illustrated in Fig. 2.54.

The Kerr effect occurs due to dependence of index of refraction on light intensity, and fiber nonlinearities belonging to this category are (i) self-phase modulation (SPM), (ii) cross-phase modulation (XPM), and (iii) four-wave mixing (FWM). The SPM is related to the single optical channel. The variation of power within the channel causes the changes in refractive index, which leads to the pulse distortion. In XPM, the refractive index changes due to variations in power not only in observed channel but also due to variation in powers of other wavelength channels leading to the pulse distortion. In FWM, several wavelength channels interact to create newly channels. This effect is dependent on both the powers of interacting channel and chromatic dispersion.

The stimulated scattering effects are caused by parametric interaction between the light and materials. There are two types of stimulated scattering effects: (i) stimulated Raman scattering (SRS) and stimulated Brillouin scattering (SBS). In the SRS, the light interacts with material through the vibrations and causes the energy transfer from short wavelength channels to longer wavelength channels, leading to the interchannel crosstalk. In the SBS, the light interacts with the matter through acoustic waves, leading to the coupling with backward propagating waves, thus limiting the available power per channel.

2.4.4.2 Effective Length and Effective Cross-Sectional Area

The nonlinear effects introduced above are function of the transmission length and cross-sectional area of the optical fiber. The nonlinear interaction is stronger for longer fibers and smaller cross-sectional area. On the other hand, the nonlinear interaction decreases along the transmission line because the signal is getting attenuated as it propagates. Therefore, the strength of the nonlinear effects may be characterized by introducing the concept of *effective length* L_{eff}, as an equivalent length so that the product of launched power P_0 and the equivalent length is the same as the area below power evolution curve $P(z) = P_0\exp(-\alpha z)$:

$$P_0 L_{eff} = \int_0^L P(z)dz = \int_0^L P_0 e^{-\alpha z} dz, \tag{2.792}$$

where α denotes the attenuation coefficient. By solving the previous equation for Leff, we obtain:

$$L_{eff} = \frac{1 - e^{-\alpha L}}{\alpha} \approx 1/\alpha, \tag{2.793}$$

and the approximation is valid for fiber which are at least several tens of kilometers in length. The affective length after M amplifiers is given by:

$$L_{eff,total} = \frac{1 - e^{-\alpha L}}{\alpha} M = \frac{1 - e^{-\alpha L}}{\alpha} \frac{L}{l}, \tag{2.794}$$

where L is the total transmission length, l is the amplifier spacing, and M is the number of amplifiers. If the amplifier spacing increases, the optical amplifier gain increases in proportion to exp.(αl), to compensate for the fiber losses. On the other hand, the increase in the power enhances the nonlinear effects. Therefore, what matters is the product of launched power P_0 and the total effective length $L_{eff,total}$ as follows: $P_0 L_{eff,total}$.

The nonlinear effects are inversely proportional to the area of the fiber core because the concentration of the optical power per unit cross-sectional area (power density) is higher for smaller cross-sectional area and vice versa. Optical power distribution across the cross-sectional area is closely related to the overall refractive index, and to characterize it, the concept of *effective cross-sectional area* (A_{eff}) is introduced [3, 4, 98]:

$$A_{eff} = \frac{\left[\int_r \int_\theta r dr d\theta |E(r,\theta)|^2 \right]^2}{\int_r \int_\theta r dr d\theta |E(r,\theta)|^4}, \tag{2.795}$$

where $E(r,\theta)$ is the distribution of electrical field in cross section and (r,θ) denotes the polar coordinates. For the Gaussian approximation $E(r,\theta) = E_0 \exp(-r^2/w^2)$, the effective cross-sectional area is simply $A_{eff} = \pi w^2$, where w is the mode radius. The Gaussian approximation is not applicable to the optical fibers having more complicated index profile, such as DSFs, NZDFs, and DCFs. In these cases, the effective cross-sectional area can be found by [4, 98]:

$$A_{eff} = c_f \pi w^2, \tag{2.796}$$

where the correction factor c_f is smaller than one for DCFs and some NZDSFs and larger than one for NZDSF with larger effective cross-sectional area.

2.4.4.3 Kerr Effect, Self-Phase Modulation, and Nonlinear Length

For relatively high power density, the index of refraction acts as the function density; the effect is known as *Kerr effect*:

$$n(P) = n_0 + n_2 \frac{P}{A_{eff}}, \tag{2.797}$$

where n_2 is the Kerr coefficient (second-order refractive index coefficient), with typical values being 2.2–3.4·10^{-20} m^2/W. The variation in refractive index due to Kerr effect yields to the propagation constant β variations, because $\beta = 2\pi n/\lambda$:

$$\beta(P) = \beta_0 + \gamma P, \quad \beta_0 = 2\pi n_0/\lambda, \quad \gamma = \frac{2\pi n_2}{\lambda A_{eff}}, \tag{2.798}$$

where γ is the nonlinear coefficient, with typical values being 0.9–2.75 W^{-1} km^{-1} at 1550 nm. The propagation constant β will vary along the duration of optical pulse, because the different points along the pulse will "see" different optical powers and the *frequency chirping* is introduced. The propagation constant associated with the leading edge of the pulse will be lower than that related to the central part of the pulse. The difference in propagation constants will cause the difference in phases associated with different portions of the pulse. The central part of the pulse will acquire phase more rapidly than the leading and trailing edges. The total nonlinear phase shift after some length L can be found by:

$$\Delta\Phi[P] = \int_0^L [\beta(P_0) - \beta]dz = \int_0^L \gamma P(z)dz = \underbrace{\gamma P_0}_{1/L_{NL}}\underbrace{\frac{1-e^{-\alpha L}}{\alpha}}_{L_{eff}} = \frac{L_{eff}}{L_{NL}}, \tag{2.799}$$

where $L_{NL} = [\gamma P_0]^{-1}$ is the nonlinear length. The frequency variation (chirp) can be found as the first derivative of the nonlinear phase shift with respect to time:

$$\Delta\omega_{SPM}(t) = \frac{d[\Delta\Phi(t)]}{dt} = \frac{2\pi}{\lambda}\frac{L_{eff}n_2}{A_{eff}}\frac{dP_0(t)}{dt}. \tag{2.800}$$

Generally speaking, the SPM introduces the pulse broadening. However, in some situations, the frequency chirp due to SPM could be opposite to that due to chromatic dispersion (such as *the anomalous dispersion region*). In such a case, SPM helps to reduce the impact of chromatic dispersion. To see this better, let us observe the propagation equation and extend it to account for the Kerr nonlinear effects as follows:

$$\frac{\partial A(z,t)}{\partial z} = -\frac{\alpha}{2}A(z,t) - \beta_1\frac{\partial A(z,t)}{\partial t} - j\frac{\beta_2}{2}\frac{\partial^2 A(z,t)}{\partial t^2} + \frac{\beta_3}{6}\frac{\partial^3 A(z,t)}{\partial t^3} + j\gamma|A(z,t)|^2 A(z,t), \tag{2.801}$$

where α is the attenuation coefficient and γ is the nonlinear coefficient introduced in (2.798). After the introduction of a new coordinate system with $T = t - \beta_1 z$, *the generalized nonlinear Schrödinger equation* (GNLSE) is obtained:

$$\frac{\partial A(z,T)}{\partial z} = -\frac{\alpha}{2}A(z,T) - j\frac{\beta_2}{2}\frac{\partial^2 A(z,t)}{\partial T^2} + \frac{\beta_3}{6}\frac{\partial^3 A(z,T)}{\partial T^3} + j\gamma|A(z,T)|^2 A(z,T). \tag{2.802}$$

Assuming that the fiber loss and the second-order GVD can be neglected, the GNLSE simplifies to:

$$\frac{\partial A(z,t)}{\partial z} = -\beta_1\frac{\partial A(z,t)}{\partial t} - j\frac{\beta_2}{2}\frac{\partial^2 A(z,t)}{\partial t^2} + j\gamma|A(z,t)|^2 A(z,t). \tag{2.803}$$

Substituting $\tau = (t - \beta_1 z)/\tau_0$, $\xi = z/L_D$ ($L_D = \tau^2{}_0/|\beta_2|$), and $U = A/\sqrt{P_0}$ ($P_0 = A^2{}_0$), we get:

$$j\frac{\partial U}{\partial \xi} - \frac{\text{sgn}(\beta_2)}{2}\frac{\partial^2 U}{\partial \tau^2} + N^2|U|^2 U = 0, \quad N^2 = L_D/L_{NL} \tag{2.804}$$

When $N << 1$, the nonlinear length is much larger than the dispersion length so that the nonlinear effects can be neglected compared to those of chromatic dispersion. *If chromatic dispersion can be neglected* (L_D- > ∞), the NLSE can be solved analytically [99]:

$$U(z,\tau) = U(0,\tau)e^{jz|U(0,\tau)|^2/L_{NL}}. \tag{2.805}$$

Interestingly, the SPM modulation causes a phase change but no change in the envelope of the pulse. Thus, SPM by itself leads only to chirping, regardless of the pulse shape. The SPM-induced chirp, however, modifies the pulse-broadening effects of chromatic dispersion.

2.4.4.4 Cross-Phase Modulation

The XPM is another effect caused by the intensity dependence of the refractive index, but it is related to the multichannel transmission and occurs during propagation of a composite signal. The nonlinear phase shift of a specific optical channel, say the n-th channel, is affected not only by the power of that channel but also by the optical power of the other channels as follows:

$$\Delta\Phi_n(t) = \gamma L_{eff} P_{0n}(t) + \gamma L_{eff} 2 \sum_{i=1,\ i\neq n}^{N} P_{0i}(t) = \frac{2\pi}{\lambda}\frac{L_{eff} n_2}{A_{eff}}\left[P_{0n}(t) + 2\sum_{i=1,\ i\neq n}^{N} P_{0i}(t)\right], \tag{2.806}$$

where $P_{0n}(t)$ denotes the pulse shape in the n-th channel, other parameters are the same as before, and N denotes the number of wavelength channels. The nonlinear phase shift is bit-pattern dependent. In *the worst-case scenario*, assuming that all wavelength channels carry the same data stream, we can write:

$$\Delta\Phi_n(t) = \frac{2\pi}{\lambda}\frac{L_{eff} n_2}{A_{eff}}(2N-1)P_{0n}(t). \tag{2.807}$$

In practice, the optical pulses from different optical channels propagate at different speeds (they have different group velocities). The phase shift given above can occur only during the overlapping time. The overlapping among neighboring channels is longer than the overlapping of channels spaced apart, and it will produce the most significant impact on the phase shift. If pulses walk through one another quickly (due to significant chromatic dispersion or when the channels are widely separated), the described effects on both pulses are diminished because the distortion done by the trailing edge is undone by the leading edge.

2.4.4.5 Four-Wave Mixing

The FWM is another effect that occurs in optical fibers during the propagation of a composite optical signal, such as the WDM signal. It gives a rise to the optical signals. The three optical signals with different carrier frequencies f_i, f_j, and f_k (i,j,k = 1,. . ., M) interact to generate the new optical signal at frequency $f_{ijk} = f_i + f_j - f_k$, providing that the *phase matching condition* is satisfied:

$$\beta_{ijk} = \beta_i + \beta_j - \beta_k, \tag{2.808}$$

where βs are corresponding propagation constants. The measure of the phase-matching condition of wavelength channels involved in the interaction is defined by:

$$\Delta\beta = \beta_i + \beta_j - \beta_k - \beta_{ijk}, \tag{2.809}$$

The FWM is effective only if $\Delta\beta$ approaches zero. Therefore, the phase-matching condition is a requirement for the *momentum conservation*. The FWM process can also be considered from quantum-mechanic point of view as the *annihilation* of two photons with energies hf_i and hf_j and *generation* of two new photons with energies hf_k and hf_{ijk}. In FWM process the indices i and j do no need to be necessarily distinct, meaning that only two channels may interact to create the new one; this case is known as *degenerate* case.

The power of the newly generated optical frequency is a function of the powers of optical signals involved in the process, the Kerr coefficient, and the degree of satisfaction of the phase-matching condition, so that we can write [3, 4, 98, 99]:

$$P_{ijk} \approx \left(\frac{2\pi f_{ijk} n_2 d_{ijk}}{3cA_{eff}}\right)^2 P_i P_j P_k L_{eff}^2, \tag{2.810}$$

where d_{ijk} is the measure of degeneracy, which takes value 3 in degenerate case or value 6 in non-degenerate case. FWM can produce significant signal degradation in WDM systems since several newly generated frequencies can coincide with any specific channel. Namely, the total number of newly generated frequencies N through the FWM process is $N = M^2(M-1)/2$. Fortunately, some of the newly generated frequencies have negligible impact. However, some of them will have significant impact especially those coinciding with already existing WDM channels.

The FWM can be reduced either by reducing the power per channel or by preventing the perfect phase-matching by increasing the chromatic dispersion or increasing the channel spacing. One option would be to use non-uniform channel spacing. On the other hand, the FWM process can be used to perform some useful functions, such as the *wavelength conversion* through the process of *phase conjugation* [2].

Another interpretation of Kerr nonlinearities can be provided as follows. Under isotropy assumption, in the presence of nonlinearities, induced polarization $\boldsymbol{P}(\boldsymbol{r},t)$ is along the same direction as the electric field $\boldsymbol{E}(\boldsymbol{r},t)$, so that corresponding vector notation can be substituted by the scalar functions and the induced polarization be written in the form:

$$P(\boldsymbol{r},t)=\varepsilon_0\int_{-\infty}^{t}\chi^{(1)}(t-t')E(\boldsymbol{r},t')dt'+\varepsilon_0\chi^{(3)}E^3(\boldsymbol{r},t) \tag{2.811}$$

where the first term corresponds to the linear part and the second term, $P_{NL}(\boldsymbol{r},t)$, to the nonlinear part. $\chi^{(1)}$ denotes the linear susceptibility, $\chi^{(3)}$ denotes the third-order susceptibility, and ε_0 is the permittivity of vacuum. (Because of the symmetry of silica molecules, only odd terms are present in Taylor expansion of $P(\boldsymbol{r},t)$.) A wavelength division multiplexing (WDM) signal can be represented as the sum of N monochromatic plane waves of angular frequency ω_i (the modulation process is ignored in this simplified analysis):

$$E(\boldsymbol{r},t)=\sum_{n=1}^{N}E_i\cos\left(\omega_n t-\beta_n z\right). \tag{2.812}$$

The nonlinear polarization can be written as after substitution of Eq. (2.812) into second term of (2.811) as follows:

$$P_{NL}(z,t)=\varepsilon_0\chi^{(3)}\sum_{i=1}^{N}\sum_{j=1}^{N}\sum_{k=1}^{N}E_i\cos\left(\omega_i t-\beta_i z\right)E_j\cos\left(\omega_j t-\beta_j z\right)E_k\cos\left(\omega_k t-\beta_k z\right) \tag{2.813}$$

which can also be rewritten as:

$$\begin{aligned}P_{NL}(z,t)&=\frac{3}{4}\varepsilon_0\chi^{(3)}\sum_{i=1}^{N}E_i^2E_i\cos\left(\omega_i t-\beta_i z\right)+\frac{3}{4}\varepsilon_0\chi^{(3)}\sum_{i=1}^{N}\sum_{j\neq i}2E_iE_jE_i\cos\left(\omega_i t-\beta_i z\right)\\&+\frac{3}{4}\varepsilon_0\chi^{(3)}\sum_{i=1}^{N}\sum_{j\neq i}E_i^2E_j\cos\left[\left(2\omega_i-\omega_j\right)t-\left(2\beta_i-\beta_j\right)z\right]\\&+\frac{6}{4}\varepsilon_0\chi^{(3)}\sum_{i=1}^{N}\sum_{j>i}\sum_{k>j}E_iE_jE_k\cos\left[\left(\omega_i+\omega_j-\omega_k\right)t-\left(\beta_i+\beta_j-\beta_k\right)z\right]+\ldots\end{aligned} \tag{2.814}$$

The first term in the first line corresponds to the SPM, the second term in the same line corresponds to the XPM, the term in the second line corresponds to the degenerate FWM case, and the last term corresponds to the non-degenerate FWM.

2.4.4.6 Raman Scattering

Raman scattering is a nonlinear effect that occurs when a propagating optical signal interacts with glass molecules in the fiber undergoing a wavelength shift. The result of interaction is a transfer of energy from some photons of the input optical signal to vibrating silica molecules and creation of new photons with lower energy than the energy of incident photons. The incident optical signal is commonly referred to as a pump, and the generated signal is called the *Stokes signal*. The difference in frequencies between the pump (ω_p) and the Stokes signal (ω_s) is known as the *Raman frequency shift* $\omega_R=\omega_p-\omega_s$. Scattered photons are not in phase with each other and do not follow the same scattering pattern, meaning that energy transfer from pump to Stokes photons *is not a uniform process*. As a result, there exists the frequency band that includes frequencies of all scattered Stokes photons. Scattered Stokes photons can take any direction that can be either forward or backward with respect to the direction of the pump, meaning that *Raman scattering is isotropic process*. If the pump power is lower than a certain *threshold value*, the Raman scattering process will have a spontaneous character, characterized by the small number of pump photons that will be scattered and converted to the Stokes photons. However, when the pump power exceeds the threshold value, Raman scattering becomes a *stimulated process*. SRS can be explained as *a positive feedback process* in the pump signal interacts with Stokes signal and creates the beat frequency $\omega_R = \omega_p - \omega_s$, and the beat frequency then serves as a stimulator of molecular oscillations, and the process is enhanced or amplified. Assuming that the Stokes signal propagates in the same direction as the pump signal, the intensity of the pump signal $I_p = P_p/A_{eff}$ and the intensity of the Stokes signal $I_s = P_s/A_{eff}$ are related by the following systems of coupled differential equations [3, 4, 98, 99]:

$$\frac{dI_p}{dz} = -g_R\left(\frac{\omega_P}{\omega_S}\right)I_pI_s - \alpha_pI_p, \quad \frac{dI_s}{dz} = g_RI_pI_s - \alpha_sI_s, \tag{2.815}$$

where α_p (α_s) is the fiber attenuation coefficient of the pump (Stokes) signal and g_R is the Raman gain coefficient. Because the SRS process is not the uniform process, the scattered Stokes photons will occupy a certain frequency band, and the Raman gain is not constant but the function of frequency. It can roughly be approximated by the *Lorentzian spectral profile* [3, 4, 98, 99]:

$$g_R(\omega_R) = \frac{g_R(\Omega_R)}{1 + (\omega_R - \Omega_R)^2T_R^2}, \tag{2.816}$$

where T_R is vibration states decay time (in the order 0.1 ps for silica based materials) and Ω_R is the Raman frequency shift corresponding the Raman gain peak. The actual gain profile extends over 40 THz (~320 nm), with a peak around 13.2 THz. The Raman gain peak $g_R(\Omega_R) = g_{R,max}$ is between 10^{-12} and 10^{-13} m/W for wavelengths above 1300 nm. The gain profile can also be approximated by a *triangle* function, which is commonly used in analytical studies of SRS [3, 4, 98, 99]:

$$g_R(\omega_R) = \frac{g_R(\Omega_R)\omega_R}{\Omega_R}. \tag{2.817}$$

The *Raman threshold*, defined as the incident power at which half of the pump power is converted to the Stokes signal, can be estimated using the system of differential equations by replacing g_R by $g_{R,max}$. The amplification of Stokes power along distance L can be determined by:

$$P_s(L) = P_{s,0}\exp\left(\frac{g_{R,\max}P_{s,0}L}{2A_{eff}}\right), \tag{2.818}$$

where $P_{s,0} = P_s(0)$. The Raman threshold can be determined by [3, 4, 98, 99]:

$$P_{R,\text{threshold}} = P_{s,0} \approx \frac{16A_{eff}}{g_{R,\max}L_{eff}} \tag{2.819}$$

and for typical SMF parameters ($A_{eff} = 50\ \mu m^2$, $L_{eff} = 20$ km, and $g_{R,max} = 7{\cdot}10^{-13}$ m/W), it is around 500 mW.

The SRS can effectively be used for optical signal amplification (*Raman amplifiers*), as we explained earlier. In WDM systems, however, the SRS effect can be quite detrimental, because the Raman gain spectrum is very broad, which enables the energy transfer from the lower to higher wavelength channels. The shortest wavelength channel within the WDM system acts as a pump for several long wavelength channels while undergoing the most intense depletion. The fraction of power coupled out of the channel 0 to all the other channels ($1, 2, \ldots, M - 1$; M is the number of wavelengths), when the Raman gain shape is approximated as a triangle, is [3, 4, 98, 99]:

$$P_0 = \sum_{n=1}^{N-1} P_0(n) = \frac{g_{R,\max}\Delta\lambda PL_{eff}}{2\Delta\lambda_C A_{eff}} \frac{N(N-1)}{2}, \tag{2.820}$$

where $\Delta\lambda$ is the channel spacing and $\Delta\lambda_c \cong 125$ nm. The energy transfer between two channels is bit-pattern dependent and occurs only if both wavelengths are synchronously loaded with 1 bits, meaning that the energy transfer will be reduced if dispersion is higher due to the *walk-off effect* (difference in velocities of different wavelength channels will reduce the time of overlapping).

2.4.4.7 Brillouin Scattering

Brillouin scattering is a physical process that occurs when an optical signal interacts with acoustical phonons, rather than the glass molecules. During this process, an incident optical signal reflects backward from the grating formed by acoustic vibrations and downshifts in frequency. The acoustic vibrations originate from the thermal effect if the power of an incident optical signal is relatively small. If the power of incident light goes up, it increases the material density through the electrostrictive effect. The change in density enhances acoustic vibrations and forces Brillouin scattering to become *stimulated*. The SBS process can also be explained as a positive feedback mechanism, in which the incident light (the pump) interacts with Stokes signal and creates the beat frequency $\omega_B = \omega_P - \omega_S$. The parametric interaction between pump, Stokes signal, and acoustical waves requires both the *energy and momentum conservation*: the energy is effectively preserved through the downshift in frequency, while the momentum conservation occurs through the backward direction of the Stokes signal. Since the SBS can be effectively suppressed by *dithering the source* (directly modulating the laser with a sinusoid at a frequency much lower than the receiver low-frequency cutoff), it is not discussed much here.

2.4.5 Generalized Nonlinear Schrödinger Equation

The GNLSE has already been introduced in previous section. Here we describe it in operator form, discus its efficient solving, and extend it to the multichannel transmission. Additionally, the GNLSE for polarization division multiplexing (PDM) is described as well. Finally, we describe the GNLSE suitable for study of propagation is spatial division multiplexing (SDM) systems.

2.4.5.1 GNLSE for Single Channel Transmission

The operator form of GNLSE is given by [3, 4, 98, 109, 110]:

$$\frac{\partial E(z,t)}{\partial z} = \left(\widehat{D} + \widehat{N}\right)E,\ \widehat{D} = -\frac{\alpha}{2} - j\frac{\beta_2}{2}\frac{\partial^2}{\partial t^2} + \frac{\beta_3}{6}\frac{\partial^3}{\partial t^3}, \widehat{N} = j\gamma\left|E^2\right|, \tag{2.821}$$

where E is the signal electric field magnitude, $\widehat{D}$ and $\widehat{N}$ denote the linear and nonlinear operators, while α, β_2, β_3, and γ represent attenuation coefficient, GVD, second-order GVD, and nonlinear coefficient, respectively. To solve the NSE, the *split-step Fourier method* is commonly used [3, 4, 98, 109, 110]. The key idea behind this method is to split the fiber into sections, each with length Δz, and perform the integration of NSE on each section, which leads to expression:

$$E(z + \Delta z, t) = \left[\exp\left(\widehat{D} + \widehat{N}\right)\Delta z\right]E(z,t). \tag{2.822}$$

In addition, the Taylor expansion can be used to present exponential term from Eq. (2.822) as:

$$\exp\left(\widehat{D} + \widehat{N}\right)\Delta z = \sum_{n=0}^{\infty}\left(\widehat{D} + \widehat{N}\right)^n \Delta z^n/n! \tag{2.823}$$

Instead of Taylor expansion, the following two approximations are commonly used in practice:

$$E(z + \Delta z, t) \approx \begin{cases} e^{\widehat{D}\Delta z/2}e^{\widehat{N}\Delta z}e^{\widehat{D}\Delta z/2}E(z,t) \\ e^{\widehat{D}\Delta z}e^{\widehat{N}\Delta z}E(z,t) \end{cases} \tag{2.824}$$

where the first method (the upper arm on the right side of above equation) is known as symmetric split-step Fourier method (SSSFM), while the second (the lower arm on the right side of above equation) is known as asymmetric split-step Fourier method (ASSFM). Clearly, these methods have similarities with corresponding methods to describe the propagation in atmospheric turbulence channels. The linear operator in either method corresponds to multiplication in frequency domain, which is:

$$\exp\left(\Delta h\widehat{D}\right)E(z,t) = FT^{-1}\left\{\exp\left[\left(-\frac{\alpha}{2}+j\frac{\beta_2}{2}\omega^2+j\frac{\beta_3}{6}\omega^3\right)\Delta h\right]FT[E(z,t)]\right\}, \tag{2.825}$$

where Δh is the step-size equal to either $\Delta z/2$ in SSSFM or Δz in ASSFM, while $FT\,(FT^{-1})$ denotes Fourier transform (inverse Fourier transform) operator. On the other side, the nonlinear operator performs the nonlinear phase "rotation" in time domain, expressed as:

$$\exp\left(\Delta z\widehat{N}\right)E(z,t) = \exp\left(j\gamma\Delta z|E(z,t)|^2\right)E(z,t). \tag{2.826}$$

It is evident that the nonlinear operator depends on electric field magnitude at location z, which is not known and should be evaluated. It was proposed in [3] to use the trapezoidal rule and express the electric field function as:

$$\begin{aligned} E(z+\Delta z,t) &= \exp\left(\widehat{D}\Delta z/2\right)\exp\left(\int_{z}^{z+\Delta z} N(z')dz'\right)\exp\left(\widehat{D}\Delta z/2\right)E(z,t) \\ &\approx \exp\left(\widehat{D}\Delta z/2\right)\exp\left[\frac{\widehat{N}(z+\Delta z)+\widehat{N}(z)}{2}\Delta z\right]\exp\left(\widehat{D}\Delta z/2\right)E(z,t). \end{aligned} \tag{2.827}$$

The iteration procedure that should be applied in accordance with Eq. (2.827) is illustrated in Fig. 2.55. Since this iteration procedure can be time-extensive, the following less accurate approximation can be considered instead:

$$E(z+\Delta z,t) \approx \begin{cases} \exp\left(\widehat{D}\Delta z/2\right)\exp\left[\widehat{N}(z+\Delta z/2)\Delta z\right]\exp\left(\widehat{D}\Delta z/2\right)E(z,t), & \text{forSSSFM} \\ \exp\left(\widehat{D}\Delta z\right)\exp\left[\widehat{N}(z)\Delta z\right]E(z,t), & \text{forASSFM} \end{cases}. \tag{2.828}$$

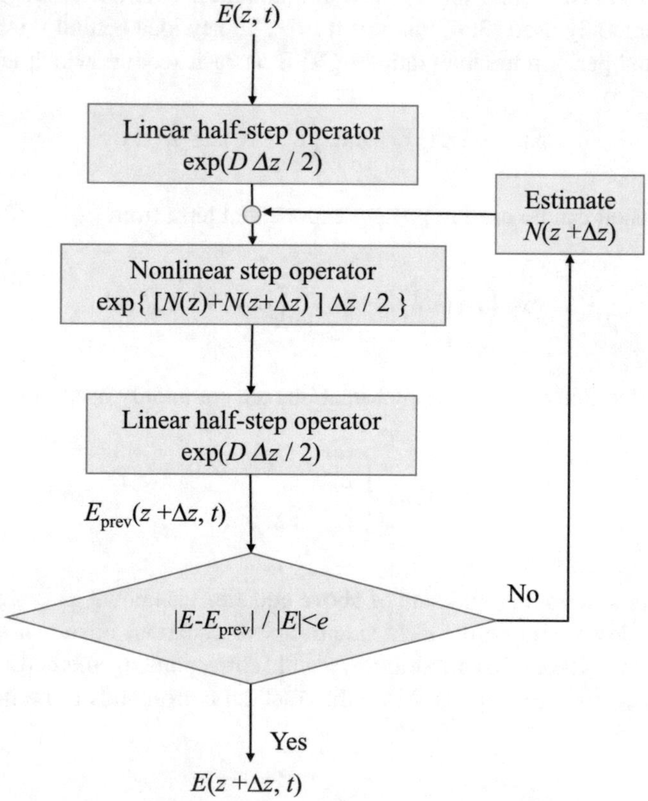

Fig. 2.55 Fiber nonlinearities classification. Solving the GNLSE by iterative symmetric split-step Fourier method

The complexity of iterative algorithms is determined by complexity of fast Fourier transform, which is proportional to ($N \cdot \log N$). To study the accuracy of split-step methods described above, we can use the Baker-Hausdorff formula [7]:

$$e^A e^B = e^{A+B+\frac{1}{2}[A,B]+\frac{1}{12}[A,[A,B]]-\frac{1}{12}[B,[A,B]]-\frac{1}{24}[B,[A,[A,B]]]+\ldots}, \tag{2.829}$$

where $[A,B]$ is the commutator of two operators A and B, defined as: $[A,B] = AB - BA$. By applying this formula to SSSFM and ASSFM, we obtain that:

$$\begin{aligned} &e^{\widehat{D}\Delta z} e^{\widehat{N}\Delta z} \cong e^{\widehat{D}\Delta z+\widehat{N}\Delta z+\frac{1}{2}[\widehat{D},\widehat{N}]\Delta z^2+\ldots}, \\ &e^{\widehat{D}\Delta z/2} e^{\widehat{N}\Delta z} e^{\widehat{D}\Delta z/2} \cong e^{\widehat{D}\Delta z+\widehat{N}\Delta z+\frac{1}{6}[\widehat{N}+\widehat{D}/2,[\widehat{N},\widehat{D}/2]]\Delta z^3+\ldots}. \end{aligned} \tag{2.830}$$

As we can see, the operator error in ASSFM is in order of (Δz^2), while the operator error in SSSFM is in order of (Δz^3), which means that SSSFM offers better accuracy.

The step-size selection is analyzed in several papers [111–113]. The step-size is traditionally chosen as a function of characteristic length over which the phase distortion reaches maximum tolerable value. For chromatic dispersion (CD) and Kerr nonlinearities (NL), the characteristic lengths are given as [109, 110]:

$$L_{\mathrm{NL}} \approx 1/\gamma P_{\mathrm{Tx}}, \quad L_{\mathrm{CD}} \approx 2/\beta_2 \omega_{\max}^2, \tag{2.831}$$

where P_{Tx} is the launched power and $\omega_{\max}$ is the highest frequency in the spectrum of modulated signal, which is:

$$\omega_{\max} \approx \begin{cases} 2\pi R_s, & \text{single-carriersystems} \\ 2\pi(N_{sc}+1)\Delta f_{sc}, & \text{OFDM systems} \end{cases} \tag{2.832}$$

with R_s being the symbol rate, N_{sc} being the number of subcarriers, and Δf_{sc} being the subcarrier spacing. If the maximum phase shifts that can be tolerated due to chromatic dispersion and nonlinearities are denoted by $\Delta\phi_{\mathrm{CD}}$ and $\Delta\phi_{\mathrm{NL}}$, respectively, we can determine the step-size as [109, 110]:

$$\Delta z = \min\left(\Delta\phi_{\mathrm{CD}} L_{\mathrm{CD}}, \Delta\phi_{\mathrm{NL}} L_{\mathrm{NL}}\right). \tag{2.833}$$

Another approach is to keep the local error $\Delta\varepsilon$ below tolerable value by choosing the step-size as [112, 113]:

$$\Delta z = \begin{cases} \left\{\Delta\varepsilon / \left[\gamma P(z)\left(|\beta_2|\omega_{\max}^2\right)^2\right]\right\}^{1/3}, & \text{forSSSFM} \\ \left\{\Delta\varepsilon / \left[\gamma P(z)\left(|\beta_2|\omega_{\max}^2\right)\right]\right\}^{1/2}, & \text{forASSFM} \end{cases} \tag{2.834}$$

where $P(z)$ is the signal power at distance z.

2.4.5.2 GNLSE for Multichannel Transmission

As for *multichannel propagation* modeling, we have two options. The first option is to express multichannel WDM signal through a composite electric field:

$$E(z,t) = \sum_l E_l(z,t) e^{-jl2\pi\Delta f}, \tag{2.835}$$

where Δf is the channel spacing. This approach can be called the "total-field propagation model." The second approach is to consider individual WDM separately and then solve the set of coupled GNLSEs:

$$\frac{\partial E_i(z,t)}{\partial z} = \left(\widehat{D}_i + \widehat{N}_i\right) E_i(z,t),$$
$$\widehat{D}_i = -\frac{\alpha}{2} - \beta_{1,i}\frac{\partial}{\partial t} - j\frac{\beta_2}{2}\frac{\partial^2}{\partial t^2} + \frac{\beta_3}{6}\frac{\partial^3}{\partial t^3}, \widehat{N}_i = j\gamma\left(|E_i|^2 + \sum_{j\neq i}|E_j|^2\right). \tag{2.836}$$

2.4.5.3 GNLSE for Polarization Division Multiplexing Systems

The NSE for PDM systems can be described as follows [109, 110]:

$$\frac{\partial \boldsymbol{E}(z,t)}{\partial z} = \left(\widehat{\boldsymbol{D}} + \widehat{\boldsymbol{N}}\right)\boldsymbol{E}(z,t),$$
$$\widehat{\boldsymbol{D}} = -\frac{\boldsymbol{\alpha}}{2} - \boldsymbol{\beta}_1\frac{\partial}{\partial t} - j\frac{\boldsymbol{\beta}_2}{2}\frac{\partial^2}{\partial t^2} + \frac{\boldsymbol{\beta}_3}{6}\frac{\partial^3}{\partial t^3}, \widehat{\boldsymbol{N}} = j\gamma\left(|\boldsymbol{E}|^2\boldsymbol{I} - \frac{1}{3}\left(\boldsymbol{E}^\dagger Y \boldsymbol{E}\right)Y\right), \tag{2.837}$$

where $\boldsymbol{E} = [E_x\ E_y]^T$ denotes the Jones vector of electric field; $\boldsymbol{\alpha}$ denotes the attenuation coefficients in x- and y-polarizations; and $\boldsymbol{\beta}_1$, $\boldsymbol{\beta}_2$, and $\boldsymbol{\beta}_3$ denote the group velocity, GVD, and second-order GVD parameters, respectively, while Y denotes the Pauli-*Y* matrix in quantum mechanics notation [7]. The Pauli matrices are defined here as:

$$X = \begin{pmatrix} 0 & 1 \\ 1 & 0 \end{pmatrix}, Y = \begin{pmatrix} 0 & -j \\ j & 0 \end{pmatrix}, Z = \begin{pmatrix} 1 & 0 \\ 0 & -1 \end{pmatrix}. \tag{2.838}$$

The $\boldsymbol{\beta}_1$ is related to the polarization rotation matrix $\boldsymbol{R}$ as follows:

$$\boldsymbol{\beta}_1(\theta_k, \varepsilon_k) = \boldsymbol{R}(\theta_k, \varepsilon_k)\left(\frac{\delta_k}{2}Z\right)\boldsymbol{R}^\dagger(\theta_k, \varepsilon_k),$$
$$\boldsymbol{R}(\theta_k, \varepsilon_k) = \begin{bmatrix} \cos\theta_k\cos\varepsilon_k - j\sin\theta_k\sin\varepsilon_k & \sin\theta_k\cos\varepsilon_k + j\cos\theta_k\sin\varepsilon_k \\ \sin\theta_k\cos\varepsilon_k + j\cos\theta_k\sin\varepsilon_k & \cos\theta_k\cos\varepsilon_k - j\sin\theta_k\sin\varepsilon_k \end{bmatrix}, \tag{2.839}$$

where $2\theta_k$ and $2\varepsilon_k$ are azimuth and ellipticity angles of *k*-th fiber section and δ_k corresponds to differential group delay (DGD) of the section. The PDM-NSE should be solved through iterative relation:

$$\boldsymbol{E}(z + \Delta z, t) = \exp\left[\Delta z\left(\widehat{\boldsymbol{D}} + \widehat{\boldsymbol{N}}\right)\right]\boldsymbol{E}(z,t), \tag{2.840}$$

where the complexity of propagation simulation is clearly higher than in the single-polarization case.

2.4.5.4 GNLSE for Spatial Division Multiplexing Systems

In the rest of this section, we describe the GNLSE for *few-mode fiber (FMF)-based mode-division multiplexing*, based on *N* modes, we introduced in [114] (see also [115]). The electric field in a group of *N*/2 degenerate spatial modes is represented by *N*-dimensional complex valued vector $\vec{\mathbf{E}}$. For example, if we consider two modes fiber (LP01 and LP11), then the $\vec{\mathbf{E}}$ can be expanded as:

$$\vec{\mathbf{E}} = [\mathbf{E}_1 \quad \mathbf{E}_2]^T = [E_{1x} \quad E_{1y} \quad E_{2x} \quad E_{2y}]^T, \tag{2.841}$$

which is constructed by stacking the Jones vectors of *N* spatial modes one on top of each other. In Eq. (2.841), $\vec{\mathbf{E}_i}$ denotes the *i*-th spatial mode. We separate the linear and nonlinear operators as follows:

$$\frac{d\vec{\mathbf{E}}}{dz} = \left(\widehat{\mathbf{D}} + \widehat{\mathbf{N}}\right)\vec{\mathbf{E}}, \tag{2.842}$$

where the $\widehat{\mathbf{D}}$ is the linear part and the $\widehat{\mathbf{N}}$ is the nonlinear part. The linear part vector form GNLSE can be written as:

$$\widehat{\mathbf{D}} = -\frac{1}{2}\overrightarrow{\boldsymbol{\alpha}} - \overrightarrow{\boldsymbol{\beta}_1}\frac{dy}{dx} - j\overrightarrow{\boldsymbol{\beta}_2}\frac{\partial^2}{\partial t^2} + \frac{1}{3!}\overrightarrow{\boldsymbol{\beta}_3}\frac{\partial^3}{\partial t^3}, \tag{2.843}$$

where in the ideal case, $\overrightarrow{\boldsymbol{\alpha}}$ is a zero matrix and $\overrightarrow{\boldsymbol{\beta}_i}$ is proportional to the identity. In general cases, the $\overrightarrow{\boldsymbol{\alpha}} = \boldsymbol{\alpha}\overrightarrow{\mathbf{I}}$ ($\overrightarrow{\mathbf{I}}$ is identity matrix), assuming that all modes are attenuated to the same level. The $\overrightarrow{\boldsymbol{\beta}_1}$ can be evaluated by:

$$\overrightarrow{\boldsymbol{\beta}_1} = \mathbf{U} \cdot \mathbf{A} \cdot \mathbf{U}^{\dagger}, \tag{2.844}$$

where **U** and **A** are random unitary Gaussian matrix [116] and different modes delays matrix, respectively. Both **U** and **A**are matrices of size 2 N-by-2 N, and $\mathbf{U}^{\dagger}$is the conjugated transpose of **U**. **A** can be further represented as:

$$\mathbf{A} = \left(\begin{matrix} \overrightarrow{\tau_1} + \frac{\tau_{PMD_1}}{2}Z & \cdots & \overrightarrow{0_{2\times 2}} \\ \vdots & \ddots & \vdots \\ \overrightarrow{0_{2\times 2}} & \cdots & \overrightarrow{\tau_N} + \frac{\tau_{PMD_N}}{2}Z \end{matrix} \right), \tag{2.845}$$

where τ_{PMD_i} is a two-by-two diagonal matrix with elements equal to differential mode group delay (DMGD):

$$\tau_{\mathrm{PMD}_i} = \frac{\mathrm{D}_{PMD_1}}{2\sqrt{2} \cdot \mathrm{L}_{\mathrm{correlation}}}, \tag{2.846}$$

where $\mathrm{L}_{\mathrm{correlation}}$ is the correlation length of the FMF. Similarly, $\overrightarrow{\boldsymbol{\alpha}}$, $\overrightarrow{\boldsymbol{\beta}_2}$, and $\overrightarrow{\boldsymbol{\beta}_3}$ can be extended to matrix form [115, 117]. The nonlinear operator $\widehat{N}_i$ can be defined as:

$$\widehat{N}_i = j\gamma\left[f_{self}\frac{8}{9}\left|\overrightarrow{\mathbf{E}_i}\right|^2\overrightarrow{\mathbf{I}} + \sum_{k\neq i}^{\mathrm{N}} f_{cross}\frac{4}{3}\left|\overrightarrow{\mathbf{E}_k}\right|^2\overrightarrow{\mathbf{I}} - \frac{1}{3}\left(\overrightarrow{\mathbf{E}_i}^{\dagger} Y\overrightarrow{\mathbf{E}_i}\right)Y\right] \quad i = 1, 2, \ldots, \mathrm{N}, \tag{2.847}$$

where $\overrightarrow{\mathbf{E}_i}$ denotes the i-th propagation mode, which includes both x- and y-polarization modes. f_{self} and f_{cross} are general self- and cross-mode coupling coefficients [115, 117]. The i-th spatial mode power $\left|\overrightarrow{\mathbf{E}_i}\right|^2$ can be calculated as $\left|\overrightarrow{\mathbf{E}_i}\right|^2 = \left|\overrightarrow{\mathbf{E}_{ix}}\right|^2 + \left|\overrightarrow{\mathbf{E}_{iy}}\right|^2$. The nonlinearity factor γ in Eq. (2.847) can be obtained as described in previous section by:

$$\gamma = \frac{2\pi \mathrm{n}_2 \mathrm{f}_{\mathrm{ref}}}{\mathrm{c}\mathrm{A}_{\mathrm{eff}}}, \tag{2.848}$$

where n_2 is the nonlinear index, $\mathrm{f}_{\mathrm{ref}}$ is the reference frequency, and $\mathrm{A}_{\mathrm{eff}}$ is the core area.

2.5 Noise Sources in Optical Channels

The total noise is a stochastic process that has both multiplicative (introduced only when the signal is present) and additive (always present) components. *Multiplicative noise components* are mode partition noise (MPN), laser intensity noise (RIN), modal noise, quantum shot noise, and avalanche shot noise. *Additive noise components* are dark current noise, thermal noise, amplified spontaneous emission (ASE) noise, and crosstalk noise.

Originators of the noise in an optical transmission system can be identified as follows. In semiconductor lasers, the laser intensity noise, laser phase noise, and mode partition noise are present. In optical cable, in fiber and splicing, the modal noise and reflection-induced noise are present. In optical amplifiers, the spontaneous emission and amplified spontaneous emission noises are present. At the receiver side, during the photodetection process, the thermal noise and quantum noise are generated.

The noise components at the receiver side are cumulative. The most relevant optical noise components at the input of optical pre-amplifier are intensity noise and spontaneous emission noise, which contains the spontaneous emission noise from the preceding in-line amplifier and accumulated ASE noise from other in-line amplifiers. The optical pre-amplifier will enhance all optical inputs in proportion to the gain, and it will also generate additional spontaneous emission noise. During the photodetection process, as already indicated above, the quantum noise and thermal noise will be generated; in addition, different beating components, such as signal-noise beating and noise-noise beating components, will be created. The most important noise sources coming from the front-end amplifiers are ASE noise and signal-noise beating component. The front end amplifier will generate additional thermal noise.

2.5.1 Mode Partition Noise

The mode partition noise (MPN) is the relative fluctuation in power between the main and side modes. The total power can remain unchanged, but the distribution of power among modes changes. MPN is present (up to certain amount) even in DFBs. MPN can affect the RIN significantly by enhancing it. It will occur even when the mode suppression ratio (MSR) is around 20 dB. Different modes will travel at different velocities due to dispersion, so that MPN can lead to the ISI.

2.5.2 Reflection-Induced Noise

The reflection-induced noise is related to the appearance of the back-reflected light due to refractive index discontinuities at optical splices, connectors, and optical fiber ends. The amount of reflected light can be estimated by the *reflection coefficient*, introduced earlier by:

$$r_{\text{ref}} = \left(\frac{n_a - n_b}{n_a + n_b}\right)^2, \tag{2.849}$$

where n_a and n_b denote the refractive index coefficients of materials facing each other.

The strongest reflection occurs at the glass-air interface, $r_{\text{ref}} \approx [(1.46–1)/(1.46 + 1)]^2 \approx 3.5\%$ (−14.56 dB). It can be reduced below 0.1% if index-matching oils or gels are used. The considerable amount of back-reflected light can come back and enter the laser cavity, negatively affecting the laser operation and leading to excess intensity noise. The RIN can be increased by up to 20 dB if the back-reflected light exceeds −30 dBm. The multiple back and forth reflections between splices and connectors can be the source of additional intensity noise.

2.5.3 Relative Intensity Noise (RIN) and Laser Phase Noise

The operating characteristics of semiconductor lasers are well described by the set of ordinary differential equations – *the rate equations* – which describe the interaction between photons and electrons inside the active region [2–4, 98]:

$$\frac{dP}{dt} = GP + R_{sp} - \frac{P}{\tau_p} + F_P(t), \tag{2.850}$$

$$\frac{dN}{dt} = \frac{I}{q} - \frac{N}{\tau_c} - GP + F_N(t) \tag{2.851}$$

and

$$\frac{d\phi}{dt} = \frac{1}{2}\alpha_{\text{chirp}}\left[G - \frac{1}{\tau_p}\right] + F_\phi(t) \tag{2.852}$$

In Eq. (2.850) the term GP denotes the increase in photon number (P) due to stimulated emission, the term R_{sp} denotes the increase in P due to spontaneous emission and can be calculated by $R_{sp} = n_{sp}G$ (n_{sp}-the spontaneous emission factor, typically $n_{sp} \approx 2$), the term $-P/\tau_p$ denotes the decrease in P due to emission through the mirrors and scattering/absorption by free carriers, and $F_p(t)$ corresponds to the noise process. In Eq. (2.851), the term I/q denotes the increase in electron numbers (N) due to injection current I, the term $-N/\tau_c$ denotes the decrease in N due to spontaneous emission and nonradiative recombination, the term $-GP$ denotes the decrease in N due to stimulated emission, and the term $F_n(t)$ corresponds to the noise process. Finally, in Eq. (2.852), the first term corresponds to dynamic chirp and the second term to the adiabatic chirp, and $F_\phi(t)$ corresponds to the noise process. The noise terms $F_p(t)$, $F_n(t)$, and $F_\phi(t)$ introduced above are known as *Langevin forces* that are commonly modeled as zero-mean Gaussian random processes. The photon lifetime τ_p is related to the internal losses (α_{int}) and mirror losses (α_{mir}) by $\tau_p = [v_g(\alpha_{int} + \alpha_{mir})]^{-1}$, with v_g being the group velocity. The carrier lifetime τ_c is related to the spontaneous recombination time (τ_{spon}) and non-radiation recombination time (τ_{nr}) by $\tau_c = \tau_{spon}\tau_{nr}/(\tau_{spon} + \tau_{nr})$. The net rate of stimulate emission G is related to the material gain g_m by $G = \Gamma v_g g_m$, where Γ is the confinement factor.

Noise in semiconductor lasers originates from two sources: (i) spontaneous emission (the dominant noise source) and (ii) electron-hole recombination shot noise. As mentioned above, the noise effects can be modeled by random driving terms in laser rate equations known as *Langevin forces*, which can be described as zero-mean Gaussian random processes with autocorrelation function (*Markoffian approximation*) [2–4, 98]:

$$\langle F_i(t)F_j(t')\rangle = 2D_{ij}\delta(t-t');\, i,j \in \{P,N,\phi\}, \tag{2.853}$$

with dominating factor being $D_{pp} = R_{sp}P$ and $D_{\phi\phi} = R_{sp}/4P$. Fluctuation in intensity can be described by *intensity autocorrelation function* [2–4, 98]:

$$C_{PP}(\tau) = \langle \delta P(t)\delta P(t+\tau)\rangle/\bar{P}^2, \bar{P} = \langle P\rangle, \delta P = P - \bar{P} \tag{2.854}$$

The RIN spectrum can be found by FT of the autocorrelation function [2–4, 98]:

$$\mathrm{RIN}(\omega) = \int_{-\infty}^{\infty} C_{pp}(\tau)e^{-j\omega t}d\tau \sim \begin{cases} 1/\bar{P}^3 \text{ at low } \bar{P} \\ 1/\bar{P} \text{ at high } \bar{P} \end{cases} \tag{2.855}$$

Finally, the SNR can be estimated by [2–4, 98]:

$$SNR = [C_{PP}(0)]^{-1/2} \sim \left(\frac{\varepsilon_{NL}}{R_{sp}\tau_p}\right)^{1/2} \bar{P} \tag{2.856}$$

where the approximation is valid for SNR above 20 dB.

Noise, especially spontaneous emission, causes phase fluctuations in lasers, leading to a nonzero spectral linewidth $\Delta\nu$. In gas- or solid-state lasers, the linewidth $\Delta\nu$ typically ranges from the subhertz to the kilohertz range. In semiconductor lasers, the linewidth $\Delta\nu$ is often much larger, up the megahertz range, because of (i) the small number of photons stored in the small cavity and (ii) the non-negligible value of the linewidth enhancement factor α_{chirp}. The spectrum of emitted light electric field $E(t) = \sqrt{(P)}\exp(j\phi)$ is related to the field autocorrelation function $\Gamma_{EE}(\tau)$ by:

$$\begin{gathered} S(\omega) = \int_{-\infty}^{\infty} \Gamma_{EE}(t)e^{-j(\omega-\omega_0)\tau}d\tau, \\ \Gamma_{EE}(t) = \langle E^*(t)E(t+\tau)\rangle = \langle \exp(\,j\Delta\phi(t))\rangle = \exp\left(-\langle\Delta\phi^2(\tau)\rangle/2\right), \end{gathered} \tag{2.857}$$

where $\Delta\phi(\tau) = \phi(t+\tau) - \phi(t)$. By describing the laser phase noise process as Wiener-Lévy process, that is, a zero-mean Gaussian process with variance $2\pi\Delta\nu|t|$ (where $\Delta\nu$ is the laser linewidth), the spectrum of emitted light is found to Lorentzian:

$$S_{EE}(\nu) = \frac{\overline{P}}{2\pi\Delta\nu}\left[\frac{1}{1+\left(\frac{2(\nu+\nu_0)}{\Delta\nu}\right)^2} + \frac{1}{1+\left(\frac{2(\nu-\nu_0)}{\Delta\nu}\right)^2}\right], \tag{2.858}$$

where ν_0 is the central frequency and other parameters are already introduced above. The phase fluctuation variance (neglecting the relaxation oscillation) is [2–4, 98]:

$$\langle \Delta\phi^2(\tau)\rangle = \frac{R_{\text{sp}}}{2\overline{P}}\left(1+\alpha_{\text{chirp}}^2\right)|\tau| = 2\pi\Delta\nu|\tau|. \tag{2.859}$$

The laser linewidth can, therefore, be evaluated by:

$$\Delta\nu = \frac{R_{sp}}{4\pi\overline{P}}\left(1+\alpha_{\text{chirp}}^2\right), \tag{2.860}$$

and we conclude that the chirp effect enhances the laser linewidth by factor $\left(1+\alpha_{\text{chirp}}^2\right)$.

2.5.4 Modal Noise

Modal noise is related to multimode optical fibers. The optical power is non-uniformly distributed among a number of modes in multimode fibers, causing the so-called speckle pattern at the receiver end containing brighter and darker sports in accordance to the mode distribution. If the speckle pattern is stable, the photodiode effectively eliminates it by registering the total power over the photodiode area. If the speckle pattern changes with time, it will induce the fluctuation in the received optical power, known as the modal noise. The modal noise is inversely proportional to the laser linewidth, so it is a good idea to use LEDs in combination with multimode fibers.

2.5.5 Thermal Noise

The photocurrent generated during photodetection process is converted to the voltage through the load resistance. The load voltage is further amplified by the front-end amplifier stage. Due to random thermal motion of electrons in load resistance, the already generated photocurrent exhibits additional noise component, also known as *Johnson noise*. The front-end amplifier enhances the thermal noise generated in the load resistance, which is described by the *amplifier noise figure* $F_{\text{n,el}}$. The *thermal noise PSD* in the load resistance R_{L} and corresponding *thermal noise power* are defined as [2–4, 98]:

$$S_{\text{thermal}} = \frac{4k_B\Theta}{R_L}, \quad \langle i^2\rangle_{\text{thermal}} = \frac{4k_B\Theta}{R_L}\Delta f \tag{2.861}$$

where Θ is the absolute temperature, k_{B} is the Boltzmann's constant, and Δf is the receiver bandwidth.

2.5.6 Spontaneous Emission Noise and Noise Beating Components

The spontaneous emission of light appears during the optical signal amplification, and it is not correlated with signal (it is additive in nature). The noise introduced by spontaneous emission has a flat frequency characterized by Gaussian probability density function (PDF), and the (double-sided) PSD in one state of polarization is given by [2–4, 98]:

$$S_{\text{sp}}(\nu) = (G-1)F_{\text{no}}h\nu/2, \tag{2.862}$$

where G is the amplifier gain and $h\nu$ is the photon energy. The *optical amplifier noise figure* F_{no}, defined as the ratio of SNRs at the input and output of the optical amplifier, is related to the *spontaneous emission* factor $n_{sp} = N_2/(N_2 - N_1)$ (for the two-level system) by [2–4, 98]:

$$F_{no} = 2n_{sp}\left(1 - \frac{1}{G}\right) \cong 2n_{sp} \geq 2. \tag{2.863}$$

In the most of practical cases, the noise figure is between 3 and 7 dB. The effective noise figure of the cascade of K amplifiers with corresponding gains G_i and noise figures $F_{no,i}$ $(i = 1,2,\ldots,K)$ [2–4, 98] is as follows:

$$F_{no} = F_{no,1} + \frac{F_{no,2}}{G_1} + \frac{F_{no,3}}{G_1 G_2} + \ldots + \frac{F_{no,K}}{G_1 G_2 \ldots G_{k-1}}. \tag{2.864}$$

The total power of the spontaneous emission noise, for an amplifier followed by an optical filter of bandwidth B_{op}, is determined by:

$$P_{sp} = 2\left|E_{sp}\right|^2 = 2S_{sp}(\nu)B_{op} = (G-1)F_{no}h\nu B_{op}, \tag{2.865}$$

where the factor 2 is used to account for both polarizations.

Optical pre-amplifiers are commonly used *to improve the receiver sensitivity* by pre-amplifying the signal before it reaches the photodetector, which is illustrated in Fig. 2.1. The gain of the optical amplifier is denoted with G, the optical filter bandwidth is denoted with B_{op}, and electrical filter bandwidth is denoted with B_{el} (Fig. 2.56).

An optical amplifier introduces the spontaneous emission noise in addition to the signal amplification:

$$E(t) = \sqrt{2P}\cos\left(2\pi f_c t + \theta\right) + n(t), \tag{2.866}$$

where $E(t)$ denotes the electrical field at the receiver input and $n(t)$ denotes the amplifier spontaneous emission noise, which is the Gaussian process with PSD given by Eq. (2.862). The photocurrent has the following form:

$$i(t) = RP(t) = RE^2(t) = 2RP\cos^2(2\pi f_c t + \theta) + 2R\sqrt{2P}n(t)\cos\left(2\pi f_c t + \theta\right) + Rn^2(t). \tag{2.867}$$

Following the procedure described [2–4, 98], we can determine the variance of different noise and beating components as follows:

$$\begin{aligned} \sigma^2 &= \int_{-B_{el}}^{B_{el}} S_i(f)df = \sigma_{sh}^2 + \sigma_{sig\text{-}sp}^2 + \sigma_{sp\text{-}sp}^2,\ \sigma_{sh}^2 = 2qR\left[GP_{in} + S_{sp}B_{op}\right]B_{el}, \\ \sigma_{sig\text{-}sp}^2 &= 4R^2GP_{in}S_{sp}B_{el},\ \sigma_{sp\text{-}sp}^2 = R^2S_{sp}^2\left(2B_{op} - B_{el}\right)B_{el} \end{aligned} \tag{2.868}$$

where we used subscripts "sh" to denote the variance of shot noise, "sp-sp" to denote the variance of spontaneous-spontaneous beat noise, and "sig-sp" to denote the variance of signal-spontaneous beating noise. S_{sp} is the PSD of spontaneous emission noise (Eq. (2.21)), P_{in} is the average power of incoming signal (see Fig. 2.1), and the components of the PSD of photocurrent $i(t)$, denoted by $S_i(f)$, are shown in Fig. 2.57.

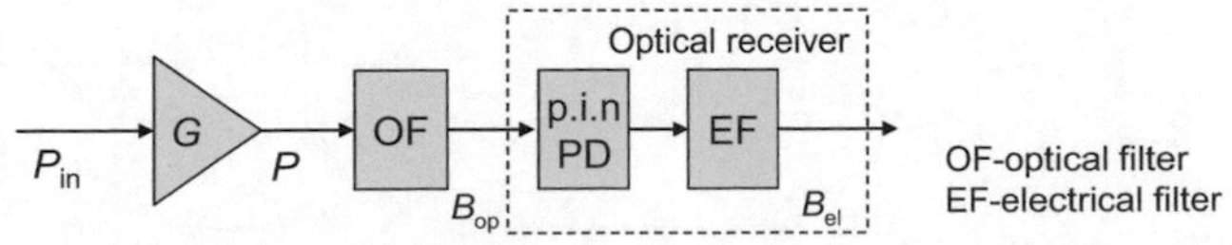

Fig. 2.56 Optical receiver with the pre-amplifier

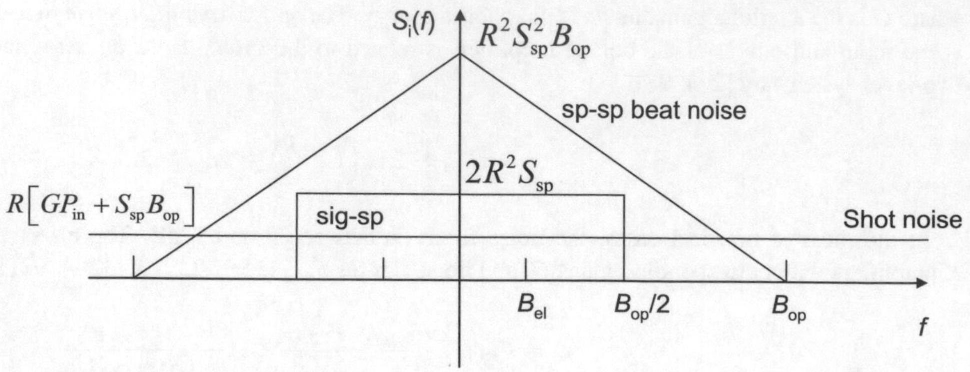

Fig. 2.57 Components of PSD of photocurrent

2.5.7 Quantum Shot Noise

The optical signal arriving at the photodetector contains a number of photons generating the electron-hole pairs through the photoelastic effect. The electron-hole pairs are effectively separated by the reversed bias voltage. However, not every photon is going to generate the electron-hole pair contributing to the total photocurrent. The probability of having n electron-hole pairs during the time interval Δt is governed by the *Poisson probability density function* (PDF):

$$p(n) = \frac{N^n e^{-N}}{n!}, \tag{2.869}$$

where N is the average number of generated electron-hole pairs. The Poisson PDF approaches Gaussian for large N. The *mean* of *photocurrent intensity*:

$$I = \langle i(t) \rangle = \frac{qN}{\Delta t} = \frac{qN}{T}, \qquad \Delta t = T \tag{2.870}$$

For Poisson distribution, the variance equals the mean $\langle (n - N)^2 \rangle = N$ so the photocurrent can be calculated by $i(t) = qn/T$. The *shot noise mean square value* can be determined by:

$$\langle i^2 \rangle_{sn} = \left\langle [i(t) - I]^2 \right\rangle = \frac{q^2 \left\langle [n - N]^2 \right\rangle}{T^2} = \frac{q^2 N}{T^2} = \frac{qI}{T} = 2qI\Delta f, \Delta f = \frac{R_b}{2} \quad \left(R_b = \frac{1}{T}\right) \tag{2.871}$$

where R_b is the bit rate. The Eq. (2.13) is derived for NRZ modulation format, but validity is more general. Therefore, the power spectral density of shot noise is determined by $S_{sn}(f) = 2qI$. To determine the APD *shot noise mean square value*, the p-i-n shot noise mean square value has to be multiplied by the excess noise factor $F(M)$ to account for the randomness of impact ionization effect:

$$\langle i^2 \rangle_{sn}^{APD} = S_{sn}^{APD}(f)\Delta f = 2qM^2 F(M) I \Delta f, F(M) = k_N M + (1 - k_N)\left[2 - \frac{1}{M}\right], \tag{2.872}$$

where M is the average APD multiplication factor and k_N is the ratio of impact ionization factors of holes and electrons, respectively. The often used excess noise factor approximation is the following one [98]:

$$F(M) \cong M^x, \quad x \in [0, 1]. \tag{2.873}$$

2.5.8 Dark Current Noise

The dark current I_d flows through the reversed-biased photodiode even in the absence of incoming light. The dark current consists of electron-hole pairs created due to thermal effects in p-n junction. The dark current noise power can be calculated as follows [2–4, 98]:

$$\langle i^2 \rangle_{\text{dcn}}^{\text{APD}} = S_{\text{dcn}}^{\text{APD}}(f)\Delta f = 2q\langle M\rangle^2 F(M) I_d \Delta f. \tag{2.874}$$

2.5.9 Crosstalk

The crosstalk effects occur in multichannel (WDM) systems and can be classified as (i) *interchannel* component (the crosstalk wavelength is sufficiently different from observed channel wavelength) and (ii) *intrachannel* (the crosstalk signal is at the same wavelength as observed channel or sufficiently close) component.

Interchannel crosstalk (also known as out-of-band or hetero-wavelength crosstalk) is introduced by either an *optical filter* or *demultiplexer* that selects desired channel and imperfectly rejects the neighboring channels, which is illustrated in Fig. 2.58 (a). Another source of interchannel crosstalk is an *optical switch*, shown in Fig. 2.58(b), in which the crosstalk arises because of imperfect isolation among different wavelength ports. The crosstalk signal behaves as noise, and it is a source of linear (non-coherent) crosstalk, so that the photocurrent at receiver can be written as [2, 4, 99]:

$$I = R_m P_m + \sum_{n \neq m} R_n X_{mn} P_n \equiv I_{ch} + I_X, \ R_m = \eta_m q / h\nu_m, \quad I_X = \sum_{n \neq m} R_n X_{mn} P_n \tag{2.875}$$

where R_m is the photodiode responsivity of mth WDM channel with average power P_m, $I_{\text{ch}} = R_m P_m$ is the photocurrent corresponding to the selected channel, I_X is the crosstalk photocurrent, and X_{mn} is the portion of nth channel power captured by the optical receiver of the mth optical channel.

The intrachannel crosstalk in transmission systems may occur due to multiple reflections. However, it is much more important in optical networks. In optical networks, it arises from cascading optical (WDM) demux and mux or from an optical switch, which is illustrated in Fig. 2.59. Receiver current $I(t) = R|E_m(t)|^2$ contains interference or beat terms, in addition to the desired signal: (i) signal-crosstalk beating terms (like $E_m E_n$) and (ii) crosstalk-crosstalk beating terms (like $E_n E_k$, $k \neq m, n \neq m$). The total electrical field E_m and receiver current $I(t)$, observing only the in-band crosstalk, can be written as [2, 4, 99]:

$$\begin{aligned} E_m(t) &= \left(E_m + \sum_{n \neq m} E_n\right) e^{\,j\omega_m t}, \\ I(t) &= R|E_m(t)|^2 \approx R P_m(t) + 2R \sum_{n \neq m}^{N} \sqrt{P_m P_n} \cos\left[\phi_m(t) - \phi_n(t)\right] \end{aligned} \tag{2.876}$$

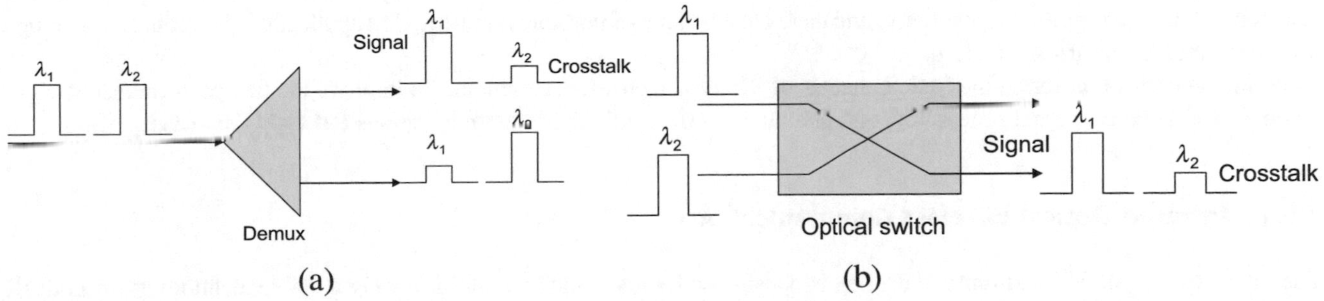

Fig. 2.58 Illustration of the sources of interchannel crosstalk: (**a**) demux and (**b**) the optical switch

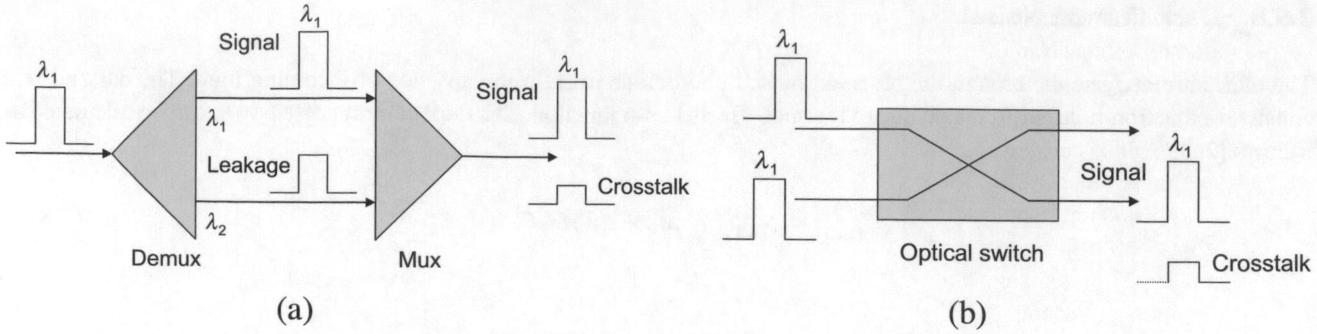

Fig. 2.59 Illustration of the sources of intrachannel crosstalk: (**a**) a demux-mux pair and (**b**) the optical switch

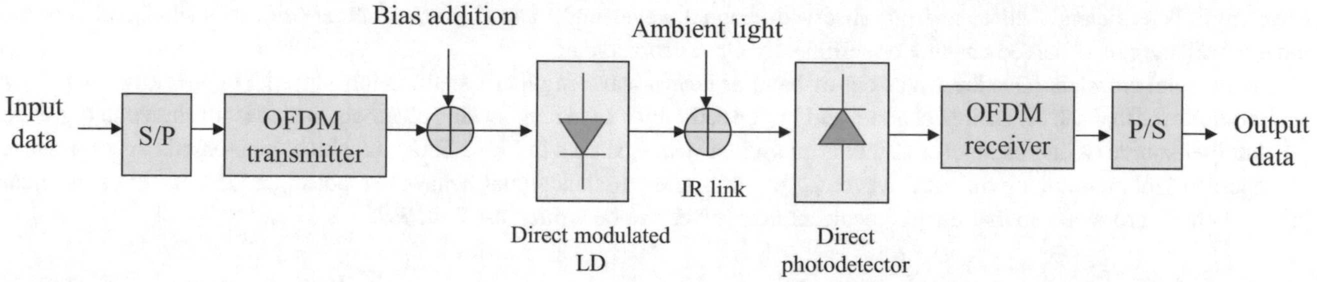

Fig. 2.60 A typical OFDM-based optical wireless communication link with direct modulation and direct detection. *LD* laser diode

where $\phi_n(t)$ denotes the phase of nth channel and R is the photodiode responsivity. Each term acts as an independent random variable, and the crosstalk effect can be observed as the intensity noise (i.e., Gaussian for large N):

$$I(t) = R(P_m + \Delta P) \quad (P_n << P_m, \ n \neq m), \tag{2.877}$$

where ΔP is the intensity noise due to intrachannel crosstalk.

2.6 Indoor Optical Wireless Communication Channels

Optical wireless communications (OWCs) [103, 118–124] represent a viable and promising supplemental technology to next-generation wireless technology (5G). OWCs offer several attractive features, compared to radio, such as [4, 125] (i) huge available unregulated bandwidth, (ii) systems that do not introduce electromagnetic interference to existing radio systems, (iii) optical wireless signal that does not penetrate through the walls so that the room represents an optical wireless cell, (iv) the same equipment that can be reused in other cells without any inter-cell interference, (v) simple design of high-capacity LANs because coordination between neighboring cells is not needed, and (vi) their inherent applicability in environments when radio systems are not desired (airplanes, hospitals, etc.).

Majority of OWCs are designed to operate in the near-infrared (IR) region at wavelength of either 850 nm or 1550 nm. The research in this area is getting momentum, and there are a number of possible point-to-point applications including short-range links described in the IrDA standard.

In this section, we describe the basic concepts of IR OWCs, different techniques to improve the link performance such as diversity and adaptive signal processing, and potential use of visible light communications (VLCs) [121–124].

2.6.1 Infrared Optical Wireless Communications

The intensity modulation and direct detection is considered by many authors, and it is only a viable technology for both IR OWCs and VLCs. One typical OFDM-based WOC link is shown in Fig. 2.60. Because of unguided propagation, the OWC links faces significant free-space loss. Similarly as wireless indoor links, the most important factor in performance degradation is the multipath effect. Namely, in addition to the LOS component, several diffuse reflection-based components arrive at receiver side. Because the direct detection is used, we do not worry about co-phasing of different multipath components. However, since high-speed signals are transmitted over an IR link, different multipath components arrive in time slots of neighboring symbols causing the ISI. To deal with ISI effect, different methods can be classified into three broad categories: (i) equalization, (ii) multicarrier modulation (such as OFDM), and (iii) spread spectrum. Multicarrier modulation can also be combined with diversity and multi-input multi-output (MIMO) techniques. The dominant noise source in IR links is the ambient light, which is responsible for the shot noise component. Artificial light sources also introduce harmonics up to 1 MHz in the transmitted signal. When the ambient light is weak, the receiver pre-amplifier is predominant noise source. Fortunately, most of the ambient noise can be removed by proper optical filtering. A typical receiver for OWC consists of an optical front end, with an optical concentrator being used to collect the incoming radiation, an optical filter (to reduce the ambient light), a photodetector, and an electrical front end. The OFDM-based OWC receiver contains an OFDM receiver. By reducing the number of subcarriers to one, the phase-modulated signals can be transmitted over direct detection links. For intensity modulation with direct detection (IM/DD), the bias addition circuit should be omitted and OFDM transmitter replaced by a NRZ/RZ/pulse-position modulator. On receiver side, for IM/DD, OFDM transmitter should be replaced by the corresponding RF demodulator.

Different link types used in OWC systems, shown in Fig. 2.61, can be classified as [120] point-to-point links (there exists a LOS component), diffuse links (transmitter emits optical power over a wide solid angle; emitted light gets reflected from ceiling so that receiver gets illuminated from diffused light), and quasi-diffuse links (the transmitter illuminates the ceiling with a series of slowly diverging beam sources). The quasi-diffuse links are also known as non-LOS links. The point-to-point links may contain both LOS component and diffused reflected multipath components (see Fig. 2.61(d)). The channel transfer function can be written as follows:

$$H(f) = H_{\mathrm{LOS}} + H_{\mathrm{diff}}(f), \tag{2.878}$$

where H_{LOS} corresponds to the LOS path amplitude gain and $H_{\mathrm{diff}}(f)$ represents the contribution of diffuse reflected multipath components, which can be approximated as a low-pass filtering function [119]. When the room is homogeneous and isotropic, the Rician factor, introduced in section on wireless channel modeling, can be used to quantify the ratio between the LOS and diffuse components as follows $K = (H_{\mathrm{LOS}}/H_{\mathrm{diff}})^2$.

When the LOS link is reasonably short, and the LED is used as the source and the non-imaging concentrator to collect the light; we can model the transfer function of LOS as follows [124, 125]:

$$H_{\mathrm{LOS}} = A_R(m+1)\cos^m\phi\, t_s(\psi)g(\psi)\cos\psi/\left(2\pi r^2\right), 0 \le \psi \le \psi_{\mathrm{FOV}}, \tag{2.879}$$

where A_{R} is the active area, ϕ is the transmit angle with respect to the maximum radiation direction, while ψ and r are the angle of irradiance and the distance to illuminated surface, respectively, as shown in Fig. 2.62. In (2.879), ψ_{FOV} denotes the field-of-view (FOV) angle. Parameter m denotes the Lambert index related to the source radiation semi-angle by $m = -1/\log_2(\cos\theta_{1/2})$, with $\theta_{1/2}$ being the IR LED semi-angle at half-power. In (2.879), we use $t_s(\psi)$ to denote the transmission of an optical filter and $g(\psi)$ to denote the gain of non-imaging concentrator $g(\psi) = n^2/\sin^2\psi_{\mathrm{FOV}}$, $0 \le \psi \le \psi_{\mathrm{FOV}}$ (n is the refractive index). Providing that $r >> \sqrt{A_R}$, the corresponding impulse response of the LOS component is given by:

$$h_{\mathrm{LOS}}(t) = A_R(m+1)\cos^m\phi\, t_s(\psi)g(\psi)\cos\psi/\left(2\pi r^2\right)\delta(t - r/c), 0 \le \psi \le \psi_{\mathrm{FOV}}, \tag{2.880}$$

where r/c is clearly the propagation delay.

For the *point-to-pint links* and *IM/DD*, the Rx signal, after the photodetection takes place, can be determined as the convolution of transmitted signal $s(t)$ and impulse response of the channel as follows:

$$r(t) = R\, s(t) * [h_{\mathrm{LOS}}(t) + h_{\mathrm{diff}}(t)] + n(t) = R\int_{-\infty}^{\infty} s(t')[h_{\mathrm{LOS}}(t-t') + h_{\mathrm{diff}}(t-t')]dt' + n(t), \tag{2.881}$$

where R is the photodiode responsivity, $h_{\mathrm{diff}}(t) = \mathrm{FT}^{-1}\{H_{\mathrm{diff}}(f)\}$ is the diffuse components impulse response, and $n(t)$ is the noise originating from ambient noise and thermal noise of transimpedance amplifier that typically follows the photodetector.

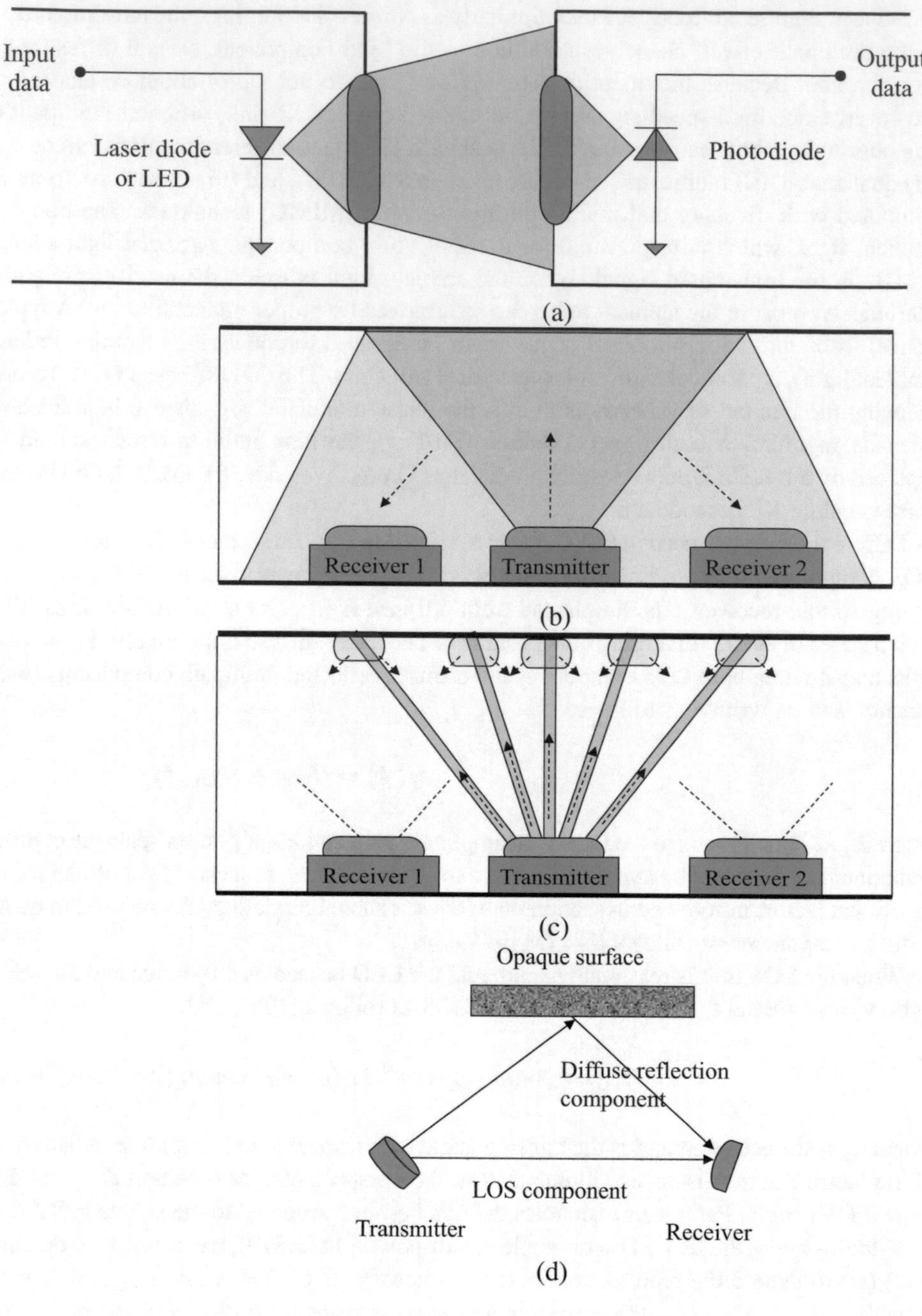

Fig. 2.61 Different link types in OWCs: (**a**) a point-to-point link, (**b**) a diffuse link, and (**c**) quasi-diffuse link (non-LOS link). (**d**) In point-to-point links, received signal may contain both LOS component and diffusion reflected components

The *i*-th reflected component of a *quasi-diffuse link*, shown in Fig. 2.63, can be considered as a generalization of the LOS propagation model from Fig. 2.62. The corresponding impulse response of the *i*-th reflected component is given by:

$$h_{\mathrm{refl},i}(t) = \frac{A_R A_{\mathrm{ref},i} r_{\mathrm{refl},i}}{2\pi r_{Tx,i}^2 r_{Rx,i}^2}(m+1)\cos^m \phi_{Tx,i} \cos\left(\psi_{Tx,i}\right) t_s(\psi_{Rx,i}) g(\psi_{Rx,i}) \cos\psi_{Rx,i}\delta\left(t - \frac{r_{Tx,i} + r_{Rx,i}}{c}\right), \tag{2.881}$$

where $r_{\mathrm{refl},i}$ is the reflection coefficient, $A_{\mathrm{ref},i}$ is the area of the reflector, and corresponding angles are shown in Fig. 2.63. The transmission and gain functions are the same as defined above.

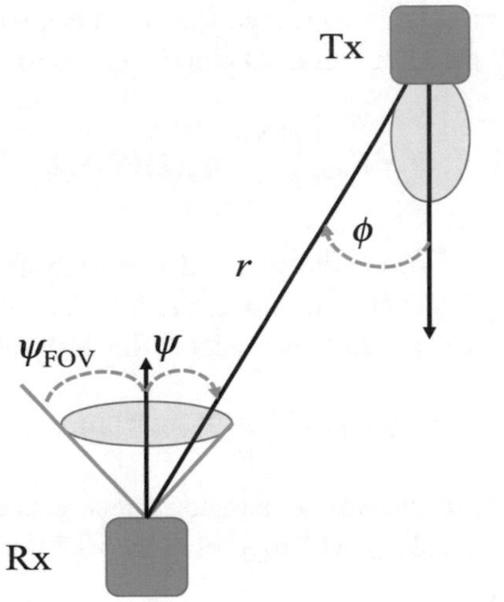

Fig. 2.62 Illustrating the LOS propagation model for IR-LED (Lambertian source)

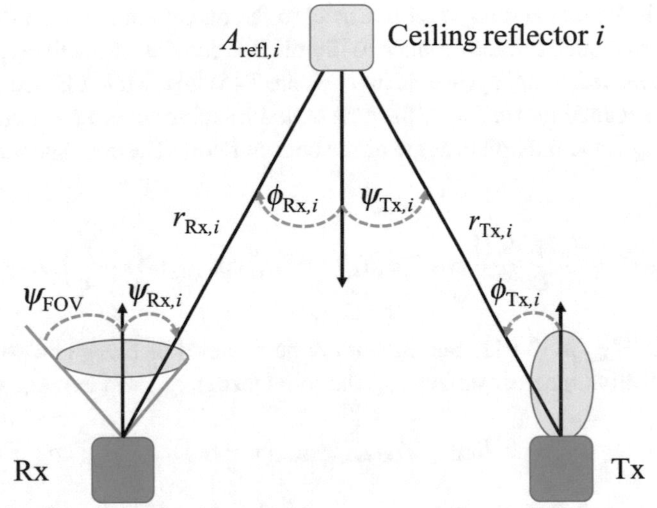

Fig. 2.63 Illustrating the *i*-th reflected component of the quasi-diffuse link

2.6.2 Visible Light Communications (VLCs)

VLCs represent a relatively new research area [121–123] with operating wavelengths of 380–720 nm by employing white-light LEDs with available bandwidth of around 20 MHz. With visible-light LEDs (VLEDs), it would be possible to broadcast broadband signals in various environments such as offices where the lighting is used already. Power line communication systems or plastic optical fiber systems can be used as feeders. VLCs can also be used in car-to-car communications by employing LED-based head and tail lights and probably in airplanes to provide high-speed internet. The VLC systems model is similar to that shown in Fig. 2.60; white-light LED is used instead of laser diode. The VLC channel model can be represented by:

$$H(f) = \sum_i H_{\mathrm{LOS}}, i \exp\left(-j2\pi f \Delta\tau_{\mathrm{LOS},i}\right) + \underbrace{H_{\mathrm{diff}} \frac{\exp\left(-j2\pi f \Delta\tau_{\mathrm{diff}}\right)}{1 + jf/f_{\mathrm{cutoff}}}}_{H_{\mathrm{diff}}(f)}, \tag{2.882}$$

where $H_{\mathrm{LOS},i}$ and H_{diff} represent the channel gain of *i*-th LOS component (coming from *i*th LED) and diffuse signal gain, respectively. The $\Delta\tau_{\ \mathrm{LOS},i}$ and $\Delta\tau_{\mathrm{diff}}$ represent the corresponding delays of *i*-th LOS component and diffuse signal components. Frequency f cutoff denotes the cutoff frequency of diffuse channel frequency response, and for medium-sized room, it is about 10 MHz [123]. The LOS gain from ith LED chip is related to the luminous flux:

$$\Phi = K_{\max} \int_{380\,\mathrm{nm}}^{780\,\mathrm{nm}} \Phi_e(\lambda) V(\lambda) d\lambda, \tag{2.883}$$

where $\Phi_e(\lambda)$ is the source power distribution, $V(\lambda)$ is the standard sensitivity function, and $K_{\max}$ is the maximum visibility (683 lm/W at 555 nm). The corresponding luminous intensity is the first derivative of the luminous flux with respect to the solid angle Ω; in other words, $I = d\Phi/d\Omega$. The radiation intensity at the desk surface is given by:

$$I(\phi) = I(\phi = 0)\cos^m(\phi), \tag{2.884}$$

where ϕ is the transmit angle with respect to the maximum radiation direction and m denotes the Lambert index related to the source radiation semi-angle by $m = -1/\log_2(\cos\theta_{1/2})$, with $\theta_{1/2}$ being the VLED semi-angle at half-power. Now, the LOS gain from the i-th white-LED is given by:

$$H_{\mathrm{LOS}}, i = A_R(m+1)\cos^m(\phi_i)t_s(\psi_i)g(\psi_i)\cos(\psi_i)/\left(2\pi r_i^2\right), \tag{2.885}$$

where A_{R} is the active area, ϕ_i is the transmit angle with respect to the maximum radiation direction of the i-th light source, while ψ_i and r_i are the angle of irradiance and the distance to illuminated surface, respectively. As expected, this expression is very similar to Eq. (2.879). Figure 2.62 is still applicable here, where Tx is now white LED. In (2.885), we use $t_s(\psi_i)$ to denote the transmission of the optical filter and $g(\psi_i)$ to denote the gain of non-imaging concentrator $g(\psi_i) = n^2/\sin^2\psi_{\mathrm{con}}$, $0 \le \psi \le \psi_{\mathrm{con}}$ (n is the refractive index and ψ_{con} is the maximum angle of the concentrator). The corresponding impulse response of the i-th LOS component is given by:

$$h_{\mathrm{LOS}}, i(t) = \frac{A_R(m+1)}{(2\pi r_i^2)}\cos^m(\phi_i)t_s(\psi_i)g(\psi_i)\cos(\psi_i)\delta\left(t - \frac{r_i}{c}\right), r_i \ge \sqrt{A_R} \tag{2.886}$$

The diffuse signal magnitude H_{diff} in (2.882) can commonly be modeled as being constant, in average, anywhere in the room, and it is related to the effective receiver surface A_{R}, the room size A_{room}, and average reflectivity r_{refl} by [123]:

$$H_{\mathrm{diff}} = A_R r_{\mathrm{refl}}/A_{\mathrm{room}}(1 - r_{\mathrm{refl}}). \tag{2.887}$$

For the *point-to-pint VLC links* and *IM/DD*, the Rx signal, after the photodetection takes place, can be determined as the convolution of transmitted signal $s(t)$ and impulse response of the VLC channel as follows:

$$\begin{aligned} r(t) &= R\, s(t) * \left[\sum\nolimits_i h_{\mathrm{LOS},i}(t) + h_{\mathrm{diff}}(t)\right] + n(t) \\ &= R\int_{-\infty}^{\infty} s(t')\left[\sum\nolimits_i h_{\mathrm{LOS},i}(t) + h_{\mathrm{diff}}(t-t')\right]dt' + n(t), \end{aligned} \tag{2.888}$$

where R is the photodiode responsivity and $h_{\mathrm{diff}}(t) = \mathrm{FT}^{-1}\{H_{\mathrm{diff}}(f)\}$.

2.7 Concluding Remarks

This chapter has been devoted to propagation effects, noise sources, and channel impairments in wireless, free-space optical, and fiber-optics communications systems. The key idea behind the chapter has been to study the propagation effects in different channel types based on Maxwell's equations. The chapter has been organized as follows. In Sect. 2.1, basic concepts from electromagnetic fields and wave equations have been described. In this section, the vector derivatives needed to study the propagation over different media have been described in generic orthogonal curvilinear coordinate system, for which Cartesian, cylindrical polar, and spherical coordinate systems have been just special cases. In Sect. 2.2.1, we have studied the vectorial nature of the light, derived Snell's law of refraction from Maxwell's equations, studied total internal reflection, and explained the light confinement in step-index optical fibers. Additionally, we have described how arbitrary polarization state can be generated by employing three basic polarizing elements, namely, polarizer, phase shifter (retarder), and rotator. In Sect. 2.2.3, we have introduced the electromagnetic potentials to simplify the study of generation of electromagnetic waves, followed by antenna theory basics. In Sect. 2.2.4, we have discuss the interference, coherence, and diffraction effects in optics, which have been used later to study the propagation effects in various wireless and optical communication channels. In Sect. 2.2.5, the detailed description of the laser beam propagation effects over the atmospheric turbulence channels has been provided, including the description of simulation models. In Sect. 2.3, the wireless communication channel propagation effects have been studied, in particular the path loss, shadowing, and multipath fading propagation effects. Both ray theory-based and statistical multipath wireless communication channel models have been described. Section 2.4 has been devoted to the signal propagation effects in optical fibers, namely, (i) fiber attenuation and insertion loss, (ii) chromatic dispersion effects, (iii) polarization mode dispersion effects, and (iv) fiber nonlinearities. Regarding the fiber nonlinearities, both Kerr nonlinear effects (self-phase modulation, cross-phase modulation) and nonlinear scattering effects, such as stimulated Raman scattering, have been described. In Sect. 2.4.5, the generalized nonlinear Schrödinger equation has been described for both single-mode and few-mode fibers, as well as the corresponding split-step Fourier algorithms to solve for it. In Sect. 2.5, the following noise sources related to optical communication have been described: mode partition noise, relative intensity noise, laser phase noise, modal noise, thermal noise, spontaneous emission noise, noise beating components, quantum shot noise, dark current noise, and crosstalk. Section 2.6 has been devoted to indoor optical wireless communications. In this section, the channel models for both infrared and visible light communications have been introduced. The set of problems is provided in next section.

2.8 Problems

1. The beam expander is composed of two lenses, the first one of short focal length f_s and the second one of long focal length f_l, with the goal to expand the beam.
   (a) Sketch the schematic of the beam expander.
   (b) How much should these two lenses be separated so that for collimated incident beam the expanded beam to be collimated as well?
   (c) Relate the expansion ratio to the focal lengths.
   (d) Design the beam expander with expansion ratio of $4\times$.
   (e) Design the corresponding beam compressor.
2. Let us observe a rectangular aperture of width $a = 1$ mm and height $b = 2$ mm that get illuminated by normally incident plane wave with wavelength 1550 nm.
   (a) Determine the distance from the aperture when Fresnel approximation is valid.
   (b) Determine the distance from the aperture when Fraunhofer approximation is valid.
3. Let us consider the double-slit aperture composed of two parallel slits of width a each and separated by distance h. Assuming the double-slit is illuminated by normally incident plane wave.
   (a) Derive the irradiance distribution function and plot corresponding Fraunhofer diffraction pattern.
   (b) Generalize this case to the multiple slits' aperture, composed of N parallel slits of width a with separation between slits being h. Derive the irradiance distribution function and plot corresponding Fraunhofer diffraction patterns in which number of slits N is used as a parameter.
4. Let a transparent object of transmittance function t_{obj} be placed against the ideal biconvex lens of focal length f. Derive the diffraction expression assuming the propagation to the back focal plane and normally incident plane wave beam.
5. Let a transparent object of transmittance function t_{obj} be placed in front of, at distance d from, the ideal biconvex lens of focal length f. Derive the diffraction expression assuming the propagation to the back focal plane and normally incident plane wave beam. For $d = 0$ your solution should reduce to that from previous problem.
6. Let a transparent object of transmittance function t_{obj} be placed behind the ideal biconvex lens of focal length f, placed at distance d-f, wherein $d < <f$. Derive the diffraction expression assuming the propagation to the back focal plane and normally incident plane wave beam.
7. In this problem you are expected to study the propagation of Gaussian and LG beams over the atmospheric turbulent channel, using the split-step algorithm. You are expected to take the diffraction effects into account. To study the FSO propagation, the following parameters of atmospheric turbulence channel should be considered: the propagation distance 1 km, the outer scale eddies $L_0 = 20$ m, the inner scale $l_0 = 1$ mm, the number of slabs at least 10, operating wavelength

1550 nm, and the beam waist at $z = 0$ being 5 mm. The phase power spectral density (PSD) to be used to generate random phase screens should be a modified version of the Kolmogorov spectrum, which includes both inner and outer scales, also known as Andrews' spectrum. *Hint*: To verify your split-step method, you can use the phase structure function. As an example, you can use S. S. Chesnokov, V. P. Kandidov, V. I. Shmalhausen, and V. V. Shuvalov, "Numerical/optical simulation of laser beam propagation through atmospheric turbulence," DTIC Document (1995).

(a) Provide the phase screen realizations after 1 km, for the following refractive structure parameters $C_n^2 = 10^{-15}, 10^{-14}$, and 10^{-13} m$^{-2/3}$. For Gaussian beam, provide also corresponding magnitude and phase characteristics for the same set of refractive structure parameters.

(b) This problem is continuation of (a). Determine the Rytov variance and scintillation index for the same set of refractive structure parameters. Determine the distributions of irradiance and compare them against gamma-gamma distribution.

(c) This problem is continuation of (a). Determine the intensity correlation function (ICF) defined as $C(\rho_1, \rho_2, L) = \langle (I_1 - \langle I_1 \rangle)(I_2 - \langle I_2 \rangle) \rangle$, where I_1 and I_2 are the captured irradiances of positions ρ_1 and ρ_2 at distance L, respectively; the operator $\langle \cdot \rangle$ denotes an ensemble average of the quantity inside the parenthesis.

(d) The field distribution of the LG beam traveling along the z-axis can be expressed in cylindrical coordinates $[\rho,\phi,z]$ (ρ denotes the radial distance from propagation axis, ϕ denotes the azimuthal angle, and z denotes the propagation distance) as:

$$u_{m,p}(\rho,\phi,z) = \sqrt{\frac{2p!}{\pi(p+|m|)!}}\frac{1}{w(z)}\left[\frac{\rho\sqrt{2}}{w(z)}\right]^{|m|} L_p^m\left(\frac{2\rho^2}{w^2(z)}\right) e^{-\frac{\rho^2}{w^2(z)}} \cdot e^{-\frac{jk\rho^2 z}{2\left(z^2+z_R^2\right)}} e^{j(2p+|m|+1)\tan^{-1}\frac{z}{z_R}} e^{-jm\phi},$$

where $w(z) = w_0[1 + (z/z_R)^2]^{1/2}$ (w_0 is the zero-order Gaussian radius at the waist), $z_R = \pi w_0^2/\lambda$ is the Rayleigh range (with λ being the wavelength), $k = 2\pi/\lambda$ is the propagation constant, and $L^m{}_p(\cdot)$ is the associated Laguerre polynomial, with p and m representing the radial and angular mode numbers (indices), respectively.

(e) **For $m = 3$ and $p = 0$**, provide the corresponding spatial irradiance and phase characteristics for the same set of refractive structure parameters as in (a).

8. In this problem we study a two-path wireless channel model with impulse response $h(t) = \alpha_1\delta(\tau) + \alpha_2\delta(\tau - 0.03\ \mu s)$. Under assumption that the reflection coefficient is -1, determine coefficients in impulse response as well as the distance between the transmitter and receiver. Assume that the carrier frequency is 1 GHz, while the transmitter and receiver are both 10 m above the ground.

9. In this problem we are concerned with four-ray wireless channel model provided in Fig. P9, composed of line-of-sight (LOS), reflected, scattered, and diffracted components.
The height of transmit (Tx) antenna measured from the road, as indicated in Fig. P9, is $h_{Tx} = 75$ m, while the receive (Rx) antenna height is $h_{Rx} = 1.5$ m. The distance between the picks of Tx and Rx antennas is $L = 350$ m. The distance

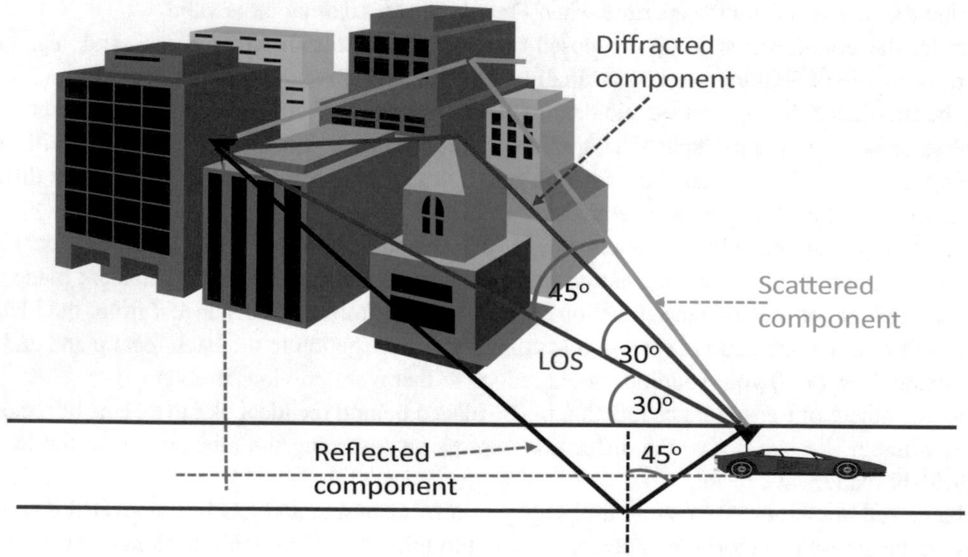

Fig. P9 Illustration of the four-ray model

from Tx antenna to the diffracting object is $d = 150$ m, while the distance from diffracting object to Rx antenna is $D = 250$ m. The distance from Tx antenna to the scattering object is $r = 200$ m, while the distance from scattering object to Rx antenna is $R = 300$ m. The scattering object radar cross section is $\sigma = 10$ dBm^2. Assume that dielectric constant of the road is $\varepsilon_r = 1.5$ and that reflected ray is TE polarized. The carrier frequency is 1 GHz.

(a) Determine the receiver-to-transmit power ratio for this four-ray problem.
(b) We declare an outage when then received power falls below −30 dBm. Assuming that the transmitted power is 500 mW and the shadowing effect with standard deviation of 3 dB is present, determine the outage probability.
(c) For the same assumptions as in (b), when the cell radius is 500 m, determine the cell coverage area.

10. Let us study a cellular system in which the received signal power is distributed according to the log-normal distribution with a mean $\mu_{\psi_{\mathrm{dB}}}$ dBm and a standard deviation $\sigma_{\psi_{\mathrm{dB}}}$ dBm. Under the assumptions that the acceptable performance is achieved for received signal power above 8 dBm, determine:
    (a) The outage probability for $\mu_{\psi_{\mathrm{dB}}} = 12$ dBm and $\sigma_{\psi_{\mathrm{dB}}} = 5$ dBm.
    (b) The value of $\mu_{\psi_{\mathrm{dB}}}$ required, for $\sigma_{\psi_{\mathrm{dB}}} = 4.5$ dBm, such that the outage probability <1%.
11. Determine the coverage area for a microcellular system with path loss that follows the simplified model with slope $s = 3$, $l_{\mathrm{ref}} = 1$ m, and $C = 1$ in the presence of log-normal shadowing with $\sigma_{\psi_{\mathrm{dB}}} = 5$ dB under the following assumptions: the cell radius is 90 m, the transmit power is 60 mW, and the minimum received power requirement is −90 dBm.
12. Let the scattering function $S(\tau, \nu)$ be nonzero over $0 \leq \tau \leq 0.1$ ms and $-0.1 \leq \nu \leq 0.1$ Hz. Assume that the distribution of power of the scattering function is approximately uniform over the nonzero range.
    (a) Determine the multipath spread and the Doppler spread of the channel.
    (b) Suppose that the input to this channel are two sinusoids with the same amplitude separated in time by Δt. Determine the minimum value of Δf for which the channel response to the first sinusoid is approximately independent of the channel response to the second sinusoid.
    (c) For two sinusoidal inputs to the channel given $u_1(t) = \sin 2\pi ft$ and $u_2(t) = \sin[2\pi f(t + \Delta t)]$, determine the minimum value of Δt for which the channel response to the first sinusoid $u_1(t)$ is approximately independent of the channel response to the second sinusoid $u_2(t)$.
    (d) For a typical voice channel with a 3 KHz bandwidth, classify the fading type.
    (e) For a cellular channel with a 30 KHz bandwidth, classify the fading type.
13. The purpose of this problem is to develop a Rayleigh fading simulator employing the Clark-Jakes model. The corresponding program should be easy to follow and well commented. The concepts to implement this simulator are already provided in Sect. 2.3.2.3.
    (a) Develop the Rayleigh fading simulator to generate the signal with Rayleigh fading amplitude in which the average received envelope is normalized to one and the Doppler frequency ν_D is used as a parameter.
    (b) Plot the received amplitude vs. time over a 5-second interval for the following Doppler frequencies $\nu_D \in \{1, 5, 10, 20, 50, 100$ Hz$\}$.
14. In this problem we are interested in approximating the Rayleigh fading simulator developed in previous problem with a finite state Markov model with sufficient number states and sufficient number of envelope regions. For each Doppler frequency listed in Problem 10(b), determine the corresponding finite state Markov model and provide the transition probabilities. By using such developed models, generate fading amplitude vs. time and report the mean-squared error compared to the simulator.
15. The fabrication of the single-mode fiber (SMF) is imperfect, and we need to associate with it certain fabrication tolerances. Under the assumptions that the tolerance in the core radius is ±3%, while the tolerance in the normalized refractive index difference Δ is ±7% (from their respective nominal values, $\Delta = 0.005$, refractive index of the core $n_{\mathrm{core}} = 1.5$), determine the largest nominal value that can specified for the core radius ensuring that the fiber operates as a single-mode for $\lambda > 1.2$ μm in the worst-case scenario of deviations of core radius and Δ.
16. A step-index multimode fiber has the cladding refractive index of 1.45 and a core diameter of 45 μm. Assuming that the intermodal dispersion of 12 ns/km can be tolerated, determine the acceptance angle. Determine the maximum bit rate for transmission over a distance of 15 km.
17. For a Gaussian pulse of peak power 10 mW and full-width half-maximum pulse duration of 20 ps, transmitted over the SMF, determine the frequency chirp due to the SPM.
18. Assuming that the step-index fiber has core radius r, cladding refractive index n_{cladding}, and core refractive index n_{core}, derive the corresponding equation for the cutoff wavelength $\lambda_{\mathrm{cutoff}}$. For a fiber of core radius of $r = 5$ μm and $\Delta = 0.004$, determine the cutoff wavelength.

19. Assuming that the fundamental mode can be approximated as a Gaussian beam, determine the confinement factor, defined as the ratio of the power in core and the total power (both in core and cladding), as a function of core radius and the beam width.
20. The broadening of the unchirped Gaussian pulse [with an envelope $A(0,t) = A_0\exp(-t^2/2\tau_0^2)$, A_0-the pulse amplitude, τ_0-the pulse half-width at 1/e intensity point] in the presence of GVD and SPM can be evaluated as the ratio of output and input RMS widths:

$$\frac{\sigma(L)}{\sigma_0} = \sqrt{\left(1 + \frac{\sqrt{2}L_{eff}L\beta_2}{2L_{NL}\sigma_0^2}\right) + \left(1 + \frac{4}{3\sqrt{3}}\frac{L_{eff}^2}{L_{NL}^2}\right)\frac{L^2\beta_2^2}{8\sigma_0^4}}, \quad \sigma_0^2 = \tau_0^2/2$$

where with L_{eff} and L_{NL} are denoted effective and nonlinear lengths.
(a) For which value of L (expressed as a multiple of dispersion length L_D) does the launched pulse attain its minimum width?
(b) Observe a RZ transmission at 10 Gb/s with infinitely large extinction ratio and 50% of duty cycle. The bit one is represented as Gaussian pulse from (a). Determine the initial pulse width that corresponds to 50% duty cycle. For such determined initial pulse width and for length determined in (a), find the maximum bit rate that can be supported. Assume that system operates at wavelength 1550 nm, for which the fiber has the dispersion coefficient $D = 17$ ps/(nm-km) and the Kerr nonlinearity coefficient $n_2 = 3 \cdot 10^{-20}$ m^2/W. The fiber attenuation is 0.2 dB/ km, and the Gaussian pulse peak power is 0 dBm.
21. This problem is related to the link design. We would like to design the system operating either at 1300 nm or 1550 nm, which supports 2.5 Gb/s data rates using NRZ modulation format. The fiber parameters are as follows. The dispersion at 1550 nm is 16.5 ps/nm-km, while the attenuation coefficient is 0.29 dB/km. The dispersion at 1300 nm is 0 ps/nm-km, while the attenuation coefficient is 0.5 dB/km.
(a) If (i) a transmitter operating at 1550 nm with spectral width of 1 nm and with launched power of 0 dBm is available and (ii) receiver with receiver sensitivity of −29 dBm is available, determine the maximum possible length of the link.
(b) If (i) a transmitter operating at 1300 nm with spectral width of 2 nm and with launched power of 3 dBm is available and (ii) receiver with receiver sensitivity of −29 dBm is available, determine the maximum possible length of the link in this case.
(c) Let's assume that transmitter from (b) is available. We are asked to achieve a signal-to-noise ratio (SNR) of 30 dB, using an APD of responsivity 10 A/W and gain of 10. The excess noise factor is 6 dB, the load resistance is 50 Ω, and an amplifier noise figure is 4 dB. Assuming the NRZ transmission, determine the maximum transmission length.
22. We would like to design a fiber-optics 160 Gb/s wavelength division multiplexing (WDM) system with $M = 4$ channels operating at 40 Gb/s employing *the hybrid Raman-EDFA* in-line optical amplifiers, as shown in Fig. P22. The total launched power is restricted by available laser sources and cannot be larger than 9 dBm, and the SMF fiber loss is $\alpha = 0.2$ dB/km. SMF dispersion coefficient is 17 ps/nm-km. Dispersion compensating module (DCM) is composed of DCF with dispersion coefficient of −95 ps/nm-km, while DCF attenuation coefficient is 0.5 dB/km. Notice that DCF length does not contribute to the total transmission length. Raman amplifier (RA) is used to compensate for SMF fiber loss, while EDFA is used to compensate for DCM insertion loss. Assuming that double Rayleigh backscattering can be neglected, the RA noise figure can be calculated by $NF_{RA} = (P_{ASE}/h\nu_s B_{op} + 1)/(G_{RA}e^{-\alpha l})$, where G_{RA} is RA gain (B_{op} is the optical filter bandwidth). The ASE noise power of RA can be calculated by $P_{ASE} = 2h\nu_s\Delta\nu_R/[1-\exp(-h\nu_s/k_B T)]$, k_B is the Boltzmann constant (1.381×10^{-23}J/K), ν_s is the signal frequency, and $\Delta\nu_R$ is the separation between signal and pump frequencies. The pump wavelength is 1450 nm. T is the absolute temperature. Assume that G_{RA} is just enough to compensate for SMF fiber loss. The EDFA noise figure (NF_{EDFA}) is 6 dB. The equivalent noise figure of both RA and EDFA can be calculated by $NF = NF_{RA} + (NF_{EDFA} - 1)/(G_{RA}e^{-\alpha l}) + 1 - \Gamma$, where Γ is loss caused by DCM. The predicted margins to achieve $BER = 10^{-12}$ allocated for chromatic dispersion, extinction ratio, polarization effects, fiber nonlinearities, component aging, and system margin are 0.5, 0.5, 1.5, 1.5, 2, and 2 dB, respectively. The operating temperature is 25 °C. WDM DEMUX can be modeled as a bank of optical filters, each of them with a bandwidth B_o equal to $0.3B$ (B-the bit rate per channel). Assuming NRZ transmission, the electrical filter bandwidth $B_e = 0.7B$, and amplifier spacing $l = 100$ km, determine the required optical signal-to-noise ratio (OSNR) to achieve BER of 10^{-12} for the central channel located at 1552.524 nm. What is the maximum possible transmission distance? Assume that the saturation of EDFAs due to ASE noise accumulation is negligible.

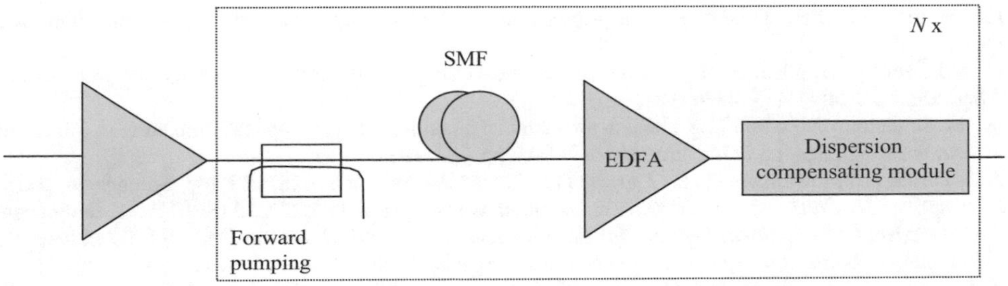

Fig. P22 hybrid Raman-EDFA concept

23. In this problem you are expected to solve the GNLSE numerically, using Fourier split-step algorithm.

(a) Let us first ignore the attenuation and fiber nonlinearities. Generate the Gaussian pulse of FWHM 25 ps and launch power 1 mW and transmit it over SMF of length L. If the dispersion coefficient is 16 ps/(nm ·km) and dispersion slope is 0.06 ps/(nm^2 ·km), determine and plot the pulse shape after 25 km, 50 km, and 100 km by solving the GNLSE. Determine the pulse spread. Operating wavelength is 1550 nm.

(b) Let us ignore the attenuation and second-order GVD, and study the soliton transmission. Assume that dispersion coefficient is 16 ps/(nm·km) and nonlinear coefficient of fiber is $\gamma = 2.5$ 1/(W·km). Chose the launch power such that dispersion length is equal to the nonlinear length. Determine the analytical expression of pulse shape. Plot the pulse shape after 1000 km and 10,000 km by solving the NLSE. The initial soliton pulse FWHM is 25 ps. Operating wavelength is again 1550 nm.

(c) Generate now a PRBS sequence of sufficient length to accurately estimate BER down to 10^{-6}. Represent the symbol 1 pulse by a Gaussian pulse of FWHM of 12.5 ps. Assume the launch power of 0 dBm, the extinction ratio of 20 dB, and data rate of 40 Gb/s. Transmit this sequence over 6000 km, with dispersion map composed of N sections of SMF of length 80 km (with dispersion and dispersion slope parameters being 16 ps/(nm·km) and 0.06 ps/(nm^2 ·km), respectively) and 40 km of DCF. The DCF is chosen to exactly compensate for residual chromatic dispersion, and dispersion map is slope matched. The nonlinear coefficient of both fibers is $\gamma = 2.5$ 1/(W·km). The EDFAs of NF of 5 dB and $BW = 3R_b$ (R_b is the bit rate) are periodically deployed every fiber section to exactly compensate for fiber loss. The SMF loss is 0.2 dB/km, and DCF loss is 0.5 dB/km. The operating wavelength is 1550 nm, and photodiode responsivity is 1 A/W. The receiver electronics can be model as Gaussian filter of 3 dB BW $=0.75\,R_b$. Determine the optical SNR (OSNR) at the receiver side as well as the BER. Show the eye diagrams and spectra at transmitter and receiver sides. Vary the launch power from −6 dBm to 6 dBm in steps of 2 dBm and plot how BER changes as a function of launch power.

References

1. G. Keiser, *Optical Fiber Communications*, 3rd edn. (McGraw-Hill, Boston, 2000)
2. G.P. Agrawal, *Fiber-Optic Communication Systems*, 4th edn. (Wiley, Hoboken, NJ, 2010)
3. G.P. Agrawal, *Nonlinear Fiber Optics*, 4th edn. (Academic Press, Oxford, UK, 2007)
4. I.B. Djordjevic, W. Ryan, B. Vasic, *Coding for Optical Channels* (Springer, New York, 2010)
5. D.M. Velickovic, *Electromagnetics – Vol. I*, 2nd edn. (Faculty of Electronic Engineering, University of Nis, Nis, 1999)
6. K.F. Riley, M.P. Hobson, S.J. Bence, *Mathematical Methods for Physics and Engineering: A Comprehensive Guide*, 3rd edn. (Cambridge University Press, Cambridge, New York, 2006)
7. I.B. Djordjevic, *Quantum Information Processing and Quantum Error Correction: An Engineering Approach* (Elsevier/Academic Press, Amsterdam, Boston, 2012)
8. C.H. Papas, *Theory of Electromagnetic Wave Propagation* (Dover, New York, 1988)
9. E. Collet, *Polarized Light for Scientists and Engineers* (The PolaWave Group, Long Branch, NJ, 2012)
10. G.R. Fowles, *Introduction to Modern Optics* (Dover, New York, 1989)
11. E. Hecht, A.R. Ganesan, *Optics* (Pearson Education, Essex, 2014)
12. N. George, *Fourier Optics* (Institute of Optics/University of Rochester, 2013)
13. J.W. Goodman, *Introduction to Fourier Optics*, 3rd edn. (Roberts/Greenwood Village, 2005)
14. M. Abramowitz, I. Stegun, *Handbook of Mathematical Functions with Formulas, Graphs, and Mathematical Tables* (Dover, New York, 1965)
15. J.L. Walsh, P.B. Ulrich, *Laser Beam Propagation in the Atmosphere* (Springer, Berlin/Heidelberg/New York, 1978)
16. R. Buckley, Diffraction by a random phase-change in screen: a numerical experiment. J. Atm. Terrestrial Phys. **37**(12), 1431–1446 (1975)

17. J.A. Fleck, J.R. Morris, M.D. Feit, Time-dependent propagation of a high-energy laser beam through the atmosphere. Appl. Phys. **10**(1), 129–139 (1976)
18. V.P. Kandinov, V.I. Ledenev, Application of the method of statistical testing for investigation of wave beam propagation in statistically random media. IVUZ Ser. Radiofiz. **24**(4), 434–442 (1981)
19. P.A. Konayev, Modification of the splitting method for numerical solution of quasi-optical problems, in Abstracts of reports at the VI conference on laser beam propagation in the atmosphere (1981), pp. 195–198
20. V.P. Lukin, B.V. Fortes, *Adaptive Beaming and Imaging in the Turbulent Atmosphere* (SPIE Press, Bellingham, 2002)
21. J.D. Schmidt, *Numerical Simulation of Optical Wave Propagation with examples in MATLAB* (SPIE Press, Bellingham, 2010)
22. J.M. Jarem, P.P. Banerjee, *Computational Methods for Electromagnetic and Optical Systems* (Marcel Dekker, New York, 2000)
23. T.-C. Poon, P.P. Banerjee, *Contemporary Optical Image Processing with Matlab* (Elsevier Science, Oxford, 2001)
24. T.-C. Poon, T. Kim, *Engineering Optics with MATLAB* (World Scientific Publishing, New Jersey/London/Singapore/Beijing, 2006)
25. J.J. Sakurai, S.F. Tuan, *Modern Quantum Mechanics*, Revised edn. (Addison-Wesley Publishing, Reading, 1994)
26. M. Nazarathy, J. Shamir, Fourier optics described by operator algebra. J. Opt. Soc. Am. **70**, 150–159 (1980)
27. L. Stankovic, *Digital Signal Processing* (Naucna Knjiga, Belgrade, 1990)
28. Fastest Fourier Transform in the West, FFTW. Available at: http://www.fftw.org/
29. S.M. Rytov, *Introduction to Statistical Radiophysics* (Nauka, Moscow, 1966)
30. L.C. Andrews, R.L. Philips, *Laser Beam Propagation Through Random Media*, 2nd edn. (SPIE Press, Bellingham, 2005)
31. H. Willebrand, B.S. Ghuman, *Free-Space Optics: Enabling Optical Connectivity in Today's Networks* (Sams Publishing, Indianapolis, 2002)
32. A. Ishimaru, *Wave Propagation and Scattering in Random Media* (Academic, NY, 1978)
33. L.C. Andrews, R.L. Phillips, C.Y. Hopen, *Laser Beam Scintillation with Applications* (SPIE Press, 2001)
34. S. Karp, R. Gagliardi, S.E. Moran, L.B. Stotts, *Optical Channels* (Plenum, New York, 1988)
35. J.A. Anguita, I.B. Djordjevic, M.A. Neifeld, B.V. Vasic, Shannon capacities and error–correction codes for optical atmospheric turbulent channels. OSA J. Opt. Netw. **4**, 586–601 (2005)
36. I.B. Djordjevic, B. Vasic, M.A. Neifeld, LDPC coded OFDM over the atmospheric turbulence channel. Opt. Express **15**(10), 6332–6346 (2007)
37. I.B. Djordjevic, J. Anguita, B. Vasic, Error-correction coded orbital-angular-momentum modulation for FSO channels affected by turbulence. IEEE/OSA J. Lightw. Technol. **30**(17), 2846–2852 (2012)
38. Shieh W, Djordjevic I (2009), OFDM for Optical Communications. Elsevier/Academic Press, Oct 2009
39. I.B. Djordjevic, Coding and modulation techniques for optical wireless channels, in *Advanced Optical Wireless Communication Systems*, ed. by S. Arnon, J. Barry, G. Karagiannidis, R. Schober, M. Uysal, (Cambridge University Press, Cambridge, 2012)
40. Z. Qu, I.B. Djordjevic, Coded orbital-angular-momentum-based free-space optical transmission, in *Wiley Encyclopedia of Electrical and Electronics Engineering*, (Wiley, Chichester, Feb. 2016)
41. A.N. Kolmogorov, Local structure of turbulence in liquid under very large Rayleigh numbers. Dok. Akad. Nauk SSR **30**(4), 299–303 (1941) (in Russian)
42. A.M. Obukhov, Energy dissipation in spectra of a turbulent stream. Dok. Akad. Nauk SSR **32**, 22–24 (1943)
43. A.N. Kolmogorov, The local structure of turbulence in incompressible viscous fluid for very large Rayleigh numbers. Proc. R. Soc. Lond. **434**, 9–13 (1991) Translated by V. Levin
44. A.M. Obukhov, Concerning the distribution of energy in the spectrum of a turbulent stream. Izvestiia Akadimii Nauk S.S.S.R. Ser. Geogr. Geofiz. **4–5** (1941)
45. A.M. Obukhov, Structure of the temperature field in turbulent flow. Izvestiia Akademii Nauk S.S.S.R. Ser. Geogr. Geofiz. **2**(1), 58–69 (1949)
46. L.D. Landau, E.M. Lifschitz, *Fluid Mechanics*, 2nd edn. (Pergamon Press, Oxford, 1982)
47. S.F. Clifford, The classical theory of wave propagation in a turbulent medium, in *Laser Beam Propagation in the Atmosphere*, ed. by J. W. Strohbehn, (Springer, New York, 1978), pp. 9–43
48. J.G. Proakis, *Digital Communications*, 4th edn. (McGraw-Hill, Boston, 2001)
49. S. Haykin, *Digital Communication Systems* (Wiley, Hoboken, 2014)
50. M.H. Hayes, *Statistical Digital Signal Processing* (Wiley, New York, 1996)
51. V.I. Tatarskii, The Effects of the Turbulent Atmosphere on Wave Propagation (Translated by Israel Program for Scientific Translations in 1971) (U.S. Dept. of Commerce/National Technical Information Service, Springfield, 1967)
52. L.C. Andrews, An analytical model for the refractive index power spectrum and its application to optical scintillation in the atmosphere. J. Mod. Opt. **39**(9), 1849–1853 (1992)
53. D.L. Fried, Statistics of a geometric representation of wavefront distortion. J. Opt. Soc. Am. **55**(11), 1427–1431 (1965)
54. R. Noll, Zernike polynomials and atmospheric turbulence. J. Opt. Soc. Am. **66**, 207–211 (1976)
55. B.M. Welsh, A Fourier phase screen generator for simulating anisoplanatic geometries and temporal evolution. Proc. SPIE **3125**, 327–338 (1997)
56. R.J. Eckert, M.E. Goda, Polar phase screens: a comparative analysis with other methods of random phase screen generation. In. Proc. SPIE **6303**, 630301-1–630301-14 (2006)
57. R.G. Lane, A. Glindemann, J.C. Dainty, Simulation of a Kolmogorov phase screen. Waves Random Media **2**, 209–224 (1992)
58. S.M. Rytov, Diffraction of light by ultrasound waves. Ivestiya SSSR Ser. Fizicheskaya **2**, 223 (1937)
59. A.M. Obukhov, Effect of weak inhomogeneities in the atmosphere on the propagation of light and sound. Izvestiya SSSR Ser. Geofiz. **2**, 155 (1953)
60. L.A. Chernov, Correlation of amplitude and phase fluctuations for wave propagation in a medium with random inhomogeneities. Akusticheskii Zhurnal **1**(1), 89 (1955)
61. L.A. Chernov, Wave propagation in a medium with random inhomogeneities. Izdatel'stvo AN SSSR (1958)
62. V.I. Tatarskii, Light propagation in a medium with random refractive index inhomogeneities in the Markov random process approximation. Sov. Phys. JETP **29**(6), 1133–1138 (1969)

63. M.A. Al-Habash, L.C. Andrews, R.L. Phillips, Mathematical model for the irradiance probability density function of a laser beam propagating through turbulent media. Opt. Eng. **40**, 1554–1562 (2001)
64. A. Goldsmith, *Wireless Communications* (Cambridge University Press, Cambridge, 2005)
65. D. Tse, P. Viswanath, *Fundamentals of Wireless Communication* (Cambridge University Press, Cambridge, 2005)
66. S. Benedetto, E. Biglieri, *Principles of Digital Transmission with Wireless Applications* (Kluwer Academic/Plenum Publishers, New York, 1999)
67. E. Biglieri, *Coding for Wireless Channels* (Springer, New York, 2005)
68. T.M. Duman, A. Ghrayeb, *Coding for MIMO Communication Systems* (Wiley, Chichester, 2007)
69. E.F. Knott, J.F. Shaeffer, M.T. Tuley, *Radar Cross Section*, 2nd edn. (SciTech Publishing, 2004)
70. M. Lanzagorta, *Quantum Radar* (Morgan and Claypool Publishers, San Rafael, 2012)
71. N. Amitay, Modeling and computer simulation of wave propagation in lineal line-of-sight microcells. IEEE Trans. Vech. Technol. **41**(4), 337–342 (1992)
72. K. Schaubach, N.J. Davis, T.S. Rappaport, A ray tracing method for predicting path loss and delay spread in microcellular environments, in Proceedings of the IEEE Vehiculat Technology Conference, vol 2,pp. 932–935 (Denver, 1992)
73. M.C. Lawton, J.P. McGeehan, The application of GTD and ray launching techniques to channel modeling for cordless radio systems, in Proceedings of the IEEE Vehiculat Technology Conference, vol 1, pp. 125–130 (Denver, 1992)
74. W.C.Y. Lee, *Mobile Communications Engineering* (McGraw-Hill, New York, 1982)
75. L. Tsang, J.A. Kong, K.-H. Ding, *Scattering of Electromagnetic Waves: Theories and Applications* (Wiley, New York, 2000)
76. S.M.I. Seidel, *Introduction to Radar Systems*, 2nd edn. (McGraw-Hill, Auckland, 1981)
77. S.Y. Seidel, T.S. Rappaport, S. Jain, M.L. Lord, R. Singh, Path loss, scattering and multipath delay statistics in four European cities for digital cellular and microcellular radiotelephone. IEEE Trans. Veh. Technol. **40**(4), 721–730 (1991)
78. P. Harley, Short distance attenuation measurements at 900 MHz and 1.8 GHz using low antenna heights for microcells. IEEE Sel. Area Comm. **7**(1), 5–11 (1989)
79. T. Okumura, E. Ohmori, K. Fukuda, Field strength and its variability in VHF and UHF land mobile services. Rev. Electr. Comm. Lab. **16**(9–10), 825–873 (1968)
80. M. Hata, Empirical formula for propagation loss in land mobile radio services. IEEE Trans. Veh. Technol. **VT-29**(3), 317–325 (1980)
81. European Cooperative in the Field of Science and Technical Research EURO-COST 213, Urban transmission loss models for mobile radio in the 900 and 1800 MHz bands. Rev., Hague (1991)
82. Hata/Davidson, A Report on technology independent methodology for the modeling, simulation and empirical verification of wireless communications system performance in noise and interference limited systems operating on frequencies between 30 and 1500 MHz. TIA TR8 Working Group, IEEE Vehicular Technology Society Propagation Committee (1997)
83. J.S. Lee, L.E. Miller, *CDMA Systems Engineering Handbook* (Artech House, Boston/London, 1998)
84. W.C. Jakes, *Microwave Mobile Communications* (Wiley, New York, 1974)
85. M. Gudmundson, Correlation model for shadow fading in mobile radio systems. Electr. Lett. **27**(23), 2145–2146 (1994)
86. J.A. Weitzen, T.J. Lowe, Measurement of angular and distance correlation properties of log-normal shadowing at 1900 MHz and its application to design of PCS systems. IEEE Trans. Veh. Technol. **51**(2), 265–273 (2002)
87. R.H. Clark, A statistical theory of mobile-radio reception. Bell Syst. Technol. J. **47**(6), 957–1000 (1968)
88. M.D.Yacoub, The $\alpha - \mu$ distribution: a general fading distribution, in Proceedings of IEEE international symposium on personal, indoor and mobile radio communications, vol 2 pp. 629–633 (2002)
89. M.D. Yacoub, The α–μ distribution: a physical fading model for the Stacy distribution. IEEE Trans. Veh. Technol. **56**(1), 27–34 (2007)
90. I.B. Djordjevic, G.T. Djordjevic, On the communication over strong atmospheric turbulence channels by adaptive modulation and coding. Opt. Express **17**(20), 18250–18262 (2009)
91. G.T. Djordjevic, I.B. Djordjevic, G.K. Karagiannidis, Partially coherent EGC reception of uncoded and LDPC-coded signals over generalized fading channels. Wirel. Pers. Commun. **66**(1), 25–39 (2012)
92. S.O. Rice, Mathematical analysis of random noise. Bell Syst. Technol. J. **23**(3), 282–333 (1944)
93. S.O. Rice, Statistical properties of a sine wave plus noise. Bell Syst. Technol. J. **27**, 109–157 (1948)
94. G.L. Stüber, *Principles of Mobile Communication*, 3rd edn. (Springer, New York, 2012)
95. G.L. Turin, Introduction to spread spectrum antimultipath techniques and their application to urban radio. Proc. IEEE **68**(3), 328–353 (1980)
96. H.S. Wang, N. Moayeri, Finite-state Markov channel-a useful model for radio communication channels. IEEE Trans. Veh. Technol. **44**(1), 163–171 (1995)
97. P.A. Bello, Characterization of randomly time-variant linear channel. IEEE Trans. Comm. Syst. **11**(4), 360–393 (1963)
98. M. Cvijetic, I.B. Djordjevic, *Advanced Optical Communication Systems and Networks* (Artech House, Boston/London, 2013)
99. R. Ramaswami, K. Sivarajan, *Optical Networks: A Practical Perspective*, 2nd edn. (Morgan Kaufman, San Francisco, 2002)
100. L. Kazovsky, S. Benedetto, A. Willner, *Optical Fiber Communication Systems* (Artech House, Boston, 1996)
101. J.C. Palais, *Fiber Optic Communications*, 5th edn. (Pearson Prentice Hall, Upper Saddle River, 2005)
102. G.P. Agrawal, *Lightwave Technology: Telecommunication Systems* (Wiley, Hoboken, NJ, 2005)
103. W. Shieh, I.B. Djordjevic, *OFDM for Optical Communications* (Elsevier/Academic Press, Amsterdam/Boston, 2009)
104. I. Kaminow, T. Li, A. Willner, *Optical Fiber Telecommunications VIB* (Elsevier/Academic Press, Oxford, 2013)
105. M. Staif, A. Mecozzi, J. Nagel, Mean square magnitude of all orders PMD and the relation with the bandwidth of principal states. IEEE Photon. Tehcnol. Lett. **12**, 53–55 (2000)
106. H. Kogelnik, L.E. Nelson, R.M. Jobson, Polarization mode dispersion, in *Optical Fiber Telecommunications*, ed. by I. P. Kaminow, T. Li, (Academic Press, San Diego, 2002)
107. G.J. Foschini, C.D. Pole, Statistical theory of polarization mode dispersion in single-mode fibers. IEEE J. Lightw. Technol. **9**(11), 1439–1456 (1991)
108. P. Ciprut, B. Gisin, N. Gisin, R. Passy, J.P. von der Weid, F. Prieto, C.W. Zimmer, Second-order PMD: impact on analog and digital transmissions. IEEE J. Lightw. Technol. **16**(5), 757–771 (1998)

109. E. Ip, J.M. Kahn, Compensation of dispersion and nonlinear impairments using digital backpropagation. IEEE J. Lightw. Technol. **26**(20), 3416–3425 (2008)
110. E. Ip, J.M. Kahn, Nonlinear Impairment Compensation using Backpropagation, in *Optical Fibre, New Developments*, ed. by C. Lethien, (In-Tech, Vienna, 2009)
111. O.V. Sinkin, R. Holzlohner, J. Zweck, C.R. Menyuk, Optimization of the split-step Fourier method in modeling optical-fiber communications systems. IEEE J. Lightw. Technol. **21**(1), 61–68 (2003)
112. A. Rieznik, T. Tolisano, F.A. Callegari, D. Grosz, H. Fragnito, Uncertainty relation for the optimization of optical-fiber transmission systems simulations. Opt. Express **13**, 3822–3834 (2005)
113. Q. Zhang, M.I. Hayee, Symmetrized split-step Fourier scheme to control global simulation accuracy in fiber-optic communication systems. IEEE J. Lightw. Technol. **26**(2), 302–316 (2008)
114. C. Lin, I.B. Djordjevic, D. Zou, Achievable information rates calculation for optical OFDM transmission over few-mode fiber long-haul transmission systems. Opt. Express **23**(13), 16846–16856 (2015)
115. C. Lin, I.B. Djordjevic, M. Cvijetic, D. Zou, Mode-multiplexed multi-Tb/s superchannel transmission with advanced multidimensional signaling in the presence of fiber nonlinearities. IEEE Trans. Comm. **62**(7), 2507–2514 (2014)
116. K.-P. Ho, J.M. Kahn, Statistics of group delays in multimode fiber with strong mode coupling. IEEE J. Lightw. Technol. **29**(21), 3119–3128 (2011)
117. S. Mumtaz, R.-J. Essiambre, G.P. Agrawal, Nonlinear propagation in multimode and multicore fibers: generalization of the Manakov equations. J. Lightw. Technol. **31**(3), 398–406 (2012)
118. K.-D. Langer, J. Grubor, Recent developments in optical wireless communications using infrared and visible Light, in Proceedings of the ICTON 2007, pp. 146–152 (2007)
119. J. Grubor, V. Jungnickel, K. Langer, Rate-adaptive multiple-subcarrrier-based transmission for broadband infrared wireless communication, in Optical fiber communication conference and exposition and the national fiber optic engineers conference, Technical Digest (CD) (Optical Society of America, 2006), paper NThG2
120. S. Hranilovic, *Wireless Optical Communication Systems* (Springer, New York, 2004)
121. Y. Tanaka et al., Indoor visible communication utilizing plural white LEDs as lighting, in Proceedings of the Personal, Indoor & Mobile Radio Communications (PIMRC) 2001, pp. F81–F85
122. T. Komine, M. Nakagawa, Fundamental analysis for visible-light communication system using LED lightings. IEEE Trans. Consum. Electron. **15**(1), 100–107 (2004)
123. J. Grubor, O.C. Gaete Jamett, J.W. Walewski, S. Randel, K.-D. Langer, High-speed wireless indoor communication via visible light, in ITG Fachbericht 198 (VDE-Verlag, Berlin und Offenbach, 2007), pp. 203–208
124. Z. Ghassemlooy, W. Popoola, S. Rajbhandari, *Optical Wireless Communications: System and Channel Modelling with MATLAB* (CRC Press, Boca Raton, 2013)
125. J.M. Kahn, J.R. Barry, Wireless infrared communications. Proc. IEEE **85**(2), 265–298 (1997)

3 Components, Modules, and Subsystems

Abstract

In this chapter, basic components, modules, and subsystems relevant in both optical and wireless communications are described. In the first half of the chapter, key optical components, modules, and subsystems are described. After describing optical communication basic, focus is changed to optical transmitters. The following classes of lasers are introduced: Fabry-Perot, distributed feedback (DFB), and distributed Bragg reflector (DBR) lasers. Regarding the external modulators, both Mach-Zehnder modulator and electro-absorption modulator are described. Regarding the external modulator for multilevel modulation schemes, phase modulator, I/Q modulator, and polar modulator are described. In subsection on optical receivers, after the photodetection principle, the p-i-n photodiode, avalanche photodiode (APD), and metal-semiconductor-metal (MSM) photodetectors are briefly described. In the same subsection, both optical receiver high-impedance front end and trans-impedance front end are described. In section on optical fibers, both single-mode fiber (SMF) and multimode fiber (MMF) are introduced. In subsection on optical amplifiers, after describing optical amplification principles, the following classes of optical amplifiers are described: semiconductor optical amplifier (SOA), Raman amplifier, and erbium-doped fiber amplifier (EDFA). In subsection on optical processing components and modules, the following components/modules are described: optical couplers, optical filters, and WDM multiplexers/demultiplexers. After that, the principles of coherent optical detection are introduced, followed by the description of optical hybrids. In subsection on coherent optical balanced detectors, both homodyne and heterodyne optical balanced detectors for both binary and multilevel (QAM, PSK, APSK) modulations are described. In the second half of the chapter, basic components and modules of relevance to modern wireless communications are described. First, the DSP basics, including various discrete-time realizations, are provided. Then the direct digital synthesizer (DDS), relevant in discrete-time (DT) wireless communications systems, is described. In subsection on multirate DSP and resampling, various upsampling, downsampling, and resampling schemes/modules are described. In the same subsection, the polyphase decomposition of the system (transfer) function is introduced and employed in resampling. In subsection on antenna array antennas, the linear array antennas are described. Finally, in subsection on automatic gain control (AGC), the AGC schemes operating in passband and baseband are described. The set of problems is provided after the concluding remarks.

3.1 Key Optical Components, Modules, and Subsystems

This section describes the basic optical components used in an optical transmission system. We start this section with optical communications basics [1–35]. Components and modules relevant in wireless communications will be covered in Sect. 3.2.

3.1.1 Optical Communications Basics

An exemplary optical network, identifying the key optical components, is shown in Fig. 3.1. The end-to-end optical transmission involves both electrical and optical signal paths. To perform conversion from electrical to optical domain, the

I. B. Djordjevic, *Advanced Optical and Wireless Communications Systems*, https://doi.org/10.1007/978-3-030-98491-5_3

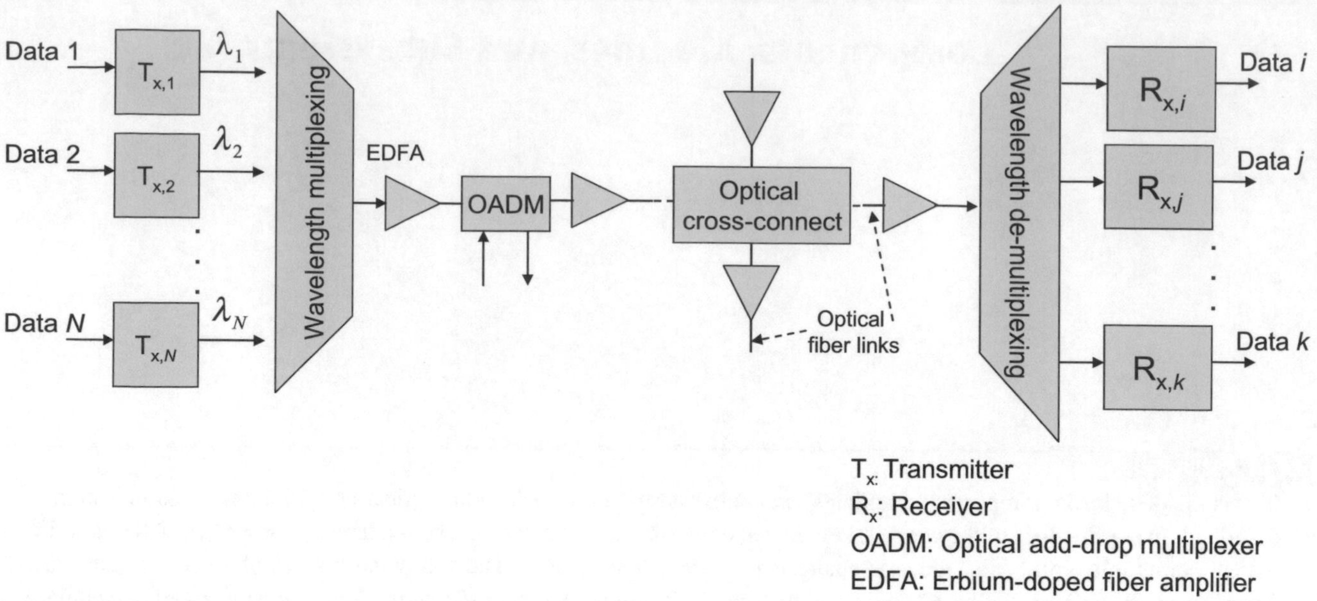

Fig. 3.1 An exemplary optical network identifying key optical components

optical transmitters are used, while to perform conversion in opposite direction (optical to electrical conversion), the optical receivers are used. The optical fibers serve as foundation of an optical transmission system because the optical fiber is used as medium to transport the optical signals from source to destination. The optical fibers attenuate the signal during transmission, and someone has to use optical amplifiers, such as erbium-doped fiber amplifiers (EDFAs), Raman amplifiers, or parametric amplifiers, to restore the signal quality [1–3, 5, 6]. However, the process of amplification is accompanied with the noise addition. The simplest optical transmission system employs only one wavelength. The wavelength division multiplexing (WDM) can be considered as an upgrade of the single-wavelength system. WDM corresponds to the scheme in which multiple optical carriers at different wavelengths are modulated by using independent electrical bit streams, as shown in Fig. 3.1, and then transmitted over the same fiber. WDM has potential of exploiting the enormous bandwidth offered by the optical fiber. During transmission of WDM signals, occasionally one or several wavelengths are to be added or dropped, which is performed by the optical component known as optical add-drop multiplexer (OADM), as illustrated in Fig. 3.1. The optical networks require the switching of information among different fibers, which is performed in optical cross-connect (OXS). To combine several distinct wavelength channels into composite channel, the wavelength multiplexers are used. On the other hand, to split the composite WDM channel into distinct wavelength channels, the wavelength demultiplexers are used. To impose the information signal, optical modulators are used. The optical modulators are commonly used in combination with semiconductor lasers.

The typical receiver configuration with direct detection is shown in Fig. 3.2 [1–3]. The main purpose of the optical receiver, terminating the lightwave path, is to convert the signal coming from single-mode fiber from optical to electrical domain and process appropriately such obtained electrical signal to recover the data being transmitted. The incoming optical signal may be pre-amplified by an optical amplifier and further processed by an optical filter to reduce the level of amplified spontaneous emission (ASE) noise or by wavelength demultiplexer to select a desired wavelength channel. The optical signal is converted into electrical domain by using a photodetector, followed by electrical post-amplifier. To deal with residual intersymbol interference (ISI), an equalizer may be used. The main purpose of clock recovery circuit is to provide timing for decision circuit by extracting the clock from the received signal. The clock recovery circuit is most commonly implemented using the phase-locked loop (PLL). Finally, the purpose of decision circuit is to provide the binary sequence being transmitted by comparing the sampled signal to a predetermined threshold. Whenever the received sample is larger than threshold, the decision circuit decides in favor of bit 1, otherwise in favor of bit 0.

The optical signal generated by semiconductor laser has to be modulated by information signal before being transmitted over the optical fiber. This can be achieved by *directly modulating* the bias current of semiconductor laser, which can be done even at high speed (even up to 40 Gb/s in certain lasers). Unfortunately, this concept although conceptually simple is rarely used in practice because of the frequency chirp introduced by direct modulation, nonuniform frequency response, and large current swing needed to provide operation. For transmitters operating at 10 Gb/s and above, instead, the semiconductor laser

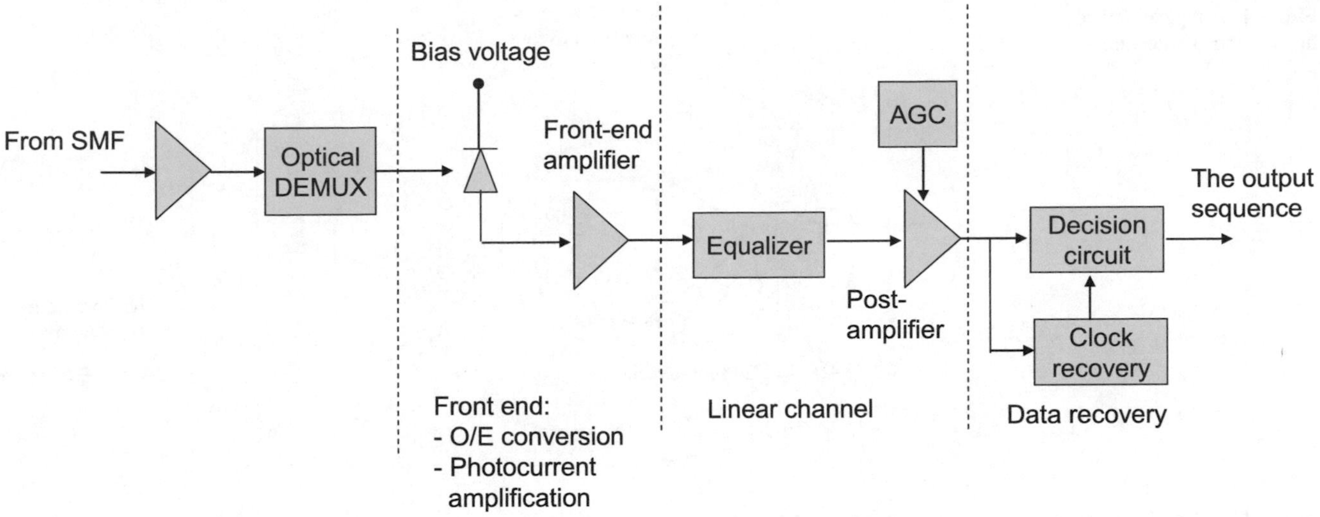

Fig. 3.2 A typical direct detection receiver architecture. O/E, optical to electrical; AGC, automatic gain control

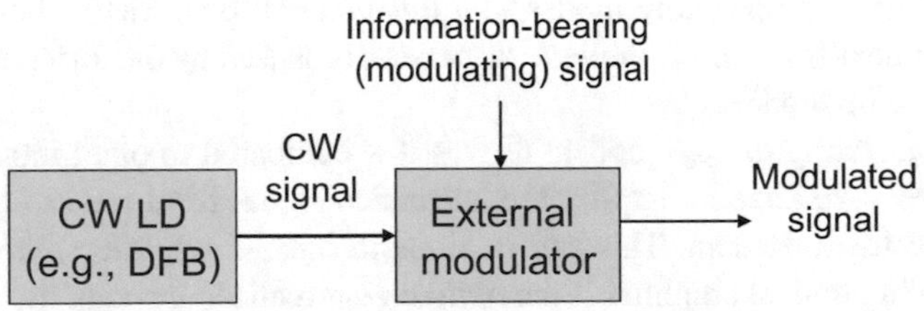

Fig. 3.3 An illustration of external modulation principle

diode (LD) is commonly biased at constant current to provide continuous wave (CW) output, and *external modulators* are used to impose the information signal to be transmitted. The most popular modulators are electro-optic optical modulators, such as Mach-Zehnder modulators, and electro-absorption modulators. The principle of the external modulator is illustrated in Fig. 3.3. Through the external modulation process, a certain parameter of the CW signal, used as a signal carrier, is varied in accordance with the information-bearing signal. For example, a monochromatic electromagnetic wave is commonly used as a carrier, and its electrical field $\boldsymbol{E}(t)$ can be represented by:

$$\boldsymbol{E}(t) = \boldsymbol{p}A\cos(\omega t + \varphi), \tag{3.1}$$

where A, ω, and φ are amplitude, frequency, and phase, respectively, while $\boldsymbol{p}$ denotes the polarization orientation. Each of those parameters can be used to carry information, and the information-bearing signal can be either CW or discrete. If the information-bearing signal is CW, corresponding modulation formats are amplitude modulation (AM), frequency modulation (FM), phase modulation (PM), and polarization modulation (PolM). On the other hand, if the information-bearing signal is digital, the corresponding modulations are amplitude-shift keying (ASK), frequency-shift keying (FSK), phase-shift keying (PSK), and polarization-shift keying (PolSK).

Optical fibers serve as foundation of an optical transmission system because they transport optical signals from source to destination. The combination of low-loss and large bandwidth allows high-speed signals to be transmitted over long distances before the regeneration is needed. A low-loss optical fiber is manufactured from several different materials; the base row material is pure silica, which is mixed with different dopants in order to adjust the refractive index of optical fiber. The optical fiber consists of two waveguide layers, the *core* and the *cladding*, protected by buffer coating. The majority of the power is concentrated in the core, although some portion can spread to the cladding. There exists a difference in refractive indices between the core and cladding, which is achieved by mixing dopants, commonly added to the fiber core. There exist two types of optical fibers: multimode fiber (MMF) and single-mode fiber (SMF). Multimode optical fibers transfer the light through a collection of spatial transversal modes. Each mode, defined through a specified combination of electrical and magnetic components, occupies a different cross section of the optical fiber core and takes a slightly distinguished path along the optical

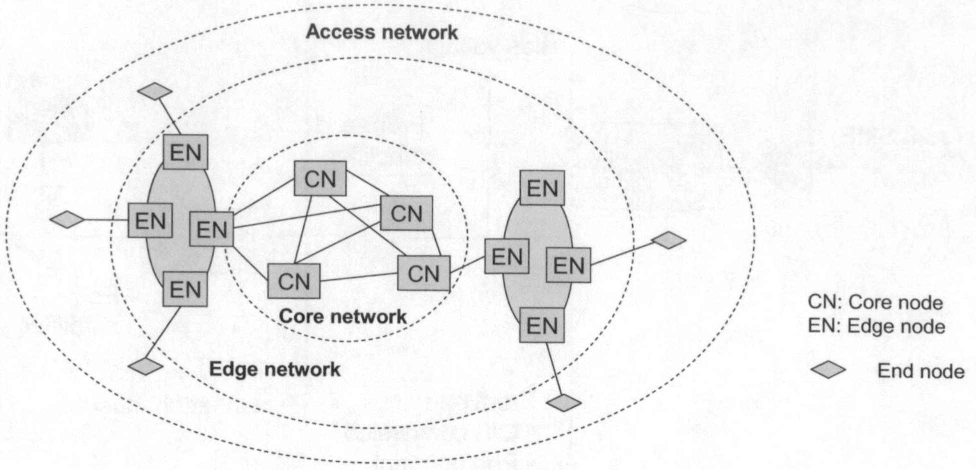

Fig. 3.4 A typical optical networking architecture

fiber. The difference in mode path lengths in multimode optical fibers produces a difference in arrival times at the receiving point. This phenomenon is known as *multimode dispersion* (or *intermodal dispersion*) and causes signal distortion and imposes the limitations in signal bandwidth. The second type of optical fibers, SMFs, effectively eliminates multimode dispersion by limiting the number of propagating modes to a *fundamental* one. SMFs, however, introduce another signal impairment known as the chromatic dispersion. *Chromatic dispersion* is caused by the difference in velocities among different spectral components within the same mode.

The attenuation of signal propagating through optical fiber is low compared to other transmission media, such as copper cables or free space. Nevertheless, we have to amplify the attenuated signal from time to time, to restore the signal level, without any conversion into electrical domain. This can be done in optical amplifiers, through the process of *stimulated emission*. The main ingredient of an optical amplifier is the *optical gain* realized through the amplifier pumping (being either electrical or optical) to achieve the so-called population inversion. The common types of optical amplifiers are semiconductor optical amplifiers (SOAs), EDFAs, and Raman amplifiers. The amplification process is commonly followed by the noise process, not related to the signal, which occurs due to spontaneous emission. The amplification process degrades the SNR, because of amplified spontaneous emission (ASE), added to the signal in every amplifier stage.

Before providing more details about basic building blocks identified in this section, let us give more global picture, by describing a typical optical network shown in Fig. 3.4. We can identify three ellipses representing the *core network*, the *edge network*, and the *access network* [1–3]. The long-haul core network interconnects big cities, major communications hubs, and even different continents by means of submarine transmission systems. The core networks are often called the wide area networks (WANs) or interchange carrier networks. The edge optical networks are deployed within smaller geographical areas and are commonly recognized as metropolitan area networks (MANs) or local exchange carrier networks. The access networks represent peripheral part of optical network and provide the last-mile access or the bandwidth distribution to the individual end users. The common access networks are local area networks (LANs) and distribution networks. The common physical network topologies are *mesh* network (often present in core networks), *ring* network (in edge networks), and *star* networks (commonly used in access networks).

Given this general description of key optical components in the rest of this section, we provide more details about basic building blocks: optical transmitters are described in Subsect. 3.1.2, optical receivers in Subsect. 3.1.3, optical fibers in Subsect. 3.1.4, and optical amplifiers in Subsect. 3.1.5, and other optical processing blocks are described, such as WDM multiplexers/demultiplexers, optical filters, and couplers, in Subsect. 3.1.6. The Subsect. 3.1.7 is devoted to the principles of coherent optical detection, followed by subsection on optical hybrids (3.1.8). In Subsect. 3.1.9, the balanced coherent optical detectors are described.

3.1.2 Optical Transmitters

The role of the optical transmitter is to generate the optical signal, impose the information-bearing signal, and launch the modulated signal into the optical fiber. This section is devoted to different lasers including Fabry-Perot, distributed feedback (DFB), and distributed Bragg reflector (DBR) lasers as well as to different external modulators including Mach-Zehnder modulator, electro-absorption modulator, phase modulator, I/Q modulator, and polar modulator.

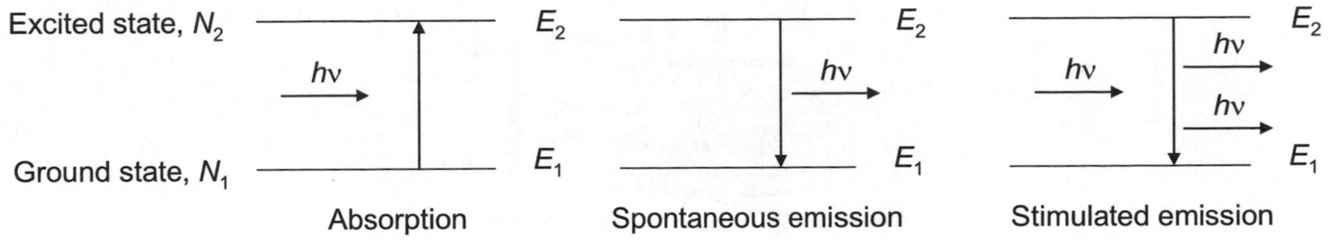

Fig. 3.5 Illustrating the interaction of the light with the matter

3.1.2.1 Lasers

The semiconductor light sources are commonly used in state-of-the-art optical communication systems. The light generation process occurs in certain semiconductor materials due to recombination of electrons and holes in p-n junctions, under direct biasing. Depending on the nature of recombination process, we can classify different semiconductor light sources as either light-emitting diodes (LEDs) in which spontaneous recombination dominates or semiconductor lasers in which the stimulated emission is a dominating mechanism. Namely, there are three basic processes in semiconductor materials, as illustrated in Fig. 3.5, by which the light interacts with matter: absorption, spontaneous emission, and stimulated emission. In normal conditions the number of electrons in ground state (with energy E_1) N_1 is larger than the number of electrons in excited state (with energy E_2) N_2, and in the thermal equilibrium, their ratio follows the Boltzmann's statistics [1–7]:

$$\frac{N_2}{N_1} = \exp\left(-\frac{h\nu}{k_B T}\right) = \exp\left(-\frac{E_2 - E_1}{k_B T}\right), \tag{3.2}$$

where $h\nu$ is a photon energy (h is the Plank's constant, ν is the optical frequency proportional to the energy difference between the energy levels E_2–E_1), k_B is the Boltzmann's constant, and T is the absolute temperature. In the same regime, the spontaneous emission rate $dN_{2,\ spon}/dt = A_{21}N_2$ (A_{21} denotes the spontaneous emission coefficient) and the stimulated emission rate $dN_{2,\ stim}/dt = B_{21}\rho(\nu)N_2$ (B_{21} denotes the stimulated emission coefficient and $\rho(\nu)$ denotes the spectral density of electromagnetic energy) are equalized with absorption rate $dN_{1,\ abs}/dt = A_{12}\rho(\nu)N_1$ (A_{12} denotes the absorption coefficient):

$$A_{21}N_2 + B_{21}\rho(\nu)N_2 = B_{12}\rho(\nu)N_1. \tag{3.3}$$

In the visible or near-infrared region ($h\nu \sim 1$ eV), the spontaneous emission always dominates over stimulated emission in thermal equilibrium at room temperature ($k_B T \approx 25$ meV) (see Eq. (3.2)). The stimulated emission rate can exceed absorption rate only when $N_2 > N_1$, the condition is referred to as *population inversion* and can never be realized for systems being in thermal equilibrium. The population inversion is a prerequisite for laser operation, and in atomic system, it is achieved by using *three- and four-level pumping schemes* (an external energy source raises the atomic population from ground to an excited state). There are three basic components required to sustain stimulated emission and to form useful laser output: the pump source, the active medium, and the feedback mirrors. The active medium can be solid (such as in semiconductor lasers), gaseous, or liquid in nature. The pump can be electrical (e.g., semiconductor lasers), optical, or chemical. The purpose of the pump is to achieve the population inversion. The basic structure of semiconductor laser of Fabry-Perot type is shown in Fig. 3.6(a), together with equivalent model. The injection (bias) current flows through the p-n junction and stimulates the recombination of electron and holes, leading to the generation of photons.

For the lasing action to be sustainable, the gain and phase matching condition should be satisfied. In the active medium, both *gain/absorption* described by $\gamma(\nu)$ and *scattering* described by α_s are present. The intensity inside the cavity can be described by the following dependence $I(z) = I_0\exp[(\gamma(\nu) - \alpha_s)z]$. The lasing is possible when collective gain is larger than loss after a round trip pass through the cavity:

$$I(2L) = I_0 R_1 R_2 \exp\left[2L(\gamma(\nu) - \alpha_s)\right] = I_0, \tag{3.4}$$

where R_1 and R_2 are facet reflectivities (see Fig. 3.6), L is the length of active medium, and I_0 and $I(2L)$ correspond to initial and round-trip intensities. The gain threshold is obtained by solving the Eq. (3.4) per γ:

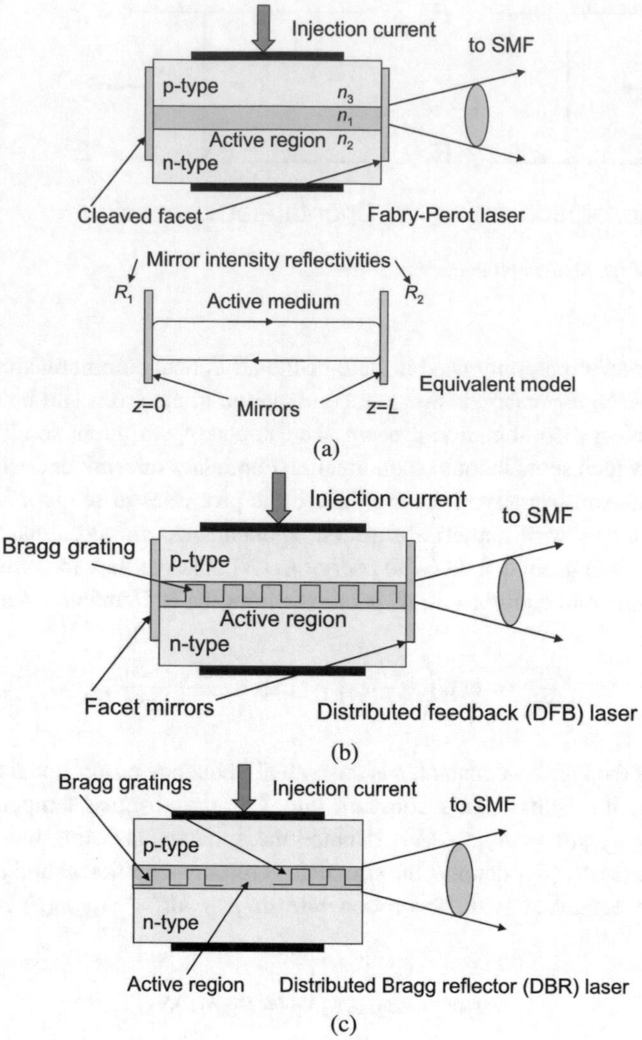

Fig. 3.6 Semiconductor lasers: (**a**) Fabry-Perot semiconductor laser, (**b**) DFB laser, and (**c**) DBR laser

$$\gamma_{th} = \alpha_s + \frac{1}{2L}\ln\left(\frac{1}{R_1R_2}\right) = \alpha_{\text{int}} + \alpha_{\text{mir}}, \tag{3.5}$$

where the internal losses (corresponding to α_s) and mirror losses ($(1/2\,L)\ln(1/R_1R_2)$) are denoted by α_{int} and α_{mir}, respectively. After the round trip, the resultant phase must be equal to the initial phase, leading to the phase matching condition:

$$\exp[-j2\beta L] = 1 \Rightarrow 2\beta L = q2\pi, \beta = 2\pi n/\lambda \tag{3.6}$$

where β denotes the propagation constant, n is the refractive index of active medium, and λ is the free-space wavelength.

The phase matching condition can be satisfied for many different integers q, representing different longitudinal modes of frequency $\nu_q = q\,c/(2\,nL)$. The separation between neighboring longitudinal modes is known as the *free spectral range*:

$$\Delta\nu = \nu_q - \nu_{q-1} = q\frac{c}{2nL} - (q-1)\frac{c}{2nL} = \frac{c}{2nL}. \tag{3.7}$$

Because of the presence of longitudinal modes, the Fabry-Perot laser belongs to the class of multimode lasers. To improve the coherence of output light and the laser modulation, speed distributed feedback (DFB) lasers, shown in Fig. 3.6(b), are used. The key idea of this laser is to effectively select one of the longitudinal modes while suppressing the remaining ones.

This is achieved by introducing the Bragg grating inside of the laser cavity. The wavelength of selected longitudinal mode can be determined from the Bragg condition:

$$2\Lambda = m \cdot \frac{\lambda}{n_{\mathrm{av}}} \Rightarrow \lambda_B = \frac{2n_{\mathrm{av}}\Lambda}{m}, \tag{3.8}$$

where Λ is the grating period, n_{av} is the average refractive index of a waveguide mode, and λ/n_{av} is the average wavelength of the light in the waveguide mode. If the grating element is put outside of active region or instead of the facet mirrors, the distributed Bragg reflector (DBR) laser is obtained, which is illustrated in Fig. 3.6(c). Both DFB and DBR lasers belong to the class of single-mode lasers. Different semiconductor lasers shown in Fig. 3.6 are edge-emitting lasers.

Another important semiconductor laser type is vertical-cavity surface-emitting laser (VCSEL), which emits the light vertical to the active layer plane [1–7]. The VCSELs are usually based on In-GaAs-P layers acting as Bragg reflectors and providing the positive feedback leading to the stimulated emission.

The spectral curve of the single-mode lasers is a result of transition between discrete energy levels and can often be represented using the Lorentzian shape [1–7]:

$$g(\nu) = \frac{\Delta\nu}{2\pi\left[(\nu - \nu_0)^2 + \left(\frac{\Delta\nu}{2}\right)^2\right]}, \tag{3.9}$$

where ν_0 is the central optical frequency and $\Delta\nu$ represents the laser linewidth [1–7]:

$$\Delta\nu = \frac{n_{\mathrm{sp}}G\left(1 + \alpha_{\mathrm{chirp}}^2\right)}{4\pi P}, \tag{3.10}$$

where n_{sp} is the spontaneous emission factor, G is the net rate of stimulated emission, P denotes the output power, and α_{chirp} is the chirp factor (representing the amplitude-phase coupling parameter).

The small-signal frequency response of the semiconductor laser is determined by [1–7]:

$$H(\omega) = \frac{\Omega_R^2 + \Gamma_{\mathrm{R}}^2}{(\Omega_{\mathrm{R}} + \omega - j\Gamma_{\mathrm{R}})(\Omega_{\mathrm{R}} - \omega + j\Gamma_{\mathrm{R}})}, \tag{3.11}$$

where Γ_{R} is the damping factor and Ω_{R} is the relaxation frequency $\Omega_R^2 \approx G_N P_b/\tau_p$ $(\Gamma_R << \Omega_R)$, with G_{N} being the net rate of stimulated emission, P_{b} being the output power corresponding to the bias current, and τ_{p} being the photon life time related to the excited energy level. The modulation bandwidth (defined as 3 dB bandwidth) is therefore determined by the relaxation frequency $\omega_{3\mathrm{dB}} = \sqrt{1 + \sqrt{2}\Omega_R}$, and for fast semiconductor lasers can be 30 GHz. Unfortunately, the direct modulation of semiconductor lasers leads to frequency chirp, which can be described by the instantaneous frequency shift from steady-state frequency ν_0 as follows [1–7]:

$$\delta\nu(t) = \frac{\alpha_{\mathrm{chirp}}}{4\pi}\left[\frac{d}{dt}\ln P(t) + \chi P(t)\right] = C_{\mathrm{dyn}}\frac{d}{dt}P(t) + C_{\mathrm{ad}}P(t), \tag{3.12}$$

where $P(t)$ is the time variation of the output power, χ is the constant (varying from zero to several tens) related to the material and design parameters, and α_{chirp} is the chirp factor defined as the ratio between the refractive index n change and gain G change with respect to the number of carriers N: $\alpha_{\mathrm{chirp}} = (dn/dN)/(dG/dN)$. The first term on the right-hand side of (3.12) represents dynamic (transient or instantaneous) chirp and the second term the adiabatic (steady state) frequency chirp. The random fluctuation in carrier density due to spontaneous emission also leads to the linewidth enhancement proportional to $(1 + \alpha^2_{\mathrm{chirp}})$. To avoid the chirp problem, the external modulation is used, while the semiconductor lasers are biased by a dc voltage to produce a continuous wave operation.

3.1.2.2 External Modulators

There are two types of external modulators commonly used in practice: Mach-Zehnder modulator (MZM) and electro-absorption modulator (EAM), whose operational principle is illustrated in Fig. 3.7. The MZM is based on electro-optic effect, the effect that in certain materials (such as $LiNbO_3$) where the refractive index n changes with respect to the voltage V applied across electrodes [1–7]:

$$\Delta n = -\frac{1}{2}\Gamma n^3 r_{33}(V/d_e) \Rightarrow \Delta \phi = \frac{2\pi}{\lambda}\Delta nL, \tag{3.13}$$

where Δn denotes the refractive index change, $\Delta\phi$ is the corresponding phase change, r_{33} is the electro-optic coefficient (~30.9 pm/V in $LiNbO_3$), d_e is the separation of electrodes, L is the electrode length, and λ is the wavelength of the light. The MZM (see Fig. 3.7(a)) is a planar waveguide structure deposited on the substrate, with two pairs of electrodes: (i) for high-speed ac voltage representing the modulation data (RF) signal and (ii) for dc bias voltage. Let $V_1(t)$ and $V_2(t)$ denote the electrical drive signals on the upper and lower electrodes, respectively. The output electrical field $E_{out}(t)$ of the second Y-branch can be related to the input electrical filed E_{in} by:

$$E_{out}(t) = \frac{1}{2}\left[\exp\left(j\frac{\pi}{V_\pi}V_1(t)\right) + \exp\left(j\frac{\pi}{V_\pi}V_2(t)\right)\right]E_{in}, \tag{3.14}$$

where V_π is now the differential drive voltage required to introduce the phase shift of π radians between two waveguide arms. With $V_1(t)$ and $V_2(t)$, we denoted the electrical drive signals applied to upper and lower electrodes, respectively. We can implement different modulation formats by using this MZM as a building block, including on-off keying (OOK), binary phase-shift keying (BPSK), QPSK, differential PSK (DPSK), differential QPSK (DQPSK), and return-to-zero (RZ) OOK. For

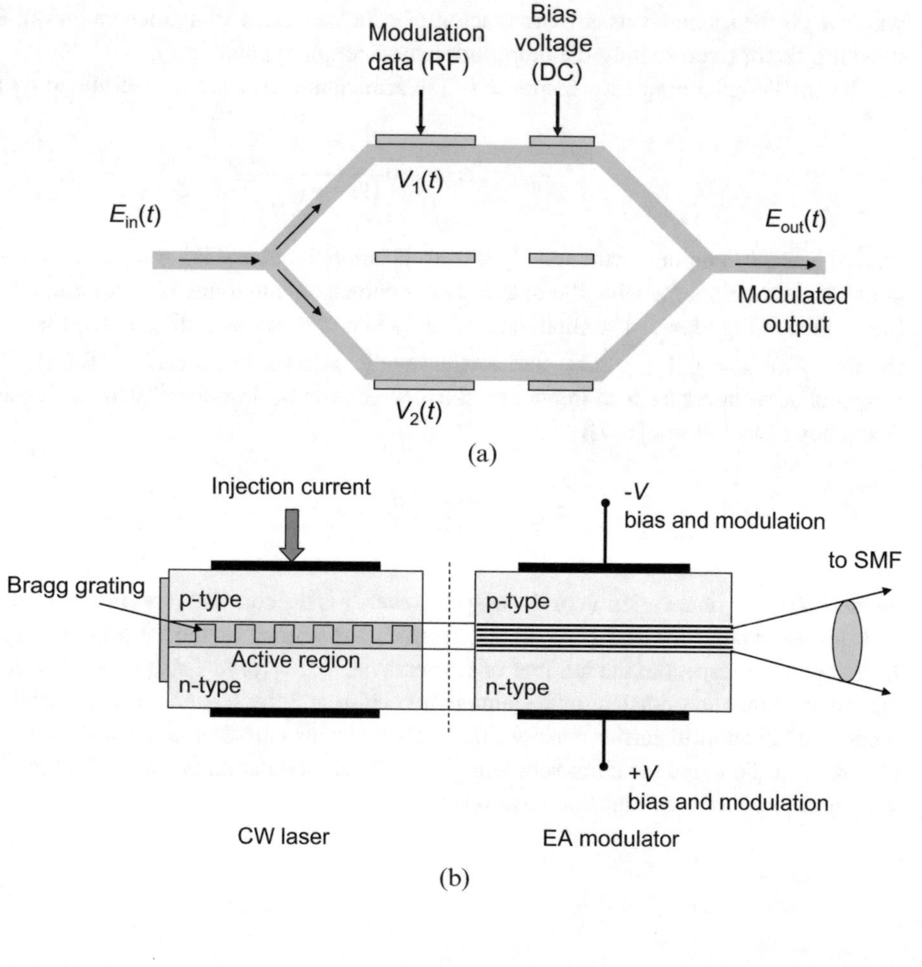

Fig. 3.7 External optical modulators: (**a**) Mach-Zehnder modulator and (**b**) electro-absorption modulator

instance, zero-chirp OOK, or BPSK, can be generated by setting $V_1(t) = V(t) - V_\pi/2$, $V_2(t) = -V(t) + V_\pi/2$, so that Eq. (3.14) becomes:

$$\frac{E_{\text{out}}(t)}{E_{\text{in}}} = \sin\left(\frac{\pi V(t)}{V_\pi}\right). \tag{3.15}$$

On the other hand, if we set $V_1(t) = V(t)/2$, $V_2(t) = -V(t)/2$, the Eq. (3.14) becomes:

$$\frac{E_{\text{out}}(t)}{E_{\text{in}}} = \cos\left(\frac{\pi V(t)}{2V_\pi}\right), \tag{3.16}$$

and we say that MZM is operated in the push-pull mode. The corresponding power transfer function of MZM is obtained as magnitude squared of electric field ratio:

$$\left|\frac{E_{\text{out}}(t)}{E_{\text{in}}}\right|^2 = \cos^2\left(\frac{\pi V(t)}{2V_\pi}\right) = \frac{1 + \cos\left(\pi V(t)/V_\pi\right)}{2}. \tag{3.17}$$

The EAM is a semiconductor-based planar waveguide composed of multiple p-type and n-type layers that form multiple quantum wells (MQWs). The basic design of EAM is similar to that of semiconductor lasers. The MQW is used to support the quantum-confined Stark effect (the absorption spectrum being a function of applied field) more effectively. Because of similarities of EAMs and semiconductor lasers design, it is possible to fabricate them on the same substrate (see Fig. 3.7(b)), providing that EAM and laser are electrically isolated. Bandgap of quantum wells is larger than photon energy, so that the light is completely transmitted in the absence of bias, which corresponds to the ON state. When the reverse bias is applied, the input signal is absorbed, which corresponds to the OFF state. The modulation speed of EAMs is typically comparable to the modulation speed of MZMs. However, the extinction ratio (the ratio of average powers corresponding to symbol 1 and symbol 0) is lower.

It is evident from Eqs. (3.15) to (3.16) that MZM can be used for both intensity modulation and phase modulation. For intensity modulation, the MZM is operated at so-called quadrature point (QP), with a DC bias of $\pm V_{\pi/2}$ and peak-to-peak modulation voltage lower than V_π. On the other hand, for BPSK, the MZM is operated at null point (NP), with a DC bias of $\pm V_\pi$ and peak-to-peak modulation voltage lower than $2V_\pi$. The operating points of MZM for amplitude and phase modulation are shown in Fig. 3.8. In addition, MZM can be used for conversion of non-return-to-zero (NRZ) digital forms to return-to-zero (RZ) form. (Otherwise, the NRZ-to-RZ conversion can be performed either in electrical or optical domain.) In optical domain, we typically use a MZM driven by a properly chosen sinusoidal signal with frequency equal to symbol rate R_s:

$$V(t) = \frac{V_\pi}{2}\sin\left(2\pi R_s t - \frac{\pi}{2}\right) - \frac{V_\pi}{2}, \tag{3.18}$$

so that RZ pulses of duty cycle 50% are being generated at the output of MZM. Accordingly, we have that:

$$\frac{E_{\text{out}}(t)}{E_{\text{in}}(t)} = \cos\left(\frac{\pi}{4}\sin\left(2\pi R_s t - \frac{\pi}{2}\right) - \frac{\pi}{4}\right). \tag{3.19}$$

Additionally, the pulse shape can further be modified by driver amplifiers.

3.1.2.3 I/Q Modulators and Polar Modulators

Another basic building block used to implement more advanced modulators is phase modulator, whose physical properties have been discussed above. The implementation of phase modulator is illustrated in Fig. 3.9.

The electrical fields at the output and input of phase modulator are related as follows:

$$\frac{E_{\text{out}}(t)}{E_{\text{in}}} = \exp\left(j\frac{\pi V(t)}{V_\pi}\right) \tag{3.20}$$

Fig. 3.8 The operating points of Mach-Zehnder modulator

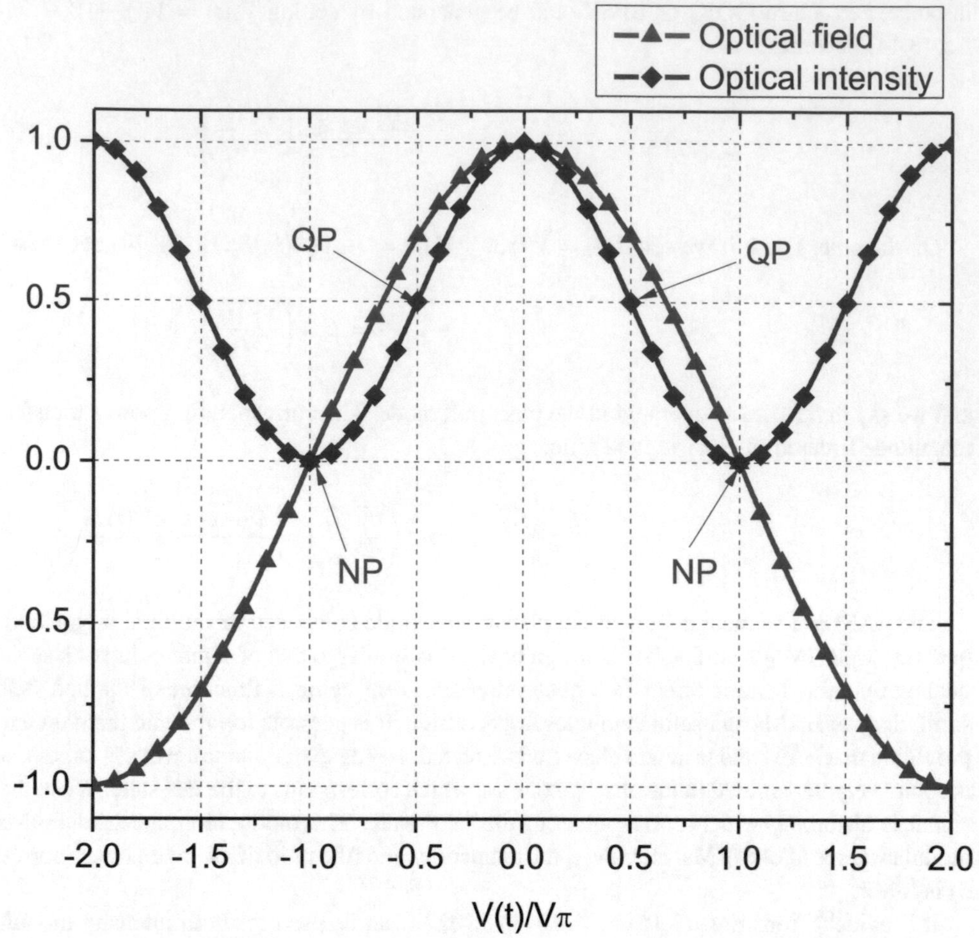

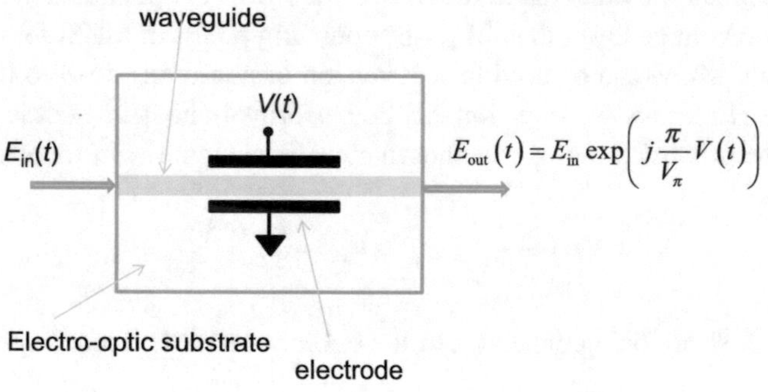

Fig. 3.9 The operation principle of phase modulator

By using two MZM modules and one PM, which introduces the π/2 phase shift, we can implement the I/Q modulator in a way shown in Fig. 3.10. It can be shown that output E_{out} and input E_{in} electrical fields of I/Q modulator are related as:

$$\begin{aligned}\frac{E_{\text{out}}(t)}{E_{\text{in}}} &= \frac{1}{2}\cos\left[\frac{1}{2}\frac{\pi V_I(t)}{V_\pi}\right] + j\frac{1}{2}\cos\left[\frac{1}{2}\frac{\pi V_Q(t)}{V_\pi}\right] \\ &= \frac{1}{2}\cos\left[\Phi_I(t)/2\right] + j\frac{1}{2}\cos\left[\Phi_Q(t)/2\right], \quad \Phi_{I\,(Q)}(t) = \pi V_{I\,(Q)}(t)/V_\pi.\end{aligned} \tag{3.21}$$

where V_I and V_Q are in-phase and quadrature RF modulating signals. When the amplitudes of V_I and V_Q are sufficiently small and in-phase/quadrature signals are chosen as follows $V_I(t) = \frac{-4}{\pi}V_\pi I(t) + V_\pi, V_Q(t) = \frac{-4}{\pi}V_\pi Q(t) + V_\pi,$ the MZM nonlinearities can be neglected and Eq. (3.21) can be rewritten as:

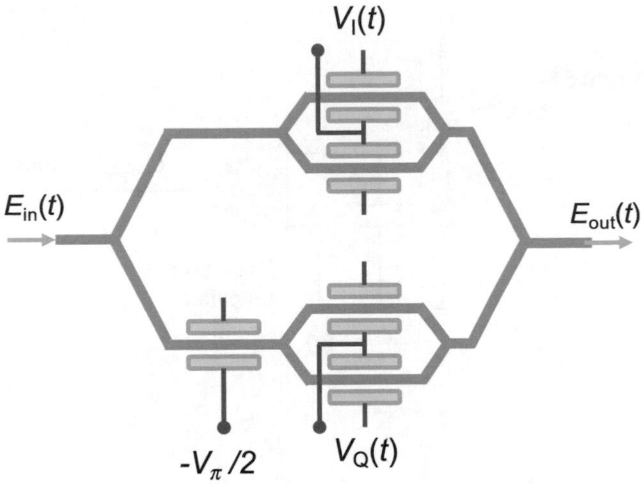

Fig. 3.10 The operation principle of I/Q modulator

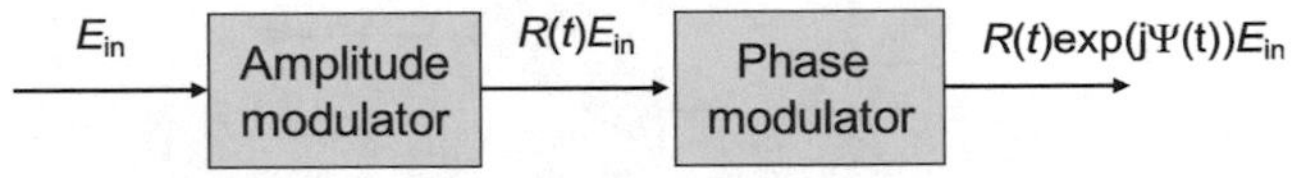

Fig. 3.11 The operation principle of polar modulator

$$\frac{E_{out}(t)}{E_{in}} \cong I(t) + jQ(t), \tag{3.22}$$

where $I(t)$ and $Q(t)$ are in-phase and quadrature components of RF signal. Thereforc, different QAM constellations can be directly applied on this scheme. Equation (3.21) can also be represented in polar coordinates in a form:

$$\frac{E_{out}(t)}{E_{in}} = R(t)\cos\left[\Psi(t)\right], \tag{3.23}$$

where the envelope $R(t)$ and phase $\Psi(t)$ are expressed as:

$$\begin{aligned} R(t) &= \left|\frac{E_{out}(t)}{E_{in}}\right| = \frac{1}{2}\left[\cos^2\left(\frac{1}{2}\frac{\pi V_I(t)}{V_\pi}\right) + \cos^2\left(\frac{1}{2}\frac{\pi V_Q(t)}{V_\pi}\right)\right]^{1/2}, \\ \Psi(t) &= \tan^{-1}\left\{\frac{\cos^2\left(\frac{1}{2}\frac{\pi V_Q(t)}{V_\pi}\right)}{\cos^2\left(\frac{1}{2}\frac{\pi V_I(t)}{V_\pi}\right)}\right\}. \end{aligned} \tag{3.24}$$

The representation given by Eq. (3.23) indicates that instead of I/Q modulator, we can use an equivalent *polar modulator* shown in Fig. 3.11. The real part of the output of polar modulator corresponds to Eq. (3.21).

3.1.3 Optical Receivers

The purpose of the optical receiver is to convert the optical signal into electrical domain and to recover the transmitted data. The typical OOK receiver configuration is already given in Fig. 3.2. We can identify three different stages: front end stage, the linear channel stage, and data recovery stage. The front end stage is composed of a photodetector and a pre-amplifier. The most commonly used front end stages are high-impedance front end and trans-impedance front end, both shown in Fig. 3.8.

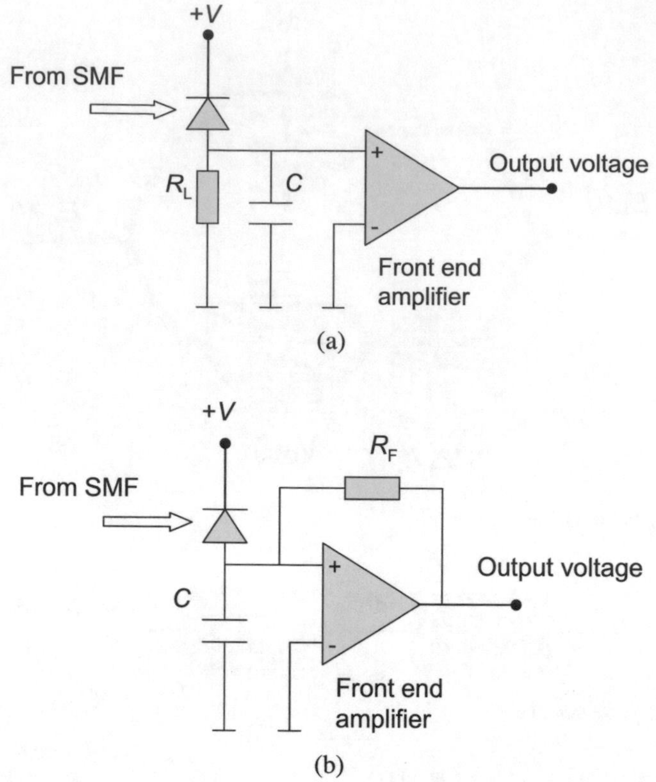

Fig. 3.12 Optical receiver front end stage schemes: (**a**) high-impedance front end and (**b**) trans-impedance front end

High-impedance front end (Fig. 3.12(a)) employs a large value load resistance to reduce the level of thermal noise and has a good receiver sensitivity. However, the bandwidth of this scheme is low because the RC constant is large. To achieve both the high receiver sensitivity and large bandwidth, the trans-impedance front end scheme, shown in Fig. 3.12(b), is used. Even though the load resistance is high, the negative feedback reduces the effective input resistance by a factor of $G - 1$, where G is the front end amplifier gain. The bandwidth is increased for the same factor compared to high-impedance front end scheme.

The photodiode is an integral part of both front end stage schemes. The key role of the photodiode is to absorb photons in incoming optical signal and convert them back to the electrical level through the process opposite to one taking place in semiconductor lasers. The common photodiodes are p-n photodiode, p-i-n photodiode, avalanche photodiode (APD), and metal-semiconductor-metal (MSM) photodetectors [1–7]. The *p-n photodiode* is based on a reverse-biased *p-n* junction. The thickness of the depletion region is often less than the absorption depth for incident light, and the photons are absorbed outside of the depletion region, leading to the slow response speed. *p-i-n photodiode* consists of an intrinsic region sandwiched between *p-* and *n*-type layers, as shown in Fig. 3.13(a). Under the reverse bias, the depletion depth can be made sufficiently thick to absorb most of the incident photons. *Avalanche photodiode*, shown in Fig. 3.13(b), is a modified *p-i-n* photodiode that is operated at very high reverse bias. Under high-field conditions, photo-generated carriers induce generation of *secondary electron-hole pairs* by the process of *impact ionization*, and this process leads to internal electrical gain. *Metal-semiconductor-metal photodetectors* employ interdigitated Schottky barrier contacts on one face of the device and are compatible with planar processing and optoelectronic integration. Depending on the device design, the device is illuminated through the p- or n-type contact. In Si, Ge, or GaAs diodes, the substrate is absorbing so that the device has to be illuminated through the top contact, as shown in Fig. 3.13(a).

On the other hand, in InGaAs or InGaAsP, the substrate is transparent, and the device can be designed to be illuminated either through the substrate or top contact. In order to increase the depletion region and to minimize the diffusion current component, an intrinsic layer (i-type) is introduced to the p-i-n photodiode structure. The p-i-n photodiode is reverse biased and has very high internal impedance, meaning that it acts as a current source generating the photocurrent proportional to the incoming optical signal power. The equivalent scheme of p-i-n photodiode is shown in Fig. 3.13(c). Typically the internal series resistance R_s is low, while the internal shunt resistance is high, so that the junction capacitance C_p dominates and can be determined by:

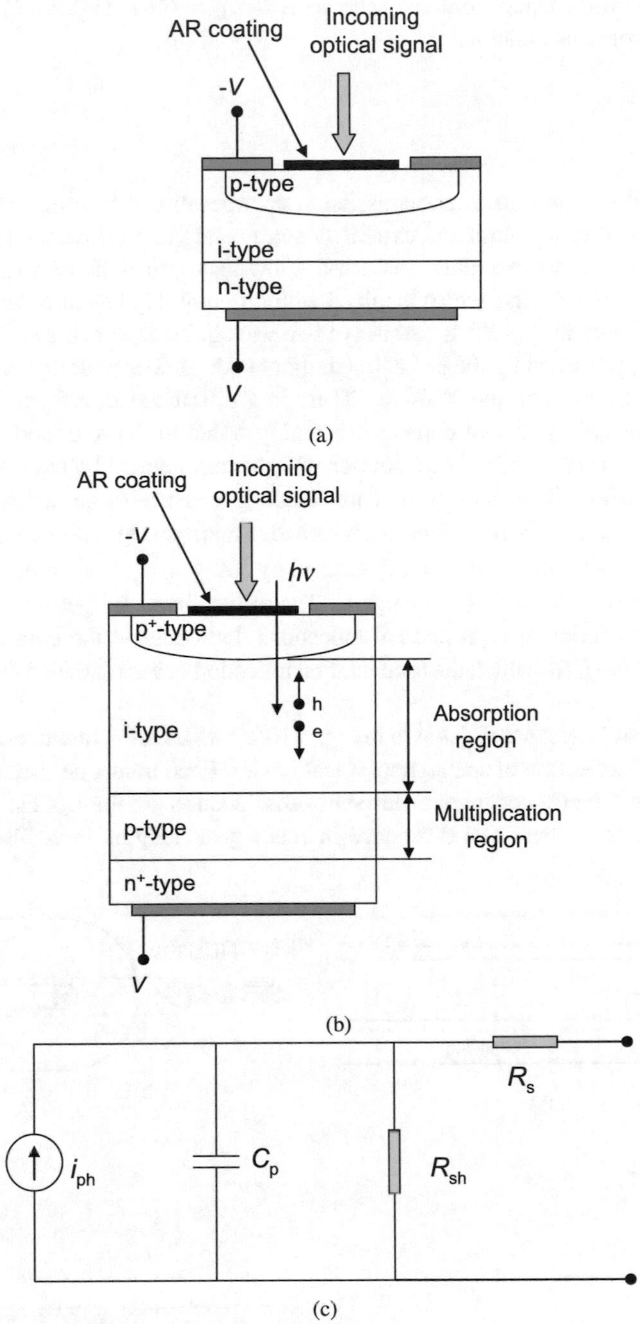

Fig. 3.13 The semiconductor photodiodes: (**a**) p-i-n photodiode and (**b**) avalanche photodiode. (**c**) The equivalent p-i-n photodiode model

$$C_p = \varepsilon_s \frac{A}{w} = \left[\frac{\varepsilon_s N_A N_D}{2(N_A - N_D)(V_0 - V_A)} \right]^{1/2}, \tag{3.25}$$

where ε_s is the semiconductor permittivity, A is the area of the space charge region (SCR), w is the width of SCR, N_A and N_D denote dopant (acceptor and donor) densities, V_0 is the built-in potential across the junction, and V_A is the applied negative voltage. The photocurrent $i_{ph}(t)$ is proportional to the power of incident light $P(t)$, that is, $i_{ph}(t) = R\,P(t)$, where R [A/W] is the photodiode responsivity. The photodiode responsivity is related to the quantum efficiency η, defined as the ratio of number of generated electrons and the number of incident photons, by $R = \eta q/h\nu$, where q is an electron charge and $h\nu$ is a photon energy. Using this model, we can determine the 3 dB bandwidth of high-impedance front end scheme as $B_{3\text{ dB}} = 1/(2\pi R_L C_p)$

and the 3 dB bandwidth of trans-impedance front end scheme as $B_{3\ \mathrm{dB}} = (G + 1)/(2\pi R_F C_p)$, which is G times higher than bandwidth of high-impedance front end scheme.

3.1.4 Optical Fibers

Optical fibers serve as foundation of an optical transmission system because they transport optical signals from source to destination [1–7]. The combination of low-loss and extremely large bandwidth allows high-speed signals to be transmitted over long distances before the regeneration becomes necessary. A low-loss optical fiber is manufactured from several different materials; the base row material is pure silica, which is mixed with different dopants in order to adjust the refractive index of optical fiber. The optical fiber, shown in Fig. 3.14, consists of two waveguide layers, the *core* (of refractive index n_1) and the *cladding* (of refractive index n_2), protected by the jacket (the buffer coating). The majority of the power is concentrated in the core, although some portion can spread to the cladding. There is a difference in refractive indices between the core and cladding ($n_1 > n_2$), which is achieved by mix of dopants commonly added to the fiber core. The refractive-index profile for step-index fiber is shown in Fig. 3.14(c), while the illustration of light confinement by the total internal reflection is shown in Fig. 3.14(d). The ray will be totally reflected from the core-cladding interface (a guided ray) if the following condition is satisfied $n_0 \sin \theta_i < \sqrt{n_1^2 - n_2^2}$, where θ_i is the angle of incidence. Max($n_0 \sin\theta_i$) defines the light-gathering capacity of an optical fiber, and it is called the *numerical aperture* (*NA*), defined by $NA = \sqrt{n_1^2 - n_2^2} \approx n_1\sqrt{2\Delta}, \quad (\Delta \ll 1)$, where Δ is the *normalized index difference* defined as $\Delta = (n_1 - n_2)/n_1$. Therefore, from the geometrical optics point of view, light propagates in optical fiber due to series of total internal reflections that occur at the core-cladding interface. The smallest angle of incidence ϕ (see Fig. 3.10(d)) for which the total internal reflection occurs is called the *critical angle* and equals $\sin^{-1} n_2/n_1$.

There exist two types of optical fibers: MMF (shown in Fig. 3.10(a)) and SMF (shown in Fig. 3.10(b)). Multimode optical fibers transfer the light through a collection of spatial transversal modes. Each mode, defined through a specified combination of electrical and magnetic components, occupies a different cross section of the optical fiber core and takes a slightly distinguished path along the optical fiber. The difference in mode path lengths in multimode optical fibers produces a

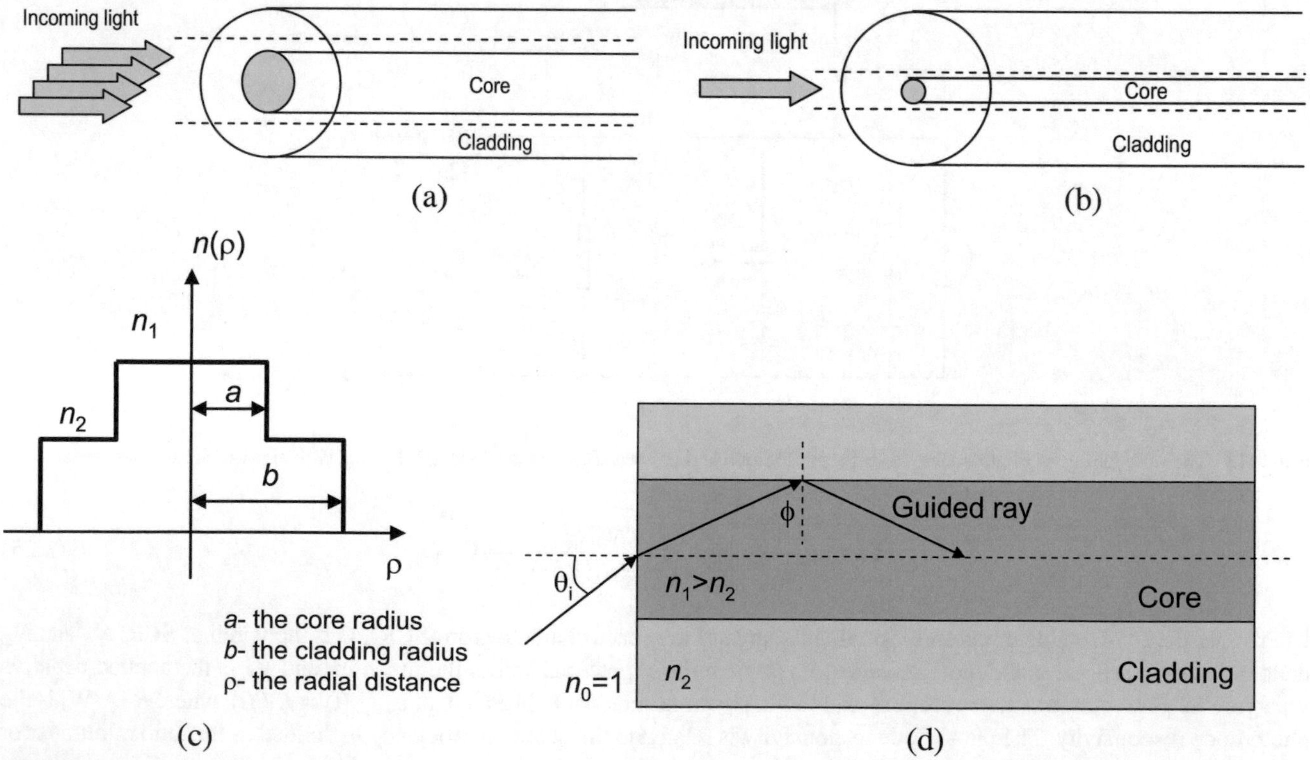

Fig. 3.14 Optical fibers: (**a**) multimode optical fiber, (**b**) single-mode optical fiber, (**c**) refractive-index profile for step-index fiber, and (**d**) the light confinement in step-index fibers through the total internal reflection

difference in arrival times at the receiving point. This phenomenon is known as *multimode dispersion* (or *intermodal dispersion*) and causes the signal distortion and imposes the limitations in signal bandwidth. The second type of optical fibers, single-mode fibers, effectively eliminates multimode dispersion by limiting the number of propagating modes to a *fundamental* one. The *fundamental mode* occupies the central portion of the optical fiber and has an energy maximum at the axis of the optical fiber core. Its radial distribution can be approximated by Gaussian curve. The number of modes (M) that can effectively propagate through an optical fiber is determined by the *normalized frequency* (*V parameter* or V number): $M \approx V^2/2$, when V is large. The normalized frequency is defined by:

$$V = \frac{2\pi a}{\lambda}\sqrt{n_1^2 - n_2^2}, \tag{3.26}$$

where a is the fiber core radius, λ is the carrier wavelength, and n_1 and n_2 are refractive indices related to the fiber core and the fiber cladding, respectively.

Each mode propagating through the fiber is characterized by its own propagation constant β. The dependence of the electric and magnetic fields on axial coordinate z is expressed through the factor exp.($-j\beta z$). The propagation constant must satisfy the following condition:

$$2\pi n_2/\lambda < \beta < 2\pi n_1/\lambda. \tag{3.27}$$

In order to evaluate the transmission characteristics of the optical fiber, the functional dependence of the mode propagation constant on the optical signal wavelength has to be known. The *normalized propagation constant b* is defined for that purpose:

$$b = \frac{\beta^2 - (2\pi n_2/\lambda)^2}{(2\pi n_1/\lambda)^2 - (2\pi n_2/\lambda)^2}. \tag{3.28}$$

The normalized propagation constant is related to the normalized frequency V by [1–11]:

$$b(V) \approx (1.1428 - 0.9960/V)^2, \quad 1.5 \le V \le 2.5 \tag{3.29}$$

The multimode dispersion can effectively be eliminated by limiting the number of propagating modes to a fundamental one: $V \le V_c = 2.405$ with V_c being the cutoff frequency. The cutoff frequency is controlled by keeping the core radius small and the normalized index difference $\Delta = (n_1 - n_2)/n_1$ between 0.2% and 0.3%.

3.1.5 Optical Amplifiers

The purpose of an optical amplifier is to restore the signal power level, reduced due to losses during propagation, without any optical to electrical conversion. The general form of an optical amplifier is given in Fig. 3.15(a). Most optical amplifiers amplify incident light through the *stimulated emission*, the same mechanism that is used in lasers but without the feedback

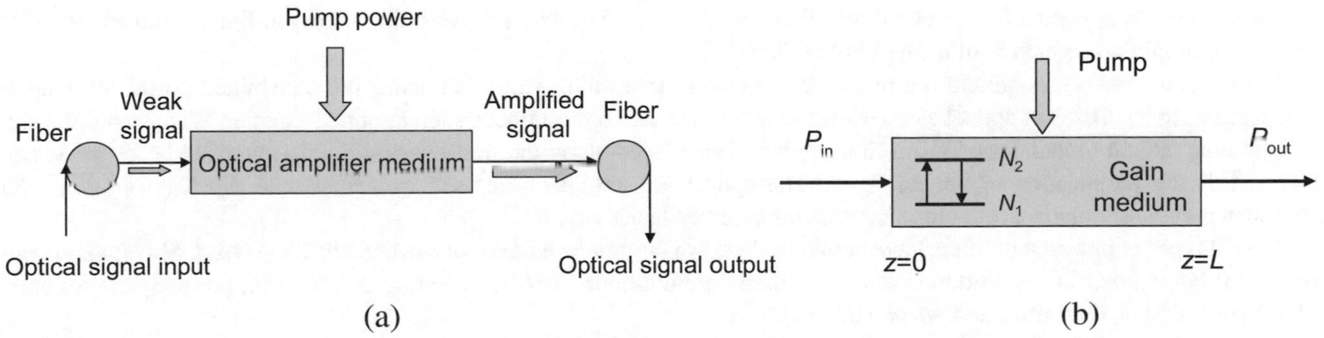

Fig. 3.15 (**a**) Optical amplifier principle, (**b**) two-level amplifier system model

mechanism. The main ingredient is the *optical gain* realized through the amplifier pumping (electrical or optical) to achieve the *population inversion*. The optical gain, generally speaking, is not only a function of frequency but also a function of local beam intensity. To illustrate the basic concepts, we consider the case in which the gain medium is modeled as *two-level system*, as shown in Fig. 3.15(b). The *amplification factor* G is defined as the ratio of amplifier output P_{out} and input P_{in} powers $G = P_{out}/P_{in}$. The amplification factor can be determined by knowing the dependence of evolution of power through the gain media [1–7]:

$$\frac{dP}{dz} = gP, \qquad g(\omega) = \frac{g_0}{1 + (\omega - \omega_0)^2 T_2^2 + P/P_s}, \tag{3.30}$$

where g is the gain coefficient, g_0 is the gain peak value, ω_0 is the atomic transition frequency, T_2 is the dipole relaxation time (<1 ps), ω is the optical frequency of incident signal, P is the incident signal power, and P_S is the saturation power.

In the unsaturated regime ($P << P_S$), the differential Eq. (3.30) can be solved by separation of variables to get the following dependence of power $P(z) = P(0)\exp.(gz)$, so that the amplification factor can be obtained by:

$$G(\omega) = \exp\left[g(\omega)L\right], \tag{3.31}$$

and corresponding full width half-maximum (FWHM) bandwidth is determined by:

$$\Delta\nu_A = \Delta\nu_g \sqrt{\frac{\ln 2}{\ln (G_0/2)}}, \qquad G_0 = \exp(g_0 L) \tag{3.32}$$

where $\Delta\nu_g$ is the FWHM gain coefficient bandwidth.

The *gain saturation* comes from the power dependence of the gain coefficient (3.30). The coefficient is reduced when the incident power P becomes comparable to the saturation power P_S. Let's assume that the incident frequency is tuned to the peak gain ($\omega = \omega_0$), and then from Eq. (3.30), we obtain:

$$\frac{dP}{dz} = \frac{g_0 P}{1 + P/P_S}. \tag{3.33}$$

By solving the differential Eq. (3.33) with respect to the *boundary conditions*, $P(0) = P_{in}$ and $P(L) = P_{out} = GP_{in}$, we get:

$$G = G_0 \exp\left[-\frac{G-1}{G}\frac{P_{out}}{P_S}\right]. \tag{3.34}$$

From Eq. (3.34), we can determine another important optical amplifier parameter, the *output saturation power* as being the optical power at which the gain G is reduced to $G_0/2$ (3 dB down):

$$P_{out}^{sat} = \frac{G_0 \ln 2}{G_0 - 2} P_s \approx (\ln 2) P_s \approx 0.69 P_s (G_0 > 20\text{dB}). \tag{3.35}$$

Three common applications of optical amplifiers are (i) power boosters (of transmitters), (ii) in-line amplifiers, and (iii) optical pre-amplifiers, which is illustrated in Fig. 3.16.

The booster (power) amplifiers are placed at the optical transmitter side to enhance the transmitted power level or to compensate for the losses of optical elements between the laser and optical fibers, such as optical coupler, WDM multiplexers, and external optical modulators. The in-line amplifiers are placed along the transmission link to compensate for the losses incurred during propagation of optical signal. The optical pre-amplifiers are used to increase the signal level before the photodetection takes place, improving therefore the receiver sensitivity.

Several types of optical amplifiers have been introduced so far: *semiconductor optical amplifiers* (SOAs), *fiber Raman* (and Brillouin) *amplifiers*, rare-earth-doped fiber amplifiers (erbium-doped *EDFA* operating at 1500 nm, praseodymium-doped PDFA operating at 1300 nm), and *parametric amplifiers*.

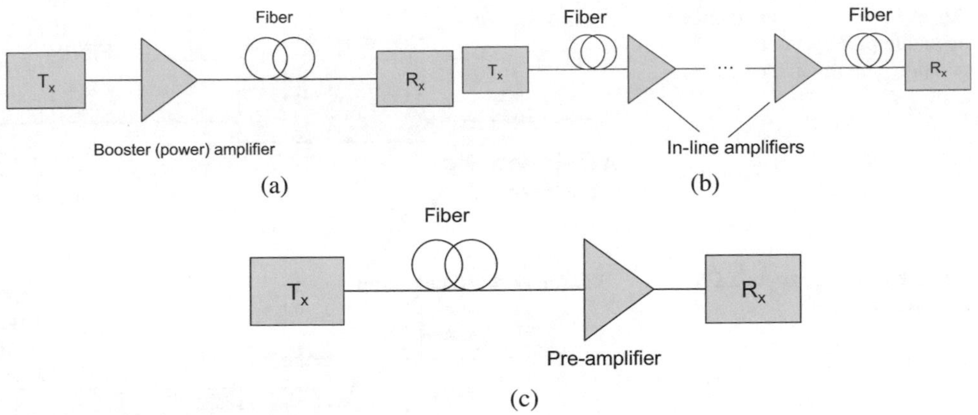

Fig. 3.16 Possible application of optical amplifiers: (**a**) booster amplifier, (**b**) in-line amplifiers, and (**c**) pre-amplifier

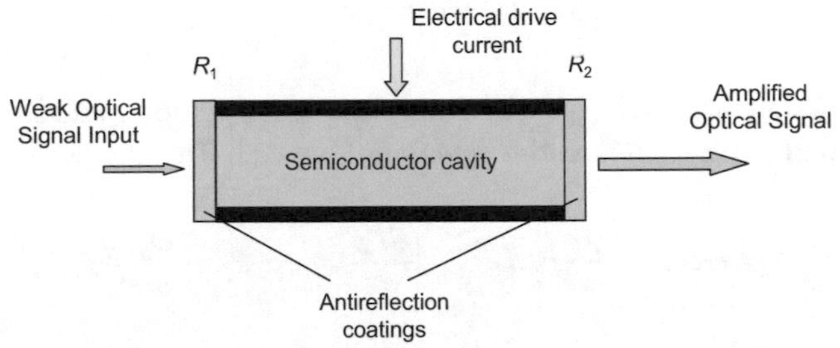

Fig. 3.17 The semiconductor optical amplifier (SOA) operation principle

3.1.5.1 Semiconductor Optical Amplifier (SOA)

Semiconductor lasers act as amplifiers before reaching the threshold. To prevent the lasing, *antireflection (AR) coatings* are used in SOAs, as shown in Fig. 3.17. Even with AR coating, the multiple reflections must be included when considering the Fabry-Perot (FP) cavity. The gain of FP amplifiers is given by [1–7]:

$$G_{\mathrm{FP}}(\nu)=\frac{(1-R_1)(1-R_2)G(\nu)}{\left(1-G(\nu)\sqrt{R_1R_2}\right)^2+4G\sqrt{R_1R_2}\sin^2[\pi(\nu-\nu_m)/\Delta\nu_L]}, \tag{3.36}$$

where R_1 and R_2 denote the facet reflectivities, $G(\nu)$ is the single-pass amplification factor, ν_m is the cavity resonance frequency, and $\Delta\nu_L$ is the free spectral range.

The FP amplifier bandwidth can be determined as follows [1–7]:

$$\Delta\nu_A=\frac{2\Delta\nu_L}{\pi}\sin^{-1}\left[\frac{1-G\sqrt{R_1R_2}}{\left(4G\sqrt{R_1R_2}\right)^{1/2}}\right]. \tag{3.37}$$

3.1.5.2 Raman Amplifiers

A *fiber-based Raman amplifier* employs the *stimulated Raman scattering* (SRS) occurring in silica fibers when an intense pump propagates through it [1–7]. *SRS fundamentally differs from stimulated emission*: in stimulated emission, an incident photon stimulates emission of another identical photon; in SRS, the incident pump photon gives up its energy to create another photon of reduced energy at a lower frequency (inelastic scattering); and the remaining energy is absorbed by the medium in the form of molecular vibrations (optical phonons). *Raman amplifiers must be pumped optically* to provide gain, as shown in Fig. 3.18.

The Raman gain coefficient g_R is related to the optical gain $g(z)$ as $g(z)=g_RI_p(z)$, where I_p is the pump intensity given by $I_p=P_p/a_p$, with P_p being the pump power and a_p being the pump cross-sectional area. Since the cross-sectional area is different for different types of fibers, the ratio g_R/a_p is the measure of the Raman gain efficiency. The DCF efficiency can even

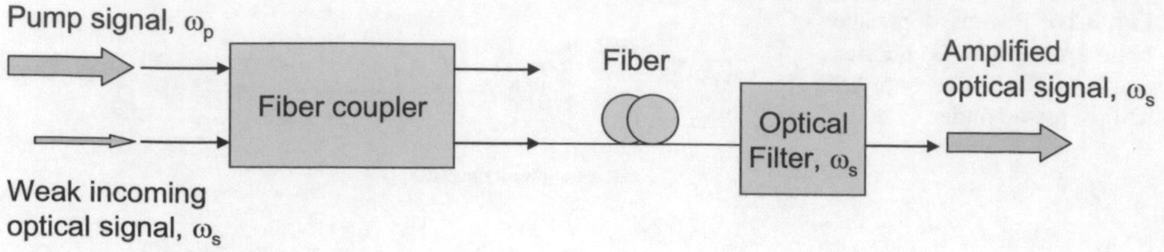

Fig. 3.18 The Raman amplifier operation principle in forward-pumping configuration

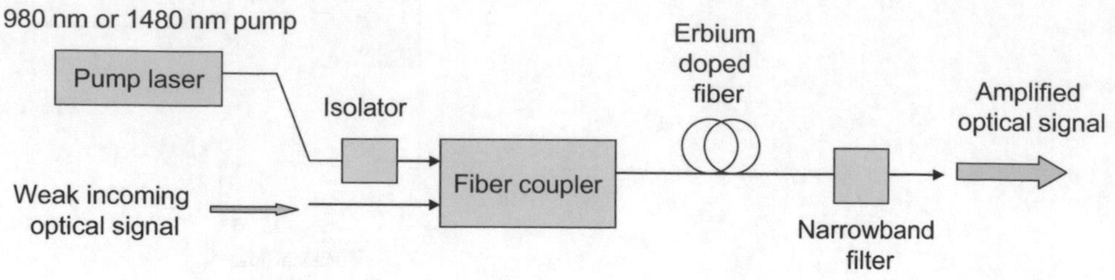

Fig. 3.19 The general EDFA configuration

be eight times better than that of a standard SMF [1–7]. The evolution of the pump P_p and signal P_s powers (in distance z) can be studied by solving the system of *coupled differential equations* below [1–7]:

$$\frac{dP_s}{dz} = -\alpha_s P_s + \frac{g_R}{a_p} P_p P_s, \qquad \frac{dP_p}{dz} = -\alpha_p P_p - \frac{\omega_p g_R}{\omega_s a_p} P_p P_s, \tag{3.38}$$

where a_s denotes the signal cross-sectional area ω_p and ω_s denote the pump and signal frequency, respectively, while other parameters are already introduced above.

In *small-signal amplification regime* (when the pump depletion can be neglected), the pump power evolution is exponential, $P_p(z) = P_p(0)\exp[-\alpha_p z]$, so that the *Raman amplifier gain* is found to be:

$$\begin{aligned} G_A &= \frac{P_s(0)\exp\left(g_R P_p(0) L_{\text{eff}}/a_p - \alpha_s L\right)}{P_s(0)\exp\left(-\alpha_s L\right)} = \exp\left(g_0 L\right), \\ g_0 &= g_R \frac{P_p(0)}{a_p}\frac{L_{\text{eff}}}{L} \approx \frac{g_R P_p(0)}{a_p \alpha_p L} \quad (\alpha_P L >> 1). \end{aligned} \tag{3.39}$$

The origin of saturation in Raman amplifiers is pump power depletion, which is quite different from that in SOAs. *Saturated amplifier gain* G_s can be determined (assuming $\alpha_p = \alpha_s$) by [1–7]:

$$G_S = \frac{1 + r_0}{r_0 + 1/G_A^{1+r_0}}, \qquad r_0 = \frac{\omega_p}{\omega_s}\frac{P_s(0)}{P_p(0)} \tag{3.40}$$

The amplifier gain is reduced down by 3 dB when $G_A r_0 \approx 1$, the condition that is satisfied when the amplified signal power becomes comparable to the input pump power $P_s(\mathrm{L}) = P_p(0)$. Typically $P_0 \sim 1$ W, and channel powers in a WDM systems are around 1 mW, meaning that *Raman amplifier operates in unsaturated or linear regime.*

3.1.5.3 Erbium-Doped Fiber Amplifier (EDFA)

The *rare-earth-doped fiber amplifiers* are finding increasing importance in optical communication systems. The most important class is erbium-doped fiber amplifiers (EDFAs) due to their ability to amplify in 1.55 μm wavelength range. The active medium consists of 10–30 m length of optical fiber highly doped with a rare-earth element, such as erbium (Er), ytterbium (Yb), neodymium (Nd), or praseodymium (Pr). The host fiber material can be pure silica, a fluoride-based glass, or a multi-component glass. General EDFA configuration is shown in Fig. 3.19.

The pumping at a suitable wavelength provides gain through *population inversion*. The gain spectrum depends on the pumping scheme as well as on the presence of other dopants, such as Ge or Al within the core. The amorphous nature of silica broadens the energy levels of Er^{3+} into the bands, as shown in Fig. 3.20.

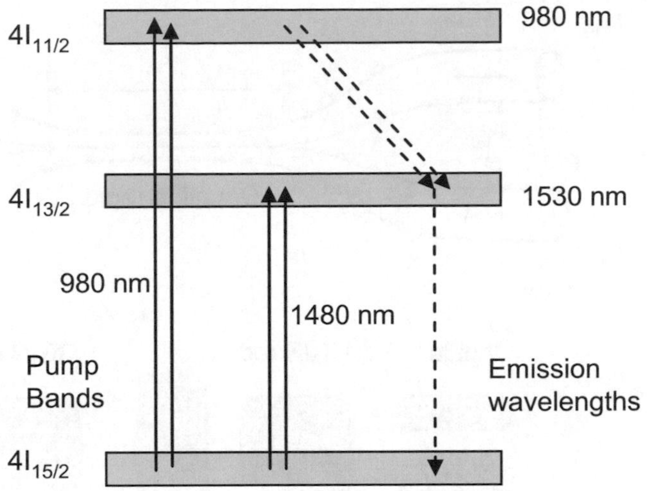

Fig. 3.20 The EDFA energy diagram

The pumping is primarily done in optical domain with the *primary pump wavelengths* at 1.48 μm and 0.98 μm. The atoms pumped to the $4I_{11/2}$ 0.98 μm decay to the primary emission transition band. The pumping with 1.48 μm is directly to the upper transition levels of the emission band.

EDFAs can be designed to operate in such a way that the pump and signal travel in opposite directions; this configuration is commonly referred to as *backward pumping*. In *bidirectional pumping*, the amplifier is pumped in both directions simultaneously by using two semiconductor lasers located at the both fiber ends.

3.1.6 Optical Processing Components and Modules

Different optical components can be classified into two broad categories depending on whether they can operate without an external electric power source or not into passive or active. Important active components are lasers, external modulators, optical amplifiers, photodiodes, optical switches, and wavelength converters. On the other hand, important passive components are optical couplers, isolators, multiplexers/demultiplexers, and filters. Some components, such as optical filters, can be either passive or active depending on operational principle. In this section, we will briefly explain some important optical components not being described in previous subsections.

3.1.6.1 Optical Couplers

The 2 × 2 optical coupler is a fundamental device that can be implemented using the fiber fusing or be based on graded-index (GRIN) rods and optical filters, as shown in Fig. 3.21. The fused optical couplers (shown in Fig. 3.21(a)) are obtained when the cladding of two optical fibers are removed; the cores are brought together and then heated and stretched. The obtained waveguide structure can exchange energy in the coupling region between the branches. If both inputs are used, 2 × 2 coupler is obtained; if only one input is used, 1 × 2 coupler is obtained. The optical couplers are recognized either as optical taps (1 × 2) couplers or directional (2 × 2) couplers. The power coupler splitting ratio depending on purpose can be different with typical values being 50%/50%, 10%/90%, 5%/95%, and 1%/99%. Directional coupler parameters (defined when only input 1 is active) are splitting ratio $P_{\text{out},1} / (P_{\text{out},1} + P_{\text{out},2})$, excess loss $10\log_{10}[P_{\text{in},1}/(P_{\text{out},1} + P_{\text{out},2})]$, insertion loss $10\log_{10}(P_{\text{in},i}/P_{\text{out},j})$, and crosstalk $10\log_{10}(P_{\text{cross}}/P_{\text{in},1})$.

The operation principle of directional coupler can be explained using coupled mode theory [1, 2, 6, 9, 10] or simple scattering (propagation) matrix $\boldsymbol{S}$ approach, assuming that coupler is lossless and reciprocal device:

$$\begin{bmatrix} E_{\text{out},1} \\ E_{\text{out},2} \end{bmatrix} = \boldsymbol{S} \begin{bmatrix} E_{\text{in},1} \\ E_{\text{in},2} \end{bmatrix} = \begin{bmatrix} s_{11} & s_{12} \\ s_{21} & s_{22} \end{bmatrix} \begin{bmatrix} E_{\text{in},1} \\ E_{\text{in},2} \end{bmatrix} = e^{-j\beta L} \begin{bmatrix} \cos(kL) & j\sin(kL) \\ j\sin(kL) & \cos(kL) \end{bmatrix} \begin{bmatrix} E_{\text{in},1} \\ E_{\text{in},2} \end{bmatrix}, \tag{3.41}$$

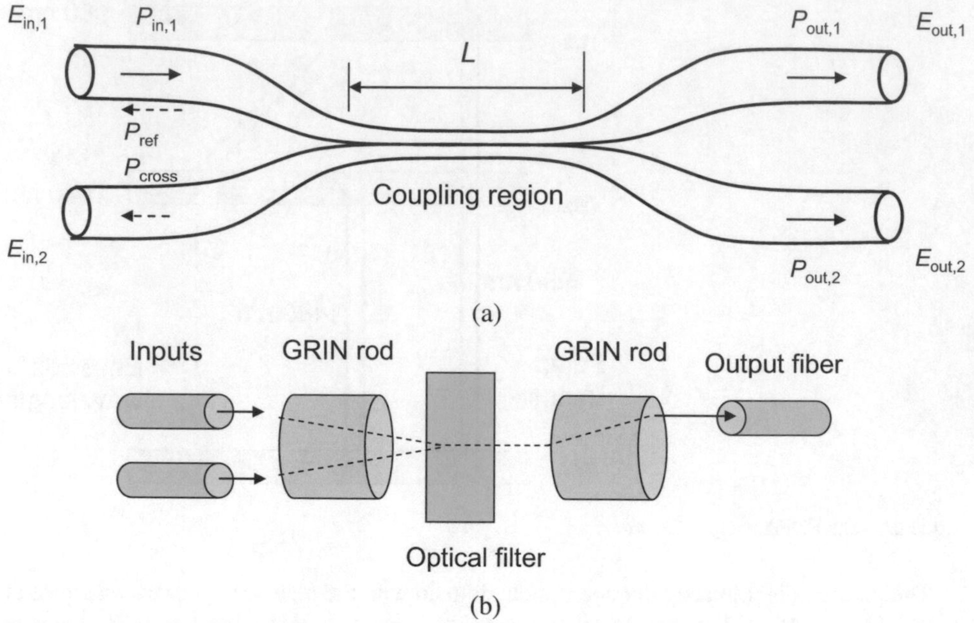

Fig. 3.21 Optical couplers: (**a**) fiber fusing based and (**b**) GRIN rode based

where β is propagation constant, k is coupling coefficient, L is the coupling region length, $E_{in,1}$ and $E_{in,2}$ are corresponding inputs electrical fields, and $E_{out,1}$ and $E_{out,2}$ are corresponding output electrical fields. Scattering matrix **S** elements are denoted with s_{ij}. For example, for *3 dB coupler*, we have to select $kL = (2\,m + 1)\pi/4$ (m is a positive integer) to get:

$$\begin{bmatrix} E_{out,1} \\ E_{out,2} \end{bmatrix} = \frac{1}{\sqrt{2}} \begin{bmatrix} 1 & j \\ j & 1 \end{bmatrix} \begin{bmatrix} E_{in,1} \\ E_{in,2} \end{bmatrix}. \tag{3.42}$$

The combination of two GRIN rods and an optical filter can effectively be used as an optical coupler, as illustrated in Fig. 3.21(b). The GRIN rods are used as collimators, to collimate the light from two input ports and deliver to the output port, while the optical filter is used to select a desired wavelength channel.

The optical couplers can be used to create more complicated optical devices such as *M*x*N* optical stars, directional optical switches, different optical filters, multiplexers, etc.

3.1.6.2 Optical Filters

An *optical filter* modifies the spectrum of incoming light and can mathematically be described by corresponding transfer function $H_{of}(\omega)$:

$$E_{out}(t) = \frac{1}{2\pi} \int_{-\infty}^{\infty} \widetilde{E}_{in}(\omega) H_{of}(\omega) e^{j\omega t} d\omega, \tag{3.43}$$

where $E_{in}(t)$ and $E_{out}(t)$ denote the input and output electrical field, respectively, and we used ~ to denote the FT as before. Depending on the operational principle, the optical filters can be classified into two broad categories as *diffraction* or *interference* filters. The important classes of optical filters are tunable optical filters, which are able to dynamically change the operating frequency to the desired wavelength channel. The *basic tunable optical filter types* include tunable 2 × 2 directional couplers, Fabry-Perot (FP) filters, Mach-Zehnder (MZ) interferometer filters, Michelson filters, and acousto-optical filters. Two basic optical filters, FP filter and MZ interferometer filter, are shown in Fig. 3.22. An FP filter is in fact a cavity between two high-reflectivity mirrors. It can act as a tunable optical filter if the cavity length is controlled, for example, by using a piezoelectric transducer. Tunable FP filters can also be made by using liquid crystals, dielectric thin films, semiconductor waveguides, etc. A *transfer function of an FP filter* whose mirrors have the same reflectivity R and the cavity length L can be written as [1–7]:

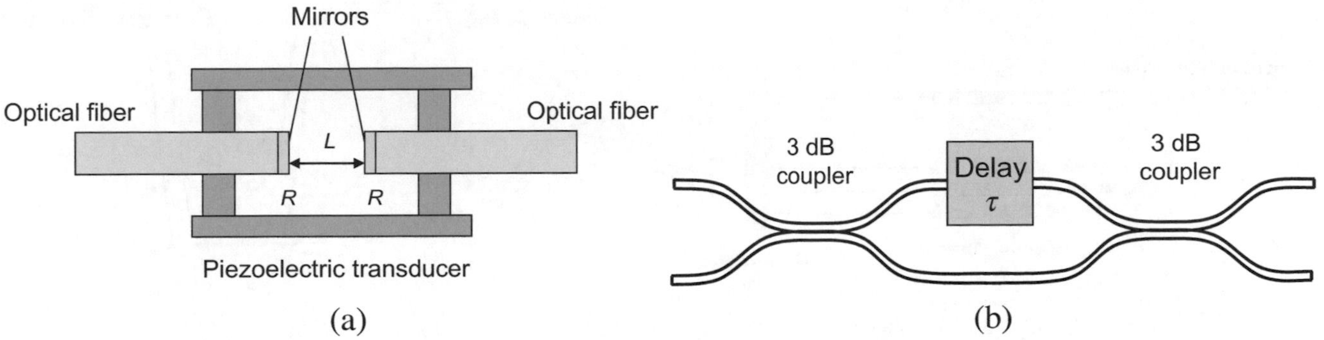

Fig. 3.22 Basic optical filters: (**a**) Fabry-Perot filter and (**b**) Mach-Zehnder interferometer

$$H_{\mathrm{FP}}(\omega)=\frac{(1-R)e^{j\pi}}{1-Re^{j\omega\tau}},\qquad \tau=2L/v_g \tag{3.44}$$

where τ is the round-trip time within the cavity and v_g is the group velocity. The transfer function of FP filter is periodic with period being the free spectral range (FSR) $\Delta\nu = v_g/(2L)$. Another FP filter important parameter is the *finesse* defined as [1–7]:

$$F=\Delta\nu\ /\Delta\nu_{\mathrm{FP}}\cong\frac{\pi\sqrt{R}}{1-R}, \tag{3.45}$$

where $\Delta\nu_{\mathrm{FP}}$ is the FP transmission peak width.

Mach-Zehnder interferometer filters, shown in Fig. 3.22(b), can also be used as tunable optical filters. The first coupler splits the signal into two equal parts, which acquire different phase shifts before they interfere at the second coupler. Several MZ interferometers can be cascaded to create an optical filter. When cross output of 3 dB coupler is used, the square magnitude of transfer function is [1–7] $|H_c(\omega)| = \cos^2(\omega\tau/2)$, so that the transfer function of M-stage MZ filter based on 3 dB couplers can be written as:

$$|H_{\mathrm{MZ}}(\omega)|^2=\prod_{m=1}^{M}\cos^2(\omega\tau_m/2), \tag{3.46}$$

where τ_m is the adjustable delay of mth ($m = 1, 2,\ldots, M$) cascade.

3.1.6.3 WDM Multiplexers and Demultiplexers

Multiplexers and demultiplexers are basic devices of a WDM system. Demultiplexers contain a *wavelength-selective element* to separate the channels of a WDM signal. Based on underlying physical principle, different demultiplexer devices can be classified as *diffraction-based demultiplexers* (based on a diffraction grating) and *interference-based demultiplexers* (based on optical filters and directional couplers).

Diffraction-based demultiplexers are based on Bragg diffraction effect and use an angular dispersive element, such as the diffraction grating. The incoming composite light signal is reflected from the grating and dispersed spatially into different wavelength components, as shown in Fig. 3.23(a). Different wavelength components are focused by lenses and sent to individual optical fibers. The same device can be used as multiplexer by switching the roles of input and output ports. This device can be implemented using either conventional or GRIN lenses. To simplify design, the concave grating can be used.

The second group of optical multiplexers is based on interference effect and employs the optical couplers and filters to combine different wavelength channels into a composite WDM signal. The multiplexers employing the interference effect include thin-film filters multiplexers and the array waveguide grating (AWG) [1–7, 17], which is shown in Fig. 3.23(b). The AWG is a highly versatile WDM device, because it can be used as a multiplexer, a demultiplexer, a drop-and-insert element, or even as a wavelength router. It consists of M_{input} and M_{output} slab waveguides and two identical focusing planar star couplers connected by N uncoupled waveguides with a propagation constant β. The length of adjacent waveguides in the central region differs by a constant value ΔL, with corresponding phase difference being $2\pi n_c\Delta L/\lambda$, where n_c is the refractive index of arrayed waveguides. Based on the phase-matching condition (see Fig. 3.23(c)) and knowing that the focusing is

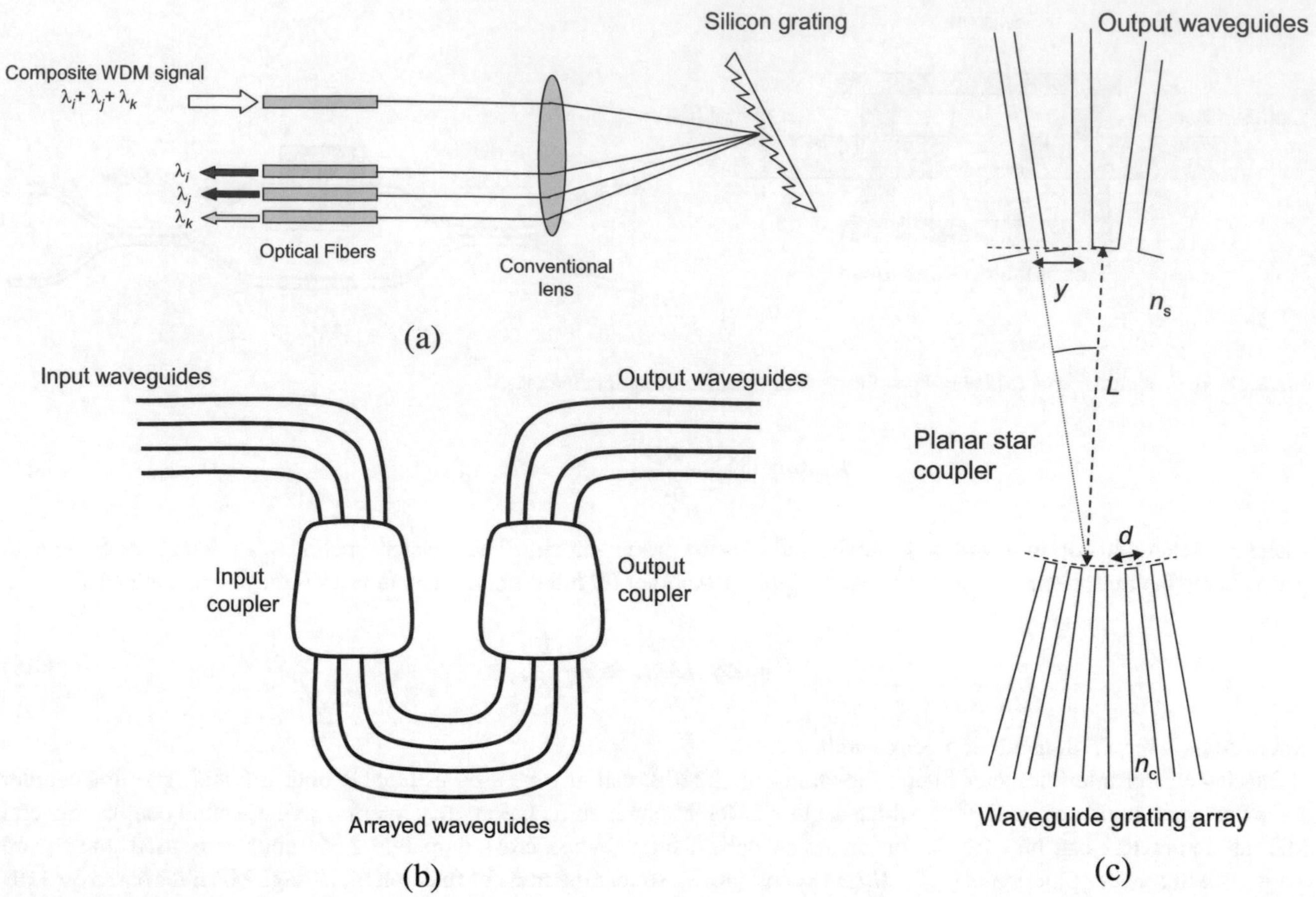

Fig. 3.23 Optical multiplexers/demultiplexers: (**a**) grating-based demultiplexer, (**b**) array waveguide grating (AWG) demultiplexer, and (**c**) illustrating the operating principle of AWG

achieved when the path-length difference ΔL between adjacent array waveguides is an integer multiple of the central design wavelength λ_c, that is, $n_c\Delta L = m\lambda_c$, we derive the following expression for the channel spacing [17]:

$$\Delta\nu = \frac{y}{L}\frac{n_s cd}{m\lambda^2}\frac{n_c}{n_g}, \tag{3.47}$$

where d is the spacing between the grating array waveguides, y is the spacing between the centers of output ports, L is the separation between center of arrayed waveguides and center of output waveguides, n_c is the refractive index of waveguides in grating array, n_s is the refractive index of star coupler, n_g the group index, and m is the diffraction order.

The free spectral range can be obtained by [17]:

$$\Delta\nu_{\mathrm{FSR}} = \frac{c}{n_g\left(\Delta L + d\sin\Theta_{\mathrm{in},i} + d\sin\Theta_{\mathrm{out},j}\right)}, \tag{3.48}$$

where the diffraction angles from the ith input $\Theta_{\mathrm{in},i}$ and jth output $\Theta_{\mathrm{out},j}$ ports (measured from the center of the array) can be found as $\Theta_{\mathrm{in},i} = iy/L$ and $\Theta_{\mathrm{out},j} = jy/L$, respectively.

3.1.7 Principles of Coherent Optical Detection

In order to exploit the enormous bandwidth potential of the optical fiber different multiplexing techniques (OTDMA, WDMA, CDMA, SCMA), modulation formats (OOK, ASK, PSK, FSK, PolSK, CPFSK, DPSK, etc.), demodulation schemes (direct

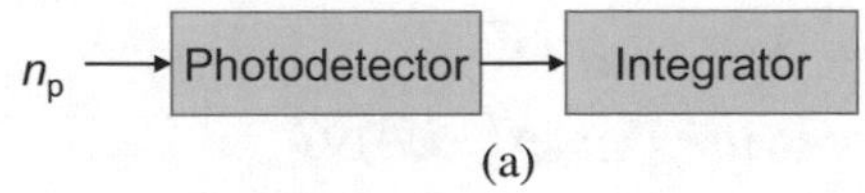

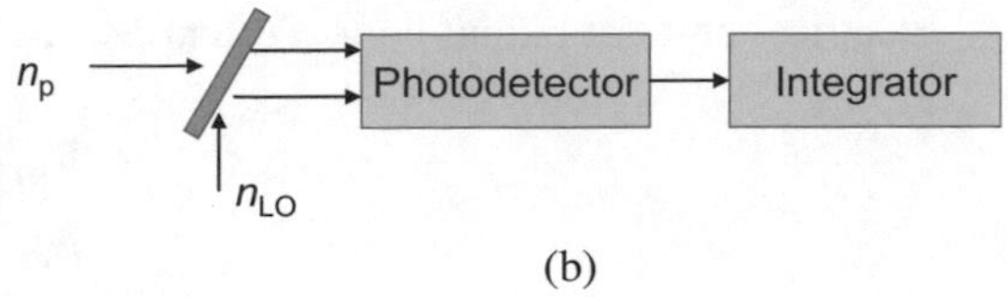

Fig. 3.24 Block diagrams of (**a**) direct detection scheme and (**b**) coherent detection scheme

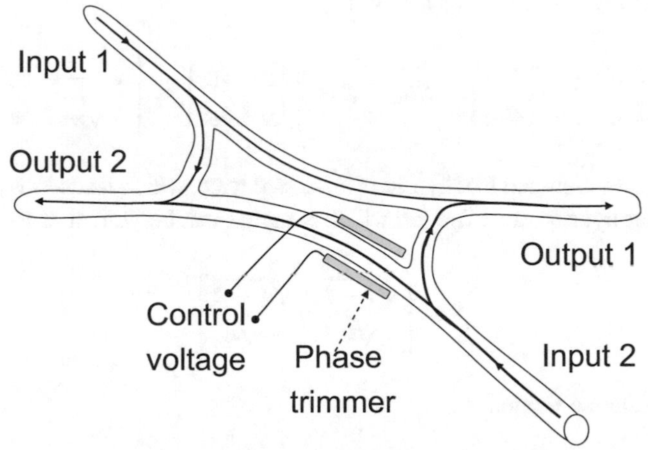

Fig. 3.25 Optical hybrid

detection or coherent optical detection) and technologies were developed, and some important aspects are discussed in this section. The coherent optical detection offers several important advantages compared to direct detection: (i) improved receiver sensitivity, (ii) better frequency selectivity, (iii) possibility of using constant amplitude modulation formats (FSK, PSK), (iv) tunable optical receivers similar to RF receivers are possible, and (v) with coherent detection the chromatic dispersion and PMD can easier be mitigated.

There is local laser, known as local oscillator laser, employed within coherent optical receiver. We can recognize several detection schemes based on the value of operating frequency of the local laser with the respect to the frequency of incoming optical signal, and they are (i) homodyne detection, in which these frequencies are identical; (ii) heterodyne detection, in which the frequency difference is larger than the signal symbol rate, so that all related signal processing upon photodetection is performed at suitable intermediate frequency (IF); and (iii) intradyne detection, in which the frequency difference is smaller than the symbol rate but higher than zero. Further on, different coherent detection schemes can be classified into following categories: (i) synchronous (PSK, FSK, ASK) schemes, (ii) asynchronous (FSK,ASK) schemes, (iii) differential detection (CPFSK, DPSK) schemes, (iv) phase diversity reception schemes, (v) polarization diversity reception schemes, and (vi) polarization-division multiplexing schemes. Synchronous detection schemes can further be categorized as residual carrier or suppressed carrier ones.

In Fig. 3.24 we show the basic difference between direct detection (Fig. 3.24(a)) and coherent detection (Fig. 3.24(b)) schemes. Coherent optical detection, in addition to the photodetector and integrator already in use for direct detection, employs a local laser whose frequency and polarization is matched to the frequency and polarization of incoming optical signal. The incoming optical signal and local laser output signal are mixed in optical domain (a mirror is used in this illustrative example; in practice, an optical hybrid, described in incoming section, is used instead).

For additional details on coherent optical systems, an interested reader is referred to [1, 3, 5, 10, 26, 28].

3.1.8 Optical Hybrids

So far we used an optical mirror to perform optical mixing before photodetection takes place. In practice this operation is performed by four-port device known as *optical hybrid*, which is shown in Fig. 3.25.

Electrical fields at output ports E_{1o} and E_{2o} are related to the electrical fields at input ports $E_{1\text{in}}$ and $E_{2\text{in}}$ as follows:

$$\begin{aligned} E_{1o} &= (E_{1i} + E_{2i})\sqrt{1-k} \\ E_{2o} &= (E_{1i} + E_{2i}\exp(-j\phi))\sqrt{k} \end{aligned} \tag{3.49}$$

where k is the power splitting ratio and ϕ is the phase shift introduced by the phase trimmer (see Fig. 3.25). The Eqs. (3.49) can also be written in terms of scattering (S-) matrix as follows:

$$\boldsymbol{E}_o = \begin{bmatrix} E_{o1} \\ E_{o2} \end{bmatrix} = \boldsymbol{S}\begin{bmatrix} E_{o1} \\ E_{o2} \end{bmatrix} = \boldsymbol{SE}_i, \quad \boldsymbol{S} = \begin{bmatrix} s_{11} & s_{12} \\ s_{21} & s_{22} \end{bmatrix} = \begin{bmatrix} \sqrt{1-k} & \sqrt{1-k} \\ \sqrt{k} & e^{-j\phi}\sqrt{k} \end{bmatrix} \tag{3.50}$$

In expressions (3.49) and (3.50), we assumed that hybrid is lossless device, which leads to $s_{11} = |s_{11}|$, $s_{12} = |s_{12}|$, $s_{21} = |s_{21}|$, and $s_{22} = |s_{22}|\exp(j\theta_{22})$. Popular hybrids are π *hybrid*, which S-matrix can be written as (by setting $\phi = \pi$ in (3.50)):

$$\boldsymbol{S} = \begin{bmatrix} \sqrt{1-k} & \sqrt{1-k} \\ \sqrt{k} & -\sqrt{k} \end{bmatrix}, \tag{3.51}$$

and $\pi/2$ *hybrid*, which S-matrix can be written as:

$$\boldsymbol{S} = \begin{bmatrix} s_{11} & s_{12} \\ s_{21} & s_{22}e^{-j\pi/2} \end{bmatrix}. \tag{3.52}$$

Well-known π hybrid is 3 dB coupler ($k = 1/2$). If $\pi/2$ hybrid is symmetric $|s_{ij}| = 1/\text{Loss}$ $(\forall i, j)$, the phase difference between the input electrical fields $E_{1i} = |E_{1i}|, E_{2i} = |E_{2i}|e^{\,j\theta_i}$ can be chosen on such a way so that total output ports power

$$\boldsymbol{E}_0^{\dagger}\boldsymbol{E}_0 = \frac{2}{\text{Loss}}\left[P_{1i} + P_{2i} + \sqrt{P_{1i}P_{2i}}(\cos\theta_i + \sin\theta_i)\right] \tag{3.53}$$

is maximized. (We use † to denote Hermitian transposition-simultaneous transposition and complex conjugation.) For equal input powers, the maximum of (3.53) is obtained when $\theta_i = \pi/4$, leading to Loss $\geq 2 + \sqrt{2}$. The corresponding loss in dB-scale is determined as $10\log_{10}(\text{Loss}/2) = 2.333$ dB. For Costas loop and decision-driven loop-based homodyne systems, there exists optimum k in corresponding S-matrix:

$$\boldsymbol{S} = \frac{1}{\sqrt{\text{Loss}}}\begin{bmatrix} \sqrt{1-k} & \sqrt{1-k} \\ \sqrt{k} & -j\sqrt{k} \end{bmatrix}. \tag{3.54}$$

3.1.9 Coherent Optical Balanced Detectors

To reduce the relative intensity noise (RIN) of transmitting laser and to eliminate the direct-detection and signal-cross-signal interferences, the most deleterious sources for multi-channel applications, the balanced coherent optical receiver, shown in Fig. 3.26, is commonly used. For homodyne detection, by ignoring the laser phase noise, the upper and lower photodetectors output currents can respectively be written as:

$$\begin{aligned} i_1(t) &= R|E_1|^2 = \frac{1}{2}R\left(\underbrace{|E_s|^2}_{P_S} + \underbrace{|E_{LO}|^2}_{P_{LO}} + 2\sqrt{P_S P_{LO}}\cos\theta_S\right) + n_1(t) \\ i_2(t) &= R|E_2|^2 = \frac{1}{2}R\left(P_S + P_{LO} - 2\sqrt{P_S P_{LO}}\cos\theta_S\right) + n_2(t) \end{aligned} \tag{3.55}$$

where θ_s is the phase of incoming optical signal and $n_i(t)$ (i = 1, 2) is the i-th photodetector shot noise process of PSD $S_{n_i} = qR|E_i|^2$. The balanced receiver output current (see Fig. 3.26) can be written as:

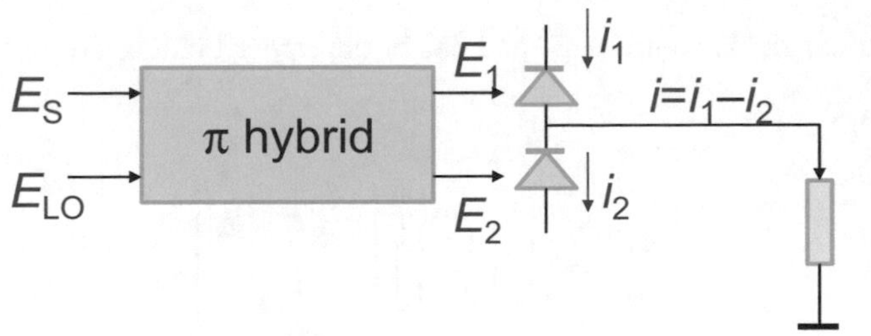

Fig. 3.26 Balanced detector

Fig. 3.27 Directional coupler-based coherent optical detection schemes for 2D signaling (PSK, QAM): (**a**) balanced coherent optical detector architecture and (**b**) single optical balanced detector based on heterodyne design. *LPF* low-pass filter

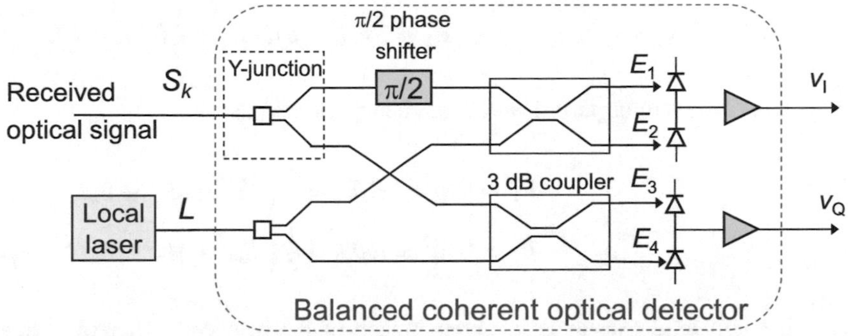

(a)

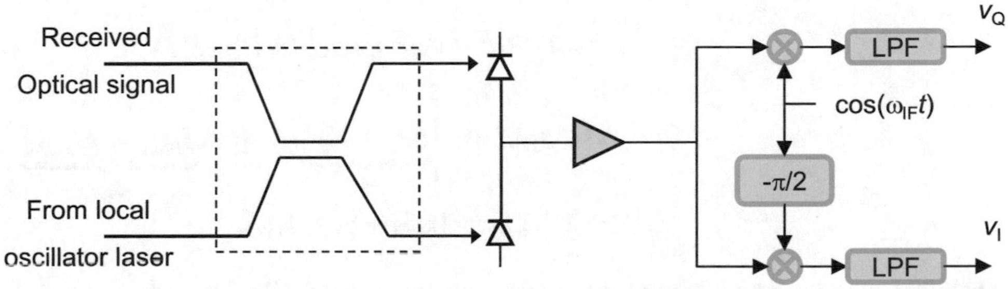

(b)

$$i(t) = i_1(t) - i_2(t) = 2R\sqrt{P_S P_{LO}} \cos\theta_S + n(t), n(t) = n_1(t) - n_2(t) \tag{3.56}$$

where $n(t)$ is a zero-mean Gaussian process of PSD $S_n = S_{n_1} + S_{n_2} = qR(P_S + P_{LO}) \cong qRP_{LO}$. For binary BPSK signaling, we have that $\theta_s = \pm\pi$.

The M-ary PSK, M-ary QAM, and M-ary DPSK all achieve the transmission of $\log_2 M = m$ bits per symbol. As an illustration, in coherent optical detection, the data phasor for M-ary PSK $\phi_k \in \{0, 2\pi/M, .., 2\pi(M-1)/M\}$ is sent at each k-th transmission interval. We should also outline that in direct detection, the modulation is differential since the data phasor $\phi_k = \phi_{k-1} + \Delta\phi_k$ is sent instead, where $\Delta\phi_k \in \{0, 2\pi/M, .., 2\pi(M-1)/M\}$ value is determined by the sequence of $\log_2 M$ input bits using an appropriate mapping rule. The mapper accepts m bits, $\boldsymbol{c} = (c_1, c_2, .., c_m)$, from the input sequence at time instance i and determines the corresponding M-ary ($M = 2^m$) constellation point $s_k = (I_k, Q_k) = |S_k| \exp(j\theta_k)$. The coordinates from the mapper, after the upsampling, are used as the RF inputs of optical I/Q modulator.

Let the input to the coherent detector be given by $S_k = |S_k| \exp[j(\omega_T t + \phi_{PN,T} + \theta_k)]$, where the subscript k is used to denote the kth symbol being transmitted, ω_T denotes the transmit laser frequency (in rad/s), $\phi_{PN,T}$ denotes the laser phase noise of transmitter laser, and $|S_k|$ and θ_k denote the amplitude and the phase of the kth symbol for M-ary QAM/PSK (or any other 2D signal constellation). Let the local laser (oscillator) output be denoted by $L = |L| \exp.[j(\omega_{LOL} t + \phi_{PN,LOL})]$, where ω_{LOL} denotes the local oscillator laser frequency and $\phi_{PN,LOL}$ denotes the laser phase noise of local laser. There exist different versions of balanced coherent optical detectors for 2D signal constellations. The directional coupler-based balanced coherent optical detector architecture is shown in Fig. 3.27(a). The scattering matrix of 3 dB directional coupler, based on Eq. (3.42), is given by $S=\frac{1}{\sqrt{2}}\begin{bmatrix}1 & j\\ j & 1\end{bmatrix}$. The electrical fields' outputs E_1 and E_2, in the upper branch, are related to input electrical fields S_k and L, in matrix form, as follows:

$$\begin{bmatrix}E_1\\ E_2\end{bmatrix}=\frac{1}{\sqrt{2}}\begin{bmatrix}1 & j\\ j & 1\end{bmatrix}\begin{bmatrix}\frac{1}{\sqrt{2}}e^{j\pi/2}S_k\\ \frac{1}{\sqrt{2}}L\end{bmatrix}, \tag{3.57}$$

while the corresponding scalar form representation is given by:

$$E_1=\frac{1}{2}j(S_k+L),\quad E_2=\frac{1}{2}(-S_k+L). \tag{3.58}$$

The powers at the output ports 1 and 2 can be represented as:

$$\begin{aligned}P_1&=|E_1|^2=E_1E_1^*=\frac{1}{4}\left(|S_k|^2+|L|^2+S_kL^*+S_k^*L\right),\\ P_2&=|E_2|^2=E_2E_2^*=\frac{1}{4}\left(|S_k|^2+|L|^2-S_kL^*-S_k^*L\right).\end{aligned} \tag{3.59}$$

Assuming the unit load resistance in upper branch of the balanced coherent optical detector, the in-phase component is obtained as (by ignoring the shot and thermal noises):

$$\begin{aligned}v_{I,k}&\cong R\left(|E_1|^2-|E_2|^2\right)=R\frac{1}{2}\left(S_kL^*+S_k^*L\right)\\ &=R|S_k||L|\cos\Bigg[\underbrace{(\omega_T-\omega_{LO})}_{\omega_{IF}}t+\theta_k+\underbrace{\phi_{PN,T}-\phi_{PN,LOL}}_{\phi_{PN}}\Bigg]\\ &=R|S_k||L|\cos\left(\omega_{IF}t+\theta_k+\phi_{PN}\right),\end{aligned} \tag{3.60}$$

where $\omega_{IF}=|\omega_T-\omega_{LOL}|$ denotes the intermediate frequency, while $\phi_{PN}=\phi_{PN,\,T}-\phi_{PN,\,LOL}$ denotes the total phase noise originating from both transmit and local oscillator lasers.

In similar fashion, the electrical fields' outputs E_3 and E_4, in the lower branch of Fig. 3.27(a), are related to input electrical fields S_k and L, in matrix form, as follows:

$$\begin{bmatrix}E_3\\ E_4\end{bmatrix}=\frac{1}{\sqrt{2}}\begin{bmatrix}1 & j\\ j & 1\end{bmatrix}\begin{bmatrix}\frac{1}{\sqrt{2}}S_k\\ \frac{1}{\sqrt{2}}L\end{bmatrix}, \tag{3.61}$$

while the corresponding scalar form representation is given by:

$$E_3=\frac{1}{2}(S_k+jL),\quad E_4=\frac{1}{2}(jS_k+L). \tag{3.62}$$

The powers at the output ports 3 and 4 can be represented as:

$$\begin{aligned}P_3&=|E_3|^2=E_3E_3^*=\frac{1}{4}\left(|S_k|^2+|L|^2-jS_kL^*+jS_k^*L\right),\\ P_4&=|E_4|^2=E_4E_4^*=\frac{1}{4}\left(|S_k|^2+|L|^2+jS_kL^*-jS_k^*L\right).\end{aligned} \tag{3.63}$$

Assuming the unit load resistance in the lower branch of the balanced coherent optical detector, the in-phase component is obtained as (by ignoring the shot and thermal noises):

$$\begin{aligned} v_{Q,k} &\cong R\left(|E_3|^2 - |E_4|^2\right) = R\frac{1}{2j}\left(S_k L^* - S_k^* L\right) \\ &= R|S_k||L| \sin\left[\underbrace{(\omega_T - \omega_{LO})}_{\omega_{IF}} t + \theta_k + \underbrace{\phi_{PN,T} - \phi_{PN,LOL}}_{\phi_{PN}}\right] \\ &= R|S_k||L| \sin\left(\omega_{IF} t + \theta_k + \phi_{PN}\right). \end{aligned} \tag{3.64}$$

To summarize, the balanced coherent optical detector outputs of I- and Q-channel branches at the time instance k can be written as:

$$\begin{aligned} v_{I,k} &\cong R|S_k||L| \cos\left(\omega_{IF} t + \theta_k + \phi_{PN}\right), \\ v_{Q,k} &\simeq R|S_k||L| \sin\left(\omega_{IF} t + \theta_k + \phi_{PN}\right), \end{aligned} \tag{3.65}$$

where R is the photodiode responsivity and $\omega_{IF} = |\omega_T - \omega_{LOL}|$ is the intermediate frequency, while ϕ_{PN} is the total laser phase noise originating from both transmit laser and LOL.

The heterodyne receiver can also be implemented based only on single balanced detector, such as lower branch of Fig. 3.27(a), as shown in Fig. 3.27(b). For ASE noise-dominated scenario, both heterodyne schemes perform comparably. However, in the shot noise-dominated scenario, the heterodyne design with single optical balanced detector performs better. In homodyne detection, we need to set the intermediate frequency to zero, $\omega_{IF} = 0$.

3.2 Modules/Components Relevant to Wireless Communications

In this section we describe the modules/components relevant to wireless communications [36–43], in particular to software-defined radio (SDR). The basic concept related to antennas has already been provided in Chap. 2. Many of the modules described here are also applicable in software-defined optical transmission (SDOT).

We first describe some basic concepts from DSP relevant in this section [44–49]; for additional details, an interested reader is referred to the Appendix. Alternatively, the readers can refer to DSP books such as [45, 47].

3.2.1 DSP Basics

We start the description of DSP basic by describing the role of the system (transfer) function as well as establishing relationship between discrete-time (DT) and continuous-time (CT) Fourier transforms.

3.2.1.1 The Role of System (Transfer) Function and Relationship Between DT and CT Fourier Transforms

The discrete-time (DT) linear time-invariant (LTI) system can be described by linear difference equation with constant coefficients:

$$a_N y(n-N) + a_{N-1} y(n-(N-1)) + \cdots + a_0 y(n) = b_M x(n-M) + \cdots + b_0 x(n), \tag{3.66}$$

where input and output to the system are denoted with $x(n)$ and $y(n)$, respectively. By applying z-transform on Eq. (3.66), we can determine the system function (also known as the transfer function) $H(z)$ as follows:

$$H(z) = \frac{Y(z)}{X(z)} = \frac{\sum_{m=0}^{M} b_m z^{-m}}{\sum_{n=0}^{N} a_n z^{-n}}, \tag{3.67}$$

where $X(z)$ $[Y(z)]$ denotes the z-transform of $x(n)$ $[y(n)]$, namely, $X(z) = Z\{x(n)\}$ $[Y(z) = Z\{y(n)\}]$. The frequency response can be obtained by substituting z with $e^{j\Omega}$:

$$H(e^{j\Omega}) = H(z)|_{z=e^{j\Omega}}. \tag{3.68}$$

The unit-sample (impulse) response can be obtained as inverse z-transform of the system function: $h(n) = Z^{-1}\{H(z)\}$. The output of the system $y(n)$ can be obtained as a linear convolution of the input to the system $x(n)$ and impulse response $h(n)$, which upon application of the convolution property yields to:

$$Y(z) = Z\{y(n)\} = Z\left\{\sum_{k=-\infty}^{\infty} x(k)h(n-k)\right\} = X(z)H(z). \tag{3.69}$$

Therefore, the system function has a central role in DT system analysis, as illustrated in Fig. 3.28.

Two important system properties of DT systems are causality and stability. The system is *causal* if there is no response before excitation. It can be shown that the system is causal if and only if $h(n) = 0$ for $n < 0$. The LTI system is causal if an only if the region of convergence (ROC) is the exterior of circle of radius $r < \infty$. On the other hand, DT system is bounded input-bounded output (*BIBO*) *stable* if every bounded input sequence $x(n)$ produces a bounded output sequence $y(n)$. It can be shown that an LTI system is BIBO stable if and only if the ROC of system function contains the unit circle. Finally, a causal LTI system is BIBO stable if and only if all the poles of the system function are located inside the unit circle.

The relationship between Fourier transforms of a continuous-time (CT) signal $x_a(t)$ and a DT signal $x(n)$ obtained by sampling (every T seconds) is illustrated in Fig. 3.29. The z-transform can be related to DT Fourier transform (DTFT) by simply substituting $z = \exp.(j\Omega)$. In other words, the DTFT is the z-transform evaluated on the unit circle. Because of the periodicity path along the unit circle, the DTFT will be periodic with a period of 2π.

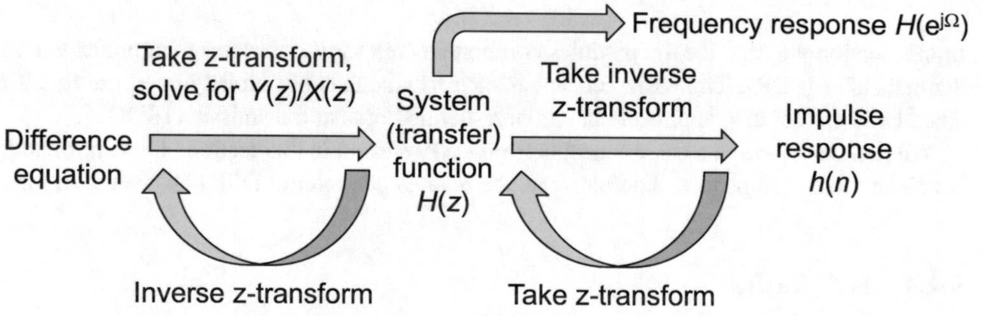

Fig. 3.28 The role of the system (transfer) function

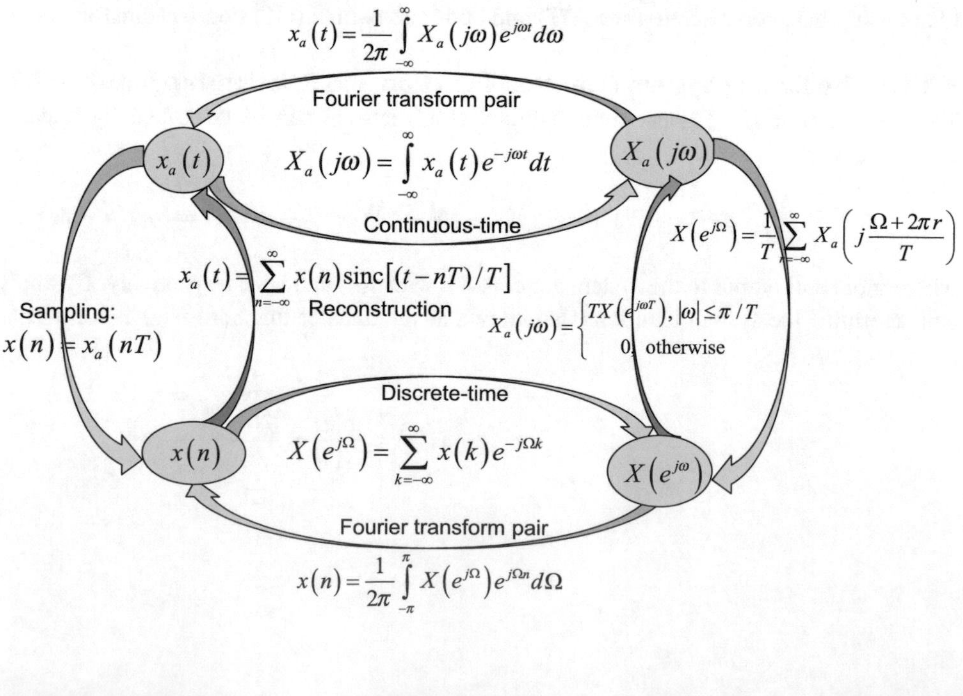

Fig. 3.29 The relationship between DT and CT Fourier transforms. We use the subscript *a* to denote the analog signals

3.2.1.2 Discrete-Time System Realizations

When $a_0 = 1$, we can express $y(n)$ as the function of previous LTI system outputs and current and previous LTI system inputs, based on Eq. (3.66), as follows:

$$y(n) = -a_1 y(n-1) - \cdots - a_N y(n-N) + b_M x(n-M) + \cdots + b_0 x(n), \tag{3.70}$$

with the corresponding implementation (realization), known as the *direct realization*, shown in Fig. 3.30. The systems for which only forward coefficients exist are commonly referred to as the *finite impulse response (FIR)* systems (filters). On the other hand, when there exist nonzero feedback coefficients, for $M < N$, the impulse response will be infinitely long, and such systems are known as the *infinite impulse response (IIR)* systems (filters).

By separating the all-zero (AZ) and all-pole (AP) subsystems' transfer functions as follows:

$$H(z) = \frac{\sum_{m=0}^{M} b_m z^{-m}}{\sum_{n=0}^{N} a_n z^{-n}} = \underbrace{\left(\sum_{m=0}^{M} b_m z^{-m}\right)}_{H_{\mathrm{AZ}}(z)} \cdot \underbrace{\frac{1}{\sum_{n=0}^{N} a_n z^{-n}}}_{H_{\mathrm{AP}}(z)} = H_{\mathrm{AZ}}(z) H_{\mathrm{AP}}(z), \tag{3.71}$$

the corresponding realization, known as the *direct canonic form I*, is shown in Fig. 3.31. Since $H_{\mathrm{AZ}}(z)H_{\mathrm{AP}}(z) = H_{\mathrm{AP}}(z)H_{\mathrm{AZ}}(z)$, we can swap the AZ and AP subsystems so that we can use the same delay elements for both subsystems, and corresponding realization, shown in Fig. 3.32, is known as *direct canonic form II*.

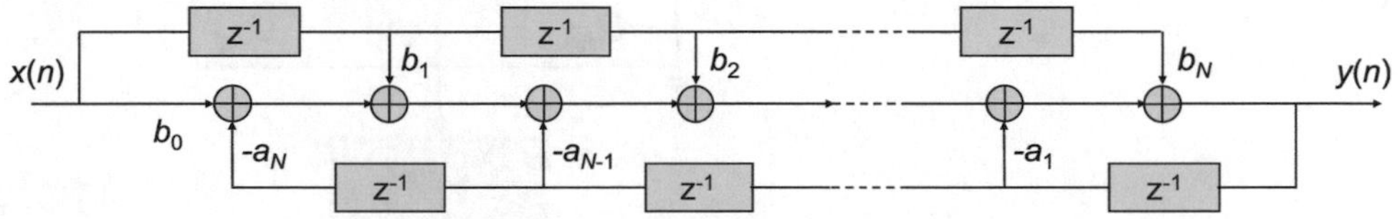

Fig. 3.30 The direct realization of DT LTI systems

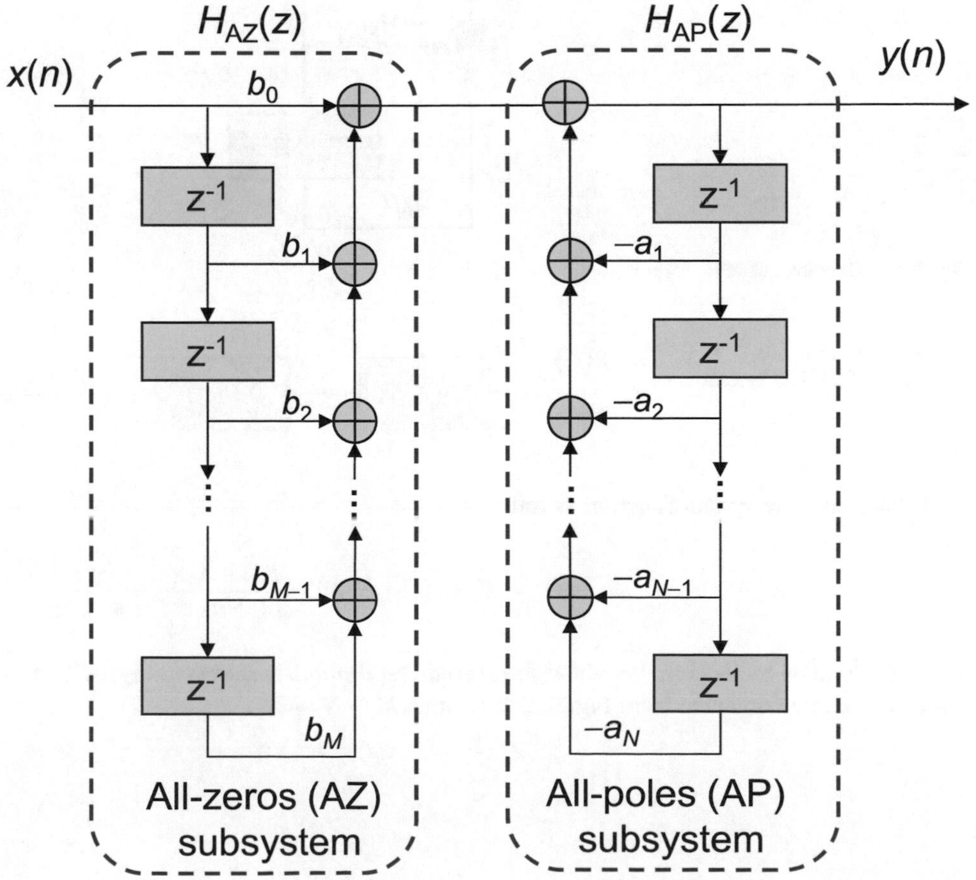

Fig. 3.31 The direct canonic form I

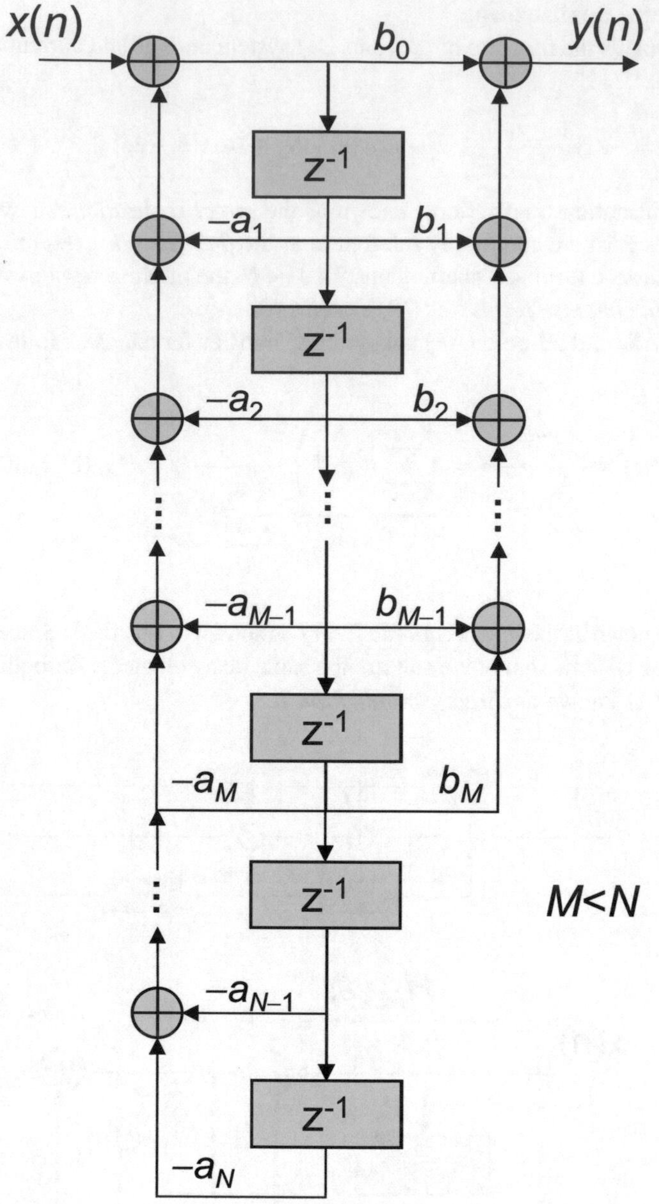

Fig. 3.32 The direct canonic form II

Fig. 3.33 The cascade realization of DT LTI system

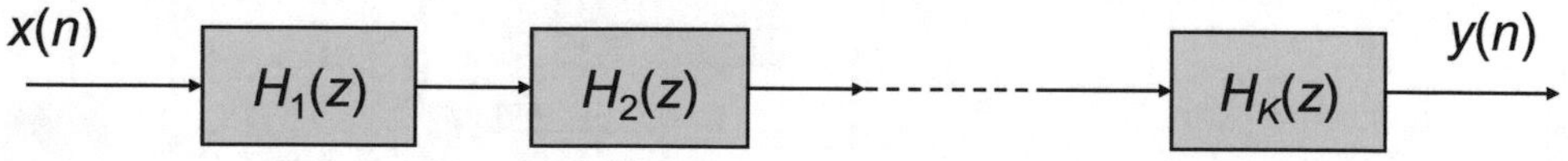

By factoring the system function as follows:

$$H(z) = \prod_{k=1}^{K} H_k(z), \quad H_k(z) = \frac{b_{k_0} + b_{k_1}z^{-1} + b_{k_2}z^{-2}}{1 + a_{k_1}z^{-1} + a_{k_2}z^{-2}}, \tag{3.72}$$

the corresponding realization, known as the *cascade realization*, is shown in Fig. 3.33. The second-order section $H_k(z)$, from Eq. (3.72), can be obtained from Fig. 3.32 by setting $M = N = 2$.

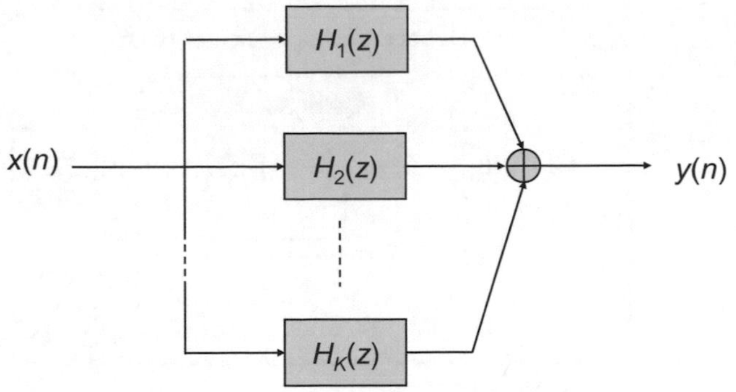

Fig. 3.34 The parallel realization of DT LTI system

By performing the partial fraction expansion, the system function can be represented as:

$$H(z) = \sum_{k=1}^{K} H_k(z), \quad H_k(z) = \frac{b_{k_0} + b_{k_1} z^{-1} + b_{k_2} z^{-2}}{1 + a_{k_1} z^{-1} + a_{k_2} z^{-2}}, \tag{3.73}$$

and the corresponding realization, known as the *parallel realization*, is shown in Fig. 3.34. Notice that some sections in parallel realization can be of the first order; in that case, we need to set $b_{k_1} = b_{k_2} = 0, a_{k_2} = 0$.

Design of IIR filters (subsystems) can be performed by designing first the analog filters and then converting to digital filter by proper mapping. Regarding the mapping from the s-domain to z-domain, the following two conditions must be satisfied: (i) imaginary axis in s-plane should map into the unit circle in z-plane, and (ii) the left half-plane of s-plane should map into inside of unit circle in the z-plane. The *bilinear transform*, given by:

$$s = \frac{2}{T} \frac{1 - z^{-1}}{1 + z^{-1}}, \tag{3.74}$$

satisfies both conditions. By using this mapping, the continuous-time frequency [rad/s] ω is mapped to the DT frequency [rad/sample] Ω by the following nonlinear mapping:

$$\Omega = 2 \tan^{-1}\left(\frac{\omega T}{2}\right). \tag{3.75}$$

3.2.2 Direct Digital Synthesizer (DDS)

The direct digital synthesizer (DDS) in discrete-time (DT) communication systems has the same role as the voltage control oscillator (VCO) in continuous-time (CT) communication systems. In VCO, the output frequency is directly proportional to the input signal $x(t)$ so that the output phase $\phi(t)$ will be the integral of the input:

$$\phi(t) = a_0 \int_{-\infty}^{t} x(t')dt', \tag{3.76}$$

where a_0 is the VCO gain [rad/V]. The VCO output signal will be the following cosinusoidal signal:

$$y(t) = \cos\left(\omega_0 t + a_0 \int_{-\infty}^{t} x(t')dt'\right). \tag{3.77}$$

In similar fashion, the output of the DDS will be:

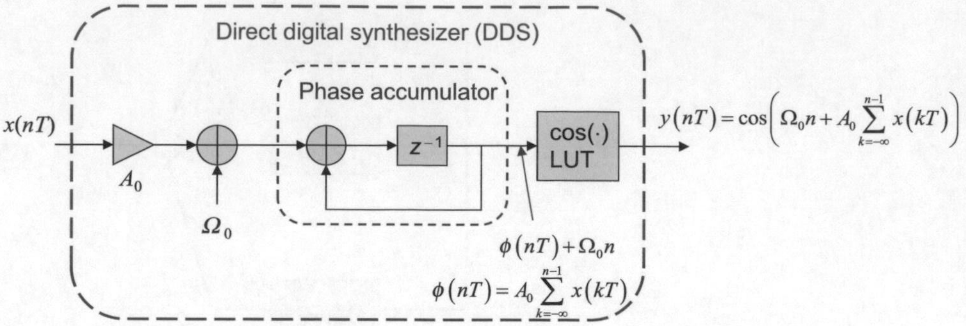

Fig. 3.35 The block diagram of direct digital synthesizer (DDS) with DSP implementation details

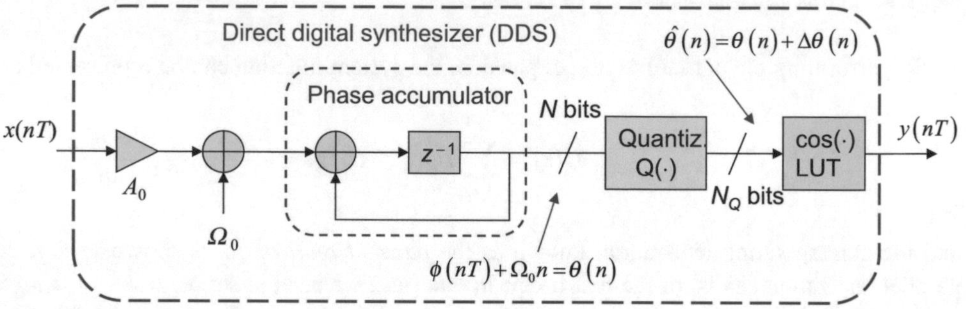

Fig. 3.36 The block diagram of the DDS with finite precision

$$y(nT) = \cos\left(\Omega_0 n + A_0 \sum_{k=-\infty}^{n-1} x(kT)\right), \Omega_0 = \omega_0 T, A_0 = a_0 T. \tag{3.78}$$

The block diagram of DDS together with corresponding DSP representation is shown in Fig. 3.28. The portion of the DDS calculating the accumulated phase is known as the phase accumulator.

The DDS shown in Fig. 3.35 does not take the finite precision of the phase accumulator representation. A realistic DDS with finite precision taken into account is shown in Fig. 3.36, where we use Q(·) to denote the quantization operation. The number of quantization bits $N_Q < N$ can be determined as $N_Q = N - N_t$, where N_t is the number of truncated least significant bits. Let us assume that two's complement integer representing the phase at the output of accumulator is given by $a_0 a_1 \cdots a_{N-1} = -a_0 2^{N-1} + a_1 2^{N-2} + \cdots + a_{N-1} 2^0$. The decimal representation of the accumulated phase would be $\theta_0 \theta_1 \cdots \theta_{N-1} = -\theta_0 2^0 + \theta_1 2^{-1} + \cdots + \theta_{N-1} 2^{-(N-1)}$. Since the sampling frequency is given by $F_s = 1/T$, the maximum frequency, according to the sampling theorem, will be $f_{\max} = F_s/2$. The minimum frequency is the resolution frequency determined by $f_{\min} = F_s/2^N$. The number of required bits for the phase accumulator content representation will be then:

$$\frac{f_{\max}}{f_{\min}} = \frac{F_s/2}{F_s/2^N} = 2^{N-1} \Rightarrow N = \left\lceil \log_2 \left[\frac{f_{\max}}{f_{\min}}\right] + 1 \right\rceil, \tag{3.79}$$

where we use $\lceil\cdot\rceil$ to denote the smallest integer larger than or equal to the enclosed quantity. As an illustration, for $f_{\max} = 400$ MHz and resolution of $f_{\min} = 0.1$ Hz, the number of required bits is $N = \left\lceil \log_2 \left[\frac{400 \cdot 10^6}{0.1}\right] + 1 \right\rceil = 33$, so that the size of LUT for cos(·) is too large, $8.589934592 \cdot 10.^9$ To reduce the size of LUT, one method is to use the truncation, mentioned above and illustrated in Fig. 3.36. The truncation method, however, will introduce the phase error $\Delta\theta(n)$ in the represented phase $\theta(n) = \widehat{\theta}(n) + \Delta\theta(n)$. The corresponding spectrum of complex exponential will contain the spurious spectral lines. The complex exponential can be represented by:

$$\begin{aligned} e^{j\theta(n)} &= e^{j[\hat{\theta}(n)+\Delta\theta(n)]} = e^{j\hat{\theta}(n)} e^{j\Delta\theta(n)} \overset{\Delta\theta(n)->0}{\cong} e^{j\hat{\theta}(n)}[1 + j\Delta\theta(n)] \\ &\cong \cos\left[\hat{\theta}(n)\right] - \Delta\theta(n) \cdot \sin\left[\hat{\theta}(n)\right] + j\left\{\sin\left[\hat{\theta}(n)\right] + \Delta\theta(n) \cdot \cos\left[\hat{\theta}(n)\right]\right\}. \end{aligned} \tag{3.80}$$

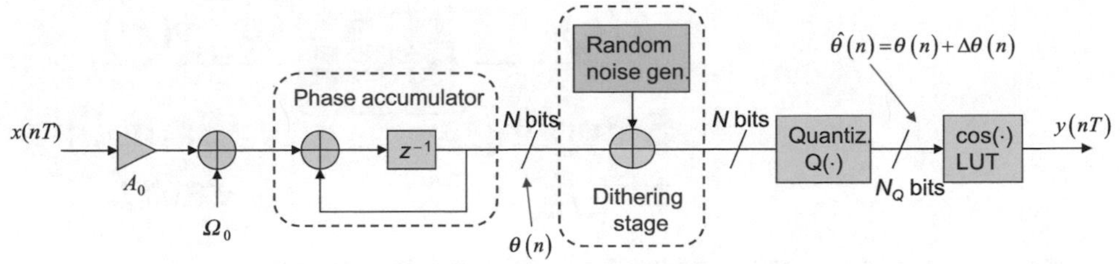

Fig. 3.37 The compensation of the spurious spectral lines by the phase dithering method

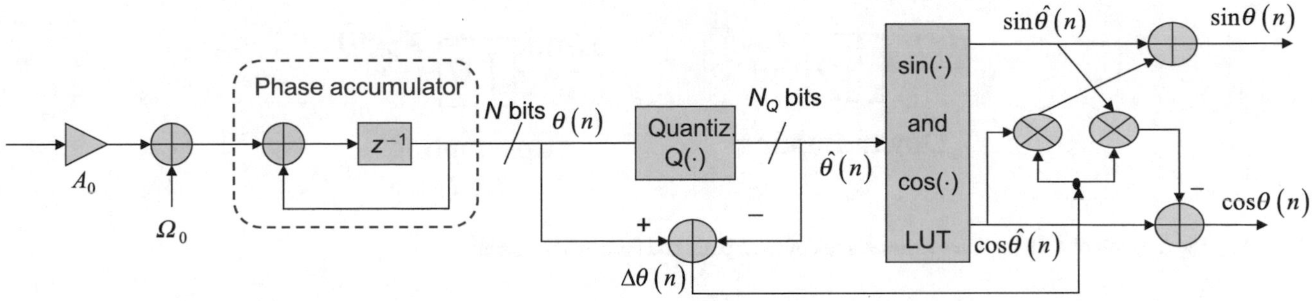

Fig. 3.38 The compensation of the phase error by the feedforward error compensation method

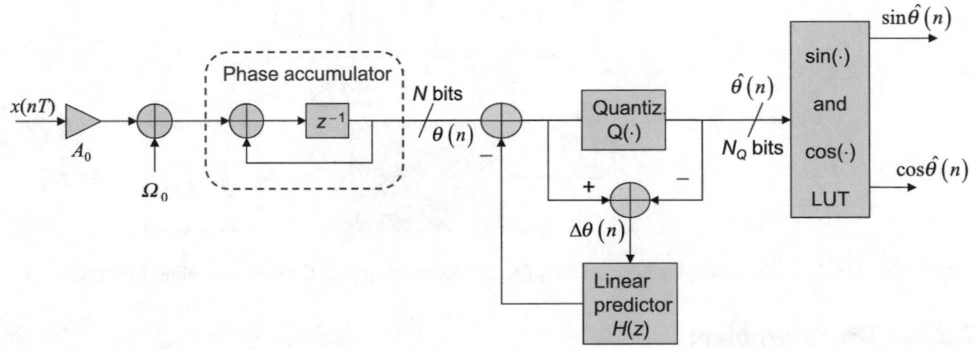

Fig. 3.39 The compensation of the phase error by the feedback error compensation method

Based on Eq. (3.80), the first line, we conclude that it is possible to average out the phase error by phase dithering, as illustrated in Fig. 3.37. The amplitude should be sufficiently small (on the order of 2^{-N}), and the samples of the random generator should follow the probability density function of the phase error, with proper tails included. The end result will be the compensation of the spurious spectral line, and the uncompensated phase error will appear as the white noise with level well below that of the desired spectral line.

Another method to compensate the phase error will be the application of the second line in Eq. (3.80), and corresponding phase error correction circuit is shown in Fig. 3.38. However, this approach requires additional multiplication and addition operations.

It is also possible to consider the phase error as the noise process and employ the linear predictive approaches to estimate the quantization error and subtract it from the phase accumulator output, as illustrated in Fig. 3.39.

3.2.3 Multirate DSP and Resampling

We are interested here in DSP modules employing multiple sampling rates, as illustrated in Fig. 3.40. In particular, the most interesting case is when the ratio of sampling rates f_y/f_x is a rational number. When this ratio $f_y/f_x = 1/D$ (D is positive integer) is smaller than 1, it is commonly referred to as *downsampling* by D or decimation by D. On the other hand, when the ratio $f_y/f_x = U$ (U is positive integer) is larger than 1, the corresponding DSP operation is called *upsampling* by U or interpolation by U. The DSP modules performing downsampling and upsampling operations are denoted as shown in Fig. 3.41.

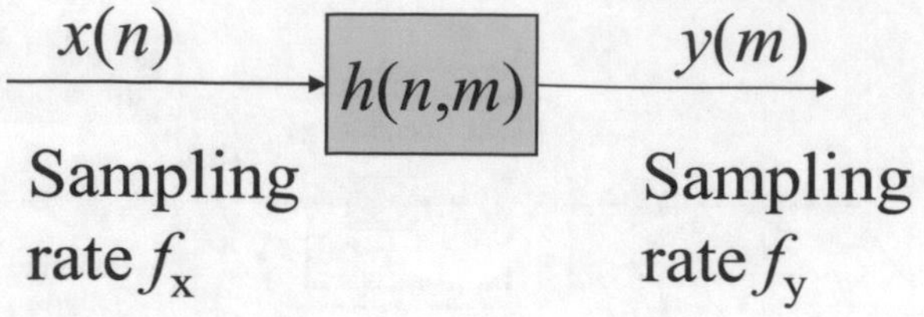

Fig. 3.40 The illustration of DSP module employing two different sampling rates

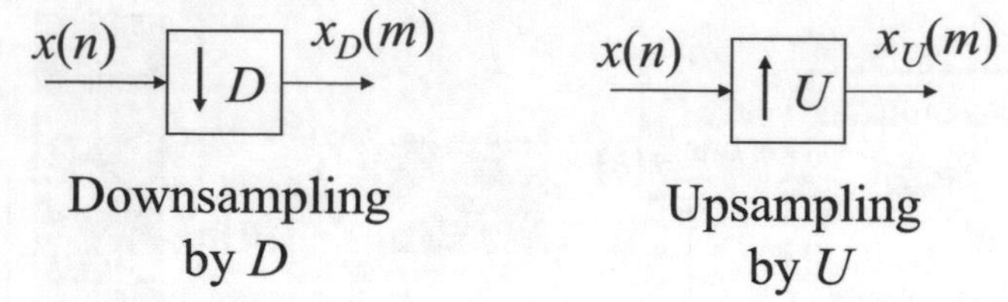

Fig. 3.41 The notation of DSP modules representing downsampling and upsampling operations

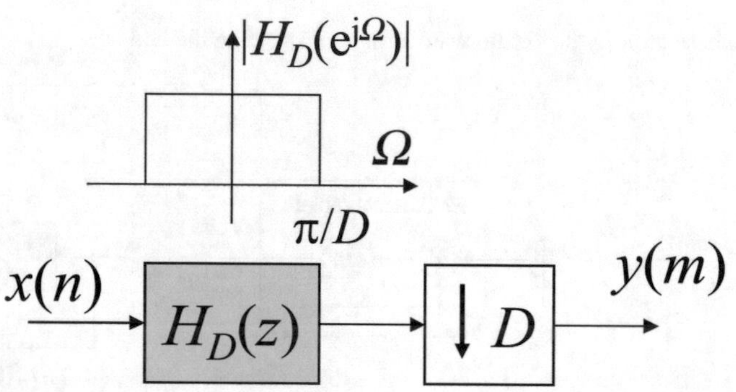

Fig. 3.42 The LPF downsampler DSP module (decimator) performing the downsampling operation

3.2.3.1 Downsampling

Downsampling the sequence $x(n)$ by D *decimation* by D represents the operation by which the sampling rate is reduced D times, and every D-th sample from original sequence is retained so that we can write:

$$x_D(m) = x(mD). \tag{3.81}$$

By reducing the sample rate D times, by downsampling, the spectrum of $X_D(\Omega) = \text{DTFT}\{x_D(m)\}$ will fold at $f_y/2 = (f_x/2)/D = f_x/(2D)$, and the aliasing will occur. Because of periodicity of DTFT, the period of $X_D(^{j\Omega})$ will be $2\pi/D$, and we can write:

$$X_D\left(e^{j\Omega}\right) = \frac{1}{D}\sum_{k=0}^{D-1} X\left(e^{j\frac{\Omega-k2\pi}{D}}\right), \tag{3.82}$$

indicating that the spectra of aliases get expanded D times. To avoid this aliasing problem, we need first to perform low-pass filtering with an LPF of cutoff frequency $f_c = f_x/(2D)$ [Hz], which is equivalent to $F_c = 1/(2D)$ [cycles/sample]. Therefore, the block diagram of DSP module for downsampling operation is a cascade of the LPF and downsampler, as shown in Fig. 3.42.

3.2.3.2 Upsampling

Upsampling the sequence $x(n)$ by U or *interpolation* by U represents the operation by which the sampling rate is increased U times, which is achieved by inserting $U-1$ zeros between neighboring samples in the sequence of samples so that we can write:

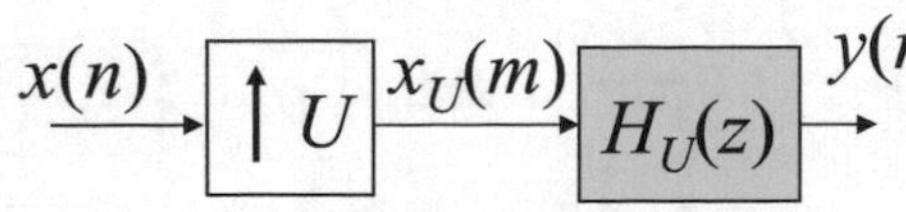

Fig. 3.43 The upsampler LPF DSP module (interpolator) performing the upsampling operation

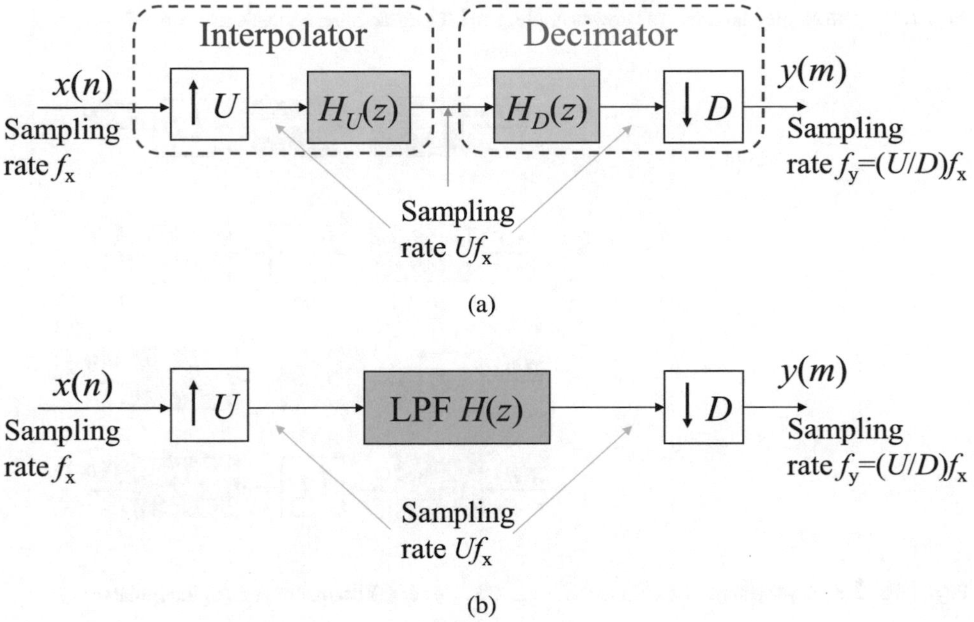

Fig. 3.44 DSP module to perform the resampling by a rational factor *U*/*D*: (**a**) cascade of interpolator and decimator and (**b**) DSP resampling module in which two low-pass filters are combined into one

$$x_U(m) = \begin{cases} x(m/U); \; m = 0, \pm U, \pm 2U, \ldots, \\ 0, \text{ else} \end{cases} \tag{3.83}$$

Given that the upsampling is the opposite operation of downsampling, we would expect that the spectra of aliases in oversampling will be compressed U times so that we can write:

$$X_U(e^{j\Omega}) = X(e^{jU\Omega}). \tag{3.84}$$

To get rid of aliases, we need the LPF with the system function:

$$H_U(e^{j\Omega_y}) = \begin{cases} C, \; 0 \le |\Omega_y| \le \pi/U \\ 0, \text{otherwise} \end{cases} \tag{3.85}$$

and corresponding DSP module for upsampling operation is a cascade of upsampler and LPF as shown in Fig. 3.43.

3.2.3.3 Resampling

To *resample* the DT signal by a rational factor *U*/*D*, or in other words to perform the rate conversion by a factor *U*/*D*, we can cascade the upsampling and downsampling modules as shown in Fig. 3.44(a). The order of interpolator and decimator is chosen on such a way so that the two low-pass filters can be combined into one, as shown in Fig. 3.4(b), with the system function:

$$H(e^{j\Omega_y}) = \begin{cases} C, \; 0 \le |\Omega_y| \le \min(\pi/U, \pi/D) \\ 0, \text{otherwise} \end{cases} \tag{3.86}$$

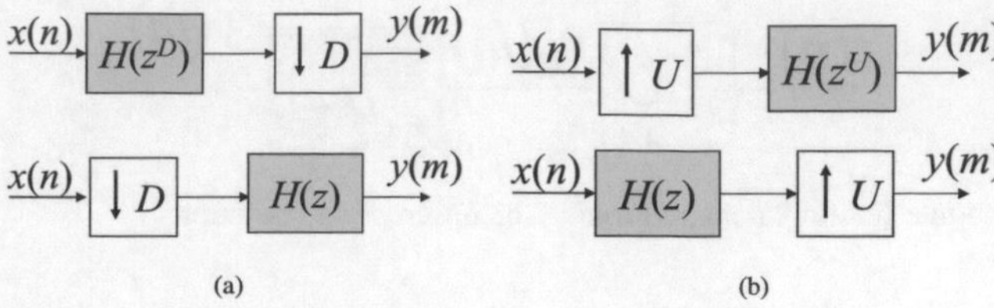

Fig. 3.45 The Noble identities: (**a**) downsampling and (**b**) upsampling equivalencies (identities)

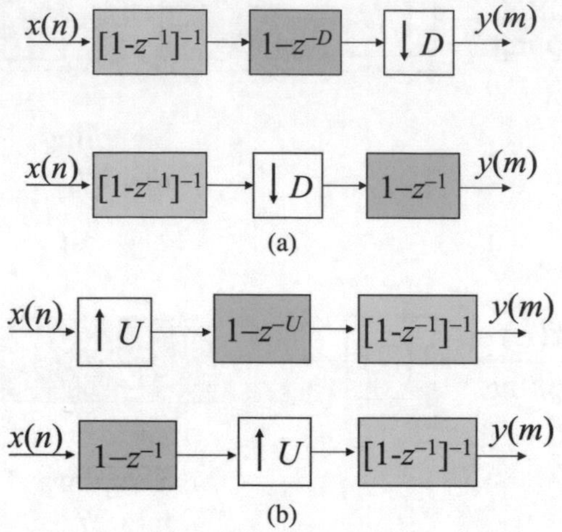

Fig. 3.46 The employment of Noble identities in CIC-based: (**a**) decimator and (**b**) interpolator

3.2.3.4 Noble Identities

Important relationships for upsampling and downsampling, known as *Noble identities*, are summarized in Fig. 3.45. These identities indicate that when the transfer function of the subsystem (filter) is polynomial in z^N (N is a positive integer), it can be swapped with corresponding upsampler/downsampler. This swapping enables us to implement the corresponding filters at lower clock rates.

3.2.3.5 Cascaded-Integrator-Comb (CIC) Filter-Based Upsampling and Downsampling

One possible implementation of LPFs required in decimator and interpolator is the *cascaded-integrator-comb (CIC) filter*. This is multiply-free LPF, because impulse response is all-one FIR:

$$h(n) = \begin{cases} 1, 0 \le n < M \\ 0, \text{ otherwise} \end{cases} \tag{3.87}$$

so that the corresponding z-transform is given by:

$$\begin{aligned} H(z) &= \sum_{n=-\infty}^{\infty} h(n)z^{-n} = \sum_{n=0}^{M-1} z^{-n} = \frac{1-z^{-M}}{1-z^{-1}} \\ &= \underbrace{\frac{1}{1-z^{-1}}}_{H_{\text{integrator}}(z)} \underbrace{\left(1-z^{-M}\right)}_{H_{\text{comb}}(z)} = H_{\text{integrator}}(z)H_{\text{comb}}(z), \end{aligned} \tag{3.88}$$

indicating that CIC filter can be represented as a cascade of an integrator and the *comb filter*, typically employed to reject periodical frequencies (or equally spaced frequencies). The CIC filter cascaded with an upsampler/downsampler is known as a *Hogenauer filter*. The CIC filter can be used either in decimator or interpolator as the LPF. By using Noble identities, the implementation of the comb filter in both decimator and interpolator can be simplified as shown in Fig. 3.46. Given that DTFT of CIC filter has the form:

$$H\left(e^{j\Omega}\right) = H(z)|_{z=e^{j\Omega}} = \left.\frac{1-z^{-M}}{1-z^{-1}}\right|_{z=e^{j\Omega}} = \frac{\sin\left(\frac{M\Omega}{2}\right)}{\sin\left(\frac{\Omega}{2}\right)} e^{-j(M-1)\Omega/2}, \tag{3.89}$$

we conclude that the magnitudes of the side-band will not be sufficiently suppressed so that the multiple integrator stages maybe used when needed.

3.2.3.6 Polyphase Decomposition

Another relevant implementation of LPFs required in decimator, interpolator, or resampling module is based on the *polyphase decomposition* of the system function. Namely, the system function can be represented in the following form:

$$\begin{aligned} H(z) &= h(0) \quad + h(1)z^{-1} \quad + h(2)z^{-2} \quad + \cdots + h(N-1)z^{-N+1} \\ &+ h(N)z^{-N} + h(N+1)z^{-N-1} + h(N+2)z^{-N-2} + \cdots + h(2N-1)z^{-2N+1} \\ &+ h(2N)z^{-2N} + h(2N+1)z^{-2N-1} + h(2N+2)z^{-2N-2} + \cdots + h(3N-1)z^{-3N+1} \\ &+ \cdots \end{aligned} \tag{3.90}$$

Let the terms in the first column of Eq. (3.90) define a sub-filter $H_0(z^N)$:

$$H_0\left(z^N\right) = h(0) + h(N)z^{-N} + h(2N)z^{-2N} + \cdots \tag{3.91}$$

Let the terms in the second column of Eq. (3.90), after removing the common factor z^{-1}, define the sub-filter $H_1(z^N)$:

$$H_1\left(z^N\right) = h(1) + h(N+1)z^{-N} + h(2N+1)z^{-2N} + \cdots \tag{3.92}$$

Further, let the terms in the third column of Eq. (3.90), after removing the common factor z^{-2}, define the sub-filter $H_2(z^N)$:

$$H_2\left(z^N\right) = h(2) + h(N+2)z^{-N} + h(2N+2)z^{-2N} + \cdots \tag{3.93}$$

Finally, let the $(N-1)$-th sub-filter be defined as:

$$H_{N-1}\left(z^N\right) = h(N-1) + h(2N-1)z^{-N} + h(3N-1)z^{-2N} + \cdots \tag{3.94}$$

We can then decompose the filter with system function (3.90) in terms of transfer functions of sub-filters as follows:

$$H(z) = H_0\left(z^N\right) + H_1\left(z^N\right)z^{-1} + H_2\left(z^N\right)z^{-2} + \cdots + H_{N-1}\left(z^N\right)z^{-(N-1)}, \tag{3.95}$$

and this representation is known as the polyphase filter decomposition/representation, and corresponding block scheme is shown in Fig. 3.47.

3.2.3.7 Polyphase Filter-Based Resampling

The polyphase filter representation can be used in decimation as shown in Fig. 3.48(a), and after application of the downsampling Noble identity, the polyphase filter can be implemented at D-times lower clock rate as shown in Fig. 3.48(b).

The use of polyphase filter structure in interpolation is shown in Fig. 3.49, both before (**a**) and after (**b**) applying the upsampling Noble identity.

The configuration of decimator can be greatly simplified, when commutator is used, as shown in Fig. 3.50. Unit-delay elements and downsamplers can be omitted in this case. This configuration is also known as the *polyphase filter bank*. In similar fashion, the configuration of the interpolator can also be greatly simplified when the commutator is used as shown in Fig. 3.51.

Finally, by employing the commutator and the DSP module to perform the resampling by a rational factor U/D shown in Fig. 3.44(b), we can implement the polyphase filter-based resampling by U/D as shown in Fig. 3.52. In this module, each input sample generates U output samples in parallel. The commutator strides through the polyphase filter bank and outputs D samples per cycle. Once the commutator exceeds the U-th polyphase filter bank branch, it rolls over per modulo U, and

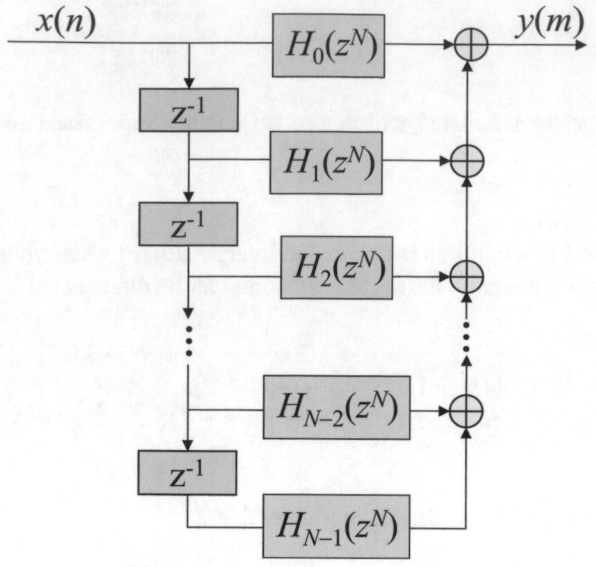

Fig. 3.47 The polyphase filter decomposition of the transfer function $H(z)$

Fig. 3.48 The use of polyphase filter structure in decimation: (**a**) before and (**b**) after the downsampling Noble identity

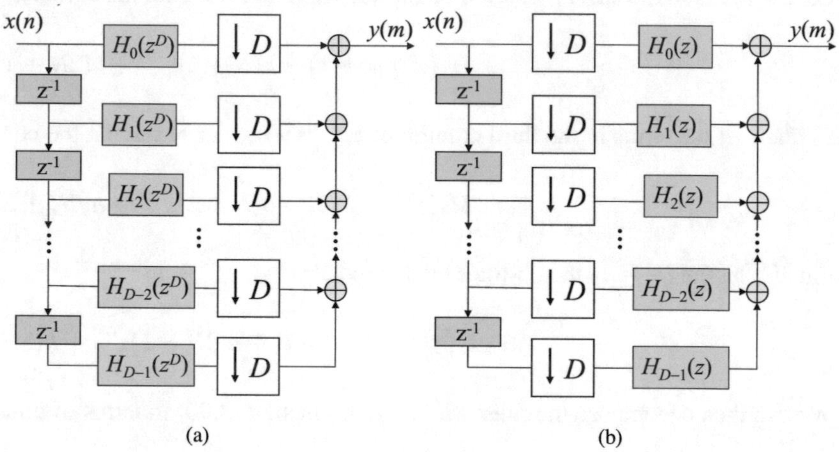

Fig. 3.49 The use of polyphase filter structure in interpolation: (**a**) before and (**b**) after the upsampling Noble identity

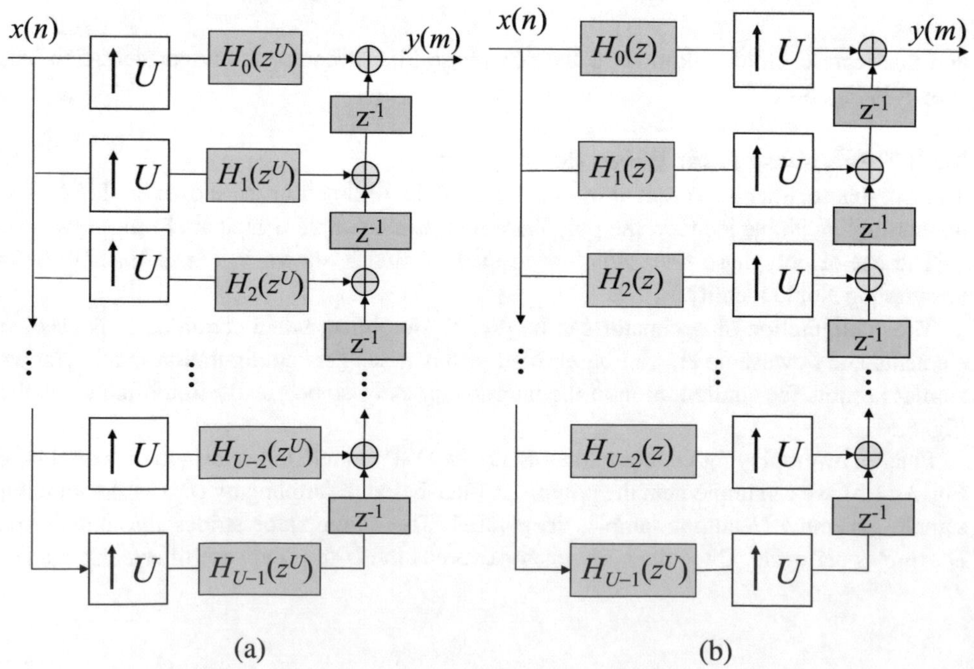

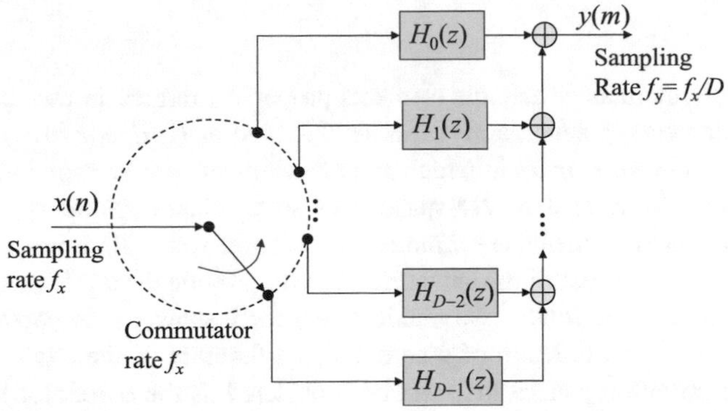

Fig. 3.50 The commutator and polyphase filter-based decimator

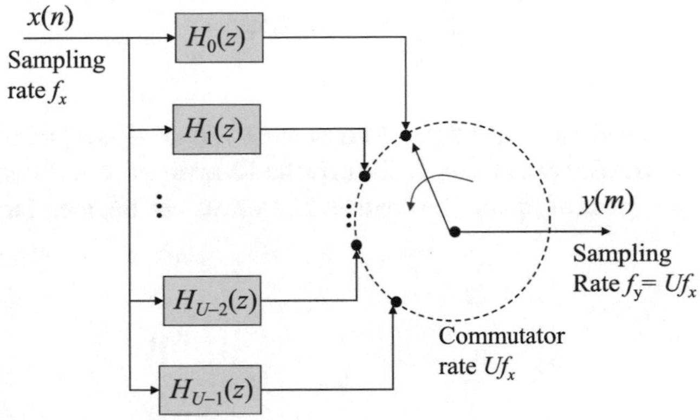

Fig. 3.51 The commutator and polyphase filter-based interpolator

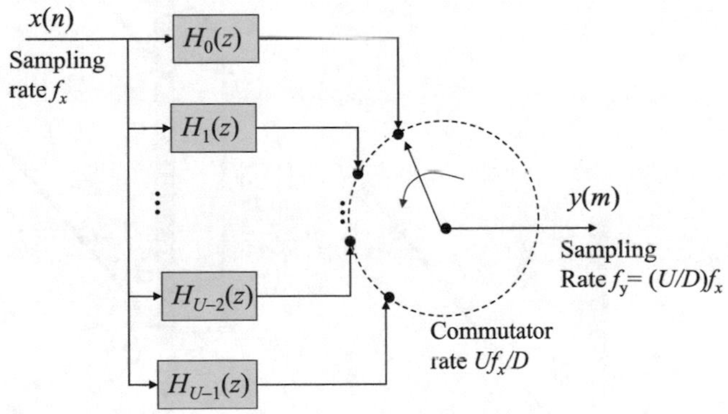

Fig. 3.52 The commutator and polyphase filter-based resampling by U/D module

the new input sample gets clocked into the polyphase filter bank. When $U > D$, then multiple outputs get generated for each input clocked into the module. On the other hand, when $U < D$, then multiple inputs get clocked into the module between two neighboring output samples. In hardware implementation, it is possible to use a single filter with coefficients being stored in corresponding memory.

The polyphase filter bank has different applications including interfacing subsystems with different rates, efficient implementation of narrowband filters, sub-band source coding, implementing precise phase shifters, digital filter banks, and quadrature-mirror filter bank, to mention few. In sub-band coding, we break the sequence $\{x(n)\}$ into different sequences representing different frequency bands. After that we code the low-frequency bands with larger number of bits because they are more important in speech processing.

3.2.4 Antenna Arrays

Antenna arrays consist of several antennas or antenna elements properly arranged in the space [50, 51]. Depending on the space arrangement of antenna elements, antenna arrays can be classified as (i) *linear array* in which antenna elements are arranged along a straight line; (ii) *circular array* in which antenna elements are arranged along a circular ring; (iii) *planar array* in which antenna elements are ranged in 2D space over some planar geometry, such as rectangular array; and (iv) *conformal array* in which antenna elements are arranged to conform some 3D object. As an illustration, two-element linear array is shown in Fig. 3.53. The purpose of the antenna array is to provide the angle diversity to deal with fading effects. The antenna array can also be used in multiuser communications, for instance, to separate different users. At any given location, the received signal is not only a function of time but also a function of the angle of arrival $r(t,\theta)$, $t\in\{-\infty,\infty\}$, $\theta\in[0,2\pi)$. The antenna array with radiation pattern $R(\theta)$ can be considered as the correlator in the space domain, so that the signal after antenna array can be represented as:

$$\frac{1}{2\pi}\int_{0}^{2\pi} r(t,\theta)R(\theta)d\theta \tag{3.96}$$

Let the two-element array be observed, see Fig. 3.53(a), with elements, denoted as A_1 and A_2, separated by d, with incident angle of incoming electromagnetic (EM) waves being θ. Clearly, the EM waves path difference is $d\cos\theta$, resulting in path delay deference of $d\cos\theta/c$ (c is the speed of light). The incident EM waves can be, therefore, represented as:

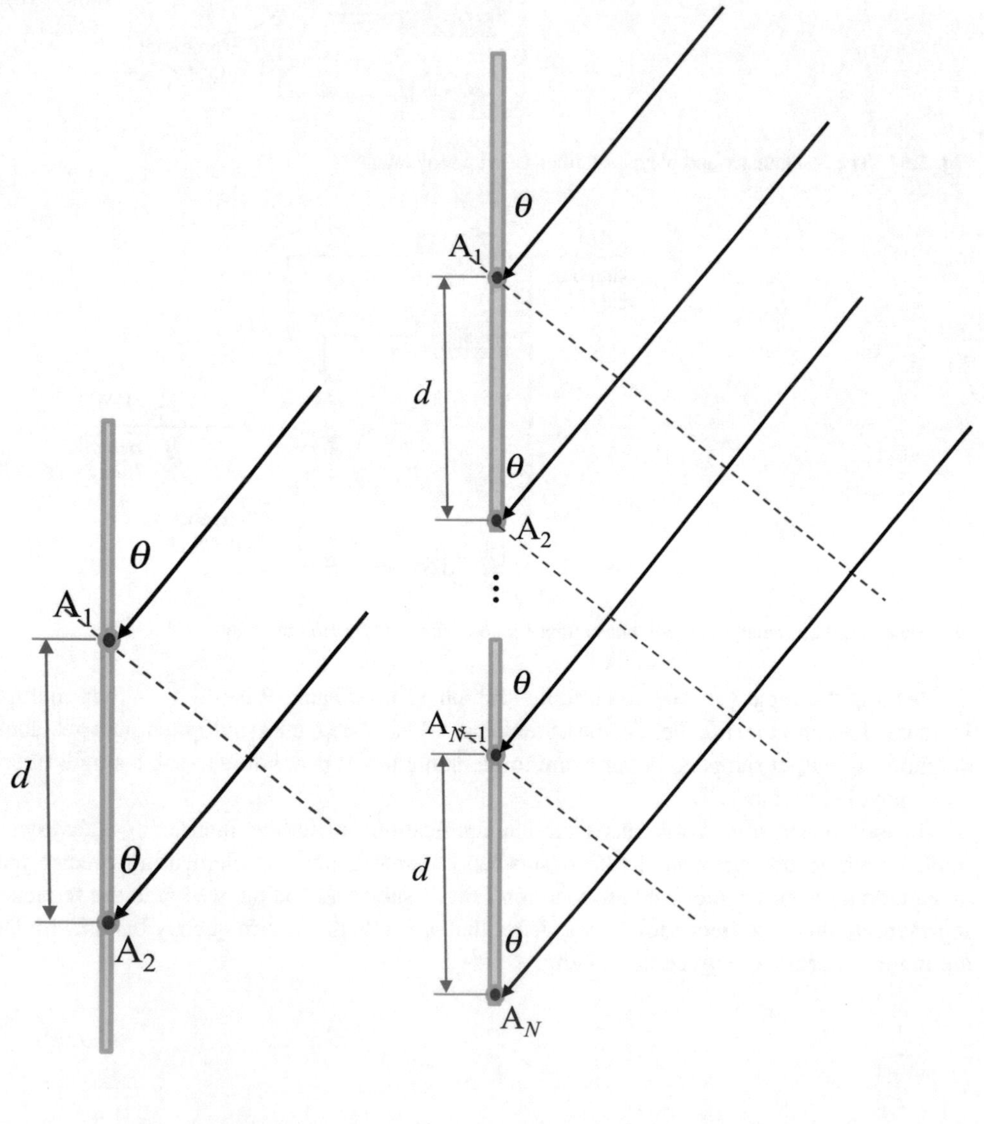

Fig. 3.53 Linear antenna arrays: (**a**) two-element linear array and (**b**) *N*-element linear array

$$x_{A_1}(t) = \exp\left(j2\pi f_c t + j\varphi\right), x_{A_2}(t) = \exp\left[j2\pi f_c\left(t - \frac{d}{c}\cos\theta\right) + j\varphi\right], \tag{3.97}$$

where f_c is the carrier frequency and φ is the carrier phase shift. Assuming that g and φ are the gain coefficient and phase shift of antenna element A_2 with respect to antenna element A_1, the resulting antenna array signal can be represented as:

$$\begin{aligned} z(t) &= x_{A_1}(t) + ge^{j\phi}x_{A_2}(t) = e^{j2\pi f_c t + j\varphi}\left(1 + ge^{j\phi - j2\pi f_c \frac{d}{c}\cos\theta}\right) \\ &= e^{j2\pi f_c t + j\varphi'}\left\{\left[1 + g\cos\left(\phi - 2\pi f_c \tfrac{d}{c}\cos\theta\right)\right]^2 + \left[g\sin\left(\phi - 2\pi f_c \tfrac{d}{c}\cos\theta\right)\right]^2\right\}^{1/2} \\ &= e^{j2\pi f_c t + j\varphi'}\left[1 + g^2 + 2g\cos\left(\phi - 2\pi f_c \tfrac{d}{c}\cos\theta\right)\right]^{1/2} = |R(\theta)|e^{j2\pi f_c t + j\varphi'}, \end{aligned} \tag{3.98}$$

where φ' is the equivalent phase shift of the carrier and $|R(\theta)|^2$ is the magnitude squared of the radiation pattern given by:

$$|R(\theta)|^2 = 1 + g^2 + 2g\cos\left(\phi - 2\pi f_c \frac{d}{c}\cos\theta\right). \tag{3.99}$$

In particular, when the distance between array elements is the half of wavelength, $d = \lambda_c/2$, the magnitude squared of the radiation pattern simplifies to:

$$\begin{aligned} |R(\theta)|^2\Big|_{d=\lambda_c/2} &= 1 + g^2 + 2g\cos\left(\phi - \pi f_c \frac{\lambda_c}{c}\cos\theta\right) \\ &\overset{\lambda_c = c/f_c}{=} 1 + g^2 + 2g\cos\left(\phi - \pi\cos\theta\right). \end{aligned} \tag{3.100}$$

Clearly, by varying the parameters ϕ and g, different radiation patterns can be obtained, as illustrated in Fig. 3.54.

Let us now observe the N-elements linear array, shown in Fig. 3.53(b), with neighboring elements separated by distance d, whose outputs are combined with following respective complex weights $\{g_1e^{j\phi_1}, g_2e^{j\phi_2}, ..., g_Ne^{j\phi_N}\}$. From Fig. 3.53(b), we conclude that the path difference of the n-th ray ($n = 1,2,\ldots,N$) with respect to the ray incident to the antenna element A_1 is equal to:

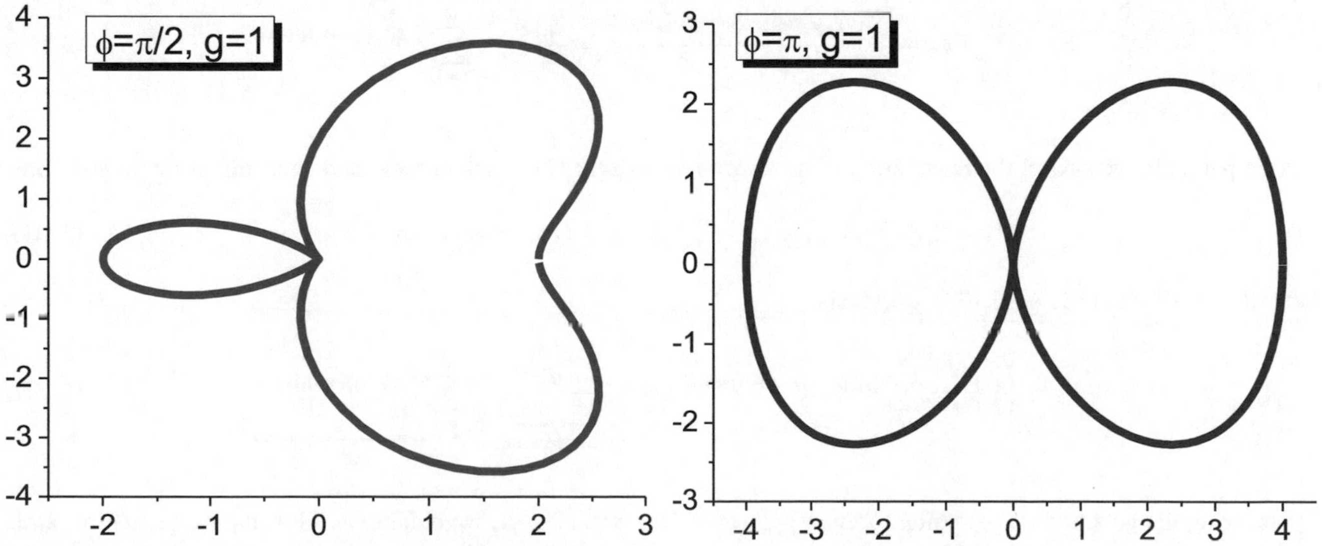

Fig. 3.54 The radiation patterns can be varied by changing the parameters g and ϕ

$$\Delta d(A_n, A_1) = (n-1)d\cos\theta; n = 1, 2, \cdots, N. \tag{3.101}$$

Based on the first line in Eq. (3.98), the radiation pattern of the N-element linear array will be:

$$R(\theta) = \sum_{n=1}^{N} g_n e^{j\phi_n - j2\pi f_c \frac{(n-1)d\cos\theta}{c}}. \tag{3.101}$$

When $d = \lambda_c/2$, the radiation pattern of N-element linear array reduces to:

$$R(\theta)|_{d=\lambda_c/2} = \sum_{n=1}^{N} g_n e^{j\phi_n - j\pi(n-1)\cos\theta}. \tag{3.102}$$

By varying the number of element in the linear array, the element spacing, amplitudes, and phases of the complex weights, it is possible to obtain different shapes of the radiation pattern.

For the transmit array antenna of identical elements, the radiation pattern can be found by applying the *pattern multiplication theorem* [51], claiming that the array pattern is the product of the array element pattern and the *array factor (AF)*. The AF is a function dependent only on the geometry of the antenna array and excitation of the elements. Since the array factor is independent of the antenna type, the isotropic radiators can be employed in the derivation of the AF. The field of an isotropic radiator located in the origin is given by [51]:

$$E_\theta = \frac{I_0}{4\pi}\frac{e^{-jkr}}{r}, \tag{3.103}$$

where r is the distance of the observation point P from the origin, k is the wave number ($k = 2\pi/\lambda$), and I_0 is the current magnitude applied. Let us now observe the transmit N-element linear array shown in Fig. 3.55. The current magnitudes of all array elements are the same I_0, while the phase of the n-th element relative to the first element is denoted as ϕ_n so that the current of the n-th element is given by $\underline{I}n = I_0\exp(j\phi_n)$. Based on Fig. 3.55, we conclude that the far field for the n-th array element is given by:

$$E_{\theta,n} \cong \frac{I_0 e^{j\phi_n}}{4\pi}\frac{e^{-jkr_n}}{r} = \frac{I_0 e^{j\phi_n}}{4\pi}\frac{e^{-jk[r-(n-1)d\cos\theta]}}{r}; n = 1, \cdots, N; \phi_1 = 0. \tag{3.104}$$

The overall N-element array far field can be determined by the superposition principle:

$$E_\theta = \sum_{n=1}^{N} E_{\theta,n} = \sum_{n=1}^{N} \frac{I_0 e^{j\phi_n}}{4\pi}\frac{e^{-jk[r-(n-1)d\cos\theta]}}{r} = \frac{I_0}{4\pi}\frac{e^{-jkr}}{r}\sum_{n=1}^{N} e^{j[\phi_n + (n-1)kd\cos\theta]}. \tag{3.105}$$

One particular version of the linear array is the *uniform linear array* in which phases follow the arithmetic progression:

$$\phi_1 = 0, \phi_2 = \phi, \phi_3 = 2\phi, \cdots, \phi_n = (n-1)\phi, \cdots, \phi_N = (N-1)\phi. \tag{3.106}$$

After the substitution in Eq. (3.105), we obtain:

$$E_\theta = \frac{I_0}{4\pi}\frac{e^{-jkr}}{r}\sum_{n=1}^{N} e^{j[(n-1)\phi + (n-1)kd\cos\theta]} = \underbrace{\frac{I_0}{4\pi}\frac{e^{-jkr}}{r}}_{arrayelementpattern} \cdot \underbrace{\sum_{n=1}^{N} e^{j(n-1)(\phi + kd\cos\theta)}}_{AF}. \tag{3.107}$$

By applying the geometric progression formula $\sum_{n=1}^{N} q^n = (1-q^n)/(1-q)$, we obtain the following expression for array factor:

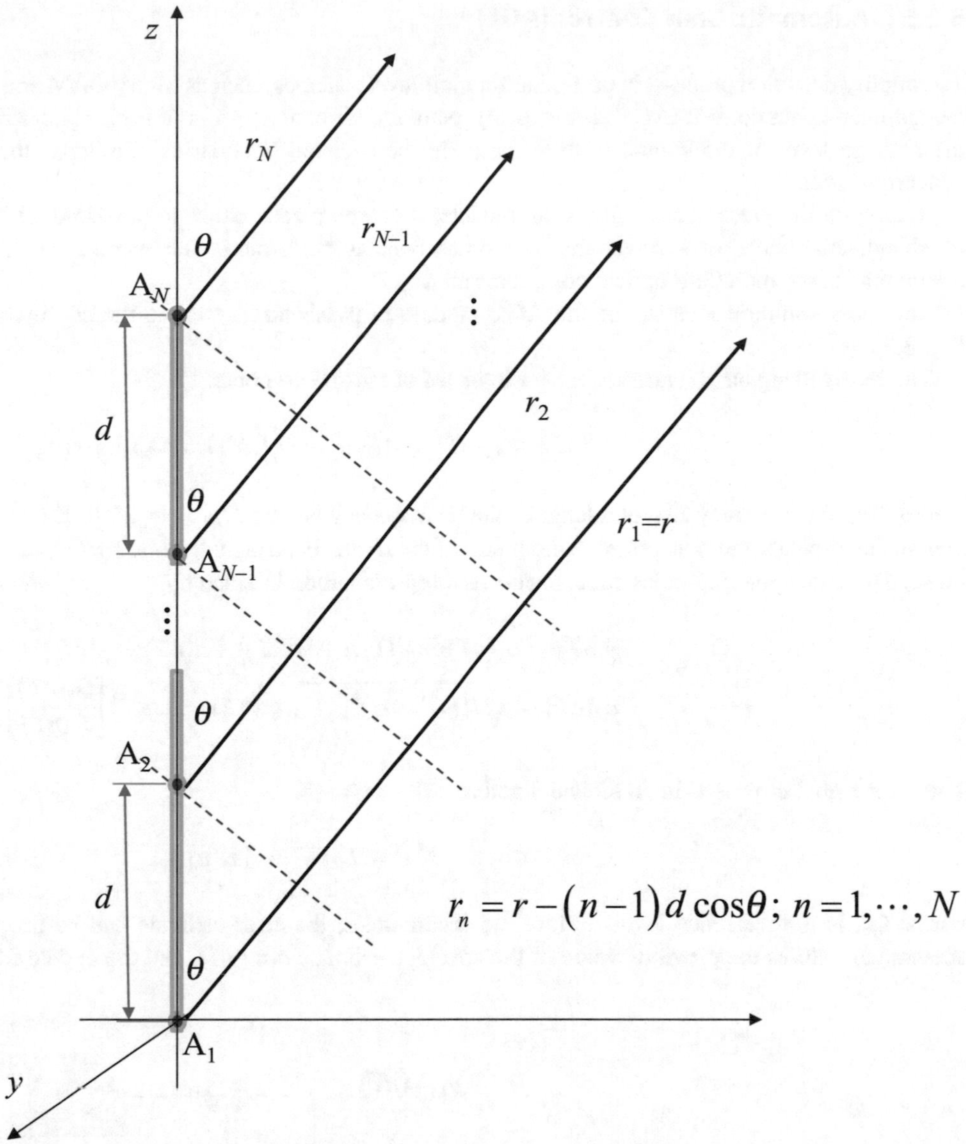

Fig. 3.55 Transmit *N*-element linear array

$$AF = \sum_{n=1}^{N} e^{\,j(n-1)(\phi+kd\cos\theta)} = \frac{e^{jN(\phi+kd\cos\theta)}-1}{e^{\,j(\phi+kd\cos\theta)}-1} = e^{\,j(N-1)\phi}\,\frac{\sin\left(\frac{N(\phi+kd\cos\theta)}{2}\right)}{\sin\left(\frac{\phi+kd\cos\theta}{2}\right)}. \tag{3.108}$$

The complex exponent term will disappear when the origin is moved to the center of the linear array so that the expression for AF simplifies to:

$$AF = \frac{\sin\left(\frac{N(\phi+kd\cos\theta)}{2}\right)}{\sin\left(\frac{\phi+kd\cos\theta}{2}\right)}, \tag{3.109}$$

The maximum of $AF_{max} = N$ is obtained for $\phi + kd\cos\theta = 0$. There are *N*-1 lobes, one main, and *N*-2 side lobes. The main lobe width is $4\pi/N$, while the minor lobe width is $2\pi/N$.

3.2.5 Automatic Gain Control (AGC)

To simplify detection process, in particular for multilevel schemes, such as M-ary QAM and M-ary APSK, the employment of the automatic gain control (AGC) is necessary. Namely, for hard decision, it is highly desired to have accurate knowledge of the average level of the signal, to draw properly the decision boundaries. Similarly, the AGC is required in PLL-based synchronization.

Regarding the placement of the AGC module, it can be placed either in passband, at intermediate frequency (IF), or in baseband, after the down-conversion to the baseband is performed. The second case is more attractive for high-speed communications, including optical communications.

The most common location of the AGC module in passband as well as the configuration of the AGC is provided in Fig. 3.56.

The received signal $y(t)$ after ADC is resampled at rate $1/T$ to obtain:

$$y(nT) = I_y(nT)\cos(\Omega_c n) - Q_y(nT)\sin(\Omega_c n) + z(nT), \tag{3.110}$$

where discrete frequency Ω_{IF} is related to the IF frequency ω_{IF} by $\Omega_{\mathrm{IF}} = \omega_{\mathrm{IF}}T$. In Eq. (3.110), we use (nT) and $Q_y(nT)$ to denote the in-phase and quadrature components of the IF passband signal, while $z(nT)$ represent the sample of the narrowband noise. The corresponding polar coordinate-based representation is given by:

$$\begin{aligned} y(nT) &= a_y(nT)\cos[\Omega_{IF}n + \psi(nT)] + z(nT), \\ a_y(nT) &= \sqrt{I_y^2(nT) + Q_y^2(nT)},\ \psi(nT) = \tan^{-1}\left[\frac{Q_y(nT)}{I_y(nT)}\right]. \end{aligned} \tag{3.111}$$

The error signal $e(n)$ used in AGC is defined as:

$$e(n) = L_{\mathrm{ref}} - a(n)a_y(n), \tag{3.112}$$

where L_{ref} is the reference level. Before the accumulator, the error estimate can be properly scaled by the factor s. The accumulator stores the previous value of the envelope estimate $a(n-1)$, and the update formula is given by:

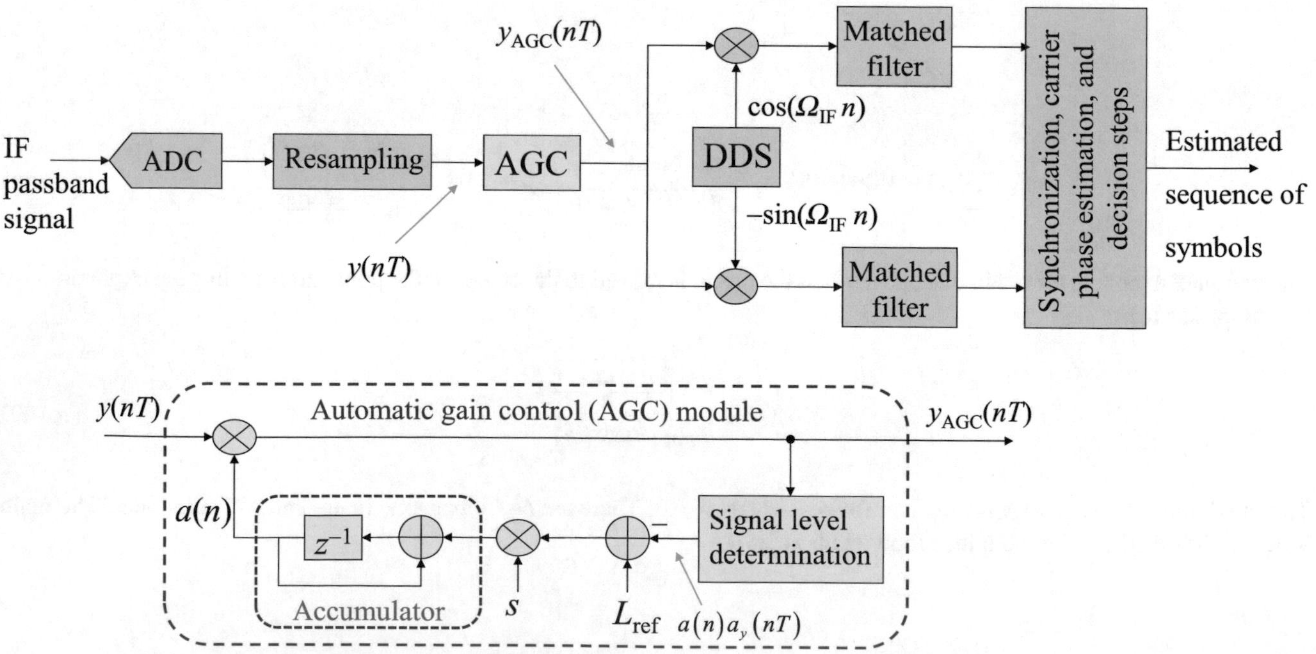

Fig. 3.56 The location of the AGC module in the passband (at IF) and the configuration of corresponding AGC module

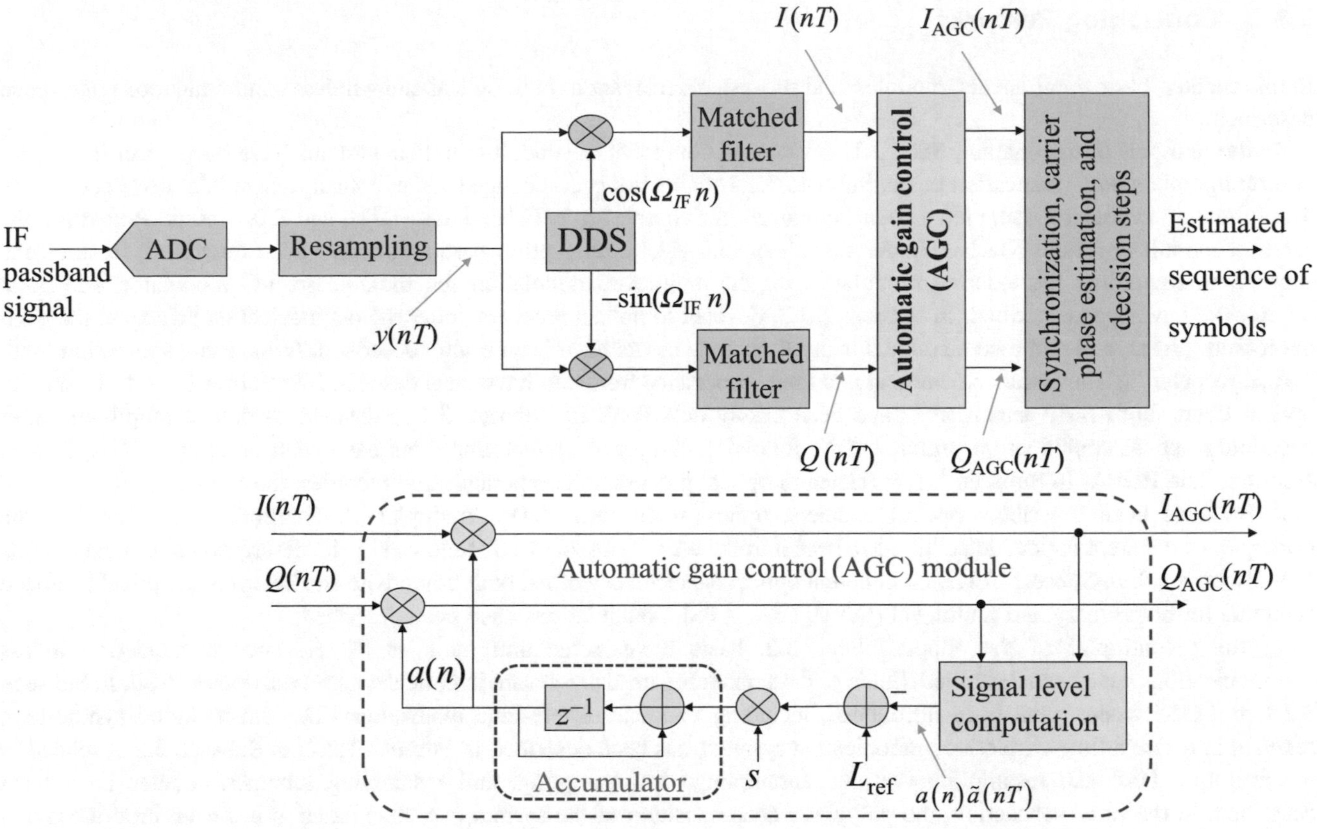

Fig. 3.57 The location of the AGC module in the baseband and the configuration of corresponding AGC module

$$a(n) = a(n-1) + s\left[L_{\text{ref}} - a(n-1)a_y((n-1)T)\right]. \tag{3.113}$$

The AGC can also be placed in the baseband domain as shown in Fig. 3.57.

The envelope after the matched filters is given by:

$$\tilde{a}(nT) = \sqrt{I^2(nT) + Q^2(nT)}. \tag{3.114}$$

The corresponding envelope update formula is very similar to Eq. (3.113) and given by:

$$a(n) = a(n-1) + s[L_{\text{ref}} - a(n-1)\tilde{a}((n-1)T)]. \tag{3.115}$$

The key difference of this AGC scheme with respect to the corresponding scheme from Fig. 3.56 is that it operates in baseband and as such it is possible to implement it at lower rates. Both envelope update expressions are nonlinear and might converge slow when the signal level is low. To overcome this problem, we can modify the update formula to operate with dB-values of the signals as follows:

$$a(n)[\text{dB}] = a(n-1)[\text{dB}] + s\{L_{\text{ref}}[\text{dB}] - a(n-1)[\text{dB}] - \tilde{a}((n-1)T)[\text{dB}]\}. \tag{3.116}$$

However, the conversion to linear scale is needed before the multiplication stage, after the accumulator. Clearly, the update formula is now linear.

3.3 Concluding Remarks

In this chapter, basic components, modules, and subsystems relevant in both optical and wireless communications have been described.

In the first half of the chapter, Sect. 3.1, key optical components, modules, and subsystems have been described. After describing optical communication basic, Subsect. 3.1.1, focus has been changed to the optical transmitters (Subsect. 3.1.2). The following classes of lasers have been introduced in Subsect. 3.1.2: Fabry-Perot, DFB, and DBR lasers. Regarding the external modulators, both Mach-Zehnder modulator and electro-absorption modulator have been described, in the same subsection. Regarding the external modulators for the multilevel modulation schemes, phase, I/Q modulator, and polar modulators have been described. In Subsect. 3.1.3, devoted to optical receivers, after the photodetection principle, the p-i-n photodiode, APD, and metal-semiconductor-metal photodetectors have been briefly described. In the same subsection, both optical receiver high-impedance front end and trans-impedance front end have been described. In Subsect. 3.1.4, devoted to optical fibers, both SMF and MMF have been briefly described. In Subsect. 3.1.5, devoted to optical amplifiers, after describing optical amplification principles, the following classes of optical amplifiers have been described: SOA, Raman amplifier, and EDFA. In Subsect. 3.1.6, related to optical processing components and modules, the following components/modules have been described: optical couplers, optical filters, and WDM multiplexers/demultiplexers. After that, the principles of coherent optical detection have been introduced, in Subsect. 3.1.7, followed by the description of optical hybrids in Subsect. 3.1.8. In Subsect. 3.1.9, on coherent optical balanced detectors, both homodyne and heterodyne optical balanced detectors for both binary and multilevel (QAM, PSK, APSK) modulations have been described.

In the second half of the chapter, Sect. 3.2, basic components and modules of relevance to modern wireless communications have been described. Some of these modules are also relevant in optical communications as well. In Subsect. 3.2.1, the DSP basics have been introduced, including various discrete-time realizations. The direct digital synthesizer, relevant in discrete-time wireless communication systems, has been described in Subsect. 3.2.2. In Subsect. 3.2.3, related to the multirate DSP and resampling, various upsampling, downsampling, and resampling schemes/modules have been described. In the same subsection, the polyphase decomposition of the system (transfer) function has been introduced and employed in resampling. In Subsect. 3.2.4, devoted to the antenna arrays, the linear array antennas have been described. Finally, in Subsect. 3.2.5, on automatic gain control, the AGC schemes operating in passband and baseband have been described. The set of problems is provided in incoming section.

3.4 Problems

1. An electric field is applied along the z-crystallographic axis of a $LiNbO_3$ modulator that operates at 1550 nm. Assuming that electrodes are separated by 9 μm, $r_{33} = 30.9$ pm/V, refractive mode index $n = 2.2$, and the overlapping between optical and electrical fields $\Gamma = 0.45$, determine the V_π-voltage for a 3.5-cm-long waveguide.

2. In this problem we study the $LiNbO_3$ (LN) modulator in the symmetric Mach-Zehnder (MZ) configuration implemented with the help of two 3 dB directional couplers, as shown in Fig. P2, for which $kl = (2m + 1)\pi/4$ (m-an integer).

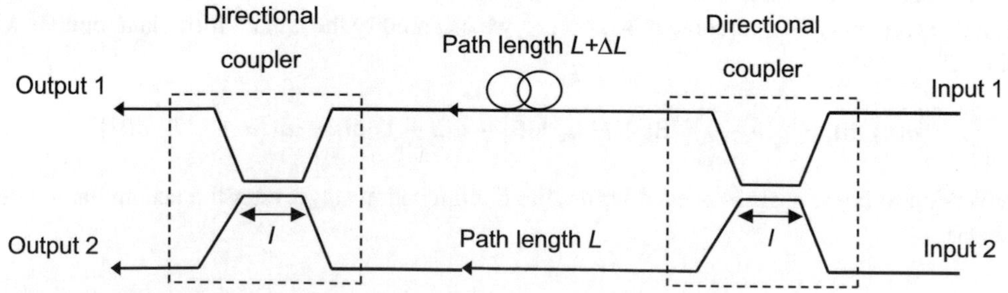

Fig. P2 LN modulator in symmetric MZ configuration

Assuming that the voltage V is applied along the arm of length $L + \Delta L$ (longer arm). By employing the transfer-matrix approach, derive corresponding expression for the power at the bar output port.
Hint. The electrical fields of output and input ports of directional coupler are connected by:

$$\begin{pmatrix} E_{o1} \\ E_{o2} \end{pmatrix} = e^{-j\beta l}\begin{pmatrix} \cos(kl) & j\sin(kl) \\ j\sin(kl) & \cos(kl) \end{pmatrix}\begin{pmatrix} E_{i1} \\ E_{i2} \end{pmatrix}$$

3. The binary sequence 00110001110101 is to be transmitted employing RZ pulses with duty cycle of 50%. For CS-RZ, RZ-DPSK, duobinary, and AMI modulation formats, determine the amplitude and phase at the output of the transmitter.
4. The RZ-DPSK receiver is implemented using Mach-Zehnder delay interferometer (MZDI) and two 3 dB couplers:

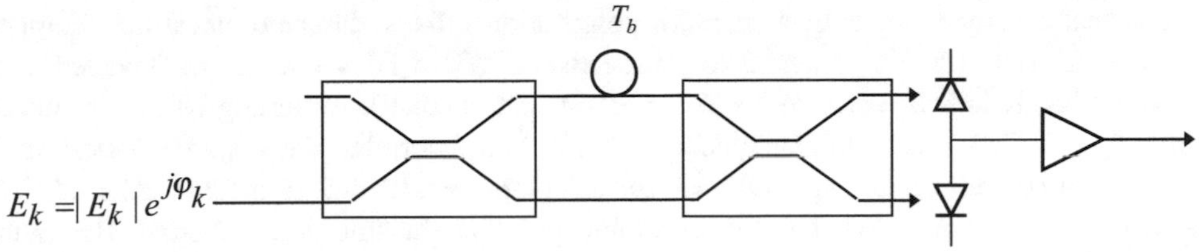

Determine the output of a balanced detector, assuming that photodiode output is related to the input electrical field E_{in} by $P_{\text{out}} = R|E_{\text{in}}|^2$, where R is the photodiode responsivity. Show that transmitted sequence in Problem 3 (corresponding to RZ-DPSK) can be recovered using this kind of receiver.
5. This problem is concerned with the InGaAsP laser.
   (a) Determine the threshold current for a 300-μm-long InGaAsP laser with an internal loss 30 cm^{-1} and operating in a single mode at 1550 nm with modal index 3.3 and group index 3.4. Assume that carrier life time is 2 ns and that the gain varies as $G = G_N(N - N_O)$, with $G_N = 8{\cdot}10^3$ s^{-1} and $N_O = 2{\cdot}10.^8$ The bias current applied is twice above threshold. Determine the emitted power from one faucet and the slope efficiency, and sketch the corresponding *P-I* curve. Assume the internal efficiency of 95%.
   (b) Assuming the laser linewidth of 10 MHz, the spontaneous emission factor $n_{\text{sp}} = 2$, and the average launched power of 2 mW, determine the frequency chirp expressed in Hz. Identify the adiabatic and dynamic terms.
   (c) Determine the effective chirp coefficient from (b). Assume that a chirped Gaussian pulse (with initial chirp equal to the effective chirp coefficient from (b)) is transmitted over DCF with dispersion coefficient of −90 ps/nm-km at 1550 nm. Assuming the NRZ transmission at 10 Gb/s, determine the transmission distance at which the pulse width is minimum. What is the maximum possible bit rate that can be supported?
6. Determine the maximum possible transmission distance of a 10Gb/s system operating at 1550 nm and employing the external modulator. Assume transmission over standard SMF with dispersion coefficient of 16 ps/nm-km. What would be the maximum transmission distance if the pre-chirping technique is employed? Determine the maximum transmission distance if the pre-chirping technique is employed together with Gires-Tournois interferometer having a 2-cm-long Fabry-Perot cavity and mirror reflectivity of 0.95.
7. A Raman amplifier is pumped *bidirectionally* using two pump laser diodes emitting 0.5 W each. Derive an expression for the amplifier output power if the input power is known. Determine the output power when 2 μW signal is injected into a 4-km-long amplifier. Assume losses of 0.2 and 0.25 dB/km at the signal (1550 nm) and pump (1450 nm) wavelengths, respectively. $A_{\text{eff}} = 50$ μm^2, and $g_R = 6 \times 10^{-14}$m/W. What is the amplifier gain in dBs? Neglect the saturation effect.
8. Consider an optically pre-amplified *pin* direct detection receiver in which the receiver thermal noise and shot noise can be neglected. Assuming the NRZ transmission, the optical filter bandwidth B_o, and electrical filter bandwidth B_e, show that the following relationship between Q-factor and optical signal-to-noise ratio (OSNR) is valid:

$$Q = \frac{2\sqrt{\frac{B_o}{B_e}}\text{OSNR}}{1 + \sqrt{1 + 4\text{OSNR}}}.$$

The OSNR is defined as the ratio between average signal power and total noise power in both polarizations. The amplifier gain is G, received optical power is P, the spontaneous emission factor is n_{sp}, and photodiode responsivity is R. Consider now an amplifier chain of N identical optical amplifiers with noise figure (NF) of 7 dB. To reduce the ASE noise, every amplifier stage has an optical filter of bandwidth $B_o = 4R_b$, with R_b being the bit-rate set to 10Gb/s. Assuming NRZ single-channel transmission with the launch power of 0 dBm, determine the maximum number of spans before bit-error rate reaches 10^{-12}. The noise figure of pre-amplifier is 6 dB, and corresponding optical filter is of bandwidth $3R_b$. The transfer function of receiver electronics can be modeled as a low-pass filter of bandwidth $B_e = 0.7R_b$. The operating wavelength is 1550 nm, and the photodiode responsivity $R = 1$ A/W.

9. We would like to design an optical 100 Gb/s WDM system with $M = 10$ channels operating at 10 Gb/s employing the in-line optical amplifiers of noise figure 5dB. The total launched power is restricted by available laser sources and cannot be larger than 12 dBm, and the fiber loss is $\alpha = 0.2$ dB/km. The predicted margins to achieve $BER = 10^{-9}$ (or $Q = 6$) allocated for chromatic dispersion, extinction ratio, polarization effects, fiber nonlinearities, component aging, and system margin are 0.5, 0.5, 1.5, 1.5, 2, and 3 dB, respectively. WDM DEMUX can be modeled as a bank of filters, each of them with a bandwidth B_o equal to $3B$ (B-the bit rate per channel). Assuming NRZ transmission, the electrical filter bandwidth $B_e = 0.65B$, and amplifier spacing $l = 100$ km, determine the required optical signal-to-noise ratio (OSNR) (expressed in 0.1 nm optical spectral analyzer resolution bandwidth) to achieve BER of 10^{-9} for the central channel located at 1552.524 nm. What is the maximum possible transmission distance? The saturation of optical amplifiers due to ASE noise accumulation is not negligible.
10. Determine the receiver sensitivity of an APD receiver with a trans-impedance amplifier (TIA) front end of noise figure F_A and feedback resistor R_F assuming (i) a finite extinction ratio r_{ex}, (ii) the thermal noise, (iii) the shot noise, (iv) the intensity noise, (v) the reflection noise, and (vi) the mode partition noise. What is the optimum APD multiplication ratio? What is the bandwidth of this receiver?
11. You are given to design the receiver for 40 Gb/s transmission that employs the Mach-Zehnder modulator with 14 dB of extinction ratio. Providing that you have available either a p-i-n photodetector or an APD with ionization factor 0.5, both of quantum efficiency 95%, a load resistance of 10 MΩ, and an electrical amplifier of noise figure 6dB, determine the receiver sensitivities of both receiver versions required to achieve the BER of 10^{-9}. Compare the signal-to-noise ratios of corresponding receivers. If an EDFA of noise figure 5dB is available for how many dBs the receiver sensitivity can be improved? Assume that the receiver bandwidth is 35 GHz and the bandwidth of optical filter that follows EDFA is 80 GHz.
12. You are given to design the system for 10 Gb/s RZ transmission that employs a very bad Mach-Zehnder modulator with extinction ratio of only 10 dB. Providing that you have available a p-i-n photodetector of quantum efficiency 0.9, an EDFA of gain 25 dB and noise figure 5dB to be used as pre-amplifier, and a load resistor R_L for high-impedance configuration, determine the receiver sensitivity required to achieve the BER of 10^{-12}. The bandwidth of an optical filter that follows EDFA is 20 GHz. The junction capacitance is 1 pF, while the input capacitance is negligible. The operating wavelength is 1550 nm.
13. You are supposed to design a transmission link operating at 160 Gb/s, employing RZ transmission of duty cycle 50%, over standard SMF with dispersion coefficient of 16 ps/nm-km and attenuation coefficient 0.2 dB/km. The effective cross-sectional area is 80 μm^2, and the second-order refractive index coefficient is $n_2 = 3.4 \times 10^{-8}$ μm^2/W. Gaussian pulses with peak power of 0 dBm are used for RZ transmission. An ideal external modulator is supposed to be used. Operating wavelength is 1550 nm.
    (a) Determine the maximum possible transmission distance.
    (b) What would be the maximum transmission distance if the SPM of fiber is used to compensate for chromatic dispersion? Assume that 50% of the broadening is allowed.
    (c) Determine the maximum transmission distance if the Gires-Tournois interferometer, having a 3-cm-long Fabry-Perot cavity and mirror reflectivity of 0.98, is used. Again, assume that 50% of the broadening is allowed.
14. An NRZ single-channel system operating at 10 Gb/s employs the Mach-Zehnder modulator (MZM) of extinction ratio of −20 dB, whose driver amplifier for any input 1-bit generates a pulse of the shape:

$$v_{RF}(t) = V_0 \begin{cases} 1, & 0 \le |t| \le (1-\alpha)T/2 \\ \frac{1}{2}\left[1 + \cos\left(\frac{\pi}{\alpha T}\left(|t| - \frac{1-\alpha}{2}T\right)\right)\right], & (1-\alpha)T/2 < |t| \le (1+\alpha)T/2, \\ 0, & |t| > (1+\alpha)T/2 \end{cases}$$

where V_0[V] is the amplitude of the pulse, T is the bit duration, and α is the so-called roll-off parameter ($0 \le \alpha \le 1$). Determine (i) the width of the eye opening, (ii) height of the eye opening, (iii) noise margin, (iv) the rise time, and (v) the decision threshold of the electrical field at MZM output. Sketch the eye diagram and determine the parameters (i)–(v) for $\alpha = 0.5$ and $\alpha = 1$.

15. You are asked to design 50 Gb/s per wavelength link between Tucson and Phoenix by using DPSK with direct detection. You have available two variable gain EDFAs with maximum gain of 18 dB each and noise figure of 5 dB each. To compensate for fiber losses of SMF (with attenuation coefficient being 0.2 dB/km at 1550 nm), you can use the available EDFAs and adjust the gains as needed. Assume that chromatic dispersion can be compensated by using appropriate equalizer. The responsivity of available PIN photodiode is 1 A/W.
    (a) Determine the equivalent noise figure of these two EDFAs in cascade.
    (b) Determine the optical SNR at receiver side.
    (c) Describe how you would determine the Q-factor.
    (d) Describe how you would measure the bit-error rate (BER).
    (e) Describe transmitter and receiver configurations for DPSK systems with direct detection.
    (f) If the following sequence is to be transmitted 0110101001100, determine the sequence of phases on transmitter side and show that original sequence can be recovered using DPSK demodulator from problem (e).
16. This problem is related to the coherent detection.
    (a) Describe key difference between direct and coherent detections.
    (b) Describe the principle of operation of optical hybrid and balanced detector.
    (c) If the input to the coherent detector is given by $S_k = |S_k|\exp.[j(\omega_T t + \phi_{PN,T} + \theta_k)]$ where the subscript k is used to denote the kth symbol being transmitted, ω_T denotes the transmitting laser frequency (in rad/s), $\phi_{PN,T}$ denotes the laser phase noise of transmitter, θ_k is the phase of the kth symbol for M-ary PSK, and the local laser (oscillator) is denoted by $L = |L|\exp.[j(\omega_{LO} t + \phi_{PN,LO})]$ (ω_{LO} denotes the local laser frequency and $\phi_{PN,LO}$ denotes the laser phase noise of local laser), determine the output of balanced receiver. Assume that both photodetectors in balanced receiver are identical with responsivity R. Assume further that corresponding hybrid is used to perform received and local laser signals mixing. Specify the phases to be used in M-ary PSK.
    (d) Describe the difference between polarization control and polarization scrambling.
    (e) Describe the operational principle of phase diversity receiver.
17. Consider the reconfigurable $M \times M$ wavelength-routing switch shown in Fig. P17.

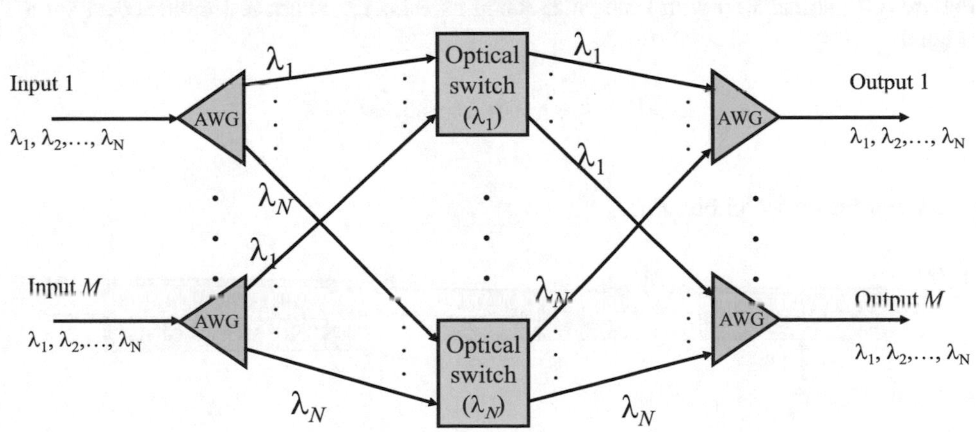

Fig. P17 Reconfigurable wavelength-routing switch

Each of the input fiber carries WDM signal with N channels, each with average power P, and extinction ratio $r = P_1/P_0$ (P_1-average power in one, P_0-average power in zero-bit). Determine out-of-band crosstalk power penalty with respect to an ideal system (with no crosstalk and ideal extinction ratio). For the extinction ratio of 10 dB, determine the maximum tolerable crosstalk level so that the introduced power penalty is 1 dB. Assume that the crosstalk from adjacent channels is dominant.

18. The analog system of the second order is shown in Fig. P18. Assuming $R/L = 8$ and $1/(LC) = 25$, determine an equivalent discrete-time system using the direct form realization.

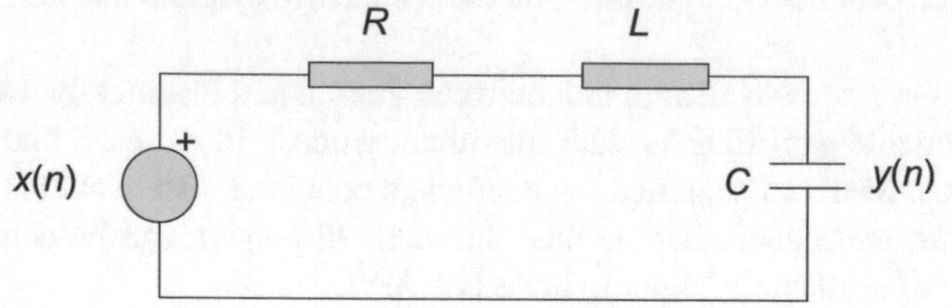

Fig. P18 An analog system of the second order

19. A digital notch filter is required to remove an undesirable 60-Hz hum associated with a power supply in ECG recording application. The sampling frequency is 500 samples/s. Design a second-order FIR notch filter. Chose the gain such that $|H(\omega)| = 1$ for $\omega = 0$.
20. Using an all-pass system, determine the causal stable system with the same magnitude response as the system with system function given by:

$$H(z) = \frac{1 + z^{-1}}{\left(1 - \frac{1}{3}z^{-1}\right)\left(1 - \frac{1}{2}z^{-1}\right)\left(1 - 2z^{-1}\right)}.$$

21. Design the discrete-time low-pass filter with passband edge frequency 10 GHz, the maximum allowable attenuation in passband being 3 dB, the stopband edge frequency being 16.5 GHz, and the minimum attenuation in stopband being 15 dB. For filter design, use the impulse invariance method. Determine the parallel realization of this system. Is this system stable?
22. The system function of a communication channel is given by: $H(z) = (1 - 0.9e^{j0.4\pi}z^{-1})(1 - 0.9e^{-j0.4\pi}z^{-1})(1 - 1.5e^{j0.6\pi}z^{-1})(1 - 1.5e^{-j0.6\pi}z^{-1})$.Determine the system function of a causal and stable compensating system so that the cascade interconnection of the two has a flat magnitude response. Sketch the zero-pole plots of subsystems.
23. A continuous-time (CT) signal $x_a(t)$ with bandwidth B and its echo $x_a(t\text{-}\tau)$ arrive simultaneously at a TV receiver. The received CT signal

$$s_a(t) = x_a(t) + \alpha x_a(t - \tau), \ |\alpha| < 1$$

is processed by the system shown in Fig. P23.

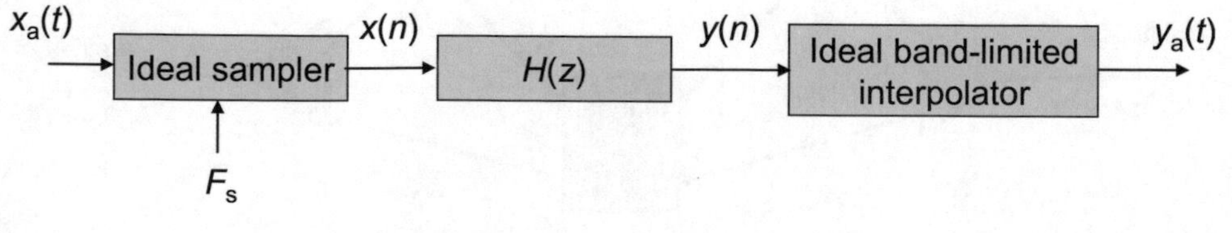

Fig. P23 The CT system under study

Specify the sampling frequency F_s and the filter system function $H(z)$ so that the "ghost" pulse $x_a(t\text{-}\tau)$ is removed from the received signal.

24. We would like to design a non-recursive decimator for $D = 8$ using the following factorization:

$$H(z) = \left[\left(1+z^{-1}\right)\left(1+z^{-2}\right)\ldots\left(1+z^{-2^{K-1}}\right)\right]^5$$

Determine the implementation of decimator using filters with system function: $H_k(z) = (1 + z^{-1})$.[5] Show that each stage of the decimator can be implemented more efficiently using the polyphase decomposition.

25. By using the bilinear transform, design a discrete low-pass filter of Butterworth type with passband being from 0 to 10 GHz, the attenuation at passband edge frequency being 3 dB, and minimum attenuation in stopband being 10 dB. The stopband edge frequency is 15 GHz, and the sampling frequency is 40 GHz.

References

1. W. Shieh, I. Djordjevic, *OFDM for Optical Communications* (Elsevier/Academic Press, Amsterdam-Boston, Oct 2009)
2. I.B. Djordjevic, W. Ryan, B. Vasic, *Coding for Optical Channels* (Springer, New York/Dordrecht/Heidelberg/London, 2010)
3. M. Cvijetic, I.B. Djordjevic, *Advanced Optical Communication Systems and Networks* (Artech House, Norwood, 2013)
4. R. Ramaswami, K. Sivarajan, *Optical Networks: A Practical Perspective* (Morgan Kaufman, 2nd, 2002)
5. G.P. Agrawal, *Fiber-Optic Communication Systems*, 3rd edn. (Wiley, Hoboken, 2002)
6. G.P. Agrawal, *Lightwave Technology: Components and Devices* (Wiley, Hoboken, 2004)
7. G.P. Agrawal, *Lightwave Technology: Telecommunication Systems* (Wiley, Hoboken, 2005)
8. G.P. Agrawal, *Nonlinear Fiber Optics*, 4th edn. (Academic Press, Oxford, 2007)
9. G. Keiser, *Optical Fiber Communications*, 3rd edn. (McGraw-Hill, 2000)
10. L. Kazovsky, S. Benedetto, A. Willner, *Optical Fiber Communication Systems* (Artech House, Boston, 1996)
11. J.C. Palais, *Fiber Optic Communications*, 5th edn. (Pearson Prentice Hall, Upper Saddle River, 2005)
12. B. Ramamurthy, Switches, wavelength routers, and wavelength converters, in *Optical WDM Networks: Principles and Practice*, ed. by K. M. Sivalingam, S. Subramaniam, (Kluwer Academic Publishers, Boston, 2000)
13. I. B. Djordjevic, ECE 430/530: Optical Communication Systems (Lecture Notes), University of Arizona, 2010/2011
14. R. Kostuk, ECE 430/530: Optical Communication Systems (Lecture Notes), University of Arizona, 2002/2004
15. I. B. Djordjevic, ECE 632: Advanced Optical Communication Systems (Lecture Notes), University of Arizona, 2007
16. A. Yariv, *Optical Electronics in Modern Communications* (Oxford University Press, 1997)
17. M. Amersfoort, Arrayed Waveguide Grating, Application note A1998003, Concept to volume, 1998
18. M. Staif, A. Mecozzi, J. Nagel, Mean square magnitude of all orders PMD and the relation with the bandwidth of principal states. IEEE Photon. Tehcnol. Lett. **12**, 53–55 (2000)
19. H. Kogelnik, L.E. Nelson, R.M. Jobson, Polarization mode dispersion, in *Optical Fiber Telecommunications*, ed. by I. P. Kaminow, T. Li, (Academic Press, San Diego, 2002)
20. G.J. Foschini, C.D. Pole, Statistical theory of polarization mode dispersion in single-mode fibers. IEEE/OSA J. Lightw. Technol. **LT-9**, 1439–1456 (1991)
21. G. Bendelli et al., Optical performance monitoring techniques, in Proc. ECOC 2000, Munich, vol. 4, pp. 213–216 (2000)
22. I.B. Djordjevic, B. Vasic, An advanced direct detection receiver model. J. Opt. Commun. **25**(1), 6–9 (Feb 2004)
23. P.J. Winzer, M. Pfennigbauer, M.M. Strasser, W.R. Leeb, Optimum filter bandwidths for optically preamplified NRZ receivers. J. Lightw. Technol. **19**, 1263–1272 (Sept 2001)
24. R. Hui et al., Technical Report: Advanced Optical Modulation Formats and Their Comparison in Fiber-Optic Systems., University of Kansas (Jan 2004)
25. V. Wongpaibool, Improvement of fiber optic system performance by synchronous phase modulation and filtering at the transmitter. Virginia Polytechnic Institute and State University, Ph.D. Dissertation (Jan 2003)
26. G. Jacobsen, *Noise in Digital Optical Transmission Systems* (Artech House, Boston, 1994)
27. I. B. Djordjevic, Analysis and optimization of coherent optical systems with the phase-locked loop. University of Nis, Serbia, Ph.D. Dissertation (1999)
28. P.W. Hooijmans, *Coherent Optical System Design* (Wiley, Chichester, 1994)
29. S.J. Savory, Digital filters for coherent optical receivers. Opt. Express **16**, 804–817 (2008)
30. I.B. Djordjevic, L. Xu, T. Wang, PMD compensation in multilevel coded-modulation schemes with coherent detection using BLAST algorithm and iterative polarization cancellation. Opt. Express **16**(19), 14845–14852 (15 Sept 2008)
31. I.B. Djordjevic, L. Xu, T. Wang, PMD compensation in coded-modulation schemes with coherent detection using Alamouti-type polarization-time coding. Opt. Express **16**(18), 14163–14172 (9 Jan 2008)
32. I.B. Djordjevic, M. Arabaci, L. Minkov, Next generation FEC for high-capacity communication in optical transport networks. IEEE/OSA J. Lightw. Technol. **27**(16), 3518–3530 (15 Aug 2009) Invited paper
33. I.B. Djordjevic, On advanced FEC and coded modulation for ultra-high-speed optical transmission. IEEE Commun. Surv. Tutor. **18**(3), 1920–1951, third quarter 2016 (19 Aug 2016). https://doi.org/10.1109/COMST.2016.2536726
34. E. Ip, A.P.T. Lau, D.J.F. Barros, J.M. Kahn, Coherent detection in optical fiber systems. Opt. Express **16**, 753–791 (2008)
35. K. Kikuchi, Fundamentals of coherent optical fiber communications. J. Lightw. Technol. **34**(1), 157–179 (2016)
36. A. Goldsmith, *Wireless Communications* (Cambridge University Press, Cambridge, 2005)
37. D. Tse, P. Viswanath, *Fundamentals of Wireless Communication* (Cambridge University Press, Cambridge, 2005)

38. S. Benedetto, E. Biglieri, *Principles of Digital Transmission with Wireless Applications* (Kluwer Academic/Plenum Publishers, New York, 1999)
39. E. Biglieri, *Coding for Wireless Channels* (Springer, New York, 2005)
40. T.M. Duman, A. Ghrayeb, *Coding for MIMO Communication Systems* (Wiley, Chichester, 2007)
41. J.R. Hampton, *Introduction to MIMO Communications* (Cambridge University Press, Cambridge, 2014)
42. G.L. Stüber, *Principles of Mobile Communication*, 3rd edn. (Springer, New York, 2012)
43. W.C.Y. Lee, *Mobile Communications Engineering* (McGraw-Hill, 1982)
44. J.G. Proakis, *Digital Communications* (McGraw-Hill, Boston, 2001)
45. J.G. Proakis, D.G. Manolakis, *Digital Signal Processing: Principles, Algorithms, and Applications*, 4th edn. (Prentice-Hall, Upper Saddle River, 2007)
46. S. Haykin, B. Van Veen, *Signals and Systems*, 2nd edn. (Wiley, 2003)
47. A.V. Oppenheim, R.W. Schafer, *Discrete-Time Signal Processing*, 3rd edn. (Pearson Higher Ed, 2010)
48. I. B. Djordjevic, ECE 429/529 Digital Signal Processing (Lecture Notes), University of Arizona, 2008
49. I. B. Djordjevic, ECE 340 Engineering Systems Analysis (Lecture Notes), University of Arizona (2012)
50. S. Verdú, *Multiuser Detection* (Cambridge University Press, New York, 1998)
51. C.A. Balanis, *Antenna Theory: Analysis and Design*, 4th edn. (Wiley, Hoboken, 2016)

Capacities of Wireless and Optical Channels

4

Abstract

This chapter is devoted to wireless channel and optical channel capacities. The chapter starts with definitions of mutual information and channel capacity, followed by the information capacity theorem. We discuss the channel capacity of discrete memoryless channels, continuous channels, and channels with memory. Regarding the wireless channels, we describe how to calculate the channel capacity of flat-fading and frequency-selective channels. We also discuss different optimum and suboptimum strategies to achieve channel capacity including the water-filling method, multiplexed coding and decoding, channel inversion, and truncated channel inversion. We also study different strategies for channel capacity calculation depending on what is known about the channel state information. Further, we explain how to model the channel with memory and describe the McMillan-Khinchin model for channel capacity evaluation. We then describe how the forward recursion of the BCJR algorithm can be used to evaluate the information capacity of channels with memory. The next topic is related to the evaluation of the information capacity of fiber-optic communication channels with coherent optical detection. Finally, we describe how to evaluate the capacity of hybrid free-space optical (FSO)-RF channels, in which FSO and RF subsystems cooperate to compensate for shortcoming of each other. The set of problems is provided after the concluding remarks.

4.1 Mutual Information, Channel Capacity, and Information Capacity Theorem

4.1.1 Mutual Information and Information Capacity

Figure 4.1 shows an example of a discrete memoryless channel (DMC), which is characterized by channel (transition) probabilities. If $X = \{x_0, x_1, \ldots, x_{I-1}\}$ and $Y = \{y_0, y_1, \ldots, y_{J-1}\}$ denote the channel input alphabet and the channel output alphabet, respectively, the channel is completely characterized by the following set of transition probabilities:

$$p(y_j|x_i) = P(Y = y_j|X = x_i), \; 0 \le p(y_j|x_i) \le 1, \tag{4.1}$$

where $i \in \{0, 1, \ldots, I-1\}$, $j \in \{0, 1, \ldots, J-1\}$, while I and J denote the sizes of input and output alphabets, respectively. The transition probability $p(y_j|x_i)$ represents the conditional probability that $Y = y_j$ for given input $X = x_i$.

One of the most important characteristics of the transmission channel is the *information capacity*, which is obtained by maximization of mutual information $I(\boldsymbol{X},\boldsymbol{Y})$ over all possible input distributions:

$$C = \max_{\{p(x_i)\}} I(\boldsymbol{X},\boldsymbol{Y}), \; I(\boldsymbol{X},\boldsymbol{Y}) = H(\boldsymbol{X}) - H(\boldsymbol{X}|\boldsymbol{Y}), \tag{4.2}$$

where $H(\mathbf{U}) = -\langle \log_2 P(\boldsymbol{U}) \rangle$ denotes the entropy of a random variable $\boldsymbol{U}$. The mutual information can be determined as

I. B. Djordjevic, *Advanced Optical and Wireless Communications Systems*, https://doi.org/10.1007/978-3-030-98491-5_4

Fig. 4.1 Discrete memoryless channel (DMC)

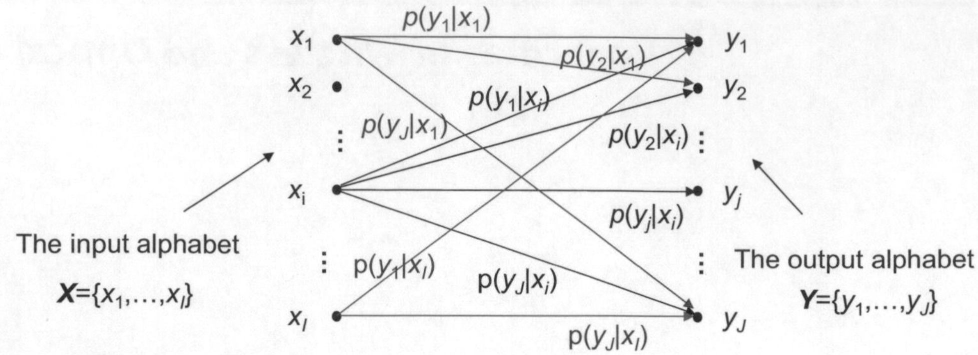

$$
\begin{aligned}
I(X;Y) &= H(X) - H(X|Y) \\
&= \sum_{i=1}^{M} p(x_i) \log_2\left[\frac{1}{p(x_i)}\right] - \sum_{j=1}^{N} p(y_j) \sum_{i=1}^{M} p(x_i|y_j) \log_2\left[\frac{1}{p(x_i|y_j)}\right].
\end{aligned}
\tag{4.3}
$$

In the above equation, $H(X)$ represents the uncertainty about the channel input before observing the channel output, also known as *entropy*, while $H(X|Y)$ denotes the conditional entropy or the amount of uncertainty remaining about the channel input after the channel output has been received. (The log function from the above equation relates to the base 2, and it will be like that throughout this chapter.) Therefore, the mutual information represents the amount of information (per symbol) that is conveyed by the channel, which represents the uncertainty about the channel input that is resolved by observing the channel output. The mutual information can be interpreted by means of Venn diagram [1–9] shown in Fig. 4.2(a). The left and right circles represent the entropy of the channel input and channel output, respectively, while the mutual information is obtained as intersection area of these two circles. Another interpretation is illustrated in Fig. 4.2(b) [8]. The mutual information, i.e., the information conveyed by the channel, is obtained as the output information minus information lost in the channel.

Since for M-ary input and M-ary output symmetric channel (MSC), we have that $p(y_j|x_i) = P_s/(M-1)$ and $p(y_j|x_j) = 1 - P_s$, where P_s is the symbol error probability, the channel capacity, in bits/symbol, can be found as

$$
C = \log_2 M + (1 - P_s)\log_2(1 - P_s) + P_s \log_2\left(\frac{P_s}{M-1}\right). \tag{4.4}
$$

The channel capacity represents an important bound on data rates achievable by any modulation and coding schemes. It can also be used in comparison of different coded modulation schemes in terms of their distance to the maximum channel capacity curve.

Now we have built enough knowledge to formulate a very important theorem, the *channel coding theorem* [1–9], which can be formulated as follows. Let a discrete memoryless source with an alphabet S have the entropy $H(S)$ and emit the symbols every T_s seconds. Let a DMC have capacity C and be used once in T_c seconds. Then, if

$$
H(S)/T_s \leq C/T_c,
$$

there exists a coding scheme for which the source output can be transmitted over the channel and reconstructed with an arbitrary small probability of error. The parameter $H(S)/T_s$ is related to the average information rate, while the parameter C/T_c is related to the channel capacity per unit time.

For binary symmetric channel ($N = M = 2$), the inequality is reduced down to $R \leq C$, where R is the code rate. Since the proof of this theorem can be found in any textbook on information theory, such as [5–8, 14], the proof of this theorem will be omitted.

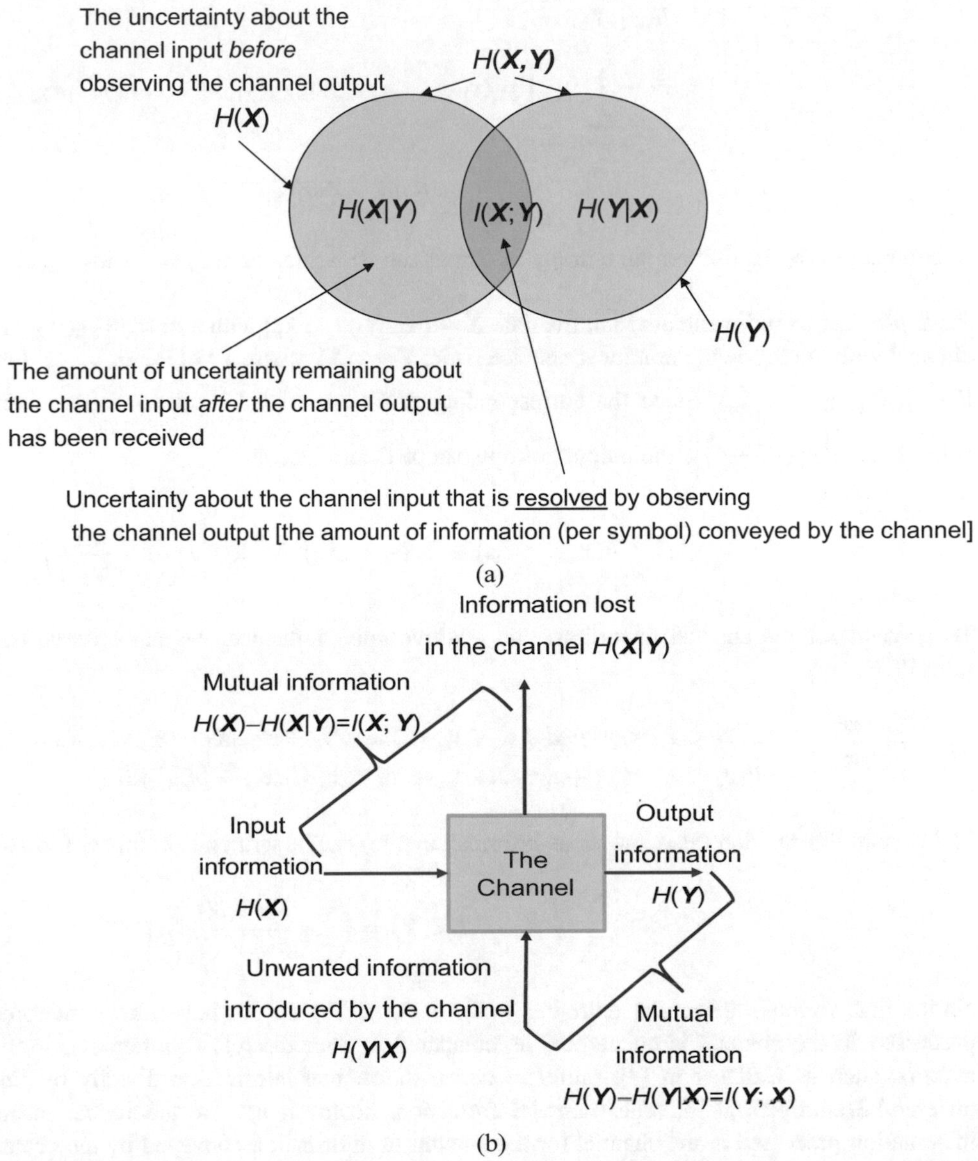

Fig. 4.2 Interpretation of the mutual information by using (**a**) Venn diagrams and (**b**) the approach due to Ingels

4.1.2 Capacity of Continuous Channels

In this section, we will discuss the channel capacity of continuous channels. Let $X = [X_1, X_2, \ldots, X_n]$ denote an n-dimensional multivariate, with a PDF $p_1(x_1, x_2, \ldots, x_n)$, representing the channel input. The corresponding *differential entropy* is defined by [2, 3, 5, 6].

$$h(X_1, X_2, \ldots, X_n) = -\underbrace{\int_{-\infty}^{\infty} \cdots \int_{-\infty}^{\infty}}_{n} p_1(x_1, x_2, \ldots, x_n) \log p_1(x_1, x_2, \ldots, x_n) dx_1 dx_2 \ldots dx_n = \langle -\log p_1(x_1, x_2, \ldots, x_n) \rangle, \quad (4.5)$$

where we use $\langle \cdot \rangle$ to denote the expectation operator. In order to simplify explanations, we will use the compact form of Eq. (4.5), namely, $h(X) = \langle -\log p_1(X) \rangle$, which was introduced in [6]. In a similar fashion, the channel output can be represented as m-dimensional random variable $Y = [Y_1, Y_2, \ldots, Y_m]$ with a PDF $p_2(y_1, y_2, \ldots, y_m)$, while the corresponding differential entropy is defined by

$$
\begin{aligned}
&h(Y_1, Y_2, \ldots, Y_m) \\
&= -\underbrace{\int_{-\infty}^{\infty} \cdots \int_{-\infty}^{\infty}}_{m} p_2(y_1, y_2, \ldots, y_m) \log p_1(y_1, y_2, \ldots, y_m) dy_1 dy_2 \ldots dy_m \\
&= \langle -\log p_2(y_1, y_2, \ldots, y_m) \rangle.
\end{aligned}
\tag{4.6}
$$

In compact form, the differential entropy of output can be written as $h(\boldsymbol{Y}) = \langle -\log p_1(\boldsymbol{Y}) \rangle$.

Example Let an n-dimensional multivariate $\boldsymbol{X} = [X_1, X_2, \ldots, X_n]$ with a PDF $p_1(x_1, x_2, \ldots, x_n)$ be applied to the nonlinear channel with the following nonlinear characteristic: $\boldsymbol{Y} = g(\boldsymbol{X})$, where $\boldsymbol{Y} = [Y_1, Y_2, \ldots, Y_n]$ represents the channel output with PDF $p_2(y_1, y_2, \ldots, y_n)$. Since the corresponding PDFs are related by the Jacobian symbol as follows, $p_2(y_1, \cdots, y_n) = p_1(x_1, \cdots, x_n) \left| J\left(\frac{X_1, \cdots, X_n}{Y_1, \cdots, Y_m}\right) \right|$, the output entropy can be determined as

$$
h(Y_1, \cdots, Y_m) \cong h(X_1, \cdots, X_n) - \left\langle \log \left| J\left(\frac{X_1, \cdots, X_n}{Y_1, \cdots, Y_m}\right) \right| \right\rangle.
$$

To account for the channel distortions and additive noise influence, we can observe the corresponding conditional and joint PDFs:

$$
\begin{aligned}
&P(y_1 < Y_1 < y_1 + dy_1, \ldots, y_m < Y_m < y_m + dy_m | X_1 = x_1, \ldots, X_n = x_n) = p(\widetilde{y}|\widetilde{x}) d\widetilde{y} \\
&P(y_1 < Y_1 < y_1 + dy_1, \ldots, x_n < X_n < x_n + dx_n) = p(\widetilde{x}, \widetilde{y}) d\widetilde{x} d\widetilde{y}.
\end{aligned}
\tag{4.7}
$$

The mutual information (also known as information rate) can be written in compact form as follows [6]:

$$
I(\boldsymbol{X}; \boldsymbol{Y}) = \left\langle \log \frac{p(\boldsymbol{X}, \boldsymbol{Y})}{p(\boldsymbol{X}) P(\boldsymbol{Y})} \right\rangle. \tag{4.8}
$$

Notice that various differential entropies $h(\boldsymbol{X})$, $h(\boldsymbol{Y})$, $h(\boldsymbol{Y}|\boldsymbol{X})$ do not have direct interpretation as far as the information processed in the channel is concerned, as compared to their discrete counterparts, from the previous subsection. Some authors, such as Gallager in [7], prefer to define the mutual information directly by Eq. (4.8), without considering the differential entropies at all. The mutual information, however, has the theoretical meaning and represents the average information processed in the channel (or the amount of information conveyed by the channel). The mutual information has the following important properties [5–8]: (i) it is symmetric, $I(X;Y) = I(Y;X)$; (ii) it is nonnegative; (iii) it is finite; (iv) it is invariant under linear transformation; (v) it can be expressed in terms of the differential entropy of channel output by $I(X;Y) = h(Y) - h(Y|X)$; and (vi) it is related to the channel input differential entropy by $I(X;Y) = h(X) - h(X|Y)$.

The *information capacity* can be obtained by maximization of Eq. (4.8) under all possible input distributions, which is

$$
C = \max I(\boldsymbol{X}; \boldsymbol{Y}). \tag{4.9}
$$

Let us now determine the mutual information of two random vectors, $\boldsymbol{X} = [X_1, X_2, \ldots, X_n]$ and $\boldsymbol{Y} = [Y_1, Y_2, \ldots, Y_m]$, which are normally distributed. Let $\boldsymbol{Z} = [\boldsymbol{X};\boldsymbol{Y}]$ be the random vector describing the joint behavior. Without loss of generality, we further assume that $\overline{X}_k = 0\ \forall k$ and $\overline{Y}_k = 0\ \forall k$. The corresponding PDF for $\boldsymbol{X}$, $\boldsymbol{Y}$, and Z are, respectively, given as [6].

$$
\begin{aligned}
p_1(\boldsymbol{x}) &= \frac{1}{(2\pi)^{n/2} (\det \boldsymbol{A})^{1/2}} \exp\left(-0.5\left(\boldsymbol{A}^{-1}\boldsymbol{x}, \boldsymbol{x}\right)\right), \\
\boldsymbol{A} &= \left[a_{ij}\right], a_{ij} = \int x_i x_j p_1(\boldsymbol{x}) d\boldsymbol{x},
\end{aligned}
\tag{4.10}
$$

$$\begin{aligned} p_2(\boldsymbol{y}) &= \frac{1}{(2\pi)^{n/2}(\det\boldsymbol{B})^{1/2}}\exp\left(-0.5\left(\boldsymbol{B}^{-1}\boldsymbol{y},\boldsymbol{y}\right)\right), \\ \boldsymbol{B} &= \left[b_{ij}\right], b_{ij} = \int y_i y_j p_2(\boldsymbol{y})d\boldsymbol{y}, \end{aligned} \tag{4.11}$$

$$\begin{aligned} p_3(\boldsymbol{z}) &= \frac{1}{(2\pi)^{(n+m)/2}(\det\boldsymbol{C})^{1/2}}\exp\left(-0.5\left(\boldsymbol{C}^{-1}\boldsymbol{z},\boldsymbol{z}\right)\right), \\ \boldsymbol{C} &= \left[c_{ij}\right], c_{ij} = \int z_i z_j p_3(\boldsymbol{z})d\boldsymbol{z}, \end{aligned} \tag{4.12}$$

where (·,·) denotes the dot product of two vectors. By substitution of Eqs. (4.10, 4.11, and 4.12) into Eq. (4.8), we obtain [6]

$$I(\boldsymbol{X};\boldsymbol{Y}) = \frac{1}{2}\log\frac{\det\boldsymbol{A}\det\boldsymbol{B}}{\det\boldsymbol{C}}. \tag{4.13}$$

The mutual information between two Gaussian random vectors can also be expressed in terms of their correlation coefficients [6]:

$$I(\boldsymbol{X};\boldsymbol{Y}) = -\frac{1}{2}\log\left[\left(1-\rho_1^2\right)\ldots\left(1-\rho_l^2\right)\right],\ l = \min(m,n), \tag{4.14}$$

where ρ_j is the correlation coefficient between X_j and Y_j.

In order to obtain the information capacity for additive Gaussian noise, we make the following assumptions: (i) the input X, output Y, and noise Z are n-dimensional random variables; (ii) $\overline{X}_k = 0, \overline{X_k^2} = \sigma_{x_k}^2\ \forall k$ and $\overline{Z}_k = 0,\ \overline{Z_k^2} = \sigma_{z_k}^2 \forall k$; and (iii) the noise is additive, $Y = X + Z$. Since we have that

$$p_{\boldsymbol{x}}(\boldsymbol{y}|\boldsymbol{x}) = p_{\boldsymbol{x}}(\boldsymbol{x}+\boldsymbol{z}|\boldsymbol{x}) = \prod_{k=1}^{n}\left[\frac{1}{(2\pi)^{1/2}\sigma_{z_k}}e^{-z_k^2/2\sigma_{z_k}^2}\right] = p(\boldsymbol{z}), \tag{4.15}$$

the conditional differential entropy can be obtained as

$$H(\boldsymbol{Y}|\boldsymbol{X}) = H(\boldsymbol{Z}) = -\int_{-\infty}^{\infty} p(\boldsymbol{z})\log p(\boldsymbol{z})d\boldsymbol{z}. \tag{4.16}$$

The mutual information is then

$$I(\boldsymbol{X};\boldsymbol{Y}) = h(\boldsymbol{Y}) - h(\boldsymbol{Y}|\boldsymbol{X}) = h(\boldsymbol{Y}) - h(\boldsymbol{Z}) = h(\boldsymbol{Y}) - \frac{1}{2}\sum_{k=1}^{n}\log 2\pi e\sigma_{z_k}^2. \tag{4.17}$$

The information capacity, expressed in bits per channel use, is therefore obtained by maximizing $h(\boldsymbol{Y})$. Because the distribution maximizing the differential entropy is Gaussian, the information capacity is obtained as

$$\begin{aligned} C(\boldsymbol{X};\boldsymbol{Y}) &= \frac{1}{2}\sum_{k=1}^{n}\log 2\pi e\sigma_{y_k}^2 - \frac{1}{2}\sum_{k=1}^{n}\log 2\pi e\sigma_{z_k}^2 = \frac{1}{2}\sum_{k=1}^{n}\log\left(\frac{\sigma_{y_k}^2}{\sigma_{z_k}^2}\right) \\ &= \frac{1}{2}\sum_{k=1}^{n}\log\left(\frac{\sigma_{x_k}^2+\sigma_{z_k}^2}{\sigma_{z_k}^2}\right) = \frac{1}{2}\sum_{k=1}^{n}\log\left(1+\frac{\sigma_x^2}{\sigma_z^2}\right). \end{aligned} \tag{4.18}$$

For $\sigma_{x_k}^2 = \sigma_x^2, \sigma_{z_k}^2 = \sigma_z^2$, we obtain the following expression for information capacity:

$$C(X;Y) = \frac{n}{2}\log\left(1 + \frac{\sigma_x^2}{\sigma_z^2}\right), \tag{4.19}$$

where σ_x^2/σ_z^2 presents the signal-to-noise ratio (SNR). The expression above represents the maximum amount of information that can be transmitted per symbol.

From a practical point of view, it is important to determine the amount of information conveyed by the channel per second, which is the information capacity per unit time, also known as the *channel capacity*. For bandwidth-limited channels and Nyquist signaling employed, there will be 2 W samples per second (W is the channel bandwidth), and the corresponding channel capacity becomes

$$C = W\log\left(1 + \frac{P}{N_o W}\right)\ [\text{bits/s}], \tag{4.20}$$

where P is the average transmitted power and $N_0/2$ is the noise power spectral density (PSD). Equation (4.20) represents the well-known *information capacity theorem*, commonly referred to as the Shannon third theorem [17].

Since the Gaussian source has the maximum entropy clearly, it will maximize the mutual information. Therefore, the equation above can be derived as follows. Let the n-dimensional multivariate $\boldsymbol{X} = [X_1, \ldots, X_n]$ represent the Gaussian channel input with samples generated from zero-mean Gaussian distribution with variance σ_x^2. Let the n-dimensional multivariate $\boldsymbol{Y} = [Y_1, \ldots, Y_n]$ represent the Gaussian channel output, with samples spaced 1/2 W apart. The channel is additive with noise samples generated from zero-mean Gaussian distribution with variance σ_z^2. Let the PDFs of input and output be denoted by $p_1(\boldsymbol{x})$ and $p_2(\boldsymbol{y})$, respectively. Finally, let the joint PDF of input and output of channel be denoted by $p(\boldsymbol{x}, \boldsymbol{y})$. The maximum mutual information can be calculated from

$$I(\boldsymbol{X};\boldsymbol{Y}) = \iint p(\boldsymbol{x},\boldsymbol{y})\log\frac{p(\boldsymbol{x},\boldsymbol{y})}{p(\boldsymbol{x})P(\boldsymbol{y})}d\boldsymbol{x}d\boldsymbol{y}. \tag{4.21}$$

By following the similar procedure to that used in Eqs. (4.15, 4.16, 4.17, 4.18, and 4.19), we obtain the following expression for Gaussian channel capacity:

$$C = W\log\left(1 + \frac{P}{N}\right), N = N_o W, \tag{4.22}$$

where P is the average signal power and N is the average noise power.

4.2 Capacity of Flat-Fading and Frequency-Selective Wireless Fading Channels

In this section, we study both flat-fading and frequency-selective fading channels [10–16] and describe how to calculate the corresponding channel capacities [17] as well as approaches to achieve these channel capacities.

4.2.1 Flat-Fading Channel Capacity

The typical wireless communication system together with an equivalent channel model is shown in Fig. 4.3. The channel power gain $g[i]$, where i is the discrete-time instance, is related to the channel coefficient $h[i]$ by $g[i] = |h[i]|^2$ and follows a given probability density function (PDF) $f(g)$. For instance, as shown in Chap. 2, for Rayleigh fading, the PDF $f(g)$ is exponential. The channel gain is commonly referred to as *channel side information* or *channel state information* (CSI). The samples of channel gain $g[i]$ change at each time instance i and could be generated from an independent identically distributed (i.i.d.) process or could be generated from a correlated source. In a block-fading channel, considered in this section, the channel gain $g[i]$ is constant over some block length T, and once this block length is over, a new realization of channel gain $g[i]$ is generated based on PDF $f(g)$.

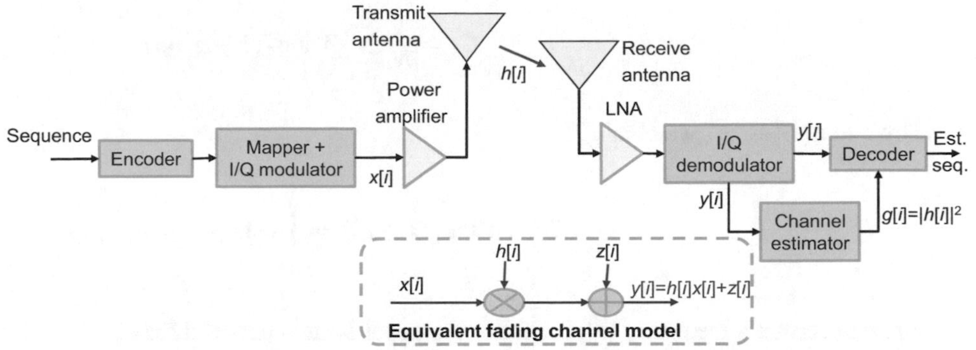

Fig. 4.3 Typical wireless communication system together with an equivalent fading channel model

The channel capacity is strongly dependent on what is known about CSI at the transmitter and/or receiver sides. In this section, three scenarios of interest will be studied: (i) *channel distribution information* (CDI), (ii) CSI is available at the receiver side (CSIR), and (iii) the CSI is available both at the transmitter and receiver sides (full CSI). In CDI scenario, the PDF of g is known to both transmitter and receiver sides. Determination of the capacity-achieving input distribution and corresponding capacity of fading channels under CDI scenario for any distribution $f(g)$, in closed form, is still an open problem, except from specific fading channel models such as i.i.d. Rayleigh fading channels and FSMCs. Other two scenarios, CSIR and full CSI, are discussed with more details below.

4.2.1.1 Channel Side Information at Receiver

Two channel capacity definitions in CSIR scenario are of interest in the system design: (i) *Shannon capacity*, also known as the *ergodic capacity*, which represents the maximum data rate that can be sent over the channel with symbol error probability P_s tending to zero; and (ii) *capacity with outage*, which represents the maximum data rate that can be transmitted over a channel with some outage probability P_{outage}, defined as the probability that SNR falls below a given SNR threshold value ρ_{tsh} corresponding to the maximum tolerable symbol error probability P_s. Clearly, in the outage, the transmission cannot be completed with negligible symbol error probability. The capacity with outage corresponds to the highest data rate that can be transmitted over the fading channel reliably except when the fading channel is in deep fade. To determine either the average symbol error probability or outage probability in the presence of fading, the distribution of SNR, $f(\rho)$, is needed. The *average symbol error probability* is then defined as

$$\overline{P}_s = \langle P_s(\gamma) \rangle = \int_0^{\infty} P_s(\rho) f(\rho) d\rho, \tag{4.23}$$

where $P_s(\rho)$ is the SNR-dependent expression for symbol error probability, for a given modulation format. The *outage probability* is defined as

$$P_{\text{outage}} = P(\rho \le \rho_{\text{tsh}}) = \int_0^{\rho_{\text{tsh}}} f(\rho) d\rho, \tag{4.24}$$

where ρ_{tsh} is the threshold SNR corresponding to the maximum tolerable symbol error probability.

Shannon (Ergodic) Capacity

Under the assumptions that the channel gain $g[i]$ originates from the flat fading, average power gain is equal to 1, and receiver knows CSI (CSIR), we study the channel capacity calculation. When fading is fast, with a certain decoding delay requirement, we transmit the signal over the time duration that contains N coherence time periods T_c, namely, NT_c, wherein $N >> 1$. In this scenario, since the transmission is sufficiently long and the fading is fast, all possible channel gain realizations $g[i]$ come to the "picture" and the channel capacity can be calculated by averaging out channel capacities for different realizations. For nth coherence time period, we assume that channel gain is constant equal to $g[n]$, so the received SNR can be estimated as $g[n]\overline{\rho}$. The corresponding channel capacity, related to nth gain realization, will be $C[n] = W \log_2(1 + g[n]\overline{\rho})$, where W is the channel bandwidth. The Shannon (ergodic) channel capacity can be obtained by averaging over N coherence time periods as follows:

$$C_N = \frac{1}{N}\sum_{n=1}^{N} W\log_2(1+g[n]\overline{\rho}). \tag{4.25a}$$

Now by letting $N \to \infty$, we obtain

$$C = \langle W\log_2(1+\rho)\rangle = \int_0^\infty W\log_2(1+\rho)f(\rho)d\rho. \tag{4.25b}$$

By applying Jensen's inequality [5], the following upper limit is obtained for C:

$$C = \langle W\log_2(1+\rho)\rangle \leq W\log_2(1+\langle\rho\rangle) = W\log_2(1+\overline{\rho}). \tag{4.26}$$

Channel Capacity with Outage

Capacity with outage is related to the *slowly varying fading channels*, where the instantaneous SNR ρ is constant for the duration of transmission of a large number of symbols, which is also known as *a transmission burst*, and changes to a new SNR value at the end of the burst period according to the given fading distribution. When the received SNR ρ is constant for the duration of the transmission burst, then the maximum possible transmission rate will be $W\log_2(1+\rho)$, with symbol error probability tending to zero. Because the transmitter does not know the exact value of SNR ρ, it will be forced to transmit at the fixed data rate corresponding to the minimum tolerable SNR ρ_{min}. The maximum data rate corresponding to SNR ρ_{min} will be $C = W\log_2(1+\rho_{min})$. As indicated earlier, the probability of outage is defined as $P_{outage} = P(\rho < \rho_{min})$. The average data rate, reliably transmitted over many transmission bursts, can be defined as

$$C_{outage} = (1 - P_{outage})W\log_2(1+\rho_{min}), \tag{4.27}$$

given that data is reliably transmitted in $1 - P_{outage}$ transmission bursts. In other words, this is the transmission rate that can be supported in $100(1 - P_{outage})\%$ of the channel realizations.

4.2.1.2 Full CSI

In full CSI scenario, both the transmitter and the receiver have CSI available, and the transmitter can adapt its transmission strategy based on this CSI. Depending on the type of fading channel, fast- or slow-varying, we can define either the Shannon capacity or capacity with outage.

Shannon Capacity

Let S denote the set of discrete memoryless wireless channels and let $p(s)$ denote the probability that the channel is in state $s \in S$. In other words, $p(s)$ is the fraction of the time for which the channel was in state s. The channel capacity, as defined by Wolfowitz [14], of the time-varying channel, in full CSI scenario (when both the transmitter and the receiver have the CSI available), is obtained by averaging the channel capacity when the channel is in state s, denoted as C_s, as follows:

$$C = \sum\nolimits_{s\in S} C_s p(s). \tag{4.28}$$

By letting the cardinality of set S to tend to infinity, $|S| \to \infty$, the summation becomes integration and $p(s)$ becomes the PDF of ρ, so that we can write

$$C = \int_0^\infty C(\rho)f(\rho)d\gamma = \int_0^\infty W\log_2(1+\rho)f(\rho)d\rho, \tag{4.29}$$

which is the same as in the CSIR case. Therefore, the channel capacity does not increase in full CSI case unless some form of the *power adaptation* is employed, which is illustrated in Fig. 4.4.

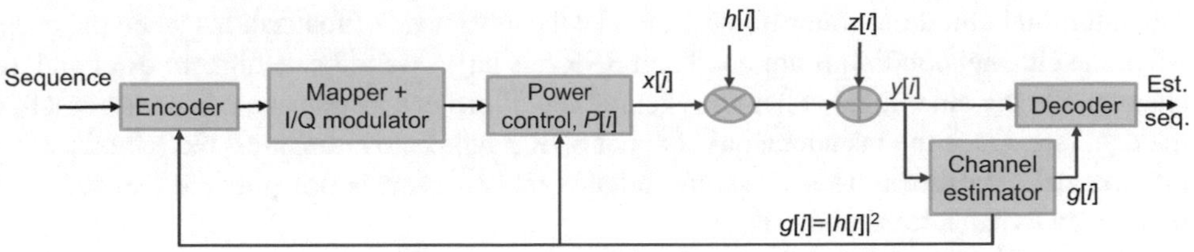

Fig. 4.4 The wireless communication system employing the power adaptation

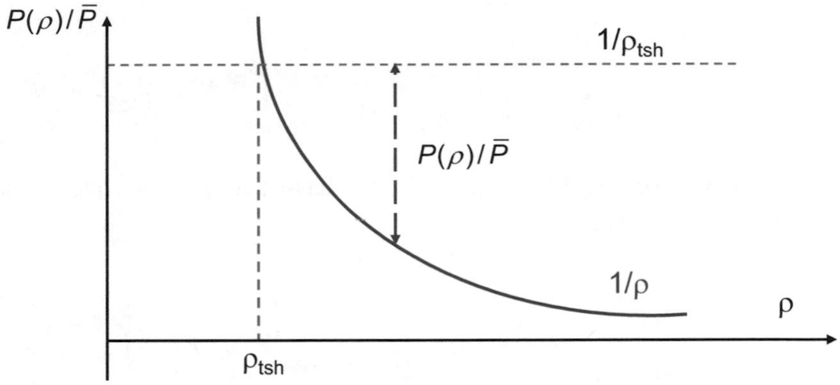

Fig. 4.5 The illustration of the optimum power allocation strategy, known as the water-filling method

The fading channel capacity, with power adaptation, is an optimization problem:

$$C = \max_{P(\rho):\int_0^\infty P(\rho)f(\rho)d\rho = \overline{P}} \int_0^\infty W\log_2\left(1+\frac{P(\rho)\rho}{\overline{P}}\right)f(\rho)d\rho, \quad \rho = \frac{\overline{P}g}{N_0 W}. \tag{4.30}$$

The optimum solution can be obtained by employing the Lagrangian method, wherein the Lagrangian $\mathcal{L}$ is defined as

$$\mathcal{L}(P(\rho)) = \int_0^\infty W\log_2\left(1+\frac{P(\rho)\rho}{\overline{P}}\right)f(\rho)d\rho - \mu\int_0^\infty P(\rho)f(\rho)d\rho, \tag{4.31}$$

which is subject to the following constraint:

$$\int_0^\infty P(\rho)f(\rho)d\rho \leq \overline{P}. \tag{4.32}$$

By finding the first derivative of Lagrangian with respect to $P(\rho)$ and setting it to zero, we obtain

$$\frac{\partial\mathcal{L}(P(\rho))}{\partial P(\rho)} = \left[\frac{W/\ln 2}{1+\rho P(\rho)/\overline{P}}\frac{\rho}{\overline{P}} - \mu\right]f(\rho) = 0. \tag{4.33}$$

By solving for $P(\rho)$ from (4.33), we obtain the optimum power allocation policy, also known as the *water-filling method*, which is illustrated in Fig. 4.5, as follows:

$$\frac{P(\rho)}{\overline{P}} = \begin{cases} 1/\rho_{tsh} - 1/\rho, & \rho \geq \rho_{tsh} \\ 0, & \rho < \rho_{tsh} \end{cases}. \tag{4.34}$$

The amount of power allocated on a transmitter side for a given SNR ρ is equal to $1/\rho_{tsh} - 1/\rho$. In other words, the amount of allocated power is proportional to the "water" filled between the bottom of the bowl ($1/\rho$) and the constant water line ($1/\rho_{tsh}$).

The intuition behind the water filling is to take the advantage of realizations when the channel conditions are good. Namely, when the channel conditions are good and SNR ρ is large, we allocate higher power and transmit at higher data rate over the channel. On the other hand, when the channel conditions deteriorate and ρ becomes small, we allocate less power and reduce the data rate. Once the instantaneous channel SNR ρ falls below the threshold (cutoff) value, we do not transmit at all. When the channel conditions are so bad, the reliable transmission is not possible and the average symbol error rate would be dominated by such transmissions.

After substituting Eq. (4.34) into (4.30), we obtain the following expression for the channel capacity corresponding to the water-filling method:

$$C = \int_{\rho_{tsh}}^{\infty} W \log_2\left(\frac{\rho}{\rho_{tsh}}\right) f(\rho) d\rho. \tag{4.35}$$

After substituting Eq. (4.34) into (4.32), we obtain the following equation that needs to be solved numerically to determine the threshold SNR, ρ_{tsh}:

$$\int_{\rho_{tsh}}^{\infty} \left(\frac{1}{\rho_{tsh}} - \frac{1}{\rho}\right) f(\rho) d\rho = 1. \tag{4.36}$$

Let us now consider the optimum allocation policy in discrete-time domain. The channel model is similar to that from Fig. 4.3:

$$y[n] = h[n]x[n] + z[n], \quad |h[n]|^2 = g[n], \tag{4.37}$$

where n is the discrete-time index. The optimization problem can be formulated as

$$\begin{aligned} &\max_{P_1, P_2, \cdots, P_N} \frac{1}{N} \sum_{n=1}^{N} \log_2\left(1 + \frac{P[n]|h[n]|^2}{N_0}\right) \\ &= \max_{P_1, P_2, \cdots, P_N} \frac{1}{N} \sum_{n=1}^{N} \log_2\left(1 + \frac{P[n]g[n]}{N_0}\right), \end{aligned} \tag{4.38}$$

which is subject to the following constraint:

$$\sum_{n=1}^{N} P[n] \leq P. \tag{4.39}$$

The optimum solution can again be determined by the Lagrangian method:

$$\begin{aligned} &\frac{P_{opt}[n]}{P} = \Big(\underbrace{\frac{1}{\mu P}}_{\frac{1}{\rho_{tsh}}} - \underbrace{\frac{N_0}{g[n]P}}_{\frac{1}{\rho}}\Big)^+ = \left(\frac{1}{\rho_{tsh}} - \frac{1}{\rho}\right)^+, \\ &(x)^+ = \max(0, x), \end{aligned} \tag{4.40}$$

and it is clearly the same allocation policy (the water-filling method) as given by Eq. (4.34), which is illustrated in Fig. 4.6. The threshold SNR equation can be obtained after substituting Eq. (4.40) into (4.39):

$$\frac{1}{N} \sum_{n=1}^{N} \left(\frac{1}{\rho_{tsh}} - \frac{1}{\rho}\right)^+ = 1. \tag{4.41}$$

As $N \to \infty$, summation becomes integral, and we obtain from Eq. (4.41) the following power allocation policy:

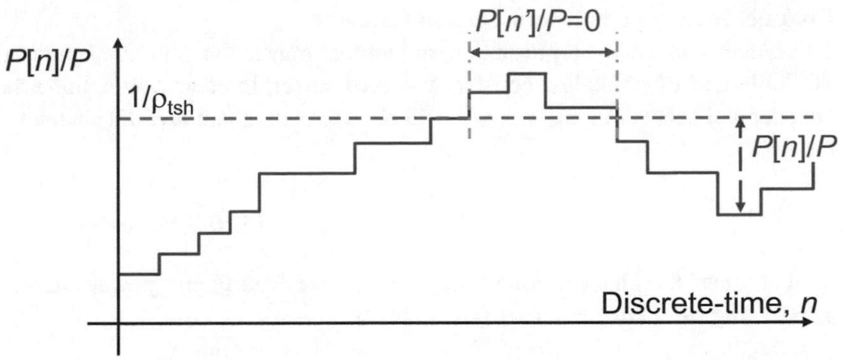

Fig. 4.6 The illustration of the optimum power allocation strategy in time domain

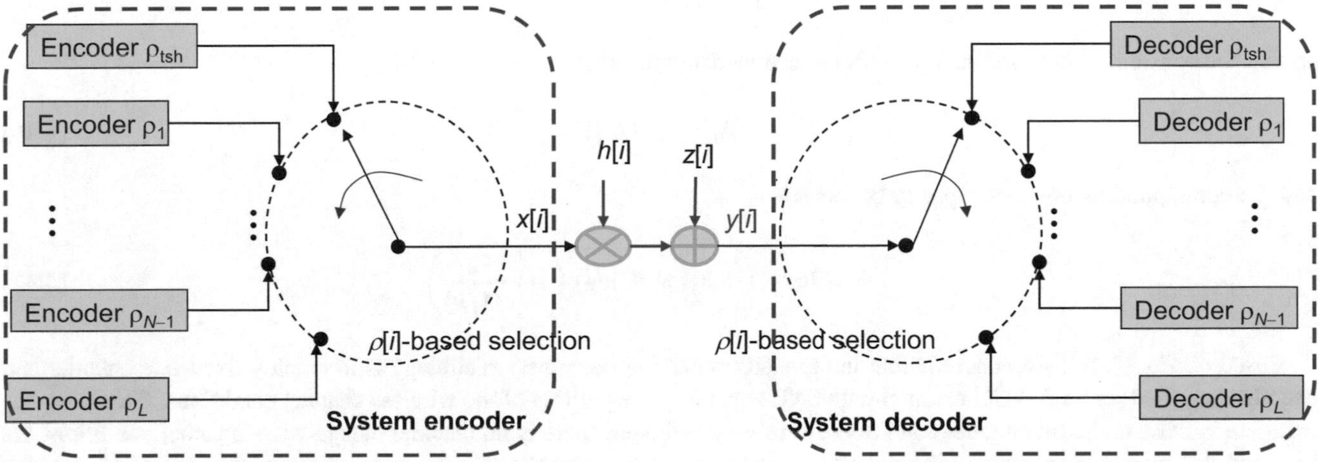

Fig. 4.7 The illustration of the multiplexed coding and decoding scheme

$$\left\langle \left(\frac{1}{\rho_{\text{tsh}}} - \frac{1}{\rho} \right)^{+} \right\rangle = 1, \tag{4.42}$$

which is identical to Eq. (4.34). The corresponding channel capacity for the optimum power adaptation strategy in time domain will be

$$C = \langle W \log_2 (1 + \frac{P_{\text{opt}}(g)}{P} \underbrace{\frac{Pg}{N_0}}_{\rho})\rangle_g = \langle W \log_2 (1 + \frac{P_{\text{opt}}(g)}{P} \rho)\rangle_g ,$$
$$\frac{P_{\text{opt}}(g)}{P} = \left(\frac{1}{\rho_{\text{tsh}}} - \frac{1}{\rho}\right)^{+}. \tag{4.43}$$

The channel capacity-achieving coding scheme, inspired by this method, is the *multiplexed coding and decoding scheme*, shown in Fig. 4.7.

The starting point is the quantization of the range of SNRs for different fading effects to a finite set $\{\rho_n \mid 1 \le n \le L\}$. For each ρ_n, we design an encoder/decoder pair for an AWGN channel with SNR ρ_n achieving the channel capacity. The output x_n of encoder ρ_n is transmitted with an average power $P(\rho_n)$, and the corresponding data rate is $R_n = C_n$, where C_n is the capacity of a time-invariant AWGN channel with received SNR $P(\rho_n)\ \rho_n/P$. These encoder/decoder pairs are selected according to the CSI, that is, SNR ρ_n. In other words, when $\rho[i] \cong \rho_n$, the corresponding pair of ports get connected through the wireless channel. Of course, we assume that CSI is perfect. Clearly, this multiplexed encoding and decoding scheme effectively replaces the time-varying channel with a set of time-invariant channels operating in parallel, wherein the nth channel is only used when $\rho[i] \cong \rho_n$. The average rate transmitted over the channel is determined by averaging, namely, $\sum_n C_n p_n$, where p_n is the probability of CSI being ρ_n—in other words, the percentage of time that the channel SNR was equal to ρ_n. Of course, the complexity and cost of this scheme are high. To reduce the system complexity and cost, the channel inversion can be used.

Channel Inversion and Zero-Outage Capacity

The *channel inversion* represents a suboptimal transmitter power adaptation strategy, where the transmitter employs the available CSI to maintain a constant received power; in other words, this scheme "inverts" the channel fading effects. The encoder and decoder in this scenario will see a time-invariant AWGN channel. The power loading is reversely proportional to the SNR:

$$P(\rho)/\overline{P} = \rho_0/\rho, \tag{4.44}$$

so that when the channel conditions are bad, we load higher power such that the receiver sees the constant power. By substituting Eq. (4.44) into (4.32), we obtain

$$\int (\rho_0/\rho) f(\rho) d\rho = 1, \tag{4.45}$$

so that the constant of proportionality (SNR) ρ_0 can be determined as

$$\rho_0 = 1/\langle 1/\rho \rangle, \tag{4.46}$$

and the corresponding channel capacity becomes

$$C = W \log_2(1+\rho_0) = W \log_2\left(1 + \frac{1}{\langle 1/\rho \rangle}\right). \tag{4.47}$$

Based on Eq. (4.47), we conclude that the capacity-achieving transmission strategy is to employ fixed-rate encoder and decoder designed for an AWGN channel with SNR equal to ρ_0, regardless of the wireless channel conditions. Given that the data rate is fixed under all channel conditions, even very bad ones, there is no channel outage when channel conditions are bad, and the corresponding channel capacity is called *zero-outage capacity*. The main drawback of this strategy is that the zero-outage capacity can exhibit a large data rate reduction, compared to the Shannon capacity in extremely bad fading conditions, when $\rho \to 0$ and subsequently $C \to 0$. To solve for this problem, the truncated channel inversion is used.

Truncated Channel Inversion and Outage Capacity

As explained above, the requirement to maintain a constant data rate regardless of the fading state results in inefficacy of this method, so that the corresponding zero-outage capacity can be significantly smaller than the Shannon capacity in a deep fade. When the fading state is very bad by not transmitting at all, we can ensure that data rates are higher in the other fading states and on such a way we significantly improve the channel capacity. The state when we do not transmit is an outage state, so that the corresponding capacity can be called the *outage capacity*, which is defined as the maximum data rate that can be kept constant, when there is no outage, times the probability of not having outage ($1 - P_{\text{outage}}$). Outage capacity is achieved with a so-called *truncated channel inversion policy* for power adaptation, in which we only transmit reversely proportional to SNR, when the SNR is larger than the threshold SNR ρ_{tsh}:

$$P(\rho)/\overline{P} = \begin{cases} \rho_0/\rho, & \rho \geq \rho_{\text{tsh}} \\ 0, & \rho < \rho_{\text{tsh}} \end{cases}. \tag{4.48}$$

The parameter ρ_0 can be determined as

$$\rho_0 = 1/\langle 1/\rho \rangle_{\rho_{\text{tsh}}}, \quad \langle 1/\rho \rangle_{\rho_{\text{tsh}}} = \int_{\rho_{\text{tsh}}}^{\infty} \frac{1}{\rho} f(\rho) d\rho. \tag{4.49}$$

For the outage probability, defined as $P_{\text{outage}} = P(\rho < \rho_{\text{tsh}})$, the corresponding channel capacity will be

$$C(P_{out}) = W\log_2\left(1 + \frac{1}{\langle 1/\rho\rangle_{\rho_{tsh}}}\right)P(\rho \geq \rho_{tsh}). \tag{4.40}$$

The channel capacity can be obtained by the following maximization with respect to ρ_{tsh}:

$$C = \max_{\rho_{tsh}} W\log_2\left(1 + \frac{1}{\langle 1/\rho\rangle_{\rho_{tsh}}}\right)(1 - P_{outage}). \tag{4.51}$$

4.2.2 Frequency-Selective Fading Channel Capacity

Two cases of interest are considered here: (i) time-invariant and (ii) time-variant channels.

4.2.2.1 Time-Invariant Channel Capacity

We consider here the time-invariant channel, as shown in Fig. 4.8, and we assume that a total transmit power cannot be larger than P. In full CSI scenario, the transfer function $H(f)$ is known to both the transmitter and the receiver. We first study the *block-fading* assumption for $H(f)$ so that the whole frequency band can be divided into subchannels each of bandwidth equal to W, wherein the transfer function is constant for each block i, namely, $H(f) = H_j$, which is illustrated in Fig. 4.9.

Therefore, we decomposed the frequency-selective fading channel into a set of the parallel AWGN channels with corresponding SNRs being $|H_i|^2P_i/(N_0W)$ on the ith channel, where P_i is the power allocated to the ith channel in this parallel set of subchannels, subject to the power constraint $\sum_i P_i \leq P$.

The *capacity* of this *parallel set of channels* is the sum of corresponding rates associated with each channel, wherein the power is optimally allocated over all subchannels:

$$C = \max_{P_i:\ \sum_i P_i \leq P} \sum W\log_2\left(1 + \underbrace{\frac{|H_i|^2P_i}{N_0W}}_{\rho_i}\right) = \max_{P_i:\ \sum_i P_i \leq P} \sum W\log_2(1 + \rho_i). \tag{4.52}$$

The optimum power allocation policy of the ith subchannel is the water filling:

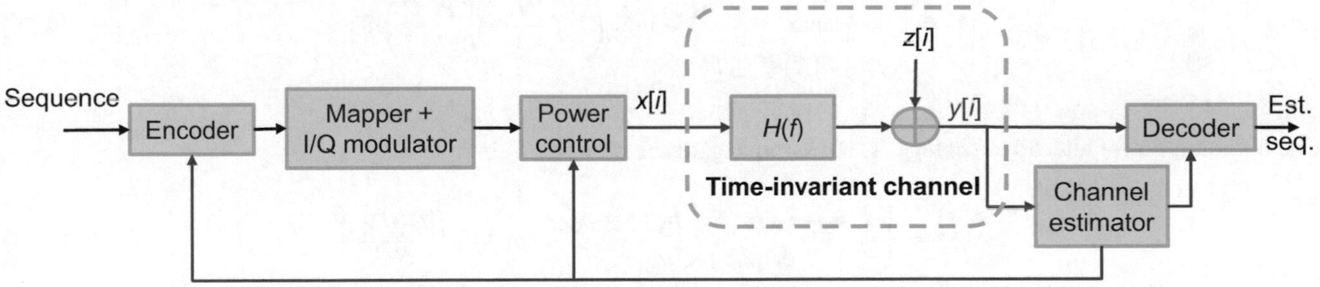

Fig. 4.8 The illustration of the wireless communication system operating over the time-invariant channel characterized by the transfer function $H(f)$

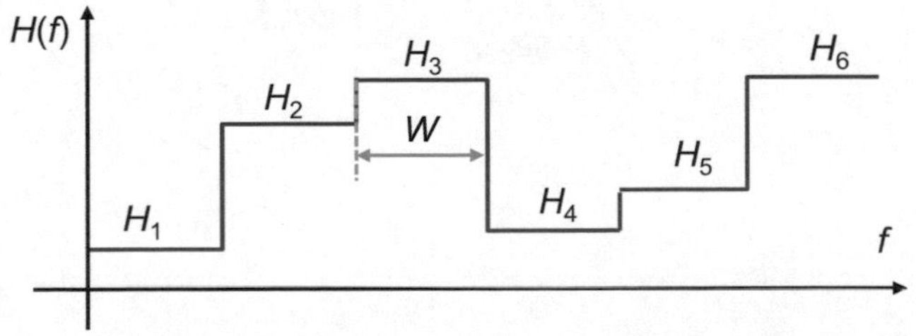

Fig. 4.9 The illustration of the block frequency-selective fading in which the channel is divided into subchannels, each of them having the bandwidth W

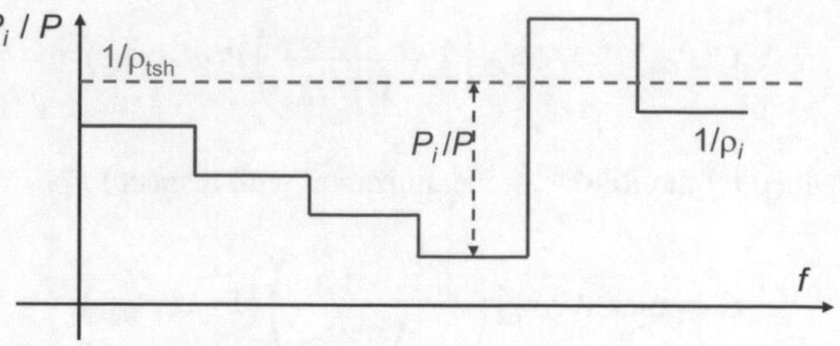

Fig. 4.10 The illustration of the water filling in the frequency-selective block-fading channels

$$\frac{P_i}{P} = \begin{cases} 1/\rho_{tsh} - 1/\rho_i, & \rho_i \geq \rho_{tsh} \\ 0, & \rho_i < \rho_{tsh} \end{cases}, \tag{4.53}$$

and the threshold SNR is obtained as the solution of the following equation:

$$\sum_i \left(\frac{1}{\rho_{tsh}} - \frac{1}{\rho_i} \right)^+ = 1. \tag{4.54}$$

The illustration of the water filling in the frequency-selective block-fading channels is provided in Fig. 4.10. The corresponding channel capacity of the optimum distribution, obtained by substituting Eq. (4.53) into (4.52), is obtained as follows:

$$C = \sum_{i:\rho_i \geq \rho_{tsh}} W \log_2 \left(\frac{\rho_i}{\rho_{tsh}} \right). \tag{4.55}$$

When the transfer function $H(f)$ is *continuous*, the capacity for the same power constraint P is similar to the case of the block-fading channel, but instead of summation, we need to perform the integration so that the corresponding channel capacity expression is given by

$$C = \max_{P(f):\ \int P(f) df \leq P} \int W \log_2 \left(1 + \frac{|H(f)|^2 P(f)}{N_0} \right) df. \tag{4.56}$$

The optimum power allocation strategy is, not surprising, again the water filling:

$$\frac{P(f)}{P} = \begin{cases} 1/\rho_{tsh} - 1/\rho(f), & \rho(f) \geq \rho_{tsh} \\ 0, & \rho(f) < \rho_{tsh} \end{cases}, \quad \rho(f) = \frac{|H(f)|^2 P}{N_0 W}. \tag{4.57}$$

After substituting Eq. (4.57) into (4.56), we obtain the following channel capacity for the optimum power allocation case:

$$C = \int_{f:\ \rho(f) \geq \rho_{tsh}} W \log_2 \left[\frac{\rho(f)}{\rho_{tsh}} \right] df. \tag{4.58}$$

Example Let us now consider the time-invariant frequency-selective block-fading channel with three subchannels of bandwidth 2 MHz, with frequency response amplitudes being 1, 2, and 3, respectively. For the transmit power 10 mW and noise power spectral density of $N_0 = 10^{-9}$ W/Hz, we are interested to determine the corresponding Shannon capacity.

The signal power-to-noise power is given by $\rho_0 = P/(N_0W) = 5$. The subchannels' SNRs can be expressed as $\rho_i = P|H_i|^2/(N_0W) = \rho_0|H_i|^2$. For $H_1 = 1$, $H_2 = 2$, and $H_3 = 3$, the corresponding subchannel SNRs are $\rho_1 = \rho_0|H_1|^2 = 5$, $\rho_2 = \rho_0|H_2|^2 = 20$, and $\rho_3 = \rho_0|H_3|^2 = 45$, respectively. Based on Eq. (4.54), the threshold SNR can be determined from the following equation:

$$\frac{3}{\rho_{\text{tsh}}} = 1 + \frac{1}{\rho_1} + \frac{1}{\rho_2} + \frac{1}{\rho_3} = 1 + \frac{1}{5} + \frac{1}{20} + \frac{1}{45},$$

as follows: $\rho_{\text{tsh}} = 2.358 < \rho_i$ for very i. The corresponding channel capacity is given by Eq. (4.55) as

$$C = \sum_{i=1}^{3} W \log_2(1 + \rho_i) = W \sum_{i=1}^{3} \log_2(1 + \rho_i) = 16.846\ \text{kb/s}.$$

4.2.2.2 Time-Variant Channel Capacity

The time-varying frequency-selective fading channel is similar to the time-invariant model, except that $H(f) = H(f, i)$, i.e., the channel varies over both frequency and time, and the corresponding channel model is given in Fig. 4.11.

We can approximate channel capacity in time-varying frequency-selective fading by taking the channel bandwidth W of interest and divide it up into subchannels of bandwidth equal to the channel coherence bandwidth B_c, as shown in Fig. 4.12. Under the assumption that each of the resulting subchannels is independent, time-varying with flat fading within the subchannel $H(f, k) = H_i\,[k]$ on the ith subchannel at time instance k, we can obtain the capacity for each of these flat-fading subchannels based on the average power P_i that we allocate to each subchannel, subject to a total power constraint P.

Since the channels are mutually independent, the *total channel capacity* is just equal to the *sum of capacities on the individual narrowband flat-fading channels*, subject to the total average power constraint, averaged over both time and frequency domains:

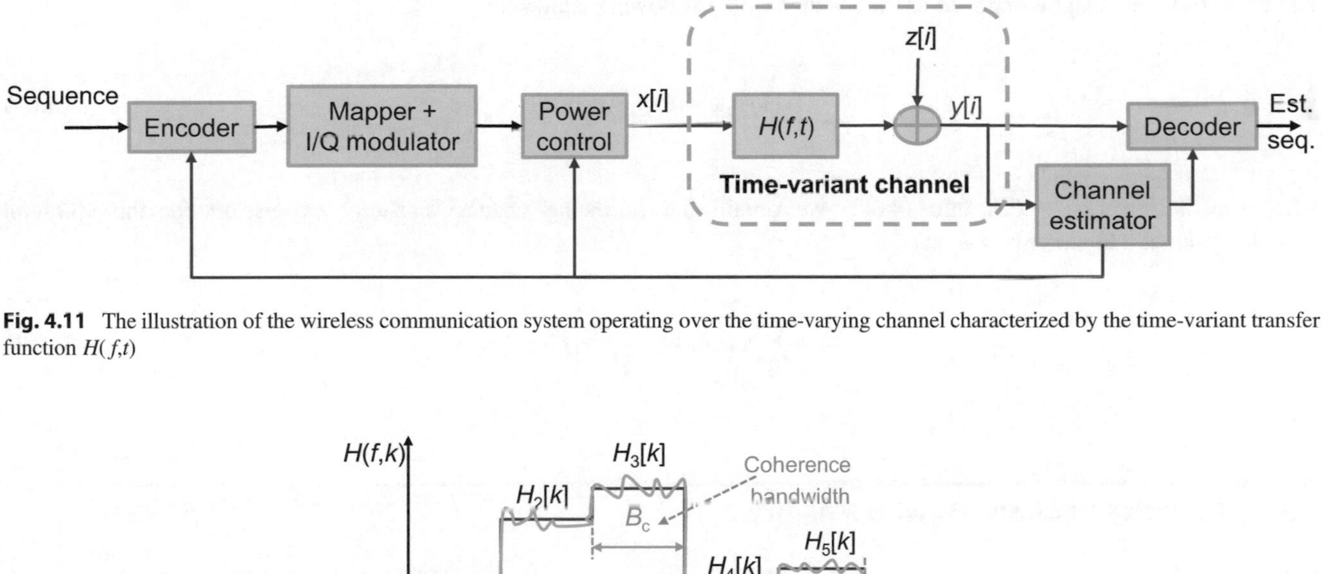

Fig. 4.11 The illustration of the wireless communication system operating over the time-varying channel characterized by the time-variant transfer function $H(f,t)$

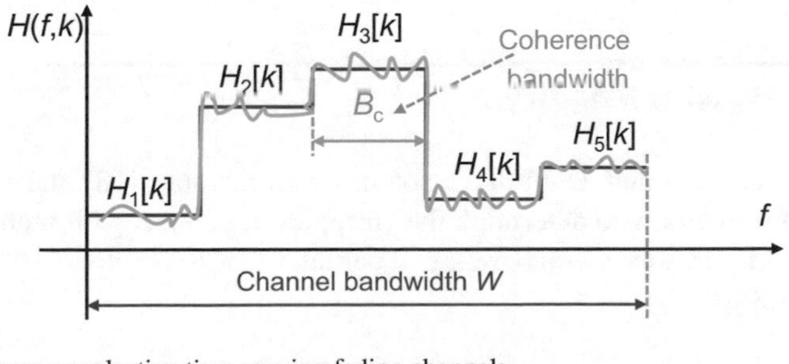

Fig. 4.12 The illustration of the frequency-selective time-varying fading channels

$$C = \max_{\{\overline{P}_i\}: \sum_i \overline{P}_i \leq \overline{P}} \sum_i C_i(\overline{P}_i), C_i = \int_{\rho_{tsh}}^{\infty} B_c \log_2\left(\frac{\rho_i}{\rho_{tsh}}\right) f(\rho_i) d\rho_i, \tag{4.59}$$

which requires the knowledge of the distribution functions of SNR for each subchannel, namely, $f(\rho_i)$.

When we fixed the average power per subchannel, the optimum power allocation policy is the water filling in the time domain. On the other hand, the optimum power allocation strategy among subchannels is again the water filling, but now in the frequency domain. Therefore, the optimum power allocation strategy for time-varying channels is the two-dimensional water filling, over both time and frequency domains. The instantaneous SNR in the *i*th subchannel at the *k*th time interval is given by $\rho_i[k] = |H_i[k]|^2 \overline{P}/(N_0 B_c)$.

The Shannon capacity assuming the perfect full CSI will be achieved by optimizing the power allocation in both time domain, by fixing $\rho_i[k] = \rho_i$, and frequency domain over subchannel indices *i* as follows:

$$C = \max_{P_i(\rho_i): \sum_i \int P_i(\rho_i) f(\rho_i) d\rho_i \leq \overline{P}} \sum_i^{\infty} \int_0 B_c \log_2\left(1 + \frac{P_i(\rho_i)}{\overline{P}}\right) f(\rho_i) d\rho_i, \tag{4.60}$$

which is subject to the following constraint:

$$\sum_i \int_0^{\infty} P_i(\rho_i) f(\rho_i) d\rho_i = \overline{P}. \tag{4.61}$$

The optimum power allocation strategy can be determined by the Lagrangian method, and the corresponding solution is given by

$$\frac{P_i(\rho_i)}{\overline{P}} = \begin{cases} 1/\rho_{tsh} - 1/\rho_i, & \rho_i \geq \rho_{tsh} \\ 0, & \rho_i < \rho_{tsh} \end{cases}. \tag{4.62}$$

The threshold SNR can be obtained as the solution of the following equation:

$$\sum_i \int_{\rho_{tsh}}^{\infty} (1/\rho_{tsh} - 1/\rho_i)^+ f(\rho_i) d\rho_i = 1. \tag{4.63}$$

After substituting Eq. (4.62) into (4.60), we obtain the following channel capacity expression for the optimum two-dimensional adaptation strategy:

$$C = \sum_i \int_{\rho_{tsh}}^{\infty} B_c \log_2\left[\frac{\rho_i}{\rho_{tsh}}\right] f(\rho_i) d\rho_i. \tag{4.64}$$

4.3 Capacity of Channels with Memory

In this section, we will describe Markov and McMillan's sources with memory [18] and the McMillan-Khinchin channel model with memory [19] and describe how to determine the entropies of sources with memory and mutual information of channels with memory [20, 22]. All this serves as a baseline in our later analysis of optical channel capacity, in particular for fiber-optic communications [23–30].

4.3.1 Markov Sources and Their Entropy

The finite Markovian chain is a commonly used model to describe both the sources and channels with memory. The Markovian stochastic process with a finite number of states $\{S\} = \{S_1, \ldots, S_n\}$ is characterized by transition probabilities π_{ij} of moving from state S_i to state S_j $(i,j = 1, \ldots, n)$. The Markov chain is the sequence of states with transitions governed by the following transition matrix:

$$\mathbf{\Pi} = \left[\pi_{ij}\right] = \begin{bmatrix} \pi_{11} & \pi_{12} & \cdots & \pi_{1n} \\ \pi_{21} & \pi_{22} & \cdots & \pi_{2n} \\ \cdots & \cdots & \ddots & \vdots \\ \pi_{n1} & \pi_{n2} & \cdots & \pi_{nn} \end{bmatrix}, \tag{4.65}$$

where $\sum_j \pi_{ij} = 1$.

Example Let us observe three-state Markov source shown in Fig. 4.13(a). The corresponding transition matrix is given by

$$\mathbf{\Pi} = \left[\pi_{ij}\right] = \begin{bmatrix} 0.6 & 0.4 & 0 \\ 0 & 1 & 0 \\ 0.3 & 0.7 & 0 \end{bmatrix}.$$

As we see, the sum in any row is equal to 1. From state S_1, we can move to either state S_2 with probability 0.4 or to stay in state S_1 with probability 0.6. Once we enter the state S_2, we stay there forever. Such a state is called absorbing, and the corresponding Markov chain is called *absorbing Markov chain*. The probability of moving from state S_3 to state S_2 in two steps can be calculated as $\pi_{32}^{(2)} = 0.3 \cdot 0.4 + 0.7 \cdot 1 = 0.82$. Another way to calculate this probability is to find the second power of transition matrix and then read out the probability of desired transition:

$$\mathbf{\Pi}^2 = \left[\pi_{ij}^{(2)}\right] = \mathbf{\Pi\Pi} = \begin{bmatrix} 0.36 & 0.64 & 0 \\ 0 & 1 & 0 \\ 0.18 & 0.82 & 0 \end{bmatrix}.$$

The probability of reaching all states from initial states after k steps can be determined by

$$\mathbf{\Pi}^{(k)} = \boldsymbol{P}^{(0)}\mathbf{\Pi}^k, \tag{4.66}$$

where $\boldsymbol{P}^{(0)}$ is the row vector containing the probabilities of initial states.

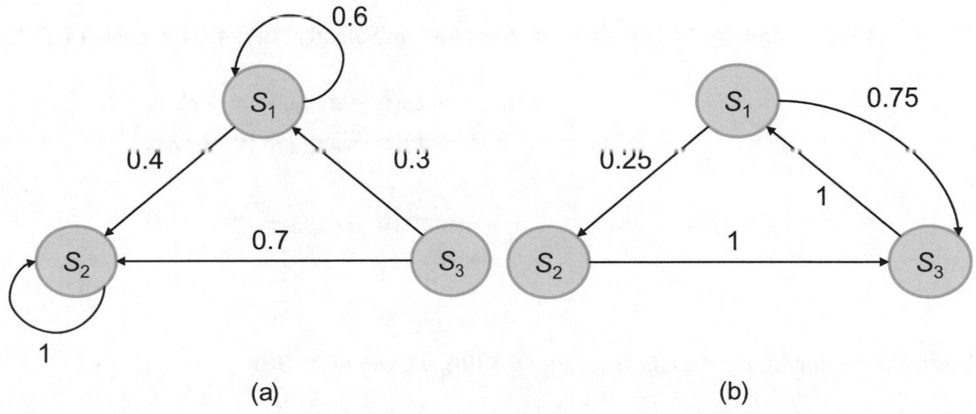

Fig. 4.13 Two three-state Markov chains: **(a)** irregular Markov chain and **(b)** regular Markov chain

If the transition matrix $\boldsymbol{\Pi}^k$ has only nonzero elements, we say that Markov chain is *regular*. Accordingly, if the k_0th power of $\boldsymbol{\Pi}$ does not have any zero entry, any kth power of $\boldsymbol{\Pi}$ for $k > k_0$ will not have any zero entry either. The so-called *ergodic Markov chains* are the most important ones from a communication system point of view. We can say that Markov chain is ergodic if it is possible to move from any specific state to any other state in a finite number of steps with nonzero probability. The Markov chain from the example above is non-ergodic. It is also interesting to notice that the transition matrix for this example has the following limit:

$$T = \lim_{k\to\infty} P^k = \begin{bmatrix} 0 & 1 & 0 \\ 0 & 1 & 0 \\ 0 & 1 & 0 \end{bmatrix}.$$

Example Let us now observe an example of regular Markov chain, which is shown in Fig. 4.13(b). The transition matrix $\boldsymbol{\Pi}$, its fourth and fifth powers, and matrix $\boldsymbol{\Pi}$ limit as $k \to \infty$ are given, respectively, as

$$\boldsymbol{\Pi} = \begin{bmatrix} 0 & 0.25 & 0.75 \\ 0 & 0 & 1 \\ 1 & 0 & 0 \end{bmatrix}, \quad \boldsymbol{\Pi}^4 = \begin{bmatrix} 0.5625 & 0.0625 & 0.3750 \\ 0.7500 & 0 & 0.2500 \\ 0.2500 & 0.1875 & 0.5625 \end{bmatrix},$$
$$\boldsymbol{\Pi}^5 = \begin{bmatrix} 0.3750 & 0.1406 & 0.4844 \\ 0.2500 & 0.1875 & 0.5625 \\ 0.5625 & 0.0625 & 0.3750 \end{bmatrix}, \quad T = \lim_{k\to\infty} \boldsymbol{\Pi}^k = \begin{bmatrix} 0.4444 & 0.1112 & 0.4444 \\ 0.4444 & 0.1112 & 0.4444 \\ 0.4444 & 0.1112 & 0.4444 \end{bmatrix}.$$

We can see that the fourth power has one zero entry, while the fifth power and all higher powers do not have zero entries. Therefore, this Markov chain is both regular and ergodic one. We can also notice that the stationary transition matrix T has identical rows.

It is evident from the example above that for regular Markov chain, the transition matrix converges to the stationary transition matrix T with all rows identical to each other:

$$T = \lim_{k\to\infty} \boldsymbol{\Pi}^k = \begin{bmatrix} t_1 & t_2 & \cdots & t_n \\ t_1 & t_2 & \cdots & t_n \\ \cdots & \cdots & \ddots & \vdots \\ t_1 & t_2 & \cdots & t_n \end{bmatrix}. \tag{4.67}$$

Additionally, the following is valid:

$$\lim_{k\to\infty} \boldsymbol{\Pi}^{(k)} = \lim_{k\to\infty} P^{(0)} \boldsymbol{\Pi}^k = P^{(0)} T = [t_1 \ t_2 \ \cdots \ t_n], \tag{4.68}$$

so we can find stationary probabilities of states (or equivalently solve for elements of T) from equations

$$\begin{aligned} t_1 &= \pi_{11} t_1 + \pi_{21} t_2 + \cdots + \pi_{n1} t_n \\ t_2 &= \pi_{12} t_1 + \pi_{22} t_2 + \cdots + \pi_{n2} t_n \\ &\vdots \\ t_n &= \pi_{1n} t_1 + \pi_{2n} t_2 + \cdots + \pi_{nn} t_n \\ \sum_{i=1}^{n} t_i &= 1. \end{aligned} \tag{4.69}$$

For instance, for Markov chain from Fig. 4.13(b), we can write that

$$t_1 = t_3,\ t_2 = 0.25t_1,\ t_3 = 0.75t_1 + t_2,\ t_1 + t_2 + t_3 = 1,$$

while the corresponding solution is given by $t_1 = t_3 = 0.4444$, $t_2 = 0.1112$.

The uncertainty of source associated with Markov source $\{S\} = \{S_1, \ldots, S_n\}$ when moving one step ahead from an initial state A_i, here denoted as $H_i^{(1)}$, can be expressed as

$$H_i^{(1)} = -\sum_{j=1}^{n} \pi_{ij} \log \pi_{ij}. \tag{4.70}$$

If the probability associated with state S_i is equal to p_i, we can obtain the entropy of Markov source by averaging over entropies associated with all states. The uncertainty of moving one step ahead becomes

$$H(X) = H^{(1)} = E\left\{H_i^{(1)}\right\} = \sum_{i=1}^{n} p_i H_i^{(1)} = -\sum_{i=1}^{n} p_i \sum_{j=1}^{n} \pi_{ij} \log \pi_{ij}. \tag{4.71}$$

In a similar fashion, the entropy of Markov source for moving k steps ahead from initial states is given by

$$H^{(k)} = E\left\{H_i^{(k)}\right\} = \sum_{i=1}^{n} p_i \underbrace{H_i^{(k)}}_{-\sum_{j=1}^{n} \pi_{ij}^{(k)} \log \pi_{ij}^{(k)}} = -\sum_{i=1}^{n} p_i \sum_{j=1}^{n} \pi_{ij}^{(k)} \log \pi_{ij}^{(k)}. \tag{4.72}$$

It can be shown that for ergodic Markov sources, there is a limit defined as $H^{(\infty)} = \lim_{k \to \infty} H^{(k)}/k$. In order to prove this, we can use the following property, which should be proved as a homework problem:

$$H^{(k+1)} = H^{(k)} + H^{(1)}. \tag{4.73}$$

By applying this property in an iterative fashion, we obtain that

$$H^{(k)} = H^{(k-1)} + H^{(1)} = H^{(k-2)} + 2H^{(1)} = \cdots = kH^{(1)} = kH(X). \tag{4.74}$$

From Eq. (4.74), it is evident that

$$\lim_{k \to \infty} \frac{H^{(k)}}{k} = H^{(1)} = H(X). \tag{4.75}$$

Equation (4.75) can be now used as an alternative definition of entropy of Markov source, which is applicable to arbitrary *stationary* source as well.

Example Let us determine the entropy of Markov source shown in Fig. 4.13(b). By using the definition from Eq. (4.71), we obtain that

$$\begin{aligned} H(X) &= -\sum_{i=1}^{n} p_i \sum_{j=1}^{n} p_{ij} \log p_{ij} \\ &= -0.4444(0.25 \log 0.25 + 0.75 \log 0.75) - 0.1111 \cdot 1 \log 1 \\ &\quad -0.4444 \cdot 1 \log 1 = 0.6605 \text{ bits}. \end{aligned}$$

4.3.2 McMillan Sources and Their Entropy

A McMillan's description [18] of the discrete source with memory is more general than that of Markov chain. In our case, we will pay attention to stationary sources. Let S represent a finite source alphabet with corresponding letters $\{s_1, s_2, \ldots, s_M\} = S$. The source emits one symbol at the time instance t_k. The transmitted sequence can be represented as $X = \{\ldots, x_{-1}, x_0, x_1, \ldots\}$, where $x_i \in S$. Among all members of ensemble $\{X\}$, we are interested just in those having specified source symbols at certain prespecified instances of time. All sequences with these properties create so-called *cylinder set*.

Example Let specified location be defined by indices -1, 0, 3, and k, with corresponding symbols at these locations equal to $x_{-1} = s_2$, $x_0 = s_5$, $x_3 = s_0$, $x_k = s_{n-1}$. The corresponding cylinder set will be given as $C_1 = \{\ldots, x_{-2}, s_2, s_5, x_1, x_2, s_0, \ldots, x_k = s_{n-1}, \ldots\}$. Since we observe the stationary processes, the statistical properties of cylinder will not change if we shift the cylinder for one time unit in either direction (either by T or by T^{-1}). For instance, the time-shifted C_1 cylinder is given by $TC_1 = \{\ldots, x_{-1}, s_2, s_5, x_2, x_3, s_0, \ldots, x_{k+1} = s_{n-1}, \ldots\}$.

The *stationary property* for arbitrary cylinder C can be expressed as

$$P\{TC\} = P\{T^{-1}C\} = P\{C\}, \tag{4.76}$$

where $P\{\cdot\}$ denotes the probability measure.

Let us now specify n letters from alphabet S to be sent at positions $k + 1, \ldots, k + n$. This sequence can be denoted as $x_{k+1}, \ldots, x_{k+n}$, and there is the total of M^n possible sequences. The entropy of all possible sequences is defined as

$$H_n = -\sum_C p_m(C) \log p_m(C), \tag{4.77}$$

where $p_m(\cdot)$ is the probability measure. The McMillan's definition of entropy of stationary discrete source is given by [18]

$$H(X) = \lim_{n\to\infty} \frac{H_n}{n}. \tag{4.78}$$

As we can see, the McMillan's definition of entropy is consistent with Eq. (4.75), which is applicable to stationary Markovian sources.

4.3.3 McMillan-Khinchin Model for Channel Capacity Evaluation

Let the input and output alphabets of the channel be finite and denoted by A and B, respectively, while the channel input and output sequences are denoted by X and Y. The noise behavior for memoryless channels is generally captured by a conditional probability matrix $P\{b_j|a_k\}$ for all $b_j \in B$ and $a_j \in A$. On the other side, in channels with finite memory (such as the optical channel), the transition probability is dependent on the transmitted sequences up to the certain prior finite instance of time. For instance, the transition matrix for channel described by the Markov process has the form $P\{Y_k = b|\ldots, X_{-1}, X_0, X_1, \ldots, X_k\} = P\{Y_k = b|X_k\}$.

Let us consider a member x of input ensemble $\{X\} = \{\ldots, x_{-2}, x_{-1}, x_0, x_1, \ldots\}$ and its corresponding channel output y from ensemble $\{Y\} = \{\ldots, y_{-2}, y_{-1}, y_0, y_1, \ldots\}$. Let $\boldsymbol{X}$ denote all possible input sequences and $\boldsymbol{Y}$ denote all possible output sequences. By fixing a particular symbol at specific location, we can obtain the *cylinder*. For instance, cylinder $x^{4,1}$ is obtained by fixing symbol a_1 to position x_4, so it is $x^{4,1} = \ldots, x_{-1}, x_0, x_1, x_2, x_3, a_1, x_5, \ldots$ The output cylinder $y^{1,2}$ is obtained by fixing the output symbol b_2 to position 1, namely, $y^{1,2} = \ldots, y_{-1}, y_0, b_2, y_2, y_3, \ldots$ In order to characterize the channel, we have to determine transition probability $P(y^{1,2}|x^{4,1})$, which is probability that cylinder $y^{1,2}$ was received if cylinder $x^{4,1}$ was transmitted. Therefore, for all possible input cylinders $S_A \subset \boldsymbol{X}$, we have to determine the probability that cylinder $S_B \subset \boldsymbol{Y}$ was received if S_A was transmitted.

The channel is completely specified by the following: (i) input alphabet A, (ii) output alphabet B, and (iii) transition probabilities $P\{S_B|S_A\} = v_x$ for all $S_A \in \boldsymbol{X}$ and $S_B \in \boldsymbol{Y}$. Accordingly, the channel is specified by the triplet $[A, v_x, B]$. If transition probabilities are invariant with respect to time shift T, which means that $v_x(TS) = v_x(S)$, the channel is said to be *stationary*. If the distribution of Y_k depends only on the statistical properties of sequence $\ldots, x_{k-1}, x_k$, we say that the channel is without *anticipation*. If, furthermore, the distribution of Y_k depends only on $x_{k-m}, \ldots, x_k$, we say that the channel has finite *memory* of *m* units.

The source and channel may be described as a new source $[C, \zeta]$, where C is the Cartesian product of input A and output B alphabets ($C = A \times B$), while ζ is a corresponding probability measure. The joint probability of symbol $(x,y) \in C$, where $x \in A$ and $y \in B$, is obtained as the product of marginal and conditional probabilities: $P(x \cap y) = P\{x\}P\{y|x\}$.

Let us further assume that both source and channel are stationary. Following the description presented in [6, 19], it is useful to describe the concatenation of a stationary source and a stationary channel as follows:

1. If the source $[A, \mu]$ (μ is the probability measure of the source alphabet) and the channel $[A, v_x, B]$ are stationary, the product source $[C, \zeta]$ will also be stationary.
2. Each stationary source has an entropy, and therefore $[A, \mu]$, $[B, \eta]$ (η is the probability measure of the output alphabet), and $[C, \zeta]$ each have the finite entropies.
3. These entropies can be determined for all n-term sequences $x_0, x_1, \ldots, x_{n-1}$ emitted by the source and transmitted over the channel as follows [19]:

$$\begin{aligned} &H_n(X) \leftarrow \{x_0, x_1, \ldots, x_{n-1}\} \\ &H_n(Y) \leftarrow \{y_0, y_1, \ldots, y_{n-1}\} \\ &H_n(X,Y) \leftarrow \{(x_0,y_0), (x_1,y_1), \ldots, (x_{n-1},y_{n-1})\} \\ &H_n(Y|X) \leftarrow \{(Y|x_0), (Y|x_1), \ldots, (Y|x_{n-1})\} \\ &H_n(X|Y) \leftarrow \{(X|y_0), (X|y_1), \ldots, (X|y_{n-1})\}. \end{aligned} \tag{4.79}$$

It can be shown that the following is valid:

$$\begin{aligned} H_n(X,Y) &= H_n(X) + H_n(Y|X) \\ H_n(X,Y) &= H_n(Y) + H_n(X|Y). \end{aligned} \tag{4.80}$$

The equations above can be rewritten in terms of entropies per symbol:

$$\begin{aligned} \frac{1}{n}H_n(X,Y) &= \frac{1}{n}H_n(X) + \frac{1}{n}H_n(Y|X) \\ \frac{1}{n}H_n(X,Y) &= \frac{1}{n}H_n(Y) + \frac{1}{n}H_n(X|Y). \end{aligned} \tag{4.81}$$

For sufficiently long sequences, the following channel entropies exist:

$$\begin{aligned} &\lim_{n\to\infty}\frac{1}{n}H_n(X,Y) = H(X,Y) \qquad \lim_{n\to\infty}\frac{1}{n}H_n(X) = H(X) \\ &\lim_{n\to\infty}\frac{1}{n}H_n(Y) = H(Y) \qquad \lim_{n\to\infty}\frac{1}{n}H_n(X|Y) = H(X|Y) \\ &\lim_{n\to\infty}\frac{1}{n}H_n(Y|X) = H(Y|X). \end{aligned} \tag{4.82}$$

The mutual information also exists, and it is defined as

$$I(X,Y) = H(X) + H(Y) - H(X,Y). \tag{4.83}$$

The *stationary information capacity* of the channel is obtained by maximization of mutual information over all possible information sources:

$$C(X,Y) = \max I(X,Y). \tag{4.84}$$

The results of analysis in this section will be applied later in the evaluation the information capacity of optical channel with memory. But, before that, we will briefly describe the adopted model for signal propagation in single-mode optical fibers, which will also be used in the evaluation of channel capacity.

4.4 Calculation of Information Capacity by the Forward Recursion of the BCJR Algorithm

We will now evaluate the channel capacity of multilevel/multidimensional modulation schemes for an i.i.d. information source, which is in literature also known as the achievable information rate [2, 3, 20–28]. The approach that will be applied was initially proposed in [20–22] to evaluate the information rate of binary modulation schemes over linear ISI channels. A similar method has been also used in [23–27] to study the information capacity of on-off keying channels in the presence of fiber nonlinearities. Finally, the method has been generalized to multilevel modulations in [28]. The i.i.d. channel capacity represents a lower bound on channel capacity. Some papers employ the mutual information definition formula to evaluate the achievable rates [29, 30]; however, this approach does not take the optical channel memory into account.

To calculate the i.i.d. channel capacity, we will model the whole transmission system as a dynamical intersymbol interference (ISI) channel, in which m previous and next m symbols have influence on the observed symbol, as shown in Fig. 4.14. Therefore, the model is a particular instance of the McMillan-Khinchin model described in Sect. 4.3.3. The optical communication system is characterized by the conditional PDF of the output N-dimensional matrix of samples $\mathbf{y} = (y_1, \ldots, y_n, \ldots)$, where $y_i = (y_{i,1}, \ldots, y_{i,D}) \in \mathbf{Y}$, for given source sequence $\mathbf{x} = (x_1, \ldots, x_n, \ldots)$, $x_i \in \mathbf{X} = \{0, 1, \ldots, M-1\}$. The set $\mathbf{X}$ represents the set of indices of constellation points in the corresponding M-ary N-dimensional signal constellation diagram, where N is the number of basis functions used to represent a given signal constellation. (For instance, N equals 2 for M-ary PSK and M-ary QAM, while it equals 3 for cube constellation.) The $\mathbf{Y}$ represents the set of all possible channel outputs, where $y_{i,j}$ $(j = 1, \ldots, N)$ corresponds to the jth coordinate of the ith sample of channel output. The state in the trellis from Fig. 4.14 is defined as $\mathbf{s}_j = (x_{j-m}, x_{j-m+1}, \ldots, x_j, x_{j+1}, \ldots, x_{j+m}) = \mathbf{x}[j-m, j+m]$, where x_k denotes the index of the symbol from the set $\mathbf{X} = \{0, 1, \ldots, M-1\}$ of possible indices. Every symbol carries $m = \log_2 M$ bits and was constructed using the appropriate mapping rule (natural, Gray, anti-Gray, etc.). The memory of the state is equal to $2m + 1$, with $2m$ being the number of *symbols* that influence the observed symbol from both sides. The trellis has M^{2m+1} states, where each of them corresponds to

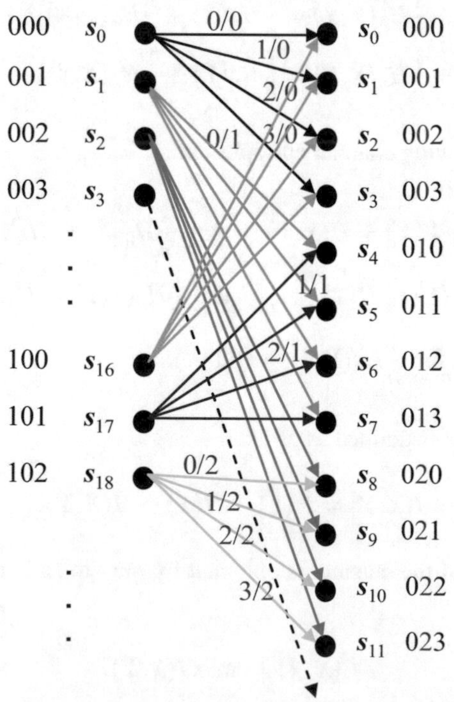

Fig. 4.14 A portion of trellis describing the model of optical channel with memory for 4-ary N-dimensional signal constellations with memory $2m + 1 = 3$

a different (2 m + 1)-symbol pattern (configuration). The state index is determined by considering (2 m + 1) symbols as digits in the numerical system with the base M.

For complete characterization of trellis, conditional PDFs are needed, and they can be evaluated by using one of the following methods: (i) estimating the histograms [20], (ii) using an instanton approach [24], or (iii) by Edgeworth expansion [25]. Notice that this channel model is consistent with the McMillan model discussed above. The information rate can be calculated by Eq. (4.83), rewritten in vector form as follows:

$$I(\mathbf{Y},\mathbf{X}) = H(\mathbf{Y}) - H(\mathbf{Y}|\mathbf{X}), \tag{4.85}$$

where $H(\mathbf{U}) = -\langle \log_2 P(\mathbf{U}) \rangle$ denotes the entropy of a random (variable) vector $\mathbf{U}$. By using the Shannon-McMillan-Breiman theorem, which states that [5]

$$\langle \log_2 P(\mathbf{Y}) \rangle = -\lim_{n\to\infty} (1/n) \log_2 P(\mathbf{y}[1,n]), \tag{4.86}$$

the information rate can be determined by calculating $\log_2(P(\mathbf{y}[1,n]))$ for the sufficiently long source sequence propagated through the channel. By substituting Eq. (4.86) into (4.85), we obtain the following expression suitable for practical calculation of information capacity:

$$I(\mathbf{Y},\mathbf{X}) = \lim_{n\to\infty} \frac{1}{n} [\sum_{i=1}^{n} \log_2 P(y_i|\mathbf{y}[1,i-1],\mathbf{x}[1,n]) - \sum_{i=1}^{n} \log_2 P(y_i|\mathbf{y}[1,i-1])]. \tag{4.87}$$

The first term on the right side of Eq. (4.87) can be straightforwardly calculated from conditional probability density functions (PDFs) $P(\mathbf{y}[j-m, j+m]|\mathbf{s})$. We can use the forward recursion of the multilevel BCJR algorithm, described in detail in Chap. 9, to calculate $\log_2 P(y_i|\mathbf{y}[1,i-1])$. We can now define the *forward metric* $\alpha_j(\mathbf{s}) = \log\{p(\mathbf{s}_j = \mathbf{s}, \mathbf{y}[1,j])\}$ ($j = 1, 2, \ldots, n$) and the *branch metric* $\gamma_j(\mathbf{s}',\mathbf{s}) = \log[p(\mathbf{s}_j = \mathbf{s}, y_j, \mathbf{s}_{j-1} = \mathbf{s}')]$ as

$$\begin{aligned} \alpha_j(\mathbf{s}) &= \max_{\mathbf{s}'}{}^*[\alpha_{j-1}(\mathbf{s}') + \gamma_j(\mathbf{s}',\mathbf{s}) - \log_2 M], \\ \gamma_j(\mathbf{s}',\mathbf{s}) &= \log\,[p(y_j|\mathbf{x}[j-m,j+m])], \end{aligned} \tag{4.88}$$

where the max* operator is defined by $\max^*(x,y) = \log(e^x + e^y) = \max(x,y) + \log[1 + \exp(-|x-y|)]$.

The ith term, $\log_2 P(y_i|\mathbf{y}[1,i-1])$, can now be calculated in an iterative fashion by

$$\log_2 P(y_i|\mathbf{y}[1,i-1]) = \max_{\mathbf{s}}{}^* \alpha_i(\mathbf{s}), \tag{4.89}$$

where the max* operator was applied for all $\mathbf{s} \in \mathbf{S}$ ($\mathbf{S}$ denotes the set of states in the trellis shown in Fig. 4.14). Information capacity is defined as

$$C = \max I(\mathbf{Y},\mathbf{X}), \tag{4.90}$$

where the maximization is performed over all possible input distributions. Since the optical channel has memory, it is natural to assume that optimum input distribution will also be the one with the memory. By considering the stationary input distributions in the form $p(x_i|x_{i-1}, x_{i-2}, \ldots) = p(x_i|x_{i-1}, x_{i-2}, \ldots, x_{i-k})$, we can determine the transition probabilities of the corresponding Markov model that would maximize the information rate by nonlinear numerical optimization [31, 32].

This method is applicable to both memoryless channels and for channels with memory. For instance, we calculated the memoryless information capacities for different signal constellation sizes and two types of QAM constellations (square QAM and star QAM) by observing a linear channel model, and results are shown in Fig. 4.15. As we can see, the information capacity can be closely approached even with a uniform information source if constellation size is sufficiently large. It is interesting to notice that star QAM outperforms the corresponding square QAM for low and medium signal-to-noise ratios (SNRs), while square QAM outperforms star QAM for high SNRs. The iterative polar quantization (IPQ)-based modulation format, introduced in [33, 34], also known as iterative polar modulation (IPM), significantly outperforms both square QAM and star QAM.

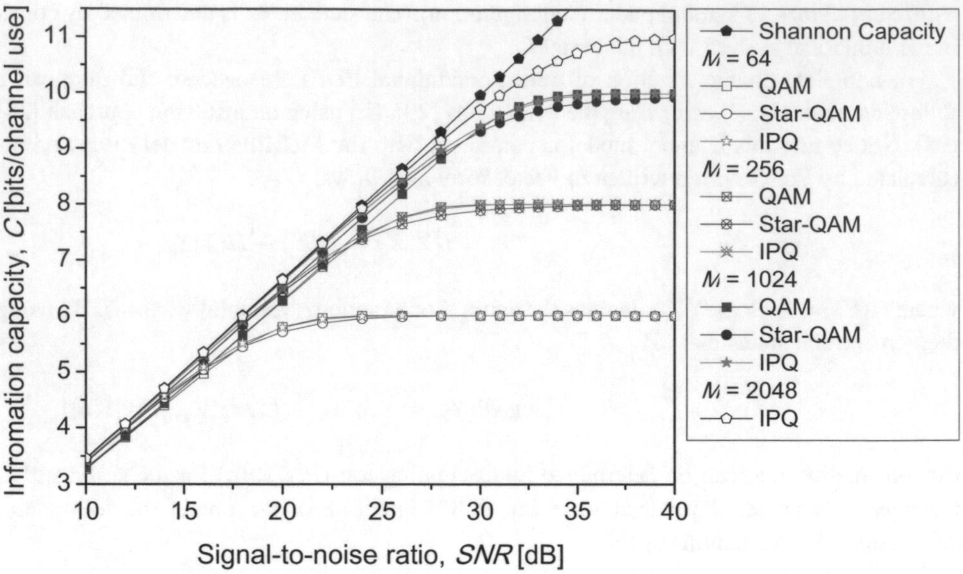

Fig. 4.15 Information capacities for linear channel model and different signal constellation sizes. (64-star QAM contains 8 rings with 8 points each, 256-star QAM contains 16 rings with 16 points, and 1024-star QAM contains 16 rings with 64 points.) SNR is defined as E_s/N_0, where E_s is the symbol energy and N_0 is the power spectral density

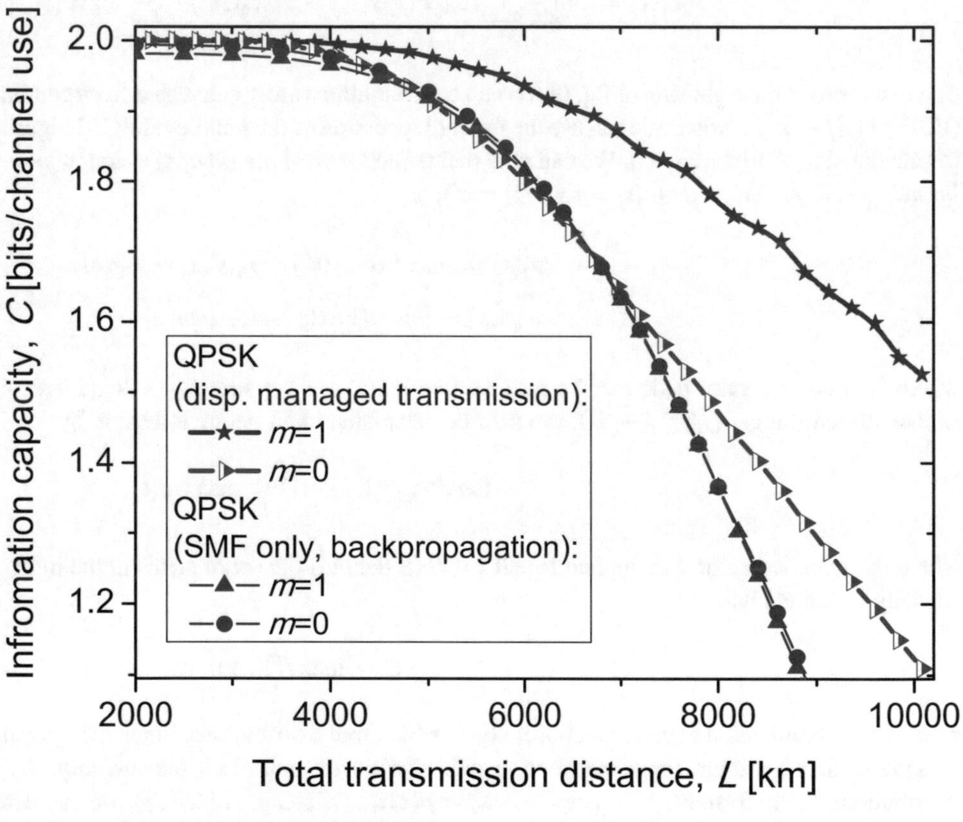

Fig. 4.16 The i.i.d. information capacity per single polarization for QPSK of aggregate data rate of 100 Gb/s against the transmission distance. The dispersion map is shown in Fig. 4.17

4.5 Information Capacity of Systems with Coherent Optical Detection

The fiber-optic channel capacity seems to be hot research topics, judging by the number of publications related to this topic [1–4, 23–30, 35–45]. In this section, we study the information capacity of optical systems with coherent detection by applying some commonly used transmission scenarios. The i.i.d. information capacity versus the number of fiber spans is shown in Fig. 4.16. It is obtained by using Monte Carlo simulations assuming the following: (i) dispersion map is the one shown in Fig. 4.17, with fiber parameters summarized in Table 4.1; (ii) QPSK modulation format of aggregate data rate 100 Gb/s per

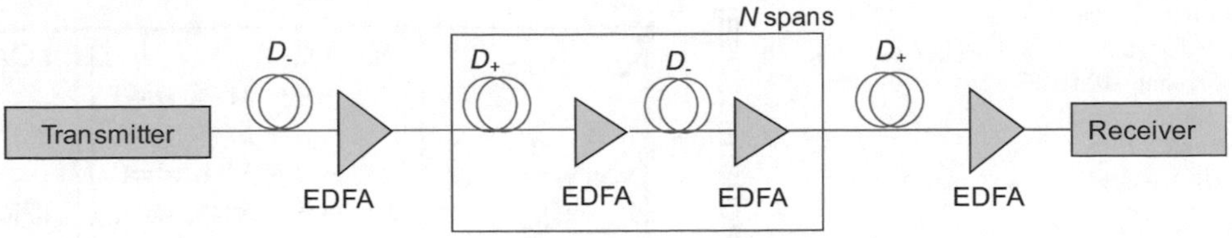

Fig. 4.17 Dispersion map under study is composed of N spans of length L = 120 km, consisting of 2 L/3 km of D_+ fiber followed by L/3 km of D_- fiber, with pre-compensation of −1600 ps/nm and corresponding post-compensation. The fiber parameters are given in Table 4.1

Table 4.1 Fiber parameters

	D_+ fiber	D_- fiber
Dispersion [ps/(nm km)]	20	−40
Dispersion slope [ps/(nm^2 km)]	0.06	−0.12
Effective cross-sectional area [μm^2]	110	50
Nonlinear refractive index [m^2/W]	$2.6{\cdot}10^{-20}$	$2.6{\cdot}10^{-20}$
Attenuation coefficient [dB/km]	0.19	0.25

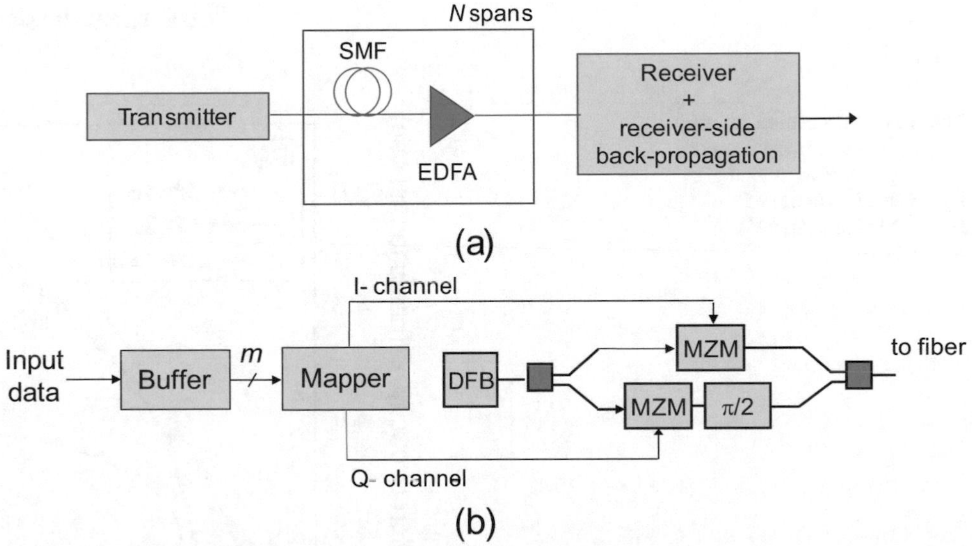

Fig. 4.18 (**a**) Dispersion map composed of SMF sections only with receiver-side digital backpropagation and (**b**) transmitter configuration for two-dimensional signal constellations

single polarization; and (iii) two channel memory settings and transmitter-receiver configurations. As we can see, by using the LDPC code of sufficient length and large girth, it is possible, in principle, to achieve the total transmission distance of ~8800 km for state memory $m = 0$ and ~ 9600 km for state memory $m = 1$. The transmission distance can further be increased by observing larger memory channel assumptions, which requires higher computational complexity of corresponding turbo equalizer. On the other hand, we can use the coarse digital backpropagation approach [28] to keep the channel memory reasonable low and then apply the method for information capacity evaluation described in this section. However, since digital backpropagation method cannot account for the nonlinear ASE noise-Kerr nonlinearity interaction, we should use the method described in the previous section in order to account for this effect in the calculation of information capacity. As a reference, the i.i.d. information capacity when digital backpropagation method is used, for fiber link composed of standard SMF with dispersion map from Fig. 4.18(a) with EDFAs with a noise figure of 6 dB deployed every 100 km, is also shown in Fig. 4.16. We can see that digital backpropagation method helps to reduce channel memory, since the improvement for $m = 1$ over $m = 0$ case is small.

In Fig. 4.19, we show the i.i.d. information capacities for three different modulation formats: (i) MPSK, (ii) star QAM (sQAM), and (iii) iterative polar modulation (IPM), obtained by employing the dispersion map from Fig. 4.17. The chosen symbol rate is 50 GS/s, while the launch power was set to 0 dBm. For a transmission distance of 5000 km, the

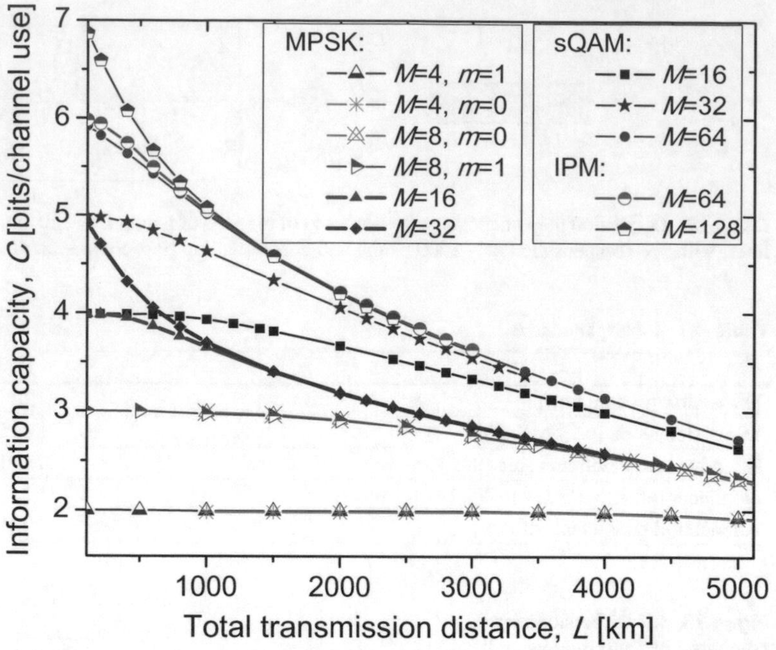

Fig. 4.19 Information capacities per single polarization for star QAM (sQAM), MPSK, and IPM for different constellation sizes. The dispersion map is shown in Fig. 4.17. EDFA's NF = 6 dB

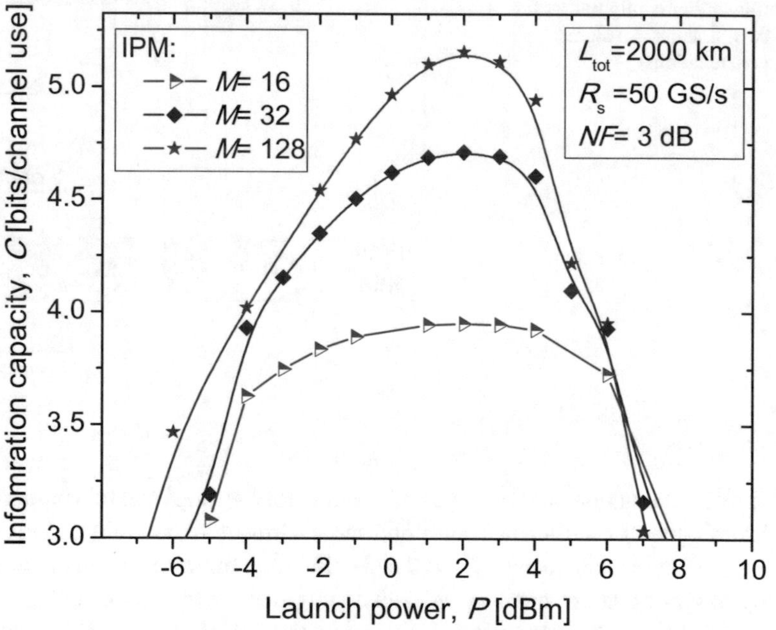

Fig. 4.20 Information capacity per single polarization against launch power P for a total transmission distance of $L_{tot} = 2000$ km. EDFA's NF = 3 dB

i.i.d. information capacity is 2.72 bits/symbol (the aggregate rate is 136 Gb/s per wavelength), for 2000 km it is 4.2 bits/ symbol (210 Gb/s), and for 1000 km the i.i.d. information capacity is 5.06 bits/symbol (253 Gb/s per wavelength).

The information capacity as a function of launch power for fixed total transmission distance $L_{tot} = 2000$ km and $NF = 3$ dB, and dispersion map from Fig. 4.18(a), is shown in Fig. 4.20. As we see, the single polarization channel capacity of $C_{opt} = 5.16$ bits/s/Hz over transmission distance to 2000 km can be achieved for optimum launch power of $P_{opt} = 2$ dBm and $M = 128$, which is very close to the result reported in [30] where star QAM of size 2048 and optimum dispersion map based on Raman amplifiers were employed (Figs. 4.19 and 4.20).

Regarding the channel capacity calculation for few-mode-fiber (FMF)-based communication systems, an interested reader is referred to [35].

4.6 Hybrid Free-Space Optical (FSO)-RF Channel Capacity

In this section, we describe the use of a hybrid FSO-RF system with adaptive modulation and coding (AMC) as an efficient way to deal with strong atmospheric turbulence. For FSO only channels' performance an interested reader is referred to [52–54]. Adaptive modulation and coding [2, 46, 47] can enable robust and spectrally efficient transmission over both α-μ (or generalized gamma) wireless fading channel and FSO channel. To enable high-speed communication over atmospheric turbulence channels, we describe a scheme in which an LDPC-coded sequence is partially over FSO channel and partially over wireless channel. The channel state information of both channels is forwarded to the transmitters by an RF feedback channel. The transmitters then adapt powers and rates so that the total channel capacity is maximized. The AMC concept has been initially introduced for wireless communications [10, 48].

4.6.1 Hybrid FSO-RF System Model Description

We first describe the hybrid system, wireless channel model, and FSO channel model. The adaptive hybrid FSO-RF communication system, shown in Fig. 4.21, consists of two parallel FSO and RF channels. The encoded data stream is partially transmitted over FSO and partially over RF channel. Operating symbol rate of FSO channel is commonly many times higher than that of RF channel. FSO channel comprises an FSO transmitter, propagation path through the atmosphere, and an FSO receiver. The optical transmitter includes a semiconductor laser of high launch power, adaptive mapper, and power control block. To reduce the system cost, the direct modulation of laser diode is used.

The modulated beam is projected toward the distant receiver by using an expanding telescope assembly. Along the propagation path through the atmosphere, the light beam experiences absorption, scattering, and atmospheric turbulence, which cause attenuation and random variations in amplitude and phase. At the receiver side, an optical system collects the incoming light and focuses it onto a detector, which generates an electrical current proportional to the incoming power. The RF channel comprises adaptive RF mapper, RF power control, RF transmitter (Tx), transmitting antenna, wireless propagation path, receiver antenna, and RF receiver (Rx).

The RF channel estimates and FSO irradiance estimates are transmitted back to transmitters using the same RF feedback channel. Because the atmospheric turbulence changes slowly, with correlation time ranging from 10 μs to 10 ms, this is a plausible scenario for FSO channels with data rates in the order of Gb/s. Notice that erbium-doped fiber amplifiers (EDFAs) cannot be used at all in this scenario because the fluorescence time is too long (about 10 ms). The semiconductor optical amplifiers (SOAs) should be used instead, if needed. The data rates and powers in both channels are varied in accordance with channel conditions. The symbol rates on both channels are kept fixed, while the signal constellation diagram sizes are varied based on channel conditions. When FSO (RF) channel condition is favorable, larger constellation size is used; when FSO (RF) channel condition is poor, smaller constellation size is used; and when the FSO (RF) channel signal-to-noise ratio (SNR) falls below threshold, the signal is not transmitted at all. Both subsystems (FSO and RF) are designed to achieve the same target bit error probability (P_b). The RF subsystem employs M-ary quadrature amplitude modulation (MQAM), while the FSO subsystem employs the M-ary pulse amplitude modulation (MPAM). MPAM is selected for the FSO subsystem because negative amplitude signals cannot be transmitted over FSO channels with direct detection. These two modulation formats are selected as an illustrative example; the proposed scheme, however, is applicable to arbitrary multilevel modulations. The optimum variable-power variable-rate policy, maximizing total channel capacity, is described in the next subsection. In the rest of this section, we describe the wireless and the FSO channel models. Both RF and FSO channels are modeled as *block-*

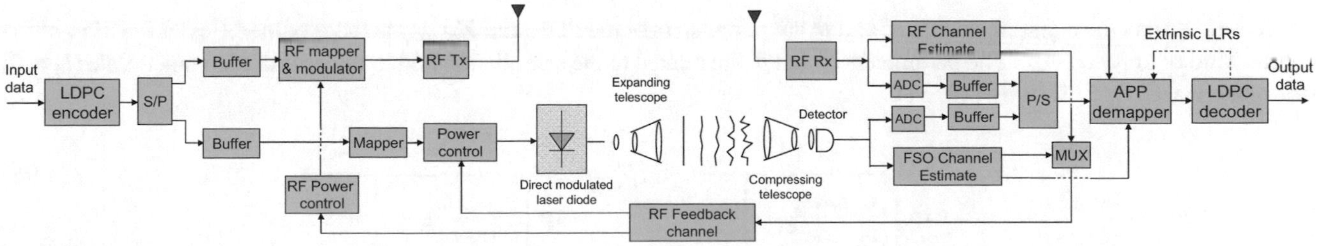

Fig. 4.21 Hybrid FSO-RF system model. *S/P* serial-to-parallel conversion, *LD* laser diode, *ADC* A/D converter, *P/S* parallel-to-serial converter, *APP* a posteriori probability

fading channels, because symbol durations in both channels are much shorter than the corresponding coherence times. The duration of the block is proportional to the coherence time of the channel, which is from 10 μs to 10 ms for FSO channel and from 10 ms to 1 s for RF channels. We therefore assume the i.i.d. fading between the blocks.

The signal at the RF receiver antenna (see Fig. 4.21), denoted by $\mathbf{s}_k$, can be written as

$$\mathbf{s}_k = r_k e^{j\gamma_k}\mathbf{c}_k + \mathbf{z}_k, \tag{4.91}$$

where r_k is the fading envelope for the kth transmission block and γ_k is the random phase shift occurred during signal transmission of the same block of symbols over a fading channel. The fading at antenna is frequency nonselective, it does not change for the duration of block (slow fading), and it is independent from block to block. The $\mathbf{c}_k$ is the vector of transmitted symbols within the kth block. The $\mathbf{z}_k$ denotes the block of zero-mean circular Gaussian noise samples corresponding to block k.

The corresponding α-μ PDF is given by [49].

$$f(r_t) = \frac{\alpha\mu^{\mu} r_t^{\alpha\mu-1}}{\widehat{r}^{\alpha\mu}\Gamma(\mu)} \exp\left(-\mu\frac{r_t^{\alpha}}{\widehat{r}^{\alpha}}\right), \tag{4.92}$$

where $\Gamma(.)$ is the gamma function. In Eq. (4.92), $\mu > 0$ denotes the inverse of the normalized variance of r_t^{α}, while $\widehat{r}$ is a α-root mean value, and they are defined, respectively, as

$$\mu = \langle r_t^{\alpha}\rangle^2 / \mathrm{Var}\{r_t^{\alpha}\},\ \widehat{r} = \sqrt[\alpha]{\langle r_t^{\alpha}\rangle}. \tag{4.93}$$

The α-μ fading model is employed because Rayleigh, Nakagami-m, exponential, Weibull, and one-sided Gaussian distribution functions are all special cases of this model. For example, by setting $\alpha = 2$ and $\mu = 1$, we obtain the Rayleigh distribution, while by setting $\alpha = 2$ and $\mu = 2$, we obtain Nakagami $m = 2$ distribution.

The FSO communication channel model is described by

$$\mathbf{y}_k = \mathcal{R} i_k \mathbf{x}_k + \mathbf{n}_k, \tag{4.94}$$

where $\mathbf{x}_k$ is the kth transmitted block, i_k is the instantaneous intensity gain, $\mathbf{y}_k$ is the kth received block of symbols, $\mathbf{n}_k$ is the vector of additive white Gaussian noise (AWGN) samples with a normal distribution $N(0,\sigma^2)$ representing the trans-impedance amplifier thermal noise, and $\mathcal{R}$ denotes the photodiode responsivity. (Without the loss of generality in the rest of the chapter, we will set $R = 1$ A/W.) (All signals in (4.94) are real valued.)

Several probability density functions (PDFs) have been proposed for the intensity variations at the receiver side of an FSO link [50, 51]. For example, Al-Habash et al. [50, 51] proposed a statistical model that factorizes the irradiance as the product of two independent random processes each with a gamma PDF. It was shown in [51] that predicted distribution matches very well the distributions obtained from numerical propagation simulations and experiments, and as such is adopted here. The PDF of the intensity fluctuation is given by [50].

$$f(i_k) = \frac{2(\alpha'\beta')^{(\alpha'+\beta')/2}}{\Gamma(\alpha')\Gamma(\beta')} i_k^{(\alpha'+\beta')/2-1} K_{\alpha'-\beta'}\left(2\sqrt{\alpha'\beta' i_k}\right), i_k > 0, \tag{4.95}$$

where i_k ($k \geq 0$) is the signal intensity, α' and β' are parameters of the PDF, and $K_{\alpha' - \beta'}$ is the modified Bessel function of the second kind of order $\alpha' - \beta'$. The parameters α' and β' are related to the scintillation and in the case of zero inner scale ($l_0 = 0$) (for plane waves) are given by [50, 51].

$$\alpha' = \frac{1}{\exp\left[\frac{0.49\sigma_R^2}{\left(1+1.11\sigma_R^{12/5}\right)^{7/6}}\right] - 1},\quad \beta' = \frac{1}{\exp\left[\frac{0.51\sigma_R^2}{\left(1+0.69\sigma_R^{12/5}\right)^{5/6}}\right] - 1}, \tag{4.96}$$

where $\sigma_R^{\ 2}$ is the Rytov variance [51].

$$\sigma_R^2 = 1.23\, C_n^2 k^{7/6} L^{11/6}, \tag{4.97}$$

where $k = 2\pi/\lambda$ (λ is the wavelength), L denotes the propagation distance, and $C_n{}^2$ is the refractive index structure parameter. The Rytov variance is an excellent parameter to be used to describe the turbulence strength, because it takes the refractive structure parameter, the propagation distance, and the operating wavelength into account. Weak fluctuations are associated with $\sigma^2{}_R < 1$, the moderate with $\sigma^2{}_R \approx 1$, the strong with $\sigma^2{}_R > 1$, and the saturation regime is defined by $\sigma^2{}_R \to \infty$ [51].

4.6.2 Adaptive Modulation and Coding (AMC) in Hybrid FSO-RF Communications

There are many parameters that can be varied at the transmitter side relative to the FSO (RF) channel conditions, including data rate, power, coding rate, and combinations of different adaptation parameters. The transmitter power adaptation can be used to compensate for SNR variation due to atmospheric turbulence/fading, with the aim to maintain a desired BER in both FSO and RF channels. The power adaptation therefore "inverts" the FSO channel scintillation and fading in wireless channel so that both FSO and RF channels behave similarly as an AWGN channel. The FSO channel, upon channel inversion, appears to the receiver as standard AWGN channel with $\mathrm{SNR}^{\mathrm{FSO}} = \Gamma_0^{\mathrm{FSO}}/\langle 1/i_t^2\rangle$, where $\Gamma_0^{\mathrm{FSO}} = E_s^{\mathrm{FSO}}/N_0$ is the signal-to-noise ratio in the absence of scintillation, with E_s^{FSO} being the symbol energy and $N_0/2$ being the double-side power spectral density of AWGN related to variance by $\sigma^2 = N_0/2$. Notice that this definition [36, 47, 51], commonly used in wireless communications [10–13], is different from [51] where SNR is defined as P_o/σ, where $\langle\, i_t \,\rangle \le P_o$. The wireless fading channel, upon channel inversion, appears to the receiver as standard AWGN channel with $\mathrm{SNR}^{\mathrm{RF}} = \Gamma_0^{\mathrm{RF}}/\langle 1/h^2\rangle$, where h is the channel coefficient and $\Gamma_0^{\mathrm{RF}} = E_s^{\mathrm{RF}}/N_0$ is the signal-to-noise ratio in the absence of fading, with E_s^{RF} being the symbol energy and $N_0/2$ being the double-side power spectral density of AWGN.

In the rest of this section, we derive the optimum power adaptation policy that maximizes the total channel capacity [2, 35, 47]. We further derive the rate adaptation policy but assuming that symbol rates in both channels are fixed, and the symbol rate in RF channel is much smaller than that in FSO channel. As an illustrative example, we assume that MQAM is used in RF channel, while MPAM is used in FSO channel, and determine the spectral efficiency. MPAM is selected because negative amplitude signals cannot be transmitted over FSO channels with direct detection. The MQAM is not power efficient for transmission over FSO channel with direct detection because it requires the addition of DC bias to convert negative amplitudes to positive ones, and as such is not considered here. Notice that M-ary pulse-position modulation (MPPM) can also be used for transmission over FSO channel. Because MPPM is highly spectrally inefficient, we restrict our attention to MPAM instead.

Before we continue with the description of different adaptation policies, we have to derive a target bit error probability equation P_b for MAPM and MQAM on an AWGN channel. In MPAM, the transmitted signal x takes values from the discrete set $X = \{0, d, \ldots, (M-1)d\}$ $(M \ge 2)$, where d is the Euclidean distance between two neighboring points. If all signal constellation points are equiprobable, the average signal energy is given by $E_s = d^2(M-1)(2M-1)/6$, and it is related to the bit energy E_b by $E_s = E_b \log_2 M$, so that signal-to-noise ratio per bit is defined by $SNR_b = E_b/N_0 = E_S/(N_0 \log_2 M) = d^2(M-1)(2M-1)/(6N_0 \log_2 M)$. Following the derivation similar to that reported in [10], the following expression for bit error probability of FSO channel has been derived in [35, 47]:

$$P_b^{\mathrm{MPAM}} \cong \frac{M-1}{M \log_2 M} \operatorname{erfc}\left(\sqrt{\frac{3\Gamma_0^{\mathrm{MPAM}}}{2(M-1)(2M-1)}}\right), \tag{4.98}$$

where the symbol SNR Γ_0^{MPAM} is the symbol SNR in the absence of scintillation and the erfc(z) function is defined by $\operatorname{erfc}(z) = (2/\sqrt{\pi})\int_z^\infty \exp(-u^2)du$. Because Eq. (4.98) is not invertible, we derive the following empirical formula which is valid in the regime of medium and high signal-to-noise ratios [35, 47]:

$$P_b^{\mathrm{MPAM}} \cong 0.2 \exp\left[-\frac{1.85\Gamma_0^{\mathrm{MPAM}}}{2^{2.19 \log_2 M} - 1}\right]. \tag{4.99}$$

The corresponding expressions for MQAM are the same as already provided in [10]. In derivations that follow, we will assume that target bit error probabilities in both channels are the same $P_b^{\mathrm{MPAM}} = P_b^{\mathrm{MQAM}} = P_b$. The total spectral efficiency R as the function of bit error probability P_b can be found as [35, 47].

$$R = \frac{1}{2.19} B_{\mathrm{FSO}} \log_2\left(1 + K_{\mathrm{FSO}}\Gamma_0^{\mathrm{FSO}}\right) + B_{\mathrm{RF}} \log_2\left(1 + K_{\mathrm{RF}}\Gamma_0^{\mathrm{RF}}\right), \tag{4.100}$$

where $K_{\mathrm{FSO}} = -1.85\backslash \ln(5P_b)$ and $K_{\mathrm{RF}} = -1.5\backslash \ln(5P_b)$. Because the total spectral efficiency changes as the channel conditions in either channel change, the spectral efficiency is a function of FSO channel irradiance i_k and RF fading coefficient h as follows [35, 47]:

$$R = \frac{1}{2.19} B_{\mathrm{FSO}} \log_2\left(1 + K_{\mathrm{FSO}}\Gamma^{\mathrm{FSO}}(i_k)\frac{P^{\mathrm{FSO}}(i_k)}{P}\right) + B_{\mathrm{RF}} \log_2\left(1 + K_{\mathrm{RF}}\Gamma^{\mathrm{RF}}(h)\frac{P^{\mathrm{RF}}(h)}{P}\right), \tag{4.101}$$

where $\Gamma^{\mathrm{FSO}}(i_k) = i_k^2\Gamma_0^{\mathrm{FSO}}$ and $\Gamma^{\mathrm{RF}}(h) = h^2\Gamma_0^{\mathrm{RF}}$. To derive the optimum power adaptation policy, subject to $P^{\mathrm{FSO}}(i_k) + P^{\mathrm{RF}}(h) \le P$, we have to define the corresponding Lagrangian, differentiate it with respect to $P^{\mathrm{FSO}}(i_k)$ and $P^{\mathrm{RF}}(h)$, and set corresponding derivatives to be equal to zero. The optimum power adaptation policy is obtained as the result of this derivation [35, 47]:

$$\begin{aligned} \frac{K_{\mathrm{FSO}}P^{\mathrm{FSO}}(i_k)}{P} &= \begin{cases} \dfrac{1}{\Gamma_{\mathrm{tsh}}} - \dfrac{1}{\Gamma^{\mathrm{FSO}}}, \Gamma^{\mathrm{FSO}} \ge \Gamma_{\mathrm{tsh}} \\ 0, \ \Gamma^{\mathrm{FSO}} < \Gamma_{\mathrm{tsh}} \end{cases}, \quad \Gamma^{\mathrm{FSO}} = \Gamma_0^{\mathrm{FSO}} i_k^2 \\ \frac{K_{\mathrm{RF}}P^{\mathrm{RF}}(h)}{P} &= \begin{cases} \dfrac{1}{\Gamma_{\mathrm{tsh}}} - \dfrac{1}{\Gamma^{\mathrm{RF}}}, \Gamma^{\mathrm{RF}} \ge \Gamma_{\mathrm{tsh}} \\ 0, \ \Gamma^{\mathrm{RF}} < \Gamma_{\mathrm{tsh}} \end{cases}, \quad \Gamma^{\mathrm{RF}} = \Gamma_0^{\mathrm{RF}} h^2 \end{aligned}, \tag{4.102}$$

where Γ_{tsh} is the threshold SNR, which is common to both channels. With this adaptation policy, more power and higher data rates are transmitted when the FSO (RF) channel conditions are good, less power and lower data rates are transmitted when FSO (RF) channel is bad, and nothing is transmitted when the SNR falls below the threshold Γ_{tsh}. The optimum threshold Γ_{tsh} can be obtained numerically by solving the following equation [35, 47]:

$$\int_{\sqrt{\Gamma_{\mathrm{tsh}}/\Gamma_0^{\mathrm{FSO}}}}^{\infty} \left(\frac{1}{K_{\mathrm{FSO}}\Gamma_{\mathrm{tsh}}} - \frac{1}{K_{\mathrm{FSO}}\Gamma_0^{\mathrm{FSO}} i_k^2}\right) f(i_k)di_k + b \int_{\sqrt{\Gamma_{\mathrm{tsh}}/\Gamma_0^{\mathrm{RF}}}}^{\infty} \left(\frac{1}{K_{\mathrm{RF}}\Gamma_{\mathrm{tsh}}} - \frac{1}{K_{\mathrm{RF}}\Gamma_0^{\mathrm{RF}} h^2}\right) f(h)dh = 1, \tag{4.103}$$

where $b = B_{\mathrm{RF}}/B_{\mathrm{FSO}}$, $f(i_t)$ is the PDF of FSO irradiance i_k given by Eq. (4.95), and $f(h)$ is the PDF of RF channel coefficient h given by Eq. (4.92). The optimum spectral efficiency, defined as data rate R over channel bandwidth B, can be evaluated by substituting Eq. (4.102) into Eq. (4.101) to obtain [35, 47].

$$\frac{R}{B} = \frac{1}{2.19} \int_{\sqrt{\Gamma_{\mathrm{tsh}}/\Gamma_0^{\mathrm{FSO}}}}^{\infty} \log_2\left(\frac{\Gamma_0^{\mathrm{FSO}} i_k^2}{\Gamma_{\mathrm{tsh}}}\right) f(i_k)di_k + b \int_{\sqrt{\Gamma_{\mathrm{tsh}}/\Gamma_0^{\mathrm{RF}}}}^{\infty} \log_2\left(\frac{\Gamma_0^{\mathrm{RF}} h^2}{\Gamma_{\mathrm{tsh}}}\right) f(h)dh \ [\mathrm{bits/s/Hz}]. \tag{4.104}$$

Although this adaptive-rate adaptive-power scheme provides excellent spectral efficiencies, the optimum threshold computation in Eq. (4.103) is time extensive. Instead, we can perform the *truncated channel inversion with fixed rate*. The truncated channel inversion adaptation can be performed by [35, 47]

$$\frac{P^{FSO}(i_k)}{P} = \begin{cases} \frac{1}{i_k^2 E_{\Gamma_{tsh}}^{FSO}[1/i_k^2]}, & \Gamma^{FSO} \geq \Gamma_{tsh} \\ 0, & \Gamma^{FSO} < i_{tsh} \end{cases},$$
$$\frac{P^{RF}(h)}{P} = \begin{cases} \frac{1}{h^2 E_{\Gamma_{tsh}}^{RF}[1/h^2]}, & \Gamma^{RF} \geq \Gamma_{tsh} \\ 0, & \Gamma^{RF} < \Gamma_{tsh} \end{cases} \tag{4.105}$$

where

$$\left\langle \frac{1}{i_k^2} \right\rangle_{\Gamma_{tsh}}^{FSO} = \int_{\sqrt{\Gamma_{tsh}/\Gamma_0^{FSO}}}^{\infty} \frac{f(i_k)}{i_k^2} di_k, \left\langle \frac{1}{h^2} \right\rangle_{\Gamma_{tsh}}^{RF} = \int_{\sqrt{\Gamma_{tsh}/\Gamma_0^{RF}}}^{\infty} \frac{f(h)}{h^2} dh. \tag{4.106}$$

The threshold Γ_{tsh} in Eq. (4.106) is obtained by maximizing the spectral efficiency as given below [35, 47]:

$$R = \max_{\Gamma_{tsh}} \left\{ \frac{1}{2.19} B_{FSO} \log_2 \left(1 + K_{FSO} \Gamma_0^{FSO} \frac{1}{\langle 1/i_k^2 \rangle_{\Gamma_{tsh}}^{FSO}} \right) P\left(i_k \geq \sqrt{\Gamma_{tsh}/\Gamma_0^{FSO}} \right) \right.$$
$$\left. + B_{RF} \log_2 \left(1 + K_{RF} \Gamma_0^{RF} \frac{1}{\langle 1/h^2 \rangle_{\Gamma_{tsh}}^{RF}} \right) P\left(h \geq \sqrt{\Gamma_{tsh}/\Gamma_0^{RF}} \right) \right\}, \tag{4.107}$$

where

$$P\left(i_k \geq \sqrt{\Gamma_{tsh}/\Gamma_0^{FSO}} \right) = \int_{\sqrt{\frac{\Gamma_{tsh}}{\Gamma_0^{FSO}}}}^{\infty} f(i_k) di_k, P\left(h \geq \sqrt{\Gamma_{tsh}/\Gamma_0^{RF}} \right) = \int_{\sqrt{\frac{\Gamma_{tsh}}{\Gamma_0^{RF}}}}^{\infty} f(h) dh. \tag{4.108}$$

In Fig. 4.22, we show the spectral efficiencies for FSO system only, which can be achieved using the optimum power and rate adaptation and MPAM for different target bit error probabilities, and both (a) weak turbulence regime ($\sigma_R = 0.2$, $\alpha' = 51.913$, $\beta' = 49.113$) and (b) strong turbulence regime ($\sigma_R = 2$, $\alpha' = 4.3407$, $\beta' = 1.3088$). For example, the spectral efficiency *R*/*B* of 2 bits/s/Hz at $P_b = 10^{-9}$ is achieved for symbol SNR of 23.3 dB in weak turbulence regime and 26.2 dB in strong turbulence regime. In the same figure, we report the spectral efficiencies that can be achieved by both channel inversion ($\Gamma_{tsh} = 0$) and truncated channel inversion ($\Gamma_{tsh} > 0$). In the weak turbulence regime (see Fig. 4.24(a)), even simple channel inversion performs comparable to an optimum adaptive-power adaptive-rate scheme. However, in the strong turbulence regime (see Fig. 4.24(b)), this scheme faces significant performance degradation. On the other hand, the truncated channel inversion scheme in strong turbulence regime faces moderate performance degradation, about 3.7 dB at $P_b = 10^{-9}$ for a spectral efficiency of 2 bits/s/Hz. The optimum adaptation policy for FSO channel at $P_b = 10^{-6}$ for a spectral efficiency of 4 bits/s/Hz provides a moderate improvement of 3.3 dB in the weak turbulence regime over a nonadaptive scheme, while the improvement in the strong turbulence regime is significant at 31.7 dB.

In Fig. 4.23, we report the spectral efficiencies for the hybrid FSO-RF system shown in Fig. 4.21, with RF subsystem fading parameters $\alpha = 3$, $\mu = 2$ in both weak turbulence regime [Fig. 4.23(a)] and strong turbulence regime [Fig. 4.23(b)]. We assume that the FSO subsystem symbol rate is ten times larger than the RF subsystem data rate, that is, $b = 0.1$. For a spectral efficiency of 2 bits/s/Hz, the hybrid FSO-RF system outperforms the FSO system by 3.39 dB at a BER of 10^{-6} and 3.49 dB at a BER of 10^{-9}. It is interesting to notice that even truncated channel inversion for the hybrid system outperforms the optimum adaptation of the FSO system by 0.8 dB at a BER of 10^{-9} and a spectral efficiency of 2 bits/s/Hz.

In Fig. 4.24, we report the spectral efficiencies for the hybrid FSO-RF system, with RF subsystem fading parameters $\alpha = 2$, $\mu = 1$ (corresponding to Rayleigh fading) in both weak turbulence regime [Fig. 4.24(a)] and strong turbulence regime [Fig. 4.24(b)]. This case corresponds to the situation where there is no line of site between transmit and receive antennas for

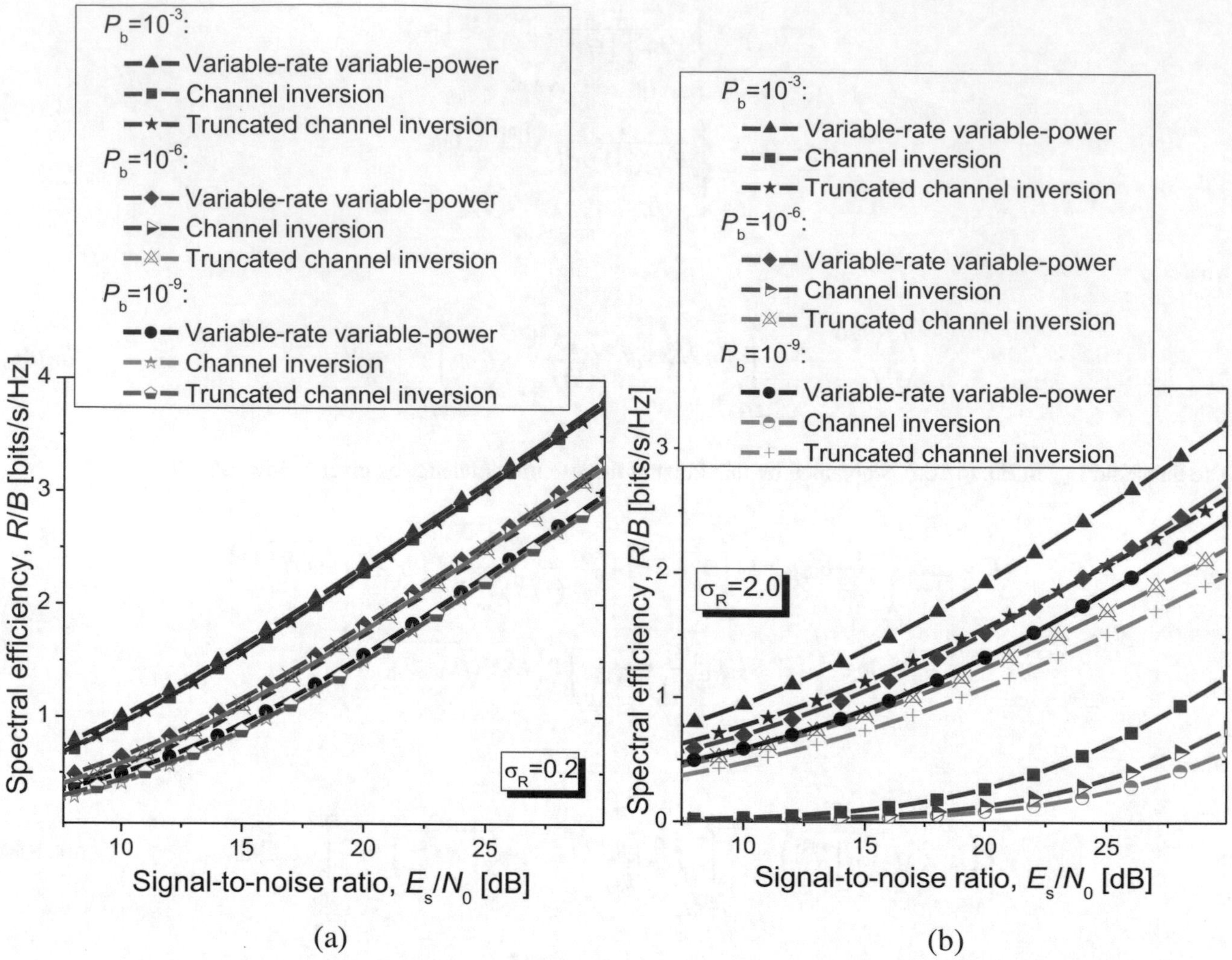

Fig. 4.22 Spectral efficiencies of an FSO system against symbol SNR for different target bit probabilities of error: (**a**) in weak turbulence regime and (**b**) in strong turbulence regime

the RF subsystem. We again assume that $b = 0.1$. For a spectral efficiency of 2 bits/s/Hz, the hybrid FSO-RF system outperforms the FSO system by 3.01 dB at a BER of 10^{-6} and 3.11 dB at a BER of 10^{-9}. The truncated channel inversion for the hybrid system performs comparable to the optimum adaptation of the FSO system.

In this section, we described two different adaptive modulation scenarios for both the FSO system with RF feedback and the hybrid FSO-RF system. In the next section, we describe the adaptive coding based on coded modulation.

By using the trellis-coded modulation (TCM) or cosset codes, we can separate the encoding and modulation process (see [10, 48] for more details). However, to keep the complexity of this approach reasonably low, the convolutional or block codes should be simple and short. Those codes are in principle of low rate and weak so that coding gains are moderate. For example, the adaptive coding scheme based on the TCM proposed in [48] is about 5 dB away from channel capacity. Instead, in this chapter, we propose to implement adaptive coding based on LDPC-coded modulation. For the FSO system, the input data are LDPC encoded and written to a buffer. Based on FSO channel irradiance, i_t, $\log_2 M(i_t)$ bits are taken at a time from a buffer and used to select the corresponding point from MPAM signal constellation. For the hybrid FSO-RF system, the LDPC-encoded sequence is split between FSO and RF subsystems (see Fig. 4.21). To facilitate the implementation, we assume that symbol rates in both subsystems are fixed, while constellation sizes and emitted powers are determined based on channel conditions in both channels using the adaptation scenarios described in the previous section. We further assume that the symbol rate in the RF subsystem is at least ten times lower than that in the FSO subsystem (e.g., 1 giga symbol/s in the FSO subsystem and 100 mega symbol/s in the RF subsystem).

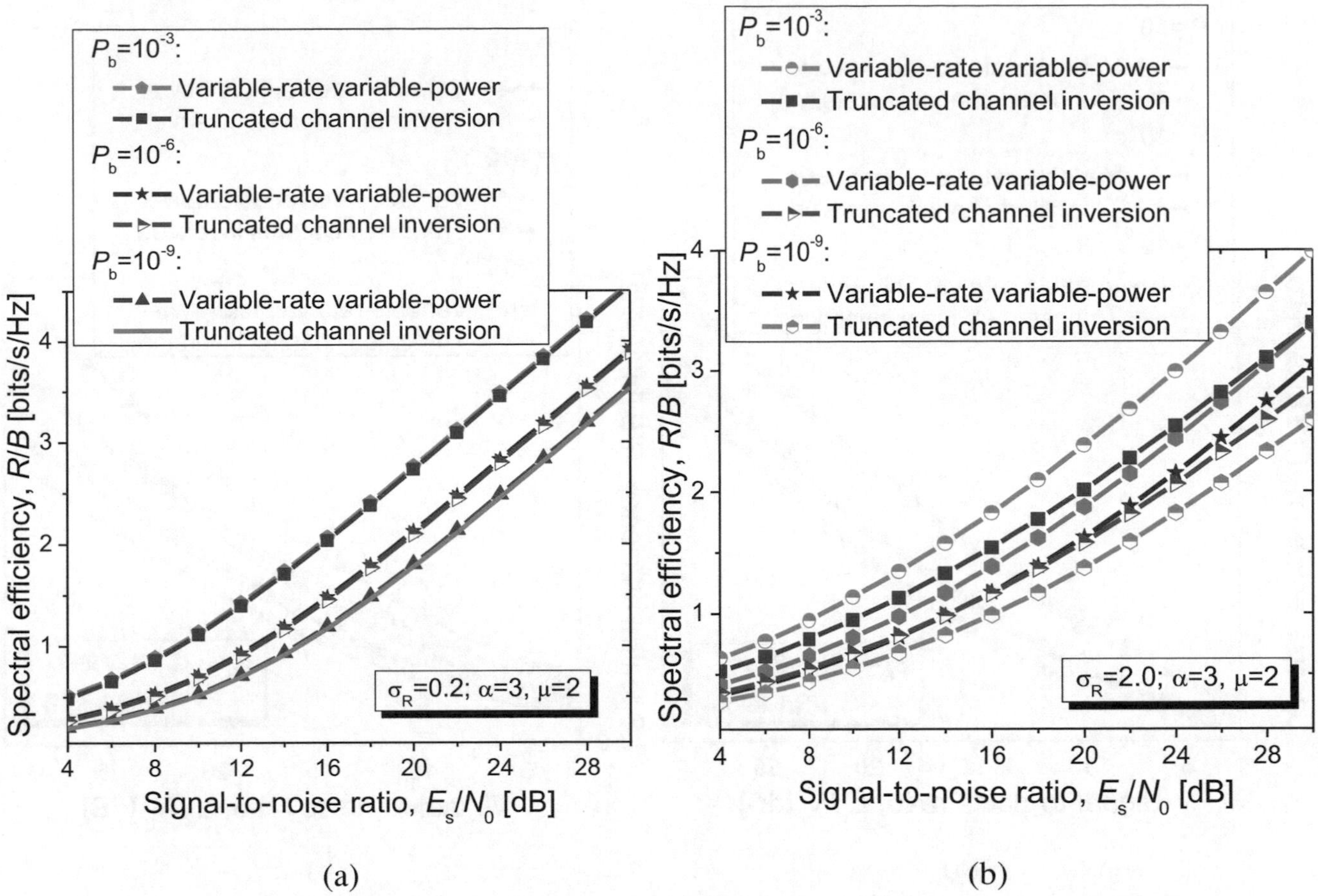

Fig. 4.23 Spectral efficiencies of a hybrid FSO-RF system with α = 3, μ = 2 fading against symbol SNR for different target bit probabilities of error: (**a**) in weak turbulence regime and (**b**) in strong turbulence regime

In Fig. 4.25(a), we show the *R*/*B* performance of the FSO system with adaptive LDPC-coded MPAM for different adaptation scenarios. Given the fact that the channel capacity of FSO channel under atmospheric turbulence is an open problem, we show in the same figure an *upper bound* in the absence of atmospheric turbulence from [54]. The coding gain over adaptive modulation at $P_b = 10^{-6}$ for $R/B = 4$ bits/s/Hz is 7.2 dB in both (weak and strong) turbulence regimes. Larger coding gains are expected at lower BERs and for higher spectral efficiencies. Further improvements can be obtained by increasing the girth of LDPC codes and employing better modulation formats. The increase in code word length to 100,515 does not improve the *R*/*B* performance that much as shown in Fig. 4.25(a). It is interesting to notice that by employing adaptive coding, the communication under saturation regime is possible, as shown in Fig. 4.25(a). Moreover, for a variable-rate variable-power scheme, there is no degradation in saturation regime compared to strong turbulence regime. Overall improvement from adaptive modulation and coding for $R/B = 4$ bits/s/Hz at $P_b = 10^{-6}$ over nonadaptive uncoded modulation ranges from 10.5 dB (3.3 dB from adaptive modulation and 7.2 dB from coding) in the weak turbulence regime to 38.9 dB in the strong turbulence regime (31.7 dB from adaptive modulation and 7.2 dB from coding). In Fig. 4.25(b), we show the *R*/*B* performance of the hybrid FSO-RF system with adaptive LDPC-coded modulation (MPAM is used in the FSO subsystem and MQAM in the RF subsystem) for different adaptation scenarios. The symbol rate in the FSO subsystem is set to be ten times larger than that in the RF subsystem ($b = 0.1$). For a spectral efficiency of 4 bits/s/Hz at a BER of 10^{-6}, the improvement of the hybrid FSO-RF system over the FSO system is 5.25 dB in Rayleigh fading (α = 2, μ = 1), 5.51 dB in Nakagami $m = 2$ fading (α = 2, μ = 2), and 5.63 dB in α =3, μ = 2 fading. For a spectral efficiency of 2 bits/s/Hz at the same BER, the improvement of the hybrid FSO-RF system over the FSO system is 3.32 dB in Rayleigh fading, 3.72 dB in Nakagami $m = 2$ fading, and 3.86 dB in α = 3, μ = 2 fading.

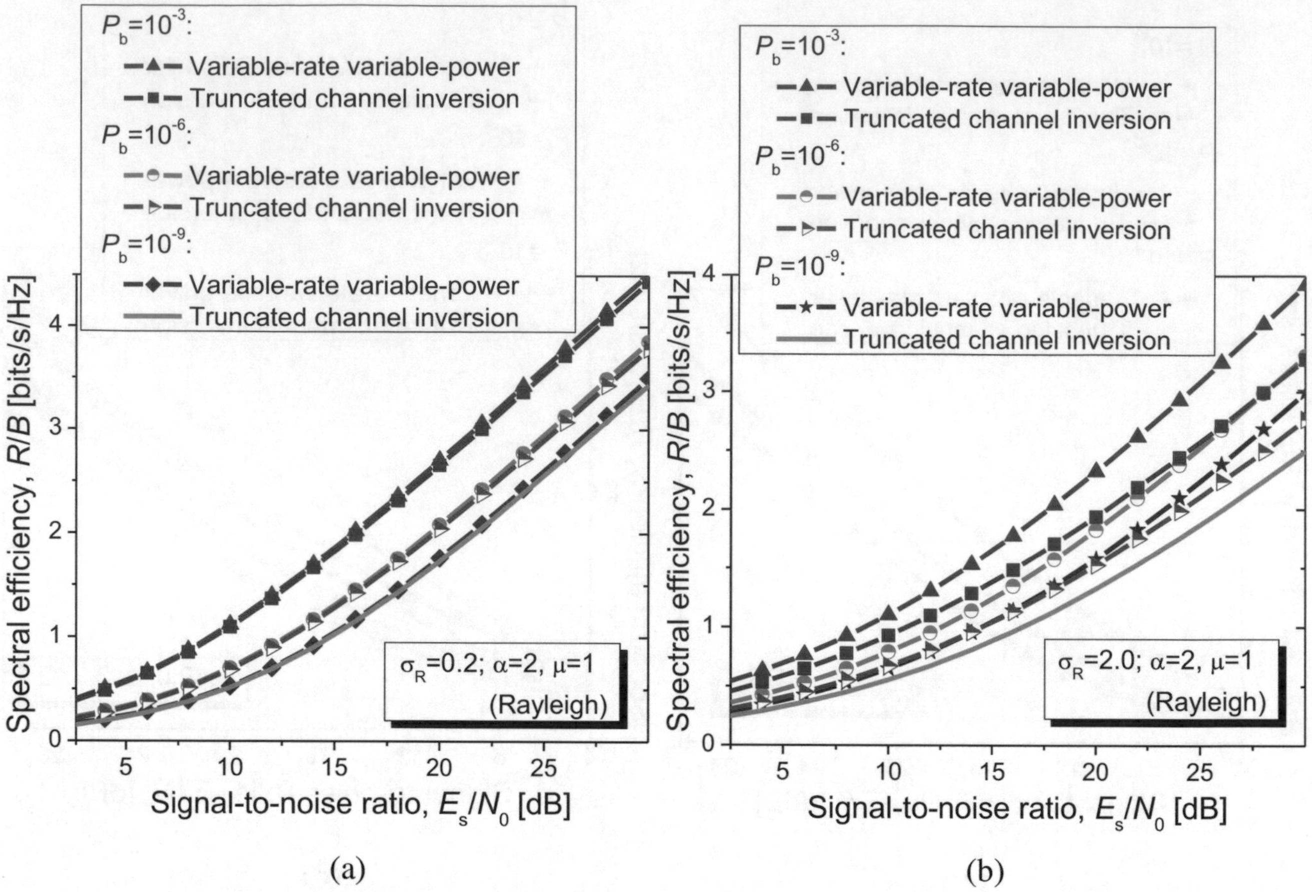

Fig. 4.24 Spectral efficiencies of a hybrid FSO-RF system with $\alpha = 2$, $\mu = 1$ (Rayleigh) fading against symbol SNR for different target bit probabilities of error: (**a**) in weak turbulence regime and (**b**) in strong turbulence regime

4.7 Concluding Remarks

This chapter has been devoted to wireless channel and optical channel capacities. The chapter starts with definitions of mutual information and channel capacity for discrete memoryless channels in Sect. 4.1.1, followed by the capacity of continuous channels and information capacity theorem in Sect. 4.1.2. Regarding the wireless channel capacity, described in Sect. 4.2, we describe how to calculate the channel capacity of flat-fading channels in Sect. 4.2.1 and frequency-selective channels in Sect. 4.2.2. We also discuss different optimum and suboptimum strategies to achieve channel capacity including the water-filling method, multiplexed coding and decoding, channel inversion, and truncated channel inversion. Depending on what is known about the channel state information (CSI), different strategies are described for channel capacity evaluation. When CSI is known on the receiver side only (CSI), we describe the Shannon (ergodic) channel capacity and capacity with outage in Sect. 4.2.1.1. On the other hand, when CSI is available on both transmitter and receiver sides (full CSI), we describe the Shannon capacity, zero-outage capacity, and outage capacity in Sect. 4.2.1.2. Regarding the frequency-selective fading channel capacity, described in Sect. 4.2.2, we describe how to calculate the channel capacity for both time-invariant channels (Sect. 4.2.2.1) and time-variant channels (Sect. 4.2.2.2). Additionally, we describe the optimum power adaption strategies in the same sections. For instance, the optimum power adaptation strategy for time-variant fading channels is the two-dimensional water-filling method in both time and frequency domains. Further, we explain how to model the channel with memory and describe how to calculate its capacity in Sect. 4.3. The key topics of this section include Markov sources (Sect. 4.3.1), McMillan sources (4.3.2), together with corresponding entropies, followed by the McMillan-Khinchin model for channel capacity evaluation (Sect. 4.3.3). We then describe in Sect. 4.4 how to employ the forward recursion of the BCJR algorithm to evaluate the information capacity. The next topic, described in Sect. 4.5, is related to the evaluation of the information capacity of fiber-optic communication channels with coherent optical detection. Finally, in Sect. 4.6, we describe how to evaluate the capacity of hybrid FSO-RF channels, with a hybrid FSO-RF system model being described in Sect. 4.6.1 and AMC for hybrid FSO-RF communication being described in Sect. 4.6.2. The set of problems is provided in the incoming section.

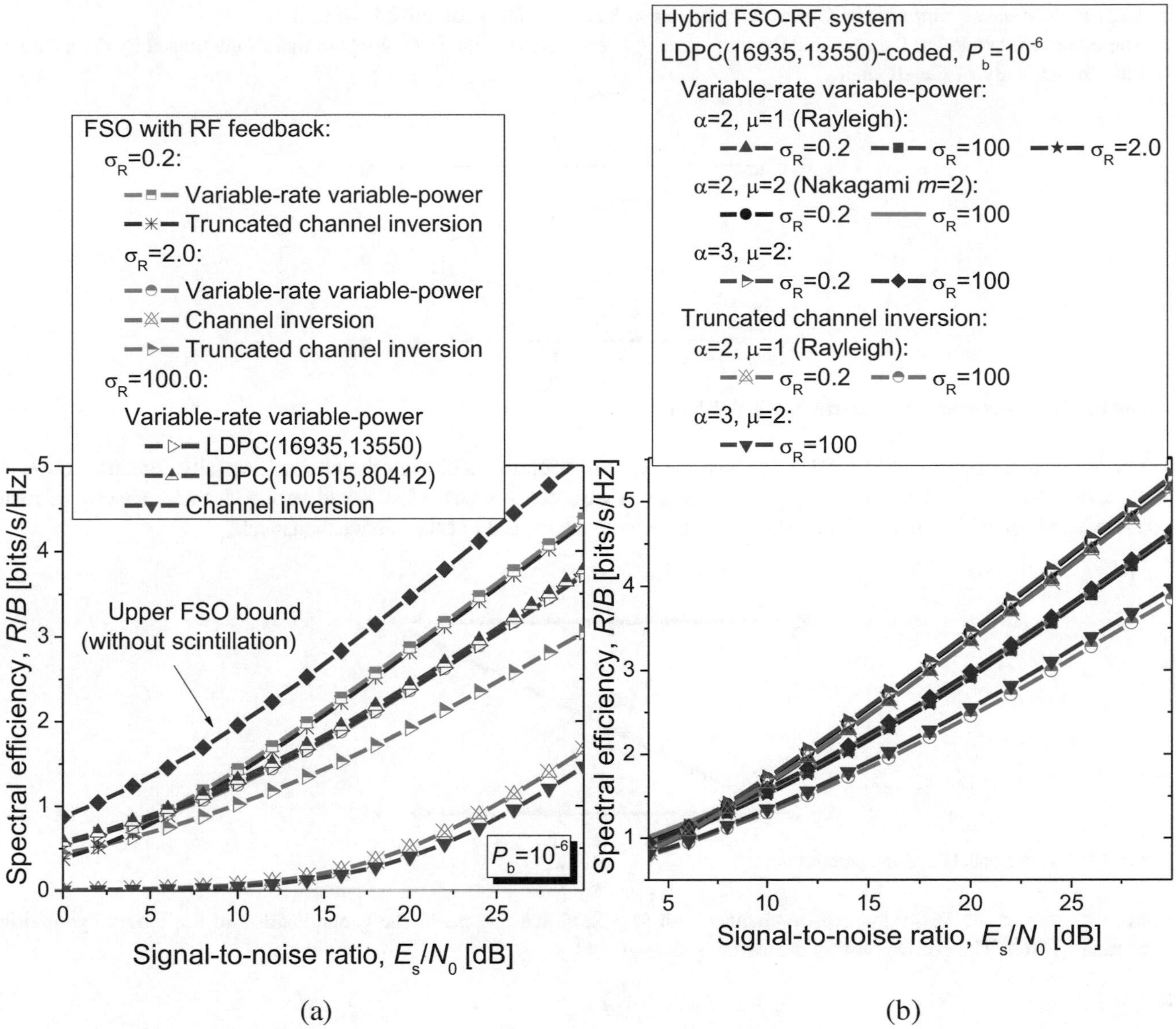

Fig. 4.25 Spectral efficiencies against symbol SNR for adaptive LDPC-coded modulation: (**a**) FSO with RF feedback only and (**b**) hybrid FSO-RF system

4.8 Problems

1. Let a 2×2 binary asymmetric channel matrix be defined as follows:

$$\Pi = \begin{bmatrix} P(y_0|x_0) & P(y_1|x_0) \\ P(y_0|x_1) & P(y_1|x_1) \end{bmatrix},$$

where x_i $(i = 0,1)$ are channel inputs and y_i $(i = 0,1)$ are channel outputs. A digital transmitter sends a binary sequence over a cascade of two binary asymmetric channels, characterized by corresponding channel matrices Π_1 and Π_2, respectively. The known elements of the first channel matrix are $P_1(0|0) = 0.8$ and $P_1(1|1) = 0.9$. On the other hand, the known elements of the second channel matrix are $P_2(0|0) = 0.7$ and $P_2(1|1) = 0.9$. The a priori probabilities for transmitting 0 and 1 are 0.6 and 0.4, respectively. Determine the overall channel matrix of the cascade of these two binary asymmetric channels. Determine a posteriori probabilities of receiving 0 and 1 at the output of the channel cascade. Determine the average error probability at the output of the channel cascade. Determine the mutual information of the channel cascade.

2. The *binary erasure channel* (BEC) has two inputs and three outputs as described in Fig. P2. The inputs are labeled as 0 and 1, and the outputs as 0, 1, and *e*. A fraction *f* of the incoming bits are erased by the channel. Find the capacity of the channel.

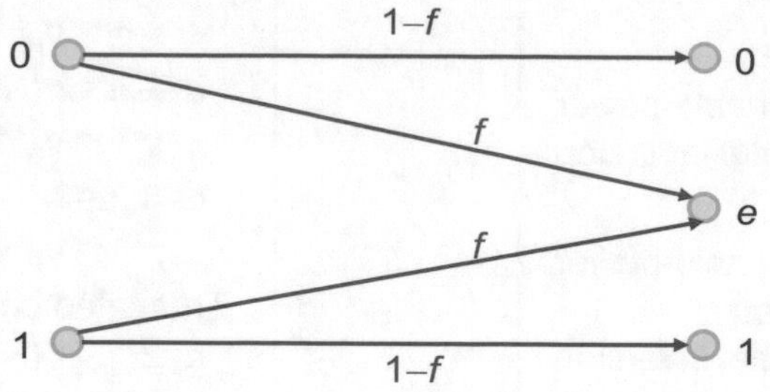

Fig. P2 The binary erasure channel transition probabilities

3. For the channel shown in Fig. P3, determine the input distribution achieving the channel capacity. Evaluate channel capacities for $\varepsilon = 0$, 0.5, and 1. When this channel is cascaded *n* times, what would be the channel capacity of such cascaded channel? Finally, in the limit $n \to \infty$, determine the capacity of the cascaded channel.

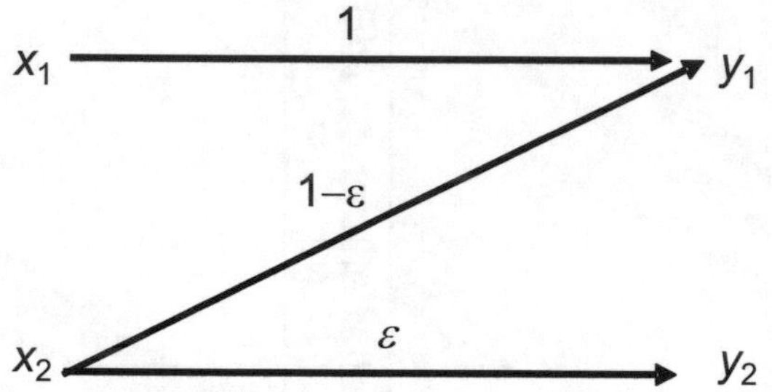

Fig. P3 The Z-channel transition probabilities

4. Let us consider the *M*-ary symmetric channel with $M > 2$, which represents the generalization of the binary symmetric channel (BSC). The corresponding channel matrix is given by.

$$P_{MSC} = \begin{bmatrix} v & p/(M-1) & p/(M-1) & \cdots & p/(M-1) \\ p/(M-1) & v & p/(M-1) & p/(M-1) & p/(M-1) \\ \vdots & \vdots & \vdots & \ddots & \vdots \\ p/(M-1) & p/(M-1) & p/(M-1) & p/(M-1) & v \end{bmatrix}, p + v = 1.$$

Determine the channel capacity as a function of *M*.

5. The binary symmetric channel (BSC) is described by the following channel matrix:

$$P_{BSC} = \begin{bmatrix} v & p \\ p & v \end{bmatrix}, p + v = 1.$$

Determine the channel matrices for two and three BSC channels serially connected (in a cascade). Determine the channel capacity for the cascade of two BSCs and cascade of three BSCs.

6. Let us consider the time-invariant frequency-selective block-fading channel with three subchannels of bandwidth 2 MHz, with frequency response amplitudes being 2, 4, and 6, respectively. For the transmit power 10 mW and noise power spectral density of $N_0 = 10^{-9}$ W/Hz, determine the corresponding Shannon capacity. What is the corresponding optimum power allocation strategy to achieve this capacity?

7. Let us consider the communication over Nakagami-m channel of bandwidth 20 MHz. Determine the distribution of SNR for this channel. Assuming that the channel state information is available at both transmitter and receiver sides (full CSI) and the SNR mean is 12 dB, solve problems (a)–(d).

(a) Determine the optimum power adaptation policy achieving the Shannon capacity.
(b) Determine the Shannon capacities for the following values of parameter $m = 0.5$, 1, and $m = (K + 1)^2/(2K + 1)$, with K being 20 dB (convert to linear scale first). Discuss the results. How far we are from the AWGN channel capacity of the same SNR for each of the cases?
(c) Repeat (b) but now assuming that only the receiver knows the CSI. Discuss the degradation compared to (b).
(d) Determine the zero-outage capacity and outage capacity, assuming that the outage probability is 0.02 for the same cases as in (b). Discuss the degradation compared to (b).

8. In this problem, we study the frequency-selective fading channel with a total available bandwidth of 20 MHz and corresponding coherence bandwidth being 5 MHz. Divide the total bandwidth into four subchannels with corresponding bandwidth being equal to the coherence bandwidth and assume that each subchannel is a Nakagami-m flat-fading channel with independent realizations for each subchannel. Assume the subchannels have average gains $\langle |H_1[i]|^2 \rangle = 1$, $\langle |H_2[i]|^2 \rangle = 0.75$, $\langle |H_3[i]|^2 \rangle = 0.5$, and $\langle |H_4[i]|^2 \rangle = 0.25$. Assume a total transmit power of 50 mW and a receiver noise spectral density of 0.002 μW/Hz. For $m = 0.5$, 1, and 2, solve for (a) and (b).
   (a) Determine the optimal two-dimensional water-filling power adaptation for this channel and the corresponding Shannon capacity, assuming both the transmitter and the receiver know the instantaneous value of $H_j[i]$, $j = 1, \ldots, 4$.
   (b) Allocate now the equal average power to each subchannel and then perform water filling for each subchannel. Compare the overall capacity against that in (a).
9. Let us now consider the time-invariant frequency-selective block-fading channel with 32 subchannels of bandwidth 2 MHz, with frequency response amplitudes being 1 for odd subchannels and $\sqrt{2}$ for even subchannels, respectively. For the transmit power 10 mW and noise power spectral density of $N_0 = 10^{-9}$ W/Hz, determine the OFDM channel capacity. What is the corresponding optimum power allocation strategy based on OFDM to achieve this channel capacity?
10. In this problem, we are concerned with *scintillation effects*, and the i.i.d. FSO channel model, in which the samples of irradiance are generated by gamma-gamma distribution.
    (a) For (binary) intensity modulation with direct detection (IM/DD) system, determine the average bit error probability when operating over turbulent FSO channel.
    (b) For coherent detection, determine the ergodic capacity, assuming that channel state information is known at either receiver (Rx) side or both (Tx and Rx) sides.
    (c) Determine the water-filling, channel inversion, and truncated channel inversion capacities, assuming that coherent detection is used.
11. The entropy of Markov source for moving k steps ahead from initial states is given as $H^{(k)} = -\sum_{i=1}^{n} p_i \sum_{j=1}^{n} p_{ij}^{(k)} \log p_{ij}^{(k)}$.

    Prove the following property: $H^{(k+1)} = H^{(k)} + H^{(1)}$.
12. For the Markov source from Fig. P10, determine:
    (a) Stationary transition and state probabilities.
    (b) Entropy of this source.

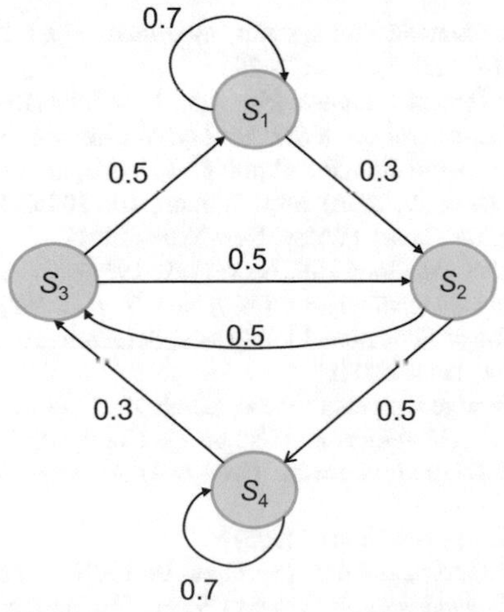

Fig. P10 A particular Markov source

13. Reproduce the results shown in Fig. 4.15 by using Monte Carlo integration.
14. Implement the algorithm described in Sect. 4.4. Reproduce the results shown in Fig. 4.15 and compare them against those from Problem 4.11.
15. Plot the i.i.d. channel capacity against optical SNR for different memories in the forward step of the BCJR algorithm, in the presence of both residual chromatic dispersion (11,200 ps/nm) and PMD with a DGD of 50 ps by observing the NRZ transmission system operating at 10 Gb/s.
16. Plot the i.i.d. channel capacity in the presence of intrachannel nonlinearities against the number of spans, for dispersion map shown in Fig. 4.17, for different memories in the forward step of the BCJR algorithm. The span length is set to $L = 120$ km, and each span consists of 2 $L/3$ km of D_+ fiber followed by $L/3$ km of D_- fiber. Pre-compensation of -1600 ps/nm and corresponding post-compensation are also applied. The parameters of D_+ and D_- fibers, used in simulations, are given in Table 4.1. Assume that the RZ modulation format of a duty cycle of 33% is observed, the extinction ratio is 14 dB, and the launched power is set to 0 dBm. EDFAs with a noise figure of 6 dB are deployed after every fiber section, the bandwidth of optical filter (modeled as super-Gaussian filter of eight order) is set to $3R_l$, and the bandwidth of electrical filter (modeled as Gaussian filter) to $0.7R_l$, with R_l being the line rate (defined as the bit rate divided by a code rate). Assume that the bit rate is 40 Gb/s.
17. Reproduce the results shown in Figs. 4.19 and 4.20.
18. Let us now study by the information capacity by using the dispersion map shown in 4.18(a), but now replacing EDFAs with a hybrid Raman/EDFA scheme described in Problem 16 of Chap. 2. Study the information capacities in a fashion similar to Fig. 4.20. Discuss the improvements with respect to Fig. 4.20.
19. Repeat Problem 18, but now by taking quantization effects into account. Observe the information capacities for the following number of bits: 8, 6, 4, and 2. Discuss the results.
20. Let us now study by the information capacity by using the dispersion map shown in Fig. 4.18(a), but now by replacing EDFAs with distributed Raman amplifiers, with typical parameters as described in Chap. 3. Study the information capacities in a fashion similar to Fig. 4.20. Discuss the improvements with respect to Fig. 4.20 and Problem 18.
21. Plot the ergodic channel capacity against SNR for 2×2 MIMO over MMF, assuming that MMF can support 500 modes. Assume that channel coefficients follow zero-mean Gaussian distribution. Study different strategies: CSIT, CSIR, and full CSI. Compare the results against the SISO system. *Hint*. You can use the MMF model described in Sect. 8.2.1 of Chap. 8.
22. Plot the ergodic channel capacity against SNR for different MIMO systems over FMF, assuming that FMF can support 2, 4, and 8 modes. Assume that channel coefficients follow zero-mean Gaussian distribution. Study different strategies (CSIT, CSIR, and full CSI). Compare the results against the SISO system. *Hint*. You can use the FMF model described in Sect. 8.2.1 of Chap. 8.

References

1. I.B. Djordjevic, Advanced coding for optical communication systems, in *Optical Fiber Telecommunications VI*, ed. by I. Kaminow, T. Li, A. Willner, (Elsevier/Academic Press, Boston, May 2013), pp. 221–296
2. I.B. Djordjevic, W. Ryan, B. Vasic, *Coding for Optical Channels* (Springer, New York/Dordrecht/Heidelberg/London, 2010)
3. M. Cvijetic, I.B. Djordjevic, *Advanced Optical Communication Systems and Networks* (Artech House, Norwood, MA, 2013)
4. I.B. Djordjevic, On advanced FEC and coded modulation for ultra-high-speed optical transmission. IEEE Communications Surveys and Tutorials **18**(3), 1920–1951, third quarter 2016 (Aug 19, 2016). https://doi.org/10.1109/COMST.2016.2536726
5. T.M. Cover, J.A. Thomas, *Elements of Information Theory* (Wiley, New York, 1991)
6. F.M. Reza, *An Introduction to Information Theory* (McGraw-Hill, New York, 1961)
7. R.G. Gallager, *Information Theory and Reliable Communication* (Wiley, New York, 1968)
8. F.M. Ingels, *Information and Coding Theory* (Intext Educational Publishers, Scranton, 1971)
9. S. Haykin, *Communication Systems* (Wiley, New York, 2004)
10. A. Goldsmith, *Wireless Communications* (Cambridge University Press, Cambridge, 2005)
11. D. Tse, P. Viswanath, *Fundamentals of Wireless Communication* (Cambridge University Press, Cambridge, 2005)
12. S. Benedetto, E. Biglieri, *Principles of Digital Transmission with Wireless Applications* (Kluwer Academic/Plenum Publishers, New York, 1999)
13. E. Biglieri, *Coding for Wireless Channels* (Springer, New York, 2005)
14. J. Wolfowitz, *Coding Theorems of Information Theory*, 2nd edn. (Springer, Berlin/New York, 1964)
15. T.M. Duman, A. Ghrayeb, *Coding for MIMO Communication Systems* (Wiley, Chichester, 2007)
16. J.G. Proakis, *Digital Communications* (McGraw-Hill, Boston, MA, 2001)
17. C.E. Shannon, A mathematical theory of communication. Bell Syst. Tech. J. **27**, 379–423. 623–656 (July, Oct 1948)
18. B. McMillan, The basic theorems of information theory. Ann. Math. Statistics **24**, 196–219 (1952)

19. A.I. Khinchin, *Mathematical Foundations of Information Theory* (Dover Publications, New York, 1957)
20. D. Arnold, A. Kavcic, H.-A. Loeliger, P. O. Vontobel, W. Zeng, Simulation-based computation of information rates: upper and lower bounds, in Proceedings of the IEEE International symposium information theory (ISIT 2003), 2003, p. 119
21. D. Arnold, H.-A. Loeliger, On the information rate of binary-input channels with memory, in Proceedings of the 2001 International conference communications, Helsinki, Finland, 11–14 June 2001, pp. 2692–2695
22. H.D. Pfitser, J. B. Soriaga, P. H. Siegel, On the achievable information rates of finite state ISI channels, in Proceedings of the Globecom 2001, San Antonio, TX, 25–29 Nov 2001, pp. 2992–2996
23. I.B. Djordjevic, B. Vasic, M. Ivkovic, I. Gabitov, Achievable information rates for high-speed long-haul optical transmission. IEEE/OSA J. Lightw. Technol. **23**, 3755–3763 (Nov 2005)
24. M.̇ Ivkovic, I.B. Djordjevic, B. Vasic, Calculation of achievable information rates of long-haul optical transmission systems using instanton approach. IEEE/OSA J. Lightw. Technol. **25**, 1163–1168 (May 2007)
25. M. Ivkovic, I. Djordjevic, P. Rajkovic, B. Vasic, Pulse energy probability density functions for long-haul optical fiber transmission systems by using instantons and Edgeworth expansion. IEEE Photon. Technol. Lett. **19**(20), 1604, 15 Oct 2007–1606
26. I.B. Djordjevic, N. Alic, G. Papen, S. Radic, Determination of achievable information rates (AIRs) of IM/DD systems and AIR loss due to chromatic dispersion and quantization. IEEE Photon. Technol. Lett. **19**(1), 12–14 (1 Jan 2007)
27. I.B. Djordjevic, L.L. Minkov, H.G. Batshon, Mitigation of linear and nonlinear impairments in high-speed optical networks by using LDPC-coded turbo equalization. IEEE J. Sel. Areas Comm. Optical Comm. Netw. **26**(6), 73–83 (Aug 2008)
28. I.B. Djordjevic, L.L. Minkov, L. Xu, T. Wang, Suppression of fiber nonlinearities and PMD in coded-modulation schemes with coherent detection by using turbo equalization. IEEE/OSA J. Opt. Comm. Netw. **1**(6), 555–564 (Nov 2009)
29. R.-J. Essiambre, G.J. Foschini, G. Kramer, P.J. Winzer, Capacity limits of information transport in fiber-optic networks. Phys. Rev. Lett. **101**, 163901-1–163901-4 (17 Oct 2008)
30. R.-J. Essiambre, G. Kramer, P.J. Winzer, G.J. Foschini, B. Goebel, Capacity limits of optical fiber networks. J. Lightw. Technol. **28**(4), 662–701 (Feb 2010)
31. D.P. Bertsekas, *Nonlinear Programming*, 2nd edn. (Athena Scientific, Belmont, 1999)
32. E.K.P. Chong, S.H. Zak, *An Introduction to Optimization*, 3rd edn. (Wiley, New York, 2008)
33. I.B. Djordjevic, H. G. Batshon, L. Xu, T. Wang, Coded polarization-multiplexed iterative polar modulation (PM-IPM) for beyond 400 Gb/s serial optical transmission, in Proceedings of the OFC/NFOEC 2010, Paper No. OMK2, San Diego, CA, 21–25 Mar 2010
34. H.G. Batshon, I.B. Djordjevic, L. Xu, T. Wang, Iterative polar quantization based modulation to achieve channel capacity in ultra-high-speed optical communication systems. IEEE Photon. J. **2**(4), 593–599 (Aug 2010)
35. C. Lin, I.B. Djordjevic, D. Zou, Achievable information rates calculation for optical OFDM transmission over few-mode fiber long-haul transmission systems. Opt. Express **23**(13), 16846–16856 (2015)
36. J.M. Kahn, K.-P. Ho, Spectral efficiency limits and modulation/detection techniques for DWDM systems. IEEE Selected Topics Quantum Electron. **10**, 259–272 (Mar/Apr 2004)
37. K.S. Turitsyn, S.A. Derevyanko, I.V. Yurkevich, S.K. Turitsyn, Information capacity of optical fiber channels with zero average dispersion. Phys. Rev. Lett. **91**(20), 203–901 (Nov 2003)
38. J. Tang, The channel capacity of a multispan DWDM system employing dispersive nonlinear optical fibers and an ideal coherent optical receiver. J. Lightw. Technol. **20**, 1095–1101 (July 2002)
39. E.E. Narimanov, P. Mitra, The channel capacity of a fiber optics communication system: perturbation theory. J. Lightw. Technol. **20**, 530–537 (Mar 2002)
40. A. Mecozzi, M. Shtaif, On the capacity of intensity modulated systems using optical amplifiers. IEEE Photon. Technol. Lett. **13**, 1029–1031 (Sept 2001)
41. P.P. Mitra, J.B. Stark, Nonlinear limits to the information capacity of optical fiber communications. Nature **411**, 1027–1030 (June 2001)
42. J. Tang, The Shannon channel capacity of dispersion-free nonlinear optical fiber transmission. J. Lightw. Technol. **19**, 1104–1109 (Aug 2001)
43. J. Tang, The multispan effects of Kerr nonlinearity and amplifier noises on Shannon channel capacity for a dispersion-free nonlinear optical fiber. J. Lightw. Technol. **19**, 1110–1115 (Aug 2001)
44. K.-P. Ho, J. M. Kahn, Channel capacity of WDM systems using constant-intensity modulation formats, in Optical Fiber Communications Conferences (OFC '02), 2002, paper ThGG85
45. J.M. Kahn, K.-P. Ho, Ultimate spectral efficiency limits in DWDM systems, in Optoelectronics communication conference, Yokohama, Japan, 8–12 July 2002
46. I.B. Djordjevic, G.T. Djordjevic, On the communication over strong atmospheric turbulence channels by adaptive modulation and coding. Opt. Express **17**(20), 18250–18262 (25 Sept 2009)
47. I.B. Djordjevic, Adaptive modulation and coding for free-space optical channels. IEEE/OSA J. Opt. Commun. Netw. **2**(5), 221–229 (May 2010)
48. A. Goldsmith, S.-G. Chua, Adaptive coded modulation for fading channels. IEEE Trans. Comm. **46**, 595–601 (May 1998)
49. M.D. Yacoub, The α-μ distribution: a physical fading model for the Stacy distribution. IEEE Trans. Vehic. Techn. **56**(1), 27–34 (Jan 2007)
50. M.A. Al-Habash, L.C. Andrews, R.L. Phillips, Mathematical model for the irradiance probability density function of a laser beam propagating through turbulent media. Opt. Eng. **40**, 1554–1562 (2001)
51. L.C. Andrews, R.L. Philips, *Laser Beam Propagation through Random Media* (SPIE Press, 2005)
52. I.B. Djordjevic, S. Denic, J. Anguita, B. Vasic, M.A. Neifeld, LDPC-coded MIMO optical communication over the atmospheric turbulence channel. IEEE J. Lightw. Technol. **26**(5), 478–487 (2008)
53. I.B. Djordjevic, Coding and modulation techniques for optical wireless channels, in *Advanced Optical Wireless Communication Systems*, ed. by S. Arnon, J. Barry, G. Karagiannidis, R. Schober, M. Uysal, (Cambridge University Press, May 2012), pp. 11–53
54. A.A. Farid, S. Hranilovic, Upper and lower bounds on the capacity of wireless optical intensity channels, in Proceedings of the ISIT 2007, pp. 2416–2420, Nice, France, 24 June–29 June 2007. A.R. Shah, R.C.J. Hsu, A.H. Sayed, B. Jalali, Coherent optical MIMO (COMIMO). *J. Lightw. Technol.*, **23**(8), 2410–2419, Aug 2005

Advanced Modulation and Multiplexing Techniques

5

Abstract

This chapter is devoted to advanced modulation and multiplexing techniques suitable for both wireless and optical communication systems. The chapter starts with signal space theory concepts applied to wireless communication systems. After the geometric representations of signals, we describe various multidimensional modulators and demodulators suitable for wireless communication applications, in particular the Euclidean distance, correlation, and matched filter-based detectors, followed by the description of frequency-shift keying (FSK). Both continuous-time (CT) and discrete-time (DT) implementations for pulse amplitude modulation schemes, suitable for wireless communications, are described as well. After that, the focus is moved to multilevel schemes suitable for both wireless and optical communication applications, including M-ary PSK, star-QAM, square-QAM, and cross-QAM. Regarding the optical communications, the transmitters for M-ary PSK, star-QAM, and square−/cross-QAM are described in detail. The next topic in the chapter is related to multicarrier modulation, including the description of multicarrier systems with both nonoverlapping and overlapping subcarriers as well as the introduction of various approaches to deal with fading effects at the subcarrier level. The concept of OFDM is also introduced; however, details are provided in Chap. 7. The MIMO fundamentals are then provided, including the description of key differences with respect to diversity scheme, as well as the introduction of array, diversity, and multiplexing gains. The parallel MIMO channel decomposition is briefly described. The details on MIMO signal processing are postponed for Chap. 8. In the section on polarization-division multiplexing (PDM) and four-dimensional (4-D) signaling, we describe key differences between PDM and 4-D signaling and describe how both types of schemes can be implemented in both wireless and optical communications. The focus is moved to the spatial-division multiplexing (SDM) and multidimensional signaling. We describe how SDM can be applied in wireless communications first. Then we describe how various degrees of freedom including amplitude, phase, frequency, polarization states, and spatial modes can be used to convey the information in the optical domain. In the same sections, the SDM concepts for fiber-optic communications are described as well. The section concludes with SDM and multidimensional signaling concepts applied to free-space optical (FSO) communications. The next topic is devoted to the signal constellation design, including iterative polar modulation (IPM), signal constellation design for circular symmetric optical channels, energy-efficient signal constellation design, and optimum signal constellation design (OSCD). The final section of the chapter is devoted to the nonuniform signaling, in which different signal constellation points are transmitted with different probabilities. The set of problems is provided after the concluding remarks.

5.1 Signal Space Theory in Wireless Communications

Before we start with a detailed description of advanced modulation schemes, we will present a generic view of optical digital communication system, explain the geometric representation of a signal, and introduce signal constellation diagrams. Additionally, we will provide generic configurations of modulators and demodulators.

I. B. Djordjevic, *Advanced Optical and Wireless Communications Systems*, https://doi.org/10.1007/978-3-030-98491-5_5

5.1.1 Geometric Representation of Signals

The key idea behind the geometric representation of the signals can be summarized as follows [1–7]: represent any set of M energy (complex-valued) signals $\{s_i(t)\}$ as a linear combination of N orthonormal basis functions $\{\Phi_j\}$, where $N \leq M$, as follows:

$$s_i(t) = \sum_{j=1}^{N} s_{ij}\Phi_j(t), \begin{cases} 0 \leq t \leq T \\ i = 1, 2, \ldots, M \end{cases}, \tag{5.1}$$

where the coefficients of expansion represent projections along the basis functions:

$$s_{ij} = \langle \Phi_j(t)|s_i(t)\rangle = \int_0^T s_i(t)\Phi_j^*(t)dt, \begin{cases} i = 1, 2, \ldots, M \\ j = 1, 2, \ldots, N \end{cases}. \tag{5.2}$$

The basis functions satisfy the *principle of orthogonality* given as [1–9]

$$\int_0^T \Phi_i(t)\Phi_j^*(t)dt = \delta_{ij} = \begin{cases} 1, i = j \\ 0, i \neq j \end{cases}. \tag{5.3}$$

The signal vector can now be represented in a vector form as

$$\mathbf{s}_i = |s_i\rangle = \begin{bmatrix} s_{i1} \\ s_{i2} \\ \cdots \\ s_{iN} \end{bmatrix}; i = 1, 2, \ldots, M, \tag{5.4}$$

where the jth component (coordinate) represents a projection along the jth basis function. The inner (dot) product of a signal vector with itself is defined by

$$\langle s_i|s_i\rangle = \|\mathbf{s}_i\|^2 = \mathbf{s}_i^\dagger \mathbf{s}_i = \sum_{j=1}^{N} |s_{ij}|^2; i = 1, 2, \ldots, M. \tag{5.5}$$

The signal energy of the ith symbol can be now expressed as

$$\begin{aligned} E_i &= \int_0^T |s_i(t)|^2 dt = \int_0^T \left[\sum_{j=1}^{N} s_{ij}\Phi_j(t)\right]\left[\sum_{k=1}^{N} s_{ik}^*\Phi_k^*(t)\right] dt \\ &= \sum_{j=1}^{N}\sum_{k=1}^{N} s_{ij}s_{ik}^* \underbrace{\int_0^T \Phi_j(t)\Phi_k^*(t)dt}_{1, k=j} = \sum_{j=1}^{N} |s_{ij}|^2 = \|\mathbf{s}_i\|^2. \end{aligned} \tag{5.6}$$

The inner product of signals $s_i(t)$ and $s_k(t)$ is related to the correlation coefficient between them as

$$\langle s_i|s_k\rangle = \int_0^T s_i^*(t)s_k(t)dt = \mathbf{s}_i^\dagger \mathbf{s}_k, \tag{5.7}$$

where we used † to denote the Hermitian conjugate. The Euclidean distance between signals $s_i(t)$ and $s_j(t)$ can be determined as

$$\|\mathbf{s}_i - \mathbf{s}_k\| = \sqrt{\sum_{j=1}^{N} |s_{ij} - s_{ik}|^2} = \sqrt{\int_0^T |s_i(t) - s_k(t)|^2 dt}. \tag{5.8}$$

On the other hand, the angle between two signal vectors can be determined as

$$\theta_{ik} = \cos^{-1}\left(\frac{\mathbf{s}_i^\dagger \mathbf{s}_k}{\|\mathbf{s}_i\|\|\mathbf{s}_k\|}\right). \tag{5.9}$$

By using Eqs. (5.6) and (5.9), we can write that

$$\cos\theta_{i,k} = \frac{\mathbf{s}_i^\dagger \mathbf{s}_k}{\|\mathbf{s}_i\|\|\mathbf{s}_k\|} = \frac{\int_{-\infty}^{\infty} s_i^*(t) s_k(t) dt}{\left[\int_{-\infty}^{\infty} |s_i(t)|^2 dt\right]^{1/2} \left[\int_{-\infty}^{\infty} |s_k(t)|^2 dt\right]^{1/2}}. \tag{5.10}$$

Since $|\cos\theta_{i,k}| \leq 1$, the following inequality can be derived from (5.10):

$$\left|\int_{-\infty}^{\infty} s_i^*(t) s_k(t) dt\right| \leq \left[\int_{-\infty}^{\infty} |s_i(t)|^2 dt\right]^{1/2} \left[\int_{-\infty}^{\infty} |s_k(t)|^2 dt\right]^{1/2}. \tag{5.11}$$

The arbitrary set of orthonormal basis functions $\{\Phi_j\}$ can be used in the geometric representation of signals. As an illustration, the geometric representation of two sets of signals, also known as *signal constellations*, is shown in Fig. 5.1. The one-dimensional (for $N = 1$) ternary signal constellation (for $M = 3$) is shown in Fig. 5.1(a), while the two-dimensional ($N = 2$) quaternary signal constellation, also known as quaternary quadrature amplitude modulation (4-QAM) or quadriphase-shift keying (QPSK), is shown in Fig. 5.1(b).

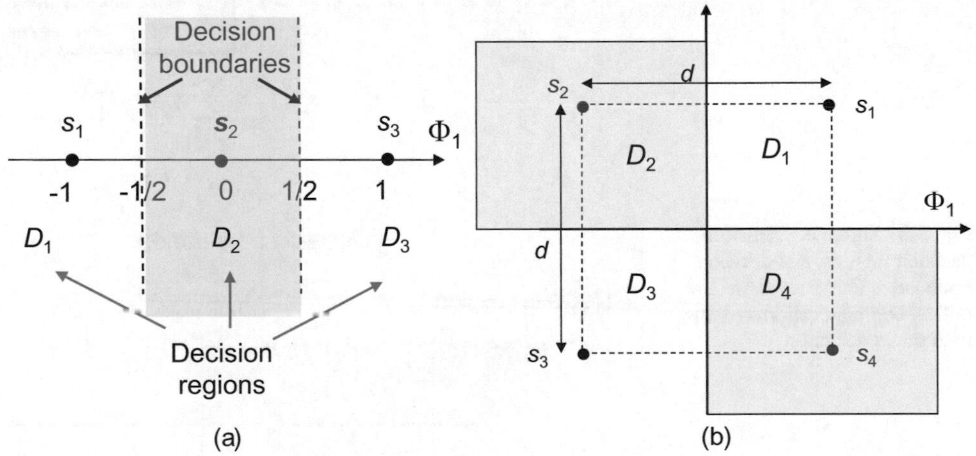

Fig. 5.1 Illustration of geometric representation of signals for (**a**) $N = 1, M = 3$ and (**b**) $N = 2, M = 4$

5.1.2 Modulators and Demodulators

To generate a given signal constellation point $\boldsymbol{s}_i$, we can use Eq. (5.4), commonly referred to as the *synthesis equation*; we can use the *modulator* (*synthesizer*) shown in Fig. 5.2, in the context of wireless communication systems. For binary sources, we need to use parallel-to-serial (P/S) converter, whose output is used as an address for the *N*-dimensional mapper, implemented as a look-up table (LUT). In this case, every M-ary symbol carries $m = \log_2 M$ bits.

At the receiver side, decision boundaries should be established in the signal space to identify constellation points being transmitted. If we assume the equal probability of signal transmission, the *decision boundaries* (see Fig. 5.1) between states can be positioned as straight lines perpendicular to lines $\boldsymbol{s}_i - \boldsymbol{s}_k$ and passing at distance $d/2$ from $\boldsymbol{s}_i$. Neighboring decision boundaries determine a given *decision region*. In such a way, the signal space is split into M decision regions, denoted as D_i. When received vector $\boldsymbol{r}$, obtained as a sum of transmitted vector $\boldsymbol{s}_j$ and noise vector $\boldsymbol{w}$, falls within the decision regionD_i, we decide in favor of signal point $\boldsymbol{s}_i$. As an illustrative example, applicable to M-ary PAM suitable for optical communication systems with direct detection, when the signals used in signal space do not have negative amplitude, we provide the signal constellation and decision boundaries in Fig. 5.3.

Another decision strategy will be to use the *Euclidean distance receiver* [1], for which the Euclidian distance square between the received signal $r(t)$ and the mth transmitted signal $s_m(t)$ is defined as

$$d_E^2(r, s_m) = \int_0^T |r(t) - s_m(t)|^2 dt. \tag{5.12a}$$

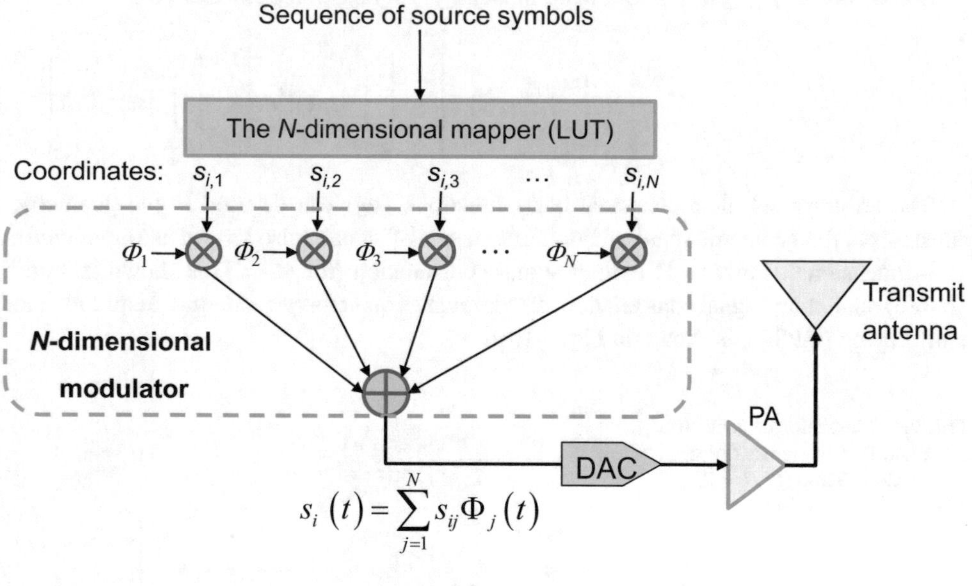

Fig. 5.2 The configuration of the M-ary *N*-dimensional modulator (synthesizer) for wireless communications. *DAC* digital-to-analog converter, *PA* power amplifier

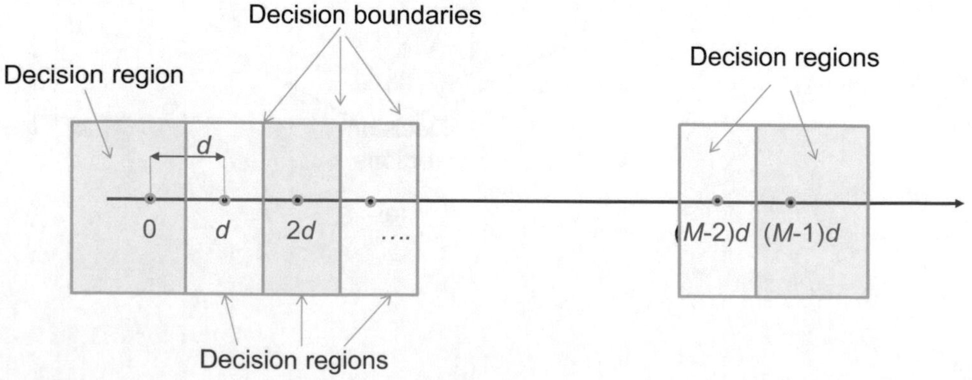

Fig. 5.3 Signal constellation, decision regions, and decision boundaries for M-ary PAM for optical communication systems with direct detection

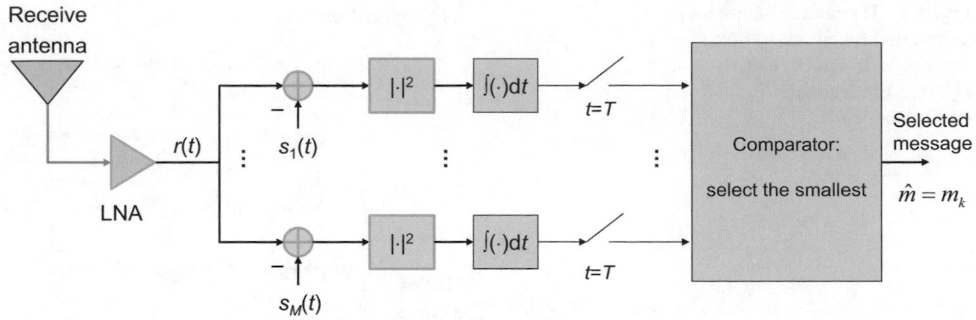

Fig. 5.4 Euclidean distance receiver for wireless communications. *LNA* low-noise amplifier

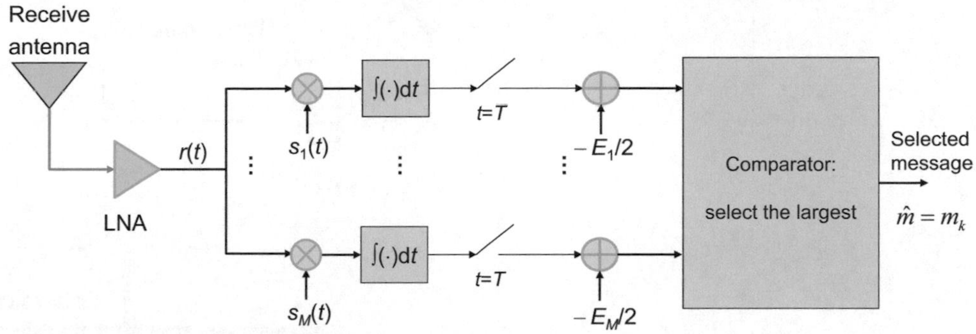

Fig. 5.5 The correlation receiver for wireless communications

In the Euclidean distance receiver, we decide in favor of signal $s_k(t)$ closest to the received signal in Euclidean distance, based on Eq. (5.12a), and the corresponding receiver is shown in Fig. 5.4. The decision strategy can also be described as follows:

$$d_E^2(r, s_m) \underset{H_n}{\overset{H_m}{\lessgtr}} d_E^2(r, s_n) \Leftrightarrow \int_0^T |r(t) - s_m(t)|^2 dt \underset{H_n}{\overset{H_m}{\lessgtr}} \int_0^T |r(t) - s_m(t)|^2 dt, \tag{5.12b}$$

where we use H_m to denote the hypothesis that signal $s_m(t)$ was transmitted.

By squaring the bracket terms, assuming that signals are real valued, and after rearranging, we obtain that

$$\int_0^T r(t)s_m(t)dt - \frac{1}{2}\underbrace{\int_0^{T_s} |s_m(t)|^2 dt}_{E_m} \underset{H_n}{\overset{H_m}{\gtrless}} \int_0^{T_s} r(t)s_n(t)dt - \frac{1}{2}\underbrace{\int_0^T |s_n(t)|^2 dt}_{E_n}. \tag{5.13}$$

The corresponding receiver, based on Eq. (5.13), is known as the *correlation receiver*, and it is shown in Fig. 5.5.

The received vector for this receiver, also known as the *observation vector*, can be obtained by projecting the received signal along different basis functions as follows (assuming that signals are complex valued):

$$\boldsymbol{r} = \begin{bmatrix} r_1 \\ r_2 \\ \vdots \\ r_N \end{bmatrix}, \quad r_n = \int_0^T r(t)\Phi_n^*(t)dt. \tag{5.14}$$

The N-dimensional demodulator (analyzer) is constructed based on Eq. (5.14) and provides the estimate of signal constellation coordinates. Based on coordinates (projections of received signal), the Euclidean distance receiver assigns the observation vector to the closest constellation point within Euclidean distance. The corresponding decision rule can be defined as

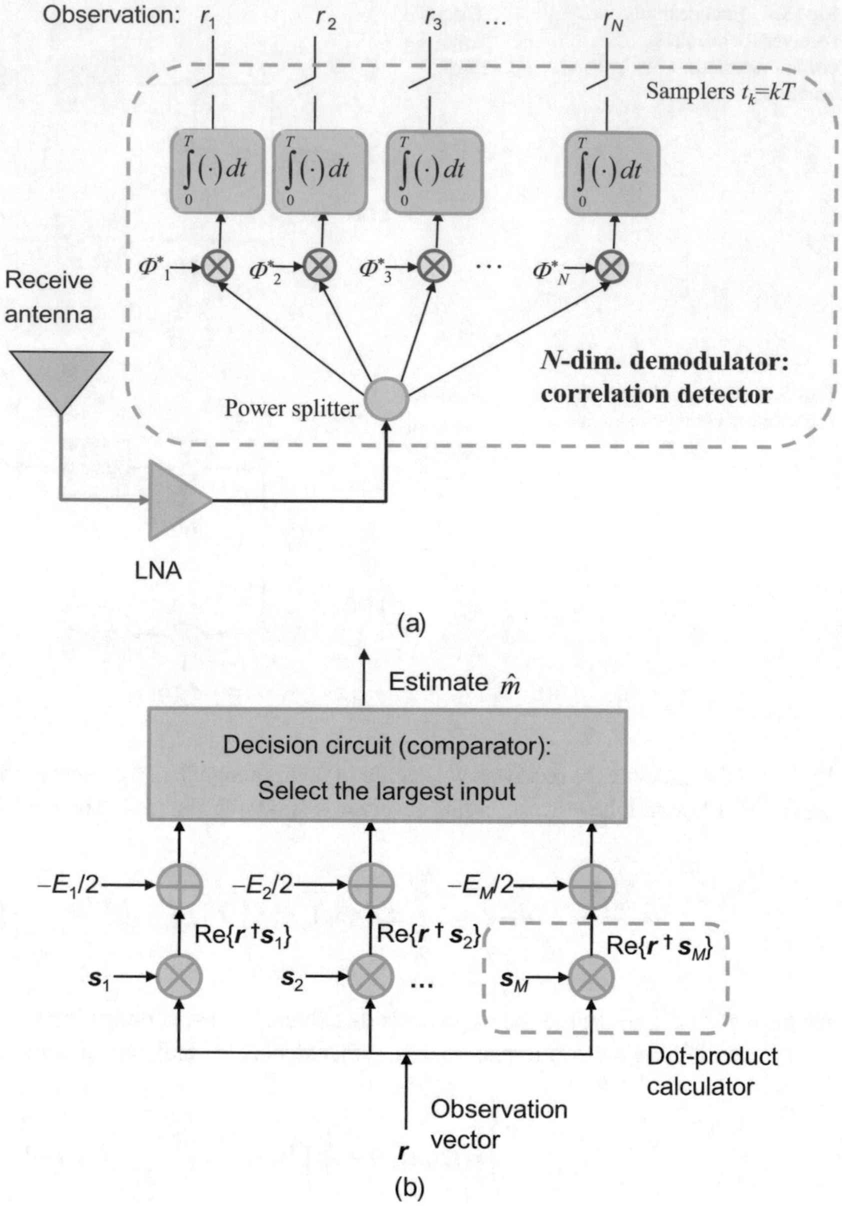

Fig. 5.6 The reduced-complexity correlation receiver: (**a**) detector/demodulator and (**b**) signal transmission decoder configurations

$$\widehat{s}_m = \arg\min_{s_k} \|\boldsymbol{r} - \boldsymbol{s}_k\|^2 = \arg\min_{s_k} \sum_{j=1}^{N} \left|r_j - s_{kj}\right|^2. \tag{5.15}$$

By rearranging Eq. (5.15), we obtain an alternative representation of the decision rule:

$$\widehat{s}_m = \arg\max_{s_k} \boldsymbol{r}^\dagger \boldsymbol{s}_k - E_k/2. \tag{5.16}$$

The corresponding correlation receiver, based on Eq. (5.16), is shown in Fig. 5.6. This scheme has lower complexity since there are $N < M$ branches in detector/demodulator, as shown in Fig. 5.6(a).

An alternative receiver to the scheme discussed above is based on the so-called *matched filter*-based demodulator/detector [1–7], as shown in Fig. 5.7. The output of the *n*th matched filter having impulse response $h_n(t) = \Phi_n^*(T-t)$ is given as

$$r_n(t) = \int_{-\infty}^{\infty} r(\tau)h_n(t-\tau)d\tau = \int_{-\infty}^{\infty} r(\tau)\Phi_n^*(T-t+\tau)d\tau. \tag{5.17}$$

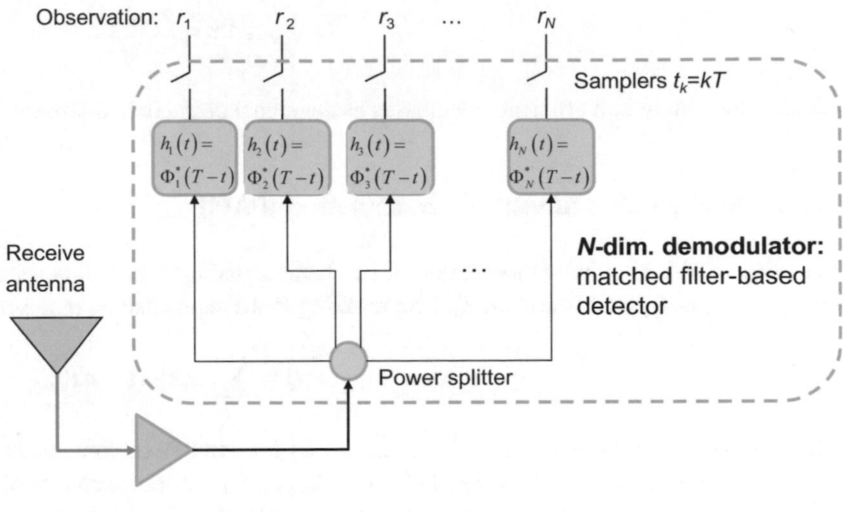

Fig. 5.7 Matched filter-based demodulator/detector configuration

By sampling at time instance $t = T$, we obtain that

$$r_n(T) = \int_{-\infty}^{\infty} r(\tau)h_n(T-\tau)d\tau = \int_0^T r(\tau)\Phi_n^*(\tau)d\tau, \tag{5.18}$$

which is the same form as the nth output of detector in the correlation receiver. Therefore, the correlation and matched filter-based receivers are equivalent to each other.

Even though in this section we are focused on wireless communications, by replacing the transmit antenna with corresponding electro-optical modulator, and receive antenna with corresponding coherent optical detector, all modulators and demodulators described here are also applicable in optical communications as well.

5.1.3 Frequency-Shift Keying (FSK)

The M-ary FSK signal set is defined as

$$s_i(t) = \sqrt{\frac{2E}{T}} \cos\left[\frac{\pi}{T}(n+i)t\right], 0 \le t \le T; i = 1, 2, \ldots, M, \tag{5.19}$$

where n is an integer. Clearly, the individual signals are mutually orthogonal since each signal contains an integer number of cycles within the symbol interval T and the number of cycles between adjacent FSK signals differs by one cycle, so that the orthogonality principle among symbols is satisfied:

$$\int_0^T s_m(t)s_n(t)dt = \delta_{mn}. \tag{5.20}$$

Since the FSK signals are mutually orthogonal, they can be used as basis functions as follows:

$$\Phi_m(t) = \frac{1}{\sqrt{E_s}} s_m(t), \quad \begin{array}{c} 0 \le t \le T \\ m = 1, 2, \ldots, M \end{array}, \tag{5.21}$$

where E_s is the symbol energy. The minimum distance between adjacent FSK signals is given by $d_{\min} = (2E_s)^{1/2}$. The adjacent frequency spacing is $1/(2\,T)$, so that the total bandwidth is given as $B = M/(2\,T)$. The corresponding bandwidth efficiency, for data rate R_b, can be expressed as

$$\rho = \frac{R_b}{B} = \frac{2\log_2 M}{M}. \tag{5.22}$$

Clearly, the bandwidth efficiency decreases as the signal constellation size increases for all values $M > 4$.

5.1.4 M-ary Pulse Amplitude Modulation (PAM)

The M-ary pulse amplitude modulation in baseband signaling is one-dimensional signal set with the basis function being a unit energy pulse $p(t)$ of duration T_s. The resulting PAM signal can be represented as a train of rectangular pulses

$$s(t) = \sum\nolimits_n s(n)p(t - nT_s), \tag{5.23}$$

where $s(n) \in \{A_i = (2i - 1 - L)d;\ i = 1, 2, \ldots, L\}$ for wireless communications and optical communications with coherent optical detection or $s(n) \in \{A_i = (i - 1)d,\ i = 1, 2, \ldots, L\}$ for optical communications with direct detection. The pulse shape $p(t)$ is assumed to have a rectangular form with an amplitude A for $-1/2 < t/T < 1/2$ (and zero value otherwise), but other pulse shapes are possible such as Gaussian pulse shape. When the number of amplitude levels is $L = 2$ and direct detection is used, the corresponding modulation scheme is known as on-off keying (OOK) or unipolar signaling. On the other hand, when the number of amplitude levels is also two but coherent detection is used, the corresponding scheme is known as the antipodal signaling scheme. Assuming an equally probable transmission of symbols, the average symbol energy for coherent detection PAM is given by $E_s = (1/M)\sum_{i=1}^{L} A_i^2 = (1/3)(L^2 - 1)d^2$. On the other hand, for the direct detection scheme with PAM, the average symbol energy is given as $E_s = (1/M)\sum_{i=1}^{L} A_i^2 = E_s = (L-1)(2L-1)d^2/6$. The corresponding bit energy E_b is related to symbol energy as $E_b = E_s/\log_2 L$. The number of bits per symbol equals $\log_2 L$. Finally, the bit rate R_b can be expressed by symbol rate $R_s = 1/T$ as $R_b = R_s \log_2 L$.

We will now turn our attention to the description of both continuous-time (CT) and discrete-time (DT) implementations of the PAM scheme, described in the context of wireless communications. The CT PAM modulator and demodulator are shown in Fig. 5.8. At the output of the serial-to-parallel (S/P) conversion block, the $\log_2 L$ bits are used to select the point from the PAM mapper, implemented as the look-up table (LUT). The output of LUT can be represented as

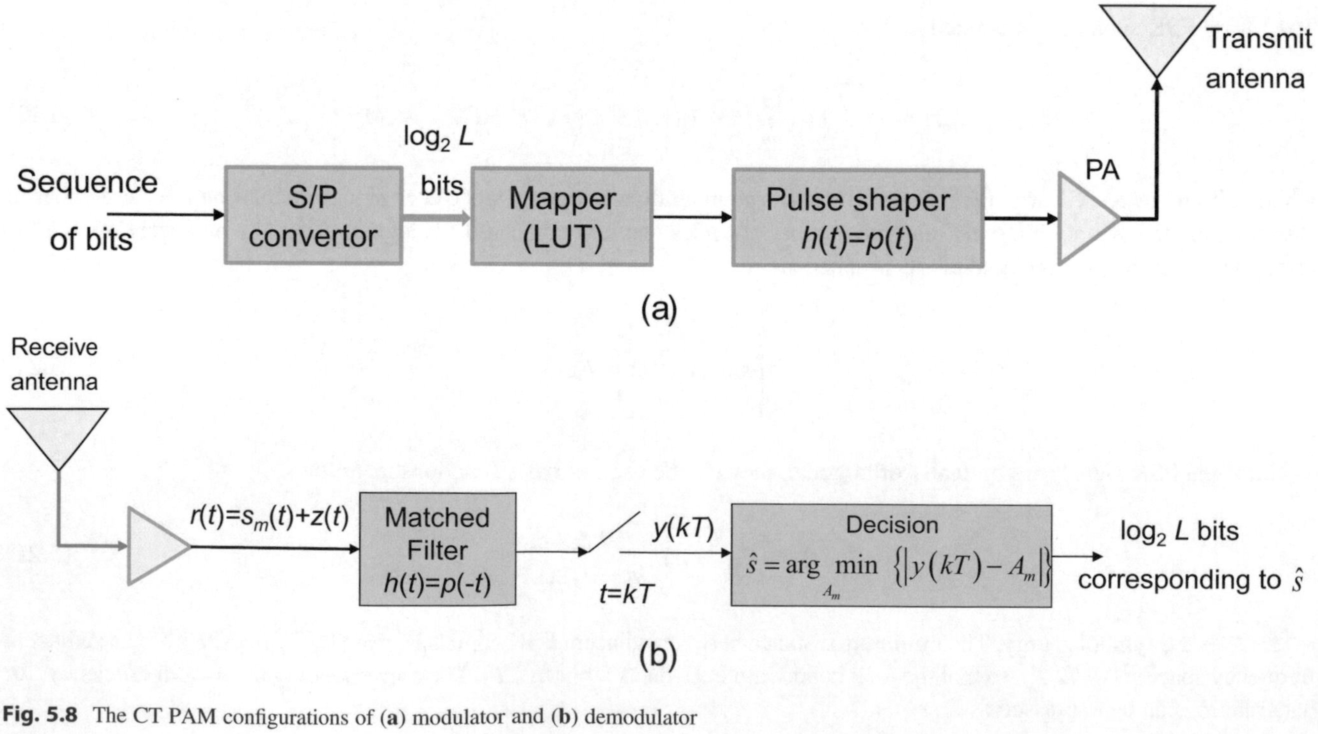

Fig. 5.8 The CT PAM configurations of (a) modulator and (b) demodulator

$$s_\delta(t) = \sum\nolimits_n s(n)\delta(t - nT), \tag{5.24}$$

where $\delta(t)$ is the Dirac delta function. The output of the pulse shaper of the impulse response $h(t) = p(t)$ is obtained as a convolution of the input $a(t)$ and the pulse shaper impulse response:

$$\begin{aligned} s(t) &= s_\delta(t) * p(t) = \textstyle\sum_n s(n)\delta(t - nT) * p(t) \\ &= \textstyle\sum_n s(n)p(t - nT), \end{aligned} \tag{5.25}$$

which is the same form as the one given by Eq. (5.23). The continuous-time correlation PAM demodulator/detector is shown in Fig. 5.8(b).

The output of the correlator, which presents the projection along the basis function $p(t)$, is used in the Euclidean distance receiver to determine the closest amplitude level A_i from the signal constellation set. For the additive channel model, where the received signal $r(t) = s(t) + z(t)$, where $z(t)$ is the additive white Gaussian noise, the output of the correlator can be written as

$$\begin{aligned} y(t) &= \int_{-\infty}^{\infty} r(\tau)p(\tau - t)d\tau \\ &= \textstyle\sum_n s(n) \underbrace{\int_{nT}^{t+nT} p(\tau - nT)p(\tau - t)d\tau}_{R_p(t-nT)} + \underbrace{\int_{nT}^{t+nT} z(\tau)p(\tau - t)d\tau}_{z'(t)}. \end{aligned} \tag{5.26}$$

which after sampling becomes

$$y(kT) = \sum\nolimits_n s(n)R_p(kT - nT) + z'(kT) = s(k) + z'(kT). \tag{5.27}$$

We assume in the equation above that the pulse shape is properly chosen in accordance to the Nyquist criterion [1] and there is no intersymbol interference at the output of the matched filter so that we can write

$$R_p(lT_s) = \begin{cases} 1, l = 0 \\ 0, l \neq 0 \end{cases}.$$

The DT implementation of the PAM modulator and demodulator is shown in Fig. 5.9. At the output of the serial-to-parallel (S/P) conversion, the $\log_2 L$ bits are used to select the point from the PAM mapper, which is implemented as the look-up table.

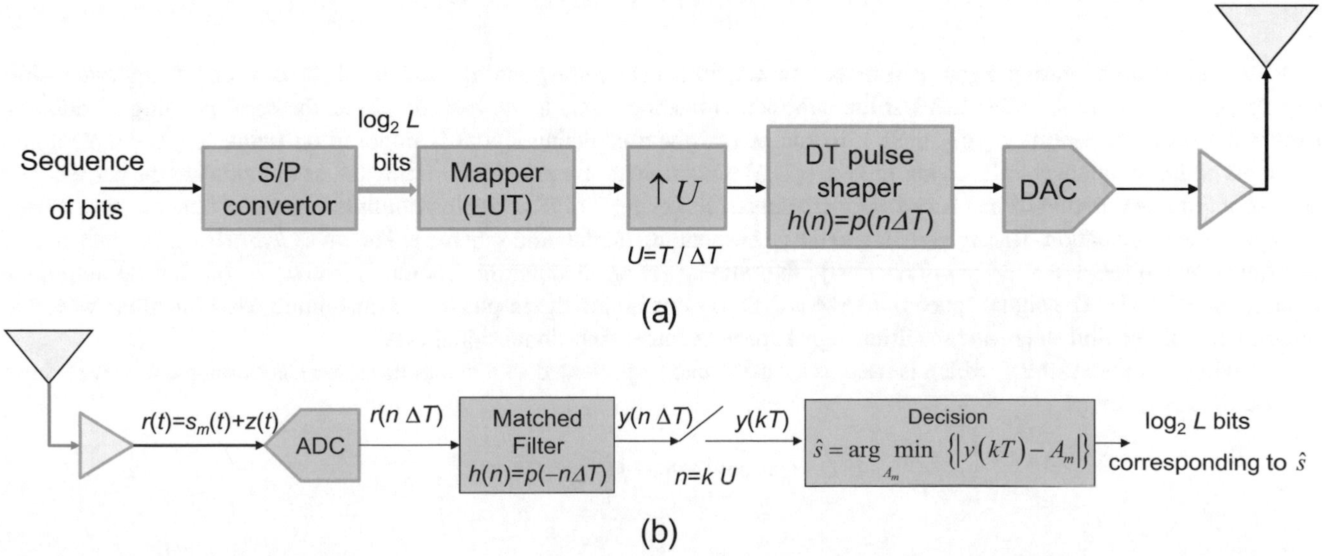

Fig. 5.9 The DT PAM configurations of (**a**) modulator and (**b**) demodulator for wireless communications

The output of a mapper is upsampled for a factor $U = T/\Delta T$, where ΔT is the sampling period. The upsampling is performed by inserting $(U - 1)$ zero samples between two neighboring samples in original sequence. The output of the pulse shaper can be obtained by a DT convolution sum of the upsampler output $\sum_k s(k)\delta(n\Delta T - kT_s)$ and the pulse shaper impulse response $h(n) = p(n\Delta T)$, which gives the following result:

$$s_\delta(n\Delta T) = \sum_k s(k)\delta(n\Delta T - kT). \tag{5.28}$$

As a summary, after digital-to-analog conversion (DAC) at the transmitter side, we obtain the modulating signal given by Eq. (5.25). On the receiver side, after performing ADC, matched filtering, and sampling, we obtain the input to the decision circuit given by Eq. (5.27). The decision circuit operates in the same fashion as in the CT case described above.

5.1.5 Passband Digital Wireless/Optical Transmission

The passband modulated signal $s(t)$ can be described in terms of its in-phase and in-quadrature (which is often referred to just as quadrature) components, in standard (or canonical) form [1, 2], as follows:

$$s(t) = s_I(t)\cos(2\pi f_c t) - s_Q(t)\sin(\omega_c t), \tag{5.29}$$

where $s_I(t)$ and $s_Q(t)$ represent the in-phase and quadrature components, respectively, while ω_c denotes the carrier frequency [rad/s]. Because the sine and cosine signals are orthogonal to each other in the kth symbol interval ($kT \leq t \leq (k + 1)T$), the corresponding signal space is two-dimensional. To facilitate further discussion, we can use the *complex envelope* (low-pass (LP)) representation of the signal $s(t)$ given by

$$\tilde{s}(t) = s_I(t) + js_Q(t). \tag{5.30}$$

In the frequency domain, we have the following relation between the power spectral densities related to standard form (denoted as PSD_s) and to complex envelope representation (denoted as PSD_{LP}):

$$PSD_s(\omega) = \frac{1}{4}[PSD_{LP}(\omega - \omega_c) + PSD_{LP}(\omega + \omega_c)], \tag{5.31}$$

where the indices "S" and "LP" stand for "standard" and "low pass" representations, respectively. The standard form can also be represented by polar coordinates in a form

$$s(t) = a(t)\cos[\omega_c t + \phi(t)]; a(t) = \left[s_I^2(t) + s_Q^2(t)\right]^{1/2}, \phi(t) = \tan^{-1}\left[\frac{s_Q(t)}{s_I(t)}\right]. \tag{5.32}$$

If the information-bearing signal is imposed on amplitude, the corresponding modulated signal is called amplitude-shift keying (ASK) one. On the other hand, if the information-bearing signal is imposed on phase, the corresponding modulating format is called phase-shift keying (PSK). In addition, if the modulating signal is imposed on frequency, the modulation format is called frequency-shift keying (FSK). It is also possible to use polarization orientation to modulate the carrier. The corresponding modulation format is called polarization shift keying (PolSK). Finally, multiple degrees of freedom can be used to impose the information-bearing signal, and the corresponding modulation scheme is known as *hybrid* one. For instance, if amplitude and phase are used simultaneously, the corresponding modulation scheme is known as quadrature amplitude modulation (QAM). The signal space of QAM is 2-D. By combining the in-phase and quadrature basis functions with two orthogonal polarization states, the resulting signal space becomes four-dimensional (4-D).

The wireless/optical signal, which is used as a carrier, can be presented as a monochromatic electromagnetic wave whose electric field is given as

$$\boldsymbol{E}(t) = \boldsymbol{p}A\cos[2\pi ft + \phi(t)], 0 \leq t < T. \tag{5.33}$$

Each of the parameters in Eq. (5.33) (amplitude A, frequency f, phase ϕ, and polarization orientation $\boldsymbol{p}$) can be utilized to carry the information. The corresponding modulation formats implemented on RF/optical carrier are known as ASK, FSK, PSK, and PolSK, respectively, as indicated already above.

As an illustration, in M-ary ASK, the signal waveforms can be represented as

$$s_m(t) = A_m p(t) \cos \omega_c t;\ \ 0 \le t < T; m = 1, 2, \cdots, M, \tag{5.34}$$

where $p(t)$ is the pulse shape, while $\{A_m | \ m = 1, 2, \ldots, M\}$ represents a set of possible amplitudes. For $M = 2$, we can set $A_1 = 0$ and $A_2 = A$. For $M > 2$, we can use the PAM representation as follows: $A_m = (2\ m - 1\text{-}M)d/2$, indicating that the distance between adjacent amplitudes is d.

Similar to signal, any linear time-invariant (LTI) passband system can be represented by its canonical impulse response:

$$h(t) = h_I(t) \cos (\omega_c t) - h_Q(t) \sin (\omega_c t), \tag{5.35}$$

where $h_I(t)$ and $h_Q(t)$ represent the in-phase and quadrature components, respectively. The complex envelope representation of LTI band-pass systems is given by

$$\widetilde{h}(t) = h_I(t) + jh_Q(t). \tag{5.36}$$

The complex envelope of signal at the output of the band-pass system is obtained by CT convolution of the complex forms:

$$\widetilde{y}(t) = \frac{1}{2}\widetilde{h}(t) * \widetilde{x}(t). \tag{5.37}$$

Finally, the canonical form of the output of the LTI band-pass system can be obtained as

$$y(t) = \ \mathrm{Re}\left[\widetilde{y}(t) \exp\left(\ j\omega_c t\right)\right]. \tag{5.38}$$

An important parameter of any modulation scheme is its *bandwidth efficiency*, defined as $S = R_b/B$[bits/s/Hz], where B is the bandwidth occupied by modulated signal, determined from $\mathrm{PSD_{LP}}$ shape (often as the width of the main lobe in the PSD). The *spectrally efficient* modulation schemes are aimed to maximize the bandwidth efficiency. In the next part, we will describe the spectrally efficient modulation schemes that can be used in both wireless and optical communication systems.

5.2 Multilevel (Two-Dimensional) Modulation Schemes

In this section, we describe the multilevel and multidimensional modulation schemes suitable for and/or in use for both wireless and optical communication systems.

5.2.1 Two-Dimensional Signal Constellations for Wireless Communications

M-ary QAM is a 2-D generalization of M-ary PAM, with two orthogonal passband basis functions given as

$$\begin{aligned} \Phi_I(t) &= \sqrt{\frac{2}{T_s}} \cos (\omega_c t), 0 \le t \le T \\ \Phi_Q(t) &= \sqrt{\frac{2}{T_s}} \sin (\omega_c t), 0 \le t \le T. \end{aligned} \tag{5.39}$$

The mth signal constellation point in signal (Φ_I, Φ_Q) space can be represented by ($I_m d_{min}/2$, $Q_m d_{min}/2$), where d_{min} is the minimum Euclidean distance among signal constellation points, which is related to the minimum energy signal constellation point E_{min} by $d_{min}/2 = E_{min}^{1/2}$. With ($I_m$,$Q_m$), we denoted the coordinates of the mth signal constellation point. The transmitted signal in the kth symbol interval can be represented as

$$s_k(t) = \sqrt{\frac{2E_{min}}{T_s}} I_k \cos(\omega_c t) - \sqrt{\frac{2E_{min}}{T_s}} Q_k \sin(\omega_c t), \quad \begin{matrix} 0 \le t \le T \\ k = 0, \pm 1, \pm 2, \ldots \end{matrix} . \tag{5.40}$$

Therefore, two phase-quadrature carriers are being employed, each modulated by the corresponding set of discrete amplitudes. All QAM constellations can be categorized into three broad categories: (i) square-QAM in which the number of bits per symbol is even (signal constellation size is a square number, $M = L^2$, where L is a positive integer); (ii) cross constellations in which the number of bits per symbol is odd; and (iii) star-QAM constellations in which signal constellation points are placed along concentric circles of different radiuses. The M-ary PSK can be considered as a special case of star-QAM with constellation points placed uniformly on a single circle.

The *square-QAMconstellations* can be interpreted as being obtained as a two-dimensional Cartesian product of 1-D-PAM, described in Subsect. 5.1.4, as follows:

$$X^2 = X \times X = \{(x_1, x_2) | x_n \in X,\ 1 \le n \le 2\}, \tag{5.41}$$

where $X = \{2l - 1 - L | \ l = 1, 2, \ldots, L\}$, while L is the size of 1-D-PAM, which is related to the size of QAM square constellation as $L^2 = M_{QAM}$. In other words, the coordinates of the kth signal constellation point for square M-ary QAM signal constellation are given by

$$\{I_k, Q_k\} = \begin{bmatrix} (-L+1, L-1) & (-L+3, L-1) & \ldots & (L-1, L-1) \\ (-L+1, L-3) & (-L+3, L-3) & \ldots & (L-1, L-3) \\ \ldots & \ldots & \ldots & \ldots \\ (-L+1, -L+1) & (-L+3, -L+1) & \ldots & (L-1, -L+1) \end{bmatrix}. \tag{5.42}$$

As an illustration, for 16-ary square constellation, the signal constellation points are given as

$$\{I_k, Q_k\} = \begin{bmatrix} (-3,3) & (-1,3) & (1,3) & (3,3) \\ (-3,1) & (-1,1) & (1,1) & (3,1) \\ (-3,-1) & (-1,-1) & (1,-1) & (3,-1) \\ (-3,-3) & (-1,-3) & (1,-3) & (3,-3) \end{bmatrix}.$$

The corresponding constellation is shown in Fig. 5.10, together with 64-QAM.

The *cross constellation* can be obtained by properly combining two square constellations; the inner square constellation carries $m - 1$ bits per symbol, while the outer constellation is split into four regions, with each region carrying m-3 bits/symbol. The symbols are placed along each side of inner constellation as shown in Fig. 5.11(a). Another way to interpret the cross signal constellation is to start with the Cartesian product of 1-D-PAM to obtain L^2 square constellation and remove the needed number of points in the corners to obtain a desired cross constellation. For example, 32-ary cross constellation can be obtained by starting from 36-ary square constellation and by removing the four points in the corners, as shown in Fig. 5.11(b).

The *star-QAM*, also known as amplitude phase-shift keying (APSK), can be obtained by placing the constellation points along the concentric rings. The M-ary phase-shift keying (PSK) is just a special case of star-QAM having only one ring. The transmitted M-ary PSK signal, which can be considered as a special case of star-QAM, for the mth symbol can be expressed as

$$s_m(t) = \begin{cases} \sqrt{\frac{2E}{T}} \cos\left(\omega_c t + (m-1)\frac{2\pi}{M}\right), 0 \le t \le T \\ 0, \quad \text{otherwise} \end{cases} ; m = 1, 2, \ldots, M. \tag{5.43}$$

As an illustration, the 8-PSK and 8-star-QAM constellation diagrams are shown in Fig. 5.12. Different mapping rules can be used to map binary sequence to corresponding constellation points including natural mapping, Gray mapping, and anti-Gray mapping. In Gray mapping rule, the neighboring signal constellation points differ in only one bit position. For additive

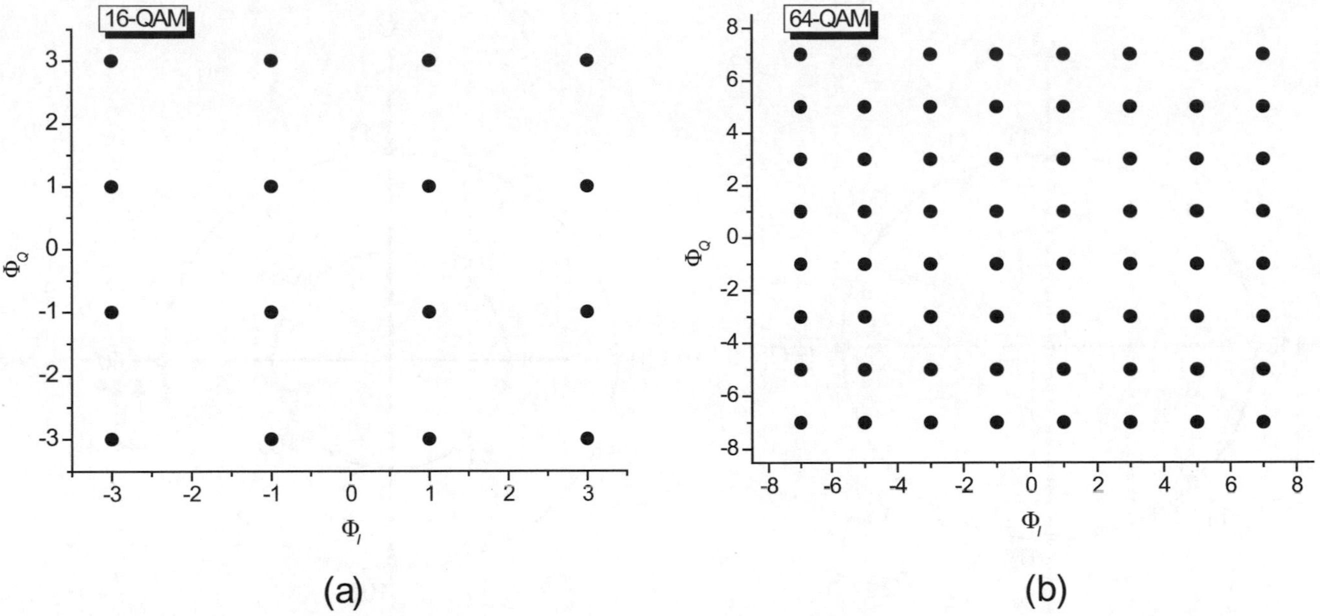

Fig. 5.10 Square-QAM constellations: (**a**) 16-QAM and (**b**) 64-QAM

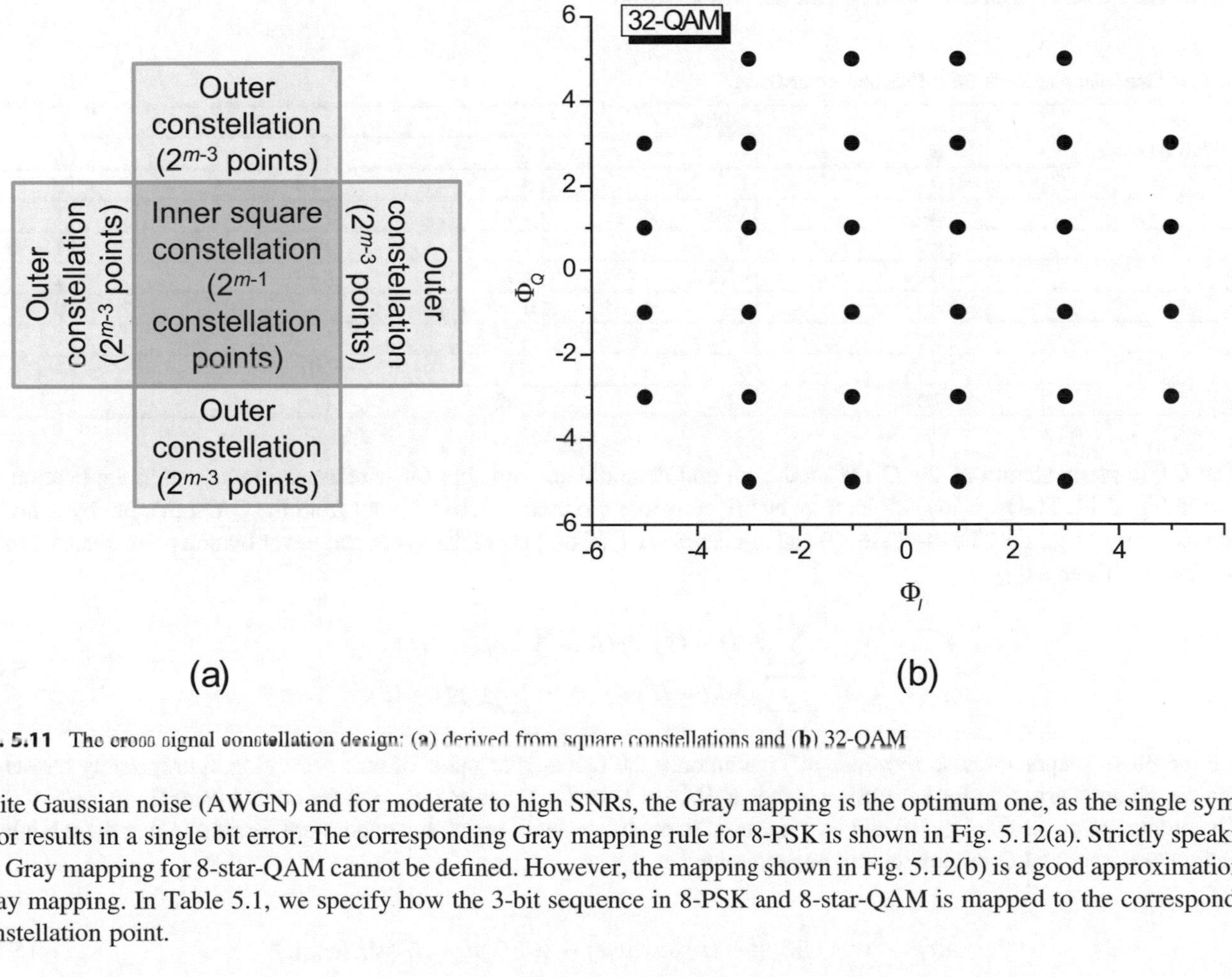

Fig. 5.11 The cross signal constellation design: (**a**) derived from square constellations and (**b**) 32-QAM

white Gaussian noise (AWGN) and for moderate to high SNRs, the Gray mapping is the optimum one, as the single symbol error results in a single bit error. The corresponding Gray mapping rule for 8-PSK is shown in Fig. 5.12(a). Strictly speaking, the Gray mapping for 8-star-QAM cannot be defined. However, the mapping shown in Fig. 5.12(b) is a good approximation of Gray mapping. In Table 5.1, we specify how the 3-bit sequence in 8-PSK and 8-star-QAM is mapped to the corresponding constellation point.

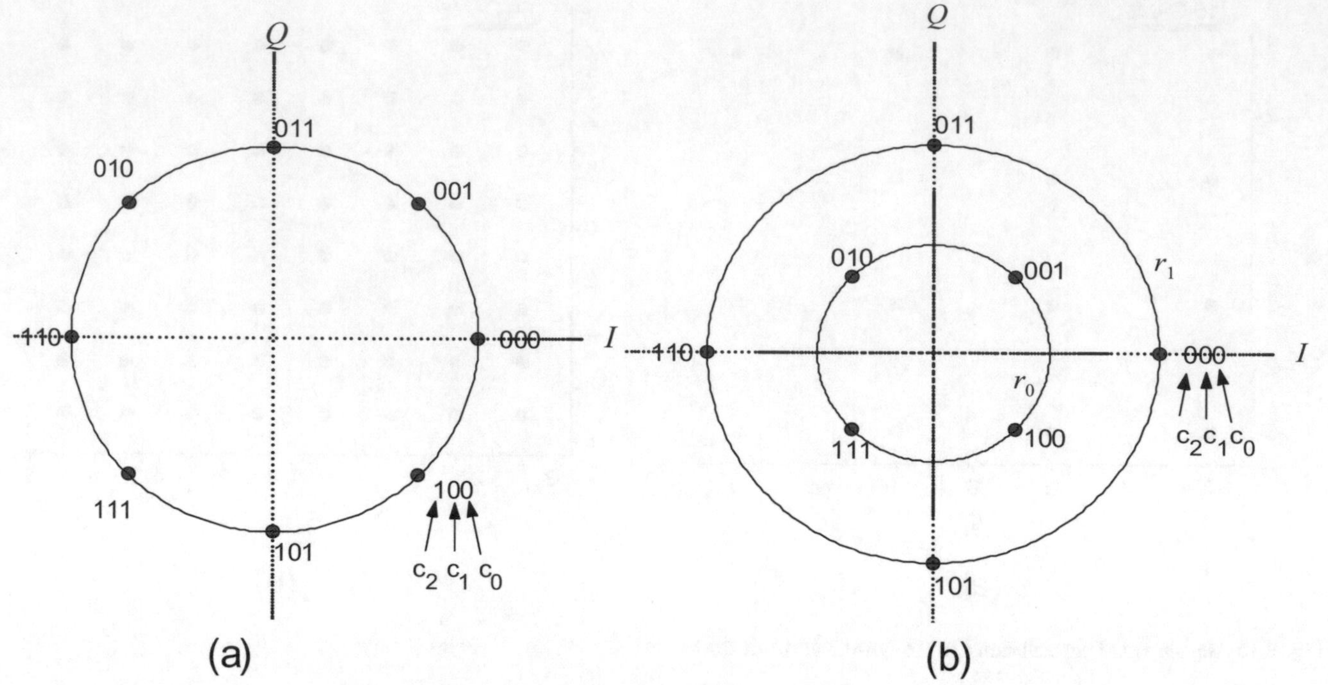

Fig. 5.12 The constellation diagrams for (**a**) 8-PSK and (**b**) 8-star-QAM

Table 5.1 Gray mapping rules for 8-PSK and 8-star-QAM

		8-PSK		8-QAM	
Input bits ($c_2c_1c_0$)	ϕ_k	I_k	Q_k	I_k	Q_k
000	0	1	0	1 + √3	0
001	π/4	√2/2	√2/2	1	1
011	π/2	0	1	0	1 + √3
010	3π/4	-√2/2	√2/2	−1	1
110	π	−1	0	− (1 + √3)	0
111	5π/4	−√2/2	−√2/2	−1	−1
101	3π/2	0	−1	0	− (1 + √3)
100	7π/4	√2/2	−√2/2	1	−1

The CT implementation of the QAM modulator and demodulator, suitable for wireless communication applications, is shown in Fig. 5.13. The $m = \log_2 M$ bits from an S/P converter are used to select a point from the QAM mapper by using the appropriate mapping rule. The in-phase $I(t)$ and in-quadrature $Q(t)$ outputs of the upper and lower branch pulse shaper can be, respectively, written as

$$\begin{aligned} I(t) &= \sum\nolimits_l I_l\delta(t - lT) * p(t) = \sum\nolimits_l I_l p(t - lT), \\ Q(t) &= \sum\nolimits_l Q_l\delta(t - lT) * p(t) = \sum\nolimits_l Q_l p(t - lT), \end{aligned} \tag{5.44}$$

where the pulse shaper impulse response $p(t)$ (commonly the rectangular pulse of unit energy) is appropriately chosen to satisfy the Nyquist criterion for zero ISI, which is $R_p(lT_s) = 1$ for $l = 0$ and $R_p(lT_s) = 0$ for other l-values.

The outputs of pulse shapers, $I(t)$ and $Q(t)$, are used to modulate the in-phase and quadrature carriers so that the modulated signal at the output of I/Q modulator can be represented as

$$s(t) = \sum\nolimits_l \sqrt{\frac{2}{T}} I_l p(t - lT) \cos(\omega_c t) - \sqrt{\frac{2}{T}} Q_l p(t - lT) \sin(\omega_c t), \tag{5.45}$$

which is consistent with Eq. (5.40). On the receiver side (see Fig. 5.13(b)), the outputs of matched filters represent projections along basis functions, which after sampling can be written as

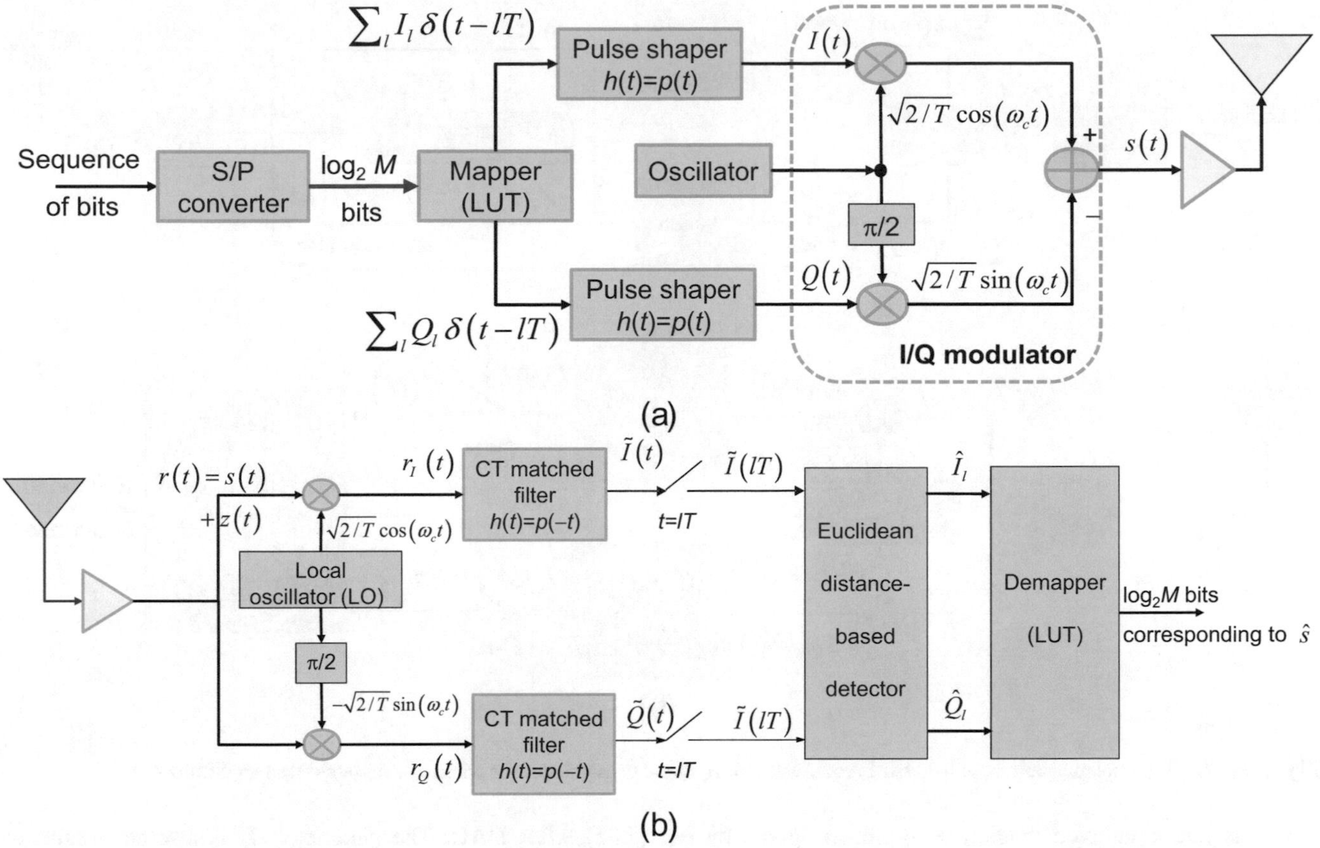

Fig. 5.13 The CT implementations of (**a**) QAM modulator and (**b**) QAM demodulator suitable for wireless communication applications

$$\begin{aligned}\widehat{I}(lT_s) &= \sum\nolimits_m I_m R_p(mT - lT) + z_I(lT),\\ \widehat{Q}(lT_s) &= \sum\nolimits_m Q_m R_p(mT - lT) + z_Q(lT),\end{aligned} \tag{5.46}$$

where z_I and z_Q represent the projection of the noise process along the in-phase and quadrature basis functions, respectively. Since $p(t)$ has the rectangular shape described above and in the absence of noise, we can perfectly recover the in-phase and quadrature coordinates of the transmitted QAM constellation point, as $R_p(mT_s) = 1$ for $m = 0$, and $R_p(mT_s) = 0$ for $m \neq 0$. However, in the presence of noise and channel distortions, we use the Euclidean distance receiver and decide in favor of a candidate constellation point closest to the received point $\left(\widehat{I}(lT), \widehat{Q}(lT)\right)$.

The corresponding DT implementation of the QAM modulator and demodulator is shown in Fig. 5.14. The $m = \log_2 M$ bits from an S/P converter are used to select a point from the QAM mapper. The outputs of a mapper are upsampled for a factor $U = T/\Delta T$, where ΔT is the sampling period.

The outputs of pulse shapers can be obtained as the DT convolution sums of inputs and impulse responses $h(i) = p(i\Delta T)$ of corresponding pulse shapers, to obtain

$$\begin{aligned}I(i\Delta T) &= \sum\nolimits_l I_l p(i\Delta T - lT),\\ Q(i\Delta T) &= \sum\nolimits_l Q_l p(i\Delta T - kT).\end{aligned} \tag{5.47}$$

These outputs are used to modulate in-phase and quadrature DT carriers, generated by direct digital synthesizer. The output of DT I/Q modulator can be represented as

$$s(iT) = \sum\nolimits_l \sqrt{\frac{2}{T_s}} I_l p(iT - lT)\cos(\Omega_c i) - \sqrt{\frac{2}{T}} Q_l p(iT - lT)\sin(\Omega_c i), \tag{5.48}$$

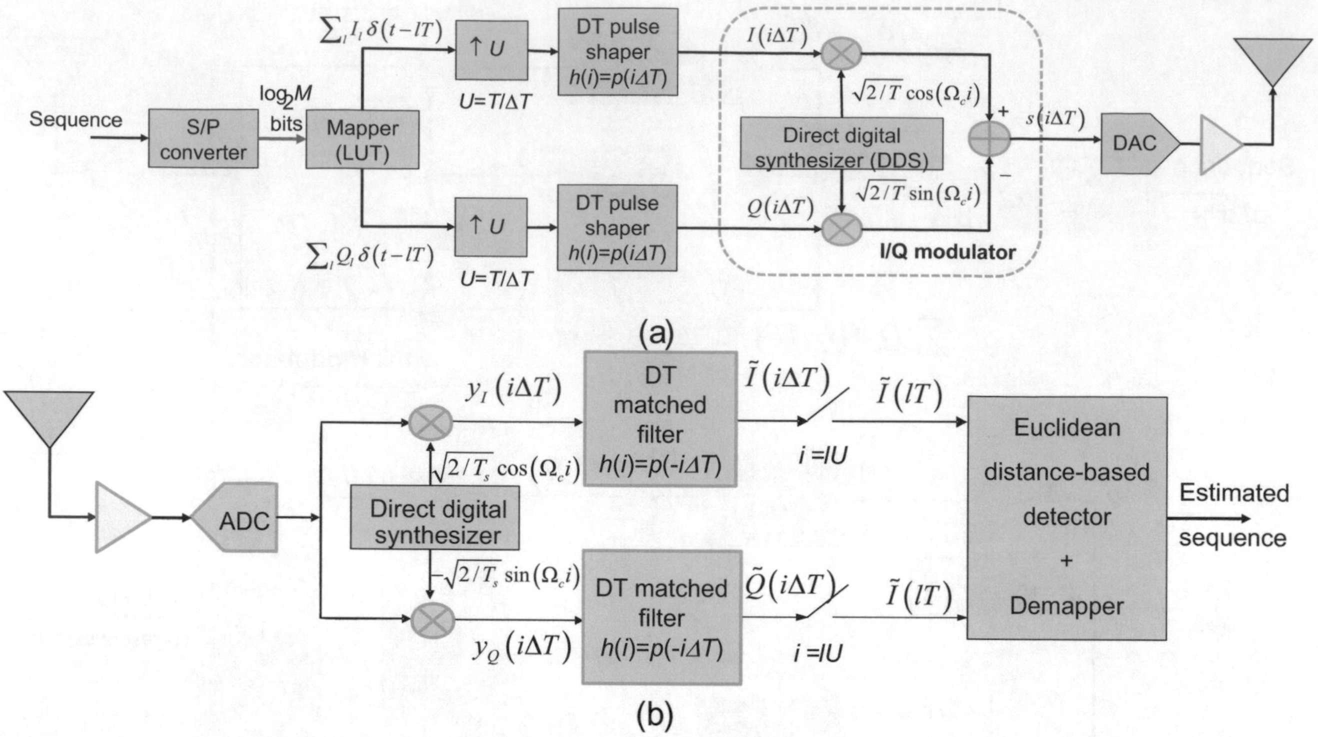

Fig. 5.14 The DT implementations of (**a**) QAM modulator and (**b**) QAM demodulator for wireless communication applications

which is just a sampled version of waveform given by Eq. (5.45); after DAC. The parameter Ω_c is discrete frequency (expressed in radians per sample) related to the carrier frequency ω_c [rad/s] by $\Omega_c = \omega_c \cdot \Delta T$, where ΔT is the sampling period. The operation of the direct digital synthesizer (DDS) is described in Chap. 3. On the receiver side, after band-pass filtering (BPF) and ADC, discrete-time projections along DT basis functions are obtained as follows:

$$\begin{aligned}\widehat{I}(iT) &= \sum\nolimits_{l} I_l R_p(lT - iT) + z_I(iT),\\ \widehat{Q}(iT) &= \sum\nolimits_{l} Q_l R_p(lT - iT) + z_Q(iT),\end{aligned} \tag{5.49}$$

where z_I and z_Q, similar to the CT case, represent the projections of the noise process along the in-phase and quadrature basis functions, respectively. After resampling the matched filters' outputs at $i = lU$, we can obtain the estimates of the transmitted QAM signal constellation point coordinates as follows:

$$\begin{aligned}\widehat{I}(lT) &= \sum\nolimits_{m} I_m R_p(mT - lT) + z_I(lT) = I_l + z_I(lT),\\ \widehat{Q}(lT) &= \sum\nolimits_{m} Q_m R_p(mT - lT) + z_Q(lT) = Q_l + z_Q(lT),\end{aligned} \tag{5.50}$$

which are identical to the corresponding CT case equations given by (5.46).

Before concluding this section, let us discuss the bandwidth efficiency of PSK and QAM formats. The power spectral density (PSD) of LP representation of M-PSK is given by

$$\mathrm{PSD}_{LP}^{\mathrm{PSK}}(f) = 2E_s \mathrm{sinc}^2(Tf) = 2E_b \log_2 M \mathrm{sinc}^2(fT_b \log_2 M), \tag{5.51}$$

where we assume that the pulse shaper impulse response is a rectangular pulse of duration T. With E_s we denoted symbol energy, E_b is the bit energy, while T_b denotes the bit duration. If we use the null-to-null bandwidth definition, corresponding to the main lobe in the PSD, we can establish relation $B = 2/T$ between the bandwidth occupied by the modulated signal B and the symbol duration T. Since $R_b = 1/T_b$, it follows that

$$B = 2R_b / \log_2 M. \tag{5.52}$$

The *bandwidth efficiency* is then obtained as

$$S = R_b / B = (\log_2 M)/2, \tag{5.53}$$

which means that it increases as the signal constellation size grows, but the dependence is logarithmic. As an illustration, we can calculate that the bandwidth efficiency for 4-QAM (QPSK) is equal to 1 bit/s/Hz. If 3 dB bandwidth definition of the spectrum is used instead, the bandwidth efficiency values will be higher.

The transmitter configurations for optical communications are more complicated than those for wireless communications [6, 7, 9]. In principle, a single electro-optical (E-O) I/Q modulator, as described in Chap. 3 (see Fig. 3.10), can be used, providing that sufficiently small amplitude variations of both in-phase and in-quadrature channels are used so that the approximation given by Eq. (3.22) can be applied. A mapper, implemented as an LUT, is then sufficient to provide Cartesian coordinates of a corresponding M-PSK/M-QAM constellation. However, large constellations are quite challenging to implement with this approach. In the incoming sections, we describe various transmitters for optical communication applications, based on [7, 9] (and references therein). The readers not familiar with external modulators for optical communications are referred to Sect. 3.1.2.3 in Chap. 3.

5.2.2 M-ary PSK Transmitters for Optical Communications

There are several ways to generate M-PSK signals by using the following: (i) concatenation (serial connection) of M E-O phase modulators (PMs), described in Fig. 3.9 of Chap. 3; (ii) combination of E-O I/Q and phase modulators; and (iii) single E-O I/Q modulator. The *serial connection of PMs* is shown in Fig. 5.15. After serial-to-parallel conversion (or 1:*m* DEMUX), *m* output bits are properly mapped into output signals as follows:

$$V_i(t) = \frac{V_\pi}{2^{i-1}} \sum_{k=-\infty}^{\infty} c_i(k) p(t - kT_s); i = 1, 2, \cdots, m. \tag{5.54}$$

If the optical transmission system is sensitive to cycle slips, then differential encoding should be used. With $p(t)$ we denoted the pulse shape, which is obtained by the convolution of the impulse response of pulse shaper and impulse response of the driver amplifier. The output amplitude of the ith pulse shaper (PS) equals $V\pi/2^{i-1}$. The phase change after the first phase modulator is equal to $\pm\pi$, thus indicating that the first PM generates a BPSK format. The phase change introduced by the second PM equals $\pm\pi/2$, which makes the resulting signal constellation after the second PM to be in a QPSK format. Next, the

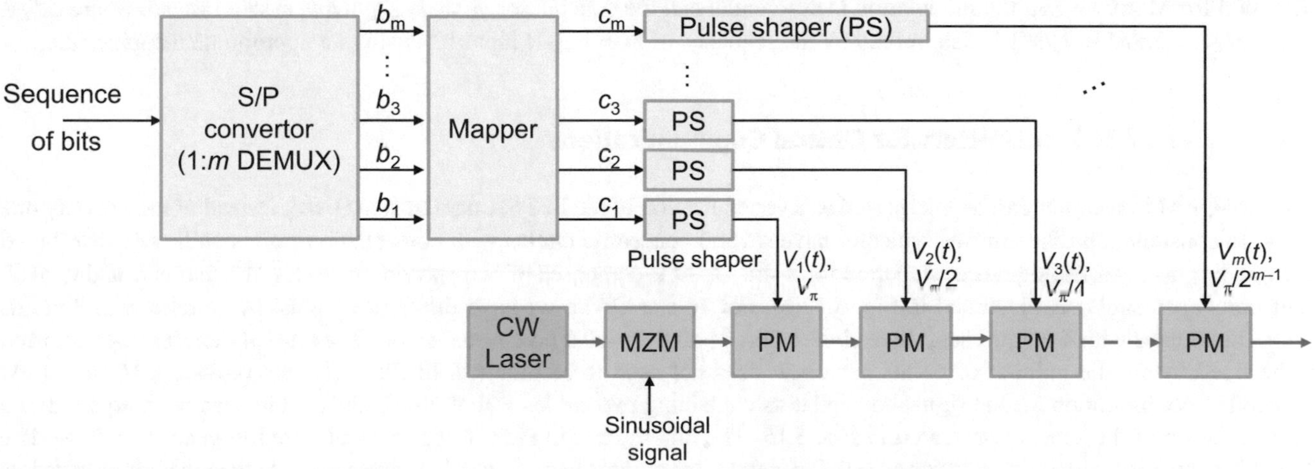

Fig. 5.15 The serial configuration of multilevel PSK transmitter for optical communications. *PM* E-O phase modulator, *MZM* Mach-Zehnder modulator, *PS* pulse shaper

third PM introduces the phase change $\pm\pi/4$, so that the signal constellation after the third PM has an 8-PSK format. In a similar fashion, it can be shown that the signal constellation after the mth E-O PM has an $M = 2^m$-PSK format.

The Mach-Zehnder modulator (MZM) is optional in an M-PSK transmitter setup; its purpose would be to perform the NRZ-to-RZ conversion. The overall electric field at the output of the serial M-PSK transmitter can be represented as

$$E_T(t) = \sqrt{P_T} e^{j(\omega_T t + \phi_{T,PN})} \exp\left(j\frac{\pi}{V_\pi} \sum_{i=1}^{m} V_i(t) \right), \tag{5.55}$$

where P_T denotes the transmitted power, ω_T is the carrier emitting frequency, while $\phi_{T,PN}$ denotes the phase noise introduced by the CW transmitter laser. Equation (5.55) also indicates that it is possible, at least in principle, to perform a summation of various signals described by Eq. (5.54) in the electrical domain while employing only a single E-O PM.

The differential encoding is an efficient way to deal with the cycle slips at the expense of SNR performance degradation. The differential PSK eliminates the need for frequency locking of the local laser oscillator at the receiving side with the transmitting laser, since it combines two basic operations at transmitter: differential encoding and phase-shift keying. For instance, in differential BPSK (DBPSK), to send bit 0 we advance in phase the current signal waveform by π radians, while to send bit 1 we leave the phase unchanged. The receiver measures the relative phase difference received during two successive bit intervals by using either the Mach-Zehnder interferometer, which is followed by either direct detection or conventional coherent difference detection. The coherent optical detection scheme is also suitable for the compensation of linear channel impairments while being insensitive to cycle slips. The transmitted signal in this case can be represented as

$$\begin{aligned} E_1(t) &= \sqrt{P_T} e^{j(\omega_T t + \phi_{T,PN})} \begin{cases} p(t), 0 \le t \le T_b \\ p(t), T_b \le t \le 2T_b \end{cases} \\ E_0(t) &= \sqrt{P_T} e^{j(\omega_T t + \phi_{T,PN})} \begin{cases} p(t), 0 \le t \le T_b \\ -p(t), T_b \le t \le 2T_b \end{cases}, \end{aligned} \tag{5.56}$$

where the subscript "T" denotes the transmitted bit. If DBPSK transmission is interpreted as symbol transmission of duration $2T_b$, it is evident from Eq. (5.56) that two symbols are orthogonal to each other. The following differential encoding rule is applied: (i) if the incoming bit at the kth time instance, denoted as b_k, is 1, leave the symbol c_k unchanged with respect to the previous bit; and (ii) if the incoming binary symbol b_k is 0, we complement the symbol c_k with respect to the previous bit. In other words, we can write that $c_k = \overline{b_k \bigoplus c_{k-1}}$, where the overbar denotes the complement operation to the summation per mod 2.

Another interpretation of M-PSK/DPSK, which requires the use of only one phase modulator but more complicated electrical signal generation, can be described as follows. In a coherent detection scheme for M-ary PSK, the data phasor $\phi_l \in \{0, 2\pi/M, \ldots, 2\pi(M-1)/M\}$ is sent in each lth transmission interval. In both direct detection and coherent difference detection for M-ary DPSK, the modulation is differential and the data phasor $\phi_l = \phi_{l-1} + \Delta\phi_l$ is sent instead, where $\Delta\phi_l \in \{0, 2\pi/M, \ldots, 2\pi(M-1)/M\}$ is determined by the sequence of $m = \log_2 M$ input bits using an appropriate mapping rule.

5.2.3 Star-QAM Transmitters for Optical Communications

The star-QAM transmitter can be interpreted as a generalization of the M-PSK one. In star-QAM, instead of having only one circle in constellation diagram, we typically have several concentric circles with constellation points uniformly distributed around. For instance, the signal constellation for 8-star-QAM is composed of two concentric circles of radiuses r_0 and r_1, with four points per circle, as illustrated in Fig. 5.12(b). For 16-star-QAM, we have different options: two circles with 8 points each, four circles with 4 points each, three circles with 6 points, 5 and 5 points, and so on. The ratio of radiuses does not need to be fixed, while the number of points per circle does not need to be uniform. In Fig. 5.16, we provide a 16-star-QAM transmitter configuration for the signal constellation containing two circles with 8 points each, which can be interpreted as a generalization of the scheme presented in Fig. 5.15. The first three E-O PMs in Fig. 5.16 are used to generate 8-PSK. The MZM is controlled with bit c_4, and for $c_4 = 0$, it generates the signal of amplitude that corresponds to the inner circle of radius r_0, while for $c_4 = 1$, it generates the signal of amplitude that corresponds to the outer circle of radius r_1. A similar strategy can be applied for an arbitrary star-QAM signal constellation.

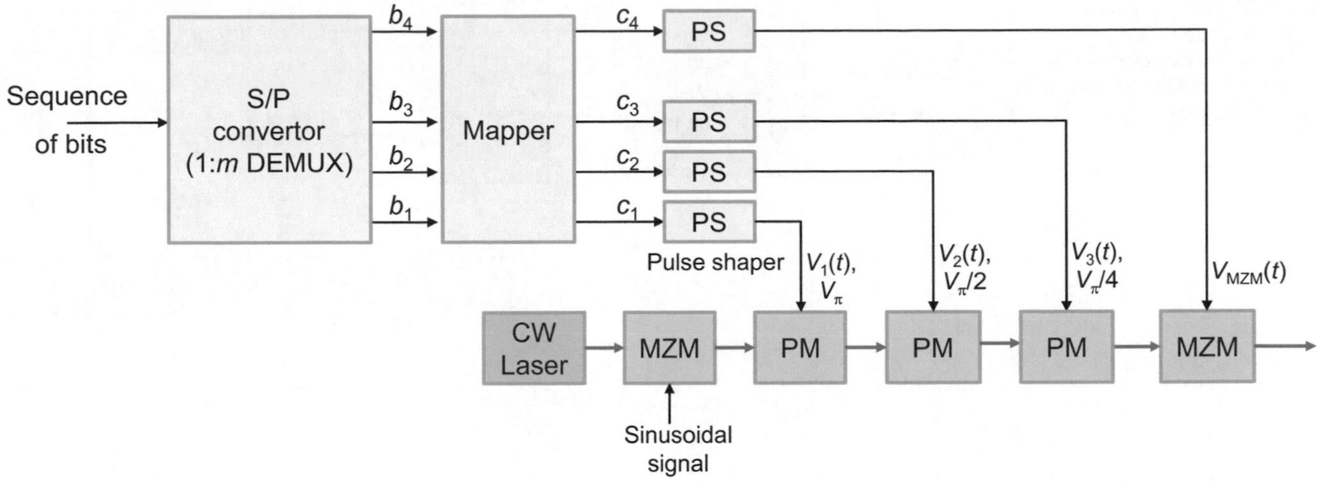

Fig. 5.16 The operation principle of a concatenated (serial) configuration of a 16-star-QAM transmitter for optical communications

Let us say that signal constellation contains M_c concentric circles with M_p points per circle. The number of required E-O PMs is $m_\mathrm{p} = \log_2 M_\mathrm{p}$, while the number of required bits to select a given circle is $m_c = \log_2 M_\mathrm{c}$. Only one MZM will be needed, with driving signal having M_c amplitude levels. The number of bits per constellation point is $m_c + m_p$, where the first m_c bits are used to select a circle, while the remaining m_p bits are used to select a point within a given circle. The set of M_p phase modulators, each driven with binary sequence, can be replaced with a single one dealing with the M_c-level signal. This configuration corresponds to the polar modulator shown in Fig. 3.11 in Chap. 3. Alternatively, a single I/Q modulator can be used, but requires the mapper implemented as an LUT followed by a DAC of sufficient resolution.

For differential encoding, the m_c bits that are used to select a given circle must stay in natural mapping form, while the remaining m_p bits can be differentially encoded. The 16-star-QAM modulator can also be implemented by concatenating one E-O I/Q modulator, one E-O PM, and one MZM. The E-O I/Q modulator is used to generate a QPSK constellation, while the E-O PM is used to introduce an additional phase shift of $\pm\pi/4$, so that 8-PSK is generated. Finally, the MZM is used to select one out of two 8-PSK constellations. Notice that the use of more than two amplitude levels in MZM degrades the overall OSNR.

The 16-ary star-QAM modulator shown in Fig. 5.16 can be generalized by using m_p of E-O PMs and one MZM. This modulator is able to generate a star-QAM constellation composed of two 2^{mp}-ary PSK constellations. The output electric field of this star-QAM modulator is given by

$$E_T(t) = \sqrt{P_T} e^{j(\omega_T t + \phi_{T,PN})} \exp\left(j\frac{\pi}{V_\pi} \sum_{i=1}^{m_p} V_i(t) \right) \cos\left(\frac{\pi V_{\mathrm{MZM}}(t)}{2V_\pi} \right), \tag{5.57}$$

where $V_i(t)$ $(i = 1,2,\ldots,m_p)$ is given by Eq. (5.54), while $V_{\mathrm{MZM}}(t)$ is given by

$$V_{\mathrm{MZM}}(t) = \frac{2V_\pi}{\pi} \cos^{-1}(r_0/r_1) \left[\sum_k c_{m_p+1}[k] p(t - kT_s) - 1 \right]. \tag{5.58}$$

5.2.4 Square- and Cross-QAM Transmitters for Optical Transmission

The square-/cross-QAM constellations can be generated in a number of different ways by using (i) single E-O I/Q modulator (Fig. 3.10), (ii) single E-O polar modulator (Fig. 3.11), (iii) improved E-O I/Q modulator having one E-O PM and one MZM per phase branch, (iv) parallel E-O I/Q modulators, (v) so-called tandem-QAM transmitter, and so on. The configuration employing a single E-O I/Q modulator is shown in Fig. 5.17. The $m = \log_2 M$ bits are used to select a QAM constellation point stored in a QAM mapper, implemented as an LUT. The in-phase and quadrature driving signals are given as

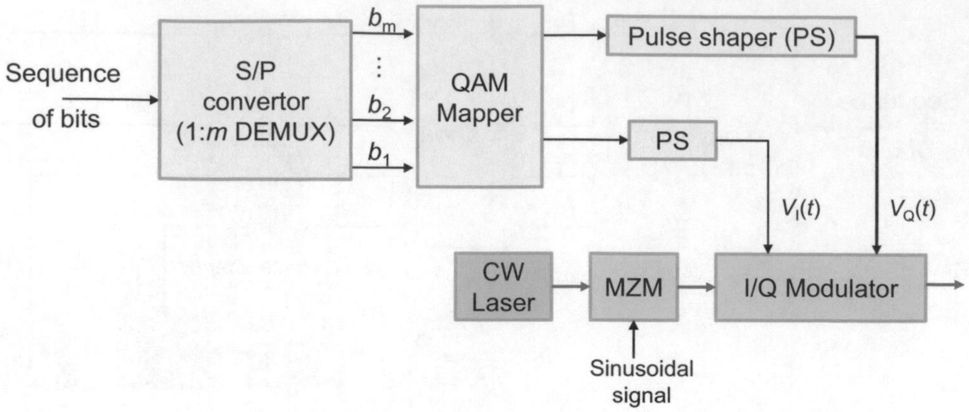

Fig. 5.17 The operation principle of an E-O I/Q modulator-based square-QAM transmitter suitable for optical communications

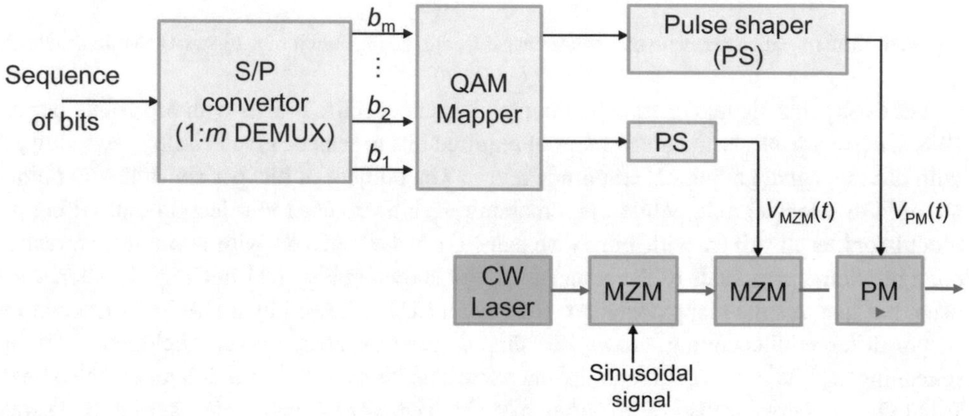

Fig. 5.18 The operation principle of an E-O polar modulator-based square-QAM transmitter for optical communications

$$
\begin{aligned}
V_I(t) &= 2\frac{V_\pi}{\pi}\left[\sum_{k=-\infty}^{\infty} \sin^{-1}(b_1(k))p(t-kT_s) - \frac{\pi}{2}\right], \\
V_Q(t) &= 2\frac{V_\pi}{\pi}\left[\sum_{k=-\infty}^{\infty} \sin^{-1}(b_2(k))p(t-kT_s) - 1\right].
\end{aligned}
\tag{5.59}
$$

It is evident, based on Eqs. (3.23) and (3.24), that polar modulator can be used instead of an E-O I/Q modulator to generate an arbitrary square/cross constellation. The polar modulator-based square-QAM transmitter is shown in Fig. 5.18. The transmitted electric field can be written as

$$
E_T(t) = \sqrt{P_T} e^{j(\omega_T t + \phi_{T,PN})} \cos\left(\frac{\pi V_{\mathrm{MZM}}(t)}{2V_\pi}\right) \exp\left(j\frac{\pi V_{\mathrm{PM}}(t)}{V_\pi}\right), \tag{5.60}
$$

where with V_{MZM} and V_{PM} we denoted the MZM and PM driving signals, respectively. However, the polar modulator-based square/cross transmitter requires the nonlinear transformation of the QAM constellation points' coordinates.

The square–/cross-QAM transmitter employing an alternative E-O I/Q modulator is shown in Fig. 5.19. In the same figure, the configuration of the alternative E-O I/Q modulator is shown. The key difference with respect to conventional E-O I/Q modulator is the introduction of additional E-O PM in an in-phase branch. Notice that, from a communication theory point of view, this additional PM is redundant. However, its introduction provides more flexibility in the sense that simpler driving signals can be used to generate a desired QAM constellation. The MZM and E-O PM driving signals in an I-branch are given by

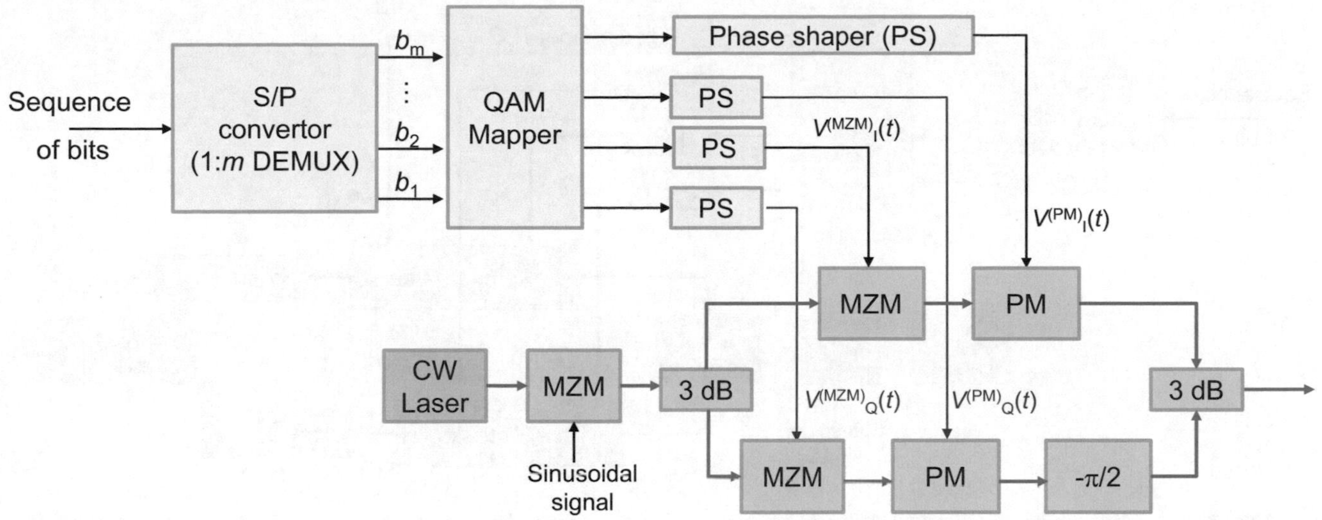

Fig. 5.19 The operation principle of an alternative I/Q modulator-based square-QAM transmitter. *PS* pulse shaper; I/Q modulator is shown in Fig. 5.17; 3 dB, 3 dB coupler

$$
\begin{aligned}
V_I^{(MZM)}(t) &= 2\frac{V_\pi}{\pi}\left[\sum_{k=-\infty}^{\infty} \sin^{-1}(|b_1(k)|)p(t-kT_s) - \frac{\pi}{2}\right], \\
V_I^{(PM)}(t) &= \frac{V_\pi}{2}\sum_{k=-\infty}^{\infty}(1-\mathrm{sign}(b_1(k)))p(t-kT_s),
\end{aligned}
\tag{5.61}
$$

while the MZM and PM driving signals in a Q-branch are given by

$$
\begin{aligned}
V_Q^{(MZM)}(t) &= 2\frac{V_\pi}{\pi}\left[\sum_{k=-\infty}^{\infty} \sin^{-1}(|b_2(k)|)p(t-kT_s) - \frac{\pi}{2}\right], \\
V_Q^{(PM)}(t) &= \frac{V_\pi}{2}\sum_{k=-\infty}^{\infty}(1-\mathrm{sign}(b_2(k)))p(t-kT_s).
\end{aligned}
\tag{5.62}
$$

The output electric field of the alternative I/Q modulator-based square−/cross-QAM transmitter can be represented in polar coordinates as

$$
E_T(t) = \sqrt{P_T}e^{j(\omega_T t+\phi_{T,PN})}R(t)\exp(\,j\Psi(t)),
\tag{5.63}
$$

where the envelope $R(t)$ and phase $\Psi(t)$ are given, respectively, as

$$
\begin{aligned}
R(t) &= \frac{1}{2}\left[R_I^2(t)+R_Q^2(t)+2R_I(t)R_Q(t)\cos(\Psi_I(t)-\Psi_Q(t))\right]^{1/2}, \\
\Psi(t) &= \tan^{-1}\left\{\frac{R_I(t)\sin\Psi_I(t)-R_Q(t)\sin\Psi_Q(t)}{R_I(t)\cos\Psi_I(t)-R_Q(t)\cos\Psi_Q(t)}\right\},
\end{aligned}
\tag{5.64}
$$

with R_I (R_Q) and Ψ_I (Ψ_Q) defined by

$$
R_{I,Q}(t) = \cos\left(\frac{1}{2}\frac{\pi V_{I,Q}^{(\mathrm{MZM})}(t)}{V_\pi}\right), \Psi_{I,Q}(t) = \frac{1}{2}\frac{\pi V_{I,Q}^{(\mathrm{PM})}(t)}{V_\pi}.
$$

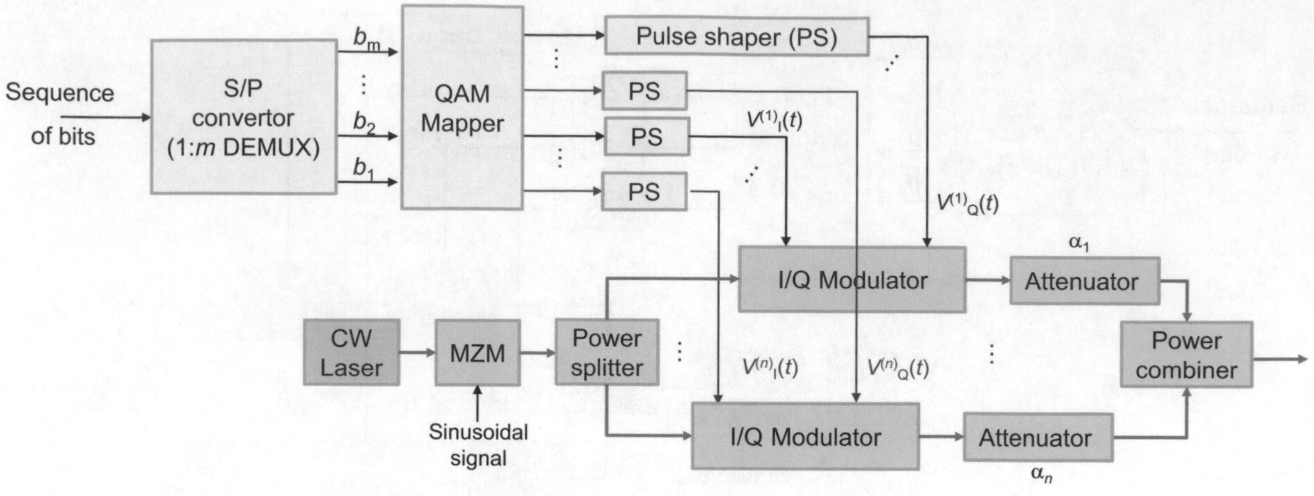

Fig. 5.20 The operation principle of a parallel I/Q modulator-based square-QAM transmitter. Attenuation coefficient of the *i*th branch is specified by $\alpha_i = (i - 1) \times 6$ dB

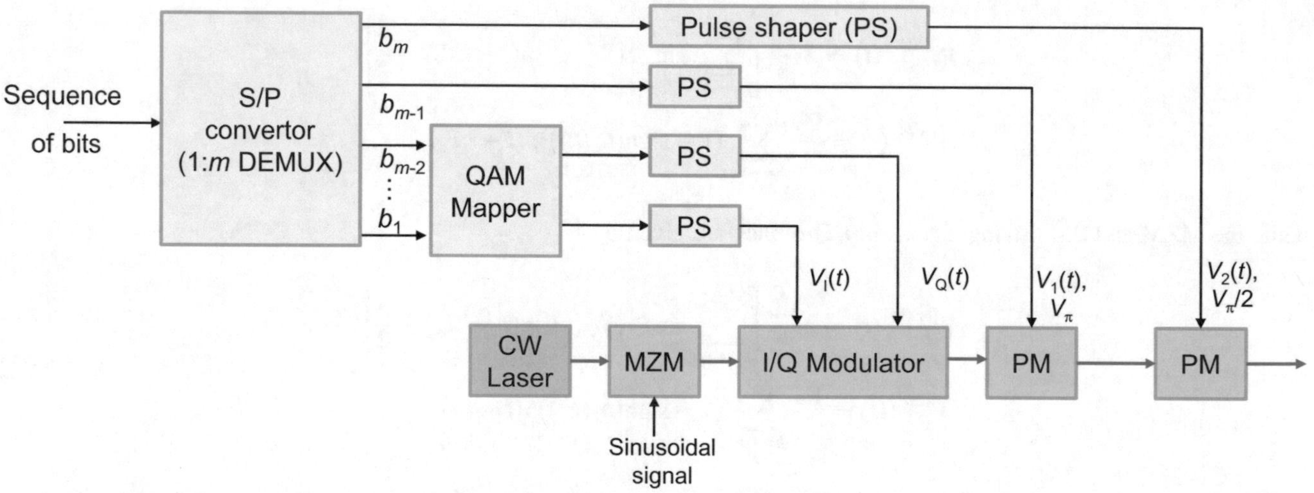

Fig. 5.21 The operation principle of a tandem-QAM transmitter for optical communications

The *parallel I/Q modulator-based square–/star-QAMtransmitter* is shown in Fig. 5.20. The CW signal is split into *n* branches, with each branch containing one I/Q modulator and attenuator. The attenuation coefficient of the *i*th branch is given as $\alpha_i = (i - 1) \times 6$ dB. The key idea behind this approach is to simplify a generation of larger constellation sizes by properly combing smaller constellation sizes. For instance, 16-QAM can be generated by summing up two QPSK signals, while one of the QPSK signals is being attenuated by 6 dB.

The *tandem-QAM transmitter* is shown in Fig. 5.21, based on [9]. The I/Q modulator is used to generate the initial QAM constellation residing in the first quadrant only. Two consecutive PM modulators are used to introduce the phase shifts of $\pi/2$ and π, respectively. With this additional manipulation, we essentially rotate the constellation diagram residing in the first quadrant for $\pm\pi/2$ and $\pm\,\pi$, respectively, thus creating the resulting QAM constellation residing in all four quadrants. The first *m*-2 bits are used to select a point from the initial signal constellation residing in the first quadrant. The bit b_{m-1} is used to control the first PM so that, after the pulse shaper (PS), a phase shift of $\pm\pi/2$ is introduced. The bit b_m is used to control the second PM so that after the PS a phase shift of $\pm\pi$ is introduced. The resulting signal constellation has 2^m constellation points located in all four quadrants.

5.3 Multicarrier Modulation

The key idea behind the multicarrier modulation is to divide the transmitted bit stream into many lower-speed substreams, which are typically orthogonal to each other, and send these over many different subchannels [1–3, 5–7]. For wireless communication applications, the number of substreams is chosen to ensure that each subchannel has a bandwidth smaller than the coherence bandwidth of the channel, so that each subchannel will experience the flat fading. This is an efficient way to deal with frequency-selective fading effects. For fiber-optics communications, the guard interval is chosen in such a way that the delay spread due to dispersion effects is smaller than the guard interval so that these dispersion effects can be successfully compensated.

The discrete implementation of multicarrier modulation, well known as orthogonal frequency division multiplexing (OFDM) [1–3, 5–7, 13–15], can be successfully used to deal with ISI introduced by various fading and dispersion effects through the use of concept of cyclic prefix. As the name suggests, the OFDM can be considered as a modulation technique, a multiplexing technique, or both. This gives the OFDM flexibility that does not exist in conventional single-carrier systems. The full details of OFDM can be found in Chap. 7; here, we are concerned with some fundamental concepts of multicarrier modulation. We describe two versions of multicarrier systems: (i) multicarrier systems with nonoverlapping subcarriers and (ii) multicarrier systems with overlapping subcarriers.

5.3.1 Multicarrier Systems with Nonoverlapping Subcarriers

The multicarrier transmitter configuration is provided in Fig. 5.22, assuming that real-valued signals are transmitted, to facilitate explanations. If we assume that N_{sc} subcarriers are used, each having an R_{sc} symbol rate, the total symbol rate R_s for multicarrier systems can be calculated as

$$R_s = N_{sc}R_{sc}. \tag{5.65}$$

Let BW denote the signal bandwidth. The corresponding bandwidth per subcarrier will be $BW_{sc} = BW/N_{sc}$. When the wireless channel is frequency selective with coherence bandwidth $B_c > BW$, for a sufficient number of subcarriers, we can ensure that the $B_c << BW_{sc}$, so that each subcarrier will experience the flat fading. In other words, the symbol duration at the subcarrier level $T_{sc} = 1/R_{sc} \cong 1/BW_{sc} >> 1/B_c \approx 1/\Delta\tau$, where $\Delta\tau$ is the multipath delay spread of the wireless channel. When raised-cosine pulses, with roll-off factor β, are used at each subcarrier, the bandwidth occupied per subcarrier will be $BW_{sc} = (1 + \beta)R_{sc}$. For the multicarrier system with nonoverlapping subcarriers, the guard band Δ/R_{sc} ($\Delta <<1$) must be introduced to limit co-channel crosstalk. By choosing the subcarrier frequencies as follows, $f_n = f_1 + (n - 1)BW_{sc}$ $(n - 1, \ldots, N_{sc})$, the subcarrier channels will be orthogonal to each other, with each subcarrier channel occupying the bandwidth BW_{sc}. The total bandwidth of the multicarrier system with nonoverlapping subcarriers will be then

$$BW = N_{sc}(1 + \beta + \Delta)R_{sc}. \tag{5.66}$$

The modulated subcarriers are multiplexed together in an FDM fashion (see Fig. 5.22), and this multicarrier multiplexing can be expressed as

$$s(t) = \sum_{n=1}^{N_{sc}} s_n p(t)\cos(2\pi f_n t + \varphi_n), \quad f_n = f_1 + (n-1)BW_{sc}\ (n = 1, \ldots, N_{sc}), \tag{5.67}$$

where s_n is the symbol transmitted by the nth subcarrier. After that, the multicarrier signal is amplified and transmitted with the help of corresponding antenna.

On the receiver side, to reconstruct the nth subcarrier, the band-pass filter of bandwidth BW_{sc}, centered at f_n, is sufficient to reject the neighboring multicarrier subchannels, as illustrated in Fig. 5.23. Following down-conversion to the baseband, the rest of the receiver per subcarrier is the same as described earlier.

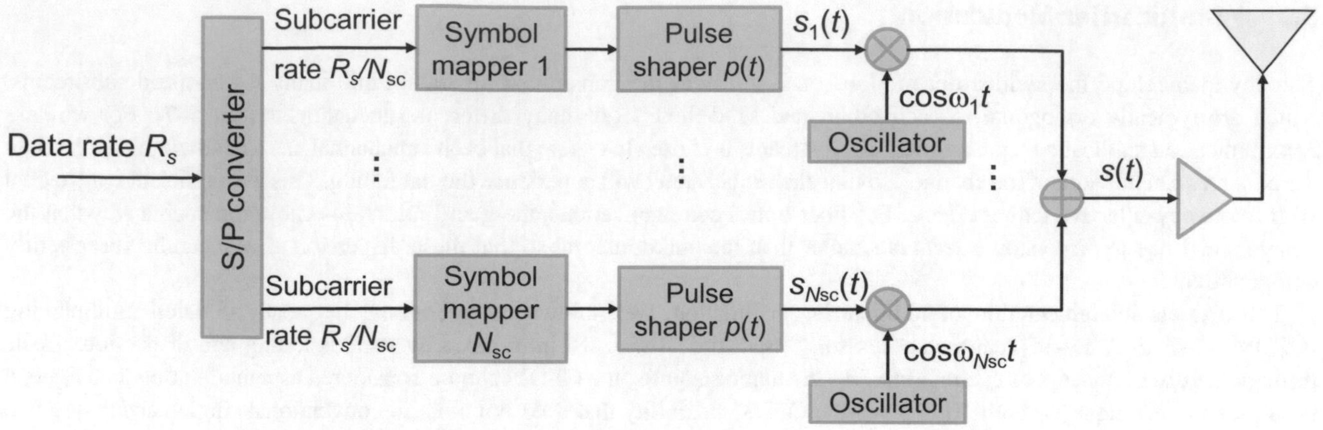

Fig. 5.22 Multicarrier transmitter configuration for wireless communications (for real-valued signals)

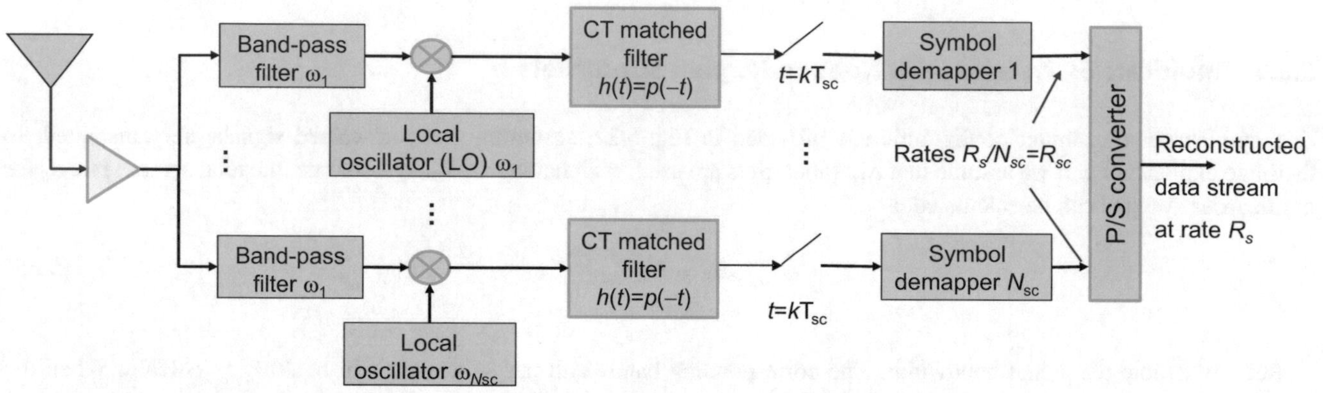

Fig. 5.23 The multicarrier receiver for nonoverlapping subcarriers (for real-valued signals)

5.3.2 Multicarrier Systems with Overlapping Subcarriers

Regarding the multicarrier receiver with overlapping subcarriers, the subcarrier frequencies are properly chosen such that the following principle of orthogonality is satisfied:

$$\begin{aligned}&\frac{1}{T_{sc}}\int_0^{T_{sc}} \cos\left[2\pi\left(f_1+\frac{m}{T_{sc}}\right)t+\varphi_i\right]\cos\left[2\pi\left(f_1+\frac{n}{T_{sc}}\right)t+\varphi_n\right]dt\\ &=\frac{1}{T_{sc}}\int_0^{T_{sc}}\cos\left[2\pi\frac{m-n}{T_N}t+\varphi_m-\varphi_n\right]dt+\ldots\cong\frac{1}{2}\delta(m-n),\end{aligned} \tag{5.68}$$

even for random phase shifts per subcarrier, providing that $f_1T_{sc}>>1$. Thanks to this orthogonality of subcarriers, we can allow for partial overlap of neighboring subcarriers, and we can choose the subcarrier frequencies as $f_n=f_1+(n-1)R_{sc}$ $(n=1,\ldots,N_{sc})$. By assuming again that raised-cosine pulse shape is used on the subcarrier level, with roll-off factor β, the total occupied bandwidth will be

$$BW=N_{sc}R_{sc}+(\beta+\Delta)R_{sc}\approx N_{sc}R_{sc}. \tag{5.69}$$

Clearly, for a sufficient number of subcarriers, the bandwidth efficiency of the multicarrier system with overlapping subcarriers is $(1+\beta+\Delta)$ times higher than that of the multicarrier system with nonoverlapping subcarriers.

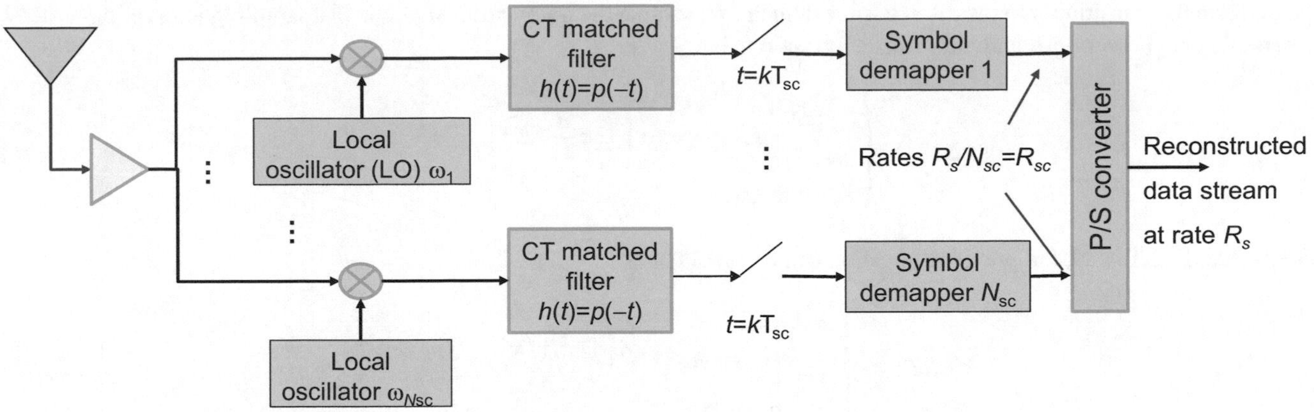

Fig. 5.24 Configuration of the multicarrier receiver with overlapping subcarriers (for real-valued signals)

The transmitter configuration of the multicarrier system with overlapping subcarriers is the same that shown in Fig. 5.22, providing that the principle of orthogonality, given by Eq. (5.68), is satisfied. On the other hand, the receiver configuration of the multicarrier system with overlapping subcarriers is shown in Fig. 5.24. So the key difference of this receiver with respect to the corresponding receiver shown in Fig. 5.23 is that the band-pass filters can be omitted, thanks to the orthogonality principle. To illustrate that subcarrier channels can be properly demultiplexed, without co-channel crosstalk, the input to the symbol mapper in the nth branch of the receiver is given by

$$\begin{aligned}\widehat{s}_n &= \int_0^{T_{sc}} \left[\sum_{m=1}^{N_{sc}} s_m p(t)\cos\left(2\pi f_m t+\varphi_m\right)\right] p(t)\cos\left(2\pi f_n t+\varphi_n\right)dt = \\ &= \sum_{m=1}^{N_{sc}} s_m \int_0^{T_{sc}} p^2(t)\cos\left(2\pi\left(f_0+\frac{m}{T_{sc}}\right)t+\phi_m\right)\cos\left(2\pi\left(f_0+\frac{n}{T_{sc}}\right)t+\varphi_n\right)dt \\ &= \sum_{m=1}^{N_{sc}} s_m\delta(m-n)=s_n,\end{aligned} \tag{5.70}$$

and clearly, the nth subcarrier symbol is properly detected. Even in the presence of selective fading, the nth subcarrier will see the flat fading with corresponding channel coefficient $H(f_n)$, and the corresponding channel model is given by

$$r_n = H(f_n)s_n + z_n;\, n=1,2,\cdots,N_{sc}, \tag{5.71}$$

where z_n is the AWGN sample with variance N_0BW_{sc}. Clearly, with multicarrier modulation, by employing the orthogonality principle, we have successfully decomposed the wireless channel into N_{sc} parallel subchannels.

For complex-valued signals, the multicarrier signal can be represented as

$$s(t)=\begin{cases}\mathrm{Re}\left\{\sum_{n=-N_s/2}^{n=N_s/2-1} \underline{s}_{n+N_{sc}/2}\exp\left[j2\pi\left(f_c-\frac{n+0.5}{T_s}\right)t\right]\right\}, & 0\le t\le T_s \\ 0, & \text{otherwise}\end{cases}, \tag{5.72}$$

where f_c is the carrier frequency and T_s is the symbol duration of the multicarrier signal. The corresponding complex envelope representation is given by

$$\widetilde{s}(t)=\begin{cases}\sum_{n=-N_s/2}^{n=N_s/2-1} \underline{s}_{n+N_{sc}/2}\exp\left[j2\pi\frac{n}{T_s}t\right], & 0\le t\le T_s \\ 0, & \text{otherwise}\end{cases}. \tag{5.73}$$

To perform the transition into the discrete-time domain, we sample the multicarrier signal with a sampling interval $\Delta T = T_s/N$, where N is the power of 2, and after sampling, we obtain

$$\tilde{s}(k\Delta T) = \begin{cases} \sum_{n=-N_s/2}^{n=N_s/2-1} \underline{s}_{n+N_{sc}/2} \exp\left[j2\pi \frac{n}{T_s} k\Delta T\right], & 0 \le t \le T_s \\ 0, & \text{otherwise} \end{cases}. \tag{5.74}$$

Given that Ts/ΔT = N, the previous equation can be rewritten as

$$\tilde{s}(k\Delta T) = \begin{cases} \sum_{n=-N_s/2}^{n=N_s/2-1} \underline{s}_{n+N_{sc}/2} \exp\left[j2\pi nk/N\right], & 0 \le t \le T_s \\ 0, & \text{otherwise} \end{cases} \tag{5.75}$$

and represents the inverse discrete Fourier transform (IDFT) of 2-D (in-phase-quadrature) sequence $\{\underline{s}_n\}$. Therefore, to efficiently implement multicarrier transmitter and receiver, the fast Fourier transform (FFT) can be used, and this version of the multicarrier system is commonly referred to as OFDM [1–3, 5–7, 13–15]. For additional details on OFDM as well as various applications, an interested reader is referred to Chap. 7, devoted to OFDM systems.

5.3.3 Dealing with Fading Effects on Subcarrier Level

As discussed above, in multicarrier modulation, each subcarrier channel is narrowband, so that each subcarrier experiences the flat fading, which mitigates the multipath effects. Nevertheless, each subchannel experiences flat fading, which can introduce the deep fade on certain subcarriers. The various methods to compensate for the flat fading effects on subcarrier channels can be categorized as follows: (i) frequency equalization, which is described in Chap. 6; (ii) precoding; (iii) adaptive loading, also described in Chap. 6; and (iv) coding with interleaving in time and frequency domains, which is described in Chap. 9. Here briefly describe the basic multicarrier modulation concepts. In frequency equalization, the flat fading $H(f_n)$ on the nth subcarrier is essentially inverted at the receiver side, the received signal is multiplied by $1/H(f_n)$, and the resulting resultant signal power is independent on flat fading since $|H(f_n)|^2P_n/|H(f_n)|^2 = P_n$. Even though the impact of flat fading effects on the signal is removed, unfortunately, in equalization, in particular for zero-forcing equalizers, the noise is enhanced because the noise power is proportional to $N_0BW_{sc}/|H(f_n)|^2$. Different equalization schemes are subject of study in Chap. 6.

Precoding employs the same idea as in frequency equalization, except that the flat fading is inverted at the transmitter side instead of the receiver. The channel state information related to the nth subcarrier needs to be fed back to the transmitter side so that transmitter transmits with power $P_n/|H(f_n)|$, so that after the flat fading channel, the receiver will see constant power since $[P_n/|H(f_n)|^2]|H(f_n)|^2 = P_n$. Because the channel inversion takes place at the transmitter side instead of the receiver, the noise power at the receiver side will be N_0BW_{sc}, and the noise enhancement problem is avoided.

Adaptive loading is based on the adaptive modulation techniques.

The key idea is to vary the data rate and power assigned to each subchannel relative to that subchannel gain.

In adaptive loading techniques, the power and rate on each subcarrier is adapted in such a way to maximize the total rate of the system by using adaptive modulation techniques, described with more details in Chap. 6. The capacity of the multicarrier system with N_{sc} independent subchannels of bandwidth BW_{sc} and subcarrier channel coefficients $\{H(f_n),\ n = 1, \ldots, N_{sc}\}$, based on Chap. 4, is given by

$$C = \max_{P_n:\ \sum P_n = P} \sum_{n=1}^{N_{sc}} BW_{sc} \log_2\left(1 + \frac{|H(f_n)|^2 P_n}{N_0 BW_{sc}}\right). \tag{5.76}$$

The optimum power allocation policy is the water filling in the frequency domain:

$$\frac{P_n}{P} = \begin{cases} 1/\rho_{\text{tsh}} - 1/\rho_n, \rho_n \ge \rho_{\text{tsh}} \\ 0, \text{ otherwise} \end{cases}, \quad \rho_n = |H(f_n)|^2 P/N_0 BW_{sc}, \tag{5.77}$$

where ρ_{tsh} is the threshold SNR. After substituting Eq. (5.77) into (5.76), we obtain

$$C = \sum_{n:\rho_n > \rho_{\rm tsh}} BW_{sc} \log_2\left(\frac{\rho_n}{\rho_{\rm tsh}}\right). \tag{5.78}$$

The key idea of coding with interleaving is to encode information bits into code words, interleave the coded bits over both time and frequency domains, and then transmit the coded bits over different subcarriers such that the coded bits within a code word all experience independent flat fading. In such a way, the channel code will be correcting random errors only, instead of both random and burst errors introduced by flat fading. For additional details on coding with interleaving, an interested reader is referred to Chap. 9.

5.4 MIMO Fundamentals

Multiple transmitters can be used either to improve the performance through array gain, related to improvement in SNR, and diversity gain, related to improving the BER slope, or to increase data rate through multiplexing of independent data streams. In multiple-input multiple-output (MIMO) systems, the transmitters and receivers can exploit the knowledge about the channel to determine the rank of so-called channel matrix $\boldsymbol{H}$, denoted as R_H, which is related to the number of independent data streams that can be simultaneously transmitted. When identical data stream is transmitted over multiple transmit antennas, the corresponding scheme is a *spatial diversity scheme*. In spatial diversity scheme, M either transmit or receive branches are properly combined in a linear combiner to improve the SNR. This improvement in SNR is known as the *array gain* and can be defined as the linear combiner average SNR $\overline{\rho}_{\oplus}$ to single branch average SNR $\overline{\rho}$ ratio:

$$AG = \frac{\overline{\rho}_{\oplus}}{\overline{\rho}} \leq M, \tag{5.79}$$

and the maximum possible array gain is equal to M. The *average symbol error probability* can be determined from SNR-dependent expression of symbol error probability $P_s(\rho)$ by averaging out over all possible values of ρ:

$$\overline{P}_s = \int_0^{\infty} P_s(\rho) f_{\rho_{\oplus}}(\rho) d\rho, \tag{5.80}$$

where $f_{\rho_{\oplus}}(\rho)$ is the PDF of the linear combiner SNR. The reduction in the symbol error rate due to diversity can be described by more rapid slope in the corresponding symbol error probability curve. For very high SNRs, the symbol error probability is reversely proportional to ρ, so that the *diversity gain* can be defined as

$$DG = -\lim_{\rho \to \infty} \frac{\log \overline{P}_s(\rho)}{\log \rho}. \tag{5.81}$$

In MIMO systems, multiple independent data streams are transmitted simultaneously. To specify improvement due to MIMO, we can define the *multiplexing gain* as the asymptotic slope in SNR of the outage capacity:

$$MG = -\lim_{\rho \to \infty} \frac{C_{{\rm outage},p}(\rho)}{\log_2 \rho}, \tag{5.82}$$

where $C_{\text{outage}, p}(\rho)$ is the p percentage outage capacity at signal-to-noise ratio ρ. The p percentage capacity is defined as the transmission rate that can be supported in (100-p)% of the channel realizations.

As an illustration, in Fig. 5.25, we provide the basic concept of wireless MIMO communication, with the number of transmit antennas denoted as $M_{\rm Tx}$ and the number of receive antennas denoted as $M_{\rm Rx}$. The 2-D symbol transmitted on the mth transmit antennas is denoted by x_m ($m = 1, 2, \ldots, M_{\rm Tx}$), while the received symbol on the nth received antenna is denoted as y_n. The channel coefficient relating the received symbol on the nth receive antenna and transmitted symbol on the mth transmit antenna is denoted as h_{mn}. By using this model, the output of the nth receive antenna can be represented as

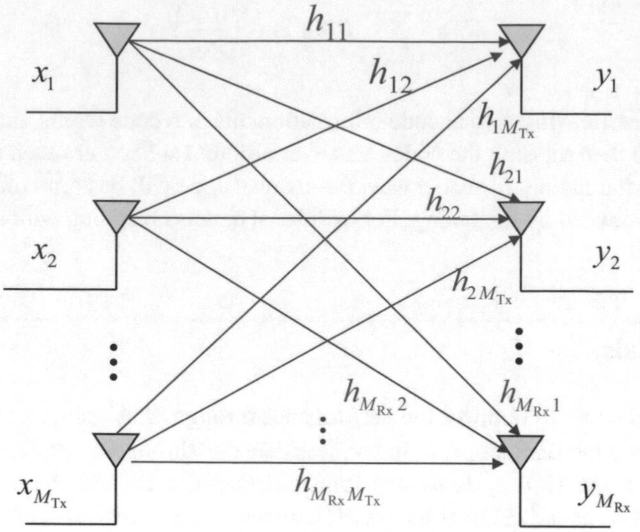

Fig. 5.25 The wireless MIMO concept

$$y_n = \sum_{m=1}^{M_{Tx}} h_{nm} x_m + z_n, \tag{5.83}$$

where z_n is the zero-mean Gaussian noise sample with variance $\sigma^2 = N_0 B$, with $N_0/2$ being the double-sided power spectral density and B being the channel bandwidth.

A narrowband *point-to-point wireless communication system* composed of M_{Tx} transmit and M_{Rx} receive antennas can be represented by the following discrete-time model, which is a matrix representation of Eq. (5.83):

$$\boldsymbol{y} = \boldsymbol{H}\boldsymbol{x} + \boldsymbol{z}, \tag{5.84}$$

where $\boldsymbol{x}$, $\boldsymbol{y}$, and $\boldsymbol{z}$ represent the transmitted, received, and noise vectors defined as

$$\boldsymbol{x} = \begin{bmatrix} x_1 \\ x_2 \\ \cdots \\ x_{M_{Tx}} \end{bmatrix}, \ \boldsymbol{y} = \begin{bmatrix} y_1 \\ y_2 \\ \cdots \\ y_{M_{Rx}} \end{bmatrix}, \ \boldsymbol{z} = \begin{bmatrix} z_1 \\ z_2 \\ \cdots \\ z_{M_{Rx}} \end{bmatrix}. \tag{5.85}$$

The channel noise samples z_n are complex Gaussian with zero mean and covariance matrix $\sigma^2 I_{M\ Rx}$, where $\sigma^2 = N_0 B$. If the noise variance is normalized to one then the transmitted power P can be normalized as follows: $P/\sigma^2 = \rho$, where ρ denotes the average SNR per receiver antenna, assuming unity channel gain. The total normalized transmitted power would be then $\sum_{m=1}^{M_{Tx}} \langle x_i x_i^* \rangle = \rho$. We use $\boldsymbol{H} = [h_{mn}]_{M_{Rx} \times M_{Tx}}$ in Eq. (5.84) to denote the so-called *channel matrix*:

$$\boldsymbol{H} = \begin{bmatrix} h_{11} & h_{12} & \cdots & h_{1M_{Tx}} \\ h_{21} & h_{22} & \cdots & h_{2M_{Tx}} \\ \cdots & \cdots & \cdots & \cdots \\ h_{M_{Rx}1} & h_{M_{Rx}2} & \cdots & h_{M_{Rx}M_{Tx}} \end{bmatrix}. \tag{5.86}$$

The multiplexing gain of a MIMO system, introduced by Eq. (5.82), originates from the fact that a MIMO channel can be decomposed into a number of parallel independent channels, so that the data rate can be increased R_H times, where R_H is the rank of the channel matrix $\boldsymbol{H}$.

Let us perform the *singular value decomposition*(SVD) of channel matrix $\boldsymbol{H}$ given by Eqs. (5.84), and (5.86). In SVD, we have the following: (i) the $\Sigma -$ matrix corresponds to scaling operation; (ii) the columns of matrix $\boldsymbol{U}$, obtained as eigenvectors

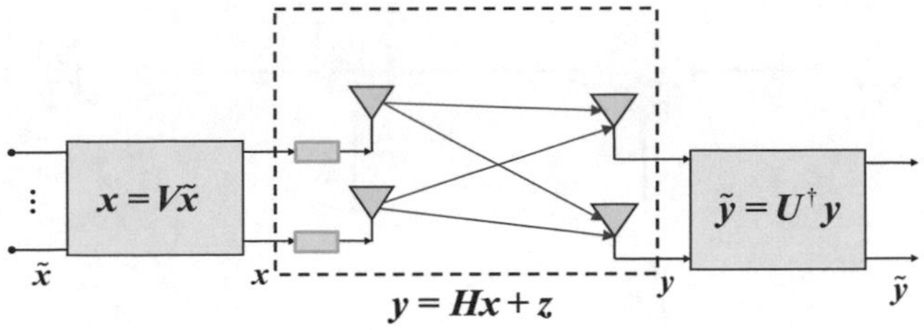

Fig. 5.26 The parallel decomposition of wireless MIMO channel

of the Wishart matrix $HH^\dagger$, correspond to the rotation operation; and (iii) the other rotation matrix V that has for columns the eigenvectors of $H^\dagger H$. The rank of matrix H corresponds to the multiplexing gain. In few-mode fiber (FMF) and few-core fiber (FCF)-based applications, we can ensure the full rank of channel matrix, which is not guaranteed when either regular multimode fiber links or wireless communication links are employed.

The *parallel decomposition* of the MIMO wireless channel, illustrated in Fig. 5.26, is obtained by introducing the following two transformations to the channel input vector x and channel output vector y: (i) *transmit precoding*, in which the input vector of symbols x to transmitter is linearly transformed by pre-multiplying with rotation matrix V; (ii) *receiver shaping*, in which the receiver vector y, upon optical coherent detection, is multiplied by rotation matrix $U^\dagger$. With these two operations, we effectively performed the following manipulation:

$$\tilde{y} = U^\dagger \left(\underbrace{H}_{U\Sigma V^\dagger} x + z \right) = U^\dagger U \Sigma V^\dagger V \tilde{x} + \underbrace{U^\dagger z}_{\tilde{z}} = \Sigma \tilde{x} + \tilde{z}, \quad (5.87)$$

and successfully decomposed the wireless MIMO channel into $R_H = \text{rank}(H)$ parallel single-input single-output (SISO) channels. The equivalent wireless/optical channel model is shown in Fig. 5.27. Therefore, with the help of precoding on the transmitter side and receiver shaping on the receiver side, we were able to decompose the MIMO channel into R_H independent parallel channels and therefore increase the data rate of the SISO system R_H times.

The readers interested in MIMO communications are referred to Chap. 8 for additional details.

5.5 Polarization-Division Multiplexing (PDM) and 4-D Signaling

The electric field of plane linearly polarized electromagnetic waves propagating in the z-direction can be presented as follows:

$$E(z,t) = E_x(z,t) + E_y(z,t) = e_x E_{0x} \cos(\omega t - kz) + e_y E_{0y} \cos(\omega t - kz + \delta), \quad (5.88)$$

where $k = 2\pi/\lambda$ is the wave propagation vector magnitude (with λ being the operating wavelength) and δ is the relative phase difference between the two orthogonal polarizations. In Eq. (5.88), E_x and E_y denote x- (horizontal) and y- (vertical) polarization states, respectively. By using the Jones vector representation of polarization wave [6, 7], we can represent the electric field as follows:

$$E(t) = \begin{bmatrix} E_x(t) \\ E_y(t) \end{bmatrix} = E \begin{bmatrix} \sqrt{1-p_r} \\ \sqrt{p_r} e^{j\delta} \end{bmatrix} e^{j(\omega t - kz)}, \quad (5.89)$$

where p_r is the power splitting ratio between states of polarizations (SOPs), while the complex phasor term is omitted. Since x- and y-polarization states are orthogonal, they can be used as the additional basis functions. Given the fact that in-phase and quadrature channels are orthogonal as well, we can conclude that the corresponding space becomes four-dimensional (4-D). This enables us to use 4-D constellations instead of 2-D ones (PSK and QAM). As compared to 2-D constellations with the same symbol energy, 4-D constellations can increase the Euclidean distance among neighboring constellation points and thus improve the SNR [10, 11]. In addition, as shown in [12], 4-D signal constellations, when applied to fiber-optic communications, are more robust to fiber nonlinearities.

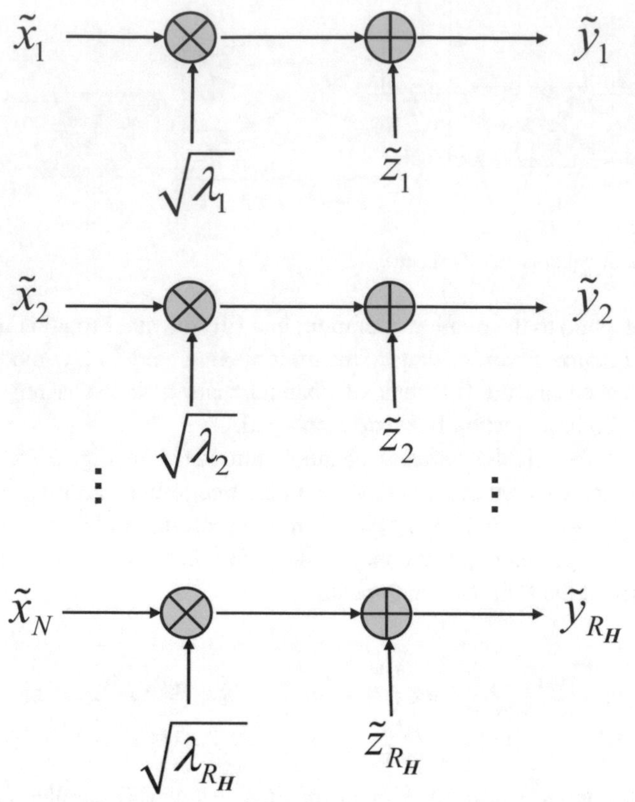

Fig. 5.27 Equivalent parallel MIMO channel. The λ_i denotes the ith eigenvalue of the Wishart matrix $\boldsymbol{HH}^\dagger$

The generic scheme suitable for wireless communication applications is provided in Fig. 5.25.

In this particular example, which employs dual-polarization transmit and receive antennas, two independent data streams are combined properly together in a polarization beam combiner. The configurations of RF transmitters corresponding to x- and y-polarization states are identical to that from Fig. 5.13(a), except from the antenna element. Due to multipath effect, the initial orthogonality of polarization states is no longer preserved on a receiver side, and we can use the channel coefficients as shown in Fig. 5.28 to describe this depolarization effect. The receiver now needs to compensate for cross polarization effects and then perform demodulation in a similar fashion as shown in Fig. 5.13(b). The polarization-division combiner could be just multiplexing two independent data streams or perform polarization-time coding, in a fashion similar to space-time coding, which is described in Chap. 8.

The 4-D electro-optical modulator is shown in Fig. 5.29. The CW laser signal is separated into two orthogonal polarizations by polarization beam splitter (PBS). Each polarization branch contains either single I/Q or polar modulator. In *polarization-division multiplexing (PDM)* applications, the QAM constellation coordinates in x- and y-polarization branches are used, after pulse shaping, as in-phase (I) and quadrature (Q) inputs of the corresponding I/Q modulator. The independent QAM streams are multiplexed together by a polarization beam combiner (PBC). The I/Q modulator can be replaced by polar modulators, as indicated in Fig. 5.29(b). Alternatively, various 2-D modulators described in Sect. 5.2 can be used in polarization branches. In 4-D modulation schemes, a single 4-D mapper, implemented as an LUT, is used to provide four coordinates for a 4-D modulator. In either case, the first two coordinates (after pulse shaping) are used as inputs to an I/Q modulator in an x-polarization branch, while the other two coordinates (after pulse shaping) are used as inputs to an I/Q modulator in a y-polarization branch. The corresponding signal constellation point of the ith symbol interval can be represented in a vector form as follows:

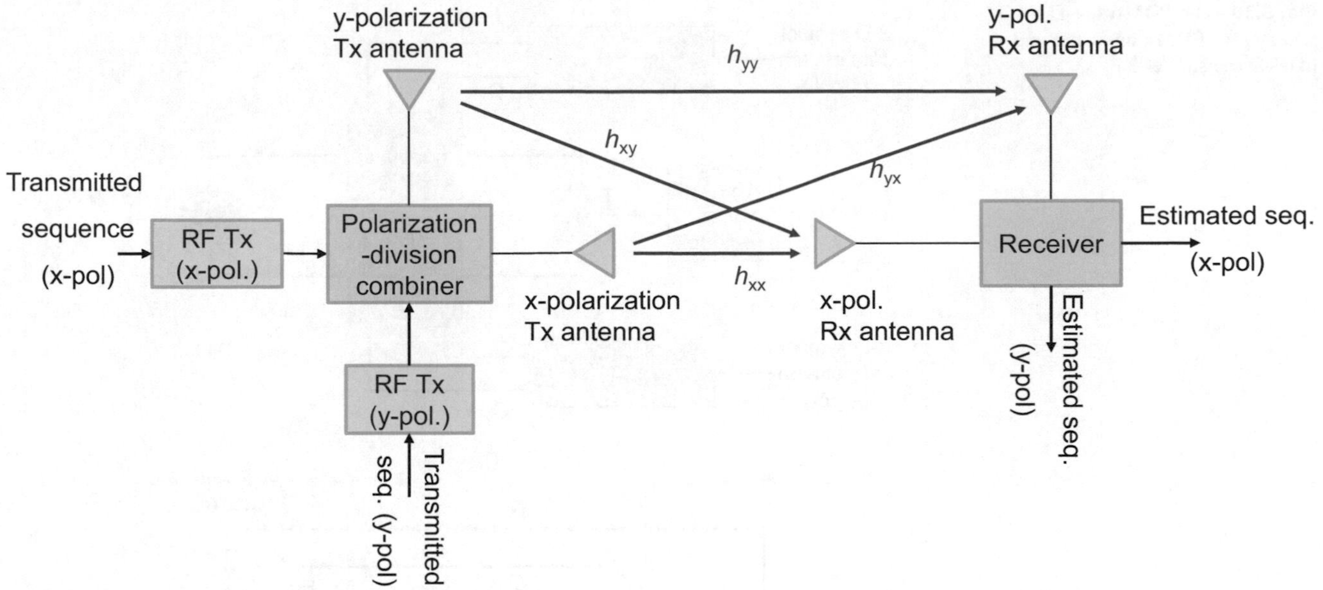

Fig. 5.28 Generic polarization-division multiplexing scheme for wireless communication applications

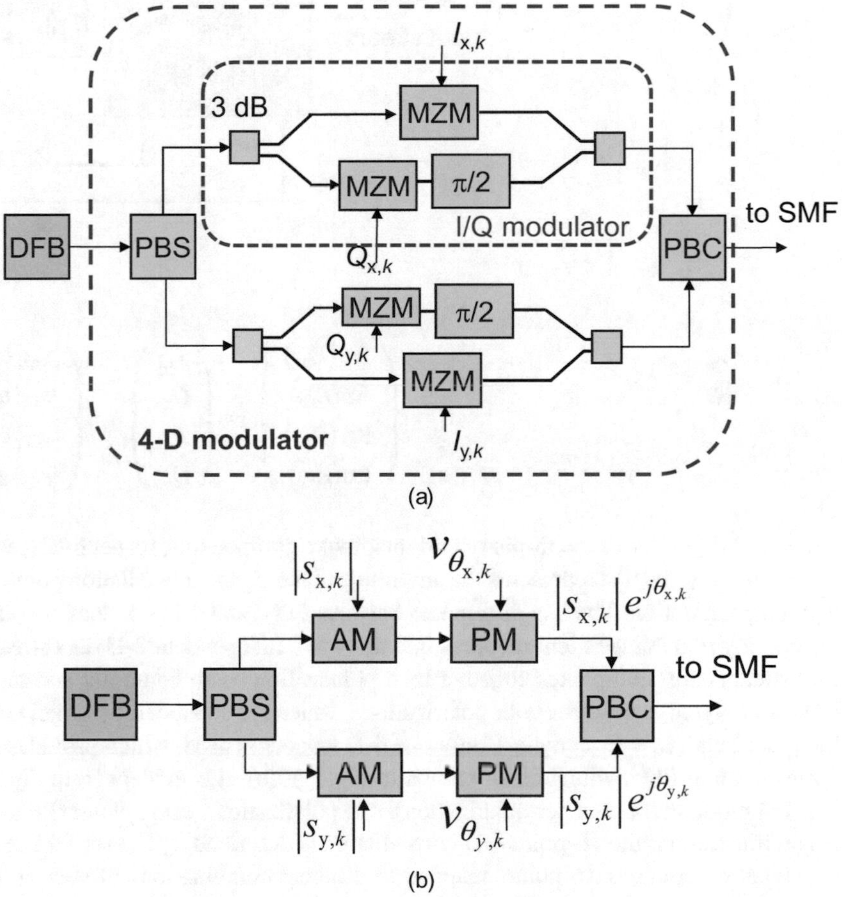

Fig. 5.29 The configuration of a 4-D modulator suitable for polarization-division multiplexing (PDM) and 4-D optical transmission: (**a**) Cartesian coordinate-based modulator and (**b**) polar coordinate-based modulator

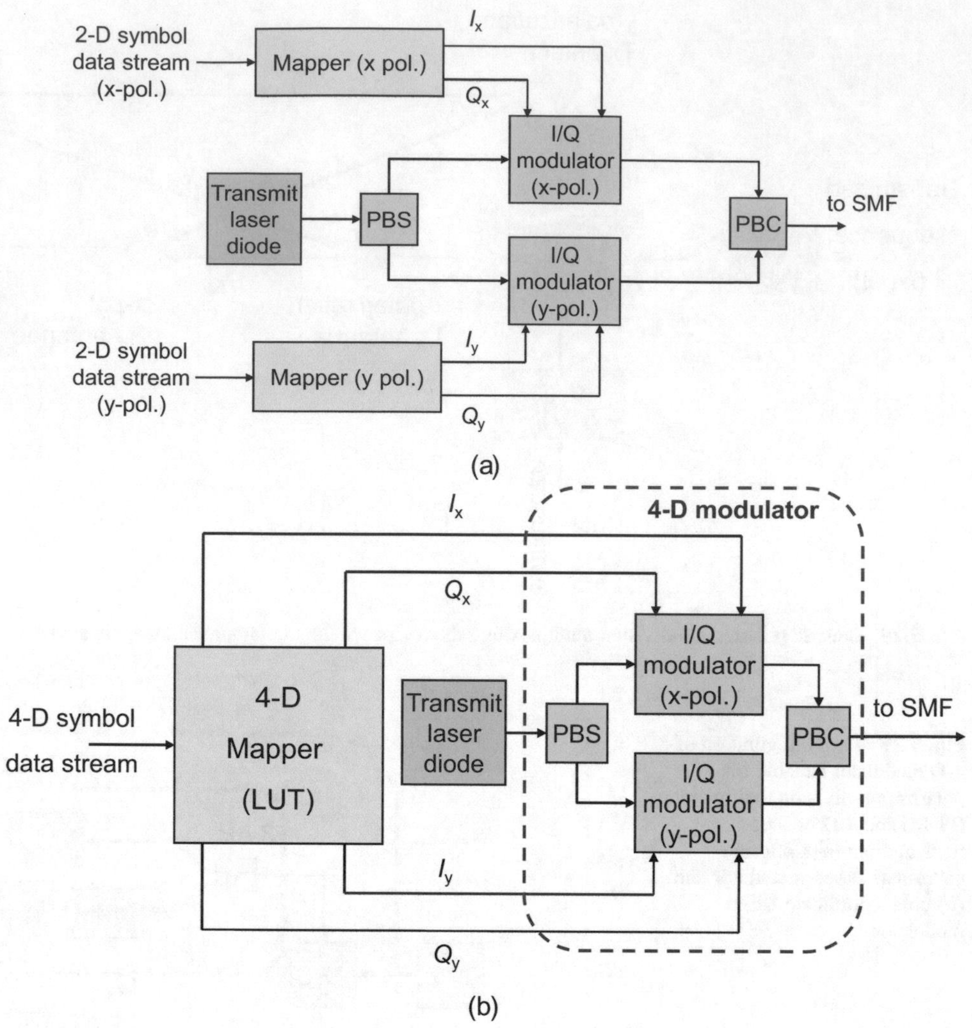

Fig. 5.30 The PDM vs. 4-D scheme: (**a**) PDM transmitter and (**b**) 4-D transmitter

$$\mathbf{s}_i = \begin{pmatrix} \mathrm{Re}\,(E_{x,i}) \\ \mathrm{Im}(E_{x,i}) \\ \mathrm{Re}\,(E_{y,i}) \\ \mathrm{Im}(E_{y,i}) \end{pmatrix} = \begin{pmatrix} I_{x,i} \\ Q_{x,i} \\ I_{y,i} \\ Q_{y,i} \end{pmatrix} = \begin{pmatrix} |s_{x,i}|\cos\theta_{x,i} \\ |s_{x,i}|\sin\theta_{x,i} \\ |s_{y,i}|\cos\theta_{y,i} \\ |s_{y,i}|\sin\theta_{y,i} \end{pmatrix}, \tag{5.90}$$

where I and Q refer to the in-phase and quadrature components, respectively, while x and y refer to the x- and y-polarization states. In Eq. (5.90), $|s|$ represents the amplitude of the signal constellation point, while θ represents the corresponding phase in a given polarization. The key differences between PDM and 4-D schemes are outlined in Fig. 5.30. In PDM (see Fig. 5.30(a)), we employ two independent mappers that drive two independent 2-D data streams through I/Q modulators. The independent 2-D streams are multiplexed together by a polarization beam combiner and transmitted over an optical fiber line. The same CW laser signal is used for both polarizations, which are separated by polarization beam splitter prior to the modulation. On the other hand, in 4-D signaling, only one 4-D mapper is used, which provides four coordinates that are used as inputs to the corresponding I/Q modulators, as shown in Fig. 5.30(b). It is evident from Fig. 5.30(b) that a 4-D modulator is composed of two I/Q modulators (one per polarization), one polarization beam splitter (PBS), and one polarization beam combiner (PBC). As an illustration, the 16-point 4-D constellation is described by the set $\{\pm1,\pm1,\pm1,\pm1\}$ [10]. At the same time, the 32-4-D constellation contains 16 points mapped to different combinations of vectors $\{\pm0.5,\pm0.5,\pm0.5,\pm0.5\}$, 4 points mapped to combinations of $\{\pm1,0,0,\pm1\}$, 4 points mapped to combinations of $\{0,\pm1,\pm1,0\}$, and 8 points mapped to the different combinations of $\{\pm1,0,0,0\}$. Even though 4-D schemes are more popular in fiber-optic communications than wireless communications, the scheme shown in Fig. 5.29 is also applicable in 4-D wireless communication-based signaling.

It also possible to transmit the same data stream in both polarizations, and we refer to this scheme as the *polarization diversity* scheme. The purpose of the polarization diversity scheme is to avoid the need for PMD compensation at the expense of deteriorating the OSNR sensitivity by 3 dB. The various PMD compensation schemes will be described in Chap. 6. Another interesting interpretation of a polarization diversity modulation is to consider it as a 2 × 2 multi-input multi-output (MIMO) scheme and employ the MIMO signal processing to deal with PMD, chromatic dispersion, polarization-dependent loss (PDL), and various filtering effects.

The spectral efficiency of 4-D modulation schemes is given by

$$S = R_s \log_2 M_{4-D}, \tag{5.91}$$

where R_s is the symbol rate and $M_{4\text{-D}}$ is the 4-D signal constellation size. On the other hand, the spectral efficiency of the polarization PDM scheme is given as

$$S = 2R_s \log_2 M_{2-D}, \tag{5.92}$$

where factor 2 accounts for polarization states, while $M_{2\text{-D}}$ is the 2-D signal constellation size used in each polarization. Signal constellation sizes for the same aggregate data rate are related by $M_{4-D} = M_{2-D}^2$.

For instance, 16-ary 4-D constellation corresponds to PDM-QPSK. Interestingly enough, in the presence of nonlinearities, 4-D modulation schemes of larger constellation sizes show better robustness compared to a PMD-QAM scheme of the same aggregate rate, as demonstrated in 0.

The 4-D signaling can also be combined with subcarrier multiplexing as described in 0. The characteristics of a dual-polarization receiver were described in Chap. 3. Another alternative is to combine 4-D signaling with OFDM 0. Also, instead of subcarrier multiplexing, we can use fiber mode multiplexing, and the corresponding scheme can be called mode-multiplexed 4-D modulation. Finally, the polarization diversity multiplexing can be used in combination with the *polarization-time* (PT) coding, which is in particular suitable for use in combination with OFDM 0. The multidimensional coded modulation (CM) studies [1, 14, 16, 17] indicate that the simultaneous employment of multiple photon degrees of freedom in an optical communication channel can enable a substantial increase in bandwidth capacity. Given the high potential of multidimensional signaling schemes, the next section will be devoted to spatial-division multiplexing and multidimensional signaling.

5.6 Spatial-Division Multiplexing and Multidimensional Signaling

In spatial-division multiplexing (SDM), we transmit multiple independent data streams over either multipath channels in wireless communications or spatial modes in SDM fiber-based optical communication systems.

5.6.1 SDM in Wireless Communications

The SDM techniques with M_{Tx} transmit antennas and M_{Rx} receiver antennas can achieve the spatial rates equal to min (M_{Tx}, M_{Rx}). The generic scheme for wireless SDM scheme is shown in Fig. 5.31.

Incoming information data stream might be first encoded by conventional channel code (not shown in the figure), serving as an outer code. The encoded data stream is forwarded to 1:L demultiplexer, where serial-to-parallel conversation takes place. The key transmitter performs simultaneously channel encoding and modulation process thus performing L to M_{Tx} mapping. The mth ($m = 1, 2, \ldots, M_{\text{Tx}}$) transmit antenna emits the symbol x_m. These M_{Tx} symbols are transmitted over wireless MIMO channel of interest such as rich-scattering channel, described by the channel matrix $\boldsymbol{H}$. The maximum number of independent streams is determined by the rank of channel matrix $R_{\boldsymbol{H}} = \text{rank}(\boldsymbol{H})$, as described in Sect. 5.4, which is upper bounded by $R_{\boldsymbol{H}} \leq \min(M_{\text{Tx}}, M_{\text{Rx}})$. In order to properly detect independent data streams, typically the number of receive antennas must be larger than the number of transmitted antennas so that the maximum possible number of independent data streams is upper bounded by M_{Tx}. Let R_{Tx} denote the code rate of both outer and inner channel codes (obtained as a product of individual code rates); then the spectral efficiency (S) of the SDM scheme is upper bounded by

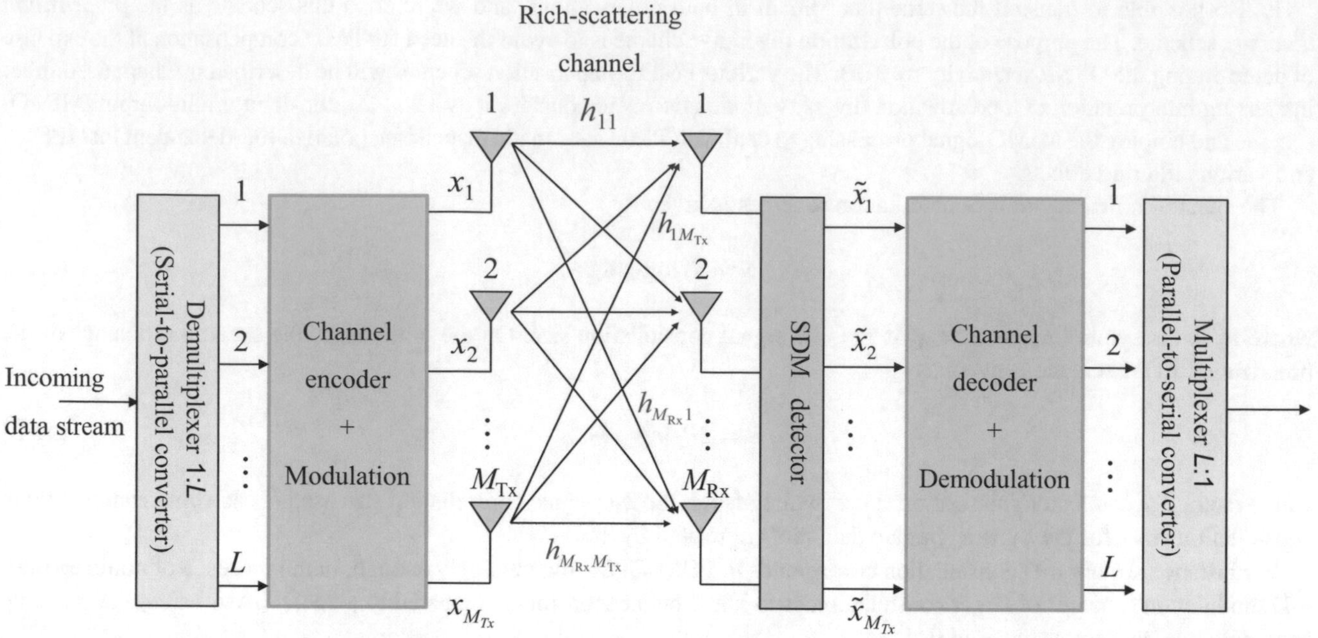

Fig. 5.31 Generic SDM scheme for wireless MIMO communications

$$S_{SDM} \leq M_{Tx} R_{Tx} \log_2(M) \text{ [bits/s/Hz]}, \tag{5.93}$$

where M is the signal constellation size (assuming that all independent data streams employ the same constellation size). At the receiver side, each receive antenna detects the complex linear combination of signal from all transmit antennas. The purpose of SDM detector is to compensate for spatial interference and provide accurate estimates of transmitted symbols. After SDM detection, demodulation takes place, followed by the channel decoding. The L estimated data streams are multiplexed together, and the corresponding output is passed to outer channel decoder, which delivers estimated information sequence to the end users. The SDM concept is also applicable to both free-space (FSO) optical channels and fiber-optic communications. Below, we describe the hybrid multidimensional coded modulation scheme, introduced in [16–21].

5.6.2 SDM and Multidimensional Signaling in Fiber-Optic Communications

The hybrid multidimensional modulation schemes can be effectively implemented with the help of SDM-based fibers; in particular, the few-mode fibers (FMFs) and few-core fibers (FCFs) are excellent candidates for this purpose. It is well known that photon can carry both spin angular momentum and orbital angular momentum (OAM) [22–29, 40, 46]. The spin angular momentum is associated with polarization, while OAM is associated with the azimuthal phase of the complex electric field. The angular momentum, $\boldsymbol{L}$, of the classical electromagnetic field can be written as [22].

$$\boldsymbol{L} = \frac{1}{4\pi c} \int_V (\boldsymbol{E} \times \boldsymbol{A}) dV + \frac{1}{4\pi c} \int_V \sum_{k=x,y,z} E_k (\boldsymbol{r} \times \nabla) A_k dV, \tag{5.94}$$

where $\boldsymbol{E}$ is the electric field intensity, $\boldsymbol{A}$ is the vector potential, and c is the speed of light. $\boldsymbol{A}$ is related to the magnetic field intensity $\boldsymbol{H}$ by $\boldsymbol{H} = \nabla \times \boldsymbol{A}$ and to the electric field intensity by $\boldsymbol{E} = -c^{-1} \partial \boldsymbol{A} / \partial t$. The second term in Eq. (5.94) is identified as the "OAM" due to the presence of the angular momentum operator $\boldsymbol{r} \times \boldsymbol{\nabla}$. In general, among various optical beams that can carry OAM, the Laguerre-Gauss (LG) vortex beams/modes can easily be implemented. The field distribution of an LG beam traveling along the z-axis can be expressed in cylindrical coordinates (r,ϕ,z) (r denotes the radial distance from propagation axis, ϕ denotes the azimuthal angle, and z denotes the propagation distance) as follows [28, 29]:

$$u_{m,p}(r,\phi,z) = \sqrt{\frac{2p!}{\pi(p+|m|)!}\frac{1}{w(z)}}\left[\frac{r\sqrt{2}}{w(z)}\right]^{|m|} L_p^m\left(\frac{2r^2}{w^2(z)}\right)\times$$
$$e^{-\frac{r^2}{w^2(z)}} e^{-\frac{jkr^2 z}{2\left(z^2+z_R^2\right)}} e^{j(2p+|m|+1)\tan^{-1}\frac{z}{z_R}} e^{-jm\phi}, \tag{5.95}$$

where $w(z) = w_0\sqrt{1+(z/z_R)^2}$ (w_0 is the zero-order Gaussian radius at the waist), $z_R = \pi w_0^2/\lambda$ is the Rayleigh range (with λ being the wavelength), $k = 2\pi/\lambda$ is the wave number, and $L^m{}_p(\cdot)$ is the associated Laguerre polynomial, with p and m representing the radial and angular mode numbers, respectively. It can be seen from Eq. (5.95) that the mth mode of the LG beam has the azimuthal angular dependence given by the term $\exp(-jm\phi)$, where m is the azimuthal mode index. In free space, for $m = 0$, $u(r,\phi,z)$ becomes a zero-order Gaussian beam known as TEM_{00} mode. For $p = 0$, $L^m{}_p(\cdot) = 1$ for all m's, so that the intensity of an LG mode is a ring of radius proportional to $|m|^{1/2}$.

It can be shown that for fixed p, the following principle of orthogonality is satisfied [22–26, 29]:

$$\begin{aligned}\langle u_{m,p}|u_{n,p}\rangle &= \int u^*_{m,p}(r,\phi,z)u_{n,p}(r,\phi,z)rdrd\phi \\ &= \begin{cases}\int\int |u_{m,p}|^2 rdrd\phi, n = m \\ 0, \quad n \neq m\end{cases}.\end{aligned} \tag{5.96}$$

Evidently, different OAM "states" corresponding to a fixed index p are all mutually orthogonal, and they can be used as the basis functions for an *OAM modulation*. Instead of OAM modes, in fiber optics, other spatial modes can be used such as LP modes [30–39]. However, on a quantum level, only SAM and OAM of the photon can be defined. Moreover, other spatial modes can be represented in terms of OAM modes.

Different signal constellations can be used in combination with OAM modulations, thus producing multidimensional hybrid modulation schemes. These schemes are called hybrid in [16–21] since they employ all available degrees of freedom: the amplitude, the phase, the polarization, and the OAM. As an illustration, for each polarization state, we can employ L electrical basis functions, such as L orthogonal subcarriers [14], so we have that

$$\Phi_l(nT) = \exp\left[j2\pi(l-1)\Delta T/T\right]; l = 1,\cdots,L, \tag{5.97}$$

where $\Phi_l(nT)$ define the basis functions, T is the symbol duration, and ΔT denotes the sampling interval that is related to the symbol duration by $\Delta T = T/U$, with U being the oversampling factor. The corresponding space described by Eq. (5.97) is 2 L-dimensional. In addition, we can apply N orthogonal either OAM states or other spatial modes and two polarization states for use in multidimensional modulation. Therefore, the corresponding signal space is $D = 4\,L{\cdot}N$-dimensional. By increasing the number of dimensions, we can also increase the aggregate data rate of the system while ensuring reliable transmission at these ultrahigh speeds using capacity-approaching LDPC codes, described in Chap. 9.

The D-dimensional space, if compared to the conventional 2-D space, can provide larger Euclidean distances between signal constellation points, resulting in improved BER performance and better tolerance to fiber nonlinearities. We should outline that modified orthogonal polynomials or any other set of orthogonal complex basis functions can be used as well, such as Slepian sequences based on [18–21]. The overall multidimensional transmitter configuration is depicted in Fig. 5.32, while the corresponding multidimensional receiver in Fig. 5.33. The D-dimensional modulator, whose configuration is shown in Fig. 5.32(a), generates the signal constellation points by

$$\mathbf{s}_i = \sum_{d=1}^{D}\phi_{i,d}\Phi_d, \tag{5.98}$$

where $\phi_{i,d}$ denotes the dth coordinate ($d = 1,\ldots, D$) of the ith signal constellation point and the set $\{\Phi_1,\ldots,\Phi_D\}$ denotes the basis functions introduced above. The transmitter architecture is provided in Fig. 5.32(a). A continuous wave laser diode, such as DFB laser, signal is split into N branches by using a power splitter to feed 4 L-dimensional electro-optical modulators, each corresponding to one out of N spatial modes, such as LP or OAM modes. The 4 L-dimensional electro-optical modulator is implemented as shown in Fig. 5.32(b). The SDMA multiplexer combines N independent data streams for transmission over

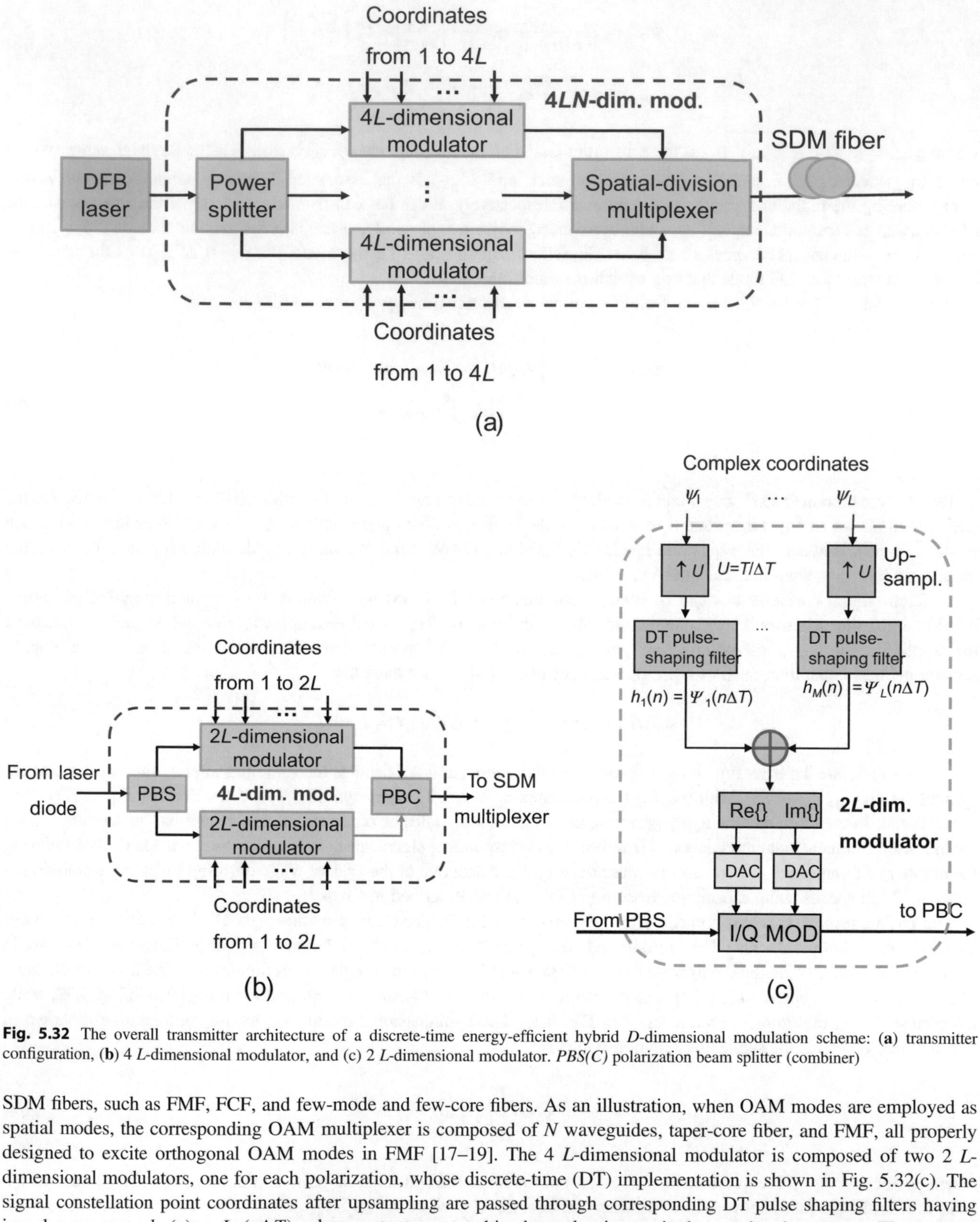

Fig. 5.32 The overall transmitter architecture of a discrete-time energy-efficient hybrid D-dimensional modulation scheme: (**a**) transmitter configuration, (**b**) 4 L-dimensional modulator, and (c) 2 L-dimensional modulator. *PBS(C)* polarization beam splitter (combiner)

SDM fibers, such as FMF, FCF, and few-mode and few-core fibers. As an illustration, when OAM modes are employed as spatial modes, the corresponding OAM multiplexer is composed of N waveguides, taper-core fiber, and FMF, all properly designed to excite orthogonal OAM modes in FMF [17–19]. The 4 L-dimensional modulator is composed of two 2 L-dimensional modulators, one for each polarization, whose discrete-time (DT) implementation is shown in Fig. 5.32(c). The signal constellation point coordinates after upsampling are passed through corresponding DT pulse shaping filters having impulse responses $h_m(n) = \Phi_m(n\Delta T)$, whose outputs are combined together into a single complex data stream. The real and imaginary parts of this complex data stream, after digital-to-analog conversion (DAC), drive RF inputs of I/Q modulator. The outputs from M-dimensional modulators are combined into a single SDM stream by the polarization beam combiner (PBC). The obtained N SDM streams are combined by the spatial-division multiplexer.

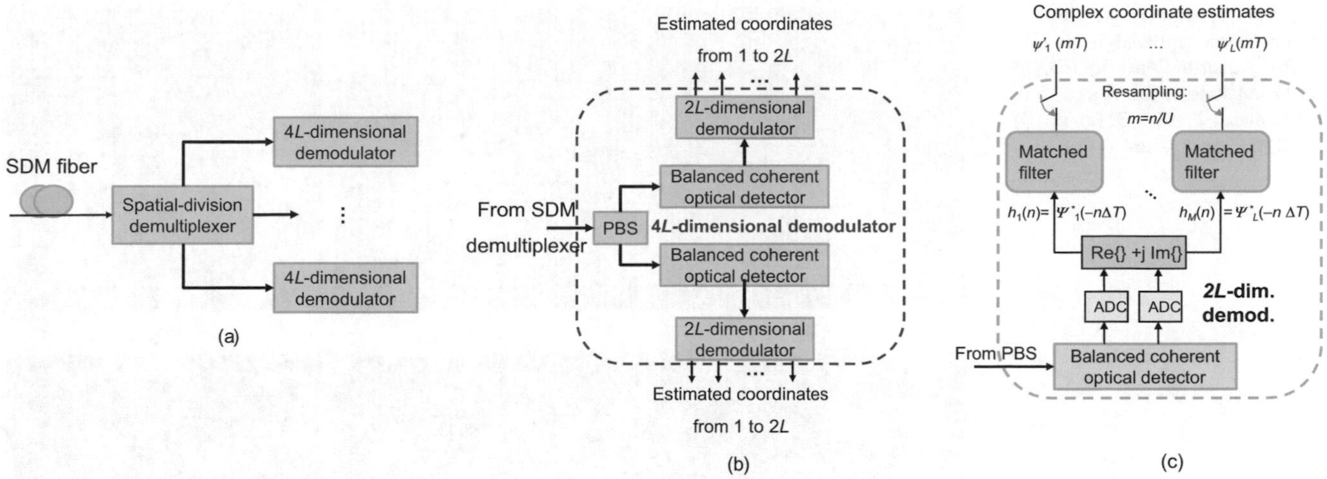

Fig. 5.33 Receiver architecture of a discrete-time energy-efficient hybrid D-dimensional modulation scheme: (**a**) receiver, (**b**) 4 *L*-dimensional demodulator, and (**c**) 2 *L*-dimensional demodulator configurations

On the other hand, on the receiver side, the 4 $L{\cdot}N$-dimensional demodulator architecture is shown in Fig. 5.33(a).

We first perform SDM/OAM mode demultiplexing in the SDM/OAM-demultiplexing block, whose outputs are 4 L-dimensional projections along N spatial modes/OAM states. Each spatial/OAM mode undergoes polarization diversity coherent detection, as shown in Fig. 5.33(b), and the corresponding outputs are forwarded to 2 L-dimensional demodulators, implemented as shown in Fig. 5.33(c). After polarization diversity detection, we recover Re and Im parts, which are after ADCs combined into a single complex data stream. The same complex data stream is applied to the inputs of L complexed matched filters with impulse responses $h_l(n) = \Phi^*_l(-nT)$. The corresponding outputs after resampling represent projections along basis functions Φ_l.

5.6.3 SDM and Multidimensional Signaling in Free-Space Optical (FSO) Communications

Thanks to the mutual orthogonality among the electric fields of OAM modes with different integer OAM states, as given by Eq. (5.96), one can in principle superimpose as many OAM modes as one wishes [24–26]. OAM state superposition using diffractive elements enables implementation of OAM-based multiplexers as an integral part of an FSO communication link. The simplest mathematical expression of an OAM mode and zero-order Gaussian beam with an incidence angle θ can be written, respectively, as [25, 26]

$$u(r, \phi, z) = \exp(-jm\phi)\exp(-jkz) \text{ and } u_0 = \exp(jkx\sin\theta)\exp(-jkz). \tag{5.99}$$

A computer-generated hologram (CGH) is a recording of the interference pattern between two incident beams, in our case a zero-order Gaussian beam and the beam we want to generate. The resulting interference pattern, assuming that $z = 0$, can be expressed as

$$I = |u(r, \phi, z = 0) + u_0|^2 = 2 + 2\cos(kx\sin\theta - m\phi). \tag{5.100}$$

Equation (5.100) is a sinusoidal grating pattern, which is easy to generate but with low diffraction efficiency [35]. It is well known that blazed grating can obtain 100% diffraction efficiency, whose interference pattern can be expressed as $I = kx\sin\theta - m\phi \bmod 2\pi$..

As an illustration, the SLM display programmed with blazed interference pattern is shown in Fig. 5.34(a)–(c). The experimentally generated (Fig. 5.34(d)–(f)) and analytical (Fig. 5.34(g)–(i)) intensity distributions of LG beams can be compared. In Fig. 5.35, the detection efficiency of individual OAM mode is illustrated in terms of intensity profile after the detection. The imperfections in experimental generation of OAM modes, visible in Figs. 5.34 and 5.35, are contributed to the finite resolution of SLMs. To our best knowledge, even though phase-only SLMs are used, it is impossible to realize in

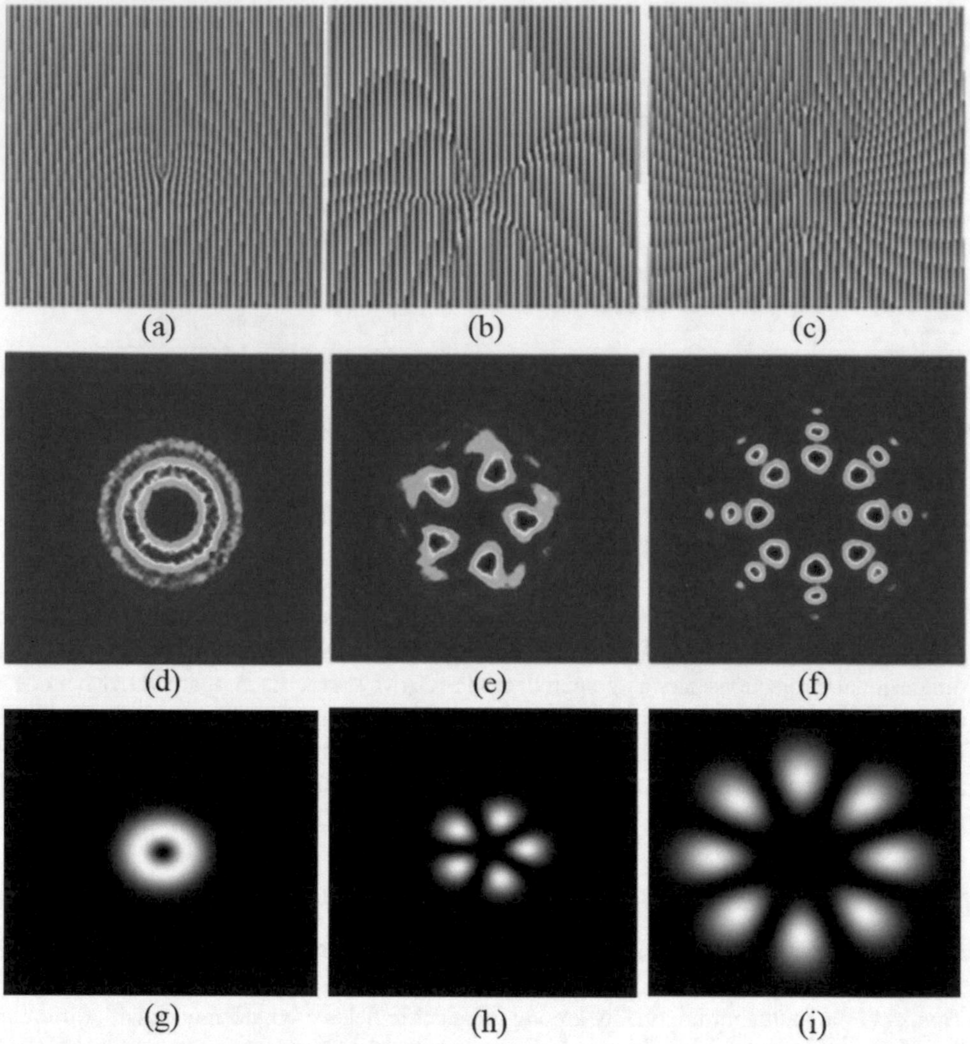

Fig. 5.34 The phase pattern, experimental, and analytical intensity distributions: (**a**), (**d**), (**g**) for OAM state 1; (**b**), (**e**), (**h**) for OAM states 2 and − 3; (**c**), (**f**), (**i**) for OAM states 4 and − 4

practice phase-only modulation without any residual amplitude modulation. This residual amplitude modulation may generate the residual ring outside the desirable OAM mode in the far field.

The overall configurations of OAM*N*-dimensional modulator/multiplexer and demultiplexer, based on spatial light modulators (SLMs), are shown in Figs. 5.36(a) and (b), respectively. A CW laser signal is split into *N* branches with the help of 1:*N* power splitter, whose every output is used as an input to the E/O MOD. After E/O optical conversion, in each branch, CGH (based on SLM) is used to impose the corresponding OAM mode. The OAM-based *N*-dimensional modulator is composed of power splitter, *N* E/O MODs, *N* SLMs, power combiner, and corresponding waveguides.

On the receiver side, the OAM multiplexed signal is demultiplexed with the help of corresponding complex-conjugate CGHs (see Fig. 5.36(b)), so that the *i*th output represents a projection along the *i*th OAM mode. The local laser is then used to coherently detect the signals transmitted in all OAM modes. To improve OAM modes' orthogonality, the MIMO signal processing approaches, described in Chap. 8, can be used.

5.7 Optimum Signal Constellation Design

Different optimization criteria can be used in optimum signal constellation design including minimum MSE (MMSE) [41, 42], minimum BER, and maximum information rate (channel capacity) [43].

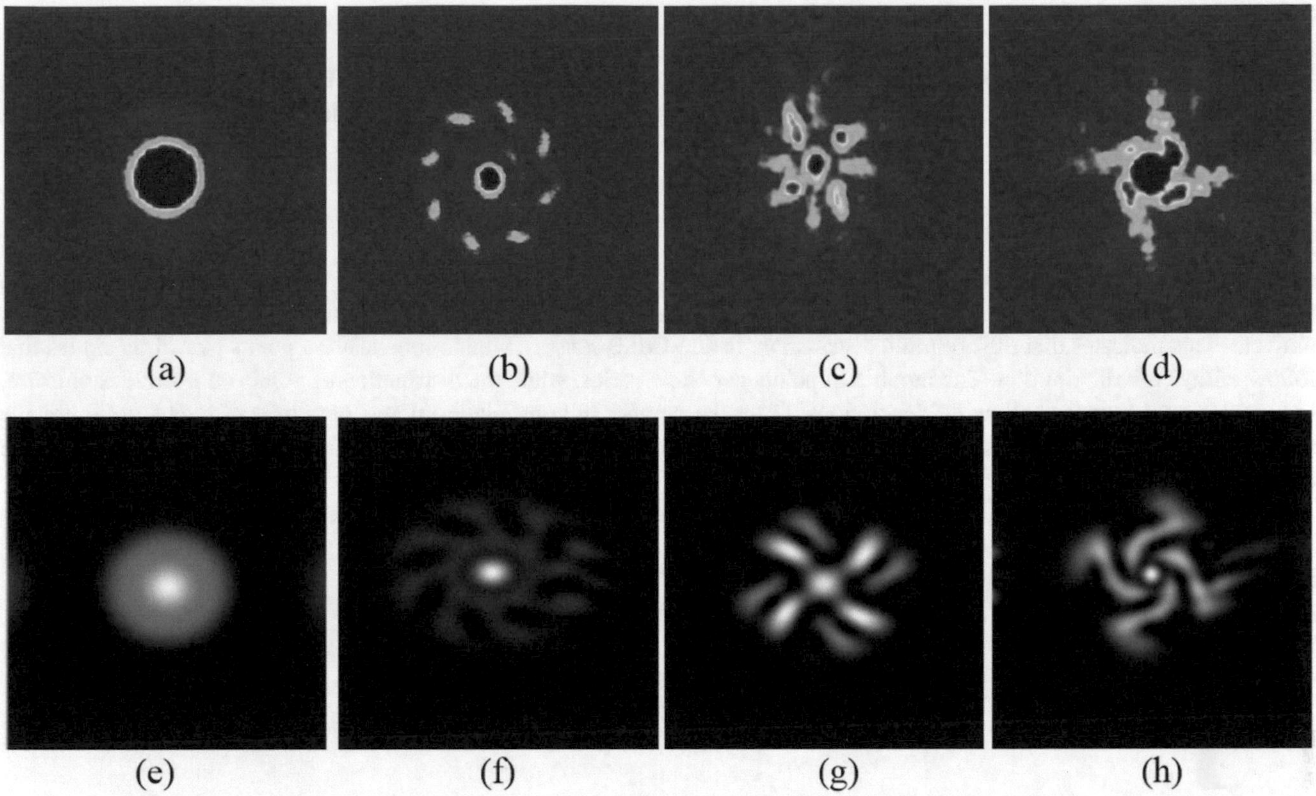

Fig. 5.35 Experimental (top) and numerical (bottom) detection of OAM states from the superimposed OAM modes: (**a**), (**e**) OAM state 1 is detected using OAM state −1; (**b**), (**f**) superposition of OAM states 4, −4 is detected using OAM state 4; (**c**), (**g**) superposition of OAM states 2, −2, 6, −6 is detected using OAM state 2; (**d**), (**h**) superposition of OAM states 2, −2, 6, −6 is detected using OAM state −6. Imperfections in experimental generation of OAM modes are contributed to finite resolution of SLMs

Fig. 5.36 SLM-based configurations of (**a**) OAM *N*-dimensional modulator and multiplexer and (**b**) OAM demultiplexer

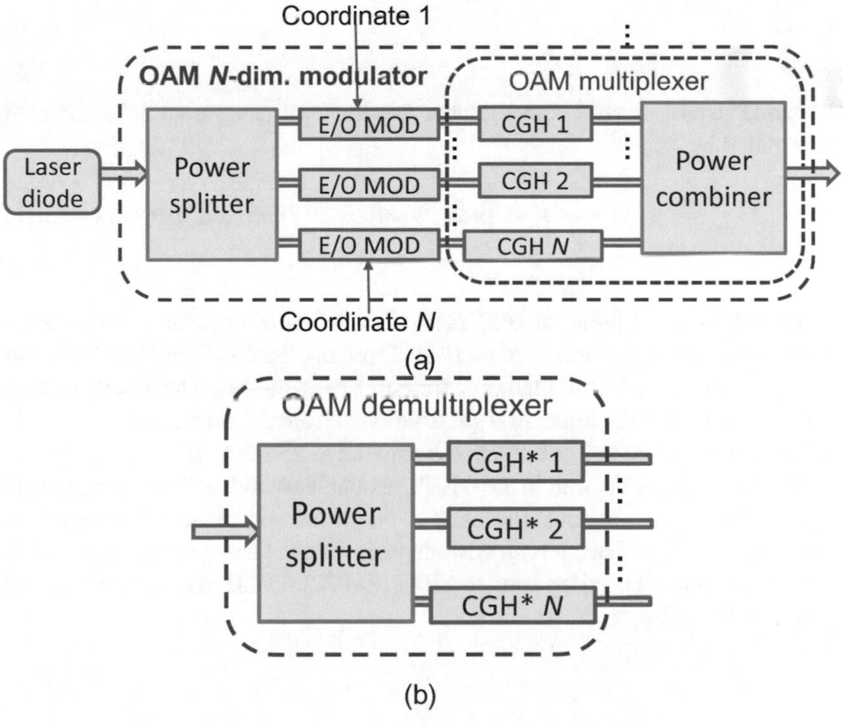

5.7.1 Iterative Polar Modulation (IPM)

We first describe the iterative polar modulation (IPM) based on iterative polar quantization (IPQ) [42]. We can assume that, for an ASE noise dominated channel model, the optimum information source is Gaussian with distribution:

$$p(I,Q) = \frac{1}{2\pi\sigma^2}\exp\left[-\frac{I^2+Q^2}{2\sigma^2}\right], \tag{5.101}$$

where (I,Q) denote in-phase and quadrature coordinates. By expressing Cartesian coordinates through polar coordinates (r,θ) as $I = r\cos\theta$ and $Q = r\sin\theta$, it can be shown that radius r will follow a Rayleigh distribution [42], while the phase stays uniform. This indicates that the optimum distribution, in an MMSE sense, contains constellation points placed on circles that follow a Rayleigh distribution. The number of points per circle varies, while the distribution of points on a circle is uniform. Let us also introduce the following denotations: L_i for the number of constellation points per circle of radius m_i; L_r for the number of circles in the constellation; Q for the total number of signal constellation points ($Q = \sum_{i=1}^{L_r} L_i$); and $p(r)$ for the Rayleigh distribution function of radius r.

The *IPMsignal constellationdesign algorithm* can be formulated through the following steps [41, 42]:

Step 0: Choose an arbitrary signal constellation of size Q for initialization.
Step 1: Determine the number of constellation points per ith circle as

$$L_i = \sqrt[3]{m_i^2\int_{r_i}^{r_{i+1}} p(r)dr} \Bigg/ \left[\sum_{i=2}^{L_r}\frac{1}{Q}\sqrt[3]{m_i^2\int_{r_i}^{r_{i+1}} p(r)dr}\right]; i = 1,2,\ldots,L_r. \tag{5.102}$$

Step 2: Determine the radius of the ith circle as

$$\begin{aligned} m_i &= \left[2\sin(\Delta\theta_i/2)\int_{r_i}^{r_{i+1}} rp(r)dr\right] \Bigg/ \left[\Delta\theta_i\int_{r_i}^{r_{i+1}} p(r)dr\right], \\ \Delta\theta_i &= 2\pi/L_i; i = 1,2,\ldots,L_r. \end{aligned} \tag{5.103}$$

Iterate steps 1–2 until convergence. The limits of integration, which correspond to decision boundaries, in steps 1–2 are determined by

$$\begin{aligned} r_i &= \left[\pi\left(m_i^2 - m_{i-1}^2\right)/2\right]/[m_iL_i\sin(\Delta\theta_i/2) - m_{i-1}L_{i-1}\sin(\Delta\theta_{i-1}/2)]; \\ i &= 1,2,\ldots,L_r. \end{aligned} \tag{5.104}$$

Therefore, in addition to IPM signal constellation in polar coordinates, this algorithm can also provide the decision boundaries. As an illustration of an IPM algorithm, the 64-IPQ/IPM signal constellation is shown in Fig. 5.37. With r_3 and r_5 we denoted the decision boundaries for circle of radius m_4. The details of this constellation are provided in Table 5.2. With index i we denoted the index of a circle counted from the center. The 256-ary IPM has been experimentally evaluated in [43], and it exhibits much better tolerance compared to 256-QAM.

We should point out that in the 64-IPQ example we did not use the signal constellation point placed in origin, and for large signal constellations it does not really matter if the point placed in origin is used or not. However, for moderate and small signal constellation sizes, it is an advantage, in terms of energy efficiency, to place the point in the origin. Such obtained signal constellations can be called centered-IPQ/IPM (CIPQ/CIPM). As an illustration, 16-ary and 32-ary CIPQ constellations are shown in Fig. 5.38.

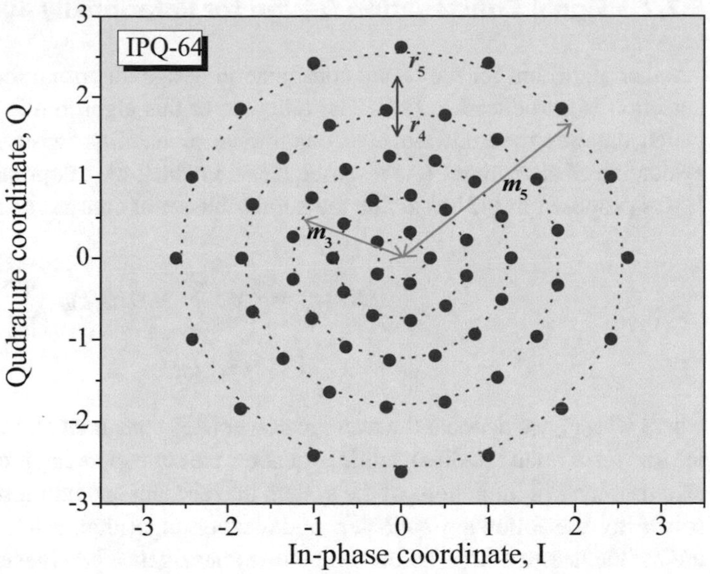

Fig. 5.37 Details of a 64-IPQ/IPM signal constellation

Table 5.2 Details of 64-IPQ/IPM constellation

i	1	2	3	4	5	6
r_i	0	0.54	1.02	1.55	2.25	4.5
m_i	0.33	0.78	1.26	1.84	2.61	
L_i	5	11	15	17	16	

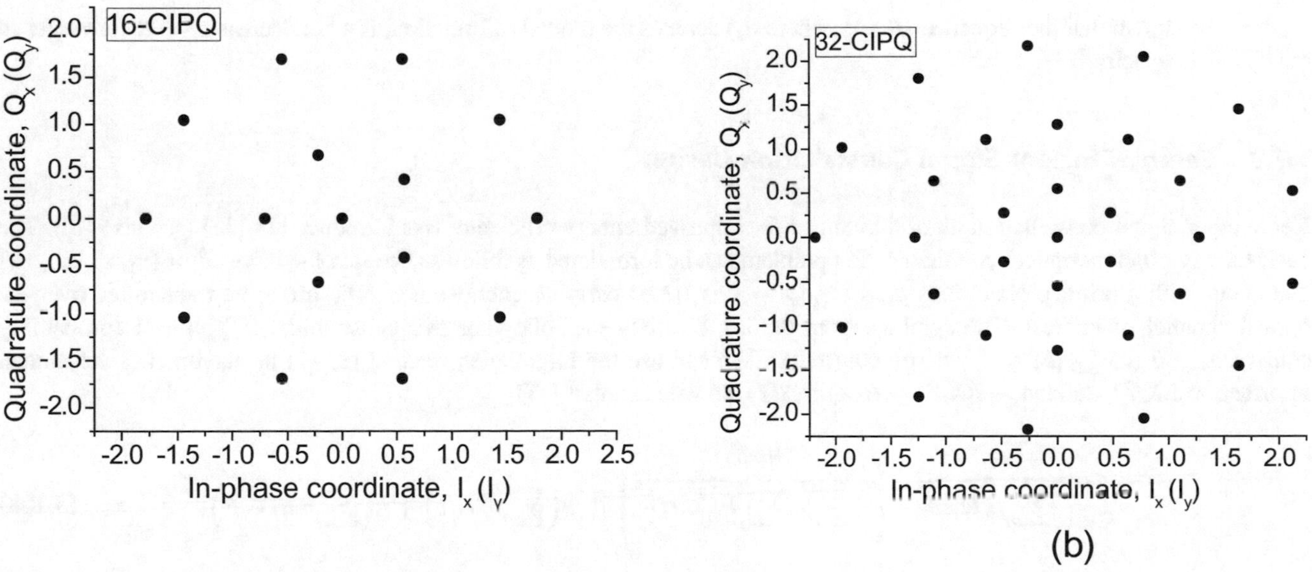

Fig. 5.38 The CIPQ-based signal constellations for (**a**) 16 points and (**b**) 32 points

5.7.2 Signal Constellation Design for Rotationally Symmetric Optical Channels

Another algorithm for the signal constellation design algorithm for "rotationally symmetric optical channels" with coherent detection is introduced in [43]. The rationale of this algorithm is dictated by the fact that the optical channel has memory which requires the knowledge of conditional probability density functions (PDFs), which are typically estimated by the evaluation of histograms. (As a consequence, gradient-based optimization methods cannot be directly applied in such a case.) It was proposed in [43] to define the admissible set of constraints as follows:

$$\Omega(p;r) = \left\{ p : \sum_{i=1}^{n} p_i r_i^2 \leq P_a,\ \sum_{i=1}^{n} p_i = 1,\ 0 \leq r_i \leq \sqrt{P_p} \right\},\quad i = 1, 2, \ldots, L_r \tag{5.105}$$

where with P_a we denoted the average power (AP) constraint and with P_p the peak power (PP) constraint. Further, r_i denotes the ith "mass point" (radius), while p_i denotes the corresponding probability of occurrence. These constraints are introduced to take the physical limitations of the system into account and minimize the effect of nonlinearities. The optimization problem is solved by the following split-step optimization algorithm. After the initialization by conventional 2-D constellation, we initiate the iterative steps below until either convergence is achieved or a predetermined number of iterations is reached [43]:

Step 1: *$p_i$update rule*:

$$p_i = \arg\max_{p} \{ I(p) : p \in \Omega(p; r_{i-1}) \}. \tag{5.106}$$

Step 2: *$r_i$update rule*:

$$r_i = \arg\max_{r} \{ I(p) : r \in \Omega(p_i; r) \}. \tag{5.107}$$

It can be shown that the sequence $\{I(p_i)\}$, where $I(\cdot)$ denotes the mutual information, is a nondecreasing and converges to the channel capacity.

5.7.3 Energy-Efficient Signal Constellation Design

Yet another signal constellation algorithm aimed for improved energy efficiency was introduced in [17] (see also [8]). The basic energy-efficient optical communication problem can be formulated as follows. The set of symbols $X = \{x_1, x_2, \ldots, x_Q\}$ that occur with a priori probabilities $p_1,\ldots,p_Q$ [$p_i = \Pr(x_i)$] and carrying energies $E_1,\ldots,E_Q$ are to be transmitted over the optical channel of interest. The symbols from the set X satisfy the following two constraints: (i) $\sum_i p_i = 1$ (probability constraint) and (ii) $\sum_i p_i E_i \leq E$ (energy constraint). We can use the Lagrangian method [8, 47] in maximizing the mutual information $I(X,Y)$, defined as $I(X,Y) = H(X)\text{-}H(X|Y)$, so we have that [17].

$$L = \overbrace{-\sum_i p_i \log p_i}^{H(X)} - \left(\overbrace{-\sum_i p_i \sum_j P_{ij} \log Q_{ji}}^{H(X|Y)} \right) + \eta\left(\sum_i p_i - 1\right) + \mu\left(\sum_i p_i E_i - E\right), \tag{5.108}$$

where $P_{ij} = \Pr(y_j|x_i)$ can be determined by the evaluation of histograms and $Q_{ji} = \Pr(x_i|y_j)$ from Bayes' rule [48]: $Q_{ji} = \Pr(x_i, y_j)/\Pr(y_j) = P_{ij}p_i/\sum_k P_{kj}p_k$. The *energy-efficient signal constellation designalgorithm (EE-SCDA)* can be now formulated as follows [17].

Step 0 (Initialization step): Choose an arbitrary auxiliary input distribution and signal constellation, with a number of constellation points M_a much larger than target signal constellation M.

Step 1: Q_{ji}update rule:

$$Q_{ji}^{(t)} = P_{ij}p_i^{(t)} / \sum\nolimits_k P_{kj}p_k^{(t)}, P_{ij} = \Pr(y_j|x_i). \tag{5.109}$$

Step 2: *$p_i$update rule*:

$$p_i^{(t+1)} = \frac{\exp\left(-\mu E_i - H^{(t)}(x_i|Y)\right)}{\sum_k \exp\left(-\mu E_k - H^{(t)}(x_k|Y)\right)},$$
$$H(x_i|Y) = -\sum\nolimits_k P_{ik} \log Q_{ki}. \tag{5.110}$$

Iterate steps 1–2 until convergence.

Step 3: Determine the constellation points of target constellation as a center of mass of closest M_a/M constellation points in the auxiliary signal constellation.

Notice that the well-known Arimoto-Blahut algorithm [48, 49] does not impose an energy constraint and yields the optimum source distribution only. Using EE-SCDA, we can obtain the optimized signal constellation while taking the energy constraint into account. Both optimum source distribution and EE signal constellation are obtained from this algorithm.

5.7.4 Optimum Signal Constellation Design (OSCD)

Finally, we will describe the *optimum signal constellation design (OSCD)algorithm*, introduced in [50] (see also [51]). This algorithm is convenient since it does not require the employment of constrained optimization software. Instead, a simple Monte Carlo simulation can be used in signal constellation design. Given that the IPM algorithm is derived under assumptions that are valid for large constellations, we can say that OSCD-generated constellations are the most favorable for small and medium sizes of signal constellations. The OSCD can be formulated as follows:

Step 0 (Initialization): Choose an arbitrary signal constellation of size Q.

Step 1: Apply the Arimoto-Blahut algorithm to determine an optimum source distribution.

Step 2: Generate a long sequence of samples from the optimum source distribution. Group the samples from this sequence into Q clusters. The membership of the cluster is determined based on a squared Euclidean distance between the sample point and the signal constellation points from previous iteration. Each sample point is assigned to the cluster with the smallest squared distance.

Step 3: Determine the signal constellation points as the center of mass for each cluster.

Repeat steps 2–3 until convergence.

As an illustration, we show 32-ary and 64-ary constellations obtained by employing this algorithm in Fig. 5.39.

5.8 Nonuniform Signaling

In the conventional data transmission schemes, each point in a given signal constellation is transmitted equally likely [1–7, 44, 45]. However, if we take the different energy costs of various constellation points into account and transmit each point in the constellation diagram with different probabilities, the resulting modulation technique is called either nonuniform signaling [5, 52, 53] or probability shaping-based signaling [54]. The basic idea of the nonuniform signaling is to transmit symbols of larger energy with lower probabilities, so that the energy efficiency of conventional modulation schemes can be improved [52]. This improvement in energy efficiency is commonly referred to as the shaping gain. In any data transmission scheme, the goal is to transmit at a higher bit rate, with higher reliability and as low transmitter power as possible. When the distribution of constellation points matches Maxwell-Boltzmann distribution, the ultimate shaping gain of 1.53 dB can be achieved [53]. Moreover, the nonuniform signaling scheme may be more suitable for optical communication because the symbols with small energies, transmitted with higher probabilities, suffer less fiber nonlinearity effects.

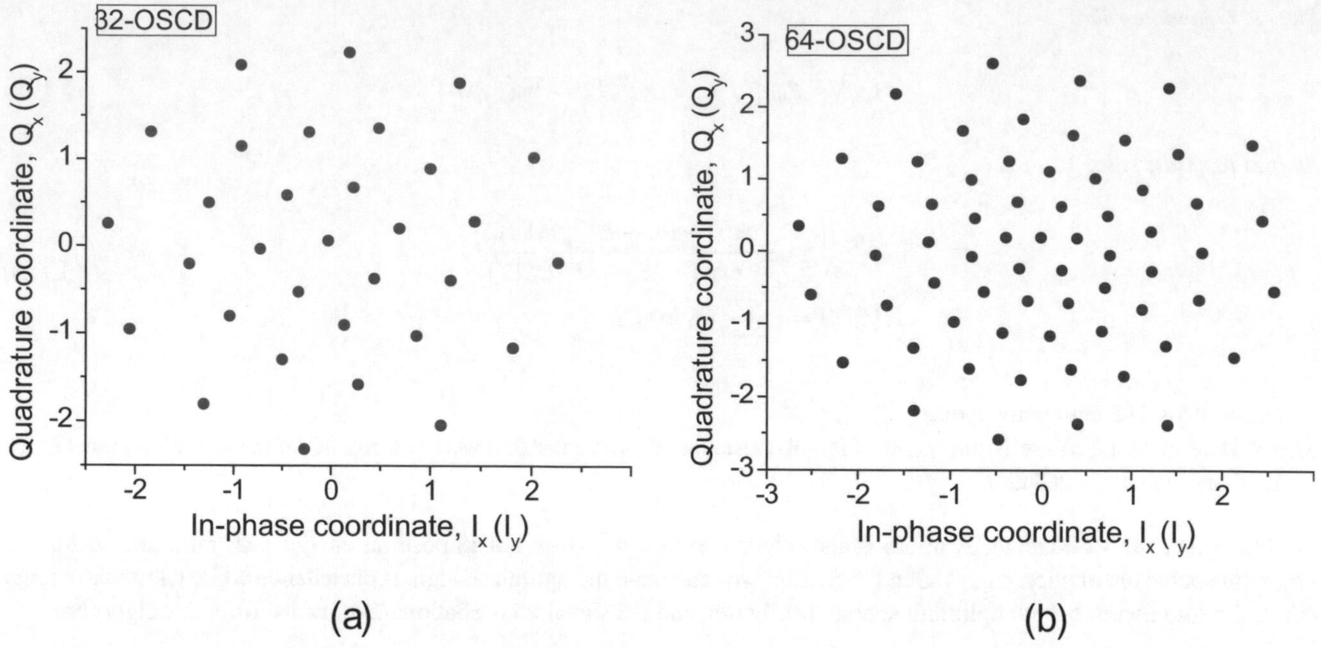

Fig. 5.39 Signal constellations obtained by an OSCD algorithm: (**a**) 32-ary signal constellation and (**b**) 64-ary signal constellation

In this section, based on our recent invited paper [52], we introduce the nonuniform signaling techniques suitable for use in coherent optical communication systems. There are two basic ways to map the uniform binary source bits to nonuniform distributed symbols. The first method is to use arithmetic code [54]. This method can be applied to existing signal constellation sets like 16-QAM, 64-QAM, and so on, to map the uniform binary source to nonuniform symbols based on arbitrary distributions, which are based on different criteria. Even though this scheme can easily fit to current optical systems, its complexity is high. Moreover, it is only applicable to square-QAM constellations.

The second method is to use the Huffman code, which is based on a binary tree structure and has much lower complexity. This method can be combined with either binary or nonbinary FEC code, but it cannot be applied to arbitrary constellation sets [55]. The structure of the constellation is determined by the Huffman code so that the design of new signal constellation set is needed. One of the corresponding constellation sets, 9-QAM, has the same spectral efficiency with traditional star-8-QAM. On the other hand, the LDPC-coded 9-QAM scheme outperforms the LDPC-coded 8-QAM case by at least 0.8 dB [55].

The fundamental method of nonuniform signaling with arithmetic code is introduced in [54] by Böcherer, and then Fehenberger and Böcherer proposed applying this approach to optical communication systems [56, 57]. In [57], the authors extended the work on nonuniform signaling and investigated the effect of constellation shaping on nonlinearities in WDM systems [57]. The block diagram of the corresponding nonuniform scheme employing arithmetic coding is shown in Fig. 5.40. As an illustration, the receiver constellation diagrams of the nonuniform signaling 16-QAM and regular 16-QAM are also shown in the same figure. The key part of the system is the distribution matcher, which is used for the nonuniform signaling based on the arithmetic code. Note that the distribution matcher has been added outside the FEC encoder and decoder, which is the reason why this scheme requires minor modifications regarding the current optical communication system. Because of the flexibility of arithmetic code, the distribution matcher can be designed based on different criteria, such as Maxwell-Boltzmann distribution and maximizing mutual information rate. According to [57], the proposed nonuniform signaling scheme increases the transmission distance by 8% with the help of binary LDPC code and bit-interleaved coded modulation. However, the use of arithmetic code increases the overall overhead of this nonuniform signaling scheme.

In the rest of this section, we investigate the method of achieving nonuniform signaling schemes for the transmission of binary data by using the Huffman code. In this section, we mainly focus on the 2-D 9-point constellation sets as an illustrative example. The conceptual diagram of the nonuniform signaling process is illustrated in Fig. 5.40. In this scheme, each symbol can carry two, three, or four bits per symbol, and symbols are selected for transmission with different probabilities. It is obvious that the center point carrying 00 bits has the largest probability of 0.25 and the symbols carrying 3 or 4 bits are transmitted with probabilities 0.125 and 0.0625, respectively. The mapping rule is determined by employing the Huffman procedure described as a tree diagram provided in Fig. 5.41. The receiving constellation diagram is shown in Fig. 5.42.

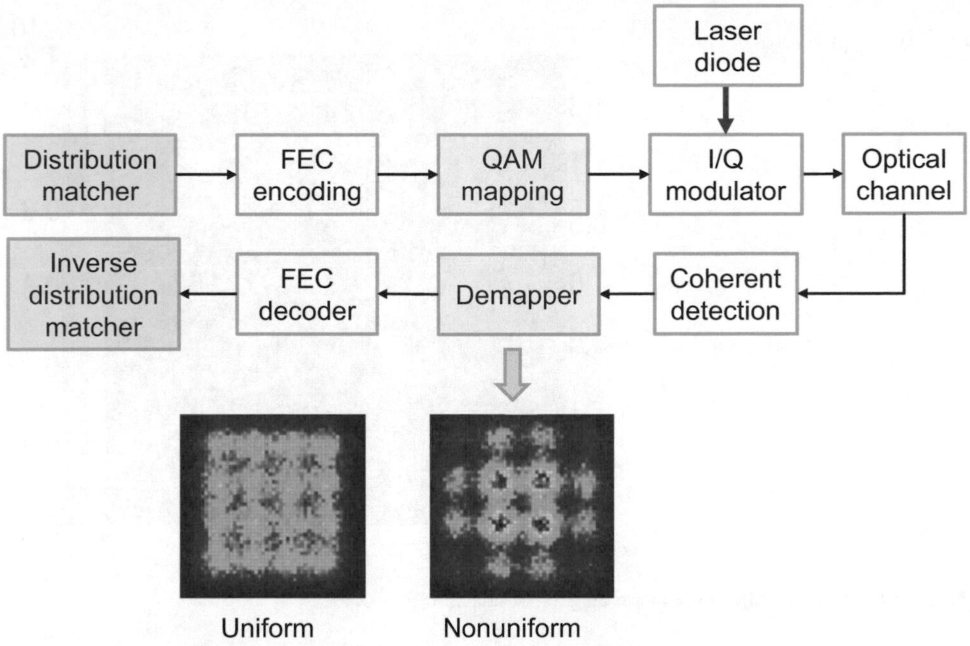

Fig. 5.40 The nonuniform signaling scheme based on an arithmetic code

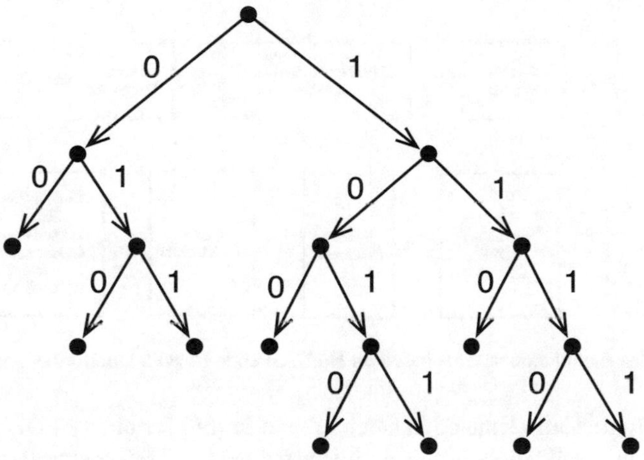

Fig. 5.41 The tree diagram of the Huffman code

It is obvious that the constellation structure of the nonuniform constellation sets is dictated by the structure of the binary tree of the Huffman code. In order to determine the signal constellation that is suitable for nonuniform signaling scheme, an overall search algorithm maximizing the constellation figure of merit (CFM) is used in signal constellation design. For additional details, an interested reader is referred to [58, 59]. The key advantage of this scheme, compared to the corresponding scheme with arithmetic coding, is the fact that overhead coming from the Huffman code is placed into larger constellation so that overall code rate is identical to the traditional uniform signaling scheme.

The nonbinary LDPC-coded modulation scheme for nonuniform signaling based on the Huffman code is shown in Fig. 5.43, based on ref. [59]. With the help of interleaver and 9-QAM [58], binary bits are mapped to the nonuniform symbols, which are then passed to the nonbinary LDPC encoder. The power distributer block will avoid the presentence of long zero energy sequences. After that, the I and Q RF signals will be used as inputs to the I/Q modulator to perform the electrical-to-optical conversion, and such obtained nonuniform signal will be transmitted over fiber-optic communication system of interests. On the receiver side, the homodyne coherent optical detection is used and followed with the compensation of linear and nonlinear impairments and carrier phase estimator (CPE). The sliding-MAP equalizer provides soft symbol LLRs, which are sent to the nonbinary LDPC decoder. Once decoding is completed, we can get information bits after

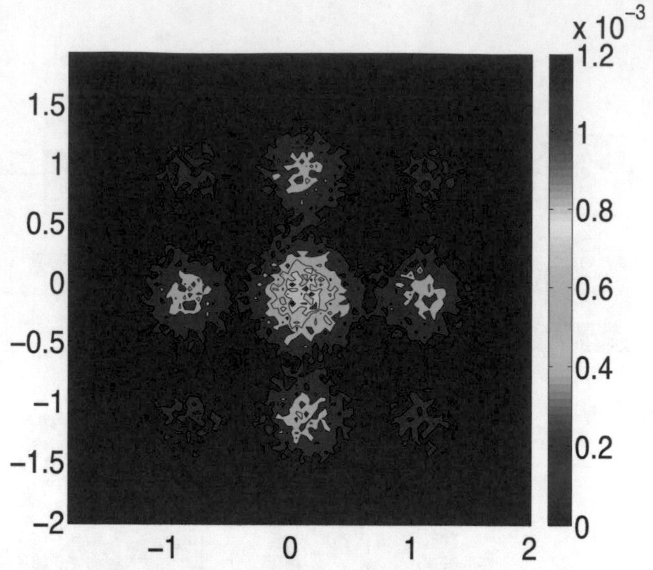

Fig. 5.42 The receiving constellation diagram of nonuniform 9-QAM

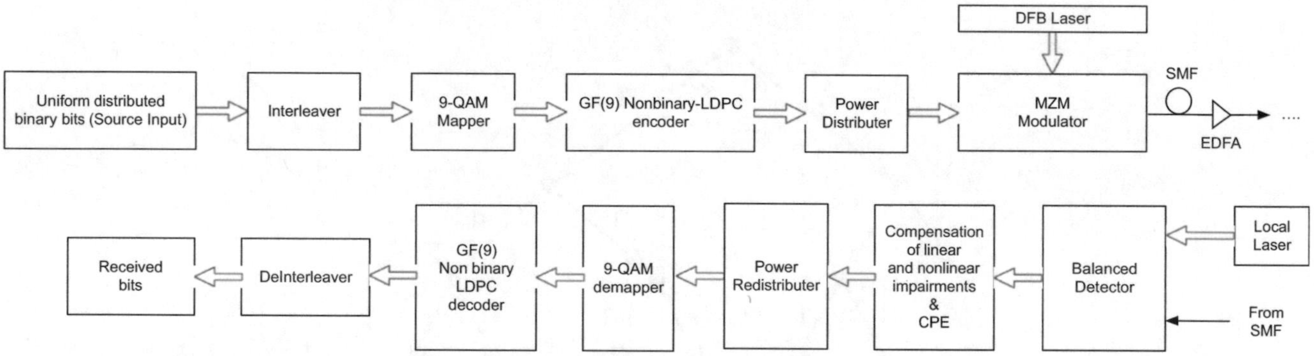

Fig. 5.43 The nonbinary LDPC-coded modulation scheme based on Huffman code-based nonuniform signaling

deinterleaving. The results of Monte Carlo simulation can be found in [58] for binary LDPC coding and [59] for nonbinary LDPC coding. For the experimental validation results, an interested reader is referred to [60].

A reader interested to study how to combine nonuniform signaling and LDPC-coded modulation, nonuniform signaling-based Shannon limits, hybrid probabilistic-geometric shaping, and combining probabilistic shaping and pulse shaping schemes is referred to refs. [61–70].

5.9 Concluding Remarks

This chapter has been devoted to advanced modulation and multiplexing techniques suitable for both wireless and optical communication systems. In Sect. 5.1, the signal space theory concepts have been applied to wireless communication systems. After the geometric representations of signals (Sect. 5.1.1), various multidimensional modulators and demodulators suitable for wireless communication applications have been described, in particular the Euclidean distance, correlation, and matched filter-based detectors, followed by the description of FSK scheme in Sect. 5.1.3. Both continuous-time (CT) and discrete-time (DT) implementations for pulse amplitude modulation schemes, suitable for wireless communications, have been described as well.

After that, the focus has been moved in Sect. 5.2 to multilevel schemes suitable for both wireless and optical communication applications, including M-ary PSK, star-QAM, square-QAM, and cross-QAM, all described in Sect. 5.2.1. Regarding the optical communications, the transmitters for M-ary PSK (Sect. 5.2.2), star-QAM (Sect. 5.2.3), and square−/cross-QAM (Sect. 5.2.4) have been described in detail.

The next topic in the chapter, in Sect. 5.3, has been related to the multicarrier modulation, including the description of multicarrier systems with both nonoverlapping (Sect. 5.3.1) and overlapping (Sect. 5.3.2) subcarriers as well as the introduction of various approaches to deal with fading effects at the subcarrier level (Sect. 5.3.3). The concept of OFDM has been introduced as well; however, details have been provided in Chap. 7 instead.

The MIMO fundamentals have been then provided in Sect. 5.4, including the description of key differences with respect to diversity scheme, as well as the introduction of array, diversity, and multiplexing gains. The parallel MIMO channel decomposition has been briefly described as well, while the details on MIMO signal processing have been postponed for Chap. 8.

In the section on polarization-division multiplexing (PDM) and 4-D signaling, Sect. 5.5, we have described key differences between PDM and 4-D signaling and have described as well how both types of schemes can be implemented in both wireless and optical communications.

The focus has been moved in Sect. 5.6 to the spatial-division multiplexing (SDM) and multidimensional signaling. We have described in Sect. 5.6.1 how SDM can be applied in wireless communications. Then we have described in Sect. 5.6.2 how various degrees of freedom including amplitude, phase, frequency, polarization states, and spatial modes can be used to convey the information in the optical domain. In the same section, the SDM concepts for fiber-optic communications have been described as well. The section concludes with SDM and multidimensional signaling concepts applied to free-space optical (FSO) communications.

The next topic in Sect. 5.7 has been devoted to the signal constellation design, including iterative polar modulation (IPM) in Sect. 5.7.1, signal constellation design for circular symmetric optical channels in Sect. 5.7.2, energy-efficient signal constellation design (Sect. 5.7.3), and optimum signal constellation design (OSCD) in Sect. 5.7.4.

Section 5.8 has been devoted to the nonuniform signaling, in which different signal constellation points have been transmitted with different probabilities.

The set of problems is provided in the incoming section.

5.10 Problems

5.1. Determine the Euclidean distance between the pair of signals provided below:
   (a) Antipodal (bipolar) signals.
   (b) Unipolar (on-off) signals.
   (c) Orthogonal signals.

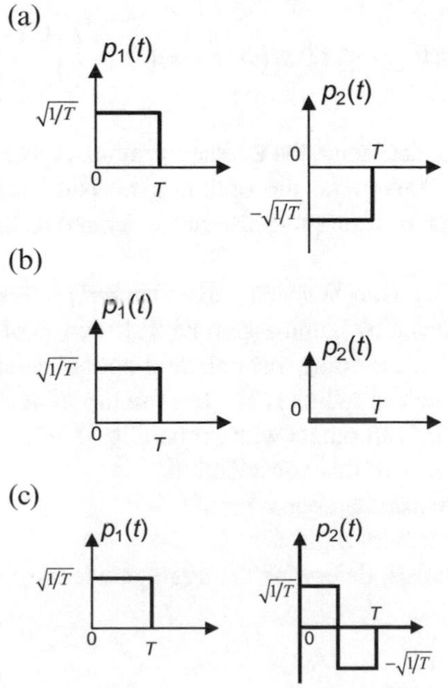

5.2. Let us observe the 4-ary digital communication system, with symbols represented by the following waveforms: $p_1(t) = 0$, $p_2(t) = \text{rect}((t - T/2)/T)$, $p_3(t) = 2\text{rect}((t - T/2)/T)$, $p_4(t) = 3\text{rect}((t - T/2)/T)$. The sequence of pair of bits is used to select the waveform to be sent, according to the following mapping rule:

$$00 \to p_1(t) \quad 01 \to p_2(t) \quad 10 \to p_3(t) \quad 11 \to p_4(t).$$

Determine the average energy per bit if the a priori probabilities are given by $P(00) = 1/16$, $P(01) = P(10) = 3/16$, $P(11) = 9/16$.

5.3. One possible implementation of QPSK modulators is given in Fig. P3, where $b_i(t) = \sum_k b_{ik}\text{rect}((t - kT)/t)$, $b_{ik} \in \{-1, 1\}, i \in \{I, Q\}$. Show that the output of this modulator can be represented as

$$s_m(t) = \sqrt{\frac{2E_s}{T_0}} \cos\left[\omega_c t - \left(\frac{\pi}{4} + \frac{m2\pi}{4}\right)\right].$$

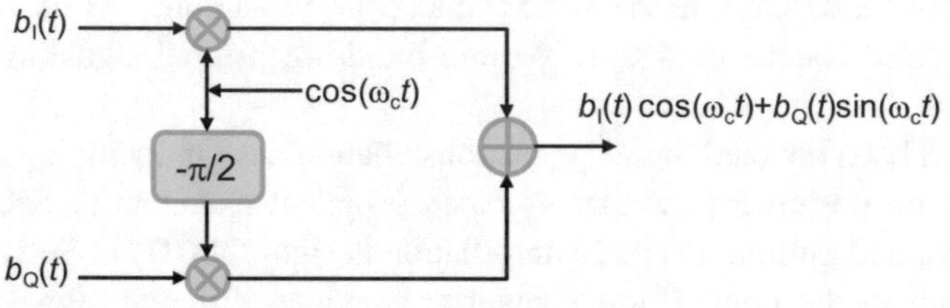

Fig. P3 The QPSK modulator

5.4. The signal constellation of a ternary signal set is obtained by placing the signal constellation points in vertices of an *equilateral triangle* with side length a. Sketch the signal constellation and determine the average signal energy as the function of a by assuming that all symbols are equally likely. Determine the optimum decision regions. Select the arbitrary basis functions $\phi_1(t)$ and $\phi_2(t)$ and express the signals from signal constellation diagram in terms of basis functions. Determine the corresponding minimum energy signal constellation.

5.5. A ternary one-dimensional signal constellation is given by {−1,0,1}. Sketch the signal constellation and decision regions and determine the average symbol energy assuming equal probable transmission. Further, determine the optimum receiver. Finally, derive the exact symbol error probability.

5.6. A ternary set of signals is given by

$$s_0(t) = 0; 0 \le t \le T; \; s_1(t) = -s_2(t) = \begin{cases} d, 0 \le t \le T \\ 0, \text{ otherwise} \end{cases}.$$

5.7. If $P(s_0) = 0.4$ and $P(s_1) = P(s_2) = 0.3$, determine the average symbol energy. Determine the basis functions and sketch corresponding constellation diagram. Determine the optimum decision regions. Derive the expression for average symbol error probability as the function of signal-to-noise ratio, defined as the average signal energy-to-power spectral density of AWGN.

5.8. Let us consider a ternary transmission by using *9-ary two-dimensional (2-D) constellation over FSO channel* containing both multiplicative component represented by gamma-gamma distribution of irradiance and additive zero-mean Gaussian noise of power spectral density $N_0/2$, assuming that coherent optical detection is used. The 8 constellation points in this 2-D constellation are placed on a circle of radius 1. The last (ninth) point is placed in the origin. Assume that the point in the origin appears with probability 0.5, all others with probability 0.5/8.

(a) Determine the average symbol energy of this constellation.
(b) How many ternary digits can be transmitted per symbol?
(c) Determine the average energy per ternary digit.
(d) If the symbol rate is 25 giga symbols/s, determine the aggregate ternary data rate.

(e) Determine the Gray mapping rule to map ternary digits to different constellation points in 9-ary 2-D constellation.
(f) Determine the average symbol error probability for this modulation format in the presence of scintillation and AWGN.
(g) Determine the optimum receiver for this 2-D constellation in the presence of scintillation and AWGN.
(h) In a fading wireless communication environment, the received signal power varies randomly over distance or time due to multipath fading. Let us assume that the same signal constellation is used to transmit messages over wireless multipath channel, when there is no line of sight. In this case, the distribution of envelope of received signal is Rayleigh. Determine the average symbol error probability for this modulation format in the presence of Rayleigh fading and AWGN. Compare the severity of fading and severity of turbulence in (g).

Note: Do not solve for integrals; just express them in terms of either Q-function or erfc-function.

5.9. Suppose we want to transmit the following data sequence: 1101000110 by binary DPSK. If $s(t) = A\cdot\cos(\omega_c t + \theta)$ represents the transmitted signal in any signaling interval of duration T, determine the phases of transmitted signal for this data sequence. Begin with the phase $\theta = 0$ rad for the phase of the first bit to be transmitted.

5.10. Suppose we want to transmit the following data sequence: 0101110011010110 by *differential* 16-PSK. Determine the phases of transmitted signal for this data sequence. Assume that the referent phase is 0 rad. Describe how the transmitted sequence can be recovered on the receiver side. Provide the block diagrams of transmitter and receiver.

5.11. Let us consider the binary transmission with a bit rate of 10 Gb/s. The bit 0 is represented by the absence of signal, while bit 1 is represented by a rectangular pulse of amplitude 1 V. The binary sequence, in which symbols 0 and 1 are equally likely, is transmitted over a zero-mean additive white Gaussian noise (AWGN) channel of variance 0.2 W. We assume that signal is attenuated during transmission in addition to being affected by AWGN. Let the received amplitude when symbol 1 is transmitted be $A = 0.8$ V. Determine the average time that elapses between two bits being in error. If the low-noise amplifier (LNA) is used, the signal amplitude get increased to $A = 1.6$ V. Determine the average time that elapses between two bits being in error, when LNA is used (ignore the noise introduced by LNA).

5.12. Suppose we want to transmit a multilevel digital signal at rate 12 giga symbols/s through a channel described by transfer function

$$H_c(f) = \begin{cases} H_0, |f| \leq 8 \text{ GHz} \\ 0, \text{otherwise} \end{cases},$$

where $H_0 \in (0,1]$ is a constant. Design the near optimum transmit and receive filters so that transmission is ISI free and probability of error is minimized.

5.13. In this problem, we are concerned with noncoherent detection of binary signals in AWGN, when the phase is unknown. In this communication system, the zero symbol is represented by the absence of signal, while the symbol one as a cosine signal of amplitude A and frequency ω_0. The transmission of both symbols is affected by the AWGN $n(t)$ of PSD $N_0/2$. The received signal $r(t)$ is represented by

$$r(t) = \begin{cases} n(t), \text{when 0 istransmitted} \\ A\cos(\omega_0 t + \theta) + n(t), \quad \text{when 1 istransmitted} \end{cases}$$

where the distribution of the phase θ is uniform.
(a) Determine the optimum receiver minimizing bit error probability.
(b) Derive the expression for bit error probability in this case.

5.14. The *cube* signal constellation is obtained by placing the signal constellation points in vertices of the cube of side length a. Sketch the signal constellation and determine the average signal energy as the function of a, assuming that all symbols are equally likely. Select the arbitrary basis functions $\phi_i(t)$ ($i = 1,2,3$) and express the signals from signal constellation diagram in terms of basis functions. Determine the corresponding minimum energy signal constellation. How many bits per symbol can be transmitted? Apply the Gray mapping and show how the bits should be mapped to symbols.

5.15. In an M-ary pulse amplitude modulation (MPAM) system, the transmitted signal x takes values from the discrete set $X = \{0,d,\ldots,(M-1)d\}$ ($M \geq 2$), where d is the Euclidean distance between two neighboring points in corresponding signal constellation diagram. If all signal constellation points are equally probable, determine the average signal energy. Derive the probability of symbol error as the function of signal-to-noise ratio per bit and signal constellation size. Sketch the signal constellation diagram, and denote the decision regions and decision boundaries. Determine an equivalent minimum energy MPAM system, and determine its average energy.

5.16. The 5-QAM signal constellation is shown in Fig. P16. We are interested in evaluating its performance against that of 5-PSK with constellation points located on a circle of radius a. Assuming that all constellation points are equally likely:
(a) Compare the energy efficiencies of these two signal constellations.
(b) Which signal constellation is more robust to phase errors? Justify your answer.
(c) Determine the decision boundaries and decision regions.

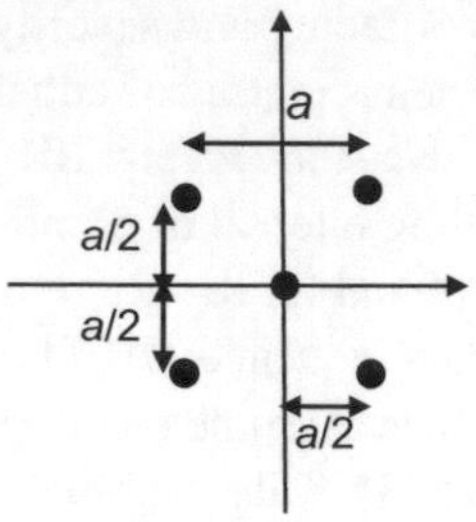

Fig. P16 The 5-QAM signal constellation diagram

5.17. Prove that the rectangular pulse shape $p(t) = \mathrm{rect}(t/T_s)$ satisfies the Nyquist criterion for zero ISI at the output of the matched filter: $R_p(lT_s) = \begin{cases} 1, l = 0 \\ 0, l \neq 0 \end{cases}$, where $R_p(\tau)$ is the autocorrelation function of $p(t)$. By using the result above and the fact that the output of I/Q modulator can be represented as

$$s(t) = \sum_k \sqrt{\frac{2}{T_s}} I_k p(t - kT_s) \cos(\omega_c t) - \sqrt{\frac{2}{T_s}} Q_k p(t - kT_s) \sin(\omega_c t),$$

prove that at the receiver side the outputs of matched filters after sampling can be written as

$$I'(kT_s) = \sum_l I_l R_p(lT_s - kT_s) + v_I(kT_s), Q'(kT_s) = \sum_l Q_l R_p(lT_s - kT_s) + v_Q(kT_s).$$

Describe the operation principle of the DT QAM modulator and demodulator and relate the output of the DT demodulator with sampled version of CT demodulator output.

5.18. Based on concatenated transmitter configuration of 16-star-QAM, provide the corresponding transmitter configuration for star-8-QAM signal constellation composed of two circles of radius 1 and r each carrying 4 points. Fully describe the operation of this transmitter. Determine r so that the average signal energy is lowest possible. Provide the Gray mapping rule.

5.19. The peak-to-average ratio (PAR) for a given waveform $s(t)$, to be used in digital communications, is defined as PAR $= \max|s(t)|^2/\mathrm{E}\{|s(t)|^2\}$, where both $\max(\cdot)$ and expectation $\mathrm{E}\{\cdot\}$ operators are defined with respect to time t. Determine the PAR of a raised-cosine pulse when the roll-off parameter is either 0 or 1. Which pulse shape has the lowest PAR? Is this pulse shape more or less sensitive to timing errors compared to others?

5.20. Explain how the multidimensional hybrid modulation scheme, described in Sect. 5.6.2, can be used as both 4 TbE and 10 TbE enabling technology by using the devices operating at a symbol rate of 40 GS/s. Discuss different alternatives and their corresponding complexities.

5.21. Provide details of 128-CIPQ constellation in a fashion similar to Table 5.2 in Sect. 5.7.1. Use the algorithm described in the same section.

5.22. Explain how the p_i update rule given by Eq. (5.110) in an energy-efficient signal constellation design algorithm is derived. Provide fully the steps in this derivation.

5.23. Provide the system parameters of the CO-OFDM system that is capable of compensating the accumulated chromatic dispersion over 250 km of SMF and corresponding PMD. The aggregate data rate should be at least 800 Gb/s, and the OFDM signal bandwidth should be either 25 GHz or 40 GHz. Use polarization-division multiplexing and square-QAM constellations. Assume that the chromatic dispersion parameter is 16 ps/(nm·km) and that the PMD parameter is

$D_p = 0.05$ ps·km$^{-1/2}$. Ignore the second-order GVD and fiber nonlinearities. Estimate the OSNR penalty due to guard interval compared to back-to-back configuration. Determine the spectral efficiency of your design. Is it possible to find the optimum number of subcarriers to maximize the spectral efficiency?

5.24. OFDM samples will follow Gaussian distribution for a sufficient number of subcarriers. The envelope will clearly follow a Rayleigh distribution, while the power will have a chi-square distribution. Determine the cumulative distribution function of envelope. Determine the probability that PAPR is below a certain threshold z_{tsh}.

5.25. In order to study the efficiency of CO-OFDM in PMD compensation, the 2 × 2 Jones matrix for the first-order PMD is typically used. Generalize this model for higher-order PMD studies.

5.26. Based on parallel decomposition of MIMO optical channel, similar to Fig. 5.27, explain how MIMO-OFDM can be used in multi-Tb/s optical transport.

References

1. J.G. Proakis, *Digital Communications*, 5th edn. (McGraw Hill, New York, 2007)
2. S. Haykin, *Communication Systems*, 4th edn. (Wiley, New York, 2001)
3. S. Hayking, *Digital Communications Systems* (Wiley, Hoboken, 2014)
4. B.P. Lathi, Z. Ding, *Modern Digital and Analog Communication Systems*, 4th edn. (Oxford University Press, Oxford, 2009)
5. A. Goldsmith, *Wireless Communications* (Cambridge University Press, Cambridge, 2005)
6. I.B. Djordjevic, W. Ryan, B. Vasic, *Coding for Optical Channels* (Springer, New York/Dordrecht/Heidelberg/London, 2010)
7. M. Cvijetic, I.B. Djordjevic, *Advanced Optical Communication Systems and Networks* (Artech House, Norwood, 2013)
8. I.B. Djordjevic, L. Xu, T. Wang, Statistical physics inspired energy-efficient coded-modulation for optical communications. Opt. Lett. **37**(8), 1340–1342 (2012)
9. M. Seimetz, *High-Order Modulation for Optical Fiber Transmission* (Springer, Berlin, 2009)
10. H.G. Batshon, I.B. Djordjevic, T. Schmidt, Ultra high speed optical transmission using subcarrier-multiplexed four-dimensional LDPC-coded modulation. Opt. Express **18**, 20546–20551 (2010)
11. I. Djordjevic, H.G. Batshon, L. Xu, T. Wang, Four-dimensional optical multiband-OFDM for beyond 1.4 Tb/s serial optical transmission. Opt. Express **19**(2), 876–882 (2011)
12. Y. Zhang, M. Arabaci, I.B. Djordjevic, Evaluation of four-dimensional nonbinary LDPC-coded modulation for next-generation long-haul optical transport networks. Opt. Express **20**(8), 9296–9301 (2012)
13. I.B. Djordjevic, L. Xu, T. Wang, PMD compensation in coded-modulation schemes with coherent detection using Alamouti-type polarization-time coding. Opt. Express **16**(18), 14163–14172 (2008)
14. I.B. Djordjevic, M. Arabaci, L. Xu, T. Wang, Generalized OFDM (GOFDM) for ultra-high-speed optical transmission. Opt. Express **19**(7), 6969–6979 (2011)
15. W. Shieh, I. Djordjevic, *OFDM for Optical Communications* (Elsevier/Academic Press, Burlington, 2009)
16. I.B. Djordjevic, M. Arabaci, L. Xu, T. Wang, Spatial-domain-based multidimensional modulation for multi-Tb/s serial optical transmission. Opt. Express **19**(7), 6845–6857 (2011)
17. I.B. Djordjevic, Energy-efficient spatial-domain-based hybrid multidimensional coded-modulations enabling multi-Tb/s optical transport. Opt. Express **19**(17), 16708–16714 (2011)
18. I.B. Djordjevic, A. Jovanovic, Z.H. Peric, T. Wang, Multidimensional optical transport based on optimized vector-quantization-inspired signal constellation design. IEEE Trans. Commun. **62**(9), 3262–3273 (Sept 2014)
19. I.B. Djordjevic, On the irregular nonbinary QC-LDPC-Coded hybrid multidimensional OSCD-modulation enabling beyond 100 Tb/s optical transport. IEEE/OSA J. Lightw. Technol. **31**(16), 2969–2975 (15 Aug 2013)
20. I.B. Djordjevic, M. Cvijetic, C. Lin, Multidimensional signaling and coding enabling multi-Tb/s optical transport and networking. IEEE Sig. Proc. Mag. **31**(2), 104–117 (Mar 2014)
21. I.B. Djordjevic, On advanced FEC and coded modulation for ultra-high-speed optical transmission. IEEE Commun. Surv. Tutor. **18**(3), 1920–1951, third quarter 2016 (19 Aug 2016). https://doi.org/10.1109/COMST.2016.2536726
22. I.B. Djordjevic, M. Arabaci, LDPC-coded orbital angular momentum (OAM) modulation for free-space optical communication. Opt. Express **18**(24), 24722–24728 (22 Nov 2010)
23. I.B. Djordjevic, J. Anguita, B. Vasic, Error-correction coded orbital-angular-momentum modulation for FSO channels affected by turbulence. IEEE/OSA J. Lightw. Technol. **30**(17), 2846–2852 (1 Sept 2012)
24. Z. Qu, I.B. Djordjevic, 500 Gb/s free-space optical transmission over strong atmospheric turbulence channels. Opt. Lett. **41**(14), 3285–3288 (15 July 2016)
25. I.B. Djordjevic, Z. Qu, Coded orbital angular momentum modulation and multiplexing enabling ultra-high-speed free-space optical transmission, in *Optical Wireless Communications – An Emerging Technology*, ed. by M. Uysal, C. Capsoni, Z. Ghassemlooy, A. Boucouvalas, E. G. Udvary, (Springer, 2016), pp. 363–385
26. I.B. Djordjevic, Z. Qu, Coded orbital-angular-momentum-based free-space optical transmission, in *Wiley Encyclopedia of Electrical and Electronics Engineering*, (Feb 2016). http://onlinelibrary.wiley.com/doi/10.1002/047134608X.W8291/abstract
27. J.D. Jackson, *Classical Electrodynamics* (Wiley, New York, 1975)
28. Z. Bouchal et al., Selective excitation of vortex fiber modes using a spatial light modulator. New J. Phys. **7**, 125 (2005)

29. J.A. Anguita, M.A. Neifeld, B.V. Vasic, Turbulence-induced channel crosstalk in an orbital angular momentum-multiplexed free-space optical link. Appl. Opt. **47**(13), 2414–2429 (2008)
30. A. Li, A. Al Amin, X. Chen, W. Shieh, Reception of mode and polarization multiplexed 107-Gb/s CO-OFDM signal over a two-mode fiber, in Proceedings of the OFC/NFOEC, Postdeadline Papers (OSA, 2011), Paper PDPB8
31. L. Gruner-Nielsen, Y. Sun, J.W. Nicholson, D. Jakobsen, R. Lingle, B. Palsdottir, Few mode transmission fiber with low DGD, low mode coupling and low loss, in Proceedings of the OFC/NFOEC, Postdeadline Papers (OSA, 2012), Paper PDP5A.1
32. R. Ryf, et al., Low-loss mode coupler for mode-multiplexed transmission in few-mode fiber, in Proceedings of the OFC/NFOEC, Postdeadline Papers (OSA, 2012), Paper PDP5B.5
33. N.K. Fontaine, et al., Space-division multiplexing and all-optical MIMO demultiplexing using a photonic integrated circuit, in Proceedings of the OFC/NFOEC, Postdeadline Papers (OSA, 2012), Paper PDP5B.1
34. X. Chen, A. Li, J. Ye, A. Al Amin, and W. Shieh, Reception of dual-LP11-mode CO-OFDM signals through few-mode compatible optical add/drop multiplexer, in Proceedings of the OFC/NFOEC, Postdeadline Papers (OSA, 2012), Paper PDP5B.4
35. S. Randel, et al., Mode-multiplexed 6×20-GBd QPSK transmission over 1200-km DGD-compensated few-mode fiber, in Proceedings of the OFC/NFOEC, Postdeadline Papers (OSA, 2012), Paper PDP5C.5
36. R. Ryf, et al., Space-division multiplexed transmission over 4200 km 3-core microstructured fiber, in Proceedings of the OFC/NFOEC, Postdeadline Papers (Optical Society of America, 2012), Paper PDP5C.5
37. B.J. Puttnam, W. Klaus, Y. Awaji, N. Wada, A. Kanno, T. Kawanishi, K. Imamura, H. Inaba, K. Mukasa, R. Sugizaki, T. Kobayashi, , and M. Watanabe, 19-core fiber transmission of 19x100x172-Gb/s SDM-WDM-PDM-QPSK signals at 305Tb/s, in Proceedings of the OFC/NFOEC, Postdeadline Papers (OSA, 2012), Paper PDP5C.1
38. P.M. Krummrich, Optical amplifiers for multi mode/multi core transmission, in Proceedings of the *OFC/NFOEC* (OSA, 2012), Paper OW1D.1
39. R. Essiambre, A. Mecozzi, Capacity limits in single mode fiber and scaling for spatial multiplexing, in Proceedings of the *OFC/NFOEC* (OSA, 2012), Paper OW3D
40. J.A. Anguita, J. Herreros, I.B. Djordjevic, Coherent multi-mode OAM superpositions for multi-dimensional modulation. IEEE Photonics Journal **6**(2), 900811 (Apr 2014)
41. Z.H. Peric, I.B. Djordjevic, S.M. Bogosavljevic, M.C. Stefanovic, Design of signal constellations for Gaussian channel by iterative polar quantization, in Proceedings of the 9th Mediterranean electrotechnical conference, vol. 2, pp. 866–869, Tel-Aviv, Israel, 18–20 May 1998
42. I.B. Djordjevic, H.G. Batshon, L. Xu, T. Wang, Coded polarization-multiplexed iterative polar modulation (PM-IPM) for beyond 400 Gb/s serial optical transmission, in Proceedings of the OFC/NFOEC 2010, Paper No. OMK2, San Diego, CA, 21–25 Mar 2010
43. X. Liu, S. Chandrasekhar, T. Lotz, P. Winzer, H. Haunstein, S. Randel, S. Corteselli, B. Zhu, Generation and FEC-decoding of a 231.5-Gb/s PDM-OFDM signal with 256-Iterative-Polar-Modulation achieving 11.15-b/s/Hz intrachannel spectral efficiency and 800-km reach, in Proceedings of the OFC/NFOEC, Postdeadline Papers (OSA, 2012), Paper PDP5B.3
44. Zhang, J., I.B. Djordjevic, Optimum signal constellation design for rotationally symmetric optical channel with coherent detection, in Proceedings of the *OFC/NFOEC 2011*, Paper No. OThO3, LA, USA, 6–10 Mar 2011
45. R.-J. Essiambre, G. Kramer, P.J. Winzer, G.J. Foschini, B. Goebel, Capacity limits of optical fiber networks. J. Lightwave Technol. **28**(4), 662–701 (2010)
46. N.J.A. Sloane, R.H. Hardin, T.S. Duff, J.H. Conway, Minimal-energy clusters of hard spheres. Discrete Comput. Geom. **14**, 237–259 (1995)
47. G. Wannier, *Statistical Physics* (Dover Publications, New York, 1987)
48. T. Cover, J. Thomas, *Elements of Information Theory* (Wiley, 1991)
49. R.E. Blahut, Computation of channel capacity and rate distortion functions. IEEE Trans. Inform. Theory **IT-18**, 460–473 (1972)
50. I.B. Djordjevic, T. Liu, L. Xu, T. Wang, Optimum signal constellation design for high-speed optical transmission, in Proceedings of the OFC/NFOEC 2012, Paper no. OW3H.2, Los Angeles, CA, USA, 6–8 Mar 2012
51. T. Liu, I.B. Djordjevic, L. Xu, T. Wang, Feedback channel capacity inspired optimum signal constellation design for high-speed optical transmission, in Proceedings of the CLEO 2012, Paper no. CTh3C.2, San Jose, CA, 6–11 May 2012
52. T. Liu, Z. Qu, C. Lin, I.B. Djordjevic, Nonuniform Signaling based LDPC-coded modulation for high-speed optical transport networks, in OSA Asia Communications and Photonics Conference (ACP) 2016, 2–5 Nov 2016, Wuhan, China. (Invited Paper)
53. F. Kschischang, S. Pasupathy, Optimal Nonuniform Signaling for Gaussian Channels. IEEE Trans. Inform. Theory **39**, 913–929 (1993)
54. T. Fehenberger, D. Lavery, R. Maher, A. Alvarado, P. Bayvel, N. Hanik, Sensitivity gains by mismatched probabilistic shaping for optical communication systems. IEEE Photo. Technol. Lett. **28**(7), 786–789 (2016)
55. T. Liu, I.B. Djordjevic, LDPC-coded BICM-ID based nonuniform signaling for ultra-high-speed optical transport, in Optical Fiber Communication Conference, OSA Technical Digest (online) (Optical Society of America, 2016), paper M3A.3
56. G. Böcherer, Probabilistic signal shaping for bit-metric decoding, in Proceedings of the 2014 IEEE International Symposium on Information Theory (ISIT 2014), (Institute of Electrical and Electronics Engineers, New York, 2014), pp. 431–435
57. T. Fehenberger, G. Böcherer, A. Alvarado, and N. Hanik, LDPC coded modulation with probabilistic shaping for optical fiber systems, in Optical Fiber Communication Conference, OSA Technical Digest (online) (Optical Society of America, 2015), paper Th2A.23
58. T. Liu, C. Lin, I.B. Djordjevic, Advanced GF(3^2) nonbinary LDPC coded modulation with non-uniform 9-QAM outperforming star 8-QAM. Opt. Express **24**(23), 13866–13874 (2016)
59. C. Lin, D. Zou, T. Liu, I.B. Djordjevic, Capacity achieving nonbinary LDPC coded non-uniform shaping modulation for adaptive optical communications. Opt. Express **24**(16), 18095–18104 (2016)
60. Z. Qu, C. Lin, T. Liu, I. Djordjevic, Experimental study of nonlinearity tolerant modulation formats based on LDPC coded non-uniform signaling, in *OFC 2017*, Paper W1G.7, 19–23 Mar 2017, Los Angeles, CA, USA
61. Z. Qu, I.B. Djordjevic, J. Anderson, Two-dimensional constellation shaping in fiber-optics communications. Appl. Sci. **9**(9), 1889 (8 May 2019)
62. Z. Qu, I.B. Djordjevic, On the probabilistic shaping and geometric shaping in optical communication systems. IEEE Access **7**, 21454–21464 (4 Feb 2019)
63. X. Han, M. Yang, I.B. Djordjevic, Y. Yue, Q. Wang, Z. Qu, J. Anderson, Joint probabilistic-nyquist pulse shaping for LDPC-coded 8-PAM signal in DWDM data center communications. Appl. Sci. **9**(23), 4996 (20 Nov 2019)

64. X. Han, Y. Yue, Z. Qu, R. Holmes, I. Djordjevic, Probabilistic shaped LDPC-coded 400G PM-64QAM DWDM transmission in 50-GHz Grid, in 26th Optoelectronic and Communications Conference (OECC), 3–7 July 2021
65. X. Han, I.B. Djordjevic, A. Jovanovic, A hybrid PGS LDPC-coded scheme for PM-16QAM modulation in 140 km DWDM metro network communication, in Proceedings of the CLEO 2021, paper SF2D.5, 9–14 May 2021
66. X. Han, Y. Yue, Z. Qu, I.B. Djordjevic, Probabilistic-shaping-enabled ADC/DAC resolution tolerance improvement in optical PAM/QAM communication systems, in SPIE OPTO, Photonics West 2021, San Francisco, CA, 6–11 Mar 2021
67. X. Han, I.B. Djordjevic, Probabilistic-shaped, Nyquist-pulse-shaped four-dimensional LDPC-coded modulation scheme for 100 km DWDM metro network transmission, in Proceedings of the 22nd international conference on transparent optical networks ICTON 2020, paper Mo.E1.7, Bari, Italy, 19–23 July 2020
68. X. Han, M. Yang, I. Djordjevic, A. Li, Probabilistically-shaped four-dimensional LDPC-coded modulation in 100 km DWDM optical transmission for metro network applications, in Asia Communications and Photonics Conference (ACP) 2019, 2–5 Nov 2019, Chengdu, Sichuan, China
69. X. Han, M. Yang, I.B. Djordjevic, Y. Yue, Q. Wang, Z. Qu, J. Anderson, Combined probabilistic shaping and nyquist pulse shaping for PAM8 signal transmission in WDM systems, in Proceedings of the *CLEO 2019*, 5–10 May 2019, San Jose, California, USA
70. X. Han, I.B. Djordjevic, Probabilistically shaped 8-PAM suitable for data centers communication, in Proceedings of the ICTON 2018, Paper Th. B3.4, Bucharest, Romania, 1–5 July 2018

Advanced Detection Techniques and Compensation of Channel Impairments

6

Abstract

This chapter is devoted to advanced detection and channel impairments compensation techniques applicable to both wireless and optical communication systems. To better understand the principles behind advanced detection and channel impairments compensation, we start the chapter with the fundamentals of detection and estimation theory including optimum receiver design in symbol error probability sense and symbol error probability derivations. In the section on wireless communication systems performance, we describe different scenarios including outage probability, average error probability, and combined outage-average error probability scenarios. We also describe the moment-generating function-based approach to average error probability calculation. We also describe how to evaluate performance in the presence of Doppler spread and fading effects. In the section on channel equalization techniques, after short introduction, we first describe how zero-forcing equalizers can be used to compensate for ISI, chromatic dispersion, polarization mode dispersion (PMD) in fiber-optics communication; ISI and multipath fading effects in wireless communications; and ISI and multipath effects in indoor optical wireless communications. After that, we describe the optimum linear equalizer design in the minimum mean-square error sense as well as Wiener filtering concepts. The most relevant post-compensation techniques, applicable to both wireless and optical communications, are described next including feedforward equalizer, decision-feedback equalizer, adaptive equalizer, blind equalizers, and maximum likelihood sequence detector (MLSD), also known as the Viterbi equalizer. The turbo equalization technique is postponed for Chap. 9. The next section is devoted to the relevant synchronization techniques. In the section on adaptive modulation techniques, we describe how to adapt the transmitter to time-varying wireless/optical channel conditions to enable reliable and high spectral-efficient transmission. Various scenarios are described including data rate, power, code rate, error probability adaptation scenarios, as well as a combination of various adaptation strategies. In particular, variable-power variable-rate modulation techniques and adaptive coded modulation techniques are described in detail. Next, the Volterra series-based equalization to deal with fiber nonlinearities and nonlinear effects in wireless communications is described. In the section on digital backpropagation, we describe how this method can be applied to deal with fiber nonlinearities, chromatic dispersion, and PMD in a simultaneous manner. In the section on coherent optical detection, we describe various balanced coherent optical detection schemes for two-dimensional modulation schemes; polarization diversity and polarization demultiplexing schemes; homodyne coherent optical detection based on phase-locked loops; phase diversity receivers; and the dominant coherent optical detection sources including laser phase noise, polarization noise, transimpedance amplifier noise, and amplifier spontaneous emission (ASE) noise. In the section on compensation of atmospheric turbulence effects, we describe various techniques to deal with atmospheric turbulence including adaptive coding, adaptive coded modulation, diversity approaches, MIMO signal processing, hybrid free-space optical (FSO)-RF communication approach, adaptive optics, and spatial light modulator (SLM)-based backpropagation method. In particular, linear adaptive optics techniques are described in detail. The set of problems is provided after the concluding remarks.

I. B. Djordjevic, *Advanced Optical and Wireless Communications Systems*, https://doi.org/10.1007/978-3-030-98491-5_6

6.1 Detection and Estimation Theory Fundamentals

To better understand the principles of advanced detection and channel impairments compensation techniques, in this section, we provide the fundamentals of detection and estimation theory.

6.1.1 Geometric Representation of Received Signals

Earlier in Sect. 5.1, we mentioned that in a digital wireless/optical communication system, there is a source of the message that at any time instance generates a symbol m_i $(i = 1, 2, \ldots, M)$ from the set of symbols $\{m_1, m_2, \ldots, m_M\}$. These symbols are generated with a priori probabilities: $p_1, p_2, \ldots, p_M$, where $p_i = P(m_i)$. A transmitter then converts the output m_i from the message source into a *distinct* real signal $s_i(t)$ of duration T, which is suitable for transmission over the wireless/optical channel. We also outlined that this set of signals $\{s_i(t)\}$ can be represented as properly weighted orthonormal basis functions $\{\Phi_n\}$ $(n = 1, \ldots, N)$ (wherein $N \leq M$), so that [1–5]

$$s_i(t) = \sum_{n=1}^{N} s_{in}\Phi_n(t), 0 \leq t < T; i = 1, 2, \ldots, M, \tag{6.1}$$

where s_{in} is the projection of the ith signal along the nth basis function. The received signal $r(t)$ can also be represented in terms of projections along the same basis functions as follows:

$$r(t) = \sum_{j=1}^{N} r_n\Phi_n(t), r_n = \int_0^T r(t)\Phi_n(t)dt \ (n = 1, 2, \ldots, N). \tag{6.2}$$

Therefore, both received and transmitted signal can be represented as vectors $\boldsymbol{r}$ and $\boldsymbol{s}_i$ in the signal space, respectively, as

$$\boldsymbol{r} = [r_1 \ r_2 \ \cdots \ r_N]^{\mathrm{T}}, \qquad \boldsymbol{s}_i = |s_i\rangle = [s_{i1} \quad s_{i2} \quad \cdots \quad s_{iN}]^{\mathrm{T}}. \tag{6.3}$$

In Chap. 5, we used the Euclidean distance receiver to decide in favor of a signal constellation point closest to the received signal vector. In the linear regime for both wireless and optical communications, we can use the *additive* (linear) *channel model* and represent the received signal as addition of transmitted signal $s_i(t)$ and accumulated additive noise $z(t)$:

$$r(t) = s_i(t) + z(t), 0 \leq t < T; i = 1, 2, \ldots, M. \tag{6.4}$$

We can represent the projection of received signal $r(t)$ along the nth basis function as follows:

$$\begin{aligned} r_n &= \int_0^T r(t)\Phi_n(t)dt = \int_0^T [s_i(t) + z(t)]\Phi_n(t)dt \\ &= \underbrace{\int_0^T s_i(t)\Phi_n(t)dt}_{s_{in}} + \underbrace{\int_0^{T_s} z(t)\Phi_n(t)dt}_{z_n} = \\ &= s_{in} + z_n; \ n = 1, \cdots, N. \end{aligned} \tag{6.5}$$

We can also rewrite Eq. (6.5) into a compact vector form as

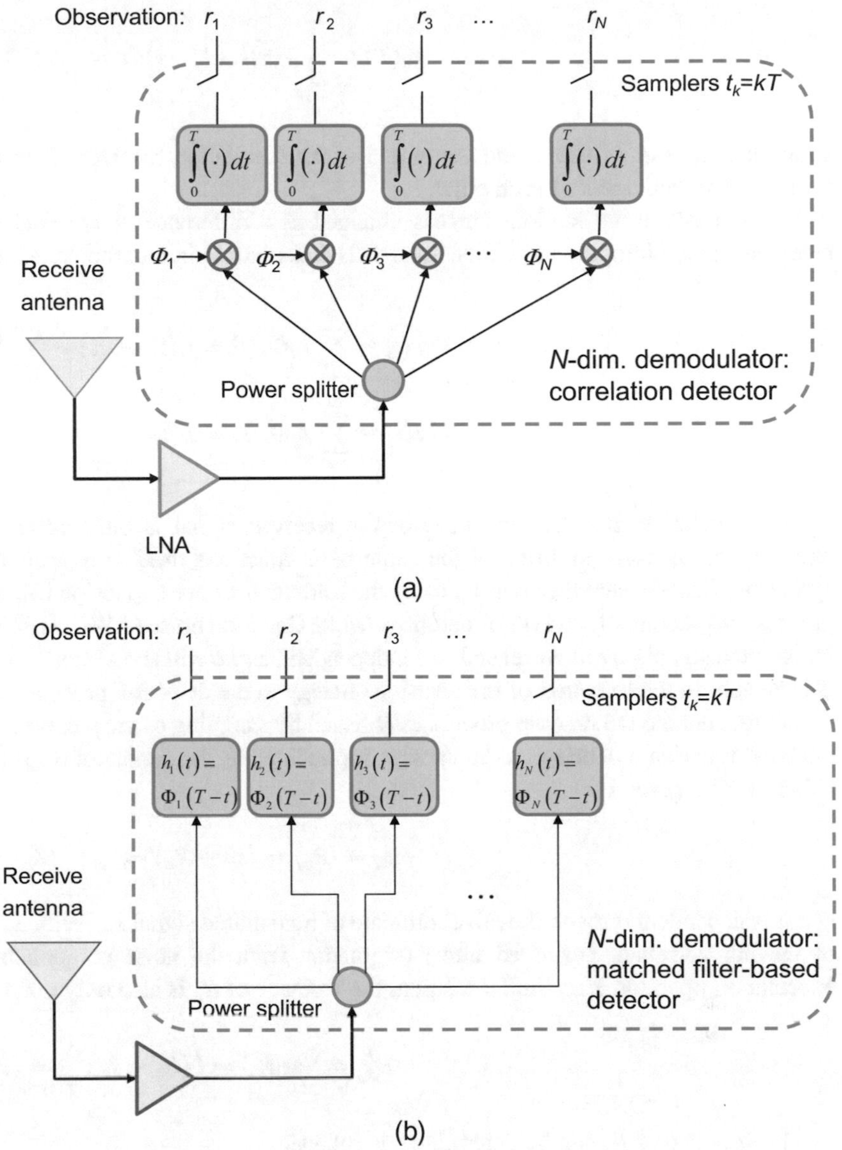

Fig. 6.1 Two equivalent implementations of detector (or demodulator) for wireless communications: (**a**) correlation detector and (**b**) matched filter-based detector

$$\boldsymbol{r} = \begin{bmatrix} r_1 \\ r_2 \\ \vdots \\ r_N \end{bmatrix} = \underbrace{\begin{bmatrix} s_{i1} \\ s_{i2} \\ \vdots \\ s_{iN} \end{bmatrix}}_{\boldsymbol{s}_i} + \underbrace{\begin{bmatrix} z_1 \\ z_2 \\ \vdots \\ z_N \end{bmatrix}}_{\boldsymbol{z}} = \boldsymbol{s}_i + \boldsymbol{z}. \tag{6.6}$$

Therefore, the received vector, also known as *observation vector*, can be represented as a sum of transmitted vector and noise vector. The observation vector can be obtained by using one of two possible alternative detection methods: matched filter-based detector and correlation detector, as summarized in Fig. 6.1.

The output of the nth matched filter having impulse response $h_n(t) = \Phi_n(T-t)$ is given as the convolution of the matched filter input and impulse response:

$$r_n(t) = \int_{-\infty}^{\infty} r(\tau)h_n(t-\tau)d\tau = \int_{-\infty}^{\infty} r(\tau)\Phi_n(T-t+\tau)d\tau. \tag{6.7}$$

By performing a sampling at the time instance $t = T$, we obtain

$$r_n(T) = \int_{-\infty}^{\infty} r(\tau)h_n(T-\tau)d\tau = \int_{0}^{T} r(\tau)\Phi_n(\tau)d\tau, \tag{6.8}$$

which has the same form as the *n*th output of the correlation receiver. Therefore, the correlation and matched filter-based receivers are equivalent to each other.

Let us observe the random process obtained as a difference of received signal and signal obtained by projecting the received signal along the basis functions. This difference is in fact the "remainder" of the received signal:

$$\begin{aligned} r'(t) &= r(t) - \sum_{n=1}^{N} r_n\Phi_n(t) = s_i(t) + z(t) - \sum_{n=1}^{N} (s_{in} + z_n)\Phi_n(t) \\ &= z(t) - \sum_{n=1}^{N} z_n\Phi_n(t) = z'(t). \end{aligned} \tag{6.9}$$

We can notice that the remainder of the received signal is only noise dependent. In all equations above, we have represented the noise in terms of the same basis functions used to represent the transmitted signal. However, it is more appropriate to represent the noise by using the Karhunen-Loève expansion [2], by taking the correlation properties of the noise process into account. However, for additive white Gaussian noise (AWGN), the noise projections along the basis functions of transmitted signals are uncorrelated and independent, as we will show shortly. Consequently, the basis used to represent noise is irrelevant in the selection of the decision strategy in the detection process.

Let $R(t)$ denote the random process established by sampling of the received signal $r(t)$. The projection R_n of $R(t)$ along the *n*th basis function will be a random variable represented by the correlator output r_n ($n = 1, 2, \ldots, N$). The corresponding mean value of R_n is given by

$$m_{R_n} = \langle R_n \rangle = \langle s_{in} + Z_n \rangle = s_{in} + \langle Z_n \rangle = s_{in}, \tag{6.10}$$

and it is dependent only on the *n*th coordinate of transmitted signal s_{in}. With Z_n we denoted the random variable at the output of the *n*th correlator (matched filter) originating from the noise component $z(t)$, and with $\langle \cdot \rangle$ being the mathematical expectation operator. In a similar fashion, the variance of R_n is also only noise dependent since it is

$$\sigma_{R_n}^2 = \mathrm{Var}[R_n] = \left\langle (R_n - s_{in})^2 \right\rangle = \langle Z_n^2 \rangle. \tag{6.11}$$

The variance of R_n can be determined as follows:

$$\begin{aligned} \sigma_{R_n}^2 = \langle Z_n^2 \rangle &= \left\langle \int_0^T Z(t)\Phi_n(t)dt \int_0^T Z(u)\Phi_n(u)du \right\rangle \\ &= \left\langle \int_0^T\int_0^T \Phi_n(t)\Phi_n(u)Z(t)Z(u)dtdu \right\rangle \\ &= \int_0^T\int_0^T \Phi_n(t)\Phi_n(u)\underbrace{\langle Z(t)Z(u)\rangle}_{R_Z(t,u)}dtdu \\ &= \int_0^T\int_0^T \Phi_n(t)\Phi_n(u)R_Z(t,u)dtdu. \end{aligned} \tag{6.12}$$

In Eq. (6.12), $R_Z(\cdot)$ presents the autocorrelation function of the noise process $Z(t)$, which is otherwise given as [1, 2]

$$R_Z(t,u) = \frac{N_0}{2}\delta(t-u). \tag{6.13}$$

Now, after substitution of autocorrelation function (6.13) into Eq. (6.12), we obtain that

$$\sigma_{R_n}^2 = \frac{N_0}{2}\int_0^T\int_0^T \Phi_n(t)\Phi_n(u)\delta(t-u)dtdu = \frac{N_0}{2}\underbrace{\int_0^T \Phi_n^2(t)dt}_{1} = \frac{N_0}{2}. \tag{6.14}$$

Therefore, all matched filter (correlator) outputs have the variance equal to the power spectral density of the noise process $Z(t)$, which is equal to $N_0/2$. The covariance between the nth and kth ($n \neq k$) outputs of the matched filters is given by

$$\begin{aligned}
\mathrm{Cov}[R_n R_k] &= \langle (R_n - s_{in})(R_k - s_{ik})\rangle = \langle R_n R_k\rangle \\
&= \left\langle \int_T^0 Z(t)\Phi_j(t)dt \int_T^0 Z(u)\Phi_k(u)dt \right\rangle \\
&= \left\langle \int_T^0\int_T^0 \Phi_n(t)\Phi_k(u)Z(t)Z(u)dtdu \right\rangle \\
&= \int_T^0\int_T^0 \Phi_n(t)\Phi_k(u)\underbrace{\langle Z(t)Z(u)\rangle}_{R_Z(t-u)}dtdu \\
&= \int_T^0\int_T^0 \Phi_n(t)\Phi_k(u)\underbrace{R_Z(t-u)}_{\frac{N_0}{2}\delta(t-u)}dtdu = \frac{N_0}{2}\int_T^0\int_T^0 \Phi_n(t)\Phi_k(u)\delta(t-u)dtdu \\
&= \frac{N_0}{2}\underbrace{\int_T^0 \Phi_n(t)\Phi_k(t)dt}_{0,\, n\neq k} = 0.
\end{aligned} \tag{6.15}$$

As a consequence, the outputs R_n of correlators (matched filters) are mutually uncorrelated. In conclusion, the outputs of correlators, given by Eq. (6.6), are Gaussian random variables. These outputs are statistically independent so that the joint conditional PDF of the observation vector $\boldsymbol{R}$ can be written as a product of conditional PDFs of its individual outputs from correlators as follows:

$$f_{\boldsymbol{R}}(\boldsymbol{r}|m_i) = \prod_{n=1}^{N} f_{R_n}(r_n|m_i); i = 1, \cdots, M. \tag{6.16}$$

The components R_n's are independent Gaussian random variables with mean values s_{in} and variance equal to $N_0/2$, so that the corresponding conditional PDF of R_n is given as

$$f_{R_n}(r_n|m_i) = \frac{1}{\sqrt{\pi N_0}}\exp\left[-\frac{1}{N_0}(r_n - s_{in})^2\right]; \begin{array}{l} n = 1, 2, \ldots, N \\ i = 1, 2, \ldots, M \end{array}. \tag{6.17}$$

By substituting the conditional PDFs into joint PDF, and after simple arrangements, we obtain

$$f_{\boldsymbol{R}}(\boldsymbol{r}|m_i) = (\pi N_0)^{-N/2}\exp\left[-\frac{1}{N_0}\sum_{n=1}^{N}(r_n - s_{in})^2\right]; i = 1, \cdots, M. \tag{6.18}$$

We can conclude that the elements of random vector given by Eq. (6.6) completely characterize the term $\sum_{n=1}^{N} r_n \Phi_n(t)$.

What left to be characterized is the remainder $z'(t)$ of the received signal given by Eq. (6.9), which is only noise dependent. It can be easily shown that the sample of the noise process $Z'(t_m)$ is statistically independent on outputs R_j of the correlators; in other words,

$$\langle R_n Z'(t_m) \rangle = 0, \begin{cases} n = 1, 2, \ldots, N \\ 0 \leq t_m \leq T_s \end{cases}. \tag{6.19}$$

Therefore, the outputs of correlators (matched filters) are only relevant in statistics of the decision-making process. In other words, as long as the signal detection in AWGN is concerned, only the projections of the noise onto basis functions used to represent the signal set $\{s_i(t)\}$ ($i = 1, \ldots, M$) affect the statistical process in the detection circuit; the remainder of the noise is irrelevant. This claim is sometimes called the *theorem of irrelevance* [1–4]. According to this theorem, the channel model given by Eq. (6.4) can be represented by an equivalent N-dimensional vector given by Eq. (6.6).

6.1.2 Correlation and Matched Filter-Based Receivers Maximizing the Signal-to-Noise Ratio

The schemes from Fig. 6.1 can serve as a base of the receiver decision circuit, as shown in Fig. 6.2. The observation vector obtained by using either correlation of matched filter detector is used as the input vector. The observation vector is used as input to M dot-product calculators. The ith product calculator evaluates the inner product between the observation vector $\boldsymbol{r}$ and the ith signal vector $\boldsymbol{s}_i$ (corresponding to transmitted message m_i) as follows:

$$\langle r|s_i \rangle = \int_0^T r(t) s_i(t) dt = \boldsymbol{r}\boldsymbol{s}_i^{\mathrm{T}} = \sum_{n=1}^{N} r_n s_{in}. \tag{6.20}$$

If it happens that different symbols have different energies after dot-product calculations, we need to remove the half of the energy of the corresponding symbol in each of the branches. Decision circuit decides in favor of the symbol with the largest input and provides the estimate to the N-dimensional demapper. For nonbinary transmission, the demapper block can be omitted. The configurations of both correlation detector and matched filter-based detector, both shown in Fig. 6.1, and the configuration of signal transmission decoder/detector, shown in Fig. 6.2, have been derived based on the Euclidean distance receiver in Chap. 5.

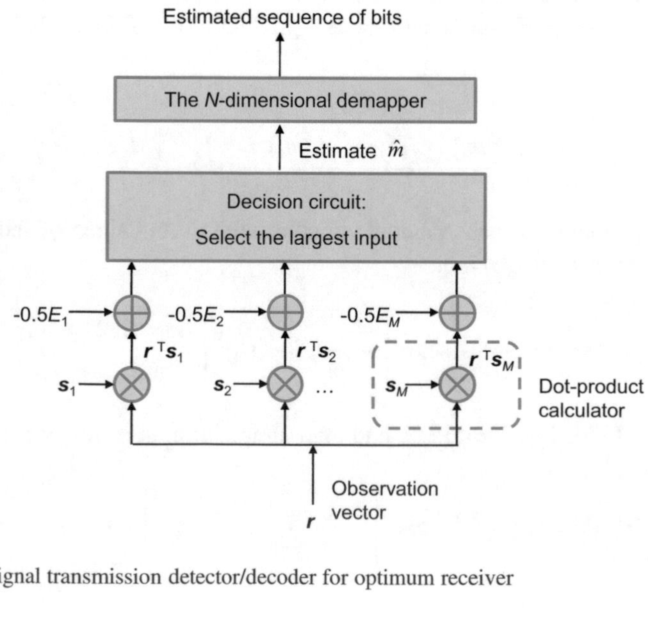

Fig. 6.2 The configuration of the signal transmission detector/decoder for optimum receiver

In the rest of this section, we will show that the matched filter is optimum from the signal-to-noise ratio (SNR) point of view. For this proof, we will use the Schwartz inequality given by Eq. (5.11). Let us observe the nth output from the matched filter shown in Fig. 6.1b, which is the convolution of received signal and impulse response of the matched filter evaluated at $t = T$:

$$\begin{aligned} r_n(T) = r(t) * h_n(t)|_{t=T} &= \underbrace{[s_n(t) * h_n(t)]|_{t=T}}_{a_n(T)} + \underbrace{[z(t) * h_n(t)]|_{t=T}}_{z_n(T)} \\ &= a_n(T) + z_n(T), \end{aligned} \tag{6.21}$$

where $a_n(T)$ represents the signal portion of the output from matched filter, while $z_n(T)$ denotes the noise portion of the matched filter output. The SNR can be defined as

$$SNR = \frac{\langle a_n^2(T) \rangle}{\langle z_n^2(T) \rangle}. \tag{6.22}$$

The variance of the noise process can be expressed as

$$\begin{aligned} \langle z_n^2(T) \rangle &= \left\langle \int_{-\infty}^{\infty} h_n(T-\tau) z(\tau) d\tau \cdot \int_0^T h_n(T-u) z(u) du \right\rangle \\ &= \int_{-\infty}^{\infty} \int_{-\infty}^{\infty} \underbrace{\langle z(\tau) z(u) \rangle}_{R_z(\tau-u)} h_n(T-\tau) h_n(T-u) d\tau du, \end{aligned} \tag{6.23}$$

where $R_z(\tau\text{-}u)$ is the autocorrelation function of AWGN introduced by Eq. (6.13). Now we have that $R_z(\tau\text{-}u) = (N_0/2)\,\delta(\tau\text{-}u)$, which upon substitution in Eq. (6.23) gives the following:

$$\begin{aligned} \langle z_n^2(T) \rangle &= \frac{N_0}{2} \int_{-\infty}^{\infty} \int_{-\infty}^{\infty} \delta(\tau-u) h_n(T-\tau) h_n(T-u) d\tau du \\ &= \frac{N_0}{2} \int_0^T h_n^2(T-u) du. \end{aligned} \tag{6.24}$$

On the other hand, the variance of the signal at the output of matched filter is given as

$$\begin{aligned} \langle a_n^2(T) \rangle &= \left\langle [s_n(t) * h_n(t)|_{t=T}]^2 \right\rangle = [s_n(t) * h_n(t)|_{t=T}]^2 = \\ &= \left[\int_{-\infty}^{\infty} h_n(T-\tau) s_n(\tau) d\tau \right]^2. \end{aligned} \tag{6.25}$$

After substitution of corresponding variances into Eq. (6.22), we obtain the following expression for SNR:

$$SNR = \frac{\left[\int_{-\infty}^{\infty} h_n(T-\tau) s_n(\tau) d\tau \right]^2}{\frac{N_0}{2} \int_0^T h_n^2(T-u) du}. \tag{6.26}$$

By applying the Schwartz inequality to numerator in Eq. (6.26), becomes upper bounded by

$$SNR = \frac{\left[\int_0^T h_n(T-\tau)s_n(\tau)d\tau\right]^2}{\frac{N_0}{2}\int_0^T h_n^2(T-\tau)d\tau} \leq \frac{\int_0^T h_n^2(T-\tau)d\tau \int_0^T s_n^2(\tau)d\tau}{\frac{N_0}{2}\int_0^T h_n^2(T-\tau)d\tau}$$
$$= \frac{2}{N_0}\int_0^{T_s} s_n^2(\tau)d\tau = \frac{2E_n}{N_0} = SNR_{n,\max}. \tag{6.27}$$

The equality sign in Eq. (6.27) is satisfied when $h_n(T-\tau) = Cs_n(\tau)$, where C is an arbitrary constant (for convenience, we can assume that $C = 1$). Accordingly, the impulse response of the matched filter can be expressed as

$$h_n(\tau) = s_n(T-\tau). \tag{6.28}$$

In conclusion, we can say that the matched filter maximizes the SNR. It can be straightforwardly shown that the output signal of a matched filter is proportional to a shifted version of the autocorrelation function of the input signal, namely,

$$r_n(t) = R_{s_n}(t-T). \tag{6.29}$$

It is also important to mention that the spectrum of the output signal of a matched filter is proportional to the spectral density of the input signal:

$$H(f) = FT\{s_n(t) * s_n(T-\tau)\} \overset{FT\{s_n(t)\}=S_n(f)}{=} \underbrace{S_n(f)S_n^*(f)}_{|S_n(f)|^2}\exp\left[-j2\pi fT\right]$$
$$= |S_n(f)|^2\exp\left[-j2\pi fT\right]. \tag{6.30}$$

Therefore, we have just shown that the Euclidean distance receiver, the correlation receiver, and matched filter-based receivers are equivalent to each other for Gaussian-like channels.

6.1.3 Optimum and LLR Receivers

In this section, we are concerned with the receivers minimizing symbol error probability, commonly referred to as the *optimum receivers*. The joint conditional PDF given by $f_{\mathbf{R}}(\mathbf{r}| m_i)$ can be used in a decision-making process. It is often called *likelihood function* and denoted as $L(m_i)$. For Gaussian channels, the likelihood function contains exponential terms that can lead to numerical instabilities. It is more convenient to work with its logarithmic version, which is known as *log-likelihood function*:

$$l(m_i) = \log L(m_i);\ i = 1, 2, \ldots, M. \tag{6.31}$$

For an AWGN channel, the log-likelihood function can be obtained by substituting Eq. (6.18) into Eq. (6.31) (by ignoring independent terms in the channel statistics) as follows:

$$l(m_i) = -\frac{1}{N_0}\sum_{j=1}^{N}(r_j - s_{ij})^2;\ i = 1, 2, \ldots, M. \tag{6.32}$$

One can notice that the log-likelihood function is similar to the Euclidean distance squared introduced earlier.

The log-likelihood function is quite useful in the decision-making procedure. Since we are particularly concerned with *optimum receiver* design, we can use the log-likelihood function as a design tool. Namely, given the observation vector $\mathbf{r}$, the optimum receiver performs the mapping from the observation vector $\boldsymbol{r}$ to the estimate value $\widehat{m}$ of the transmitted symbol (say, m_i) so that the symbol error probability is minimized. The probability P_e of errorness decision, in other words the probability of detecting $\widehat{m} = m_i$ when it was not transmitted, can be calculated by the following relation:

$$P_e(m_i|\boldsymbol{r}) = P(m_i \text{ isnot sent}|\boldsymbol{r}) = 1 - P(m_i \text{ issent}|\boldsymbol{r}), \tag{6.33}$$

where $P(\cdot|\cdot)$ denotes the conditional probability.

The optimum decision strategy, also known as the *maximum a posteriori probability (MAP) rule*, can be formulated as the following decision rule:

$$\begin{aligned} &\text{Set } \widehat{m} = m_i \text{ if} \\ &P(m_i \text{ issent}|\boldsymbol{r}) \geq P(m_k \text{ issent}|\boldsymbol{r}), \forall k \neq i. \end{aligned} \tag{6.34}$$

At this point, it is useful to provide the geometric interpretation of optimum detection problem. Let D denote the N-dimensional signal space of all possible observation vectors (D is also known as the *observation space*). The total observation space is partitioned into M nonoverlapping N-dimensional decision regions D_1, D_2, . . ., D_M, which are defined as

$$D_i = \{\boldsymbol{r} : P(m_i \text{ sent}|\boldsymbol{r}) > P(m_k \text{ sent}|\boldsymbol{r}) \forall k \neq i\}. \tag{6.35}$$

In other words, the ith decision region D_i is defined as the set of all observation vectors for which the probability that m_i is sent is larger than any other probability. As an illustration, in Fig. 6.3, we show decision regions for QPSK (also known as 4-QAM), assuming an equal probability transmission. Whenever the observation vector falls within the decision region D_i, we can decide in favor of symbol m_i.

The optimum decision strategy given by Eq. (6.35) is not quite suitable from a practical point of view. Let us consider an alternative practical representation for $M = 2$. The average error probability for $M = 2$ is given by

$$P_e = p_1 \int_{\overline{D}_1} f_{\boldsymbol{R}}(\boldsymbol{r}|m_1) d\boldsymbol{r} + p_2 \int_{\overline{D}_2} f_{\boldsymbol{R}}(\boldsymbol{r}|m_2) d\boldsymbol{r}, \tag{6.36}$$

where $p_i = P(m_i)$ represents the a priori probability that symbol m_i is transmitted, while $\overline{D}_i = D - D_i$ denotes the complement of D_i ("not D_i decision region").

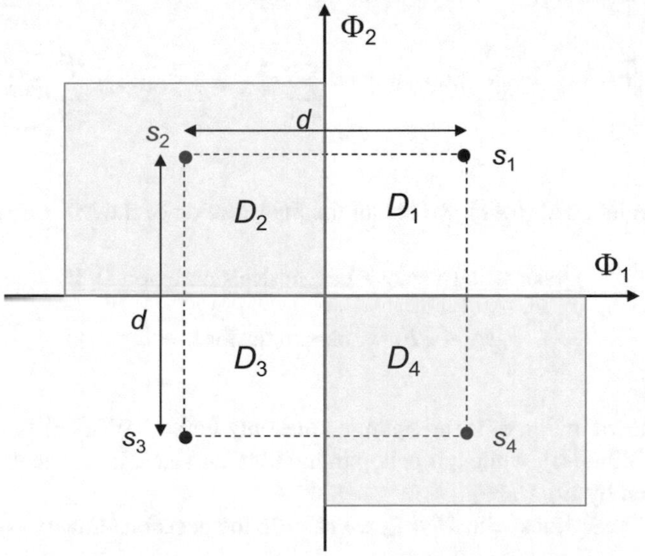

Fig. 6.3 Decision regions for QPSK

Since $D = D_1 + D_2$, we can rewrite Eq. (6.36) as

$$\begin{aligned} P_e &= p_1 \int_{D-D_1} f_{\boldsymbol{R}}(\boldsymbol{r}|m_1)d\boldsymbol{r} + p_2 \int_{D_1} f_{\boldsymbol{R}}(\boldsymbol{r}|m_2)d\boldsymbol{r} = \\ &= p_1 + \int_{D_1} [p_2 f_{\boldsymbol{R}}(\boldsymbol{r}|m_2) - p_1 f_{\boldsymbol{R}}(\boldsymbol{r}|m_1)]d\boldsymbol{r}. \end{aligned} \tag{6.37}$$

In order to minimize the average error probability given by Eq. (6.37), we can apply the following decision rule: if $p_1 f_{\boldsymbol{R}}(\boldsymbol{r}|m_1)$ is greater than $p_2 f_{\boldsymbol{R}}(\boldsymbol{r}|m_2)$, then assign $\boldsymbol{r}$ to D_1 and accordingly decide in favor of m_1. Otherwise, we assign $\boldsymbol{r}$ to D_2 and decide in favor of m_2. This decision strategy can be generalized for $M > 2$ as follows:

$$\begin{aligned} &\text{Set } \widehat{m} = m_i \text{ if} \\ &p_k f_{\boldsymbol{R}}(\boldsymbol{r}|m_k) \text{ is maximum for } k = i. \end{aligned} \tag{6.38}$$

When all symbols occur with the same probability $p_i = 1/M$, then the corresponding decision rule is known as the *maximum likelihood (ML) rule* and can be formulated as

$$\begin{aligned} &\text{Set } \widehat{m} = m_i \text{ if} \\ &l(m_k) \text{ is maximum for } k = i. \end{aligned} \tag{6.39}$$

In terms of the observation space, the ML rule can also be formulated as

$$\begin{aligned} &\text{Observationvector } \boldsymbol{r} \text{ liesindecisionregion } D_i \text{ when} \\ &l(m_k) \text{ is maximum for } k = i. \end{aligned} \tag{6.40}$$

For an AWGN channel, by using the likelihood function given by Eq. (6.32) (which is related to the Euclidean distance squared), we can formulate the ML rule as

$$\begin{aligned} &\text{Observation vector } \boldsymbol{r} \text{ lies in region } D_i \text{ if} \\ &\text{the Euclidean distance } \|\boldsymbol{r} - \boldsymbol{s}_k\| \text{ is minimum for } k = i. \end{aligned} \tag{6.41}$$

From Euclidean distance in signal space (see Chap. 5), we know that

$$\|\boldsymbol{r} - \boldsymbol{s}_k\|^2 = \sum_{n=1}^{N} |r_n - s_{kn}|^2 = \sum_{n=1}^{N} r_n^2 - 2\sum_{n=1}^{N} r_n s_{kn} + \underbrace{\sum_{n=1}^{N} s_{kn}^2}_{E_k}, \tag{6.42}$$

and by substituting this equation into Eq. (6.41), we obtain the final version of the ML rule as follows:

$$\begin{aligned} &\text{Observation vector } \boldsymbol{r} \text{ lies in decisionregion } \mathrm{D}_i \text{ if} \\ &\sum_{n=1}^{N} r_n s_{kn} - \frac{1}{2} E_k \text{ is maximum for } k = i. \end{aligned} \tag{6.43}$$

The receiver configuration shown in Fig. 6.2 is an optimum one only for an AWGN channel (such as wireless channel and optical channel dominated by ASE noise), while it is suboptimum for Gaussian-like channels. As for non-Gaussian channels, we need to use the ML rule given by Eq. (6.39).

As an example, for modulation schemes with $M = 2$, we can use the decision strategy based on the following rule:

$$LLR(\boldsymbol{r}) = \log\left[\frac{f_{\boldsymbol{R}}(\boldsymbol{r}|m_1)}{f_{\boldsymbol{R}}(\boldsymbol{r}|m_2)}\right] \underset{m_2}{\overset{m_1}{\gtrless}} \log\left(\frac{p_2}{p_1}\right), \tag{6.44}$$

Fig. 6.4 The log-likelihood ratio receiver

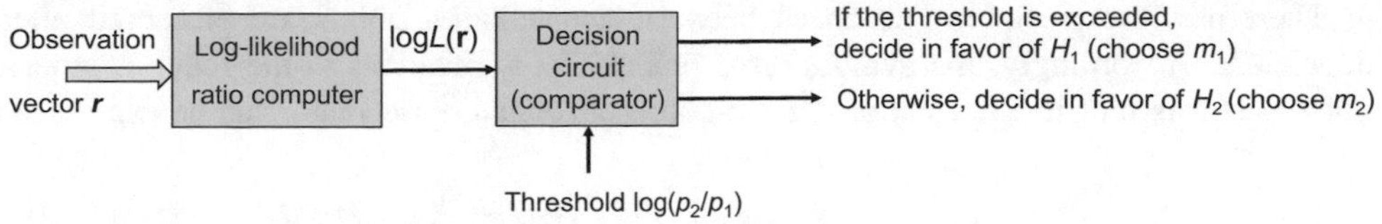

where $LLR(\cdot)$ denotes the log-likelihood ratio. Equation (6.44) is also known as *Bayes' test* [2], while the factor $\log(p_2/p_1)$ is known as a threshold of the test. The scheme of the corresponding log-likelihood ratio receiver is shown in Fig. 6.4. When $p_1 = p_2 = 1/2$, the threshold of the test is 0, while the corresponding receiver is the ML one.

6.1.4 Symbol Error Probability Calculation

In this section, we describe how to calculate symbol error probability and provide two illustrative examples: (i) calculation of bit error probabilities for binary signal constellations and (ii) calculation of symbol error probability for M-ary QAM. For large signal constellations, when it becomes quite difficult to calculate symbol error probabilities, we describe a union-bound approximation.

6.1.4.1 Calculation of Symbol Error Probability

The *average symbol error probability* can be calculated as

$$P_e = \sum_{i=1}^{M} p_i P(\boldsymbol{r} \text{ does not lie in } D_i | m_i \text{ sent}) = \sum_{i=1}^{M} p_i \int_{\overline{D}_i} f_{\boldsymbol{R}}(\boldsymbol{r}|m_i) d\boldsymbol{r}, \tag{6.45}$$

where $p_i = P(m_i)$ represents the a priori probability that symbol m_i is transmitted, while $\overline{D}_i = D - D_i$ denotes the complement of D_i (not D_i decision region). By using this definition of the complement of decision region, Eq. (6.45) can be rewritten as

$$\begin{aligned} P_e &= \sum_{i=1}^{M} p_i \int_{\overline{D}_i} f_{\boldsymbol{R}}(\boldsymbol{r}|m_i) d\boldsymbol{r} = \sum_{i=1}^{M} p_i \int_{D-D_i} f_{\boldsymbol{R}}(\boldsymbol{r}|m_i) d\boldsymbol{r} \\ &= 1 - \sum_{i=1}^{M} p_i \int_{D_i} f_{\boldsymbol{R}}(\boldsymbol{r}|m_i) d\boldsymbol{r}. \end{aligned} \tag{6.46}$$

For an equal probability transmission of symbols ($p_i = 1/M$), the average error probability can be calculated as

$$P_e = 1 - \frac{1}{M} \sum_{i=1}^{M} \int_{D_i} f_{\boldsymbol{R}}(\boldsymbol{r}|m_i) d\boldsymbol{r}. \tag{6.47}$$

In the ML detection, the probability P_e of the symbol error depends only on the relative Euclidean distances between the message points in the constellation. Since AWGN is spherically symmetric in all directions in the signal space, the changes in the orientation (with respect to both coordinate axes and the origin) do not affect the probability of the symbol error. Therefore, the rotation and translation of signal constellation produce another signal constellation that has the same symbol error probability as the one observed against the minimum Euclidean distance.

Let us first observe the *rotation operation* and denote the original complex-valued signal constellation by $\{\boldsymbol{s}_i\}$. The rotation matrix $\boldsymbol{U}$ is a Hermitian matrix ($\boldsymbol{U}\boldsymbol{U}^\dagger = \boldsymbol{I}$, where $\boldsymbol{I}$ is the identity matrix). The corresponding constellation obtained after application of rotation matrix is given by $\{\boldsymbol{U}\boldsymbol{s}_i\}$. The Euclidean distance between the observation vector $\boldsymbol{r}$ and a constellation point from rotated signal constellation, say, $\boldsymbol{U}\boldsymbol{s}_i$, is given as

$$\|\boldsymbol{r}_{\text{rotate}} - \boldsymbol{s}_{i,\text{rotate}}\| = \|\boldsymbol{U}\boldsymbol{s}_i + \boldsymbol{z} - \boldsymbol{U}\boldsymbol{s}_i\| = \|\boldsymbol{z}\| = \|\boldsymbol{r} - \boldsymbol{s}_i\|. \tag{6.48}$$

Therefore, the Euclidean distances between constellation points are preserved after rotation since it is only noise dependent. Accordingly, the symbol error probability is invariant to the rotation process. We need also to verify if the noise vector is sensitive to rotation. The variance of rotated noise vector can be expressed as

$$\begin{aligned}\left\langle z_{\text{rotate}} z_{\text{rotate}}^{\dagger}\right\rangle &= \left\langle \boldsymbol{Uz}(\boldsymbol{Uz})^{\dagger}\right\rangle = \langle \boldsymbol{Uzz}^{\dagger}\boldsymbol{U}^{\dagger}\rangle = \boldsymbol{U}\underbrace{\langle \boldsymbol{zz}^{\dagger}\rangle}_{(N_0/2)\boldsymbol{I}}\boldsymbol{U}^{\dagger} \\ &= \frac{N_0}{2}\boldsymbol{UU}^{\dagger} = \frac{N_0}{2}\boldsymbol{I} = \langle \boldsymbol{zz}^{\dagger}\rangle,\end{aligned} \tag{6.49}$$

which means that variance does not change during rotation. The mean value of the rotated noise vector is still equal to zero, i.e.,

$$\langle z_{\text{rotate}}\rangle = \langle \boldsymbol{Uz}\rangle = \boldsymbol{U}\langle z\rangle = \boldsymbol{0}. \tag{6.50}$$

Therefore, the AWGN exhibits spherical symmetric property.

Regarding the *translation operation*, let all constellation points be translated for the same translation vector $\{\boldsymbol{s}_i\text{-}\boldsymbol{a}\}$. The Euclidean distance between the observation vector and transmitted point $\boldsymbol{s}_i$ is given as

$$\|\boldsymbol{r}_{\text{translate}} - \boldsymbol{s}_{i,\text{translate}}\| = \|\boldsymbol{s}_i - \boldsymbol{a} + z - (\boldsymbol{s}_i - \boldsymbol{a})\| = \|z\| = \|\boldsymbol{r} - \boldsymbol{s}_i\|, \tag{6.51}$$

and does not change with translation. Accordingly, the symbol error probability, as a function of the minimum Euclidean distance, does not change either. However, the average symbol energy changes with translation, and upon translation it becomes

$$E_{\text{translate}} = \sum_{i=1}^{M} \|\boldsymbol{s}_i - \boldsymbol{a}\|^2 p_i. \tag{6.52}$$

Since $\|\boldsymbol{s}_i - \boldsymbol{a}\|^2 = \|\boldsymbol{s}_i\|^2 - 2\boldsymbol{a}^{\dagger}\boldsymbol{s}_i + \|\boldsymbol{a}\|^2$, we can rewrite Eq. (6.52) as follows:

$$E_{\text{translate}} = \underbrace{\sum_{i=1}^{M} \|\boldsymbol{s}_i\|^2 p_i}_{E} - 2\boldsymbol{a}^{\dagger}\underbrace{\sum_{i=1}^{M} \boldsymbol{s}_i p_i}_{\langle \boldsymbol{s}\rangle} + \|\boldsymbol{a}\|^2 \sum_{i=1}^{M} p_i = E - 2\boldsymbol{a}^{\dagger}\langle \boldsymbol{s}\rangle + \|\boldsymbol{a}\|^2. \tag{6.53}$$

The minimum energy constellation can be obtained by differentiating $E_{\text{translate}}$ with respect to $\boldsymbol{a}$, by setting this result to be zero, and by solving for the optimum translate value, which leads to $\boldsymbol{a}_{\min} = \langle \boldsymbol{s}\rangle$. The minimum energy signal constellation is then obtained as

$$\boldsymbol{s}_{i,\text{translate}} = \boldsymbol{s}_i - \langle \boldsymbol{s}\rangle. \tag{6.54}$$

The average symbol energy that corresponds to optimum constellation is derived after substitution of Eq. (6.54) into Eq. (6.52) to obtain $E_{\text{translate, min}} = E - \|\langle \boldsymbol{s}\rangle\|^2$. Therefore, we can rotate and translate the signal constellation as needed to simplify the symbol error probability calculations while not affecting the symbol error probability expressed as the function of minimum distance in the signal constellation.

At this point, it is convenient to introduce the Q-function, defined as a probability that zero-mean Gaussian random variable X of unit variance is larger than u:

$$Q(u) = P(X > u) = \int_u^{\infty} \frac{1}{\sqrt{2\pi}} e^{-x^2/2} dx = \frac{1}{2}\text{erfc}\left(\frac{u}{\sqrt{2}}\right), \tag{6.55}$$

wherein the complementary error function (erfc-function) is defined by

$$\text{erfc}(u) = \frac{2}{\sqrt{\pi}} \int_u^{\infty} e^{-x^2} dx.$$

6.1.4.2 Calculation of Bit Error Probability of Binary Signal Constellations

Let us now consider *binary signal constellations*. We know that the output of matched filter (correlator) r is a Gaussian random variable since the input is a Gaussian random process. In any linear filter, the Gaussian input produces the output that is a Gaussian random process as well, but with different mean value and variance [1–4]. The statistics of this Gaussian random variable, r, changes depending on whether the transmitted signal was s_1 (corresponding to message m_1) or s_2 (corresponding to message m_2). If s_1 was sent, then r is a Gaussian random variable with the mean value μ_1 and variance σ^2 given by

$$\mu_1 = \langle r|m_1\rangle = E_1 - \rho_{12}, \sigma^2 = \frac{N_0}{2}(E_1 + E_2 - 2\rho_{12}), \tag{6.56}$$

where ρ_{12} is the correlation coefficient $\rho_{12} = \int_0^T s_1(t)s_2(t)dt$.

If s_2 was sent, then r has mean value and variance given by

$$\mu_2 = \langle r|m_2\rangle = -E_2 + \rho_{12}, \sigma^2 = \frac{N_0}{2}(E_1 + E_2 - 2\rho_{12}). \tag{6.57}$$

For an equal probability of symbol transmission, the optimum decision thresholds can be determined in intersection of corresponding conditional probability density functions (PDFs). We can now use Eqs. (6.45), (6.56), and (6.57) to express the error probability as

$$P_e\{D_1|m_2\} = P_e\{D_2|m_1\} = Q\left(\sqrt{\frac{E_1 + E_2 - 2\rho_{12}}{2N_0}}\right). \tag{6.58}$$

The average bit error probability, for an equal probability transmission, is obtained as

$$P_b = \frac{1}{2}(P_e\{D_1|m_2\} + P_e\{D_2|m_1\}) = Q\left(\sqrt{\frac{E_1 + E_2 - 2\rho_{12}}{2N_0}}\right). \tag{6.58}$$

We now pay attention to three representative examples of binary communication systems known as antipodal signaling, orthogonal signaling, and unipolar signaling. The average symbol energy is given by $E_s = \sum_{k=1}^{M} E_k p_k$, which for equally likely transmission of binary symbols becomes $E_s = 0.5E_1 + 0.5E_2$. For binary transmission, the bit energy E_b is the same as average symbol energy, i.e., $E_b = E_s/\log_2 M = E_s$.

In *antipodal signaling*, such as BPSK, the transmitted symbols are $s_1(t) = +s(t)$ and $s_2(t) = -s(t)$. In this case, $E_1 = E_2 = E_b$, and $\rho_{12} = -E$, so that the average error probability is given by $P_e = Q\left(\sqrt{2E_b/N_0}\right)$.

In *orthogonal signaling*, such as FSK, we have that correlation coefficient $\rho_{12} = 0$ and $E_1 = E_2 = E$. In this case, the average probability of error probability is $P_e = Q\left(\sqrt{\frac{E}{N_0}}\right) = Q\left(\sqrt{\frac{E_b}{N_0}}\right)$, which is about 3 dB worse than in antipodal signaling.

In *unipolar signaling*, such as on-off keying, transmitted symbols are represented as $s_1(t) = s(t)$ and $s_2(t) = 0$. In this case, $E_1 = E$, $E_2 = 0$, $E_b = E/2$, and $\rho_{12} = 0$, so that the average error probability is given by $P_e = Q\left(\sqrt{\frac{E}{2N_0}}\right) = Q\left(\sqrt{\frac{E_b}{N_0}}\right)$, which is about 3 dB worse than in antipodal signaling.

Let us now establish the relationship between the average error probability and the Euclidean distance. From the definition of Euclidean distance, we have that

$$\begin{aligned} d_E^2(s_1, s_2) &= \int_0^T [s_1(t) - s_2(t)]^2 dt = \int_0^T s_1^2(t) - 2s_1(t)s_2(t) + s_2^2(t)dt \\ &= E_1 + E_2 - 2\rho_{12}, \end{aligned} \tag{6.59}$$

so that the average probability of error can be written as

$$P_b = Q\left(\frac{d_E(s_1, s_2)}{\sqrt{2N_0}}\right). \tag{6.60}$$

It can be shown that when $E_1 = E_2 = E$, $d_E(s_1, s_2)$ is maximized (while P_e is minimized) when $s_1(t) = -s_2(t)$, which corresponds to antipodal signaling. For antipodal signaling, $\rho_{12} = -E$, the Euclidean distance is $d_E(s_1, s_2) = 2\sqrt{E}$ and the corresponding error probability is $P_e = Q\left(\sqrt{2E/N_0}\right)$, which is consistent with the expression obtained above.

6.1.4.3 Calculation of Symbol Error Probability of M-ary QAM

From Chap. 5, we know that M-ary square-QAM signal constellation can be obtained as two-dimensional Cartesian product of $L = \sqrt{M}$-ary pulse amplitude modulation (PAM) signals:

$$X \times X = \{(x_1, x_2) | x_i \in X\}, X = \{(2i - 1 - L)d; i = 1, 2, \cdots, L\}. \tag{6.61}$$

The error probability for M-QAM can be expressed as

$$P_e = 1 - P_{e,\text{PAM}}^2, \tag{6.62}$$

wherein for in-phase and quadrature channels we have two independent PAM constellations. This assumption is correct for medium and high SNRs. The neighboring constellation points in PAM are separated by $2d$. In order to derive the expression of average PAM symbol error probability $P_{e,\text{ PAM}}$, we should first determine the probability of error for the following two symbols: (i) the edge symbol having only one neighbor and (ii) the inner symbol which has two neighbors. The inner symbol error probability can be expressed as

$$P_e(m_i) = 2P(\|z\| > d) = 2Q\left(\sqrt{2d^2/N_0}\right); i = 2, 3, \cdots, L - 1. \tag{6.63}$$

On the other hand, the outer symbol error probability is just half of the inner symbol error probability. Accordingly, the average symbol error probability of L-ary PAM can be expressed as

$$P_{e,\text{PAM}} = \frac{2}{L} Q\left(\sqrt{\frac{2d^2}{N_0}}\right) + \frac{L-2}{L} 2Q\left(\sqrt{\frac{2d^2}{N_0}}\right) = \frac{2(L-1)}{L} Q\left(\sqrt{\frac{2d^2}{N_0}}\right). \tag{6.64}$$

The average symbol energy of PAM constellations is given by

$$E_{s,\text{PAM}} = \frac{1}{L} \sum_{i=1}^{L} (2i - L)^2 d^2 = \frac{1}{3}(L^2 - 1)d^2. \tag{6.65}$$

By expressing d^2 in terms of the average energy, we can write the average symbol error probability of PAM as

$$\begin{aligned} P_{e,\text{PAM}} &= \frac{2(L-1)}{L} Q\left(\sqrt{\frac{6}{L^2 - 1} \frac{E_{s,\text{PAM}}}{N_0}}\right) \\ &= \frac{L-1}{L} \operatorname{erfc}\left(\sqrt{\frac{3}{L^2 - 1} \frac{E_{s,\text{PAM}}}{N_0}}\right). \end{aligned} \tag{6.66}$$

By assuming that all L amplitude levels of in-phase or quadrature channels are equally likely, the average symbol energy of M-ary QAM is simply

$$E_s = 2E_{s,\text{PAM}} = \frac{2}{3}(L^2 - 1)d^2 = \frac{2}{3}(M - 1)d^2. \tag{6.67}$$

Similarly, by expressing d^2 in terms of the average energy, we can write the average symbol error probability of QAM as

$$P_e = 1 - \left\{1 - 2\frac{\sqrt{M}-1}{\sqrt{M}} Q\left(\sqrt{\frac{3}{M-1}\frac{E_s}{N_0}}\right)\right\}^2 = 1 - \left\{1 - \frac{\sqrt{M}-1}{\sqrt{M}} \operatorname{erfc}\left(\sqrt{\frac{3}{2(M-1)}\frac{E_s}{N_0}}\right)\right\}^2. \tag{6.68}$$

For sufficiently high SNRs, we can neglect the $\operatorname{erfc}^2\{\cdot\}$ term in the previous equation so that we can write that

$$P_e \cong 2\frac{\sqrt{M}-1}{\sqrt{M}} \operatorname{erfc}\left(\sqrt{\frac{3}{2(M-1)}\frac{E_s}{N_0}}\right). \tag{6.69}$$

6.1.4.4 Union-Bound Approximation

Although the equations derived above allow us to accurately calculate average symbol error probabilities, it is quite challenging to find a closed form solution for large signal constellations. In these situations, we use the *union-bound* approximation [1, 4], which gives us an expression that is only function of Euclidean distances among signal constellation points. This approximation is accurate only for sufficiently high SNRs. Let A_{ik} denote the event that the observation vector $\boldsymbol{r}$ is closer to the signal vector $\boldsymbol{s}_k$ than to $\boldsymbol{s}_i$, when the symbol m_i (vector $\boldsymbol{s}_i$) is sent, which means that $\|\boldsymbol{r} - \boldsymbol{s}_k\| < \|\boldsymbol{r} - \boldsymbol{s}_i\|$. The constellation point $\boldsymbol{s}_i$ is detected correctly if $\|\boldsymbol{r} - \boldsymbol{s}_i\| < \|\boldsymbol{r} - \boldsymbol{s}_k\| \; \forall\, k \neq i$, which means that the average error probability can be upper bounded as

$$P_e(m_i \text{ sent}) = P\left(\bigcup_{k=1, k\neq i}^{M} A_{ik}\right) \leq \sum_{k=1, k\neq i}^{M} P(A_{ik}). \tag{6.69}$$

The probability of event A_{ik} can be evaluated as

$$P(A_{ik}) = P(\|\boldsymbol{r} - \boldsymbol{s}_k\| < \|\boldsymbol{r} - \boldsymbol{s}_i\| | m_i) = P(\|\boldsymbol{s}_i - \boldsymbol{s}_k + \boldsymbol{z}\| < \|\boldsymbol{z}\|). \tag{6.70}$$

In other words, $P(A_{ik})$ corresponds to the probability that the noise vector is closer to the vector $\boldsymbol{s}_i$-$\boldsymbol{s}_k$ than to the origin. Given the fact that the ith noise component is a zero-mean Gaussian random variable of variance $N_0/2$, what matters is just a projection n of the noise vector on line containing vector $\boldsymbol{s}_i$-$\boldsymbol{s}_k$. This projection has zero mean and variance equal to $N_0/2$, so that $P(A_{ik})$ can be expressed as

$$\begin{aligned} P(A_{ik}) &= P\left(z > \underbrace{\|\boldsymbol{s}_i - \boldsymbol{s}_k\|}_{d_{ik}}/2\right) = \int_{d_{ik}/2}^{\infty} \frac{1}{\sqrt{\pi N_0}} e^{-u^2/2} du \\ &= Q\left(\frac{d_{ik}}{\sqrt{2N_0}}\right) = \frac{1}{2}\operatorname{erfc}\left(\frac{d_{ik}}{2\sqrt{N_0}}\right). \end{aligned} \tag{6.71}$$

The error probability related to symbol m_i can be now upper bounded as

$$P_e(m_i) < \frac{1}{2} \sum_{k=1, k\neq i}^{M} \operatorname{erfc}\left(\frac{d_{ik}}{2\sqrt{N_0}}\right); i = 1, \cdots, M. \tag{6.72}$$

Finally, the average symbol error probability is upper bounded as

$$P_e = \sum_{i=1}^{M} p_i P_e(m_i) \leq \frac{1}{2} \sum_{i=1}^{M} p_i \sum_{k=1, k\neq i}^{M} \operatorname{erfc}\left(\frac{d_{ik}}{2\sqrt{N_0}}\right), \tag{6.73}$$

and this inequality is known as *union bound*. For an equal probability transmission of symbols, $p_i = 1/M$, the union bound simplifies to

$$P_e \le \frac{1}{2M}\sum_{i=1}^{M}\sum_{k=1,k\neq i}^{M}\text{erfc}\left(\frac{d_{ik}}{2\sqrt{N_0}}\right). \tag{6.74}$$

By defining the minimum distance of constellation as $d_{\min} = \min_{i,\ k} d_{ik}$, we can obtain a looser bound as

$$P_e \le \frac{M-1}{2}\text{erfc}\left(\frac{d_{\min}}{2\sqrt{N_0}}\right). \tag{6.75}$$

For circularly symmetric signal constellation, we can use the following bound:

$$P_e \le \frac{1}{2}\sum_{k=1,k\neq i}^{M}\text{erfc}\left(\frac{d_{ik}}{2\sqrt{N_0}}\right). \tag{6.76}$$

Finally, the *nearest neighbor approximation* is given as

$$P_e \approx \frac{M_{d_{\min}}}{2}\text{erfc}\left(\frac{d_{\min}}{2\sqrt{N_0}}\right), \tag{6.77}$$

where $M_{d_{\min}}$ is the number of neighbors at distance d_{min} from observed constellation point. Regarding the bit error probability P_b, when Gray-like mapping rule [1] is used for sufficiently high SNRs, we can use the following approximation:

$$P_b \approx P_e / \log_2 M. \tag{6.78}$$

If we would like to compare various signal constellation sizes, the bit energy-to-power spectral density ratio is typically used, E_b/N_0. In *optical communications*, very often we use the optical SNR (OSNR), introduced by

$$OSNR = \frac{P_s}{P_{\text{ASE}}} = \frac{P_s}{2n_{sp}h\nu(G-1)B_{op}}, \tag{6.79}$$

instead of electrical SNR. In Eq. (6.79), P_s denotes the signal power, while P_{ASE} is the amplified spontaneous emission (ASE) noise power determined by $P_{ASE} = 2n_{sp}h\nu(G-1)B_{op}$, where n_{sp} is the spontaneous emission factor of an erbium-doped fiber amplifier (EDFA), G is the EDFA gain, B_{op} is the bandwidth of optical passband filter that follows EDFA, and $h\nu$ is a single photon energy. To compare different modulation schemes with different signal constellation sizes, it is convenient to introduce OSNR per bit and per a single-polarization state and define it as [11]

$$OSNR_b = \frac{OSNR}{\log_2 M} = \frac{R_s}{2B_{\text{op}}}SNR_b, \tag{6.80}$$

where M is the signal constellation size, R_s is the symbol rate, and B_{op} is the referent optical bandwidth (it is commonly assumed that $B_{\text{op}} = 12.5$ GHz, which corresponds to a 0.1 nm spectral width around a carrier wavelength of 1550 nm). The parameter $\text{SNR}_\text{b} = E_\text{b}/N_0$ denotes the signal-to-noise ratio per bit of information, where E_b is the bit energy and N_0 is a power spectral density (PSD) originating from ASE noise. For convenience, it is assumed that photodiode responsivity is 1 A/W so that $N_0 = N_{\text{ASE}}$, where N_{ASE} is the PSD of ASE noise in a single-polarization state. The OSNR per single polarization is commonly defined as $OSNR = E/(2N_{\text{ASE}}B_{\text{op}})$, where E is the symbol energy. Additionally, in optical communications, it is very common to use the Q-factor as the figure of merit instead of SNR (that has nothing to do with resonators'/oscillators' Q-factor), which is related to the bit-error rate (BER) on an AWGN as follows [11]:

$$BER = \frac{1}{2}\text{erfc}\left(\frac{Q}{\sqrt{2}}\right). \tag{6.81}$$

Finally, for a binary modulation scheme, the Q-factor in optical communications is defined as

$$Q = \frac{|\mu_2 - \mu_1|}{\sigma_2 + \sigma_1}, \tag{6.82}$$

where μ_i and σ_i denote the mean value and standard deviation of symbol s_i (message m_i), wherein $i = 1, 2$.

6.1.5 Estimation Theory Fundamentals

So far, our focus was on detection of signal in the presence of additive noise, such as ASE or Gaussian noise. In this section, we will pay attention to the estimation of a certain signal parameter, such as phase, frequency offset, and so on, important from the detection point of view. Let the received signal be denoted by $r(t)$ while the transmitted signal is $s(t,p)$, where p is the parameter to be estimated. The received signal in the presence of additive noise $z(t)$ can be written as

$$r(t) = s(t,p) + z(t),\ 0 \le t \le T, \tag{6.83}$$

where T is the observation interval (symbol duration for modulation schemes). The operation of assigning a value $\widehat{p}$ to an unknown parameter p is known as parameter estimation; the value assigned is called an estimate, and the algorithm used to perform this operation is called an estimator.

There are different criteria used to evaluate the estimator, including minimum mean-square (MMS) estimate, maximum a posteriori probability (MAP) estimate, and maximum likelihood (ML) estimate. We can define the cost function in MMS estimate (MMSE) as an integral of quadratic form [2]

$$C_{\mathrm{MMSE}}(\varepsilon) = \int \underbrace{(p - \widehat{p})^2}_{\varepsilon^2} f_P(p|\mathbf{r})dp, \tag{6.84}$$

where $\varepsilon = \widehat{p} - p$, while $f_P(p|\mathbf{r})$ is the *a posteriori* probability density function (PDF) of random variable P. The observation vector $\mathbf{r}$ in Eq. (6.84) is defined by its r_i components, where the ith component is determined by

$$r_i = \int_0^T r(t)\Phi_i(t)dt;\ i = 1, 2, \cdots, N \tag{6.85}$$

$\{\Phi_i(t)\}$ denotes the set of orthonormal basis functions. The estimate that minimizes the average cost function given by Eq. (6.84) is known as Bayes' MMS estimate [2, 3]. The estimate that maximizes the a posteriori probability density function $f_P(p|\mathbf{r})$ is known as the MAP estimate. Finally, the estimate which maximizes the conditional PDF $f_r(\mathbf{r}|p)$, which is also known as the likelihood function, is called the ML estimate. The MAP estimate can be obtained as a special case of Bayes' estimation by using the uniform cost function $C(\varepsilon) = (1/\delta)[1\text{-rect}(\varepsilon/\delta)]$ as follows:

$$\begin{aligned}
\langle C(\varepsilon)\rangle &= \int C(\varepsilon)\, f_P(p|\mathbf{r})dp = \frac{1}{\delta}\left[\int_{-\infty}^{\hat{p}-\delta/2} + \int_{\hat{p}+\delta/2}^{\infty}\right] f_P(p|\mathbf{r})dp \\
&= \frac{1}{\delta}\left[1 - \int_{\hat{p}-\delta/2}^{\hat{p}+\delta/2} f_P(p|\mathbf{r})dp\right] \cong (1/\delta)[1 - \delta\, f_P(p|\mathbf{r})] \\
&= \frac{1}{\delta} - f_P(p|\mathbf{r})
\end{aligned} \tag{6.86}$$

where the approximation above is valid for sufficiently small δ. The quality of an estimate is typically evaluated in terms of expected mean value $\langle\widehat{p}\rangle$ and variance of the estimation error. We can recognize several cases with respect to the expected value. If the expected value of the estimate equals the true value of the parameter, $\langle\widehat{p}\rangle = p$, we say that the estimate is *unbiased*. If the expected value of estimate differs from true parameter by fixed value b, i.e., if $\langle\widehat{p}\rangle = p + b$, we say that the estimate has a *known bias*. Finally, if the expected value of estimate is different from true parameter by variable amount $b(p)$, i.e., if $\langle \widehat{p} \rangle = p + b(p)$, we say that the estimate has a *variable bias*. The estimation error ε is defined as the difference between true value p and estimate $\widehat{p}$, namely, $\varepsilon = p - \widehat{p}$. A good estimate is one that simultaneously provides a small bias and a small variance of estimation error. The lower bound on the variance of estimation error can be obtained from the Cramér-Rao inequality [1, 2] which is

$$\left\langle (p - \widehat{p})^2 \right\rangle \geq \frac{-1}{\left\langle \frac{\partial^2 \log f_r(\mathbf{r}|p)}{\partial p^2} \right\rangle}, \tag{6.87}$$

where we assumed that the second partial derivative of conditional PDF exists and that it is absolutely integrable. Any estimate that satisfies the Cramér-Rao bound defined by the right side of Eq. (6.87) is known as an *efficient estimate*.

In order to explain the ML estimation, we can assume that the transmitted signal $s(t,p)$ can also be presented as a vector $\mathbf{s}$ in signal space, with the nth component given as

$$s_n(p) = \int_0^T s(t,p)\Phi_n(t)dt; n = 1, 2, \cdots, N. \tag{6.88}$$

The noise vector $\mathbf{z}$ can be presented in a similar fashion

$$\mathbf{z} = \begin{bmatrix} z_1 \\ z_2 \\ \vdots \\ z_N \end{bmatrix}, \quad z_n = \int_0^T z(t)\Phi_n(t)dt; n = 1, 2, \cdots, N. \tag{6.89}$$

Since the components z_n of the noise are zero-mean Gaussian with PSD equal to $N_0/2$, the components of observation vector will also be Gaussian with mean values $s_n(p)$. The joint conditional PDF for a given parameter p will be

$$f_{\mathbf{R}}(\mathbf{r}|p) = \frac{1}{(\pi N_0)^{N/2}} \prod_{n=1}^{N} e^{-[r_n - s_n(p)]^2/N_0}. \tag{6.90}$$

The likelihood function can be defined as

$$L[r(t), p] = \lim_{N\to\infty} \frac{f_r(\mathbf{r}|p)}{f_r(\mathbf{r})} = \frac{2}{N_0}\int_0^T r(t)s(t,p)dt - \frac{1}{N_0}\int_0^T s^2(t,p)dt. \tag{6.91}$$

The ML estimate is derived by maximizing the likelihood function, which is done by differentiating the likelihood function with respect to the estimate $\widehat{p}$ and setting the result to zero, so we obtain [3]

$$\int_T^0 [r(t) - s(t,\hat{p})] \frac{\partial s(t,\hat{p})}{\partial \hat{p}} dt = 0, \tag{6.92}$$

and this condition is known as the *likelihood equation*.

Example. Let us consider the following sinusoidal signal whose amplitude a and frequency ω are known, while the phase is unknown: a cos(ωt+θ). Based on the likelihood equation (6.92), we obtain the following ML estimate of the phase in the presence of AWGN:

$$\hat{\theta} = \tan^{-1}\left[\frac{\int_0^T r(t)\cos(\omega t)dt}{\int_0^T r(t)\sin(\omega t)dt}\right].$$

6.2 Wireless Communication Systems Performance

In a fading environment, as shown in Chap. 2, the envelope of the signal varies with the propagation distance as the environment and consequently multipath and shadowing effects vary. Accordingly, the signal-to-noise ratio ρ also varies with propagation distance, and these variations can be described with the probability density function $f(\rho)$. Given that SNR becomes random variable, the symbol error probability $P_s(\rho)$ becomes random [5, 8, 14, 15] as well, and we can define different scenarios to characterize these variations:

(i) The *outage probability*, denoted as P_{outage}, represents the probability that SNR falls below a given threshold value ρ_{tsh}, corresponding to the maximum tolerable symbol error probability.
(ii) The *average error probability* corresponds to the symbol error probability averaged over SNR distribution.
(iii) *Combined outage and average error probability scenario*, when shadowing and multipath fading effects are simultaneously present.

6.2.1 Outage Probability Scenario

The outage probability scenario is applicable when the symbol duration T_s is comparable to the coherence time T_c so that the signal fade level is roughly constant for the duration of the symbol. As indicated above, the outage probability P_{outage} represents the probability that SNR falls below a given threshold value ρ_{tsh}:

$$P_{\text{outage}} = P(\rho \le \rho_{\text{tsh}}) = \int_0^{\rho_{\text{tsh}}} f(\rho)d\rho, \tag{6.93}$$

where $f(\rho)$ is the PDF of SNR ρ. For Rayleigh fading, the corresponding outage probability becomes

$$P_{\text{outage}} = P(\rho \le \rho_{\text{tsh}}) = \int_0^{\rho_{\text{tsh}}} \frac{1}{\overline{\rho}} e^{-\rho/\overline{\rho}} d\rho = 1 - e^{-\rho_{\text{tsh}}/\overline{\rho}}. \tag{6.94}$$

The required SNR to achieve a target outage probability P_{outage} is given by

$$\overline{\rho} = \frac{\rho_{\text{tsh}}}{-\log_{10}(1 - P_{\text{outage}})}. \tag{6.95}$$

From (6.95), we can determine the *fade margin* (FM) defined as SNR increase required to maintain acceptable performance more than $(1-P_{\text{outage}}) \cdot 100\%$ of the time as follows:

$$FM = -10 \log_{10}\left[-\log_{10}(1 - P_{\text{outage}})\right]. \tag{6.96}$$

6.2.2 Average Error Probability Scenario

The average error probability scenario is applicable when the symbol duration T is comparable to wireless channel coherence time T_c, and it can be computed by averaging the AWGN error probability for different fading realizations; in other words,

$$\overline{P}_s = \langle P_s(\rho) \rangle = \int_0^{\infty} P_s(\rho) f(\rho) d\rho, \tag{6.97}$$

where $P_s(\rho)$ is the AWGN symbol error probability, with ρ being the symbol SNR. The generic expression applicable to many 2-D and 1-D signal constellations is given by

$$P_s(\rho) \simeq C_M Q\left(\sqrt{D_M \rho}\right), \tag{6.98}$$

where C_M and D_M are the modulation format-dependent parameters, with M being the signal constellation size. For instance, for QPSK, $C_4 = 2$ and $D_4 = 1$. For M-PSK, the corresponding parameters are $C_M = 2$ and $D_M = 2\sin^2(\pi/M)$. On the other hand, the corresponding parameters for M-QAM are $C_M = 4(1\text{-}1/\sqrt{M})$ and $D_M = 3/(M\text{-}1)$. Finally, for M-PAM, the corresponding parameters are $C_M = 2(M\text{-}1)/M$ and $D_M = 6/(M^2\text{-}1)$. For Gray mapping rule, to estimate the bit error probability, we need to divide symbol error probability by $\log_2 M$. Since symbol SNR ρ and bit SNR ρ_b are related by $\rho = \rho_b \log_2 M$, the corresponding expression for bit error probability will be

$$P_b(\rho_b) \approx \frac{C_M}{\log_2 M} Q\left(\sqrt{D_M \rho_b \log_2 M}\right). \tag{6.99}$$

For Rayleigh fading, the expression for average symbol error probability can be written as

$$\overline{P}_s = \langle P_s(\rho)\rangle \simeq \int_0^\infty C_M Q\left(\sqrt{D_M \rho}\right) \frac{1}{\overline{\rho}} e^{-\rho/\overline{\rho}} d\rho \simeq \frac{C_M}{2}\left[1 - \sqrt{\frac{D_M \overline{\rho}/2}{1 + D_M \overline{\rho}/2}}\right]. \tag{6.100}$$

For very high SNR, the following asymptotic expression can be used: $\overline{P}_s \underset{\rho \to \infty}{\to} C_M/(2D_M\overline{\rho})$.

As an illustration, we show in Fig. 6.5 bit error probability vs. average SNR per bit for different constellation sizes for AWGN and Rayleigh fading channels, by employing approximate formulas (6.98) for an AWGN channel and (6.100) for a Rayleigh fading channel. Clearly, significant degradation in P_b can be found in the presence of Rayleigh fading.

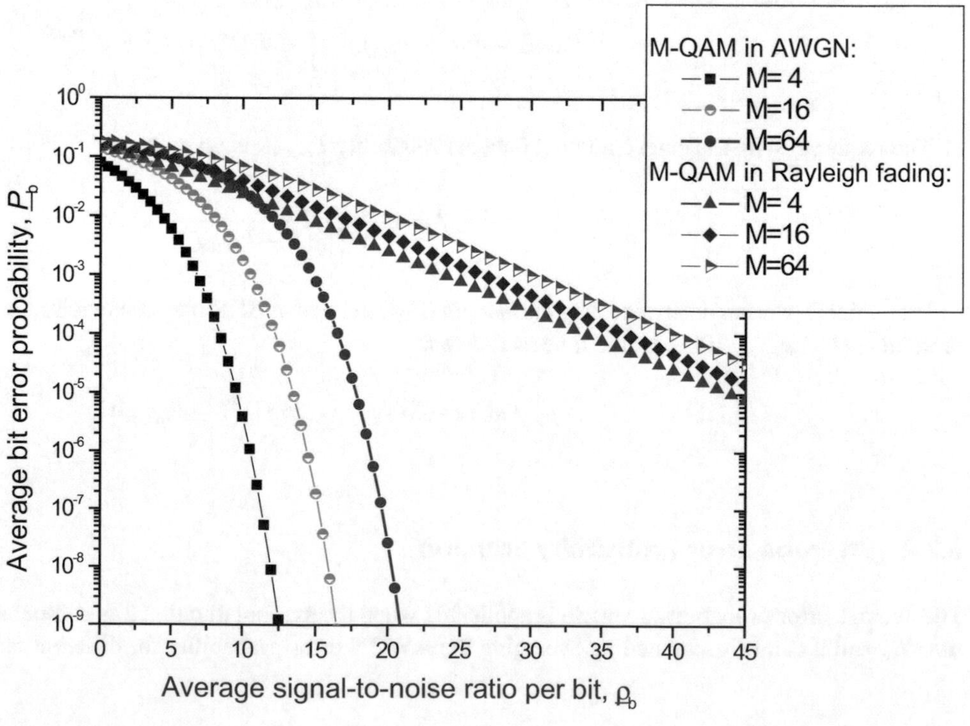

Fig. 6.5 Average bit error probability vs. average SNR per bit for AWGN and Rayleigh fading channels

6.2.3 Combined Outage and Average Error Probability Scenario

The combined outage and average error probability scenario is applicable when the fading environment can be represented as a superposition of both fast fading (shadowing effects) and slow fading (such as Rayleigh fading). In this scenario, we can define the average error probability with some probability $1 - P_{\text{outage}}$. Therefore, when not in outage, in other words, the symbol SNR is averaged over fast fading for fixed path loss and random shadowing $\overline{\rho} \geq \overline{\rho}_{\text{tsh}}$ (where $\overline{\rho}_{\text{tsh}}$ is the average threshold SNR), the average probability of error is obtained by averaging over the conditional PDF of the fast fading as follows:

$$\overline{P}_s = \int_0^\infty P_s(\rho) f(\rho|\overline{\rho}) d\rho. \tag{6.101}$$

The average threshold SNR $\overline{\rho}_{\text{tsh}}$ is obtained from Eq. (6.101) by setting $\overline{P}_s = \overline{P}_{s,\text{tsh}}$, where $\overline{P}_{s,\text{tsh}}$ is the maximum tolerable average symbol error probability, and solving for $\overline{\rho}$.

6.2.4 Moment-Generating Function (MGF)-Based Approach to Average Error Probability Calculation

The moment-generating function (MGF) for a nonnegative random variable ρ with PDF $f_\rho(\rho)$ is defined as

$$MGF_\rho(s) = \int_0^\infty e^{s\rho} f_\rho(\rho) d\rho. \tag{6.102}$$

Clearly, the MGF is related to the Laplace transform (LT) of PDF by

$$LT\{ f_\rho(\rho)\} = MGF_\rho(-s).$$

As the name suggests, the MGF is related to the moments $\langle \rho^n \rangle$ by

$$\langle \rho^n \rangle = \frac{\partial^n}{\partial s^n} MGF_\rho(s) \bigg|_{s=0}. \tag{6.103}$$

The MGF for Nakagami-m fading is given by [15]

$$MGF_\rho(s) = \frac{1}{(1 - s\overline{\rho}/m)^m}. \tag{6.104}$$

Rayleigh fading is a special case of Nakagami-m fading with $m = 1$. On the other hand, the MGF for Rician fading with Rician factor K is given by [15]

$$MGF_\rho(s) = \frac{1+K}{1+K-s\overline{\rho}} \exp\left[\frac{Ks\overline{\rho}}{1+K-s\overline{\rho}}\right]. \tag{6.105}$$

As an illustration of this method, let us consider symbol error probability given by Eq. (6.98). For convenience, let us use an alternative definition of Q-function [14, 15]:

$$Q(z) = \frac{1}{\pi} \int_0^{\pi/2} \exp\left[-\frac{z^2}{2\sin^2\phi}\right] d\phi. \tag{6.106}$$

Equation (6.98) can be now rewritten as

$$P_s(\rho) \simeq C_M Q\left(\sqrt{D_M \rho}\right) = \frac{C_M}{\pi} \int_0^{\pi/2} \exp\left[-\frac{D_M \rho}{2\sin^2\phi}\right] d\phi. \tag{6.107}$$

The average error probability in a fading environment will be then

$$\begin{aligned} \overline{P}_s &\cong \int_0^{\infty} C_M Q(\sqrt{D_M \rho}) f(\rho) d\rho \\ &= \frac{C_M}{\pi} \int_0^{\pi/2} \underbrace{\int_0^{\infty} \left\{ \exp\left[-\frac{D_M \rho}{2\sin^2\phi}\right] f(\rho) d\rho \right\}}_{MGF_\rho\left[-\frac{D_M}{2\sin^2\phi}\right]} d\phi \Rightarrow \\ &= \frac{C_M}{\pi} \int_0^{\pi/2} MGF_\rho\left[-\frac{D_M}{2\sin^2\phi}\right] d\phi. \end{aligned} \tag{6.108}$$

Equation (6.108) still requires numerical integration, but the corresponding integral is easier to evaluate.

6.2.5 Performance Evaluation in the Presence of Doppler Spread and Fading

When differential detection is used, the Doppler spread introduces an error floor phenomenon caused by the correlation of symbols, which is a function of Doppler frequency ν_D and symbol duration T. The Doppler spread is accounted by the correlation coefficient for different Doppler power spectrum models, defined as $\rho_C = R(T)/R(0)$, and for the Clark-Jakes model, described in Chap. 2, we have that

$$\rho_C = R(T)/R(0) = J_0(2\pi B_D T), \tag{6.109}$$

where B_D is the Doppler spread.

The average probability of error for DPSK in the presence of Doppler spread and Rician fading is given by

$$\overline{P}_b^{(\mathrm{DPSK})} = \frac{1}{2} \frac{1 + K + \overline{\rho}_b(1 - \rho_C)}{1 + K + \overline{\rho}_b} \exp\left[-\frac{K\overline{\rho}_b}{1 + K + \overline{\rho}_b}\right]. \tag{6.110}$$

For Rayleigh fading, the corresponding average probability of error is obtained from Eq. (6.110), by setting $K = 0$ to obtain $\overline{P}_b = \frac{1}{2} \frac{1+\overline{\rho}_b(1-\rho_C)}{1+\overline{\rho}_b}$. The error floor, in Rician fading, can be estimated as the following limit:

$$\overline{P}_{b,\mathrm{floor}}^{(\mathrm{DPSK})} = \overline{P}_b^{(\mathrm{DPSK})}\Big|_{\rho\to\infty} = \frac{(1 - \rho_C)e^{-K}}{2}. \tag{6.111}$$

The average probability of error for DQPSK in the presence of Doppler spread and in Rician fading is upper bounded by

$$\begin{aligned} \overline{P}_b^{(\mathrm{DQPSK})} &\leq \frac{1}{2}\left(1 - \sqrt{\frac{(\rho_C \overline{\rho}/\sqrt{2})^2}{(1 + \overline{\rho})^2 - (\rho_C \overline{\rho}/\sqrt{2})^2}}\right) \cdot \\ &\exp\left[-\frac{(2 - \sqrt{2})K\overline{\rho}/2}{(1 + \overline{\rho}_s) - (\rho_C \overline{\rho}/\sqrt{2})}\right]. \end{aligned} \tag{6.112}$$

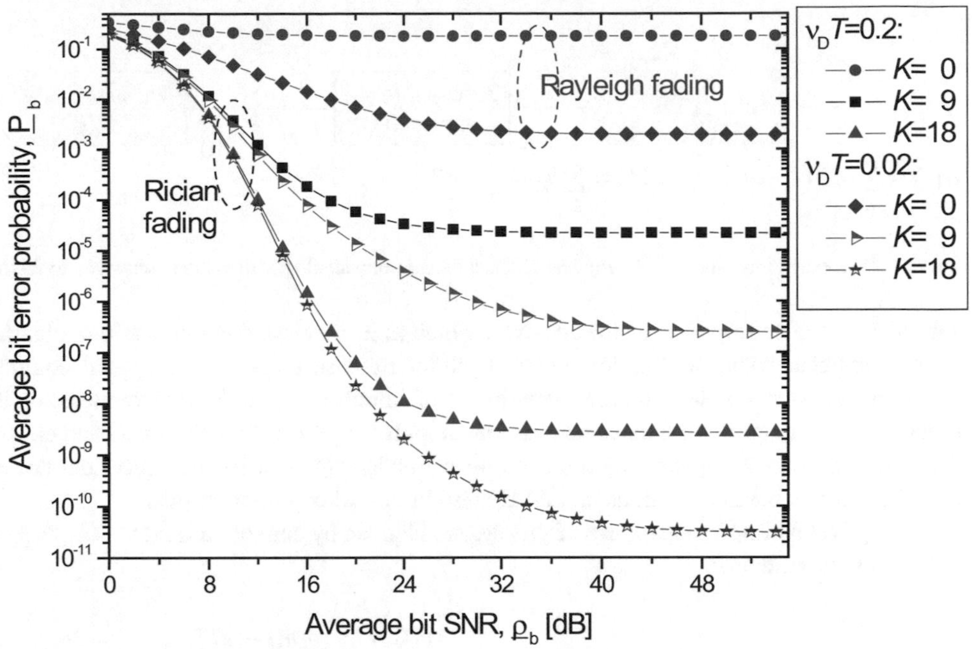

Fig. 6.6 Average bit error probability vs. average bit SNR in DPSK for different Rician parameters and different Doppler frequencies under the Clark-Jakes (uniform scattering) model

The corresponding error floor is obtained as the following limit:

$$\overline{P}_{b,\text{floor}}^{(\text{DQPSK})}=\overline{P}_{b}^{(\text{DQPSK})}\Big|_{\rho\to\infty}=\frac{1}{2}\left(1-\sqrt{\frac{\left(\rho_C/\sqrt{2}\right)^2}{1-\left(\rho_C/\sqrt{2}\right)^2}}\right)\exp\left[-\frac{(2-\sqrt{2})K/2}{1-\left(\rho_C/\sqrt{2}\right)}\right]. \tag{6.113}$$

The DPSK error floor for the Clark-Jakes (uniform scattering) model in the presence of Doppler spread and Rayleigh fading can be obtained upon substitution of Eq. (6.109) into Eq. (6.113) as follows:

$$\overline{P}_{b,\text{floor}}^{(\text{DQPSK})}=\frac{1-J_0(2\pi\nu_D T)}{2}, \tag{6.114}$$

where ν_D is the Doppler frequency.

As an illustration, in Fig. 6.6, we provide the average DPSK bit error probabilities for different Rician fading parameters $K\in\{0, 9, 18\}$ and normalized Doppler frequencies $\nu_D T\in\{0.2, 0.02\}$ under the Clark-Jakes (uniform scattering) model. Clearly, in the presence of Rayleigh fading and Doppler spread, the DPSK system is facing significant performance degradation and very strong error floor phenomenon.

6.3 Channel Equalization Techniques

In this section, we describe various equalization techniques suitable to deal with different channel impairments in both wireless and optical communication systems.

6.3.1 ISI-Free Digital Transmission and Partial-Response Signaling

We will start with the equivalent wireless/optical channel model by observing only one polarization state (either x- or y-polarization state), which is shown in Fig. 6.7. The function $h_{\text{Tx}}(t)$ denotes the equivalent impulse response of transmitter, which is in wireless communications the convolution of pulse shaper and power amplifier. In optical communications, $h_{\text{Tx}}(t)$ is obtained as the convolution of pulse shaper impulse response, driver amplifier, and Mach-Zehnder modulator (MZM) impulse responses. In the same figure, $h_c(t)$ denotes the impulse response of wireless/optical channel. $h_{\text{Rx}}(t)$ denotes the

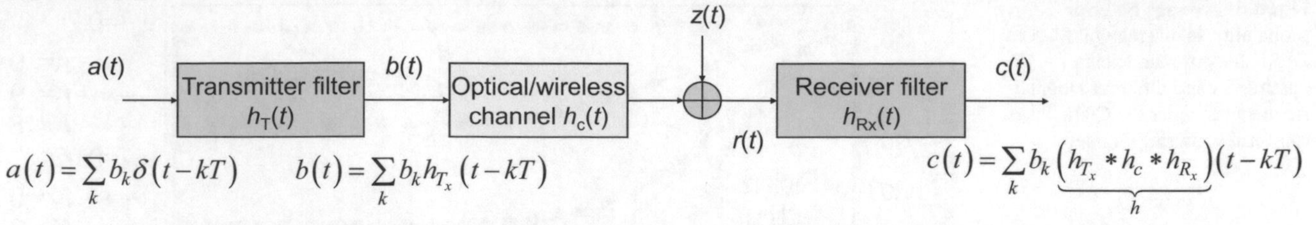

Fig. 6.7 Equivalent low-pass wireless/optical channel model (only single polarization is observed to facilitate explanations)

equivalent impulse response of the receiver, which is in wireless communications obtained as the convolution of receive antenna impulse response and low-noise amplifier impulse response. In optical communications with coherent optical detection, $h_{Rx}(t)$ is obtained as the convolution of impulse responses of photodetector impulse response, front end with transimpedance amplifier impulse response, and impulse responses of both optical and electrical filters that are employed. The electrical filter can be replaced by matched filter. Finally, $z(t)$ denotes the equivalent noise process dominated by Gaussian noise in wireless communications and ASE noise in optical communications.

The equivalent impulse response of this system is given by convolution $h(t) = (h_{Tx} * h_c * h_{Rx})(t)$, so that the output of the system can be written as

$$c(t) = \sum_k b_k h(t - kT), \tag{6.115}$$

where b_k is the transmitted sequence of symbols, while T denotes the symbol duration. By sampling the system output, we obtain

$$\begin{aligned} c(mT) &= \sum_k b_k h(mT - kT) = b_m h(0) + \sum_{k \neq m} b_k h(mT - kT) \\ &= b_m h(0) + \underbrace{\sum_{k \neq 0} b_k h(kT)}_{\text{ISI term}}, \end{aligned} \tag{6.116}$$

where the term $b_m h(0)$ corresponds to the transmitted symbol, while the second term represents the intersymbol interference (ISI) term. The second term can disappear if ISI-free optical transmission is achieved, which is possible when transmitter and receiver are properly designed such that the equivalent impulse response satisfies the following condition:

$$h(kT) = \begin{cases} h_0 = 1, k = 0 \\ 0, k \neq 0 \end{cases}. \tag{6.117}$$

If the frequency response of equivalent system function $H(f) = \text{FT}[h(t)]$, where FT stands for the Fourier transform, satisfies the following condition

$$\sum_{k=-\infty}^{\infty} H\left(f + \frac{k}{T}\right) = T, \tag{6.118}$$

then there will not be ISI at sampling time instances. This claim is known as *Nyquist's criterion for zero-ISI* and can be used in the design of all channel equalizers.

The above claim is quite straightforward to prove. Let us start with inverse Fourier transform definition and let us split the integration interval into frequency bands of bandwidth $1/T$, so we have that

$$h(t) = FT^{-1}\{H(f)\} = \int_{-\infty}^{\infty} H(f) e^{j2\pi ft} df = \sum_{k=-\infty}^{\infty} \int_{(2k-1)/2T}^{(2k+1)/2T} H(f) e^{j2\pi ft} df. \tag{6.119}$$

By sampling the equivalent impulse response from Eq. (6.119), we obtain

$$h(mT) = \sum_{k=-\infty}^{\infty} \int_{(2k-1)/2T}^{(2k+1)/2T} H(f) e^{j2\pi fmT} df. \tag{6.120}$$

Now, by substituting $f' = f - k/T$, Eq. (6.120) becomes

$$\begin{aligned} h(mT) &= \sum_{k=-\infty}^{\infty} \int_{-1/2T}^{1/2T} H(f' + k/T) e^{j2\pi f' mT} df' = \int_{-1/2T}^{1/2T} \overbrace{\sum_{k=-\infty}^{\infty} H(f' + k/T)}^{T} e^{j2\pi f' mT} df' \\ &= \int_{-1/2T}^{1/2T} T e^{j2\pi f' mT} df' = \frac{\sin(m\pi)}{m\pi} = \begin{cases} 1 & m = 0 \\ 0 & m \neq 0 \end{cases}. \end{aligned} \tag{6.121}$$

We can see from Eq. (6.121) that the ISI term disappeared, which proves that Nyquist's criterion indeed yields to zero-ISI among symbols. The channel with frequency response satisfying Eq. (6.118) with a single term is known as the *ideal Nyquist channel*. The transfer function of the ideal Nyquist channel has a rectangular form given as $H(f) = (1/R_s)\mathrm{rect}(f/R_s)$ (R_s is the symbol rate defined as $R_s = 1/T$), while the corresponding impulse response is given as

$$h(t) = \frac{\sin(\pi t/T)}{\pi t/T} = \mathrm{sinc}(t/T), \tag{6.122}$$

where the sinc-function is defined as $\mathrm{sinc}(u) = \sin(\pi u)/(\pi u)$.

The ideal rectangular frequency response has sharp transition edges, which is difficult to implement in practice. The frequency response (spectrum) satisfying Eq. (6.118) with additional terms ($k = -1, 0, 1$) is known as the *raised-cosine* (RC) spectrum given by

$$H_{RC}(f) = \begin{cases} T, & 0 \leq |f| \leq (1-\beta)/2T \\ \frac{T}{2}\{1 - \sin[\pi T(|f| - \frac{1}{2T})/\beta]\}, & (1-\beta)/2T \leq |f| \leq (1+\beta)/2T \end{cases} \tag{6.123}$$

The impulse response of the raised-cosine spectrum is given as

$$h_{RC}(t) = \frac{\sin(\pi t/T)}{\pi t/T} \frac{\cos(\beta\pi t/T)}{1 - 4\beta^2 t^2/T^2}. \tag{6.124}$$

The parameter β from Eqs. (6.123) and (6.124) is known as the *roll-off factor* and determines the bandwidth required to accommodate the raised-cosine spectrum, which is given by $BW = (1+\beta)/(2T)$. The raised-cosine spectrum reduces to the ideal Nyquist channel if $\beta = 0$. It is evident from the raised cosine filter impulse response that in addition to zero-crossings at $\pm T$, $\pm 2T$, . . ., there are additional zero-crossings at points $\pm 3T/2$, $\pm 5T/2$, ... The RC impulse responses and spectra for several roll-off factors are illustrated in Fig. 6.8.

In fiber-optics communications, when different wavelength responses are sinc-functions, the corresponding scheme is commonly referred to as the Nyquist-WDM scheme. In this scheme, the pulse shapes in the time domain are rectangular. When the Nyquist-WDM channel responses are of raised-cosine type in wavelength domain, the corresponding WDM channel spacing will be $(1+\beta)R_s$.

In Sect. 6.1, we studied different approaches to minimize symbol error probability and learned that employment of the matched filter is optimum for Gaussian noise-dominated channels. On the other hand, in this section, we learned that for ISI-free transmission, the receive filter should be chosen in such a way that the overall system function has the raised-cosine

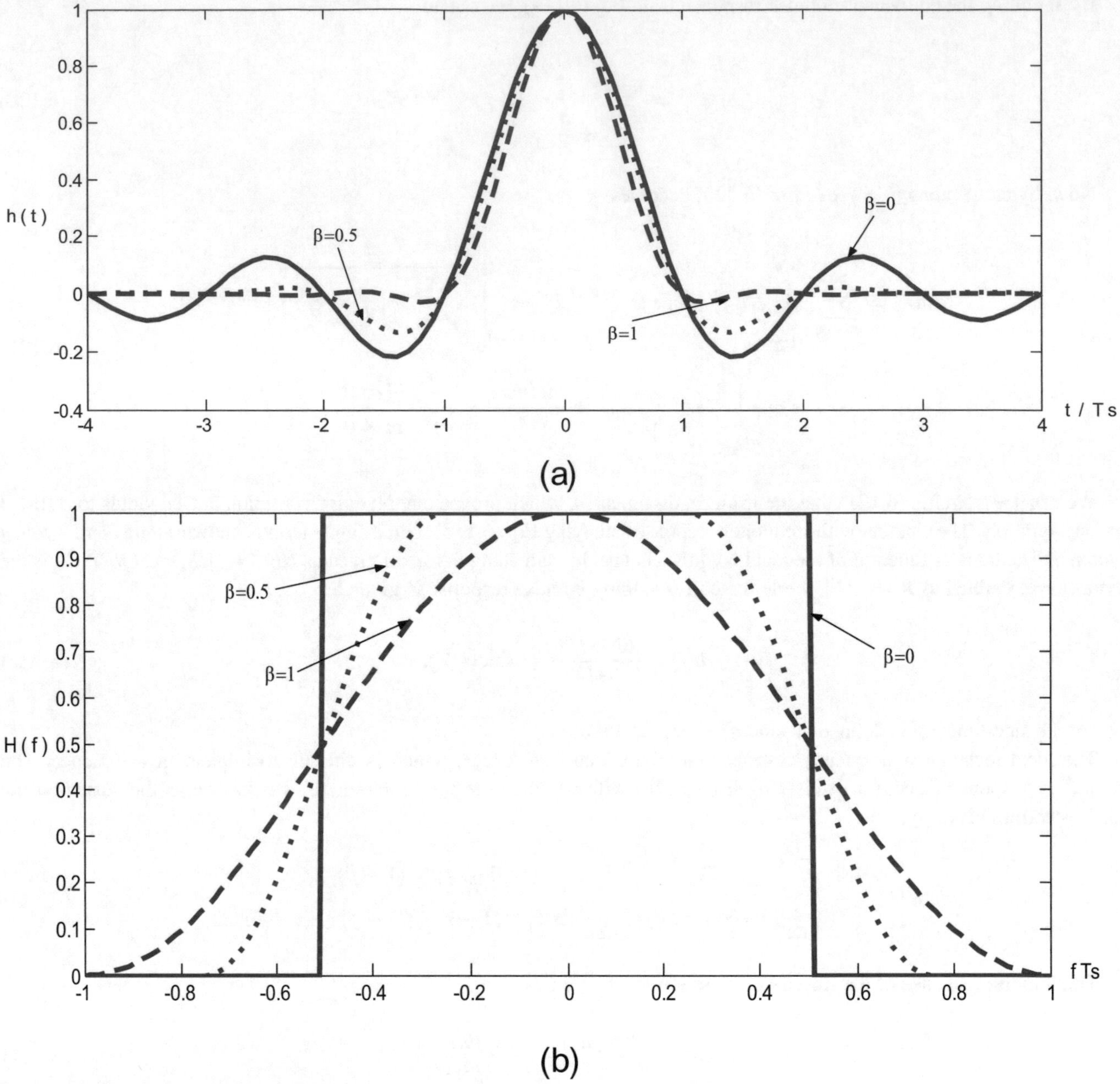

Fig. 6.8 Raised-cosine function: (**a**) impulse response and (**b**) frequency response

form. By combining these two approaches, we can come up with a near-optimum transmit and receive filters. Namely, from the zero-ISI condition, we require that

$$h(t) = (h_{Tx} * h_c * h_{Rx})(t) = h_{RC}(t), \tag{6.125}$$

or in the frequency domain

$$|H_{RC}| = |H_{Tx} H_c H_{Rx}|. \tag{6.126}$$

The condition for matching filter can be written as

Table 6.1 Common partial-response signaling schemes

Type of class	*N*	w_0	w_1	w_2	w_3	w_4	*Class is also known as:*
I	2	1	1				Duobinary
II	3	1	2	1			
III	3	2	1	−1			
IV	3	1	0	−1			Modified duobinary
V	5	−1	0	2	0	−1	

$$|H_{Rx}(f)| = |H_{Tx}(f)H_c(f)|. \tag{6.127}$$

When these two criteria are combined, we obtain that the following is valid:

$$|H_{Tx}| = \frac{\sqrt{H_{RC}}}{|H_c|}, \quad |H_{Rx}| = \sqrt{H_{RC}}. \tag{6.128}$$

Notice that this still represents an approximate solution, as two problems get solved independently and then get combined.

In our discussion above, we considered the ISI as a detrimental effect. However, by inserting the ISI in a controllable manner, it is possible to transmit signal at a Nyquist rate of 2*BW*, where *BW* is the bandwidth of the channel. The systems using this strategy are known as *partial-response (PR) signaling schemes*, and they are widely used in magnetic recording and various digital communications [1, 32, 33]. The impulse response of PR systems is given as

$$h(t) = \sum_{n=0}^{N-1} w_n \text{sinc}(t/T - n). \tag{6.129}$$

The common PR signaling schemes are listed in Table 6.1. For instance, the scheme with $N = 2$ is known as a duobinary scheme, while the scheme with $N = 3$ and weight coefficients 1, 0, −1 is known as a modified duobinary scheme.

6.3.2 Zero-Forcing Equalizers

To compensate for linear distortion effects, such as chromatic dispersion in fiber-optics communications and multipath effects in wireless communications, we may use the circuit known as equalizer, connected in cascade with the system in question, as shown in Fig. 6.9a. The overall frequency response is equal to $H_c(\omega)\cdot H_{eq}(\omega)$, where $H_c(\omega)$ is the system transfer function, while $H_{eq}(\omega)$ is the equalizer transfer function. We assume that balanced coherent detection is used. For distortionless transmission, we require that

$$H_c(\omega)H_{eq}(\omega) = e^{-j\omega t_0}, \tag{6.130}$$

so that the frequency response of the equalizer is inversely proportional to the system transfer function, as given below:

$$H_{eq}(\omega) = \frac{e^{-j\omega t_0}}{H_c(\omega)}. \tag{6.131}$$

A functional block that is well suited for equalization in general is *tapped-delay-line (transversal) equalizer*, or the finite impulse response (FIR) filter, which is shown in Fig. 6.9b. The total number of filter taps is equal to *N*+1.

The FIR equalizer design can be done by using some of the methods described in [26], which are (i) the symmetry method, (ii) the window method, (iii) the frequency sampling method, and (iv) the Chebyshev approximation method. The FIR filters obtained by symmetry method have a linear phase so that their discrete impulse response $h[n]$ and system function $H(z)$ satisfy the following symmetry property [26]:

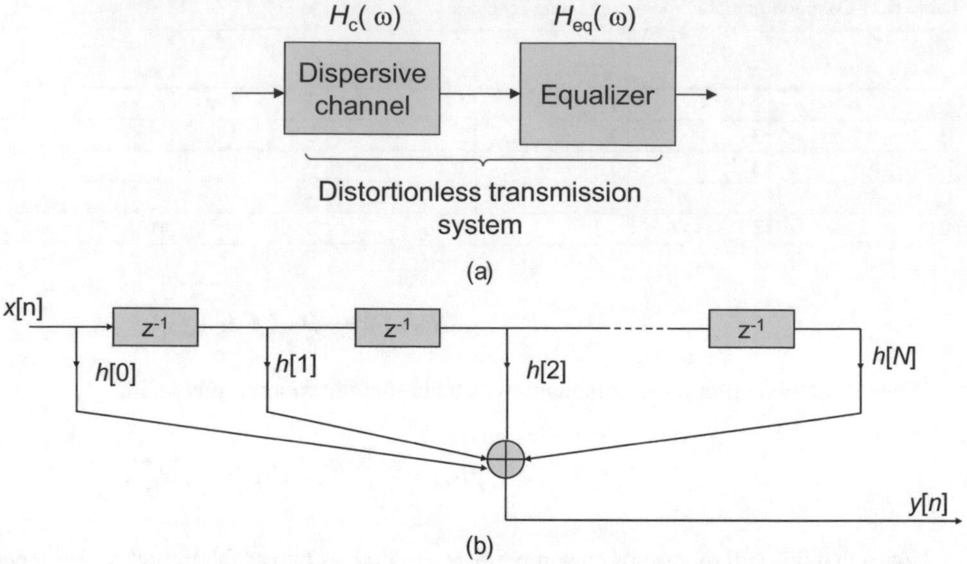

Fig. 6.9 (**a**) Equalization principle and (**b**) tapped-delay-line equalizer (FIR filter)

$$h[n] = \pm h[N-1-n]; n = 0, 1, \cdots, N-1$$
$$H(z) = Z\{h[n]\} = \pm z^{-(N-1)} H(z^{-1}), \tag{6.132}$$

where $Z\{\cdot\}$ presents the z-transform of impulse response $h[n]$, as defined in Chap. 3. From the symmetry property (6.132), we conclude that the roots of $H(z)$ must be the same as the roots of $H(z^{-1})$, which indicates that the roots of $H(z)$ must occur in reciprocal pairs: z_k, z_k^{-1}. Moreover, for filter coefficients to be real valued, the roots must appear in complex-conjugate pairs: z_k, z_k^{-1}, z^*_k, z^{*-1}_k.

The *windowing method* starts with a specified filter order N (N is typically an even integer), and for a given sampling interval T_s, the following steps are taken: (i) set the constant time delay $t_0 = (N/2)/T_s$; (ii) take the inverse Fourier transform of $H_{eq}(\omega)$ to obtain a desired impulse response $h_{eq}(t)$; and (iii) set $h[n] = w[n]h_{eq}[nT_s]$, where $w[n]$ is a window of length $(N+1)$. The simplest windowing function is a "rectangular window" given by

$$w(n) = \begin{cases} 1, & n = 0, 1, \ldots, N-1 \\ 0, & \text{otherwise} \end{cases}. \tag{6.133}$$

Therefore, the frequency response of smoothed equalizer can be obtained as the convolution $H(\Omega) = H_{eq}(\Omega) * W(\Omega)$, where $\Omega = \omega \cdot T_s$ is the discrete frequency (expressed in rad/sample). For large filter order having rectangular window, the Gibbs phenomenon in $H(\Omega)$ can be noticed [26], which comes from abrupt truncation of $h_{eq}[n]$. For better results, the tapered windows can be used. The commonly used tapered windowing functions include Blackman, Hamming, Hanning, Kaiser, Lanczos, and Tukey windows [26]. For instance, the Hanning window is given by the function

$$w[n] = \frac{1}{2}\left(1 - \cos\frac{2\pi n}{N-1}\right). \tag{6.134}$$

Tapered windows reduce the Gibbs ringing but increase the transition bandwidth (they provide better smoothing but less sharp transition).

In fiber-optics communications, the first-order polarization mode dispersion (PMD) can be compensated by using four FIR filters. The output symbols in x-polarization can be determined by [21]

$$x[k] = \boldsymbol{h}_{xx}^T \boldsymbol{x}' + \boldsymbol{h}_{xy}^T \boldsymbol{y}' = \sum_{n=0}^{N-1} \{h_{xx}[n]x'[k-n] + h_{xy}[n]y'[k-n]\}, \tag{6.135}$$

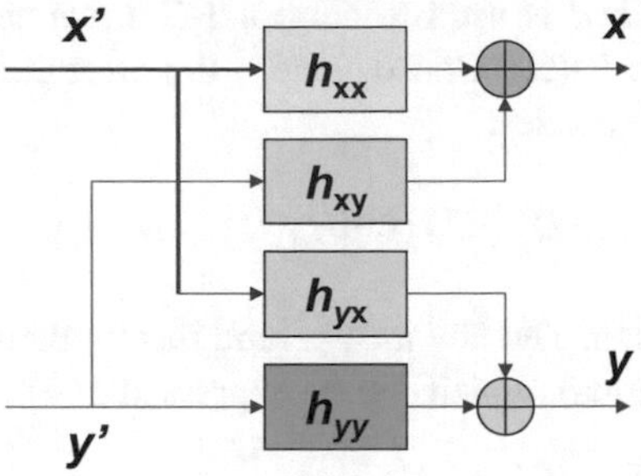

Fig. 6.10 The compensation of polarization-dependent impairments in fiber-optics communications by FIR filters

where $\boldsymbol{h}_{ij}$ ($i,j \in \{x,y\}$) are responses of FIR filters, each with N taps. The corresponding equation to determine the symbols in y-polarization can be written in a similar fashion, and the PDM compensation procedure can be summarized as shown in Fig. 6.10.

The design of an FIR equalizer to compensate for chromatic dispersion in fiber-optics communications is a straightforward procedure when the windowing method is used. As shown in Chap. 2, the impact of chromatic dispersion can be described by the following transfer function:

$$H_c(\omega) = \exp\left[j\left(\beta_2\omega^2/2 + \beta_3\omega^3/6\right)L\right], \tag{6.136}$$

where β_2 and β_3 are the group-velocity dispersion (GVD) and second-order GVD parameters, respectively, while L is the fiber length. In particular, when the second-order GVD parameter can be neglected, by taking inverse Fourier transform of Eq. (6.136), and applying the windowing method, the FIR equalizer coefficients become

$$h[n] = \sqrt{\frac{jcT^2}{D\lambda^2 L}}\exp\left(-j\frac{\pi cT^2}{D\lambda^2 L}n^2\right),\ -\left\lfloor\frac{N}{2}\right\rfloor \le n \le \left\lfloor\frac{N}{2}\right\rfloor, N = 2\left\lfloor\frac{|D|\lambda^2 L}{2cT^2}\right\rfloor + 1, \tag{6.137}$$

where D is the dispersion parameter, related to GVD parameter β_2 by $D = -(2\pi c/\lambda^2)\beta_2$.

The *frequency sampling method* is based on sampling of frequency response of the equalizer as follows:

$$H(k) = H(\Omega)|_{\Omega_k = 2\pi k/N}. \tag{6.138}$$

The function $H(\Omega)$ can be expressed by discrete-time Fourier transform as

$$H(\Omega) = \sum_{n=0}^{N-1} h[n]e^{-j\Omega n}, \tag{6.139}$$

and, since the frequency response is related to z-transform by setting $z = \exp(j\Omega)$, we can easily determine the system function $H(z)$. This is particularly simple to do when the system function is decomposable as follows:

$$H(z) = \underbrace{\frac{1 - z^{-N}}{N}}_{H_{\text{comb}}(z)}\underbrace{\sum_{k=0}^{N-1}\frac{H(k)}{1 - e^{j2\pi k/N}z^{-1}}}_{H_{\text{bank}}(z)} = H_{\text{comb}}(z)H_{\text{bank}}(z), \tag{6.140}$$

where $H_{\text{comb}}(z)$ represents the all-zero system or a comb filter, with equally spaced zeros located at the unit circle ($z_k = e^{j2\pi(k+\alpha)/N}$, with $\alpha = 0$ or 0.5), while $H_{\text{bank}}(z)$ consists of a parallel bank of single-pole filters with resonant frequencies equal to $p_k = e^{j2\pi(k+\alpha)/N}$. The unit-sample impulse response $h[n]$ needs now to be determined from $H(z)$. The main drawback of this method is nonexistence of control of $H(\Omega)$ at frequencies in between points Ω_k.

In practice, the *optimum equiripple method* is used to control $H(\Omega)$. Namely, the frequency response $H(\Omega)$ of actual equalizer is different from desired function $H_d(\Omega)$. Consequently, the error function $E(\Omega)$ can be found as the difference between the desired and actual frequency responses:

$$E(\Omega) = W(\Omega)[H_d(\Omega) - H(\Omega)], \tag{6.141}$$

where $W(\Omega)$ is known as a weighting function. The key idea behind this method is to determine the $h[k]$ by minimizing the maximum absolute error (the minimax criterion), which can be expressed as

$$\widehat{h}[n] = \arg \min_{\text{over } h(n)} \left[\max_{\Omega \in S} |E(\Omega)| \right], \tag{6.142}$$

where S denotes the disjoint union of frequency bands over which the optimization is performed. For linear phase FIR filter design, it is possible to use the Remez algorithm [26]. On the other side, for problems involving both quadratic and cubic phase terms, the Chebyshev approximation method can be used instead.

We can conclude from Eq. (6.137) that chromatic dispersion compensator is essentially an all-pass filter with constant magnitude of the response, $|H(\Omega)|=1$, $0 \le \Omega \le \pi$. An important property of all-pass filters is that zero and poles are reciprocals of one another; in other words,

$$H(z) = z^{-N} \frac{A(z^{-1})}{A(z)}, A(z) = 1 + \sum_{n=1}^{N} a[n] z^{-n} \tag{6.143}$$

with response magnitudes $|H(\omega)|^2 = H(z)H(z^{-1})\big|_{z=e^{j\Omega}} = 1$. Upon factorization, the system function of an all-pass filter is given by [26]

$$H(z) = \prod_{k=1}^{N_r} \frac{z^{-1} - \alpha_k}{1 - \alpha_k z^{-1}} \prod_{k=1}^{N_c} \frac{(z^{-1} - \beta_k)(z^{-1} - \beta_k^*)}{(1 - \beta_k z^{-1})(1 - \beta_k^* z^{-1})}; \quad |\alpha_k| < 1, |\beta_k| < 1, \tag{6.144}$$

where N_r is the number of sections with simple poles, while N_c is the number of sections with complex-conjugate poles. By interpreting the chromatic dispersion compensator design as the all-pass filter design, it is expected that the number of taps will be smaller than that needed in FIR filter design (for which the number of taps is a linear function of fiber length). It is evident from Eq. (6.144) that the all-pass filter is essentially infinite impulse response (IIR) filter. Additional details on IIR all-pass filter design can be found in [27] and [28].

The function of FIR/IIR equalizers described above is to ensure the zero-ISI at all instances $t = nT$, except for $n = 0$. Because of that, they are also known as the *zero-forcing equalizers*. However, the problem in their employment is that the Gaussian noise, chromatic dispersion, and PMD in fiber-optics communications all act together, similar to multipath fading and Gaussian noise in wireless communication, affecting the behavior of a transmission system in a combined manner, while transversal equalizers ignore the effect of channel noise. This leads to the *noise enhancement phenomenon*, which can be explained as follows. Let us again consider the model shown in Fig. 6.9, but now with the impulse response of receive filter being decomposed into matched filter impulse response and equalizer impulse response:

$$h_{Rx}(t) = h_{Tx}(-t) * h_{\text{eq}}(t). \tag{6.145}$$

The Fourier transform of system output, assuming $H_{\text{eq}}(f) = \text{FT}\{h_{\text{eq}}(t)\} = 1/H_c(f)$, is given by

$$\begin{aligned} C(f) &= [H_T(f)H_c(f) + N(f)]H_T^*(f)H_{eq}(f) \\ &= |H_T(f)|^2 + \underbrace{\frac{N(f)H_T^*(f)}{H_c(f)}}_{N'(f)}. \end{aligned} \tag{6.146}$$

The corresponding power spectral density of colored Gaussian noise $N'(f)$ is given by

$$PSD_{N'}(f) = \frac{N_0}{2}\frac{|H_T(f)|^2}{|H_c(f)|^2}. \tag{6.147}$$

We can see that the ISI is compensated for, but the noise is enhanced unless we have that $|H_{Tx}(f)|=|H_c(f)|$.

6.3.3 Optimum Linear Equalizer in MMSE Sense

The better approach for receiver design, which avoids the noise enhancement phenomenon, would be to use the *minimum mean-square error* (MMSE) criterion to determine the equalizer coefficients, which provides a balanced solution to the problem by taking the effects of both channel noise and ISI into account. Let $h_{Rx}(t)$ be the receiver filter impulse response and $x(t)$ be the channel output determined by

$$x(t) = \sum_k s_k q(t - kT) + z(t), \quad q(t) = h_{Tx}(t) * h_c(t), \tag{6.148}$$

where $h_{Tx}(t)$ is the transmit filter impulse response, $h_c(t)$ is the channel impulse response, s_k is the transmitted symbol at the kth time instance, and $z(t)$ is the additive channel noise. (As before, T denotes the symbol duration.) The receive filter output can be determined by the convolution of receive filter impulse response and corresponding input as

$$y(t) = \int_{-\infty}^{\infty} h_{Rx}(\tau)x(t - \tau)d\tau. \tag{6.149}$$

By sampling at $t = nT$, we obtain the following:

$$\begin{aligned} y(nT) &= \zeta_n + z_n, \quad \zeta_n = \sum_k s_k \int_{-\infty}^{\infty} h_{Rx}(\tau)q(nT - kT - \tau)d\tau, \\ z_n &= \int_{-\infty}^{\infty} h_{Rx}(\tau)z(nT - \tau)d\tau. \end{aligned} \tag{6.150}$$

The signal error can be defined as a difference between receive sample and transmitted symbol as follows:

$$e_n = y(nT) - s_n = \zeta_n + z_n - s_n. \tag{6.151}$$

The corresponding mean-square error is

$$MSE = \frac{1}{2}\langle e_n^2 \rangle = \frac{1}{2}\langle \zeta_n^2 \rangle + \frac{1}{2}\langle z_n^2 \rangle + \frac{1}{2}\underbrace{\langle s_n^2 \rangle}_{=1} + \underbrace{\langle \zeta_n z_n \rangle}_{=0} - \underbrace{\langle \zeta_n s_n \rangle}_{\int_{-\infty}^{\infty} h_{Rx}(\tau)q(-\tau)d\tau} - \underbrace{\langle z_n s_n \rangle}_{=0}. \tag{6.152}$$

Assuming the stationary environment, the equation above can be rewritten as

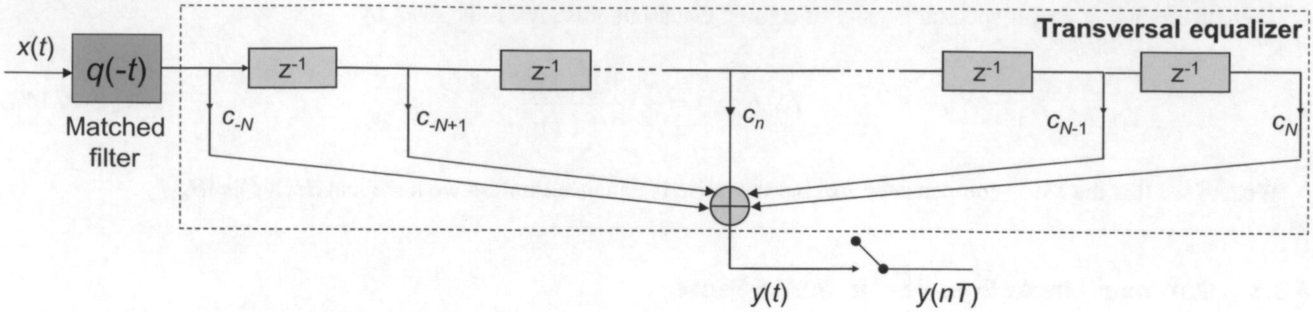

Fig. 6.11 Optimum linear equalizer (in the MMSE sense)

$$MSE = \frac{1}{2} + \frac{1}{2} \int_{-\infty}^{\infty} \int_{-\infty}^{\infty} \left[R_q(t - \tau) + \frac{N_0}{2} \delta(t - \tau) \right] h_{Rx}(t) h_{Rx}(\tau) dt d\tau \\ - \int_{-\infty}^{\infty} h_{Rx}(t) q(-t) dt, \tag{6.153}$$

where $R_q(\tau)$ is the autocorrelation function of $q(t)$ and N_0 is the power spectral density of Gaussian noise. In order to determine the optimum filter in the MMSE sense, we have to find the derivative of MSE with respect to h_{Rz}(t) and set it to zero, which yields to

$$\int_{-\infty}^{\infty} \left[R_q(t - \tau) + \frac{N_0}{2} \delta(t - \tau) \right] h_{Rx}(\tau) d\tau = q(-t). \tag{6.154}$$

By applying the Fourier transform (FT) of the equation above and solving it for receive filter transfer function $H_{Rx}(f) =$ FT $[h_{Rx}(t)]$, we obtain the following:

$$H_{Rx}(f) = \frac{Q^*(f)}{S_q(f) + \frac{N_0}{2}} = Q^*(f) \underbrace{\frac{1}{S_q(f) + \frac{N_0}{2}}}_{H_{eq}(f)}, \tag{6.155}$$

where the impulse response of equalizer is obtained as $h_{eq}(t) = \mathrm{FT}^{-1}[H_{eq}(f)]$, followed by the windowing method described above. Therefore, the optimum linear receiver, in the MMSE sense, consists of the cascade connection of matched filter and transversal equalizer, as shown in Fig. 6.11.

If delay T_d is equal to the symbol duration T, the corresponding equalizer is known as symbol-spaced equalizer. If the symbol rate ($R_s = 1/T$) is smaller than $2BW$ (BW is the channel bandwidth), the equalizer will need to compensate for both channel distortions and aliasing effect. On the other hand, when T_d is chosen to satisfy the condition $1/T_d \geq 2BW > R_s$, aliasing problem is avoided, and equalizer needs to compensate for channel distortions only. This type of equalizer is known as *fractionally spaced equalizer*, and the most common case is the one for which $T_d = T/2$.

6.3.4 Wiener Filtering

We have already discussed in Sect. 6.1.5 how to estimate a certain signal parameter in the presence of additive noise. In this section, we extend that study to the estimation of random vector $\boldsymbol{s}$ (of length N) from the received vector $\boldsymbol{r}$. We are specifically interested in linear (unbiased) minimum error variance estimation [2]. Namely, we would like our estimate to be a linear function of a received vector, which can be expressed as

$$\hat{s} = Ar + b, \tag{6.156}$$

where A and b are the matrix and vector, respectively, that need to be determined. The vector b can be determined from the requirement for estimate to be unbiased as follows:

$$\langle \hat{s} \rangle = \underbrace{m_s}_{\langle s \rangle} = A \underbrace{\langle r \rangle}_{m_r} + b = Am_r + b, \tag{6.157}$$

and by solving for b, we obtain

$$b = m_s - Am_r. \tag{6.158}$$

The error made by unbiased estimator will be then defined as

$$\varepsilon = \hat{s} - s = \underbrace{(\hat{s}\text{-}m_s)}_{A(r-m_r)} - (s\text{-}m_s) = A(r - m_r) - (s\text{-}m_s). \tag{6.159}$$

The matrix A can be determined from the condition that trace of covariance matrix, defined as

$$\mathrm{Cov}(\varepsilon) = \langle \varepsilon \varepsilon^{\mathrm{T}} \rangle, \tag{6.160}$$

is minimized. By setting the first derivative of trace of covariance matrix with respect to matrix A from Eq. (6.160) to be zero, we derive the following *orthogonality principle*:

$$\langle (r - m_r)\varepsilon^{\mathrm{T}} \rangle = 0. \tag{6.161}$$

The orthogonality principle indicates that error vector is independent of the data. By substituting Eq. (6.159) into Eq. (6.161) and by solving for A, we obtain

$$A = C_{rs}^{\mathrm{T}} C_r^{-1};\ C_{rs} = E\{(r - m_r)(x - m_s)^{\mathrm{T}}\}, C_r = E\{(r - m_r)(r \quad m_r)^{\mathrm{T}}\}, \tag{6.162}$$

where C_{rs} and C_r are the corresponding covariance matrices, and $E\{\cdot\}$=<·> denotes the expectation operator. Now, by substituting Eqs. (6.162) and (6.158) into Eq. (6.156), we obtain the following form of the estimated transmitted vector:

$$\hat{s} = C_{rs}^{\mathrm{T}} C_r^{-1}(r - m_r) + m_s, \tag{6.163}$$

which represents the most general form of the *Wiener filter*. The minimum error variance of this estimator is given by

$$\langle \varepsilon \varepsilon^{\mathrm{T}} \rangle = C_s - C_{sr} C_r^{-1} C_{sr}^{\mathrm{T}}. \tag{6.164}$$

The Wiener filtering is also applicable in polarization-division multiplexed (PDM) systems. In PDM, the transmitted vector s can be represented as $s = [s_{\mathrm{x}}\ s_{\mathrm{y}}]^{\mathrm{T}}$, where the subscripts x and y are used to denote x- and y-polarization states. In a similar fashion, the received vector r can be represented as $r = [r_{\mathrm{x}}\ r_{\mathrm{y}}]^{\mathrm{T}}$. The only difference in Wiener filtering problem is just in dimensionality of vectors s and r, which in PDM have twice more components ($2N$). The dimensionality of matrix A is also higher, namely, $2N \times 2N$.

6.3.5 Adaptive Equalization

So far, we have assumed that different channel impairments in either wireless or optical communication systems are time invariant, which is not completely true, especially with respect to multipath fading in wireless communications as well as the

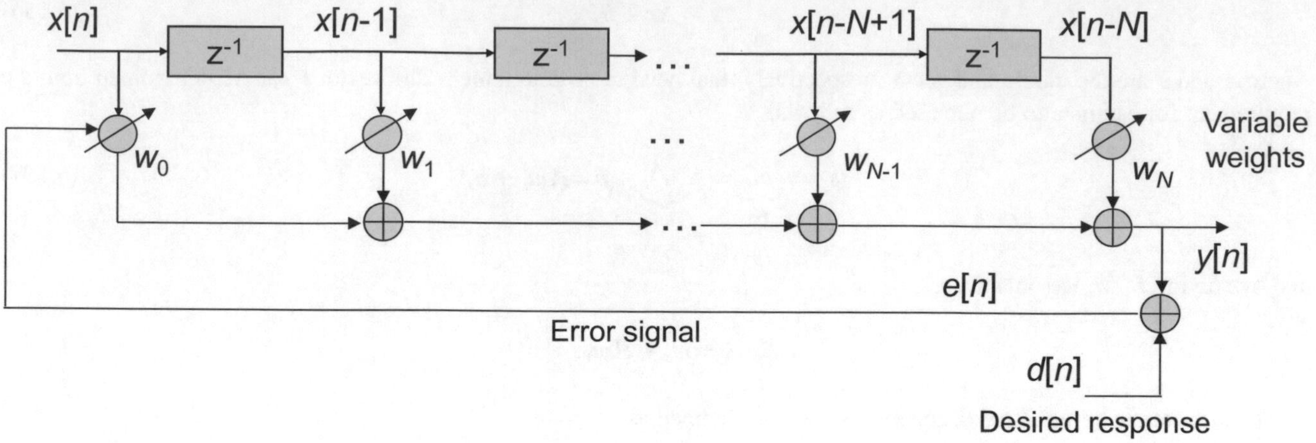

Fig. 6.12 Adaptive equalizer

PMD and polarization-dependent losses (PDL) in fiber-optics communications. The *adaptive filtering* offers an attractive solution to address the channel equalization [1, 3, 4, 26]. An adaptive filter has a set of adjustable filter coefficients, as shown in Fig. 6.12, which are adjusted based on the algorithms described below. The widely used adaptive filter algorithms are the steepest descent algorithm and the least-mean-square (LMS) algorithm [3, 4, 26]. These algorithms can be used to determine the coefficients of transversal equalizer.

According to the *steepest descent algorithm*, we update the kth filter coefficient w_k, shown in Fig. 6.12, by making correction of the present value in the direction opposite to the gradient ∇_k, in other words, in the direction of the steepest descent on the error-performance surface [3], as follows:

$$w_k[n+1] = w_k[n] + \frac{1}{2}\mu\{-\nabla_k[n]\}; \quad k = 0, 1, \ldots, N, \tag{6.165}$$

where the real-valued parameter μ determines the convergence speed.

When the error signal $e[n]$ is defined as the difference of desired signal (commonly the received sequence of a training sequence in a back-to-back configuration) and the output $y[n]$ of corresponding FIR filter output, we can write

$$e[n] = d[n] - \sum_{k=0}^{N} w_k[n]x[n-k]. \tag{6.166}$$

The gradient ∇_k can be determined as

$$\nabla_k[n] = \frac{\partial}{\partial w_k[n]}\langle e^2[n]\rangle = -2\langle e[n]x[n-k]\rangle = -2R_{EX}[k]; \quad k = 0, 1, \ldots, N, \tag{6.167}$$

where $R_{EX}[k]$ is the cross-correlation function between the error signal and the adaptive filter input. By substituting Eq. (6.167) into Eq. (6.165), we obtain the following form of steepest descent algorithm:

$$w_k[n+1] = w_k[n] + \mu R_{EX}[k]; \; k = 0, 1, \ldots, N. \tag{6.168}$$

It can be shown that for this algorithm to converge the following condition for the parameter μ should be satisfied:

$$0 < \mu < \mu_{\max}, \tag{6.169}$$

where $\mu_{\max}$ is the largest eigenvalue of the correlation matrix of the filter input $\boldsymbol{R}_X$:

Fig. 6.13 Decision-feedback equalizer (DFE)

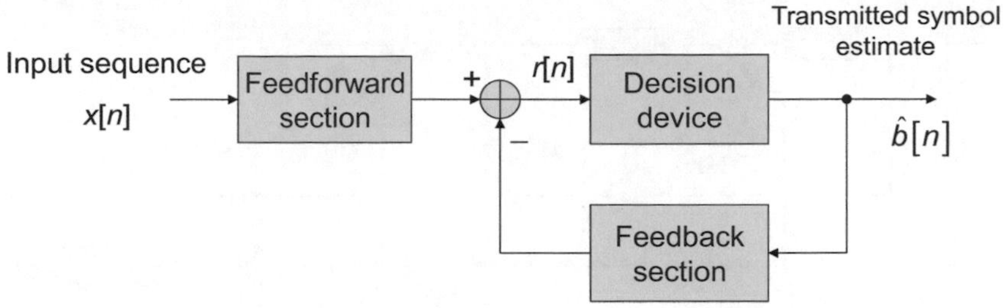

$$R_X = \begin{bmatrix} R_X(0) & R_X(1) & \dots & R_X(N) \\ R_X(1) & R_X(0) & \dots & R_X(N-1) \\ \dots & \dots & \dots & \dots \\ R_X(N) & R_X(N-1) & \dots & R_X(0) \end{bmatrix}, \tag{6.170}$$

$$R_X(l) = \langle x[n]x[n-l] \rangle.$$

The main drawback of the steepest descent algorithm is that it requires the knowledge of gradient ∇_k in each iteration. On the other side, the key idea in *LMS algorithm* is to approximate the operator of averaging $\langle \cdot \rangle$ in Eq. (6.167) by its instantaneous value $e[n]x[n-k]$, so that the LMS weight update rule is simply

$$w_k^{(\mathrm{LMS})}[n+1] = w_k^{(\mathrm{LMS})}[n] + \mu e[n]x[n-k]; \quad k = 0, 1, \dots, N. \tag{6.171}$$

6.3.6 Decision-Feedback Equalization

Another relevant equalizer is the *decision-feedback equalizer* (DFE), which is shown in Fig. 6.13. The key idea of DFE is to use the decision made on the basis of precursors of the channel impulse response in order to take care of post-cursors. The assumption in DFE is that decisions in decision circuit are correct. Let the channel impulse response in discrete form be denoted by $h_c[n]$. The response of the channel, in the absence of noise, to an input sequence $x[n]$ is given by

$$\begin{aligned} y[n] &= \sum_k h_c[k]x[n-k] \\ &= h_c[0]x[n] + \underbrace{\sum_{k<0} h_c[k]x[n-k]}_{\text{precursors}} + \underbrace{\sum_{k>0} h_c[k]x[n-k]}_{\text{postcursors}}. \end{aligned} \tag{6.172}$$

The first term on the right side $h[0]x[n]$ represents the desired data symbol; the second term is a function of previous samples only (where the channel coefficients are known as precursors), while the third term is a function of incoming samples, with channel coefficients being known as post-cursors. The DFE is composed of a feedforward equalizer (FFE) section, a feedback equalizer (FBE) section, and a decision device, as shown in Fig. 6.13.

The feedforward and feedback sections can be implemented as FIR filters (transversal equalizers) and can also be adaptive. The input to the detector can be written as

$$r[m] = \sum_{n=1}^{N_{\mathrm{FFE}}} f[n]x[mT - nT] - \sum_{n=1}^{N_{\mathrm{FBE}}} g[n]\hat{b}[m-n], \tag{6.173}$$

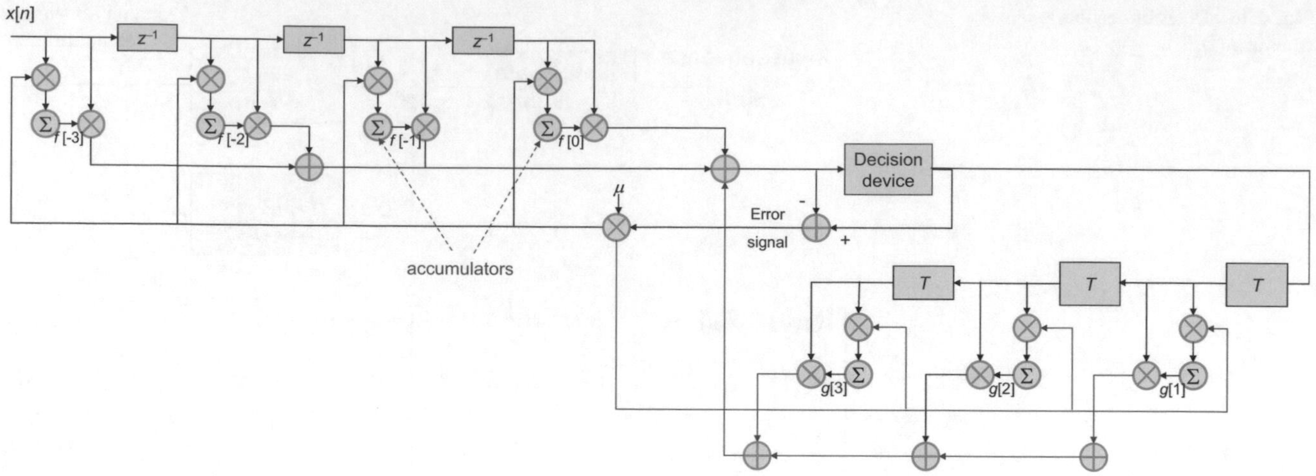

Fig. 6.14 An adaptive decision-feedback equalizer example

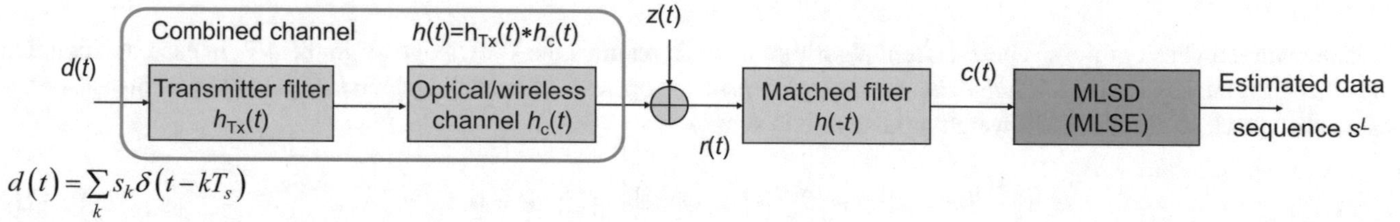

Fig. 6.15 Equivalent system model to study MLSD (to facilitate explanations only single-polarization case is studied)

where $f[n]$ and $g[n]$ are values of adjustable taps in feedforward and feedback sections, respectively. The N_{FFE} (N_{FBE}) denotes the number of taps in feedforward (feedback) section.

We can see that the feedforward section is a fractionally spaced equalizer, which operates with rate equal to an integer multiple of the symbol rate ($1/T_d = m/T$, m is an integer). The taps in feedforward and feedback sections are typically chosen based on MSE criterion, by employing the LMS algorithm. An example of the adaptive decision-feedback equalizer is shown in Fig. 6.14. Since the transmitted symbol estimate is dependent on previous decisions, this equalizer is a nonlinear functional module. If previous decisions are in error, the error propagation effect will occur. However, the errors do not persist indefinitely, but rather occur in bursts.

6.3.7 MLSD (MLSE) or Viterbi Equalization

Another very important equalizer is based on *maximum likelihood sequence detection (estimation)* (MLSD or MLSE) [34–36]. Since this method estimates the sequence of transmitted symbols, it avoids the problems of noise enhancement and error propagation. The MLSE chooses an input sequence of transmitted symbols $\{s_k\}$ in a way to maximize the likelihood function of received signal $r(t)$. The equivalent system model that illustrates MLSD is shown in Fig. 6.15.

The received signal $r(t)$ can be expressed in terms of a complete set of orthonormal basis functions $\{\Phi_n(t)\}$ as follows:

$$r(t) = \sum_{n=1}^{N} r_n \Phi_n(t), \tag{6.174}$$

where

$$\begin{aligned}r_n &= \sum_{k=-\infty}^{\infty} s_k h_{nk} + z_n = \sum_{k=0}^{L} s_k h_{nk} + z_n;\\ h_{nk} &= \int_0^{LT} h(t-kT)\Phi_n^*(t)dt;\ z_n = \int_0^{LT} z(t)\Phi_n^*(t)dt;\ h(t) = h_{Tx}(t) * h_c(t).\end{aligned} \tag{6.175}$$

The impulse responses $h_{\mathrm{Tx}}(t)$ and $h_c(t)$ used in the equation above denote (as before) impulse responses of transmitter and channel, respectively. The N denotes the cardinality of the basis set, L is the channel memory expressed in the number of symbols, while $h(t)$ is the combined channel impulse response (the convolution of transmit and channel impulse responses). Since z_n are Gaussian random variables, the distribution of $\boldsymbol{r}_{[1,N]} = [r_1\ r_2\ \ldots\ r_N]^{\mathrm{T}}$ is a multivariate Gaussian:

$$f(\boldsymbol{r}[1,N]|\boldsymbol{s}[1,L], h(t)) = \prod_{n=1}^{N}\left\{\frac{1}{\pi N_0}\exp\left[-\frac{1}{N_0}\left|r_n - \sum_{k=0}^{L} s_k h_{nk}\right|^2\right]\right\}, \tag{6.176}$$

where N_0 is the power spectral density of the Gaussian noise. The MLSD decides in favor of the symbol sequence $\boldsymbol{s}[1,L] = [s_0\ s_1\ \ldots\ s_L]$ that maximizes the likelihood function given by Eq. (6.176), so that we can write

$$\begin{aligned}\widehat{\boldsymbol{s}}[0,L] &= \underset{\boldsymbol{s}[0,L]}{\arg\max} f(\boldsymbol{y}[1,N]|\boldsymbol{s}[0,L], h(t)) = \underset{\boldsymbol{s}[0,L]}{\arg\max}\left[-\sum_{n=1}^{N}\left|r_n - \sum_{k=0}^{L} s_k h_{nk}\right|^2\right]\\ &= \underset{\boldsymbol{s}[0,L]}{\arg\max}\left\{2\,\mathrm{Re}\left[\sum_k s_k^* \sum_{n=1}^{N} r_n h_{nk}^*\right] - \sum_k\sum_m s_k s_m^* \sum_{n=1}^{N} h_{nk} h_{nm}^*\right\}.\end{aligned} \tag{6.177}$$

By using substitution $u(t) = h(t) * h^*(-t)$ and noticing that the following is valid:

$$\begin{aligned}\sum_{n=1}^{N} r_n h_{nk}^* &= \int_{-\infty}^{\infty} r(\tau) h^*(\tau - kT)d\tau = c[k],\\ \sum_{n=1}^{N} h_{nk} h_{nm}^* &= \int_{-\infty}^{\infty} h(\tau - kT) h^*(\tau - mT)d\tau = u(mT - kT) = u[m-k]\end{aligned} \tag{6.178}$$

Eq. (6.177) can be simplified to

$$\widehat{\boldsymbol{s}}[0,L] = \underset{\boldsymbol{s}[0,L]}{\arg\max}\left\{2\,\mathrm{Re}\left[\sum_k s_k^* c[k]\right] - \sum_k\sum_m s_k s_m^* u[m-k]\right\}. \tag{6.179}$$

Equation (6.179) can efficiently be calculated by the Viterbi algorithm [1] (see also Chap. 9). However, the Viterbi algorithm provides the hard decisions and as such is not suitable for use with soft decision decoding/detection schemes. To fully exploit the advantages of soft decoding, the soft reliabilities are needed. These reliabilities can be obtained by soft-output Viterbi algorithm (SOVA) [37], BCJR algorithm [38], or by Monte Carlo-based equalization [39]. To further improve the BER performance, we can perform the iteration of extrinsic information between soft equalizer and soft decoder, the procedure known as *turbo equalization* [40–42], which will be described in Chap. 9. The turbo equalization scheme can be used in fiber-optics communications to simultaneously compensate for chromatic dispersion, PMD, PDL, and fiber nonlinearities, as shown in [43].

If Gaussian-like approximation is not valid, which is the case with the presence of strong fiber nonlinearities in fiber-optics communications or in satellite communications, the conditional PDF $f(\boldsymbol{r}[1, N]\|\ \boldsymbol{s}[0, L], h(t))$ should be estimated from histograms, by propagating the sufficiently long training sequence.

6.3.8 Blind Equalization

In MMSE equalizer, the use of training sequences is needed for initial adjustment of equalizer coefficients. However, the adjustment of equalizer coefficients can be done without using the training sequences, and this approach is known as blind equalization (or self-recovery equalization) [44–50]. All blind equalizers can be classified into three broad categories [1]: (i) steepest descent algorithm based, (ii) high-order statistics based, and (ii) maximum likelihood approach based.

To facilitate explanation of blind equalization, let us observe a signal polarization state. The output of optical fiber channel upon balanced coherent detection, assuming only either x- or y-polarization, can be expressed as

$$r[n] = \sum_{k=0}^{K} h[k]a[n-k] + z[n], \tag{6.180}$$

where $h[n]$ represents the optical channel impulse response (in either x- or y-polarization), $a[n]$ is the transmitted sequence of symbols, while $z[n]$ are samples of additive Gaussian noise. The joint PDF of received sequence $\boldsymbol{r}[1,L] = [r_1\, r_2 \ldots r_L]^{\mathrm{T}}$, under Gaussian noise assumption and for given impulse response vector $\boldsymbol{h}[0,K] = [h_0\, h_1 \ldots h_K]^{\mathrm{T}}$ and transmitted sequence $\boldsymbol{a}[1,L] = [a_1 \ldots a_L]$, can be written as

$$f_{\boldsymbol{R}}(\boldsymbol{r}|\boldsymbol{h},\boldsymbol{a}) = (\pi N_0)^{-L}\exp\left[-\frac{1}{N_0}\|\boldsymbol{r}-\boldsymbol{A}\boldsymbol{h}\|^2\right], \boldsymbol{A} = \begin{bmatrix} a_1 & 0 & 0 & \cdots & 0 \\ a_2 & a_1 & 0 & \cdots & 0 \\ a_3 & a_2 & a_1 & & 0 \\ \vdots & \vdots & \vdots & \ddots & \vdots \\ a_L & a_{L-1} & a_{L-2} & \cdots & a_{L-K} \end{bmatrix}. \tag{6.181}$$

When training sequence is used and data vector $\boldsymbol{a}$ is known, the ML estimate of channel impulse response can be obtained by maximizing the likelihood function (6.181) to obtain

$$\boldsymbol{h}_{ML}(\boldsymbol{a}) = \left(\boldsymbol{A}^\dagger \boldsymbol{A}\right)^{-1}\boldsymbol{A}^T\boldsymbol{r}. \tag{6.182}$$

On the other hand, when the channel impulse response is known, we can use the Viterbi equalizer, described in the section above, to detect the most likely transmitted sequence $\boldsymbol{a}$. However, when neither $\boldsymbol{a}$ nor $\boldsymbol{h}$ is known, we will need to determine them in such a way that the likelihood function from Eq. (6.181) is maximized. As an alternative solution, we can estimate $\boldsymbol{h}$ from $f_{\boldsymbol{R}}(\boldsymbol{r}|\boldsymbol{h})$, which is obtained by averaging over all possible data sequences (M^K in total, where M is the signal constellation size). In other words, the conditional PDF can be determined as

$$f_{\boldsymbol{R}}(\boldsymbol{r}|\boldsymbol{h}) = \sum_{k=1}^{M^K} f_{\boldsymbol{R}}(\boldsymbol{r}|\boldsymbol{h},\boldsymbol{a})P\left(\boldsymbol{a}^{(k)}\right). \tag{6.183}$$

The estimate of $\boldsymbol{h}$ that maximizes this new likelihood function from Eq. (6.183) can be obtained from the first derivative of $f_{\boldsymbol{R}}(\boldsymbol{r}|\boldsymbol{h})$ with respect to $\boldsymbol{h}$, which is set to zero, to obtain

$$\boldsymbol{h}^{(i+1)} = \left[\sum_k \boldsymbol{A}^{(k)\dagger}\boldsymbol{A}^{(k)}e^{-\left\|\boldsymbol{r}-\boldsymbol{A}^{(k)}\boldsymbol{h}^{(i)}\right\|^2/N_0}P\left(\boldsymbol{a}^{(k)}\right)\right]^{-1}\sum_k \boldsymbol{A}^{(k)\dagger}\boldsymbol{r}e^{-\left\|\boldsymbol{r}-\boldsymbol{A}^{(k)}\boldsymbol{h}^{(i)}\right\|^2/N_0}P\left(\boldsymbol{a}^{(k)}\right). \tag{6.184}$$

Given that the corresponding equation is transcendental, we can use an iterative procedure to determine $\boldsymbol{h}$ recursively. Once $\boldsymbol{h}$ is determined, the transmitted sequence $\boldsymbol{a}$ can be estimated by the Viterbi algorithm that maximizes the likelihood function from Eq. (6.181) or minimizes the Euclidean distance $\|\boldsymbol{r}-\boldsymbol{A}\boldsymbol{h}_{\mathrm{ML}}\|^2$:

$$\hat{\boldsymbol{a}} = \arg\min_{\boldsymbol{a}} \|\boldsymbol{r} - \boldsymbol{A}\boldsymbol{h}_{\mathrm{ML}}\|^2. \tag{6.185}$$

It is clear that this approach is computationally extensive. Moreover, since $\boldsymbol{h}$ is estimated from average conditional PDF, its estimate is not going to be that accurate as compared to the case when training sequences are used. The better strategy in this case will be to perform channel and data estimation jointly.

In *joint channel and data estimation*, the process is done in several stages. In the first stage, the corresponding ML estimate of $\boldsymbol{h}$ is determined for each candidate data sequence $\boldsymbol{a}^{(k)}$, which is

$$\boldsymbol{h}_{ML}\left(\boldsymbol{a}^{(k)}\right) = \left(\boldsymbol{A}^{(k)\dagger}\boldsymbol{A}^{(k)}\right)^{-1}\boldsymbol{A}^{(k)\mathrm{T}}\boldsymbol{r}; k = 1, \cdots, M^K. \tag{6.186}$$

In the second stage, we select the data sequence that minimizes the Euclidean distance for all channel estimates:

$$\hat{\boldsymbol{a}} = \arg\min_{\boldsymbol{a}^{(k)}} \left\|\boldsymbol{r} - \boldsymbol{A}^{(k)}\boldsymbol{h}_{\mathrm{ML}}\left(\boldsymbol{a}^{(k)}\right)\right\|^2. \tag{6.187}$$

For efficient implementation of joint channel and data estimation, a generalized Viterbi algorithm (GVA) in which the best B ($\geq$1) estimates of transmitted sequence are preserved is proposed in [47]. In this algorithm, the conventional Viterbi algorithm up to the Kth stage is applied, which performs an exhaustive search. After that, only B surviving sequences are retained. This algorithm performs well for $B = 4$ and for medium SNR values, as shown in [47].

In stochastic-gradient blind equalization algorithms, a memoryless nonlinear assumption is typically used. The most popular from this class of algorithms is *Godard's algorithm* [48], also known as the *constant modulus algorithm* (CMA). In fiber-optics communications, this algorithm has been used for PMD compensation [21, 29, 31], which is performed in addition to compensation of I/Q imbalance and other channel impairments. In conventional adaptive equalization, we use a training sequence as desired sequence, so it is $d[n] = a[n]$. However, in blind equalization, we have to generate a desired sequence $d[n]$ from the observed equalizer output based on a certain nonlinear function:

$$d[n] = \begin{cases} g(\hat{a}[n]), \text{memorylesscase} \\ g(\hat{a}[n], \hat{a}[n-1], \cdots, \hat{a}[n-m]), \text{ memoryoforder } m \end{cases}. \tag{6.188}$$

The commonly used nonlinear functions are as follows:

(1) Godard's function [48] given as

$$g(\hat{a}[n]) = \frac{\hat{a}[n]}{|\hat{a}[n]|}\left(|\hat{a}[n]| + \frac{\left\langle|\hat{a}[n]|^4\right\rangle}{\left\langle|\hat{a}[n]|^2\right\rangle}|\hat{a}[n]| - |\hat{a}[n]|^3\right), \tag{6.189}$$

(2) Sato's function [49] given as

$$g(\hat{a}[n]) = \mathrm{sign}(\hat{a}[n])\frac{\left\langle[\mathrm{Re}\,(a[n])]^2\right\rangle}{\langle|\mathrm{Re}\,(a[n])|\rangle}, \tag{6.190}$$

and

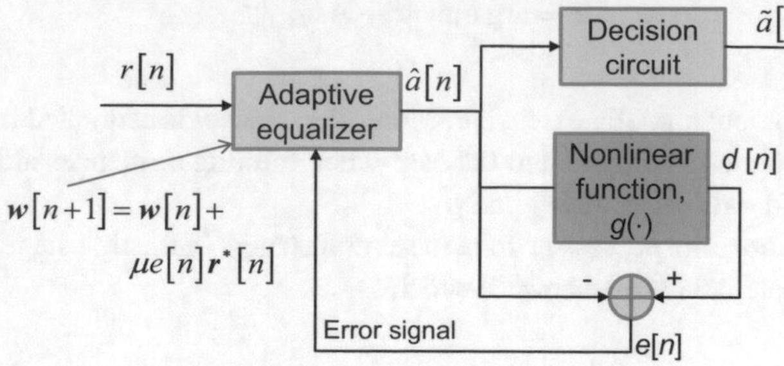

Fig. 6.16 The description of generic stochastic-gradient blind equalization algorithm

Fig. 6.17 Godard's scheme for simultaneous blind equalization and carrier phase tracking

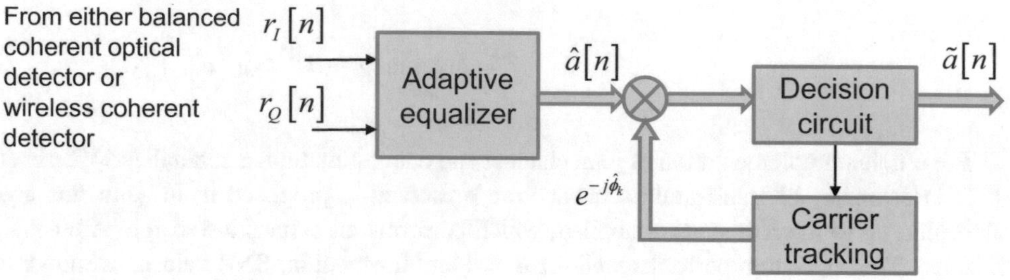

(3) Benveniste-Goursat's function [50] given as

$$g(\widehat{a}[n]) = \widehat{a}[n] + c_1(\widehat{a}[n] - a[n]) + c_2|\widehat{a}[n] - \widetilde{a}[n]|\left[\operatorname{sign}(\widehat{a}[n])\frac{\left\langle [\operatorname{Re}(a[n])]^2\right\rangle}{\langle|\operatorname{Re}(a[n])|\rangle} - \widetilde{a}[n]\right], \tag{6.191}$$

where c_1 and c_2 are the properly chosen positive constants, while $\widetilde{a}[n]$ is the decision circuit output as shown in Fig. 6.16. In the same figure, we also provide the update rule for adaptive equalizer tap coefficients, which is applicable to all three nonlinear functions described above.

Godard's algorithm belongs to the class of steepest descent algorithms, and it is widely used when the training sequence is not available. Godard's scheme is suitable for simultaneous blind adaptive equalization and carrier phase tracking, as shown in Fig. 6.17. With $r_I[n]$ and $r_Q[n]$ we denoted the in-phase and quadrature components of the received signal, after either wireless coherent detection or optical coherent detection. The equalizer output is represented by discrete-time convolution of the input complex sequence $r[n] = (r_I[n], r_Q[n])$ and the equalizer tap coefficients $w[n]$ given by

$$\widehat{a}[k] = \sum_{n=-K}^{K} w[n]r[k-n]. \tag{6.192}$$

This output is multiplied with $\exp\left(-j\widehat{\phi}_k\right)$, where $\widehat{\phi}_k$ is the carrier phase estimate at the kth time instance (symbol interval). The error signal can be defined as

$$e[k] = a[k] - \exp\left(-j\widehat{\phi}_k\right)\widehat{a}[k], \tag{6.193}$$

which assumes that $a[k]$ were known. The MSE can be minimized with respect to equalizer tap coefficients and carrier phase estimate to become

$$\left(\widehat{\phi}_k, \boldsymbol{w}\right) = \underset{\widehat{\phi}_k, \boldsymbol{w}}{\arg\min}\, E\left\{\left|a[k] - \exp\left(-j\widehat{\phi}_k\right)\widehat{a}[k]\right|^2\right\}. \tag{6.194}$$

The LMS-like algorithm can be used to determine the carrier phase estimate and equalizer tap coefficients, respectively, as

$$\begin{aligned} \widehat{\phi}_k &= \widehat{\phi}_k + \mu_\phi \mathrm{Im}\left\{a[k]\widehat{a}^*[k]\exp\left(j\widehat{\phi}_k\right)\right\}, \\ \widehat{\boldsymbol{w}}_{k+1} &= \widehat{\boldsymbol{w}}_k + \mu_w\left(a[k] - \widehat{a}[k]\exp\left(-j\widehat{\phi}_k\right)\right). \end{aligned} \tag{6.195}$$

However, since the desired sequence $a[n]$ is not known, the algorithm above will not converge. To solve this problem, we can use the cost function that is independent of the carrier phase, defined as

$$C(p) = \left\langle [|\widehat{a}[k]|^p - |a[k]|^p]^2 \right\rangle, \tag{6.196}$$

where p is a positive integer ($p = 2$ is the most used value). However, only signal amplitude will be equalized in this case. Another more general cost function was introduced for Godard's algorithm, defined as [48]

$$\mathcal{D}(p) = \left\langle \left[|\widehat{a}[k]|^p - R_p\right]^2 \right\rangle, \tag{6.197}$$

where R_p is a positive real number to be determined. The minimization of cost function from Eq. (6.197) with respect to equalizer tap coefficients can be done by using the steepest descent algorithm, with the weight coefficient update rule as follows:

$$\boldsymbol{w}_{i+1} = \boldsymbol{w}_i - \mu_p \frac{d\mathcal{D}(p)}{d\boldsymbol{w}_i}. \tag{6.198}$$

After performing a derivation of variance and omitting the expectation operator $\langle\cdot\rangle$, the following LMS-like adaptation algorithm can be used:

$$\boldsymbol{w}_{k+1} = \boldsymbol{w}_k + \mu_p \boldsymbol{r}_k^* \widehat{a}[k]|\widehat{a}[k]|^{p-2}\left(R_p - |\widehat{a}[k]|^p\right),\ R_p = \frac{\left\langle |\widehat{a}[k]|^{2p} \right\rangle}{\langle |\widehat{a}[k]|^p \rangle}. \tag{6.199}$$

As we can see, the knowledge of carrier phase is not needed to determine equalizer tap coefficients. The algorithm is particularly simple for $p = 2$, since $|\widehat{a}[k]|^{p-2} = 1$. The algorithm from Eq. (6.199) will converge if all tap coefficients have been initialized to zero, but with exception of the center tap that was chosen to satisfy the following inequality:

$$|w_0|^2 > \frac{\left\langle |a[k]|^4 \right\rangle}{2|h_{\max}|\left\{\left\langle |a[k]|^2 \right\rangle\right\}^2}, \tag{6.200}$$

where $h_{\max}$ is the channel impulse response sample with the largest amplitude. The inequality from Eq. (6.200) represents the necessary but not the sufficient condition for the convergence.

6.4 Synchronization Techniques

There are two basic forms of synchronization employed in digital optical receivers: (i) carrier synchronization, in which the frequency and phase of optical carrier are estimated; and (ii) symbol synchronization, in which instants of time at which modulation changes its state are estimated. The synchronization problem can be interpreted as an estimation problem with

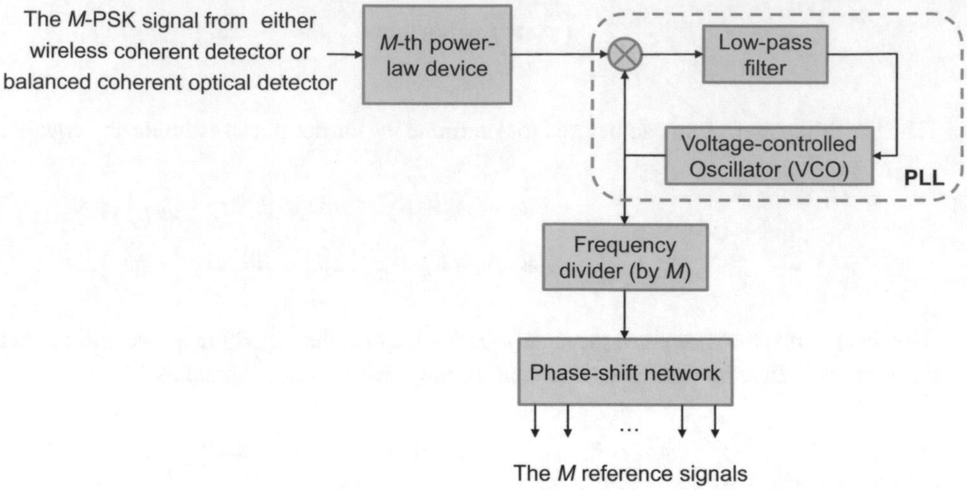

Fig. 6.18 The Mth power loop for phase recovery (in M-ary PSK systems)

respect to statistical parameter. For instance, maximum likelihood (ML) estimation can be straightforwardly applied here. We can classify the various synchronization methods as data-aided and nondata-aided ones, depending on whether a preamble is used for synchronization. In this section, we will describe nondata-aided synchronization methods. These methods can further be classified as (i) conventional, phase-locked loop (PLL)-based approaches and (ii) algorithmic approaches.

As for M-ary PSK format, we can use a variant of the Costas loop of the Mth power loop, as illustrated in Fig. 6.18. This method can easily be generalized and extended to star-QAM constellations. However, it suffers from phase ambiguities within the $[0,2\pi]$ interval, as well as the cyclic slips. Both of these problems can be solved by differential encoding, but at the expense of SNR degradation.

The *algorithmic approaches* are based on specially developed DSP algorithms. The M-ary PSK signal after wireless coherent detector or balanced coherent optical detection can be represented as

$$r(t) = A\cos\left(2\pi f_{IF}t + \theta + \phi_k\right)p\left(t - t_g\right) + z(t);\phi_k = k2\pi/M\ (k = 0, \cdots, M-1), \tag{6.201}$$

where A is an amplitude, θ is the random phase due to either laser phase noise in optical communications or multipath delay in wireless communications, t_g is the group delay, $p(t)$ is the pulse shape used on the transmitter side, $z(t)$ is the equivalent Gaussian noise, while f_{IF} denotes the intermediate frequency, which is the difference in carrier frequencies of received signal and local oscillator. The basis functions to be used at the receiver side are given by

$$\begin{aligned}\Phi_1(t) &= \sqrt{\frac{2}{T}}\cos\left(2\pi f_{IF}t\right), t_g \le t \le t_g + T\\ \Phi_2(t) &= \sqrt{\frac{2}{T}}\sin\left(2\pi f_{IF}t\right), t_g \le t \le t_g + T.\end{aligned} \tag{6.202}$$

The received signal can now be represented as a two-dimensional vector:

$$\boldsymbol{r}\left(t_g\right) = \begin{bmatrix} r_1\left(t_g\right) \\ r_2\left(t_g\right)\end{bmatrix}, r_{1,2}\left(t_g\right) = \int_{t_g}^{t_g+T} r(t)\Phi_{1,2}(t)dt. \tag{6.203}$$

In a similar fashion, the transmitted signal portion of $\boldsymbol{r}(t_g)$ can be represented as

$$\begin{aligned}\boldsymbol{s}\left(a_k, \theta, t_g\right) &= \begin{bmatrix} s_1\left(a_k, \theta, t_g\right) \\ s_2\left(a_k, \theta, t_g\right)\end{bmatrix},\\ s_{1,2}\left(a_k, \theta, t_g\right) &= \int_{t_g}^{t_g+T} A\cos\left(2\pi f_{IF}t + \theta + \phi_k\right)\Phi_{1,2}(t)dt\end{aligned} \tag{6.204}$$

where a_k is a transmitted symbol, which is equal to $\exp(j\phi_k)$ for M-ary PSK. By choosing the frequencies of transmit laser and local laser oscillator so that f_{IF} is an integer multiple of symbol rate (1/*T*), Eq. (6.204) can be simplified to

$$s(a_k, \theta, t_g) \approx \begin{bmatrix} \sqrt{E}\cos(\theta + \phi_k) \\ -\sqrt{E}\sin(\theta + \phi_k) \end{bmatrix}. \tag{6.205}$$

The noise signal *z*(*t*) can also be represented in a vector form as

$$z = \begin{bmatrix} z_1 \\ z_2 \end{bmatrix}, \quad z_{1,2} = \int_{t_g}^{t_g+T} z(t)\Phi_{1,2}(t)dt \tag{6.206}$$

so that the equivalent channel model in a vector form is given by

$$\boldsymbol{r}_k(t_g) = \boldsymbol{s}(a_k, \theta, t_g) + z. \tag{6.207}$$

In fiber-optics communications, when the chromatic dispersion, PMD, and PDL effects are properly compensated for, the resulting noise process is Gaussian-like with the power spectral density of N_0 so that the corresponding conditional probability density function is Gaussian. Similarly, after compensation of multipath fading effects in wireless communications, the resulting distribution is Gaussian as well:

$$f_{\boldsymbol{R}}(\boldsymbol{r}|a_k, \theta, t_g) = \frac{1}{\pi N_0} e^{-\left\|\boldsymbol{r}_k(t_g) - \boldsymbol{s}(a_k, \theta, t_g)\right\|^2}. \tag{6.208}$$

For non-Gaussian channels, we will need to use the method of histograms to estimate the conditional probability density function $f_{\boldsymbol{R}}(\boldsymbol{r}|\, a_k, \theta, t_g)$. For convenience, we can use the likelihood function defined as

$$L(a_k, \theta, t_g) = \frac{f_{\boldsymbol{R}}(\boldsymbol{r}|a_k, \theta, t_g)}{f_{\boldsymbol{R}}(\boldsymbol{r}|a_k = 0)} = e^{\frac{2}{N_0}\boldsymbol{r}_k^{\mathrm{T}}(t_g)\boldsymbol{s}(a_k, \theta, t_g) - \frac{1}{N_0}\left\|\boldsymbol{s}(a_k, \theta, t_g)\right\|^2}. \tag{6.209}$$

In M-ary PSK, the term $\|s(a_k, \theta, t_g)\|^2$ is constant and can be ignored, so that the previous equation becomes

$$L(a_k, \theta, t_g) = \exp\left[\frac{2}{N_0}\boldsymbol{r}^{k\mathrm{T}}(t_g)\boldsymbol{s}(a_k, \theta, t_g)\right]. \tag{6.210}$$

If the sequence of *L* statistically independent symbols, $\boldsymbol{a} = [a_0 \ldots a_{L-1}]^{\mathrm{T}}$, is transmitted, the corresponding likelihood function will be

$$L(\boldsymbol{a}, \theta, t_g) = \prod_{l=0}^{L-1} \exp\left[\frac{2}{N_0}\boldsymbol{r}_l^{\mathrm{T}}(t_g)\boldsymbol{s}(a_l, \theta, t_g)\right]. \tag{6.211}$$

To avoid numerical overflow problems, the log-likelihood function should be used instead, so that we can write

$$l(\boldsymbol{a}, \theta, t_g) = \log L(\boldsymbol{a}, \theta, t_g) = \sum_{l=0}^{L-1} \frac{2}{N_0}\boldsymbol{r}_l^{\mathrm{T}}(t_g)\boldsymbol{s}(a_l, \theta, t_g). \tag{6.212}$$

The first derivative of log-likelihood function with respect to θ is given by

$$\begin{aligned} &\frac{\partial l(\boldsymbol{a}, \theta, t_g)}{\theta} \cong \frac{2E^{-1/2}}{N_0}\sum_{k=0}^{L-1} \mathrm{Im}\,\{\widehat{a}_k^* \breve{r}_k e^{-j\theta}\}; \breve{r}_k = r_{1,k} + jr_{2,k}, \\ &r_{1,k} = E^{-1/2}\cos(\widehat{\phi}_k + \theta), r_{2,k} = E^{-1/2}\sin(\widehat{\phi}_k + \theta), \end{aligned} \tag{6.213}$$

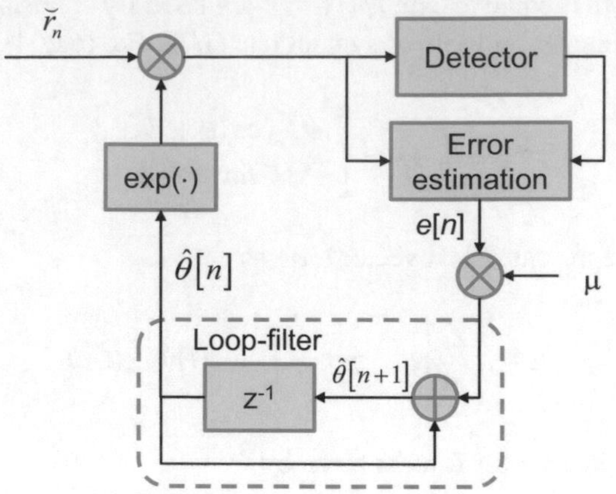

Fig. 6.19 The recursive implementation of the Costas loop

where $\widehat{a}_k$ is the estimate of a_k, while $\widehat{\phi}_k$ is the estimate of ϕ_k. Now we can formulate an LMS-like algorithm that can be used to estimate θ. Namely, the error signal in the current iteration can be estimated as

$$e[n] = \mathrm{Im}\left\{\widehat{a}_k^* \breve{r}_k e^{-j\widehat{\theta}[n]}\right\}, \tag{6.214}$$

where $\widehat{\theta}[n]$ is the previous estimate of a random phase, while an updated estimate can be obtained as

$$\widehat{\theta}[n+1] = \widehat{\theta}[n] + \mu e[n], \tag{6.215}$$

where μ is the step-size parameter. This algorithm essentially implements the recursive Costas loop as depicted in Fig. 6.19, where the loop filter is in fact the first-order filter. As a consequence, this loop can have a static error, which can be avoided if the second-order filter is used as a loop filter.

For *timing synchronization*, we need to know the intermediate frequency f_{IF}. The starting point in developing an algorithm for timing synchronization is again the likelihood function given by Eq. (6.211). However, the likelihood function is dependent on random phase θ so that it needs to be averaged since we know that distribution of the frequency offset. In optical communications, the Gaussian distribution can be used to describe the laser phase noise, so that the average likelihood function can be defined as

$$\begin{aligned}\overline{L}(a_k, t_g) &= \int L(a_k, \theta, t_g)\frac{1}{\sigma_\theta\sqrt{2\pi}}e^{-\theta^2/2\sigma_\theta^2}d\theta \\ &= \int \exp\left[\frac{2}{N_0}\boldsymbol{r}^{k\mathrm{T}}(t_g)\boldsymbol{s}(a_k, \theta, t_g)\right]\frac{1}{\sigma_\theta\sqrt{2\pi}}e^{-\theta^2/2\sigma_\theta^2}d\theta,\end{aligned} \tag{6.216}$$

where the variance of the phase noise is given by $\sigma_\theta^2 = 2\pi\Delta f T$, with Δf being the frequency offset. For laser phase noise, we can set $\Delta f = \Delta\nu$, where $\Delta\nu$ is the sum of linewidths of transmit laser and local laser oscillator. The integral above can be solved numerically or by Monte Carlo integration. The further procedure is similar as for carrier synchronization since we have to first create the likelihood function for the sequence of L statistically independent symbols, $\boldsymbol{a}=[a_0 \ldots a_{L-1}]^{\mathrm{T}}$, given by

$$\overline{L}(\boldsymbol{a}, t_g) = \prod_{l=0}^{L-1} \overline{L}(a_k, t_g). \tag{6.217}$$

The next step is to create the log-likelihood function by $\overline{l}(\boldsymbol{a}, t_g) = \log \overline{L}(\boldsymbol{a}, t_g)$. Finally, we can properly choose the error signal and create an LMS-like algorithm for timing synchronization from the derivative of log-likelihood function with respect to group delay, namely, from $\partial\overline{l}(\boldsymbol{a}, t_g)/\partial t_g$.

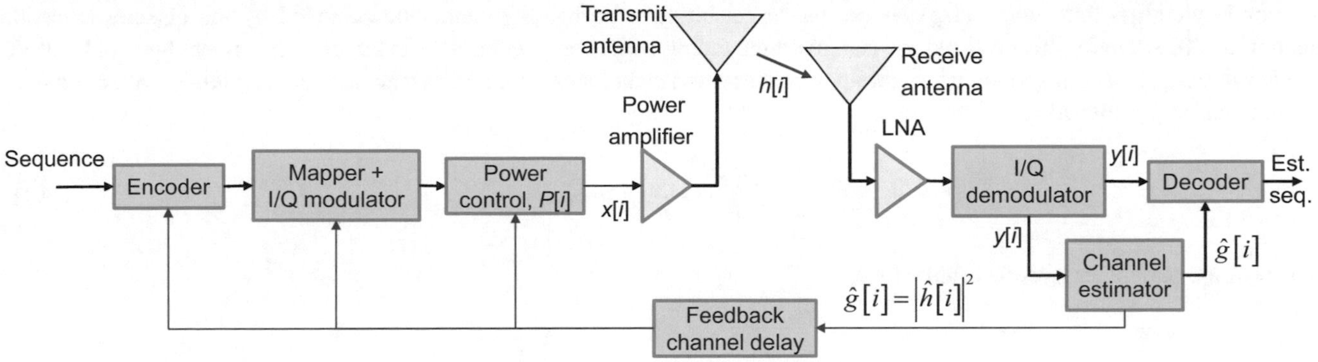

Fig. 6.20 Adaptive wireless communication system model

6.5 Adaptive Modulation Techniques

Adaptive modulation techniques [5, 11, 14, 51–56] enable spectrally efficient and reliable transmission over time-varying wireless and optical channels. As illustrated in Fig. 6.20, the basic premise is to estimate the channel state information (CSI) at the receiver side and feedback this CSI estimate to the transmitter, which can be adapted relative to the time-varying channel conditions. The nonadaptive modulation techniques are designed for the worst-case scenario and require a fixed system margin to maintain acceptable performance when the channel quality is poor.

There are many parameters that can be varied at the transmitter side depending on the channel SNR. The most common parameters to adapt include data rate, launch power, error correction strength (code rate), error probability, as well as combinations of various adaptive techniques. In this section, we describe the adaptive modulation techniques in the context of wireless communication systems. However, these concepts can easily be modified for optical communication applications [53–56]. For this purpose, for discrete-time interval i, the SNR can be defined as $\rho[i] = Pg[i]/N_0BW$, where P is the average power, $g[i] = |h[i]|^2$ is the channel gain, N_0 is the additive noise power spectral density, and BW is the channel bandwidth. The *data rate* and *spectral efficiency* of M-ary modulation scheme are defined, respectively, as $R[i] = \log_2 M[i]/T = BW \log_2 M\,[i]$ [bits/s] and $R[i]/BW = \log_2 M[i]$ [bits/s/Hz], where we assumed that the channel bandwidth is equal to the symbol rate, namely, $BW = 1/T$.

In *variable-rate modulation technique*, the data rate $R[\rho]$ is varied relative to the channel SNR ρ, which can be done by (1) fixing the symbol rate to $R_s = 1/T$ and vary signal constellation sizes for the same modulation format or employing multiple modulation formats and (2) fixing the modulation format while the symbol rate is varied. Symbol rate variation is difficult to implement in wireless communications because of the bandwidth sharing problem. However, this is typically not a problem in optical communication systems.

The *variable-power modulation technique* is used to compensate for SNR variation due to fading effects, to enable constant SNR at the receiver side, which is equivalent to the constant receiver side bit error probability. Therefore, with variable-power adaptation technique, we effectively invert the channel fading effects so that the channel appears as an AWGN channel at the receiver side, as described in Chap. 4. The *power adaptation policy for channel inversion* is given by

$$\frac{P(\rho)}{\overline{P}} = \frac{\rho_0}{\rho}, \tag{6.218}$$

wherein the constant ρ_0 is determined from the following condition:

$$\int \frac{P(\rho)}{\overline{P}} f(\rho) d\rho = \int \frac{\rho_0}{\rho} f(\rho) d\rho = 1, \tag{6.219}$$

with $f(\rho)$ being the PDF of SNR, as follows:

$$\rho_0 = 1/\langle 1/\rho \rangle. \tag{6.220}$$

For Rayleigh fading, when $\langle 1/\rho \rangle \to \infty$, the target bit error probability cannot be achieved by the channel inversion technique. To solve for this problem, we can invert the fading effects only when SNR is larger than a given threshold (cutoff) SNR value ρ_{tsh}, and corresponding variable-power adaptation techniques are known as the *truncated channel inversion policy*, which can be formulated as

$$\frac{P(\rho)}{\overline{P}} = \frac{\rho_0}{\rho}, \rho > \rho_{tsh} \tag{6.221}$$

wherein the parameter ρ_0 is determined from

$$\rho_0 = 1/\langle 1/\rho \rangle_{\rho_{tsh}}, \ \langle 1/\rho \rangle_{\rho_{tsh}} = \int_{\rho_{tsh}}^{\infty} \frac{1}{\rho} f(\rho) d\rho. \tag{6.222}$$

In *variable-error probability technique*, we vary the signal constellation size and/or modulation format so that the average bit error probability, defined as $\overline{P}_b = \int P_b(\rho) f(\rho) d\rho$, is smaller than target P_b.

In *adaptive coding*, a stronger channel code is used when channel conditions (or SNR) are (is) bad, while a weaker channel code is used when channel conditions are favorable (or SNR is high). Various adaptive coding approaches include (i) code multiplexing, in which different channel codes are used for different SNR ranges; (ii) rate-compatible punctured convolutional (RCPC) codes or rate-compatible punctured LDPC codes, in which error correction strength is reduced by puncturing a certain number of parity bits; (iii) shortening approach, in which the smaller number of information bits is transmitted for the same number of parity bits to improve error correction strength; (iv) re-encoding approach, in which a certain number of bits/symbols that form the previous codeword are re-encoded with incoming information bits to enable almost a continuous tuning of error correction strength (code rate) [11]; and (v) *fountain codes* [57], in particular *Raptor codes* [58], in which error correction strength is improved by increasing the codeword length.

In *hybrid adaptive modulation techniques*, we adapt multiple parameters of the wireless/optical communication transmission system including code rate, power, data rate, and bit error probability. To satisfy a given performance metric, the joint optimization of the different modulation techniques is employed. For instance, the data rate adaptation and power adaptation can be simultaneously performed to maximize the spectral efficiency. In adaptive modulation and coding (AMC), the modulation format/constellation size is combined with code rate adaptation.

In the incoming sections, we describe variable-power variable-rate modulation techniques and adaptive coded modulation with more details.

6.5.1 Variable-Power Variable-Rate Modulation Techniques

In variable-power variable-rate modulation technique, we simultaneously adapt the transmitted power and data rate. For illustrative purposes, we study a family of M-ary QAM signal constellations, but the described concepts are also applicable to any other M-ary (M – the number of points in corresponding signal constellation) modulation scheme. We assume that the symbol duration T is related to the channel bandwidth BW by $T = 1/BW$ based on the ideal Nyquist pulse shaping. Let P, N_0, and $\rho = Pg/N_0 BW$ denote the average power, the PSD of AWN, and the average SNR, respectively. The average symbol SNR is given by the following equation:

$$\frac{\overline{E}_s}{N_0} = \frac{PT}{N_0} = \overline{\rho}. \tag{6.223}$$

We use the following bit error probability bound, which is accurate within 1 dB penalty for $M \geq 4$ [5]:

$$P_b \leq 0.2 \exp\left[-1.5\rho/(M-1)\right], \ 0 \leq \rho \leq 30 \text{ dB}, \ M \geq 4. \tag{6.224}$$

Let us now adapt the transmit power $P(\rho)$ depending on symbol SNR ρ, subject to the average power constraint P and an instantaneous probability of error constraint $P_b(\rho) = P_b$. Given that transmitted power is adapted, the received SNR will be $\rho P(\rho)/P$, while the corresponding P_b bound can be written as

$$P_b \leq 0.2\exp\left[\frac{-1.5\rho}{M(\rho)-1}\frac{P(\rho)}{\bar{P}}\right],\quad 0 \leq \rho \leq 30\,\text{dB}, M \geq 4. \tag{6.225}$$

By solving Eq. (6.225) for $M(\rho)$, we obtain

$$M(\rho) = 1 + \underbrace{\frac{1.5}{-\log\,(5P_b)}}_{C_f}\rho\frac{P(\rho)}{\bar{P}} = 1 + C_f\rho\frac{P(\rho)}{\bar{P}},\quad 0 \leq \rho \leq 30\,\text{dB}, M \geq 4, \tag{6.226}$$

where $C_f = 1.5/(-\log(5P_b))$ is the correction factor, which is used to account for the fact that non-optimized signal constellation is used. The spectral efficiency is maximized by optimizing the $\langle \log_2 M \rangle$ [bits/symbol], subject to (s.t.) the corresponding power constraint as follows:

$$\begin{aligned}\langle \log_2 M \rangle &= \max \int_0^{+\infty} \log_2\left[1 + C_f\rho\frac{P(\rho)}{\bar{P}}\right] f(\rho)d\rho,\\ \text{s.t.} &\int_0^{+\infty} P(\rho)f(\rho)d\rho = \bar{P}.\end{aligned} \tag{6.227}$$

Based on Chap. 4, as expected, the optimum power adaptation policy is the water-filling method:

$$\begin{aligned}\frac{P(\rho)}{\bar{P}} &= \begin{cases} 1/\rho_{\text{tsh}} - 1/\rho C_f, & \rho \geq \rho_{\text{tsh}}/C_f \\ 0, & \rho < \rho_{\text{tsh}}/C_f \end{cases} \Leftrightarrow\\ \frac{C_f P(\rho)}{\bar{P}} &= \begin{cases} 1/\rho_{C_f} - 1/\rho, & \rho \geq \rho_{C_f} \\ 0, & \rho < \rho_{C_f} \end{cases}, \rho_{C_f} = \rho_{\text{tsh}}/C_f.\end{aligned} \tag{6.228}$$

After substituting the optimum power allocation into power constraint, we obtain the following equation:

$$\int_{\rho_{C_f}}^{+\infty} \left(\frac{1}{\rho_{C_f}} - \frac{1}{\rho}\right) f(\rho)d\rho = C_f, \tag{6.229}$$

which can be used to determine ρ_{Cf}. On the other hand, after substituting Eq. (6.228) into Eq. (6.227), we obtain the following expression for instantaneous data rate [bits/symbol]:

$$\log_2 M(\rho) = \log_2\left(\frac{\rho}{\rho_{C_f}}\right). \tag{6.230}$$

The corresponding expression for spectral efficiency is obtained by averaging out the instantaneous data rate as follows:

$$\frac{R}{BW} = \langle \log_2 M(\rho) \rangle = \left\langle \log_2\left(\frac{\rho}{\rho_{C_f}}\right) \right\rangle = \int_{\rho_{C_f}}^{\infty} \log_2\left(\frac{\rho}{\rho_{C_f}}\right) f(\rho)d\rho. \tag{6.231}$$

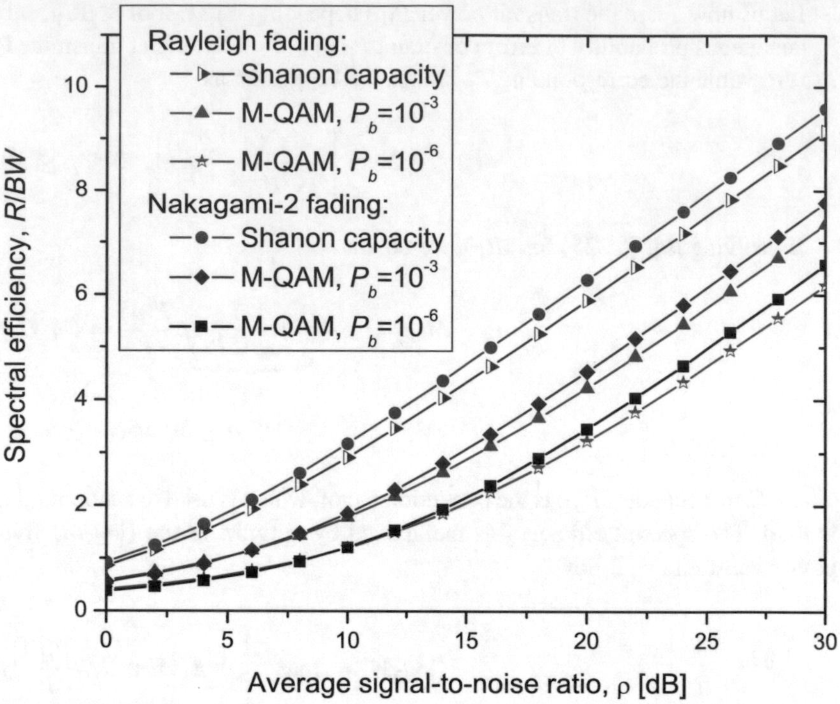

Fig. 6.21 Average spectral efficiencies for adaptive M-QAM in Rayleigh fading and Nakagami-2 fading

As an illustration, in Fig. 6.21, we provide the spectral efficiencies vs. average SNR for Rayleigh fading and Nakagami-2 fading for adaptive M-ary QAM. Because the M-ary QAM is not capacity modulation format, there is the gap with respect to Shannon capacity, related to the constant C_f.

The optimization given by Eq. (6.229) can be computationally extensive. To solve for this problem, we can apply *channel inversion-based power adaptation*. With channel inversion, the receiver will experience the constant SNR, and the transmitter can employ a single fixed data rate M-QAM modulation archiving the target P_b. The constellation size M that satisfies the target P_b can be obtained by substituting the channel inversion power adaptation given $P(\rho)/P = \rho_0 / \rho$, $\rho_0 = 1/\langle 1/\rho\rangle$. The corresponding spectral efficiency for the channel inversion-based power adaptation is given as follows:

$$\frac{R}{BW} = \log_2 M = \log_2\left[1 + C_f\frac{P(\rho)}{\bar{P}}\right] = \log_2\left[1 + C_f\frac{1}{\langle 1/\rho\rangle}\right],$$
$$C_f = \begin{cases} \dfrac{1.5\rho}{-\log(5P_b)}, M \geq 4 \\ \dfrac{1.5}{-\log(0.5P_b)}, M < 4 \end{cases}. \tag{6.232}$$

To avoid the problem associated with small SNRs tending to zero, the *truncated channel inversion-based power adaptation* can be used in which we do not transmit when SNR falls below the threshold ρ_{tsh}, which is determined to maximize the spectral efficiency below:

$$\frac{R}{BW} = \max_{\rho_{\text{tsh}}} \log_2\left[1 + C_f\frac{1}{\langle 1/\rho\rangle_{\rho_{\text{tsh}}}}\right] P(\rho > \rho_{\text{tsh}}). \tag{6.233}$$

As an illustration, in Fig. 6.22, we provided the average spectral efficiencies for various power adaptation strategies related to M-QAM in Rayleigh fading. Clearly, the truncated channel inversion, although not optimal, performs comparable to water-filling method in low- and medium-SNR regimes. On the other hand, the channel inversion strategy faces significant performance degradation in the spectral efficiency sense. The corresponding spectral efficiency plots related to Nakagami-2 fading are provided in Fig. 6.23.

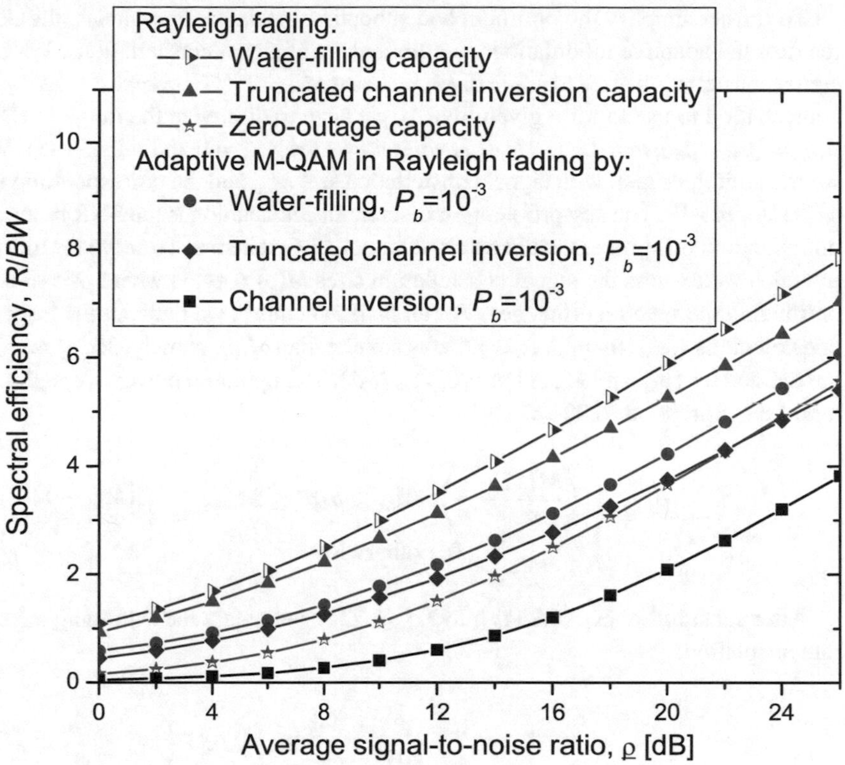

Fig. 6.22 Average spectral efficiencies for various power adaptation strategies related to M-QAM in Rayleigh fading

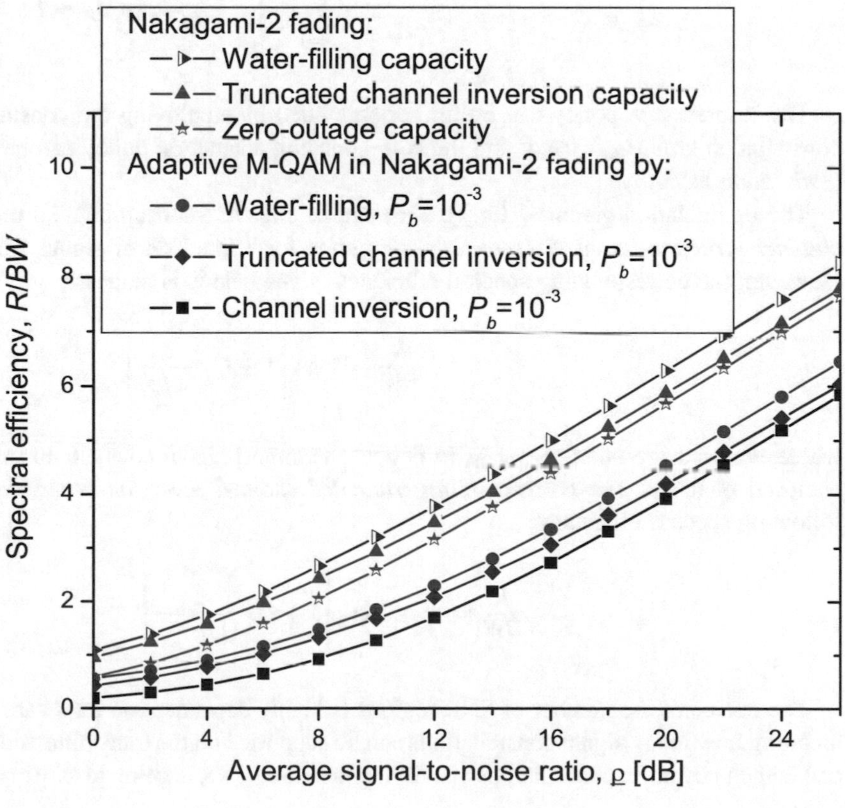

Fig. 6.23 Average spectral efficiencies for various power adaptation strategies related to M-QAM in Nakagami-2 fading

To further simplify the optimum and suboptimum adaptation policies, the *discrete-rate adaptation* should be used, which restricts the adaptive modulations to a limited set of signal constellations. For instance, for square-QAM, we can restrict the signal constellation sizes to $M_0 = 0$, $M_1 = 2$, and $M_m = 2^{2(m-1)}$, $m = 2, \ldots, N-1$ (for some $N > 2$). To determine which signal constellation to use to for a given SNR ρ, we have to discretize the range of SNRs due to fading effects into N SNR regions $\Delta\rho_m = [\rho_{m-1}, \rho_m)$, $m = 0, \ldots, N-1$, wherein we set $\rho_{-1} = 0$ and $\rho_{N-1} \to \infty$. Whenever the signal SNR falls into region R_m, we transmit the signal with signal constellation size M_m, and the corresponding spectral efficiency in this case is $\log_2 M_m$ [bits/s/Hz] (for $m > 0$). The key problem for discrete-rate adaptation is the SNR boundaries of the $\Delta\rho_m$ regions. Even with numerical optimization, this process could be computationally extensive. To solve for this problem, we can use a suboptimum technique in which we express the signal constellation sizes $M(\rho) = \rho/\rho^*$, where $\rho^* > 0$ is a single SNR parameter to be determined by optimizing the spectral efficiency. For an instantaneous ρ, we transmit the largest signal constellation from the set of possible constellations $\{M_m \mid m = 0, \ldots, N\}$ that is smaller than $M(\rho)$, namely, $M_m \leq M(\rho) < M_{m+1}$. The corresponding SNR boundaries are defined by $\{\rho_m = \rho^* M_{m+1} \mid m = 0, \ldots, N-2\}$. The optimum power adaptation policy is again water filling, but with slightly modified expression (6.228):

$$\frac{C_f P(\rho)}{\overline{P}} = \begin{cases} \left(\frac{M(\rho)}{\rho} - \frac{1}{\rho}\right), & M_m < \rho/\rho^* \leq M_{m+1} \\ 0, & \text{otherwise} \end{cases} = \begin{cases} (M_m - 1)(1/\rho), & M_m < \rho/\rho^* \leq M_{m+1} \\ 0, & \text{otherwise} \end{cases}. \tag{6.234}$$

After substituting Eq. (6.234) into Eq. (6.227), we obtain the following expression for the spectral efficiency for discrete-rate adaptation:

$$\frac{R}{BW} = \sum_{m=1}^{N-1} \log_2(M_m) P\left(M_m \leq \frac{\rho}{\rho^*} < M_{m+1}\right). \tag{6.235}$$

The parameter ρ^* is determined from the power constraint condition:

$$\sum_{m=1}^{N-1} \int_{\rho^* M_m}^{\rho^* M_{m+1}} \frac{P_m(\rho)}{\overline{P}} f(\rho) d\rho = 1. \tag{6.236}$$

The discrete-rate policy can be further simplified by employing the constant power for each SNR region, when signal constellation size M_m is used; and the corresponding adaptation policy can be called *discrete-power discrete-rate adaptive modulation technique.*

The optimization given by Eq. (6.236) can be still time-consuming. To reduce the complexity further, we can use the *channel inversion-based discrete-rate adaptation* for a fixed set of signal constellation sizes $\{M_m \mid m = 0, 1, \ldots, N-1\}$. However, the corresponding spectral efficiency, given below, is reduced:

$$\frac{R}{BW} = \log_2 \left\lfloor 1 + C_f \frac{1}{\langle 1/\rho \rangle} \right\rfloor_{\{M_m | m=0,1,\cdots,N-1\}}, \tag{6.237}$$

where we used the notation $\lfloor \cdot \rfloor_{Mm}$ to denote the largest signal constellation in the set $\{M_m\}$ smaller than or equal to the enclosed quantity. The corresponding *truncated channel inversion-based discrete-rate adaptation* policy will have the following spectral efficiency:

$$\frac{R}{BW} = \max_{\rho_{tsh}} \log_2 \left\lfloor 1 + C_f \frac{1}{\langle 1/\rho \rangle_{\rho_{tsh}}} \right\rfloor_{\{M_m | m=0,1,\cdots,N-1\}} P(\rho > \rho_{tsh}). \tag{6.238}$$

The choice of the number of SNR regions is highly dependent on hardware constraints and how fast the fading changes, dictating how many signal constellations can be supported on the transmitter side and at what rate to adapt modulation format and launch power. In particular, by specifying how long SNR is going to stay in a given SNR region gives us an information at which rate to adapt and how many SNR regions to use. This information is provided by the *average fade duration* concept introduced in Chap. 2. We use T_m to denote the average fade region duration that SNR ρ stays within the mth SNR region. The finite-state Markov model, also described in Chap. 2, provides a reasonable complexity model as it assumes that the SNR ρ remains within one SNR region for the duration of symbol period and that from the given SNR region the transition is possible only to neighboring regions. Under these assumptions, the transition probabilities between the regions are given by [60]

$$p_{m,m+1} = \frac{L_{m+1}T}{\pi_m}, \; p_{m,m-1} = \frac{L_m T}{\pi_m}, \quad p_{m,m} = 1 - p_{m,m+1} - p_{m,m-1}, \tag{6.239}$$

where π_m is the probability of being in the mth SNR region, namely, $\pi_m = P(\Delta\rho_m \le \rho \le \Delta\rho_{m+1})$. With L_m we denoted the mth SNR level ($\Delta\rho_m$) crossing rate, which is for Rician fading given by [13]

$$\begin{aligned} L_m &= \sqrt{2\pi(K+1)}\nu_D\left(\frac{\rho_m}{\overline{\rho}}\right)^{1/2} \\ &\exp\left(-K-(K+1)\frac{\rho_m}{\overline{\rho}}\right) I_0\left(2\sqrt{K(K+1)\frac{\rho_m}{\overline{\rho}}}\right), \end{aligned} \tag{6.240}$$

where K is the Rician parameter and ν_D is the Doppler frequency. For Rayleigh fading, by setting $K = 2$, the previous equation simplifies to

$$L_m^{(\text{Rayleigh})} = \sqrt{2\pi}\nu_D\left(\frac{\rho_m}{\overline{\rho}}\right)^{1/2} \exp\left(-\frac{\rho_m}{\overline{\rho}}\right). \tag{6.241}$$

The average fade duration for the mth SNR region can be determined as

$$\overline{T}_m = \frac{T}{p_{m,m+1} + p_{m,m-1}} = \frac{\pi_m}{L_{m+1} + L_m}. \tag{6.242}$$

Clearly, T_m is reversely proportional to the Doppler frequency.

6.5.2 Adaptive Coded Modulation

When compared to adaptive modulation alone, the additional coding gain can be achieved by superimposing either trellis codes or coset codes on top of the adaptive modulation scheme. The key idea behind this adaptive coded modulation is to exploit the separability of coding stage and constellation design inherent to the coset codes [61, 62], which are described in Chap. 9. The instantaneous SNR varies with time, which will change the distance in the received signal constellation $d_0(t)$, and, therefore, the corresponding minimum distance between coset sequences d_s and the minimum distance between coset points d_c will be time varying as well. By using adaptive modulation in tandem with the coset codes, we can ensure that distances ds and dc do not vary as channel conditions change by adapting the signal constellation size $M(\rho)$, transmit power $P(\rho)$, and/or symbol rate $R_s(\rho)$ of the transmitted signal constellation relative to the SNR, subject to an average transmit power $\overline{P}$ constraint on SNR-dependent transmit power $P(\rho)$. As minimum distance does not vary as channel conditions change $d_{\min}(t) = \min\{d_s(t), d_c(t)\} = \underline{d_{\min}}$, the adaptive coded modulation will exhibit the same coding gain as coded modulation developed for a Gaussian-like channel with corresponding minimum codeword distance $d_{\min}$. Therefore, this adaptive coded modulation scheme requires small changes in the generic coded modulation scheme described above, as shown in Fig. 6.24. Clearly, the coding stage does not change at all. The number of cosets does not change as well. On the other hand, the number of points within the coset is channel SNR dependent. When channel condition get improved, we enlarge the coset size. In such a way, we transmit more uncoded bits while not affecting the distance of the channel code. Additionally, this approach is hardware friendly.

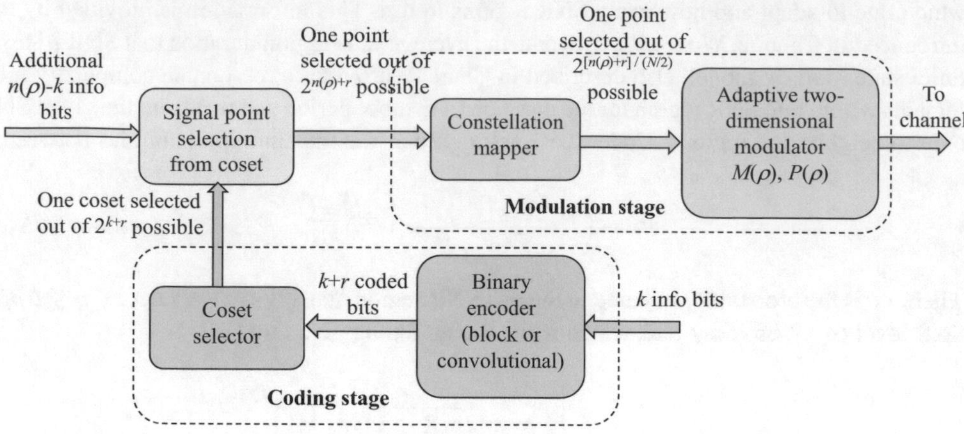

Fig. 6.24 Generic adaptive coded modulation scheme

The following approximation for bit error probability for M-ary QAM systems can be used:

$$P_b \leq 0.2 \exp\left[\frac{-1.5\rho \cdot G_c}{M(\rho) - 1}\frac{P(\rho)}{\overline{P}}\right], \quad M \geq 4, 0 \leq \rho\,[\mathrm{dB}] \leq 30\ \mathrm{dB}, \tag{6.243}$$

where G_c is the coding gain. We adjust the signal constellation size $M(\rho)$ depending on channel SNR and target BER as follows:

$$M(\rho) = 1 + \underbrace{\frac{1.5}{-\log(5P_b)}}_{C_f}\rho \cdot G_c\frac{P(\rho)}{\overline{P}} = 1 + C_f\rho \cdot G_c\frac{P(\rho)}{\overline{P}}, \tag{6.244}$$

where C_f is again the correction factor, defined as $C_f = -1.5/\log(5P_b)$, which is used to account for the fact that non-optimized signal constellation is used. The number of uncoded bits needed to select a point within the coset is determined by

$$n(\rho) - 2k/N = \log_2 M(\rho) - 2(k + r)/N. \tag{6.245}$$

The normalized transmission rate is given by $\log_2 M(\rho)$, while the normalized adaptive data rate is determined by $\log_2 M(\rho)$-$2r/N$. To determine the optimum launch power, we need to optimize $\langle \log_2 M(\rho) \rangle$, subject to the average power constraint, and not surprisingly the optimum power adaptation strategy is the water-filling law:

$$\frac{P(\rho)}{\overline{P}} = \begin{cases} 1/\rho_0 - 1/\rho_c, & \rho \geq \underbrace{\rho_0/C_c}_{\rho_c} \\ 0, & \rho < \rho_0/C_c \end{cases}, \quad C_c = C_f G_c, \tag{6.246}$$

where ρ_0 is the threshold signal-to-noise ratio for an AWGN channel. After substituting Eq. (6.246) into Eq. (6.244), we obtain the following equation for adaptive signal constellation size:

$$M(\rho) = \rho/\rho_c. \tag{6.247}$$

The corresponding spectral efficiency is given by

$$\frac{R}{BW} = \langle \log_2 M(\rho) \rangle = \int_{\rho_c}^{\infty} \log_2\left(\frac{\rho}{\rho_c}\right) f(\rho) d\rho. \tag{6.248}$$

6.6 Volterra Series-Based Equalization

One important application of the Volterra series is in compensation of nonlinear effects in fiber-optics communications [63–70] or to deal with nonlinearities on satellite channels. The nonlinear compensation can be performed either in time domain [63, 70] or frequency domain [67, 91]. The frequency domain is particularly attractive, as the theory behind it is well developed [67, 68]. Expansion in the Volterra series is given as [64]

$$\begin{aligned} y(t) &= \sum_{n=0}^{\infty} y_n(t), \\ y_n(t) &= \int_{-\infty}^{\infty} \cdots \int_{-\infty}^{\infty} h_n(\tau_1, \tau_2, \cdots \tau_n) x(t-\tau_1) x(t-\tau_2) \cdots x(t-\tau_n) d\tau_1 d\tau_2 \cdots d\tau_n, \end{aligned} \tag{6.249}$$

where $x(t)$ is the input to the nonlinear system, while the nth term $y_n(t)$ represents the n-fold convolution of the input and the n-fold impulse response $h_n(\tau_1, \cdots \tau_n)$. The set $\{h_0, h_1(\tau_1), h_2(\tau_1, \tau_2), \cdots, h_n(\tau_1, \tau_2, \cdots \tau_n), \cdots\}$ of impulse responses is known as the Volterra kernel of the system. The *0*th order corresponds to d.c. component, which is not represented here, while the first-order term corresponds to impulse response of the linear system. The first-order impulse response can be expressed in analytical form, where the second-order GVD can be neglected. The frequency domain representation of the Volterra series is given by [67]

$$Y(\omega) = \sum_{n=1}^{\infty} \int_{-\infty}^{\infty} \cdots \int_{-\infty}^{\infty} H_n(\omega_1, \cdots \omega_n) X(\omega_1) \cdots X(\omega_n) X(\omega - \omega_1 - \cdots \omega_n) d\omega_1 \cdots d\omega_n. \tag{6.250}$$

As an example, if H_3 denotes the three-dimensional Fourier transform of kernel h_3, it can be estimated as [64]

$$\begin{aligned} H_3(f,g,h) &= \frac{S_{xxxy}(f,g,h)}{6 S_{xx}(f) S_{xx}(g) S_{xx}(h)}, \\ S_{xx}(f) &= \frac{1}{T} E\left\{ |X_i(f)|^2 \right\}, \ S_{xxxy}(f,g,h) = E\left\{ X_i^*(f) X_i^*(g) X_i^*(h) Y_i(f+g+h) \right\}, \end{aligned} \tag{6.251}$$

where $X(f)$ and $Y(f)$ are the Fourier transforms of $x(t)$ and $y(t)$, T is the observation interval, while the subscript i denotes the index of record of duration T.

The propagation over the SMF is governed by the nonlinear Schrödinger equation (NLSE), as described in Chap. 2. Since we are concerned here with compensation of chromatic dispersion and nonlinearities, let us observe the following inverse version of NLSE that can be obtained from original NLSE by changing the sign of the terms:

$$\begin{aligned} \frac{\partial E(z,t)}{\partial z} &= (-\hat{D} - \hat{N}) E(z,t), \\ \hat{D} &= -\frac{\alpha}{2} - j\frac{\beta_2}{2}\frac{\partial^2}{\partial t^2} + \frac{\beta_3}{6}\frac{\partial^3}{\partial t^3}, hatN = j\gamma |E^2|, \end{aligned} \tag{6.252}$$

where E is the signal electric field, $\hat{D}$ and $\hat{N}$ denote the linear and nonlinear operators, while α, β_2, β_3, and γ represent the attenuation coefficient, GVD, second-order GVD, and nonlinear coefficient, respectively. By setting that $X(\omega) = E(\omega,z) = \text{FT}\{E(z,t)\}$, the electric field from Eq. (2.252) can be presented as [69]

$$E(\omega,0) \approx H_1(\omega,z) E(\omega,z) + \iint H_3(\omega_1, \omega_2, \omega - \omega_1 + \omega_2, z) E(\omega_1, z) E^*(\omega_2, z) E(\omega - \omega_1 + \omega_2, z) d\omega_1 \omega_2, \tag{6.253}$$

where the linear kernel in the frequency domain is given as

$$H_1(\omega, z) = \exp\left[\alpha z/2 - j\beta_2\omega^2 z/2\right], \tag{6.254}$$

while the second-order GVD is neglected. The third-order kernel in the frequency domain is given by [69]

$$H_3(\omega_1, \omega_2, \omega - \omega_1 + \omega_2, z) = -j\gamma H_1(\omega, z)\frac{1 - \exp\left[\alpha z - j\beta_2(\omega_1 - \omega)(\omega_1 - \omega_2)z\right]}{-\alpha + j\beta_2(\omega_1 - \omega)(\omega_1 - \omega_2)}. \tag{6.255}$$

As we see in Eq. (6.255), fiber parameters of opposite sign are used in the design of the nonlinear equalizer based on the Volterra series. For ultra-long-haul transmission systems' studies, the second-order GVD must be included. Also, the fifth-order kernel term should be calculated as well. Regarding the complexity of this method, the following rule of thumb can be used. The number of operations related for the nth kernel is approximately equal to the nth power of the number of operations needed for the first-order kernel.

6.7 Digital Backpropagation in Fiber-Optics Communications

Another method to simultaneously compensate for chromatic dispersion and fiber nonlinearities is known as the digital backpropagation compensation [72, 73]. The key idea of this method is to assume that received signal in the digital domain can now propagate backward through the same fiber. This virtual backpropagation is done by using the fiber with parameters just opposite in sign to real fiber parameters (applied in forward propagation). By this approach, we will be able in principle to compensate for fiber nonlinearities if there was not signal-to-noise nonlinear interaction. This virtual backpropagation (BP) can be performed either on transmitter side or receiver side, as illustrated in Fig. 6.25. In the absence of noise, those two approaches are equivalent to each other.

The backpropagation is governed by inverse NLSE, given by Eq. (6.252). The backpropagation method operates on the signal electric field, and it is, therefore, universal and independent on modulation format. It uses a split-step Fourier method of reasonable high complexity to solve it. Both asymmetric and symmetric split-step Fourier methods [74] can be used. As an illustration, we briefly describe the iterative symmetric split-step Fourier method. Additional details can be found in Chap. 2. The key idea of this method is to apply the linear and nonlinear operators in an iterative fashion.

The linear operator, which corresponds to multiplication in the frequency domain, is expressed as

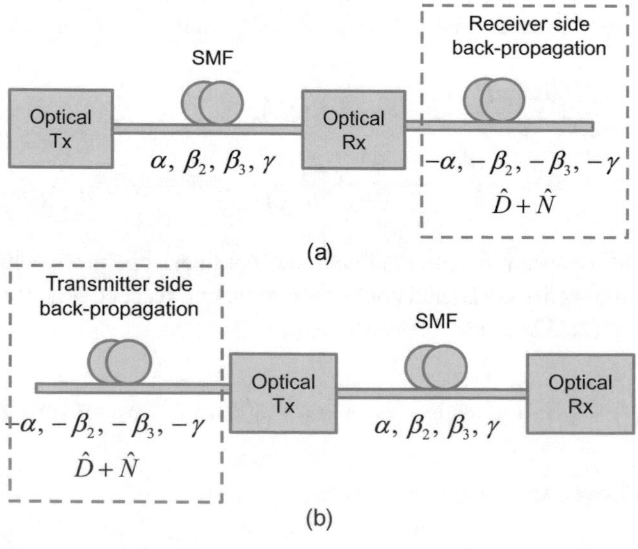

Fig. 6.25 Digital backpropagation (BP) method illustration: (**a**) receiver side BP and (**b**) transmitter side BP

$$\exp\left[-\Delta z\widehat{D}\right]E(z+\Delta z,t)=FT^{-1}\left\{\exp\left[-\left(-\frac{\alpha}{2}+j\frac{\beta_2}{2}\omega^2+j\frac{\beta_3}{6}\omega^3\right)\Delta z\right]FT[E(z+\Delta z,t)]\right\}, \tag{6.256}$$

where Δz is the step size, while FT (FT^{-1}) denotes the Fourier transform (inverse Fourier transform). On the other hand, the nonlinear operator, which performs the nonlinear phase "rotation" in the time domain, is expressed as

$$\exp\left(-\Delta z\xi\hat{N}\right)E(z+\Delta z,t)=\exp\left(-j\xi\gamma\Delta z|E(z+\Delta z,t)|^2\right)E(z+\Delta z,t), \tag{6.257}$$

where $0\leq\xi\leq1$ is the correction factor, which is needed to account for ASE noise-signal nonlinear interaction during propagation.

It is evident that the nonlinear operator depends on the magnitude of the electric field at location z, which should be determined. The electric field can be found by using the trapezoidal rule, as proposed in [74], so that we can write

$$\begin{aligned}E(z,t)&=\exp\left(-\hat{D}\Delta z/2\right)\exp\left(-\xi\int_{z+\Delta z}^{z}N(z')dz'\right)\exp\left(-\hat{D}\Delta z/2\right)E(z+\Delta z,t)\\&\approx\exp\left(-\hat{D}\Delta z/2\right)\exp\left[-\xi\frac{\hat{N}(z+\Delta z)+\hat{N}(z)}{2}\Delta z\right]\exp\left(-\hat{D}\Delta z/2\right)E(z+\Delta z,t).\end{aligned} \tag{6.258}$$

Since the operator $\widehat{N}(z)=j\gamma|E(z,t)|^2$ is dependent on the output, Eq. (6.258) can be solved in an iterative fashion, as illustrated by the flowchart in Fig. 6.26.

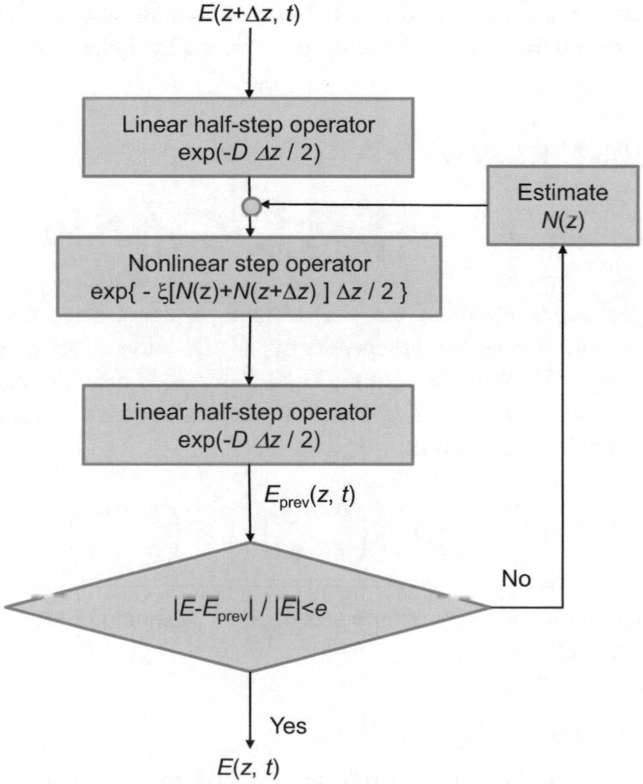

Fig. 6.26 The iterative symmetric split-step Fourier method to solve the inverse nonlinear Schrödinger equation

The operating frequency range of interest in BP method must be smaller than the sampling rate, and therefore, the oversampling is needed. Regarding the complexity issues, FFT complexity scales with the factor $N{\cdot}\log N$, while the filtering and nonlinear phase rotation operations have complexity proportional to N.

As for the *multichannel BP*, there are two options that can be applied. The first one is to combine different WDM channels together so that the electric field can be written as

$$E(z,t)=\sum\nolimits_n E_n(z,t)e^{-jn2\pi\Delta f}, \tag{6.259}$$

where Δf is the channel spacing. This approach can be called the "total-field BP" [73]. The second approach is to consider individual WDM separately and then solve the set of coupled inverse NLSE [74]:

$$\begin{aligned}\frac{\partial E_i(z,t)}{\partial z}&=\left(-\widehat{D}_i-\xi\widehat{N}_i\right)E_i(z,t),\\ \widehat{D}_i&=-\frac{\alpha}{2}-\beta_{1,i}\frac{\partial}{\partial t}-j\frac{\beta_2}{2}\frac{\partial^2}{\partial t^2}+\frac{\beta_3}{6}\frac{\partial^3}{\partial t^3},\widehat{N}_i=j\gamma\left(|E_i|^2+\sum_{j\neq i}|E_j|^2\right).\end{aligned} \tag{6.260}$$

Although this method appears to be more extensive, the synchronization of all WDM channels is not required. Both methods are suitable for point-to-point application. A different situation is in mesh networks, since some wavelength channels might be dropped and added several times along the route. As a consequence, if the exact situation along the route is not known at the receiver side, the application of multiband BP might even lead to system performance degradation.

The BP method can also be used in polarization-division multiplexed (PDM) systems. Since the PMD rotation matrix changes in a random fashion from fiber section to fiber section, it is quite challenging to predict the individual rotation matrices. As a consequence, it is expected that the BP method will not be that efficient in simultaneous compensation of nonlinear effects and PMD. In addition, the complexity of BP method to address PMD is higher as now the electric field scalars used in single-channel single-polarization BP should be replaced by Jones matrices. The inverse NSE for PDM systems will be now

$$\begin{aligned}\frac{\partial \boldsymbol{E}(z,t)}{\partial z}&=\left(-\widehat{\boldsymbol{D}}-\xi\widehat{\boldsymbol{N}}\right)\boldsymbol{E}(z,t),\\ \widehat{\boldsymbol{D}}&=-\frac{\boldsymbol{\alpha}}{2}-\boldsymbol{\beta}_1\frac{\partial}{\partial t}-j\frac{\boldsymbol{\beta}_2}{2}\frac{\partial^2}{\partial t^2}+\frac{\boldsymbol{\beta}_3}{6}\frac{\partial^3}{\partial t^3},\widehat{\boldsymbol{N}}=j\gamma\left(|\boldsymbol{E}|^2\boldsymbol{I}\text{-}\frac{1}{3}\left(\boldsymbol{E}^\dagger Y\boldsymbol{E}\right)Y\right),\end{aligned} \tag{6.261}$$

where $\boldsymbol{E}=[E_x\ E_y]^T$ denotes the Jones vector of the electric field, $\boldsymbol{\alpha}$ denotes the attenuation coefficients in x- and y-polarizations, while $\boldsymbol{\beta}_1$, $\boldsymbol{\beta}_2$, and $\boldsymbol{\beta}_3$ denote the group-velocity, GVD, and second-order GVD parameters in x- and y-polarizations, respectively. Finally, with Y we denoted the Pauli-Y matrix in quantum mechanics notation [75]. Namely, the Pauli matrices are different in quantum mechanics than in representation used in papers and books related to optics [73, 74]. In our case, the Pauli matrices are defined as

$$X=\begin{pmatrix}0&1\\1&0\end{pmatrix},Y=\begin{pmatrix}0&-j\\j&0\end{pmatrix},\ Z=\begin{pmatrix}1&0\\0&-1\end{pmatrix}. \tag{6.262}$$

If there is no possibility to track the Jones vector of each section, then $\boldsymbol{\beta}_1$ should be set to zero. Otherwise, $\boldsymbol{\beta}_1$ is related to the polarization rotation matrix $\boldsymbol{R}$ as follows:

$$\begin{aligned}\boldsymbol{\beta}_1(\theta_k,\varepsilon_k)&=\boldsymbol{R}(\theta_k,\varepsilon_k)\left(\frac{\delta_k}{2}Z\right)\boldsymbol{R}^\dagger(\theta_k,\varepsilon_k),\\ \boldsymbol{R}(\theta_k,\varepsilon_k)&=\begin{bmatrix}\cos\theta_k\cos\varepsilon_k-j\sin\theta_k\sin\varepsilon_k & \sin\theta_k\cos\varepsilon_k+j\cos\theta_k\sin\varepsilon_k\\ \sin\theta_k\cos\varepsilon_k+j\cos\theta_k\sin\varepsilon_k & \cos\theta_k\cos\varepsilon_k-j\sin\theta_k\sin\varepsilon_k\end{bmatrix},\end{aligned} \tag{6.263}$$

where $2\theta_k$ and $2\varepsilon_k$ are the azimuth and ellipticity angles of the kth fiber section and δ_k corresponds to the differential group delay (DGD) of the section. The solution of Eq. (6.236) should be obtained by the split-step Fourier method

$$\boldsymbol{E}(z,t) = \exp\left[-\Delta z\left(\widehat{\boldsymbol{D}} + \xi\widehat{\boldsymbol{N}}\right)\right]\boldsymbol{E}(z+\Delta z,t), \tag{6.264}$$

where the complexity of the iteration process is higher than in the single-polarization case.

6.8 Optical Communication Systems with Coherent Optical Detection

The principles of coherent optical detection have been already introduced in Chap. 3. In particular, in Sect. 3.1.8, we described balanced coherent optical detectors for 1-D and 2-D modulation schemes. The balanced coherent detector suitable for homodyne, intradyne, and heterodyne detections based on two Y-junctions and two 3 dB couplers was described. In this section, we describe other alternatives for coherent balanced detection, in particular 2 × 4 optical hybrid-based balanced coherent optical detector, polarization-division demultiplexing scheme, homodyne coherent optical detection based on phased-locked loop (PLL), phase diversity concepts, as well as dominant coherent optical detection noise sources.

6.8.1 Balanced Coherent Optical Detection for 2-D Modulation Schemes

Various alternatives suitable for balanced coherent optical detector implementation include (i) 2 × 4 optical hybrid, (ii) 3 dB couplers and phase shifter, (iii) 3 dB coupler and polarization beam splitter (PBS), (iv) 3 × 3 coupler, (v) 4 × 4 multimode interference (MMI) coupler, and (vi) Y-junction, 3 dB coupler, and phase shifter-based solutions, to mention a few. The option (v) has already been described in Sect. 3.1.9. In Fig. 6.27, we provide the description of basic 2 × 2 optical hybrid, with two phase trimmers, suitable for implementation of 2 × 4 optical hybrid. Let the input to the coherent detector be given by $S_n = |S_n|\exp[j(\omega_T t + \phi_{PN,T} + \theta_n)]$, where the subscript n is used to denote the nth symbol being transmitted, ω_T denotes the transmit laser frequency (in rad/s), $\phi_{PN,T}$ denotes the laser phase noise of transmitter laser, and $|S_n|$ and θ_n denote the amplitude and the phase of the kth symbol for M-ary QAM/PSK (or any other 2-D signal constellation). Let the local laser (oscillator) output be denoted by $L = |L|\exp[j(\omega_{LOL} t + \phi_{PN,LOL})]$, where ω_{LOL} denotes the local oscillator laser frequency and $\phi_{PN,LOL}$ denotes the laser phase noise of local laser. The output port 1 electric field, denoted as $E_{o,1}$, is clearly related to the input port fields S_n and L by

$$E_{o,1} = \left(S_n e^{j\phi_1} + L\right)\sqrt{1-\kappa}, \tag{6.265}$$

where κ is the power splitting ratio and ϕ_1 is the phase shift introduced by the control voltage 1 (phase trimmer 1). On the other hand, the output port 2 electric field, denoted as $E_{o,2}$, is related to the input port fields S_n and L by

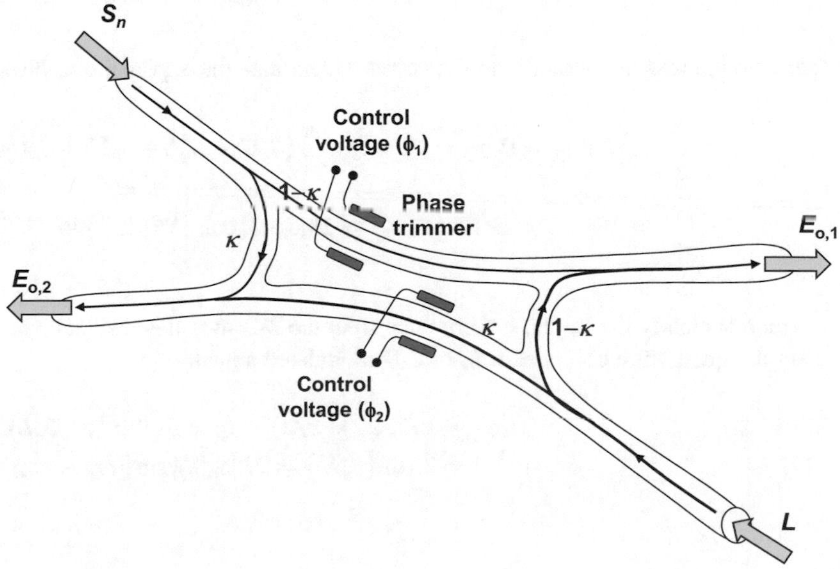

Fig. 6.27 The basic 2 × 2 optical hybrid with two phase trimmers

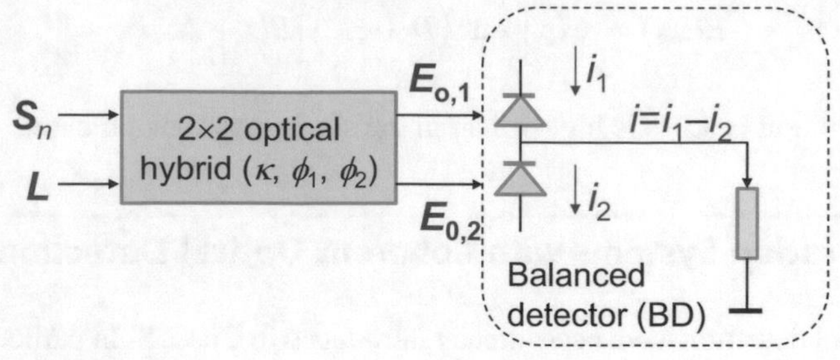

Fig. 6.28 The coherent balanced optical detector based on 2 × 2 optical hybrid

$$E_{o,2} = \left(S_n + e^{\,j\phi_2}L\right)\sqrt{\kappa}, \tag{6.266}$$

where ϕ_2 is the phase shift introduced by the control voltage 2 (phase trimmer 2). The previous two equations can be represented in the matrix form by

$$\begin{bmatrix} E_{o,1} \\ E_{o,2} \end{bmatrix} = \underbrace{\begin{bmatrix} e^{\,j\phi_1}\sqrt{1-\kappa} & \sqrt{1-\kappa} \\ \sqrt{\kappa} & e^{\,j\phi_1}\sqrt{\kappa} \end{bmatrix}}_{H} \begin{bmatrix} S_n \\ L \end{bmatrix} = H \begin{bmatrix} S_n \\ L \end{bmatrix},$$
$$H = \begin{bmatrix} e^{\,j\phi_1}\sqrt{1-\kappa} & \sqrt{1-\kappa} \\ \sqrt{\kappa} & e^{\,j\phi_1}\sqrt{\kappa} \end{bmatrix}, \tag{6.267}$$

where H is the scattering matrix. For power splitting ratio $\kappa = 1/2$, the scattering matrix simplifies to

$$H|_{\kappa=1/2} = \frac{1}{\sqrt{2}}\begin{bmatrix} e^{\,j\phi_1} & 1 \\ 1 & e^{\,j\phi_2} \end{bmatrix}. \tag{6.268}$$

Let us now observe the balanced coherent optical detector shown in Fig. 6.28, which is based on 2 × 2 optical hybrid shown in Fig. 6.27, characterized by power splitting ration κ and phase shifts ϕ_1 and ϕ_2. Assuming that both photodetectors have the same responsivity R and the power splitting ratio is $\kappa = 1/2$, the resulting photocurrent $i = i_1$-i_2 can be written as

$$\begin{aligned} i &= i_1 - i_2 = R|E_{o,1}|^2 - R|E_{o,2}|^2 \\ &= \frac{R}{2}\left(S_n e^{\,j\phi_1}L^* + S_n^* e^{-j\phi_1}L - S_n L^* e^{-j\phi_2} - S_n^* L e^{\,j\phi_2}\right), \end{aligned} \tag{6.269}$$

where we ignored the photodiode shot noise to facilitate the explanations. Now by setting $\phi_1 = 0$ and $\phi_2 = \pi$, we obtain

$$\begin{aligned} i(\phi_1 = 0, \phi_2 = \pi) &= i_I = \frac{R}{2}\left(S_n L^* + S_n^* L + S_n L^* + S_n^* L\right) = R\left(S_n L^* + S_n^* L\right) \\ &= 2R\,\mathrm{Re}\{S_n L^*\} = 2R|S_n||L|\cos\left[(\omega_T - \omega_{LO})t + \underbrace{\phi_{T,PN} - \phi_{LO,PN}}_{\phi_{PN}} + \theta_n\right], \end{aligned} \tag{6.270}$$

which is clearly the in-phase component i_I of the 2-D modulated signal. On the other hand, by $\phi_1 = \phi_2 = \pi/2$, we clearly detect the quadrature component i_Q of 2-D modulated signal:

$$\begin{aligned} i(\phi_1 = \pi/2, \phi_2 = \pi/2) &= i_Q = R\left(jS_n L^* - jS_n^* L\right) \\ &= 2R\mathrm{Im}\{S_n L^*\} = 2R|S_n||L|\sin\left[(\omega_T - \omega_{LO})t + \phi_{PN} + \theta_n\right]. \end{aligned} \tag{6.271}$$

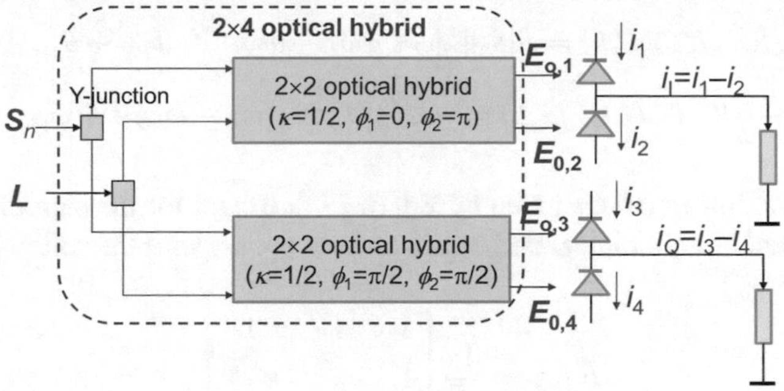

Fig. 6.29 The coherent balanced optical detector for demodulation of 2-D signals based on 2 × 4 optical hybrid

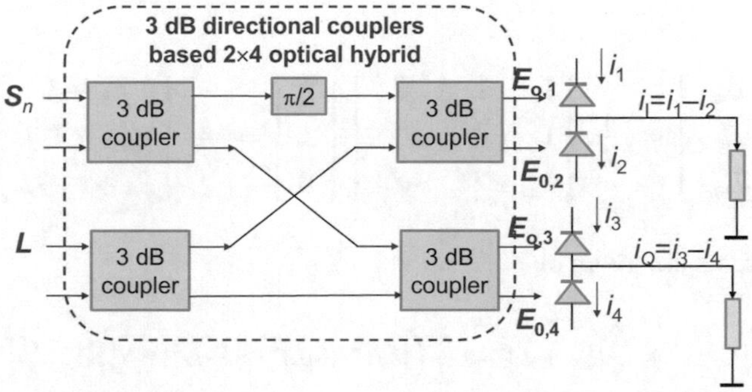

Fig. 6.30 The coherent balanced optical detector for demodulation of 2-D signals based on 2×4 optical hybrid implemented with the help of four 3 dB directional couplers. The scattering matrix of 3 dB directional coupler is given by $\boldsymbol{H}_{3\text{ dB}} = \frac{1}{\sqrt{2}}\begin{bmatrix} 1 & j \\ j & 1 \end{bmatrix}$

By combining these two balanced coherent detectors, we obtain the balanced coherent detector suitable of demodulation of 2-D modulated signals as shown in Fig. 6.29. The configuration of corresponding 2 × 4 optical hybrid is provided as well. To take the presence of Y-junctions into account, the corresponding expressions for detected in-phase i_I and quadrature i_Q components become, respectively,

$$\begin{aligned} i_I &= \frac{1}{2}R\left(S_n L^* + S_n^* L\right) = R|S_n||L| \cos\left[(\omega_T - \omega_{LO})t + \phi_{PN} + \theta_n\right], \\ i_Q &= \frac{1}{2}R\left(jS_n L^* - jS_n^* L\right) = R|S_n||L| \sin\left[(\omega_T - \omega_{LO})t + \phi_{PN} + \theta_n\right]. \end{aligned} \tag{6.272}$$

This scheme is applicable to (i) *homodyne detection* by setting $\omega_T = \omega_{LO}$; (ii) *heterodyne detection* by setting $\omega_T = \omega_{LO} + \omega_{IF}$ wherein ω_{IF} is the intermediate frequency larger than the symbol rate R_s, that is, $\omega_{IF} > 2\pi R_s$; and (iii) *intradyne detection* by properly choosing the frequency difference as follows. $|\omega_T - \omega_{LO}| \leq 2\pi R_s$.

The *3 dB directional coupler-based 2×4 optical hybrid* and corresponding balanced coherent optical detectors are provided in Fig. 6.30. By using the basic matrix multiplication, the output electric fields $E_{o,i}$ are related to the received optical field S_n and local oscillator laser output field L as follows:

$$\begin{aligned} E_{o,1} &= \frac{1}{2}(jS_n + jL), E_{o,2} = \frac{1}{2}(-S_n + L), E_{o,3} = \frac{1}{2}(jS_n - L), \\ E_{o,4} &= \frac{1}{2}(-S_n + jL). \end{aligned} \tag{6.273}$$

The corresponding in-phase i_I and quadrature i_Q photocurrents of balanced detector branches are equal to

$$\begin{aligned}i_I &= \frac{1}{2}R(S_nL^* + S_n^*L) = R|S_n||L|\cos\left[(\omega_T - \omega_{LO})t + \phi_{PN} + \theta_n\right],\\ i_Q &= -\frac{1}{2}R(\,jS_nL^* - jS_n^*L) = -R|S_n||L|\sin\left[(\omega_T - \omega_{LO})t + \phi_{PN} + \theta_n\right],\end{aligned} \tag{6.274}$$

and are identical to the corresponding equations given by Eq. (6.272), except for the sign of quadrature component.

The 3×3 coupler, implemented in either integrated optics or fiber optics, with the following scattering matrix [76]

$$\boldsymbol{H}_{3\times 3} = \frac{1}{\sqrt{3}}\begin{bmatrix}1 & 1 & 1\\ 1 & e^{j\frac{2\pi}{3}} & e^{-j\frac{2\pi}{3}}\\ 1 & e^{-j\frac{2\pi}{3}} & e^{j\frac{2\pi}{3}}\end{bmatrix}, \tag{6.275}$$

can also be used in coherent optical detection by using S_n and L as the first two inputs. The corresponding output port electric fields are

$$\begin{bmatrix}E_{o,1}\\ E_{o,2}\\ E_{o,3}\end{bmatrix} = \frac{1}{\sqrt{3}}\begin{bmatrix}1 & 1 & 1\\ 1 & e^{j\frac{2\pi}{3}} & e^{-j\frac{2\pi}{3}}\\ 1 & e^{-j\frac{2\pi}{3}} & e^{j\frac{2\pi}{3}}\end{bmatrix}\begin{bmatrix}S_n\\ L\\ 0\end{bmatrix} = \frac{1}{\sqrt{3}}\begin{bmatrix}S_n + L\\ S_n + e^{j\frac{2\pi}{3}}L\\ S_n + e^{-j\frac{2\pi}{3}}L\end{bmatrix}. \tag{6.276}$$

The corresponding photocurrents are given by

$$\begin{aligned}i_{o,1} &= R|E_{o,1}|^2 = \frac{R}{3}|S_n + L|^2 = \frac{R}{3}\left(|S_n|^2 + |L|^2 + S_nL^* + S_n^*L\right),\\ i_{o,2} &= R|E_{o,2}|^2 = \frac{R}{3}\left|S_n + e^{j\frac{2\pi}{3}}L\right|^2 = \frac{R}{3}\left(|S_n|^2 + |L|^2 + S_ne^{-j\frac{2\pi}{3}}L^* + S_n^*e^{j\frac{2\pi}{3}}L\right),\\ i_{o,3} &= R|E_{o,3}|^2 = \frac{R}{3}\left|S_n + e^{-j\frac{2\pi}{3}}L\right|^2 = \frac{R}{3}\left(|S_n|^2 + |L|^2 + S_ne^{j\frac{2\pi}{3}}L^* + S_n^*e^{-j\frac{2\pi}{3}}L\right).\end{aligned} \tag{6.277}$$

Now by subtracting the first two photocurrents, we obtain

$$\begin{aligned}i_{o,1} - i_{o,2} &= \frac{R}{3}\left(S_nL^* + S_n^*L - S_ne^{-j\frac{2\pi}{3}}L^* - S_n^*e^{j\frac{2\pi}{3}}L\right)\\ &= \frac{2R}{\sqrt{3}}|S_n||L|\cos\left[(\omega_T - \omega_{LO})t + \phi_{PN} + \theta_n - \pi/6\right]\\ &= \frac{2R}{\sqrt{3}}|S_n||L|\sin\left[(\omega_T - \omega_{LO})t + \phi_{PN} + \theta_n + \pi/6\right] \doteq i_Q,\end{aligned} \tag{6.278}$$

and this expression represents the quadrature component. On the other hand, the in-phase component can be obtained as follows:

$$i_{o,1} + i_{o,2} - 2i_{o,3} = \frac{2R}{\sqrt{3}}|S_n||L|\cos\left[(\omega_T - \omega_{LO})t + \phi_{PN} + \theta_n + \pi/6\right] \doteq i_I. \tag{6.279}$$

The 3 dB directional coupler together with two PBS can also be used to implement 2×4 optical hybrid required in 2-D modulation schemes, as illustrated in Fig. 6.31. We use $|\nearrow\rangle$ and $|R\rangle$ to denote the +π/4 and right-circular polarization states, defined, respectively, as

$$|\nearrow\rangle = \frac{1}{\sqrt{2}}\begin{bmatrix}1\\ 1\end{bmatrix} = \frac{1}{\sqrt{2}}(|x\rangle + |y\rangle),\ |R\rangle = \frac{1}{\sqrt{2}}\begin{bmatrix}1\\ j\end{bmatrix} = \frac{1}{\sqrt{2}}(|x\rangle + j|y\rangle), \tag{6.280}$$

where $|x\rangle$ and $|y\rangle$ are the x- and y-polarization states.

Therefore, the inputs to the 3 dB directional coupler are given by

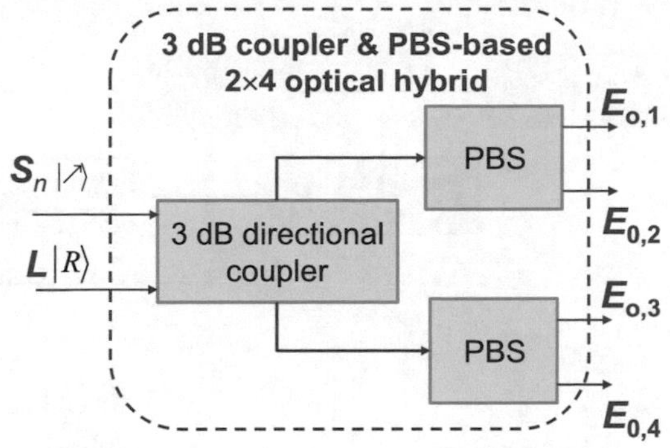

Fig. 6.31 The 2×4 optical hybrid implemented with the help of one 3 dB directional coupler and two polarization beam splitters

$$\boldsymbol{E}_{\mathrm{i},1} = S_n|\nearrow\rangle = \frac{S_n}{\sqrt{2}}(|x\rangle + |y\rangle), \boldsymbol{E}_{\mathrm{i},2} = L|R\rangle = \frac{L}{\sqrt{2}}(|x\rangle + j|y\rangle). \tag{6.281}$$

The corresponding output port electric fields are given, respectively, by

$$\begin{aligned} \boldsymbol{E}_{\mathrm{o},1} &= \frac{1}{2}(S_n + jL)|x\rangle, \boldsymbol{E}_{\mathrm{o},2} = \frac{1}{2}(S_n - L)|y\rangle, \\ \boldsymbol{E}_{\mathrm{o},3} &= \frac{1}{2}(jS_n + L)|x\rangle, \boldsymbol{E}_{\mathrm{o},4} = \frac{1}{2}(jS_n + jL)|y\rangle. \end{aligned} \tag{6.282}$$

The in-phase component can be obtained by placing balanced detector in front of output ports 2 and 4 to obtain

$$\begin{aligned} i_{\mathrm{o},2} - i_{\mathrm{o},4} &= \frac{R}{4}(|S_n - L|^2 - |S_n + L|^2) \\ &= -\frac{R}{2}(S_n L^* + S_n^* L) \\ &= -R|S_n||L|\cos\left[(\omega_T - \omega_{LO})t + \phi_{PN} + \theta_n\right] \doteq i_I. \end{aligned} \tag{6.283}$$

On the other hand, the quadrature component can be obtained by placing balanced detector in front of output ports 1 and 3 to obtain

$$\begin{aligned} i_{\mathrm{o},1} - i_{\mathrm{o},3} &= \frac{R}{4}\left(|S_n + jL|^2 - |jS_n + L|^2\right) \\ &= \frac{R}{2}\left(-jS_n L^* + jS_n^* L\right) = -R|S_n||L|\sin\left[(\omega_T - \omega_{LO})t + \phi_{PN} + \theta_n\right] \doteq i_Q. \end{aligned} \tag{6.284}$$

Regarding the implementation of the 2×4 optical hybrid with the help of 4×4 MMI, to [77, 78] an interested reader is referred.

6.8.2 Polarization Diversity and Polarization-Division Demultiplexing

The coherent optical receivers require matching the state of polarization (SOP) of the local laser with that of the received optical signal. In practice, only the SOP of local laser can be controlled; one possible configuration of the *polarization control receiver* is shown in Fig. 6.32. Polarization controller from Fig. 6.32 is commonly implemented by using four squeezers as described in [20]. Therefore, by polarization transformer, the SOP of local laser signal is changed until it matches the SOP of received optical signal.

Receiver resilience with respect to polarization fluctuations can be improved if the receiver derives two demodulated signals from two orthogonal polarizations of the received signal, as illustrated in Fig. 6.33. This scheme is known as a *polarization diversity receiver*, and it has been derived from Fig. 6.31. In polarization diversity schemes, however, the same data sequence is transmitted over both polarization states, which means that spectral efficiency is the same as in receivers

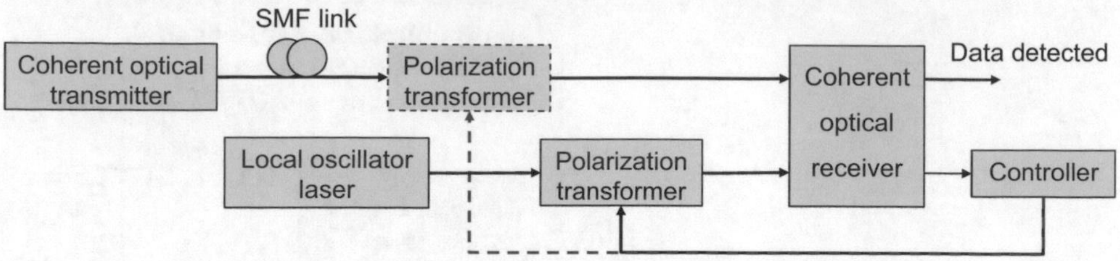

Fig. 6.32 The configuration of the polarization control receiver

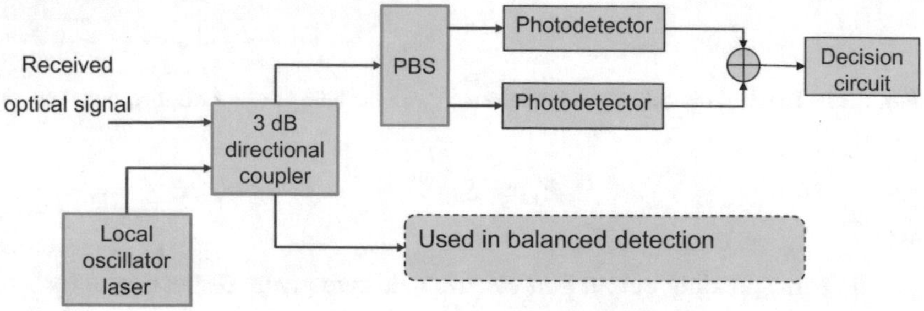

Fig. 6.33 The configuration of polarization diversity receiver. PBS, polarization beam splitter

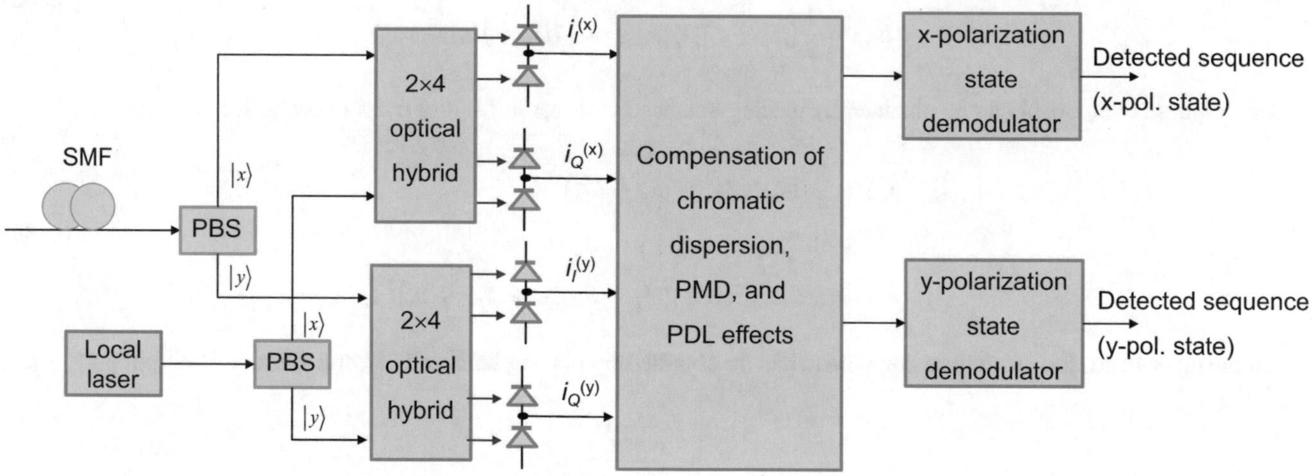

Fig. 6.34 The polarization-division demultiplexing-based receiver configuration

without polarization diversity. To double the spectral efficiency of polarization diversity schemes, the polarization-division multiplexing can be applied [22, 23].

In polarization-division multiplexing (PDM) [22, 23] both polarizations carry independent multilevel modulated streams, as described in Chap. 5, Sect. 5.5 , which doubles the spectral efficiency. As indicated in Chap. 5, the *M*-ary PSK, *M*-ary QAM, and *M*-ary 2-D modulation schemes, in general, achieve the transmission of $\log_2 M = m$ bits per symbol. In M-ary PSK, the data phasor $\theta_n \in \{0, 2\pi/M, \ldots, 2\pi(M\text{-}1)/M\}$ is sent at each *n*th transmission interval. In M-ary QAM, the mapper is typically used that takes *m* bits at time instance *n* to determine the corresponding *M*-ary ($M = 2^m$) constellation point $s_n = (I_n, Q_n) = |s_n|\exp(j\theta_n)$, and corresponding coordinates after DAC are used in RF inputs of electro-optical I/Q modulator. In PDM, two I/Q modulators (with configuration provided in Chap. 3) and two 2-D mappers are needed, one for each polarization. The outputs of I/Q modulators are combined using the polarization beam combiner (PBC). The same transmit DFB laser is used as CW source, with x- and y-polarizations being separated by a polarization beam splitter (PBS).

The corresponding coherent optical detector receiver architecture is shown in Fig. 6.34, while the balanced coherent optical detector architecture is described already above. The balanced output photocurrents of I- and Q-channel branches for x-polarization at the time instance *n* can be written as

$$\begin{aligned} i_{I,n}^{(x)} &\simeq R|S_n^{(x)}||L^{(x)}|\cos\left(\omega_{IF}t+\theta_n^{(x)}+\underbrace{\phi_{T,PN}^{(x)}-\phi_{LO,PN}^{(x)}}_{\phi_{PN}^{(x)}}\right), \\ i_{Q,n}^{(x)} &\simeq R|S_n^{(x)}||L^{(x)}|\sin\left(\omega_{IF}t+\theta_n^{(x)}+\phi_{PN}^{(x)}\right), \\ \phi_{PN}^{(x)} &= \phi_{T,PN}^{(x)}-\phi_{LO,PN}^{(x)}, \end{aligned} \tag{6.285}$$

where R is the photodiode responsivity, while $\phi_{T,PN}$ and $\phi_{LO,PN}$ represent the laser phase noise of transmitting and local laser oscillator, respectively. $\omega_{IF} = |\omega_T - \omega_{LO}|$ denotes the intermediate frequency, while $S_n^{(x)}$ and $L^{(x)}$ represent the incoming signal (at the nth symbol interval) in x-polarization and x-polarization output from local laser oscillator, respectively. Similar expressions hold for y-polarization state:

$$\begin{aligned} i_{I,n}^{(y)} &\simeq R\left|S_n^{(y)}\right|\left|L^{(y)}\right|\cos\left(\omega_{IF}t+\theta_n^{(y)}+\phi_{PN}^{(y)}\right), \\ i_{Q,n}^{(y)} &\simeq R\left|S_n^{(y)}\right|\left|L^{(y)}\right|\sin\left(\omega_{IF}t+\theta_n^{(y)}+\phi_{PN}^{(y)}\right). \end{aligned} \tag{6.286}$$

The heterodyne receiver can also be implemented based only on single balanced detector as described in Chap. 3. For ASE noise-dominated scenario, both heterodyne schemes perform comparably. However, in shot noise-dominated scenario, the heterodyne design with single balanced detector performs better. In homodyne detection, it is set that $\omega_{IF} = 0$. After the in-phase and quadrature signals are detected in both polarization states, we perform ADC and compensate for various dispersion effects including chromatic dispersion, polarization mode dispersion (PMD), and polarization-dependent losses (PDL). After timing recovery and phase estimation steps in each polarization state, the demodulation procedure takes place.

6.8.3 Homodyne Coherent Optical Detection Based on PLLs

Homodyne coherent optical detection receivers based on PLL can be classified as either residual carrier receivers or suppressed carrier receivers [14]. Corresponding PLLs used in optical communications are commonly referred to as optical PLLs (OPLLs). In systems with residual carrier, shown in Fig. 6.35, the phase deviation between the mark- and space-state bits in BPSK is less then π/2 rad, so that the part of transmitted signal power is used for the non-modulated carrier transmission and as consequence some power penalty occurs. Local oscillator used in various OPLL schemes is tunable.

The Costas loop and decision-driven loop (DDL) for BPSK [18], shown in Fig. 6.36, are two alternatives to the receivers with residual carrier. Both these alternatives employ a fully suppressed carrier transmission, in which the entire transmitted power is used for data transmission only. However, at the receiver side, a part of the power is used for the carrier extraction, so some power penalty is incurred with this approach too. The corresponding decision-directed OPLL for QPSK is shown in Fig. 6.37, with details provided in [79].

The coherent optical detection receivers with OPLL are very sensitive to laser linewidth, as discussed in [18–20]. Another big challenge is to implement the optical voltage-controlled oscillator (OVCO). To avoid this problem, typically DSP approaches, such as those described in Sect. 6.4, are used. Alternatively, the *subcarrier-based OPLL*, proposed in [80, 81], described in Fig. 6.38 can be used, which replaces the loop filter and local laser from Fig. 6.37. In OVCO, external modulator,

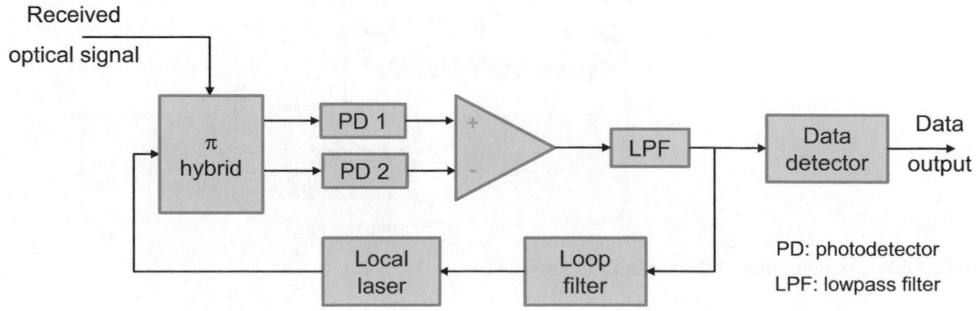

Fig. 6.35 Balanced loop-based receiver for BPSK

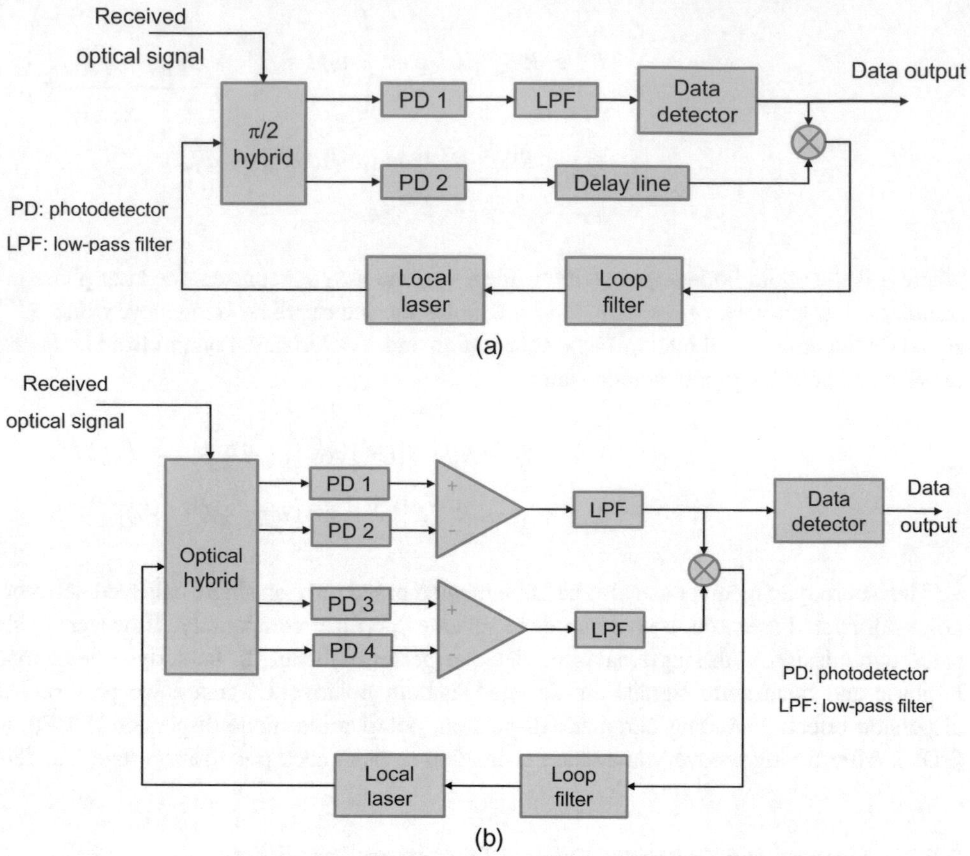

Fig. 6.36 BPSK coherent optical detection receivers with (**a**) decision-driven (decision-directed) loop and (**b**) Costas loop. *PD* photodetector, *LPF* low-pass filter

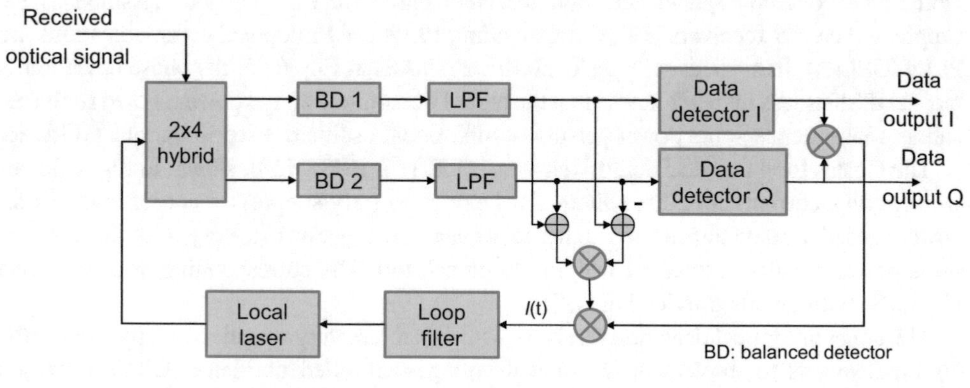

Fig. 6.37 Decision-directed OPLL for QPSK. The configuration of balanced detector (BD) is shown in Fig. 6.28

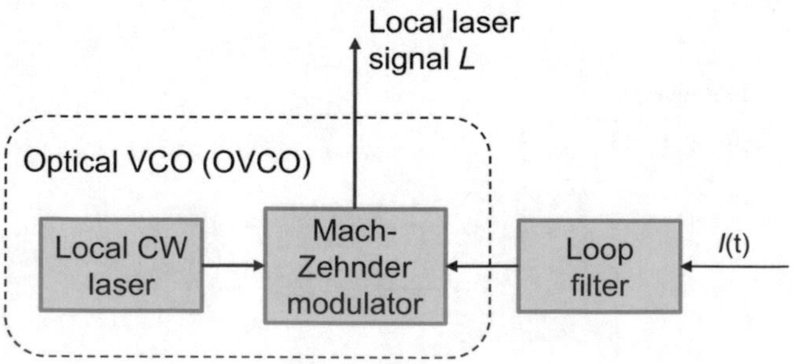

Fig. 6.38 Configuration of the subcarrier-based OPLL

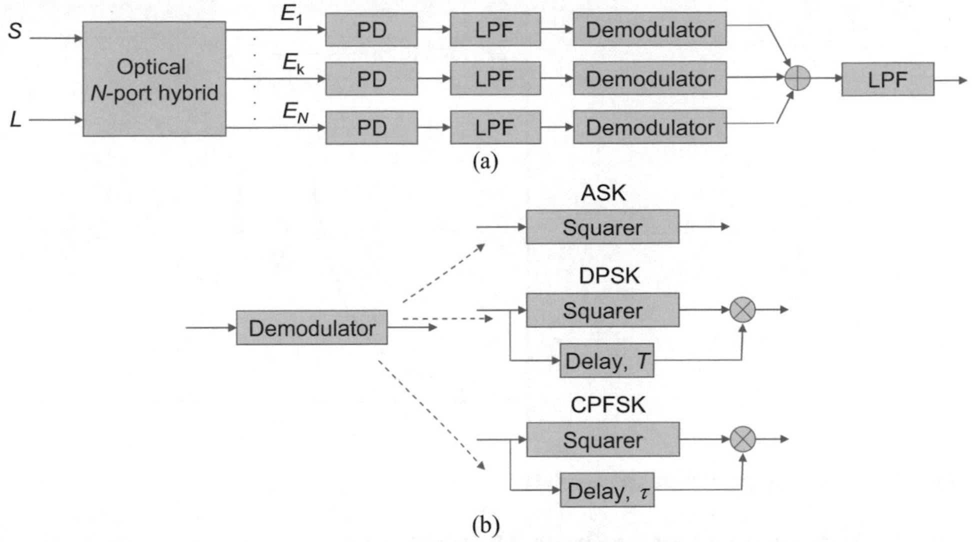

Fig. 6.39 Phase diversity receivers (**a**) and demodulator configurations (**b**)

such as the Mach-Zehnder modulator, is used to change the OVCO output frequency by $\omega_{CW} \pm \omega_{VCO}$, where ω_{CW} is the angular frequency of continuous-wave (CW) laser and ω_{VCO} is the angular frequency of the electrical VCO.

6.8.4 Phase Diversity Receivers

To avoid the problems associated with OPLL-based receivers, the phase diversity receivers can be used. The general architecture of a multiport homodyne receiver is shown in Fig. 6.39(a) [9, 20].

The incoming optical signal and local laser output signal can be written as

$$S(t) = aE_s e^{j[\omega_c t + \phi_S(t)]}, \qquad L(t) = E_{LO} e^{j[\omega_c t + \phi_{LO}(t)]}, \tag{6.287}$$

where the information is imposed either in amplitude a or phase ϕ_s. Both incoming optical signal S and local laser output signal L are used as inputs to N output ports of an optical hybrid, which introduces fixed phase difference $k(2\pi/N)$ ($k = 0, 1, \ldots, N\text{-}1$) between the ports, so that the output electric fields can be written as

$$E_k(t) = \frac{1}{\sqrt{N}}\left[S(t)e^{jk\frac{2\pi}{N}} + L(t)\right]. \tag{6.288}$$

The corresponding photodetector outputs are as follows:

$$\begin{aligned} i_k(t) &= R|E_k(t)|^2 + I_{nk} \\ &= \frac{R}{N}\left\{|E_{LO}|^2 + a|E_S|^2 + 2a|E_S||E_{LO}|\cos\left[\phi_S(t) - \phi_{LO}(t) + k\frac{2\pi}{N}\right]\right\} + I_{sn,k}, \end{aligned} \tag{6.289}$$

where $I_{sn,k}$ is the kth photodetector shot noise sample. Different versions of demodulators for ASK, DPSK, and CPFSK are shown in Fig. 6.39(b). For ASK we have simply to square photodetector outputs and add them together to get

$$y = \sum_{k=1}^{N} I_k^2. \tag{6.290}$$

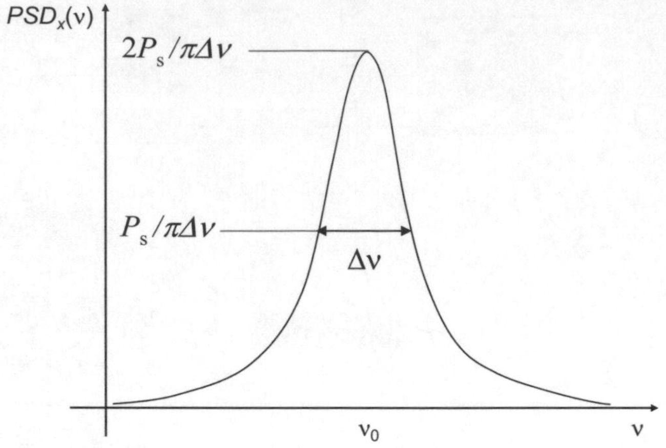

Fig. 6.40 Lorentzian spectrum of the laser diode

6.8.5 Dominant Coherent Optical Detection Noise Sources

In this section, we will outline the noise processes due to phase, polarization, and intensity variations in coherent optical detector; a more detailed description of other optical relevant channel impairments has been already provided in Chap. 2.

6.8.5.1 Laser Phase Noise

Noise due to spontaneous emission causes phase fluctuations in lasers, leading to a nonzero spectral linewidth $\Delta\nu$. In semiconductor lasers, the linewidth $\Delta\nu$ is dependent on the cavity length and the value of the linewidth enhancement factor and can vary from 10 KHz up to the megahertz range. The output from the semiconductor laser can be expressed as

$$x(t) = \sqrt{P_S} e^{\,j[\omega_0 t + \phi_n(t)\,]}, \tag{6.291}$$

where $\phi_n(t)$ is the laser phase noise process. We can also assume that the corresponding PSD of x(t) can be expressed as the Lorentzian curve. For this purpose, we can express the PSD of Eq. (6.292) in a simplified form as

$$PSD_x(\nu) = \frac{2P_S}{\pi\ \Delta\nu}\left[1 + \left(2\frac{\nu - \nu_0}{\Delta\nu}\right)^2\right]^{-1}, \tag{6.292}$$

which is shown in Fig. 6.40. The laser phase noise, expressed through the spectral linewidth, will cause bit error probability degradation.

As an illustration, the impact of the laser phase noise to BPSK performance is illustrated in Fig. 6.41. As we can see, the effect of phase noise on BER curves is twofold: (i) the BER curves are shifted to the right, and (ii) BER floor appears. In the design of advanced coherent detection schemes, the impact of the laser phase noise is mitigated by digital signal processing that takes place after photodetection, and it does impose a serious limitation if spectral linewidth is measured by hundreds of kilohertz. Moreover, for higher multilevel modulation schemes, such as M-ary QAM, the laser phase noise can cause larger performance degradation.

6.8.5.2 Polarization Noise

Polarization noise, which comes from discrepancies between the SOP of incoming optical signal and local laser oscillator, is another factor that causes performance degradation in coherent receiver schemes. In order to mitigate its impact, several avoidance techniques have been considered. These techniques can be classified as follows: (i) polarization control, (ii) employment of polarization-maintaining fibers, (iii) polarization scrambling, (iv) polarization diversity, and (v) polarization-division multiplexing. The employment of polarization control, polarization diversity, and polarization-division demultiplexing has already been discussed above.

The polarized electromagnetic field launched into the fiber can be represented as

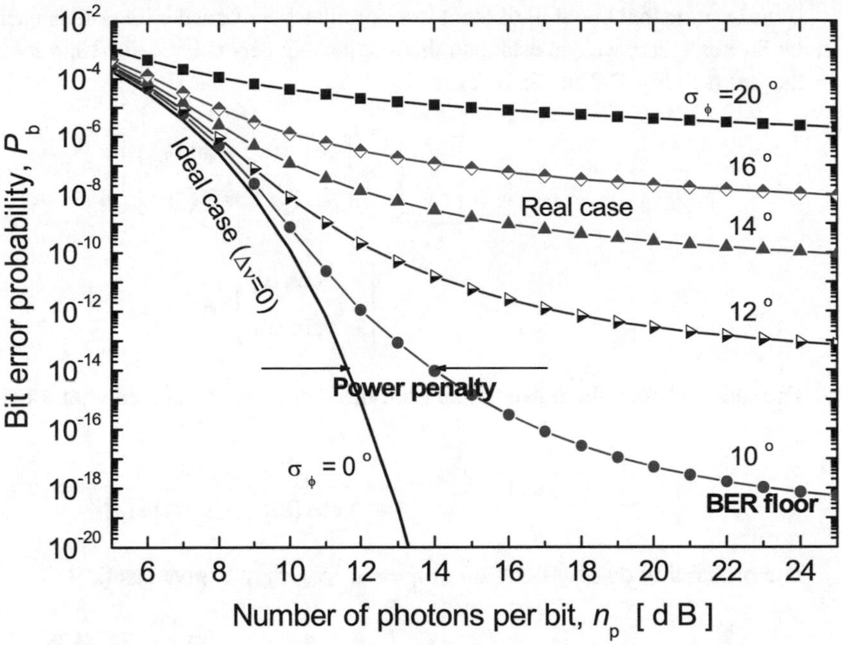

Fig. 6.41 Probability of error vs. number of photons per bit in the presence of the laser phase noise for homodyne BPSK signaling

$$E(t) = \begin{bmatrix} e_x(t) \\ e_y(t) \end{bmatrix} e^{j\omega_c t}, \tag{6.293}$$

where e_x and e_y represent two orthogonal SOP components and ω_c is the optical carrier frequency. The received field can be represented by

$$E_S(t) = H' \begin{bmatrix} e_x(t) \\ e_x(t) \end{bmatrix} e^{j\omega_c t}, \tag{6.294}$$

where H' is the Jones matrix of birefringence, introduced in Chap. 2. In order to match the SOPs of local laser with that of incoming optical signal, an additional transformation is needed, and it can be represented with transformation matrix H'', so we have that

$$E_s'(t) = \underbrace{H''H'}_{H} \begin{bmatrix} e_x(t) \\ e_y(t) \end{bmatrix} e^{j\omega_c t} = H \begin{bmatrix} e_x(t) \\ e_y(t) \end{bmatrix} e^{j\omega_c t}, \quad H = H''H'. \tag{6.295}$$

The SOP of local laser oscillator can be represented by Stokes coordinates as $S_{LO} = (S_{1,LO}\ S_{2,LO}\ S_{3,LO})$, as introduced in Chap. 2. The full heterodyne mix is possible only if $S_{LO} = S_R$, where S_R is the SOP of the received optical signal. The action of birefringence corresponds to rotating the vector, which represents the launched SOP, on the surface of the Poincaré sphere. This rotation can also be represented in terms of rotation matrices R_i ($i = 1, 2, 3$) for angles α_i around axes s_i. These matrices can be represented as [75]

$$\begin{aligned}
R_1(\alpha_1) &= \begin{bmatrix} \cos(\alpha_1/2) & -j\sin(\alpha_1/2) \\ -j\sin(\alpha_1/2) & \cos(\alpha_1/2) \end{bmatrix}, \\
R_2(\alpha_2) &= \begin{bmatrix} \cos(\alpha_2/2) & -\sin(\alpha_2/2) \\ \sin(\alpha_2/2) & \cos(\alpha_2/2) \end{bmatrix}, \\
R_3(\alpha_3) &= \begin{bmatrix} e^{-j\alpha_3/2} & 0 \\ 0 & e^{j\alpha_3/2} \end{bmatrix}.
\end{aligned} \tag{6.296}$$

If we assume that the SOP of local laser oscillator is aligned with s_1-axis (whose Jones vector is given by $\boldsymbol{E}_{LO} = [e_x(t)\ 0]^T$) in the Stokes space, we can establish the relationship between received and local laser SOPs in terms of spherical coordinates $(r, 2\varepsilon, 2\eta)$ (see Fig. 2.2 in Chap. 2) as

$$\underbrace{\begin{bmatrix} e'_x(t) \\ e'_y(t) \end{bmatrix}}_{\boldsymbol{E}_S} = \begin{bmatrix} \cos(\varepsilon) & -\sin(\varepsilon) \\ \sin(\varepsilon) & \cos(\varepsilon) \end{bmatrix} \begin{bmatrix} e^{-j\eta} & 0 \\ 0 & e^{j\eta} \end{bmatrix} \underbrace{\begin{bmatrix} e_x(t) \\ 0 \end{bmatrix}}_{\boldsymbol{E}_{LO}} = \begin{bmatrix} e^{-j\eta}\cos(\varepsilon) \\ e^{-j\eta}\sin(\varepsilon) \end{bmatrix} e_x(t). \tag{6.297}$$

The ratio between the power of heterodyned component (aligned with local laser) and the total power is given by

$$p_{het}(\eta, \varepsilon) = \frac{P_{het}}{P_{tot}} = \frac{|e^{-j\eta}\cos(\varepsilon)|^2}{|e^{-j\eta}\cos(\varepsilon)|^2 + |e^{-j\eta}\sin(\varepsilon)|^2} = |\cos\varepsilon|^2 = \frac{1}{2}[1 + \cos(2\varepsilon)]. \tag{6.298}$$

The probability density function of $\eta = \measuredangle(\boldsymbol{S}_R, \boldsymbol{S}_{LO})$ is given as [20]

$$f(\eta) = \frac{\sin(2\eta)}{2} e^{-\frac{A^2}{4\sigma^2}[1-\cos(2\psi)]} \left\{ 1 + \frac{4R^2 P_S P_{LO}}{4qRP_{LO}} [1 + \cos(2\eta)] \right\}, \quad \eta \in [0, \pi]. \tag{6.299}$$

6.8.5.3 Transimpedance Amplifier Noise

The typical optical receiver is based on transimpedance FET stage, as illustrated in Fig. 6.42. The PSD of receiver noise can be written as [19]

$$PSD_{rec}(f) = \frac{4k_B T_a}{R_f} + 4k_B T_a g_m \Gamma \left(\frac{f}{f_{T,eff}} \right)^2, \quad f_{T,eff} = g_m / 2\pi C_T, \ C_T = C_i + C_{PIN}, \tag{6.300}$$

where g_m is the transconductance, Γ is the FET channel-noise factor, and C_T is the total capacitance (FET input capacitance C_i plus PIN photodetector capacitance C_{PIN}). (R_L is the load resistor, R_f is the feedback resistor, T_a is the absolute temperature, and k_B is the Boltzmann constant.) This amplifier stage is popular because it has large gain-bandwidth product (GBW_i) defined as [19]

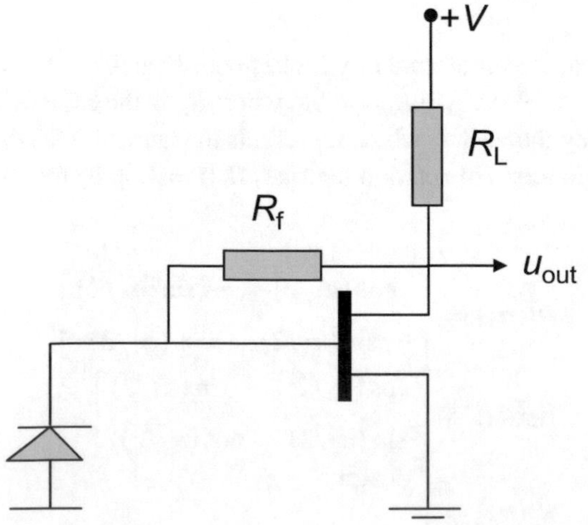

Fig. 6.42 Transimpedance FET receiver

$$GBW_{\mathrm{i}} = \left| -R_f \frac{g_m R_L}{1 + g_m R_L} \right| \frac{1 + g_m R_L}{2\pi C_T (R_f + R_i)} = f_{\mathrm{T,eff}} \left(R_f \middle\| R_L \right). \tag{6.301}$$

The intensity noise comes from the variation of optical power of transmitting laser $P = \langle P \rangle + \Delta P$ ($\langle \cdot \rangle$ denotes, as before, the statistical averaging). The relative intensity noise is defined as $RIN = \langle \Delta P^2 \rangle / \langle P \rangle^2$. Because the power of transmitting laser fluctuates, the photocurrent fluctuates as well, $I = \langle I \rangle + \Delta I$. The corresponding shot noise PSD can be determined as $2q\langle I \rangle = 2qR\langle P \rangle$. The intensity noise PSD is simply $\langle \Delta I^2 \rangle = R^2 RIN \langle P \rangle^2$. The SNR in the presence of shot, receiver, and RIN can be determined as signal power over total noise power:

$$SNR = \frac{2R^2 P_S P_{LO}}{2qRP_{LO}BW + RIN \cdot R^2 P_{LO}^2 BW + PSD_{\mathrm{rec}} BW}, \tag{6.302}$$

where BW is the receiver bandwidth and PSD_{rec} is introduced by Eq. (6.300). The SNR for balanced coherent optical reception can be written as

$$SNR = \frac{2R^2 P_S P_{LO}}{2qRP_{LO}B + \frac{RIN}{CMRR} R^2 P_{LO}^2 B + 2PSD_{\mathrm{rec}} B}, \tag{6.303}$$

where CMRR is the common mode rejection ratio of FET. Therefore, the RIN of balanced receiver is significantly reduced by balanced detection.

6.8.5.4 Coherent Optical Detection in the Presence of ASE Noise

In the rest of this section, we describe the coherent optical detection in the presence of amplified spontaneous emission (ASE) noise. To facilitate explanations, we return to the basic balanced coherent optical detector, shown in Fig. 6.28. The 2×2 optical hybrid output port electric field for $\kappa = 1/2$, $\phi_1 = 0$, $\phi_2 = \pi$ can be written as follows:

$$E_{\mathrm{o},1} = \frac{1}{\sqrt{2}}(S_n + Z + L), E_{\mathrm{o},2} = \frac{1}{\sqrt{2}}(S_n + Z - L), \tag{6.304}$$

where we use Z to denote the ASE noise vector, which can be represented as $\mathrm{Z} = \mathrm{z}\exp(\mathrm{j}\omega_{\mathrm{T}} t)$, where ω_{T} is the angular frequency of the transmit laser. Further, z can be decomposed in terms of x- and y-polarization states by $z = z_x |x\rangle + z_y |y\rangle$. The photocurrent i_1 now becomes

$$\begin{aligned} i_1 &= R|E_{\mathrm{o},1}|^2 = \frac{R}{2}(S_n + Z + L)(S_n^* + Z^* + L^*) + I_{1,\mathrm{sh}} = \\ &= \underbrace{\frac{R}{2}|S_n|^2}_{\text{Signalpower}} + \underbrace{\frac{R}{2}|L|^2}_{\text{LO laserpower}} \\ &+ \underbrace{\frac{R}{2}|Z|^2}_{\text{ASE-ASE noise}} + \underbrace{\frac{R}{2}(S_n L^* + S_n^* L)}_{\text{Signal-LO laserbeating}} + \underbrace{\frac{R}{2}(S_n Z^* + S_n^* Z)}_{\text{Signal-ASE noisebeating}} + \underbrace{\frac{R}{2}(LZ^* + L^* Z)}_{\text{LO laser-ASE noisebeating}} + I_{1,\mathrm{sh}}, \end{aligned} \tag{6.305}$$

where we have identified various beating terms as ASE-ASE noise interaction, signal-LO laser beating, signal-ASE noise beating, and LO laser-ASE noise beating components. With $I_{1,\mathrm{sh}}$ we denoted the shot noise of photodiode 1. In a similar fashion, the photocurrent i_2 can be expressed as

$$
\begin{aligned}
i_2 &= R|E_{o,2}|^2 = \frac{R}{2}(S_n + Z - L)\left(S_n^* + Z^* - L^*\right) + I_{2,\mathrm{sh}} = \\
&= \underbrace{\frac{R}{2}|S_n|^2}_{\text{Signalpower}} + \underbrace{\frac{R}{2}|L|^2}_{\text{LO laserpower}} \\
&+ \underbrace{\frac{R}{2}|Z|^2}_{\text{ASE-ASE noise}} + \underbrace{\frac{R}{2}\left(-S_nL^* - S_n^*L\right)}_{\text{Signal-LO laserbeating}} + \underbrace{\frac{R}{2}\left(S_nZ^* + S_n^*Z\right)}_{\text{Signal-ASE noisebeating}} + \underbrace{\frac{R}{2}\left(-LZ^* - L^*Z\right)}_{\text{LO laser-ASE noisebeating}} + I_{2,\mathrm{sh}}.
\end{aligned}
\tag{6.306}
$$

The balanced detector photocurrent will be now

$$
i = i_1 - i_2 = R\left(S_nL^* + S_n^*L\right) + R\left(LZ^* + L^*Z\right) + I_{1,\mathrm{sh}} - I_{2,\mathrm{sh}}. \tag{6.307}
$$

Based on Eq. (6.307), we conclude that in this balanced detector, ASE-ASE noise components and signal-ASE noise component cancel each other, providing that photodetectors in balanced detector have identical parameters. Therefore, the LO laser signal-ASE noise beating is the dominant noise source, which can be modeled as zero-mean Gaussian with PSD given by

$$
PSD_{\text{LO-ASE}} = 2R^2|L|^2N_0, \tag{6.308}
$$

where N_0 is the PSD of ASE noise, as described in Chap. 2.

6.9 Compensation of Atmospheric Turbulence Effects

The relevant approaches to deal with atmospheric turbulence effects include adaptive coding and adaptive coded modulation [5, 9, 44–53], hybrid RF-FSO communications [9, 53], diversity techniques, MIMO signal processing [5, 8, 9, 12, 16], adaptive optics approaches [82–88], and spatial light modulator (SLM)-based backpropagation method [88], to mention a few.

In adaptive coding approaches, the error correction strength (code rate) is adapted based on time-varying channel conditions. When channel conditions are bad, a stronger channel code is used; on the other hand, when channel conditions are favorable, a weaker channel code is used. In adaptive modulation and coding, both modulation format and code rate are simultaneously adapted. These schemes have been already described in Sect. 6.5 (see also Chaps. 9 and 10 in ref. [9]).

The key idea behind the diversity-combining techniques [5, 8, 9, 12, 16] is to use several independent realizations of that same transmitted signal to improve tolerance to channel impairments. The independent signal copies transmitted over different paths or degrees of freedom will have a low probability of experiencing the same level of channel distortion. In multiple-input multiple-output (MIMO) systems, the transmitters and receivers can exploit the knowledge about the channel to determine the rank of so-called channel matrix $\boldsymbol{H}$, which is related to the number of independent data streams that can be simultaneously transmitted. The diversity and MIMO techniques are subject of study in Chap. 8.

In hybrid FSO-RF communications, FSO and RF subsystems cooperate to overcome the shortcomings of each other [9, 53]. The atmospheric turbulence, as described in Chap. 2, is one of the most important factors that degrade the performance of an FSO link, which is present even in clear sky conditions. Another relevant factor that impacts FSO transmissions is the Mie scattering, which occurs in the near-infrared wavelength region when particle sizes due to fog, haze, and pollution (aerosols) become comparable to the operating wavelength. Fog itself is the most detrimental factor that impacts the reliability of an FSO link. In contrast, RF signals are not impacted by these problems, but are affected by other issues, particularly rain and snow. This suggests that the two transmission media (FSO and RF) can be operated in a complementary fashion, depending on the predominant weather and atmospheric conditions. During certain channel conditions, FSO and RF link performance becomes comparable, when it makes sense to simultaneously and cooperatively communicate over the two channels so as to maximize the overall link performance. For additional details, an interested reader is referred to Chap. 4 (see Sect. 4.6).

Atmospheric turbulence effects can also be handled through the use of the azimuthal phase correction method, similar to the Gerchberg-Saxton (GS) phase retrieval algorithm [82, 83]. This method can be described as follows [89]. In the first stage, we determine the azimuthal phase of a given spatial mode, $\varphi(x,y)$, in the absence of turbulence, based on the magnitude response of the Fourier transform of $\exp[j\varphi(x,y)]$. In the second stage, we determine the azimuthal phase of spatial modes in the presence of turbulence. The distortions introduced by atmospheric turbulence can be compensated for by using the GS phase retrieval algorithm. However, there is no evidence that this method can be used to deal with strong turbulence effects.

In the remainder of this chapter, we describe the usage of adaptive optics and SLM-based backpropagation method to deal with atmospheric turbulence effects.

6.9.1 Adaptive Optics Techniques

A linear adaptive optics system consists of three subsystems: (i) wavefront sensor to detect the atmospheric distortions, (ii) wavefront corrector to compensate for turbulence effects, and (iii) the control processor to monitor the wavefront sensor information and control the wavefront corrector.

Before we describe each individual subsystem, we provide the Zernike representation of the atmospheric turbulence.

6.9.1.1 Zernike Representation of Atmospheric Turbulence

The polynomial expansion of the arbitrary wavefront $\varphi(\boldsymbol{r})$ over the circular aperture of radius R can be represented as [85, 86]:

$$\varphi(\boldsymbol{r}) = \sum_{l=1}^{\infty} a_l Z_l(\boldsymbol{r}), \tag{6.309}$$

where a_l are the coefficients of expansion. Zernike polynomials $Z_l(\boldsymbol{r})$ form a set of orthogonal polynomials expressed as product of radial R_n^m and azimuth V_{ml} components as follows [85]:

$$Z_l(\boldsymbol{r}) = \underbrace{\sqrt{2^{\delta_{m0}}}\sqrt{n+1}}_{c_{mn}} R_n^m(r/R)V_{ml}(\theta) = c_{mn}R_n^m(r/R)V_{ml}(\theta), \tag{6.310}$$

where radial components R_n^m are defined as

$$R_n^m(\rho) = \sum_{s=0}^{(n-m)/2} \frac{(-1)^s(n-m)!}{s!\left(\frac{n+m}{2}-s\right)!\left(\frac{n-m}{2}-s\right)!}\rho^{n-2s} = \sum_{s=0}^{(n-m)/2} (-1)^s \binom{n-s}{s}\binom{n-2s}{\frac{n-m}{2}-s}\rho^{n-2s}, \tag{6.311}$$

while azimuth components V_{ml} as

$$V_{ml} = \begin{cases} \cos(m\theta), \ l = 0, 2, 4, \cdots (l \,\text{iseven}) \\ \sin(m\theta), l = 1, 3, 5, \cdots (l \text{ is odd}) \end{cases}. \tag{6.312}$$

Zernike polynomials are orthogonal within a circular aperture of radius R:

$$\iint_{r\le R} Z_l(\boldsymbol{r})Z_k(\boldsymbol{r})d^2r = c_l\delta_{lk}. \tag{6.313}$$

By employing the orthogonality principle, from Eq. (6.309), we can write

$$\iint_{r\le R} \varphi(\boldsymbol{r})Z_l(\boldsymbol{r})d^2r = \sum_{k=1}^{\infty} a_k \underbrace{\iint_{r\le R} Z_l(\boldsymbol{r})Z_k(\boldsymbol{r})d^2r}_{c_l\delta_{lk}}, \tag{6.314}$$

and determine the coefficient a_l as follows:

Table 6.2 Zernike polynomials sorted by Noll index l

Noll index l	Radial degree n	Azimuthal frequency m	Z_l	Typical name
1	0	0	1	Piston
2	1	1	$2\rho \cos \theta$	Tip, lateral position, tilt-x
3	1	−1	$2\rho \sin \theta$	Tilt, lateral position, tilt-y
4	2	0	$\sqrt{3}(2\rho^2 - 1)$	Defocus, longitudinal position
5	2	−2	$\sqrt{6}\rho^2 \sin 2\theta$	Oblique astigmatism
6	2	2	$\sqrt{6}\rho^2 \cos 2\theta$	Vertical astigmatism
7	3	−1	$\sqrt{8}(3\rho^2 - 2\rho) \sin \theta$	Vertical coma
8	3	1	$\sqrt{8}(3\rho^2 - 2\rho) \cos \theta$	Horizontal coma
9	3	−3	$\sqrt{8}\rho^3 \sin 3\theta$	Vertical trefoil
10	3	3	$\sqrt{8}\rho^3 \cos 3\theta$	Oblique trefoil
11	4	0	$\sqrt{5}(6\rho^4 - 6\rho^2 + 1)$	Primary spherical
12	4	2	$\sqrt{10}(4\rho^4 - 3\rho^2) \cos 2\theta$	Vertical secondary astigmatism
13	4	−2	$\sqrt{10}(4\rho^4 - 3\rho^2) \sin 2\theta$	Oblique secondary astigmatism
14	4	4	$\sqrt{10}\rho^4 \cos 4\theta$	Oblique quadrafoil
15	4	−4	$\sqrt{10}\rho^4 \sin 4\theta$	Vertical quadrafoil

$$a_l = \frac{1}{\pi R^2} \iint_{r \le R} \varphi(\mathbf{r}) Z_l(\mathbf{r}) d^2 r. \tag{6.315}$$

During propagation over the atmospheric turbulence channels, the wavefront $\varphi(\mathbf{r})$ get randomly distorted with samples generated as described in Chap. 2, so that we can use the Gaussian approximation to describe the random phase screen variations. To characterize this Gaussian process, we need to determine the correlation matrix with corresponding correlation coefficients as follows:

$$\langle a_l a_k \rangle = \frac{1}{(\pi R^2)^2} \iint_{r \le R} Z_l(\mathbf{r}) \iint_{r' \le R} \langle \varphi(\mathbf{r}) \varphi(\mathbf{r}') \rangle Z_l(\mathbf{r}') d^2 \mathbf{r}' d^2 \mathbf{r}. \tag{6.316}$$

The first several Zernike polynomials, ordered by Noll index l, are summarized in Table 6.2, based on [90]. The following recurrence relation can be used to calculate all other radial terms:

$$R_n^m = \{2(n-1)[2n(n-2)\rho^2 - m^2 - n(n-2)]R_{n-2}^m(\rho) - n(n+m-2)(n-m-2)R_{n-4}^m(\rho)\}/[(n+m)(n-m)(n-2)]. \tag{6.317}$$

The convenience of the Zernike polynomials is that we can determine the power associated with each Zernike term. Once a certain number of Zernike terms get compensated for, we can determine the residual aberration as follows:

$$\Delta_L = \iint_{r<R} |\varphi(\mathbf{r}) - \varphi_L(\mathbf{r})|^2 d^2 r, \tag{6.318}$$

where $\varphi_L(\mathbf{r})$ is the approximation with the first L terms:

$$\varphi_L(\mathbf{r}) = \sum_{l=1}^{L} a_l Z_l(\mathbf{r}). \tag{6.319}$$

When $L > 10$, the following approximation due to Noll can be used [90]:

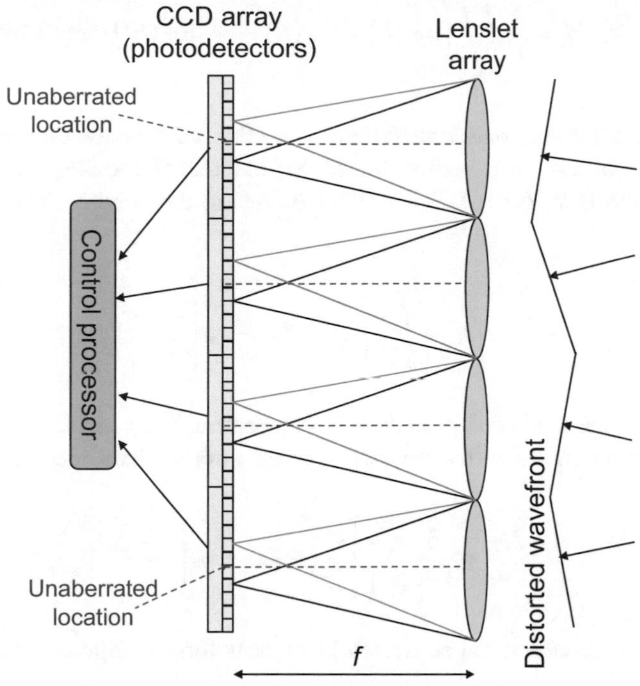

Fig. 6.43 Illustration of the Shack-Hartmann sensor

$$\Delta_L \approx 0.2944 L^{-\sqrt{3}/2} \left(\frac{D}{r_0}\right)^{5/3} \left[\text{rad}^2\right], \tag{6.320}$$

where r_0 is the Fried parameter introduced in Chap. 2.

6.9.1.2 Shack-Hartmann Wavefront Sensor

The most widely used wavefront sensor is the *Shack-Hartmann sensor*, whose basic layout is illustrated in Fig. 6.43.

This wavefront sensor consists of an array of collecting lenslets, in whose focal plan an array of photodetectors is placed. The subapertures (lenslets) can be arranged in rows, the hexagon packing can be used, etc. Each subaperture focuses the corresponding portion of the light beam to the photodetector plane. The signals from different photodetectors are used to determine the displacement of every sub-beam from the optical axis of the corresponding subaperture, which is proportional to the mean tilt of the wavefront within the subaperture. When sufficient number of photodetectors in the array is used, such as CCD array, the displacement of the focal spot is determined as the displacement of its centroid. The focal spot centroid vector within subaperture A can be determined as [90] (see also [85])

$$\boldsymbol{r}_c = \frac{kf}{P} \iint_A I(\boldsymbol{r}) \nabla \varphi d^2 \boldsymbol{r}, \tag{6.321}$$

where f is the focal length of the lenslet and P is the total radiation power incident to the subaperture A. The normalized phase gradient can be expressed in terms of the gradients of the real and imaginary parts of the complex amplitude E as follows [85]:

$$I(\boldsymbol{r}) \nabla \varphi = |E|^2 \nabla \tan^{-1} \left[\frac{\text{Im}\,(E)}{\text{Re}\,(E)}\right] = \text{Re}\,(E) \nabla \text{Im}\,(E) - \nabla \left[\text{Re}\,(E)\right] \text{Im}\,(E). \tag{6.322}$$

After substituting Eq. (6.322) into Eq. (6.321), we obtain the following equation for the local tilt:

$$\boldsymbol{s} = \boldsymbol{r}_c/kf = \frac{1}{P}\iint_A \{\text{Re}\,(E)\nabla\text{Im}\,(E) - \nabla[\text{Re}\,(E)]\text{Im}\,(E)\}d^2\boldsymbol{r}. \tag{6.323}$$

The wavefront modal reconstruction of the sensor aperture can be obtained from the estimates of local tilts, as described in [92] (see also [85]). The starting point is the approximation of $\varphi(\boldsymbol{r})$ in terms of Zernike polynomials Z_l, given by Eq. (6.319). The local tilts of the mth subaperture $\boldsymbol{s}_m$ ($m = 1, 2, \ldots, M$) are fitted to the estimates obtained in the wavefront sensor as follows:

$$\varepsilon_L = \sum_{m=1}^{M}\left[\sum_{l=1}^{L} a_l \boldsymbol{Z}_{lm} - \boldsymbol{s}_m\right]^2, \tag{6.324}$$

where $\boldsymbol{Z}_{lm}$ is the local tilt of the lth basis function Z_l on the mth subaperture. The variation method can be applied to determine the expansion coefficients a_k, by setting the partial derivatives of the error ε with respect to a_k to zero [85]:

$$\frac{\partial \varepsilon_L}{\partial a_k} = \frac{\partial}{\partial a_k}\sum_{m=1}^{M}\left[\sum_{l=1}^{L} a_l \boldsymbol{Z}_{lm} - \boldsymbol{s}_m\right]^2 = 0. \tag{6.325}$$

The corresponding set of linear equations can be written in a matrix form as follows:

$$\boldsymbol{A}\boldsymbol{a} = \boldsymbol{b}, A_{kl} = \sum_{m=1}^{M} \boldsymbol{Z}_{lm}^{\mathrm{T}}\boldsymbol{Z}_{km},\; b_k = \sum_{m=1}^{M} \boldsymbol{s}_m \boldsymbol{Z}_{km}. \tag{6.326}$$

The solution of expansion coefficient can be obtained by the following matrix inversion and multiplication [85]:

$$\boldsymbol{a} = \boldsymbol{A}^{-1}\boldsymbol{b}. \tag{6.327}$$

To determine the local tilts according to Eq. (6.323) would require quadrature sensors, which is not possible by the Shack-Hartmann sensor, and this equation can only be used for modeling purposes. With the Shack-Hartmann sensor we measure the displacements of the focal spots, denoted as $\Delta\boldsymbol{r}_m$, with respect to reference positions. The local tilts are directly proportional to the displacement and reversely proportional to the focal length f, that is, $\boldsymbol{s}_m = \Delta\boldsymbol{r}_m/f$. Local tilts are then substituted in Eq. (6.236) to determine the coefficients of the expansion a_l. Assuming the intensity of aperture incident beam $I(x,y)$, the position of the image spot in each small region of the grid is determined by evaluating the centroid:

$$\boldsymbol{r} = (x, y); x = \frac{\sum_{i=1}^{N_x}\sum_{j=1}^{N_y} x_i I_{ij}}{\sum_{i=1}^{N_x}\sum_{j=1}^{N_y} I_{ij}}, y = \frac{\sum_{i=1}^{N_x}\sum_{j=1}^{N_y} y_i I_{ij}}{\sum_{i=1}^{N_x}\sum_{j=1}^{N_y} I_{ij}}, \tag{6.328}$$

where N_x (N_y) is the number of focal spots along the x-direction (y-direction).

6.9.1.3 Wavefront Correctors

Regarding the *wavefront correctors*, the most popular are *adaptive mirrors*, which can be classified as either deformable mirrors or segmented mirrors. In modal corrector, the residual aberration error, based on Eq. (6.318), can be defined as

$$\Delta_L = \iint W(\boldsymbol{\rho})\left|\varphi(\boldsymbol{\rho}) - \sum_{l=1}^{L} a_l Z_l(\boldsymbol{\rho})\right|^2 d^2\boldsymbol{\rho}, \tag{6.329}$$

where the aperture function $W(\boldsymbol{r})$ corresponds to the circle of the radius R:

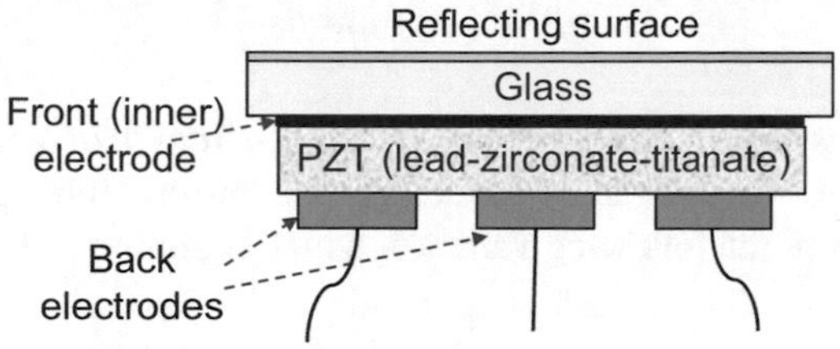

Fig. 6.44 Illustrating the cross section of the biomorphic deformable mirror

$$W(\boldsymbol{\rho}) = \begin{cases} 1, & \rho \le R \\ 0, & \rho > R \end{cases}. \tag{6.330}$$

By employing the orthogonality principle, the expansion coefficients can be determined as

$$a_l = \frac{\iint\limits_{\rho \le R} \varphi(\boldsymbol{\rho}) Z_l(\boldsymbol{\rho}) d^2\boldsymbol{\rho}}{\iint\limits_{\rho \le R} Z_l^2(\boldsymbol{\rho}) d^2\boldsymbol{\rho}}. \tag{6.331}$$

Alternatively, the variational method, similar to that given by Eqs. (6.325, 6.326 and 6.327), can be used. The only difference is that the elements of matrix $\boldsymbol{A}$ and vector $\boldsymbol{b}$ are determined, respectively, as

$$A_{kl} = \int W(\boldsymbol{\rho}) Z_k(\boldsymbol{\rho}) Z_l(\boldsymbol{\rho}) d^2\boldsymbol{\rho}, \; b_k = \int W(\boldsymbol{\rho}) Z_k(\boldsymbol{\rho}) \varphi(\boldsymbol{\rho}) d^2\boldsymbol{\rho}. \tag{6.332}$$

The most commonly used *deformable mirrors* are those with discretely arranged points of force (force moment). The simplest and most popular method is to control the mirror surface with the help of actuators. The actuators, performing the deformation of the deformable mirror, could be the discrete ones, acting either perpendicular to the surface or affecting the bending moments for edge actuators. The actuators can also be continuous such as in a membrane mirror or a bimorph mirror. The bimorph deformable mirror, whose cross section is illustrated in Fig. 6.44, is composed of either glass with reflecting surface or metal mirror faceplate bonded to a crystal of piezoelectric ceramics, such as lead-zirconate-titanate (PZT)-based ceramics, which is polarized normal to its surface. The bonded (glued) interface between glass and PZT ceramics contains inner electrode. The back electrodes are attached to the PZT-based ceramics. The glass is polished and coated to create the surface of the deformable mirror. The wires are attached to the electrodes.

By applying the voltage V across front and back electrodes, the corresponding dimension of the PZT-based ceramics changes. The curvature R of the deformable mirror is directly proportional to the voltage applied and reversely proportional to the thickness t of the PZT-based ceramics, that is [94],

$$R = d_{13} \frac{V}{t^2}, \tag{6.333}$$

where d_{13} is the piezoelectric tensor coefficient. The actuators are organized in either square or hexagonal array. Square arrays come in the following sizes: 4, 9, 16, 21 (5×5-4 on corners), 37 (7×7-12 on corners), 69 (9×9-24), 97 (5×5-4), etc. On the other hand, hexagonal arrays come in the following sizes: 3, 7, 19, 37, 61, 91, 127, etc., with generic formula being $1+6\sum_n n$, where n is a positive integer. For additional details, an interested reader is referred to [93, 94].

The mirror surface can be described as the weighted sum of the response functions f_l [85]:

$$S(\boldsymbol{\rho}) = \sum_{l=1}^{L} a_l f_l(\boldsymbol{\rho} - \boldsymbol{\rho}_l), \tag{6.334}$$

wherein the response functions are determined experimentally, based on the system of mechanical equations. To simplify the problem, the Gaussian response approximation is used in numerous papers, such as [85, 93]:

$$f_l(\boldsymbol{\rho}) = \exp\left(-\rho^2/w_l^2\right), \tag{6.335}$$

where the halfwidth w_l is determined experimentally and typically ranges from $0.7d$ to $0.8d$, with d being the spacing between actuators. The approximation of the surface required by the deformable mirror, with a given set of response functions, can be obtained as the minimization problem of the following variance, which is similar to Eq. (6.329):

$$\sigma_L^2 = \iint W(\boldsymbol{\rho})\left|\varphi(\boldsymbol{\rho}) - \sum_{l=1}^{L} a_l f_l(\boldsymbol{\rho} - \boldsymbol{\rho}_l)\right|^2 d^2\boldsymbol{\rho}. \tag{6.336}$$

To reduce the computation time, the truncated Gaussian response function can be used:

$$f_l(\boldsymbol{\rho}) = \begin{cases} \exp\left(-\rho^2/w_l^2\right), \rho \le w_l \\ 0, \rho > w_l \end{cases} \tag{6.337}$$

Now we can apply the variation method given by Eqs. (6.325, 6.326 and 6.327), wherein the elements of matrix $\boldsymbol{A}$ and vector $\boldsymbol{b}$ are determined, respectively, as

$$\begin{aligned} A_{kl} &= \sum_{\left|\boldsymbol{\rho}_{i,j}-\boldsymbol{\rho}_k\right|<2w} W(\boldsymbol{\rho}_{i,j}) f_k(\boldsymbol{\rho}_{i,j} - \boldsymbol{\rho}_k) f_l(\boldsymbol{\rho}_{i,j} - \boldsymbol{\rho}_k), \\ b_k &= \sum_k W(\boldsymbol{\rho}_{i,j}) f_k(\boldsymbol{\rho}_{i,j} - \boldsymbol{r}_k)\varphi(\boldsymbol{\rho}_{i,j}), \end{aligned} \tag{6.338}$$

where we used the subscripts i, j to denote the grid ($i = 1, \ldots, I; j = 1, \ldots, J$). The actuator fastening points are typically placed on an equidistant grid with node coordinates (x_l, y_l) being determined by [85]

$$x_l = -R + \left[l/\sqrt{L}\right]\sqrt{L}d, \; y_l = -R + \left\{l/\sqrt{L}\right\}\sqrt{L}d, \tag{6.339}$$

where the brackets $[\cdot]$ are used to denote the integer part of enclosed entity, while the brackets $\{\cdot\}$ are used to denote the fractional part.

Now we have all elements required for an adaptive optics system, which can be placed at either transmitter or receiver side. The receiver side adaptive optics system is illustrated in Fig. 6.45. The wavefront errors are measured with the help of the Shack-Hartmann wavefront sensor, and to correct for the aberration errors, the deformable mirror is used in a closed loop, controlled by the control subsystem. The pilot beam is used to estimate the distortions introduced by atmospheric turbulence channels and deformable mirror. Once the residual aberrations are minimized, the signal beam is reflected from deformable mirrors and the atmospheric turbulence-induced wavefront errors are corrected for, providing that resolution is sufficiently good, and time needed for deformable mirror adjustment is shorter than the channel coherence time.

The size of aperture in front of photodetector is properly chosen, depending on FSO link length. For signal spatial mode transmission, a single photodetector might be sufficient. Alternatively, a photodetector array can be used to exploit spatial diversity, by properly selecting the compression telescope, collimating lens, the aperture size in front of photodetector array, and separation of photodetectors so that the different optical paths will exhibit different turbulence effects. In such a way, we combine the adaptive optics with spatial diversity to deal with strong turbulence effects. Alternatively, the low-cost wavefront sensorless adaptive optics can be used to reduce the effects due to turbulence, and to deal with remaining turbulence effects, the adaptive LDPC coding can be used, as described in [88]. By employing the multiple spatial laser sources, the MIMO signal processing can be combined with adaptive optics.

6.9.2 SLM-Based Backpropagation Method

The receiver side adaptive optics can be used to deal with atmospheric turbulence effects in both single-mode and multimode free-space optical communication systems including classical, QKD, and quantum communication systems. However, it has

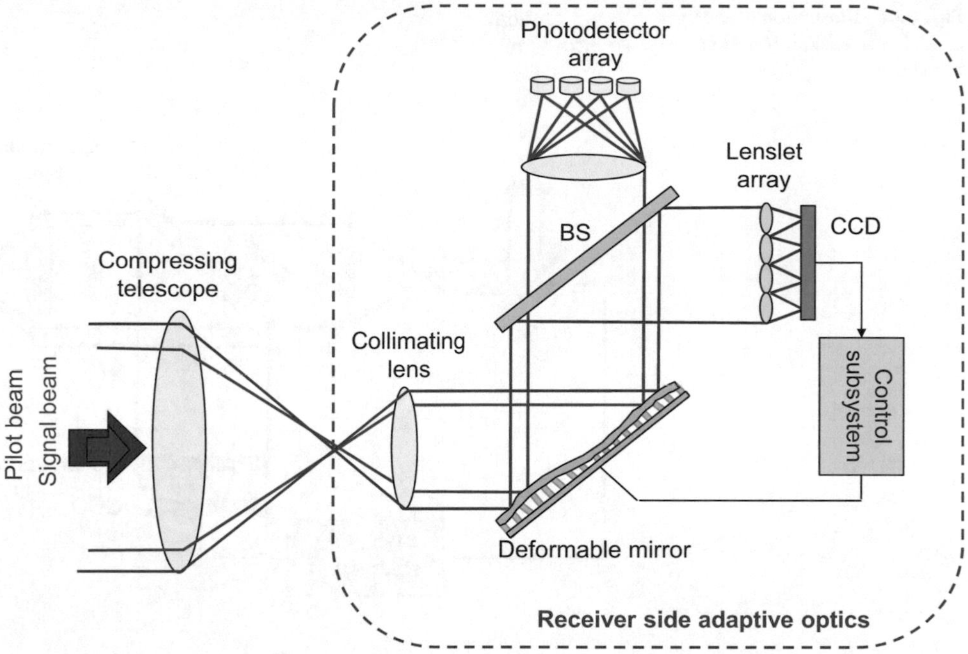

Fig. 6.45 Typical receiver side adaptive optics system. BS, beam splitter

been shown in [87] that the residual wavefront error is limited by the finite resolution of the deformable mirror (DM). To deal with strong turbulence effects, the resolution of existing DMs is not sufficient. To solve for this problem, a liquid crystal spatial light modulator (LC-SLM) can be used as a promising alternative to the DM [87]. As an illustration, an LC-SLM with a 1000×1000 pixel array in a 10×10 mm^2 area has been reported in [94]. The SLMs can also be used to emulate atmospheric turbulence effects [95]. However, because the LC-SLM is polarization sensitive, we need two SLMs, one for each orthogonal polarization state. Alternatively, SLM can be segmented into two areas corresponding to each SOP. To deal with strong turbulence effects, the SLM-based backpropagation mode (BPM) method has been proposed in [87]. In the SLM-based BPM scheme, the pilot beam is propagated backward from the receiver to the transmitter to probe the atmospheric channel. The distorted wavefront is now determined at the transmitter side with the help of wavefront sensor, as illustrated in Fig. 6.46.

For the signal beam, SLM (SLMs) is (are) used to transmit the data sequence into specific spatial mode, which is the conjugation of the determined wavefront. Even though there exist amplitude-modulating liquid crystal on silicon (LCOS) devices, it has been shown in [87] that the low-cost phase-only SLMs to modulate the signal beams in the SLM-based BPM scheme are sufficient to deal with turbulence effects in a strong atmospheric turbulence regime. Compared to receiver side adaptive optics scheme, in which the segmented DM applies modal correction with superposition Zernike modes, the SLM is programed to generate the conjugated phase of the piecewise wavefront measured by the wavefront sensor directly [87]. Essentially, SLM operates as a zonal corrector to take advantage of its high spatial resolution. Therefore, the wavefront of the signal beam is pre-shaped so that after propagation the light will focus at a predetermined position, i.e., the target detector. This scheme is applicable in both single transmitter-single receiver and MIMO systems. In multi-spatial mode scenario, when we use one probing source for each parallel channel, the same number of wavefront sensors will be needed, which result in high complexity of the transmitter optics. To reduce the complexity, fewer probing sources can be used with linear phase interpolation, as proposed in [87]. When the number of probe sources is smaller than the number of spatial mode-based transmitters, a linear phase term is needed to modify the conjugation wavefront to provide the necessary signal displacement to the position of the target detector at the receiver side. Therefore, the pre-shaped wavefront U_{shaped} is given by [87]

$$U_{\text{shaped}} = U^{\dagger}_{\text{back-propagation}} \exp\left\{-j2\pi \frac{x(x_{det} - x_{prob}) + y(y_{det} - y_{prob})}{\lambda f}\right\}, \tag{6.340}$$

where (x_{det}, y_{det}) is the position of a target detector, while (x_{prob}, y_{prob}) is the position of the closest probing source to the target detector, and f is the focal length of the receiver (compressing) telescope. Additional details of SLM-based BPM can be found in [87].

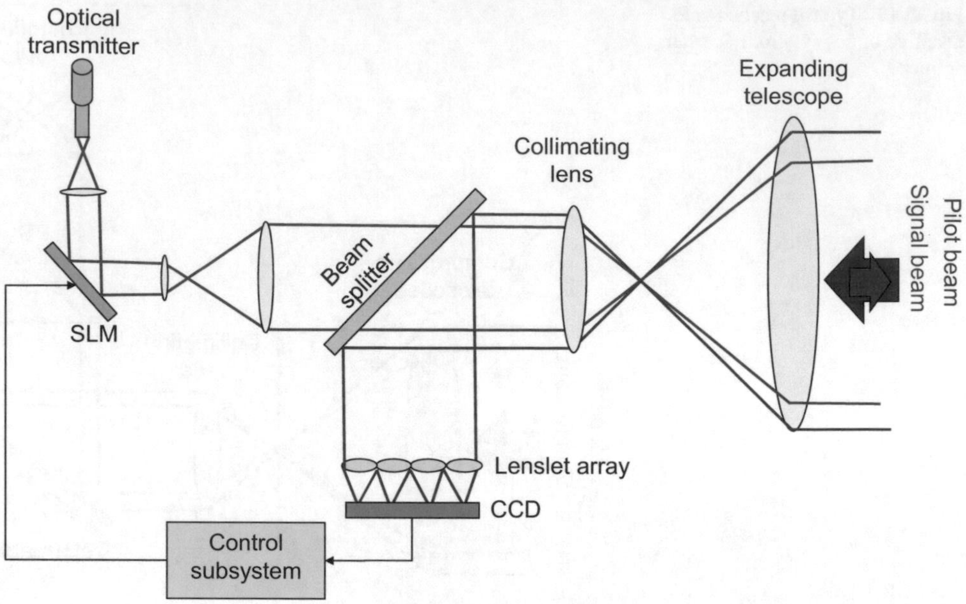

Fig. 6.46 Illustration of the SLM-based backpropagation method

6.10 Concluding Remarks

This chapter has been devoted to the advanced detection and channel impairments compensation techniques applicable to both wireless and optical communication systems. To better understand the principles behind advanced detection and channel impairments compensation, the fundamentals of detection and estimation theory have been described in Sect. 6.1, including geometric representation of received signals (Sect. 6.1.1), optimum receiver design (Sects. 6.1.2 and 6.1.3) in either maximum SNR sense (Sect. 6.1.2) or symbol error probability sense (Sect. 6.1.3), and symbol error probability derivations (Sect. 6.1.4). Fundaments of estimation theory have been provided in Sect. 6.1.5.

In the section on wireless communication systems performance, Sect. 6.2, different scenarios have been described including outage probability scenario (Sect. 6.2.1), average error probability scenario (Sect. 6.2.2), and combined outage-average error probability scenario (Sect. 6.2.3). The moment-generating function-based approach to average error probability calculation has been described in Sect. 6.2.4. The performance calculation in the presence of Doppler spread and fading effects has been described in Sect. 6.2.5.

In the section on channel equalization techniques, Sect. 6.3, after introductory remarks on ISI-free transmission and partial-response signaling (Sect. 6.3.1), the zero-forcing equalizers suitable for compensation of ISI, chromatic dispersion, PMD in fiber-optics communication; ISI and multipath fading effects in wireless communications; and ISI and multipath effects in indoor optical wireless communications have been described. The optimum linear equalizer design in the minimum mean-square error sense has been provided in Sect. 6.3.3, while the Wiener filtering concepts have been introduced in Sect. 6.3.4. The most relevant post-compensation techniques, applicable to both wireless and optical communications, have been described next including adaptive equalizer (Sect. 6.3.5), decision-feedback equalizer (Sect. 6.3.6), MLSD or Viterbi equalizer (Sect. 6.3.7), and blind equalization techniques (6.3.8). Relevant synchronization techniques have been described in Sect. 6.4.

In the section on adaptive modulation techniques, Sect. 6.5, various techniques to adapt the transmitter to time-varying wireless/optical channel conditions, enabling reliable and high spectral-efficient transmission, have been described. Various scenarios have been described including data rate, power, code rate, error probability adaptation scenarios, as well as a combination of various adaptation strategies. Various variable-power variable-rate modulation techniques have been described in Sect. 6.5.1 including water-filling method, channel inversion, and truncated channel inversion. The discrete-rate and discrete-power discrete-rate approaches have been discussed as well. The adaptive coded modulation techniques have been described in Sect. 6.5.2.

The Volterra series-based equalization to deal with fiber nonlinearities and nonlinear effects in wireless communications has been described in Sect. 6.6. In Sect. 6.7, the digital backpropagation has been described as a method to deal with fiber nonlinearities, chromatic dispersion, and PMD in a simultaneous manner.

Section 6.8 has been devoted to the coherent optical detection. The following topics have been covered: various balanced coherent optical detection schemes for two-dimensional modulation schemes in Sect. 6.8.1; polarization diversity and polarization demultiplexing schemes in Sect. 6.8.2; homodyne coherent optical detection based on phase-locked loops in Sect. 6.8.3; phase diversity receivers in Sect. 6.8.4; and the dominant coherent optical detection sources, such as laser phase noise, polarization noise, transimpedance amplifier noise, and ASE noise, in Sect. 6.8.5.

In the section on compensation of atmospheric turbulence effects, Sect. 6.9, various techniques to deal with atmospheric turbulence have been described including adaptive coding, adaptive coded modulation, diversity approaches, MIMO signal processing, and hybrid FSO-RF communication approach. In Sect. 6.9.1, the adaptive optics approaches to deal with turbulence effects have been described in detail. The SLM-based backpropagation method has been described in Sect. 6.9.2.

The set of problems is provided in the incoming section to help the reader gain deeper understanding of material described in this chapter.

6.11 Problems

6.1. A low-pass, zero-mean, Gaussian random process *X*(*t*) has a PSD of

$$G_X(f) = \begin{cases} A & |f| < B \\ 0 & |f| \geq B \end{cases}.$$

Let $Y = (1/T)\int_{-T/2}^{T/2} X(t)dt$. Assuming that $T >> 1/B$, calculate σ_Y^2 and compare it against σ_X^2.

6.2. The AWGN signal is passed through the system whose transfer function can be modeled as in Fig. P6.1. What is the output noise power? Determine the autocorrelation function of the random noise process at the output of the system. Assume that $R/L = 8$ and $1/(LC) = 25$.

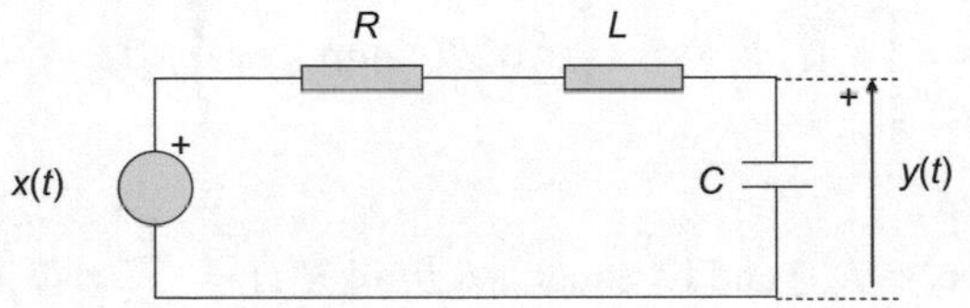

Fig. P6.1 The circuit under study

6.3. Determine the average energy per symbol for 4-ASK, assuming that each of the four symbols is equally likely. Derive the expression for symbol error probability. Compare it against QPSK.

6.4. The signal constellation is obtained by placing the signal constellation points in vertices of a rhombus (rhomb) of side length *a*. Sketch the signal constellation and determine the average signal energy as the function of *a*, assuming that all symbols are equally likely. Determine the optimum decision regions. Select the arbitrary basis functions $\phi_1(t)$ and $\phi_2(t)$ and express the four signals (from signal constellation diagram) in terms of basis functions. Determine the optimum receiver.

6.5. In this problem, we study the performance of 5-QAM signal constellation shown in Fig. P6.2 against that of 5-PSK with constellation points located on a circle of radius *a*. Assuming that all constellation points are equally likely:

(a) Determine the decision boundaries and decision regions for both 5-QAM and 5-PSK.
(b) Derive the expression for average error probability of 5-QAM.
(c) Derive the expression for average error probability of 5-PSK and compare it against that of 5-QAM.

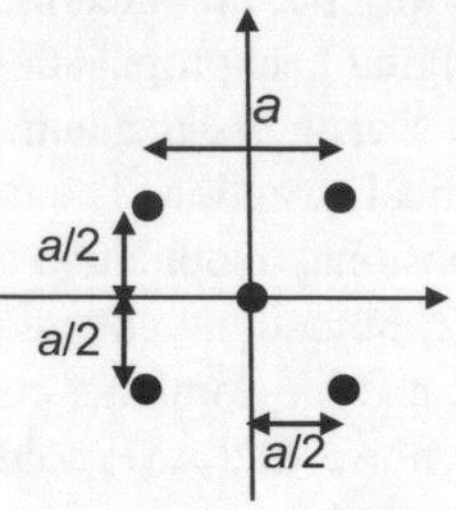

Fig. P6.2 The 5-QAM signal constellation diagram

6.6. The *cube* signal constellation is obtained by placing the signal constellation points in vertices of the cube of side length a. Assuming that all symbols are equally likely, derive the expression for average symbol error probability. Compare it against 8-star-QAM constellation shown in Fig. 5.12 (see Chap. 3).

6.7. Let $r(T_s)$ (T_s is the symbol duration) be the output of the optimum binary receiver. Prove that the following is valid:

$$m_1 = E\{r|m_1\} = E_1 - \rho_{12},$$

$$m_2 = E\{r|S_2\} = -E_2 + \rho_{12},$$

$$\sigma^2 = \mathrm{Var}\{r|m_1\} = \mathrm{Var}\{r|m_2\} = \mathrm{Var}\left\{\int_0^T n(t)[s_1(t) - s_2(t)]dt\right\}$$

$$= \frac{N_0}{2}(E_1 + E_2 - 2\rho_{12})$$

where

$$E_1 = \int_0^{T_s} s_1^2(t)dt, E_2 = \int_0^{T_s} s_2^2(t)dt, \text{and} \rho_{12} = \int_0^{T_s} s_1(t)s_2(t)dt.$$

6.8. Let us consider the random process $R(t)$ and its expanded form $R(t) = \sum_{j=1}^{D} r_j\Phi_j(t) + Z(t),\ 0 \le t \le T_s,$ where Z(t) is the remainder of noise term. $\{\Phi_j(t)\}$ represents the set of orthonormal basis functions, and r_j is the projection of $R(t)$ along the jth basis function. Let $Z(t_m)$ be the random variable obtained by observing $Z(t)$ at $t = t_m$. Show that the sample of the noise process $Z(t_m)$ is statistically independent on correlators' outputs $\{R_j\}$:$E[R_jZ(t_m)] = 0,\ \begin{cases} j = 1, 2, \ldots, D \\ 0 \le t_m \le T_s \end{cases}.$

6.9. An orthogonal set of signals is characterized by the property that the dot product of any pair of signals is zero. Provide a pair of signals that satisfies this condition. Construct the corresponding signal constellations.

6.10. A set of $2M$ biorthogonal signals can be derived from the set of orthogonal signals by augmenting it with the negative sign signals from orthogonal set. Does the biorthogonal signal construction increase the dimensionality of original signal set? The construction of the biorthogonal signal constellation for orthogonal signal constellation is developed in Problem 6.9.

6.11. Formulate the signal constellations for the following lines codes: unipolar NRZ, polar NRZ, unipolar RZ, and Manchester line code. For Manchester line code, derive the expression for error probability when the ML rule is applied in an AWGN channel.

6.12. The simplex signals are equally likely and highly correlated signals. When derived from M orthogonal signals, the correlation coefficient between any two pairs of signals in the set is given by

$$\rho_{kl} = \begin{cases} -1/(M-1), & k \neq l \\ 1, & k = l \end{cases}.$$

The simplest way to construct the simplex signal set is to start with a set of orthogonal signal set, with each element from the set having the same energy, and then create minimum energy signal set. Consider now the signal set with constellation points placed on equilateral triangle vertices. Prove that these three signals represent a simplex signal set.

6.13. Let us observe the M-ary cross-constellation described in Chap. 5 (see Fig. 5.11). Prove that for sufficiently high SNRs, the average symbol error probability can be estimated as $P_e \cong 2\left(1-1/\sqrt{2M}\right)\ \mathrm{erfc}\left(\sqrt{E_0/E_0}\right)$, where E_0 is the smallest symbol energy in constellation. Determine also the union bound of this constellation and discuss its accuracy. Finally, determine the nearest neighbor approximation.

6.14. Let us observe M-ary PSK signal constellation. Derive the expression for the average symbol error probability. Determine also the union-bound approximation. Finally, determine the nearest neighbor approximation. Discuss the accuracy of these approximations against the exact expression for average symbol error probability.

6.15. Let us consider a signal constellation that is symmetric with respect to the origin. Assume that there are M signal constellation points in the set, which are equally likely. Using the upper bound approach, determine the corresponding average symbol error of this signal constellation.

6.16. Let us reconsider the example from Sect. 6.1.5. Show how the equation in the example of this section is derived. By using it, deduce how the phase-locked loop can be implemented.

6.17. Derive the orthogonality principle given by Eq. (6.161). By using it, derive the most general form of the Wiener filter (6.163). Finally, derive the variance error of the Wiener filter. Show the steps of your derivations.

6.18. This problem is useful in MIMO signal processing and filter design. Let $\boldsymbol{x}$ be transmitted, $\boldsymbol{r}$ be received, and $\boldsymbol{n}$ be noise vectors, which are related by $r = Hx + n$, where $\boldsymbol{H}$ is the channel matrix. Derive the Wiener estimate and determine its minimum variance of error.

6.19. In this problem, we study the carrier phase recovery by data-aided method. In data-aided method, the receiver has the knowledge of preamble of length L, namely, $\{a_l\}_{l=0}^{L-1}$. Determine the ML estimate of the carrier phase. Provide the block diagram of this ML phase estimator.

6.20. Let us consider heterodyne coherent detection designs with two balanced detectors and single balanced detector, as shown in Fig. P6.3. Derive v_I and v_Q outputs in both designs, assuming that the incoming optical signal is given by $S_i = \sqrt{P_s}|a_i|e^{j(\omega_s t+\phi_i+\phi_{S,PN})}$, where ω_s is the carrier frequency of transmitting laser, (a_i,ϕ_i) are the polar coordinates of two-dimensional signal constellation being transmitted at the ith time instance (symbol interval), while $\phi_{s,PN}$ is the laser

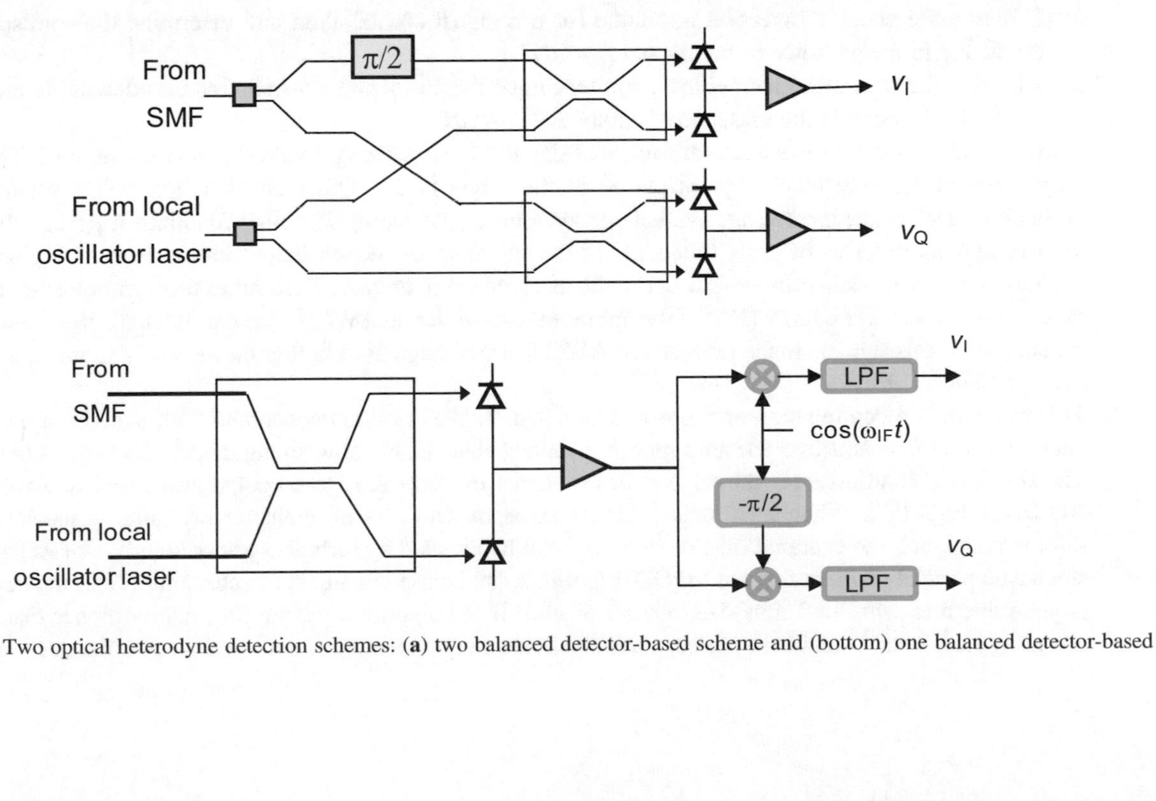

Fig. P6.3 Two optical heterodyne detection schemes: **(a)** two balanced detector-based scheme and (bottom) one balanced detector-based scheme

phase noise of transmitting laser. Assume that the local oscillator (LO) laser signal is given by $L = \sqrt{P_{LO}}e^{j(\omega_{LO}t+\phi_{LO,PN})}$, where ω_{LO} is the carrier frequency of local laser oscillator and $\phi_{LO,PN}$ is the corresponding laser phase noise. Compare the receiver sensitivities of these two schemes in the following two regimes: (i) ASE noise-dominated scenario and (ii) shot noise-dominated scenario.

6.21. The signal constellation of a ternary signal set is obtained by placing the signal constellation points in vertices of an *equilateral triangle* of side length a.
   (a) Sketch the signal constellation and determine the average signal energy as the function of a, assuming that all symbols are equally likely. Determine the optimum decision regions.
   (b) Select the arbitrary basis functions $\phi_1(t)$ and $\phi_2(t)$ and express the signals from signal constellation diagram in terms of basis functions.
   (c) Determine the corresponding minimum energy signal constellation.
   (d) What would be the optimum receiver in an AWGN channel?
   (e) Derive an expression for average symbol error probability against average symbol energy for an AWGN channel.
   (f) In a *fading environment*, the received signal power varies randomly over distance or time due to shadowing and/or multipath fading. Determine the average symbol error probability when in Nakagami $m = 2$ fading.
   (g) Let us now assume that we would like to transmit video signal, by employing the same signal constellation, with symbol rate larger than coherence bandwidth over the Nakagami-m fading. What would be the outage probability?
   (h) Determine the required signal-to-noise ratio corresponding to the outage probability of 1%, for the outage probability expression derived in (g). Do not evaluate exact value; just write down the corresponding equation to be solved numerically. Now, in special case for $m = 1$, determine the fade margin.

6.22. A four-dimensional (4-D) signal constellation is described by the following set of points $\{\pm a, \pm a, \pm a, \pm a\}$ $(a > 0)$. Different signal constellation points are used with the same probability. Determine the average symbol energy. Choose the appropriate basis functions $\{\Phi_1(t), \Phi_2(t), \Phi_3(t), \Phi_4(t)\}$ and express the signal constellation points in terms of basis functions. For an AWGN channel, determine the average symbol error probability as a function of the average energy (E_{av}). Evaluate now the average symbol error probability in the presence of Rayleigh fading.

6.23. This problem is related to the *adaptive modulation applied to 9-ary 2-D constellation* described below, transmitted over wireless channel in the presence of Nakagami fading and AWGN. The eight constellation points in this 2-D constellation are placed on a circle of radius 1. The last (ninth) point is placed in the origin. Assume that the point in the origin appears with probability 0.5, all others with probability 0.5/8.
   (a) Determine the average symbol error probability for this modulation format in the presence of fading and AWGN.
   (b) Determine the optimum power adaptation strategy as well as the corresponding spectral efficiency of this scheme in the presence of fading and AWGN.
   (c) Describe the channel inversion technique for this signal constellation and determine the corresponding spectral efficiency in the presence of fading and AWGN.
   (d) Describe the truncated channel inversion technique for this signal constellation and determine the corresponding spectral efficiency in the presence of fading and AWGN.

6.24. A five-signal configuration in a two-dimensional signal space has been provided already in Fig. P6.2. The probability of occurrence of the constellation point placed in the origin is 2/5. Other constellation points occur with the same probability. Determine the average symbol energy. Choose the appropriate basis functions $\Phi_1(t)$ and $\Phi_2(t)$ and express the five signals in terms of basis functions. In the signal space, sketch the optimum decision regions, assuming an AWGN channel model. Provide the optimum receiver configuration. Determine the symbol error probability as a function of the average energy (E_{av}) of the optimum receiver for an AWGN channel. Evaluate the average symbol error probability of this scheme in the presence of AWGN and Nakagami-m fading for $m = 0.5$, 1, and 2, assuming that the average SNR is 12 dB.

6.25. This problem is a continuation of Problem 1 in Chap. 2. We are again concerned with a study of the propagation of Gaussian and LG beams over the atmospheric turbulent channel, but now taking the adaptive optics (AO) into account. The simulation conditions are the same as in Problem 1 of Chap. 2. Figure P6.4 shows a typical wavefront sensorless AO-based FSO link. When a collimated Gaussian beam with a beam diameter of 5 mm is transmitted over 1 km turbulent channel, the captured on-axis intensity will be affected by turbulence accordingly. An AO system based on stochastic parallel gradient descent (SPGD) algorithm can be used to adjust the phase profile of the deformable mirror (square aperture with 8×8 mm size, 50×50 pixels). If the algorithm converging rate is higher than the turbulence changing rate, a stable and near-perfect Gaussian beam can be then recovered, in a weak turbulence regime.

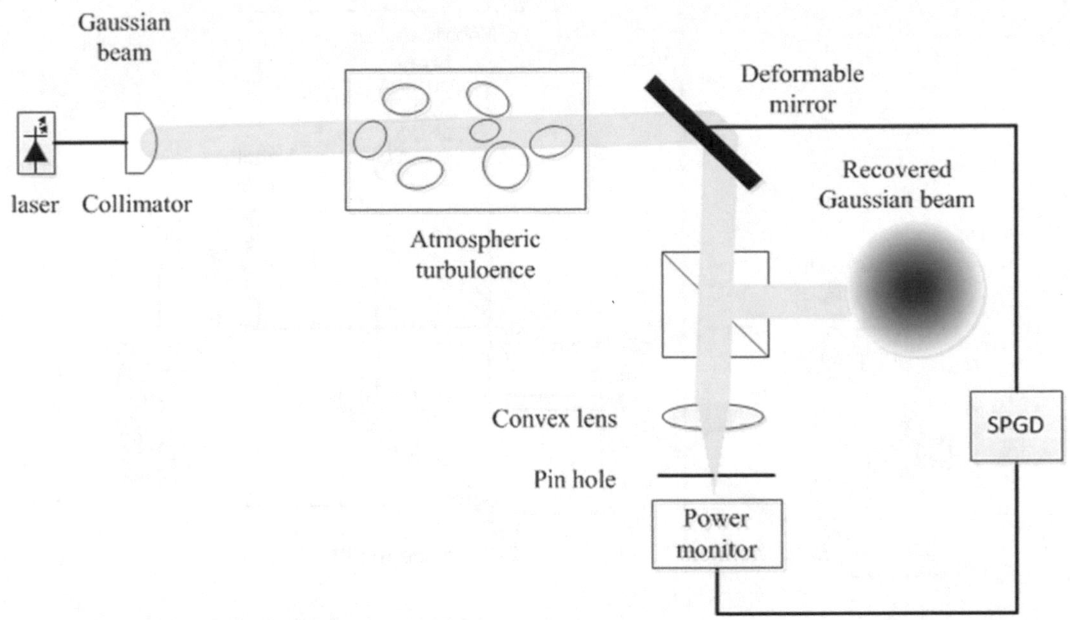

Fig. P6.4 Wavefront sensorless AO-based turbulence mitigation

Simulate the whole AO process for the following refractive structure parameters: $C_n^{\ 2} = 10^{-15}$, 10^{-14}, and 10^{-13} $\mathrm{m}^{-2/3}$.

(a) Plot the monitored intensity evolution over each SPGD algorithm step.
(b) When the monitored intensity converges to a certain level, plot the intensity profiles of the distorted Gaussian beam affected by the turbulence and recovered Gaussian beam after the turbulence mitigation.
(c) Plot the intensity probability density function and compare it with the case when no AO is used.

6.26. In this problem, which is a continuation of the previous problem, you are expected to simulate the wavefront sensor (WFS)-based AO system, shown in Fig. P6.5. Assume that WFS is based on 2-D Shack-Hartmann sensor. To represent the atmospheric turbulence effects, you are suggested to observe at least 56 Zernike terms. To determine the local tilts, you can use either of the equations below:

$$\begin{aligned} \boldsymbol{s} = \boldsymbol{r}_c/kf &= \frac{1}{P}\iint\limits_A I(\boldsymbol{r})\nabla\varphi d^2\boldsymbol{r} \\ &= \frac{1}{P}\iint\limits_A \{\mathrm{Re}\,(E)\nabla\mathrm{Im}\,(E) - \nabla[\mathrm{Re}\,(E)]\mathrm{Im}\,(E)\}d^2\boldsymbol{r}, \end{aligned}$$

where E is the reference beam complex amplitude of the electric field and I is the corresponding intensity. The surface of deformable mirror $S(\boldsymbol{\rho})$ can be represented as

$$S(\boldsymbol{\rho}) = \sum_{n=1}^{N} A_n f_n(\boldsymbol{\rho} - \boldsymbol{\rho}_n),$$

where A_n is the deflection of the mirror surface at the point of fastening of the nth actuator $\boldsymbol{\rho}_n$ and $f_n(\cdot)$ are the response functions $f_n(\boldsymbol{\rho}) = \exp\left(-\rho^2/w_n^2\right)$, with w_n being the halfwidth, ranging from $0.7d$ to $0.8d$, where d denotes the spacing between actuators.

By setting $S(\boldsymbol{\rho}) = \varphi(\boldsymbol{\rho})$, repeat steps (a), (b), and (c) from Problem 6.25 and discuss the improvements with respect to the wavefront sensorless AO. Also discuss the improvement for a different number of Zernike terms in WFS.

6.27. Repeat Problem 6.26 for LG mode with azimuthal mode index $m = 3$ and radial index $p = 0$.

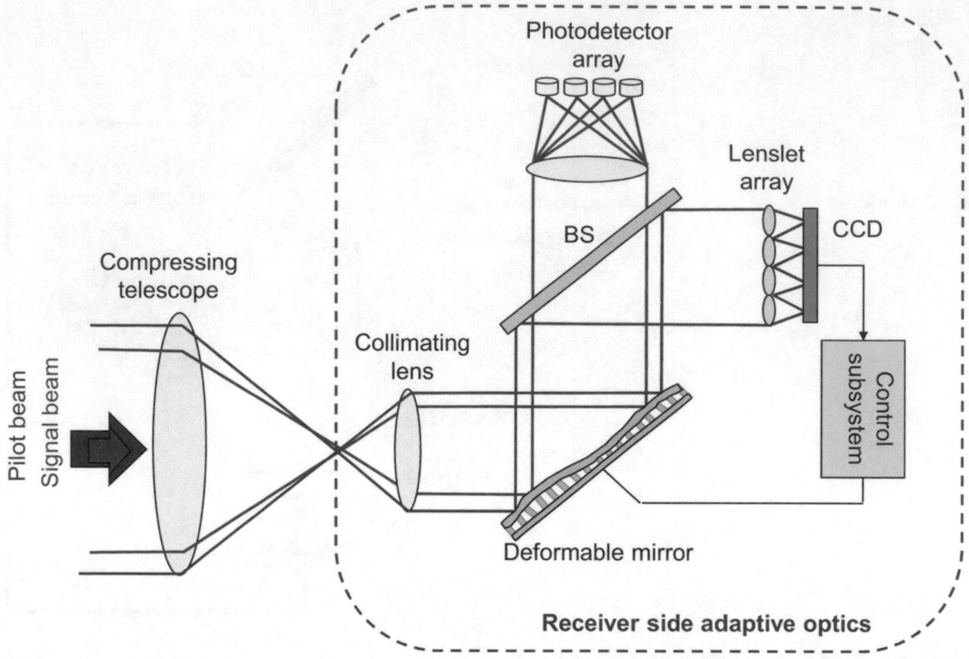

Fig. P6.5 Receiver side AO system

References

1. J.G. Proakis, *Digital Communications* (McGraw-Hill, Boston, 2001)
2. R.N. McDonough, A.D. Whalen, *Detection of Signals in Noise*, 2nd edn. (Academic Press, San Diego, 1995)
3. S. Haykin, *Digital Communications* (John Wiley & Sons, 1988)
4. S. Haykin, *Communication Systems*, 4th edn. (John Wiley & Sons, 2001)
5. A. Goldsmith, *Wireless Communications* (Cambridge University Press, 2005)
6. D. Tse, P. Viswanath, *Fundamentals of Wireless Communication* (Cambridge University Press, Cambridge, 2005)
7. S. Benedetto, E. Biglieri, *Principles of Digital Transmission with Wireless Applications* (Kluwer Academic/Plenum Publishers, New York, 1999)
8. E. Biglieri, *Coding for Wireless Channels* (Springer, New York, 2005)
9. I. Djordjevic, W. Ryan, B. Vasic, *Coding for Optical Channels* (Springer, 2010)
10. W. Shieh, W. Djordjevic, *OFDM for Optical Communications* (Elsevier, 2009)
11. I.B. Djordjevic, On Advanced FEC and Coded Modulation for Ultra-High-Speed Optical Transmission. IEEE Communications Surveys and Tutorials **18**(3), 1920–1951, third quarter 2016, August 19, 2016 (2016). https://doi.org/10.1109/COMST.2016.2536726
12. T.M. Duman, A. Ghrayeb, *Coding for MIMO Communication Systems* (John Wiley & Sons, Chichester, 2007)
13. G.L. Stüber, *Principles of Mobile Communication*, 3rd edn. (Springer, New York, 2012)
14. M.K. Simon, M.-S. Alouini, A unified approach to the performance analysis of digital communication over generalized fading channels. *Proceedings of the IEEE* **86**(9), 1860–1877 (1998)
15. M.K. Simon, M.-S. Alouini, *Digital Communication over Fading Channels: A Unified Approach to Performance Analysis* (John Wiley & Sons, New York, 2000)
16. M. Cvijetic, I.B. Djordjevic, *Advanced Optical Communication Systems and Networks* (Boston-London, Artech House, 2013)
17. G. Jacobsen, *Noise in Digital Optical Transmission Systems* (Artech House, Boston, 1994)
18. Djordjevic, I. B., and Stefanovic, M. C., "Performance of optical heterodyne PSK systems with Costas loop in multichannel environment for nonlinear second order PLL model," *IEEE J. Lightw. Technol.*, vol. 17, no. 12, pp. 2470-2479, Dec. 1999.
19. P.W. Hooijmans, *Coherent Optical System Design*. J. Wiley and Sons (1994)
20. L. Kazovsky, S. Benedetto, A. Willner, *Optical Fiber Communication Systems* (Artech House, Boston, 1996)
21. S.J. Savory, Digital filters for coherent optical receivers. *Opt. Express* **16**, 804–817 (2008)
22. I.B. Djordjevic, L. Xu, T. Wang, PMD compensation in multilevel coded-modulation schemes with coherent detection using BLAST algorithm and iterative polarization cancellation. *Opt. Express* **16**(19), 14845–14852, Sept. 15 (2008)
23. I.B. Djordjevic, L. Xu, T. Wang, PMD compensation in coded-modulation schemes with coherent detection using Alamouti-type polarization-time coding. *Opt. Express* **16**(18), 14163–14172, 09/01/2008 (2008)

24. I.B. Djordjevic, M. Arabaci, L. Minkov, Next generation FEC for high-capacity communication in optical transport networks. *IEEE/OSA J. Lightw. Technol.* accepted for publication. (Invited Paper.)
25. I.B. Djordjevic, L. Xu, T. Wang, Beyond 100 Gb/s optical transmission based on polarization multiplexed coded-OFDM with coherent detection. *IEEE/OSA J. Opt. Commun. Netw.* **1**(1), 50–56, June 2009 (2009)
26. J.G. Proakis, D.G. Manolakis, *Digital Signal Processing: Principles, Algorithms, and Applications*, 4th edn. (Prentice-Hall, 2007)
27. C.-C. Tseng, Design of IIR digital all-pass filters using least *p*th phase error criterion. *IEEE Trans. Circuits Syst. II, Analog Digit. Signal Process* **50**(9), 653–656, Sep. 2003 (2003)
28. G. Goldfarb, G. Li, Chromatic dispersion compensation using digital IIR filtering with coherent detection. *IEEE Photon. Technol. Lett* **19**(13), 969–971, July 1, 2007 (2007)
29. E. Ip, A. Pak, T. Lau, D.J.F. Barros, J.M. Kahn, Coherent detection in optical fiber systems. *Opt. Express* **16**(2), 753–791, 21 January 2008 (2008)
30. E. Ip, J.M. Kahn, Digital equalization of chromatic dispersion and polarization mode dispersion. *J. Lightw. Technol.* **25**, 2033–2043, Aug. 2007 (2007)
31. M. Kuschnerov, F.N. Hauske, K. Piyawanno, B. Spinnler, M.S. Alfiad, A. Napoli, B. Lankl, DSP for coherent single-carrier receivers. *J. Lightw. Technol* **27**(16), 3614–3622, Aug. 15, 2009 (2009)
32. J.G. Proakis, Partial response equalization with application to high density magnetic recording channels, in *Coding and Signal Processing for Magnetic Recording Systems*, ed. by B. Vasic, E. M. Kurtas, (CRC Press, Boca Raton, 2005)
33. I. Lyubomirsky, Optical duobinary systems for high-speed transmission, in *Advanced Technologies for High-Speed Optical Communications*, ed. by L. Xu, (Research Signpost, 2007)
34. N. Alic, G.C. Papen, R.E. Saperstein, L.B. Milstein, Y. Fainman, Signal statistics and maximum likelihood sequence estimation in intensity modulated fiber optic links containing a single optical preamplifier. *Opt. Express* **13**, 4568–4579, 13 June 2005 (2005)
35. G. Colavolpe, T. Foggi, E. Forestieri, G. Prati, Multilevel optical systems with MLSD receivers insensitive to GVD and PMD. *IEEE/OSA J. of Lightw. Technol.* **26**, 1263–1273 (2008)
36. M. Ivkovic, I.B. Djordjevic, B. Vasic, Hard decision error correcting scheme based on LDPC codes for long-haul optical transmission, in *Proc. Optical Transmission Systems and Equipment for Networking V-SPIE Optics East Conference*, vol. 6388, (Boston, 2006), pp. 63880F.1–63880F.7, 1–4 October 2006
37. Hagenauer, J., and Hoeher, P., "A Viterbi algorithm with soft-decision outputs and its applications," in Proc. IEEE Globecom Conf., Dallas, pp. 1680-1686, Nov. 1989.
38. L.R. Bahl, J. Cocke, F. Jelinek, J. Raviv, Optimal decoding of linear codes for minimizing symbol error rate. *IEEE Trans. Inform. Theory* **IT-20**, 284–287 (Mar. 1974)
39. H. Wymeersch, M.Z. Win, Soft electrical equalization for optical channels. Proc. ICC'08, 548–552, 19–23 May 2008 (2008)
40. C. Douillard, M. Jézéquel, C. Berrou, A. Picart, P. Didier, A. Glavieux, Iterative correction of intersymbol interference: turbo equalization'. *Eur. Trans. Telecommun.* **6**, 507–511 (1995)
41. M. Tüchler, R. Koetter, A.C. Singer, Turbo equalization: principles and new results. *IEEE Trans. Commun.* **50**(5), 754–767 (May 2002)
42. M. Jäger, T. Rankl, J. Spiedel, H. Bulöw, F. Buchali, Performance of turbo equalizers for optical PMD channels. *IEEE/OSA J. Lightw. Technol.* **24**(3), 1226–1236 (Mar 2006)
43. I.B. Djordjevic, L.L. Minkov, L. Xu, T. Wang, Suppression of fiber nonlinearities and PMD in coded-modulation schemes with coherent detection by using turbo equalization. *IEEE/OSA J. Opt. Commun. Netw.* **1**(6), 555–564 (Nov 2009)
44. M.H. Hayes, *Statistical Digital Signal Processing and Modeling* (John Wiley & Sons, Inc, 1996)
45. A. Haykin, *Adaptive Filter Theory*, 4th edn. (Pearson Education, 2003)
46. D.G. Manolakis, V.K. Ingle, S.M. Kogon, *Statistical and Adaptive Signal Processing* (McGraw-Hill, Boston, 2000)
47. N. Seshadri, Joint data and channel estimation using fast blind trellis search techniques. IEEE Trans. Comm., col. COMM-42, 1000–1011, Feb/Mar/Apr (1994)
48. D.N. Godard, Self-recovering equalization and carrier tracking in two-dimensional data communication systems. *IEEE Trans. Comm* **28**(11), 1867–1875 (Nov. 1980)
49. Y. Sato, A method of self-recovering equalization for multilevel amplitude-modulation systems. *IEEE Tran. Comm.* **23**(6), 679–682 (Jun 1975)
50. A. Benveniste, M. Goursat, G. Ruget, Robust identification of a nonminimum phase system. *IEEE Trans. Auto. Control* **AC-25**, 385–399 (June 1980)
51. A. Goldsmith, S.-G. Chua, "Adaptive coded modulation for fading channels," *IEEE Trans. Comm.*, vol. 46, pp. 595-601, May 1998.
52. I.B. Djordjevic, Adaptive modulation and coding for free-space optical channels. *IEEE/OSA J. Opt. Commun. Netw.* **2**(5), 221–229 (May 2010)
53. I.B. Djordjevic, G.T. Djordjevic, On the communication over strong atmospheric turbulence channels by adaptive modulation and coding. *Optics Express* **17**(20), 18250–18262, September 25, 2009 (2009)
54. I.B. Djordjevic, On the Irregular Nonbinary QC-LDPC-Coded Hybrid Multidimensional OSCD-Modulation Enabling Beyond 100 Tb/s Optical Transport. *IEEE/OSA J. Lightwave Technol* **31**(16), 2969–2975, Aug 15, 2013 (2013)
55. D. Zou, I.B. Djordjevic, FPGA-based rate-adaptive LDPC-coded modulation for the next generation of optical communication systems. *Optics Express* **24**(18), 21159–21166 (2016)
56. D. Zou, I.B. Djordjevic, FPGA-based rate-compatible LDPC codes for the next generation of optical transmission systems. *IEEE Photonics Journal* **8**(5), 7906008-1–7906008-8 (Oct. 2016). https://doi.org/10.1109/JPHOT.2016.2602063
57. D.J.C. MacKay, Fountain codes. *IEE Proc.-Commun* **152**, 1062–1068 (Dec 2005)
58. A. Shokrollahi, Raptor codes. *IEEE Inf. Theory* **52**, 2551–2567 (June 2006)
59. I.B. Djordjevic, Spatial-domain-based hybrid multidimensional coded-modulation schemes enabling multi-Tb/s optical transport. *IEEE/OSA J. Lightw. Technol* **30**(14), 2315–2328, July 15, 2012 (2012)
60. H.S. Wang, N. Moayeri, Finite-state Markov channel-a useful model for radio communication channels. IEEE Trans. Veh. Technol. **44**(1), 163–171 (1995)
61. G.D. Forney, R.G. Gallager, G.R. Lang, F.M. Longstaff, S.U. Qureshi, Efficient modulation for band-limited channels. IEEE J. Sel. Areas Comm **SAC-2**(5), 632–647 (1984)

62. G.D. Forney, Coset codes-part I: introduction and geometrical classification. IEEE Trans. Infrom. Theory **34**(5), 1123–1151 (Sept 1988)
63. R.D. Nowak, B.D. Van Veen, Volterra filter equalization: a fixed point approach. *IEEE Tran. Sig. Proc.* **45**(2), 377–388 (1997)
64. M.C. Jeruchim, P. Balaban, K.S. Shanmugan, *Simulation of Communication systems: Modeling, Methodology, and Techniques*, 2nd edn. (Kluwer Academic/Plenum Pub, N. York, 2000)
65. M. Nazarathy, R. Weidenfeld, Nonlinear impairments in coherent optical OFDM and their mitigation, in *Impact of Nonlinearities on Fiber Optic Communication*, ed. by S. Kumar, (Springer, Mar. 2011), pp. 87–175
66. F.P. Guiomar, J.D. Reis, A.L. Teixeira, A.N. Pinto, Mitigation of intra-channel nonlinearities using a frequency-domain Volterra series equalizer, in *Proc. ECOC 2011*, (Paper Tu.6.B.1, 18–22 September 2011, Geneva, 2011)
67. K.V. Peddanarappagari, M. Brandt-Pearce, Volterra series transfer function of single-mode fibers. *J. Lightw. Technol.* **15**, 2232–2241 (Dec 1997)
68. B. Xu, M. Brandt-Pearce, Modified Volterra series transfer function method. *IEEE Photon. Technol. Lett.* **14**(1), 47–49 (2002)
69. F.P. Guiomar, J.D. Reis, A.L. Teixeira, A.N. Pinto, Digital post-compensation using Volterra series transfer function. *IEEE Photon. Technol. Lett* **23**(19), 1412–1414, October 1, 2011 (2011)
70. Y. Gao, F. Zgang, L. Dou, Z. Chen, A. Xu, Intra-channel nonlinearities mitigation in pseudo-linear coherent QPSK transmission systems via nonlinear electrical equalizer. *Opt. Commun.* **282**, 2421–2425 (2009)
71. L. Liu, L. Li, Y. Huang, K. Cui, Q. Xiong, F.N. Hauske, C. Xie, Y. Cai, Intrachannel nonlinearity compensation by inverse Volterra series transfer function. *J. Lightw. Technol* **30**(3), 310–316, Feb. 1, 2012 (2012)
72. E. Ip, J.M. Kahn, Compensation of dispersion and nonlinear impairments using digital backpropagation. *J. Lightw. Technol.* **26**(20), 3416–3425 (2008)
73. E. Ip, J.M. Kahn, Nonlinear Impairment Compensation using Backpropagation, in *Optical Fibre, New Developments*, ed. by C. Lethien, (In-Tech, Vienna Austria, Dec. 2009)
74. G.P. Agrawal, *Nonlinear Fiber Optics*, 3rd edn. (Academic Press, San Diego)
75. I.B. Djordjevic, *Quantum Information Processing and Quantum Error Correction: An Engineering Approach* (Elsevier/Academic Press, Apr. 2012)
76. R.G. Priest, Analysis of fiber interferometer utilizing 3 × 3 fiber coupler. IEEE Journal of Quantum Electronics **QE-18(10)**, 1601–1603 (1982)
77. M.R. Paiam, R.I. MacDonald, Design of phased-array wavelength division multiplexers using multimode interference couplers. Applied Optics **36**(21), 5097–5108 (1997)
78. M. Seimetz, C.M. Weinert, Options, feasibility and availability of 2×4 90° hybrids for coherent optical systems. IEEE Journal of Lightwave Technology **24**(3), 1317–1322 (2006)
79. S. Norimatsu et al., An 8 Gb/s QPSK optical homodyne detection experiment using external cavity laser diodes. IEEE Photonics Technology Letters **4**(7), 765–767 (1992)
80. S. Camatel et al., Optical phase-locked loop for coherent detection optical receiver. Electronics Letters **40**(6), 384–385 (2004)
81. S. Camatel et al., 2-PSK homodyne receiver based on a decision driven architecture and a sub-carrier optical PLL, in *Proceedings of Optical Fiber Communication Conference (OFC)*, (Paper OTuI3, 2006)
82. A. Jesacher, A. Schwaighofer, S. Frhapter, C. Maurer, S. Bernet, M. Ritsch-Marte, Wavefront correction of spatial light modulators using an optical vortex image. *Optics Express* **15**(5), 5801–5808 (Apr. 2007)
83. R.W. Gerchberg, W.O. Saxton, A practical algorithm for the determination of phase from image and diffraction plane pictures. *Optik* **35**, 237–246 (1972)
84. N. Chandrasekaran and J. H. Shapiro, "Turbulence-induced crosstalk in multiple-spatial-mode optical communication," in Proc. *CLEO 2012*, San Jose, CA, May 6–11, 2012, Paper CF3I.6.
85. V.P. Lukin, B.V. Fortes, *Adaptive Beaming and Imaging in the Turbulent Atmosphere* (SPIE Press, Bellingham, Washington, 2002)
86. A. Glindemann, S. Hippler, T. Berkfeld, W. Hackenberg, Adaptive optics on large telescopes. Experimental Astronomy **10**(1), 5–47 (2000)
87. X. Sun, I.B. Djordjevic, M.A. Neifeld, Multiple spatial modes based QKD over marine free-space optical channels in the presence of atmospheric turbulence. Optics Express **24**(24), 27663–27673, 28 Nov 2016 (2016)
88. Z. Qu, I.B. Djordjevic, 500 Gb/s Free-Space Optical Transmission over Strong Atmospheric Turbulence Channels. *Optics Letters* **41**(14), 3285–3288, July 15 2016 (2016)
89. I.B. Djordjevic, J. Anguita, B. Vasic, Error-Correction Coded Orbital-Angular-Momentum Modulation for FSO Channels Affected by Turbulence. IEEE/OSA J. Lightw. Technol **30**(17), 2846–2852, September 1 2012 (2012)
90. R.J. Noll, Zernike polynomials and atmospheric turbulence. J. Opt. Soc. Am. **66**(3), 207–211 (1976)
91. Tatarskii VI (1967), The Effects of the Turbulent Atmosphere on Wave Propagation (Translated by Israel Program for Scientific Translations in 1971), U.S. Dept. of Commerce, National Technical Information Service, Springfield.
92. R. Cubalchini, Modal wave-front estimation from phase derivative measurements. J. Opt. Soc. Am. **69**, 972–977 (1979)
93. M.A. Ealey, J.A. Wellman, *Deformable mirrors: design fundamentals, key performance specifications, and parametric trades* (Proc. SPIE 1543, Active and Adaptive Optical Components, 1992), pp. 36–51. https://doi.org/10.1117/12.51167. January 13, 1992
94. R.K. Tyson, *Principles of Adaptive Optics* (CRC Press, Boca Raton, 2015)
95. Z. Qu, I.B. Djordjevic, M.A. Neifeld, RF-subcarrier-assisted Four-state Continuous-variable QKD Based on Coherent Detection. Optics Letters **41**(23), 5507–5510, Dec. 1, 2016 (2016)

OFDM for Wireless and Optical Communications 7

Abstract

This chapter is devoted to OFDM fundamentals and applications to wireless and optical communications. The chapter starts with the basic of OFDM including the generation of OFDM signal by inverse FFT. After that several basic manipulations on OFDM symbols, required to deal with multipath effects in wireless communications and dispersion effects in fiber-optics communications, including guard time and cyclic extension as well as the windowing operation, are described. Next, the bandwidth efficiency of OFDM is discussed. The next topic in the chapter is devoted to the OFDM system design, description of basic OFDM blocks, and OFDM-based parallel channel decomposition, including multipath fading radio and dispersion-dominated channel decomposition. Following the description of coherent optical OFDM (CO-OFDM) principles, the basic estimation and compensation steps, common to both wireless and optical communication systems, are described including DFT windowing, frequency synchronization, phase estimation, and channel estimation. Regarding the channel estimation, both pilot-aided and data-aided channel estimation techniques are described. Following the differential detection in OFDM description, the focus is on various applications of OFDM in wireless communications including digital audio broadcasting (DAB), digital video broadcasting (DVB), Wi-Fi, LTE, WiMAX, and ultra-wideband (UWB) communication. Next, optical OFDM applications are discussed starting with description of basic OFDM system types such as CO-OFDM and direct detection optical OFDM (DDO-OFDM) systems. Regarding the high-speed spectrally efficient CO-OFDM systems, the polarization-division multiplexed OFDM systems and OFDM-based superchannel optical transmission systems are described. Before, the concluding remarks, the application of OFDM in multimode fiber links is described. The set of problems is provided after the concluding remarks.

7.1 Introduction, OFDM Basics, and Generation of Subcarriers Using Inverse FFT

7.1.1 Introduction to OFDM

Orthogonal frequency division multiplexing (OFDM) belongs to the multicarrier transmission, where a single data stream is transmitted over a number of lower-rate orthogonal subcarriers. As the name suggests, the OFDM can be considered as a modulation technique, a multiplexing technique, or both. This gives the OFDM flexibility that does not exist in conventional single carrier systems. The OFDM has been introduced by Chang [1] in his seminal paper in 1966, followed by his patent in 1970 [2]. The OFDM is already a part of numerous standards for mobile wireless systems, wireless LANs, asymmetric digital subscriber lines, and coax cables [3, 4]. In optical communications [5–7], the OFDM occurred around 1996 [8], and the paper on optical OFDM was sporadic [9, 10], until recent resurgence [11–16]. The OFDM is in these days intensively studied as enabler of ultrahigh speed optical communications [15–19]. Regarding the optical communications, there are two basic classes of OFDM systems, namely, direct detection optical (DDO) OFDM [11, 13] and coherent optical (CO) OFDM [12], each of them with a number of variants.

There are several fundamental advantages of OFDM as compared to single-carrier systems, such as robustness against various multipath effects in wireless communication, infrared (IR) indoor and visible light communication (VLC) channels,

I. B. Djordjevic, *Advanced Optical and Wireless Communications Systems*, https://doi.org/10.1007/978-3-030-98491-5_7

robustness to dispersion effects (chromatic dispersion, PMD, multimode dispersion) and polarization-dependent loss (PDL) for fiber-optics communications, adaptivity to time-varying channel conditions, flexibility for software-defined radio (SDR) and software-defined optical transport, straightforward channel estimation, and low complexity compared to conventional equalizer schemes. However, it is sensitive to frequency offset and phase noise, and it exhibits large peak-to-average power ratio. Since the OFDM experts have already figured out how to deal with these drawbacks, the OFDM remains to be a hot research topic, in particular for fiber-optics communications. In addition, given the resurgence of optical fibers that support spatial multiplexing (FMF and FCF), the OFDM can be seen as a key enabling technology for next-generation optical transport [20–25].

7.1.2 OFDM Basics

An OFDM signal can be generated as a sum of orthogonal subcarriers that are modulated using conventional two-dimensional (2D) signal constellations such as PSK and QAM. The word orthogonality indicates that there exists a precise mathematical relationship among the frequencies of the carriers in the OFDM system. Namely, each subcarrier has exactly an integer number of cycles within the symbol interval T_s, and the number of cycles between adjacent subcarriers differs by exactly one, so that k-th subcarrier, denoted as f_k, and the l-th one, denoted as f_l, are related as follows:

$$f_k - f_l = n/T_s, \tag{7.1}$$

where n is an integer. The k-th subcarrier waveform can be represented by

$$s_k(t) = \text{rect}(t/T_s)e^{j2\pi f_k t}, \text{rect}(t) = \begin{cases} 1, & -1/2 < t \le 1/2 \\ 0, & \text{otherwise} \end{cases} \tag{7.2}$$

The following principle of orthogonality between k-th and l-th ($k \neq l$) subcarrier waveforms is valid only when the subcarriers are related by Eq. (7.1):

$$\begin{aligned} \langle s_l, s_k \rangle &= \frac{1}{T_s} \int_{-T_s/2}^{T_s/2} s_k(t) s_l^*(t) dt = \frac{\sin(\pi(f_k - f_l)T_s)}{\pi(f_k - f_l)T_s} \\ &= \text{sinc}((f_k - f_l)T)|_{f_k - f_l = n/T_s} = 0, \end{aligned} \tag{7.3}$$

where sinc(x) = sin(πx)/(πx). Therefore, in this case the orthogonal subcarriers present the basis functions. And, as common, an arbitrary signal can be represented in terms of projections along basis functions. In other words, by using PSK or QAM constellation points as projections along basis functions, we can generate an arbitrary OFDM waveform as

$$s(t) = \begin{cases} \sum\limits_{k=-N_{sc}/2}^{k=N_{sc}/2-1} X_{k+N_{sc}/2} \exp\left[j2\pi \frac{k}{T_s} t \right], & -T_s/2 < t \le T_s/2 \\ 0, & \text{otherwise} \end{cases} \tag{7.4}$$

where with X_k we denoted the QAM/PSK symbol imposed on k-th subcarrier and N_{sc} denotes the number of subcarriers. Based on this interpretation, we can represent the OFDM modulator as shown in Fig. 7.1a. The QAM serial data stream is demultiplexed into N_{sc} parallel streams, which are multiplied with different subcarrier basis functions to generate the OFDM waveform, shown in Fig. 7.1b. The real and imaginary part of this OFDM signal after digital-to-analog conversions can be used as RF inputs of I/Q modulator. Therefore, this OFDM modulator is consistent with synthesizer, described in Chap. 5. In this particular configuration, the OFDM principle is employed to perform modulation. On the other hand, if N_{sc} independent QAM streams for different sources are multiplied with corresponding subcarrier basis functions before being combined together, the OFDM synthesizer will serve as multiplexer.

The demodulation of l-th subcarrier QAM symbol can be performed by down-conversion by l/T_s, in other words by correlation receiver from Chap. 5 as follows:

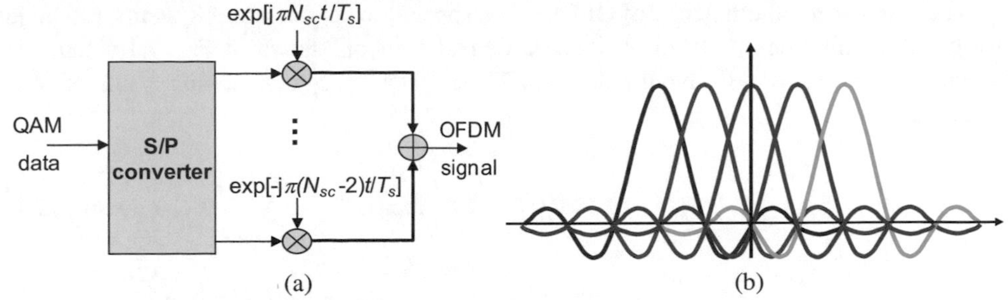

Fig. 7.1 The OFDM signal: (**a**) waveform generation and (**b**) Nyquist overlapping subcarriers

$$
\begin{aligned}
&\frac{1}{T_s}\int\limits_{-T_s/2}^{T_s/2} \exp\left[-j2\pi\frac{l}{T}t\right]\sum_{k=-N_{sc}/2}^{N_{sc}/2-1} X_{k+N_s/2}\exp\left[j2\pi\frac{k}{T_s}t\right]dt \\
&=\frac{1}{T_s}\sum_{k=-N_{sc}/2}^{N_{sc}/2-1} X_{k+N_{sc}/2}\underbrace{\int\limits_{-T_s/2}^{T_s/2} \exp\left[j2\pi\frac{k-l}{T}t\right]dt}_{T_s,\,k=l\ (0,\,k\neq l)}= X_{l+N_s/2}
\end{aligned}
\tag{7.5}
$$

Therefore, thanks to the orthogonality principle, the QAM symbol transmitted over l-th subcarrier has been demodulated. However, in the presence frequency offset, the initial orthogonality of subcarriers will be violated, and there will be a crosstalk among basis functions (subcarriers), commonly referred to as inter-carrier interference (ICI).

7.1.3 Generation of OFDM Signals

In order to perform transition in DT domain, let us sample the OFDM CT signal, given by Eq. (7.4), with a sampling interval $T = T_s/N$, where N is an integer of the form 2^n ($n > 1$). The samples can be expressed as

$$
s(mT) = \begin{cases} \sum\limits_{k=-N_{sc}/2}^{k=N_{sc}/2-1} X_{k+N_{sc}/2}\exp\left[j2\pi\frac{k}{T_s}mT\right], & T_s/2 < t < T_s/2 \\ 0, \quad \text{otherwise} \end{cases}
\tag{7.6}
$$

Since $T_s/T = N$, we can rewrite the previous equation as

$$
s(mT) = \begin{cases} \sum\limits_{k=-N_{sc}/2}^{k-N_{sc}/2-1} X_{k+N_{sc}/2}\exp\left[j\frac{2\pi}{N}km\right], & T_s/2 < t \le T_s/2 \\ 0, \quad \text{otherwise} \end{cases}
\tag{7.7}
$$

which represents the inverse discrete Fourier transform (IDFT) of input QAM sequence. The IDFT can be efficiently implemented by inverse fast Fourier transform (IFFT). This discrete-time implementation is an important reason of high appealing of OFDM scheme. The DT implementation is very flexible as several subcarriers can be used for channel estimation. In addition, the radix-2 and radix-4 algorithms can be used to reduce the complexity of IDFT. Namely, in radix-2 algorithm an N-point IFFT requires $(N/2)\log_2 N$ complex multiplications, while radix-4 algorithm requires only $(3\,N/8)(\log_2 N\text{-}2)$ complex multiplications or phase rotations. On the other hand, the regular IDFT requires N^2 complex multiplications. Unfortunately, the samples used to calculate IFFT are not sufficient for practical implementation since the use of digital-to-analog conversion (DAC) is needed. To avoid aliasing effect, the oversampling must be used. This is typically performed by zero padding. The zero samples should be added in the middle of QAM data sequence rather than appending them at the edges. With this approach we will ensure that data QAM points are mapped around d.c. component, while zero samples will be mapped close to plus/minus sampling rate. After the IFFT calculation and parallel-to-serial conversion, DAC process is performed.

The better spectral efficiency of OFDM, as compared to single-carrier systems, can be justified as follows. From Eq. (7.4) we conclude that the spectrum of a single OFDM symbol, shown in Fig. 7.1b, can be obtained as a convolution of the spectrum of a square pulse having duration T_s and given by sinc(fT_s), with a sum of Dirac pulses located at the subcarrier frequencies:

$$\begin{aligned} S(f) &= FT\{s(t)\} = FT\left\{\operatorname{rect}(t/T_s)\sum_{k=-N_{sc}/2}^{k=N_{sc}/2-1} X_{k+N_{sc}/2}\exp\left[j2\pi\frac{k}{T_s}t\right]\right\} \\ &= \operatorname{sinc}\,(fT_s)*\sum_{k=-N_{sc}/2}^{k=N_{sc}/2-1} X_{k+N_{sc}/2}\delta\left(f-k\frac{1}{T_s}\right) \\ &= \sum_{k=-N_{sc}/2}^{k=N_{sc}/2-1} X_{k+N_{sc}/2}\operatorname{sinc}\left(\left(f-k\frac{1}{T_s}\right)T_s\right). \end{aligned} \tag{7.8}$$

where FT denotes Fourier transform operator. Therefore, the overlapping sinc spectra of individual subcarriers (OFDM spectrum) satisfy the Nyquist criterion for an intersymbol interference (ISI)-free signaling [26, 27]. Even though the neighboring subcarrier pulses partially overlap at the observed OFDM subcarrier location, the neighboring subcarrier pulse shapes cross zero at that location so that there is no inter-carrier interference. The key difference as compared to zero-ISI signaling is that the pulse shape of OFDM subcarrier is observed in the frequency-domain rather than in the time-domain. Since the partial overlapping of neighboring subcarrier pulses is allowed, the bandwidth efficiency can be improved.

7.2 Guard Time, Cyclic Extension, and Windowing

In this section we describe several basic manipulations on OFDM symbol required to deal with dispersion effects in optical communications and multipath effects in wireless communications (including indoor optical wireless and IR communications), as well as out-of-band spectrum reduction.

7.2.1 Guard Time and Cyclic Extension

One of the key advantages of OFDM is that it can straightforwardly be used to deal with various dispersion and multipath effects. To reduce the ISI among OFDM symbols, the *guard time* is introduced for every OFDM symbol. The guard time is typically chosen to be longer than the expected dispersion/multipath widening so that dispersion/multipath components from one symbol are not able to interfere with the neighboring symbols. The guard time could be just an empty time interval; however, the problem of inter-carrier interference (ICI) will arise in that case as the subcarrier components cease to be orthogonal. In order to eliminate the ICI introduced by dispersive optical and multipath wireless channels, the OFDM symbol is commonly *cyclically extended* in the guard time to restore the orthogonality of the subcarrier components. The symbol duration, after cyclic extension, is determined as $T_s = T_G + T_{FFT}$, where T_G is the guard interval duration, while T_{FFT} is the duration of effective portion of OFDM symbol (FFT part) related to subcarrier spacing Δf_{sc} by $T_{FFT} = 1/\Delta f_{sc}$. The cyclic extension can be performed by repeating the last $N_G/2$ samples (corresponding to $T_G/2$) of the FFT frame (of duration T_{FFT} with N_{FFT} samples) as a prefix and repeating the first $N_G/2$ samples (out of N_{FFT}) as a suffix, which is illustrated in Fig. 7.2. Another alternative is to use only either prefix or suffix of duration T_G.

In *fiber-optics communications*, for complete elimination of ISI, the total delay spread Δt due to chromatic dispersion and differential group delay (DGD) should be shorter than the guard interval [5]:

$$\Delta t = |\beta_2|L_{tot}\Delta\omega + DGD_{max} = \frac{c}{f^2}|D_t|N_{FFT}\Delta f_{sc} + DGD_{max} \le T_G, \tag{7.9}$$

where D_t is the accumulated dispersion, c is the speed of the light, f is the central frequency (typically set to 193.1 THz), L_{tot} is the total transmission distance, DGD_{max} is the maximum DGD, and β_2 is the second-order group velocity dispersion (GVD)

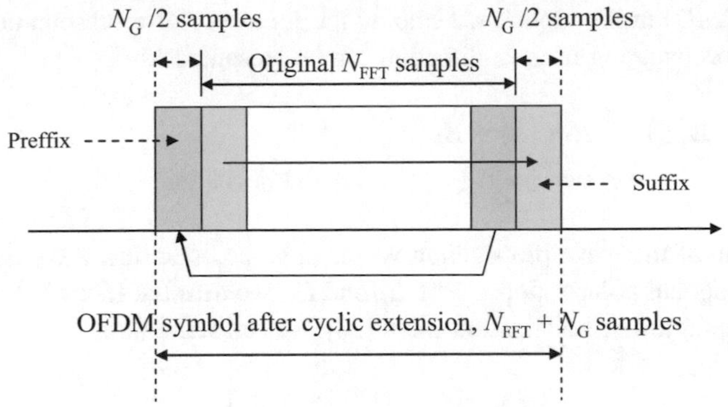

Fig. 7.2 The illustration of cyclic extension procedure

parameter related to chromatic dispersion parameter D by $\beta_2 = -(2\pi c/\lambda^2)D$, with λ being the operating wavelength. The received electrical field, in the presence of chromatic dispersion, can be written as

$$E_s = \exp\left[j(2\pi f_{LD}t + \phi_{PN,S})\right]\sum_{k=1}^{N_{FFT}} X_k e^{j2\pi f_k t} e^{j\phi_{CD}(k)}, \tag{7.10}$$

where $\phi_{CD}(k)$ denotes the phase factor of k-th subcarrier

$$\phi_{CD}(k) = \frac{\omega_k^2|\beta_2|L}{2} = \frac{4\pi^2 f_k^2}{2}\frac{\lambda_{LD}^2}{2\pi c}DL = \frac{\pi c}{f_{LD}^2}D_t f_k^2, \tag{7.11}$$

with f_k being the k-th subcarrier frequency (corresponding angular frequency is $\omega_k = 2\pi f_k$), f_{LD} being the transmitting laser frequency (the corresponding wavelength is λ_k), D_t being the total dispersion ($D_t = D{\cdot}L$, D dispersion coefficient, L the fiber length), and c being the speed of light. The output of the OFDM balanced coherent optical detector can be represented as

$$v(t) \simeq R_{PIN}R_L e^{\left\{j\left[2\pi(f_{LD}-f_{LO})t+\phi_{PN,S}-\phi_{PN,LO}\right]\right\}}\sum_{k=1}^{N_{FFT}} X_k e^{j2\pi f_k t} e^{j\phi_{CD}(k)} + N(t), \tag{7.12}$$

where f_{LO} denotes the carrier frequency of the local laser oscillator, R_{PIN} is the photodetector responsivity, R_L is feedback resistance in trans-impedance amplifier configuration, while $\phi_{S,PN}$ and $\phi_{L,PN}$ represent the laser phase noise of the transmitting laser and local laser oscillator, respectively. These two noise sources are commonly modeled as Wiener-Lévy process [7], which is a zero-mean Gaussian process with variance $2\pi(\Delta\nu_S + \Delta\nu_L)|t|$, where $\Delta\nu_S$ and $\Delta\nu_L$ are the laser linewidths of transmitting laser and local oscillator, respectively, while $N=N_I\text{-}jN_Q$ represent the noise process, mostly dominated by ASE noise. The received symbol Y_k of k-th subcarrier can be represented as

$$Y_k = X_k \exp\left[j(\phi_{PN,S} - \phi_{PN,LO})\right]\exp\left[j\phi_{CD}(k)\right] + N_k, \qquad N_k = N_{k,I} + N_{k,Q} \tag{7.13}$$

where N_k represents a circular Gaussian process due to ASE noise expressed in a complex form. The estimation of the transmitted symbol on k-th subcarrier can be done by relation

$$\widetilde{X}_k = Y_k \exp\left[-j(\phi_{PN,S} - \phi_{PN,LO})\right]\exp\left[-j\phi_{CD}(k)\right], \tag{7.14}$$

where the chromatic dispersion phase factor of k-th subcarrier $\phi_{CD}(k)$ is estimated by training-based channel estimation while the phase factor $\phi_{PN,\,S} - \phi_{PN,\,LO}$ is determined by pilot-aided channel estimation. The additional details of various OFDM channel estimation techniques can be found in Sec. 7.5. Therefore, with proper guard interval selection, we can decompose the dispersive channel into N_{FFT} parallel channels, which will be described later with additional details. To include the polarization mode dispersion (PMD) in the model, we employ the Jones vector representation. Namely, the electric field of plane linearly polarized waves propagating in the z-direction can be presented as

$$\begin{aligned}\mathbf{E}(z,t) &= \mathbf{E}_H(z,t) + \mathbf{E}_V(z,t)\\ &= \mathbf{e}_H E_{0H}\cos(\omega t - kz) + \mathbf{e}_V E_{0V}\cos(\omega t - kz + \delta),\end{aligned} \tag{7.15}$$

where $k = 2\pi/\lambda$ is the magnitude of the wave propagation vector (λ is the operating wavelength) and δ is the relative phase difference between the two orthogonal polarizations. With E_H and E_V, we denoted H- and V-polarization states, respectively. If using the Jones vector representation of polarization wave [28], we can write that

$$\mathbf{E}(t) = \begin{bmatrix} E_H(t) \\ E_V(t) \end{bmatrix} = E\begin{bmatrix} \sqrt{1-p_r} \\ \sqrt{p_r}e^{j\delta} \end{bmatrix} e^{\,j(\omega t - kz)}, \tag{7.16}$$

where p_r is the power splitting ratio between states of polarizations (SOPs) while the complex phasor term is typically omitted. From Eq. (7.16), the equivalent channel model can be expressed as

$$\begin{bmatrix} Y_k^{(H)} \\ Y_k^{(V)} \end{bmatrix}_k = U(f_k)\begin{bmatrix} X_k^{(H)} \\ X_k^{(V)} \end{bmatrix} \exp\left[j(\phi_{PN,S} - \phi_{PN,LO})\right]\exp\left[j\phi_{CD}(k)\right] + \begin{bmatrix} N_k^{(H)} \\ N_k^{(V)} \end{bmatrix}, \tag{7.17}$$

where $U(f_k)$ denotes the Jones matrix of fiber link, defined as

$$U(f_k) = \begin{bmatrix} U_{HH}(f_k) & U_{HV}(f_k) \\ U_{VH}(f_k) & U_{VV}(f_k) \end{bmatrix}. \tag{7.18}$$

In *wireless communications*, to deal with multipath effects, the guard interval must be two to four times longer than the root mean square (RMS) of the multipath delay spread.

We will now provide the interpretation of cyclic extension from digital signal processing (DSP) [7, 29, 30] point of view. The N-point discrete Fourier transform (DFT) of discrete-time sequence $x(n)$ $(0 \le n \le N\text{-}1)$ and corresponding inverse DFT (IDFT) are defined as

$$\begin{aligned}\mathrm{DFT}\{x(n)\} &= X_k = \sum_{n=0}^{N-1} x[n]e^{-j2\pi nk/N},\quad 0 \le k \le N-1\\ \mathrm{IDFT}\{X_k\} &= x(n) = \frac{1}{N}\sum_{k=0}^{N-1} X_k e^{\,j2\pi nk/N},\quad 0 \le n \le N-1\end{aligned} \tag{7.19}$$

The output $y(n)$ of a linear time-invariant (LTI) system can be obtained by convolution of the input $x(n)$ and LTI dispersive channel impulse response $h(n)$:

$$y(n) = x(n) * h(n) = h(n) * x(n) = \sum_k h(k)x(n-k). \tag{7.20}$$

On the other hand, the circular convolution of $x(n)$ and $h(n)$ is defined as

$$\begin{aligned}y(n) &= x(n)\boxed{*}h(n) = \sum_k h(k)x(n-k)_N,\\ (n-k)_N &\equiv (n-k)\bmod N,\end{aligned} \tag{7.21}$$

where with $(n\text{-}k)_N$ we denoted the periodic extension of $x(n\text{-}k)$, with period N. Therefore, the resulting sequence obtained by circular convolution is periodic with a period of N. DFT of circular convolution of $x(n)$ and $h(n)$ is product of corresponding DFTs of individual terms:

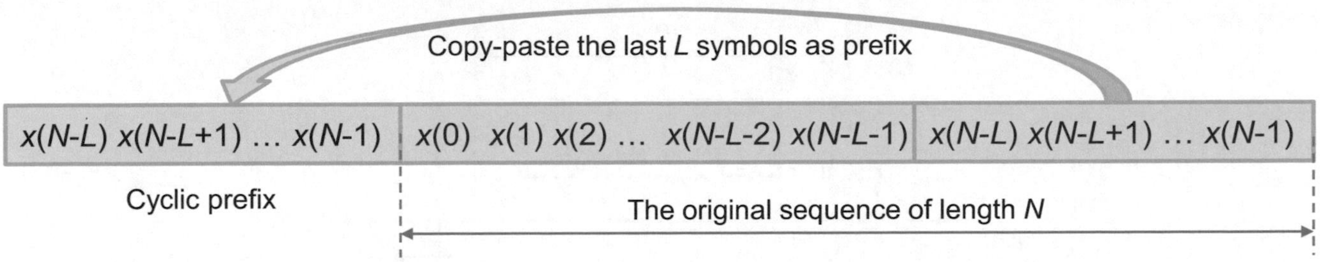

Fig. 7.3 The illustration of cyclic prefix creation for FFT-based OFDM

$$\text{DFT}\{y(n) = x(n)\boxed{*}h(n)\} = X_k H_k,\ \ 0 \le k \le N-1. \tag{7.22}$$

In this case, the channel output is not a circular but rather a linear convolution of the channel input and its impulse response. However, the linear convolution can be converted to circular by adding a cyclic extension. The cyclic extension can be inserted as a prefix or suffix or can be split into prefix and suffix portions. The cyclic extension of $\{x(n)\}$, inserted as a prefix, is defined by $\{x(N\text{-}L),\ldots,x(N\text{-}1)\}$, where L presents the last L samples of sequence, as illustrated in Fig. 7.3.

When the cyclic extension is appended as a prefix (to the beginning of the sequence $x(n)$), the convolution of the resulting sequence and impulse response $h(n)$ yields to the circular convolution:

$$\begin{aligned} y(n) &= x_{\text{circular}}(n) * h(n) = \sum_{k=0}^{L} h(k) x_{\text{circular}}(n-k) = \sum_{k=0}^{L} h(k) x(n-k)_N \\ &= x(n)\boxed{*}h(n), \end{aligned} \tag{7.23}$$

where with $x_{\text{circular}}(n)$ we denoted the cyclic extended sequence. The DFT of circular convolution of $x(n)$ and $h(n)$ is given as

$$\text{DFT}\{y(n) = x(n)\boxed{*}h(n)\} = X_k H_k,\ \ 0 \le k \le N-1. \tag{7.24}$$

The estimation of the input sequence, denoted here as $\tilde{x}(n)$, can be done by

$$\tilde{x}(n) = \text{IDFT}\left\{\frac{Y_k}{H_k}\right\} = \text{IDFT}\left\{\frac{\text{DFT}\{y(n)\}}{\text{DFT}\{h(n)\}}\right\}, \tag{7.25}$$

where H_k can be obtained by pilot-aided channel estimation [5]. The function $y(n)$, used in Eq. (7.25), has the length $N + L$. However, the first L samples in $y(n)$, (where $n = -L,\ldots,-1$) are not required to recover the original sequence $x(n)$ (where $n = 0,1,\ldots,N\text{-}1$). By adding the cyclic prefix, we have solved the problem of circular convolution, but we have introduced an overhead of L/N, while the overall data rate has been reduced by $N/(L + N)$ times. Instead of using DFT/IDFT in OFDM, we can apply FFT/IFFT. The length of sequence N in discussion above will correspond to N_{FFT}. In OFDM, we add the cyclic prefix to every OFDM symbol. As a summary, with the help of cyclic prefix, we can simultaneously address the problems of both ISI and circular convolution.

7.2.2 Windowing

Another important procedure with respect to OFDM is the *windowing*, which is illustrated in Fig. 7.4. The purpose of the windowing is to reduce the out-of band spectrum, which is important for multiband coherent optical (CO) OFDM applications, often known as superchannel optical OFDM structure. The windowing is also important in wireless communication to reduce out-of-band spectrum so that the neighboring wireless channels can be placed closer. A commonly used window type has the raised cosine shape:

$$w(t) = \begin{cases} \frac{1}{2}[1 - \cos \pi(t + T_{\text{win}} + T_G/2)/T_{\text{win}}], & -T_{\text{win}} - T_G/2 \le t < -T_G/2 \\ 1.0, & -T_G/2 \le t < T_{\text{FFT}} + T_G/2 \\ \frac{1}{2}[1 - \cos \pi(t - T_{\text{FFT}})/T_{\text{win}}], & T_{\text{FFT}} + T_G/2 \le t < T_{\text{FFT}} + T_G/2 + T_{\text{win}} \end{cases} \tag{7.26}$$

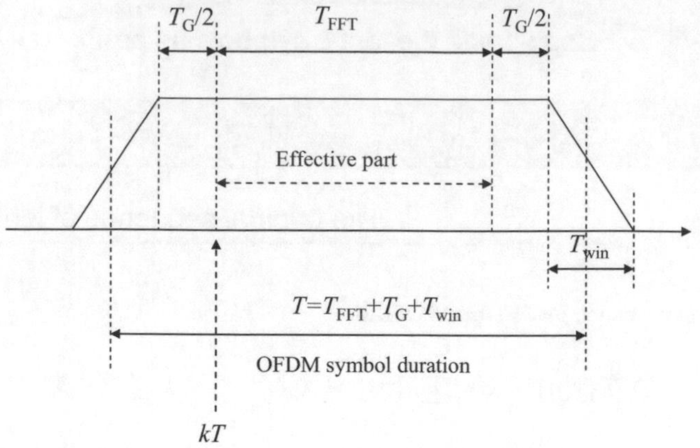

Fig. 7.4 The illustration of the windowing operation

where T_G is the guard interval duration and T_{win} is the duration of windowing interval. The window shape expressed above can be represented in digital signal processing by various digital functions (also known just as windows), such as Hamming, Hanning, Bartlett, etc. [28]. The windowing interval needs to be carefully chosen not to violate the orthogonality principle since a partial overlap of neighboring OFDM is allowed in practice. Please notice that digital filtering techniques can be used instead of windowing, to reduce the out-of-band spectrum. However, the digital filtering requires at least several multiplications per sample, while windowing needs several multiplications per symbol, which means that the windowing is simpler to implement. The OFDM signal after windowing can be represented as

$$s_B(t) = \sum_{n=-\infty}^{\infty} w(t - nT_s) \sum_{k=-N_{FFT}/2}^{N_{FFT}/2-1} X_{n,k} \cdot e^{j2\pi \frac{k}{T_{FFT}} \cdot (t - nT_s)}, \tag{7.27}$$

where we used the index n to denote the n-th OFDM symbol while index k denotes the k-th OFDM subcarrier. For single-band OFDM systems, the windowing operation can be omitted. (In Eq. (7.27), subscript "B" stands for the baseband signal.)

In summary, the OFDM signal can be generated as follows. The N_{QAM} input QAM symbols are zero-padded to obtain N_{FFT} input samples used to calculate the IFFT. The last $N_G/2$ samples of the IFFT output are inserted at the start of the OFDM symbol, while the first $N_G/2$ samples are appended at the end. The OFDM symbol is multiplied by corresponding windowing function. After that, the real and imaginary parts are separated and digital-to-analog conversion performed on both the real and imaginary parts. Finally, after amplification in the driver, the real and imaginary parts are used as inputs to I and Q ports of an I/Q modulator.

7.3 Bandwidth Efficiency of OFDM

If we assume that N_{sc} subcarriers are used, each having R_{sc} symbol rate, the total symbol rate R_s for CO-OFDM and wireless OFDM systems can be calculated as

$$R_s = N_{sc} R_{sc}. \tag{7.28}$$

The bandwidth occupied by OFDM symbol, without employing the windowing operation, can be expressed as

$$BW_{OFDM} = \frac{N_{sc} - 1}{T_{FFT}} + 2R_{sc}, \tag{7.29}$$

where T_{FFT} is the effective portion of OFDM symbol (carrying data), as shown in Fig. 5.39. Since the subcarrier spacing is given by $\Delta f_{sc} = 1/T_{FFT}$, the bandwidth occupied by subcarriers equals $(N_{sc}\text{-}1)\Delta f_{sc}$. On the other hand, the bandwidth portion

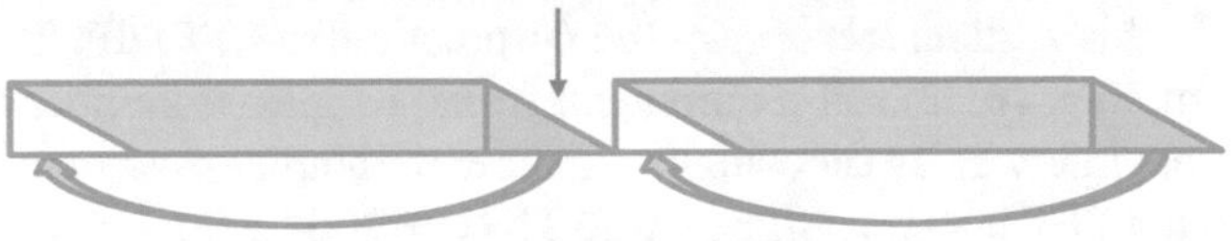

Leave the guard interval empty on transmitter side
Copy-paste the samples of received trail as prefix on receiver side

Fig. 7.5 The illustration of the zero-prefix OFDM

due to finite rise and falling edges of the first and last subcarriers is given by $2R_{sc}$. Therefore, the overall bandwidth efficiency of CO-OFDM systems can be expressed as

$$\rho = 2\frac{bN_{sc}R_{sc}}{BW_{\text{OFDM}}} = 2\frac{bN_{sc}R_{sc}}{2R_{sc} + (N_{sc} - 1)/T_{\text{FFT}}}, \tag{7.30}$$

where b denotes the number of bits per QAM symbol while factor 2 originates from polarization-division multiplexing (PDM). In wireless communications, typically, the PDM is not used so that the factor 2 in (7.30) should be omitted. The subcarrier spacing can be determined as $\Delta f_{sc} = 1/(T_s\text{-}T_G)$, and it is system design dependent. The guard interval is strongly dependent on the amount of dispersion/multipath fading the system needs to compensate for. Therefore, the true bandwidth efficiency can be determined only if all details of OFDM system design are known. Nevertheless, the following rule of thumb can be used for an initial study. Typically, the ratio $R_s\ /B_{\text{CO-OFDM}}$ is around 8/9 so that the bandwidth efficiency is about $(1.778{\cdot}b)$ bits/s per Hz [5]. As an illustration, when all subcarriers carry 16-QAM symbols, the bandwidth efficiency is ~7.112 bits/s/Hz. Notice that transmission of cyclic prefix, in addition of reducing the bandwidth efficiency, also reduces the power efficiency.

The power efficiency of OFDM can be improved by using *all-zero prefix*, which is illustrated in Fig. 7.5. Namely, on the transmitter side we leave the guard intervals empty. The optical channel will introduce the trails in guard intervals due to chromatic dispersion and PMD. These trails need to be "copy-pasted" as prefix of OFDM symbols before OFDM demodulation takes place. In principle, there is no difference in dispersion compensation capabilities of this scheme as compared to cyclic prefix in CO-OFDM systems. However, the power efficiency can be improved $(N + L)/N$ times. The drawback of this scheme is that ASE noise gets approximately doubled, as the noise from trail is transferred to already noisy prefix portion of OFDM symbol. On the other hand, the bandwidth efficiency can be improved through the so-called no-guard interval OFDM [17, 18]. The no-guard interval OFDM, however, requires the use of conventional linear equalizer to compensate for distortions introduced by dispersion effects.

7.4 OFDM Parameter Selection, OFDM Building Blocks, and Parallel Channel Decomposition

7.4.1 OFDM Parameter Selection

The selection of various OFDM parameters is a trade-off between various, very often conflicting requirements. In wireless communications, the main requirements to start with include the bandwidth, bit rate, and multipath delay spread. The multipath delay spread directly dictates the guard time: two to four times RMS of delay spread. The symbol duration is typically chosen to be at least five times the guard time, to ensure 1 dB SNR loss because of the guard time [3, 4]. The number of subcarriers can be determined as the ratio of the required – 3 dB bandwidth and the subcarrier spacing. The subcarrier spacing is determined as the inverse of symbol duration minus the guard time, i.e., $\Delta f_{sc} = 1/(T_s\text{-}T_G)$. Alternatively, the number of subcarriers may be determined as the ratio of required bit rate and the bit rate per subcarrier [3, 4].

Example. Let us design the OFDM system, for wireless communication applications, with required bit rate of 40 Mbps and tolerable multipath delay spread of 100 ns. Let us assume that the available channel bandwidth is $\leq$25 MHz. The guard time is chosen to be four times the multipath delay spread, which is 400 ns. The OFDM symbol duration can be chosen to be 6 $\times$ the guard time, which is 2.4 μs. The subcarrier spacing is determined as 1/(2.4–0.4) μs = 500 kHz. The number of bits per OFDM symbol is determined as the required bit rate divided by the OFDM symbol rate = 40 Mb/s / (1/2.4 μs) = 96 bits per OFDM symbol. When 8-QAM is used with 2/3 rate coding, the required number of subcarriers will be 48, with 3 $\times$ 2/3 = 2 information bits per subcarrier symbol. The bandwidth occupied will be 48 × 500 kHz = 24 MHz, which is smaller than the available bandwidth (25 MHz). In this particular example, the 64-point radix-4 FFT/IFFT can be used, leaving 16 unused subcarriers to provide the oversampling. Additional requirement is that an integer number of samples must be used within FFT/IFFT interval and in the symbol interval. By choosing the number of samples per symbol to be 72, the sampling rate will be 72/2.4 μs = 30 MHz. Finally, the FFT interval will be 64/30 MHz = 2.134 μs.

7.4.2 OFDM Building Blocks

The generic OFDM system suitable for wireless applications is shown in Fig. 7.6. The binary data sequence is first LDPC or turbo encoded, depending on application of interest, and the encoder output is written into interleaver. For turbo coding s-random interleaver can be used, while for LDPC coding the block-interleaver can be used. For block-interleaver, $\log_2(M)$, where M is signal constellation size, codewords are written in row-wise fashion interleaver. Notice that arbitrary two-dimensional (2D) constellation can be used. Then $\log_2(M)$ bits are taken into column-wise fashion from the interleaver and used to select a point from 2D constellation such as QAM, by applying appropriate mapping rule. The QAM symbol data stream is converted from serial to parallel form, and with the help of subcarrier mapper, different subcarriers are loaded with different QAM symbols. The inverse FFT is then applied, followed by the cyclic extension and windowing operations. After digital-to-analog conversion (DAC), the corresponding RF signal is up-converted, amplified, and transmitted by using a transmit antenna. On the receiver side, after low-noise amplification, down-conversion, and analog-to-digital conversion (ADC), the timing and frequency-offset compensation are performed. On such obtained symbol sequence, with proper timing and frequency-offset compensated, the cyclic removal step is applied, followed by serial-to-parallel (S/P) conversion. Further, FFT demodulation is performed followed by symbol detection and QAM demapping. Further, deinterleaving is preformed, followed by turbo or LDPC decoding. Deinterleaving and turbo/LDPC decoding are typically jointly performed.

On the other hand, the generic CO-OFDM system architecture (suitable for optical communications) is shown in Fig. 7.7. The QAM data stream, obtained after encoding of binary data stream and performing the corresponding binary-to-QAM mapping, arrives as serial symbol flow to OFDM RF transmitter, as shown in Fig. 7.7a. The symbols in QAM stream belong to the set of constellation points {0,1,. . ., M-1}, where M is the QAM signal constellation size. When nonbinary coding is used, the codeword symbols can be interpreted as QAM symbols, thus providing that encoder nonbinary field size is the same as the size of the QAM constellation. The serial QAM stream is converted into parallel stream in P/S converter (or DEMUX), with N_{sc} symbols provided to the subcarrier mapper. The subcarrier mappers serve as a look-up table (LUT), which assigns the in-phase and quadrature coordinates to every input symbol, and create the corresponding complex numbers passed to IFFT OFDM modulator in parallel fashion. After the FFT portion of OFDM symbol is created, the cyclic extension is performed. The cyclically extended OFDM symbol undergoes P/S conversion and up-sampling operation, before being passed to the DACs. The real and imaginary outputs of DACs are (after pulse shaping and amplification, denoted as low-pass filtering

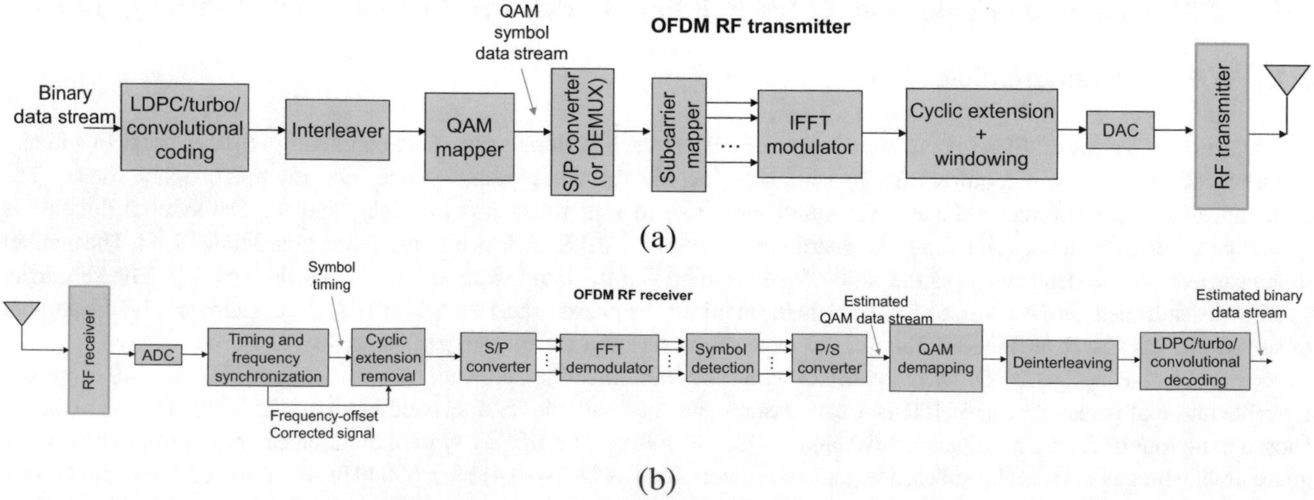

Fig. 7.6 The block diagram of OFDM transceiver suitable for wireless applications: (**a**) OFDM transmitter configuration and (**b**) OFDM receiver configuration

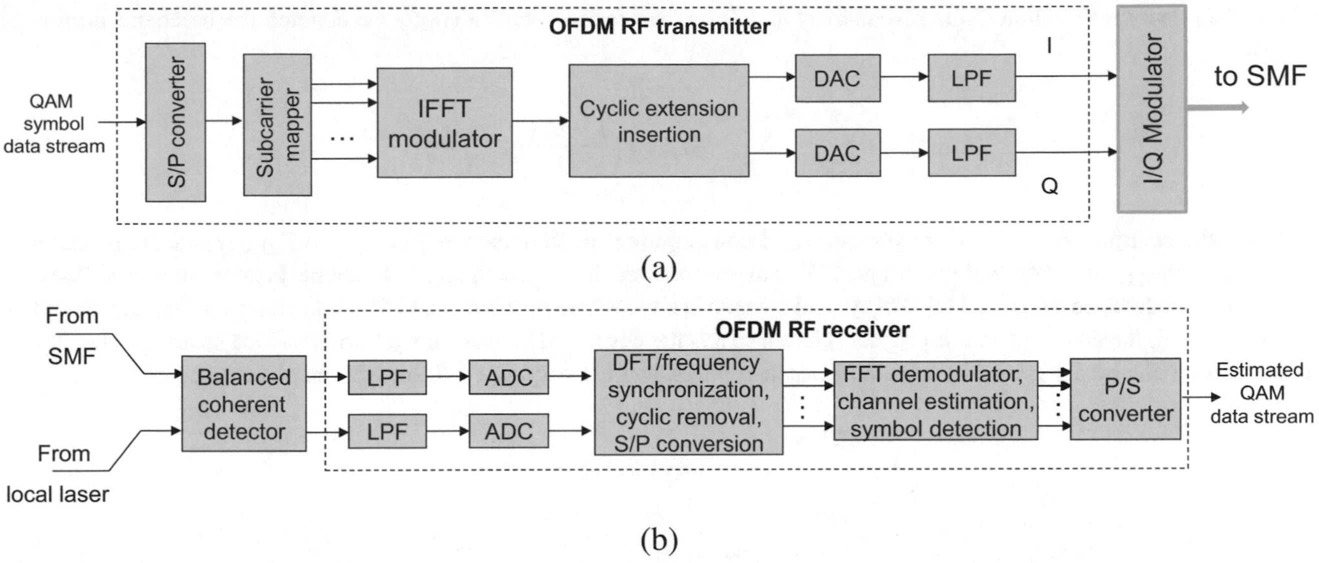

Fig. 7.7 The CO-OFDM system architecture: (**a**) OFDM transmitter configuration, and (**b**) CO-OFDM receiver configuration

operation) used as inputs of an optical I/Q modulator. The optical OFDM data stream is transmitted over optical communication system of interest. On the receiver side [see Fig. 7.7b], the received optical signal is mixed with local laser signal in balanced coherent detector. The balanced coherent detector provides the projections along in-phase and quadrature basis functions, denoted as v_I and v_Q in Fig. 7.7b, which are after amplification and ADC interpreted as real and imaginary parts of corresponding complex number. Upon the S/P conversion, the FFT demodulation is performed, followed by QAM symbol detection. The QAM symbol stream is after P/S conversion delivered to the end user. In OFDM RF receiver from Fig. 7.7b, the OFDM signal after DACs undergoes the following synchronization steps: (i) FFT window synchronization, in which OFDM symbols are properly aligned to avoid any ISI, and (ii) frequency synchronization, in which the frequency offset is estimated and compensated. Additionally, after FFT demodulation the subcarrier recovery is performed, where the channel coefficients at subcarrier level are estimated and compensated for. Full details of these steps, together with various channel estimation techniques, will be provided in incoming sections.

The OFDM signal at the output of transmitter shown in Fig. 7.7a can be expressed as

$$s(t) = e^{j[2\pi f_{LD}t+\phi_{PN,S}]} \times \sum_{n=-\infty}^{\infty} \sum_{k=-N_{FFT}/2}^{N_{FFT}/2-1} X_{n,k}\mathrm{rect}\left(\frac{t-nT_s}{T_s}\right) e^{j2\pi\frac{k}{T_{FFT}}\cdot(t-nT_s)}, \tag{7.31}$$

where we omitted the windowing function. In Eq. (7.31), f_{LD} denotes the optical carrier frequency of transmitter laser diode, while $\phi_{PN,S}$ denotes the stochastic laser phase noise process. The other parameters in Eq. (7.31) are already described in previous sections.

7.4.3 OFDM Parallel Channel Decomposition and Channel Modelling

The transmitted optical OFDM signal propagates over optical medium of impulse response $h(t,\tau)$, where the parameter τ is used to denote that the impulse response is time-variant, so that the received signal can be represented as

$$r(t) = s(t) * h(t,\tau) + w(t), \tag{7.32}$$

where $w(t)$ denotes the ASE noise process that originates from EDFAs periodically deployed to compensate for fiber loss. Upon the balanced coherent detection, the corresponding signal can be written as

$$v(t) \simeq R_{PIN}R_L e^{j\{2\pi(f_{LD}-f_{LO})t+\phi_{PN,S}-\phi_{PN,LO}\}} s_B(t) * h(t,\tau) + w(t), \tag{7.33}$$

where R_{PIN} denotes the photodiode responsivity and R_L is load resistance. With $s_B(t)$ we denoted the baseband portion of (7.33):

$$s_B(t) = \sum_{n=-\infty}^{\infty} \sum_{k=-N_{FFT}/2}^{N_{FFT}/2-1} X_{n,k} \mathrm{rect}\left(\frac{t-nT_s}{T_s}\right) e^{j2\pi \frac{k}{T_{FFT}} \cdot (t-nT_s)}. \tag{7.34}$$

It is evident from Eq. (7.33) that estimation and compensation of frequency offset $\Delta f_{\mathrm{off}} = f_{LD}\text{-}f_{LO}$ and the phase error $\Delta\phi = \phi_{PN,S}\text{-}\phi_{PN,LO}$ are of crucial importance. We can now assume the following: (i) The channel is quasi-static and does not change for the duration of several OFDM symbols, which is always true for typical CO-OFDM systems, so we can write that $h(t,\tau) \cong h(t)$; (ii) the guard interval is properly chosen so that there is no ISI between neighboring OFDM symbols. The OFDM demodulating can be performed by a bank of matched filters, and the output of k-th branch can be obtained as

$$\begin{aligned} Y_{n,k} &= \frac{1}{T_{FFT}} \int_{t=kT_s}^{kT_s+T_{FFT}} v(t) e^{-j2\pi i(t-kT_s)/T_{FFT}} dt \\ &= \frac{1}{T_{FFT}} \int_{t=kT_s}^{kT_s+T_{FFT}} \left[\int_0^{\Delta t} h_k(\tau) s_B(t-\tau) d\tau + w(t) \right] e^{-j2\pi n \frac{t-kT_s}{T_{FFT}}} dt, \end{aligned} \tag{7.35}$$

where for simplicity of presentation we omitted $R_{PIN} \cdot R_L$ term and assumed that frequency offset and phase error were compensated for.

By substituting $s_B(t)$ with expression given by Eq. (7.34), we obtain

$$\begin{aligned} Y_{n,k} &= \sum_{n'=-N/2}^{N/2-1} X_{n',k} \frac{1}{T_{FFT}} \int_{\sigma=0}^{T_{FFT}} \left[\int_0^{\Delta t} h_k(\tau) e^{-j2\pi n'(\sigma-\tau)/T_{FFT}} d\tau \right] e^{-j2\pi n\sigma/T_{FFT}} d\sigma + w_{n,k} \\ &= \sum_{n'=-N/2}^{N/2-1} X_{n',k} \frac{1}{T_{FFT}} \int_{\sigma=0}^{T_{FFT}} \underbrace{\int_0^{\Delta t} h_k(\tau) e^{-j2\pi n'\tau/T_{FFT}} d\tau}_{H_{n',k}} e^{-j2\pi(n-n')\sigma/T_{FFT}} d\sigma + w_{n,k} \end{aligned} \tag{7.36}$$

where $H_{n,k}$ denotes the transfer function term of k-th subcarrier in the n-th OFDM symbol. This term is not a function of outer integral and can be moved out so that the equation above becomes

$$Y_{n,k} = \sum_{n'=-N/2}^{N/2-1} X_{n',k} H_{n',k} \overbrace{\frac{1}{T_{FFT}} \int_{\sigma=0}^{T_{FFT}} e^{-j2\pi(n-n')\sigma/T_{FFT}} d\sigma}^{1,\; n=n'} + w_{n,k}, \tag{7.37}$$

By using the orthogonality principle, we obtain

$$Y_{n,k} = H_{n,k} X_{n,k} + w_{n,k}, \tag{7.38}$$

which represents the parallel decomposition of an optical dispersive channel in frequency-domain, as illustrated in Fig. 7.8.

If the phase error in coherent optical system is not compensated for, the corresponding parallel optical channel decomposition is given by

$$Y_{n,k} = X_{n,k} H_{n,k} e^{j(\phi_{PN,S}-\phi_{PN,LO})} + w_{n,k}. \tag{7.39}$$

The channel coefficient $H_{n,k}$ accounts for chromatic dispersion, filtering effects, and other linear distortion effects. The chromatic dispersion portion has the form $\exp[j\phi_{CD}(k)]$, where $\phi_{CD}(k)$ is defined earlier. We can now substitute $H_{n,k} = H'_{n,k} \exp[j\phi_{CD}(k)]$ to obtain

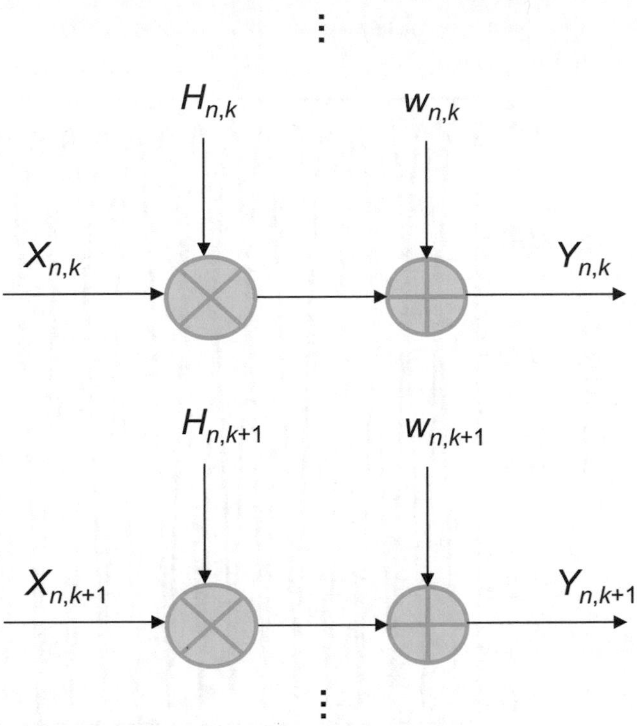

Fig. 7.8 The parallel dispersive optical channel decomposition in frequency-domain

$$Y_{n,k} = X_{n,k}H'_{n,k}e^{j(\phi_{PN,S}-\phi_{PN,LO})}e^{j\phi_{CD}(k)} + w_{n,k}. \tag{7.40}$$

The generalization of Equation (7.40) to PDM systems can be done by introducing the Jones matrix of fiber link $\boldsymbol{U}(f_k)$, so we obtain the following channel model:

$$\boldsymbol{Y}_{n,k} = \boldsymbol{U}(f_k)H'_{n,k}e^{j(\phi_{PN,S}-\phi_{PN,LO})}e^{j\phi_{CD}(k)}\boldsymbol{X}_{n,k} + \boldsymbol{W}_{n,k}, \tag{7.41}$$

where the corresponding symbols get replaced with two-component column vectors, with the first component representing x-polarization state and the second component representing the y-polarization state. This model can be used to study CO-OFDM systems and various channel distortion compensation methods. For instance, to study the impact of first-order PMD, we can use the following Jones matrix [5, 47]:

$$\begin{aligned}\boldsymbol{U} &= \begin{bmatrix} U_{xx}(\omega) & U_{xy}(\omega) \\ U_{yx}(\omega) & U_{yy}(\omega) \end{bmatrix} = \boldsymbol{R}\boldsymbol{P}(\omega)\boldsymbol{R}^{-1}, \\ \boldsymbol{P}(\omega) &= \begin{bmatrix} e^{-j\omega\tau/2} & 0 \\ 0 & e^{j\omega\tau/2} \end{bmatrix},\end{aligned} \tag{7.42}$$

where τ is DGD, ω is the angular frequency, and $R = R(\theta,\varepsilon)$ is the rotational matrix

$$\boldsymbol{R} = \begin{bmatrix} \cos\left(\frac{\theta}{2}\right)e^{j\varepsilon/2} & \sin\left(\frac{\theta}{2}\right)e^{-j\varepsilon/2} \\ -\sin\left(\frac{\theta}{2}\right)e^{j\varepsilon/2} & \cos\left(\frac{\theta}{2}\right)e^{-j\varepsilon/2} \end{bmatrix} \tag{7.43}$$

with θ being the polar angle, while ε is the azimuthal angle. The magnitude of U_{xx} and U_{xy} coefficients of Jones channel matrix against normalized frequency $f\tau$ (the frequency is normalized with DGD τ so that the conclusions are independent on the data rate) is shown in Fig. 7.9 for two different cases: (i) $\theta = \pi/2$ and $\varepsilon = 0$ and (ii) $\theta = \pi/3$ and $\varepsilon = 0$. In the first case,

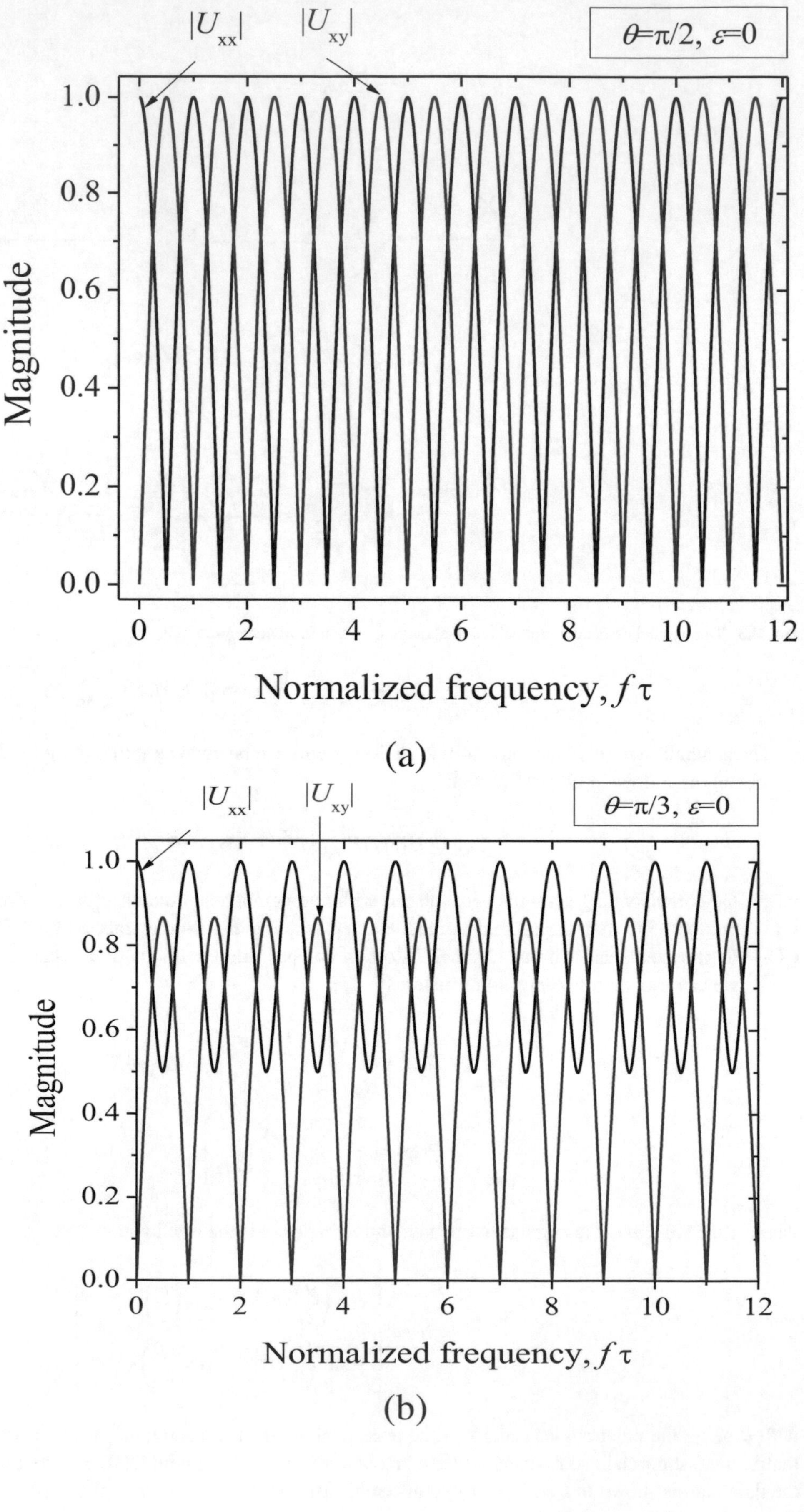

Fig. 7.9 Magnitude response of U_{xx} and U_{xy} Jones matrix coefficients against the normalized frequency for: (**a**) $\theta = \pi/2$ and $\varepsilon = 0$ and (**b**) $\theta = \pi/3$ and $\varepsilon = 0$

channel coefficient U_{xx} completely fades away for certain frequencies, while in the second case, it never completely fades away, thus suggesting that the first case represents the worst-case scenario.

For *multipath fading radio channel*, we can denote the impulse response as $h(t,\tau)$, where the parameter τ is used to denote that the impulse response is time-variant. Therefore, Eq. (7.32) is also applicable here. By assuming that the guard interval is longer than the maximum delay spread, while that the wireless channel is quasi-static for at least the duration of one OFDM symbol, we can perform the multipath fading channel decomposition in similar fashion as given by Eqs. (7.35)–(7.38). The time-varying impulse response, as already shown in Chap. 2, can be represented as a sum of multipath components arriving at the receiver side:

$$h(\tau,t) = \sum\nolimits_n h_n(t)e^{-j\theta_n(t)}\delta[\tau - \tau_n(t)], \tag{7.44}$$

where h_n, θ_n, and τ_n represent the n-th propagation path amplitude, phase, and delay, respectively. The Fourier transform (FT) of the time-varying impulse response with respect to propagation path delay τ will give us time-varying transfer function:

$$H(f,t) = \int_{-\infty}^{\infty} h(\tau,t)e^{-j2\pi f\tau} = \sum\nolimits_n h_n(t)e^{-j[2\pi f\tau_n(t)+\theta_n(t)]}, \tag{7.45}$$

which changes rapidly in both time and frequency, exhibiting therefore both time- and frequency-selective fading. Given that the transfer function is time-varying, it makes sense to define the autocorrelation function of the transfer function:

$$R_H(f_1, f_2; t_1, t_2) = \langle H^*(f_1, t_1)H(f_2, t_2)\rangle. \tag{7.46}$$

This autocorrelation function gets simplified under wide-sense stationary-uniform scattering (WSS-US) assumption (see Sect. 2.3 of Chap. 2 for additional details):

$$R_H(\Delta f; \Delta t) = \langle H^*(f,t)H(f+\Delta f, t+\Delta t)\rangle. \tag{7.47}$$

The ranges in time and frequency over which the transfer function exhibits significant correlation are known as coherence time and coherence bandwidth. The coherence time is defined as the time range when the spaced-time correlation function $R_H(\Delta t) = R_H(0;\Delta t)$ drops below 0.9. On the other hand, the coherence bandwidth is defined as the frequency range when the spaced frequency correlation function $R_H(\Delta f) = R_H(\Delta f;0)$ drops below 0.9. The *delay power spectrum (DPS)*, denoted as $R_h(\tau)$, is the inverse Fourier transform of the space-frequency correlation function $R_H(\Delta f) = R_H(\Delta f;0)$. The following simple model can be used to study OFDM systems for wireless applications [4]:

$$R_h(\tau) = \begin{cases} \rho^2\delta(\tau), \tau = 0 \\ \Pi, 0 < \tau \le \tau_c \\ \Pi e^{-\gamma(\tau-\tau_c)}, \tau > \tau_c \end{cases} \tag{7.48}$$

where ρ^2 is the normalized power of the line-of-site (LOS) component, Π is the normalized power density of the constant-level portion of DPS, τ_c is the duration of the DPS constant level, and γ is the decay exponent of the exponentially decreasing portion of the DPS. Given that the magnitude $R = |H(f,t)|$ of time-varying transfer function $H(f,t)$ for WSS US model follows the Rician distribution,

$$\begin{aligned} f_R(r) &= \frac{r}{\sigma^2}\exp\left(-\frac{r^2+s^2}{2\sigma^2}\right)I_0\left(\frac{rs}{\sigma^2}\right), \\ I_0(x) &= \frac{1}{2\pi}\int_0^{2\pi} \exp\left(x\cos\psi\right)d\psi, \end{aligned} \tag{7.49}$$

where $I_0(x)$ is the modified Bessel function of the first kind and zero order, $s = |\langle H(f,t)\rangle|$, and σ^2 is the variance of the real and imaginary components of the complex Gaussian process. As already discussed in Chap. 2, the total power in multipath components different from the LOS component is $2\sigma^2 = \sum_{n,n\neq 0}\langle h_n^2\rangle$, while the total power is given by

$$\overline{P}_{Rx} = \int_0^\infty r^2 f_R(r)dr = s^2 + 2\sigma^2. \tag{7.50}$$

The Rician distribution can also be expressed in terms of *Rician parameter K*, defined by $K = s^2/(2\sigma^2)$, as follows:

$$f_R(r) = \frac{2r(K+1)}{\overline{P}_{Rx}} \exp\left(-K - \frac{(K+1)r^2}{\overline{P}_{Rx}}\right) I_0\left(2r\sqrt{\frac{K(K+1)}{\overline{P}_{Rx}}}\right), r \geq 0. \tag{7.51}$$

The Rician parameter can be used as a measure of the fading severity: the small K implies severe fading, a large K implies mild fading, and when K tends to infinity, there is no fading at all. The parameters of the model given by Eq. (7.48) can easily be related to K, $\overline{P}_{Rx}$, and root mean square (RMS) delay spread, defined as $\tau_{RMS} = \sqrt{\langle\tau^2\rangle - \langle\tau\rangle^2}$, as described in [4]. Based on this model, we can build the frequency-domain channel simulator, based on Gaussian random generator, with details found in [4], which agrees very well with the measurement results.

7.5 CO-OFDM Principles, DFT Windowing, Frequency Synchronization, Phase Estimation, and Channel Estimation

7.5.1 Principles of Coherent Optical OFDM (CO-OFDM)

The operation principles for wireless communications have already been provided in Sect. 7.4.2. The operation principle of CO-OFDM system is illustrated in Fig. 7.10. Several basic building blocks can be identified including RF OFDM transmitter, optical I/Q modulator (to perform electro-optical conversion), dispersion map of interest, balanced coherent detector, and RF OFDM receiver. Upon the balanced coherent detection, the received signal can be expressed as

$$r(t) \simeq A\exp\left\{j\left[\underbrace{(\omega_{\mathrm{Tx}} - \omega_{\mathrm{LO}})}_{\Delta\omega}t + \underbrace{\phi_{PN,\mathrm{Tx}} - \phi_{PN,\mathrm{LO}}}_{\Delta\phi}\right]\right\} \times s_{\mathrm{BB}}(t) * h(t) + w(t), \tag{7.52}$$

where A is the balanced coherent detection constant, $h(t)$ is optical channel impulse response, $w(t)$ is additive noise process (dominated by ASE noise), while $s_{\mathrm{BB}}(t)$ denotes the baseband (BB) portion of the signal:

$$s_{\mathrm{BB}}(t) = \sum_{n=-\infty}^{\infty} \sum_{k=-N_{\mathrm{FFT}}/2}^{N_{\mathrm{FFT}}/2-1} X_{n,k}\mathrm{rect}\left(\frac{t-nT_s}{T_s}\right)e^{j2\pi\frac{k}{T_{FFT}}\cdot(t-nT_s)}. \tag{7.53}$$

The $X_{n,k}$ denotes a two-dimensional signal constellation (such as QAM) point transmitted on the k-th subcarrier in the n-th OFDM symbol, T_s is symbol duration, and T_{FFT} is the effective FFT part of OFDM symbol. The ω_{Tx} and ω_{LO} denote the carrier frequencies of transmit laser and local laser oscillator, respectively. Finally, $\phi_{\mathrm{PN,Tx}}$ and $\phi_{\mathrm{PN,LO}}$ denote the transmit laser and local oscillator laser phase noise, respectively. It is evident from Eqs. (7.52) and (7.53) that estimation and compensation of frequency offset $\Delta\omega = \omega_{\mathrm{Tx}} - \omega_{LO}$ and the phase error $\Delta\phi = \phi_{PN,\mathrm{Tx}} - \phi_{PN,\mathrm{LO}}$ are of crucial importance.

After this review of the optical OFDM signal processing, at conceptual level, we describe the following RF OFDM signal processing steps: DFT window synchronization, frequency synchronization, channel estimation, and phase noise estimation. In DFT window synchronization, the OFDM symbols are properly aligned to avoid ISI. In frequency synchronization, we estimate the frequency offset between transmit laser and local oscillator laser frequency in order to compensate for it. In subcarrier recovery, we estimate the channel coefficients for each subcarrier and compensate for channel distortion. Finally, in phase noise estimation, we are trying to estimate the phase noise originating from laser phase noise of transmit and local lasers and trying to compensate for it. (The reader is advised to look at references [5, 30–37] if needed.)

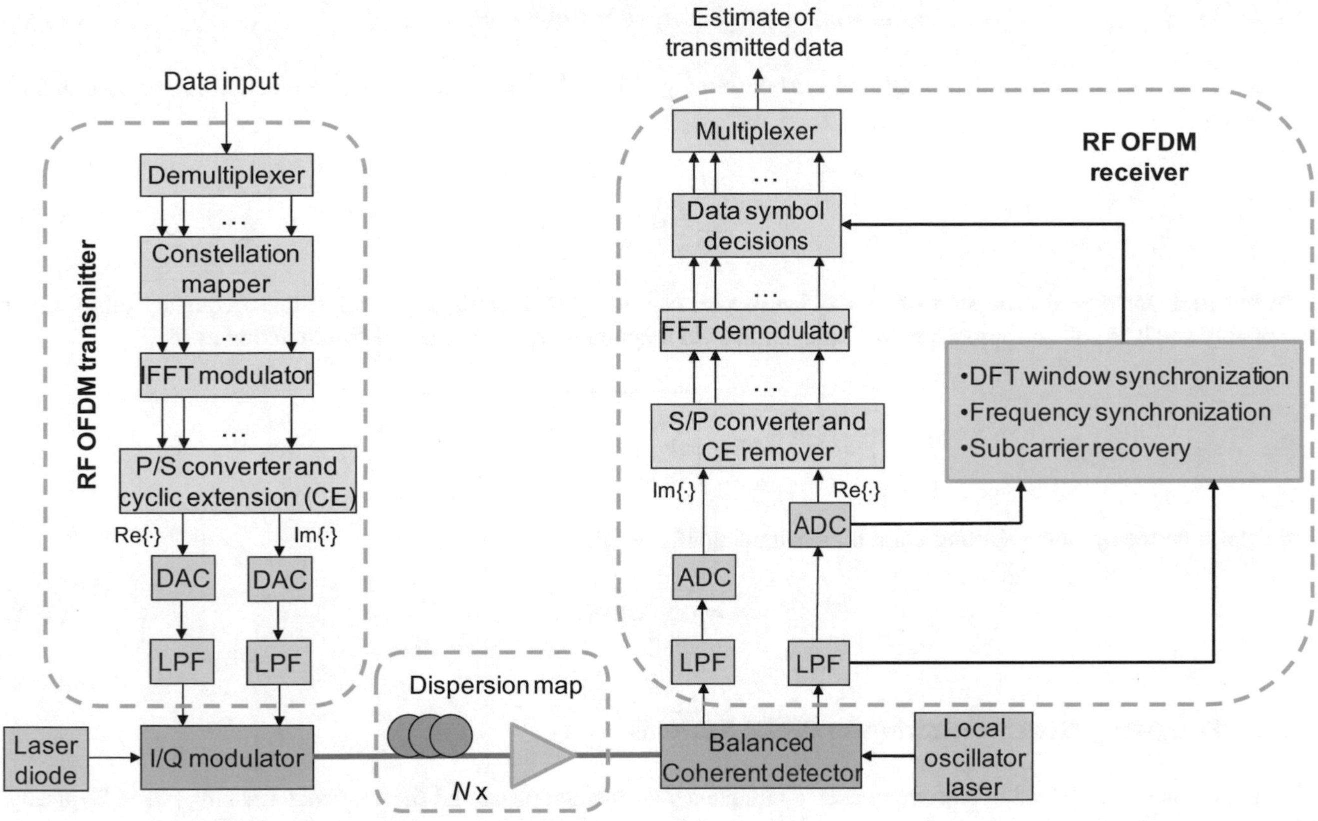

Fig. 7.10 The operation principle of CO-OFDM system

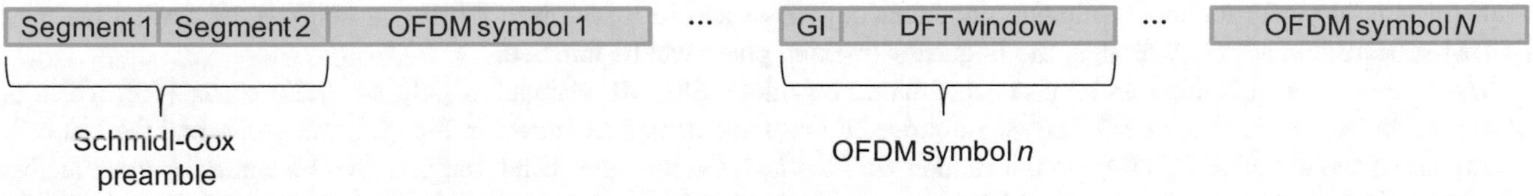

Fig. 7.11 The organization of OFDM frame showing Schmidl-Cox preamble for DFT windowing

In the rest of this section, we describe the receiver operations common to both wireless OFDM communication and coherent optical OFDM communication including DFT window synchronization, frequency synchronization, phase estimation, and channel estimation.

7.5.2 DFT Window Synchronization

DFT windowing [32], in which a preamble of pilot symbols consisting of two identical segments is transmitted, has become very attractive from the application perspective. This method is illustrated in Fig. 7.11. The autocorrelation method is used to identify the start of DFT window. It is also possible to use cyclic extended guard interval and autocorrelation method to determine the starting point of DFT window, as discussed in [11]. However, the guard interval is typically shorter than the preamble used in DFT synchronization, which might result in inferior performance. The DFT synchronization preamble can be described as [5, 32]

$$p_m = p_m + N_{sc}/2; m = 1, 2, \cdots, N_{sc}/2 \tag{7.54}$$

where p_m denotes the m-th pilot symbol and N_{sc} denotes the number of subcarriers. By sampling the received signal, based on parallel channel decomposition model described earlier, we obtain

$$r_m = r(mT_{\mathrm{FFT}}/N_{sc}) = p_m e^{j(\Delta\omega m T_{\mathrm{FFT}}/N_{sc} + \Delta\phi)} + w_m, \tag{7.55}$$

where T_{FFT} is the effective portion of OFDM symbol (see Fig. 7.11). The following correlation function can be used in DFT windowing:

$$R_d = \sum_{m=1}^{N_{sc}/2} r^*_{m+d} r_{m+d+N_{sc}/2}, \tag{7.56}$$

with the peak for $d = 0$. The start of FFT effective part can easily be identified, providing that frequency offset $\Delta\omega$ is reasonably small. Another approach is to define normalized magnitude squared of correlation function as [5]

$$M_d = |R_d/S_d|^2, S_d = \sqrt{\sum_{m=1}^{N_{sc}/2} |r_{m+d}|^2 \sum_{m=1}^{N_{sc}/2} |r_{m+d+N_{sc}/2}|^2}, \tag{7.57}$$

and determine the optimum starting point by maximizing M_d, which is

$$\widehat{d} = \arg\max_{d} M_d. \tag{7.58}$$

7.5.3 Frequency Synchronization in OFDM Systems

The frequency synchronization has two phases: (i) frequency acquisition phase and (ii) frequency tracking phase. Typically, telecommunication lasers are locked to references in accordance with ITU-T-defined frequency grid. Regarding the optical communications, a wavelength locker that is commonly used has accuracy of ~2.5 GHz. Any frequency offset between −5 GHz and 5GHz will cause serious problems for OFDM system operation, as discussed in [5]. The purpose of frequency acquisition is to perform course estimation of frequency offset and to reduce it to fall into a range equal several times of OFDM subcarrier spacing. After that, the frequency tracking phase can be initiated.

The frequency acquisition can be performed based on pilots [38], ML estimation [39], or cyclic prefix [40], which is illustrated in Fig. 7.12. If the cyclic extension procedure is implemented as shown in Fig. 7.2, the prefix and the last *NG*/2 samples of the effective OFDM symbol portion are identical. On the other hand, the first *NG*/2 samples of the effective OFDM symbol portion and the suffix, which are separated by *T*FFT seconds, are also identical. We are able to estimate the frequency offset and compensate for it by cross-correlating the corresponding segments. At the same time, correlation peaks obtained after every OFDM symbol can be used for timing. However, this approach, given the duration of guard interval, is inferior to Schmidl-Cox timing approach proposed in [32].

The frequency offset that remained upon frequency acquisition can be estimated by using Schmidl-Cox approach. By substituting Eq. (7.55) into Eq. (7.56), we can rewrite R_d in terms of the frequency offset by

$$\begin{aligned} R_d &= \sum_{m=1}^{N_{sc}/2} |r_{m+d}|^2 e^{j\left(\overbrace{\Delta\omega}^{2\pi f_{\mathrm{offset}}} \underbrace{T_{\mathrm{FFT}}}_{1/\Delta f_{sc}}/2 + \Delta\phi\right)} + n_d \\ &= \sum_{m=1}^{N_{sc}/2} |r_{m+d}|^2 e^{j(\pi f_{\mathrm{offset}}/\Delta f_{sc} + \Delta\phi)} + n_d, \end{aligned} \tag{7.59}$$

Fig. 7.12 The frequency acquisition by using the cyclic prefix

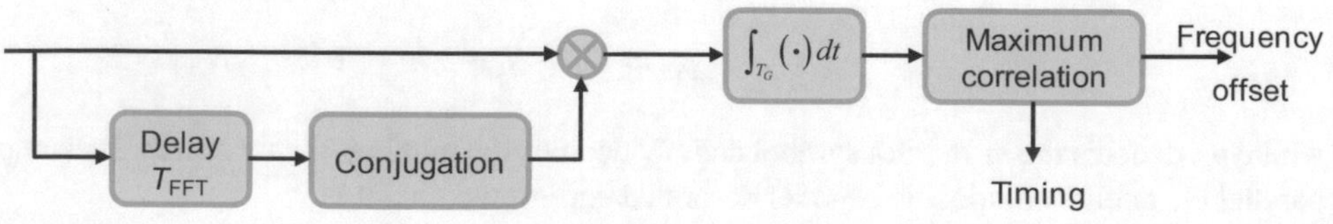

where n_d is the equivalent noise term. By ignoring the phase noise error $\Delta\phi$, we can estimate the frequency offset f_{offset} as follows [5, 32]:

$$\widehat{f}_{\text{offset}} = \frac{\Delta f_{sc}}{\pi} \measuredangle R_{\widehat{d}}, \tag{7.60}$$

where $\measuredangle R_{\widehat{d}}$ is the angle of correlation function (in radians) while $\widehat{d}$ is obtained from Eq. (7.58). Once the frequency offset is estimated, it can be compensated from received signal by

$$r_{\text{comp}}(t) = e^{-j2\pi \widehat{f}_{\text{offset}}} r(t). \tag{7.61}$$

After the time synchronization and frequency synchronization were completed, the *subcarrier recovery* takes place. The subcarrier recovery involves phase estimation and channel estimation.

7.5.4 Phase Estimation in OFDM Systems

In the equivalent OFDM channel model, described earlier, the received symbol vector of the k-th subcarrier in the n-th OFDM symbol, denoted as $r_{n,\,k}$, can be represented as

$$r_{n,k} = H_k s_{n,k} e^{\,j\phi_i} + w_{n,k}, \tag{7.62}$$

where $s_{n,\,k}$ denotes the transmitted symbol of the k-th subcarrier in the i-th OFDM symbol, H_k is the channel coefficient of k-th subcarrier, and $\phi_n = (1/N_{sc})\sum_{k=1}^{N_{sc}} \phi_{nk}$ is the common phase error (CPE). Parameter $w_{n,k}$ denotes the additive noise term (again dominated by AWGN). The likelihood function in this case is given as

$$f(r_{i1}, r_{i2}, \cdots, r_{i,N_{sc}} | H_1, H_2, \cdots, H_{N_{sc}}; \phi_i) = C \exp\left(-\sum_{k=1}^{N_{sc}} \frac{\left| r_{i,k} - H_k s_{i,k} e^{\,j\phi_i} \right|^2}{2\sigma_k^2} \right), \tag{7.63}$$

where C is the normalization constant while σ_k^2 is the variance of noise. The corresponding log-likelihood function is given by

$$l(r_{i1}, \cdots, r_{iN} | H_1, \cdots, H_{N_{sc}}; \phi_i) = -\sum_{k=1}^{N_{sc}} \frac{\left| r_{i,k} - H_k s_{i,k} e^{\,j\phi_i} \right|^2}{2\sigma_k^2}. \tag{7.64}$$

By minimizing the log-likelihood function, we can obtain the following solution for the common phase error [5]:

$$\hat{\phi}_n = \arg\left\{ \sum_{k=1}^{N_{sc}} r_{n,k} H_k^* s_{n,k} / \sigma_k^2 \right\}. \tag{7.65}$$

7.5.5 Channel Estimation in OFDM Systems

There exist a number of algorithms for channel estimation proposed so far. They could be time-domain based, frequency-domain based, or combination of the two. The channel estimation techniques can be classified as either data-aided or pilot-aided. The data-aided channel estimation techniques are prone to the error propagation effect and as such will not be discussed here with much details.

7.5.5.1 Pilot-Aided Channel Estimation

In pilot-aided channel estimation, a portion of subcarriers are allocated for channel estimation. There are two types of pilot arrangements in one-dimensional pilot-aided channel estimation [35–37], block type and comb type, as illustrated in Fig. 7.13. The block-type arrangement is suitable when channel conditions change with slow pace, which is the case with chromatic dispersion effect in fiber-optics communications. The comb-type arrangement is suitable when channel conditions change fast, like in the case when PMD is considered, in fiber-optics communication. In this case, the pilots are inserted into a certain number of subcarriers of each OFDM symbol. It is possible to use the two-dimensional (2D) channel estimation as well, which is illustrated in Fig. 7.14. For instance, the preamble can be used for chromatic dispersion compensation, while the comb-pilot arrangement is used for PMD compensation. In 2D channel estimation, the pilot symbol spacings in time-domain (TD) and frequency-domain (FD) must satisfy the following constraints: (1) the TD-symbol spacing $\Delta s_{TD} < 1/B_D$, where B_D is the Doppler spread, and (2) the FD-symbol spacing $\Delta s_{FD} < 1/\tau_{max}$, where τ_{max} is the maximum delay spread.

Various channel estimators typically belong to the class of either ML estimators or MMSE estimators. Let us again consider the model expressed by Eq. (7.62), but now we will set the common phase error to zero and omit the subscript corresponding to the index of OFDM symbol (the assumption is that optical channel does not change too fast). The likelihood function, assuming Gaussian noise-dominated scenario, is given by

$$f(r_1, r_2, \cdots, r_{N_{sc}} | H_1, H_2, \cdots, H_{N_{sc}}) = C \exp\left(-\sum_{k=1}^{N_{sc}} \frac{|r_k - H_k s_k|^2}{2\sigma^2} \right), \tag{7.66}$$

where C is the normalization constant and σ^2 is the variance of noise. The corresponding log-likelihood function is given as

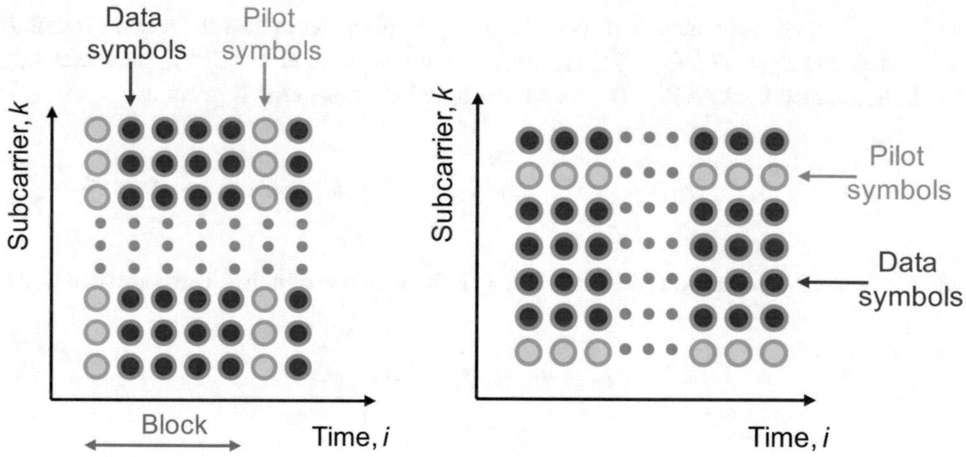

Fig. 7.13 Pilot arrangements for OFDM channel estimation: (left) block-type arrangement and (right) comb-type arrangement

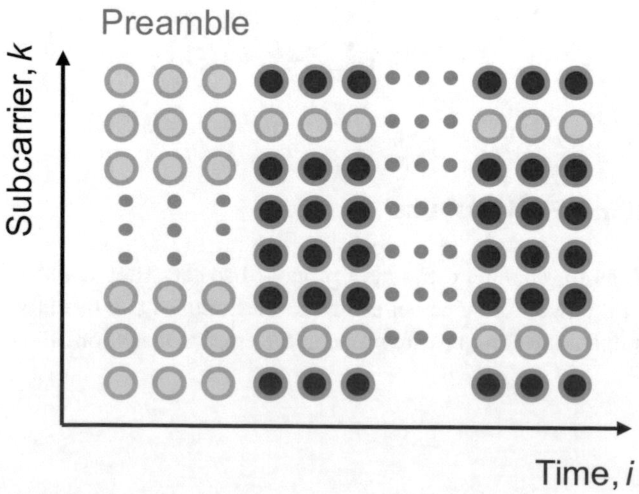

Fig. 7.14 Pilot arrangements for two-dimensional OFDM channel estimation

$$\begin{aligned} l(r_1, r_2, \cdots, r_{N_{sc}} | H_1, H_2, \cdots, H_{N_{sc}}) &= -\sum_{k=1}^{N_{sc}} \frac{|r_k - H_k s_k|^2}{2\sigma^2} \\ &= -\sum_{k=1}^{N_{sc}} \frac{(r_k - H_k s_k)(r_k - H_k s_k)^*}{2\sigma^2} \end{aligned} \tag{7.67}$$

To determine the channel coefficients, someone needs to perform maximization of log-likelihood function with respect to all H_k's, which needs an extensive computation time. Instead, by setting the first derivative of likelihood function with respect to H_k to be zero, and by solving it for H_k, we will obtain the following solution:

$$H_{k,LS} = \frac{r_k}{s_k} = H_k + \frac{w_k}{s_k}. \tag{7.68}$$

The channel estimate obtained by Eq. (7.68) is commonly referred to as the least square (LS) solution. The estimated information symbols by LS estimator are given as

$$s_k = \frac{r_k}{H_{k,LS}} = \frac{H_k s_k + w_k}{H_k + \frac{w'_k}{s'_k}} \approx s_k + \frac{w_k}{H_k} - \frac{w'_k}{H_k} \frac{s_k}{s'_k}, \tag{7.69}$$

where we used s'_k and w'_k to denote the transmitted symbol and noise sample generated when the LS estimate of H_k is obtained. Since there are two additive terms, it is evident that noise gets enhanced, indicating that this scheme represents an instance of zero-forcing estimator.

In *MMSE channel estimation*, we intend to linearly transform the LS solution by pre-multiplying $\boldsymbol{H}_{\mathrm{LS}}$ with a properly chosen matrix $\boldsymbol{A}$ as

$$\boldsymbol{H}_{\mathrm{MMSE}} = \boldsymbol{A}\boldsymbol{H}_{LS}, \tag{7.70}$$

which will minimize the MSE. The estimator error can be defined as

$$\boldsymbol{\varepsilon} = \boldsymbol{A}\boldsymbol{H}_{LS} - \boldsymbol{H}. \tag{7.71}$$

By employing the orthogonality principle from Sect. 6.3.4 in Chap. 6, and by claiming that the error matrix is independent on "data," we obtain

$$\langle(\boldsymbol{A}\boldsymbol{H}_{\mathrm{LS}} - \boldsymbol{H})\boldsymbol{H}^\dagger\rangle = 0 \Leftrightarrow \boldsymbol{A}\underbrace{\langle\boldsymbol{H}_{\mathrm{LS}}\boldsymbol{H}^\dagger\rangle}_{\boldsymbol{R}_{\boldsymbol{H},\boldsymbol{H}_{\mathrm{LS}}}} - \underbrace{\langle\boldsymbol{H}\boldsymbol{H}^\dagger\rangle}_{\boldsymbol{R}_{\boldsymbol{H}}} = 0, \tag{7.72}$$

where $\boldsymbol{R}_{\boldsymbol{H}}$ denotes the correlation matrix of $\boldsymbol{H}$ while $\boldsymbol{R}_{\boldsymbol{H},\boldsymbol{H}_{\mathrm{LS}}}$ denotes the cross-correlation matrix between $\boldsymbol{H}$ and $\boldsymbol{H}_{\mathrm{LS}}$. By solving Eq. (7.72) for $\boldsymbol{A}$, we obtain

$$\boldsymbol{A} = \boldsymbol{R}_{\boldsymbol{H}}\boldsymbol{R}_{\boldsymbol{H},\boldsymbol{H}_{LS}}^{-1}, \tag{7.73}$$

and after substituting it in Eq. (7.70), we derive the MMSE channel estimation as

$$\boldsymbol{H}_{\mathrm{MMSE}} = \boldsymbol{R}_{\boldsymbol{H}}\boldsymbol{R}_{\boldsymbol{H},\boldsymbol{H}_{\mathrm{LS}}}^{-1}\boldsymbol{H}_{\mathrm{LS}}, \tag{7.74}$$

which resembles a quite general Wiener filter solution. Based on Eq. (7.68), the cross-correlation matrix $\boldsymbol{R}_{\boldsymbol{H},\boldsymbol{H}_{\mathrm{LS}}} \cong \boldsymbol{R}_{\boldsymbol{H},\boldsymbol{H}_{\mathrm{LS}}} + \sigma_n^2\langle\boldsymbol{S}\boldsymbol{S}^\dagger\rangle^{-1}$, where σ_n^2 is the variance of the noise and $\boldsymbol{S}$ is the diagonal pilot matrix. Now, upon substitution into Eq. (7.74), we obtain a simpler equation for MMSE estimation of $\boldsymbol{H}$:

$$\boldsymbol{H}_{\text{MMSE}} = \boldsymbol{R_H}\left\{\boldsymbol{R}_{\boldsymbol{H},\boldsymbol{H}_{\text{LS}}} + \sigma_n^2\langle \boldsymbol{SS}^\dagger\rangle^{-1}\right\}^{-1}\boldsymbol{H}_{\text{LS}}. \tag{7.75}$$

The discussion above, related to MMSE estimation, is applicable to *block-type* pilot arrangements. To reduce the complexity and lower the overhead, the combo-type and 2D pilot arrangements should be used instead, as already shown in Fig. 7.14. The expressions for MMSE estimates are still applicable, but we need just to replace every $\boldsymbol{H}$ with $\boldsymbol{H}_\text{p}$, with subscript p referring to the pilots. For instance, Eq. (7.74) for combo-type/2D arrangements can be rewritten as

$$\widehat{\boldsymbol{H}}_{p,\text{MMSE}} = \boldsymbol{R}_{\boldsymbol{H}_p}\boldsymbol{R}^{-1}_{\boldsymbol{H}_p,\boldsymbol{H}_{p,\text{LS}}}\widehat{\boldsymbol{H}}_{p,\text{LS}}, \tag{7.76}$$

where $\boldsymbol{H}_{\text{p, LS}}$ is given by [36, 37]

$$\widehat{\boldsymbol{H}}_{p,\text{LS}} = \left[H_{p,\text{LS}}(1)\cdots H_{p,\text{LS}}(N_\text{p})\right]^\text{T},\ H_{p,LS}(i) = \frac{Y_p(i)}{X_p(i)}; i = 1,\cdots,N_p \tag{7.77}$$

In Eq. (7.77), we use N_p to denote the number of pilot subcarriers, while $X_p(i)$ and $Y_p(i)$ denote the i-th transmitted and received pilot symbols ($i = 1,\ldots,N_p$), respectively. Once the pilots have been determined in combo-type and 2D arrangements, we need to perform the interpolation to estimate the channel coefficients for data subcarriers. Various interpolation methods, such as linear interpolation, second-order interpolation, low-pass interpolation, spline cubic interpolation, and time-domain interpolation, have been proposed so far. For instance, in *linear interpolation* the channel estimation for k-th subcarrier is given by [35]

$$\begin{gathered}\widehat{H}(k) = \left[H_p(m+1) - H_p(m)\right]\frac{l}{N_{sc}/N_p} + H_p(m) \\ m\frac{N_{sc}}{N_p} < k < (m+1)\frac{N_{sc}}{N_p};\ 0 < l < \frac{N_{sc}}{N_p}\end{gathered} \tag{7.78}$$

where we assumed the uniform spacing of pilots, $X(mN_{sc}/N_p + l) = X_p(m)$, with $X(i)$, $i = 1,\ldots, N_{sc}$, being the symbol carried by i-th subcarrier. The *time-domain interpolation method* [35, 48] represents a high-resolution method based on zero-padding FFT/IFFT. The IFFT is applied to $\widehat{\boldsymbol{H}}_{p,\text{LS}}$ in order to perform conversion into time-domain [35]

$$h_p(n) = \sum_{k=0}^{N_p-1} H_{p,LS}e^{j2\pi kn/N_p}; n = 0, 1, \cdots, N_p - 1 \tag{7.79}$$

and interpolate to N_{sc} points as follows [35]:

$$\widehat{h}(n) = \begin{cases} h_p(n), 0 \le n < \frac{N_p}{2} - 1 \\ 0,\ \frac{N_p}{2} \le n < N_{sc} - \frac{N_p}{2} + 1 \\ h_p\left(n - N_{sc} + N_p + 1\right), N_{sc} - \frac{N_p}{2} + 1 \le n \le N_{sc} - 1 \end{cases} \tag{7.80}$$

Then the FFT is then applied to convert N_{sc}-sequence to frequency-domain:

$$\widehat{H}(k) = \sum_{n=0}^{N_{sc}-1} \widehat{h}(n)e^{-j2\pi kn/N_{sc}}; k = 0, 1, \cdots, N_{sc} - 1 \tag{7.81}$$

and such obtained channel coefficient is used to compensate for channel impairments.

The calculation of channel coefficients by using Eq. (7.74) is computationally extensive. The following approximation has been shown to be effective for both multi-Tb/s superchannel CO-OFDM optical transport [36, 37] and multi-Tb/s CO-OFDM transport over few-mode fibers [41]:

$$\hat{\boldsymbol{H}}_{p,\mathrm{MMSE}} = \boldsymbol{R}_{\boldsymbol{H}_p}\left(\boldsymbol{R}_{\boldsymbol{H}_p} + \frac{\alpha(M)}{SNR}\right)^{-1}\hat{\boldsymbol{H}}_{p,\mathrm{LS}}, \tag{7.82}$$

where parameter $\alpha(M)$ is defined as $\alpha(M) = \left\langle |X_p(m)|^2 \right\rangle \left\langle |X_p^{-1}(m)|^2 \right\rangle$, and it is dependent on both the signal constellation type and the size. The signal-to-noise ratio is defined as $SNR = \left\langle |X_p(m)|^2 \right\rangle / \sigma_n^2$ (σ_n^2 is the noise variance). This method is also effective in dealing with multipath fading effects.

7.5.5.2 Data-Aided Channel Estimation

So far the pilot-aided channel estimation techniques suitable for both wireless coherent and optical coherent detections have been studied. The use of pilots reduces the overall spectral efficiency. In other words, the power efficiency can be improved if the pilot subcarriers are not transmitted. In data-aided channel estimation, the first OFDM symbol must be known at the receiver side to start the detection process. The subcarrier symbols from previous OFDM symbol detection are used to predict the channel for the current OFDM symbol. Namely, when the channel is slowly time-varying, there will be a large correlation of current OFDM symbol with previous OFDM symbol, so that the channel estimation based on previous OFDM symbol will be reliable. Otherwise, the error propagation effect will occur. For additional details on data-aided channel estimation, the interested reader is refereed to [50, 51].

7.6 Differential Detection in OFDM Systems

The key idea behind various channel estimation techniques described so far, based on coherent detection, is to determine accurate magnitude and phase estimates of the subcarriers' channel coefficients. These estimates are then used to compensate for channel distortions. On the other hand, the differential detection does not perform any channel estimation, since the information is written in the phase difference between the current and previous sub-symbol. The overall complexity and latency have been reduced at the cost of about 3 dB penalty in signal-to-noise ratio. The configuration of OFDM receiver with differential detection, suitable for wireless applications, is shown in Fig. 7.15.

This scheme can easily be adopted in direct-detection OFDM, to be described later. The absolute phase reference is not needed, as the differential detector compares the current sub-symbol with previous sub-symbol. Given that the OFDM is a two-dimensional scheme, the differential encoding can be applied in either time-domain or frequency-domain.

In *time-domain differential encoding*, the information is imposed in phase difference between two sub-symbols of the current and previous OFDM symbols within the same subcarrier. Therefore, the differential encoding is applied on subcarrier level. For M-ary PSK (M-PSK), the differentially encoded phase related to the k-th subcarrier of the n-th OFDM symbol, denoted as $\theta_{n,k}$, can be written as

$$\theta_{n,k} = \sum_{m=0}^{n} \vartheta_{m,k} \bmod 2\pi, \tag{7.83}$$

where we use $\vartheta_{n,\,k}$ to denote the corresponding absolute phase. The corresponding transmitted PSK symbol is $X_{n,k} = e^{\,j\theta_{n,k}}$. At the receiver side, based on parallel channel decomposition, described by Eq. (7.38), the received sub-symbol n in k-th subcarrier can be represented as

$$Y_{n,k} = \underbrace{|H_{n,k}|e^{\,j\phi_{n,k}}}_{H_{n,k}} \cdot \underbrace{e^{\,j\theta_{n,k}}}_{X_{n,k}} + w_{n,k}, \tag{7.84}$$

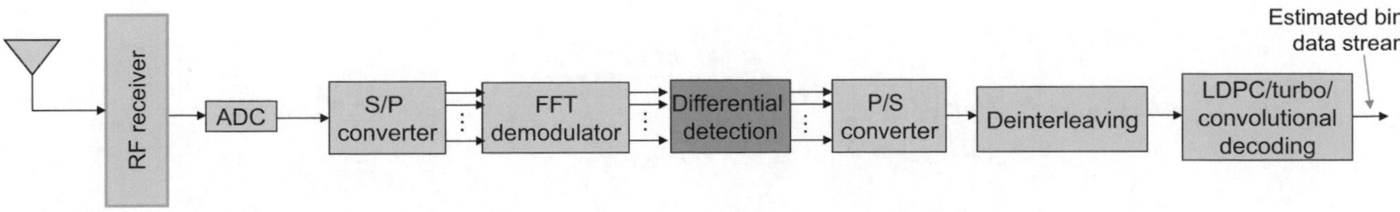

Fig. 7.15 The simplified configuration of OFDM receiver with differential detection, suitable for wireless applications

with $w_{n,k}$ representing an AWGN noise sample, as before. The n-th phase differential detection can be performed by

$$Y_{n,k}Y_{n-1,k}^{*} = |H_{n,k}||H_{n-1,k}^{*}|e^{j(\phi_{n,k}-\phi_{n-1,k})}e^{\overbrace{j(\theta_{n,k}-\theta_{n-1,k})}^{\vartheta_{n,k}}} + |H_{n,k}|e^{j\phi_{n,k}}w_{n-1,k}^{*} + |H_{n-1,k}^{*}|e^{-j\phi_{n-1,k}}w_{n,k} + w_{n,k}w_{n-1,k}^{*} \tag{7.85}$$

Even though that the phase $\vartheta_{n,\,k}$ is properly detected, there is the phase disturbance term $\phi_{n,k}$-$\phi_{n-1,k}$, which is introduced by the time-selective fading. When OFDM symbol duration is properly chosen to small relative to the phase coherence time, this phase disturbance can be neglected. On the other hand, the signal level gets affected by fading, while the additive noise gets doubled, resulting in at least 3 dB degradation in SNR compared to the coherent detection case. On the other hand, when pilot-aided channel estimation is imperfect, the differential detection penalty is actually lower than 3 dB. Based on *Clark-Jakes model* described in Chap. 2, the normalized correlation coefficient corresponding to the n-th OFMD symbol would be

$$R(n) = R(nT_s)/P_{Rx} = J_0(2\pi\nu_{D,\max}nT_s), \tag{7.86}$$

where T_s if OFDM symbol duration while $\nu_{D,\max}$ is the maximum Doppler frequency. This correlation coefficient can be used to estimate the maximum tolerable Doppler frequency. The difference between received sub-symbols can be related to the correlation coefficient as follows [3]:

$$Y_{n,k} - Y_{n-1,k} = [1 - R(1)]Y_{n-1,k} + \sqrt{1 - R^2(1)}z_{n,k}, \tag{7.87}$$

where $z_{n,k}$ is the unit variance noise sample. When $y_{n-1,k}$ and $z_{n,k}$ are uncorrelated, the signal-to-distortion ratio (SDR) can be estimated as [3]

$$SDR = \frac{1}{2(1 - R(1))}, \tag{7.88}$$

and must much larger than SNR not to affect the BER performance. For a SNR penalty of 1 dB, the SDR should be at least 10 dB, so that the corresponding correlation coefficient is $R(1) \approx 0.95$, corresponding to Doppler frequency×OFDM symbol duration product of $\nu_{D,\max}T_s \approx 0.07$ or, equivalently, the maximum tolerable Doppler frequency of $\nu_{D,\max} \approx 0.07/T_s$. Since the maximum Doppler frequency is related to the user speed u by $\nu_{D,\max} = f_c u/c$ (f_c is the carrier frequency), the maximum tolerable user velocity would be

$$u_{\max} = c\frac{\nu_{D,\max}}{f_c} \approx c\frac{0.07}{f_c T_s}. \tag{7.89}$$

As an illustration, for carrier frequency of 3 GHz and OFDM symbol duration 125 μs, the maximum tolerable car speed would be 56 m/s.

In *frequency-domain differential encoding*, the information is imposed in phase difference between two neighboring subcarrier symbols of the same OFDM symbol. The differentially encoded phase $\theta_{n,k}$ is related to the absolute phase $\vartheta_{n,\,k}$ by

$$\theta_{n,k} = \sum_{l=0}^{k} \vartheta_{n,l} \bmod 2\pi. \tag{7.90}$$

The k-th subcarrier differential detection can be performed by

$$Y_{n,k}Y_{n,k-1}^{*} = |H_{n,k}||H_{n,k-1}^{*}|e^{j(\phi_{n,k}-\phi_{n,k-1})}e^{\overbrace{j(\theta_{n,k}-\theta_{n,k-1})}^{\vartheta_{n,k}}} + |H_{n,k}|e^{j\phi_{n,k}}w_{n,k-1}^{*} + |H_{n,k-1}^{*}|e^{-j\phi_{n,k-1}}w_{n,k} + w_{n,k}w_{n,k-1}^{*} \tag{7.91}$$

Not surprisingly, this equation has similar form as Eq. (7.85). The phase distortion is now related to the phase difference between the n-th and (n-1)-th subcarrier. In other words, this phase disturbance is now introduced by the frequency-selective fading. For the exponentially decreasing multipath power delay profile, the frequency-domain (FD) normalized correlation function is given by [3]

$$R_{FD}(k) = R(k/T)/P_{Rx} = \frac{1}{1 + j2\pi\tau_{RMS}k/T}, \tag{7.92}$$

τRMS is the RMS multipath delay spread, and the subcarrier channel spacing is given by $1/T$. The corresponding frequency-domain SDR for $k = 1$ would be

$$SDR_{FD} = \frac{1}{2(1 - R_{FD}(1))}. \tag{7.93}$$

For the FD SDR of 10 dB, the corresponding correlation will be 0.95, which results in normalized subcarrier spacing $(1/T)/(1/\tau_{RMS})$ of 0.03. In other words, the maximum tolerable RMS delay spread is 3% of the FFT period T. Clearly, in frequency-domain differential detection, the robustness to multipath fading is FFT period dependent and, therefore, significantly worse than coherent detection where the tolerance to multipath fading is guard interval dependent.

7.7 OFDM Applications in Wireless Communications

The OFDM has found various applications in wireless communications including digital audio broadcasting (DAB) [51], digital video broadcasting (DVB) [52, 53], wireless LAN (WLAN) [3], WiMAX, Wi-Fi, WiGig, and ultra-wideband (UWB) communications [54], to mention few.

7.7.1 OFDM in Digital Audio Broadcasting (DAB)

The digital audio broadcasting has several advantages compared to FM including the improved sound quality (comparable to that of a compact disk), higher spectral efficiency, it enables single-frequency network (SFN) operation, it offers new data services, and it provides an unequal error protection. The DAB standard is described in ETSI standard ETS300401, where the data structure, the FEC, and coded-OFDM modulation are described. There are four selectable transmission modes in DAB, as summarized in Table 7.1 [3, 49]; the first three are related to the different frequency bands, while the fourth one is introduced to provide a better coverage range for mode II. The SFN capability can significantly improve the spectrum efficiency. For a receiver at a given location, it is possible to receive the signal from two different DAB transmitters located at distances d_1 and d_2. When the difference in arrival times $\Delta\tau = |d_1\text{-}d_2|/c$ is smaller than the guard interval duration T_G, the signals coming from different DAB transmitters can be interpreted as the transmit diversity scheme to improve SNR and reduce the BER. The audio signals in DAB are encoded into either MPEG-1 (frame duration of 24 ms) or MPEG-2 (48 ms) frames. During the

Table 7.1 Different DAB modes

Mode	Mode I	Mode II	Mode III	Mode IV
Number of subcarriers	1536	384	192	768
Subcarrier spacing	1 kHz	4 kHz	8 kHz	2 kHz
Symbol duration (no guard interval)	1 ms	250 μs	125 μs	500 μs
Guard interval duration	246 μs	62 μs	31 μs	123 μs
Frame length	96 ms (76 symbols)	24 ms (76 symbols)	24 ms (153 symbols)	48 ms (76 symbols)
Frequency range, carrier frequency	Band III VHF, < 375 MHz	L band, < 1.5 GHz	L band, <3 GHz	L band, <1.5 GHz
Application	Single-frequency networks (SFN)	Multi-frequency network (MFN)	Satellite	SFN
Maximum propagation path difference	100 km	25 km	12.5 km	50 km

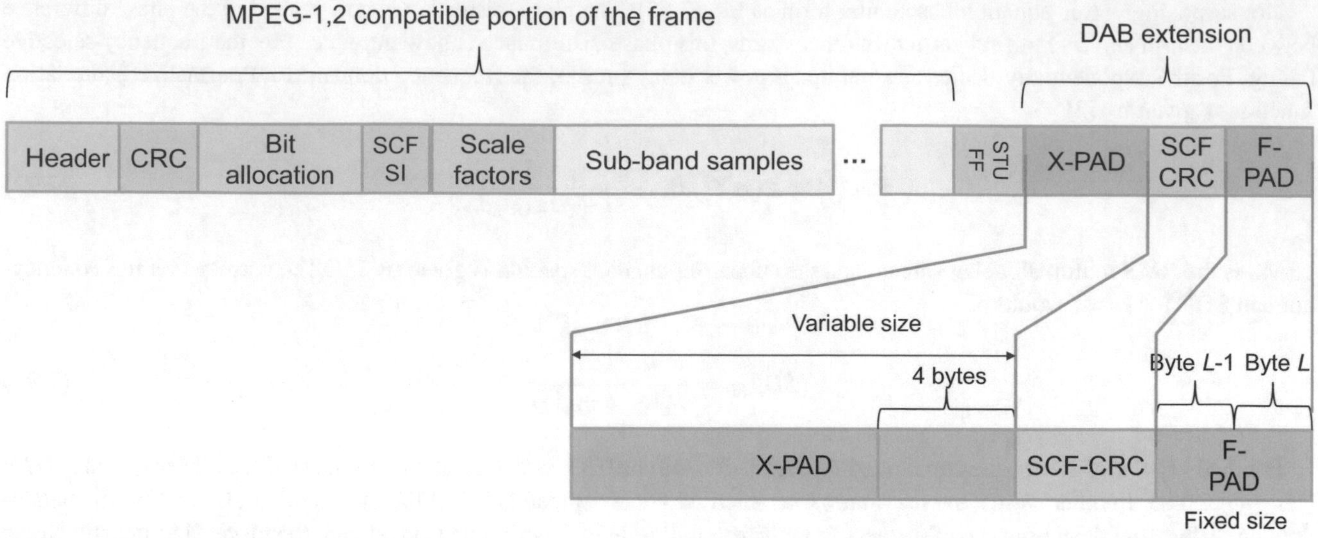

Fig. 7.16 DAB audio frame including DAB extension

compression, from 1.5 Mb/s down to 64–384 kb/s, the perceptual coding is used in which the audio components that cannot be detected by human earing system are omitted. MPEG-compatible DAB audio frame is illustrated in Fig. 7.16.

The header is 32 bits long, while the CRC checksum is 16 bits long. The bit allocation field, scale factors, and sub-band samples follow the CRC field. The audio sampling rate in MPEG-1 is maximum 48 kHz, which is higher than 44.1 kHz used in the standard CD. For mono sound the data rates range from 32 to 192 kb/s, while for stereo, dual sound, and joint stereo, the data rates range between 64 kb/s and 384 kb/s. The MPEG-2 frame represents the MPEG-1 frame supplemented with MPEG-2 extension. The sampling rate in MPEG-2 is 24 kHz; thus, the frame duration is twice longer (48 ms). The sub-band coding of audio signals is used. After the staffing bytes field, the DAB extension is used in DAB standard. In DAB extension, the program-associated data (PAD) are transmitted, with possible versions extended-PAD (X-PAD), which is variable in size with minimum size of 4 bytes, and fixed-PAD (FPAD), which is of fixed size (2 bytes). The PAD contains the information related to the music/voice identifier, program-related text, and additional scale factor-CRC bits (SCF-CRC). The DAB audio rates used in the UK are 256 kb/s for classical music, 128 kb/s for popular music, and 64 kb/s for the voice signal.

The coded-OFDM frame, in DAB, consists of 77 coded-OFDM symbols, the null symbol is used to denote the start of DAB frame, the first symbol is used as the phase reference for phase/frequency synchronization on the receiver side, the data OFDM symbols start with the symbol number 2. The coded-OFDM symbol duration ranges from 125 μs, for mode III, to 1 ms, for mode I. To deal with multipath fading effects, the guard interval is appended to the OFDM symbol, with duration of guard interval being approximately 1/4 of duration of OFDM symbol. The first several symbols of total rate 96 kb/s compose the so-called fast information channel (FIC), with a number of OFDM symbols dictated by the DAB mode (sampling rate). The remaining data OFDM symbols of rate 2.304 Mb/s compose the so-called main service channel (MSC). The number of subcarrier per OFDM symbol is dictated by DAB mode and ranges from 192 for mode III to 1536 for mode I. Each subcarrier carries $\pi/4$-shift-DQPSK symbols, with corresponding constellation diagram shown in Fig. 7.17a. The carrier phase gets shifted for $\pi/4$ from symbol to symbol. On such a way, we move from QPSK to rotated QPKS for $\pi/4$ rad from symbol to symbol. From Fig. 7.17a, it is evident that with $\pm\pi/4$ phase shift, we move to 00 and 10 correction symbols (states), while with $\pm 3\pi/4$ shifts, we move to 01 and 11 states. Clearly there is no transition of π rad. For comparison purposes, the conventional DQPSK signal constellation is shown as well. The transition from 00 state to 11 state results in the phase shift of π rad, which changes amplitude of the carrier causing additional symbol errors in detection stage, due to filtering effects. After the computation of projections of received signal along the basis functions $\phi_1(t)$ and $\phi_2(t)$, to get the in-phase and quadrature estimates, the corresponding detector is composed of the following blocks: (i) arctangent computation block to determine the absolute phase, (ii) the phase-difference block, and (iii) modulo-2π correction block to correct the errors due to phase angles wrapping around the $\phi_1(t)$-basis function.

The bandwidth of DAB signal is 1.536 MHz, and together with the guard band, it occupies in total 1.75 MHz, which is 1/4 of the 7 MHz channel available in band III (174–240 MHz). Therefore, four DAB channels get frequency multiplexed into one VHF channel.

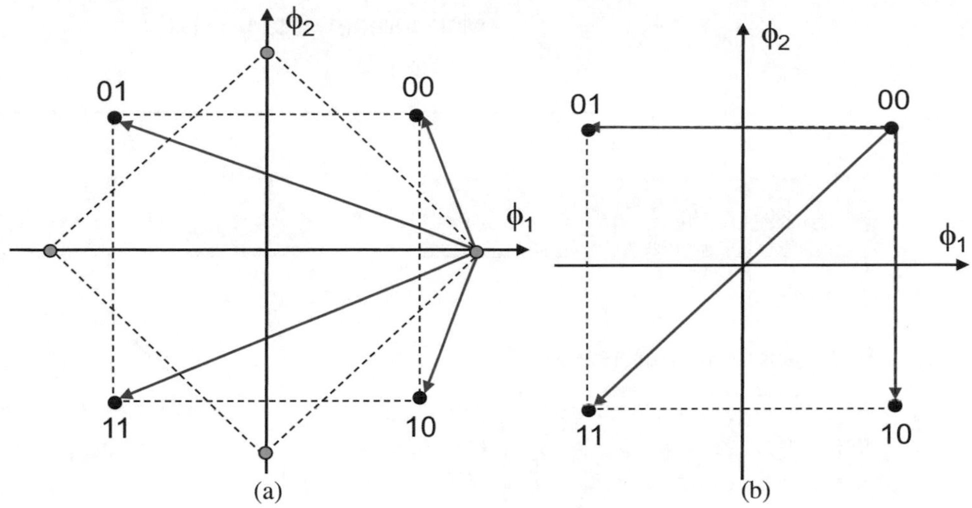

Fig. 7.17 The constellation diagrams for: (**a**) $\pi/4$-shift-DQPSK and (**b**) DQPSK

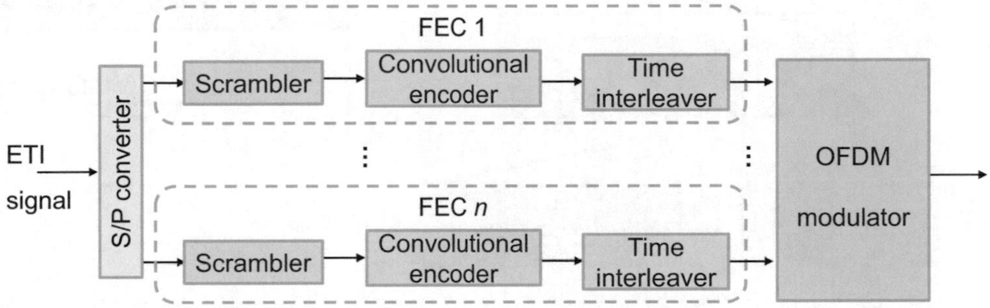

Fig. 7.18 The configuration of DAB modulator

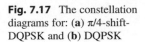

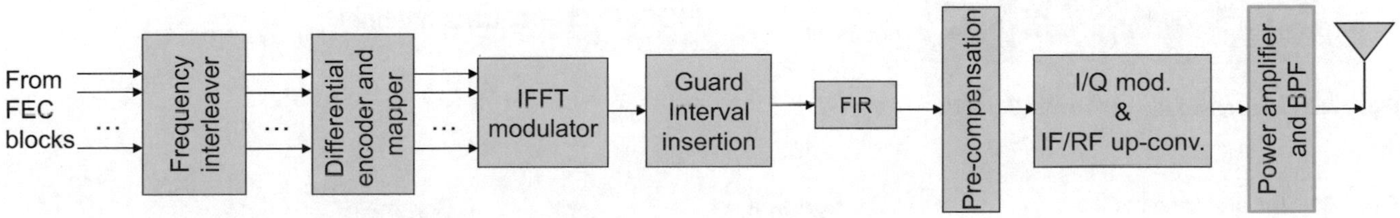

Fig. 7.19 The configuration of OFDM modulator for DAB

The DAB signal is composed of up to 64 subchannels carrying audio, data, or a combination of the two signals, multiplexed together into the common interleaved frame, with the help of the *ensemble multiplexer*. The data rate per subchannel is a multiple of 8 KHz. Such obtained ensemble transfer interface (ETI) signal is used as the input to the DAB modulator, shown in Fig. 7.18, composed of *n* FEC schemes, one per subchannel, providing unequal error protection (UEP), and an OFDM modulator. Clearly, the DAB modulator is in fact the coded-OFDM modulator. The configuration of OFDM modulator is shown in Fig. 7.19. The frequency interleaver is used to deal with frequency-selective fading effects. The differential encoding and mapping is already described earlier; the pre-compensation block is used to partially pre-compensate for the nonlinearities introduced by the amplifier. After I/Q modulation and intermediate frequency (IF)/RF up-conversion, the OFDM signal is amplified, followed by the band-pass filtering (BPF), and broadcasted by a transmit antenna. The OFDM demodulator is similar to that shown in Fig. 7.15. The organization of DAB packet is shown in Fig. 7.20, which is self-explanatory.

To avoid synchronization problems on the receiver, originating from sequences of all zeroes/ones, the *scrambler* shown in Fig. 7.21 is used. The scrambler has been reset every 24 ms by the all-ones pattern into the shift register.

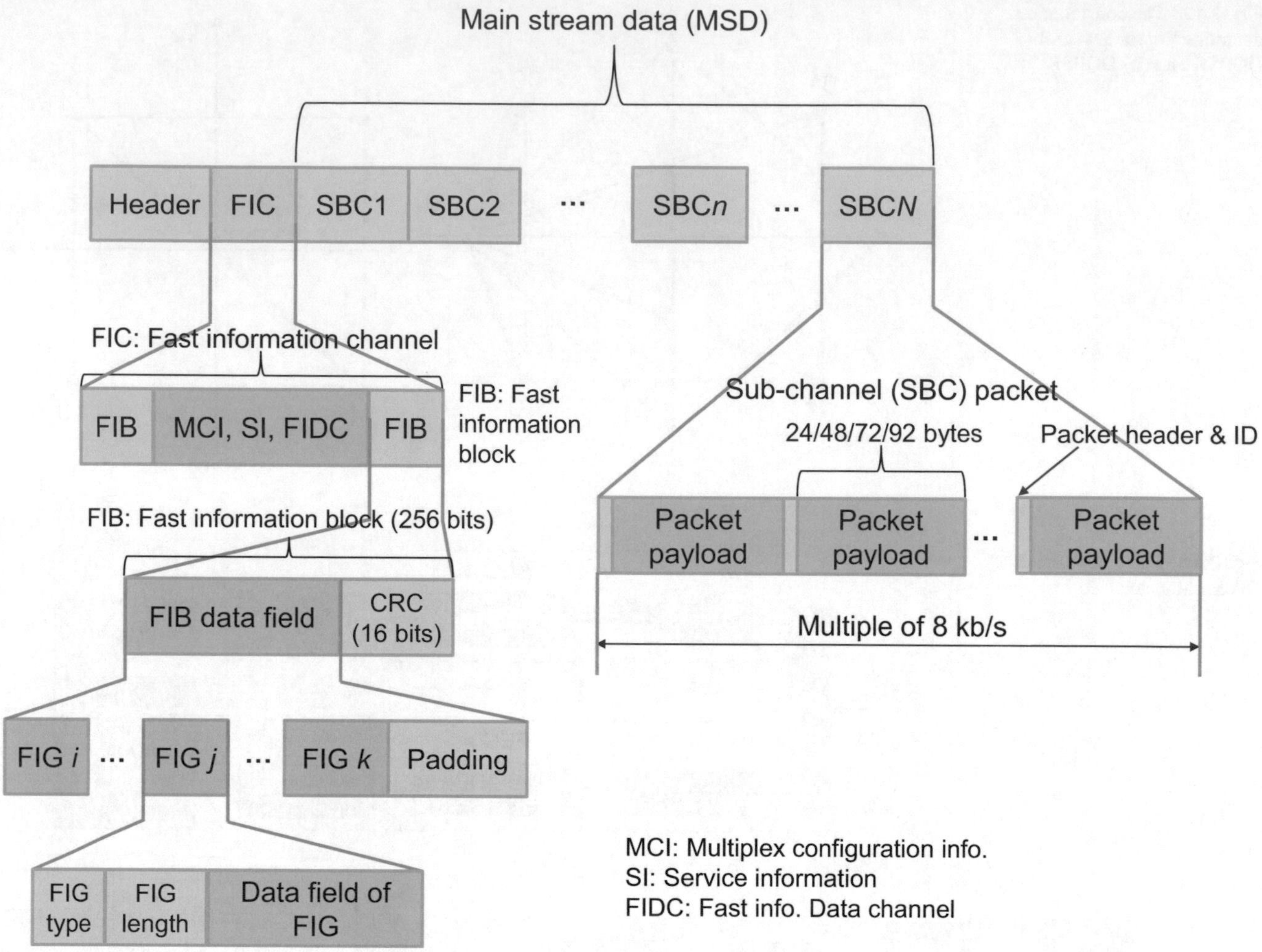

Fig. 7.20 The organization of the DAB packet

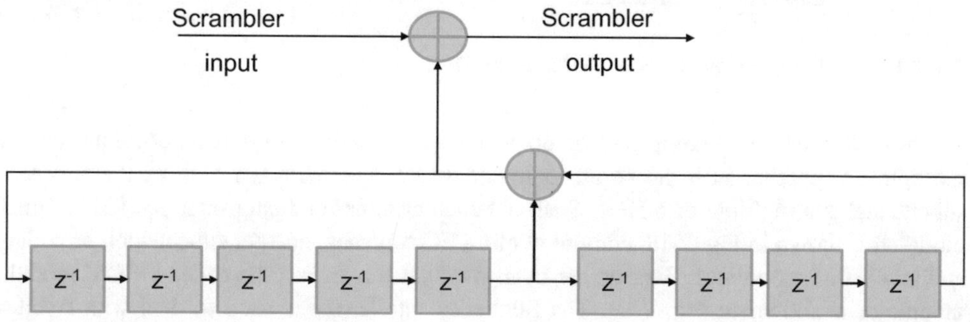

Fig. 7.21 The description of the scrambler

The FEC is based on convolutional coding, as illustrated in Fig. 7.22. The convolutional codes have been described in Chap. 9; here we briefly describe the puncturing mechanism. To achieve the unequal error protection, the code rate R is varied by changing the puncturing index (PI) according to the following equation:

$$R = \frac{8}{8 + PI}, PI \in \{1, 2, \cdots, 24\}. \tag{7.94}$$

Therefore, the code rate ranges from 8/9 to 8/32. Typically, the FIC is protected by code rate 1/3, while the code rates related to the data payload are chosen from the set {2/8, 3/8, 4/8, 6/8}, depending on the priority level.

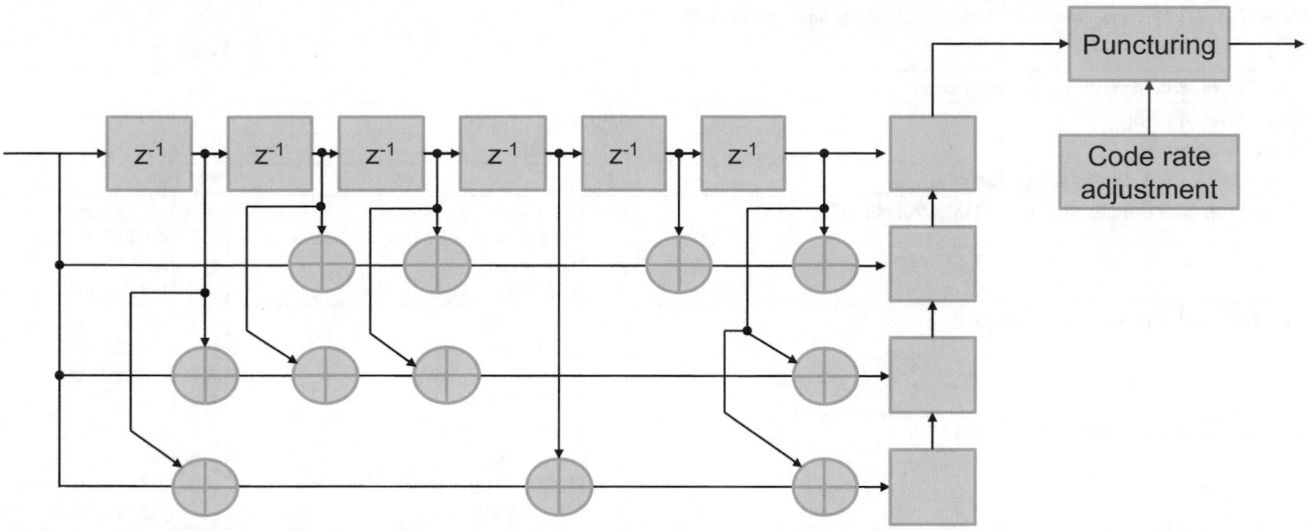

Fig. 7.22 The convolutional encoder configuration and puncturing mechanism. The code rate is determined by $R = 8/(8 + PI)$, where PI is the puncturing index, which is dependent on the protection level. The set of code rates of this punctured convolutional code is {8/9, 8/10, . . ., 8/31, 8/32}

7.7.2 Coded-OFDM in Digital Video Broadcasting (DVB)

Today there exist numerous standards related to DVB [52, 53]. DVB-S, DVB-C, and DVB-T represent the DVB standards for the first-generation satellite, cable, and terrestrial transmission, respectively. On the other hand, DVB-S2, DVB-C2, and DVB-T2 represent the second-generation standards for satellite, cable, and terrestrial transmission, respectively. DVB-H standard represents an extension of DVB-T for handheld mobiles. DVB-SH standard describes the hybrid method for handheld mobiles via satellite and terrestrial transmission systems. DVB-IP is the standard for video stream transmission over the IP networks. DVB-SNG represents the satellite transmission standard for professional applications. DVB-T, DVB-C2, and DVB-T2 standards are coded-OFDM based, while DVB-S, DVB-C, DVB-SNG, and DVB-S2 are single-carrier-based standards.

In the remainder of this section, we briefly describe the basics of DVB-T and DVB-T2. DVB-T channel has a bandwidth of 6, 7, or 8 MHz. There are two operating modes in DVB-T: 2 K mode employing 2046-point-based IFFT and 8 K mode employing 8192-point-based IFFT. The basic system parameter in DVB-T is the IFFT sampling frequency for 8 MHz bandwidth, denoted as $f_{s\ \mathrm{IFFT,\ 8MHz}}$ and determined by

$$f_{s\ \mathrm{IFFT,8MHZ}} = 64/7\ \mathrm{MHz} = 9.142857143\ \mathrm{MHz}. \tag{7.95}$$

The corresponding sampling frequencies for 7 MHz and 6 MHz channels are given respectively by

$$\begin{aligned} f_{s\ \mathrm{IFFT,7\,MHz}} &= \frac{64}{7} \cdot \frac{7}{8} \mathrm{MHz} = 8\ \mathrm{MHz} \\ f_{s\ \mathrm{IFFT,6\,MHz}} &= \frac{64}{7} \cdot \frac{6}{8} \mathrm{MHz} = 6.857142857\ \mathrm{MHz} \end{aligned} \tag{7.96}$$

The *subcarrier spacing* can be obtained by dividing the sampling frequency with the number of subcarriers N:

$$\begin{aligned} \Delta f_{sc} &= f_{s\ \mathrm{IFFT,8/7/6\,MHz}}/N, \\ \Delta f_{2K} = f_{s\ \mathrm{IFFT,8/7/6\,MHz}}/2048,\ &\ \Delta f_{8K} = f_{s\ \mathrm{IFFT,8/7/6\,MHz}}/8192. \end{aligned} \tag{7.97}$$

The *symbol duration* (duration of IFFT portion) can be determined as the reciprocal of the subcarrier spacing:

Table 7.2 DVB-T system parameters for 8 MHz channel bandwidth

Mode	2 K	8 K
Number of subcarriers N_{sc} actually used	1705	6817
Subcarrier spacing	4.464285714 kHz	1.116071429 kHz
Signal bandwidth	7.61161 MHz	7.608259 MHz
Symbol duration Δt (no guard interval)	224 μs	896 ms
Guard interval durations $g\Delta t$, $g\in(1/4,1/8,1/16,1/32)$	$g = 1/4$: 56 μs $g = 1/8$: 28 μs $g = 1/16$: 14 μs $g = 1/32$: 7 μs	$g = 1/4$: 224 ms $g = 1/8$: 112 ms $g = 1/16$: 56 ms $g = 1/32$: 28 ms
Total symbol durations, $(1 + g)\ \Delta t$	$g = 1/4$: 280 μs $g = 1/8$: 252 μs $g = 1/16$: 238 μs $g = 1/32$: 231 μs	$g = 1/4$: 1120 ms $g = 1/8$: 1008 ms $g = 1/16$: 952 ms $g = 1/32$: 924 ms
Symbol rates (per single subcarrier)	$g = 1/4$: 3.571429 kS/s $g = 1/8$: 3.968254 kS/s $g = 1/16$: 4.201681 kS/s $g = 1/32$: 4.329004 kS/s	$g = 1/4$: 0.8928571 kS/s $g = 1/8$: 0.9920635 kS/s $g = 1/16$: 1.05042 kS/s $g = 1/32$: 1.082251 kS/s

$$\Delta t = 1/\Delta f_{sc},\ \Delta t_{2K} = 1/\Delta f_{2K}, \Delta t_{8K} = 1/\Delta f_{8K}. \tag{7.98}$$

The *guard interval* T_G is chosen to be 1/4, 1/8, 1/16, or 1/32 part of the symbol duration. The *total symbol duration* (*OFDM symbol duration*) is then given by

$$T_s = \underbrace{T_G}_{g\Delta t} + \Delta t = g\Delta t + \Delta t = \frac{1+g}{\Delta f_{sc}}, g \in \{1/4, 1/8, 1/16, 1/32\}. \tag{7.99}$$

The subcarrier symbol rate would be then

$$R_{\text{subcarrier}} = 1/T_s = \frac{\Delta f_{sc}}{1+g}, g \in \{1/4, 1/8, 1/16, 1/32\}. \tag{7.100}$$

The number of subcarriers actually used in 2 K mode is $N_{sc} = 1705$, while the number of subcarriers used in 8 K mode is $N_{sc} = 6817$. Therefore, the bandwidth occupied $BW_{\text{DVB-T}}$ would be then

$$BW_{\text{DVB-T}} \simeq N_{sc}\Delta f_{sc}. \tag{7.101}$$

Out of the 1705 subcarriers used in 2 K mode, there are 1512 payload subcarriers, carrying QPSK, 16-QAM, or 64-QAM symbols, 142 scattered pilots (whose position changes), 45 continuous pilots, and 17 transmission parameter signaling (TPS) subcarriers. On the other hand, out of the 6817 subcarriers used in 8 K mode, there are 6048 payload subcarriers, 568 scattered pilots, 177 continuous pilots, and 68 TPC carriers.

The DVB-T system parameters are summarized in Table 7.2, for 8 MHz channel bandwidth. By using Eq. (7.96), it is easy to calculate the corresponding parameters for 7 MHz and 6 MHz channel bandwidths.

The FEC used in DVB-T, DVB-C, and DVB-S is illustrated in Fig. 7.23. For completeness of the presentation, the source coding and MPEG-2 multiplexing is described as well. The energy dissipation is used to decorrelate the byte TS sequence. Reed Solomon RS (204,188) code, capable of correcting up to 8 bytes in error, is used as an outer code. The convolution interleaver is used to deal with burst errors introduced by deep fade (please refer to Chap. 9 for additional details). The convolutional encoder is used as an inner code. The puncturing block is used to puncture parity bits to deal with time-varying channel conditions.

The code rates of punctured convolutional code are selected from the following set {1/2, 2/3, 3/4, 5/6, 7/8}. Two different MPEG-TS streams can be transmitted simultaneously through the so-called hierarchical modulation. The high-priority (HP) path is used to transmit a low-data rate stream, lower-quality signal with higher compression ratio, but with stronger FEC code and more robust modulation format such as QPSK. On the other hand, the low-priority (LP) path is used to transmit MPEG-TS with a higher data rate, employing a weaker FEC scheme, but higher modulation format order, such as 16-QAM

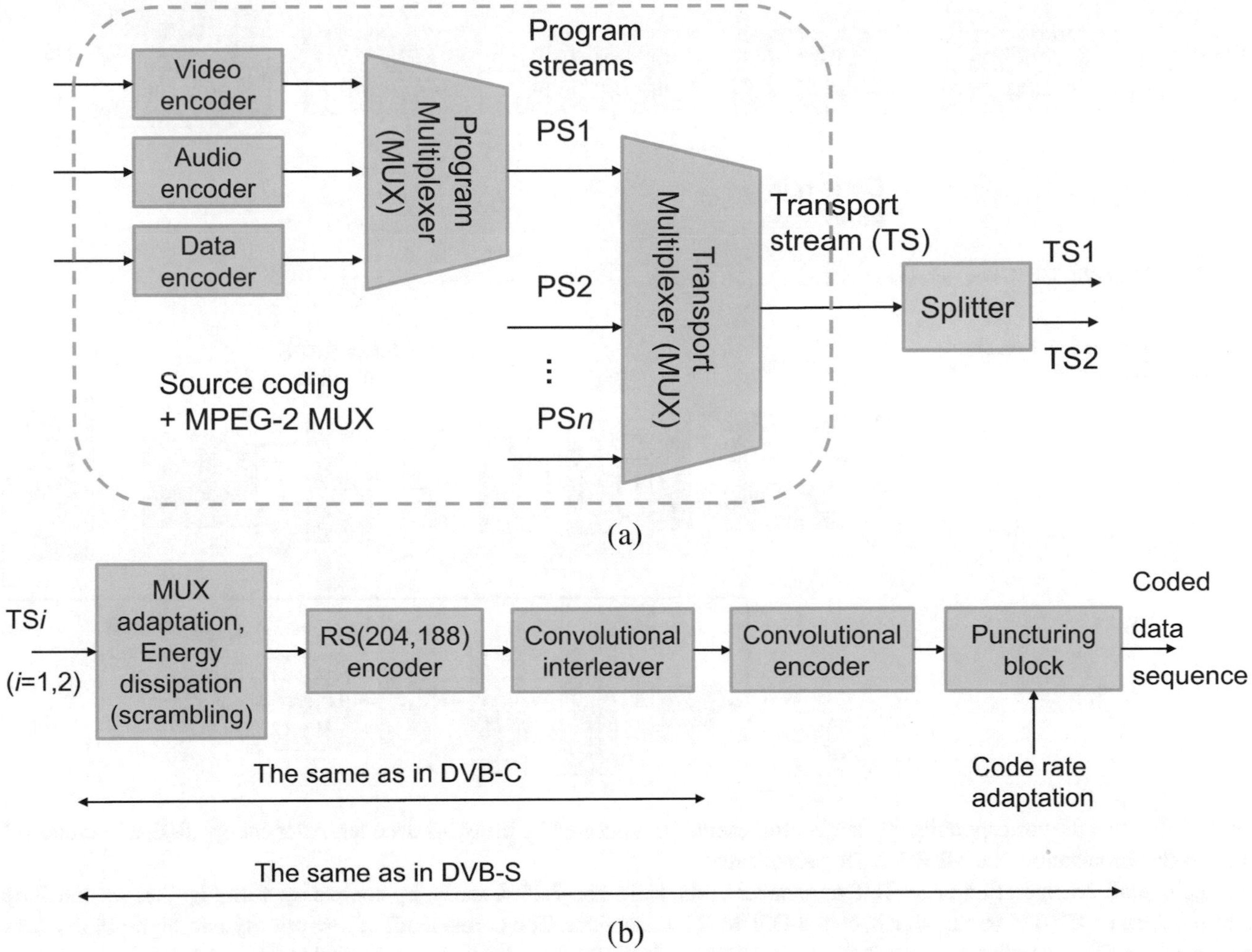

Fig. 7.23 The source coding and FEC in DVB-T: (**a**) source coding and MPEG-2 multiplexing and (**b**) FEC in DVB-T, DVB-C, and DVB-S

and 64-QAM. At the receiver side, the TV set can decide which mode to select (LP or HP), depending on the received signal quality.

The aggregate data information rate (net data rate), denoted as R_D, can be determined as

$$R_D = N_{sc}\frac{\Delta f_{sc}}{1+g}\frac{188}{204}R_{PCC}\log_2 M \cdot cf_{BW}, g \in \{1/4, 1/8, 1/16, 1/32\}, \tag{7.102}$$

where R_{PCC} is the punctured convolutional code rate, M is the signal constellation size, and cf_{BW} is channel bandwidth-dependent correction factor, which is based on Eq. (7.96), determined as

$$cf_{BW} = \begin{cases} 1, & BW = 8\text{ MHz} \\ 7/8, & BW = 7\text{ MHz}. \\ 6/8, & BW = 6\text{ MHz} \end{cases}$$

The DVB-T FEC decoder configuration is shown in Fig. 7.24. After symbol/bit deinterleaving, the Viterbi decoder inserts dummy bits on positions where the parity bits get punctured on the transmitter side. After the Viterbi decoding, the convolutional deinterleaver breaks the burst errors, so that RS (204,188) decoder deals only with random errors and short bursts. If the number of errors exceeds the error correction capability of the RS decoder, the transport error indicator gets

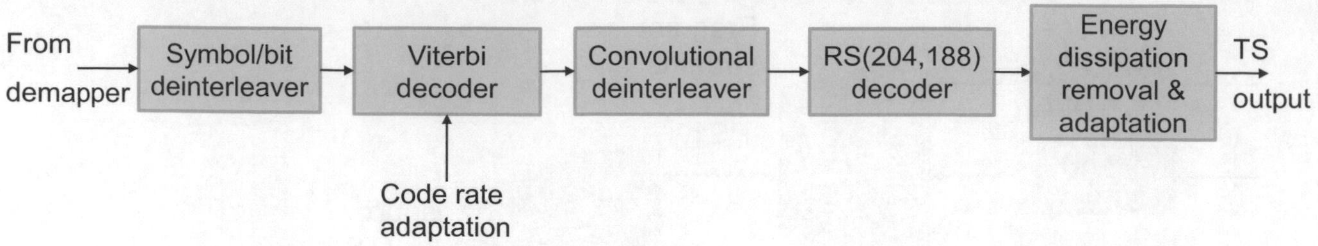

Fig. 7.24 The DVB-T FEC decoder

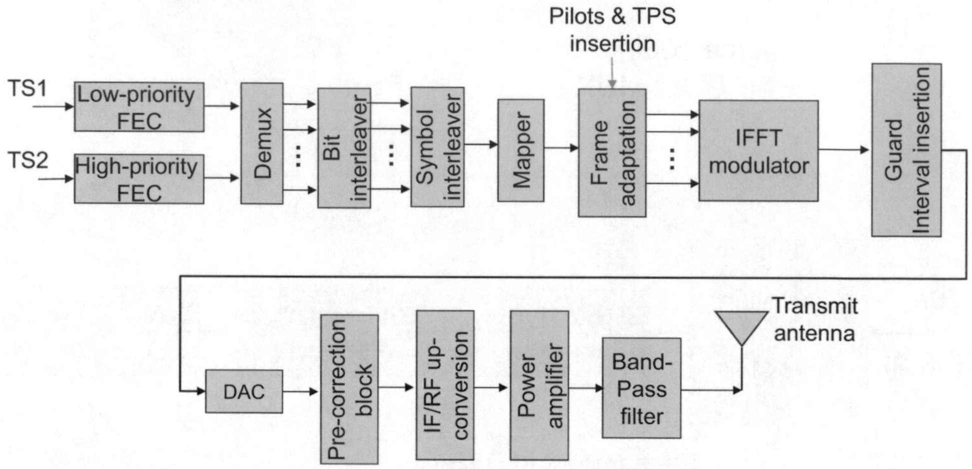

Fig. 7.25 The configuration of the DVT-B transmitter

activated so that the corresponding packet in error cannot be processed by MPEG-2 decoder. After energy dissipation removal and re-synchronization, the MPEG-2 TS get recreated.

The overall configuration of DVB-T transmitter is shown in Fig. 7.25. Clearly, by comparing with Fig. 7.6, we conclude that the core of DVB-T transmitter is coded-OFDM RF transmitter. The corresponding low-priority and high-priority data streams after FEC encoding, as described in Fig. 7.23b, undergo the demultiplexing, followed by bit and symbol interleaving steps. After subcarrier QAM mapping, the frame adaptation is performed by appropriately inserting the pilots used on the receiver side for channel estimation and TPS subcarrier symbols.

The remaining steps are identical to those described in Fig. 7.6a and include the guard interval insertion and DAC, while the pre-correction step to deal with amplifier nonlinearities is optional. After the IF/RF up-conversion, power amplification, and band-pass filtering, the corresponding DVB-T signal is transmitted over a transmit antenna.

The TPS subcarriers contain information about the mode (2 K or 8 K), duration of the guard interval (1/4, 1/8, 1/16, or 1/32 of IFFT part), convolutional code rate (1/2, 2/3, 3/4, 5/6, 7/8), the type of modulation (QPSK, 16-QAM, 64-QAM), and the employment of the hierarchical modulation. To transmit this information, the differential BPSK (DBPSK) is used. As indicated above, 17 subcarriers in 2 K mode and 68 subcarriers in 8 K mode are used as TPC subcarriers. All TPC subcarriers within the same OFDM symbol carry the same bit. On the receiver side, the decision rule is a majority voting rule. We can interpret this approach as a repetition code. The TPS information sequence is transmitted over 68 OFDM symbols, with the first one serving as a reference, meaning that the corresponding TPS sequence length is 67 bits. Positions 1–16 represent a synchronization word, while bits 25 and 26 describe signal constellation size as follows: 00 denotes QPSK, 01 denotes 16-QAM, and 10 denotes 64-QAM. Further, bits 27–29 are related to the hierarchical modulation information, with 000 denoting non-hierarchical modulation (conventional QAM), while 001, 010, and 011 correspond to hierarchical modulation with expansion factor $ef = 1$, 2, and 4, respectively. Bits 36 and 27 represent the (normalized) guard interval duration as follows: 00 represents 1/32, 01 represents 1/16, 10 represents 1/8, and 11 represents 1/4. The transmission mode is transmitted by bits 38 and 39, with 00 representing 2 K mode and 01 representing 8 K mode. Bits 33–35 are used to represent the convolutional code rate of the LP mode, with 000 denoting 1/2 rate, 001 representing 2/3 rate, 010 denoting 3/4 rate, 011 representing 5/6 rate, and 100 denoting the 7/8 code rate. In similar fashion, bits 3–32 are used to represent the code of the HP stream.

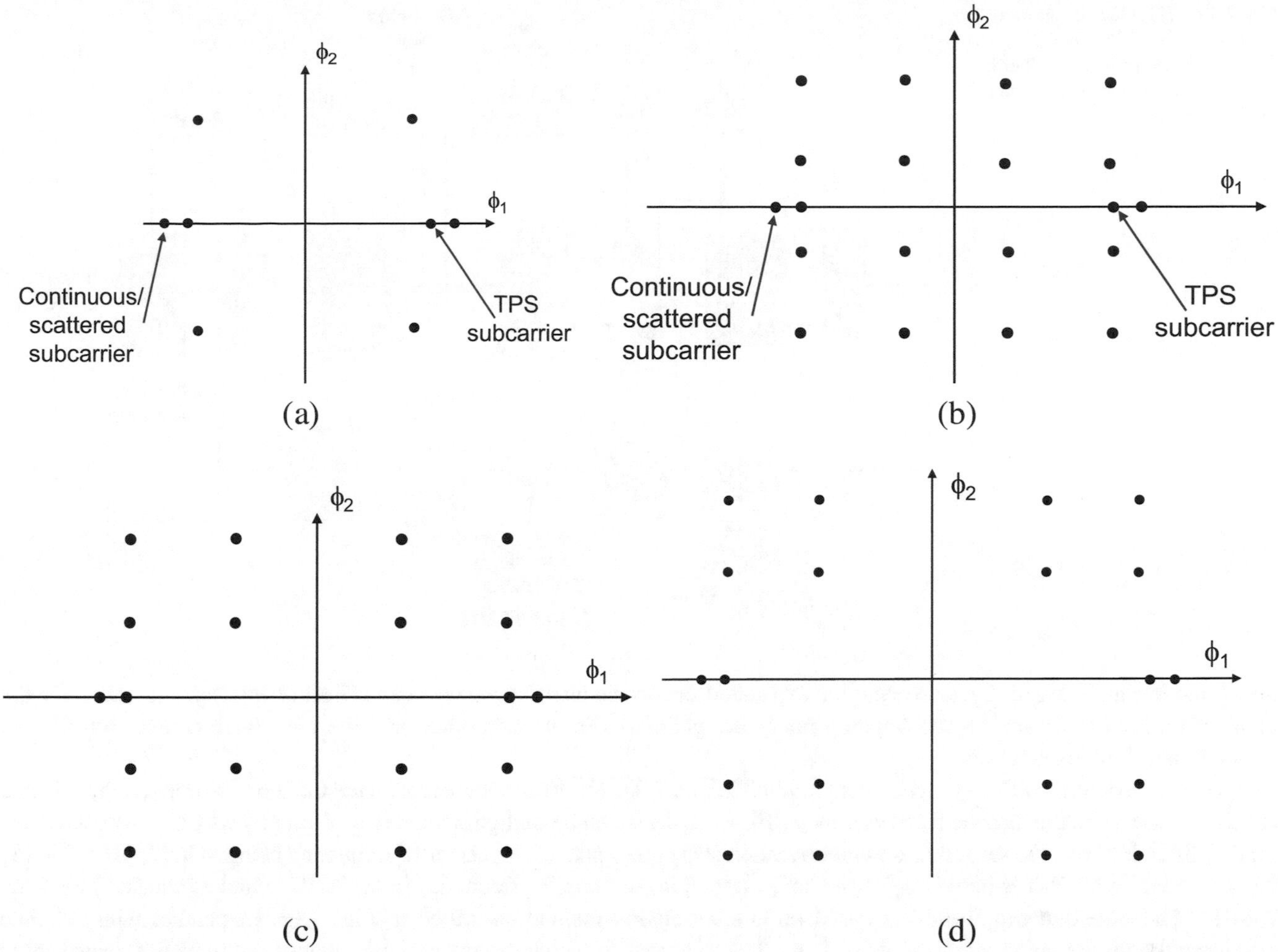

Fig. 7.26 DVT-B signal constellation diagrams for (**a**) QPSK and (**b**) 16-QAM (for $ef = 1$). Hierarchical 16-QAM modulation for expansion factors: (**c**) $ef = 2$ and (**d**) $ef = 4$

The *continuous* (*continual*) *pilots* are based on BPSK and are transmitted with 3 dB higher power compared to the average transmit power, and they are used on the receiver side for phase reference in automatic frequency control (AFC) block.

Within the OFDM symbol, every 12th subcarrier is the *scattered pilot* used in channel estimation. In the next OFDM symbol, the positions of the scattered pilots are increased for three positions. Therefore, it is possible that the position of the scattered pilot coincides with the position of the continuous pilot.

The payload subcarriers carry QPSK, 16-QAM, or 64-QAM symbols. To illustrate the coordinates of the TPS and continuous/scattered subcarriers relative to payload subcarriers signal constellation, in Fig. 7.26 QPSK and 16-QAM signal constellations are provided.

To ensure that reliable transmission is still possible, when the received signal is weak (SNR is small), the hierarchical modulation can be used as an option in DVB-T. In hierarchical modulation, there are two modulation paths in DVB-T transmitter, as mentioned earlier, the LP and HP modulation paths. The HP path employs smaller constellation size (QPKS) and lower code rate, such as 1/2, to guarantee the reliable transmission. In principle, these two modulation paths can carry the same video signal with different compression ratios and consequently different data rates. On the other hand, HP and LP modulation paths can carry different video signals. The HP path represents an embedded QPSK in either 16-QAM or 64-QAM. As an illustration, in Fig. 7.26a, b, the four constellation points in each quadrant can be interpreted as points in QPSK for HP modulation path. The quadrant information carries 2 bits, so the remaining 2 bits in 16-QAM can be used to identify the point within the quadrant. In 64-QAM, we can decompose the constellation into 4 quadrants each with 16 points. The quadrant information can be used for HP stream, while with 4 bits we can select 1 of the 16 points in each quadrant, which corresponds to the LP stream. The expansion factor increases the distance among center-of-mass points for each quadrant, as

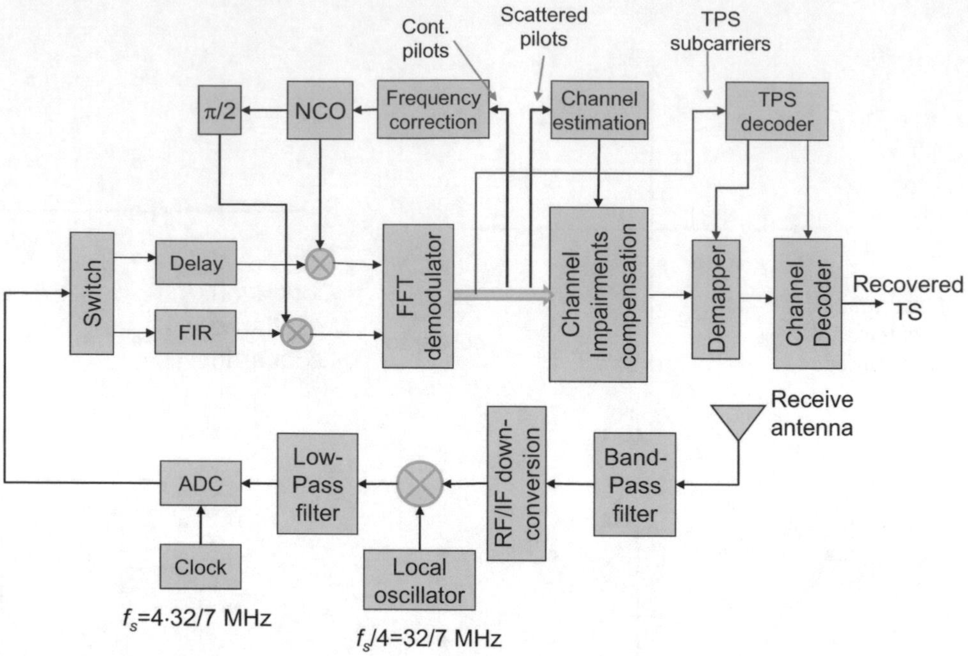

Fig. 7.27 The block diagram of the DVB-T receiver. NCO: numerically controlled oscillator. The channel decoder is described in Fig. 7.24

illustrated in Fig. 7.26c, d. By increasing the expansion factor, the interference between HP and LP streams is lower. On the other hand, for LP stream the constellation points are getting closer to each other, and the LP stream is more sensitive to crosstalk and low SNR values.

The configuration of DVB-T receiver is provided in Fig. 7.27. The first module represents the tuner, to convert the selected RF channel down to the intermediate frequency (IF). At IF the signal is band-pass filtered, typically by a surface acoustic wave (SAW) filter. Further, the second IF down-conversion stage takes place, with center IF frequency being at 32/7 MHz. The I/Q demodulation in DVB-T is based on $f_s/4$-method, where f_s is the sampling frequency in the ADC, which is equal to $f_s = 4{\cdot}32/7$ MHz. The time synchronization is based on autocorrelation method described in Fig. 7.12. In parallel with the time synchronization, the signal after the ADC is used as the input to the changeover switch, whose outputs get selected in an alternative fashion. Therefore, the corresponding branch data rate is $f_s/2$. There is the offset between two signal branches for the half of the sampling cycle. To compensate for this, the FIR filter is used in the lower branch and delay line in the upper branch (to make sure that delays in I and Q channels are properly aligned in time-domain). The output data streams of the upper and lower branches are used as the input of the numerically controlled oscillator (NCO). The NCO and mixer are essentially used as a digital PLL. However, since the oscillators are not of very high precision (to ensure that the TV set is inexpensive), the AFC approach is used by employing the information conveyed by the continuous pilots after FFT demodulation. When transmit and receive frequencies differ, there will be either clockwise or anti-clockwise rotation of the constellation diagram. The phase difference between continuous pilots from symbol to symbol is used as a control variable of the NCO, which changes the output in the direction of reducing the difference between incoming frequency and digital PLL frequency. Once this difference becomes zero, the rotation of constellation diagram is compensated for, and we employ the scattered pilots to compensate for the channel impairments. The detection of TPS subcarriers is differential and therefore independent on the phase error. The repetition coding and additional BCH coding is used to ensure that the TPS sequence is error-free. After that, the demapping takes place followed by channel decoding, with details of channel decoder provided already in Fig. 7.24. The output of decoder represents the recovered TS MPEG-2 stream.

The fs/4 method for I/Q demodulation is illustrated in Fig. 7.28. The QAM signal with carrier frequency f_{IF} is used as input to ADC clocked at $f_s = 4f_{IF}$. On the other hand, the 1-to-2 switch is clocked at twice lower frequency ($f_s/2 = 2f_{IF}$) so that two samples per I and two samples per Q channel are obtained at the upper and lower output branches of the switch, respectively. Since the second sample in either branch is of opposite sign that is needed, we flip the signs with rate f_{IF}. The FIR is used in the lower branch to ensure the proper synchronization of the upper and lower branches. The outputs of the upper and lower branches represent the estimates of in-phase and quadrature samples, used at the input of the demodulator.

In order to increase the overall data rates of DVB-T, the DVB-T2 standard is introduced [52, 55]. In addition to channel bandwidths already in use in DVB-T, the 10 MHz bandwidth is introduced. Stronger FEC is used in DVB-T2, employing

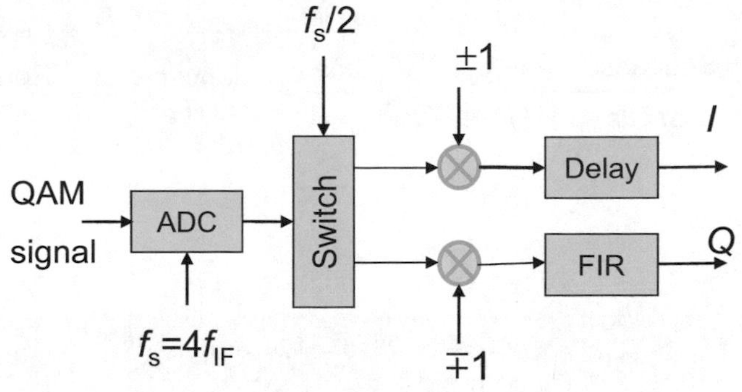

Fig. 7.28 The $f_s/4$ method based I/Q demodulation

Fig. 7.29 The FEC used in DVB-T2 is based on concatenated code in which LDPC code is used as the inner code and BCH code is used as the outer code

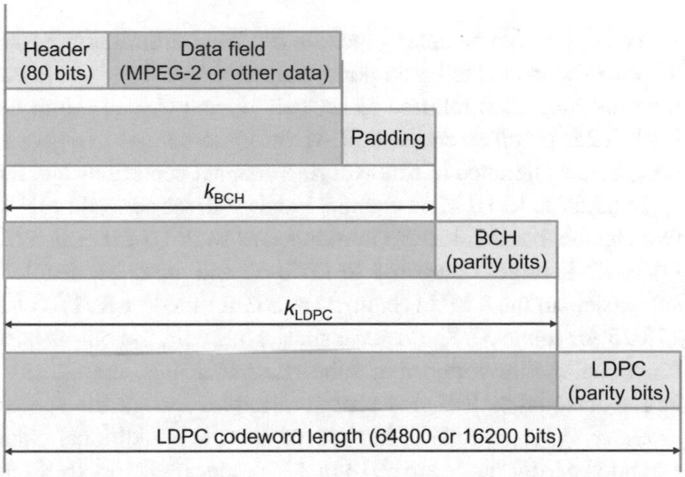

Fig. 7.30 The FEC frame used in DVB-T2

LDPC coded as the inner code and BCH code as the outer code, which is illustrated in Fig. 7.29. The following code rates are supported: 1/2, 3/5, 2/3, 4/5, and 5/6. The corresponding FEC frame is shown in Fig. 7.30. Both BCH and LDPC codes belong to the class of systematic codes, in which information portion remains unchanged, while generalized parity bits are added to the information word to form a codeword. The number of parity bits in BCH code is 16 t, where t is the error-correcting capability of BCH code (8, 10, or 12). The LDPC codeword length is either 64,800 or 16,200 bits. The information word length for the LDPC code, denoted as k_{LPDC}, is determined from the target code rate and FEC frame length. Moreover, the LDPC code is adaptive; the FEC strength can be increased (code rate decreased) when channel conditions get worse. The adaptive coding can be combined with adaptive modulation in the so-called adaptive modulation and coding (AMC) [6, 7, 30].

The following modulation formats are supported in DVB-T2: QPSK, 16-QAM, 64-QAM, and 256-QAM. In DVB-T2 Gray mapping and coherent detection are used. For 8 MHz channel bandwidth, the signal bandwidth is 7.61 MHz. Therefore, the date rates for guard interval fraction $g = 1/128$ range from 4 Mb/s for QPSK with code rate $R = 1/2$ to 50.32 Mb/s for 256-QAM with $R = 5/6$ [52]. The coordinates for these normal QAM constellations can be found in textbooks [26]. Another type of signal constellations used in DVB-T2 belongs to the class of the so-called rotated Q-coordinate-delayed signal constellations, whose operation principles are illustrated in Fig. 7.31. The $m = \log_2 M$ bits from input sequence are used to select the point from M-ary QAM, by employing the Gray mapping rule, with the corresponding coordinates stored in a look-up table (LUT), with coordinates being I'_k and Q'_k. The rotator block rotates the signal constellation by ϕ [deg] as follows:

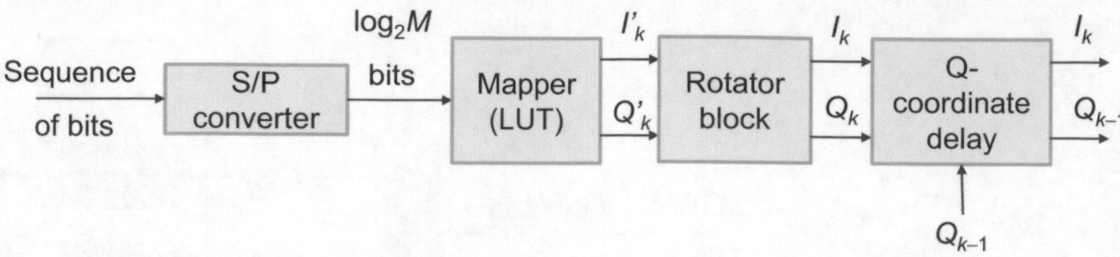

Fig. 7.31 The illustration of rotated Q-coordinate delayed signal constellation principle

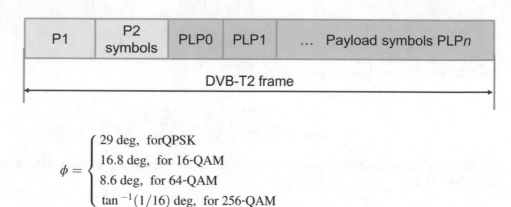

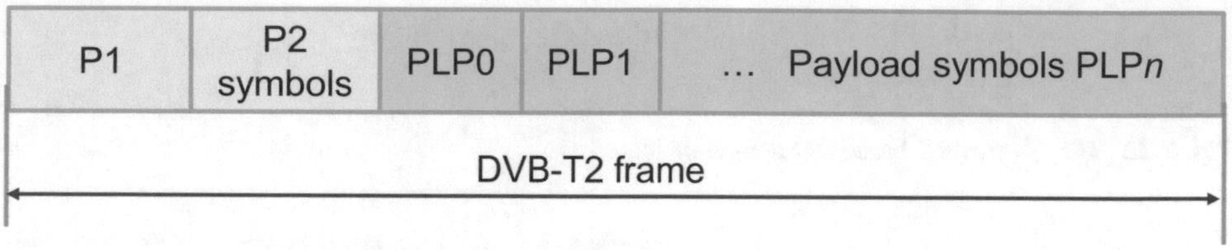

Fig. 7.32 The organization of the DV-T2 frame

$$\phi = \begin{cases} 29\ \text{deg},\ \text{forQPSK} \\ 16.8\ \text{deg},\ \text{for 16-QAM} \\ 8.6\ \text{deg},\ \text{for 64-QAM} \\ \tan^{-1}(1/16)\ \text{deg},\ \text{for 256-QAM} \end{cases}$$

The Q-coordinate delay block, in the k-th transmission instance, transmits the following coordinates (I_k, Q_{k-1}). Therefore, the k-th signal constellation point of rotated QAM (I_k, Q_k) is transmitted into two consecutive instances (I_k, Q_{k-1}) in the k-th time instance, also referred to as "cell" k, and (I_{k+1}, Q_k) in the (k + 1)-st time instance, also referred as cell k + 1. Since, DVB-T2 is based on coded-OFDM, the Q-coordinate of a given signal constellation point is on the next subcarrier. On such a way, better tolerance to time-varying channel conditions and fading effects is obtained.

In addition to 10 MHz channel bandwidth introduced in DVB-T2, for cellular (mobile) telephony applications, additional two channel bandwidths are introduced as well: 1.7 MH and 5 MHz. The number of different FFT modes for coded-OFDM in DVB-T2 is larger compared to DVB-T, and these are denoted as 1 K, 2 K, 4 K, 8 K, 16 K, and 32 K. The numbers of subcarriers in the 8 MHz channel are 853 for mode 1 K, 1705 for mode 2 K, 3409 for 4 K, 6817 for 8 K, 13633 for 16 K, and 27,625 for mode 32 K. Because sinc(x) has tails, the out-of-band spectrum is typically filtered by corresponding windowing functions. As the number of subcarriers increases, the transition in frequency-domain is sharper. Nevertheless, for modes larger or equal to 8 K, the spectrum moves toward the spectrum of neighboring channel. To solve for this problem, the *extended carrier mode* is introduced in which additional channel bandwidth is allocated. The numbers of subcarriers in extended carrier mode are 6913 in 8 K mode, 13,921 in 16 K mode, and 27,861 in 32 K mode. As before, the script K is used to denote the size of FFT block of 1024. For instance, the size of FFT block in 32 K is 32,768.

The organization of DVB-T2 frame is shown in Fig. 7.32. The symbol P1 is used for synchronization and frame detection purposes. The symbols P2 are used for signaling purposes on the receiver side to determine the parameters of OFDM frame, and the number of P2 symbol is a function of the FFT mode: 16 P2 symbols are used in 1 K mode, 8 in 2 K, 4 in 4 K, 2 in 8 K mode, and 1 in 16 K and 32 K modes. The payload symbols that follow after P2 symbols are organized into the so-called physical layer pipes (PLPs), and up to 255 input data streams are possible. The length of DVB-T2 frame in terms of the number of OFDM symbols is dependent on FFT mode and guard interval fraction g and, for 8 MHz channel and g = 1/16, ranges from 64 in 32 K mode to 2098 in 1 K mode.

Before the IFFT modulation takes place, different types of interleavers are used: bit interleaver is used just after FEC encoder, cell interleaver as described in Fig. 7.31, and the time interleaver is used after the cell interleaver. The symbol streams from different PLPs are frequency interleaved before the IFFT modulation, as shown in Fig. 7.33.

The coded-OFDM portion of transmitter is very similar to generic coded-OFDM transmitter shown in Fig. 7.6a, except from P1 symbol insertion. To deal with time-selective fading effects, the time interleaver is used. To deal with frequency-selective fading effects, the frequency interleaver is used. To deal with multipath fading, the multiple-input single-output (MISO) processing is used, based on Alamouti coding Scheme [56, 57]. Additional details about Alamouti coding scheme can be found in Chap. 8.

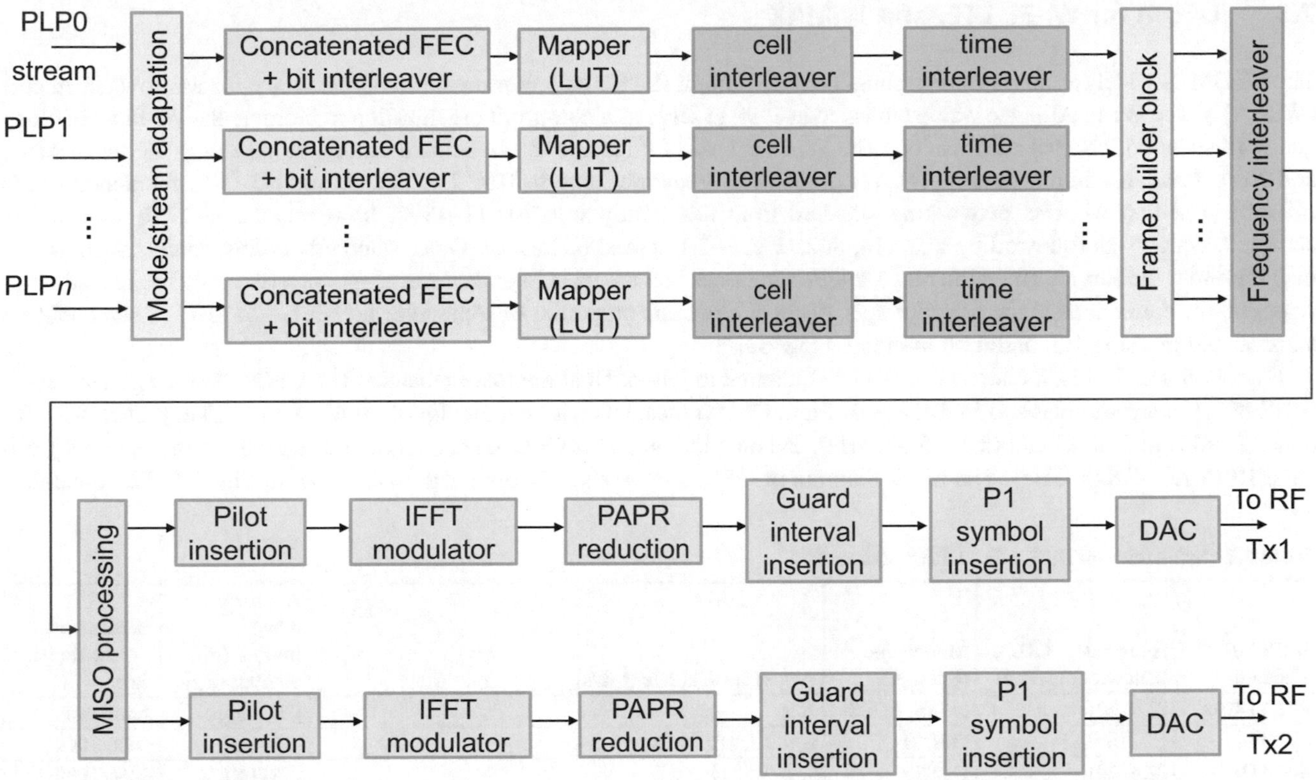

Fig. 7.33 The configuration of DVB-T2 transmitter based on coded-OFDM and multiple input single output (MISO)

Since in OFDM different subcarriers are modulated independently, some of the subcarriers can add constructively leading to high peak-to-average power ratio (PAPR), defined as

$$\mathrm{PAPR} = 10\log_{10}\left(\frac{\max\limits_{0\leq t<T_{\mathrm{OFDM}}} |s_{\mathrm{OFDM}}(t)|^2}{\left\langle |s_{\mathrm{OFDM}}(t)|^2 \right\rangle}\right)\ [\mathrm{dB}], \tag{7.103}$$

where $s_{\mathrm{OFDM}}(t)$ represents the OFDM frame of duration T_{OFDM}. High PAPR can introduce nonlinearities in driver amplifier and DAC and ADC blocks. To solve this problem, either tone reservation or active constellation extension (ACE) is used in DVB-T2. In tone reservation, certain subcarriers are planed for payload symbols, which are not used as pilots, but reserved to reduce the PAPR. These tones are not used unless the PAPR is too high, when they get activated to reduce the PAPR. In ACE the outmost constellation points are displaced further away from the original constellation in an effort to reduce the PAPR.

The pilots used in DVB-Tw can be categorized as edge pilots (at the beginning and end of spectrum), continuous pilots, scattered pilots, P2 pilots, and frame-closing pilots. The continuous and scattered pilots have the same role as in DVB-T, for AFC and channel estimation, respectively. On the other hand, the P2 symbols have the same role as TPS subcarriers in DVB-T. There exist eight pilot patterns (PPs) in DVB-T2 depending on the distance between neighboring subcarriers in time domain and frequency-domain. Let d_1 and d_2 denote the distance between scattered pilot subcarrier positions and distance between scattered pilots within the same symbol, respectively. Let d3 denote the number of symbols involved in the pilot sequence. Then the PPs can be specified by the following triple (d_1, d_2, d_3). The PP1 is specified by the triple (3,12,4), PP2 by (6,12,2), PP3 by (6,24,4), PP4 by (12,24,2), PP5 by (12,48,4), PP6 by (24, 48, 2), PP7 by (24, 96, 4), and PP8 by the triple (6, 96, 16). The scattered pilot patterns PP1 and PP2 are transmitted with normalized amplitude of 4/3, PP3 and PP4 with 7/4, while PP5-PP7 with normalized amplitude 7/3. As an illustration, the pilot patterns used in MISO mode for 32 K with $g = 1/16$ are PP2 and PP8. The corresponding PPs used in single-input single-output (SISO) mode are PP2, PP8, and PP4.

7.8 OFDM for Wi-Fi, LTE, and WiMAX

The OFDM is widely used in Wi-Fi, Long-Term Evolution (LTE), and Worldwide Interoperability for Microwave Access (WiMAX). The Wi-Fi Alliance was established in 1999 to serve as a non-profit organization to promote the Wi-Fi technology and corresponding product certification. The Wi-Fi is based on IEEE 802.11 standard family, representing the set of MAC and PHY standards implementing WLAN computer communications in 0.9, 2.4, 3.6, 5, and 60 GHz frequency bands [58–60]. The first wireless networking standard from this family was 802.11–1997; however, the 802.11b was widely accepted, which was followed by 802.11a, 802.11 g, 802.11n, and 802.11 ac. Other standards in this family (such as c–f, h, j) represent the service amendments to the existing ones that are used to extend the scope, or eventually the amendments represent the corrections to the previous specification. The basic properties of some relevant IEEE 802.11 PHY standards are summarized in Table 7.3, based on references [58–63].

For additional details, the interested reader is referred to [58–63] and references therein. The OFDM system parameters in IEEE 802.11a are summarized in Table 7.4. Eight OFDM channels with bandwidth of 20 MHz can be supported in UNIII bands I and II of 5GHz band (5.15–5.35 GHz), and an additional four OFDM channels can be supported in the upper UNII band III (5.725–5.825 GHz). The block diagram of OFDM transceiver is similar to that shown in Fig. 7.6. The preamble,

Table 7.3 Some relevant IEEE 802.11 PHY standards

IEEE 802.11 standard	[Frequency (GHz), bandwidth (MHz)]	Modulation technique	Aggregate data rate (Mb/s)	Approximate range [indoor (m), outdoor (m)]	Release date/ expected release date
802.11–1997	[2.4, 22]	DSSS, FHSS	1, 2	[20, 100]	Jun. 1997
802.11a	[5, 20]	OFDM	6, 9, 12, 18, 24, 36, 48, 54	[35, 120]	Sept. 1999
802.11b	[2.4, 22]	DSSS	1, 2, 5.5, 11	[35, 140]	Sept. 1999
802.11 g	[2.4, 20]	OFDM	6, 9, 12, 18, 24, 36, 48, 54	[38, 140]	Jun. 2003
802.11n	[2.4/5, 20/40]	MIMO-OFDM	For 400 ns guard interval (GI) and bandwidth (BW) of 20 MHz: 7.2, 14.4, 21.7,28.9, 43.3, 57.8, 65, 72.2 GI = 800 ns, BW = 20 MHz: 6.5, 13, 19.5, 26, 39, 52, 58.5, 65	[70, 250]	Oct. 2009
802.11 ac	[5, 20/40/80/160]	MIMO-OFDM	GI = 400 ns, BW = 20 MHz: 7.2, 14.4, 21.7, 28.9, 43.3, 57.8, 65, 72.2, 86.7, 96.3 GI = 800 ns, BW = 20 MHz: 6.5, 13, 19.5, 26, 39, 52, 58.5, 65, 78, 86.7 GI = 400 ns, BW = 160 MHz: 65, 130, 195, 260, 390, 520, 585, 650, 780, 866.7 GI = 800 ns, BW = 160 MHz: 58.5, 117, 175.5, 234, 351, 468, 702, 780	[35, –]	Dec. 2013
802.11ad	[60, 2160]	OFDM, single-carrier	≤6912	[60, 100]	Dec. 2012
802.11ay	[60, 8000]	OFDM, single-carrier	≤100,000	[60, 1000]	2017

Table 7.4 OFDM system parameters in IEEE 802.11a PHY standard

OFDM symbol duration	4 μs
Number of subcarriers actually used/number of pilots	52/4
Subcarrier spacing	312.5 kHz
Signal bandwidth	16.66 MHz
Guard interval duration	800 ns
Channel spacing	20 MHz
Modulation formats	BPSK, QPSK, 16-QAM, 64-QAM
Code rates	½, 9/16, 2/3, 3/4
Aggregate data rates	6, 9, 12, 18, 36, 48, 54 Mb/s

which precedes the OFDM packed, is composed of three sections of durations 8 μs, 8 μs, and 4 μs, respectively. The first section is used in automatic gain control (AGC) and coarse frequency offset estimation, and it is composed of ten training symbols with duration of 800 ns each. These training symbols are generated by employing only nonzero QAM symbols, with located subcarrier indices being a multiple of 4, namely, $\{-24, -20, -16, -12, -8, -4, 4, 8, 12, 16, 20, 24\}$, wherein the subcarrier indices range from -26 to 26. Given that for the digital signal with a symbol duration of T the shortest measurable frequency offset is $1/(2\,T)$, we conclude that the offset up to $1/(2 \times 800\text{ ns}) = 625$ kHz can be estimated. If the whole FFT duration is used for estimation, namely, 4 μs–0.8 μs = 3.2 μs, the frequency offset of 156 kHz can be estimated, so that the corresponding relative frequency error of 26.89 ppm at carrier frequency 5.8 GHz can be measured. The sequence of short training symbols is also used for frame detection by correlating to neighboring short symbols (pulses), and the frame will be detected once correlator outputs exceed a desired threshold. The AGC circuit can adjust operation once the frame is detected.

The sequence of short training symbols is followed by a long training symbol (with duration of 8 μs), composed for 52 QPSK-modulated subcarriers, which is used for fine-frequency offset estimation, timing, and channel estimation, with corresponding concepts explained already above. Given the duration of this long training symbol, it is possible to place 8 μs/3.2 μs = 2.5 FFT frames and therefore perform the cyclic extension. This first portion with duration of 1.6 μs is used as guard interval and contains the copy of the last 1.6 μs of FFT part. By measuring the frequency drift of two samples separated by 3.2 μs, the fine-frequency offset estimation is possible. Moreover, the long training symbol is used to estimate the amplitude for coherent (synchronous) detection. The training symbols are properly selected so that the PAPR of 3 dB is obtained, significantly lower than in data carrying OFDM symbols, thus avoiding any nonlinear distortion due to power amplifier.

The last 4 μs of preamble contains the information about modulation type, code rate employed, and the duration of the packet length.

Even after the fine-frequency offset compensation, there will be the uncompensated frequency drift causing the common phase error (drift) for all subcarriers, and to compensate for the common phase error, 4 out of 52 subcarriers are used as pilots, located at subcarrier indices $\{-21, -7, 7, 21\}$. The pilots are scrambled by a pseudorandom sequence of length 127 to avoid the spectral lines in power spectral density of the OFDM spectrum.

To adjust the error correction strength to time-varying channel conditions, the punctured convolutional code is used with possible code rates already specified in Table 7.4. Clearly, the lowest data rate would be $[(52–4)/4{\cdot}10^{-6}] \times 1/2$ b/s = 6 Mb/s. On the other hand, the highest possible data rate in this standard would be $(48/4{\cdot}10^{-6}) \times (\log_2 64) \times 3/4$ b/s = 54 Mb/s.

The IEEE 802.11 g PHY standard is very similar to 801.11a, except that it is related to the 2.4 GHz unlicensed ISM band (2.4–2.497 GHz).

To increase the aggregated data rates in IEEE 802.11n and 802.11 ac PHY standards, the MIMO-OFDM concept, explained in Chap. 8, is used. Namely, through MIMO concept, several independent data streams can be simultaneously transmitted increasing the aggregate data rates. The IEEE 802.11n PHY standard is applicable to both 2.4 and 5 GHz bands while 802.11 ac to 5 GHz band only. In IEEE 802.11n standard, the available channel bandwidths are {20 MHz, 40 MHz}, while in 802.11 ac the available channel bandwidths are {20, 40, 80, 160} MHz. The highest data rate in 802.11 ac standard, for 160 MHz bandwidth and GI of 400 ns, is 780 Mb/s.

Recent 802.11ax (Wi-Fi 6E) standard, also known as a high-efficiency Wi-Fi, adopted in 2020, is designed to operate in license-exempt bands (1–7.125 GHz), including commonly used 2.4 and 5 GHz as well as the 6 GHz band (5.925–7.125 GHz in the USA) [64], and it is based on OFDMA and multi-user (MU)-MIMO. The purpose of the Wi-Fi 6E standard is to enhance throughput/area for high-density scenarios including shopping malls, corporate offices, and dense residential areas. The modulation formats range from BPSK to 1024-QAM, with code rates from 1/2 to 5/6. The data rates range from 135 Mb/s (for 20 MHz channels) to 1.201 Gb/s (for 120 MHz channels). The guard intervals are extended to 0.8 μs, 1.6 μs, or 3.2 μs so that the corresponding systems are more tolerant to the delay spread. Compared to the 802.11 ac, the OFDM symbol duration is increased four times (extend to 12.8 μs), while the subcarrier spacing is reduced four times.

The third Generation Partnership Project (3GPP) represents an international standardization body working on 3G Universal Terrestrial Access Network (UTRAN) and GSM specifications. This standardization body developed the 3G radio access widely known as Long-Term Evolution (LTE) or Evolved UTRAN as well as the evolved packet access core network in the System Architecture Evolution (SAE) [60]. The 3GPP standards are reported as so-called Releases. For instance, Release 8 (2008 Q4) represents the first LTE release and all-IP SAE, based on OFDMA, frequency-domain equalization (FDE), and MIMO-based interface, not compatible with previous CDMA-based interfaces. Release 10 (2011 Q1) describes the LTE Advanced standard satisfying the International Mobile Telecommunications-Advanced (IMT-Advanced) 4G requirements and being backward compatible with Release 8. The LTE Advanced can use up to 8 × 8 MIMO signal processing and 128-QAM in downlink direction, and with 100 MHz in aggregated bandwidth, it provides almost 3.3 Gb/s in peak download rates per sector of the base station under perfect channel conditions. LTE Advanced Pro represents the 3GPP Releases 13 and

Table 7.5 OFDM system parameters and other relevant parameters in WiMAX IEEE 802.16e standard

Channel bandwidth	1.25 MHz	5 MHz	7 MHz	8.75 MHz	10 MHz	20 MHz
Sampling frequency	1.4	5.6 MSa/s	8 MSa/s	10 MSs/s	11.2 MSa/s	22.4 MSa/s
The FFT size	128	512	1024	1024	1024	2048
Number of subchannels	2	8	16	16	16	32
Subcarrier spacing (Δf)	10.94 kHz	10.94 kHz	7.81 kHz	9.77 kHz	10.94 kHz	10.94 kHz
Symbol duration ($T = 1/\Delta f$)	91.4 μs	91.4 μs	128 μs	102.4 μs	91.4 μs	91.4 μs
Guard interval duration ($T_G = T/8$)	11.425 μs	11.425 μs	16 μs	12.8 μs	11.425 μs	11.425 μs
OFDM symbol duration ($T_s = T + T_G$)	102.825 μs	102.825 μs	144 μs	115.2 μs	102.825 μs	102.825 μs
Number of OFDMA symbols per frame	48	48	34	43	48	48
Frame duration	4.94 ms ≅ 5 ms	4.94 ms ≅ 5 ms	4.89 ms ≅ 5 ms	4.95 ms ≅ 5 ms	4.94 ms ≅ 5 ms	4.94 ms ≅ 5 ms
Modulation	QPSK, 16-QAM, 64-QAM					
Channel Coding Rates	Convolutional coding (CC): 1/2, 2/3,3/4 Block turbo coding (BTC): 1/2, 3/4 LDPC coding: 1/2, 2/3,3/4, 5/6 Convolutional turbo coding (CTC) in both downlink (DL) and uplink (UL): For QPSK and 16-QAM: 1/2, 3/4; for 64-QAM: 1/2, 2/3,3/4, 5/6					

14. Another relevant 3GPP standard-air interface for the 5G networks is the 5G NR (New Radio), also known as Release 15 [65]. The 5G NR employs two frequency bands: frequency range 1 (FR1) 410 MHz–7.125 GHz and frequency range 2 (FR2) 24.25–52.6 GHz. Early FR1 deployments employ 5G NR software on 4G hardware, which are slightly better than newer 4G systems, and essentially represent 4.5G technology. The 5G-Advanced standard is currently under development by 3GPP, known as Release 18 [66].

WiMAX represents one of the most popular broadband wireless (BWA) technologies, aiming to provide high-speed bandwidth access for wireless MANs [60, 69–71]. WiMAX is based on IEEE 802.16 set of standards, in particular IEEE 802.16-2004 (fixed-WiMAX) and IEEE 802.16e-2005 (mobile-WiMAX) standards, which provide multiple PHY and MAC options. WiMAX represents the last mile wireless broadband access alternative to the cable and DSL services, and it is a direct competitor of the LTE Advanced standard. The IEEE 802.16e-2005 provides several improvements compared to the IEEE 802.16-2004 including [69–71] (1) providing support for mobility by enabling the soft and hard handover between base stations; (2) introducing the concept of scalable OFDM (SOFDM) by scaling the FFT to the channel bandwidth to ensure that the carrier spacing constant across different channel bandwidths, ranging from 1.25 MHz through 5 MHz and 10 MHz all way to 20 MHz, and thus improving the spectral efficiency; (3) introducing the advanced antenna diversity schemes and hybrid automatic repeat-request (HARQ); (4) introducing the adaptive antenna systems (AAS) and MIMO signal processing; (5) providing the denser sub-channelization and therefore improving the indoor penetration; (6) introducing downlink subchannelization, in order to trade the coverage for capacity or vice versa; (7) introducing the LDPC coding to improve the error correction strength; and (8) introducing an extra QoS class for VoIP applications. The basic OFDM parameters and other relevant system parameters in IEEE 802.16e-2005 are summarized in Table 7.5.

For OFDM PHY, the following FEC schemes are used: (1) mandatory concatenated RS-convolutional code (RS), in which RS code is used as the outer code and rate-compatible CC code as the inner code; (2) optional block turbo coding (BTC); and (3) optional convolutional turbo coding (CTC). The RS-CC concatenated code has already been discussed above; for additional detail the interested reader is referred to Chap. 9. The BTC and CTC schemes are fully described in Chap. 9. In the burst mode, when we are concerned with correcting both burst and random errors, the following FEC schemes are used: (1) mandatory tail-biting CC code as specified in IEEE 802.16 standard or zero-tailing CC in WiMAX profiles; (2) CTC, which is optional in IEEE 802.16 standards, but mandatory in WiMAX profiles; (3) optional TC; and (4) optional LDPC coding. In September 2018 the IEEE 802.16-2017 standard was released, relevant for combined fixed and mobile point-to-multipoint BWA systems providing multiple services [67]. The IEEE 802.16 Working Group ceased their activities on 9 March 2018 [68].

7.9 OFDM in Ultra-Wideband Communication (UWC)

Ultra-wideband (UWB) communication represents a fast-emerging RF technology enabling short-range high-bandwidth communications with high energy efficiency while occupying a large RF bandwidth [72–75]. In the USA, UWB radio communication links operate in the range from 3.1 GHz up to 10.6 GHz, satisfying FCC 15.517(b,c) requirements (for indoor applications). The available bandwidth is 7.5 GHz, and the maximum allowable output power is −41.3 dBm/MHz. There are two major UWB technologies: (1) impulse radio UWB (IR-UWB) and (2) multiband OFDM (MB-OFDM). The UWB technologies are fully described in Chap. 10; here we briefly introduce the MB-OFDM concept [72, 73, 75].

The multiband modulation represents approach to modulate information in UWB systems, by splitting the 7.5 GHz band into multiple smaller frequency bands. The key idea is to efficiently utilize UWB spectrum by simultaneously transmitting multiple UWB signals. The MB-OFDM system can be interpreted as a combination of OFDM and frequency hopping. With this approach, the information bits are "spread" across the whole UWB spectrum, exploiting, therefore, frequency diversity and providing the robustness against multipath fading and narrowband interference, originating from Wi-Fi channels. The 7.5 GHz UWB spectrum can be divided into 14 bands, each with bandwidth of $\Delta f = 528$ MHz, as illustrated in Fig. 7.34. The center frequency of the nth band $f_{c,n}$, the low boundary frequency $f_{l,n}$, and the upper boundary frequency $f_{h,n}$ can be determined, respectively, as

$$f_{c,n} = f_0 + n\Delta f, \quad f_{l,n} = f_{c,n} - \Delta f/2, \quad f_{h,n} = f_{c,n} + \Delta f/2, \\ f_0 = 2904 \text{ MHz}. \tag{7.104}$$

The MB-OFDM transmitter configuration is provided in Fig. 7.35. Clearly, the transmitter configuration is very similar to conventional OFDM transmitter, shown in Fig. 7.6a, except from the time-frequency coding block. The time-frequency code (TFC) denotes which band within the band group is occupied at a given transmission interval. As an illustration, the TFCs of length 6 for band group 1 are given as [72]: {[1 2 3 1 2 3], [1 3 2 1 3 2], [1 1 2 2 3 3], [1 1 3 3 2 2], [1 2 1 2 1 2], [1 1 1 2 2 2]}. When TFC code 1 is used, the first OFDM symbol is transmitted on band 1, the second OFDM symbol on band 2, the third OFDM symbol on band 3, the fourth OFDM symbol on band 1, the firth OFDM symbol on band 2, and the sixth OFDM symbol on band 3; and this sequence of bands within the band group is repeated until the TFC is changed in time-frequency coding block.

The binary data bits are scrambled and then encoded employing a convolutional code of rate 1/3, with constraint length $K = 7$, and generating polynomials $g_0 = 133_8$, $g_1 = 165_8$, and $g_2 = 171_8$ (the subscript denotes octal number system). Higher-rate convolutional codes (of rate 11/32, 1/2, 5/8, and 3/4) can be obtained by puncturing. The number of subcarriers used in MB-OFDM is 128, 100 of which are used to transmit QPSK data, 12 are pilots, and 10 are guard tones (the remained subcarriers are set to 0). The pilots are used for channel estimation and carrier-phase tracking on the receiver side. To improve the performance of MB-OFDM scheme, LDPC-coded MB-OFDM can be used instead. It has been shown [76, 77] that the transmission distance over wireless channel of LDPC-coded MB-OFDM can be increased by 29%–73% compared to that of MB-OFDM with convolutional codes. After pilot insertion, serial-to-parallel conversion (S/P), inverse FFT (IFFT) operation, and parallel-to-serial (P/S) conversion, cyclic extension and windowing are performed. After D/A conversion (DAC) and time-frequency coding, the RF signal is transmitted over transmit (Tx) antenna. The transmitted MB-OFDM signal can be written as [5, 73]

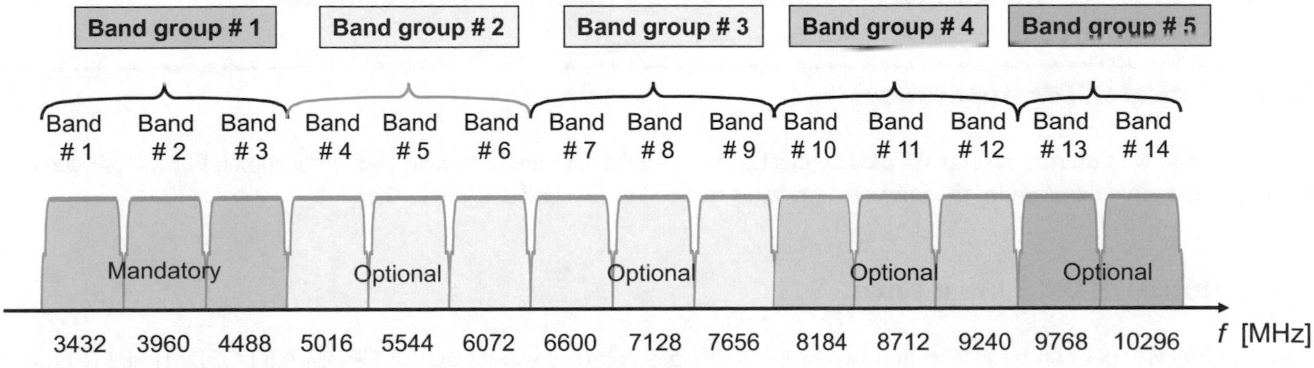

Fig. 7.34 The frequency allocation in MB-OFDM for UWB communications

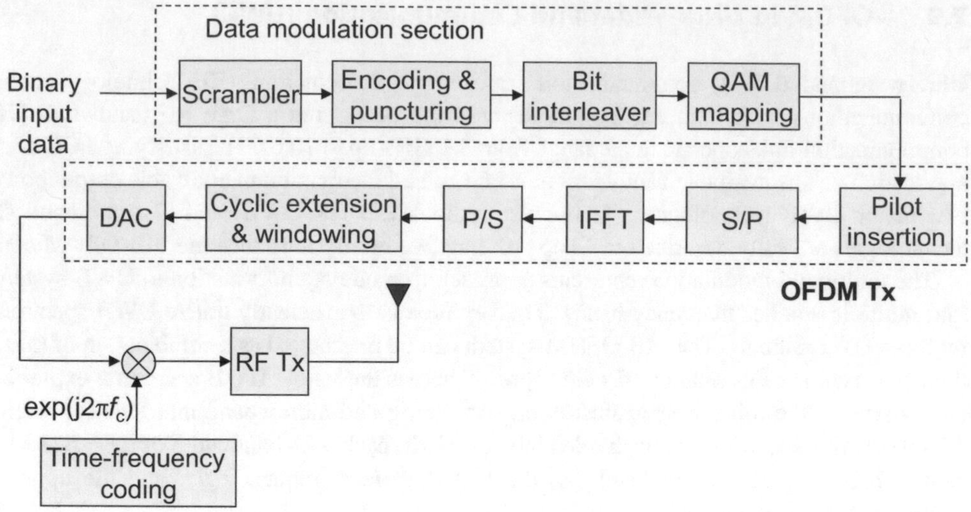

Fig. 7.35 The MB-OFDM transmitter configuration for UWB communications

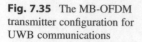

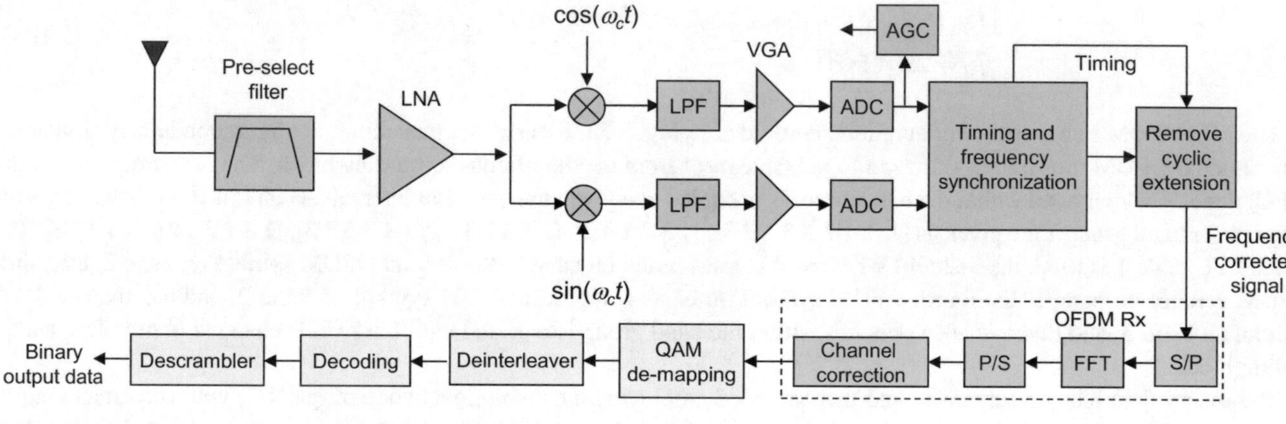

Fig. 7.36 The MB-OFDM receiver configuration for UWB communications

$$s_{\text{MB-OFDM}}(t) = \text{Re}\left\{\sum_{n=0}^{N-1} s_{\text{OFDM,n}}(t - nT_{\text{OFDM}}) \exp\left[j2\pi f_{(n \bmod 6)} t\right]\right\}, \tag{7.105}$$

where N is the number of OFDM symbols transmitted while $s_{\text{OFDM},\,n}(t)$ denotes the nth OFDM symbol of duration T_{OFDM} transmitted on carrier frequency $f_{n \bmod 6}$. As discussed above, the carrier frequency changes over three frequencies assigned to the band group, organized in sequences of length 6 (TFCs). The nth OFDM symbol is generated in similar fashion as we described in Sect. 7.1.2.

The receiver configuration, shown in Fig. 7.36, is very similar to conventional OFDM receiver [Fig. 7.6b], except from pre-select filter that is used to select OFDM signal of 528 MHz bandwidth at a given carrier frequency.

7.10 Optical OFDM Applications

In this section, we describe several selected applications of OFDM relevant in optical communications. For a more detailed and comprehensive description, the interested reader is referred to [5] (and references therein).

7.10.1 Optical OFDM System Types

Various OFDM types can be placed into two broad categories [5]: (i) coherent optical OFDM (CO-OFDM) and (ii) direct detection-based optical OFDM (DDO-OFDM).

7.10.1.1 Coherent Optical OFDM (CO-OFDM) Systems

The basic principles of coherent optical OFDM (CO-OFDM) have already been introduced in Fig. 7.10. Here we provide some additional details. As indicated in [5], the synergy between OFDM and coherent optical detection is twofold: (i) coherent optical detection brings the required linearity of RF-to-optical up-conversion and optical-to-RF down-conversion and (ii) brings the efficiency for channel and phase noise estimation. Given that basic operation principles of CO-OFDM have already been described, here we provide more details of various modules within CO-OFDM system. The basic modules for CO-OFDM systems are (1) RF OFDM transmitter, (2) RF OFDM receiver, (3) RF-to-optical (RF2O) up-converter based on either MZM or optical I/Q modulator, and (4) optical-to-RF (O2RF) down-converter based on coherent optical detection. The details of RF OFDM transmitter and receiver have been provided already in Figs. 7.7 and 7.10, together with the description of operation principles. There are two options for RF-to-optical up-conversion, as indicated in [5], namely, direct RF2O up-conversion and RF2O up-conversion through RF intermediate frequency, with both options shown in Figs. 7.37 and 7.38. The detailed description of operation principles of balanced coherent optical detector, Mach-Zehnder modulator (MZM), and optical I/Q modulator has been provided already in Chap. 3. To perform RF2O up-conversion in direct up-conversion configuration, the optical I/Q modulator is used [Fig. 7.37a]; and to perform O2RF down-conversion, the balanced coherent optical detector is used [Fig. 7.37b]. On the other hand, to implement CO-OFDM transmitter in RF intermediate-frequency-based configuration, RF up-conversion is combined with optical RF-to-optical converter based on single MZM followed by an optical band-pass filter (OBPF), as shown in Fig. 7.38a. To implement CO-OFDM receiver in RF intermediate-frequency-based configuration, the optical-to-RF down-converter, implemented with the help of an OBPF and an optical π-hybrid-based balanced optical detector, is combined with an RF down-converter, as shown in Fig. 7.38b.

The advantages of direct up-configuration are twofold [5]: (1) the required electrical bandwidth in transmitter and receiver is significantly lower, and (2) there is no need for the image rejection filters (BPF and OBPF) in transmitter and receiver. The MZMs in either configuration are operated at null point (see Chap. 3 for additional details). It is also possible to use direct RF-to-optical up-conversion on the transmitter side and RF intermediate-frequency-assisted configuration on the receiver side and vice versa.

By omitting the distortions due to imperfect channel and phase estimations, the receiver sensitivity will be identical to that of single-carrier in ASE noise-dominated scenario. For instance, the BER performance of CO-OFDM with QPSK can be described as

Fig. 7.37 The block diagram of CO-OFDM system in direct optical-to-RF up-conversion configuration. Architectures of (**a**) transmitter and (**b**) receiver for CO-OFDM

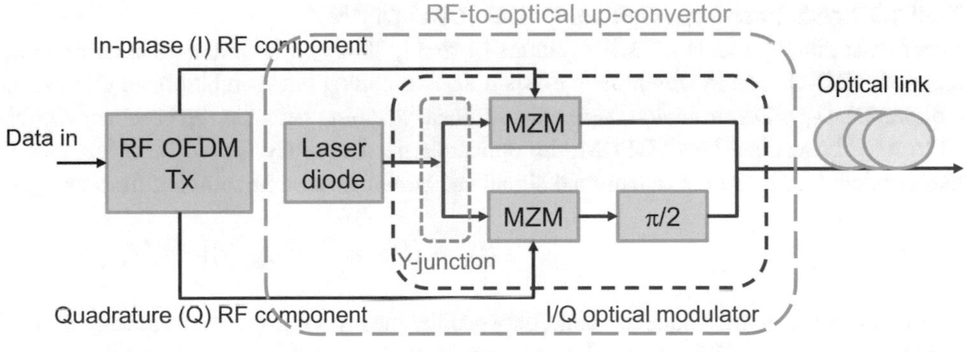

(a)

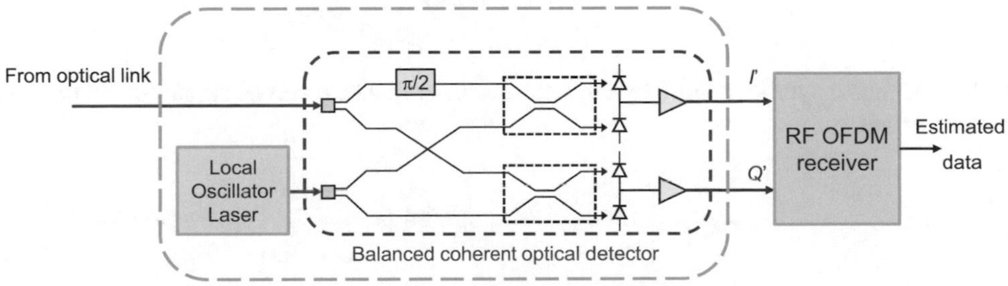

(b)

Fig. 7.38 The block diagram of CO-OFDM system in optical-to-RF up-conversion through RF intermediate frequency. Architectures of (**a**) transmitter and (**b**) receiver for CO-OFDM

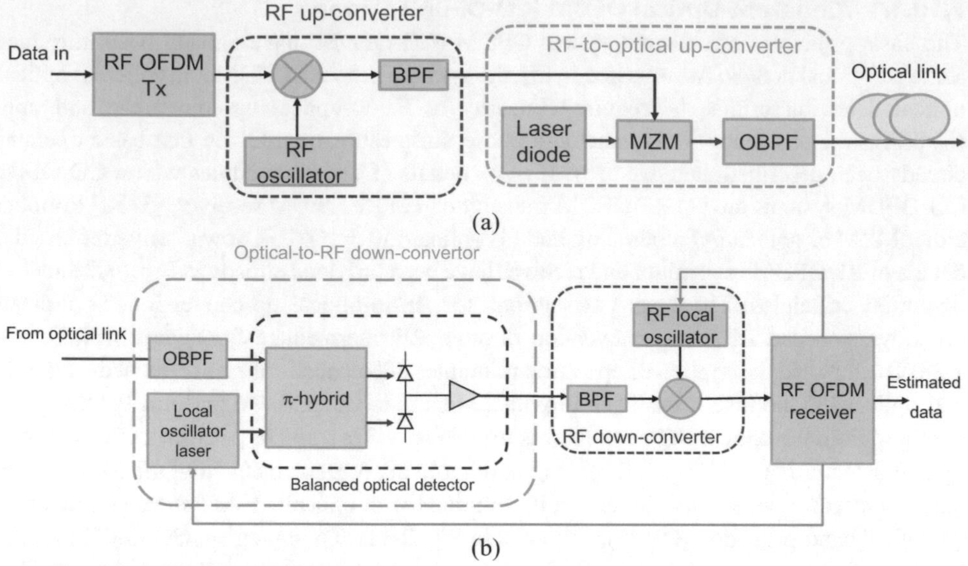

$$BER \simeq \frac{1}{2}\operatorname{erfc}\left(\sqrt{\rho/2}\right), \rho = \frac{\sigma_s^2}{\sigma_{wn}^2} \tag{7.106}$$

where ρ is signal-to-noise ratio, defined as the ratio of the signal variance σ_s^2 and the white noise variance σ_{wn}^2. The symbol SNR (for QPSK) is related to OSNR, number of subcarriers N_{sc}, and subcarrier spacing Δf by [5, 78]

$$\rho \cong 2OSNR\frac{B_{\text{ref}}}{N_{sc}\Delta f}, \tag{7.107}$$

where B_{ref} is the reference bandwidth (12.5 GHz in 0.1 nm at 1550 nm).

7.10.1.2 Direct Detection Optical OFDM (DDO-OFDM)

There exist many variants of DDO systems [5, 8–11, 79–81], which can be classified into broad categories [5]: (i) linearly mapped DDO-OFDM in which there exists a linear mapping between baseband OFDM and optical field and (ii) nonlinearly mapped DDO-OFDM in which there exists a linear mapping between the baseband OFDM and optical intensity instead.

In linearly mapped DDO-OFDM, the optical carrier frequency should be at least the OFDM signal bandwidth from the main optical carrier, so the transmitted signal can be represented in complex from as

$$\underline{s}(t) = \left[1 + ae^{j2\pi\Delta f_G}\underline{s}_{BB}(t)\right]e^{j2\pi f_o t}, \tag{7.108}$$

where Δf_G is the guard frequency band between the main optical carrier frequency f_o and OFDM signal, a is the coefficient related to the ratio of OFDM signal power and the main optical carrier, while the baseband OFDM signal is defined by

$$\underline{s}_{BB}(t) = \sum_{k=-N_{sc}/2+1}^{N_{sc}/2} x_k e^{j2\pi f_k t}, \tag{7.109}$$

with x_k being the symbol carried by the k-th subcarrier f_k. The received signal, in the presence of chromatic dispersion, can be represented as

$$\underline{r}(t) = \left[e^{j\Phi_{CD}(-\Delta f_G)} + ae^{j2\pi\Delta f_G}\sum_{k=-N_{sc}/2+1}^{N_{sc}/2} x_k e^{j2\pi f_k t + j\Phi_{CD}(f_k)}\right]e^{j2\pi f_o t + j\phi(t)} \tag{7.110}$$

where $\Phi_{CD}(f_k)$ is the phase distortion of the k-th subcarrier introduced by chromatic dispersion and $\phi(t)$ is the random phase shift of the main optical carrier. When the second-order GVD can be neglected, the phase distortion is quadratic [5]:

$$\Phi_D(f_k) = \pi c D_{tot}(f_k/f_o)^2, \tag{7.111}$$

where D_{tot} denotes the accumulated chromatic dispersion (expressed in ps/pm) and c is the speed of light. After the PIN photodetector, the photo current $I(t)$ will be proportional to the magnitudes squared of the received optical field:

$$\begin{aligned} I(t) \propto |r(t)|^2 &= \left| e^{j\Phi_{CD}(-\Delta f_G)} + a e^{j2\pi\Delta f_G t} \sum_{k=-N_{sc}/2+1}^{N_{sc}/2} x_k e^{j2\pi f_k t + j\Phi_{CD}(f_k)} \right|^2 \\ &= 1 + 2a\,\mathrm{Re}\left\{ e^{j2\pi\Delta f_G t} \sum_{k=-N_{sc}/2+1}^{N_{sc}/2} x_k e^{j2\pi f_k t + j[\Phi_{CD}(f_k) - \Phi_{CD}(-\Delta f_G)]} \right\} \\ &\quad + |a|^2 \sum_{k_1,k_2=-N_{sc}/2+1}^{N_{sc}/2} x_{k_1} x_{k_2}^* e^{j2\pi(f_{k_1} - f_{k_2}) + j[\Phi_{CD}(f_{k_1}) - \Phi_{CD}(f_{k_2})]} \end{aligned} \tag{7.112}$$

The DC component can easily be removed by DC blocker, while the third-term represents the second-order nonlinearity that must be properly canceled out. The second-term contains the desired OFDM signal. To improve the spectral efficiency, the single-side band (SSB) transmission is advocated in [11, 13]. To remove one of the bands (either the upper or lower band), the optical filtering approach is used in [11, 13]. Alternatively, the Hilbert transform approach for SSB generation can be used at the expense of increased complexity in digital domain on the transmitter side [26, 27]. Namely, the quadrature signal will represent the OFDM baseband signal passed through the Hilbert transform filter, and in passband representation, the quadrature signal will represent an interference canceling the desired upper/lower band of the double-side band (DSB) signal. The quadratic term is then filtered out in electrical domain (after the photodetection takes place). Alternatively, the parameter a in Eq. (7.108) should be chosen sufficiently low so that the quadratic term becomes negligible. It is also possible to estimate the quadratic term in an iterative fashion, by using the estimate linear term to cancel the quadratic term [82]. It is also possible to transmit QAM/PSK symbols only on odd subcarrier; the second-order term will introduce the crosstalk only on even subcarrier, not carrying any information [83]. Each of these approaches to deal with the second-order distortion has advantages and disadvantages as discussed in [5].

As indicated earlier, in nonlinearly mapped DDO-OFDM, we establish the linear mapping between the baseband OFDM signal, defined by (7.109), and optical intensity by employing the direct modulation of the laser diode as follows:

$$\underline{E}(t) = |E(t)| e^{j2\pi f_o t}, |E(t)| = E_0\sqrt{1 + a\,\mathrm{Re}\{e^{j2\pi f_{IF} t} s_{BB}(t)\}}, \tag{7.113}$$

where $\underline{E}(t)$ is the complex representation of laser diode output field and f_{IF} is the RF intermediate frequency (IF). By ignoring the dispersion effects, the PIN photodetector photocurrent would be proportional the OFDM baseband signal:

$$\begin{aligned} I(t) \propto |\underline{E}(t)|^2 &= E_0^2\{1 + a\,\mathrm{Re}\{e^{j2\pi f_{IF} t} s_{BB}(t)\}\} \\ &= E_0^2\left\{1 + a\,\mathrm{Re}\left\{e^{j2\pi f_{IF} t} \sum_{k=-N_{sc}/2+1}^{N_{sc}/2} x_k e^{j2\pi f_k t}\right\}\right\}. \end{aligned} \tag{7.114}$$

For additional details on schemes belonging to this type, the interested reader is referred to [5] (and references therein).

7.10.2 High-Speed Spectrally Efficient CO-OFDM Systems

Among various alternatives considered in the previous section, the CO-OFDM offers better spectral efficiency, receiver sensitivity, and tolerance to chromatic dispersion. By employing polarization states, by either performing the polarization-division multiplexing (PDM) of two independent optical data streams or four-dimensional (4D) signaling, the spectral efficiency can be doubled.

7.10.2.1 Polarization-Division Multiplexed OFDM

In conventional polarization-division multiplexed CO-OFDM scheme, two independent two-dimensional (2D) signal constellation data streams are used as inputs of two RF OFDM transmitters, which is similar to the OFDM transmitter configuration described earlier (see Figs. 7.7 and 7.10). One transmit laser diode is used for both polarizations, with output being split by polarization beam splitter (PBS). Also, two I/Q modulators are used, one for each polarization, as shown in Fig. 7.39(a). The 2D (QAM data streams) are processed in RF OFDM transmitter in the same fashion as described in Sects. 7.4.2, 7.4.3, and 7.5. The outputs of RF OFDM transmitter present the real and imaginary parts of OFDM signal, which are used as I and Q inputs to corresponding I/Q modulator. Upon electro-optical conversion by I/Q modulator, two independent optical signals corresponding to x- and y-polarizations are combined together by polarization beam combiner (PBC) and transmitted over optical transmission system in question. At the receiver side, the incoming optical signal is split into two orthogonal polarizations by PBS, which is also done with the signal coming from local laser oscillator. The x-polarization outputs of PBSs are used as inputs to the upper branch of the balanced coherent detector, while the y-polarization outputs are used as inputs to the down-branch of the balanced coherent detector, as shown in Fig. 7.39b. The configuration of coherent balanced detector has been provided already in Sec. 7.10.1.1 (see also Chap. 3). The outputs of balanced coherent detectors represent estimates of real and imaginary parts of corresponding RF OFDM receiver, whose configuration is shown in Figs. 7.7 and 7.10. After FFT demodulations in x- and y-polarization branches, chromatic dispersion compensation, PMD/PDL compensation, and symbol detection have been performed. In the presence of polarization-dependent loss (PDL), we also need to perform the matrix inversion by using the parallel decomposition (see Fig. 7.8 and corresponding text related to it). This operation is not computationally extensive given the fact that the channel matrix size is 2 × 2.

Another alternative is to use *polarization-time (PT) coding*. One such scheme is shown in Fig. 7.40. The scheme, which is based on Alamouti code described later in Chap. 8 (see also [56]), was proposed in [57] (see also ref. [5]). The operations of all blocks on the transmitter side, except that of the PT encoder, are similar to those shown in Fig. 7.39. The functional blocks at the receiver side are identical to the blocks shown in Fig. 7.39, except from symbol detection that is based on the Alamouti-type combiner. The operation principle of the encoder from Fig. 7.40 can be described as follows: (i) in the first half of *i*-th time instance ("the first channel use"), symbol s_x is sent by using x-polarization channel, while symbol s_y is sent by using y-polarization channel; (ii) in the second half of *i*-th time instance ("the second channel use"), symbol $-s^*_y$ is sent by using x-polarization channel, while symbol s^*_x is sent by using y-polarization.

The received symbol vectors for the first and second channel use can be written as

$$\boldsymbol{r}_{i,k}^{(m)} = \boldsymbol{U}(k)\boldsymbol{s}_{i,k}^{(m)} e^{\,j(\phi_T-\phi_{LO})} + \boldsymbol{z}_{i,k}^{(m)}, \quad m = 1,2 \tag{7.115}$$

where the Jones (channel) matrix $\boldsymbol{U}(k)$ has been already introduced by Eqs. (7.18) and (7.42) (we use again index k to denote the frequency ω_k of k-th subcarrier), $\boldsymbol{r}_{i,k}^{(m)} = \left[r_{x,i,k}^{(m)}, r_{y,i,k}^{(m)}\right]^T$ denotes the received symbol vector in the m-th ($m = 1,2$) channel use of i-th OFDM symbol and k-th subcarrier, while $z_{i,k}^{(m)} = \left[z_{x,i,k}^{(m)}, n_{y,i,k}^{(m)}\right]^T$ denotes the corresponding noise vector. We also use $s_{i,k}^{(1)} = \left[s_{x,i,k}, s_{y,i,k}\right]^T$ to denote the symbol transmitted in the first channel use and $s_{i,k}^{(2)} = \left[-s_{y,i,k}^*, s_{x,i,k}^*\right]^T$ to denote the symbol transmitted in the second channel use (of the same symbol interval). Because the symbol vectors transmitted in the first and the second channel use of i-th time instance are orthogonal, Eq. (7.115) can be rewritten by grouping separately x- and y-polarizations, so we have that

$$\begin{bmatrix} r_{x,i,k}^{(1)} \\ r*_{x,i,k}^{(2)} \end{bmatrix} = \begin{bmatrix} U_{xx}e^{\,j\phi_{PN}} & U_{xy}e^{\,j\phi_{PN}} \\ U_{xy}^{*}e^{-j\phi_{PN}} & -U_{xx}^{*}e^{-j\phi_{PN}} \end{bmatrix} \begin{bmatrix} s_{x,i,k} \\ s_{y,i,k} \end{bmatrix} + \begin{bmatrix} z_{x,i,k}^{(1)} \\ z_{x,i,k}^{*(2)} \end{bmatrix}, \tag{7.116}$$

$$\begin{bmatrix} r_{y,i,k}^{(1)} \\ r_{y,i,k}^{*(2)} \end{bmatrix} = \begin{bmatrix} U_{yx}e^{\,j\phi_{PN}} & U_{yy}e^{\,j\phi_{PN}} \\ U_{yy}^{*}e^{-j\phi_{PN}} & -U_{yx}^{*}e^{-j\phi_{PN}} \end{bmatrix} \begin{bmatrix} s_{x,i,k} \\ s_{y,i,k} \end{bmatrix} + \begin{bmatrix} z_{y,i,k}^{(1)} \\ z_{y,i,k}^{*(2)} \end{bmatrix}. \tag{7.117}$$

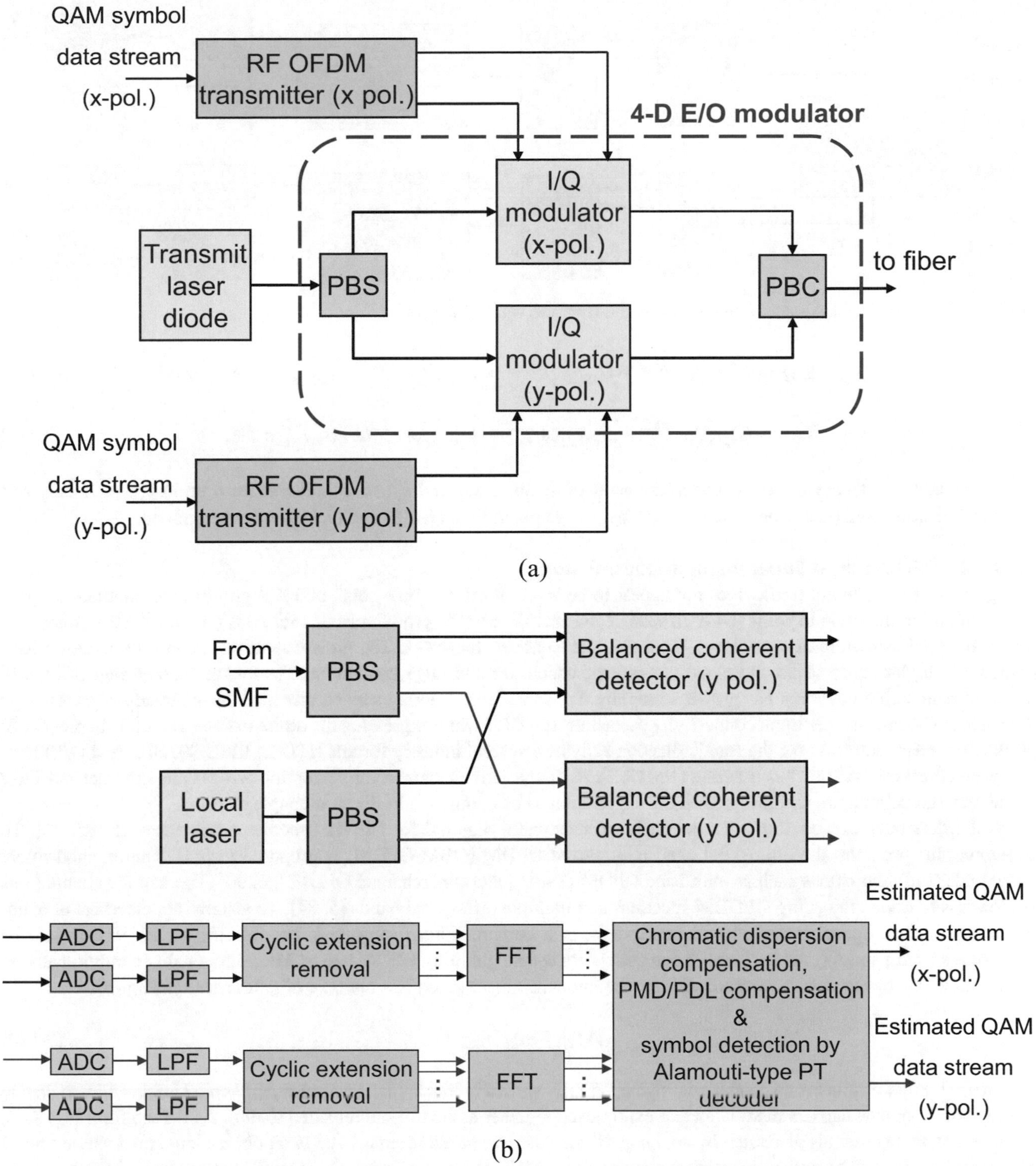

Fig. 7.39 Polarization-division multiplexed CO-OFDM scheme: (**a**) transmitter architecture and (**b**) receiver architecture. *PBS(C)* polarization beam splitter (combiner)

where $\phi_{PN} = \phi_T - \phi_{LO}$. If only one polarization is used, we could solve either Eq. (7.116) or Eq. (7.117). However, the use of only one polarization would result in 3 dB penalty with respect to the case when both polarizations are employed. Following the derivation similar to that described in the previous section, it can be shown that the estimates of transmitted symbols at the output of PT decoder (for ASE noise-dominated scenario, neglecting fiber nonlinearities) can be obtained as

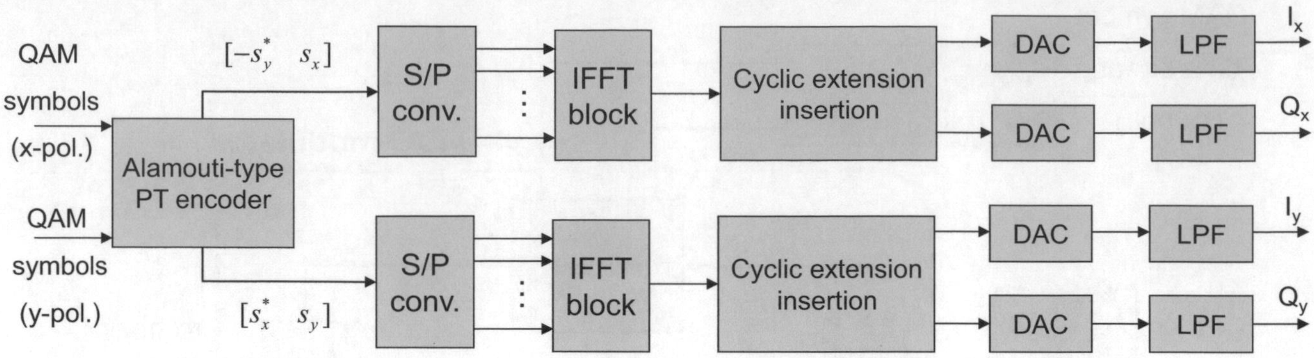

Fig. 7.40 The transmitter architecture of PT coding scheme combined with CO-OFDM

$$\widetilde{s}_{x,i,k} = U_{xx}^* r_{x,i,k}^{(1)} e^{-j\phi_{PN}} + U_{xy} r_{x,i,k}^{*(2)} e^{j\phi_{PN}} + U_{yx}^* r_{y,i,k}^{(1)} e^{-j\phi_{PN}} + U_{yy} r_{y,i,k}^{*(2)} e^{j\phi_{PN}}, \tag{7.118}$$

$$\widetilde{s}_{y,i,k} = U_{xy}^* r_{x,i,k}^{(1)} e^{-j\phi_{PN}} - U_{xx} r_{x,i,k}^{*(2)} e^{j\phi_{PN}} + U_{yy}^* r_{y,i,k}^{(1)} e^{-j\phi_{PN}} - U_{yx} r_{y,i,k}^{*(2)} e^{j\phi_{PN}}, \tag{7.119}$$

where $\widetilde{s}_{x,i}$ and $\widetilde{s}_{y,i}$ denote the PT decoder estimates of symbols $s_{x,i}$ and $s_{y,i}$ transmitted in ith time instance. If only one polarization were used (i.e., x-polarization), the last two terms in Eqs. (7.118) and (7.119) must be omitted.

7.10.2.2 OFDM-Based Superchannel Transmission

The growth of the Internet traffic does not appear to be leveling off any time soon, and it is projected to continue to grow exponentially in the years to come [84–90]. This exponential-like traffic growth places enormous pressure on the underlying information infrastructure at every level, from the core to access networks. The growth is driven mainly by new services, particularly high-quality video and cloud computing, which demand huge bandwidths. As for the market demand, the IP traffic in nationwide networks has grown so rapidly that some large Internet service providers have already reported huge demands in router-to-router trunk connectivity exceeding 100 Gb/s while expressing the desire to have Terabit Ethernet (TbE) ports in the near future. As for the standardization activities, several industry forums (ITU-T, IEEE 802.3ba, and OIF) have completed the work on 100 Gb/s Ethernet (100 GbE), while the activities for an early adoption of 400 Gb/s Ethernet and 1 Tb/s Ethernet and beyond are underway, with the expectation to be completed in the next couple of years.

As the data rates exceed 100 Gb/s, the electrical bandwidth required for OFDM becomes a bottleneck [5, 85, 86]. To overcome this problem, the orthogonal band multiplexed OFDM (OBM-OFDM) is proposed in [84]. This method might come under different names such as multiband OFDM [88, 89] and superchannel OFDM [86, 90]. The key idea behind this approach is to divide the entire OFODM spectrum into multiple orthogonal bands [5, 84]. To achieve so, electrical domain-based approaches, optical domain-based approaches, or a combination of the two is possible [5, 84, 91–93]. Let Δf be subcarrier spacing and Δf_G the frequency band guard between neighboring OFDM bands. The orthogonality condition among different OFDM bands would be satisfied when the frequency band guard is a multiple of subcarrier spacing [5]:

$$\Delta f_G = n\Delta f, n \geq 1. \tag{7.120}$$

A generic superchannel transponder employing WDM is shown in Fig. 7.41. The term *superchannel* is used here to denote a collection of optical carriers modulated and multiplexed together at the transmitter side [86, 90]. As we see from Fig. 7.41, the superchannel content is generated by applying 4D signals (two phase channels and two polarizations) on a single optical carrier (or better to say an optical subcarrier) by using 4D modulator, being composed of two I/Q modulators, one polarization beam splitter and one polarization beam combiner, as shown in Fig. 7.39a. This 4D modulator can be used either for PDM of independent OFDM data streams or fully 4 D signaling. The OFDM signals of all optical subcarriers within the superchannel are spectrally multiplexed, in an all-optical OFDM fashion, to create the superchannel. The superchannels are then spectrally multiplexed together to create even larger entities, called here *superchannel band groups*. Finally, the superchannel band groups are further multiplexed together by employing a frequency-locked WDM multiplexer.

The process at the receiving side is just opposite, and it is composed of three-step spectral demultiplexing, which all results in the selection of individual optical subcarriers loaded with 4D electrical signals. It is also possible to use the anti-aliasing

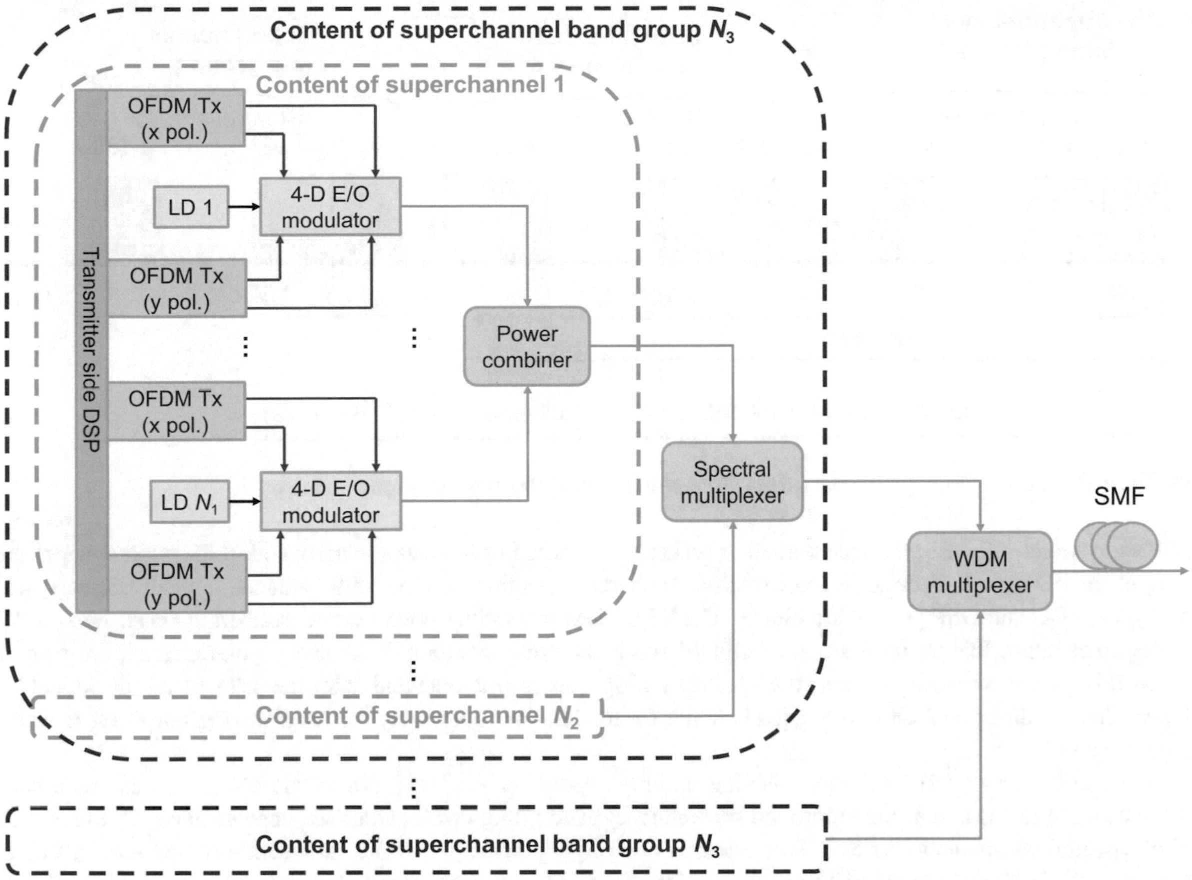

(a)

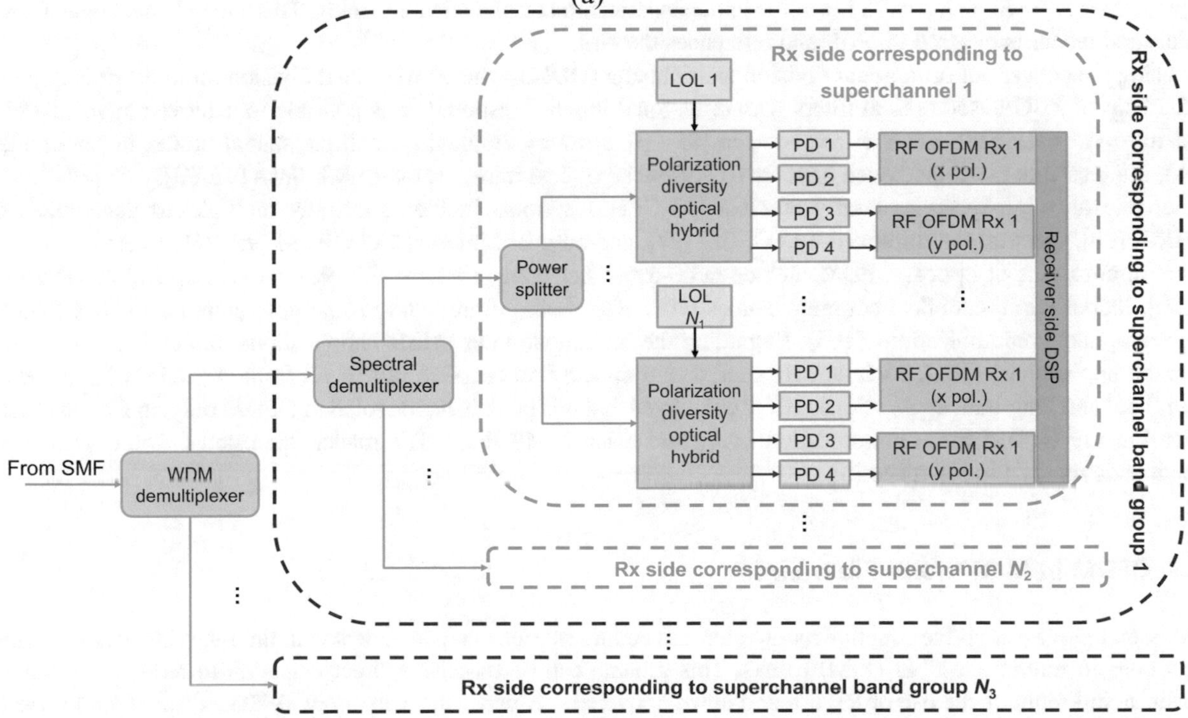

(b)

Fig. 7.41 The generic superchannel OBM-OFDM enabling ultrahigh-speed optical transport: (**a**) transmitter side and (**b**) receiver side architectures. N_1 denotes the number of frequency-locked laser signal within the superchannel, N_2 denotes the number of superchannels within the superchannel band group, and N3 denotes the number of superchannel band groups. *LD* laser diode, *LOL* local oscillator laser, *PD* photodetector

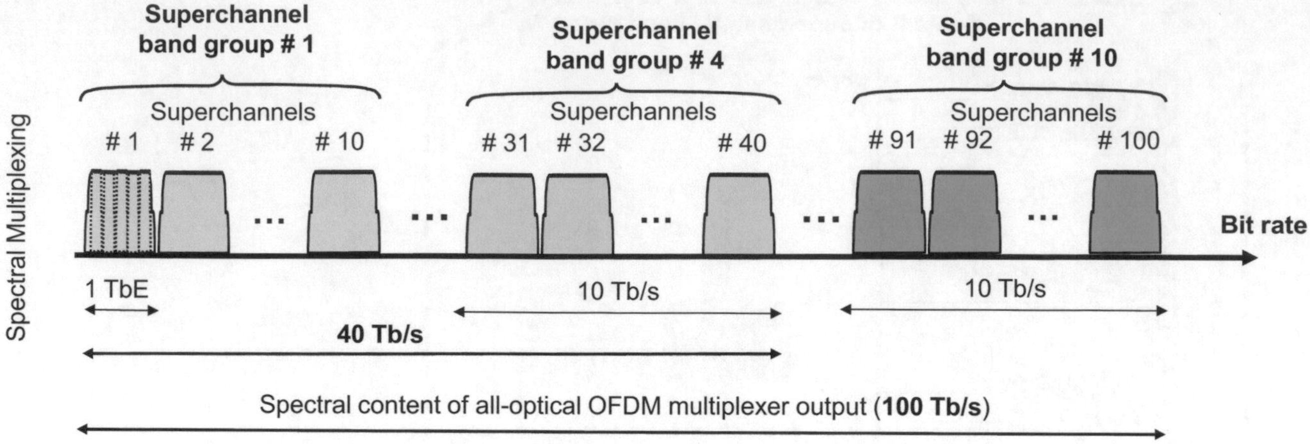

Fig. 7.42 The digital hierarchy of superchannel OBM-OFDM enabling up to 100 Tb/s serial optical transport

optical filters to select the desired superchannel as well as the desired optical supercarrier within the superchannel. Further processing of the 4D signals is done in a conventional coherent detection scheme with balanced optical receivers, which is followed by the ADC and DSP processing blocks. The RF OFDM transmitter operates as described in Sects. 7.4.2, 7.4.3, and 7.5. On the other hand, the configurations of OFDM receivers, corresponding to x- and y-polarizations, are provided in Fig. 7.7. In this particular configuration, the spectral multiplexer is implemented with the help of an all-optical OFDM multiplexer. Clearly, all-optical OFDM is used here not for modulation, but purely for multiplexing purposes, to reduce the overall system cost.

Now we describe the digital hierarchy enabling ultrahigh-speed optical transport. In the illustrative example shown in Fig. 7.42, the signal frame is organized into ten superchannel band groups with center frequencies being orthogonal to each other. Each spectral component carries 1 Tb/s Ethernet (1 TbE) superchannel, while each superchannel band group carries 10 TbE traffic. With 10 superchannel band groups, 100 Tb/s serial optical transport is possible over SMF links. In this example, we set $N_1 = N_2 = N_3 = 10$. For some initial experimental results related to multi-Tb/s superchannel optical transport, the interested reader is referred to [90] (and references therein).

By using the concept of orthogonal division multiplexing (ODM), where orthogonal division multiplexer is implemented with the help of FBG-based optical filters with orthogonal impulse response, it is possible to achieve beyond 1 Pb/s serial optical transport over SMF links as proposed in [91–93]. Now by employing multiple spatial modes in few-mode fibers (FMFs), it is possible to achieve even beyond 10 Pb/s serial optical transport over FMF links [92, 93].

Other relevant high-speed spectrally efficient CO-OFDM systems include spectrally multiplexed generalized OFDM (GOFDM) [88], spectrally multiplexed 4-D OFDM [94], and no-guard interval CO-OFDM [95, 96] systems.

Other applications of optical OFDM include radio-over-fiber (RoF) systems [97, 98], passive optical networks (PONs) [99, 100], indoor wireless optical communications [101, 102], both infrared and visible light communications (VLCs), and free-space optical communications [103]. Regarding the multimode fiber (MMF) applications, the OFDM can be used for short-reach applications [104], such as data centers and optical broadcasting in MIMO fashion [105, 106], as well as for medium-haul and long-haul applications [107, 108]. Given that this book is not devoted to OFDM only, in the next section we describe the use OFDM for communication over short-reach MMF links. The reader interested on other optical OFDM applications is referred to reference [5].

7.10.3 OFDM in Multimode Fiber Links

OFDM as technique can utilize adaptive modulation and coding as a mechanism to deal with time-varying channel conditions, such as ones in multimode fiber (MMF) links. This scheme can be based on direct detection to reduce the system cost. However, it still requires the use of RF up- and down-converters. A particular version of OFDM, discrete multitone (DMT) modulation, is considered as a cost-effective solution for transmission over multimode fiber links in LAN applications [18]. Although the same sequence is transmitted twice if DMT modulation is applied, which results in reduced data rate, DMT offers the following two advantages as compared to traditional single-carrier approaches: (i) it has the ability to maximize the

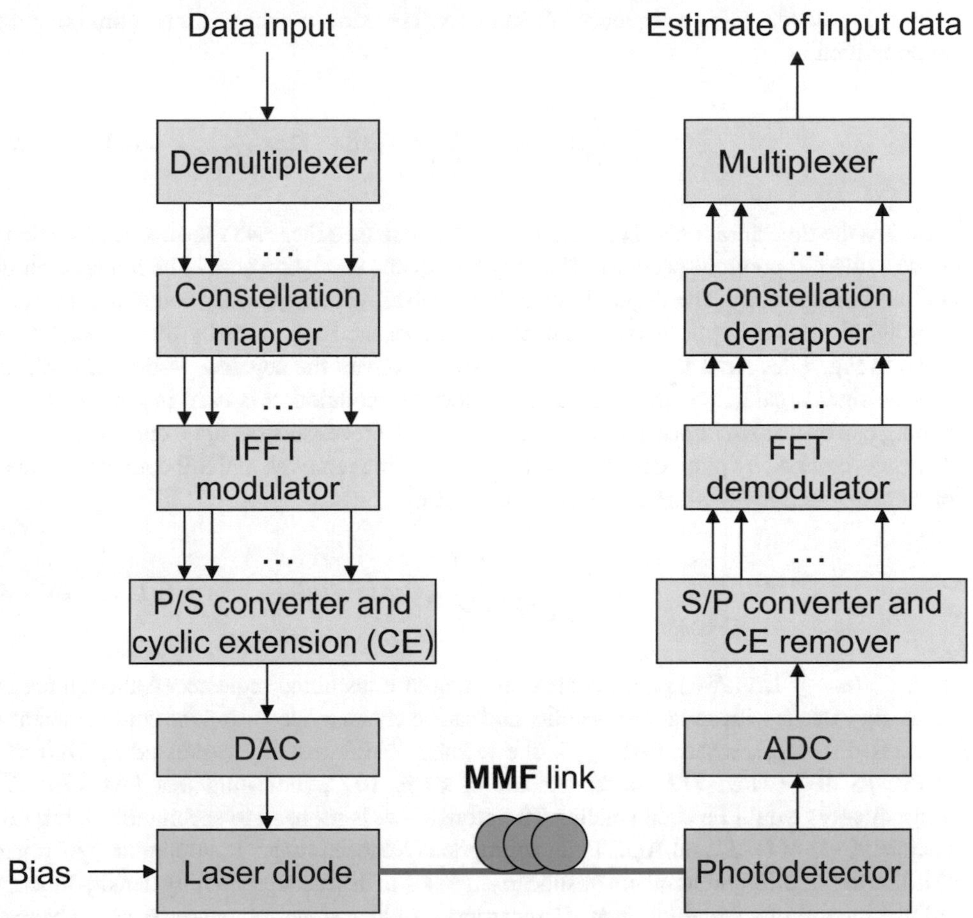

Fig. 7.43 The DMT system for transmission over MMF. *P/S* parallel-to-serial, *S/P* serial-to-parallel, *DAC* digital-to-analog converter, *ADC* analog-to-digital converter

information rate by tailoring the information-bearing distribution across the channel in accordance with the channel conditions; and (ii) it can adapt to time-varying channel conditions, which is achieved by the virtue that original data stream is divided among many subcarriers, while the subcarriers affected by the channel conditions can be avoided through adaptive modulation. The block diagram of DMT system for transmission over MMF links is shown in Fig. 7.43. The demultiplexer converts incoming information data stream into parallel form. Constellation mapper maps parallel data into N subcarriers using QAM. The modulation is performed by applying the IFFT, which transforms frequency-domain parallel data into time-domain parallel data. The complex-valued time-domain sequence s_k after IFFT can be expressed as

$$s_k = \frac{1}{\sqrt{N}} \sum_{n=0}^{N-1} C_n \exp\left(j2\pi k \frac{n}{N} \right); \; k = 0, 1, \ldots, N-1 \tag{7.121}$$

where C_n ($n = 0,1,\ldots,N$-1) is frequency-domain input QAM sequence of symbols. In DMT, the time-domain sequence is real-valued, which is achieved by employing 2 N-IFFT instead, with input values satisfying the Hermitian symmetry property [6]:

$$C_{2N-n} = C_n^*; \; n = 1, 2, \ldots, N-1; \; \mathrm{Im}\{C_0\} = \mathrm{Im}\{C_N\} = 0 \tag{7.122}$$

meaning that the second half of input sequence (going to the IFFT block) is complex conjugate of the first half. In addition, 0-th and N-th subcarriers must be real-valued as given in Eq. (7.122). Therefore, the 2 N-point IFFT in DMT is obtained as

$$s_k = \frac{1}{\sqrt{2N}} \sum_{n=0}^{2N-1} C_n \exp\left(j2\pi k \frac{n}{2N} \right); \; k = 0, 1, \ldots, 2N-1 \tag{7.123}$$

where s_k is now real-valued sequence of length 2 N. The corresponding, discrete-time signal upon parallel-to-serial conversion can be written as

$$s\left(k\frac{T}{2N}\right)=\frac{1}{\sqrt{2N}}\sum_{n=0}^{2N-1}C_n\exp\left(j2\pi n\frac{kT}{2N}\right);\ k=0,1,\ldots,2N-1 \tag{7.124}$$

where T is the time duration of DMT frame. In P/S block (see Fig. 7.43), the cyclic extension is performed in similar fashion as one described in previous sections. The length of cyclic extension should be longer than the maximum pulse spread due to multimode dispersion. The digital-to-analog converter, which typically contains a transmit filter, performs the conversion from digital into analog domain. Direct modulation of the laser diode by DMT signal is used to reduce the system cost, as shown in Fig. 7.43. Also, the bias voltage is used to convert the negative portion of DMT signal into a positive one since the negative signals cannot be transmitted if an intensity modulation is used in combination with direct detection. The signal clipping can be used to improve power efficiency. At the receiver side, upon optical-to-electrical conversion by photodetector, DC bias is blocked by photodetection, while ADC, cyclic removal, and S/P conversion have been performed. The demodulation is performed by 2 N-point FFT with the output

$$\widehat{C}_n=\frac{1}{\sqrt{2N}}\sum_{k=0}^{2N-1}c_k\exp\left(-j2\pi k\frac{n}{2N}\right);\ n=0,1,\ldots,2N-1 \tag{7.125}$$

where $\widehat{C}_n$ ($n = 0,1,\ldots,N$-1) presents the estimation of transmitted sequence. Although the DMT frames can have high PAPR values, they occur with certain probability and can be characterized using the complementary cumulative distribution function Pr(PAPR>PAPR$_{ref}$), where PAPR$_{ref}$ is the referent PAPR (usually expressed in dBs). It has been shown in [18] that Pr (PAPR>15 dB) (when 512 subcarriers are used) is 10^{-4}, indicating that Pr(PAPR>27 dB) (when all subcarriers add constructively) would be even smaller. Therefore, there is no need to accommodate the full dynamic range. By limiting the dynamic range of DAC and ADC to an appropriately chosen range, the optimum performance can be obtained, as shown in [42]. The use of different numbers of subcarriers leads to different probability density functions of PAPR. It has been shown in [18] that the use of a larger number of subcarriers provides better robustness to poor channel response by allowing for longer guard intervals, for the same overhead.

The adjustment of the dynamic range of DAC and ADC can simply be achieved by clipping, as given below

$$s_{\mathrm{clipped}}(t)=\begin{cases}s_{\mathrm{DMT}}(t), & |s_{\mathrm{DMT}}(t)|\le A\\ 0, & \text{otherwise}\end{cases} \tag{7.126}$$

where $s_{\mathrm{clipped}}(t)$ denotes the clipped DMT signal while A denotes the maximum tolerable amplitude before clipping. We can determine the optimum clipping level that minimizes the BER if we measure the BER against the clipping level C_L, based on Eq. (). As an example, it was shown in [18] that for 256 subcarriers and 64-QAM, the optimum clipping ratio for different ADC/DAC resolutions is between 8.5 and 9.5 dB. On the other side, the clipping introduces distortion to DMT signals. It is possible to reduce the PAPR by using distortionless PAPR reduction [43–45]. Another approach to reduce the PAPR would be through coding [46]. Some of these distortionless methods have complexity significantly higher than ones allowing clipping. This is the reason why medium-level complexity methods, such as selective mapping [43, 44], have been proposed in [18]. The selective mapping is based on dual use of real and imaginary parts of a complex FFT in a procedure described below. Let us first observe the complex sequence D_n ($n = 0,1,\ldots,N$-1) of length N. Next, assume the sequence of complex numbers of length 2 N satisfies the symmetry property

$$D_{2N-n}=-D_n^*;\ n=1,2,\ldots,N-1;\ \mathrm{Im}\{D_0\}=\mathrm{Im}\{D_N\}=0 \tag{7.127}$$

The sequence given by Eq. (7.127) has purely imaginary values for the 2 N-point IFFT. The IFFT of the sequence $X_n = C_n + jD_n$ (C_n is the transmitted sequence as defined above) is now given as

$$\begin{aligned} s_k &= \frac{1}{\sqrt{2N}} \sum_{n=0}^{2N-1} X_n \exp\left(j2\pi k \frac{n}{2N}\right); \; k = 0, 1, \ldots, 2N-1 \\ &= \frac{1}{\sqrt{2N}} \left[\sum_{n=0}^{2N-1} C_n \exp\left(j2\pi k \frac{n}{2N}\right) + j \sum_{n=0}^{2N-1} D_n \exp\left(j2\pi k \frac{n}{2N}\right) \right] \\ &= \mathrm{Re}\{s_k\} + j\mathrm{Im}\{s_k\} \end{aligned} \tag{7.128}$$

We have just created a two-channel input-output modulator with a single 2 N-point IFFT. By using a gray mapping for C_n, anti-gray mapping for D_n, and applying the same input sequence $D_n = C_n$, we will get two DMT frames with different PAPR values. We can then transmit the DMT frame with smaller PAPR value. Receiver has to know which sequence was used at the transmitter side. It can be shown that the same probability Pr(PAPR>PAPR$_{\mathrm{ref}}$) of 10^{-4} is now obtained for PAPR$_{\mathrm{ref}} \sim 13.2$ dB (for 512 subcarriers and 64-QAM) [18].

The capacity of the DMT system, having N independent subcarriers with bandwidth B_N and subcarrier gain $\{g_i, i = 0, \ldots, N-1\}$, can be expressed as

$$C = \max_{P_i: \sum P_i = P} \sum_{i=0}^{N-1} B_N \log_2\left(1 + \frac{g_i(P_i + P_{\mathrm{bias}})}{N_0 B_N}\right), \tag{7.129}$$

where P_i is the optical power allocated to i-th subcarrier, N_0 is the power spectral density of transimpedance amplifier, while P_{bias} represents the corresponding power used to transmit the bias. The signal bias is used to convert the negative values of DMT signal into positive ones. The subcarrier gain g_i can be calculated as $g_i = R^2|H_i|^2$, with R being the photodiode responsivity and H_i being the MMF transfer function amplitude of the i-th subcarrier. Since the direct detection is used, Eq. (7.129) represents essentially a lower bound on channel capacity. Nevertheless, it can be used as an initial figure of merit. By using the Lagrangian method, it can be shown that [5]

$$\frac{P_i + P_{\mathrm{bias}}}{P} = \begin{cases} 1/\rho_c - 1/\rho_i, \rho_i \geq \rho_c \\ 0, \text{ otherwise} \end{cases}, \quad \rho_i = g_i(P + P_{\mathrm{bias}})/N_0 B_N \tag{7.130}$$

where ρ_i is the signal-to-noise ratio of i-th subcarrier while γ_c is the threshold SNR. By substituting Eq. (7.128) into Eq. (7.127), the following expression for channel capacity can be obtained:

$$C = \sum_{i:\rho_i > \rho_c} B_N \log_2\left(\frac{\rho_i}{\rho_c}\right). \tag{7.131}$$

The i-th subcarrier is used when the corresponding SNR is above the threshold. The number of bits per i-th subcarrier is determined by $m_i = \lfloor B_N \log_2(\rho_i/\rho_c) \rfloor$, where $\lfloor \cdot \rfloor$ denotes the largest integer smaller than the enclosed number. Therefore, the signal constellation size and power per subcarrier are determined based on MMF channel coefficients. When subcarrier's SNR is high, larger constellation sizes are used, and power per subcarrier is chosen in accordance with Eq. (7.130); smaller constellation sizes are used when the subcarrier's SNR is low; and nothing is transmitted when the subcarrier SNR falls below a certain threshold value. An example of adaptive QAM is shown in Fig. 7.44.

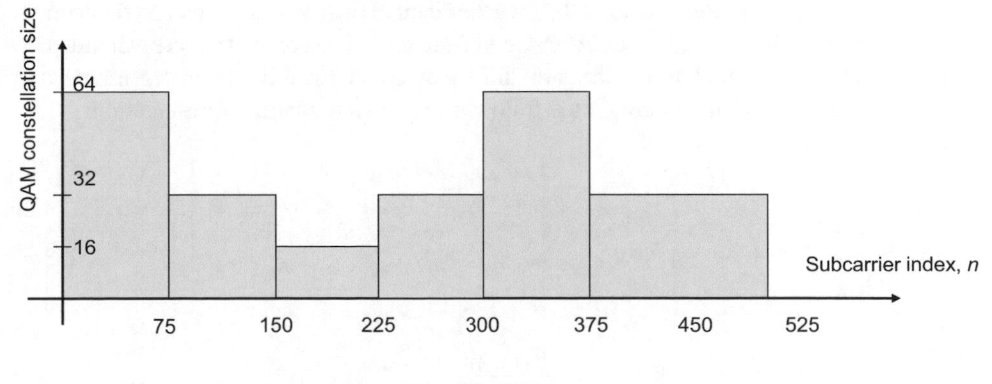

Fig. 7.44 Illustration of adaptive QAM-based subcarrier mapping

7.11 Concluding Remarks

The main subject of this chapter has been on the OFDM fundamentals and OFDM application to wireless and optical communications. The chapter starts with Sect. 7.1, where the basics of OFDM, including the generation of OFDM signal by inverse FFT, have been described. After that, in Sect. 7.2, several basic manipulations on OFDM symbols, required to deal with multipath effects in wireless communications and dispersion effects in fiber-optics communications, including guard time and cyclic extension as well as the windowing operation, have been described. Next, the bandwidth efficiency of OFDM has been discussed in Sect. 7.3. The next topic in the chapter, Sect. 7.4, has been devoted to the OFDM system design, description of basic OFDM blocks, and OFDM-based parallel channel decomposition, including multipath fading radio and dispersion-dominated channel decomposition. In Sect. 7.5, following the description of CO-OFDM principles (Sect. 7.5.1), the basic estimation and compensation steps, common to both wireless and optical communication systems, have been described including DFT windowing (Sect. 7.5.2), frequency synchronization (Sect. 7.5.3), phase estimation (Sect. 7.5.4), and channel estimation (Sect. 7.5.5). Regarding the channel estimation, both pilot-aided and data-aided channel estimation techniques have been described. Following the differential detection in OFDM description (Sect. 7.6), the focus has moved in Sect. 7.7 on various applications of OFDM in wireless communications including DAB (Sect. 7.7.1) and DVB (Sect. 7.7.2). Wi-Fi, LTE, and WiMAX have been described in Sect. 7.8, while the OFDM application in UWB has been described in Sect. 7.9. Next, the optical OFDM applications have been discussed in Sect. 7.10, starting with description of basic OFDM system types, Sect. 7.10.1, such as CO-OFDM systems (Sect. 7.10.1.1) and DDO-OFDM systems (Sect. 7.10.1.2). Regarding the high-speed spectrally efficient CO-OFDM systems, described in Sect. 7.10.2, the polarization-division multiplexed OFDM systems (Sect. 7.10.2.1) and OFDM-based superchannel optical transmission systems (Sect. 7.10.2) have been described. Section 7.10.3 has been devoted to the application of OFDM in multimode fiber links. In Sect. 7.11 concluding remarks have been provided. The set of problems is provided in incoming section.

7.12 Problems

1. Design the OFDM system for wireless communications with the required bit rate of 20 Mbps, tolerable delay spread of 200 ns, and bandwidth occupied being less than 16 MHz. Discuss different options. Determine corresponding bandwidth efficiencies.
2. OFDM is the basis for the 5 GHz standard selected by the IEEE 802.11 standardization group in July 1998 to support data rates ranging from 6 to 54 Mbps. Describe the corresponding OFDM system design. Which modulation formats and coding rates are used for 6 and 54 Mbps?
3. An alternative description of OFDM is based on matrix representation:

$$\underbrace{\begin{bmatrix} y_{N-1} \\ y_{N-2} \\ \vdots \\ y_0 \end{bmatrix}}_{\boldsymbol{y}} = \underbrace{\begin{bmatrix} h_0 & h_1 & \cdots & h_m & 0 & \cdots & 0 \\ 0 & h_0 & \cdots & h_{m-1} & h_m & \cdots & 0 \\ \vdots & \vdots & \ddots & \ddots & \ddots & \ddots & \vdots \\ 0 & \cdots & 0 & h_0 & \cdots & h_{m-1} & h_m \end{bmatrix}}_{\boldsymbol{H}} \begin{bmatrix} x_{N-1} \\ \vdots \\ x_0 \\ x_{-1} \\ \vdots \\ x_{-m} \end{bmatrix} + \underbrace{\begin{bmatrix} z_{N-1} \\ z_{N-2} \\ \vdots \\ z_0 \end{bmatrix}}_{\boldsymbol{z}},$$

where the last m samples from signal vector $\boldsymbol{x} = [x_{N-1} \ldots x_0]^T$ correspond to the cyclic prefix: $x_{-1} = x_{N-1}, x_{-2} = x_{N-2}, \ldots, x_{-\mu} = x_{N-m}$. In the equation above, $\boldsymbol{H}$ denotes the channel matrix, $\boldsymbol{y}$ is the received vector, and $\mathbf{z}$ is the noise vector. Often, the received symbols $y_{-1}, \ldots, y_{-\mu}$ get disregarded from consideration on the receiver side because they are not really needed to recover the input signal and at the same time they are affected by the intersymbol interference (ISI). By re-arranging the above matrix equation, we obtain the following equivalent matrix representation:

$$\begin{bmatrix} y_{N-1} \\ y_{N-2} \\ \vdots \\ y_0 \end{bmatrix} = \underbrace{\begin{bmatrix} h_0 & h_1 & \cdots & h_m & 0 & \cdots & 0 \\ 0 & h_0 & \cdots & h_{m-1} & h_m & \cdots & 0 \\ \vdots & \vdots & \ddots & \ddots & \ddots & \ddots & \vdots \\ 0 & \cdots & 0 & h_0 & \cdots & h_{m-1} & h_\mu \\ \vdots & \vdots & \ddots & \ddots & \ddots & \ddots & \vdots \\ h_2 & h_3 & \cdots & hm & \cdots & h_0 & h_1 \\ h_1 & h_2 & \cdots & h_{m-1} & \cdots & 0 & h_0 \end{bmatrix}}_{\tilde{\boldsymbol{H}}} \begin{bmatrix} x_{N-1} \\ x_{N-1} \\ \vdots \\ x_0 \end{bmatrix} + \begin{bmatrix} z_{N-1} \\ z_{N-2} \\ \vdots \\ z_0 \end{bmatrix}$$

Clearly, the inserted cyclic prefix allows us to model the channel as a circulant convolution matrix over the N samples of interest. By using the corresponding eigenvalue decomposition of $\widetilde{\boldsymbol{H}}$, describe how the inter-carrier interference (ICI) between subcarriers can be compensated for.

4. By using the singular value decomposition (SVM) of the channel matrix $\boldsymbol{H}$, describe how the concept from parallel decomposition of wireless MIMO channel, described in Sect. 5.4 of Chap. 5, can be used to compensate for the ICI among the subcarriers. This approach allows the guard intervals to be empty, thus improving the power efficiency. Describe drawbacks of this approach compared to the conventional OFDM described in this chapter.
5. The frequency acquisition can be performed based on pilot tones, maximum likelihood estimation, or cyclic prefix. Describe the details of these schemes. Discuss advantages and disadvantages of each scheme against each other.
6. Let us consider 16-QAM transmission for 64-subcarriers OFDM over two-ray channel model. Assume that 16 subcarriers are used as pilots for LS channel estimation. In two-ray channel model, reflection coefficient is -1, and attenuation of reflected ray is 6 dB. The OFDM system design is similar to Problem 1. Show how the reconstructed 16-QAM constellation will look like for the following delays:
   (a) Delay is one-half of guard interval.
   (b) Delay exceeds the FFT interval by 2.5%.
   (c) Delay exceeds the FFT interval by 10%.
7. Repeat Problem 6 assuming that there is the second reflected component of attenuation 4 dB with delay being 1/2 of that of the first reflected component. The reflection coefficient for the second reflected ray is 0.5.
8. Let us study OFDM channel estimation for DAB employing mode III, with parameters provided in Table 7.1. Assuming a two-ray model between the transmitter and receiver wherein the reflected ray is 6 dB lower than the LOS ray, by using 16-QAM compare the LS and MMSE channel estimation efficiencies for different SNRs. Assume that the reflection coefficient is -1. Use the delay relative to the guard interval as a parameter.
9. The small-scale wireless channel models are valid within small local areas, ranging from 5λ to 40λ. To describe the channel's behavior in the local area, the following three average parameters can be used: Rician factor K, RMS deal spread τ_{RMS}, and normalized received power P_{n}. The channel impulse response (IR) can be modeled as the sum of discrete multipath components arriving at the receiver $h(\tau,t)=\sum_i \alpha_i(t)e^{-j\theta_i(t)}\delta(\tau-\tau_i(t))$, where α_i is the i-th propagation path amplitude and τ_i is the corresponding propagation path delay. θ_i denotes the i-th propagation path phase. The IR is time-varying. Power delay profile (PDP) represents the distribution of power among the multipath components $PDP(\tau)=\sum_i \alpha_i^2\delta(\tau-\tau_i)$. The normalized received power is simply $P_n=\sum_i \alpha_i^2$. The Rician factor is the ratio of the dominant's power to the power in the scattered multipath components $K=\frac{\alpha_{\max}^2}{P_n-\alpha_{\max}^2}$, $\alpha_{\max}=\max_i\{\alpha_i\}$. Channel transfer function (TF) is just the Fourier transform (FT) of the IR $H(f,t)=\int_{-\infty}^{\infty} h(\tau,t)e^{-j2\pi f\tau}d\tau=\sum_i \alpha_i(t)e^{-j[2\pi f\tau_i(t)+\theta_i(t)]}$. Under the Gaussian wide-sense stationary uncorrelated scattering (WSSUS) model, assume that different multipath components are uncorrelated, and for a sufficient number of rays with similar amplitudes (except for the line-of-sight ray), the TF will have a complex Gaussian distribution. The corresponding magnitude $R=|H(f,t)|$ will have *Rician PDF*. Establish the connection between small-scale model parameters and Rician distribution.
10. The channel correlation function for small-scale model described in Problem 9, under WSS assumption, is defined as $R_H(\Delta f,\Delta t)=\langle H^*(f,t)H(f+\Delta f,t+\Delta t)\rangle$. *Delay power spectrum* (DPS), denoted as $R_h(\tau)$, is defined as the inverse FT of the spaced-frequency correlation function $R_H(\Delta f)=R_H(\Delta f,0)$. For the small-scale model, we can use the approximation of the DPS given by Eq. (7.48).

$$R_h(\tau)=\begin{cases}\rho^2\delta(\tau),\tau=0\\ \Pi,0<\tau\le\tau_c\\ \Pi e^{-\gamma(\tau-\tau_c)},\tau>\tau_c\end{cases}$$

where ρ^2 is the normalized power of the line-of-site (LOS) component, Π is the normalized power density of the constant-level portion of DPS, τ_c is the duration of the DPS constant level, and γ is the decay exponent of the exponentially decreasing portion of the DPS. Establish the connection between this model and the actual channel parameters. By using this model, develop the corresponding simulator, and evaluate it for indoor use.

11. Apply the model developed in Problem 10 to the OFDM system for use in Wi-Fi according to the IEEE 802.11a PHY standard, with OFDM system parameters provided in Table 7.4. Provide BER results for different modulation formats and different aggregate data rates.
12. Let us now consider the time-invariant frequency-selective block-fading channel with 32 subchannels of bandwidth 2 MHz, with frequency response amplitudes being 1 for odd subchannels and √2 for even subchannels, respectively. For the transmit power of 20 mW and noise power-spectral density of $N_0 = 10^{-9}$ W/Hz, determine the OFDM channel capacity. What is the corresponding optimum power allocation strategy based on OFDM to achieve this channel capacity? Provide the corresponding OFDM transmitter and receiver configurations.
13. Provide the system parameters of CO-OFDM system that is capable of compensating the accumulated chromatic dispersion over 1000 km of SMF and corresponding PMD. The aggregate data rate should be at least 400 Gb/s, and the OFDM signal bandwidth should be either 25 GHz or 40 GHz. Use polarization-division multiplexing and square-QAM constellations. Assume that chromatic dispersion parameter is 16 ps/(nm·km) and that PMD parameter is $D_p = 0.05$ ps·km$^{-1/2}$. Ignore the second-order GVD and fiber nonlinearities. Estimate the OSNR penalty due to guard interval compared to back-to-back configuration. Determine the spectral efficiency of your design. Is it possible to find the optimum number of subcarriers to maximize the spectral efficiency?
14. OFDM samples at the output of OFDM transmitter follow Gaussian distribution for a sufficient number of subcarriers. The envelope will clearly follow Rayleigh distribution, while the power will have chi-square distribution. Determine the cumulative distribution function of the envelope. Determine the probability that PAPR is below certain threshold z_{tsh}. Plot the logarithm of this probability against PAPR [dB] for the following number of subcarriers: 32, 64, 128, 256, and 512. Finally, discuss possible approaches to reduce the PAPR.
15. This problem is related to direct detection OFDM (DD-OFDM). Let us generalize the OFDM equation by introducing the RF frequency term exp.($j\omega_{RF}t$) as follows:

$$s(t) = \begin{cases} \sum_{k=-N_{sc}/2}^{k=N_{sc}/2-1} X_{k+N_{sc}/2} e^{j2\pi \frac{k}{T_s} t} e^{j\omega_{RF} t}, & -T_s/2 < t \leq T_s/2 \\ 0, & \text{otherwise} \end{cases}$$

16. The real part of this signal, after appropriate bias addition, is used as input of an MZM. On the receiver side, only one photodetector is used to perform optical-to-electrical conversion. Provide the corresponding block diagram of this DD-OFDM system, and describe the operation principle of each block. Explain why the addition of bias is needed. This scheme is known as double-side band (DSB) DD-OFDM scheme. Describe how the DSB scheme can be converted to single-side band (SSB) scheme by using either time-domain or frequency-domain approach. Provide the block diagrams of corresponding SSB schemes. Discuss the implementation complexity of these two SSB schemes.
17. Eq. (7.33) describes the OFDM signal upon coherent detection in the presence of frequency offset and phase noise. In this problem we study the influence of frequency offset on BER performance of QPSK-OFDM system. The frequency offset introduces the inter-carrier interference (ICI). By using a sufficient number of subcarriers, the ICI can be approximated as Gaussian process by invoking the central limit theorem. The BER of QPSK in the absence of ICI is given by $BER = 0.5\text{erfc}\left(\sqrt{\rho/2}\right)$, where ρ is signal-to-noise ratio (SNR) and erfc(·) is the complementary error function. In the presence of ICI, we need to increase the SNR to achieve the same BER. Plot the BER against SNR [dB] for different values of normalized frequency offset $\delta = \Delta f_{off}/(1/T_s)$, where Δf_{off} is frequency offset and T_s is symbol duration.
18. Repeat the previous problem but now in terms of phase error. Plot the BER for QPSK-OFDM against SNR [dB] for different values of normalized laser linewidth $\Delta\nu/(1/T_s)$, where T_s is the symbol duration.
19. In this problem we study the efficiency of DFT window synchronization for CO-OFDM systems. Implement the CO-OFDM system in Matlab or C/C++ to perform the following Monte Carlo simulations. The OFDM system parameters should be set as follows: the symbol period of 25.6 ns, the guard time 3.2 ns, the number of subcarriers 256, the aggregate data rate of 10 Gb/s with BPSK format being used, and the laser linewidths of transmit and local oscillator lasers equal to 100 kHz. Plot the timing metric against the timing offset (in samples) for accumulated chromatic dispersion of 34,000 ps/nm and OSNR of 6 dB. Repeat the simulation when QPSK is used instead. Finally, repeat the simulation when the first-order PMD is present for different DGD values: 100 ps, 500 ps, and 1000 ps.
20. Here we study the efficiency of Alamouti-type polarization-time (PT) coding in CO-OFDM systems. Alamouti code is described in Sect. 8.2.4.1 of Chap. 8. Design a CO-OFDM system, based on Alamouti-type PT coding, capable of

100 Gb/s serial optical transmission over 6500 km of SMF. Perform Monte Carlo simulations in linear regime, by taking the following effects into account: chromatic dispersion, the first-order PMD, and laser linewidth of 100 kHz (for both transmitter and local lasers). Use the typical fiber parameters given in previous chapters. Compare this scheme with an equivalent polarization-division multiplexed (PDM) scheme, based on LS channel estimation, of the same aggregate data rate. Compare and discuss the results. What are the advantages and disadvantages of PT coding with respect to PDM?

References

1. R.W. Chang, Synthesis of band-limited orthogonal signals for multichannel data transmission. Bell Syst. Tech. J. **45**, 1775–1796 (1966)
2. R. W. Chang, Orthogonal frequency division multiplexing, US Patent no. 3488445 (1970)
3. R. Van Nee, R. Prasad, *OFDM wireless multimedia communications* (Artech House, Boston-London, 2000)
4. R. Prasad, *OFDM for wireless communications systems* (Artech House, Boston-London, 2004)
5. W. Shieh, I. Djordjevic, *OFDM for optical communications* (Elsevier, Oct. 2009)
6. I. Djordjevic, W. Ryan, B. Vasic, *Coding for optical channels* (Springer, 2010)
7. M. Cvijetic, I.B. Djordjevic, *Advanced optical communications and networks* (Artech House, 2013)
8. Q. Pan, R.J. Green, Bit-error-rate performance of lightwave hybrid AM/OFDM systems with comparision with AM/QAM systems in the presence of clipping noise. IEEE Photon. Technol. Lett. **8**, 278–280 (1996)
9. B.J. Dixon, R.D. Pollard, S. Iezekiel, Orthogonal frequency-division multiplexing in wireless communication systems with multimode fiber feeds. IEEE Tran. Microw. Theory Techn. **49**(8), 1404–1409 (Aug. 2001)
10. A.J. Lowery, J. Armstrong, 10 Gbit/s multimode fiber link using power-efficient orthogonal-frequency-division multiplexing. Opt. Express **13**(25), 10003–10009 (Dec. 2005)
11. I.B. Djordjevic, B. Vasic, Orthogonal frequency division multiplexing for high-speed optical transmission. *Opt. Express* **14**, 3767–3775 (May 1, 2006)
12. W. Shieh, C. Athaudage, Coherent optical frequency division multiplexing. Electron. Lett. **42**, 587–589 (2006)
13. A. J. Lowery, L. Du, J. Armstrong, Orthogonal frequency division multiplexing for adaptive dispersion compensation in long haul WDM systems, in Proc. *OFC Postdeadline Papers*, Paper no. PDP39, (Mar. 2006)
14. I.B. Djordjevic, L. Xu, T. Wang, Beyond 100 Gb/s optical transmission based on polarization multiplexed coded-OFDM with coherent detection. IEEE/OSA J. Opt. Commun. Netw. **1**(1), 50–56 (June 2009)
15. Y. Ma, Q. Yang, Y. Tang, S. Chen, W. Shieh, 1-Tb/s single-channel coherent optical OFDM transmission over 600-km SSMF fiber with subwavelength bandwidth access. Opt. Express **17**, 9421–9427 (2009)
16. X. Liu, S. Chandrasekhar, Beyond 1-Tb/s superchannel transmission, in Proc. *IEEE Photonics Conference (PHO)* 2011, pp. 893–894, 9–13 Oct. 2011
17. S. Sano et al., No-guard-interval coherent optical OFDM for 100-Gb/s long-haul WDM transmission. J. Lightw. Technol. **27**(16), 3705–3713 (2009)
18. B. Zhu, X. Liu, S. Chandrasekhar, D.W. Peckham, R. Lingle, Ultra-long-haul transmission of 1.2-Tb/s multicarrier no-guard-interval CO-OFDM superchannel using ultra-large-area fiber. IEEE Photon. Technol. Lett. **22**(11), 826–828 (2010)
19. S.C.J. Lee, F. Breyer, S. Randel, H.P.A. Van den Boom, A.M.J. Koonen, High-speed transmission over multimode fiber using discrete multitone modulation [invited]. J. Opt. Netw. **7**(2), 183–196 (Feb. 2008)
20. I.B. Djordjevic, M. Arabaci, L. Xu, T. Wang, Spatial-domain-based multidimensional modulation for multi-Tb/s serial optical transmission. Opt. Express **19**(7), 6845–6857 (2011)
21. A. Li, A. Al Amin, X. Chen, W. Shieh, Reception of mode and polarization multiplexed 107-Gb/s CO-OFDM signal over a two-mode fiber, in Proc. *OFC/NFOEC, Postdeadline Papers* (OSA, 2011), Paper PDPB8
22. L. Gruner-Nielsen, Y. Sun, J. W. Nicholson, D. Jakobsen, R. Lingle, B. Palsdottir, Few mode transmission fiber with low DGD, low mode coupling and low loss, in Proc. *OFC/NFOEC, Postdeadline Papers* (OSA, 2012), Paper PDP5A.1
23. Ryf, R., et al., "Low-loss mode coupler for mode-multiplexed transmission in few-mode fiber," in Proc. *OFC/NFOEC, Postdeadline Papers* (OSA, 2012), Paper PDP5B.5
24. Fontaine, N. K., et al., Space-division multiplexing and all-optical MIMO demultiplexing using a photonic integrated circuit, in Proc. *OFC/NFOEC, Postdeadline Papers* (OSA, 2012), Paper PDP5B.1
25. X. Chen, A. Li, J. Ye, A. Al Amin, W. Shieh, Reception of dual-LP11-mode CO-OFDM signals through few-mode compatible optical add/drop multiplexer, in Proc. *OFC/NFOEC, Postdeadline Papers* (OSA, 2012), Paper PDP5B.4
26. J.G. Proakis, *Digital communications*, 5th edn. (McGraw Hill, New York, 2007)
27. S. Haykin, *Communication Systems*, 4th edn. (John Wiley & Sons, 2001)
28. E. Collett, *Polarized light in fiber optics* (SPIE Press, 2003)
29. J.G. Proakis, D.G. Manolakis, *Digital signal processing: principles, algorithms, and applications*, 4th edn. (Prentice-Hall, 2007)
30. A. Goldsmith, *Wireless communications* (Cambridge University Press, Cambridge, 2005)
31. W. Shieh, X. Yi, Y. Ma, Q. Yang, Coherent optical OFDM: Has its time come? [invited]. J. Opt. Netw. **7**, 234–255 (2008)
32. T.M. Schmidl, D.C. Cox, Robust frequency and time synchronization for OFDM. IEEE Trans. Commun. **45**, 1613–1621 (1997 Dec)
33. Minn, H., Bhargava, V. K., and Letaief, K. B., "A robust timing and frequency synchronization for OFDM systems," IEEE Trans Wireless Comm., vol. 2, pp. 822–839, July 2003
34. T. Pollet, M. Van Bladel, M. Moeneclaey, BER sensitivity of OFDM systems to carrier frequency offset and Wiener phase noise, *IEEE Trans Comm.*, vol. 43, pp. 191–193, Feb./Mar./Apr. 1995

35. S. Coleri, M. Ergen, A. Puri, A. Bahai, Channel estimation techniques based on pilot arrangement in OFDM systems. IEEE. Trans. Broadcasting **48**(3), 223–229 (2002 Sept)
36. D. Zou, I. B. Djordjevic, Multi-Tb/s optical transmission based on polarization-multiplexed LDPC-coded multi-band OFDM, in Proc. *13th International Conference on Transparent Optical Networks (ICTON)* 2011, Paper Th.B3.3, 26–30 June 2011, Stockholm
37. D. Zou, I. B. Djordjevic, Beyond 1Tb/s superchannel optical transmission based on polarization multiplexed coded-OFDM over 2300 km of SSMF, in Proc. 2012 *Signal Processing in Photonics Communications (SPPCom)*, Paper SpTu2A.6, 17–21 June, 2012, Colorado Springs
38. H. Sari, G. Karam, I. Jeanclaude, Transmission techniques for digital terrestrial TV broadcasting. IEEE Commun. Mag. **33**(2), 100–109 (1995)
39. P. Moose, A technique for orthogonal frequency division multiplexing frequency offset correction. IEEE Trans. Commun. **42**, 2908–2914 (1994)
40. M. Sandell, J. J. Van de Beek, P. O. Börjesson, Timing and frequency synchronization in OFDM systems using cyclic prefix, in Proc. *Int. Symp. Synchron.*, pp. 16–19, Saalbau, Essen, Germany, Dec. 1995
41. D. Zou, C. Lin, I. B. Djordjevic, LDPC-coded mode-multiplexed CO-OFDM over 1000 km of few-mode fiber, in Proc. *CLEO 2012*, Paper no. CF3I.3, San Jose, CA, 6–11 May 2012
42. X.Q. Jin, J.M. Tang, P.S. Spencer, K.A. Shore, Optimization of adaptively modulated optical OFDM modems for multimode fiber-based local area networks [invited]. J. Opt. Netw. **7**(3), 198–214 (2008 Mar)
43. D.J.G. Mestdagh, P.M.P. Spruyt, A method to reduce the probability of clipping in DMT-based transceivers. IEEE Trans. Commun. **44**, 1234–1238 (1996)
44. P.V. Eetvelt, G. Wade, M. Thompson, Peak to average power reduction for OFDM schemes by selected scrambling. IEE Electron. Lett. **32**, 1963–1964 (1996)
45. M. Friese, Multicarrier modulation with low peak-to-average power ratio. IEE Electron. Lett. **32**, 712–713 (1996)
46. J.A. Davis, J. Jedwab, Peak-to-mean power control in OFDM, Golay complementary sequences, and Reed-Muller codes. IEEE Trans. Inf. Theory **45**, 2397–2417 (1999)
47. I. B. Djordjevic, L. Xu, T. Wang, PMD Compensation in Multilevel Coded-Modulation Schemes with Coherent Detection using BLAST Algorithm and Iterative Polarization Cancellation, *Optics Express*, vol. 16, no. 19, pp. 14845–14852, Sept. 15, 2008
48. Y. Zhao, A. Huang, A novel channel estimation method for OFDM mobile communication systems based on pilot signals and transform-domain processing, in Proc. *IEEE 47th Vehicular Technology Conference,* 1997, vol. 3, pp. 2089–2093, Phoenix, AZ, 1997
49. P.K. Frenger, N. Arne, B. Svensson, Decision-directed coherent detection in multicarrier systems on Rayleigh fading channels. IEEE Trans. Veh. Technol. **48**(2), 490–498 (Mar 1999)
50. T. Kella, Decision-directed channel estimation for supporting higher terminal velocities in OFDM based WLANs, in Proc. *IEEE Global Telecommunications Conference* 2003 *(GLOBECOM '03), vol. 3*, pp. 1306–1310, 1–5 Dec. 2003
51. W. Hoeg, T. Lauterbach, *Digital audio broadcasting: principles and applications of digital radio* (Wiley, Chichester, 2003)
52. W. Fischer, *Digital video and audio broadcasting technology: a practical engineering guide*, 3rd edn. (Springer, Heidelberg-Dordrecht-London-New York, 2010)
53. ETSI Standard: EN 300 744 V1.5.1, Digital Video Broadcasting (DVB); Framing structure, channel coding and modulation for digital terrestrial television
54. M. Ghavami, L.B. Michael, R. Kohno, *Ultra wideband signal and systems in communication engineering*, 2nd edn. (Wiley, Chichester, 2007)
55. ETSI Standard: EN 302755, digital video broadcasting (DVB), frame structure, channel coding and modulation for a second generation digital terrestrial video broadcasting system, ETSI, 2009
56. S.M. Alamouti, A simple transmit diversity technique for wireless communications. IEEE J. Sel. Areas Commun. **16**(8), 1451–1458 (1998)
57. I.B. Djordjevic, L. Xu, T. Wang, PMD compensation in coded-modulation schemes with coherent detection using Alamouti-type polarization-time coding. Opt. Express **16**(18), 14163–14172 (09/01/2008)
58. IEEE Standard for Information technology—Telecommunications and information exchange between systems Local and metropolitan area networks—Specific requirements Part 11: Wireless LAN Medium Access Control (MAC) and Physical Layer (PHY) Specifications, in *IEEE Std 802.11–2012 (Revision of IEEE Std 802.11–2007)*, pp. 1–2793, March 29 2012, https://doi.org/10.1109/IEEESTD.2012.6178212
59. J. Heiskala, J. Terry, *OFDM wireless LANs: a theoretical and practical guide* (Sams Publishing, 2002)
60. L. Hanzo, Y. Akhtman, L. Wang, M. Jiang, *MIMO-OFDM for LTE, WiFi, and WiMAX* (IEEE/Wiley, 2011)
61. M.S. Gast, *802.11 wireless networks: the definitive guide*, 2nd edn. (O'Reilly Media, Inc, Sebastopol, 2005)
62. M.S. Gast, *802.11ac: A Survival Guide* (O'Reilly Media, Inc, Sebastopol, 2013)
63. M.S. Gast, *802.11an: A Survival Guide* (O'Reilly Media, Inc, Sebastopol, 2012)
64. Unlicensed Use of the 6 GHz Band; Expanding Flexible Use in Mid-Band Spectrum Between 3.7 and 24 GHz, available at: https://docs.fcc.gov/public/attachments/FCC-20-51A1.pdf
65. 5G NR (Rel-15), available at: https://www.3gpp.org/lte-2
66. 3GPP Release 18, available at: https://www.3gpp.org/release18
67. IEEE Standard for Air Interface for Broadband Wireless Access Systems. IEEE Std 802.16–2017 (Revision of IEEE Std 802.16–2012), pp.1–2726, 2 March 2018
68. IEEE 802.16 Working Group on Broadband Wireless Access Standards, see https://grouper.ieee.org/groups/802/16/
69. L. Nuaymi, *WiMAX: technology for broadband wireless access* (Wiley, Chichester, 2007)
70. IEEE 802.16–2004, IEEE Standard for Local and Metropolitan Area Networks, Air Interface for Fixed Broadband Wireless Access Systems, October 2004
71. IEEE 802.16e, IEEE Standard for Local and Metropolitan Area Networks, Air Interface for Fixed Broadband Wireless Access Systems, Amendment 2: Physical and Medium Access Control Layers for Combined Fixed and Mobile Operation in Licensed Bands and Corrigendum 1, February 2006
72. M. Ghavami, L.B. Michael, R. Kohno, *Ultra wideband signals and systems in communication engineering* (Wiley, 2007)
73. A. Batra, J. Balakrishnan, A. Dabak, R. Gharpurey, J. Lin, P. Fontaine, J.-M. Ho, S. Lee, M. Frechette, S. March, H. Yamaguchi, "Multi-band OFDM physical layer proposal for IEEE 802.15 Task Group 3a," Doc. IEEE P802.15-03/268r3, Mar. 2004
74. H. Nikookar, R. Prasad, *Introduction to Ultra Wideband for Wireless Communications* (Springer Science+Business Media B.V, 2009)

75. W.P. Siriwongpairat, K.J.R. Liu, *Ultra-wideband communications systems: multiband OFDM approach* (Wiley, Hoboken, 2007)
76. P. K. Boon, X. Peng, F. Chin, Performance studies of a multi-band OFDM system using a simplified LDPC code, in Proc. 2004 International Workshop on Ultra Wideband Systems & 2004 Conference on Ultrawideband Systems and Technologies (UWBST & IWUWBS), pp. 376–380, 18–21 May 2004
77. S.-M. Kim, J. Tang, K. K. Parhi, Quasi-cyclic low-density parity-check coded multiband-OFDM UWB systems, in Proc. *2005 IEEE International Symposium on Circuits and Systems*, pp. 65–68, 23–26 May 2005
78. W. Shieh, R.S. Tucker, W. Chen, X. Yi, G. Pendock, Optical performance monitoring in coherent optical OFDM systems. Opt. Express **15**, 350–356 (2007)
79. J.M. Tang, K.A. Shore, Maximizing the transmission performance of adaptively modulated optical OFDM signals in multimode-fiber links by optimizing analog-to-digital converters. J. Lightwave Technol. **25**, 787–798 (2007)
80. B.J.C. Schmidt, A.J. Lowery, J. Armstrong, Experimental demonstrations of 20 Gbit/s direct-detection optical OFDM and 12 Gbit/s with a colorless transmitter. In Opt. Fiber Commun. Conf. 2007, paper no. PDP18, San Diego (2007)
81. M. Schuster, S. Randel, C.A. Bunge, et al., Spectrally efficient compatible single-sideband modulation for OFDM transmission with direct detection. IEEE Photon. Technol. Lett. **20**, 670–6722 (2008)
82. W. Peng, X. Wu, V.R. Arbab, et al., *Experimental demonstration of 340 km SSMF transmission using a virtual single sideband OFDM signal that employs carrier suppressed and iterative detection techniques* (Opt. Fiber Commun. Conf. 2008, paper no. OMU1, San Diego, 2008)
83. W.R. Peng, X. Wu, V.R. Arbab, et al., *Experimental demonstration of a coherently modulated and directly detected optical OFDM system using an RF-tone insertion* (In: Opt. Fiber Commun. Conf. 2008, paper no. OMU2, San Diego, 2008)
84. W. Shieh, Q. Yang, Y. Ma, 107 Gb/s coherent optical OFDM transmission over 1000-km SSMF fiber using orthogonal band multiplexing. Opt. Express **16**, 6378–6386 (2008)
85. I. B. Djordjevic, W. Shieh, X. Liu, M. Nazarathy, Advanced Digital Signal Processing and Coding for Multi-Tb/s Optical Transport [From the Guest Editors], *IEEE Sig. Proc. Mag.*, vol. 31, no. 2, pages 15 and 142, Mar. 2014
86. I.B. Djordjevic, M. Cvijetic, C. Lin, Multidimensional signaling and coding enabling multi-Tb/s optical transport and networking. IEEE Sig. Proc. Mag. **31**(2), 104–117 (Mar. 2014)
87. I. B. Djordjevic, "On the irregular nonbinary QC-LDPC-coded hybrid multidimensional OSCD-modulation enabling beyond 100 Tb/s optical transport," *IEEE/OSA J. Lightwave Technol.*, vol. 31, no. 16, pp. 2969–2975, Aug.15, 2013
88. I.B. Djordjevic, M. Arabaci, L. Xu, T. Wang, Generalized OFDM (GOFDM) for ultra-high-speed optical transmission. Opt. Express **19**(7), 6969–6979 (2011)
89. I. B. Djordjevic, Spatial-domain-based hybrid multidimensional coded-modulation schemes enabling multi-Tb/s optical transport, *J. Lightwave Technol.*, vol. 30, no. 14, pp. 2315–2328, July 15, 2012
90. X. Liu, S. Chandrasekhar, P.J. Winzer, Digital signal processing techniques enabling multi-Tb/s superchannel transmission: An overview of recent advances in DSP-enabled superchannels. IEEE Signal Process. Mag. **31**(2), 16–24 (March 2014)
91. I. B. Djordjevic, A. H. Saleh, F. Küppers, Design of DPSS based fiber Bragg gratings and their application in all-optical encryption, OCDMA, optical steganography, and orthogonal-division multiplexing, *Opt. Express*, vol. 22, no. 9, pp. 10882–10897, 5 May 2014
92. I. B. Djordjevic, A. Jovanovic, Z. H. Peric, T. Wang, "Multidimensional optical transport based on optimized vector-quantization-inspired signal constellation design," IEEE Trans. Commun., vol. 62, no. 9, pp. 3262–3273, Sept. 2014
93. I. B. Djordjevic, On advanced FEC and coded modulation for ultra-high-speed optical transmission, *IEEE Communications Surveys and Tutorials*, vol. 18, no. 3, pp. 1920–1951, third quarter 2016, August 19, 2016, https://doi.org/10.1109/COMST.2016.2536726
94. I. Djordjevic, H.G. Batshon, L. Xu, T. Wang, Four-dimensional optical multiband-OFDM for beyond 1.4 Tb/s serial optical transmission. Opt. Express **19**(2), 876–882 (01/17/2011)
95. A. Sano, E. Yoshida, H. Masuda, et al., *30 × 100-Gb/s all-optical OFDM transmission over 1300 km SMF with 10 ROADM nodes* (Eur. Conf. Opt. Commun. 2007 (ECOC 2007), paper no. PD 1.7, Berlin, 2007)
96. E. Yamada, A. Sano, H. Masuda, et al., *Novel no-guard interval PDM CO-OFDM transmission in 4.1 Tb/s (50 88.8-Gb/s) DWDM link over 800 km SMF including 50-GHz spaced ROADM nodes* (Opt. Fiber Commun. Conf. 2008 (OFC 2008), paper no. PDP8, San Diego, 2008)
97. G Singh, A Alphones. OFDM modulation study for a radio-over-fiber system for wireless LAN (IEEE 802.11a). In: Proc. ICICS-PCM. Singapore. p. 1460–4 (2003)
98. J.B. Song, A.H.M.R. Islam, Distortion of OFDM signals on radio-over-fiber links integrated with an RF amplifier and active/passive electroabsorption modulators. J. Lightwave Technol. **26**(5), 467–477 (2008)
99. L. Xu, D. Qian, J. Hu, W. Wei, T. Wang, *OFDMA-based passive optical networks (PON)* (2008 Digest IEEE LEOS Summer Topical Meetings, 2008), pp. 159–160
100. A. Emsia, M. Malekizandi, T. Q. Le, I. B. Djordjevic, F. Kueppers, 1 Tb/s WDM-OFDM-PON power budget extension techniques, in Proc. *IEEE Photonics Conference 2013 (IPC 2013)*, Paper WG3.3, 8–12 September 2013, Seattle, USA
101. J. Grubor, V. Jungnickel, K.-D. Langer, *Capacity analysis in indoor wireless infrared communication using adaptive multiple subcarrier transmission* (Proc ICTON, 2005), pp. 171–174
102. J. Grubor, O.C. Gaete Jamett, J.W. Walewski, S. Randel, K.-D. Langer, High-speed wireless indoor communication via visible light, in *ITG Fachbericht 198*, (VDE Verlag, Berlin, 2007), pp. 203–208
103. I.B. Djordjevic, B. Vasic, M.A. Neifeld, LDPC coded OFDM over the atmospheric turbulence channel. Opt. Express **15**(10), 6332–6346 (May 2007)
104. I.B. Djordjevic, LDPC-coded OFDM transmission over graded-index plastic optical fiber links. IEEE Photon. Technol. Lett. **19**(12), 871–873 (2007)
105. A.R. Shah, R.C.J. Hsu, A.H. Sayed, B. Jalali, Coherent optical MIMO (COMIMO). J. Lightwave Technol. **23**(8), 2410–2419 (2005)
106. R.C.J. Hsu, A. Tarighat, A. Shah, A.H. Sayed, B. Jalali, Capacity enhancement in coherent optical MIMO (COMIMO) multimode fiber links. J. Lightwave Technol. **23**(8), 2410–2419 (2005)

107. Z. Tong, Q. Yang, Y. Ma, W. Shieh. 21.4 Gb/s coherent optical OFDM transmission over multimode fiber, post-deadline papers technical digest. 13th Optoelectronics and Communications Conference (OECC) and 33rd Australian Conference on Optical Fibre Technology (ACOFT), Paper No. PDP-5. (2008)
108. F. Yaman, S. Zhang, Y.-K. Huang, E. Ip, J. D. Downie, W. A. Wood, A. Zakharian, S. Mishra, J. Hurley, Y. Zhang, I. B. Djordjevic, M.-F. Huang, E. Mateo, K. Nakamura, T. Inoue, Y. Inada, T. Ogata, First Quasi-Single-Mode Transmission over Transoceanic Distance using Few-mode Fibers, in Proc. *OFC Postdeadline Papers*, Paper Th5C.7, 22–26 March 2015, Los Angeles

Diversity and MIMO Techniques

8

Abstract

The purpose of this chapter is to describe various diversity and MIMO techniques capable of improving single-input single-output system performance. Through MIMO concept it is also possible to improve the aggregate data rate. The diversity and MIMO techniques described are applicable to wireless MIMO communications as well as free-space optical and fiber-optic communications with coherent optical detection. Various diversity schemes are described including polarization diversity, spatial diversity, and frequency diversity schemes. The following receiver diversity schemes are described: selection combining, threshold combining, maximum-ratio combining, and equal-gain combining schemes. Various transmit diversity schemes are described, depending on the availability of channel state information on transmitter side. Most of the chapter is devoted to wireless and optical MIMO techniques. After description of various wireless and optical MIMO models, we describe the parallel decomposition of MIMO channels, followed by space-time coding (STC) principles. The maximum likelihood (ML) decoding for STC is described together with various design criteria. The following classes of STC are described: Alamouti code, orthogonal designs, linear space-time block codes, and space-time trellis codes. The corresponding STC decoding algorithms are described as well. After that we move our attention to spatial division multiplexing (SDM) principles and describe various BLAST encoding architectures as well as multi-group space-time coded modulation. The following classes of linear and feedback MIMO receiver for uncoded signals are subsequently described: zero-forcing, linear minimum MSE, and decision-feedback receivers. The next topic in the chapter is related to suboptimum MIMO receivers for coded signals. Within this topic we first describe linear zero-forcing (ZF) and linear MMSE receivers (interfaces) but in the context of STC. Within decision-feedback and BLAST receivers, we describe various horizontal/vertical (H/V) BLAST architecture including ZF and MMSE ones. The topic on suboptimum receivers is concluded with diagonal (D)-BLAST and iterative receiver interfaces. The section on iterative MIMO receivers starts with brief introduction of concept of factor graphs, followed with the description of factor graphs for MIMO channel and channels with memory. We then describe the sum-product algorithm (SPA) operating on factor graphs, followed by description of SPA for channels with memory. The following iterative MIMO receivers for uncoded signals are described: ZF, MMSE, ZF H/V-BLAST, and LMMSE H/V-BLAST receivers. After a brief description of factor graphs for linear block and trellis codes, we describe several iterative MIMO receivers for space-time coded signals. The section on broadband MIMO describes how frequency-selectivity can be used as an additional degree of freedom, followed by description of MIMO-OFDM and space-frequency block-coding principles. The focus is then moved to MIMO channel capacity calculations for various MIMO channel models including deterministic, ergodic, and non-ergodic random channels, as well as correlated channel models. The concepts of ergodic and outage channel capacity are described. The section on MIMO channel capacity ends with MIMO-OFDM channel capacity description. In MIMO channel estimation section, the following MIMO channel estimation techniques are described: ML, least squares (LS), and linear minimum MSE MIMO channel estimation techniques. After the conclusion section, the set of problems is provided for deeper understanding of the chapter material.

I. B. Djordjevic, *Advanced Optical and Wireless Communications Systems*, https://doi.org/10.1007/978-3-030-98491-5_8

8.1 Diversity Techniques

The key idea behind the diversity-combining techniques [1–3] is to use several independent realizations of that same transmitted signal to improve tolerance to channel impairments. The independent signal copies transmitted over different paths or degrees of freedom will have a low probability of experiencing the same level of channel distortion. In wireless communications, the probability that all independent fading paths are in deep fade will be low. The independent realizations of the same signal will be properly combined to improve overall reliability of transmission.

8.1.1 Basic Diversity Schemes

There are many different ways of transmitting the independent signal paths. In *polarization diversity*, for optical communications, the laser signal can be split into two polarizations by the polarization beam splitter (PBS), and the same data signal can be imposed on two orthogonal polarization states. In wireless communications, two transmit or receive antennas with horizontal and vertical polarizations can be used to implement the polarization diversity. The coherent optical receivers require matching the state of polarization (SOP) of the local oscillator laser (LOL) with that of the received optical signal. In practice, only the SOP of local laser can be controlled, and one possible polarization control receiver configuration is shown in Fig. 8.1. Polarization controller is commonly implemented by using four squeezers [4, 5]. Insensitivity with respect to polarization fluctuations is possible if the receiver derives two demodulated signals from two orthogonal polarizations of the received signal, which is illustrated in Fig. 8.2. This scheme is known as polarization diversity receiver. There two main issues with respect to this approach: (i) there only two polarization diversity branches and (ii) since the same signal (s) used in both polarizations, the spectral efficiency of this scheme is twice lower than that of the polarization division multiplexing (PDM) scheme.

In *frequency diversity*, the same narrowband signal is transmitted over different subcarriers, while the frequency channel spacing is larger than the coherence bandwidth. In *wavelength diversity*, similarly, several copies of the same signal are transmitted over different wavelength channels. The main drawback of this scheme is poor spectral efficiency.

In *time diversity*, the same signal is transmitted over different time -slot, while the slot separation is larger than the coherence time or reciprocal of the channel Doppler spread. This scheme also suffers from the poor spectral efficiency. To overcome this problem, the multidimensional signaling [6, 7] can be employed, namely, the multiple degrees of freedom, such as time or frequency slots, can be used as dimensions for multidimensional signaling. Now instead of using this degrees of freedom to multiplex independent data streams, we can define the signal constellations in multidimensional space, as discussed in a chapter on advanced modulation schemes. If some of coordinates are affected by deep fade, for instance, the other will not and the signal constellation point will not be completely destroyed, which will improve tolerance to channel impairments. In another approach, it is possible to combine the coding with interleaving.

In *directional/angle diversity*, for wireless applications, the receive antenna beamwidth is restricted to a given angle increment. If the angle increment is sufficiently small, it is possible to ensure that only one multipath array reaches the antenna's beamwidth. On such a way, we can efficiently deal with the multipath fading. On the other hand, the complexity of such receiver diversity scheme will be high. Thanks to high directivity of laser beams, this method might be suitable for free-

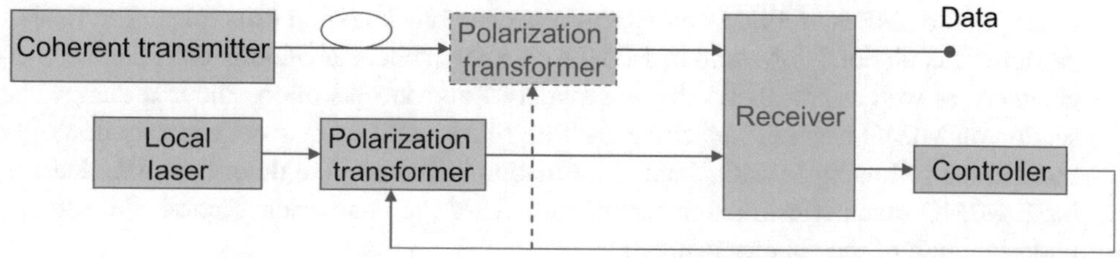

Fig. 8.1 Polarization control receiver configuration for fiber-optic communications

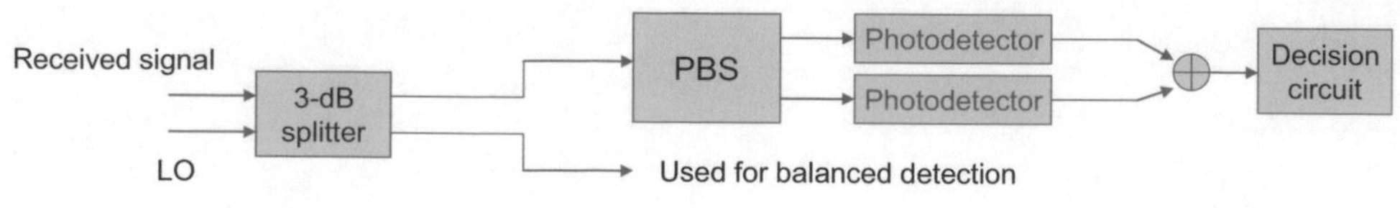

Fig. 8.2 Polarization diversity receiver configuration in fiber-optic communications. *PBS* polarization beam splitter

space optical communications to deal with atmospheric turbulence effects. However, such system has a sensitive beam wandering effect. The so-called smart antennas are built from an array of directional antenna elements, wherein the phase of each element can be independently adjusted. By steering the antenna element to the direction of the strongest multipath component, we can effectively deal with the multipath fading.

In *space/spatial diversity*, for wireless applications, multiple transmit or receive antennas are used, and the antennas are sufficiently separated so that the corresponding fading paths are uncorrelated and independent of each other. In free-space optical communications, the space diversity can be achieved through multiple laser diodes on the transmitter side and multiple photodetectors on the receiver side. In fiber-optic communications, the space diversity can be achieved through multiple fiber; however, the total system capacity will be reduced. In spatial division multiplexing (SDM) systems, in multicore fibers, different cores can be used to transmit the same data to achieve the spatial diversity. In few-mode fibers (FMFs), the different spatial modes can be used to enable the spatial diversity. Through space/spatial diversity, the overall signal-to-noise ratio (SNR) can be improved, and simultaneously the symbol error probability (P_s) can be reduced. Therefore, we can define two parameters specifying the improvement of various diversity techniques against the single-branch system, namely, array gain and diversity gain. The *array gain* is defined as the improvement in the signal-to-noise ratio when diversity is used. On the other hand, the *diversity gain* is defined as the change in the slope of symbol error probability curve.

8.1.2 Receiver Diversity

In receiver (Rx) diversity, the independent propagation paths after reception are properly combined, and such obtained signal is demodulated. We are mostly concerned with the *linear combiners*, shown in Fig. 8.3, in which multipath propagation paths are properly weighted according to certain criterion. In wireless applications, the *i*-th receiver Rx_i is represented by the receive antenna followed by a low-noise amplifier. In free-space optical communications with direct detection, the *i*-th receiver corresponds to the photodetector, be it a PIN or APD. In SDM systems, the *i*-th receiver is essentially a coherent optical detector. The *i*-th receiver output can be represented as $r_i \exp(j\phi_i)\underline{s}$, where ϕ_i is the random phase shift introduced over the *i*-th propagation path, r_i is the corresponding attenuation coefficient, and $\underline{s}$ denotes the transmitted signal constellation point. The linear combiner requires the *co-phasing*, after the coherent detection, which can be performed by multiplying by $w_i\exp(-j\phi_i)$ for some real-valued weights (w_i). The output of linear combiner is then $r = \Sigma_i w_i r_i \underline{s}$. Assuming that all receivers are identical with equivalent power spectral density N_0, the SNR after the linear combiner, denoted as $\rho_{\oplus}$, can be estimated as:

$$\rho_{\oplus} = \frac{\left(\sum_{i=1}^{M} w_i r_i\right)^2}{N_0 \sum_{i=1}^{M} w_i^2}, \tag{8.1}$$

where the nominator represents the signal power, while the denominator represents the total noise power in all branches. To determine the maximum possible SNR at the output of the linear combiner, let us ignore the distortion introduced during the

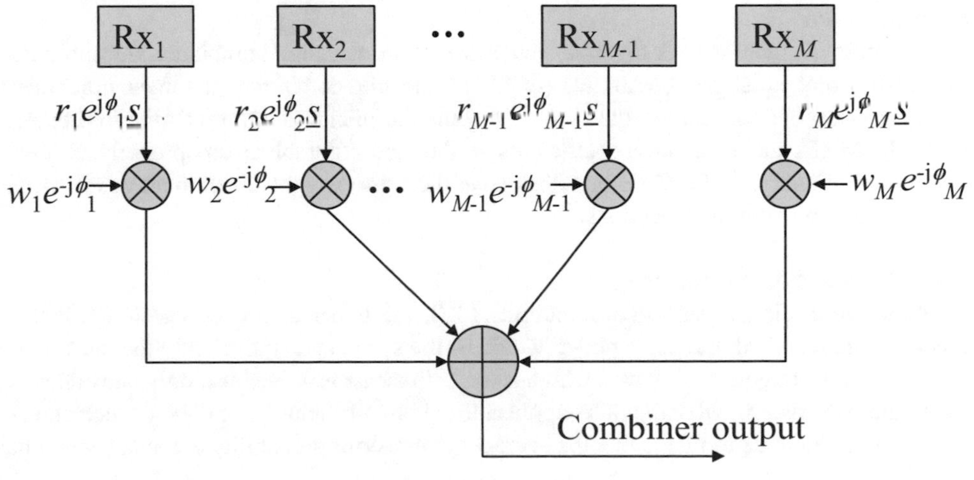

Fig. 8.3 The generic linear combiner configuration in receiver diversity

propagation, such as the fading in wireless channels, so that the output of the matched filter in the i-th branch is proportional to $\sqrt{E_s}$, where E_s is the average symbol energy and the SNR of linear combiner can be upper bounded by:

$$\rho_{\oplus} \leq \frac{\left(\sum_{i=1}^{M} w_i \sqrt{E_s}\right)^2}{N_0 \sum_{i=1}^{M} w_i^2} \leq M \underbrace{\frac{E_s}{N_0}}_{\rho} = M\rho, \tag{8.2}$$

where ρ is the average SNR of any branch. Therefore, the *array gain* can be now defined as the linear combiner SNR to single-branch SNR:

$$AG = \frac{\overline{\rho}_{\oplus}}{\overline{\rho}} \leq M, \tag{8.3}$$

and the maximum possible array gain is M. The *average symbol error probability* can be determined from SNR-dependent expression of symbol error probability $P_s(\rho)$ by averaging out over all possible values of ρ:

$$\overline{P}_s = \int_0^{\infty} P_s(\rho) f_{\rho_{\oplus}}(\rho) d\rho, \tag{8.4}$$

where $f_{\rho_{\oplus}}(\rho)$ is the PDF of linear combiner SNR. The *outage probability* can be derived from the cumulative distribution function (CDF):

$$P_{\text{outage}} = \Pr(\rho_{\oplus} \leq \rho_0) = \int_0^{\rho_0} f_{\rho_{\oplus}}(\rho) d\rho, \tag{8.5}$$

where ρ_0 is the target SNR. The linear diversity combining helps getting a more favorable distribution of combine SNR, $f_{\rho_{\oplus}}(\rho)$; the average symbol rate and outage probability get reduced. This reduction in the symbol error rate can be described by more rapid slope in the corresponding symbol error probability curve. We know that for very high SNRs, the symbol error probability is reversely proportional to ρ, so the *diversity gain* can be defined as:

$$DG = -\lim_{\rho \to \infty} \frac{\log \overline{P}_s(\rho)}{\log \rho}. \tag{8.6a}$$

For some diversity schemes, symbol error probability is proportional to ρ^{-M}, and based on Eq. (8.6a), the diversity gain for such schemes is $DG = M$. Clearly, the maximum diversity order is M, and when achieved it is commonly referred to as the *full diversity order*.

The most popular receiver diversity schemes include selection combining, threshold combining, maximum-ratio combining (MRC), and equal gain combining (EGC). In selection combining, the linear combiner selects the branch with the largest SNR. In threshold combining, the combiner scans the branches and select the first branch with SNR larger than threshold SNR. In MRC, the weighting coefficients in the linear combiner are properly chosen such that the combiner SNR is maximized. Finally, in EGC all branches have the same weight. Additional details of various receiver diversity schemes are provided in subsequent sections.

8.1.2.1 Selection Combining

As indicated earlier, in selection combining (SC), the linear combiner selects the branch with the maximum SNR. If all receivers are identical, the noise power would be the same, and equivalently the linear combiner in SC can instead select the branch with the largest total power, which is easier to measure. Given that only one branch is used in this case, there is no need to use the co-phasing so that the SC is applicable to both differential and coherent detection cases. As discussed in the previous section, in order to determine either the average symbol error probability or the outage probability, the first step is to determine the distribution of SNR of the combiner, which can be obtained as the first derivative of CDF with respect to SNR. The CDF is related to the probability that the combiner SNR $\rho_\oplus$ is smaller than a given SNR ρ, which is equivalent to the probability that the maximum of SNRs from all branches is smaller than ρ. When the maximum SNR is smaller than a given ρ, then SNRs from other branches will be smaller than ρ and we can write:

$$CDF_{\rho_\oplus}(\rho) = \Pr(\rho_\oplus < \rho) = \Pr(\max[\rho_1, \rho_1, \ldots, \rho_M] < \rho) = \prod_{i=1}^{M} \Pr(\rho_i < \rho). \tag{8.6b}$$

The PDF of combiner SNR is then the first derivative of CDF with respect to ρ:

$$f_{\rho_\oplus}(\rho) = \frac{dCDF_{\rho_\oplus}(\rho)}{d\rho}. \tag{8.7}$$

Based on Eqs. (8.4) and (8.5), the average symbol error probability and the outage probability are expressed with respect to CDF as follows:

$$\overline{P}_s = \int_0^\infty P_s(\rho)\frac{dCDF_{\rho_\oplus}(\rho)}{d\rho}d\rho, \; P_{\text{outage}} = CDF_{\rho_\oplus}(\rho_0), \tag{8.8}$$

where ρ_0 is the target SNR. The PDF of combiner SNR is particularly easy to determine for Rayleigh fading. It has been shown earlier that the PDF of SNR in the i-th branch is given by:

$$f(\rho_i) = \frac{1}{\overline{\rho}_i}e^{-\rho_i/\overline{\rho}_i}, \tag{8.9}$$

while the corresponding CDF of the i-th branch is:

$$CDF_{\rho_i}(\rho) = 1 - e^{-\rho/\overline{\rho}_i}. \tag{8.10}$$

By using Eq. (8.6b), the CDF of combiner SNR is given by:

$$CDF_{\rho_\oplus}(\rho) = \prod_{i=1}^{M} \Pr(\rho_i < \rho) = \prod_{i=1}^{M} CDF_{\rho_i}(\rho) = \prod_{i=1}^{M}\left(1 - e^{-\rho/\overline{\rho}_i}\right). \tag{8.11}$$

Assuming that all branches are identical with the same SNR $\overline{\rho}_i = \overline{\rho}$, the CDF of SC SNR becomes:

$$CDF_{\rho_\oplus}(\rho) = \prod_{i=1}^{M}\left(1 - e^{-\rho/\overline{\rho}}\right) = \left(1 - e^{-\rho/\overline{\rho}}\right)^M. \tag{8.12}$$

The PDF of SC SNR is then:

$$f_{\rho_\oplus}(\rho) = \frac{dCDF_{\rho_\oplus}(\rho)}{d\rho} = \frac{d}{d\rho}\left(1 - e^{-\rho/\overline{\rho}}\right)^M = \frac{M}{\overline{\rho}}\left(1 - e^{-\rho/\overline{\rho}}\right)^{M-1}e^{-\rho/\overline{\rho}}. \tag{8.13}$$

The average SC SNR is given by:

$$\overline{\rho}_\oplus = \int_0^\infty \rho f_{\rho_\oplus}(\rho)d\rho = \int_0^\infty \rho\frac{M}{\overline{\rho}}\left(1 - e^{-\rho/\overline{\rho}}\right)^{M-1}e^{-\rho/\overline{\rho}}d\gamma = \overline{\rho}\sum_{i=1}^{M}\frac{1}{i}, \tag{8.14}$$

and clearly the array gain $AG = \sum_{i=1}^{M} \frac{1}{i}$ increases with the number of branches, but not in a linear fashion. The largest improvement is when we move from the case of single-branch (no diversity at all) to two-branch case diversity. The outage probability is obtained from Eq. (8.12) by $P_{\text{outage}} = CDF_{\rho_{\oplus}}(\rho_0) = \prod_{i=1}^{M} \left(1 - e^{-\rho_0/\overline{\rho}_i}\right)$, while the average symbol error probability for different modulation formats can be obtained by substituting (8.12) into (8.8). For instance, for BPSK the average bit error probability, in Rayleigh fading, for SC is given by:

$$\overline{P}_b(\text{BPSK}) = \int_0^{\infty} Q\left(\sqrt{2\rho}\right) f_{\rho_{\oplus}}(\rho) d\rho = \int_0^{\infty} Q\left(\sqrt{2\rho}\right) \frac{M}{\overline{\rho}} \left(1 - e^{-\rho/\overline{\rho}}\right)^{M-1} e^{-\rho/\overline{\rho}} d\rho. \tag{8.15}$$

For DPSK, in Rayleigh fading, the average bit error probability can be found in closed form:

$$\overline{P}_b(\text{DPSK}) = \int_0^{\infty} \underbrace{\frac{1}{2} e^{-\rho}}_{\overline{P}_b(\rho)} \underbrace{\frac{M}{\overline{\rho}} \left(1 - e^{-\rho/\overline{\rho}}\right)^{M-1} e^{-\rho/\overline{\rho}}}_{f_{\rho_{\oplus}}(\rho)} d\rho = \frac{M}{2} \sum_{m=0}^{M-1} (-1)^m \frac{\binom{M-1}{m}}{1 + m + \overline{\rho}}. \tag{8.16a}$$

In Fig. 8.4 we show the outage probability vs. normalized SNR, defined as $\overline{\rho}/\rho_0$, of the selection combining scheme for different numbers of diversity branches M in Rayleigh fading. Clearly, the largest improvement is obtained by moving from single-branch case (no diversity at all) to two-branch-based SC. On the other hand, in Fig. 8.5 we show the average bit error probability against average SNR $\overline{\rho}$ for BPSK in Rayleigh fading when SC is used for different numbers of diversity branches. The results are obtained by numerical evaluation of integral in Eq. (8.15). The largest improvements over single diversity branch are obtained for $M = 2$ and $M = 4$ diversity branches. Going above $M = 8$ diversity branches does not make sense, as the further increasing number of branches provides diminishing improvements.

Fig. 8.4 The outage probability vs. normalized SNR ($\overline{\rho}/\rho_0$) of selection combining scheme for different number of diversity branches in Rayleigh fading

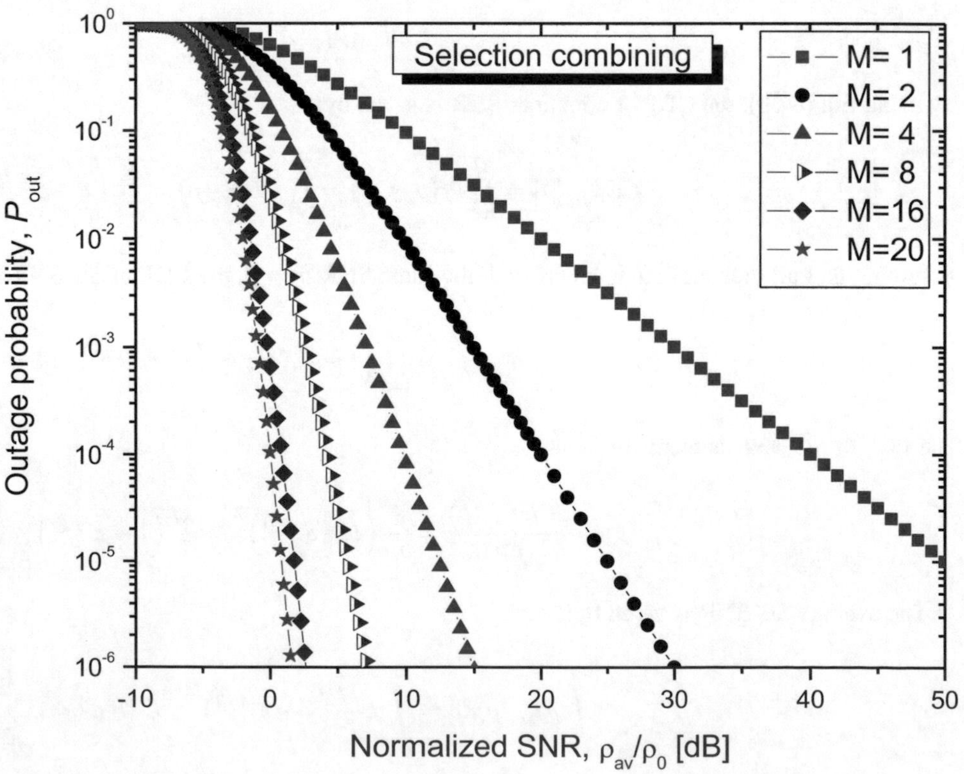

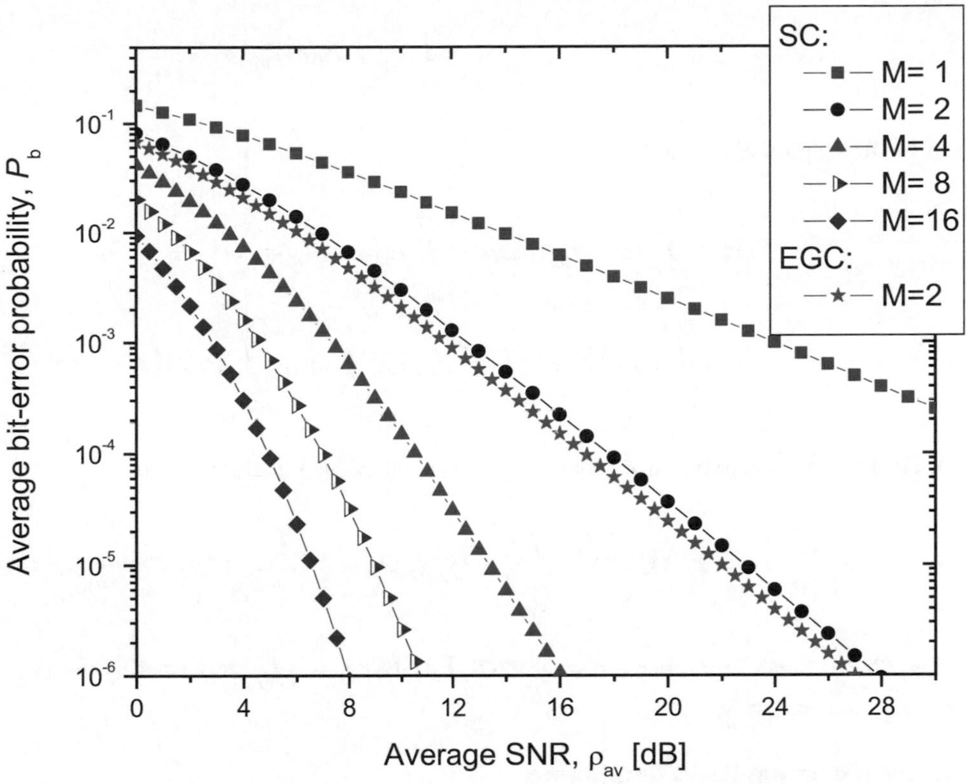

Fig. 8.5 The average bit error probability vs. average SNR ($\overline{\rho}$) of selection combining scheme for BPSK for different numbers of diversity branches in Rayleigh fading

8.1.2.2 Threshold Combining

In threshold combining (TC), the combiner scans different branches and select the first branch with SNR larger than threshold SNR, denoted as ρ_{Tsh}. The TC outputs the signal from this branch as long as the SNR is larger than threshold SNR. Once the SNR falls below ρ_{Tsh}, the TC scans sequentially other branches until it finds another branch with SNR larger than ρ_{Tsh}. The TC with only two branches is commonly referred to as switch-and-stay combiner (SSC). The CDF for SSC can be determined as [1, 9]:

$$CDF_{\rho_\oplus}(\rho) = \begin{cases} CDF_{\rho_1}(\rho_{\text{Tsh}})CDF_{\rho_2}(\rho), & \rho < \rho_{\text{Tsh}} \\ \Pr(\rho_{\text{Tsh}} \le \rho_1 \le \rho) + CDF_{\rho_1}(\rho_{\text{Tsh}})CDF_{\rho_2}(\rho), & \rho \ge \rho_{\text{Tsh}} \end{cases} \tag{8.16b}$$

The first case in (8.16b), $\rho < \rho_{\text{Tsh}}$, corresponds to the case that when the SNR in branch 1, ρ_1, dropped below the threshold, we switched to branch 2; however, the SNR on branch 2, ρ_2, is also below the threshold. The second case in (8.16b), $\rho \ge \rho_{\text{Tsh}}$, corresponds to the situation when either SNR on the first branch ρ_1 is larger than threshold but smaller than ρ or ρ_1 dropped below the threshold and we moved to branch 2, while the SNR on branch two was above the threshold.

As an illustration, the CDF for Rayleigh fading, providing that $\overline{\rho}_i = \overline{\rho}$, becomes:

$$CDF_{\rho_\oplus}(\rho) = \begin{cases} 1 - e^{-\rho_{\text{Tsh}}/\overline{\rho}} - e^{-\rho/\overline{\rho}} + e^{-(\rho_{\text{Tsh}}+\rho)/\overline{\rho}}, & \rho < \rho_{\text{Tsh}} \\ 1 - 2e^{-\rho/\overline{\rho}} + e^{-(\rho_{\text{Tsh}}+\rho)/\rho}, & \rho \ge \rho_{\text{Tsh}} \end{cases} \tag{8.17}$$

The PDF of SSC in Rayleigh fading can be found as the first derivative with respect to ρ:

$$f_{\rho_\oplus}(\rho) = \frac{dCDF_{\rho_\oplus}(\rho)}{d\rho} = \begin{cases} \left(1 - e^{-\rho_{\text{Tsh}}/\overline{\rho}}\right)(1/\overline{\rho})e^{-\rho/\overline{\rho}}, & \rho < \rho_{\text{Tsh}} \\ \left(2 - e^{-\rho_{\text{Tsh}}/\overline{\rho}}\right)(1/\overline{\rho})e^{-\rho/\overline{\rho}}, & \rho \ge \rho_{\text{Tsh}} \end{cases} \tag{8.18}$$

The outage probability can be obtained from Eq. (8.17), by setting $\rho = \rho_0$, namely, $P_{\text{outage}} = CDF_{\rho_\oplus}(\rho_0)$. The average symbol error probability can be expressed as follows:

$$\overline{P}_s = \int_0^\infty P_s(\rho) f_{\rho_\oplus}(\rho) d\rho, \tag{8.19}$$

and for Rayleigh fading, we can write:

$$\begin{aligned}\overline{P}_s &= \int_0^{\rho_{\text{Tsh}}} P_s(\rho) f_{\rho_\oplus}(\rho) d\rho + \int_{\rho_{\text{Tsh}}}^\infty P_s(\rho) f_{\rho_\oplus}(\rho) d\rho = \\ &= \int_0^{\rho_{\text{Tsh}}} P_s(\rho)\left(1 - e^{-\rho_{\text{Tsh}}/\overline{\rho}}\right)(1/\overline{\rho})e^{-\rho/\overline{\rho}} d\rho + \int_{\rho_{\text{Tsh}}}^\infty P_s(\rho)\left(2 - e^{-\rho_{\text{Tsh}}/\overline{\rho}}\right)(1/\overline{\rho})e^{-\rho/\overline{\rho}} d\rho.\end{aligned} \tag{8.20}$$

For DPSK, the average bit error probability can be determined in closed form:

$$\overline{P}_b(\text{DPSK}) = \int_0^\infty \frac{1}{2} e^{-\rho} f_{\rho_\oplus}(\rho) d\rho = \frac{1}{2(1+\overline{\rho})}\left(1 - e^{-\rho_{\text{Tsh}}/\overline{\rho}} + e^{-\rho_{\text{Tsh}}} e^{-\rho_{\text{Tsh}}/\overline{\rho}}\right). \tag{8.21}$$

When SNR is much larger than threshold SNR, $\overline{P}_b(\text{DPSK}) \to \frac{1}{2(1+\overline{\rho})}$, representing the single-branch result. Therefore, SSC is inferior compared to SC.

8.1.2.3 Maximum-Ratio Combining

In maximum-ratio combining (MRC), the outputs of different diversity branches, shown in Fig. 8.3, are properly weighted so that the combiner SNR is maximized. Assuming that all branches are composed of identical devices so that the noise PSD N_0 is the same, the combiner output is given by $r = \Sigma_i w_i r_i \underline{s}$, where $\underline{s}$ is the transmitted 2D symbol of energy normalized to one. Further assuming that the channel bandwidth × symbol duration product is one, the total noise power on all branches will be $\sum_{i=1}^{M} r_i^2 N_0$, and the combiner SNR will be upper bounded by:

$$\rho_\oplus = \frac{\left(\sum_{i=1}^{M} w_i r_i\right)^2}{N_0 \sum_{i=1}^{M} w_i^2} \le \frac{\sum_{i=1}^{M} w_i^2 \sum_{i=1}^{M} r_i^2}{N_0 \sum_{i=1}^{M} w_i^2} = \sum_{i=1}^{M} \rho_i. \tag{8.22}$$

The equality sign is achieved when the squared weight of the i-th branch is proportional to the i-th branch SNR, namely, $w_i^2 \simeq r_i^2/N_0$. Therefore, in MRC the combiner SNR is equal to the sum of SNRs of individual branches.

The PDF of $\rho_\oplus$ can be obtained as "inverse" Laplace transform of moment-generating functions (MGFs) of SNRs for individual branches. The PDF in Rayleigh fading is χ^2 with 2 M degrees of freedom, expected value $M\overline{\rho}$, and variance $2M\overline{\rho}$:

$$f_{\rho_\oplus}(\rho) = \frac{\rho^{M-1} e^{-\rho/\overline{\rho}}}{\overline{\rho}^M (M-1)!}, \quad \rho \ge 0 \tag{8.23}$$

The outage probability in this case can be determined in closed form:

$$P_{\text{outage}} = \Pr(\rho_\oplus \le \rho_0) = \int_0^{\rho_0} f_{\rho_\oplus}(\rho) d\rho = 1 - e^{-\rho_0/\overline{\rho}} \sum_{m=1}^{M} \frac{(\rho_0/\overline{\rho})^{m-1}}{(m-1)!}. \tag{8.24}$$

The average symbol error probability for Rayleigh fading can be determined as:

$$\overline{P}_s = \int_0^{\infty} P_s(\rho)\underbrace{\frac{\rho^{M-1}e^{-\rho/\overline{\rho}}}{\overline{\rho}^M(M-1)!}}_{f_{\rho_\oplus}(\rho)}d\rho, \tag{8.25}$$

and for BPSK the average bit error probability can be found in closed form:

$$\begin{aligned}\overline{P}_b(\text{BPSK}) &= \int_0^{\rho_0} Q\left(\sqrt{2\rho}\right)\underbrace{\frac{\rho^{M-1}e^{-\rho/\overline{\rho}}}{\overline{\rho}^M(M-1)!}}_{f_{\rho_\oplus}(\rho)}d\rho \\ &= \left(\frac{1-\rho_n}{2}\right)^M \sum_{k=1}^{M}\binom{M-1+m}{m}\left(\frac{1+\rho_n}{2}\right)^m, \quad \rho_n = \sqrt{\overline{\rho}/(1+\overline{\rho})}\end{aligned} \tag{8.26}$$

The upper bound on symbol error probability for M-ary QAM, obtained by employing the Chernoff bound [$Q(x) \leq \exp(-x^2/2)$], is given by:

$$P_s = \alpha_M Q\left(\sqrt{\beta_M \rho_\oplus}\right) \leq \alpha_M \exp\left(-\beta_M \rho_\oplus/2\right) = \alpha_M \exp\left(-\beta_M[\rho_1 + \ldots + \rho_M]/2\right) \tag{8.27}$$

so that the average symbol error probability can be upper bounded by:

$$\overline{P}_s = \int_0^{\infty} P_s(\rho)f_{\rho_\oplus}(\rho)d\rho \leq \alpha_M \prod_{i=1}^{M} \frac{1}{1+\beta_M\overline{\rho}_i/2}. \tag{8.28}$$

When all branches in MRC scheme are identical, for high SNR, the upper bound is simply:

$$\overline{P}_s \leq \alpha_M(\beta_M\overline{\rho}/2)^{-M}, \tag{8.29}$$

indicating that the diversity gain is given by:

$$DG = -\lim_{\rho\to\infty}\frac{\log \overline{P}_s}{\log \rho} = M, \tag{8.30}$$

and the MRC schemes achieves the full diversity.

For other fading distributions, the MGF approach is more suitable. The MGF for a nonnegative random variable ρ with PDF $f_\rho(\rho)$ is defined as:

$$MGF_\rho(s) = \int_0^{\infty} e^{s\rho} f_\rho(\rho)d\rho. \tag{8.31}$$

Clearly, the MGF is related to the Laplace transform (LT) of $f_\rho(\rho)$ by $LT\{f_\rho(\rho)\} = MGF_\rho(-s)$. As the name suggests, the moments of ρ^n, denoted as $E[\rho^n]$, can be determined as:

$$E[\rho^n] = \frac{\partial^n}{\partial s^n}MGF_\rho(s)\bigg|_{s=0}. \tag{8.32}$$

The MGF for Nakagami-m fading is given by:

$$MGF_\rho(s) = \frac{1}{(1 - s\overline{\rho}/m)^m}, \tag{8.33}$$

while for the Rician fading is given as:

$$MGF_\rho(s) = \frac{1+K}{1+K-s\overline{\rho}} \exp\left[\frac{Ks\overline{\rho}}{1+K-s\overline{\rho}}\right], \tag{8.34}$$

with K being the Rician parameter K, defined by $K = s^2/(2\sigma^2)$ (s^2 is the power in line of sight and $2\sigma^2$ is the power in multipath components). Based on Eq. (8.22), we concluded that in MRC, the combiner SNR is equal to the sum of SNRs of individual branches, $\rho_\oplus = \sum_i \rho_i$. Assuming that SNRs at different branches are independent of each other, the joint PDF can be represented as the product of PDFs:

$$f_{\rho_1,\ldots,\rho_M}(\rho_1,\ldots,\rho_M) = f_{\rho_1}(\rho_1) f_{\rho_2}(\rho_2)\cdot\ldots\cdot f_{\rho_M}(\rho_M). \tag{8.35}$$

When the symbol error probability is of the form:

$$P_s(\rho) = \alpha \exp(-\beta\rho), \tag{8.36}$$

where α and β are constants, such as for DPSK and M-ary QAM for high SNRs (see Eq. (8.27)); the average symbol error probability can be related to MGF as follows:

$$\begin{aligned}
\overline{P}_s &= \int_0^\infty\int_0^\infty\cdots\int_0^\infty \alpha \exp[-\beta(\rho_1+\ldots+\rho_M)] f_{\rho_1}(\rho_1)\ldots f_{\rho_M}(\rho_M) d\rho_1\ldots d\rho_M \\
&= \alpha \int_0^\infty\int_0^\infty\cdots\int_0^\infty \prod_{i=1}^M \exp[-\beta\rho_i] f_{\rho_i}(\rho_i) d\rho_i = \alpha \prod_{i=1}^M \underbrace{\int_0^\infty \exp[-\beta\rho_i] f_{\rho_i}(\rho_i) d\rho_i}_{MGF_{\rho_i}(-\beta)} \\
&= \alpha \prod_{i=1}^M MGF_{\rho_i}[-\beta].
\end{aligned} \tag{8.37}$$

As an illustration, the average bit error probability for DPSK will be obtained from (8.32) by setting $\alpha = 1/2$ and $\beta = 1$ to obtain $\overline{P}_b = \frac{1}{2}\prod_{i=1}^M MGF_{\rho_i}[-1]$.

Another relevant case is when the symbol error probability is of the form [1]:

$$P_s(\rho) = \int_c^d \alpha \exp(-\beta(u)\rho) du, \tag{8.38}$$

which is applicable to M-ary PSK and M-ary QAM, namely, an alternative expression for M-ary PSK is due to Craig [8]:

$$P_s(\rho) = \frac{1}{\pi} \int_0^{\pi-\pi/M} \exp\left(-\frac{\sin^2(\pi/M)}{\sin^2\phi}\rho\right) d\phi, \tag{8.39}$$

and can be related to (8.38) by $\alpha = 1/\pi$, $\beta(\phi) = \sin^2(\pi/M)/\sin^2\phi$, $c = 0$, and $d = \pi - \pi/M$. Assuming again that SNRs at different branches are independent of each other, the average symbol error probability is as follows:

$$\bar{P}_s = \int_{\infty}^{0}\int_{\infty}^{0}\cdots\int_{\infty}^{0}\int_{d}^{c} \alpha\exp(-\beta(x)\rho)dxf_{\rho_{\oplus}}(\rho)d\rho = \int_{\infty}^{0}\int_{\infty}^{0}\cdots\int_{\infty}^{0}\int_{d}^{c} \prod_{i=1}^{M}\alpha\exp(-\beta(u)\rho_i)f_{\rho_i}(\rho_i)dud\rho$$
$$= \int_{d}^{c}\prod_{i=1}^{M}\int_{0}^{\infty}\alpha\exp(-\beta(u)\rho_i)f_{\rho_i}(\rho_i)dud\rho = \alpha\int_{c}^{d}\prod_{i=1}^{M} MGF_{\rho_i}(-\beta(u))du \tag{8.40}$$

The average symbol error probability for M-ary PSK will be then:

$$\overline{P}_s(\text{MPSK}) = \frac{1}{\pi}\int_{0}^{\pi-\pi/M}\prod_{i=1}^{M} MGF_{\rho_i}\left(-\frac{\sin^2(\pi/M)}{\sin^2\phi}\right)d\phi. \tag{8.41}$$

This MGF-based method is suitable only for MRC schemes, since the combiner SNR in this case is equal to the sum of SNRs of individual branches, which is not the case for other diversity combining schemes.

In Fig. 8.6 we show the outage probability vs. normalized SNR, also defined as $\overline{\rho}/\rho_0$, of the maximum-ratio combining scheme for different numbers of diversity branches M in Rayleigh fading. By comparing Figs. 8.6 and 8.4, we conclude that MRC significantly outperforms the SC scheme.

On the other hand, in Fig. 8.7 we show the average bit error probability against average SNR $\overline{\rho}$ for BPSK in Rayleigh fading when MRC is used for different numbers of diversity branches. Similarly as in SC, the largest improvements over single diversity branch are obtained for $M = 2$ and $M = 4$ diversity branches. Going above $M = 8$ diversity branches does not make sense, as the further increasing in number of branches provides diminishing improvements. By comparing Figs. 8.7 and 8.5, we conclude that MRC significantly outperforms SC. As an illustration, for $M = 4$ and BPSK, the improvement of MRC over SC is about 4.8 dB at average bit error probability of 10^{-6}.

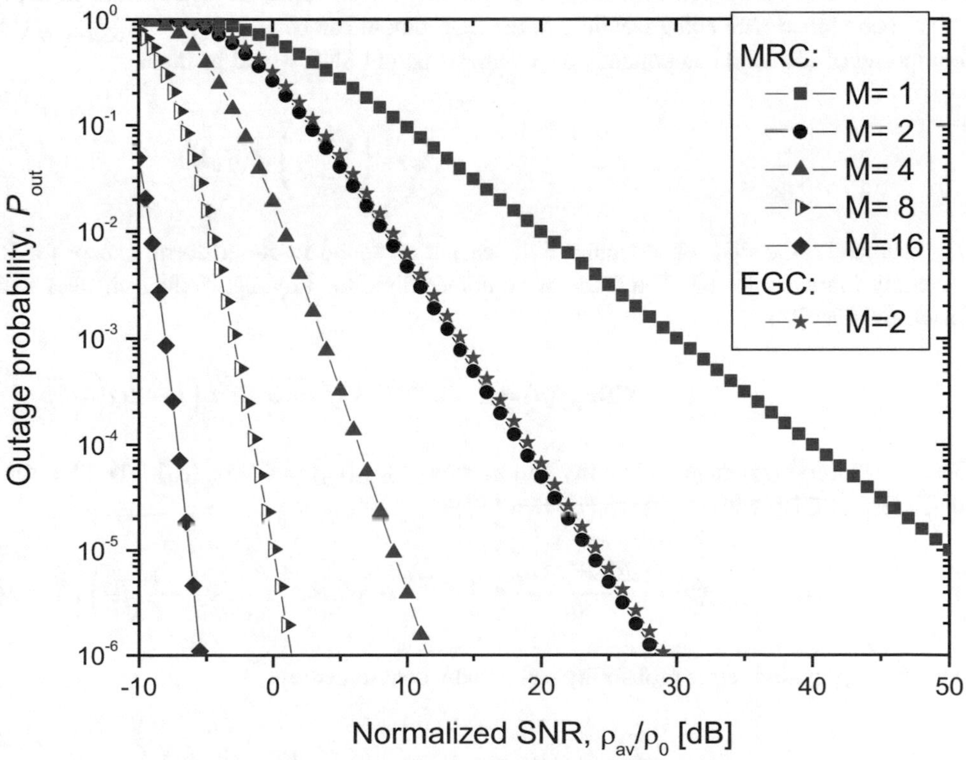

Fig. 8.6 The outage probability vs. normalized SNR ($\overline{\rho}/\rho_0$) of maximum-ratio combining scheme for different numbers of diversity branches in Rayleigh fading as well as for EGC with two diversity branches

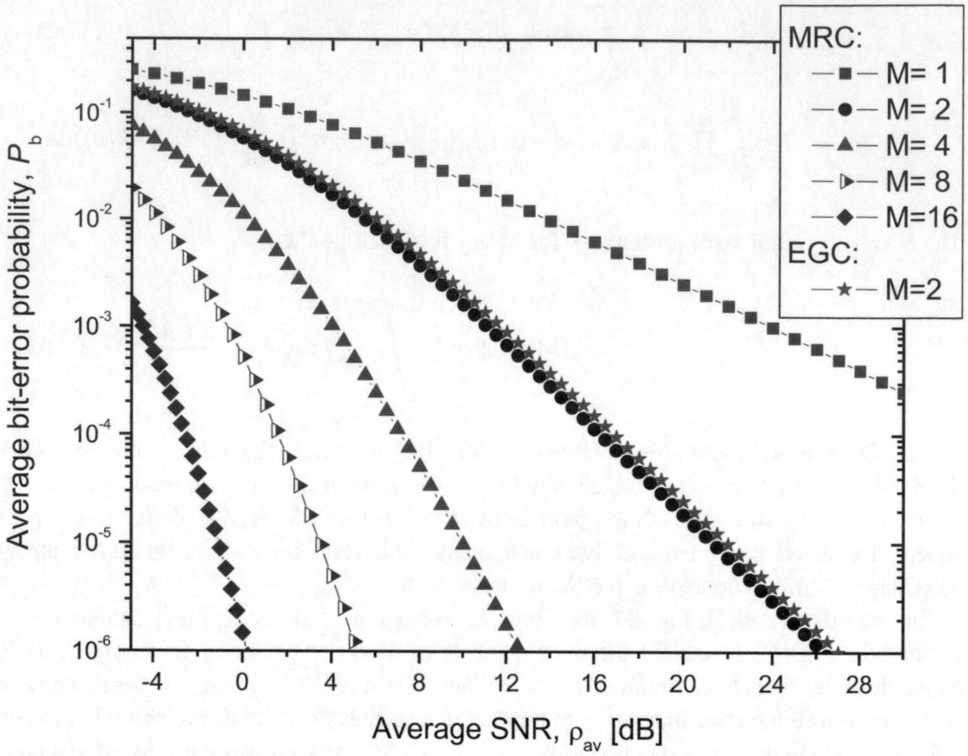

Fig. 8.7 The average bit error probability vs. averaged SNR ($\overline{\rho}$) of maximum-ratio combining scheme for different number of diversity branches in Rayleigh fading as well as for EGC with two diversity branches

8.1.2.4 Equal-Gain Combining

In equal-gain combining (EGC), once the co-phasing is performed, the signals from different diversity branches r_i ($i = 1,2,\ldots,M$) are combined with equal weights, so the EGC output can be written as $r = \sum_{i=1}^{M} r_i/\sqrt{M}$. Assuming that all branches are composed of identical components, the combiner output SNR would be then:

$$\rho_{\oplus} = \left(\sum_{i=1}^{M} r_i\right)^2 /(N_0 M). \tag{8.42}$$

Unfortunately, the PDF of combiner SNR cannot be found in closed form, except for Rayleigh fading channel with two diversity branches [1, 9]. The CDF of combiner SNR for Rayleigh fading channel with two diversity branches can be determined as [9]:

$$CDF_{\rho_{\oplus}}(\rho) = 1 - e^{-2\rho/\overline{\rho}} - \sqrt{\pi\rho/\overline{\rho}}\, e^{-\rho/\overline{\rho}}\left(1 - 2Q\left(\sqrt{2\rho/\overline{\rho}}\right)\right). \tag{8.43}$$

The corresponding outage probability will be then $P_{\text{outage}}(\rho_0) = CDF_{\rho_{\oplus}}(\rho_0)$. The PDF for this case can be found as the first derivative of CDF with respect to combiner SNR:

$$f_{\rho_{\oplus}}(\rho) = \frac{dCDF_{\rho_{\oplus}}(\rho)}{d\rho} = \frac{1}{\overline{\rho}}e^{-2\rho/\overline{\rho}} - \sqrt{\pi}e^{-\rho/\overline{\rho}}\left(\frac{1}{\sqrt{4\rho\overline{\rho}}} - \frac{1}{\overline{\rho}}\sqrt{\frac{\rho}{\overline{\rho}}}\right)\left(1 - 2Q\left(\sqrt{2\frac{\rho}{\overline{\rho}}}\right)\right). \tag{8.44}$$

The average symbol error probability can be now determined by:

$$\overline{P}_s = \int_0^{\infty} P_s(\rho)\underbrace{\left[\frac{1}{\overline{\rho}}e^{-2\rho/\overline{\rho}} - \sqrt{\pi}e^{-\rho/\overline{\rho}}\left(\frac{1}{\sqrt{4\rho\overline{\rho}}} - \frac{1}{\overline{\rho}}\sqrt{\frac{\rho}{\overline{\rho}}}\right)\left(1 - 2Q\left(\sqrt{2\frac{\rho}{\overline{\rho}}}\right)\right)\right]}_{f_{\rho_{\oplus}}(\rho)}d\rho. \tag{8.45}$$

For BPSK, the average error probability expression can be derived in closed form:

$$P_b(\text{BPSK}) = \int_0^\infty Q\left(\sqrt{2\rho}\right) f_{\rho_\oplus}(\rho)d\rho = \frac{1}{2}\left[1 - \sqrt{1 - \left(\frac{1}{1+\overline{\rho}}\right)^2}\right] \tag{8.46}$$

Based on Fig. 8.5, we conclude that the power penalty of EGC with respect to MGC, for medium SNRs, when the outage probability is concerned for $M = 2$ case, is about 0.6 dB at P_{out} of 10^{-6}. Interestingly enough, based on Fig. 8.6, the power penalty of EGC for $M = 2$ and BPSK, in Rayleigh fading, is also about 0.6 dB. On the other hand, as in Fig. 8.5, we conclude that EGC scheme for $M = 2$ (with BPSK) outperforms the corresponding SC scheme for almost 0.9 dB at average bit error probability of 10^{-6}.

8.1.3 Transmitter Diversity

In transmitter diversity multiple transmitters are used, wherein the transmit power is divided among these transmitters. In optical communications, these transmitters are either laser diodes or LEDs, while in wireless communications, these are transmit antennas. Two relevant cases of transmitter diversity include (i) the case when the channel state information (CSI) is available on transmitter side and (ii) when the CSI is not available on transmitter side, the Alamouti-type Scheme [10] is used in this case.

8.1.3.1 Transmitter Diversity when CSI Is Available on Transmitter Side

Let us consider a transmit diversity system with M transmitters and one receiver. For optical communications with coherent optical detection and wireless communications, the CSI associated with the i-th transmitter (laser diode, antenna), given by $r_i\exp(j\theta_i)$, is known at the transmitter side. The transmitted signal constellation point $\underline{s}$ is pre-multiplied by a complex channel coefficient $w_i\exp(-j\theta_i)$ ($0 \leq w_i \leq 1$), before being sent with the help of the i-th transmitter (laser diode, antenna). The corresponding weight coefficients (in electrical domain) satisfy the condition $\sum_i w_i^2 = 1$. Clearly, the co-phasing is performed on a transmitter side. The weights w_i maximizing the SNR on a receiver side are given by [1, 9]:

$$w_i = r_i / \left(\sum_{i=1}^{M} r_i^2\right)^{1/2}. \tag{8.47}$$

The receiver side SNR can be expressed now as:

$$\rho_\oplus = \frac{E_s}{N_0}\sum_{i=1}^{M} r_i^2 = \sum_{i=1}^{M} \rho_i. \tag{8.48}$$

Clearly, the reviver side SNR is equal to the sum of SNRs originating from different transmit branches, the result identical to that in MRC, indicating that the corresponding combiner PDF for Rayleigh fading is given by Eq. (8.23). When the samples originating from different transmitters are identical, $r_i = r$, the overall SNR is $\rho_\oplus = Mr^2E_s/N_0$, and the corresponding array gain is $AG = M$.

The upper bound on symbol error probability for M-ary QAM, obtained by employing the Chernoff bound, is given by:

$$P_s(\rho_\oplus) = \alpha_M Q\left(\sqrt{\beta_M \rho_\oplus}\right) \leq \alpha_M \exp\left(-\beta_M \rho_\oplus/2\right) = \alpha_M \exp\left(-\beta_M \sum_{i=1}^{M} \rho_i/2\right). \tag{8.49}$$

The average error probability can be determined by averaging over different SNR-values:

$$\overline{P}_s = \int_0^\infty P_s(\rho) f_{\rho_\oplus}(\rho)d\rho \leq \alpha_M \prod_{i=1}^{M} \frac{1}{1+\beta_M \overline{\rho}_i/2}. \tag{8.50}$$

When the average SNR values originating from different transmitter branches are identical, $\overline{\rho}_i = \overline{\rho}$, the average error probability is upper bounded by:

$$\overline{P}_s \leq \alpha_M (\beta_M \overline{\rho}/2)^{-M}, \tag{8.51}$$

and the corresponding diversity gain is given by $DG = M$, indicating that the full diversity is possible, the same as for MRC scheme.

8.1.3.2 Transmitter Diversity when CSI Is Not Available: Alamouti Scheme

When the CSI is not available on the transmitter side, we can employ the Alamouti scheme [10], shown in Fig. 8.7, illustrated for wireless communication applications. The *code matrix of Alamouti code* is given by:

$$\boldsymbol{X} = \begin{bmatrix} \underline{x}_1 & -\underline{x}_2^* \\ \underline{x}_2 & \underline{x}_1^* \end{bmatrix} \tag{8.52}$$

where the columns correspond to 2D signal constellation points transmitted on antennas Tx_1 and Tx_2, while the rows correspond to two channel uses. Therefore, the symbols transmitted on antennas Tx_1 and Tx_2 in the first channel use are $\underline{x}_1$ and $-\underline{x}_2^*$, respectively. On the other hand, the symbols transmitted on antennas Tx_1 and Tx_2 in the second channel use are $\underline{x}_2$ and $\underline{x}_1^*$, respectively

The signals received in two consecutive time slots, denoted as $\underline{y}_1$ and $\underline{y}_2$, are given by:

$$\underline{y}_1 = h_1\underline{x}_1 + h_2\underline{x}_2 + \underline{z}_1,\ \underline{y}_2 = -h_1\underline{x}_2^* + h_2\underline{x}_1^* + \underline{z}_2, \tag{8.53}$$

where h_1 is the channel coefficient between receiver antenna and transmitter antenna Tx_1, while h_2 is the channel coefficient between receiver antenna and transmitter antenna Tx_2. We use $\underline{z}_1$ and $\underline{z}_2$ to denote the zero-mean circularly symmetric Gaussian noise samples at two consecutive channel uses.

The combiner, as shown in Fig. 8.8, operates as follows:

$$\tilde{x}_1 = h_1^*\underline{y}_1 + h_2\underline{y}_2^*,\ \tilde{x}_2 = h_2^*\underline{y}_1 - h_1\underline{y}_2^*, \tag{8.54}$$

where $\tilde{x}_1$ and $\tilde{x}_2$ are estimations of transmitted symbols $\underline{x}_1$ and $\underline{x}_2$. After the substitutions of Eq. (8.53) into (8.54), we obtain:

$$\begin{aligned} \tilde{x}_1 &= h_1^*(h_1\underline{x}_1 + h_2\underline{x}_2 + \underline{z}_1) + h_2(-h_1^*\underline{x}_2 + h_2^*\underline{x}_1 + \underline{z}_2^*) \\ &= (|h_1|^2 + |h_2|^2)\underline{x}_1 + (h_1^*\underline{z}_1 + h_2\underline{z}_2^*) \\ \tilde{x}_2 &= (|h_1|^2 + |h_2|^2)\underline{x}_2 + (h_2^*\underline{z}_1 - h_1\underline{z}_2^*). \end{aligned} \tag{8.55}$$

Clearly, after the combiner, the spatial interference is compensated for, while the noise power is doubled. The receiver SNR is given by:

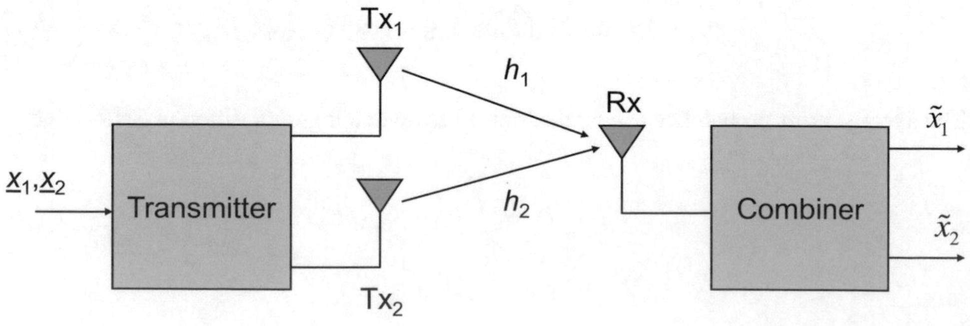

Fig. 8.8 The Alamouti scheme for two transmit and one receive antennas

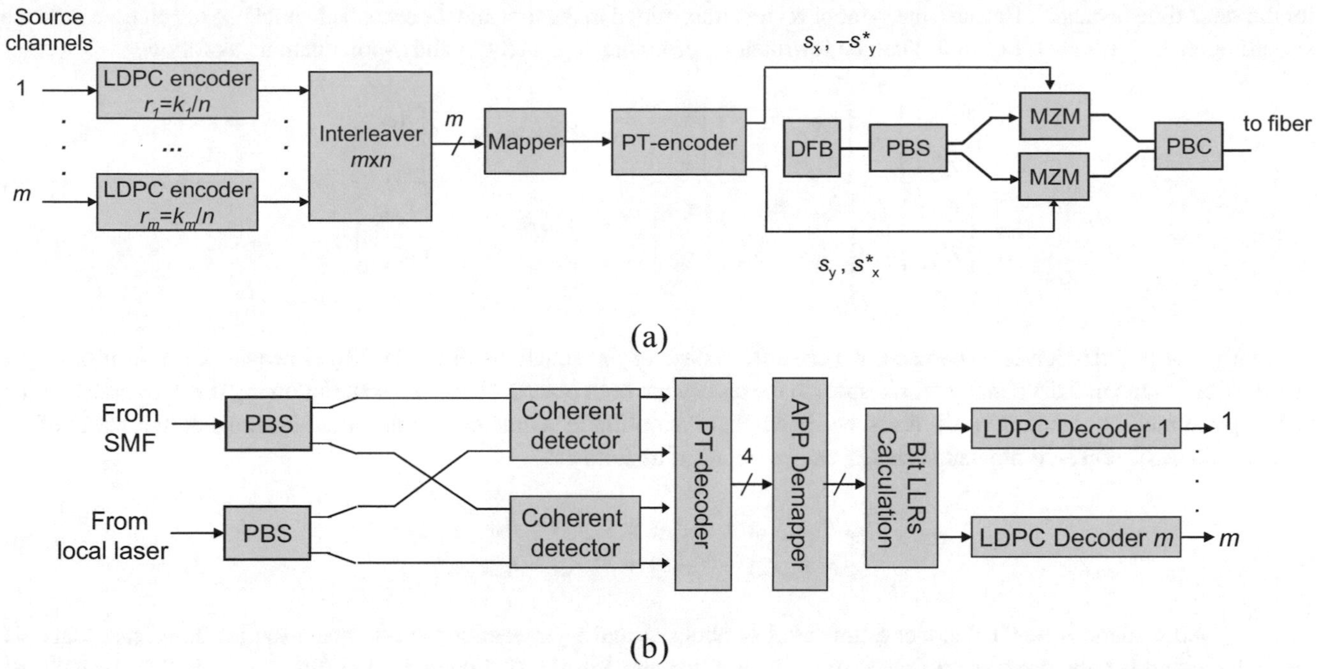

Fig. 8.9 The architecture of PT-coding scheme concatenated with LDPC coding for multilevel modulations: (**a**) transmitter architecture and (**b**) receiver architecture

$$\rho = \frac{\left(|h_1|^2 + |h_2|^2\right)}{2} \frac{E_s}{N_0}, \tag{8.56}$$

and represents 1/2 of the sum of SNRs from both transmitted branches. Clearly, the array gain is 1. Since the spatial interference is compensated for (see Eq. (8.55)), the Alamouti scheme achieves the full diversity ($DG = 2$).

In the rest of this section, we describe how this scheme can be used to compensate for polarization mode dispersion (PMD) and polarization-dependent loss (PDL) in fiber-optic communications [11] (see also [5]). The polarization time (PT) encoder operates as follows [11]. In the first half of *i*-th time instance ("the first channel use"), it sends the symbol s_x to be transmitted using x-polarization channel and the symbol s_y to be transmitted using y-polarization channel, as illustrated in Fig. 8.9a. In the second half of *i*-th time instance ("the second channel use"), it sends the symbol $-s_\mathrm{y}^*$ to be transmitted using x-polarization channel and the symbol $-s_\mathrm{x}^*$ to be transmitted using y-polarization. Therefore, the PT-coding procedure is similar to the Alamouti scheme [10]. Notice that Alamouti-type PT-coding scheme has the spectral efficiency comparable to coherent OFDM with polarization diversity scheme. When the channel is used twice during the same symbol period, the spectral efficiency of this scheme is twice higher than that of polarization diversity OFDM. Notice that the hardware complexity of PT encoder/decoder is trivial compared to that of OFDM. The transmitter complexity is slightly higher than that that of polarization diversity receiver, as it requires additional PT encoder, a polarization beam splitter (PBS), and a polarization beam combiner (PBC). On the receiver side, we have the option to use only one polarization or to use both polarizations. The receiver architecture employing both polarizations is shown in Fig. 8.9b. The received symbol vectors in the *i*-th time instance for the first ($\boldsymbol{r}_{i,1}$) and second channel use ($\boldsymbol{r}_{i,2}$) can be written, respectively, as follows:

$$\boldsymbol{r}_{i,1} = \boldsymbol{H}\boldsymbol{s}_{i,1}e^{\,j(\phi_T-\phi_{LO})+j\phi_{\mathrm{CD}}(k)} + \boldsymbol{n}_{i,1}, \;\; \boldsymbol{r}_{i,2} = \boldsymbol{H}\boldsymbol{s}_{i,2}e^{\,j(\phi_T-\phi_{LO})+j\phi_{\mathrm{CD}}(k)} + \boldsymbol{n}_{i,2}, \tag{8.57}$$

where $\boldsymbol{H}$ is the Jones (channel) matrix, $\boldsymbol{r}_{i,1(2)} = [r_{\mathrm{x},i,1(2)}\; r_{\mathrm{y},i,1(2)}]^\mathrm{T}$ denotes the received symbol vector in the first (second) channel use of *i*-th time instance, while $\boldsymbol{n}_{i,1(2)} = [n_{\mathrm{x},i,1(2)}\; n_{\mathrm{y},i,1(2)}]^\mathrm{T}$ is the amplified spontaneous emission (ASE) noise vector corresponding the first (second) channel use in the *i*-th time instance. We use $\boldsymbol{s}_{i,1} = [s_{\mathrm{x},i}\; s_{\mathrm{y},i}]^\mathrm{T}$ to denote the symbol transmitted in the first channel use of *i*-th time instance and $\boldsymbol{s}_{i,2} = \left[-s_{\mathrm{y},i}^* s_{\mathrm{x},i}^*\right]^\mathrm{T}$ to denote the symbol transmitted in the second channel use (of the same time instance). Because the symbol vectors transmitted in the first and the second channel use of i-th time instance are orthogonal, $\boldsymbol{s}_{i,1}^{+}\boldsymbol{s}_{i,2} = \boldsymbol{0}$, Eq. (8.57) can be rewritten by grouping separately x- and y-polarizations as follows:

$$\begin{bmatrix} r_{x,i,1} \\ r_{x,i,2}^{*} \end{bmatrix} = \begin{bmatrix} h_{xx} & h_{xy} \\ h_{xy}^{*} & -h_{xx}^{*} \end{bmatrix} \begin{bmatrix} s_{x,i} \\ s_{y,i} \end{bmatrix} e^{j(\phi_T - \phi_{LO}) + j\phi_{CD}(k)} + \begin{bmatrix} n_{x,i,1} \\ n_{x,i,2}^{*} \end{bmatrix},$$
$$\begin{bmatrix} r_{y,i,1} \\ r_{y,i,2}^{*} \end{bmatrix} = \begin{bmatrix} h_{yx} & h_{yy} \\ h_{yy}^{*} & -h_{yx}^{*} \end{bmatrix} \begin{bmatrix} s_{x,i} \\ s_{y,i} \end{bmatrix} e^{j(\phi_T - \phi_{LO}) + j\phi_{CD}(k)} + \begin{bmatrix} n_{y,i,1} \\ n_{y,i,2}^{*} \end{bmatrix}. \tag{8.58}$$

If only one polarization is to be used, we can solve either of the equations in Eq. (8.58). However, the use of only one polarization results in 3 dB penalty with respect to the case when both polarizations are used. Following the derivation similar to that performed by Alamouti [10], it can be shown that the optimum estimates of transmitted symbols at the output of PT decoder, for ASE noise-dominated scenario, can be obtained as follows:

$$\begin{aligned} \widetilde{s}_{x,i} &= h_{xx}^{*} r_{x,1} + h_{xy} r_{x,2}^{*} + h_{yx}^{*} r_{y,1} + h_{yy} r_{y,2}^{*}, \\ \widetilde{s}_{y,i} &= h_{xy}^{*} r_{x,1} - h_{xx} r_{x,2}^{*} + h_{yy}^{*} r_{y,1} - h_{yx} r_{y,2}^{*}, \end{aligned} \tag{8.59}$$

where $\widetilde{s}_{x,i}$ and $\widetilde{s}_{y,i}$ denote the PT-decoder estimates of symbols $s_{x,i}$ and $s_{y,i}$ transmitted in i-th time instance. In the case that only one polarization is to be used, say x-polarization, the last two terms in Eq. (8.59) are to be omitted. For additional details, an interested reader is referred to [11].

8.2 MIMO Optical and Wireless Techniques

Multiple transmitters can be used either to improve the performance through array and diversity gains or to increase data rate through multiplexing of independent data streams. In multiple-input multiple-output (MIMO) systems, the transmitters and receivers can exploit the knowledge about the channel to determine the rank of the so-called channel matrix $\boldsymbol{H}$, denoted as $R_{\boldsymbol{H}}$, which is related to the number of independent data streams that can be simultaneously transmitted. In MIMO systems, we can define the *multiplexing gain* as the asymptotic slope in SNR of the outage capacity:

$$MG = -\lim_{\rho \to \infty} \frac{C_{\text{outage},p}(\rho)}{\log_2 \rho}, \tag{8.60}$$

where $C_{\text{outage},\, p}\,(\rho)$ is the p percentage outage capacity at signal-to-noise ratio ρ. The p percentage capacity is defined as the transmission rate that can be supported in (100-p)% of the channel realizations.

As an illustration, in Fig. 8.10 we provide the basic concept of wireless MIMO communication, with the number of transmit antennas denoted as M_{Tx} and the number of receive antennas denoted as M_{Rx}. The 2D symbol transmitted on the m-th transmitter antennas is denoted by x_m ($m = 1, 2,\ldots, M_{\text{Tx}}$), while the received symbol on the n-th receiver antenna is denoted as y_n. The channel coefficient relating the received symbol on n-th receiver antenna and transmitted symbol on the m-th transmitter antenna is denoted as h_{mn}. By using this model, the output of the n-th receiver antenna can be represented as:

$$y_n = \sum_{m=1}^{M_{\text{Tx}}} h_{nm} x_m + z_n, \tag{8.61}$$

where z_n is the zero-mean Gaussian noise sample with variance $\sigma^2 = N_0 B$, with $N_0/2$ being the double-sided power-spectral density and B being the channel bandwidth.

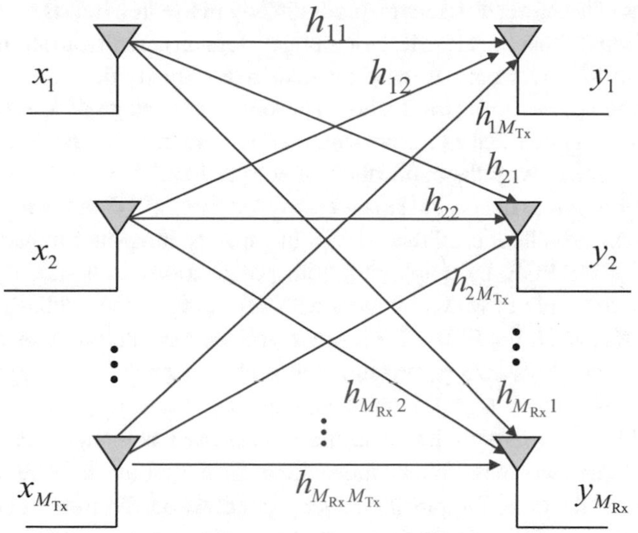

Fig. 8.10 The wireless MIMO concept

8.2.1 MIMO Wireless and Optical Channel Models

A narrowband *point-to-point wireless communication system* composed of M_{Tx} transmitter and M_{Rx} receiver antennas can be represented by the following discrete-time model, which is a matrix representation of Eq. (8.61):

$$\boldsymbol{y} = \boldsymbol{H}\boldsymbol{x} + \boldsymbol{z}, \tag{8.62}$$

where $\boldsymbol{x}$, $\boldsymbol{y}$, and $\boldsymbol{z}$ represent the transmitted, received, and noise vectors defined as:

$$\boldsymbol{x} = \begin{bmatrix} x_1 \\ x_2 \\ \cdots \\ x_{M_{Tx}} \end{bmatrix}, \quad \boldsymbol{y} = \begin{bmatrix} y_1 \\ y_2 \\ \cdots \\ y_{M_{Rx}} \end{bmatrix}, \quad \boldsymbol{z} = \begin{bmatrix} z_1 \\ z_2 \\ \cdots \\ z_{M_{Rx}} \end{bmatrix}, \tag{8.63}$$

As indicated above channel noise samples (z_n) are complex Gaussian with zero-mean and covariance matrix $\sigma^2 I_{Mr}$, where $\sigma^2 = N_0 B$. If the noise variance is normalized to one that the transmitted power P can be normalized as follows $P/\sigma^2 = \rho$, where ρ denotes the average SNR per receiver antenna, assuming unity channel gain, the total normalized transmitted power would be then $\sum_{m=1}^{M_{Tx}} \langle x_i x_i^* \rangle = \rho$. The correlation matrix of transmitted vector $\boldsymbol{x}$ is given by $\boldsymbol{R}_x = \langle \boldsymbol{x}\boldsymbol{x}^\dagger \rangle$. The trace of correlation matrix is clearly $\mathrm{Tr}(\boldsymbol{R}_x) = \rho$. We use $\boldsymbol{H} = [h_{mn}]_{M_{Tx} \times M_{Rx}}$ in Eq. (8.63) to denote the so-called channel matrix:

$$\boldsymbol{H} = \begin{bmatrix} h_{11} & h_{12} & \cdots & h_{1M_{Tx}} \\ h_{21} & h_{22} & \cdots & h_{2M_{Tx}} \\ \cdots & \cdots & \cdots & \cdots \\ h_{M_{Rx}1} & h_{M_{Rx}2} & \cdots & h_{M_{Rx}M_{Tx}} \end{bmatrix}. \tag{8.64}$$

The concept of *MIMO optical communications* has already been introduced in a section on PDM, namely, the polarization-division multiplexing in which two orthogonal polarizations are used to transmit two independent QAM streams. Therefore, the PDM represents an example of 2 × 2 MIMO communication system. On the other hand, in multimode fibers the light propagates in spatial modes, each with different propagation constant and group velocity. The spatial mode excitement at the fiber entrance depends on the launch conditions, while the conditions during propagation depend on the initial conditions and the mode coupling, as analyzed in Sect. 3.6.1. Since a single pulse of light excites a number of spatial modes, it arrives at the output of the multimode fiber as a sum of multiple pulses due to the modal dispersion, an effect that resembles the multipath effect in wireless communications. Therefore, different methods already in use in wireless communications are also applicable in this case including equalization (either in electrical or optical domain), multicarrier modulation and OFDM, spread spectrum technique, diversity, and MIMO. Some of those methods have already been studied for use in both point-to-point links and broadcasting/multicasting applications [14–21, 36]. The key disadvantage of MMF-based MIMO is the problem of coupling between too many modes, which makes it very complex to perform all calculations needed when MIMO signal processing is applied. This is the reason why the application of conventional MMFs is limited to short-reach applications. However, by using the so-called few-mode fibers (FMFs) and few-core fibers (FCFs), the mode-coupling related calculations became simpler, which makes them excellent candidates for high-capacity long-haul transmission. An example of OFDM-MIMO signal transmission over FMF/FCF, by employing both polarizations in coherent detection scheme, is shown in Fig. 8.11. Figure 8.11a shows an example of MIMO system with M_{Tx} transmitters and M_{Rx} receivers, while for simplicity sake, we can assume that $M_{\mathrm{Tx}} = M_{\mathrm{Rx}} = M$. For FMF/FCF applications, we can interpret this MIMO scheme as one employing M OAM modes. Since this system employs both polarizations, it is able to carry 2 M independent two-dimensional (QAM) data streams. The laser diode CW output signal is split among M_{Rx} branches, each containing a transmitter, whose configuration is given in Fig. 8.11b. Two independent data streams are used as inputs to each branch OFDM transmitter. In fact, we can assume that there exist two independent transmitters in each branch (denoted as x and y), one aimed for transmission over x-polarization and the other for transition over y-polarization. To convert OFDM signals to optical domain, I/Q modulators are used. The optical OFDM signals from both polarizations are combined by polarization beam combiner. The PDM data streams are further mode-multiplexed, as illustrated in Fig. 8.11a, and transmitted over FMF/FCF-based system of interest. At the receiver side, M different outputs from FMF/FCF are used as inputs to M coherent receivers, after mode demultiplexing is done. The basic configuration of a coherent detector scheme is shown in Fig. 8.11c, with additional

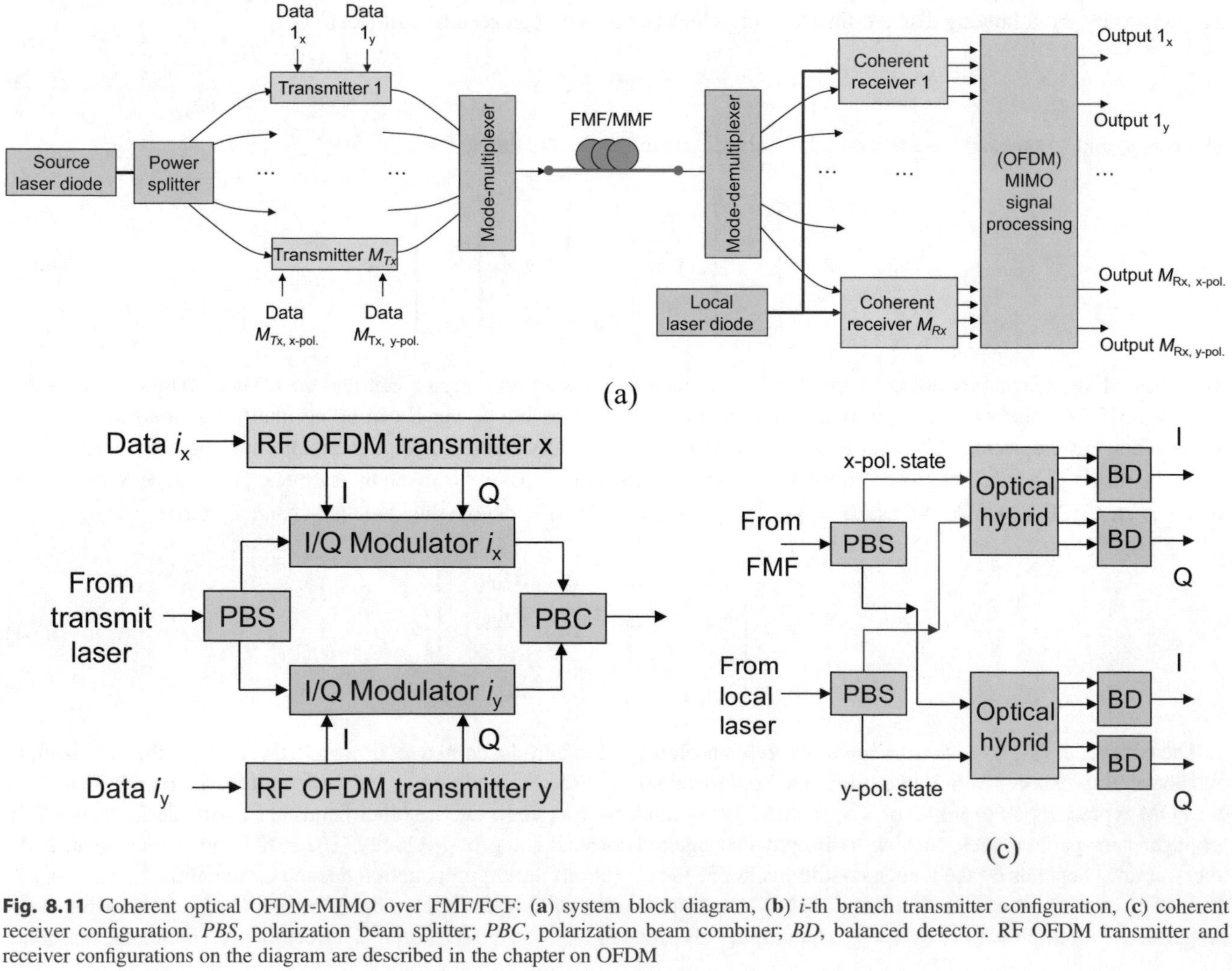

Fig. 8.11 Coherent optical OFDM-MIMO over FMF/FCF: (**a**) system block diagram, (**b**) *i*-th branch transmitter configuration, (**c**) coherent receiver configuration. *PBS*, polarization beam splitter; *PBC*, polarization beam combiner; *BD*, balanced detector. RF OFDM transmitter and receiver configurations on the diagram are described in the chapter on OFDM

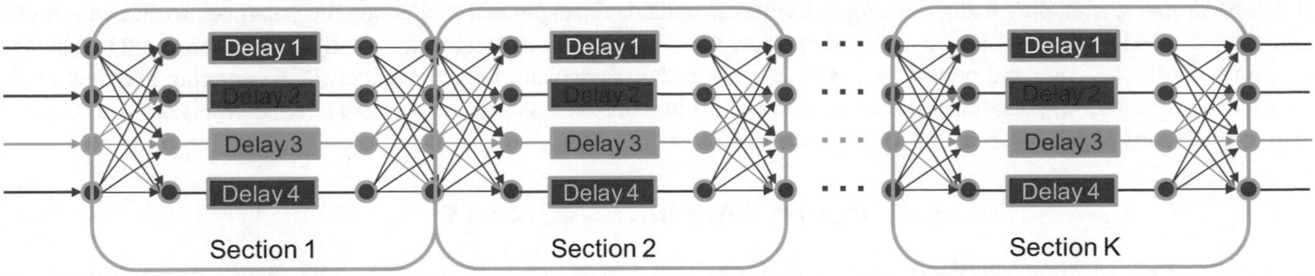

Fig. 8.12 Conceptual model of FMF in the strong coupling regime

details to be provided in OFDM chapter. One local laser oscillator is used for all coherent receivers, to reduce the system cost. The in-phase (I) and quadrature (Q) channel outputs from both polarizations are used as inputs to OFDM-MIMO signal processing block. The use of OFDM elevates simultaneously the impacts of chromatic dispersion, mode-coupling effects, and PMD. The scheme shown in Fig. 8.11 can be used in both point-to-point and multipoint-to-multipoint applications. If optical MIMO system over FMF/FCF has M_{Tx} transmitters and M_{Rx} receivers, the corresponding input-output relation can be expressed as:

$$y_n(t) = \sum_{m=1}^{M_{\mathrm{Tx}}} \sum_{k=1}^{P} h_{mn}(k) e^{\,j\omega_c\left(t-\tau_{p,k}\right)} x_m\left(t-\tau_{g,k}\right) + z_n(t), \tag{8.65}$$

where $y_n(t)$ is the signal received by n-th receiver ($n = 1,2,..,M_{\mathrm{Rx}}$), $h_{mn}(k)$ is the channel gain from m-th transmitter through the k-th mode, while $\tau_{p,k}$ and $\tau_{g,k}$ denote the phase and group delay associated with k-th mode, respectively. Also, ω_{c} is the optical carrier frequency, P is the number of spatial modes, z_n presents the noise process, while $x_m(t)$ denotes the transmitted signal originating from m-th transmitter. We should notice that Eq. (8.65) deals with a single polarization state. When the group delay spread $\Delta\tau_g = \tau_{g,P} - \tau_{g,1}$ is small as compared to symbol duration, we can assume that $x(t - \tau_{g,k}) \approx x(t - \tau_g)$, meaning that all paths arrive at approximately the same time, which is valid as distance does not exceed a certain critical value. The sampled baseband equivalent of function given by Eq. (8.65), with sampling interval T_s, can be written in a matrix form as follows:

$$\boldsymbol{y}(l) = \boldsymbol{Hx}(l) + z(l), \tag{8.66}$$

where:

$$\boldsymbol{x}(l) = \begin{bmatrix} x_1(lT_s) \\ \cdots \\ x_{M_{\mathrm{Tx}}}(lT_s) \end{bmatrix}, \boldsymbol{y}(l) = \begin{bmatrix} y_1(lT_s) \\ \cdots \\ y_{M_{\mathrm{Rx}}}(lT_s) \end{bmatrix},\ z(l) = \begin{bmatrix} z_1(lT_s) \\ \cdots \\ z_{M_R}(lT_s) \end{bmatrix}, [\boldsymbol{H}]_{nm} = \sum_{k=1}^{P} h_{nm}(k) e^{-j\omega_c\tau_{p,k}}. \tag{8.67}$$

We can assume that, based on previous results presented in [15], the number of modes is large so that each transmitter/receiver launch/sample from sufficiently different groups of modes and that the product of carrier frequency and phase-delay spread is large enough to satisfy relationship $\omega_c\tau_{p,k} \gg 2\pi$. In such a case, each element in $\boldsymbol{H}$ structure can be considered as having a uniform distribution of phase. Accordingly, the elements of $\boldsymbol{H}$ will have a complex Gaussian distribution, while their amplitudes follow Rayleigh distribution [1]. Accordingly, it is feasible sending the estimated channel state information (CSI) back to the transmitter since the multimode channel transfer function varies at relatively slower rate as compared to data rate that is being transmitted.

The group delays (GDs) among the spatial modes in FMFs are proportional to the square root of link length if there is a strong coupling among the spatial modes, which is similar to dependence of the first-order PMD on the length of fiber link [22, 23]. Accordingly, the model with optical fiber sections, used in analysis of PMD coupling, can also be applied to FMF by presenting the total link length by the number of independent fiber sections. By applying such a model, joint probability density function (PDF) of GDs in FMF has been derived in [23]. The FMF is divided into K sections, as shown in Sect. 8.12, and the coarse-step modeling method is adopted, which is a standard technique for modeling the polarization effects in single-mode fibers. The length of each section should be larger than the fiber correlation length. Ideally, the section lengths should be randomly distributed, to avoid periodic artifacts in the wavelength domain. Let us consider FMF operating in strong mode coupling regime. The propagation characteristics of k-th section of FMF having D modes can be modeled by a $D \times D$ channel matrix with terms $M_k(\omega)$ given as [23, 24] (see also [7]):

$$\boldsymbol{H}_k(\omega) = \boldsymbol{V}_k \boldsymbol{\Lambda}_k(\omega) \boldsymbol{U}_k^{\dagger}, \quad k = 1, 2, \cdots, K \tag{8.68}$$

where $\boldsymbol{V}_k$ and $\boldsymbol{U}_k$ are frequency-independent random complex unitary matrices representing the strong random coupling of modes in neighboring sections in the absence of mode-independent gain or loss; $\boldsymbol{\Lambda}_k(\omega) = \text{diag}[e^{-j\omega\tau_1}, e^{-j\omega\tau_2}, \ldots, e^{-j\omega\tau_D}]$ is the diagonal matrix describing the uncoupled modal group delays; while τ_i is the group delay of the i-th mode. (K denotes the number of fiber sections.) The number of sections (K) depends on the length of the fiber and correlation length, so the total frequency domain response can be represented by concatenating the individual sections, as follows:

$$\boldsymbol{H}_{\text{total}}(\omega) = \boldsymbol{H}_1(\omega) \cdots \boldsymbol{H}_k(\omega) \cdots \boldsymbol{H}_K(\omega). \tag{8.69}$$

The corresponding optical MIMO scheme for FSO channels is illustrated in Fig. 8.13. We assume that the beam spots on the receiver side are sufficiently wide to illuminate a whole photodetector array, which simplifies the transmitter-receiver pointing problem. The transmitter side is composed of M_{Tx} optical transmitters, with configuration of the m-th transmitter shown in Fig. 8.13b. To facilitate explanations, only the single polarization is shown. The external modulator is used to impose the data on CW laser beam signal. Alternatively, signal CW laser can be used for all optical transmitters with the help of power splitter. For direct detection applications, the external modulator is typically Mach-Zehnder modulator, imposing on-off keying (OOK) signal. For coherent detection, the external modulator is either PSK or I/Q modulator. After amplifying the modulated optical signal, by an optical amplifier, such as erbium-doped fiber amplifier (EDFA), the signal is expanded by an expanding telescope, properly designed so that the whole optical receiver array is illuminated. After transmission through atmospheric turbulence channel, each optical receiver after the compressing telescope is converted into electrical domain by using either direct or coherent optical detection. For direct detection, the n-th receiver is composed of a compressing telescope and PIN photodetector, followed by a trans-impedance amplifier (TIA). For coherent optical detection, as shown in Fig. 8.13c, after compressing telescope, the incoming optical signal is mixed with local oscillator (LO) laser signal in an optical hybrid, followed by two balanced detectors (BDs) to determine I- and Q-channels' estimates. Alternatively, a single LO laser can be used for the whole receiver array, with the help of a power splitter.

The channel coefficients for direct detection are real numbers. In the case of the *repetition* MIMO, the n-th optical receiver output, for OOK, can be represented by [5, 25]:

$$y_n(l) = x(l) \sum_{m=1}^{M_{\text{Tx}}} I_{nm} + z_n(l); \quad n = 1, \ldots, M_{\text{Rx}}; x(l) \in \{0, A\} \tag{8.70}$$

where A denotes the intensity of the pulse in the absence of scintillation and $x(l)$ denotes data symbol at the l-th time slot. I_{nm} represents the intensity channel coefficient between the n-th photodetector ($n = 1,2,\ldots,N$) and the m-th ($m = 1,2,\ldots,M$) optical source, which is described by the Gamma-Gamma probability density function. The optical transmitters and receivers are positioned in such a way that different transmitted symbols experience different atmospheric turbulence conditions. In Eq. (8.70), z_n denotes the n-th receiver TIA thermal noise that is modeled as a zero-mean Gaussian process with double-sided power spectral density $N_0/2$.

8.2.2 Parallel Decomposition of Optical and Wireless MIMO Channels

The multiplexing gain of MIMO system, introduced by Eq. (8.60), originates from the fact that a MIMO channel can be decomposed into a number of parallel independent channels, so that the data rate can be increased R_H-times, where R_H is the rank of the channel matrix $\boldsymbol{H}$. Let us perform the singular value decomposition of channel matrix $\boldsymbol{H}$ given by Eqs. (8.64), (8.67), and (8.69). In singular value decomposition, we have the following: (i) the Σ − matrix corresponds to scaling operation, (ii) the columns of matrix $\boldsymbol{U}$, obtained as eigenvectors of the Wishart matrix $\boldsymbol{H}\boldsymbol{H}^{\dagger}$, correspond to the rotation

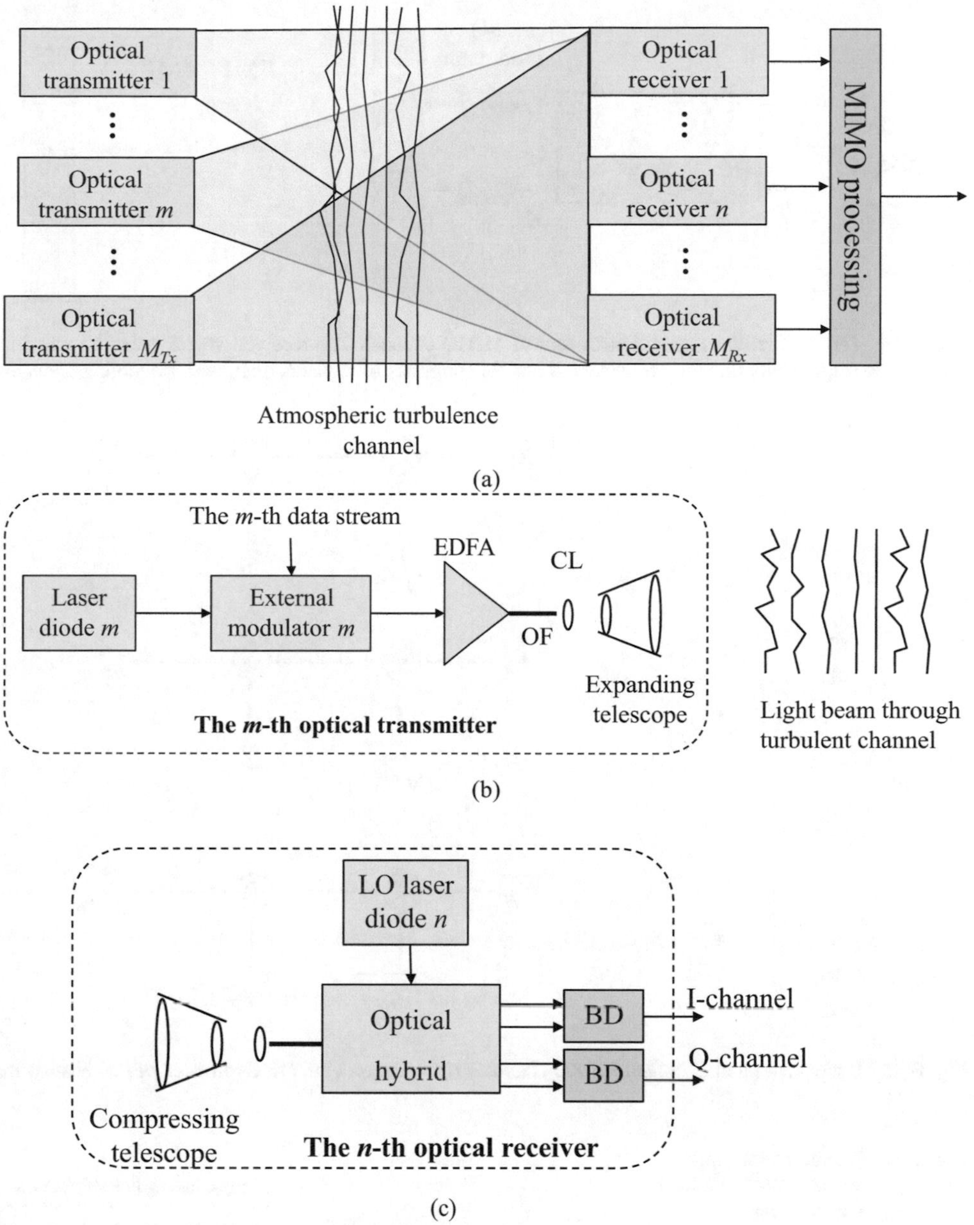

Fig. 8.13 Optical MIMO scheme for FSO channels: (**a**) generic MIMO scheme, (**b**) m-th transmitter configuration, and (**c**) n-th receiver configuration for coherent optical detection. *EDFA*, erbium-doped fiber amplifier; *CL*, compressing lenses; *OF*, optical fiber; *BD*, balanced detector

operation, and (iii) the other rotation matrix $\boldsymbol{V}$ that has for columns the eigenvectors of $\boldsymbol{H}^{\dagger}\boldsymbol{H}$. The rank of matrix $\boldsymbol{H}$ corresponds to the multiplexing gain. In FMF/FCF we can ensure the full rank of channel matrix all the time by employing OAM eigenstates, which is not guaranteed if regular multimode fiber links are employed.

The *parallel decomposition* of the optical channel, illustrated in Fig. 8.14, is obtained by introducing the following two transformations to the channel input vector $\boldsymbol{x}$ and channel output vector $\boldsymbol{y}$: (i) *transmit precoding*, in which the input vector of symbol $\boldsymbol{x}$ to transmitter is linearly transformed by pre-multiplying with rotation matrix $\boldsymbol{V}$, and (ii) *receiver shaping*, in which the receiver vector $\boldsymbol{y}$, upon optical coherent detection, is multiplied by rotation matrix $\boldsymbol{U}^{\dagger}$. With these two operations, we effectively performed the following manipulation:

$$\tilde{\boldsymbol{y}} = \boldsymbol{U}^{\dagger}\left(\underbrace{\boldsymbol{H}}_{\boldsymbol{U\Sigma V}^{\dagger}}\boldsymbol{x} + \boldsymbol{z}\right) = \boldsymbol{U}^{\dagger}\boldsymbol{U}\boldsymbol{\Sigma}\boldsymbol{V}^{\dagger}\boldsymbol{V}\tilde{\boldsymbol{x}} + \underbrace{\boldsymbol{U}^{\dagger}\boldsymbol{z}}_{\tilde{\boldsymbol{z}}} = \boldsymbol{\Sigma}\tilde{\boldsymbol{x}} + \tilde{\boldsymbol{z}}, \tag{8.71}$$

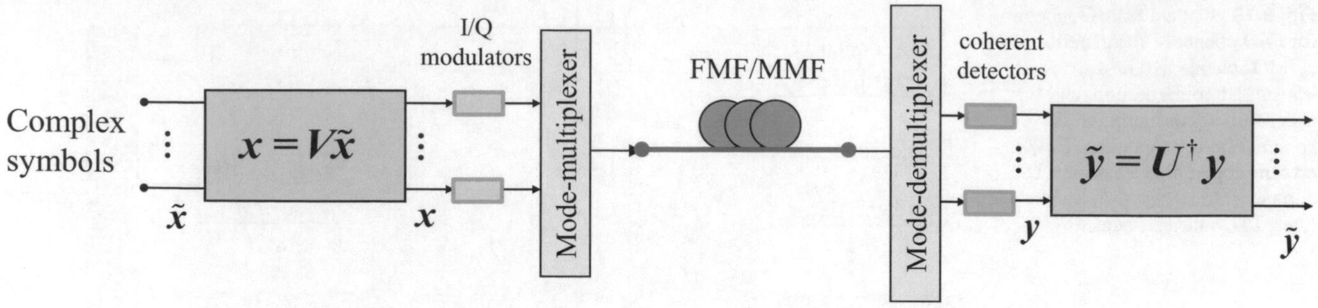

Fig. 8.14 The parallel decomposition of optical MIMO channel. The received vector $\boldsymbol{y}$ can be written as $\boldsymbol{y} = \boldsymbol{Hx} + z$ (where we omitted the photodiode responsivity and load resistance terms for simplicity of explanations), with z denoting the equivalent noise process

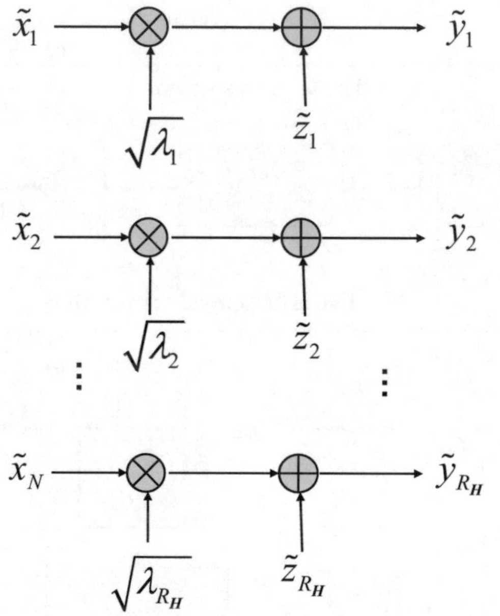

Fig. 8.15 Equivalent parallel optical MIMO channel. The λ_i denotes the i-th eigenvalue of the Wishart matrix $\boldsymbol{HH}^{\dagger}$

Fig. 8.16 The beam-forming strategy in optical MIMO. Optical transmitter and receiver configurations are similar to those shown in Fig. 8.13b, c

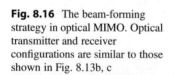

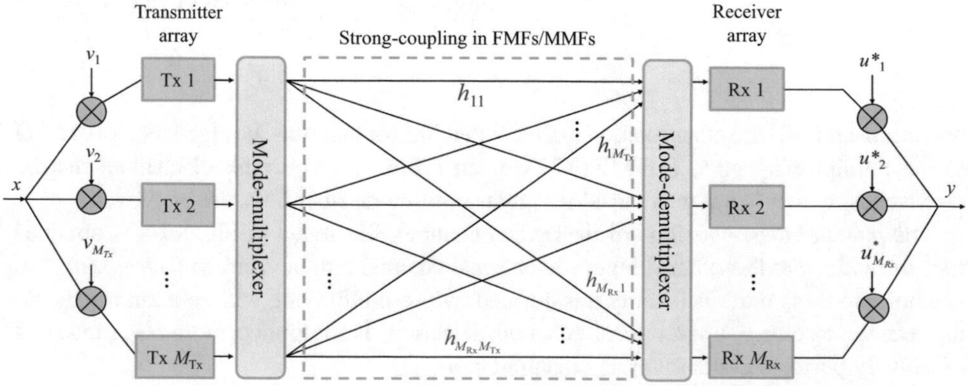

and successfully decomposed the optical MIMO optical channel into $R_{\boldsymbol{H}} = \text{rank}(\boldsymbol{H})$ parallel single-input single-output (SISO) channels. The equivalent optical channel model is shown in Fig. 8.15.

If there is no concern about multiplexing gain, a simpler scheme known in wireless communication literature as *beam-forming* [1–3] can be used instead. In this scheme, as illustrated in Fig. 8.16, the same symbol x, weighted by a complex scale

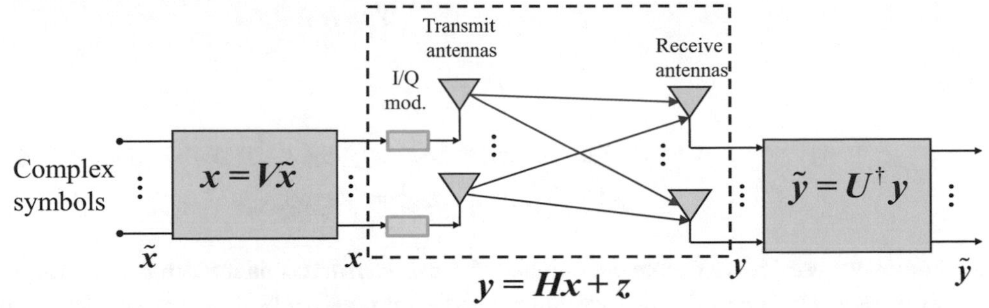

Fig. 8.17 The parallel decomposition of wireless MIMO channel

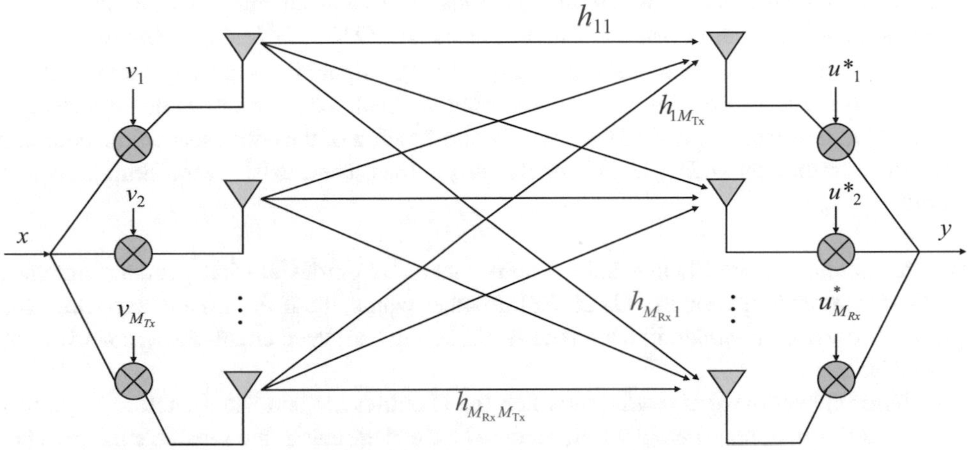

Fig. 8.18 The beam-forming strategy in wireless MIMO

factor, is sent over all spatial modes. Therefore, by applying the beam-forming strategy, the precoding and receiver matrices become just column vectors, so it is $\boldsymbol{V} = \boldsymbol{v}$ and $\boldsymbol{U} = \boldsymbol{u}$. The resulting received signal is now given by:

$$y = \boldsymbol{u}^{\dagger}\boldsymbol{H}\boldsymbol{v}x + \boldsymbol{u}^{\dagger}z, \;\; \|\boldsymbol{u}\| = \|\boldsymbol{v}\| = 1, \tag{8.72}$$

where with ||·|| we denoted the norm of a vector. When the optical channel matrix $\boldsymbol{H}$ is known at the receiver side, the received SNR can be optimized by choosing $\boldsymbol{u}$ and $\boldsymbol{v}$ as the principal left and right singular vectors of the channel matrix $\boldsymbol{H}$, while the corresponding SNR is given by $\sigma^2_{\max}\rho$, where $\sigma_{\max}$ is the largest eigenvalue of the Wishart matrix $\boldsymbol{HH}^{\dagger}$. The channel capacity beam-forming scheme can be expressed as $C = B\log^2\left(1 + \sigma^2_{\max}\mathrm{SNR}\right)$, which corresponds to the capacity of a single SISO channel having the equal channel power gain $\sigma^2_{\max}$. On the other hand, when the channel matrix is not known, we will need to do maximization only with respect to $\boldsymbol{u}$. The array gain of beam-forming diversity is between max(M_{Tx},M_{Rx}) and $M_{\mathrm{Tx}}M_{\mathrm{Rx}}$, while the diversity gain equals $M_{\mathrm{Tx}}M_{\mathrm{Rx}}$:

The wireless MIMO channel can be parallelized in a fashion similar to that shown in Fig. 8.14, and this parallelization is illustrated in Fig. 8.17. Clearly, the parallel wireless MIMO channel model is identical to that shown in Fig. 8.15. The corresponding beam-forming strategy for wireless MIMO is shown in Fig. 8.18.

8.2.3 Space-Time Coding: ML Detection, Rank Determinant, and Euclidean Distance Criteria

The symbol transmitted over the coherent MIMO optical channel with spatial modes or over the wireless MIMO channel is rather a vector than a scalar since channel has the input-output relationship given as $\boldsymbol{y} = \boldsymbol{Hx} + z$. In the case when the signal design extends over both space coordinate (by multiple spatial modes in FMF/MMF or multiple transmit antennas) and time coordinate (by multiple symbol intervals), we can refer to this design as the *space-time coding*. Under the assumption that space-mode channel is static for the duration of N_s symbols, its input and output become matrices, with dimensions corresponding to space coordinate (spatial modes) and time coordinate (the symbol intervals). The output matrix is then given as:

$$Y = HX + Z, \tag{8.73}$$

where:

$$X = [x_1, x_2, \ldots, x_{N_s}] = (X_{ij})_{M_{Tx} \times N_s}, Y = [y_1, y_2, \ldots, y_{N_s}] = (Y_{ij})_{M_{Rx} \times N_s},$$
$$Z = [z_1, z_2, \ldots, z_{N_s}] = (Z_{ij})_{M_{Rx} \times N_s}. \tag{8.74}$$

As before, we use M_{Tx} to denote the number of transmitters/transmitter antennas and M_{Rx} to denote the number of receivers/receiver antennas. Therefore, the transmitted matrix X is obtained by concatenating transmitted column vectors in N_s channel uses. On the other hand, the received matrix Y is obtained by concatenating received column vectors in N_s symbol intervals. Finally, the Z matrix entries are circularly symmetric zero-mean complex Gaussian random variables with variance N_0. The corresponding correlation matrix is given by $R_Z = \langle ZZ^\dagger \rangle = N_s N_0 I_{M_{\mathrm{Rx}}}$, where $I_{M_{\mathrm{Rx}}}$ is the identity matrix of size M_{Rx}, indicating that the additive noise affecting the received signal is spatially and temporally independent.

Let us observe a case with space-time coding in which the receiver has the knowledge of the channel matrix H. We also assume that the channel matrix H is fixed for the duration of the entire space-time codeword. To analyze such a channel, we need to know the PDF of $M_{\mathrm{Tx}}M_{\mathrm{Rx}}$ channel matrix entries. There exist several simplified models suitable for MIMO processing studies:

(i) *Rich scattering model*, in which channel matrix H entries are independent circularly symmetric complex zero-mean Gaussian random variables [2, 26, 28]. In other words, the H entries are zero-mean spatially white (ZMSW) [1].
(ii) *Fully correlated model*, in which the H entries are correlated circularly symmetric complex zero-mean Gaussian random variables.
(iii) *Separately correlated model*, in which the H entries are correlated circularly symmetric complex zero-mean Gaussian random variables, wherein the correlation between the entries are separable into two factors, corresponding to receive and transmit correlations:

$$\langle [H]_{mn} [H]^*_{m'n'} \rangle = [R]_{mm'} [T]_{nn'}, \tag{8.75}$$

where R ($M_{\mathrm{Rx}} \times M_{\mathrm{Rx}}$) and T ($M_{\mathrm{Tx}} \times M_{\mathrm{Tx}}$) are Hermitian nonnegative definite matrices. This model is applicable when the surrounding objects on transmitter and receiver sides introduce the local antenna-element correlations while not having influence on the other end of the link [2, 26, 30, 37]. The channel matrix for separately correlated model can be decomposed as:

$$H = R^{1/2} H_{\mathrm{ZMSW}} T^{1/2}, \tag{8.76}$$

where H_{ZMSW} is the matrix of circularly symmetric complex zero-mean Gaussian random variables with unit variance.

Unless otherwise stated we will adopt the rich scattering model. As an illustration, the simplest space-time code, the *delay diversity* Scheme [2, 38, 39], is essentially rate-$1/M_{\mathrm{Tx}}$ repetition code, whose transmitted code matrix for $M_{\mathrm{Tx}} = 2$ is given by:

$$X = \begin{bmatrix} x_0 & x_1 & x_2 & \cdots \\ 0 & x_0 & x_1 & \cdots \end{bmatrix}. \tag{8.77}$$

Given that each transmitted symbol transverses in total $M_{\mathrm{Tx}}M_{\mathrm{Rx}}$ independent paths, this scheme achieves the full diversity with the cost of very low data rate, only 1 symbol/channel use.

Under maximum likelihood (ML) detection, the optimum transmit matrix of the space-time code is obtained as the result of the following minimization problem [1, 2, 7, 26]:

$$\begin{aligned} \hat{X} &= \arg \min_{X \in \mathcal{X}^{M_{Tx} \times N_s}} \|Y - HX\|_F^2 = \arg \min_{X \in \mathcal{X}^{M_{Tx} \times N_s}} \sum_{i=1}^{N_s} \|y_i - Hx_i\|^2 \\ &= \arg \min_{X \in \mathcal{X}^{M_{Tx} \times N_s}} \sum_{n=1}^{M_{Rx}} \sum_{i=1}^{N_s} \left| y_{ni} - \sum_{m=1}^{M_{Tx}} h_{nm} x_{mi} \right|^2, \end{aligned} \tag{8.78}$$

where the minimization is performed over all possible space-time input matrices $\boldsymbol{X}$. In Eq. (8.78), we use the operation $\|\boldsymbol{A}\|_F$ to denote the *Frobenius (Euclidean) norm* of matrix $\boldsymbol{A} = [\boldsymbol{A}]_{mn}$ defined as:

$$\|\boldsymbol{A}\|_F \doteq \sqrt{(\boldsymbol{A},\boldsymbol{A})} = \sqrt{\mathrm{Tr}(\boldsymbol{A}\boldsymbol{A}^\dagger)} = \sqrt{\sum\nolimits_m \sum\nolimits_n |a_{mn}|^2}, \tag{8.79}$$

where we use $(\boldsymbol{A},\boldsymbol{B})$ to denote the scalar product of matrices $\boldsymbol{A} = [\boldsymbol{A}]_{mn}$ and $\boldsymbol{B} = [\boldsymbol{B}]_{mn}$, defined as $(\boldsymbol{A},\boldsymbol{B}) = \sum_m \sum_n a_{mn} b_{mn}^*$. Because it is extremely difficult to calculate the exact error probability, we employ the union bound [2]:

$$P_e \leq \frac{1}{|\mathcal{X}|} \sum\nolimits_{\mathcal{X}} \sum\nolimits_{\boldsymbol{X} \neq \boldsymbol{X}} P\left(\boldsymbol{X} \to \hat{\boldsymbol{X}}\right), \tag{8.80}$$

where the piece-wise error probability (PEP) is given by:

$$\begin{aligned} P\left(\boldsymbol{X} \to \widehat{\boldsymbol{X}}\right) &= \Pr\left(\left\|\boldsymbol{Y} - \boldsymbol{H}\widehat{\boldsymbol{X}}\right\|_F^2 < \|\boldsymbol{Y} - \boldsymbol{H}\boldsymbol{X}\|_F^2\right) \overset{\boldsymbol{Y}=\boldsymbol{H}\boldsymbol{X}+\boldsymbol{Z}}{=} \\ &= \Pr\left(\left\|\boldsymbol{H}\boldsymbol{X} + \boldsymbol{Z} - \boldsymbol{H}\widehat{\boldsymbol{X}}\right\|_F^2 < \|\boldsymbol{H}\boldsymbol{X} + \boldsymbol{Z} - \boldsymbol{H}\boldsymbol{X}\|_F^2\right) \\ &= \Pr\left(\left\|\boldsymbol{H}\underbrace{\left(\boldsymbol{X} - \widehat{\boldsymbol{X}}\right)}_{\boldsymbol{D}} + \boldsymbol{Z}\right\|_F^2 < \|\boldsymbol{Z}\|_F^2\right) = \Pr\left(\|\boldsymbol{H}\boldsymbol{D} + \boldsymbol{Z}\|_F^2 < \|\boldsymbol{Z}\|_F^2\right), \end{aligned} \tag{8.81}$$

where $\boldsymbol{D}$ is defined as $\boldsymbol{D} = \boldsymbol{X} - \widehat{\boldsymbol{X}}$. By expressing the Frobenius norm in terms of scalar product, we can rewrite the previous expression as follows:

$$\begin{aligned} P\left(\boldsymbol{X} \to \widehat{\boldsymbol{X}}\right) &= \Pr((\boldsymbol{H}\boldsymbol{D} + \boldsymbol{Z}, \boldsymbol{H}\boldsymbol{D} + \boldsymbol{Z}) < (\boldsymbol{Z},\boldsymbol{Z})) \\ &= \Pr\left(\mathrm{Tr}\left[(\boldsymbol{H}\boldsymbol{D} + \boldsymbol{Z})(\boldsymbol{H}^\dagger\boldsymbol{D}^\dagger + \boldsymbol{Z}^\dagger) - \boldsymbol{Z}\boldsymbol{Z}^\dagger\right] < 0\right) \\ &= \Pr\left(\|\boldsymbol{H}\boldsymbol{D}\|_F^2 + 2(\boldsymbol{H},\boldsymbol{Z}) < 0\right). \end{aligned} \tag{8.82}$$

The variance of Gaussian random variable $(\boldsymbol{H}, \boldsymbol{Z})$ is given by:

$$\begin{aligned} \left\langle (\boldsymbol{H},\boldsymbol{Z})^2 \right\rangle &= \left\langle \left[\mathrm{Tr}\left(\underbrace{\boldsymbol{H}}_{\boldsymbol{H}_I + j\boldsymbol{H}_Q} \underbrace{\boldsymbol{Z}^\dagger}_{\boldsymbol{Z}_I - j\boldsymbol{Z}_Q}\right)\right]^2 \right\rangle = \left\langle [\mathrm{Tr}(\boldsymbol{H}_I\boldsymbol{Z}_I + \boldsymbol{H}_Q\boldsymbol{Z}_Q)]^2 \right\rangle \\ &= \left\langle \left\{\sum\nolimits_m \sum\nolimits_n [\boldsymbol{H}_I]_{mn}[\boldsymbol{Z}_I]_{nm} + [\boldsymbol{H}_Q]_{mn}[\boldsymbol{Z}_Q]_{nm}\right\}^2 \right\rangle \\ &= \sum_m \sum_n [\boldsymbol{H}_I]_{mn}^2 \underbrace{\left\langle [\boldsymbol{Z}_I]_{nm}^2 \right\rangle}_{N_0/2} + [\boldsymbol{H}_Q]_{nm}^2 \underbrace{\left\langle [\boldsymbol{Z}_Q]_{nm}^2 \right\rangle}_{N_0/2} \\ &= \frac{N_0}{2} \underbrace{\sum_m \sum_n [\boldsymbol{H}_I]_{mn}^2 + [\boldsymbol{H}_Q]_{nm}^2}_{\|\boldsymbol{H}\|_E^2} \\ &= \frac{N_0}{2} \|\boldsymbol{H}\|_E^2 \end{aligned} \tag{8.83}$$

The PEP can now be written as:

$$P\left(X \rightarrow \widehat{X}\right) = \left\langle Q\left(\sqrt{\frac{\|HD\|_F^2}{2N_0}}\right)\right\rangle. \tag{8.84}$$

By representing $\|HD\|_F^2$ as $\|HD\|_F^2 = \mathrm{Tr}\left(HDD^\dagger H^\dagger\right) = \mathrm{Tr}\left(H^\dagger HDD^\dagger\right)$, we can conclude that after the spatial interference cancellation $H^\dagger H = I_{M_{\mathrm{Tx}}}$, and the PEP is upper bounded by:

$$P\left(X \rightarrow \widehat{X}\right) \leq Q\left(\sqrt{\frac{\|D\|_F^2}{2N_0}}\right). \tag{8.85}$$

By using the following approximation for Q-function $Q(x) \leq \exp(-x^2/2)$, we can upper bound the Q-function in Eq. (8.84) by:

$$Q\left(\sqrt{\frac{\|HD\|_F^2}{2N_0}}\right) \leq \exp\left[-\frac{\|HD\|_F^2}{4N_0}\right]. \tag{8.86}$$

Under Rayleigh fading assumption, in other words assuming that channel coefficients are i.i.d. complex Gaussian $\mathcal{N}_c(0,1)$, the PEP will be upper bounded by:

$$P\left(X \rightarrow \widehat{X}\right) \leq \left(\frac{1}{\det\left[I_{M_{Tx}} + \frac{DD^\dagger}{4N_0}\right]}\right)^{M_{Rx}}. \tag{8.87}$$

The PDF of i.i.d. complex Gaussian entries of a random matrix A, of size $M \times N$, is given by [2, 40]:

$$\pi^{-MN} \exp\left[\mathrm{Tr}\left(-AA^\dagger\right)\right], \tag{8.88}$$

This result can be used to prove the following theorem [2, 41].

Theorem If A is a given $M \times M$ Hermitian matrix such that $I_M + A > 0$, and B is an $M \times N$ matrix whose entries are i.i.d. with $\mathcal{N}_c(0,1)$, then:

$$\left\langle \exp\left[\mathrm{Tr}\left(-ABB^\dagger\right)\right]\right\rangle = \det(I_M + A)^{-N}. \tag{8.89}$$

To prove this theorem, let us represent matrix B as follows: $B = [b_1, b_2, \ldots, b_N]$, where b_n denotes the n-th column of B. The expected value of $\exp[\mathrm{Tr}(-ABB^\dagger)]$ can be determined with the help of Eq. (8.88) as follows:

$$\begin{aligned}\left\langle \exp\left[\mathrm{Tr}\left(-ABB^\dagger\right)\right]\right\rangle &= \pi^{-MN}\int \exp\left[\mathrm{Tr}\left(-ABB^\dagger - BB^\dagger\right)\right]dB \\ &= \left\{\pi^{-M}\int \exp\left[\mathrm{Tr}\left(-(A + I_M)bb^\dagger\right)\right]db\right\}^N.\end{aligned} \tag{8.90}$$

After the following transformation $b = (A + I_M)^{-1/2}c$, since the Jacobian of this transformation is $\det(I_M + A)^{-1}$, we arrive at Eq. (8.89).

Eq. (8.84) can be now upper bounded by:

$$P\left(X \rightarrow \widehat{X}\right) \leq \left\langle \exp\left[-\frac{\mathrm{Tr}\left(H^\dagger HDD^\dagger\right)}{4N_0}\right]\right\rangle, \tag{8.91}$$

and by directly applying (8.89), we derive Eq. (8.87). Because the determinant of a matrix can be determined as the product of its eigenvalues, we can rewrite Eq. (8.87) as:

$$P\left(X \to \widehat{X}\right) \leq \prod_{k=1}^{M_{Tx}} \left(\frac{1}{1 + \frac{\lambda_k}{4N_0}}\right)^{M_{Rx}}, \tag{8.92}$$

where λ_k ($k = 1,2,\ldots,M_{Tx}$) are eigenvalues for $\boldsymbol{DD}^{\dagger}$. Let us use $\mathcal{K}$ to denote the set of nonzero eigenvalues, with the size of set corresponding to rank of $\boldsymbol{DD}^{\dagger}$. For sufficiently high SNR, defined as $\gamma = 1/4N_0$, the inequality (8.92) simplifies to:

$$P\left(X \to \hat{X}\right) \leq \left(\prod_{k=1}^{|\mathcal{K}|} \lambda_k\right)^{-M_{Rx}} \gamma^{-M_{Rx}|\mathcal{K}|}. \tag{8.93}$$

The first term in PEP upper bound is independent on SNR, while dependent of product of eigenvalues of $\boldsymbol{DD}^{\dagger}$ as well as power M_{Rx}. Because this term is independent of SNR, it does not change the slope of PEP but rather shifts the PEP curve to the left; in a similar fashion as a channel code, it can be related to the *coding gain* of the space-time code. To improve the coding gain, we need to maximize the minimum of the determinant of $\boldsymbol{DD}^{\dagger}$ over all possible pairs $\boldsymbol{X}, \widehat{\boldsymbol{X}}$; we refer to this criterion as the *determinant criterion*. On the other hand, the second term is SNR dependent, and the cardinality of $\mathcal{K}$ affects the slope of PEP curve indicating that diversity gain is given by $M_{Rx}|\mathcal{K}|$. Thus to maximize the diversity gain, we have to design space-time code on such a way that the matrix $\boldsymbol{DD}^{\dagger}$ is of full rank equal to M_{Tx}, and we refer to this design criterion as the *rank criterion*. When $\boldsymbol{DD}^{\dagger}$ is of the full rank, we can write:

$$\prod_{k=1}^{M_{Tx}} \lambda_k = \det\left(\boldsymbol{DD}^{\dagger}\right), \tag{8.94}$$

and the necessary condition for matrix $\boldsymbol{DD}^{\dagger}$ to be of the full rank is that space-time code block length must be equal or higher than the number of transmit antennas.

The determinant in Eq. (8.87) can be written as follows:

$$\det\left[\boldsymbol{I}_{M_{Tx}} + \gamma \boldsymbol{DD}^{\dagger}\right] = 1 + \gamma \mathrm{Tr}\left(\boldsymbol{DD}^{\dagger}\right) + \cdots + \gamma^{M_{Tx}} \det\left(\boldsymbol{DD}^{\dagger}\right). \tag{8.95}$$

Clearly, for small SNRs with $\gamma \ll 1$ the determinant and consequently PEP depend predominantly on Tr($\boldsymbol{DD}^{\dagger}$), in other words on Euclidean distance squared between $\boldsymbol{X}$ and $\widehat{\boldsymbol{X}}$. This is the reason why the space-time code design criterion based on Eq. (8.95) is commonly referred to as the *Euclidean distance criterion*. When SNR is high with $\gamma \gg 1$, the PEP is dominated by det($\boldsymbol{DD}^{\dagger}$), in other words the product of eigenvalues of $\boldsymbol{DD}^{\dagger}$, and the rank determinant criterion is relevant here. Therefore, when SNR is small, the Euclidean distance among codewords is relevant, in a similar fashion as in a channel code, and the PEP performance is dictated by the additive noise rather than fading. For high SNRs, the rank of $\boldsymbol{DD}^{\dagger}$ dictates the PEP performance, and fading effects are relevant.

To study the error probability in waterfall region, we can allow the number of received antennas to tend to infinity, for a fixed number of transmitted antennas. In that case, the spatial interference get cancelled as $\boldsymbol{H}^{\dagger}\boldsymbol{H} \to \boldsymbol{I}_{M_{Tx}}$, and the PEP, based on inequality (8.85), tends to AWGN case:

$$P\left(X \to \widehat{X}\right) \underset{M_{Tx}<\infty}{\overset{M_{Rx}\to\infty}{\longrightarrow}} Q\left(\sqrt{\frac{\|\boldsymbol{D}\|_F^2}{2N_0}}\right), \tag{8.96}$$

indicating that union bound on error probability is only dependent on Euclidean distances among space-time codewords. In this particular case, the PEP is not affected by the number of transmitted antennas, but only the multiplexing gain that grows linearly with the number of transmitted antennas.

When the number of received antennas tends to infinity, we need to change the variance of entries in $\boldsymbol{H}$ from 1 to $1/M_{Rx}$ to avoid the divergence problems. The PEP needs to be modified as follows:

$$P\left(X \to \widehat{X}\right) \leq \left(\frac{1}{\det\left[\boldsymbol{I}_{M_{Tx}} + \frac{\boldsymbol{DD}^{\dagger}}{4M_{Rx}N_0}\right]}\right)^{M_{Rx}}. \tag{8.97}$$

The corresponding determinant in (8.97) can be represented now as:

$$\det\left[\boldsymbol{I}_{M_{Tx}}+\frac{\gamma}{M_{Rx}}\boldsymbol{D}\boldsymbol{D}^{\dagger}\right]=1+\frac{\gamma}{M_{Rx}}\mathrm{Tr}\left(\boldsymbol{D}\boldsymbol{D}^{\dagger}\right)+\cdots+\left(\frac{\gamma}{M_{Rx}}\right)^{M_{Tx}}\det\left(\boldsymbol{D}\boldsymbol{D}^{\dagger}\right). \tag{8.98}$$

When the SNR increases at the same rate as the number of received antennas, the rank and determinant criteria are applicable. On the other hand, when SNR is fixed and finite, the Euclidean distance criterion should be used instead.

8.2.4 Relevant Classes of Space-Time Codes (STCs)

In this section we describe several relevant classes of space-time codes including Alamouti code, orthogonal designs, linear space-time block codes, and trellis space-time codes.

8.2.4.1 Alamouti Code (Revisited) and Orthogonal Designs

The Alamouti code is given by the following coding matrix [10]:

$$\boldsymbol{X}=\begin{bmatrix} x_1 & -x_2^* \\ x_2 & x_1^* \end{bmatrix}, \tag{8.99}$$

in which, during the first time interval, symbol x_1 is transmitted using spatial mode (antenna 1) 1, while symbol x_2 occupies the spatial mode (antenna) 2. In the second time interval, the symbol $-x_2^*$ is transmitted by spatial mode (antenna) 1, while the symbol x_1^* goes over spatial mode (antenna) 2. The corresponding spatial mode/antenna channel matrix for two transmitted modes (antennas) and two received modes (antennas), as illustrated in Fig. 8.19, can be written as:

$$\boldsymbol{H}=\begin{bmatrix} h_{11} & h_{12} \\ h_{21} & h_{22} \end{bmatrix}. \tag{8.100}$$

The received current signals after coherent optical detection as well as the receive antenna outputs can be organized in matrix form as follows:

$$\begin{aligned}\begin{bmatrix} y_{11} & y_{12} \\ y_{21} & y_{22} \end{bmatrix}&=\begin{bmatrix} h_{11} & h_{12} \\ h_{21} & h_{22} \end{bmatrix}\begin{bmatrix} x_1 & -x_2^* \\ x_2 & x_1^* \end{bmatrix}+\begin{bmatrix} z_{11} & z_{12} \\ z_{21} & z_{22} \end{bmatrix}\\ &=\begin{bmatrix} h_{11}x_1+h_{12}x_2+z_{11} & -h_{11}x_2^*+h_{12}x_1^*+z_{12} \\ h_{21}x_1+h_{22}x_2+z_{21} & h_{21}x_2^*+h_{22}x_1^*+z_{22} \end{bmatrix}\end{aligned} \tag{8.101}$$

The combiner outputs, described in [10], can be obtained as follows:

$$\begin{aligned}\tilde{x}_1&=h_{11}^*y_{11}+h_{12}y_{12}^*+h_{21}^*y_{21}+h_{22}y_{22}^*=\left(|h_{11}|^2+|h_{12}|^2+|h_{21}|^2+|h_{22}|^2\right)x_1+\text{noise}\\ \tilde{x}_2&=h_{12}^*y_{11}-h_{11}y_{12}^*+h_{22}^*y_{21}-h_{21}y_{22}^*=\left(|h_{11}|^2+|h_{12}|^2+|h_{21}|^2+|h_{22}|^2\right)x_2+\text{noise}\end{aligned} \tag{8.102}$$

indicating that the spatial interference is cancelled.

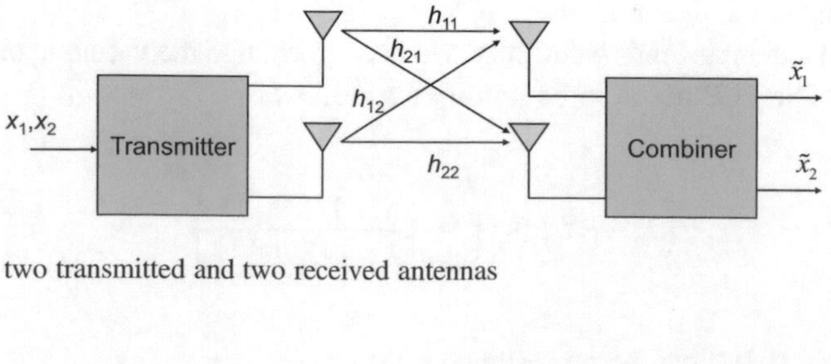

Fig. 8.19 The Alamouti scheme for two transmitted and two received antennas

The general Alamouti code with two transmit antennas and arbitrary *number of* receive antennas can be obtained by generalization.

The received signal in the Alamouti code with two transmitters and one receiver can be rewritten in the following form:

$$\begin{aligned} y_1 &= h_1 x_1 + h_2 x_2 + z_1, \\ y_2^* &= -h_1^* x_2 + h_2^* x_1 + z_2^*. \end{aligned} \tag{8.103}$$

The corresponding matrix representation is given by:

$$\begin{bmatrix} y_1 \\ y_2^* \end{bmatrix} = \underbrace{\begin{bmatrix} h_1 & h_2 \\ h_2^* & -h_1^* \end{bmatrix}}_{\widetilde{\boldsymbol{H}}} \underbrace{\begin{bmatrix} x_1 \\ x_2 \end{bmatrix}}_{\boldsymbol{x}} + \underbrace{\begin{bmatrix} z_1 \\ z_2^* \end{bmatrix}}_{\widetilde{\boldsymbol{z}}} = \widetilde{\boldsymbol{H}}\boldsymbol{x} + \widetilde{\boldsymbol{z}}. \tag{8.104}$$

By multiplying both sides of MIMO equation $\boldsymbol{y} = \boldsymbol{H}\boldsymbol{x} + \boldsymbol{z}$ by $\widetilde{\boldsymbol{H}}^{\dagger}$ from the left, we obtain:

$$\widetilde{\boldsymbol{H}}^{\dagger}\boldsymbol{y} = \underbrace{\widetilde{\boldsymbol{H}}^{\dagger}\boldsymbol{H}}_{\left(|h_1|^2+|h_2|^2\right)\boldsymbol{I}_2} \boldsymbol{x} + \widetilde{\boldsymbol{H}}^{\dagger}\boldsymbol{z} = \left(|h_1|^2 + |h_2|^2\right)\boldsymbol{x} + \widetilde{\boldsymbol{H}}^{\dagger}\boldsymbol{z}, \tag{8.105}$$

and the spatial interference is successfully compensated for. By using the Alamouti design, we were able to determine the modified channel matrix $\widetilde{\boldsymbol{H}}$ orthogonal to the channel matrix $\boldsymbol{H}$ up to the multiplication constant $|h_1|^2 + |h_2|^2$. This is the reason why the Alamouti code belongs to the class of *orthogonal designs*.

Another interesting example of orthogonal design for $M_{\mathrm{Tx}} = 3$, $M_{\mathrm{Rx}} = 1$, $N_s = 4$ is given by the code matrix:

$$\boldsymbol{X} = \begin{bmatrix} x_1 & -x_2^* & -x_3^* & 0 \\ x_2 & x_1^* & 0 & -x_3^* \\ x_3 & 0 & x_1^* & x_2^* \end{bmatrix}. \tag{8.106}$$

The corresponding MIMO model of this system is given by:

$$\boldsymbol{y} = \begin{bmatrix} y_1 \\ y_2 \\ y_3 \\ y_4 \end{bmatrix} = \boldsymbol{H}\boldsymbol{X} + \boldsymbol{z} = [h_1 \; h_2 \; h_3] \begin{bmatrix} x_1 & -x_2^* & -x_3^* & 0 \\ x_2 & x_1^* & 0 & -x_3^* \\ x_3 & 0 & x_1^* & x_2^* \end{bmatrix} + \begin{bmatrix} z_1 \\ z_2 \\ z_3 \\ z_4 \end{bmatrix}. \tag{8.107}$$

Let us rewrite the MIMO model as follows:

$$\begin{aligned} \begin{bmatrix} y_1 \\ y_2^* \\ y_3^* \\ y_4^* \end{bmatrix} &= \begin{bmatrix} h_1 x_1 + h_2 x_2 + h_3 x_3 \\ -h_1^* x_2 + h_2^* x_1 \\ -h_1^* x_3 + h_3^* x_1 \\ -h_2^* x_3 + h_3^* x_2 \end{bmatrix} + \begin{bmatrix} z_1 \\ z_2^* \\ z_3^* \\ z_4^* \end{bmatrix} = \underbrace{\begin{bmatrix} h_1 & h_2 & h_3 \\ h_2^* & -h_1^* & 0 \\ h_3^* & 0 & -h_1^* \\ 0 & h_3^* & -h_2^* \end{bmatrix}}_{\widetilde{\boldsymbol{H}}} \underbrace{\begin{bmatrix} x_1 \\ x_2 \\ x_3 \end{bmatrix}}_{\boldsymbol{x}} + \underbrace{\begin{bmatrix} z_1 \\ z_2^* \\ z_3^* \\ z_4^* \end{bmatrix}}_{\widetilde{\boldsymbol{z}}} \\ &= \widetilde{\boldsymbol{H}}\boldsymbol{x} + \widetilde{\boldsymbol{z}}. \end{aligned} \tag{8.108a}$$

By multiplying both sides of MIMO Eq. (8.107) by $\widetilde{\boldsymbol{H}}^{\dagger}$ from the left, we obtain:

$$\widetilde{\boldsymbol{H}}^{\dagger}\boldsymbol{y} = \underbrace{\widetilde{\boldsymbol{H}}^{\dagger}\boldsymbol{H}}_{\left(|h_1|^2+|h_2|^2+|h_3|^2\right)\boldsymbol{I}_3} \boldsymbol{x} + \widetilde{\boldsymbol{H}}^{\dagger}\boldsymbol{z} = \left(|h_1|^2 + |h_2|^2 + |h_3|^2\right)\boldsymbol{x} + \widetilde{\boldsymbol{H}}^{\dagger}\boldsymbol{z}, \tag{8.108b}$$

and the spatial interference is compensated for. Because three symbols are transmitted in four symbol intervals, the code rate of this space-time code is 3/4 signal per channel use, which is smaller than in Alamouti scheme whose rate is 1. Moreover, it has been shown in [42] that the code rate of orthogonal designs with the number of transmit antennas larger than two cannot exceed 3/4. However, by concatenating a linear encoder with a layered space-time mapper, it is possible simultaneously to achieve the full-diversity and full-rate for any number of transmitted and received antennas as shown in [43].

8.2.4.2 Linear Space-Time Block Codes

The Alamouti code belongs to the class of *linear space-time block codes* [12]. In linear space-time codes, the L symbols x_1, $x_2,..,x_L$ are transmitted by using M_{Tx} transmitters in N_s time intervals, and their code matrix has the form [2]:

$$\boldsymbol{X} = \sum_{l=1}^{L} (a_l \boldsymbol{A}_l + \ jb_l \boldsymbol{B}_l), \ \ a_l = \text{Re}\{x_l\}, b_l = \text{Im}\{x_l\} \tag{8.109}$$

where $\boldsymbol{A}_l$ and $\boldsymbol{B}_l$ are the complex matrices of dimension $M_{\text{Tx}} \times N_s$. For instance, the Alamouti code can be expressed in the following form:

$$\begin{aligned} \boldsymbol{X} = \begin{bmatrix} x_1 & -x_2^* \\ x_2 & x_1^* \end{bmatrix} &= \begin{bmatrix} a_1 + jb_1 & -a_2 + jb_2 \\ a_2 + jb_2 & a_1 - jb_1 \end{bmatrix} \\ &= a_1 \begin{bmatrix} 1 & 0 \\ 0 & 1 \end{bmatrix} + jb_1 \begin{bmatrix} 1 & 0 \\ 0 & -1 \end{bmatrix} + a_2 \begin{bmatrix} 0 & -1 \\ 1 & 0 \end{bmatrix} + jb_2 \begin{bmatrix} 0 & 1 \\ 1 & 0 \end{bmatrix} \end{aligned} \tag{8.110}$$

Let us define the transmitted vector as follows:

$$\widetilde{\boldsymbol{x}} = [a_1 \, b_1 \ \ldots \ a_L \, b_L]^T. \tag{8.111}$$

Let the noise vector $\widetilde{z}$ be obtained by stacking the columns of the noise matrix $\boldsymbol{Z}$ on the top of each other, and we denote this manipulation by $\widetilde{z} = \text{vect}(\boldsymbol{Z})$. By using the same manipulation, the received matrix $\boldsymbol{Y}$ can be then converted to received vector $\widetilde{\boldsymbol{y}}$ as follows:

$$\widetilde{\boldsymbol{y}} = \text{vec}(\boldsymbol{Y}) = \text{vec}(\boldsymbol{HX} + \boldsymbol{Z}) = \sum_{l=1}^{L} (a_l \text{vec}(\boldsymbol{HA}_l) + b_l \text{vec}(\text{j}\boldsymbol{HB}_l)) = \widetilde{\boldsymbol{H}}\widetilde{\boldsymbol{x}} + \widetilde{z}, \tag{8.112}$$

wherein $L \leq N_s M_{\text{Rx}}$ and the modified channel matrix $\widetilde{\boldsymbol{H}}$ is defined as:

$$\widetilde{\boldsymbol{H}} = [\text{vec}(\boldsymbol{HA}_1) \ \text{vec}(\text{j}\boldsymbol{HB}_1) \ \ldots \ \text{vec}(\boldsymbol{HA}_L) \ \text{vec}(\text{j}\boldsymbol{HB}_L)]. \tag{8.113}$$

We now perform the QR factorization of $\widetilde{\boldsymbol{H}}$:

$$\widetilde{\boldsymbol{H}} = \widetilde{\boldsymbol{Q}}\widetilde{\boldsymbol{R}}, \tag{8.114}$$

with $\widetilde{\boldsymbol{Q}}$ being the unitary matrix and $\widetilde{\boldsymbol{R}}$ being the upper triangular matrix. By multiplying both sides of MIMO Eq. (8.112) by $\widetilde{\boldsymbol{Q}}^\dagger$ from the left, we obtain:

$$\widetilde{\boldsymbol{Q}}^\dagger \widetilde{\boldsymbol{y}} = \widetilde{\boldsymbol{R}}\widetilde{\boldsymbol{x}} + \widetilde{\boldsymbol{Q}}^\dagger \widetilde{z}, \tag{8.115}$$

wherein the last entry in $\widetilde{\boldsymbol{R}}\widetilde{\boldsymbol{x}}$ is proportional to b_L and can be detected. The next-to-last entry is a linear combination of a_L and b_L; since b_L has already been detected, the a_L can be detected as well. The third from the last entry is a linear combination of b_{L-1}, a_L, and b_L, so that b_{L-1} can be detected. This iterative procedure is conducted until all L symbols get detected. For the best performance, it is a good idea to rearrange the received symbols according to SNRs, so that the received symbol with the highest SNR is placed at the bottom. This approach has certain similarities with the so-called Bell Labs Layered Space-Time (BLAST) architecture [28].

8.2.4.3 Space-Time Trellis Codes (STTCs)

Another class of space-time codes that are applicable in wireless communication as well as SDM systems are *trellis space-time codes* [1, 2, 12]. These codes are similar to trellis-coded modulation (TCM) Schemes [2], in which transition among states, described by a trellis branch, is labeled by M_{Tx} signals, each associated with one transmitter. At each time instant t, depending on the state of the encoder and the input bits, a different transition branch is chosen. If the label of this branch is $x_t^1 x_t^2 \ldots x_t^{M_{Tx}}$, then transmitter m is used to send signal constellation point x_t^m, and all these transmission processes are simultaneous. The spectral efficiency is determined by the modulation format size M and it is $\log_2(M)$ bits/symbol. As an illustration, a QPSK trellis space-time coding schemes with $M_{\mathrm{Tx}} = 2$ are shown in Fig. 8.20. Four-state and eight-state STTCs' examples are provided. QPSK symbols are denoted as {0, 1, 2, 3}, each symbol carrying two bits from the following set {00, 01, 10, 11}. The input symbols select a branch from a given state. The branches are numerated by {0,1,. . .,M-1}. The states are denoted by the circles. The left column of states denotes the current time instance t, while the right column of states

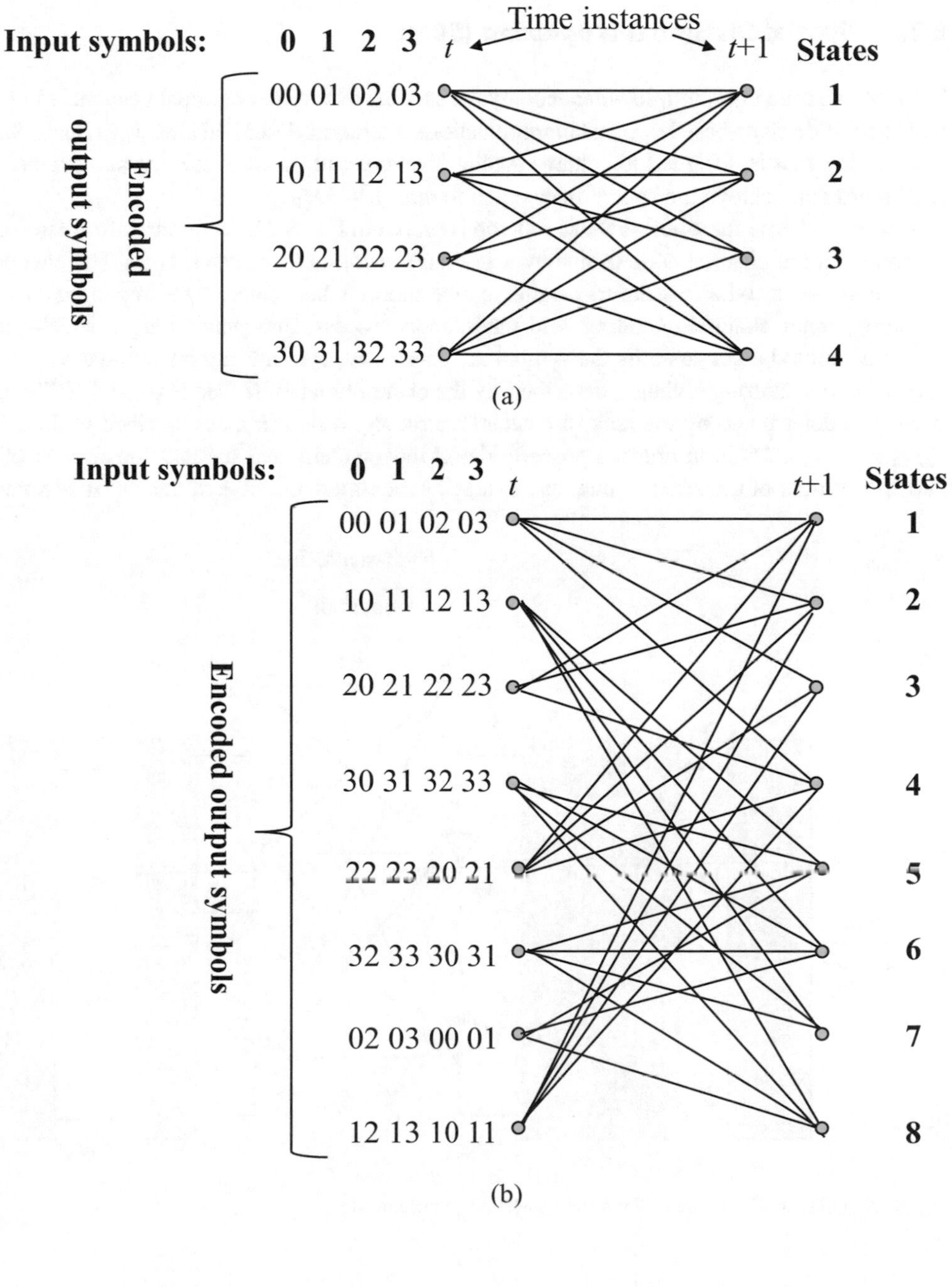

Fig. 8.20 QPSK trellis space-time code with $M_{\mathrm{Tx}} = 2$ and diversity of $2M_{\mathrm{Rx}}$: (**a**) four-state STTC and (**b**) eight-state STTC

denotes the next time instance $t + 1$. The initial state and incoming symbol dictate the encoded symbols. The first encoded symbol is transmitted on the first antenna and the second symbol on the second antenna. The trellis with larger number of states for the same modulation format and the transmit antennas represents a stronger STTC in BER sense.

Assuming an ideal channel state information (CSI) and that the path gains h_{nm} ($m = 1,\ldots,M_{\mathrm{Tx}}$; $n = 1,\ldots,M_{\mathrm{Rx}}$) are known to the detector, the *branch metric* for a transition labeled $x_t^1 x_t^2 \ldots x_t^{M_{Tx}}$ is given as:

$$\sum_{n=1}^{M_{Rx}} \left| y_t^n - \sum_{m=1}^{M_{Tx}} h_{nm} x_t^m \right|^2, \tag{8.116}$$

where y_t^n is the received signal at n-th receiver at the time instance t. The Viterbi algorithm [27] is further used to compute the path with the lowest accumulated metric. Clearly, the trellis space-time codes can achieve higher code rates than orthogonal designs and other linear space-time block codes; however, the receiver complexity grows exponentially as the number of transmit antennas increases.

8.2.5 Spatial Division Multiplexing (SDM)

In SDM, we transmit multiple independent data streams over either multipath channels in wireless communications or spatial modes in SDM fiber-based optical communication systems, as described already in Sects. 8.2.1 and 8.2.2. Compared to space-time coding that is designed to achieve spatial diversity and transmit at most one symbol per modulation period, the SDM techniques can achieve the spatial rates equal to min (M_{Tx}, M_{Rx}).

Generic scheme for wireless SDM scheme is shown in Fig. 8.21. Incoming information data stream might be first encoded by conventional channel code (not shown in figure), serving as an outer code. The encoded data stream is forwarded to 1: L demultiplexer, where serial-to-parallel conversation takes place. The key transmitter performs simultaneously STC encoding, inner channel encoding, and modulation process, thus performing L to M_{Tx} mapping. The m-th ($m = 1,2,\ldots, M_{\mathrm{Tx}}$) transmitted antenna emits the symbol x_m. These M_{Tx} symbols are transmitted over wireless MIMO channel of interest such as rich-scattering channel, described by the channel matrix $\boldsymbol{H}$ (see Eq. (8.64)). The maximum number of independent streams is determined by the rank of channel matrix $R_{\boldsymbol{H}} = \mathrm{rank}(\boldsymbol{H})$, as described in Sect. 8.2.2, which is upper bounded by $R_{\boldsymbol{H}} \le \min(M_{\mathrm{Tx}}, M_{\mathrm{Rx}})$. In order to properly detect independent data streams, the number of received antennas must be larger than the number of transmitted antennas so that the maximum possible number of independent data streams is upper bounded

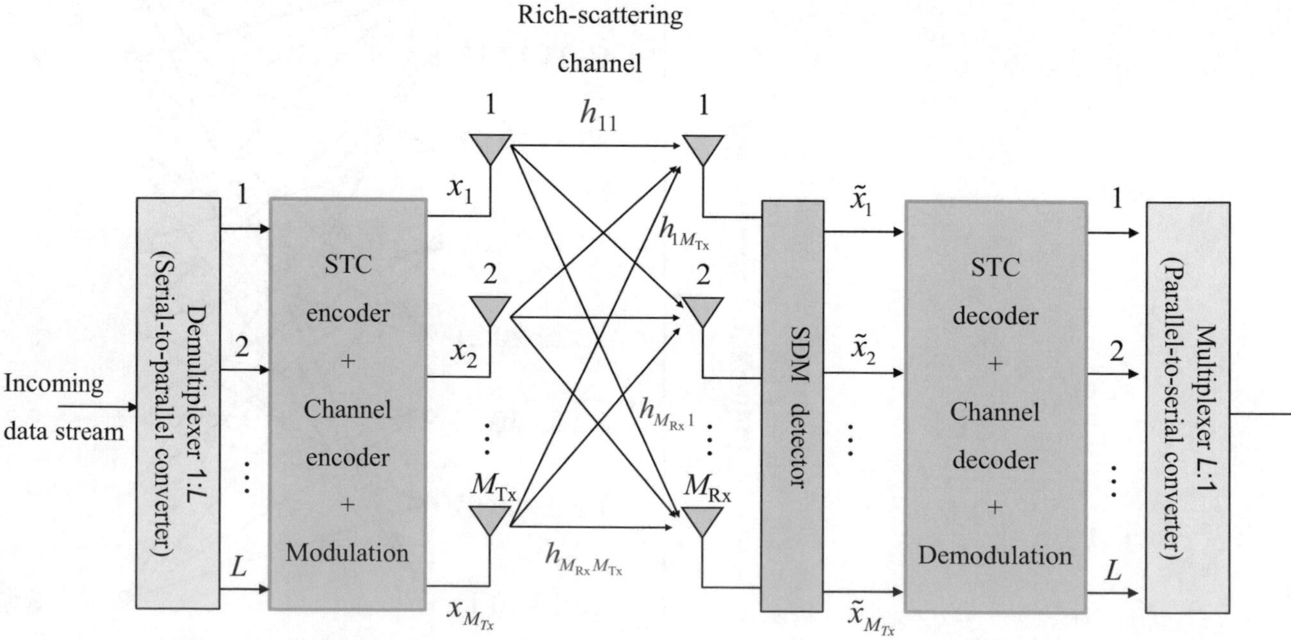

Fig. 8.21 Generic SDM scheme for wireless MIMO communications

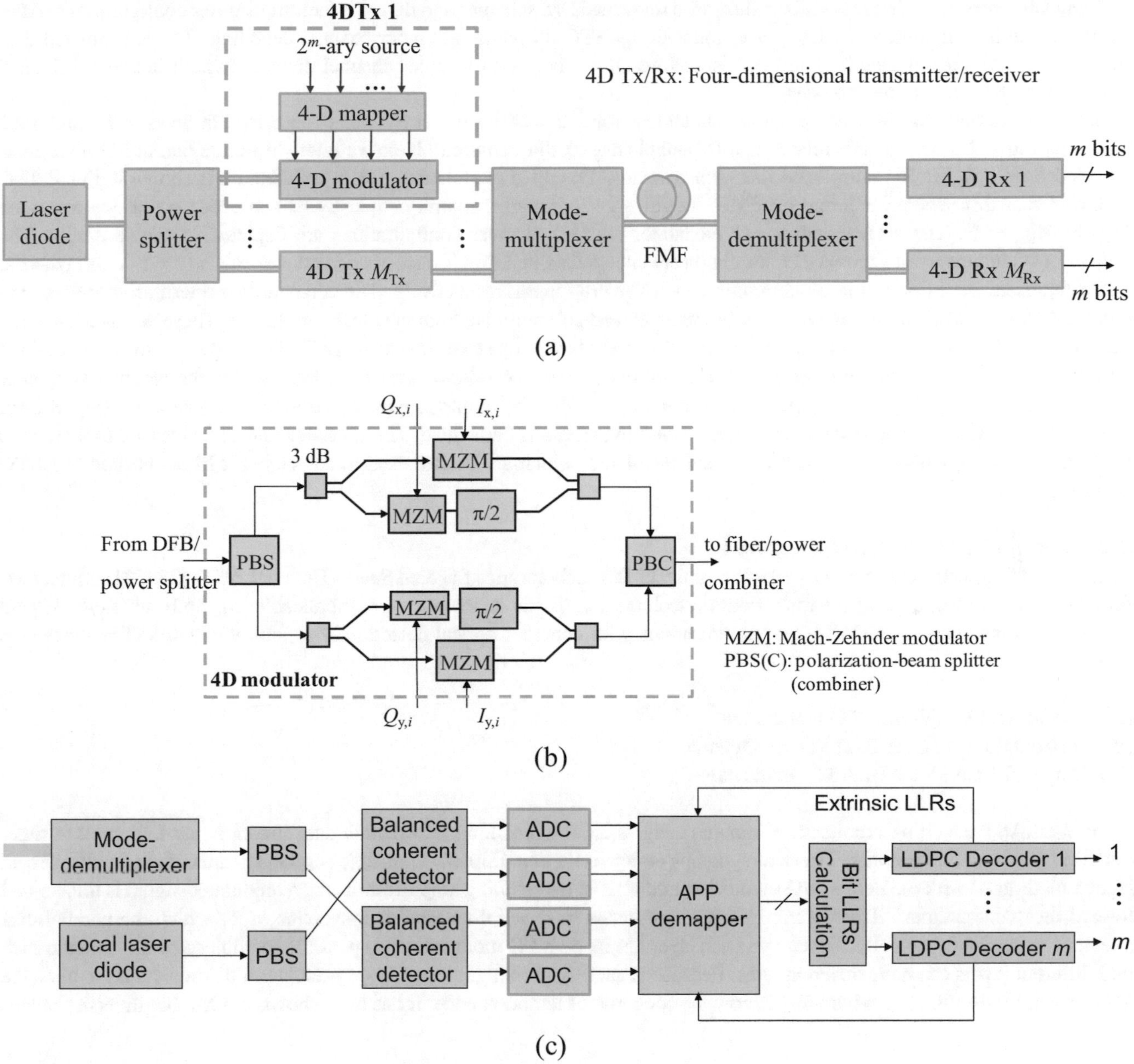

Fig. 8.22 Mode-multiplexed 4D coded modulation scheme for FMF-based optical communication systems: (**a**) overall transmitter configuration, (**b**) 4D modulator configuration, and (**c**) receiver configuration

by M_{Tx}. Let R_{Tx} denote the code rate of both outer and inner channel codes (obtained as a product of individual code rates); then the spectral efficiency (SE) of the SDM scheme is upper bounded by:

$$SE_{SDM} \leq M_{Tx}R_{Tx}\log_2(M) \text{ [bits/s/Hz]}, \quad (8.117)$$

where M is the signal constellation size (assuming that all independent data streams employ the same constellation size). At the receiver side, each receive antenna detects the complex linear combination of signal from all transmit antennas, as illustrated by Eq. (8.61). The purpose of SDM detector is to compensate for spatial interference and provides accurate estimates of transmitted symbols. The SDM detector configurations are subject of study in incoming sections, which are more general and applicable not only SDM scheme but also STC and uncoded MIMO schemes.

The main purpose of this section is to introduce relevant SDM schemes and describe their transmitter configurations. After SDM detection, demodulation takes place, followed by STC decoding and inner channel decoding. The L estimated data streams are multiplexed together, and corresponding output is passed to outer channel decoder, which delivers estimated information sequence to the end users.

The corresponding SDM scheme relevant in optical communications has already been described in Sects. 8.2.1 and 8.2.2 (see, e.g., Fig. 8.11, with details related to STC and channel coding omitted). Here we briefly describe one of SDM schemes introduced in [28]. Mode-multiplexed four-dimensional (4D) coded modulation system architecture is shown in Fig. 8.22a.

Given that this scheme is based on FMF, the full rank of channel matrix can easily be achieved so that we can write $M_{Tx} = M_{Rx} = N$. The corresponding 4D modulator and 4D receiver configurations are depicted in Fig. 8.22b, c. The $m = \log_2(M)$ independent streams of information are encoded using LDPC codes of code rates $r_i = k_i/n$ $(i = 1,\ldots,m)$ where k_i and n represent the information block length and codeword length, respectively. The codewords are written row-wise into block interleaver, while m bits at every symbol interval read column-wise from the block interleaver. These m-bits are used to determine four coordinates of corresponding signal constellation point stored inside the LUT. The N 4D streams are combined together using the mode multiplexer and transmitted over FMF transmission system of interest. On the receiver side, after mode demultiplexing (Fig. 8.22a), the 4D detection is performed by employing polarization diversity receiver (Fig. 8.22c), followed by ADCs. Then compensation of mode-coupling effects is performed. The 4D coordinate estimates are forwarded to the APP demapper where the symbol LLRs are calculated, which are further used to calculate bit LLRs needed in LDPC decoding.

8.2.5.1 BLAST Encoding Architectures

In the rest of this section, we will describe the family of Bell Laboratories Layered Space-Time (BLAST) [29–32] architecture, followed by multi-group space-time coding techniques. These techniques are applicable to both wireless MIMO communications and optical MIMO communications with coherent optical detection. The family of BLAST architectures includes:

(i) Vertical BLAST (V-BLAST) architecture.
(ii) Horizontal BLAST (H-BLAST) architecture.
(iii) Diagonal BLAST (D-BLAST) architecture.

In V-BLAST encoding architecture, shown in Fig. 8.23, the incoming information data stream is used as input to time-domain (TD)/temporal encoder. The corresponding codeword is demultiplexed into M_{Tx} parallel streams. Each parallel stream is then modulated with the help of I/Q modulator, and after corresponding amplification, such modulated signal is transmitted toward received antennas. The keyword parallel originates from serial-to-parallel conversion, after which the parallel data streams are depicted vertically. Given that each layer has its own I/Q modulator, it is possible to implement this scheme such that different layers can have different rates. Because at each symbolling interval M_{Tx} symbols can be transmitted in total, the SE is given by Eq. (8.117), where R_{Tx} is now the code rate of temporal code. It has been shown in [33] that diversity ranges

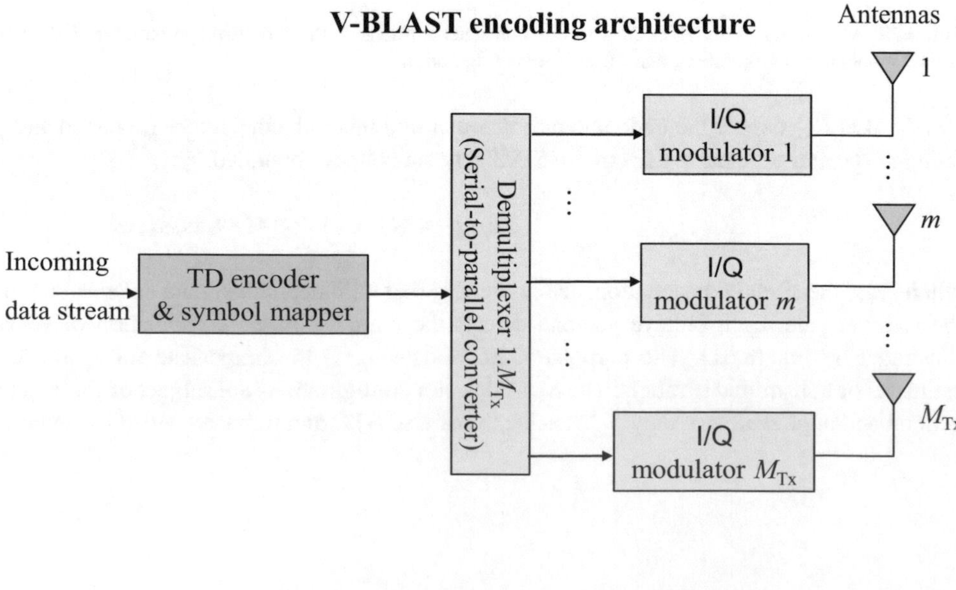

Fig. 8.23 V-BLAST encoding principles. TD: time-domain (temporal)

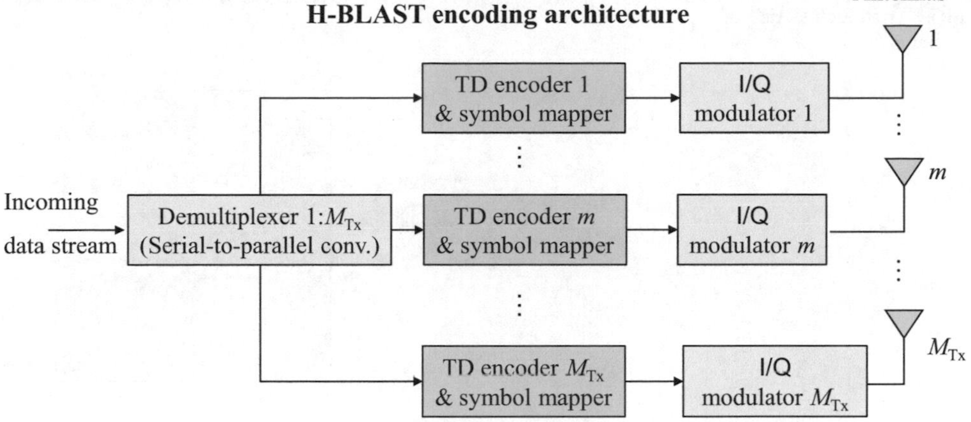

Fig. 8.24 H-BLAST encoding principles

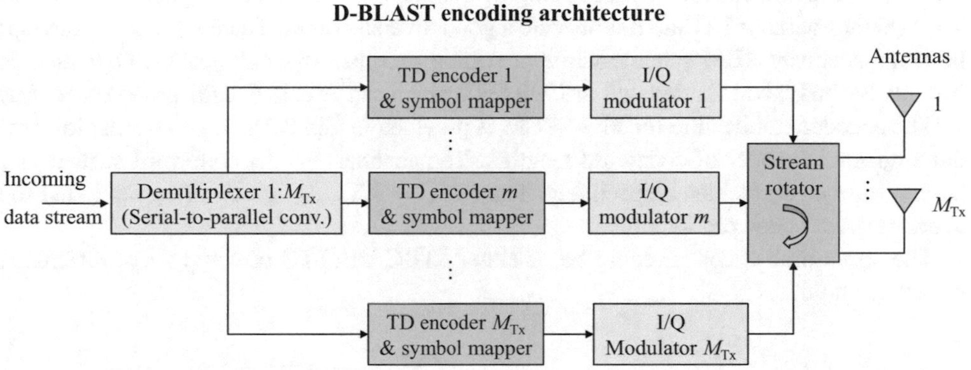

Fig. 8.25 The illustration of D-BLAST architecture of encoder. *TD* time-domain (temporal)

from $M_{\mathrm{Rx}} - M_{\mathrm{Tx}} + 1$ to M_{Rx}, depending on the layer detected. Given that overall performance is dominated by the smallest spatial diversity level, the maximum diversity gain is about $M_{\mathrm{Rx}} - M_{\mathrm{Tx}} + 1$.

The encoding process in H-BLAST, illustrated in Fig. 8.24, can be interpreted as follows. Each of the parallel sub-streams undergoes TD/temporal block encoding, with codeword length N_s. Because every coded symbol is detected by all M_{Rx} received antennas, H-BLAST can achieve at most a diversity order of M_{Rx}.

To restore the full diversity order $M_{\mathrm{Tx}}M_{\mathrm{Rx}}$, we need to modify the H-BLAST by inserting the *stream rotator* [1, 45]. In this scheme, instead of transmitting the independent space-time codewords on separate antennas, by stream rotator the codeword symbols from a given space-time codeword are rotated across transmit antennas, so that a codeword is effectively spread over all M_{Tx} antennas, as illustrated in Fig. 8.25. Such obtained BLAST scheme is commonly referred to as diagonal-BLAST or D-BLAST Scheme [1, 2, 26, 45]. Let the mth time-domain (TD)/temporal encoder codeword be denoted by $\boldsymbol{x}_m = [x_{m1} \quad x_{m2} \quad \cdots \quad x_{mN_s}]$. The stream rotator transmits n-th codeword symbol on n-th antenna, where $n = 1,2,\ldots,M_{\mathrm{Tx}}$. When the TD codeword length exceeds M_{Tx}, then the rotation begins again from the first antenna. The corresponding code matrix can be described as:

$$\boldsymbol{X} = \begin{bmatrix} x_{11} & x_{21} & x_{31} & \cdots & x_{N_s1} & 0 & \cdots \\ 0 & x_{12} & x_{22} & x_{32} & \cdots & x_{N_s2} & \cdots \\ 0 & 0 & \ddots & \ddots & \ddots & \ddots & \cdots \\ 0 & 0 & 0 & x_{1M_{Tx}} & x_{2M_{Tx}} & x_{3M_{Tx}} & \cdots \end{bmatrix}. \tag{8.118}$$

Even though that D-BLAST scheme archives full diversity, for sufficiently long codewords, there are many unused space-time dimensions, denoted by zeros in Eq. (8.118). This problem can be solved by the so-called *threaded space-time architecture*, in which the symbols are distributed in the codeword matrix on such a way to achieve full spatial span M_{Tx} (to guarantee desired spatial diversity order) and full temporal span N_s (to guarantee desired temporal diversity order to deal with fast fading effects). The key idea is to properly fill symbols from previous and incoming codewords so that there are no zero symbols in $\boldsymbol{X}$.

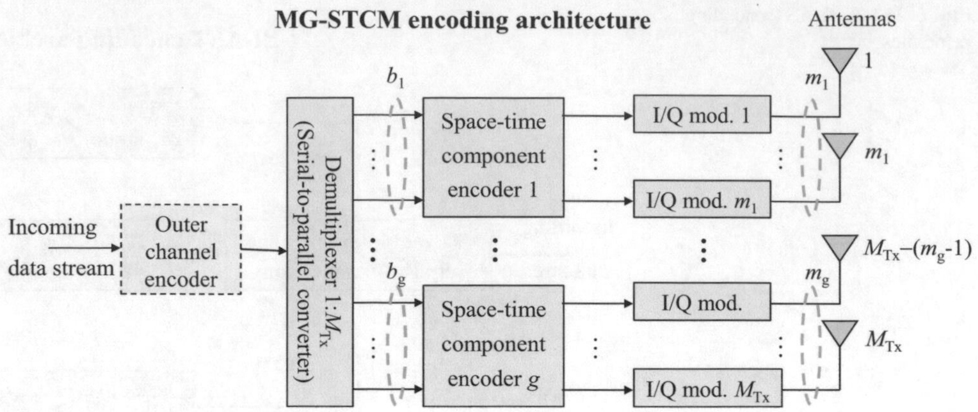

Fig. 8.26 The illustration of MG-STCM architecture of encoder

8.2.5.2 Multi-group Space-Time Coded Modulation (MG-STCM)

The BLAST architectures, mentioned in previous subsections, do not employ any STC but instead represent SDM schemes. The schemes without STC do not provide a good diversity order. There were many attempts to develop a scheme that would be simultaneously SDM scheme while providing excellent diversity orders. One such scheme get introduced by Tarokh, Naguib, Seshadri, and Calderbank in 1999 [34], which can be called multi-group space-time coded modulation (MG-STCM).

The encoder architecture for MG-STCM is provided in Fig. 8.26. Input information sequence is optionally encoded by an outer channel encoder of codeword length b. The channel encoder codeword sequences is split into g groups of bits each having b_1, b_2, …, b_g bits, respectively. The i-th ($i = 1,2,\ldots,g$) group of b_i bits is fed to space-time component encoder i, generating the codeword length m_i.

The space-time encoders could be of either STBC or STTC or could be of different types. Clearly, the following two conditions hold:

$$\sum_{i=1}^{g} b_i = b, \quad \sum_{i=1}^{g} m_i = M_{Tx}. \tag{8.119}$$

Similarly to BLAST schemes suffering from interlayer spatial interference, this scheme suffers from inter-group spatial interference. The interference cancellation must be performed on a group-by-group basis rather than on a layer-by-layer basis. The outputs of space-time encoders are mapped to corresponding constellation points with the help of I/Q modulators whose corresponding outputs drive transmit antennas. The codeword matrix of MG-STCM encoder can be represented as:

$$\mathbf{X} = \begin{bmatrix} x_{11} & \cdots & x_{1N_s} \\ \vdots & & \vdots \\ x_{m_1 1} & \cdots & x_{m_1 N_s} \\ \vdots & & \vdots \\ x_{M_{Tx}-m_g+1,1} & \cdots & x_{M_{Tx}-m_g+1,N_s} \\ \vdots & & \vdots \\ x_{M_{Tx}1} & \cdots & x_{M_{Tx}N_s} \end{bmatrix} = \begin{bmatrix} \mathbf{X}_{C_1} \\ \mathbf{X}_{C_2} \\ \vdots \\ \mathbf{X}_{C_g} \end{bmatrix}, \tag{8.120}$$

where the i-th ($i = 1,2,\ldots,g$) code C_i codeword matrix is given by:

$$\mathbf{X}_{C_i} = \begin{bmatrix} x_{m_1+\cdots m_{i-1}+1,1} & \cdots & x_{m_1+\cdots m_{i-1}+1,N_s} \\ \vdots & & \vdots \\ x_{m_1+\cdots m_{i-1}+m_i,1} & \cdots & x_{m_1+\cdots m_{i-1}+m_i,N_s} \end{bmatrix}. \tag{8.121}$$

The channel matrix for MG-STCM case be represented as:

$$\begin{aligned}\boldsymbol{H}&=\begin{bmatrix}h_{11} & \cdots & h_{1m_1} & \cdots & x_{1,M_{Tx}-m_g+1} & \cdots & h_{1M_{Tx}}\\ \vdots & & & \vdots & & & \vdots\\ h_{M_{Rx}1} & \cdots & h_{M_{Rx}m_1} & \cdots & x_{M_{Rx},M_{Tx}-m_g+1} & \cdots & h_{M_{Rx}M_{Tx}}\end{bmatrix}\\ &=\begin{bmatrix}\boldsymbol{H}_{C_1} & \boldsymbol{H}_{C_2} & \cdots & \boldsymbol{H}_{C_g}\end{bmatrix},\end{aligned} \tag{8.122}$$

where $\boldsymbol{H}_{C_i}$ is the submatrix corresponding to the i-th space-time component code of dimensions $M_{\text{Rx}} \times m_i$. The corresponding MIMO channel model is given by $\boldsymbol{Y} = \boldsymbol{H}\boldsymbol{X} + \boldsymbol{Z}$, as for STCs. The *truncated channel matrix* in which the first $m_1 + \ldots + m_i$ columns are omitted can be defined as:

$$\boldsymbol{H}_{C-C_{1,2,\ldots,i}} = \begin{bmatrix}\boldsymbol{H}_{C_{i+1}} & \boldsymbol{H}_{C_{i+2}} & \cdots & \boldsymbol{H}_{C_g}\end{bmatrix}. \tag{8.123}$$

Based on Fig. 8.26, we conclude that the rank of $\boldsymbol{H}_{C-C_1}$ is upper bounded by $\text{rank}(\boldsymbol{H}_{C_1}) \le M_{TX} - m_1$. The dimensionality of the null space of $\boldsymbol{H}_{C-C_1}$, $Null(\boldsymbol{H}_{C-C_1})$, and rank of $\boldsymbol{H}_{C_1}$ is related by [34, 35]:

$$\dim[Null(\boldsymbol{H}_{C-C_1})] + \text{rank}(\boldsymbol{H}_C) = M_{Rx}. \tag{8.124}$$

Therefore, the dimensionality of the null space of $\boldsymbol{H}_{C-C_1}$ is lower bounded by:

$$\dim[Null(\boldsymbol{H}_{C-C_1})] \ge M_{Rx} - (M_{TX} - m_1). \tag{8.125}$$

Let the set of orthonormal vectors spanning the null space of $\boldsymbol{H}_{C-C_1}$ be denoted as $\{\boldsymbol{v}_1, \boldsymbol{v}_2, \cdots, \boldsymbol{v}_{M_{Rx}-(M_{TX}-m_1)}\}$. Let us arrange the vector as the rows of the Hermitian matrix $\boldsymbol{\Theta}$, which reminds us to the parity-check matrix of an linear block code:

$$\boldsymbol{\Theta}_{C-C_1} \doteq \begin{bmatrix}\boldsymbol{v}_1\\ \boldsymbol{v}_2\\ \vdots\\ \boldsymbol{v}_{M_{Rx}-(M_{TX}-m_1)}\end{bmatrix}. \tag{8.126}$$

Because vectors $\boldsymbol{v}_i$ span the null space $\boldsymbol{H}_{C-C_1}$, they are orthogonal to $\boldsymbol{H}_{C-C_1}$ and we can write:

$$\boldsymbol{\Theta}_{C-C_1}\boldsymbol{H}_{C-C_1} = \boldsymbol{0}. \tag{8.127}$$

Decoding at the receiver side is performed in an iterative fashion. The key idea of decoding process is to decode each code separately while suppressing the spatial interference from other component codes. Because space-time code C_1 employs m_1 antennas, the signals from remaining $M_{\text{Tx}} - m_1$ antennas represent the spatial interference in decoding process for C_1. The zero-forcing (ZF) spatial interference approaches described below can be employed. Let us multiply the received matrix from the left by $\boldsymbol{\Theta}_{C-C_1}$ to obtain:

$$\begin{aligned}\boldsymbol{\Theta}_{C-C_1}\boldsymbol{Y} &= \boldsymbol{\Theta}_{C-C_1}\left(\underbrace{\begin{bmatrix}\boldsymbol{H}_{C_1} & \boldsymbol{H}_{C-C_1}\end{bmatrix}}_{\boldsymbol{H}}\underbrace{\begin{bmatrix}\boldsymbol{X}_{C_1}\\ \boldsymbol{X}_{C-C_1}\end{bmatrix}}_{\boldsymbol{X}} + \boldsymbol{Z}\right) = \boldsymbol{\Theta}_{C-C_1}(\boldsymbol{H}\boldsymbol{X} + \boldsymbol{Z})\\ &= \boldsymbol{\Theta}_{C-C_1}\boldsymbol{H}_{C_1}\boldsymbol{X}_{C_1} + \underbrace{\boldsymbol{\Theta}_{C-C_1}\boldsymbol{H}_{C-C_1}}_{=\boldsymbol{0}}\boldsymbol{X}_{C-C_1} + \underbrace{\boldsymbol{\Theta}_{C-C_1}\boldsymbol{Z}}_{\widetilde{\boldsymbol{Z}}}\\ &= \boldsymbol{\Theta}_{C-C_1}\boldsymbol{H}_{C_1}\boldsymbol{X}_{C_1} + \widetilde{\boldsymbol{Z}},\end{aligned} \tag{8.128}$$

and we have been able to successfully compensate for the spatial interference from other groups of STCs, in other words the inter-group spatial interference. We can denoted this result by $\widetilde{\boldsymbol{Y}}_{C_1} \doteq \boldsymbol{\Theta}_{C-C_1}\boldsymbol{Y}$. The transmitted block can now be estimated to minimize the Euclidean distance squared derived from (8.128):

$$\widehat{X}_{C_1} = \underset{\{X_{C_1}\}}{\arg\min} \left\| \underbrace{\Theta_{C-C_1} Y}_{\widetilde{Y}_{C_1}} - \Theta_{C-C_1} H_{C_1} X_{C_1} \right\|_F^2 . \tag{8.129}$$

Once the estimate of codeword matrix for C_1 is obtained, $\widehat{X}_{C_1}$, it can be used to cancel the spatial interference from C_1 as follows:

$$Y_{C_1} = Y - H_{C_1} \widehat{X}_{C_1}. \tag{8.130}$$

In a similar fashion as for C_1, we can now multiply Y_{C_1} by another matrix whose rows span null space of $H_{C-C_{1,2}}$, denoted as Θ_{C-C_1}, to obtain:

$$\begin{aligned}
\widetilde{Y}_{C_2} &\doteq \Theta_{C-C_{1,2}} Y_{C_2} = \Theta_{C-C_{1,2}} \left(\underbrace{Y}_{HX+Z} - H_{C_1} \widehat{X}_{C_1} \right) = \Theta_{C-C_{1,2}} \left(\overbrace{\begin{bmatrix} H_{C_1} & H_{C_2} & H_{C-C_{1,2}} \end{bmatrix}}^{H} \overbrace{\begin{bmatrix} X_{C_1} \\ X_{C_2} \\ X_{C-C_{1,2}} \end{bmatrix}}^{X} + Z - H_{C_1} \widehat{X}_{C_1} \right) \\
&= \Theta_{C-C_{1,2}} H_{C_1} X_{C_1} + \Theta_{C-C_{1,2}} H_{C_2} X_{C_2} + \underbrace{\Theta_{C-C_{1,2}} H_{C-C_{1,2}}}_{=0} X_{C-C_{1,2}} + \Theta_{C-C_{1,2}} Z - \Theta_{C-C_{1,2}} H_{C_1} \widehat{X}_{C_1} \\
&= \Theta_{C-C_{1,2}} H_{C_1} \left(X_{C_1} - \widehat{X}_{C_1} \right) + \Theta_{C-C_{1,2}} H_{C_2} X_{C_2} + \Theta_{C-C_{1,2}} Z,
\end{aligned} \tag{8.131}$$

indicating that the spatial interference from other groups is cancelled providing that $X_{C_1} \simeq \widehat{X}_{C_1}$. The ML estimate of $\widehat{X}_{C_2}$ can be determined as follows:

$$\widehat{X}_{C_2} = \underset{\{X_{C_2}\}}{\arg\min} \left\| \widetilde{Y}_{C_2} - \Theta_{C-C_{1,2}} H_{C_2} X_{C_2} \right\|_F^2 . \tag{8.132}$$

The i-th codeword ($i > 2$) is estimated by first cancelling the spatial interference originating from the (i − 1)-st STC:

$$Y_{C_{i-1}} = Y_{C_{i-2}} - H_{C_{i-1}} \widehat{X}_{C_{i-1}}. \tag{8.133}$$

Then we need to determine the orthonormal basis vectors that span $Null\left(H_{C-C_{1,2,\ldots,i}}\right)$, which are arranged as the rows of the matrix denoted by $\Theta_{C-C_{1,2,\ldots,i}}$. We now multiply $Y_{C_{i-1}}$ with $\Theta_{C-C_{1,2,\ldots,i}}$ from the left side to obtain:

$$\begin{aligned}
\widetilde{Y}_{C_i} &\doteq \Theta_{C-C_{1,2,\ldots,i}} Y_{C_{i-1}} = \Theta_{C-C_{1,2,\ldots,i}} \left(Y_{C_{i-2}} - H_{C_{i-1}} \widehat{X}_{C_{i-1}} \right) = \\
&= \Theta_{C-C_{1,2,\ldots,i}} H_{C_i} X_{C_i} + \sum_{j=1}^{i-1} \Theta_{C-C_{1,2,\ldots,i}} H_{C_j} \left(X_{C_j} - \widehat{X}_{C_j} \right) + \Theta_{C-C_{1,2,\ldots,i}} Z,
\end{aligned} \tag{8.134}$$

which indicates that the spatial interference from all other previous groups (j = 1,2,. . .,i − 1) is cancelled out providing that the following is valid $X_{C_j} - \widehat{X}_{C_j}$. The ML estimate of $\widehat{X}_{C_i}$ can be determined as follows:

$$\widehat{X}_{C_i} = \underset{\{X_{C_i}\}}{\arg\min} \left\| \widetilde{Y}_{C_i} - \Theta_{C-C_{1,2,\ldots,i}} H_{C_i} X_{C_i} \right\|_F^2 ; i = 2, 3, \cdots, g \tag{8.135}$$

8.2.6 Linear and Decision-Feedback MIMO Receivers for Uncoded Signals

We will now discuss several suboptimum strategies [2, 7, 26], in order to reduce the ML detector complexity. We will start our discussion with linear and decision-feedback receivers, applied to the schemes without space-time coding. The key idea behind linear receivers is to perform certain linear processing operation on received vector so that the transformed channel matrix does not have off-diagonal elements, which is:

$$\tilde{\boldsymbol{y}} = \boldsymbol{A}(\boldsymbol{H})\boldsymbol{y} = \boldsymbol{A}(\boldsymbol{H})\boldsymbol{H}\boldsymbol{x} + \boldsymbol{A}(\boldsymbol{H})\boldsymbol{z}. \tag{8.136}$$

The union bound for the error probability in this case is given by:

$$P_e \le \frac{1}{M}\sum\nolimits_{\boldsymbol{x}}\sum\nolimits_{\widehat{\boldsymbol{x}}} P(\boldsymbol{x} \to \widehat{\boldsymbol{x}}), \tag{8.137}$$

where the conditional PEP for Euclidean metric is expressed as:

$$\begin{aligned} P(\boldsymbol{x} \to \widehat{\boldsymbol{x}}|\boldsymbol{H}) &= P\left(\|\boldsymbol{A}\boldsymbol{y} - \widehat{\boldsymbol{x}}\|^2 < \|\boldsymbol{A}\boldsymbol{y} - \boldsymbol{x}\|^2|\boldsymbol{H}\right) \\ &= P\left(\|\boldsymbol{A}\boldsymbol{H}\boldsymbol{x} - \widehat{\boldsymbol{x}} + \boldsymbol{A}\boldsymbol{z}\|^2 < \|\boldsymbol{A}\boldsymbol{H}\boldsymbol{x} - \boldsymbol{x} + \boldsymbol{A}\boldsymbol{z}\|^2|\boldsymbol{H}\right) \\ &= P\left(\|\boldsymbol{d}\|^2 + 2((\boldsymbol{A}\boldsymbol{H}\text{-}\boldsymbol{I})\boldsymbol{x}, \boldsymbol{d}) + 2(\boldsymbol{A}\boldsymbol{z}, \boldsymbol{d}) < 0|\boldsymbol{H}\right), \end{aligned} \tag{8.138}$$

where $\boldsymbol{d} = \boldsymbol{x} - \widehat{\boldsymbol{x}}$, while $(\boldsymbol{a},\boldsymbol{b})$ denotes the dot product of vectors $\boldsymbol{a}$ and $\boldsymbol{b}$. Since $\mathrm{Var}\{(\boldsymbol{A}\boldsymbol{z}, \boldsymbol{d})\} = N_0\|\boldsymbol{A}^\dagger \boldsymbol{d}\|^2$, the PEP becomes:

$$P\left(\boldsymbol{X} \to \widehat{\boldsymbol{X}}\right) = \left\langle Q\left(\sqrt{\frac{\|\boldsymbol{d}\|^2 + 2((\boldsymbol{A}\boldsymbol{H} - \boldsymbol{I})\boldsymbol{x}, \boldsymbol{d})}{2N_0\|\boldsymbol{A}^\dagger \boldsymbol{d}\|^2}}\right)\right\rangle. \tag{8.139}$$

In the *zero-forcing receiver*, the linear matrix $\boldsymbol{A}$ is chosen to have the form:

$$\begin{aligned} &\boldsymbol{A} = \boldsymbol{H}^+ = \begin{cases} \boldsymbol{V}\begin{bmatrix}\boldsymbol{\Sigma}^+ \\ \boldsymbol{0}\end{bmatrix}\boldsymbol{U}^\dagger, & M_{\mathrm{Rx}} \le M_{\mathrm{Tx}} \\ \boldsymbol{V}[\boldsymbol{\Sigma}^+ \quad \boldsymbol{0}]\boldsymbol{U}^\dagger, & M_{\mathrm{Rx}} > M_{\mathrm{Tx}} \end{cases}, \quad \boldsymbol{\Sigma}^+ = \mathrm{diag}\left(\sigma_1^{-1}, \sigma_2^{-1}, \ldots, \sigma_p^{-1}, 0, \ldots, 0\right) \\ &\boldsymbol{H} = \boldsymbol{U}[\boldsymbol{\Sigma} \quad \boldsymbol{0}]\boldsymbol{V}^\dagger;\ \sigma_1 \ge \sigma_2 \ge \ldots \ge 0, \sigma_i = \sqrt{\lambda_i},\ \boldsymbol{\Sigma} = \mathrm{diag}(\sigma_1, \sigma_2, \ldots,) \\ &\boldsymbol{U} = \mathrm{eigenvectors}\left(\boldsymbol{H}\boldsymbol{H}^\dagger\right),\ \boldsymbol{V} = \mathrm{eigenvectors}\left(\boldsymbol{H}^\dagger\boldsymbol{H}\right) \end{aligned} \tag{8.140}$$

where we use the superscript + to denote the Moore-Penrose pseudoinverse of the matrix.

Since, for $M_{\mathrm{Rx}} \ge M_{\mathrm{Tx}}$, the matrix $\boldsymbol{H}^+\boldsymbol{H}$ is invertible:

$$\boldsymbol{H}^+(\boldsymbol{H}^+)^\dagger = \left(\boldsymbol{H}^\dagger\boldsymbol{H}\right)^{-1} \Rightarrow \boldsymbol{H}^+ = \left(\boldsymbol{H}^\dagger\boldsymbol{H}\right)^{-1}\boldsymbol{H}^\dagger, \tag{8.141}$$

we can compensate for spatial interference in wireless links or mode coupling FMF/MMF links by pre-multiplying the MIMO equation with $\boldsymbol{H}^+$ from the left:

$$\boldsymbol{H}^+\boldsymbol{y} = \boldsymbol{x} + \boldsymbol{H}^+\boldsymbol{z}. \tag{8.142}$$

The variance for PEP expression is given by:

$$\mathrm{Var}\{(\boldsymbol{H}^+\boldsymbol{z}, \boldsymbol{d})\} = \frac{N_0}{2}\mathrm{Tr}\left\{\boldsymbol{d}^\dagger\boldsymbol{H}^+(\boldsymbol{H}^+)^\dagger\boldsymbol{d}\right\} = \frac{N_0}{2}\mathrm{Tr}\left\{\boldsymbol{d}^\dagger(\boldsymbol{H}^+\boldsymbol{H})^{-1}\boldsymbol{d}\right\}, \tag{8.143}$$

indicating that the price to be paid during zero-forcing of spatial interference/mode coupling is the noise enhancement effect.

An alternative strategy would be to minimize jointly off-diagonal elements of transformed channel matrix $\boldsymbol{AH}$ and colored noise $\boldsymbol{Az}$. In this case, the receiver is known as the *minimum mean square error (MMSE)* one, with $\boldsymbol{A}$-matrix chosen as:

$$\boldsymbol{A}_{\text{MMSE}} = \left(\boldsymbol{H}^{\dagger}\boldsymbol{H} + \boldsymbol{I}/SNR\right)^{-1}\boldsymbol{H}^{\dagger}, \tag{8.144}$$

where SNR $= E/N_0$ is the signal to noise ratio for the average energy E of one component of vector $\boldsymbol{x}$.

The last method to be described in this subsection is related to *decision-feedback receivers*, which is similar to that described in the section on linear space-time block codes. The key idea is to perform preprocessing of $\boldsymbol{H}$ by QR factorization $\boldsymbol{H} = \boldsymbol{QR},\ \boldsymbol{Q}^{\dagger}\boldsymbol{Q} = \boldsymbol{I}_{M_{\text{Tx}}},$ so that we can transform the received vector as:

$$\tilde{\boldsymbol{y}} = \boldsymbol{Q}^{\dagger}\boldsymbol{y} = \boldsymbol{Rx} + \boldsymbol{Q}^{\dagger}\boldsymbol{z}, \tag{8.145}$$

wherein the transformed noise vector retains the properties of $\boldsymbol{z}$. Since $\boldsymbol{R}$ is an upper-triangular matrix, we can perform the following optimization:

$$\hat{\boldsymbol{x}} = \arg\min_{\boldsymbol{x}} \|\boldsymbol{y} - \boldsymbol{Rx}\|_F. \tag{8.146}$$

The optimization starts with detection of $x_{M_{\text{Tx}}}$ by minimizing the $\left|\tilde{y}_{M_{\text{Tx}}} - R_{M_{\text{Tx}},M_{\text{Tx}}}x_{M_{\text{Tx}}}\right|^2$; next, decision $\hat{x}_{M_{\text{Tx}}}$ is used to detect $x_{M_{\text{Tx}}-1}$ by minimizing $\left|\tilde{y}_{M_{\text{Tx}}-1} - R_{M_{\text{Tx}}-1,M_{\text{Tx}}-1}x_{M_{\text{Tx}}-1} - R_{M_{\text{Tx}}-1,M_{\text{Tx}}}\hat{x}_{M_{\text{Tx}}}\right|^2 + \left|\tilde{y}_{M_{\text{Tx}}} - R_{M_{\text{Tx}},M_{\text{Tx}}}\hat{x}_{M_{\text{Tx}}}\right|^2$ and so on. The problem with this approach is related to the error propagation.

8.2.7 Suboptimum MIMO Receivers for Coded Signals

In this section we study linear detector strategies, such as zero-forcing and linear MMSE (LMMSE) interfaces [2, 7, 26], but described in the context of space-time coding.

8.2.7.1 Linear and Zero-Forcing Receivers (Interfaces)

Similarly as in previous section, the key idea behind linear receivers is to perform linear transformation $\boldsymbol{A}(\boldsymbol{H})$ on received matrix so that the spatial interference is cancelled:

$$\tilde{\boldsymbol{Y}} = \boldsymbol{A}(\boldsymbol{H})\boldsymbol{Y} = \boldsymbol{A}(\boldsymbol{H})\boldsymbol{HX} + \boldsymbol{A}(\boldsymbol{H})\boldsymbol{Z}. \tag{8.147}$$

The conditional PEP is given by, obtained by generalization of (8.139):

$$P\left(\boldsymbol{X} \to \hat{\boldsymbol{X}}\right) = \left\langle Q\left(\sqrt{\frac{\|\boldsymbol{\Delta}\|^2 + 2((\boldsymbol{AH}\text{-}\boldsymbol{I})\boldsymbol{X},\boldsymbol{D})}{2N_0\|\boldsymbol{A}^{\dagger}\boldsymbol{D}\|^2}}\right)\right\rangle, \boldsymbol{D} = \boldsymbol{X} - \hat{\boldsymbol{X}}. \tag{8.148}$$

In *zero-forcing interface* we chose the linear transformation matrix $\boldsymbol{A}$ to be pseudo-inverse of $\boldsymbol{H}$, i.e., $\boldsymbol{A} = \boldsymbol{H}^{+}$, in a fashion similar to one from Eq. (8.142), so that spatial interference in MIMO wireless communication and mode coupling effect in FMF/MMF links can be compensated by:

$$\boldsymbol{H}^{+}\boldsymbol{Y} = \boldsymbol{X} + \boldsymbol{H}^{+}\boldsymbol{Z} \tag{8.149}$$

The Euclidean metrics is given by $\|\boldsymbol{H}^{+}\boldsymbol{Y} - \boldsymbol{X}\|_F^2$, and based on Eq. (8.141), we conclude that the conditional PEP is given by:

$$P(\boldsymbol{X} \to \boldsymbol{X}|\boldsymbol{H}) = Q\left(\sqrt{\frac{\|\boldsymbol{D}\|^2}{2\sigma^2}}\right), \tag{8.150}$$

where the variance is given by, obtained by generalization of (8.143):

$$\mathrm{Var}\{(\boldsymbol{H}^{+}\boldsymbol{Z},\boldsymbol{D})\} = \frac{N_0}{2}\mathrm{Tr}\left\{\boldsymbol{D}^{\dagger}(\boldsymbol{H}^{+}\boldsymbol{H})^{-1}\boldsymbol{D}\right\}. \tag{8.151}$$

Clearly, the price to be paid for compensation of spatial interference by zero-forcing is the noise enhancement effect.

8.2.7.2 Linear MMSE Receiver (Interface)

In *LMMSE interface (receiver)*, the linear transformation matrix $\boldsymbol{A}$ is chosen in such a way to minimize the MSE, defined by:

$$\begin{aligned} MSE(\boldsymbol{A}) &= \langle \|\boldsymbol{AY}-\boldsymbol{X}\|_F^2 \rangle \\ &= \langle \mathrm{Tr}[((\boldsymbol{AH}-\boldsymbol{I}_{M_{\mathrm{Tx}}})\boldsymbol{X}+\boldsymbol{AZ})((\boldsymbol{AH}-\boldsymbol{I}_{M_{\mathrm{Tx}}})\boldsymbol{X}+\boldsymbol{AZ})^{\dagger}]\rangle \\ &= \mathrm{Tr}[\rho(\boldsymbol{AH}-\boldsymbol{I}_{M_{\mathrm{Tx}}})(\boldsymbol{AH}-\boldsymbol{I}_{M_{\mathrm{Tx}}})^{\dagger}+N_0\boldsymbol{AA}^{\dagger}], \end{aligned} \tag{8.152}$$

where the third line in the above equation originates from the i.i.d. zero-mean assumption of $\boldsymbol{X}$, whose second moment is denoted by ρ. By setting the MSE variation with respect to $\boldsymbol{A}$ to be zero [2], i.e.:

$$\delta(MSE) = \mathrm{Tr}\left\{\delta\boldsymbol{A}\left[\rho\boldsymbol{H}(\boldsymbol{AH}-\boldsymbol{I}_{M_{\mathrm{Tx}}})^{\dagger}+N_0\boldsymbol{A}^{\dagger}\right]+\left[\rho(\boldsymbol{AH}-\boldsymbol{I}_{M_{\mathrm{Tx}}})\boldsymbol{H}^{\dagger}+N_0\boldsymbol{A}\right]\delta\boldsymbol{A}^{\dagger}\right\}=0, \tag{8.153}$$

we obtain the following solution for $\boldsymbol{A}$:

$$\boldsymbol{A}_{\mathrm{MMSE}} = \boldsymbol{H}^{\dagger}\left(\boldsymbol{HH}^{\dagger}+I_{M_{\mathrm{Rx}}}/SNR\right)^{-1} = \left(\boldsymbol{H}^{\dagger}\boldsymbol{H}+I_{M_{\mathrm{Tx}}}/SNR\right)^{-1}\boldsymbol{H}^{\dagger}, \tag{8.154}$$

where the equivalent SNR is defend as $SNR = \rho/N_0$.

The corresponding piece-wise error probability, needed for estimation of upper bound of error probability, is now given as:

$$P(\boldsymbol{X}\to\boldsymbol{X}) = \left\langle Q\left(\sqrt{\frac{\|\boldsymbol{D}\|^2+2([(\boldsymbol{H}^{\dagger}\boldsymbol{H}+\frac{\boldsymbol{I}_{M_{\mathrm{Tx}}}}{SNR})^{-1}\boldsymbol{H}^{\dagger}\boldsymbol{H}-\boldsymbol{I}_{M_{\mathrm{Tx}}}]\boldsymbol{X},\boldsymbol{D})}{2N_0\left\|\boldsymbol{H}(\boldsymbol{H}^{\dagger}\boldsymbol{H}+\frac{\boldsymbol{I}_{M_{\mathrm{Tx}}}}{SNR})^{-1}\boldsymbol{D}\right\|_F^2}}\right)\right\rangle. \tag{8.155}$$

When SNR tends to plus infinity, we obtain the same result as in zero-forcing receiver, as it was expected.

When the number of transmitters is fixed and finite, while the number of receiver tends to plus infinity, as discussed earlier, the spatial interference/mode coupling is compensated for and we can write $\boldsymbol{H}^{\dagger}\boldsymbol{H}\to\boldsymbol{I}_{M_{\mathrm{Tx}}}$. Therefore, in this case there is no performance loss with respect to ML detector. On the other hand, when the number of both transmitters and receivers tends to plus infinity by their ratio as constant $M_{\mathrm{Tx}}/M_{\mathrm{Rx}}\to c>0$, the corresponding PEP tends to [2]:

$$P\left(\boldsymbol{X}\to\widehat{\boldsymbol{X}}\right) \underset{M_{Tx}/M_{Rx}\to c}{\overset{M_{Tx}\to\infty,M_{Rx}\to\infty}{\longrightarrow}} Q\left(\sqrt{|1-c|\frac{M_{Rx}\|\boldsymbol{D}\|^2}{2N_0}}\right), \tag{8.156}$$

indicating that there is a performance loss with respect to ML detector. For the same PEP the SNR loss is $10\log_{10}|1-c|^{-1}$ [dB].

8.2.7.3 Decision-Feedback and BLAST Receivers (Interfaces)

In the *linear interface with nonlinear decision-feedback* method, illustrated in Fig. 8.21, we first process the received signal $\boldsymbol{Y}$ linearly by matrix $\boldsymbol{A}$ and then subtract the estimate of the spatial interference/mode-coupling $\boldsymbol{B}\widehat{\boldsymbol{X}}$ where $\widehat{\boldsymbol{X}}$ is a preliminary decision on transmitted codeword. The preliminary decision has been made by using the following metric $\left\|\widetilde{\boldsymbol{Y}}-\boldsymbol{X}\right\|$, where $\widetilde{\boldsymbol{Y}}=\boldsymbol{AY}-\boldsymbol{B}\widehat{\boldsymbol{X}}$. It is important that the diagonal entries of matrix $\boldsymbol{B}$ are set to zero, so that spatial interference and coupling effects are effectively removed. We should notice that this strategy has already been used to deal with PMD effect in polarization division multiplexed coherent optical systems [44]. This scheme is essentially a generalization of decision-feedback equalizers, used to compensate for ISI.

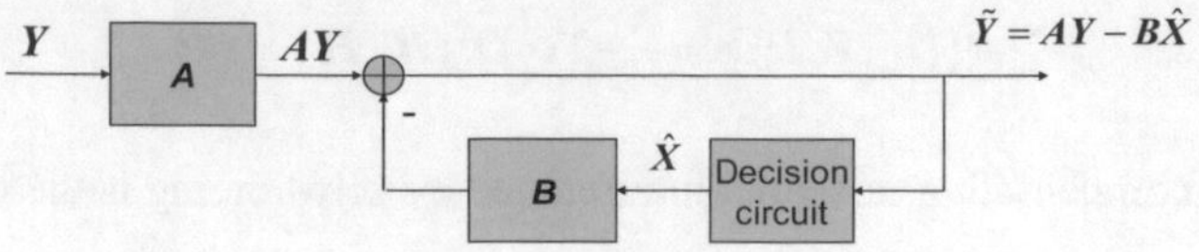

Fig. 8.27 The illustration of linear interface with nonlinear decision feedback

We will describe two attractive versions of the BLAST interface: (i) minimizing MSE in the absence of noise and (ii) minimizing the MSE in the presence of noise. The first scheme is also known as the zero-forcing vertical/horizontal-BLAST (ZF V/H-BLAST), while the second scheme is known as MMSE V-BLAST.

In *ZF V/H-BLAST*, we first perform QR factorization of channel matrix, $\boldsymbol{H} = \boldsymbol{QR}$, and then determine $\boldsymbol{A}$ and $\boldsymbol{B}$ matrices from Fig. 8.27 so that the MSE is set to zero. Because in zero-forcing receiver we ignore the presence of additive noise, the MSE is defined as:

$$MSE(\boldsymbol{A},\boldsymbol{B}) = \left\langle \left\| \widetilde{\boldsymbol{Y}} - \boldsymbol{X} \right\|_F^2 \right\rangle = \left\langle \left\| \boldsymbol{AHX} - \boldsymbol{B}\widehat{\boldsymbol{X}} - \boldsymbol{X} \right\|^2 \right\rangle. \tag{8.157}$$

For sufficiently high SNRs and when the level of spatial interference/mode coupling is not too high, the following is valid $\widehat{\boldsymbol{X}} \approx \boldsymbol{X}$, so that we can use the following two approximations $\left\langle \boldsymbol{X}\widehat{\boldsymbol{X}}^\dagger \right\rangle \approx \langle \boldsymbol{XX}^\dagger \rangle$, $\left\langle \widehat{\boldsymbol{X}}\widehat{\boldsymbol{X}}^\dagger \right\rangle \approx \langle \boldsymbol{XX}^\dagger \rangle$, and after employing them in (8.157), we obtain:

$$\begin{aligned} MSE(\boldsymbol{A},\boldsymbol{B}) &= \left\langle \|(\boldsymbol{AH} - \boldsymbol{B} - \boldsymbol{I}_{M_{Tx}})\boldsymbol{X}\|_F^2 \right\rangle \\ &= \left\langle \mathrm{Tr}\left\{ (\boldsymbol{AH} - \boldsymbol{B} - \boldsymbol{I}_{M_{Tx}})^\dagger (\boldsymbol{AH} - \boldsymbol{B} - \boldsymbol{I}_{M_{Tx}})\boldsymbol{XX}^\dagger \right\} \right\rangle \\ &\overset{\langle \boldsymbol{XX}^\dagger \rangle = N_s\rho \boldsymbol{I}_{M_{Tx}}}{=} N_s\rho \left\langle \|\boldsymbol{AH} - \boldsymbol{B} - \boldsymbol{I}_{M_{Tx}}\|_F^2 \right\rangle. \end{aligned} \tag{8.158}$$

By representing the channel matrix as $\boldsymbol{H} = \boldsymbol{QR}$, the MSE becomes for the following values for matrices $\boldsymbol{A}$ and $\boldsymbol{B}$:

$$\boldsymbol{A}_{\text{ZF V-BLAST}} = \mathrm{diag}^{-1}(\boldsymbol{R})\boldsymbol{Q}^\dagger, \ \ \boldsymbol{B}_{\text{ZF V-BLAST}} = \mathrm{diag}^{-1}(\boldsymbol{R})\boldsymbol{R} - \boldsymbol{I}_{M_{\text{Tx}}}, \tag{8.159}$$

where matrix $\boldsymbol{B}$ is strictly upper-triangular ($[\boldsymbol{B}]_{mn} = 0,\ m \geq n$). The soft estimate of this receiver, based on Fig. 8.26, is given as:

$$\begin{aligned} \widetilde{\boldsymbol{Y}} &= \mathrm{diag}^{-1}\{\boldsymbol{R}\}\boldsymbol{Q}^\dagger \boldsymbol{Y} - \left[\mathrm{diag}^{-1}\{\boldsymbol{R}\}\boldsymbol{R}^\dagger - \boldsymbol{I}_{M_{\text{Tx}}}\right]\widehat{\boldsymbol{X}} \\ &= \boldsymbol{X} + \left[\mathrm{diag}^{-1}\{\boldsymbol{R}\}\boldsymbol{R}^\dagger - \boldsymbol{I}_{M_{\text{Tx}}}\right]\boldsymbol{D} + \mathrm{diag}^{-1}\{\boldsymbol{R}\}\boldsymbol{Q}^\dagger \boldsymbol{Z}, \end{aligned} \tag{8.160}$$

where the first term is the useful term, the second term corresponds to remain mode-coupling fractions from other modes to the observed mode, while the last term is the colored Gaussian noise.

In *MMSE V/H-BLAST*, we minimize the MSE by taking into account the presence of additive noise:

$$\begin{aligned} MSE(\boldsymbol{A},\boldsymbol{B}) &= \left\langle \left\| \widetilde{\boldsymbol{Y}} - \boldsymbol{X} \right\|_F^2 \right\rangle = \left\langle \|\boldsymbol{AY} - \boldsymbol{BX} - \boldsymbol{X}\|_F^2 \right\rangle \\ &= E\left\{ \|(\boldsymbol{AH} - \boldsymbol{B} - \boldsymbol{I}_{M_{Tx}})\boldsymbol{X} + \boldsymbol{AZ}\|^2 \right\} \\ &= N_s\rho \left\{ \|\boldsymbol{AH} - \boldsymbol{BX} - \boldsymbol{I}_{M_{Tx}}\|^2 + SNR^{-1}\|\boldsymbol{A}\|^2 \right\}, \ \ SNR = \rho/N_0 \end{aligned} \tag{8.161}$$

The minimization of the MSE is performed into two steps. We first minimize MSE with respect to $\boldsymbol{A}$ to obtain:

$$\boldsymbol{A}_{\text{MMSE}} = \left(\boldsymbol{B} + \boldsymbol{I}_{M_{Tx}}\right)\left(\boldsymbol{H}^{\dagger}\boldsymbol{H} + SNR^{-1}\boldsymbol{I}_{M_{Tx}}\right)^{-1}\boldsymbol{H}^{\dagger}. \tag{8.162}$$

The residual MSE becomes:

$$MSE(\boldsymbol{B}) = N_s N_0 \text{Tr}\left[\left(\boldsymbol{B} + \boldsymbol{I}_{M_{Tx}}\right)\left(\boldsymbol{H}^{\dagger}\boldsymbol{H} + SNR^{-1}\boldsymbol{I}_{M_{Tx}}\right)^{-1}\left(\boldsymbol{B} + \boldsymbol{I}_{M_{Tx}}\right)^{\dagger}\right]. \tag{8.163}$$

In the second step, the remaining MSE is then minimized over $M_{\text{Tx}} \times M_{\text{Tx}}$ strictly upper-triangular matrices ($[\boldsymbol{B}]_{mn} = 0$, $m \geq n$) by Cholesky factorization [2, 26] of:

$$\boldsymbol{H}^{\dagger}\boldsymbol{H} + SNR^{-1}\boldsymbol{I}_{M_{\text{Tx}}} = \boldsymbol{S}^{\dagger}\boldsymbol{S}, \tag{8.164}$$

where $\boldsymbol{S}$ is an upper-triangular matrix. The remaining MSE is now:

$$\begin{aligned} MSE(\boldsymbol{B}) &= N_s N_0 \left\|\left(\boldsymbol{B} + \boldsymbol{I}_{M_{Tx}}\right)\boldsymbol{S}^{-1}\right\|^2 \geq N_s N_0 \left\|\text{diag}\left[\left(\boldsymbol{B} + \boldsymbol{I}_{M_{Tx}}\right)\boldsymbol{S}^{-1}\right]\right\|^2 \\ &= N_s N_0 \left\|\text{diag}\left[\boldsymbol{S}^{-1}\right]\right\|^2 = N_s N_0 \sum_{m=1}^{M_{Tx}} \left|[\boldsymbol{S}]_{m,m}\right|^{-2} \end{aligned} \tag{8.165}$$

The minimum is obtained when $\left(\boldsymbol{B} + \boldsymbol{I}_{M_{Tx}}\right)\boldsymbol{S}^{-1} = \text{diag}\left[\boldsymbol{S}^{-1}\right]$. In conclusion, the matrices $\boldsymbol{A}$ and $\boldsymbol{B}$ minimizing the MSE in the presence of noise are given as follows:

$$\boldsymbol{A}_{\text{MMSE V-BLAST}} = \text{diag}^{-1}(\boldsymbol{S})\boldsymbol{S}^{-\dagger}\boldsymbol{H}^{\dagger}, \ \boldsymbol{B}_{\text{MMSE V-BLAST}} = \text{diag}^{-1}(\boldsymbol{S})\boldsymbol{S} - \boldsymbol{I}_{M_{\text{Tx}}}. \tag{8.166}$$

The soft estimate of this receiver, based on Fig. 8.27, is given by:

$$\begin{aligned} \widetilde{\boldsymbol{Y}} &= \text{diag}^{-1}\{\boldsymbol{S}\}\boldsymbol{S}^{-\dagger}\boldsymbol{H}^{\dagger}\boldsymbol{Y} - \left[\text{diag}^{-1}\{\boldsymbol{S}\}\boldsymbol{S} - \boldsymbol{I}_{M_{Tx}}\right]\widehat{\boldsymbol{X}} \\ &= \left[\boldsymbol{I}_{M_{Tx}} - \text{diag}^{-1}\{\boldsymbol{S}\}\boldsymbol{S}^{-\dagger}\right]\boldsymbol{X} + \left[\text{diag}^{-1}\{\boldsymbol{S}\}\boldsymbol{S} - \boldsymbol{I}_{M_{Tx}}\right]\boldsymbol{D} + \text{diag}^{-1}\{\boldsymbol{S}\}\boldsymbol{S}^{-\dagger}\boldsymbol{H}^{\dagger}\boldsymbol{Z} \end{aligned} \tag{8.167}$$

where the first term is a useful term, the second term corresponds to residual mode-coupling from other modes to the observed mode, and the last term is the colored Gaussian noise.

Because in both cases the matrix $\boldsymbol{B}$ is strictly upper diagonal, the steps in V-BLAST algorithms can be summarized as follows [2]. We use $[\boldsymbol{A}]_m$ to denote the m-th row in $\boldsymbol{A}$ and $[\boldsymbol{A}]_{mn}$ to denote the m-th row and n-th column. We first solve for the M_{Tx}-th row in $\widehat{\boldsymbol{X}}$ from $\left[\widetilde{\boldsymbol{Y}}\right]_{M_{Tx}} = [\boldsymbol{AY}]_{M_{Tx}}$. And then we solve ($M_{\text{Tx}}$-1)-th row in $\widehat{\boldsymbol{X}}$ from $\left[\widetilde{\boldsymbol{Y}}\right]_{M_{Tx}-1} = [\boldsymbol{AY}]_{M_{Tx}-1} - [\boldsymbol{B}]_{M_{Tx}-1,M_{Tx}}\left[\widehat{\boldsymbol{X}}\right]_{M_{Tx}}$. Since the M_{Tx}-th and ($M_{\text{Tx}} - 1$)-th rows are known, the ($M_{\text{Tx}} - 2$)-th row in $\widehat{\boldsymbol{X}}$ can now be determined from $\left[\widetilde{\boldsymbol{Y}}\right]_{M_{Tx}-2} = [\boldsymbol{AY}]_{M_{Tx}-2} - [\boldsymbol{B}]_{M_{Tx}-2,M_{Tx}}\left[\widehat{\boldsymbol{X}}\right]_{M_t} - [\boldsymbol{B}]_{M_{Tx}-2,M_{Tx}-1}\left[\widehat{\boldsymbol{X}}\right]_{M_{Tx}-1}$. We continue this iterative procedure, until we reach the first row in $\widehat{\boldsymbol{X}}$, which is determined from $\left[\widetilde{\boldsymbol{Y}}\right]_1 = [\boldsymbol{AY}]_1 - [\boldsymbol{B}]_{1,M_{Tx}}\left[\widehat{\boldsymbol{X}}\right]_{M_{Tx}} - \ldots - [\boldsymbol{B}]_{1,2}\left(\widehat{\boldsymbol{X}}\right)_2$.

The *receiver in D-BLAST architecture* first detects x_{11} symbol that does not suffer from spatial interference. And then the symbol x_{12} is detected, which suffers from spatial interference from the symbol x_{21}, which can be compensated for by a zero-forcing equalizer. This procedure is continued until all symbols for codeword $\boldsymbol{x}_1$ are detected. Once all symbols of codeword $\boldsymbol{x}_1$ get detected, their contribution get subtracted from the next received word corresponding to codeword $\boldsymbol{x}_2$ and so forth.

8.2.7.4 Spatial Interference Cancellation by Iterative Receiver (Interface)

Yet another scheme of canceling out the spatial interference in wireless MIMO systems and mode coupling FMF/MMF links in an iterative fashion is illustrated in Fig. 8.28. This scheme is suitable for compensation of spatial interference in fading channels. A similar scheme has been proposed in [44] to deal with PMD effects in polarization division multiplexing coherent optical systems. The key idea behind this scheme is to cancel the spatial interference/mode coupling in an iterative fashion. The iterative mode-coupling cancellation algorithm can be formulated as follows:

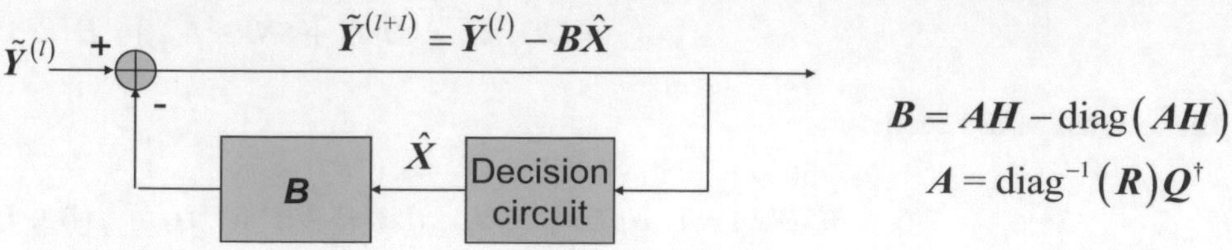

Fig. 8.28 The operation principle of iterative spatial interference/mode-coupling cancellation scheme

Step 0 (initialization): Set $\boldsymbol{A} = \text{diag}^{-1}(\boldsymbol{R})\boldsymbol{Q}^{\dagger}, \widetilde{\boldsymbol{X}}^{(0)} = 0, \widetilde{\boldsymbol{Y}}^{(0)} = \boldsymbol{Y}$.

Step 1 (horizontal update rule): $\widehat{\boldsymbol{X}}^{(k)} = \arg\min_{\boldsymbol{X}} \left\| \widetilde{\boldsymbol{Y}}^{(k)} - \boldsymbol{X} \right\|^2$.

Step 2 (vertical update rule): $\widetilde{\boldsymbol{Y}}^{(k+1)} = \widetilde{\boldsymbol{Y}} - (\boldsymbol{AH} - \text{diag}(\boldsymbol{AH}))\widehat{\boldsymbol{X}}^{(k)}$.

We have to iterate steps 1 and 2 until the convergence.

It can be shown that in the absence of nonlinearities and for moderate SNRs, $\widehat{\boldsymbol{X}}^{(k)}$ converges to transmitted $\boldsymbol{X}$.

8.3 Iterative MIMO Receivers

In this section we study iterative MIMO processing of received signals by employing the *factor graph theory* [2, 26, 46–49].

8.3.1 Factor Graph Fundamentals

Factor graphs are suitable to describe the factorization of a global function as the product of local function, with each local function being a product of subset of the variables. As an illustration let us consider maximum a posteriori probability (MAP) detection/decoding problem. Let x denote the transmitted sequence of length N and y the received sequence. The *optimum receiver* (in symbol error probability sense) assigns $\widehat{x}_n$ to the value x from the signal set X that maximizes the APP $P(x_n = x|y)$ given the received sequence y as follows:

$$\widehat{x}_n = \arg\max_{x \in X} f(x_n = x|\boldsymbol{y}). \tag{8.168}$$

This maximization is easy to perform once the APP functions $f(x_n|\boldsymbol{y})$ are known. Assuming that conditional PDF $f(\boldsymbol{x}|\boldsymbol{y})$ is known, the APP functions $f(x_n|\boldsymbol{y})$ can be determined as follows:

$$f(x_n|\boldsymbol{y}) = \sum_{x_1} \sum_{x_2} \cdots \sum_{x_{n-1}} \sum_{x_{n+1}} \cdots \sum_{x_N} f(\boldsymbol{x}|\boldsymbol{y}). \tag{8.169}$$

Clearly, the summation in Eq. (8.169) is performed over all variables except variable x_n, and we can denote this summation as $\sim x_n$ so that we can rewrite Eq. (8.169) as:

$$f(x_n|\boldsymbol{y}) = \sum_{\tilde{x}_n} f(\boldsymbol{x}|\boldsymbol{y}). \tag{8.170}$$

This operation is known as *marginalization*, and once performed we can make a decision based on maximization given by Eq. (8.168).

Evidently, the complexity of marginalization operation grows exponentially with sequence length N. The marginalization operation can be greatly simplified if the function to be marginalized (f) can be factored as a product of local functions, each function of the subset of variables. As an illustration let us consider the function $f(x_1,\ldots,x_4)$, which can be factorized as follows:

$$f(x_1, \cdots, x_4) = g_1(x_1, x_2)g_2(x_1, x_3)g_3(x_1, x_4). \tag{8.171}$$

The marginal $f_1(x_1)$ can be computed as:

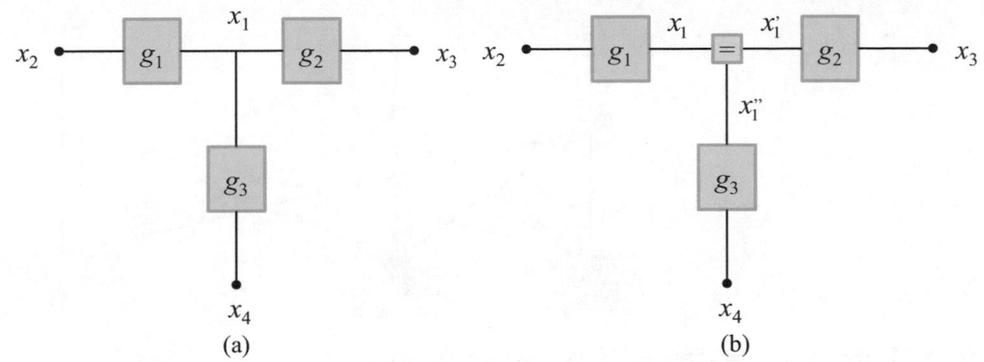

Fig. 8.29 The factor graph of $f(x_1, \cdots x_4) = g_1(x_1, x_2)g_2(x_1, x_3) g_3(x_1, x_4)$: (**a**) non-normal form and (**b**) normal form

$$\begin{aligned} f_1(x_1) &= \sum_{\tilde{x}_1} g_1(x_1,x_2)g_2(x_1,x_3)g_3(x_1,x_4) \\ &= \sum_{x_2}\sum_{x_3} g_1(x_1,x_2)g_2(x_1,x_3)g_3(x_1,x_4) \\ &= \sum_{x_2} g_1(x_1,x_2)\sum_{x_3} g_2(x_1,x_3)\sum_{x_4} g_3(x_1,x_4). \end{aligned} \tag{8.172}$$

The corresponding factor graph describing this marginalization operation is shown in Fig. 8.29a. Each node represents a processor performing a given function with arguments being the labels of incoming edges, while each edge serves as a link over which the nodes exchange the messages. The edge connected to only one node is called the half-edge, and it is terminated by the filled circle. The node g_1 performs the following calculation locally: $\sum_{x_2} g_1(x_1, x_2)$. On the other hand, the nodes g_2 and g_3 perform the summations $\sum_{x_3} g_2(x_1, x_3)$ and $\sum_{x_4} g_3(x_1, x_4)$, respectively.

The factor graph composed only of nodes, edges and half-edges is known as *normal graph*. The normal graph for graph shown in Fig. 8.29a is shown in Fig. 8.29b. This normal graph contains the *repetition node* $\boxed{=}$, which repeats the same variable multiple times. This repetition node can be also represented in terms of the so-called Iverson function, denoted by [*P*] with value 1 when proposition *P* is true, and zero otherwise:

$$[P] \doteq \begin{cases} 1, P \text{ is true} \\ 0, P \text{ is false} \end{cases} \tag{8.173}$$

The repetition function for three arguments, denoted as $f_=$, can be now defined as:

$$f_=(x_1, x_1'', x_1'') \doteq [x_1 = x_1'' = x_1''] . \tag{8.174}$$

If multiple propositions P_n ($n = 1,2,\ldots,N$) need to be simultaneously satisfied, we can represent this as follows:

$$[P_1 \wedge P_2 \wedge \cdots \wedge P_N] = [P_1][P_1]\cdots[P_N]. \tag{8.175}$$

8.3.2 Factor Graphs for MIMO Channels and Channels with Memory

Let us now observe the MIMO channel, given by Eq. (8.62), namely, $\boldsymbol{y} = \boldsymbol{Hx} + \boldsymbol{z}$. Providing that the number of transmit antennas (optical transmitters) is M_{Tx} and the number of receive antennas (optical receivers) is M_{Rx}, the output-input relationship can be described in terms of conditional PDF:

$$\begin{aligned} f(\boldsymbol{y}|\boldsymbol{x}) &= C\exp\left[-\|\boldsymbol{y} - \boldsymbol{Hx}\|_F^2/N_0\right] = C\prod_{n=1}^{M_{Rx}} \underbrace{\exp\left(-|y_n - \boldsymbol{h}_n\boldsymbol{x}|^2\right)}_{f(y_n|\boldsymbol{x})} \\ &= C\prod_{n=1}^{M_{Rx}} f(y_n|\boldsymbol{x}), \end{aligned} \tag{8.176}$$

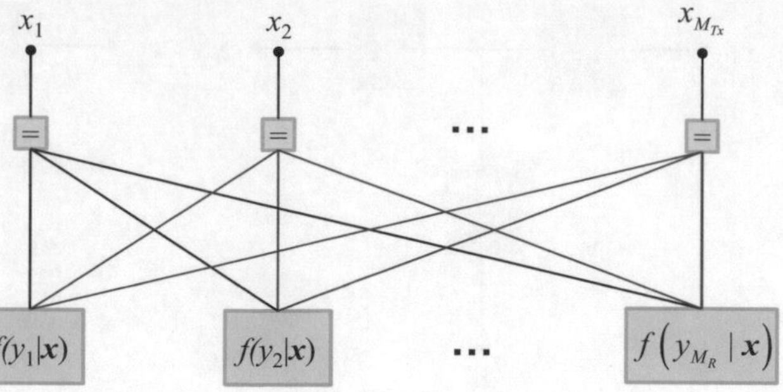

Fig. 8.30 The factor graph of the optical/wireless MIMO channel

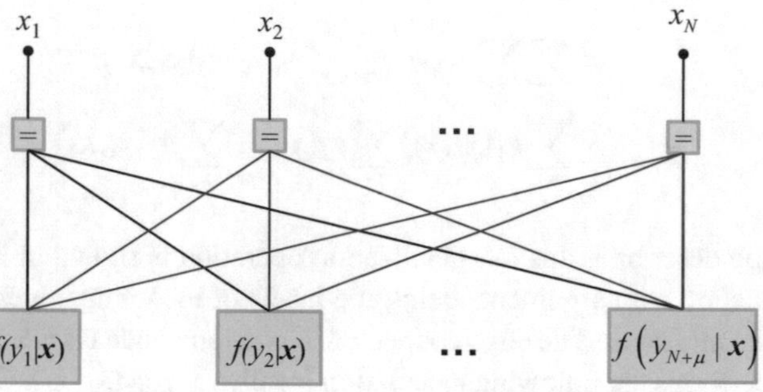

Fig. 8.31 The factor graph of the linear ISI channel (linear channel with memory)

where C is the normalization constant and we use $\boldsymbol{h}_n$ to denote the n-th row of $\boldsymbol{H}$, namely, $\boldsymbol{h}_n = [\boldsymbol{H}]_n$. The corresponding factor graph of this MIMO channel is shown in Fig. 8.30.

In linear *intersymbol interference (ISI) channel*, the output (N+μ)-component vector $\boldsymbol{y}$ and the input N-component vector $\boldsymbol{x}$ are related by the discrete-time convolution:

$$y_n = \sum_{k=0}^{\mu} h_k x_{n-k} + z_n, \tag{8.177}$$

where μ is the channel memory and h_0, h_1,..,h_μ are channel coefficients; while z_n is a zero-mean Gaussian noise sample. The matrix representation of Eq. (8.177) is given by:

$$\boldsymbol{y} = \boldsymbol{H}\boldsymbol{x} + \boldsymbol{z}, \tag{8.178}$$

where $\boldsymbol{H}$ is the corresponding channel matrix of format $(N + \mu) \times N$. The conditional PDF, describing output-input relationship, is then:

$$\begin{aligned} f(\boldsymbol{y}|\boldsymbol{x}) &= C\exp\left[-\|\boldsymbol{y} - \boldsymbol{H}\boldsymbol{x}\|_F^2/N_0\right] = C\prod_{n=1}^{N+\mu} \underbrace{\exp\left(-\left|y_n - \sum_{k=1}^{N} x_k h_{n-k}\right|^2\right)}_{f(y_n|\boldsymbol{x})} \\ &= C\prod_{n=1}^{N+\mu} f(y_n|\boldsymbol{x}), \end{aligned} \tag{8.179}$$

which is very similar to Eq. (8.176), and not surprisingly the corresponding factor graph shown in Fig. 8.31 is very similar to the one shown in Fig. 8.30.

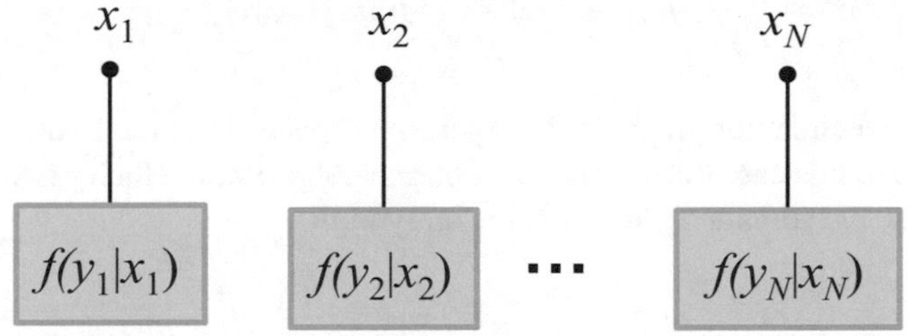

Fig. 8.32 The factor graph of a channel without memory (memoryless channel)

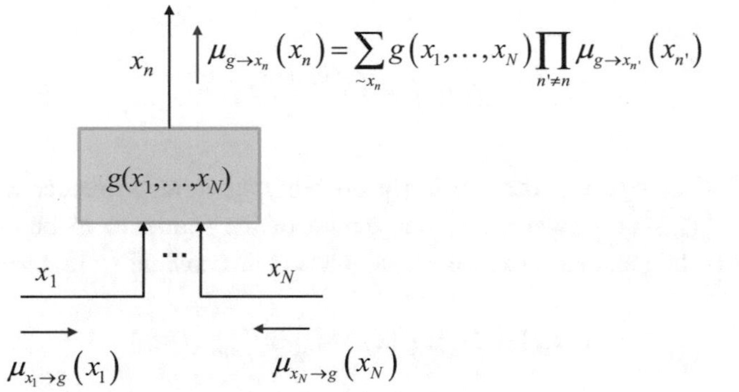

Fig. 8.33 The illustration of the sum-product rule

When the channel model is *memoryless*, $\mu = 0$, $y_n = h_0 x_n + z_n$, the conditional PDF is simply:

$$f(\mathbf{y}|\mathbf{x}) = C\prod_{n=1}^{N} f(y_n|x_n). \tag{8.180}$$

The factor graph of the memoryless channel is shown in Fig. 8.32, and it is composed of N disconnected subgraphs.

8.3.3 Sum-Product Algorithm

To compute marginal functions efficiently, the *sum-product algorithm* (SPA) [2, 5, 7, 26, 47, 50, 51] is used, which is suitable when the factor graph is free of cycles. In SPA, with each edge we associate two types of messages being the functions of variable x_n, denoted as $\mu(x_n)$, one for each direction. These messages are vectors, with components calculated for different values of x_n. For binary signaling, the message has the form $\mu(x_n) = (\mu_0, \mu_1) = \mu_0(1, \mu_1/\mu_0)$. Alternatively the logarithm can be used so that the message is then $\mu'(x_n) = \log\mu(x_n) = (0,\log(\mu_1/\mu_0))$, and clearly the first component can be omitted. Let u observe the node function $g(x_1,\ldots,x_N)$, as illustrated in Fig. 8.33. The message to be sent along the edge x_n from node g, denoted as $\mu_{g\to x_n}(x_n)$, can be calculated as:

$$\mu_{g\to x_n}(x_n) = \sum_{\sim x_n} g(x_1,\ \ldots, x_N)\prod_{n'\neq n}\mu_{g\to x'_n}(x_{n'}), \tag{8.181}$$

and this rule is commonly referred to as the *sum-product rule*. In other words, $\mu_{g\to x_n}(x_n)$ is the product of the local function g and all messages associated with incoming edges to g, except x_n, summed over all variables except x_n. The half-edge message is considered to be 1, while for only one variable the product in (8.181) does not exist and we can write $\mu_{g\to x_n}(x_n) = g(x_n)$. The *message originating from the repetition node* sent over edge x_n is simply the product of incoming messages, except x_n:

$$\mu_{f=\to x_n}(x_n) = \prod_{n'\neq n} \mu_{g\to x_{n'}}(x_{n'}). \tag{8.182}$$

The SPA can also be applied when factor graph has the cycles, by specifying initial values of local nodes and a stopping criterion [2, 5, 50, 51]. However, in this case the SPA is a suboptimum algorithm, which performs well when the length of the shortest cycle in the factor graph, the so-called girth, is sufficiently high.

8.3.4 Sum-Product Algorithm for Channels with Memory

We return our attention now to the *MAP detection problem*, given by Eq. (8.168). To calculate the $f(\mathbf{x}|\mathbf{y})$, we invoke Bayes' rule:

$$f(\boldsymbol{x}|\boldsymbol{y}) = \frac{f(\boldsymbol{y}|\boldsymbol{x})f(\boldsymbol{x})}{f(\boldsymbol{y})}, \tag{8.183}$$

where $f(\boldsymbol{y}|\boldsymbol{x})$ is conditional PDF and $f(\boldsymbol{x})$ is the a priori probability of input sequence $\boldsymbol{x}$, which when the symbols are independent factors as $f(\boldsymbol{x}) = \prod_{n=1}^{N} f(x_n)$, where N is the length of the sequence to be detected. Given that the PDF of received sequence of symbols is independent of $\boldsymbol{x}$, we consider 1/f(y) as constant so that we can rewrite Eq. (8.183) as:

$$f(\boldsymbol{x}|\boldsymbol{y}) \approx f(\boldsymbol{y}|\boldsymbol{x})f(\boldsymbol{x}) = f(\boldsymbol{y}|\boldsymbol{x})\prod_{n=1}^{N} f(x_n). \tag{8.184}$$

Let us define the prior information as the intrinsic message $\mu^i(x_n) = f(x_n)$ and the extrinsic message (denoted by superscript *e*), from symbol demapper point of view, related to edge symbol x_n by:

$$\mu^e(x_n) = \sum_{\tilde{x}_n} f(\boldsymbol{y}|\boldsymbol{x})\prod_{n'\neq n} \underbrace{f(x_{n'})}_{\mu^i(x_{n'})} = \sum_{\tilde{x}_n} f(\boldsymbol{y}|\boldsymbol{x})\prod_{n'\neq n}\mu^i(x_{n'}), \tag{8.185}$$

And then the corresponding sum-product for general channel with memory (including ISI channel) is illustrated in Fig. 8.34.

8.3.5 Iterative MIMO Receivers for Uncoded Signals

Based on similarities of MAP detection problem for channel with memory and MIMO channels, $f(\boldsymbol{x}|\boldsymbol{y})$ for *MIMO channels* can be factorized as follows:

$$\begin{aligned} f(\boldsymbol{x}|\boldsymbol{y}) &\approx f(\boldsymbol{y}|\boldsymbol{x})f(\boldsymbol{x}) = \underbrace{f\left(\boldsymbol{y}|\overbrace{x_1,\cdots,x_{M_{Tx}}}^{\boldsymbol{x}}\right)}_{\prod_{n=1}^{M_{Rx}} f(y_n|\boldsymbol{x})}\prod_{m=1}^{M_{Tx}} f(x_m), \\ &= \prod_{n=1}^{M_{Rx}} f(y_n|\boldsymbol{x})\prod_{m=1}^{M_{Tx}} f(x_m),\ f(y_n|\boldsymbol{x}) = \exp\left(-|y_n - \boldsymbol{h}_n\boldsymbol{x}|^2\right). \end{aligned} \tag{8.186}$$

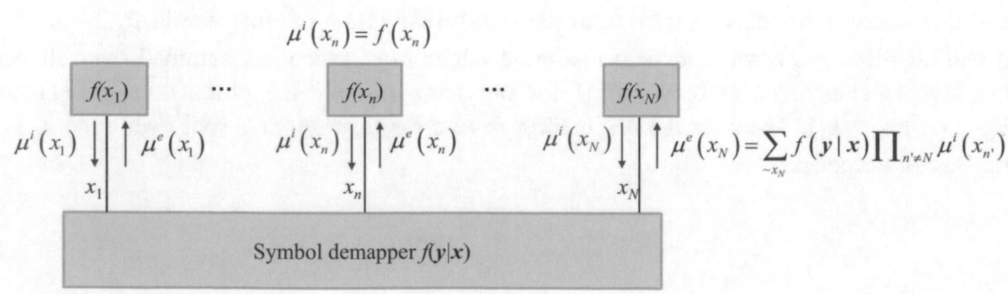

Fig. 8.34 The illustration of the sum-product algorithm for MAP detection ("equalization") in channels with memory

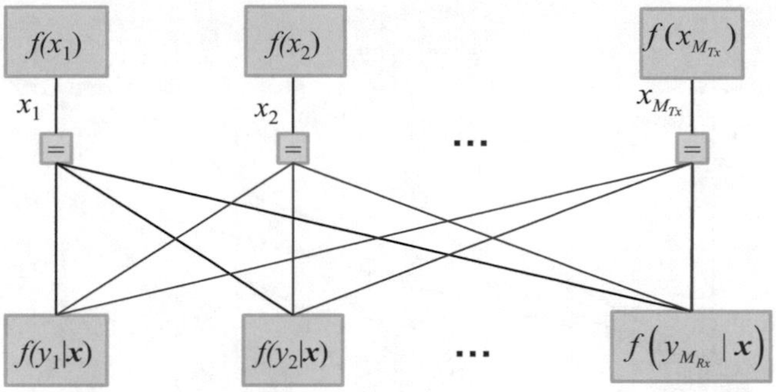

Fig. 8.35 The illustration of factor graph for MAP detection of uncoded signals for MIMO channels

Fig. 8.36 The illustration of factor graphs for MIMO receivers with (**a**) ZF and (**b**) LMMSE interfaces

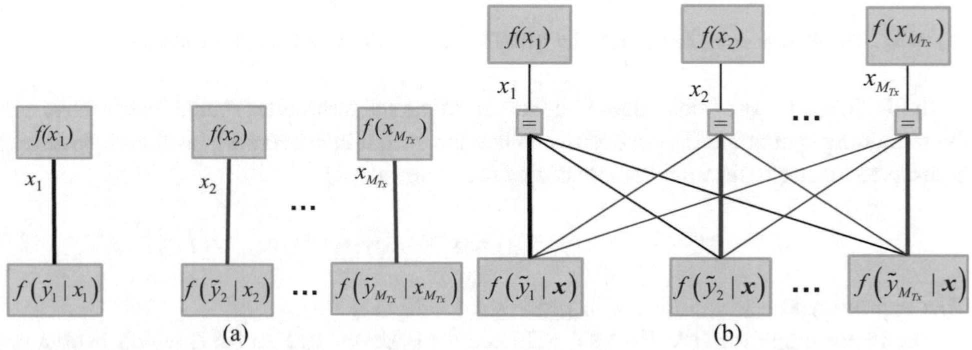

The corresponding factor graph for MAP detection of uncoded signals at the output of MIMO channel is shown in Fig. 8.35. Because of the existence of cycles, the corresponding SPA is suboptimum. By increasing the number of antennas sufficiently, the factor graph girth can be increased, and performance of suboptimum iterative MIMO receiver based on SPA can be improved. However, the complexity of computation would be high.

To reduce the complexity, we can perform pre-compensation of spatial interference by employing suboptimum linear receivers (interfaces) described in Sects. 8.2.6 and 8.2.7. The key idea, given by Eq. (8.136), is to pre-multiply the received vector $\boldsymbol{y}$ by a matrix being function of channel matrix $\boldsymbol{H}$, denoted as $\boldsymbol{A}(\boldsymbol{H})$, to obtain $\tilde{\boldsymbol{y}} = \boldsymbol{A}(\boldsymbol{H})\boldsymbol{y}$, so that the spatial interference is either compensated for or reduced.

The $\boldsymbol{A}(\boldsymbol{H})$ matrix can be chosen as follows, depending on the corresponding interface:

- In *zero-forcing interface* as the Moore-Penrose pseudoinverse of $\boldsymbol{H}$: $\boldsymbol{A}_{\mathrm{ZF}}(\boldsymbol{H}) = \boldsymbol{H}^{+}$.
- In *linear MMSE (LMMSE) interface*, $\boldsymbol{A}(\boldsymbol{H})$ is chosen on such a way to minimize MSE:

$$\boldsymbol{A}_{\mathrm{MMSE}} = \left(\boldsymbol{H}^{\dagger}\boldsymbol{H} + \boldsymbol{I}/SNR\right)^{-1}\boldsymbol{H}^{\dagger}$$

In *zero-forcing H/V-BLAST architecture*, after QR factorization of $\boldsymbol{H}$, the transform matrix is chosen as:

$$\boldsymbol{A}_{\mathrm{ZF\ V\text{-}BLAST}} = \mathrm{diag}^{-1}(\boldsymbol{R})\boldsymbol{Q}^{\dagger}$$

In *linear minimum MSE H/V-BLAST architecture*, after Cholesky factorization of $\boldsymbol{H}^{\dagger}\boldsymbol{H} + SNR^{-1}\boldsymbol{I}_{M_{\mathrm{Tx}}} = \boldsymbol{S}^{\dagger}\boldsymbol{S}$, the transform matrix is chosen to minimize MSE:

$$\boldsymbol{A}_{\mathrm{MMSE\ V\text{-}BLAST}} = \mathrm{diag}^{-1}(\boldsymbol{S})\boldsymbol{S}^{-\dagger}\boldsymbol{H}^{\dagger}.$$

After the initial compensation of spatial interference, the corresponding factor graphs for ZF and LMMSE interfaces are shown in Fig. 8.36.

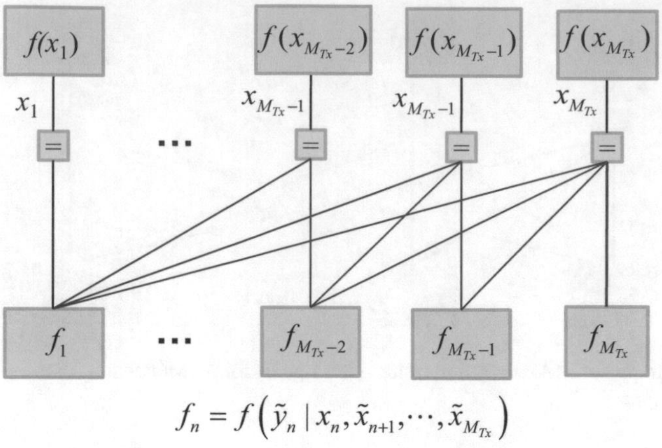

Fig. 8.37 The illustration of factor graph for MIMO receivers with H/V-BLAST architecture

In Fig. 8.36b we use thick edges to denote the diagonal terms after spatial interference cancellation and thin edges to denote the remaining spatial interference terms. When the spatial interference is well compensated, the following approximation can be used to further reduce the complexity of the factor graph:

$$f(\tilde{y}_n|\boldsymbol{x}) = f(\tilde{y}_n|x_1, \cdots, x_{M_{Tx}}) \simeq f(\tilde{y}_n|x_n). \tag{8.187}$$

This approximation is valid when $M_{\mathrm{Rx}} \gg M_{\mathrm{Tx}}$ [2, 52–54].

The factor graph for H/V-BLAST architectures is shown in Fig. 8.37, which is obtained by simplifying the factor graph shown in Fig. 8.35 by using approximation given by Eq. (8.187). The detection algorithm outputs the solution after the following steps are executed:

$$\begin{aligned} \widehat{x}_{M_{Tx}} &= \arg\max_{x \in X} f\left(\tilde{y}_{M_{Tx}}|x_{M_{Tx}}\right), \\ \widehat{x}_n &= \arg\max_{x \in X} f(\tilde{y}_n|x_n, \tilde{x}_{n+1}, \cdots, \tilde{x}_{M_{Tx}}); n = M_{Tx} - 1, M_{Tx} - 2, \cdots 1. \end{aligned} \tag{8.188}$$

Before moving to the description of iterative MIMO receivers for space-time coded signals, we briefly describe the factor graphs for linear block and trellis channel codes.

8.3.6 Factor Graphs for Linear Block and Trellis Channel Codes

The linear block codes (LBCs) can be described as a K-dimensional subspace C of the N-dimensional vector space of N-tuples, F_2^N, over the binary field F_2 [55]. For this K-dimensional subspace, we can find the basis $B = \{\boldsymbol{g}_0, \boldsymbol{g}_1, \ldots, \boldsymbol{g}_{K-1}\}$ that spans C so that every codeword $\boldsymbol{v}$ can be written as a linear combination of basis vectors $\boldsymbol{v} = m_0\boldsymbol{g}_0 + m_1\boldsymbol{g}_1 + \ldots + m_{K-1}\boldsymbol{g}_{K-1}$ for message vector $\boldsymbol{m} = (m_0, m_1, \ldots, m_{K-1})$; or in compact form we can write $\boldsymbol{v} = \boldsymbol{mG}$, where $\boldsymbol{G}$ is the so-called generator matrix with the i-th row being $\boldsymbol{g}_i$. The (N-K)-dimensional null space $C^{\perp}$ of $\boldsymbol{G}$ comprises all vectors $\boldsymbol{x}$ from F_2^N such that $\boldsymbol{xG}^{\mathrm{T}} = \boldsymbol{0}$, and it is spanned by the basis $B^{\perp} = \{\boldsymbol{h}_0, \boldsymbol{h}_1, \ldots, \boldsymbol{h}_{n-k-1}\}$. Therefore, for each codeword $\boldsymbol{v}$ from C, $\boldsymbol{vh}_i^{\mathrm{T}} = 0$ for every i; or in compact form, we can write $\boldsymbol{vH}^{\mathrm{T}} = \boldsymbol{0}$, where $\boldsymbol{H}$ is the so-called parity-check matrix whose i-th row is $\boldsymbol{h}_i$. Therefore, the parity-check matrix of LBC can be written as:

$$\boldsymbol{H} = \begin{bmatrix} \boldsymbol{h}_0 \\ \boldsymbol{h}_1 \\ \vdots \\ \boldsymbol{h}_{N-K-1} \end{bmatrix}. \tag{8.189}$$

The codeword must satisfy all parity-check equations as discussed above, and this condition can be represented with the help of Iverson function as follows:

$$[\boldsymbol{v} \in C] = \left[\boldsymbol{v}\boldsymbol{h}_0^T = 0, \cdots, \boldsymbol{v}\boldsymbol{h}_{N-K-1}^T = 0\right] = \prod_{n=0}^{N-K-1} \left[\boldsymbol{v}\boldsymbol{h}_n^T = 0\right]. \tag{8.190}$$

The n-th factor in right-hand side of Eq. (8.190) can be represented as mod-2 adder $\oplus$, connecting the edges involved in the n-th parity-check equation. A *bipartite (Tanner) graph* is a graph whose nodes may be separated into two classes (*variable* and *check* nodes) and where *undirected edges* may only connect two nodes not residing in the same class. The Tanner graph of a code is drawn according to the following rule: check (function) node c is connected to variable (bit) node v whenever element h_{cv} in a parity-check matrix $\boldsymbol{H}$ is a 1. In an $(N-K)\times N$ parity-check matrix, there are N-K check nodes and N variable nodes. As an illustrative example, consider the H matrix of the Hamming (7,4) code:

$$\boldsymbol{H} = \begin{bmatrix} 1 & 0 & 0 & 1 & 0 & 1 & 1 \\ 0 & 1 & 0 & 1 & 1 & 1 & 0 \\ 0 & 0 & 1 & 0 & 1 & 1 & 1 \end{bmatrix}$$

For any valid codeword $\boldsymbol{v} = [v_0\ v_1 \ldots v_{n-1}]$, the checks used to decode the codeword are written as:

- Equation (c_0): $v_0 + v_3 + v_5 + v_6 = 0 \pmod 2$
- Equation (c_1): $v_1 + v_3 + v_4 + v_5 = 0 \pmod 2$
- Equation (c_2): $v_2 + v_4 + v_5 + v_6 = 0 \pmod 2$

The bipartite graph (Tanner graph) representation of this code is shown in Fig. 8.38. Both non-normal and normal versions of Tanner graph have been shown.

The circles represent the bit (variable) nodes while $\oplus$-nodes represent the check (function) nodes. For example, the variable nodes v_0, v_3, v_5, and v_6 are involved in (c_0) and therefore connected to the check node c_0. A closed path in a bipartite graph comprising l edges that close back on itself is called a *cycle* of length l. The shortest cycle in the bipartite graph is called the *girth*. The shortest cycle in Tanner graph in Hamming (7,4) code is 4; therefore, the girth is 4.

Codes described by trellis can also be described using the concept of normal factor graphs. A *trellis* can be interpreted as a set of triples (s_{n-1}, v_n, s_n), describing the transition from state s_{n-1} to state s_n driven by the incoming channel symbol v_n at

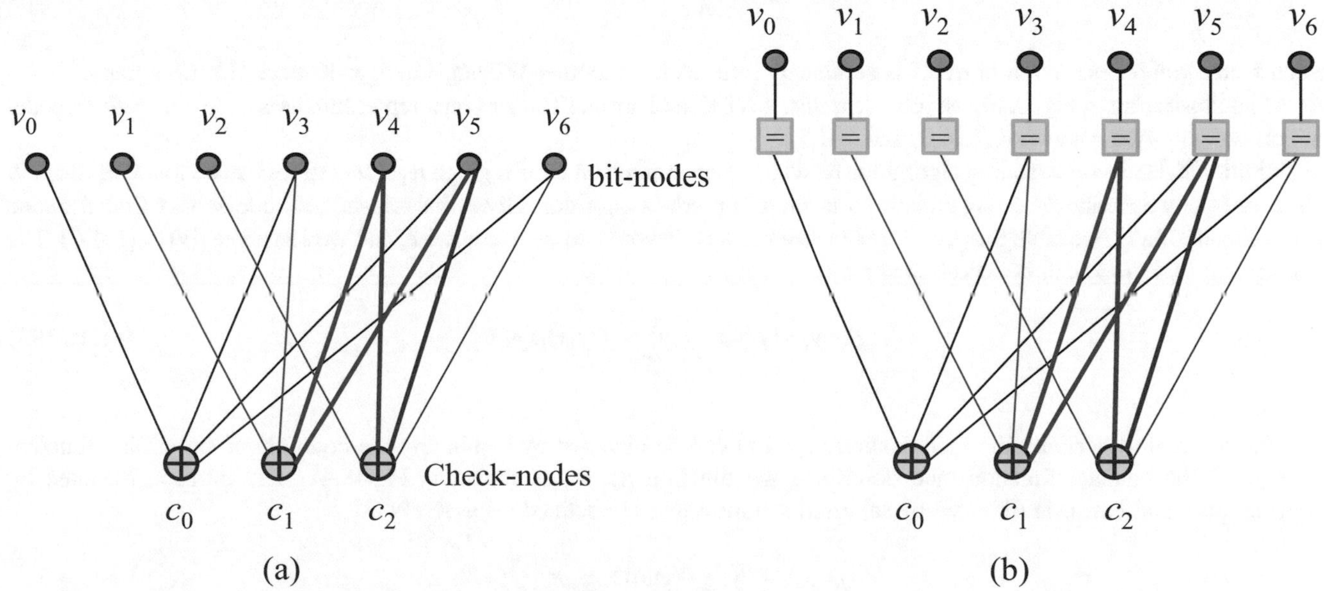

Fig. 8.38 Tanner graph of (7,4) Hamming code: (**a**) non-normal and (**b**) normal versions

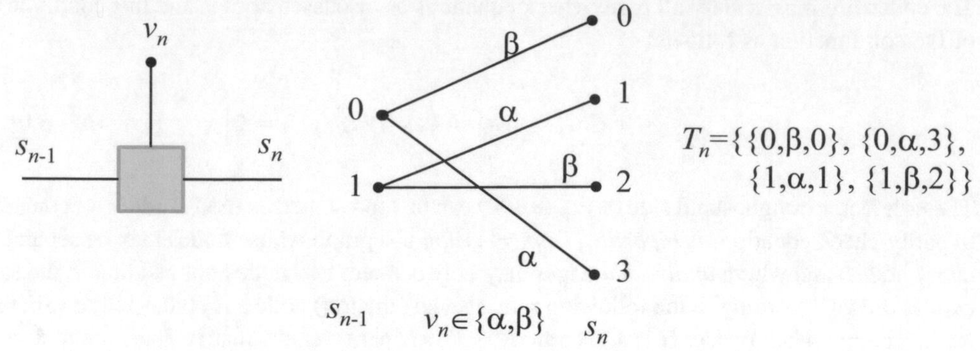

Fig. 8.39 A node in TWLK graph representing a given trellis section

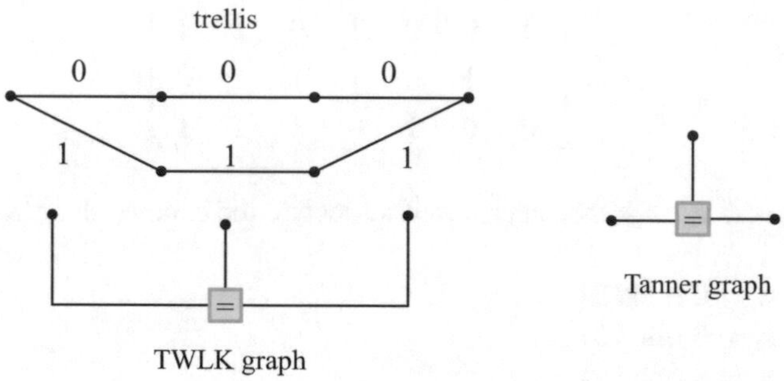

Fig. 8.40 Trellis, TWLK graph, and normal Tanner graph representations of repetition (3,1) code

time instance $n-1$ ($n=1,2,\ldots,N$). Let the set of branch labels connecting the states s_{n-1} and s_n be denoted by T_n. And then the local function corresponding to the n-th trellis section, shown in Fig. 8.39, is given by:

$$[(s_{n-1}, v_i, s_n) \in T_n], \tag{8.191}$$

and the whole trellis can be described as a product of corresponding Iverson functions of individual trellis sections:

$$[\boldsymbol{v} \in C] = \prod_{n=1}^{N} [(s_{n-1}, v_i, s_n) \in T_n]. \tag{8.192}$$

Such factor graph description of trellis is commonly referred to as Tanner-Wilberg-Loeliger-Koetter (TWLK) graph.

As an illustration in Fig. 8.40, we provide trellis, TWLK, and normal Tanner graph representations of (3,1) repetition code, which has only two codewords (0,0,0) and (1,1,1).

Additional details on decoding algorithms based on Tanner graph and trellis graph representations can be found in Chap. 9. Here we briefly describe decoding process based on sum-product algorithm. Given that transmitted codewords $\boldsymbol{v}$ from the code C are equally likely, we have that $f(\boldsymbol{v}) = 1/|C|$ when $\boldsymbol{v}$ is a codeword and zero otherwise, and we can write $f(\boldsymbol{v}) \approx [\boldsymbol{v} \in C]$. The conditional PDF $f(\boldsymbol{v}|\boldsymbol{y})$ can be represented as:

$$f(\boldsymbol{v}|\boldsymbol{y}) \approx f(\boldsymbol{y}|\boldsymbol{v}) \underbrace{f(\boldsymbol{v})}_{[\boldsymbol{v}\in C]} = f(\boldsymbol{y}|\boldsymbol{v})[\boldsymbol{v} \in C]. \tag{8.193}$$

The factor graph describing the factorization of $f(\boldsymbol{v}|\,\boldsymbol{y})$ can be obtained by combining the code graph, describing function [v∈C] and the channel function, and describing the function $f(\boldsymbol{y}|\,\boldsymbol{v})$, as shown in Fig. 8.41. The intrinsic (denoted by superscript e) and extrinsic (denoted by superscript i) messages are defined respectively as:

$$\begin{aligned} \mu^i(v_n) &= \sum\nolimits_{\sim v_n} f(\boldsymbol{y}|\boldsymbol{v}) \prod\nolimits_{n' \neq n} \mu^e(v_{n'}), \\ \mu^e(v_n) &= \sum\nolimits_{\sim v_n} [\boldsymbol{v} \in C] \prod\nolimits_{n' \neq n} \mu^i(v_{n'}). \end{aligned} \tag{8.194}$$

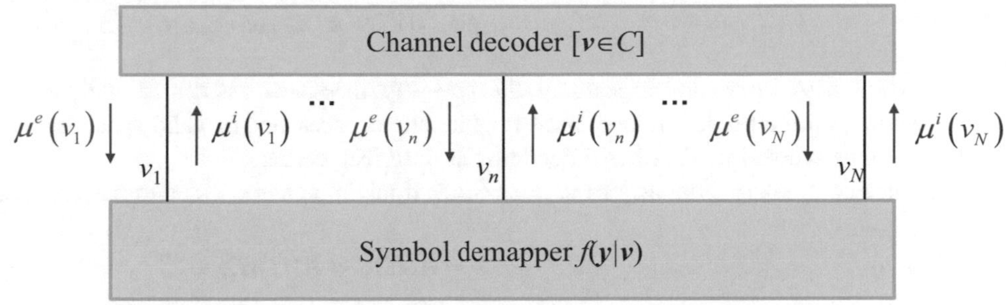

Fig. 8.41 The SPA illustrating decoding over an arbitrary channel

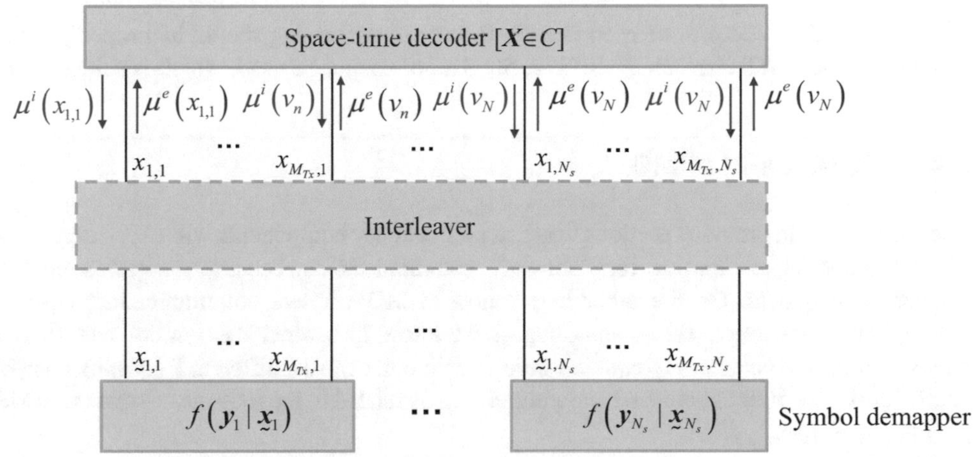

Fig. 8.42 The factor graph for space-time coded MIMO (with interleaver)

8.3.7 Iterative MIMO Receivers for Space-Time Coded Signals

A codeword generated by a space-time encoder, as already described in Sect. 8.2.3, can be described as $M_{\mathrm{Tx}} \times N_s$ matrix $\boldsymbol{X} = [\boldsymbol{x}_1, \ldots, \boldsymbol{x}_{Ns}]$, where $\boldsymbol{x}_n$ ($n = 1,..,N_s$) denotes the n-th column of $\boldsymbol{X}$, and the length of vector $\boldsymbol{x}_n$ is M_{Tx}. Clearly, rows correspond to transmit antennas (space dimensions) while columns to time intervals. The received signal is represented as $M_{\mathrm{Rx}} \times N_s$ matrix $\boldsymbol{Y} = [\boldsymbol{y}_1, \ldots, \boldsymbol{y}_{Ns}]$. The received and transmitted signals are related by $\boldsymbol{Y}=\boldsymbol{HX} + \boldsymbol{Z}$, where $\boldsymbol{H}$ is the channel matrix and Z is a noise matrix of independent zero-mean symmetric complex Gaussian random variables. (For additional details on STC-based MIMO system, please refer to Sect. 8.2.3.) In analogy with Eq. (8.193), the conditional PDF $f(\boldsymbol{X}|\boldsymbol{Y})$ can be factorized as follows:

$$f(\boldsymbol{X}|\boldsymbol{Y}) \approx f(\boldsymbol{Y}|\boldsymbol{X})[\boldsymbol{X} \in C] = [\boldsymbol{X} \in C] \prod_{n=1}^{N_s} f(\boldsymbol{y}_n|\boldsymbol{x}_n), \tag{8.195}$$

and the corresponding factor graph describing this factorization is shown in Fig. 8.42. The purpose of the interleaver (which is optional) is to increase the girth of the factor graph. (We use tilde below variable to denote that the corresponding variable is obtained after interleaving.) To facilitate the explanations, below we omit the interleaver.

From *space-time decoder point of view*, the extrinsic massage is calculated by:

$$\mu^e(x_{m,n}) = \sum\nolimits_{\tilde{x}_{m,n}} [\boldsymbol{X} \in C] \prod\nolimits_{(m',n') \neq (m,n)} \mu^i(x_{m',n'}), \tag{8.196}$$

where $\mu^i(x_{m,\,n})$ are messages output by the functional nodes $f(\boldsymbol{y}_m|\boldsymbol{x}_m)$ computed as:

$$\mu^i(x_{m,n}) = \sum\nolimits_{\tilde{x}_{m,n}} f(\boldsymbol{y}_n|\boldsymbol{x}_n) \prod\nolimits_{m' \neq m} \mu^e(x_{m,n}). \tag{8.197}$$

Once the above messages converge to exact values, the APPs can be computed as follows:

$$f(x_{m,n}|\boldsymbol{Y}) \approx \mu^i(x_{m,n})\mu^e(x_{m,n}). \tag{8.198}$$

This iterative SPA computes repeatedly the two-way messages associated with the edges, after proper interleaving/deinterleaving, until a predetermined number of iterations is reached or valid space-time codeword is obtained, after which APPs are computed and used in MAP decoding as described earlier.

To reduce complexity, similarly as in Sect. 8.3.5, it might be a good idea to perform preprocessing as follows:

$$\widetilde{\boldsymbol{Y}} = \boldsymbol{A}(\boldsymbol{H})\boldsymbol{Y} = \boldsymbol{A}(\boldsymbol{H})\boldsymbol{H}\boldsymbol{X} + \boldsymbol{Z}, \tag{8.199}$$

in a similar fashion as described in Sect. 8.2.7.1, to compensate for spatial interference. When the spatial interference is properly compensated, the factor graph corresponding to each function $f(\boldsymbol{y}_n|\underset{\sim}{\boldsymbol{x}}_n)$ get disconnected, which simplifies the SPA.

The space-time codes to be used in MIMO receiver processing should be properly matched to the corresponding symbol demapper. The EXIT chart analysis [2, 5, 50, 51, 56] should be used, which is subject of study in the next chapter.

8.4 Broadband MIMO

The assumption in previous sections was that all frequency components will experience the same channel coefficients. This is never the correct assumption for fiber-optic communications, because dispersion effects and mode-coupling effects are frequency-dependent. On the other hand, most MIMO wireless communications operate at data rates when flat fading assumption is incorrect. Based on Chap. 2 we know that when the symbol rate R_s is larger than (channel) coherence bandwidth B_c, the channel is *frequency-selective*; in other words, different frequency components will have different channel coefficients. The 50% coherence bandwidth $B_{c,\ 50\%}$ is related to the root-mean-squared (RMS) multipath delay spread $\sigma_{\tau_{MDS}}$ by [57] $B_{c,50\%} \simeq (5\sigma_{\tau_{MDS}})^{-1}$.

We have to modify the MIMO channel model to take into account the number of symbols over which the channel coefficients can be considered static. If we divide the RMS multipath delay spread $\sigma_{\tau_{MDS}}$ with the symbol duration T_s, we can determine the number of symbols over which the channel coefficients can be considered fixed, denoted as L, namely, $L = \sigma_{\tau_{MDS}}/T_s$. The quasi-static MIMO channel model can be now described as [58]:

$$y_n(k) = \sum_{m=1}^{M_{Tx}} \sum_{l=0}^{L-1} h_{nm}^{(l)} x_m(k-l); n = 1, \cdots, M_{Rx}; k = 1, \cdots, N_{sf} \tag{8.200}$$

where N_{sf} is number of symbols in frame (for which the channel can be considered fixed).

To deal with frequency-selective fading, different techniques can be classified as either single-carrier or multicarrier. The single-carrier systems typically employ the equalization to deal with frequency selectivity, which was subject of investigation in Chap. 6. On the other hand, the multicarrier techniques are typically based on OFDM, which is a subject of study in Chap. 7. In this section, we mainly deal with relevant MIMO topics omitted in previous sections and Chaps. 6 and 7, which are related on MIMO-OFDM and space-frequency block coding (SFBC) [58–60].

The frequency selectivity can be used as an additional degree of freedom, *frequency diversity*, as discussed at the beginning of this chapter. The maximum number of possible diversity channels, denoted as $M_{D,\max}$, can be summarized as follows:

$$M_{D,\max} = \begin{cases} M_{Tx}M_{Rx}, \text{for flat fading channels} \\ M_{Tx}M_{Rx}L, \text{for frequency-selective channels} \end{cases} \tag{8.201}$$

This equation suggests that multiplexing gain of frequency-selective channels is higher than that of flat fading channels, which was confirmed in [58]. This paper inspired the design of space-time codes to exploit the diversity related to frequency-selective fading [59, 60].

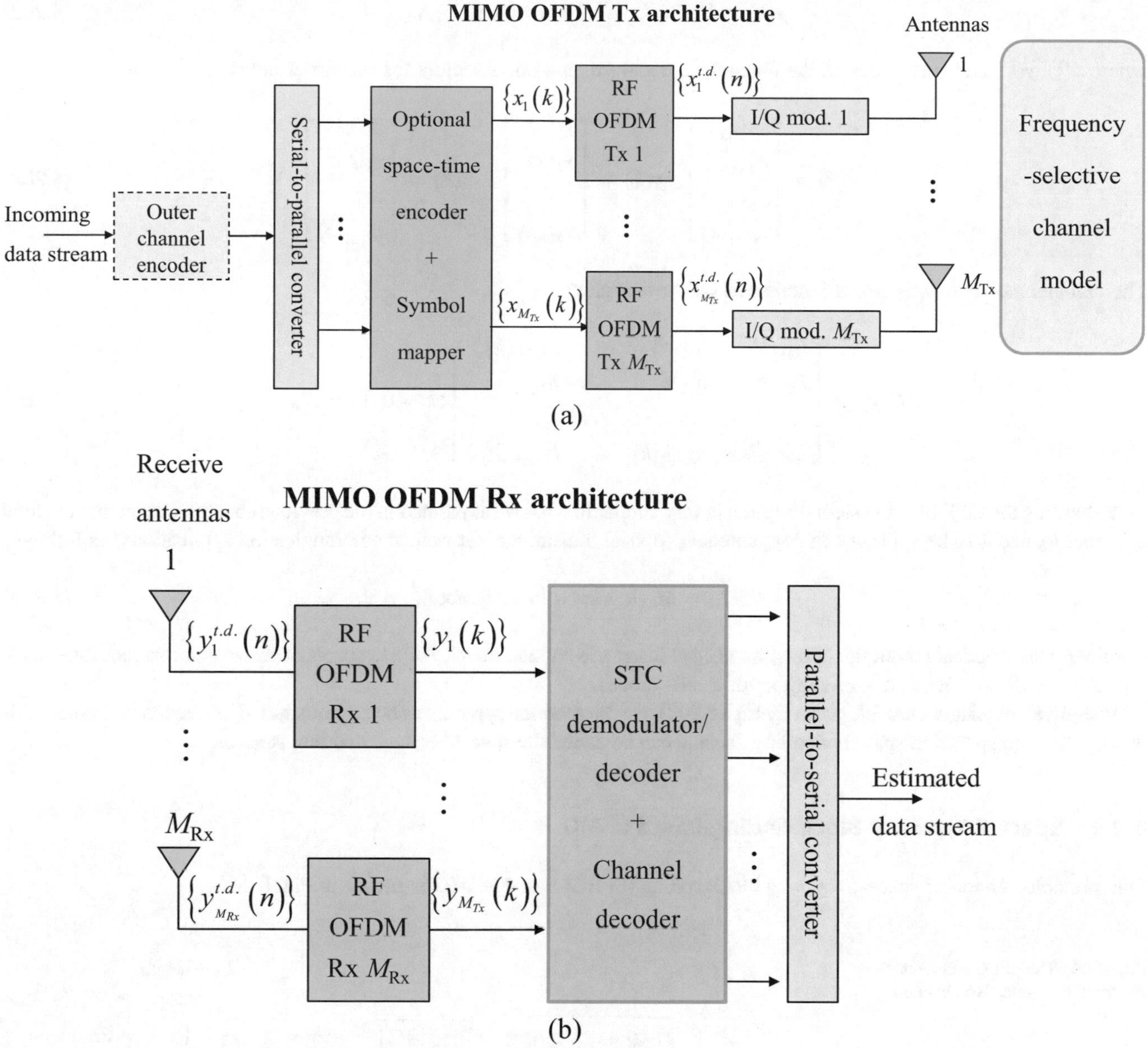

Fig. 8.43 MIMO-OFDM system architecture for frequency-selective channels: (**a**) transmitter and (**b**) receiver architectures. *t.d.* time domain

8.4.1 MIMO-OFDM Scheme

The concept of MIMO-OFDM is shown in Fig. 8.43. After serial-to-parallel (S/P) conversion, we perform symbol mapping, and the corresponding symbols are mapped to different subcarriers in conventional OFDM, as described in the previous chapter. Optionally the space-time coding can be used as well as the channel coding. The OFDM concept is used to deal with frequency-selective fading, so that each subcarrier will see flat fading. The multiple transmitter and multiple receiver antennas are used to improve multiplexing and diversity gains as in traditional MIMO systems. The space-time coding can help exploit the frequency diversity gain.

In a chapter on OFDM, we have shown that when the guard interval is larger than the RMS multipath delay spread (together with perfect time and carrier synchronization), we can decompose the wireless fading channel into N_{sc} parallel channels, where N_{sc} is the number of subcarriers used in OFDM. For SMF-based optical communications, as shown in [61], the guard interval (cyclic prefix duration) must be longer than RMS due to chromatic dispersion, plus RMS due to differential group delay of two polarization states. Finally, in FMF-based optical applications, we need to account for RMS of mode group delay. Further assuming that the number of subcarriers is larger than or equal to the number of resolvable paths due to selective fading (L), the corresponding OFDM parallelization model we developed can be extended as follows:

$$\boldsymbol{y}(k)=\boldsymbol{H}(k)\boldsymbol{x}(k)+\boldsymbol{z}(k);\ k=0,1,\cdots,N_{sc}-1 \tag{8.202}$$

where $\boldsymbol{x}(k)$, $\boldsymbol{y}(k)$, and $z(k)$ represent the transmitted, received, and noise vectors for subcarrier index k:

$$\boldsymbol{x}(k)=\begin{bmatrix} x_1(k) \\ x_2(k) \\ \dots \\ x_{M_{Tx}}(k) \end{bmatrix},\ \boldsymbol{y}(k)=\begin{bmatrix} y_1(k) \\ y_2(k) \\ \dots \\ y_{M_{Rx}}(k) \end{bmatrix},\ \ z(k)=\begin{bmatrix} z_1(k) \\ z_2(k) \\ \dots \\ z_{M_{Rx}}(k) \end{bmatrix}, \tag{8.203}$$

The *channel matrix* now is also a function of subcarrier index:

$$\boldsymbol{H}(k)=\begin{bmatrix} h_{11}(k) & h_{12}(k) & \dots & h_{1M_{Tx}}(k) \\ h_{21}(k) & h_{22}(k) & \dots & h_{2M_{Tx}}(k) \\ \dots & \dots & \dots & \dots \\ h_{M_{Rx}1}(k) & h_{M_{Rx}2}(k) & \dots & h_{M_{Rx}M_{Tx}}(k) \end{bmatrix};k=0,1,\cdots,N_{sc}-1. \tag{8.204}$$

Regarding the OFDM subsystem design, it is very similar to what we explained in the previous chapter, except that the total data rate R_b needs to be split among M_{Tx} antennas (optical transmitters for optical communication applications) as follows:

$$R_b=M_{Tx}R_s\log_2 M\ [\text{bits/symbol}]. \tag{8.205}$$

For FMF-based optical communications, we need to insert a factor 2 in Eq. (8.205), to account for two polarization states used, while M_{Tx} will now account for a number of spatial modes.

In equivalent channel model, given by Eq. (8.202), we interpret the symbols as being transmitted in frequency domain; this is why the corresponding spatial encoding scheme can be called the space-frequency coding scheme.

8.4.2 Space-Frequency Block Coding-Based MIMO

One possible version of space-frequency block coding (SFBC)-based MIMO is shown in Fig. 8.44.

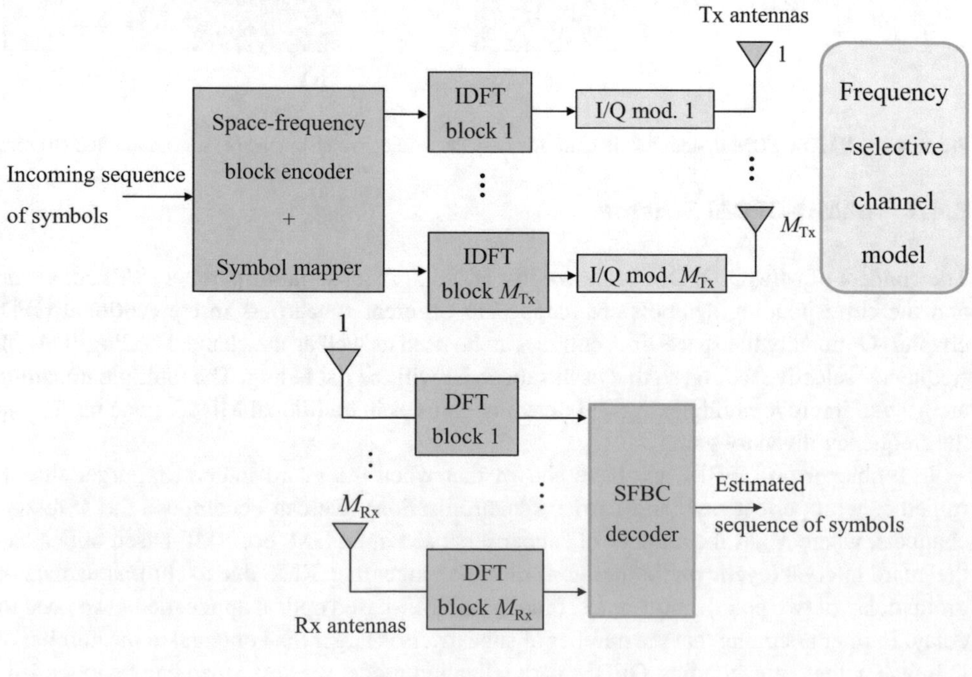

Fig. 8.44 The SFBC architecture for frequency-selective channels

A block of N_s information symbols is arriving at the input of SFBC encoder. In principle, any STC can be used. The STC generates M_{Tx} encoded outputs. Each output is used as an input of inverse DFT (IDFT) block, composed of S/P converter, IFFT, and P/S converter. In principle, RF OFDM transmitters can be used in a similar fashion as shown in Fig. 8.43. On the receiver side, the n-th ($n = 1,2,\ldots,M_{\mathrm{Rx}}$) receiver antenna signal after A/D conversion undergoes S/P conversion, FFT, and P/S conversion. And then SFBC decoding takes place, and the estimated sequence of symbols is delivered to the end user. There are different options; single SFBC decoder can be used or M_{Rx} SFBC decoders can be used operating in parallel. The SFBC principles can be combined with various BLAST architectures.

For additional details on SFBC systems, an interested reader is referred to [59, 60].

8.5 MIMO Channel Capacity

The channel capacity of single-input single-output (SISO) channels has already been studied in Chap. 4. Here we are concerned with MIMO channel capacity, where the channel model is defined by Eq. (8.62), namely, $\boldsymbol{y} = \boldsymbol{Hx} + \boldsymbol{z}$, where $\boldsymbol{x}$, $\boldsymbol{y}$, and $\boldsymbol{z}$ denote the transmitted, received, and noise vectors, respectively.

MIMO channel capacity evaluation depends on what is known about the channel matrix or its distribution at the transmitter and/or receiver sides. We assume that channel state information at the receiver (CSIR) side is always available. We consider the following relevant cases [62]:

- The channel matrix $\boldsymbol{H}$ is deterministic (static).
- The channel matrix $\boldsymbol{H}$ is random, and at every channel use an independent realization of $\boldsymbol{H}$ is generated; in other words the channel is *ergodic*.
- The channel matrix is random but fixed for the duration of transmission; in other words, the channel is *non-ergodic*.

8.5.1 Capacity of Deterministic (Static) MIMO Channels

The MIMO channel capacity for deterministic channel can be defined as the following maximization problem:

$$C = \max_{f_X\left(x_1,\cdots,x_{M_{Tx}}\right)} [H(\boldsymbol{y}) - H(\boldsymbol{y}|\boldsymbol{x})], \tag{8.206}$$

where the maximization is performed over the multivariate PDF $f_X(\boldsymbol{x})$. Given that the components of noise vector are independent on $\boldsymbol{x}$ and zero-mean Gaussian, we can simplify the problem of estimation of entropy $H(\boldsymbol{y}|\,\boldsymbol{x}) = H(\boldsymbol{Hx} + \boldsymbol{z}|\,\boldsymbol{x}) = H(\boldsymbol{z})$, so that the previous equation can be rewritten as:

$$C = \max_{f_X\left(x_1,\cdots,x_{M_{Tx}}\right)} [H(\boldsymbol{y}) - H(\boldsymbol{z})]. \tag{8.207}$$

Given that components of noise vector are i.i.d. and complex circular symmetric Gaussian, the corresponding distribution of $\boldsymbol{z}$ is a multivariate Gaussian [62].

$$f_Z(\boldsymbol{z}) = \frac{1}{\det(\pi \boldsymbol{R}_z)} e^{-\left(\boldsymbol{z}-\boldsymbol{\mu}_z\right)^\dagger \boldsymbol{R}_z^{-1}\left(\boldsymbol{z}-\boldsymbol{\mu}_z\right)}, \tag{8.208}$$

where $\boldsymbol{\mu}_z \doteq \langle \boldsymbol{z}\rangle$ and $\boldsymbol{R}_z$ is the covariance matrix of $\boldsymbol{z}$, defined as $\boldsymbol{R}_z \doteq \left\langle \left(\boldsymbol{z}-\boldsymbol{\mu}_z\right)\left(\boldsymbol{z}-\boldsymbol{\mu}_z\right)^\dagger \right\rangle = \langle \boldsymbol{z}\boldsymbol{z}^\dagger\rangle - \boldsymbol{\mu}_z\boldsymbol{\mu}_z^\dagger$. The corresponding differential entropy can be determined as:

$$\begin{aligned}
H(z) &= \langle -\log_2 f_Z(z) \rangle = \log_2 \det(\pi \boldsymbol{R}_z) + \left\langle \overbrace{(z-\boldsymbol{\mu}_z)^\dagger \boldsymbol{R}_z^{-1}(z-\boldsymbol{\mu}_z)}^{\sum_m \sum_n (z_m^* - \mu_{z_m}^*)[\boldsymbol{R}_z^{-1}]_{mn}(z_n - \mu_{z_n})} \right\rangle \log_2 e \\
&= \log_2 \det(\pi \boldsymbol{R}_z) + \log_2 e \cdot \sum_m \sum_n \underbrace{\left\langle \left(z_m^* - \mu_{z_m}^*\right)\left(z_n - \mu_{z_n}\right) \right\rangle}_{[\boldsymbol{R}_z]_{mn}} \left[\boldsymbol{R}_z^{-1}\right]_{mn} \\
&= \log_2 \det(\pi \boldsymbol{R}_z) + \log_2 e \cdot \mathrm{Tr}(\boldsymbol{I}_{M_{Rx}}) \\
&= \log_2 \det(e\pi \boldsymbol{R}_z).
\end{aligned} \tag{8.209}$$

This result is applicable to any complex Gaussian multivariate. For i.i.d. zero-mean complex Gaussian multivariate, the following are valid $\mu_z = 0, \boldsymbol{R}_z = \sigma^2 \boldsymbol{I}_{M_{Rx}}$, so that the differential entropy becomes:

$$H(z) = \log_2 \det\left(e\pi\sigma^2 \boldsymbol{I}_{M_{Rx}}\right). \tag{8.210}$$

The correlation matrix of $\boldsymbol{y}$ can be determined as:

$$\boldsymbol{R}_y = \left\langle \overbrace{\boldsymbol{y}}^{\boldsymbol{Hx}+z} \boldsymbol{y}^\dagger \right\rangle = \boldsymbol{H}\boldsymbol{R}_x\boldsymbol{H}^\dagger + \sigma^2 \boldsymbol{I}_{M_{Rx}}. \tag{8.211}$$

The differential entropy of $\boldsymbol{y}$ can be obtained by substituting $\boldsymbol{R}_y$ into (8.209) to obtain:

$$H(\boldsymbol{y}) = \log_2 \det(e\pi \boldsymbol{R}_y) = \log_2 \det\left[e\pi\left(\boldsymbol{H}\boldsymbol{R}_x\boldsymbol{H}^\dagger + \sigma^2 \boldsymbol{I}_{M_{Rx}}\right)\right]. \tag{8.212}$$

After substitution of Eqs. (8.210) and (8.212) into (8.207), we obtain:

$$\begin{aligned}
C &= \max_{f_X(x_1,\cdots,x_{M_{Tx}})} \left\{ \log_2 \det\left[e\pi\left(\boldsymbol{H}\boldsymbol{R}_x\boldsymbol{H}^\dagger + \sigma^2 \boldsymbol{I}_{M_{Rx}}\right)\right] - \log_2 \det\left(e\pi\sigma^2 \boldsymbol{I}_{M_{Rx}}\right) \right\} \\
&= \max_{f_X(x_1,\cdots,x_{M_{Tx}})} \log_2 \det\left(\frac{\boldsymbol{H}\boldsymbol{R}_x\boldsymbol{H}^\dagger + \sigma^2 \boldsymbol{I}_{M_{Rx}}}{\sigma^2 \boldsymbol{I}_{M_{Rx}}} \right) \\
&= \max_{\boldsymbol{R}_x : \mathrm{Tr}(\boldsymbol{R}_x) = P} \log_2 \det\left(\boldsymbol{I}_{M_{Rx}} + \frac{1}{\sigma^2} \boldsymbol{H}\boldsymbol{R}_x\boldsymbol{H}^\dagger \right) \text{ [bits/channel use]},
\end{aligned} \tag{8.213}$$

where the transmit power is constrained by $\sum_{i=1}^{M_{Tx}} \langle \boldsymbol{x}_i \boldsymbol{x}_i^* \rangle \le P$, so that the maximization is performed over all input covariance matrices satisfying this power constraint. If we want to convert units into bit(s), we need to multiply (8.213) by the channel bandwidth B.

This optimization process is dependent on the availability of CSI on the transmitter side. As in Chap. 4, we know that when *CSI is available on transmitter (CSIT) side*, the optimum adaptation strategy is the *water-filling*. As already discussed in Sect. 8.2.2, by performing the SVD of the channel matrix we obtain:

$$\begin{aligned}
&\boldsymbol{H} = \boldsymbol{U}\boldsymbol{\Sigma}\boldsymbol{V}^\dagger;\ \boldsymbol{U}^\dagger\boldsymbol{U} = \boldsymbol{I}_{M_{Rx}};\ \boldsymbol{V}^\dagger\boldsymbol{V} = \boldsymbol{I}_{M_{Tx}};\ \boldsymbol{\Sigma} = \mathrm{diag}(\sigma_i), \\
&\sigma_i = \sqrt{\lambda_i},
\end{aligned} \tag{8.214}$$

where λ_i are eigenvalues of the Wishart matrix $\boldsymbol{H}\boldsymbol{H}^\dagger$. As discussed in Sect. 8.2.2, we can now decompose the MIMO channel into $R_{\boldsymbol{H}}$ parallel channels, where $R_{\boldsymbol{H}}$ is the rank of channel matrix. The MIMO channel capacity would be then:

$$C = \max_{P_i : \sum_i P_i \le P} \sum_{i=1}^{R_H} \log_2\left(1 + \frac{P_i}{P} SNR_i\right),\quad SNR_i = \frac{\sigma_i^2 P}{\sigma^2}. \tag{8.215}$$

The water-filling power allocation strategy achieves the channel capacity:

$$\frac{P_i}{P} = \begin{cases} 1/SNR_c - 1/SNR_i, SNR_i \ge SNR_c \\ 0,\ \text{otherwise} \end{cases} \tag{8.216}$$

where SNR_c is the SNR cutoff value (we do not transmit when SNR is below the cutoff value). By substituting (8.216) into (8.215), we obtain:

$$C = \sum_{i:SNR_i > SNR_c} \log_2\left(\frac{SNR_i}{SNR_c}\right) \tag{8.217}$$

When the *CSI is not known on the transmitter side*, there is no reason to transmit on one antenna with higher launch power than on the others, and there is no need to introduce the correlation among transmit antennas. Therefore, it makes sense to apply the *uniform power allocation* strategy. Assuming the rich scattering scenario, the input covariance matrix is a scaled version of the identity matrix $\boldsymbol{R_x} = (P/M_{T_x})\boldsymbol{I}_{M_{T_x}}$, and after substitution into (8.213) we obtain:

$$I(\boldsymbol{x},\boldsymbol{y}) = \log_2 \det\left(\boldsymbol{I}_{M_{Tx}} + \frac{P}{M_{Tx}\sigma^2}\boldsymbol{HH}^\dagger\right). \tag{8.218}$$

In the limit, as the number of transmit antennas tends to plus infinity, we have that $\boldsymbol{HH}^\dagger/M_{Tx} \to \boldsymbol{I}_{M_{Tx}}$, so that the mutual information tends to $C = M_{R_x}\log_2(1 + P/\sigma^2)$, and the capacity is a linear function in $M = \min(M_{\mathrm{Tx}}, M_{\mathrm{Rx}})$. By performing the SVD of channel matrix, Eq. (8.218) can also be written in the following form:

$$I(\boldsymbol{x},\boldsymbol{y}) = \sum_{i=1}^{R_H} \log_2\left(1 + \frac{\sigma_i^2 P/\sigma^2}{M_{Tx}}\right), R_H = \mathrm{rank}(\boldsymbol{H}) \le \min(M_{Tx}, M_{Rx}) \tag{8.219}$$

When the transmitter does not know the channel matrix, it can select a fixed data rate R to transmit, which might exceed the mutual information and an outage will occur. The probability that the actual data rate R exceeds mutual information $I(\boldsymbol{x},\boldsymbol{y})$ is known as the *outage probability*, defined as:

$$P_{\mathrm{outage}}(R) = \Pr\left(\log_2\left[\det\left(\boldsymbol{I}_{M_{Rx}} + \frac{P/\sigma^2}{M_{Tx}}\boldsymbol{HH}^\dagger\right)\right] < R\right). \tag{8.220}$$

The maximum data rate that can be supported at a given outage probability is commonly referred to as the *outage capacity*.

8.5.2 Capacity of Random MIMO Channels

So far we assumed that the channel matrix is static, which is rarely true in optical and wireless communication channels. The channel matrix describing the chromatic dispersion is frequency selective, but it is closest to the deterministic assumption. Other fiber-optic channel impairments such as PMD in SMFs and mode coupling in FMFs are time varying and random. The FSO channel matrix is time varying with the correlation time ranging typically from 10 μs to 10 ms. The wireless channel matrix is time varying due to the dynamic nature of the channel and movement through the spatially varying field. As discussed earlier, we consider two scenarios: (i) the ergodic channel capacity, in which for each channel use the new realization of channel matrix is drawn, and (ii) non-ergodic channel capacity, in which the channel matrix $\boldsymbol{H}$ is randomly chosen and kept fixed for the duration of transmission.

8.5.2.1 Ergodic Capacity

When the receiver has the perfect knowledge of the CSI, the mutual information between channel input $\boldsymbol{x}$ and channel output $\boldsymbol{y}$, given the channel matrix $\boldsymbol{H}$, can be determined as [2, 63]:

$$\begin{aligned} I(\boldsymbol{x};\boldsymbol{y},\boldsymbol{H}) &= \underbrace{I(\boldsymbol{x};\boldsymbol{H})}_{=0} + I(\boldsymbol{x};\boldsymbol{y}|\boldsymbol{H}) = I(\boldsymbol{x};\boldsymbol{y}|\boldsymbol{H}) \\ &= \langle I(\boldsymbol{x};\boldsymbol{y}|\boldsymbol{H} = \boldsymbol{H}')\rangle_{\boldsymbol{H}'}, \end{aligned} \tag{8.221}$$

where we use the subscript $\boldsymbol{H}'$ in expectation operator to denote different realizations $\boldsymbol{H}'$ of the random channel matrix $\boldsymbol{H}$. The term $I(\boldsymbol{x};\,\boldsymbol{H}) = 0$ as the input vector and the channel matrix are independent of each other.

Based on Eqs. (8.213) and (8.221), we can write the following expression for ergodic capacity:

$$C = \left\langle \max_{\boldsymbol{R}_x:\mathrm{Tr}(\boldsymbol{R}_x)=P} \log_2 \det\left(\boldsymbol{I}_{M_{Rx}} + \frac{1}{\sigma^2}\boldsymbol{H}\boldsymbol{R}_x\boldsymbol{H}^\dagger\right)\right\rangle_{\boldsymbol{H}}. \tag{8.222}$$

When the CSI is known on the transmitter side, assuming that the short-term power constraint $\overline{P}$ is satisfied, we perform the SVD of the channel matrix, so that based on (8.215) and (8.222) we obtain:

$$C = \left\langle \max_{P_i:\sum_i P_i \le P} \sum_{i=1}^{R_H} \log_2\left(1 + \frac{P_i}{\overline{P}} SNR_i\right)\right\rangle_{\boldsymbol{H}}, \quad SNR_i = \frac{\sigma_i^2 \overline{P}}{\sigma^2}. \tag{8.223}$$

Clearly, the optimum strategy is water-filling in space domain.

When we are concerned with the less restrictive long-term power constraint [1].

$$\left\langle \overbrace{\sum_{i=1}^{M_{Tx}} \langle \boldsymbol{x}_i(\boldsymbol{H})\boldsymbol{x}_i^*(\boldsymbol{H})\rangle}^{P_H} \right\rangle_{\boldsymbol{H}} \le \overline{P} \tag{8.224}$$

at each realization, we can use channel matrix-dependent power $P_{\boldsymbol{H}}$ so that the ergodic channel capacity is given by [1]:

$$C = \max_{P_H:\langle P_H\rangle \le \overline{P}} \left\langle \max_{P_i:\sum_i P_i \le P_H} \sum_{i=1}^{R_H} \log_2\left(1 + \frac{P_i}{P_H} SNR_i\right)\right\rangle_{\boldsymbol{H}}, \quad SNR_i = \frac{\sigma_i^2 P_H}{\sigma^2}. \tag{8.225}$$

In this case, the optimum strategy is two-dimensional water-filling in space and time.

When transmitter does not know the CSI, it assumes that the elements of the channel matrix are i.i.d. complex symmetric circular Gaussian; in other words the envelope of entries follow Rayleigh distribution while the phase is uniform, so that the optimization problem is to maximize ergodic capacity subject to transmit power constraint [1]:

$$C = \max_{\boldsymbol{R}_x:\mathrm{Tr}(\boldsymbol{R}_x)=P} \left\langle \log_2\left[\det\left(\boldsymbol{I}_{M_{Tx}} + \frac{1}{\sigma^2}\boldsymbol{H}\boldsymbol{R}_x\boldsymbol{H}^\dagger\right)\right]\right\rangle_{\boldsymbol{H}}. \tag{8.226}$$

The optimum covariance function that maximizes ergodic capacity is a properly scaled identity matrix so that the ergodic channel capacity expression simplifies to:

$$C = \left\langle \log_2\left[\det\left(\boldsymbol{I}_{M_{Rx}} + \frac{P/\sigma^2}{M_{Tx}}\boldsymbol{H}\boldsymbol{H}^\dagger\right)\right]\right\rangle_{\boldsymbol{H}}. \tag{8.227}$$

For a fixed number of receive antennas, and as the number of transmit antennas tends to infinity, we have that:

$$\frac{1}{M_{Tx}}\boldsymbol{H}\boldsymbol{H}^\dagger \underset{M_{Tx}\to\infty}{\to} \boldsymbol{I}_{M_{Rx}} \Rightarrow \log_2\det\left(\boldsymbol{I}_{M_{Rx}} + \frac{P}{\sigma^2}\boldsymbol{I}_{M_{Rx}}\right) = \log_2\left(1 + \frac{P}{\sigma^2}\right)^{M_{Rx}} = M_{Rx}\log_2\left(1 + \frac{P}{\sigma^2}\right), \tag{8.228}$$

indicating that ergodic channel capacity grows linearly with M_{Rx}.

The upper bound can be determined by applying Jensen's inequality [63] on Eq. (8.227) as follows:

$$\begin{aligned} C &= \left\langle \log_2\left[\det\left(\boldsymbol{I}_{M_{Rx}} + \frac{P/\sigma^2}{M_{Tx}}\boldsymbol{H}\boldsymbol{H}^\dagger\right)\right]\right\rangle_{\boldsymbol{H}} \\ &\le \log_2\left[\det\left(\boldsymbol{I}_{M_{Rx}} + \frac{P/\sigma^2}{M_{Tx}}\langle \boldsymbol{H}\boldsymbol{H}^\dagger\rangle_{\boldsymbol{H}}\right)\right] = M_{Rx}\log_2\left(1 + P/\sigma^2\right). \end{aligned} \tag{8.229}$$

By recalling that $\boldsymbol{H}\boldsymbol{H}^\dagger$ and $\boldsymbol{H}^\dagger\boldsymbol{H}$ have the same spectrum (the set of eigenvalues), then the following is valid:

$$\det\left(\boldsymbol{I}_{M_{Rx}} + \frac{P/\sigma^2}{M_{Tx}}\boldsymbol{H}\boldsymbol{H}^\dagger\right) = \det\left(\boldsymbol{I}_{M_{Rx}} + \frac{P/\sigma^2}{M_{Tx}}\boldsymbol{H}^\dagger\boldsymbol{H}\right), \tag{8.230}$$

and after substitution of (8.230) into (8.227), Jensen's inequality yields to:

$$\begin{aligned} C &= \left\langle \log_2\left[\det\left(\boldsymbol{I}_{M_{Rx}} + \frac{P/\sigma^2}{M_{Tx}}\boldsymbol{H}^\dagger\boldsymbol{H}\right)\right]\right\rangle_{\boldsymbol{H}} \\ &\le \log_2\left[\det\left(\boldsymbol{I}_{M_{Rx}} + \frac{P/\sigma^2}{M_{Tx}}\langle\boldsymbol{H}^\dagger\boldsymbol{H}\rangle_{\boldsymbol{H}}\right)\right] = M_{Tx}\log_2\left(1 + \frac{M_{Rx}}{M_{Tx}}P/\sigma^2\right). \end{aligned} \tag{8.231}$$

By combining these two inequalities into one, we obtained the following upper bound:

$$C \le \min\left\{M_{Rx}\log_2(1 + P/\sigma^2), M_{Tx}\log_2\left(1 + \frac{M_{Rx}}{M_{Tx}}P/\sigma^2\right)\right\}. \tag{8.232}$$

For the exact computation of (8.227), we can use Monte Carlo integration approach. For semi-analytical computation, an interested reader is referred to [2].

8.5.2.2 Non-ergodic Capacity

As discussed at the beginning of MIMO channel capacity section, in non-ergodic channel, the channel matrix is generated at random but kept fixed for the duration of the transmission. Since the average channel capacity does not have any meaning here, we can define the instantaneous channel capacity as a function of the channel matrix:

$$C(\boldsymbol{H}) = \log_2\det\left(\boldsymbol{I}_{M_{Rx}} + \frac{P/\sigma^2}{M_{Tx}}\boldsymbol{H}\boldsymbol{H}^\dagger\right), \tag{8.233}$$

and determine the *outage probability* as the probability that the transmission rate exceeds instantaneous channel capacity $C(\boldsymbol{H})$ [2]:

$$P_{\text{outage}} = \Pr\{C(\boldsymbol{H}) < R\}. \tag{8.234}$$

The maximum rate that can be supported by this channel, for the outage probability P_{outage}, is called the *outage capacity*. More formally, the *p-percent outage capacity* is defined as the data capacity that is guaranteed in $(1 - p) \times 100\%$ of channel realizations [45]:

$$\Pr\{C(\boldsymbol{H}) \le C_{\text{outage},p}\} = p. \tag{8.235}$$

Given that $C_{\text{outage},p}$ occurs in $(1 - p)\%$ of channel uses, the *throughput* of the system associated with, denoted as R_p, can be determined as [45]:

$$R_p = (1 - p)C_{\text{outage},p}. \tag{8.236}$$

An interesting special case of non-ergodic channels is the so-called *block-fading* channel model. In this channel model, the channel matrix is selected randomly and kept fixed of duration of N_s symbols. After N_s symbols a new channel matrix realization is drawn. The *B block-fading channel model* can be described as:

$$\boldsymbol{y}_b[n] = \boldsymbol{H}_b\boldsymbol{x}_b[n] + \boldsymbol{z}_b[n]; b = 1, \cdots, B; n = 1, \cdots, N_s \tag{8.237}$$

where b is the index of the block and n is the index of the symbol within that block. Other assumptions for input, output, and noise vectors are the same as before. Again, we can perform the SVD of channel matrix $\boldsymbol{H}_b$ in a similar fashion to Eq. (8.214):

$$H_b = U_b \Sigma_b V_b^\dagger, \tag{8.238}$$

and by pre-multiplying (8.237) by $U_b^\dagger$, the output-input relationship becomes:

$$\begin{aligned} U_b^\dagger y_b[n] &= U_b^\dagger \underbrace{U_b \Sigma_b V_b^\dagger}_{H_b} x_b[n] + U_b^\dagger z_b[n] \\ &= \Sigma_b V_b^\dagger x_b[n] + U_b^\dagger z_b[n]. \end{aligned} \tag{8.239}$$

when transmitted vector $x_b[n]$ get preprocessed on transmitter side, $V_b x_b[n]$, the output-input relationship (8.239) becomes free of spatial interference. As the number of blocks tends to plus infinity, the block-fading model becomes ergodic, and in the case of CSIR, the channel capacity can be evaluated as:

$$C = \left\langle \sum\nolimits_i \log_2 \left(1 + \frac{P/\sigma^2}{M_{Tx}} \sigma_i^2 \right) \right\rangle, \tag{8.240}$$

and this equation is similar to (8.223), so in the case of CSIT, the optimum power allocation strategy is water-filling.

When the number of blocks in block-fading model is finite, the channel capacity calculation is more involved; therefore, an interested reader is referred to specialized books on MIMO communication principles, such as [2, 26].

8.5.2.3 Correlated Fading Channel Capacity

In a *separately correlated MIMO model*, introduced earlier, the H entries are correlated circularly symmetric complex zero-mean Gaussian random variables, and the channel matrix can be decomposed as:

$$H = R^{1/2} H_{\text{ZMSW}} T^{1/2}, \tag{8.241}$$

where H_{ZMSW} is the matrix of circularly symmetric complex zero-mean Gaussian random variables with unit variance. The CSIR ergodic channel capacity, based on Eq. (8.227), is given by:

$$\begin{aligned} C &= \left\langle \log_2 \left[\det \left(I_{M_{Rx}} + \frac{P/\sigma^2}{M_{Tx}} H H^\dagger \right) \right] \right\rangle \\ &= \left\langle \log_2 \left[\det \left(I_{M_{Rx}} + \frac{P/\sigma^2}{M_{Tx}} H_{\text{ZMSW}} T H_{\text{ZMSW}}^\dagger R \right) \right] \right\rangle. \end{aligned} \tag{8.242}$$

For high SNRs, when $M_{\text{Tx}} = M_{\text{Rx}} = M$, the following asymptotic approximation is valid [64]:

$$C \approx M \log_2 \left(\frac{P/\sigma^2}{M} \right) + \log(M!) - |\log_2 \det(TR)|, \tag{8.243}$$

with the last term representing the loss due to local spatial correlation. Clearly, the spatial correlation degrades the system performance, while the linear growth with respect to the minimum number of transmit/receive antennas is preserved.

8.5.2.4 MIMO-OFDM Channel Capacity

The channel matrix for MIMO-OFDM system can be generalized as a block diagonal matrix with i-th block diagonal element corresponding to $M_{\text{Tx}} \times M_{\text{Rx}}$ MIMO channel [13]:

$$H = \begin{bmatrix} H(0) & \dots & 0 \\ \dots & \dots & \dots \\ 0 & \dots & H(N_{sc} - 1) \end{bmatrix}, \tag{8.244}$$

where N_{sc} is the number of subcarriers. The corresponding ergodic channel capacity expressions given above can be generalized to OFDM MIMO systems with N_{sc} subcarriers as follows [13]:

$$C = \left\langle \frac{1}{N_{sc}} \max_{\mathrm{Tr}(\boldsymbol{\Sigma}) \leq P} \log_2 \left[\det\left(\boldsymbol{I}_{M_{Rx}N_{sc}} + \frac{1}{\sigma^2} \boldsymbol{H}\boldsymbol{\Sigma}\boldsymbol{H}^\dagger \right) \right] \right\rangle_{\boldsymbol{H}}, \tag{8.245}$$

where $\boldsymbol{\Sigma}$ is the covariance matrix of the Gaussian input vector, defined as:

$$\boldsymbol{\Sigma} = \mathrm{diag}\{\boldsymbol{\Sigma}_k\}_{k=0}^{N_c-1},\ \boldsymbol{\Sigma}_k = \frac{P}{M_{Rx}N_{xc}} \boldsymbol{I}_{M_{Rx}};\ k = 0, 1, \ldots, N_{sc} - 1 \tag{8.246}$$

With *uniform power allocation*, the OFDM MIMO channel capacity is given by [13]:

$$C = \left\langle \frac{1}{N_{sc}} \sum_{k=0}^{N_{sc}-1} \log_2 \left[\det\left(\boldsymbol{I}_{M_{Rx}} + \frac{P/\sigma^2}{M_{Rx}N_{sc}} \boldsymbol{H}(k)\boldsymbol{H}^\dagger(k) \right) \right] \right\rangle_{\boldsymbol{H}}. \tag{8.245}$$

Various MIMO channel capacity scenarios described above in addition to wireless MIMO communications are also applicable to free-space optical communications with coherent detection. Regarding fiber-optic communications, these approaches can be used only in linear regime. Once the fiber nonlinearities become relevant, the generalized nonlinear Schrödinger equation should be used to study the propagation effects, nonlinear signal-ASE noise interaction, nonlinear interaction between polarization states, as well as nonlinear interaction among spatial modes in FMF links. By modeling such nonlinear fiber-optic channel as nonlinear channel with memory, the method proposed in [67] can be used to evaluate the achievable information rates.

8.6 MIMO Channel Estimation

So far we assumed that CSI was perfect. In this section we discuss how to estimate the MIMO channel coefficients. We have seen above that parallel MIMO channel decomposition has similarities with parallel decomposition of OFDM channels. So it is not surprising that similar methods can be used for channel estimation. Here we describe three techniques to estimate MIMO channel matrix:

(i) Maximum likelihood (ML) MIMO channel estimation.
(ii) Least squares (LS) MIMO channel estimation.
(iii) Linear minimum MSE (LMMSE) MIMO channel estimation.

The common denominator to all three channel estimation techniques is the assumption of using a sequence of N_p training symbols as a preamble. The transmitted pilot matrix $\boldsymbol{X}_p$ has the format $M_{\mathrm{Tx}} \times N_{\mathrm{p}}$, the received matrix has $\boldsymbol{Y}_{\mathrm{p}}$, and the noise matrix have $M_{\mathrm{Rx}} \times N_{\mathrm{p}}$. The channel matrix $\boldsymbol{H}$ dimensionality is the same as before ($M_{\mathrm{Tx}} \times M_{\mathrm{Rx}}$). We also assume that channel is quasi-static; in other words the $\boldsymbol{H}$ does not change during transmission of pilot symbols. The corresponding output-input relationship is given by:

$$\boldsymbol{Y}_p = \boldsymbol{H}\boldsymbol{X}_p + \boldsymbol{Z}. \tag{8.246}$$

8.6.1 Maximum Likelihood (ML) MIMO Channel Estimation

Given that $\boldsymbol{Y}_p$ is a random function of the channel matrix $\boldsymbol{H}$, the conditional multivariate PDF $f(\boldsymbol{Y}_p | \boldsymbol{H})$ can be used for ML estimation of $\boldsymbol{H}$ as follows:

$$\widehat{\boldsymbol{H}}_{ML} = \arg\max_{\boldsymbol{H}} f\left(\boldsymbol{Y}_p | \boldsymbol{H}\right). \tag{8.247}$$

Because the entries of noise matrix $\boldsymbol{Z}$ are ZMSW, from detection theory [65], with basics provided in Chap. 6, we know that maximizing ML is equivalent to minimizing the Euclidean (Frobenius) norm, and the optimization problem can be simplified to:

$$\widehat{\boldsymbol{H}}_{ML} = \arg\min_{\boldsymbol{H}} \left\| \boldsymbol{Y}_p - \boldsymbol{H}\boldsymbol{X}_p \right\|_F^2. \tag{8.248}$$

Defining now the MSE by:

$$\begin{aligned} MSE(\boldsymbol{H}) &= \left\| \boldsymbol{Y}_p - \boldsymbol{H}\boldsymbol{X}_p \right\|_F^2 = \mathrm{Tr}\left[\left(\boldsymbol{Y}_p - \boldsymbol{H}\boldsymbol{X}_p \right) \left(\boldsymbol{Y}_p - \boldsymbol{H}\boldsymbol{X}_p \right)^\dagger \right] \\ &= \mathrm{Tr}\left[\boldsymbol{Y}_p \boldsymbol{Y}_p^\dagger - \boldsymbol{Y}_p \boldsymbol{X}_p^\dagger \boldsymbol{H}^\dagger - \boldsymbol{H}\boldsymbol{X}_p \boldsymbol{Y}_p^\dagger + \boldsymbol{H}\boldsymbol{X}_p \boldsymbol{X}_p^\dagger \boldsymbol{H}^\dagger \right], \end{aligned} \tag{8.249}$$

we can determine $\widehat{\boldsymbol{H}}_{ML}$ by solving the equation $\partial MSE(\boldsymbol{H})/\partial \boldsymbol{H} = 0$. By using the following property of matrices $\partial \mathrm{Tr}(\boldsymbol{AB})/\partial \boldsymbol{A} = \boldsymbol{B}^T$, $\partial \mathrm{Tr}(\boldsymbol{AB}^\dagger)/\partial \boldsymbol{B} = \boldsymbol{0}$, we obtain the following equation for $\partial MSE(\boldsymbol{H})/\partial \boldsymbol{H}$:

$$\frac{\partial MSE(\boldsymbol{H})}{\partial \boldsymbol{H}} = -\left(\boldsymbol{X}_p \boldsymbol{Y}_p^\dagger \right)^T + \left(\boldsymbol{X}_p \boldsymbol{X}_p^\dagger \boldsymbol{H}^\dagger \right)^T = 0. \tag{8.250}$$

From Eq. (8.250), we obtain the following solution for $\boldsymbol{H}^\dagger$:

$$\boldsymbol{H}^\dagger = \left(\boldsymbol{X}_p \boldsymbol{X}_p^\dagger \right)^{-1} \boldsymbol{X}_p \boldsymbol{Y}_p^\dagger, \tag{8.251}$$

or equivalently, after applying the †-operation, we arrive at the following solution for $\boldsymbol{H}$:

$$\widehat{\boldsymbol{H}}_{\mathrm{ML}} = \boldsymbol{Y}_p \boldsymbol{X}_p^\dagger \left(\boldsymbol{X}_p \boldsymbol{X}_p^\dagger \right)^{-1}. \tag{8.252}$$

From matrix theory we know that in order for $\left(\boldsymbol{X}_p \boldsymbol{X}_p^\dagger \right)^{-1}$ to exist, the following condition most be satisfied: $N_p \geq M_{Tx}$.

8.6.2 Least Squares (LS) MIMO Channel Estimation

In LS MIMO estimation, we determine the channel matrix $\boldsymbol{H}$ that minimizes the squared error between the actual received signal $\boldsymbol{Y}_p$ and the estimated received signal $\widehat{\boldsymbol{H}}\boldsymbol{X}_p$; in other words:

$$\widehat{\boldsymbol{H}}_{LS} = \arg\min_{\widehat{\boldsymbol{H}}} \left\| \boldsymbol{Y}_p - \widehat{\boldsymbol{H}}\boldsymbol{X}_p \right\|_F^2. \tag{8.253}$$

Clearly, this equation is almost identical to Eq. (8.248), and not surprisingly, the same solution for $\widehat{\boldsymbol{H}}$ is obtained after solving the equation $\partial \left\| \boldsymbol{Y}_p - \widehat{\boldsymbol{H}}\boldsymbol{X}_p \right\|_F^2 / \partial \widehat{\boldsymbol{H}} = 0$, which is:

$$\widehat{\boldsymbol{H}}_{\mathrm{LS}} = \boldsymbol{Y}_p \boldsymbol{X}_p^\dagger \left(\boldsymbol{X}_p \boldsymbol{X}_p^\dagger \right)^{-1}. \tag{8.254}$$

The LS estimation method can be justified as follows. For reasonably high SNRs, the MIMO channel model is given by $\boldsymbol{Y}_p \simeq \boldsymbol{H}\boldsymbol{X}_p$. The $\boldsymbol{X}_p$ is not a square matrix and thus it is not invertible. However, if we pre-multiply this MIMO equation by $\boldsymbol{X}_p^\dagger$ from the right, we obtain the modified MIMO model $\boldsymbol{Y}_p \boldsymbol{X}_p^\dagger \simeq \boldsymbol{H}\boldsymbol{X}_p \boldsymbol{X}_p^\dagger$, and $\boldsymbol{X}_p \boldsymbol{X}_p^\dagger$ is an invertible square matrix. By multiplying now the modified MIMO model by $\left(\boldsymbol{X}_p \boldsymbol{X}_p^\dagger \right)^{-1}$ from the right, we obtain the following solution for the channel matrix $\boldsymbol{H} \simeq \boldsymbol{Y}_p \boldsymbol{X}_p^\dagger \left(\boldsymbol{X}_p \boldsymbol{X}_p^\dagger \right)^{-1}$, which is the same as the LS solution given by Eq. (8.254).

8.6.3 Linear Minimum Mean Square Error (LMMSE) MIMO Channel Estimation

In LMMSE MIMO channel estimation, we try to determine the channel matrix minimizing the MSE between true channel matrix $\boldsymbol{H}$ and channel matrix estimate $\widehat{\boldsymbol{H}}$:

$$\widehat{\boldsymbol{H}}_{\mathrm{LMMSE}} = \underset{\widehat{\boldsymbol{H}}}{\arg\min} \left\langle \left\| \boldsymbol{H} - \widehat{\boldsymbol{H}} \right\|_F^2 \right\rangle. \tag{8.255}$$

To simplify the solution, we assume that channel matrix estimate $\widehat{\boldsymbol{H}}$ is the linear transformation of received signal $\boldsymbol{Y}_p$; in other words:

$$\widehat{\boldsymbol{H}} = \boldsymbol{Y}_p \boldsymbol{A}, \tag{8.256}$$

where $\boldsymbol{A}$ is the transformation matrix to be determined. After substitution of (8.256) into (8.255), the equivalent optimization problem becomes:

$$\widehat{\boldsymbol{H}}_{MMSE} = \underset{\boldsymbol{A}}{\arg\min} \left\langle \left\| \boldsymbol{H} - \boldsymbol{Y}_p \boldsymbol{A} \right\|_F^2 \right\rangle = \underset{\boldsymbol{A}}{\arg\min} \left\langle \mathrm{Tr}\left[(\boldsymbol{H} - \boldsymbol{Y}_p \boldsymbol{A})(\boldsymbol{H} - \boldsymbol{Y}_p \boldsymbol{A})^\dagger \right] \right\rangle. \tag{8.257}$$

As Telatar has shown [62], the expectation and trace operations commute, so we can rewrite the previous equation as follows:

$$\begin{aligned} \widehat{\boldsymbol{H}}_{MMSE} &= \underset{\boldsymbol{A}}{\arg\min}\, \mathrm{Tr}\left[\left\langle (\boldsymbol{H} - \boldsymbol{Y}_p \boldsymbol{A})(\boldsymbol{H} - \boldsymbol{Y}_p \boldsymbol{A})^\dagger \right\rangle \right] \\ &= \underset{\boldsymbol{A}}{\arg\min}\, \mathrm{Tr}\left[\underbrace{\langle \boldsymbol{H}\boldsymbol{H}^\dagger \rangle}_{\boldsymbol{R}_H} - \left\langle \boldsymbol{H}\boldsymbol{A}^\dagger \boldsymbol{Y}_p^\dagger \right\rangle - \langle \boldsymbol{Y}_p \boldsymbol{A} \boldsymbol{H}^\dagger \rangle + \left\langle \boldsymbol{Y}_p \boldsymbol{A}\boldsymbol{A}^\dagger \boldsymbol{Y}_p^\dagger \right\rangle \right], \end{aligned} \tag{8.258}$$

where we used $\boldsymbol{R}_H$ to denote the correlation matrix of $\boldsymbol{H}$, namely, $\boldsymbol{R}_H = \langle \boldsymbol{H}\boldsymbol{H}^\dagger \rangle$. Let us define the MSE by:

$$MSE(\boldsymbol{A}) = \boldsymbol{R}_H - \left\langle \boldsymbol{H}\boldsymbol{A}^\dagger \boldsymbol{Y}_p^\dagger \right\rangle - \langle \boldsymbol{Y}_p \boldsymbol{A} \boldsymbol{H}^\dagger \rangle + \left\langle \boldsymbol{Y}_p \boldsymbol{A}\boldsymbol{A}^\dagger \boldsymbol{Y}_p^\dagger \right\rangle. \tag{8.259}$$

By assuming that $\langle \boldsymbol{H} \rangle = \boldsymbol{0}$, $\langle \boldsymbol{Z} \rangle = \boldsymbol{0}$ after substitution of $\boldsymbol{Y}_p = \boldsymbol{H}\boldsymbol{X}_p + \boldsymbol{Z}$ into the previous equation, we obtain:

$$MSE(\boldsymbol{A}) = \boldsymbol{R}_H - \boldsymbol{A}^\dagger \boldsymbol{X}_p^\dagger \boldsymbol{R}_H - \boldsymbol{X}_p \boldsymbol{A} \boldsymbol{R}_H + \boldsymbol{X}_p \boldsymbol{A}\boldsymbol{A}^\dagger \boldsymbol{X}_p^\dagger \boldsymbol{R}_H + \boldsymbol{A}\boldsymbol{A}^\dagger \boldsymbol{R}_Z, \tag{8.260}$$

where we used $\boldsymbol{R}_Z$ to denote the correlation matrix of $\boldsymbol{Z}$, namely, $\boldsymbol{R}_Z = \langle \boldsymbol{Z}\boldsymbol{Z}^\dagger \rangle = N_p \sigma^2 \boldsymbol{I}_{N_p}$. By taking the partial derivative with respect to $\boldsymbol{A}$, we obtain:

$$\frac{\partial MSE(\boldsymbol{A})}{\partial \boldsymbol{A}} = -\boldsymbol{X}_p \boldsymbol{R}_H + \boldsymbol{X}_p \boldsymbol{A}^\dagger \boldsymbol{X}_p^\dagger \boldsymbol{R}_H + \boldsymbol{A}^\dagger \boldsymbol{R}_Z = 0, \tag{8.261}$$

and after solving for $\boldsymbol{A}$, we obtained the optimum solution for transformation matrix as:

$$\boldsymbol{A}_{\mathrm{optimum}} = \left(\boldsymbol{R}_H \boldsymbol{X}_p^\dagger \boldsymbol{X}_p + \boldsymbol{R}_Z \right)^{-1} \boldsymbol{R}_H \boldsymbol{X}_p^\dagger. \tag{8.262}$$

Based on Eq. (8.256), LMMSE estimate of the channel matrix becomes:

$$\widehat{\boldsymbol{H}}_{\mathrm{LMMSE}} = \boldsymbol{Y}_p \left(\boldsymbol{R}_H \boldsymbol{X}_p^\dagger \boldsymbol{X}_p + \boldsymbol{R}_Z \right)^{-1} \boldsymbol{R}_H \boldsymbol{X}_p^\dagger. \tag{8.263}$$

Clearly, this result is applicable to any random channel matrix, not necessarily ZMSW model as long as the following is valid $\langle \boldsymbol{H} \rangle = \boldsymbol{0}$.

8.6.4 Selection of Pilot Signals

So far we have not put any restriction on pilot matrix, except for the corresponding invertible matrix which we required that $N_p \geq M_{Tx}$. If we choose $X_p X_p^{\dagger}$ to be a scaled identity matrix:

$$\boldsymbol{X}_p \boldsymbol{X}_p^{\dagger} = \frac{N_p}{M_{Tx}} \boldsymbol{I}_{M_{Tx}}, \tag{8.264}$$

the LS solution becomes:

$$\widehat{\boldsymbol{H}}_{\mathrm{LS}} = \frac{M_{Tx}}{N_p} \underbrace{(\boldsymbol{H}\boldsymbol{X}_p + \boldsymbol{Z})}_{\boldsymbol{Y}_p} \boldsymbol{X}_p^{\dagger} = \boldsymbol{H} + \frac{M_{Tx}}{N_p} \boldsymbol{Z}\boldsymbol{X}_p^{\dagger}, \tag{8.265}$$

and we can draw the following conclusions:

- The estimate error increases as the number of transmit antennas increases.
- The estimate error decreases as the pilot sequence length increases.
- The estimate error decreases as SNR increases.
- The LS estimate is unbiased when the noise samples are zero-mean Gaussian.

Interestingly enough, the authors in [66] used information theoretic approach and concluded that rows in training matrix should be orthogonal among each other, which represents just the generalization of Hadamard matrices, namely, Hadamard matrix, which is defined as a square matrix with entries being either 1 or − 1, whose rows are mutually orthogonal. Hadamard matrices can be defined recursively as follows:

$$\boldsymbol{H}_1 = [1], \boldsymbol{H}_2 = \begin{bmatrix} 1 & 1 \\ 1 & -1 \end{bmatrix}, \boldsymbol{H}_{2^k} = \boldsymbol{H}_2 \otimes \boldsymbol{H}_{2^{k-1}}, \tag{8.266}$$

where $\otimes$ denotes the Kronecker product. Therefore, the simplest way to design the pilot matrices satisfying Eq. (8.264), when $M_{\mathrm{Rx}} > M_{\mathrm{Tx}}$, is to start with Hadamard matrix of size $M_{\mathrm{Tx}} \times M_{\mathrm{Tx}}$ and add additional $M_{\mathrm{Rx}} - M_{\mathrm{Tx}}$ rows that are orthogonal to already existing rows in Hadamard matrix.

8.7 Massive MIMO

The massive MIMO is a generalization of multiuser MIMO in which the base station (BS) antenna array has hundreds of antenna elements, supporting tens of single-antenna users [68–75].

8.7.1 Massive MIMO Concepts

The massive MIMO concept is illustrated in Fig. 8.45, in which the number of antenna elements in BS M is much larger (could be orders of magnitude) than the number of active users' devices N. The large number of antenna elements allows highly directive links between the BS and user equipment; thus more energy can be concentrated, and consequently the spectral efficiency and throughput can be improved. Thanks to higher directivity, the required power per link can be reduced, and thus a lower cost power amplifier per antenna element can be used. The use of a large number of antenna elements smoothens the frequency response dependences so that the power control to improve the spectral efficiency can be simplified. To come close to the channel capacity, thanks to a large number of antenna elements, low complexity precoder and decoder can be used.

For uplink transmission N single-antenna users send either data or pilots for channel estimation toward the base station. After decoding the base station can detect data sequences belonging to different users, and this process is illustrated on Fig. 8.46a. Let the transmitted vector originating from N users be denoted by $\boldsymbol{x} \in C^N$, where C is the set of complex numbers.

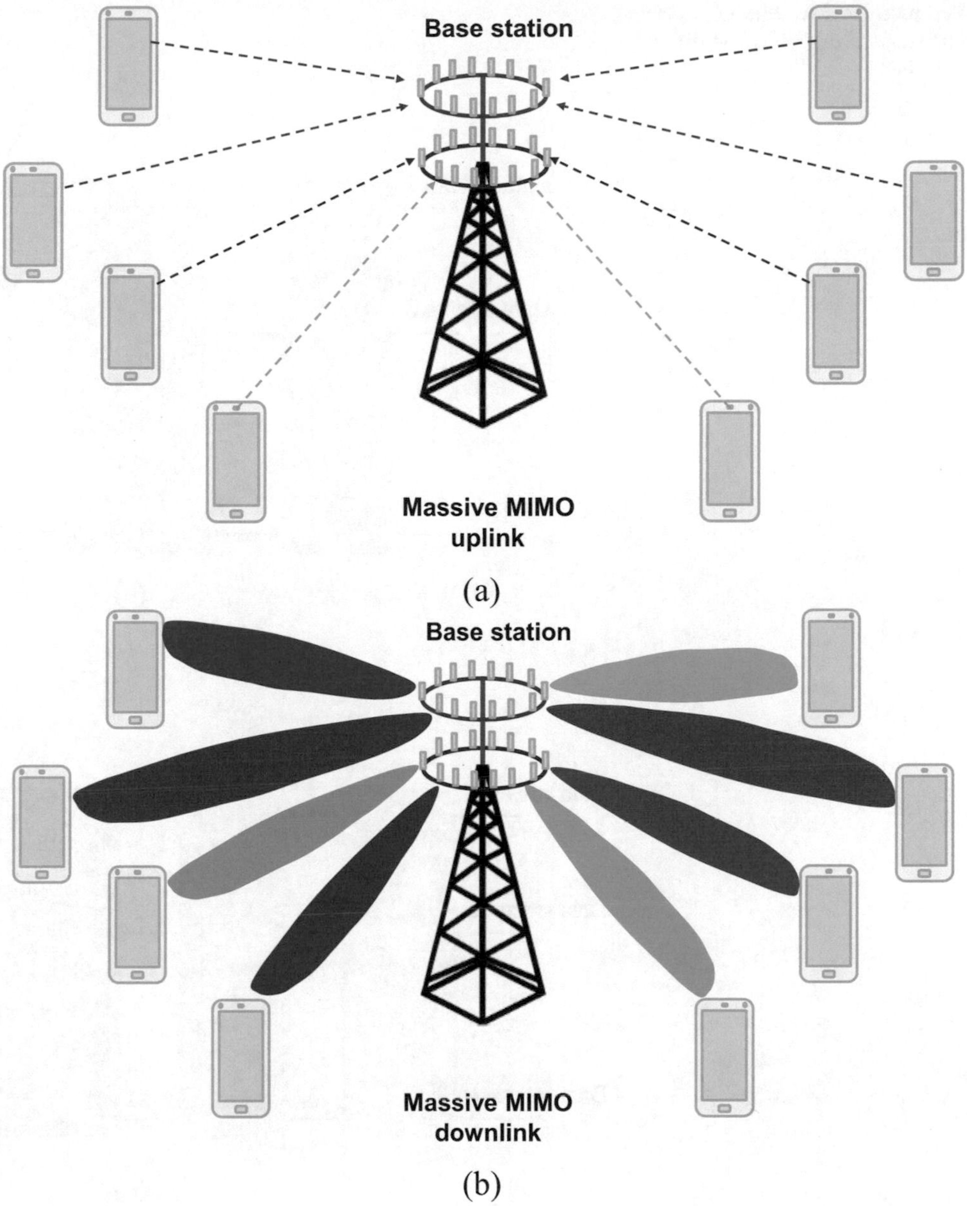

Fig. 8.45 Illustrating massive MIMO concept: (**a**) uplink and (**b**) downlink operations

The received vector at the base station is denoted by $\boldsymbol{y} \in C^M$. Let the channel matrix of size $M \times N$ be denoted by $\boldsymbol{H}$. A simple channel mode for uplink communication will be then:

$$\boldsymbol{y} = \boldsymbol{Hx} + \boldsymbol{i}_{\text{uplink}} + \boldsymbol{n}, \tag{8.267}$$

where $\boldsymbol{n}$ is a noise vector and $\boldsymbol{i}_{\text{uplink}}$ describes the interference from simultaneous transmissions. In the simplest approximation, the elements of the channel matrix can be assumed to zero-mean spatially white.

The downlink channel is used to send the data to different users or to estimate the channel. After the training phase is completed, the channel state information (CSI) is used in the precoder to transmit the data streams for different users as illustrated in Fig. 8.46b. The simplified downlink channel model from the BS to the n-th user can be represented by:

$$\boldsymbol{y}_n = \boldsymbol{h}_n \boldsymbol{x}_n + \boldsymbol{i}_{\text{downlink}} + \boldsymbol{n}, \tag{8.268}$$

where $\boldsymbol{x}_n \in C^M$ is the signal vector transmitted by the BS intended for the n-th user, $\boldsymbol{h}_n$ is the row channel vector between the BS and the n-th user, and $\boldsymbol{y}_n$ is the n-th user received vector. As before, we use $\boldsymbol{n}$ to denote the noise vector and $\boldsymbol{i}_{\text{downlink}}$ to describe the interference coming from other users.

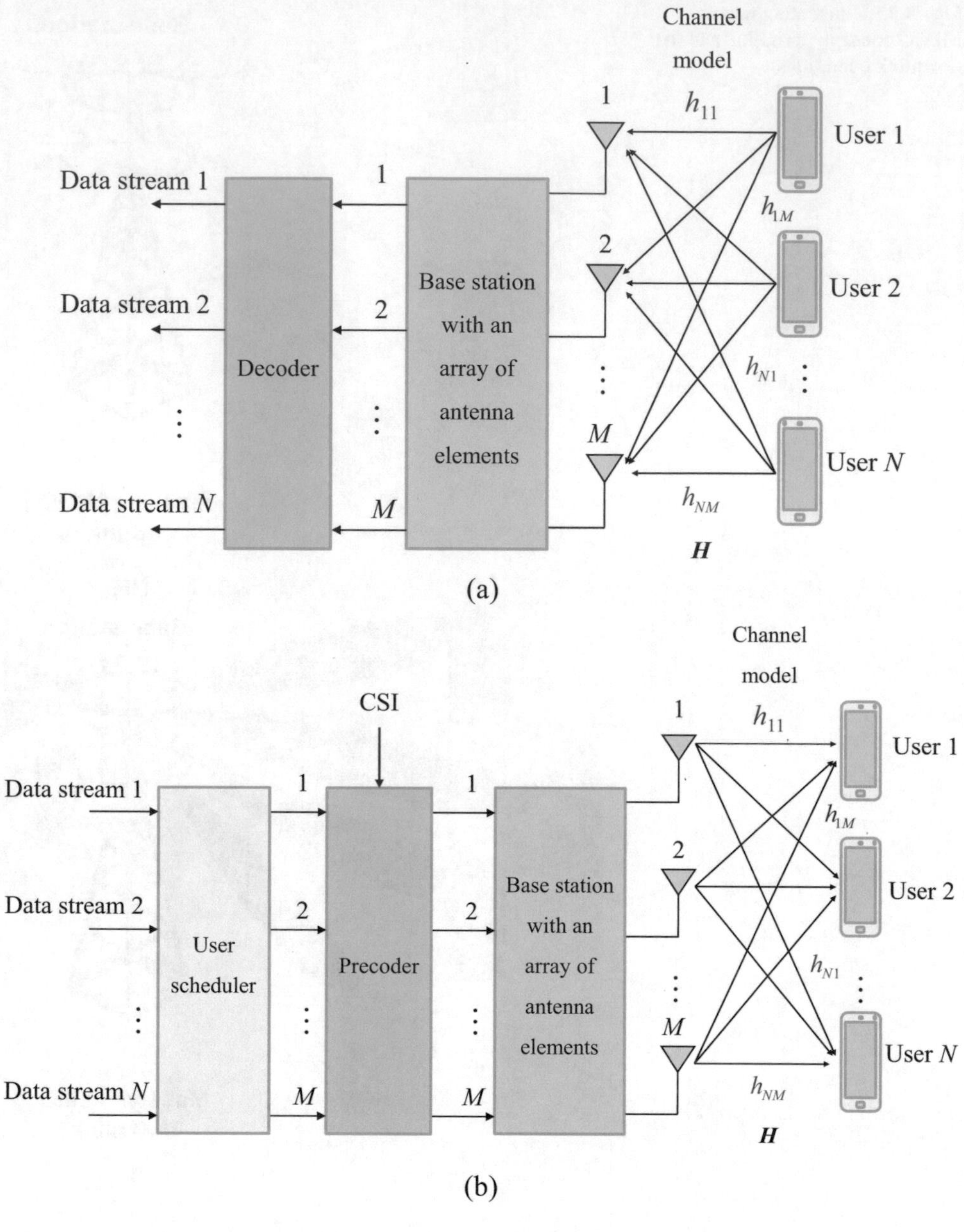

Fig. 8.46 Massive MIMO operations of (**a**) uplink and (**b**) downlink

Even though that the massive MIMO has many promises, there are some challenges. As an illustration, given that the wireless channel coherence time is finite, the number of possible orthogonal pilots is limited so that the pilot sequences must be reused in the neighboring cells resulting in the *pilot contamination* problem, which is illustrated in Fig. 8.47. In this example, two single-antenna users, belonging to neighboring cells, employ the same orthogonal pilots to estimate the channels which results in interference. The pilot contamination affects the *channel estimation*, resulting in imperfect CSI. The BS needs the accurate CSI to take the advantages of the massive MIMO concepts. To focus the spatial radiation patterns to respective users, as illustrated in Fig. 8.45b, the beamforming is used enabled by *precoding* as described in Sect. 8.2.2.

The CSI needs to be accurate enough and not outdated for precoding to be effective in spatial interference cancelation and be able to improve the spectral efficiency and thus the pilot contamination affects the precoding as well. Finally, various *detection techniques* described in Sect. 8.2 require the CSI so that the pilot contamination affects the detection as well. Other challenges for massive MIMO include *hardware impairments* and user scheduling. Massive MIMO employs a large number of antenna elements at the BS, which increase the overall system complexity and cost. To reduce the cost and complexity, lower-complexity algorithms and low-cost hardware should be used. Unfortunately, the low-cost hardware introduces various impairments including the phase noise, I/Q imbalance, and amplifier distortions, to mention a few. When the number of users increases, the multi-access interference grows, and we need to do proper users' scheduling, which is illustrated in Fig. 8.48.

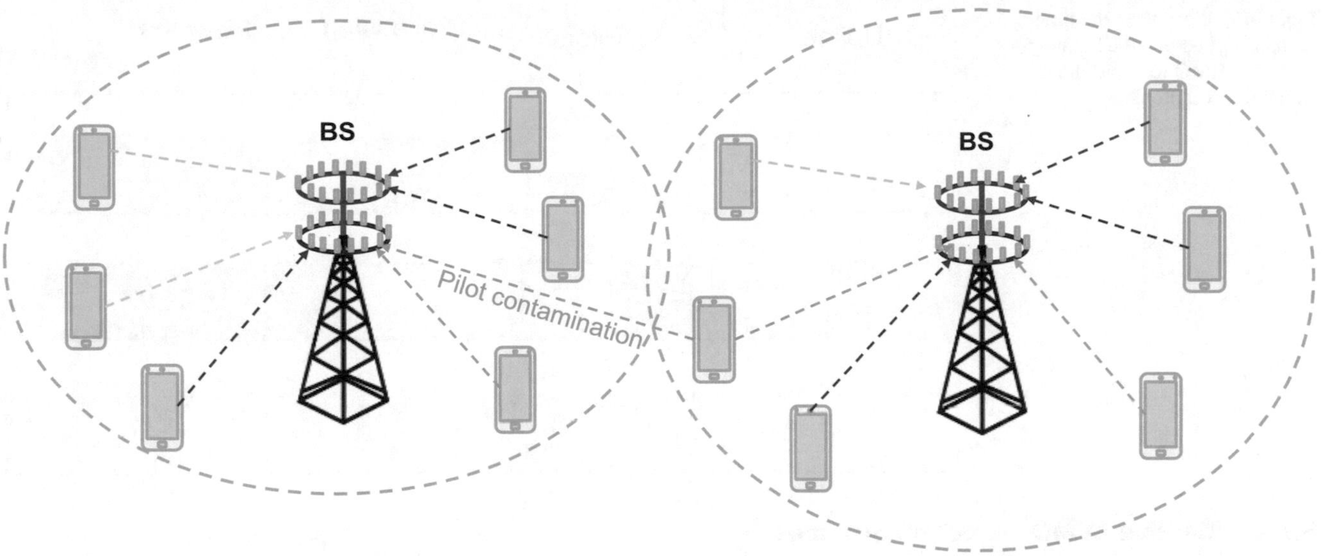

Fig. 8.47 Illustrating the pilot contamination problem

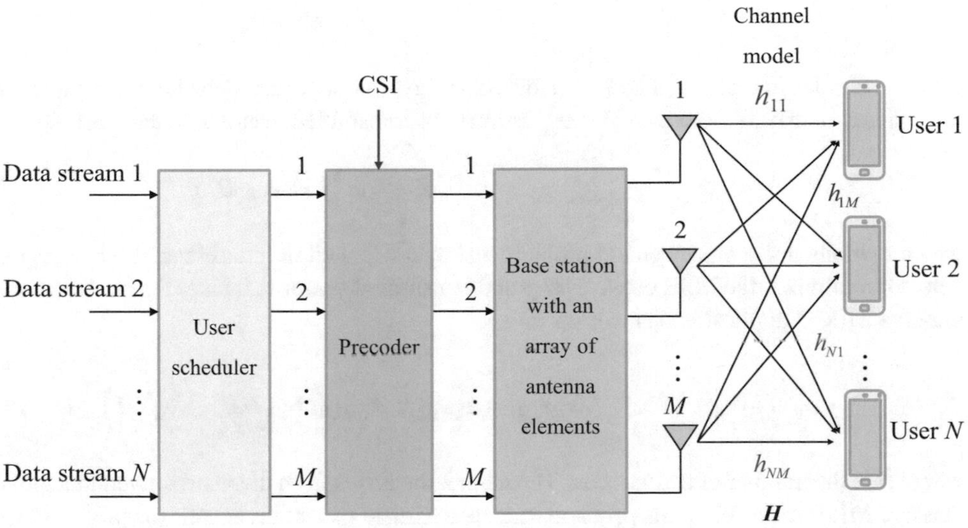

Fig. 8.48 Illustrating massive MIMO user scheduling

To solve for the pilot contamination problem, different approaches have been proposed [74–78]. The channel estimation can be based either on OFDM channel estimation described in Sect. 7.5.5 or MIMO channel estimation described in Sect. 8.6. Alternatively, the combination of these two approaches is relevant too. In particular, the two-dimensional OFDM channel estimation, described in Fig. 7.14, is applicable here. In this scheme, multiple subcarriers within the OFDM symbol and multiple OFDM symbols are used for the channel estimation. Since the number of pilots is limited, they need to be reused in neighboring cells. The simplest approach is to place them in different time and frequency slots so that there is no overlap. For the time-division duplexing (TDD), the authors in [74] proposed to use time-shifted pilot scheme in which cells are divided into several groups and for each group the pilots are placed in non-overlapping time slots. In this scenario, neighboring cells should belong to different groups to avoid the pilot contamination, which is illustrated in Fig. 8.49.

In this example, we assumed that three time slots are used for uplink transmission, three time slots for pilots, one time slot to process pilots and obtain CSI, and four time lots for downlink transmission. Various OFDM- and MIMO-based channel estimation techniques have already been described in respective section; we move our attention to lower-complexity detection schemes suitable for massive MIMO applications.

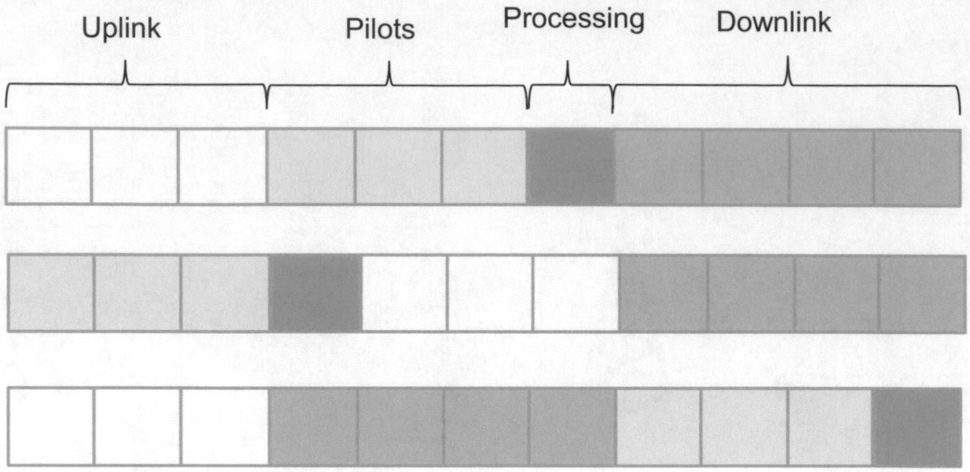

Fig. 8.49 Illustrating the time-domain shifted pilot technique for massive MIMO to avoid the pilot contamination problem

8.7.2 Massive MIMO Detection Schemes

Let us consider the channel model given by Eq. (8.267) in which the noise and interference vectors get combined into a z vector:

$$y = Hx + z, \tag{8.269}$$

which is similar to that given by Eq. (8.62). The zero-forcing linear detector performs the Moore-Penrose pseudo-inverse of the channel matrix H^+ and provides the estimate of transmitted vector x by Eq. (8.142), that is:

$$\hat{x}_{\mathrm{ZF}} = H^+ y = x + H^+ z, \tag{8.270}$$

which exhibits noise enhancement problem and operates well in interference-limited regime. On the other hand, the MMSE detector minimizes the MSE between $H^\dagger y$ and transmitted vector x; it takes the noise into account but has high complexity for massive MIMO applications since:

$$\hat{x}_{\mathrm{MMSE}} = A^\dagger_{\mathrm{MMSE}} y, A_{\mathrm{MMSE}} = \left(HH^\dagger + \frac{N}{SNR} I\right)^{-1} H^\dagger, \tag{8.271}$$

where N is the number of active users. The main issue arises from the matrix inversion above, which is of huge complexity for massive MIMO. To solve this problem, the approximate matrix inversion was proposed in [79].

When the number of users is small compared to the number of antenna elements in the BS ($N \ll M$), the small-scale fading effects dominate, and for sufficiently high SNR, our problem reduces to the matrix inversion of the *Gram matrix* $G = H^\dagger H$. In the *Neumann series* (NS) approach, we decompose the Gram matrix in terms of diagonal matrix D and off-diagonal matrix O, that is, $G = D + O$. We then perform the following NS expansion [80]:

$$G^{-1} = \sum_{k=0}^{\infty} \left(-D^{-1} O\right)^k D^{-1}. \tag{8.272}$$

The diagonal matrix inversion is trivial, and for sufficiently large k, we have that $-D^{-1}O \to 0$ as $k \to \infty$.

In the *Gauss-Seidel (GS) method*, we further decompose the off-diagonal matrix into strictly upper U and strictly lower L triangular matrices, that is, $O=U + L$, and detect the transmitted vector in an iterative fashion [81]:

$$x^{(k)} = (D + L)^{-1} \left(H^\dagger y - U x^{(k-1)}\right); k = 1, 2, \cdots \tag{8.273}$$

In the *successive over-relaxation (SOR) method*, generalize the GS method as follows [82]:

$$\boldsymbol{x}^{(k)} = \left(r^{-1}\boldsymbol{D} + \boldsymbol{L}\right)^{-1}\left[\boldsymbol{H}^{\dagger}\boldsymbol{y} + \left(r^{-1} - 1\right)\boldsymbol{D} - \boldsymbol{U}\boldsymbol{x}^{(k-1)}\right]; k = 1, 2, \cdots \tag{8.274}$$

where the relaxation parameter r, affecting the convergence rate, is selected as $r \in (0,2)$. Clearly, for $r = 1$ the SOR method reduces down to the GS method.

In the *Jacobi method*, we apply the following iterative formula to detect the transmitted signal [83]:

$$\boldsymbol{x}^{(k)} = \boldsymbol{D}^{-1}\left[\boldsymbol{H}^{\dagger}\boldsymbol{y} + (\boldsymbol{D} - \boldsymbol{A})\boldsymbol{x}^{(k-1)}\right]; k = 0, 1, \cdots \tag{8.275}$$

which is applicable when $\boldsymbol{I}\text{-}\boldsymbol{D}^{-1}\boldsymbol{A} \to \boldsymbol{0}$ as $k \to \infty$. We use matrix $\boldsymbol{A}$ to denote the matrix $\boldsymbol{G}\boldsymbol{\Lambda} + N_0\boldsymbol{I}$, where $\boldsymbol{\Lambda}$ is the signal power matrix and N_0 is the PSD of the noise. The starting point is $\widehat{\boldsymbol{x}}^{(0)} = \boldsymbol{D}^{-1}\boldsymbol{H}^{\dagger}\boldsymbol{y}$.

In the *Richardson method*, we apply the following iterative relation [84]:

$$\boldsymbol{x}^{(k)} = \boldsymbol{x}^{(k)} + r\left(\boldsymbol{y} - \boldsymbol{H}\boldsymbol{x}^{(k)}\right); k = 1, 2, \cdots \tag{8.276}$$

where the relaxation parameter is chosen as $r \in (0, 2/\lambda_{\max})$ with λmax being the largest eigenvalue of the symmetric positive definite matrix $\boldsymbol{H}$. $\boldsymbol{x}^{(0)}$ is all zeros vector.

In the *Newton iteration method*, the matrix inversion of the Gram matrix $\boldsymbol{G}$ is obtained by the following recursive relation [85]:

$$\boldsymbol{X}_k^{-1} = \boldsymbol{X}_{k-1}^{-1}\left(2\boldsymbol{I} - \boldsymbol{G}\boldsymbol{X}_{k-1}^{-1}\right), \tag{8.277}$$

which converges quadratically providing that the following is valid $\left\|\boldsymbol{I} - \boldsymbol{G}\boldsymbol{X}_0^{-1}\right\| < 1$.

A large number of antenna elements in massive MIMO increase the number of degrees of freedom, so many elements in the channel matrix are either zero or negligible. This gives an idea to transform the channel matrix $\boldsymbol{H}$ into a sparse matrix $\boldsymbol{H}_s$ with the help of two unitary matrices $\boldsymbol{U}$ and $\boldsymbol{V}$ as follows [86, 87]:

$$\boldsymbol{H} = \boldsymbol{U}\boldsymbol{H}_s\boldsymbol{V}. \tag{8.278}$$

Now by exploiting the properties of this sparse matrix, we can estimate the channel matrix by employing the *compressive sensing* (CS) concepts [87]. In CS we reconstruct the sparse signals $\boldsymbol{s}$ from the compressed measurements $\boldsymbol{y} = \boldsymbol{\Phi}\boldsymbol{s}$, where $\boldsymbol{y} \in C^m$, through a convex programming, where $\boldsymbol{\Phi}$ is the sensing matrix. If the original signal $\boldsymbol{x} \in C^n$, $m < n$, is not sparse, we need to sparsify it by $\boldsymbol{s} = \boldsymbol{\Psi}\boldsymbol{x}$, where $\boldsymbol{\Psi}$ is the sparsifying matrix. Therefore, $\boldsymbol{y}$ and $\boldsymbol{x}$ are connected by $\boldsymbol{y} = \boldsymbol{\Phi}\boldsymbol{\Psi}\boldsymbol{x} = \boldsymbol{A}\boldsymbol{x}$ with $\boldsymbol{A} = \boldsymbol{\Phi}\boldsymbol{\Psi}$ being the measurement matrix. The solution for $\boldsymbol{x}$ will be $\boldsymbol{x} = (\boldsymbol{A}^{\mathrm{T}}\boldsymbol{A})^{-1}\boldsymbol{A}^{\mathrm{T}}\boldsymbol{y}$. Unfortunately, the system of equations is underdetermined, that is, the number of equations is smaller than the number of unknowns. To reconstruct $\boldsymbol{x}$ we can apply the following combinatorial optimization problem:

$$\widehat{\boldsymbol{x}} = \underset{\boldsymbol{x} \in C^m}{\arg\min} \|\boldsymbol{x}\|_0 \text{ subject to } \boldsymbol{A}\boldsymbol{x} = \boldsymbol{y}, \tag{8.279}$$

where $\|\cdot\|_0$ denotes the l_0-norm (the number of nonzero components). Unfortunately, this problem in general is NP-hard. A very popular method to solve this problem is the basis pursuit or l_1-minimization:

$$\widehat{\boldsymbol{x}} = \underset{\boldsymbol{x} \in C^m}{\arg\min} \|\boldsymbol{x}\|_1 \text{ subject to } \boldsymbol{A}\boldsymbol{x} = \boldsymbol{y}, \tag{8.280}$$

Because the l_1-norm is convex, the efficient methods from convex optimization are applicable. For instance, the greedy methods such as the orthogonal matching pursuit (OMP) can be employed [88]. This optimization is applicable even in the presence of noise, that is, when $\boldsymbol{y} = \boldsymbol{A}\boldsymbol{x} + \boldsymbol{z}$, which is similar to the massive MIMO channel model. As an illustration, a multipath matching pursuit (MMP) detector, employing the sphere decoding (SD) detector, is used in ref. [89]. The SD search is performed by [89, 90]:

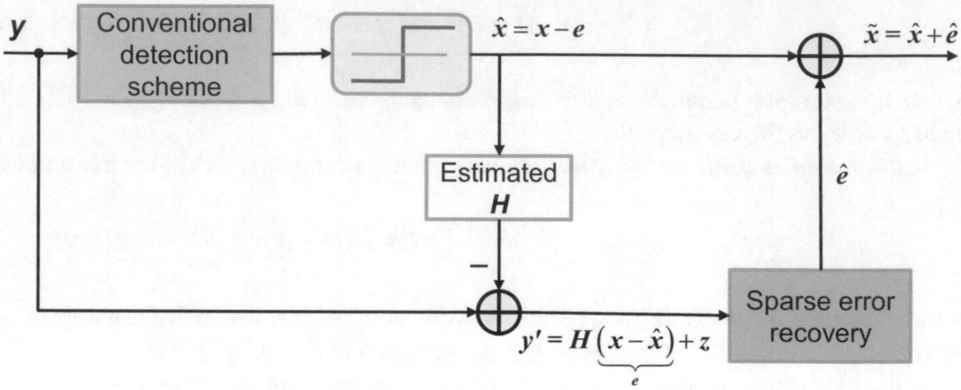

Fig. 8.50 Illustrating the post-detection sparse error recovery (PDSR) scheme

$$\widehat{\boldsymbol{x}} = \arg\min_{x \in C^m} \left\{ \|\boldsymbol{y} - \boldsymbol{R}\boldsymbol{x}\|^2 < d^2 \right\}, \tag{8.281}$$

where $\boldsymbol{R}$ is the upper triangular matrix in QR factorization of the channel matrix, that is, $\boldsymbol{H} = \boldsymbol{Q}\boldsymbol{R}$. The distance d is properly chosen to get a near-ML solution. In [91] the CS is used to improve the performance of the conventional receiver, with the corresponding post-detection sparse error recovery (PDSR) scheme provided in Fig. 8.50. This scheme assumes that the CSI is perfect. We first perform the conventional massive MIMO detection to get $\widehat{\boldsymbol{x}}$. Since the error vector $\boldsymbol{e} = \boldsymbol{x} - \widehat{\boldsymbol{x}}$ is sparse, we can apply the CS techniques, such as MMP or OMP, to estimate it. The estimate error is added to the detected vector to get the improved detected vector $\widetilde{\boldsymbol{x}} = \widehat{\boldsymbol{x}} + \boldsymbol{e}$.

Various multiuser detection techniques described in Chap. 10 (see Sect. 10.3) are also applicable in massive MIMO applications. The decision-feedback and BLAST receivers described in Sect. 8.2.7.3 are applicable as well. Spatial interference cancelation scheme described in Sect. 8.2.7.4 can also be used in massive MIMO. The successive interference cancelation (SIC) Schemes [71, 92] belong to this category.

The performance of linear detectors can be improved by employing the *lattice reduction-based (LRB) algorithms* [71, 93, 94]. The complex-valued lattice of rank m in n-dimensional complex space C^n can be defined as follows:

$$\mathcal{L} = \{\boldsymbol{y} | \boldsymbol{y} = \boldsymbol{H}\boldsymbol{x}\},\ \boldsymbol{x} = [x_1 \cdots x_n]^T, \tag{8.282}$$

where $\boldsymbol{H}$ is the basis of the lattice (with columns representing the basis vectors). In other words, $\mathcal{L}$ is the set of received signal points not perturbed by the noise. The column vectors in the channel matrix are non-orthogonal. The goal of LRB algorithms is to find an improved basis and then perform detection in this improved basis. By multiplying the channel matrix with a unimodular matrix $\boldsymbol{T}$ (a square integer matrix whose determinant is either +1 or −1), we obtain the improved basis, that is, $\widetilde{\boldsymbol{H}} = \boldsymbol{T}\boldsymbol{H}$. The decision regions in modified basis are the parallelograms around the lattice points, while the Voronoi regions in original space could be of highly irregular shape. We can modify the MIMO channel model given by Eq. (8.269) as follows:

$$\boldsymbol{y} = \underbrace{\boldsymbol{H}\boldsymbol{T}}_{\widetilde{H}}\underbrace{\boldsymbol{T}^{-1}\boldsymbol{x}}_{\widetilde{x}} + z = \widetilde{\boldsymbol{H}}\widetilde{\boldsymbol{x}} + z, \tag{8.283}$$

and then perform the detection of $\widetilde{\boldsymbol{x}}$. As an illustration, the LRB zero-forcing detector will be:

$$\widehat{\boldsymbol{x}}_{\mathrm{ZF}} = \widetilde{\boldsymbol{H}}^{+}\boldsymbol{y} = \widetilde{\boldsymbol{x}} + \widetilde{\boldsymbol{H}}^{+}z. \tag{8.284}$$

Various LRB algorithms are reviewed in [93].

Various *machine learning algorithms* (MLAs) can be also used in massive MIMO communications, in particular to simplify the optimizations. As an illustration, MLAs have been already studied for use in channel estimation, detection, MIMO beamforming, load balancing, and optimization of available spectrum [71, 72, 95–98].

8.8 Concluding Remarks

The purpose of this chapter has been to describe various diversity and MIMO techniques capable of improving SISO system performance as well as the system throughput. The various diversity and MIMO techniques that have been described can be used not only in wireless MIMO communications but also in free-space optical and fiber-optic communications with coherent optical detection.

Various diversity schemes including polarization diversity, spatial diversity, and frequency diversity schemes are summarized in Sect. 8.1.1. The following receiver diversity schemes have been described in Sect. 8.1.2: selection combining, threshold combining, maximum-ratio combining, and equal-gain combining schemes. In Sect. 8.1.3, various transmit diversity schemes have been described, depending on the availability of channel state information on the transmitter side.

Most of the chapter has been devoted to wireless and optical MIMO techniques. After description of various wireless and optical MIMO models in Sect. 8.2.1, we describe the parallel decomposition of MIMO channels in Sect. 8.2.2, followed by space-time coding (STC) principles in Sect. 8.2.3. The maximum likelihood (ML) decoding for STC is described together with various design criteria. The following relevant classes of STC have been described in Sect. 8.2.4: Alamouti code, orthogonal designs, linear space-time block codes, and space-time trellis codes. The corresponding STC decoding algorithms have been described as well in the same section.

After that the attention has moved to SDM principles in Sect. 8.2.5, where various BLAST encoding architectures (Sect. 8.2.5.1) as well as multi-group space-time coded modulation (Sect. 8.2.5.2) have been introduced. The following classes of linear and feedback MIMO receiver for uncoded signals have been subsequently described in Sect. 8.2.6: zero-forcing, linear minimum MSE, and decision-feedback receivers. The next topic in the chapter has been related to the suboptimum MIMO receivers for coded signals (see Sect. 8.2.7). Within this topic, the linear zero-forcing (ZF) and linear MMSE receivers (interfaces) have been described in the context of STC. Within decision-feedback and BLAST receivers, various H/V-BLAST architectures, including ZF and MMSE ones, have been introduced. The topic on suboptimum receivers has been concluded with diagonal (D)-BLAST and iterative receiver interfaces.

The section on iterative MIMO receivers (Sect. 8.3) starts with a brief introduction of concept of factor graphs (Sect. 8.3.1), followed by the description of factor graphs for MIMO channel and channels with memory (Sect. 8.3.2). And then the sum-product algorithm (SPA) operating on factor graphs has been described in Sect. 8.3.3, followed by description of SPA for channels with memory (Sect. 8.3.4). The following iterative MIMO receivers for uncoded signals have been described in Sect. 8.3.5: ZF, MMSE, ZF H/V-BLAST, and LMMSE H/V-BLAST receivers. After a brief description of factor graphs for linear block and trellis codes in Sect. 8.3.6, several iterative MIMO receivers for space-time coded signals have been described in Sect. 8.3.7.

The section on broadband MIMO (Sect. 8.4) has been devoted to the description on how frequency selectivity can be used as an additional degree of freedom, followed by a description of MIMO-OFDM in Sect. 8.4.1 and space-frequency block-coding principles in Sect. 8.4.2.

The focus has been then moved to MIMO channel capacity calculations in Sect. 8.5 for various MIMO channel models including deterministic (Sect. 8.5.1), ergodic and non-ergodic random channels, as well as correlated channel models, described in Sect. 8.5.2. The concepts of ergodic and outage channel capacity have been introduced. The section on MIMO channel capacity ends with MIMO-OFDM channel capacity description (Sect. 8.5.2.4).

In MIMO channel estimation section (Sect. 8.6), the following MIMO channel estimation techniques have been described: ML MIMO channel estimation (Sect. 8.6.1), least squares (LS) MIMO channel estimation (Sect. 8.6.2), and linear minimum MSE MIMO channel estimation (Sect. 8.6.3). The massive MIMO concepts have been described in Sect. 8.7.

The set of problems is provided in incoming section to help reader gain deeper understanding of material described in this chapter.

8.9 Problems

8.1. A four-dimensional (4D) signal constellation is described by the following set of points $\{\pm a, \pm a, \pm a, \pm a\}$ ($a > 0$). Different signal constellation points are used with the same probability. Determine the average symbol energy. Choose the appropriate basis functions $\{\Phi_1(t), \Phi_2(t), \Phi_3(t), \Phi_4(t)\}$ and express the signal constellation points in terms of basis functions. For AWGN channel, determine the average symbol error probability as a function of the average energy (E_{av}). Evaluate now the average symbol error probability in the presence of Rayleigh fading. Finally, evaluate the symbol error probability in the presence of Rayleigh fading when the *maximum-ratio combining diversity* is used.

8.2. In this problem, we study the receiver diversity based on equal gain combining (EGC). Let us assume that either BPSK or QPSK is used to transmit messages over a generalized fading wireless channel. According to the generalized fading channel model, the signal envelope at the i-th receiver branch ($i = 1,2,\ldots,L$) follows α-μ distribution with probability density function (PDF) given by:

$$f_{R_i}(r_i) = \frac{\alpha_i \mu_i^{\mu_i} r_i^{\alpha_i \mu_i - 1}}{\widehat{r}_i^{\alpha_i \mu_i} \Gamma(\mu_i)} \exp\left(\mu_i \frac{r_i^{\alpha_i}}{\widehat{r}_i^{\alpha_i}}\right),$$

where $\alpha_i > 0$ is the parameter of nonlinearity, $\Gamma(.)$ is the Gamma function, and $\mu_i > 0$ is the inverse of the normalized variance of $r_i^{\alpha_i}$, namely, $\mu_i = \mathrm{E}^2\{r_i^{\alpha_i}\}/\mathrm{Var}\{r_i^{\alpha_i}\}$, where E{.} is the expectation operator and $\widehat{r}_i$ is a α_i-root mean value $\widehat{r}_i = \sqrt[\alpha_i]{\mathrm{E}\{r_i^{\alpha_i}\}}$. If the phase estimation is done from unmodulated carrier by using phase-locked loop (PLL) and if only the Gaussian noise is present in the phase-locked loop circuit, then the PDF of this phase error is:

$p_{\varphi_i}(\varphi_i) = \frac{1}{2\pi} \frac{\exp(\varsigma_i \cos(\varphi_i))}{\mathrm{I}_0(\varsigma_i)}, \ -\pi < \varphi_i \leq \pi$, where $I_0(x)$ is the modified Bessel function of the first kind and zero order for the argument x, ζ_i is the SNR in the PLL circuit at the i-th receiver branch.

Write the expressions for the average bit error probability of uncoded BPSK and QPSK signal detections in the presence of phase error and α-μ fading assuming that EGC receiver is used.

8.3. Based on parallel decomposition of MIMO optical channel, explain how MIMO-OFDM can be used in multi-Tb optical transport.

8.4. In the section on space-time coding for MIMO communications, we described the Alamouti scheme employing two transmitters and two receivers. Describe how this Alamouti scheme can be generalized to a number of receivers larger than two. Provide the corresponding combiner rule. Prove it by mathematical induction.

8.5. Discuss the advantages and disadvantages of Alamouti-type polarization-time (PT) coding scheme against that of conventional PDM-OFDM scheme.

8.6. Provide the star-8-QAM trellis space-time coding scheme with two transmitters and diversity of $2M_{\mathrm{Rx}}$, where M_{Rx} is the number of transmitters.

8.7. The space-time code for a number of transmit antennas $M_{\mathrm{Tx}} = 4$, number of receive antennas $M_{\mathrm{Rx}} = 1$, and number of channel's uses $N = 4$ is described by the following signaling matrix:

$$\boldsymbol{X} = \begin{bmatrix} x_1 & -x_2^* & -x_3^* & 0 \\ x_2 & x_1^* & 0 & x_3^* \\ x_3 & 0 & x_1^* & -x_2^* \\ 0 & -x_3^* & x_2^* & x_1^* \end{bmatrix}$$

(a) Can this space-time block code be used to design the receiver that will result in spatial interference cancellation? If so describe the corresponding receiver configuration. What is the code rate of this space-time code?

(b) Represent the signaling matrix $\boldsymbol{X}$ in the same fashion as for linear space-time (LST) codes. If the channel matrix is given by $\boldsymbol{H} = [h_1\ h_2\ h_3\ h_4]$ using detection approach for LST, determine the equations for estimated symbols $\widehat{x}_i$.

8.8. Figure 8.P8 depicts a 2 × 2 polarization-time coding MIMO scheme, which employs dual-polarization transmitter and receiver antennas. Due to multipath effect, the initial orthogonality of polarization states is no longer preserved on a receiver side, and we can use the channel coefficients as shown in Fig. 8.P8 to describe this depolarization effect. Show that Alamouti 2 × 2 scheme can be used to deal with depolarization effect. Determine the array, diversity, and multiplexing gains of this scheme. How would you determine the channel capacity of this scheme? Consider now a MIMO scheme employing two dual-polarization Tx antennas and two dual-polarization Rx antennas. How would you approach this system to deal simultaneously with depolarization and multipath effects? How would you determine the channel capacity of this scheme?

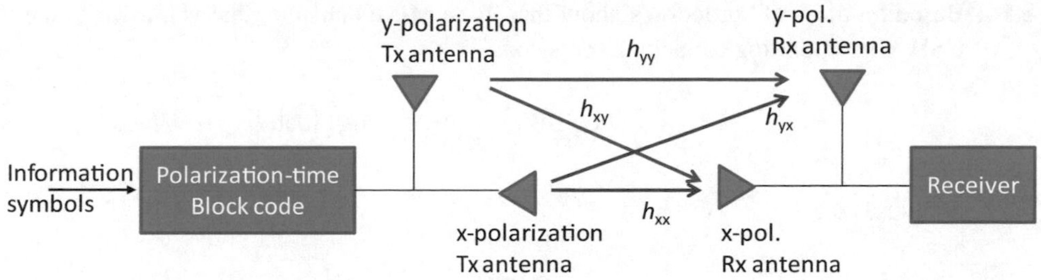

Fig. 8.P8 A 2 × 2 polarization-time coding MIMO scheme

8.9. This problem is related to the derivation of $\boldsymbol{A}_{\text{MMSE}}$ matrix given by Eq. (8.144). Let us define the MSE by:

$$MSE(\boldsymbol{A}) = \left\langle \|\boldsymbol{A}\boldsymbol{y} - \boldsymbol{x}\|^2 \right\rangle = \left\langle \mathrm{Tr}\left[((\boldsymbol{A}\boldsymbol{H} - \boldsymbol{I}_{M_{\text{Tx}}})\boldsymbol{x} + \boldsymbol{A}\boldsymbol{z})((\boldsymbol{A}\boldsymbol{H} - \boldsymbol{I}_{M_{\text{Tx}}})\boldsymbol{x} + \boldsymbol{A}\boldsymbol{z})^\dagger \right] \right\rangle.$$

By assuming that components of $\boldsymbol{x}$ are i.i.d. zero-mean with the second moment corresponding to the average energy, determine the optimum (in MMSE sense) linear transformation $\boldsymbol{A}$. Derive the corresponding PEP. Discuss its asymptotic behavior for (a) finite M_T, M_R- > ∞ and (b) M_T- > ∞, M_R- > ∞, while M_T/M_R- > a.

8.10. In this problem we deal with the derivation of matrices $\boldsymbol{A}$ and $\boldsymbol{B}$ in ZF V-BLAST. The MSE in the absence of noise is given by $MSE(\boldsymbol{A}, \boldsymbol{B}) = \left\langle \left\| \widetilde{\boldsymbol{Y}} - \boldsymbol{X} \right\|^2 \right\rangle = \left\langle \left\| \boldsymbol{A}\boldsymbol{H}\boldsymbol{X} - \boldsymbol{B}\widehat{\boldsymbol{X}} - \boldsymbol{X} \right\|^2 \right\rangle$. By using the high SNR approximation and QR factorization of channel matrix, determine the matrices $\boldsymbol{A}$ and $\boldsymbol{B}$ so that the MSE is minimized. Once you determine these matrices, demonstrate that mode coupling is indeed canceled out. Fully describe the detection procedure.

8.11. Here we deal with the derivation of matrices $\boldsymbol{A}$ and $\boldsymbol{B}$ in MMSE V-BLAST. The MSE in the presence of noise is given by: $MSE(\boldsymbol{A}, \boldsymbol{B}) = \left\langle \left\| \widetilde{\boldsymbol{Y}} - \boldsymbol{X} \right\|^2 \right\rangle = \left\langle \|\boldsymbol{A}\boldsymbol{Y} - \boldsymbol{B}\boldsymbol{X} - \boldsymbol{X}\|^2 \right\rangle = \left\langle \|(\boldsymbol{A}\boldsymbol{H} - \boldsymbol{L} - \boldsymbol{I}_{M_{\text{Tx}}})\boldsymbol{X} + \boldsymbol{A}\boldsymbol{Z}\|^2 \right\rangle$. By using the Cholesky factorization of $\boldsymbol{H}^\dagger\boldsymbol{H} + SNR^{-1}\boldsymbol{I}_{M_{\text{Tx}}} = \boldsymbol{S}^\dagger\boldsymbol{S}$, determine the matrices $\boldsymbol{A}$ and $\boldsymbol{B}$ so that the MSE is minimized. What is the remaining MSE error? Once you determine these matrices, demonstrate that mode coupling is indeed canceled out.

8.12. Let us study a MIMO system with the following channel matrix:"

$$\boldsymbol{H} = \begin{bmatrix} 0.1 & 0.2 & 0.3 \\ 0.3 & 0.4 & 0.5 \\ 0.7 & 0.6 & 0.5 \end{bmatrix}.$$

(a) Write the channel matrix in terms of its singular value decomposition (SVD) $\boldsymbol{H} = \boldsymbol{U}\boldsymbol{\Sigma}\boldsymbol{V}^\dagger$.

(b) Based on SVD develop an equivalent MIMO systems composed of $R_{\boldsymbol{H}} = \text{rank}(\boldsymbol{H})$ independent channels. After determining $R_{\boldsymbol{H}}$, devise pre-coding and receiver shaping matrices suitable to transform the original system into an equivalent system.

(c) Find optimum power allocation P_i (i = 1,2,3) across the $R_{\boldsymbol{H}}$ channels found in (b), and find the corresponding total capacity of the equivalent system assuming SNR of 22 dB and a system bandwidth of 1 MHz.

(d) When the channel is not known to the transmitter, the equal power is allocated to each antenna. Compare the corresponding capacity part (c) with that for equal power allocation.

8.13. Consider the channel with the following channel matrix:

$$\boldsymbol{H} = \begin{bmatrix} 0.1 & 0.1 & 0.4 \\ 0.2 & 0.2 & 0.5 \\ 0.6 & 0.3 & 0.7 \end{bmatrix}.$$

Under assumption that ρ = 12 dB, determine the output SNR when beamforming is used over the channel with equal weights on each transmit antenna and optimal weights at the receiver side. Compare the SNR with the corresponding SNR under beamforming when optimum weights are applied at both sides.

8.14. Based on the SVD properties show that, for a MIMO channel that is known to the transmitter and receiver side (full-CSI), corresponding capacity expression:

$$C/BW = \max_{\boldsymbol{R}_x:\mathrm{Tr}(\boldsymbol{R}_x)=\rho} \log_2\left[\det\left(\boldsymbol{I}_{M_{Rx}} + \boldsymbol{H}\boldsymbol{R}_x\boldsymbol{H}^\dagger\right)\right],$$

simplifies to:

$$C/BW = \max_{\rho_i:\sum_i \rho_i \le \rho} \sum_i \log_2\left(1 + \sigma_i^2 \rho_i\right),$$

for singular values $\{\sigma_i\}$ and SNR values ρ_i.

8.15. Let us consider 16 × 16, 64 × 64, and 256 × 256 massive MIMO systems for which entries of the channel matrix are generated using the simulator developed in Problem 13 of Chap. 2 for Doppler frequency of 20 Hz. Estimate the channel by using both LS and LMMSE channel estimators. Evaluate BER performance of Neumann series, Gauss-Seidel, successive over-relaxation, Jacobi, and Richardson detection methods for different SNRs. Which detection scheme performs the best? Discuss the complexity of this scheme.

8.16. Repeat Problem 15 but now using the lattice reduction-based approach applied to ZF and MMSE detectors. Evaluate the BER performance against different schemes discussed in Problem 16.

8.17. Here we study the efficiency of V-BLAST scheme in CO-OFDM systems. The V-BLAST scheme is described in Sect. 8.2.7.3. Design a CO-OFDM system, capable of 100 Gb(s) serial optical transmission over 10,000 km of SMF. Perform Monte Carlo simulations in linear regime, by taking the following effects into account: chromatic dispersion, the first-order PMD, and laser linewidth of 100 kHz (for both transmitter and local lasers). Use the typical fiber parameters given in previous chapters. To compensate for PDM and chromatic dispersion, use the V-BLAST approach. Compare this scheme with an equivalent PDM scheme, based on LS channel estimation, of the same aggregate data rate. Compare and discuss the results. What are the advantages and disadvantages of V-BLAST detection scheme with respect to PDM?

8.18. Repeat Problem 17, but now in the nonlinear regime, by solving the *generalized nonlinear Schrödinger equation* (GNLSE) for both polarizations. Study the BER performance as the function of total transmission distance.

8.19. In this problem we are interested in improving the BER performance and nonlinearity tolerance of V-BLAST-based and conventional PDM CO-OFDM systems by using MMSE channel estimation. Plot BER vs. total transmission distance for several cases: (i) V-BLAST-LS estimation, (ii) conventional PDM-LS estimation, and (iii) V-BLAST-MMSE estimation, conventional PDM-MMSE estimation. For MMSE estimation use the first- and second-order interpolation. Discuss results.

8.20. In this problem we are concerned with 1 Tb optical transport over 1500 km of few-mode fiber (FMF) by using mode-multiplexed PDM CO-OFDM. For modeling of FMF, use the model described in Sect. 8.2.1 of this chapter (see Fig. 8.12). Design a CO-OFDM system with aggregate data rate exciding 1 Tb by employing two spatial and two polarization modes. Use five orthogonal OFDM bands for simulations, with 200 Gb in aggregate data rate per band. Perform Monte Carlo simulations for the following cases of interest: (i) V-BLAST-LS estimation, (ii) conventional PDM-LS estimation, and (iii) V-BLAST-MMSE estimation, conventional PDM-MMSE estimation. For MMSE estimation use the first- and second-order interpolation. Discuss results.

8.21. For FSO channels for IM/DD, the repetition MIMO outperforms any space-time coding scheme. Explain this counterintuitive claim.

8.22. Let us consider 3 × 3 FSO MIMO system, in which three independent data streams are transmitted by sufficiently separated FSO transmitters. Assume that each FSO transmitter illuminates both FSO receivers. Assume that coherent detection is used and that the simulation parameters are the same as in Problem 1 of Chap. 2.

(a) For 3 × 3 FSO MIMO system with either zero-forcing (ZF) or minimum MSE (MMSE) receiver, for the refractive structure parameter of $C_n^2 = 10^{-15}$, perform Monte Carlo simulations for $M = 2$, 4, and 16. Plot BERs versus optical SNR per information bit. Determine the coding gains at BER of 10^{-5}. Assume that channel matrix $\boldsymbol{H}$ is available on the receiver side.

(b) Repeat (a) but now for different values of the refractive structure parameter. At BER of 10^{-6}, for $M = 2$, 4, 16, determine improvements when MMSE receiver is used compared to the ZF receiver case.

References

1. A. Goldsmith, *Wireless Communications* (Cambridge University Press, Cambridge, 2005)
2. E. Biglieri, *Coding for Wireless Channels* (Springer, New York, 2005)
3. D. Tse, P. Viswanath, *Fundamentals of Wireless Communication* (Cambridge University Press, Cambridge, 2005)
4. L. Kazovsky, S. Benedetto, A. Willner, *Optical Fiber Communication Systems* (Artech House, Boston, 1996)
5. I. Djordjevic, W. Ryan, B. Vasic, *Coding for Optical Channels* (Springer, New York, Mar. 2010)
6. I.B. Djordjevic, M. Cvijetic, C. Lin, Multidimensional signaling and coding enabling multi-Tb/s optical transport and networking. IEEE Sig. Proc. Mag. **31**(2), 104–117 (Mar. 2014)
7. M. Cvijetic, I.B. Djordjevic, *Advanced Optical Communications and Networks* (Artech House, Norwood, Jan. 2013)
8. J.W. Craig, A new, simple and exact result for calculating the probability of error for two-dimensional signal constellations, in *Proceedings of the IEEE Military Communications Conference 1991 (MILCOM"91)*, McLean, VA, vol 2, pp. 571–575, 4–7 Nov. 1991
9. G.L. Stüber, *Principles of Mobile Communication*, 3rd edn. (Springer, New York, 2012)
10. S.M. Alamouti, A simple transmit diversity technique for wireless communications. IEEE J. Sel. Areas Commun. **16**(8), 1451–1458 (1998)
11. I.B. Djordjevic, L. Xu, T. Wang, PMD compensation in coded-modulation schemes with coherent detection using Alamouti-type polarization-time coding. Opt. Express **16**(18), 14163–14172 (5 Sep. 2008)
12. V. Tarokh, N. Seshadri, A.R. Calderbank, Space–time codes for high data rate wireless communication: performance criterion and code construction. IEEE Trans. Inf. Theory **44**(2), 744–765 (Mar. 1998)
13. H. Bölcskei, D. Gesbert, A.J. Paulraj, On the capacity of wireless systems employing OFDM-based spatial multiplexing. IEEE Trans. Commun. **50**, 225–234 (Feb. 2002)
14. A.R. Shah, R.C.J. Hsu, A.H. Sayed, B. Jalali, Coherent optical MIMO (COMIMO). J. Lightwave Technol. **23**(8), 2410–2419 (Aug. 2005)
15. R.C.J. Hsu, A. Tarighat, A. Shah, A.H. Sayed, B. Jalali, Capacity enhancement in coherent optical MIMO (COMIMO) multimode fiber links. J. Lightwave Technol. **23**(8), 2410–2419 (Aug. 2005)
16. A. Tarighat, R.C.J. Hsu, A.H. Sayed, B. Jalali, Fundamentals and challenges of optical multiple-input multiple output multimode fiber links. IEEE Commuun. Mag. **45**, 57–63 (May 2007)
17. N.W. Bikhazi, M.A. Jensen, A.L. Anderson, MIMO signaling over the MMF optical broadcast channel with square-law detection. IEEE Trans. Commun. **57**(3), 614–617 (Mar. 2009)
18. A. Agmon, M. Nazarathy, Broadcast MIMO over multimode optical interconnects by modal beamforming. Opt. Express **15**(20), 13123–13128 (26 Sept. 2007)
19. E. Alon, V. Stojanovic, J.M. Kahn, S.P. Boyd, M. Horowitz, Equalization of modal dispersion in multimode fibers using spatial light modulators, in *Proceedings of the IEEE Global Telecommunication Conference* 2004, Dallas, TX, Nov. 29–Dec. 3, 2004
20. R.A. Panicker, J.M. Kahn, S.P. Boyd, Compensation of multimode fiber dispersion using adaptive optics via convex optimization. J. Lightwave Technol. **26**(10), 1295–1303 (15 May 2008)
21. S. Fan, J.M. Kahn, Principal modes in multi-mode waveguides. Opt. Lett. **30**(2), 135–137 (15 Jan. 2005)
22. H. Kogelnik, R.M. Jopson, L.E. Nelson, Polarization-mode dispersion, in *Optical Fiber Telecommunications IVB: Systems and Impairments*, ed. by I. Kaminow, T. Li, (Academic, San Diego, 2002)
23. K.-P. Ho, J.M. Kahn, Statistics of group delays in multimode fiber with strong mode coupling. J. Lightwave Technol. **29**(21), 3119–3128 (2011)
24. Zou, D., Lin, C., Djordjevic, I.B., LDPC-coded mode-multiplexed CO-OFDM over 1000 km of few-mode fiber, in *Proceedings of CLEO 2012*, Paper no. CF3I.3, San Jose, CA, 6–11 May 2012
25. I.B. Djordjevic, S. Denic, J. Anguita, B. Vasic, M.A. Neifeld, LDPC-coded MIMO optical communication over the atmospheric turbulence channel. IEEE/OSA J. Lightwave Technol. **26**(5), 478–487 (1 Mar. 2008)
26. E. Biglieri, R. Calderbank, A. Constantinides, A. Goldsmith, A. Paulraj, H.V. Poor, *MIMO Wireless Communications* (Cambridge University Press, Cambridge, 2007)
27. J.G. Proakis, *Digital Communications*, 5th edn. (McGraw Hill, New York, 2007)
28. I.B. Djordjevic, Spatial-domain-based hybrid multidimensional coded-modulation schemes enabling multi-Tb/s optical transport. IEEE/OSA J. Lightwave Technol. **30**(14), 2315–2328 (15 July 2012)
29. G.J. Foschini, Layered space-time architecture for wireless communication in a fading environment when using multi-element antennas. Bell Labs Tech. J. **1**, 41–59 (1996)
30. P.W. Wolniansky, G.J. Foschini, G.D. Golden, R.A. Valenzuela, V-BLAST: an architecture for realizing very high data rates over the rich-scattering wireless channel, in *Proceedings of 1998 URSI International Symposium on Signals, Systems, and Electronics (ISSSE 98)*, Pisa, pp. 295–300, 29 Sept.–2 Oct. 1998
31. X. Li, H. Huang, G.J. Foschini, R.A. Valenzuela, Effects of iterative detection and decoding on the performance of BLAST, in *Proceedings of the IEEE Global Telecommunications Conference 2000. (GLOBECOM '00)*, vol 2, San Francisco, CA, pp. 1061–1066, 2000
32. G.J. Foschini, D. Chizhik, M.J. Gans, C. Papadias, R.A. Valenzuela, Analysis and performance of some basic space-time architectures. IEEE J. Sel. Areas Commun. **21**(3), 303–320 (Apr. 2003)
33. S. Loyka, F. Gagnon, Performance analysis of the V-BLAST algorithm: An analytical approach. IEEE Trans. Wirel. Commun. **3**(4), 1326–1337 (July 2004)
34. V. Tarokh, A. Naguib, N. Seshadri, A.R. Calderbank, Combined array processing and space-time coding. IEEE Trans. Inf. Theory **45**(4), 1121–1128 (May 1999)
35. R.A. Horn, C.R. Johnson, *Matrix Analysis* (Cambridge University Press, New York, 1988)
36. A.L. Moustakas, H.U. Baranger, L. Balents, A.M. Sengupta, S.H. Simon, Communication through a diffusive medium: coherence and capacity. Science **287**, 287–290 (14 Jan. 2000)
37. S. Da-Shan, G.J. Foschini, M.J. Gans, J.M. Kahn, Fading correlation and its effect on the capacity of multielement antenna systems. IEEE Trans. Commun. **48**(3), 502–513 (Mar. 2000)

38. A. Wittneben, Basestation modulation diversity for digital simulcast, *41st IEEE Vehicular Technology Conference, 1991. Gateway to the Future Technology in Motion,* pp. 848–853, St. Louis, MO, 19–22 May 1991
39. N. Seshadri, J.H. Winters, Two signaling schemes for improving the error performance of frequency-division-duplex (FDD) transmission systems using transmitter antenna diversity, *43rd IEEE Vehicular Technology Conference, 1993*, pp. 508–511, Secaucus, NJ, 18–20 May 1993
40. A. Edelman, Eigen Values and Condition Numbers of Random Matrices. PhD Dissertation, MIT, Cambridge, May 1989
41. A.T. James, Distributions of matrix variates and latent roots derived from normal samples. Ann. Math. Stat. **35**(2), 475–501 (1964)
42. H. Wang, X.-G. Xia, Upper bounds of rates of complex orthogonal space-time block codes. IEEE Trans. Inf. Theory **49**(10), 2788–2796 (Oct. 2003)
43. X. Ma, G.B. Giannakis, Full-diversity full-rate complex-field space-time coding. IEEE Trans. Signal Process. **51**(11), 2917–2930 (Nov. 2003)
44. I.B. Djordjevic, L. Xu, T. Wang, PMD compensation in multilevel coded-modulation schemes with coherent detection using BLAST algorithm and iterative polarization cancellation. Opt. Express **16**(19), 14845–14852 (15 Sept. 2008)
45. J.R. Hampton, *Introduction to MIMO Communications* (Cambridge University Press, Cambridge, 2014)
46. N. Wiberg, H.A. Loeliger, R. Kotter, Codes and iterative decoding on general graphs, in *Proceedings of 1995 IEEE International Symposium on Information Theory*, p. 468, Whistler, BC, 1995
47. F.R. Kschischang, B.J. Frey, H.A. Loeliger, Factor graphs and the sum-product algorithm. IEEE Trans. Inf. Theory **47**(2), 498–519 (Feb. 2001)
48. H.A. Loeliger, An introduction to factor graphs. IEEE Signal Process. Mag. **21**(1), 28–41 (Jan. 2004)
49. B.J. Frey, F.R. Kschischang, H.-A. Loeliger, N. Wiberg, Factor graphs and algorithms, in *Proceedings of 35th Allerton Conference on Communications, Control, and Computing*, Allerton House, Monticello, IL, pp. 666–680, 29 Sept.–1 Oct. 1997
50. S. Lin, D.J. Costello, *Error Control Coding: Fundamentals and Applications*, 2nd edn. (Pearson Prentice Hall, Upper Saddle River, 2004)
51. W.E. Ryan, S. Lin, *Channel Codes: Classical and Modern* (Cambridge University Press, Cambridge, 2009)
52. E. Biglieri, G. Taricco, A. Tulino, Performance of space-time codes for a large number of antennas. IEEE Trans. Inf. Theory **48**(7), 1794–1803 (July 2002)
53. E. Biglieri, G. Taricco, A. Tulino, Decoding space-time codes with BLAST architectures. IEEE Trans. Signal Process. **50**(10), 2547–2552 (Oct. 2002)
54. A. Pulraj, R. Nabar, D. Gore, *Introduction to Space-Time Wireless Communications* (Cambridge University Press, Cambridge, 2003)
55. I.B. Djordjevic, On advanced FEC and coded modulation for ultra-high-speed optical transmission, IEEE Commun. Surv. Tutorials **99**, 1–31, 1 Mar. 2016. https://doi.org/10.1109/COMST.2016.2536726
56. S. Ten Brink, Convergence behavior of iteratively decoded parallel concatenated codes. IEEE Trans. Commun. **49**(10), 1727–1737 (2001)
57. T.S. Rappaport, *Wireless Communications: Principles and Practice*, 2nd edn. (Prentice Hall, Hoboken, 2002)
58. Z. Zheng, T.M. Duman, E.M. Kurtas, Achievable information rates and coding for MIMO systems over ISI channels and frequency-selective fading channels. IEEE Trans. Commun. **52**(10), 1698–1710 (Oct. 2004)
59. R. Schober, W.H. Gerstacker, L.H.J. Lampe, Performance analysis and design of STBCs for frequency-selective fading channels. IEEE Trans. Wirel. Commun. **3**(3), 734–744 (May 2004)
60. W. Zhang, X.G. Xia, P.C. Ching, High-rate full-diversity space–time–frequency codes for broadband MIMO block-fading channels. IEEE Trans. Commun. **55**(1), 25–34 (Jan. 2007)
61. W. Siheh, I.B. Djordjevic, *OFDM for Optical Communications* (Elsevier/Academic Press, Amsterdam/Boston, 2009)
62. E. Telatar, Capacity of multi-antenna Gaussian channels. Eur. Trans. Telecommun. **10**, 585–595 (1999)
63. T.M. Cover, J.A. Thomas, *Elements of Information Theory* (Wiley, New York, 1991)
64. H. Shin, J.H. Lee, Capacity of multiple-antenna fading channels: Spatial fading correlation, double scattering, and keyhole. IEEE Trans. Inf. Theory **49**(10), 2636–2647 (Oct. 2003)
65. R.N. McDonough, A.D. Whalen, *Detection of Signals in Noise*, 2nd edn. (Academic Press, San Diego, 1995)
66. B. Hassibi, B.M. Hochwald, How much training is needed in multiple-antenna wireless links? IEEE Trans. Inf. Theory **49**(4), 951–963 (Apr. 2003)
67. C. Lin, I.B. Djordjevic, D. Zou, Achievable information rates calculation for optical OFDM transmission over few-mode fiber long-haul transmission systems. Opt. Express **23**(13), 16846–16856 (2015)
68. E.G. Larsson, O. Edfors, F. Tufvesson, T.L. Marzetta, Massive MIMO for next generation wireless systems. IEEE Commun. Mag. **52**(2), 186–195 (2014)
69. T. Marzetta, E.G. Larsson, H. Yang, H.Q. Ngo, *Fundamentals of Massive MIMO* (Cambridge University Press, Cambridge, 2016)
70. H.H. Yang, T.Q.S. Quek, *Massive MIMO Meets Small Cell: Backhaul and Cooperation* (Springer Nature, Cham, 2017)
71. M.A. Albreem, M. Juntti, S. Shahabuddin, Massive MIMO detection techniques: A survey. IEEE Commun. Surv. Tutorials **21**(4), 3109–3132 (2019)
72. R. Chataut, R. Akl, Massive MIMO systems for 5G and beyond networks-overview, recent trends, challenges, and future research direction. Sensors (Basel) **20**(10), 2753 (2020)
73. S.A. Khwandah, J.P. Cosmas, P.I. Lazaridis, et al., Massive MIMO systems for 5G communications. Wireless Pers. Commun. **120**, 2101–2115 (2021)
74. F. Fernandes, A. Ashikhmin, T.L. Marzetta, Inter-cell interference in noncooperative TDD large scale antenna systems. IEEE J. Sel. Areas Commun. **31**(2), 192–201 (2013)
75. J.L.L. Morales, S. Roy, Channel estimation using time-shifted pilot sequences in non-cooperative cellular TDD networks with large antenna arrays. In *Proceedings of 2013 Asilomar Conference on Signals, Systems and Computers*, pp. 1258–1262, 2013
76. V. Saxena, G. Fodor, E. Karipidis, Mitigating pilot contamination by pilot reuse and power control schemes for massive MIMO systems. In *Proceedings of the IEEE VTC Spring*, Glasgow, UK, 11–14 May 2015, pp. 1–6, 2015
77. R. Muller, L. Cottatellucci, M. Vehkapera, Blind pilot decontamination. IEEE J. Sel. Top. Signal Proc. **8**, 773–786 (2014)
78. O. Elijah, C.Y. Leow, T.A. Rahman, S. Nunoo, S.Z. Iliya, A comprehensive survey of pilot contamination in massive MIMO-5G system. IEEE Commun. Surv. Tutorials **18**, 905–923 (2016)
79. M. Wu, B. Yin, A. Vosoughi, C. Studer, J.R. Cavallaro, C. Dick, Approximate matrix inversion for high-throughput data detection in the large-scale MIMO uplink, in *Proceedings of 2013 IEEE International Symposium on Circuits and Systems (ISCAS)*, pp. 2155–2158, 2013

80. B. Kang, J.-H. Yoon, J. Park, Low complexity massive MIMO detection architecture based on Neumann method, in *Proceedings of 2015 International SoC Design Conference (ISOCC)*, pp. 293–294, 2015
81. Z. Wu, Y. Xue, X. You, C. Zhang, Hardware efficient detection for massive MIMO uplink with parallel Gauss-Seidel method, in *Proceedings of 2017 22nd International Conference on Digital Signal Processing (DSP)*, pp. 1–5, 2017
82. X. Gao, L. Dai, Y. Hu, Z. Wang, Z. Wang, Matrix inversion-less signal detection using SOR method for uplink large-scale MIMO systems, in *Proceedings of 2014 IEEE Global Communications Conference*, pp. 3291–3295, 2014
83. B.Y. Kong, I.-C. Park, Low-complexity symbol detection for massive MIMO uplink based on Jacobi method, in *Proceedings of 2016 IEEE 27th Annual International Symposium on Personal, Indoor, and Mobile Radio Communications (PIMRC)*, pp. 1–5, 2016
84. X. Gao, L. Dai, Y. Ma, Z. Wang, Low-complexity near-optimal signal detection for uplink large-scale MIMO systems. IEEE Electron. Lett. **50**(18), 1326–1328 (2014)
85. M. Ylinen, A. Burian, J. Takala, Direct versus iterative methods for fixed-point implementation of matrix inversion, in *Proceedings of 2004 IEEE International Symposium on Circuits and Systems (IEEE Cat. No.04CH37512)*, pp. III–225, 2004
86. G. Wunder, H. Boche, T. Strohmer, P. Jung, Sparse signal processing concepts for efficient 5G system design. IEEE Access **3**, 195–208 (2015)
87. M. Masood, L.H. Afify, T.Y. Al-Naffouri, Efficient coordinated recovery of sparse channels in massive MIMO. IEEE Trans. Signal Process. **63**(1), 104–118 (2015)
88. S. Foucart, H. Rauhut, *A Mathematical Introduction to Compressive Sensing* (Springer, New York, 2013)
89. A.K. Sah, A.K. Chaturvedi, An MMP-based approach for detection in large MIMO systems using sphere decoding. IEEE Wirel. Commun. Lett. **6**(2), 158–161 (2017)
90. M.A.M. Albreem, M.F.M. Salleh, Regularized lattice sphere decoding for block data transmission systems. Wirel. Pers. Commun. **82**(3), 1833–1850 (2015)
91. J.W. Choi, B. Shim, New approach for massive MIMO detection using sparse error recovery, in *Proceedings of 2014 IEEE Global Communications Conference*, pp. 3754–3759, 2014
92. Y. Jiang, M.K. Varanasi, J. Li, Performance analysis of ZF and MMSE equalizers for MIMO systems: an in-depth study of the high SNR regime. IEEE Trans. Inf. Theory **57**(4), 2008–2026 (2011)
93. D. Wübben, D. Seethaler, J. Jaldén, G. Matz, Lattice reduction. IEEE Signal Process. Mag. **28**(3), 70–91 (2011)
94. S. Shahabuddin, J. Janhunen, Z. Khan, M. Juntti, A. Ghazi, A customized lattice reduction multiprocessor for MIMO detection, in *Proceedings of 2015 IEEE International Symposium on Circuits and Systems (ISCAS)*, pp. 2976–2979, 2015
95. J. Liu, R. Deng, S. Zhou, Z. Niu, Seeing the unobservable: CHANNEL learning for wireless communication networks, in *Proceedings of 2015 IEEE Global Communications Conference (GLOBECOM)*, San Diego, CA, USA, pp. 1–6, 2015
96. M. Koller, C. Hellings, M. Knödlseder, T. Wiese, D. Neumann, W. Utschick, Machine learning for channel estimation from compressed measurements, in *Proceedings of 2018 15th International Symposium on Wireless Communication Systems (ISWCS)*, Lisbon, Portugal, pp. 1–5, 2018
97. D. Neumann, T. Wiese, W. Utschick, Learning the MMSE channel estimator. IEEE Trans. Signal Process. **66**, 2905–2917 (2018)
98. A. Zappone, L. Sanguinetti, M. Debbah, User association and load balancing for massive MIMO through deep learning, in *Proceedings of 2018 Asilomar Conference on Signals, Systems and Computers*, Pacific Grove, CA, USA, pp. 1262–1266 , 2018

Advanced Coding and Coded Modulation Techniques

9

Abstract

This chapter is devoted to advanced channel coding and coded modulation techniques. After linear and BCH codes fundamentals, we provide the trellis description of linear block codes and describe the corresponding Viterbi decoding algorithm. After describing the fundamentals of convolutional, RS, concatenated, and product codes, we describe coding with interleaving as an efficient way to deal with burst of errors and fading effects. Significant space in the chapter is devoted to codes on graphs, in particular turbo, turbo product, and LDPC codes are described together with corresponding decoding algorithms. Regarding LDPC codes, both binary and nonbinary (NB) LDPC codes are introduced, and corresponding decoding algorithms as well as their FPGA implementation are described. Additionally, LDPC code design procedures are described, followed by rate adaptation. Rate-adaptive FPGA implementations of LDPC and generalized LDPC codes are described as well. The next portion of the chapter is devoted to coded modulation (CM) and unequal error protection (UEP). After providing the coded modulation fundamentals, we describe trellis-coded modulation (TCM), multilevel coding and UEP, bit-interleaved coded modulation (BICM), turbo TCM, and various hybrid multidimensional coded modulation schemes suitable for ultrahigh-speed optical transmission including multilevel nonbinary LDPC-coded modulation. After coded modulation sections, the focus of the chapter is on multidimensional turbo equalization. The following topics are described: nonlinear channels with memory, nonbinary MAP detection, sliding-window multidimensional turbo equalization, simulation study of multidimensional turbo equalization, and several experimental demonstrations including time-domain 4D-NB-LDPC-CM and quasi-single-mode transmission over transoceanic distances. In section on optimized signal constellation design and optimized bit-to-symbol mappings-based coded modulation, we describe multidimensional optimized signal constellation design (OSCD), EXIT chart analysis of OSCD mapping rules, nonlinear OSCD-based coded modulation, and transoceanic multi-Tb/s transmission experiments enabled by OSCD. Finally, in adaptive coding and adaptive coded modulation section, we describe adaptive coded modulation, adaptive nonbinary LDPC-coded multidimensional modulation suitable for high-speed optical communications, and adaptive hybrid free-space optical (FSO)-RF-coded modulation. For better understanding, the set of problems is provided after concluding remarks.

9.1 Linear Block Codes Fundamentals

The *linear block code* (n,k) satisfies a linearity property, which means that a linear combination of arbitrary two codewords results in another codeword. If we use the terminology of vector spaces, it can be defined as a subspace of a vector space over finite (Galois) field, denoted as GF(q), with q being the prime power. Every space is described by its *basis* – a set of linearly independent vectors. The number of vectors in the basis determines the dimension of the space. Therefore, for an (n,k) linear block code, the dimension of the space is n, and the dimension of the code subspace is k.

Let $\boldsymbol{m} = (m_0\, m_1 \ldots m_{k-1})$ denote the k-bit message vector. Any codeword $\boldsymbol{x} = (x_0\, x_1 \ldots x_{n-1})$ from the (n,k) linear block code can be represented as a linear combination of k basis vectors $\boldsymbol{g}_i$ $(i = 0,1,..,k-1)$ as follows:

I. B. Djordjevic, *Advanced Optical and Wireless Communications Systems*, https://doi.org/10.1007/978-3-030-98491-5_9

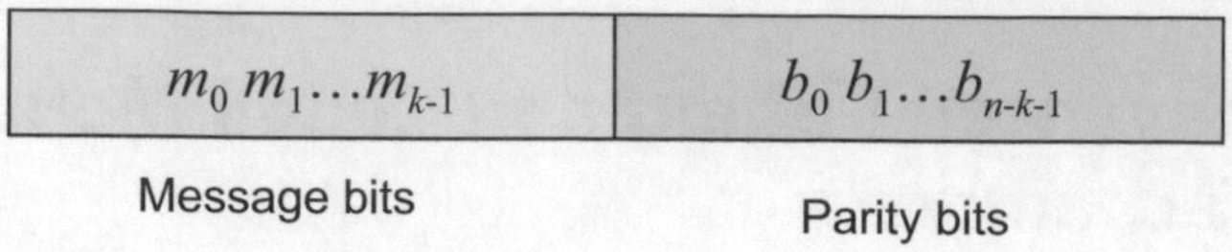

Fig. 9.1 Systematic codeword structure

$$\boldsymbol{x} = m_0\boldsymbol{g}_0 + m_1\boldsymbol{g}_1 + \cdots + m_{k-1}\boldsymbol{g}_{k-1} = \boldsymbol{m}\begin{bmatrix}\boldsymbol{g}_0\\ \boldsymbol{g}_1\\ \ldots\\ \boldsymbol{g}_{k-1}\end{bmatrix} = \boldsymbol{mG};\ \boldsymbol{G} = \begin{bmatrix}\boldsymbol{g}_0\\ \boldsymbol{g}_1\\ \ldots\\ \boldsymbol{g}_{k-1}\end{bmatrix}, \tag{9.1}$$

where $\boldsymbol{G}$ is the generator matrix (of dimensions $k \times n$), in which every row represents basis vector from the coding subspace. Therefore, in order to be encoded, the message vector $\boldsymbol{m}$ has to be multiplied with a generator matrix $\boldsymbol{G}$ to get the codeword, namely, $\boldsymbol{x} = \boldsymbol{mG}$.

The code may be transformed into a systematic form by elementary operations on rows in the generator matrix, i.e.,

$$\boldsymbol{G}_s = [\boldsymbol{I}_k|\boldsymbol{P}], \tag{9.2}$$

where $\boldsymbol{I}_k$ is the unity matrix of dimensions $k \times k$ and $\boldsymbol{P}$ is the matrix of dimensions $k \times (n - k)$ with columns denoting the positions of parity checks:

$$\boldsymbol{P} = \begin{bmatrix} p_{00} & p_{01} & \cdots & p_{0,n-k-1}\\ p_{10} & p_{11} & \cdots & p_{1,n-k-1}\\ \ldots & & \ldots & \ldots\\ p_{k-1,0} & p_{k-1,1} & \cdots & p_{k-1,n-k-1}\end{bmatrix}.$$

The codeword of a *systematic code* is obtained as

$$\boldsymbol{x} = [\boldsymbol{m}|\boldsymbol{b}] = \boldsymbol{m}[\boldsymbol{I}_k|\boldsymbol{P}] = \boldsymbol{mG},\quad \boldsymbol{G} = [\boldsymbol{I}_k|\boldsymbol{P}], \tag{9.3}$$

and has the structure as shown in Fig. 9.1. The message vector stays unaffected during systematic encoding, while the vector of parity checks $\boldsymbol{b}$ is appended having the bits that are algebraically related to the message bits as follows:

$$b_i = p_{0i}m_0 + p_{1i}m_1 + \ldots + p_{k-1,i}m_{k-1}, \tag{9.4}$$

where

$$p_{ij} = \begin{cases} 1, & \text{if } b_i \text{ depends on } m_j\\ 0, & \text{otherwise}\end{cases}.$$

The optical/wireless channel introduces the errors during transmission, and the received vector $\boldsymbol{r}$ can be represented as $\boldsymbol{r} = \boldsymbol{x} + \boldsymbol{e}$, where $\boldsymbol{e}$ is the error vector (error pattern) whose components are determined by

$$e_i = \begin{cases} 1, & \text{if an error occurred in the } i\text{th location}\\ 0, & \text{otherwise}\end{cases}$$

In order to verify if the received vector $\boldsymbol{r}$ is a codeword one, in incoming subsection we will introduce the concept of a *parity-check matrix* as another useful matrix associated with the linear block codes.

Let us expand the matrix equation $\boldsymbol{x} = \boldsymbol{mG}$ in a scalar form as follows:

$$\begin{aligned}
x_0 &= m_0 \\
x_1 &= m_1 \\
&\ldots \\
x_{k-1} &= m_{k-1} \\
x_k &= m_0 p_{00} + m_1 p_{10} + \ldots + m_{k-1} p_{k-1,0} \\
x_{k+1} &= m_0 p_{01} + m_1 p_{11} + \ldots + m_{k-1} p_{k-1,1} \\
&\ldots \\
x_{n-1} &= m_0 p_{0,n-k-1} + m_1 p_{1,n-k-1} + \ldots + m_{k-1} p_{k-1,n-k-1}
\end{aligned} \tag{9.5}$$

By using the first k equalities, the last $n - k$ equations can be rewritten in terms of the codeword elements as follows:

$$\begin{aligned}
&x_0 p_{00} + x_1 p_{10} + \ldots + x_{k-1} p_{k-1,0} + x_k = 0 \\
&x_0 p_{01} + x_1 p_{11} + \ldots + x_{k-1} p_{k-1,0} + x_{k+1} = 0 \\
&\ldots \\
&x_0 p_{0,n-k+1} + x_1 p_{1,n-k-1} + \ldots + x_{k-1} p_{k-1,n-k+1} + x_{n-1} = 0
\end{aligned} \tag{9.6}$$

The equations presented above can be rewritten through matrix representation as

$$\underbrace{\begin{bmatrix} x_0 & x_1 & \ldots & x_{n-1} \end{bmatrix}}_{\boldsymbol{x}} \underbrace{\begin{bmatrix} p_{00} & p_{10} & \cdots & p_{k-1,0} & 1 & 0 & \ldots & 0 \\ p_{01} & p_{11} & \cdots & p_{k-1,1} & 0 & 1 & \ldots & 0 \\ \ldots & & \cdots & & & \ldots & & \ldots \\ p_{0,n-k-1} & p_{1,n-k-1} & \cdots & p_{k-1,n-k-1} & 0 & 0 & \ldots & 1 \end{bmatrix}^T}_{\boldsymbol{H}^T} = \boldsymbol{0}$$
$$\Leftrightarrow \boldsymbol{xH}^T = \boldsymbol{0}, \quad \boldsymbol{H} = \begin{bmatrix} \boldsymbol{P}^T & \boldsymbol{I}_{n-k} \end{bmatrix}_{(n-k)xn} \tag{9.7}$$

The $\boldsymbol{H}$-matrix in Eq. (9.7) is known as *the parity-check* one. We can easily verify that $\boldsymbol{G}$ and $\boldsymbol{H}$ matrices satisfy equation

$$\boldsymbol{GH}^T = \begin{bmatrix} \boldsymbol{I}_k & \boldsymbol{P} \end{bmatrix} \begin{bmatrix} \boldsymbol{P} \\ \boldsymbol{I}_{n-k} \end{bmatrix} = \boldsymbol{P} + \boldsymbol{P} = \boldsymbol{0}, \tag{9.8}$$

meaning that the parity-check matrix H of an (n,k) linear block code has rank n-k and dimensions $(n - k) \times n$ whose null space is k-dimensional vector with basis forming the generator matrix $\boldsymbol{G}$.

Every (n,k) linear block code with generator matrix $\boldsymbol{G}$ and parity-check matrix $\boldsymbol{H}$ has a dual code, this time having generator matrix $\boldsymbol{H}$ and parity-check matrix $\boldsymbol{G}$. As an example, $(n,1)$ repetition code and $(n,n - 1)$ single-parity-check code are dual ones.

In order to determine the error correction capability of the linear block code, we have to introduce the concepts of Hamming distance and Hamming weight [1, 2]. The Hamming distance $d(\boldsymbol{x}_1,\boldsymbol{x}_2)$ between two codewords $\boldsymbol{x}_1$ and $\boldsymbol{x}_2$ is defined as the number of locations in which these two vectors differ. The Hamming weight $wt(\boldsymbol{x})$ of a codeword vector $\boldsymbol{x}$ is defined as the number of nonzero elements in the vector. The minimum distance d_{min} of a linear block code is defined as the smallest Hamming distance between any pair of code vectors in the code space. Since the zero vector is also a codeword, the minimum distance of a linear block code can be defined as the smallest Hamming weight of the nonzero code vectors in the code.

We can write the parity-check matrix in a form $\boldsymbol{H} = [\boldsymbol{h}_1\, \boldsymbol{h}_2 \ldots \boldsymbol{h}_n]$, where $\boldsymbol{h}_i$ presents the i-th column in the matrix structure. Since every codeword $\boldsymbol{x}$ must satisfy the syndrome equation $\boldsymbol{xH}^{\mathrm{T}} = \boldsymbol{0}$, the minimum distance of a linear block code is determined by the minimum number of columns in the $\boldsymbol{H}$-matrix whose sum is equal to zero vector. As an example, (7,4) Hamming code discussed above has a minimum distance $d_{\text{min}} = 3$ since the sum of first, fifth, and sixth columns leads to zero vector.

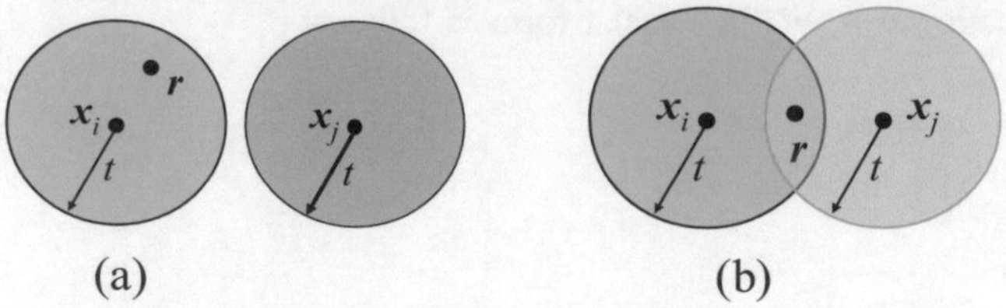

(a) (b)

Fig. 9.2 The illustration of Hamming distance: **(a)** $d(\boldsymbol{x}_i,\boldsymbol{x}_j) \geq 2\,t + 1$ and **(b)** $d(\boldsymbol{x}_i,\boldsymbol{x}_j) < 2\,t + 1$

The codewords can be represented as points in n-dimensional space, as shown in Fig. 9.2. Decoding process can be visualized by creating the spheres of radius t around codeword points. The received word vector $\boldsymbol{r}$ in Fig. 9.2a will be decoded as a codeword $\boldsymbol{x}_i$ because its Hamming distance $d(\boldsymbol{x}_i,\boldsymbol{r}) \leq t$ is closest to the codeword $\boldsymbol{x}_i$. On the other hand, in example shown in Fig. 9.2b, the Hamming distance satisfies relation $d(\boldsymbol{x}_i,\boldsymbol{x}_j) \leq 2\,t$, and the received vector $\boldsymbol{r}$ that falls in intersection area of two spheres cannot be uniquely decoded. Therefore, (n,k) linear block code of minimum distance d_{min} can correct up to t errors if, and only if, $t \leq \lfloor 1/2(d_{\mathrm{min}} - 1)\rfloor$ or $d_{\mathrm{min}} \geq 2\,t + 1$ (where $\lfloor \cdot \rfloor$ denotes the largest integer smaller or equal to the enclosed quantity). If we are only interested in detecting e_{d} errors, then the minimum distance should be $d_{\mathrm{min}} \geq e_{\mathrm{d}} + 1$. However, if we are interested in detecting e_{d} errors and correcting e_{c} errors, then the minimum distance should be $d_{\mathrm{min}} \geq e_{\mathrm{d}} + e_{\mathrm{c}} + 1$. Accordingly, the Hamming (7,4) code is a single error correcting and double error detecting code. More generally, Hamming codes are (n,k) linear block codes with the following parameters:

- Block length: $n = 2^m - 1$
- Number of message bits: $k = 2^m - m - 1$
- Number of parity bits: $n - k = m$
- $d_{\mathrm{min}} = 3$

where $m \geq 3$. Hamming codes belong to the class of perfect codes, the codes that satisfy the Hamming inequality given as [1–7]

$$2^{n-k} \geq \sum_{i=0}^{t} \binom{n}{i}. \tag{9.9}$$

This bound gives how many errors t can be corrected with a specific (n,k) linear block code.

The coding gain a linear (n,k) block is defined as the relative saving in the energy per information bit for a given bit error probability when coding is applied. Since the total information word energy kE_{b} must be the same as the total codeword energy nE_{c} (where E_{c} is the transmitted bit energy, while E_{b} is the information bit energy), we can establish the following relationship:

$$E_{\mathrm{c}} = (k/n)E_{\mathrm{b}} = RE_{\mathrm{b}} \tag{9.10}$$

where R defines the coding rate. As an example, the probability of error for BPSK on an AWGN channel, when coherent hard-decision (bit-by-bit) demodulator is used, can be obtained as

$$p = \frac{1}{2}\mathrm{erfc}\left(\sqrt{\frac{E_c}{N_0}}\right) = \frac{1}{2}\mathrm{erfc}\left(\sqrt{\frac{RE_b}{N_0}}\right), \tag{9.11}$$

where erfc(x) function is defined in Chap. 2. By using the Chernoff bound [3], we obtain the following expression for the coding gain when hard decision is applied in decoding process:

$$\frac{(E_b/N_0)_{\mathrm{uncoded}}}{(E_b/N_0)_{\mathrm{coded}}} \approx R(t+1), \tag{9.12}$$

where t is the error correction capability of the code. The corresponding soft-decision coding gain can be estimated by [2, 3, 6].

$$\frac{(E_b/N_0)_{\text{uncoded}}}{(E_b/N_0)_{\text{coded}}} \approx Rd_{\min}. \tag{9.13}$$

By comparing Eqs. (9.12) and (9.13), we can see that the soft-decision coding gain is about 3 dB better than hard-decision coding gain since the minimum distance $d_{\min} \geq 2t + 1$.

In fiber-optics communications, it is common to use the expression for BER on an AWGN for on-off keying and express both the coding gain (CG) and net coding gain (NCG) as follows [1, 2, 6–12]:

$$\text{CG} = 20\log_{10}\left[\text{erfc}^{-1}(2\text{BER}_{\text{out,t}})\right] - 20\log_{10}\left[\text{erfc}^{-1}(2\text{BER}_{\text{in}})\right] \ \ [\text{dB}] \tag{9.14}$$

$$\text{NCG} = \text{CG} + 10\log_{10}R \ \ [\text{dB}] \tag{9.15}$$

where BER_{in} denotes the BER at the input of FEC decoder, BER_{out} denotes the BER at the output of FEC decoder, while BER_{t} denotes the target BER (which is typically 10^{-15}).

In the rest of this section, we describe several important coding bounds including Hamming, Plotkin, Gilbert-Varshamov, and Singleton ones [2, 3]. The *Hamming* bound has already been introduced for binary linear block codes (LBC) and was expressed by Eq. (9.9). The Hamming bound for q-ary (n,k) LBC is given as

$$\left[1 + (q-1)\binom{n}{1} + (q-1)^2\binom{n}{2} + \ldots + (q-1)^i\binom{n}{i} + \ldots + (q-1)^t\binom{n}{t}\right]q^k \leq q^n, \tag{9.16}$$

where t is the error correction capability and $(q-1)^i\binom{n}{i}$ is the number of received words that differ from a given codeword in i symbols. The codes satisfying the Hamming bound with equality sign are known as *perfect codes*. Hamming codes are perfect codes since $n = 2^{n-k} - 1$, which is equivalent to $(1 + n)2^k = 2^n$, so that the relationship between the right and left side of Eq. (9.16) is expressed by equality sign. The $(n,1)$ repetition code is also an example of a perfect code.

The *Plotkin* bound is defined by the following relation for minimum distance:

$$d_{\min} \leq \frac{n2^{k-1}}{2^k - 1}. \tag{9.17}$$

Namely, if all codewords are written as the rows of a $2^k \times n$ matrix, each column will contain 2^{k-1} "zeros" and 2^{k-1} "ones," with the total weight of all codewords being equal to $n2^{k-1}$.

Gilbert-Varshamov bound is based on the property that the minimum distance $d_{\min}$ of a linear (n,k) block code and can be determined as the minimum number of columns in $\boldsymbol{H}$-matrix that adds to zero, i.e.,

$$\binom{n-1}{1} + \binom{n-1}{2} + \ldots + \binom{n-1}{d_{\min}-2} < 2^{n-k} - 1. \tag{9.18}$$

The *Singleton bound* is defined by inequality:

$$d_{\min} \leq n - k + 1. \tag{9.19}$$

This bound is straightforward to prove. Let only one bit of value 1 be present in information vector. If it is involved in n-k parity checks, the total number of ones in codeword cannot be larger than $n - k + 1$. The codes satisfying the Singleton bound with equality sign are known as the *maximum-distance separable* (MDS) codes (e.g., RS codes are MDS codes).

9.2 BCH Codes Fundamentals

The BCH codes, the most famous cyclic codes, were discovered by Hocquenghem in 1959 and by Bose and Chaudhuri in 1960 [2–5]. Among many different decoding algorithms, the most important are Massey-Berlekamp algorithm and Chien's search algorithm. An important subclass of BCH is a class of Reed-Solomon codes proposed in 1960.

Equipped we the knowledge of Galois fields from Appendix, we can continue our description of structure of BCH codes. Let the finite field GF(q) (*symbol field*) and extension field GF(q^m) (*locator field*), $m \geq 1$, be given. For every m_0 ($m_0 \geq 1$) and Hamming distance d, there exists a BCH code with the generating polynomial $g(x)$, if and only if it is of smallest degree with coefficients from GF(q) and the roots from the extension field GF(q^m) as follows [2, 5]:

$$\alpha^{m_0}, \alpha^{m_0+1}, \cdots, \alpha^{m_0+d-2} \tag{9.20}$$

where α is from GF(q^m). The codeword length is determined as the least common multiple of orders of roots. (The order of an element β from finite field is the smallest positive integer j such that $\beta^j = 1$.)

It can be shown that for any positive integer m ($m \geq 3$) and t ($t < 2^{m-1}$), there exists a *binary* BCH code having the following properties:

- Codeword length: $n = 2^m - 1$
- Number of parity bits: $n - k \leq mt$
- Minimum Hamming distance: $d \geq 2t + 1$

This code is able to correct up to t errors. The generator polynomial can be found as the LCM of the minimal polynomials of α^i:

$$g(x) = \mathrm{LCM}[P_{\alpha^1}(x), P_{\alpha^3}(x), \ldots, P_{\alpha^{2t-1}}(x)] \tag{9.21}$$

where α is a primitive element in GF(2^m) and $P_{\alpha^i}(x)$ is the minimal polynomial of α^i.

Let $c(x) = c_0 + c_1x + c_2x^2 + \ldots + c_{n-1}x^{n-1}$ be the codeword polynomial, and let the roots of generator polynomial be α, $\alpha^2, \ldots, \alpha^{2t}$, where t is the error correction capability of BCH code. Because the generator polynomial $g(x)$ is the factor of codeword polynomial $c(x)$, the roots of $g(x)$ must also be the roots of $c(x)$:

$$c(\alpha^i) = c_0 + c_1\alpha^i + \ldots + c_{n-1}\alpha^{(n-1)i} = 0; 1 \leq i \leq 2t \tag{9.22}$$

This equation can also be written as inner (scalar) product of codeword vector $\boldsymbol{c} = [c_0\, c_1 \ldots c_{n-1}]$ and the following vector $[1\ \alpha^i\ \alpha^{2i} \ldots \alpha^{2(n-1)i}]$:

$$[c_0\, c_1 \ldots c_{n-1}] \begin{bmatrix} 1 \\ \alpha^i \\ \ldots \\ \alpha^{(n-1)i} \end{bmatrix} = 0; 1 \leq i \leq 2t \tag{9.23}$$

Equation (9.23) can also be written in the following matrix form:

$$[c_0\, c_1 \ldots c_{n-1}] \underbrace{\begin{bmatrix} \alpha^{n-1} & \alpha^{n-2} & \ldots & \alpha & 1 \\ (\alpha^2)^{n-1} & (\alpha^2)^{n-2} & \ldots & \alpha^2 & 1 \\ (\alpha^3)^{n-1} & (\alpha^3)^{n-2} & \ldots & \alpha^3 & 1 \\ \ldots & \ldots & \ldots & \ldots & \ldots \\ (\alpha^{2t})^{n-1} & (\alpha^{2t})^{n-2} & \ldots & \alpha^{2t} & 1 \end{bmatrix}^T}_{\boldsymbol{H}^T} = \boldsymbol{c}\boldsymbol{H}^T = \boldsymbol{0}, \tag{9.24}$$

where $\boldsymbol{H}$ is the parity-check matrix of BCH code. Using the property 2 of GF(q) from Appendix, we conclude that α^i and α^{2i} are the roots of the same minimum polynomial, so that the even rows in $\boldsymbol{H}$ can be omitted to get the final version of the parity-check matrix of BCH codes:

$$\boldsymbol{H}=\begin{bmatrix}\alpha^{n-1} & \alpha^{n-2} & \dots & \alpha & 1\\ (\alpha^3)^{n-1} & (\alpha^3)^{n-2} & \dots & \alpha^3 & 1\\ (\alpha^5)^{n-1} & (\alpha^5)^{n-2} & \dots & \alpha^5 & 1\\ \dots & \dots & \dots & \dots & \dots\\ (\alpha^{2t-1})^{n-1} & (\alpha^{2t-1})^{n-2} & \dots & \alpha^{2t-1} & 1\end{bmatrix} \tag{9.25}$$

For example, (15,7) 2-error correcting BCH code has the generator polynomial [6]

$$\begin{aligned}g(x)&=LCM[P_\alpha(x),P_{\alpha^3}(x)]\\&=LCM\left[x^4+x+1,\left(x+\alpha^3\right)\left(x+\alpha^6\right)\left(x+\alpha^9\right)\left(x+\alpha^{12}\right)\right]\\&=x^8+x^7+x^6+x^4+1,\end{aligned}$$

wherein the primitive polynomial used to design this code is $p(x)=x^4+x+1$. The corresponding parity-check matrix is given by

$$\boldsymbol{H}=\begin{bmatrix}\alpha^{14} & \alpha^{13} & \alpha^{12} & \cdots & \alpha & 1\\ \alpha^{42} & \alpha^{39} & \alpha^{36} & \cdots & \alpha^3 & 1\end{bmatrix}.$$

By using the GF(2^4) property $\alpha^{15}=1$, we can rewrite the parity-check matrix as

$$\boldsymbol{H}=\begin{bmatrix}\alpha^{14} & \alpha^{13} & \alpha^{12} & \alpha^{11} & \alpha^{10} & \alpha^9 & \alpha^8 & \alpha^7 & \alpha^6 & \alpha^5 & \alpha^4 & \alpha^3 & \alpha^2 & \alpha & 1\\ \alpha^{12} & \alpha^9 & \alpha^6 & \alpha^3 & 1 & \alpha^{12} & \alpha^9 & \alpha^6 & \alpha^3 & 1 & \alpha^{12} & \alpha^9 & \alpha^6 & \alpha^3 & 1\end{bmatrix}.$$

The BCH code is able to correct up to t errors. Let us assume that error polynomial $e(x)$ does not have more than t errors, which can then be written as

$$e(x)=e_{j_1}x^{j_1}+e_{j_2}x^{j_2}+\ldots+e_{j_l}x^{j_l}+\ldots+e_{j_v}x^{j_v};\;\;0\le v\le t \tag{9.26}$$

e_{j_l} is the error magnitude, α^{j_l} is the error-location number, while j_l is the error location. Notice that the error magnitudes are from symbol field, while the error-location numbers are from extension field. The corresponding syndrome components can be obtained from (9.26) and (9.24) as follows:

$$S_i=e_{j_1}\left(\alpha^i\right)^{j_1}+e_{j_2}\left(\alpha^i\right)^{j_2}+\ldots+e_{j_l}\left(\alpha^i\right)^{j_l}+\ldots+e_{j_v}\left(\alpha^i\right)^{j_v};\;\;0\le v\le t \tag{9.27}$$

In order to avoid the double indexing, let us introduce the following notation $X_l=\alpha^{j_l}$, $Y_l=e_{j_l}$. The pair (X_l,Y_l) completely identifies the errors ($l\in[1,v]$). We have to solve the following set of equations:

$$\begin{aligned}S_1&=Y_1X_1+Y_2X_2+\ldots+Y_vX_v\\S_2&=Y_1X_1^2+Y_2X_2^2+\ldots+Y_vX_v^2\\&\dots\\S_{2t}&=Y_1X_1^{2t}+Y_2X_2^{2t}+\ldots+Y_vX_v^{2t}.\end{aligned} \tag{9.28}$$

The procedure to solve this system of equations represents the corresponding decoding algorithm. Direct solution of this system of equations is impractical. There exists many different algorithms to solve the system of Eq. (9.28), ranging from iterative to Euclidean algorithms [3–7]. The very popular decoding algorithm of BCH codes is *Massey-Berlekamp*

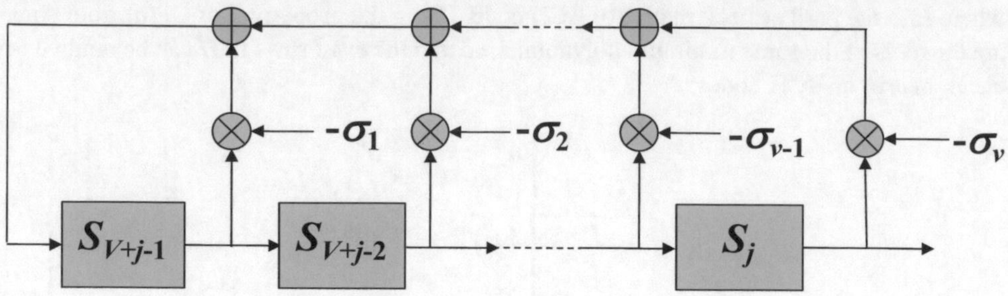

Fig. 9.3 A shift register that generates the syndromes S_j

algorithm [2–7]. In this algorithm the BCH decoding is observed as shift register synthesis problem: given the syndromes S_i, we have to find the minimal length shift register that generates the syndromes. Once we determine the coefficients of this shift register, we construct the *error locator* polynomial [2–7]:

$$\sigma(x) = \prod_{i=1}^{v}(1 + X_i x) = \sigma_v x^v + \sigma_{v-1}x^{v-1} + \ldots + \sigma_1 x + 1, \tag{9.29}$$

where the σ_i's, also known as elementary symmetric functions, are given by Viète's formulas:

$$\begin{aligned} \sigma_1 &= X_1 + X_2 + \ldots + X_v \\ \sigma_2 &= \sum_{i<j} X_i X_j \\ \sigma_3 &= \sum_{i<j<k} X_i X_j X_k \\ &\ldots \\ \sigma_v &= X_1 X_2 \ldots X_v \end{aligned} \tag{9.30}$$

Because $\{X_l\}$ are the inverses of the roots of $\sigma(x)$, $\sigma(1/X_l) = 0 \ \forall\ l$, and we can write [2–7]

$$X_l^v \sigma\left(X_l^{-1}\right) = X_l^v + \sigma_1 X_l^{v-1} + \ldots + \sigma_v. \tag{9.31}$$

By multiplying the previous equation by X_l^j and performing summation over l for fixed j, we obtain

$$S_{v+j} + \sigma_1 S_{v+j-1} + \ldots + \sigma_v S_j = 0;\ j = 1, 2, \ldots, v \tag{9.32}$$

This equation can be rewritten as follows:

$$S_{v+j} = -\sum_{i=1}^{v} \sigma_i S_{v+j-i};\ j = 1, 2, \ldots, v \tag{9.33}$$

where S_{v+j} represents the output of shift register as shown in Fig. 9.3. The Massey-Berlekamp algorithm is summarized by flowchart shown in Fig. 9.4, which is self-explanatory.

Once the locator polynomial is determined, we have to find the roots and invert them to obtain the error locators. To determine the error magnitudes, we have to define another polynomial, known as the *error-evaluator* polynomial, defined as follows [6]:

$$\varsigma(x) = 1 + (S_1 + \sigma_1)x + (S_2 + \sigma_1 S_1 + \sigma_2)x^2 + \ldots + (S_v + \sigma_1 S_{v-1} + \ldots + \sigma_v)x^v. \tag{9.34}$$

The error magnitudes are then obtained from [5]

$$Y_l = \frac{\varsigma\left(X_l^{-1}\right)}{\prod_{i=1,\, i\neq l}^{v}\left(1 + X_i X_l^{-1}\right)}. \tag{9.35}$$

For binary BCH codessince $Y_l = 1$ we do not need to evaluate the error magnitudes.

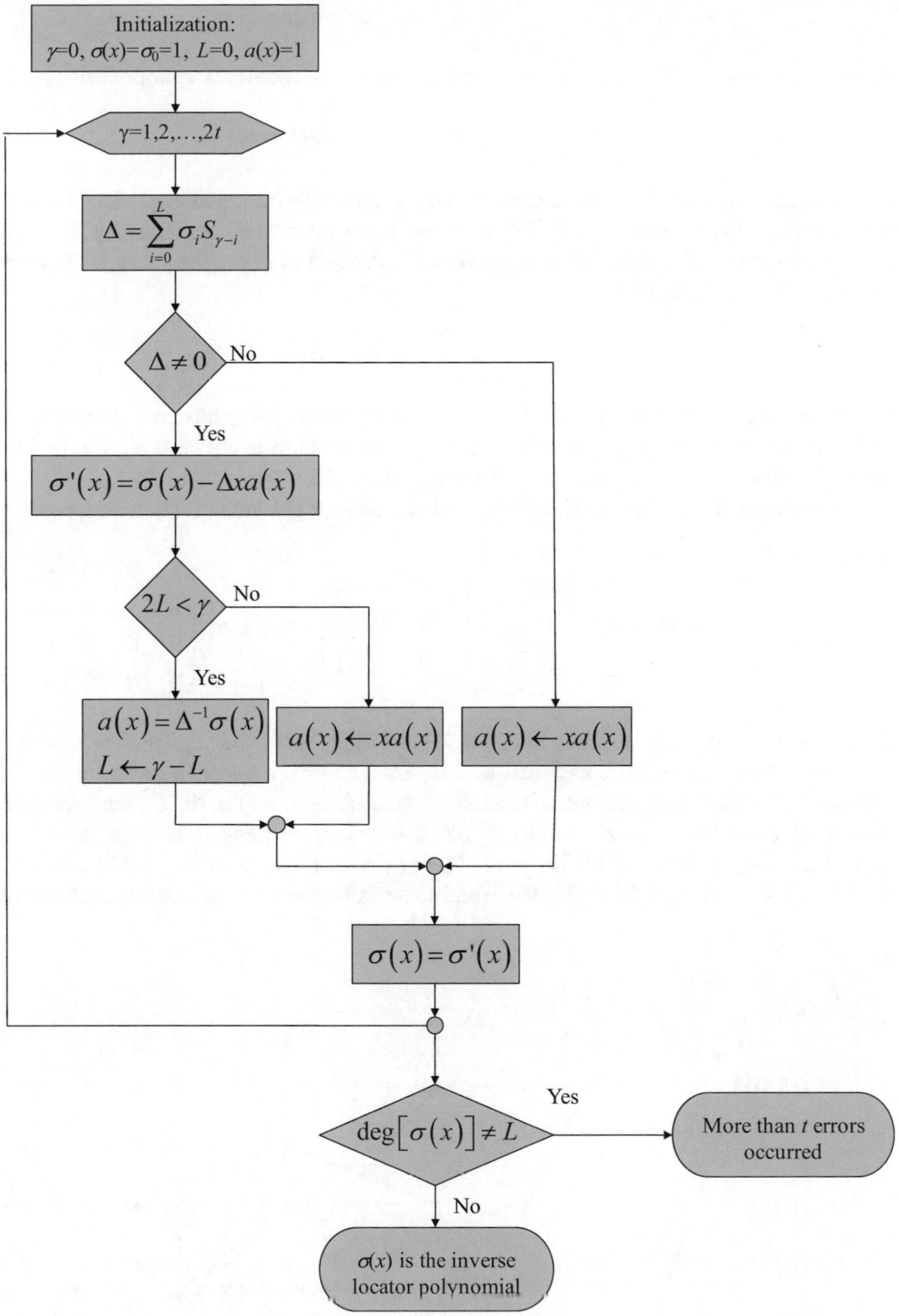

Fig. 9.4 Flowchart of Massey-Berlekamp algorithm. Δ denotes the error (discrepancy) between the syndrome and the shift register output; $a(x)$ stores the content of shift register (normalized by Δ^{-1}) prior the lengthening

9.3 Trellis Description of Linear Block Codes and Viterbi Decoding Algorithm

Since the bits in codeword are statistically dependent, the encoding process can be described by the finite state machine, with precisely determined initial and terminal states. We can further describe the encoding by means of trellis [13] and perform maximum likelihood decoding using the Viterbi algorithm [2–7].

Let linear block code (n,k) be defined over GF(2). The corresponding parity-check matrix of dimension $(n-k)\times n$ can be written as

$$\boldsymbol{H} = [\boldsymbol{h}_1 \boldsymbol{h}_2 \ldots \boldsymbol{h}_n], \tag{9.36}$$

where $\boldsymbol{h}_i$ denotes the i-th column of parity-check matrix. The codeword vector $\boldsymbol{c} = (c_1 c_2 \ldots c_n)$ satisfies the following equation:

$$\boldsymbol{c}\boldsymbol{H}^{\mathrm{T}} = 0. \tag{9.37}$$

Let the code be systematic with the first k bits representing the information bits and last $n - k$ bits representing the parity bits. The state at depth j is denoted by $\boldsymbol{S}_j$. The number of states in trellis is determined by 2^{n-k}, each represented in binary form as a vector of length $n - k$. Let the information vector be denoted by $\boldsymbol{m} = (m_1 m_2 \ldots m_k)$. The new state at depth j, $\boldsymbol{S}_j$, is related to the previous state $\boldsymbol{S}_{j-1}$ by [13]

$$\boldsymbol{S}_j = \boldsymbol{S}_{j-1} + c_j \boldsymbol{h}_j, \tag{9.38}$$

where $c_j = m_j$ (for $j \leq k$). For $j = k + 1, \ldots, n$ the corresponding parity bits are obtained by generalized parity checks according to the generator matrix $\boldsymbol{G}$. For terminal state $\boldsymbol{S}_n$ to be equal to initial state $\boldsymbol{S}_0$, Eq. (9.37) is to be satisfied. Different paths throughout the trellis correspond to 2^k different codewords.

For example, the parity-check and generator matrices of Hamming (7,4) code are given, respectively, as

$$\boldsymbol{H} = \begin{bmatrix} 1 & 1 & 0 & 1 & 1 & 0 & 0 \\ 1 & 0 & 1 & 1 & 0 & 1 & 0 \\ 1 & 1 & 1 & 0 & 0 & 0 & 1 \end{bmatrix} \qquad \boldsymbol{G} = \begin{bmatrix} 1 & 0 & 0 & 0 & 1 & 1 & 1 \\ 0 & 1 & 0 & 0 & 1 & 0 & 1 \\ 0 & 0 & 1 & 0 & 0 & 1 & 1 \\ 0 & 0 & 0 & 1 & 1 & 1 & 0 \end{bmatrix}$$

The corresponding trellis is shown in Fig. 9.5. The solid edges correspond to $c_j = 1$, while dashed edges to $c_j = 0$. There are $2^{n-k} = 8$ states in trellis represented as 3-tuples. The total number of paths through trellis is $2^k = 16$, and these paths correspond to different codewords. This trellis is created based on Eq. (9.38). For example, for $j = 1$ and $c_1 = 0$, the terminal state is obtained by $\boldsymbol{S}_1 = \boldsymbol{S}_0 + 0{\cdot}\boldsymbol{h}_1 = (000) + 0{\cdot}(111) = (000)$, while for $c_1 = 1$ the terminal state is obtained by $\boldsymbol{S}_1 = \boldsymbol{S}_0 + 1{\cdot}\boldsymbol{h}_1 = (000) + 0{\cdot}(111) = (111)$. For $j = 2$, initial state $\boldsymbol{S}_1 = (111)$, and $c_2 = 1$, we arrive at terminal state $\boldsymbol{S}_2 = \boldsymbol{S}_1 + 1{\cdot}\boldsymbol{h}_2 = (111) + 1{\cdot}(101) = (010)$ and so on. After $j = 3$ we can see that trellis is fully developed; there exist 16 paths

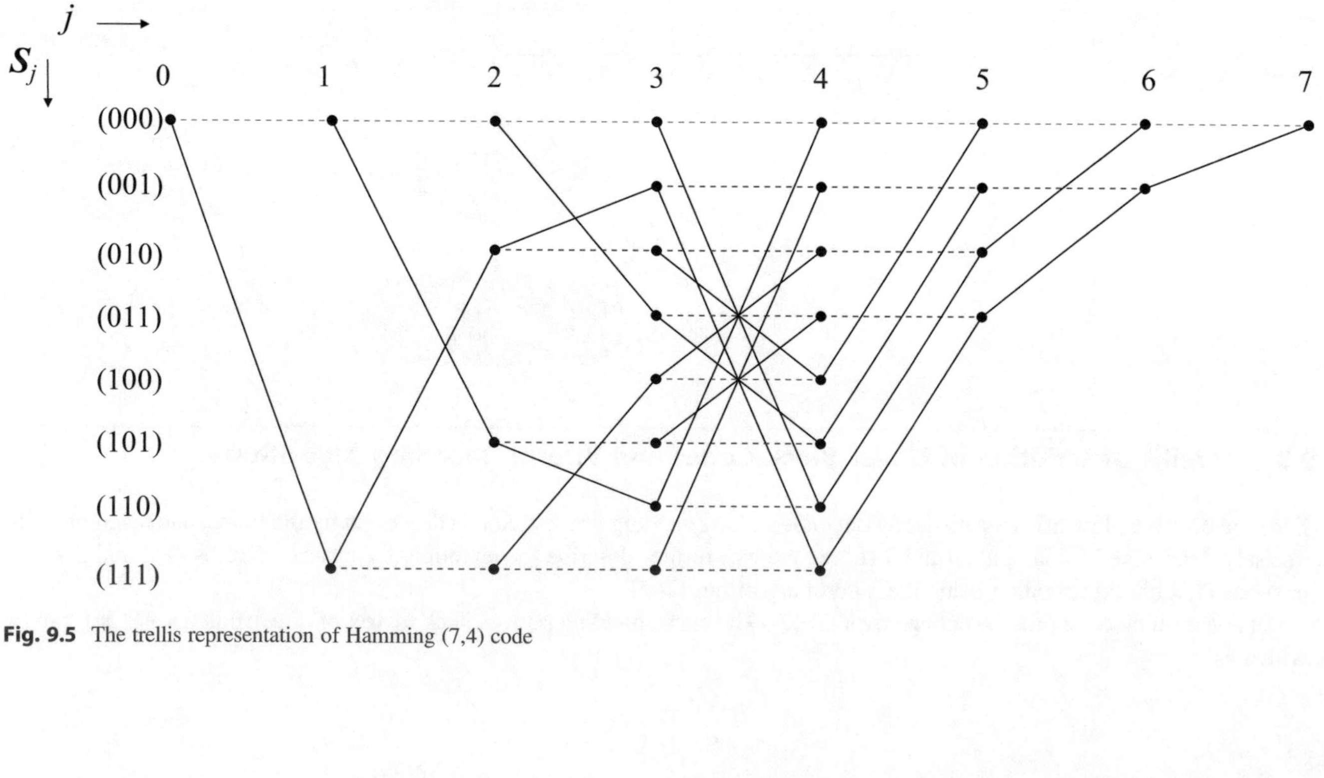

Fig. 9.5 The trellis representation of Hamming (7,4) code

leading to 8 nodes (with 2 edges reaching the terminal state). These 16 paths correspond to 16 possible information sequences of length 4. The number of possible transitions after $j = k$ reduces dramatically because the remaining bits c_5, c_6, and c_7 are parity bits, which are algebraically related to the previous information bits. For q-ary linear block codes, once the trellis is fully developed, every node will have q outgoing and q incoming edges. It is interesting to notice that the trellis for liner block codes is *time-variant*.

The decoding rule for choosing an estimate of the code vector $\boldsymbol{C}$, given the received vector $\boldsymbol{r}$, is optimum when the probability of decoding error is minimized. The *maximum likelihood (ML) decoder decision rule* can be formulated as follows:

Choose the estimate of $\boldsymbol{C}$ if the log-likelihood function $\log p_{\boldsymbol{R} \mid \boldsymbol{C}}(\boldsymbol{r} \mid \boldsymbol{c})$ is maximum.

The log-likelihood function, given the transmitted codeword $\boldsymbol{c} = (c_1 c_2 \ldots c_n)$, assuming that components of receiver word $\boldsymbol{r} = (r_1 r_2 \ldots r_n)$ are statistically independent and noise is additive can be written as follows:

$$\log p_{\boldsymbol{R}|\boldsymbol{C}}(\boldsymbol{r}|\boldsymbol{c}) = \sum_{i=1}^{n} \log \left[p_{N_i}(r_i|c_i)\right] = -\sum_{i=1}^{n} M_i(r_i, c_i) \triangleq -\mathrm{M}(\boldsymbol{C}), \tag{9.39}$$

where $p_{\boldsymbol{R}|\boldsymbol{C}}(\boldsymbol{r}|\boldsymbol{c})$ is the joint probability density function (PDF), $p_{N_i}(r_i|c_i)$ is the noise PDF, and M_i is the branch metric. The ML decoder provides the codeword $\boldsymbol{C}$ that minimizes the path metric $M(\boldsymbol{C})$. Consider BPSK coherent optical communication system with homodyne balanced detection and amplified spontaneous emission (ASE) noise dominated scenario. The corresponding PDF function will be

$$p_{N_i}(r_i|c_i) = \frac{1}{\sigma\sqrt{2\pi}} e^{-\frac{\left(r_i - c_i 2R_{PD}\sqrt{P_S P_{LO}}\right)^2}{2\sigma^2}}, \tag{9.40}$$

where R_{PD} is the photodiode responsivity; P_s and P_LO are the average powers of incoming and local laser, respectively; and σ^2 is the variance of equivalent noise process dominated by ASE noise. The corresponding log-likelihood function can be obtained by

$$\begin{aligned} \log p_{\boldsymbol{R}|\boldsymbol{C}}(\boldsymbol{r}|\boldsymbol{c}) &= -\sum_{i=1}^{n} \frac{\left(r_i - c_i 2R_{PD}\sqrt{P_S P_{LO}}\right)^2}{2\sigma^2} - \frac{n}{2}\log\left(\pi\sigma^2\right) \\ &= -A\sum_{i=1}^{n} r_i c_i + B = -A(\boldsymbol{r}\cdot\boldsymbol{c}) + B, \end{aligned} \tag{9.41}$$

where A and B are constants independent on transmitted codeword. The path metric is therefore the inner (scalar) product of the received word and transmitted codeword.

The brute force decoding method would mean to try all possible codewords 2^k and select one that minimizes the path metric $M(\boldsymbol{C}) = \boldsymbol{r}\cdot\boldsymbol{c}$. The better option would be to use *Viterbi algorithm* [2–7, 14], which is a recursive algorithm, whereby many codewords can be discarded from consideration in determination of $\boldsymbol{c}$ that minimizes the path metric. Alternatively, we can use the square Euclidean metric, which is equivalent to the correlation metric as shown by Eq. (9.41). The square Euclidean metric can also be used for ASK transmission, wherein $c_i \in \{0,1\}$. Below we provide an illustrative example that employs the squared Euclidean metric and ASK. To keep the exposition simple, we perform the following normalization, $2\sigma^2 = 1$ and $2R_{PD}\sqrt{(P_s P_{LO})} = 1$, and assume that Hamming (7,4) code (with corresponding trellis shown in Fig. 9.5) is used. Let the received word be as follows $\boldsymbol{r} = (0.4, 0.2, 0.5, 0.1, 0.2, 0.1, 0.2)$, while the all-zero codeword was transmitted. The first four samples correspond to information bits, and last three to the parity bits. The decoding starts when all branches merge together for the first time, which occurs at trellis depth $j = 4$. The cumulative metric for (000)- > (000) path at depth $j = 4$, corresponding to the information sequence 0000, is $(0.4 - 0)^2 + (0.2 - 0)^2 + (0.5 - 0)^2 + (0.1 - 0)^2 = 0.46$; while the cumulative metric for (000)- > (000) path at depth $j = 4$, corresponding to the information sequence 0111, is $(0.4 - 0)^2 + (0.2 - 1)^2 + (0.5 - 1)^2 + (0.1 - 1)^2 = 1.86$. Out of two possible paths, the path with lower accumulated metric (0.46) is preserved and is commonly referred to as the *survivor* path. The other path is discarded from further consideration. The similar procedure is repeated for all state nodes at level $j = 4$, and the survivor paths are denoted as bolded letters in Table 9.1. In step 2 ($j = 5$), we repeat the similar procedure using the accumulated metric from survived paths as the starting

Table 9.1 Illustrating the soft Viterbi decoding of Hamming (7,4) code

	Step 1	Step 2	Step 3	Step 4
Received samples → Paths ↓	0.4, 0.2, 0.5, 0.1	0.2	0.1	0.2
(000) → (000)	0000: **0.46** 0111: 1.86	0: 0.46 + (0.2–0)^2 = **0.5** 1: 0.66 + (0.2–1)^2 = 1.3	0: 0.5 + (0.1–0)^2 = **0.51** 1: 1.3 + (0.1–1)^2 = 2.11	0: 0.51 + (0.2–0)^2 = **0.55** 1: 1.31 + (0.2–1)^2 = 1.95
(000) → (001)	1110: **1.26** 1001: 1.46	0: 1.26 + (0.2–0)^2 = **1.3** 1: 1.06 + (0.2–1)^2 = 1.7	0: 1.3 + (0.1–0)^2 = 1.31 1: 0.5 + (0.1–1)^2 = **1.31**	
(000) → (010)	1100: **1.26** 1011: 1.46	0: 1.26 + (0.2–0)^2 = **1.3** 1: 1.06 + (0.2–1)^2 = 1.7		
(000) → (011)	0010: **0.46** 0101: 1.86	0: 0.46 + (0.2–0)^2 = **0.5** 1: 0.66 + (0.2–1)^2 = 1.3		
(000) → (100)	1101: 2.06 1010: **0.66**			
(000) → (101)	0011: 1.26 0100: **1.06**			
(000) → (110)	0001: 1.26 0110: **1.06**			
(000) → (111)	1111: 2.06 1000: **0.66**			

Adopted from [5]

point. When two paths have the same accumulated metric, we flip the coin to determine which one to choose without affecting the final result. We see that in step 5 the only survived path is all-zeros path, and the transmitted codeword is properly decoded. We can also introduce a single error, for example, by setting 0.5 sample to 0.6, and repeat the decoding procedure. The all-zero codeword will be detected again. However, when two errors are introduced, the Viterbi decoder will fail to determine the correct codeword. On the other hand, by checking the syndrome equation, we can detect the double error.

If the probability of error is independent of the transmitted bit value, the corresponding channel is called the binary symmetric channel (BSC). This channel is completely described by transition (crossover) probability: $P(r_i = 1| c_i = 0) = P(r_i = 0| c_i = 1) = p$. Because

$$p(r_i|c_i) = \begin{cases} p, & r_i \neq c_i \\ 1-p, & r_i = c_i \end{cases},$$

the corresponding log-likelihood function can be written as follows:

$$\begin{aligned} \log p(\mathbf{r}|\mathbf{c}) &= d_H \log p + (n - d_H)\log(1-p) \\ &= d_H \log\left(\frac{p}{1-p}\right) + n\log(1-p), \end{aligned} \tag{9.42}$$

where d_H is the Hamming distance between the received word $\boldsymbol{r}$ and transmitted codeword $\mathbf{c}$. Therefore, for $p < 1/2$ the *ML rule for BSC* can be formulated as:

Choose the estimate $\boldsymbol{C}$ that minimizes the Hamming distance between the received vector $\boldsymbol{r}$ and codeword $\mathbf{c}$.

For intensity modulation with direct detection (IM/DD), it was found in [15] that distribution of PDF tails is exponential rather than Gaussian, so that square root metric is more suitable than Euclidean metric. Some other metrics suitable for IM/DD systems are discussed in [16]. For high-speed long-haul optical communication system, such as those operating at 40 Gb/s and above, the influence of self-phase modulation (intrachannel four-wave mixing in systems with direct detection and nonlinear phase noise in systems with coherent detection) is important, and we have to use the metric given by Eq. (9.42), providing that interleaving is used so that neighboring symbols are independent of each other. The PDF has to be estimated by histogram methods. An alternative would be to use the edgeworth expansion method as we described in [17].

So far we were concerned with different classes of linear block codes. Another important class of codes is based on convolutional codes. Convolutional codes of practical importance are commonly $(n,1)$ codes, whose code rate is low $1/n$, and as such are not suitable for fiber-optics communication, because the penalty due to chromatic dispersion is highly severe at high-data rates. The code rate can be increased by puncturing, at the expense of performance degradation. On the other hand, the recursive systematic convolutional (RSC) codes are usually employed as component codes for turbo codes that are used in wireless and deep-space communications.

9.4 Convolutional Codes

In a systematic (n,k) linear block code, information bits get unchanged during the encoding process; only $n - k$ parity-check bits are added that are algebraically related to the information bits. If the information bits are statistically independent, the parity (redundant) bits will only be dependent on the information block of current codeword, so that the codewords are statistically independent. On the other hand, in an (n,k,M) convolutional code, the parity bits are not only the function of current information k-tuple but also the function of previous m information k-tuples. The statistical dependence is therefore introduced in a window of $K = n(M + 1)$ symbols, which is known as the *constraint length*. The constrained length represents the number of coded symbols influenced by a single-message symbol. The general architecture of a convolutional code is shown in Fig. 9.6. The k information bits are taken during encoding process from a serial-to-parallel converter (S/P) that follows an information buffer as the input of the encoder memory of size k-by-M. The same bits are also used as the input to the combinational logic, which determines the n-outputs based on current k-tuple and previous Mk bits. The logic outputs are written in parallel to the output shift register, and the codeword is transmitted in a serial fashion over the channel.

An example of nonsystematic convolutional (2,1,2) code is shown in Fig. 9.7. This encoder has the code rate $R = 1/2$ and has $M = 2$ flip-flops as memory elements. Let the input sequence to the encoder be denoted by $\boldsymbol{m} = (m_0\ m_1\ \ldots\ m_l\ \ldots)$.

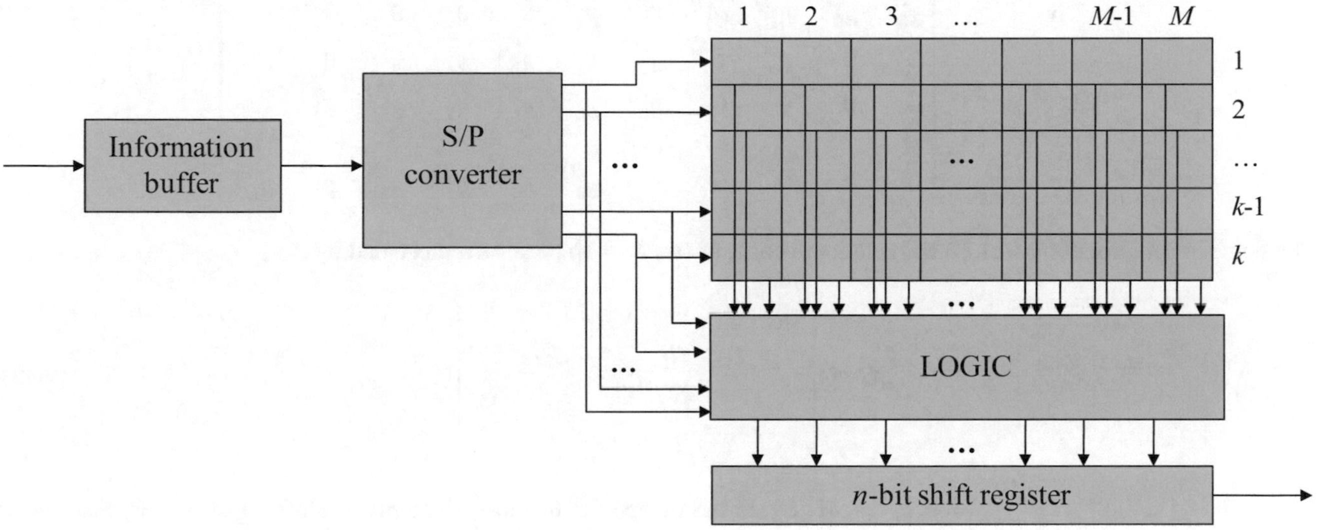

Fig. 9.6 Illustrating the operational principle of convolutional codes

Fig. 9.7 Nonsystematic convolutional (2,1,1) encoder

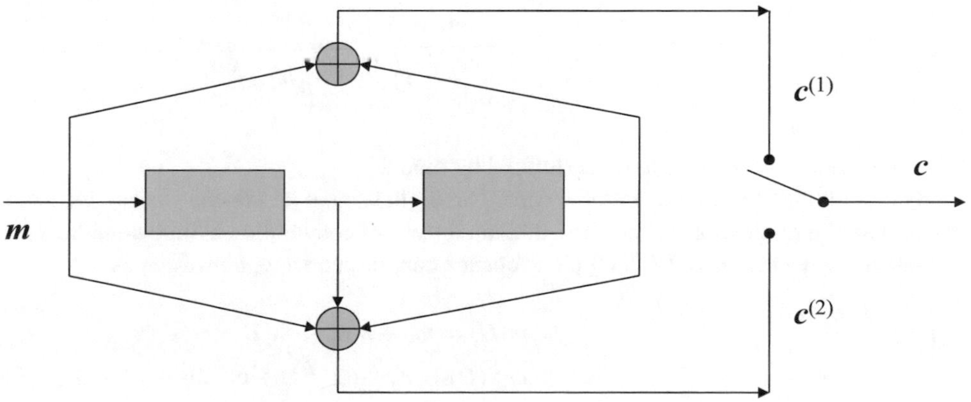

The corresponding output sequences are denoted with $\boldsymbol{c}^{(1)} = \left(c_0^{(1)} c_1^{(1)} \ldots c_l^{(1)} \ldots\right)$ and $\boldsymbol{c}^{(2)} = \left(c_0^{(2)} c_1^{(2)} \ldots c_l^{(2)} \ldots\right)$, respectively. The encoder sequence is obtained by multiplexing two output sequences as follows: $\boldsymbol{c} = \left(c_0^{(1)} c_0^{(2)} c_1^{(1)} c_1^{(2)} \ldots c_l^{(1)} c_l^{(2)} \ldots\right)$. For convolutional (2,1,2) code shown in Fig. 9.7, the output sequences are related to the input sequence by

$$c_l^{(1)} = m_l + m_{l-2}, \quad c_l^{(2)} = m_l + m_{l-1} + m_{l-2}.$$

Another approach to describe the encoding operation is by convolution of the input sequence and corresponding *generator sequences* $\boldsymbol{g}^{(1)} = (101)$ and $\boldsymbol{g}^{(2)} = (111)$ as follows:

$$\boldsymbol{c}^{(1)} = \boldsymbol{m} * \boldsymbol{g}^{(1)}, \quad \boldsymbol{c}^{(2)} = \boldsymbol{m} * \boldsymbol{g}^{(2)},$$

where the generator sequences are obtained as the response to the unit step sequence (100...). The generator sequence, representing the impulse response of corresponding input-output path can also be determined as the sequence of connections from the shift register to the pertinent adder, with a 1 representing the existence of connection and a 0 representing the absence of connection.

The encoding process in convolutional encoder can also be represented in similar fashion to that of the block codes, as matrix product of the input message vector $\boldsymbol{m}$ and generator matrix $\boldsymbol{G}$:

$$\boldsymbol{c} = \boldsymbol{m}\boldsymbol{G}, \ \boldsymbol{G} = \begin{bmatrix} g_0^{(1)} & g_0^{(2)} & g_1^{(1)} & g_1^{(2)} & g_2^{(1)} & g_2^{(2)} & 0 & 0 & 0 & 0 & \ldots \\ 0 & 0 & g_0^{(1)} & g_0^{(2)} & g_1^{(1)} & g_1^{(2)} & g_2^{(1)} & g_2^{(2)} & 0 & 0 & \ldots \\ 0 & 0 & 0 & 0 & g_0^{(1)} & g_0^{(2)} & g_1^{(1)} & g_1^{(2)} & g_2^{(1)} & g_2^{(2)} & \ldots \\ \vdots & & \vdots & & \ddots & & \ddots & & \vdots & & \vdots \\ 0 & 0 & 0 & 0 & \cdots & g_0^{(1)} & g_0^{(2)} & g_1^{(1)} & g_1^{(2)} & g_2^{(1)} & g_2^{(2)} \end{bmatrix}.$$

The generator matrix of an $(n,1,m)$ convolutional code, of rate $R = 1/n$, can be represented by

$$\boldsymbol{G} = \begin{bmatrix} \boldsymbol{G}_0 & \boldsymbol{G}_1 & \cdots & \boldsymbol{G}_M & & & \\ & \boldsymbol{G}_0 & \boldsymbol{G}_1 & \cdots & \boldsymbol{G}_M & & \\ & & \boldsymbol{G}_0 & \boldsymbol{G}_1 & \cdots & \boldsymbol{G}_M & \\ & & & \ddots & & & \end{bmatrix}, \tag{9.43}$$

where $\boldsymbol{G}_l = \left(g_l^{(1)} g_l^{(2)} \ldots g_l^{(n)}\right); l = 0, 1, \ldots, \boldsymbol{M}$. Since the number of memory elements in shift register is M, there exist n generator sequences $\boldsymbol{g}^{(i)}$ ($i = 1,2,\ldots,n$) of length $M + 1$, and n output sequences $\boldsymbol{c}^{(l)}$ ($l = 1,\ldots,n$) are multiplexed before being transmitted over the channel. The j-th output symbol and l-th input symbol are related by the following convolution relationship:

$$c_l^{(j)} = \sum_{i=0}^{M} m_{l-i} g_i^{(j)}. \tag{9.44}$$

(The summation in (9.44) is to be performed by mod 2.)

By employing the delay operator D (or z^{-1} in digital signal processing books), we can simplify the encoding process and perform the multiplication in transform domain instead of convolution in time domain. The transform-domain representations of message sequence $\boldsymbol{m}$ and j-th output sequence can, respectively, be written as

$$\begin{aligned} m(D) &= m_0 + m_1 D + m_2 D^2 + \ldots \\ c^{(j)}(D) &= c_0^{(j)} + c_1^{(j)} D + c_2^{(j)} D^2 + \ldots \end{aligned} \tag{9.45}$$

For (2,1,2) convolutional encoder shown in Fig. 9.7, the corresponding transform-domain representations of generator polynomials can be written as

$$g_1(D) = 1 + D^2 \qquad g_2(D) = 1 + D + D^2.$$

The corresponding transform-domain generator matrix $\boldsymbol{G}(D)$ will be

$$\boldsymbol{G}(D) = [\, g_1(D) \quad g_2(D) \,] = \left[\, 1 + D^2 \quad 1 + D + D^2 \,\right].$$

The output sequences in transform domain can be written as

$$c^{(1)}(D) = m(D)g_1(D) \qquad c^{(2)}(D) = m(D)g_2(D).$$

The multiplexed sequence in transform domain is as follows:

$$c(D) = c^{(1)}(D) + Dc^{(2)}(D).$$

For an arbitrary $(n,1,m)$ convolutional code, the corresponding generator matrix in transform domain can be represented as

$$\boldsymbol{G}(D) = [\, g_1(D) \quad g_2(D) \quad \cdots \quad g_n(D) \,], \tag{9.46}$$

while the transform domain of output sequence can be represented by

$$c(D) = m(D)\boldsymbol{G}(D), \qquad c(D) = \left[c^{(1)}(D)\ c^{(2)}(D)\ \cdots\ c^{(n)}(D) \right]. \tag{9.47a}$$

For an arbitrary (n,k,M) convolutional code, we have to split the input sequence into k subsequences $\boldsymbol{m} = [\boldsymbol{m}_1\, \boldsymbol{m}_2 \ldots \boldsymbol{m}_k]$ and define the generator matrix as

$$\boldsymbol{G} = \begin{bmatrix} \boldsymbol{G}_0 & \boldsymbol{G}_1 & \cdots & \boldsymbol{G}_M & & & \\ & \boldsymbol{G}_0 & \boldsymbol{G}_1 & \cdots & \boldsymbol{G}_M & & \\ & & \boldsymbol{G}_0 & \boldsymbol{G}_1 & \cdots & \boldsymbol{G}_M & \\ & & & \ddots & & & \end{bmatrix}, \tag{9.47b}$$

where

$$\boldsymbol{G}_l = \begin{bmatrix} g_l^{(11)} & g_l^{(12)} & \cdots & g_l^{(1n)} \\ g_l^{(21)} & g_l^{(22)} & \cdots & g_l^{(2n)} \\ \vdots & \vdots & \ddots & \vdots \\ g_l^{(k1)} & g_l^{(k2)} & \cdots & g_l^{(kn)} \end{bmatrix},$$

and proceed in similar fashion as above.

The convolutional codes can also be put in systematic form, by transforming the generator matrix and representing it into the following form:

$$\boldsymbol{G}(D) = [\, \boldsymbol{I} \quad \boldsymbol{P}(D) \,], \tag{9.48}$$

where $\boldsymbol{I}$ is $k \times k$ identity matrix and $\boldsymbol{P}(D)$ is $kx(n-k)$ matrix of polynomials. The nonsystematic generator matrix of linear block codes can easily be transformed in systematic form by elementary operations per rows, while this operation is not always possible to do for the convolutional codes. Namely, the two generator matrices are equivalent if they generate the same convolutional code. Let us again observe the (2,1,2) convolutional code shown in Fig. 9.7, whose generator matrix is $\boldsymbol{G}(D) = [1 + D^2\ 1 + D + D^2]$. We can perform the following transformation:

$$\begin{aligned} c(D) &= m(D)\boldsymbol{G}(D) = m(D)\left[\, 1 + D^2 \quad 1 + D + D^2 \,\right] = m(D)\left(1 + D^2\right)\left[\, 1 \quad \frac{1 + D + D^2}{1 + D^2} \,\right] \\ &= m'(D)\boldsymbol{G}'(D), \end{aligned}$$

where

$$m'(D) = m(D)\boldsymbol{T}(D);\ \boldsymbol{T}[D] = \left[1 + D^2\right],\ \boldsymbol{G}'(D) = \left[\, 1 \quad \frac{1 + D + D^2}{1 + D^2} \,\right].$$

Because both generator matrices, $\boldsymbol{G}$(D) and $\boldsymbol{G}'(D)$, will generate the same output sequence for the same input sequence, we can say that these two matrices are equivalent, which represents a necessary condition. For two generator matrices to be equivalent, the $\boldsymbol{T}$(D) matrix should be invertible, which represents the sufficient condition. In previous example, m(D) can be obtained from $m'(D)$ by

$$m(D) = m'(D)\boldsymbol{T}^{-1}(D);\ \boldsymbol{T}^{-1}[D] = \left[\frac{1}{1 + D^2}\right].$$

Therefore, two generator matrices $\boldsymbol{G}(\boldsymbol{D})$ and $\boldsymbol{G}'(D) = \boldsymbol{T}(D)\boldsymbol{G}(D)$ are equivalent if the matrix $\boldsymbol{T}(D)$ is invertible. In the example above, $\boldsymbol{G}'(\boldsymbol{D})$ is in systematic form; unfortunately its elements are not polynomials but rather rational functions. The parity sequence can be generated by first multiplying the input sequence by $1 + D + D^2$ and then dividing by $1 + D^2$. One possible implementation is shown in Fig. 9.8.

Similar to the linear block codes, we can define the parity-check matrix in transform domain $\boldsymbol{H}(D)$ (of size $(n\text{-}k)$xk) that satisfies the following equation:

$$\boldsymbol{G}(D)\boldsymbol{H}^T(D) = \boldsymbol{0}. \tag{9.49}$$

Every codeword $c(D)$ satisfies the following equation:

$$\boldsymbol{c}(D)\boldsymbol{H}^T(D) = \boldsymbol{0}. \tag{9.50}$$

If $\boldsymbol{G}(\boldsymbol{D})$ is the generator matrix of a systematic convolutional code, given by Eq. (9.48), the corresponding parity-check matrix can be written in the following format:

$$\boldsymbol{H}(D) = \left[\boldsymbol{P}^T(D) \quad \boldsymbol{I}\right], \tag{9.51}$$

where $\boldsymbol{I}$ is $(n - k) \times (n - k)$ identity matrix. For example, shown in Fig. 9.8, the parity-check matrix in transform domain can be written as

$$\boldsymbol{H}(D) = \left[\frac{1 + D + D^2}{1 + D^2} \quad 1\right].$$

For nonsystematic codes, the code sequence on receiver $c(D)$ should be multiplied by $\boldsymbol{G}^{-1}(D)$ to determine the information sequence:

$$c(D)\boldsymbol{G}^{-1}(D) = m(D)\boldsymbol{G}(D)\boldsymbol{G}^{-1}(D) = m(D)D^l, \tag{9.52}$$

which can be done if the following is valid:

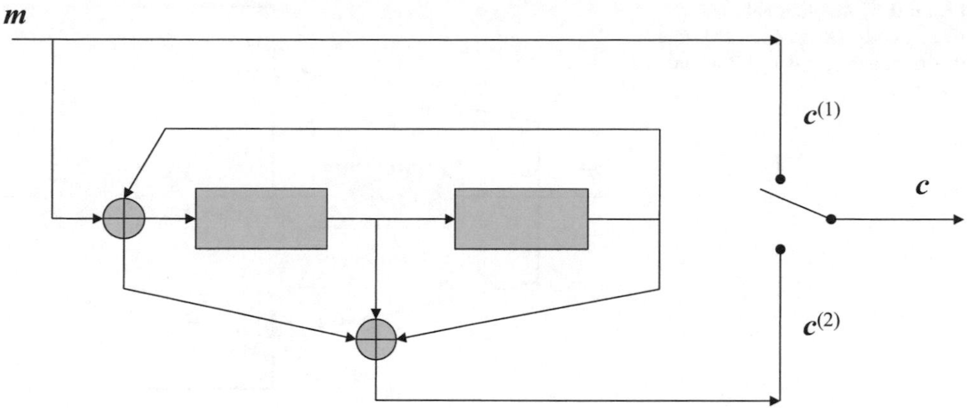

Fig. 9.8 The systematic encoder equivalent to the nonsystematic encoder shown in Fig. 9.7

$$\boldsymbol{G}(D)\boldsymbol{G}^{-1}(D) = \boldsymbol{I}D^l, l \geq 0 \tag{9.53}$$

where the multiplication with D^l corresponds to the delay. Therefore, the generator matrix must be invertible. If not, the corresponding convolutional code is *catastrophic*, that is, the finite number of channel errors will introduce infinite number of decoding errors. The sufficient condition for the convolutional code to be non-catastrophic is that the greatest common divisor (GCD) of all submatrices of size *kxk* be equal to D^l. (The number of these submatrices is $\binom{n}{k}$.) The systematic codes cannot be catastrophic because their generator matrix is invertible. Any non-catastrophic convolutional code can be transformed into systematic equivalent. For our example shown in Fig. 9.8, the GCD can be found as

$$\mathrm{GCD}\left(1 + D^2, 1 + D + D^2\right) = 1 = D^0.$$

Therefore, for the $(n,1,M)$ codes, it is sufficient not to have any common roots among generator polynomials except probable one of the form D^l.

The convolutional codes can be considered as being a subclass of so-called trellis codes. The trellis codes can be described as finite state machines, with the output being the function of *Mk* previous bits and *k* current bits, whose output can be described in the form of trellis. If the trellis encoder is time-invariant and linear, the corresponding code is called the convolutional code.

Similarly as in block codes where the error correction capability is related to the minimum Hamming distance and decoding algorithm, the performance of convolutional codes is determined by decoding algorithm and so-called *free distance* of the code, $d_{\text{free.}}$ The free distance of convolutional codes is defined as the minimum Hamming distance between any two codewords in the code. Because the convolutional codes contain all-zero codeword and belong to the class of linear codes, the free distance can be determined as the minimum weight output codeword sequence caused by a certain nonzero input sequence. The following three algorithms are commonly used in decoding of convolutional codes: (i) majority logic decoding, (ii) sequential decoding, and (iii) decoding by Viterbi algorithm. The Viterbi algorithm is the most popular and it is already explained in the previous section. The only difference is in trellis, which is in case of convolutional codes time-invariant, so that Viterbi decoder of convolutional codes is simpler to implement than that of block codes. The description of other two decoding algorithms can be found in any textbook on channel coding [3–7]. The calculation of free distance of convolutional codes is described below.

The calculation of free distance of convolutional codes is closely related to the *generating function* of a convolutional code [3–7], i.e., the transfer function of the encoder with respect to state transitions. We have shown in Fig. 9.6 that convolutional codes can be represented using the finite state machine (FSM) approach, and its operation can be described using the state diagram. The state of convolutional encoder of rate $1/n$ is defined by the sequence of M most recent bits ($m_{j\text{-}M}, m_{j-M+1}, \ldots, m_{j-1}, m_j$) moved into encoder's shift register, and it is denoted by ($m_{j-1}, \ldots, m_{j-M}$), where m_j denotes the current bit. For example, for convolutional encoder shown in Fig. 9.9a with $M = 2$ flip-flops, there are $2^M = 4$ possible states denoted by *a* (00), *b* (10), c (01), and *d* (11). The corresponding state diagram of this encoder is shown in Fig. 9.9b, c. The FSM states correspond to the 2-bit states of shift register. The pair of labels above edges, in Fig. 9.9b, represents the outputs of encoder for a given input. For example, if the content of shift register is 10 (we are in state 10) and incoming bit is 1, the generated output sequence is 01, and in the next clock cycle, the content of shift register will be 11, meaning that the terminal state is 11.

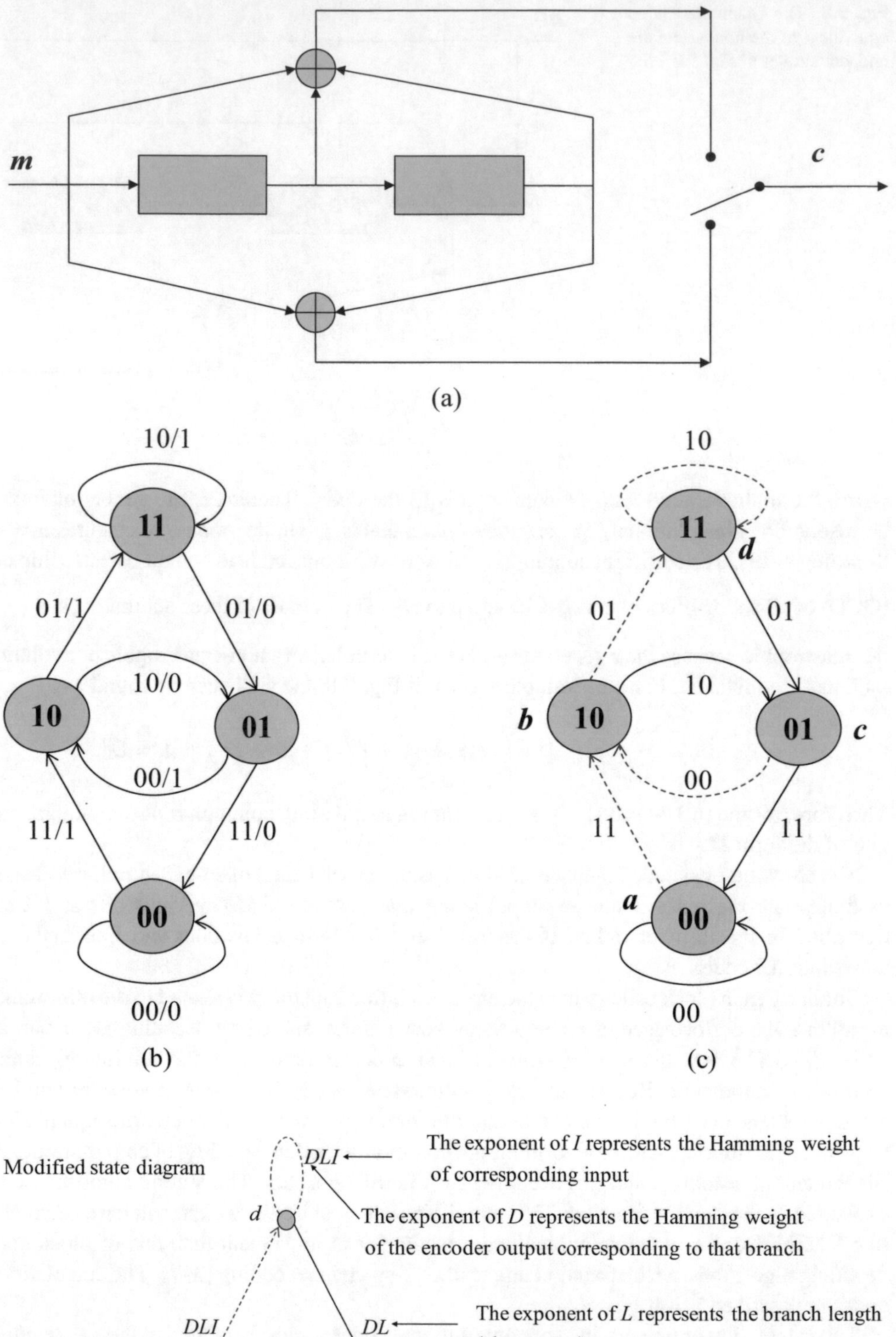

Fig. 9.9 Convolutional (2,1,2) code: (**a**) encoder, (**b**) state diagram, and (**c**) version 2 of state diagram

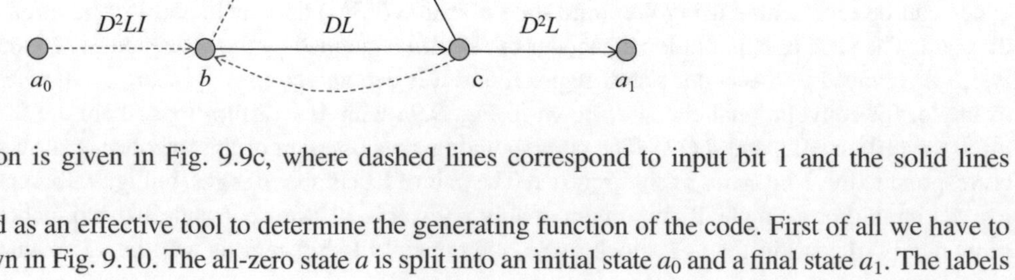

Fig. 9.10 The modified state diagram of convolutional (2,1,2) code from Fig. 9.9a

Another equivalent representation is given in Fig. 9.9c, where dashed lines correspond to input bit 1 and the solid lines correspond to input bit 0.

The state diagram can be used as an effective tool to determine the generating function of the code. First of all we have to modify the state diagram as shown in Fig. 9.10. The all-zero state a is split into an initial state a_0 and a final state a_1. The labels

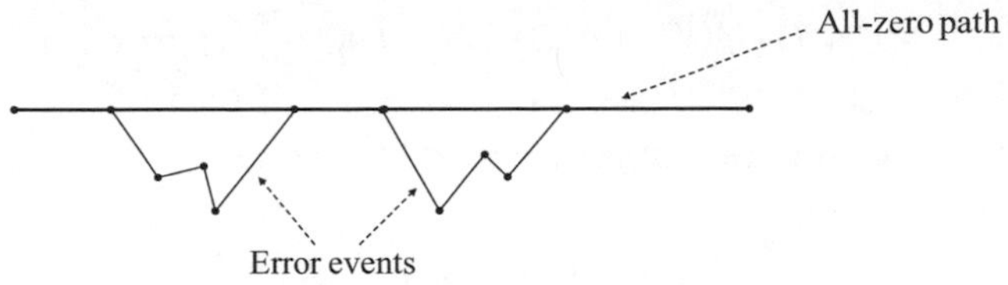

Fig. 9.11 Introducing the concept of error event

of edges are changed as well. The label of each edge is composed of three letters with corresponding exponents. The exponent of I represents the Hamming weight of input (for bit zero we obtain $I^0 = 1$), the exponent of L is the branch length, and the exponent of D represents the Hamming weight of the encoder output corresponding to the input of observed edge. With different paths from initial to final state in modified state diagram, we associate the corresponding label by concatenating the edge labels. For example, the path a_0bdca_1 has the label $D^6L^4I^3$. The path starting at a_0 and ending at a_1 is called the fundamental path. If the number of paths starting at a_0 and ending at a_1 with label $D^dL^lI^i$ is denoted by $T_{d,l,i}$, the following generating function will represent the *complete path enumerator* [2–7, 18]:

$$T(D,L,I) = \sum_{d=1}^{\infty}\sum_{l=1}^{\infty}\sum_{i=1}^{\infty} T_{d,l,i}D^dL^lI^i. \tag{9.54}$$

For example, shown in Fig. 9.10, we can determine the $T(D,L,I)$ by using Mason's gain formula [19] from signal-flaw graph theory to obtain [18]

$$T(D,L,I) = \frac{D^5L^3I}{1 - DLI(1+L)} = D^5L^3I + D^6L^4I^2(1+L) + D^7L^5I^3(1+L)^2 + \ldots \tag{9.55}$$

From Eq. (9.55) we conclude that there are no fundamental paths at distance 0, 1, 2, or 3 from the all-zero path; and there is a single fundamental path at distance 5 from all-zero path meaning that the free distance of this convolutional code is $d_{\text{free}} = 5$. This implies that Viterbi decoder can correct up to two errors in decoded sequence.

The modified state diagram can also be used to estimate the bit error ratios (BERs). Without loss of generality, for an AWGN channel model, we will assume that the all-zero codeword was transmitted ($\mathbf{c} = \mathbf{0}$). We will say that an *error event* has occurred if a nonzero codeword survives at state $\mathbf{0}$, therefore eliminating the correct path (all-zero codeword), which is illustrated in Fig. 9.11.

The probability that the first error event E of weight d occurs is given by [6]

$$P(E,d) = \begin{cases} \sum_{i=\frac{d+1}{2}}^{d} \binom{d}{i} p^i(1-p)^{d-i}, & d \text{ is odd} \\ \frac{1}{2}\binom{d}{d/2} p^{d/2}(1-p)^{d/2} + \sum_{i=\frac{d}{2}+1}^{d} \binom{d}{i} p^i(1-p)^{d-i}, & d \text{ is even} \end{cases} \tag{9.56}$$

which is always smaller than $P(E, d) < 2^d p^{d/2}(1 - p)^{d/2}$, where p is the crossover probability of BSC. The event error probability upper bound can be determined by [6].

$$P_E < \sum_{d=d_{\text{free}}}^{\infty} T_d P(E,d), \tag{9.57}$$

where T_{d} denotes the number of codewords of weight d. Combining (9.56) and (9.57), we obtain

$$P_E < T_d[4p(1-p)]^{d/2} = T(D,I)|_{I=1,D=\sqrt{4p(1-p)}}. \tag{9.58}$$

The BER of a convolutional code of rate $r = k/n$ is upper bounded by [6]

$$BER \leq \frac{1}{k}\sum_{d=d_{\text{free}}}^{\infty} \beta_d P(E,d) \tag{9.59}$$

where β_d denotes the number of nonzero data bits carried by all T_d paths of weight d. By setting $L = 1$ in (9.55), we obtain

$$T(D,I) = \sum_{d=d_{\text{free}}}^{\infty}\sum_{i=1}^{\infty} T_{d,i}D^d I^i. \tag{9.60}$$

By differentiating $T(D,I)$ with respect to I and by setting $I = 1$ upon differentiation, we obtain

$$\left.\frac{\partial T(D,I)}{\partial I}\right|_{I=1} = \sum_{d=d_{\text{free}}}^{\infty}\sum_{i=1}^{\infty} iT_{d,i}D^d = \sum_{d=d_{\text{free}}}^{\infty}\beta_d D^d, \quad \beta_d = \sum_{i=1}^{\infty} iT_{d,i} \tag{9.61}$$

Finally, by comparing (9.59) and (9.61), we conclude that BER is upper bounded by

$$BER \leq \frac{1}{k}\left.\frac{\partial T(D,I)}{\partial I}\right|_{I=1} = \frac{1}{k}\sum_{d=d_{\text{free}}}^{\infty}\beta_d D^d. \tag{9.62}$$

For binary symmetric channel, the crossover probability is

$$p = \frac{1}{2}\text{erfc}\left(\sqrt{\frac{E_b}{N_0}}\right) \approx \frac{1}{2\sqrt{\pi}}\exp\left(-\frac{E_b}{N_0}\right), \tag{9.63}$$

where E_b/N_0 is electrical bit energy per power spectral density ratio, and the parameter D from (9.62) can be estimated as

$$D = \sqrt{4p(1-p)} \approx 1.06\exp\left(-\frac{E_b}{N_0}\right). \tag{9.64a}$$

By substituting (9.64a) into (9.62), we obtain

$$BER \leq \frac{1}{k}\sum_{d=d_{\text{free}}}^{\infty}\beta_d D^d \approx \frac{1}{k}\sum_{d=d_{\text{free}}}^{\infty} 1.06^d \beta_d \exp\left(-\frac{dRE_b}{2N_0}\right), \tag{9.64b}$$

where R is the code rate. For high SNRs only the first term in summation above dominates:

$$BER \approx \frac{1}{k}1.06^{d_{\text{free}}}\beta_d \exp\left(-\frac{d_{\text{free}}RE_b}{2N_0}\right) \quad \left(\text{for large } \frac{E_b}{N_0}\right) \tag{9.65}$$

The corresponding expression for uncoded BER at high SNR can be estimated by

$$BER_{\text{unc}} \approx 0.282\exp\left(-\frac{E_b}{N_0}\right). \tag{9.66}$$

By comparing the exponents in (9.65) and (9.66), we conclude that argument in coded case is $d_{\text{free}}R/2$ times larger, so that corresponding hard-decision asymptotic coding gain is

$$G_h = 10\log_{10}\left(\frac{d_{\text{free}}R}{2}\right) \quad [\text{dB}]. \tag{9.67}$$

For binary input AWGN channel, we can repeat the similar procedure and obtain the following asymptotic coding gain for soft decoding:

$$G_s = 10\log_{10}(d_{\text{free}}R) \quad [\text{dB}], \tag{9.68}$$

which is 3 dB better than that for hard-decision decoding.

9.5 RS Codes, Concatenated Codes, and Product Codes

The Reed-Solomon (RS) codes were discovered in 1960 and represent a special class of nonbinary BCH codes [2–7, 20]. RS codes represent the most commonly used nonbinary codes. Both the code symbols and the roots of generating polynomial are from the locator field. In other words, the symbol field and locator field are the same. The codeword length of RS codes is determined by $n = q^m - 1 = q - 1$, so that these codes are relatively short codes. The minimum polynomial for some element β is $P_\beta(x) = x - \beta$. If α is the primitive element of GF(q) (q is a prime or prime power), the generator polynomial for t-error correcting Reed-Solomon code is given by

$$g(x) = (x - \alpha)(x - \alpha^2)\cdots(x - \alpha^{2t}). \tag{9.69}$$

The generator polynomial degree is $2t$ and it is the same as the number of parity symbols $n - k = 2t$, while the block length of the code is $n = q - 1$. Since the minimum distance of BCH codes is $2t + 1$, the minimum distance of RS codes is $d_{\min} = n - k + 1$, satisfying therefore the Singleton bound ($d_{\min} \leq n - k + 1$) with equality, thus belonging to the class of *maximum-distance separable* (MDS) *codes*. When $q = 2^m$, the RS code parameters are $n = m(2^m - 1)$, $n - k = 2mt$, and $d_{\min} = 2mt + 1$. Therefore, the minimum distance of RS codes, when observed as binary codes, is large. The RS codes may be considered as burst error correcting codes and as such are suitable for bursty-error-prone channels. This binary code is able to correct up to t bursts of length m. Equivalently, this binary code will be able to correct a single burst of length $(t - 1)m + 1$.

The weight distribution of RS codes can be determined as [2, 5, 14]

$$A_i = \binom{n}{i}(q-1)\sum_{j=0}^{i-d_{\min}}(-1)^j\binom{i-1}{j}q^{i-d_{\min}-j}, \tag{9.70}$$

and by using this expression, we can evaluate the undetected error probability as [2, 5, 14]

$$P_u(e) = \sum_{i=1}^{n} A_i p^i (1-p)^{n-i}, \tag{9.71}$$

where p is the crossover probability and A_i denotes the number of codewords of weight i. The codeword weights can be determined by *McWilliams identity* [2, 5, 7, 14].

Example Let the GF(4) be generated by $1 + x + x^2$. The symbols of GF(4) are 0, 1, α, and α^2. The generator polynomial for RS(3,2) code is given by $g(x) = x - \alpha$. The corresponding codewords are 000, 101, $\alpha 0 \alpha$, $\alpha^2 0 \alpha^2$, 011, 110, $\alpha 1 \alpha^2$, $\alpha^2 1 \alpha$, $0\alpha\alpha$, $1\alpha\alpha^2$, $\alpha\alpha 0$, $\alpha^2\alpha 1$, $0\alpha^2\alpha^2$, $1\alpha^2\alpha$, $\alpha\alpha^2 1$, and $\alpha^2\alpha^2 0$. This code is essentially the even parity-check code ($\alpha^2 + \alpha + 1 = 0$). The generator polynomial for RS(3,1) is $g(x) = (x - \alpha)(x - \alpha^2) = x^2 + x + 1$, while the corresponding codewords are 000, 111, $\alpha\alpha\alpha$, and $\alpha^2\alpha^2\alpha^2$. Therefore, this code is in fact the repetition code.

Since RS codes are special class of nonbinary BCH codes, they can be decoded by using the same decoding algorithm as one already explained in the previous section.

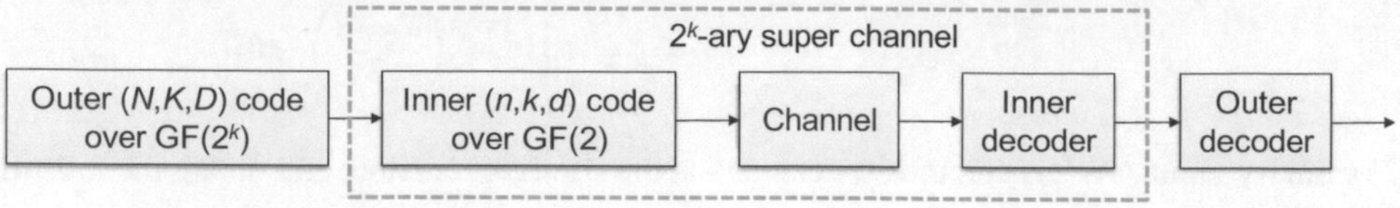

Fig. 9.12 The concatenated (Nn, Kk,$\geq Dd$) code

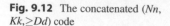

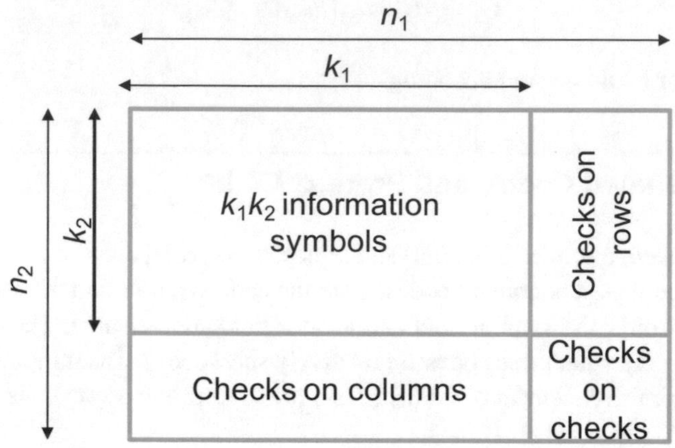

Fig. 9.13 The structure of a codeword of a product code

To further improve the burst error correction capability of RS codes, RS code can be combined with an inner binary block code in a *concatenation* scheme as shown in Fig. 9.12. The key idea behind the concatenation scheme can be explained as follows. Consider the codeword generated by inner (n,k,d) code (with d being the minimum distance of the code) and transmitted over the bursty channel. The decoder processes the erroneously received codeword and decodes it correctly. However, occasionally the received codeword is decoded incorrectly. Therefore, the inner encoder, the channel, and the inner decoder may be considered as a superchannel whose input and output alphabets belong to GF(2^k). The outer encoder (N,K,D) (D – the minimum distance of outer code) encodes input K symbols and generates N output symbols transmitted over the superchannel. The length of each symbol is k information digits. The resulting scheme is known as concatenated code [2, 5, 7, 14]. This scheme is an (Nn,Kk,$\geq Dd$) code with the minimum distance of at least Dd. For instance, RS(255,239,8) code can be combined with the (12,8,3) single-parity-check code in the concatenation Scheme ($12{\cdot}255$,$239{\cdot}8$,≥ 24). The concatenated scheme from Fig. 9.12 can be generalized to q-ary channels, where the inner code operates over GF(q) and outer over GF(q^k).

Another way to deal with burst errors is to arrange two RS codes in a *product* manner as shown in Fig. 9.13. A new product code is an (n_1n_2,k_1k_2,d_1d_2) scheme in which codewords form an n_1 x n_2 array such that each row is a codeword from C_1 (n_1,k_1,d_1) code, while each column is a codeword from another C_2 (n_2,k_2,d_2) code, where n_i, k_i, and d_i ($i = 1,2$) are the codeword length, dimension, and minimum distance of i-th component code, respectively. Turbo-product codes were originally proposed in [29]. Both binary (such as binary BCH codes) and nonbinary codes (such as RS codes) may be arranged in a product code manner. It is possible to show that the minimum distance of a product codes is the product of minimum distances of component codes. Accordingly, it can be shown that the product code is able to correct the burst error of length $b = \max(n_1b_2,n_2b_1)$, where b_i is the burst error capability of component code $i = 1,2$.

The results of Monte Carlo simulations for several RS concatenation schemes on an optical on-off keying AWGN channel are shown in Fig. 9.14. The net coding gain is measured by difference in Q-values for a specific BER. As we can see, the concatenation Scheme RS(255,239) + RS(255,223) of code rate $R = 0.82$ outperforms all other combinations.

9.6 Coding with Interleaving

An *interleaved code* is obtained by taking L codewords (of length n) of a given code $\boldsymbol{x}_i = (x_{i1},x_{i2},\ldots,x_{in})$ ($i = 1,2,\ldots,L$) and forming the new codeword by interleaving the L codewords as $\boldsymbol{y}_i = (x_{11},x_{21},\ldots,x_{L1}, x_{12},x_{22},\ldots,x_{L2},\ldots,x_{1n},x_{2n},\ldots,x_{Ln})$. The process of interleaving can be visualized as the process of forming an $L \times n$ matrix of L codewords written row by row and transmitting the matrix column by column, as given below:

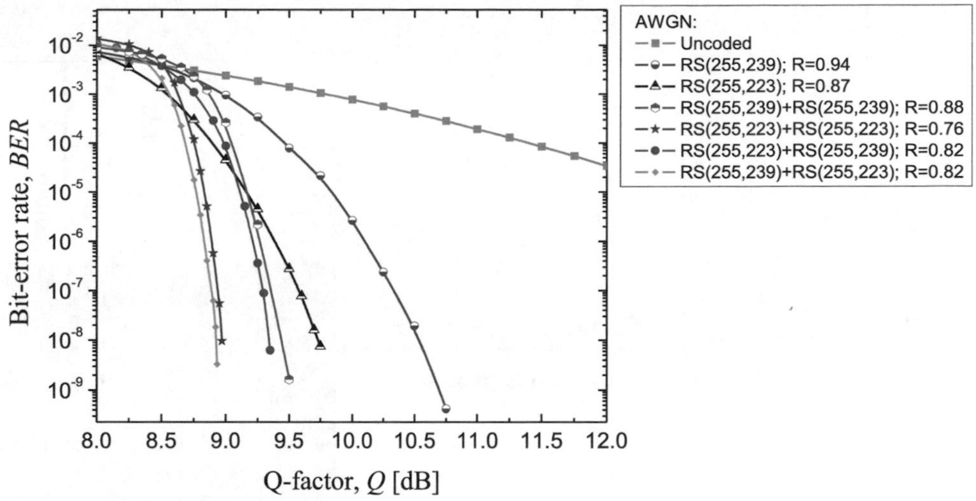

Fig. 9.14 BER performance of concatenated RS codes on an optical on-off keying channel

$$\begin{array}{cccc} x_{11} & x_{12} & \cdots & x_{1n} \\ x_{21} & x_{22} & \cdots & x_{2n} \\ \cdots & \cdots & \cdots & \cdots \\ x_{L1} & x_{L2} & \cdots & x_{Ln} \end{array}$$

The parameter L is known as the interleaving *degree*. The transmission must be delayed until L codewords are collected. To be able to transmit a column whenever a new codeword becomes available, the codewords should be arranged down diagonals as shown below. This interleaving scheme is known as the delayed interleaving (1-frame delayed interleaving):

$$\begin{array}{ccccccccc} x_{i-(n-1),1} & \cdots & x_{i-2,1} & x_{i-1,1} & x_{i,1} & & & & \\ & x_{i-(n-1),2} & \cdots & x_{i-2,2} & x_{i-1,2} & x_{i,2} & & & \\ & & \cdots & \cdots & \cdots & \cdots & \cdots & & \\ & & & x_{i-(n-1),n-1} & x_{i-(n-2),n-1} & & & & \\ & & & & x_{i-(n-1),n} & x_{i-(n-2),n} & & & \end{array}$$

Each new codeword completes one column of this array. In the example above, the codeword $\boldsymbol{x}_i$ completes the column (frame) $x_{i,1}, x_{i-1,2}, \ldots, x_{i-(N-1),n}$. A generalization of this scheme, in which the components of i-th codeword $\boldsymbol{x}_i$ (say $x_{i,j}$ and $x_{i,j+1}$) are spaced λ frames apart, is known as λ-frame delayed interleaving.

Interleaving is an efficient way to deal with error bursts introduced by *fading effects* in wireless communication channels. In the presence of burst of errors, the code designed for random error needs to deal with both random errors and burst errors. By using coding with interleaving, we spread error bursts due to deep fades over many codewords such that each received codeword only exhibits at most a few simultaneous symbol errors, which can be corrected for. The interleaver depth must be large enough so that fading is independent across a received codeword. Slowly fading channels require large interleavers, which can introduce the large decoding delays. In the block interleaver discussed above, during transmission, two neighboring codeword symbols, originating the same codeword, are separated by $L - 1$ symbols from other codewords, so that the symbols from the same codeword will experience approximately independent fading conditions providing that their separation in time is greater than the channel coherence time T_c:

$$LT_s \geq T_c \approx 1/B_D, \tag{9.72}$$

where T_s is the symbol duration and B_D is the Doppler spread. The block coding with interleaving is illustrated in Fig. 9.15.

The codewords from block encoder are written into block interleaver in column-wise fashion. After L codewords are written, the modulator reads bits from the interleaver in column-wise fashion, starting from the first column. On receiver side,

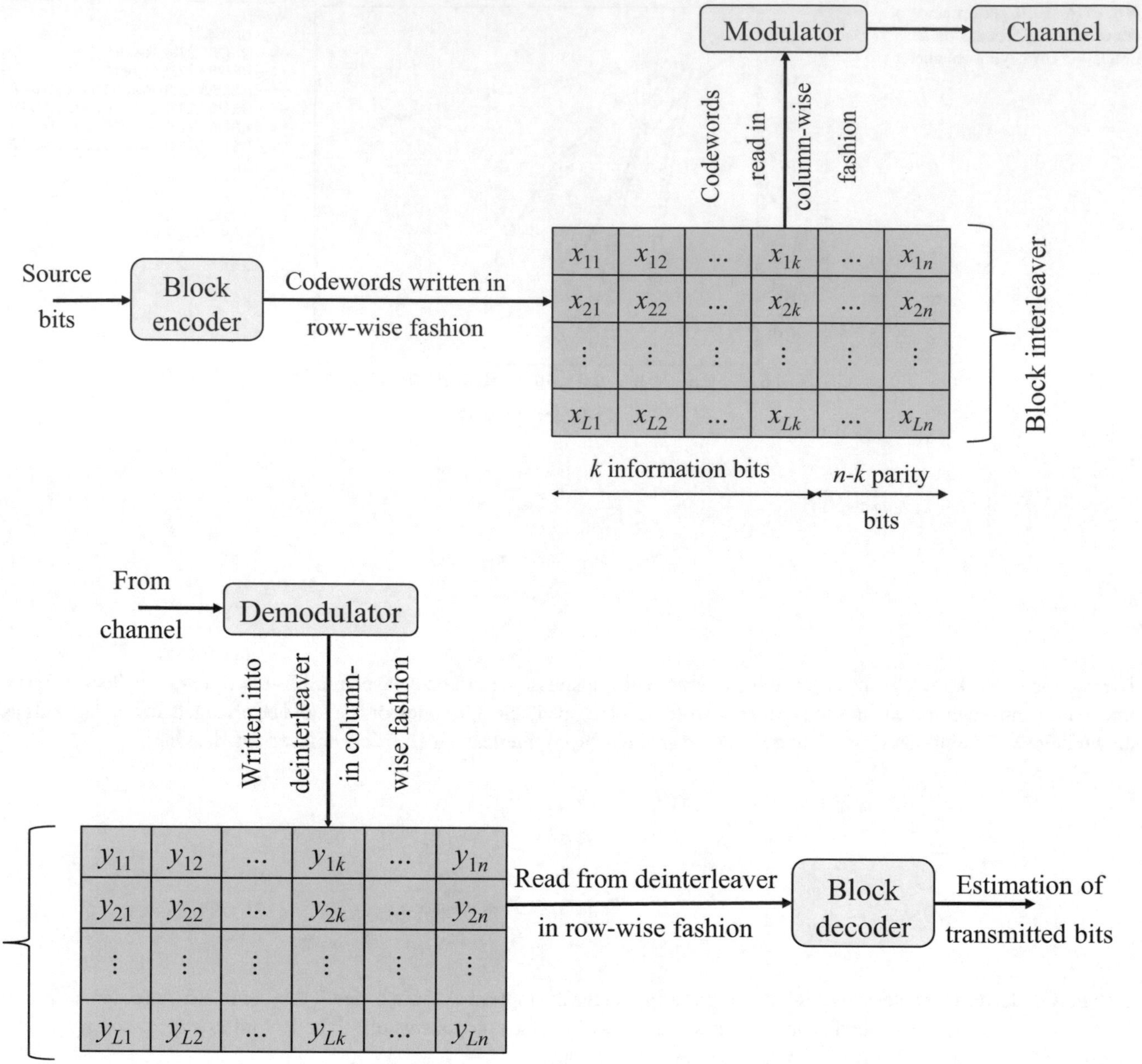

Fig. 9.15 Illustrating block coding with interleaving and deinterleaving principles

after demodulation, for hard decision the bits upon hard decision are written into deinterleaver in column-wise fashion. Once the deinterleaver is filled, the block decoder reads the content of deinterleaver in row-wise fashion, and decoding process takes place. On the other hand, for soft-decision decoding, corresponding samples are written into deinterleaver in column-wise fashion. For two-dimensional signaling schemes, the two samples, corresponding to in-phase and quadrature channel, are written in column-wise fashion. The *i*-th row and *j*-th element of deinterleaver are then represented as $y_{ij} = (y_{ij,\mathrm{I}}, y_{ij,\mathrm{Q}})$, where the first component corresponds to the sample in in-phase channel, while the second component corresponds to the sample in quadrature channel. These channel samples are read in row-wise fashion by soft-decision block decoder.

When convolutional code is used instead of block code, the corresponding *convolutional interleaver* is provided in Fig. 9.16. The convolutional encoder output is multiplexed into the stage of L buffers of increasing size, from zero buffering size to the buffer of size $(L - 1)$. We use z^{-1} to denote one unit delay of T_s seconds. The buffer stage output is multiplexed into the channel in a similar fashion. On receiver side, the reverse operation is performed. Clearly, any two code symbols are separated by LT_s, and providing that the buffering delay is larger than the coherence time T_c, the interleaver is efficient in dealing with fading effects. The total number of memory elements per buffering stage is 0.5 $L(L - 1)$.

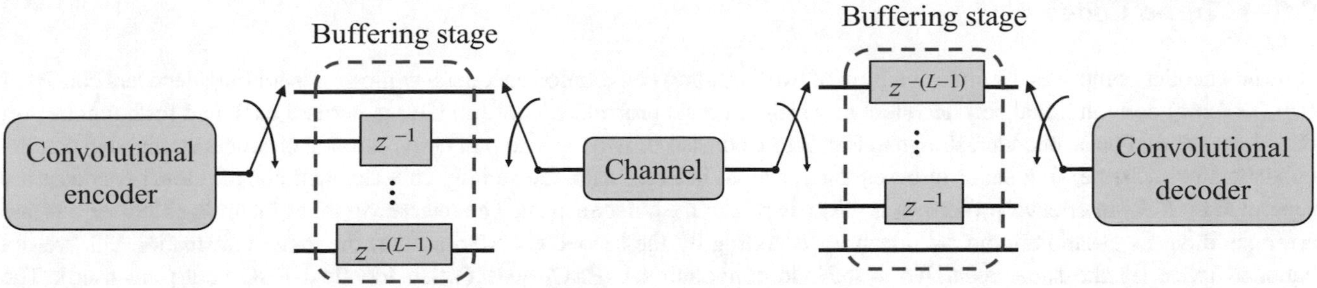

Fig. 9.16 Illustration of convolutional coding with interleaving and deinterleaving principles

Fig. 9.17 Bipartite graph examples: (**a**) bipartite graph for a generalized LDPC (GLDPC) code and (**b**) bipartite graph for parallel concatenation of codes. The constraint node could be any linear code

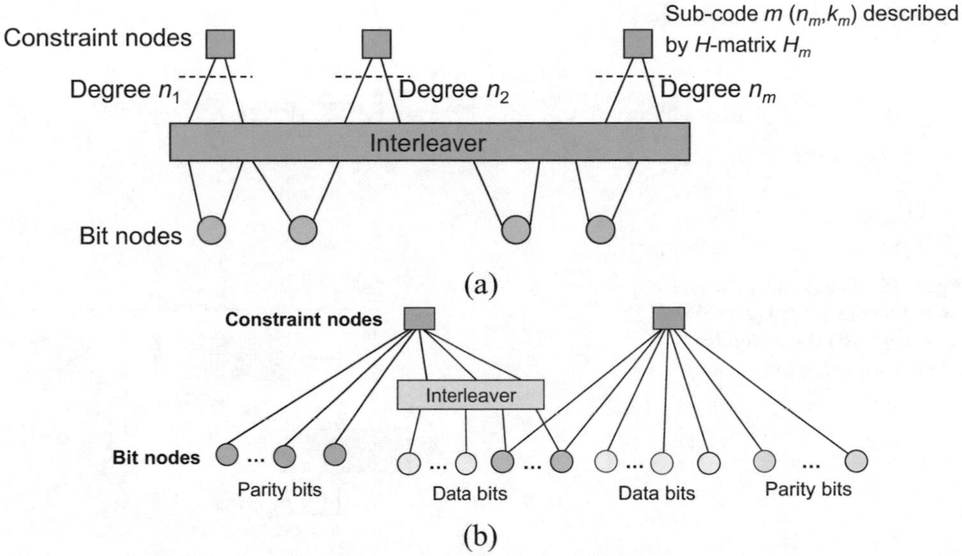

9.7 Codes on Graphs Basics

A graph is typically defined as a set of nodes connected by edges. There are many ways to describe a code by using graphical approach including the factor graph approach [21]; one such approach is described below. We associate two types of nodes to a given code: bit (variable) nodes and constraint (function) nodes. The bit nodes can only be connected to the function nodes and vice versa. Such a graph, which allows only the nodes of different type to be connected, is known as a *bipartite* graph [1, 2, 4, 7, 22]. Two examples of bipartite graphs, corresponding to generalized LDPC (GLDPC) codes and parallel concatenation of codes, are shown in Fig. 9.17. In GLDPC coding, shown in Fig. 9.17a, each constraint node, also known as subcode, represents an (n_s, k_s) $(s = 1,\ldots,m)$ linear block code such as single-parity-check (SPC) code, BCH code (including Hamming code), or Reed-Müller (RM) code. There is one-to-one correspondence between the set of variable nodes and the set of all bits in a codeword. However, we might allow certain number of variable nodes to be *punctured* (corresponding bits are not transmitted), to adapt to time-varying optical channel conditions. In parallel concatenation of codes (Fig. 9.17b), constraint nodes are typically convolutional codes. Very often only two constraint nodes (constituent codes) are used in conventional turbo codes [20, 23–27].

The codes on graphs of interest in wireless and optical communications include turbo codes [20, 23–28], turbo-product codes [2, 8, 9, 28, 30], and LDPC codes [1, 2, 4, 5, 7, 10–12, 31–51]. As fiber-optics communications are affected by fiber nonlinearities, the code rate should be kept sufficiently high ($\geq$0.75).

9.8 Turbo Codes

A turbo encoder comprises the concatenation of two (or more) convolution encoders, while corresponding decoders consist of two (or more) convolutional soft decoders in which extrinsic probabilistic information is iterated back and forth among soft decoders. Parallel turbo encoder, shown in Fig. 9.18a, consists of two rate one-half convolutional encoders arranged in parallel concatenation scheme. In a serial turbo encoder, shown in Fig. 9.18b, the serially concatenated convolutional encoders are separated by K/R_0-interleaver, where R_0 is the code rate of the outer encoder. The interleaver takes incoming block of bits and arranges them in pseudorandom fashion prior encoding by the second encoder, so that the same information bits are not encoded twice by the same recursive systematic convolutional (RSC) code (when identical RSC codes are used). The generator matrix of an RSC code can be written as $G_{\text{RSC}}(x) = [1 \quad g_2(x)/g_1(x)]$, where $g_2(x)/g_1(x)$ denotes the transfer function of the parity branch of encoder (see Fig. 9.18a). As an example, the RSC encoder described by $G_{\text{RSC}}(x) = [1 \quad (1+x^4)/(1+x+x^2+x^3+x^4)]$ is shown in Fig. 9.18c. The output of the RSC encoder in parallel turbo code, the codeword, can be represented as

$$\boldsymbol{x} = \left[x_1 x_2 \cdots \underbrace{x_k}_{\left[x_k^{(\text{info})}, x_k^{(\text{parity})}\right]} \cdots x_K \right] = \left[x_1^{(\text{info})}, x_1^{(\text{parity})}, x_2^{(\text{info})}, x_2^{(\text{parity})}, \ldots, u_K^{(\text{info})}, x_K^{(\text{parity})} \right], \tag{9.73}$$

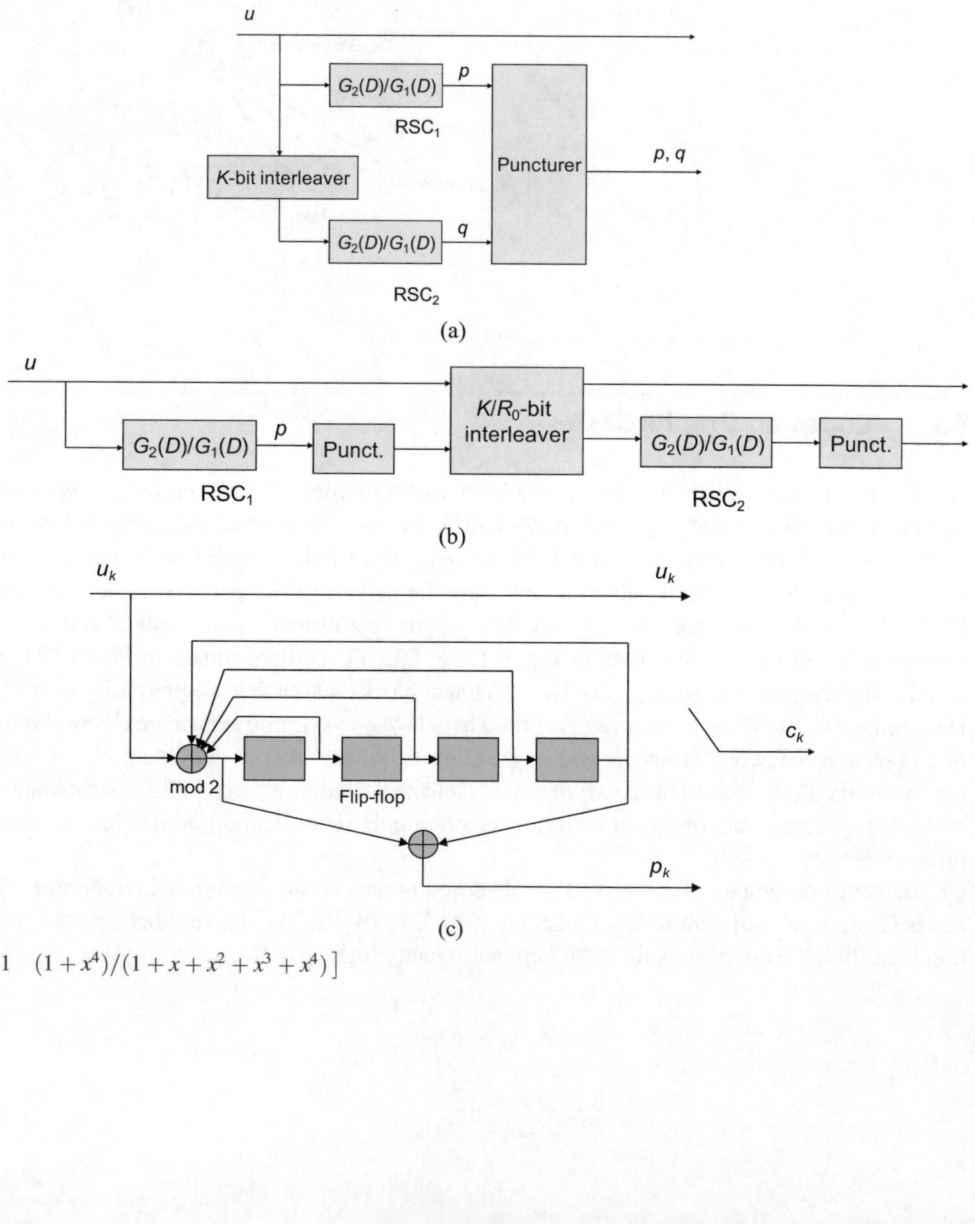

Fig. 9.18 Turbo codes encoder configurations for **(a)** parallel and **(b)** serial turbo codes. **(c)** RSC encoder with $G_{\text{RSC}}(x) = [1 \quad (1+x^4)/(1+x+x^2+x^3+x^4)]$

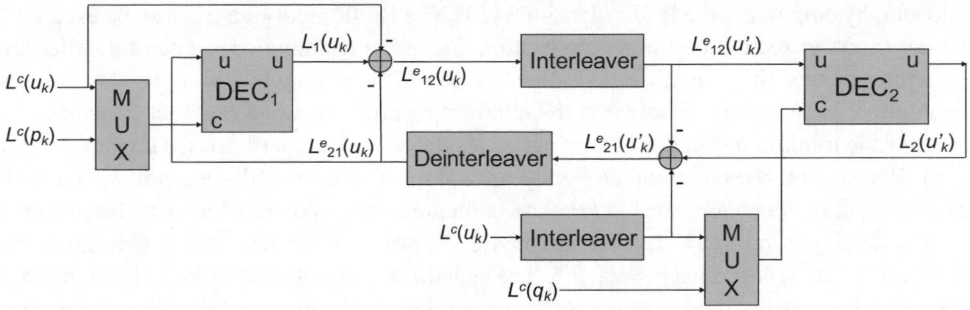

Fig. 9.19 Parallel turbo decoder configuration. *MUX* multiplexer, *DEC* decoder

where $x_k^{(\text{info})}$ is k-th information bit ($k = 1,2,\ldots,K$) and $x_k^{(\text{parity})}$ is the k-th parity bit. For antipodal signaling, we can represent the modulation process as $u_k = (-1)^{x_k^{(\text{info})}}, p_k = (-1)^{x_k^{(\text{parity})}}$ so that the transmitted word $\boldsymbol{c}$ is represented as

$$\boldsymbol{c} = [c_1, c_2, \ldots, c_K] = [u_1, p_1, u_2, p_2, \ldots, u_K, p_K], \quad c_k = [u_k, p_k] \tag{9.74}$$

The iterative (turbo) decoder, shown in Fig. 9.19, interleaves two soft-input/soft-output (SISO) decoders, exchanging the extrinsic information iteratively and cooperatively. The role of iterative decoder is to iteratively estimate the a posteriori probabilities (APPs) $\Pr(u_k|\boldsymbol{y})$, where $\boldsymbol{y}$ is a received codeword plus noise $\boldsymbol{n}$:

$$\boldsymbol{y} = \boldsymbol{c} + \boldsymbol{n} = [y_1, y_2, \ldots, y_K] = [y_1^u, y_1^p, y_2^u, y_2^p, \ldots, y_K^u, y_K^p], \quad y_k = [y_k^u, y_k^p] \tag{9.75}$$

with y_k^u being the received sample corresponding to k-th data bit, while y_k^p being the sample corresponding to the k-th parity bit.

In iterative decoders, each component decoder receives extrinsic or soft information for each u_k from its companion decoder, which provides "the prior" information, as illustrated in Fig. 7.20. The key idea behind extrinsic information is that decoder DEC$_2$ provides the soft information to DEC$_1$ for each u_k by using only information that is not available to DEC$_1$. The knowledge of APPs allows the optimum decisions for information bits u_k to be made by the maximum a posteriori probability (MAP) rule:

$$\frac{P(u_k = +1|\boldsymbol{y})}{P(u_k = -1|\boldsymbol{y})} \underset{-1}{\overset{+1}{\gtrless}} 1 \Leftrightarrow u_k = \text{sign}\,[L(u_k)], L(u_k) = \log\left[\frac{P(u_k = +1|\boldsymbol{y})}{P(u_k = -1|\boldsymbol{y})}\right], \tag{9.76}$$

where $L(u_k)$ is the log-APP also known as log-likelihood ratio (LLR). By applying Bayes' rule, we can write that

$$\begin{aligned} L(b_k) &= \log\left[\frac{P(b_k = +1|\boldsymbol{y})\,p(\boldsymbol{y})}{P(b_k = -1|\boldsymbol{y})\,p(\boldsymbol{y})}\right] = \log\left[\frac{P(\boldsymbol{y}|b_k = +1)\,P(b_k = +1)}{P(\boldsymbol{y}|b_k = -1)\,P(b_k = -1)}\right] \\ &= \log\left[\frac{P(\boldsymbol{y}|b_k = +1)}{P(\boldsymbol{y}|b_k = -1)}\right] + \log\left[\frac{P(b_k = +1)}{P(b_k = -1)}\right], \end{aligned} \tag{9.77}$$

where $b_k \in \{u_k, p_k\}$ and the second term represents *a priori* information. The first term in Eq. (9.77) represents the *extrinsic* information for iterative decoders obtained from corresponding companion decoder. Notice that for conventional decoders we typically have $P(b_k = +1) = P(b_k = -1)$.

The key idea behind extrinsic information is to provide only soft information about b_k (that is not already available) to the companion decoder. Therefore, although initially the prior information is zero, it becomes nonzero after the first iteration. As shown in Fig. 7.20, the extrinsic information to be sent from DEC$_1$ to DEC$_2$, which is denoted as $L_{12}^{\text{e}}(u_k)$, can be calculated by subtracting from DEC$_1$ output LLR $L_1(u_k)$ both the channel reliability $L^{\text{c}}(u_k)$ and extrinsic information $L_{21}^{\text{e}}(u_k)$ already received from DEC$_2$. On the other hand, the extrinsic information $L_{21}^{\text{e}}(u_k)$, to be sent from DEC$_2$ to DEC$_1$, is obtained by subtracting both the interleaved channel reliability $L^{\text{c}}(u_k)$ and interleaved extrinsic information $L_{12}^{\text{e}}(u_k')$ (already received from DEC$_1$) from DEC$_2$'s output LLR $L_2(u_k)$. Since the DEC$_2$ operates on interleaved sequence, this extrinsic information needs to be deinterleaved before being sent to DEC$_1$. This exchange of extrinsic information is performed until the successful decoding is done or until the predetermined number of iterations is reached. Decoders DEC$_i$ ($i = 1,2$) operate on trellis description of encoder, by employing the BCJR algorithm [13]. The BCJR algorithm can also be used for MAP detection, and as such it will be described in section on turbo equalization. Since the resulting code rate of parallel turbo encoder is low ($R = 1/3$), the puncturer deletes the selected bits in order to reduce the coding overhead, as shown in Fig. 9.18a. The resulting code rate becomes $R = K/(K + P)$, where P is the number of parity bits remained after puncturing. On the other hand, the puncturing reduces the minimum distance of original code, which leads to performance degradation and an early error floor as shown in [28]. Because of low-code rate and early error floor, the turbo codes are not widely used in fiber-optics communications. However, they are widely used in wireless communication systems as well as deep-space optical communications.

The details of the *BCJR decoder* are provided below. We assume that RSC is represented using the trellis description, in similar fashion as described in Sect. 9.3. The key difference with respect to the linear block codes is that trellis of RSC is time-invariant. Let s' denote one of the states in current stage of the trellis, while s denote the corresponding state in the next stage of trellis. The transition from the state s' to state s can be represented as the pair (s', s). Let U^+ (U^-) denote the set of pairs (s',s) for the state transitions $(s_{k-1} = s') \rightarrow (s_k = s)$ corresponding to $u_k = +1$ ($u_k = -1$). The LLR of the k-th data bit can be represented as

$$L(u_k) = \log \frac{\sum_{U^+} p(s_{k-1} = s', s_k = s, \mathbf{y})}{\sum_{U^-} p(s_{k-1} = s', s_k = s, \mathbf{y})}. \tag{9.78}$$

Therefore, we have to compute $p(s',s,\mathbf{y})$ for all state transitions and then sum up over appropriate transitions in numerator and denominator. The BCJR algorithm is composed of two recursions, forward and backward recursion as illustrated in Fig. 9.21. To efficiently compute LLRs, we factor $p(s',s,\mathbf{y})$ as follows:

$$p(s', s, \mathbf{y}) = \alpha_{k-1}(s') \cdot \gamma_k(s', s) \cdot \beta_k(s), \tag{9.79}$$

where the state probabilities in forward recursion $\alpha_k(s)$, the transition probabilities, $\gamma_k(s',s)$, and the state probabilities in backward recursion $\beta_k(s)$ are defined, respectively, as

$$\alpha_k(s) = p\left(s_k = s, \mathbf{y}_1^k\right), \gamma_k(s', s) = p\left(s_k = s, \mathbf{y}_k | s_{k-1} = s'\right), \beta_k(s) = p\left(\mathbf{y}_{k+1}^K | s_k = s\right) \tag{9.80}$$

where we used $\mathbf{y}_a^b$ to denote $[y_a, y_{a+1}, \ldots, y_b]$. In the *forward recursion*, we calculate the state probabilities for the next trellis as follows:

$$\alpha_k(s) = \sum_{s'} \alpha_{k-1}(s') \gamma_k(s', s), \tag{9.81}$$

where the summation is performed over all edges merging into state s, as illustrated in Fig. 9.20a. The *transition probabilities* $\gamma_k(s',s)$ are determined as

$$\begin{aligned} \gamma_k(s', s) = p(s, \mathbf{y}_k | s') &= \frac{p(s, \mathbf{y}_k | s') p(s')}{p(s')} = \frac{p(s', s, \mathbf{y}_k)}{p(s')} \frac{p(s', s)}{p(s', s)} = \frac{p(s', s)}{p(s')} \frac{p(s', s, \mathbf{y}_k)}{p(s', s)} \\ &= p(s|s') p(\mathbf{y}_k | s', s) = p(u_k)\, p(\mathbf{y}_k | s', s), \end{aligned} \tag{9.82}$$

where $p(s|s')$ equals 1/2 for a valid (existing) transition, while $p(\mathbf{y}_k|s',s)$ is determined from channel conditional PDF. For an AWGN channel and antipodal signaling, the transition probabilities are simply

$$\gamma_k(s', s) = p(u_k) \frac{1}{\sigma\sqrt{2\pi}} e^{-\frac{\|\mathbf{y}_k - \mathbf{c}_k\|^2}{2\sigma_k^2}} = p(u_k) \frac{1}{\sigma\sqrt{2\pi}} e^{-\frac{\left(y_k^u - u_k\right)^2 + \left(y_k^p - p_k\right)^2}{2\sigma_k^2}}. \tag{9.83}$$

In the *backward recursion*, the state probabilities in the previous trellis stage are determined by, as illustrated in Fig. 9.20b,

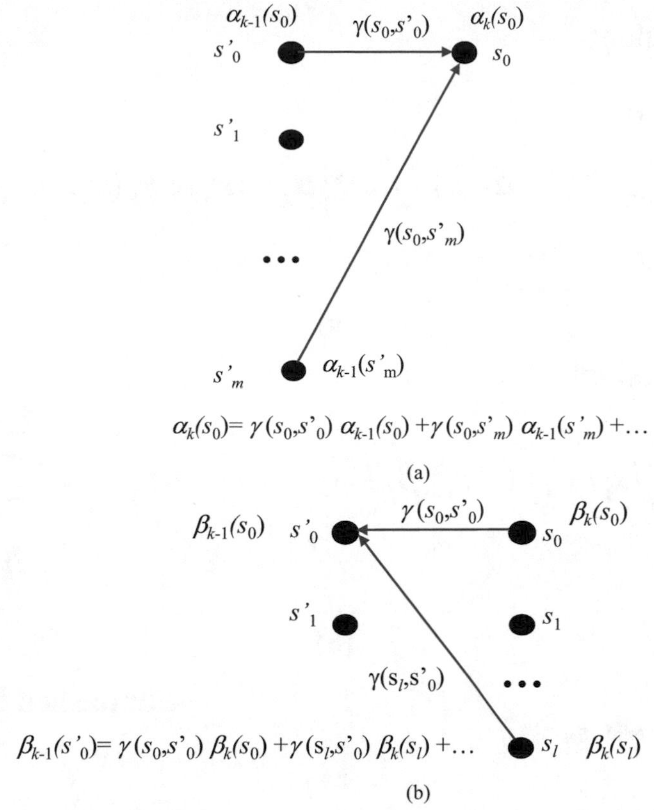

Fig. 9.20 Illustration of forward (**a**) and backward (**b**) recursions in BCJR algorithm

$$\beta_{k-1}(s') = \sum_{s} \gamma_k(s', s)\beta_k(s), \tag{9.84}$$

where the summation is performed over all incoming edges to state s' in backward direction. Clearly, the BCJR algorithm is composed of two Viterbi-like algorithms running in opposite directions. Now we have to calculate the LLRs, as follows:

$$L(u_k) = \log \frac{\sum_{U^+} \alpha_{k-1}(s') \cdot \gamma_k(s', s) \cdot \beta_k(s)}{\sum_{U^-} \alpha_{k-1}(s') \cdot \gamma_k(s', s) \cdot \beta_k(s)}, \tag{9.85}$$

and this expression is obtained by substitution of Eq. (9.79) into (9.78).

The factorization claim (9.79) can be proved by applying conditional probability formula two times and removing redundant variables.

$$\begin{aligned} p(s', s, \mathbf{y}) &= p\left(s', s, \mathbf{y}_1^{k-1}, y_k, \mathbf{y}_{k+1}^K\right) = p\left(\mathbf{y}_{k+1}^K | s', s, \mathbf{y}_1^{k-1}, y_k\right) \cdot p\left(s', s, \mathbf{y}_1^{k-1}, y_k\right) \\ &= p\left(\mathbf{y}_{k+1}^K | s', s, \mathbf{y}_1^{k-1}, y_k\right) \cdot p\left(s, y_k | s', \mathbf{y}_1^{k-1}\right) \cdot p\left(s', \mathbf{y}_1^{k-1}\right) \\ &= \underbrace{p\left(\mathbf{y}_{k+1}^K | s'\right)}_{\beta_k(s)} \cdot \underbrace{p\left(s, y_k | s'\right)}_{\gamma_k(s', s)} \cdot \underbrace{p\left(s', \mathbf{y}_1^{k-1}\right)}_{\alpha_{k-1}(s')} \\ &= \alpha_{k-1}(s') \cdot \gamma_k(s', s) \cdot \beta_k(s). \end{aligned} \tag{9.86}$$

On the other hand, the forward recursion step (9.81) has been derived as follows:

$$\begin{aligned} \alpha_k(s) &= p(s_k = s, \mathbf{y}_1^k) = p(s, \mathbf{y}_1^k) = \sum_{s'} p(s', s, \mathbf{y}_1^k) = \sum_{s'} \underbrace{p(s, y_k | s', \mathbf{y}_1^{k-1})}_{\gamma_k(s', s)} \underbrace{p(s', \mathbf{y}_1^{k-1})}_{\alpha_{k-1}(s')} \\ &= \sum_{s'} \alpha_{k-1}(s')\gamma_k(s', s). \end{aligned} \tag{9.89}$$

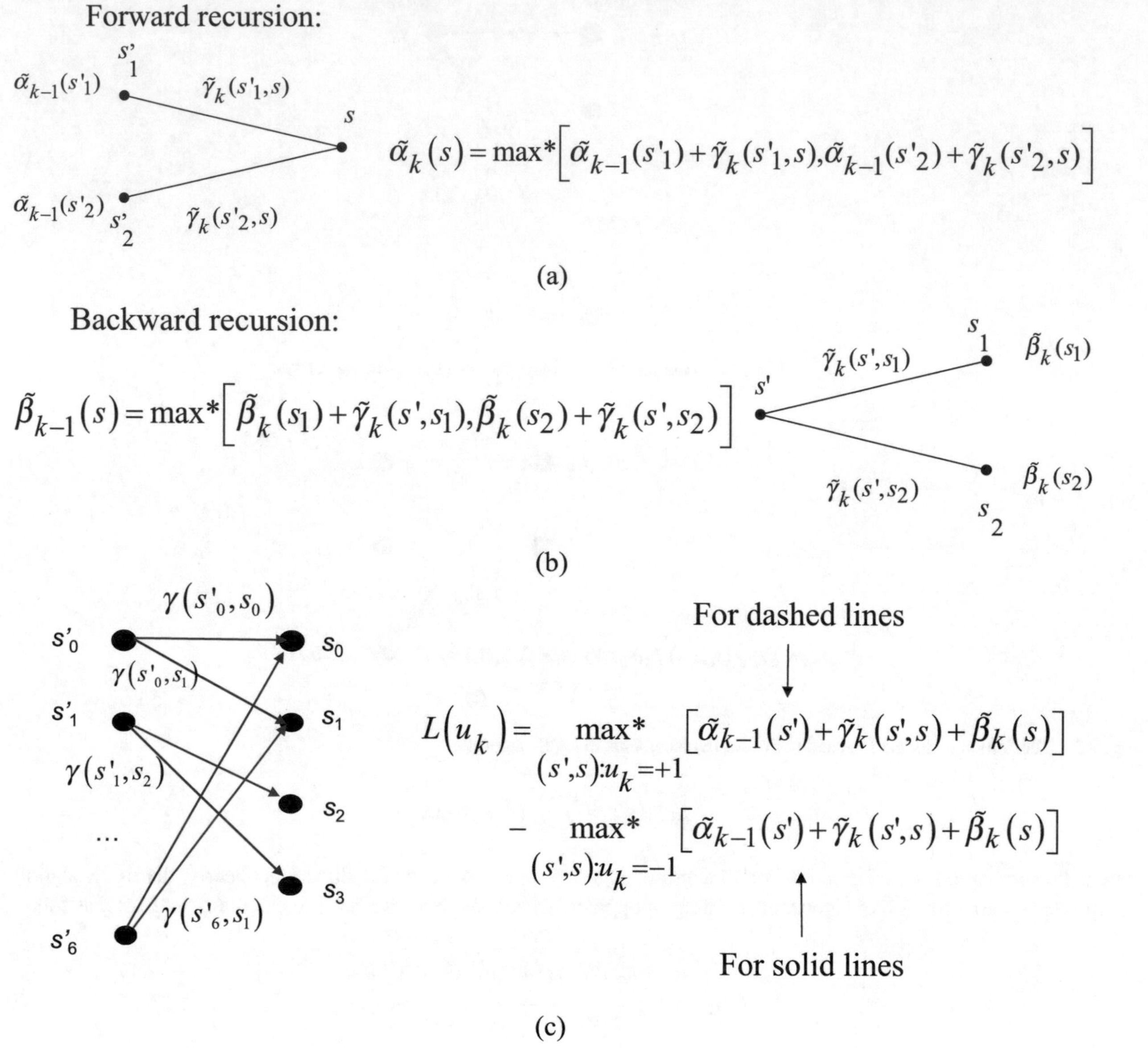

Forward recursion:

$$\tilde{\alpha}_k(s) = \max{}^* \left[\tilde{\alpha}_{k-1}(s'1) + \tilde{\gamma}_k(s'1,s), \tilde{\alpha}_{k-1}(s'2) + \tilde{\gamma}_k(s'2,s) \right]$$

(a)

Backward recursion:

$$\tilde{\beta}_{k-1}(s) = \max{}^* \left[\tilde{\beta}_k(s_1) + \tilde{\gamma}_k(s',s_1), \tilde{\beta}_k(s_2) + \tilde{\gamma}_k(s',s_2) \right]$$

(b)

For dashed lines
\downarrow

$$L(u_k) = \max_{(s',s):u_k=+1}{}^* \left[\tilde{\alpha}_{k-1}(s') + \tilde{\gamma}_k(s',s) + \tilde{\beta}_k(s) \right]$$

$$- \max_{(s',s):u_k=-1}{}^* \left[\tilde{\alpha}_{k-1}(s') + \tilde{\gamma}_k(s',s) + \tilde{\beta}_k(s) \right]$$

\uparrow
For solid lines

(c)

Fig. 9.21 Log-domain BCJR algorithm steps: (**a**) forward recursion, (**b**) backward recursion, (**c**) LLR calculation step

In similar fashion, the backward recursion step (9.84) is derived as follows:

$$\begin{aligned}
\beta_{k-1}(s') &= p(\mathbf{y}_k^K | s') = \sum_s p(\mathbf{y}_k^K, s | s') \\
&= \sum_s p(\mathbf{y}_{k+1}^K | s', s, y_k) p(s, y_k | s') = \sum_s \underbrace{p(\mathbf{y}_{k+1}^K | s)}_{\beta_k(s)} \underbrace{p(s, y_k | s')}_{\gamma_k(s', s)} \\
&= \sum_s \beta_k(s) \gamma_k(s', s).
\end{aligned} \tag{9.90}$$

The initialization can be performed as

$$\alpha_0(s) = \begin{cases} 1, & s = 0 \\ 0, & s \neq 0 \end{cases} \qquad \beta_K(s) = \begin{cases} 1, & s = 0 \\ 0, & s \neq 0 \end{cases} \tag{9.91}$$

The calculation of state probabilities as well as LLRs involves a product of small probabilities, which can cause numerical problems, and to solve for this problem, the *log-domain BCJR algorithm* can be formulated as follows. The *forward metric* in log domain is introduced by taking the logarithm of probability-domain forward metric by

$$\tilde{\alpha}_k(s) = \log\left[\alpha_k(s)\right] = \log\left[\sum_{s'}\alpha_{k-1}(s')\gamma_k(s',s)\right] = \log\left[\sum_{s'}\exp\left(\tilde{\alpha}_{k-1}(s') + \tilde{\gamma}_k(s',s)\right)\right] \tag{9.92}$$

On the other hand, the *backward metric* in log domain is introduced by taking the logarithm of probability-domain backward metric by

$$\begin{aligned}\tilde{\beta}_{k-1}(s') &= \log\left[\beta_{k-1}(s')\right] = \log\left[\sum_{s'}\gamma_k(s',s)\beta_k(s)\right] \\ &= \log\left[\sum_{s'}\exp\left(\tilde{\gamma}_k(s',s) + \tilde{\beta}_k(s)\right)\right].\end{aligned} \tag{9.93}$$

The *branch metric* in log domain is obtained by taking the logarithm of corresponding transition probability as

$$\begin{aligned}\tilde{\gamma}_k(s',s) &= \log\left[\gamma_k(s',s)\right] = \log\left[P(u_k)p(y_k|u_k)\right] \\ &= \log\left[P(u_k)\right] + \log\left[p(y_k|u_k)\right].\end{aligned} \tag{9.94}$$

For an AWGN channel and antipodal signaling, the branch metric gets simplified to

$$\tilde{\gamma}_k(s',s) = -\log\left(2\sigma\sqrt{2\pi}\right) - \frac{\|y_k - c_k\|^2}{2\sigma^2}. \tag{9.95}$$

The LLRs can be expressed in terms of forward, backward, and branch metrics as follows:

$$\begin{aligned}L(u_k) &= \log\frac{\sum_{U^+}\alpha_{k-1}(s')\gamma_k(s',s)\beta_k(s)}{\sum_{U^-}\alpha_{k-1}(s')\gamma_k(s',s)\beta_k(s)} \\ &= \log\left\{\sum_{U^+}\underbrace{\alpha_{k-1}(s')}_{\exp\left[\tilde{\alpha}_{k-1}(s')\right]}\underbrace{\gamma_k(s',s)}_{\exp\left[\tilde{\gamma}_k(s',s)\right]}\underbrace{\beta_k(s)}_{\exp\left[\tilde{\beta}_k(s)\right]}\right\} - \\ \log&\left\{\sum_{U^-}\underbrace{\alpha_{k-1}(s')}_{\exp\left[\tilde{\alpha}_{k-1}(s')\right]}\underbrace{\gamma_k(s',s)}_{\exp\left[\tilde{\gamma}_k(s',s)\right]}\underbrace{\beta_k(s)}_{\exp\left[\tilde{\beta}_k(s)\right]}\right\},\end{aligned} \tag{9.96}$$

which can be written in more compact form as

$$\begin{aligned}L(u_k) = &\log\left\{\sum_{U^+}\exp\left[\tilde{\alpha}_{k-1}(s') + \tilde{\gamma}_k(s',s) + \tilde{\beta}_k(s)\right]\right\} \\ &-\log\left\{\sum_{U^-}\exp\left[\tilde{\alpha}_{k-1}(s') + \tilde{\gamma}_k(s',s) + \tilde{\beta}_k(s)\right]\right\}.\end{aligned} \tag{9.97}$$

Clearly, in (9.35) we recognize the following operation $\log(e^x + e^y)$, which we can denote as $\max^*(x,y)$. The max*-operation can be efficiently calculated as

$$\log\left(e^x + e^y\right) = \max * (x,y) = \max(x,y) + \log\left[1 + e^{-|x-y|}\right], \tag{9.98}$$

where the second term is a correction term and can efficiently be implemented with the help of a look-up table (LUT) in hardware. The max*-operator for more than two terms can be applied recursively, to terms at the time. For instance, for three entries x, y, and z, the max*-operator is applied as

$$\max*(x, y, z) = \max*(\max*(x, y), z). \tag{9.99}$$

To prove the claim (9.98), we can express the max(x,y) as follows:

$$\max(x, y) \stackrel{?}{=} \log\left(\frac{e^x + e^y}{1 + e^{-|x-y|}}\right). \tag{9.100}$$

Let $x > y$, then the left side of equation is clearly x, while the right side of previous equation becomes

$$\log\left(\frac{e^x + e^y}{1 + e^{-|x-y|}}\right) = \log\left(\frac{e^x\left(1 + e^{-(x-y)}\right)}{1 + e^{-(x-y)}}\right) = \log(e^x) = x, \tag{9.101}$$

which proves the correctness of claim (9.98). With the introduction of max*-operation, the forward recursion (metric) simplifies to

$$\widetilde{\alpha}_k(s) = \log\left[\sum_{s'} \exp\left(\widetilde{\alpha}_{k-1}(s') + \widetilde{\gamma}_k(s', s)\right)\right] = \max_{s'}*[\widetilde{\alpha}_{k-1}(s') + \widetilde{\gamma}_k(s', s)], \tag{9.102}$$

and it is illustrated in Fig. 9.21a. On the other hand, the backward recursion (metric) becomes

$$\widetilde{\beta}_{k-1}(s') = \log\left[\sum_{s'} \exp\left(\widetilde{\gamma}_k(s', s) + \widetilde{\beta}_k(s)\right)\right] = \max_{s}*\left[\widetilde{\beta}_k(s) + \widetilde{\gamma}_k(s', s)\right], \tag{9.103}$$

and it is illustrated in Fig. 9.21b. Finally, the LLR computation step simplifies to

$$L(u_k) = \max_{(s',s):\ u_k=0}*\left[\widetilde{\alpha}_{k-1}(s') + \widetilde{\gamma}_k(s', s) + \widetilde{\beta}_k(s)\right] - \max_{(s',s):\ u_k=1}*\left[\widetilde{\alpha}_{k-1}(s') + \widetilde{\gamma}_k(s', s) + \widetilde{\beta}_k(s)\right] \tag{9.104}$$

and it is illustrated in Fig. 9.21c. The initialization step is obtained by taking logarithm of (9.91)

$$\widetilde{\alpha}_0(s) = \log[\alpha_0(s)] = \begin{cases} 0, & s = 0 \\ -\infty, & s \neq 0 \end{cases} \qquad \widetilde{\beta}_K(s) = \log[\beta_K(s)] = \begin{cases} 0, & s = 0 \\ -\infty, & s \neq 0 \end{cases} \tag{9.105}$$

The log-domain BCJR algorithm can be summarized as follows:

0. *Initialization step*:

$$\widetilde{\alpha}_0(s) = \begin{cases} 0, & s = 0 \\ -\infty, & s \neq 0 \end{cases} \quad \widetilde{\beta}_K(s) = \begin{cases} 0, & s = 0 \\ -\infty, & s \neq 0 \end{cases}$$

1. For $k = 1,2,\ldots,K$, take the samples from the channel $y_k = \left[y_k^u, y_k^p\right]$, and compute the *branch metric* $\gamma_k(s', s) = \log[p(y_k | u_k)] + \log[P(u_k)]$.

2. For $k = 1,2,\ldots,K$, apply the *forward recursion*:

$$\widetilde{\alpha}_k(s) = \max_{s'} * [\widetilde{\alpha}_{k-1}(s') + \widetilde{\gamma}_k(s', s)].$$

3. For $k = K, K-1,\ldots,2$, apply the *backward recursion*:

$$\widetilde{\beta}_{k-1}(s) = \max_{s} * [\widetilde{\beta}_k(s) + \widetilde{\gamma}_k(s', s)].$$

4. For $k = 1,2,\ldots,K$, *compute LLRs*:

$$L(u_k) = \max_{(s',s):\ u_k=+1} * \left[\widetilde{\alpha}_{k-1}(s') + \widetilde{\gamma}_k(s', s) + \widetilde{\beta}_k(s)\right] - \max_{(s',s):\ u_k=-1} * \left[\widetilde{\alpha}_{k-1}(s') + \widetilde{\gamma}_k(s', s) + \widetilde{\beta}_k(s)\right]$$

5. Make decisions for $k = 1,2,\ldots,K$:

$$\widehat{u}_k = \text{sign}[L(u_k)].$$

If a valid codeword is obtained or maximum number of iterations is reached, then stop; otherwise, go to step 1 to update the branch metric with new extrinsic information $\log[P(u_k)]$.

The BCJR algorithm is a basic building block for implementing DEC_1 and DEC_2 in Fig. 9.20.

9.9 Turbo-Product Codes

A turbo-product code (TPC), which is also known as block turbo code (BTC) [2, 8, 9, 29], is an (n_1n_2,k_1k_2,d_1d_2) code. The n_i, k_i, and d_i $(i = 1,2)$ denote the codeword length, dimension, and minimum distance, respectively, of the i-th component code $(i = 1,2)$. A TPC codeword, shown in Fig. 9.13, forms an $n_1 \times n_2$ array in which each row represents a codeword from an (n_1, k_1,d_1) code C_1, while each column represents a codeword from an (n_2,k_2,d_2) code C_2. The code rate of TPC is given by $R = R_1R_2 = (k_1k_2)/(n_1n_2)$, where R_i is the code rate of the i-th $(i = 1,2)$ component code, namely, $R_i = k_i/n_i$. The corresponding overhead (OH) of each component code is defined as $OH_i = (1/R_i - 1)\cdot 100\%$. The TPC can be considered as a particular instance of serial concatenated block codes. In this interpretation, rows belong to the outer code and columns to the inner code, and the interleaver is a deterministic column-row interleaver. The corresponding TPC decoder is shown in Fig. 9.22.

In soft-decision decoding, the extrinsic reliabilities are iterated between soft-input/soft-output (SISO) decoders for C_1 and C_2. The extrinsic information in iterative decoders is obtained from corresponding companion decoder. The key idea behind extrinsic information is to provide to the companion decoder only with soft information not already available to it. In fiber-optics communications, TPCs based on extended BCH (including extended Hamming) component codes were intensively studied, e.g., [8, 9, 30]. The product codes were proposed by Elias [52], but the term "turbo" is used when two SISO decoders exchange the extrinsic information. As already mentioned above, the minimum distance of a product code is the product of minimum distances of component codes, $d = d_1d_2$. The constituent codes are typically extended BCH codes, because with extended BCH codes, we can increase the minimum distance for $d_1 + d_2 + 1$ compared to the nominal BCH codes. The

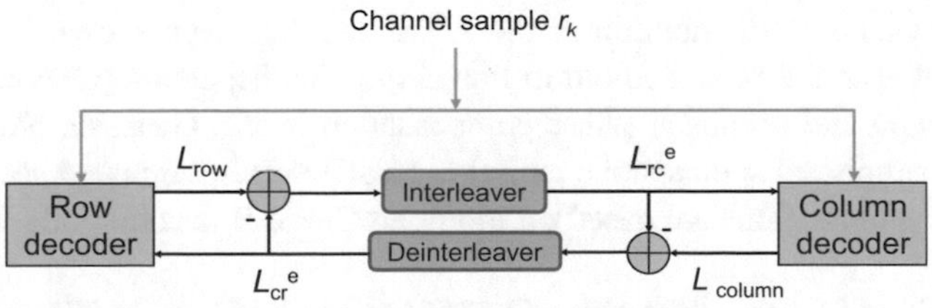

Fig. 9.22 The decoder of a turbo-product code

expressions for bit error probability P_b and codeword error probability P_{cw} under maximum likelihood (ML) decoding, for additive white Gaussian noise (AWGN) channel, can be estimated as follows [2]:

$$P_b \simeq \frac{w_{\min}}{2k_1k_2} \operatorname{erfc}\left(\sqrt{\frac{RdE_b}{N_0}}\right), \; P_{cw} \simeq \frac{A_{\min}}{2} \operatorname{erfc}\left(\sqrt{\frac{RdE_b}{N_0}}\right), \tag{9.106}$$

where $w_{\min}$ is the minimum weight of all $A_{\min}$ TPC codewords at the minimum distance d. Notice that $w_{\min}$ and $A_{\min}$ are quite large, resulting in excellent bit error rate (BER) performance of BTCs. Unfortunately, due to high complexity, maximum likelihood (ML) decoding is typically not used in BTC decoding, but simple *Chase II-like decoding algorithms* have been used instead [2, 29]. One such algorithm, which is independent of the channel model [2], is described below. Let v_i be i-th bit in a codeword $\mathbf{v} = [v_1 \ldots v_n]$ and y_i be the corresponding received sample in a received vector of samples $\mathbf{y} = [y_1 \ldots y_n]$. The role of decoder is to iteratively estimate the a posteriori probabilities (APPs) $\Pr(v_i|y_i)$. The knowledge of APPs allows for optimum decisions on the bits v_i by MAP rule

$$\frac{P(v_i = 0|\mathbf{y})}{P(v_i = 1|\mathbf{y})} \underset{1}{\overset{0}{\gtrless}} 1, \tag{9.107}$$

or, more conveniently, we can write

$$\widehat{v}_i = \frac{1 - \operatorname{sign}[L(v_i)]}{2}, \; \operatorname{sign}[x] = \begin{cases} +1, x \geq 0 \\ -1, x < 0 \end{cases} \tag{9.108}$$

where $L(v_i)$ is the log a posteriori probability (log-APP), commonly referred to as the log-likelihood ratio (LLR), defined as

$$L(v_i) = \log \left[\frac{P(v_i = 0|\mathbf{y})}{P(v_i = 1|\mathbf{y})}\right]. \tag{9.109}$$

The component SISO decoders for C_1 and C_2 calculate $L(v_i)$ and exchange extrinsic information as explained above. The initial/channel bit LLRs, to be used in SISO decoding including Chase II algorithm, can be calculated by

$$L_{\text{ch}}(v_i) = \log \frac{P(v_i = 0|y_i)}{P(v_i = 1|y_i)}, \tag{9.110}$$

where the conditional probability $P(v_i|y_i)$ is determined by

$$P(v_i|y_i) = \frac{P(y_i|v_i)P(v_i)}{P(y_i|v_i = 0)P(v_i = 0) + P(y_i|v_i = 1)P(v_i = 1)}, \tag{9.111}$$

In the presence of fiber nonlinearities, the conditional $P(y_i|\cdot)$ can be evaluated by estimation of histograms. On the other hand, the channel LLRs for AWGN channel can be calculated as follows:

$$L_{\text{ch}}(v_i) = 2y_i/\sigma^2, \text{for binary input AWGN} \tag{9.112}$$

where σ^2 is the variance of the Gaussian distribution of the AWGN. It has been shown in experimental study for beyond 100 Gb/s transmission [53, 54] that in the links without in-line chromatic dispersion (CD) compensation, the distribution of samples upon compensation of CD and nonlinear phase compensation is still Gaussian-like, which justifies the use of the Gaussian assumption (9.112) in amplified spontaneous emission (ASE) noise-dominated scenario.

The constituent *SISO decoding* algorithms are based on modified Chase II decoding algorithm [2]:

1. Determine p least reliable positions starting from either (9.110) or (9.112). Generate 2^p test patterns to be added to the hard-decision word obtained after (3).

2. Determine the i-th ($i = 1,\ldots,2^p$) perturbed sequence by adding (modulo-2) the test pattern to the hard-decision word (on least reliable positions).
3. Perform the algebraic or hard decoding to create the list of candidate codewords. Simple syndrome decoding is suitable for high-speed transmission.
4. Calculate the k-th candidate codeword $\boldsymbol{c}_k$ LLRs by

$$L[\boldsymbol{v}_k = (v_k(1)\cdots v_k(n))] = \sum_{i=1}^{n} \log\left[\frac{e^{(1-v_k(i))L(v_k(i))}}{1+e^{L(v_k(i))}}\right], \tag{9.113}$$

where $v_k(i)$ denotes the i-th bit in the k-th candidate codeword and $L(\cdot)$ is a corresponding LLR.

5. Calculate the extrinsic bit reliabilities, denoted as $L_e(\cdot)$, for the next decoding stage using

$$L_e(v_i) = L'(v_i) - L(v_i), \tag{9.114}$$

where

$$L'(v_i) = \log\left[\frac{\sum_{v_k(i)=0} L(\boldsymbol{v}_k)}{\sum_{v_k(i)=1} L(\boldsymbol{v}_k)}\right]. \tag{9.115}$$

In (9.115), summation in numerator (denominator) is performed over all candidate codewords having 0 (1) at position i.

6. Set $L(v_i) = L'(v_i)$ and move to step 1.

The following "max-star" operator can be applied to (9.113) and (9.115) recursively as follows:

$$\max * (x,y) \triangleq \log\left(\mathrm{e}^x + \mathrm{e}^y\right) = \max\left(\mathrm{x},\mathrm{y}\right) + \log\left[1+\mathrm{e}^{-|\mathrm{x}-\mathrm{y}|}\right]. \tag{9.116}$$

Given this description of SISO constituent decoding algorithm, the TPC decoder, shown in Fig. 9.22, operates as follows. Let $L^{\mathrm{e}}_{\mathrm{rc},j}$ denote the extrinsic information to be passed from row to column decoder, and let $L^{\mathrm{e}}_{\mathrm{cr},j}$ denote the extrinsic information to be passed in opposite direction. Then, assuming that the column decoder operates first, the *TPC decoder* performs the following steps:

0. Initialization: $L^{\mathrm{e}}_{\mathrm{cr},i} = L^{\mathrm{e}}_{\mathrm{rc},i} = 0$ for all i.
1. Column decoder: Run the SISO decoding algorithm described above with the following inputs $L(v_j) + L^{\mathrm{e}}_{\mathrm{rc},j}$ to obtain $\{L_{\mathrm{column}}(v_i)\}$ and $\{L^{\mathrm{e}}_{\mathrm{cr},i}\}$, as shown in Fig. 9.22. The extrinsic information is calculated by (9.114). Pass the extrinsic information $\{L^{\mathrm{e}}_{\mathrm{cr},j}\}$ to companion row decoder.
2. Row decoder: Run the SISO decoding algorithm with the following inputs $L(v_i) + L^{\mathrm{e}}_{\mathrm{cr},i}$ to obtain $\{L_{\mathrm{row}}(v_i)\}$ and $\{L^{\mathrm{e}}_{\mathrm{rc},i}\}$. Pass the extrinsic information $\{L^{\mathrm{e}}_{\mathrm{rc},i}\}$ to companion column decoder.
3. Bit decisions: Repeat the steps 1 to 2 until a valid codeword is generated or a predetermined number of iterations have been reached. Make the decisions on bits by sign$[L_{\mathrm{row}}(u_k)]$.

9.10 LDPC Codes

9.10.1 Binary LDPC Codes

Because LDPC codes belong to the class of linear block codes, they can be described as a k-dimensional subspace C of the n-dimensional vector space of n-tuples, F_2^n, over the binary field $F_2 = \mathrm{GF}(2)$. For this k-dimensional subspace, we can find the basis $B = \{g_0, g_1, \ldots, g_{k-1}\}$ that spans C so that every codeword v can be written as a linear combination of basis vectors $v = m_0 g_0 + m_1 g_1 + \ldots + m_{k-1} g_{k-1}$ for message vector $m = (m_0, m_1, \ldots, m_{k-1})$; or in compact form, we can write $v = mG$, where G is so-called generator matrix with the i-th row being g_i. The $(n-k)$-dimensional null space $C^{\perp}$ of G comprises all vectors x from F_2^n such that $xG^T = 0$, and it is spanned by the basis $B^\perp = \{h_0, h_1, \ldots, h_{n-k-1}\}$. Therefore, for each codeword v from C, $vh_i^T = 0$ for every i; or in compact form, we can write $vH^T = 0$, where H is so-called parity-check matrix whose i-th row is h_i.

An LDPC code can now be defined as an (n, k) linear block code whose parity-check matrix $\boldsymbol{H}$ has a low density of 1's. A *regular LDPC code* is a linear block code whose $\boldsymbol{H}$-matrix contains exactly w_c 1's in each column and exactly $w_r = w_c n/(n-k)$ 1's in each row, where $w_c \ll n - k$. The code rate of the regular LDPC code is determined by $R = k/n = 1 - w_c/w_r$. The graphical representation of LDPC codes, known as bipartite (Tanner) graph representation, is helpful in efficient description of LDPC decoding algorithms. A *bipartite (Tanner) graph* is a graph whose nodes may be separated into two classes (*variable* and *check* nodes) and where *undirected edges* may only connect two nodes not residing in the same class. The Tanner graph of a code is drawn according to the following rule: check (function) node c is connected to variable (bit) node v whenever element h_{cv} in a parity-check matrix $\boldsymbol{H}$ is a 1. In an $m \times n$ parity-check matrix, there are $m = n - k$ check nodes and n variable nodes. LDPC codes were invented by Robert Gallager (from MIT) in 1960, in his PhD dissertation [31], but received no attention from the coding community until the 1990s [32]. As an illustrative example, consider the $\boldsymbol{H}$-matrix of the Hamming (7,4) code:

$$\boldsymbol{H} = \begin{bmatrix} 1 & 0 & 0 & 1 & 0 & 1 & 1 \\ 0 & 1 & 0 & 1 & 1 & 1 & 0 \\ 0 & 0 & 1 & 0 & 1 & 1 & 1 \end{bmatrix}$$

For any valid codeword $\boldsymbol{v} = [v_0\ v_1 \ldots v_{n-1}]$, the checks used to decode the codeword are written as

- Equation (c_0): $v_0 + v_3 + v_5 + v_6 = 0 \pmod 2$
- Equation (c_1): $v_1 + v_3 + v_4 + v_5 = 0 \pmod 2$
- Equation (c_2): $v_2 + v_4 + v_5 + v_6 = 0 \pmod 2$

The bipartite graph (Tanner graph) representation of this code is given in Fig. 9.23a.

The circles represent the bit (variable) nodes, while squares represent the check (function) nodes. For example, the variable nodes v_0, v_3, v_5, and v_6 are involved in (c_0) and therefore connected to the check node c_0. A closed path in a bipartite graph comprising l edges that closes back on itself is called a *cycle* of length l. The shortest cycle in the bipartite graph is called the *girth*. The girth influences the minimum distance of LDPC codes, correlates the extrinsic LLRs, and therefore affects the decoding performance. The use of large (high) girth LDPC codes is preferable because the large girth increases the minimum distance and prevents early correlations in the extrinsic information during decoding process. To improve the iterative decoding performance, we have to avoid cycles of lengths 4 and 6, which are illustrated in Fig. 9.23b, c, respectively.

The code description can also be done by the *degree distribution polynomials* $\mu(x)$ and $\rho(x)$, for the variable node (v-node) and the check node (c-node), respectively [2]:

$$\mu(x) = \sum_{d=1}^{d_v} \mu_d x^{d-1}, \qquad \rho(x) = \sum_{d=1}^{d_c} \rho_d x^{d-1}, \tag{9.117}$$

where μ_d and ρ_d denote the fraction of the edges that are connected to degree-d v-nodes and c-nodes, respectively, and d_v and d_c denote the maximum v-node and c-node degrees, respectively.

The most obvious way to design LDPC codes is to construct a low-density parity-check matrix with prescribed properties. Some important LDPC designs, among others, include Gallager codes (semi-random construction) [31], MacKay codes (semi-random construction) [32], combinatorial design-based LDPC codes [34], finite-geometry-based LDPC codes [68–70], and array [also known as quasi-cyclic (QC)] LDPC codes [41], to mention a few. The generator matrix of a QC-LDPC code can be represented as an array of circulant submatrices of the same size B indicating that QC-LDPC codes can be encoded in linear time using simple shift-register-based architectures [35]. A QC-LDPC code can be defined as an LDPC code for which every sectional cyclic shift to the right (or left) for $l \in [0, B-1]$ places of a codeword $\boldsymbol{v} = [\boldsymbol{v}_0\ \boldsymbol{v}_1 \ldots \boldsymbol{v}_{B-1}]$ (each section $\boldsymbol{v}_i$ contains B elements) results in another codeword. Additional details on LDPC code designs are postponed for Sect. 9.10.5.

Fig. 9.23 (**a**) Bipartite graph of Hamming (7,4) code described by ***H***-matrix above. Cycles in a Tanner graph: (**b**) cycle of length 4 and (**c**) cycle of length 6

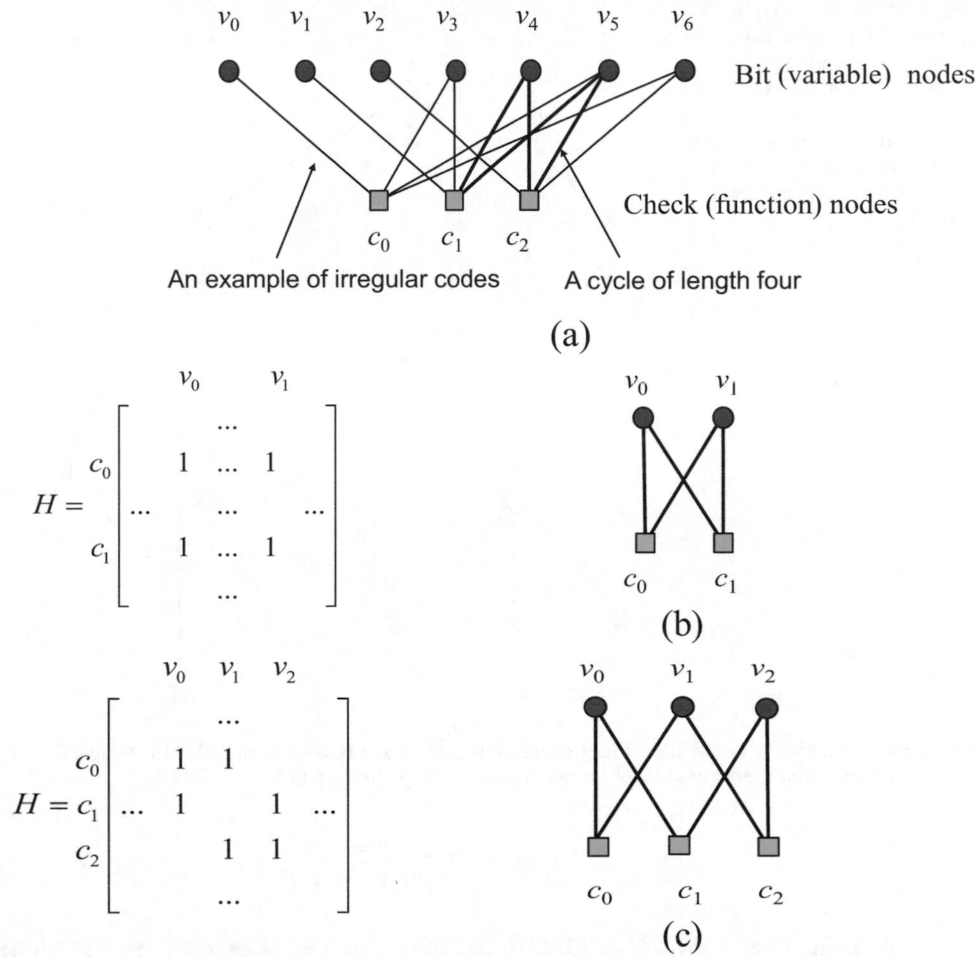

9.10.2 Decoding of Binary LDPC Codes

The sum-product algorithm (SPA) is an iterative LDPC decoding algorithm in which extrinsic probabilities are iterated forward and back between variable and check nodes of bipartite (Tanner) graph representation of a parity-check matrix [1, 2, 4, 5, 7, 36, 37, 65]. To facilitate the explanation of the various versions of the SPA, we use $N(v)$ [$N(c)$] to denote the neighborhood of v-node v (c-node c) and introduce the following notations:

- $N(c) = \{v\text{-nodes connected to } c\text{-node } c\}$
- $N(c)\backslash\{v\} = \{v\text{-nodes connected to } c\text{-node } c \text{ except } v\text{-node } v\}$
- $N(v) = \{c\text{-nodes connected to } v\text{-node } v\}$
- $N(v)\backslash\{c\} = \{c\text{-nodes connected to } v\text{-node } v \text{ except } c\text{-node } c\}$
- $P_v = \Pr(v = 1|y)$
- $E(v)$: the event that the check equations involving variable node v are satisfied
- $M(c')\backslash\{c\} = \{\text{messages from all } c'\text{-nodes except node } c\}$
- $q_{vc}(b) = \Pr(v = b \mid E(v), y, M(c')\backslash\{c\})$
- $M(v')\backslash\{v\} = \{\text{messages from all } v'\text{-nodes except node } v\}$
- $r_{cv}(b) = \Pr(\text{check equation } c \text{ is satisfied} \mid v = b, M(v')\backslash\{v\})$
- L_{vc}: the extrinsic likelihood to be sent from v-node v to c-node c
- L_{cv}: the extrinsic likelihood to be sent from c-node c to v-node v

We are interested in computing the APP that a given bit in a transmitted codeword $\boldsymbol{v} = [v_0\ v_1\ \ldots\ v_{n-1}]$ equals 1, given the received word $\boldsymbol{y} = [y_0\ y_1\ \ldots\ y_{n-1}]$. Let us focus on decoding of the variable node v and we are concerned in computing LLR:

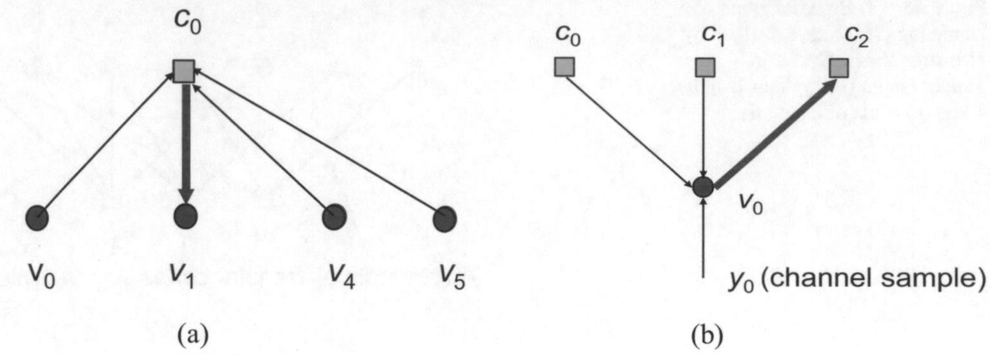

Fig. 9.24 Illustration of half-iterations of the SPA: (**a**) subgraph of bipartite graph corresponding to the H-matrix with the 0-th row [1 1 0 0 1 1 0. . .0], (**b**) subgraph of bipartite graph corresponding to the *H*-matrix with the 0-th column [1 1 1 0. . .0]

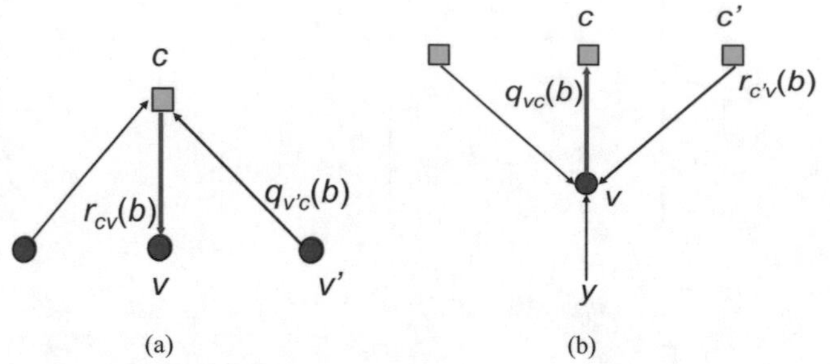

Fig. 9.25 Illustration of calculation of (**a**) extrinsic information to be passed from node *c* to variable node *v*, (**b**) extrinsic information (message) to be passed from variable node *v* to check node *c* regarding the probability that $v = b$, $b \in \{0,1\}$

$$L(v) = \log\left[\frac{\Pr(v = 0|\mathbf{y})}{\Pr(v = 1|\mathbf{y})}\right], \quad v \in \{v_0, \cdots, v_{n-1}\}. \tag{9.118}$$

The SPA, as indicated above, is an iterative decoding algorithm based on Tanner graph description of an LDPC code. We interpret the v-nodes as one type of processors and c-nodes as another type of processors, while the edges as the message paths for LLRs.

The subgraph illustrating the passing of messages (extrinsic information) from c-node to v-nodes in the check-node update half-iteration is shown in Fig. 9.24a. The information passed is the probability that parity-check equation c_0 is satisfied. The information passed from node c_0 to node v_1 represents the extrinsic information it had received from nodes v_0, v_4, and v_5 on the previous half-iteration. We are concerned in calculating the APP that a given bit *v* equals 1, given the received word $\mathbf{y}$ and the fact that all parity-check equations involving bit v are satisfied; with this APP is denoted as $\Pr(v = 1|\, \mathbf{y}, E(v))$. Instead of APP we can use log-APP or LLR defined as $\log\left[\frac{\Pr(v=0|\mathbf{y}, E(v))}{\Pr(v=1|\mathbf{y}, E(v))}\right]$.

In the v-node update half-iteration, the messages are passed in opposite direction – from v-nodes to c-nodes, as depicted in subgraph of Fig. 9.24b. The information passed concerns $\log[\Pr(v_0 = 0|\mathbf{y})/\Pr(v_0 = 1|\mathbf{y})]$. The information being passed from node v_0 to node c_2 is information from the channel (via y_0) and extrinsic information node v_0 had received from nodes c_0 and c_1 on a previous half-iteration. This procedure is performed for all v-node/c-node pairs.

The calculation of the probability that *c*-th parity-check equation is satisfied given $v = b$, $b \in \{0,1\}$, denoted as $r_{cv}(b)$, is illustrated in Fig. 9.25a. On the other hand, the calculation of the probability that $v = b$ given extrinsic information from all check nodes, except node *c*, and given channel sample *y*, denoted as $q_{vc}(b)$, is illustrated in Fig. 9.25b.

In the first half-iteration, we calculate the extrinsic LLR to be passed from node *c* to variable node *v*, denoted as $L_{vc} = L(r_{cv})$, as follows:

$$L_{cv} = \left(\prod_{v'} \alpha_{v'c}\right) \phi\left[\sum_{v' \in N(c)\backslash v} \phi(\beta_{v'c})\right]; \alpha_{vc} = \text{sign}\,[L_{vc}], \beta_{vc} = |L_{vc}| \tag{9.119}$$

where with $\phi(x)$ we denoted the following function:

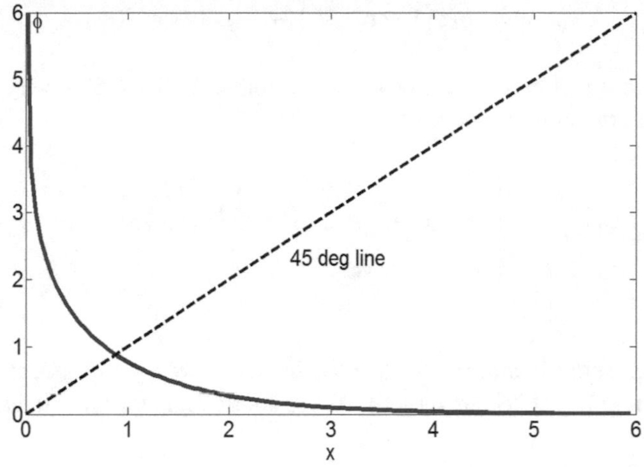

Fig. 9.26 The plot of function $\phi(x)$

$$\phi(x) = -\log \tanh\left(\frac{x}{2}\right) = \log\left[\frac{e^x+1}{e^x-1}\right], \tag{9.120}$$

which is plotted in Fig. 9.26. In the second half-iteration, we calculate the extrinsic LLR to be passed from variable node v to function node c regarding the probability that $v = b$, denoted as $L_{vc} = L(q_{vc})$, as

$$L_{vc} = L(q_{vc}) = L_{\mathrm{ch}}(v) + \sum_{c' \in N(v)\backslash c} L(r_{c'v}). \tag{9.121}$$

For the derivation of Eqs. (9.119) and (9.121), an interested reader is referred to [37].

Now we can summarize the *log-domain SPA* as follows:

0. *Initialization*: For $v = 0,1,\ldots,n-1$, initialize the messages L_{vc} to be sent from v-node v to c-node c to channel LLRs $L_{\mathrm{ch}}(v)$, namely, $L_{vc} = L_{\mathrm{ch}}(v)$.
1. *c-node update rule*: For $c = 0,1,\ldots,n-k-1$, compute $L_{cv} = \boxplus_{N(c)\backslash\{v\}} L_{vc}$. The box-plus operator is defined by

$$L_1 \boxplus L_2 = \prod_{k=1}^{2} \mathrm{sign}(L_k) \cdot \phi\left(\sum_{k=1}^{2} \phi(|L_k|)\right),$$

where $\phi(x) = -\log \tanh(x/2)$. The box operator for $|N(c)\backslash\{v\}|$ components is obtained by recursively applying two-component version defined above.

2. *v-node update rule*: For $v = 0,1,\ldots,n-1$, set $L_{vc} = L_{\mathrm{ch}}(v) + \sum_{N(v)\backslash\{c\}} L_{cv}$ for all c-nodes for which $h_{cv} = 1$.
3. *Bit decisions*. Update $L(v)$ $(v = 0,\ldots,n-1)$ by $L(v) = L_{\mathrm{ch}}(v) + \sum_{N(v)} L_{cv}$ and set $\hat{v} = 1$ when $L(v) < 0$ (otherwise, $\hat{v} = 0$).

If $\hat{\boldsymbol{v}}\boldsymbol{H}^T = \boldsymbol{0}$ or predetermined number of iterations has been reached, then stop; otherwise go to step 1.

Because the c-node update rule involves log and tanh functions, it is computationally intensive, and there exist many approximations. The very popular is the *min-sum-plus-correction-term approximation* [39]. Namely, it can be shown that "box-plus" operator $\boxplus$ can also be calculated by

$$L_1 \boxplus L_2 = \prod_{k=1}^{2} \mathrm{sign}(L_k) \cdot \min(|L_1|, |L_2|) + c(x,y), \tag{9.122}$$

where $c(x,y)$ denotes the correction factor defined by

$$c(x,y) = \log\left(1 + e^{-|x+y|}\right) - \log\left(1 + e^{-|x-y|}\right), \tag{9.123}$$

commonly implemented as a look-up table (LUT). Given the fact that $|c(x,y)| < 0.693$, very often this term can be ignored. Alternatively, the following approximation can be used:

$$c(x,y) \simeq \begin{cases} -d, & |x-y| < 2 \cap |x+y| > 2|x-y| \\ d, & |x+y| < 2 \cap |x-y| > 2|x+y| \\ 0, & \text{otherwise} \end{cases} \tag{9.124}$$

with typical d being 0.5.

Another popular decoding algorithm is *min-sum algorithm* in which we simply ignore the correction term in (9.123). Namely, the shape of $\phi(x)$, shown in Fig. 9.26, suggests that the smallest β_{vc} in the summation (9.119) dominates and we can write:

$$\phi\left[\sum_{v' \in N(c) \backslash v} \phi(\beta_{v'c})\right] \cong \phi\left(\phi\left(\min_{v'} \beta_{v'c}\right)\right) = \min_{v'} \beta_{v'c}. \tag{9.125}$$

Therefore, the min-sum algorithm is thus the log-domain algorithm with step 1 replaced by

$$L_{cv} = \left(\prod_{v' \in N(c) \backslash v} \alpha_{c'v}\right) \cdot \min_{v'} \beta_{v'c}. \tag{9.126}$$

9.10.3 FPGA Implementation of Binary LDPC Decoders

In this section, we present an FPGA-based partial-parallel implementation of QC-LDPC decoder. The decoder of girth-10 QC-LDPC (34635, 27708) with column weight $W_c = 3$, row weight $W_r = 15$, and submatrix size $B = 2309$ has been implemented by using BEEcube system BEE4. This rapid prototyping FPGA system contains four fully integrated Xilinx Virtex-6 FPGAs. Four LDPC decoders have been implemented per single FPGA, resulting in total 16 LDPC decoders per BEE4 system.

As shown in Fig. 9.27a, the implementation consists of three types of memories: *Mem*, *MemInit*, and *MemBit*. The $W_c \times W_r$ *Mem* units store the messages to be interchanged between variable nodes and check nodes, W_r *MemInit* units store the initial LLRs from the channel, and W_r *MemBit* units store the decoded bits. There are W_r variable node processors (VNPs) and W_c check node processors (CNPs) in our implementation. While the VNPs read the extrinsic information from *Mem* and prior information from *MemInit* and write the updated information to the corresponding addresses of *Mem* and *MemBit* in the first half-iteration, the CNPs read the information from *Mem* and *MemBit* and write the updated extrinsic information to the corresponding addresses of *Mem* in the second half-iteration. Decoding stops when either the predefined maximum number of iterations has been reached or decoder outputs a codeword.

With above described architecture, the lowest decoding throughput per decoder and FPGA can be expressed as follows:

$$\text{Throughput} = \frac{N \times f_{\text{clock}}}{B + (B + B + \delta) \times Iter} \simeq 143\,\text{Mb/s}$$

where $N = 34635$ is the codeword length, $f_{\text{clock}} = 200$ MHz is the clock frequency, $B = 2309$ is the size of the submatrix, $Iter = 20$ is the maximum number of iterations, and $\delta = 6$ is the critical path delay of the processing unit. The first B clock cycles used to initialize the *MemInit*, W_r VNPs need B clock cycles to process $W_r \times B$ variable nodes, while W_c CNPs need B clock cycles to process $W_c \times B$ check nodes in each iteration. The aggregate minimum throughput of 16 LDPC decoders

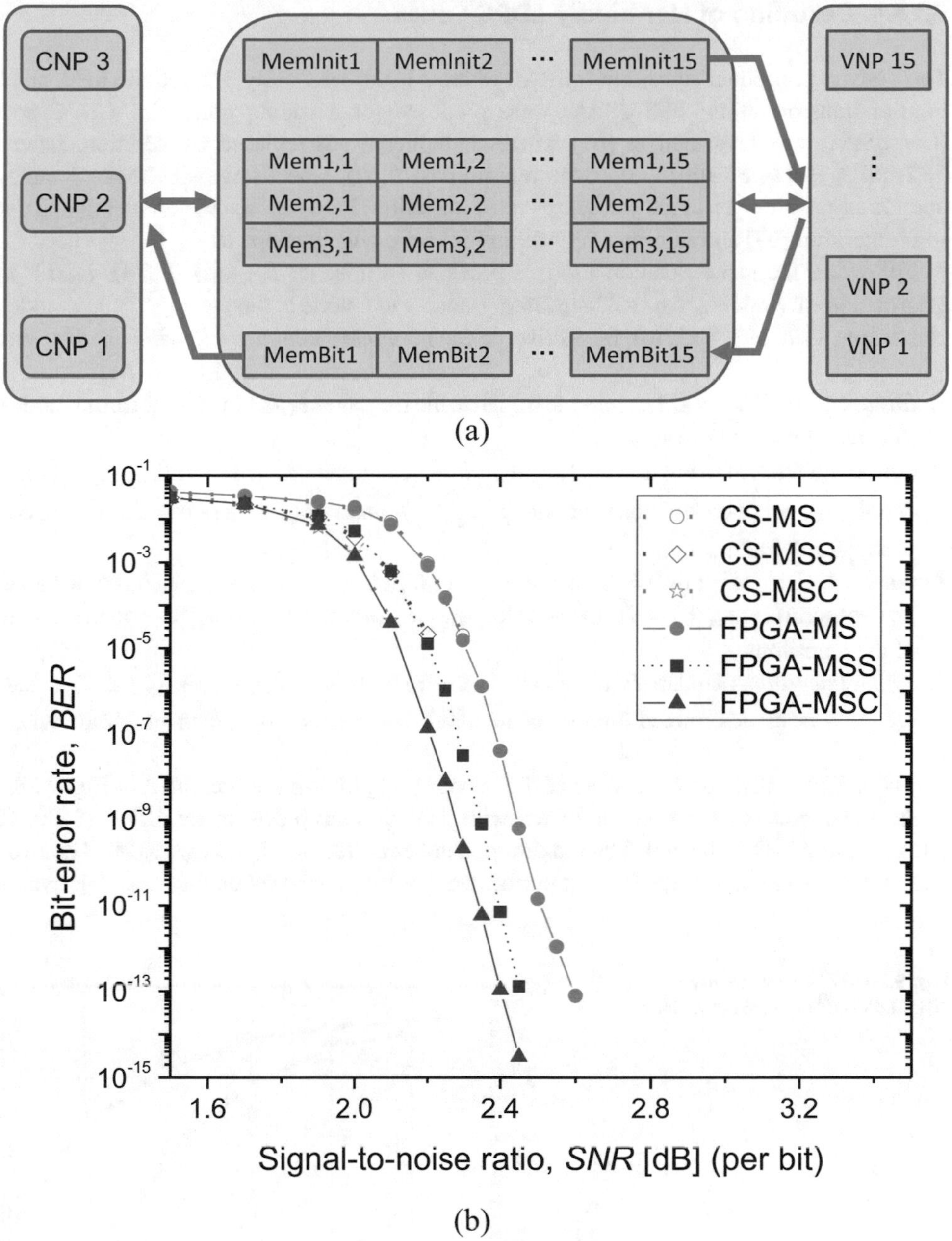

Fig. 9.27 (**a**) Implementation architecture of FPGA-based partial-parallel LDPC decoder and (**b**) BER performance comparison of simulated and FPGA-based LDPC decoder. *CS* computer simulations, *FPGA* FPGA results, *MS* min-sum algorithm, *MSC* min-sum with correction term algorithm, *MSS* min-sum algorithm with scaling

implemented on BEE4 system is ≅143 Mb/s × 4 × 4 = 2.288 Gb/s. Since the average number of iterations until a valid codeword is reached is 8.8, the average total throughput is actually higher.

Figure 9.27b presents the BER performances of the FPGA-based min-sum algorithm (MS), min-sum with scaling factor (MSS), and min-sum with correction term (MSC). The computer simulation (CS) results are provided as well for verification. In practice, it is also possible to implement other decoding algorithms to reduce the complexity, described above. Here, the input LLRs and extrinsic information are represented by 8 soft-decision bits (1 sign bit, 5 integer bits, and 2 fractional bits). It can be observed that there is no error floor down to 10^{-15}, which is consistent with ref. [55], which claims that girth-10 LDPC codes do not exhibit error floor phenomenon in the region of interest for optical communications (down to 10^{-15}). The net coding gain (NCG) at BER of 10^{-15} is 11.8 dB. The good BER performance of our QC-LDPC code is contributed to the large minimum distance and large girth. For short LDPC codes, error floor can be suppressed by (i) BCH with corresponding interleaver or RS codes [56, 57] as outer code in concatenation fashion or (ii) post-processing techniques [58].

9.10.4 Decoding of Nonbinary LDPC Codes

The q-ary sum-product algorithm (qSPA), where q is the cardinality of the finite field, and a low-complexity version via fast Fourier transformation (FFT-qSPA) were proposed for decoding nonbinary LDPC codes [59, 60]. Log-domain qSPA (Log-qSPA) was presented in [61] where multiplication is replaced by addition. Later, a mixed-domain version of the FFT-qSPA (MD-FFT-qSPA) decoder was proposed [62], where large-size look-up tables are required for the purpose of domain conversion. To avoid the complicated operations involved in the aforementioned decoding algorithms, extended min-sum algorithm [63] and min-max algorithm [64] were widely adopted.

Following the same notation used in previous section, let $L_{ch,\ v}(a)$, $L_{vc}(a)$, $L_{cv}(a)$, $L_{cv}^{k,l}$ and $L_v(a)$ represent the prior information of v-node v, the message from v-node v to c-node c, the message from c-node c to v-node v, and the APP value concerning symbol $a \in \mathrm{GF}(q)$, respectively. The *nonbinary min-max algorithm* can be summarized as follows:

0. *Initialization*: For $v = 0,1,\ldots,n-1$, initialize the messages $L_{vc}(a)$ to be sent from v-node v to c-node c to channel LLRs $L_{ch,\ v}(a)$, namely,$L_{vc}(a) = L_{ch,\ v}(a)$.
1. *c-node update rule*: For $c = 0,1,\ldots,n-k-1$, compute $L_{cv}(a) = \min_{a'_v \in I(c|a_v)} \left[\max_{v' \in N(c)\backslash v} L_{vc}(a'_{v'}) \right]$, where $I(c|\ a_v)$ denotes the set of codewords such that $\sum_{v' \in N(c)\backslash v} h_{cv'} a'_{v'} = h_{cv} a_v$. Here h_{cv} denotes the c-th row, v-th column element of the parity-check matrix.
2. *v-node update rule*: For $v = 0,1,\ldots,n-1$, set $L_{vc}(a) = L_{\mathrm{ch},\ v}(a) + \sum_{N(v)\backslash c} L_{cv}(a)$ for all c-nodes for which $h_{cv} \neq 0$.
3. *Post-processing*: $L_{vc}(a) = L_{vc}(a) - \min_{a' \in \mathrm{GF}(q)} L_{vc}(a')$, it is necessary for numerical reasons to ensure the non-divergence of the algorithm.
4. *Symbol decisions*: Update $L_v(a)$ $(v = 0,\ldots,n-1)$ by $L_v(a) = L_{\mathrm{ch},\ v}(a) + \sum_{N(v)} L_{cv}(a)$, and set $\widehat{v} = \arg\min_{a \in GF(q)} (L_v(a))$. If $\widehat{\boldsymbol{v}}\boldsymbol{H}^T = \mathbf{0}$ or predetermined number of iterations has been reached, then stop; otherwise go to step 1.

The BER vs. SNR performances of the decoding algorithms are presented in Fig. 9.28.

All the decoders have been simulated with floating point precision for GF(4) (8430, 6744) nonbinary LDPC code over binary input AWGN channel. The maximum number of iteration $I_{\max}$ is set to 20. The two decoding algorithms, Log-QSPA and MD-FFT-qSPA, achieve the best performance as they are based on theoretical derivations, while the reduced-complexity

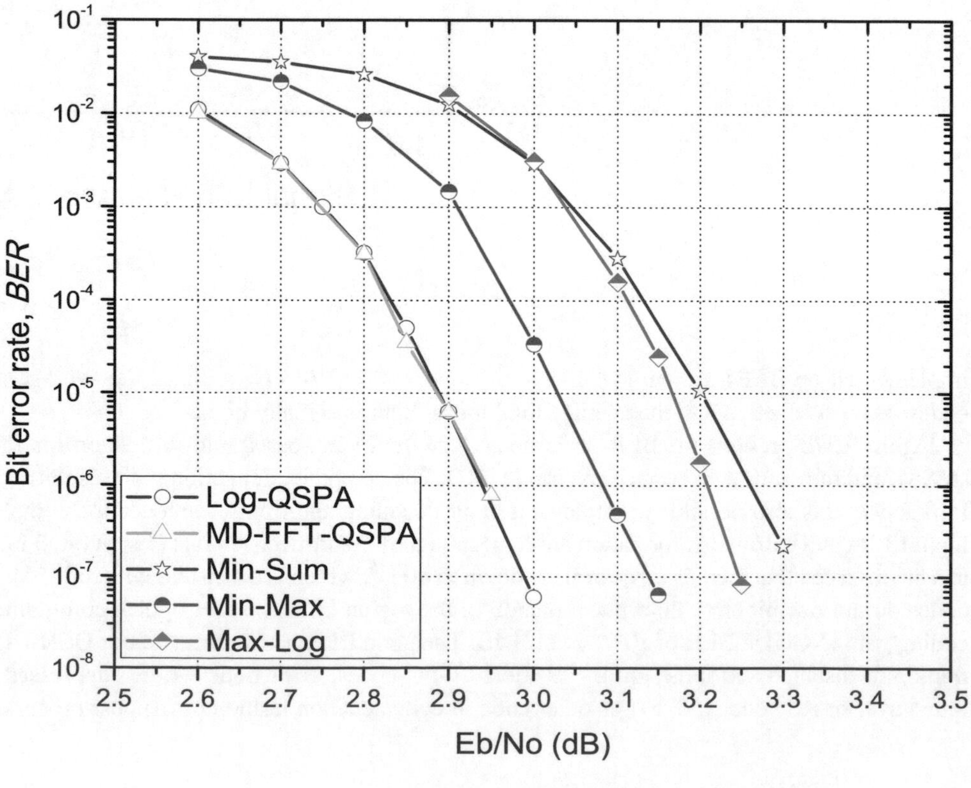

Fig. 9.28 BER performance of NB-LDPC code over BI-AWGN

decoding algorithms, min-max, max-log, and min-sum algorithms, face the performance degradation at BER of 10^{-7} by ~0.13 dB, ~0.27 dB, and ~ 0.31 dB, respectively.

The details of corresponding FPGA implementation of nonbinary LDPC decoders can be found in [66, 67]. In the same papers, the FPGA resource utilization has been discussed in detail. For instance, the reduced-complexity nonbinary decoding algorithms and finite precision introduce the error floor phenomenon, which has been successfully eliminated by an outer RS code.

9.10.5 Design of LDPC Codes

The most obvious way to design LDPC codes is to construct a low-density parity-check matrix with prescribed properties. Some important designs among others include (i) Gallager codes (semi-random construction) [31], (ii) MacKay codes (semi-random construction) [32], (iii) finite-geometry-based LDPC codes [68–70], (iv) combinatorial-design-based LDPC codes [34], (v) quasi-cyclic (array, block-circulant) LDPC codes [2, 41], (vi) irregular LDPC codes [71], and (vii) Tanner codes [22]. In this section we describe briefly several classes of LDPC codes including Gallager, Tanner, MacKay, and quasi-cyclic LDPC codes.

9.10.5.1 Gallager Codes

The $\boldsymbol{H}$-matrix for Gallager code has the following general form:

$$\boldsymbol{H} = [\boldsymbol{H}_1 \quad \boldsymbol{H}_2 \quad \cdots \quad \boldsymbol{H}_{w_c}]^{\mathrm{T}} \tag{9.127a}$$

where $\boldsymbol{H}_1$ is $pxp{\cdot}w_r$ matrix of row weight w_r, and $\boldsymbol{H}_i$ are column-permuted versions of $\boldsymbol{H}_1$-submatrix. The row weight of $\boldsymbol{H}$ is w_r, and column weight is w_c. The permutations are carefully chosen to avoid the cycles of length 4. The $\boldsymbol{H}$-matrix is obtained by computer search.

9.10.5.2 Tanner Codes and Generalized LDPC (GLDPC) Codes

In Tanner codes [22], each bit node is associated with a code bit, and each check node is associated with a subcode whose length is equal to the degree of node, which is illustrated in Fig. 9.29. Notice that so-called generalized LDPC codes [72–78] were inspired by Tanner codes.

Let us consider the following example:

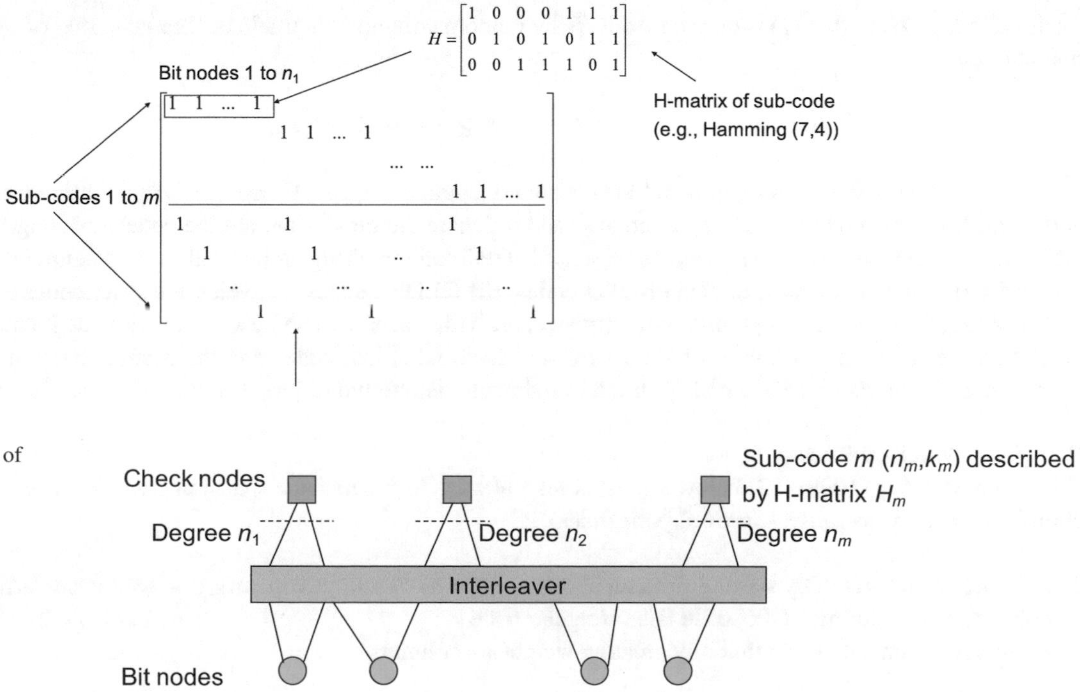

Fig. 9.29 An illustration of Tanner codes

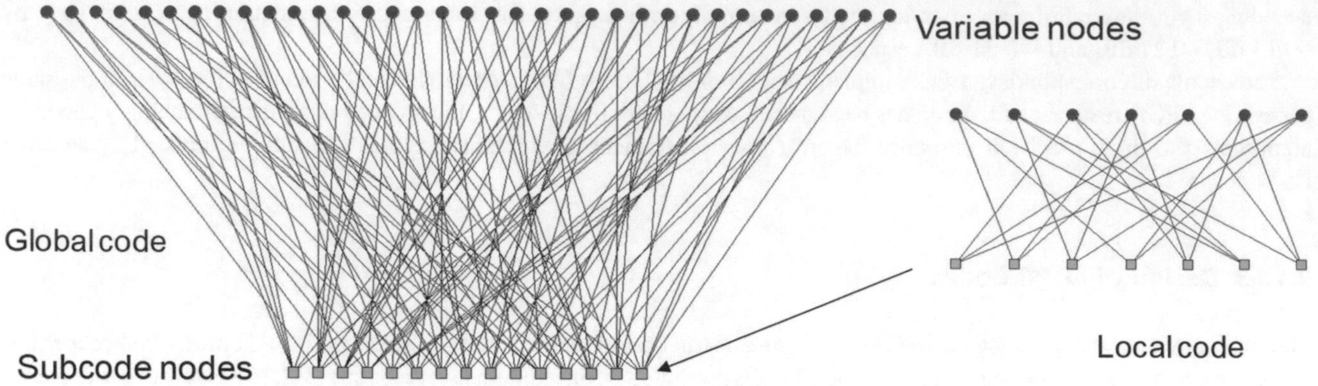

Fig. 9.30 Illustration of construction of LZ-GLDPC codes

The Tanner code design in this example is performed in two stages. We first design the global code by starting from an identity matrix $I_{m/2}$ and replace every nonzero element with n_1 ones and every zero element by n_1 zeros. The lower submatrix is obtained by concatenating the identity matrices In_1. In the second stage, we substitute all-one row vectors (of length n_1) by the parity-check matrix of local linear block code (n_1,k_1), such as Hamming, BCH, or RM code. The resulting parity-check matrix is used as the parity-check matrix of an LDPC code.

The GLDPC codes can be constructed in similar fashion. To construct a GLDPC code, one can replace each single parity-check equation of a global LDPC code by the parity-check matrix of a simple linear block code, known as the constituent (local) code, and this construction is proposed by Lentmaier and Zigangirov [77], and we will refer to this construction as LZ-GLDPC code construction. One illustrative example is shown in Fig. 9.30. In another construction proposed by Boutros et al. in [76], referred here as B-GLDPC code construction, the parity-check matrix, $\boldsymbol{H}$, is a sparse matrix partitioned into W submatrices $\boldsymbol{H}_1$,. . .,$\boldsymbol{H}_W$. The $\boldsymbol{H}_1$-submatrix is a block-diagonal matrix generated from an identity matrix by replacing the ones by a parity-check matrix $\boldsymbol{H}_0$ of a local code of codeword length n and dimension k:

$$\boldsymbol{H} = [\boldsymbol{H}_1^T \ \boldsymbol{H}_2^T \ \ldots \ \boldsymbol{H}_W^T]^T, \boldsymbol{H}_1 = \begin{bmatrix} \boldsymbol{H}_0 & & & 0 \\ & \boldsymbol{H}_0 & & \\ & & \cdots & \\ 0 & & & \boldsymbol{H}_0 \end{bmatrix} \tag{9.127b}$$

Each submatrix $\boldsymbol{H}_j$ in (9.127) is derived from $\boldsymbol{H}_1$ by random column permutations. The code rate of a GLDPC code is lower bounded by

$$R = K/N \geq 1 - W(1 - k/n), \tag{9.128}$$

where K and N denote the dimension and the codeword length of a GLDPC code, W is the column weight of a global LDPC code, and k/n is the code rate of a local code (k and n denote the dimension and the codeword length of a local code). The GLDPC codes can be classified as follows [72]: (i) GLDPC codes with algebraic local codes of short length, such as Hamming codes, BCH codes, RS codes, or Reed-Muller codes, (ii) GLDPC codes for which the local codes are high-rate regular or irregular LDPC codes with large minimum distance, and (iii) fractal GLDPC codes in which the local code is in fact another GLDPC code. For more details on LZ-GLDPC and B-GLDPC like codes and their generalization-fractal GLDPC codes (a local code is another GLDPC code), an interested reader is referred to [33, 37] (and references therein).

9.10.5.3 MacKay Codes

Following MacKay (1999) [32] below are listed several ways to generate the sparse matrices in order of increasing algorithm complexity (not necessarily improved performance):

1. $\boldsymbol{H}$-matrix is generated by starting from an all-zero matrix and randomly inverting w_c (not necessarily distinct bits) in each column. The resulting LDPC code is an irregular code.
2. $\boldsymbol{H}$-matrix is generated by randomly creating weight-w_c columns.

3. $\boldsymbol{H}$-matrix is generated with weight-w_c columns and uniform row weight (as near as possible).
4. $\boldsymbol{H}$-matrix is generated with weight-w_c columns, weight-w_r rows, and no two columns having overlap larger than one.
5. $\boldsymbol{H}$-matrix is generated as in (4), and short cycles are avoided.
6. $\boldsymbol{H}$-matrix is generated as in (5) and can be represented $\boldsymbol{H} = [\boldsymbol{H}_1 | \boldsymbol{H}_2]$, where $\boldsymbol{H}_2$ is invertible or at least has a full rank.

The construction via (5) may lead to an $\boldsymbol{H}$-matrix that is not of full rank. Nevertheless, it can be put in the following form by column swapping and Gauss-Jordan elimination:

$$\boldsymbol{H} = \begin{bmatrix} \boldsymbol{P}^T & \boldsymbol{I} \\ \boldsymbol{0} & \boldsymbol{0} \end{bmatrix}, \tag{9.129}$$

and by eliminating the all-zero submatrix, we obtain the parity-check matrix of systematic LDPC code:

$$\widetilde{\boldsymbol{H}} = \begin{bmatrix} \boldsymbol{P}^T & \boldsymbol{I} \end{bmatrix}. \tag{9.130}$$

Among various construction algorithms listed above, the construction (5) will be described with more details.

Outline of Construction Algorithm (5)

1. Choose code parameters n, k, w_c, w_r, and g (minimum cycle length girth). The resulting H-matrix will be an m x n ($m = n - k$) matrix with w_c ones per column and w_r ones per row.

2. Set the column counter to $i_c = 0$.
3. Generate a weight-w_c column vector and place it in i_c-th column of H-matrix.
4. If the weight of each row $\leq w_r$, the overlap between any two columns ≤ 1, and all cycle lengths are $\geq g$, then increment the counter $i_c = i_c + 1$.
5. If $i_c = n$, stop; else go to step 3.

This algorithm could take hours to run with no guarantee of regularity of $\boldsymbol{H}$-matrix. Moreover, it may not finish at all, and we need to restart the search with another set of parameters. Richard and Urbanke (in 2001) proposed a linear complexity in length technique based on $\boldsymbol{H}$-matrix [33].

An alternative approach to simplify encoding is to design the codes via algebraic, geometric, or combinatoric methods. The $\boldsymbol{H}$-matrix of those designs can be put in cyclic or quasi-cyclic form, leading to implementations based on shift registers and mod-2 adders.

9.10.5.4 Large-Girth Quasi-Cyclic (QC) Binary LDPC Codes

Based on Tanner's bound for the minimum distance of an LDPC code [22]

$$d \geq \begin{cases} 1 + \dfrac{w_c}{w_c - 2}\left((w_c - 1)^{\lfloor (g-2)/4 \rfloor} - 1 \right), g/2 = 2m + 1 \\ 1 + \dfrac{w_c}{w_c - 2}\left((w_c - 1)^{\lfloor (g-2)/4 \rfloor} - 1 \right) + (w_c - 1)^{\lfloor (g-2)/4 \rfloor}, g/2 = 2m \end{cases} \tag{9.131}$$

(where g and w_c denote the girth of the code graph and the column weight, respectively, and where d stands for the minimum distance of the code), it follows that large girth leads to an exponential increase in the minimum distance, provided that the column weight is at least 3. ($\lfloor \cdot \rfloor$ denotes the largest integer less than or equal to the enclosed quantity.) For example, the minimum distance of girth-10 codes with column weight $r = 3$ is at least 10. The parity-check matrix of regular QC LDPC codes [2, 10, 12] can be represented by

$$
\boldsymbol{H} = \begin{bmatrix} I & I & I & \cdots & I \\ P^{S[0]} & P^{S[1]} & P^{S[2]} & \cdots & P^{S[c-1]} \\ P^{2S[0]} & P^{2S[1]} & P^{2S[2]} & \cdots & P^{2S[c-1]} \\ \cdots & \cdots & \cdots & \cdots & \cdots \\ P^{(r-1)S[0]} & P^{(r-1)S[1]} & P^{(r-1)S[2]} & \cdots & P^{(r-1)S[c-1]} \end{bmatrix}, \tag{9.132}
$$

where I is $B \times B$ (B is a prime number) identity matrix, P is $B \times B$ permutation matrix given by $P = (p_{ij})_{B \times B}$, $p_{i,i+1} = p_{B,1} = 1$ (zero otherwise), and r and c represent the number of block rows and block columns in (9.132), respectively. The set of integers S are to be carefully chosen from the set $\{0,1,\ldots,B-1\}$ so that the cycles of short length, in the corresponding Tanner (bipartite) graph representation of (9.132), are avoided. We have to avoid the cycles of length $2k$ ($k = 3$ or 4) defined by the following equation:

$$
S[i_1]\, j_1 + S[i_2]\, j_2 + \cdots + S[i_k]\, j_k = S[i_1]\, j_2 + S[i_2]\, j_3 + \cdots + S[i_k]\, j_1 \bmod B, \tag{9.133}
$$

where the closed path is defined by (i_1,j_1), (i_1,j_2), (i_2,j_2), (i_2,j_3), …, (i_k,j_k), (i_k,j_1) with the pair of indices denoting row-column indices of permutation blocks in (2) such that $l_m \neq l_{m+1}$, $l_k \neq l_1$ ($m = 1,2,..,k$; $l \in \{i,j\}$). Therefore, we have to identify the sequence of integers $S[i] \in \{0,1,\ldots,B\text{-}1\}$ ($i = 0,1,\ldots,r\text{-}1$; $r < B$) not satisfying Eq. (9.133), which can be done either by computer search or in a combinatorial fashion. We add an integer at a time from the set $\{0,1,\ldots,B-1\}$ (not used before) to the initial set S and check if Eq. (9.133) is satisfied. If Eq. (9.133) is satisfied, we remove that integer from the set S and continue our search with another integer from set $\{0,1,\ldots,B-1\}$ until we exploit all the elements from $\{0,1,\ldots,B-1\}$. The code rate of these QC codes, R, is lower bounded by

$$
R \geq \frac{|S|B - rB}{|S|B} = 1 - r/|S|, \tag{9.134}
$$

and the codeword length is $|S|B$, where $|S|$ denotes the cardinality of set S. For a given code rate R_0, the number of elements from S to be used is $\lfloor r/(1 - R_0) \rfloor$. With this algorithm, LDPC codes of arbitrary rate can be designed.

As an illustration, by setting $B = 2311$, the set of integers to be used in (2) is obtained as $S = \{1, 2, 7, 14, 30, 51, 78, 104, 129, 212, 223, 318, 427, 600, 808\}$. The corresponding LDPC code has rate $R_0 = 1\text{–}3/15 = 0.8$, column weight 3, girth 10, and length $|S|B = 15 \cdot 2311 = 34{,}665$. In the example above, the initial set of integers was $S = \{1,2,7\}$, and the set of row to be used in (6.25) is $\{1,3,6\}$. The use of a different initial set will result in a different set from that obtained above.

9.10.6 Rate-Adaptive LDPC Coding

In this section, we first describe the nonbinary (NB) irregular QC-LDPC code design derived from PBDs, introduced in [48], defined over GF(q). Let a set V of size v represent a set of elements (points), with any subset being called a block. Then, a pairwise balanced design PBD(v,K,δ) is defined as a collection of blocks of different sizes taken from set K, such that every pair of points is contained in δ of the blocks. For additional details on PBDs, an interested reader is referred to [79]. The parity-check matrix of irregular NB-QC-LDPC codes based on PBDs is given by [48]

$$
\boldsymbol{H} = \begin{bmatrix} \alpha^0 I(b_{00}) I & \alpha^1 I(b_{10}) I & \cdots & \alpha^{c-1} I(b_{c-1,0}) I \\ \alpha^{c-1} I(b_{01}) P^{S[0]} & \alpha^0 I(b_{11}) P^{S[1]} & \cdots & \alpha^{c-2} I(b_{c-1,1}) P^{S[c-1]} \\ \alpha^{c-2} I(b_{02}) P^{2S[0]} & \alpha^{c-1} I(b_{12}) P^{2S[1]} & \cdots & \alpha^{c-3} I(b_{c-1,2}) P^{2S[c-1]} \\ \vdots & \vdots & \ddots & \vdots \\ \alpha^{c-(r-1)} I(b_{0,r-1}) P^{(r-1)S[0]} & \alpha^{c-r+2} I(b_{1,r-1}) P^{(r-1)S[1]} & \cdots & \alpha^{c-r} I(b_{c-1,r-1}) P^{(r-1)S[c-1]} \end{bmatrix}, \tag{9.135}
$$

where I is $B \times B$ (B is a prime number) identity matrix, P is $B \times B$ permutation matrix given by $P = (p_{ij})_{B \times B}$, $p_{i,i+1} = p_{B,1} = 1$ (zero otherwise), and r and c represent the number of block rows and block columns in (9.135), respectively. In (9.135), $\{b_{ij}\}$ are points of the i-th block in PBD(r,K,δ), with the largest size of block k in set of sizes K satisfying the inequality $k \leq r$. α^i are nonzero elements of Galois field of size q, denoted as GF(q). Finally, the $I(b_{ij})$ denotes the indicator function, which has the value 1 for the existing point within the i-th block and 0 for the non-existing point. Therefore, only those submatrices for which indicator function is 1 will be preserved from template, regular, QC-LDPC code design. Notice the repetition of PBD blocks in (9.135) is allowed. Given the fact that PBDs have regular mathematical structure that can be algebraically described, the irregular NB-QC-LDPC codes derived from PBDs have the complexity comparable or lower to that of regular NB-QC-LDPC code design. As an illustration, the irregular NB-QC-LDPC code derived from PBD(5, {3,2,1}, 1) = {{0,1,3}, {1,2,4}, {1,2}, {0,4}, {3,4}, {0}, {1}} has the following form:

$$\boldsymbol{H}=\begin{bmatrix} \alpha^0 I & 0 & 0 & \alpha^3 I & 0 & \alpha^5 I & 0 \\ \alpha^6 I & \alpha^0 P^{S[1]} & \alpha^1 P^{S[2]} & 0 & 0 & 0 & \alpha^5 P^{S[5]} \\ 0 & \alpha^6 P^{2S[1]} & \alpha^0 P^{2S[2]} & 0 & 0 & 0 & 0 \\ \alpha^4 I & 0 & 0 & 0 & \alpha^1 P^{3S[4]} & 0 & 0 \\ 0 & \alpha^4 P^{4S[1]} & 0 & \alpha^6 P^{4S[3]} & \alpha^0 P^{4S[4]} & 0 & 0 \end{bmatrix}.$$

Since both the identity matrix and the power of permutation matrix have a single 1 per row, the block size of the i-th block from PBD determines i-th block-column weight. In the example above, the first two block columns have column weight 3; the third, fourth, and fifth have the column weight 2; and the last two block columns have weight 1. Notice that for GF (4) = $\{0,1,\alpha,\alpha^2\}$, we have that $\alpha^3 = 1$, $\alpha^4 = \alpha$, $\alpha^5 = \alpha^2$, and $\alpha^6 = \alpha^3$.

We now describe how to derive different classes of codes proposed for next generation of optical networks from (9.135). Let the template $\boldsymbol{H}$-matrix of (4,8) irregular QC code be given as

$$\boldsymbol{H}=\begin{bmatrix} \alpha^0 I(b_{00})I & \alpha^1 I(b_{10})I & \cdots & \alpha^7 I(b_{70})I \\ \alpha^7 I(b_{01})I & \alpha^0 I(b_{11})P^{S[1]} & \cdots & \alpha^6 I(b_{71})P^{S[7]} \\ \alpha^6 I(b_{02})I & \alpha^7 I(b_{12})P^{2S[1]} & \cdots & \alpha^5 I(b_{72})P^{2S[7]} \\ \alpha^5 I(b_{03})I & \alpha^6 I(b_{13})P^{3S[1]} & \cdots & \alpha^4 I(b_{73})P^{3S[7]} \end{bmatrix}.$$

By employing the following PBD, {{0}, {0,1}, {0,1,2}, {0,1,2,3},{1,2,3},{2,3},{3}}, we obtain the NB staircase LDPC code:

$$\boldsymbol{H}=\begin{bmatrix} I & \alpha^1 I & \alpha^2 I & \alpha^3 I & \alpha^4 I & & \\ & P^{S[1]} & \alpha^1 P^{S[2]} & \alpha^2 P^{S[3]} & \alpha^3 P^{S[4]} & \alpha^4 P^{S[5]} & \\ & & P^{2S[2]} & \alpha^1 P^{2S[3]} & \alpha^2 P^{2S[4]} & \alpha^3 P^{2S[5]} & \alpha^4 P^{2S[6]} \\ & & & P^{3S[3]} & \alpha^1 P^{3S[4]} & \alpha^2 P^{3S[5]} & \alpha^3 P^{3S[6]} \quad \alpha^4 P^{3S[7]} \end{bmatrix},$$

which represents the generalization of binary staircase-like codes. The corresponding *spatially coupled LDPC code*, derived from the following PBD {{0,1,2},{1,2,3}}, is given as

$$\boldsymbol{H}=\begin{bmatrix} I & \alpha I & \alpha^2 I & \alpha^3 I & & & & \\ \alpha^7 I & P^{S[1]} & \alpha P^{S[2]} & \alpha^2 P^{S[3]} & \alpha^3 P^{S[4]} & \alpha^4 P^{S[5]} & \alpha^5 P^{S[6]} & \alpha^6 P^{S[7]} \\ \alpha^6 I & \alpha^7 P^{2S[1]} & P^{2S[2]} & \alpha P^{2S[3]} & \alpha^2 P^{2S[4]} & \alpha^3 P^{2S[5]} & \alpha^4 P^{2S[6]} & \alpha^5 P^{2S[7]} \\ & & & & \alpha^1 P^{3S[4]} & \alpha^2 P^{3S[5]} & \alpha^3 P^{3S[6]} & \alpha^4 P^{3S[7]} \end{bmatrix},$$

the

The convolutional code can be obtained from the template code as follows:

$$\boldsymbol{H}=\begin{bmatrix} I & & & & & \\ \alpha^7 I & P^{S[1]} & & & & \\ \alpha^6 I & \alpha^7 P^{2S[1]} & P^{2S[2]} & & & \\ \alpha^5 I & \alpha^6 P^{3S[1]} & \alpha^7 P^{3S[2]} & P^{3S[3]} & & \\ & \alpha^5 P^{4S[1]} & \alpha^6 P^{4S[2]} & \alpha^7 P^{4S[3]} & P^{4S[4]} & \\ & & \alpha^5 P^{5S[2]} & \alpha^6 P^{5S[3]} & \alpha^7 P^{5S[4]} & P^{5S[5]} \\ & & \ddots & & \ddots & \quad \ddots \end{bmatrix}.$$

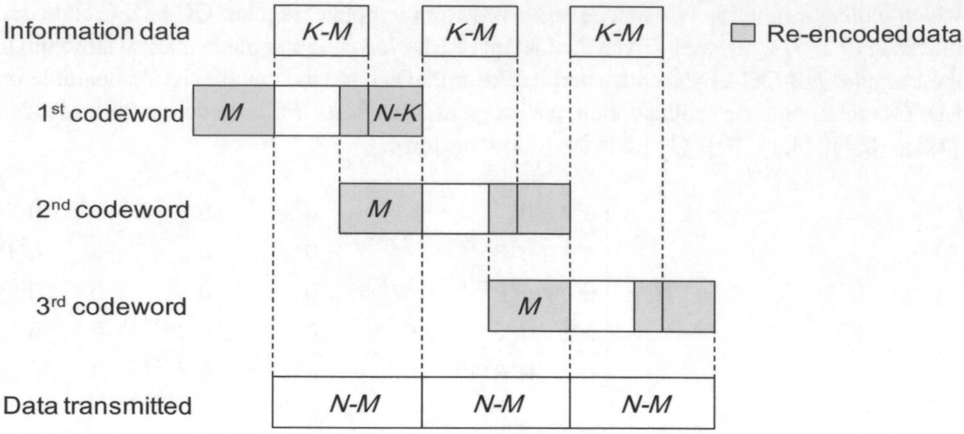

Fig. 9.31 The principles of rate-adaptive LDPC coding with re-encoding

Therefore, the staircase-like, spatially coupled-like, and some convolutional LDPC codes can be interpreted as different arrangements derived from irregular QC-LDPC codes.

In the rest of this section, we discuss different strategies for rate adaptation. The code rate adaptation is performed by *partial reconfiguration* of decoder based on (9.135), for a fixed codeword length, by changing the size of the permutation matrix P while keeping the number of block rows constant, and/or by varying the number of employed block rows while keeping P fixed. Another alternative to change code rate is to use the same regular parity-check matrix as template but employ the PBDs corresponding to different code rates. It is also possible to perform the puncturing of parity symbols in original code (n,k,d) to obtain a linear block code $(n - p,k,d_p)$, $d_p \leq d$, where p is the number of removed parity symbols. Notice that this approach when applied to LDPC codes can introduce an early error floor phenomenon. This is the reason why the re-encoding approach has been introduced in [12, 80] instead of using the conventional shortening or puncturing approach [81]. As shown in Fig. 9.31, to encode, the adaptive LDPC encoder encapsulates the last M symbols of the proceeding codeword and the incoming $K - M$ information symbols into a K-symbol vector. In other words, each codeword is generated continuously by re-encoding the last M symbols of its preceding codeword. (The first M symbols in the first codeword are set as known since no proceeding codeword exists and the first M re-encoded symbols of each codeword are not transmitted.) Therefore, the actual code rate is $R' = (K - M)/(N - M)$, where $0 < M < K$ and can be tuned easily in the range (0, R] (R is the code rate of template code) by adjusting the re-encoded data size M. Notice that the template code should be systematic to leverage the advantage of re-encoding in decoding.

The decoder employs the aforementioned nonbinary LDPC decoding algorithm [82]. Correspondingly, decoding of each codeword utilizes both the LLRs of the M re-encoded symbols produced by the decoding of its preceding codeword and the initial LLRs of the received K-M symbols. As LLRs of the M re-encoded symbols are converged LLRs and thus of higher confidence, it helps improve the error correction capability of the LDPC code. With higher number of symbols re-encoded, the lower code rate R' or equivalently the stronger error correction capability can be obtained.

As an illustration, we provide Monte Carlo simulation results for an adaptive LDPC code based on the quasi-cyclic nonbinary LDPC (12128, 10991) code [82] of girth 10 over GF(8) and code rate $R = K/N = 0.906$, for polarization-division multiplexed (PDM) 8-QAM of aggregate date information rate over 100 Gb/s. The FFT-based 8-ary MD-FFT-SPA [82] is employed in simulations as the decoding algorithm with 50 iterations. By choosing the proper parameter M, we have constructed the LDPC codes of rate 0.875 and 0.833, the BER vs. OSNR performances of which along with that of the mother code are shown in Fig. 9.32. By increasing the LDPC overhead from 10.3% to 14.3% and 20%, LDPC coding gain can be raised by 0.45 and 0.85 dB, respectively, at BER$=10^{-8}$. The rate-adaptive coding has been used in Section IX.C to deal with wavelength-dependent Q-factor degradation.

9.10.7 Rate-Adaptive Coding Implementations in FPGA

9.10.7.1 Rate-Adaptive LDPC Coding in FPGA

The basic rate adaptation via either shortening or puncturing is widely used everywhere in communication systems and can be introduced in both block and convolutional codes. In this subsection, we use shortening to achieve rate-adaptive LDPC coding since it can allow a wide range of rate adjustment with unified decoder architecture through a set of reconfigurable registers in

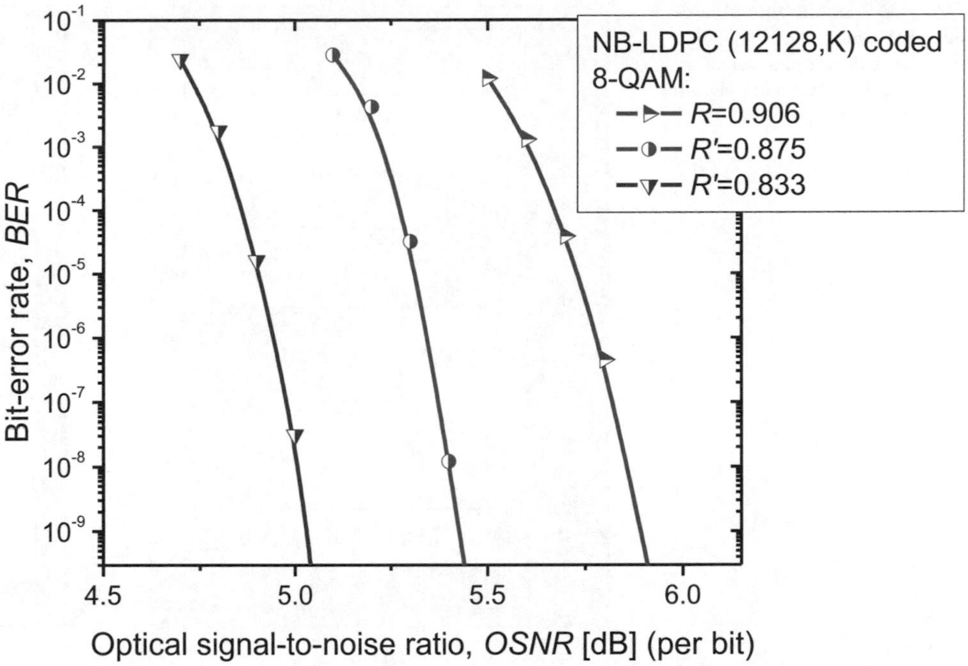

Fig. 9.32 BER performance of polarization-division multiplexed rate-adaptive NB-LDPC codes

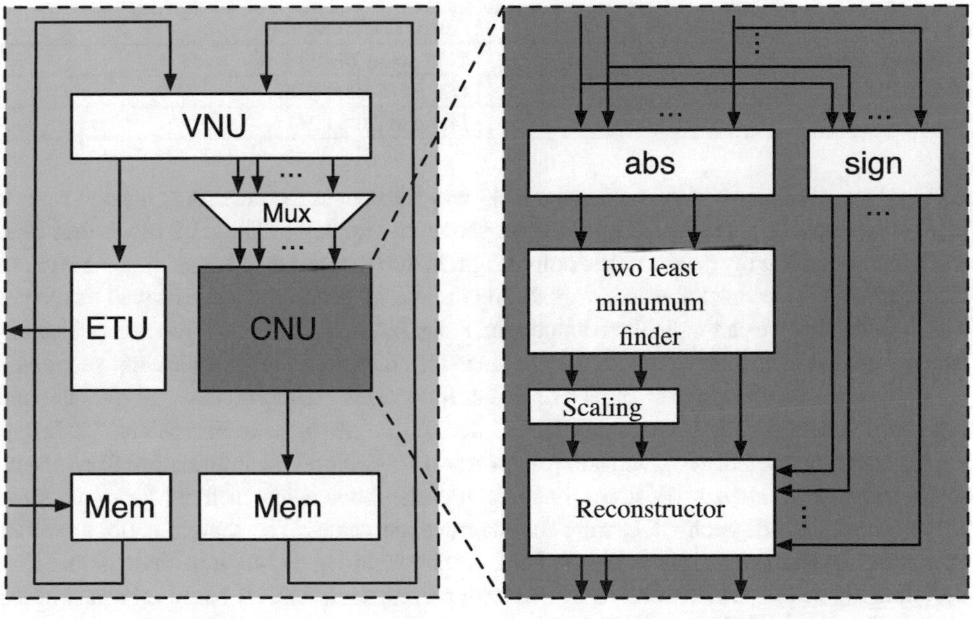

Fig. 9.33 Architecture of rate-adaptive binary LDPC decoder: (**a**) overall schematic diagram, (**b**) schematic diagram of CNU

FPGA [67]. Because of the quasi-cyclic structure of our nonbinary LDPC codes, we shorten the entire sub-block by adding the least number of logics' blocks. For example, we start from a (3,15)-regular nonbinary LDPC codes with rate of 0.8, and we can obtain a class of shortened regular nonbinary LDPC codes with column weight and row weight of {(3,14), (3,13), (3,12), (3,11), (3,10)}, which correspond to code rates of {0.786, 0.77, 0.75, 0.727, 0.7}.

As shown in Fig. 9.33a, the binary LDPC decoder consists of two major memory blocks (one stores channel LLR and another stores a posteriori probability (APP) messages), two processing blocks (variable node unit (VNU) and check node unit (CNU)), an early termination unit (ETU), and a number of mux blocks, wherein its selection of output signal can be software reconfigurable to adjust the shortening length. The memory consumption is dominated by LLR message with size of $n \times W_L$

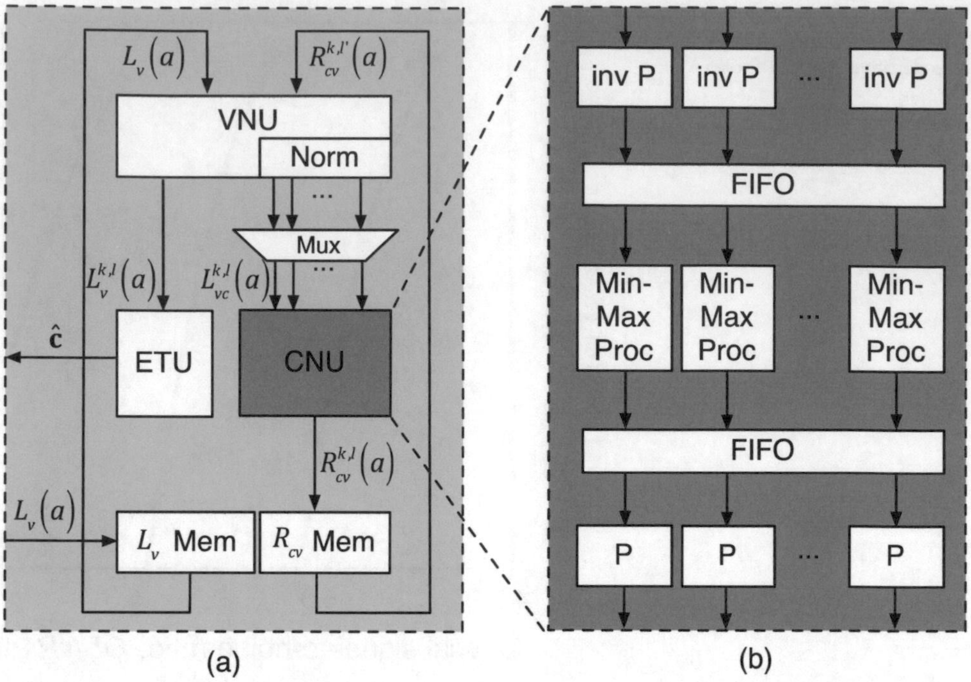

Fig. 9.34 Architecture of rate-adaptive nonbinary LDPC decoder: (**a**) overall schematic diagram, (**b**) schematic diagram of CNU

Table 9.2 Logic resource utilization

Resources	Binary LDPC decoder	Nonbinary LDPC decoder
Occupied slices	2969 out of 74,400 (3%)	16,842 out of 74,400 (22%)
RAMB36E1	113 out of 1064 (10%)	338 out of 1064 (31%)
RAMB18E1	14 out of 2128 (1%)	221 out of 2128 (9%)

APP messages of size $c \times n \times W_R$, where W_L and W_R denote the precisions used to represent LLR and APP messages. The logic consumption is dominated by CNU, as shown in Fig. 9.33b. The ABS block first takes the absolute value of the inputs, and the sign XOR array produces the output sign. In the two least minimums' finder block, we find the first minimum value via binary tree and trace back the survivors to find the second minimum value as well as the position of the first minimum value. This implies that we can write three values and r sign bits back to the APP memories instead of r values. However, we will not take advantage of memory reduction techniques for comparison in the following sections.

Similar to the rate-adaptive binary LDPC decoder architecture discussed above, the architecture of the layered min-max algorithm (LMMA)-based nonbinary LDPC decoder is presented in Fig. 9.34a. There are four types of memories used in implementation: memory $R_{c,v}$ with size of $c \times n \times q \times W_R$ stores the information from check nodes to variable nodes, memory for L_v with size of $n \times q \times W_L$ stores the initial log-likelihood ratios, memory for $\widehat{c}$ with size $n \times \log_2 q$ stores the decoded bits, and memories inside each CNU store the intermediate values. The same notations are borrowed from previous subsection except that q denotes the size of Galois field. As shown in Fig. 9.34b, it is obvious that CNU is the most complex part of the decoding algorithm, which consists of r inverse permutators, r BCJR-based min-max processors and r permutators, and two types of the first-in-first-out (FIFO) registers. The inverse permutator block shifts the incoming message vector cyclically. The first FIFO is used to perform the parallel-to-serial conversion. After the min-max processor, which is implemented by low-latency bidirectional BCJR algorithm, the processed data is fed into another FIFO block performing serial-to-parallel conversion, followed by the permutator block. Because of high complexity of CNU design and high memory requirements of nonbinary decoder than that of binary decoder, reduced-complexity architectures and selective version of MMA can be further exploited.

We compare three rate-adaptive schemes based on binary LDPC codes and nonbinary LDPC codes. These three architectures can be software-defined by initializing configurable registers in FPGA. The resource utilization is summarized in Table 9.2. One can clearly notice that the LMMA-based nonbinary LDPC codes consume 3.6 times larger memory size than the binary one because of large field size and high quantization precision, while the occupied number of slices is five times larger than that in binary case because of higher complexity of CNU.

Fig. 9.35 BER performance of rate-adaptive binary LDPC decoder implemented in FPGA

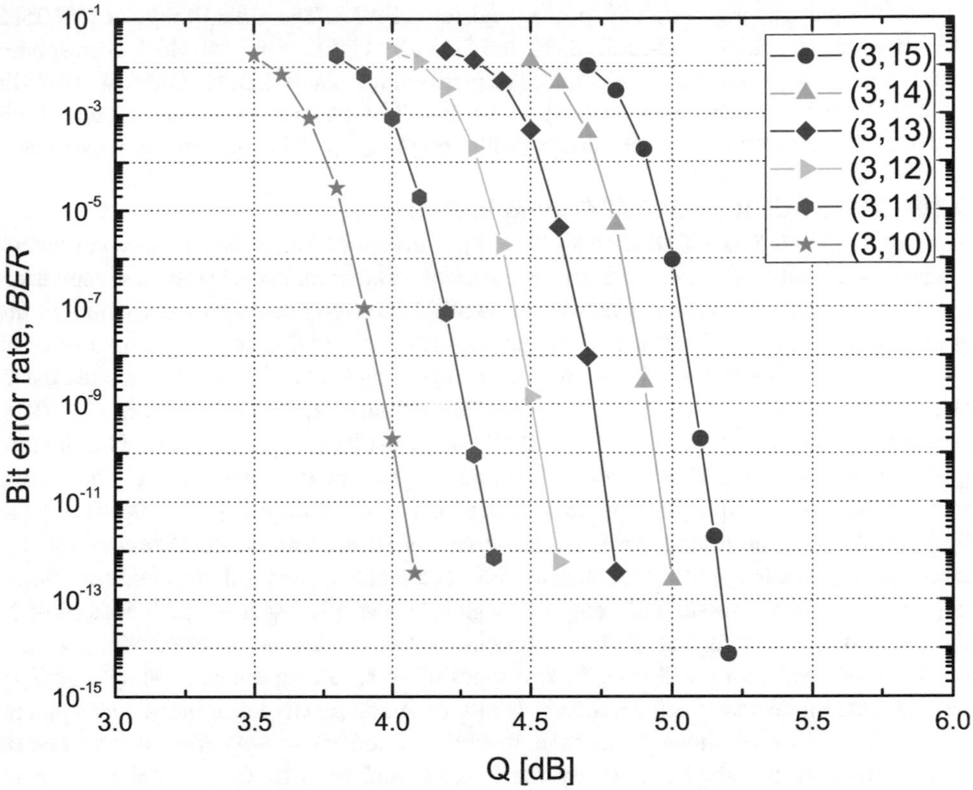

Fig. 9.36 BER performance of rate-adaptive nonbinary LDPC decoder implemented in FPGA

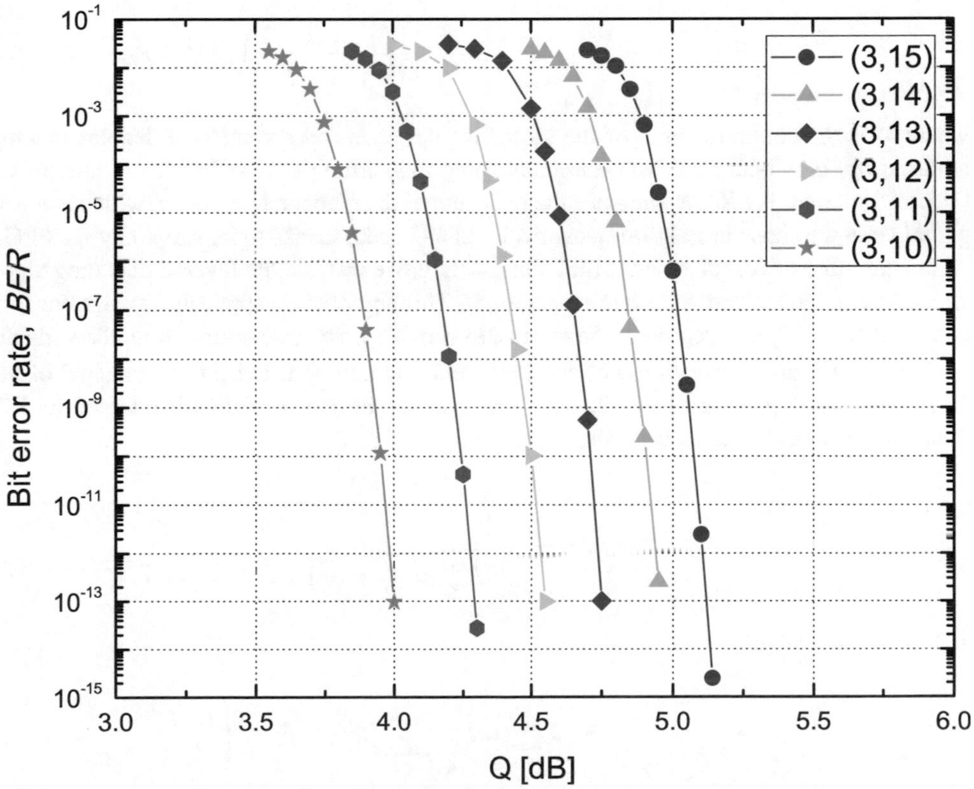

The BER vs. Q-factor performances of the rate-adaptive binary and nonbinary LDPC code are presented in Figs. 9.35 and 9.36. The FPGA-based emulation was conducted over binary (BI)-AWGN channel, and 6 and 8 bits precision are used in binary and nonbinary LDPC decoder, respectively. A set of column weight and row weight configurations of {(3, 15), (3, 14), (3, 13), (3, 12), (3, 11), (3, 10)}, which corresponds to the code rates of {0.8, 0.786, 0.77, 0.75, 0.727, 0.7}, can be achieved by software-based reconfiguration of specific register in FPGA. The girth-10 regular (34635, 27710, 0.8) binary and nonbinary mother code can achieve a Q-limit of 5.2 and 5.14 dB at BER of 10^{-15}, which corresponds to NCG of 11.83 and 11.89 dB. The rate-adaptive nonbinary LDPC codes outperform the binary LDPC codes by approximated 0.06 dB in all range of rate from 0.7 ~ 0.8. In addition, we believe this gap will be larger when combined with higher modulation schemes enabling 100Gbits/s (with QPSK) and 400Gbits/s (with 16-QAM) optical communication systems.

9.10.7.2 Rate-Adaptive GLDPC Coding in FPGA

The key idea behind the GLDPC codes [72–78], as discussed earlier, is to replace the parity-check equations in a parity-check matrix of a *global* LDPC code by a linear block code. The decoding is based on several low-complexity soft-input/soft-output (SISO) linear block decoders operating in parallel. The SISO decoders are commonly implemented based on maximum a posteriori probability (MAP) decoders, and as such they provide accurate estimates of bit reliabilities for a global LDPC decoder after small number of iterations. Due to high complexity of the BCJR decoder, the GLDPC coding is limited to simple linear block component codes such as Hamming, binary BCH, and Reed-Muller (RM) codes. To keep the code rate reasonably high, the global code of GLDPC code must be of column weight (variable node degree) 2. Unfortunately, the global codes with parity-check matrix of column weight 2 are known as weak LDPC codes as they exhibit an early error floor phenomenon. In principle, multiple local codes can be used instead of parity checks of a global code. The proper selection of these local codes can lead to further improvement of BER performance. However, the high-code rate requirement problem important for certain applications, such as optical communications and magnetic recording, will not be solved. As solution to high-code rate requirement, while employing global codes of column weight 3, instead of 2 used in [72, 73], here we describe the possibility not to replace all parity checks by local codes, but only certain, selected, number of parity checks. Such obtained GLDPC codes will contain two types of so-called super-nodes, simple parity-check equations for which MAP decoders are trivial and selected number of reasonable-complexity linear block codes (not necessary identical ones) [74]. This type of GLDPC codes shows remarkable flexibility for code rate adaptation as it will be shown later in the subsection.

For instance, if only single-parity-check codes and (n,k) local code are used in regular GLDPC code design, the corresponding code rate R can be bounded as follows [74]:

$$1-\frac{W}{n}-\frac{W}{d}\left(1-\frac{k+1}{n}\right)\leq R\leq R_G, \tag{9.136}$$

where W is the column weight of the global regular code and parameter d denotes that every d-th row in a global code is replaced by (n,k) local code, while the remaining rows from the global code are interpreted as single-parity-check codes in GLDPC code. In (9.136), R_G denotes the code rate of the global regular code, which is $R_G = 1 - W/n$. In this subsection, for global code we chose large-girth quasi-cyclic LDPC code, thanks to its simplicity for FPGA implementation.

For the sake of completeness of the presentation, we provide the layered decoding algorithm of the GLDPC codes [75], which is suitable for hardware implementations. This algorithm is applicable to any linear block code to be used as the local code. Let $R_{c,v}^{(k,l)}, L_{v,c}^{(k,l)}, L_v$ represent the check c to variable v extrinsic information, the variable v to check c at k-th iteration and l-th layer extrinsic information, and the log-likelihood ratio (LLR) from the channel of variable (bit) node v, respectively, where $k = 1,2,\ldots,I_{\max}$ and $l = 1,2,\ldots,c$. The layered min-max algorithm is adopted for FPGA implementation, whose data flow can be summarized as follows:

0. Initialization step:

$$R_{cv}^{(k,l)}=0,\ k=1;l=1,2,\ \ldots,c \tag{9.137}$$

1. Bit decision step:

$$L_v^{(k,l)}=L_v+\sum\nolimits_{l'}R_{cv}^{(k,l')},\ \widehat{c}_v=\begin{cases}0, & L_v^{(k,l)}\geq 0\\ 1, & L_v^{(k,l)}<0\end{cases} \tag{9.138}$$

2. Variable node processing rule:

$$L_{vc}^{(k,l)} = L_v + \sum\nolimits_{l' \neq l} R_{cv}^{(k',l')}, k' = \begin{cases} k-1, l < l' \\ k, l \geq l' \end{cases} \tag{9.139}$$

3. Check node processing rule:
   - If it is a local code: BCJR-based APP updating rule is applied:

$$R_{cv}^{(k,l)} = \text{BCJR}\left(L_{vc}^{(k,l)}\right). \tag{9.140}$$

   - Else: scaled min-sum updating rule is applied:

$$R_{c,v}^{(k,l)} = a \prod\nolimits_{v' \neq v} \text{sign}\left(L_{v',c}^{(k,l)}\right) \min_{v' \neq v} \left| L_{v',c}^{(k,l)} \right|, \tag{9.141}$$

where a is the attenuation factor.

The criterion of early termination is achieved when decoded bits converge to transmitted codeword, since only $1/c$ portion of check nodes involved in each layer.

The implementation of BCJR-based MAP decoder from Eq. (9.140) can be divided into three parts. As shown in Fig. 9.37, the first part calculates forward and backward recursion likelihoods, the second part corresponds to the memories storing the intermediate data α and β, and the third part is a combiner calculating the output. Since the trellis derived from a block code is time-variant, it implies that a selection signals should be pre-stored in ROM so that we can select an appropriate feedback output to the input of forward and backward recursion blocks. Aiming to keep reasonable complexity and latency, we replace the max-star operation by max operation and adopt a bidirectional recursion scheme, which minimizes the memory sizes for α and β as well as the latency. To be more specific, for illustrative purposes, Fig. 9.38 shows the timing diagram of decoding of

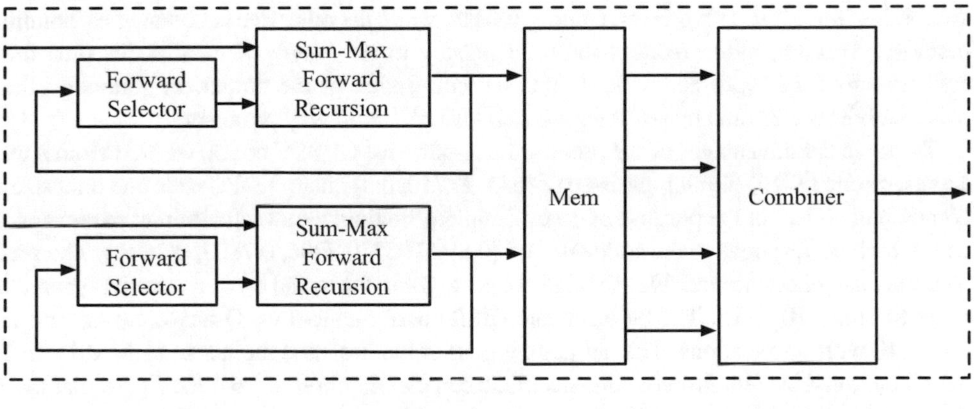

Fig. 9.37 BCJR-based APP decoder architecture

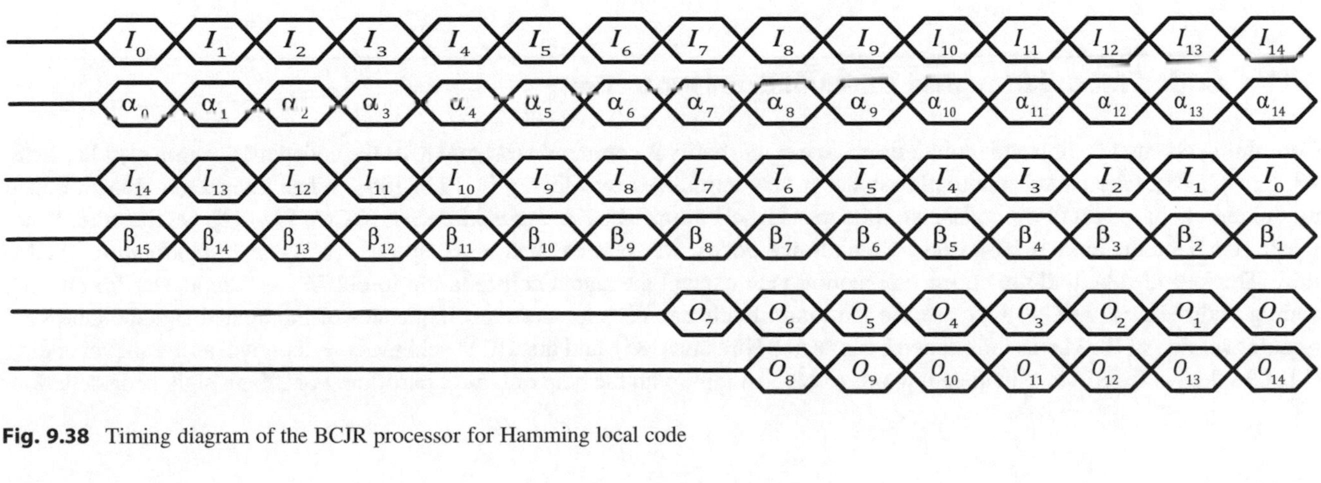

Fig. 9.38 Timing diagram of the BCJR processor for Hamming local code

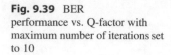

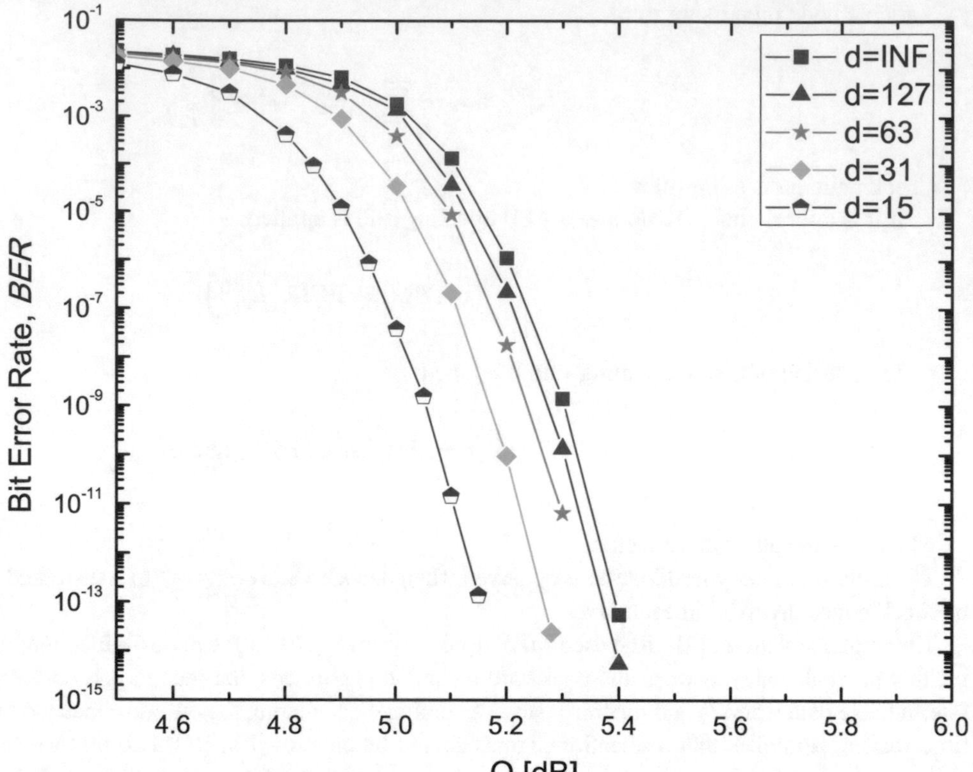

Fig. 9.39 BER performance vs. Q-factor with maximum number of iterations set to 10

the (15, 11) local Hamming code. The first half of the forward and backward recursion is updated and stored in memories, and current α_7 and current β_8 are combined to the output O_7. After that, each cycle will generate two outputs, one is obtained by combining current α with β from memory blocks, while the other one is obtained by combining the current β with α from the memory. This technique reduces the total latency to the length of local codes plus the latency due to parallel-to-serial conversion of the input and serial-to-parallel conversion of the output. In summary, the complexity of MAP decoder is reasonably low and thus makes the proposed GLDPC code very promising.

To prove the advantages of the proposed rate-adaptive GLDPC codes, we start from a well-designed LDPC code, which is a quasi-cyclic (3, 15)-regular, girth-10 (34635, 27710, 0.8) binary LDPC code and choose a simple (15, 11) Hamming code as component code. For the purpose of ease of implementation, we sweep different parameters d in the range of $\{\infty, 127, 63, 31, 15\}$, which corresponds to the code rates of $\{0.8, 0.7953, 0.7906, 0.7807, 0.7601\}$. The precision of LLRs, variable-to-check message, and check-to-variable message are set to 5-bit, 5-bit, and 6-bit, respectively; and the maximum number of iterations is set to either 10 or 15. The bit error rate (BER) performances vs. Q-factor are present in Fig. 9.39 with 10 iterations and Fig. 9.40 with 15 iterations. The net coding gain of the designed mother is 11.61 dB and 11.71 dB for 10 and 15 iterations, which demonstrates its fast convergence. One can clearly observe from the figure that the BER performance is enhanced as d decreases; thus fine-tuning of code rate can be achieved.

9.11 Coded Modulation and Unequal Error Protection

Combining channel coding and modulation into a single entity is commonly referred to as the coded modulation [1, 11, 12, 48, 49, 51, 82, 83]. The coded modulation concept was introduced by Massey in 1974 [84]. When channel coding in coded modulation is based on block codes, we refer to such scheme as *block-coded modulation* (BCM) [85–92]. On the other hand, when coded modulation is based on convolutional codes, we refer to such a scheme as *trellis-coded modulation* (TCM) [93]. The key idea behind the coded modulation is to expand the signal constellation to obtain the redundancy for channel coding and then use FEC to increase the minimum Euclidean distance between sequences of modulated signals. One very popular version of BCM was introduced by Imai and Hirakawa [85], and this BCM scheme is well known as multilevel coding (MLC) scheme. In MLC we map codewords to bit positions, with the help of binary partition. For 2^m-ary signal constellation,

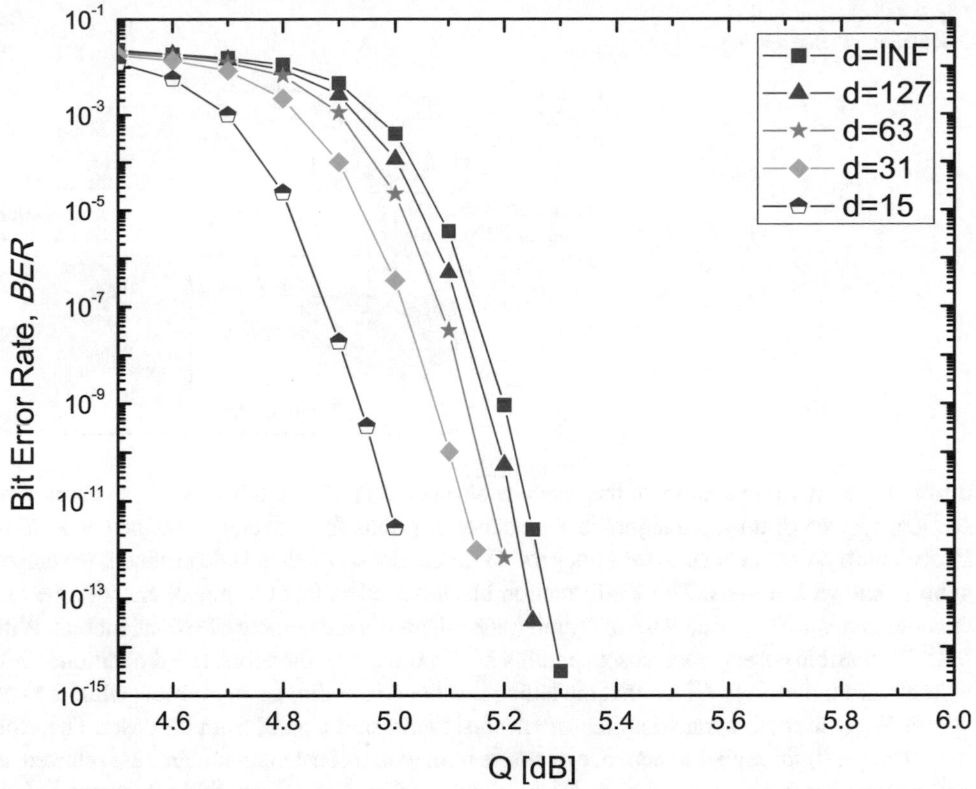

Fig. 9.40 BER performance vs. Q-factor with maximum number of iterations set to 15

we use *m* different error correcting codes, one per each label. On receiver side, we perform so-called multistage decoding, in which the LLRs from higher decoding stage, corresponding to stronger code, are forwarded to lower decoding stage, corresponding to a weaker code. When the same code is used on all levels, we refer to such coded modulation scheme as bit-interleaved coded modulation (BICM) scheme, which was introduced by Caire et al. [89]. In TCM, introduced by Ungerboeck in [93], we apply a natural mapping of bits to symbols through so-called set partitioning. Given a trellis description of corresponding code, we assign symbol sequences to the trellis paths. On receiver side, we perform the Viterbi decoding of trellis describing the trellis code. It is also possible to trade the constellation expansion with bandwidth expansion in so-called *hybrid coded modulation* Scheme [12, 50].

9.11.1 Coded Modulation Fundamentals

Let d_{unc} denote the minimum Euclidean distance in original, uncoded signal constellation diagram, and let $d_{\min}$ denote the minimum distance in sequence of symbols after coding. When the uncoded normalized average energy, denoted as E_{unc}, is different from the normalized average energy after coding, denoted as E_{coded}, the *asymptotic coding gain* can be defined as follows:

$$\begin{aligned} G_a &= \frac{E_{\text{unc}}/d_{\text{unc}}^2}{E_{\text{coded}}/d_{\min}^2} = \underbrace{\frac{E_{\text{unc}}}{E_{\text{coded}}}}_{G_C} \cdot \underbrace{\frac{d_{\min}^2}{d_{\text{unc}}^2}}_{G_D} \\ &= G_C \cdot G_D, \end{aligned} \tag{9.142}$$

where G_C is the *constellation expansion factor* and G_D is the *increased distance gain*. The generic coded modulation scheme, applicable to both BCM and TCM, is illustrated in Fig. 9.41, which is based on Forney's interpretation [94, 95] (see also [96, 98]). Two-dimensional (2-D) modulator such as M-ary QAM (I/Q) modulator or M-ray PSK modulator is used to impose the sequence of symbols to be transmitted. The *N*-dimensional lattice is used for code design. A *lattice* represents a discrete set of vectors in real Euclidean *N*-dimensional space, which forms a group under ordinary vector addition, so the sum or

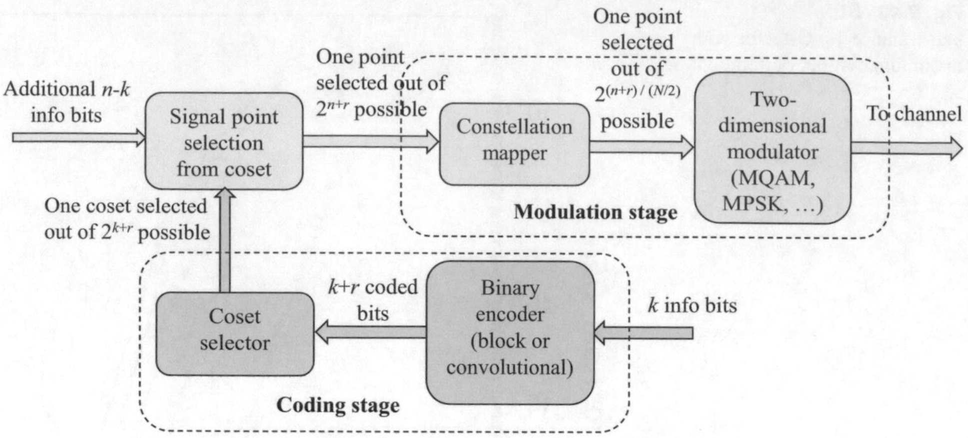

Fig. 9.41 Generic coded modulation scheme

difference of any two vectors in the lattice is also in the lattice. A sub-lattice is a subset of a lattice that is itself a lattice. The sequence space of uncoded signal is a sequence of points from *N*-cube, obtained as a Cartesian product of 2-D rectangular lattices with points located at odd integers. Once the densest lattice is determined, we create its decomposition into partition subsets known as *cosets*. The *k* information bits are used as input to binary encoder, be block or convolutional. The binary encoder generates the codeword of length $k + r$, where *r* is the number of redundant bits. With $k + r$ bits, we can select one out of 2^{k+r} possible cosets. Each coset contains 2^{n-k} points, and therefore, $n - k$ additional information bits are needed to select a point within the coset. Given that the lattice is defined in *N*-dimensional space, while 2-D modulator imposes the data in 2-D space, *N*/2 consecutive transmissions are needed to transmit a point from the coset. Therefore, information bits are conveyed in two ways: (i) through the sequence of cosets from which constellation points are selected and (ii) through the points selected within each coset. Let d_{points} denote the minimum distance of points within the coset and d_{seq} denote the minimum distance among coset sequences; then the minimum distance of the code can be determined as $d_{\min} = \min(d_{\text{points}}, d_{\text{seq}})$. Since the fundamental volume per *N* dimensions is 2^r, where *r* is the number of redundant bits, the normalized volume per two dimensions would be $\sqrt[N/2]{2^r} = 2^{2r/N}$; the coding gain of this coded modulation scheme can be estimated as

$$G = \frac{d_{\min}^2}{2^{2r/N}} = 2^{-2r/N} d_{\min}^2. \tag{9.143}$$

Clearly, the normalized redundancy per two dimensions is equal to $r/(N/2) = 2r/N$. When the constellation is chosen to be *N*-sphere-like to reduce the average energy, we need to account for an additional gain, often called the *shaping gain*, denoted with G_s, which measures the reduction in energy. Calderbank and Sloane have shown that the shaping gain of an *N*-sphere over *N*-cube, when *N* is even, is given by [97]

$$G_s = \frac{\pi}{6} \frac{(N/2) + 1}{[(N/2)!]^{1/(N/2)}}. \tag{9.144}$$

For instance, for $N = 2$ and 4, the shaping gains are $\pi/3$ (0.2 dB) and $\pi/2^{3/2}$ (0.456 dB), respectively. The shaping gain in limiting case, as *N* tends to infinity, is $\pi e/6$ (1.533 dB).

Based on Forney's guidelines [94], it is possible to generate maximum-density *N*-dimensional lattices by using a simple partition of a 2-D lattice with corresponding block/convolutional code, when $N = 4$, 8, 16, and 24, by applying the subset partition rules as shown in Fig. 9.42. The *4-D lattice* is obtained by taking all sequences of pair of points from the same subset, such as (*A*,*A*) or (*B*,*B*). The *8-D lattice* is obtained by taking all sequences of four points of either type A or type B. Each point in subset has a single subscript. Within each of four-point subset, the point subscripts satisfy the parity-check equation $i_1 + i_2 + i_3 + i_4 = 0$, so that the sequence subscripts must be codewords in the (4,3) parity-check code, which has a minimum Hamming distance of two. Therefore, three information bits and one parity bit are used to determine the lattice subset. The *16-D lattice* is obtained by taking all sequences of eight points of either type A or type B. Each point has two subscripts. The subscripts in the subset belong to a codeword from extended Hamming code (16,11) of min Hamming distance 4. Finally, the *24-D lattice* is obtained by taking all sequences of 12 points of either type A or type B. Each point has three subscripts (*i*,*j*,*k*). The subscripts (*i*,*j*) in 12-point subset form a codeword from Golay (24,12) code of minimum Hamming distance 8. The third

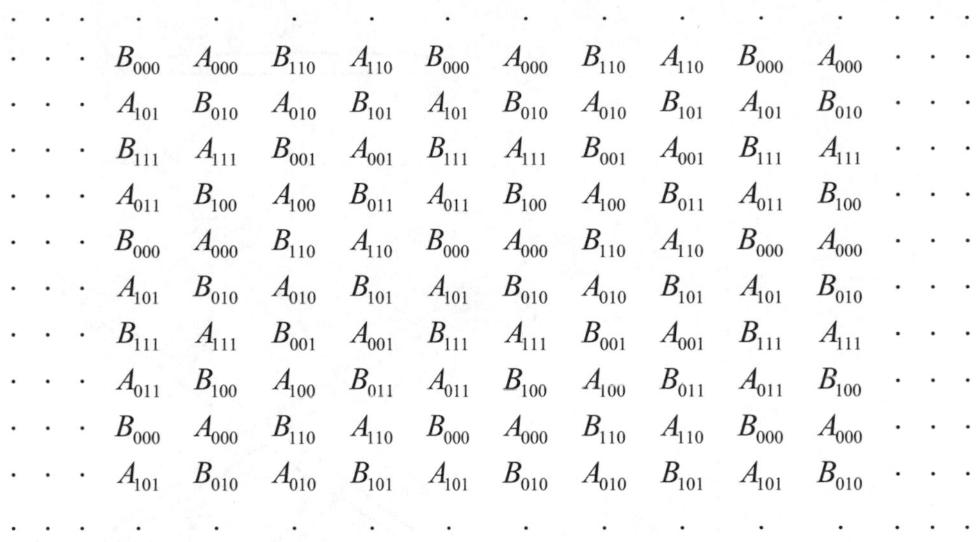

Fig. 9.42 Subset partition for up to 24-dimensional lattice

B_{000}	A_{000}	B_{110}	A_{110}	B_{000}	A_{000}	B_{110}	A_{110}	B_{000}	A_{000}
A_{101}	B_{010}	A_{010}	B_{101}	A_{101}	B_{010}	A_{010}	B_{101}	A_{101}	B_{010}
B_{111}	A_{111}	B_{001}	A_{001}	B_{111}	A_{111}	B_{001}	A_{001}	B_{111}	A_{111}
A_{011}	B_{100}	A_{100}	B_{011}	A_{011}	B_{100}	A_{100}	B_{011}	A_{011}	B_{100}
B_{000}	A_{000}	B_{110}	A_{110}	B_{000}	A_{000}	B_{110}	A_{110}	B_{000}	A_{000}
A_{101}	B_{010}	A_{010}	B_{101}	A_{101}	B_{010}	A_{010}	B_{101}	A_{101}	B_{010}
B_{111}	A_{111}	B_{001}	A_{001}	B_{111}	A_{111}	B_{001}	A_{001}	B_{111}	A_{111}
A_{011}	B_{100}	A_{100}	B_{011}	A_{011}	B_{100}	A_{100}	B_{011}	A_{011}	B_{100}
B_{000}	A_{000}	B_{110}	A_{110}	B_{000}	A_{000}	B_{110}	A_{110}	B_{000}	A_{000}
A_{101}	B_{010}	A_{010}	B_{101}	A_{101}	B_{010}	A_{010}	B_{101}	A_{101}	B_{010}

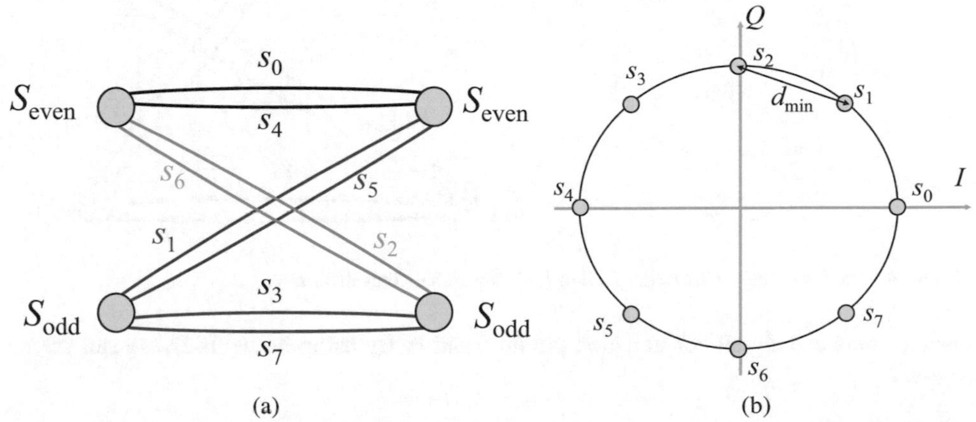

Fig. 9.43 Illustration of the two-state TCM trellis section for 8-ary signal constellation $S = (s_0, s_1,...,s_7)$ (**a**) and corresponding 8-PSK signal constellation (**b**)

subscript k represents overall parity check; for B-points even parity check is used and for A-points odd-parity check is used (or vice versa). The corresponding coding gains are 1.5 dB, 3 dB, 4.5 dB, and 6 dB for lattice codes of dimensionalities 3, 8, 16, and 24, respectively [94].

9.11.2 Trellis-Coded Modulation

In TCM, introduced by Ungerboeck [93], the redundancy is introduced by expanding the signal constellation into a larger signal constellation. Typically, if the size of original constellation is M, the size of extended constellation would be $2M$. The TCM can be described as follows. We define the trellis and then assign signal constellation points to different branches in trellis, as illustrated in Fig. 9.43a. Unlike convolutional codes where only one edge is allowed between two states, in TCM parallel transitions are allowed. This particular example is applicable to both 8-QAM and 8-PSK (shown in Fig. 9.43b). This trellis has two states, denoted as S_{even} and S_{odd}. When the encoder is in state S_{even}, only a signal constellation point s with even subscript can be generated, in other words $s \in \{s_0, s_2, s_4, s_6\}$. On the other hand, when the encoder is in state S_{odd}, only a signal constellation point s with odd subscript can be generated, in other words $s \in \{s_1, s_3, s_5, s_7\}$. The original constellation is QPSK, defined as $S_{\text{QPSK}} = \{s_0, s_2, s_4, s_6\}$. The Euclidean distance square for QPSK is $d^2_{\text{unc}} = 2E_{\text{unc}}, E_{\text{unc}} = 1$. On the other hand, the Euclidean distance in 8-PSK is $d^2_{\text{8-PSK}} = 4\sin^2(\pi/8)E_{\text{coded}}, E_{\text{coded}} = 1$. As the generalization of the free distance of convolutional codes, the *free distance* of TCM can be defined as the minimum Euclidean distance between any two codewords in the code. The path corresponding to the minimum distance from (s_0, s_0)-path is (s_2, s_1) path. Therefore, the free distance of this TCM scheme is $d_{\text{free}} = \sqrt{d^2_E(s_2, s_0) + d^2_E(s_1, s_0)} = \sqrt{2 + 4\sin^2(\pi/8)}$, where $d_E(s_i, s_j)$ is the Euclidean

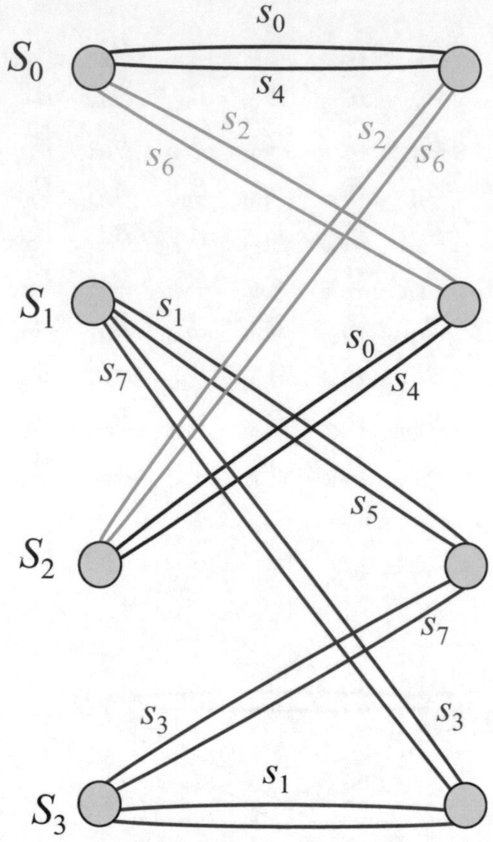

Fig. 9.44 The four-state TCM trellis section for 8-ary signal constellation

distance between signal constellation points s_i and s_j. By using Eq. (9.142), we can calculate the asymptotic coding gain as follows:

$$G_a = \underbrace{\frac{E_{\text{unc}}}{E_{\text{coded}}}}_{=1} \frac{d_{\min}^2}{d_{\text{unc}}^2} = \frac{d_{\text{free}}^2}{d_{\text{unc}}^2} = \frac{2 + 4\sin^2(\pi/8)}{2} = 1 + 2\sin^2(\pi/8)$$
$$= 1.29289\ (1.1156\ \text{dB}).$$

To increase the coding gain, we can use trellis with larger number of states. For instance, the four-state TCM trellis is shown in Fig. 9.44. Two shortest paths require three steps to return back to the state s_0. The shortest path with respect to the all-zero path is (s_4,s_0,s_0)-path and the corresponding Euclidean distance square is 4. The asymptotic coding gain for four-state TCM scheme is determined as follows: $G_a = d_{\text{free}}^2/d_{\text{unc}}^2 = 4/2 = 2\ (3\ \text{dB})$.

The mapping of constellation points should be achieved on such a way to maximize the asymptotic coding gain, in other words to maximize the free Euclidean distance. One popular way to achieve so is through so-called set partitioning. A 2^m-ary signal constellation set is partitioned into m-levels. At the i-th partition level ($0 \le i \le m - 1$), the signal S_i is split into two subsets $S_i(0)$ and $S_i(1)$ so that the intra-set squared Euclidean distance d_i^2 is maximized. With each subset choice $S_i(c_i)$, we associate a bit label c_i. The end result of this process is unique m-label $c_0c_1 \ldots c_{m-1}$ associated with each constellation point $s(c_0c_1 \ldots c_{m-1})$. By employing this Ungerboeck (standard) set partitioning of 2^m-ary signal constellation, the intra-set distances per levels are arranged in nondecreasing order as level index increases; in other words:

$$d_0^2 \le d_1^2 \le \cdots d_{m-1}^2.$$

Fig. 9.45 The set partitioning of 8-PSK with natural mapping rule

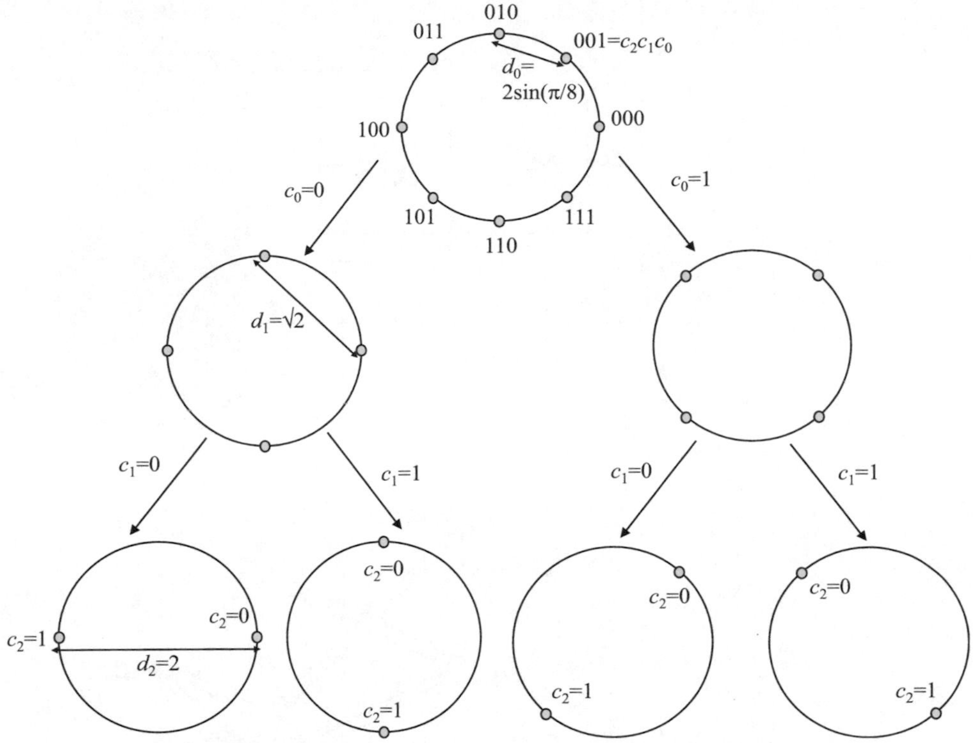

As an illustration, the set partitioning of 8-PSK with natural mapping rule is shown in Fig. 9.45. Clearly, in level 3 there are four subsets (cosets) with each subset containing two points. To select a given coset, two bits are needed. Rate ½ convolutional code can be used for this purpose as illustrated in Fig. 9.46a. To select a point within a given subset (coset), one information bit is needed. The overall code rate of this trellis-coded 8-PSK modulation scheme is ¾. The corresponding trellis diagram is shown in Fig. 9.46b. The current state is defined by the content of shift register (D_0D_1), while the next state by (u_0D_1). The first two bits (c_0c_1) are used to select the coset, while the information bit u_1 (code bit c_2) is used to select a point within that coset.

The general encoder for $(m\quad 1)/m$ TCM scheme is shown in Fig. 9.47. The k information bits are used as input to rate $k/(k+1)$ convolutional code, whose codeword ($c_0c_1\ldots c_k$) is used to select a coset (signal constellation subset). Additional $m-(k+1)$ information bits are used to select a point within the coset.

To decode the most likely TCM sequence, the *Viterbi algorithm* can be used, in similar fashion as described in Sect. 9.3. For the branch metric, either correlation metric or Euclidean distance can be used. The survivor path memory must include the $m-(k+1)$ information (uncoded) bits. Because of the parallel edges, the computation complexity of the corresponding Viterbi algorithm is higher than that for either convolutional or linear block codes.

The symbol error rates of TCM schemes can be evaluated in similar fashion as for convolutional codes, with the help of weight-enumerated sequence obtained from the state diagram of trellis code. The key difference is the replacement of Hamming distance with Euclidean distance. Another relevant difference is that in TCM we have parallel branches in trellis diagram.

9.11.3 Multilevel Coded Modulation and Unequal Error Protection

Multilevel coding (MLC) is initially proposed by Imai and Hirakawa in 1977 [85]. The key idea behind the MLC is to protect individual bits using different binary codes and use M-ary signal constellations, as illustrated in Fig. 9.48. The i-th component code, denoted as C_i, is $(n,k_i,d_i^{(H)})$ $(i=1,2,\ldots,m)$ linear block code of Hamming distance $d_i^{(H)}$. The length of information word $\boldsymbol{u}_i$ for the i-th component code is k_i, while the corresponding codeword length n is the same for all codes. The bits-to-symbol mapper can be implemented with the help of block interleaver, in which the codewords from different component codes are written in row-wise fashion into interleaver and read out m bits at the time in column-wise fashion. With this interpretation, the symbol at j-th time instance $(j=1,2,\ldots,n)$ is obtained after the mapping $s_j=s(\boldsymbol{c}_j)$, where $\boldsymbol{c}_j=(c_{i,j}\, c_{i,2}\, \ldots\, c_{i,m})$ and $s(\cdot)$ is the

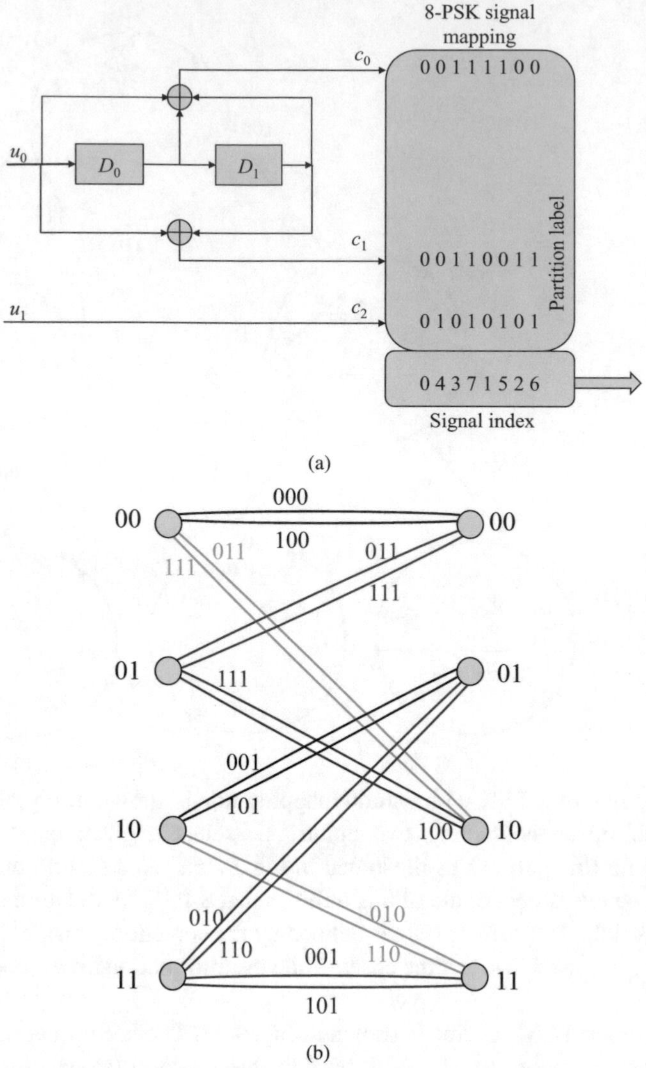

Fig. 9.46 Encoder for four-state rate-2/3 trellis-coded 8-PSK modulation (**a**) and corresponding trellis diagram (**b**)

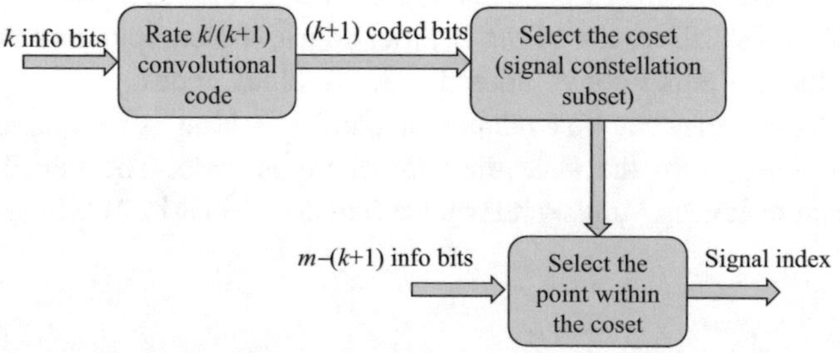

Fig. 9.47 General encoder for a rate $(m-1)/m$ TCM scheme

corresponding mapping rule. The 2^m-ary 2-D modulator, such as I/Q and PSK modulators, is used to impose the selected signal constellation point $\underline{s}_i = (s_{i,I}, s_{i,Q})$, where $s_{i,\ I}$ and $s_{i,\ Q}$ are corresponding in-phase and quadrature components, on corresponding carrier signal. For instance, the signal index can be used as the address for look-up table (LUT) where the coordinates are stored. Once the coordinates are selected, the in-phase and quadrature components are used as inputs of corresponding pulse shapers, followed by driver amplifiers. For optical communications, the optical I/Q modulator is further used to perform electrical-to-optical conversion. For wireless applications, after the I/Q modulator, the corresponding signal is after power amplifier used as input to the transmit antenna.

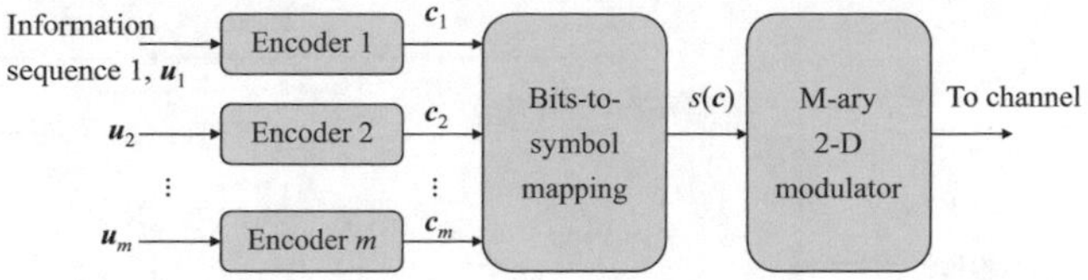

Fig. 9.48 Multilevel encoder for MLC/MCM scheme

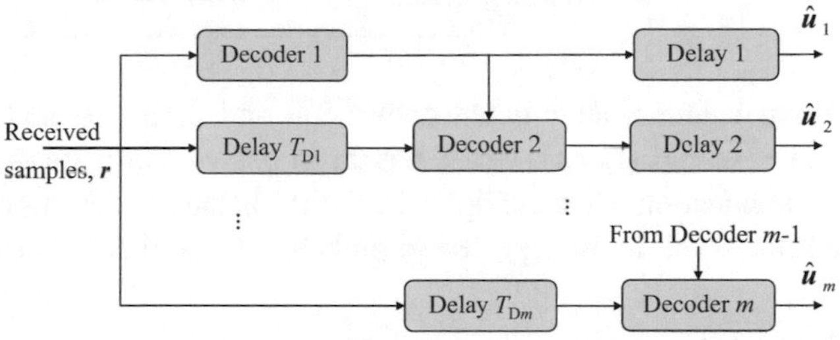

Fig. 9.49 Multistage decoder for MLC/MCM scheme

The overall code rate (spectral efficiency) of this MCM scheme, expressed in bits/symbol, is given by

$$R = \sum_{i=1}^{m} R_i = \sum_{i=1}^{m} k_i/n = \frac{\sum_{i=1}^{m} k_i}{n}. \tag{9.145}$$

The minimum squared Euclidean distance (MSED) of this scheme is lower bounded by [85].

$$MSED \geq \min_{1 \leq i \leq m} \left\{ d_i^{(H)} d_i \right\}, \tag{9.146}$$

where d_i is the Euclidean distance at the i-th level of set partitioning. As an illustration, the Euclidean distances at i-th level of set partitioning ($i = 1,2,3$) have been already provided in Fig. 9.45. When the component codes are selected as follows, (8,1,8), (8,7,2), (8,8,1), the MSED for coded 8-PSK is lower bounded by $MSED \geq \min \left\{ 0.7654^2 \cdot 8, \left(\sqrt{2}\right)^2 \cdot 2, 2^2 \cdot 1 \right\} = 4$, indicating that the asymptotic coding gain with respect to uncoded QPSK is at least 3 dB. When properly designed LDPC codes are used as component codes, the capacity-achieving performance can be obtained [99, 100].

Each component code can be represented as trellis diagram as described in Sect. 9.3. By taking the Cartesian product of individual trellises and then performing mapping based on set partitioning, we obtain a trellis description of this MCM scheme. Now we can apply either Viterbi algorithm or BCJR algorithm to perform decoding on such trellis. Unfortunately, the complexity of such decoding is too high to be of practical importance. Instead, the decoding is typically based on so-called *multistage decoding* (MSD) algorithm [7] in which the decisions from prior (lower) decoding stage are passed to next (higher) stage, which is illustrated in Fig. 9.49.

The i-th decoder operation ($i = 2,\ldots,m$) is delayed until the ($i - 1$)-th decoder completes the decoding algorithm. Further, after the decoding, the i-th decoder ($i = 1,\ldots,m - 1$) output is delayed until the m-th decoder decoding algorithm is completed. Clearly, even though the complexity of MSD algorithm is lower compared to Viterbi/BCJR algorithm operating on MLC trellis diagram, since the i-th component decoder operates on trellis of C_i code, the overall decoding latency is still high. Moreover, when the component codes are weak codes, the error not corrected in prior decoding stage will affect the next stage and result in error multiplicity. On the other hand, when component codes are properly designed LDPC codes, this problem can be avoided [99, 100]. Moreover, it has been shown in [100], when the code rates and degree distribution are optimized for a given mapping rule, BER performance degradation when component LDPC decoders operate independently can be made arbitrary small compared to MSD algorithm. This decoding algorithm in which all component decoders operate independently of each other is commonly referred to as *parallel independent decoding* (PID) [99, 100]. The PID decoding architecture is illustrated in Fig. 9.50, assuming that properly designed LDPC codes are used as component codes.

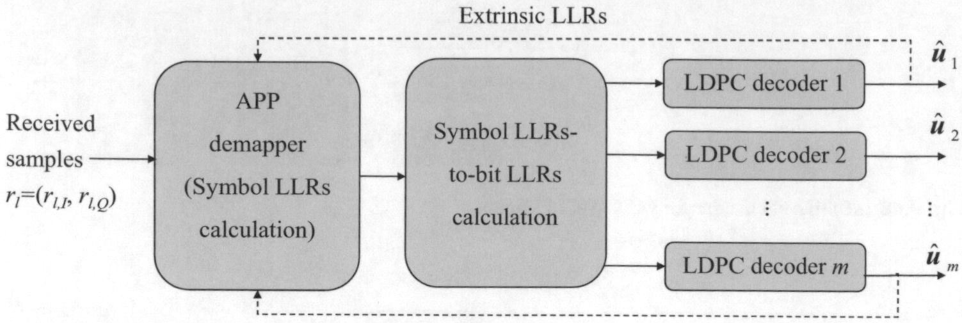

Fig. 9.50 Parallel independent decoding (PID) for MLC/MCM scheme

The received signal from receive antenna, after down conversion and digital-to-analog conversion in the observed symbolling interval, can be represented as $\underline{r} = (r_I, r_Q)$, where r_I is the in-phase coordinate and r_Q is the quadrature coordinate of received symbol. In optical communication, after the optical coherent balanced detection, as shown earlier, the received symbol can be represented in the same form. In the a posteriori probability (APP) demapper, the symbols' $\underline{s}_i$ ($i = 1,2,..,2^m$) LLRs are calculated as follows:

$$L(\underline{s}_i) = \log \frac{P(\underline{s}_i|\underline{r})}{P(\underline{s}_0|\underline{r})}, \tag{9.147}$$

where $P(\underline{s}|\underline{r})$ is determined by using Bayes' rule:

$$P(\underline{s}_i|\underline{r}) = \frac{P(\underline{r}|\underline{s}_i)P(\underline{s}_i)}{\sum_{\underline{s}'}P(\underline{r}|\underline{s}')P(\underline{s}')}. \tag{9.148}$$

In the presence of fiber nonlinearities, $P(\underline{r}|\underline{s}_i)$ in (9.148) needs to be estimated evaluating the histograms, by propagating a sufficiently long training sequence. With $P(\underline{s}_i)$ we denote the a priori probability of symbol $\underline{s}_i$, while $\underline{s}_0$ is a reference symbol. By substituting (9.148) into (9.147), we obtain

$$\begin{aligned} L(\underline{s}_i) &= \log\left[\frac{P(\underline{r}|\underline{s}_i)P(\underline{s}_i)}{P(\underline{r}|\underline{s}_0)P(\underline{s}_0)}\right] = \log\left[\frac{P(\underline{r}|\underline{s}_i)}{P(\underline{r}|\underline{s}_0)}\right] + \log\left[\frac{P(\underline{s}_i)}{P(\underline{s}_0)}\right] \\ &= \log\left[\frac{P(\underline{r}|\underline{s}_i)}{P(\underline{r}|\underline{s}_0)}\right] + L_a(\underline{s}_i), L_a(\underline{s}_i) = \log\left[\frac{P(\underline{s}_i)}{P(\underline{s}_0)}\right], \end{aligned} \tag{9.149}$$

where with $L_a(\underline{s}_i)$ we denote the prior reliability of symbol $\underline{s}_i$. Let us denote by $\underline{s}_{ij}$ the j-th level of the transmitted symbol $\underline{s}_i$ represented as $\underline{s}_i = (s_{i1},\ldots,s_{im})$, where $s_{ij}\in\{0,1\}$. The prior symbol LLRs for the next iteration are determined by

$$\begin{aligned} L_a(\underline{s}_i) &= \log\frac{P(\underline{s}_i)}{P(\underline{s}_0)} = \log\frac{\prod_{j=1}^{m} P(\underline{s}_{ij})}{\prod_{j=1}^{m} P(\underline{s}_{ij}=0)} = \\ &= \sum_{j=1}^{m} \log\frac{P(s_{ij})}{P(s_{ij}=0)} = \sum_{j=1}^{m} L_a(s_{ij}), L_a(s_{ij}) = \log\frac{P(s_{ij})}{P(s_{ij}=0)} \end{aligned} \tag{9.150}$$

where we assumed that reference symbol is $\underline{s}_0 = (0\ldots0)$.

The prior symbol estimate for the next outer iteration can be obtained from

$$L_a(\hat{\underline{s}}_i) = \sum_{j=1}^{m} L_{D,e}\left(\hat{\underline{s}}_{ij}\right), \tag{9.151}$$

where

$$L_{D,e}\left(\widehat{\underline{s}}_{ij}\right) = L\left(\underline{s}_{ij}^{(t)}\right) - L\left(\underline{s}_{ij}^{(t-1)}\right). \tag{9.152}$$

In (9.152), we use $L\left(s_{ij}^{(t)}\right)$ to denote the LDPC decoder output in current iteration (iteration t). The symbol LLRs corresponding to the j-th decoder $L(\underline{s}_{ij})$ are determined from symbol LLRs (9.149) by

$$L\left(\widehat{\underline{s}}_{ij}\right) = \log \frac{\sum_{\underline{s}_i:s_{ij}=1} \exp\left[L(\underline{s}_i)\right] \exp\left(\sum_{\underline{s}_i:s_{ik}=0,k\neq j} L_a(\underline{s}_{ik})\right)}{\sum_{\underline{s}_i:s_{ij}=0} \exp\left[L(\underline{s}_i)\right] \exp\left(\sum_{\underline{s}_i:s_{ik}=0,k\neq j} L_a(\underline{s}_{ik})\right)}. \tag{9.153}$$

Therefore, the j-th position reliability in (9.153) is calculated as the logarithm of the ratio of a probability that $\underline{s}_{ij} = 1$ and probability that $\underline{s}_{ij} = 0$. In the nominator, the summation is done over all symbols $\boldsymbol{s}_i$ having 1 at the position j, while in the denominator over all symbols $\boldsymbol{s}_i$ having 0 at the position j. With $L_a(\underline{s}_{ik})$ we denoted the prior (extrinsic) information determined from the APP demapper. The inner summation in (9.153) is performed over all positions of symbol $\underline{s}_i$, selected in the outer summation, for which $\underline{s}_{ik} = 0$, $k \neq j$. The j-th position LLRs are forwarded to corresponding binary LDPC decoders. The iteration between the APP demapper and LDPC decoders is performed until the maximum number of iterations is reached or the valid codewords are obtained.

Clearly, in MLC different bits transmitted over the channel are differently protected, and such the MLC is suitable for unequal error protection (UEP). The MLC channel encoder consists of m different binary error-correcting codes $C_1, \ldots, C_m$ with decreasing codeword distances. The i-th priority bit stream is used as the input to the i-th encoder, which generates the coded bits c_i. The rest is the same as described above. As an illustration, in image compression, one type of channel code can be used for bits corresponding to the low-resolution reproduction of the image, whereas high-resolution bits that simply refine the image can be protected by a different channel code. This is in particular important in JPEG 2000 image compression standard, in which the image is decomposed into a multiple-resolution representation. The JPEG 2000 provides efficient code-stream organizations that are progressive by pixel accuracy and by image resolution. Thanks to such organization, after a smaller portion of the whole file has been received, the viewer can see a lower-quality version of the picture. The quality of the picture then gets improved progressively through downloading additional data bits from the source.

For UEP, the *nested coding* can also be used. Nested coding can be interpreted as follows. Let us consider m different information vectors $\boldsymbol{u}_i$ ($i = 1, \ldots, m$) of length k_m. We would like to jointly encode these information vectors on such a way that each information vector is associated with a codeword from a different subcode. The i-th subcode C_i is represented by the generator matrix $\boldsymbol{G}_i$ of rate $R_i = k_i/n$. The overall generator matrix is given by

$$\boldsymbol{G} = \begin{bmatrix} \boldsymbol{G}_1 \\ \boldsymbol{G}_2 \\ \vdots \\ \boldsymbol{G}_m \end{bmatrix}. \tag{9.154}$$

The overall codeword $\boldsymbol{c}$ can be obtained as follows:

$$\begin{aligned} \boldsymbol{c}^T &= \left[\boldsymbol{u}_1^T \ \boldsymbol{u}_2^T \ \ldots \ \boldsymbol{u}_m^T\right]\boldsymbol{G} = \left[\boldsymbol{u}_1^T \ \ \boldsymbol{u}_2^T \ \ \ldots \ \ \boldsymbol{u}_m^T\right] \begin{bmatrix} \boldsymbol{G}_1 \\ \boldsymbol{G}_2 \\ \vdots \\ \boldsymbol{G}_m \end{bmatrix} \\ &= \boldsymbol{u}_1^T\boldsymbol{G}_1 \oplus \boldsymbol{u}_2^T\boldsymbol{G}_2 \oplus \cdots \oplus \boldsymbol{u}_m^T\boldsymbol{G}_m, \end{aligned} \tag{9.155}$$

where we use $\oplus$ to denote the bitwise XOR operation. If we are interested in unequal error correction, by setting $\boldsymbol{u}_i = \boldsymbol{u}$, by varying the number of generator matrices $\boldsymbol{G}_i$, we can achieve different levels of protection. The lowest level of protection would be to use only one generator matrix. The highest level of protection, corresponding to high-priority bits, will be achieved by encoding the same information vector $\boldsymbol{u}$ m-times.

To improve the spectral efficiency of MLC scheme, the *time-multiplexed coded modulation* can be used. The key idea is to use different coded modulation schemes for different priority classes of information bits. Let T_i be the fraction of time in which the i-th priority code C_i is employed, and let $R_{s,i}$ denote the corresponding symbol rate. The overall symbol rate R_s of this scheme would be

$$R_s = \frac{\sum_i R_{s,i} T_i}{\sum_i T_i}. \tag{9.156}$$

When all component codes in MCM scheme have the same rate, the corresponding scheme in [12], and related papers, is called as *block-interleaved* coded modulation, as it contains the block interleaver. This particular version, when combined with optimum signal constellation and optimum mapping rule designs [12, 123–125], performs comparable to MLC, but has even lower complexity. At the same time, it is suitable for implementation in hardware at high speed, as m LDPC decoders operate in parallel at bit rate equal to symbol rate R_s. In conventional bit-interleaved coded modulation (BICM) Scheme [89], described in the next section, a single LDPC decoder is required operating at rate $m\,R_s$, which can exceed the speed of existing electronics for fiber-optics communications, where the information symbol rate is typically 25 GS/s.

For convergence behavior analysis of the block-interleaved coded modulation for optical communication applications, the EXIT chart analysis should be performed. In order to determine the mutual information (MI) transfer characteristics of the demapper, we model the a priori input LLR, $L_{\mathrm{M,a}}$, as a conditional Gaussian random variable [101]. The MI between c and $L_{\mathrm{M,a}}$ is determined numerically as explained in [101] (see also [90, 91, 99, 100, 102, 103]). Similarly, the MI $I_{L\ M,e}$ between c and $L_{\mathrm{M,e}}$ is calculated numerically, but with the p.d.f. of c and $L_{\mathrm{M,e}}$ determined from histogram obtained by Monte Carlo simulation, as explained in [101]. By observing the $I_{L\ M,e}$ as a function of the MI of $I_{L\ M,a}$ and optical signal-to-noise ratio, *OSNR*, in dB, (*SNR* in wireless applications) the demapper EXIT characteristic (denoted as T_{M}) is given by

$$I_{L\ M,e} = T_M(I_{L\ M,a}, OSNR). \tag{9.157}$$

The EXIT characteristic of LDPC decoder (denoted as T_{D}) is defined in a similar fashion as

$$I_{L\ D,e} = T_D(I_{L\ D,a}). \tag{9.158}$$

The "turbo" demapping-based receiver operates by passing extrinsic LLRs between demapper and LDPC decoder. The iterative process starts with an initial demapping in which $L_{\mathrm{M,a}}$ is set to zero, and as a consequence, $I_{L\ M,a}$ becomes zero as well. The demapper output LLRs, described by

$$I_{L\ M,e} = I_{L\ D,a}. \tag{9.159}$$

are fed to LDPC decoder. The LDPC decoder output LLRs, described by

$$I_{L\ D,e} = I_{L\ M,a} \tag{9.160}$$

are fed to the APP demapper. The iterative procedure is repeated until the convergence or the maximum number of iterations has been reached. This procedure is illustrated in Fig. 9.51, where the APP demapper and LDPC decoder EXIT charts are shown together on the same graph. Three modulation formats (8-PSK, 16-PSK, 16-QAM) are observed, as well as the following mappings: natural, Gray, and anti-Gray. The EXIT curves have different slopes for different mappings. The existence of "tunnel" between corresponding demapping and decoder curves indicates that iteration between demapper and decoder will be successful. The smallest OSNR, at which iterative scheme starts to converge, is known as threshold (pinch-off) limit [101]. The threshold limit in the case of 16-PSK (Fig. 9.51b) is about 3 dB worse as compared to 8-PSK (Fig. 9.51a). The 16-QAM mapping curve is well above the 16-PSK curve (see Fig. 9.1b), indicating that 16-QAM scheme is going to significantly outperform the 16-PSK one.

Fig. 9.51 EXIT charts for different mappings and modulation formats

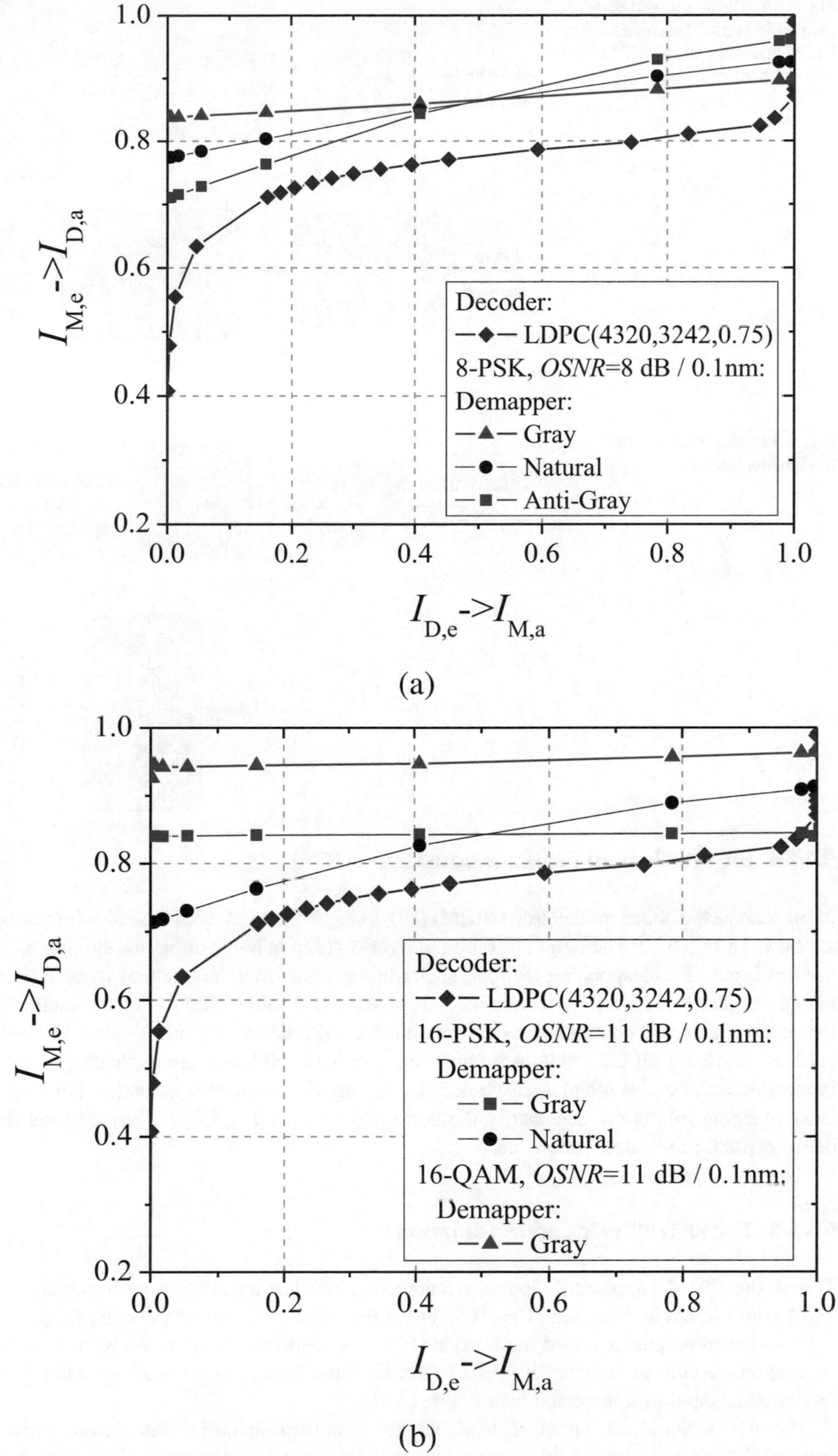

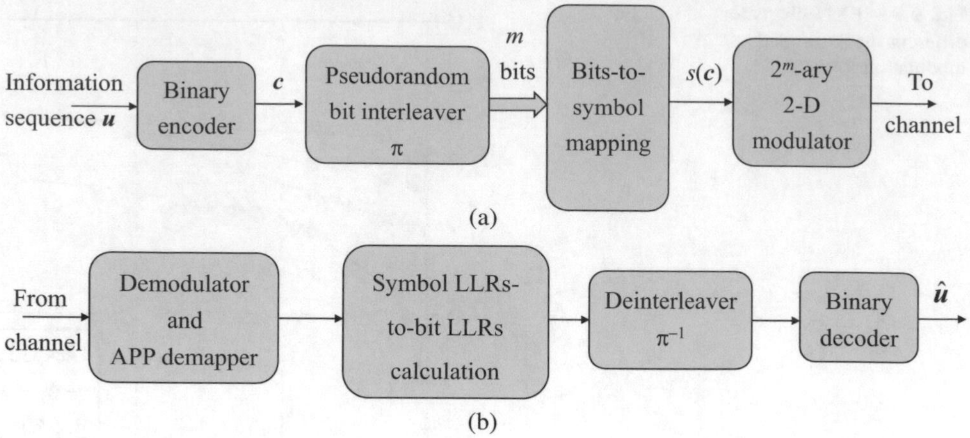

Fig. 9.52 Bit-interleaved coded modulation (BICM) scheme: (**a**) BICM encoder and (**b**) BICM decoder configurations

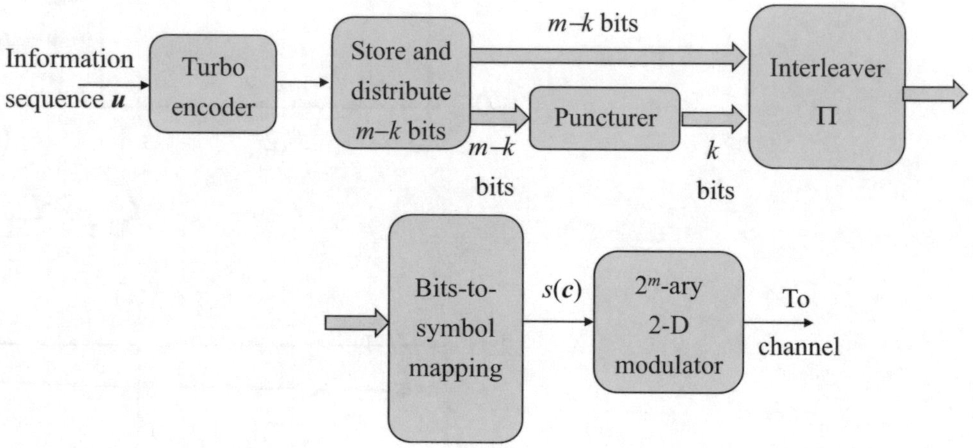

Fig. 9.53 Pragmatic coded modulation scheme

9.11.4 Bit-Interleaved Coded Modulation (BICM)

In bit-interleaved coded modulation (BICM) [89], a single binary encoder is used followed by a pseudorandom bit interleaver, as shown in Fig. 9.52. The output of bit interleaver is group in block of m bits, and during signaling interval these m bits are used as input of corresponding mapper, such as Gray, anti-Gray, and natural mapping, and after the mapper, m bits from mapper are used to select a point from a 2^m-ary signal constellation. The rest of transmitter is similar to that for MLC scheme. On receiver side, the configurations of demodulator, APP demapper, and symbol LLRs-to-bit LLRs calculation block are similar to those for MLC scheme. After the deinterleaving of bit LLRs, corresponding bit reliabilities are forwarded to a single binary decoder. For this scheme, the choice of mapping rule is of high importance. For iterative decoding and demapping, the Gray mapping rule is not necessarily the best mapping rule. The EXIT chart analysis should be used to match the APP demapper and LDPC decoder choice.

9.11.5 Turbo Trellis-Coded Modulation

There exist different approaches to coded modulation based on turbo codes such as pragmatic coded modulation [104], turbo TCM with symbol interleaving [105, 106], and turbo TCM with beat interleaving [88].

In so-called pragmatic coded modulation [104], encoders and decoders are binary, as shown in Fig. 9.53. Therefore, this scheme has certain similarities with BICM scheme. Turbo encoder sequence is split into blocks of $m - k$ bits each, which are then used as input to storage and distribution block.

The $m - k$ bits at the output of this block are forwarded directly to interleaver, while at the same time $m - k$ bits are punctured down to k bits and then forwarded to the interleaver Π. Therefore, the interleaver collects m bits at the time. After interleaving, m bits are taken from interleaver to select a point from corresponding 2^m-ary signal constellation. The rest of transmitter is similar to that of BICM scheme.

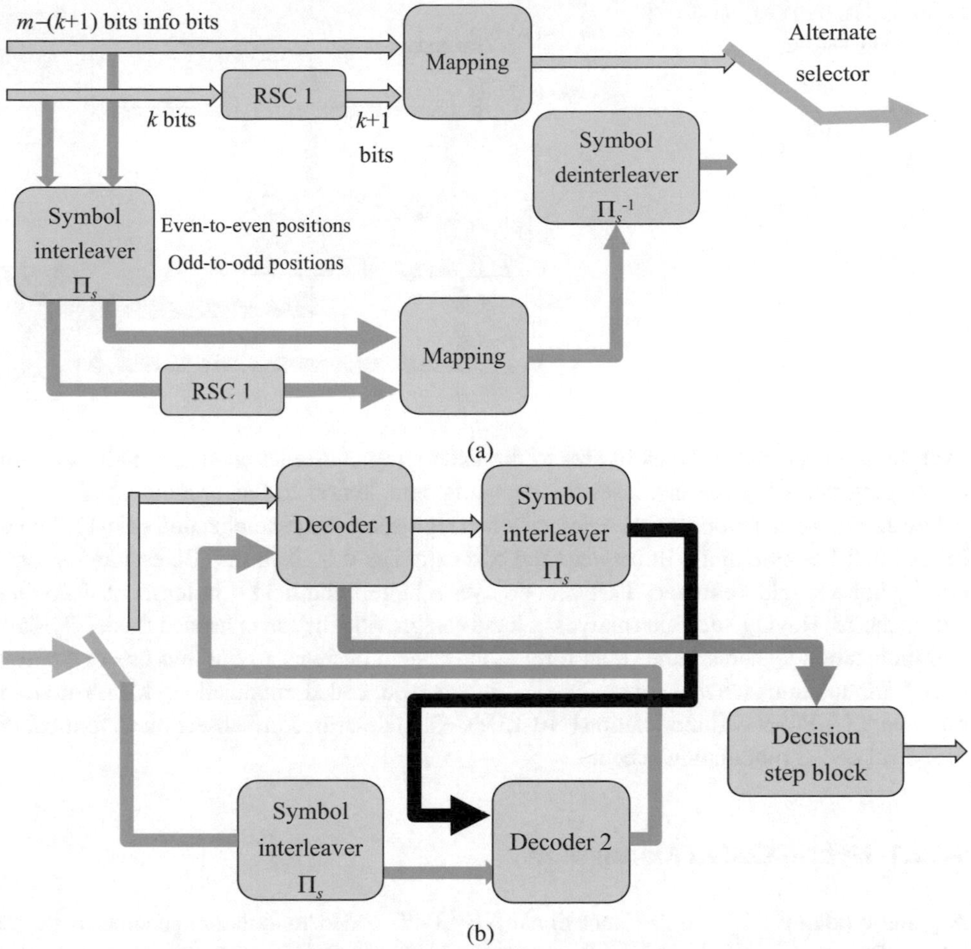

Fig. 9.54 Symbol-interleaved turbo TCM: (**a**) encoder and (**b**) decoder architectures

The *symbol interleaving TCM* scheme is shown in Fig. 9.54. The encoder architecture is shown in Fig. 9.54a. Clearly, the interleaver operates on symbol level, each symbol contains *m* bits, even positions in interleaver get mapped to even positions, and odd positions get mapped to odd positions. Two TCM encoders are coupled by the symbol interleaver. The RSC codes have the rate $k/(k + 1)$. The alternate selector selects the upper and lower branches in an alternative fashion. The m-$(k + 1)$ uncoded bits and k + 1 coded bits from RSC are used to select a point in a signal constellation diagram. Given the existence of two alternative modulation branches, the careful puncturing of the redundant symbols is needed.

The symbol-interleaved TCM decoder architecture is shown in Fig. 9.54b. Clearly, the symbol-interleaved TCM decoding principle is very similar to that of turbo decoding principle, and the key difference is that interleavers operate on a symbol level. Additional details on this scheme can be found in [105, 106].

The bit-interleaved turbo TCM, shown in Fig. 9.55, was introduced by Benedetto et al. in 1996 [88]. In this scheme, instead of symbol interleaving and puncturing of redundant symbols, multiple bit interleavers are employed. After RSC1 we select $m/2$ systematic bits and puncture the remaining ones. In RSC2 we select the bit positions punctured in RSC1 as the output and puncture the remaining ones. The permutation block Π_1 permutes the bits selected by RSC1. On the other hand, the permutation block Π_2 permutes the bits punctured by RSC1. The remaining $m/2$ bits from RSC1 and $m/2$ bits from RSC2 are used to select a point from 2^m-ary signal constellation.

9.12 Hybrid Multidimensional Coded Modulation Scheme for High-Speed Optical Transmission

It has been shown that multilevel coding with parallel independent decoding (MLC/PID) and iterative detection and decoding (IDD) perform very well under diverse optical fiber transmission scenarios [2, 10–12]. Recently, nonbinary LDPC-coded modulation (NB-LDPC-CM) has been shown to outperform its corresponding MLC/PID+IDD scheme with performance gaps increasing as the underlying signal constellation size increases [82]. NB-LDPC-CM employs a single component code defined

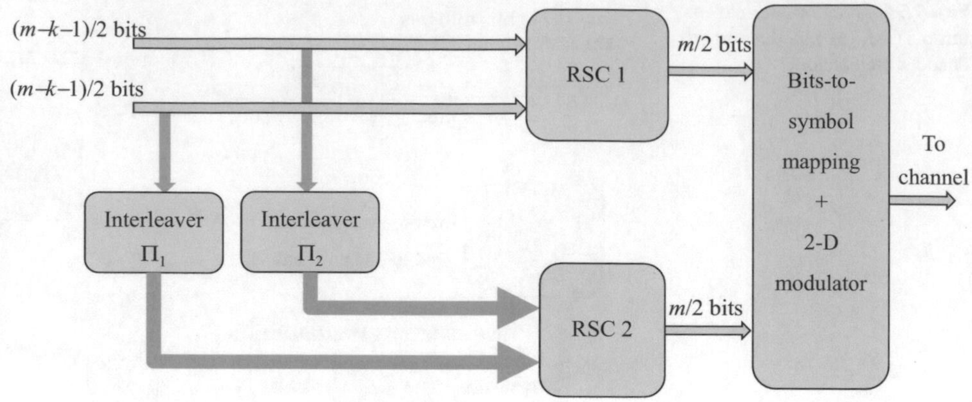

Fig. 9.55 Turbo TCM encoder with bit interleaving

over an alphabet that matches in size to the underlying constellation size, which in return eliminates the need for iterative demapping-decoding but increases the decoding complexity. In this section, we described alternative, *hybrid multidimensional LDPC-coded modulation scheme*, which represents the generalization of hybrid coded modulation scheme introduced in [50] that lies essentially in between the two extremes described in [10], employing only binary LDPC codes, and [82] – employing a single nonbinary LDPC code over a large finite field – in terms of decoding complexity and error correction performance. Having such alternatives at hand can provide the much needed flexibility during link budget analysis of optical communication systems rather than forcing the system designer to opt into one of the two extremes. Additionally, various coded modulation schemes including bit-interleaved coded modulation, MLC, nonbinary LDPC CM, and multilevel nonbinary LDPC-coded modulation (NB-LDPC-CM) described in Subsection C, to mention few, are just particular instances of hybrid coded modulation scheme.

9.12.1 Hybrid Coded Modulation

A generic adaptive hybrid D-dimensional NB-LDPC-coded modulation scheme to be used in combination with irregular NB-LDPC coding, discussed in Sec. 9.10.4, is shown in Fig. 9.56. To facilitate explanations, the details related to compensation of dispersion and nonlinear effects, carrier frequency, and carrier phase estimations are omitted. It is applicable to various spatial division multiplexing (SDM) fibers: few-mode fibers (FMFs), few-core fibers (FCFs), and few-mode-few-core fibers (FMFCFs). It is also applicable to single-mode fibers (SMFs), which can be considered as SDM fibers with only one fundamental mode and two polarization states. Both electrical and optical basis functions are applicable. Electrical basis functions include modified orthogonal polynomials and orthogonal prolate spheroidal wave functions (also known as Slepian sequences in discrete-time domain), while optical basis functions include spatial modes and polarization states. This scheme is very flexible and can be used in various configurations ranging from spatial multiplexing of various multidimensional signals to fully D-dimensional signaling.

Generic Tx configuration for the hybrid multidimensional LDPC-coded modulation scheme is presented in Fig. 9.56a. The MLC scheme, shown in Fig. 9.57a, is a degenerate case of the hybrid CM scheme and occurs when source channels are binary, while LDPC codes at different levels have different code rates. When all LDPC codes shown in Fig. 9.57a have the same code rate, we refer to this scheme as the *block-interleaved coded modulation* (BICM). The block interleaver can be replaced by more advanced interleavers [20, 24]. The conventional bit-interleaved coded modulation is a serial concatenation of channel encoder, interleaver, and multilevel modulator [89]. This scheme is based on a single LDPC code, which operates at rate mR_s, where m is the number of bits per symbol and R_s is the symbol rate. For currently available fiber-optics communication systems operating at 25 Gs/s, this scheme will require the implementation of LDPC encoder and decoder operating at rates $m \times 25$ Gb/s, which would be quite challenging to implement with currently available commercial electronics. To avoid this problem, m parallel LDPC encoders and decoders should be used instead in fiber-optics communications, with one possible implementation shown in Fig. 9.57a (see also Fig. 9.56a), as indicated above. Alternatively, the source channels shown in Fig. 9.57a may originate from 1:m demultiplexer. Moreover, the block interleaver can be replaced with m interleavers, one per LDPC encoder, decomposing, therefore, the BICM scheme from Fig. 9.57a into m parallel conventional bit-interleaved coded modulation schemes due to Caire et al. [89]. The corresponding schemes perform comparable under AWGN channel assumption.

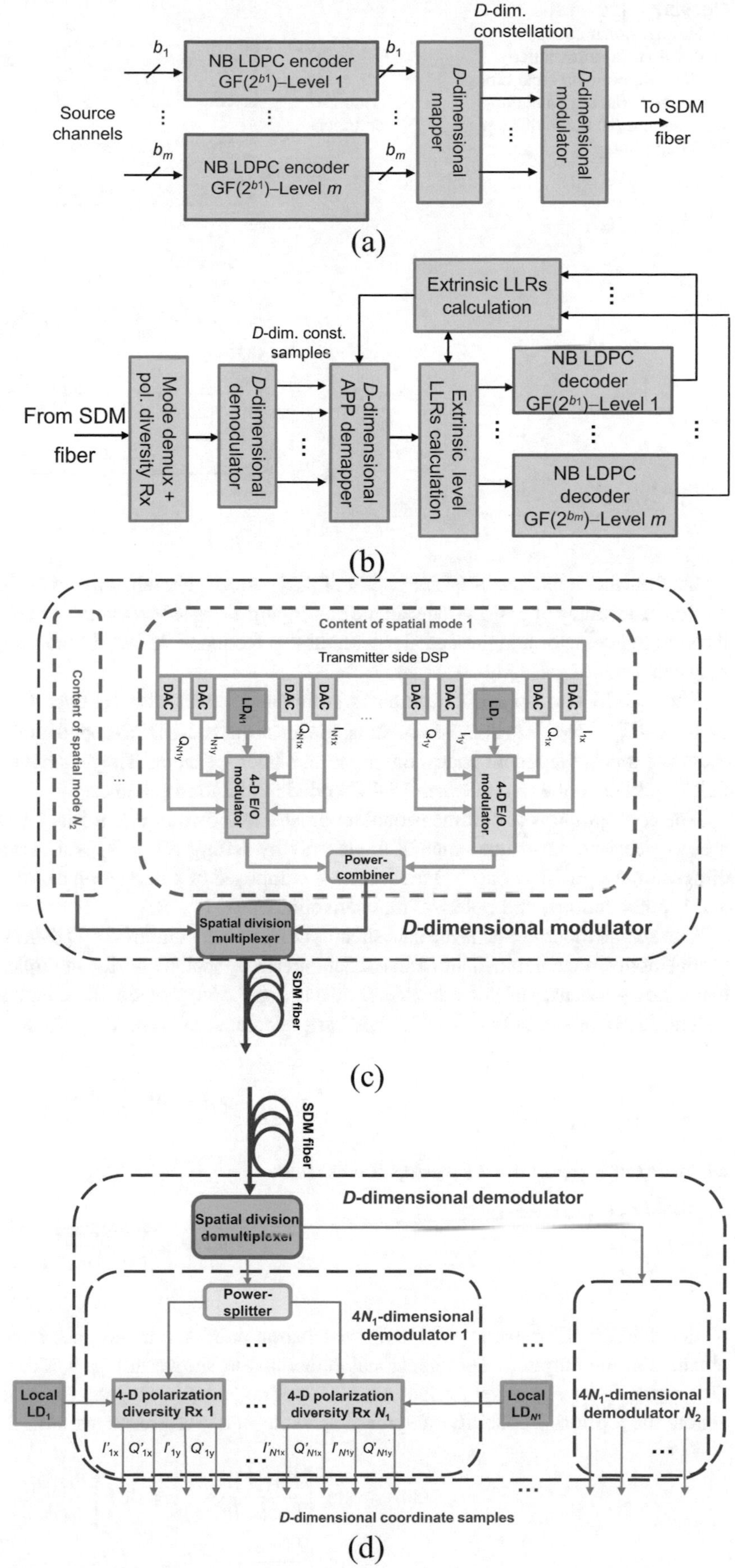

Fig. 9.56 Hybrid multidimensional coded modulation: (**a**) transmitter configuration, (**b**) receiver configuration, (**c**) configuration of D-dimensional modulator, and (**d**) D-dimensional demodulator configuration. *SDM* spatial division multiplexing. SDM fibers could be FMF, FCF, FMFCF, or SMF. The corresponding signal space is $D = 4N_1N_2$-dimensional, where N_1 is the number of sub-bands in a superchannel with orthogonal center frequencies and N_2 is the number of spatial modes. The factor 4 originates from two polarization states and in-phase and quadrature channels

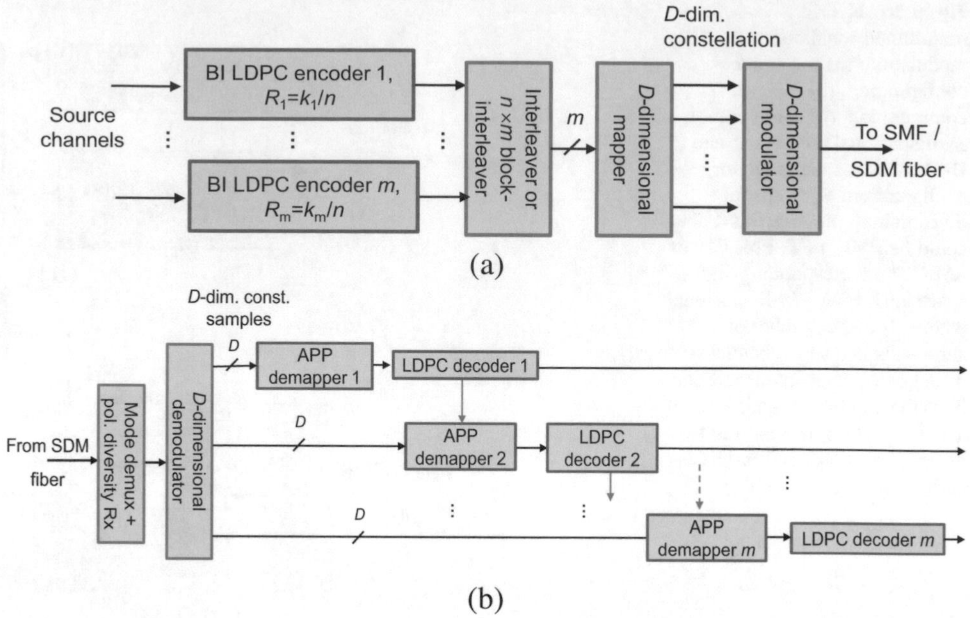

Fig. 9.57 MLC-based multidimensional coded modulation: (**a**) transmitter configuration of MLC and block-interleaved coded modulation schemes and (**b**) the multistage decoding principle

On the other hand, the NB-LDPC-CM of [82], whose generalization to D-dimensional space is shown later in Fig. 9.87, is a particular instance of hybrid CM scheme employing a single level with an M-ary LDPC code and performing just a single pass through the detector and the decoder without any feedback. In hybrid multidimensional CM scheme, the number of levels is between 1 and $\log_2(M)$.

The hybrid CM approach combines component codes defined over $GF(2^{b_i}), 1 \leq i \leq m$ in a way that the condition $b_1 + \ldots + b_m = \log_2(M)$ is satisfied. Thus, the resulting NB-LDPC-coded modulation schemes are not homogenous in the field orders of their component codes, but are rather heterogeneous. They enable the interaction of component codes over various fields and hence the name *hybrid* LDPC-coded modulation scheme.

The configuration of D-dimensional modulator is shown in Fig. 9.56c. Clearly, conventional PDM scheme is just a special case corresponding to fundamental mode only by setting $N_1 = N_2 = 1$. Transmitter side DSP provides coordinates for D-dimensional signaling. The 4-D modulator is composed of polarization beam splitter, two I/Q modulators corresponding to x- and y-polarizations, and polarization beam combiners.

One spatial mode contains the data streams of N_1 power combined 4-D data streams. The N_2 data streams, corresponding to N_2 spatial modes, are combined into a signal data stream by spatial division multiplexer and transmitted over SDM fiber. On the other hand, the D-dimensional demodulator, shown in Fig. 9.56d, provides the samples corresponding to signal constellation estimates.

The LLRs of symbols $\boldsymbol{s}_i$ ($i = 1,\ldots,M$) are calculated in an APP demapper block as follows:

$$L(\boldsymbol{s}_i) = \log \frac{P(\boldsymbol{s}_i|\boldsymbol{r})}{P(\boldsymbol{s}_0|\boldsymbol{r})}, \tag{9.161}$$

where $P(\boldsymbol{s}|\boldsymbol{r})$ is determined by using Bayes' rule:

$$P(\boldsymbol{s}_i|\boldsymbol{r}) = \frac{P(\boldsymbol{r}|\boldsymbol{s}_i)P(\boldsymbol{s}_i)}{\sum_{\boldsymbol{s}'} P(\boldsymbol{r}|\boldsymbol{s}')P(\boldsymbol{s}')}. \tag{9.162}$$

Notice that $\boldsymbol{s}_i$ is the transmitted signal constellation point at time instance i, and $\boldsymbol{r}_{ij}$ ($j = 1,2,\ldots,D$) are corresponding samples obtained at the output of D-dimensional demodulator, shown in Fig. 9.56d. In the presence of fiber nonlinearities, $P(\boldsymbol{r}|\boldsymbol{s}_i)$ in (36) needs to be estimated evaluating the histograms, by propagating a sufficiently long training sequence. With $P(\boldsymbol{s}_i)$ we denote the a priori probability of symbol $\boldsymbol{s}_i$, while $\boldsymbol{s}_0$ is a reference symbol. By substituting (9.162) into (9.161), we obtain

$$\begin{aligned} L(\boldsymbol{s}_i) &= \log\left[\frac{P(\boldsymbol{r}_i|\boldsymbol{s}_i)P(\boldsymbol{s}_i)}{P(\boldsymbol{r}_i|\boldsymbol{s}_0)P(\boldsymbol{s}_0)}\right] = \log\left[\frac{P(\boldsymbol{r}_i|\boldsymbol{s}_i)}{P(\boldsymbol{r}_i|\boldsymbol{s}_0)}\right] + \log\left[\frac{P(\boldsymbol{s}_i)}{P(\boldsymbol{s}_0)}\right] \\ &= \log\left[\frac{P(\boldsymbol{r}_i|\boldsymbol{s}_i)}{P(\boldsymbol{r}_i|\boldsymbol{s}_0)}\right] + L_a(\boldsymbol{s}_i), L_a(\boldsymbol{s}_i) = \log\left[\frac{P(\boldsymbol{s}_i)}{P(\boldsymbol{s}_0)}\right], \end{aligned} \tag{9.163}$$

where with $L_a(\boldsymbol{s}_i)$ we denote the prior reliability of symbol $\boldsymbol{s}_i$. Let us denote by $\boldsymbol{s}_{ij}$ the j-th component of the transmitted symbol $\boldsymbol{s}_i$ represented as $\boldsymbol{s}_i = (s_{i1},\ldots,s_{im})$, where $s_{ij}\in\mathrm{GF}(2^{m_i})$. The prior symbol LLRs for the next iteration are determined by

$$L_a(\boldsymbol{s}_i) = \log\frac{P(\boldsymbol{s}_i)}{P(\boldsymbol{s}_0)} = \log\frac{\prod_{j=1}^{m} P(\boldsymbol{s}_{ij})}{\prod_{j=1}^{m} P(\boldsymbol{s}_{ij}=0)} = \\ = \sum_{j=1}^{m}\log\frac{P(\boldsymbol{s}_{ij})}{P(\boldsymbol{s}_{ij}=0)} = \sum_{j=1}^{m} L_a(\boldsymbol{s}_{ij}), L_a(\boldsymbol{s}_{ij}) = \log\frac{P(\boldsymbol{s}_{ij})}{P(\boldsymbol{s}_{ij}=0)} \tag{9.164}$$

where we assumed that reference symbol is $\boldsymbol{s}_0 = (0\ldots0)$.

Finally, the prior symbol estimate can be obtained from

$$L_a(\widehat{\boldsymbol{s}}_i) = \sum_{j=1}^{m} L_{D,e}(\widehat{\boldsymbol{s}}_{ij}), \tag{9.165}$$

where

$$L_{D,e}(\widehat{\boldsymbol{s}}_{ij}) = L\left(\boldsymbol{s}_{ij}^{(t)}\right) - L\left(\boldsymbol{s}_{ij}^{(t-1)}\right) \tag{9.166}$$

In (9.166), we use $L\left(\boldsymbol{s}_{ij}^{(t)}\right)$ to denote the LDPC decoder output in current iteration (iteration t).

The symbol LLRs corresponding to the j-th decoder $L(\boldsymbol{s}_{ij})$ are determined from symbol LLRs (9.163) by

$$L(\widehat{\boldsymbol{s}}_{ij}) = \log\frac{\sum_{\boldsymbol{s}_i:s_{ij}=s}\exp\left[L(\boldsymbol{s}_i)\right]\exp\left(\sum_{\boldsymbol{s}_i:s_{ik}=0,k\neq j}L_a(\boldsymbol{s}_{ik})\right)}{\sum_{\boldsymbol{s}_i:s_{ij}=0}\exp\left[L(\boldsymbol{s}_i)\right]\exp\left(\sum_{\boldsymbol{s}_i:s_{ik}=0,k\neq j}L_a(\boldsymbol{s}_{ik})\right)}. \tag{9.167}$$

Therefore, the j-th position reliability in (9.167) is calculated as the logarithm of the ratio of a probability that $\boldsymbol{s}_{ij} = s$, $s\in\mathrm{GF}(2^{m_i})\backslash\{0\}$ and probability that $\boldsymbol{s}_{ij} = 0$. In the numerator, the summation is done over all symbols $\boldsymbol{s}_i$ having $s\in\mathrm{GF}(2^{m_i})\backslash\{0\}$ at the position j, while in the denominator over all symbols $\boldsymbol{s}_i$ having 0 at the position j.

With $L_a(\boldsymbol{s}_{ik})$ we denoted the prior (extrinsic) information determined from the APP demapper. The inner summation in (9.167) is performed over all positions of symbol $\boldsymbol{s}_i$, selected in the outer summation, for which $\boldsymbol{s}_{ik} = 0$, $k \neq j$. The j-th position LLRs are forwarded to corresponding LDPC decoders operating over $\mathrm{GF}(2^{m_j})$. The iteration between the D-dimensional APP demapper and NB-LDPC decoders is performed until the maximum number of iterations is reached or the valid codewords are obtained.

Instead of parallel independent decoding (PID) fashion, described above, iterative demapping and decoding can be performed in a multistage decoding fashion, in which the reliabilities from the higher i-th ($i = 1,2,\ldots,m-1$) decoding stage are forwarded to the lower ($i+1$)-th decoding stage, as shown in Fig. 9.57b. In this scheme the code rates of LDPC codes are arranged in increasing order: $R_1 < R_2 < \ldots < R_m$. However, this decoding scheme exhibits higher decoding complexity and higher latency compared to PID.

9.12.2 Multilevel Nonbinary LDPC-Coded Modulation (ML-NB-LDPC-CM) for High-Speed Optical Transmissions

Still we restrict our analysis of coded modulation schemes over a predetermined modulation format of constellation size $M = 2^m$. We compare multilevel NB-LDPC-CM with block-interleaved LDPC-coded modulation (BI-LDPC-CM) and NB-LDPC-CM as presented in Fig. 9.58. The transmitter/receiver architectures of BI-LDPC-CM are depicted in Fig. 9.58a. The source information entering the transmitter is composed of m parallel binary information streams of length K bits each. The m parallel bit streams are separately encoded into m codewords through m binary LDPC encoders of identical

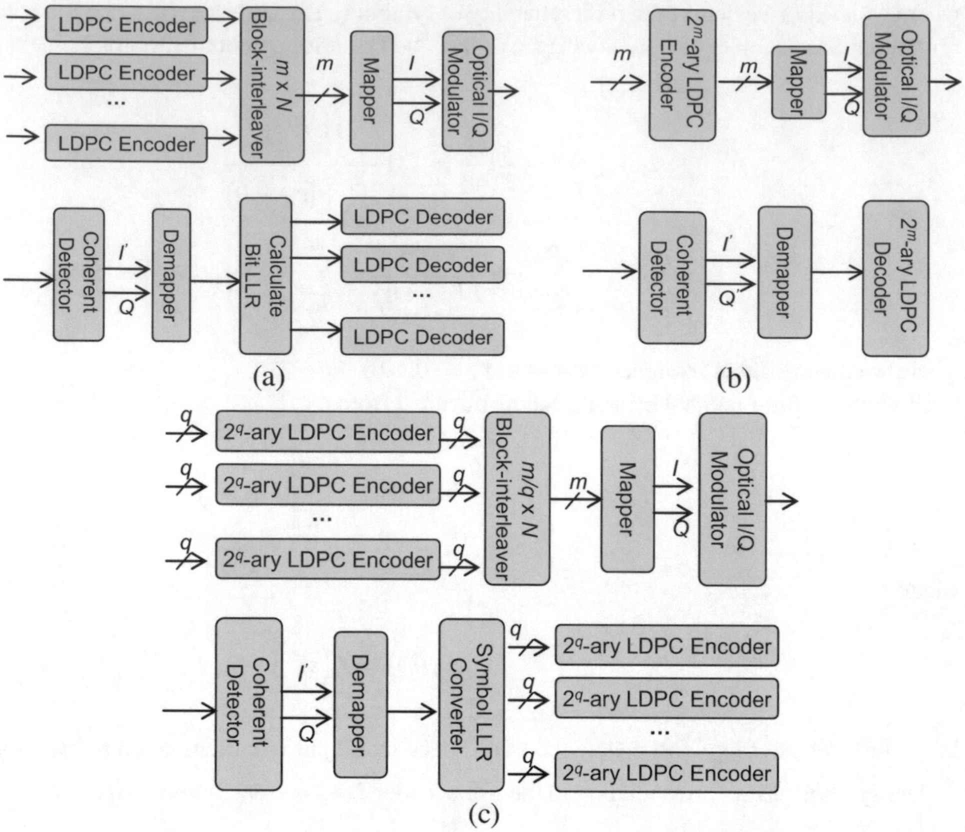

Fig. 9.58 Transmitter and receiver structures of (**a**) BI-LDPC-CM, (**b**) NB-LDPC-CM, and (**c**) multilevel NB-LDPC-CM. These are particular instances of hybrid multidimensional coded modulation scheme shown in Fig. 9.56 applied to 2-D signaling schemes

code rate $R = K/N$. The codewords are then placed in an $m \times N$ block interleaver row-wise. The mapper reads the data out of the interleaver column-wise and maps the m-bit tuple to a 2^m-ary constellation symbol (I_i, Q_i), $i = 1,2,...,2^m$, corresponding to the in-phase and quadrature components which are further modulated onto the optical carrier through the optical I/Q modulator. For NB-LDPC-CM in Fig. 9.58b, the m parallel binary information streams can be regarded as a single nonbinary stream over an alphabet of $Q = 2^m$ symbols. This Q-ary information stream of length K is encapsulated into a codeword of length N symbols by a single Q-ary LDPC encoder of code rate $R = K/N$. The mapper then maps each Q-ary codeword symbol to a Q-ary (or 2^m-ary) constellation point $\mathbf{s} = (I_x, Q_x, I_y, Q_y)$. The following optical modulator operates in the same way as described before. As shown in Fig. 9.58c, different from NB-LDPC-CM, the multilevel NB-LDPC-CM employs nonbinary LDPC code with smaller Q, i.e., $Q = 2^q$, while $1 < q < \mathrm{m}$. The m parallel information bit streams are first grouped into $L = m/q$ sets of bit streams, while L represents the number of decoding levels.

Each set of bit streams is considered as a stream of 2^q-ary symbols. The L 2^q-ary information streams of length K are then encoded into codewords of length N separately by L 2^q-ary LDPC encoders and later are block-interleaved through an $L \times N$ block interleaver, similar to that for BI-LDPC-CM. Then each tuple of L 2^q-ary symbols is mapped to a 2^m-ary constellation point followed with the optical I/Q modulation. We have noticed that BI-LDPC-CM and NB-LDPC-CM are in fact two special cases of the proposed multilevel NB-LDPC-CM for $q = 1$ and $q = m$, respectively. Obviously, the selection of q becomes a tradeoff between performance and complexity.

At the receiver, the demapper first calculates the symbol LLRs $L(s)$ for each received symbol with $s \in \{0,1,...,M\text{-}1\}$ indicating the log-likelihood of sending a symbol on the constellation map. Since L-level Q-ary symbols have been used, it is natural to represent a symbol s by a tuple of L Q-ary symbols as $\mathbf{S} = \{s_0, s_1, ..., s_{L-1}\}$ with $s_i \in \{0,1,...,Q-1\}$ for any $i \in \{0,1,...,L-1\}$. Corresponding to the configuration of the transmitter, $L(s)$ with $s \in \{0,1,...,M-1\}$ are then converted into L parallel Q-ary symbol LLRs $L(s_i)$ with $s_i \in \{0,1,...,Q-1\}$ and $i \in \{0,1,...,L-1\}$ according to the following equation:

$$L(s_i) = \log \left(\sum_{s=0,\, s_i \in S}^{M-1} e^{L(s)} \right) \tag{9.168}$$

Each GF(Q) decoder then performs decoding separately after which L Q-ary symbol estimates are converted into the one M-ary symbol and later m bits to count bit errors. As noticed in our previous analysis, the BI-LDPC-CM and NB-LDPC-CM schemes serve as two special cases of multilevel NB-LDPC-CM; the above receiver side processing can also be applied to BI-LDPC-CM and NB-LDPC-CM by setting $q = 1$ and $q = m$, respectively. Notice that for BI-LDPC-CM, iterative decoding between demapper and LDPC decoder can be applied.

Coded modulation with the adoption of nonbinary LDPC codes achieves promising net coding gain (NCG) performance at the expense of very high complexity in both hardware and computation. As hardware complexity analysis is affected by many practical tradeoff factors like throughput and parallelization and hence not easy to get a clear expression, analysis of decoding computation is mostly executed instead. Particularly, an expression of the total number of additions involved in decoding a quasi-cyclic LDPC codeword based on MD-FFT-SPA (which is already known as hardware-friendly) is given as [82]

$$2\rho Q(N-K)\left(q+1-\frac{1}{2\rho}\right), \tag{9.169a}$$

where ρ is the number of nonzero elements in the row of the parity-check matrix. For detailed derivation of the decoding complexity computation, we refer readers to [82]. We adopt the expression to analyze the complexity reduction of the proposed scheme over the conventional NB-LDPC-CM. Hence, the complexity ratio (CR) of the conventional NB-LDPC-CM over that of the proposed scheme is given as

$$CR = \frac{2\rho 2^m (N-K)\left(m+1-\frac{1}{2\rho}\right)}{\frac{m}{q}2\rho 2^q (N-K)\left(q+1-\frac{1}{2\rho}\right)}. \tag{9.169b}$$

Figure 9.59 plots the complexity ratio with $\rho = 24$, $q = 2$, $m = 2,3,\ldots,7$ and shows the potential of the ML-NB-LDPC-CM scheme in reducing computation complexity. Notice that we use $q = 2$ for the ML-NB-LDPC-CM scheme for the purpose of revealing the potential of complexity reduction, it may not be the optimal choice in terms of performance, and the selection of q should also take the performance into account for practical applications.

Figure 9.60 presents the simulated 25-Gbaud 16-QAM optical transmission system in the ASE noise-dominated scenario, based on which we evaluate the aforementioned three coded modulation schemes. We have applied identical coded modulation scheme to both polarizations. Parameters for the adopted quasi-cyclic LDPC codes are $N = 13{,}680$, $K = 11{,}970$, $R = 0.875$, and girth 8. The simulated optical transmission data rate is $4 \times 2 \times 25 \times 0.875 = 175$ Gb/s. For BI-LDPC-CM, we use 25 iterations in binary LDPC decoders based on the sum-product algorithm and perform 3 outer iterations between the demapper and the decoders as iterative decoding is adopted. For both NB-LDPC-CM and multilevel NB-LDPC-CM, we use 50 iterations for the nonbinary LDPC decoders based on MD-FFT-SPA [82]. For NB-LDPC-CM, we use 16-ary LDPC code to match the constellation size. For multilevel NB-LDPC-CM, we set $q = 2$ and employ a couple of

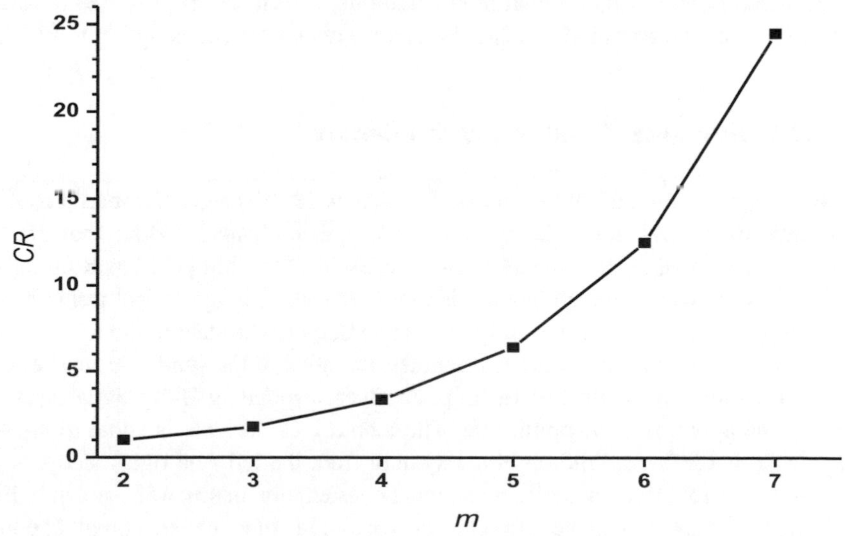

Fig. 9.59 Computation complexity ratio of the conventional NB-LDPC-CM over multilevel NB-LDPC-CM with $q = 2$

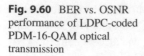

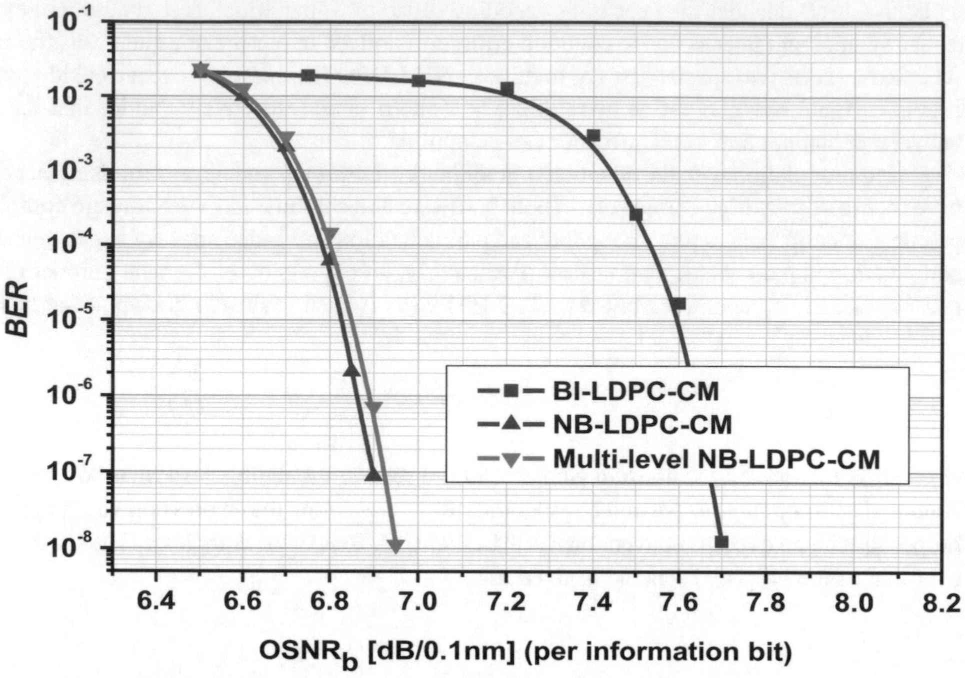

Fig. 9.60 BER vs. OSNR performance of LDPC-coded PDM-16-QAM optical transmission

identical 4-ary LDPC codes. Figure 9.60 shows the performances of LDPC-coded PDM 16-QAM optical transmission system in terms of BER vs. $OSNR_b$. As shown in Fig. 9.60, the ML-NB-LDPC-CM outperforms BICM-LDPC-CM by ~0.65 dB, while achieving nearly identical performance as the conventional NB-LDPC-CM. The results have indicated that the ML-NB-LDPC-CM scheme has reduced the complexity by adopting low-complexity nonbinary LDPC codes over smaller finite field while keeping the performance unaffected.

9.13 Multidimensional Turbo Equalization

In this section we describe turbo equalization [107–112] with the employment of LDPC codes named as LDPC-coded turbo equalization (TE) [2, 10, 12, 107, 108], as a universal scheme that can be used to deal with the imperfectly compensated linear effects and nonlinear effects. This scheme was initially introduced to deal with ISI in wireless channels [112]. However, given its exponential complexity, this scheme is more suitable to deal with nonlinear effects rather than linear ones. In fiber-optics communications, it can be used to deal with imperfectly compensated: (i) fiber nonlinearities, (ii) residual polarization-mode dispersion (PMD), (iii) residual polarization-dependent loss (PDL), (iii) residual CD, (iv) residual I/Q-imbalance effects in multilevel coded modulation schemes, and (v) residual mode coupling in FMFs.

9.13.1 Nonlinear Channels with Memory

The multidimensional Bahl-Cocke-Jelinek-Raviv (BCJR) algorithm-based equalizer [12] to be described below operates on a discrete dynamical trellis description of the optical channel. Notice that this equalizer is universal and applicable to any multidimensional signal constellation. The BCJR algorithm [13] has been used earlier in this chapter as a MAP decoding algorithm; however, its application is more general. This dynamical trellis is uniquely defined by the following triplet: the previous state, the next state, and the channel output. The state in the trellis is defined as $s_j = (x_{j-m}, x_{j-m+1}, \ldots, x_j, x_{j+1}, \ldots, x_{j+m}) = \boldsymbol{x}[j-m, j+m]$, where x_k denotes the index of the symbol from the following set of possible indices $\boldsymbol{X} = \{0, 1, \ldots, M-1\}$, with M being the number of points in corresponding M-ary signal constellation. Every symbol carries $b = \log_2 M$ bits, using the appropriate mapping rule. The memory of the state is equal to $m_1 + m_2 + 1$, with m_1 (m_2) being the number of *symbols* that influence the observed symbol from the left (the right) side. An example of such dynamic trellis of memory $m_1 + m_2 + 1$ for arbitrary multidimensional constellation of size M is shown in Fig. 9.61. The trellis has $M^{m_1+m_2+1}$ states, each of which corresponds to the different $(m_1 + m_2 + 1)$-symbol patterns (symbol configurations). The state index is determined by

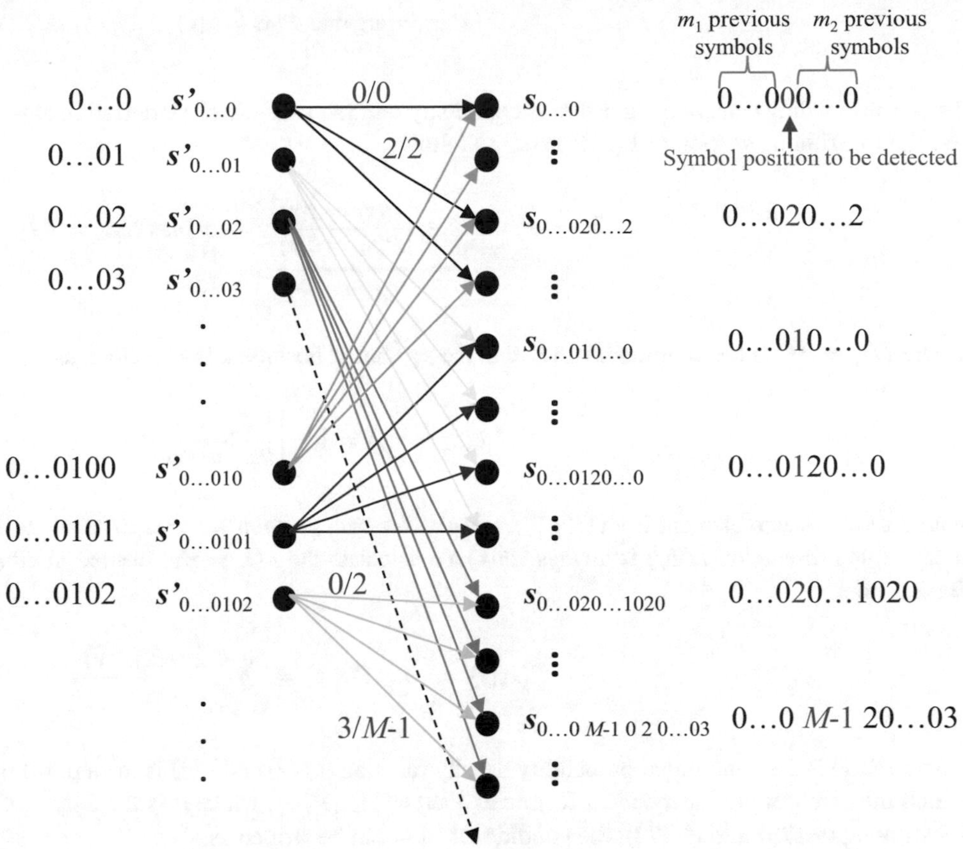

Fig. 9.61 A portion of trellis to be used in multidimensional BCJR detector with memory $m_1 + m_2 + 1$. The m_1 previous symbols and m_2 next symbols relative to the observed symbol define the memory of the detector

considering $(m_1 + m_2 + 1)$ symbols as digits in numerical system with the base M. The left column in dynamic trellis represents the current states, and the right column denotes the terminal states. The branches are labeled by two symbols, the input symbol is the last symbol in initial state (the blue symbol), and the output symbol is the central symbol of terminal state (the red symbol). Therefore, the current symbol is affected by both previous and incoming symbols.

The dynamic trellis shown in Fig. 9.61 is generated by assuming that the symbols are coming from the right and that the size of window of symbols is $m_1 + m_2 + 1$.

For instance, let us observe the symbol 10. . .0 ***M*-3** 0. . .023, with the center of the window denoted in bold. For the same window size, for incoming symbol 5, we will move to the state 0. . .0 *M*-3 **0**. . .0235. By using this strategy, the trellis portion shown in Fig. 9.61 is generated. During detection process instead of detecting whole sequence of length n, we instead detect the subsequence (the window) of length $w > m_1 + m_2 + 1$, with each window containing m_2 samples from previous window and m_1 samples from the next window. Since the detection process is performed by splitting the sequence to be detected into n/w subsequences containing w symbols each, the detection algorithm is called the *sliding window*. This approach is suitable for parallelization as n/w BCJR processors can perform detection in parallel.

For the complete description of the dynamical trellis, the transition PDFs $p(\mathbf{y}_j|x_j) = p(\mathbf{y}_j|s)$, $s \in S$ are needed, where S is the set of states in the trellis and $\mathbf{y}_j$ is the vector of samples (corresponding to the transmitted symbol index x_j). The conditional PDFs can be determined from *collected histograms* or by using *instanton Edgeworth expansion* method [17]. The number of edges originating in any of the left-column states is M, and the number of merging edges in arbitrary terminal state is also M.

9.13.2 Nonbinary MAP Detection

Before we describe the LDPC-coded turbo equalization scheme, we provide the basic concepts of optimum detection of nonbinary signaling in minimum probability of error sense [2, 12, 109]. Let x denote the transmitted sequence and y the received sequence. The *optimum receiver* (in symbol error probability sense) assigns $\hat{x}_k$ to the value $x \in GF(2^m)$ that maximizes the APP $P(x_k = x|y)$ given the received sequence y as follows:

$$\widehat{x}_k = \arg\max_{x \in GF(2^m)} P(x_k = x|\mathbf{y}). \tag{9.170}$$

The corresponding detection algorithm is commonly referred to as the MAP detection algorithm. In practice, it is common to use the logarithmic version of Eq. (9.170a) as follows:

$$\widehat{x}_k = \underbrace{\arg\max_{x \in GF(2^m)} \log \left[\frac{P(x_k = x|\mathbf{y})}{P(x_k = 0|\mathbf{y})}\right]}_{L(x_k = x|\mathbf{y})} = \arg\max_{x \in GF(2^m)} L(x_k = x|\mathbf{y}), \tag{9.171}$$

where $L(x_k = x|\mathbf{y})$ is the conditional LLR that the symbol at position k is x, defined as

$$L(x_k = x|\mathbf{y}) = \log \left[\frac{P(x_k = x|\mathbf{y})}{P(x_k = 0|\mathbf{y})}\right], \tag{9.172}$$

where x is a nonzero element from GF(2^m). Clearly, for each position we need to calculate $2^m - 1$ conditional LLRs. (Notice that in this formulation $L(0|\mathbf{y})$ is always zero.) To calculate the $P(x_k = x|\mathbf{y})$ needed in either (9.171) or (9.172), we invoke Bayes' rule:

$$P(x_k|\mathbf{y}) = \sum_{\forall \mathbf{x}: x_k = x} P(\mathbf{x}|\mathbf{y}) = \sum_{\forall \mathbf{x}: x_k = x} \frac{P(\mathbf{y}|\mathbf{x})P(\mathbf{x})}{P(\mathbf{y})}, \tag{9.173}$$

where $P(\mathbf{y}|\mathbf{x})$ is the conditional probability density function (PDF) and $P(\mathbf{x})$ is the a priori probability of input sequence $\boldsymbol{x}$, in which the symbols are independent factors as $P(\boldsymbol{x}) = \prod_{i=1}^{n} P(x_i)$, where n is the length of the sequence to be detected. By substituting (9.173) into (9.172), the conditional LLR can be written as

$$\begin{aligned} L(x_k = x|\mathbf{y}) &= \log \left[\frac{\sum_{\forall \mathbf{x}: x_k = x} p(\mathbf{y}|\mathbf{x}) \prod_{i=1}^{n} P(x_i)}{\sum_{\forall \mathbf{x}: x_k = 0} p(\mathbf{y}|\mathbf{x}) \prod_{i=1}^{n} P(x_i)}\right] \\ &= L_{\text{ext}}(x_k = x|\mathbf{y}) + L_a(x_k = x), \end{aligned} \tag{9.174}$$

where the extrinsic information about x_k contained in y $L_{\text{ext}}(x_k = x|\mathbf{y})$ and the *a priori* LLR $L_a(x_k = x)$ are defined, respectively, as

$$\begin{aligned} L_{\text{ext}}(x_k = x|\mathbf{y}) &= \log \left[\frac{\sum_{\forall \mathbf{x}: x_k = x} p(\mathbf{y}|\mathbf{x}) \prod_{i=1, i \neq k}^{n} P(x_i)}{\sum_{\forall \mathbf{x}: x_k = 0} p(\mathbf{y}|\mathbf{x}) \prod_{i=1, i \neq k}^{n} P(x_i)}\right], \\ L_a(x_k = x) &= \log \left[\frac{P(x_k = x)}{P(x_k = 0)}\right]. \end{aligned} \tag{9.175}$$

From (9.175), it is clear that computation of conditional LLRs can be computationally intensive. One possible computation is based on BCJR algorithm, with log-domain version when applied to the multilevel/multidimensional modulation schemes being described in the following section.

9.13.3 Sliding-Window Multidimensional Turbo Equalization

The multidimensional LDPC-coded turbo equalizer is composed of two ingredients: (i) the MAP detector based on multilevel/multidimensional BCJR detection algorithm and (ii) the LDPC decoder. The multidimensional transmitter configuration is

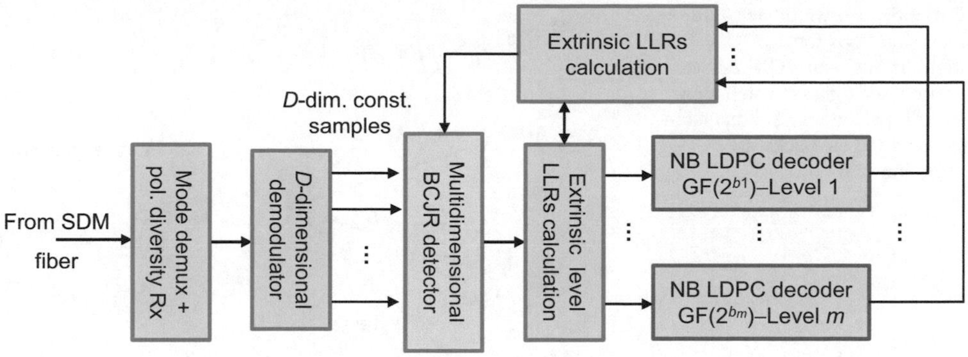

Fig. 9.62 Multidimensional LDPC-coded turbo equalization scheme architecture

identical to one shown in Fig. 9.56a. The receiver configuration of multidimensional LDPC-coded turbo equalizer is shown in Fig. 9.62.

Clearly, the *D*-dimensional APP demapper from Fig. 9.56b is replaced by multidimensional BCJR detector, whose operation principle is described next, and represents the generalization of [2, 107, 108]. The *forward metric* is defined as $\alpha_j(\boldsymbol{s}) = \log\{p(\boldsymbol{s}_j = \boldsymbol{s},\boldsymbol{y}[1,j])\}$ $(j = 1,2,\ldots,n)$; the *backward metric* is defined as $\beta_j(\boldsymbol{s}) = \log\{p(\boldsymbol{y}[j+1,n]|\boldsymbol{s}_j = \boldsymbol{s})\}$; and the *branch metric* is defined as $\gamma_j(\boldsymbol{s}',\boldsymbol{s}) = \log[p(\boldsymbol{s}_j = \boldsymbol{s},y_j,\boldsymbol{s}_{j-1} = \boldsymbol{s}')]$. The corresponding metrics can be calculated iteratively as follows [2, 107, 108]:

$$\alpha_j(\boldsymbol{s}) = \max_{\boldsymbol{s}'} * \left[\alpha_{j-1}(\boldsymbol{s}') + \gamma_j(\boldsymbol{s}',\boldsymbol{s})\right], \tag{9.176}$$

$$\beta_{j-1}(\boldsymbol{s}') = \max_{\boldsymbol{s}} * \left[\beta_j(\boldsymbol{s}) + \gamma_j(\boldsymbol{s}',\boldsymbol{s})\right], \tag{9.177}$$

$$\gamma_j(\boldsymbol{s}',\boldsymbol{s}) = \log\left[p\left(y_j|\boldsymbol{x}[j-m,j+m]\right)P\left(x_j\right)\right]. \tag{9.178}$$

The max*-operator used in (9.176) and (9.177) can be calculated efficiently according to (9.98), while the correction factor $\log[1+\exp(-|x-y|)]$ is commonly approximated or implemented using a look-up table. $p(y_j|\boldsymbol{x}[j-m_1,j+m_2])$ is obtained, by either collecting the histograms or by instanton-Edgeworth expansion method, and $P(x_j)$ represents a priori probability of transmitted symbol x_j.

In the first outer iteration, $P(x_j)$ is set to either $1/M$ (because equally probable transmission is observed) for an existing transition from trellis given in Fig. 9.61 or to zero for a non-existing transition. The outer iteration is defined as the calculation of symbol LLRs in multidimensional BCJR detector block, the calculation of corresponding LLRs needed for LDPC decoding, the LDPC decoding, and the calculation of extrinsic symbol LLRs needed for the next iteration. The iterations within LDPC decoder are called here inner iterations.

The initial forward and backward metrics values are set to [2, 107, 108]

$$\alpha_0(\boldsymbol{s}) = \begin{cases} 0, & \boldsymbol{s} = \boldsymbol{s}_0 \\ -\infty, & \boldsymbol{s} \neq \boldsymbol{s}_0 \end{cases}, \quad \beta_n(\boldsymbol{s}) = \begin{cases} 0, & \boldsymbol{s} = \boldsymbol{s}_0 \\ -\infty, & \boldsymbol{s} \neq \boldsymbol{s}_0 \end{cases}, \tag{9.179}$$

where $\boldsymbol{s}_0$ is an initial state.

Let $\boldsymbol{s}' = \boldsymbol{x}[j-m_1-1,j+m_2-1]$ represent the previous state, $\boldsymbol{s} = \boldsymbol{x}[j-m,j+m]$ the present state, $\boldsymbol{x} = (x_1,x_2,\ldots,x_n)$ the transmitted word of symbols, and $\boldsymbol{y} = (y_1,y_2,\ldots,y_n)$ the received sequence of samples. The LLR, denoting the reliability, of symbol $x_j = \delta$ $(j = 1,2,\ldots,n)$ can be calculated by [2, 107, 108].

$$\begin{aligned} L(x_j=\delta) = & \max_{(\boldsymbol{s}',\boldsymbol{s}):x_j=\delta} * \left[\alpha_{j-1}(\boldsymbol{s}') + \gamma_j(\boldsymbol{s}',\boldsymbol{s}) + \beta_j(\boldsymbol{s})\right] \\ & - \max_{(\boldsymbol{s}',\boldsymbol{s}):x_j=\delta_0} * \left[\alpha_{j-1}(\boldsymbol{s}') + \gamma_j(\boldsymbol{s}',\boldsymbol{s}) + \beta_j(\boldsymbol{s})\right], \end{aligned} \tag{9.180}$$

where δ represents the observed symbol ($\delta \in \{0,1,\ldots,M-1\} \backslash \{\delta_0\}$) and δ_0 is the reference symbol. The forward and backward metrics are calculated using the (9.176) and (9.177). The forward and backward recursion steps of multidimensional BCJR

Fig. 9.63 Forward/backward recursion steps for multidimensional BCJR detector of arbitrary signal constellation size M: (**a**) the forward recursion step and (**b**) the backward recursion step

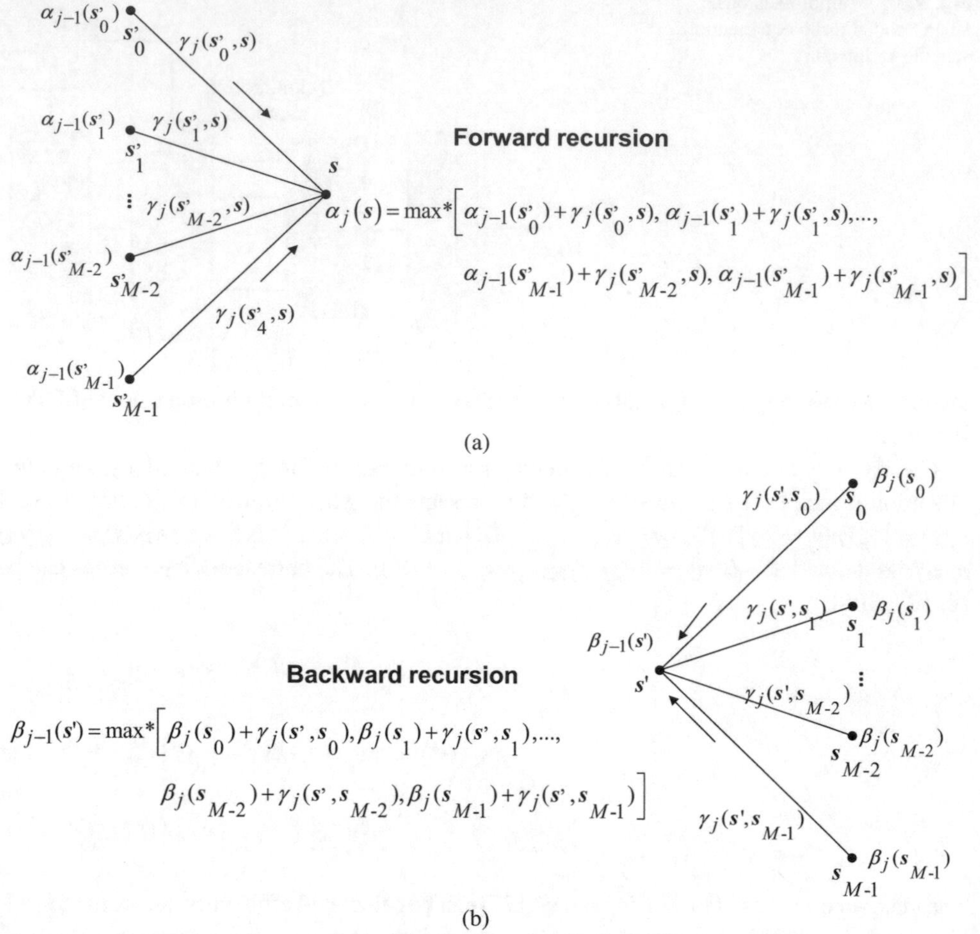

MAP detector for signal constellation size M are illustrated in Fig. 9.63a, b, respectively. In Fig. 9.63a, s denotes an arbitrary terminal state, which has M edges originating from corresponding initial states, denoted as $s'_0, s'_1, s'_2, \ldots, s'_{M-1}$. Notice that the first term in branch metric is calculated only once, before the detection/decoding takes place, and stored. The second term, log $(P(x_j))$, is recalculated in every outer iteration. The forward metric of state s in the j-th step ($j = 1,2,\ldots,n$) is updated by preserving the maximum term (in max*-sense) $\alpha_{j-1}(s'_k) + \gamma_j(s, s'_k)$ ($k = 0,1,\ldots,M-1$). The procedure is repeated for every state in column of terminal states of j-th step. The similar procedure is used to calculate the backward metric of state s', $\beta_{j-1}(s')$ (in $(j-1)$-th step), as shown in Fig. 9.63b, but now proceeding in backward direction ($j = n, n-1, \ldots, 1$).

To improve the overall performance of LDPC-coded turbo equalizer, we perform the iteration of *extrinsic* LLRs between LDPC decoder and multidimensional BCJR equalizer.

To reduce the complexity of sliding-window BCJR detector for large constellation sizes and/or larger memories, the reduced-complexity BCJR algorithm should be used instead. For instance, we do not need to memorize all branch metrics but several largest ones. In forward/backward metrics' update, we need to update only the metrics of those states connected to the edges with dominant branch metrics. Moreover, when $\max^*(x,y) = \max(x,y) + \log[1 + \exp(-|x - y|)]$ operation, required in forward and backward recursion steps, is approximated by $\max(x,y)$ operation, the forward and backward BCJR steps become the forward and backward Viterbi algorithms, respectively. In systems with digital back-propagation [113, 114], since the channel memory is reduced before BCJR detection takes place, we can set m_2 to zero as shown in [115]. Moreover, after digital back-propagation, the statistics of samples is Gaussian-like as discussed earlier in this chapter, and instead of histogram approach to estimate the conditional PDFs, the multivariate Gaussian assumption can be used instead without too much degradation in performance as shown in [115]. Instead of a single BCJR detector operating on whole sequence to be detected, several sliding-window BCJR detectors can be operated in parallel.

9.13.4 Nonlinear Propagation Simulation Study of Turbo Equalization

In this section we provide the propagation simulation study results for nonbinary LDPC-coded turbo equalization applied to rate-adaptive four-dimensional signaling [116]. The transmitter/receiver structures of the 4D nonbinary LDPC-coded modulation (4D-NB-LDPC-CM) are depicted in Fig. 9.64. As introduced before, the source signal entering the transmitter is indeed composed of m parallel binary information streams of length K bits each. These m parallel bit streams are considered as a single nonbinary stream over an alphabet of $Q = 2^m$ symbols. This Q-ary information stream of length K is encapsulated into a codeword of length N symbols by a Q-ary LDPC encoder of code rate $R = K/N$. The mapper maps each Q-ary codeword symbol to a Q-ary 4D constellation point $s = (I_x, Q_x, I_y, Q_y)$, where I_x and Q_x correspond to the in-phase and quadrature components in x-polarization, while I_y and Q_y correspond to those in y-polarization. Rate adaptation is achieved by using software to configure the mapper which then selects the proper 4D modulation format (tuning the number of bits per symbol). The output of the mapper is used to drive the 4D modulator, composed of a distributed feedback (DFB) laser, a polarization beam splitter (PBS), two I/Q modulators, and a polarization beam combiner (PBC). At the Rx side, the received optical signal is first split into two polarizations using the PBS, and the resulting signals are fed to two balanced coherent detectors. Outputs of the coherent detectors are sampled at the symbol rate to obtain the estimates on the coordinates corresponding to the transmitted symbols and then passed to the equalizer employing BCJR algorithm. Compared to the conventional 2D schemes where each polarization branch is equipped with a separate BCJR equalizer block, the corresponding 4D scheme uses a single BCJR equalizer which handles both polarizations simultaneously, and hence, it can compensate for the nonlinear crosstalk between polarizations, whereas the conventional 2D schemes lack such capability. The symbol LLRs at the output of the BCJR equalizer are finally forwarded to a Q-ary LDPC decoder matched to the encoder at the transmitter side. The estimates of the m information sequences sent by the transmitter are then passed to the sink.

The 4D-NB-LDPC-CM scheme is evaluated in the optical transmission system depicted in Fig. 9.64 by using VPI Transmission Maker v8.5, for dispersion map shown in Fig. 9.65 [116], based on the nonlinear propagation model described in Section III. Our fiber model takes into account dispersion effects, ASE noise, and impairments due to nonlinear phase noise, Kerr nonlinearities, and stimulated Raman scattering. To explore the fundamental limits on the transmission distances that could be achieved by the 4D-LDPC-coded modulation schemes (both binary and nonbinary) and compare them against those achievable by its counterpart polarization-division multiplexing 2D-LDPC-coded modulation schemes, perfect CD and PMD compensation are assumed. In order to achieve this in our simulations, we perform inline CD compensation via dispersion-compensating fibers (DCFs) and adopt polarization tracking. As shown in Fig. 9.65, each span in the dispersion map consists

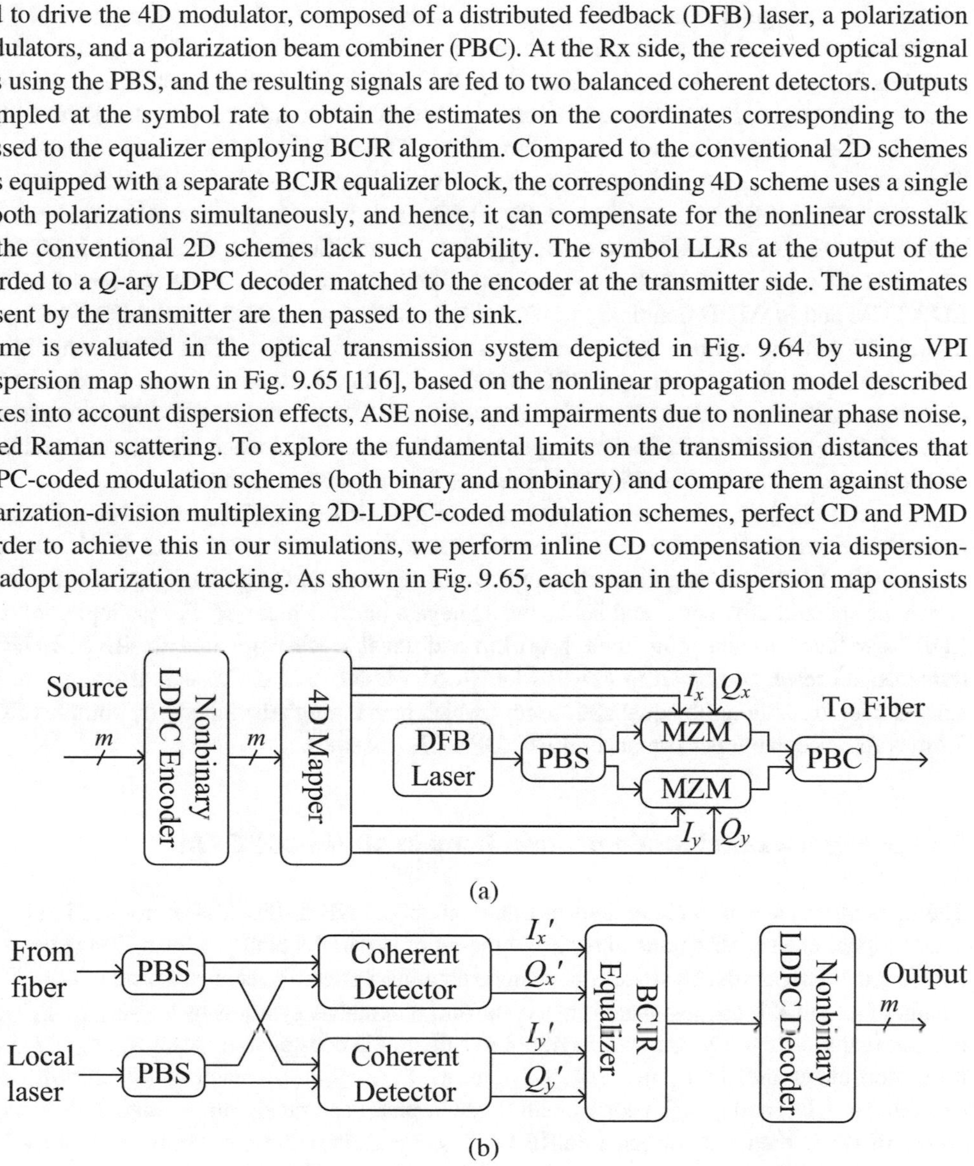

Fig. 9.64 (**a**) Transmitter and (**b**) receiver configurations of 4D-NB-LDPC-CM

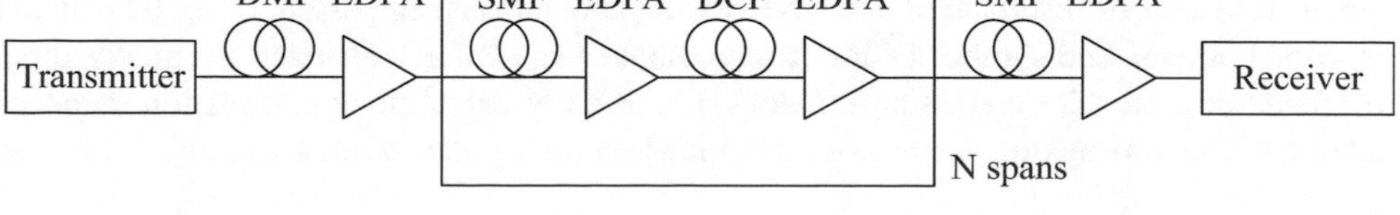

Fig. 9.65 Dispersion map of the 4D-LDPC-CM transmission system under test

of a 40-km-long SMF (dispersion 16 ps/nm·km, dispersion slope 0.08 ps/nm^2·km, effective cross-sectional area 80 μm^2, attenuation 0.22 dB/km) and a 7.11-km-long DCF (dispersion −90 ps/nm·km, dispersion slope − 0.45 ps/nm^2·km, effective cross-sectional area 30 μm^2, attenuation 0.5 dB/km) each followed by an erbium-doped fiber amplifier (EDFA). For both SMF and DCF, we set nonlinear refractive index as 2.6×10^{-20} m^2/W and PMD coefficient 0.1×10^{-12} s/km$^{1/2}$. We also applied pre-compensation of −320 ps/nm and the corresponding post-compensation in our transmission link. The optimal launch power of the system is found to be near −5 dBm. The operating wavelength and the laser linewidth are set to 1552 nm and 100 KHz, respectively. Bandwidths of the optical and electrical filters are set to $3R_s$ and $0.75R_s$, respectively, where R_s is the symbol rate which is defined as the information bit rate $R_i = 40$ Gb/s divided by the FEC code rate R, which is set to $R = 0.8$, corresponding to 25% overhead. For efficient encoding and hardware-friendly decoding, we employ regular and QC-LDPC codes. The software-configured 4D signal constellations are of sizes 16, 32, and 64, denoted as 16-, 32-, and 64-4D, respectively. The 16-4D constellation is described by the set $\{\pm1,\pm1,\pm1,\pm1\}$, which is indeed the set of vertices of a tesseract (i.e., a regular octachoron). Also, each EDFA in Fig. 9.65 has a noise figure of 5 dB.

The coherent receiver is composed of an optical filter, two π-hybrids, eight photodiodes, four electrical filters, and four samplers followed by a BCJR equalizer and LDPC decoder(s). In Fig. 9.66a, we present first the BER against transmission distance performances of 4D block-interleaved LDPC-coded modulation (4D-BI-LDPC-CM) and the corresponding PDM 2D block-interleaved LDPC-coded modulation (PDM-2D-BI-LDPC-CM) schemes under the dispersion map given in Fig. 24. Our comparisons are performed against conventional modulation formats, i.e., 16-4D vs. PDM-QPSK and 64-4D vs. PDM-8-QAM. As shown in Fig. 9.66a, we achieve more than 8000 km with 16-4D at 160 Gb/s aggregate bit rate and 2800 km with 64-4D at 240 Gb/s aggregate bit rate. The corresponding PDM-QPSK and 8-QAM distances are limited to around 6000 km and 2000 km, respectively. Similar observation can be obtained from the comparison between 4D-NB-LDPC-CM and PDM 2D nonbinary LDPC-coded modulation (PDM-2D-NB-LDPC-CM) in Fig. 9.66b. As revealed by both figures, 4D scheme exhibits solid advantage over the corresponding PDM-2D scheme. Specifically, such advantage is even more significant when nonbinary LDPC-coded modulation is adopted. To put it numerically, 4D-NB-LDPC-CM maintains more than 2000 km additional transmission reach over PDM-2D-NB-LDPC-CM during transmission at aggregate bit rates of both 160 Gb/s and 240 Gb/s. Such an improvement in transmission reach is just a well-grounded justification why 4D schemes should be adopted rather than PDM-2D schemes for long-haul transmission.

In order to further manifest the superiority of 4D-NB-LDPC-CM over 4D-BI-LDPC-CM, we performed comparisons between the two for three different 4D modulation formats, i.e., 16-, 32-, and 64-4D. Our results are presented in Fig. 9.67. The 4D-NB-LDPC-CM significantly outperforms 4D-BI-LDPC-CM especially as the underlying constellation size, and hence the spectral efficiency, and hence the aggregate bit rates increase. For example, for transmission at 240 Gb/s, 4D-NB-LDPC-CM can provide more than 1500 km additional reach compared to 4D-BI-LDPC-CM. In addition to improving transmission reach compared to 4D-BI-LDPC-CM, 4D-NB-LDPC-CM can also improve the latency at the receiving ends since it avoids costly turbo equalization steps which pass through the inherently complex BCJR equalization block about 3 to 5 times for acceptable performance in BI-LDPC-CM schemes.

9.13.5 Experimental Study of Time-Domain 4D-NB-LDPC-CM

The transmitter/receiver architectures of time-domain 4D-NB-LDPC-CM, introduced in [117], are depicted in Fig. 9.68. The source signal entering the transmitter is composed of m parallel binary information streams of length K bits each. Again, we consider these m parallel bit streams as a single nonbinary stream over an alphabet of $Q = 2^m$ symbols. This Q-ary information stream of length K is encapsulated into a codeword of length N symbols by a Q-ary LDPC encoder of code rate $R = K/N$. The mapper maps each Q-ary codeword symbol to a Q-ary 4D constellation point $(I_{x,t}, Q_{x,t}, I_{x,t+1}, Q_{x,t+1})$. We then serialize the four coordinates into two pairs, $(I_{x,t}, Q_{x,t})$ and $(I_{x,t+1}, Q_{x,t+1})$, each of which will occupy half symbol duration. The coordinates I_x (I_y) and Q_x (Q_y) correspond to the in-phase and quadrature components in x-polarization (y-polarization). The output of P/S is then used to generate RF I and Q signals in both time instances t and $t + 1$ to drive the I/Q modulator.

A typical coherent receiver and standard offline DSP algorithm were used for recovering the original transmitted signals, I'_x, Q'_x, I'_y, and Q'_y. Corresponding to the transmitter, each continuous two pairs of (I'_x, Q'_x) is grouped as $(I'_{x,t}, Q'_{x,t}, I'_{x,t+1}, Q'_{x,t+1})$, which represents the estimate of a received 4D symbol and is then passed to the 4D equalizer employing BCJR algorithm, as described above. The symbol LLRs at the output of the BCJR equalizer are finally forwarded to a Q-ary LDPC decoder matched to the encoder at the transmitter side. The estimates of the m information sequences sent by the transmitter are then obtained. The information sequences transmitted on the y polarization are recovered in a similar fashion.

Fig. 9.66 BER vs. transmission distance comparisons between (**a**) 4D-BI-LDPC-CM and PDM-2D-BI-LDPC-CM and (**b**) 4D-NB-LDPC-CM and PDM-2D-NB-LDPC-CM

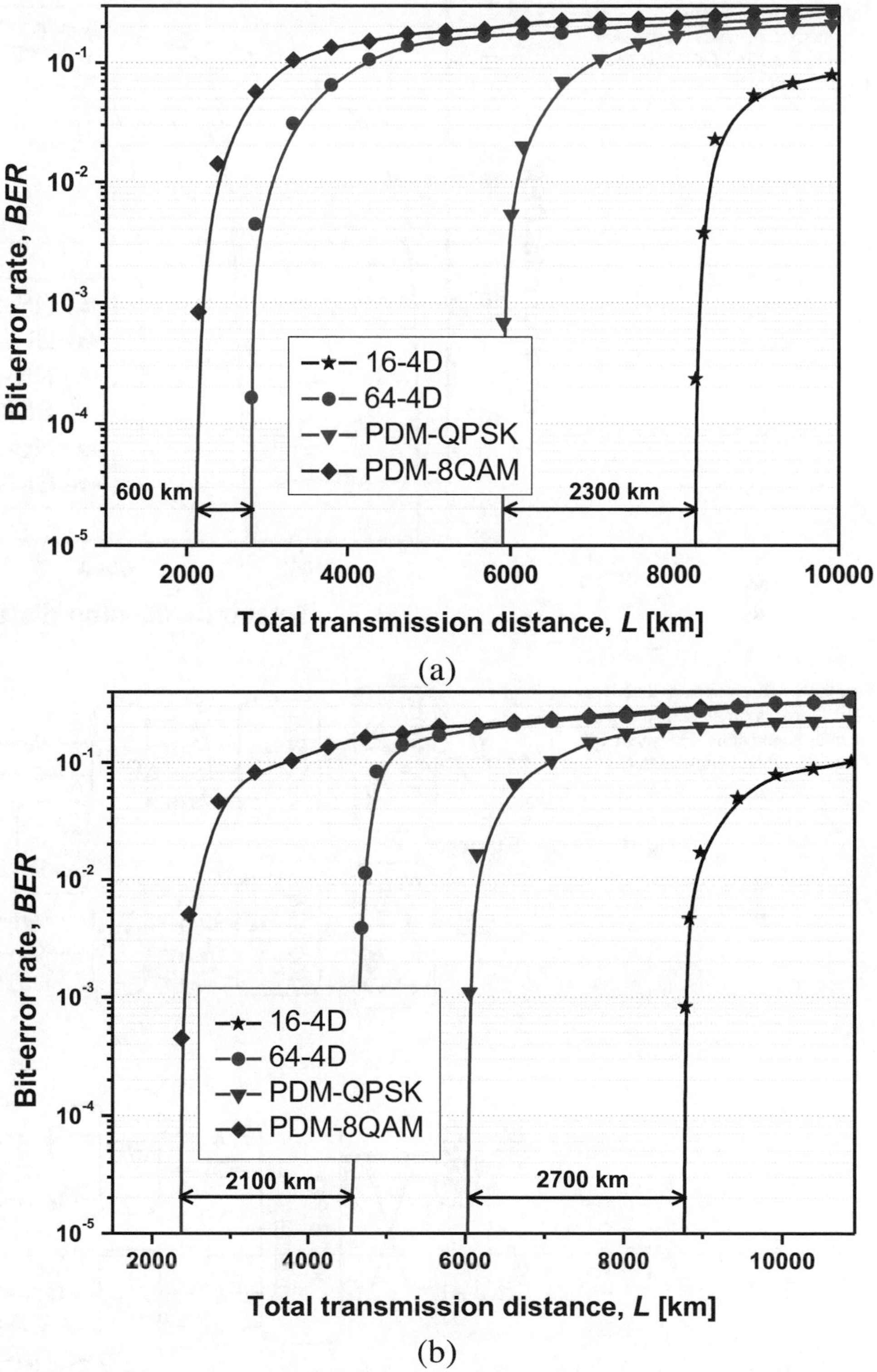

The counterpart of the time-domain 4D-LDPC-CM scheme is PDM nonbinary LDPC-coded modulation, which employs the conventional 2D QAM. The PDM-QPSK system operates using two 2D equalizers one for each time interval under consideration. On the other hand, the 16-4D system operates with one 4D equalizer. Both 2D and 4D equalizers operate on trellis description of the optical channel with memory. Therefore, the 4D equalization is able to compensate for the intrachannel nonlinear effects, which expand over the time domain.

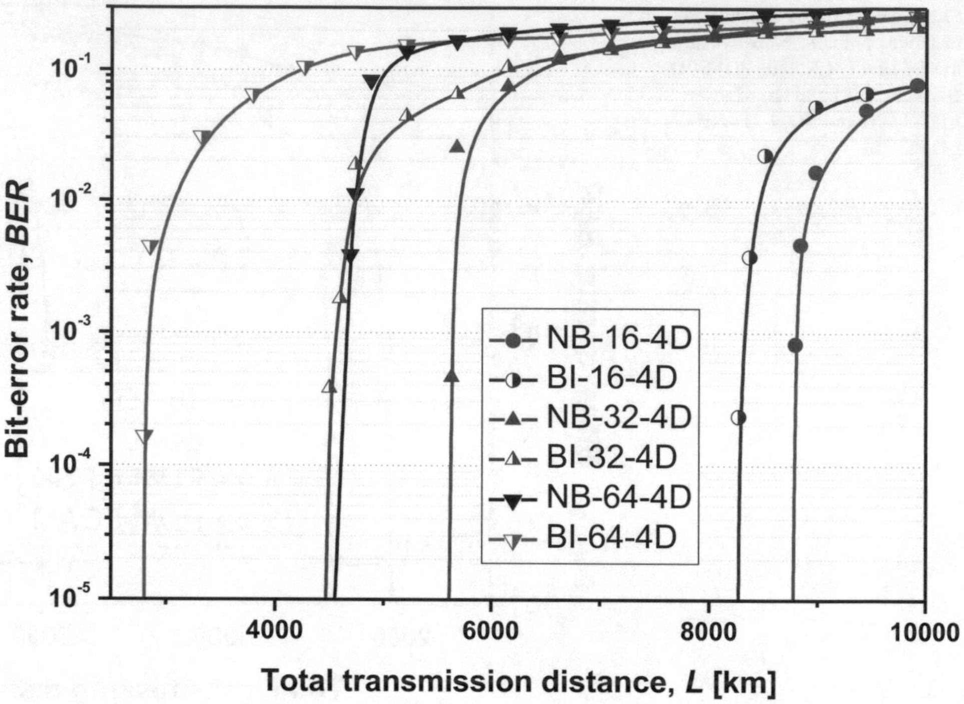

Fig. 9.67 BER against transmission distance comparison between 4D-NB-LDPC-CM and 4D-BI-LDPC-CM

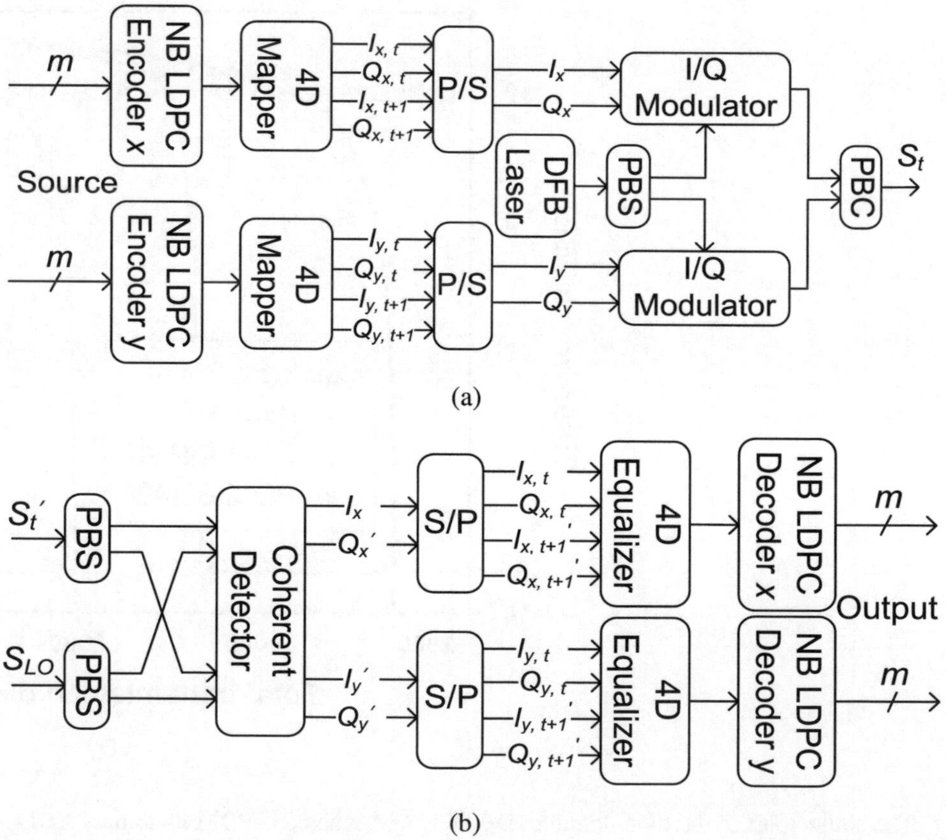

Fig. 9.68 Time-domain 4D-NB-LDPC-CM: (**a**) transmitter and (**b**) receiver architecture. *P/S* parallel-to-serial, *S/P* serial-to-parallel

Figure 9.69 shows the experiment setup of the eight-channel DWDM optical transmission system at 50 GHz channel spacing as well as the optical spectrum at the transmitter and receiver sides. The eight 100-kHz line-width external cavity lasers with $\lambda_1 = 1550.12$ nm and $\lambda_8 = 1550.40$ nm are divided into odd and even groups, each of which is modulated with 16 Gbaud 4D signals. The 4D signals are generated by digital-to-analog converters (DACs) using coordinates that are produced according to the time-domain 4D-LDPC-CM described before.

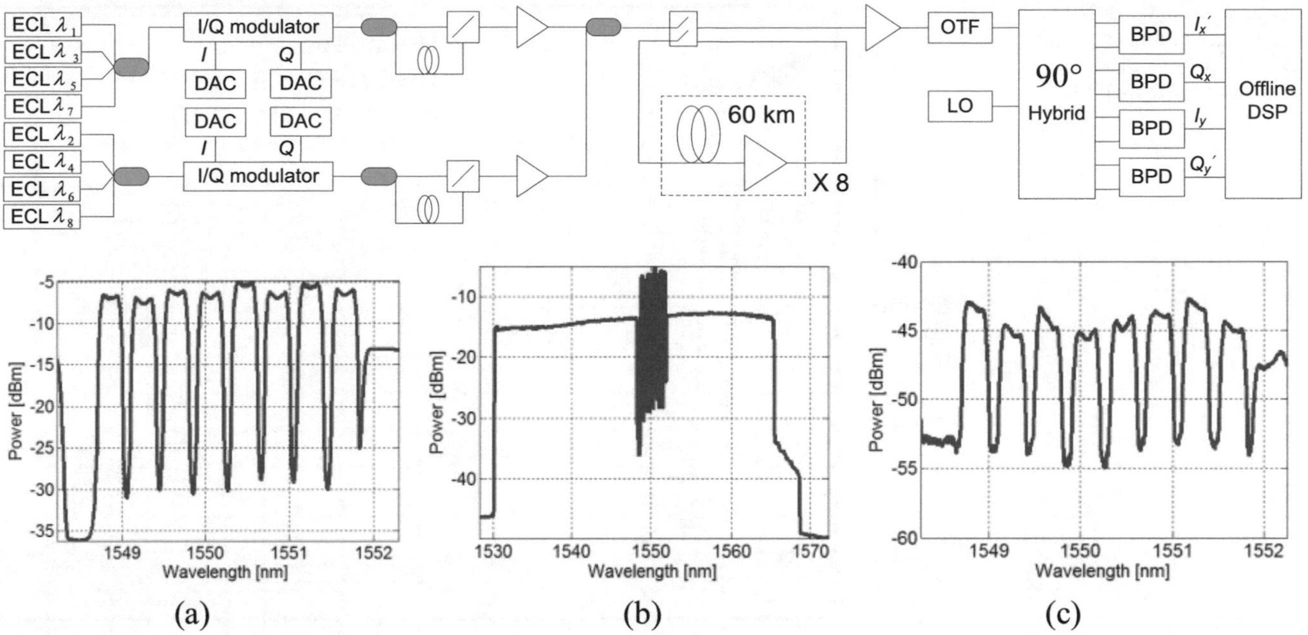

Fig. 9.69 Experimental setup (top figure) and optical spectra (at 0.1 nm RBW): (**a**) Tx output WDM signal, (**b**) Tx output signal in C-band, (**c**) WDM signal after 7680 km. *ECL* external cavity laser, *DAC* digital-to-analog converter, *OTF* optical tunable filter, *LO* local oscillator laser, *BPD* balanced photodetector

A de-correlated copy of signal is combined with its original signal for emulating PDM signals. The odd and even channels are then combined and launched into the fiber recirculating loop that consists of 8 × 60-km spans of DMF with an inline EDFA after each span. Signal loading and circulation are controlled by a mechanical transmitter switch and loop switch. At the receiver side, the desired channel is extracted by an optical tunable filter and down-converted to electrical baseband domain by combining it with a local laser centered at the proper frequency using a polarization-diversity 90° optical hybrid followed by balanced photodetectors. The equalized symbols are forwarded to the 4D equalizer to compensate the intrachannel nonlinearities and later decoded by the nonbinary LDPC decoder. In this experiment, the 4D modulation format of constellation size 16 (16-4D) and 16-ary (16935, 13550) LDPC code achieve $16 \times 10^9 \times 4 \times 2 \times 0.8 = 102.4$ Gb/s per channel and 2.05 b/s/Hz. The adopted decoding algorithm is MD-FFT-SPA [82]. For comparison, we also transmit the counterpart 32-Gbaud PDM-QPSK signals with NB-LDPC-CM, which adopts conventional 2D equalizer and 4-ary (16935, 13550) LDPC code at the same data rate and spectral efficiency.

Figure 9.70 shows the back-to-back BER performance of time-domain 4D-LDPC-CM of 16-4D and NB-LDPC-CM of PDM QPSK, in which the former is ~0.4 dB better, at *BER* of 10^{-4}, than the latter in terms of OSNR due to the adoption of 16-ary LDPC code instead of 4-ary. The result confirms that nonbinary LDPC codes in a larger Galois Field exhibit stronger error correction capability.

To further illustrate the advantage of time-domain 4D-LDPC-CM in compensating intrachannel nonlinear effects, we present the transmission performance in terms of Q-factor (derived from BER) vs. transmission distance in Fig. 9.71. We show Q-factor values of the received 16-4D and PDM-QPSK signals before and after the corresponding equalization. For PDM-QPSK with NB-LDPC-CM, the 2D equalizer does not help improve Q-factor, while for 16-4D with 4D-LDPC-CM, Q-factor is improved by ~0.8 dB after 4D equalization as the intrachannel nonlinear effects are compensated by the 4D equalizer. Namely, because 4D equalizer operates in 4D space, with basis functions being two neighboring symbols and in-phase and quadrature channels, it can compensate for the nonlinear interaction between neighboring symbols. In this work, we collect the samples of 4×10^6 symbols after every six spans. No countable error is found after transmitting 7680 km at the output of LDPC decoder for the 4D scheme. In contrast, the NB-LDPC-CM with PDM QPSK can only reach 6240 km. Therefore, the optical transmission reach has been extended by 1440 km by employing the time-domain 4D-LDPC-CM.

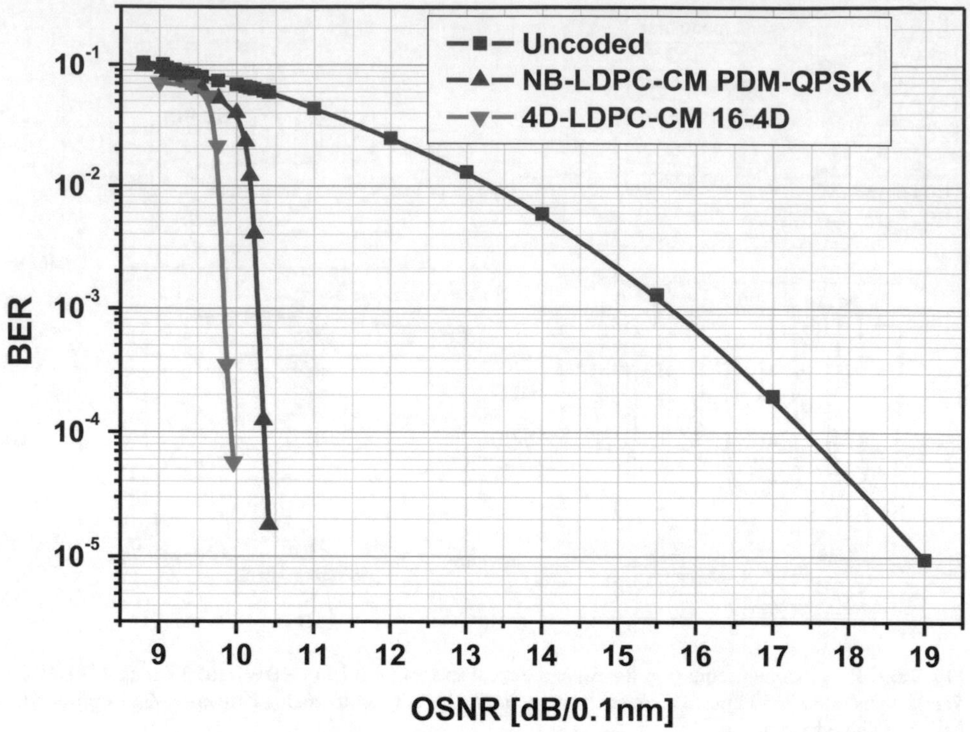

Fig. 9.70 BER vs. OSNR performance of time-domain 4D-LDPC-CM of 16-4D and NB-LDPC-CM of PDM-QPSK

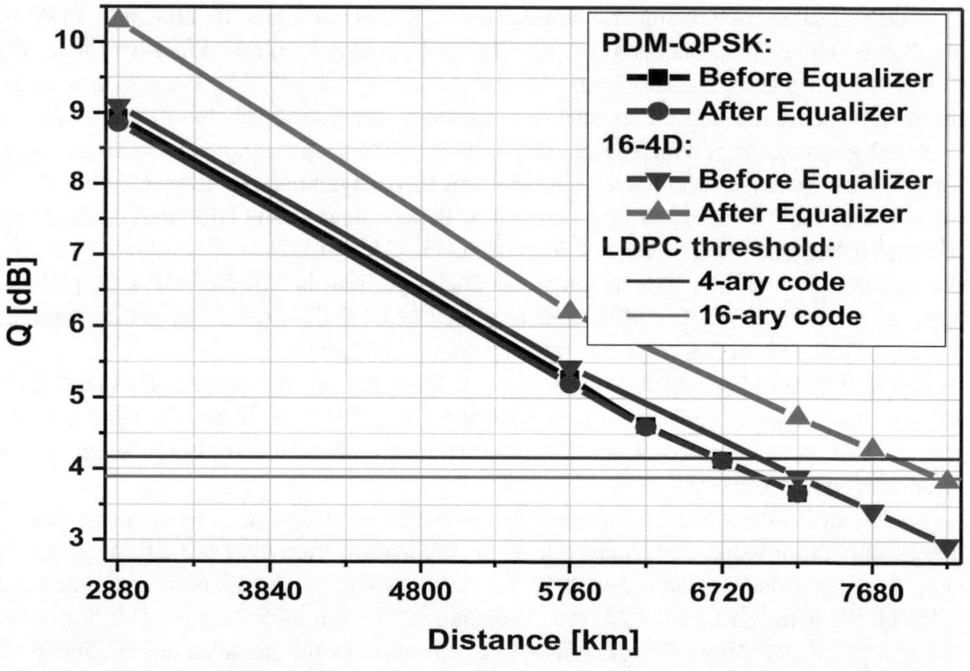

Fig. 9.71 Q-factor vs. distance performance of time-domain 16-4D and PDM-QPSK before and after BCJR equalization

9.13.6 Quasi-Single-Mode Transmission Over Transoceanic Distances Using Few-Mode Fibers

High-speed optical systems rely solely on single-mode fibers for transmission beyond a few kilometers reach, and state-of-the art single-mode fibers are required to achieve spectral efficiencies beyond 6 b/s/Hz over transoceanic distances. Recently, there has been a renewed interest in transmission over FMFs to increase the capacity over a single fiber strand by the multitude of the spatial modes. However, such an approach requires a radical transformation of optical communication infrastructure which has been built on the premise of transmission over a single spatial mode. It was proposed that FMFs can be integrated into the otherwise single-mode infrastructure by transmitting only in the fundamental mode, while taking advantage of their significantly larger effective area. The feasibility of using FMFs was demonstrated in a recent transmission experiment over 2600 km. In our article [118], 101.6-km-long hybrid spans consisting of low-loss silica-core FMFs and single-mode fibers are used to achieve a new record spectral efficiency of 6.5 b/s/Hz in unidirectional transmission configuration over the transoceanic distance of 6600 km using EDFAs only. To the best of our knowledge, this is the first time FMFs are used to outperform state-of-the-art single-mode fibers in a transoceanic transmission experiment. To make the FMFs compatible with single-mode transmission, single-mode fiber jumpers are spliced to each end. This excites only the fundamental mode of the FMF at launch and also strips any high-order modes excited during transmission at the exit splice. Such systems may suffer from multipath interference (MPI) even with inherently low-mode-coupling FMFs. In effect, by using FMFs the nonlinear fiber impairments are traded for linear impairments from MPI which are easier to manage. While the improvements possible from mitigating nonlinear impairments by DSP are fundamentally limited in scope and computationally costly, it was demonstrated recently that most of the penalty from the MPI can be mitigated with much less computational complexity by including an additional decision-directed least mean-square (DD-LMS) equalizer to the standard digital coherent DSP. The differential modal group delay (DMGD) of a FMF can be of the order of 1 ns/km, and the required number of taps in the DD-LMS equalizer may be several hundreds to thousands, which not only makes them less stable and computationally demanding but also degrades its performance. In this work, instead of using standard single-carrier modulation (SCM), a digital multi-subcarrier modulation (MSCM) signal with 32 subcarriers at 1 Gbaud is used to reduce the number of equalizer taps per subcarrier by more than an order of magnitude while increasing the temporal extent of MPI that can be mitigated. To further reduce MPI, a hybrid FMF-single-mode fiber span design is used for the first time with the FMF at the beginning of the span to reduce nonlinearity and the single-mode fiber at the end to avoid excessive MPI. Hybrid spans also reduce the DSP complexity as they have smaller modal group delay per span compared to all FMF spans.

The experimental setup is shown in Fig. 9.72. At the transmitter side, digital-to-analog converters (DACs) with a sampling rate of 64 GHz are used to generate WDM channels tuned to a 33 GHz grid. WDM channels are prepared in two groups. The first group consists of six neighboring channels generated with tunable ECLs. These six channels are tuned together across the C-band, and only the center channel which uses a narrow linewidth (<1 kHz) laser is measured. Each of the even and odd subgroups of the six tunable channels is modulated separately by four independent streams of data for all in-phase and quadrature (I/Q) rails in both polarizations. The 111 loading channels are modulated with only independent I/Q rails followed by polarization multiplexing emulators. The tunable channels and the dummy channels are combined using a wavelength selective switch (WSS) with a 1-GHz grid resolution and a coupler. Using the DACs, either a 32 Gbaud, Nyquist-shaped SCM 16-QAM, or a MSCM signal of 32 × 1Gbaud 16-QAM subcarriers with 10 MHz guard-band is generated per wavelength. The 2^{18}–1 PRBS binary data are encoded by 16-ary irregular quasi-cyclic nonbinary LDPC encoder of girth 10 to generate codewords at three overheads 14.3%, 20%, and 25%, to maximize the capacity across C-band. The codeword length is 13,680 for 14.3% and 20% OHs and 74,985 for 25% OH. Encoder output is uniformly distributed into 32 subcarriers of MSCM signal.

At the receiver side, the WDM channel under test is filtered and captured with a standard offline coherent receiver, using a narrow-linewidth (<1 kHz) laser as the local oscillator and a real-time 80 GSa/s sampling scope. In the case of a SCM signal, after resampling and chromatic dispersion compensation (CDC), the signal is fed into a multi-modulus algorithm for initial convergence of polarization demultiplexing followed by carrier phase recovery. Subsequently a second stage of T-spaced equalizer using DD-LMS is applied to the 32Gbaud SCM signal to mitigate MPI. In the case of MSCM, after frequency offset estimation and CDC, each subcarrier is digitally filtered and processed individually in the same way as the single-carrier signal. The data of all subcarriers is combined and is further improved by the BCJR equalization and then sent to nonbinary LDPC decoder for detecting post-FEC bit errors.

The circulating loop test bed is composed of five hybrid spans obtained by splicing 51.3-km-long FMFs with 50.3-km-long Vascade® EX3000 fiber with average attenuations of 0.157 dB/km and 0.153 dB/km, respectively. The FMFs support the LP11 mode albeit with a slightly higher attenuation, and the differential modal delay is estimated to be 0.95 ns/km with the LP01 lagging. The average effective areas of the single-mode fiber and the FMF portions are 151 μm^2 and an estimated 200 μm^2, respectively. A short piece of Vascade® EX3000 is used as a bridge fiber between the input end of the FMF and the standard-single-mode fiber pigtail of the EDFA output. The average span loss is 16.48 dB corresponding to an additional 0.7 dB incurred by the four splices and the two connectors. The mid-band chromatic dispersion is approximately 21 ps/nm/km. Span loss is compensated by C-band EDFAs, and the accumulated amplifier gain tilt is compensated by a WSS each loop.

The Q-factor is measured across the C-band at 6600 km for a total of 111 WDM channels, from 1534.5 nm to 1563.6 nm. Q-factors averaged over 32 carriers, both polarizations, and over 10 separate measurements are plotted in Fig. 9.73 along with the received spectrum. The Q-factor varies from 4.5 dB to 5.7 dB from the shorter to the longer wavelengths. To maximize the

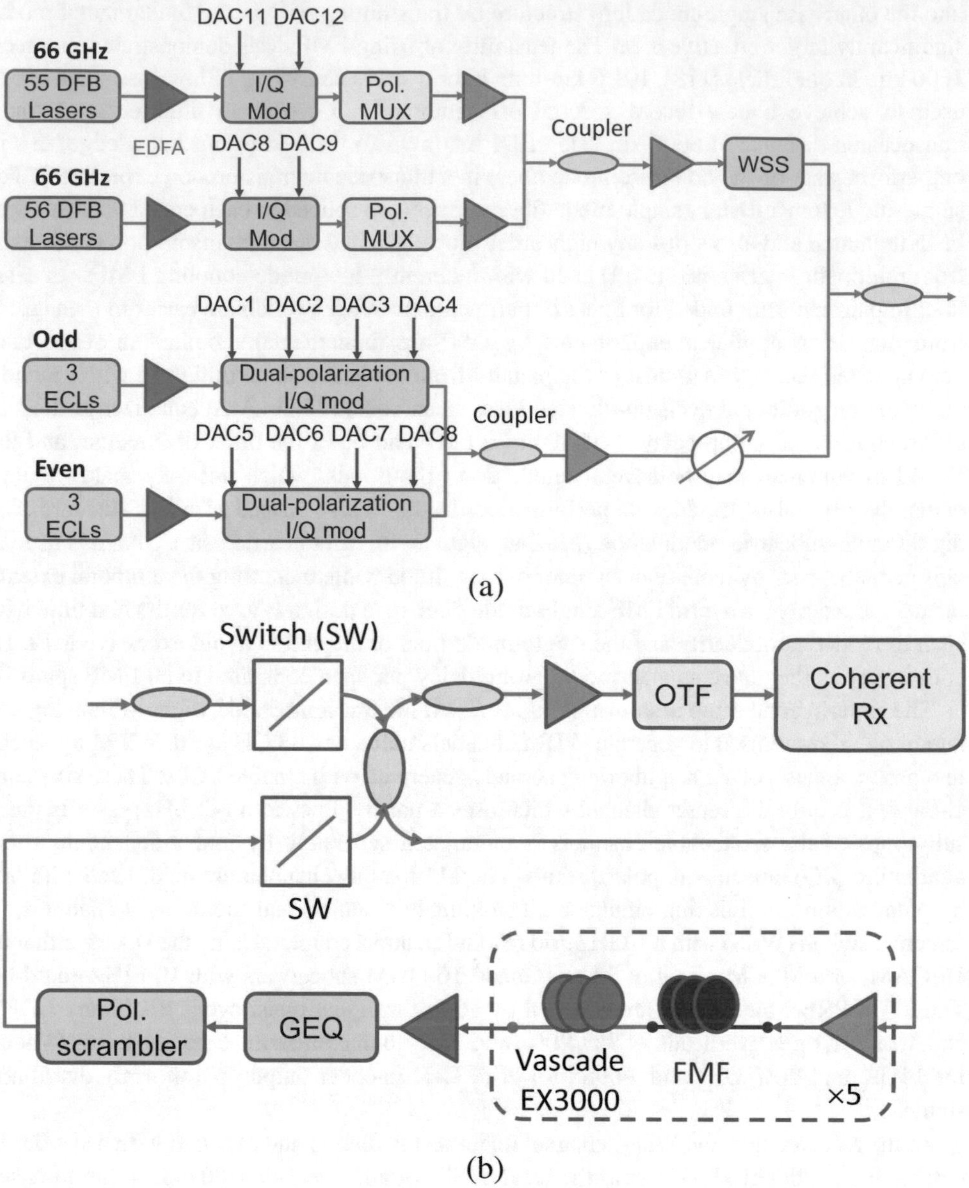

Fig. 9.72 System setup: (**a**) transmitter with 6 tunable channels and 111 loading channels, on a 33 GHz grid; (**b**) recirculating loop with 5 × 101.6-km-long hybrid spans of FMF and Vascade EX3000

capacity with such a variation of Q-factor, three different FEC OHs at 14.3%, 20%, and 25% are used for 28, 64, and 19 WDM channels across the C-band, as shown in Fig. 9.73, resulting in an average spectral efficiency of 6.5 b/s/Hz. For each channel, more than 25 million bits were processed for LDPC decoding, and all the channels were decoded error-free in the LDPC decoder assisted with BCJR equalizer. The coding performance is further evaluated in the BTB configuration as a function of pre-FEC Q-factor derived from bit error counting as shown in Fig. 9.74. To achieve sufficient confidence level, over 100 million bits are considered for each BER point, and exponential fitting is used for extrapolating the measured BER data and thus estimating the FEC limit as 5.5 dB, 4.9 dB, and 4.35 dB Q-factor for 14.3%, 20%, and 25% OH, respectively.

To make direct performance comparisons, we next constructed three comparable loops from either all single-mode fiber, all FMF, or hybrid spans. Four of the hybrid spans are broken up and re-spliced to obtain either two spans of all-single-mode fiber spans or all-FMF spans with almost identical span lengths and losses. For the hybrid case, three spans are removed from the loop. Figure 9.75 shows the comparison after 10 and 20 circulations using the same MSCM signals and the same receiver DSP with 101-tap-long CMA-based filer. Because the FMFs have larger effective area, the optimum OSNR is larger by about 1 dB compared to the all-SMF case. However, some of this improvement is lost because the impact of MPI cannot be removed altogether which is more evident in portions of the curves dominated by the linear noise, especially when all-FMF spans are used. This is most clear at 4060 km in the low OSNR regime. However, for the hybrid span case, the nonlinear improvement from the large effective area more than compensates for the penalty due to residual MPI affording 0.4 dB improvement over the all-single-mode fiber configuration. For additional details on this experiment, an interested reader is referred to [118].

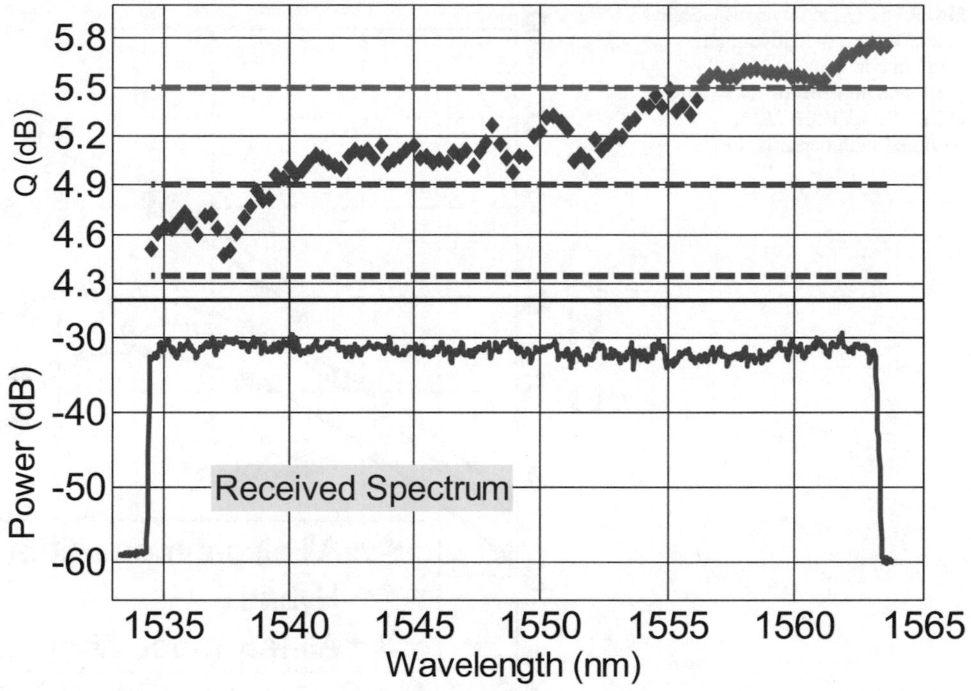

Fig. 9.73 (Top) Q-factor vs. wavelengths with 14.3% (green), 20% (blue), and 25% (red) FEC OH after 6600 km. (Bottom) Received spectrum. (After ref. [118]; © IEEE 2015; reprinted with permission)

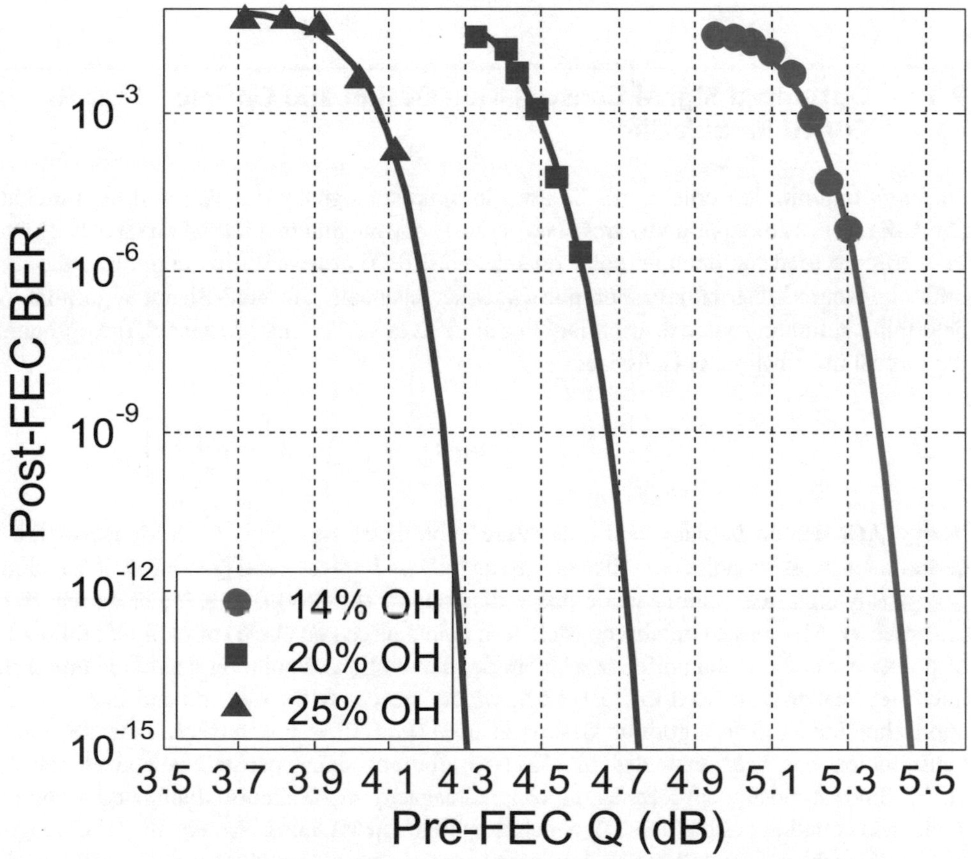

Fig. 9.74 Back-to-back FEC coding performance. (After ref. [118]; © IEEE 2015; reprinted with permission)

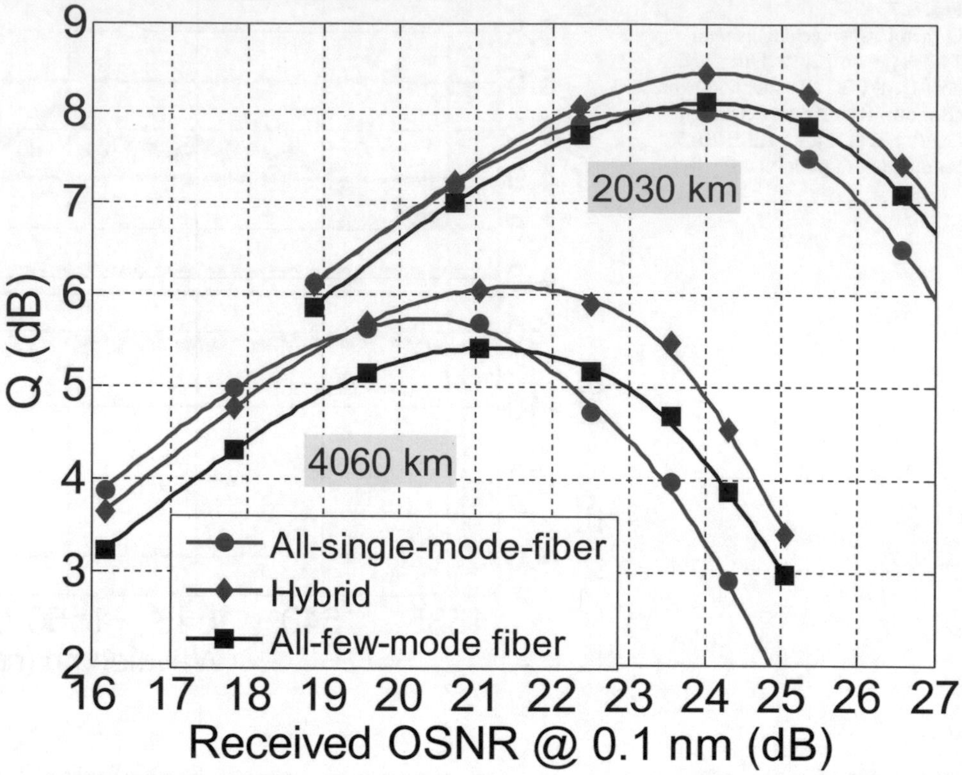

Fig. 9.75 Q-factor versus OSNR at 2030 km and 4060 km, all-single-mode fiber (blue), hybrid (red), and all FMF (black). (After ref. [118]; © IEEE 2015; reprinted with permission)

9.14 Optimized Signal Constellation Design and Optimized Bit-to-Symbol Mapping-Based Coded Modulation

Different optimization criteria can be used in optimum signal constellation design including minimum mean-square error (MMSE) [119, 120], minimum symbol error rate [121], maximum information rate [122], and maximum LLR [123]. Recently, in [120], we used the iterative polar quantization (IPQ) procedure that minimizes the mean-squared quantization error of optimum source distribution. For non-Gaussian channels, Arimoto-Blahut algorithm (ABA) [124] should be used to determine optimum source distribution. For an ASE noise-dominated channel, the optimum information source, maximizing the mutual information, is Gaussian:

$$p(I,Q)=\frac{1}{2\pi\sigma^2}\exp\left[-\frac{I^2+Q^2}{2\sigma_s^2}\right], \tag{9.181}$$

where (I,Q) denote in-phase and quadrature coordinates and σ_s^2 is the variance of the source. By expressing Cartesian coordinates by using polar coordinates (r,θ) as follows $I=r\cos\theta$ and $Q=r\sin\theta$, it is straightforward to show that radius will follow Rayleigh distribution, while phase distribution is uniform. This indicates that the optimum distribution, in mutual information (MI) sense, contains constellation points placed on circles of radii that follow Rayleigh distribution. The number of points per circle is nonuniform, wherein the distribution of point on a circle is uniform. For corresponding algorithm an interested reader is referred to [119, 120], which is accurate for medium and large signal constellation sizes. Namely, the approximation used in algorithm described in [119, 120] is not very accurate for small constellation sizes. This signal constellation has been extended in [122] to arbitrary signal constellation sizes and to multidimensional signal space [125]. Since it closely approaches the channel capacity in ASE noise-dominated scenario, it has been called the optimum signal constellation design (OSCD), which is optimum in MI sense. Namely, the OSCD algorithm is formulated on such a way to maximize the MI for ASE noise-dominated scenario, which justifies the name introduced. The OSCD has been extended in [126, 127] to the phase noise-dominated channels. The optimum mapping rule design for OSCDs is the subject of ref. [125]. Finally, in [128] the vector quantization concepts have been used to large signal constellation design for multidimensional signaling.

9.14.1 Multidimensional Optimized Signal Constellation Design

Given the space limitations, here we briefly describe the OSCD for multidimensional signaling, the optimum mapping rule determination, as well as the EXIT chart analysis [125] of OSCDs. In Sect. 9.14.3, we describe the nonlinear optimized signal constellation design (NL-OSCD), suitable for SPM-dominated channels. After the initialization stage by signal constellation obtained by sphere-packing method, we generate the multidimensional training sequences from the optimum source distribution and split them into the clusters of points according to the Euclidean distance squared from constellation obtained in previous iteration. New constellation points are obtained as the center of mass of such obtained clusters. This procedure is repeated until convergence or until a predetermined number of iterations have been reached.

As an illustration in Fig. 9.76, we provide the 8-ary and 16-ary 2D signal constellations obtained by using the OSCD algorithm for ASE noise-dominated scenario as well as the optimized mapping rules. Corresponding 3D constellations and mapping rules are provided in Fig. 9.78. The 3D constellations are relevant in SDM-based applications.

9.14.2 EXIT Chart Analysis of OSCD Mapping Rules

For the convergence behavior analysis, the EXIT chart analysis should be performed [101]. In order to determine the MI transfer characteristics of the demapper, we model the a priori input LLR, $L_{M,a}$, as a conditional Gaussian random variable. The MI between c (c denotes the initial bits) and $L_{M,a}$ is determined numerically as explained in [101]. Similarly, the MI between c and demapper extrinsic LLR, denoted as $L_{M,e}$, is also calculated numerically, but with the PDF of c and $L_{M,e}$ determined from histogram obtained by Monte Carlo simulation. By observing the MI of demapper extrinsic likelihoods $I_{LM,e}$ as a function of the MI of the prior LLRs $L_{M,a}$, denoted as $I_{LM,a}$, and OSNR, in dB, the demapper EXIT characteristic (denoted as T_M) is given by

$$I_{L_{M,e}} = T_M\left(I_{L_{M,a}}, OSNR\right). \tag{9.182}$$

The EXIT characteristic of LDPC decoder (denoted as T_D) is defined in a similar fashion as

$$I_{L_{D,e}} = T_D\left(I_{L_{D,a}}\right), \tag{9.183}$$

where we use $I_{LD,e}$ to denote MI of decoder extrinsic likelihoods and $I_{LD,a}$ to denote the MI of decoder the prior LLRs.

The EXIT chart analysis of optimized mapping, optimized as described in [125] (and references therein), for 2D 8-ary OSCD is shown in Fig. 9.77a. Namely, the optimization criterion for these mapping rules can be summarized as follows [125]: find the mapping rule in such a way that every two symbols that differ in only one bit position has larger distance than the minimum Euclidean distance of the constellation. The channel OSNR (for symbol rate of 25 GS/s) equals 5.0 dB, and we use a QC-LDPC (16935, 13550) decoder. It is obvious to see that in this case only 8-ary OSCD constellation can operate with the turbo decoding process. Other mapping rules, for this OSNR, will result in turbo decoding errors. In Fig. 9.78a, we can see the tradeoff performance for MI in optimized mapping. The MI at the output of the demapper increases over iterations when using the optimized mapping. The corresponding EXIT chart for 8-ary 3D OSCD is shown in Fig. 9.78b.

The optimized mappings for OSCDs are studied for use in adaptive polarization-division multiplexed/mode-division multiplexed LDPC-coded modulation schemes to enable both 400GbE and 1 TbE. The channel symbol rate was set to 31.25 GS/s, and QC-LDPC (16935, 13550) code of girth 8 and column weight 3 was used in simulations. All results are obtained for 20 LDPC decoder (inner) iterations and 3 APP demapper-LDPC decoder (outer) iterations and summarized in Fig. 9.79. Clearly, the OSCD constellations outperform QAM constellations in both 16-ary and 8-ary cases. Also, the optimized mapping rules outperform randomly selected mapping for OSCDs, and the gap, at *BER* of 10^{-8}, is almost 0.5 dB in the 8-ary case and 0.6 dB in the 16-ary case. Note that we used natural mapping for 8-QAM and 16-QAM constellations.

Fig. 9.76 The optimum signal constellations and mapping rules for (**a**) 8-ary and (**b**) 16-ary 2D signal space

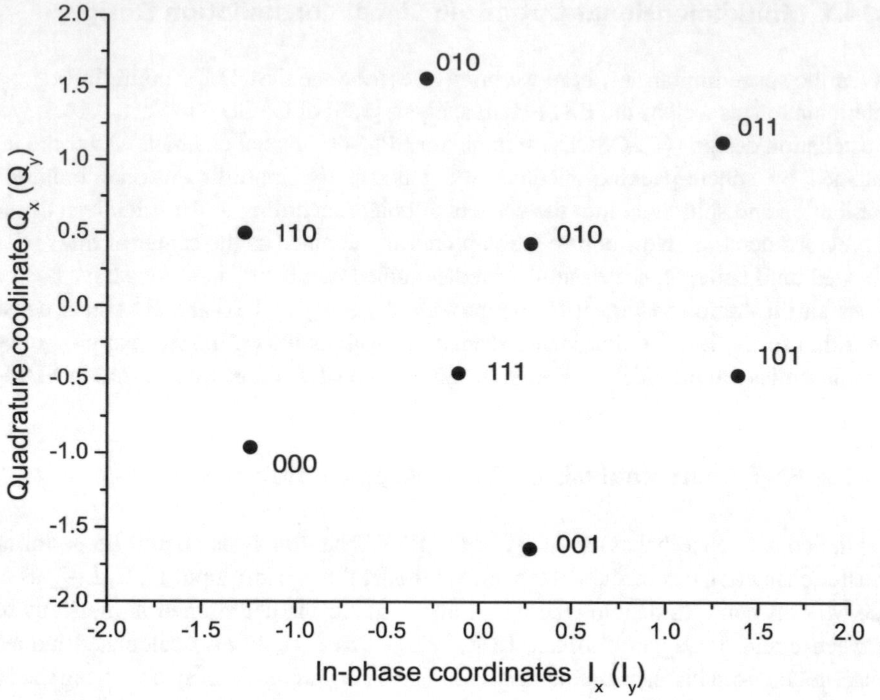

(a)

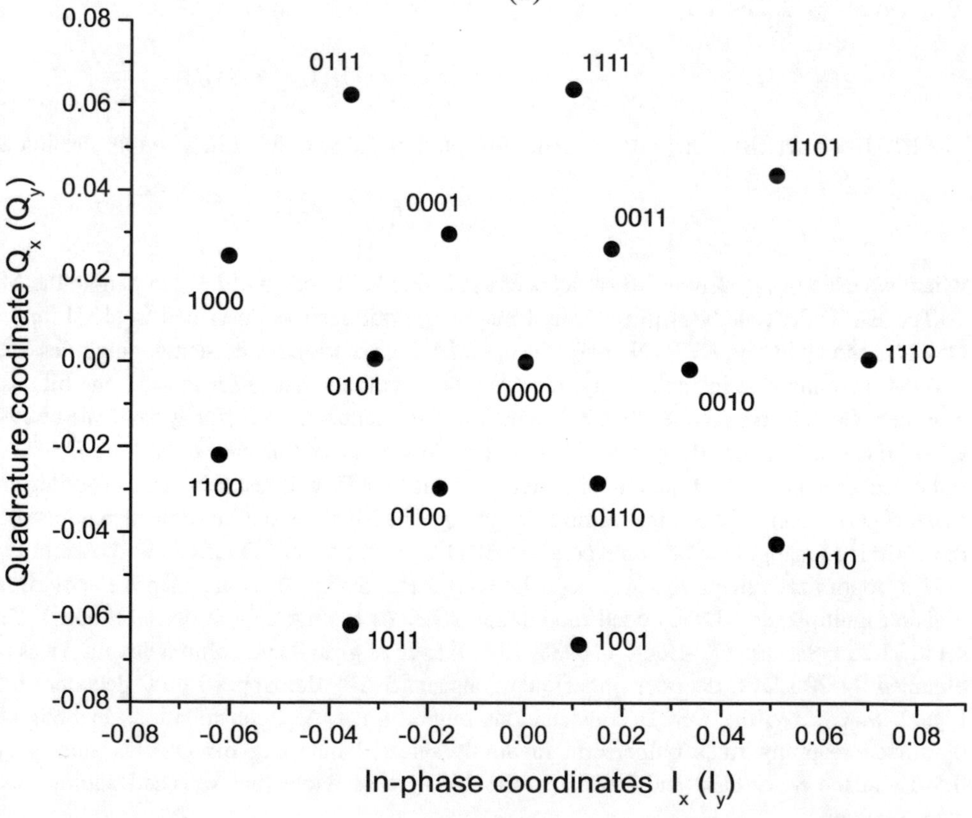

(b)

Fig. 9.77 3D (**a**) 8-ary and (**b**) 16-ary optimum signal constellations and mapping rules. *X*, *Y*, *Z* are used to denote three either electrical (Slepian sequences) or optical (spatial modes) basis functions to be used in SDM applications

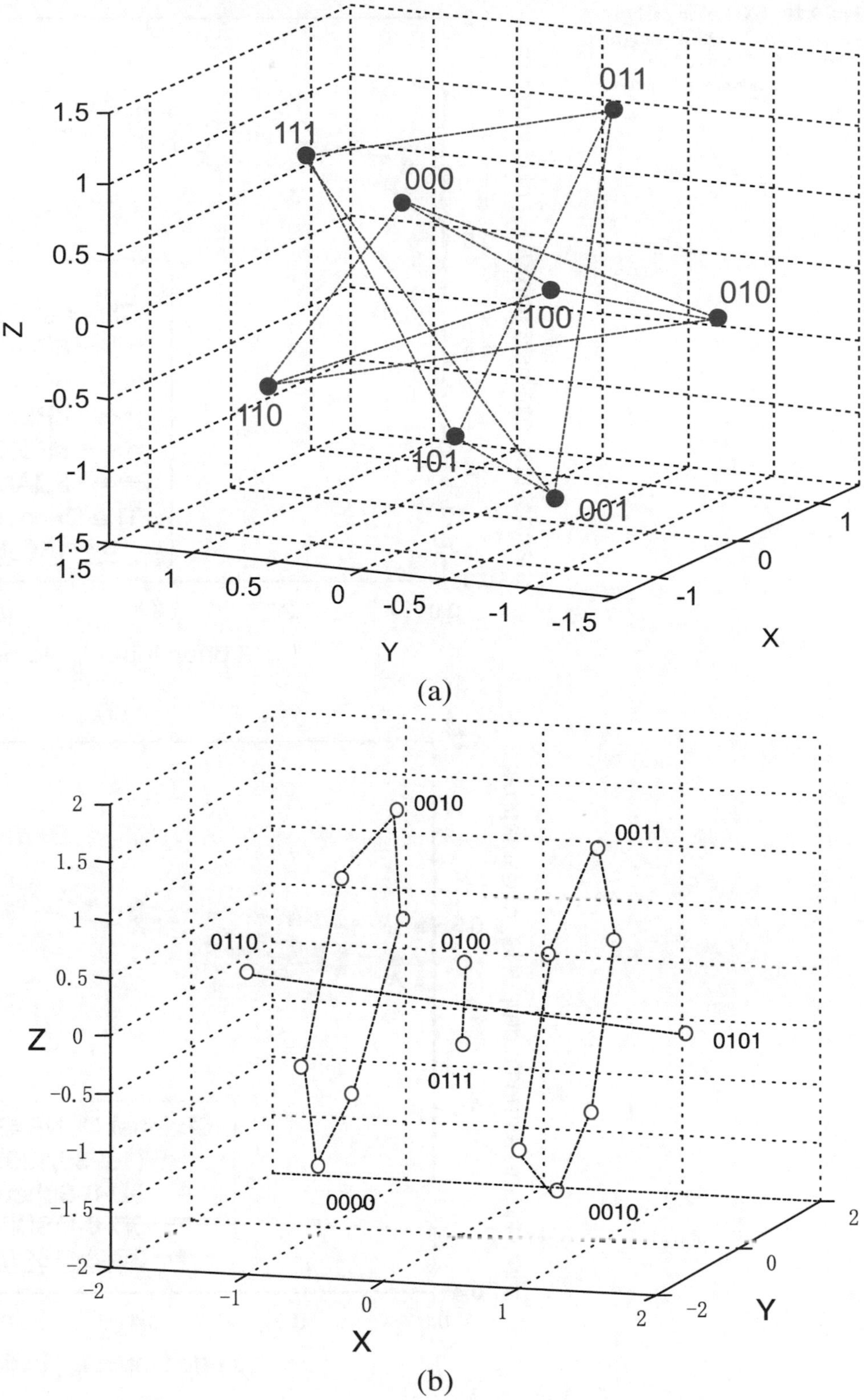

Fig. 9.78 EXIT chart analysis for (**a**) 8-ary 2D OSCD and (**b**) 8-ary 3D OSCD. Random mapping: randomly selected mapping rule

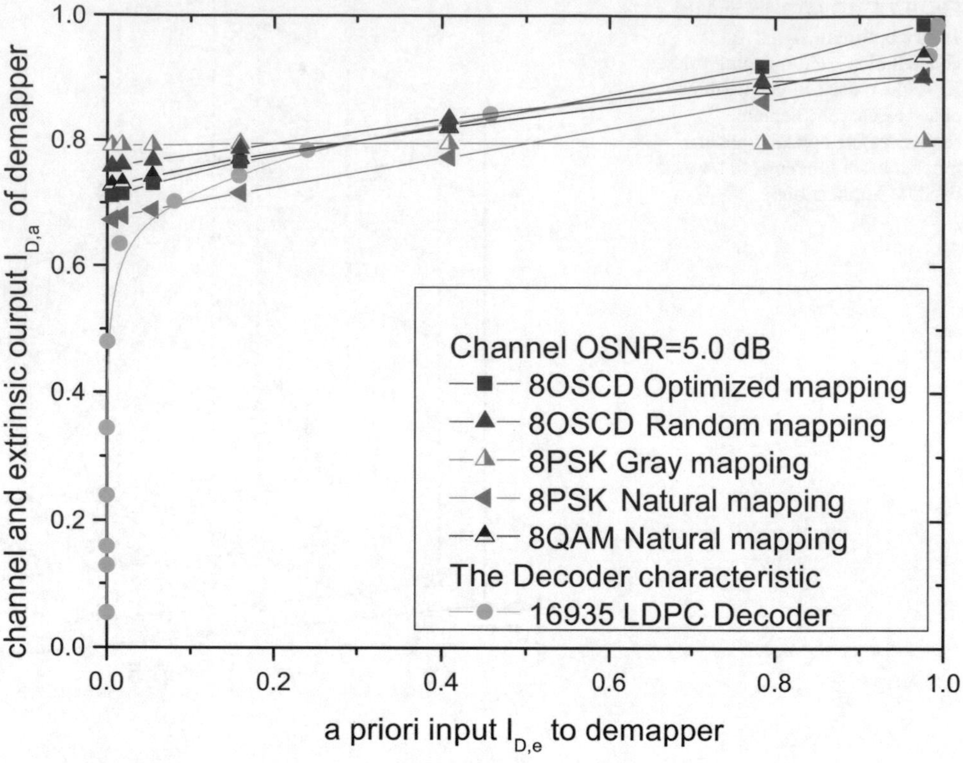

(a)

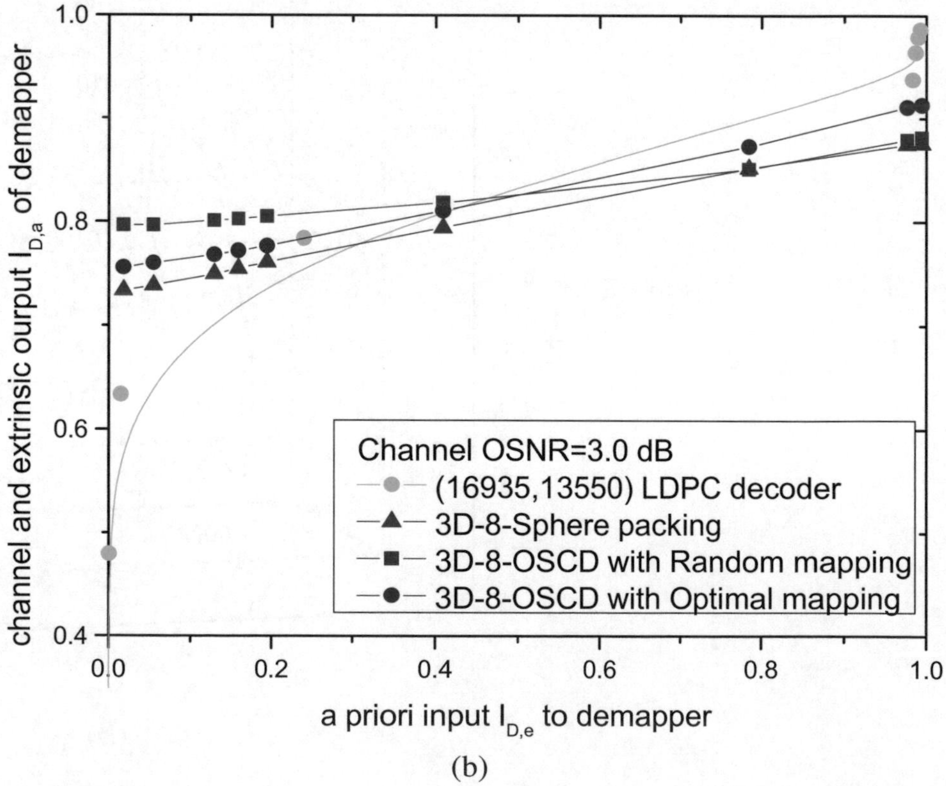

(b)

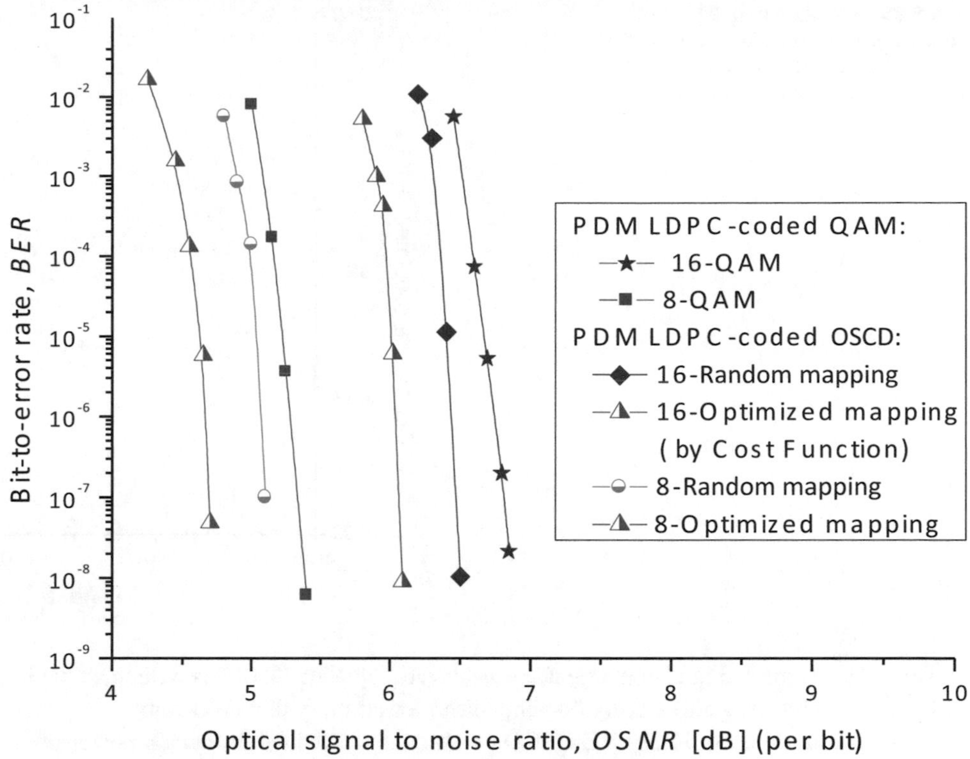

Fig. 9.79 BER performance for optimal mapping rules of LDPC-coded PDM 2D-OSCDs. PDM, polarization-division multiplexing

9.14.3 Nonlinear Optimized Signal Constellation Design-Based Coded Modulation

In this subsection, we describe NL-OSCD algorithm suitable for SPM dominated channel. The *NL-OSCD algorithm* can be summarized as follows [127]:

0. *Initialization*: Choose the signal constellation of size M that will be used for initialization.
1. Generate the training sequence from the optimum source distribution, denoted as $\{x_j = (x_{j,I}, x_{j,Q}) | j = 0,\ldots,n\text{-}1\}$.
2. Group the samples from this sequence into M clusters. The membership to the cluster is determined based on LLR of sample point and candidate signal constellation points from previous iteration. Each sample point is assigned to the cluster with the largest log-likelihood (LL) function, defined below. Given the m-th cluster (subset) $\widehat{A}_m = \{y_j = (y_{j,I}, y_{j,Q}) | j = 0, 1, \ldots\}$, find the LL function of the m-th cluster as follows:

$$LL_m = n^{-1} \sum_{k=0}^{n-1} \max_{y \in \widehat{A}_m} L(x_k, y), \tag{9.184}$$

where $L(x_j, y)$ is the nonlinear phase noise-averaged LLR defined as

$$L(x_k, y) = -\frac{1}{N_s} \sum_{i=0}^{N_s-1} \left\{ \frac{\left[x_{k,I} - \mathrm{Re}\left\{(y_I + y_Q j) e^{-j\Phi_{NL,i}}\right\}\right]^2}{2\delta^2} + \frac{\left[x_{k,Q} - \mathrm{Im}\left\{(y_I + y_Q j) e^{-j\Phi_{NL,i}}\right\}\right]^2}{2\delta^2} \right\} \tag{9.185}$$

where N_s denotes the number of nonlinear phase noise samples $\Phi_{NL,\,i}$.

3. Determine the new signal constellation points as center of mass for each cluster.

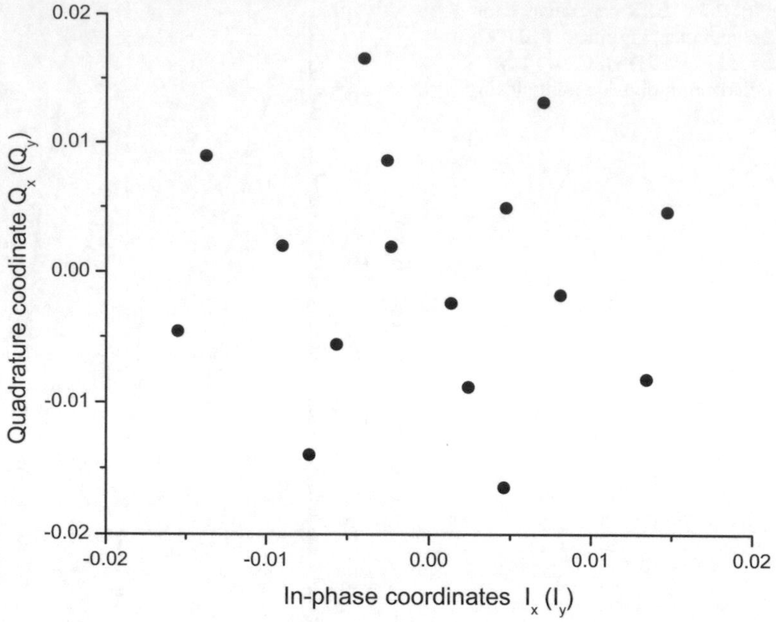

Fig. 9.80 The optimized 2D 16-ary signal constellations for the SPM-dominated scenario (after 2000 km of SMF and for launch power of – 6 dBm)

Repeat the steps 2–3 until convergence is achieved. Equation (9.185) is written by having in mind a 2D signal constellation design, but it can straightforwardly be generalized for arbitrary dimensionality.

As an illustration, the 16-ary NL-OSCD constellation, obtained for launch power of −6 dB and transmission distance of 2000 km, is shown in Fig. 9.80. The optimal launch power (in uncoded BER sense) for 16-ary NL-OSCD is 4 dBm, as shown in Fig. 9.81a, and the total transmission distance for LDPC-coded 16-NL-OSCD exceeds 4400 km, as shown in Fig. 9.81b. Figure 9.81 indicates that 16-ary NL-OSCD has better nonlinear tolerance than 16-QAM. On the other hand, Fig. 9.81b indicates that LDPC-coded 16-ary NL-OSCD outperforms 16-QAM by almost 1000 km. The signal constellations obtained by using the NL-OSCD algorithm clearly have better nonlinear tolerance and can significantly extend the total transmission distance of QAM. The main reason for this improvement is due to the fact that during the signal constellation design, the nonlinear nature of the channel is taken into account.

9.14.4 40 Tb/s Transoceanic Transmission Enabled by 8-ary OSCD

By using 8-ary OSCD, we have demonstrated in [129] a record to date 714 Gb/s capacity-per-amplifier (40 Tb/s in aggregated data rate) transoceanic transmission over 6787 km is demonstrated. The system is based on dual-carrier 400G transmission with C + L-band EDFA amplification only (at 121 km span length). In this subsection, we have briefly described this experimental verification.

Figure 9.82 outlines the experimental setup. At the transmitter side, a total of 109 C-band DFB lasers spanning from 1529.88 to 1564.55 nm and 91 L-band lasers spanning from 1572.40 to 1602.81 nm at 40.2GHz spacing are used and independently modulated by the data bar outputs of 4 64Gs/s DACs. For the measurement, instead of DFB lasers, eight external cavity lasers (ECLs) are separated into odd and even channels and are modulated separately by four 64Gs/s digital-to-analog converters (DACs) driving two I/Q modulators to produce 40Gbaud Nyquist-shaped NB-LDPC-coded 8-ary OSCD signals. First, a PRBS of order 2^{18}–1 was encoded by the 8-ary irregular quasi-cyclic LDPC (81126, 67605) encoder of girth 10 to generate codewords at 20% overhead and then mapped into 8-ary OSCD symbols via optimal mapping described above. For additional details on experimental setup including recirculating loop, an interested reader is referred to [129].

The offline DSP algorithm digitally compensates for the coarse frequency offset between the transmitter and LO lasers and then filters out the testing carrier and resamples the signals back to two samples per symbol. After resampling, either CD compensation (CDC) is performed in single stage in frequency domain or nonlinear compensation (NLC) is applied using digital back-propagation with one step per span [130]. Received signal is then fed into multi-modulus algorithm for initial convergence of polarization demultiplexing. Due to the non-regular constellation layout for the 8-ary OSCD, the blind phase searching method [131, 132] is applied for carrier phase recovery in the stage of DD-LMS algorithm. The recovered symbols are then sent to nonbinary LDPC decoder for detecting post-FEC bit errors. MD-FFT-SPA is employed with 50 iterations or less.

Fig. 9.81 (**a**) Uncoded BER vs. launch power for 16-ary NL-OSCD and 16-QAM, (**b**) BER vs. transmission length plot for LDPC-coded 16-ary NL-OSCD and 16-QAM. LLR-OSCD: constellation obtained for optimization in the presence of linear phase noise. The parameters (44000) mean that launch power was 4 dBm and target distance for optimization was 4000 km

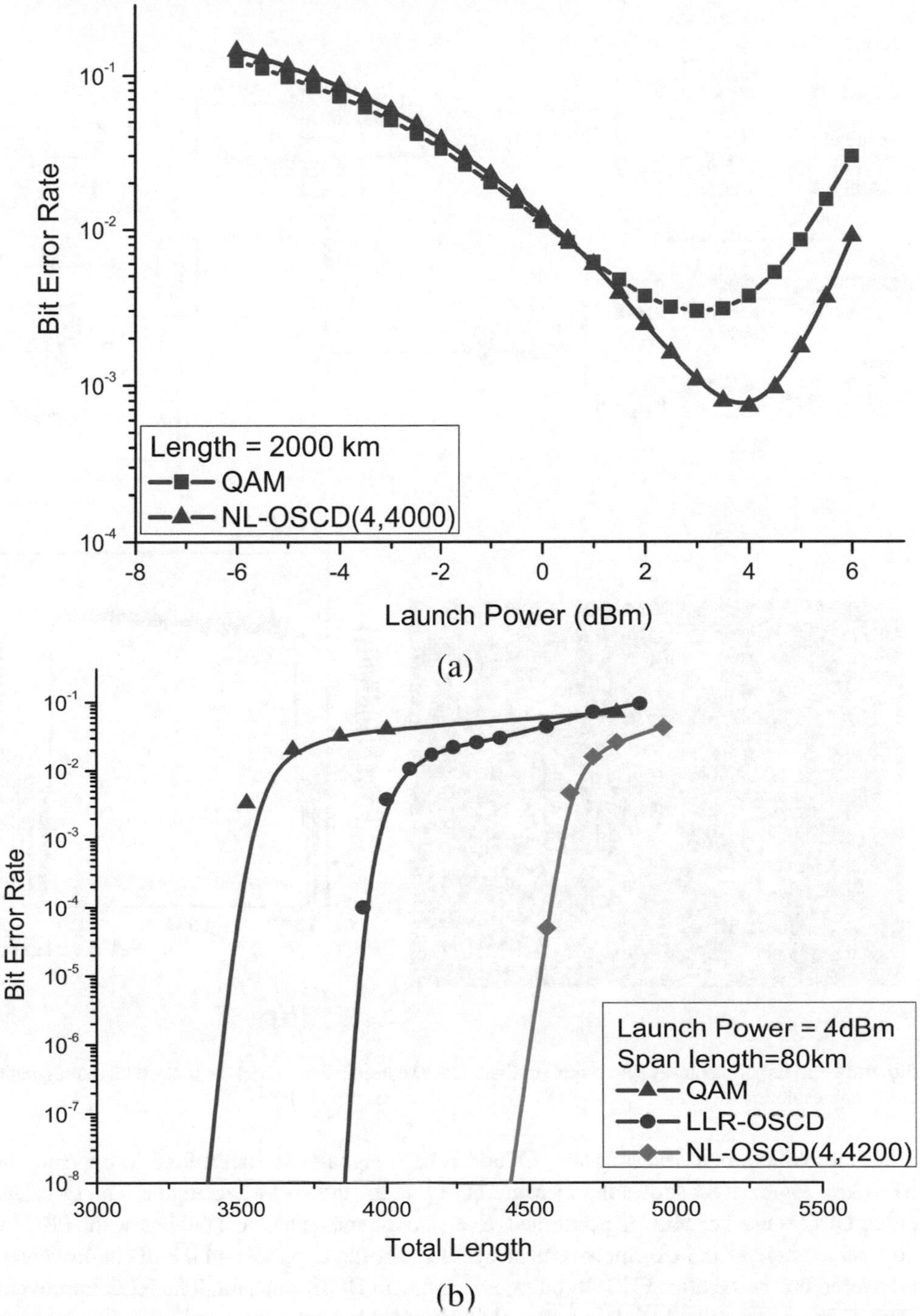

The coding performance of using regular 8-QAM and 8-ary OSCD modulation formats is compared in the back-to-back (BTB) scenario as a function of pre-FEC Q-factor derived from *BER* counting. The regular 8-QAM constellations can be found in [133]. To achieve sufficient confidence level, over 100 million bits are considered for each BER point, and exponential fitting is used for extrapolating the measured BER data and thus estimating the FEC limit. As observed in Fig. 9.83, with the same LDPC codewords transmitted, the FEC limit achieved by using optimal 8-ary OSCD is 4.05 dB at 20% overhead, whereas 4.6 dB for regular 8-QAM. Conforming to the findings in [48], there is ~0.55 dB improvement when utilizing optimal nonbinary LDPC-coded 8-QAM due to the joint coded modulation involving bit mapping, coding, and modulation format.

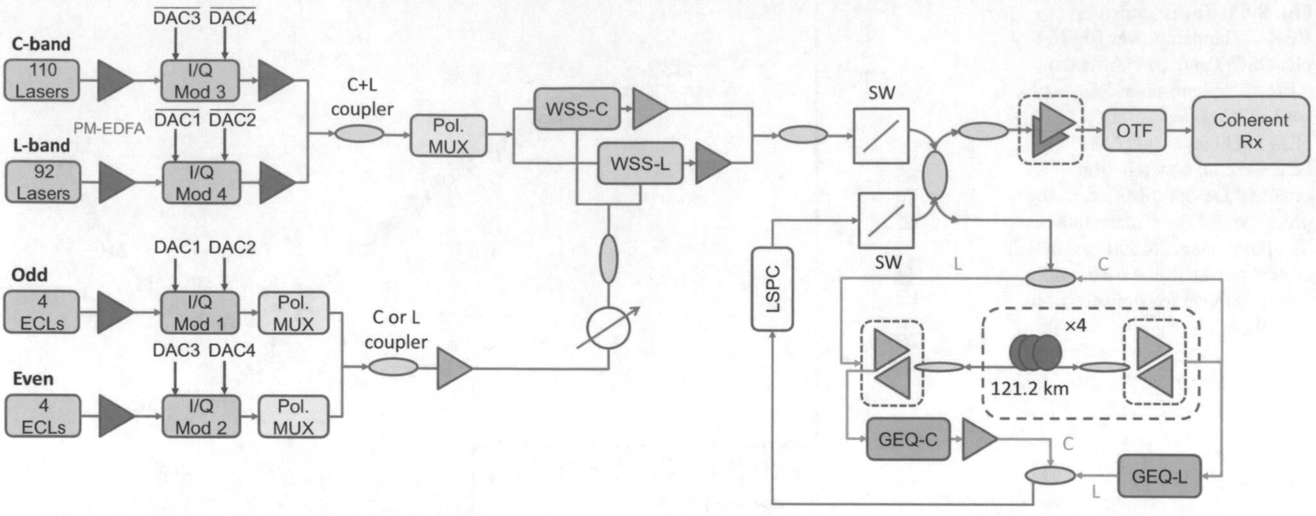

(a)

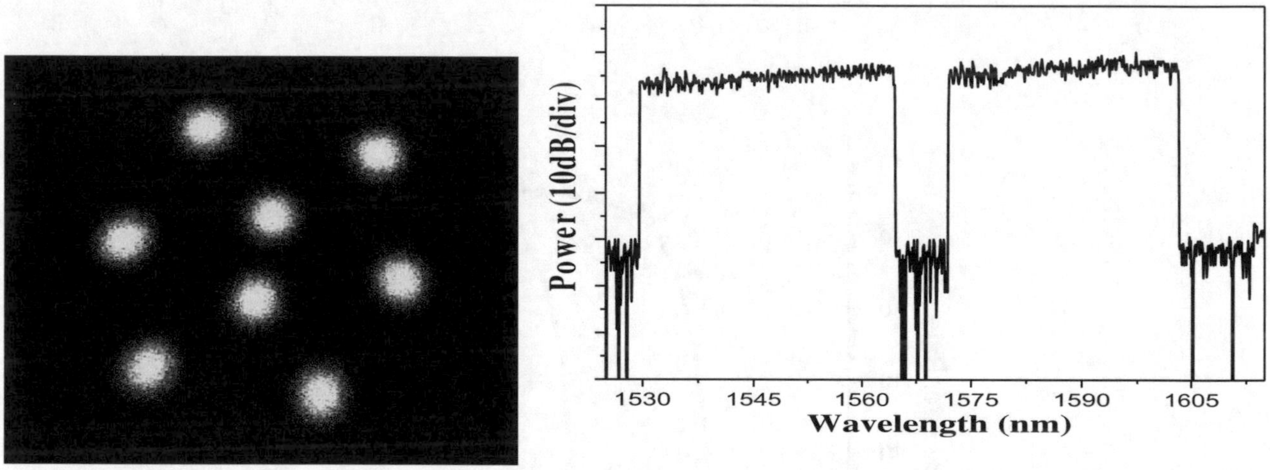

(b)

Fig. 9.82 (**a**) Experimental system setup, (**b**) 8-ary OSCD constellation in back-to-back (BTB) configuration and received spectrum after 6787 km at 0.1 nm resolution

In the bidirectional transmission, C- and L-band signals are transmitted in opposite directions over the same fiber over 6780 km. Figure 9.84 shows the measured Q-value for the 200 wavelengths. The Q-values are shown for both cases where either CDC is used or NLC is performed. Even though most channels fall below the FEC limit when only CDC is performed, after an average of 0.4 dB improvement by NLC, for the C-band, and 0.8 dB improvement for the L-band, all channels are recovered error-free after FEC decoding assisted with BCJR detector. The NLC improvement is different for the two bands since C-band is operated slightly below the optimum channel power and L-band is operated at a slightly above the optimum power to avoid excessive Raman interaction between bands and also to minimize gain tilt at the L-band.

Figure 9.85 shows the Q-value vs. received OSNR at 1551.2 nm as an example, where CDC only and NLC are compared as the channel power is swept freely. [The Q-factor is obtained from measured BERs]. The lower two figures in Fig. 9.85 depict the constellation density after CDC and NLC, respectively. These results show that NLC not only increases system margin but provides additional flexibility for system design and power budget allocation. In addition, the coded 8-OSCD could benefit more from the NLC since the coded modulation has lower FEC limit and the Q-factor improvement from NLC is sufficient to let the Q-factor move away (to the right) from the LDPC waterfall region, thus avoiding bit errors. Further experiments employing OSCD-based constellations can be found in [134, 135].

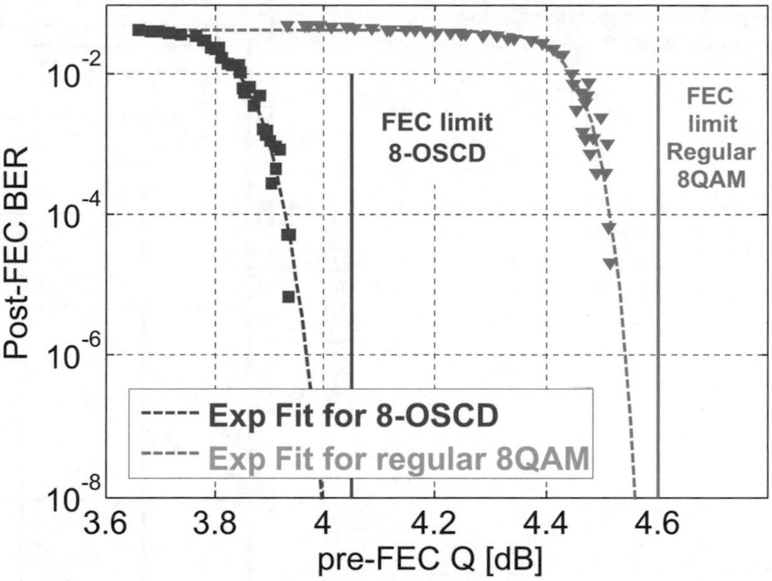

Fig. 9.83 Experimental comparison of the coding performance between regular 8-QAM and 8-OSCD

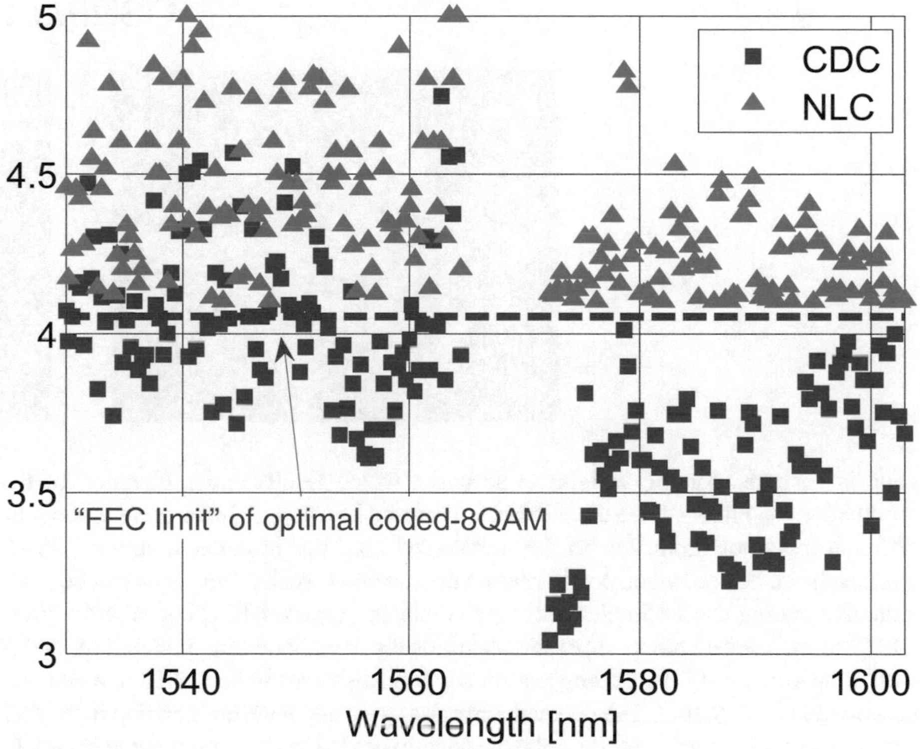

Fig. 9.84 Measured transmission performance for all channels with CDC and NLC after 6780 km. We use the term "FEC limit" to denote that not any countable error is found in experiment after LDPC decoder for this particular Q-value

9.15 Adaptive Coding and Adaptive Coded Modulation

As discussed in Chap. 5, adaptive modulation and adaptive coding enable both robust and spectrally efficient transmission over time-varying wireless and optical channels [67, 75, 96, 136–140]. The key idea is to estimate the channel state at the receiver and feed this estimate back to the transmitter side, so that the transmitter can adapt its parameters relative to the channel characteristics. In case when modulation format and channel code are fixed, the transmission system margin must be taken into consideration, to enable the proper system operation when the channel conditions are poor. Additionally, the component aging effects should be taken into account. Clearly, the system with fixed modulation format and fixed channel

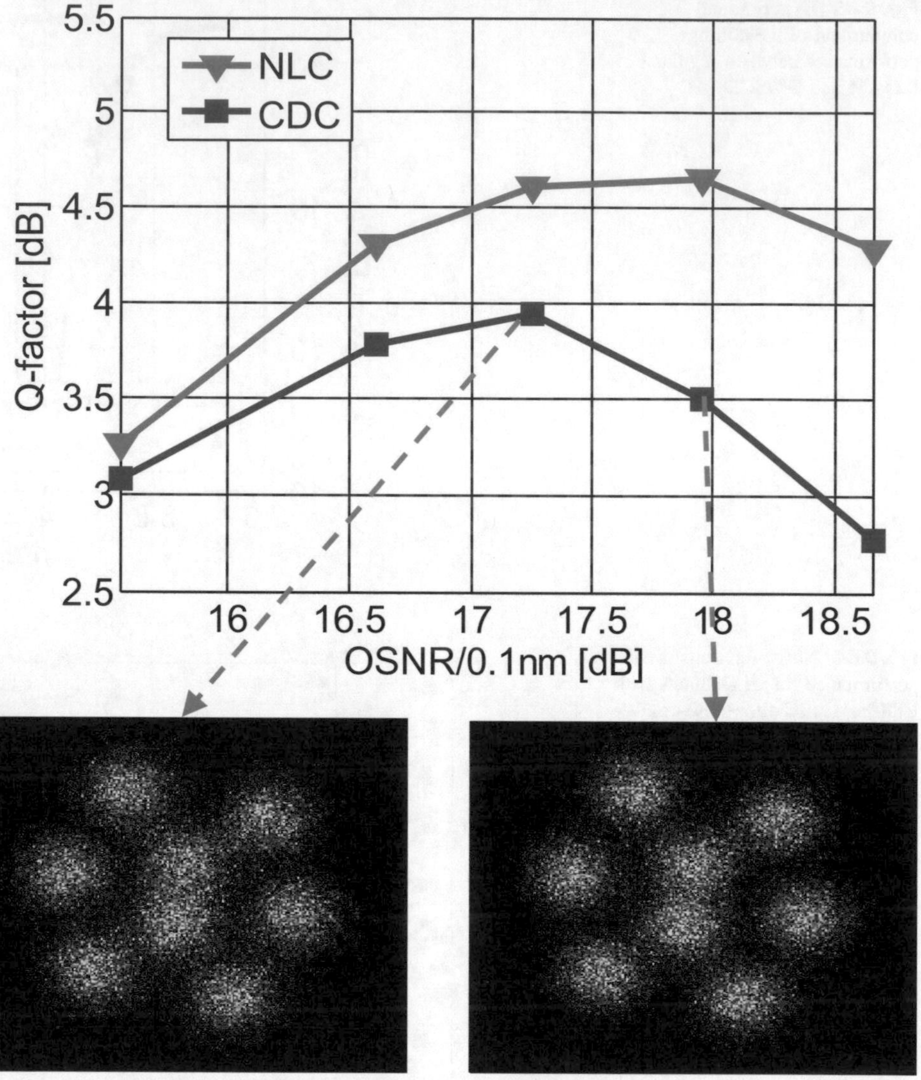

Fig. 9.85 Measured transmission performance for the channel at 1551.2 nm with CDC and NLC as a function of received OSNR

code is designed for the worst-case scenario, which results underutilization of the system most of the time. Adaptive modulation techniques have already been discussed in Chap. 5. In adaptive coding, different FEC codes are used to provide different levels of protection to the transmitted bits. For instance, a stronger FEC code with smaller coder rate (larger overhead) can be used when conditions are poor, while a weaker FEC code can be used when the channel conditions are good. Adaptive coding can be implemented by combining together FEC codes with different error correction capabilities, using UEP and multilevel coding approach. Additionally, the shortening, puncturing, and re-encoding approaches can be used as well. Rate-adaptive LDPC coding has already been discussed in Sect. 9.10.6, while corresponding FPGA implementations are described in Sect. 9.10.7. The unequal error protection and multilevel coding have already been described in Sect. 9.11. In the remainder of this chapter, we describe the adaptive coded modulation scheme in which trellis or coset codes are superimposed on the top of adaptive modulation scheme (Sect. 9.15.1), the adaptive modulation and coding scheme in which both the signal constellation size and error correction strength (code rate) are simultaneously adapted to come as close as possible to the channel capacity (Sect. 9.15.2), and hybrid adaptive modulation and coding scheme in which FSO and RF channels are operated in a cooperative fashion compensating for shortcomings of each other (Sect. 9.15.2).

9.15.1 Adaptive Coded Modulation

Compared to adaptive modulation alone, the additional coding gain can be achieved by superimposing either trellis codes or coset codes on top of the adaptive modulation scheme. The key idea behind this adaptive coded modulation is to exploit the

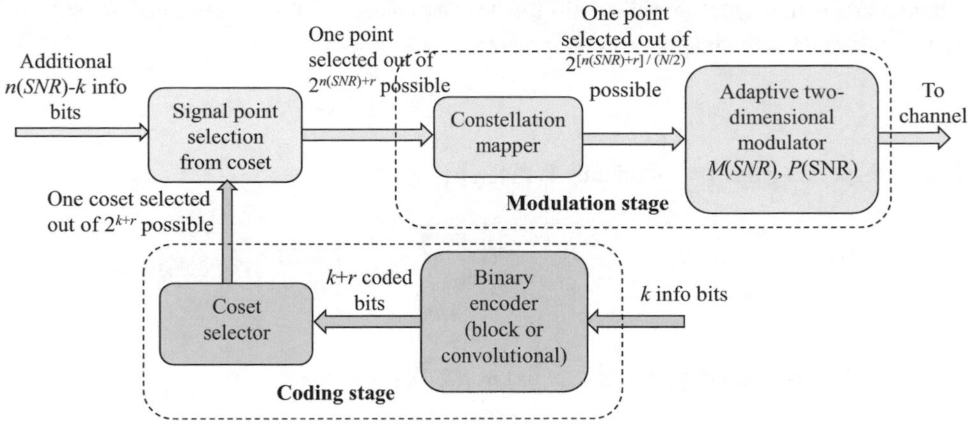

Fig. 9.86 Generic adaptive coded modulation scheme

separability of coding stage and constellation design inherent to the coset codes. The instantaneous SNR or OSNR varies with time, which will change the distance in the received signal constellation $d_0(t)$, and, therefore, the corresponding minimum distance between coset sequences d_s and the minimum distance between coset points d_c will be time-varying as well. By using adaptive modulation in a tandem with the coset codes, we can ensure that distances ds and dc do not vary as channel conditions change by adapting the signal constellation size $M(SNR)$, transmit power $P(SNR)$, and/or symbol rate $R_s(SNR)$ of the transmitted signal constellation relative to the SNR, subject to an average transmit power $\overline{P}$ constraint on SNR-dependent transmit power $P(SNR)$. Minimum distance does not vary as channel conditions change $d_{\min}(t) = \min\{d_s(t), d_c(t)\} = \underline{d_{\min}}$; the adaptive coded modulation will exhibit the same coding gain as coded modulation developed for a Gaussian-like channel with corresponding minimum codeword distance $d_{\min}$. Therefore, this adaptive coded modulation scheme requires small changes in generic coded modulation scheme described in Sect. 9.11.1, as shown in Fig. 9.86. Clearly, the coding stage does not change at all. The number of cosets does not change as well. On the other hand, the number of points within the coset is channel SNR dependent. When channel condition gets improved, we enlarge the coset size. On such a way, we transmit more uncoded bits, while not affecting the distance of the channel code. Additionally, this approach is hardware-friendly.

The following approximation for bit error probability for M-ary QAM systems can be used:

$$P_b \leq 0.2 \exp\left[\frac{-1.5SNR \cdot G_c}{M(SNR)-1}\frac{P(SNR)}{\overline{P}}\right], \quad M \geq 4, 0 \leq SNR\,[\mathrm{dB}] \leq 30\ \mathrm{dB}, \tag{9.186}$$

where G_c is the coding gain. We adjust the signal constellation size $M(SNR)$ depending on channel SNR and target BER as follows:

$$M(SNR) = 1 + \underbrace{\frac{1.5}{-\log(5P_b)}}_{C_f} SNR \cdot G_c \frac{P(SNR)}{\overline{P}} = 1 + C_f SNR \cdot G_c \frac{P(\gamma)}{\overline{P}}, \tag{9.187}$$

where C_f is the correction factor, defined as $C_f = -1.5SNR/\log(5P_b)$, which is used to account for the fact that non-optimized signal constellation is used. The number of uncoded bits needed to select a point within the coset is determined by

$$n(\gamma) - 2k/N = \log_2 M(\gamma) - 2(k+r)/N. \tag{9.188}$$

The normalized transmission rate is given by $\log_2 M(SNR)$, while the normalized adaptive data rate is determined by $\log_2 M(SNR) - 2r/N$. To determine the optimum launch power, we need to optimize $\langle \log_2 M(SNR) \rangle$, subject to the average power constraint, and not surprisingly the optimum power adaptation strategy is the water-filling law:

$$\frac{P(SNR)}{\overline{P}} = \begin{cases} 1/SNR_0 - 1/SNR_c, & SNR \geq \underbrace{SNR_0/C_c}_{SNR_c} \\ 0, & SNR < SNR_0/C_c \end{cases}, \quad C_c = C_f G_c \tag{9.189}$$

where SNR_0 is the signal-to-noise ratio for AWGN channel. After substituting (9.189) into (9.187), we obtain the following equation for adaptive signal constellation size:

$$M(SNR) = SNR/SNR_c. \tag{9.190}$$

The corresponding spectral efficiency is given by

$$\frac{R}{B} = \int_{SNR_c}^{\infty} \log_2\left(\frac{SNR}{SNR_c}\right) f(SNR) dSNR, \tag{9.191}$$

for which the distribution of SNR, denoted as *f*(*SNR*), is needed.

9.15.2 Adaptive Nonbinary LDPC-Coded Multidimensional Modulation for High-Speed Optical Communications

In this scheme, which employs a single NB-LDPC encoder on Tx side and a single NB-LDPC decoder operating over GF of order matching the signal constellation size on Rx side, as depicted in Fig. 9.87, the signal constellation size $M = 2^m$ and code rate R adaptation have been performed such that $mR \leq C$, where C is the multidimensional optical channel capacity.

The simulation results for SDM NB irregular QC-LDPC-coded 2D constellations are summarized in Fig. 9.88, for information symbol rate $R_{s,\text{ info}} = 25$ GS/s and aggregate data rate of $2mNRR_s$, where m is the number of bits/symbol, N is the number of spatial modes, R is the code rate, and R_s is the channel symbol rate ($R_{s,\text{info}}/R$). (To compensate for mode-coupling effects, the MIMO signal processing is used as described in [1].) The corresponding signal constellation size and code rate have been chosen based on the estimate of channel OSNR per bit. Both conventional QAM and constellations from optimized signal constellation design (OSCD) have been observed (see chapter on advanced modulation formats for additional details on OSCD constellations). In Fig. 9.88 fixed irregular (34635, 27708,0.8) code is observed. The results for adaptive NB-QC-LDPC coded modulation, for code rates ranging from 0.667 to 0.8 are summarized in Fig. 9.89.

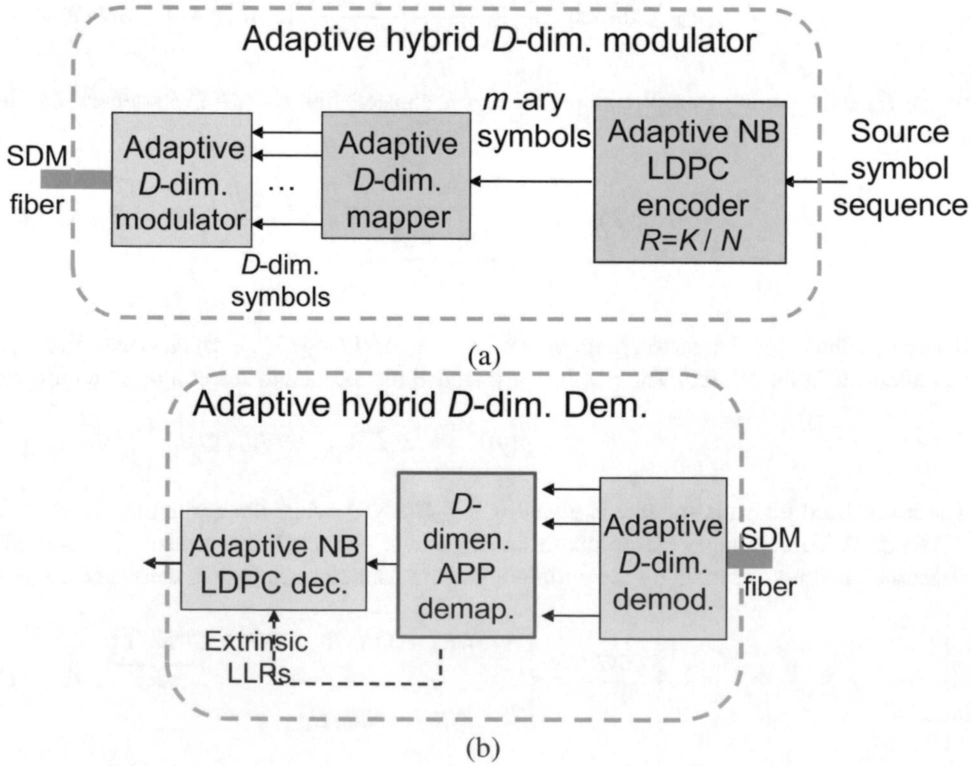

Fig. 9.87 A generic adaptive hybrid *D*-dimensional irregular NB-QC-LDPC-coded modulation scheme employing single LDPC encoder/decoder operating over GF of order matching the signal constellation size

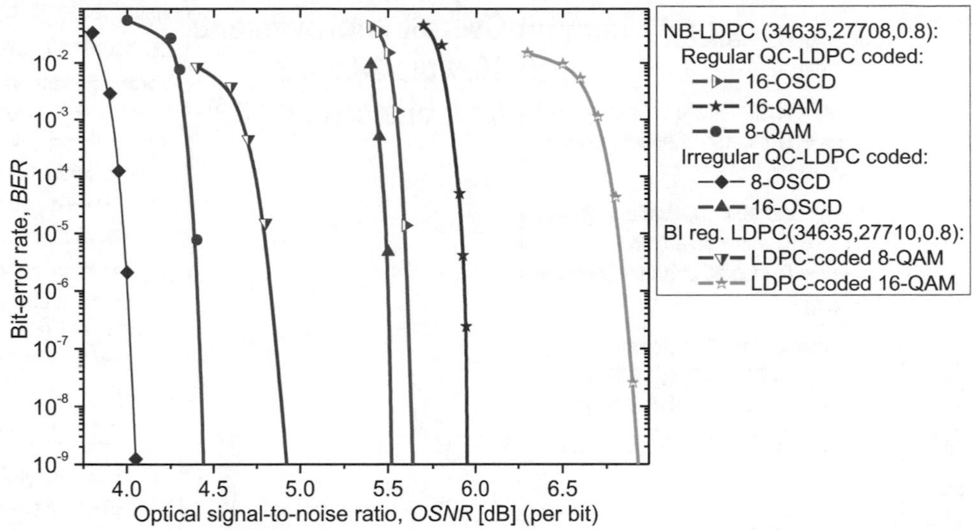

Fig. 9.88 BER performance of irregular NB-QC-LDPC (34635, 27708,0.8)-coded 2D constellations

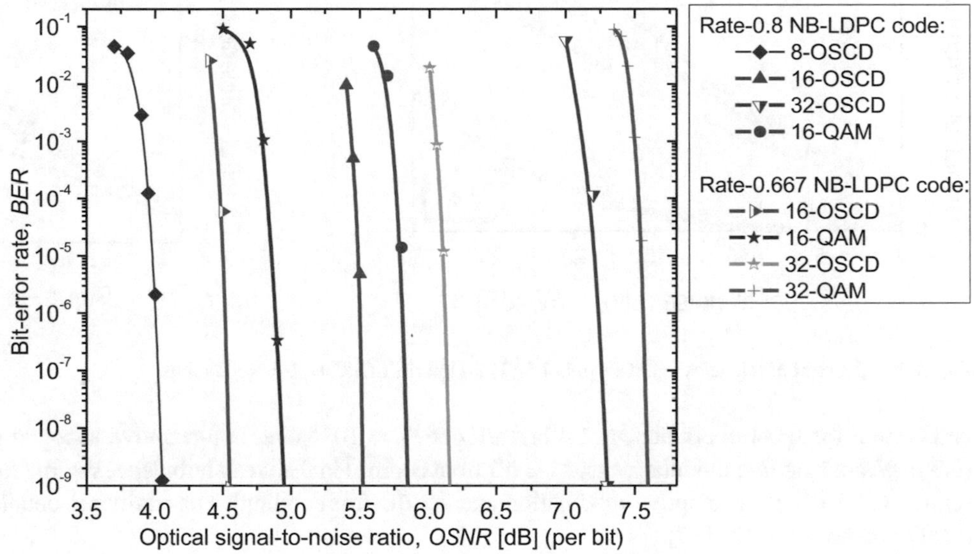

Fig. 9.89 BER performance of irregular NB-QC-LDPC-coded 2D constellations for code rates ranging from 0.667 to 0.8

9.15.3 Adaptive Hybrid FSO-RF-Coded Modulation

In a hybrid FSO/RF system, FSO and RF channels exhibit drastically different dynamics and transmission speeds, so they naturally require different coding and modulation techniques. Determining the optimum split of information bits between the two is quite challenging. Rate-adaptive LDPC-coded modulation can be used for this purpose by adjusting the error correction strength according to the time-varying conditions of the two channels [137]. The LDPC code design has been performed based on the channel capacity of the hybrid link, with some of the symbols transmitted over FSO and the rest over the RF channel. In one possible solution, one LDPC decoder has been used for both subsystems.

To illustrate high potential of hybrid techniques, combined together with adaptive hybrid LDPC-coded modulation, in Fig. 9.90 we provide spectral efficiency results per single hybrid Tx-Rx pair. The PAM is used in FSO subsystem with direct detection, while QAM is used in RF subsystem. The hybrid coded modulation scenario is compared against corresponding FSO system with RF feedback. The coding gain over adaptive modulation at bit error probability $P_b = 10^{-6}$ for spectral efficiency $R/B = 4$ bits/s/Hz is 7.2 dB in both (weak and strong) turbulence regimes. Further improvements can be obtained by increasing the girth of LDPC codes and employing better modulation formats. The increase in codeword length does not improve spectral efficiency performance that much. It is interesting to notice that by employing adaptive hybrid coded modulation, the communication under saturation regime is possible. Moreover, for variable-rate variable-power scheme, there is no degradation in saturation regime compared to strong turbulence regime. Overall improvement from adaptive modulation

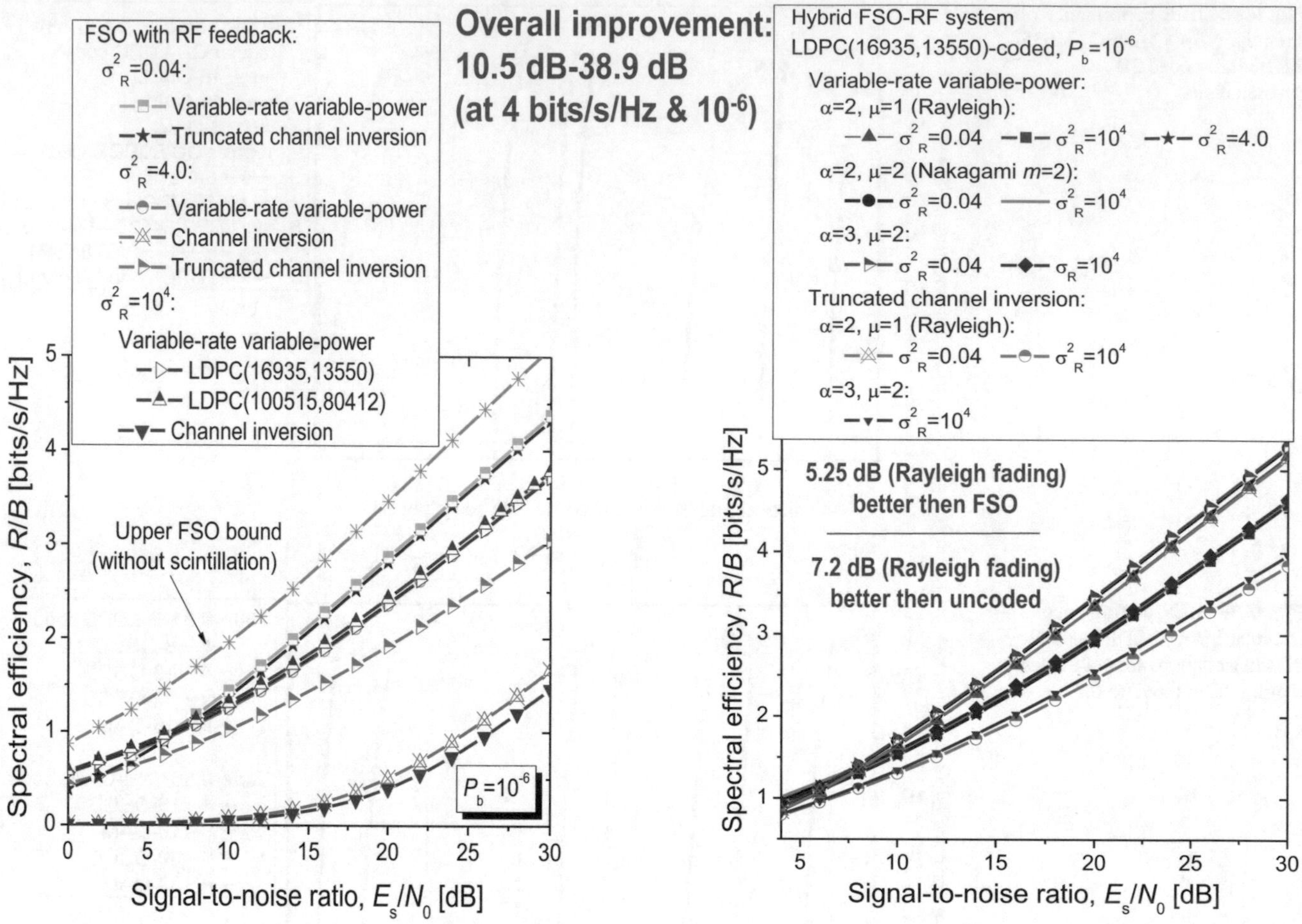

Fig. 9.90 Spectral efficiencies against symbol SNR for hybrid LDPC-coded modulation

and coding for spectral efficiency of 4 bits/s/Hz at $P_b = 10^{-6}$ over non-adaptive uncoded modulation ranges from 10.5 dB (3.3 dB from adaptive modulation and 7.2 dB from coding) in the weak turbulence regime to 38.9 dB in the strong turbulence regime (31.7 dB from adaptive modulation and 7.2 dB from coding). For additional details on this approach, an interested reader is referred to [2, 137].

9.16 Concluding Remarks

This chapter has been devoted to advanced channel coding and coded modulation techniques. After linear block codes fundamentals described in Sect. 9.1, BCH codes fundamentals have been described in Sect. 9.2, including Massey-Berlekamp algorithm. The trellis description of linear block codes and the corresponding Viterbi decoding algorithm have been described in Sect. 9.3. The fundamentals of convolutional codes have been provided in Sect. 9.4. The RS, concatenated, and product codes have been described in Sect. 9.5. Section 9.6 has been devoted to the description of coding with interleaving as an efficient way to deal with burst of errors and fading effects.

Significant space in the chapter has devoted to codes on graphs, starting with codes on graph fundamentals in Sect. 9.7. The turbo codes and BCJR algorithm-based decoding procedure have been described in Sect. 9.8, while turbo-product codes have been described in Sect. 9.9. LDPC codes and corresponding decoding algorithms have been described in Sect. 9.10. Regarding LDPC codes, both binary and nonbinary (NB) LDPC codes have been introduced, and corresponding decoding algorithms as well as their FPGA implementation have been described. Section on LDPC codes has been organized as follows: binary LDPC codes have been described in Sect. 9.10.1, decoding of binary LDPC codes in Sect. 9.10.2, FPA implementation of binary LDPC decoders in Sect. 9.10.3, decoding of NB-LDPC codes in Sect. 9.10.4, design of LDPC codes in Sect. 9.10.5, rate-adaptive LDPC coding in Sect. 9.10.6, and rate-adaptive coding implementations in FPGA in Sect. 9.10.7. The next section in the chapter, Sect. 9.11, has been devoted to coded modulation and unequal error protection. After providing the coded modulation fundamentals (Sect. 9.11.1), TCM has been described in Sect. 9.11.2, multilevel coding and UEP in Sect. 9.11.3, BICM in Sect. 9.11.4, and turbo TCM in Sect. 9.11.5. Various hybrid multidimensional coded modulation schemes suitable for ultrahigh-speed optical transmission have been described in Sect. 9.12. Hybrid coded modulation has been described in Sect. 9.12.1, while multilevel NB-LDPC-coded modulation in Sect. 9.12.2.

After coded modulation sections, the focus of the chapter is on multidimensional turbo equalization, Sect. 9.13. The following topics have been described: nonlinear channels with memory in Sect. 9.13.1, nonbinary MAP detection in Sect. 9.13.2, sliding-window multidimensional turbo equalization in Sect. 9.13.3, simulation study of multidimensional turbo equalization in Sect. 9.13.4, and several experimental demonstrations including time-domain 4D-NB-LDPC-CM and quasi-single-mode transmission over transoceanic distances in Sects. 9.13.5 and 9.13.6, respectively. In section on optimized signal constellation design and optimized bit-to-symbol mapping-based coded modulation, Sect. 9.14, multidimensional OSCD (Sect. 9.14.1), EXIT chart analysis of OSCD mapping rules (Sec. 9.14.2), nonlinear OSCD-based coded modulation (Sect. 9.14.3), and transoceanic multi-Tb/s transmission experiments enabled by OSCD (Sect. 9.14.4) have been described.

Finally, in adaptive coding and adaptive coded modulation section (Sect. 9.15) the following coded modulation schemes are described: adaptive coded modulation (Sect. 9.15.1), adaptive nonbinary LDPC-coded multidimensional modulation suitable for high-speed optical communications (Sect. 9.15.2), and adaptive hybrid FSO-RF coded modulation (Sect. 9.15.3).

The set of problems is provided in incoming section to help reader gain deeper understanding of material described in this chapter.

9.17 Problems

9.1. The binary symmetric channel (BSC) is described by the following channel matrix:

$$\boldsymbol{P}_{\mathrm{BSC}} = \begin{bmatrix} q & p \\ p & q \end{bmatrix}, \quad p + q = 1.$$

(a) Determine the channel capacity of this channel, and plot its dependence against the crossover probability p. Observe now the channel that is obtained by cascading several BSC channels.
(b) Determine the channel matrix after cascading two and three BSC channels.
(c) Plot the channel capacity of cascaded channels against p. Discuss the results.

9.2. For the channel shown in Fig. 9.P2, with the prior probabilities $P(x_1) = p$, $P(x_2) = q$, and the transition probabilities being different $p_1 \neq p_2$, determine the channel capacity. This channel is known as the binary asymmetric channel (BAC), and the optical channel for IM/DD belongs to this category.

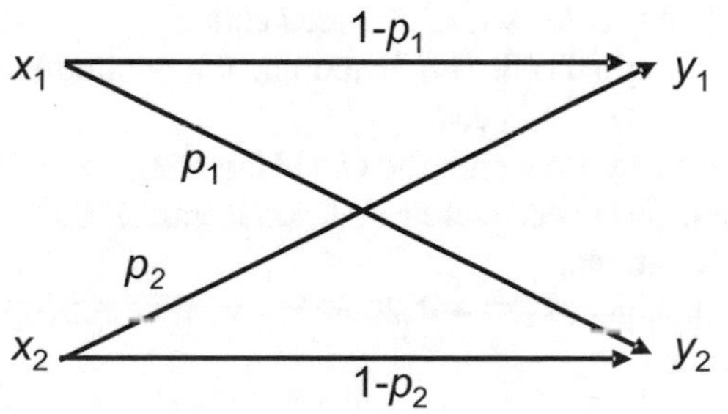

Fig. 9.P2 The binary asymmetric channel

9.3. An (n,k) linear block code is described by the following parity-check matrix:

$$\boldsymbol{H} = \begin{bmatrix} 0 & 0 & 0 & 0 & 0 & 0 & 0 & 1 & 1 & 1 & 1 & 1 & 1 & 1 & 1 \\ 0 & 0 & 0 & 1 & 1 & 1 & 1 & 0 & 0 & 0 & 0 & 1 & 1 & 1 & 1 \\ 0 & 1 & 1 & 0 & 0 & 1 & 1 & 0 & 0 & 1 & 1 & 0 & 0 & 1 & 1 \\ 1 & 0 & 1 & 0 & 1 & 0 & 1 & 0 & 1 & 0 & 1 & 0 & 1 & 0 & 1 \end{bmatrix}.$$

(a) Determine the code parameters: codeword length, number of information bits, code rate, overhead, minimum distance, and error correction capability.
(b) Represent the $\boldsymbol{H}$-matrix in systematic form, and determine the generator matrix $\boldsymbol{G}$ of the corresponding systematic code.

9.4. Consider a (7,4) code with generator matrix:

$$\boldsymbol{G} = \begin{bmatrix} 0 & 1 & 0 & 1 & 1 & 0 & 0 \\ 1 & 0 & 1 & 0 & 1 & 0 & 0 \\ 0 & 1 & 1 & 0 & 0 & 1 & 0 \\ 1 & 1 & 0 & 0 & 0 & 0 & 1 \end{bmatrix}$$

(a) Find all the codewords of the code.
(b) What is the minimum distance of the code?
(c) Determine the parity-check matrix of the code.
(d) Determine the syndrome for the received vector [1101011].
(e) Assuming that an information bit sequence of all 0 s is transmitted, find all minimum weight error patterns $\boldsymbol{e}$ that result in a valid codeword that is not the all-zero codeword.
(f) Use row and column operations to transform $\boldsymbol{G}$ to systematic form, and find its corresponding parity-check matrix. Sketch a shift register implementation of this systematic code.
(g) All Hamming codes have a minimum distance of 3. What is the error correction and error detection capability of a Hamming code?

9.5. Consider the following set:

$$C = \{(000000), (001011), (010101), (011110), (100111), (101100), (110010), (111001)\}$$

(a) Show that C is vector space over GF(2).
(b) Determine the dimension of this set.
(c) Determine the set of basis vectors.
If the set C is considered as the code book of an LBC:
(d) Determine the parameters of this code.
(e) Determine the generator and parity-check matrices in systematic form.
(f) Determine the minimum distance and error correction capability of this code.

9.6. Let C be an (n,k) LBC of minimum Hamming distance d. Define a new code C_e by adding an additional overall parity-check equation. Such obtained code is known as *extended* code:
(a) If the minimum distance of original code is d, determine the minimum distance of extended code.
(b) Show that extended code is an $(n + 1,k)$ code.

9.7. Determine an (8,4) code obtained by extending the (7,4) Hamming code. Determine its minimum distance and error correction capability. Show that (8,4) code and its dual are identical. Such codes are called self-dual and are of high importance in quantum error correction.

9.8. The maximum length C^m_{dual} is the dual of $(2^m - 1, 2^m - 1 - m, 3)$ of corresponding Hamming code. Let $m = 3$:
(a) List all codewords of this code.
(b) Show that C^3_{dual} is an (7,3,4) LBC.

9.9. Let $\boldsymbol{G} = [\boldsymbol{I}_k \mid \boldsymbol{P}]$ be the generator matrix of a systematic (n,k) LBC. Show that this code is self-dual if and only if $\boldsymbol{P}$ is a square and $\boldsymbol{P}\boldsymbol{P}^{\mathrm{T}} = \boldsymbol{I}$. Show that for dual codes, $n = 2\,k$. Finally, design self-dual codes for $n = 4$, 6, and 8.

9.10. Prove the following properties of standard array:
(a) All n-tuples of row are distinct.
(b) Each n-tuple appears exactly once in the standard array.
(c) There are exactly 2^{n-k} rows in the standard array.
(d) For perfect codes (satisfying the Hamming bound with equality sign), all n-tuples of weight $t = \text{int}[(d_{\min} - 1)/2]$ or less appear as coset leaders (int[x] is the integer part of x).
(e) For quasi-perfect codes, in addition to all n-tuples of weight t or less, some but not all n-tuples of weight $t + 1$ appear as coset leaders.

(f) All elements in the same row (coset) have the same syndrome.
(g) Elements in different rows have different syndromes.
(h) There are 2^{n-k} different syndromes corresponding to 2^{n-k} rows.

9.11. For extended code from Problem 9.7, determine the standard array and syndrome decoding table. If the received word was (10101010), what would be the result of standard array decoding? Determine the weight distribution of coset leaders and word error probability for crossover probability of BSC being 10^{-4}. Finally, determine the coding gain at target BER of 10^{-6}.

9.12. The (n,k) LBC can be *shortened* by deleting the information symbols, which is equivalent to the deleting of parity columns in the parity-check matrix $\boldsymbol{H} = [\boldsymbol{P}^{\mathrm{T}}|\boldsymbol{I}_{n-k}]$. Let us remove all columns of even weight in $\boldsymbol{P}^{\mathrm{T}}$ of the $(2^m - 1, 2^m - 1 - m)$ Hamming code. Derive the new code parameters in terms of the original code, and determine the minimum distance of the shortened code. What is the error correction capability of this shortened code?

9.13. Prove that an LBC can correct all error patterns having e_{c} or fewer errors, and simultaneously detect all error patterns containing e_{d} ($e_{\mathrm{d}} \geq e_{\mathrm{c}}$) or fewer errors if the minimum distance $d_{\min}$ satisfies the following inequality: $d_{\min} \geq e_{\mathrm{d}} + e_{\mathrm{c}} + 1$.

9.14. Let the polynomial $x^7 + 1$ over GF(2) be given. Determine the generator and parity-check polynomials of cyclic (7,4) code. By using these polynomials, create the corresponding generator and parity-check matrices. Encode the information sequence (1101) using this cyclic code. Provide the encoder and syndrome circuits.

9.15. Consider RS (31,15) code:
(a) Determine the number of bits per symbol in the code and determine the block length in bits.
(b) Determine the minimum distance of the code and the error correction capability.

9.16. Let us observe RS (n,k) code:
(a) Prove that minimum distance is given by $d_{\min} = n - k + 1$.
(b) Prove that the weight distribution of RS codes can be determined by

$$A_i = \binom{n}{i}(q-1)\sum_{j=0}^{i-d_{\min}}(-1)^j\binom{i-1}{j}q^{i-d_{\min}-j}.$$

9.17. Design a concatenated code using (15,9) 3-symbol-error-correcting RS code as the outer code and (7,4) binary Hamming code as the inner code. Determine the overall codeword length and the number of binary information bits contained in the codeword. Determine the error correction capability of this code.

9.18. This problem is related to *coded modulation*:
(a) Describe how to determine the symbol log-likelihood ratios (LLRs): (i) assuming Gaussian approximation and (ii) assuming that conditional PDF is obtained by collecting the histograms.
(b) Describe the transmitter and receiver configuration for polarization-multiplexed bit-interleaved LDPC-coded modulation.
(c) Describe the transmitter and receiver configuration for multilevel coding (MLC) scheme with LDPC codes as channel codes.
(d) Providing that symbol LLRs are determined as in (a), describe how to determine the bit LLRs required for LDPC decoding. Describe how to determine the extrinsic information at the output of LDPC decoder for APP demapper. Describe how to determine the prior symbol LLRs for next APP-LDPC decoder iteration step.

9.19. This problem is related to the *sum-product algorithm*. Let $q_{ij}(b)$ denote the extrinsic information (message) to be passed from variable node v_i to function node f_j regarding the probability that $c_i = b$, $b \in \{0,1\}$, as illustrated in Fig. 9.P19 on the left. This message is concerned about probability that $c_i = b$, given extrinsic information from all check nodes, except

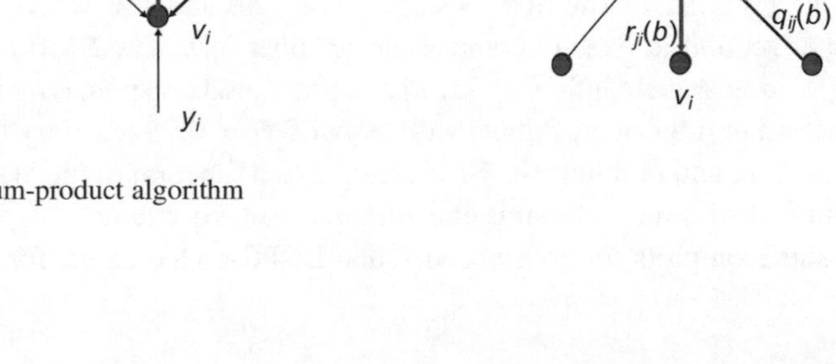

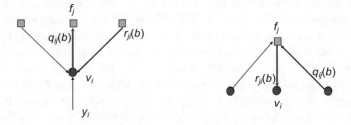

Fig. 9.P19 Illustration of the sum-product algorithm

node f_j, and given the channel sample y_j. Let $r_{ji}(b)$ denote extrinsic information to be passed from node f_j to node v_i, as illustrated in figure below on the right. This message is concerned about probability that j-th parity-check equation is satisfied given that $c_i = b$, while other bits have separable distribution given by $\{q_{ij'}\}_{j' \neq j}$.

– Show that $r_{ji}(0)$ and $r_{ji}(1)$ can be calculated from $q_{i'j}(0)$ and $q_{i'j}(1)$ as follows:

$$r_{ji}(0) = \frac{1}{2} + \frac{1}{2} \prod_{i' \in V_j \backslash i} (1 - 2q_{i'j}(1)), \quad r_{ji}(1) = \frac{1}{2} - \frac{1}{2} \prod_{i' \in V_j \backslash i} (1 - 2q_{i'j}(1))$$

– Show that $q_{ji}(0)$ and $q_{ji}(1)$ can be calculated as follows:

$$q_{ij}(0) = (1 - P_i) \prod_{j' \in C_i \backslash j} r_{j'i}(0), \quad q_{ij}(1) = P_i \prod_{j' \in C_i \backslash j} r_{j'i}(1)$$

9.20. The probability domain sum-product algorithm has the following disadvantages: (i) multiplications are involved (additions are less costly to implement) and (ii) the calculation of many product probabilities could become numerically unstable (especially for long codes with more than 50 iterations). The log-domain version of sum-product algorithm is a preferable option. Instead of beliefs in log domain, we use the log-likelihood ratios (LLRs), which for codeword bits c_i $(i = 2,\ldots,n)$ are defined by

$$L(c_i) = \log\left[\frac{\Pr(c_i = 0|y_i)}{\Pr(c_i = 1|y_i)}\right],$$

where y_i is a noisy sample corresponding to c_i. The LLRs corresponding to r_{ji} and q_{ij} are defined as follows:

$$L(r_{ji}) = \log\left[\frac{r_{ji}(0)}{r_{ji}(1)}\right], \quad L(q_{ij}) = \log\left[\frac{q_{ij}(0)}{q_{ij}(1)}\right].$$

By using $\tanh[\log(p_0/p_1)/2] = p_0 - p_1 = 1 - 2p_1$, show that

$$L(r_{ji}) = 2\tanh^{-1}\{\prod_{i' \in V_j \backslash i} \tanh\left[\frac{1}{2}L(q_{i'j})\right]\} \text{ and } L(q_{ij}) = L(c_i) + \sum_{j' \in C_i \backslash j} L(r_{j'i}).$$

9.21. This problem is related to the LDPC decoder implementation using the min-sum-with-correction term algorithm:

(a) Implement the LDPC decoder using the min-sum-with-correction-term algorithm. By assuming that zero codeword was transmitted over an additive white Gaussian noise (AWGN) channel for the LDPC code described in example of Sect. 9.10.5.4, show BER dependence against Q-factor for on-off keying. Determine the coding gains at BER of 10^{-6} and 10^{-9}. Discuss the results.

(b) Implement now corresponding encoder, use sufficiently long pseudorandom binary sequence, encode it, transmit over the AWGN channel, and repeat the simulation from (a). Discuss the differences.

(c) You are asked now to transmit an LDPC encoded sequence over the realistic optical channel. Generate the PRBS sequence of sufficient length to accurately estimate BER down to 10^{-6}. Encode this sequence using the LDPC encoder you developed in problem (b), and transmit over the optical fiber transmission systems as described below. Represent the symbol 1 pulse by a Gaussian pulse with FWHM of 12.5 ps. Assume the launch power of 0 dBm, the extinction ratio of 20 dB, and data rate of 40 Gb/s. Transmission system is 6000 km in length, with dispersion map composed of N sections of SMF of length 80 km (with dispersion and dispersion slope parameters being 16 ps/(nm·km) and 0.06 ps/(nm^2·km), respectively) and 40 km of DCF. The DCF is chosen to exactly compensate for residual chromatic dispersion, and dispersion map slope is matched. The nonlinear coefficient of both fibers is $\gamma = 2.5$ (W·km)$^{-1}$. The EDFAs with NF of 5 dB and $BW = 3R_b$ (R_b is the bit rate) are periodically deployed every fiber section to exactly compensate for fiber loss. The SMF loss is 0.2 dB/km and DCF loss is 0.5 dB/km. The operating wavelength is 1550 nm and photodiode responsivity is 1 A/W. The receiver electronics can be model as Gaussian filter of 3 dB bandwidth equal BW = 0.75 R_b. Vary the launch power from −6 dBm to 6 dBm in steps of 2 dBm, and plot how the BER changes as a function of the launch power before and after LDPC decoding. Determine how much transmission distance can be extended by using the LDPC code. Provide BER vs. total transmission plots for both uncoded and LDPC-coded cases, for different launch powers.

9.22. This problem is related to the QAM with coherent detection in a back-to-back configuration. The transmitter and receiver are shown in Fig. 9.P22a, b, respectively. Determine by simulation both bit error rate P_b and symbol error rate P_s against optical SNR per bit for the following signal constellation sizes: $M = 4$, 16, and 64.

9.23. This problem is related again to the QAM with coherent detection, but now in a long-haul optical transmission system with dispersion map as shown in Fig. 9.P22c. The transmitter and receiver configurations are identical to those shown in Fig. 9.P22a, b.

Assume the average symbol launch power of 0 dBm, the symbol rate is 25 Giga symbol/s (25 GS/s), and that either QPSK or 16-QAM is used. Transmission system is composed of N sections of SMF of length 80 km (with dispersion and dispersion slope parameters being 16 ps/(nm·km) and 0.06 ps/(nm^2·km), respectively) and 40 km of DCF. The DCF is chosen to exactly compensate for residual chromatic dispersion, and dispersion map is slope matched. The nonlinear coefficient of both fibers is $\gamma = 2.5\ (\text{W·km})^{-1}$. The EDFAs of NF of 5 dB and $BW = 3R_b$ (R_b is the bit rate) are periodically deployed every fiber section to exactly compensate for fiber loss. The SMF loss is 0.2 dB/km and DCF loss is 0.5 dB/km. The operating wavelength is 1550 nm and photodiode responsivity is 1 A/W. The receiver electronics can be model as Gaussian filter of 3 dB bandwidth equal BW $= 0.75\ R_b$. Vary the launch power from −6 dBm to 6 dBm in steps of 2 dBm, and plot how the BER changes as a function of total transmission length by using the launch power as a parameter. Show both bit error and symbol error rates with both natural and Gray mappings.

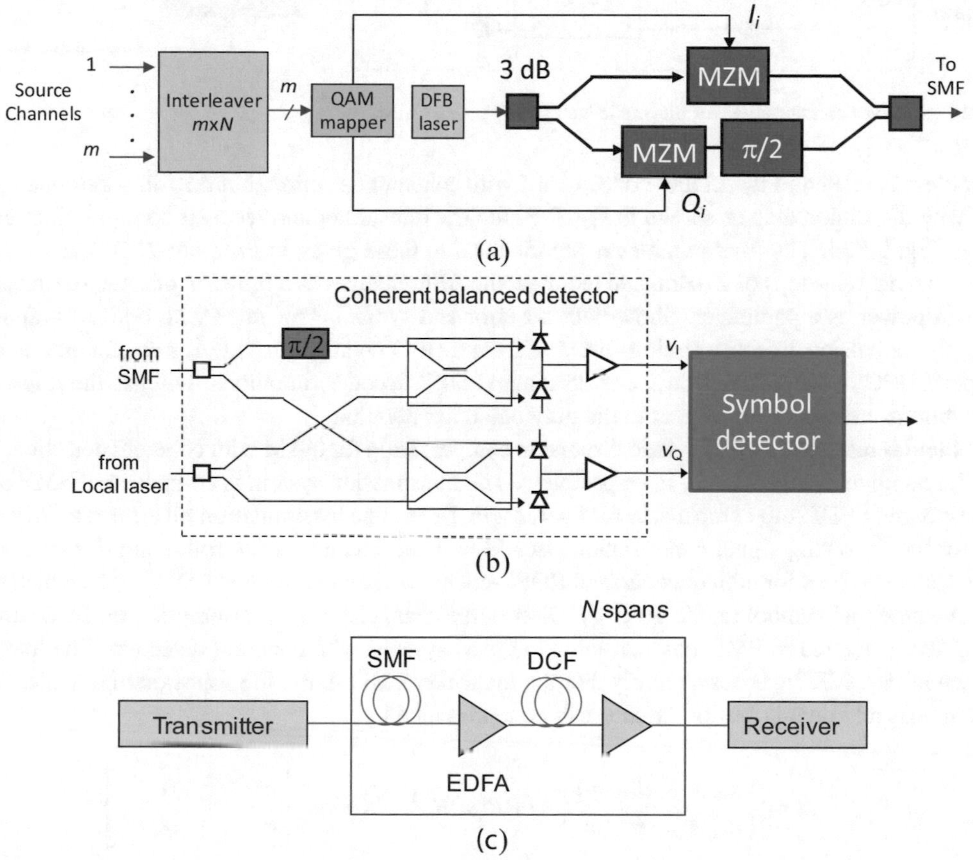

Fig. 9.P22 QAM with coherent detection: (**a**) transmitter configuration, (**b**) receiver configuration, and (**c**) dispersion map. *3dB* 3dB coupler, *MZM* Mach-Zehnder modulator

9.24. This problem is related to the coded modulation with coherent detection in a back-to-back configuration. Transmitter configuration is shown in Fig. 9.P22a, and receiver configuration is shown in Fig. 9.P24. Let us observe again QAM, but this time in combination with LDPC codes. The outputs of LDPC encoders are written in corresponding block interleaver by block row-wise order, as shown in Fig. 9.P22a. The $m = \log_2 M$ bits are taken from block interleaver column-wise to select the corresponding constellation point using the Gray mapping rule. The upper branch at the input of modulator represents the in-phase channel (I) and lower the quadrature channel (Q). The a posteriori probability (APP) block determines the symbol log-likelihood ratios (LLRs). Bit LLRs block calculates the bit LLRs needed for LDPC decoding:

(a) Derive the symbol LLRs. How to calculate bit LLRs based on symbol LLRs?
(b) Perform Monte Carlo simulations for $M = 2$, 4, and 16 for both uncoded case and LDPC-coded case. Plot BERs versus optical SNR per information bit.
(c) Determine the coding gains at BER of 10^{-6}.
(d) Iterate extrinsic information between LDPC decoders and APP demapper 3 times, and provide BER plots vs. optical SNR per bit, down to 10^{-6}. Determine the OSNR improvement with respect to (c).

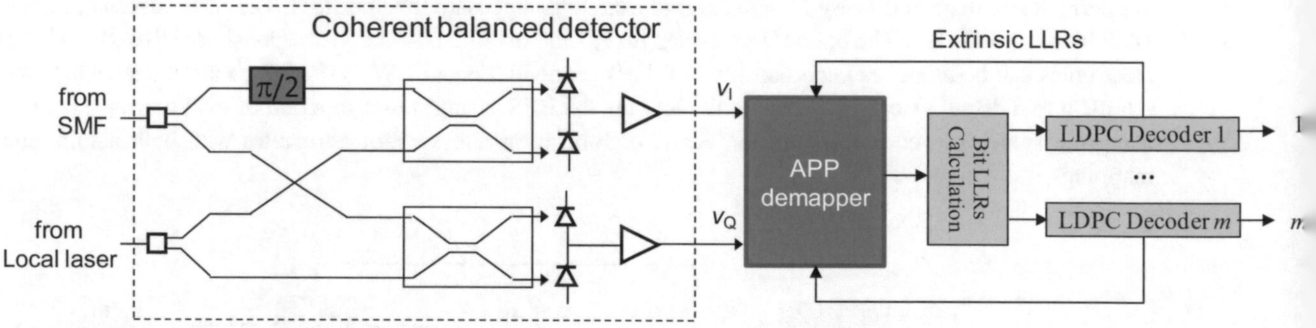

Fig. 9.P24 Receiver configuration for bit-interleaved LDPC-coded QAM

9.25. This problem is related to the LDPC-coded QAM with coherent detection, but now in a long-haul optical transmission system with dispersion map as shown in Fig. 9.P22c. The transmitter and receiver configurations are identical to those shown in Fig. 9.P24. The fiber parameters are identical to those given in Problem 9.23. Vary the launch power from −6 dBm to 6 dBm in steps of 2 dBm, and plot how the BER changes as a function of total transmission length by using the launch power as a parameter. Show both bit error and symbol error rates with both natural and Gray mappings. Identify the extension in transmission distance from LDPC coding. Provide results for one and three outer (APP demapper-LDPC decoder) iterations. Use 25 inner (LDPC decoder) iterations. Discuss the improvement when three outer iterations are used with respect to the only one outer iteration.

9.26. This problem is related to the chromatic dispersion compensation for QAM with coherent detection. The transmitter and receiver are shown in Fig. 9.P26a, b, respectively. The transmission system is composed of SMF of length L followed by corresponding EDFA to compensate for fiber losses. Determine by simulation bit error rate *BER* against optical SNR per bit for the following signal constellation sizes $M = 4$ and 16 and for the following distances $L = 40$, 60, 80, and 100 km. Calculate *BER* for both uncoded and LDPC-coded cases. Use the same LDPC code, which was provided to you earlier. Assume that symbol rate is 40 GS/s. Discuss the channel memory assumption on BER results.

9.27. This problem is related to PMD compensation in QAM systems with coherent detection. The transmitter and receiver are shown in Fig. 9.P27a, b, respectively. For the first-order PMD study, the Jones matrix, neglecting the polarization-dependent loss and depolarization effects, can be represented by

$$\boldsymbol{H} = \begin{bmatrix} h_{xx}(\omega) & h_{xy}(\omega) \\ h_{yx}(\omega) & h_{yy}(\omega) \end{bmatrix} = \boldsymbol{RP}(\omega)\boldsymbol{R}^{-1},\ \boldsymbol{P}(\omega) = \begin{bmatrix} e^{-j\omega\tau/2} & 0 \\ 0 & e^{j\omega\tau/2} \end{bmatrix},$$

where τ denotes DGD, ω is the angular frequency, and $R = R(\theta,\varepsilon)$ is the rotational matrix:

$$\boldsymbol{R} = \begin{bmatrix} \cos\left(\frac{\theta}{2}\right)e^{j\varepsilon/2} & \sin\left(\frac{\theta}{2}\right)e^{-j\varepsilon/2} \\ -\sin\left(\frac{\theta}{2}\right)e^{j\varepsilon/2} & \cos\left(\frac{\theta}{2}\right)e^{-j\varepsilon/2} \end{bmatrix},$$

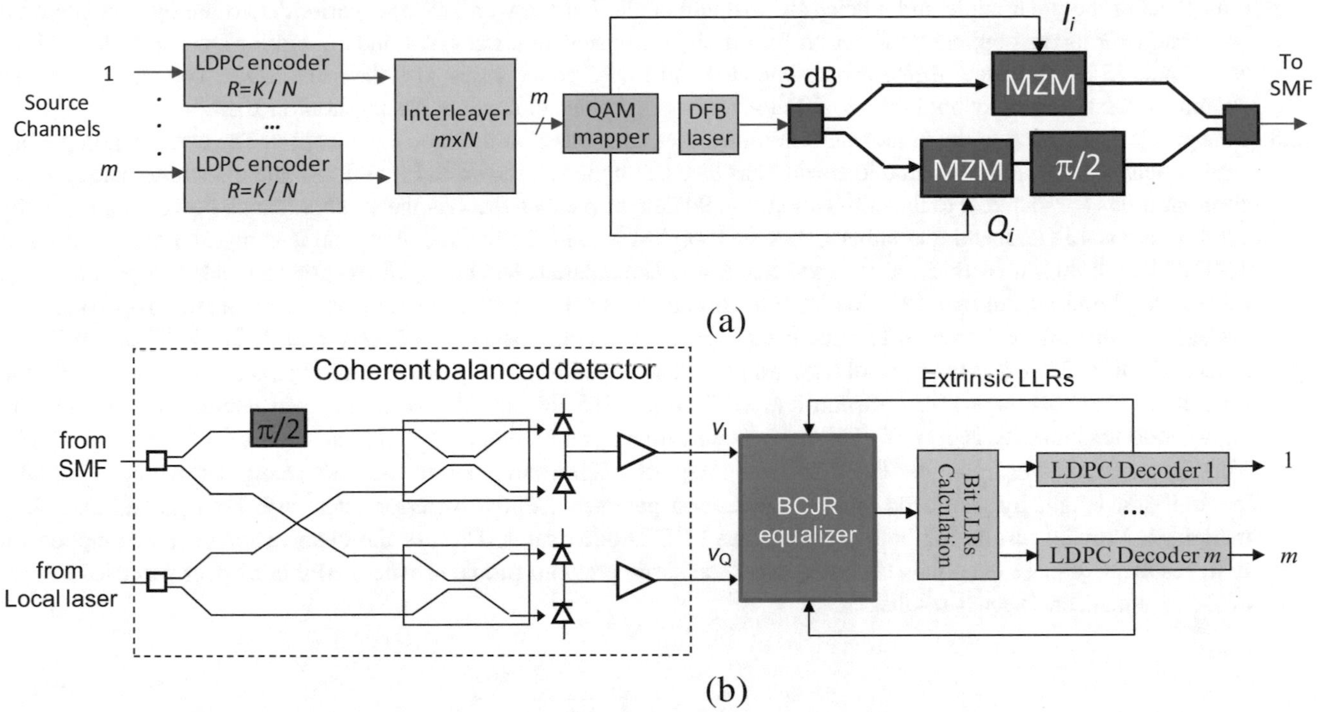

Fig. 9.P26 Bit-interleaved LDPC-coded QAM with turbo equalization: (**a**) Tx and (**b**) Rx configurations

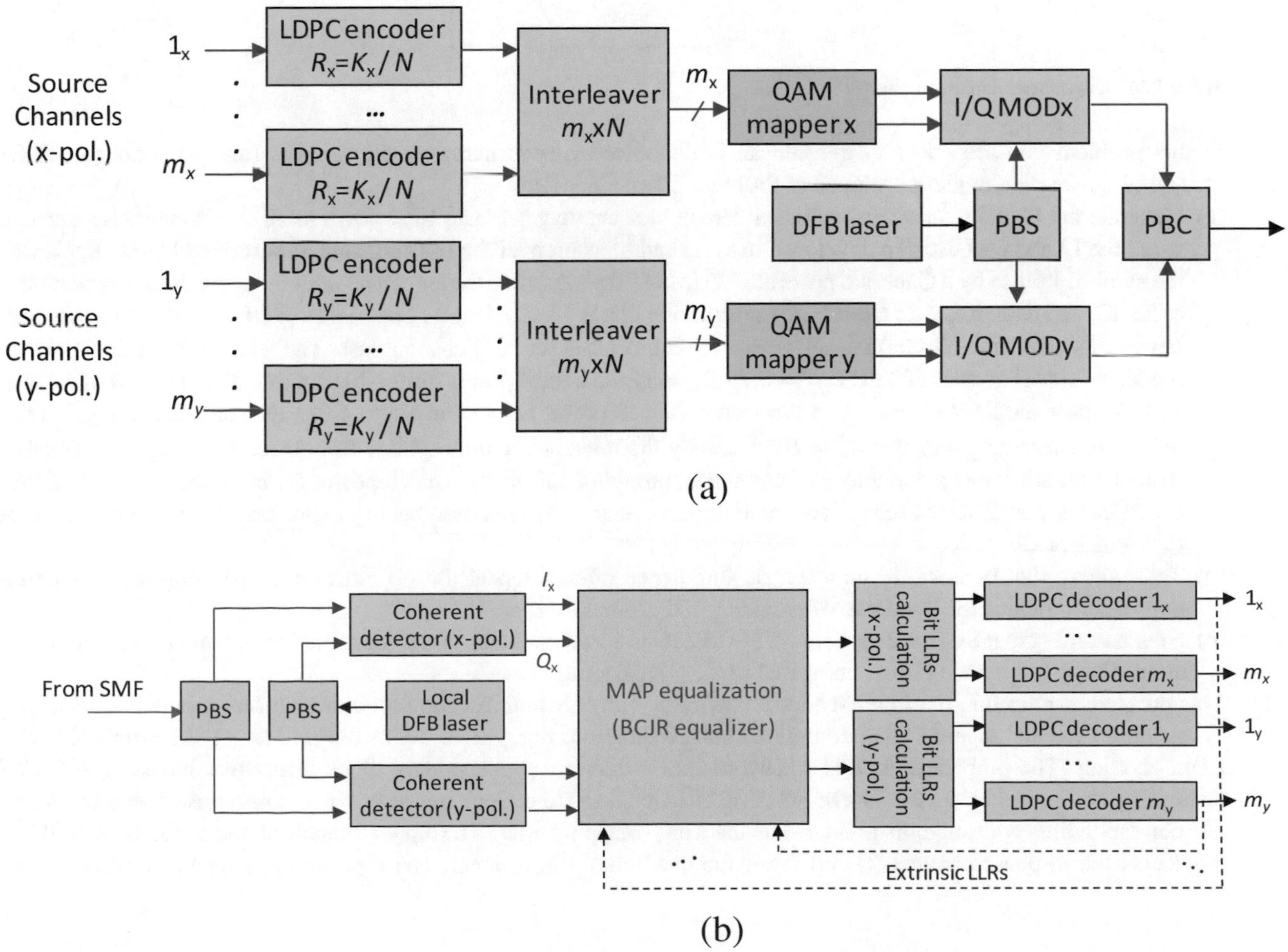

Fig. 9.P27 Polarization-multiplexed bit-interleaved LDPC-coded QAM with turbo equalization: (**a**) Tx and (**b**) Rx configurations

with θ being the polar angle and ε being the azimuth angle. For the worst-case scenario, determine by simulation bit error rate BER against optical SNR per bit for the signal constellation sizes of 4 and 16 and for DGD values of 25 ps, 50 ps, and 75 ps. Calculate BER for both uncoded and LDPC-coded cases. Use the same LDPC code as in previous examples. Assume that symbol rate is 40 GS/s. Discuss the channel memory assumption on BER.

9.28. This problem is related to the intrachannel nonlinearities compensation by turbo equalization. The corresponding long-haul optical transmission system under study has dispersion map as shown in Fig. 9.P28. The transmitter and receiver configurations are identical to those shown in Fig. 9.P22a, b. Assume the average symbol launch power of 0 dBm, the symbol rate is 40 GS/s, and that either QPSK or 16-QAM is used. Transmission system is composed of N sections of SMF of length 80 km (with dispersion and dispersion slope parameters being 16 ps/(nm·km) and 0.06 ps/(nm^2 km), respectively) and 40 km of DCF. The DCF is chosen to exactly compensate for residual chromatic dispersion, and dispersion map is slope matched. The nonlinear coefficient of both fibers is $\gamma = 2.5\ (\mathrm{W{\cdot}km})^{-1}$. The EDFAs with NF of 5 dB and $BW = 3R_s$ (R_s is the symbol rate) are periodically deployed every fiber section to exactly compensate for the fiber loss. The SMF loss is 0.2 dB/km and DCF loss is 0.5 dB/km. The operating wavelength is 1550 nm and photodiode responsivity is 1 A/W. The receiver electronics can be model as Gaussian filter of 3 dB BW $= 0.75R_s$. Vary the launch power from −6 dBm to 6 dBm in steps of 2 dBm, and plot how the BER changes as a function of total transmission length by using the launch power as a parameter. Show bit error rates with both natural and Gray mappings. Provide results for both uncoded and LDPC-coded cases. Discuss the channel memory assumption on BER results. Compare the obtained results with those from previous problems where APP demapper was used instead of BCJR equalizer. Discuss results.

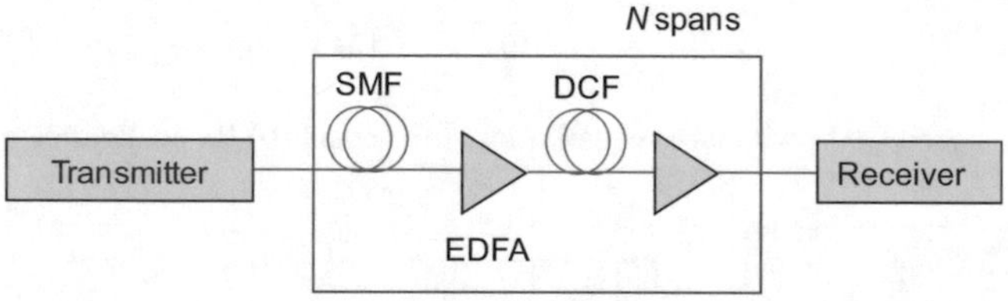

Fig. 9.P28 Dispersion map under study

9.29. In this problem you are asked to transmit an LDPC encoded sequence over the realistic free-space optical (FSO) channel, by using the model developed in Problem 2.7 of Chap. 2:

(a) Generate the PRBS sequence of sufficient length to accurately estimate BER down to 10^{-6}. Encode this sequence using the LDPC encoder you developed in 9.21, and transmit over the FSO channel as described below. Represent the symbol 1 pulse by a Gaussian pulse of FWHM of 20 ps. Assume the launch power of 0 dBm, the extinction ratio of 25 dB, and data rate of 25 Gb/s. FSO transmission link is 1 km in length. The EDFA of NF of 5 dB and bandwidth $BW = 3R_b$ (R_b is the bit rate) is used to exactly compensate for FSO channel loss. The receiver electronics can be model as Gaussian filter of 3 dB BW $= 0.75\ R_b$. Vary the launch power from −6 dBm to 6 dBm in steps of 1 dBm, and plot how the BER changes as a function of launch power before and after LDPC decoding, assuming that the refractive structure parameter $C_n^2 = 10^{-15}$. Study the tolerance to atmospheric turbulence by using the refractive structure parameter as a variable. To summarize, provide BER vs. the launch power for both uncoded and LDPC-coded cases, for different refractive structure parameters. The Gaussian beam parameters should be the same in Problem 1 of Chap. 2.

(b) To improve the tolerance to atmospheric turbulence effects, repeat the (a) but now performing the shortening approach to reduce the code rate of provided LDPC code.

(c) Now assuming that the 2 × 2 repetition FSO MIMO is used, repeat (a), and discuss the improvements in tolerance to atmospheric turbulence effects compared to adaptive coding.

9.30. This problem is related to the LDPC-coded modulation with coherent detection for communication over FSO channel. Transmitter (Tx) and receiver (Rx) configurations are shown in Fig. 9.P26. Let us observe the QAM, combined with LDPC coding. The outputs of LDPC codewords are written in corresponding block-interleaver block in row-wise fashion, as shown in Fig. 9.P26. The $m = \log_2 M$ bits are taken from block interleaver in column-wise fashion to select the corresponding constellation point using the Gray mapping rule. The upper branch at the input of modulator represents the in-phase channel (I) and lower the quadrature (Q) channel. The a posteriori probability (APP) block determines the symbol log-likelihood ratios (LLRs). Bit LLRs block calculates the bit LLRs needed for LDPC decoding:

(a) Assuming that the refractive structure parameter is $C_n^2 = 10^{-15}$, perform the Monte Carlo simulations for $M = 2$, 4, and 16 for both uncoded case and LDPC-coded case. Plot BERs versus optical SNR per information bit. Determine the coding gains at BER of 10^{-6}. The simulation conditions are identical to those described in Problem 9.29(a).

(b) Iterate extrinsic information between LDPC decoders and APP demapper 3 times, and provide BER plots vs. optical SNR per bit, down to 10^{-6}. Determine the OSNR improvement with respect to (a).

(c) Perform adaptive coding with shortening approach, and show improvements by adaptive coding for $M = 2$, 4, and 16 for different turbulence strengths.

9.31. This problem is continuation of Problem 9.30, but now in MIMO environment. Let us consider 2 × 2 FSO MIMO system, in which two independent data streams are transmitted by two sufficiently separated FSO transmitters. Assume that each FSO transmitter illuminates both FSO receivers:

(a) For 2 × 2 FSO MIMO system with either zero-forcing (ZF) or minimum MSE (MMSE) receiver, for the refractive structure parameter of $C_n^2 = 10^{-15}$, perform Monte Carlo simulations for $M = 2$, 4, and 16 for both uncoded case and LDPC-coded case. Plot BERs versus optical SNR per information bit. Determine the coding gains at BER of 10^{-6}. Assume that channel matrix $\boldsymbol{H}$ is available on receiver side. The simulation conditions are identical to those described in Problem 9.29(a).

(b) Repeat (a) but now for different values of the refractive structure parameter. At BER of 10^{-6}, for LDPC-coded QAM, determine improvements when MMSE receiver is used compared to the ZF receiver case.

(c) Discuss the improvements of LDPC-coded QAM when MMSE receiver is used with respect to the adaptive coding case discussed in Problem 9.31(c) for different turbulence strengths.

9.32. This problem is related to *coded modulation applied to 9-ary 2-D constellation* described in Problem 5.8 of Chap. 5, in the presence of scintillation and AWGN:

(a) Describe the transmitter and receiver configuration for polarization-multiplexed 9-ary LDPC-coded modulation.

(b) Determine the symbol LLRs required for 3-ary LDPC decoding. Describe how to determine the extrinsic information at the output of 3-ary LDPC decoder for APP demapper. Describe how to determine the prior symbol LLRs for next APP-LDPC decoder iteration step.

9.33. This problem is related to the *adaptive modulation and coding applied to 9-ary 2-D constellation* described in Problem 5.8 of Chap. 5, in the presence of scintillation and AWGN. First of all derive an error probability bound for 9-ary 2-D constellation in the presence AWGN to be used throughout the problem:

(a) Determine the optimum power adaptation strategy as well as the corresponding spectral efficiency of this scheme, in the presence of scintillation and AWGN.

(b) Describe the channel inversion technique for this signal constellation, and determine the corresponding spectral efficiency, in the presence of scintillation and AWGN.

9.34. This problem is related to the adaptive modulation and coding applied to 9-ary 2-D constellation described below, transmitted over *wireless channel in the presence of Nakagami fading and AWGN*. The eight constellation points in this 2-D constellation are placed on a circle of radius 1. The last (ninth) point is placed in the origin. Assume that the point in the origin appears with probability 0.25, all others with probability 0.75/8. For channel coding use the LDPC code from Problem 9.21:

(a) Determine the average symbol error probability for this modulation format in the presence of fading and AWGN.

(b) Estimate the coding gain for this LDPC code and use it in incoming (c)–(e) problems.

(c) Determine the optimum power adaptation strategy as well as the corresponding spectral efficiency of this scheme, in the presence of fading and AWGN.

(d) Describe the channel inversion technique for this signal constellation, and determine the corresponding spectral efficiency, in the presence of fading and AWGN.

(e) Describe the truncated channel inversion technique for this signal constellation, and determine the corresponding spectral efficiency, in the presence of fading and AWGN.

9.35. Implement the parallel (1152, 1024) turbo decoder. The turbo code should be of rate $R = 8/9$. Use two identical RSC codes for encoding, whose encoder in octal form is given by $(g_1,g_2) = (7,5)$. To achieve the desired code rate, perform the puncturing (e.g., you can save only 1 parity bit from block of 16 parity bits). Perform Monte Carlo simulations on an additive white Gaussian channel (AWGN) model by employing BPSK. Plot BER against signal-to-noise ratio (SNR) (in dB scale) for both uncoded case and turbo-coded case. Determine the coding gain at BER of 10^{-5}.

9.36. This step is related to the coded modulation. The transmitter and receiver are shown in Fig. 9.P36a, b, respectively. The noise from different amplifiers at receiver side can be modeled as AWGN.

Let us assume that M-ary QAM is used. The output of parallel turbo encoder is written into a buffer. The $m = \log_2 M$ bits are taken from the buffer and used to select the corresponding constellation point using the Gray mapping rule. After D/A conversion (DAC), M-QAM modulator is used. The upper branch at the input of modulator represents the in-phase channel (I) and lower the quadrature channel (Q). After up-conversion and amplification signal is emitted by transmit antenna. Assume that there exists a line of sight (LOS) between transmitter and receiver, and ignore the multipath fading and shadowing effect. On the receiver side, the signal after antenna is amplified, the I- and Q-channel signals obtained upon demodulation are converted into digital form (assume double precision). The a posteriori probability (APP) block determines the symbol log-likelihood ratios (LLRs). Bit LLRs block calculates the bit LLRs needed for turbo decoding:

(a) Derive the symbol LLRs. How to calculate bit LLRs based on symbol LLRs?
(b) Perform Monte Carlo simulations for $M = 2$, 4, and 16 for both uncoded case and turbo-coded case. Plot BERs versus SNR per information bit.
(c) Determine the coding gains at BER of 10^{-5}.

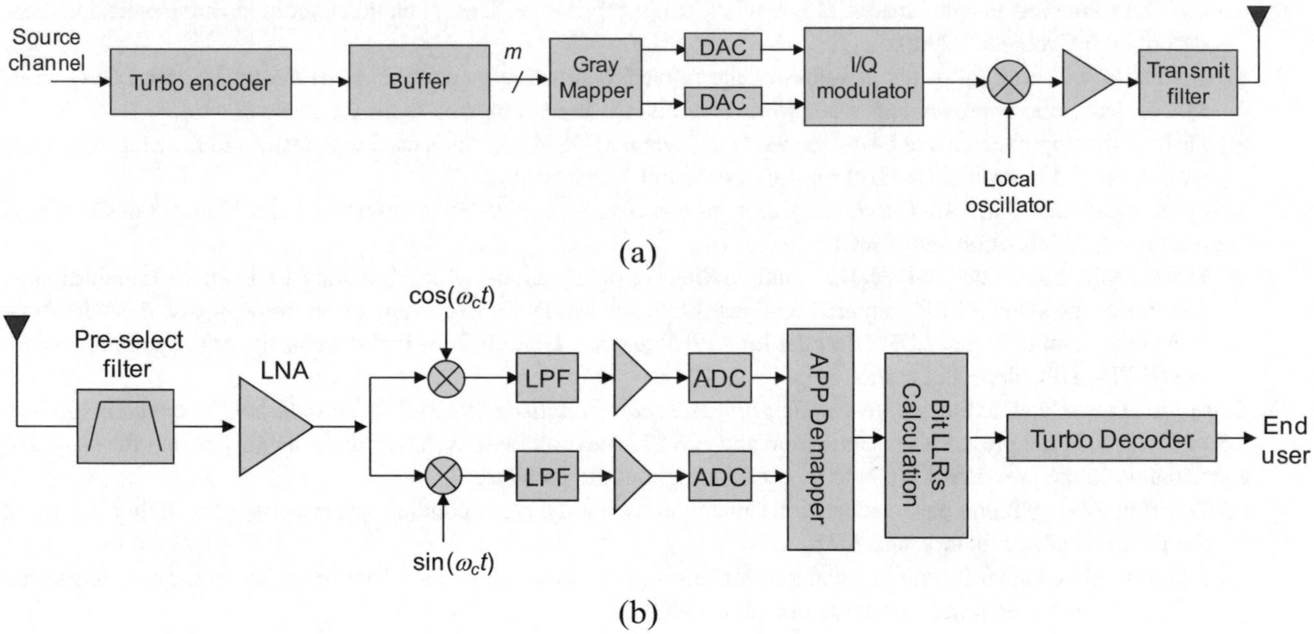

Fig. 9.P36 Bit-interleaved LDPC-coded modulation-based wireless system: (**a**) transmitter configuration and (**b**) receiver configuration. *DAC* digital-to-analog converter, *LNA* low-noise amplifier, *LPF* low-pass filter, *ADC* analog-to-digital converter

9.37. Repeat the Problem 9.36 by modeling the wireless channel using the Rayleigh fading simulator you have developed in Problem 2.13 of Chap. 2. In addition to questions (a)–(c), determine the penalty coming from wireless channel at BER of 10^{-5}.

9.38. This problem is related to the adaptive modulation. We are concerned with adaptive MQAM transmission over Rayleigh fading channel, as shown in Fig. 9.P38. The channel estimate from receiver side is transmitted back to transmitter that adapts its power and signal constellation size according to the channel conditions. When channel conditions are favorable, more data is transmitted. When channel conditions are poor, less data is transmitted; and when SNR falls below threshold, data are not transmitted at all. Derive the optimum power adaptation policy. Determine the spectral efficiency for optimum power adaptation policy. Write a MATLAB program that determines the spectral efficiency vs. SNR for target bit error probability of 10^{-5}. For spectral efficiency of 2 bits/s/Hz, determine how far away this adaptation policy is from the channel capacity. When the truncated channel inversion is used, determine SNR degradation for spectral efficiency of 2 bits/s/Hz with respect to the optimum adaptation strategy. The noise from different amplifiers at receiver side can be modeled as AWGN.

9.39. Repeat the Problem 9.38 but now observing the Nakagami $m = 2$ fading instead. Compare the conclusions between Problems 9.38 and 9.39.

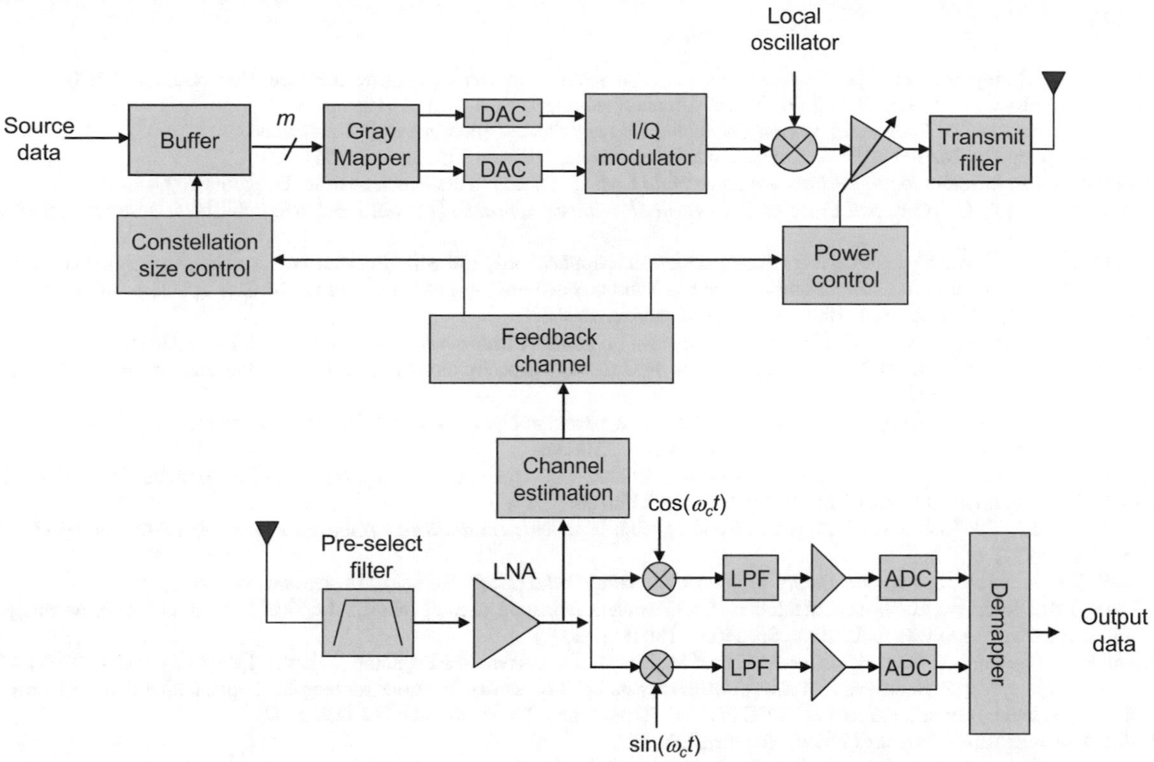

Fig. 9.P38 The variable-power variable-rate adaptation system. *DAC* digital-to-analog converter, *LNA* low-noise amplifier, *LPF* low-pass filter, *ADC* analog-to-digital converter

9.40. We now observe the adaptive turbo-coded modulation. Corresponding system configuration is shown in Fig. 9.P40. For turbo code you developed in Problem 9.35 and for spectral efficiency of 2 bits/s/Hz, determine how far away this adaptive coding scheme is from the channel capacity. When the truncated channel inversion is used, determine SNR degradation for spectral efficiency of 2 bits/s/Hz with respect to the optimum adaptation strategy. Write a MATLAB or C/C++ program that plots the spectral efficiency against SNR when turbo coding is used. Determine the improvement due to adaptive coding.

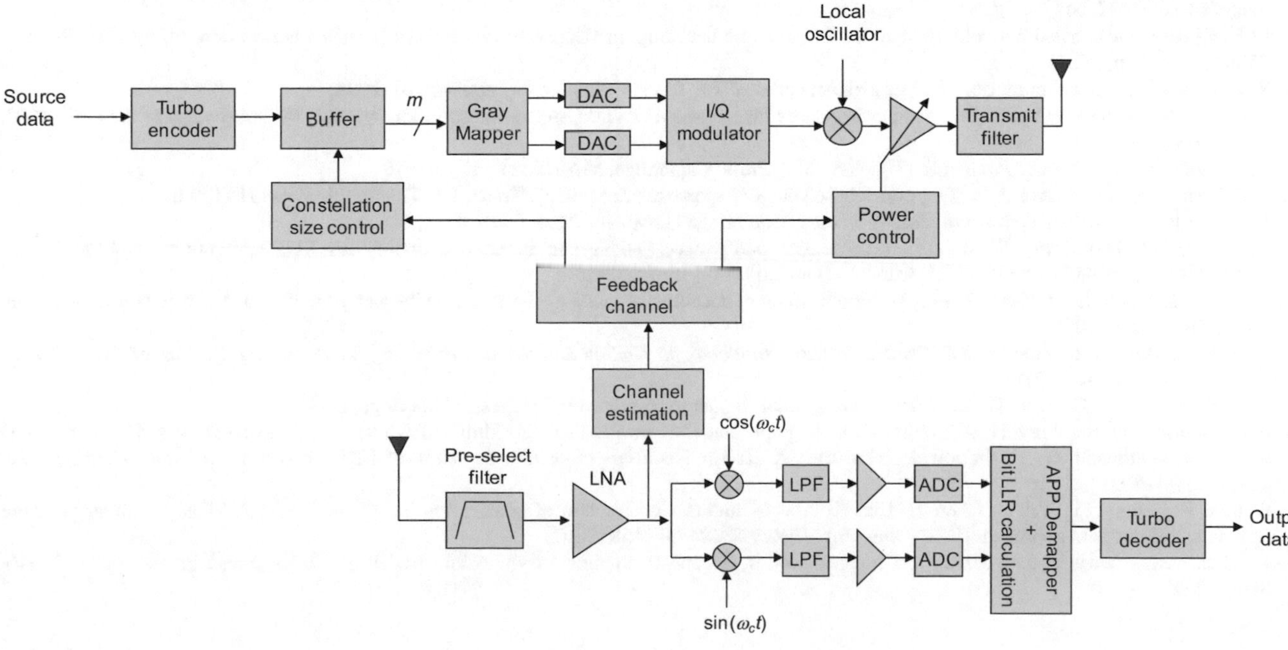

Fig. 9.P40 The turbo-coded variable-power variable-rate adaptation system. *DAC* digital-to-analog converter, *LNA* low-noise amplifier, *LPF* low-pass filter, *ADC* analog-to-digital converter, *APP*, a posteriori probability

References

1. M. Cvijetic, I.B. Djordjevic, *Advanced Optical Communication Systems and Networks* (Artech House, Norwood, Jan. 2013)
2. I. Djordjevic, W. Ryan, B. Vasic, *Coding for Optical Channels* (Springer, New York, 2010)
3. S.B. Wicker, *Error Control Systems for Digital Communication and Storage* (Prentice Hall, Englewood Cliffs, 1995)
4. R.H. Morelos-Zaragoza, *The Art of Error Correcting Coding*, 2nd edn. (Wiley, Chichester, 2006)
5. D. Drajic, P. Ivanis, *Introduction to Information Theory and Coding*, 3rd edn. (Akademska Misao, Belgrade, 2009)
6. J.B. Anderson, S. Mohan, *Source and Channel Coding: An Algorithmic Approach* (Kluwer Academic Publishers, Boston/Dordrecht/London, 1991)
7. S. Lin, D.J. Costello, *Error Control Coding: Fundamentals and Applications*, 2nd edn. (Pearson Prentice Hall, Upper Saddle River, 2004)
8. T. Mizuochi et al., Forward error correction based on block turbo code with 3-bit soft decision for 10 Gb/s optical communication systems. IEEE J. Sel. Top. Quantum Electron. **10**(2), 376–386 (Mar./Apr. 2004)
9. T. Mizuochi, et al., Next generation FEC for optical transmission systems, in *Proceedings of OFC 2003*, Paper ThN1
10. I.B. Djordjevic, M. Arabaci, L. Minkov, Next generation FEC for high-capacity communication in optical transport networks. J. Lightwave Technol. **27**, 3518–3530 (2009)
11. I.B. Djordjevic, Advanced coding for optical communication systems, in *Optical Fiber Telecommunications VI*, ed. by I. Kaminow, T. Li, A. Willner, (Elsevier/Academic Press, Boston, May 2013), pp. 221–296
12. I.B. Djordjevic, On advanced FEC and coded modulation for ultra-high-speed optical transmission. IEEE Commun. Surv. Tutorials **18**(99), 1–31 (1 Mar. 2016). https://doi.org/10.1109/COMST.2016.2536726
13. L.R. Bahl, J. Cocke, F. Jelinek, J. Raviv, Optimal decoding of linear codes for minimizing symbol error rate. IEEE Trans. Inf. Theory **IT-20**(2), 284–287 (Mar. 1974)
14. F.J. MacWilliams, N.J.A. Sloane, *The Theory of Error-Correcting Codes* (North Holland, Amsterdam, 1977)
15. M. Ivkovic, I.B. Djordjevic, B. Vasic, Calculation of achievable information rates of long-haul optical transmission systems using instanton approach. IEEE/OSA J. Lightwave Technol. **25**, 1163–1168 (May 2007)
16. G. Bosco, P. Poggiolini, Long-distance effectiveness of MLSE IMDD receivers. IEEE Photon. Tehcnol. Lett. **18**(9), 1037–1039 (1 May 2006)
17. M. Ivkovic, I. Djordjevic, P. Rajkovic, B. Vasic, Pulse energy probability density functions for long-haul optical fiber transmission systems by using instantons and edgeworth expansion. IEEE Photon. Technol. Lett. **19**, 1604–1606 (15 Oct. 2007)
18. S. Haykin, *Communication Systems* (Wiley, Hoboken, 2004)
19. M.E. van Valkenburg, *Network Analysis*, 3rd edn. (Prentice-Hall, Englewood Cliffs, 1974)
20. B. Vucetic, J. Yuan, *Turbo Codes-Principles and Applications* (Kluwer Academic Publishers, Boston, 2000)
21. B.P. Smith, F.R. Kschischang, Future prospects for FEC in fiber-optic communications. IEEE J. Sel. Top. Quantum Electron. **16**(5), 1245–1257 (Sept.–Oct. 2010)
22. R.M. Tanner, A recursive approach to low complexity codes. IEEE Tans. Inf. Theory **IT-27**, 533–547 (1981)
23. C. Berrou, A. Glavieux, P. Thitimajshima, Near Shannon limit error-correcting coding and decoding: turbo-codes, in *Proceedings of the IEEE International Conference on Communications 1993 (ICC '93)*, vol 2, pp. 1064–1070, 23–26 May 1993
24. C. Heegard, S. Wicker, *Turbo Codes* (Kluwer Academic Press, Boston, 1999)
25. B. Moision, J. Hamkins, Deep space optical communications downlink budget: Modulation and coding, Jet Propulsion Laboratory, Pasadena, CA, Interplanetary Netw. Progr. Rep., pp. 42–154, Apr.–Jun. 2003
26. Available at: http://ipnpr.jpl.nasa.gov/progress_report/42-154/154K.pdf
27. B. Moision, J. Hamkins, Coded modulation for the deep space optical channel: serially concatenated pulse-position modulation, Jet Propulsion Laboratory, Pasadena, CA, Interplanetary Netw. Progr. Rep. 42–161, pp. 1–25, 15 May 2005. Available at: http://ipnpr.jpl.nasa.gov/progress_report/42-161/161T.pdf
28. W.E. Ryan, Concatenated convolutional codes and iterative decoding, in *Wiley Encyclopedia in Telecommunications*, ed. by J. G. Proakis, (Wiley, Hoboken, 2003)
29. R.M. Pyndiah, Near optimum decoding of product codes. IEEE Trans. Commun. **46**, 1003–1010 (1998)
30. O.A. Sab, V. Lemarie, Block turbo code performances for long-haul DWDM optical transmission systems, in *Proceedings of the OFC 2001*, Paper ThS5
31. R.G. Gallager, *Low Density Parity Check Codes* (MIT Press, Cambridge, MA, 1963)
32. D.J.C. MacKay, Good error correcting codes based on very sparse matrices. IEEE Trans. Inf. Theory **45**, 399–431 (1999)
33. T. Richardson, R. Urbanke, *Modern Coding Theory* (Cambridge University Press, Cambridge, 2008)
34. B. Vasic, I.B. Djordjevic, R. Kostuk, Low-density parity check codes and iterative decoding for long haul optical communication systems. J. Lightwave Technol. **21**(2), 438–446 (Feb. 2003)
35. Z.-W. Li, L. Chen, L.-Q. Zeng, S. Lin, W. Fong, Efficient encoding of quasi-cyclic low-density parity-check codes. IEEE Trans. Commun. **54**(1), 71–81 (Jan. 2006)
36. W.E. Ryan, An introduction to LDPC codes, in *CRC Handbook for Coding and Signal Processing for Recording Systems*, ed. by B. Vasic, (CRC Press, London, 2004)
37. W.E. Ryan, S. Lin, *Channel Codes: Classical and Modern* (Cambridge University Press, Cambridge, 2009)
38. F.R. Kschischang, B.J. Frey, H.-A. Loeliger, Factor graphs and the sum-product algorithm. IEEE Trans. Inf. Theory **47**, 498–519 (Feb. 2001)
39. J. Chen, A. Dholakia, E. Eleftheriou, M. Fossorier, X.-Y. Hu, Reduced-complexity decoding of LDPC codes. IEEE Trans. Commun. **53**, 1288–1299 (2005)
40. L. Lan, L. Zeng, Y.Y. Tai, L. Chen, S. Lin, K. Abdel-Ghaffar, Construction of quasi-cyclic LDPC codes for AWGN and binary erasure channels: a finite field approach. IEEE Trans. Inf. Theory **53**, 2429–2458 (2007)
41. J.L. Fan, Array-codes as low-density parity-check codes, Second International Symposium on Turbo Codes, Brest, France, pp. 543–546, Sept. 2000

42. S. Kudekar, T.J. Richardson, R.L. Urbanke, Threshold saturation via spatial coupling: why convolutional LDPC ensembles perform so well over the BEC. IEEE Trans. Inf. Theory **57**(2), 803–834 (Feb. 2011)
43. M. Lentmaier, G. Fettweis, K.S. Zigangirov, D. Costello, Approaching capacity with asymptotically regular LDPC codes, in *Proceedings of the Information Theory and Applications 2009*, San Diego, CA, USA, pp. 173–177, Feb. 2009
44. K. Sugihara, et al., A spatially-coupled type LDPC code with an NCG of 12 dB for optical transmission beyond 100 Gb/s, in *Proceedings of the OFC/NFOEC 2013*, Paper OM2B.4
45. D. Chang, et al., LDPC convolutional codes using layered decoding algorithm for high speed coherent optical transmission, in *Proceedings of OFC/NFOEC 2012*, Paper OW1H.4
46. Y. Zhang, I.B. Djordjevic, Staircase rate-adaptive LDPC-coded modulation for high-speed intelligent optical transmission, in *Proceedings of the OFC/NFOEC 2014*, Paper M3A.6
47. A.J. Felström, K.S. Zigangirov, Time-varying periodic convolutional codes with low-density parity-check matrix. IEEE Trans. Inf. Theory **45**(6), 2181–2191 (1999)
48. I.B. Djordjevic, On the irregular nonbinary QC-LDPC-coded hybrid multidimensional OSCD-modulation enabling beyond 100 Tb/s optical transport. J. Lightwave Technol. **31**(16), 2969–2975 (15 Aug. 2013)
49. M. Arabaci, I.B. Djordjevic, L. Xu, T. Wang, Nonbinary LDPC-coded modulation for rate-adaptive optical fiber communication without bandwidth expansion. IEEE Photon. Technol. Lett. **23**(18), 1280–1282 (15 Sept. 2012)
50. M. Arabaci, I. B. Djordjevic, L. Xu, T. Wang, Hybrid LDPC-coded modulation schemes for optical communication systems, in *Proceedings of the CLEO 2012*, Paper CTh3C.3
51. M. Arabaci, I.B. Djordjevic, R. Saunders, R.M. Marcoccia, High-rate non-binary regular quasi-cyclic LDPC codes for optical communications. J. Lightwave Technol. **27**(23), 5261–5267 (1 Dec. 2009)
52. P. Elias, Error-free coding. IRE Trans. Inf. Theory **IT-4**, 29–37 (Sept. 1954)
53. Y. Zhao, J. Qi, F.N. Hauske, C. Xie, D. Pflueger, G. Bauch, Beyond 100G optical channel noise modeling for optimized soft-decision FEC performance, in *Proceedings of the OFC/NFOEC 2012*, Paper OW1H.3
54. F. Vacondio, et al., Experimental characterization of Gaussian-distributed nonlinear distortions, in *Proceedings of the ECOC 2011*, Paper We.7.B.1
55. F. Chang, K. Onohara, T. Mizuochi, Forward error correction for 100 G transport networks. IEEE Commun. Mag. **48**(3), S48–S55 (Mar. 2010)
56. Y. Miyata, K. Sugihara, W. Matsumoto, K. Onohara, T. Sugihara, K. Kubo, H. Yoshida, T. Mizuochi, A triple-concatenated FEC using soft-decision decoding for 100 Gb/s optical transmission, in *Proceedings of the OFC/NFOEC 2010*, Paper OThL3
57. T. Mizuochi et al., Experimental demonstration of concatenated LDPC and RS code by FPGAs emulation. IEEE Photon. Technol. Lett. **21**(18), 1302–1304 (Sept. 2009)
58. Z. Zhang, et al., Lowering LDPC error floors by postprocessing, in *Proceedings of the IEEE GLOBECOM*, 2008
59. G.J. Byers, F. Takawira, Fourier transform decoding of non-binary LDPC codes, in *Proceedings Southern African Telecommunication Networks and Applications Conference (SATNAC)*, Sept. 2004
60. M. Davey, D.J.C. MacKay, Low-density parity check codes over GF(q). IEEE Commun. Lett. **2**(6), 165–167 (1998)
61. L. Barnault, D. Declercq, Fast decoding algorithm for LDPC over GF(2^q), in *Proceedings of the ITW2003*, pp. 70–73
62. C. Spagnol, W. Marnane, E. Popovici, FPGA implementations of LDPC over GF(2^m) decoders. IEEE Workshop on Signal Processing Systems, Shanghai, China, pp. 273–278, 2007
63. D. Declercq, M. Fossorier, Decoding algorithms for nonbinary LDPC codes over GF(q). IEEE Trans. Commun. **55**(4), 633–643 (2007)
64. V. Savin, Min-max decoding for non-binary LDPC codes, in *Proceedings of the ISIT 2008*, Toronto, ON, Canada, pp. 960–964, 6–11 July 2008
65. A. Voicila, F. Verdier, D. Declercq, M. Fossorier, P. Urard, Architecture of a low-complexity non-binary LDPC decoder for high order fields, in *Proceedings of the ISIT* 2007, pp. 1201–1206
66. D. Zou, I.B. Djordjevic, FPGA implementation of concatenated non-binary QC-LDPC codes for high-speed optical transport. Opt. Express **23**(10), 14501–14509 (18 May 2015)
67. D. Zou, I. B. Djordjevic, FPGA implementation of advanced FEC schemes for intelligent aggregation networks, in *Proceedings of the PhotonicsWest 2016, OPTO, Optical Metro Networks and Short-Haul Systems VIII, SPIE*, San Francisco, CA, USA, vol. 9773, pp. 977309-1–977309-7, 13–18 Feb. 2016. (Invited Paper)
68. I.B. Djordjevic, S. Sankaranarayanan, B. Vasic, Projective plane iteratively decodable block codes for WDM high-speed long-haul transmission systems. IEEE/OSA J. Lightwave Technol. **22**, 695–702 (Mar. 2004)
69. I.B. Djordjevic, B. Vasic, Performance of affine geometry low-density parity-check codes in long-haul optical communications. Eur. Trans. Telecommun. **15**, 477–483 (2004)
70. I.B. Djordjevic, B. Vasic, Projective geometry low-density parity-check codes for ultra-long haul WDM high-speed transmission. IEEE Photon. Technol. Lett. **15**, 784–786 (May 2003)
71. I.B. Djordjevic, S. Sankaranarayanan, B. Vasic, Irregular low-density parity-check codes for long haul optical communications. IEEE Photon. Technol. Lett. **16**, 338–340 (Jan. 2004)
72. I.B. Djordjevic, O. Milenkovic, B. Vasic, Generalized low-density parity-check codes for optical communication systems. J. Lightwave Technol. **23**(5), 1939–1946 (2005)
73. I.B. Djordjevic, L. Xu, T. Wang, M. Cvijetic, GLDPC codes with Reed-Muller component codes suitable for optical communications. IEEE Commun. Lett. **12**(9), 684–686 (2008)
74. I.B. Djordjevic, T. Wang, Multiple component codes based generalized LDPC codes for high-speed optical transport. Opt. Express **22**(14), 16694–16705 (2014)
75. D. Zou, I. B. Djordjevic, An FPGA design of generalized low-density parity-check codes for rate-adaptive optical transport networks. In *Proceedings of the SPIE PhotonicsWest 2016, optical metro networks and short-haul systems VIII, SPIE*, San Francisco, CA, USA, vol. 9773, pp. 97730M-1–97730M-6, 13–18 Feb. 2016
76. J. Boutros, O. Pothier, G. Zemor, Generalized low density (Tanner) codes, in *Proceedings of the IEEE Intternational Conference on Communication (ICC'99),* pp. 441–445, 1999

77. M. Lentmaier, K.S. Zigangirov, On generalized low-density parity-check codes based on Hamming component codes. IEEE Commun. Lett. **3**(8), 248–250 (1999)
78. V.A. Zyablov, R. Johannesson, M. Loncar, Low-complexity error correction of Hamming-code-based LDPC codes. Probl. Inf. Transm. **45**(2), 95–109 (2009)
79. I. Andersen, Combinatorial designs: construction methods, in *Mathematics and Its Applications*, (Ellis Horwood, Chichester, 1990)
80. M.-F. Huang, A. Tanaka, E. Ip, Y.-K. Huang, D. Qian, Y. Zhang, S. Zhang, P.N. Ji, I.B. Djordjevic, T. Wang, Y. Aono, S. Murakami, T. Tajima, T.J. Xia, G.A. Wellbrock, Terabit/s Nyquist superchannels in high capacity fiber field trials using DP-16QAM and DP-8QAM modulation formats. J. Lightwave Technol. **32**(4), 776–782 (15 Feb. 2014)
81. G.-H. Gho, J.M. Kahn, Rate-adaptive modulation and low-density parity-check coding for optical fiber transmission systems. J. Opt. Commun. Netw. **4**(10), 760–768 (Oct. 2012)
82. M. Arabaci, I.B. Djordjevic, R. Saunders, R.M. Marcoccia, Polarization-multiplexed rate-adaptive non-binary-LDPC-coded multilevel modulation with coherent detection for optical transport networks. Opt. Express **18**(3), 1820–1832 (2010)
83. L. Beygi, E. Agrell, P. Johannisson, M. Karlsson, A novel multilevel coded modulation scheme for fiber optical channel with nonlinear phase noise, in *Proceedings of the IEEE GLOBECOM*, 2010
84. J.L. Massey, Coding and modulation in digital communications, in *Proceedings of the International Zurich Seminar on Digital Communications*, Zurich, Switzerland, pp. E2(1)–E2(4), 1974
85. H. Imai, S. Hirakawa, A new multilevel coding method using error correcting codes. IEEE Trans. Inf. Theory **IT-23**(3), 371–377 (May 1977)
86. U. Wachsmann, R.F.H. Fischer, J.B. Huber, Multilevel codes: theoretical concepts and practical design rules. IEEE Trans. Inf. Theory **45**(5), 1361–1391 (July 1999)
87. K.R. Narayanan, J. Li, Bandwidth efficient low density parity check codes using multilevel coding and iterative multistage decoding, in *Proceedings of the 2nd Symposium on Turbo Codes and Related Topics*, Brest, France, pp. 165–168, 2000
88. S. Benedetto, D. Divsalar, G. Montorsi, H. Pollara, Parallel concatenated trellis coded modulation, in *Proceedings of the IEEE International Conference on Communications*, Dallas, TX, USA, pp. 974–978, 1996
89. G. Caire, G. Taricco, E. Biglieri, Bit-interleaved coded modulation. IEEE Trans. Inf. Theory **44**(3), 927–946 (May 1998)
90. J. Tan, G.L. Stüber, Analysis and design of symbol mappers for iteratively decoded BICM. IEEE Trans. Wirel. Comm. **4**(2), 662–672 (Mar. 2005)
91. J. Hou, P.H. Siegel, L.B. Milstein, H.D. Pfitser, Capacity-approaching bandwidth-efficient coded modulation schemes based on low-density parity-check codes. IEEE Trans. Inf. Theory **49**(9), 2141–2155 (2003)
92. L. Beygi et al., Coded modulation for fiber-optic networks. IEEE Sig. Proc. Mag. **31**(2), 93–103 (Mar. 2014)
93. G. Ungerboeck, Channel coding with multilevel/phase signals. IEEE Trans. Inf. Theory **IT-28**, 55–67 (1982)
94. G.D. Forney, R.G. Gallager, G.R. Lang, F.M. Longstaff, S.U. Qureshi, Efficient modulation for band-limited channels. IEEE J. Sel. Areas Comm. **SAC-2**(5), 632–647 (1984)
95. G.D. Forney, Coset codes-part I: introduction and geometrical classification. IEEE Trans. Inf.. Theory **34**(5), 1123–1151 (Sept. 1988)
96. A. Goldsmith, *Wireless Communications* (Cambridge University Press, Cambridge, 2005)
97. A.R. Calderbank, N.J.A. Sloane, New trellis codes based on lattices and cosets. IEEE Trans. Inf. Theory **IT-33**, 177–195 (1987)
98. G.D. Forney, Trellis shaping. IEEE Trans. Inf. Theory **38**(2), 281–300 (Mar. 1992)
99. I.B. Djordjevic, B. Vasic, Multilevel coding in *M*-ary DPSK/differential QAM high-speed optical transmission with direct detection. IEEE/OSA J. Lightwave Technol. **24**(1), 420–428 (Jan. 2006)
100. J. Hou, P.H. Siegel, L.B. Milstein, H.D. Pfitser, Capacity approaching bandwidth-efficient coded modulation schemes based on low-density parity-check codes. IEEE Trans. Inf. Theory **49**(9), 2141–2155 (Sept. 2003)
101. S. ten Brink, Designing iterative decoding schemes with the extrinsic information transfer chart. Intern. J. Electron. Commun. **54**, 389–398 (Dec. 2000)
102. J. Hou, P.H. Siegel, L.B. Milstein, Design of multi-input multi-output systems based on low-density parity-check codes. IEEE Trans. Commun. **53**, 601–611 (Apr. 2005)
103. I.B. Djordjevic, M. Cvijetic, L. Xu, T. Wang, Using LDPC-coded modulation and coherent detection for ultra high-speed optical transmission. IEEE/OSA J. Lightwave Technol. **25**, 3619–3625 (Nov. 2007)
104. S. Le Goff, A. Glavieux, C. Berrou, Turbo-codes and high-spectral efficiency modulation, *Proceedings of the 1994 IEEE International Conference on Communications (ICC'94)*, New Orleans, LA, USA, pp. 645–649, 1994
105. P. Robertson, T. Wörz, Coded modulation scheme employing turbo codes. Electron. Lett. **31**, 1546–1547 (Aug. 1995)
106. P. Robertson, T. Wörz, Bandwidth-efficient turbo trellis-coded modulation using punctured component codes. IEEE J. Sel. Areas Commun. **16**(2), 206–218 (Feb. 1998)
107. I.B. Djordjevic, B. Vasic, Nonlinear BCJR equalizer for suppression of intrachannel nonlinearities in 40 Gb/S optical communications systems. Opt. Express **14**, 4625–4635 (29 May 2006)
108. I.B. Djordjevic, L.L. Minkov, L. Xu, T. Wang, Suppression of fiber nonlinearities and PMD in coded-modulation schemes with coherent detection by using turbo equalization. IEEE/OSA J. Opt. Commun. Netw. **1**(6), 555–564 (Nov. 2009)
109. R. Koetter, A.C. Singer, M. Tüchler, Turbo equalization. IEEE Sig. Proc. Mag. **21**(1), 67–80 (Jan. 2004)
110. C. Duan, et al., A low-complexity sliding-window turbo equalizer for nonlinearity compensation, in *Proceedings of the OFC 2012*, Paper JW2A.59
111. T. Fujimori, et al., A study on the effectiveness of turbo equalization with FEC for nonlinearity compensation in coherent WDM transmissions, in *Proceedings OECC/PS 2013*, Paper OM2B
112. C. Douillard, M. Jézéquel, C. Berrou, A. Picart, P. Didier, A. Glavieux, Iterative correction of intersymbol interference: turbo equalization. Eur. Trans. Telecommun. **6**, 507–511 (1995)
113. E. Ip, J.M. Kahn, Compensation of dispersion and nonlinear impairments using digital backpropagation. J. Lightwave Technol. **26**(20), 3416–3425 (2008)
114. E. Ip, J.M. Kahn, Nonlinear impairment compensation using backpropagation, in *Optical Fibre, New Developments*, ed. by C. Lethien, (In-Tech, Vienna, Dec. 2009) (Invited Chapter)

115. Y. Zhang, S. Zhang, I.B. Djordjevic, On the pragmatic turbo equalizer for optical communications, in *Proceedings of the CLEO-PR&OECC/PS 2013*, Paper ThR2-4
116. Y. Zhang, M. Arabaci, I.B. Djordjevic, Evaluation of four-dimensional nonbinary LDPC-coded modulation for next-generation long-haul optical transport networks. Opt. Express **20**(8), 9296–9301 (9 Apr. 2012)
117. Y. Zhang, S. Zhang, I.B. Djordjevic, F. Yaman, T. Wang, Extending 100G transatlantic optical transmission over legacy DMF fibers using time-domain four-dimensional nonbinary LDPC-coded modulation, In *Proceedings of ECOC* 2013, Paper P.4.18
118. F. Yaman, S. Zhang, Y.-K. Huang, E. Ip, J.D. Downie, W.A. Wood, A. Zakharian, S. Mishra, J. Hurley, Y. Zhang, I.B. Djordjevic, M.-F. Huang, E. Mateo, K. Nakamura, T. Inoue, Y. Inada, T. Ogata, First quasi-single-mode transmission over transoceanic distance using few-mode fibers, in *Proceedings of the OFC Postdeadline Papers*, Paper Th5C.7, Los Angeles, CA, USA, pp. 22–26, Mar. 2015
119. Z.H. Peric, I.B. Djordjevic, S.M. Bogosavljevic, M.C. Stefanovic, Design of signal constellations for Gaussian channel by iterative polar quantization, in *Proceedings of the 9th Mediterranean Electrotechnical Conference*, Tel-Aviv, Israel, vol. 2, pp. 866–869, 18–20 May, 1998
120. I.B. Djordjevic, H.G. Batshon, L. Xu, T. Wang, Coded polarization-multiplexed iterative polar modulation (PM-IPM) for beyond 400 Gb/s serial optical transmission, in *Proceedings of the OFC/NFOEC* 2010, Paper OMK2
121. A.P.T. Lau, J.M. Kahn, Signal design and detection in presence of nonlinear phase noise. J. Lightwave Technol. **25**(10), 3008–3016 (Oct. 2007)
122. T. Liu, I.B. Djordjevic, On the optimum signal constellation design for high-speed optical transport networks. Opt. Express **20**(18), 20396–20406 (27 Aug. 2012)
123. T. Liu, I.B. Djordjevic, Optimal signal constellation design for ultra-high-speed optical transport in the presence of nonlinear phase noise. Opt. Express **22**(26), 32188–32198 (29 Dec. 2014)
124. R.E. Blahut, Computation of channel capacity and rate distortion functions. IEEE Trans. Inf. Theory **IT-18**, 460–473 (1972)
125. T. Liu, I.B. Djordjevic, Multi-dimensional optimal signal constellation sets and symbol mappings for block-interleaved coded-modulation enabling ultra-high-speed optical transport. IEEE Photonics J. **6**(4), Paper 5500714 (Aug. 2014)
126. T. Liu, I.B. Djordjevic, T. Wang, Optimal signal constellation design for ultra-high-speed optical transport in the presence of phase noise, In *Proceedings of the CLEO* 2014, Paper STu3J.7
127. T. Liu, I.B. Djordjevic, M. Li, Multidimensional signal constellation design for channel dominated with nonlinear phase noise, in *Proceedings of the 12th International Conference on Telecommunications in Modern Satellite, Cable and Broadcasting Services – TELSIKS 2015*, Nis, Serbia, pp. 133–136, 14–17 Oct. , 2015
128. I.B. Djordjevic, A. Jovanovic, M. Cvijetic, Z.H. Peric, Multidimensional vector quantization-based signal constellation design enabling beyond 1 Pb/s serial optical transport networks. IEEE Photonics J. **5**(4), Paper 7901312 (Aug. 2013)
129. S. Zhang, F. Yaman, M.-F. Huang, Y.-K. Huang, Y. Zhang, I. B. Djordjevic, E. Mateo, 40Tb/s transoceanic transmission with span length of 121km and EDFA-only amplification, in *Proceedings of the OECC/ACOFT* 2014, Paper THPDP1-3
130. S. Chandrasekhar, X. Liu. B. Zhu, D.W. Peckham, Transmission of a 1.2-Tb/s 24-carrier no-guard-interval coherent OFDM superchannel over 7200-km of ultra-large-area fiber, in *Proceedings of the ECOC* 2009, Paper PD2.6
131. T. Pfau, S. Hoffmann, R. Noé, Hardware-efficient coherent digital receiver concept with feedforward carrier recovery for M-QAM constellations. J. Lightwave Technol. **27**, 989–999 (2009)
132. X. Liu, S. Chandrasekhar, et al., Digital coherent superposition for performance improvement of spatially multiplexed coherent optical OFDM superchannels. Opt. Express **20**, B595–B600 (2012)
133. J.G. Proakis, *Digital Communications*, 4th edn. (McGraw-Hill, New York, 2001)
134. X. Liu, S. Chandrasekhar, P.J. Winzer, Digital signal processing techniques enabling multi-Tb/s superchannel transmission. IEEE. Sig. Proc. Mag. **31**(2), 16–24 (Mar. 2014)
135. T.H. Lotz et al., Coded PDM-OFDM transmission with shaped 256-iterative-polar modulation achieving 11.15-b/s/Hz intrachannel spectral efficiency and 800-km reach. J. Lightwave Technol. **31**, 538–545 (2013)
136. A. Goldsmith, S.-G. Chua, Adaptive coded modulation for fading channels. IEEE Trans. Commun. **46**, 595–601 (May 1998)
137. I.B. Djordjevic, G.T. Djordjevic, On the communication over strong atmospheric turbulence channels by adaptive modulation and coding. Opt. Express **17**(20), 18250–18262 (25 Sept. 2009)
138. I.B. Djordjevic, Adaptive modulation and coding for free-space optical channels. IEEE/OSA J. Opt. Commun. Netw. **2**(5), 221–229 (May 2010)
139. I.B. Djordjevic, Adaptive modulation and coding for communication over the atmospheric turbulence channels, in *Proceedings of the IEEE Photonics Society Summer Topicals 2009*, Paper TuD3. 3, Newport Beach, CA, USA, 20–22 July 2009
140. I.B. Djordjevic, G.T. Djordjevic, Adaptive modulation and coding for generalized fading channels, in *Proceedings of the IEEE TELSIKS 2009*, Nis, Serbia, pp. 418–422, 7–9 Oct. 2009

Spread Spectrum, CDMA, and Ultra-Wideband Communications 10

Abstract

The subject of this chapter is devoted to spread spectrum (SS), CDMA, and ultra-wideband (UWB) communication systems. After the introduction of SS systems, both direct sequence-spread spectrum (DS-SS) and frequency-hopping-spread spectrum (FH-SS) systems are described in detail. Regarding the DS-SS systems, after typical transmitter and receiver configurations are described, we describe in detail the synchronization process including acquisition and tracking stages. To deal with the multipath fading effects, the use of RAKE receiver is described. The section on DS-SS systems concludes with the spreading sequences description, in particular pseudo-noise (PN) sequences. In this section, after introduction of the basic randomness criterions, we introduce maximal-length sequences (m-sequences) and describe their basic properties. In the same section, the concept of nonlinear PN sequences is introduced as well. Regarding the FH-SS systems, both slow-frequency hopping (SFH) and fast-frequency hopping (FFH) systems are introduced. We provide the detailed description of FH-SS systems, including both transmitter and demodulator as well as various time acquisition schemes. In the same section we describe how to generate the FH sequences. After the introduction of CDMA systems, we describe various signature waveforms suitable for use in DS-CDMA systems including Gold, Kasami, and Walsh sequences. We then describe relevant synchronous and asynchronous CMDA models, with special attention devoted to discrete-time CDMA models. In the section on multiuser detection (MUD), we describe how to deal with interference introduced by various CDMA users. The conventional single-user detector is used as the reference case. The various MUD schemes are described then, with special attention being paid to the jointly optimum MUD. The complexity of jointly optimum MUD might be prohibitively high for certain applications, and for such applications we provide the description of the decorrelating receiver, which can compensate for multiuser interference but enhances the noise effect. To solve for this problem, the linear minimum mean-square error (MMSE) receiver is introduced. The section on MUD concludes with description of the nonlinear MUD schemes employing decisions on bits from other users (be preliminary or final), with description of successive cancelation and multistage detection schemes. In the section on optical CDMA (OCDMA), we describe both incoherent and coherent optical detection-based OCDMA schemes. Regarding the incoherent optical detection-based schemes, we first describe how to design various unipolar signature sequences, known as optical orthogonal codes (OOC), followed by the description of basic incoherent OCDMA schemes including time spreading, spectral amplitude coding (SAC), and wavelength hopping (WH) schemes. Related to coherent optical detection-based OCDMA systems, we describe both pulse laser-based and CW laser-based OCDMA systems. Further, the hybrid OFDM-CDMA systems are described, taking the advantages of both OFDM and CDMA concepts into account, particularly in dealing with frequency selective fading and narrowband interference, while at the same time supporting multiple users. The following broad classes of hybrid OFDM-CDMA systems are described: multicarrier CDMA (MC-CDMA), OFDM-CDMA, and OFDM-CDMA-FH systems. In the section on UWB communications, we describe the concepts and requirements as well as transmitter and receiver configurations. Further, various modulation formats, suitable for UWB communications, are described and categorized into time-domain category (pulse-position nodulation and pulse-duration modulation) and shape-based category [on-off keying, pulse-amplitude modulation, bi-phase modulation (BPM) or BPSK, and orthogonal pulse modulation]. Regarding the pulse shapes suitable for orthogonal pulse modulation (OPM), the modified Hermite polynomials, Legendre polynomials, and orthogonal prolate spheroidal wave functions (OPSWF) are

I. B. Djordjevic, *Advanced Optical and Wireless Communications Systems*, https://doi.org/10.1007/978-3-030-98491-5_10

described. Regarding the UWB channel models suitable for indoor applications, the model due to Saleh and Valenzuela is described. Both types of UWB systems are described: the impulse radio UWB (IR-UWB), typically DS-CDMA-based, and multiband OFDM (MB-OFDM). After the conclusion section, the set of problems is provided for deeper understanding of the chapter material.

10.1 Spread Spectrum Systems

The main issues in digital communications are related to the efficient use of the bandwidth and power. However, in certain situations, in a very hostile environment, in the presence of eavesdroppers and jammers, we may want to scarify these efficiencies in order to enable the secure and reliable transmission. One way to achieve this goal is through the *spread spectrum* communications [1–7]. Signals belonging to the spread spectrum class must simultaneously satisfy the following *spreading* and *despreading criterions*. The generic spread spectrum digital communication system is illustrated in Fig. 10.1. The *spreading criterion* indicates that information-bearing signal is spread in frequency domain, on a transmitter side, as shown in Fig. 10.1, with the help of the *spreading signal* so that the resulted signal spectrum is much larger than the bandwidth of the original information-bearing signal. On the other hand, the *despreading criterion* indicates that the recovery of information-bearing signal from noisy spread spectrum received signal is achieved by correlating the received signal with the synchronized replica of the spreading signal, as illustrated in Fig. 10.1. The channel encoder and decoder, the basic ingredients of a digital communication system, are shown as well. If there is no any interference/jammer present, an exact copy (replica) of the original signal can be recovered on the receiver side, and the spread spectrum system performance is transparent to the spreading-despreading process.

In the presence of the narrowband interference signal, introduced during transmission, during the despreading process, the bandwidth of the interference signal will be increased by the receiver side spreading signal. Since the filter that follows despreading matches the information-bearing signal, the interfering spread signal will be significantly filtered, and the output signal-to-interference ratio (SIR), often called signal-to-noise ratio (SNR), will be improved. This improvement in SIR due to spread spectrum processing is commonly referred to as the *processing gain* (PG) and will be formally introduced below.

10.1.1 Spread Spectrum Systems Fundamentals

In a typical digital communication problem, we require the dimensionality of the system D to be smaller than or equal to the number of signaling waveforms M. In spread spectrum systems, on the other hand, we require the dimensionality N to be much larger than M. Let us observe M linearly independent signaling real-valued waveforms $\{s_m(t)\}$ $\{s_m(t); 0 \leq t < T\}_{m=1}^{M}$. At every signaling interval T, one signaling waveform is selected at random, indicating that the data rate is $\log_2 M/T$. The dimensionality $D \leq M$ is related to the signal set bandwidth B_D by $D \cong 2B_D T$. We are concerned with embedding this signal set into much larger signal space of dimensionality $N >> D$. The required bandwidth of the spread spectrum (SS) signal, denoted as B_{SS}, can be determined from the following relationship $N \cong 2B_{SS}T$. We can now represent the m-th transmitted signaling waveform in terms of orthonormal basis functions $\{\phi_n(t); 0 \leq t < T\}_{n=1}^{N}$ as follows:

$$s_m(t) = \sum_{n=1}^{N} s_{mn}\phi_n; m = 1, 2, \cdots, M; 0 \leq t < T, \tag{10.1}$$

where

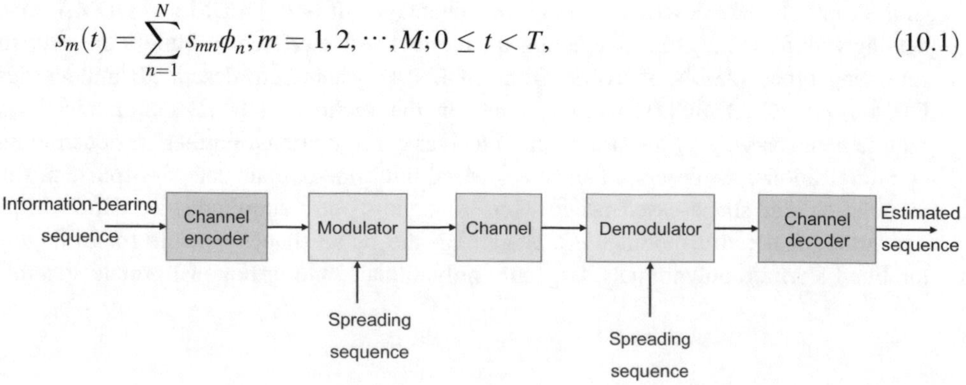

Fig. 10.1 The generic spread spectrum digital communication system. The spreading sequences on transmitter and receiver sides are synchronized

$$s_{mn} = \int_0^T s_m(t)\phi_n(t)dt. \tag{10.2}$$

The basis functions satisfy the orthogonality principle:

$$\int_0^T \phi_n(t)\phi_k(t)dt = \delta_{nk} = \begin{cases} 1, n = k \\ 0, n \neq k \end{cases}. \tag{10.3}$$

We also assume that all signaling waveforms have the same energy E_s, in other words:

$$\int_0^T \langle s_m^2(t)\rangle dt = \sum_{n=1}^{N} \langle s_{mn}^2\rangle \triangleq \mathrm{E_s}, \tag{10.4}$$

where we use the notation $\langle\cdot\rangle$ to denote the ensemble averaging. To better hide the signaling waveform into N-dimensional space, we chose to select the coefficients of expansion s_{mn} independently so that they have zero mean and the correlation given by [6]:

$$\langle s_{mn}s_{mk}\rangle = \frac{E_s}{N}\delta_{nk}. \tag{10.5}$$

Now we can represent narrowband jamming (interference) signal in terms of basis functions as:

$$J(t) = \sum_{n=1}^{N} J_n\phi_n, 0 \leq t < T, \tag{10.6}$$

with corresponding energy, denoted as E_J, being:

$$\int_0^T J^2(t)dt = \sum_{n=1}^{N} J_n^2 \triangleq \mathrm{E_J}. \tag{10.7}$$

When the jamming signal is additive to transmitted signal, and by ignoring the noise, the received signal is given by:

$$r(t) = s_m(t) + J(t) = \sum_{n=1}^{N} (s_{mn} + J_n)\phi_n; m = 1, 2, \cdots, M; 0 \leq t < T. \tag{10.8}$$

The output of the m-th branch of correlator receiver can be determined as:

$$r_m = \int_0^T r(t)s_m(t)dt = \int_0^T [s_m(t) + J(t)]s_m(t)dt = \sum_{n=1}^{N} \left(s_{mn}^2 + s_{mn}J_n\right). \tag{10.9}$$

The mean value of r_m given s_m will be then:

$$\langle r_m|s_m\rangle = \sum_{n=1}^{N}\left(\langle s_{mn}^2\rangle + \underbrace{\langle s_{mn}\rangle\langle J_n\rangle}_{=0}\right) = \sum_{n=1}^{N}\langle s_{mn}^2\rangle = E_s, \tag{10.10}$$

while the mean value of r_m, assuming equal probable transmission:

$$\langle r_m \rangle = \underbrace{\langle r_m | s_m \rangle}_{E_s} \underbrace{P(s_m)}_{1/M} = E_s/M. \tag{10.11a}$$

The variance of r_m given s_m can be determined now by:

$$\begin{aligned}\left\langle (r_m - \langle r_m|s_m\rangle)^2 | s_m \right\rangle &= \left\langle (r_m - E_s)^2 | s_m \right\rangle = \sum_{n=1}^{N} \left\langle \left(s_{mn}^2 + s_{mn}J_n - s_{mn}^2\right)^2 \right\rangle \\ &= \sum_{n=1}^{N} \langle s_{mn}^2 \rangle \underbrace{\langle J_n^2 \rangle}_{E_J/N} = E_s E_J / N.\end{aligned} \tag{10.11b}$$

The variance of r_m is then:

$$\left\langle (r_m - E_s)^2 \right\rangle = \frac{E_s E_J}{N} P(s_m) = \frac{E_s E_J}{MN}. \tag{10.12}$$

The SIR can be then determined as:

$$SIR = \frac{\langle r_m \rangle^2}{\left\langle (r_m - E_s)^2 \right\rangle} = \frac{\left(\frac{E_s}{M}\right)^2}{\frac{E_s E_J}{MN}} = \frac{E_s}{E_J} \underbrace{\frac{N}{M}}_{PG} = \frac{E_s}{E_J} \cdot PG, \tag{10.13}$$

where PG is the *processing gain*, also known as the spreading factor. By assuming that $M = D$, we can express the PG as follows:

$$PG = \frac{N}{D} \cong \frac{2B_{SS}T}{2B_D T} = \frac{B_{SS}}{B_D}, \tag{10.14}$$

indicating that PG is in fact the ratio of spread spectrum bandwidth to information signal bandwidth. In practice, the PG ranges from 10 to 1000.

Depending on how the spread spectrum signals have been generated, the SS systems can be classified into two broad categories: (i) *direct sequence-spread spectrum* (*DS-SS*) and (ii) *frequency hopping-spread spectrum* (*FH-SS*).

In *DS-SS* modulation process, the modulated signal $s(t)$ is multiplied by a wideband *spreading signal*, also known as spreading code/sequence, denoted by $s_c(t)$. Typically, the spreading code has a constant value + or − 1 over a time duration T_c, which is much shorter than the symbol duration T_s. The bits in spreading code are commonly referred to as *chips*, while the duration of the bit in spreading code T_c is known as the *chip time*. Finally, the rate $1/T_c$ is called the *chip rate*. The bandwidth of spreading signal, denoted as B_c, is approximately $1/T_c$, so that the processing gain can be estimated as $PG \cong B_{SS}/B_D \cong (1/T_c)/(1/T_s) = T_s/T_c$. The multiplication of $s(t)$ and $s_c(t)$ in time domain, namely, $s(t)s_c(t)$, can be represented as the convolution in frequency domain $S(f)*S_c(f)$, where $S(f)$ is the Fourier transform (FT) of $s(t)$, while $S_c(f)$ is the FT of $s_c(t)$. The corresponding bandwidth of the resulting signal is approximately equal to $B_c + B_D$. In the presence of additive white Gaussian noise (AWGN) process $z(t)$, the received SS signal is given by $r(t) = s(t)s_c(t) + z(t)$. After despreading, the resulting signal becomes:

$$s_c(t)r(t) = s(t)\underbrace{s_c^2(t)}_{=1} + \underbrace{s_c(t)z(t)}_{z'(t)} = s(t) + z'(t). \tag{10.15}$$

Since $z'(t)$ still has the white Gaussian statistics, the despreading does not have effect on signal transmitted over AWGN. When the jamming (interference) signal is wideband, it can be approximated by an equivalent AWGN process, and the end effect will be reduction of overall SNR. On the other hand, when the jamming signal is narrowband, in frequency domain after despreading it can be represented as $S_c(f)*J(f)$, where $J(f) = \text{FT}\{J(t)\}$. Therefore, the power of the jamming signal gets distributed over SS bandwidth, and the corresponding jamming signal power after the matched filter is approximately

$B_{SS}/B_D \cong PG$ times lower. This indicates that SS approach is effective in dealing with narrowband interference. We will now show that SS approach is also effective in dealing with multipath fading-induced intersymbol interference (ISI). For simplicity, let us observe the two-ray multipath model, introduced in Chap. 2, whose impulse response is given by $h(t) = a\delta(t) + b\delta(t - \tau)$, where the first component corresponds to the line of sight (LOS) component, while the second component is the reflected component, which arrives at receiver side with delay τ. The corresponding frequency response is given by $H(f) = a + b\ \exp.(-j2\pi f\tau)$. The received signal, in the absence of noise, after two-ray multipath channel, can be represented in frequency domain as $[S_c(f)*S(f)]H(f)$. The corresponding time domain representation is given by $as_c(t)s(t) + bs_c(t\text{-}\tau)s(t\text{-}\tau)$. After despreading we can write $as(t) + b\ s_c(t)s_c(t\text{-}\tau)s(t\text{-}\tau)$. Since the reflected component is asynchronous with respect to the spreading signal, it will not be despread, and the power of reflected component after the matched filter will be reduced approximately PG times. This indicates that DS-SS systems are effective in dealing with multipath fading-induced ISI.

In *FH-SS*, we occupy a large bandwidth by *randomly hopping* input data-modulated carrier form one frequency to another. Therefore, the spectrum is spread sequentially rather than instantaneously. The term sequentially is used to denote that frequency hops are arranged in pseudo-randomly ordered code sequence $s_c(t)$. The chip time T_c dictates the time elapsed between frequency hops. The FH-SS is suitable to combine with M-ary frequency-shift keying (MFSK). When N different carrier frequencies are used in FH-SS systems, the bandwidth occupied is given by NB_D, where B_D is the bandwidth of original information-bearing signal, as before. The FH-SS systems are attractive for defense/military applications. Depending on the rate at which frequency hops occur, we can classify the FH-SS systems into two broad categories:

- The *slow-frequency hopping* (*SFH*), in which the symbol rate of MFSK $R_s = 1/T_s$ is an integer multiple of the hop rate $R_c = 1/T_c$. In other words, several symbols are transmitted with the same carrier frequency (for each frequency hop).
- The *fast-frequency hopping* (*FFH*), in which the hop rate R_c is an integer multiple of the MFSK symbol rate R_s. In other words, the carrier frequency gets changed several times during the transmission of the same symbol. Clearly, in multipath fading environment, FFH scheme provides inherent frequency diversity.

In the presence of AWGN, the FH does not have influence on AWGN performance. Let us consider the narrowband interference at carrier frequency f_n of bandwidth $B=B_D$. When the FH frequency coincides with the interference carrier, the particular symbol transmitted will be affected by the interference. However, this occurs $1/N$ fraction of time. The power of interference is effectively reduced approximately $1/N$ times. However, compared to DS-SS systems, where the power of narrowband interference is reduced all the time, the FH-SS systems are affected by the full power of the jammer a fraction of the time. In FFH systems, the interference affects the portion of symbol only, and this problem can be solved by coding, to correct for occasional errors. On the other hand, in SFH systems many symbols are affected by the interference, so that the coding with interleaving is required to solve for this problem.

Regarding the influence of multipath fading-induced ISI, for two-ray channel, on FH-SS systems, whenever $\tau > T_c$, the reflected component will exhibit a different carrier frequency than LOS component, and will not affect the detection of LOS component. However, when $\tau < T_c$, the FFH system will exhibit the flat fading, while the SFH system will exhibit either slow-varying flat fading for interference bandwidth $B < 1/\tau$ or slow-varying frequency-selective fading for $B > 1/\tau$.

10.1.2 Direct Sequence-Spread Spectrum (DS-SS) Systems

As we discussed in the previous section, in *DS-SS* systems, the modulated signal $s(t)$ is multiplied by a wideband *spreading signal* $s_c(t)$, very often derived from *pseudo-noise (PN) sequences* [1–6], which possess noise-like properties which are as follows: (i) the mean value is approximately zero, and (ii) the autocorrelation function is approximately δ-function. Additional details on PN sequences will be provided later in this section. More generally, DS refers to the specific way to generate SS signals satisfying the following three conditions [5]:

(i) The chip waveform of duration T_c, denoted as p_{Tc}, is chosen such that:

$$\int_{-\infty}^{\infty} p_{T_c}(t)p_{T_c}(t - nT_c)dt = 0, \forall n = 1, 2, \cdots \tag{10.16}$$

(ii) The number of chips is N.
(iii) The binary spreading sequence is of length N, namely, $\boldsymbol{b} = (b_1, b_2, \ldots, b_N)$, and does not carry any information.

The DS-SS waveform (signal) can be then generated as:

$$s_c(t) = A_c \sum_{n=1}^{N} (2b_i - 1) p_{T_c}[t - (n-1)T_c], \tag{10.17}$$

where A_c is the normalization constant. Very often the rectangular pulses of duration T_c are used as the chip waveform:

$$p_{T_c}(t) = \mathrm{rect}(t/T_c),\ \mathrm{rect}(t) = \begin{cases} 1, 0 \le t < 1 \\ 0, \text{otherwise} \end{cases}. \tag{10.18}$$

However, there exist significant spectral components outside the spectral null $1/T_c$. When someone is concerned with spectral efficiency, in particular for CDMA systems, the following waveform can be used for the chip waveform [5]:

$$p_{T_c}(t) = \mathrm{sinc}(2t/T_c - 1),\ \mathrm{sinc}(t) = \sin(\pi t)/(\pi t). \tag{10.19}$$

10.1.2.1 DS-SS Transmitters and Receivers

To implement DS-SS system, we can start from either spreading sequence generator or spreading signal generator. The DS-SS modulator for QPSK, based on PN sequence generator, is shown in Fig. 10.2a.

The spreading sequence generators used for in-phase and quadrature channels are typically identical and generate sequence $\boldsymbol{b} = (b_1 b_2 \ldots b_N)$. The block channel encoder (n,k) generates the codeword $\boldsymbol{c} = (c_1 c_2 \ldots c_n)$. The sequence after mod-2 adder can be obtained by $b_i \oplus c_i$. The in-phase component before product modulator can be represented as:

Fig. 10.2 DS-SS QPSK signal generation based on (**a**) binary spreading sequence generator and (**b**) spreading signal generator

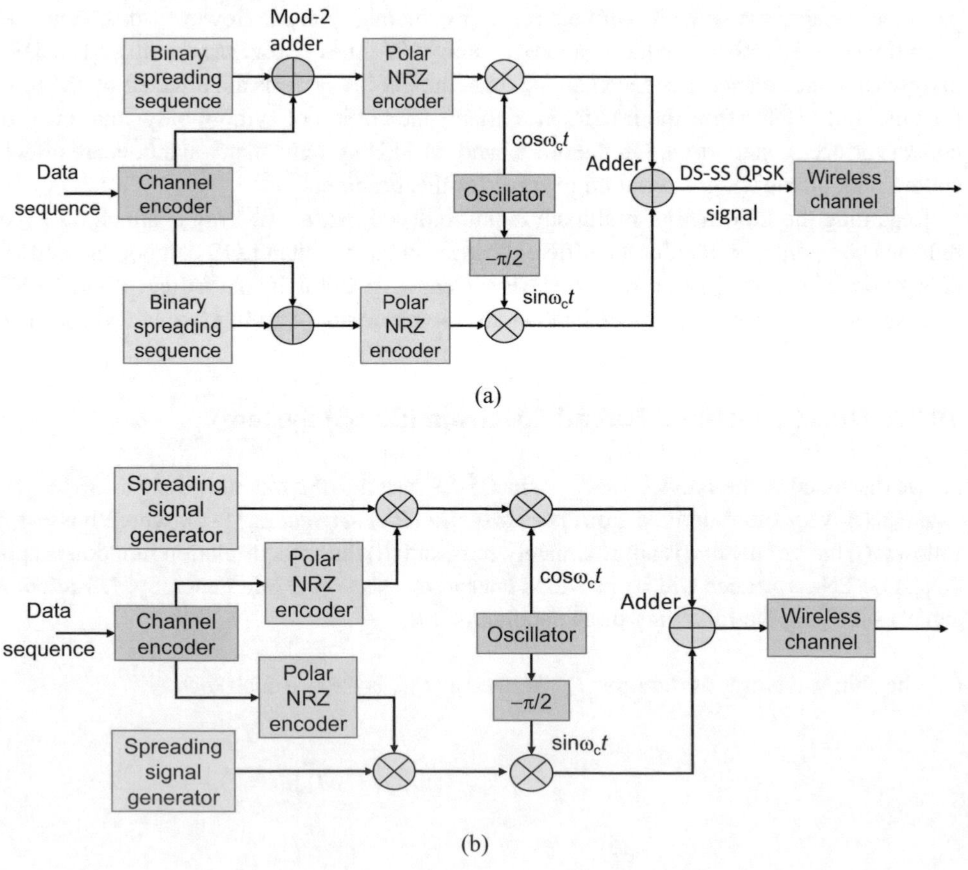

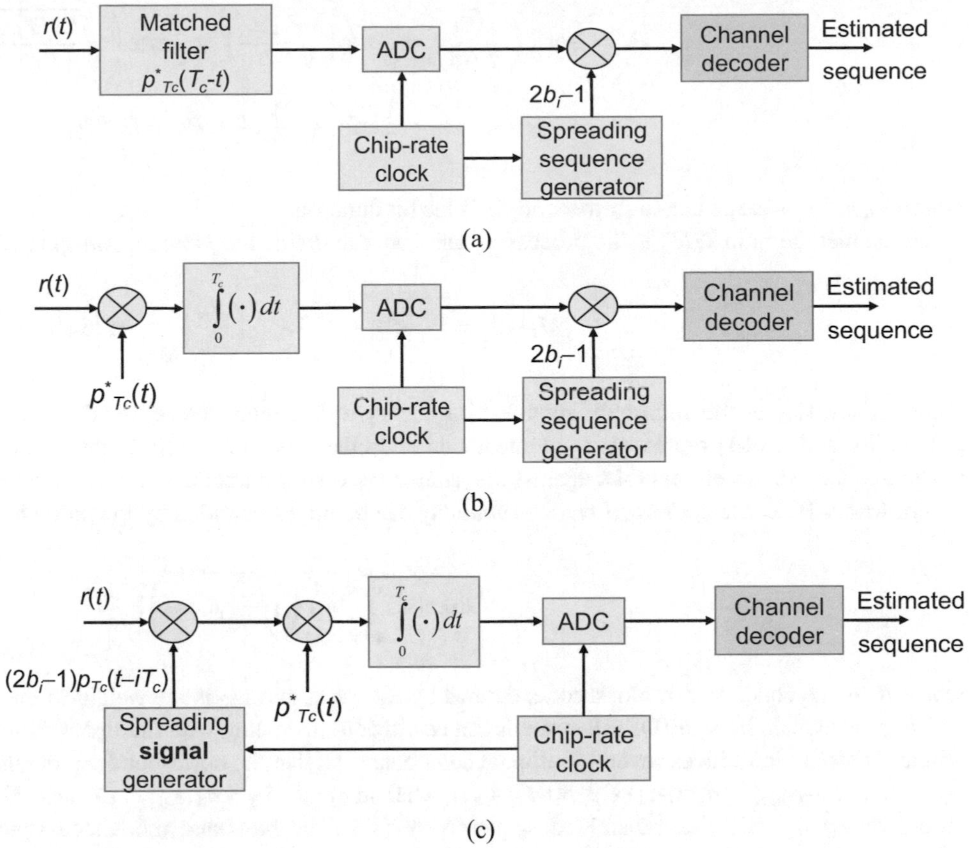

Fig. 10.3 Various in-phase signal demodulator architectures for DS-SS signals: (**a**) matched-filter and spreading sequence-based demodulator, (**b**) correlator and spreading sequence-based demodulator, and (**c**) spread signal-based demodulator with correlator

$$I_i(t) = (2b_i - 1)(2c_i - 1)p_{T_c}(t - iT_c). \tag{10.20}$$

Similar equation holds for the quadrature channel. After product modulation, the in-phase and quadrature SS signals are added together and transmitted toward remote destination by a transmit antenna. The spreading signal generator-based configuration of DD-SS QPSK modulator is shown in Fig. 10.2b. The spreading signal generator generates the signal according to Eq. (10.17). The mod-2 adder version is easier to implement instead of waveform multiplication.

On the receiver side, we can also employ either spreading sequence generator or spreading signal generator to perform despreading. Additionally, either matched-filter receiver or correlation receiver configurations can be used. To facilitate explanations, only in-phase demodulator structures are summarized in Fig. 10.3. The quadrature demodulator structure is very similar to the one shown in Fig. 10.3.

The received *in-phase equivalent low-pass signal* can be represented by:

$$r(t) = (2b_i - 1)(2c_i - 1)p_{T_c}(t - iT_c) + z(t), \tag{10.21}$$

where $z(t)$ originates from the jamming signal and/or Gaussian noise. When $z(t)$ is Gaussian noise or when interference can be modeled as equivalent zero mean Gaussian process with power spectral density (PSD) $N_0/2 = P_J T_c/2$, where P_J is the average power of the jamming (interference) signal, we can use either matched-filter or correlator-based receiver, with corresponding spreading sequence-based configurations provided in Fig. 10.3a, b. The corresponding spreading signal generator-based demodulator with correlator is shown in Fig. 10.3c. Given that $(2b_i - 1)^2 = 1$ in either case, we have successfully performed the despreading operation. Based on Chap. 6, we conclude that bit-error probability P_b for *uncoded* binary PSK (BPSK) can be estimated as:

$$\begin{aligned} P_b &= \frac{1}{2}\operatorname{erfc}\left(\sqrt{\frac{E_b}{N_0}}\right) = \frac{1}{2}\operatorname{erfc}\left(\sqrt{\frac{E_b}{P_J T_c}}\right) = \frac{1}{2}\operatorname{erfc}\left(\sqrt{\frac{E_b/T_b}{P_J}\frac{T_b}{T_c}}\right) \\ &= \frac{1}{2}\operatorname{erfc}\left(\sqrt{\frac{P_b}{P_J}\frac{T_b}{T_c}}\right), P_b = E_b/T_b, \end{aligned} \tag{10.22}$$

where P_b is the average signal power and T_b is the bit duration.

Given that the ratio T_b/T_c is the processing gain, we can define the *jamming margin* [dBs], denoted as M_J, as follows:

$$M_J[\mathrm{dB}] = PG[\mathrm{dB}] - 10\log_{10}\left(\frac{E_b}{N_0}\right)_{\min} - L[\mathrm{dB}], \tag{10.23}$$

where $(E_b/N_0)_{\min}$ is the minimum required bit energy-to-jamming "noise" PSD ratio to support a prescribed bit error probability and L [dB] represents implementation and other losses. Therefore, the jamming margin describes the residual advantage that SS system exhibits against the jammer once we subtract $(E_b/N_0)_{\min}$ [dB] and L [dB] from PG [dB].

For M-ary PSK, the *codeword error probability* can be upper-bounded by the union bound as follows [3]:

$$P_M \le \frac{1}{2}\sum_{m=2}^{M}\operatorname{erfc}\left(\sqrt{\frac{E_b}{N_0}R_c w_m}\right), \tag{10.24}$$

where R_c is the coder rate of block code, defined by $R_c = k/n$, and w_m is the weight of the codeword $\boldsymbol{C}_m$.

We now explain how the DS-SS systems can be efficient in dealing with multipath fading-induced ISI. Let us assume that wireless channel introduces several multipath components so that the corresponding impulse response can be represented by $h(t) = a_{\mathrm{LOS}}\delta(t-\tau_{\mathrm{LOS}}) + a_1\delta(t-\tau_1) + a_2\delta(t-\tau_2) + \ldots$, wherein $a_{\mathrm{LOS}} > a_1 > a_2 > \ldots$. Let us further denote the spreading sequence (code) by $s_c(t)$, while the transmitted sequence by $\{s_k\}$. The baseband modulated signal can be represented by $\tilde{s}(t) = \sum_k s_k g(t - kT_s)$, where $g(t)$ is the baseband modulator impulse response, which is assumed to be rectangular $g(t) = \sqrt{2/T_s}, 0 \le t < T_s$. We also assume that the chip waveform is rectangular given by Eq. (10.18). For receiver shown in Fig. 10.3(c), the input to the correlator can be written as:

$$\begin{aligned} \hat{\tilde{s}}(t) &= \{[\tilde{s}(t)s_c(t)\cos(\omega_c t)] * h(t)\}s_c(t-\tau)\cos(\omega_c t + \phi) \\ &+ z(t)s_c(t-\tau)\cos(\omega_c t + \phi). \end{aligned} \tag{10.25}$$

Now by further assuming that multipath components a_m, $m = 2,3,\ldots$ can be neglected, while the receiver locks to the LOS component ($\tau_{\mathrm{LOS}} = 0$, $\phi = 0$), we can rewrite Eq. (10.25), after substitution for impulse response $h(t)$, as follows:

$$\begin{aligned} \hat{\tilde{s}}(t) &= a_{LOS}\tilde{s}(t)\underbrace{s_c^2(t)}_{=1}\cos^2(\omega_c t) + a_1\tilde{s}(t-\tau_1)s_c(t-\tau_1)\cos[\omega_c(t-\tau_1)]s_c(t)\cos(\omega_c t) \\ &+ z(t)s_c(t)\cos(\omega_c t). \end{aligned} \tag{10.26}$$

We are concerned in demodulation of the m-th symbol, and assuming that multipath fading-induced ISI originates from the transmission of the (m-k)-th symbol, we can set $\tau_1 = kT_s$ to obtain:

$$\hat{s}_m = \int_0^{T_s} \hat{s}(t) g^*(T_s - t) dt = a_{LOS} \frac{2}{T_s} \int_0^{T_s} s_m \underbrace{\cos^2(\omega_c t)}_{\frac{1}{2}[1+\cos(2\omega_c t)]} dt$$
$$+ a_1 \frac{2}{T_s} \int_0^{T_s} s_{m-k} s_c(t) s_c(t-\tau_1) \underbrace{\cos(\omega_c t) \cos[\omega_c(t-\tau_1)]}_{\frac{1}{2}[\cos(\omega_c\tau_1)+\cos(2\omega_c t - \omega_c \tau_1)]} dt \tag{10.27}$$
$$\underbrace{+ \sqrt{\frac{2}{T_s}} \int_0^{T_s} z(t) s_c(t) \cos(\omega_c t) dt.}_{z_m}$$

Given that the double-frequency terms in (10.27) will not survive the low-pass filtering (integration), the previous equation simplifies to:

$$\hat{s}_m \cong a_{LOS} s_m + a_1 s_{m-k} \cos(\omega_c \tau_1) \underbrace{\frac{1}{T_s} \int_0^{T_s} s_c(t) s_c(t-\tau_1) dt}_{\rho_c(t, t-\tau_1)} + z_m, \tag{10.28}$$

where $\rho_c(t, t - \tau_1)$ is the *autocorrelation function* of the spreading sequence (code), defined as:

$$\rho_c(t, t-\tau) \triangleq \frac{1}{T_s} \int_0^{T_s} s_c(t) s_c(t-\tau) dt = \frac{1}{N} \sum_{n=1}^{N} s_c(nT_c) s_c(nT_c - \tau), N = T/T_c. \tag{10.29}$$

When the autocorrelation of the spreading sequence is periodic in T, then autocorrelation becomes only function in τ:

$$\rho_c(t, t-\tau) = \frac{1}{T_s} \int_0^{T_s} s_c(t) s_c(t-\tau) dt = \rho_c(\tau), \tag{10.30}$$

and we can rewrite (10.28) as:

$$\hat{s}_m \cong a_{LOS} s_m + a_1 s_{m-k} \rho_c(\tau_1) \cos(\omega_c \tau_1) + z_m. \tag{10.31}$$

Clearly, when $T = T_s$ and $\rho_c(\tau_1) = \delta(\tau_1)$, the multipath fading-induced ISI gets completely eliminated by the despreading procedure. So, one of the key topics in DS-SS systems is to properly design the spreading sequence (code) of finite length, whose autocorrelation over T_s is approximately δ-function.

In the discussion above, we assumed that the receiver is perfectly locked to the LOS, which is not always the case in practice, and we need to study the *synchronization* process of synchronizing the locally generated spreading sequence with the received SS signal.

10.1.2.2 Acquisition and Tracking

The synchronization process is generally accomplished in two steps [2, 8]: (i) *acquisition* or coarse synchronization step and (ii) *tracking* or fine synchronization step. In the acquisition step, we are coarsely aligning the timing of local spreading sequence with incoming SS signal. Once the acquisition step is completed, we perform the fine timing alignment with a help of phase-locked loop (PLL)-based approaches.

The DS *acquisition* can be performed in either serial or parallel search fashion. The *serial search acquisition*, illustrated in Fig. 10.4, represents a low-cost solution, in which we use a single correlator or matched filter to serially search for the correct phase of DS spread sequence. The locally generated spreading sequence is correlated with incoming spreading signal. The

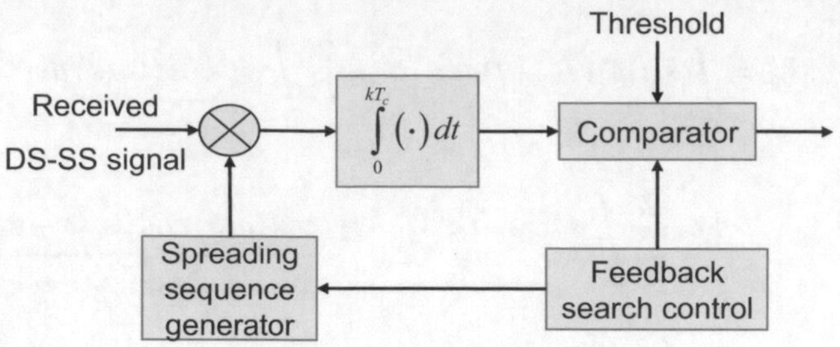

Fig. 10.4 The DS-SS serial search acquisition

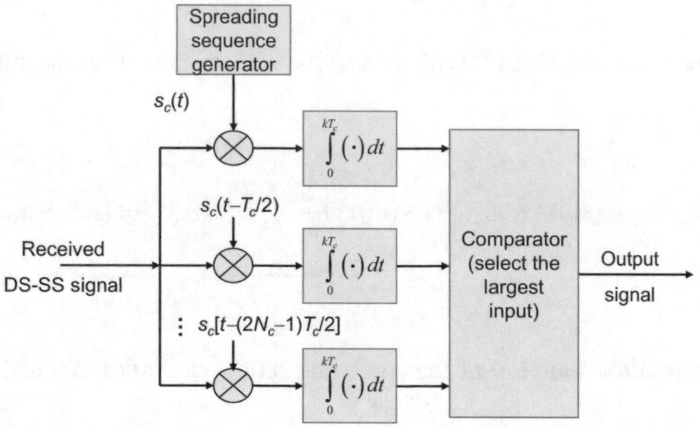

Fig. 10.5 The DS-SS parallel search acquisition

incoming signal is compared with the preset threshold at search dwell time intervals kT_c, where $k> > 1$. If the comparator output is below the threshold value, the delay of locally generated spreading sequence is incremented by $0.5T_c$ and correlation is reexamined.

Once the threshold is exceeded the acquisition step will be completed, while the tracking step will be initiated. Assuming that the uncertainty between the local spreading sequence and received one is N_c chips, the maximum acquisition time to perform a fully serial DS search is:

$$(T_{\text{acquisition}})_{\max} = 2N_c kT_c. \tag{10.32}$$

The mean acquisition time for a serial DS search system has been found in [9] to be:

$$\langle T_{\text{acquisition}} \rangle = \frac{(2-P_D)(1+KP_{FA})}{P_D} N_c kT_c, \tag{10.33}$$

where P_D and P_{FA} are probabilities of the correct detection and false alarm, respectively. We assumed that the time required to verify detection is $N_c kT_c$.

To reduce the acquisition time of DS serial search, the *parallel search acquisition*, illustrated in Fig. 10.5, can be used. In this acquisition scheme, the locally generated spreading sequence s_c(t) is available together with $2N_c$ delayed versions, spaced $T_c/2$ apart. Clearly, $2N_c$ correlators (matched filters are needed), and the comparator after kT_c seconds select the largest input. As the number of chips k increases, the probability of synchronization error decreases. When for each k the correlator outputs are examined to determine the largest one, the maximum time interval required for a fully parallel search is:

$$(T_{\text{acquisition}})_{\max} = kT_c. \tag{10.34}$$

If an incorrect output is chosen, additional k chips will be required to determine the correct output, so that the average acquisition time can be estimated as [10, 11]:

$$\begin{aligned}\langle T_{\text{acquisition}}\rangle &= kT_cP_D + 2kT_cP_D(1-P_D) + 3kT_cP_D(1-P_D)^2 + \cdots \\ &= \frac{kT_c}{P_D},\end{aligned} \tag{10.35}$$

where P_D is the probability of detection. The complexity of fully parallel acquisition is high; however, the average acquisition time is much shorter compared to the serial search acquisition.

Another acquisition technique, which has rapid acquisition time, but employs sequential estimation, is shown in Fig. 10.6. It is commonly referred to as *rapid acquisition by sequential estimation* (*RASE*) search technique [12]. With switch in position 1, the first n chips are detected and stored inside n-stage PN sequence generator. The PN generator has a unique property that the next combination of the register states depends only on the present states, so, when the first n chips are correctly estimated, all next chips from a local PN generator will be correctly generated, and the switch can be moved to position 2. Now we correlate the output from n-stage PN sequence generator with that of incoming SS signal. Once the threshold is exceeded after kT_c seconds, we assume that correct coarse synchronization is achieved. On the other hand, when the output of correlator is below the threshold, we move the switch in position 1, and the register is reloaded with the estimates of next n received chips, and the procedure is repeated. Clearly, when the noise level is sufficiently low, the minimum acquisition time will be nT_c. The RASE provides rapid acquisition time; however, it is sensitive to noise and jamming signals. Other approaches to acquisition can be found in [10, 11].

Once the acquisition step is completed, we move to the *tracking step*. There are two types of tracking spreading sequence (code) loops, noncoherent loops and coherent ones. In coherent loops, the carrier frequency and phase are perfectly known so that the loop can operate on baseband level. Given that the carrier frequency and phase are not known due to Doppler and various multipath effects, the noncoherent loops to track PN sequences are more common. The popular noncoherent tracking loops include (i) the *delay-locking loop* (DLL), also known as full-time early-late tracking loop, and (ii) the *tau-dither loop* (TDL), also known as time-shared early-late tracking loop. The DLL for BPSK signals is provided in Fig. 10.7. Given that the

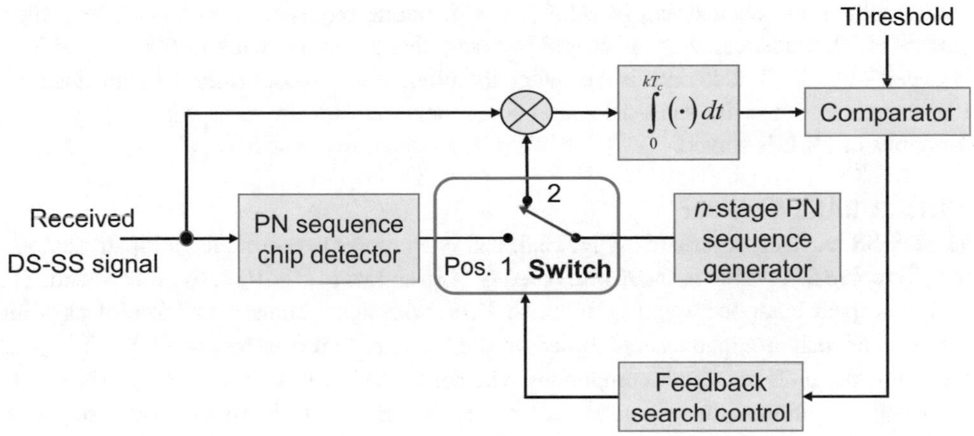

Fig. 10.6 The rapid acquisition by sequential estimation (RASE) search technique

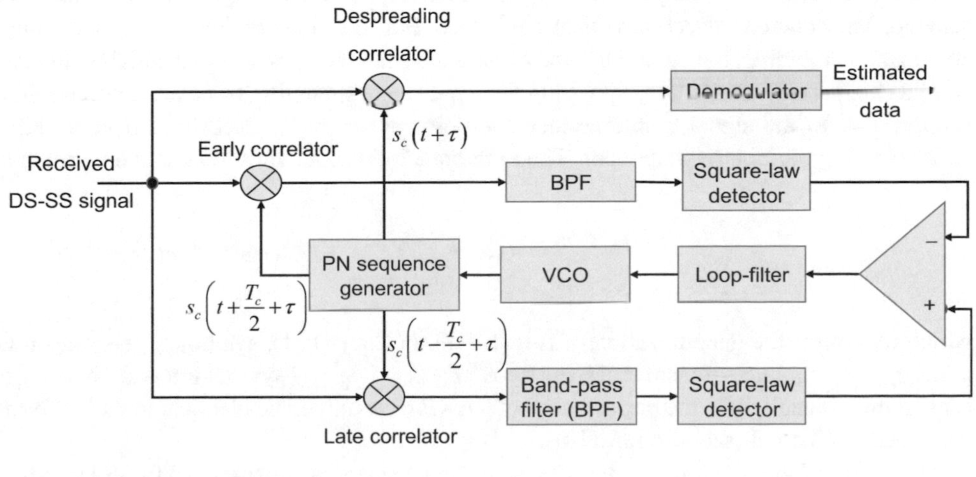

Fig. 10.7 The delay-locked loop (DLL) for DS-SS signal tracking

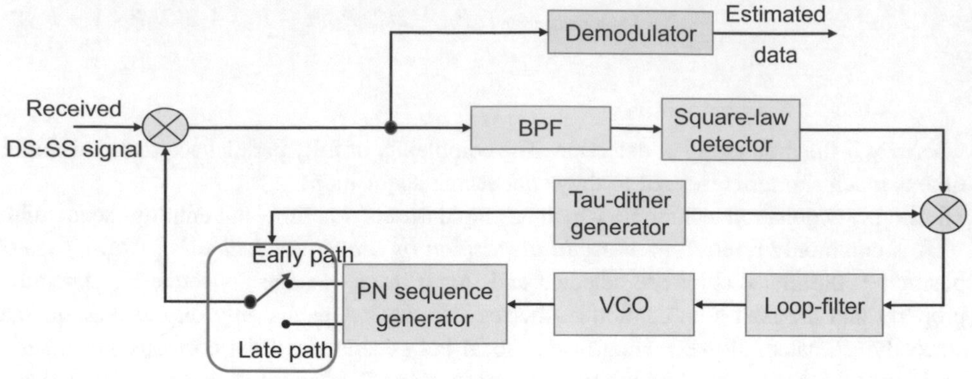

Fig. 10.8 The tau-dither loop (TDL) for DS-SS signal tracking

tracking step follows the acquisition step, we assume that local PN sequence generator is offset from the incoming spreading sequence by $\tau < T_c/2$. The local PN generator provides two PN sequences: $s_c(t + T_c/2 + \tau)$ for early correlator and $s_c(t-T_c/2 + \tau)$ for late correlator. The square-law detector is applied to erase the modulation data, since $|\pm 1|^2 = 1$. The feedback signal at the output of comparator indicates whether the frequency of voltage control oscillator (VCO) should be increased, when the late correlator output is larger in magnitude, or decreased. The increase in VCO frequency is proportional to $1/\tau$, namely, when VCO frequency is increased, τ is decreased and vice versa. On such a way, the τ is getting smaller and smaller, and in limit we have that $s_c(t)s_c(t+\tau) \underset{\tau \to 0}{\to} 1$. For sufficiently small τ, the dispreading correlator output signal represents the dispread signal that is further demodulated in BPSK demodulator. The main problem DLL is facing is related to imperfect gain balancing of early and late correlator branches, which can cause the offset in feedback signal, when magnitudes are expected to be the same, resulting in nonzero error signal.

To avoid this problem, the *tau-dither loop*, shown in Fig. 10.8, can be used, in which only one correlator is used for both code tracking and dispreading functions. The incoming sequence is correlated with either early or late version of locally generated PN sequence, which is dictated by tau-dither generator. The tau-dither signal is a square wave signal that alternates between +1 and − 1 and controls the switch that alternatively passes early path (advanced) and late path (delayed) versions of PN sequence. Since the same BPF and square-law detector are used with both paths, the problem of identical transfer functions of DLL is solved.

10.1.2.3 RAKE Receiver

The DS-SS receiver, described in Fig. 10.3, will synchronize to one of the multipath components, not necessarily the strongest one. To solve this problem, the *RAKE receiver* [13], shown in Fig. 10.9, should be used. The assumption is here that there are M-1 multipath components, in addition to LOS, with delay being a multiple of chip interval T_c. Therefore, the impulse response of such multipart system, based on Chap. 2, can be represented as $h(t) = \sum_{m=1}^{M} a_m \delta(t - mT_c)$, where a_m is the gain related to the m-th multipath component. The component with $m = 0$ corresponds to the LOS component. The spreading sequence and different delayed versions serve as referent signals. It is also possible to perform the down-conversion first prior to the multiplication with spreading signal. We also assume that co-phasing has been performed after dispreading, to ensure that different multipath components add constructively, and the weights w_m are chosen according to the desired diversity scheme. The coherent detection is then performed. For selection combining, we select the branch with strongest gain. With equal gain combining, the demodulators' outputs are summed up with equal weights. In maximal ratio combining (MRC), the weights are proportional to corresponding pertinent signal strengths. To be more precise, in MRC the weight in each branch is proportional to the signal-to-interference-and-noise power ratio. Each branch in RAKE receiver gets synchronized to a corresponding multipath component. The m-th branch demodulator output can be written as:

$$\widehat{s_k^{(m)}} \cong a_m s_k + \sum_{m'=1,\, m' \neq m}^{M} a_{m'} s_{k,m'} \rho_c[(m - m')T_c] + z_k^{(m)}, \tag{10.36}$$

which represents the generalization of Eq. (10.31). In Eq. (10.31), symbol s_k gets transmitted over the k-th symbol interval, while $s_{k,m'}$ symbol gets transmitted over $[(k\text{-}m')T_s, (k + 1\text{-}m')T_s]$ symbol interval. When $\rho_c(\tau) \cong 0$, for $|\tau| > T_c$, and the ISI in each branch is negligible, the performance of RAKE receiver will be identical to the performance of various diversity schemes described in Chap. 8 (see also ref. [14]).

In the rest of this section, we describe spreading sequences suitable for DS-SS systems [15–17] and beyond.

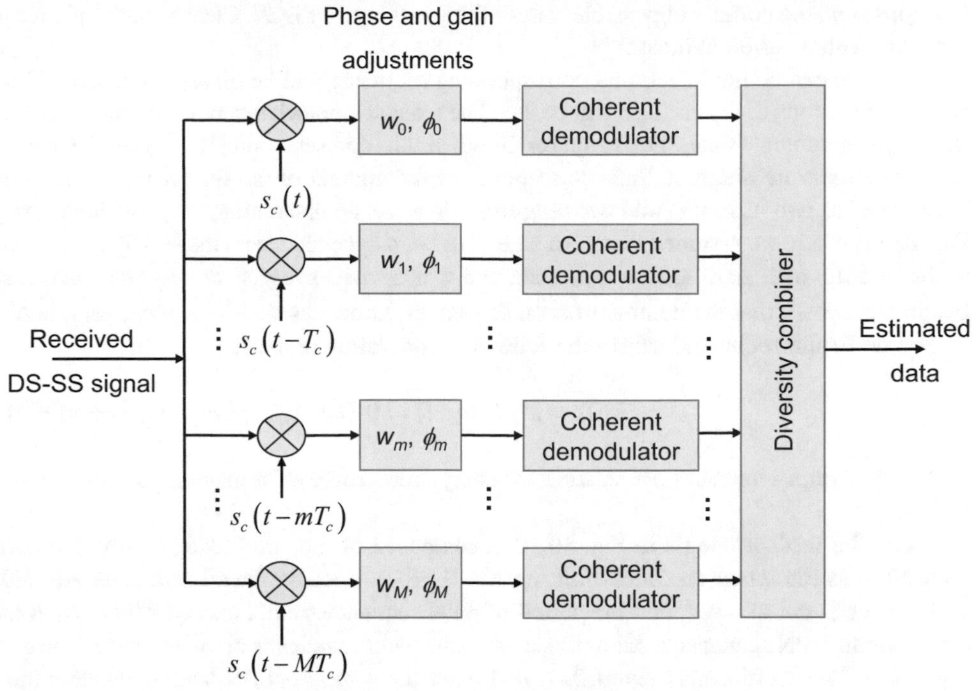

Fig. 10.9 The generic configuration of RAKE receiver

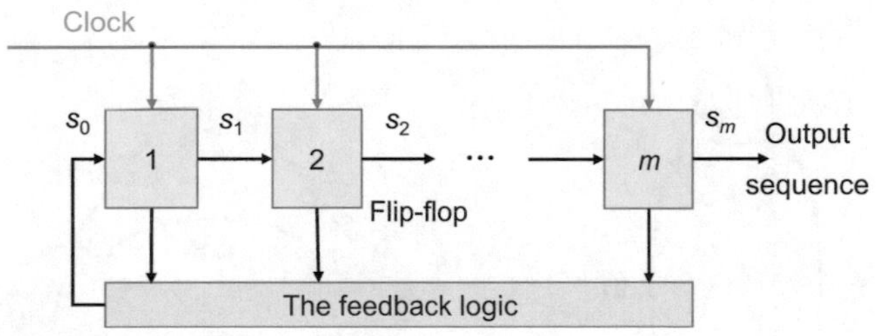

Fig. 10.10 The generic block scheme of the feedback shift register

10.1.2.4 Spreading Sequences

A pseudo-noise (PN) sequence is a binary sequence, typically generated by means of a *feedback shift register*, having a noise-like waveform. The most popular PN sequences are maximal-length or m-sequences, Gold sequences, Walsh sequences, and Kasami sequences. The pseudo-random sequences of 1 and − 1 must satisfy the following randomness criterions [16, 17]:

(i) *Balance property* indicates that the number of binary symbols of one type must be approximately equal to the number of symbols of other type (more precisely, the difference should not be larger than one).
(ii) *Run property* indicates that shorter successive runs of symbols of the same type are more probable. More precisely, one-half of runs has length 1, one-fourth of runs has length 2, one-eighth of runs has length 3, and so on.
(iii) *Correlation property* indicates that autocorrelation function must have a large peak in the origin and to decay fast as separation index from origin increases. The most relevant is the pseudo-random sequence whose autocorrelation function has two values:

$$\rho_c(k) = \begin{cases} 1, k = 0 \\ \text{const} < 1, 0 < |k| < L \end{cases}, \tag{10.37}$$

where L is the sequence length. The pseudo-random sequences satisfying Eq. (10.37) are commonly referred to as the *perfect sequences*.

The generic block scheme of the feedback shift register is shown in Fig. 10.10. Let $s_i(k)$ denote the state of the i-th flip-flop after the k-th clock pulse. Then the state of the shift register can be described as $\mathbf{s} = s_1(k)s_2(k)\ldots s_m(k)$. Given that the number flip-flops is m, the number of possible states of the shift register is 2^m. Clearly, the sequence generated by shift register must be periodic, with a period at most 2^m.

The logic can be nonlinear, and corresponding sequence will be binary nonlinear PN sequence. The possible number of Boolean functions of m variables will be 2^{2^m}. The feedback branch symbol is obtained as the arbitrary Boolean function of the shift register content (state). De Bruijn (or Good) graph representation [18, 19] can be used to describe this calculation, which represents the state diagram. State corresponds to the content of the shift register, and edges represent possible transitions. Each state has two incoming and two outgoing edges. As an illustration, we provide de Bruijn graphs for $m = 1$ and $m = 2$ in Fig. 10.11, where we assumed that the next bit has been appended from the left. The maximal length (2^m) sequence is obtained by finding the path starting from one state and which visits every state only once and reading out the output symbols. De Bruijn has shown that the number of such sequences, known as *de Bruijn sequences*, is $N_m = 2^{2^{m-1}-m}$.

The de Bruijn sequences satisfy the following correlation property:

$$R(0) = 2^n/L, R(\pm i) = 0\ (i = 1, 2, \cdots, m-1), R(\pm m) \neq 0. \tag{10.38}$$

The de Bruijn sequences are suitable for encryption. There exist many algorithms to generate the Bruijn sequences, such as [20].

When the feedback logic in Fig. 10.10 is composed of modulo-2 adders only, the corresponding register is commonly referred to as the *linear feedback-shift register* (LFSR). Because the all-zero state will always generate all-zeros sequence, such a state is not allowed, and the period of a PN sequence cannot exceed 2^m-1. When the period is exactly $L = 2^m$-1, the corresponding PN sequence is known as a *maximal-length sequence* or *m-sequence*. The generic LFSR register is provided in Fig. 10.12. The coefficients f_i ($i = 1,2,\ldots,m$-1) have the value 1 or 0, depending whether the corresponding switch is closed or

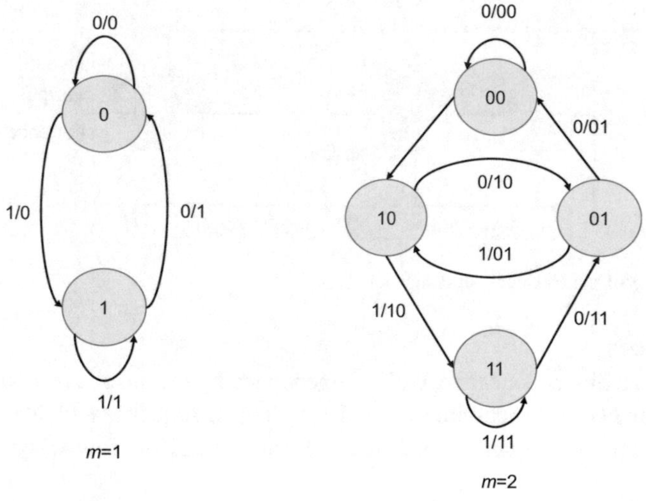

Fig. 10.11 The de Bruijn graphs for $m = 1$ and $m = 2$

Fig. 10.12 The generic block scheme of the linear-feedback shift register (LFSR)

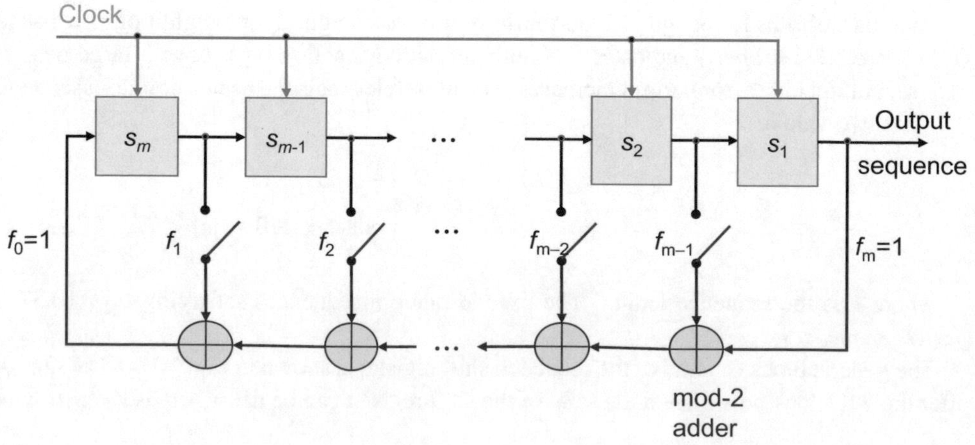

open. The switches f_0 and f_m are always closed ($f_0 = f_m = 1$). Assuming that initial content of the shift register $s_m s_{m-1} \ldots s_2 s_1$ is different from all-zero word, the i-th feedback output symbol can be determined by:

$$s_i = f_1 s_{i-1} \oplus f_2 s_{i-2} \oplus \cdots \oplus f_m s_{i-m} = \sum_{j=1}^{m} f_j s_{i-j}. \tag{10.39}$$

After certain number of clocks, the initial content of register will repeat. The switches should be properly chosen such that period is maximum possible $L = 2^m$-1.

By introducing the delay operator $x = z^{-1}$, we can define the *generating function* as follows:

$$G(x) = \sum_{i=0}^{\infty} s_i x^i, \tag{10.40}$$

where s_0 is the very first bit being generated. Based on recurrent relation (10.39), we can represent the generating function as:

$$G(x) = \frac{\sum_{i=1}^{m} f_i x^i \left(s_i x^{-i} + s_{i-1} x^{-(i-1)} + \cdots + s_{-1} x^{-1}\right)}{1 - \sum_{i=1}^{m} f_i x^i}. \tag{10.41}$$

Since the initial content of the shift register can be any sequence of length m different from all-zero word, we can set $s_{-1} = s_{-2} = \ldots = s_{-(m-1)} = 0$, $s_{-m} = 1$ (similar to unit response) to obtain:

$$G(x) = \frac{1}{\underbrace{1 - \sum_{i=1}^{m} f_i x^i}_{f(x)}} = \frac{1}{f(x)}, \tag{10.42}$$

where $f(x)$ is the *characteristic polynomial*. It can be shown that when $f(x)$ is irreducible (not divisible with any polynomial of lower degree), the period L will be independent of initial conditions, and L can be found as the smallest positive integer for which 1-x^L is divisible with the characteristic polynomial $f(x)$ [16]. In other words, the characteristic polynomial is one of factors of 1-x^L. Let us assume that the characteristic polynomial is the product of two irreducible polynomials, that is, $f(x) = f(x_1) f(x_2)$. Then we can perform the partial fraction expansion of $1/f(x)$ as follows:

$$\frac{1}{f(x)} = \frac{a_1(x)}{f_1(x)} + \frac{a_2(x)}{f_2(x)}; \deg[a_i(x)] < \deg[f_i(x)]; i = 1, 2 \tag{10.43}$$

The maximum possible period will be the least common multiplier of possible periods $2^{m_1+m_2} = 2^m, 2^{m_1}, 2^{m_2}$. Since

$$(2^{m_1} - 1)(2^{m_2} - 1) = 2^{m_1+m_2} - 2^{m_1} - 2^{m_2} + 1 < 2^m - 2 - 2 + 1 < 2^m - 1, \tag{10.44}$$

we conclude that the *necessary condition* to obtain the m-sequence is to make sure that the characteristic polynomial is irreducible. However, this is not a sufficient condition. The sufficient condition will be satisfied when the characteristic polynomial is *primitive*. The primitive polynomial is an irreducible polynomial with corresponding roots being the primitive elements of finite field GF(2^m). Finally, a primitive element is a field element that generates all nonzero elements of the field as its successive powers.

As an illustration, to get m-sequence of length $L = 2^3$–$1 = 7$, we need to factorize $x^7 - 1$ as follows:

$$x^7 - 1 = (x - 1)\left(x^3 + x + 1\right)\left(x^3 + x^2 + 1\right).$$

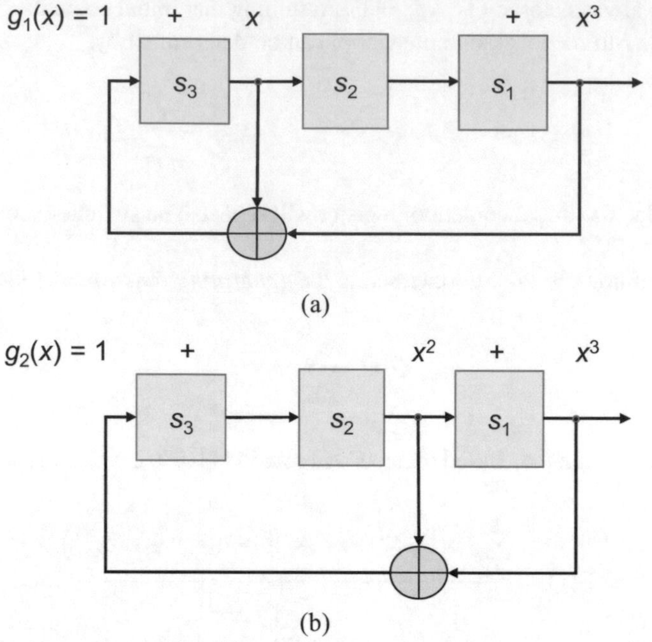

Fig. 10.13 Two different ways to generate *m*-sequences of length 7

Both polynomials $g_1(x) = x^3 + x + 1$ and $g_2(x) = x^3 + x^2 + 1$ are irreducible and primitive and as such can be used to generate *m*-sequences of length 7. The corresponding LSFRs are provided in Fig. 10.13. Assuming that initial content of shift register is 001, the corresponding *m*-sequences are 0111001 and 1101001, respectively.

The generation of *m*-sequences can also be described using a matrix approach as follows [16]. The transformation matrix corresponding to $g_1(x) = x^3 + x + 1$ is given by:

$$\boldsymbol{M} = \begin{bmatrix} 1 & 1 & 0 \\ 0 & 0 & 1 \\ 1 & 0 & 0 \end{bmatrix}.$$

Let s_i be the content of register in the *i*-th step. The content of shift register in the next step will be $\boldsymbol{s}_{i+1} = \boldsymbol{s}_i\boldsymbol{M}$. In general, the transformation matrix has the form:

$$\boldsymbol{M} = \begin{bmatrix} f_1 & 1 & 0 & \cdots & 0 \\ f_2 & 0 & 1 & \cdots & 0 \\ \vdots & \vdots & \vdots & \ddots & \vdots \\ f_{m-1} & 0 & 0 & \cdots & 1 \\ f_m & 0 & 0 & \cdots & 0 \end{bmatrix}, \tag{10.45}$$

where the first column are coefficients of the characteristic polynomial, corresponding to the feedback. The rest of the matrix is an (*m*-1) × (*m*-1) identity submatrix, while the elements in the last row, except the first element, are zeros. It can be shown that the characteristic equation for this matrix is:

$$p(\lambda) = \det(\boldsymbol{M} - \lambda\boldsymbol{I}) = (-1)^{m+1}\left[1 - \underbrace{\left(f_1x + f_2x^2 + \cdots + f_mx_m\right)}_{f(x)}\right]/x^m, \tag{10.46}$$

where $x = 1/\lambda$. (Notice that, in GF(2), − sign can be replaced with + sign.)

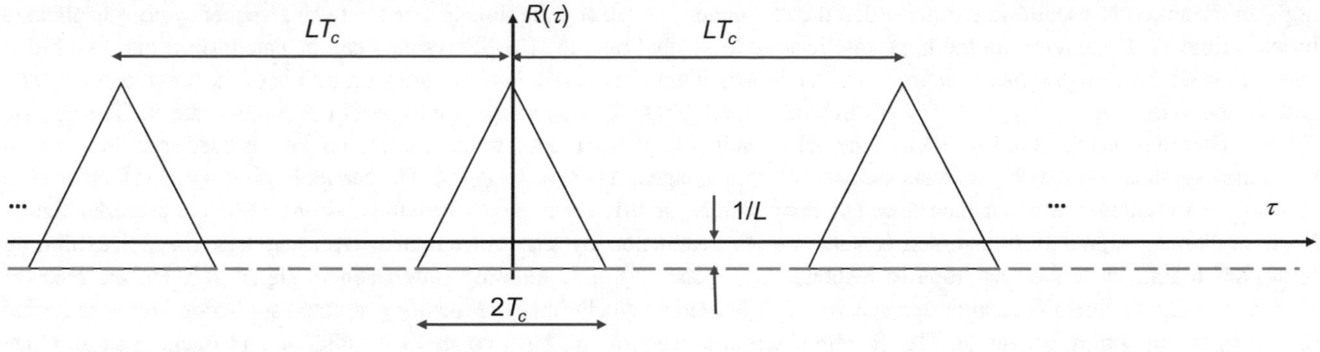

Fig. 10.14 Illustration of the autocorrelation function of the *m*-sequence

The period of *m*-sequence can be determined as the smallest power of *M* yielding to the identity matrix:

$$\boldsymbol{M}^L = \boldsymbol{I}. \tag{10.47}$$

Since every matrix satisfies its characteristic equation $p(\boldsymbol{M}) = \boldsymbol{0}$, if $p(x)$ is the factor of $1\text{-}x^L$, then $\boldsymbol{M}$ will be the root of eq. $\boldsymbol{I}\text{-}\boldsymbol{M}^L$, given that $\boldsymbol{M}$ is the root of $p(\boldsymbol{M})$. Therefore, when $p(x)$ divides $1\text{-}x^L$, then $\boldsymbol{M}^L = \boldsymbol{I}$.

The *autocorrelation function* of the *m*-sequence is dependent on the chip waveform. When the chip waveform is a rectangular pulse of duration T_c, given by Eq. (10.18), the continuous autocorrelation function will have a triangle shape around the origin of duration $2T_c$, in other words:

$$R_{m-\text{sequence}}(\tau) = \begin{cases} 1 - \frac{L+1}{L}\frac{|\tau|}{T_c}, |\tau| < T_c \\ -\frac{1}{L}, |\tau| > T_c \end{cases}. \tag{10.48}$$

On the other hand, the autocorrelation function of random binary sequence of symbols of duration T_c and amplitude either +1 or − 1 is given by:

$$R_{\text{random}}(\tau) = \begin{cases} 1 - \frac{|\tau|}{T_c}, |\tau| < T_c \\ 0, |\tau| > T_c \end{cases}. \tag{10.49}$$

Clearly, Eq. (10.49) can be derived from (10.48) by letting *L* to tend to plus infinity. Given that *m*-sequence is periodic with period LT_c, the corresponding autocorrelation function will be periodic as well and can be sketched as shown in Fig. 10.14. The corresponding power spectral density (PSD) can be obtained by calculating the Fourier transform (FT) of the autocorrelation function. From FT theory, we know that the FT of the triangle function is the sinc^2 function of frequency, so we expect the continuous PSD to be sinc^2 function, while the discrete portion of PSD will contain the series of δ-functions spaced $1/(NT_c)$ apart, with magnitude dictated by the $\text{sinc}^2(fT)$ function. The magnitude of the D.C. component of PSD can be determined by calculating the area below the autocorrelation curve, shown in Fig. 10.14, which is $1/L^2$. Therefore, the PSD of *m*-sequence will be:

$$PSD_{m-\text{sequence}}(\tau) = \frac{1}{L^2}\delta(f) + \frac{L+1}{L^2}\sum_{l=-\infty,\, l\neq 0}^{\infty} \text{sinc}^2\left(\frac{l}{L}\right)\delta\left(f - \frac{l}{LT_c}\right). \tag{10.50}$$

10.1.3 Frequency Hopping-Spread Spectrum (FH-SS) Systems

As discussed in Sect. 10.1.1, in *FH-SS* systems, the large bandwidth occupation is achieved by *randomly hopping* input data-modulated carrier from one frequency to another. Clearly, the spectrum is spread sequentially; in other words, the frequency hops are arranged in pseudo-randomly ordered code sequence $s_c(t)$, and the time elapsed between frequency hops is dictated by chip time T_c. Depending on the hope rate relative to symbol rate, the FH-SS systems can be categorized into two broad categories: (i) the *slow-frequency hopping* (*SFH*), in which the symbol rate R_s is an integer multiple of the hop rate $R_c = 1/T_c$, and (ii) the *fast-frequency hopping* (*FFH*), in which the hop rate R_c is an integer multiple of the symbol rate R_s. The generic FH-SS scheme is provided in Fig. 10.15. The FH spreading sequence (code), to be described below, is used as an input to the frequency synthesizer, which generates the carrier hopping signal $A\cos[\omega_k t + \phi_k(t)]$. The carrier hoping signal is then used as input to the modulator, which can be coherent, noncoherent, or differential, to perform the up-conversion to the carrier signal. Such modulated signal is transmitted toward remote destination by transmit antenna, over wireless channel exhibiting multipath fading, described by impulse response $h(t)$, noise $z(t)$, and jamming (interference) signal $J(t)$. On the receiver side, following the antenna reception, a *synchronizer* is used to synchronize the locally generated spreading sequence (code) to that of the incoming SS signal. The synchronization procedure also has two steps, acquisition and tracking steps. Once synchronization is achieved, the demodulation process takes place. When either differential or noncoherent modulation is employed, it is not required to synchronize phases of receive and transmit signals.

For FFH-MFSK, we provide additional details of noncoherent demodulator in Fig. 10.16. For completeness of presentation, the configuration of FH-MSFK system is shown as well. After frequency dehopping, we perform filtering to select the

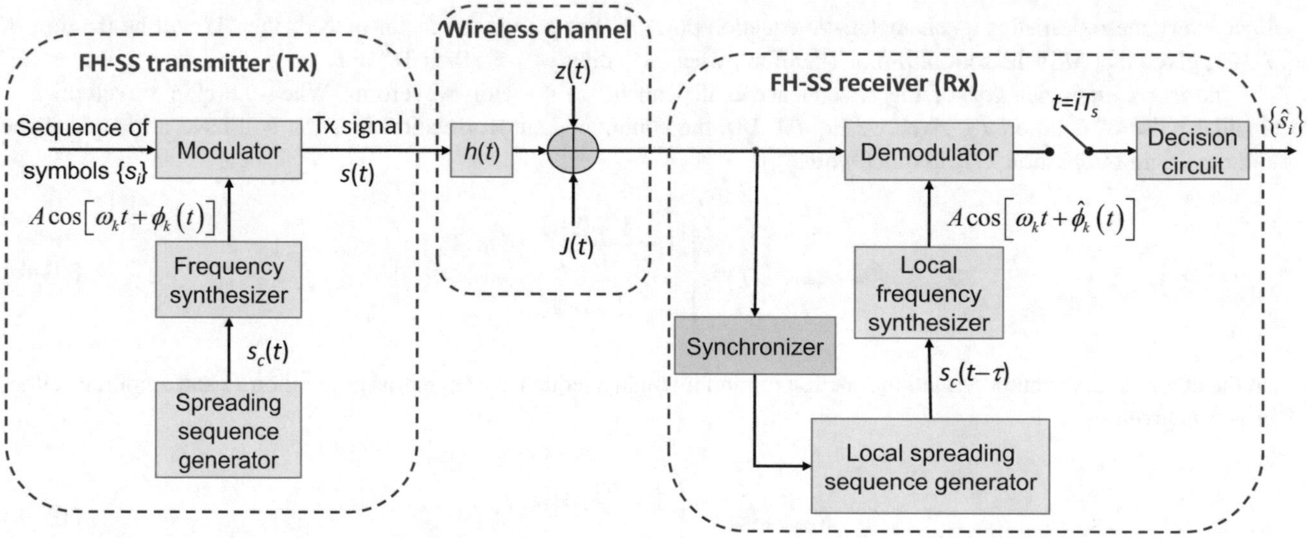

Fig. 10.15 Generic FH-SS system architecture

Fig. 10.16 Configurations of (**a**) FH-MSFK system and (**b**) FFH-MFSK demodulator

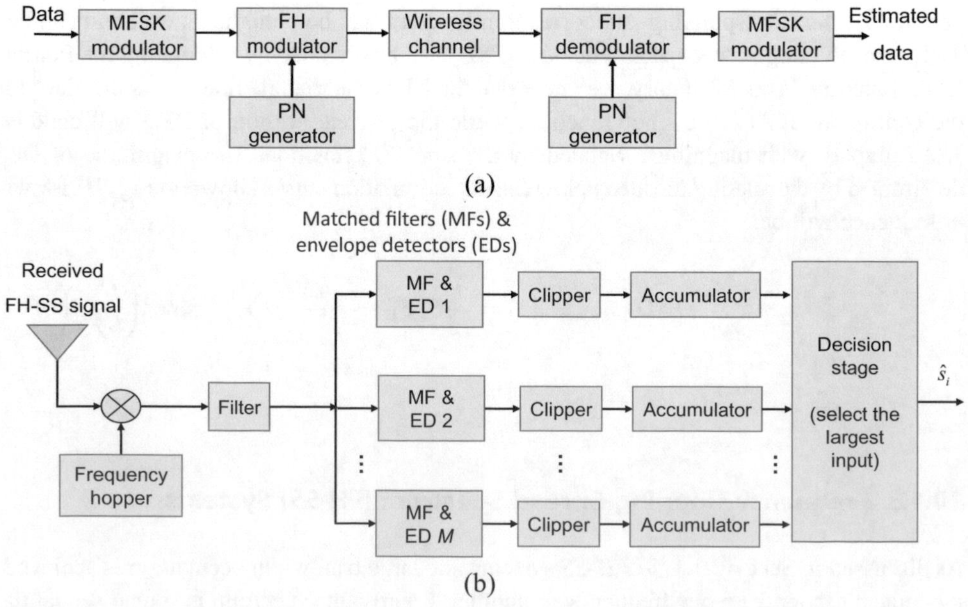

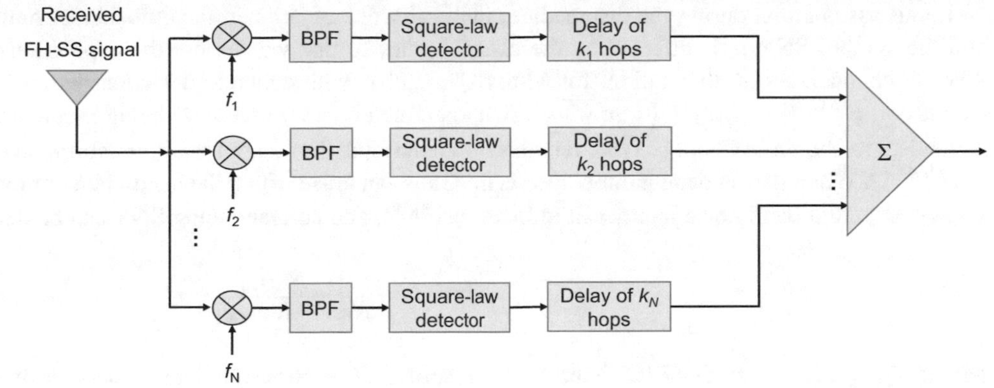

Fig. 10.17 Parallel FH acquisition scheme

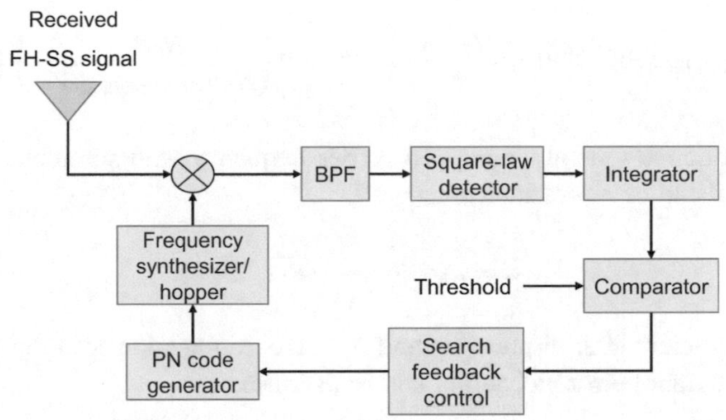

Fig. 10.18 FH serial search acquisition scheme

desired bandwidth of data signal only. After that, signal is split into M branches, one for each frequency f_k ($k = 1,\ldots,M$). The k-th branch is composed of matched filter for frequency f_k, envelope detector, clipper, and accumulator. The clipper is needed in the presence of jamming signal. Demodulator does not make a decision on a chip level but rather on a symbol level. The accumulator sums up the energies after all chips. In the decision stage, the accumulator output of highest cumulative energy is selected.

The parallel FH acquisition scheme is shown in Fig. 10.17. The received FH-SS signal is split into N branches, each corresponding to one of possible hopping frequencies. Correlators can be replaced by matched filters. After match filtering, the square-low detection takes place, and in each branch the corresponding delays are introduced, depending on the spreading sequence. Once the delays are properly chosen, the strong peak gets generated at the output, indicating that the local spreading sequence matches that of the received signal.

To reduce the complexity of this acquisition scheme, single correlator/matched filter can be used, combined with frequency synthesizer/hopper, as shown in Fig. 10.18. The PN generators control the frequency synthesizer/hopper. Once the correct frequency hopping sequence is found, the comparator exceeds the threshold, and the acquisition indication is generated. The tracking step is very similar to that of DS-SS systems.

We have discussed the impact of multipath fading on FH-SS systems already in Sect. 10.1.1. When multipath effects cause either flat or frequency-selective fading, in the absence of interference signal, the performances are similar to those of not-FH systems. In the presence of interference, let us assume that SFH is used and that interference (jamming) signal occupies K out of N frequency bands. The symbol error probability P_s can be estimated as follows:

$$\begin{aligned} P_s = &\Pr(\text{symbol error}|\text{no jammer})\underbrace{\Pr(\text{no jammer})}_{\frac{N-K}{N}} \\ &+\Pr(\text{symbol error}|\text{jammer})\underbrace{\Pr(\text{jammer})}_{\frac{K}{N}}. \end{aligned} \tag{10.51}$$

Let us assume that during the dispreading, in the absence of jammer, the influence of multipath components is mitigated, and the symbol SNR ρ_s is affected by the channel gain, so that we can use the expression of symbol error probability for AWGN channel. As an illustration, for M-ary QAM/PSK with coherent detection, we can use the following approximate expression $P_s^{(\mathrm{AWGN})} \simeq a(M)Q\left(\sqrt{b(M)\rho_s}\right)$, with parameters $a(M)$ and $b(M)$ being modulation and signal constellation size dependent as shown in Chap. 6. What remains is to determine the symbol error probability in the presence of jammer, denoted as $P_s^{(\mathrm{Jammer})}$. When narrowband jammer inserts the Gaussian noise within the bandwidth of modulated signal B, such a jammer, known as partial band noise jammer, it reduces the SNR, and corresponding SNR can be determined as:

$$\rho_s^{(\mathrm{Jammer})} = \frac{N_0 B}{N_0 B + N_{\mathrm{Jammer}} B}\rho_s, \tag{10.52}$$

where N_{Jammer} is the PSD of the jammer noise source. The corresponding symbol error probability will be $P_s^{(\mathrm{Jammer})} = P_s^{(\mathrm{AWGN})}\left(\rho_s^{(\mathrm{Jammer})}\right)$.The overall symbol error probability, based on Eq. (10.51), can be determined as:

$$P_s(\rho_s) = P_s^{(\mathrm{AWGN})}(\rho_s)\frac{N-K}{N} + P_s^{(\mathrm{Jammer})}\left(\frac{N_0 B}{N_0 B + N_{\mathrm{Jammer}} B}\rho_s\right)\frac{K}{N}. \tag{10.53}$$

Let now the jamming signal be the tone at the hopped carrier frequency with some phase offset $\delta\phi_i$. The demodulator output at the i-th symbol interval can be represented as:

$$\widehat{s}_i = a_i s_i + z_i + \sqrt{P_J} e^{\,j\delta\phi_i}, \tag{10.54}$$

where a_i is the channel gain coefficient after dispreading and P_J is the average power of the jamming tone. For M-ary PSK, assuming that $\mathrm{Arg}\{s_i\} = 0$, the symbol error probability can be estimated by:

$$P_s = 1 - \Pr\left(\left|\mathrm{Arg}\left\{a_i s_i + z_i + \sqrt{P_J} e^{\,j\delta\phi_i}\right\}\right| \le \pi/M\right), \tag{10.55}$$

with additional details to be found in [10, 11].

In the rest of this section, we describe how to generate sequences for FH-SS systems. When we are concerned with synthesis of sequences for multiuser FS-SS systems, the following parameters are relevant [17, 21]: number of available frequency slots Q, sequence period L, cardinality of sequence set M, and maximum Hamming cross-correlation $H_{\max}$. The Hamming cross-correlation for two sequences $\{x\}$ and $\{y\}$ of length L each is defined as:

$$H_{xy}(i) = \sum_{l=0}^{L-1} h\left(x_l y_{l+i}\right), \tag{10.56}$$

where $h(a,b)$ is the Hamming metrics defined as $h(a,b) = \begin{cases} 1, & a = b \\ 0, & a \neq b \end{cases}$. The hopping pattern can be defined as L-tuple:

$$f_L = \left(f_{j_1}, f_{j_2}, \cdots, f_{j_L} \right), \tag{10.57}$$

where f_{j_l} is the symbol selected from a finite set of Q elements. The maximum number of patterns, for synchronous multiuser case, with no more than k hits between any two patterns, is Q^{k+1} [21]. For asynchronous transmission, the maximum number of patterns reduces down to Q^{k+1}/L. The upper limit for maximum sequence length is given by [21]:

$$L \le \frac{Q^{k+1} - 1}{Q^k - 1}, \tag{10.58}$$

which indicates that for $k = 1$ we have that $L_{\max} = Q + 1$, while for $k > 1$, the $L_{\max} = kQ$. Compared to the Singleton bound for maximum distance separable (MDS) codes, this bound is not tight bound. The Singleton bound, as shown in Chap. 9, for codeword length n and number of information bits k_{info}, the Singleton bound is given by:

$$H_{\max} \leq n - k_{\text{info}} + 1. \tag{10.59}$$

The set of patterns with k hits is equivalent to the MDS code with $k_{\text{info}} = k + 1$. So only for $k = 1$, these two bounds are in agreement. The lower bound for $H_{\max}$ is given by [22]:

$$H_{\max} \geq \frac{(L-b)(L+b+Q)}{Q(L-1)}, \quad b = L \bmod Q. \tag{10.60}$$

It has been shown in Chap. 9 that Reed-Solomon (RS) codes satisfy the Singleton bound with equality sign and belong to the class of MDS codes. If we are concerned with asynchronous multiuser transmission, and sequences of length $L = Q$-1, we can generate Q^k sequences as follows [17, 23–25]. Let Q be a prime power for the finite field GF(Q) defined, and let α be a primitive element in GF(Q). We need to determine the polynomial:

$$P(x) = \sum_{i=0}^{k} p_i x^i, p_i \neq 0, \tag{10.61}$$

such that the set of sequences:

$$\{a_i\} = P(\alpha^i); i = 0, 1, 2, \cdots, Q-2 \tag{10.62}$$

have autocorrelation with maximum of k-1. In particular, for one-coincidence sequences ($k = 1$), the polynomial $P(x)$ is simple:

$$P(x) = p_0 + p_1 x, p_0 \in GF(Q), p_1 = \text{const} \in GF(Q). \tag{10.63}$$

The LSFR in the so-called parallel configuration can be used to determine different powers of primitive element as the content of shift register. Now by adding p_0 to the shift register and reading out the content as integer mod Q, we obtain the desired frequency hopping sequence.

As an illustration, let us observe the GF(2) and extended field GF($Q = 2^3$), which can be obtained from primitive polynomial $g(x) = 1 + x + x^3$, whose parallel implementation is shown in Fig. 10.19. The representation of GF(8) elements in GF(2) is given by:

α^i	GF(2) representation	Decimal representation
0	000	0
1	001	1
α	010	2
α^2	100	4
$\alpha^3 = \alpha + 1$	011	3
$\alpha^4 = \alpha^2 + \alpha$	110	6
$\alpha^5 = \alpha^2 + \alpha + 1$	111	7
$\alpha^6 = \alpha^2 + 1$	101	5

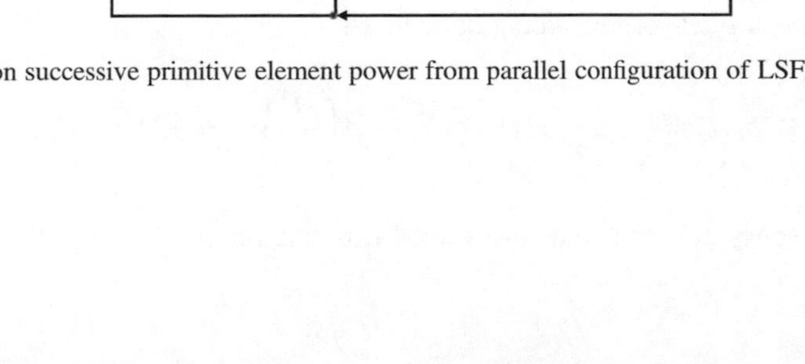

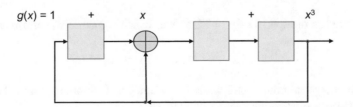

Fig. 10.19 Illustration of generation successive primitive element power from parallel configuration of LSFR

Now by calculating the polynomial $P(x) = x + p_0$, $p_0 \in GF(8)$, we obtained the hopping sequences as follows [17, 25]:

Polynomial	Hopping sequence
$x + 0$	1,243,675
$x + 1$	0352764
$x + \alpha$	3,061,457
$x + \alpha^2$	5,607,231
$x + \alpha^3$	2,170,546
$x + \alpha^4$	7,425,013
$x + \alpha^5$	6,534,102
$x + \alpha^6$	4,716,320

10.2 Code-Division Multiple Access (CDMA) Systems

The spread spectrum systems can be used in *multiaccess communications*, in which multiple users share the spectrum. Multiaccess communications suffer from *multiaccess interference*, which can be both intentional and unintentional, such as crosstalk or multipath fading. Moreover, the system design, such as in *code-division multiple access* (CDMA), can introduce the crosstalk among the users. When spread spectrum principles are used in multiaccess communication, sometimes we refer to such scheme as *spread spectrum multiple access* (SSMA), which is one particular instance of CDMA. To deal with multiuser interference, originating from multiple interfering signals, we use the *multiuser detection (demodulation)* [5], which is also known as interference cancelation or co-channel interference suppression, and it is subject of investigation in Sect. 10.3. The performances of multiaccess communication are also dependent on the fact if the multiaccess communication link is a *downlink channel* (one transmitter to many receivers scenario) or an *uplink channel* (many transmitters to one receiver scenario). The downlink channel is also known as the forward or broadcast channel as well as point-to-multipoint channel, and all signals are typically synchronous as they originate from the same transmitter. In contrast, in the uplink channel, also known as multiaccess channel or reverse link as well as multipoint-to-point channel, the signals are typically asynchronous as they originate from different users at different locations. In frequency-division multiple access (FDMA), different non-overlapped frequency bands are allocated to different users. On the other hand, in time-division multiple access (TDMA), the time scale is partitioned into time slots, which are shared among different users in a round-robin fashion. The TDMA requires precise time synchronism. Therefore, in both FDMA and TDMA, users are required not to overlap each other in either time domain or frequency domain. However, it is possible for users' signals to be orthogonal while simultaneously overlapping in both time and frequency domains, such as the one illustrated in Fig. 10.20a, and we refer to such multiaccess scheme as CDMA. Clearly, the signature waveforms $p_1(t)$ and $p_2(t)$, assigned to two different users, are orthogonal to each other since:

$$\langle p_2 | p_1 \rangle = \int_0^T p_1(t) p_2^*(t) dt = 0. \tag{10.64}$$

Now let both users apply antipodal signaling with $a_{ik} \in \{-1,+1\}$, $i = 1,2$. Then modulated signals can be represented as:

$$s_1(t) = \sum_i a_{1i} p_1(t - iT), s_2(t) = \sum_i a_{2i} p_2(t - iT), \langle s_2 | s_1 \rangle = 0. \tag{10.65}$$

The corresponding CDMA system model is provided in Fig. 10.20(b). Because the signature waveforms ("codes") $p_1(t)$ and $p_2(t)$ are orthogonal to each other, and perfect synchronism is assumed, the optimum receiver will be matched filter or correlation receiver.

The FDMA can be considered as the spatial case of CDMA in which signature waveforms are defined as:

$$p_k(t) = g(t) \cos(2\pi f_k t), | f_k - f_m | = l/T, \tag{10.66}$$

where $g(t)$ is the pulse of unit energy and duration T while l is an integer. In similar fashion, TDMA can be considered as a special case of CDMA with signature waveforms being defined as:

$$p_k(t) = \frac{1}{\sqrt{T/K}} \text{rect}\left(\frac{t - (k-1)T/K}{T/K}\right); k = 1, 2, \cdots, K \tag{10.67}$$

In the next subsection, we discuss different signature waveforms suitable for use in CDMA systems.

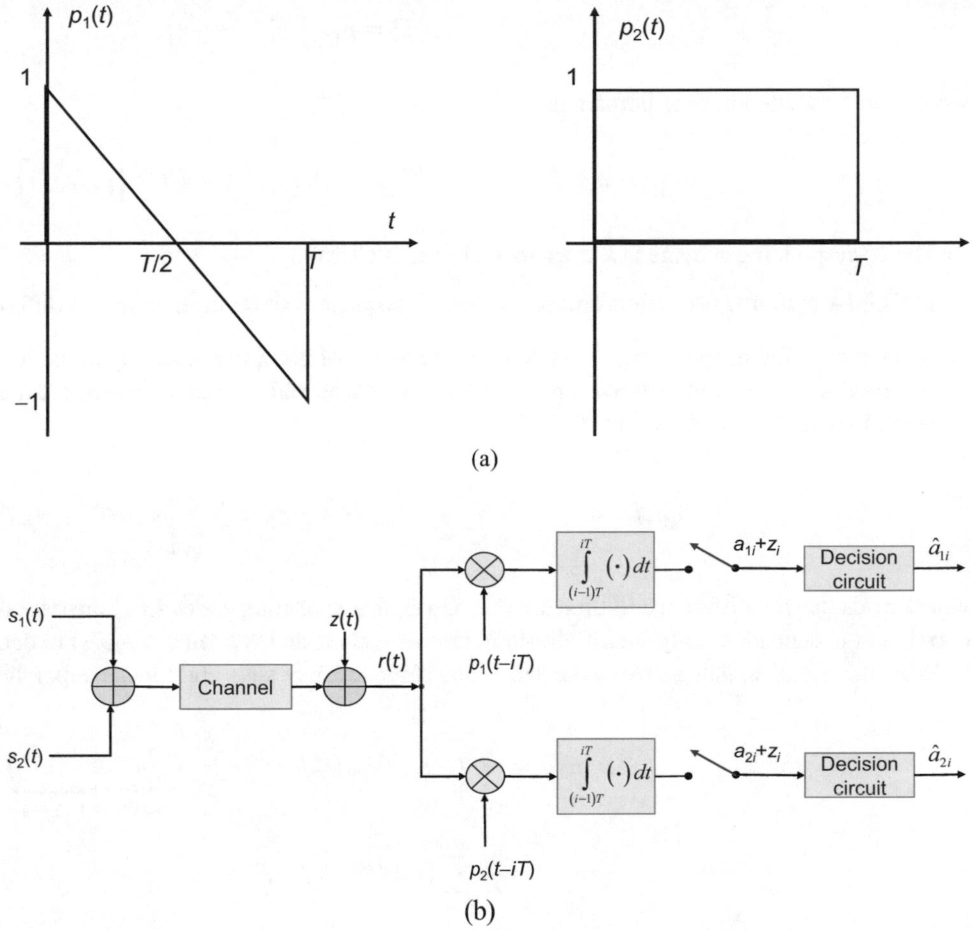

Fig. 10.20 (**a**) Two orthogonal signature waveforms assigned to two different users. (**b**) Two users CDMA system model

10.2.1 Signature Waveforms

The spreading sequences suitable for DS-SS systems have already been discussed in Sect. 10.1.2, in particular in Sect. 10.1.2.4. On the other hand, the spreading sequences suitable for both FH-SS- and FH-based CDMA systems have been discussed in Sect. 10.1.3. The DS-SS waveform given by Eq. (10.17) can be generalized as follows [5]:

$$s_c(t) = A_c \sum_{n=1}^{N} \alpha_n \phi_n(t), \tag{10.68}$$

where $[\alpha_1,\ldots, \alpha_n,\ldots, \alpha_N]$ is the spreading code with components not necessarily binary and the basis functions $\{\phi_i(t)\}$ satisfy the orthogonality conditions:

$$\langle \phi_m | \phi_n \rangle = \delta_{nm}.$$

For DS-SS systems, the basis functions are chip waveforms given by $\phi_n(t) = p_{T_c}[t - (n-1)T_c]$ and spreading code components $\alpha_n \in \{-1,+1\}$. In FH-SS, the chips are modulated in frequency domain:

$$s_c^{(\mathrm{FH})}(t) = A_c \sum_{n=1}^{N} \phi_n(t, f_n),\ \ \phi_n(t, f_n) = \begin{cases} \cos(2\pi f_n t), (n-1)T_c \le t < nT_c \\ 0, \text{otherwise} \end{cases} \tag{10.69}$$

The dual for DS chip waveforms $\phi_n(t) = p_{T_c}[t - (n-1)T_c]$ can be defined in frequency domain as:

$$\Phi_n(f) = p_{1/T}\left(f - \frac{n-1}{T}\right), \tag{10.70}$$

which can be written in time domain as:

$$\phi_n(f) = e^{j2\pi(n-1)t/T} P_{1/T}(t), P_{1/T}(t) = FT^{-1}\left\{p_{1/T}(f)\right\}. \tag{10.71}$$

The corresponding scheme is known as multicarrier CDMA.

In CDMA systems, we assign different signature sequences (spreading codes) to different users $\{s_{c_k}\}_{k=1}^{K}$. The cross-correlation of different spreading codes defines the amount of interference among users. In asynchronous CDMA, the signals for different users arrive at different times at the receiver side, and we can define the cross-correlation of the spreading codes assigned to user k and k' as follows:

$$\rho_{kk'}(t, t-\tau) = \frac{1}{T_s}\int_0^{T_s} s_{c_k}(t) s_{c_{k'}}(t-\tau)dt = \frac{1}{N}\sum_{n=1}^{N} s_{c_k}(nT_c) s_{c_{k'}}(nT_c - \tau), \tag{10.72}$$

where τ denotes the difference in time arrivals. Given that spreading codes $\{s_{c_k}\}$ are periodic with the period T_s, then cross-correlation is dependent only on difference in time arrivals τ, and we can use $\rho_{kk'}(\tau)$ to denote $\rho_{kk'}(t, t-\tau)$. In synchronous CDMA, the users' signals arrive at the same time at the receiver side, and the corresponding cross-correlation becomes:

$$\begin{aligned} \rho_{kk'}(0) &= \frac{1}{T_s}\int_0^{T_s} s_{c_k}(t) s_{c_{k'}}(t)dt = \frac{1}{N}\sum_{n=1}^{N} \underbrace{s_{c_k}(nT_c)}_{(-1)^{c_{kn}}} \underbrace{s_{c_{k'}}(nT_c)}_{(-1)^{c_{k'n}}} \\ &= \frac{1}{N}\sum_{n=1}^{N} (-1)^{c_{kn}}(-1)^{c_{k'n}}, \end{aligned} \tag{10.73}$$

where we used $(-1)^{c_{kn}}$ to denote $s_{c_k}(nT_c)$. Clearly, the negative logic is used here: -1 to denote binary 1 and + 1 to denote binary 0. Ideally, we would like in synchronous (asynchronous) CDMA to have $\rho_{kk'}(0) = 0$ [$\rho_{kk'}(\tau) = 0$] whenever $k \neq k'$. The codes set satisfying this cross-correlation property can be called as *orthogonal spreading code set*, otherwise non-orthogonal spreading code set. The literature on PN sequences is quite rich [10, 11, 16–26]. Some examples of pseudo-random sequences include Gold codes, Kasami sequences [10, 17], Barker sequences, etc. The cross-correlation between two synchronous signature sequences can be determined as:

$$\begin{aligned} \rho_{kk'}(0) &= \frac{1}{N}\sum_{n=1}^{N} (-1)^{c_{kn}}(-1)^{c_{k'n}} = \frac{1}{N}(\text{\#agreements} - \text{\#disagreements}) \\ &= \frac{1}{N}[\text{\#agreements} - (N - \text{\#agreements})] \\ &= -1 + \frac{2}{N}(\text{\#agreements}) \\ &= -1 + \frac{2}{N}\sum_{n=1}^{n} [c_{kn} = c_{k'n}], \end{aligned} \tag{10.74}$$

where we used $[P]$ to denote the *Inversion function*, which has value 1 when proposition P is true and zero otherwise:

$$[P] \doteq \begin{cases} 1, P \text{ is true} \\ 0, P \text{ is false} \end{cases} \tag{10.75}$$

For an m-sequence, defined as $(c_1, c_2, \ldots, c_N)$, there are 2^{m-1} zeros and 2^{m-1} ones in the sequence. The sequence length is $N = 2^m$-1, and when we let the m-sequence to repeat, the corresponding correlation function is given by:

$$R(l) = \frac{1}{N}\sum_{n=1}^{N}(-1)^{c_n}(-1)^{c_{n+l}} = \begin{cases} 1, & l = 0, N, 2N, \cdots \\ -\frac{1}{N}, & \text{otherwise} \end{cases} \tag{10.76}$$

The spreading codes for asynchronous CDMA must satisfy the following two constraints: (i) the out-of-phase autocorrelation must be low and (ii) the cross-correlation must be low as well. The m-sequences do satisfy the constraint (i); however, their cross-correlation constraint is not that good. The Welch has shown that the maximum cross-correlation for any two spreading (periodic) binary sequences of length N, taken from the set of size M, are lower bounded as follows [27]:

$$R(k) \geq N\left(\frac{M-1}{MN-1}\right)^{1/2} \approx \sqrt{N}. \tag{10.77}$$

The *Gold sequences (codes)* [28, 29] have better cross-correlation properties than m-sequence but worse autocorrelation properties. At this point it is convenient to introduce the concept of *preferred pairs of m-sequences*, which represent a pair of m-sequences $\{a\}$ and $\{b\}$ of length N with a periodic cross-correlation function taking three possible values:

$$\rho_{a,b} = \begin{cases} -1/N \\ -t(n)/N \\ [t(n)-2]/N \end{cases}, t(n) = \begin{cases} 1 + 2^{(n+1)/2}, & n - \text{odd} \\ 1 + 2^{(n+2)/2}, & n - \text{even} \end{cases} \tag{10.78}$$

Let us observe two sequences of length $N = 2^m$-1; the second is obtained from the first $\{a_n\}$ by decimation (taking every q-th element): $\{b_n\} = \{a_{qn}\}$. When N and q are co-prime, with either $q = 2^{k+1}$ or $q = 2^{2k}–2^k + 1$, and when $m = 0 \bmod 4$ with the greatest common divider (GCD) for m and k satisfying the following constraint $GCD(m,k) = \begin{cases} 1, & m - \text{odd} \\ 2, & m - \text{even} \end{cases}$, then it can be shown that cross-correlation between these two sequences will have three possible values, as specified by Eq. (10.78). Therefore, $\{a_n\}$ and $\{b_n\} = \{a_{qn}\}$ are preferred pair of sequences. They can be used to create the Gold sequences:

$$\{a\}, \{b\}, \{a+b\}, \{a+Db\}, \{a+D^2b\}, \cdots, \{a+D^{N-1}b\}, \tag{10.79}$$

where $D = z^{-1}$ is a delay operator, having trivalent cross-correlation (10.78). The first two sequences are m-sequences, while the remaining N are not. For sufficiently large N, the maximum cross-correlation is either $\sqrt{2}$ (m-odd) or two times (m-even) larger than that predicted by Welch bound. As an illustration, in Fig. 10.21, we describe how to generate Gold sequences of length $N = 31$ starting from polynomials $g_1(x) = 1 + x^2 + x^5$ and $g_2(x) = 1 + x + x^2 + x^4 + x^5$.

Kasami sequences are also derived from m-sequences but have better cross-correlation properties than Gold codes [2, 10, 17]. There two different sets of Kasami sequences, used to generate Kasami codes, namely, the large set and the small set. The *small set* is obtained with the help of the following generating polynomial: $f(x)f^1(x)$, where $f(x)$ corresponds to m-sequence $\{a\}$ of length $L = 2^m$-1 (m-even) and $f^1(x)$ corresponds to an $\{a^1\}$ sequence obtained from $\{a\}$ by decimation with $q = 2^{m/2} + 1$. The new sequence is shorter $L = 2^{m/2}–1$, and there are $2^{m/2}$ different sequences. The cross-correlation values are 3-valued as well $\{-1,-(2^{m/2} + 1), -(2^{m/2}–1)\}/N$, and the cross-correlation values satisfy the Welch bound with equality sign. The *large set* is obtained with the help of generating polynomial: $f(x)f^1(x)f^2(x)$, where $f^2(x)$ corresponds to an $\{a^2\}$ sequence obtained from $\{a\}$ by decimation with $q = 2^{(m+2)/2} + 1$. The cross-correlation has five different values:

$$\rho_{a,b} = \begin{cases} -1/N \\ \left(-1 \pm 2^{m/2}\right)/N \\ \left[-1 \pm \left(2^{m/2}+1\right)\right]/N \end{cases}, \tag{10.80}$$

and the number of sequences is large $2^{3m/2}$ for $m = 0 \bmod 4$ and $2^{3m/2} + 2^{m/2}$ for $m = 2 \bmod 4$.

For synchronous CDMA applications, the *orthogonal sequences* can be used. The orthogonality of two sequences, represented as vectors $|a\rangle = [a_1,\ldots,a_N]^T$ and $|b\rangle = [b_1 \ldots b_N]^T$, can be defined in similar fashion to orthogonality principle for signals [see Eq. (10.64)] as follows:

$$\langle b|a\rangle = \sum_{n=1}^{N} a_n b_n^* = 0.$$

The very popular orthogonal sequences are *Walsh functions (sequences)* [30–32], the set of bipolar sequences of size $N = 2^m$, defined recursively as follows:

$$\begin{aligned} &x_1 = \{(+1)\}, x_2 = \{(+1,+1),(+1,-1)\}, \\ &x_3 = \{(+1,+1,+1,+1),(+1,+1,-1,-1),(+1,-1,+1,-1),(+1,-1,-1,+1)\}, \cdots \end{aligned} \tag{10.81}$$

An alternative way to obtain the Walsh sequences is to interpret them as the rows (columns) of the *Hadamard matrices*, defined recursively as:

$$\begin{aligned} &\boldsymbol{H}_1 = [+1], \boldsymbol{H}_2 = \begin{bmatrix} +1 & +1 \\ +1 & -1 \end{bmatrix}, \boldsymbol{H}_4 = \boldsymbol{H}_2^{\otimes 2} = \begin{bmatrix} +1 & +1 & +1 & +1 \\ +1 & +1 & -1 & -1 \\ +1 & -1 & +1 & -1 \\ +1 & -1 & -1 & +1 \end{bmatrix}, \cdots, \\ &\boldsymbol{H}_{2n} = \boldsymbol{H}_n^{\otimes 2} = \begin{bmatrix} \boldsymbol{H}_n & \boldsymbol{H}_n \\ \boldsymbol{H}_n & -\boldsymbol{H}_n \end{bmatrix}, \end{aligned} \tag{10.82}$$

where we use $(\cdot)^{\otimes}$ to denote the Kronecker product of matrices. Clearly, the Hadamard matrix is symmetric and has equal number of +1 and − 1 in every row (column), except the first one. Additionally, any two rows (columns) are orthogonal to each other. In conclusion, the rows (columns) of Hadamard matrix can be used as Walsh sequences.

10.2.2 Synchronous and Asynchronous CDMA Models

We first describe the basic models, followed by discrete-time models.

10.2.2.1 Basic CDMA Models

In *synchronous CDMA*, because the signals originating from different users are perfectly time-aligned, it is sufficient to consider a single symbol interval [0, *T*]. To facilitate explanations, let us denote the *k*-th ($k = 1,2,\ldots,K$) user antipodally modulated signature waveform of unit energy with $s_k(t)$ instead of $s_{c_k}(t)$used before. As the generalization of Fig. 10.20 to *K* users, the received signal can be represented as:

$$r(t) = \sum_{k=1}^{K} A_k b_k s_k(t) + \sigma z(t), \tag{10.83}$$

where A_k is the amplitude of the *k*-th user at receiver side, $b_k \in \{+1, -1\}$ is the *k*-th user information bit, and *z*(*t*) is a zero-mean Gaussian noise of unit variance. When the real-valued signature waveforms are orthogonal to each other, the performance of

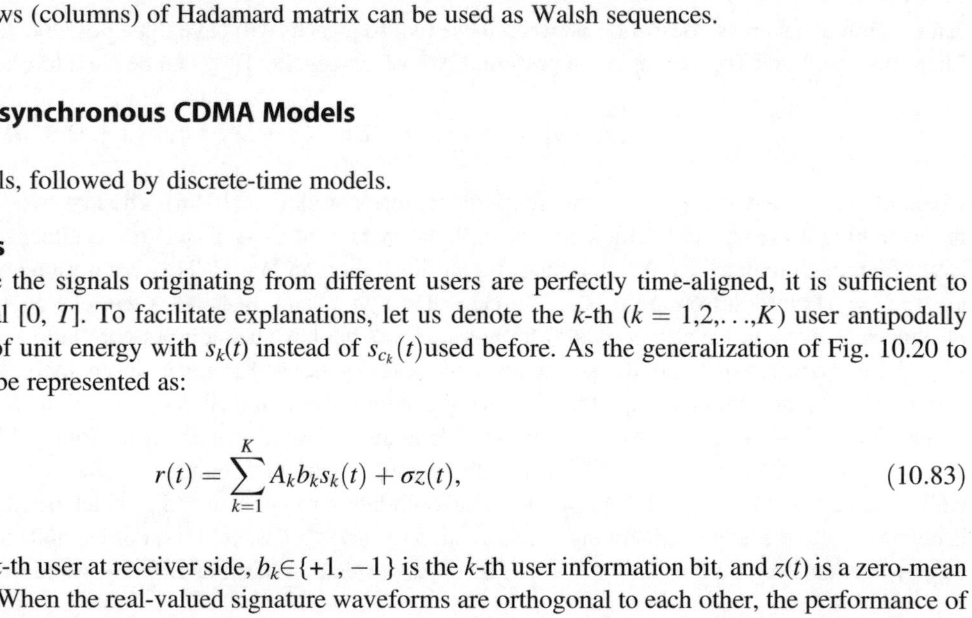

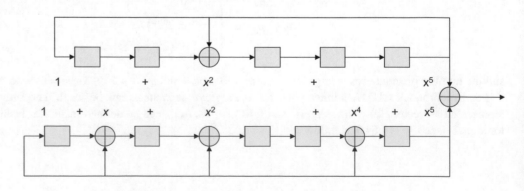

Fig. 10.21 Generation of one set of Gold sequences form polynomials $g_1(x) = 1 + x^2 + x^5$ and $g_2(x) = 1 + x + x^2 + x^4 + x^5$

the k-th user is dependent on signal-to-noise ratio A_k/σ, where σ signal is the standard deviation of Gaussian noise process. When the signature waveforms are quasi-orthogonal (or even non-orthogonal), we can use the cross-correlation to quantify the orthogonality:

$$\rho_{kk'} = \langle s_{k'} | s_k \rangle = \int_0^T s_k(t) s_{k'}^*(t) dt, \tag{10.84}$$

and since the waveform signatures are real-valued, the complex-conjugate operation can be omitted. By applying the Bunjakowski-Cauchy-Schwartz inequality, the following upper bound on magnitude of cross-correlation is obtained:

$$|\rho_{kk'}| = \left| \int_T^0 s_k(t) s_{k'}^*(t) dt \right| \leq \|s_k\| \|s_{k'}\| = 1, \ \|s_k\|^2 = \int_0^T |s_k(t)|^2 dt = 1. \tag{10.85}$$

The cross-correlation matrix $\boldsymbol{R}$ can be defined as:

$$\boldsymbol{R} = \begin{bmatrix} \rho_{11} & \rho_{12} & \cdots & \rho_{1K} \\ \rho_{21} & \rho_{22} & \cdots & \rho_{2K} \\ \vdots & \vdots & \ddots & \vdots \\ \rho_{K1} & \rho_{2K} & \cdots & \rho_{KK} \end{bmatrix}; \ \rho_{kk} = 1, k = 1, \cdots, K \tag{10.86}$$

It is straightforward to show that correlation matrix is non-negative definite, since for arbitrary vector $|a\rangle = [a_1 \ldots a_N]^{\mathrm{T}}$ we have that:

$$\begin{aligned} \langle a | \boldsymbol{R} | a \rangle &= \underbrace{[a_1^* a_2^* \cdots a_K^*]}_{\langle a|} \begin{bmatrix} \rho_{11} & \rho_{12} & \cdots & \rho_{1K} \\ \rho_{21} & \rho_{22} & \cdots & \rho_{2K} \\ \vdots & \vdots & \ddots & \vdots \\ \rho_{K1} & \rho_{2K} & \cdots & \rho_{KK} \end{bmatrix} \underbrace{\begin{bmatrix} a_1 \\ a_2 \\ \vdots \\ a_K \end{bmatrix}}_{|a\rangle} = \sum_{k=1}^{K} \sum_{k'=1}^{K} a_k a_{k'}^* \rho_{kk'} \\ &= \sum_{k=1}^{K} \sum_{k'=1}^{K} a_k a_{k'}^* \int_0^T s_k(t) s_{k'}^*(t) dt = \int_0^T \left(\sum_{k=1}^{K} a_k s_k(t) \right) \left(\sum_{k'=1}^{K} a_{k'} s_{k'}(t) \right)^* dt \\ &= \left\| \sum_{k=1}^{K} a_k s_k(t) \right\|^2 \geq 0. \end{aligned} \tag{10.87}$$

In *asynchronous CDMA*, we need to specify delays in arrival for each user $\tau_k \in [0,T]$; $k = 1,2,\ldots,K$. Because different users' signals arrive at different times at the receiver side, it is not sufficient to observe one symbol but block of symbols. Let us denote the block of symbols for the k-th user containing $2M + 1$ symbols as follows [5]: $\boldsymbol{b}_k = [b_k[-M], \ldots, b_k[0], \ldots, b_k[M]]$. The basic asynchronous CDMA model can be now represented as:

$$r(t) = \sum_{k=1}^{K} \sum_{i=-M}^{M} A_k b_k[i] s_k(t - iT - \tau_k) + \sigma z(t). \tag{10.88}$$

We further assume that information block length $\boldsymbol{b}_k$ is sufficiently large so that edge effects are not relevant, so that we can observe delays modulo-T. Clearly, when signals from all users arrive at the same time ($\tau_k = \tau$), the asynchronous case reduces down to the synchronous chase. Another interesting special case is when all amplitudes are the same ($A_k = A$) and all users use the same signature waveform, namely, $s_k(t) = s(t)$. If we further assume that delays can be represented as $\tau_k = (k\text{-}1)T/K$, like in TDMA, clearly, Eq. (10.88) reduces down to ISI case:

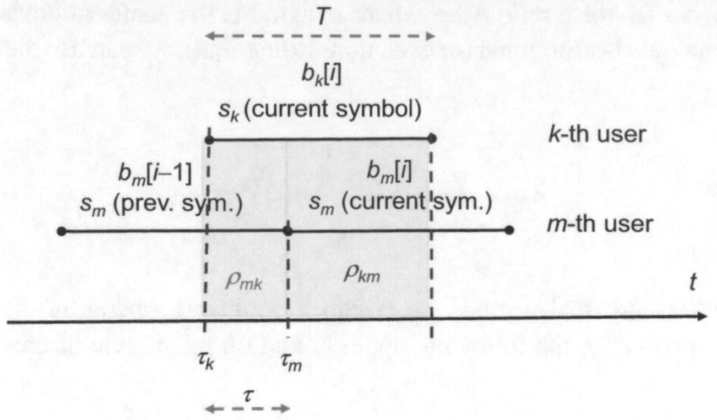

Fig. 10.22 Illustration of asynchronous cross-correlations' definitions, assuming that $\tau_k < \tau_m$

$$
\begin{aligned}
r(t) &= \sum_{k=1}^{K} \sum_{i=-M}^{M} Ab_k[i]s\left(t - iT - \frac{k-1}{K}T\right) + \sigma z(t) \\
&= \sum_{k=1}^{K} \sum_{i=-M}^{M} \underbrace{A\mathbf{b}[iK + k - 1]}_{b_k[i]} s\left(t - iT - \frac{k-1}{K}T\right) + \sigma z(t).
\end{aligned}
\tag{10.89}
$$

Because the signals from users k and m have different delays, τ_k and τ_m, we have to specify two cross-correlations, assuming that $\tau = \tau_m$-$\tau_k > 0$ (the signal from the k-th user arrives earlier than the signal from the m-th user for τ seconds), as illustrated in Fig. 10.22. The cross-correlation ρ_{km} is related to cross-correlation between the current symbol of k-th user and current symbol of the m-th user, and it is defined as:

$$
\rho_{km} = \int_{\tau}^{T} s_k(t)s_m(t-\tau)dt. \tag{10.90}
$$

On the other hand, the cross-correlation ρ_{mk} is related to cross-correlation between the current symbol of k-th user and previous symbol of the m-th user, and it is defined as:

$$
\rho_{mk} = \int_{0}^{\tau} s_k(t)s_m(t+T-\tau)dt. \tag{10.91}
$$

10.2.2.2 Discrete-Time CDMA Models

Given that most of the signal processing in CDMA systems is performed in the discrete-time (DT) domain, it makes sense to develop the DT CDMA models.

Similar to the single-user case, the samples from correlators (matched filters) multiuser *synchronous CDMA* case, as shown in Fig. 10.23, represent a sufficient statistics. The k-th correlator output can be represented as:

$$
\begin{aligned}
r_k &= \int_0^T r(t)s_k(t)dt = \int_0^T \left[\sum_{k=1}^{K} A_k b_k s_k(t) + \sigma z(t)\right] s_k(t)dt \\
&= A_k b_k + \sum_{m=1,\, m\neq k}^{K} A_m b_m \underbrace{\int_0^T s_m(t)s_k(t)dt}_{\rho_{mk}} + \underbrace{\int_0^T \sigma z(t)s_k(t)dt}_{z_k} \\
&= A_k b_k + \sum_{m=1,\, m\neq k}^{K} A_m b_m \rho_{mk} + z_k;\ k = 1, 2, \cdots, K.
\end{aligned}
\tag{10.92}
$$

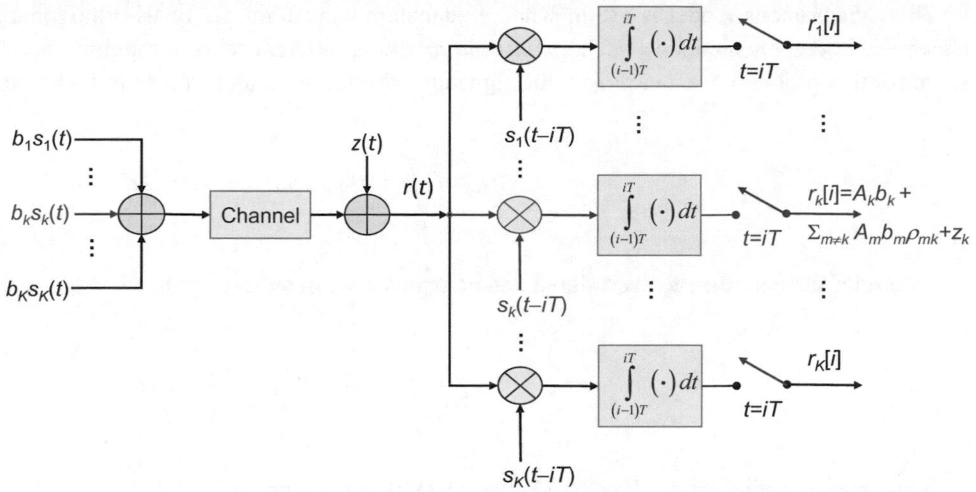

Fig. 10.23 Illustration of DT synchronous CDMA model

This set of K-equations can also be written in more compact matrix form:

$$\begin{aligned}|r\rangle &= \begin{bmatrix} r_1 \\ r_2 \\ \vdots \\ r_K \end{bmatrix} = \boldsymbol{R}\,\underbrace{\mathrm{diag}(A_1,\cdots,A_K)}_{\boldsymbol{A}}\underbrace{\begin{bmatrix} b_1 \\ b_2 \\ \vdots \\ b_K \end{bmatrix}}_{|b\rangle} + \underbrace{\begin{bmatrix} z_1 \\ z_2 \\ \vdots \\ z_K \end{bmatrix}}_{|z\rangle} \\ &= \boldsymbol{RA}\underbrace{|b\rangle}_{\boldsymbol{b}} + \underbrace{|z\rangle}_{\boldsymbol{z}} \\ &= \boldsymbol{RA}\,\boldsymbol{b} + \boldsymbol{z}.\end{aligned} \tag{10.93}$$

It can easily be shown that the following is valid:

$$\langle \boldsymbol{z}\rangle = \boldsymbol{0}, \mathrm{Cov}(\boldsymbol{z}) = \langle \boldsymbol{z}\boldsymbol{z}^T\rangle = \sigma^2\boldsymbol{R}, \tag{10.94}$$

indicating that components in $|z\rangle$ are correlated, and in analogy with single-user ISI channel, we can perform whitening by *Cholesky factorization* of the cross-correlation matrix [3, 33, 34], assuming that signals are real-valued:

$$\boldsymbol{R} = \boldsymbol{F}^T\boldsymbol{F}, \tag{10.95}$$

where $\boldsymbol{F}$ is the lower triangular matrix with positive diagonal elements. Now, by multiplying the Eq. (10.93) by $(\boldsymbol{F}^T)^{-1}$ from the left side, we obtain:

$$\begin{aligned}|r'\rangle &= (\boldsymbol{F}^T)^{-1}|r\rangle = \underbrace{(\boldsymbol{F}^T)^{-1}\boldsymbol{F}^T}_{\boldsymbol{I}}\boldsymbol{F}\;\boldsymbol{A}\,\boldsymbol{b} + \underbrace{(\boldsymbol{F}^T)^{-1}\boldsymbol{z}}_{z'} \\ &= \boldsymbol{F}\;\boldsymbol{A}\,\boldsymbol{b} + z'.\end{aligned} \tag{10.96}$$

Given that:

$$\begin{aligned}\langle z'\rangle &= \langle(\boldsymbol{F}^T)^{-1}z\rangle = (\boldsymbol{F}^T)^{-1}\langle z\rangle = \boldsymbol{0}, \\ \mathrm{Cov}(z') &= \langle z'z'^T\rangle = \langle(\boldsymbol{F}^T)^{-1}zz^T\boldsymbol{F}^{-1}\rangle = (\boldsymbol{F}^T)^{-1}\underbrace{\langle zz^T\rangle}_{\sigma^2\boldsymbol{R}}\boldsymbol{F}^{-1} \\ &= (\boldsymbol{F}^T)^{-1}\sigma^2\underbrace{\boldsymbol{R}}_{\boldsymbol{F}^T\boldsymbol{F}}\boldsymbol{F}^{-1} = \sigma^2\underbrace{(\boldsymbol{F}^T)^{-1}\boldsymbol{F}^T}_{\boldsymbol{I}}\underbrace{\boldsymbol{F}\boldsymbol{F}^{-1}}_{\boldsymbol{I}} = \sigma^2\boldsymbol{I},\end{aligned} \tag{10.97}$$

we conclude that components of $|z'\rangle$ are uncorrelated, and the whitening of the colored Gaussian noise has been successfully conducted.

This synchronous model is useful when K signature waveforms are quasi-orthogonal and known on the receiver side. However, in certain situations, such as *antenna arrays*, a different set of orthogonal functions is used, say $\{\phi_d(t)\}_{d=1}^{D}$. To solve for this problem, we can project the signature waveforms along this new basis functions as follows:

$$s_{kd} = \int_0^T s_k(t)\phi_d(t)dt; d = 1, \cdots, D. \tag{10.98}$$

In similar fashion, the received signal can be represented in terms of projection as:

$$r_d = \int_0^T r(t)\phi_d(t)dt; d = 1, \cdots, D. \tag{10.99}$$

Now, after substituting the Eq. (10.83) into (10.99), we obtain:

$$\begin{aligned} r_d &= \int_0^T \underbrace{r(t)}_{\sum_{k=1}^{K} A_k b_k s_k(t) + \sigma z(t)} \phi_d(t)dt = \int_0^T \left[\sum_{k=1}^{K} A_k b_k s_k(t) + \sigma z(t)\right]\phi_d(t)dt \\ &= \sum_{k=1}^{K} A_k b_k \underbrace{\int_0^T s_k(t)\phi_d(t)dt}_{s_{kd}} + \underbrace{\int_0^T \sigma z(t)\phi_d(t)dt}_{z_d} = \sum_{k=1}^{K} A_k b_k s_{kd} + z_d; d = 1, \cdots, D. \end{aligned} \tag{10.100}$$

The corresponding matrix representation will be:

$$\vec{r} = \begin{bmatrix} r_1 \\ r_2 \\ \vdots \\ r_D \end{bmatrix} = \sum_{k=1}^{K} A_k b_k \underbrace{\begin{bmatrix} s_{k1} \\ s_{k2} \\ \vdots \\ s_{kD} \end{bmatrix}}_{\vec{s}_k} + \underbrace{\begin{bmatrix} z_1 \\ z_2 \\ \vdots \\ z_D \end{bmatrix}}_{\vec{z}} = \sum_{k=1}^{K} A_k b_k \vec{s}_k + \vec{z}. \tag{10.101}$$

Let us now group the signature vectors into a signature matrix:

$$\mathbf{S} = \left[\vec{s}_1 \; \vec{s}_2 \; \cdots \; \vec{s}_K\right], \tag{10.102}$$

which is clearly of dimensionality $D \times K$. By using this signature matrix, with the k-th column corresponding to the k-th user signature $\vec{s}_k$, we can rewrite Eq. (10.101) as follows:

$$\vec{r} = \boldsymbol{SAb} + \vec{z}, \tag{10.103}$$

where matrix $\boldsymbol{A}$ and vector $\boldsymbol{b}$ are the same as introduced before. Let us now determine the covariance matrix of $\vec{r}$ as follows:

$$\begin{aligned} \mathrm{Cov}\left(\vec{r}\right) &= \left\langle \vec{r}\vec{r}^T \right\rangle = \left\langle \left(\boldsymbol{SAb} + \vec{z}\right)\left(\boldsymbol{SAb} + \vec{z}\right)^T \right\rangle = \left\langle \boldsymbol{SA}\underbrace{\boldsymbol{b}(\boldsymbol{SAb})^T}_{\boldsymbol{b}^T\boldsymbol{A}^T\boldsymbol{S}^T} \right\rangle + \underbrace{\left\langle \vec{z}\vec{z}^T \right\rangle}_{\sigma^2 \boldsymbol{I}} \\ &= \boldsymbol{SA}\underbrace{\left\langle \boldsymbol{bb}^T \right\rangle}_{\boldsymbol{I}}\boldsymbol{A}^T\boldsymbol{S}^T + \sigma^2\boldsymbol{I}, \\ &= \boldsymbol{SAA}^T\boldsymbol{S}^T + \sigma^2\boldsymbol{I}, \end{aligned} \tag{10.104}$$

where we used in the second line the fact that users' bits are independent. Clearly, the noise samples are independent and Eq. (10.103) is valid regardless of whether the orthogonal basis $\{\phi_d(t)\}_{d=1}^{D}$ span the signature waveforms $\{s_k(t)\}_{k=1}^{K}$ or not.

Regarding the *DT representation of asynchronous CDMA system*, without loss of generality, we assume that delays can be arranged as $\tau_1 \leq \tau_2 \leq \cdots \leq \tau_K$. The correlators' outputs can be written as:

$$\begin{aligned} r_k[i] &= \int_{\tau_k+iT}^{\tau_k+(i+1)T} r(t)s_k(t)dt \\ &= \sum_{m=1}^{K}\sum_{i=-M}^{M} A_m b_m[i] \int_{\tau_k+iT}^{\tau_k+(i+1)T} s_m(t-iT-\tau_k)s_k(t)dt + \underbrace{\int_{\tau_k+iT}^{\tau_k+(i+1)T} \sigma z(t)dt}_{z_k[i]} \\ &= A_k b_k[i] + \sum_{m=1,m\neq k}^{K}\sum_{i=-M}^{M} A_m b_m[i] \int_{\tau_k+iT}^{\tau_k+(i+1)T} s_m(t-iT-\tau_k)s_k(t)dt + z_k[i]. \end{aligned} \tag{10.105}$$

Now we have to group the crosstalk terms, relative to the k-th user signal, into two categories, as illustrated in Fig. 10.24: (i) the crosstalk terms for which $\tau_m > \tau_k$ and (ii) crosstalk terms for which $\tau_l < \tau_k$. By taking the asynchronous cross-correlations' definitions, given by (10.90 and 10.91), into account, we can rewrite the equation above as follows:

$$\begin{aligned} r_k[i] = A_k b_k[i] &+ \sum_{m>k}^{K} A_m b_m[i]\rho_{km} + \sum_{m>k}^{K} A_m b_m[i-1]\rho_{mk} \\ &+ z_k[i] + \sum_{l<k}^{K} A_l b_l[i+1]\rho_{kl} + \sum_{l<k}^{K} A_l b_l[i]\rho_{lk} \end{aligned} \tag{10.106}$$

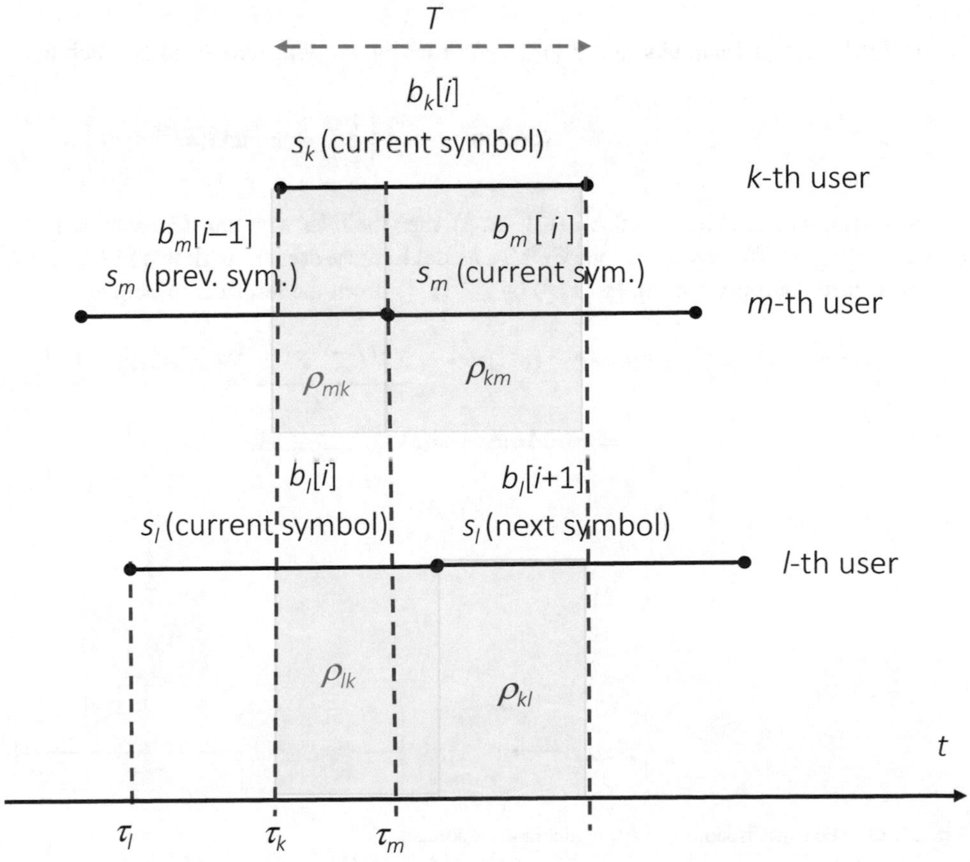

Fig. 10.24 Illustration of classification of crosstalk terms relative to channel k

Let the elements of correlation matrices $\boldsymbol{R}[0]$ and $\boldsymbol{R}[1]$ be defined as [5]:

$$R_{mk}[0] = \begin{cases} 1, m = k \\ \rho_{mk}, m < k, \\ \rho_{km}, m > k \end{cases} \quad R_{mk}[1] = \begin{cases} 0, m \geq k \\ \rho_{km}, m < k \end{cases} \tag{10.107}$$

We can now represent the K-equations from (10.106) in matrix form by:

$$\boldsymbol{r}[i] = \boldsymbol{R}^T[1]\boldsymbol{A}\boldsymbol{b}[i+1] + \boldsymbol{R}^T[0]\boldsymbol{A}\boldsymbol{b}[i] + \boldsymbol{R}[1]\boldsymbol{A}\boldsymbol{b}[i-1] + \boldsymbol{z}[i]. \tag{10.108}$$

The noise vector $z[i]$ is a zero-mean Gaussian with covariance matrix:

$$\langle \boldsymbol{z}[i]\boldsymbol{z}[j]\rangle = \begin{cases} \sigma^2\boldsymbol{R}^T[1], j = i+1 \\ \sigma^2\boldsymbol{R}^T[0], j = i \\ \sigma^2\boldsymbol{R}^T[1], j = i-1 \\ \boldsymbol{0}, \text{ otherwise} \end{cases} \tag{10.109}$$

Let us now apply the z-transform on Eq. (10.108) to obtain:

$$\begin{aligned} \boldsymbol{r}(z) &= \underbrace{\left(\boldsymbol{R}^T[1]z + \boldsymbol{R}^T[0] + \boldsymbol{R}[1]z^{-1}\right)}_{\boldsymbol{S}(z)}\boldsymbol{A}\boldsymbol{b}(z) + \boldsymbol{z}(z) \\ &= \boldsymbol{S}(z)\boldsymbol{A}\boldsymbol{b}(z) + \boldsymbol{z}(z). \end{aligned} \tag{10.110}$$

Similar to synchronous case, let us factorize $\boldsymbol{S}(z)$ as follows [5, 33]:

$$\boldsymbol{S}(z) = \boldsymbol{F}^T\left(z^{-1}\right)\boldsymbol{F}(z), \ \boldsymbol{F}(z) = \boldsymbol{F}[0] + \boldsymbol{F}[1]z^{-1}, \tag{10.111}$$

where $\boldsymbol{F}[0]$ is lower triangular and $\boldsymbol{F}$ [1] is upper triangular with zero diagonal, such that [5, 33]:

$$\det\boldsymbol{F}[0] = \exp\left\{\frac{1}{2}\int_0^1 \log\left[\det S\left(e^{j2\pi f}\right)\right]df\right\}. \tag{10.112}$$

Moreover, when $\det S(e^{j\omega}) > 0, \forall\, \omega \in [-\pi, \pi]$, then the filter with transfer function $\boldsymbol{F}^{-1}[z]$ is *causal and stable* [5, 33]. The corresponding MIMO asynchronous CDMA model in the z-domain is shown in Fig. 10.25.

Let us now multiply the Eq. (10.110) by $\boldsymbol{F}^{-T}(z^{-1})$, from the left side, to obtain:

$$\begin{aligned} \boldsymbol{r}'(z) &= \boldsymbol{F}^{-T}(z^{-1})\boldsymbol{r}(z) = \underbrace{\boldsymbol{F}^{-T}(z^{-1})\boldsymbol{F}^T(z^{-1})}_{\boldsymbol{I}}\boldsymbol{F}(z)\boldsymbol{A}\boldsymbol{b}(z) + \underbrace{\boldsymbol{F}^{-T}(z^{-1})\boldsymbol{z}(z)}_{\boldsymbol{w}(z)} \\ &= \boldsymbol{F}(z)\boldsymbol{A}\boldsymbol{b}(z) + \boldsymbol{w}(z). \end{aligned} \tag{10.113}$$

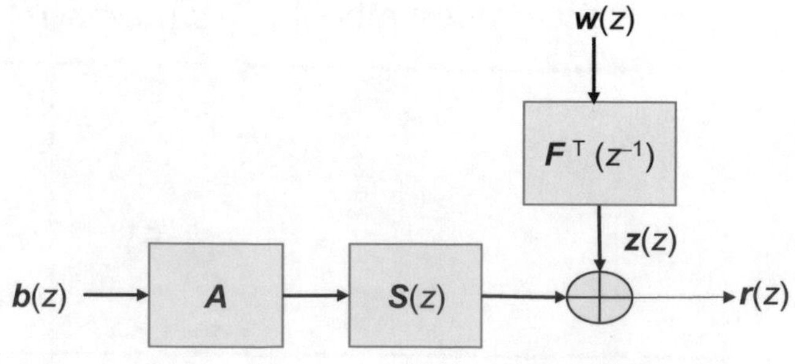

Fig. 10.25 The asynchronous CDMA model in the z-domain

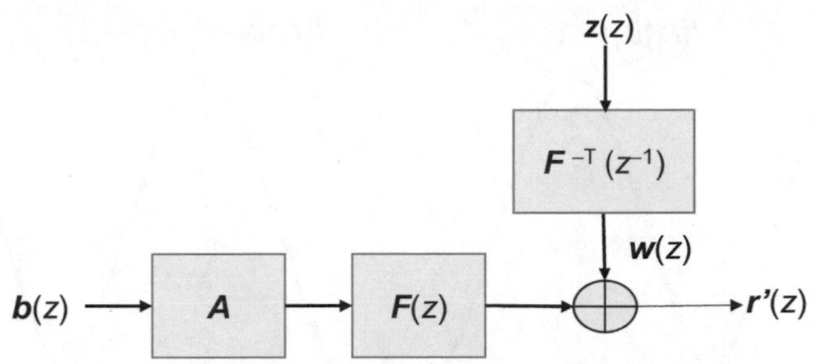

Fig. 10.26 Equivalent asynchronous CDMA model in the z-domain with MIMO whitening filter

It can be shown that the samples of noise process $\boldsymbol{w}(z)$ are uncorrelated, indicating that the filter with transfer function $\boldsymbol{F}^{-T}(z^{-1})$is in fact MIMO whitening filter. The corresponding equivalent asynchronous CDMA model in z-domain with MIMO whitening filter, based on Eq. (10.113), is shown in Fig. 10.26.

The time-domain representation of asynchronous CDMA system can be obtained by inverse z-transform of Eq. (10.113), which yields to:

$$\boldsymbol{r}'[i] = \boldsymbol{F}[0]\boldsymbol{A}\boldsymbol{b}[i] + \boldsymbol{F}[1]\boldsymbol{A}\boldsymbol{b}[i-1] + \boldsymbol{w}[i]. \tag{10.114}$$

Finally, from Eqs. (10.110) and (10.111), we can establish the following relationships between $\boldsymbol{R}[0]$, $\boldsymbol{R}[1]$ and $\boldsymbol{F}[0]$, $\boldsymbol{F}[1]$:

$$\boldsymbol{R}[0] = \boldsymbol{F}^T[0]\boldsymbol{F}[0] + \boldsymbol{F}^T[1]\boldsymbol{F}[1], \boldsymbol{R}[1] = \boldsymbol{F}^T[0]\boldsymbol{F}[1]. \tag{10.115}$$

10.3 Multiuser Detection

The interference originating from multiple users should not be treated as a noise process. Instead, the knowledge of signature waveforms should be exploited to mitigate multiaccess interference. In particular, when all users are detected simultaneously, the interference from other users can be estimated and cancelled out. The CDMA receiver can also exploit the structure of multiuser interference in detection of desired signal, which is commonly referred to as the multiuser detection. The multiuser detection was pioneered by Sergio Verdú, in a series of publications [5, 35–42].

We start our discussion with conventional single-user detection, when other users are considered as interference channels, which is illustrated in Fig. 10.23. This case is suboptimum, because the output of the k-th correlator (matched filter) does not represent the sufficient statistics for making a decision on bit transmitted b_k.

10.3.1 Conventional Single-User Correlator Receiver

Suppose that signature sequences in synchronous CDMA system are orthogonal to each other. In this case, the k-th user can be operated independently on other users, and the optimum receiver will be conventional matched filter receiver, and the probability of error for the k-th user will be $P_{1,k} = 0.5\text{erfc}(A_k/\sigma\sqrt{2})$, where erfc function is introduced in Chap. 4 as $\text{erfc}(z) = (2/\sqrt{\pi})\int_z^\infty \exp(-u^2)du$. When signature sequences are quasi-orthogonal or non-orthogonal, we need to determine cross-correlations among different users and apply the CDMA model provided in Fig. 10.23. Even though this receiver is suboptimum, it can be used for reference. Based on Eq. (10.92), we know that the output of the k-th matched filter can be represented as $r_k = A_k b_k + \sum_{m=1,m\neq k}^{K} A_m b_m \rho_{mk} + z_k, k = 1, \cdots, K$, and corresponding CDMA model for two-user case $(K = 2)$ is simply:

$$\begin{aligned} r_1 &= A_1 b_1 + A_2 b_2 \rho + z_1 \\ r_2 &= A_2 b_2 + A_1 b_1 \rho + z_2, \end{aligned} \tag{10.116}$$

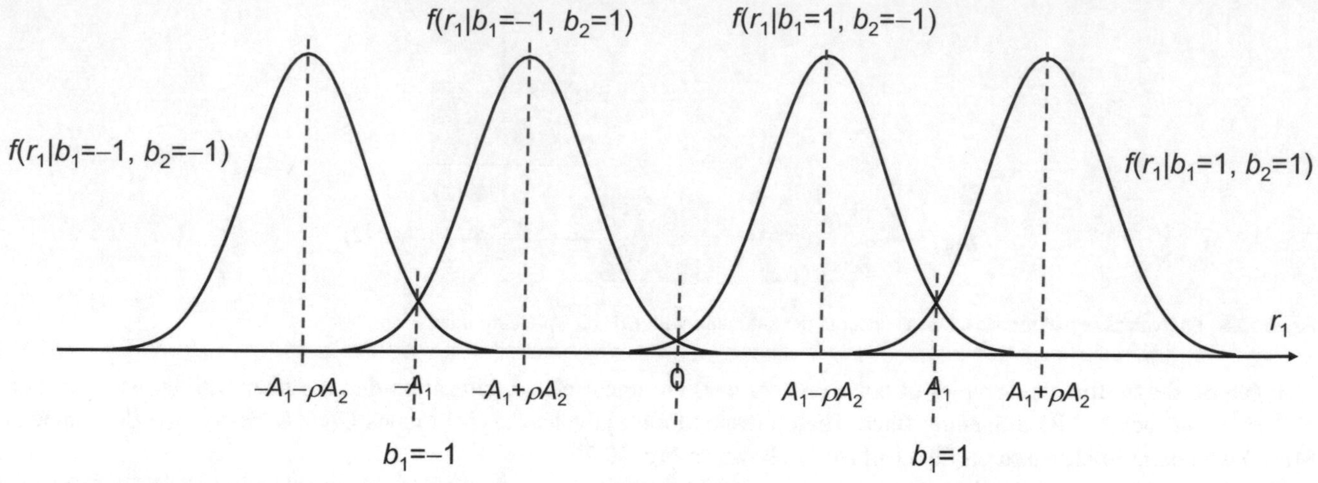

Fig. 10.27 Probability density functions for user 1 bit, b_1, by taking into account the bit originating from user 2, b_2

where for simplicity we have written $\rho_{12} = \rho_{21} = \rho$. Given that noise samples are zero-mean Gaussian with the same variance σ^2, we sketched in Fig. 10.27 the probability density functions (PDFs) for user 1 when transmitted bit b_1 was either +1 or – 1, which takes into account the bit b_2 transmitted by user 2. The probability of error for user 1 can be determined as:

$$\begin{aligned} P_{e,1} &= \Pr\left\{\widehat{b_1} \neq b_1\right\} \\ &= \Pr\{r_1 < 0|b_1 = +1\}\Pr\{b_1 = +1\} + \Pr\{r_1 > 0|b_1 = -1\}\Pr\{b_1 = -1\}. \end{aligned} \tag{10.117}$$

However, r_1 conditioned only on b_1 is not Gaussian, since it is also dependent on b_2. So we need to condition on b_2 as well, so that the last term can be written as:

$$\begin{aligned} \Pr\{r_1 > 0|b_1 = -1\} &= \Pr\left\{\underbrace{r_1 > 0}_{z_1 > A_1 + A_2\rho} |b_1 = -1, b_2 = -1\right\}\Pr\{b_2 = -1\} \\ &+ \Pr\left\{\underbrace{r_1 > 0}_{z_1 > A_1 - A_2\rho} |b_1 = -1, b_2 = +1\right\}\Pr\{b_2 = +1\} \end{aligned} \tag{10.118}$$

Based on Fig. 10.27, we conclude that:

$$\begin{aligned} \Pr\{z_1 > A_1 + A_2\rho|b_1 = -1, b_2 = -1\} &= \frac{1}{2}\operatorname{erfc}\left(\frac{A_1 + A_2\rho}{\sigma\sqrt{2}}\right) \\ \Pr\{z_1 > A_1 - A_2\rho|b_1 = -1, b_2 = +1\} &= \frac{1}{2}\operatorname{erfc}\left(\frac{A_1 - A_2\rho}{\sigma\sqrt{2}}\right), \end{aligned} \tag{10.119}$$

and after substitution in (10.118), assuming $\Pr\{b_2 = -1\} = \Pr\{b_2 = +1\} = 1/2$, we obtain:

$$\Pr\{r_1 > 0|b_1 = -1\} = \frac{1}{4}\operatorname{erfc}\left(\frac{A_1 + A_2\rho}{\sigma\sqrt{2}}\right) + \frac{1}{4}\operatorname{erfc}\left(\frac{A_1 - A_2\rho}{\sigma\sqrt{2}}\right). \tag{10.120}$$

Because of symmetry as shown in Fig. 10.27, we can write $\Pr\{r_1 < 0| \ b_1 = +1\} = \Pr\ \{r_1 > 0| \ b_1 = -1\}$, so that the expression for probability of error for user 1 becomes:

$$P_{e,1} = \frac{1}{4}\operatorname{erfc}\left(\frac{A_1 + A_2\rho}{\sigma\sqrt{2}}\right) + \frac{1}{4}\operatorname{erfc}\left(\frac{A_1 - A_2\rho}{\sigma\sqrt{2}}\right), \tag{10.121}$$

while based on Eq. (10.116), we conclude that probability of error for user 2 will be the same as probability of error for user 1, namely, $P_{e,2} = P_{e,1}$. Given that the first term in (10.121) cannot be smaller than the second term, the probability of error can be upper-bounded by:

$$P_{e,1} = P_{e,2} \leq \frac{1}{2}\mathrm{erfc}\left(\frac{A_1 - A_2|\rho|}{\sigma\sqrt{2}}\right), \tag{10.122}$$

indicating that probability of error will be smaller than ½ ($P_{e,1} < 1/2$) when $A_1 > A_2|\rho|$, and this condition is known as the *open eye condition* (when observed on the scope, the eye diagram will be open). In this case, the useful term is dominant over crosstalk term, and error probability can be approximated as:

$$P_{e,1} = P_{e,2} \simeq \frac{1}{2}\mathrm{erfc}\left(\frac{A_1 - A_2|\rho|}{\sigma\sqrt{2}}\right),$$

and this expression is similar to single-user link, except that SNR is lower due to the presence of crosstalk. In opposite situation, when $A_1 < A_2|\rho|$, the eye diagram will be completely closed, and the crosstalk will dominate over useful signal, which reminds to the near-far crosstalk problem. Since the reception will be dominated by interference, the receiver 1 will detect the bit transmitted by user 2, and probability of error will be close to ½. When $A_1 = A_2\rho$, from Eq. (10.121), we obtain the following expression for error probability:

$$P_{e,1} = \frac{1}{4}\mathrm{erfc}\left(\frac{2A_1}{\sigma\sqrt{2}}\right) + \frac{1}{4}, \tag{10.123}$$

indicating that with probability ½ the interference dominates so that probability of error is ½ in that case, while with probability of ½ the b_2 doubles the amplitude for channel 1 resulting in improvement in performance. As an illustration of error probability performance of conventional detector, in Fig. 10.28, we provide P_e versus signal-to-noise ratio (SNR) in channel 1, wherein the ratio A_2/A_1 is used as a parameter for cross-correlation of $\rho = 0.25$. As expected, when the crosstalk

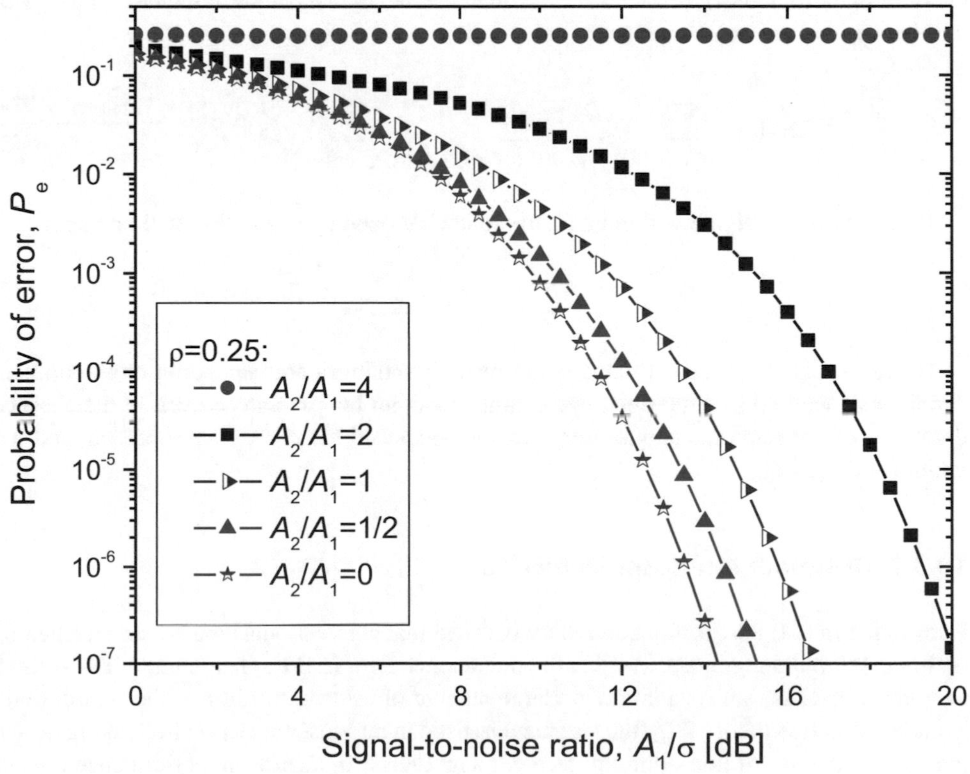

Fig. 10.28 Probability of error in channel 1, for conventional single-user correlator receiver, vs. SNR for different A_2/A_1 ratios, assuming the cross-correlation of 0.25

dominates, it is not possible to detect the transmitted bits correctly. On the other hand, when useful signal dominates over crosstalk, it is possible to detect the transmitted bits correctly, without any error floor phenomenon.

By using analogy, we can generalize two-user error probability calculation to K-user case, with details provided in [5], as follows:

$$\begin{aligned}P_{e,k} &= \Pr\{r_k < 0|b_k = +1\}\Pr\{b_k = +1\} + \Pr\{r_k > 0|b_k = -1\}\Pr\{b_k = -1\}\\ &= \frac{1}{2}\Pr\left\{z_k < A_k - \sum\nolimits_{m\neq k} A_m b_m \rho_{mk}\right\} + \frac{1}{2}\Pr\left\{z_k > A_k - \sum\nolimits_{m\neq k} A_m b_m \rho_{mk}\right\}\\ &= \Pr\left\{z_k > A_k - \sum\nolimits_{m\neq k} A_m b_m \rho_{mk}\right\}\\ &= \frac{1}{2^{K-1}}\sum\nolimits_{\alpha_1\in\{-1,+1\}}\cdots\sum\nolimits_{\alpha_m\in\{-1,+1\}}\cdots\sum\nolimits_{\alpha_K\in\{-1,+1\}}\frac{1}{2}\operatorname{erfc}\left(\frac{A_k - \sum_{m\neq k} A_m \alpha_m \rho_{mk}}{\sigma\sqrt{2}}\right),\end{aligned} \tag{10.124}$$

where we performed conditioning over all possible interfering bits. Again, for the open eye condition, the following upper bound can be used:

$$P_{e,k} \leq \frac{1}{2}\operatorname{erfc}\left(\frac{A_k - \sum_{m\neq k} A_m |\rho_{mk}|}{\sigma\sqrt{2}}\right),\ A_k > \sum\nolimits_{m\neq k} A_m \rho_{mk}. \tag{10.125}$$

For sufficient number of crosstalk terms, we can use the Gaussian approximation, according to central limit theorem, for $\sum_{m\neq k} A_m \rho_{mk}$, and corresponding expression for error probability will be:

$$\widetilde{P}_{e,k} \simeq \frac{1}{2}\operatorname{erfc}\left(\frac{A_k}{\sqrt{\sigma^2 + \sum_{m\neq k} A_m^2 \rho_{mk}^2}\sqrt{2}}\right), \tag{10.126}$$

which gives more pessimistic results, in high SNR regime, as shown in [5].

The probability of error analysis for *asynchronous CDMA* is similar, except that we need to account for: (i) left and right crosscorrelations and (ii) the fact that each bit is affected by 2 K-2 interfering bits. The resulting probability of error expression will have increased number of terms twice [5]:

$$P_{e,k} = \frac{1}{4^{K-1}}\sum_{\alpha_1,\beta_1\in\{-1,+1\}^2}\cdots\sum_{\alpha_m,\beta_m\in\{-1,+1\}^2}\cdots\sum_{\alpha_K,\beta_K\in\{-1,+1\}^2}\frac{1}{2}\operatorname{erfc}\left(\frac{A_k - \sum_{m\neq k} A_m(\alpha_m\rho_{mk} + \beta_m\rho_{km})}{\sigma\sqrt{2}}\right). \tag{10.127}$$

When the user k signal is dominant, the following open eye condition will be valid:

$$A_k > \sum\nolimits_{m\neq k} A_m(|\rho_{mk}| + |\rho_{km}|). \tag{10.128}$$

However, since the arrival times τ_k are random variables, corresponding cross-correlations ρ_{km} and ρ_{mk} will be random variables as well so that the open eye condition cannot be guaranteed even in the absence of noise. This represents a key drawback of conventional single-user correlator (matched filter) receiver approach, and we move the focus to the optimum multiuser detection.

10.3.2 Optimum Multiuser Detection

Prior to the mid-1980s, it was commonly believed that conventional single-user matched filter receiver was optimal or near-optimal. The main argument was that the interference term in DT synchronous CDMA model is binomial for a finite number of users K or Gaussian for infinitely a large number of users, according to the central limit theorem. However, as we know from the detection theory [33], the sufficient statistic in multiuser detection for making decisions on $(b_1,\ldots,b_K)$ is $(r_1,..,r_K)$, not just r_k, and the use of non-optimum receiver might result in significant performance degradation, in particular for asynchronous CDMA in the presence of multipath fading.

In our study on optimum multiuser detection for synchronous CDMA, we assume that the receiver has full knowledge of signature waveforms of all users and received amplitudes A_k for all users, bit-timing information is known, and the variance of the noise σ^2 is known as well. We apply DT synchronous CDMA model, given by Eq. (10.83), namely, $r(t) = \sum_k A_k b_k s_k(t) + \sigma z(t)$.We start our study with a *two-user case*, in similar fashion as we have done before:

$$r(t) = A_1 b_1 s_1(t) + A_2 b_2 s_2(t) + \sigma z(t), \ t \in [0, T). \tag{10.129}$$

From detection theory [33] (see also Chap. 6), we know that the maximum a posteriori probability (MAP) decision rule minimizes the decision error probability, and the *individually optimum decision rules* can be formulated as:

$$\begin{aligned} \widehat{b}_1 &= \underset{b_1 \in \{-1,+1\}}{\arg\max} \ \Pr\{r_1 | r(t)\}, \\ \widehat{b}_2 &= \underset{b_2 \in \{-1,+1\}}{\arg\max} \ \Pr\{r_2 | r(t)\}. \end{aligned} \tag{10.130}$$

We are more interested in the *jointly optimum decision rule*, formulated by:

$$\left(\widehat{b}_1, \widehat{b}_2\right) = \underset{(b_1, b_2) \in \{-1,+1\}^2}{\arg\max} \quad \Pr\{(b_1, b_2) | r(t)\}. \tag{10.131}$$

For sufficiently high SNR, we would expect these two decision rules to agree. When each of the possibilities for (b_1, b_2) is equally likely, the MAP decision rule becomes the *maximum likelihood* (*ML*) decision rule:

$$\begin{aligned} \left(\widehat{b}_1, \widehat{b}_2\right) &= \underset{(b_1, b_2)}{\arg\max} f(r(t)) \mid (b_1, b_2) \\ &= \underset{(b_1, b_2)}{\arg\max} \ e^{-\frac{1}{2\sigma^2} \int_0^T [r(t) - A_1 b_1 s_1(t) - A_2 b_2 s_2(t)]^2 dt} \\ &= \underset{(b_1, b_2)}{\arg\min} \frac{1}{2\sigma^2} \underbrace{\int_0^T [r(t) - A_1 b_1 s_1(t) - A_2 b_2 s_2(t)]^2 dt}_{d_E^2[r(t), A_1 b_1 s_1(t) + A_2 b_2 s_2(t)]} \\ &= \underset{(b_1, b_2)}{\arg\min} \frac{1}{2\sigma^2} d_E^2[r(t), A_1 b_1 s_1(t) + A_2 b_2 s_2(t)]. \end{aligned} \tag{10.132}$$

Therefore, when the variance is not known, the Euclidean distance receiver is equivalent to ML decision rule for AWGN, which is consistent with Chap. 6. Instead of using waveforms, it will be much more convenient to work with matched filter (MF) outputs r_1 and r_2, representing projections of received signal $r(t)$ onto signatures $s_1(t)$ and $s_2(t)$, as illustrated in Fig. 10.23. The corresponding ML decision rule will be now:

$$\begin{aligned} \left(\widehat{b}_1, \widehat{b}_2\right) &= \underset{(b_1, b_2)}{\arg\max} \Pr\{r_1, r_2 | (b_1, b_2)\} \\ &= \underset{(b_1, b_2)}{\arg\max} \ e^{-\frac{1}{2\sigma^2}(\boldsymbol{r}-\boldsymbol{m})^T \boldsymbol{C}^{-1}(\boldsymbol{r}-\boldsymbol{m})} \\ &= \underset{(b_1, b_2)}{\arg\min} (\boldsymbol{r} - \boldsymbol{m})^T \boldsymbol{C}^{-1} (\boldsymbol{r} - \boldsymbol{m}), \end{aligned} \tag{10.133}$$

where $\boldsymbol{r} = [r_1 \ r_2]^{\mathrm{T}}$, $\boldsymbol{m} = [A_1 b_1 + A_2 b_2 \rho \ A_2 b_2 + A_1 b_1 \rho]^{\mathrm{T}}$ and $\boldsymbol{C}$ is the covariance matrix:

$$\boldsymbol{C} = \mathrm{Cov}(r_1, r_2) = \sigma^2 \begin{bmatrix} 1 & \rho \\ \rho & 1 \end{bmatrix}.$$

The performance of such ML decision rule can be improved if we project to orthogonal basis functions $\{\phi_1(t), \phi_2(t)\}$, instead of non-orthogonal signal set $\{s_1(t), s_2(t)\}$, obtained by Gram-Schmidt procedure:

$$\begin{aligned}\phi_1(t) &= s_2(t),\\ \phi_2(t) &= \frac{s_2(t) - \rho s_1(t)}{\sqrt{1-\rho^2}}.\end{aligned} \tag{10.134}$$

These new projections, denoted as $(\widetilde{r}_1, \widetilde{r}_2)$, will have the mean and covariance as follows:

$$\widetilde{\boldsymbol{m}} = \begin{bmatrix} A_1 b_1 + A_1 b_1 \rho \\ A_2 b_2 \sqrt{1-\rho^2} \end{bmatrix}, \ \widetilde{\boldsymbol{C}} = \mathrm{Cov}(\widetilde{r}_1, \widetilde{r}_2) = \sigma^2 \begin{bmatrix} 1 & 0 \\ 0 & 1 \end{bmatrix}. \tag{10.135}$$

Corresponding ML decision rule becomes:

$$\left(\widehat{b}_1, \widehat{b}_2\right) = \underset{(b_1, b_2)}{\arg\min}\, (\widetilde{\boldsymbol{r}} - \widetilde{\boldsymbol{m}})^T \widetilde{\boldsymbol{C}}^{-1} (\widetilde{\boldsymbol{r}} - \widetilde{\boldsymbol{m}}). \tag{10.136}$$

Since the covariance matrix is diagonal, we can simplify the decision rule to:

$$\left(\widehat{b}_1, \widehat{b}_2\right) = \underset{(b_1, b_2)}{\arg\min}\, \|\widetilde{\boldsymbol{r}} - \widetilde{\boldsymbol{m}}\|^2, \tag{10.137}$$

and with this decision rule, we can split the signal space into decision regions, similarly in *K*-hypothesis decision theory.

The following bounds for error probability for ML decision for two-user channel have been derived in [5]:

$$\begin{aligned}&\max\left\{\frac{1}{2}\mathrm{erfc}\left(\frac{A_1}{\sigma\sqrt{2}}\right), \frac{1}{4}\mathrm{erfc}\left(\frac{\sqrt{A_1^2 + A_2^2 - 2A_1A_2|\rho|}}{\sigma\sqrt{2}}\right)\right\} \le P_e\\ &P_e \le \frac{1}{2}\mathrm{erfc}\left(\frac{A_1}{\sigma\sqrt{2}}\right) + \frac{1}{4}\mathrm{erfc}\left(\frac{\sqrt{A_1^2 + A_2^2 - 2A_1A_2|\rho|}}{\sigma\sqrt{2}}\right).\end{aligned} \tag{10.138}$$

In Fig. 10.29, we plotted these probabilities of error ML bounds and compared against conventional receiver, assuming $A_2/A_1 = 1.5$ and $\rho = 0.45$. First of all we notice that these ML bounds agree very well. Secondly, the ML receiver significantly outperforms the conventional receiver.

The ML decision rule (10.137) can further be written as:

$$\begin{aligned}\left(\widehat{b}_1, \widehat{b}_2\right) &= \underset{(b_1, b_2)}{\arg\min}\left[(\widetilde{r}_1 - A_1 b_1 - A_2 b_2 \rho)^2 + \left(\widetilde{r}_2 - A_2 b_2 \sqrt{1-\rho^2}\right)^2\right]\\ &= \underset{(b_1, b_2)}{\arg\max}\left(\widetilde{r}_1 A_1 b_1 + \widetilde{r}_1 A_2 b_2 \rho + \widetilde{r}_2 A_2 b_2 \sqrt{1-\rho^2} - A_1 b_1 A_2 b_2 \rho\right).\end{aligned} \tag{10.139}$$

From (10.134), we have the following relationship between new and old coordinates:

$$\widetilde{r}_1 = r_1, \widetilde{r}_2 = (r_2 - \rho r_1)/\sqrt{1-\rho^2}, \tag{10.140}$$

and after substitution we obtained the following ML decision rule:

$$\left(\widehat{b}_1, \widehat{b}_2\right) = \underset{(b_1, b_2)}{\arg\max}\, (r_1 A_1 b_1 + r_2 A_2 b_2 - A_1 b_1 A_2 b_2 \rho). \tag{10.141}$$

It has been shown in [5] that this expression can be expressed equivalently as:

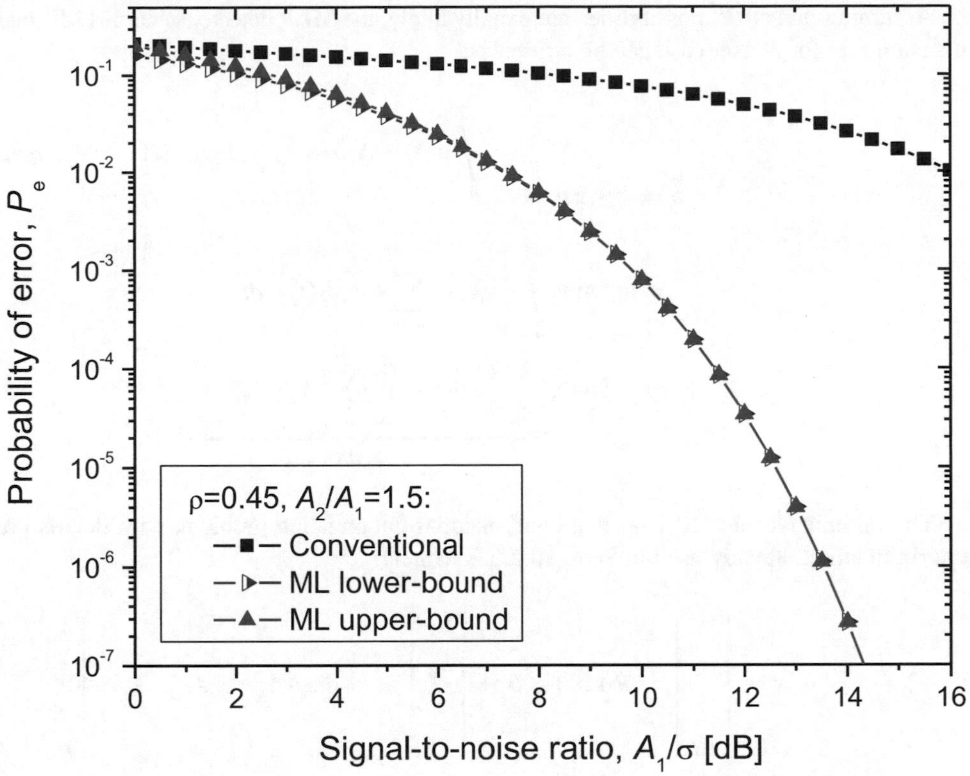

Fig. 10.29 Probability of error in a two-user channel for ML against conventional receiver for $A_2/A_1 = 1.5$ and $\rho = 0.25$

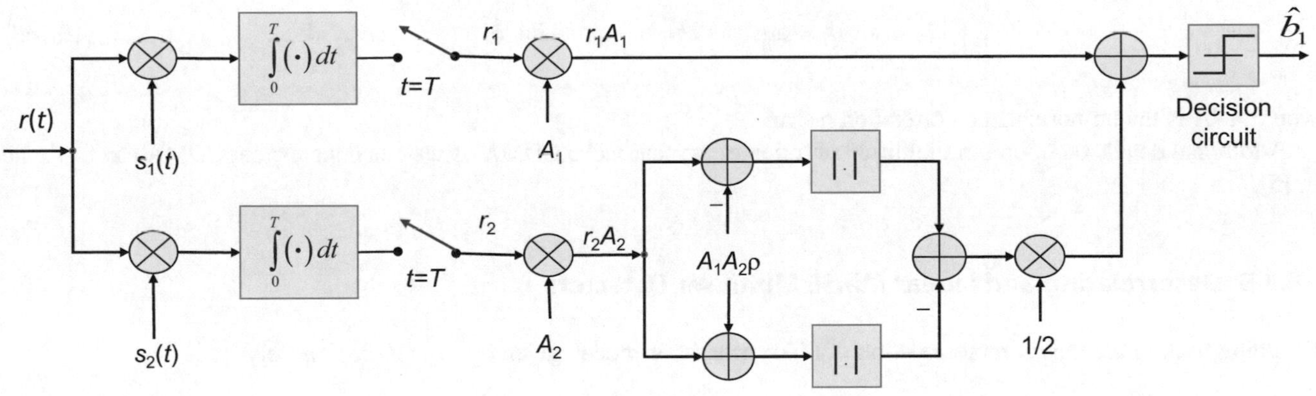

Fig. 10.30 Implementation of jointly optimum multiuser detector for user 1 in a two-user system

$$\begin{aligned}\hat{b}_1 &= \operatorname{sgn}\left(r_1A_1 + \frac{1}{2}|r_2A_2 - A_1A_2\rho| - \frac{1}{2}|r_2A_2 + A_1A_2\rho|\right)\\ \hat{b}_2 &= \operatorname{sgn}\left(r_2A_2 + \frac{1}{2}|r_1A_1 - A_1A_2\rho| - \frac{1}{2}|r_1A_1 + A_1A_2\rho|\right),\end{aligned} \tag{10.142}$$

and corresponding implementation for decision on b_1 is shown in Fig. 10.30.

Clearly, the jointly optimum multiuser detector requires sufficient statistics (r_1, r_2), amplitudes A_1 and A_2, as well as cross-correlation ρ.

In this rest of this section, we describe the *jointly optimum detector for K-user synchronous CDMA* case, described by Eq. (10.83), $r(t) = \sum_k A_k b_k s_k(t) + \sigma z(t)$. The aim is to choose the information vector $\boldsymbol{b} = [b_1, \ldots, b_K]$ maximizing the likelihood function $f(r(t)|\boldsymbol{b})$, namely:

$$\hat{\boldsymbol{b}} = \arg\max f(\boldsymbol{r}(t)|\boldsymbol{b}). \tag{10.143}$$

Assuming that all 2^K possibilities are equally likely, the MAP detector becomes ML one. Similar to Eq. (10.132), the ML decision rule for K-user case can be written as:

$$\begin{aligned}\hat{\boldsymbol{b}} &= \arg\max_{\boldsymbol{b}} e^{-\frac{1}{2\sigma^2}\int_0^T [r(t) - A_1 b_1 s_1(t) - A_2 b_2 s_2(t) - \cdots A_K b_K s_K(t)]^2 dt} \\ &= \arg\min_{\boldsymbol{b}} \int_0^T \left[r(t) - \sum_{k=1}^K A_k b_k s_k(t) \right]^2 dt \\ &= \arg\max_{\boldsymbol{b}} \underbrace{2\sum_{k=1}^K A_k b_k - \sum_{k=1}^K \sum_{m=1}^K A_k b_k A_m b_m \rho_{km}}_{L_K(\boldsymbol{b})}.\end{aligned} \tag{10.144}$$

The optimization in (10.144) is a combinatorial optimization problem, with details provided in [5]. By introducing the matrix notation, already used in Sect. 10.2.2.2, namely:

$$\boldsymbol{r} = \begin{bmatrix} r_1 \\ r_2 \\ \vdots \\ r_K \end{bmatrix} = \boldsymbol{RA}\,\boldsymbol{b} + \boldsymbol{z}, \boldsymbol{b} = \begin{bmatrix} b_1 \\ b_2 \\ \vdots \\ b_K \end{bmatrix}, \boldsymbol{A} = \operatorname{diag}(A_1, \cdots, A_K),\ \boldsymbol{R} = \begin{bmatrix} 1 & \rho_{12} & \cdots & \rho_{1K} \\ \rho_{21} & 1 & \cdots & \rho_{2K} \\ \vdots & \vdots & \ddots & \vdots \\ \rho_{K1} & \rho_{2K} & \cdots & 1 \end{bmatrix}$$

we can rewrite the last line in Eq. (10.144) in matrix form as follows:

$$\hat{\boldsymbol{b}} = \arg\max_{\boldsymbol{b}} 2\boldsymbol{b}^T \boldsymbol{A}\boldsymbol{r} - \boldsymbol{b}^T(\boldsymbol{ARA})\boldsymbol{b}, \tag{10.145}$$

where $\boldsymbol{ARA}$ is the un-normalized correlation matrix.

Additional details on optimum multiuser detection of asynchronous CDMA as well as noncoherent CDMA can be found in [5].

10.3.3 Decorrelating and Linear MMSE Multiuser Detectors

By going back to the matrix representation of DT synchronous model, given by Eq. (10.93), namely:

$$\boldsymbol{r} = \boldsymbol{RA}\,\boldsymbol{b} + \boldsymbol{z},\ \langle \boldsymbol{z} \rangle = \boldsymbol{0}, \operatorname{Cov}(\boldsymbol{z}) = \langle \boldsymbol{z}\boldsymbol{z}^T \rangle = \sigma^2 \boldsymbol{R}, \tag{10.146}$$

Assuming that covariance matrix $\boldsymbol{R}$ is invertible, we can multiply (10.146) by $\boldsymbol{R}^{-1}$ from the left side to obtain:

$$\boldsymbol{R}^{-1}\boldsymbol{r} = \underbrace{\boldsymbol{R}^{-1}\boldsymbol{R}}_{\boldsymbol{I}}\boldsymbol{A}\,\boldsymbol{b} + \boldsymbol{R}^{-1}\boldsymbol{z} = \boldsymbol{A}\,\boldsymbol{b} + \boldsymbol{R}^{-1}\boldsymbol{z}, \tag{10.147}$$

we conclude that multiuser interference is compensated for, and the decision can be made by simply looking for the signs of $\boldsymbol{R}^{-1}\boldsymbol{r}$ as follows:

$$\hat{b}_k = \operatorname{sign}\left(\boldsymbol{R}^{-1}\boldsymbol{r}\right)_k; k = 1, \cdots, K; \tag{10.148}$$

where we used the notation $(\cdot)_k$ to denote the k-th element of the vector. Clearly, in the absence of noise, $\hat{b}_k = \operatorname{sign}(\boldsymbol{Ab})_k = b_k, A_k > 0$, providing that amplitudes are non-negative. In the presence of noise, from equalization (Chap. 6) and spatial

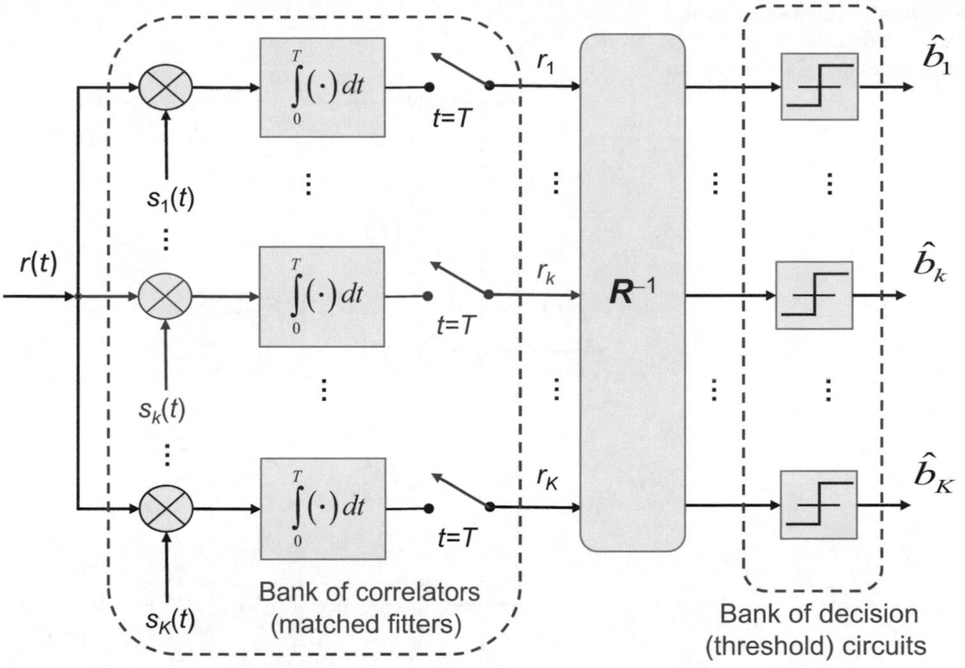

Fig. 10.31 Decorrelating detector for synchronous CDMA systems

interference cancelation in MIMO systems (Chap. 8), we know that the noise process described by $\boldsymbol{R}^{-1}z$ gets enhanced. The configuration of decorrelating detector for synchronous CDMA system is provided in Fig. 10.31. The covariance matrix inversion can be computationally intensive for a large number of users. Two interesting properties of decorrelating receiver are as follows: (i) the knowledge of amplitudes A_k is not needed, as evident from (10.148), and (ii) the demodulation process can be parallelized by operating with transformed set of basis functions in bank of decelerators (matched filters). Namely, the k-th element of $\boldsymbol{R}^{-1}\boldsymbol{r}$ can be rewritten as:

$$\begin{aligned}\left(\boldsymbol{R}^{-1}\boldsymbol{r}\right)_k &= \sum_{m=1}^{K}\left(\boldsymbol{R}^{-1}\right)_{km}\underbrace{r_m}_{\langle s_m|r\rangle} = \Bigg\langle\underbrace{\sum_{m=1}^{K}\left(\boldsymbol{R}^{-1}\right)_{km}s_m}_{\widetilde{s}_k}\Big|r\Bigg\rangle \\ &= \langle\widetilde{s}_k|r\rangle, \widetilde{s}_k = \sum_{m=1}^{K}\left(\boldsymbol{R}^{-1}\right)_{km}s_m.\end{aligned} \tag{10.149}$$

Clearly by using the set $\{\widetilde{s}_k\}$ instead of $\{s_k\}$ to implement the bank of correlators, we can greatly simplify the receiver architecture, as shown in Fig. 10.32. As long as the set of signals $\{s_k\}$ is linearly independent, the decorrelating set is a unique one. The new signal set has a unit dot product with the original signal set since:

$$\langle\widetilde{s}_k|s_k\rangle = \int_0^T\sum_{m=1}^{K}\left(\boldsymbol{R}^{-1}\right)_{km}s_m(t)s_k(t)dt = \left(\boldsymbol{R}^{-1}\boldsymbol{R}\right)_{kk} = 1. \tag{10.150}$$

However, when the cross-correlations are not known a priori, due to channel distortion, we would need to generate them locally on receiver side, as well as to calculate elements of $\boldsymbol{R}^{-1}$.

The decorrelating linear transformation can be related to the whitening matched filter approach. Namely, we can perform the Cholesky factorization $\boldsymbol{R} = \boldsymbol{F}^T\boldsymbol{F}$, and apply approach similar to Eq. (10.96), with additional details provided in [5].

We mentioned earlier that the decorrelation receiver does not require the knowledge of amplitudes. Assuming equal probable transmission, the ML decision will be:

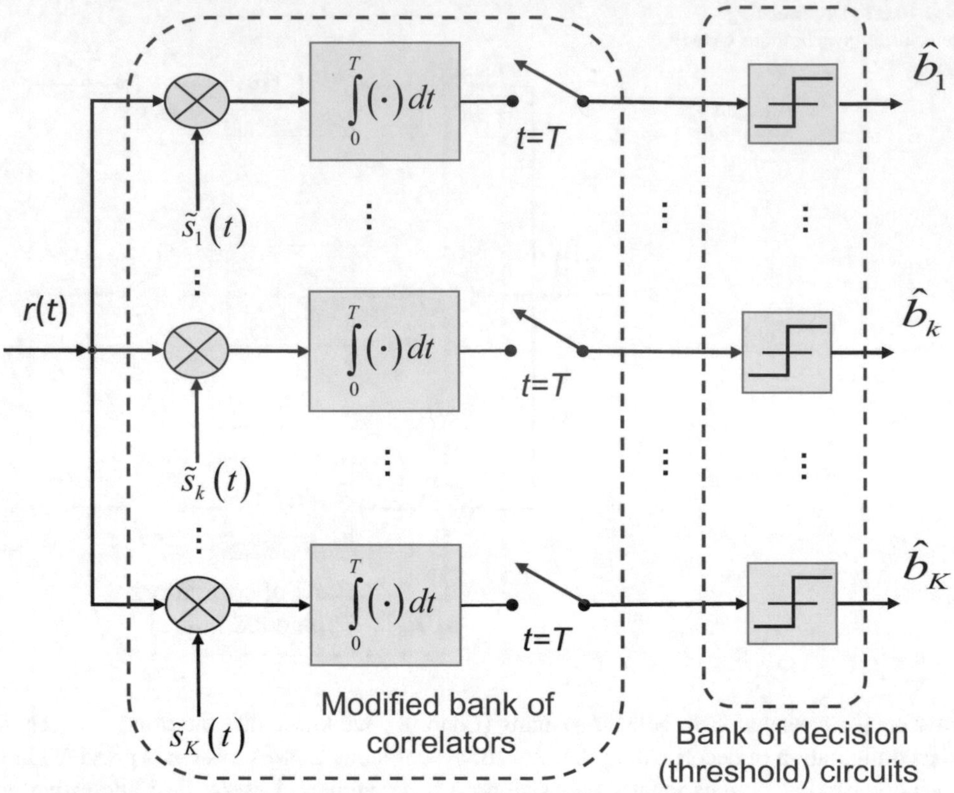

Fig. 10.32 Modified correlator bank-based decorrelating detector for synchronous CDMA systems

$$\begin{aligned}\left(\hat{\boldsymbol{b}},\hat{\boldsymbol{A}}\right)_{ML} &= \underset{(\boldsymbol{b},\boldsymbol{A})}{\arg\max} f(\boldsymbol{r}|\boldsymbol{b},\boldsymbol{A}) \\ &= \underset{(\boldsymbol{b},\boldsymbol{A})}{\arg\min}\,\|\boldsymbol{r}-\boldsymbol{R}\boldsymbol{A}\boldsymbol{b}\|^2 = \underset{(\boldsymbol{b},\boldsymbol{A})}{\arg\min}\,(\boldsymbol{r}-\boldsymbol{R}\boldsymbol{A}\boldsymbol{b})^T(\boldsymbol{r}-\boldsymbol{R}\boldsymbol{A}\boldsymbol{b}) \\ &= \underset{(\boldsymbol{b},\boldsymbol{A})}{\arg\max}\,(\boldsymbol{R}\boldsymbol{A}\boldsymbol{b})^T\boldsymbol{r}+\boldsymbol{r}^T(\boldsymbol{R}\boldsymbol{A}\boldsymbol{b})-(\boldsymbol{R}\boldsymbol{A}\boldsymbol{b})^T\boldsymbol{R}\boldsymbol{A}\boldsymbol{b} \\ &= \underset{(\boldsymbol{b},\boldsymbol{A})}{\arg\max}\,(\boldsymbol{R}\boldsymbol{A}\boldsymbol{b})^T(\boldsymbol{r}-\boldsymbol{R}\boldsymbol{A}\boldsymbol{b})+\boldsymbol{r}^T(\boldsymbol{R}\boldsymbol{A}\boldsymbol{b}),\end{aligned} \tag{10.151}$$

and the maximum is obtained by setting $\boldsymbol{r}-\boldsymbol{R}\boldsymbol{A}\boldsymbol{b}=\boldsymbol{0}$ or equivalently $(\boldsymbol{A}\boldsymbol{b})_{\text{optimum}}=\boldsymbol{R}^{-1}\boldsymbol{r}$. Therefore, the decision rule given by Eq. (10.148) is optimum when no information about amplitudes $\{\boldsymbol{A}_k\}$ is available.

As indicated above, one of the main drawbacks of the decorrelating detector is the noise enhancement problem due to transformation $\boldsymbol{R}^{-1}\boldsymbol{z}$. To avoid this problem, we will determine the transformation ("weight") matrix $\boldsymbol{W}$ applied on the received vector $\boldsymbol{r}$ on such a way to minimize the mean squared error (MSE). We are in particular interested in linear (or affine) transformations. The corresponding multiuser detection is commonly referred to as the *linear minimum MSE (MMSE) multiuser detection*. Instead of minimizing the MSE related to the k-th user, namely,

$$\min_{\boldsymbol{w}_k}\left\langle\left(b_k-\tilde{b}_k\right)\right\rangle=\min_{\boldsymbol{w}_k}\left\langle\left(b_k-\boldsymbol{w}_k^T\boldsymbol{r}\right)\right\rangle,$$

we will minimize the sum of these MSE expectations simultaneously:

$$\min_{\{w_k\}} \sum_{k=1}^{K} \left\langle \left(b_k - w_k^T r\right) \right\rangle = \min_{\{w_k\}} \left\langle \sum_{k=1}^{K} \left(b_k - w_k^T r\right) \right\rangle$$
$$= \min_{W} \left\langle \|b - Wr\|^2 \right\rangle, W = \begin{bmatrix} w_1^T \\ w_2^T \\ \vdots \\ w_K^T \end{bmatrix}. \tag{10.152}$$

We again use the matrix representation of DT synchronous model, given by (10.146), and invoke the *orthogonality principle* from detection theory [33], claiming that error vector is independent on the data vector; see also Sect. 6.3.4 in Chap. 6, so that we can write:

$$\left\langle (b - Wr)r^T \right\rangle = \mathbf{0} \Leftrightarrow \left\langle br^T \right\rangle = W\left\langle rr^T \right\rangle. \tag{10.153}$$

Form one side, the expectation of br^{T} can be represented as:

$$\left\langle br^T \right\rangle = \left\langle b(RAb + z)^T \right\rangle = \left\langle b(RAb)^T \right\rangle + \underbrace{\left\langle bz^T \right\rangle}_{\mathbf{0}} = \left\langle bb^T AR \right\rangle = \underbrace{\left\langle bb^T \right\rangle}_{I} AR = AR. \tag{10.154}$$

From the other side, we can write:

$$\begin{aligned} W\left\langle rr^T \right\rangle &= W\left\langle (RAb + z)(RAb + z)^T \right\rangle \\ &= W\left\langle RAbb^T AR \right\rangle + W\underbrace{\left\langle RAbz^T \right\rangle}_{\mathbf{0}} + W\underbrace{\left\langle z(RAb)^T \right\rangle}_{\mathbf{0}} + W\underbrace{\left\langle zz^T \right\rangle}_{\sigma^2 R} \\ &= WRA\underbrace{\left\langle bb^T \right\rangle}_{I} AR + W\sigma^2 R \\ &= W\left(RA^2 R + \sigma^2 R\right). \end{aligned} \tag{10.155}$$

Since (10.153) and (10.155) represent the left and right sides of Eq. (10.153), by equating them, we obtain:

$$AR = W\left(RA^2 R + \sigma^2 R\right), \tag{10.156}$$

and by multiplying (10.156) by R^{-1} from the right, we arrive at:

$$A = W\left(RA^2 + \sigma^2 I\right). \tag{10.157}$$

By solving for W, we obtain the optimum solution as:

$$W_{\text{optimum}} = A\left(RA^2 + \sigma^2 I\right)^{-1}, \tag{10.158}$$

and the optimal joint linear MMSE multiuser detector performs the following operations:

$$\tilde{b} = W_{\text{optimum}} r = A\left(RA^2 + \sigma^2 I\right)^{-1} r. \tag{10.159}$$

The k-th user decision is given by:

$$\tilde{b}_k = \operatorname{sign}\left\{ \left[A\left(RA^2 + \sigma^2 I\right)^{-1} r \right]_k \right\}; k = 1, \cdots, K. \tag{10.160}$$

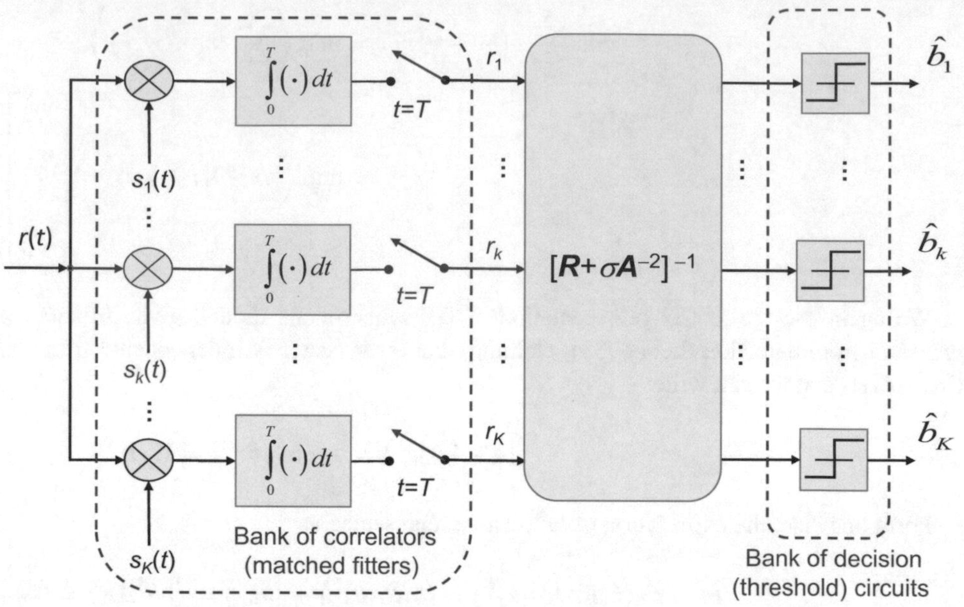

Fig. 10.33 The linear MMSE multiuser detector for synchronous CDMA systems

The sign of $\widehat{b}_k$ will not be affected if we factor out matrix $\boldsymbol{A} = \text{diag}(A_1,\ldots,A_K)$, since attenuation coefficients are nonnegative, so that the final decision for the k-th user is made according to:

$$\widehat{b}_k = \text{sign}\left\{\left[\left(\boldsymbol{R} + \sigma^2\boldsymbol{A}^{-2}\right)^{-1}\boldsymbol{r}\right]_k\right\}; k = 1, \cdots, K. \tag{10.161}$$

The corresponding linear MMSE multiuser detector is provided in Fig. 10.33. Compared to decorrelating receiver from Fig. 10.31, we replaced linear transformation $\boldsymbol{R}^{-1}$ of decorrelating detector by $(\boldsymbol{R} + \sigma^2\boldsymbol{A}^{-2})^{-1}$, wherein

$$\sigma^2\boldsymbol{A}^{-2} = \text{diag}\left\{\frac{\sigma^2}{A_1^2}, \cdots, \frac{\sigma^2}{A_k^2}, \cdots \frac{\sigma^2}{A_K^2}\right\},$$

and we do not require anymore the covariance matrix to be nonsingular. When the variance tends to zero, the decision reduces to:

$$\widehat{b}_k \underset{\sigma^2 \to 0}{\to} \text{sign}\left\{\left[\boldsymbol{R}^{-1}\boldsymbol{r}\right]_k\right\},$$

and we can say that MMSE detector becomes decorrelating detector, when noise variance tends to be zero (or equivalently SNR tends to be plus infinity).

The linear MMSE multiuser detector has certain similarities with linear MMSE interface described in Sect. 8.2.6 (in Chap. 8). By denoting $\boldsymbol{RA}$ as $\boldsymbol{H}$, our DT synchronous CDMA model becomes MIMO model $\boldsymbol{r} = \boldsymbol{Hb} + z$, wherein the channel matrix is quadratic and of full rank. Therefore, different MIMO techniques, but of lower complexity, are applicable here as well. Moreover, the iterative interface described in Sect. 8.2.7.4, in which spatial interference is cancelled in an iterative fashion is also applicable here.

10.3.4 Decision-Driven Multiuser Detectors

Nonlinear multiuser detection scheme can employ decisions on bits from other users, to estimate the interference and remove it from the received signal [5, 43–53]. Both tentatives of final decision on other users' bits may be employed. These decision-driven multiuser detectors can be classified into several categories [5, 43–53]: (i) successive cancellation Schemes [45–47], (ii) multistage detection [48, 51], (iii) continuous-time tentative decoding [5], and (iv) decision-feedback multiuser detection Schemes [52, 53]. The first two schemes will be briefly described here.

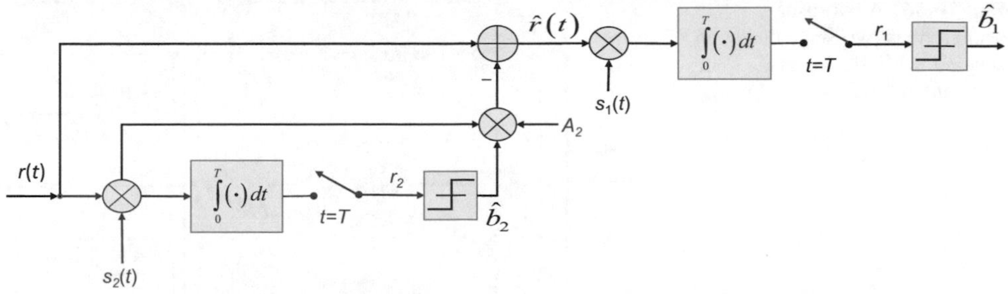

Fig. 10.34 Two-user synchronous successive cancellation scheme, employing conventional correlators

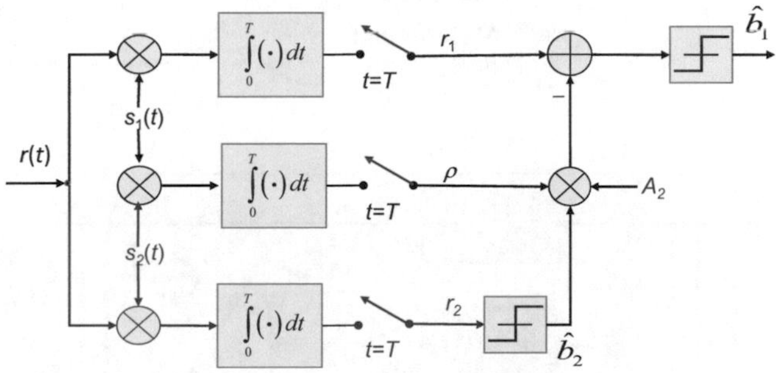

Fig. 10.35 An alternative implementation of two-user successive cancellation scheme

In *successive cancellation scheme*, the symbol of a given user is detected by observing the signals for other users as noise. Such obtained decision is then used to re-modulate the signal from that user and subtract it from the received signal. This procedure repeated in an iterative fashion until the transmitted bits from all users are demodulated. Before the successive cancelation takes place, the users are arranged with the decreased average energies at the outputs of matched filters, for synchronous CDMA defined as $\left\langle \left[\int_0^T r(t)s_k(t)dt\right]^2 \right\rangle = \sigma^2 + A_k^2 + \sum_{m\neq k} A_m^2\rho_{mk}^2$. However, this is not the best strategy since the cross-correlation is not taken into account. As an illustration, let us observe *two-user synchronous successive cancellation* scheme, employing conventional correlators (matched filters), with corresponding scheme being provided in Fig. 10.34. Assuming that $\langle r_2^2 \rangle$ is higher, we first make decision on b_2 by:

$$\widehat{b}_2 = \text{sign}\{r_2\}. \tag{10.162}$$

We used this decision to estimate the crosstalk originating from user 2 by:

$$r(t) - A_2\widehat{b}_2 s_2(t) = \widehat{r}(t), \tag{10.163}$$

and after subtracting the crosstalk from user 2, we perform match filtering for user 1 and perform decision by:

$$\begin{aligned} \hat{b}_1 &= \text{sign}\left\{ \int_0^T [r(t) - A_2\hat{b}_2 s_2(t)]s_1(t)dt \right\} \\ &= \text{sign}\left(\underbrace{r_1}_{A_1b_1+\rho A_2b_2+\sigma\langle s_1|z\rangle} - A_2 \underbrace{\hat{b}_2}_{\text{sign}(r_2)} \rho \right) = \text{sign}(A_1b_1 + \rho A_2(b_2 - \hat{b}_2) + \sigma\langle s_1|z\rangle), \end{aligned} \tag{10.164}$$

with corresponding implementation being shown in Fig. 10.35.

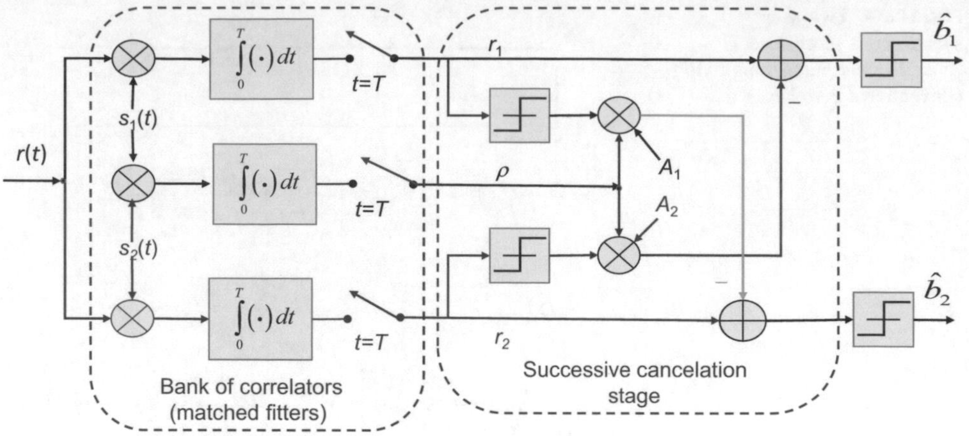

Fig. 10.36 A two-user synchronous two-stage detection scheme, with the first stage being conventional receiver

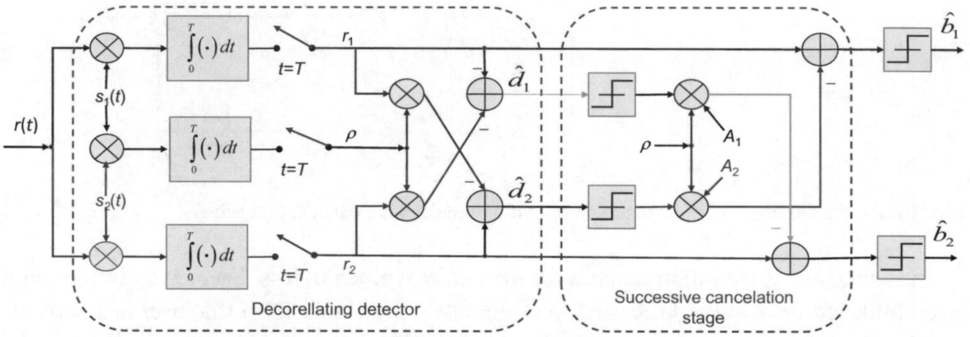

Fig. 10.37 A two-user synchronous two-stage detection scheme, with the first stage being the decorrelating detector

In the *K-user successive cancelation scheme*, assuming that decisions on bits related to the users $k + 1,\ldots,K$ are correct, the decision on the bit related to the k-th user, while ignoring the presence of other users, will be:

$$\widehat{b}_k = \text{sign}\left\{ r_k - \sum_{m=k+1}^{K} A_m \rho_{mk} \widehat{b}_m \right\}. \tag{10.165}$$

In the K-user successive cancelation scheme, the decorrelating detectors can be used as well.

In *multistage detection* scheme, the various decisions get generated in different stages. In one illustrative two-user example, shown in Fig. 10.36, the first stage is the conventional bank of correlators (matched filters), followed by the successive cancelation stage for both users.

The successive cancelation stage, based on Eq. (10.165), can be repeated as many times as needed. In another interesting example, shown in Fig. 10.37, the first stage is the decorrelating detector, while the second stage is the successive cancelation stage for both users. For the K-user case, the outputs of decorrelation stage are given by:

$$\widehat{d}_k = \text{sign}\left\{ \left(\boldsymbol{R}^{-1} \boldsymbol{r} \right)_k \right\}. \tag{10.166}$$

These decisions are then used as tentative decisions used in the successive cancelation stage, operating as described by Eq. (10.165), so that the corresponding decisions in successive cancelation will be made now as:

$$\widehat{b}_k = \text{sign}\left\{ r_k - \sum_{m=k+1}^{K} A_m \rho_{mk} \widehat{d}_m \right\}. \tag{10.167}$$

For two-user case, the corresponding decisions are made based on:

$$\begin{aligned} \widehat{b}_1 &= \text{sign}\{ r_1 - A_2 \rho \, \text{sign}(r_2 - \rho r_1) \}, \\ \widehat{b}_2 &= \text{sign}\{ r_2 - A_1 \rho \, \text{sign}(r_1 - \rho r_2) \}. \end{aligned} \tag{10.168}$$

10.4 Optical CDMA Systems

In this section, we briefly describe various optical CDMA Schemes [54–75]. Similar to wireless communications, optical CDMA allows users to share the same bandwidth at the same time using the *unique codes* or *signature sequences*. Some unique advantages of optical CDMA include the following [66–69]: (i) privacy and inherent security in transmission, (ii) simplified network control (no centralized network control is required), (iii) scalability of network, (iv) multimedia applications support, and (v) providing provision for quality of service (QoS). Optical CDMA can be either direct detection-based [56–64] or coherent optical detection-based [65]. The basic intensity modulation/direct detection (IM/DD) optical CDMA scheme, suitable for LAN applications, is provided in Fig. 10.38.

The optical source could be either laser diode or LED. The light beam gets modulated by electro-optical modulator, which can be either electro-absorption (EA) or Mach-Zehnder modulator (MZM). Different users get assigned with unique signature codes, typically referred to *optical orthogonal codes* (*OOCs*), which are strictly speaking quasi-orthogonal. The OOCs must satisfy the following two constraints [56–58, 66, 68, 69]: (i) each sequence (code) must be easily distinguishable from a shifted version of itself and (ii) each code must be easily be distinguished from any other sequence in the OOC family (and possibly corresponding shifted version as well). The OOC can be defined in either time-domain, typically called time-spreading encoding, or in spectral (wavelength)-domain, commonly referred to as spectral amplitude coding (SAC). Moreover, the frequency hopping (coding) can also be employed, in similar fashion as already employed in wireless FH-CDMA systems. The scheme shown in Fig. 10.38 is suitable for both time-spreading encoding, typically based on pulse laser, and SAC, wherein the optical source is an LED. The corresponding OOCs are imposed optically in either time-domain or spectral-domain. Independent signals are after time-domain/spectral-domain encoding combined together with the help of $N \times N$ optical coupler, such as optical star coupler. On the receiver side, the signal for the intended user is decoded and converted into electrical domain by photodetector.

The optical orthogonal codes, denoted as OOC(v,k,λ_a,λ_c), are unipolar sequences of length v, weight k, autocorrelation constraint λ_a, and crosscorrelation constraint λ_c [56, 58, 63, 66–69]. For unipolar sequences $\{x_n\}_{n=0}^{v-1}$ and $\{y_n\}_{n=0}^{v-1}$, the autocorrelation and cross-correlation constraints are defined as [56, 66]:

- *Autocorrelation constraint*:

$$\sum_{n=0}^{v-1} x_n x_{n+m} \geq \lambda_a; m = 0, 1, \cdots, v-1 \tag{10.169}$$

- *Cross-correlation constraint*:

$$\sum_{n=0}^{v-1} x_n y_{n+m} \geq \lambda_c; m = 0, 1, \cdots, v-1. \tag{10.170}$$

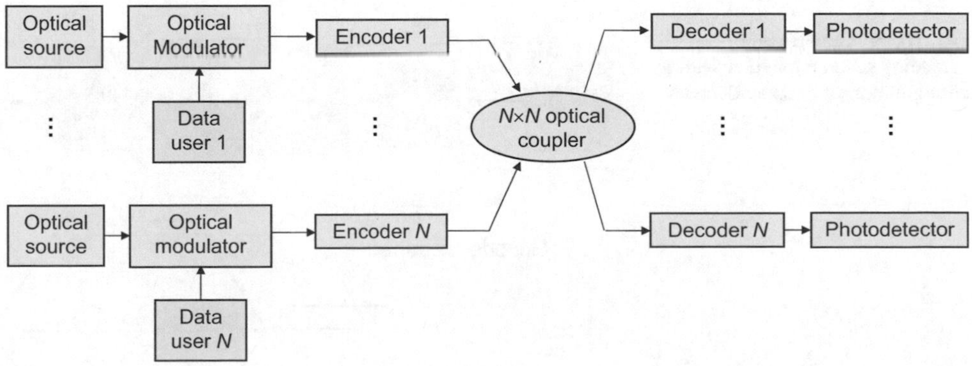

Fig. 10.38 Optical CDMA-based LAN

The strict orthogonality would mean that $\lambda_a = \lambda_c = 0$; however, when sequences are unipolar, the cardinality of set of unipolar sequences will be too small. In practice, we typically require $\lambda_a = \lambda_c = 1$. There are different ways how to design OOCs including algebraic constructions [58, 61], combinatorial constructions [66, 69], and finite geometries [68], to mention a few. The weight of codewords in OOC do not need to be of fixed weight; the variable weight can be used as a way to provide different QoS values for different user types [71, 72]. To improve the cardinality of OOCs, the two-dimensional (2-D) wavelength-time OOCs have been advocated in [70, 73, 74]. The 2-D wavelength-time OOCs ($m \times n$, k, λ_a,λ_c) are 2-D unipolar sequences, described as $m \times n$ arrays, with weight k, and autocorrelation and cross-correlation constraints being λ_a and λ_c. For 2-D time-wavelength OOCs $\{x_{i,\,j}\}$ and $\{y_{ij}\}$, the autocorrelation and cross-correlation constraints are defined, respectively, as [70, 73, 74]:

- *Autocorrelation constraint*:

$$\sum_{i=0}^{m-1}\sum_{j=0}^{n-1} x_{i,j}x_{i,(j+l)\bmod n} = \begin{cases} k, & l=0 \\ \le \lambda_a, & 1 \le l \le n-1 \end{cases} \tag{10.171}$$

- *Cross-correlation constraint*:

$$\sum_{i=0}^{m-1}\sum_{j=0}^{n-1} x_{i,j}y_{i,(j+l)\bmod n} \le \lambda_c, 0 \le l \le n-1 \tag{10.172}$$

The *incoherent optical CDMA* systems can be categorized into three categories: time-spreading encoding, spectral-amplitude-coding, and wavelength-hopping optical CDMA systems. The generic time-spreading encoding scheme, together with the configuration of encoded/decoder, is provided in Fig. 10.39. Clearly, the pulse laser is needed with pulse duration corresponding to the chip duration. The pulse laser is directly modulated by data sequence to be transmitted. The optical encoder is based on 1-D OOC sequence and composed of 1:n power splitter (star coupler), n optical delay lines (ODLs), and n:1 power combiner (gain optical star). The ODLs are properly chosen so that the pulse appears on chip positions according to the OOC sequence, and after power combining, a desired unipolar sequence gets generated. Decoder operates with complementary ODLs so that pulses get aligned when encoder and decoders are serially connected.

The spectral-amplitude coding employs a similar encoding principle, but in spectral (wavelength) domain, as illustrated in Fig. 10.40. The SACs are typically synchronous schemes, with OOCs characterized by in-phase cross-correlation. For two unipolar sequences $\{a\}$ and $\{b\}$, the in-phase cross-correlation is defined by $R_{\boldsymbol{a},\boldsymbol{b}} = \Sigma_n a_n b_n$. The balanced optical detector shown in Fig. 10.40 has been used to deal with intensity fluctuations introduced by broadband source, and for this scheme, typically the OOCs with fixed in-phase cross-correlations are used. The splitting ratio 1:a is properly chosen to maximize the eye opening upon balanced optical detector, described in Chap. 3. The broadband source spectrum is sliced into n frequency bands, and nonzero positions in OOC determine which frequency bands are used. We denoted this encoding process by $A(f)$. The spectrum complementary to $A(f)$ is denoted by $\hat{A}(f)$. The fiber Bragg grating (FBG) represents an ideal device to perform the spectral amplitude encoding.

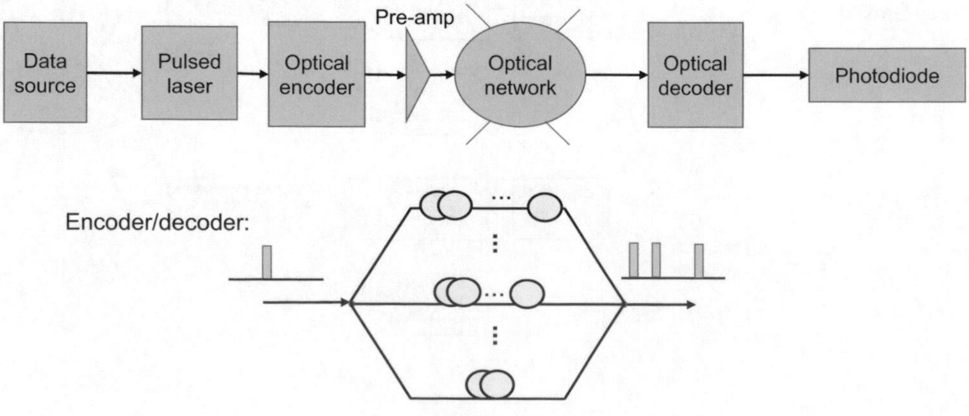

Fig. 10.39 Generic time-spreading scheme together with a configuration of encoder/decoder

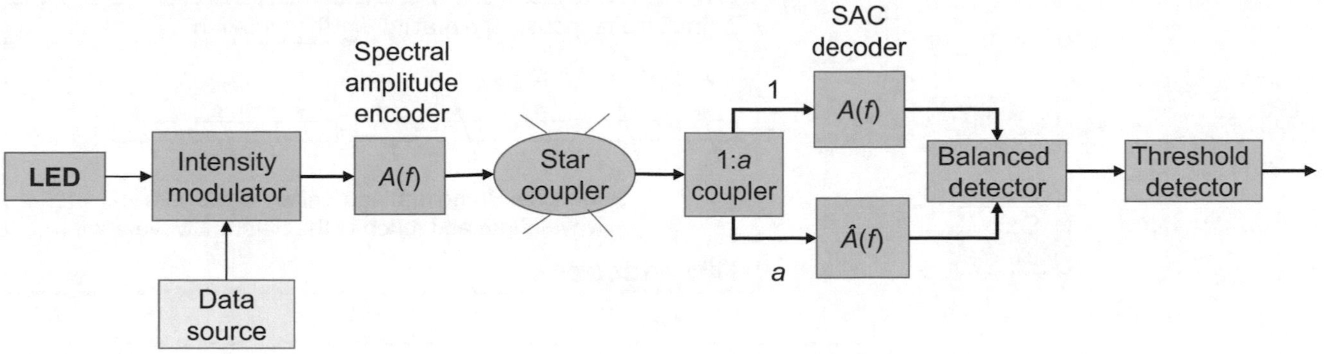

Fig. 10.40 Generic spectral amplitude coding (SAC) scheme with balanced optical detector

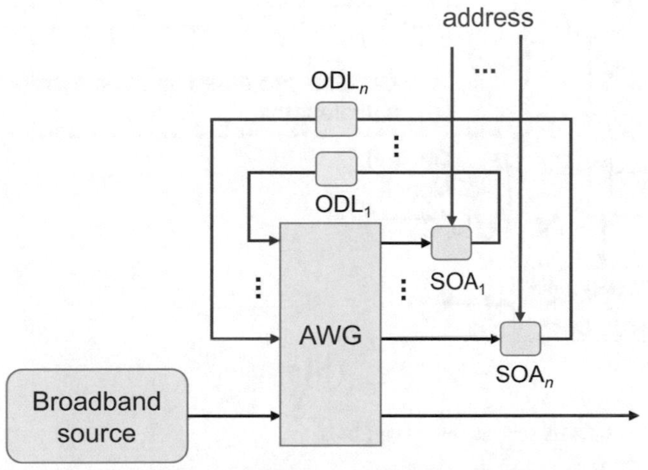

Fig. 10.41 Reconfigurable SAC encoder/decoder based on AWG and semiconductor optical amplifiers (SOAs)

In particular, for SAC decoding process, the reflected component from FBG will represent the decoded signal, while the transparent will perform the complementary function $\hat{A}(f)$. It is also possible to implement SAC schemes with non-fixed in-phase cross-correlations [71]; however, SAC decoder complexity is higher. Moreover, the SAC schemes for multimedia applications require the use of constant-length variable-weight OOCs. Instead of FBGs, arrayed waveguide gratings (AWGs) can be used in SAC Schemes [72], as illustrated in Fig. 10.41. The signal from broadband source is demultiplexed into different wavelength channels $\lambda_1, \ldots, \lambda_n$. The active wavelengths are determined by the OOC, and only those semiconductor optical amplifiers (SOAs) are turned on. Different ODLs are used to ensure that all wavelength channels experience the same delay. Clearly, this SAC encoder/decoder is reconfigurable. Moreover, ODLs and SOAs can be used to implement 2-D wavelength-time encoders/decoders.

The wavelength hopping (WH) CDMA schemes, such as the one shown in Fig. 10.42, are conceptually very similar to frequency hopping (FH)-CDMA schemes used in wireless communications and are described above. Therefore, the same spreading sequences are applicable. The pulse from broadband source is narrow and will generate a narrow impulse response from Bragg grating. Identical Bragg gratings are written to the same wavelength and are tuned to the desired wavelength [see Fig. 10.42a]. The target wavelength can be changed in either slow or fast fashion so that we can also define slow and fast WH. Either one or multiple wavelengths can be activated during the symbol duration. At the receiver side, we need to tune the wavelengths in opposite order to the WH encoder, to ensure that wavelengths are properly aligned at the output of WH decoder.

Regarding the coherent optical CDMA systems, they can be classified into two broad categories [65]: (i) pulse laser-based and (ii) CW laser-based. The pulse laser-based coherent optical CDMA systems, shown in Fig. 10.43a, are conceptually similar to the time-spreading incoherent schemes.

Namely, the incoming pulse from the laser is split into n branches by an 1:n power splitter. Each branch contains an ODL and a phase shifter (PS). The ODLs are used to choose the proper locations of chips, while phase shifters are used to introduce

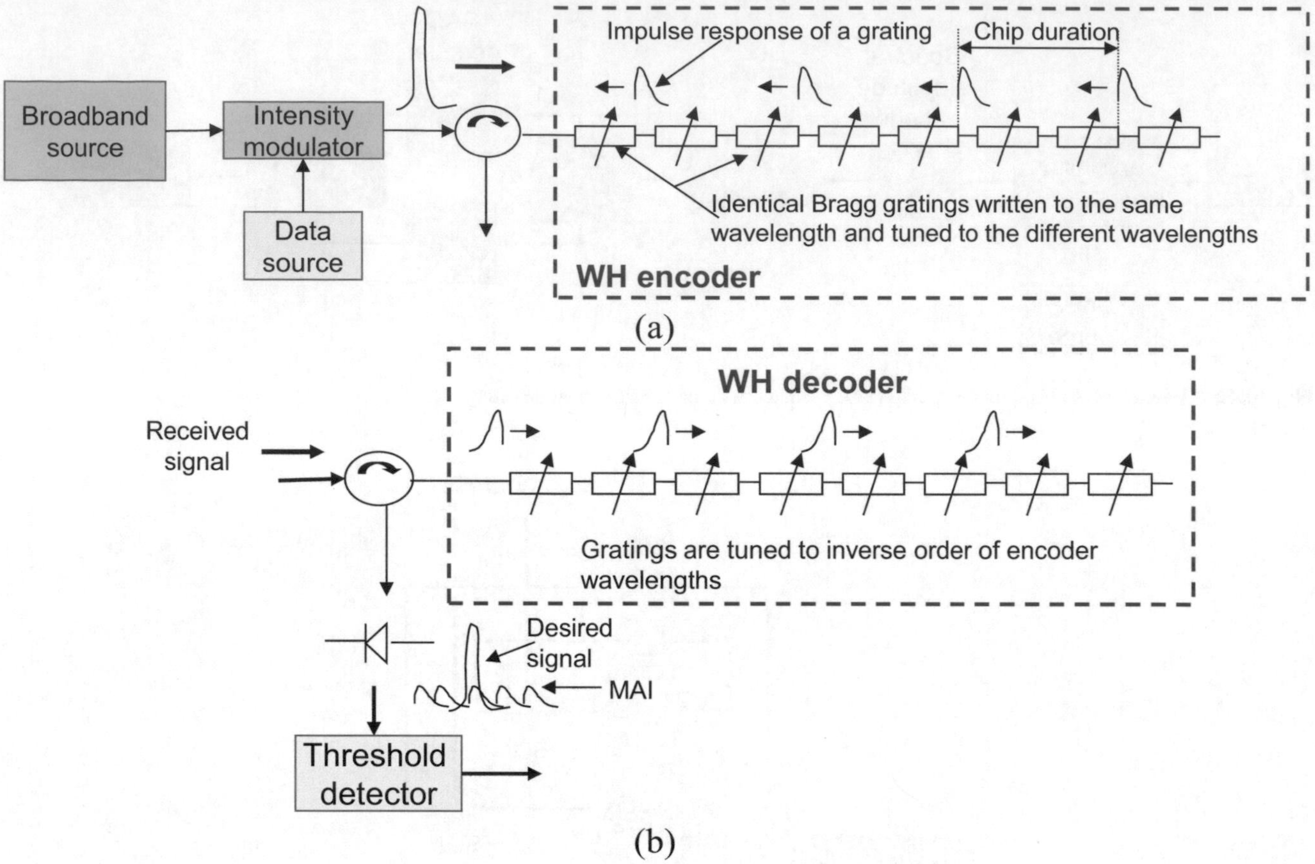

Fig. 10.42 Wavelength hopping (WH) CDMA schemes based on FBGs

the appropriate phase shifts. Various carrier synchronization algorithms described in Chap. 6 can be implemented in DSP. Another alternative scheme based on optical PLL is described in [65]. On the other hand, the CW laser-based coherent optical scheme, shown in Fig. 10.43b, is similar to DS-CDMA schemes from wireless communications. After the spread spectrum signal is generated by DS approach, such signal is converted into optical domain by electro-optical (E/O) modulator. For BPSK, either MZM or optical phase modulator can be used, while for QPSK and higher-order modulation schemes, the E/O I/Q modulator should be used. On the receiver side, we perform coherent optical balanced detection as described in Chap. 3, and the remaining steps are the same as in DS wireless CDMA systems, including the dispreading and demodulation.

The readers interested in optical CDMA are referred to reach literature, including [54–75], in particular to books specialized in optical CDMA, such as [55].

10.5 Hybrid OFDM-CDMA Systems

The hybrid OFDM-CDMA systems [76–80] are trying to take advantages of both OFDM and CDMA concepts, particularly in dealing with frequency selective fading and narrowband interference, while at the same time supporting multiple users. They can be categorized into three broad classes: (i) multicarrier CDMA (MC-CDMA), (ii) OFDM-CDMA, and (iii) OFDM-CDMA-FH systems.

In *multicarrier CDMA* [79, 81–83], on the transmitter side, shown in Fig. 10.44a, the data sequence is coming in serial fashion, in blocks of length P each, and after serial-to-parallel (S/P) conversion, each bit (symbol) of the k-th user, say b_{kp}, is multiplied by a spreading sequence of the k-th user $[s_{k1} \ldots s_{kN}]$ to obtain $b_{kp}[s_{k1} \ldots s_{kN}]$. Such frequency spread $[b_{kp} s_{k1} \ldots b_{kp} s_{kN}]$ data is then used as the input of inverse FFT (IFFT) block of OFDM subsystem. Clearly, the required number of subcarriers is NP. The remaining steps on the transmitter side are the same as described in Chap. 8 (this is why full details of OFDM transmitter are not provided here). Clearly, the spreading is applied here in frequency-domain rather than time-domain in DS-CDMA systems. Because of that, each user benefits from frequency diversity since the same user bit (symbol) gets transmitted over multiple subcarriers. The transmitted signal of the k-th user ($k = 1,\ldots,K$) can be represented as:

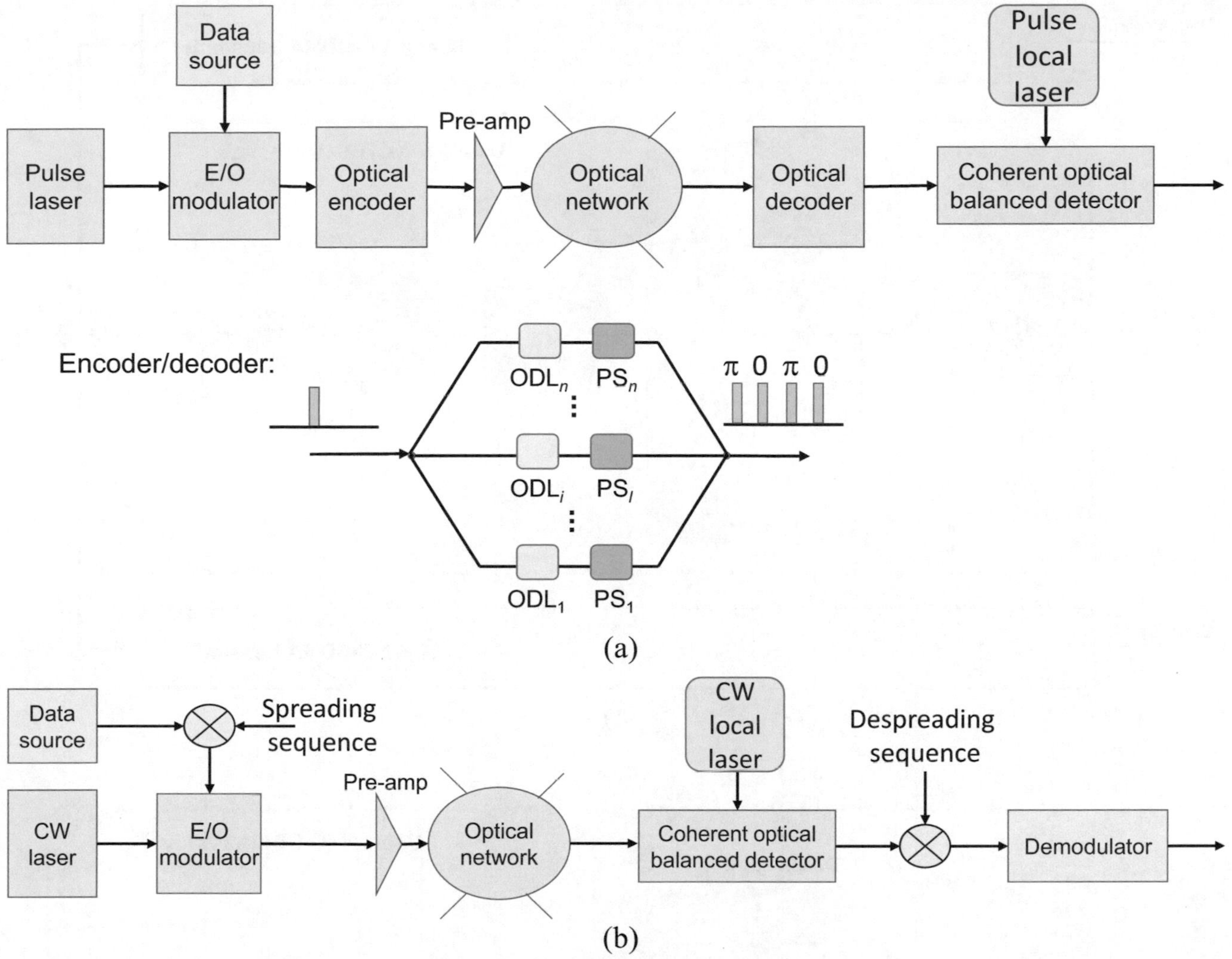

Fig. 10.43 Coherent optical detection CDMA systems: (**a**) pulse laser-based and (**b**) CW laser-based schemes

$$s_k^{(\text{MC}-\text{CDMA})}(t) = \sum_{i=-\infty}^{\infty} \sum_{p=0}^{P-1} \sum_{n=0}^{N-1} b_{kp} s_{kn} p(t - iT_s) e^{j2\pi(nP+p)\Delta f(t-iT_s)},$$
$$p(t) = \begin{cases} 1, -T_G \le t < T_s - T_G \\ 0, \text{otherwise} \end{cases} \tag{10.173}$$

where Δf is the subcarrier spacing, determined by $\Delta f = 1/(T_s - T_G)$, with T_s being OFDM symbol duration related to original symbol period by $T_s = PT$, while T_G is the guard interval duration. Clearly, the bandwidth occupied by K-user MC-CDMA signal is:

$$B_{MC-CDMA} = (NP - 1)/(T_s - T_G) + 2/T_s. \tag{10.174}$$

The received signal can be represented as:

$$r^{(\text{MC-CDMA})}(t) = \sum_{i=-\infty}^{\infty} \sum_{p=0}^{P-1} \sum_{n=0}^{N-1} \sum_{k=1}^{K} H_{np}^{(k)} b_{kp} s_{kn} p(t - iT_s) e^{j2\pi(nP+p)\Delta f(t-iT_s)} + z(t), \tag{10.175}$$

where $H_{np}^{(k)}$ is the channel coefficient corresponding to the $(nP + p)$-th subcarrier of the k-th user and $z(t)$ is the noise process. The configuration of the k-th user receiver is shown in Fig. 10.44b. After the down-conversion and analog-to-digital

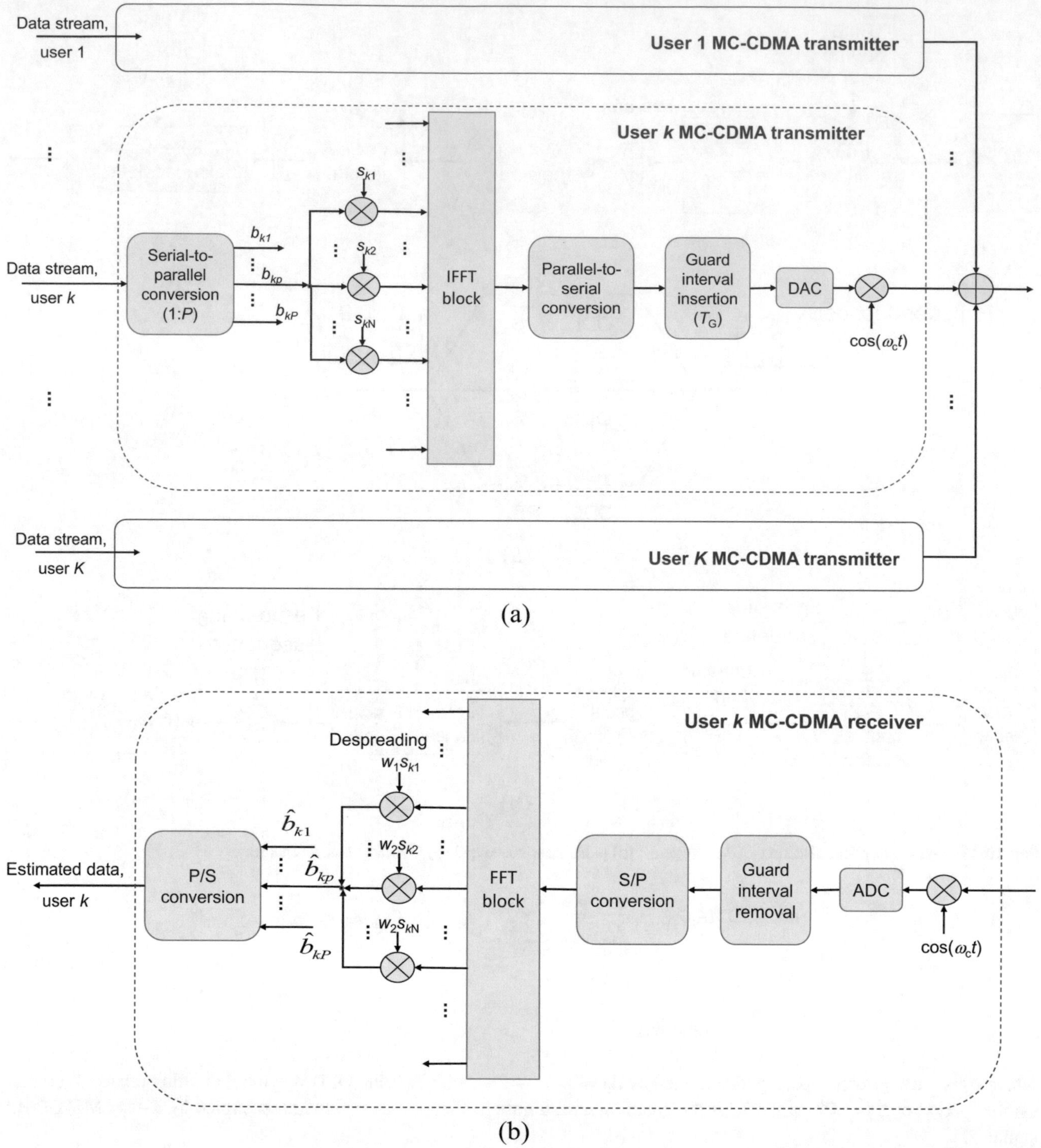

Fig. 10.44 Generic K-user MC-CDMA system: (**a**) transmitter configuration and (**b**) k-th user receiver configuration

conversion (DAC), we perform timing recovery and frequency synchronization (not shown in the figure), followed by cyclic extension removal. We further perform S/P conversion, followed by FFT. The next step is to perform dispreading together with a proper weighting, according to some of diversity schemes, such as maximum ratio combining or MMMSE combining.

The generic *hybrid OFDM-CDMA-FH Scheme* [76–78, 80] is provided in Fig. 10.45. The transmitter, shown in Fig. 10.45a, is composed of four stages: (i) data modulation stage, (ii) DS-CDMA stage, (iii) OFDM stage, and (iv) FH stage. When FH stage gets replaced with up-convertor, the corresponding scheme can be called hybrid OFDM-CDMA scheme. On the other hand, when DS-CDMA stage is omitted, the corresponding system will be hybrid OFDM-FH scheme.

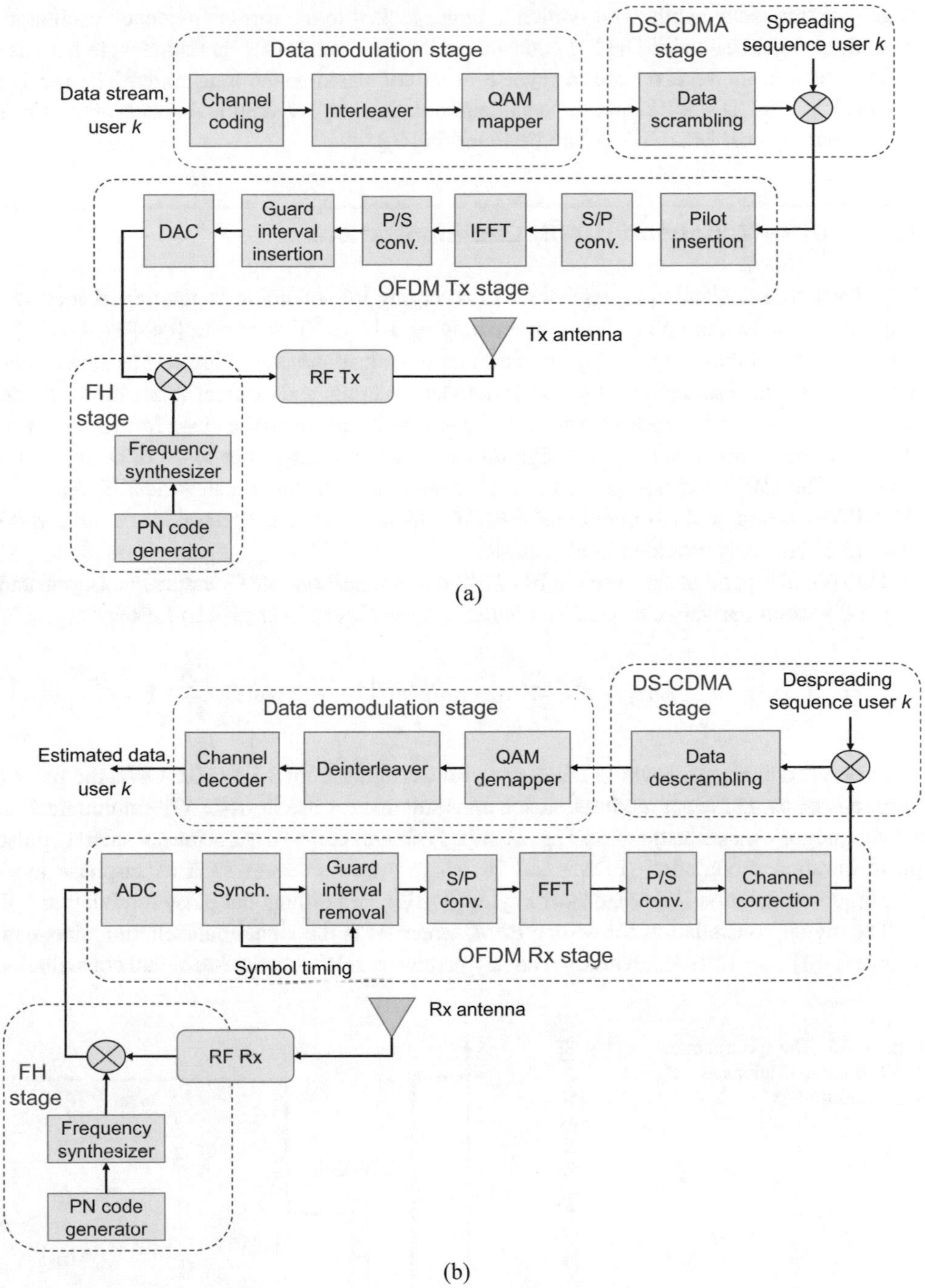

Fig. 10.45 Generic hybrid OFDM-CDMA-FH scheme related to the k-th user: (**a**) transmitter and (**b**) receiver configurations

The binary input sequence gets first encoded using one of the following alternatives: (1) convolutional code, (2) concatenated convolutional-RS code, (3) turbo code, or (4) LDPC code. The interleaver is employed to randomize the burst errors due to the deep fades on some subcarriers. After that, the I/Q modulator is used followed by QAM mapper. After that, the scrambling is used, typically based on PN sequences. After that, the spreading is achieved by multiplying the scrambled signal with a spreading sequence.

The PN sequence is not suitable for spreading since the scheme needs to be combined with OFDM and the period is not sufficient to ensure good autocorrelation and cross-correlation properties. Instead, the Walsh sequences should be used. After that, the spread signal is used as the input to the OFDM transmitter, which starts with pilot insertion, which is used on the receiver side for channel estimation to compensate for channel impairments as well as for synchronization purposes. Further, after pilot insertion, we perform S/P conversion, IFFT, P/S conversion, and guard interval insertion, followed by DAC. In FH stage, the frequency synthesizer, which is phase-locked to the carrier frequency oscillator in RF Tx, changes the frequency according to the FH code. The FH codes, described in Sect. 10.1.3, are applicable here as well.

The steps in the receiver are in reverse order compared to the transmitter. The receiver also has four stages: FH stage, OFDM Rx, DS-CDMA Rx, and the data demodulation stage. Additional details about this scheme as well as its performance comparison against MC-CDMA can be found in [80].

10.6 Ultra-Wideband (UWB) Communications

The UWB communication represents a fast emerging RF technology operating at very low energy levels, suitable for short-range high-bandwidth applications, by occupying a large RF spectrum [84–93]. UWB technology can be considered as a complement to other longer range wireless communication/networking technologies such as WiFi, WiMAX, and cellular networks. It is also suitable for localization, radar, imaging, and medical applications. In the USA, UWB radio transmissions, satisfying the transmitted power constraints, can legally operate in the range from 3.1 GHz up to 10.6 GHz [84–89]. Therefore, UWB communication can support high data rates at short range (up to 10 m or so) with limited interference to other wire devices. The UWB technologies can be classified into two broad categories: (i) *impulse radio UWB (IR-UWB)*, typically DS-CDMA-based, and (ii) *multiband OFDM (MB-OFDM)*. The spectral mask for UWB communications, as regulated by FCC 15.517(b,c), is provided in Fig. 10.46.

Two popular pulse shapes used in IR-UWB communications are Gaussian *monocycle* and *doublet*, which are derived as the first and second derivative of Gaussian pulse $g_1(t) = C_1 \exp[(-t/\tau)^2]$, as follows:

$$g_2(t) = C_2 \frac{-2t}{\tau^2} \exp\left[-(t/\tau)^2\right], \quad g_3(t) = C_3 \frac{-2}{\tau^2}\left(1 - \frac{2t^2}{\tau}\right) \exp\left[-(t/\tau)^2\right]. \tag{10.176}$$

In order to occupy the whole UWB spectrum, ranging from 3.1 to 10.6 GHz, the pulse duration should be in the order of hundreds of ps. Different modulation schemes suitable for use in IR-UWB communications can further be classified in two broad categories, as illustrated in Fig. 10.47: (i) *time-domain-based schemes* such as pulse position modulation (PPM) and pulse duration modulation (PDM) and (ii) *shape-based schemes* such as bi-phase modulation (BPM) or BPSK, pulse-amplitude modulation (PAM), on-off keying (OOK), and orthogonal pulse modulation (OPM).

The signal constellation for *M-ary PPM*, where M is the signal constellation size, can be represented by the signal set $\{s_i = p(t\text{-}\tau_i)\}$ $(i = 1,\ldots,M)$, where τ_i is delay parameter related to the i-th signal constellation point and $p(t)$ is the pulse shape.

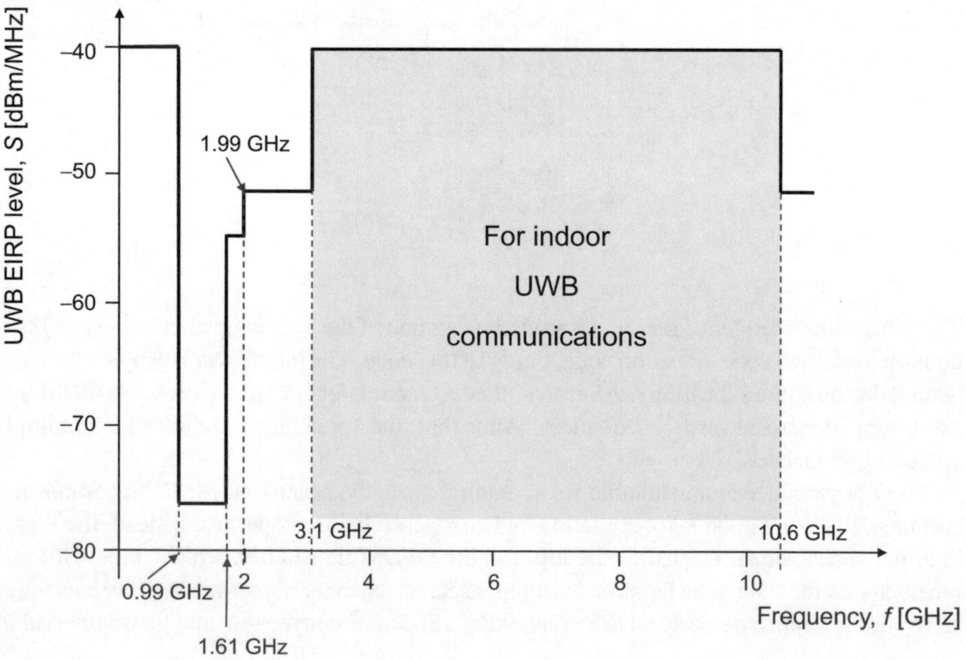

Fig. 10.46 The spectral mask for UWB communications as mandated by FCC

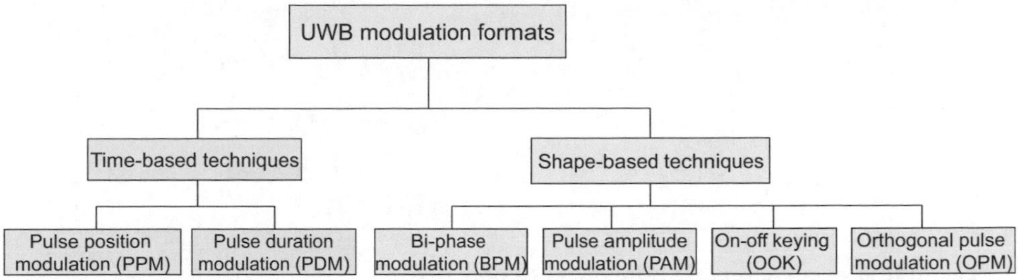

Fig. 10.47 Classification of modulation formats for UWB communications

The signal constellation for BPM (or BPSK) is represented by the following binary signal set $\{s_i = b_i p(t)\}$ $(i = 1,2;\ b_i \in \{-1,1\})$. The signal constellation for OOK is represented by the signal set $\{s_i = b_i p(t)\}$ $(i = 1,2;\ b_i \in \{0,1\})$. Finally, the signal constellation for M-ary OPM is represented by the signal set $\{p_i(t)\}$ $(i = 1,\ldots,M)$, wherein the waveforms $p_i(t)$ are mutually orthogonal. As an illustration, the *Modified Hermite polynomials* are mutually orthogonal and can be used as symbols for M-OPM as follows [84, 85]:

$$h_n(t) = k_n e^{-t^2/4\tau^2} \tau^2 \frac{d^n}{dt^n}\left(e^{-t^2/2\tau^2}\right) = k_n e^{-t^2/4\tau^2} \tau^2 \sum_{i=0}^{\lfloor n/2 \rfloor} \left(-\frac{1}{2}\right)^i \frac{(t/\tau)^{n-2i}}{(n-2i)!i!}, \tag{10.177}$$

where the coefficient k_n is related to the pulse energy E_n of h_n by:

$$k_n = \sqrt{\frac{E_n}{\tau n! \sqrt{2\pi}}}.$$

Typically, a UWB signal is in fact a sequence of sub-nanosecond pulses of low duty cycle (about 1%), with the bandwidth being in the order of several GHz. Because the spectrum of UWB signal is spread over a wide range of frequencies, its power spectral density (PSD) is low, so that the interference it causes to other existing wireless systems is extremely low.

Another class of orthogonal polynomials suitable for UWB applications are *Legendre polynomials* [95], which can be obtained by applying the Rodrigues' formula:

$$P_n(x) = \frac{1}{2^n n!} \frac{d^n}{dx^n}\left[\left(x^2 - 1\right)^n\right], x = t/\tau, \tag{10.178}$$

where τ is the scaling factor. It is trivial to show that the 0-th and the first terms are $P_0(x) = 1$ and $P_1(x) = x$. The higher-order terms can be obtained by applying the following recurrent relation:

$$(n+1)P_{n+1}(x) = (2n+1)xP_n(x) - nP_{n-1}(x). \tag{10.179}$$

The Legendre polynomials satisfy the following orthogonality principle:

$$\int_{-1}^{1} P_n(x)P_m(x)dx = \frac{1}{2n+1}\delta_{nm}. \tag{10.180}$$

The final class of functions to be considered here, to be used in OPM, are *orthogonal prolate spheroidal wave functions* (*OPSWF*) [96]. The OPSW functions are simultaneously time-limited to symbol duration T_s and bandwidth-limited to band Ω and can be obtained as solutions of the following integral equation [96, 97]:

$$\int_{-T_s/2}^{T_s/2} \Psi_n(u) \frac{\sin \Omega(t-u)}{\pi(t-u)} du = \eta_n \Psi_n(t), \quad \eta_n \in (0,1] \tag{10.181}$$

where the coefficient η_n is related to the energy concentration in the interval $[-T_s/2, T_s/2]$, and it is defined by:

$$\eta_n = \frac{\int_{-T_s/2}^{T_s/2} \Psi_n(t)dt}{\int_{-\infty}^{\infty} \Psi_n(t)dt}. \tag{10.182}$$

The OSPWs satisfy double-orthogonality principle:

$$\int_{-T_s/2}^{T_s/2} \Psi_n(u)\Psi_m(u)du = \eta_n \delta_{nm}, \quad \int_{-\infty}^{\infty} \Psi_n(u)\Psi_m(u)du = \delta_{nm}, \tag{10.183}$$

and as such these functions are ideal for both UWB [85] and optical telecommunication [97] applications. The second orthogonality condition indicates that these systems are more tolerant to intersymbol interference (ISI), since they stay orthogonal outside of symbol duration. The following unique properties make OPSWFs ideal for UWB and optical communications [85, 97]: the pulse duration is independent on index n, the pulse bandwidth does not change much with index n, they satisfy orthogonality and double-orthogonality principles, for sufficiently large index n the D.C. component tends to zero, and the pulse duration and bandwidth occupied can be tuned simultaneously.

Similarly to CDMA systems, to each user in UWB communications, we allocate a unique *PN sequence* of length N that is used to encode the pulses in either position (PPM) or polarity (BPM). Let the PN sequence be denoted by $\{s\}$ and binary sequence by $\{b\}$. Then the *PPM-based UWB* signal can be represented as:

$$s_{\text{PPM-UWB}}(t) = A \sum_{i=-\infty}^{\infty} \sum_{n=0}^{N-1} p(t - iNT_s - nT_s - c_n T_c - \tau b_i), \tag{10.184}$$

where T_s is the symbol duration, while T_c is the chip duration. With $p(t)$ we denoted the arbitrary pulse shape, with some of the pulse shapes being discussed already. Clearly, for PPM-based UWB communication, the PN sequence modulates the positions in increments/decrements of multiple of T_c, while the information bit b_m introduces additional time shift $b_m\tau$. In *BPM (BPSK)-based UWB* communications, the corresponding UWB signal can be represented as:

$$s_{\text{BPM-UWB}}(t) = A \sum_{i=-\infty}^{\infty} \sum_{n=0}^{N-1} b_i c_n p(t - iNT_s - nT_s), b_i \in \{-1, 1\}, \tag{10.185}$$

which is the same as in DS-SS systems. In other words, in BPM-based UWB communication, the PN sequence modulates the polarity of the pulse within each symbol of duration T_s, while the information bit b_n changes the polarity of the whole block of N pulses simultaneously.

The block diagram of UWB receiver and demodulator of UWB signals is provided in Fig. 10.48, with operation principle very similar to DS-SS systems, while the operation of automatic gain control (AGC) is described in Chap. 3. For PPM-based UWB system, the signal corresponding to symbols 0 and 1, to be used in correlator shown in Fig. 10.48, is given by:

$$s_0(t) = \sum_{n=0}^{N-1} p(t - nT_s - c_n T_c + \tau), s_1(t) = \sum_{n=0}^{N-1} p(t - nT_s - c_n T_c - \tau). \tag{10.186}$$

On the other hand, for BPM-based UWB system, the signal corresponding to symbols 0 and 1 is given by:

$$s_1(t) = \sum_{n=0}^{N-1} c_n p(t - nT_s), s_0(t) = -s_1(t). \tag{10.187}$$

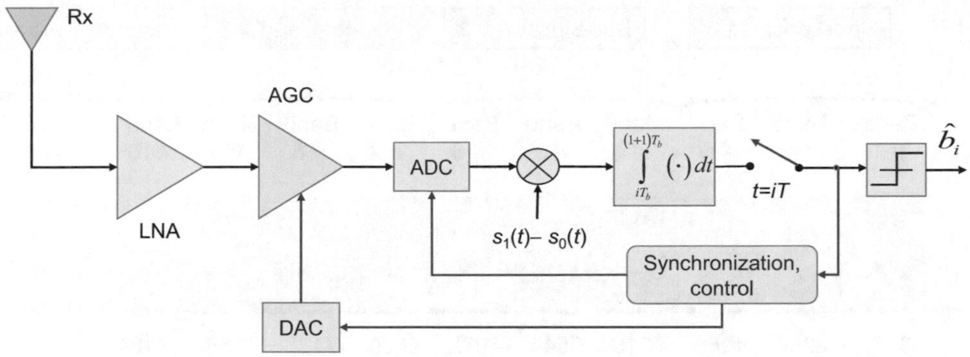

Fig. 10.48 A typical UWB receiver configuration. *LNA* low-noise amplifier, *AGC* automatic gain control, *ADC* analog-to-digital conversion, *DAC* digital-to-analog converter

Regarding the UWB channel models suitable for indoor applications, the model due to Saleh and Valenzuela [94] has been adopted in IEEE 802.15.3a. According to Saleh-Valenzuela model, the multipath components arrive in clusters, while each cluster is composed of several multipath rays. Clearly, the model has similarities with wideband multipath channel model described in Chap. 2. The impulse response can be represented as:

$$h(t) = \sum_c \sum_r a_{c,r}\delta(t - \tau_c - \tau_{c,r}), \tag{10.188}$$

where $a_{c,r}$ is normalized multipath gain related to the ray r within the cluster c. In (10.188), τ_c denotes the arrival time of the first ray within the cluster c, while $\tau_{c,r}$ denotes the arrival time of the ray r within the cluster c. The magnitude distribution in IEEE 802.15.3a has been assumed to be log-normal. The arrival time of the c-th cluster τ_c, given $\tau_{c\text{-}1}$, follows the exponential distribution:

$$\Pr\{\tau_c|\tau_{c-1}\} = \Lambda e^{-\Lambda(\tau_c - \tau_{c-1})}, \tag{10.189}$$

where Λ is the cluster arrival rate. On the other hand, the arrival time of the r-th ray within the c-th cluster follows as well the exponential distribution:

$$\Pr\{\tau_{c,r}|\tau_{c,r-1}\} = \lambda e^{-\lambda(\tau_{c,r} - \tau_{c,r-1})}, \tag{10.190}$$

where now λ is the arrival rate of the rays within the cluster. The power delay profile (PDP) is defined by double negative exponential decaying functions:

$$P_{c,r} = P_{0,0} e^{-\frac{\tau_c}{\Gamma}} e^{-\frac{\tau_{c,r}}{\gamma}}, \tag{10.191}$$

where Γ and γ denote the cluster and ray time decay constants, respectively. For additional details of this model, an interested reader is referred to [85, 94].

The *multiband modulation* represents approach to modulate information in UWB systems, by splitting the 7.5 GHz band into multiple smaller frequency bands. The key idea is to efficiently utilize UWB spectrum by simultaneously transmitting multiple UWB signals. This is applicable to both pulsed multiband and OFDM. To deal with multipath effects, mentioned above, in IR-UWB, we can use the RAKE receiver, described in Sect. 10.1.2.3. On the other hand, in MB-OFDM, to deal with multipath effects, we can use approaches described in Chap. 7.

The MB-OFDM system can be interpreted as a form of hybrid OFDM-FH system, similar to the one described in previous section, when the DS-CDMA section is omitted. The key idea of MB-OFDM is to "spread" the information bits across the whole UWB spectrum, exploiting, therefore, frequency diversity and providing the robustness against multipath fading and narrowband interference, originating from WiFi channels. The 7.5 GHz UWB spectrum can be divided into 14 bands, each of bandwidth of $\Delta f = 528$ MHz, as illustrated in Fig. 10.49. The center frequency of the nth band $f_{c,n}$, the low boundary frequency $f_{l,n}$, and upper boundary frequency $f_{h,n}$ can be determined, respectively, as:

$$f_{c,n} = f_0 + n\Delta f, \quad f_{l,n} = f_{c,n} - \Delta f/2, \quad f_{h,n} = f_{c,n} + \Delta f/2, \quad f_0 = 2904 \text{ MHz}.$$

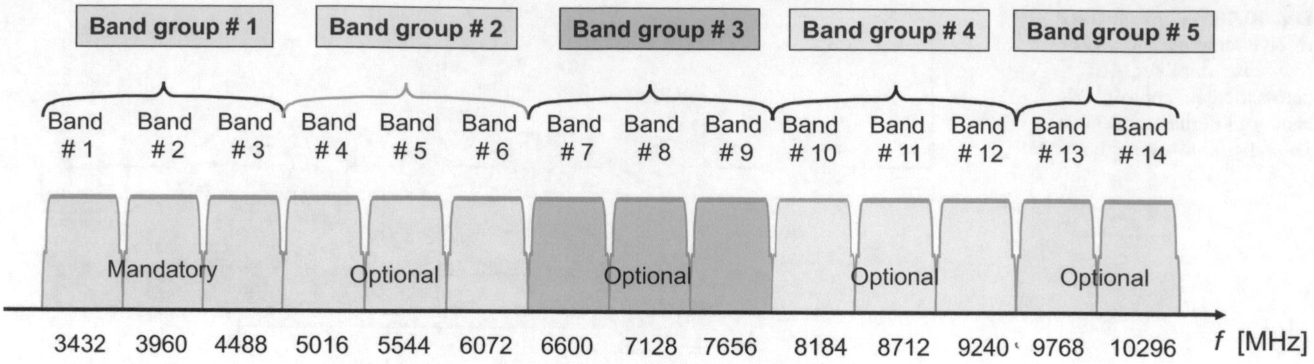

Fig. 10.49 The frequency allocation in MB-OFDM for UWB communications

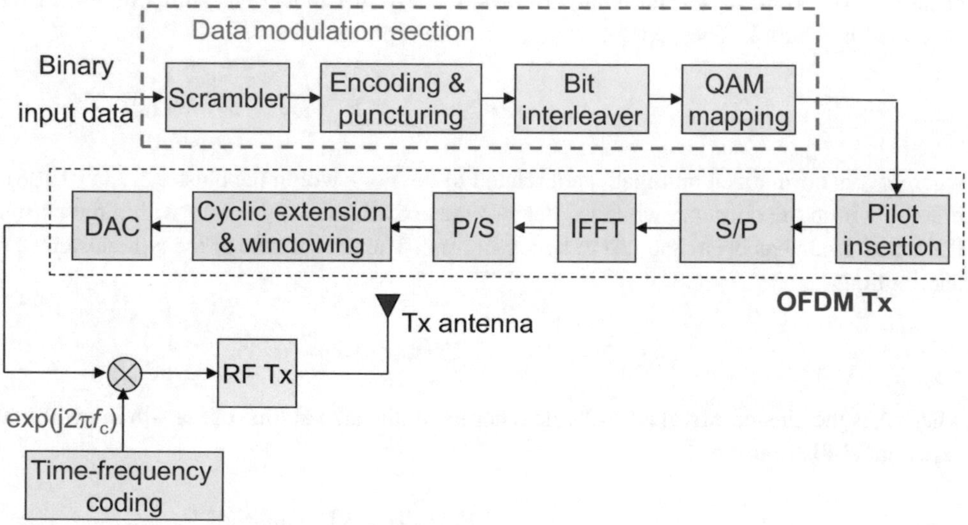

Fig. 10.50 The MB-OFDM transmitter configuration for UWB communications

The MB-OFDM transmitter configuration is provided in Fig. 10.50, and as expected it is very similar to conventional OFDM transmitter, described in Chap. 7 [see Fig. 7.6a], except from the time-frequency coding block. The time-frequency code (TFC) denotes which band within the band group is occupied at given transmission interval. As an illustration, the TFCs of length 6 for band group 1 are given as [85] {[1 2 3 1 2 3], [1 3 2 1 3 2], [1 1 2 2 3 3], [1 1 3 3 2 2], [1 2 1 2 1 2], [1 1 1 2 2 2]}. When TFC code 1 is used, the first OFDM symbol is transmitted on band 1, the second OFDM symbol on band 2, the third OFDM symbol on band 3, the fourth OFDM symbol on band 1, the firth OFDM symbol on band 2, and the sixth OFDM symbol on band 3; and this sequence of bands within the band group is repeated until the TFC is changed in time-frequency coding block. The FH codes described in Sect. 10.1.3 can be also used as TFC codes in MB-OFDM-based UWB communications. The binary data bits are scrambled, and then encoded employing a convolutional code of rate 1/3, with constraint length $K = 7$, and generating polynomials $g_0 = 133_8$, $g_1 = 165_8$, and $g_2 = 171_8$ (the subscript denotes octal number system) [93]. Higher-rate convolutional codes (of rate 11/32, 1/2, 5/8, and 3/4) can be obtained by puncturing. The number of subcarriers used in MB-OFDM is 128, 100 of which are used to transmit QPSK data, 12 are pilots, and 10 are guard tones (the remained subcarriers are set to zero). The pilots are used for channel estimation and carrier-phase tracking on receiver side. To improve performance of MB-OFDM scheme, LDPC-coded MB-OFDM can be used instead. It has been shown in [92, 93] that the transmission distance over wireless channel of LDPC-coded MB-OFDM can be increased by 29–73% compared to that of MB-OFDM with convolutional codes. After pilot insertion, and serial-to-parallel conversion (S/P), inverse FFT (IFFT) operation, and parallel-to-serial (P/S) conversion, cyclic extension and windowing are performed. After DAC and time-frequency coding, the RF signal is transmitted over transmit (Tx) antenna. The transmitted MB-OFDM-based UWB signal can be written as:

$$s_{\text{MB-OFDM-UWB}}(t) = \text{Re}\left\{\sum_{n=0}^{N-1} s_{\text{OFDM,n}}(t - nT_{\text{OFDM}}) \exp\left[j2\pi f_{(n \bmod 6)} t\right]\right\}, \tag{10.192}$$

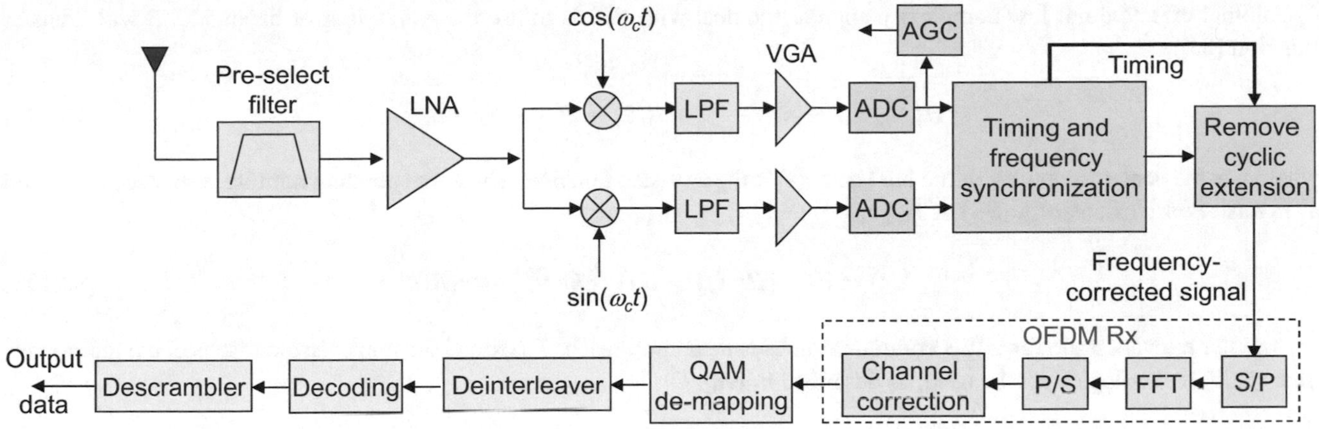

Fig. 10.51 The MB-OFDM receiver configuration for UWB communications

where N is the number of OFDM symbols transmitted, while $s_{\mathrm{OFDM},\,n}(t)$ denotes the nth OFDM symbol of duration T_{OFDM} transmitted on carrier frequency $f_{n \bmod 6}$. As discussed above, the carrier frequency changes over three frequencies assigned to the band group, organized in sequences of length 6 (TFCs). The n-th OFDM symbol is generated in similar fashion as we described in Chap. 7.

The receiver configuration, shown in Fig. 10.51, is very similar to conventional OFDM receiver [Fig. 7.6b], except from preselected filter that is used to select OFDM signal of 528 MHz bandwidth at given carrier frequency.

One of the important issues related to the UWB communications is the coexistence with other wireless communication systems and the narrowband interference (NBI) influence on UWB system performance. Even though that NBI bandwidth is low compared to UWB signal bandwidth, the power of NBI is significantly higher in that frequency band. To deal with multipath effects, we can use the RAKE receiver, shown in Fig. 10.9, where different outputs of correlators are properly weighted with weight vector $\boldsymbol{w}$, so that the decision statistics will be based on the following:

$$\boldsymbol{w}^{\dagger}\boldsymbol{r}, \boldsymbol{r} = \boldsymbol{s} + \boldsymbol{i} + \boldsymbol{z} \tag{10.193}$$

where $\boldsymbol{s}$ is the desired signal vector, $\boldsymbol{i}$ is the interference vector, and $\boldsymbol{z}$ is the noise vector. The receiver output signal-to-interference-plus-noise ratio, denoted by $SINR$, when NBI is ignored, by using maximum ratio combining (MRC), is given by [85]:

$$SINR_{MRC} = \frac{N_s E_s \left|\boldsymbol{h}\boldsymbol{h}^{\dagger}\right|}{\boldsymbol{h}^{\dagger}(\boldsymbol{R}_i + \boldsymbol{R}_z)\boldsymbol{h}}; \quad \boldsymbol{R}_i = \langle \boldsymbol{i}\boldsymbol{i}^{\dagger}\rangle, \quad \boldsymbol{R}_z = \langle \boldsymbol{z}\boldsymbol{z}^{\dagger}\rangle, \tag{10.194}$$

where E_s is the symbol energy, $\boldsymbol{h}$ is channel fading coefficient, and N_s is the number of bits per symbol. ($\boldsymbol{R}_i$ and $\boldsymbol{R}_z$ denote the covariance matrices of interference and noise, respectively). In MRC, however, the influence of interference is ignored, representing sub-optimum solution. The optimum weighting must be determined by minimizing the MSE, to obtain the following optimum weight vector for the minimum MSE combiner (MMSEC) [85]:

$$\boldsymbol{w} = \frac{N_s E_s (\boldsymbol{R}_i + \boldsymbol{R}_z)^{-1}\boldsymbol{h}}{1 + N_s E_s \boldsymbol{h}^{\dagger}(\boldsymbol{R}_i + \boldsymbol{R}_z)^{-1}\boldsymbol{h}}, \tag{10.195}$$

and the corresponding SINR is given by:

$$SINR_{MMSEC} = N_s E_s \boldsymbol{h}^{\dagger}(\boldsymbol{R}_i + \boldsymbol{R}_z)^{-1}\boldsymbol{h}. \tag{10.196}$$

It has been shown in [85] that MRC results in significant performance degradation compared to MMSEC, because it ignores the presence of NBI (due to WLAN device).

Another effective but low-complexity approach to deal with NBI is to use the *notch filter* of bandwidth B with transfer function [85]:

$$H(f) = 1 - [u(f - f_i + B/2) - u(f - f_i - B/2)], \tag{10.197}$$

where f_i is the center frequency of the NBI and $u(x)$ is the unit-step function. The corresponding impulse response is obtained as inverse Fourier transform (FT) of $H(f)$:

$$h(t) = FT^{-1}\{H(f)\} = \delta(t) - Be^{j2\pi f_i t}\text{sinc}(Bt). \tag{10.198}$$

Another relevant issue for UWB communications is their short reach. To extend the transmission distance, the radio-over-fiber (RoF) technologies can be used, as described in [98].

10.7 Concluding Remarks

The subject of this chapter has been devoted to spread spectrum (SS) communications systems, described in Sect. 10.1; CDMA systems, described in Sect. 10.2; multiuser detection schemes, described in Sect. 10.3; optical CDMA systems, described in Sect. 10.4; hybrid OFDM-CDMA systems, introduced in Sect. 10.5; and UWB communication systems, described in Sect. 10.6.

In the section on SS systems (Sect. 10.1), after the introduction of SS systems (Sect. 10.1.1), both DS-SS and FH-SS systems have been described in detail. Regarding the DS-SS systems (Sect. 10.1.2), after typical transmitter and receiver configurations have been described in Sect. 10.12.1, the synchronization process, including acquisition and tracking stages, has been provided in Sect. 10.1.2.2. To deal with the multipath fading effects, the use of RAKE receiver has been described in Sect. 10.1.2.3. In the subsection on spreading sequences (Sect. 10.1.2.4), after the introduction of PN sequences as well as the basic randomness criterions, we have introduced maximal-length sequences (m-sequences), including their basic properties. In the same subsection, the concept of nonlinear PN sequences has been introduced as well. In Sect. 10.1.3, devoted to the FH-SS systems, we have provided the detailed description of FH-SS systems, including both transmitter and demodulator as well as various time acquisition schemes. In the same section, we have described how to generate the FH sequences.

In Sect. 10.2, after the basic introduction of CDMA systems, we have introduced the various signature waveforms (Sect. 10.2.1.) suitable for use in DS-CDMA systems including Gold, Kasami, and Walsh sequences. Further, the relevant synchronous and asynchronous CDMA models have been described in Sect. 10.2.2, with special attention devoted to discrete-time CDMA models.

In the section on multiuser detection (MUD), Sect. 10.3, we have described how to deal with interference introduced by various CDMA users. The conventional single-user detector, introduced in Sect. 10.3.1, has been used as the reference case. The various MUD schemes have been described then on Sect. 10.3.2, with special attention being paid to the jointly optimum MUD. The complexity of jointly optimum MUD might be prohibitively high for certain application, and for such application, we have introduced decorrelating receiver in Sect. 10.3.3, which can compensate for multiuser interference but enhances the noise effect. To solve for this problem, the linear MMSE receiver has been introduced as well in the same section capable of dealing simultaneously with both multiuser interference and noise effects. In Sect. 10.3.4, we have described the successive cancelation and multistage detection schemes, taking either preliminary or final decisions from other users in detection process.

In the section on optical CDMA (OCDMA), Sect. 10.4, we have introduced both incoherent and coherent optical detection-based OCDMA schemes. Regarding the incoherent optical detection-based schemes, we first have described how to design various unipolar signature sequences, optical orthogonal codes (OOC), followed by the description of basic incoherent OCDMA schemes including time-spreading, spectral amplitude coding (SAC), and wavelength hopping (WH) schemes. Related to coherent optical detection-based OCDMA systems, we have described both pulse laser-based and CW laser-based OCDMA systems.

In Sect. 10.5, the hybrid OFDM-CDMA systems have been described, taking the advantages of both OFDM and CDMA concepts into account, particularly in dealing with frequency selective fading and narrowband interference, while at the same time supporting multiple users. The following three classes of hybrid OFDM-CDMA systems have been introduced: multicarrier CDMA (MC-CDMA), OFDM-CDMA, and OFDM-CDMA-FH systems.

In the section on UWB communications, Sect. 10.6, we have described the concepts and requirements for UWB communication as well as transmitter and receiver configurations. In the same section, various modulation formats suitable for UWB communications have been described. These have been categorized into time-domain category (pulse-position nodulation and pulse-duration modulation) and shape-based category (on-off keying, pulse-amplitude modulation, bi-phase modulation (BPM) or BPSK, and orthogonal pulse modulation). Regarding the pulse shapes suitable for orthogonal pulse modulation (OPM), the modified Hermite polynomials, Legendre polynomials, and orthogonal prolate spheroidal wave functions (OPSWF) have been introduced. Further, the Saleh-Valenzuela UWB channel model has been described, suitable for indoor applications. Both types of UWB systems have been described in detail: the impulse radio UWB (IR-UWB), typically DS-CDMA-based, and multiband OFDM (MB-OFDM).

The set of problems is provided in the following section to help the reader gain deeper understanding of the material described in this chapter.

10.8 Problems

1. The jammer signal is composed of three tones: $J(t) = J \cos(2\pi f_c t + \theta) + J_- \cos(2\pi(f_c - \Delta f)t + \theta) + J_+ \cos(2\pi(f_c + \Delta f)t + \theta)$,and applied to the input of DS-SS with QPSK. Determine SIR and PG as a function of Δf.
2. Assuming that in M-ary PSK DS-SS the block code of rate $R = k/n$ and codeword weight distribution $\{w_m\}$ is used, derive the upper bound for the codeword error probability.
3. Let us assume that the DS-SS system is based on m-sequence with m being 5 with the chip rate of 10^6 chips/s.
   (a) Determine PN sequence length, period, and chip duration.
   (b) Sketch the autocorrelation function and PSD of this PN sequence.
   (c) By using (a), determine the average acquisition time for a serial DS search under assumptions that the detection probability is 0.95 and maximum tolerable false alarm probability is 0.1.
4. Let us consider the following two m-sequences' generators for shift-registers of length 8: (i) the feedback taps are taken after the flip-flops {8,6,5,1} and (ii) the feedback taps are taken after flip-flops {8,6,4,3,2,1}.
   (a) Calculate and plot autocorrelation and cross-correlation functions.
   (b) Calculate and plot the corresponding PSDs.
5. Let us consider the example to generate the Gold sequences from polynomials $g_1(x) = 1 + x^2 + x^5$ and $g_2(x) = 1 + x + x^2 + x^4 + x^5$, provided in Fig. 10.21. Generate the Gold sequences using Eq. (10.79).
   (a) Calculate and plot autocorrelation and cross correlation functions.
   (b) Calculate and plot the corresponding PSDs.
6. Let us consider a QPSK DS-SS in which m-sequence has parameter $m = 23$. Under the requirement that the BER caused by the interfering jammer cannot be larger than 10^{-6}, determine the processing gain and antijam margin.
7. The multipath channel is composed of the LOS component of amplitude a_0 and two reflected components with amplitudes a_1 and a_2 with corresponding delays with respect to the LOS being τ_1 and τ_2, respectively. The amplitudes follow Nakagami-m fading with $m = 0.5$ and 1 each occurring with probability 0.5. The corresponding average powers of the LOS related to $m = 0.5$ and $m = 1$ are 10 and 20. The average powers of the first reflected components related to $m = 0.5$ and $m = 1$ are 1 and 5. Finally, the average powers of the second reflected components related to $m = 0.5$ and $m = 1$ are 5 and 10. Let us assume that the SS receiver get locked to the i-th multipath component in the absence of other components with corresponding SNR being a_i^2 (thus normalized $N_0 = 1$). The bit duration of the BPSK scheme is T_b. The m-sequence is used to implement the DS-SS system.
   (a) When the receiver gets locked to the LOS component, determine τ_i ($i = 1, 2$) so that the corresponding multipath components are attenuated by $-1/N$ (N-the number of chips).
   (b) By observing a single-branch SS receiver locked to the LOS component, determine the outage probability corresponding to the instantaneous probability of error P_b of 10^{-4}.
   (c) For the same instantaneous probability and three-branch RAKE receiver, each locked to one multipath component, determine the outage probability when selection combining is used. Compare the result against (a).
   (d) Repeat (c) when the maximum ratio combining is used instead.
8. In this problem, we are interested in FH based on M-ary FSK. The bandwidth allocated to our system is $BW = 100$ MHz. Let us assume that the multipath channel is a two-ray model channel, wherein the reflected ray is 6 dB lower than the LOS ray. Assume that the reflection coefficient is -0.2. Let us use the delay relative to the $1/BW$ as a parameter.

(a) Evaluate the BER performance for the following values of delay of reflected component relative to the LSO $\tau \in \{0.25, 0.5, 1, 2, 4\}/BW$ for the following values of $M \in \{2, 4, 8, 16\}$. Select the chip interval T_c so that the reflected component will not affect the detection of the LOS component that much.
(b) Repeat (a) but selecting T_c such the system will exhibit slow-varying flat fading.
(c) Repeat (a) but selecting T_c such the system will exhibit slow-varying frequency-selective fading.

9. In this problem we are interested in evaluation performance of FH-SS system with M-ary PSK in the presence jammer. The bandwidth of modulated signal is $B = 20$ MHz. The jammer can vary the power P_J, center frequency, and the bandwidth of his jamming signal. Evaluate the performance of this system in the presence of AWGN and jammer activity as describe above vs. SNR. Use the P_J and various jammer strategies as parameters.
10. Let us consider the DS-SS waveform given by $s_c(t) = A_c \sum_{n=1}^{N} \alpha_n p_{T_c}[t - (n-1)T_c]$, where $[\alpha_1, \ldots, \alpha_n, \ldots, \alpha_N]$ is the *nonbinary* spreading code (real-valued components) with A being the normalization amplitude such that $\|s_c\| = 1$. If the maximum tolerable cross-correlation between any two signature sequences is $\rho_{\max}$, prove that the maximum number of users that can be supported with N chips per symbol is upper-bounded by [5, 99]:

$$K_{\max} \leq \min_k \frac{(1-\rho_{\max}^2)N(N+2)\cdots(N+2k)}{\left[(2k+1)-(N+2k)\rho_{\max}^2\right][1\cdot 3\cdot \ldots \cdot (2k-1)]},$$

with minimization being performed over all k_s for which the denominator is positive. Specify the particular case for which $\rho_{\max}^2 = 1/N$.

11. The OFDM can be considered as a special case of the synchronous CDMA model in which the signature waveforms are given by $s_i(t) = T^{-1/2} \exp\left[j2\pi(i-1-0.5(K-1))t/T\right]$, where T is the symbol duration and K is number of users.
(a) Show that any two signature waveforms are orthogonal.
(b) In the absence of noise, the received signal is given by $r(t) = A\sum_i b_i s_i(t)$. Show that the K-point DFT given by $[r(0)\ r(T/K) \ldots r((K-1)T/K)]$ reconstructs the transmitted sequence.
12. This problem is related to diversity in which we receive M copies of transmitted signal $s(t)$ affected by noise $\sigma z(t)$ and Rayleigh fading coefficients R_i as follows: $r_i(t) = AR_i s(t) + \sigma z(t)$, where $i = 1,\ldots,M$.
(a) In selection combining, we select the branch with maximum SNR (equivalently maximum R_i^2). Determine the corresponding cumulative distribution function of the resulting SNR.
(b) In equal-gain combining, all diversity branches will have the same weight in resulting signal. Determine the corresponding moment-generating function of the SNR.
(c) In maximum-ratio combining, we weigh the output of the i-th branch by R_i before summing them up. Determine the corresponding moment-generating function of the SNR.
13. In this problem we are concerned with the decomposition of cross-correlation matrix by $\boldsymbol{R} = \boldsymbol{F}^{\mathrm{T}}\boldsymbol{F}$. Show that the i-th column in the lower triangular matrix $\boldsymbol{F}$ depends only on $s_i,\ldots,s_K$.
14. Solve the equations in (10.115) for $\boldsymbol{F}[0]$ and $\boldsymbol{F}[1]$ in a two-user asynchronous channel model.
15. Let us consider K-user synchronous channel under assumption that any two users' cross-correlation is the same and equal to ρ and the amplitudes for all users are the same ($A_k = A$). Determine the range of cross-correlation for which the correlation matrix is positive-definite.
(a) Determine the bit error probability when single-user matched filter is used.
(b) Apply the Gaussian approximation given by Eq. (10.126) and discuss the accuracy of this approximation for $K = 2$.
16. The optimum two-user detection scheme is one that performs the following maximization:

$$\left(\widehat{b}_1, \widehat{b}_2\right) = \underset{(b_1, b_2)}{\arg\max}\,(r_1A_1b_1 + r_2A_2b_2 - A_1b_1A_2b_2\rho).$$

Show that this expression can be expressed equivalently as:

$$\widehat{b}_1 = \mathrm{sgn}\left(r_1A_1 + \frac{1}{2}|r_2A_2 - A_1A_2\rho| - \frac{1}{2}|r_2A_2 + A_1A_2\rho|\right)$$
$$\widehat{b}_2 = \mathrm{sgn}\left(r_2A_2 + \frac{1}{2}|r_1A_1 - A_1A_2\rho| - \frac{1}{2}|r_1A_1 + A_1A_2\rho|\right),$$

17. Let us consider a three-user synchronous case in which the signature waveforms s_1 and s_2 are orthogonal, while s_3 is the superposition $s_3 = s_1 + s_2/\sqrt{2}$. Sketch the decision regions when the jointly optimum detection is used.
18. The ML decision rule for K-users' synchronous case is given by.
$\hat{\boldsymbol{b}} = \arg\max_{\boldsymbol{b}} 2\boldsymbol{b}^T\boldsymbol{Ar} - \boldsymbol{b}^T(\boldsymbol{ARA})\boldsymbol{b}$ where $\boldsymbol{ARA}$ is the un-normalized correlation matrix. Show that equivalently the optimization can be represented by:

$$\hat{\boldsymbol{x}} = \arg\min_{\boldsymbol{x}\in\{0,1\}^K} \boldsymbol{x}^T\boldsymbol{Bx} + \boldsymbol{W}^T\boldsymbol{x},$$

where $\boldsymbol{B} = \boldsymbol{ARA}\text{-}\boldsymbol{V}$ and $\boldsymbol{W}=\boldsymbol{V}\text{-}\boldsymbol{ARA}[1\ldots1]^{\mathrm{T}}\text{-}\boldsymbol{Ar}$, with $\boldsymbol{V}$ being arbitrary diagonal matrix $\boldsymbol{V} = \text{diag}\,(v_1, \ldots, v_K)$. See [100] for additional details of this representation.
19. Let us go back to the K-user synchronous channel model with M diversity branches, described in Problem 12, namely:

$$r_m(t) = \sum\nolimits_k A_{mk}b_k s_k(t) + \sigma z_m(t);\, m = 1, \ldots, M$$

Show that the following set of variables $\{r_{mk}\}$ with $\int_0^T r_m(t)s_k^*(t)dt$ represents the sufficient statistics.
20. Let us study the K-user asynchronous channel. Prove that, when the open eye condition $|r_k| > \sum_{i=1,i\neq k}^K A_i(|\rho_{ik}| + |\rho_{ki}|)$ is satisfied, the jointly optimum detector is simply $\hat{b}_k = \text{sgn}\,(r_k)$.
21. Let us observe a two-user synchronous channel model. Determine ρ, A_1/σ, and A_2/σ in such a way that the decorrelation detector outperforms single-user matched filter (in BER sense), while the signal-to-interference ratio (SIR) in decorrelator is lower than that in the single-used matched detector.
22. Let us consider the discrete time synchronous model introduced by Eq. (10.103), namely, $\vec{r} = \boldsymbol{SAb} + \vec{z}$. Show that $\boldsymbol{S}^+\vec{r}$, where $\boldsymbol{S}^+$ is the pseudo-inverse operation, is the solution of the following minimization problem $\min_{\boldsymbol{S}^+\vec{r}\in R^K} \left(\boldsymbol{SS}^+\vec{r} - \vec{r}\right)^T \boldsymbol{C}^{-1}\left(\boldsymbol{SS}^+\vec{r} - \vec{r}\right)$, where $\boldsymbol{C}$ is the covariance matrix of $\vec{z}$.
23. Determine the coefficients a_k in the following equation: $[\boldsymbol{R} + \sigma^2\boldsymbol{I}]^{-1} = \sum_{k=1}^K a_k\boldsymbol{R}^k$.
24. Derive the implementations of the multistage detectors with conventional and decorrelating two stages in terms of the model discussed in Problem 22.
25. Describe key advantages and disadvantages of incoherent optical CDMA systems with respect to corresponding coherent detection schemes. Discuss possible applications of incoherent CDMA systems.
26. Discuss the advantages of the hybrid OFDM-CDMA systems compared to CDMA-only and OFDM-only systems.
27. Let us consider the system model given by Eq. (10.193) in which multipath fading is compensated by RAKE receiver, but when in addition to AWGN noise the narrowband interference is present. Derive optimum weights minimizing the MSE in maximum-ratio combining receiver in this scenario.

References

1. S. Glisic, B. Vucetic, *Spread Spectrum CDMA Systems for Wireless Communications* (Artech House, Inc, Boston/London, 1997)
2. A. Goldsmith, *Wireless Communications* (Cambridge University Press, Cambridge/New York, 2005)
3. J.G. Proakis, *Digital Communications*, 4th edn. (McGraw-Hill, Boston, 2001)
4. S. Haykin, *Digital Communication Systems* (Wiley, Hoboken, 2014)
5. S. Verdú, *Multiuser Detection* (Cambridge University Press, Cambridge/New York, 1998)
6. R. Pickholtz, D. Schilling, L. Milstein, Theory of spread-Spectrum communications – a tutorial. IEEE Trans. Commun. **30**(5), 855–884 (1982)
7. D. Tse, P. Viswanath, *Fundamentals of Wireless Communication* (Cambridge University Press, Cambridge, 2005)
8. B. Sklar, *Digital Communications: Fundamentals and Applications*, 2nd edn. (Prentice-Hall, Inc, Upper Saddle River, 2001)
9. J.K. Holmes, C.C. Chen, Acquisition time performance of PN spread-spectrum systems. IEEE Trans. Commun. **25**(8), 778–784 (1977)
10. M.K. Simon, J.K. Omura, R.A. Scholtz, B.K. Levit, *Spread Spectrum Communications* (Computer Science Press, Inc., Rockville, 1985)
11. D. Torrieri, *Principles of Spread-Spectrum Communication Systems*, 3rd edn. (Springer International Publishing Switzerland, 2015)

12. R.B. Ward, Acquisition of pseudonoise signals by sequential estimation. IEEE Trans. Commun. Technol. **13**(4), 475–483 (1965)
13. R. Price, P.E. Green, A communication technique for multipath channels. Proc. IRE **46**(3), 555–570 (1958)
14. G.L. Turin, Introduction to spread spectrum antimultipath techniques and their application in urban digital radio. Proc. IEEE **68**(3), 328–353 (1980)
15. D.V. Sarwate, M.B. Pursley, Crosscorrelation properties of pseudorandom and related sequences. Proc. IEEE **68**(5), 593–619 (1980)
16. S.W. Golomb, *Shift Register Sequences: Secure and Limited-Access Code Generators, Efficiency Code Generators, Prescribed Property Generators, Mathematical Models*, 3rd Revised edn. (World Scientific Publishing Co., Singapore, 2017)
17. D.B. Drajic, *Introduction to Statistical Theory of Telecommunications* (Akademska Misao, Belgrade, 2003). (In Serbian)
18. N.G. De Bruijn, A combinatorial problem. Proc. Sect. Sci. Koninklijke Nederlands Akademie wan Wetenschappen **49**(2), 758–764 (1946)
19. I.J. Good, Normal recurring decimals. J. London Math. Soc. **21**(3), 169–172 (1946)
20. T. Etzion, A. Lempel, Algorithms for the generation of full-length shift-register sequences. IEEE Trans. Inform. Theory **30**(3), 480–484 (1984)
21. T.S. Seay, Hopping patterns for bounded mutual interference in frequency hopping multiple access, in *Proceedings of the IEEE Military Communications Conference – Progress in Spread Spectrum Communications (MILCOM '82)*, (Boston, 1982), pp. 22.3-1–22.3-6
22. A. Lempel, H. Greenberger, Families of sequences with optimal Hamming correlation properties. IEEE Trans. Inform. Theory **20**(1), 90–94 (1974)
23. I.S. Reed, K-th order near-orthogonal codes. IEEE Trans. Inform. Theory **17**(1), 116–117 (1971)
24. G. Solomon, Optimal frequency hopping sequences for multiple access. Proc. Syp. Spread Spectrum Commun. **1**, 33–35 (1973)
25. B.M. Popovic, *Comparative Analysis of Sequences for Asynchronous Code-division Multiplex with Frequency Hopping*, Master Thesis, University of Belgrade, Serbia. (in Serbian) (1989)
26. A.A. Shaar, P.A. Dvaiees, A survey of one-coincidence sequences for frequency-hopped spread-spectrum systems. IEEE Proc. Commun. Radar Signal Proc. Part F **131**(7), 719–724 (1984)
27. L.R. Welch, Lower bounds on the maximum cross correlation of signals (Corresp.). IEEE Trans. Inform. Theory **20**(3), 397–399 (1974)
28. R. Gold, Optimal binary sequences for spread spectrum multiplexing (Corresp.). IEEE Trans. Inform. Theory **13**(4), 619–621 (1967)
29. R. Gold, Maximal recursive sequences with 3-valued recursive cross-correlation functions (Corresp.). IEEE Trans. Inform. Theory **14**(1), 154–156 (1968)
30. J.L. Walsh, A closed set of normal orthogonal functions. Am. J. Math. **45**(1), 5–24 (1923)
31. H.F. Harmuth, Applications of Walsh functions in communications. IEEE Spectr. **6**(11), 82–91 (1969)
32. K.G. Beauchamp, *Walsh Functions and Their Applications* (Academic, London, 1975)
33. R.N. McDonough, A.D. Whalen, *Detection of Signals in Noise*, 2nd edn. (Academic, San Diego, 1995)
34. Vasic B, Digital Communications II, Lecture Notes, 2005
35. S. Verdú, *Optimum Multiuser Signal Detection*, PhD dissertation, University of Illinois, Urbana-Champaign, 1984)
36. S. Verdú, Minimum probability of error for asynchronous Gaussian multiple-access channels. IEEE Trans. Inform. Theory **32**(1), 85–96 (1986)
37. R. Lupas, S. Verdú, Linear multiuser detectors for synchronous code-division multiple-access channels. IEEE Trans. Inform. Theory **35**(1), 123–136 (1989)
38. S. Verdú, Multiuser detection, in *Advances in Statistical Signal Processing: Signal Detection*, ed. by H. V. Poor, J. B. Thomas, (JAI Press, 1993), pp. 369–410
39. S. Verdú, Computational complexity of optimum multiuser detection. Algorithmica **4**, 303–312 (1989)
40. S. Verdú, Optimum multiuser asymptotic efficiency. IEEE Trans. Commun. **34**, 890–897 (1986)
41. S. Verdú, Multiple-access channels with point-process observations: Optimum demodulation. IEEE Trans. Inform. Theory **32**, 642–651 (1986)
42. S. Verdú, Adaptive multiuser detection, in *Code Division Multiple Access Communications*, ed. by S. G. Glisic, P. A. Leppanen, (Kluwer Academic, 1995)
43. A. Duel-Hallen, J. Holtzman, Z. Zvonar, Multiuser detection for CDMA systems. IEEE Pers. Commun. **2**(2), 46–58 (1995)
44. J.G. Andrews, Interference cancellation for cellular systems: A contemporary overview. IEEE Wirel. Commun. **12**(2), 19–29 (2005)
45. Y.C. Yoon, R. Kohno, H. Imai, A spread-spectrum multiaccess system with cochannel interference cancellation for multipath fading channels. IEEE J. Sel. Areas Commun. **11**(7), 1067–1075 (1993)
46. P. Patel, J. Holtzman, Analysis of a simple successive interference cancellation scheme in a DS/CDMA system. IEEE J. Sel. Areas Commun. **12**(5), 796–807 (1994)
47. D. Divsalar, M.K. Simon, D. Raphaeli, Improved parallel interference cancellation for CDMA. IEEE Trans. Commun. **46**(2), 258–268 (1998)
48. M.K. Varanasi, B. Aazhang, Near-optimum detection in synchronous code-division multiple access systems. IEEE Trans. Commun. **39**, 725–736 (1991)
49. M.K. Varanasi, B. Aazhang, Multistage detection in asynchronous code-division multiple-access communications. IEEE Trans. Commun. **38**(4), 509–519 (1990)
50. J. Luo, K.R. Pattipati, P.K. Willett, F. Hasegawa, Near-optimal multiuser detection in synchronous CDMA using probabilistic data association. IEEE Commun. Lett. **5**(9), 361–363 (2001)
51. X. Wang, H.V. Poor, *Wireless Communication Systems: Advanced Techniques for Signal Reception* (Prentice-Hall, Emglewood Cliffs, 2004)
52. A. Duel-Hallen, Decorrelating decision-feedback multiuser detector for synchronous code-division multiple-access channel. IEEE Trans. Commun. **41**(2), 285–290 (1993)
53. A. Duel-Hallen, A family of multiuser decision-feedback detectors for asynchronous code-division multiple-access channels. IEEE Trans. Commun. **43**(2/3/4), 421–434 (1995)
54. P. Prucnal, M. Santoro, T. Fan, Spread spectrum fiber-optic local area network using optical processing. J. Lightw. Technol. **4**(5), 547–554 (1986)
55. P.R. Prucnal, *Optical Code Division Multiple Access: Fundamentals and Applications* (CRC Press, Boca Raton, 2006)
56. J.A. Salehi, Code division multiple-access techniques in optical fiber networks-part I: Fundamental principles. IEEE Trans. Commun. **37**(8), 824–833 (1989)
57. Fathallah H, Rusch LA, and S. LaRochelle S (1999), Passive optical fast frequency-hop CDMA communication system. J. Lightwave Technol. 17: 397–405

58. J.A. Salehi, F.R.K. Chung, V.K. Wei, Optical orthogonal codes: Design, analysis, and applications. IEEE Trans. Inform. Theory **35**(3), 595–605 (1989)
59. R. Petrovic, Orthogonal codes for CDMA optical fiber LAN's with variable bit interval. IEE Electron. Lett. **26**(10), 662–664 (1990)
60. S.V. Maric, A new family of optical code sequences for use in spread-spectrum fiber-optic local area networks. IEEE Trans. Commun. **41**(8), 1217–1221 (1993)
61. S.V. Maric, New family of algebraically designed optical orthogonal codes for use in CDMA fiber-optic networks. IEE Electron. Lett. **29**(6), 538–539 (1993)
62. G.-C. Yang, W.C. Kwong, Performance analysis of optical CDMA with prime codes. IEE Electron. Lett. **31**(7), 569–570 (1995)
63. Z. Wei, H. Ghafouri-Shiraz, Proposal of a novel code for spectral amplitude-coding optical CDMA systems. IEEE Photon Technol. Lett. **14**, 414–416 (2002)
64. S. Yengnarayanan, A.S. Bhushan, B. Jalali, Fast wavelength-hopping time-spreading encoding/decoding for optical CDMA. IEEE Photon. Technol. Lett. **12**(5), 573–575 (2000)
65. W. Huang, M.H.M. Nizam, I. Andonovic, M. Tur, Coherent optical CDMA (OCDMA) systems used for high-capacity optical fiber networks-system description, OTDMA comparison, and OCDMA/WDMA networking. J. Lightw. Technol. **18**(6), 765–778 (2000)
66. I.B. Djordjevic, B. Vasic, Novel combinatorial constructions of optical orthogonal codes for incoherent optical CDMA systems. J. Lightw. Technol. **21**(9), 1869–1875 (2003)
67. I.B. Djordjevic, D.F. Geraghty, A. Patil, C.H. Chen, R. Kostuk, B. Vasic, Demonstration of spectral-amplitude encoding/decoding for multimedia optical CDMA applications. J. Opt. Commun. **27**(2), 121–124 (2006)
68. I.B. Djordjevic, B. Vasic, Unipolar codes for spectral-amplitude-coding optical CDMA systems based on projective geometries. IEEE Photon. Technol. Lett. **15**(9), 1318–1320 (2003)
69. I.B. Djordjevic, B. Vasic, Combinatorial constructions of optical orthogonal codes for OCDMA systems. IEEE Commun. Lett. **8**(6), 391–393 (2004)
70. A. Patil, I.B. Djordjevic, D.F. Geraghty, Performance Assessment of Optical CDMA Systems based on Wavelength-Time Codes from Balanced Incomplete Block Designs, in *Proceedings of the 7th International Conference on Telecommunications in Modern Satellite, Cable and Broadcasting Services-TELSIKS 2005*, (Nis, Serbia and Montenegro, 2005), pp. 295–298, September 28–30
71. I.B. Djordjevic, B. Vasic, J. Rorison, Multi-weight unipolar codes for spectral-amplitude-coding optical CDMA systems. IEEE Commun. Lett. **8**(4), 259–261 (2004)
72. I.B. Djordjevic, B. Vasic, J. Rorison, Design of multi-weight unipolar codes for multimedia optical CDMA applications based on pairwise balanced designs. IEEE/OSA J. Lightw. Technol. **21**(9), 1850–1856 (2003)
73. E.S. Shivaleela, K.N. Sivarajan, A. Selvarajan, Design of a new family of two-dimensional codes for fiber-optic CDMA networks. J. Lightw. Technol. **16**(4), 501–508 (1998)
74. E.S. Shivaleela, A. Selvarajan, T. Srinivas, Two-dimensional optical orthogonal codes for fiber-optic CDMA networks. J. Lightw. Technol. **23**(2), 647–654 (2005)
75. L. Tancevski, I. Andonovic, J. Budin, Secure optical network architectures utilizing wavelength hopping/time spreading codes. IEEE Photon. Technol. Lett. **7**(5), 573–575 (1995)
76. M. Jankiraman, R. Prasad, Hybrid CDMA/OFDM/SFH: a novel solution for wideband multimedia communications, in *Proceedings of the 4th ACTS Mobile Communication Summit '99*, (Sorento, Italy, 1999), pp. 1–7
77. M. Jankiraman, R. Prasad, A novel solution to wireless multimedia application: the hybrid OFDM/CDMA/SFH approach, in *Proceedings of the 11th IEEE International Symposium on Personal Indoor and Mobile Radio Communications (PIMRC 2000), 2*, (London, UK, 2000), pp. 1368–1374
78. M. Jankiraman, R. Prasad, Wireless multimedia application: the hybrid OFDM/CDMA/SFH approach, in *Proceedings of the 2000 IEEE Sixth International Symposium on Spread Spectrum Techniques and Applications (ISSTA 2000), 2*, (Parsippany, NJ, 2000), pp. 387–393
79. R. Van Nee, R. Prasad, *OFDM Wireless Multimedia Communications* (Artech House, Boston/London, 2000)
80. R. Prasad, *OFDM for Wireless Communications Systems* (Artech House, Boston/London, 2004)
81. N. Yee, J.-P. Linnartz, G. Fettweis, Multi-carrier CDMA in indoor wireless networks, in *Proceedings of the Fourth International Symposium on Personal, Indoor and Mobile Radio Communications (PIMRC'93)*, (Yokohama, Japan, 1993), pp. 109–113
82. A. Chouly, A. Brajal, S. Jourdan, Orthogonal multicarrier techniques applied to direct sequence spread spectrum CDMA systems, in *Proceedings of the IEEE Global Telecommunications Conference 1993 (GLOBECOM '93), including a Communications Theory Mini-Conference, 3*, (Houston, TX, 1993), pp. 1723–1728
83. K. Fazel, Performance of CDMA/OFDM for mobile communication system, in *Proceedings of the 2nd IEEE International Conference on Universal Personal Communications, 2*, (Ottawa, Canada, 1993), pp. 975–979
84. H. Nikookar, R. Prasad, *Introduction to Ultra Wideband for Wireless Communications* (Springer ScienceþBusiness Media B.V, 2009)
85. M. Ghavami, L.B. Michael, R. Kohno, *Ultra Wideband Signals and Systems in Communication Engineering* (Wiley, Chichester, 2007)
86. K. Siwiak, D. McKeown, *Ultra-Wideband Radio Technology* (Wiley, Chichester, 2004)
87. F. Nekoogar, *Ultra-Wideband Communications: Fundamentals and Applications* (Prentice Hall Press, Upper Saddle River, 2005)
88. A. Batra, J. Balakrishnan, A. Dabak, R. Gharpurey, J. Lin, P. Fontaine, J.-M. Ho, S. Lee, M. Frechette, Y.H. March, *Multi-Band OFDM Physical Layer Proposal for IEEE 802.15 Task Group 3a, Doc* (IEEE P802.15-03/268r3, 2004)
89. A. Batra, J. Balakrishnan, G.R. Aiello, J.R. Foerster, A. Dabak, Design of a multiband OFDM system for realistic UWB channel environments. IEEE Trans. Microwave Theory Tech. **52**(9), 2123–2138 (2004)
90. O.-O. Shin, S.S. Ghazzemzadeh, L.J. Greenstein, V. Tarokh, Performance evaluation of MB-OFDM and DS-UWB systems for wireless personal area networks, in *Proceedings of the 2005 IEEE International Conference on Ultra-Wideband, 1*, (Zurich, Switzerland, 2005), pp. 214–219
91. M.Z. Win, R.A. Scholtz, Ultra-wide bandwidth time-hopping spread-spectrum impulse radio for wireless multiple–access communications. IEEE Trans. Commun. **48**(4), 679–691 (2000)

92. K.-B. Png, X. Peng, F. Chin, Performance studies of a multi-band OFDM system using a simplified LDPC code, in *Proceedings of the 2004 International Workshop on Ultra Wideband Systems & 2004 Conference on Ultrawideband Systems and Technologies (UWBST & IWUWBS), 1*, (Kyoto, Japan, 2004), pp. 376–380
93. S.-M. Kim, J. Tang, K.K. Parhi, Quasi-cyclic low-density parity-check coded multiband-OFDM UWB systems, in *Proceedings of the 2005 IEEE International Symposium on Circuits and Systems, 1*, (Kobe, Japan, 2005), pp. 65–68
94. A.A.M. Saleh, R. Valenzuela, A statistical model for indoor multipath propagation. IEEE J. Sel. Areas Commun. **5**(2), 128–137 (1987)
95. G.B. Arfken, H.J. Weber, *Mathematical Methods for Physicists*, 6th edn. (Elsevier/Academic, Amsterdam/Boston, 2005)
96. D. Slepian, Prolate spheroidal wave functions, Fourier analysis and uncertainty V: The discrete case. Bell Syst. Tech. J. **57**, 1371–1430 (1978)
97. I.B. Djordjevic, On the irregular nonbinary QC-LDPC-coded hybrid multidimensional OSCD-modulation enabling beyond 100 Tb/s optical transport. IEEE/OSA J. Lightw. Technol. **31**(16), 2969–2975 (2013)
98. W. Shieh, I. Djordjevic, *OFDM for Optical Communications* (Elsevier/Academic, Amsterdam/Boston, 2010)
99. J.E. Mazo, Some theoretical observations on spread-spectrum communications. Bell Syst. Tech. J. **58**(9), 2013–2023 (1979)
100. Z.-L. Shi, W. Du, P.F. Driessen, A new multistage detector for synchronous CDMA communications. IEEE Trans. Commun. **44**(5), 538–541 (1996)

11 Physical-Layer Security for Wireless and Optical Channels

Abstract

This chapter is devoted to the physical-layer security. The chapter starts with discussion on security issues, followed by the introduction of information-theoretic security and comparison against the computational security. In the same section, various information-theoretic security measures are introduced, including strong secrecy and weak secrecy conditions. After that, Wyner's wiretap channel model, also known as the degraded wiretap channel model, is introduced. In the same section, the concept of secrecy capacity is introduced as well as the nested wiretap coding. Further, the broadcast channel with confidential messages is introduced, and the secrecy capacity definition is generalized. The focus is then moved to the secret-key generation (agreement), the source- and channel-type models are introduced, and corresponding secret-key generation protocols are described. The next section is devoted to the coding for the physical-layer security systems, including both coding for weak and strong secrecy systems. Regarding the coding for weak secrecy systems, the special attention is devoted to two-edge-type LDPC coding, punctured LDPC coding, and polar codes. Regarding the coding for strong secrecy systems, the focus is on coset coding with dual of LDPC codes and hash functions/extractor-based coding. The attention is then moved to information reconciliation and privacy amplification. In wireless channels physical-layer security (PLS) section, the following topics are covered: wireless MIMO PLS and secret-key generation in wireless networks. In section on optical channels PLS, both PLS for spatial division multiplexing (SDM)-fibers-based systems and free-space optical (FSO) systems are discussed. For better understanding, the set of problems is provided.

11.1 Security Issues

Public key cryptography has several serious drawbacks such as the following: it is difficult to implement it in devices with low memory and low process constraints, Internet is becoming more and more mobile, security schemes are based on unproven assumptions of intractability of certain functions, and the assumption of limiting computing resources of the Eve is not always applicable, to mention few. The open system interconnection (OSI) reference model defines seven layers. However, only five layers relevant to security issues are provided in Fig. 11.1. The original OSI model does not specify the security issues at all. The security issues are addressed in X.800 standard (security architecture for OSI) [1]. Even though the physical-layer security (PLS) is not discussed in this standard, the services specified in these five layers can be enhanced by employing the PLS as discussed in [2–5]. The PLS scheme can also operate independently. The distortions and noise effects introduced by channel can be exploited to reduce the number of bits extracted by Eve. Compared to conventional cryptographic approaches where strong error control coding (ECC) schemes are used to provide reliable communication, the transmission in PLS scenario needs to be simultaneously reliable and secure. This indicates that different classes of ECC must be developed, and these codes will be discussed later in this chapter. Alternatively, similarly to QKD, the randomness of the channel can be exploited to generate the key, and this approach is commonly referred to as the *secret-key agreement* [3–6], and this concept is described in Fig. 11.2, inspired by [3, 4, 7].

Alice and Bob monitor the Alice-Bob channel capacity (also known as the capacity of the main channel) C_M and the secrecy capacity C_S, defined as a difference between main channel capacity and eavesdropping channel capacity C_E. When the

I. B. Djordjevic, *Advanced Optical and Wireless Communications Systems*, https://doi.org/10.1007/978-3-030-98491-5_11

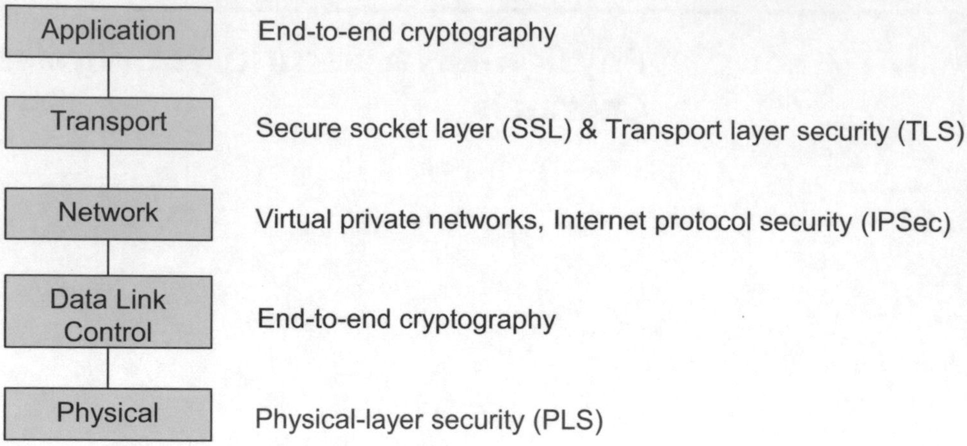

Fig. 11.1 Security mechanisms at different layers in OSI model. (Only security-relevant layers are shown)

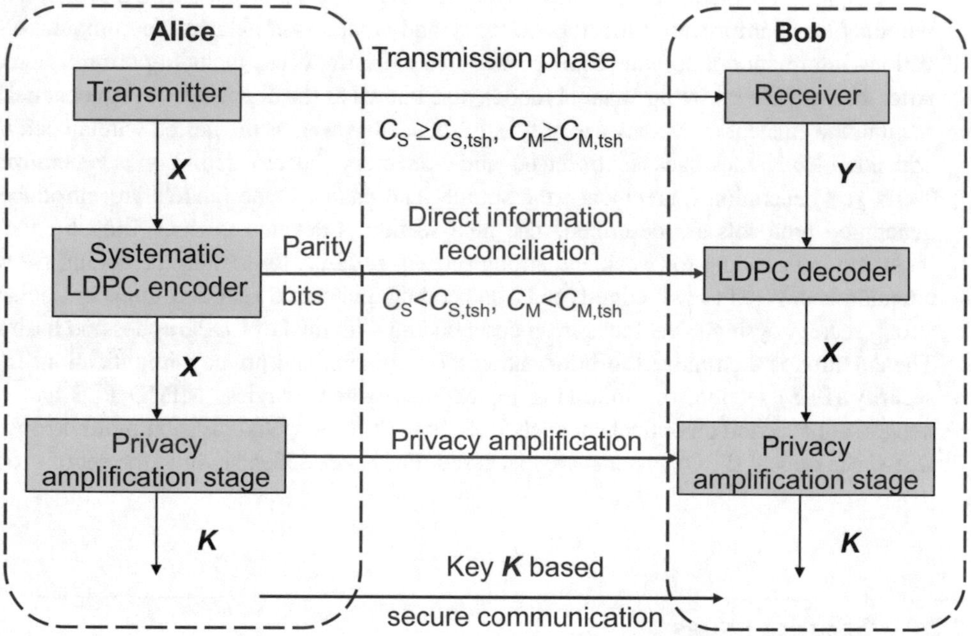

Fig. 11.2 Secret-key generation (agreement) protocol suitable for wireless communications

secrecy capacity is sufficiently above the threshold value $C_{S,\text{tsh}}$ and the main channel capacity is well above threshold value $C_{M,\text{tsh}}$, Alice transmits Gaussian-shaped symbols $\boldsymbol{X}$ to Bob. When the secrecy capacity and main channel capacity are both below corresponding thresholds due to deep fading, Alice and Bob perform *information reconciliation* of previously transmitted symbols, which is based on strong ECC scheme to ensure that errors introduced by either channel or Eve can be corrected for. Similar to QKD Schemes [8–15], a systematic low-density parity-check (LDPC) code can be used (that does not affect information bits but generates the parity-check bit algebraically related to the information bits) to generate the parity bits and transmit them over an authenticated public channel. There exist direct and reverse information reconciliation schemes. In *direct reconciliation*, shown in Fig. 11.2, Alice performs LDPC encoding and sends the parity bits to Bob. The Bob performs the LDPC decoding to get the correct key $\boldsymbol{X}$. In *reverse reconciliation*, Bob performs LDPC encoding instead. *Privacy amplification* is then performed between Alice and Bob to distil from $\boldsymbol{X}$ a smaller set of bits $\boldsymbol{K}$ (secure key), whose correlation with Eve's string is below a desired threshold. One way to accomplish privacy amplification is through the use of *universal hash functions* G [10, 13, 15, 16], which map the set of n-bit strings $\boldsymbol{X}$ to the set of m-bit strings $\boldsymbol{K}$ such that for any distinct $\boldsymbol{X}_1$ and $\boldsymbol{X}_2$ from the set of corrected keys, when the mapping g is chosen uniformly at random from G, the probability of having $g(\boldsymbol{X}_1) = g(\boldsymbol{X}_2)$ is very low.

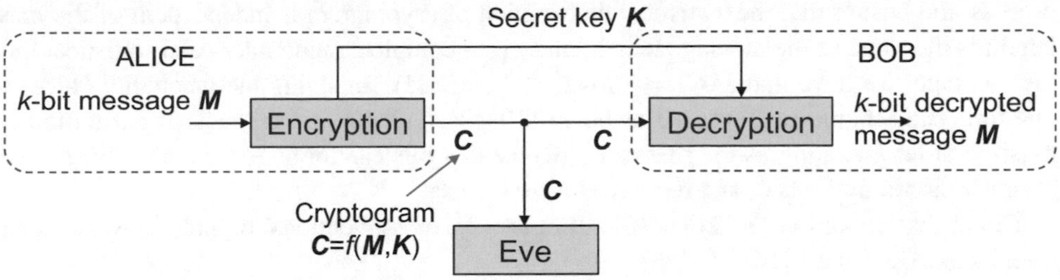

Fig. 11.3 Shannon's model of secrecy system

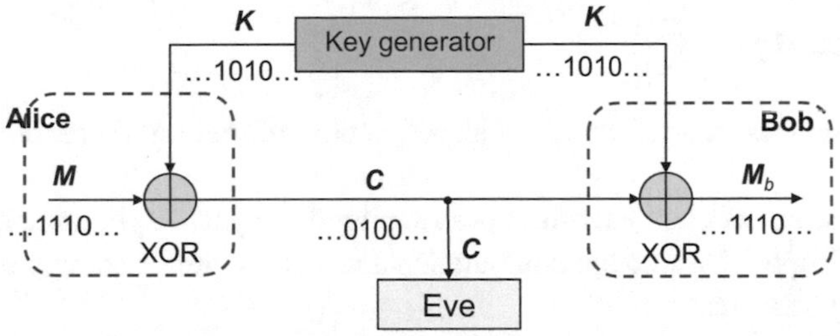

Fig. 11.4 One-time pad scheme

11.2 Information-Theoretic vs. Computational Security

Shannon's model of secrecy system [17] is depicted in Fig. 11.3. The purpose of this system is to reliably transmit the message $\boldsymbol{M}$, which is at the same time secret for Eve's point of view. To do so, Alice and Bob have access to the random key $\boldsymbol{K}$, which is not known to Eve, and it is used by Alice to encrypt the message into the cryptogram $\boldsymbol{C}$. On receiver side Bob decrypts the cryptogram $\boldsymbol{C}$ with the help of key $\boldsymbol{K}$ and determines the transmitted message $\boldsymbol{M}$.

11.2.1 Information-Theoretic (Perfect) Security

We say that the message is transmitted with *perfect security*, if given Eve's knowledge about the cryptogram it does not help her deciphering the message. In other words, $\Pr(\boldsymbol{M} \mid \text{Eve's knowledge}) = \Pr(\boldsymbol{M})$. Clearly, in perfect security sense, the codeword $\boldsymbol{C}$ is statistically independent of the message $\boldsymbol{M}$, and we can write $H(\boldsymbol{M}|\boldsymbol{C}) = H(\boldsymbol{M})$. Equivalently, the mutual information between the codeword $\boldsymbol{C}$ and the message $\boldsymbol{M}$ is zero, that is, $I(\boldsymbol{M};\boldsymbol{C}) = H(\boldsymbol{M}|\boldsymbol{C})\text{-}H(\boldsymbol{M}) = 0$. In this scenario, the best strategy for the Eve is to guess the message, and the probability of success is 2^{-k} and decreases exponentially as the key length k increases. By employing the chain rule from information theory [2, 3], we can write:

$$\begin{aligned} H(\boldsymbol{M}) &= H(\boldsymbol{M}|\boldsymbol{C}) \\ &\leq H(\boldsymbol{M},\boldsymbol{K}|\boldsymbol{C}) = \\ &= H(\boldsymbol{K}|\boldsymbol{C}) + \underbrace{H(\boldsymbol{M}|\boldsymbol{C},\boldsymbol{K})}_{=0} \\ &= H(\boldsymbol{K}|\boldsymbol{C}) = H(\boldsymbol{K}), \end{aligned} \tag{11.1}$$

where we used the fact that message is completely specified when both the cryptogram and the key are known, that is, $H(\boldsymbol{M}|\boldsymbol{K},\boldsymbol{C}) = 0$. Therefore, the *perfect secrecy condition* is given by:

$$H(\boldsymbol{M}) \leq H(\boldsymbol{K}). \tag{11.2}$$

In conclusion, the entropy (uncertainty) of the key cannot be lower than the entropy of the message, for an encryption scheme to be perfectly secure.

One-time pad encryption scheme (Vernam cipher) [18, 19], shown in Fig. 11.4, operates by adding mod 2 message $\boldsymbol{M}$ bits and uniform random key bits $\boldsymbol{K}$ and satisfies the perfect security condition [20]. Namely, the key bits randomize the encoding process and ensure that the statistical distribution of cryptogram is independent of the message. On the other hand, the key length is the same as the message length, and given complete randomness and statistical independence of the cryptogram and the message, we have that $H(\boldsymbol{C}) = -\log 2^k = k = H(\boldsymbol{M})$, satisfying the inequality (11.2) with equality sign. Unfortunately, one-time pad scheme has several drawbacks [10, 14]: (i) it requires the secure distribution of the key, (ii) the length of the key must be at least as long as the message, (iii) the key bits cannot be reused, and (iv) the keys must be delivered in advance, securely stored until used, and destroyed after the use.

Given that condition (11.2) is difficult to satisfy, in conventional cryptography, instead of perfect security the *computational security* is used [15, 21]-[25].

11.2.2 Computational Security

The computational security introduces two relaxations with respect to information-theoretic security [22--24]:

- Security is guaranteed against an efficient eavesdropper running the cryptanalytic attacks for certain limited amount of time. Of course, when eavesdropper has enough computational resources and/or enough time, he/she will be able to break the security of the encryption scheme.
- Eavesdroppers can be successful in breaking the security protocols, but with small success probability.

There are two general approaches to precisely define the computational relaxations [22]:

- The *concrete approach*, in which an eavesdropper running the eavesdropping algorithm for time interval no more than T seconds can be successful with probability at most ε. We could call this approach the concept of (T,ε)-security.
- The *asymptotic approach*, in which a probabilistic polynomial-time eavesdropper caring out an eavesdropping strategy (cryptanalytic attack), for every polynomial $p(n)$ there exists an integer N such that when $n > N$, will have the success attack probability that is smaller than $1/p(n)$.

A reader interested to learn more about computational security is referred to an excellent book due to Katz and Lindell [22].

11.2.3 Information-Theoretic Secrecy Metrics

Before concluding this section, we describe several relevant information-theoretic secrecy metrics [4, 5, 26]. The generic metric $D(\cdot,\cdot)$, measuring the distance between the joint distribution of $\boldsymbol{M}$ and $\boldsymbol{C} = [C_1,C_2,\ldots,C_k]$, denoted as $f_{\boldsymbol{M},\boldsymbol{C}}$, and the product of independent distributions for $\boldsymbol{M}$ and $\boldsymbol{C}$, denoted as $f_{\boldsymbol{M}}$ and $f_{\boldsymbol{C}}$, can be used to define the secrecy requirement as follows:

$$\lim_{k\to\infty} D\left(f_{\boldsymbol{M},\boldsymbol{C}}, f_{\boldsymbol{M}} f_{\boldsymbol{C}}\right) = 0. \tag{11.3}$$

As an illustration, the Kullback-Leibler (KL) distance [27], representing the measure of inefficiency of assuming that distribution is $f_{\boldsymbol{M}} f_{\boldsymbol{C}}$ when true distribution is $f_{\boldsymbol{M},\boldsymbol{C}}$, leads to the well-known *mutual information*:

$$D\left(f_{\boldsymbol{M},\boldsymbol{C}} \middle\| f_{\boldsymbol{M}} f_{\boldsymbol{C}}\right) = E_{f_{\boldsymbol{M},\boldsymbol{C}}} \log\left[\frac{f_{\boldsymbol{M},\boldsymbol{C}}}{f_{\boldsymbol{M}} f_{\boldsymbol{C}}}\right] = I(\boldsymbol{M},\boldsymbol{C}), \tag{11.4}$$

where E denotes an expectation operator. The L_1-distance between the joint distribution $f_{\boldsymbol{M},\boldsymbol{C}}$ and the product of independent distribution $f_{\boldsymbol{M}} f_{\boldsymbol{C}}$, known as the *variational distance* [5], can also be used as the secrecy metric:

$$V\left(f_{\boldsymbol{M},\boldsymbol{C}}, f_{\boldsymbol{M}} f_{\boldsymbol{C}}\right) = E_{f_{\boldsymbol{M},\boldsymbol{C}}} \left| f_{\boldsymbol{M},\boldsymbol{C}} - f_{\boldsymbol{M}} f_{\boldsymbol{C}}\right|. \tag{11.5}$$

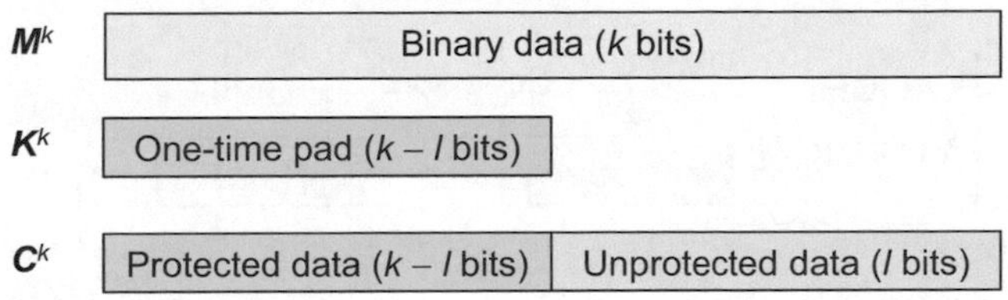

Fig. 11.5 Illustration of the weak secrecy concept

Given that mutual information $I(\mathbf{M},\mathbf{C})$ measures the average amount of information about message $\boldsymbol{M}$ leaked in $\boldsymbol{C}$, the following requirement:

$$\lim_{k\to\infty} I(\boldsymbol{M},\boldsymbol{C}) = 0 \tag{11.6}$$

is commonly referred to as the *strong secrecy condition*. From practical point of view, given that the strong secrecy condition is difficult to satisfy, instead of requesting the mutual information to vanish, we can soften the requirement and request that the *rate* of information leaked to Eve tends to zero:

$$\lim_{k\to\infty} \frac{1}{k} I(\boldsymbol{M},\boldsymbol{C}) = 0 \tag{11.7}$$

This average information rate about the message $\boldsymbol{M}$ leaked to $\boldsymbol{C}$ is well known as the *weak secrecy condition*. To illustrate the difference, let us consider the following illustrative example, provided in Fig. 11.5. The message $\boldsymbol{M}$ of length k, denoted as $\boldsymbol{M}^k$, is to be encrypted. To encrypt the message, we apply the one-time pad on first k-l bits, while the remaining l bits are unprotected, where l is fixed number. The strong secrecy metric can be expressed as:

$$I\left(\boldsymbol{M},\boldsymbol{C}^k\right) = H(\boldsymbol{M}) - H\left(\boldsymbol{M}|\boldsymbol{C}^k\right) = k - (k - l) = l,$$

and since the mutual information does not tend to zero as $k \to \infty$, the strong secrecy condition is not satisfied. On the other hand, the weak secrecy metric will be:

$$\frac{1}{k} I\left(\boldsymbol{M},\boldsymbol{C}^k\right) = \frac{1}{k}\left[H(\boldsymbol{M}) - H\left(\boldsymbol{M}|\boldsymbol{C}^k\right)\right] = \frac{k-(k-l)}{k} = \frac{l}{k},$$

and clearly tends to zero as $k \to \infty$, indicating that weak secrecy condition is satisfied.

From this example we conclude that some secrecy metrics are mathematically stronger than others. For instance, the mutual information metric is stronger than variational distance secrecy metric, while variational metric is stronger than weak secrecy metric as shown in [28].

11.3 Wyner's Wiretap Channel

Shannon's model is too pessimistic as it assumes that no noise has been introduced during transmission. Wyner introduced the so-called wiretap channel [29], now also known as a *degraded wiretap channel model*, in which Eve's channel is the degraded version of the Alice-to-Bob channel (main channel), as indicated in Fig. 11.6. Alice encodes the message $\boldsymbol{M}$ into a codeword $\boldsymbol{X}^n$ of length n and sends it over the noisy channel, represented by conditional probability density function (PDF) $f(y|x)$ toward the Bob. On the other hand, Eve observes the noisy version of the signal available to Bob. Therefore, the wiretap channel is a degraded channel represented by the conditional PDF $f(z|y)$. More formally, the discrete memoryless degraded wiretap channel is specified by input alphabet $\mathcal{X}$; two output alphabets $\mathcal{Y}$, $\mathcal{Z}$; and transition probabilities $f(y|x)$ and $f(z|y)$ such that joint distributions for main and wiretap channels are independent, that is, we can write:

$$f_{\boldsymbol{YZ}}(\boldsymbol{y}^n\boldsymbol{z}^n|\boldsymbol{x}^n) = \prod_{m=1}^{n} f(y_m|x_m) \prod_{m=1}^{n} f(z_m|y_m). \tag{11.8}$$

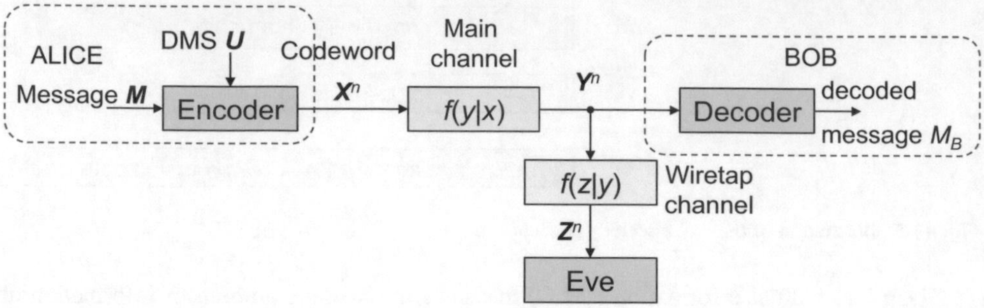

Fig. 11.6 Wyner's wiretap channel model. DMS, discrete memoryless source

Wyner suggested to use the *equivocation rate*, defined as $(1/n)H(M|\mathbf{Z}^n)$, instead of the entropy of the message $H(\mathbf{M})$. So the *secrecy condition* in Wyner's sense will be:

$$\frac{1}{n}H(\mathbf{M}) - \frac{1}{n}H(\mathbf{M}|\mathbf{Z}^n) = \frac{1}{n}I(\mathbf{M},\mathbf{Z}^n) \underset{n\to\infty}{\to} 0, \tag{11.9}$$

which is clearly the weak secrecy condition. In addition to secrecy condition, the *reliability condition* must be satisfied as well:

$$\Pr(\mathbf{M}_B \neq \mathbf{M}|\mathbf{Y}^n) \underset{n\to\infty}{\to} 0. \tag{11.10}$$

In other words, the probability that Bob's message is different from the message sent by Alice tends to zero as the codeword length $n \to \infty$. The channel codes to be used in this scenario must satisfy both reliability and secrecy conditions and are sometimes referred to as the *wiretap codes* [30]. For instance, LDPC, polar, and lattice codes can be used to design the wiretap codes. The (n,k) wiretap code $\mathcal{C}_n$ of rate $R = k/n$ is specified by [4, 30]: (i) the set of messages $\mathcal{M}$ of size 2^{nR}, (ii) the local random source $\mathcal{U}$ with distribution $f_{\mathcal{U}}$, (iii) the encoder performing the mapping $\mathcal{M} \times \mathcal{U} \to \mathcal{X}^n$ (mapping the message and a random realization of the local source into a codeword), and (iv) the decoder performing the following mapping $\mathcal{Y}^n \to \mathcal{M}\cup\{\text{Er}\}$ (mapping the received word into a message or an error Er).

The transmission rate R-equivocation rate R_{eq} pair (R,R_{eq}) is *achievable* in Wyner's sense for the sequence of codes $\{\mathcal{C}_n\}$ of code rate R if both reliability and secrecy conditions are satisfied, defined, respectively, as:

$$\begin{aligned} &\Pr(\mathbf{M}_B \neq \mathbf{M}|\mathcal{C}_n) = P_e(\mathcal{C}_n) \underset{n\to\infty}{\to} 0, \\ &\underbrace{\lim_{n\to\infty}\frac{1}{n}H(M|\mathbf{Z}^n\mathcal{C}_n)}_{ER(\mathcal{C}_n)} = \lim_{n\to\infty} ER(\mathcal{C}_n) \geq R_{eq}, \end{aligned} \tag{11.11}$$

where the first line represents the probability of error when wiretap code $\mathcal{C}_n$ is used, while the second term represents the equivocation rate (ER). The largest transmission rate at which both reliability and secrecy conditions are simultaneously satisfied is commonly referred to as the *secrecy capacity*. For any distribution f_x of $\mathcal{X}$ from set of distributions $\mathcal{P}(R \geq 0)$ for which $I(X,Y) \geq R$, Wyner has defined the merit function, which can be called a *secrecy rate*, as follows:

$$SR(R) = \sup_{f_x \in \mathcal{P}(R)} I(X,Y|Z). \tag{11.12}$$

Since random variables X, Y, Z form the Markov chain, we can apply the chain rule to obtain:

$$\begin{aligned} I(X,Y|Z) &= H(X|Z) - \underbrace{H(X|Y,Z)}_{H(X|Y)} = \\ &= I(X,Y) - I(X,Z), \end{aligned} \tag{11.13}$$

so that the alternative expression for $C(R)$ is obtained by:

$$SR(R) = \sup_{f_x \in \mathcal{P}(R)} [I(X,Y) - I(X,Z)]. \tag{11.14}$$

Wyner further proved the following *lemma* [29]: The secrecy rate *SR*(*R*), $0 \leq R \leq C_m$ (C_m is the capacity of the main channel) satisfies the following properties:

- The supremum in definitions above is in fact maximum, i.e., for each *R* there exists distribution from $\mathcal{P}(R)$ such that *I*(*X*,*Y*| *Z*) = *SR*(*R*).
- *SR*(*R*) is concave and continuous function of *R*.
- *SR*(*R*) is nonincreasing in *R*.
- *SR*(*R*) is upper bounded by C_m and lower bounded by C_m-C_e, where C_e is the capacity of the main-wiretap channel cascade; that is, C_m-$C_e \leq SR(R) \leq C_m$.

Wyner also proved the following *theorem* [29]: The achievable region $\mathcal{R}$ is determined by:

$$\mathcal{R} = \left\{ (R, R_{eq}) : 0 \leq R \leq C_m, 0 \leq ER \leq H(\boldsymbol{M}), RR_{eq} \leq SR(R)H(\boldsymbol{M}) \right\}. \tag{11.15}$$

The typical achievable region is provided in Fig. 11.7. As expected from the lemma, the achievable region is not convex, and there are two critical corner points: (C_m,H_m = *SR*(*R*)*H*(***M***)/C_m) and (C_M, *H*(***M***)). The *secrecy capacity* can now be defined as:

$$C_S = \max_{(R, H(\boldsymbol{M})) \in \mathcal{R}} R. \tag{11.16}$$

Leung-Yan-Cheong has shown in [31] that the secrecy capacity is equal to the lower bound in lemma, that is, $C_S = C_m$-C_e, when both main and wiretap channels are *weakly symmetric*. The channel $\left(\mathcal{X}, \mathcal{Y}, f_{\mathcal{Y}|\mathcal{X}}\right)$ is weakly symmetric when the rows of the channel transition matrix are permutations of each other, while the summations per column are independent of *y*. The binary symmetric channel (BSC) is a weakly symmetric channel. As an illustration, let us assume that both main and wiretap channels can be both modeled as binary symmetric channels, as shown in Fig. 11.8, but with two different error probabilities (*p* and *q*). From Chap. 4 we know that the capacity of BSC is given by C_m = 1-*h*(*p*), where *h*(*p*) is the binary entropy function $h(p) = -p \log p$-(1-*p*)log *p*. Given that cascade of two BSCs is also a BSC with equivalent error probability of *p* + *q*-2*pq*, the secrecy capacity for this case will be:

$$\begin{aligned} C_S &= C_m - C_e = 1 - h(p) - [1 - h(p + q - 2pq)] \\ &= h(p + q - 2pq) - h(p). \end{aligned}$$

If we instead use the *strong secrecy condition*, as it was done in [4], defined as follows:

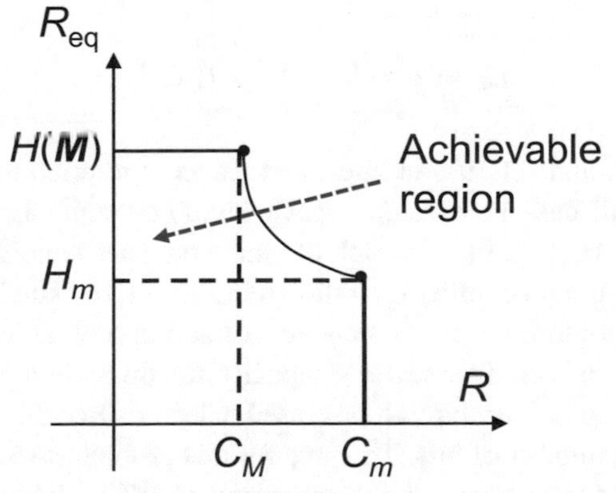

Fig. 11.7 Typical achievable region

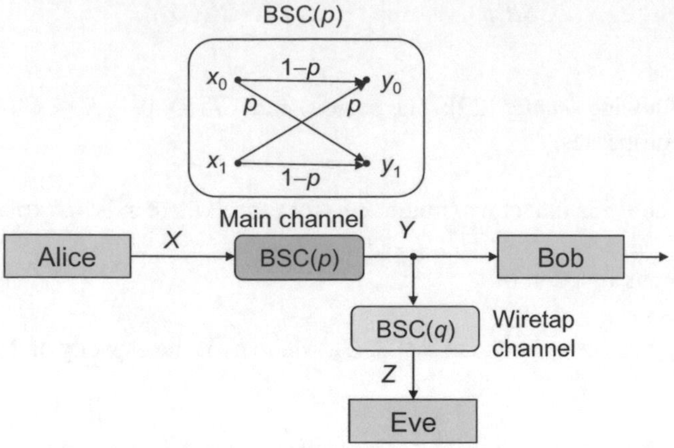

Fig. 11.8 The wiretap channel model when both main and wiretap channels are BSCs

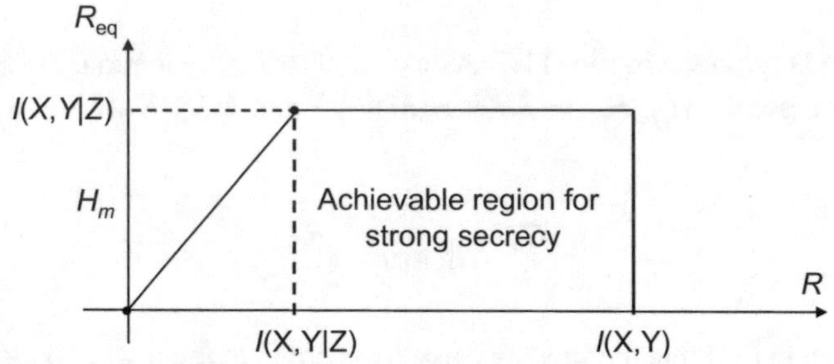

Fig. 11.9 The typical rate-equivocation achievable region in strong secrecy scenario

$$\lim_{n\to\infty}\left[H(M|\mathbf{Z}^n C_n) - nR_{eq}\right] \geq 0, \tag{11.17}$$

together with the same reliability condition, the corresponding achievable rate-equivocation region $\mathcal{R}'$ becomes convex and can be defined as a sequence of (n,nR) codes $\{C_n\}$ satisfying both strong secrecy constraint and reliability constraint and be determined by [4, 32]:

$$\mathcal{R}' = \left\{\left(R, R_{eq}\right) : 0 \leq R_{eq} \leq R \leq I(X,Y), 0 \leq R_{eq} \leq I(X,Y|Z)\right\}. \tag{11.18}$$

The typical shape of achievable region in strong secrecy scenario is provided in Fig. 11.9. As long as we are below the $I(X,Y|Z)$, it is possible to find codes satisfying both full transmission and equivocation rates. However, above $I(X,Y|Z)$ the equivocation rate saturates, while the transmission rate can be increased up to $I(X,Y)$. The secrecy capacity in strong secrecy scenario can be determined by [4]:

$$C_S = \max_{f_x}\left[I(X,Y) - I(X,Z)\right]. \tag{11.19}$$

For corresponding proofs of (11.18) and (11.19), an interested reader is referred to [4].

Before completing this section, we discuss the *wiretap codes* with more details. Let us observe an example of the wiretap channel model, mentioned in Wyner's paper [29], in which the main channel is noiseless, while the wiretap channel is the binary erasure channel (BEC) with erasure probability α as shown in Fig. 11.10. The BEC has two inputs x_0 and x_1 and three outputs y_0, y_1, and Er. The output symbol Er is called the erasure. A fraction α of the incoming bits get erased by the channel. The channel capacity of such channel is 1-α. The secrecy capacity for this wiretap channel model will be then C_S = 1-(1-α) = α. Therefore, out of n transmitted bits, $n\alpha$ bit will be erased. Clearly since the secrecy capacity is lower than 1, Eve is capable to learn the content of relevant number of bits. To solve for this problem, Alice can apply the *stochastic encoding* as follows. Let us assume that Alice wants to transmit a binary message $M\in\{0,1\}$, by encoding the message zero as codeword $\mathbf{x}_0^n = (x_{00}, x_{01}, \cdots, x_{0,n-1})$, while encoding the message one to codeword $\mathbf{x}_1^n = (x_{10}, x_{11}, \cdots, x_{1,n-1})$. Since these codewords get transmitted over the noiseless channel, Bob will receive the codeword correctly and will be able to conclude which message

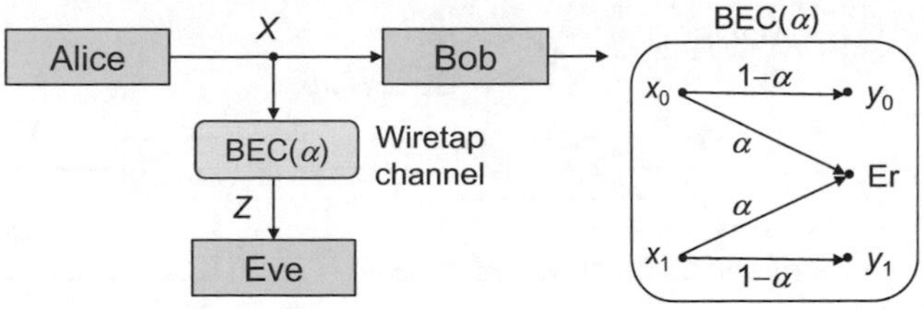

Fig. 11.10 An example of wiretap channel model in which the main channel is noiseless, while the wiretap channel is a BEC

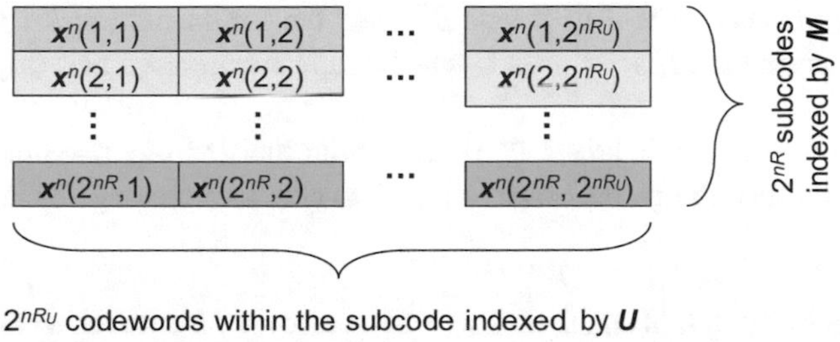

Fig. 11.11 Illustrating the nested structure of a wiretap code

was transmitted by simply taking the parity check. However, since Eve's channel is BEC and in her received word there will be $n\alpha$ erasures, so she will not be able to determine what was transmitted by taking the parity check. So the conditional entropy will be lower bounded by:

$$H(M|Z^n) \geq 1 - (1 - \alpha)^n,$$

and tends to the entropy of the message 1 as $n \to \infty$ (for sufficiently small erasure probability). On the other hand, the code rate of such stochastic encoding scheme is only $1/n$. However, this concept can be generalized as shown in Fig. 11.11. For each of 2^{nR} possible messages indexed by $\boldsymbol{M}$, there exist 2^{nR_U} codewords indexed by a local random generator $\boldsymbol{U}$. The set of codewords corresponding to a message forms a subcode of the wiretap code. Given that the main channel is noiseless, we need to ensure that the total code rate $R + R_U \leq 1$.

11.4 Broadcast Channel with Confidential Messages and Wireless Channel Secrecy Capacity

We first define the broadcast channel with confidential messages.

11.4.1 Broadcast Channel with Confidential Messages

Wyner's wiretap channel gets generalized and refined by Csiszár and Körner [32], and the corresponding model, now known as the *broadcast channel with confidential messages* (BCC), is provided in Fig. 11.12. The broadcast channel is assumed to be discrete and memoryless and characterized by input alphabet $\mathcal{X}$, output alphabets $\mathcal{Y}$ and $\mathcal{Z}$ (corresponding to Bob and Eve, respectively), and transition PDF $f(yz|x)$. So, the channel itself is modeled by a joint PDF for Bob's and Eve's observations, $f(\boldsymbol{yz}|\boldsymbol{x})$, given the channel input. Since the BCC is a memoryless channel, the PDF can be decomposed as:

$$f(\boldsymbol{y}^n\boldsymbol{z}^n|\boldsymbol{x}^n) = \prod_{m=1}^{n} f_{YZ|X}(y_m z_m|x_m). \tag{11.20}$$

In this scenario, Alice wishes to broadcast a common message $\boldsymbol{M}_c$ to both Bob and Eve and a confidential message $\boldsymbol{M}$ to Bob. The corresponding stochastic code $\mathcal{C}_n$ of codeword length n is composed of:

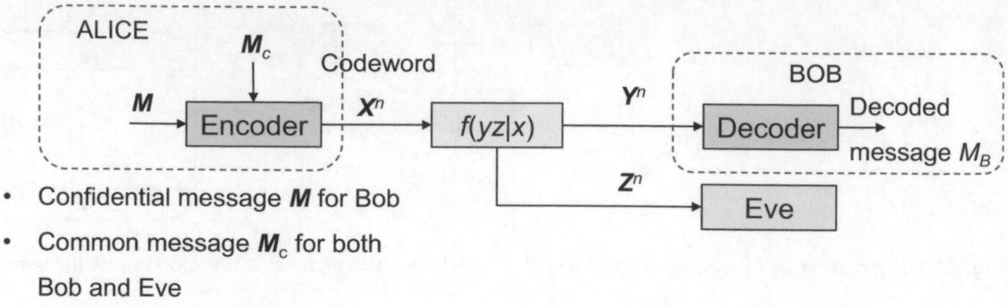

Fig. 11.12 The broadcast channel model with confidential messages (BCC)

- Two message sets: the common message set $\mathcal{M}_c$ of size 2^{nR_c} and the confidential message set $\mathcal{M}$ of size 2^{nR}.
- The encoding (stochastic) function e_n: $\mathcal{M}_c \times \mathcal{M} \to \mathcal{X}^n$, which maps the message pair $(\boldsymbol{m}_c,\boldsymbol{m}) \in \mathcal{M}_c \times \mathcal{M}$ into a codeword $\boldsymbol{x}^n \in \mathcal{X}^n$.
- The decoding functions d_n: $\mathcal{Y}^n \to \mathcal{M}_c \times \mathcal{M}$, $d_{E,n}$: $\mathcal{Z}^n \to \mathcal{M}_c$, with the first one mapping the observation vector $\boldsymbol{y}^n$ to a message pair $(\widehat{\boldsymbol{m}}_c, \widehat{\boldsymbol{m}})$, while the second one the observation z^n to common message estimate $\widehat{\boldsymbol{m}}_c$.

Similarly to Wyner's approach, the confidential message $\boldsymbol{M}$ secrecy with respect to Eve is measured in terms of the *equivocation rate* as follows:

$$ER(C_n) = \frac{1}{n} H(\boldsymbol{M}|\boldsymbol{Z}^n C_n). \tag{11.21}$$

The probability of error is defined in similar fashion as for Wyner's approach:

$$P_e(C_n) = \Pr\left(d_n(\boldsymbol{Y}^n) = \left(\widehat{\boldsymbol{M}}_c, \widehat{\boldsymbol{M}}\right) \neq (\boldsymbol{M}_c, \boldsymbol{M}) \text{or}\, d_{E,n}(\boldsymbol{Y}^n) = \widehat{\boldsymbol{M}}_c \neq \boldsymbol{M}_c \, |C_n\right) \tag{11.22}$$

We further request that the (stochastic) code C_n must satisfy both *reliability* and *secrecy constraints*, specified, respectively, as:

$$\lim_{n\to\infty} P_e(C_n) = 0, \lim_{n\to\infty} ER(C_n) \geq R_{eq}. \tag{11.23}$$

The rate tuple (R_c,R,R_{eq}) for the BCC is *achievable* if and only if there exists a (stochastic) wiretap code C_n of length n satisfying simultaneously both reliability and security constraints, defined above.

Csiszár and Körner proved the following *theorem* [32] (see also [4]): For the joint distribution f_{UVX} on $U \times V \times X$ that factorizes as $f_U f_{U|V} f_{X|V}$, the region of achievable rate tuples (R_c,R,R_{eq}), denoted as $\mathcal{R}$, is determined by:

$$\mathcal{R} = \bigcup_{f_U f_{V|U} f_{X|V}} \left\{ (R_c, R, R_{eq}) : \begin{array}{l} 0 \leq R_{eq} \leq R \\ R_{eq} \leq I(V,Y|U) - I(V,Z|U) \\ R + R_c \leq I(V,Y|U) + \min\left[I(U,Y), I(U,Z)\right] \\ 0 \leq R_c \leq \min\left[I(U,Y), I(U,Z)\right] \end{array} \right. \tag{11.24}$$

U and V are auxiliary random variables so that random variables U, V, X, and YZ form the Markov chain $U \to V \to X \to YZ$. Finally, the cardinalities of U and V maybe limited to $|\mathcal{U}| \leq |\mathcal{X}| + 3$ and $|\mathcal{V}| \leq |\mathcal{X}|^2 + 4|\mathcal{X}| + 3$.

Csiszár and Körner provided the following *corollary* [32] (see also [4]): *Secrecy capacity* is determined as the difference of mutual information for Alice-Bob and Alice-Eve links, when the rate of the common message is set to zero, that is:

$$C_s = \max_{\substack{f_{VX} \\ V\to X\to YZ}} \left[I(V,Y) - I(V,Z)\right], \tag{11.25}$$

where the maximization is performed over all possible joint distributions $f_{VX}(v,x)$ and V, X, and YZ form a Markov chain $V \to X \to YZ$.

Clearly, the secrecy capacity is strictly positive when Bob's channel is less noisy than Eve's channel, i.e., $I(X;Y) > I(X;Z)$. Namely, by setting $V = X$, the secrecy capacity expression becomes $C_s = \max_{f_X} [I(X, Y) - I(X, Z)]$, which is clearly strictly positive when $I(X;Y) > I(X;Z)$.

Similarly as in Wyner's wiretap channel case, the secrecy metric can be replaced with strong secrecy metric.

11.4.2 Wireless Channel Secrecy Capacity

Wireless communication scenario is illustrated in Fig. 11.13. Due to fading effects [33–35], the signal-to-noise ratios (SNRs) in both main and Eve's channels are time-varying. For certain time periods, the SNR in main channel is higher than that in Eve's channel and vice versa. Moreover, if Alice and Bob have access to Eve's channel state information (CSI), they can exploit this knowledge to get an advantage over Eve.

As indicated in [36, 37], the general goal is to maximize the transmission rate between Alice and Bob $R = H(M^k)/n$, while minimizing the Eve's equivocation rate $ER = H(M^k| Z^n)/H(M^k)$, wherein the error probability tends to zero. Secrecy capacity is defined as the maximum transmission rate R for which Eve's equivocation rate is equal to 1. The generic fading wireless BCC model is provided in Fig. 11.14. Clearly, both main and Eve's channels are simultaneously additive and multiplicative. The additive noise is a zero-mean Gaussian, with variance in main (Alice-Bob) channel being σ_m^2 and variance in Eve's channel being σ_e^2.

The multiplicative components are described by (i) main channel coefficients $\boldsymbol{h}_m{}^n = [h_m[1], \ldots, h_m[n]]$ and (ii) Eve's channel coefficients $\boldsymbol{h}_e{}^n = [h_e[1], \ldots, h_e[n]]$. For *ergodic-fading* channel model, channel coefficients change every channel use, that is, $h_m[i]$ and $h_e[i]$ $(i = 1,\ldots,n)$ are mutually independent. For *quasi-static* channel model, employed in the rest of this section, the channel coefficients do not change for duration of whole codeword, that is, $h_m = h_m[i]$, $h_e = h_e[i]$. The *block-fading* channel model describes the scenario between the two cases just described, in which the channel coefficients do not change for certain number of symbols, but still change many times during the codeword transmission. For additional details on wireless channel models, an interested reader is referred to [33–35]. The SNRs for main channel, denoted as ρ_m, and Eve's channel, denoted as ρ_e, assuming quasi-static fading channel are simply:

$$\rho_m = \underbrace{|h_m|^2}_{g_m} P/\sigma_m^2 = g_m P/\sigma_m^2,\ \rho_e = \underbrace{|h_e|^2}_{g_e} P/\sigma_e^2 = g_e P/\sigma_e^2, \tag{11.26}$$

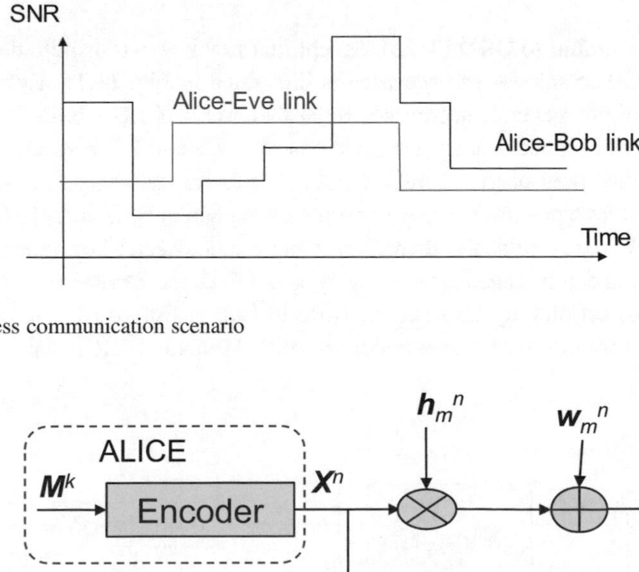

Fig. 11.13 Illustration of the wireless communication scenario

Fig. 11.14 The generic wireless BCC

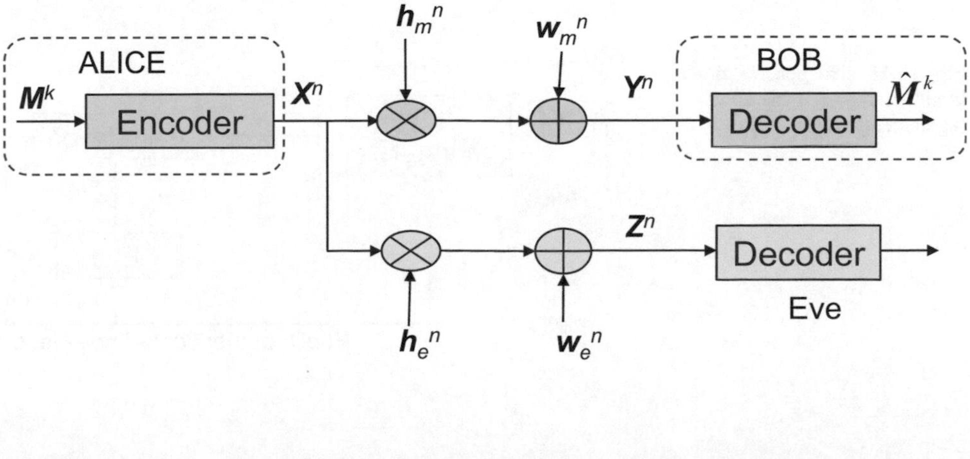

wherein P is the transmit power, g_m is the gain coefficient in main channel ($g_m = |h_m|^2$), and g_e is the gain coefficient in Eve's channel ($g_e = |h_e|^2$). Based on discussion related to (11.25), the instantaneous secrecy capacity for quasi-static fading channel is determined by:

$$C_S = \begin{cases} \log(1+\rho_m) - \log(1+\rho_e), \rho_m > \rho_e \\ 0, \text{ otherwise} \end{cases} \tag{11.27}$$

The *average secrecy capacity* for quasi-static channel can be obtained by averaging out the gain coefficients as follows:

$$\bar{C}_S = E_{g_m, g_e}\{C_s(g_m, g_e)\},$$
$$C_s(g_m, g_e) = \left[\log\left(1+\frac{g_m P}{\sigma_m^2}\right) - \log\left(1+\frac{g_e P}{\sigma_e^2}\right)\right]^+, \tag{11.28}$$

where $[x]^+ = \max(0,x)$. For formal proof, an interested reader is referred to [36].

Another relevant figure of merit related to secrecy, important in wireless communications, is the *probability of positive secrecy capacity* defined as:

$$P_S^+ = \Pr(C_S > 0). \tag{11.29}$$

11.5 Secret-Key Generation (Agreement) Protocols

Alice and Bob can employ the noisy channel to generate correlated random sequences and subsequently use a public authenticated error-free feedback channel to agree on a secret key [38–41], and this approach is commonly referred to as the *secret-key agreement* [3–6, 42]. Two types of models are typically considered for secret-key agreement [39]:

- *Source-type model*, in which terminals observe the correlated output of the source of randomness without having control of it.
- *Channel-type model*, in which one terminal transmits random symbols to other terminals using a broadcast channel. This scenario is similar to the wiretap channel model with feedback channel, which is an authenticated public noiseless channel.

Both of these models are very similar to QKD [8–15], except that raw key is transmitted over classical channel, while in QKD over the quantum channel. The *source-type scenario* is illustrated in Fig. 11.15. Even if the feedback message $\boldsymbol{F}$ is available to Eve, it is still possible to generate secure key such that $H(\boldsymbol{K}|\boldsymbol{Z}^n\boldsymbol{F})$ is arbitrarily close to $H(\boldsymbol{K})$. In this specific example, Alice, Bob, and Eve get correlated noisy observations, $\boldsymbol{X}^n$, $\boldsymbol{Y}^n$, and $\boldsymbol{Z}^n$, respectively. Alice and Bob are able to generate the common key $\boldsymbol{K}$ based on their observations and a set of feedback messages $\boldsymbol{F}$. Even if Eve's channel is better than Bob's channel, it is possible to achieve positive secrecy capacity, as explained by Maurer [38]. The corresponding procedure that Alice and Bob need to follow to determine the secret key, can be called secret-key agreement (generation) protocol. The rate at which secret key is generated can be called the same way as in QKD, the *secret-key rate* (SKR). If the protocols exploit the public messages sent in one direction only (from either Alice to Bob or Bob to Alice), the corresponding secret-key rate (SKR) is said to be achievable with *one-way communication*; otherwise, the SKR is said to be achievable with *two-way communication*.

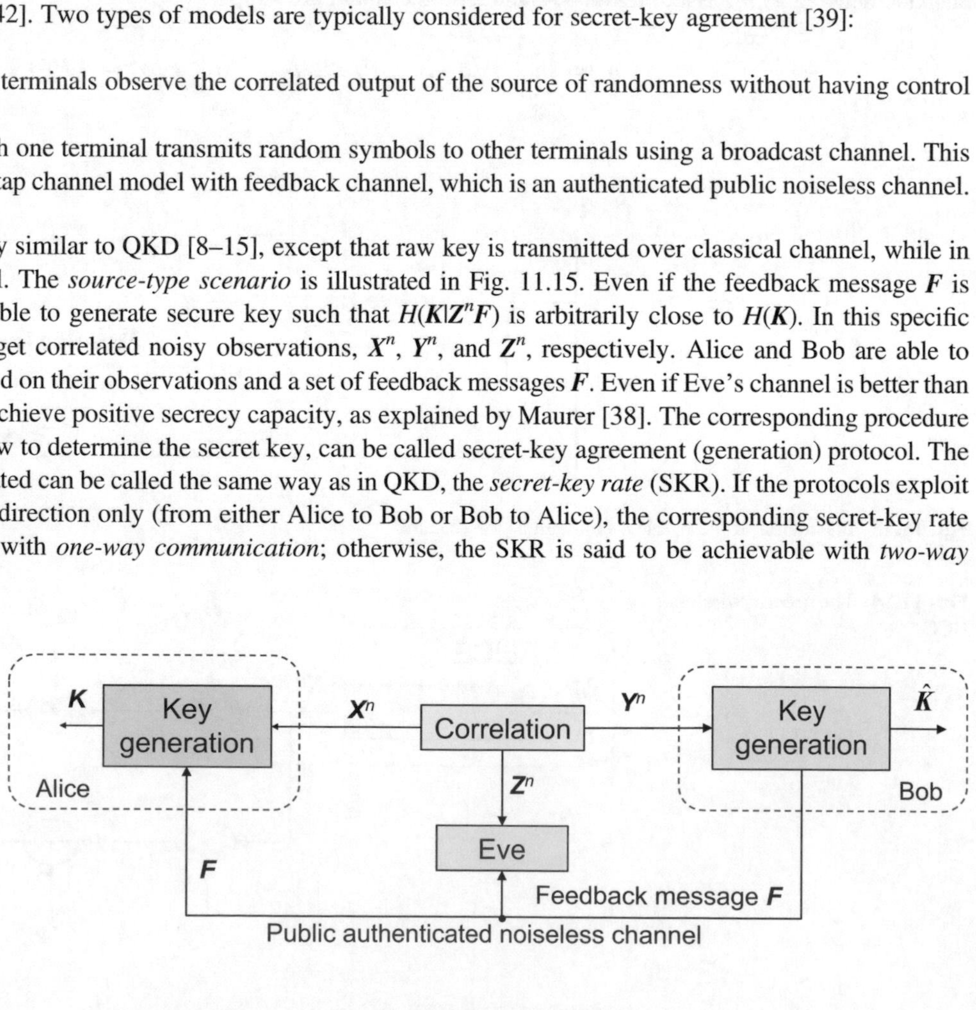

Fig. 11.15 The generic source model secret-key generation (agreement) scenario

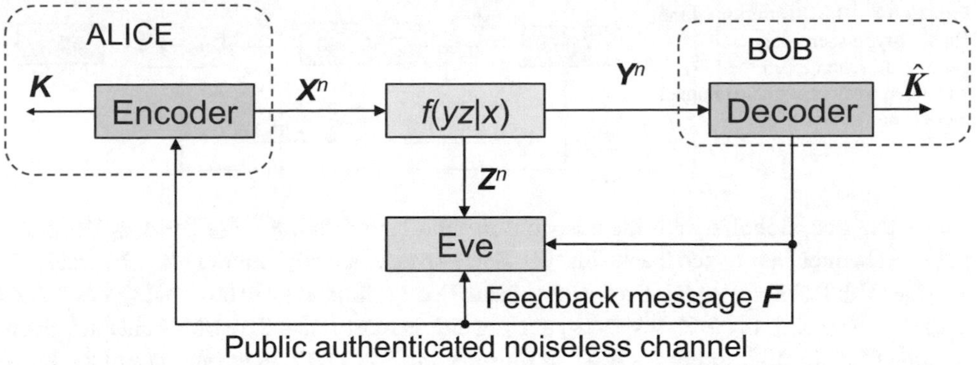

Fig. 11.16 The generic channel model secret-key generation (agreement) scenario

The generic *channel-type scenario* is described in Fig. 11.16, which is clearly the generalization (extension) of the Csiszár-Körner wiretap model. The channel is characterized by conditional PDF *f*(*yz*|*x*), and its input is controlled by Alice by generating *n* symbols $\boldsymbol{X}^n$. Bob and Eve observe the channel outputs and get $\boldsymbol{Y}^n$ and $\boldsymbol{Z}^n$ realizations. Additionally, Alice and Bob communicate over two-way authenticated noiseless channel of unlimited capacity. By exchanging the random symbols over the channel and feedback symbols $\boldsymbol{F}$ over the public channel, Alice and Bob are able to generate the secret key not known to Eve. To get the secret key, Alice and Bob apply a set of steps that can also be called the secret-key generation protocol.

This protocol should be properly designed such that in addition to *reliability condition*, defined as:

$$P_e = \Pr\left(\boldsymbol{K} \neq \widehat{\boldsymbol{K}}\right) \underset{n\to\infty}{\to} 0, \tag{11.30}$$

and *secrecy condition*, defined as:

$$L = I(\boldsymbol{K}, \boldsymbol{Z}^n \boldsymbol{F}) \underset{n\to\infty}{\to} 0, \tag{11.31}$$

the *uniformity condition* is also satisfied, which is defined as:

$$U = \left|H(\boldsymbol{K}) - \log\left\lceil 2^{nR}\right\rceil\right| \underset{n\to\infty}{\to} 0, \tag{11.32}$$

where *R* is the SKR. As before, the reliability condition tells us that the Alice and Bob must agree on secret key with the high probability. The security condition is important to ensure the key Alice and Bob finally agreed is indeed secret with respect to the Eve. Finally, the uniformity condition ensures that the secret key is uniformly distributed within the corresponding keyspace. The corresponding parameters defined above are known as *probability of error* (P_e), the *leakage of information* (*L*), and *uniformity of the keys* (*U*), respectively. We say that the secret-key rate *R* is *achievable* if there exist a sequence of secret-key generation protocols satisfying all three constraints as $n \to \infty$. The supremum of achievable SKRs is commonly referred to as the *secret-key capacity*, denoted here as C_{SK}. Given two-way communication over the authenticated public channel, it is difficult to derive an exact expression for C_{SK}; however based on [38, 39], it can be bounded from both sides as follows:

$$\max_{f_X} \max\left[I(X,Y) - I(X,Z), I(Y,X) - I(Y,Z)\right] \leq C_{SK} \leq \max_{f_X}\left[I(X,Y|Z)\right] \tag{11.33}$$

The upper-bound term indicates the secret-key capacity when Bob has access to Eve's observations. The lower-bound term max[*I*(*X*,*Y*)-*I*(*X*,*Z*)] indicates that direct reconciliation is employed, while the lower-bound term max[*I*(*Y*,*X*)-*I*(*Y*,*Z*)] indicates that reverse reconciliation is employed instead.

Let us now provide the following *example* due to Maurer [38] to illustrate how the noiseless feedback channel can help getting positive secrecy capacity even when Eve's channel is less noisy. We assume that both main and Eve's channels are BSCs with bit error probabilities p_m and p_e, respectively. Alice transmits bit *X* over BSC(p_m) to Bob, and this transmission was received by Eve over BSC(p_e). The errors introduced by channels can be represented by corresponding Bernoulli random variables E_m and E_e, added mod 2 to *X* to get $Y = X \oplus E_m$ at Bob's side and $Z = X \oplus E_e$ on Eve's side. Let us assume Bob's channel is noisier than Eve's channel, that is, $\Pr(E_m = 1) > \Pr(E_e = 1)$. Bob employs the feedback channel as follows. To send

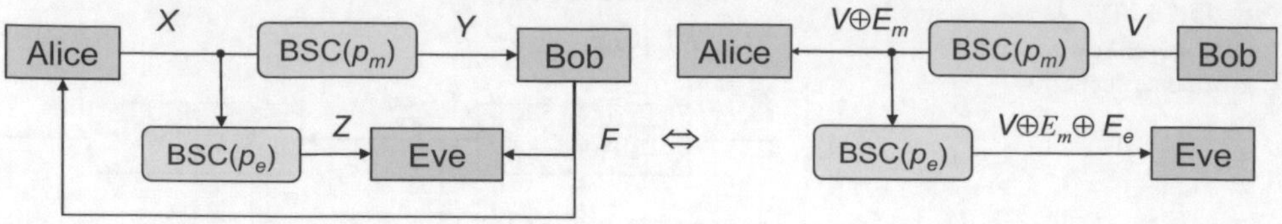

Fig. 11.17 The illustration of the channel-type secret-key generation scheme (on the left) and an equivalent wiretap channel model (on the right)

a bit V to Alice, Bob first adds the noisy observation Y, to obtain $F{=}V \oplus Y = V \oplus X \oplus E_m$, and sends it over the authenticated public channel as a feedback bit F. Since Alice already knows X, she adds it to the received F to obtain $F \oplus X = V \oplus Y \oplus X = V \oplus E_m$. On the other hand, Eve's estimation will be $F \oplus Z = V \oplus Y \oplus Z = V \oplus X \oplus E_m \oplus Z = V \oplus E_m \oplus E_e$, $\mathrm{Y} \oplus \mathrm{Z} = \mathrm{V} \oplus \mathrm{X} \oplus \mathrm{E_m} \oplus \mathrm{Z} = \mathrm{V} \oplus \mathrm{E_m} \oplus \mathrm{E_e}$, which is more noisy than Bob's channel given that equivalently V was sent over cascade of main BSC and Eve's BSC, as illustrated in Fig. 11.17. In the absence of feedback, the schemes become the wiretap channel model, and we have already shown that secrecy capacity of such channel will be:

$$C_S = [C_m - C_e]^+ = [1 - h(p_m) - (1 - h(p_e))]^+ = [h(p_e) - h(p_m)]^+. \tag{11.34}$$

So if $p_m > p_e$, clearly, the secrecy capacity will be zero. On the other hand, when feedback is used, based on Fig. 11.17 (right), we conclude that corresponding secrecy capacity will be:

$$\begin{aligned} C_S^{(F)} &= \left[C_m^{(F)} - C_e^{(F)}\right]^+ = [1 - h(p_m) - (1 - h(p_m + p_e - 2p_m p_e))]^+ \\ &= [h(p_m + p_e - 2p_m p_e) - h(p_m)]^+, \end{aligned} \tag{11.35}$$

which is strictly positive unless $p_m = 1/2$, $p_e = 0$.

To distill for a shorter shared secret key, Bennet [16] (see also [6, 10–15]) proposed the following three-phase approach:

- *Advantage distillation* phase, in which Alice and Bob out of a set of correlated observations select the positions over which they have an advantage over the Eve. In QKD this step is also known as a sifting procedure.
- *Information reconciliation* phase, in which these "advantage" observations are further processed to reconcile the discrepancies by employing the error correction.
- *Privacy amplification* phase, in which Alice and Bob perform further processing of corrected key to remove redundancy and get a uniformly distributed key sequence with no leakage to Eve, and this is done with the help of hash functions, as discussed in introductory part of this chapter.

We can extend these three phases with the 0th phase representing the *raw key transmission*, which in the source-type model can be called the *randomness sharing* [4]. Now we provide more details on both source-type and channel-type secret-key agreement scenarios.

11.5.1 Source-Type Secret-Key Generation

The *source model for secret-key agreement scenario* is provided in Fig. 11.18. Here we use the interpretation due to Bloch and Barros [4], which is slightly different than originally proposed in [39], but it has some similarities with [16]. This protocol is very similar to the *entanglement-assisted QKD* [13–15]. The main point of this scenario is discrete memoryless source (DMS), not controlled by involved parties, described by joint PDF f_{XYZ}, with three output ports $\mathcal{X}, \mathcal{Y}, \mathcal{Z}$ corresponding to Alice, Bob, and Eve, respectively. In other words, Alice, Bob, and Eve have access to correlated n-component realizations $\boldsymbol{X}^n$, $\boldsymbol{Y}^n$, and $\boldsymbol{Z}^n$, governed by the joint PDF. Additionally, Alice and Bob communicate over two-way public authenticated noiseless channel of unlimited capacity. This side channel does not contribute to the secret key, but helps Alice and Bob to distill for the shorter secure key. As mentioned above, this feedback channel is authenticated to prevent Eve tampering with the key. Finally, Alice and Bob employ the local sources of randomness, denoted as $\mathcal{R}_x$ and $\mathcal{R}_y$, respectively, to randomize the message they transmit. These sources of randomness are described by corresponding PDFs f_{Rx} and f_{Ry}. As already indicated above, the set of steps Alice and Bob need to follow constitute the *secret-key generation protocol*, which can also be called the key-distillation protocol (strategy) [4].

Fig. 11.18 The source model for secret-key generation (agreement) scenario

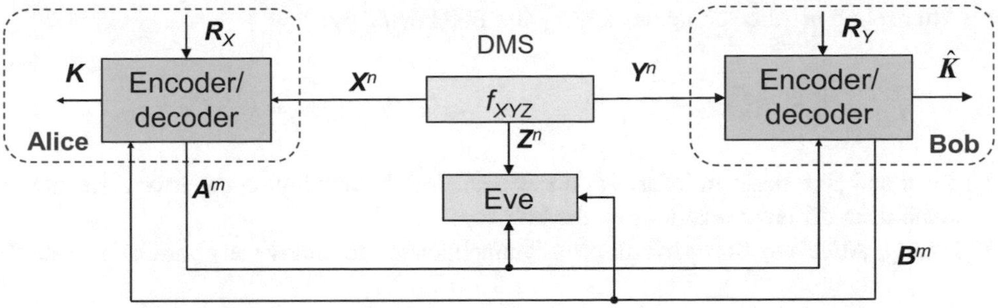

The generic *key-generation (key-distillation) protocol* P_n for *DMS*, denoted as $(\mathcal{XYZ}, f_{XYZ})$, of secret-key rate R and codeword length n, with m rounds over the authenticated public channel, employs:

The following *alphabets*:

- A key alphabet K of cardinality 2^{nR}.
- Alice alphabet A used to communicate over the authenticated public channel.
- Bob alphabet B used to communicate over the authenticated public channel.

The following sources of the local randomness:

- A source of local randomness for Alice $(\mathcal{R}_X, f_{RY})$.
- A source of local randomness for Bob $(\mathcal{R}_Y, f_{RY})$.

The following *encoding functions*:

- The m Alice encoding functions $e_i^{(A)}$:

$$e_i^{(A)} : \boldsymbol{X}^n \times \boldsymbol{B}^{i-1} \times \boldsymbol{R}_X \to \boldsymbol{A}, i \in \{1, 2, \cdots, m\} \tag{11.36}$$

- The m Bob encoding functions $e_i^{(B)}$:

$$e_i^{(B)} : \boldsymbol{Y}^n \times \boldsymbol{A}^{i-1} \times \boldsymbol{R}_Y \to \boldsymbol{B}, i \in \{1, 2, \cdots, m\} \tag{11.37}$$

The following *key-distillation functions*:

- Alice key-distillation function k_A, defined as the following mapping:

$$k_A : \boldsymbol{X}^n \times \boldsymbol{B}^m \times \boldsymbol{R}_X \to \boldsymbol{K} \tag{11.38}$$

- Bob key-distillation function k_B, defined as the following mapping:

$$k_B : \boldsymbol{Y}^n \times \boldsymbol{A}^m \times \boldsymbol{R}_Y \to \boldsymbol{K} \tag{11.39}$$

The key-distillation protocol P_n is composed of the following *steps*:

- Alice observes n realizations of the source $\boldsymbol{x}^n$, while Bob observes $\boldsymbol{y}^n$.
- Alice generates local random realization r_x, while Bob generates r_y.
- In round $i \in [1,m]$, Alice transmits a_i and Bob b_i by:

$$a_i = e_i^{(A)}\left(\boldsymbol{x}^n, \boldsymbol{b}^{i-1}, r_x\right), b_i = e_i^{(B)}\left(\boldsymbol{y}^n, \boldsymbol{a}^{i-1}, r_y\right); \tag{11.40}$$

- After round m Alice computes key k_A and Bob key k_B by:

$$\boldsymbol{k} = k_A(\boldsymbol{x}^n, \boldsymbol{b}^m, r_x), \hat{\boldsymbol{k}} = k_B(\boldsymbol{y}^n, \boldsymbol{a}^m, r_y). \tag{11.41}$$

- Alice and Bob perform information reconciliation, by employing the error correction, to correct for the discrepancies in symbols at different positions on the raw keys.
- Finally, Alice and Bob perform privacy amplification to remove any correlation with the Eve.

At the end of protocol P_n, Alice and Bob end with the common key $\boldsymbol{k}_c = \widehat{\boldsymbol{k}}_c$, which is secret with respect to the Eve. This source-type key-generation protocol P_n is evaluated in terms of:

- The *average probability of error*:

$$P_e(P_n) = \Pr\left(\boldsymbol{K} \neq \widehat{\boldsymbol{K}}|P_n\right), \tag{11.42}$$

- The *information leakage to the Eve*:

$$L(P_n) \triangleq \mathrm{I}(\boldsymbol{K}; \boldsymbol{Z}^{\mathrm{n}}\boldsymbol{A}^{\mathrm{r}}\boldsymbol{B}^{\mathrm{r}}|\mathrm{P_n}), \tag{11.43}$$

- The *uniformity of the key*:

$$U(P_n) \triangleq \log\left\lceil 2^{nR} \right\rceil - \mathrm{H}(\boldsymbol{K}|\mathrm{P_n}). \tag{11.44}$$

Similarly as before, we can define both strong and weak SKR. A *strong secret-key rate R* is achievable for a source/channel model if there exist a sequence of key-distillation protocols $\{P_n\}$ of length $n = 1,2,\ldots$ such that the reliability, secrecy, and uniformity conditions for $\{P_n\}$ are simultaneously satisfied; in other words the following constraints are valid:

$$P_e(P_n) \underset{n\to\infty}{\to} 0, L(P_n) \underset{n\to\infty}{\to} 0, U(P_n) \underset{n\to\infty}{\to} 0. \tag{11.45}$$

On the other hand, a *weak secret-key rate R* is achievable for a source/channel model if there exist a sequence of key-distillation protocols $\{P_n\}$ of length $n = 1,2,..$ such that the reliability constraint, *weak secrecy* constraint, and *weak uniformity constraint* are simultaneously satisfied. The reliability constraint is the same as in strong SKR case, while weak secrecy and uniformity constraints are defined, respectively, as follows:

$$\frac{L(P_n)}{n} \underset{n\to\infty}{\to} 0, \quad \frac{U(P_n)}{n} \underset{n\to\infty}{\to} 0. \tag{11.46}$$

The corresponding keys are weak secret keys. Given that we introduced both weak and strong secrecy constraints, we can also define the weak and strong secret-key capacity. The *weak secret-key capacity* of a source-type model with joint PDF f_{XYZ} is defined as:

$$C_{SK}^{(\mathrm{weak})} \triangleq \sup\{\mathrm{R} : \mathrm{R\ is\ an\ achievable\ weak\ SKR}\} \tag{11.47}$$

On the other hand, the *strong secret-key capacity* of a source-type model with joint PDF f_{XYZ} is defined as:

$$C_{SK}^{(\mathrm{strong})} \triangleq \sup\{\mathrm{R} : \mathrm{R\ is\ an\ achievable\ strong\ SKR}\} \tag{11.48}$$

The following *theorem* has been proved *by Maurer*, *Ahlswede*, and *Csiszár* [38, 39] (see also [4]): The weak secret-key capacity of a source-type model (XYZ, f_{XYZ}) satisfies the following lower and upper bounds:

$$\max\left(I(X;Y)-I(X;Z), I(Y;X)-I(Y;Z)\right) \leq C_{SK}^{(\text{weak})} \leq \min\left(I(X;Y), I(X,Y|Z)\right) \tag{11.49}$$

The lower bound can also be rewritten as $I(X;Y) - \min(I(X;Z), I(Y;Z))$. Moreover, for the one-way communication, the following SKR is achievable [38, 39]:

$$C_{SK}^{(\text{one-way communication})} = \max\left(I(X;Y)-I(X;Z), I(Y;X)-I(Y;Z)\right) \tag{11.50}$$

Proof. As discussed above, the lower bound corresponds to the Csiszár-Körner wiretap channel model with a feedback. We have also shown an example related to Fig. 11.17 that this channel model can be equivalently represented as Wyner's channel model, by following the similar procedure explained in text related to Fig. 11.17, which starts with Alice observing the channel realization x and adding uniformly generated integer u mod-$|\mathcal{X}|$. We know that the secrecy capacity of Wyner's channel model is higher than $I(X,Y)$-$I(X,Z)$, that is, we can write:

$$C_{SK}^{(\text{weak})} \geq I(X;Y) - I(X;Z). \tag{11.51}$$

Since the communication over authenticated public channel is two-way, the roles of Alice and Bob can be reversed by applying the reverse reconciliation so that the secret-key capacity is now lower limited by:

$$C_{SK}^{(\text{weak})} \geq I(Y;X) - I(Y;Z). \tag{11.52}$$

By combining these two inequalities, we obtain the lower bound of secret-key capacity as follows:

$$C_{SK}^{(\text{weak})} \geq \max\left(I(X;Y)-I(X;Z), I(Y;X)-I(Y;Z)\right). \tag{11.53}$$

To determine the upper bound, we observe the scenario in which Bob has access to Eve's observations and can create new random variably $Y' = YZ$. Since X, Y', and Y form the Markov chain, we can apply the chain rule to obtain:

$$\begin{aligned} I(X,Y|Z) &= \underbrace{I(X,YZ)}_{I(X,Y)+I(X,Z|Y)} - I(X,Z) = I(X,Y) + \underbrace{I(X,Z|Y)}_{0} - I(X,Z) \\ &= I(X,Y) - I(X,Z), \end{aligned} \tag{11.54}$$

which is clearly the upper bound of secrecy capacity. Moreover, from this equation we conclude that in this scenario $I(X,Y|Z) \leq I(X,Y)$. However, since the participants do not have control of DMS, there are realizations for which $I(X,Y) > I(X,Y|Z)$. Let us take the following example due to Maurer [38]. When X, Y are binary independent with $\Pr(X=0) = \Pr(Y=0) = 1/2$ and Z is formed by $Z = X \oplus Y$, we have that $I(X,Y) = 0$ and $I(X,Y|Z) = H(Z) = 1$. So to account for both scenarios, our upper bound becomes:

$$C_{SK}^{(\text{weak})} \leq \min\left(I(X,Y), I(X,Y|Z)\right). \tag{11.55}$$

11.5.2 Channel-Type Secret-Key Generation

The *channel model for secret-key agreement scenario* is provided in Fig. 11.19. Here we again use the interpretation due to Bloch and Barros [4] for consistency. In this scenario, Alice provides the input X to the discrete memoryless channel (DMC) described by the conditional PDF $f_{YZ|X}$. Bob and Eve do not have any control of the channel outputs; they can just observe random realizations at corresponding output ports. We again assume that Alice and Bob can communicate over authenticated

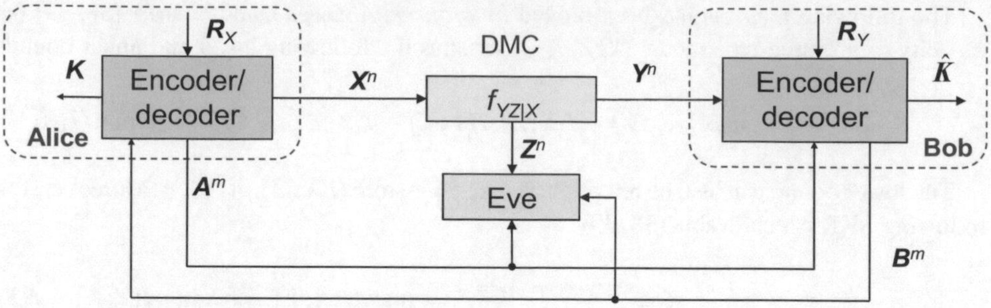

Fig. 11.19 The channel model for secret-key generation (agreement) scenario

public noiseless channel to which Eve can have access. As before, we assume that Alice and Bob employ the local sources of randomness, denoted as $\mathcal{R}_x$ and $\mathcal{R}_y$, respectively, to randomize their communication.

In analogy with the key generation protocols for DMS, we can define the corresponding generic protocol for DMC, which is briefly described in text related to Fig. 11.1. This scenario is very similar to *weak coherent states-based QKD* [13–15]. The generic *key-generation (key-distillation) protocol P_n for DMC*, described by the conditional PDF $f_{YZ|X}$, of secret-key rate R and codeword length n, with m rounds over the authenticated public noiseless channel and n transmissions over DMC, employs:

The following *alphabets*:

- A key alphabet K of cardinality 2^{nR}.
- Alice alphabet A used to communicate over the authenticated public channel.
- Bob alphabet B used to communicate over the authenticated public channel.

The following sources of the local randomness:

- A source of local randomness for Alice ($\mathcal{R}_X, f_{RY}$).
- A source of local randomness for Bob (w, f_{RY}).

The following *encoding functions* to be used over the public channel:

- The m Alice encoding functions $e_i^{(A)}$:

$$e_i^{(A)} : \boldsymbol{B}^{i-1} \times \boldsymbol{R}_X \to \boldsymbol{A}, i \in \{1, 2, \cdots, m\} \tag{11.56}$$

- The m Bob encoding functions $e_i^{(B)}$:

$$e_i^{(B)} : \boldsymbol{Y}^j \times \boldsymbol{A}^{i-1} \times \boldsymbol{R}_Y \to \boldsymbol{B}, i \in \{1, 2, \cdots, m\} \tag{11.57}$$

The index j is used to denote the number of available DMC channel outputs up to now.

The following *encoding functions* to be used over the DMC:

$$e_l^{(DMC)} : \boldsymbol{B}^k \times \boldsymbol{R}_X \to \boldsymbol{X}, l \in \{1, 2, \cdots, n\} \tag{11.58}$$

The index k is used to denote the number of bits received from Bob over the public channel up to now.

The following *key-distillation functions*:

- Alice key-distillation function k_A, defined as the following mapping:

$$k_A : \boldsymbol{X}^n \times \boldsymbol{B}^m \times \boldsymbol{R}_X \to \boldsymbol{K} \tag{11.59}$$

- Bob key-distillation function k_B, defined as the following mapping:

$$k_B : \boldsymbol{Y}^n \times \boldsymbol{A}^m \times \boldsymbol{R}_Y \to \boldsymbol{K} \tag{11.60}$$

The key-distillation protocol P_n is composed of the following *steps*:

- Alice generates local random realization r_x, while Bob generates r_y.
- In round $i \in [1,m]$, Alice transmits a_i and Bob b_i by:

$$a_i = e_i^{(A)}\left(\boldsymbol{b}^{i-1}, r_x\right), b_i = e_i^{(B)}\left(\boldsymbol{a}^{i-1}, r_y\right); \tag{11.61}$$

- In the l-th DMC transmission interval, Alice transmits symbol x_l over DMC:

$$x_l = e_l^{(DMC)}\left(\boldsymbol{b}^k, r_x\right), \tag{11.62}$$

Bob receives y_l and Eve receives z_l. (Once more, k denotes the number of bits received from Bob over the public channel up to now.) The transmission over DMC takes place when channel conditions are favorable.

- After all DMC and public channel transmissions are completed, Alice computes key k_A and Bob key k_B by:

$$\boldsymbol{k} = k_B(\boldsymbol{x}^n, \boldsymbol{b}^m, r_x), \widehat{\boldsymbol{k}} = k_B\left(\boldsymbol{y}^n, \boldsymbol{a}^m, r_y\right). \tag{11.63}$$

- Alice and Bob perform information reconciliation, by employing the error correction to correct for the errors introduced by the DMC and Eve.
- Finally, Alice and Bob perform information reconciliation to remove any correlation with the Eve.

At the end of protocol P_n, Alice and Bob end up with the common key $\boldsymbol{k}_c = \widehat{\boldsymbol{k}}_c$, which is not known to the Eve. This channel-type key-generation protocol P_n is evaluated in terms of probability of error, information leakage to the Eve, and uniformity of the key.

The *weak secret-key capacity for DMC-type scenario* can be defined as follows:

$$C_{SK}^{(\text{weak})} \triangleq \sup\{\text{R} : \text{R isanachievableweak SKR}\} \tag{11.64}$$

On the other hand, the *strong secret-key capacity* of a channel-type model is defined as:

$$C_{SK}^{(\text{strong})} \triangleq \sup\{\text{R} : \text{R isanachievablestrong SKR}\} \tag{11.65}$$

The following *theorem* has been proved *by Ahlswede* and *Csiszár* in [39] (see also [4]): The weak secret-key capacity of a channel type model satisfies the following lower and upper bounds:

$$\begin{aligned} &\max\left\{\max_{f_X}[I(X;Y) - I(X;Z)], \max_{f_X}[I(Y;X) - I(Y;Z)]\right\} \leq \\ &\leq C_{SK}^{(\text{weak})} \leq \max_{f_X} \min\left(I(X;Y), I(X, Y|Z)\right) \end{aligned} \tag{11.66}$$

By employing the information reconciliation and privacy amplification steps, the strong secrecy condition can be obtained from the weak secrecy condition.

11.6 Coding for Physical-Layer Security Systems

We have already discussed the concept of *nested code* in Sect. 11.3. In this coding scenario, the message $\boldsymbol{M}$ is used to select the subcode (n, nR). After that the source of local randomness $\boldsymbol{U}$ is used to select the codeword within the subcode. For each message there are 2^{nRU} possible codewords selected at random. In this section we discuss (i) coding for weak secrecy, (ii) coding for strong secrecy, and (iii) information reconciliation.

11.6.1 Coding for Weak Secrecy Systems

Ozarow and Wyner considered in [43] the wiretap channel model in which the main channel is noiseless, while the wiretap (Eve's) channel is noisy. Clearly, assuming the binary transmission, the secrecy of this wiretap channel model is 1-C_w, where C_w is the capacity of the wiretap channel. Let the binary linear block code of code rate R_0 be denoted by $\mathcal{C}_0$. For the parity-check matrix $\boldsymbol{H}$, each codeword $\boldsymbol{x}^n$ satisfies the parity-check equations: $\boldsymbol{Hx}^n = \mathbf{0}$. For each message $\boldsymbol{m}$, we define the *coset* $C_{\boldsymbol{m}}$ as follows:

$$C_{\boldsymbol{m}} = \{\boldsymbol{x}^n : \boldsymbol{Hx}^n = \boldsymbol{m}\}. \tag{11.67}$$

To encode Alice randomly chooses the n-tuple $\boldsymbol{x}^n$ from the coset $C_{\boldsymbol{m}}$. Since there are 2^n n-tuples and the cardinality of the coset $C_{\boldsymbol{m}}$ is $2nR_0$, the number of possible cosets is 2^n/ $2nR_0$t> = $2n(1\text{-}R_0)$, indicating that the code rate of this scheme is 1-R_0. Clearly, when $\mathcal{C}_0$ is the capacity-achieving code, the equivocation and wiretap rate can be achieved arbitrarily close to the secrecy capacity. The nested codes get extended to multiterminal scenarios in [44].

The *nested wiretap code idea* gets extended to noisy main channel in [45] as follows. Let $\boldsymbol{H}$ be an (1-R_{12})$n \times n$ parity-check matrix of full rank representing the code $\mathcal{C}$, composed of two submatrices $\boldsymbol{H}_1$ and $\boldsymbol{H}_2$:

$$\boldsymbol{H} = \begin{bmatrix} \boldsymbol{H}_1 \\ \boldsymbol{H}_2 \end{bmatrix}, \tag{11.68}$$

where $\boldsymbol{H}_1$ is (1-R_1)$n \times n$ submatrix representing the code $\mathcal{C}_1$, with $R_1 > R_{12}$, and $\boldsymbol{H}_2$ is $Rn\times n$ submatrix. Clearly, (1-R_1) n + Rn = (1-R_{12})n so that we can write $R = R_1$-R_{12}. The linear block code (LBC) $\mathcal{C}$ is a subcode of the LBC $\mathcal{C}_1$, while the distinct cosets of $\mathcal{C}$ in $\mathcal{C}_1$ represent the partition of $\mathcal{C}$. Alice wants to send the nR-bit message $\boldsymbol{M}$ to Bob. To do so she randomly selects n-tuple $\boldsymbol{X}^n$ from the following coset:

$$\mathcal{C}_{\boldsymbol{M}} = \left\{\boldsymbol{X}^n : \begin{bmatrix} \boldsymbol{H}_1 \\ \boldsymbol{H}_2 \end{bmatrix} \boldsymbol{X}^n = \begin{bmatrix} \mathbf{0} \\ \boldsymbol{M} \end{bmatrix}\right\}. \tag{11.69}$$

To decode transmitted message, Bob performs the syndrome decoding as follows. By using the observation vector $\boldsymbol{Y}^n$, he obtains the estimate of transmitted n-tuple, denoted as $\widehat{\boldsymbol{X}}^n$, by running the corresponding decoding algorithm to satisfy the parity-check equations $\boldsymbol{H}_1\widehat{\boldsymbol{X}}^n = \mathbf{0}$. Finally, Bob computes the message estimate by $\boldsymbol{H}_2\widehat{\boldsymbol{X}}^n = \widehat{\boldsymbol{M}}$. The wiretap code just described can be denoted by $\mathcal{C}_n$.

Clearly, we can partition $\mathcal{C}_1$, using the cosets of $\mathcal{C}$ into 2^{nR} disjoint subsets, and thus represent the generalization of Wyner's construction, since $\mathcal{C}_1$ is equivalent to $\mathcal{C}_0$. Assuming that $\mathcal{C}_1$ comes from the capacity-achieving channel code over the main channel and $\mathcal{C}$ forms the channel capacity-achieving code over the wiretap channel, authors in [45] (see also [46]) were able to show that this coset encoding scheme satisfies both the reliability condition ($P_e \rightarrow 0$ as $n \rightarrow \infty$) and the secrecy capacity condition $(1/n)I(\boldsymbol{M},\boldsymbol{Z}^n) \rightarrow 0$ as $n \rightarrow \infty$. Because both codes $\mathcal{C}_1$ and $\mathcal{C}$ are capacity achieving, the probability of error condition is satisfied. To prove the weak secrecy requirement, let us apply the chain rule for $I(\boldsymbol{M},\boldsymbol{X}^n,\boldsymbol{Z}^n)$ in two different ways [27]:

$$I(\boldsymbol{M},\boldsymbol{Z}^n) + I(\boldsymbol{X}^n,\boldsymbol{Z}^n|\boldsymbol{M}) = I(\boldsymbol{X}^n,\boldsymbol{Z}^n) + \underbrace{I(\boldsymbol{M},\boldsymbol{Z}^n|\boldsymbol{X}^n)}_{=0} = I(\boldsymbol{X}^n,\boldsymbol{Z}^n), \tag{11.70}$$

wherein $I(\boldsymbol{M},\boldsymbol{Z}^n|\boldsymbol{X}^n) = 0$, since $\boldsymbol{M} \rightarrow \boldsymbol{X}^n \rightarrow \boldsymbol{Z}^n$ is a Markov chain. We can rewrite the previous equation as:

$$\begin{aligned} I(M, Z^n) &= I(X^n, Z^n) - \underbrace{I(X^n, Z^n|M)}_{H(X^n|M)-H(X^n|M,Z^n)} = \\ &= I(X^n, Z^n) - H(X^n|M) + H(X^n|M, Z^n). \end{aligned} \tag{11.71}$$

Because $H(X^n|M) = nR_{12}$, $I(X^n,Z^n) \leq nC_w$, and from Fano's inequality [27, 46] $H(X^n| Z^n, M) \leq h(P_e) + P_e nR_{12}$, the previous equation becomes the following inequality:

$$I(M, Z^n) \leq nC_w - nR_{12} + h(P_e) + P_e nR_{12}, \tag{11.72}$$

and from the weak secrecy requirement definition, we obtain:

$$\lim_{n\to\infty} \frac{1}{n} I(M, Z^n) = \lim_{n\to\infty} \left[C_w - R_{12} + \frac{1}{n} h(P_e) + P_e R_{12} \right] = 0, \tag{11.73}$$

where we employed the reliability condition ($P_e \to 0$ as $n \to \infty$).

In incoming subsections, we describe several codes' designs capable of achieving the weak secrecy capacity.

11.6.1.1 Two-Edge-Type LDPC Coding

The two-edge-type LDPC coding [47] is the most natural way to implement the nested coding. It can be considered as an instance of generalized LDPC coding [48–54]. As discussed in Chap. 9, we can describe an LDPC code in terms of the *degree distribution polynomials* $\mu(x)$ and $\rho(x)$, for the variable node (v-node) and the check node (c-node), respectively, as follows:

$$\mu(x) = \sum_{d=1}^{d_v} \mu_d x^{d-1}, \qquad \rho(x) = \sum_{d=1}^{d_c} \rho_d x^{d-1}, \tag{11.74}$$

where μ_d and ρ_d denote the fraction of the edges that are connected to degree-d v-nodes and c-nodes, respectively, and d_v and d_c denote the maximum v-node and c-node degrees, respectively. For sufficiently large codeword length n, the most of parity-check matrices in a $\{\mu,\rho\}$ ensemble of LDPC codes of the same girth exhibit similar performance under the log-domain sum-product algorithm (SPA) described in Chap. 9. Now we can define the *two-edge LDPC code* ensemble as a $\{\mu_1, \mu_2, \rho_1, \rho_2\}$ two-edge-type LDPC coding ensemble of codeword length n, wherein the variable nodes are connected to μ_1 (μ_2) check nodes of type 1 (2) and all type 1 (2) check nodes have degree ρ_1 (ρ_2) [47]. The parity-check matrix of two-edge-type LDPC codes has the form introduced above, namely, $\boldsymbol{H} = \begin{bmatrix} \boldsymbol{H}_1 \\ \boldsymbol{H}_2 \end{bmatrix}$, where $\boldsymbol{H}_1$ ($\boldsymbol{H}_2$) define connections between the variable nodes and the check nodes of type 1 (2). Since the two-edge LPDC codes represent the instances of LBC codes described above, they achieve the weak secrecy capacity. Let $\boldsymbol{G}$ be the generator matrix that corresponds to the parity-check matrix $\boldsymbol{H}$. As discussed above, each coset C_m of C in C_1 contains solutions of $\boldsymbol{xH}^{\mathrm{T}} = [\boldsymbol{xH}_1^{\mathrm{T}}\ \boldsymbol{xH}_2^{\mathrm{T}}] = [\boldsymbol{0}\ \boldsymbol{m}]$ for message $\boldsymbol{m}$. Let $\boldsymbol{G}^* = [\boldsymbol{g}_1^{\mathrm{T}}\ \boldsymbol{g}_2^{\mathrm{T}} \ldots \boldsymbol{g}_{nR}^{\mathrm{T}}]^{\mathrm{T}}$ be the submatrix with rows being linearly independent and when augmented with $\boldsymbol{G}$ to form the basis for code with parity-check matrix $\boldsymbol{H}_1$. To encode nR-bit message $\boldsymbol{m}$, Alice randomly chooses nR_{12}-bit auxiliary message $\boldsymbol{m}'$. By calculating:

$$\boldsymbol{x} = [\boldsymbol{m}\ \boldsymbol{m}'] \begin{bmatrix} \boldsymbol{G}^* \\ \boldsymbol{G} \end{bmatrix}, \tag{11.75}$$

Alice randomly chooses $\boldsymbol{x}$ among all possible solutions of $\boldsymbol{xH}^{\mathrm{T}} = [\boldsymbol{0}\ \boldsymbol{m}]$. Bob applies the decoding procedure described in the paragraph below Eq. (11.69) to estimate the transmitted message.

11.6.1.2 Punctured LDPC Coding

The puncturing in communication systems is typically used to change the code rate of original code. The punctured bits are not transmitted. So the key idea of using puncturing in secrecy coding is to assign the punctured bits to message bits, denoted by $\boldsymbol{m}$ [55, 56], which is illustrated in Fig. 11.20.

Further auxiliary dummy bits get generated, denoted as $\boldsymbol{m}'$, and then parity bits get generated by LDPC encoding to get $\boldsymbol{p}$. Alice then sends the codeword $\boldsymbol{x}' = [\boldsymbol{m}'\ \boldsymbol{p}]$ to Bob over the noisy channel, and Bob performs log-domain SPA trying to

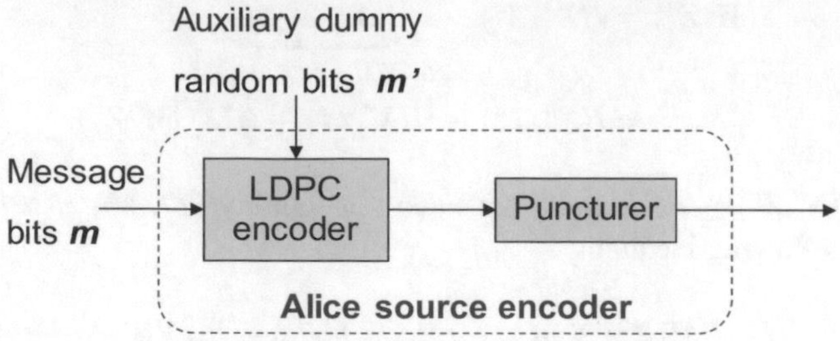

Fig. 11.20 The punctured LDPC encoding for wiretap channels

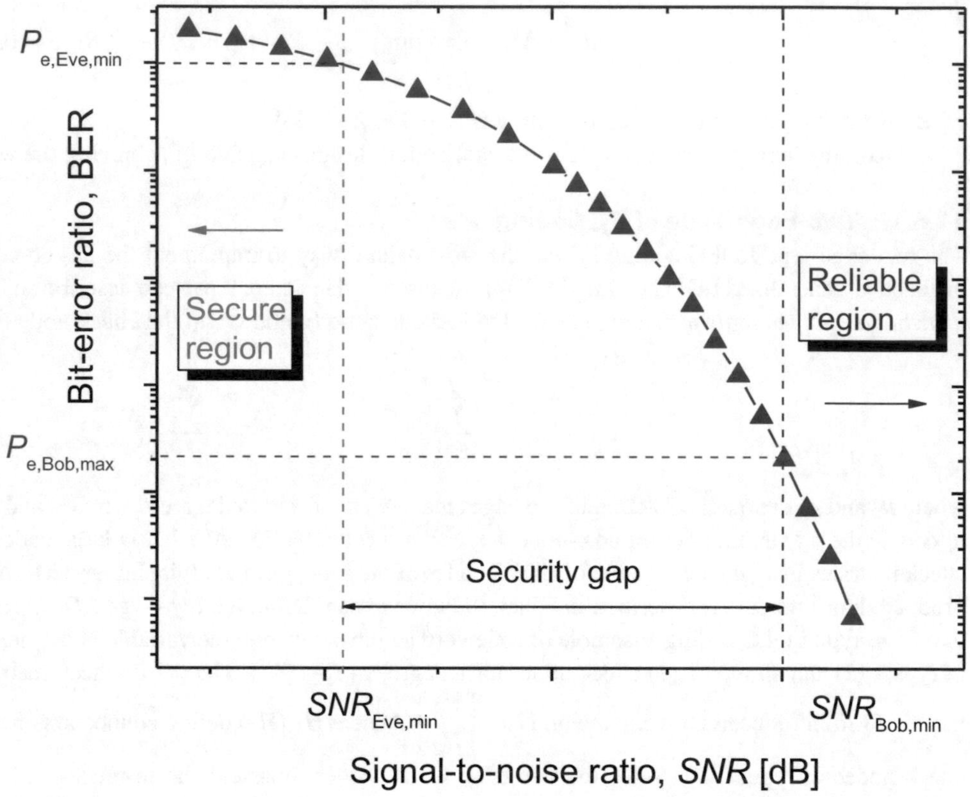

Fig. 11.21 Defining the secrecy gap

reconstruct the punctured (message) bits. Clearly, the number of message bits k_m is smaller than the number of information bits k used in LDPC encoder of rate $R = k/n$. The number of dummy bits is k-k_m. Therefore, the secrecy rate $R_s = k_m/n$ is lower than the LDPC code rate R. The key problem here is to determine the optimized puncturing distribution, which is subject of investigation in [56]. For Gaussian wiretap channel, the *reliability region* is defined by the smallest signal-to-noise ratio ($SNR_{\text{Bob,min}}$) for Bob for which the probability of error ($P_{\text{e,Bob,max}}$) is maximally tolerable, as illustrated in Fig. 11.21. The *secure region* is defined as the highest SNR of Eve ($SNR_{\text{Eve,min}}$) for which the probability of error ($P_{\text{e,Eve,min}}$) is minimum that can be tolerated (it is close to 0.5). The security gap is defined as a difference (in dB scale), $SNR_{\text{Bob,min}}$-$SNR_{\text{Eve,min}}$. The goal is to optimize the puncturing function to minimize the security gap. For additional details an interested reader is referred to [56].

Another approach advocated in [55] was to choose which bits to puncture in an LDPC codeword such that the codewords with the same punctured bits form the subcodebook, forming thus the nested structure. Let us consider (n',k') LPDC code $\mathcal{C}'$ whose parity-check matrix can be represented in the form $\boldsymbol{H} = [\boldsymbol{H}_1\ \boldsymbol{H}_2]$, where $\boldsymbol{H}_2$ is a lower-triangular matrix of size (n'-k') × (n'-k'). The codewords for $\mathcal{C}'$ are represented in the form $\boldsymbol{x} = [\boldsymbol{m}\ \boldsymbol{m}'\ \boldsymbol{p}]$, where $\boldsymbol{m}$ is the message vector of length $k < k'$, $\boldsymbol{m}'$ is the axillary dummy message vector of length k'-k, and $\boldsymbol{p}$ is the vector of parity checks of length n'-k'. The (n,k) punctured code $\mathcal{C}$ is composed of punctured codewords $\boldsymbol{x}' = [\boldsymbol{m}'\ \boldsymbol{p}]$, which get transmitted toward Bob. The partition of code $\mathcal{C}$ can be done based on punctured bits' values in the message $\boldsymbol{m}$. Bob based on the received word performs LDPC decoding to determine the punctured bits.

11.6.1.3 Polar Codes

The polar codes, introduced by Arikan in [57], are low-complexity linear block codes being capacity achieving over binary input symmetric channels (rows in transition channel matrix are permutations of each other). Let W: $\{0,1\} \rightarrow \mathcal{Y}$, $(x \rightarrow y)$, be a generic binary input discrete memoryless channel (BI-DMC) characterized with transition probabilities $W(y|x)$. For the block length $n = 2^m$ (m is a positive integer), we define the $n \times n$ matrix $\boldsymbol{G}_n$ as follows:

$$\boldsymbol{G}_n = \boldsymbol{G}_2^{\otimes m}, \boldsymbol{G}_2 \doteq \begin{bmatrix} 1 & 0 \\ 1 & 1 \end{bmatrix}, \tag{11.76}$$

where $\otimes$ denotes the Kronecker product. Let us consider the sequence $\boldsymbol{u}$ of n-tuples from the binary field $\mathcal{F}_2$, that is, $\boldsymbol{u} \in \mathcal{F}_2^n$. Each bit x_i in a codeword $\boldsymbol{x} = \boldsymbol{u}\boldsymbol{G}_n$ is transmitted over BI-DMC with transition probability $W(y|x)$, and the resulting output vector is denoted by $\boldsymbol{y}$. For each bit u_i in $\boldsymbol{u}$, we define the bit channel $W^{(i)}$: $\mathcal{U} \rightarrow \mathcal{Y}^n \times \mathcal{U}^{i-1}$, indicating that the bit u_i can be decoded by using the received sequence $\boldsymbol{y}$ and previously decoded bits $u_1, u_2, \ldots, u_{i-1}$. The i-th bit channel function is defined by:

$$W^{(i)}\left(\boldsymbol{y}, \boldsymbol{u}^{i-1} | u_i\right) = \sum_{\boldsymbol{u}_{i+1}^N \in \{0,1\}^{n-i}} W^n(\boldsymbol{y} | \boldsymbol{u}\boldsymbol{G}_n), \tag{11.77}$$

whereas before we use the notation $\boldsymbol{u}^{i-1}$ to denote the sequence $[u_1\, u_2 \ldots u_{i-1}]$. Similarly we use the notation $\boldsymbol{u}_{i+1}^n$ to denote the sequence $[u_{i+1}\, u_{i+2} \ldots u_n]$. Since the channel is memoryless, the $W^n(\boldsymbol{y}|\boldsymbol{x})$ term above can be written as product of individual transition probabilities, that is, $W^n(\boldsymbol{y}|\boldsymbol{x}) = \prod_{i=1}^n W(y_i|x_i)$. Arikan has shown that these bit channels "polarize" in the sense that they become either almost noiseless or almost completely noisy. This "polarization" effect can be described in terms of the *Bhattacharya parameter* $Z(W)$ of the binary input channel W, which is defined as:

$$Z(W) = \sum\nolimits_y \sqrt{W(y|0)W(y|1)} \in [0,1], \tag{11.78}$$

with $Z(W) \rightarrow 0$ meaning that the channel is noiseless, while for $Z(W) \rightarrow 1$ meaning that the channel is totally noisy.

Let $C(W)$ denote the channel capacity of channel W. Arikan has shown in [57] that for any fixed $\delta \in (0,1)$ as $n \rightarrow \infty$, the fraction of indices $i \in \{1,2,\ldots,n\}$ for which $Z(W^{(i)}) \in (1-\delta, 1]$ tends to $C(W)$, while the fraction of indices for which $Z(W^{(i)}) \in [0,\delta)$ tends to $1\text{-}C(W)$. The bit channels of the first type (noiseless) can be called good bit channels and denoted by $\mathcal{G}$, while the bit channels of the second type can be called bad bit channels denoted by $\mathcal{B} = \{1,2,\ldots,n\} \backslash \mathcal{G}$. This polarization phenomenon can be employed to design the polar codes. Given that bad bit channels are too noisy, they should not be used to transmit any reliable information; instead they should be frozen to a fixed value. On the other hand, the good bit channels are almost noise-free and reliable for transmission of information. So, the main task in the polar code design is to identify then indices of good bit channels $\mathcal{G}$. By using the Bhattacharya parameter, we can select the indices of the bit channels with the k least Bhattacharya parameter to design an (n,k) polar code.

Regarding the decoding procedures of polar codes, we can use either sum-product algorithm or sequential decoding. The *sequential decoding* operates as follows. For indices belonging to the good bit channels, namely, $i \in \mathcal{G}$, we apply the following decoding rule:

$$\widehat{u}_i = \begin{cases} 1, & \dfrac{W^{(i)}\left(\boldsymbol{y}^n, \widehat{\boldsymbol{u}}^{i-1} | 1\right)}{W^{(i)}\left(\boldsymbol{y}^n, \widehat{\boldsymbol{u}}^{i-1} | 0\right)} > 1 \\ 0, & \text{otherwise} \end{cases} \tag{11.79}$$

A typical polar code n-combiner architecture for $n \geq 4$, as proposed by Arikan [57] (see also [30]), is shown in Fig. 11.22. Clearly, the combining process proceeds in two steps. In step 1, we pair $\boldsymbol{u}^n$ into sequence of pairs $\left(\left(\boldsymbol{u}_1^2, \boldsymbol{u}_3^4, \cdots, \boldsymbol{u}_{n-1}^n\right)\right)$ and then apply $\boldsymbol{G}_2$ to each pair. This step can be described by the following transformation [30]:

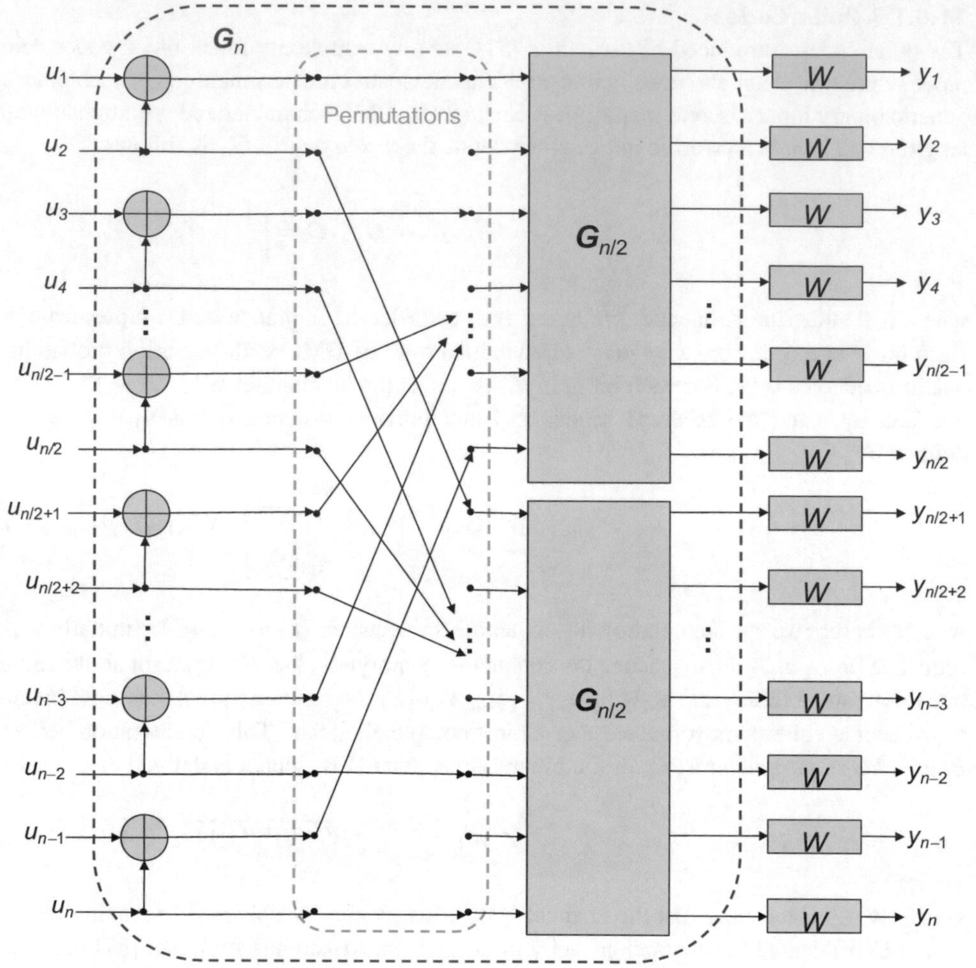

Fig. 11.22 Description of Arikan's combiner in the polar code

$$I_{n/2} \otimes G_2 = \begin{bmatrix} G_2 & \mathbf{0} & \mathbf{0} \\ \mathbf{0} & \ddots & \mathbf{0} \\ \mathbf{0} & \mathbf{0} & G_2 \end{bmatrix}. \tag{11.80}$$

In step 2, we feed even bits $(u_2, u_4, \ldots, u_n)$ to the odd positions of the bottom $n/2$-stage of the combiner. At the same time, we feed $n/2$ sum bits $(u_1 \oplus u_2, u_2 \oplus u_3, \ldots, u_{n-1} \oplus u_n)$ to the even positions of the top $n/2$-stage of the combiner. This step can be described as $\boldsymbol{R}_n(\boldsymbol{I}_2 \otimes \boldsymbol{G}_{n/2})$, where $\boldsymbol{R}_n$ is the permutation matrix separating the even bits and the sum bits.

If Eve's channel is degraded, it can be shown that Eve's set of good bits, denoted as $\mathcal{G}_{\mathrm{E}}$, is the subset of Bob's set of good bits $\mathcal{G}_{\mathrm{B}}$ [58, 59]. Clearly, in this scenario, Eve will be able to decode properly only a subset of bits that Bob is able to decode. So we can group the bit channels into three groups: (i) bit channels decodable by Bob only, denoted as $\mathcal{G}_{\mathrm{B}} \backslash \mathcal{G}_{\mathrm{E}}$; (ii) bit channels decodable by both Bob and Eve, denoted as $\mathcal{G}_{\mathrm{BE}}$; and (iii) Bob's set of bad bits denoted as $\mathcal{B}_{\mathrm{B}}$. We can now reorder information word as follows $\left[\boldsymbol{u}_{\mathcal{G}_{\mathrm{B}} \backslash \mathcal{G}_{\mathrm{E}}} \quad \boldsymbol{u}_{\mathcal{G}_{\mathrm{BE}}} \quad \boldsymbol{u}_{\mathcal{B}_{\mathrm{B}}}\right]$, wherein Bob's bad bits are frozen (fixed). The codewords $\left[\boldsymbol{u}_{\mathcal{G}_{\mathrm{B}} \backslash \mathcal{G}_{\mathrm{E}}} \quad \boldsymbol{u}_{\mathcal{G}_{\mathrm{BE}}} \quad \boldsymbol{u}_{\mathcal{B}_{\mathrm{B}}}\right]\boldsymbol{G}_n$ can now be partitioned into different cosets characterized by different values of the $\boldsymbol{u}_{\mathcal{G}_{\mathrm{B}} \backslash \mathcal{G}_{\mathrm{E}}}$, thus representing the nested structure. To encode Alice will place the message bits $\boldsymbol{m}$ into $\boldsymbol{u}_{\mathcal{G}_{\mathrm{B}} \backslash \mathcal{G}_{\mathrm{E}}}$ information bits block and auxiliary dummy bits $\boldsymbol{m}'$ into $\boldsymbol{u}_{\mathcal{G}_{\mathrm{BE}}}$ block and transmit the codeword $[\boldsymbol{m} \quad \boldsymbol{m}' \quad \boldsymbol{u}_{\mathcal{B}_{\mathrm{B}}}]\boldsymbol{G}_n$. On receiver side, Bob will perform the sequential decoding as described above.

Some other classes of codes suitable for use in physical-layer security are provided in [60–62].

11.6.2 Coding for Strong Secrecy Systems

This subsection is devoted to the coding for strong secrecy [63–66], and we describe two classes of codes capable of achieving the strong secrecy requirement.

11.6.2.1 Coset Coding with Dual of LDPC Codes

Let us consider an *(n,k)* LDPC code $\mathcal{C}$, with the parity-check matrix $\boldsymbol{H}$ belonging to the ensemble of codes with distribution (μ,λ) and with the BEC threshold $\alpha^*(\mu,\lambda)$ [63, 64], suitable for the binary erasure wiretap channel model shown in Fig. 11.10. (As a reminder, for BEC all errors with error probability α below the threshold α^* are correctable with the sum-product (belief-propagation) algorithm.) Let $\mathcal{C}^{\perp}$ be the dual of LDPC code C, in other words the *(n,n-k)* low-density generator matrix (LDGM) code with generator matrix being $\boldsymbol{H}$. The 2^k cosets of $\mathcal{C}^{\perp}$ in *n*-tuple space form a partition, thus representing the nested structure. To encode a *k*-bit message $\boldsymbol{m}$, the encoder chooses the coset with syndrome $\boldsymbol{m} = \boldsymbol{x}\boldsymbol{G}^{\mathrm{T}}$ and then randomly selects a codeword from that coset. To simplify the codeword selection process, we can proceed as follows. The linearly independent rows of $\boldsymbol{H}$ (representing the basis for $\mathcal{C}^{\perp}$ code) can be augmented with *k* linearly independent vectors $\{g_1, g_2, \ldots, g_k\}$ to form the basis for the *n*-tuple space (in $\{0,1\}^n$), representing the matrix $\boldsymbol{G}^* = \left[g_1^T \cdots g_k^T\right]^T$, and the encoding process can be represented as follows:

$$\boldsymbol{x} = \begin{bmatrix} \boldsymbol{m} & \boldsymbol{m}' \end{bmatrix} \begin{bmatrix} \boldsymbol{G} \\ \boldsymbol{H} \end{bmatrix}, \tag{11.81}$$

where $\boldsymbol{m}'$ is the dummy auxiliary vector. The decoding process is of a reasonable complexity as discussed in [63].

Let $P_e^{(\mathcal{C},BEC(1-\alpha))}$ denote the block error probability of the code $\mathcal{C}$ over the BEC with erasure probability 1-α. It can be shown that the following inequality holds [65]:

$$\frac{1}{n} I(M, Z^n) \leq \frac{k}{n} P_e^{(\mathcal{C},BEC(1-\alpha))} \tag{11.82}$$

indicating that for $1\text{-}\alpha < \alpha^*(\mu,\lambda)$ this construction guarantees the weak secrecy. However, it has been shown in [64], when an ensemble of large-girth LDPC codes is used, whose girth grows exponentially with the codeword length *n*, the probability of error has the desired decay for $1\text{-}\alpha < \alpha^*(\mu,\lambda)$, and strong secrecy can be achieved.

When the main channel is noiseless, while the wiretap channel is binary additive, it has been shown in [67] that coset coding with parity-check matrix generated uniformly at random can also achieve the strong secrecy.

11.6.2.2 Hash Functions and Extractor-Based Coding

The code constructions based on hash functions and extractors can achieve the strong secrecy as discussed in [68–70]. The key idea behind these code constructions is to employ the universal$_2$ families of hash functions [16, 68–71]. We say that the family of functions $\mathcal{F}$: $(0,1)^r \rightarrow (0,1)^k$ is a *universal$_2$ family of hash functions* if for any two distinct $\boldsymbol{x}_1$ and $\boldsymbol{x}_2$ from $(0,1)^r$, the probability that $f(\boldsymbol{x}_1) = f(\boldsymbol{x}_2)$ when *f* is chosen uniformly at random from $\mathcal{F}$ is smaller than 2^{-k}. Let us now consider *(n,r)* reliability code $\mathcal{C}$ and choose *f* uniformly at random from $\mathcal{F}$. To encode a secret message $\boldsymbol{m}$ from $(0,1)^k$, Alice selects the sequence $\boldsymbol{u}$ uniformly at random from $(0,1)^r$ in the set $f^{-1}(\boldsymbol{m})$ and encodes it using the code $\mathcal{C}$. It has been shown in [68] that any choice of *f* guarantees the strong secrecy when the code rate *k*/*n* is lower than secrecy capacity. The problem with this approach is that function $f(\boldsymbol{x})$ is not invertible or it is very difficult to invert. To solve for this problem, the use of invertible extractors is advocated in [69, 70].

11.6.3 Information Reconciliation

We turn our attention now to the source-type secret-key generation (agreement) protocol, depicted in Fig. 11.18 and described in Sect. 11.5.1. In this model, Alice, Bob, and Eve have access to the respective output ports $\mathcal{X}, \mathcal{Y}, \mathcal{Z}$ of the source characterized by the joint PDF f_{XYZ}. Each of them sees the i.i.d. realizations of the source denoted, respectively, as $\boldsymbol{X}^n$, $\boldsymbol{Y}^n$, and $\boldsymbol{Z}^n$. Given that Alice's and Bob's realizations are not perfectly correlated, they need to correct the discrepancies' errors. This step in secret-key generation as well as in QKD is commonly referred to as the *information reconciliation* (or just reconciliation). The reconciliation process can be considered as a special case of the *source coding with the side information* [27, 72]. To encode the source $\mathcal{X}$, we know from Shannon's source coding theorem that the rate R_x must be as small as possible but not lower than the entropy of the source, in other words $R_x > H(X)$. To jointly encode the source $(\mathcal{X}\mathcal{Y}, f_{XY})$, we need the rate $R > H(X,Y)$. If we separately encode the sources, the required rate would be $R > H(X) + H(Y)$. However, Slepian and Wolf have shown in [72] that the rate $R = H(X,Y)$ is still sufficient even for separate encoding of correlated sources. This claim is

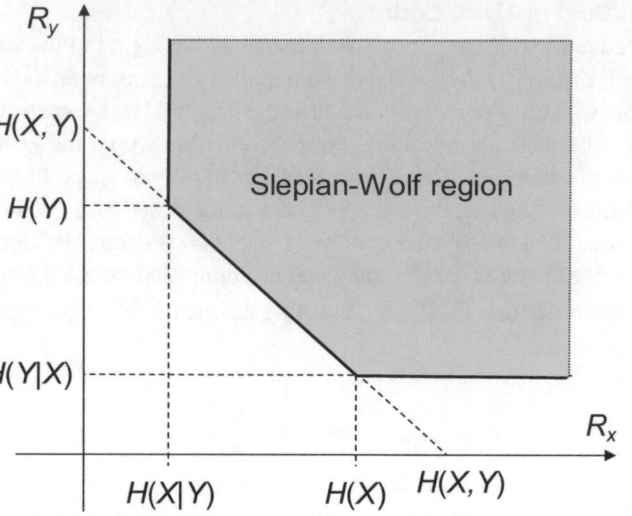

Fig. 11.23 The typical Slepian-Wolf achievable rate region for a correlated source $(\mathcal{XY}, f_{XY})$

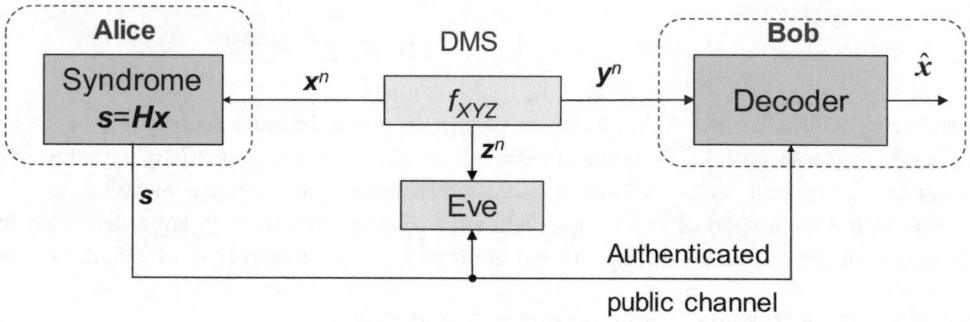

Fig. 11.24 The reconciliation problem represented as a source coding with syndrome-based side information

commonly referred to as the *Slepian-Wolf theorem* and can be formulated as follows [27, 72]: The *achievable rate region* $\mathcal{R}$ (for which the error probability tends to zero) for the separate encoding of the correlated source $(\mathcal{XY}, f_{XY})$ is given by:

$$\mathcal{R}\doteq\left\{ (R_x, R_y) : \begin{array}{c} R_x \geq H(X|Y) \\ R_y \geq H(Y|X) \\ R_x + R_y \geq H(X, Y) \end{array} \right\} \tag{11.83}$$

The typical shape of the Slepian-Wolf region is provided in Fig. 11.23. Now if we assume that (X,f_X) should be compressed when (Y,f_Y) is available as the side information, then $H(X|Y)$ is sufficient to describe X.

Going back to our reconciliation problem, Alice compresses her observations $\boldsymbol{X}^n$, and Bob decodes them with the help of correlated side information $\boldsymbol{Y}^n$, which is illustrated in Fig. 11.24. The source encoder can be derived from capacity-achieving channel code. For an (n,k) LDPC code, the parity-check matrix $\boldsymbol{H}$ of size $(n\text{-}k) \times n$ can be used to generate the syndrome by $\boldsymbol{s} = \boldsymbol{Hx}$, where $\boldsymbol{x}$ is Alice's DMS observation vector of length n. Since $\boldsymbol{x}$ is not a codeword, the syndrome vector is different from all-zeros vector. The syndrome vector is of length $n\text{-}k$ and can be represented as $\boldsymbol{s} = [s_1\ s_2\ \ldots\ s_{n\text{-}k}]^{\mathrm{T}}$. Clearly, the syndrome vector is transmitted toward Bob over the (error-free) authenticated public channel.

Once Bob receives the syndrome vector $\boldsymbol{s}$, his decoder tries to determine the vector $\boldsymbol{x}$ based on $\boldsymbol{s}$ and Bob's observation vector $\boldsymbol{y}$, which maximizes the posterior probability:

$$\widehat{\boldsymbol{x}} = \max P(\boldsymbol{y}|\boldsymbol{x}, \boldsymbol{s}), \tag{11.84}$$

and clearly, the decoding problem is similar to the maximum a posteriori probability (MAP) decoding. The compression rate is $(n\text{-}k)/n = 1\text{-}k/n$, while the linear code rate is k/n. The number of syndrome bits required is:

$$n-k \geq H(X^n|Y^n) = nH(X|Y). \tag{11.85}$$

In practice, the practical reconciliation algorithms introduce the overhead $OH > 0$, so that the number of transmitted bits over the public channel is $nH(X|Y)(1 + OH)$. Given that Alice and Bob at the end of the reconciliation step share with high probability the common sequence $\boldsymbol{X}^n$ with entropy $nH(X)$, the *reconciliation efficiency* β will be then:

$$\beta = \frac{nH(X) - nH(X|Y)(1+OH)}{nI(X,Y)} = 1 - OH\frac{H(X|Y)}{I(X,Y)} \leq 1. \tag{11.86}$$

The MAP decoding complexity is prohibitively high when LDPC coding is used; instead the low-complexity sum-product algorithm (SPA) (also known as belief-propagation algorithm) should be used as described in [73]. Given that Alice observation $\boldsymbol{x}$ is not a codeword, the syndrome bits are nonzero. To account for this problem, we need to change the sign of log-likelihood to be sent from the c-th check node to the v-th variable node when the corresponding syndrome bit s_c is 1. Based on Chap. 2 and ref. [74], we can summarize the *log-domain SPA* below. To improve the clarity of presentation, we use the index $v = 1,\ldots,n$ to denote variable nodes and index $c = 1,\ldots,n\text{-}k$ to denote the check nodes. Additionally, we use the following notation (introduced in Chap. 9): (i) $N(c)$ to denote $\{v\text{-nodes connected to } c\text{-node } c\}$, (ii) $N(c)\backslash\{v\}$ denoting $\{v\text{-nodes connected to } c\text{-node } c \text{ except } v\text{-node } v\}$, (iii) $N(v)$ to denote $\{c\text{-nodes connected to } v\text{-node } v\}$, and (iv) $N(v)\backslash\{c\}$ representing $\{c\text{-nodes connected to } v\text{-node } v \text{ except } c\text{-node } c\}$.

The formulation of the *log-domain SPA for source coding with the side information* is provided now:

1. *Initialization*: For the variable nodes $v = 1,2,\ldots,n$, initialize the extrinsic messages L_{vc} to be sent from variable node (v-node) v to the check node (c-node) c to the channel log-likelihood ratios (LLRs), denoted as $L_{\mathrm{ch}}(v)$, namely, $L_{vc} = L_{\mathrm{ch}}(v)$.
2. *The check-node (c-node) update rule*: For the check nodes $c = 1,2,\ldots,n\text{-}k$, compute $L_{cv} = (1-2s_c) \underset{N(c)\backslash\{v\}}{\boxplus} L_{vc}$. The box-plus operator is defined by:

$$L_1 \boxplus L_2 = \prod_{k=1}^{2} \mathrm{sign}(L_k) \cdot \phi\left(\sum_{k=1}^{2} \phi(|L_k|)\right),$$

where $\phi(x) = -\log\tanh(x/2)$. The box operator for $|N(c)\backslash\{v\}|$ components is obtained by recursively applying two-component version defined above.
3. *The variable-node (v-node) update rule*: For $v = 1,2,\ldots,n$, set $L_{vc} = L_{\mathrm{ch}}(v) + \sum_{N(v)\backslash\{c\}} L_{cv}$ for all c-nodes for which the c-th row/v-th column element of the parity-check matrix $\boldsymbol{H}$ is one, that is, $h_{cv} = 1$.
4. *Bit decisions*: Update $L(v)$ $(v = 1,\ldots,n)$ by $L(v) = L_{\mathrm{ch}}(v) + \sum_{N(v)} L_{cv}$ and make decisions as follows $\widehat{x}_v = 1$ when $L(v) < 0$ (otherwise set $\widehat{x}_v = 0$).

So, the only difference with respect to conventional log-domain SPA is the introduction of the multiplication factor $(1\text{-}2s_c)$ in the c-node update rule. Moreover, the channel likelihoods $L_{\mathrm{ch}}(v)$ should be calculated from the joint distribution f_{XY} by $L_{\mathrm{ch}}(v) = \log[f_{XY}(x_v = 0|\, y_v)/f_{XY}(x_v = 1|\, y_v)]$. Because the c-node update rule involves log and tanh functions, it can be computationally intensive, and to lower the complexity, the reduced complexity approximations should be used as described in Chap. 9.

11.7 Privacy Amplification

The purpose of the privacy amplification is to extract the secret key from the reconciled (corrected) sequence. The key idea is to apply a well-chosen compression function, Alice and Bob have agreed on, to the reconciled binary sequence of length n_r as follows g: $(0,1)^n \rightarrow (0,1)^k$, where $k < n$, so that Eve gets negligible information about the secret key [3, 4, 15, 16, 75]. Typically, g is selected at random from the set of compression functions $\mathcal{G}$ so that Eve does not know which g has been used. In practice, the set of functions $\mathcal{G}$ is based on the family of universal hash functions, introduced by Carter and Wegman [76, 77].

We have already introduced the concept of hash functions but repeat the definition here for completeness of the section. We say that the family of functions $\mathcal{G}$: $(0,1)^n \to (0,1)^k$ is a *universal$_2$ family of hash functions* if for any two distinct $\boldsymbol{x}_1$ and $\boldsymbol{x}_2$ from $(0,1)^n$, the probability that $g(\boldsymbol{x}_1) = g(\boldsymbol{x}_2)$ when g is chosen uniformly at random from $\mathcal{G}$ is smaller than 2^{-k}.

As an illustrative example, let us consider the multiplication in GF(2^n) [16]. Let a and x be two elements from GF(2^n). Let the g function be defined as the mapping $(0,1)^n \to (0,1)^k$ that assigns to argument x the first k bits of the product $ax \in$GF(2^n). It can be shown that the set of such functions is a universal class of functions for $k \in [1,n]$.

The analysis of privacy amplification does not rely on Shannon entropy, but Rényi entropy [78–80]. For discrete random variable X, taking the values from the sample space $\{x_1,\ldots,x_n\}$ and distributed according to $P_X = \{p_1,\ldots,p_n\}$, the *Rényi entropy of X of order α* is defined by [79]:

$$R_\alpha(X) = \frac{1}{1-\alpha}\log_2\left\{\sum\nolimits_x [P_X(x)]^\alpha\right\}. \tag{11.87}$$

The Rényi entropy of order two is also known as the *collision entropy*. From Jensen's inequality [27], we conclude that the collision entropy is upper limited by the Shannon entropy, that is, $R_2(X) \leq H(X)$. In limit when $\alpha \to \infty$, the Rényi entropy becomes the *min-entropy*:

$$\lim_{\alpha\to\infty} R_\alpha(X) = -\log_2 \max_{x\in X} P_X(x) = H_\infty(X). \tag{11.88}$$

Finally, by applying l'Hôpital's rule as $\alpha \to 1$, the Rényi entropy becomes the Shannon entropy:

$$\lim_{\alpha\to 1} R_\alpha(X) = -\sum\nolimits_x P_X(x)\log_2 P_X(x) = H(X). \tag{11.89}$$

The properties of Rényi entropy are listed in [79]. Regarding the conditional Rényi entropy, there exist different definitions; three of them have been discussed in detail in [80]. We adopt the first one. Let the joint distribution for (X, Y) be denoted as $P_{X,Y}$, and the distribution for X denoted as P_X. Further, let the Rényi entropy of the conditional random variables $X|Y = y$, distributed according to $P_{X|Y=y}$, be denoted as $R_\alpha(X|Y = y)$. Then the conditional Rényi entropy can be defined by:

$$\begin{aligned} R_\alpha(X|Y) &= \sum\nolimits_y P_Y(y) R_\alpha(X|Y=y) \\ &= \frac{1}{1-\alpha}\sum\nolimits_y P_Y(y)\log_2\sum\nolimits_x P_{X|Y}(x|y)^\alpha. \end{aligned} \tag{11.90}$$

The mutual Rényi information can be defined then as:

$$I_\alpha(X,Y) = R_\alpha(X) - R_\alpha(X|Y), \tag{11.91}$$

and it is not symmetric to $R_\alpha(Y)$-$R_\alpha(Y|X)$. Moreover, it can even be negative as discussed in [79, 80].

Bennett et al. have proved the following *privacy amplification theorem* in [16] (see theorem 3): Let X be a random variable with distribution P_X and Rényi entropy of order two $R_2(X)$. Further, let G be the random variable representing uniform selection at random of a member of universal hash functions $(0,1)^n \to (0,1)^k$. Then the following inequalities are satisfied [16]:

$$\begin{aligned} H(G(X)|G) \geq R_2(G(X)|G) &\geq k - \log_2\left[1 + 2^{k-R_2(X)}\right] \\ &\geq k - \frac{2^{k-R_2(X)}}{\ln 2}. \end{aligned} \tag{11.92}$$

Bennett et al. have also proved the following *corollary of the privacy amplification theorem* [16] (see also [4]): Let $S \in \{0,1\}^n$ be a random variable representing the sequence shared between Alice and Bob, and let E be a random variable representing the knowledge Eve was able to get about the S, with a particular realization of E denoted by e. If the conditional Rényi entropy of order two $R_2(S|E = e)$ is known to be at least r_2 and Alice and Bob choose the secret key by $K = G(S)$, where G is a hash function chosen uniformly at random from the universl$_2$ family of hash functions $(0,1)^n \to (0,1)^k$, then the following is valid:

$$H(K|G,E=e) \geq k - \frac{2^{k-r_2}}{\ln 2}. \tag{11.93}$$

Cachin has proved the following Rényi entropy *lemma* in [81] (see also [82]): Let X and Q be two random variables and $s > 0$. Then with probability of at least $1–2^{-s}$, the following is valid [81, 82]:

$$R_2(X) - R_2(X|Q=q) \leq \log|Q| + 2s + 2 \tag{11.94}$$

The total information available to Eve is composed of her observations $\mathbf{Z}^n$ from the source of common randomness and additional bits exchanged in information reconciliation phase over the authenticated public channel, represented by random variable Q. Based on Cachin's lemma, we can write:

$$R_2(S|\mathbf{Z}^n = z^n) - R_2(S|\mathbf{Z}^n = z^n, Q=q) \leq \log|Q| + 2s + 2 \text{ with probability } 1-2^{-s} \tag{11.95}$$

We know that $R_2(X) \leq H(X)$, so we conclude that $R_2(S|\mathbf{Z}^n = z^n) \leq H(S|\mathbf{Z}^n = z^n)$ and therefore we can upper bound $R_2(S|\mathbf{Z}^n = z^n)$ as follows:

$$R_2(S|\mathbf{Z}^n = z^n) \leq nH(X|Z). \tag{11.96}$$

Based on (11.95), we can lower bound $R_2(S|\mathbf{Z}^n = z^n, Q = q)$ by:

$$R_2(S|\mathbf{Z}^n = z^n, Q=q) \geq nH(X|Z) - \log|Q| - 2s - 2 \text{ with probability } 1-2^{-s} \tag{11.97}$$

Since the number of bits exchanged over the public channel $\log|Q|$ is approximately $nH(X|Y)(1 + OH)$, for sufficiently large n, the previous inequality becomes:

$$R_2(S|\mathbf{Z}^n = z^n, Q=q) \geq \underbrace{nH(X|Z) - nH(X|Y)(1+OH) - 2s - 2}_{r_2} \quad \text{with probability } 1-2^{-s} \tag{11.98}$$

From Eq. (11.93) we conclude that $k\text{-}r_2 = -k_2 < 0$, which guarantees that for $k_2 > 0$ Eve's uncertainty of the key is lower bounded by:

$$H(K|E) \geq k - 2^{-k_2}/\ln 2 \text{ with probability } 1-2^{-s} \tag{11.99}$$

11.8 Wireless Channels Physical-Layer Security

The wireless channel secrecy capacity has been already discussed in Sect. 11.4.2, see also [85, 86], while the corresponding coding schemes have been described in Sect. 11.6 as well as in [87]. Here we concentrate on (i) PLS for MIMO channels and (ii) key generation techniques for wireless channels. We first introduce the wireless MIMO systems fundamentals.

11.8.1 PLS for Wireless MIMO Channels

The MIMO wiretap channel consists of Alice's M_{Tx} antennas, Bob's M_{Rx} antennas, and Eve's M_{E} receive antennas. The matrix representation of the main (Alice-to-Bob) channel is given by:

$$\boldsymbol{y}_B = \boldsymbol{H}_B \boldsymbol{x}_A + \boldsymbol{z}_B, \tag{11.100}$$

while the corresponding matrix representation of the Alice-to-Eve's channel is:

$$\boldsymbol{y}_E = \boldsymbol{H}_E \boldsymbol{x}_A + \boldsymbol{z}_E, \tag{11.101}$$

where $\boldsymbol{x}_A$ is the Alice transmitted vector, with covariance $\boldsymbol{R}_x = E\left(\boldsymbol{x}_A \boldsymbol{x}_A^\dagger\right)$, $\boldsymbol{y}_B$ is Bob's received vector, $\boldsymbol{H}_B$ is Alice-to-Bob channel matrix of size $M_{Rx} \times M_{Tx}$, and $\boldsymbol{H}_E$ is Alice-to-Eve channel matrix of size $M_E \times M_{Tx}$. When the additive noise is Gaussian, the optimum source to achieve secrecy capacity is Gaussian, and the corresponding secrecy capacity expression under the average power constraint $\text{Tr}(\boldsymbol{R}_x) \leq P$ will be [88]:

$$C_s = \max_{\boldsymbol{R}_x,\ \text{Tr}(\boldsymbol{R}_x) \leq P} [I(\boldsymbol{X}_A, \boldsymbol{Y}_B) - I(\boldsymbol{X}_A, \boldsymbol{Y}_E)]. \tag{11.102}$$

For Gaussian source, if the noise is spatially white with unit variance, we can write:

$$C_s = \max_{\boldsymbol{R}_x,\ \text{Tr}(\boldsymbol{R}_x) \leq P} \left[\log_2 \det\left(\boldsymbol{I} + \boldsymbol{H}_B \boldsymbol{R}_X \boldsymbol{H}_B^\dagger\right) - \log_2 \det\left(\boldsymbol{I} + \boldsymbol{H}_E \boldsymbol{R}_X \boldsymbol{H}_E^\dagger\right)\right]. \tag{11.103}$$

From equation above we conclude that we still need to determine the transmit covariance matrix $\boldsymbol{R}_x$ maximizing the secrecy rate. Now we consider two scenarios, when the CSI is perfectly and partially known.

Assuming that transmitter has the *perfect CSI* for both Bob and Eve, for the multiple-input single-output multiple-Eve (MISOME) case, the optimal solution is transmit beamforming [88]:

$$\boldsymbol{R}_x = P|\lambda_m\rangle\langle\lambda_m|,\ \langle\lambda_m| = (|\lambda_m\rangle)^\dagger, \tag{11.104}$$

where the unit-norm column-eigenvector $|\lambda_m\rangle$ corresponding to the eigenvalue λ_m is obtained as the solution to eigenvalue equation [88]:

$$\left(\boldsymbol{I} + P\boldsymbol{h}_B^\dagger \boldsymbol{h}_B\right)|\lambda_m\rangle = \lambda_m \left(\boldsymbol{I} + P\boldsymbol{H}_E^\dagger \boldsymbol{H}_E\right)|\lambda_m\rangle, \tag{11.105}$$

where $\boldsymbol{h}_B$ is Bob's channel vector. The solution for the secrecy rate maximization problem of Eq. (11.103) is given by [89]:

$$C_S = \frac{1}{2}\log_2\left[\lambda_{\max}\left(\boldsymbol{I} + P\boldsymbol{h}_B^\dagger \boldsymbol{h}_B, \boldsymbol{I} + P\boldsymbol{H}_E^\dagger \boldsymbol{H}_E\right)\right], \tag{11.106}$$

where $\lambda_{\max}$ is the largest generalized eigenvalue of the following two matrices $\boldsymbol{I} + P\boldsymbol{h}_B^\dagger \boldsymbol{h}_B$ and $\boldsymbol{I} + P\boldsymbol{H}_E^\dagger \boldsymbol{H}_E$. Under the matrix power covariance constraint $\boldsymbol{R}_x \leq \boldsymbol{S}$, where "$\leq$" denotes less or equal to in the positive semidefinite partial ordering sense of the symmetric matrices, the MIMO secrecy capacity is given by [90]:

$$C_S(\boldsymbol{S}) = \sum_i \log_2 \lambda_i \left(\boldsymbol{I} + \boldsymbol{S}^{1/2}\boldsymbol{H}_B^\dagger \boldsymbol{H}_B \boldsymbol{S}^{1/2}, \boldsymbol{I} + \boldsymbol{S}^{1/2}\boldsymbol{H}_E^\dagger \boldsymbol{H}_E \boldsymbol{S}^{1/2}\right), \tag{11.107}$$

where λ_i are greater than one eigenvalue of matrices $\boldsymbol{I} + \boldsymbol{S}^{1/2}\boldsymbol{H}_B^\dagger \boldsymbol{H}_B \boldsymbol{S}^{1/2}$ and $\boldsymbol{I} + \boldsymbol{S}^{1/2}\boldsymbol{H}_E^\dagger \boldsymbol{H}_E \boldsymbol{S}^{1/2}$. The *average power constraint* is much less restrictive than the matrix power constraint, and the secrecy capacity of the MIMO wiretap channel under the average power constraint can be found by the exhaustive search over the set $\{\boldsymbol{S} | \boldsymbol{S} \geq \boldsymbol{0}),\ \text{Tr}(\boldsymbol{S}) \leq P\}$; in other words, we can write [88, 91]:

$$C_S(P_t) = \max_{\boldsymbol{S} \geq \boldsymbol{0},\ \text{Tr}(\boldsymbol{S}) \leq P_t} C_S(\boldsymbol{S}). \tag{11.108}$$

In the practical schemes, due to limited feedback capabilities, we must deal with the imperfect CSI or *partial CSI* [92, 93]. A realistic model will be to assume that actual channel realizations are not known to Alice and Bob, but are known to lie in an uncertainty set of possible channels [92, 93], and these channels are known as *compound channels* [92–95]. Now we have to perform the reliability and security studies over all possible realizations in the uncertainty sets. The uncertainty sets for Bob and Eve channel matrices' estimates can be defined, respectively, as [88, 93]:

$$\mathcal{H}_B = \left\{ \boldsymbol{H}_B | \ \boldsymbol{H}_B = \boldsymbol{H} + \Delta\boldsymbol{H}, \|\Delta\boldsymbol{H}\|_F \le \varepsilon \right\}, \tag{11.109}$$

$$\mathcal{H}_E = \left\{ \boldsymbol{H}_E | \ \|\Delta\boldsymbol{H}_E\|_F \le \varepsilon \right\}, \tag{11.110}$$

where $\|\cdot\|_F$ is the Frobenius norm, $\boldsymbol{H}$ is the estimate of Bob's channel matrix, and $\Delta\boldsymbol{H}$ is the estimation error not larger than ε. Clearly, we assume that the estimate of Eve's channel matrix is not known, only the maximum estimation error. This scenario is applicable when Eve cannot approach the transmitter beyond certain protection distance. In other words, this compound model is applicable when Eve is outside of the exclusion zone. Interestingly enough, the secrecy capacity for such compound model can be determined for the nondegraded case for Gaussian MIMO channels [93]. Based on (11.103), (11.109), and (11.110), we conclude that the compound MIMO wiretap secrecy capacity can be determined by:

$$C_s = \max_{\mathrm{Tr}(\boldsymbol{R}_x) \le P} \left[\min_{\boldsymbol{H}_B \in \mathcal{H}_B} \log_2 \det\left(\boldsymbol{I} + \boldsymbol{H}_B \boldsymbol{R}_X \boldsymbol{H}_B^\dagger \right) - \max_{\boldsymbol{H}_E \in \mathcal{H}_E} \log_2 \det\left(\boldsymbol{I} + \boldsymbol{H}_E \boldsymbol{R}_X \boldsymbol{H}_E^\dagger \right) \right]. \tag{11.111}$$

Clearly, the maximum secrecy rate is limited by the worst channel to Bob and the best channel to Eve.

The concept of *broadcast channel with confidential messages* has been already introduced in Sect. 11.4.1. In broadcast scenario one user sends the message to multiple receivers, and this scenario is very common in downlink phase of cellular communication, in which the base station sends information to multiple mobile users. In a broadcast channel with confidential messages, Alice sends confidential message to Bob, which should be kept as secret as possible from other users/eavesdroppers. The broadcast channel with parallel independent subchannels is considered in [96], and the corresponding optimal source allocation, achieving the boundary of secrecy capacity region, is determined. The transmission of two confidential messages over discrete memoryless broadcast channel is studied in [97], wherein each receiver serves as an eavesdropper of the other. The general MIMO Gaussian case over the matrix power constraint is studied in [98]. For two-user broadcast channel, with each user receiving the corresponding confidential message, it has been shown that the secrecy capacity region (R_1,R_2) is rectangular for the matrix power constraint $\boldsymbol{C}_x \le \boldsymbol{S}$. Assuming so-called secret dirty-paper coding [98], the corner point rate ($R^*{}_1$,$R^*{}_2$) for secrecy capacity is given by [88, 98]:

$$\begin{aligned} R_1^* &= \sum\nolimits_i \log_2 \lambda_i, \ \lambda_i = \mathrm{eigenvalues}\left(\boldsymbol{I} + \boldsymbol{S}^{1/2} \boldsymbol{H}_B^\dagger \boldsymbol{H}_B \boldsymbol{S}^{1/2} \right) \\ R_2^* &= -\sum\nolimits_j \log_2 \lambda'_j, \lambda'_j = \mathrm{eigenvalues}\left(\boldsymbol{I} + \boldsymbol{S}^{1/2} \boldsymbol{H}_E^\dagger \boldsymbol{H}_E \boldsymbol{S}^{1/2} \right) \end{aligned} \tag{11.112}$$

wherein λ_i are larger than one eigenvalue of $\boldsymbol{I} + \boldsymbol{S}^{1/2} \boldsymbol{H}_B^\dagger \boldsymbol{H}_B \boldsymbol{S}^{1/2}$, while λ'_j are smaller than one eigenvalue of $\boldsymbol{I} + \boldsymbol{S}^{1/2} \boldsymbol{H}_E^\dagger \boldsymbol{H}_E \boldsymbol{S}^{1/2}$. Some more recent results on MIMO broadcast channels with confidential messages can be found in [99].

In a *multiaccess channel*, multiple users send information toward the same receiver, and this situation occurs in cellular communications when multiple mobile users transmit data toward the base station. *In multiaccess channel with confidential messages*, multiple senders Alice 1, Alice 2, etc. transmit confidential messages $\boldsymbol{M}_1$, $\boldsymbol{M}_2$, ... to a single receiver Bob. Each confidential message should be decodable by Bob, but without any leakage to other transmitters. This problem is considered in [100], wherein the inner and outer bounds for the region of secret rates have been determined, while the secrecy capacity region has been left unsolved.

The interference channel describes a scenario in which multiple transmitter-receiver communication links are simultaneously active introducing the crosstalk to each other. The *interference channel with confidential messages* represents the situation in which multiple transmitters want to transmit confidential message to the respective receivers in such a way to keep those messages secure from the counterpart receivers [93, 97, 101, 102]. As an illustration, in Fig. 11.29 the system model for two-user MIMO interference channel with confidential messages is provided. The $\boldsymbol{H}_i$ is used to denote the channel matrix of the *i*-th ($i = 1,2$) direct channel, while $\boldsymbol{H}_{ji}$ to denote the channel matrix of the crosstalk channel *j*. The figure of merit suitable to study the scaling behavior of the aggregate secrecy rate $R_\oplus$ in multiuser networks with signal-to-noise ratio at transmitter side, denoted as ρ, is the *secret multiplexing gain*, denoted as *SMG*, and can be defined as [88, 101]:

$$SMG = \lim_{\rho \to \infty} \frac{R_\oplus}{\log \rho}. \tag{11.113}$$

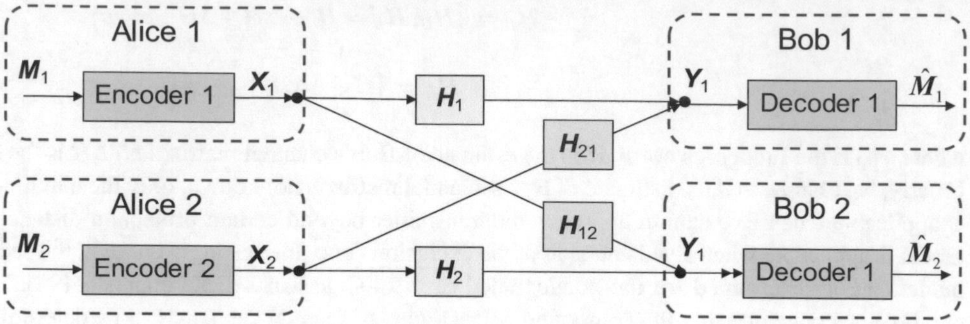

Fig. 11.25 The model for two-user MIMO interference channel with confidential messages

The SMG, also known as the *secure degree of freedom* (DoF), has been studied in [101, 102] for the *K*-users Gaussian interference channel, where $K \geq 3$, assuming that multiuser network is composed of *K* transmitter-receiver pairs, employing *F* frequency bands, and it has been found that SMG = (*K*-2)/(2 *K*-2) secure DoFs per frequency-time slot per user are almost surely achievable [101]. On the other hand, when an external eavesdropper is present, each user can achieve SMG = (*K*-1)/(2 *K*) secure DoFs per frequency-time slot [101].

The two-user MIMO interference channel, with each node having arbitrary number of antennas, has been studied in [103, 104], and the corresponding interference channel model illustrated in Fig. 11.25 can be represented as follows:

$$\begin{aligned} y_1 &= H_1 x_1 + H_{21} x_2 + z_1, \\ y_2 &= H_2 x_2 + H_{12} x_1 + z_2. \end{aligned} \tag{11.114}$$

The achievable secrecy regions have been derived in [103, 104] under different CSI assumptions and various noncooperative and cooperative scenarios. In noncooperative scenarios, it is assumed that both transmitters have perfect CSI of both the direct and crosstalk channels. The generalized singular value decomposition (GSVD) can be employed, as explained in [103, 104]. Given the direct and crosstalk channel matrices $\boldsymbol{H}_1$ and $\boldsymbol{H}_{12}$, the GSVD procedure gives unitary matrices $\boldsymbol{U}_{\mathrm{Rx1}}$ and $\boldsymbol{U}_{\mathrm{E1}}$, positive semidefinite diagonal matrices $\boldsymbol{\Lambda}_1$ and $\boldsymbol{D}_1$, and $\boldsymbol{A}_1$ matrix of size $M_{\mathrm{Rx1}} \times d$, where $d = \min(M_{\mathrm{Rx1}}, M_{\mathrm{Tx1}} + M_{\mathrm{Tx2}})$, such that the following relationships are valid [103, 104]:

$$\begin{aligned} \boldsymbol{H}_1 \boldsymbol{A}_1 &= \boldsymbol{U}_{Rx_1} \boldsymbol{\Lambda}_1, \\ \boldsymbol{H}_{12} \boldsymbol{A}_1 &= \boldsymbol{U}_{E_1} \boldsymbol{D}_1. \end{aligned} \tag{11.115}$$

The nonzero elements in $\boldsymbol{\Lambda}_1$ are ordered in ascending order, while in decreasing order in $\boldsymbol{D}_1$; and these two diagonal matrices are related by $\boldsymbol{\Lambda}_1^T \boldsymbol{\Lambda}_1 + \boldsymbol{D}_1^T \boldsymbol{D}_1 = \boldsymbol{I}$. The transmitter Tx_1 then precodes $\boldsymbol{x}_1$ as follows:

$$\boldsymbol{x}_1 = \boldsymbol{A}_1 \boldsymbol{s}_1, \;\; \boldsymbol{s}_1 \sim \mathcal{CN}(\boldsymbol{0}, \boldsymbol{\Sigma}_1), \tag{11.116}$$

where $\mathcal{CN}$ is the multivariate complex circular symmetric zero-mean Gaussian (normal) distribution with covariance matrix $\boldsymbol{\Sigma}_1$ being positive semidefinite diagonal matrix. The diagonal elements in $\boldsymbol{\Sigma}_1$ represent the powers allocated to different antennas. The transmitter Tx_2 applies an equivalent procedure. The achievable secrecy rate for Tx_1 is given by [103, 104]:

$$\begin{aligned} 0 \leq R_1 \leq \log &\left| \boldsymbol{I} + \boldsymbol{U}_{Rx_1} \boldsymbol{\Lambda}_1 \boldsymbol{\Sigma}_1 \boldsymbol{\Lambda}_1^\dagger \boldsymbol{U}_{Rx_1}^\dagger + \boldsymbol{U}_{E_2} \boldsymbol{D}_2 \boldsymbol{\Sigma}_2 \boldsymbol{D}_2^\dagger \boldsymbol{U}_{E_2}^\dagger \right| \\ &- \log \left| \boldsymbol{I} + \boldsymbol{U}_{E_2} \boldsymbol{D}_2 \boldsymbol{\Sigma}_2 \boldsymbol{D}_2^\dagger \boldsymbol{U}_{E_2}^\dagger \right| - \log \left| \boldsymbol{I} + \boldsymbol{U}_{E_1} \boldsymbol{D}_1 \boldsymbol{\Sigma}_1 \boldsymbol{D}_1^\dagger \boldsymbol{U}_{E_1}^\dagger \right|. \end{aligned} \tag{11.117}$$

The similar achievable secrecy rate inequality holds for Tx_2. A reader interested in cooperative scenario is referred to refs [103, 104].

The relay is used to support communication between transmitter and receiver, by extending the range/coverage or enabling higher transmission rate. The *relay channel with confidential messages* [88, 93, 105–107] represents scenario in which Alice wants to deliver the confidential message to Bob with the help of the relay. Given that relay can be trusted or untrusted, it is possible to classify the various MIMO wiretap networks into two broad categories: *trusted* and *untrusted* relay wiretap networks. The cooperative jamming strategies enabling secure communication without employing the external helpers have

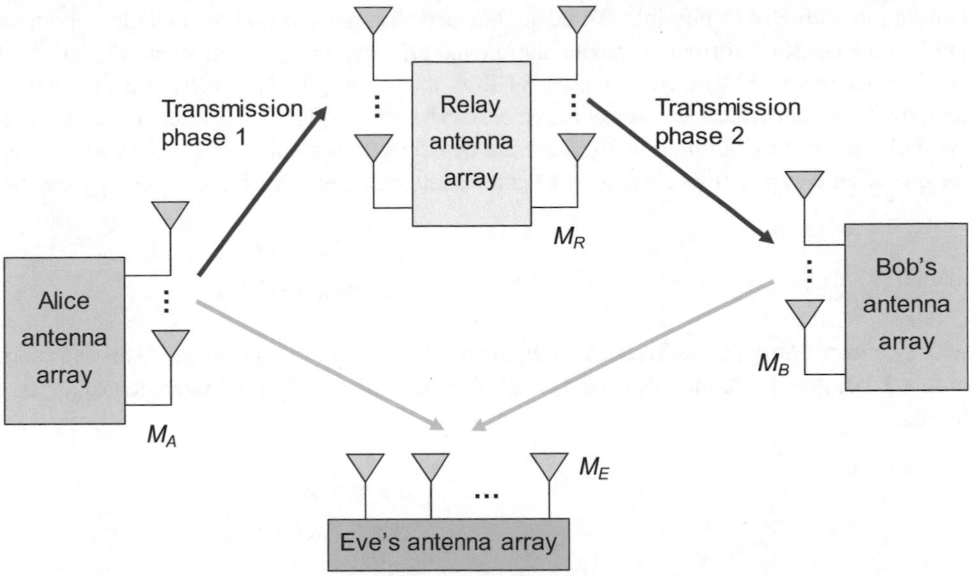

Fig. 11.26 The two-hop MIMO relay system architecture with an external eavesdropper

been discussed in [107], with corresponding two-hop MIMO relay network architecture provided in Fig. 11.26. The number of antennas available to Alice, Bob, relay, and Eve is denoted, respectively, as M_A, M_B, M_R, and M_E. In cooperative jamming strategies, Alice and Bob transmit jamming signals in stages when they do not typically transmit any data. The relay retransmits the signal received from Alice. Eve has access to both Alice and Bob's transmissions over wireless channel. The signals received by relay and Eve in transmission phase 1 can be represented as [107]:

$$\begin{aligned} y_R &= \boldsymbol{H}_{AR}(\boldsymbol{T}_A\boldsymbol{x}_A + \boldsymbol{T}'_A z_A) + \boldsymbol{H}_{BR}\boldsymbol{T}'_B z_B + z_R, \\ y_{E_1} &= \boldsymbol{H}_{AE}(\boldsymbol{T}_A\boldsymbol{x}_A + \boldsymbol{T}'_A z_A) + \boldsymbol{H}_{BE}\boldsymbol{T}'_B z_B + z_{E_1}, \end{aligned} \tag{11.118}$$

where $\boldsymbol{x}_A$ is the data signal transmitted by Alice toward the relay, while z_A and z_B are the jamming signals transmitted by Alice and Bob, respectively. (The relay and Eve's antennas noise sources are denoted by z_R and z_{E1}, respectively.) We use $\boldsymbol{T}_A$ and $\boldsymbol{T'}_B$ to denote the beamformers corresponding to Alice and Bob. On the other hand, the signals received by Bob and Eve in transmission phase 2 can be represented as [107]:

$$\begin{aligned} y_R &= \boldsymbol{H}_{RB}(\boldsymbol{T}_R\boldsymbol{x}_R + \boldsymbol{T}'_R z_R) + \boldsymbol{H}_{AB}\boldsymbol{T}'_{A2} z_{A2} + z_B, \\ y_{E_1} &= \boldsymbol{H}_{AE}(\boldsymbol{T}_R\boldsymbol{x}_R + \boldsymbol{T}'_R z_R) + \boldsymbol{H}_{AE}\boldsymbol{T}'_{A2} z_{A2} + z_{E_1}, \end{aligned} \tag{11.119}$$

where $\boldsymbol{x}_R$ is the data signal transmitted by relay and $\boldsymbol{T}_R$ is the corresponding beamformer. We use the z_R and z_{A2} to denote the jamming signals transmitted by relay and Alice, respectively. ($\boldsymbol{T'}_R$ and $\boldsymbol{T'}_{A2}$ are corresponding jammers' beamformers.) When both Alice and Bob's jammer signals are different from zero, that is, $z_A \neq 0$ and $z_B \neq 0$ in transmit phase 1, the corresponding scheme is called the *fully cooperating jamming* (FCJ) scheme. Otherwise, if one of them is zero, the corresponding scheme is called the *partially cooperating jamming* (PCJ) scheme. As expected, in terms of secrecy rate, the FCJ scheme significantly outperforms both PCJ scheme and the scheme without the jamming. Moreover, the FCJ does not exhibit the secret rate saturation effect for very high SNR_B. For full details on cooperative jamming schemes for secure communication in relay MIMO networks, an interested reader is referred to [107].

11.8.2 Secret-Key Generation in Wireless Networks

The secret-key generation (SKG) protocols have already been described in Sect. 11.5, where source- and channel-type SKG protocols are introduced and analyzed. Here we concentrate on SKG protocols in context of wireless networks. The wireless channels themselves can serve as a source of common randomness as discussed in [93, 108–111]. The SKG protocols for wireless channels are similar to those considered in Sect. 11.5; namely, Alice and Bob want to generate the secret with as little correlation with Eve as possible. To do so they perform transmission over wireless fading channel and employ authenticated public channel for information reconciliation and privacy amplification steps. The key idea is to exploit the *fading channel coefficients reciprocity* between Alice and Bob, that is, $h_{AB} = h_{BA} = h_m$, for the channel coherence time T_c. Even if the reciprocity is not perfect, the occasional errors can be corrected for by Slepian-Wolf-like coding. In training phase, Alice sends symbol sequence $\boldsymbol{x}_A$ of duration T_A, followed by Bob's transmission of symbol sequence $\boldsymbol{x}_B$ of duration T_c-T_A. During Bob's transmission Bob's and Eve's received vectors, denoted, respectively, as y_B and y_E, can be represented as:

$$\begin{aligned} y_B &= h_m \boldsymbol{x}_A + z_B, \\ y_E &= h_{AE} \boldsymbol{x}_A + z_E, \end{aligned} \tag{11.120}$$

with z_B and z_E being the multivariate complex circular symmetric zero-mean Gaussian (normal) distributions with covariance matrix $\boldsymbol{\Sigma}$ being $\sigma^2 \boldsymbol{I}$. On the other hand, when Bob transmits the symbol sequence $\boldsymbol{x}_B$, Alice and Eve's received vectors can be written as:

$$\begin{aligned} y_A &= h_m \boldsymbol{x}_B + z_A, \\ y_E &= h_{BE} \boldsymbol{x}_B + z_E. \end{aligned} \tag{11.121}$$

Let us assume that the main (Alice-to-Bob) channel coefficient is generated from a zero-mean Gaussian random variable of variance σ_m^2. The Alice and Bob's channel coefficient estimates will be:

$$\begin{aligned} h_m^{(A)} &= h_m + \frac{\boldsymbol{x}_B^\dagger}{\|\boldsymbol{x}_B\|^2} z_A, \\ h_m^{(B)} &= h_m + \frac{\boldsymbol{x}_A^\dagger}{\|\boldsymbol{x}_A\|^2} z_B. \end{aligned} \tag{11.122}$$

Clearly, the Alice (Bob) channel coefficient is a zero-mean Gaussian random variable with variance $\left(\sigma_m^{(A)}\right)^2 = \sigma_m^2 + \sigma^2/\|\boldsymbol{x}_B^2\|$ $[(\sigma_m^{(B)})^2 = \sigma_m^2 + \sigma^2/\|\boldsymbol{x}_A^2\|]$. Assuming that both Alice and Bob's average transmit powers are the same and equal to P, we have that $\|\boldsymbol{x}_A^2\| = T_A P$, $\|\boldsymbol{x}_B^2\| = (T_c - T_A)P$. The secret-key rate (SKR) can be estimated by [110]:

$$SKR = \frac{1}{T_c} I\left(h_m^{(A)}, h_m^{(B)}\right) = \frac{1}{T_c} \log_2 \left[\frac{\left(\sigma^2 + \sigma_m^2 T_A P\right)\left(\sigma^2 + \sigma_m^2 (T_c - T_A) P\right)}{\sigma^4 + \sigma^2 \sigma_m^2 T_c P} \right]. \tag{11.123}$$

Obviously, the optimum frame duration for Alice is $T_A = T_c/2$, so that the previous expression simplifies to:

$$SKR = \frac{1}{T_c} \log_2 \left[1 + \frac{\sigma_m^2 (T_c P)^2}{4\left(\sigma^4 + \sigma^2 \sigma_m^2 T_c P\right)} \right]. \tag{11.124}$$

From Eq. (11.124) we conclude that SKR increases as the average power P increases and decreases as the coherence time increases. Therefore, the fast fading is beneficial, while slow varying fading results in low SKR.

To improve further the SKRs, the *joint source-channel secret-key generation* was proposed in [111]. In this scheme, the key generation protocol, illustrated in Fig. 11.27, is composed of two phases: (i) the source-type training phase of duration T_s, which is used to generate the source-type key and estimate the main (Alice-Bob) channel coefficient at the same time and

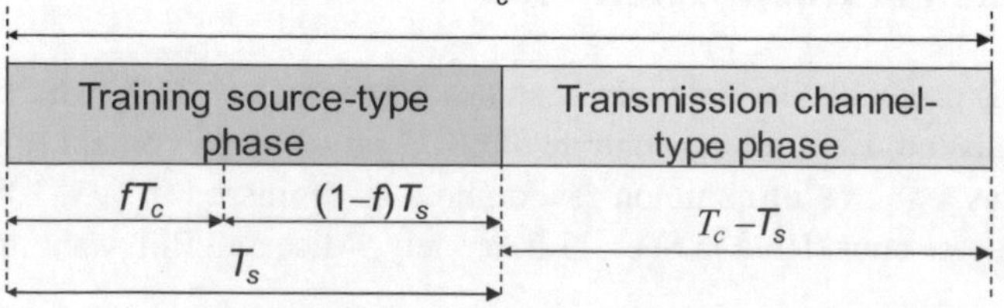

Fig. 11.27 The illustration of joint source-channel key generation scheme

(ii) the transmission of duration T_c-T_s that is used to generate the channel-type secret key. The total SKR will be then a summation of source- and channel-type SKRs. The fraction $0 < f < 1$ of training interval T_s is used by Alice to transmit the training sequence $\boldsymbol{x}_A$, while the portion of training interval $(1\text{-}f)T_s$ is used by Bob to transmit the training symbol sequence $\boldsymbol{x}_B$. Therefore, this training source-type phase is very similar to the fading channel reciprocity-based protocol discussed above. The source-type SKR can be then calculated using an expression similar to Eq. (11.123) as follows:

$$SKR_s = \frac{1}{T_c} I\left(h_m^{(A)}, h_m^{(B)}\right) = \frac{1}{T_c}\log_2\left[\frac{\left(\sigma^2 + \sigma_m^2 f T_s P_s\right)\left(\sigma^2 + \sigma_m^2 (1-f) T_s P_s\right)}{\sigma^4 + \sigma^2 \sigma_m^2 T_c P}\right], \tag{11.125}$$

where P_s is the average power used during the training source-type phase. Given that the training interval T_s is shorter than coherence time, it is a good idea to use the minimum mean square error (MMSE) channel estimation approach [33, 112] instead of the least square method discussed above. Bob's MMSE estimate of the main channel coefficient will be [111]:

$$\widetilde{h}_m^{(B)} = \frac{\sigma_m^2}{\sigma^2 + f P_s T_s \sigma_m^2} \boldsymbol{x}_A^\dagger \boldsymbol{y}_B. \tag{11.126}$$

The true channel coefficient value can be represented as:

$$h_m = \widetilde{h}_m^{(B)} + \Delta h_m, \tag{11.127}$$

where the Δh_m is the estimation error, which is a zero-mean Gaussian random variable of variance $\sigma_m^2/\left(\sigma^2 + f P_s T_s \sigma_m^2\right)$..

In second phase authors in [111] consider scheme in which Alice does not perform both power and rate adaptation to decouple the first and the second phases, since Eve can learn the Alice channel coefficient estimate for the phase 1. So, Alice transmits the symbol sequence to Bob with constant power, and the SKR for channel-type phase can be determined by [112]:

$$\begin{aligned} SKR_{ch} &= \frac{T_c - T_s}{T_c}\left[I\left(X_A, Y_B | \widetilde{h}_m^{(B)}\right) - I(X_A, Y_E | h_{AE})\right]^+ \\ &= \frac{T_c - T_s}{T_c}\left\{E\left[\log_2\left(1 + \frac{\left(\widetilde{h}_m^{(A)}\right)^2 P_{ch}}{\sigma^2 + \dfrac{\sigma_m^2 P_{ch}}{\sigma^2 + f P_s T_s \sigma_m^2}}\right)\right] - \log_2\left(1 + \frac{h_{AE}^2 P_{ch}}{\sigma^2}\right)\right\}, \end{aligned} \tag{11.128}$$

where the second term is an upper bound for Eve's mutual information, while for the first term, we need to perform the averaging for different channel coefficient estimates. The following overall SKR for this two-phase-based secret-key generation protocol is achievable [110]:

$$\begin{aligned} SKR &= \max_{f, P_s, T_s} \left(SKR_s + SKR_{ch}\right) \\ &s.t. P_s T_s + P_{ch}(T_c - T_s) \le T_c P, \end{aligned} \tag{11.129}$$

where the maximization is performed for all possible parameters f, P_s, and T_s.

For additional details on joint source-channel key generation protocols and *relay-assisted key generation*, an interested reader is referred to [110] and references therein.

11.9 Optical Channels Physical-Layer Security

Thanks to its flexibility, security, immunity to interference, high-beam directivity, and energy efficiency, the free-space optical (FSO) technology represents an excellent candidate for high-performance secure communications. Despite these advantages, large-scale deployment of FSO systems has so far been hampered by reliability and availability issues due to atmospheric turbulence in clear weather (*scintillation*), low visibility in foggy conditions, and high sensitivity to misalignment [33, 113]. Because of high directivity of optical beams, the FSO systems are much more challenging to intercept compared to

RF systems. Nevertheless, the eavesdropper can still apply the beam splitter on transmitter side, the blocking attack, or exploit beam divergence at the receiver side. The research on FSO physical-layer security is getting momentum, which can be judged based on increased number of recent papers related to this topic [83, 84, 114–117]. Most of papers on the physical-layer security for FSO communications are based on direct detection and employ wiretap channel approach introduced by Wyner [29].

In our recent papers [83, 84, 116, 117], we introduced a different strategy. It is well known that we can associate with a photon both spin angular momentum (SAM), related to polarization, and OAM, related to azimuthal dependence of the complex electric field [118–120]. Given that OAM eigenstates are orthogonal, these additional degrees of freedom can be utilized to improve both spectral efficiency and the physical-layer security in optical networks [12, 33, 74, 83, 84, 118–120]. Given that the spatial modes in spatial division multiplexing (SDM) fibers, such as few-mode fibers (FMFs), few-core fibers (FCFs), and few-mode-few-core fibers (FMFCFs), can be decomposed in terms of OAM eigenkets, the OAM can be used to enable the physical-layer security in both FSO and fiber-optics-based optical networks. Because OAM states provide an infinite basis state, while SAM states are two-dimensional only, the OAM can also be used to increase the security of QKD as we described in [124]. Therefore, the OAM eigenkets can be employed to provide physical-layer security on classical, semiclassical, and quantum levels.

11.9.1 SDM-Fibers-Based Physical-Layer Security

Three types of physical-layer security schemes based on spatial modes are possible, classical, semiclassical, and QKD schemes, depending on the desired level of security. It is well known that classical protocols rely on the computational difficulty of reversing the one-way functions and in principle cannot provide any indication of Eve's presence at any point in the communication process. However, the optical communication links can be operated at a desired margin from the receiver sensitivity, and for known channel conditions, Eve's beam-splitting attack can be detected as it will cause sudden decrease in *secrecy capacity* C_S, defined as:

$$C_S = C_{AB} - C_{AE}, \tag{11.130}$$

where C_{AB} is the instantaneous capacity of Alice-to-Bob channel and C_{AE} is the instantaneous capacity of Alice-to-Eve channel. Another relevant probabilistic measure that will be used in this paper to characterize the security of optical communication link is the *probability of strictly positive secrecy capacity*, defined as:

$$P_S^+ = \Pr(C_S > 0). \tag{11.131}$$

Our assumption for DWDM applications is that we are concerned with the physical-layer security of a particular WDM channel and that we cannot manipulate other WDM channels. From recent studies of spatial division multiplexing (SDM) systems, such as [121–123], we have learned that channel capacity can be increased linearly with number of spatial modes N, rather than logarithmically with signal-to-noise ratio for conventional 2-D schemes. These observations motivate us to employ the spatial modes to dramatically improve secrecy capacity when compared to conventional 2-D schemes. The use of SDM schemes to increase the secret-key rates is always sensitive to the crosstalk among spatial modes, and potential eavesdropper can compromise the security by relying on spatial coupling, without being detected by Alice and Bob. To solve for this problem, in addition to compensating for spatial modes coupling effects, it is possible to employ the multidimensional signaling. In multidimensional signaling, the spatial modes are used as bases functions, and by detecting the signal in any particular spatial mode, Eve will not be able to compromise security as only a single coordinate will be detected. Since the multidimensional signaling based on spatial modes has been already described in [33, 74, 121–123], here we just briefly describe the corresponding multidimensional scheme to be used for raw key transmission, which is shown in Fig. 11.28.

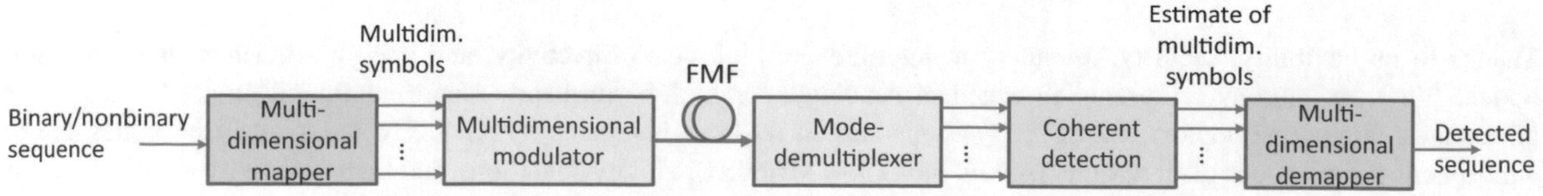

Fig. 11.28 The physical-layer security scheme employing mode-multiplexing-based multidimensional signaling

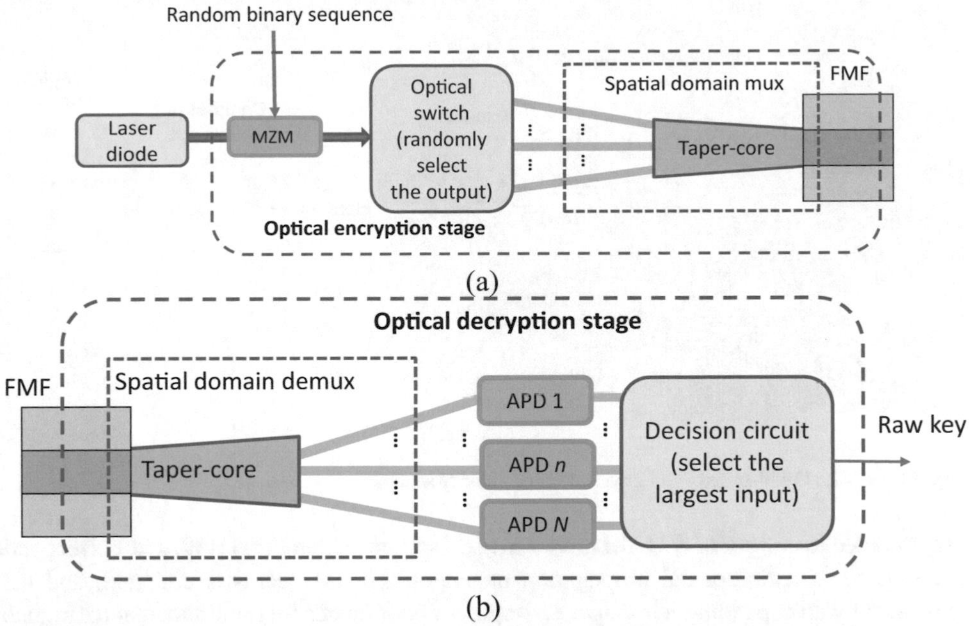

Fig. 11.29 SPML-semiclassical physical-layer security scheme: (**a**) optical encryption stage and (**b**) optical decryption stage. APD, avalanche photodiode

The configurations of spatial-modes-based multidimensional modulator and mode-demultiplexer are already provided in [122, 123]. Alice generates the binary sequence randomly. The multidimensional mapper can be implemented as a look-up table (LUT). For signal constellation size M, the $\log_2 M$ bits are used to find the coordinates of multidimensional signal constellation, obtained as described in [123]. The multidimensional coordinates are used as the inputs to corresponding Mach-Zehnder modulators (MZMs) of multidimensional modulator. After the mode-multiplexing, the signal is transmitted over SDM system of interest. On receiver side, after mode-demultiplexing and coherent detection, the estimated multidimensional coordinates are used as inputs of multidimensional a posteriori probability (APP) demapper, which provides the most probable symbol being transmitted, and the detected sequence is delivered to Bob. After that information reconciliation, based on systematic LDPC coding, is performed in similar fashion as already proposed for QKD applications [124]. To distill from the generated key a smaller set of bits whose correlation with Eve's string falls below the desired threshold, the privacy amplification is performed with the help of the *universal hash functions*.

In the rest of this subsection, we describe the corresponding *semiclassical physical-layer security* scheme. The simplest version, based on spatial position modulation-like (SPML) approach, is illustrated in Fig. 11.29.

Alice generates a random binary sequence, which is with the help of MZM converted into optical domain. The MZM's output drives the optical switch (OS) with ns switching speed, such as one introduced in [125]. The OS output is randomly selected. The OS outputs are used as inputs of spatial-domain multiplexer, whose configuration is provided in [122, 123]. Such encrypted signal is sent over FMF system to Bob. On receiver side, in the simplest (direct detection) version, Bob applies spatial-domain demultiplexer, whose each output branch drives an avalanche photodiode (APD). In decision, circuit, the largest output is selected as transmitted symbol. For M_s spatial modes, $\log_2 M_s$ bits of raw key are transmitted. In this simple scheme, MZM is used for framing purpose only. If coherent detection is used instead and MZM is replaced by I/Q modulator (with two RF inputs), we can transmit $\log_2 M + \log_2 M_s$ bits of raw key per channel use, where M is the size of I/Q constellation. The polarization state (not shown in figure) has not been used for raw key transmission, but to detect the presence of Eve, since the semiclassical encryption scheme is operated close to the quantum limit. After raw key transmission is completed, Alice encodes the raw key by employing multilevel nonbinary LDPC-coded modulation (ML-NB-LDPC-CM) [75]-based information reconciliation. The privacy amplification is then performed to distill for the shorter key with negligible correlation with Eve. This key is then used for secure communication, based on one-time pad or any symmetric cipher.

11.9.2 FSO Physical-Layer Security

For classical OAM-based physical-layer security scheme, the N raw key-carrying TEM_{00} modes are illuminated on a series of computer-generated holograms (CGHs), implemented, for instance, with the help of the spatial light modulators (SLMs), each

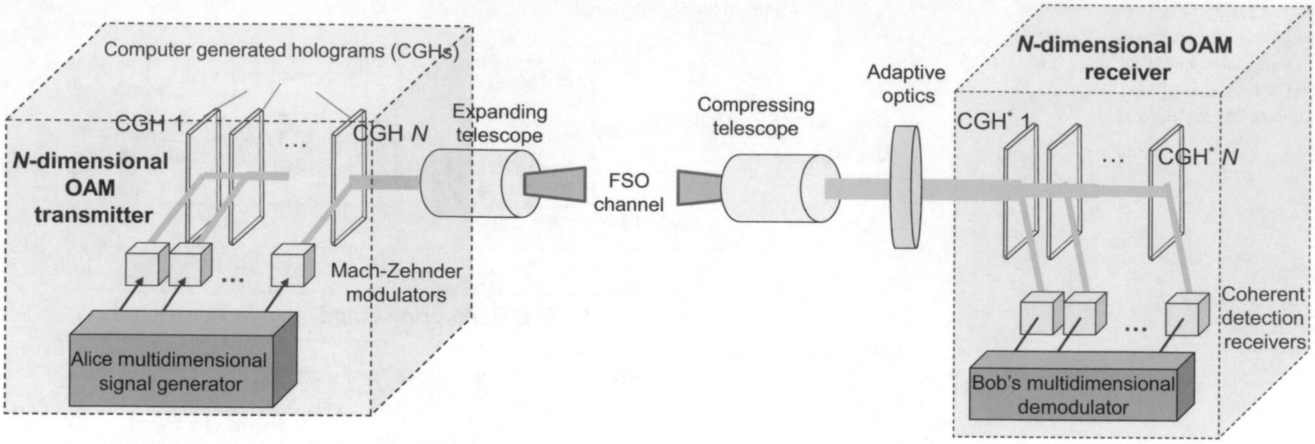

Fig. 11.30 The OAM-based FSO physical-layer security scheme

programmed to one out of N OAM modes in use, as illustrated in Fig. 11.30. The corresponding diffraction angles are properly adjusted so that the coaxial propagation of outgoing OAM beams is obtained, and the resulting superposition beam is expanded by an expanding telescope. To impose N coordinates for multidimensional signaling, a series of MZMs is used on a transmitter side. The number of bits required to select a point from multidimensional signal constellation of size M is given by $\log_2 M$. Therefore, $\log_2 M$ bits are used to select a point from multidimensional signal constellation, whose coordinates are stored in N-dimensional mapper, which can be implemented with a help of an arbitrary waveform generator (AWG). On receiver side, after compressing telescope, we pass the signal to adaptive optics subsystem to compensate for atmospheric turbulence effects. After that a series of conjugate volume holograms recorded on SLMs are used to determine the projections along corresponding OAM modes. These OAM projections (in optical domain) are used as inputs of corresponding coherent detectors to estimate the coordinates of transmitted multidimensional signal. For efficient implementation, one local laser is used for all coherent detectors. After coherent detection, the corresponding analog-to-digital converters (ADCs) outputs are passed to an N-dimensional APP demapper, in which symbol LLRs are calculated. After raw key transmission is completed, Alice encodes the raw key by employing ML-NB-LDPC-CM-based information reconciliation. The privacy amplification is further performed, to distill for the shorter key with negligible correlation with Eve. This key is then used for secure FSO communication.

To demonstrate high potential of the OAM-based physical-layer security scheme, the propagation of OAM modes at 1550 nm with the azimuthal index l from -20 to 20 and radial index p set to 0 is simulated in [83] by performing split-step propagation method [113]. The FSO link of length 1 km is observed. To model the strong turbulence effects, the use of 11 random phase screens is sufficient [83]. The phase power spectral density (PSD) used to generate random phase screens is a modified version of the Kolmogorov spectrum which includes both inner and outer scales and is given as [113]:

$$\Phi_n(\kappa) = 0.033C_n^2\left[1 + 1.802(\kappa/\kappa_l) - 0.254(\kappa/\kappa_l)^{7/6}\right]\frac{e^{-\kappa^2/\kappa_l^2}}{\left(\kappa^2 + \kappa_0^2\right)^{11/6}}, \tag{11.132}$$

where κ is the spatial frequency, $\kappa_l = 3.3/l_0$, $\kappa_0 = 1/L_0$, and C_n^2 is the refractive index structure parameter that indicates the turbulence strength; l_0 and L_0 are the inner and outer scale of the turbulence, respectively. The propagation of Laguerre-Gaussian (LG) beams with $n \in S = \{-20, \ldots, -1, 0, 1, \ldots, 20\}$ of $N = 41$ OAM modes in total is simulated [83] for three OAM subsets: $S' = \{-18, \ldots, -3, 0, 3, \ldots, 18\}$, $S'' = \{-20, \ldots, -5, 0, 5, \ldots, 20\}$, and $S''' = \{-14, -7, 0, 7, 14\}$ with spacing 2, 4, and 6 between selected modes, respectively, following by the calculation of the aggregate secrecy rate. The results of calculation are summarized in Fig. 11.31 [83]. The worst-case scenario, when Eve is located on a transmitter side and taps the portion r_e of the transmitted optical power, is considered. As we can see, in the weak turbulence regime, the total secrecy capacity can be improved to close to two orders of magnitude by using OAM multiplexed beam. However, as the refractive index structure parameter C_n^2 increases, the secrecy capacity decreases due to orthogonality loss caused by turbulence effects. Comparing with the secrecy capacity of using single channel (denoted as LG_{00} mode), the use of OAM multiplexing is not beneficial in very strong turbulence regime without adaptive optics.

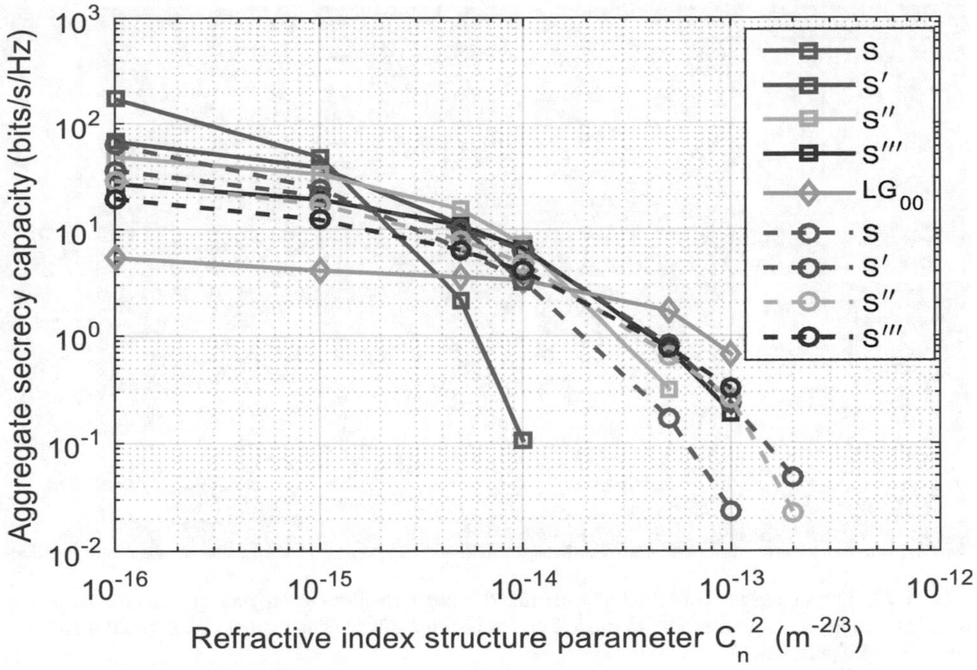

Fig. 11.31 The aggregate secrecy capacity vs. refractive index structure parameter. Solid lines: equal power per channel. Dashed lines: fixed system power (fixed transmitted power equally divided among OAM channels). The portion of transmitted power taped by Eve is set to 0.01. Signal-to-noise ratio is set to 20 dB. (After ref. [83]; © IEEE 2016; reprinted with permission)

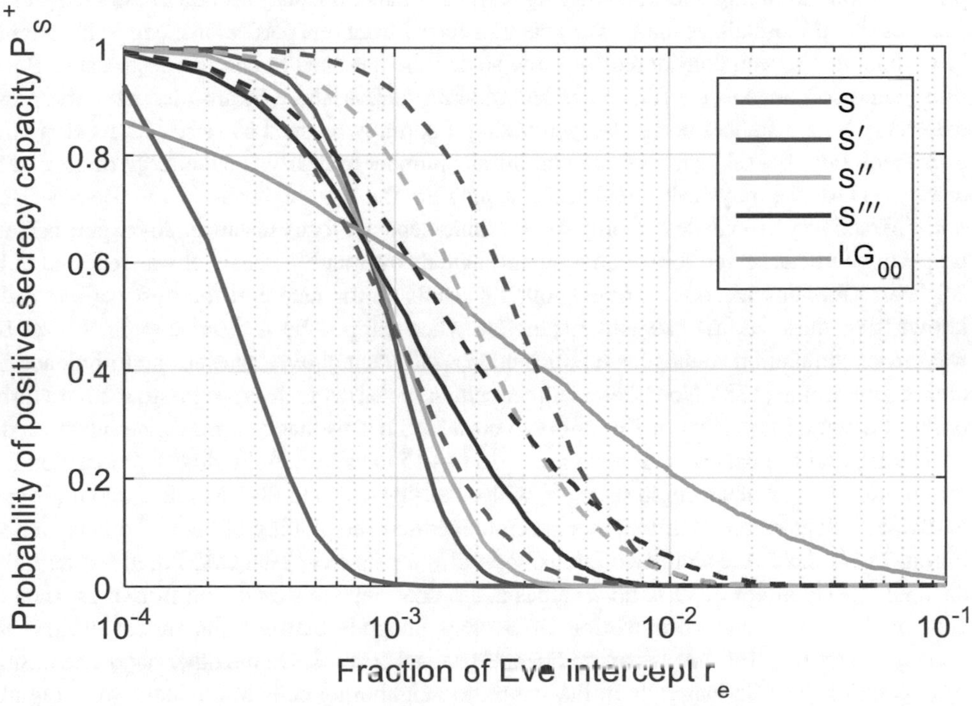

Fig. 11.32 The probability of positive secrecy capacity vs. Eve's interception fraction. Solid lines, equal power per channel; dashed lines, fixed system power. Rytov variance is set to 4 (corresponding to strong turbulence). (After ref. [83]; © IEEE 2016; reprinted with permission)

We can also use the Rytov variance $\sigma_R^2 = 1.23C_n^2 k^{7/6} L^{11/6}$ to represent the turbulence strength, which takes the propagation distance L, the operating wavelength, and the refractive structure parameter into account. For our FSO system under study (a path of 1 km and wavelength at 1550 nm), the range of σ_R^2 spans from 0.002 to 10, while for C_n^2 from 10^{-16} to 5×10^{-13}. As an illustration, the *probability of strictly positive secrecy capacity* P_S^+ by Eq. (11.131) is calculated, which represents a probabilistic metric for characterization the secrecy for the communication. The results for P_S^+ corresponding to strong turbulence regime are summarized in Fig. 11.32 [83].

In an ideal homogeneous medium, the vorticity of OAM modes is preserved as they propagate and their wavefronts remain orthogonal even after undergoing diffraction. However, in an atmospheric turbulence channel, their orthogonality is no longer

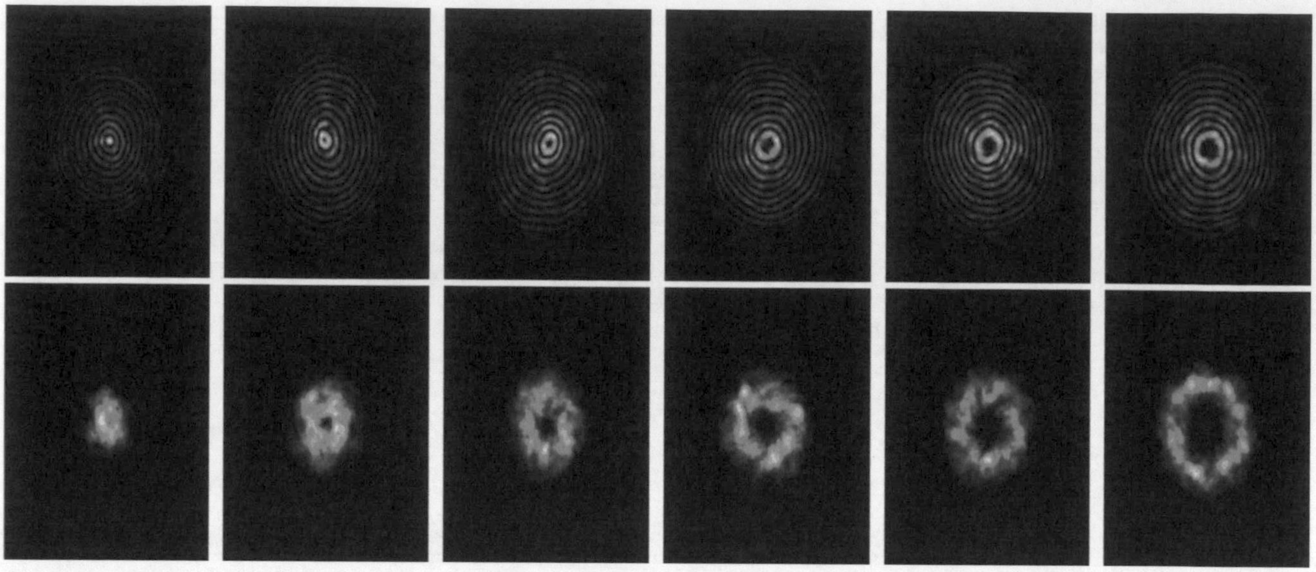

Fig. 11.33 The simulated OAM modes from order 0 to order +5 after 1 km of propagation distance through $C_n^2 = 10^{-14}$ turbulence. The top row contains sample beam profiles for BS modes, and the bottom row contains sample beam profiles for LG modes. (After ref. [84]; © IEEE 2016; reprinted with permission)

preserved due to fading and time-varying refractive index fluctuations that cause intensity scintillations in the beam profiles. This results in crosstalk as the power gets transferred from one particular mode to its neighbors. An intriguing possibility for FSO links that's been considered for quite some time has been to communicate using Bessel beams. In addition to forming orthogonal solutions to the free-space Helmholtz equation, these beams also have the property of being non-diffracting for extremely long distances and self-regenerating if partially blocked by obstructions along their propagation axis [126]. However theoretical Bessel beams contain an infinite number of nearly equal energy rings, and by definition they have an infinite energy. Thus the physically realizable analog to the Bessel beam is the Bessel-Gaussian (BS) beam that can also non-diffractively propagate but only over a finite depth of focus distance. An experimental demonstration of this promising property being harnessed for use in transmitting data over an FSO channel was reported in [127]. However, the robustness of BS beams remains tenuous and very much depends on the nature of the obstructions that the beams encounter. While it's known that these beams can still regenerate when clipped by a discrete point object blocking a small section of their wavefront, simulation studies have indicated that they cannot overcome strong turbulence-induced phase changes that distort their entire profile [128]. Nonetheless under certain scenarios such as a weak to medium turbulence channel of distance on the order of several kilometers, it's possible to create BS beams that can propagate intact, as demonstrated in [84].

The BS and LG beams of orders from −15 to +15 have been individually propagated in consecutive manner over a 1 km turbulence channel of specified strength, with details provided in [84]. Monte Carlo trials are done for each C_n^2 value with the particular values being intentionally chosen to facilitate the plotting of the C_n^2 points on a semi-logarithmic horizontal scale. Figure 11.33 shows the simulated intensity profiles for several of the OAM modes after traveling through a strong turbulence channel. The complex electric field values at the final step are stored, and from these data the channel crosstalk matrices are generated by pair-wise computation of overlap integrals between the modal electric fields. An identity matrix would correspond to the ideal case where no channel crosstalk occurs. On the other hand, the turbulence would cause power leakage from each of the diagonal cells in the matrix to neighboring cells in the same row. Figure 11.34 contains crosstalk matrix results arranged in increasing turbulence strength for several C_n^2 values. A visual inspection shows slightly more power leakage among the LG modes than among the BS modes when the C_n^2 value is below 10^{-15} $\mathrm{m}^{-2/3}$. To quantify these effects, the aggregate secrecy capacity is calculated over the entire set of modes and corresponding crosstalk matrices using the information-theoretic security formulas, with the results being summarized in Fig. 11.35 [84]. Two curves, one for the case of Eve near the transmitter and one for Eve near the receiver, are shown. As expected, the aggregate C_S values decrease with increasing turbulence strength. For reference, the maximum achievable aggregate secrecy capacity is 175 bits/sec/Hz when using consecutive orders from −15 to +15 in a perfect channel having no crosstalk regardless of whether Eve is on the transmit or receive sides.

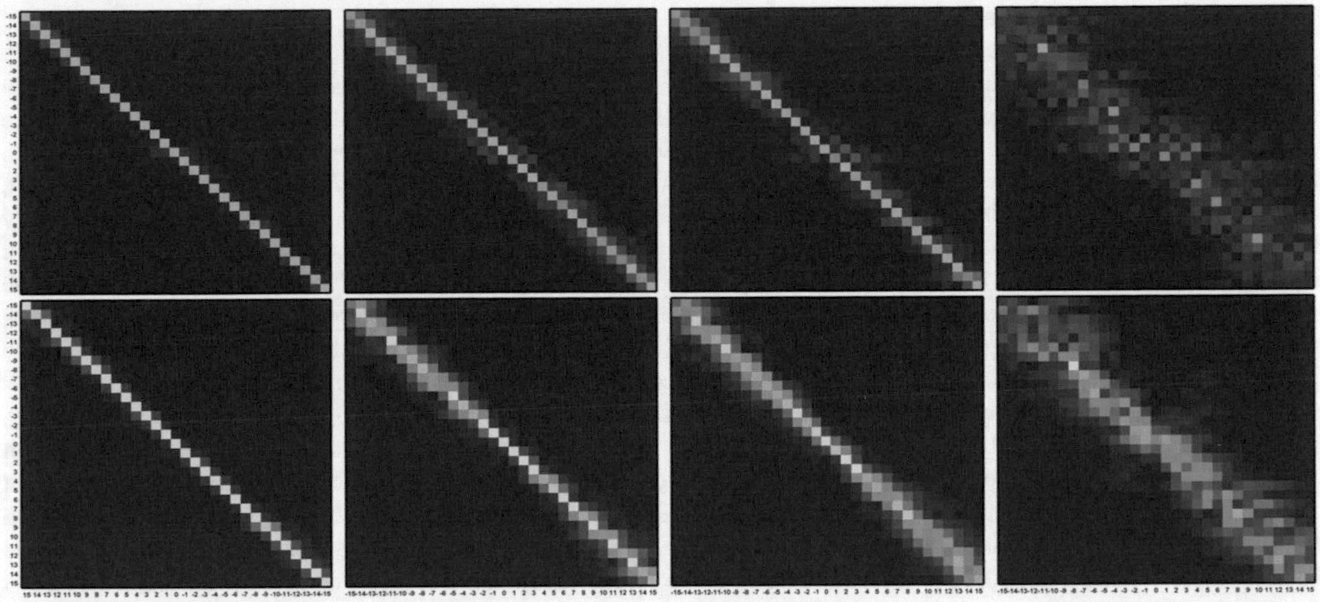

Fig. 11.34 The simulation crosstalk matrices for BS modes (top) and LG modes (bottom) for $C_n^2 = 10^{-17}$, 10^{-16}, $1.75\text{x}10^{-16}$, $1.75\text{x}10^{-15}$ (going left to right). Transmit orders are labeled along the vertical axis and the orders used to receive them are labeled along the horizontal axis. (After ref. [84]; © IEEE 2016; reprinted with permission)

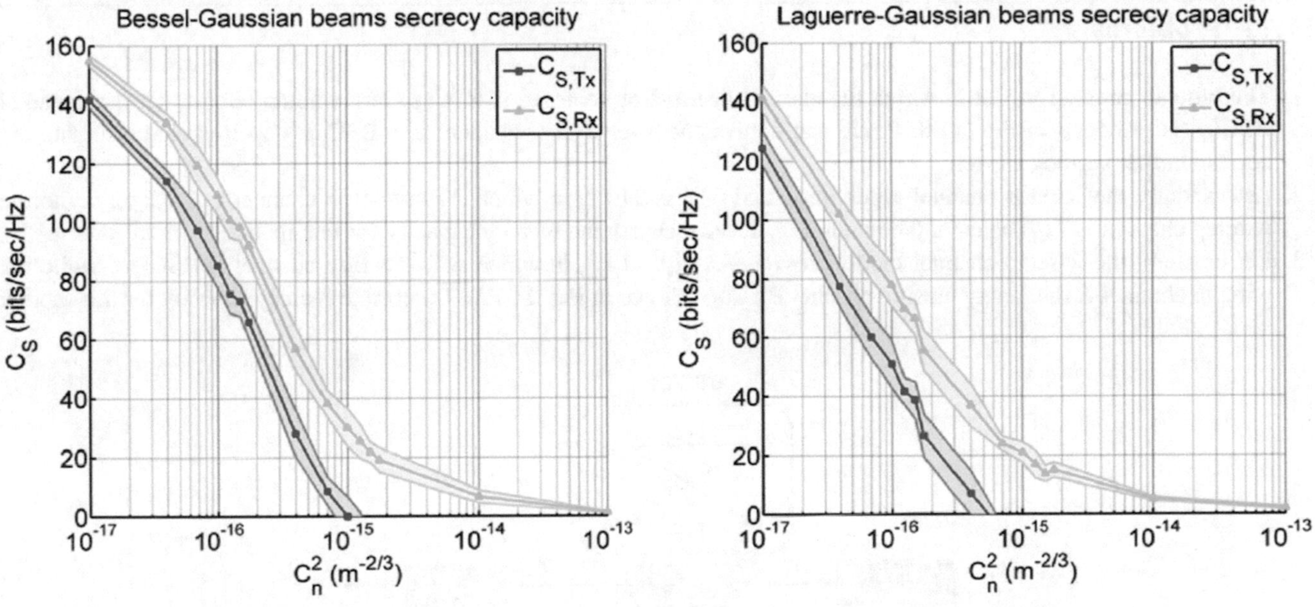

Fig. 11.35 The plots of aggregate secrecy capacity calculated from crosstalk matrices over orders −15 to +15 as a function of turbulence strength. The shaded bands surrounding the data points represent error bars over the Monte Carlo runs. (After ref. [84]; © IEEE 2016; reprinted with permission)

Comparing the data points in the BS and LG curves, we find that the C_S values are typically higher by 10 to 30 bits/sec/Hz when using BS beams, which indicates better PLS performance than when using LG beams. However, this improvement is mostly negated in strong turbulence when the C_n^2 value is above 10^{-15} $m^{-2/3}$. Under these channel conditions, we observe the capacity approaches 0 bit/sec/Hz if the Eve is located near the receiver and becomes negative if Eve is located near transmitter. For corresponding experimental results, an interested reader is referred to [117].

11.10 Concluding Remarks

This chapter has been devoted to the physical-layer security fundamentals. The chapter starts with discussion on security issues (Sect. 11.1), followed by Sect. 11.2 where the information-theoretic security is introduced and comparison against the computational security has been performed. In the same section, various information-theoretic security measures have been introduced, including strong secrecy and weak secrecy conditions. After that, in Sect. 11.3 Wyner's wiretap channel model, also known as the degraded wiretap channel model, has been introduced. In the same section, the concept of secrecy capacity has been introduced as well as the nested wiretap coding. Further, in Sect. 11.4, the broadcast channel with confidential messages has been introduced, and the secrecy capacity definition has been generalized. The focus has been then moved to the secret-key generation (agreement) protocols in Sect. 11.5, where the source- and channel-type models have been introduced, and corresponding secret-key generation protocols have been described. Section 11.6 has been devoted to the coding for the physical-layer security systems, including both coding for weak and strong secrecy systems. Regarding the coding for weak secrecy systems, described in Subsection 11.6.1, the special attention has been paid to two-edge-type LDPC coding, punctured LDPC coding, and polar coding. Regarding the coding for strong secrecy systems, described in Subsection 11.6.2, the focus has been on coset coding with dual of LDPC codes and hash functions/extractor-based coding. In Subsection 11.6.3, the information reconciliation has been introduced and described. In Sect. 11.7, the concept of privacy amplification has been introduced and described. In wireless channels physical-layer security section, Sect. 11.8, the following topics have been covered: wireless MIMO PLS (Subsection 11.8.1) and secret-key generation in wireless networks (Subsection 11.8.2). In Sect. 11.9, related to the optical channels PLS, both PLS for spatial division multiplexing (SDM)-fibers-based systems (subsection 11.9.1) and FSO systems (Subsection 11.9.2) have been discussed. The set of problems is provided in incoming section to help reader gain deeper understanding of material described in this chapter.

11.11 Problems

1. The wiretap channel model in which the main (Alice-to-Bob) channel is BSC(p) and wiretap (Bob-to-Eve) channel is BSC(q) is provided in Fig. 11.8. Prove that equivalent Alice-to-Eve channel is a BSC(p + q-2pq), and calculate the corresponding secrecy capacity.
2. Let us study the wiretap channel model provided in Fig. 11.P2, in which Alice-to-Bob channel is BSC(p), while the wiretap channel is Z(q) channel (shown in Fig. 11.P2). Determine secrecy capacity for this model.
3. Let us study the wiretap channel model provided in Fig. 11.P3, in which Alice-to-Bob channel is BSC(p), while the wiretap channel is the binary erasure channel BEC(q) (shown in Fig. 11.P2). Determine secrecy capacity for this model.

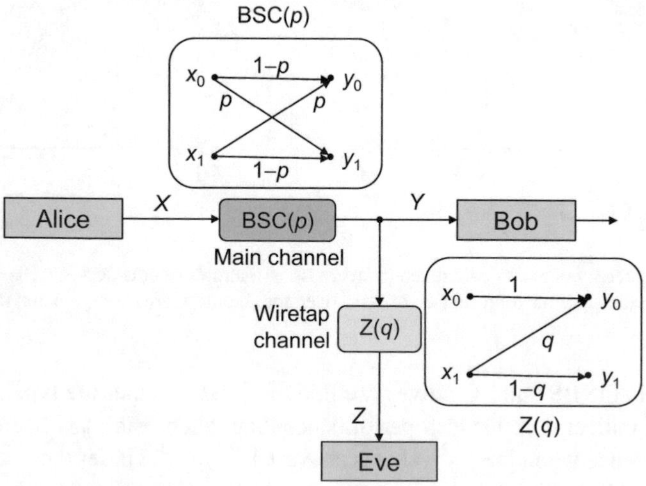

Fig. 11.P2 The wiretap channel model under study

4. Describe how the wiretap code, described in Sect. 11.3, should be modified for the wiretap channel model provided in Fig. 11.P3.

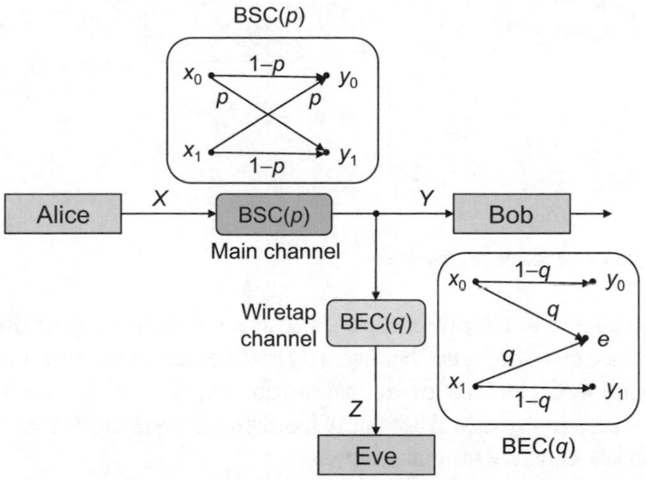

Fig. 11.P3 The wiretap channel model in which main channel is BSC and wiretap channel is BEC

5. Prove the Csiszár-Körner corollary for a broadcast channel with confidential messages, given by Eq. (11.25).
6. Calculate the FSO secrecy capacity for IM/DD in the weak turbulence regime where the distribution of irradiance is governed by the log-normal distribution, and thermal noise of transimpedance front-end is the dominant noise source.
7. Let us consider indoor infrared communication with IM/DD in which LED is used a Lambertian source and it is placed on a ceiling. Bob's receiver is placed directly below the LED. The impulse response of the main channel is composed of the LOS and four dominant reflections from the walls. What would be the best strategy for Eve to locate her receiver? Describe how to calculate the secrecy capacity for various Eve's scenarios.
8. This problem is related to physical-layer security (PLS) *applied to 9-ary 2-D constellation* described in Problem 8 of Chap. 5, in the presence of scintillation and AWGN. By employing Wyner's wiretap PLS model, determine the secrecy capacity assuming:
   (a) Eve is located on transmitter side and employs a beam-splitting attack.
   (b) Eve is located on receive side and employs the beam divergence attack.
   (c) Eve is located in the middle of the FSO link and exploits Mie scattering effects.
9. In this problem we explore the simulator developed in Problem 1 of Chap. 2. Assume that Eve employs any of three strategies from Problem 8 and that QPSK is used in the main channel, which is imposed on Gaussian beam by the I/Q modulator:
   (a) Determine the secrecy capacity for different Eve's scenario assuming that operating wavelength in the main channel is 1550 nm for different turbulence strengths (as specified in Problem 1 of Chap. 2).
   (b) Let us assume that the symbol rate per wavelength is 20 GS/s. Assume further dense WDM is used with channel spacing of 25 GHz. Calculate the crosstalk matrix for different turbulence strengths (as specified in Problem 1 of Chap. 2), wherein the number of wavelength channels is used as a parameter.
   (c) By using the crosstalk matrix developed in (b), calculate overall secrecy capacity for different number of channels and different Eve's strategies.
10. This problem is a continuation of the previous problem. Assume now that adaptive optics (AO) is used in main channel. Under assumption that Eve does not use the AO in her channel, repeat the steps (a)–(c) from the previous problem, and discuss the improvements. The AO system should be based on your solution from Problem 26 from Chap. 6.
11. In ref. [129], the atmospheric turbulence is used, among others, as the source of randomness for secret-key generation (SKG). Here we consider one possible application of this scenario. In Fig. 11.P11 a simplified version of the source-type SKG scheme is provided. Alice and Bob simultaneously send the CW laser beams with the help of compressing telescopes over the time-varying atmospheric turbulence channel. On receiver side, Alice and Bob after the compressing telescope and beam splitter detect the received optical signals by using either direct detection or coherent detection receiver. With direct detection, the fluctuations in intensity (scintillation) can be used as the common source of

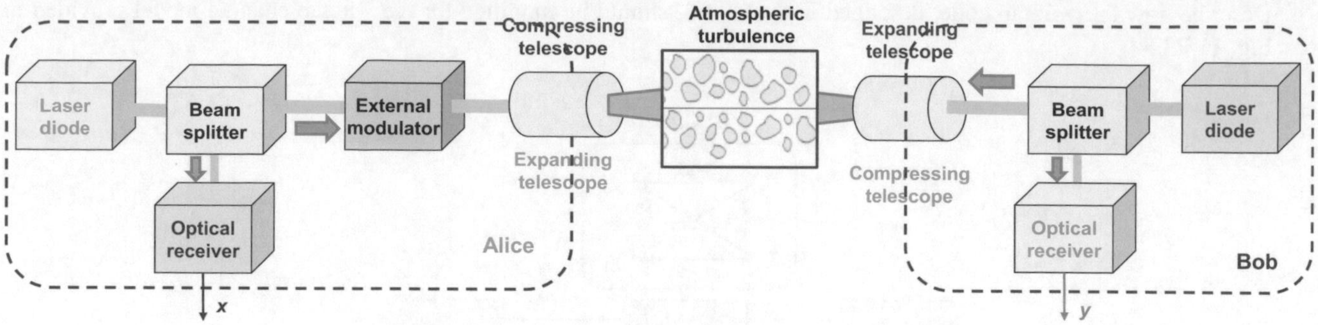

Fig. 11.P11 Atmospheric turbulence-based secret-key generation

randomness. They can quantize the intensity samples, because the reciprocity of the channel are the same. Eve will experience different turbulence conditions, and her intensity samples are uncorrelated with Alice and Bob's ones. Assume that turbulence conditions are weak, and the log-normal distribution of irradiance can be used. Assuming that both Alice and Bob employ identical transimpedance front-ends, the thermal noise from electrical amplifiers will be Gaussian. Determine the weak secret-key capacity for this scheme.

12. In this problem we are concerned with a particular coding scheme for the weak secrecy systems, the punctured LDPC coding, which is described in Sect. 11.6.1.2. In this scheme, the message bits $\boldsymbol{m}$ and the dummy random bits $\boldsymbol{m}'$ are combined before the LDCP coding takes place. In punctured FEC schemes, we typically puncture the parity bits, and these punctured bits are not transmitted. In this scheme, however, we puncture the message bits $\boldsymbol{m}$ instead. Please refer to the section just cited for additional details. Let us observe an LDPC code example provided in Sect. 9.10.5.4, which is of rate 0.8 and codeword length 34,665. Assuming that the target BER for Bob is $\leq 10^{-6}$ for the AWGN channel, determine the *security gap* if the minim BER of Eve that can be tolerated is 0.3.
13. This problem is a continuation of the previous problem in the presence of atmospheric turbulence, modeled with the gamma-gamma distribution of irradiance for different turbulence strength: weak turbulence with Rytov variance 0.01, medium turbulence with Rytov variance 1, and strong turbulence with Rytov variance of 20. Determine the security gap for different turbulence strengths with other parameters being specified in the previous problem.
14. Design a polar code with similar code rate and codeword length as the LDPC code used in Problem 12. Under the same assumptions as in Problem 12, determine the security gap, and compare it against that obtained in Problem 12.
15. Repeat Problem 13 but now employing the polar code designed in Problem 14. Discuss the security gaps for polar and LDPC codes.
16. Let the LDPC and polar codes used in Problems 12 and 14, respectively, be used in information reconciliation as described in Sect. 11.6.3. Determine reconciliation efficiencies of these two codes and discuss which one is better.
17. Let us consider 2×2 MIMO wireless system with elements of channel matrix being spatially white. Assuming the perfect CSI for Bob and Eve for multiple-input single-output multiple-Eve (MISOME) case, determine the MIMO secrecy capacity.
18. Let us now consider the FSO channel with coherent detection with distribution of irradiance following the gamma-gamma distribution and distribution of variance being Gaussian. For different turbulence strengths and MISOME case, determine the corresponding secrecy capacities.
19. Let us consider the *K*-user wireless communication channel. For the same conditions as in Problem 17, determine the secret multiplexing gain.
20. Let us consider the *K*-user FSO communication channel. For the same conditions as in Problem 18, determine the secret multiplexing gain.

References

1. X.800: Security architecture for Open Systems Interconnection for CCITT applications, Recommendation X.800 (03/91), available at: https://www.itu.int/rec/T-REC-X.800-199103-I
2. I.B. Djordjevic, *Physical-Layer Security and Quantum Key Distribution* (Springer, Cham, 2019)
3. M. Bloch, Physical-Layer Security. PhD dissertation, School of Electrical and Computer Engineering, Georgia Institute of Technology (2008)

4. M. Bloch, J. Barros, *Physical-Layer Security: From Information Theory to Security Engineering* (Cambridge University Press, Cambridge, 2011)
5. M. Bloch, Fundamentals of physical layer security, in *Physical Layer Security in Wireless Communications*, ed. by X. Zhou, L. Song, Y. Zhang, (CRC Press, Boca Raton/London/New York, 2014), pp. 1–16
6. A. Chorti et al., Physical layer security: A paradigm shift in data confidentiality, in *Physical and Data-Link Security Techniques for Future Communications Systems*, Lecture notes in electrical engineering 358, (Springer, 2016), pp. 1–15
7. M. Bloch, J. Barros, M.R.D. Rodrigues, S.W. McLaughlin, Wireless information-theoretic security. IEEE Trans. Inform. Theory **54**(6), 2515–2534 (2008)
8. C.H. Bennet, G. Brassard, Quantum cryptography: Public key distribution and coin tossing, in *Proceedings of IEEE International Conference on Computers, Systems, and Signal Processing*, (Bangalore, 1984), pp. 175–179
9. C.H. Bennett, Quantum cryptography: Uncertainty in the service of privacy. Science **257**, 752–753 (1992)
10. I.B. Djordjevic, *Quantum Information Processing and Quantum Error Correction: An Engineering Approach* (Elsevier/Academic Press, Amsterdam-Boston, 2012)
11. X. Sun, I.B. Djordjevic, M.A. Neifeld, Multiple spatial modes based QKD over marine free-space optical channels in the presence of atmospheric turbulence. Opt. Express **24**(24), 27663–27673 (2016)
12. Z. Qu, I.B. Djordjevic, Four-dimensionally multiplexed eight-state continuous-variable quantum key distribution over turbulent channels. IEEE Photonics Journal **9**(6), 7600408-1-7600408-8 (2017)
13. I.B. Djordjevic, FBG-based weak coherent state and entanglement assisted multidimensional QKD. IEEE Photonics Journal **10**(4), 7600512-1-7600512-12 (2018)
14. M.A. Neilsen, I.L. Chuang, *Quantum Computation and Quantum Information* (Cambridge University Press, Cambridge, 2000)
15. G. Van Assche, *Quantum Cryptography and Secrete-Key Distillation* (Cambridge University Press, Cambridge/New York, 2006)
16. C.H. Bennett, G. Brassard, C. Crepeau, U. Maurer, Generalized privacy amplification. IEEE Inform. Theory **41**(6), 1915–1923 (1995)
17. C.E. Shannon, Communication theory of secrecy systems. Bell Sys. Tech. J. **28**(4), 656–715 (1949)
18. G.S. Vernam, Cipher printing telegraph systems for secret wire and radio telegraphic communications. Trans. Am. Inst. Electr. Eng. **1**, 295–301 (1926)
19. D. Kahn, *The Codebreakers: The Story of Secret Writing* (Macmillan Publishing Co., Ney York, 1967)
20. G.D. Forney, On the role of MMSE estimation in approaching the information theoretic limits of linear Gaussian channels: Shannon meets Wiener, in *Proceedings of 41st Annual Allerton Conference on Communication, Control, and Computing*, (Monticello, 2003), pp. 430–439
21. B. Schneier, *Applied Cryptography, Second Edition: Protocols, Algorithms, and Source Code in C* (Wiley, Indianapolis, 2015)
22. J. Katz, Y. Lindell, *Introduction to Modern Cryptography*, 2nd edn. (CRC Press, Boca Raton, 2015)
23. J.-P. Aumasson, *Serious Cryptography: A Practical Introduction to Modern Encryption* (No Starch Press, San Francisco, 2018)
24. J. Sebbery, J. Pieprzyk, *Cryptography: An Introduction to Computer Security* (Prentice Hall, New York, 1989)
25. H. Delfs, H. Knebl, *Introduction to Cryptography: Principles and Applications (Information Security and Cryptography)*, 3rd edn. (Springer, Heidelberg/New York, 2015)
26. M. Bloch, M. Hayashi, A. Thangaraj, Error-control coding for physical-layer secrecy. Proc. IEEE **103**(10), 1725–1746 (2015)
27. T.M. Cover, J.A. Thomas, *Elements of Information Theory* (Wiley, New York, 1991)
28. M. Bloch, J.N. Laneman, On the secrecy capacity of arbitrary wiretap channels, in *Proceedings of 46th Annual Allerton Conference on Communication, Control, and Computing*, (Monticello, 2008), pp. 818–825
29. A.D. Wyner, The wire-tap channel. Bell Syst. Tech. J. **54**(8), 1355 1387 (1975)
30. F. Lin, F. Oggier, Coding for wiretap channels, in *Physical Layer Security in Wireless Communications*, ed. by X. Zhou, L. Song, Y. Zhang, (CRC Press, Boca Raton/London/New York, 2014), pp. 17–32
31. S.K. Leung-Yan-Cheong, Multi-User and Wiretap Channels Including Feedback. PhD dissertation, Stanford University (1976)
32. I. Csiszár, J. Körner, Broadcast channels with confidential messages. IEEE Trans. Inf. Theory **24**(3), 339–348 (1978)
33. I.B. Djordjevic, *Advanced Optical and Wireless Communications Systems* (Springer, Cham, 2017)
34. A. Goldsmith, *Wireless Communications* (Cambridge University Press, Cambridge, 2005)
35. D. Tse, P. Viswanath, *Fundamentals of Wireless Communication* (Cambridge University Press, Cambridge, 2005)
36. J. Barros, M.R.D. Rodrigues, Secrecy capacity of wireless channels, in *Proceedings of IEEE International Symposium on Information Theory 2006 (ISIT 2006)*, (Seattle, 2006), pp. 356–360
37. J. Barros, M. Bloch, Strong secrecy for wireless channels, in *Information-Theoretic Security*, Lecture Notes in Computer Science, vol. 5155, (Springer, Berlin, 2008), pp. 40–53
38. U.M. Maurer, Secret key agreement by public discussion from common information. IEEE Trans. Inf. Theory **39**(3), 733–742 (1993)
39. R. Ahlswede, I. Csiszár, Common randomness in information theory and cryptography-part I: Secret sharing. IEEE Trans. Inf. Theory **39**(4), 1121–1132 (1993)
40. U.M. Maurer, S. Wolf, Unconditionally secure key agreement and intrinsic conditional information. IEEE Trans. Inf. Theory **45**(2), 499–514 (1999)
41. I. Csiszár, P. Narayan, Secrecy capacities for multiple terminals. IEEE Trans. Inf. Theory **50**(12), 3047–3061 (2004)
42. X. Zhou, L. Song, Y. Zhang (eds.), *Physical layer security in wireless communications* (CRC Press, Boca Raton/London/New York, 2014)
43. L.H. Ozarow, A.D. Wyner, Wire-tap channel II. AT&T Bell Lab. Tech. J. **63**(10), 2135–2157 (1984)
44. R. Zamir, S. Shamai, U. Erez, Nested linear/lattice codes for structured multiterminal binning. IEEE Trans. Inf. Theory **48**(6), 1250–1276 (2002)
45. A. Thangaraj, S. Dihidar, A.R. Calderbank, S.W. McLaughlin, J. Merolla, Applications of LDPC codes to the Wiretap Channel. IEEE Trans. Inform. Theory **53**(8), 2933–2945 (2007)
46. M. Andersson, *Coding for Wiretap Channels*. Thesis, School of Electrical Engineering, Royal Institute of Technology (KTH), Sweden (2011)
47. V. Rathi, R. Urbanke, M. Andersson, M. Skoglund, Rate-equivocation optimal spatially coupled LDPC codes for the BEC wiretap channel, in *Proceedings of 2011 IEEE International Symposium on Information Theory Proceedings*, (St. Petersburg, 2011), pp. 2393–2397
48. R.M. Tanner, A recursive approach to low complexity codes. IEEE Tans. Inf. Theory **27**, 533–547 (1981)

49. J. Boutros, O. Pothier, G. Zemor, Generalized low density (Tanner) codes. In Proc. IEEE Int. Conf. Comm. **99**, 441–445 (1999)
50. M. Lentmaier, K. Zigangirov, On generalized low-density parity-check codes based on hamming component codes. IEEE Comm. Lett. **3**(8), 248–250 (1999)
51. I.B. Djordjevic, O. Milenkovic, B. Vasic, Generalized low-density parity-check codes for optical communication systems. J. Lightw. Technol. **23**(5), 1939–1946 (2005)
52. I.B. Djordjevic, L. Xu, T. Wang, M. Cvijetic, GLDPC codes with reed-Muller component codes suitable for optical communications. IEEE Comm. Lett. **12**(9), 684–686 (2008)
53. I.B. Djordjevic, T. Wang, Multiple component codes based generalized LDPC codes for high-speed optical transport. Opt. Express **22**(14), 16694–16705 (2014)
54. D. Zou, I.B. Djordjevic, An FPGA design of generalized low-density parity-check codes for rate-adaptive optical transport networks, in *Proceedings of SPIE PhotonicsWest 2016, Optical Metro Networks and Short-Haul Systems VIII, SPIE*, vol. 9773, (San Francisco, 2016), p. 97730M-1-97730M-6
55. C.W. Wong, T. Wong, J. Shea, Secret-sharing LDPC codes for the BPSK constrained Gaussian wiretap channel. IEEE Trans. Inform. Forensics Sec. **6**(3), 551–564 (2011)
56. D. Klinc, J. Ha, S.W. McLaughlin, J. Barros, B. Kwak, LDPC codes for the Gaussian Wiretap Channel. IEEE Trans. Inform. Forensics and Security **6**(3), 532–540 (2011)
57. E. Arikan, Channel polarization: A method for constructing capacity-achieving codes for symmetric binary-input memoryless channels. IEEE Trans. Inform. Theory **55**(7), 3051–3073 (2009)
58. H. Mahdavifar, A. Vardy, Achieving the secrecy capacity of wiretap channels using polar codes. IEEE Trans. Inform. Theory **57**(10), 6428–6443 (2011)
59. O.O. Koyluoglu, H. El Gamal, Polar coding for secure transmission and key agreement. IEEE Trans. Inform. Forensics and Security **7**(5), 1472–1483 (2012)
60. R. Liu, Y. Liang, H.V. Poor, P. Spasojevic, Secure nested codes for type II wiretap channels, in *Proceedings of 2007 IEEE Information Theory Workshop*, (Tahoe City, 2007), pp. 337–342
61. J. Belfiore, F. Oggier, Secrecy gain: A wiretap lattice code design, in *Proceedings of 2010 International Symposium On Information Theory & Its Applications*, (Taichung, 2010), pp. 174–178
62. F. Oggier, P. Solé, J. Belfiore, Lattice codes for the wiretap Gaussian Channel: Construction and analysis. IEEE Trans. Inform. Theory **62**(10), 5690–5708 (2016)
63. A. Thangaraj, S. Dihidar, A.R. Calderbank, S.W. McLaughlin, J.-M. Merolla, Applications of LDPC codes to the wiretap channel. IEEE Trans. Inform. Theory **53**(8), 2933–2945 (2007)
64. A. Subramanian, A. Thangaraj, M. Bloch, S.W. McLaughlin, Strong secrecy on the binary erasure wiretap channel using large-girth LDPC codes. IEEE Trans. Inform. Forensics Sec. **6**(3), 585–594 (2011)
65. A.T. Suresh, A. Subramanian, A. Thangaraj, M. Bloch, S.W. McLaughlin, Strong secrecy for erasure wiretap channels, in *Proceedings of IEEE Information Theory Workshop 2010 (ITW2010)*, (Dublin, 2010)
66. A. Thangaraj, Coding for wiretap channels: Channel resolvability and semantic security, in *Proceedings of 2014 IEEE Information Theory Workshop (ITW 2014)*, (Hobart, 2014), pp. 232–236
67. G. Cohen, G. Zémor, Syndrome-coding for the wiretap channel revisited, in *Proceedings of IEEE Information Theory Workshop 2006*, (Chengdu, 2006), pp. 33–36
68. M. Hayashi, R. Matsumoto, Construction of wiretap codes from ordinary channel codes, in *Proceedings of IEEE International Symposium Information Theory (ISIT) 2010*, (Austin, 2010), pp. 2538–2542
69. M. Bellare, S. Tessaro, A. Vardy, Semantic security for the wiretap channel, in *Advances in Cryptology—CRYPTO 2012*, Lecture Notes in Computer Science, vol. 7417, (Springer, Berlin/Heidelberg), pp. 294–311
70. M. Cheraghchi, F. Didier, A. Shokrollahi, Invertible extractors and wiretap protocols. IEEE Trans. Inform. Theory **58**(2), 1254–1274 (2012)
71. W.K. Harrison, J. Almeida, M.R. Bloch, S.W. McLaughlin, J. Barros, Coding for secrecy: An overview of error-control coding techniques for physical-layer security. IEEE Signal Proc. Mag. **30**(5), 41–50 (2013)
72. D. Slepian, J.K. Wolf, Noiseless coding of correlated information sources. IEEE Trans. Inf. Theory **19**(4), 471–480 (1973)
73. A.D. Liveris, Z. Xiong, C.N. Georghiades, Compression of binary sources with side information using low-density parity-check codes, in *Proceedings of IEEE Global Telecommunications Conference 2002 (GLOBECOM '02)*, vol. 2, (Taipei, 2002), pp. 1300–1304
74. I.B. Djordjevic, On advanced FEC and coded modulation for ultra-high-speed optical transmission. IEEE Commun. Surveys and Tutorials **18**(3), 1920–1951 (2016). https://doi.org/10.1109/COMST.2016.2536726
75. C. Cachin, U.M. Maurer, Linking information reconciliation and privacy amplification. J. Cryptol. **10**, 97–110 (1997)
76. J.L. Carter, M.N. Wegman, Universal classes of hash functions. J. Comput. Syst. Sci. **18**(2), 143–154 (1979)
77. M.N. Wegman, J. Carter, New hash functions and their use in authentication and set equality. J. Comp. Sci. Syst. **22**, 265–279 (1981)
78. A. Rényi, On measures of entropy and information, in *Proceedings of 4th Berkeley Symposium on Mathematical Statistics and Probability*, vol. 1, (University of California Press, 1961), pp. 547–561
79. I. Ilic, I.B. Djordjevic, M. Stankovic, On a general definition of conditional Rényi entropies, in *Proceedings of 4th International Electronic Conference on Entropy and Its Applications*, vol. 4, (Sciforum Electronic Conference Series, 2017). https://doi.org/10.3390/ecea-4-05030
80. A. Teixeira, A. Matos, L. Antunes, Conditional Rényi entropies. IEEE Trans. Information Theory **58**(7), 4273–4277 (2012)
81. C. Cachin, *Entropy Measures and Unconditional Security in Cryptography*. PhD dissertation, ETH Zurich, Hartung-Gorre Verlag, Konstanz (1997)
82. U. Maurer, S. Wolf, Information-Theoretic Key Agreement: From Weak to Strong Secrecy for Free, in *Advances in Cryptology — EUROCRYPT 2000. EUROCRYPT 2000*, Lecture Notes in Computer Science, ed. by B. Preneel, vol. 1807, (Springer, Berlin/Heidelberg, 2000)
83. X. Sun, I.B. Djordjevic, Physical-layer security in orbital angular momentum multiplexing free-space optical communications. IEEE Photon. J. **8**(1), paper 7901110 (2016)

84. T.-L. Wang, J. Gariano, I.B. Djordjevic, Employing Bessel-Gaussian beams to improve physical-layer security in free-space optical communications. IEEE Photon. J. **10**(5), paper 7907113 (2018). https://doi.org/10.1109/JPHOT.2018.2867173
85. E. Telatar, Capacity of multi-antenna Gaussian channels. Eur. Trans. Telecomm. **10**, 585–595 (1999)
86. E. Biglieri, *Coding for Wireless Channels* (Springer, New York, 2005)
87. J.R. Hampton, *Introduction to MIMO Communications* (Cambridge University Press, Cambridge, 2014)
88. A. Mukherjee, S.A.A. Fakoorian, J. Huang, L. Swindlehurst, MIMO signal processing algorithms for enhanced physical layer security, in *Physical Layer Security in Wireless Communications*, ed. by X. Zhou, L. Song, Y. Zhang, (CRC Press, Boca Raton/London/New York, 2014), pp. 93–114
89. A. Khisti, G.W. Wornell, Secure transmission with multiple antennas I: The MISOME wiretap channel/part II: The MIMOME wiretap channel. IEEE Trans. Inf. Theory **56**(7), 3088-3104/5515–5532 (2010)
90. R. Bustin, R. Liu, H.V. Poor, S. Shamai(Shitz), An MMSE approach to the secrecy capacity of the MIMO Gaussian wiretap channel. EURASIP J. Wirel. Commun. Netw. **2009**, 370970 (2009). https://doi.org/10.1155/2009/370970
91. P. Narayan, H. Tyagi, Multiterminal secrecy by public discussion. Found. Trends Commun. Inf. Theory **13**, 129–275 (2016)
92. R.F. Schaefer, H. Boche, H.V. Poor, Secure communication under channel uncertainty and adversarial attacks. Proc. IEEE **103**(10), 1796–1813 (2015)
93. H.V. Poor, R.F. Schaefer, Wireless physical layer security. PNAS **114**(1), 19–26 (2017)
94. I. Bjelakovic, H. Boche, J. Sommerfeld, Secrecy results for compound wiretap channels. Probl. Inf. Transm. **49**(1), 73–98 (2013)
95. R.F. Schaefer, S. Loyka, The secrecy capacity of compound MIMO Gaussian channels. IEEE Trans. Inf. Theory **61**(10), 5535–5552 (2015)
96. H.D. Ly, T. Liu, Y. Liang, Multiple-input multiple-output Gaussian broadcast channels with common and confidential messages. IEEE Trans. Inform. Theory **56**(11), 5477–5487 (2010)
97. R. Liu, I. Maric, P. Spasojevic, R.D. Yates, Discrete memoryless interference and broadcast channels with confidential messages: Secrecy rate regions. IEEE Trans. Inf. Theory **54**(6), 2493–2507 (2008)
98. R. Liu, T. Liu, H.V. Poor, S. Shamai (Shitz), Multiple-input multiple-output Gaussian broadcast channels with confidential messages. IEEE Trans. Inform. Theory **56**(9), 4215–4227 (2010)
99. R. Liu, T. Liu, H.V. Poor, S. Shamai (Shitz), New results on multiple-input multiple-output broadcast channels with confidential messages. IEEE Trans. Inf. Theory **59**(3), 1346–1359 (2013)
100. Y. Liang, H.V. Poor, Multiple-access channels with confidential messages. IEEE Trans. Inf. Theory **54**(3), 972–1002 (2008)
101. O.O. Koyluoglu, H. El Gamal, L. Lai, H.V. Poor, Interference alignment for secrecy. IEEE Trans. Inf. Theory **57**(6), 3323–3332 (2011)
102. X. He, A. Aylin Yener, K-user interference channels: Achievable secrecy rate and degrees of freedom, in *Proceedings of 2009 IEEE Information Theory Workshop on Networking and Information Theory*, (Volos, 2009), pp. 336–340
103. S.A.A. Fakoorian, A.L. Swindlehurst, MIMO interference channel with confidential messages: game theoretic beamforming designs, in *2010 Conference Record of the Forty Fourth Asilomar Conference on Signals, Systems and Computers*, (Pacific Grove, 2010)
104. S.A.A. Fakoorian, A.L. Swindlehurst, MIMO Interference Channel with confidential messages: Achievable secrecy rates and Precoder design. IEEE Trans. Inform. Forensics and Security **6**(3), 640–649 (2011)
105. Y. Oohama, Capacity Theorems for Relay Channels with Confidential Messages, in *Proceedings of 2007 IEEE International Symposium on Information Theory*, (Nice, 2007), pp. 926–930
106. X. He, A. Yener, Cooperation with an untrusted relay: A secrecy perspective. IEEE Trans. Inf. Theory **56**(8), 3801–3827 (2010)
107. J. Huang, A.L. Swindlehurst, Cooperative jamming for secure communications in MIMO relay networks. IEEE Trans. Signal Process. **59**(10), 4871–4884 (2011)
108. R. Wilson, D. Tse, R.A. Scholtz, Channel identification: Secret sharing using reciprocity in ultrawideband channels. IEEE Trans. Inf. Forensics and Security **2**(3), 364–375 (2007)
109. C. Ye, S. Mathur, A. Reznik, W. Trappe, N. Mandayam, Information theoretic key generation from wireless channels. IEEE Trans Inf. Forensics and Security **5**(2), 240–254 (2010)
110. L. Lai, Y. Liang, H.V. Poor, W. Du, Key generation from wireless channels, in *Physical Layer Security in Wireless Communications*, ed. by X. Zhou, L. Song, Y. Zhang, (CRC Press, Boca Raton/London/New York, 2014), pp. 47–68
111. L. Lai, Y. Liang, H.V. Poor, A unified framework for key agreement over wireless fading channels. IEEE Trans. Inf. Forensics and Security **7**(2), 480–490 (2012)
112. P.K. Gopala, L. Lai, H. El Gamal, On the secrecy capacity of fading channels. IEEE Trans. Inf. Theory **54**(10), 4687–4698 (2008)
113. L.C. Andrews, R.L. Philips, *Laser Beam Propagation through Random Media* (SPIE Press, Bellingham, 2005)
114. V.G. Sidorovich, Optical countermeasures and security of free-space optical communication links. In *Proceedings of European Symposium on Optics Photonics Defence Security 2004*, pp. 97–108, Int. Soc. opt. Photon (2004)
115. F.J. Lopez-Martinez, G. Gomez, J.M. Garrido-Balsells, Physical-layer security in free-space optical communications. IEEE Photonics J. **7**(2), paper 7901014 (2015)
116. IB Djordjevic, X. Sun, Spatial modes-based physical-layer security. In *Proceedings of IEEE ICTON 2016, Paper Mo.C1.3*, Trento, Italy, (Invited Paper) (2016)
117. T.-L. Wang, I.B. Djordjevic, Physical-layer security of a binary data sequence transmitted with Bessel-Gaussian beams over an optical Wiretap Channel. IEEE Photon. J. **10**(6), 7908611 (2018)
118. I.B. Djordjevic, M. Arabaci, LDPC-coded orbital angular momentum (OAM) modulation for free-space optical communication. Opt. Express **18**(24), 24722–24728 (2010)
119. I.B. Djordjevic, A. Anguita, B. Vasic, Error-correction coded orbital-angular-momentum modulation for FSO channels affected by turbulence. IEEE/OSA J. Lightw. Technol. **30**(17), 2846–2852 (2012)
120. I.B. Djordjevic, Z. Qu, Coded orbital-angular-momentum-based free-space optical transmission. In *Wiley encyclopedia of electrical and electronics engineering* (J. G. Webster, Ed., (2016)), pp. 1–12, https://doi.org/10.1002/047134608X.W8291
121. I.B. Djordjevic, Spatial-domain-based hybrid multidimensional coded-modulation schemes enabling multi-Tb/s optical transport. IEEE/OSA J. Lightw. Technol. **30** (14): 2315–2328 (2012). (Invited Paper)

122. I.B. Djordjevic, M. Cvijetic, C. Lin, Multidimensional signaling and coding enabling multi-Tb/s optical transport and networking. IEEE Sig. Proc. Mag. **31**(2), 104–117 (2014)
123. I.B. Djordjevic, A. Jovanovic, Z.H. Peric, T. Wang, Multidimensional optical transport based on optimized vector-quantization-inspired signal constellation design. IEEE Trans. Comm. **62**(9), 3262–3273 (2014)
124. I.B. Djordjevic, Integrated optics modules based proposal for quantum information processing, teleportation, QKD, and quantum error correction employing photon angular momentum. IEEE Photonics Journal **8**(1), paper 6600212 (2016) https://ieeexplore.ieee.org/document/7393447
125. I.B. Djordjevic, R. Varrazza, M. Hill, S. Yu, Packet switching performance at 10Gb/s across a 4×4 optical Crosspoint switch matrix. IEEE Photon. Technol. Lett. **16**(1), 102–104 (2004)
126. D. McGloin, K. Dholakia, Bessel beams: Diffraction in a new light. Contemp. Phys. **46**(1), 15–28 (2005)
127. S. Chen, S. Li, Y. Zhao, J. Liu, L. Zhu, A. Wang, J. Du, L. Shen, J. Wang, Demonstration of 20-Gbit/s high-speed Bessel beam encoding/decoding link with adaptive turbulence compensation. Opt. Lett. **41**(20), 4680–4683 (2016)
128. W. Nelson, J.P. Palastro, C.C. Davis, P. Sprangle, Propagation of Bessel and airy beams through atmospheric turbulence. J. Opt. Soc. Am. A **31**(3), 603–609 (2014)
129. I.B. Djordjevic, Atmospheric turbulence-controlled cryptosystems. IEEE Photon. J. **13**(1), 7900909 (2021)

Appendix

This chapter is mostly devoted to the abstract algebra fundamentals [1–15], needed in Chap. 9. Only the most important topics needed for better understanding of the error correction are covered. The chapter is organized as follows. In Sect. A.1, we introduce the concept of groups and provide their basic properties. In Sect. A.2, we introduce the concept of fields, while in Sect. A.3, the concept of vector spaces. Further, Sect. A.4 is devoted to algebra of finite fields, which is of high importance in forward error correction. Section A.5 is related to the pulse-position modulation (PPM), relevant in UWB communications and deep-space optical communications. The fundamentals of the z-transform, needed in Chaps. 3 and 6, are provided in Sect. A.6.

A.1 Groups

Definition D1 A group is the set G that together with an operation, denoted by "+," satisfies the following axioms:

1. *Closure*: $\forall\, a, b \in G \Rightarrow a + b \in G$.
2. *Associative law*: $\forall\, a, b, c \in G \Rightarrow a + (b + c) = (a + b) + c$.
3. *Identity element*: $\exists\, e \in G$ such that $a + e = e + a = a \;\forall\, a \in G$.
4. Inverse element: $\exists\, a^{-1} \in G \;\forall\, a \in G$, such that $a + a^{-1} = a^{-1} + a = e$.

We call a group *Abelian* or a *commutative* group, if the operation "+" is also commutative: $\forall\, a, b \subset G \Rightarrow a + b = b + a$.

Theorem T1 The identity element in a group is the unique one, and each element in the group has a unique inverse element.

Proof If the identity element e is not a unique one, then there exists another one e': $e' = e' + e = e$, implying that e' and e are identical. In order to show that a group element a has a unique inverse a^{-1}, let us assume that it has another inverse element ${a_1}^{-1}: {a_1}^{-1} = {a_1}^{-1} + e = {a_1}^{-1} + (a + a^{-1}) = ({a_1}^{-1} + a) + a^{-1} = e + a^{-1} = a^{-1}$, thus implying that the inverse element in the group is unique.

Examples

- $F_2 = \{0,1\}$ and "+" operation is in fact the modulo-2 addition, defined by $0 + 0 = 0$, $0 + 1 = 1$, $1 + 0 = 1$, $1 + 1 = 0$. The closure property is satisfied, 0 is identity element, 0 and 1 are their own inverses, and operation "+" is associative. The set F_2 with modulo-2 addition forms, therefore, a group.
- The set of integers forms a group under the usual addition operation on which 0 is identity element, and for any integer n, $-n$ is its inverse.
- Consider the set of code words of **binary linear (*N*,*K*) block code**. Any two code words added per modulo-2 form another code word (according to the definition of the linear block code). All-zeros code word is identity element, and each code word is its own inverse. The associative property also holds. Therefore, all group properties are satisfied, and the set of code words form a group, and this is the reason why this is called the *group code*.

I. B. Djordjevic, *Advanced Optical and Wireless Communications Systems*, https://doi.org/10.1007/978-3-030-98491-5

The number of elements in the group is typically called the *order* of the group. If the group has a finite number of elements, it is called the *finite* group.

Definition D2 Let H be a subset of elements of group G. We call H *subgroup* of G if the elements of H themselves form the group under the same operation as that defined on elements from G.

In order to verify that the subset S is the subgroup, it is sufficient to check the closure property and that an inverse element exists for every element of subgroup.

Theorem T2 (Lagrange Theorem) The order of a finite group is an integer multiple of any of its subgroup order.

Example Consider the set Y of N-tuples (received words on a receiver side of a communication system), each element of the N-tuple being 0 or 1. It can easily be verified that elements of Y form a group under modulo-2 addition. Consider a subset C of Y with elements being the code words of a binary (N,K) block code. Then C forms a subgroup of Y, and the order of group Y is divisible by the order of subgroup C.

There exist groups, whose all elements can be obtained as power of some element, say a. Such a group G is called the *cyclic group*, and the corresponding element a is called the *generator* of the group. The cyclic group can be denoted by $G = <a>$.

Example Consider an element α of a finite group G. Let S be the set of elements $S = \{\alpha, \alpha^2, \alpha^3, \ldots, \alpha^i, \ldots\}$. Because G is finite, S must be finite as well, and therefore, not all powers of α are distinct. There must be some l,m $(m > l)$ such that $\alpha^m = \alpha^l$ so that $\alpha^m \alpha^{-l} = \alpha^l \alpha^{-l} = 1$. Let k be the smallest such power of α for which $\alpha^k = 1$, meaning that $\alpha, \alpha^2, \ldots, \alpha^k$ are all distinct. We can verify now that set $S = \{\alpha, \alpha^2, \ldots, \alpha^k = 1\}$ is a subgroup of G. S contains the identity element, and for any element α^i, the element α^{k-i} is its inverse. Given that any two elements α^i, $\alpha^j \in S$ the corresponding element obtained as their product, $\alpha^{i+j} \in S$ if $i + j \leq k$. If $i + j > k$, then $\alpha^{i+j} \cdot 1 = \alpha^{i+j} \cdot \alpha^{-k}$, and because $i + j - k \leq k$, the closure property is clearly satisfied. S is, therefore, the subgroup of group G. Given the definition of cyclic group, the subgroup S is also cyclic. The set of code words of a cyclic (N,K) block code can be obtained as a cyclic subgroup of the group of all N-tuples.

Theorem T3 Let G be a group and $\{H_i | i \in I\}$ be a nonempty collection of subgroups with index set I. The intersection $\cap_{i \in I} H_i$ is a subgroup.

Definition D3 Let G be a group and X be a subset of G. Let $\{H_i | i \in I\}$ be the collection of subgroups of G that contain X. Then the intersection $\cap_{i \in I} H_i$ is called the *subgroup of G generated by* X and is denoted by $<X>$.

Theorem T4 Let G be a group and X nonempty subset of G with elements $\{x_i \mid i \in I\}$ (I is the index set). The subgroup of G generated by X, $<X>$, consists of all finite product of the x_i. The x_i elements are known as generators.

Definition D4 Let G be a group and H be a subgroup of G. For any element $a \in G$, the set $aH = \{ah | h \in H\}$ is called the *left coset* of H in G. Similarly, the set $Ha = \{ha | h \in H\}$ is called the *right coset* of H in G.

Theorem T5 Let G be a group and H be a subgroup of G. The collection of right cosets of H, $Ha = \{ha | h \in H\}$ forms a partition of G.

Instead of formal proof, we provide the following justification. Let us create the following table. In the first row, we list all elements of subgroup H, beginning with the identity element e. The second column is obtained by selecting an arbitrary element from G, not used in the first row, as the leading element of the second row. We then complete the second row by "multiplying" this element with all elements of the first row from the right. Out of not previously used elements, we arbitrarily select an element as the leading element of the third row. We then complete the third row by multiplying this element with all elements of the first row from the right. We continue this procedure until we exploit all elements from G. The resulting table is as follows:

$h_1 = e$	h_2	...	h_{m-1}	h_m
g_2	$h_2 g_2$	...	$h_{m-1}\ g_2$	$h_m g_2$
g_3	$h_2 g_3$	...	$h_{m-1}\ g_3$	$h_m g_3$
...	...	...	...	...
g_n	$h_2 g_n$	...	$h_{m-1} g_n$	$h_m g_n$

Each row in this table represents a coset, and the first element in each row is a coset leader. The number of cosets of H in G is in fact the number of rows, and it is called the *index of H in G*, typically denoted by $[G{:}H]$. It follows from table above $|H| = m$, $[G{:}H] = n$, and $|G| = nm = [G{:}H]|H|$, which can be used as the proof of Lagrange's theorem. In other words, $[G{:}H] = |G|/|H|$.

Definition D5 Let G be a group and H be a subgroup of G. H is *normal subgroup* of G if it is invariant under conjugation, that is, $\forall\ h \in H$ and $g \in G$, the element $ghg^{-1} \in H$.

In other words, H is *fixed* under conjugation by the elements from G, namely, $gHg^{-1} = H$ for any $g \in G$. Therefore, the left and the right cosets of H in G coincide: $\forall\ g \in G,\ gH = Hg$.

Theorem T6 Let G be a group and H a normal subgroup of G. If $G|H$ denotes the set of cosets of H in G, then the set $G|H$ with coset multiplication forms a group, known as the *quotient group* of G by H.

The coset multiplication of aH and bH is defined as $aH * bH = abH$. It follows from Lagrange's theorem that $|G/H| = [G{:}H]$. This theorem can straightforwardly be proved by using the table above.

A.2 Fields

Definition D6 A *field* is a set of elements F with two operations, addition "+" and multiplication "·", such that:

1. F is an Abelian group under addition operation, with 0 being the identity element.
2. The nonzero elements of F form an Abelian group under the multiplication operation, with 1 being the identity element.
3. The *multiplication operation is distributive over the addition operation*:

$$\forall a, b, c \in F \Rightarrow a \cdot (b + c) = a \cdot b + a \cdot c$$

Examples

- The set of real numbers, with operation + as ordinary addition, and operation · as ordinary multiplication, satisfy above three properties, and it is therefore a field.
- The set consisting of two elements {0,1}, with modulo-2 multiplication and addition, given in the table below, constitutes a field known as **Galois field** and is denoted by GF(2).

+	0	1
0	0	1
1	1	0

·	0	1
0	0	0
1	0	1

- The set of integers modulo-p, with modulo-p addition and multiplication, forms a field with p elements, denoted by GF(p), providing that p is a prime.
- For any q that is an integer power of prime number p ($q = p^m$, m - an integer), there exists a field with q elements, denoted as GF(q). (The arithmetic is not modulo-q arithmetic, except when $m = 1$.) GF(p^m) contains GF(p) as a subfield.

Addition and multiplication in GF(3) are defined as follows:

+	0	1	2
0	0	1	2
1	1	2	0
2	2	0	1

·	0	1	2
0	0	0	0
1	0	1	2
2	0	2	1

Addition and multiplication in $GF(2^2)$:

+	0	1	2	3
0	0	1	2	3
1	1	0	3	2
2	2	3	0	1
3	3	2	1	0

·	0	1	2	3
0	0	0	0	0
1	0	1	2	3
2	0	2	3	1
3	0	3	1	2

Theorem T7 Let Z_p denote the set of integers $\{0,1,\ldots,p\text{-}1\}$, with addition and multiplication defined as ordinary addition and multiplication modulo-p. Then Z_p is a field if and only if p is a prime.

A.3 Vector Spaces

Definition D7

Let V be a set of elements with a binary operation "+" and let F be a field. Further, let an operation "·" be defined between the elements of V and the elements of F. Then V is said to be a *vector space* over F if the following conditions are satisfied $\forall\, a, b \in F$ and $\forall\, \boldsymbol{x}, \boldsymbol{y} \in V$:

1. V is an Abelian group under the addition operation.
2. $\forall\, a \in F$ and $\forall\, \boldsymbol{x} \in V$, then $a \cdot \boldsymbol{x} \in V$.
3. *Distributive law*:

$$a \cdot (x + y) = a \cdot x + a \cdot y$$

$$(a + b) \cdot x = a \cdot x + b \cdot x$$

4. *Associative law*:

$$(a \cdot b) \cdot x = a \cdot (b \cdot x)$$

If 1 denotes the identity element of F, then $1 \cdot \boldsymbol{x} = \boldsymbol{x}$. Let 0 denote the zero element (identity element under +) in F, and $\mathbf{0}$ the additive element in V, -1 the additive inverse of 1, the multiplicative identity in F. It can easily be shown that the following two properties hold:

$$0 \cdot x = 0$$
$$x + (-1) \cdot x = 0$$

Examples

- Consider the set V, whose elements are n-tuples of the form $\boldsymbol{v} = (v_0, v_1, \ldots, v_{n\text{-}1})$, $v_i \in F$. Let us define the addition of any two n-tuples as another n-tuple obtained by component-wise addition and multiplication of n-tuple by an element from F. Then V forms vector space over F, and it is commonly denoted by F^n. If $F = R$ (R – the field of real number), then R^n is called the Euclidean n-dimensional space.
- The set of n-tuples whose elements are from GF(2), again with component-wise addition and multiplication by an element from GF(2), forms a vector space over GF(2).
- Consider the set V of polynomials whose coefficients are from GF(q). Addition of two polynomials is the usual polynomial addition, addition being performed in GF(q). Let the field F be GF(q). Scalar multiplication of a polynomial by a field element from GF(q) corresponds to the multiplication of each polynomial coefficient by the field element, carried out in GF (q). V is then a vector space over GF(q).

Consider a set V that forms a vector space over a field F. Let $v_1, v_2, \ldots, v_k$ be vectors from V, and $a_1, a_2, \ldots, a_k$ be field elements from F. The *linear combination* of the vectors $v_1, v_2, \ldots, v_k$ is defined by:

$$a_1 v_1 + a_2 v_2 + \ldots + a_k v_k$$

The set of vectors $\{v_1, v_2, \ldots, v_k\}$ is said to be *linearly independent* if there does not exist a set of field elements $a_1, a_2, \ldots, a_k$, not all $a_i = 0$, such that:

$$a_1 v_1 + a_2 v_2 + \ldots + a_k v_k = \mathbf{0}$$

Example The vectors (0 0 1), (0 1 0), and (1 0 0) (from F^3) are linearly independent. However, the vectors (0 0 2), (1 1 0), and (2 2 1) are linearly dependent over GF(3) because they sum to zero vector.

Let V be a vector space and S be subset of the vectors in V. If S is itself a vector space over F under the same vector addition and scalar multiplication operations applicable to V and F, then S is said to be *a subspace* of V.

Theorem T7 Let $\{v_1, v_2, \ldots, v_k\}$ be a set of vectors from a vector space V over a field F. Then the set consisting of all linear combinations of $\{v_1, v_2, \ldots, v_k\}$ forms a vector space over F and is, therefore, a subspace of V.

Example Consider the vector space V over GF(2) given by the set {(0 0 0), (0 0 1), (0 1 0), (0 1 1), (1 0 0), (1 0 1), (1 1 0), (1 1 1)}. The subset S = {(0 0 0), (1 0 0), (0 1 0), (1 1 0)} is a subspace of V over GF(2).

Example Consider the vector space V over GF(2) given by the set {(0 0 0), (0 0 1), (0 1 0), (0 1 1), (1 0 0), (1 0 1), (1 1 0), (1 1 1)}. For the subset B = {(0 1 0), (1 0 0)}, the set of all linear combinations is given by S = {(0 0 0), (1 0 0), (0 1 0), (1 1 0)} and forms a subspace of V over F. The set of vectors B is said to *span S*.

Definition D8 A *basis* of a vector space V is set of linearly independent vectors that spans the space. The number of vectors in a basis is called the *dimension* of the vector space.

In example above, the set {(0 0 1), (0 1 0), (1 0 0)} is the basis of vector space V and has dimension 3.

A.4 Algebra of Finite Fields

A *ring* is a set of elements R with two operations, addition "+" and multiplication "·", such that:

(i) R is an Abelian group under addition operation, (ii) multiplication operation is associative, and (iii) multiplication is associative over addition.

The quantity <u>ais</u> said to be *congruent to b to modulus n*, denoted as a ° b (mod n), if a-b is divisible by n. If x ° a (mod n), then a is called a *residue* to x to modulus n. *A class of residues to modulus n* is the class of all integers congruent to a given residue (mod n), and every member of the class is called a representative of the class. There are n classes, represented by (0), (1), (2), . . ., (n-1), and the representative of these classes is called *a complete system of incongruent residues* to modulus n. If i and j are two members of a complete system of incongruent residues to modulus n, then addition and multiplication between i and j are defined by:

$$i + j = (i + j)(\text{mod } n) \text{ and } i \cdot j = (i \cdot j)(\text{mod } n)$$

A complete system of residues (mod n) forms a commutative ring with unity element. Let s be a nonzero element of these residues. Then *s posses inverse if and only if n is prime, p*. Therefore, if p is a prime, a complete system of residues (mod p) forms a *Galois (or finite) field* and is denoted by GF(p).

Let P(x) be any given polynomial in x of degree m with coefficients belonging to GF(p), and let F(x) be any polynomial in x with integral coefficients. Then F(x) may be expressed as:

$$F(x) = f(x) + p \cdot q(x) + P(x) \cdot Q(x), \text{where } f(x) = a_0 + a_1x + a_2x^2 + \cdots + a_{m-1}x^{m-1}, a_i \in \mathrm{GF}(p).$$

This relationship may be written as:

$F(x)°f(x)$ mod $\{p,P(x)\}$, and we say that *f(x) is the residue of F(x) modulus p and P(x)*. If p and $P(x)$ are kept fixed but $f(x)$ varied, p^m classes may be formed, because each coefficient of $f(x)$ may take p values of GF(p). The classes defined by $f(x)$ form a commutative ring, which will be a field if and only if $P(x)$ is *irreducible* over GF(p) (not divisible with any other polynomial of degree m^{-1} or less).

The finite field formed by p^m classes of residues is called *a Galois field of order* p^m and is denoted by GF(p^m). The function $P(x)$ is said to be *minimum polynomial* for generating the elements of GF(p^m) (the smallest-degree polynomial over GF(p) having a field element $\beta \in$GF(p^m) as a root). The nonzero elements of GF(p^m) can be represented as polynomials of degree at most m-1 or as powers of a *primitive root* α such that:

$$\alpha^{p^m-1} = 1, \quad \alpha^d \neq 1 \quad (\text{for } d \text{ dividing } p^m - 1)$$

A *primitive element* is a field element that generates all nonzero field elements as its successive powers. *A primitive polynomial* is irreducible polynomial that has a primitive element as its root.

Theorem T8 *Two important properties of GF(q),* $q = p^m$ are:

1. The roots of polynomial x^{q-1}-1 are all nonzero elements of GF(q).
2. Let $P(x)$ be an irreducible polynomial of degree m with coefficients from GF(p) and β be a root from the extended field GF ($q = p^m$). Then all the m roots of $P(x)$ are:

$$\beta, \beta^p, \beta^{p^2}, \ldots, \beta^{p^{m-1}}$$

To obtain a minimum polynomial, we divide x^q-1 ($q = p^m$) by the least common multiple of all factors like x^d-1, where d is a divisor of p^m-1. Then we get the **cyclotomic equation** – the equation having for its roots all primitive roots of equation x^{q-1}-1 = 0. The order of this equation is ϕ (p^m-1), where ϕ (k) is the number of positive integers less than k and relatively prime to it. By replacing each coefficient in this equation by least nonzero residue to modulus p, we get the *cyclotomic polynomial* of order ϕ (p^m-1). Let $P(x)$ be an irreducible factor of this polynomial; then $P(x)$ is a minimum polynomial, which is in general not the unique one.

Example Let us determine the minimum polynomial for generating the elements of GF(2^3). The cyclotomic polynomial is:

$$(x^7 - 1)/(x - 1) = x^6 + x^5 + x^4 + x^3 + x^2 + x + 1 = (x^3 + x^2 + 1)(x^3 + x + 1)$$

Hence, $P(x)$ can be either $x^3 + x^2 + 1$ or $x^3 + x + 1$. Let us choose $P(x) = x^3 + x^2 + 1$.

$$\phi\ (7) = 6, \deg[P(x)] = 3$$

Three different representations of GF(2^3) are given in the table below.

Power of α	Polynomial	3-tuple
0	0	000
α^0	1	001
α^1	α	010
α^2	α^2	100
α^3	$\alpha^2 + 1$	101
α^4	$\alpha^2 + \alpha + 1$	111
α^5	$\alpha + 1$	011
α^6	$\alpha^2 + \alpha$	110
α^7	1	001

A.5 Pulse-Position Modulation (PPM)

Pulse-position modulation (PPM) is a standard modulation technique that has been considered for different purposes [16–21]. In PPM, we employ M pulse-position basis functions defined as follows:

$$\Phi_j(t) = \frac{1}{\sqrt{T_s/M}} \mathrm{rect}\left[\frac{t-(j-1)T_s/M}{T_s/M}\right]; j = 1, \cdots, M;$$

where T_s is a symbol duration and rect(t) is defined as:

$$\mathrm{rect}(t) = \begin{cases} 1, 0 \le t < 1 \\ 0, \text{otherwise} \end{cases}.$$

The signal space of M-ary PPM is therefore M-dimensional, and constellation points are located on axes. Only two amplitude levels per basis function are used. The time-domain representation of M-ary PPM signal intensity is given by [17]:

$$s(t) = \sum_{k=-\infty}^{\infty} MP\sqrt{\frac{T_s}{M}}\Phi_{a_k}(t - kT_s),$$

where $\{a_k\}$ represents a sequence of symbols being transmitted, while P denotes the average power. The symbol error probability for equal probability of transmitted symbols is given as [17]:

$$P_s \simeq (M-1)\mathrm{erfc}\left(\frac{P}{2}\sqrt{\frac{M}{2R_s\sigma^2}}\right),$$

where erfc(·) is the complementary error function introduced earlier, R_s is the symbol rate related to bit rate by R_b by $R_s = R_b/\log_2 M$, and σ^2 is the noise variance. Let us assume that all symbol errors are equally likely, and therefore, they occur with probability $P_s/(M\text{-}1)$. There are $2^{\log_2 M - 1}$ cases in which i-th bit in a given symbol is in error. Therefore, the bit-error probability can be estimated as:

$$BER \simeq \frac{2^{\log_2 M - 1}}{M-1}P_s = \frac{M/2}{M-1}P_s.$$

Bandwidth occupied by PPM signal is $B = 1/(T_s/M)$ since there are M possible time slots (also known as chips), each with duration T_s/M. The bandwidth efficiency of M-ary PPM is given by:

$$\rho = \frac{R_b}{B} = \frac{R_b}{1/(T_s/M)} = \frac{R_b}{MR_s} = \frac{\log_2 M}{M},$$

which is lower than the efficiency of M-ary QAM. The low-bandwidth efficiency of M-ary PPM can be improved by multi-pulse PPM [19], where w ($w > 1$) out of M possible time slots are used. The symbol interval T_s is now related to bit interval T_b by:

$$T_s = T_b \log_2 \binom{M}{w}.$$

The bandwidth efficiency of multi-pulse PPM is given by:

$$\rho = \frac{R_b}{B} = \frac{R_b}{1/(T_s/M)} = \frac{R_b}{1/\left[T_b \log_2 \binom{M}{w}/M\right]} = \frac{\log_2 \binom{M}{w}}{M}.$$

Another alternative approach to improve the bandwidth efficiency of PPM is based on differential PPM (DPPM) [20], also known as truncated PPM (TPPM), in which the new PPM symbol begins as soon as the slot containing the pulse is over. This scheme, however, suffers from variable-rate code design problem and catastrophic error propagation.

Finally, to improve bandwidth efficiency of PPM, so-called multidimensional PPM [21] can be used. In this scheme, the pulse positions are used as basis functions, as given above, and then L amplitude levels are imposed on each basis function. The constellation points, in this scheme, occupy all axes not just one as in ordinary PPM. This scheme can carry $\log_2(M^L)$ bits per symbol, thus improving bandwidth efficiency by L times.

A.6 The z-Transform

The z-transform plays the same role in discrete time (DT) systems as the Fourier and Laplace transforms play in continuous time (CT) systems. It converts the difference equations into algebraic ones in z-plane. By using the z-transform, the convolution sum in time domain can be converted to product of z-transforms of input and impulse response. This section is based on references [22–24].

A.6.1 Bilateral z-Transform

The *bilateral z-transform* of DT signal $x(n)$ is defined by:

$$Z\{x(n)\} = \sum_{n=-\infty}^{\infty} x(n)z^{-n} = X(z).$$

The set of values for which the summation above converges is called the *region of convergence (ROC)*. Generally speaking, the ROC is an annular region of the entire complex plane defined by $r_1 < |z| < r_2$. Two important types of sequences are left-handed and right-handed ones. A *right-handed* sequence is one for which $x(n) = 0$ for all $n < n_0$, where n_0 is a positive or negative finite integer. If $n_0 \geq 0$, the resulting sequence is *causal* or a *positive-time* sequence. A *left-handed* sequence is one for which $x(n) = 0$ for all $n \geq n_0$, where n_0 is positive or negative finite integer. If $n_0 \leq 0$, the resulting sequence is *anticausal* or a *negative-time* sequence.

Suppose that a causal sequence $x(n)$ can be written as sum of complex exponentials:

$$x(n) = \sum_{i=1}^{N} (a_i)^n u(n).$$

By applying the definition equation and geometric progression formulas, the corresponding z-transform can be written as:

$$X(z) = \sum_{i=1}^{N} \frac{z}{z - a_i},$$

where the ROC of i-th term is given as the exterior of a circle with radius $|a_i|$: $R_i = \{z: |z| > |a_i|\}$. The overall ROC is obtained as intersection of ROCs R_i as:

$$\text{ROC}: \ R = \cap_{i=1}^{N} R_i = \{z : |z| > \max(|a_i|)\}.$$

Suppose now that an anticausal sequence $x(n)$ can be written as a sum of time-shifted complex exponentials:

$$x(n) = \sum_{i=1}^{N} -b_i^n u(-n-1).$$

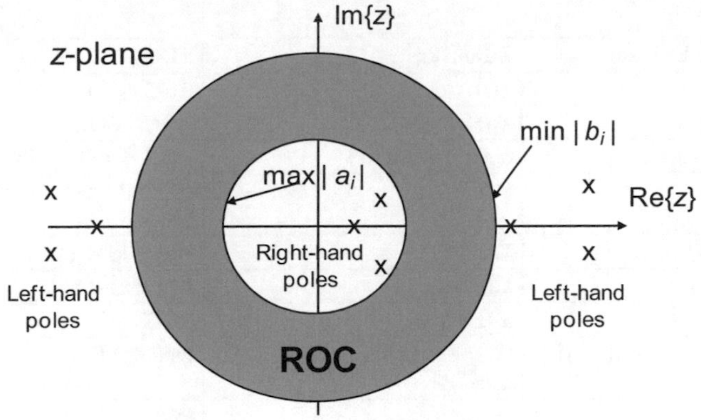

Fig. A1 Determination of the ROC

By applying again the definition equation and geometric progression formulas, the corresponding z-transform becomes:

$$X(z) = \sum_{i=1}^{N} \frac{z}{z - b_i},$$

where the ROC of i-th term is given as the interior of a circle of radius $|b_i|$: $L_i = \{z: |z| < |b_i|\}$. The overall ROC is obtained as intersection of ROCs R_i by:

$$\text{ROC}: \ L = \cap_{i=1}^{N} L_i = \{z : |z| < \min(|b_i|)\}.$$

Finally, suppose now that a sequence $x(n)$ contains both right- and left-handed sub-sequences. Its ROC is determined by intersection of right- and left-handed subsequences' ROCs as follows:

$$\text{ROC}_{\text{total}} = R \cap L = \{z : \max(|a_i|) < |z| < \min(|b_i|)\},$$

and corresponding illustration is shown in Fig. A1.

A.6.2 Properties of z-Transform and Common z-Transform Pairs

The basic operations on sequences include the addition, the time-shifting (or translation), multiplication, and convolution. The corresponding properties are summarized in Table A1. Some common z-transform pairs are summarized in Table A2.

A.6.3 The Inversion of the z-Transform

The inverse z-transform is defined as:

$$x(n) = \frac{1}{2\pi j} \oint_C X(z) z^{n-1} dz,$$

where C is the closed contour (observed counterclockwise, encircling the origin) within the ROC. The methods commonly used to evaluate the inverse z-transform are (i) contour integration, (ii) partial-fraction expansion, and (iii) division. The contour integration method is based on the Cauchy's formula:

Table A1 The z-transform properties

Property	Time Domain	Z-Domain	ROC
Notation	$x(n)$	$X(z)$	ROC:$r_2 < \|z\| < r_1$
	$x_1(n)$, $x_2(n)$	$X_1(z)$, $X_2(z)$	ROC_1, ROC_2
Linearity	$a_1x_1(n) + a_2x_2(n)$	$a_1X_1(z) + a_2X_2(z)$	$ROC = ROC_1 \cap ROC_2$
Time shifting	$x(n\text{-}k)$	$z^{-k}X(z)$	ROC: the same as $X(z)$ except $z = 0$ if $k > 0$, or $z\text{-} > \infty$ if $k < 0$
Scaling in the z-domain	$a^n x(n)$	$X(a^{-1}z)$	$\|a\|r_2 < \|z\| < \|a\|r_1$
Time reversal	$x(-n)$	$X(z^{-1})$	$1/r_1 < \|z\| < 1/r_2$
Conjugation	$x^*(n)$	$X^*(z^*)$	ROC
Real and imaginary parts	Re{$x(n)$}, Im{$x(n)$}	$[X(z) + X^*(z^*)]/2$, $j[X(z)\text{-}X^*(z^*)]/2$	Includes ROC
Differentiation in the z-domain	$nx(n)$	$-zdX(z)/dz$	ROC
Convolution	$x_1(n)*x_2(n)$	$X_1(z)X_2(z)$	At least $ROC_1 \cap ROC_2$
Correlation	$r_{x1,x2}(l) = x_1(l)^* x_2(-l)$	$R_{x1,x2}(z) = X_1(z)X_2(z^{-1})$	At least intersection of ROCs for $X_1(z)$ and $X_2(z^{-1})$
Initial value theorem	If $x(n)$ is causal	$x(0) = \lim_{z \to \infty} X(z)$	
Multiplication	$x_1(n)x_2(n)$	$\frac{1}{2\pi j}\oint_C X_1(v)X_2\left(\frac{z}{v}\right)v^{-1}dv$	At least: $r_{1l}r_{2l} < \|z\| < r_{1u}r_{2u}$
Parseval's relation	$\sum_{n=-\infty}^{\infty} x_1(n)x_2^*(n) = \frac{1}{2\pi j}\oint_C X_1(v)X_2^*\left(\frac{1}{v^*}\right)v^{-1}dv$		

Table A2 The common z-transform pairs

Signal, $x(n)$	Z-transform, $X(z)$	ROC
(n)	1	All z
$u(n)$	$1/(1\text{-}z^{-1})$	$\|z\| > 1$
$a^n u(n)$	$1/(1\text{-}az^{-1})$	$\|z\| > \|a\|$
$na^n u(n)$	$az^{-1}/(1\text{-}az^{-1})^2$	$\|z\| > \|a\|$
$-a^n u(-n\text{-}1)$	$1/(1\text{-}az^{-1})$	$\|z\| < \|a\|$
$-na^n u(-n\text{-}1)$	$az^{-1}/(1\text{-}az^{-1})^2$	$\|z\| < \|a\|$
$\cos(\omega n)u(n)$	$(1\text{-}z^{-1}\cos\omega)/(1\text{-}2z^{-1}\cos\omega + z^{-2})$	$\|z\| > 1$
$\sin(\omega n)u(n)$	$z^{-1}\sin\omega/(1\text{-}2z^{-1}\cos\omega + z^{-2})$	$\|z\| > 1$
$a^n\cos(\omega n)u(n)$	$(1\text{-}az^{-1}\cos\omega)/(1\text{-}2az^{-1}\cos\omega + a^2z^{-2})$	$\|z\| > \|a\|$
$a^n\sin(\omega n)u(n)$	$az^{-1}\sin\omega/(1\text{-}2az^{-1}\cos\omega + a^2z^{-2})$	$\|z\| > \|a\|$

$$\frac{1}{2\pi j}\oint_C \frac{f(z)}{z - z_0}dz = \begin{cases} f(z_0), & z_0 \text{ inside of } C \\ 0, & z_0 \text{ outside of } C \end{cases},$$

where C is a closed path, while $f'(z)$ exists on and inside of C. If there are $(k + 1)$-th order derivatives of $f(z)$, for poles of multiplicity k that are enclosed by the contour C, while $f(z)$ does not have the poles outside C, then the following is valid:

$$\frac{1}{2\pi j}\oint_C \frac{f(z)}{(z - z_0)^k}dz = \begin{cases} \frac{1}{(k-1)!}\frac{d^{k-1}}{dz^{k-1}}[f(z)]\big|_{z=z_0}, & z_0 \text{ inside of } C \\ 0, & z_0 \text{ outside of } C \end{cases}$$

The partial-fraction method can be used to determine the inverse z-transform when $X(z)$ is expressed as a rational function of z. The key idea behind this method is to express $X(z)$ as a sum:

$$X(z) = \alpha_1X_1(z) + \alpha_2X_2(z) + \ldots + \alpha_KX_K(z),$$

where the inverse z-transforms of i-th terms ($i = 1,\ldots,K$) are available in the table of z-transform pairs (see Table A2).

References

1. I.B. Djordjevic, *Quantum Information Processing and Quantum Error Correction: An Engineering Approach* (Elsevier/Academic, Amsterdam/ Boston, 2012)
2. I. Djordjevic, W. Ryan, B. Vasic, *Coding for Optical Channels* (Springer Science+Business Media, New York, 2010)
3. W.E. Ryan, S. Lin, *Channel Codes: Classical and Modern* (Cambridge University Press, Cambridge/New York, 2009)
4. S. Lin, D.J. Costello, *Error Control Coding: Fundamentals and Applications*, 2nd edn. (Pearson Prentice Hall, Upper Saddle River, 2004)
5. C.C. Pinter, *A Book of Abstract Algebra* (Dover Publications, New York, 2010). (reprint)
6. J.B. Anderson, S. Mohan, *Source and Channel Coding: An Algorithmic Approach* (Kluwer Academic Publishers, 1991)
7. P.A. Grillet, *Abstract Algebra* (Springer, 2007)
8. A. Chambert-Loir, *A Field Guide to Algebra* (Springer, 2005)
9. S. Lang, *Algebra* (Addison-Wesley Publishing Company, Reading, 1993)
10. F. Gaitan, *Quantum Error Correction and Fault Tolerant Quantum Computing* (CRC Press, Boca Raton, 2008)
11. D. Raghavarao, *Constructions and Combinatorial Problems in Design of Experiments* (Dover Publications, Inc., New York, 1988). (Reprint)
12. D.B. Drajic, P.N. Ivanis, *Introduction to Information Theory and Coding* (Akademska Misao, Belgrade, 2009). (in Serbian)
13. M. Cvijetic, I.B. Djordjevic, *Advanced Optical Communications and Networks* (Artech House, Boston/London, 2013)
14. S.B. Wicker, *Error Control Systems for Digital Communication and Storage* (Prentice Hall, Englewood Cliffs, 1995)
15. B. Vucetic, J. Yuan, *Turbo Codes-Principles and Applications* (Kluwer Academic Publishers, Boston, 2000)
16. Infrared Data Association. Infrared Data Association serial infrared physical layer specification. Version 1.4, available at: www.irda.org, 2001
17. S. Hranilovic, *Wireless Optical Communication Systems* (Springer, Boston, 2005)
18. H. Hemmati, A. Biswas, I.B. Djordjevic, Deep-space optical communications: Future perspectives and applications. Proc. IEEE **99**(11), 2020–2039 (2011)
19. S.G. Wilson, M. Brandt-Pearce, Q. Cao, M. Baedke, Optical repetition MIMO transmission with multipulse PPM. IEEE Sel. Areas Commun. **23**(9), 1901–1910 (2005)
20. S.J. Dolinar, J. Hamkins, B.E. Moision, V.A. Vilnrotter, in *Optical Modulation and Coding, in Deep Space Optical Communications*, ed. by H. Hemmati, (Wiley, 2006)
21. I.B. Djordjevic, Multidimensional pulse-position coded-modulation for deep-space optical communication. IEEE Photon. Technol. Lett. **23**(18), 1355–1357 (2011)
22. S. Haykin, B. Van Veen, *Signals and Systems*, 2nd edn. (Wiley, 2003)
23. J.G. Proakis, D.G. Manolakis, *Digital Signal Processing: Principles, Algorithms, and Applications*, 4th edn. (Prentice-Hall, 2007)
24. L.L. Ludeman, *Fundamentals of Digital Signal Processing* (Harper & Row, Publishers, Inc., New York, 1986)

Index

I. B. Djordjevic, *Advanced Optical and Wireless Communications Systems*, https://doi.org/10.1007/978-3-030-98491-5

J

K

L

M

N

O

T

U